Rotational Spectroscopy of Diatomic Molecules

Written to be the definitive text on the rotational spectroscopy of diatomic molecules, this book develops the theory behind the energy levels of diatomic molecules and then summarises the many experimental methods used to study the spectra of these molecules in the gaseous state.

After a general introduction, the methods used to separate nuclear and electronic motions are described. Brown and Carrington then show how the fundamental Dirac and Breit equations may be developed to provide comprehensive descriptions of the kinetic and potential energy terms which govern the behaviour of the electrons. One chapter is devoted solely to angular momentum theory and another describes the development of the so-called effective Hamiltonian used to analyse and understand the experimental spectra of diatomic molecules. The remainder of the book concentrates on experimental methods.

This book will be of interest to graduate students and researchers interested in the rotational spectroscopy of diatomic molecules.

CAMBRIDGE MOLECULAR SCIENCE SERIES

Series Editors:

Richard J. Saykally, *Department of Chemistry, University of California, Berkeley*

Ahmed H. Zewail, *Arthur Amos Noyes Laboratory of Chemical Physics, California Institute of Technology*

David A. King, *Department of Chemistry, University of Cambridge*

Rotational Spectroscopy of Diatomic Molecules

JOHN M. BROWN
Professor of Chemistry, University of Oxford
Fellow of Exeter College, Oxford

ALAN CARRINGTON
Former Royal Society Research Professor, Department of Chemistry,
University of Southampton
Honorary Fellow of Downing College, Cambridge

PUBLISHED BY THE PRESS SYNDICATE OF THE UNIVERSITY OF CAMBRIDGE
The Pitt Building, Trumpington Street, Cambridge, United Kingdom

CAMBRIDGE UNIVERSITY PRESS
The Edinburgh Building, Cambridge CB2 2RU, UK
40 West 20th Street, New York, NY 10011-4211, USA
477 Williamstown Road, Port Melbourne, VIC 3207, Australia
Ruiz de Alarcón 13, 28014 Madrid, Spain
Dock House, The Waterfront, Cape Town 8001, South Africa

http://www.cambridge.org

© John Brown and Alan Carrington

This book is in copyright. Subject to statutory exception
and to the provisions of relevant collective licensing agreements,
no reproduction of any part may take place without
the written permission of Cambridge University Press.

First published 2003

Printed in the United Kingdom at the University Press, Cambridge

Typeface Times New Roman 10/13 pt and Scala Sans   *System* LATEX 2_ε [TB]

A catalogue record for this book is available from the British Library

Library of Congress Cataloguing in Publication data

Brown, John M.
Rotational spectroscopy of diatomic molecules/John M. Brown, Alan Carrington.
   p. cm. – (Cambridge molecular science)
Includes bibliographical references and index.
ISBN 0 521 81009 4 – ISBN 0 521 53078 4 (pb.)
1. Molecular spectroscopy.   I. Carrington, Alan.   II. Title.   III. Series.
QC454.M6 B76   2003
539'.6'0287–dc21   2002073930

ISBN 0 521 81009 4 hardback
ISBN 0 521 53078 4 paperback

The publisher has used its best endeavours to ensure that the URLs for external websites referred to in this book are correct and active at the time of going to press. However, the publisher has no responsibility for the websites and can make no guarantee that a site will remain live or that the content is or will remain appropriate.

Contents

Preface	*page* xv
Summary of notation	xix
Figure acknowledgements	xxiii

1 General introduction	1
1.1 Electromagnetic spectrum	1
1.2 Electromagnetic radiation	3
1.3 Intramolecular nuclear and electronic dynamics	5
1.4 Rotational levels	9
1.5 Historical perspectives	12
1.6 Fine structure and hyperfine structure of rotational levels	14
1.6.1 Introduction	14
1.6.2 $^1\Sigma^+$ states	15
1.6.3 Open shell Σ states	21
1.6.4 Open shell states with both spin and orbital angular momentum	26
1.7 The effective Hamiltonian	29
1.8 Bibliography	32
Appendix 1.1 Maxwell's equations	33
Appendix 1.2 Electromagnetic radiation	35
References	36
2 The separation of nuclear and electronic motion	38
2.1 Introduction	38
2.2 Electronic and nuclear kinetic energy	40
2.2.1 Introduction	40
2.2.2 Origin at centre of mass of molecule	41
2.2.3 Origin at centre of mass of nuclei	43
2.2.4 Origin at geometrical centre of the nuclei	44
2.3 The total Hamiltonian in field-free space	44
2.4 The nuclear kinetic energy operator	45
2.5 Transformation of the electronic coordinates to molecule-fixed axes	51
2.5.1 Introduction	51
2.5.2 Space transformations	52
2.5.3 Spin transformations	54
2.6 Schrödinger equation for the total wave function	59
2.7 The Born–Oppenheimer and Born adiabatic approximations	60

2.8 Separation of the vibrational and rotational wave equations 61
2.9 The vibrational wave equation 63
2.10 Rotational Hamiltonian for space-quantised electron spin 67
2.11 Non-adiabatic terms 67
2.12 Effects of external electric and magnetic fields 68
Appendix 2.1 Derivation of the momentum operator 71
References 72

3 The electronic Hamiltonian 73
3.1 The Dirac equation 73
3.2 Solutions of the Dirac equation in field-free space 76
3.3 Electron spin magnetic moment and angular momentum 77
3.4 The Foldy–Wouthuysen transformation 80
3.5 The Foldy–Wouthuysen and Dirac representations for a free particle 85
3.6 Derivation of the many-electron Hamiltonian 89
3.7 Effects of applied static magnetic and electric fields 94
3.8 Retarded electromagnetic interaction between electrons 97
    3.8.1 Introduction 97
    3.8.2 Lorentz transformation 98
    3.8.3 Electromagnetic potentials due to a moving electron 99
    3.8.4 Gauge invariance 101
    3.8.5 Classical Lagrangian and Hamiltonian 103
3.9 The Breit Hamiltonian 104
    3.9.1 Introduction 104
    3.9.2 Reduction of the Breit Hamiltonian to non-relativistic form 105
3.10 Electronic interactions in the nuclear Hamiltonian 109
3.11 Transformation of coordinates in the field-free total Hamiltonian 110
3.12 Transformation of coordinates for the Zeeman and Stark terms in the total Hamiltonian 114
3.13 Conclusions 118
Appendix 3.1 Power series expansion of the transformed Hamiltonian 121
References 122

4 Interactions arising from nuclear magnetic and electric moments 123
4.1 Nuclear spins and magnetic moments 123
4.2 Derivation of nuclear spin magnetic interactions through the magnetic vector potential 125
4.3 Derivation of nuclear spin interactions from the Breit equation 130
4.4 Nuclear electric quadrupole interactions 131
    4.4.1 Spherical tensor form of the Hamiltonian operator 131
    4.4.2 Cartesian form of the Hamiltonian operator 133
    4.4.3 Matrix elements of the quadrupole Hamiltonian 134
4.5 Transformation of coordinates for the nuclear magnetic dipole and electric quadrupole terms 136
References 138

5 Angular momentum theory and spherical tensor algebra 139
5.1 Introduction 139
5.2 Rotation operators 140
    5.2.1 Introduction 140
    5.2.2 Decomposition of rotational operators 142
    5.2.3 Commutation relations 142
    5.2.4 Representations of the rotation group 143
    5.2.5 Orbital angular momentum and spherical harmonics 144
5.3 Rotations of a rigid body 146
    5.3.1 Introduction 146
    5.3.2 Rotation matrices 148
    5.3.3 Spin 1/2 systems 150
    5.3.4 Symmetric top wave functions 150
5.4 Addition of angular momenta 152
    5.4.1 Introduction 152
    5.4.2 Wigner 3-j symbols 154
    5.4.3 Coupling of three or more angular momenta: Racah algebra, Wigner 6-j and 9-j symbols 155
    5.4.4 Clebsch–Gordan series 157
    5.4.5 Integrals over products of rotation matrices 158
5.5 Irreducible spherical tensor operators 159
    5.5.1 Introduction 159
    5.5.2 Examples of spherical tensor operators 160
    5.5.3 Matrix elements of spherical tensor operators: the Wigner–Eckart theorem 163
    5.5.4 Matrix elements for composite systems 165
    5.5.5 Relationship between operators in space-fixed and molecule-fixed coordinate systems 167
    5.5.6 Treatment of the anomalous commutation relationships of rotational angular momenta by spherical tensor methods 168
Appendix 5.1 Summary of standard results from spherical tensor algebra 171
References 175

6 Electronic and vibrational states 177
6.1 Introduction 177
6.2 Atomic structure and atomic orbitals 178
    6.2.1 The hydrogen atom 178
    6.2.2 Many-electron atoms 181
    6.2.3 Russell–Saunders coupling 184
    6.2.4 Wave functions for the helium atom 187
    6.2.5 Many-electron wave functions: the Hartree–Fock equation 190
    6.2.6 Atomic orbital basis set 194
    6.2.7 Configuration interaction 196
6.3 Molecular orbital theory 197

6.4 Correlation of molecular and atomic electronic states	203
6.5 Calculation of molecular electronic wave functions and energies	206
6.5.1 Introduction	206
6.5.2 Electronic wave function for the H_2^+ molecular ion	207
6.5.3 Electronic wave function for the H_2 molecule	208
6.5.4 Many-electron molecular wave functions	212
6.6 Corrections to Born–Oppenheimer calculations for H_2^+ and H_2	219
6.7 Coupling of electronic and rotational motion: Hund's coupling cases	224
6.7.1 Introduction	224
6.7.2 Hund's coupling case (a)	225
6.7.3 Hund's coupling case (b)	226
6.7.4 Hund's coupling case (c)	228
6.7.5 Hund's coupling case (d)	228
6.7.6 Hund's coupling case (e)	229
6.7.7 Intermediate coupling	230
6.7.8 Nuclear spin coupling cases	232
6.8 Rotations and vibrations of the diatomic molecule	233
6.8.1 The rigid rotor	233
6.8.2 The harmonic oscillator	235
6.8.3 The anharmonic oscillator	238
6.8.4 The non-rigid rotor	242
6.8.5 The vibrating rotor	243
6.9 Inversion symmetry of rotational levels	244
6.9.1 The space-fixed inversion operator	244
6.9.2 The effect of space-fixed inversion on the Euler angles and on molecule-fixed coordinates	245
6.9.3 The transformation of general Hund's case (a) and case (b) functions under space-fixed inversion	246
6.9.4 Parity combinations of basis functions	251
6.10 Permutation symmetry of rotational levels	251
6.10.1 The nuclear permutation operator for a homonuclear diatomic molecule	251
6.10.2 The transformation of general Hund's case (a) and case (b) functions under nuclear permutation P_{12}	252
6.10.3 Nuclear statistical weights	254
6.11 Theory of transition probabilities	256
6.11.1 Time-dependent perturbation theory	256
6.11.2 The Einstein transition probabilities	258
6.11.3 Einstein transition probabilities for electric dipole transitions	261
6.11.4 Rotational transition probabilities	263
6.11.5 Vibrational transition probabilities	266
6.11.6 Electronic transition probabilities	267
6.11.7 Magnetic dipole transition probabilities	269

6.12 Line widths and spectroscopic resolution	273
6.12.1 Natural line width	273
6.12.2 Transit time broadening	273
6.12.3 Doppler broadening	274
6.12.4 Collision broadening	275
6.13 Relationships between potential functions and the vibration–rotation levels	276
6.13.1 Introduction	276
6.13.2 The JWKB semiclassical method	277
6.13.3 Inversion of experimental data to calculate the potential function (RKR)	280
6.14 Long-range near-dissociation interactions	282
6.15 Predissociation	286
Appendix 6.1 Calculation of the Born–Oppenheimer potential for the H_2^+ ion	289
References	298
7 Derivation of the effective Hamiltonian	**302**
7.1 Introduction	302
7.2 Derivation of the effective Hamiltonian by degenerate perturbation theory: general principles	303
7.3 The Van Vleck and contact transformations	312
7.4 Effective Hamiltonian for a diatomic molecule in a given electronic state	316
7.4.1 Introduction	316
7.4.2 The rotational Hamiltonian	319
7.4.3 Hougen's isomorphic Hamiltonian	320
7.4.4 Fine structure terms: spin–orbit, spin–spin and spin–rotation operators	323
7.4.5 Λ-doubling terms for a Π electronic state	328
7.4.6 Nuclear hyperfine terms	331
7.4.7 Higher-order fine structure terms	335
7.5 Effective Hamiltonian for a single vibrational level	338
7.5.1 Vibrational averaging and centrifugal distortion corrections	338
7.5.2 The form of the effective Hamiltonian	341
7.5.3 The N^2 formulation of the effective Hamiltonian	343
7.5.4 The isotopic dependence of parameters in the effective Hamiltonian	344
7.6 Effective Zeeman Hamiltonian	347
7.7 Indeterminacies: rotational contact transformations	352
7.8 Estimates and interpretation of parameters in the effective Hamiltonian	356
7.8.1 Introduction	356
7.8.2 Rotational constant	356
7.8.3 Spin–orbit coupling constant, A	357
7.8.4 Spin–spin and spin–rotation parameters, λ and γ	360

7.8.5 Λ-doubling parameters 362
7.8.6 Magnetic hyperfine interactions 363
7.8.7 Electric quadrupole hyperfine interaction 365
Appendix 7.1 Molecular parameters or constants 368
References 369

8 Molecular beam magnetic and electric resonance 371
8.1 Introduction 371
8.2 Molecular beam magnetic resonance of closed shell molecules 372
    8.2.1 H_2, D_2 and HD in their $X\,^1\Sigma^+$ ground states 372
    8.2.2 Theory of Zeeman interactions in $^1\Sigma^+$ states 390
    8.2.3 Na_2 in the $X\,^1\Sigma_g^+$ ground state: optical state selection and detection 416
    8.2.4 Other $^1\Sigma^+$ molecules 421
8.3 Molecular beam magnetic resonance of electronically excited molecules 422
    8.3.1 H_2 in the $c\,^3\Pi_u$ state 422
    8.3.2 N_2 in the $A\,^3\Sigma_u^+$ state 446
8.4 Molecular beam electric resonance of closed shell molecules 463
    8.4.1 Principles of electric resonance methods 463
    8.4.2 CsF in the $X\,^1\Sigma^+$ ground state 465
    8.4.3 LiBr in the $X\,^1\Sigma^+$ ground state 483
    8.4.4 Alkaline earth and group IV oxides 487
    8.4.5 HF in the $X\,^1\Sigma^+$ ground state 489
    8.4.6 HCl in the $X\,^1\Sigma^+$ ground state 500
8.5 Molecular beam electric resonance of open shell molecules 508
    8.5.1 Introduction 508
    8.5.2 LiO in the $X\,^2\Pi$ ground state 509
    8.5.3 NO in the $X\,^2\Pi$ ground state 526
    8.5.4 OH in the $X\,^2\Pi$ ground state 538
    8.5.5 CO in the $a\,^3\Pi$ state 552
Appendix 8.1 Nuclear spin dipolar interaction 558
Appendix 8.2 Relationship between the cartesian and spherical tensor forms of the electron spin–nuclear spin dipolar interaction 561
Appendix 8.3 Electron spin–electron spin dipolar interaction 563
Appendix 8.4 Matrix elements of the quadrupole Hamiltonian 568
Appendix 8.5 Magnetic hyperfine Hamiltonian and hyperfine constants 573
References 574

9 Microwave and far-infrared magnetic resonance 579
9.1 Introduction 579
9.2 Experimental methods 579
    9.2.1 Microwave magnetic resonance 579
    9.2.2 Far-infrared laser magnetic resonance 584
9.3 $^1\Delta$ states 587
    9.3.1 SO in the $a\,^1\Delta$ state 587
    9.3.2 NF in the $a\,^1\Delta$ state 591

9.4 $^2\Pi$ states	596
9.4.1 Introduction	596
9.4.2 ClO in the $X\,^2\Pi$ ground state	597
9.4.3 OH in the $X\,^2\Pi$ ground state	613
9.4.4 Far-infrared laser magnetic resonance of CH in the $X\,^2\Pi$ ground state	624
9.5 $^2\Sigma$ states	633
9.5.1 Introduction	633
9.5.2 CN in the $X\,^2\Sigma^+$ ground state	633
9.6 $^3\Sigma$ states	641
9.6.1 SO in the $X\,^3\Sigma^-$ ground state	641
9.6.2 SeO in the $X\,^3\Sigma^-$ ground state	649
9.6.3 NH in the $X\,^3\Sigma^-$ ground state	652
9.7 $^3\Pi$ states	655
9.7.1 CO in the $a\,^3\Pi$ state	655
9.8 $^4\Sigma$ states	661
9.8.1 CH in the $a\,^4\Sigma^-$ state	661
9.9 $^4\Delta$, $^3\Phi$, $^2\Delta$ and $^6\Sigma^+$ states	665
9.9.1 Introduction	665
9.9.2 CrH in the $X\,^6\Sigma^+$ ground state	666
9.9.3 FeH in the $X\,^4\Delta$ ground state	669
9.9.4 CoH in the $X\,^3\Phi$ ground state	669
9.9.5 NiH in the $X\,^2\Delta$ ground state	674
Appendix 9.1 Evaluation of the reduced matrix element of $T^3(\boldsymbol{S},\boldsymbol{S},\boldsymbol{S})$	678
References	680
10 Pure rotational spectroscopy	**683**
10.1 Introduction and experimental methods	683
10.1.1 Simple absorption spectrograph	683
10.1.2 Microwave radiation sources	685
10.1.3 Modulation spectrometers	688
10.1.4 Superheterodyne detection	701
10.1.5 Fourier transform spectrometer	703
10.1.6 Radio telescopes and radio astronomy	713
10.1.7 Terahertz (far-infrared) spectrometers	723
10.1.8 Ion beam techniques	728
10.2 $^1\Sigma^+$ states	732
10.2.1 CO in the $X\,^1\Sigma^+$ ground state	732
10.2.2 HeH$^+$ in the $X\,^1\Sigma^+$ ground state	736
10.2.3 CuCl and CuBr in their $X\,^1\Sigma^+$ ground states	738
10.2.4 SO, NF and NCl in their $b\,^1\Sigma^+$ states	741
10.2.5 Hydrides (LiH, NaH, KH, CuH, AlH, AgH) in their $X\,^1\Sigma^+$ ground states	743

10.3 $^2\Sigma$ states	745
10.3.1 CO$^+$ in the $X\,^2\Sigma^+$ ground state	745
10.3.2 CN in the $X\,^2\Sigma^+$ ground state	749
10.4 $^3\Sigma$ states	752
10.4.1 Introduction	752
10.4.2 O$_2$ in its $X\,^3\Sigma_g^-$ ground state	754
10.4.3 SO, S$_2$ and NiO in their $X\,^3\Sigma^-$ ground states	759
10.4.4 PF, NCl, NBr and NI in their $X\,^3\Sigma^-$ ground states	763
10.5 $^1\Delta$ states	776
10.5.1 O$_2$ in its $a\,^1\Delta_g$ state	776
10.5.2 SO and NCl in their $a\,^1\Delta$ states	779
10.6 $^2\Pi$ states	782
10.6.1 NO in the $X\,^2\Pi$ ground state	782
10.6.2 OH in the $X\,^2\Pi$ ground state	788
10.6.3 CH in the $X\,^2\Pi$ ground state	794
10.6.4 CF, SiF, GeF in their $X\,^2\Pi$ ground states	810
10.6.5 Other free radicals with $^2\Pi$ ground states	811
10.7 Case (c) doublet state molecules	813
10.7.1 Studies of the HeAr$^+$ ion	813
10.7.2 Studies of the HeKr$^+$ ion	832
10.8 Higher spin/orbital states	834
10.8.1 CO in the $a\,^3\Pi$ state	834
10.8.2 SiC in the $X\,^3\Pi$ ground state	836
10.8.3 FeC in the $X\,^3\Delta$ ground state	841
10.8.4 VO and NbO in their $X\,^4\Sigma^-$ ground states	841
10.8.5 FeF and FeCl in their $X\,^6\Delta$ ground states	845
10.8.6 CrF, CrCl and MnO in their $X\,^6\Sigma^+$ ground states	850
10.8.7 FeO in the $X\,^5\Delta$ ground state	853
10.8.8 TiCl in the $X\,^4\Phi$ ground state	854
10.9 Observation of a pure rotational transition in the H$_2^+$ molecular ion	856
References	862
11 Double resonance spectroscopy	**870**
11.1 Introduction	870
11.2 Radiofrequency and microwave studies of CN in its excited electronic states	871
11.3 Early radiofrequency or microwave/optical double resonance studies	876
11.3.1 Radiofrequency/optical double resonance of CS in its excited $A\,^1\Pi$ state	876
11.3.2 Radiofrequency/optical double resonance of OH in its excited $A\,^2\Sigma^+$ state	880
11.3.3 Microwave/optical double resonance of BaO in its ground $X\,^1\Sigma^+$ and excited $A\,^1\Sigma^+$ states	883

11.4 Microwave/optical magnetic resonance studies of electronically excited H_2	885
11.4.1 Introduction	885
11.4.2 H_2 in the $G\,^1\Sigma_g^+$ state	885
11.4.3 H_2 in the $d\,^3\Pi_u$ state	892
11.4.4 H_2 in the $k\,^3\Pi_u$ state	900
11.5 Radiofrequency or microwave/optical double resonance of alkaline earth molecules	902
11.5.1 Introduction	902
11.5.2 SrF, CaF and CaCl in their $X\,^2\Sigma^+$ ground states	902
11.6 Radiofrequency or microwave/optical double resonance of transition metal molecules	906
11.6.1 Introduction	906
11.6.2 FeO in the $X\,^5\Delta$ ground state	909
11.6.3 CuF in the $b\,^3\Pi$ excited state	913
11.6.4 CuO in the $X\,^2\Pi$ ground state	917
11.6.5 ScO in the $X\,^2\Sigma^+$ ground state	919
11.6.6 TiO in the $X\,^3\Delta$ ground state and TiN in the $X\,^2\Sigma^+$ ground state	922
11.6.7 CrN and MoN in their $X\,^4\Sigma^-$ ground states	924
11.6.8 NiH in the $X\,^2\Delta$ ground state	927
11.6.9 4d transition metal molecules: YF in the $X\,^1\Sigma^+$ ground state, YO and YS in their $X\,^2\Sigma^+$ ground states	930
11.7 Microwave/optical double resonance of rare earth molecules	936
11.7.1 Radiofrequency/optical double resonance of YbF in its $X\,^2\Sigma^+$ ground state	936
11.7.2 Radiofrequency/optical double resonance of LaO in its $X\,^2\Sigma^+$ and $B\,^2\Sigma^+$ states	938
11.8 Double resonance spectroscopy of molecular ion beams	942
11.8.1 Radiofrequency and microwave/infrared double resonance of HD^+ in the $X\,^2\Sigma^+$ ground state	942
11.8.2 Radiofrequency/optical double resonance of N_2^+ in the $X\,^2\Sigma_g^+$ ground state	953
11.8.3 Microwave/optical double resonance of CO^+ in the $X\,^2\Sigma^+$ ground state	958
11.9 Quadrupole trap radiofrequency spectroscopy of the H_2^+ ion	960
11.9.1 Introduction	960
11.9.2 Principles of photo-alignment	960
11.9.3 Experimental methods and results	962
11.9.4 Analysis of the spectra	964
11.9.5 Quantitative interpretation of the molecular parameters	972
References	974
General appendices	978
Appendix A Values of the fundamental constants	978

Appendix B Selected set of nuclear properties for naturally occurring
           isotopes   979
Appendix C Compilation of Wigner 3-j symbols   987
Appendix D Compilation of Wigner 6-j symbols   991
Appendix E Relationships between cgs and SI units   993

Author index   994
Subject index   1004

Preface

A book whose title refers to the spectroscopy of diatomic molecules is, inevitably, going to be compared with the classic book written by G. Herzberg under the title *Spectroscopy of Diatomic Molecules*. This book was published in 1950, and it dealt almost entirely with electronic spectroscopy in the gas phase, studied by the classic spectrographic techniques employing photographic plates. The spectroscopic resolution at that time was limited to around 0.1 cm^{-1} by the Doppler effect; this meant that the vibrational and rotational structure of electronic absorption or emission band systems could be easily resolved in most systems. The diatomic molecules studied by 1950 included conventional closed shell systems, and a large number of open shell electronic states of molecules in both their ground and excited states. Herzberg presented a beautiful and detailed summary of the principles underlying the analysis of such spectra. The theory of the rotational levels of both closed and open shell diatomic molecules was already well developed by 1950, and the correlation of experimental and theoretical results was one of the major achievements of Herzberg's book. It is a matter of deep regret to us both that we cannot present our book to 'GH' for, hopefully, his approval. On the other hand, we were both privileged to spend time working in the laboratory in Ottawa directed by GH, and to have known him as a colleague, mentor and friend.

Accepting, therefore, the possible and perhaps likely comparison with Herzberg's book, we should say at the outset that almost everything described in our book relates to work published after 1950, and the philosophy and approach of our book is different from that of Herzberg, as it surely should be over 50 years on. The Doppler width of 0.1 cm^{-1} characteristic of conventional visible and ultraviolet electronic spectra, corresponding to 3000 MHz in frequency units, conceals much of what is most interesting and fundamental to the spectroscopic and electronic properties of diatomic molecules. Our book deals with the experimental and theoretical study of these details, revealed by measurement of either transitions *between* rotational levels, or transitions *within* a single rotational level, occurring between the fine or hyperfine components. This branch of spectroscopy is often called rotational spectroscopy, and it involves much lower frequency regions of the electromagnetic spectrum than those arising in conventional electronic or vibrational spectroscopy. The experimental work described in this book ranges from the far-infrared, through the microwave, to the radiofrequency regions of the spectrum; the intrinsic Doppler width is small because of the lower frequency, and special techniques, particularly those involving molecular beams, sometimes result in very high spectroscopic resolution. Molecules in open shell electronic states possess a number of subtle intramolecular magnetic and electric interactions, revealed by these high-resolution studies. Additional studies involving the effects of applied magnetic or electric fields provide further information, particularly about electron and nuclear spin magnetic moments. All of the experimental work described in this book involves molecules in the gas phase. Consequently we include descriptions of conventional microwave and millimetre wave rotational spectroscopy, subjects which would be familiar to most molecular spectroscopists. However, we give equal prominence to the molecular beam magnetic resonance studies of the small magnetic interactions arising from the presence of magnetic nuclei in closed shell molecules. These classic studies formed the basis for subsequent nuclear magnetic resonance studies of condensed phases; similarly the magnetic interactions studied through condensed phase electron spin resonance experiments were first understood through high-resolution gas phase investigations described in this book. These are subjects which, more often than not, do not appear in the same book as rotational spectroscopy, but they should.

The important threads which link these different branches of gas phase rotational spectroscopy are, of course, those arising from the theory. We have tried to make clear the distinction between two different types of theory. A spectroscopist analyses a spectrum by using algebraic expressions for transition frequencies which involve appropriate quantum numbers and 'molecular constants'. These expressions arise from the use of an *effective* Hamiltonian, which summarises the relevant intramolecular dynamics and interactions, and is expressed in terms of molecular parameters and operators, usually angular momentum operators. A central theme of our book is the construction and use of the effective Hamiltonian, and through it some more precise definitions of the molecular constants or parameters. We show, at length, how the effective Hamiltonian is derived from a consideration of the fundamental *true* Hamiltonian, although the word 'true' must be used with caution and some respect. We take as our foundations the Dirac equation for one electron, and the Breit equation for two electrons. We show how the 'true' Hamiltonian for a molecule, in the presence of external fields, is derived, and show how this may be applied to the derivation of an 'effective' Hamiltonian appropriate for any particular molecular system or spectroscopic study. We have made a compromise in our analysis; we do not delve into quantum electrodynamics! One of the lessons in the life of a serious spectroscopist is that there is always a level of understanding deeper than that being employed, and we all have to compromise somewhere.

Chapters 2 to 7 deal with the essentials of the theory, starting with the separation of nuclear and electronic motion, and finishing with the derivation of effective Hamiltonians. An important aspect of diatomic molecules is their high symmetry, and the various angular momenta which can arise. Angular momentum theory is summarised in chapter 5 where we show the importance of rotational symmetry by introducing spherical tensors to describe the angular momenta and their interactions, both with each other and with applied fields. Spherical tensor methods are used throughout the book; we have used them to describe the analysis of particular spectra even if the original work used cartesian tensors. Spherical tensor, or irreducible tensor methods bring out the links between different parts of the subject; they make maximum use of symmetry and, to our minds at least, are simpler and more reliable in their use than the older cartesian methods. This is particularly true of problems which involve transformations from space to molecule-fixed axes, for example, the effects of applied magnetic or electric fields.

Chapters 8 to 11 describe the details and results of experimental studies. Chapter 8 deals with molecular beam magnetic and electric resonance, chapter 9 with magnetic resonance of open shell molecules in the bulk gas phase, chapter 10 with pure rotational spectroscopy in the bulk gas phase, and chapter 11 with double resonance studies. Of course, these topics overlap and some molecules, the OH radical for example, appear in all four chapters. We have deliberately allowed some repetition in our discussion, because in the process of following a complicated analysis, it is very annoying to be forced to jump to other parts of a book for some essential details. Our overriding philosophy has been to choose particularly important examples which illustrate the details for particular types of electronic state, and to work through the theory and analysis in considerable detail. Although there is a substantial amount of experimental data in our book, we have not intended to be comprehensive in this respect. Computerised data bases, and the various encyclopaedic assemblies are the places to seek for data on specific molecules. As mentioned earlier, we have analysed the experimental data using spherical tensor methods, even if the original work used cartesian methods, as was often the case with the earlier studies.

The question of units always poses a problem for anyone writing a book in our field. Most authors from North America use cgs units, and most of the work described in this book originated in the USA. Authors from the UK and Europe, on the other hand, have largely been converted to using SI units. There is no doubt that the SI system is the more logical, and that numerical calculations using SI units are more easily accomplished. Nevertheless since so many spectra are still assigned and analysed using cgs units, we have had to seek a compromise solution. The fundamental theory describing the electronic Hamiltonian, presented in chapter 3, uses SI units. Similarly we use SI units in describing the theory of nuclear hyperfine interactions in chapter 4. However, chapters 8 to 11, which deal with the analysis of spectra, are written in terms of both cgs and SI units, so that direct comparisons with the original literature can be made. A comparison of the cgs and SI units is presented in General Appendix D. To complicate matters even further, the use of *atomic* units, which is common in *ab initio* electronic structure calculations because of the simplifications introduced, is described in chapter 6.

The gestation period for this book has been particularly long, work on it having started around 1970 when we were both members of the Department of Chemistry at Southampton University. Research was going rather slowly at the time and we had a keen desire to understand the foundations of our subject properly. We worked through the various aspects together, and put the material in writing. At first we had only the other members of the group in mind but, as things developed, we started to write for a wider audience. We were encouraged and greatly helped at the time by our colleague, Dr Richard Moss, who gave an outstanding post-graduate course on relativistic quantum mechanics. Chapters 2, 3 and 4 of the present book were essentially written at that time. The writing process, however, eventually gave way to other things, particularly research, and it seemed that the unfinished book, like so many others, was destined for the scrap heap. There it remained until one of us (AC), conscious of approaching enforced 'retirement', decided to revive the project as an antidote to possible vegetation. The dusty old manuscript was scanned into a computer, revised, and over a period of four years developed into the book now published. The passage of some thirty years between the two phases of writing has undoubtedly had some benefits. In particular, it has allowed the time for important new technical developments to take place, and for the subject (and the authors) to mature generally.

The manuscript for this book was produced using MSWord text and equation editor, with MATHTYPE used to control equation numbering and cross-referencing. Those figures which include an experimental spectrum were produced using SigmaPlot, each spectrum being obtained in XY array form by a digitising scan of the original paper. All other figures were produced using CoralDraw.

We are grateful to several friends and colleagues who have read parts of the book and given us their comments. In particular we thank Professors B.J. Howard, T.A. Miller, T.C. Steimle, M. McCarthy, M.S. Child and Dr I.R. McNab. We will always be glad to receive comments from readers, kind, helpful, or otherwise! Alan Carrington would like to thank the Leverhulme Trust for an Emeritus Fellowship which has enabled him to keep in close touch with the subject through attendances at conferences.

This book is dedicated to the memory of Bill Flygare, Harry Radford and Ken Evenson.

Alan Carrington
John M. Brown
April 2002

Summary of notation

Throughout this book we have used, at different times, space-fixed or molecule-fixed axis systems, with arbitrary origin, origin at the molecular centre of mass, origin at the nuclear centre of mass, or origin at the geometrical centre of the nuclei. We use CAPITAL letters for SPACE-FIXED axes, and lower case letters for molecule-fixed axes. The various origins are denoted by primes as follows.

(i) Space-fixed axes: arbitrary origin.
  R_α = position vector of nucleus α
  P_α = momentum conjugate to R_α
  R_i = position vector of ith electron
  P_i = momentum conjugate to R_i
  S_i = spin of ith electron
  X, Y, Z = space-fixed axes

(ii) Space-fixed axes: origin at molecular centre of mass.
  $R'_\alpha, P'_\alpha, R'_i, P'_i$, defined by analogy with R_α, etc.,
  R_O = position vector of molecular centre of mass with respect to the arbitrary origin
  P_O = momentum conjugate to R_O, i.e., translational momentum
  R = internuclear vector = $R_2 - R_1 = R'_2 - R'_1$
  S_i = spin of ith electron

(iii) Space-fixed axes: origin at nuclear centre of mass.
  $P''_\alpha, R''_\alpha, P''_i, R''_i, R, S_i$

(iv) Space-fixed axes: origin at geometrical centre of nuclei.
  $P'''_\alpha, R'''_\alpha, P'''_i, R'''_i$

(v) Molecule-fixed axes: origin at nuclear centre of mass.
  r_i, p_i, s_i

When dealing with components of vector quantities we usually use subscripts X, Y, Z or x, y, z for space-fixed or molecule-fixed components, the origin of coordinates usually being denoted in the primary subscripted symbol. For the electron spin we use capital S_i for space-fixed axes and small s_i for molecule-fixed; it is not necessary to distinguish the origin of coordinates. A difficulty with this notation is that, in conformity with common practice, we also use the symbol S to denote the total spin ($\Sigma_i s_i$ or $\Sigma_i S_i$). We hope to avoid confusion in the appropriate text. We use M or M_s to denote the component of $\boldsymbol{S}$ in the space-fixed Z direction, and m or m_s to denote the component of $\boldsymbol{s}$ in the molecule-fixed z direction, i.e. along the internuclear axis. We shall also sometimes use Σ to denote m_s. Hence ψ_M denotes a spinor in the space-fixed axis system and ψ_m refers to the molecule-fixed axes.

Other symbols used are as follows:

$\boldsymbol{i}', \boldsymbol{j}', \boldsymbol{k}'$ = unit vectors along X, Y, Z
$\boldsymbol{i}, \boldsymbol{j}, \boldsymbol{k}$ = unit vectors along x, y, z
$\boldsymbol{\varepsilon}_i$ = electric field strength at electron i arising from other electrons and nuclei
$\boldsymbol{E}_i$ = applied electric field strength at electron i
$\boldsymbol{B}_i$ = applied magnetic flux density at electron i
$\boldsymbol{A}_i$ = total magnetic vector potential at electron i
$\boldsymbol{A}_i^e$ = contribution to $\boldsymbol{A}_i$ from other electrons
$\boldsymbol{A}_i^B$ = contribution to $\boldsymbol{A}_i$ from external magnetic field
ϕ_i = total electric potential at electron i
$\boldsymbol{A}_\alpha$ = total magnetic vector potential at nucleus α
$\boldsymbol{A}_\alpha^e$ = contribution to $\boldsymbol{A}_\alpha$ from electrons
$\boldsymbol{A}_\alpha^B$ = contribution to $\boldsymbol{A}_\alpha$ from external magnetic field
ϕ_α = total electric potential at nucleus α
$\boldsymbol{B}_\alpha$ = applied magnetic flux density at nucleus α
$\boldsymbol{E}_\alpha$ = applied electric field strength at nucleus α
$\boldsymbol{B} = \boldsymbol{B}_i = \boldsymbol{B}_\alpha$ for homogeneous magnetic field
$\boldsymbol{E} = \boldsymbol{E}_i = \boldsymbol{E}_\alpha$ for homogeneous electric field
Λ = projection of $\boldsymbol{L}$ along internuclear axis
Σ, m_s = projection of $\boldsymbol{S}$ along internuclear axis
$\Omega = |\Lambda + \Sigma|$ = projection of total electronic angular momentum along internuclear axis
$\mathfrak{D}_{M,m}^{(1/2)}(\phi, \theta, \chi)$ = rotational matrix for spin transformation
$\boldsymbol{V}$ or $\boldsymbol{v}$ = classical velocity vector
$\mathfrak{L}$ = Lagrangian
E = energy
t = time
$\boldsymbol{\sigma}'$ = Pauli spin vector
$\boldsymbol{\sigma}$ = Dirac spin vector
$\boldsymbol{\mu}_S$ = electron spin magnetic moment
$\boldsymbol{\mu}_I$ = nuclear spin magnetic moment
$\boldsymbol{\mu}_r$ or $\boldsymbol{\mu}_J$ = rotational magnetic moment
$\boldsymbol{\mu}_e$ = electric dipole moment
$\bar{S}$ = Foldy–Wouthuysen operator
∇ = gradient operator
∇^2 = Laplacian
$\delta(\ )$ = Dirac delta function

$\pi =$ pi, 3.141 592 653...
$M_\alpha =$ mass of nucleus α
m_i (or m) $=$ mass of electron
$M_p =$ proton mass
$\mu =$ reduced nuclear mass (possible confusion here with magnetic moment)
$\mu_\alpha = M_1 M_2/(M_1 - M_2)$
$M =$ total molecular mass
$h =$ the Planck constant
$\hbar = h/2\pi$
$\mu_B =$ electron Bohr magneton $= e\hbar/2m$
$\mu_N =$ nuclear Bohr magneton $= e\hbar/2M_p$
$c =$ speed of light
$e =$ elementary unit of charge (defined to be positive)
$-e =$ electron charge
$Z_\alpha e =$ nuclear charge
α, β denote m_s or $M_s = +1/2, -1/2$
$\pi =$ mechanical momentum in presence of electromagnetic fields
$\boldsymbol{\alpha} =$ Dirac momentum operator
$\beta =$ Dirac matrix
$\boldsymbol{L} =$ orbital angular momentum
$\boldsymbol{S} =$ electron spin angular momentum
$\boldsymbol{P} =$ total electronic angular momentum
$\boldsymbol{J} =$ total angular momentum excluding nuclear spin
$\boldsymbol{N} =$ total angular momentum excluding electron and nuclear spin
$\boldsymbol{R} =$ rotational angular momentum of the bare nuclei
$\boldsymbol{I} =$ nuclear spin angular momentum
$\boldsymbol{F} =$ grand total angular momentum including electron and nuclear spin
$R =$ internuclear distance
$\phi, \theta, \chi =$ Euler angles
$\tilde{\boldsymbol{R}} =$ mean position operator in the Dirac representation
$\boldsymbol{R}'' =$ position operator in the F–W representation (confusion)
$\boldsymbol{I}_\alpha =$ spin of nucleus α
$g_S =$ electron g factor: value $= 2$ in the Dirac theory, 2.002 32 from quantum electrodynamics
$g_N =$ nuclear g factor
$g_L =$ orbital g factor
g_r or $g_J =$ rotational g-factor
$i = \sqrt{-1}$

Some additional notes

Vector quantities are denoted by bold font. Although the square of a vector, i.e. the scalar product of the vector with itself, is a scalar quantity, we have followed the commonest convention of also denoting the vector squares in bold font.

An applied magnetic field is denoted B_Z throughout this book; we use the alternative B_0, to denote the rotational constant for the $v = 0$ level.

Additional molecular parameters which arise in effective Hamiltonians are listed in Appendix 7.1.

Figure acknowledgements

The figures in this book are of two different types. Figures which are solely line drawings were produced using CoralDraw; in some cases the drawings are similar to figures published elsewhere, and appropriate acknowledgements are given below. Other figures contain reproductions of experimentally recorded spectra. In these cases the literature spectrum was first photocopied, and the copy then digitally scanned to produce the data in the form of a numerical XY array. The spectrum was regenerated from the XY array using SigmaPlot, and appropriate annotation added as required. Acknowledgements to the original sources of the spectra are listed below.

Figure 1.2. After figure 9.7 of E.M. Purcell, *Electricity and Magnetism*, McGraw-Hill Book Company, Singapore, 1985.

Figure 2.2. After figure 3.1 of R.N. Zare, *Angular Momentum*, John Wiley and Sons, New York, 1988.

Figure 3.1. After figure 4.1 of R.E. Moss, *Advanced Molecular Quantum Mechanics*, Chapman and Hall, London, 1973.

Figure 5.3. After figure 3.1 of R.N. Zare, *Angular Momentum*, John Wiley and Sons, New York, 1988.

Figure 6.3. After figure 2.10 of T.P. Softley, *Atomic Spectra*, Oxford University Press, Oxford, 1994.
Figure 6.4. After figure 3.7 of T.P. Softley, *Atomic Spectra*, Oxford University Press, Oxford, 1994.
Figure 6.5. After figure 1.22 of W.G. Richards and P.R. Scott, *Energy Levels in Atoms and Molecules*, Oxford University Press, Oxford, 1994.
Figure 6.6. After figures 2.2, 2.4, 2.5, 2.9 and 2.10 of W.G. Richards and P.R. Scott, *Energy Levels in Atoms and Molecules*, Oxford University Press, Oxford, 1994.
Figure 6.8. After figure 157 of G. Herzberg, *Spectra of Diatomic Molecules*, D. Van Nostrand Company, Inc., Princeton, 1950.

Figure 6.9. After figure 156 of G. Herzberg, *Spectra of Diatomic Molecules*, D. Van Nostrand Company, Inc., Princeton, 1950.

Figure 6.10. After figure 151 of G. Herzberg, *Spectra of Diatomic Molecules*, D. Van Nostrand Company, Inc., Princeton, 1950.

Figure 6.13. After figure 97 of G. Herzberg, *Spectra of Diatomic Molecules*, D. Van Nostrand Company, Inc., Princeton, 1950.

Figure 6.14. After figure 100 of G. Herzberg, *Spectra of Diatomic Molecules*, D. Van Nostrand Company, Inc., Princeton, 1950.

Figure 6.15. After figure 104 of G. Herzberg, *Spectra of Diatomic Molecules*, D. Van Nostrand Company, Inc., Princeton, 1950.

Figure 6.18. After figure 110 of G. Herzberg, *Spectra of Diatomic Molecules*, D. Van Nostrand Company, Inc., Princeton, 1950.

Figure 8.1. After figure VI.5 of N.F. Ramsey, *Molecular Beams*, Oxford University Press, 1956.

Figure 8.2. After figure 1 of H.G. Kolsky, T.E. Phipps, N.F. Ramsey and H.B. Silsbee, *Phys. Rev.*, **87**, 395 (1952).

Figure 8.3. After figure 1 of N.F. Ramsey, *Phys. Rev.*, **85**, 60 (1952).

Figure 8.4. After figure 3 of H.G. Kolsky, T.E. Phipps, N.F. Ramsey and H.B. Silsbee, *Phys. Rev.*, **87**, 395 (1952).

Figure 8.5. figure 2 of N.F. Ramsey, *Phys. Rev.*, **85**, 60 (1952).

Figure 8.7. After figure 5 of J.M.B. Kellogg, I.I. Rabi, N.F. Ramsey and J.R. Zacharias, *Phys. Rev.*, **56**, 728 (1939).

Figure 8.8. After figure 4 of J.M.B. Kellogg, I.I. Rabi, N.F. Ramsey and J.R. Zacharias, *Phys. Rev.*, **57**, 677 (1940).

Figure 8.11. After figure 2.1 of W.H. Flygare, *Molecular Structure and Dynamics*, Prentice-Hall, Inc., New Jersey, 1978.

Figure 8.12. After figure 6.12 of W.H. Flygare, *Molecular Structure and Dynamics*, Prentice-Hall, Inc., New Jersey, 1978.

Figure 8.13. After figure 1 of S.D. Rosner, R.A. Holt and T.D. Gaily, *Phys. Rev. Lett.*, **35**, 785 (1975).

Figure 8.15. After figure 160 of G. Herzberg, *Spectra of Diatomic Molecules*, D. Van Nostrand Company, Inc., Princeton, 1950.

Figure 8.16. After figure 160 of G. Herzberg, *Spectra of Diatomic Molecules*, D. Van Nostrand Company, Inc., Princeton, 1950.

Figure 8.17. After figure 2 of W. Lichten, *Phys. Rev.*, **120**, 848 (1960).

Figure 8.18. After figure 2 of P.R. Fontana, *Phys. Rev.*, **125**, 220 (1962).

Figure 8.20. After figure 2 of W. Lichten, *Phys. Rev.*, **126**, 1020 (1962).

Figure 8.22. After figure 2 of R.S. Freund, T.A. Miller, D. De Santis and A. Lurio, *J. Chem. Phys.*, **53**, 2290 (1970).

Figure 8.23. After figure 3 of R.S. Freund, T.A. Miller, D. De Santis and A. Lurio, *J. Chem. Phys.*, **53**, 2290 (1970).

Figure 8.24. After figure 3 of D. De Santis, A. Lurio, T.A. Miller and R.S. Freund, *J. Chem. Phys.*, **58**, 4625 (1973).

Figure 8.25. After figure 1 of H.K. Hughes, *Phys. Rev.*, **72**, 614 (1947).
Figure 8.27. After figure 2 of H.K. Hughes, *Phys. Rev.*, **72**, 614 (1947).
Figure 8.28. After figure 3 of H.K. Hughes, *Phys. Rev.*, **72**, 614 (1947).
Figure 8.29. After figure 3 of J.W. Trischka, *Phys. Rev.*, **74**, 718 (1948).
Figure 8.30. After figure 4 of J.W. Trischka, *Phys. Rev.*, **74**, 718 (1948).
Figure 8.31. After figure 1 of T.C. English and J.C. Zorn, *J. Chem. Phys.*, **47**, 3896 (1967).
Figure 8.33. After figure 2 of S.M. Freund, G.A. Fisk, D.R. Herschbach and W. Klemperer, *J. Chem. Phys.*, **54**, 2510 (1971).
Figure 8.34. After figure 1 of R.C. Hilborn, T.F. Gallagher and N.F. Ramsey, *J. Chem. Phys.*, **56**, 855 (1972).
Figure 8.35. After figure 2 of R.C. Hilborn, T.F. Gallagher and N.F. Ramsey, *J. Chem. Phys.*, **56**, 855 (1972).
Figure 8.37. After figure 1 of J.S. Muenter and W. Klemperer, *J. Chem. Phys.*, **52**, 6033 (1970).
Figure 8.39. After figure 2 of F.H. de Leeuw and A. Dymanus, *J. Mol. Spectrosc.*, **48**, 427 (1973).
Figure 8.41. After figure 1 of S.M. Freund, E. Herbst, R.P. Mariella and W. Klemperer, *J. Chem. Phys.*, **56**, 1467 (1972).
Figure 8.47. After figure 1 of K.I. Peterson, G.T. Fraser and W. Klemperer, *Can. J. Phys.*, **62**, 1502 (1984).
Figure 8.48. After figure 3 of R.S. Freund and W. Klemperer, *J. Chem. Phys.*, **43**, 2422 (1965).
Figure 8.49. After figure 1 of R.C. Stern, R.H. Gammon, M.E. Lesk, R.S. Freund and W. Klemperer, *J. Chem. Phys.*, **52**, 3467 (1970).
Figure 8.50. After figure 2 of R.C. Stern, R.H. Gammon, M.E. Lesk, R.S. Freund and W. Klemperer, *J. Chem. Phys.*, **52**, 3467 (1970).
Figure 8.51. After figure 1 of R.H. Gammon, R.C. Stern and W. Klemperer, *J. Chem. Phys.*, **54**, 2151 (1971).

Figure 9.3. After figure 1 of A. Carrington, D.H. Levy and T.A. Miller, *Rev. Sci. Instr.*, **38**, 1183 (1967).
Figure 9.4. After figure 1 of K.M. Evenson, *Disc. Faraday Soc.*, **71**, 7 (1981).
Figure 9.7. After figure 1 of A.H. Curran, R.G. MacDonald, A.J. Stone and B.A. Thrush, *Proc. R. Soc. Lond.*, **A332**, 355 (1973).
Figure 9.9. After figure IV.26 of A. Carrington, *Microwave spectroscopy of Free Radicals*, Academic Press, London, 1974.
Figure 9.10. After figure 3 of R.J. Saykally and K.M. Evenson, *Phys. Rev. Lett.*, **43**, 515 (1979).
Figure 9.11. After figure 1 of D.C. Hovde, E. Schafer, S.E. Strahan, C.A. Ferrari, D. Ray, K.G. Lubic and R.J. Saykally, *Mol. Phys.*, **52**, 245 (1984).
Figure 9.12. After figure 1 of R.L. Brown and H.E. Radford, *Phys. Rev.*, **147**, 6 (1966).
Figure 9.14. After figure 1 of H.E. Radford, *Phys. Rev.*, **122**, 114 (1961).

Figure 9.15. After figure 2 of J.M. Brown, M. Kaise, C.M.L. Kerr and D.J. Milton, *Mol. Phys.*, **36**, 553 (1978).

Figure 9.16. After figure 2 of H.E. Radford, *Phys. Rev.*, **126**, 1035 (1962).

Figure 9.17. After figure 2 of J.M. Brown, C.M.L. Kerr, F.D. Wayne, K.M. Evenson and H.E. Radford, *J. Mol. Spectrosc.*, **86**, 544 (1981).

Figure 9.19. After figure 1 of K.M. Evenson, H.E. Radford and M.M. Moran, *App. Phys. Lett.*, **18**, 426 (1971).

Figure 9.20. After figure 1 of J.M. Brown and K.M. Evenson, *Astrophys. J.*, **269**, L51 (1983).

Figure 9.21. After figure 2 of J.M. Brown and K.M. Evenson, *J. Mol. Spectrosc.*, **98**, 392 (1983).

Figure 9.26. After figure 2 of A. Carrington, D.H. Levy and T.A. Miller, *Proc. R. Soc. Lond.*, **A298**, 340 (1967).

Figure 9.27. After figure 113(b) of G. Herzberg, *Spectra of Diatomic Molecules*, D. Van Nostrand Company, Inc., Princeton, 1950.

Figure 9.28. After figure 2 of F.D. Wayne and H.E. Radford, *Mol. Phys.*, **32**, 1407 (1976).

Figure 9.29. After figure 5 of F.D. Wayne and H.E. Radford, *Mol. Phys.*, **32**, 1407 (1976).

Figure 9.31. After figure 1 of R.J. Saykally, K.M. Evenson, E.R. Comben and J.M. Brown, *Mol. Phys.*, **58**, 735 (1986).

Figure 9.32. After figure 2 of R.J. Saykally, K.M. Evenson, E.R. Comben and J.M. Brown, *Mol. Phys.*, **58**, 735 (1986).

Figure 9.33. After figure 1 of T. Nelis, J.M. Brown and K.M. Evenson, *J. Chem. Phys.*, **92**, 4067 (1990).

Figure 9.34. After figure 3 of T. Nelis, J.M. Brown and K.M. Evenson, *J. Chem. Phys.*, **92**, 4067 (1990).

Figure 9.35. After figure 1 of S.M. Corkery, J.M. Brown, S.P. Beaton and K.M. Evenson, *J. Mol. Spectrosc.*, **149**, 257 (1991).

Figure 9.36. After figure 3 of S.M. Corkery, J.M. Brown, S.P. Beaton and K.M. Evenson, *J. Mol. Spectrosc.*, **149**, 257 (1991).

Figure 9.37. After figure 2 of S.P. Beaton, K.M. Evenson and J.M. Brown, *J. Mol. Spectrosc.*, **164**, 395 (1994).

Figure 9.38. After figure 4 of S.P. Beaton, K.M. Evenson and J.M. Brown, *J. Mol. Spectrosc.*, **164**, 395 (1994).

Figure 9.39. After figure 1 of S.P. Beaton, K.M. Evenson and J.M. Brown, *J. Mol. Spectrosc.*, **164**, 395 (1994).

Figure 9.40. After figure 1 of T. Nelis, S.P. Beaton, K.M. Evenson and J.M. Brown, *J. Mol. Spectrosc.*, **148**, 462 (1991).

Figure 10.8. After figure 1 of D.R. Johnson, F.X. Powell and W.H. Kirchhoff, *J. Mol. Spectrosc.*, **39**, 136 (1971).

Figure 10.9. After figure 1 of G.C. Dousmanis, T.M. Sanders and C.H. Townes, *Phys. Rev.*, **100**, 1735 (1955).

Figure 10.10. After figure 1 of R.C. Woods, *Rev. Sci. Instr.*, **44**, 282 (1973).

Figure 10.11. After figure 1 of T.A. Dixon and R.C. Woods, *Phys. Rev. Lett.*, **34**, 61 (1975).

Figure 10.12. After figure 1 of C. Yamada, M. Fujitake and E. Hirota, *J. Chem. Phys.*, **90**, 3033 (1989).

Figure 10.13. After figure 2 of C. Yamada, M. Fujitake and E. Hirota, *J. Chem. Phys.*, **90**, 3033 (1989).

Figure 10.14. After figure 1 of C.S. Gudeman, M.H. Begemann, J. Pfaff and R.J. Saykally, *Phys. Rev. Lett.*, **50**, 727 (1983).

Figure 10.15. After figure 3 of H.E. Radford, *Rev. Sci. Instr.*, **39**, 1687 (1968).

Figure 10.16. After figure 1 of A.C. Legon, *Ann. Rev. Phys. Chem.*, **34**, 275 (1983).

Figure 10.17. After figure 3.7 of J.M. Hollas, *Modern Spectroscopy*, John Wiley and Sons, Chichester, 1996, 3rd edn.

Figure 10.20. After figure 3.11 of J.M. Hollas, *Modern Spectroscopy*, John Wiley and Sons, Chichester, 1996, 3rd edn.

Figure 10.21. After figure 1 of J. Strong and G.A. Vanasse, *J. Opt. Soc. Amer.*, **50**, 113 (1960).

Figure 10.23. After figure 6.7 of K. Rohlfs and T.L. Wilson, *Tools of Radio Astronomy*, Springer-Verlag, Berlin, 1999, 3rd edn.

Figure 10.24. After figure 7.2 of J.D. Kraus, *Radio Astronomy*, McGraw-Hill Book Company, New York, 1966.

Figure 10.25. After figure 7.6 of J.D. Kraus, *Radio Astronomy*, McGraw-Hill Book Company, New York, 1966.

Figure 10.26. After figure 1 of J.W.V. Storey, D.M. Watson and C.H. Townes, *Int. J. IR and MM waves*, **1**, 15 (1980).

Figure 10.28. After figure 1 of G. Winnewisser, A.F. Krupnov, M. Yu. Tretyakov, M. Liedtke, F. Lewen, A.H. Salek, R. Schieder, A.P. Shkaev and S.V. Volokhov, *J. Mol. Spectrosc.*, **165**, 294 (1994).

Figure 10.29. After figure 1 of F. Lewen, R. Gendriesch, I. Pak, D.G. Paveliev, M. Hepp, R. Schieder and G. Winnewisser, *Rev. Sci. Instr.*, **69**, 32 (1998).

Figure 10.30. After figure 1 of T. Amano, *Astrophys. J.*, **531**, L161 (2000).

Figure 10.31. After figure 3 of T. Amano, *Astrophys. J.*, **531**, L161 (2000).

Figure 10.32. After figure 1 of K.M. Evenson, D.A. Jennings and M.D. Vanek, *Frontiers of Laser Spectroscopy*, Kluwer Academic Publishers, 1987.

Figure 10.34. After figure 2 of E.V. Loewenstein, *J. Opt. Soc. Amer.*, **50**, 1163 (1960).

Figure 10.35. After figure 1 of K.M. Evenson, D.A. Jennings and F.R. Petersen, *App. Phys. Lett.*, **44**, 576 (1984).

Figure 10.36. After figure 1 of R.D. Suenram, F.J. Lovas, G.T. Fraser and K. Matsumura, *J. Chem. Phys.*, **92**, 4724 (1990).

Figure 10.37. After figures 1 and 2 of R.J. Low, T.D. Varberg, J.P. Connelly, A.R. Auty, B.J. Howard and J.M. Brown, *J. Mol. Spectrosc.*, **161**, 499 (1993).

Figure 10.38. After figure 1 of K.A. Walker and M.C.L. Gerry, *J. Mol. Spectrosc.*, **182**, 178 (1997).

Figure 10.40. After figure 1 of A.A. Penzias, R.W. Wilson and K.B. Jefferts, *Phys. Rev. Lett.*, **32**, 701 (1974).

Figure 10.41. After figure 2 of A.A. Penzias, R.W. Wilson and K.B. Jefferts, *Phys. Rev. Lett.*, **32**, 701 (1974).

Figure 10.45. After figure 2 of Th. Klaus, A.H. Saleck, S.P. Belov, G. Winnewisser, Y. Hirahara, M. Hayashi, E. Kagi and K. Kawaguchi, *J. Mol. Spectrosc.*, **180**, 197 (1996).

Figure 10.46. After figure 2 of K. Namiki and S. Saito, *Chem. Phys. Lett.*, **252**, 343 (1996).

Figure 10.48. After figure 1 of S. Saito, Y. Endo and E. Hirota, *J. Chem. Phys.*, **82**, 2947 (1985).

Figure 10.49. After figure 2 of C. Yamada, Y. Endo and E. Hirota, *J. Chem. Phys.*, **79**, 4159 (1983).

Figure 10.50. After figure 1 of C. Yamada, Y. Endo and E. Hirota, *J. Chem. Phys.*, **79**, 4159 (1983).

Figure 10.52. After figure 1 of T. Sakamaki, T. Okabayashi and M. Tanimoto, *J. Chem. Phys.*, **109**, 7169 (1998), and figure 1 of T. Sakamaki, T. Okabayashi and M. Tanimoto, *J. Chem. Phys.*, **111**, 6345 (1999).

Figure 10.54. After figure 1 of E. Klisch, S.P. Belov, R. Schieder, G. Winnewisser and E. Herbst, *Mol. Phys.*, **97**, 65 (1999).

Figure 10.58. After figure 2 of J.M. Brown, L.R. Zink, D.A. Jennings, K.M. Evenson, A. Hinz and I.G. Nolt, *Astrophys. J.*, **307**, 410 (1986).

Figure 10.59. After figure 1 of J.M. Brown, L.R. Zink, D.A. Jennings, K.M. Evenson, A. Hinz and I.G. Nolt, *Astrophys. J.*, **307**, 410 (1986).

Figure 10.60. After figure 1 of D.M. Watson, R. Genzel, C.H. Townes and J.W.V. Storey, *Astrophys. J.*, **298**, 316 (1985), and figure 2 of J.W.V. Storey, D.M. Watson and C.H. Townes, *Astrophys. J.*, **244**, L27 (1981).

Figure 10.63. After figure 1 of C.R. Brazier and J.M. Brown, *J. Chem. Phys.*, **78**, 1608 (1983).

Figure 10.64. After figure 1(b) of T.C. Steimle, D.R. Woodward and J.M. Brown, *Astrophys. J.*, **294**, L59 (1985).

Figure 10.65. After figure 1(c) of T.C. Steimle, D.R. Woodward and J.M. Brown, *Astrophys. J.*, **294**, L59 (1985).

Figure 10.66. After figures 3 and 4 of T. Amano, *Astrophys. J.*, **531**, L161 (2000).

Figure 10.67. After figure 2 of T. Amano, *Astrophys. J.*, **531**, L161 (2000).

Figure 10.70. After figure 1 of T. Amano, S. Saito, E. Hirota, Y. Morino, D.R. Johnson and F.X. Powell, *J. Mol. Spectrosc.*, **30**, 275 (1969).

Figure 10.71. After figure 1 of A. Carrington, C.A. Leach, A.J. Marr, A.M. Shaw, M.R. Viant, J.M. Hutson and M.M. Law, *J. Chem. Phys.*, **102**, 2379 (1995).

Figure 10.72. After figure 3 of A. Carrington, C.A. Leach, A.J. Marr, A.M. Shaw, M.R. Viant, J.M. Hutson and M.M. Law, *J. Chem. Phys.*, **102**, 2379 (1995).

Figure 10.73. After figure 4 of A. Carrington, C.A. Leach, A.J. Marr, A.M. Shaw, M.R. Viant, J.M. Hutson and M.M. Law, *J. Chem. Phys.*, **102**, 2379 (1995).

Figure 10.74. After figure 7 of A. Carrington, C.A. Leach, A.J. Marr, A.M. Shaw, M.R. Viant, J.M. Hutson and M.M. Law, *J. Chem. Phys.*, **102**, 2379 (1995).

Figure 10.76. After figure 8 of A. Carrington, C.A. Leach, A.J. Marr, A.M. Shaw, M.R. Viant, J.M. Hutson and M.M. Law, *J. Chem. Phys.*, **102**, 2379 (1995).

Figure 10.77. After figure 5(a) of A. Carrington, C.H. Pyne, A.M. Shaw, S.M. Taylor, J.M. Hutson and M.M. Law, *J. Chem. Phys.*, **105**, 8602 (1996).

Figure 10.79. After figure 1 of J. Cernicharo, C.A. Gottlieb, M. Guelin, P. Thaddeus and J.M. Vrtilek, *Astrophys. J.*, **341**, L25 (1989).

Figure 10.80. After figure 3 of R. Mollaaghababa, C.A. Gottlieb and P. Thaddeus, *J. Chem. Phys.*, **98**, 968 (1993).

Figure 10.81. After figure 2 of J. Cernicharo, C.A. Gottlieb, M. Guelin, P. Thaddeus and J.M. Vrtilek, *Astrophys. J.*, **341**, L25 (1989).

Figure 10.83. After figure 1 of M.D. Allen, T.C. Pesch and L.M. Ziurys, *Astrophys. J.*, **472**, L57 (1996).

Figure 10.84. After figure 3 of R.D. Suenram, G.T. Fraser, F.J. Lovas and C.W. Gillies, *J. Mol. Spectrosc.*, **148**, 114 (1991).

Figure 10.85. After figure 7 of J.M. Brom, C.H. Durham and W. Weltner, *J. Chem. Phys.*, **61**, 970 (1974).

Figure 10.87. After figure 2 of M.D. Allen, B.Z. Li and L.M. Ziurys, *Chem. Phys. Lett.*, **270**, 517 (1997).

Figure 10.88. After figure 1 of M.D. Allen and L.M. Ziurys, *Astrophys. J.*, **479**, 1237 (1996).

Figure 10.89. After figure 6 of M.D. Allen and L.M. Ziurys, *J. Chem. Phys.*, **106**, 3494 (1997).

Figure 10.91. After figure 2 of K. Namiki and S. Saito, *J. Chem. Phys.*, **107**, 8848 (1997).

Figure 10.93. After figures 1 and 2 of M.D. Allen, L.M. Ziurys and J.M. Brown, *Chem. Phys. Lett.*, **257**, 130 (1996).

Figure 10.97. After figure 2 of A.D.J. Critchely, A.N. Hughes and I.R. McNab, *Phys. Rev. Lett.*, **86**, 1725 (2001).

Figure 11.2. After figure 1 of K.M. Evenson, J.L. Dunn and H.P. Broida, *Phys. Rev.*, **136**, A1566 (1964).

Figure 11.3. After figure 2 of K.M. Evenson, J.L. Dunn and H.P. Broida, *Phys. Rev.*, **136**, A1566 (1964).

Figure 11.4. After figure 3 of K.M. Evenson, J.L. Dunn and H.P. Broida, *Phys. Rev.*, **136**, A1566 (1964).

Figure 11.5. After figure 1 of S.J. Silvers, T.H. Bergeman and W. Klemperer, *J. Chem. Phys.*, **52**, 4385 (1970).

Figure 11.6. After figure 4 of S.J. Silvers, T.H. Bergeman and W. Klemperer, *J. Chem. Phys.*, **52**, 4385 (1970).

Figure 11.7. After figures 9 and 10 of S.J. Silvers, T.H. Bergeman and W. Klemperer, *J. Chem. Phys.*, **52**, 4385 (1970).

Figure 11.9. After figure 1 of K.R. German and R.N. Zare, *Phys. Rev. Lett.*, **23**, 1207 (1969).

Figure 11.10. After figure 1 of R.W. Field, A.D. English, T. Tanaka, D.O. Harris and D.A. Jennings, *J. Chem. Phys.*, **59**, 2191 (1973).

Figure 11.12. After figure 1 of R.S. Freund and T.A. Miller, *J. Chem. Phys.*, **56**, 2211 (1972).

Figure 11.13. After figure 2 of R.S. Freund and T.A. Miller, *J. Chem. Phys.*, **56**, 2211 (1972).

Figure 11.15. After figure 2 of R.S. Freund and T.A. Miller, *J. Chem. Phys.*, **58**, 3565 (1973).

Figure 11.16. After figure 1 of R.S. Freund and T.A. Miller, *J. Chem. Phys.*, **58**, 3565 (1973).

Figure 11.17. After figure 1 of T.A. Miller and R.S. Freund, *J. Chem. Phys.*, **58**, 2345 (1973).

Figure 11.19. After figure 3 of R.S. Freund and T.A. Miller, *J. Chem. Phys.*, **59**, 5770 (1973).

Figure 11.20. After figure 1 of T.A. Miller, R.S. Freund and B.R. Zegarski, *J. Chem. Phys.*, **60**, 3195 (1974).

Figure 11.21. After figure 1 of P.J. Domaille, T.C. Steimle and D.O. Harris, *J. Mol. Spectrosc.*, **68**, 146 (1977).

Figure 11.22. After figure 1 of W.E. Ernst and T. Törring, *Phys. Rev.*, **A25**, 1236 (1982).

Figure 11.23. After figure 1 of W.E. Ernst, *Appl. Phys.*, **B30**, 105 (1983).

Figure 11.24. After figure 1 of W.E. Ernst and S. Kindt, *Appl. Phys.*, **B31**, 79 (1983).

Figure 11.25. After figure 1 of T. Kröckertskothen, H. Knöckel and E. Tiemann, *Mol. Phys.*, **62**, 1031 (1987).

Figure 11.27. After figure 2 of T. Kröckertskothen, H. Knöckel and E. Tiemann, *Mol. Phys.*, **62**, 1031 (1987).

Figure 11.28. After figure 3 of T.C. Steimle, C.R. Brazier and J.M. Brown, *J. Mol. Spectrosc.*, **110**, 39 (1985).

Figure 11.29. After figure 4 of T.C. Steimle, C.R. Brazier and J.M. Brown, *J. Mol. Spectrosc.*, **110**, 39 (1985).

Figure 11.30. After figure 3 of M.C.L. Gerry, A.J. Merer, U. Sassenberg and T.C. Steimle, *J. Chem. Phys.*, **86**, 4754 (1987).

Figure 11.31. After figure 3 of W.J. Childs and T.C. Steimle, *J. Chem. Phys.*, **88**, 6168 (1988).

Figure 11.32. After figure 4 of W.J. Childs and T.C. Steimle, *J. Chem. Phys.*, **88**, 6168 (1988).

Figure 11.34. After figure 3 of K. Namiki and T.C. Steimle, *J. Chem. Phys.*, **111**, 6385 (1999).

Figure 11.35. After figures 4 and 5 of K. Namiki and T.C. Steimle, *J. Chem. Phys.*, **111**, 6385 (1999).

Figure 11.36. After figure 2 of T.C. Steimle, D.F. Nachman, J.E. Shirley, D.A. Fletcher and J.M. Brown, *Mol. Phys.*, **69**, 923 (1990).

Figure 11.37. After figure 4 of T.C. Steimle, D.F. Nachman, J.E. Shirley, D.A. Fletcher and J.M. Brown, *Mol. Phys.*, **69**, 923 (1990).

Figure 11.38. After figure 1 of D.A. Fletcher, K.Y. Jung, C.T. Scurlock and T.C. Steimle, *J. Chem. Phys.*, **98**, 1837 (1993).

Figure 11.39. After figure 3 of D.A. Fletcher, K.Y. Jung, C.T. Scurlock and T.C. Steimle, *J. Chem. Phys.*, **98**, 1837 (1993).

Figure 11.41. After figure 2 of Y. Azuma and W.J. Childs, *J. Chem. Phys.*, **93**, 8415 (1990).

Figure 11.43. After figure 1 of W.J. Childs, G.L. Goodman, L.S. Goodman and L. Young, *J. Mol. Spectrosc.*, **119**, 166 (1986).

Figure 11.44. After figure 3 of W.J. Childs, G.L. Goodman, L.S. Goodman and L. Young, *J. Mol. Spectrosc.*, **119**, 166 (1986).

Figure 11.45. After figure 3 of A. Carrington, I.R. McNab and C.A. Montgomerie, *Mol. Phys.*, **66**, 519 (1989).

Figure 11.46. After figure 2 of A. Carrington, I.R. McNab and C.A. Montgomerie, *Mol. Phys.*, **66**, 519 (1989).

Figure 11.47. After figure 1 of A. Carrington, I.R. McNab and C.A. Montgomerie, *Mol. Phys.*, **66**, 519 (1989).

Figure 11.48. After figure 1 of A. Carrington, I.R. McNab, C.A. Montgomerie and J.M. Brown, *Mol. Phys.*, **66**, 1279 (1989).

Figure 11.49. After figure 2 of A. Carrington, I.R. McNab, C.A. Montgomerie and J.M. Brown, *Mol. Phys.*, **66**, 1279 (1989).

Figure 11.50. After figure 3 of A. Carrington, I.R. McNab, C.A. Montgomerie and J.M. Brown, *Mol. Phys.*, **66**, 1279 (1989).

Figure 11.51. After figure 3 of N. Berrah Mansour, C. Kurtz, T.C. Steimle, G.L. Goodman, L. Young, T.J. Scholl, S.D. Rosner and R.A. Holt, *Phys. Rev.*, **A44**, 4418 (1991).

Figure 11.52. After figure 2 of N. Berrah Mansour, C. Kurtz, T.C. Steimle, G.L. Goodman, L. Young, T.J. Scholl, S.D. Rosner and R.A. Holt, *Phys. Rev.*, **A44**, 4418 (1991).

Figure 11.53. After figure 4 of N. Berrah Mansour, C. Kurtz, T.C. Steimle, G.L. Goodman, L. Young, T.J. Scholl, S.D. Rosner and R.A. Holt, *Phys. Rev.*, **A44**, 4418 (1991).

Figure 11.54. After figure 4 of M.A. Johnson, M.L. Alexander, I. Hertel and W.C. Lineberger, *Chem. Phys. Lett.*, **105**, 374 (1984).

Figure 11.56. After figure 2 of K.B. Jefferts, *Phys. Rev. Lett.*, **20**, 39 (1967).

Figure 11.57. After figure 3 of K.B. Jefferts, *Phys. Rev. Lett.*, **20**, 39 (1967).

1 General introduction

1.1. Electromagnetic spectrum

Molecular spectroscopy involves the study of the absorption or emission of electromagnetic radiation by matter; the radiation may be detected directly, or indirectly through its effects on other molecular properties. The primary purpose of spectroscopic studies is to understand the nature of the nuclear and electronic motions within a molecule.

The different branches of spectroscopy may be classified either in terms of the wavelength, or frequency, of the electromagnetic radiation, or in terms of the type of intramolecular dynamic motion primarily involved. Historically the first method has been the most common, with different regions of the electromagnetic spectrum classified as shown in figure 1.1. In the figure we show four different ways of describing these regions. They may be classified according to the wavelength, in ångström units ($1\text{Å} = 10^{-8}$ cm), or the frequency in Hz; wavelength (λ) and frequency (ν) are related by the equation,

$$\nu = c/\lambda, \tag{1.1}$$

where c is the speed of light. Very often the wavenumber unit, cm^{-1}, is used; we denote this by the symbol $\tilde{\nu}$. Clearly the wavelength and wavenumber are related in the simple way

$$\tilde{\nu} = 1/\lambda, \tag{1.2}$$

with λ expressed in cm. Although offensive to the purist, the wavenumber is often taken as a unit of energy, according to the Planck relationship

$$E = h\nu = hc\tilde{\nu}, \tag{1.3}$$

where h is Planck's constant. From the values of the fundamental constants given in General Appendix A, we find that 1 cm^{-1} corresponds to $1.986\,445 \times 10^{-23}$ J molecule^{-1}. A further unit of energy which is often used, and which will appear in this book, is the electronvolt, eV; this is the kinetic energy of an electron which has been accelerated through a potential difference of 1 V; 1 eV is equal to 8065.545 cm^{-1}.

In the classical theory of electrodynamics, electromagnetic radiation is emitted when an electron moves in its orbit but, according to the Bohr theory of the atom,

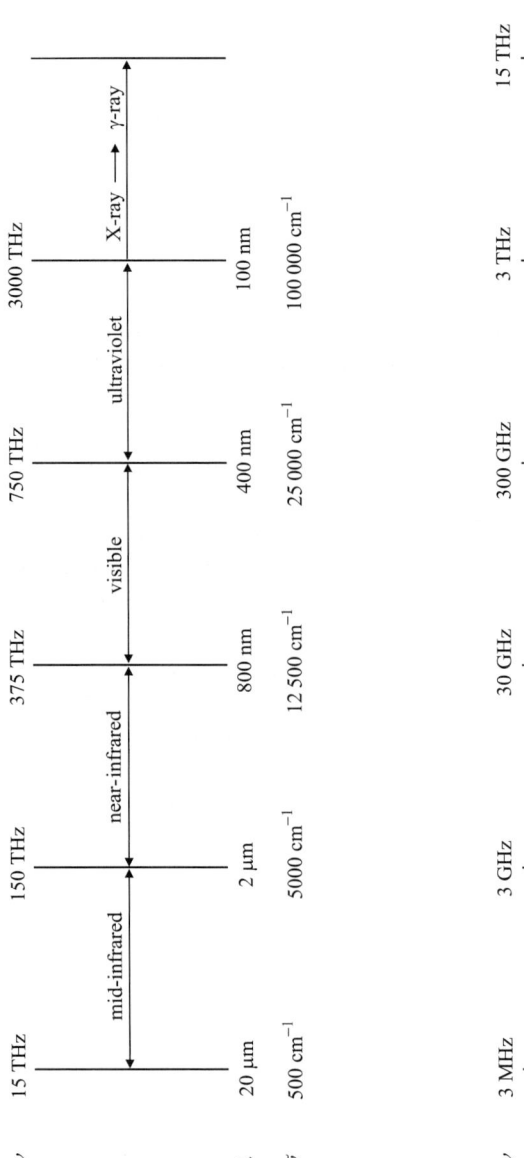
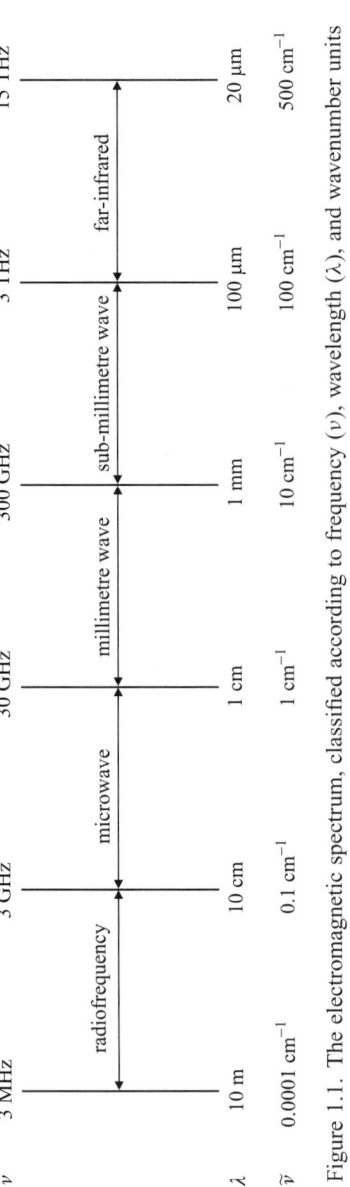

Figure 1.1. The electromagnetic spectrum, classified according to frequency (ν), wavelength (λ), and wavenumber units ($\tilde{\nu}$). There is no established convention for the division of the spectrum into different regions; we show our convention.

emission of radiation occurs only when an electron goes from a higher energy orbit E_2 to an orbit of lower energy E_1. The emitted energy is a photon of energy $h\nu$, given by

$$h\nu = E_2 - E_1, \tag{1.4}$$

an equation known as the Bohr frequency condition. The reverse process, a transition from E_1 to E_2, requires the absorption of a quantum of energy $h\nu$. The range of frequencies (or energies) which constitutes the electromagnetic spectrum is shown in figure 1.1. Molecular spectroscopy covers a nominal energy range from 0.0001 cm^{-1} to 100 000 cm^{-1}, that is, nine decades in energy, frequency or wavelength. The spectroscopy described in this book, which we term *rotational* spectroscopy for reasons to be given later, is concerned with the range 0.0001 cm^{-1} to 100 cm^{-1}. Surprisingly, therefore, it covers six of the nine decades shown in figure 1.1, very much the major portion of the molecular spectrum! Indeed our low frequency cut-off at 3 MHz is somewhat arbitrary, since molecular beam magnetic resonance studies at even lower frequencies have been described. As we shall see, the experimental techniques employed over the full range given in figure 1.1 vary a great deal. We also note here that the spectroscopy discussed in this book is concerned solely with molecules in the gas phase. Again the reasons for this discrimination will become apparent later in this chapter.

So far as the classification of the type of spectroscopy performed is concerned, the characterisation of the dynamical motions of the nuclei and electrons within a molecule is more important than the region of the electromagnetic spectrum in which the corresponding transitions occur. However, before we come to this in more detail, a brief discussion of the nature of electromagnetic radiation is necessary. This is actually a huge subject which, if tackled properly, takes us deeply into the details of classical and semiclassical electromagnetism, and even further into quantum electrodynamics. The basic foundations of the subject are Maxwell's equations, which we describe in appendix 1.1. We will make use of the results of these equations in the next section, referring the reader to the appendix if more detail is required.

1.2. Electromagnetic radiation

Electromagnetic radiation consists of both an electric and a magnetic component, which for plane-polarised (or linearly-polarised) radiation, travelling along the Y axis, may be represented as shown in figure 1.2. Each of the three diagrams represents the electric and magnetic fields at different instants of time as indicated. The electric field ($\boldsymbol{E}$) is in the YZ plane parallel to the Z axis, and the magnetic field ($\boldsymbol{B}$) is everywhere perpendicular to the electric field, and therefore in the XY plane. Consideration of Maxwell's equations [1] shows that, as time progresses, the entire field pattern shifts to the right along the Y axis, with a velocity c. The wavelength of the radiation, λ, shown in the figure, is related to the frequency ν by the simple expression $\nu = c/\lambda$. At every point in the wave at any instant of time, the electric and magnetic field strengths

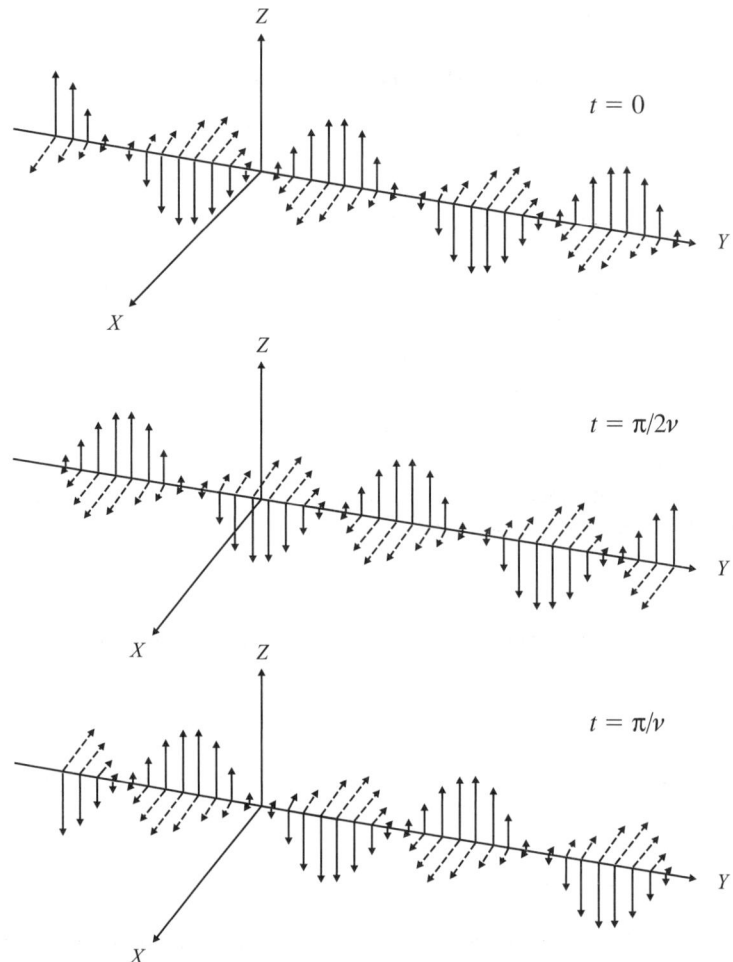

Figure 1.2. Schematic representation of plane-polarised radiation projected along the Y axis at three different instants of time. The solid arrows denote the amplitude of the electric field (E), and the dashed arrows denote the perpendicular magnetic field (B).

are equal; this means that, in cgs units, if the electric field strength is $10\,\mathrm{V\,cm^{-1}}$ the magnetic field strength is 10 G.

Although it is simplest to describe and represent graphically the example of plane polarised radiation, it is also instructive to consider the more general case [2]. For propagation of the radiation along the Y axis, the electric field E can be decomposed into components along the Z and X axes. The electric field vector in the XZ plane is then given by

$$E = i'E_X + k'E_Z \qquad (1.5)$$

where i' and k' are unit vectors along the X and Z axes. The components in equation (1.5) are given by

$$E_X = E_X^0 \cos(k^*Y - \omega t + \alpha_X),$$
$$E_Z = E_Z^0 \cos(k^*Y - \omega t + \alpha_Z), \qquad (1.6)$$
$$\alpha = \alpha_X - \alpha_Z.$$

Here $\omega = 2\pi\nu$, ω is the angular frequency in units of rad s^{-1}, ν is the frequency in Hz, and $\boldsymbol{k}^*$ is called the propagation vector with units of inverse length. In a vacuum $\boldsymbol{k}^*$ has a magnitude equal to $2\pi/\lambda_0$ where λ_0 is the vacuum wavelength of the radiation. Finally, α is the difference in phase between the X and Z components of $\boldsymbol{E}$.

Plane-polarised radiation is obtained when the phase factor α is equal to 0 or π and $E_X^0 = E_Z^0$. When $\alpha = 0$, E_X and E_Z are in phase, whilst for $\alpha = \pi$ they are out-of-phase by π. The special case illustrated in figure 1.2 corresponds to $E_X^0 = 0$. Other forms of polarisation can be obtained from equations (1.6). For elliptically-polarised radiation we set $\alpha = \pm\pi/2$ so that equations (1.6) become

$$E_X = E_X^0 \cos(k^*Y - \omega t),$$
$$E_Z = E_Z^0 \cos(k^*Y - \omega t \pm \pi/2) = \pm E_Z^0 \sin(k^*Y - \omega t),$$
$$\boldsymbol{E}_\pm = \boldsymbol{i}' E_X \pm \boldsymbol{k}' E_Z$$
$$= \boldsymbol{i}' E_X^0 \cos(k^*Y - \omega t) \pm \boldsymbol{k}' E_Z^0 \sin(k^*Y - \omega t). \qquad (1.7)$$

If $E_X^0 = E_Z^0 = \mathcal{E}$ for $\alpha = \pm\pi/2$, we have circularly-polarised radiation given by the expression

$$\boldsymbol{E}_\pm = \mathcal{E}[\boldsymbol{i}' \cos(k^*Y - \omega t) \pm \boldsymbol{k}' \sin(k^*Y - \omega t)]. \qquad (1.8)$$

When viewed looking back along the Y axis towards the radiation source, the field rotates clockwise or counter clockwise about the Y axis. When $\alpha = +\pi/2$ which corresponds to E_+, the field appears to rotate counter clockwise about Y.

Conventional sources of electromagnetic radiation are incoherent, which means that the waves associated with any two photons of the same wavelength are, in general, out-of-phase and have a random phase relation with each other. Laser radiation, however, has both spatial and temporal coherence, which gives it special importance for many applications.

1.3. Intramolecular nuclear and electronic dynamics

In order to understand molecular energy levels, it is helpful to partition the kinetic energies of the nuclei and electrons in a molecule into parts which, if possible, separately represent the electronic, vibrational and rotational motions of the molecule. The details of the processes by which this partitioning is achieved are presented in chapter 2. Here we give a summary of the main procedures and results.

We start by writing a general expression which represents the kinetic energies of the nuclei (α) and electrons (i) in a molecule:

$$T = \sum_{\alpha} \frac{1}{2M_{\alpha}} P_{\alpha}^2 + \sum_{i} \frac{1}{2m} P_i^2, \qquad (1.9)$$

where M_{α} and m are the masses of the nuclei and electrons respectively. The momenta P_{α} and P_i are vector quantities, which are defined by

$$P_i = -i\hbar \frac{\partial}{\partial R_i},$$

$$P_{\alpha} = -i\hbar \frac{\partial}{\partial R_{\alpha}}, \qquad (1.10)$$

expressed in a space-fixed axis system (X, Y, Z) of arbitrary origin. R_{α} gives the position of nucleus α within this coordinate system. The partial derivative ($\partial/\partial R_{\alpha}$) is a shorthand notation for the three components of the gradient operator,

$$\frac{\partial}{\partial R_{\alpha}} \equiv \left(\frac{\partial}{\partial R_X}\right)_{\alpha} i' + \left(\frac{\partial}{\partial R_Y}\right)_{\alpha} j' + \left(\frac{\partial}{\partial R_Z}\right)_{\alpha} k', \qquad (1.11)$$

where i', j', k' are unit vectors along the space-fixed axes X, Y, Z.

It is by no means obvious that (1.9) contains the vibrational and rotational motion of the nuclei, as well as the electron kinetic energies, but a series of origin and axis transformations shows that this is the case. First, we transform from the arbitrary origin to an origin at the centre of mass of the molecule, and then to the centre of mass of the nuclei. As we show in chapter 2, these transformations convert (1.9) into the expression

$$T = \frac{1}{2M} P_O^2 + \frac{1}{2\mu} P_R^2 + \frac{1}{2m} \sum_i P_i''^2 + \frac{1}{2(M_1 + M_2)} \sum_{i,j} P_i'' \cdot P_j''. \qquad (1.12)$$

The first term in (1.12) represents the kinetic energy due to translation of the whole molecule through space; this motion can be separated off rigorously in the absence of external fields. In the second term, μ is the reduced nuclear mass, $M_1 M_2/(M_1 + M_2)$, and this term represents the kinetic energy of the nuclei. The third term describes the kinetic energy of the electrons and the last term is a correction term, known as the mass polarisation term. The transformation is described in detail in chapter 2 and appendix 2.1. An alternative expression equivalent to (1.12) is obtained by writing the momentum operators in terms of the Laplace operators,

$$T = -\frac{\hbar^2}{2M} \nabla^2 - \frac{\hbar^2}{2\mu} \nabla_R^2 - \frac{\hbar^2}{2m} \sum_i \nabla_i''^2 - \frac{\hbar^2}{2(M_1 + M_2)} \sum_{i,j} \nabla_i'' \cdot \nabla_j''. \qquad (1.13)$$

The next step is to add terms representing the potential energy, the electron spin interactions and the nuclear spin interactions. The total Hamiltonian $\mathcal{H}_T$ can then be subdivided into electronic and nuclear Hamiltonians,

$$\mathcal{H}_T = \mathcal{H}_{el} + \mathcal{H}_{nucl}, \qquad (1.14)$$

where

$$\mathcal{H}_{el} = -\frac{\hbar^2}{2m}\sum_i \nabla_i^2 - \frac{\hbar^2}{2M_N}\sum_{i,j}\nabla_i \cdot \nabla_j + \sum_{i<j}\frac{e^2}{4\pi\varepsilon_0 R_{ij}} - \sum_{\alpha,i}\frac{Z_\alpha e^2}{4\pi\varepsilon_0 R_{i\alpha}}$$
$$+ \mathcal{H}(\boldsymbol{S}_i) + \mathcal{H}(\boldsymbol{I}_\alpha), \tag{1.15}$$

$$\mathcal{H}_{nucl} = -\frac{\hbar^2}{2\mu}\nabla_R^2 + \sum_{\alpha,\beta}\frac{Z_\alpha Z_\beta e^2}{4\pi\varepsilon_0 R}. \tag{1.16}$$

The third and fourth terms in (1.15) represent the potential energy contributions (in SI units, see General Appendix E) arising from the electron–electron and electron–nuclear interactions, whilst the second term in (1.16) describes the nuclear repulsion term between nuclei with charges $Z_\alpha e$ and $Z_\beta e$. The electron and nuclear spin Hamiltonians introduced into (1.15) are described in detail later.

The total nuclear kinetic energy is contained within the first term in equation (1.16) and we now introduce a further transformation from the axes translating with the molecule but with fixed orientation to molecule-fixed axes gyrating with the nuclei. In chapter 2 the two axis systems are related by Euler angles, ϕ, θ and χ, although for diatomic molecules the angle χ is redundant. We may use a simpler transformation to spherical polar coordinates R, θ, ϕ as defined in figure 1.3. With this transformation the space-fixed coordinates are given by

$$\begin{aligned} X &= R\sin\theta\cos\phi, \\ Y &= R\sin\theta\sin\phi, \\ Z &= R\cos\theta. \end{aligned} \tag{1.17}$$

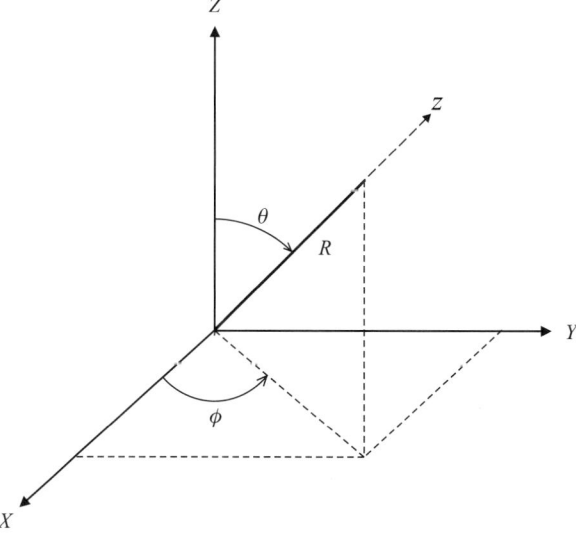

Figure 1.3. Transformation from space-fixed axes X, Y, Z to molecule-fixed axes using the spherical polar coordinates R, θ, ϕ, defined in the figure.

We proceed to show, in chapter 2, that this transformation of the axes leads to the nuclear kinetic energy term being converted into a new expression:

$$\frac{1}{2\mu} P_R^2 = -\frac{\hbar^2}{2\mu} \nabla_R^2$$
$$= -\frac{\hbar^2}{2\mu} \left\{ \frac{1}{R^2} \frac{\partial}{\partial R} \left(R^2 \frac{\partial}{\partial R} \right) + \frac{1}{R^2 \sin\theta} \frac{\partial}{\partial \theta} \left(\sin\theta \frac{\partial}{\partial \theta} \right) + \frac{1}{R^2 \sin^2\theta} \frac{\partial^2}{\partial \phi^2} \right\}. \tag{1.18}$$

This is a very important result because the first term describes the vibrational kinetic energy of the nuclei, whilst the second and third terms represent the rotational kinetic energy. The transformation is straightforward provided one takes proper note of the non-commutation of the operator products which arise.

The transformation of terms representing the kinetic energies of all the particles into terms representing, separately, the electronic, vibrational and rotational kinetic energies is clearly very important. The nuclear kinetic energy Hamiltonian, (1.18), is relatively simple when the spherical polar coordinate transformation (1.17) is used. When the Euler angle transformation is used, it is a little more complicated, containing terms which include the third angle χ:

$$\mathcal{H}_{\text{nucl}} = -\frac{\hbar^2}{2\mu R^2} \left\{ \frac{\partial}{\partial R} \left(R^2 \frac{\partial}{\partial R} \right) + \operatorname{cosec}\theta \frac{\partial}{\partial \theta} \left(\sin\theta \frac{\partial}{\partial \theta} \right) \right.$$
$$\left. + \operatorname{cosec}^2\theta \left[\frac{\partial^2}{\partial \phi^2} + \frac{\partial^2}{\partial \chi^2} - 2\cos\theta \frac{\partial^2}{\partial \phi \partial \chi} \right] \right\} + V_{\text{nucl}}(R). \tag{1.19}$$

We show in chapter 2 that when the transformation of the electronic coordinates, including electron spin, into the rotating molecule-fixed axes system is taken into account, equation (1.19) takes the much simpler form

$$\mathcal{H}_{\text{nucl}} = -\frac{\hbar^2}{2\mu R^2} \frac{\partial}{\partial R} \left(R^2 \frac{\partial}{\partial R} \right) + \frac{\hbar^2}{2\mu R^2} (\boldsymbol{J} - \boldsymbol{P})^2 + V_{\text{nucl}}(R), \tag{1.20}$$

where $\boldsymbol{J}$ is the total angular momentum and $\boldsymbol{P}$ is the total electronic angular momentum, equal to $\boldsymbol{L} + \boldsymbol{S}$. Hence although the electronic Hamiltonian is free of terms involving the motion of the nuclei, the nuclear Hamiltonian (1.20) contains terms involving the operators P_x, P_y and P_z which operate on the electronic part of the total wave function. The Schrödinger equation for the total wave function is written as

$$(\mathcal{H}_{\text{el}} + \mathcal{H}_{\text{nucl}}) \Psi_{\text{rve}} = E_{\text{rve}} \Psi_{\text{rve}}, \tag{1.21}$$

and, as we show in chapter 2, the Born approximation allows us to assume total wave functions of the form

$$\Psi_{\text{rve}}^0 = \psi_{\text{e}}^{\text{n}}(r_i) \phi_{\text{rv}}^{\text{n}}(R, \phi, \theta). \tag{1.22}$$

The matrix elements of the nuclear Hamiltonian that mix different electronic states are then neglected; the electronic wave function is taken to be dependent upon nuclear coordinates, but not nuclear momenta. If the first-order contributions of the nuclear kinetic energy are taken into account, we have the Born adiabatic approximation; if they are neglected, we have the Born–Oppenheimer approximation. This approximation occupies a central position in molecular quantum mechanics; in most situations it is a good approximation, and allows us to proceed with concepts like the potential energy curve or surface, molecular shapes and geometry, etc. Those special cases, usually involving electronic orbital degeneracy, where the Born–Oppenheimer approximation breaks down, can often be treated by perturbation methods.

In chapter 2 we show how a separation of the vibrational and rotational wave functions can be achieved by using the product functions

$$\phi_{rv}^n = \chi^n(R)e^{iM_J\phi}\Theta^n(\theta)e^{ik\chi}, \tag{1.23}$$

where M_J and k are constants taking integral or half-odd values. We show that in the Born approximation, the wave equation for the nuclear wave functions can be expressed in terms of two equations describing the vibrational and rotational motion separately. Ultimately we obtain the wave equation of the vibrating rotator,

$$\frac{\hbar^2}{2\mu R^2}\frac{\partial}{\partial R}R^2\frac{\partial \chi^n(R)}{\partial R} + \left\{E_{rve} - V - \frac{\hbar^2}{2\mu R^2}J(J+1)\right\}\chi^n(R) = 0. \tag{1.24}$$

The main problem with this equation is the description of the potential energy term (V). As we shall see, insertion of a restricted form of the potential allows one to express data on the ro-vibrational levels in terms of semi-empirical constants. If the Morse potential is used, the ro-vibrational energies are given by the expression

$$E_{v,J} = \omega_e(v+1/2) - \omega_e x_e(v+1/2)^2 + B_e J(J+1) - D_e J^2(J+1)^2$$
$$- \alpha_e(v+1/2)J(J+1). \tag{1.25}$$

The first two terms describe the vibrational energy, the next two the rotational energy, and the final term describes the vibration–rotation interaction.

1.4. Rotational levels

This book is concerned primarily with the rotational levels of diatomic molecules. The spectroscopic transitions described arise either from transitions between different rotational levels, usually adjacent rotational levels, or from transitions between the fine or hyperfine components of a single rotational level. The electronic and vibrational quantum numbers play a different role. In the majority of cases the rotational levels studied belong to the lowest vibrational level of the ground electronic state. The detailed nature of the rotational levels, and the transitions between them, depends critically upon the type of electronic state involved. Consequently we will be deeply concerned with the many different types of electronic state which arise for diatomic molecules, and the molecular interactions which determine the nature and structure of the rotational levels. We will not, in general, be concerned with transitions between different electronic states, except for the double resonance studies described in the final chapter. The vibrational states of diatomic molecules are, in a sense, relatively uninteresting.

The detailed rotational structure and sub-structure does not usually depend upon the vibrational quantum number, except for the magnitudes of the molecular parameters. Furthermore, we will not be concerned with transitions between different vibrational levels.

Rotational level spacings, and hence the frequencies of transitions between rotational levels, depend upon the values of the rotational constant, B_v, and the rotational quantum number J, according to equation (1.25). The largest known rotational constant, for the lightest molecule (H_2), is about 60 cm^{-1}, so that rotational transitions in this and similar molecules will occur in the far-infrared region of the spectrum. As the molecular mass increases, rotational transition frequencies decrease, and rotational spectroscopy for most molecules occurs in the millimetre wave and microwave regions of the electromagnetic spectrum.

The fine and hyperfine splittings within a rotational level, and the transition frequencies between components, depend largely on whether the molecular species has a closed or open shell electronic structure. We will discuss these matters in more detail in section 1.6. For a closed shell molecule, that is, one in a $^1\Sigma^+$ state, intramolecular interactions are in general very small. They depend almost entirely on the presence of nuclei with spin magnetic moments, or with electric quadrupole moments. If both nuclei in a diatomic molecule have spin magnetic moments, there will be a magnetic interaction between them which leads to splitting of a rotational level. The interaction may occur as a through-space dipolar interaction, or it may arise through an isotropic scalar coupling brought about by the electrons. Dipolar interactions are much larger than the scalar spin–spin couplings, but even so only produce splittings of a few kHz in the most favourable cases. A molecule also possesses a magnetic moment by virtue of its rotational motion, which can interact with any nuclear spin magnetic moments present in the molecule. Nuclear and rotational magnetic moments interact with an applied magnetic field, and these interactions are at the heart of the molecular beam magnetic resonance studies described in chapter 8. The pioneering experiments in this field were carried out in the period 1935 to 1955; they are capable of exceptionally high spectroscopic resolution, with line widths sometimes only a fraction of a kHz, and they form the foundations of what came to be known as nuclear magnetic resonance [3]. Nuclear electric quadrupole moments, where present, interact with the electric field gradient caused by the other charges (nuclei and electrons) in a molecule and the resulting interaction, called the nuclear electric quadrupole interaction, can in certain cases be quite large (i.e. several GHz). This interaction may be studied through molecular beam magnetic resonance experiments, but it can also be important in conventional microwave absorption studies, as we describe in chapter 10. Magnetic resonance studies require the presence of a magnetic moment, but in the closely related technique of molecular beam electric resonance, the interaction between a molecular electric dipole moment and an applied electric field is used. These experiments are also described in detail in chapter 8. The magnetic resonance studies of closed shell molecules almost always involve transitions between components of a rotational level, and usually occur in the radiofrequency region of the spectrum. Electric resonance experiments, on the other hand, often deal with electric dipole transitions between rotational levels, and occur in the millimetre wave and microwave regions of the spectrum. Molecular beam electric resonance experiments are closely related to conventional absorption experiments.

Molecules with open shell electronic states, which are often highly reactive transient species called free radicals, introduce a range of new intramolecular interactions. The largest of these, which occurs in molecules with both spin and orbital angular momentum, is spin–orbit coupling. Spin–orbit interactions range from a few cm^{-1} to several thousand cm^{-1} and determine the overall pattern of the rotational levels and their associated spectroscopy. Molecules in $^2\Pi$ states are particularly important and will appear frequently in this book; the OH and CH radicals, in particular, are principal players who will make many appearances. If orbital angular momentum is not present, spin–orbit coupling is less important (though not completely absent). However, the magnetic moment due to electron spin is large and will interact with nuclear spin magnetic moments, to give nuclear hyperfine structure, and also with the rotational magnetic moment, giving rise to the so-called spin–rotation interaction. As important, however, is the strong interaction which occurs with an applied magnetic field. This interaction leads to magnetic resonance studies with bulk samples, performed at frequencies in the microwave region, or even in the far-infrared. The Zeeman interaction is used to tune spectroscopic transitions into resonance with fixed-frequency radiation; these experiments are described in detail in chapter 9. For various reasons they are capable of exceptionally high sensitivity, and consequently have been extremely important in the study of short-lived free radicals. It is, perhaps, important at this point to appreciate the difference between the molecular beam magnetic resonance experiments described in chapter 8, and the bulk studies described in chapter 9. In most of the molecular beam experiments the Zeeman interactions are used to control the molecular trajectories through the apparatus, and to produce state selectivity. Spectroscopic transitions, which may or may not involve Zeeman components, are detected through their effects on detected beam intensities. No attempt is made to detect the absorption or emission of electromagnetic radiation directly. Conversely, in the bulk magnetic resonance experiments, direct detection of the radiation is involved and the Zeeman effect is used to tune spectroscopic transitions into resonance with the radiation. Later in this chapter we will give a little more detail about electron spin and hyperfine interactions, as well as the Zeeman effect in open shell systems.

The final, but very important, point to be made in this section is that all of the experiments described and discussed in this book involve molecules in the gas phase. Moreover the gas pressures involved are sufficiently low that the molecular rotational motion is conserved. Just as importantly, quantised electronic orbital motion is not quenched by molecular collisions, as it would be at higher pressures. Of course, condensed phase studies are important in their own right, but they are different in a number of fundamental ways. In condensed phases rotational motion and electronic orbital angular momentum are both quenched. Anisotropic interactions, such as the dipolar interactions involving electron or nuclear spins, or both, can be studied in regularly oriented solids like single crystals, but are averaged in randomly oriented solids, like glasses. In isotropic liquids they drive time-dependent relaxation processes through a combination of the anisotropy and the tumbling Brownian motion of the molecules. It should also be remembered that the strong *intermolecular* interactions that occur in solids can substantially change the magnitudes of the *intramolecular* interactions, like hyperfine interactions.

1.5. Historical perspectives

A major reference point in the history of diatomic molecule spectroscopy was the publication of a classic book by Herzberg in 1950 [4]; this book was, in fact, an extensively revised and enlarged version of one published earlier in 1939. Herzberg's book was entitled *Spectra of Diatomic Molecules*, and it deals almost entirely with electronic spectroscopy. In the years leading up to and beyond 1950, spectrographic techniques using photographic plates were almost universally employed. They covered a wide wavelength range, from the far-ultraviolet to the near-infrared, and at their best presented a comprehensive view of the complete rovibronic band system of one or more electronic transitions. In Herzberg's hands these techniques were indeed presented at their best, and his book gives masterly descriptions of the methods used to obtain and analyse these beautiful spectra. For both diatomic and polyatomic molecules, most of what we now know and understand about molecular shapes, geometry, structure, dynamics, and electronic structure, has come from spectrographic studies of the type described by Herzberg. One could not improve on his exposition of the rules leading to our comprehension of these spectra, and there is no need to attempt to do so. It is, however, a rather sad fact that the classic spectrographic techniques seem now to be regarded as obsolete; most of the magnificent instruments which were used have been scrapped. The main thrust now is to use lasers to probe intimate details with much greater sensitivity, specificity and resolution, but such studies would not be possible without the foundations provided by the classic techniques. Perhaps one day they will, of necessity, return.

Almost all of the spectroscopy described in our book involves techniques which have been developed since the publication of Herzberg's book. Rotational energy levels were very well understood in 1950, and the analysis of rotational structure in electronic spectra was a major part of the subject. The major disadvantage of the experimental methods used was, however, the fact that the resolution was limited by Doppler broadening. The Doppler line width depends upon the spectroscopic wavelength, the molecular mass, the effective translational temperature, and other factors. However, a ballpark figure for the Doppler line width of 0.1 cm^{-1} would not be far out in most cases. Concealed within that 0.1 cm^{-1} are many subtle and fascinating details of molecular structure which are major parts of the subject of this book.

In 1950, microwave and molecular beam methods were just beginning to be developed, and they are mentioned briefly by Herzberg in his book. Microwave spectroscopy was given a boost by war-time research on radar, with the development of suitable radiation sources and transmission components; an early review of the subject was given by Gordy [5], one of its pioneers. Cooley and Rohrbaugh [6] observed the first three rotational transitions of HI in 1945, whilst Weidner [7] and Townes, Merritt and Wright [8] observed microwave transitions of the ICl molecule. Because of the much reduced Doppler width at the long wavelengths in the microwave region, nuclear hyperfine effects were observed. Such effects were already known in atomic spectroscopy, but not in molecular electronic spectra apart from some observations on HgH. Microwave transitions in the O_2 molecule were observed by Beringer [9] in 1946, and Beringer and Castle [10] in 1949 observed transitions between the Zeeman components of the rotational levels in O_2 and NO, the first examples of magnetic resonance in open shell molecules. Chapter 9 in this book is devoted to the now large and important subject of magnetic resonance spectroscopy in bulk gaseous samples.

The molecular beam radiofrequency magnetic resonance spectrum of H_2 was first observed by Kellogg, Rabi, Ramsey and Zacharias [11] in 1939, and was further developed in the post-war years. An analogous radiofrequency electric resonance spectrum of CsF was described by Hughes [12] in 1947, and again the technique underwent extensive development in the next thirty years. These molecular beam experiments, which had important precursors in atomic beam spectroscopy, are very different from the traditional spectroscopic experiments described by Herzberg in his book. They are capable of very high spectroscopic resolution, partly because they usually involve radio- or microwave frequencies, partly because of the absence of collisional effects, and partly because residual Doppler effects can be removed by appropriate relative spatial alignment of the molecular beam and the electromagnetic radiation. All of these matters are discussed in great detail in chapter 8. Finally in this brief review of the techniques that were developed after Herzberg's book, we should mention the laser, which now dominates electronic spectroscopy, and much of vibrational spectroscopy as well. Laser spectroscopy as such is not an important part of this book, apart from far-infrared magnetic resonance studies, but the use of lasers, both visible and infrared, in double resonance experiments is an important aspect of chapter 11. Lasers have made it possible to apply the techniques of radiofrequency and microwave spectroscopy to excited electronic states, an aspect of the subject which is likely to be developed much further.

Herzberg's book was therefore perfectly timed. The electronic spectroscopy of diatomic molecules was well developed and understood, and continues to be important [13]. Hopefully our book is also well timed; the molecular beam magnetic and electric resonance experiments are becoming less common, and may now almost be regarded as classic techniques! Magnetic resonance experiments on bulk gaseous samples are likely to continue to be important in the study of free radicals, particularly because of their very high sensitivity. Double resonance is important, in the study of excited states, but also in the route it provides towards the study of much heavier molecules where sensitivity considerations become increasingly important. Finally, pure rotational spectroscopy has assumed even greater importance because of its relationship with radioastronomy and the study of interstellar molecules, and because of its applications in the study of atmospheric chemistry.

1.6. Fine structure and hyperfine structure of rotational levels

1.6.1. Introduction

We outlined in section 1.4 the coordinate transformations which enable us to separate the rotational motion of a diatomic molecule from the electronic and vibrational motions. We pointed out that the spectroscopy described in this book involves either transitions *between* different rotational levels, or transitions between the various sub-components *within* a single rotational level; additional effects arising from applied electric or magnetic fields may or may not be present. We now outline very briefly the origin and nature of the sub-structure which is possible for a single rotational level in different electronic states. All of the topics mentioned in this section will be developed in considerable depth elsewhere in the book, but we hope that an elementary introduction will be useful, especially for the reader approaching the subject for the first time. As we will see, the detailed sub-structure of a rotational level depends upon the nature of the electronic state being considered. We can divide the electronic states into three different types, namely, closed shell states without electronic angular momentum, open shell states with electron spin angular momentum, and open shell states with both orbital and spin angular momentum. There is also a small number of cases where an electronic state has orbital but not spin angular momentum.

We will present the *effective* Hamiltonian terms which describe the interactions considered, sometimes using cartesian methods but mainly using spherical tensor methods for describing the components. These subjects are discussed extensively in chapters 5 and 7, and at this stage we merely quote important results without justification. We will use the symbol T to denote a spherical tensor, with the particular operator involved shown in brackets. The *rank* of the tensor is indicated as a post-superscript, and the *component* as a post-subscript. For example, the electron spin vector S is a first-rank tensor, $T^1(S)$, and its three spherical components are related to cartesian components in the following way:

$$\begin{aligned} T^1_0(S) &= S_z, \\ T^1_1(S) &= -(1/\sqrt{2})(S_x + iS_y), \\ T^1_{-1}(S) &= (1/\sqrt{2})(S_x - iS_y). \end{aligned} \quad (1.26)$$

The components may be expressed in either a space-fixed axis system (p) or a molecule-fixed system (q). The early literature used cartesian coordinate systems, but for the past fifty years spherical tensors have become increasingly common. They have many advantages, chief of which is that they make maximum use of molecular symmetry. As we shall see, the rotational eigenfunctions are essentially spherical harmonics; we will also find that transformations between space- and molecule-fixed axes systems, which arise when external fields are involved, are very much simpler using rotation matrices rather than direction cosines involving cartesian components.

1.6.2. $^1\Sigma^+$ states

In a diatomic, or linear polyatomic molecule, the energies of the rotational levels within a vibrational level v are given by

$$E(v, J) = B_v J(J+1) - D_v J^2(J+1)^2 + H_v J^3(J+1)^3 + \cdots, \quad (1.27)$$

where the rotational quantum number, J, takes integral values 0, 1, 2, etc. Provided the molecule is heteronuclear, with an electric dipole moment, rotational transitions between adjacent rotational levels ($\Delta J = \pm 1$) are electric-dipole allowed. The extent of the spectrum depends upon how many rotational levels are populated in the gaseous sample, which is determined by the Boltzman distribution law for a system in thermal equilibrium. The rotational transition frequencies increase as J increases, as (1.27) shows.

Any additional complications depend entirely on the nature of the nuclei involved. General Appendix B presents a list of the naturally occurring isotopes, with their spins, magnetic moments and electric quadrupole moments. Magnetic and electric interactions involving these moments can and will occur, the most important in a $^1\Sigma$ state being the electric quadrupole interaction between the nuclear quadrupole moment and an electric field gradient at the nucleus. Nuclei possessing a quadrupole moment must also have a spin I equal to 1 or more, and the extent of the quadrupole splitting of a rotational level depends upon the value of the nuclear spin. One of the most important quadrupolar nuclei is the deuteron, and quadrupole effects were probably first observed and analysed in the molecular beam magnetic resonance spectra of HD and D_2. In describing the energy levels we will often use a hyperfine-coupled representation, written as a ket $|\eta, J, I, F\rangle$, where the symbol η represents all other quantum numbers not specified, particularly those describing the electronic and vibrational state. For any given rotational level J, the total angular momentum F takes all values $J+I, J+I-1, \ldots, |J-I|$, so that there can be splitting into a maximum of $2I+1$ hyperfine levels for a single quadrupolar nucleus provided $J \geq I$. Such a case is shown schematically in figure 1.4 for the AlF molecule [14]; the ^{27}Al nucleus has a spin I of 5/2 and a large quadrupole moment. The $J = 0$ rotational level has no quadrupole splitting but $J = 1$ is split into three components as shown. An electric dipole $J = 1 \leftarrow 0$ rotational transition between adjacent rotational levels will exhibit a quadrupole splitting, as indicated. Alternatively, a spectrum arising from transitions within a single rotational level is possible, as indicated for CsF in figure 1.5. In this case [12] the ^{133}Cs nucleus has a spin of 7/2, and there is also an additional doublet splitting from the ^{19}F nucleus, arising from its magnetic dipole moment, which we will discuss shortly. There are other subtle aspects of this spectrum, one of them being that if the spectrum is recorded in the presence of a weak electric field, the transitions shown, which would be expected to have magnetic dipole intensity only, acquire electric dipole intensity. The full details are given in chapter 8.

The essential features of the electric quadrupole interaction can, hopefully, be appreciated with the aid of figure 1.6. The Z direction defines the direction of the

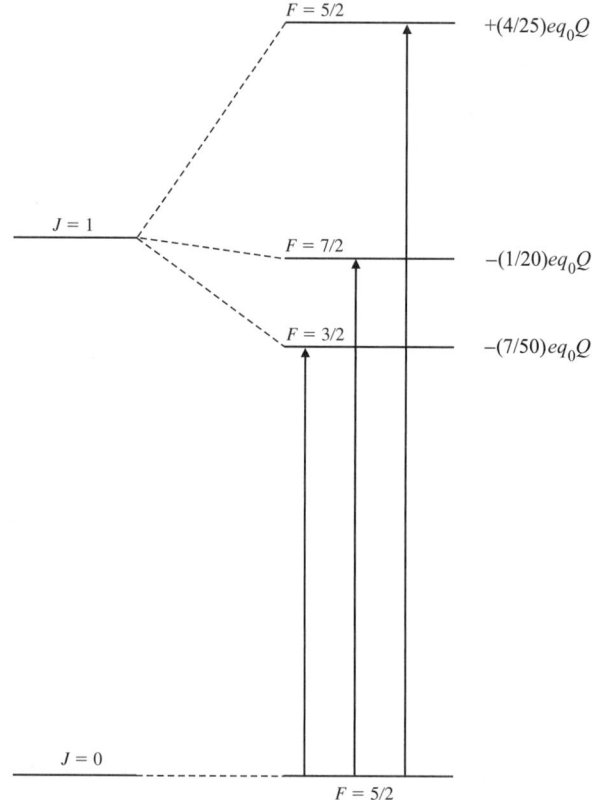

Figure 1.4. Splitting of the $J = 1$ rotational level of $^{27}\text{Al}^{19}\text{F}$ arising from the ^{27}Al quadrupole interaction with spin $I = 5/2$, and the resulting hyperfine splitting of the rotational transition. The magnetic interactions involving the ^{19}F nucleus are too small to be observed in this case.

electric field gradient, produced mainly by the electrons in the molecule. The total charge distribution of the nucleus may be decomposed into the sum of monopole, quadrupole, hexadecapole moments; the quadrupole distribution may be represented as a cigar-shaped distribution of charge having cylindrical symmetry about a principal axis fixed in the nucleus, which we define as the nuclear z axis. The quadrupolar charge distribution may be appreciated by considering the nuclear charge distribution at symmetrically disposed points on the $+z, -z, +x, -x$ axes. As we see from figure 1.6. the nuclear charge is $\delta-$ at the $\pm x$ points and $\delta+$ at the $\pm z$ points.

For a nucleus of spin $I = 1$ there are three allowed spatial orientations of the spin; in figure 1.6 these three orientations may be identified with those in which the nuclear z axis is coincident with Z, perpendicular to Z, and antiparallel to Z. These three orientations correspond to projection quantum numbers $M_I = +1$, 0 and -1 respectively, and it is clear from the figure that the state with $M_I = 0$ has a different electrostatic energy from the states with $M_I = \pm 1$. This 'quadrupole splitting' depends upon the sizes of the nuclear quadrupole moment and the electric field gradient.

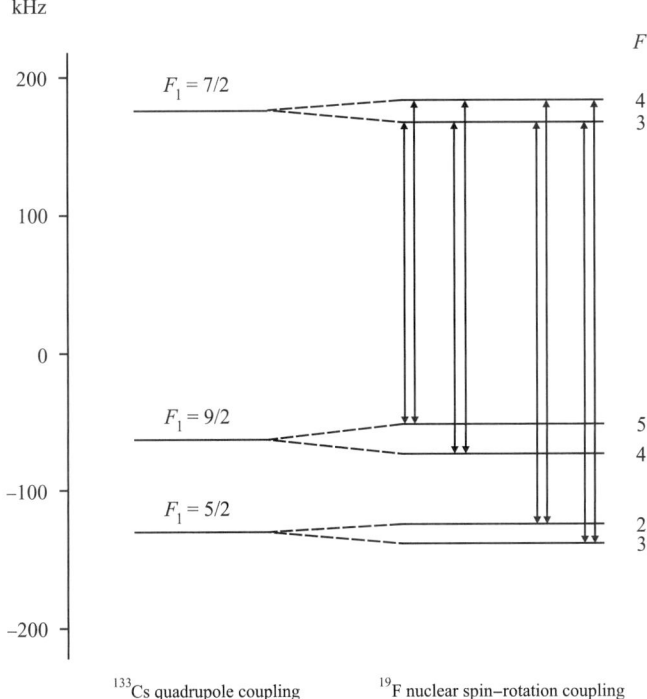

Figure 1.5. Nuclear hyperfine splitting of the $J=1$ rotational level of CsF. The major splitting is the result of the ^{133}Cs quadrupole interaction, and the smaller doublet splitting is caused by the ^{19}F interaction (see text).

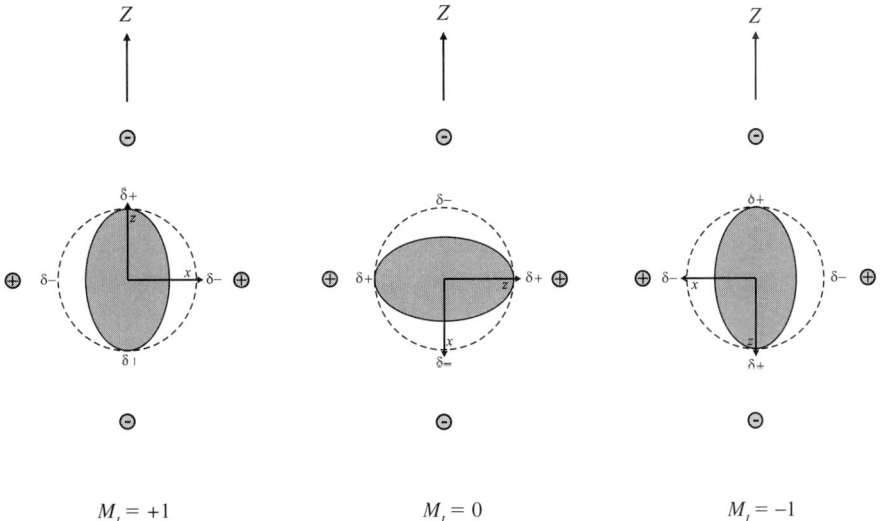

Figure 1.6. Orientation of a nucleus ($I=1$) with an electric quadrupole moment in an electric field gradient.

We show elsewhere in this book that the quadrupole interaction may be represented as the scalar product of two second-rank spherical tensors,

$$\mathcal{H}_Q = -eT^2(\nabla E) \cdot T^2(Q), \tag{1.28}$$

where the details of the electric field gradient are contained within the first tensor in (1.28) and the nuclear quadrupole moment is contained within the second tensor. We show elsewhere (chapter 8, for example) that the diagonal quadrupole energy obtained from (1.28) is given by

$$E_Q = -\frac{eq_0 Q}{2I(2I-1)(2J-1)(2J+3)}\{(3/4)C(C+1) - I(I+1)J(J+1)\}, \tag{1.29}$$

where $C = F(F+1) - I(I+1) - F(F+1)$. The quantity $eq_0 Q$ in (1.29) is called the quadrupole coupling constant, q_0 being the electric field gradient (actually its negative) and eQ the quadrupole moment of the ^{133}Cs nucleus. The value of $eq_0 Q$ for ^{133}Cs in CsF is 1.237 MHz.

The quadrupole coupling is very much the most important nuclear hyperfine interaction in $^1\Sigma^+$ states, and it takes the same form in open shell states as in closed shells. We turn now to the much smaller interactions involving magnetic dipole moments, two types of which may be present. A nuclear spin I gives rise to a magnetic moment μ_I,

$$\mu_I = g_N \mu_N I, \tag{1.30}$$

where g_N is the g-factor for the particular nucleus in question and μ_N is the nuclear magneton. In addition, the rotation of the nuclei and electrons gives rise to a rotational magnetic moment, whose value depends upon the rotational quantum number,

$$\mu_J = \mu_N J. \tag{1.31}$$

The magnetic moments given above will interact with an applied magnetic field, and these interactions are discussed extensively in chapter 8. In some diatomic molecules both nuclei have non-zero spin and an associated magnetic moment. The magnetic interactions which then occur are the nuclear spin–rotation interactions, represented by the operator

$$\mathcal{H}_{nsr} = \sum_{\alpha=1,2} c_\alpha T^1(J) \cdot T^1(I_\alpha), \tag{1.32}$$

and the nuclear spin–spin interactions. Here two different interactions are possible. The largest and most important is the through-space dipolar interaction, which in its classical form is represented by the operator

$$\mathcal{H}_{dip} = g_1 g_2 \mu_N^2 (\mu_0/4\pi) \left\{ \frac{I_1 \cdot I_2}{R^3} - \frac{3(I_1 \cdot R)(I_2 \cdot R)}{R^5} \right\}. \tag{1.33}$$

Here I_1, I_2 and g_1, g_2 are the spins and g-factors of nuclei 1 and 2 and R is the distance between them. In spherical tensor form the interaction may be written

$$\mathcal{H}_{dip} = -g_1 g_2 \mu_N^2 (\mu_0/4\pi) \sqrt{6} T^2(C) \cdot T^2(I_1, I_2), \tag{1.34}$$

where the second-rank tensors are defined as follows:

$$T_p^2(\boldsymbol{I}_1, \boldsymbol{I}_2) = (-1)^p \sqrt{5} \sum_{p_1, p_2} \begin{pmatrix} 1 & 1 & 2 \\ p_1 & p_2 & -p \end{pmatrix} T_{p_1}^1(\boldsymbol{I}_1) T_{p_2}^1(\boldsymbol{I}_2), \quad (1.35)$$

$$T_q^2(\boldsymbol{C}) = \langle C_q^2(\theta, \phi) R^{-3} \rangle. \quad (1.36)$$

These expressions require some detailed explanation, and the reader might wish to advance to chapter 5 at this point. First, here and elsewhere, the subscripts p and q refer to space-fixed and molecule-fixed axes respectively. Equation (1.35) which describes the construction of a second-rank tensor from two first-rank tensors contains a vector coupling coefficient called a Wigner 3-j symbol. Equation (1.36) contains a spherical harmonic function which gives the necessary geometric information. The equivalence of (1.34) and (1.33) is demonstrated in appendix 8.1, which also introduces another spherical tensor form for the dipolar interaction. The most important feature is, of course, the R^{-3} dependence of the interaction. In the H_2 molecule the proton–proton dipolar coupling is about 60 kHz, which is readily determinable in the high-resolution molecular beam magnetic resonance studies.

The second interaction between two nuclear spins in a diatomic molecule is a scalar coupling,

$$\mathcal{H}_{\text{scalar}} = c_s T^1(\boldsymbol{I}_1) \cdot T^1(\boldsymbol{I}_2), \quad (1.37)$$

which is often described as the electron-coupled spin–spin interaction because the mechanism involves the transmission of nuclear spin orientation through the intervening electrons (see section 1.7). This coupling is very small compared with the dipolar interaction, and is usually negligible in gas phase studies. It is, however, extremely important in liquid phase nuclear magnetic resonance because, unlike the dipolar coupling, it is not averaged to zero by the tumbling motion of the molecules.

The remaining important type of magnetic interaction is that between the rotational magnetic moment and any nuclear spin magnetic moments, given in equation (1.32). In the case of H_2 the constant c has the value 113.9 kHz. The doublet splitting in the spectrum of CsF, shown in figure 1.5, is due to the ^{19}F nuclear spin–rotation interaction. Note also that in this case the hyperfine basis kets take the form $|\eta, J, I_1, F_1; I_2, F\rangle$ where I_1 is the spin of ^{133}Cs (value 7/2) and I_2 is the spin of ^{19}F value 1/2. Hence for $J = 1$, F_1 can take the values 9/2, 7/2 and 5/2 as shown, and F takes values $F_1 \pm 1/2$. Other possible magnetic interactions in CsF are too small to be observed.

The remaining important magnetic interactions to be considered are those which arise when a static magnetic field $\boldsymbol{B}$ is applied. The Zeeman interaction with a nuclear spin magnetic moment is represented by the Hamiltonian term

$$\mathcal{H}_Z = -\sum_{\alpha=1,2} g_N^\alpha \mu_N T^1(\boldsymbol{B}) \cdot T^1(\boldsymbol{I}_\alpha), \quad (1.38)$$

and since the direction of the magnetic field is usually taken to define the space-fixed

Z or $p=0$ direction, the scalar product in (1.38) contracts to

$$\mathcal{H}_Z = -\sum_{\alpha=1,2} g_N^\alpha \mu_N \mathrm{T}_0^1(\boldsymbol{B})\mathrm{T}_0^1(\boldsymbol{I}_\alpha). \tag{1.39}$$

The nuclear spin Zeeman levels then have energies given by

$$E_Z = -\sum_{\alpha=1,2} g_N^\alpha \mu_N B_Z M_{I_\alpha}, \tag{1.40}$$

where the projection quantum number M_I takes the $2I+1$ values from $-I$ to $+I$. The nuclear spin Zeeman interaction in discussed extensively in chapter 8. In molecular beam experiments it is used for magnetic state selection, and the radiofrequency transitions studied are usually those with the selection rule $\Delta M_I = \pm 1$ observed in the presence of an applied magnetic field. We will also see, in chapter 8, that the simple expression (1.38) is modified by the inclusion of a screening factor,

$$\mathcal{H}_Z = -\sum_{\alpha=1,2} g_N^\alpha \mu_N \mathrm{T}^1(\boldsymbol{B}) \cdot \mathrm{T}^1(\boldsymbol{I}_\alpha)\{1 - \sigma_\alpha(\boldsymbol{J})\}, \tag{1.41}$$

arising mainly because of the diamagnetic circulation of the electrons in the presence of the magnetic field. In liquid phase nuclear magnetic resonance this screening gives rise to what is known as the 'chemical shift'.

The rotational magnetic moment also interacts with an applied magnetic field, the interaction term being very similar to (1.41) above, i.e.

$$\mathcal{H}_{JZ} = -g_r \mu_N \mathrm{T}^1(\boldsymbol{B}) \cdot \mathrm{T}^1(\boldsymbol{J})\{1 - \sigma(\boldsymbol{J})\}, \tag{1.42}$$

where g_r is the rotational g-factor. In a molecule where there are no nuclear spins present, the rotational Zeeman interaction can be used for selection of M_J states.

Finally in this section on $^1\Sigma^+$ states we must include the Stark interaction which occurs when an electric field ($\boldsymbol{E}$) is applied to a molecule possessing a permanent electric dipole moment ($\boldsymbol{\mu}_e$):

$$\mathcal{H}_E = -\mathrm{T}^1(\boldsymbol{\mu}_e) \cdot \mathrm{T}^1(\boldsymbol{E}). \tag{1.43}$$

As with the Zeeman interaction discussed earlier, (1.43) is usually contracted to the space-fixed $p=0$ component. An extremely important difference, however, is that in contrast to the nuclear spin Zeeman effect, the Stark effect in a $^1\Sigma$ state is second-order, which means that the electric field *mixes* different rotational levels. This aspect is thoroughly discussed in the second half of chapter 8; the second-order Stark effect is the engine of molecular beam electric resonance studies, and the spectra, such as that of CsF discussed earlier, are usually recorded in the presence of an applied electric field.

Whilst the most important examples of Zeeman and Stark effects in $^1\Sigma$ states are found in molecular beam studies, they can also be important in conventional absorption microwave rotational spectroscopy, as we describe in chapter 10. The use of the Stark effect to determine molecular dipole moments is a very important example.

1.6.3. Open shell Σ states

We now proceed to consider the magnetic interactions involving the electron spin $\boldsymbol{S}$ in Σ states with open shell electronic structures. The magnetic dipole moment arising from electron spin is

$$\boldsymbol{\mu}_S = -g_S \mu_B \boldsymbol{S}, \tag{1.44}$$

where g_S is the free electron g-factor, with the value 2.0023, and μ_B is the electron Bohr magneton; μ_B is almost two thousand times larger than the nuclear magneton, μ_N, so we see at once that magnetic interactions from electron spin are very much larger than those involving nuclear spin, considered in the previous sub-section.

With the introduction of electronic angular momentum, we have to consider how the spin might be coupled to the rotational motion of the molecule. This question becomes even more important when electronic orbital angular momentum is involved. The various coupling schemes give rise to what are known as Hund's coupling cases; they are discussed in detail in chapter 6, and many practical examples will be encountered elsewhere in this book. If only electron spin is involved, the important question is whether it is quantised in a space-fixed axis system, or molecule-fixed. In this section we confine ourselves to space quantisation, which corresponds to Hund's case (b).

We deal first with molecules containing one unpaired electron ($S = 1/2$) where magnetic nuclei are not present. The electron spin magnetic moment then interacts with the magnetic moment due to molecular rotation, the interaction being represented by the Hamiltonian term

$$\mathcal{H}_{\mathrm{sr}} = \gamma \mathbf{T}^1(\boldsymbol{S}) \cdot \mathbf{T}^1(\boldsymbol{N}), \tag{1.45}$$

in which γ is the spin–rotation coupling constant. As was originally shown by Hund [15] and Van Vleck [16], each rotational level in a given vibrational level (v) of a $^2\Sigma$ state is split into a spin doublet, with energies

$$F_1(N) = B_v N(N+1) + (1/2)\gamma_v N,$$
$$F_2(N) = B_v N(N+1) - (1/2)\gamma_v (N+1). \tag{1.46}$$

The F_1 levels correspond to $J = N + 1/2$ and the F_2 levels to $J = N - 1/2$. A typical rotational energy level diagram is shown in figure 1.7(a); each rotational transition ($\Delta N = \pm 1$) is split into a doublet (with $\Delta J = \pm 1$) and a weaker satellite ($\Delta J = 0$). This seems a simple conclusion, except that Van Vleck [16] showed that the spin splitting of each rotation level is only partly the result of the rotational magnetic moment in the direction of $\boldsymbol{N}$. The other part comes from electronic orbital angular momentum in the Σ state which precesses at right angles about the internuclear axis; in other words, although the expectation value of $\boldsymbol{L}$ is zero in a pure Σ state, the spin–orbit coupling operator mixes the Σ state with excited Π states. This introduces an additional non-zero magnetic moment in the direction of $\boldsymbol{N}$, which contributes to the spin–rotation coupling. We will return to this important subject in the next section; it represents

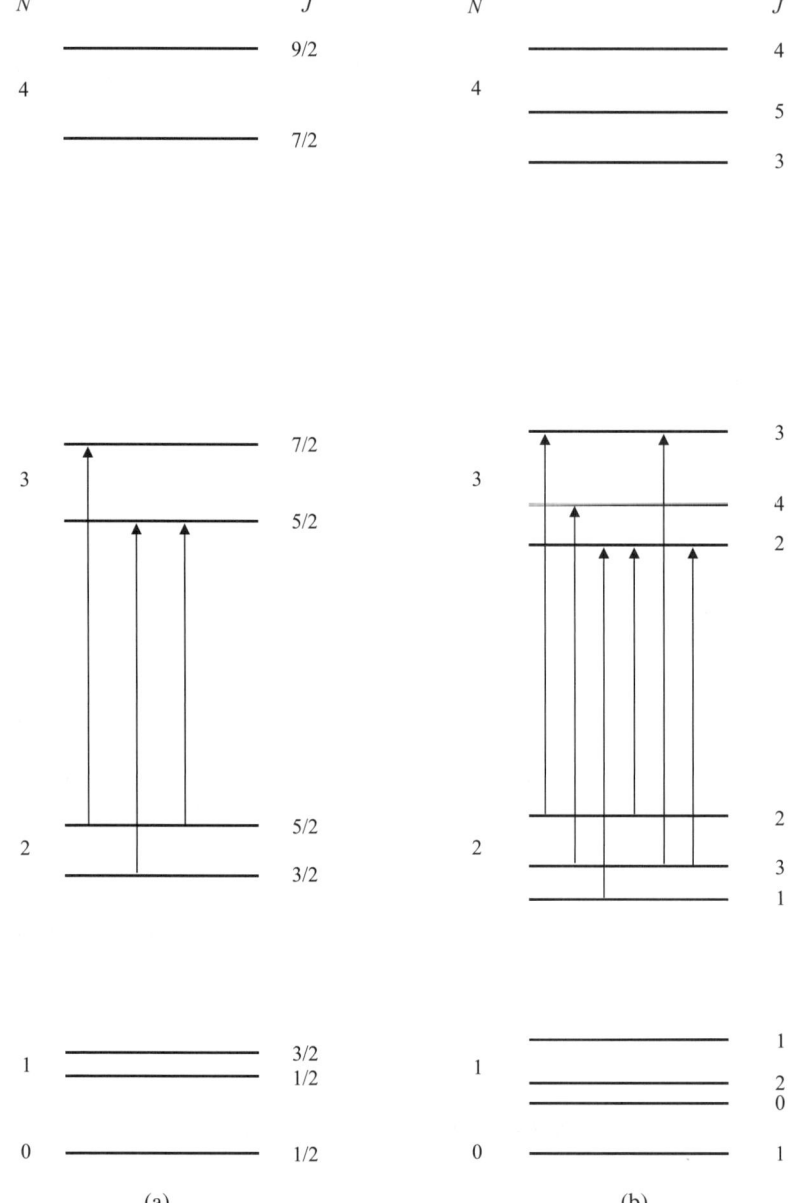

Figure 1.7. (a) Lower rotational levels and transitions in a case (b) $^2\Sigma$ state, showing the spin splitting of a rotational transition. (b) Lower rotational levels and transitions in a case (b) $^3\Sigma$ state, showing the spin splitting of a rotational transition.

our first encounter with the very important concept of the *effective* Hamiltonian. What looks like a spin–rotation interaction is not entirely what it seems!

The lower rotational levels for a case (b) $^3\Sigma$ state are shown in figure 1.7(b). The spin–rotation interaction takes the same form as for a $^2\Sigma$ state, given in equation (1.45), but in addition there is an important interaction between the spins of the two unpaired electrons, called the electron spin–spin interaction; this is usually larger than the spin–rotation interaction. The spin–spin interaction can be represented in a number of different ways, depending upon the molecule under investigation. Initially we might regard the interaction as being analogous to the classical interaction between two magnetic dipole moments so that, following equation (1.33) for nuclear spins, we write the interaction as

$$\mathcal{H}_{ss} = g_S^2 \mu_B^2 (\mu_0/4\pi) \left\{ \frac{\mathbf{S}_1 \cdot \mathbf{S}_2}{r^3} - \frac{3(\mathbf{S}_1 \cdot \mathbf{r})(\mathbf{S}_2 \cdot \mathbf{r})}{r^5} \right\}, \quad (1.47)$$

where $\mathbf{S}_1$ and $\mathbf{S}_2$ are the spins of the individual electrons, and r is the distance between them. Of course, the electrons are not point charges, so that r is an average distance which can be calculated from a suitable electronic wave function. Again, by analogy with our previous treatment of nuclear spins, the electron spin dipolar interaction can be represented in spherical tensor form by the operator

$$\mathcal{H}_{ss} = -g_S^2 \mu_B^2 (\mu_0/4\pi) \sqrt{6} \mathbf{T}^2(\mathbf{C}) \cdot \mathbf{T}^2(\mathbf{S}_1, \mathbf{S}_2), \quad (1.48)$$

where, as before, $\mathbf{T}^2(\mathbf{C})$ represents the spherical harmonic functions, the $q = 0$ component being given by

$$T_0^2(\mathbf{C}) = C_0^2(\theta, \phi)(r^{-3}) = \left(\frac{4\pi}{5} \right)^{1/2} Y_{2,0}(\theta, \phi)(r^{-3}) = \frac{1}{2}(2z^2 - x^2 - y^2)(r^{-5}). \quad (1.49)$$

In appendix 8.3 we show that (1.48) with $q = 0$ leads to the simple expression,

$$\mathcal{H}_{ss} = \frac{2}{3}\lambda \left(3S_z^2 - \mathbf{S}^2 \right), \quad (1.50)$$

where z is the internuclear axis and λ is called the spin–spin coupling constant. Provided λ is not too large compared with the rotational constant, Kramers [17] showed that each rotational level is split into a spin triplet, with relative component energies

$$F_1(N) = B_v N(N+1) - \frac{2\lambda(N+1)}{(2N+3)} + \gamma_v(N+1),$$
$$F_2(N) = B_v N(N+1), \quad (1.51)$$
$$F_3(N) = B_v N(N+1) - \frac{2\lambda N}{(2N-1)} - \gamma_v N.$$

where F_1, F_2, F_3 refer to levels with $J = N + 1$, N and $N - 1$. More accurate formulae were given by Schlapp [18] and, neglecting the small vibrational dependence of λ and γ, these are

$$F_1(N) = B_v N(N+1) + (2N+3)B_v - \lambda - \left\{ (2N+3)^2 B_v^2 + \lambda^2 - 2\lambda B_v \right\}^{1/2} + \gamma_v(N+1),$$
$$F_2(N) = B_v N(N+1), \quad (1.52)$$
$$F_3(N) = B_v N(N+1) - (2N-1)B_v - \lambda + \left\{ (2N-1)^2 B_v^2 + \lambda^2 - 2\lambda B_v \right\}^{1/2} - \gamma_v N.$$

The molecule O_2 in its $^3\Sigma_g^-$ ground state is a good example of a case (b) molecule, and the triplet energies agree with (1.52), the values of the constants being $B_0 = 1.43777 \text{ cm}^{-1}$, $\lambda = 1.984 \text{ cm}^{-1}$, $\gamma_0 = -0.0084 \text{ cm}^{-1}$. Note that yet another spherical tensor form for the dipolar interaction which is sometimes used is

$$\mathcal{H}_{ss} = 2\lambda T^2(S, S) \cdot T^2(n, n), \tag{1.53}$$

where n is a unit vector along the internuclear axis and S is the total spin of 1. Again, the relationship of this form to the others is described in appendix 8.3. A typical pattern of rotational levels for a $^3\Sigma$ state with the spin splitting is shown in figure 1.7(b), together with the allowed rotational transitions. Once again, spin–orbit coupling can mix a $^3\Sigma$ state with nearby Π states and contribute to the value of the constant λ. We show in chapter 9 that in the SeO molecule the spin–orbit coupling is so strong that the case (b) pattern of rotational levels no longer holds, and a case (a) coupling scheme is more appropriate. The formulae given above are then not applicable.

The remaining important interactions which can occur for a $^2\Sigma$ or $^3\Sigma$ molecule involve the presence of nuclear spin. Interactions between the electron spin and nuclear spin magnetic moments are called 'hyperfine' interactions, and there are two important ones. The first is called the Fermi contact interaction, and if both nuclei have non-zero spin, each interaction is represented by the Hamiltonian term

$$\mathcal{H}_F = b_F T^1(S) \cdot T^1(I). \tag{1.54}$$

The Fermi contact constant b_F is given by

$$b_F = \left(\frac{2}{3}\right) g_S \mu_B g_N \mu_N \mu_0 \int \psi^2(r) \delta(r) \, dr, \tag{1.55}$$

where the function $\delta(r)$, called the Dirac delta function, imposes the condition that $r = 0$ when we integrate over the probability density of the wave function of the unpaired electron. Hence the contact interaction can only occur when the unpaired electron has a finite probability density at the nucleus, which means that the wave function must have some s-orbital character (i.e. $\psi(0)^2 \neq 0$).

The second important hyperfine interaction is the dipolar interaction and by analogy with equations (1.34) and (1.48) it may be expressed in spherical tensor form by the expression

$$\mathcal{H}_{dip} = \sqrt{6} g_S \mu_B g_N \mu_N (\mu_0/4\pi) T^2(C) \cdot T^2(S, I). \tag{1.56}$$

There are some situations when this is the most convenient representation of the dipolar coupling, for example, when S and I are very strongly coupled to each other but weakly coupled to the molecular rotation, as in the H_2^+ ion. However, an alternative form which is often more suitable is

$$\mathcal{H}_{dip} = -\sqrt{10} g_S \mu_B g_N \mu_N (\mu_0/4\pi) T^1(I) \cdot T^1(S, C^2). \tag{1.57}$$

The spherical components of the new first-rank tensor in (1.57) are defined, in the molecule-fixed axes system, by

$$T_q^1(S, C^2) = \sqrt{3} \sum_{q_1, q_2} (-1)^q T_{q_1}^1(S) T_{q_2}^2(C) \begin{pmatrix} 1 & 2 & 1 \\ q_1 & q_2 & -q \end{pmatrix}, \tag{1.58}$$

where, as before,

$$T_{q_2}^2(\boldsymbol{C}) = C_{q_2}^2(\theta, \phi)(r^{-3}). \tag{1.59}$$

The relationships between the various forms of the dipolar Hamiltonian are explained in appendix 8.2. As we see from (1.59), the dipolar interaction has various components in the molecule-fixed axis system but the most important one, and often the only one to be determined from experiment, is $T_0^2(\boldsymbol{C})$. This leads us to define a constant t_0, the axial dipolar hyperfine component, given in SI units by,

$$t_0 = g_S \mu_B g_N \mu_N (\mu_0/4\pi) T_0^2(\boldsymbol{C}) = \frac{1}{2} g_S \mu_B g_N \mu_N (\mu_0/4\pi) \left\langle \frac{(3\cos^2\theta - 1)}{r^3} \right\rangle. \tag{1.60}$$

The most important examples of $^2\Sigma$ states to be described in this book are CO$^+$, where there is no nuclear hyperfine coupling in the main isotopomer, CN, which has ^{14}N hyperfine interaction, and the H$_2^+$ ion. A number of different $^3\Sigma$ states are described, with and without hyperfine coupling. A particularly important and interesting example is N$_2$ in its $A\,^3\Sigma_u^+$ excited state, studied by De Santis, Lurio, Miller and Freund [19] using molecular beam magnetic resonance. The details are described in chapter 8; the only aspect to be mentioned here is that in a homonuclear molecule like N$_2$, the individual nuclear spins ($I = 1$ for ^{14}N) are coupled to form a total spin, I_T, which in this case takes the values 2, 1 and 0. The hyperfine Hamiltonian terms are then written in terms of the appropriate value of I_T. As we have already mentioned, the presence of one or more quadrupolar nuclei will give rise to electric quadrupole hyperfine interaction; the theory is essentially the same as that already presented for $^1\Sigma^+$ states.

Finally we note that the interaction with an applied magnetic field is important because of the large magnetic moment arising from the presence of electron spin (see (1.44)). The Zeeman interaction is represented by the Hamiltonian term

$$\mathcal{H}_Z = g_S \mu_B T^1(\boldsymbol{B}) \cdot T^1(\boldsymbol{S}) = g_S \mu_B B_{p=0} T_{p=0}^1(\boldsymbol{S}) = g_S \mu_B B_Z T_0^1(\boldsymbol{S}), \tag{1.61}$$

which, as we show, may again be contracted to a single $p = 0$ space-fixed component. As we will see, the Zeeman interaction is central to magnetic resonance studies, either with molecular beams as described in chapter 8 where radiofrequency spectroscopy is involved, or with bulk gases (chapter 9) where microwave or far-infrared radiation is employed. The magnetic resonance studies are, in general, of two kinds. For magnetic fields which are readily accessible in the laboratory, the Zeeman splitting of different M_S (or M_J) levels often corresponds to a microwave frequency. In many studies, therefore, the transitions studied obey a selection rule $\Delta M_S = \pm 1$ or $\Delta M_J = \pm 1$, and take place between levels which are otherwise degenerate in the absence of a magnetic field. There are, however, very important experiments where the transitions occur between levels which are already well separated in zero field; fixed frequency radiation is then used, with the transition energy mismatch being tuned to zero with an applied field. Far-infrared laser magnetic resonance studies are of this type. As we will see, the theoretical problem which must be solved concerns the competition between the Zeeman interaction, which tends to decouple the electron spin from the molecular framework, and intramolecular interactions like the electron spin dipolar coupling which tends to couple the spin orientation to the molecular orientation.

1.6.4. Open shell states with both spin and orbital angular momentum

Many free radicals in their electronic ground states, and also many excited electronic states of molecules with closed shell ground states, have electronic structures in which both electronic orbital and electronic spin angular momentum is present. The precession of electronic angular momentum, $\boldsymbol{L}$, around the internuclear axis in a diatomic molecule usually leads to defined components, Λ, along the axis, and states with $|\Lambda| = 0, 1, 2, 3$, etc., are called Σ, Π, Δ, Φ, etc., states. In most cases there is also spin angular momentum $\boldsymbol{S}$, and the electronic state is then labelled $^{2S+1}\Pi$, $^{2S+1}\Delta$, etc.

Questions arise immediately concerning the coupling of $\boldsymbol{L}$, $\boldsymbol{S}$ and the nuclear rotation, $\boldsymbol{R}$. The possible coupling cases, first outlined by Hund, are discussed in detail in chapter 6. Here we will adopt case (a), which is the one most commonly encountered in practice. The most important characteristic of case (a) is that Λ, the component of $\boldsymbol{L}$ along the internuclear axis, is indeed defined and we can use the labels Σ, Π, Δ, etc., as described above. The spin–orbit coupling can be represented in a simplified form by the Hamiltonian term

$$\mathcal{H}_{so} = A \mathbf{T}^1(\boldsymbol{L}) \cdot \mathbf{T}^1(\boldsymbol{S}) = A \sum_q (-1)^q \mathbf{T}^1_q(\boldsymbol{L}) \mathbf{T}^1_{-q}(\boldsymbol{S}), \tag{1.62}$$

expanded in the molecule-fixed axis system as shown. The $q = 0$ term gives a diagonal energy $A \Lambda \Sigma$, where Σ is the component of the electron spin ($\boldsymbol{S}$) along the internuclear axis. The component of total electronic angular momentum along the internuclear axis is called Ω; it is given by $\Omega = \Lambda + \Sigma$.

If we are dealing with a $^2\Pi$ state, the possible values of the projection quantum numbers are as follows:

$$\begin{aligned}
\Lambda &= +1, & \Sigma &= +1/2, & \Omega &= +3/2; \\
\Lambda &= -1, & \Sigma &= -1/2, & \Omega &= -3/2; \\
\Lambda &= +1, & \Sigma &= -1/2, & \Omega &= +1/2; \\
\Lambda &= -1, & \Sigma &= +1/2, & \Omega &= -1/2.
\end{aligned} \tag{1.63}$$

The occurrence of $\Lambda = \pm 1$ is called Λ-doubling or Λ-degeneracy; in addition, the spin coupling gives rise to an additional two-fold doubling. The states with $|\Omega| = 3/2$ or $1/2$ are called fine-structure states, with spin–orbit energies $+A/2$ and $-A/2$ respectively; the value of $|\Omega|$ is written as a subscript in the state label. Hence we have $^2\Pi_{3/2}$ and $^2\Pi_{1/2}$ fine-structure components; if A is negative the $^2\Pi_{3/2}$ state is the lower in energy, and we have an 'inverted' doublet, the opposite case being called a 'regular' doublet. The NO molecule has a $^2\Pi_{1/2}$ ground state (regular), whilst the OH radical has a $^2\Pi_{3/2}$ ground state (inverted).

The rigid body rotational Hamiltonian can be written in the form

$$\mathcal{H}_{\text{rot}} = B\mathbf{R}^2 = B(\mathbf{J} - \mathbf{L} - \mathbf{S})^2$$
$$= B(\mathbf{J}^2 + \mathbf{L}^2 + \mathbf{S}^2 - 2\mathbf{J} \cdot \mathbf{L} - 2\mathbf{J} \cdot \mathbf{S} + 2\mathbf{L} \cdot \mathbf{S}). \tag{1.64}$$

The expansion of (1.64) is discussed in detail in chapter 8, and elsewhere, so we present only a brief and simplified summary here. Expanded in the molecule-fixed axis system, the diagonal part of the expression gives the result:

$$E_{\text{rot}}(J) = B\{J(J+1) + S(S+1) + 2\Lambda\Sigma + \Lambda^2 - 2\Omega^2\}. \tag{1.65}$$

There is, therefore, a sequence of rotational levels, characterised by their J values, for each fine-structure state. According to the discussion above, each J level has a two-fold degeneracy, forming what are called Ω-doublets or Λ-doublets. The off-diagonal ($q = \pm 1$) terms from (1.64), together with the off-diagonal components of the spin–orbit coupling operator (1.62), remove the degeneracy of the Λ-doublets. The resulting pattern of the lower rotational levels for the OH radical is shown in figure 1.8, which is discussed in more detail in chapters 8 and 9. Transitions between the rotational levels, shown in the diagram, have been observed by far-infrared laser magnetic resonance, and transitions between the Λ-doublet components of the same rotational level have been observed by microwave rotational spectroscopy, by microwave magnetic resonance, by molecular beam maser spectroscopy, and by radio-astronomers studying interstellar gas clouds.

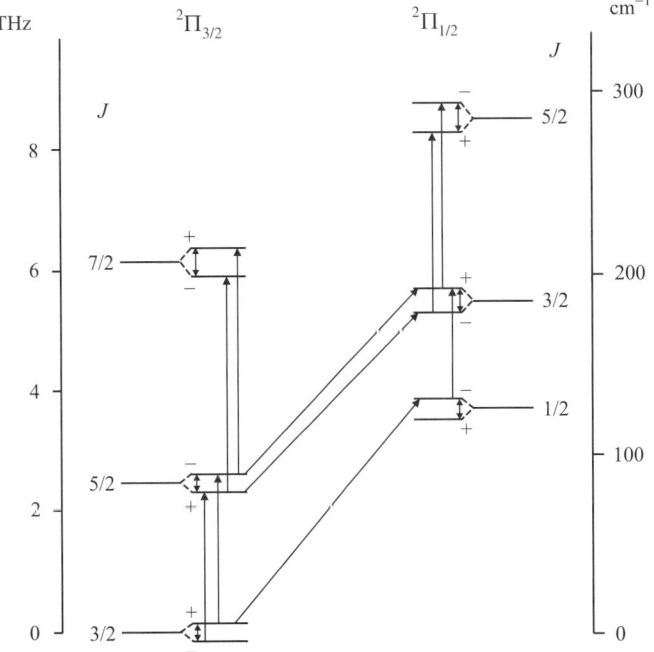

Figure 1.8. Lower rotational levels of the OH radical, and some of the transitions that have been observed. The size of the Λ-doublet splitting is exaggerated for the sake of clarity.

Interactions with an applied magnetic field are particularly important for open shell free radicals, many with $^2\Pi$ ground states having been studied by magnetic resonance methods. The Zeeman Hamiltonian may be written as the sum of four terms:

$$\mathcal{H}_Z = g_L \mu_B \mathbf{T}^1(\boldsymbol{B}) \cdot \mathbf{T}^1(\boldsymbol{L}) + g_S \mu_B \mathbf{T}^1(\boldsymbol{B}) \cdot \mathbf{T}^1(\boldsymbol{S}) - g_N \mu_N \mathbf{T}^1(\boldsymbol{B}) \cdot \mathbf{T}^1(\boldsymbol{I}) \\ - g_r \mu_B \mathbf{T}^1(\boldsymbol{B}) \cdot \{\mathbf{T}^1(\boldsymbol{J}) - \mathbf{T}^1(\boldsymbol{L}) - \mathbf{T}^1(\boldsymbol{S})\}. \qquad (1.66)$$

All of these terms must be included in an accurate analysis and their effects are described in detail in chapter 9. The most important terms, however, are the first two. Putting the orbital g-factor, g_L, equal to 1 one can show that for a good case (a) molecule the effective g-value for the rotational level J is

$$g_J = \frac{(\Lambda + \Sigma)(\Lambda + g_S \Sigma)}{J(J+1)}. \qquad (1.67)$$

If we put $g_S = 2$, we find that for the lowest rotational level of the $^2\Pi_{3/2}$ state, $J = 3/2$, the g-factor is 4/5. For any rotational level of the $^2\Pi_{1/2}$ state, however, (1.67) predicts a g-factor of zero. For a perfect case (a) molecule, therefore, we cannot use magnetic resonance methods to study $^2\Pi_{1/2}$ states. Fortunately perhaps, most molecules are intermediate between case (a) and case (b) so that both fine-structure states are magnetic to some extent. The other point to notice from (1.67) is that the g-factor decreases rapidly as J increases.

We will see elsewhere is this book many examples of the spectra of $^2\Pi$ molecules. We will see also that although our discussion above is based upon a case (a) coupling scheme for the various angular momenta, case (b) is often just as appropriate and, as we have already noted, many molecules are really intermediate between case (a) and case (b). We will also meet electronic states with higher spin and orbital multiplicity. For $S \geq 1$, the terms describing the interaction between electron spins play much the same role in Π and Δ states as they do for Σ states. Nuclear hyperfine interactions are also similar to those described already, with the addition of an orbital hyperfine term which may be written in the form

$$\mathcal{H}_{IL} = a \mathbf{T}^1(\boldsymbol{I}) \cdot \mathbf{T}^1(\boldsymbol{L}), \qquad (1.68)$$

where the orbital hyperfine constant is given by

$$a = 2\mu_B g_N \mu_N (\mu_0/4\pi) \langle r^{-3} \rangle; \qquad (1.69)$$

r is the distance between the nucleus and the orbiting electron, with the average calculated from a suitable electronic wave function.

The purpose of this section has been to introduce the complexity in the sub-structure of rotational levels, and the richness of the consequent spectroscopy which is revealed by high-resolution techniques. Understanding the origin and details of this structure also takes us very deeply into molecular quantum mechanics, as we show in chapters 2 to 7.

1.7. The effective Hamiltonian

The process of analysing a complex diatomic molecule spectrum with electron spin, nuclear hyperfine and external field interactions has several stages. We need to derive expressions for the energies of the levels involved, which means choosing a suitable basis set and a suitable 'effective Hamiltonian'. The best basis set is that particular Hund's case which seems the nearest or most convenient approximation to the 'truth'. The effective Hamiltonian is a sum of terms representing the various interactions within the molecule; each term contains angular momentum operators and 'molecular parameters'. Our choice of effective Hamiltonian is also determined by the basis set chosen. The procedure is then to set up a matrix of the effective Hamiltonian operating within the chosen basis. The matrix is often truncated artificially, and we then diagonalise the matrix to obtain the energies of the levels and the effective wave functions. Armed with this information we attempt to assign the lines in the spectrum. The spectral frequencies are expressed in terms of the molecular parameters, and usually a first set of values is determined. If the assignment is correct, a program designed to minimise the differences between calculated and measured transition frequencies is employed. The final best values of the molecular parameters may then be used for comparison with the predictions of electronic structure calculations. In this way we hope to develop a better description of the electronic structure of the molecule.

The choice of the effective Hamiltonian is often far from straightforward; indeed we have devoted a whole chapter to this subject (chapter 7). In this section we give a gentle introduction to the problems involved, and show that the definition of a particular 'molecular parameter' is not always simple. The problem we face is not difficult to understand. We are usually concerned with the sub-structure of one or two rotational levels at most, and we aim to determine the values of the important parameters relating to those levels. However, these parameters may involve the participation of other vibrational and electronic states. We do *not* want an effective Hamiltonian which refers to other electronic states explicitly, because it would be very large, cumbersome and essentially unusable. We want to analyse our spectrum with an effective Hamiltonian involving *only* the quantum numbers that arise directly in the spectrum. The effects of all other states, and their quantum numbers, are to be absorbed into the definition and values of the 'molecular parameters'. The way in which we do this is outlined briefly here, and thoroughly in chapter 7.

The development of the effective Hamiltonian has been due to many authors. In condensed phase electron spin magnetic resonance the so-called 'spin Hamiltonian' [20, 21] is an example of an effective Hamiltonian, as is the 'nuclear spin Hamiltonian' [22] used in liquid phase nuclear magnetic resonance. In gas phase studies, the first investigation of a free radical by microwave spectroscopy [23] introduced the ideas of the effective Hamiltonian, as also did the first microwave magnetic resonance study [24]. Miller [25] was one of the first to develop the more formal aspects of the subject, particularly so far as gas phase studies are concerned, and Carrington, Levy and Miller [26] have reviewed the theory of microwave magnetic resonance, and the use of the effective Hamiltonian.

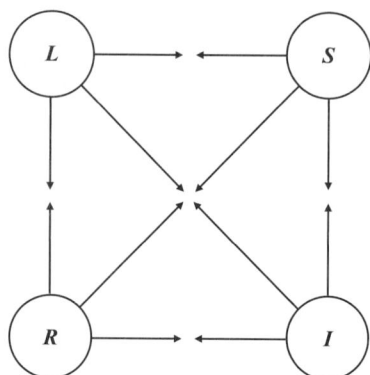

Figure 1.9. Possible pairwise interactions of **L**, **S**, **R** and **I**.

As a simple introduction to the subject [27], let us consider the four angular momentum vectors illustrated in figure 1.9. They are as follows:

R: the rotational angular momentum of the bare nuclei,
L: the electronic orbital angular momentum,
S: the electronic spin angular momentum,
I: the nuclear spin angular momentum.

Each angular momentum can interact with the other three, and figure 1.9 draws attention to the following pairwise interactions:

(**L**)(**S**): spin–orbit coupling,
(**L**)(**R**): rotational–electronic interaction,
(**L**)(**I**): hyperfine interaction between the electron orbital and nuclear spin magnetic moments,
(**S**)(**I**): hyperfine interaction between the electron and nuclear spin magnetic moments,
(**S**)(**R**): interaction between the electron spin and rotational magnetic moments,
(**I**)(**R**): interaction between the nuclear spin and rotational magnetic moments.

The direct interactions listed above and illustrated in figure 1.9 can occur in the effective Hamiltonian, but figure 1.10 shows how the effective Hamiltonian can also contain similar terms which arise indirectly. In figure 1.10(a) we illustrate the interaction of **R** with **L**, which in turn couples with the spin **S**. Consequently the effective Hamiltonian may contain a term of the form (**R**)(**S**), part of which arises from the direct coupling shown in figure 1.9, but with the remaining part coming from the indirect coupling via **L**. If we are dealing with a diatomic molecule in a Σ state, there is no first-order orbital angular momentum, but the spin–orbit coupling can mix the ground state with one or more excited Π states, thereby generating some orbital angular momentum in the ground state [28]. Consequently the spin–rotation constant γ comprises a first-order direct contribution, plus a second-order contribution arising from admixture of excited states. In all but the lightest molecules, this second-order contribution is the largest in magnitude.

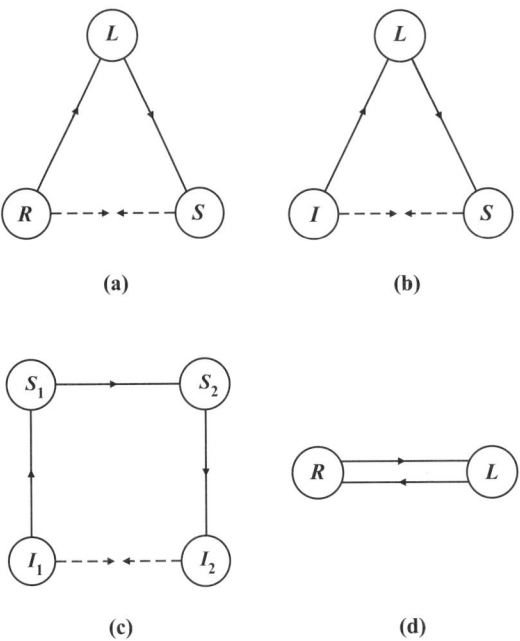

Figure 1.10. (a) Second-order spin–rotation interaction occurring via L. (b) Second-order pseudo-contact hyperfine interaction occurring via L. (c) Electron coupled nuclear spin–spin interaction. (d) Second-order interaction of R and L.

Another example is the pseudo-contact hyperfine interaction illustrated in figure 1.10(b); spin–orbit mixing of excited Π states with a Σ ground state generates orbital angular momentum in the ground state, which interacts with the nuclear spin magnetic moment. Overall, therefore, the interaction looks somewhat like a direct Fermi contact interaction, $S \cdot I$. A third example, illustrated in figure 1.10(c), involves two electron spins (S_1 and S_2) and two nuclear spins (I_1 and I_2). The nuclear spin I_1 interacts with the electron spin S_1; S_1 is coupled with S_2, which in turn interacts with I_2. The net result is an interaction which is represented in the effective Hamiltonian by a term of the form $I_1 \cdot I_2$. This interaction is called the 'electron-coupled nuclear spin–spin interaction', and it is the origin of the spin–spin splittings observed in liquid phase nuclear magnetic resonance spectra. Note that it is not necessary for the total spin $S = S_1 + S_2$ to be non-zero; the interaction can and does occur in closed shell molecules.

Our final example, illustrated in figure 1.10(d), involves the rotational angular momentum R and the orbital angular momentum L. The second-order effect of the coupling for a ground $^1\Sigma$ state, operating through admixture of excited states, involves the product of matrix elements containing the operator products $(R \cdot L)(R \cdot L)$. The net effect is a term in the effective Hamiltonian which contains the operator R^2. Remembering that the rotational angular momentum of the nuclei is also represented in the effective Hamiltonian by a term BR^2, we see that the rotational constant B must be interpreted with some care, because it contains a very small contribution from the electrons, added to the major contribution of the nuclei.

Many different procedures for reducing the complete Hamiltonian to a suitable effective Hamiltonian have been devised. These are reviewed in detail in chapter 7; we will see that the methods involve different forms of perturbation theory [29].

1.8. Bibliography

The subject matter of this book is scattered between many other books, and much of it does not yet appear in any book. The most important book which deals exclusively with diatomic molecules is that by Herzberg [4], combined with the later supplement by Huber and Herzberg [13] which lists data up to 1979. This takes us into the era of computerised data bases (see below), which are the best sources for numerical data, and are the reason why we have made no attempt at a comprehensive data coverage in this book. A further important book is that by Lefebvre-Brion and Field [29]; the title of this book suggests a rather specialised treatment but it is actually both wide and deep in its coverage. Other books which deal specifically with theoretical aspects of diatomic molecules are those by Judd [30], dealing with angular momentum theory, Kovács [31] and Mizushima [32]. Angular momentum theory occupies a central position in understanding the energy levels of both diatomic and polyatomic molecules. In this book we use the methods and conventions of Edmonds [33], but have also benefited from the reader-friendly accounts provided by Rose [34], Zare [35] and Brink and Satchler [36]. Quantum mechanics is the fundamental theory which must be mastered if molecular spectroscopy is to be understood. This is not the place for a comprehensive listing of the many books on this subject, but we have found the books by Flygare [2], Moss [37] and Hannabuss [38] to be helpful; in particular, our treatment of relativistic quantum mechanics in chapters 3 and 4 owes much to those by Moss and Hannabuss. The book by Bunker and Jensen [39] is our standard source for problems involving symmetry and group theory.

Molecular beams are important in this book. For the early work the book by Ramsey [3] is indispensable, and a recent two-volume comprehensive survey edited by Scoles [40] covers recent developments in the technology. However, books dealing with microwave, millimetre wave or far-infrared spectroscopy, whether using beams or not, are scarce. The early books of Townes and Schawlow [41], Kroto [42] and Carrington [27] still have some value, and more recently Hirota [43] has described spectroscopic work (mainly Japanese) on transient molecules. There is, however, a vast amount of published original work on the high-resolution spectroscopy of transient species, using far-infrared or lower radiation frequencies. This book is devoted to a description of this type of work applied to diatomic molecules. More general books on molecular spectroscopy, including diatomic molecules, are those by Hollas [44], Demtröder [45] and Bernath [46]. In the field of radio astronomy we have found the book by Rohlfs and Wilson [47] to be most helpful.

There are, of course, a number of review articles from which we select three, all written by Hirota and all dealing with free radicals and molecular ions. The first of these covers the period up to 1992 [48] and deals with diatomic and polyatomic species. It is supplemented by further reviews published in 1994 [49] and 2000 [50]. Finally we should draw attention to a computer data base covering all types of spectroscopy of diatomic molecules, produced by Bernath and McLeod [51]. This is available free of charge on the Internet and may be seen at <http://diref.uwaterloo.ca>. It will be maintained for the indefinite future.

Appendix 1.1. Maxwell's equations

An important connection between optical and electromagnetic phenomena was first discovered by Faraday in 1846. He observed that when plane-polarised radiation passes through certain materials exposed to a magnetic field that is parallel to the propagation direction of the radiation, the plane of polarisation is rotated. The degree of rotation depends upon the nature of the material and the strength of the magnetic field. The union of optical and electromagnetic properties was subsequently put on firm foundations by Maxwell in the form of his wave theory of electromagnetic interactions. As we shall see, Maxwell's equations also provide the explanation for optical properties like dispersion and refraction. The nature of electromagnetic radiation, which is central to almost everything in this book, was described earlier in this chapter, but without much justification. Maxwell's equations, which form the basis for understanding electromagnetic radiation, will now be described. There are, in fact, four equations that connect macroscopic electric and magnetic phenomena, and two further equations that describe the response of a material medium to electric and magnetic fields.

(i) The first equation is

$$\nabla \wedge \boldsymbol{E} + \frac{1}{c}\frac{\partial \boldsymbol{B}}{\partial t} = 0 \quad \text{(in cgs units)},$$
$$\nabla \wedge \boldsymbol{E} + \frac{\partial \boldsymbol{B}}{\partial t} = 0 \quad \text{(in SI units)}. \tag{1.70}$$

∇ is the vector operator given by

$$\nabla = \boldsymbol{i}\frac{\partial}{\partial x} + \boldsymbol{j}\frac{\partial}{\partial y} + \boldsymbol{k}\frac{\partial}{\partial z}, \tag{1.71}$$

where $\boldsymbol{i}, \boldsymbol{j}, \boldsymbol{k}$ are orthogonal unit vectors. When ∇ operates on a scalar ϕ the resulting vector $\nabla \phi$ is called the gradient of ϕ (i.e. grad ϕ). When ∇ operates on a vector $\boldsymbol{A}$ there are two possibilities. The scalar product, $\nabla \cdot \boldsymbol{A}$, results in a new scalar, and is known as the divergence of $\boldsymbol{A}$ (i.e. div $\boldsymbol{A}$). The vector product, $\nabla \wedge \boldsymbol{A}$, is a vector called the curl of $\boldsymbol{A}$; c, as elsewhere, is the speed of light.

Equation (1.70) is Faraday's law of electromagnetic induction; it shows that a time-dependent magnetic flux density, $\boldsymbol{B}$, gives rise to an electric field, $\boldsymbol{E}$, in a direction perpendicular to the original magnetic field. Equations (1.70) are often written in the abbreviated form

$$\operatorname{curl} E + \frac{1}{c}\frac{\partial B}{\partial t} = 0 \quad \text{(in cgs units)},$$
$$\operatorname{curl} E + \frac{\partial B}{\partial t} = 0 \quad \text{(in SI units)}. \tag{1.72}$$

(ii) The second equation is

$$\nabla \wedge H - \frac{1}{c}\frac{\partial D}{\partial t} = \frac{4\pi}{c} J \quad \text{(in cgs units)},$$
$$\nabla \wedge H - \frac{\partial D}{\partial t} = J \quad \text{(in SI units)}. \tag{1.73}$$

H is the magnetic field vector and D is called the electric induction or displacement field. This equation is known as the Ampere–Oersted law and shows that a magnetic field will exist near an electric current density J. The displacement field, D, is necessary to propagate electromagnetic energy through space. J has units charge $\cdot$ area$^{-1}\cdot$ t^{-1}

(iii) The third equation is

$$\nabla \cdot D = 4\pi\bar{\rho} \quad \text{(in cgs units)},$$
$$\nabla \cdot D = \bar{\rho} \quad \text{(in SI units)}. \tag{1.74}$$

$\nabla \cdot$ is called the div and $\bar{\rho}$ is the electric charge density with units charge $\cdot$ volume^{-1}. There is a relationship between J and $\bar{\rho}$, given by

$$J = \bar{\rho}v, \tag{1.75}$$

where v is the velocity of the charge distribution. J and E are also related by

$$J = \sigma \cdot E, \tag{1.76}$$

where σ is the conductivity. Equation (1.74) is actually the Coulomb law in electrostatics.

(iv) The fourth equation is the same in both cgs and SI units, and is

$$\nabla \cdot B = 0. \tag{1.77}$$

This equation states that there are no sources of magnetic field except currents; in other words, there are no free magnetic poles.

The remaining two equations both relate to properties of the medium.

(v) The fifth equation may be written

$$D = \varepsilon \cdot E \quad \text{(in cgs units)},$$
$$D = \varepsilon_0 \varepsilon \cdot E \quad \text{(in SI units)}. \tag{1.78}$$

ε is the relative electric permittivity, or dielectric constant, of the medium, expressed in general as a tensor, and ε_0 is the permittivity of a vacuum.

(vi) The sixth equation is the magnetic analogue of the fifth:

$$\begin{aligned} \boldsymbol{B} &= \mu_p \cdot \boldsymbol{H} \quad \text{(in cgs units)}, \\ \boldsymbol{B} &= \mu_0 \mu_p \cdot \boldsymbol{H} \quad \text{(in SI units)}. \end{aligned} \tag{1.79}$$

$\boldsymbol{B}$ is the magnetic induction or magnetic flux density, and μ_p is the relative magnetic permeability of the medium, also expressed in general as a tensor. For an anisotropic medium the scalars ε and μ_p are used; their values are unity for a vacuum (so that $\boldsymbol{B}$ and $\boldsymbol{H}$ are then equivalent). ε has a wide range of values for different substances, but μ_p is usually close to unity. If μ_p is less than 1.0 the substance is diamagnetic, and if it is greater than 1.0 the substance is paramagnetic.

The permittivity of a vacuum is

$$\varepsilon_0 = 8.854\,187\,818 \times 10^{-12}\,\text{s}^4\,\text{A}^2\,\text{kg}^{-1}\,\text{m}^{-3}, \tag{1.80}$$

and the permeability of free space is

$$\mu_0 = 4\pi \times 10^{-7}\,\text{kg}\,\text{m}\,\text{s}^{-2}\,\text{A}^{-2}. \tag{1.81}$$

It also follows from the above equations that

$$(1/\varepsilon_0 \mu_0)^{1/2} = c. \tag{1.82}$$

Appendix 1.2. Electromagnetic radiation

The oscillating electric and magnetic fields of a plane wave, shown in figure 1.2, may be represented by the following simple equations:

$$\begin{aligned} \boldsymbol{E} &= \boldsymbol{k} E_0 \sin(Y - vt), \\ \boldsymbol{B} &= \boldsymbol{i} B_0 \sin(Y - vt), \end{aligned} \tag{1.83}$$

in which E_0, B_0 and v are simply constants. We now show that this electromagnetic field satisfies Maxwell's equations provided certain conditions are met. We find the following results:

$$\begin{aligned} \text{div}\,\boldsymbol{E} &= 0, \\ \text{curl}\,\boldsymbol{E} &= \boldsymbol{i}\frac{\partial E_Z}{\partial Y} = \boldsymbol{i} E_0 \cos(Y - vt) \\ \frac{\partial \boldsymbol{E}}{\partial t} &= -v\boldsymbol{k} E_0 \cos(Y - vt). \end{aligned} \tag{1.84}$$

$$\begin{aligned} \text{div}\,\boldsymbol{B} &= 0, \\ \text{curl}\,\boldsymbol{B} &= -\boldsymbol{k}\frac{\partial B_X}{\partial Y} = -\boldsymbol{k} B_0 \cos(Y - vt), \\ \frac{\partial \boldsymbol{B}}{\partial t} &= -v\boldsymbol{i} B_0 \cos(Y - vt). \end{aligned} \tag{1.85}$$

We note also that $\boldsymbol{J}=0$ in empty space. If these results are combined with (1.72) and (1.73), the conditions that must be satisfied are, in SI units,

$$E_0 = \text{v} B_0, \qquad B_0 = \mu_0 \varepsilon_0 \text{v} E_0. \tag{1.86}$$

Taken together, these equations require that

$$\text{v} = \pm c, \qquad E_0 = c B_0. \tag{1.87}$$

In the old cgs units, the second relationship is even simpler:

$$E_0 = B_0. \tag{1.88}$$

We have therefore established three important features of the electromagnetic radiation. The first is that the field pattern travels with the speed of light, c. The second is that at every point in the wave at any instant of time, the electric and magnetic field strengths are directly related to each other. The third is that the electric and magnetic fields are perpendicular to one another, and to the direction of travel.

References

[1] E.M. Purcell, *Electricity and Magnetism*, McGraw-Hill Book Company, Singapore, 1985.
[2] W.H. Flygare, *Molecular Structure and Dynamics*, Prentice-Hall, Inc., New Jersey, 1978.
[3] N.F. Ramsey, *Molecular Beams*, Oxford University Press, Oxford, 1956.
[4] G. Herzberg, *Spectra of Diatomic Molecules*, D. Van Nostrand Company, Inc., Princeton, 1950.
[5] W. Gordy, *Rev. Mod. Phys.*, **20**, 668 (1948).
[6] J.P. Cooley and J.H. Rohrbaugh, *Phys. Rev.*, **67**, 296 (1945).
[7] R.T. Weidner, *Phys. Rev.*, **72**, 1268 (1947); **73**, 254 (1948).
[8] C.H. Townes, F.R. Merritt and B.D. Wright, *Phys. Rev.*, **73**, 1334 (1948).
[9] R. Beringer, *Phys. Rev.*, **70**, 53 (1946).
[10] R. Beringer and J.G. Castle, *Phys. Rev.*, **75**, 1963 (1949); **76**, 868 (1949).
[11] J.M.B. Kellogg, I.I. Rabi, N.F. Ramsey and J.R. Zacharias, *Phys. Rev.*, **56**, 728 (1939).
[12] H.K. Hughes, *Phys. Rev.*, **72**, 614 (1947).
[13] K.P. Huber and G. Herzberg, *Constants of Diatomic Molecules*, Van Nostrand Reinhold Company, New York, 1979.
[14] D.R. Lide, *J. Chem. Phys.*, **38**, 2027 (1963).
[15] F. Hund, *Z. Physik*, **42**, 93 (1927).
[16] J.H. Van Vleck, *Phys. Rev.*, **33**, 467 (1929).
[17] H.A. Kramers, *Z. Physik*, **53**, 422 (1929).
[18] R. Schlapp, *Phys. Rev.*, **51**, 342 (1937).
[19] D. De Santis, A. Lurio, T.A. Miller and R.S. Freund, *J. Chem. Phys.*, **58**, 4625 (1973).
[20] A. Abragam and B. Bleaney, *Electron Paramagnetic Resonance of Transition Ions*, Clarendon Press, Oxford, 1970.
[21] A. Carrington and A.D. McLachlan, *Introduction to Magnetic Resonance*, Harper and Row, New York, 1967.

[22] J.A. Pople, W.G. Schneider and H.J. Bernstein, *High-resolution Nuclear Magnetic Resonance*, McGraw-Hill Book Company, Inc., New York, 1959.
[23] G.C. Dousmanis, T.M. Sanders and C.H. Townes, *Phys. Rev.*, **100**, 1735 (1955).
[24] H.E. Radford, *Phys. Rev.*, **126**, 1035 (1962).
[25] T.A. Miller, *Mol. Phys.*, **16**, 105 (1969).
[26] A. Carrington, D.H. Levy and T.A. Miller, *Adv. Chem. Phys.*, **18**, 149 (1970).
[27] A. Carrington, *Microwave Spectroscopy of Free Radicals*, Academic Press, London, 1974.
[28] J.H. Van Vleck, *Rev. Mod. Phys.*, **23**, 213 (1951).
[29] H. Lefebvre-Brion and R.W. Field, *Perturbations in the Spectra of Diatomic Molecules*, Academic Press, Inc., Orlando, 1986.
[30] B.R. Judd, *Angular Momentum Theory for Diatomic Molecules*, Academic Press, Inc., New York, 1975.
[31] I. Kovács, *Rotational Structure in the Spectra of Diatomic Molecules*, Adam Hilger Ltd., London, 1969.
[32] M. Mizushima, *The Theory of Rotating Diatomic Molecules*, John Wiley and Sons, New York, 1975.
[33] A.R. Edmonds, *Angular Momentum in Quantum Mechanics*, Princeton University Press, Princeton, 1960.
[34] M.E. Rose, *Elementary Theory of Angular Momentum*, John Wiley and Sons, Inc., New York, 1957.
[35] R.N. Zare, *Angular Momentum*, John Wiley and Sons, New York, 1988.
[36] D.M. Brink and G.R. Satchler, *Angular Momentum*, Clarendon Press, Oxford, 1962.
[37] R.E. Moss, *Advanced Molecular Quantum Mechanics*, Chapman and Hall, London, 1973.
[38] K. Hannabuss, *An Introduction to Quantum Theory*, Oxford University Press, 1997.
[39] P.R. Bunker and P. Jensen, *Molecular Symmetry and Spectroscopy*, NRC Research Press, Ottawa, 1998.
[40] *Atomic and Molecular Beam Methods*, ed. G. Scoles, Oxford University Press, New York, 1988.
[41] C.H. Townes and A.L. Schawlow, *Microwave Spectroscopy*, McGraw-Hill Book Company, New York, 1955.
[42] H.W. Kroto, *Molecular Rotation Spectra*, John Wiley and Sons, London, 1975.
[43] E. Hirota, *High-resolution Spectroscopy of Transient Molecules*, Springer-Verlag, Berlin, 1985.
[44] J.M. Hollas, *Modern Spectroscopy*, John Wiley and Sons, Ltd., Chichester, 1996.
[45] W. Demtröder, *Laser Spectroscopy*, Springer-Verlag, Berlin, 1982.
[46] P.F. Bernath, *Spectra of Atoms and Molecules*, Oxford University Press, New York, 1995.
[47] K. Rohlfs and T.L. Wilson, *Tools of Radio Astronomy (Third Edition)*, Springer-Verlag, Berlin, 2000.
[48] E. Hirota, *Chem. Rev.*, **92**, 141 (1992).
[49] E. Hirota, *Ann. Rep. Prog. Chem., Sect. C*, **91**, 3 (1994).
[50] E. Hirota, *Ann. Rep. Prog. Chem., Sect. C*, **96**, 95 (2000).
[51] P.F. Bernath and S.McLeod, *DiRef – A Database of References Associated with the Spectra of Diatomic Molecules*, University of Waterloo Chemical Physics report.

2 The separation of nuclear and electronic motion

2.1. Introduction

A molecule is an assembly of positively charged nuclei and negatively charged electrons that forms a stable entity through the electrostatic forces which hold it all together. Since all the particles which make up the molecule are moving relative to each other, a full mechanical description of the molecule is very complicated, even when treated classically. Fortunately, the overall motion of the molecule can be broken down into various types of motion, namely, translational, rotational, vibrational and electronic. To a good approximation, each of these motions can be considered on its own. The basis of this classification was established in a ground-breaking paper written by Born and Oppenheimer [1] in 1927, just one year after the introduction of wave mechanics. The main objective of their paper was the separation of electronic and nuclear motions in a molecule. The physical basis of this separation is quite simple. Both electrons and nuclei experience similar forces in a molecular system, since they arise from a mutual electrostatic interaction. However, the mass of the electron, m, is about four orders-of-magnitude smaller than the mass of the nucleus M. Consequently, the electrons are accelerated at a much greater rate and move much more quickly than the nuclei. On a very short time scale (less than 10^{-16} s), the electrons will move but the nuclei will barely do so; as a first approximation, the nuclei can be regarded as being fixed in space.

Born and Oppenheimer expanded the molecular Hamiltonian in terms of a parameter κ given by the ratio of a typical nuclear displacement to the internuclear distance R. Simple order-of-magnitude arguments showed that

$$\kappa = (m/M)^{1/4} \tag{2.1}$$

which has a value close to $1/10$. Born and Oppenheimer went on to show that, in a typical situation,

$$\Delta E_{\text{nucl}}/\Delta E_{\text{el}} \approx \kappa^2, \tag{2.2}$$

where ΔE_{el} is the separation between electronic energy levels and ΔE_{nucl} similarly for nuclear (i.e. vibrational) energy levels. An extension of these ideas revealed a further separation of nuclear motion into vibrational and rotational parts for which

$$\Delta E_{\text{rot}}/\Delta E_{\text{vib}} \approx \kappa^2. \tag{2.3}$$

In terms of κ, the electronic energy is of zeroth order, the vibrational energy of second order and the rotational energy of fourth order; the first- and third-order energies vanish. This general approach to the classification of molecular energy levels is known as the *Born–Oppenheimer separation* (or the Born–Oppenheimer approximation). For the vast majority of situations, particularly for molecules in their closed shell ground states, it is extremely reliable. It forms a reassuringly sound foundation for molecular quantum mechanics and is very robust when subjected to quantitative test.

In this chapter we show how the separation of the quantum mechanical problem into translational, rotational, vibrational and electronic parts can be achieved. The basis of our approach is to define coordinates which describe the various motions and then attempt to express the wave function as a product of factors, each of which depends only on a small sub-set of coordinates, along the lines:

$$\psi(X, x) = R(X)S(x). \tag{2.4}$$

Such a clean separation cannot generally be achieved in practice. One has instead to settle for second best, getting as close to the separation as possible. There is one exception to this statement, however, the case of translational motion of the molecule as a whole through three-dimensional space. In the absence of external electric and magnetic fields, space is isotropic. Translation of the body from one point to another (or translation of the coordinate system which is used to describe it) is a symmetry operation. For this reason, translational motion can be separated *rigorously* from the other three types of motion. In the detailed description of a molecular system, this motion is separated off by moving the origin of our arbitrary space-fixed coordinate system to one with its origin fixed at the molecular centre of mass. As the molecule moves through laboratory space, this coordinate system moves with it.

In this chapter we describe the various stages of the factorisation process. Following the separation of translational motion by reference of the particles' coordinates to the molecular centre of mass, we separate off the rotational motion by referring coordinates to an axis system which rotates with the molecule (the so-called molecule-fixed axis system). Finally, we separate off the electronic motion to the best of our ability by invoking the Born–Oppenheimer approximation when the electronic wave function is obtained on the assumption that the nuclei are at a fixed separation R. Some empirical discussion of the involvement of electron spin, in either Hund's case (a) or (b), is also included. In conclusion we consider how the effects of external electric or magnetic fields are modified by the various transformations.

2.2. Electronic and nuclear kinetic energy

2.2.1. Introduction

The kinetic energy of the nuclei and electrons in field-free space may be written in the form

$$T = -\hbar^2 \sum_{\alpha=1}^{2} \frac{1}{2M_\alpha} \nabla_\alpha^2 - \hbar^2 \sum_{i=1}^{n} \frac{1}{2m_i} \nabla_i^2, \tag{2.5}$$

where α and i sum over the nuclei and electrons of masses M_α and m_i respectively. This operator may also be written in terms of the momenta,

$$T = \sum_\alpha \frac{1}{2M_\alpha} P_\alpha^2 + \sum_i \frac{1}{2m_i} P_i^2, \tag{2.6}$$

where $P_i = -i\hbar(\partial/\partial R_i)$ and $P_\alpha = -i\hbar(\partial/\partial R_\alpha)$ are expressed in a space-fixed axis system of arbitrary origin; $\partial/\partial R_\alpha$ is a shorthand notation for the three components of the gradient operator in a particular coordinate system:

$$\frac{\partial}{\partial R_\alpha} \equiv \left(\frac{\partial}{\partial X_\alpha}\right) i' + \left(\frac{\partial}{\partial Y_\alpha}\right) j' + \left(\frac{\partial}{\partial Z_\alpha}\right) k' \tag{2.7}$$

where i', j' and k' are unit vectors along the X, Y and Z axes.

In order to discuss the spectroscopic properties of diatomic molecules it is useful to transform the kinetic energy operators (2.5) or (2.6) so that the translational, rotational, vibrational, and electronic motions are separated, or at least partly separated. In this section we shall discuss transformations of the *origin* of the space-fixed axis system; the following choices of origin have been discussed by various authors (see figure 2.1):

(a) centre of mass of the molecule,
(b) centre of mass of the two nuclei,
(c) geometrical centre of the two nuclei,
(d) the position of one nucleus.

Transformation from an arbitrary origin to (a) allows the translational motion of the whole molecule to be separated. Transformation to (b), first discussed by Kronig [2] and by Van Vleck [3], is a useful starting point for examination of the coupling of electronic and nuclear motions. Transformation (c) to the geometrical centre of the nuclei is independent of nuclear mass and so is useful in the discussion of electronic isotope effects (Kolos and Wolniewicz [4], Bunker [5]). Transformation (d) has been discussed by Pack and Hirschfelder [6], particularly in connection with long-range interactions between two atoms. All of these various coordinate systems have the same orientation in space as the original space-fixed axis system.

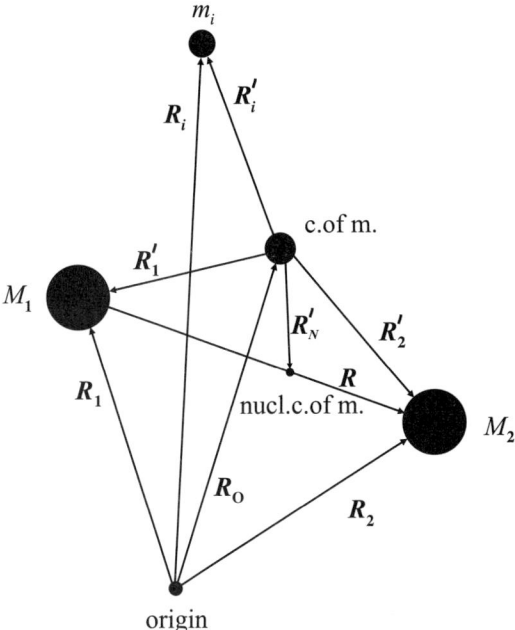

Figure 2.1. Axis systems for origin transformations.

2.2.2. Origin at centre of mass of molecule

We transform the origin of the space-fixed coordinates through R_O to the centre of mass of the molecule; the vector R_O is thus given by

$$R_O = \frac{1}{M}\left\{m\sum_i R_i + \sum_\alpha M_\alpha R_\alpha\right\} \tag{2.8}$$

where M is the total mass of all particles,

$$M = \sum_i m_i + \sum_\alpha M_\alpha. \tag{2.9}$$

The new coordinates R'_i and R'_α are given by the relations

$$R'_i = R_i - R_O, \tag{2.10}$$
$$R'_\alpha = R_\alpha - R_O, \tag{2.11}$$

and it follows from the definitions of R'_i and R'_α that

$$\sum_i m_i R'_i + \sum_\alpha M_\alpha R'_\alpha = 0. \tag{2.12}$$

The state of the system consisting of n electrons and two nuclei is now specified in terms of $3n + 6$ spatial coordinates referred to the centre of mass plus three centre-of-mass coordinates. Three of the spatial coordinates are redundant and can therefore be eliminated. Various choices of redundant coordinates can be made; we choose to introduce the internuclear vector $\boldsymbol{R}$ defined by

$$\boldsymbol{R} = \boldsymbol{R}'_2 - \boldsymbol{R}'_1 = \boldsymbol{R}_2 - \boldsymbol{R}_1. \tag{2.13}$$

The transformation therefore involves a change from the coordinates $\boldsymbol{R}_i$, $\boldsymbol{R}_\alpha$ of equation (2.6) to the new coordinates, $\boldsymbol{R}'_i$, $\boldsymbol{R}_O$ and $\boldsymbol{R}$, and we must rewrite $\boldsymbol{P}_i^2$, $\boldsymbol{P}_1^2$, and $\boldsymbol{P}_2^2$ in terms of $\boldsymbol{P}'_i$, $\boldsymbol{P}_R$ and $\boldsymbol{P}_O$, conjugate to $\boldsymbol{R}'_i$, $\boldsymbol{R}$ and $\boldsymbol{R}_O$ respectively. Using the chain rule we find (see appendix 2.1)

$$\boldsymbol{P}_i = \left(\frac{\partial \boldsymbol{R}_O}{\partial \boldsymbol{R}_i}\right)\boldsymbol{P}_O + \left(\frac{\partial \boldsymbol{R}}{\partial \boldsymbol{R}_i}\right)\boldsymbol{P}_R + \sum_j \left(\frac{\partial \boldsymbol{R}'_j}{\partial \boldsymbol{R}_i}\right)\boldsymbol{P}'_j. \tag{2.14}$$

Similar equations for $\boldsymbol{P}_1$ and $\boldsymbol{P}_2$ yield the required results,

$$\frac{1}{2m}\sum_i \boldsymbol{P}_i^2 = \frac{nm}{2M^2}\boldsymbol{P}_O^2 + \frac{1}{2m}\sum_i \boldsymbol{P}'^2_i + \frac{nm}{2M^2}\left(\sum_i \boldsymbol{P}'_i\right)^2$$

$$+ \frac{1}{M}\boldsymbol{P}_O \cdot \sum_i \boldsymbol{P}'_i - \frac{nm}{M^2}\boldsymbol{P}_O \cdot \sum_i \boldsymbol{P}'_i - \frac{1}{M}\left(\sum_i \boldsymbol{P}'_i\right)^2, \tag{2.15}$$

$$\frac{1}{2M_1}\boldsymbol{P}_1^2 = \frac{M_1}{2M^2}\boldsymbol{P}_O^2 + \frac{1}{2M_1}\boldsymbol{P}_R^2 + \frac{M_1}{2M^2}\left(\sum_i \boldsymbol{P}'_i\right)^2 - \frac{1}{M}\boldsymbol{P}_O \cdot \boldsymbol{P}_R$$

$$- \frac{M_1}{M^2}\boldsymbol{P}_O \cdot \sum_i \boldsymbol{P}'_i + \frac{1}{M}\boldsymbol{P}_R \cdot \sum_i \boldsymbol{P}'_i, \tag{2.16}$$

$$\frac{1}{2M_2}\boldsymbol{P}_2^2 = \frac{M_2}{2M^2}\boldsymbol{P}_O^2 + \frac{1}{2M_2}\boldsymbol{P}_R^2 + \frac{M_2}{2M^2}\left(\sum_i \boldsymbol{P}'_i\right)^2 + \frac{1}{M}\boldsymbol{P}_O \cdot \boldsymbol{P}_R$$

$$- \frac{M_2}{M^2}\boldsymbol{P}_O \cdot \sum_i \boldsymbol{P}'_i - \frac{1}{M}\boldsymbol{P}_R \cdot \sum_i \boldsymbol{P}'_i. \tag{2.17}$$

Taking the sum of these last three equations, we obtain the required operator

$$T = \frac{1}{2M}\boldsymbol{P}_O^2 + \frac{1}{2\mu}\boldsymbol{P}_R^2 - \frac{1}{2M}\sum_{i,j=1}^n \boldsymbol{P}'_i \cdot \boldsymbol{P}'_j + \frac{1}{2m}\sum_{i=1}^n \boldsymbol{P}'^2_i \tag{2.18}$$

where μ is the reduced nuclear mass, $M_1 M_2/(M_1 + M_2)$. The first term in equation (2.18) represents the kinetic energy due to the translational motion of the whole molecule and in field-free space it can be omitted at this stage. Symmetry arguments show that translational motion can be separated rigorously from the other molecular motions in the absence of external fields. The third term on the right-hand side of equation (2.18) is called the mass polarisation term. It describes the small fluctuations in the position of the centre of mass as the electrons move around within the molecule.

The kinetic energy expression can also be written in terms of the Laplace operators:

$$T = -\frac{\hbar^2}{2M}\nabla_O^2 - \frac{\hbar^2}{2\mu}\nabla_R^2 + \frac{\hbar^2}{2M}\sum_{i,j=1}^n \nabla'_i \cdot \nabla'_j - \frac{\hbar^2}{2m}\sum_{i=1}^n \nabla'^2_i. \tag{2.19}$$

2.2.3. Origin at centre of mass of nuclei

We now transform from the centre of mass of the molecule to the centre of mass of the nuclei which is located at

$$\bm{R}'_N = \frac{1}{(M_1 + M_2)} \sum_\alpha M_\alpha \bm{R}'_\alpha \qquad (2.20)$$

in the (X', Y', Z') coordinate system. The new coordinates are $\bm{R}_O$, $\bm{R}$ and $\bm{R}''_i$, where $\bm{R}''_i$ is defined by

$$\bm{R}''_i = \bm{R}'_i - \frac{1}{M_1 + M_2} \sum_\alpha M_\alpha \bm{R}'_\alpha \qquad (2.21)$$

$$= \bm{R}'_i + \frac{1}{M_1 + M_2} \sum_i m_i \bm{R}'_i. \qquad (2.22)$$

Equation (2.22) follows from (2.21) because of (2.12). Hence we can rewrite equation (2.18) by expressing $\bm{P}'_i$ in terms of $\bm{P}''_i$, the momentum conjugate to the new coordinate $\bm{R}''_i$. By the chain rule

$$\bm{P}'_i = \sum_j \frac{\partial \bm{R}''_j}{\partial \bm{R}'_i} \bm{P}''_j \qquad (2.23)$$

$$= \bm{P}''_i + \frac{m}{M_1 + M_2} \sum_i \bm{P}''_i, \qquad (2.24)$$

from which we obtain the results

$$-\frac{1}{2M} \sum_{i,j} \bm{P}'_i \cdot \bm{P}'_j = -\frac{1}{2M} \left\{ 1 + \frac{2nm}{M_1 + M_2} + \frac{n^2 m^2}{(M_1 + M_2)^2} \right\} \sum_{i,j} \bm{P}''_i \cdot \bm{P}''_j, \qquad (2.25)$$

$$\frac{1}{2m} \sum_i \bm{P}'^2_i = \frac{1}{2m} \sum_i \bm{P}''^2_i + \frac{nm}{2(M_1 + M_2)^2} \sum_{i,j} \bm{P}''_i \cdot \bm{P}''_j$$

$$+ \frac{1}{M_1 + M_2} \sum_{i,j} \bm{P}''_i \cdot \bm{P}''_j. \qquad (2.26)$$

Substituting into equation (2.18) we obtain the new operator,

$$T = \frac{1}{2M} \bm{P}^2_O + \frac{1}{2\mu} \bm{P}^2_R + \frac{1}{2m} \sum_i \bm{P}''^2_i + \frac{1}{2(M_1 + M_2)} \sum_{i,j} \bm{P}''_i \cdot \bm{P}''_j. \qquad (2.27)$$

As before the kinetic energy may also be written in terms of the Laplace operators:

$$T = -\frac{\hbar^2}{2M} \nabla^2_O - \frac{\hbar^2}{2\mu} \nabla^2_R - \frac{\hbar^2}{2m} \sum_i \nabla''^2_i - \frac{\hbar^2}{2(M_1 + M_2)} \sum_{i,j} \nabla''_i \cdot \nabla''_j. \qquad (2.28)$$

The above terms represent the kinetic energy due to translation, the kinetic energy of the nuclei, the kinetic energy of the electrons, and finally a correction term, commonly known as the mass polarisation term.

2.2.4. Origin at geometrical centre of the nuclei

Starting from equation (2.6) with arbitrary origin we first transform to the molecular centre of mass (which ensures that the translational motion is separable), and then to the geometrical centre of the nuclei. The total transformation from arbitrary origin to the new origin is represented by

$$\boldsymbol{R} = \boldsymbol{R}_2 - \boldsymbol{R}_1, \tag{2.29}$$

$$\boldsymbol{R}_i''' = \boldsymbol{R}_i - (1/2)(\boldsymbol{R}_1 + \boldsymbol{R}_2), \tag{2.30}$$

$$\boldsymbol{R}_O = \frac{1}{M}\left\{m\sum_i \boldsymbol{R}_i + \sum_\alpha M_\alpha \boldsymbol{R}_\alpha\right\}. \tag{2.31}$$

Hence the momenta $\boldsymbol{P}_i$, $\boldsymbol{P}_1$ and $\boldsymbol{P}_2$ of equation (2.6) are now given by

$$\boldsymbol{P}_i = \frac{m}{M}\boldsymbol{P}_O + \sum_i \boldsymbol{P}_i''', \tag{2.32}$$

$$\boldsymbol{P}_1 = \frac{M_1}{M}\boldsymbol{P}_O - \boldsymbol{P}_R - (1/2)\sum_i \boldsymbol{P}_i''', \tag{2.33}$$

$$\boldsymbol{P}_2 = \frac{M_2}{M}\boldsymbol{P}_O + \boldsymbol{P}_R - (1/2)\sum_i \boldsymbol{P}_i'''. \tag{2.34}$$

Substituting in equation (2.6) we obtain the new Hamiltonian

$$T = \frac{1}{2M}\boldsymbol{P}_O^2 + \frac{1}{2\mu}\boldsymbol{P}_R^2 + \frac{1}{2m}\sum_i \boldsymbol{P}_i'''^2 + \frac{1}{8\mu}\sum_{i,j}\boldsymbol{P}_i'''\cdot\boldsymbol{P}_j''' - \frac{1}{2\mu_\alpha}\boldsymbol{P}_R\cdot\sum_i \boldsymbol{P}_i''' \tag{2.35}$$

where $\mu_\alpha = M_1 M_2/(M_1 - M_2)$.

2.3. The total Hamiltonian in field-free space

In our subsequent development we shall take the origin of coordinates to be at the centre of mass of the two nuclei, although we could equally well have chosen the molecular centre of mass as origin. Setting aside the translational motion of the molecule, we use equation (2.28) to represent the kinetic energy of the electrons and nuclei. To this we add terms representing the potential energy, electron spin interactions, and nuclear spin interactions. We subdivide the total Hamiltonian $\mathcal{H}_T$ into electronic and nuclear Hamiltonians,

$$\mathcal{H}_{\text{el}} = -\frac{\hbar^2}{2m}\sum_i \nabla_i^2 - \frac{\hbar^2}{2M_N}\sum_{i,j}\nabla_i\cdot\nabla_j + \sum_{i<j}\frac{e^2}{4\pi\varepsilon_0 R_{ij}}$$

$$- \sum_{\alpha,i}\frac{Z_\alpha e^2}{4\pi\varepsilon_0 R_{i\alpha}} + \mathcal{H}(\boldsymbol{S}_i) + \mathcal{H}(\boldsymbol{I}_\alpha), \tag{2.36}$$

$$\mathcal{H}_{\text{nucl}} = -\frac{\hbar^2}{2\mu}\nabla_R^2 + \sum_{\alpha,\beta}\frac{Z_\alpha Z_\beta e^2}{4\pi\varepsilon_0 R}. \tag{2.37}$$

$\mathcal{H}(\boldsymbol{I}_\alpha)$ is included in the electronic Hamiltonian since, as we shall see, its most important effects arise from interactions involving electronic motions. The interactions which arise from electron spin, $\mathcal{H}(\boldsymbol{S}_i)$, will be derived later from relativistic quantum mechanics; for the moment electron spin is introduced in a purely phenomenological manner. The electron–electron and electron–nuclear potential energies are included in equation (2.36) and the purely nuclear electrostatic repulsion is in equation (2.37). The double prime superscripts have been dropped for the sake of simplicity. We remind ourselves that μ in equation (2.37) is the reduced nuclear mass, $M_1 M_2/(M_1 + M_2)$.

We wish to divide $\mathcal{H}_T$ into a part describing the nuclear motion and a part describing the electronic motion in a fixed nuclear configuration, as far as possible. Equations (2.36) and (2.37) do not themselves represent such a separation because $\mathcal{H}_{el}$ is still a function of R, ϕ and θ and cannot therefore commute with $\mathcal{H}_{nucl}$ which, as we shall see, involves partial differential operators with respect to these coordinates. The obvious way to remove the effects of nuclear motion from $\mathcal{H}_{el}$ is by transforming from space-fixed axes to molecule-fixed axes gyrating with the nuclei.

In the Born–Oppenheimer approximation the basis set for $\mathcal{H}_{el}$ would consist of products of electronic space and spin functions. Transformation to the gyrating axis system may involve transformation of both space and spin variables, leading to a Hamiltonian in which the spin is quantised in the molecule-fixed axis system (as, for example, in a Hund's case (a) coupling scheme) or transformation of spatial variables only, in which case spatially quantised spin is implied (for example, Hund's case (b)). We will deal in detail with the former transformation and subsequently summarise the results appropriate to spatially quantised spin.

2.4. The nuclear kinetic energy operator

The rotational and vibrational kinetic energies of the nuclei are represented by the term $-(\hbar^2/2\mu)\nabla_R^2$ in equation (2.37); we now seek its explicit form and the relation between the momentum operators $\boldsymbol{P}_R$ and $\boldsymbol{P}_\alpha$ in equation (2.6). If we take components of $\boldsymbol{P}_R$ in a space-fixed frame, we have the straightforward relationship:

$$\boldsymbol{P}_R = -i\hbar \left\{ \frac{\partial}{\partial R_X} \boldsymbol{i}' + \frac{\partial}{\partial R_Y} \boldsymbol{j}' + \frac{\partial}{\partial R_Z} \boldsymbol{k}' \right\}. \tag{2.38}$$

However, it is more convenient to use curvilinear cartesian coordinates to describe the rotational motion of the nuclei. To this end we relate a set of rotating, molecule-fixed axes to the space-fixed axes by the three Euler rotations. In our experience the Euler angles and the rotations based upon them are not easily visualised; Zare [7] has given as good a description as any. Figure 2.2 defines the Euler angles ϕ, θ, χ, and the operations involved are as follows:

(i) a rotation about the initial Z axis through an angle $\phi (0 < \phi \leq 2\pi)$,
(ii) a subsequent rotation about the resultant Y axis through an angle $\theta (0 \leq \theta \leq \pi)$,
(iii) a final rotation about the resultant Z axis through an angle $\chi (0 \leq \chi \leq 2\pi)$.

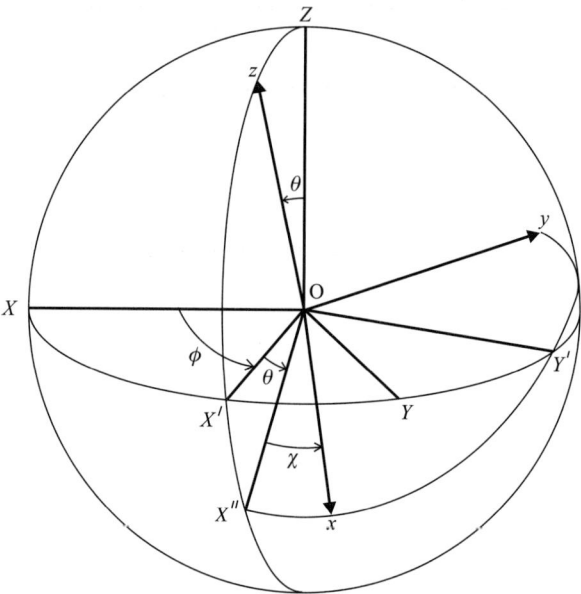

Figure 2.2. Definition of the Euler angles used in the transformation from space-fixed to molecule-fixed axes.

These rotations are performed sequentially and a rotation which takes one along an axis in the sense of a right-handed screw is defined as being positive. The nuclei are labelled so that the molecule-fixed z axis points from nucleus 1 to nucleus 2. It must be appreciated that this rotating coordinate system is a completely new one; it was not mentioned in section 2.3 where all the various coordinate systems have a fixed orientation in laboratory space.

The transformation between the space- and molecule-fixed coordinate systems is thus expressed by

$$\begin{bmatrix} X \\ Y \\ Z \end{bmatrix} = \mathfrak{M} \begin{bmatrix} x \\ y \\ z \end{bmatrix}, \tag{2.39}$$

where $\mathfrak{M}$ is the following unitary matrix:

$$\begin{bmatrix} \cos\phi\cos\theta\cos\chi - \sin\phi\sin\chi & -\sin\phi\cos\chi - \cos\phi\cos\theta\sin\chi & \cos\phi\sin\theta \\ \sin\phi\cos\theta\cos\chi + \cos\phi\sin\chi & \cos\phi\cos\chi - \sin\phi\cos\theta\sin\chi & \sin\phi\sin\theta \\ -\sin\theta\cos\chi & \sin\theta\sin\chi & \cos\theta \end{bmatrix}.$$

$$\tag{2.40}$$

We require only three nuclear coordinates to define the nuclear motion and we choose these to be R, the internuclear distance, ϕ and θ; the third Euler angle χ is a redundant coordinate. In fact, because there are no nuclei lying off-axis in a diatomic molecule, χ is undefineable; it is, however, expedient to retain it because of simplification in the final form of the rotational Hamiltonian. We shall examine this point in more detail in chapter 7. It is a simple matter to relate the components of $\boldsymbol{R}$ to R, θ, ϕ:

$$R_X = X_2 - X_1 = R\cos\phi\sin\theta, \tag{2.41}$$
$$R_Y = Y_2 - Y_1 = R\sin\phi\sin\theta, \tag{2.42}$$
$$R_Z = Z_2 - Z_1 = R\cos\theta. \tag{2.43}$$

These equations do not involve the redundant coordinate χ and can thus be inverted to give:

$$R^2 = R_X^2 + R_Y^2 + R_Z^2, \tag{2.44}$$
$$\tan\phi = R_Y/R_X, \tag{2.45}$$
$$\tan^2\theta = (R_X^2 + R_Y^2)/R_Z^2. \tag{2.46}$$

We wish to evaluate $\boldsymbol{P}_R$ from equation (2.38) and therefore need the following partial differentials:

$$\begin{aligned}
\frac{\partial}{\partial R_X} &= \frac{\partial R}{\partial R_X}\frac{\partial}{\partial R} + \frac{\partial \phi}{\partial R_X}\frac{\partial}{\partial \phi} + \frac{\partial \theta}{\partial R_X}\frac{\partial}{\partial \theta}, \\
\frac{\partial}{\partial R_Y} &= \frac{\partial R}{\partial R_Y}\frac{\partial}{\partial R} + \frac{\partial \phi}{\partial R_Y}\frac{\partial}{\partial \phi} + \frac{\partial \theta}{\partial R_Y}\frac{\partial}{\partial \theta}, \\
\frac{\partial}{\partial R_Z} &= \frac{\partial R}{\partial R_Z}\frac{\partial}{\partial R} + \frac{\partial \phi}{\partial R_Z}\frac{\partial}{\partial \phi} + \frac{\partial \theta}{\partial R_Z}\frac{\partial}{\partial \theta}.
\end{aligned} \tag{2.47}$$

We use equations (2.44), (2.45) and (2.46) to evaluate the partial differentials, and obtain the results given below. First the partial differentials of R with respect to R_X, R_Y and R_Z using (2.44):

$$\begin{aligned}
2R\frac{\partial R}{\partial R_X} &= 2R_X = 2R\sin\theta\cos\phi: & \text{hence } \frac{\partial R}{\partial R_X} &= \sin\theta\cos\phi, \\
2R\frac{\partial R}{\partial R_Y} &= 2R_Y = 2R\sin\theta\sin\phi: & \text{hence } \frac{\partial R}{\partial R_Y} &= \sin\theta\sin\phi, \\
2R\frac{\partial R}{\partial R_Z} &= 2R_Z = 2R\cos\theta: & \text{hence } \frac{\partial R}{\partial R_Z} &= \cos\theta.
\end{aligned} \tag{2.48}$$

Next, the partial differentials of θ using (2.46):

$$\begin{aligned}
2\tan\theta\sec^2\theta\frac{\partial \theta}{\partial R_X} &= \frac{2R_X}{R_Z^2} = \frac{2\sin\theta\cos\phi}{R\cos^2\theta}: & \text{hence } \frac{\partial \theta}{\partial R_X} &= \frac{\cos\theta\cos\phi}{R}, \\
2\tan\theta\sec^2\theta\frac{\partial \theta}{\partial R_Y} &= \frac{2R_Y}{R_Z^2} = \frac{2\sin\theta\sin\phi}{R\cos^2\theta}: & \text{hence } \frac{\partial \theta}{\partial R_Y} &= \frac{\cos\theta\sin\phi}{R}, \\
2\tan\theta\sec^2\theta\frac{\partial \theta}{\partial R_Z} &= -\frac{2(R_X^2+R_Y^2)}{R_Z^3} = -\frac{2\sin^2\theta}{R\cos^3\theta}: & \text{hence } \frac{\partial \theta}{\partial R_Z} &= -\frac{\sin\theta}{R}.
\end{aligned} \tag{2.49}$$

Thirdly, the partial differentials of ϕ using (2.45):

$$\sec^2\phi \frac{\partial \phi}{\partial R_X} = -\frac{R_Y}{R_X^2} = -\frac{\sin\phi}{R\sin\theta \cos^2\phi}: \quad \text{hence} \quad \frac{\partial \phi}{\partial R_X} = -\frac{\sin\phi}{R\sin\theta},$$

$$\sec^2\phi \frac{\partial \phi}{\partial R_Y} = \frac{1}{R_X} = \frac{1}{R\sin\theta \cos\phi}: \quad \text{hence} \quad \frac{\partial \phi}{\partial R_Y} = \frac{\cos\phi}{R\sin\theta}, \quad (2.50)$$

$$\frac{\partial \phi}{\partial R_Z} = 0.$$

We may now substitute in (2.38) to obtain

$$\begin{aligned}\boldsymbol{P}_R = -i\hbar\Bigg\{&\left(\sin\theta\cos\phi\frac{\partial}{\partial R} - \frac{\sin\phi}{R\sin\theta}\frac{\partial}{\partial \phi} + \frac{\cos\theta\cos\phi}{R}\frac{\partial}{\partial \theta}\right)\boldsymbol{i}' \\ &+ \left(\sin\theta\sin\phi\frac{\partial}{\partial R} + \frac{\cos\phi}{R\sin\theta}\frac{\partial}{\partial \phi} + \frac{\cos\theta\sin\phi}{R}\frac{\partial}{\partial \theta}\right)\boldsymbol{j}' \\ &+ \left(\cos\theta\frac{\partial}{\partial R} - \frac{\sin\theta}{R}\frac{\partial}{\partial \theta}\right)\boldsymbol{k}'\Bigg\}. \end{aligned} \quad (2.51)$$

We now note that the nuclear kinetic energy operator in equations (2.35) and (2.37) is

$$\frac{1}{2\mu}\boldsymbol{P}_R^2 = -\frac{\hbar^2}{2\mu}\nabla_R^2. \quad (2.52)$$

The required expression for the nuclear kinetic energy operator is therefore

$$\frac{1}{2\mu}\boldsymbol{P}_R^2 = -\frac{\hbar^2}{2\mu}\left\{\frac{1}{R^2}\frac{\partial}{\partial R}\left(R^2\frac{\partial}{\partial R}\right) + \frac{1}{R^2\sin\theta}\frac{\partial}{\partial \theta}\left(\sin\theta\frac{\partial}{\partial \theta}\right) + \frac{1}{R^2\sin^2\theta}\frac{\partial^2}{\partial \phi^2}\right\}, \quad (2.53)$$

or equivalently,

$$\frac{1}{2\mu}\boldsymbol{P}_R^2 = -\frac{\hbar^2}{2\mu}\left\{\frac{1}{R^2}\frac{\partial}{\partial R}\left(R^2\frac{\partial}{\partial R}\right) + \frac{\operatorname{cosec}\theta}{R^2}\frac{\partial}{\partial \theta}\left(\sin\theta\frac{\partial}{\partial \theta}\right) + \frac{\operatorname{cosec}^2\theta}{R^2}\frac{\partial^2}{\partial \phi^2}\right\}. \quad (2.54)$$

The partial derivatives in (2.53) or (2.54) are performed with the electronic coordinates in the space-fixed axis system held constant. The first term, which involves the internuclear distance only, represents the vibrational motion of the nuclei, whilst the second and third terms describe the rotational motion. We will examine them in more detail in due course.

We now investigate the relationship between $\boldsymbol{P}_R$ and the original nuclear momenta $\boldsymbol{P}_\alpha$ in more detail. We have introduced a redundant coordinate χ in the transformation matrix $\mathfrak{M}$ (2.40). We therefore need to define our original $(3n+6)$ coordinates in terms of the $(3n+7)$ final coordinates. To do this, we note that the position of the molecular centre of mass in the (X'', Y'', Z'') coordinate system is

$$\boldsymbol{R}_O'' = \frac{1}{M}\left\{\sum_\alpha M_\alpha \boldsymbol{R}_\alpha'' + m\sum_i \boldsymbol{R}_i''\right\} = \frac{m}{M}\sum_i \boldsymbol{R}_i''. \quad (2.55)$$

Thus

$$\boldsymbol{R}_\alpha = \boldsymbol{R}_\mathrm{O} - \boldsymbol{R}_\mathrm{O}'' + \boldsymbol{R}_\alpha'' = \boldsymbol{R}_\mathrm{O} - \frac{m}{M}\sum_i \boldsymbol{R}_i'' \mp \frac{\mu}{M_\alpha}\boldsymbol{R}, \qquad (2.56)$$

and

$$\boldsymbol{R}_i = \boldsymbol{R}_\mathrm{O} - \boldsymbol{R}_\mathrm{O}'' + \boldsymbol{R}_i'' = \boldsymbol{R}_\mathrm{O} - \frac{m}{M}\sum_j \boldsymbol{R}_j'' + \boldsymbol{R}_i''. \qquad (2.57)$$

In equation (2.56) the upper sign is taken for $\alpha = 1$ and the lower for $\alpha = 2$. From (2.56) we see that differentials such as $(\partial \boldsymbol{R}_\alpha/\partial \phi)$ are uniquely defined so that we can express $\boldsymbol{P}_R$ in terms of $\boldsymbol{P}_\alpha$ (and $\boldsymbol{P}_i$) but again we cannot in general write down the inverse relationships. We see that

$$\boldsymbol{P}_R = \sum_\alpha \frac{\partial \boldsymbol{R}_\alpha}{\partial \boldsymbol{R}}\boldsymbol{P}_\alpha + \sum_i \frac{\partial \boldsymbol{R}_i}{\partial \boldsymbol{R}}\boldsymbol{P}_i \qquad (2.58)$$

$$= \sum_\alpha \mp \frac{\mu}{M_\alpha}\boldsymbol{P}_\alpha. \qquad (2.59)$$

The analogous expressions for $\boldsymbol{P}_\mathrm{O}$ and $\boldsymbol{p}_i$ are

$$\boldsymbol{P}_\mathrm{O} = \sum_\alpha \boldsymbol{P}_\alpha + \sum_i \boldsymbol{P}_i, \qquad (2.60)$$

$$\boldsymbol{p}_i = \boldsymbol{P}_i - \frac{m}{M}\left\{\sum_\alpha \boldsymbol{P}_\alpha + \sum_j \boldsymbol{P}_j\right\}. \qquad (2.61)$$

These expressions will be useful in a later section; for the moment we turn our attention to the explicit form of (2.59). If we substitute equations (2.41), (2.42) and (2.43) in (2.56) we obtain

$$X_\alpha = \mp \frac{\mu}{M_\alpha} R \sin\theta \cos\phi + X_\mathrm{O} - \frac{m}{M}\sum_i X_i'', \qquad (2.62)$$

$$Y_\alpha = \mp \frac{\mu}{M_\alpha} R \sin\theta \sin\phi + Y_\mathrm{O} - \frac{m}{M}\sum_i Y_i'', \qquad (2.63)$$

$$Z_\alpha = \mp \frac{\mu}{M_\alpha} R \cos\theta + Z_\mathrm{O} - \frac{m}{M}\sum_i Z_i''. \qquad (2.64)$$

Therefore we have

$$\frac{\partial}{\partial R} = \mp \frac{\mu}{M_\alpha}\left(\sin\theta\cos\phi\frac{\partial}{\partial X_\alpha} + \sin\theta\sin\phi\frac{\partial}{\partial Y_\alpha} + \cos\theta\frac{\partial}{\partial Z_\alpha}\right), \qquad (2.65)$$

$$\frac{\partial}{\partial \phi} = \mp \frac{\mu}{M_\alpha}R\left(-\sin\theta\sin\phi\frac{\partial}{\partial X_\alpha} + \sin\theta\cos\phi\frac{\partial}{\partial Y_\alpha}\right), \qquad (2.66)$$

$$\frac{\partial}{\partial \theta} = \mp \frac{\mu}{M_\alpha}R\left(\cos\theta\cos\phi\frac{\partial}{\partial X_\alpha} + \cos\theta\sin\phi\frac{\partial}{\partial Y_\alpha} - \sin\theta\frac{\partial}{\partial Z_\alpha}\right), \qquad (2.67)$$

and

$$\left(\frac{\partial}{\partial \chi}\right)_{X_i'', Y_i'', Z_i''} = 0. \tag{2.68}$$

We have included the double-primed subscripts in (2.68) to emphasise that the differentiation is performed with *space-fixed* nuclear centre-of-mass electronic coordinates held constant. Equation (2.68) can be appreciated when we realise that the total Hamiltonian is independent of χ so that we can take the eigenfunctions Ψ_{rve} to be independent of χ also. This relationship provides a crucial restriction on the redundant coordinates; its form is such that we could, if we wished, write down the inverses of equations (2.59), (2.60) and (2.61).

From (2.68) we see that we can add selected terms in $\partial/\partial \chi$ to our expression for $\boldsymbol{P}_R$ in (2.51) and hence to the nuclear Hamiltonian, without altering the values of any of the physical observables. We choose these terms so that the rotational Hamiltonian has the same form as the rotational Hamiltonian of a spherical top molecule. We shall see later that with this choice for the rotational Hamiltonian, we can make use of the very powerful techniques of angular momentum theory, in particular, irreducible tensor methods, which would otherwise be denied to us. Accordingly, we modify equation (2.51) to be

$$\begin{aligned}
\boldsymbol{P}_R = -i\hbar \bigg\{ &\left(\sin\theta \cos\phi \frac{\partial}{\partial R} - \frac{\sin\phi}{R\sin\theta} \frac{\partial}{\partial \phi} + \frac{\cos\theta\cos\phi}{R} \frac{\partial}{\partial \theta} + \frac{\cot\theta\sin\phi}{R} \frac{\partial}{\partial \chi} \right) \boldsymbol{i}' \\
+ &\left(\sin\theta \sin\phi \frac{\partial}{\partial R} + \frac{\cos\phi}{R\sin\theta} \frac{\partial}{\partial \phi} + \frac{\cos\theta\sin\phi}{R} \frac{\partial}{\partial \theta} - \frac{\cot\theta\cos\phi}{R} \frac{\partial}{\partial \chi} \right) \boldsymbol{j}' \\
+ &\left(\cos\theta \frac{\partial}{\partial R} - \frac{\sin\theta}{R} \frac{\partial}{\partial \theta} \right) \boldsymbol{k}' \bigg\}. \tag{2.69}
\end{aligned}$$

We can simplify this expression by introducing the components of the angular momentum operator $\boldsymbol{N}$ of the rotating coordinate system. These components are defined by

$$N_X = i \left\{ \cot\theta \cos\phi \frac{\partial}{\partial \phi} + \sin\phi \frac{\partial}{\partial \theta} - \operatorname{cosec}\theta \cos\phi \frac{\partial}{\partial \chi} \right\}, \tag{2.70}$$

$$N_Y = i \left\{ \cot\theta \sin\phi \frac{\partial}{\partial \phi} - \cos\phi \frac{\partial}{\partial \theta} - \operatorname{cosec}\theta \sin\phi \frac{\partial}{\partial \chi} \right\}, \tag{2.71}$$

$$N_Z = -i \frac{\partial}{\partial \phi}. \tag{2.72}$$

Hence we can write

$$\boldsymbol{P}_R = \hbar \left\{ -\frac{1}{R}(\boldsymbol{k} \wedge \boldsymbol{N}) - i\boldsymbol{k} \frac{\partial}{\partial R} \right\} \tag{2.73}$$

where $\boldsymbol{k}$ is the unit vector along the molecule-fixed z axis:

$$\boldsymbol{k} = \sin\theta \cos\phi \, \boldsymbol{i}' + \sin\theta \sin\phi \, \boldsymbol{j}' + \cos\theta \, \boldsymbol{k}'. \tag{2.74}$$

Note that the two terms in (2.73) do not commute since $\mathbf{k}$ is a function of the Euler angles. We can substitute either (2.69) or (2.73) into (2.38) to obtain the modified form of (2.53),

$$\frac{1}{2\mu}\mathbf{P}_R^2 = -\frac{\hbar^2}{2\mu R^2}\frac{\partial}{\partial R}\left(R^2\frac{\partial}{\partial R}\right) - \frac{\hbar^2}{2\mu R^2}\left\{\operatorname{cosec}\theta\frac{\partial}{\partial\theta}\left(\sin\theta\frac{\partial}{\partial\theta}\right)\right.$$
$$\left. + \operatorname{cosec}^2\theta\left[\frac{\partial^2}{\partial\phi^2} + \frac{\partial^2}{\partial\chi^2} - 2\cos\theta\frac{\partial^2}{\partial\phi\partial\chi}\right]\right\} \quad (2.75)$$

$$= -\frac{\hbar^2}{2\mu R^2}\frac{\partial}{\partial R}\left(R^2\frac{\partial}{\partial R}\right) + \frac{\hbar^2}{2\mu R^2}\mathbf{N}^2. \quad (2.76)$$

Along with the introduction of the redundant coordinate χ, we define the normalisation condition as

$$\int\int\int\int\int \Psi_{\text{rve}}^*\Psi_{\text{rve}}R^2\sin\theta\,\mathrm{d}\mathbf{R}_i''\,\mathrm{d}R\,\mathrm{d}\phi\,\mathrm{d}\theta\,\mathrm{d}\chi = 1, \quad (2.77)$$

where $\mathbf{R}_i''$ represents the $3n$ cartesian coordinates of the electrons. The modified form of $\mathcal{H}_{\text{rot}}$ in equation (2.75) is equivalent to the following expression for the original nuclear momenta:

$$\mathbf{P}_\alpha = -\mathrm{i}\hbar\left\{\frac{M_\alpha}{M}\left[\mathbf{i}'\frac{\partial}{\partial X_0} + \mathbf{j}'\frac{\partial}{\partial Y_0} + \mathbf{k}'\frac{\partial}{\partial Z_0}\right] - \frac{M_\alpha}{M_N}\sum_i\left[\mathbf{i}'\frac{\partial}{\partial X_i''} + \mathbf{j}'\frac{\partial}{\partial Y_i''} + \mathbf{k}'\frac{\partial}{\partial Z_i''}\right]\right.$$
$$\mp\left[\mathbf{i}'\left(\cos\phi\sin\theta\frac{\partial}{\partial R} - \frac{\sin\phi\operatorname{cosec}\theta}{R}\frac{\partial}{\partial\phi} + \frac{\cos\phi\cos\theta}{R}\frac{\partial}{\partial\theta} + \frac{\sin\phi\cot\theta}{R}\frac{\partial}{\partial\chi}\right)\right.$$
$$+ \mathbf{j}'\left(\sin\phi\sin\theta\frac{\partial}{\partial R} + \frac{\cos\phi\operatorname{cosec}\theta}{R}\frac{\partial}{\partial\phi} + \frac{\sin\phi\cos\theta}{R}\frac{\partial}{\partial\theta} - \frac{\cos\phi\cot\theta}{R}\frac{\partial}{\partial\chi}\right)$$
$$\left.\left. + \mathbf{k}'\left(\cos\theta\frac{\partial}{\partial R} - \frac{\sin\theta}{R}\frac{\partial}{\partial\theta}\right)\right]\right\}. \quad (2.78)$$

2.5. Transformation of the electronic coordinates to molecule-fixed axes

2.5.1. Introduction

We have derived the total Hamiltonian expressed in a space-fixed (i.e. non-rotating) coordinate system in (2.36), (2.37) and (2.75). We can now simplify the electronic Hamiltonian $\mathcal{H}_{\text{el}}$ by transforming the electronic coordinates to the molecule-fixed axis system defined by (2.40) because the Coulombic potential term, when expressed as a function of these new coordinates, is independent of θ, ϕ and χ. From a physical standpoint it is obviously sensible to transform the electronic coordinates in this way because under the influence of the electrostatic interactions, the electrons rotate in space with the nuclei. We shall take the opportunity to refer the electron spins to the molecule-fixed axis system in this section also, and leave discussion of the alternative scheme of space quantisation to a later section. Since we assume the electron spin wave function to be completely separable from the spatial (i.e. orbital) wave function, the decision as to whether the spin is quantised in the space- or molecule-fixed axis system can be made independently of the choice of axis system for the electronic spatial coordinates, and will depend on the particular molecule that one is trying to describe. In addition we can deal with each transformation separately.

We shall later be seeking solutions $\Psi_{rve}(x_1, y_1, z_1, \ldots, z_n; R, \theta, \phi; \Sigma_1, \ldots, \Sigma_n)$ to the total Hamiltonian where this eigenfunction is to be considered as one component of a 2^n-rank spinor and Σ_i can be either α_i or β_i, corresponding to quantisation of the spin parallel or antiparallel to the molecule-fixed z axis. When we perform partial differentiation with respect to R, ϕ, θ or χ in the nuclear Hamiltonian (2.75), we must take care to include both the explicit functional dependence of Ψ_{rve} on these coordinates and the implicit dependence that arises because x_i, y_i, z_i and Σ_i are functions of the three Euler angles (see (2.39), for example). It is desirable to rewrite the Hamiltonian in such a form that, when we operate on Ψ_{rve} with $\partial/\partial\theta$, $\partial/\partial\phi$ or $\partial/\partial\chi$, only the explicit dependence of Ψ_{rve} on the Euler angles is to be considered; in other words, we can ignore the implicit effects resulting from the θ, ϕ and χ dependence of the transformations of electronic spatial and spin coordinates from the fixed to the gyrating axis system in our final Hamiltonian. In the case that Ψ_{rve} can be expressed as a product function $\psi_e(x_i, y_i, z_i; \Sigma_i)\psi_{rv}(R, \theta, \phi)$, this means that $\partial/\partial\theta$, $\partial/\partial\phi$, and $\delta/\delta\chi$ operate only on the factor ψ_{rv}.

2.5.2. Space transformations

The operators in (2.36) are easily re-expressed in the molecule-fixed coordinate system since the ∇_i'' operator merely becomes the ∇_i operator in the new coordinate system and, as mentioned earlier, $V_{el,nucl}$ becomes independent of the Euler angles. We must also consider the transformation of the partial differential operators $\partial/\partial\phi$, $\partial/\partial\theta$ and $\partial/\partial\chi$ from space to molecule-fixed axes. Thus

$$\left(\frac{\partial}{\partial\theta}\right)_s = \left(\frac{\partial}{\partial\theta}\right)_m + \sum_i \left\{\left(\frac{\partial x_i}{\partial\theta}\right)_s \left(\frac{\partial}{\partial x_i}\right)_m + \left(\frac{\partial y_i}{\partial\theta}\right)_s \left(\frac{\partial}{\partial y_i}\right)_m + \left(\frac{\partial z_i}{\partial\theta}\right)_s \left(\frac{\partial}{\partial z_i}\right)_m\right\},$$

$$\left(\frac{\partial}{\partial\phi}\right)_s = \left(\frac{\partial}{\partial\phi}\right)_m + \sum_i \left\{\left(\frac{\partial x_i}{\partial\phi}\right)_s \left(\frac{\partial}{\partial x_i}\right)_m + \left(\frac{\partial y_i}{\partial\phi}\right)_s \left(\frac{\partial}{\partial y_i}\right)_m + \left(\frac{\partial z_i}{\partial\phi}\right)_s \left(\frac{\partial}{\partial z_i}\right)_m\right\},$$

$$\left(\frac{\partial}{\partial\chi}\right)_s = \left(\frac{\partial}{\partial\chi}\right)_m + \sum_i \left\{\left(\frac{\partial x_i}{\partial\chi}\right)_s \left(\frac{\partial}{\partial x_i}\right)_m + \left(\frac{\partial y_i}{\partial\chi}\right)_s \left(\frac{\partial}{\partial y_i}\right)_m + \left(\frac{\partial z_i}{\partial\chi}\right)_s \left(\frac{\partial}{\partial z_i}\right)_m\right\}.$$

(2.79)

The subscripts 's' and 'm' denote that space or molecule-fixed electron coordinates are held constant; for example, $(\partial/\partial\theta)_s$ means $(\partial/\partial\theta)_{R,\phi,\chi,X_i'',Y_i'',Z_i''}$. The desirability of rewriting the total Hamiltonian in terms of $(\partial/\partial\theta)_m$, $(\partial/\partial\phi)_m$ and $(\partial/\partial\chi)_m$ is thus evident; $(\partial/\partial\theta)_s$ corresponds to the effect on the wavefunction of an infinitesimal rotation of the nuclei alone, with the electrons held fixed in space, whereas $(\partial/\partial\theta)_m$ corresponds to the effect of an infinitesimal rotation of the molecule-fixed axis system and all the particles with it.

In order to carry out the differentiations in (2.79), we use the transformation matrix between the two coordinate systems given in equation (2.40) and obtain

$$\left(\frac{\partial}{\partial \theta}\right)_s = \left(\frac{\partial}{\partial \theta}\right)_m + \sum_i \left\{ [-\cos\phi \sin\theta \cos\chi X_i'' \right.$$
$$\left. - \sin\phi \sin\theta \cos\chi Y_i'' - \cos\theta \cos\chi Z_i''] \left(\frac{\partial}{\partial x_i}\right)_m \right.$$
$$\left. + [\cos\phi \sin\theta \sin\chi X_i'' + \sin\phi \sin\theta \sin\chi Y_i'' + \cos\theta \sin\chi Z_i''] \left(\frac{\partial}{\partial y_i}\right)_m \right.$$
$$\left. + [\cos\phi \cos\theta X_i'' + \sin\theta \cos\theta Y_i'' - \sin\theta Z_i''] \left(\frac{\partial}{\partial z_i}\right)_m \right\}, \quad (2.80)$$

$$\left(\frac{\partial}{\partial \phi}\right)_s = \left(\frac{\partial}{\partial \phi}\right)_m + \sum_i \left\{ [(-\sin\phi \cos\theta \cos\chi - \cos\phi \sin\chi) X_i'' \right.$$
$$\left. + (\cos\phi \cos\theta \cos\chi - \sin\phi \sin\chi) Y_i''] \left(\frac{\partial}{\partial x_i}\right)_m \right.$$
$$\left. + [(-\cos\phi \cos\chi + \sin\phi \cos\theta \sin\chi) X_i'' \right.$$
$$\left. + (-\sin\phi \cos\chi - \cos\phi \cos\theta \sin\chi) Y_i''] \left(\frac{\partial}{\partial y_i}\right)_m \right.$$
$$\left. + [-\sin\phi \sin\theta X_i'' + \cos\phi \sin\theta Y_i''] \left(\frac{\partial}{\partial z_i}\right)_m \right\}, \quad (2.81)$$

$$\left(\frac{\partial}{\partial \chi}\right)_s = \left(\frac{\partial}{\partial \chi}\right)_m + \sum_i \left\{ [(-\cos\phi \cos\theta \sin\chi - \sin\phi \cos\chi) X_i'' \right.$$
$$\left. + (-\sin\phi \cos\theta \sin\chi + \cos\phi \cos\chi) Y_i'' + \sin\theta \sin\chi Z_i''] \left(\frac{\partial}{\partial x_i}\right)_m \right.$$
$$\left. + [(\sin\phi \sin\chi - \cos\phi \cos\theta \cos\chi) X_i'' \right.$$
$$\left. + (-\cos\phi \sin\chi - \sin\phi \cos\theta \cos\chi) Y_i'' + (\sin\theta \cos\chi) Z_i''] \left(\frac{\partial}{\partial y_i}\right)_m \right\}. \quad (2.82)$$

We now revert to molecule-fixed coordinates, obtaining

$$\left(\frac{\partial}{\partial \theta}\right)_s = \left(\frac{\partial}{\partial \theta}\right)_m + \sum_i \left\{ -\cos\chi z_i \left(\frac{\partial}{\partial x_i}\right)_m + \sin\chi z_i \left(\frac{\partial}{\partial y_i}\right)_m \right.$$
$$\left. + (\cos\chi x_i - \sin\chi y_i) \left(\frac{\partial}{\partial z_i}\right)_m \right\}, \quad (2.83)$$

$$\left(\frac{\partial}{\partial \phi}\right)_s = \left(\frac{\partial}{\partial \phi}\right)_m + \sum_i \left\{ (\cos\theta y_i - \sin\theta \sin\chi z_i) \left(\frac{\partial}{\partial x_i}\right)_m \right.$$
$$\left. + (-\cos\theta x_i - \sin\theta \cos\chi z_i) \left(\frac{\partial}{\partial y_i}\right)_m \right.$$
$$\left. + (\sin\theta \sin\chi x_i + \sin\theta \cos\chi y_i) \left(\frac{\partial}{\partial z_i}\right)_m \right\} \quad (2.84)$$

$$\left(\frac{\partial}{\partial \chi}\right)_{\text{s}} = \left(\frac{\partial}{\partial \chi}\right)_{\text{m}} + \sum_i \left\{ y_i \left(\frac{\partial}{\partial x_i}\right)_{\text{m}} - x_i \left(\frac{\partial}{\partial y_i}\right)_{\text{m}} \right\}. \tag{2.85}$$

Now the orbital angular momentum operator $\hbar \boldsymbol{L}$ has components in the space-fixed axis system which are defined as

$$L_X = -\mathrm{i} \sum_j \left\{ Y_j'' \left(\frac{\partial}{\partial Z_j''}\right)_{\text{s}} - Z_j'' \left(\frac{\partial}{\partial Y_j''}\right)_{\text{s}} \right\}, \tag{2.86}$$

$$L_Y = -\mathrm{i} \sum_j \left\{ Z_j'' \left(\frac{\partial}{\partial X_j''}\right)_{\text{s}} - X_j'' \left(\frac{\partial}{\partial Z_j''}\right)_{\text{s}} \right\}, \tag{2.87}$$

$$L_Z = -\mathrm{i} \sum_j \left\{ X_j'' \left(\frac{\partial}{\partial Y_j''}\right)_{\text{s}} - Y_j'' \left(\frac{\partial}{\partial X_j''}\right)_{\text{s}} \right\}. \tag{2.88}$$

When we refer these components to the gyrating axis system, we find that the resultant operators L_x, L_y and L_z are the same functions of the molecule-fixed coordinates and their conjugate momenta as L_X, L_Y and L_Z are of the space-fixed electronic coordinates and their conjugate momenta. Thus L_x, L_y and L_z are given by expressions of the same form as (2.86), (2.87) and (2.88) respectively. However, since the partial differentiations are performed with θ, ϕ, and χ held constant, these components are not the components of the orbital angular momentum measured in the gyrating axis system but are the components measured in a space-fixed axis system which is instantaneously coincident with the moving axis system. Similar remarks apply to the components of the spin angular momentum operator which will be introduced in section 2.5.3.

We thus have the final results:

$$\left(\frac{\partial}{\partial \theta}\right)_{\text{s}} = \left(\frac{\partial}{\partial \theta}\right)_{\text{m}} - \mathrm{i} \sin \chi L_x - \mathrm{i} \cos \chi L_y, \tag{2.89}$$

$$\left(\frac{\partial}{\partial \phi}\right)_{\text{s}} = \left(\frac{\partial}{\partial \phi}\right)_{\text{m}} + \mathrm{i} \sin \theta \cos \chi L_x - \mathrm{i} \sin \theta \sin \chi L_y - \mathrm{i} \cos \theta L_z, \tag{2.90}$$

$$\left(\frac{\partial}{\partial \chi}\right)_{\text{s}} = \left(\frac{\partial}{\partial \chi}\right)_{\text{m}} - \mathrm{i} L_z. \tag{2.91}$$

It is interesting to note that since $(\partial/\partial \chi)_{\text{s}}$ is zero for a diatomic molecule, $(\partial/\partial \chi)_{\text{m}}$ is, by (2.91), equivalent to $\mathrm{i} L_z$.

2.5.3. Spin transformations

We now consider some aspects of the theory of electronic spin angular momentum. What follows here is a relatively brief and simple exposition; we will return to a comprehensive description of the details of electron spin theory in chapter 3. For a single electron, the spin vector $\boldsymbol{S}$ is set equal to $(1/2)\boldsymbol{\sigma}'$ and in the representation where S_Z (or σ_Z') is diagonal, the components of the vector $\boldsymbol{\sigma}'$ may be represented by 2×2

matrices, first introduced by Pauli [8],

$$\sigma'_X = \begin{bmatrix} 0 & 1 \\ 1 & 0 \end{bmatrix}, \quad \sigma'_Y = \begin{bmatrix} 0 & -i \\ i & 0 \end{bmatrix}, \quad \sigma'_Z = \begin{bmatrix} 1 & 0 \\ 0 & -1 \end{bmatrix}. \tag{2.92}$$

For $S = 1/2$, functions ψ_M are eigenfunctions of the operators S^2 and S_Z if they satisfy the equations

$$S^2 \psi_M = S(S+1)\psi_M = (3/4)\psi_M, \tag{2.93}$$

$$S_Z \psi_M = M\psi_M, \quad \text{where } M = \pm 1/2. \tag{2.94}$$

There are therefore two eigenfunctions $\psi_{M=\pm 1/2}$ and from the matrix representation of S^2 and S_Z each ψ_M must be a two-component function,

$$\psi_{+1/2} = \begin{bmatrix} 1 \\ 0 \end{bmatrix}, \quad \psi_{-1/2} = \begin{bmatrix} 0 \\ 1 \end{bmatrix}. \tag{2.95}$$

These functions are called *spinors* and any other two-component function (spinor) can be written as a linear combination of them,

$$\begin{bmatrix} a \\ b \end{bmatrix} = a\psi_{1/2} + b\psi_{-1/2}. \tag{2.96}$$

Let us now consider a representation in which the component of $\boldsymbol{\sigma}'$ in any direction $\mathbf{n}$ (i.e. $\boldsymbol{\sigma}' \cdot \mathbf{n}$) is diagonal. Then

$$(\boldsymbol{\sigma}' \cdot \mathbf{n})_m = \lambda \psi_m \quad \text{where } \psi_m = \sum_M a_M \psi_M. \tag{2.97}$$

Since $(\boldsymbol{\sigma}' \cdot \mathbf{n})^2 = 1$ it follows that $\lambda = \pm 1$. A rotation from the first representation (which we now identify as space-fixed) to the second (instantaneous molecule-fixed) can be represented by a rotational matrix, i.e.,

$$\psi_m(\mathbf{S}) = \sum_M \mathcal{D}^{1/2}_{M,m}(\phi, \theta, \chi) \psi_M(\mathbf{S}) \tag{2.98}$$

where the rotational matrix is given by

$$\mathcal{D}^{1/2}(\phi, \theta, \chi) = \begin{bmatrix} e^{-i\phi/2} \cos(\theta/2) e^{-i\chi/2} & -e^{-i\phi/2} \sin(\theta/2) e^{i\chi/2} \\ e^{i\phi/2} \sin(\theta/2) e^{-i\chi/2} & e^{i\phi/2} \cos(\theta/2) e^{i\chi/2} \end{bmatrix}. \tag{2.99}$$

We are now in a position to investigate the effects of $\partial/\partial \phi$, $\partial/\partial \theta$ and $\partial/\partial \chi$ on the electron spin functions. When the electron spins are quantised in the molecule-fixed axis system, we see that each component of the 2^n-rank spinor is an implicit function of ϕ, θ and χ through its dependence on the transformation matrix (2.99). The total spinor $\psi(\mathbf{S})$ may be expressed as a product of one-electron spinors,

$$\psi(\mathbf{S}) = \prod_i \psi_m(s_i) \tag{2.100}$$

where m equals either $+1/2$ or $-1/2$; we therefore first consider the effects of operating with $\partial/\partial \phi$, $\partial/\partial \theta$ and $\partial/\partial \chi$ on the one-electron spinors (for simplicity we drop the subscript i):

$$\frac{\partial}{\partial \phi}\psi_{m=\pm 1/2}(s) = \sum_M \frac{\partial}{\partial \phi}\mathcal{D}^{1/2}_{M,\pm 1/2}(\phi,\theta,\chi)\psi_M(s)$$

$$= \sum_M \frac{i}{2}\{\mp\cos\theta\mathcal{D}^{1/2}_{M,\pm 1/2}(\phi,\theta,\chi) + \sin\theta e^{\mp i\chi}\mathcal{D}^{1/2}_{M,\mp 1/2}(\phi,\theta,\chi)\}\psi_M(s)$$

$$= \frac{i}{2}\{\mp\cos\theta\,\psi_{\pm 1/2}(s) + \sin\theta e^{\mp i\chi}\psi_{\mp 1/2}(s)\}, \tag{2.101}$$

$$\frac{\partial}{\partial \theta}\psi_{m=\pm 1/2}(s) = \sum_M \pm(1/2)e^{\mp i\chi}\mathcal{D}^{1/2}_{M,\mp 1/2}(\phi,\theta,\chi)\psi_M(s)$$

$$= \pm(1/2)e^{\mp i\chi}\psi_{\mp 1/2}(s), \tag{2.102}$$

$$\frac{\partial}{\partial \chi}\psi_{m=\pm 1/2}(s) = \sum_M \mp(i/2)\mathcal{D}^{1/2}_{M,\pm 1/2}(\phi,\theta,\chi)\psi_M(s)$$

$$= \mp(i/2)\psi_{\pm 1/2}(s). \tag{2.103}$$

We next use the Pauli matrix representations of the spin angular momentum operator components in the instantaneous molecule-fixed axis system from equation (2.92) to rewrite the above relationships:

$$\frac{\partial}{\partial \phi}\psi_{m=\pm 1/2}(s) = -i\{\cos\theta s_z + \sin\theta(\sin\chi s_y - \cos\chi s_x)\}\psi_{m=\pm 1/2}(s), \tag{2.104}$$

$$\frac{\partial}{\partial \theta}\psi_{m=\pm 1/2}(s) = -i\{\sin\chi s_x + \cos\chi s_y\}\psi_{m=\pm 1/2}(s), \tag{2.105}$$

$$\frac{\partial}{\partial \chi}\psi_{m=\pm 1/2}(s) = -is_z\psi_{m=\pm 1/2}(s). \tag{2.106}$$

We are interested in the result for the total spinor $\psi(S)$ and we find, for example,

$$\frac{\partial}{\partial \theta}\psi(S) = \sum_j \prod_{i\neq j}\psi_m(s_i)\frac{\partial}{\partial \phi}\psi_m(s_j)$$

$$= \sum_j \prod_{i\neq j}\psi_m(s_i)\{-i[\sin\chi(s_j)_x + \cos\chi(s_j)_y]\}\psi_m(s_j)$$

$$= \sum_i [-i\sin\chi(s_i)_x - i\cos\chi(s_i)_y]\prod_i \psi_m(s_i)$$

$$= (-\sin\chi S_x - i\cos\chi S_y)\psi(S), \tag{2.107}$$

where S_x and S_y are molecule-fixed components of the total spin $S = \sum_i s_i$. Similarly:

$$\frac{\partial}{\partial \phi}\psi(S) = (-i\cos\theta S_z + i\sin\theta\cos\chi S_x - i\sin\theta\sin\chi S_y)\psi(S), \tag{2.108}$$

$$\frac{\partial}{\partial \chi}\psi(S) = -iS_z\psi(S). \tag{2.109}$$

In addition to these modifications on transforming the electron spin to molecule-fixed quantisation, we must rewrite the electron spin interactions $\mathcal{H}(S_i)$ as

$$\mathcal{H}(s_i) = \mathfrak{U}\mathcal{H}(S_i)\mathfrak{U}^{-1} \tag{2.110}$$

where the unitary transformation matrix $\mathcal{U}$ is the product of the transposed rotation matrices

$$\mathcal{U} = \prod_i \mathcal{U}(s_i; S_i) \tag{2.111}$$

and

$$[\mathcal{U}(s_i; S_i)]_{m,M} = \left[\mathcal{D}_{M,m}^{1/2}(\phi, \theta, \chi)\right]^{\text{trans}}. \tag{2.112}$$

We have now derived the results necessary to write the total Hamiltonian $\mathcal{H}_T$ with the electronic coordinates expressed in the molecule-fixed axis system. The electronic Hamiltonian (2.36) becomes

$$\mathcal{H}_T = -\frac{\hbar^2}{2m}\sum_i \nabla_i^2 - \frac{\hbar^2}{2M_N}\sum_{i,j}\nabla_i \cdot \nabla_j + V_{\text{el,nucl}}(r_i, R) + \sum_i \mathcal{H}(s_i) + \mathcal{H}(I_\alpha). \tag{2.113}$$

We have not, in fact, transformed the nuclear spin term but we leave discussion of this term until a later chapter.

The nuclear Hamiltonian obtained earlier is, in the space-fixed coordinate system, given by (2.75)

$$\mathcal{H}_{\text{nucl}} = -\frac{\hbar^2}{2\mu R^2}\left\{\frac{\partial}{\partial R}\left(R^2 \frac{\partial}{\partial R}\right) + \operatorname{cosec}\theta \frac{\partial}{\partial \theta}\left(\sin\theta \frac{\partial}{\partial \theta}\right)\right.$$
$$\left. + \operatorname{cosec}^2\theta\left[\frac{\partial^2}{\partial \phi^2} + \frac{\partial^2}{\partial \chi^2} - 2\cos\theta \frac{\partial^2}{\partial \phi \partial \chi}\right]\right\} + V_{\text{nucl}}(R), \tag{2.114}$$

and it may now be rewritten in order to take account of the effects of the transformation of electronic coordinates by using the results given in equations (2.83), (2.84), (2.85), (2.107), (2.108) and (2.109). If we define the total electronic (orbital and spin) angular momentum $\hbar P$ by

$$P = L + S, \tag{2.115}$$

the complete effect of the transformation on the partial differential operators may be expressed as

$$\left(\frac{\partial}{\partial \phi}\right)_s = \left(\frac{\partial}{\partial \phi}\right)_m - i\cos\theta P_z + i\sin\theta \cos\chi P_x - i\sin\theta \sin\chi P_y, \tag{2.116}$$

$$\left(\frac{\partial}{\partial \theta}\right)_s = \left(\frac{\partial}{\partial \theta}\right)_m - i\sin\chi P_x - i\cos\chi P_y, \tag{2.117}$$

$$\left(\frac{\partial}{\partial \chi}\right)_s = \left(\frac{\partial}{\partial \chi}\right)_m - iP_z. \tag{2.118}$$

Substitution in equation (2.114) gives

$$
\begin{aligned}
\mathcal{H}_T = -\frac{\hbar^2}{2\mu R^2} & \left\{ \frac{\partial}{\partial R}\left(R^2 \frac{\partial}{\partial R}\right) + \operatorname{cosec}\theta \left[\left(\frac{\partial}{\partial \theta}\right)_m - \mathrm{i}\sin\chi\, P_x - \mathrm{i}\cos\chi\, P_y \right] \sin\theta \right. \\
& \times \left[\left(\frac{\partial}{\partial \theta}\right)_m - \mathrm{i}\sin\chi\, P_x - \mathrm{i}\cos\chi\, P_y \right] + \operatorname{cosec}^2\theta \left[\left(\frac{\partial}{\partial \phi}\right)_m + \mathrm{i}\sin\theta\cos\chi\, P_x \right. \\
& \left. - \mathrm{i}\sin\theta\sin\chi\, P_y - \mathrm{i}\cos\theta\, P_z \right]^2 + \operatorname{cosec}^2\theta \left[\left(\frac{\partial}{\partial \chi}\right)_m - \mathrm{i}P_z \right]^2 - 2\cot\theta\operatorname{cosec}\theta \\
& \times \left[\left(\frac{\partial}{\partial \phi}\right)_m + \mathrm{i}\sin\theta\cos\chi\, P_x - \mathrm{i}\sin\theta\sin\chi\, P_y - \mathrm{i}\cos\theta\, P_z \right] \\
& \left. \times \left[\left(\frac{\partial}{\partial \chi}\right)_m - \mathrm{i}P_z \right] \right\} + V_{\text{nucl}}(R). \quad (2.119)
\end{aligned}
$$

This may be rewritten in the simpler form

$$\mathcal{H}_T = -\frac{\hbar^2}{2\mu R^2}\frac{\partial}{\partial R}\left(R^2 \frac{\partial}{\partial R}\right) + \frac{\hbar^2}{2\mu R^2}(\boldsymbol{J}-\boldsymbol{P})^2 + V_{\text{nucl}}(R) \quad (2.120)$$

where the molecule-fixed components of $\boldsymbol{J}$ are defined [7] by

$$J_x = -\mathrm{i}\left\{\cos\chi\left[\cot\theta\left(\frac{\partial}{\partial\chi}\right)_m - \operatorname{cosec}\theta\left(\frac{\partial}{\partial\phi}\right)_m\right] + \sin\chi\left(\frac{\partial}{\partial\theta}\right)_m\right\}, \quad (2.121)$$

$$J_y = -\mathrm{i}\left\{-\sin\chi\left[\cot\theta\left(\frac{\partial}{\partial\chi}\right)_m - \operatorname{cosec}\theta\left(\frac{\partial}{\partial\phi}\right)_m\right] + \cos\chi\left(\frac{\partial}{\partial\theta}\right)_m\right\}, \quad (2.122)$$

$$J_z = -\mathrm{i}\left(\frac{\partial}{\partial\chi}\right)_m. \quad (2.123)$$

We present a detailed description of angular momentum theory in chapter 5, and the reader may wish to examine the results given there at this stage. It emerges that the angular momentum operator $\boldsymbol{J}$ commutes with $\boldsymbol{L}$ and $\boldsymbol{S}$ in this axis system and its molecule-fixed components obey the usual commutation relations for angular momentum operators provided that the anomalous sign of i is used,

$$[J_i, J_j] = -\mathrm{i}\epsilon_{ijk}J_k, \quad (2.124)$$

where ϵ_{ijk} is equal to $+1$ or -1 depending on whether ijk form a cyclic permutation of x, y, z or not, and equal to 0 if any two of ijk are identical. By loose analogy with classical mechanics, J_x, J_y and J_z are the instantaneous components in the molecule-fixed axis system of the total angular momentum of the system of nuclei and electrons. $\boldsymbol{J}-\boldsymbol{P}$ is therefore just the nuclear rotational angular momentum expressed in the same coordinate system.

We have thus achieved our aim of eliminating the effects of nuclear motion from $\mathcal{H}_{\text{el}}$; on the other hand, $\mathcal{H}_{\text{nucl}}$ now contains the operators P_x, P_y and P_z which operate on the electronic part of the total wave function.

2.6. Schrödinger equation for the total wave function

The total Hamiltonian with the electronic coordinates expressed in the molecule-fixed axis system is given by (2.120). If the total wave function for electronic and nuclear motion is written as $\Psi_{rve}(r_i, s_i, R, \phi, \theta)$ the Schrödinger equation can be expressed as

$$(\mathcal{H}_{el} + \mathcal{H}_{nucl})\Psi_{rve} = E_{rve}\Psi_{rve}. \tag{2.125}$$

In order to develop this equation we assume that Ψ_{rve} can be expanded as a complete set of electronic functions multiplied by nuclear functions

$$\Psi_{rve} = \sum_n \psi_e^n(r_i, R)\psi_{rv}^n(R, \phi, \theta) \tag{2.126}$$

where the electronic functions $\psi_e^n(r_i, R)$ are eigenfunctions of the electronic Hamiltonian

$$\mathcal{H}_{el}\psi_e^n(r_i, R) = E_e^n(R)\psi_e^n(r_i, R). \tag{2.127}$$

The electronic functions $\psi_e^n(r_i, R)$ are also eigenfunctions of the operator P_z, since P_z commutes with $\mathcal{H}_{el}$,

$$P_z\psi_e^n(r_i, R) = \Omega_n\psi_e^n(r_i, R). \tag{2.128}$$

This exact form of the molecular wavefunction Ψ_{rve} was first introduced by Born and Huang [9]. The equation to be solved, equation (2.125), may therefore be written

$$\{\mathcal{H}_{el} + \mathcal{H}_{nucl} - E_{rve}\}\left\{\sum_n \psi_e^n(r_i, R)\psi_{rv}^n(R, \phi, \theta)\right\} = 0. \tag{2.129}$$

Multiplication on the left by the complex conjugate function $\psi_e^{n'*}$ and expansion yields

$$\psi_e^{n'*}\mathcal{H}_{el}\sum_n \psi_e^n(r_i, R)\psi_{rv}^n(R, \phi, \theta) + \psi_e^{n'*}\mathcal{H}_{nucl}\sum_n \psi_e^n(r_i, R)\psi_{rv}^n(R, \phi, \theta)$$
$$- \psi_e^{n'*}E_{rve}\sum_n \psi_e^n(r_i, R)\psi_{rv}^n(R, \phi, \theta) = 0. \tag{2.130}$$

The first and third terms in this expression immediately simplify on integration over the electronic coordinates, and this allows us to express the eigenvalue problem in the more compact form,

$$\{E_e^{n'}(R) - E_{rve}\}\psi_{rv}^{n'} + \sum_n C_{n',n}\psi_{rv}^n = 0, \tag{2.131}$$

where

$$C_{n',n}\psi_{rv}^n = \int \psi_e^{n'*}\left(-\frac{\hbar^2}{2\mu}\nabla_R^2\psi_e^n\,dr_i\right)\phi_{rv}^n + \frac{\hbar^2}{2\mu}\nabla_R^2\phi_{rv}^n\delta_{n,n'}. \tag{2.132}$$

2.7. The Born–Oppenheimer and Born adiabatic approximations

The complete solution of the eigenvalue problem would require that we solve equation (2.127) for the infinite complete set of functions ψ_e^n and then solve the infinite number of coupled equations (2.131) in a self-consistent manner with the infinite number of functions ψ_{rv}^n. This is clearly an impossible task and our procedure is to find approximate solutions to Ψ_{rve} which can then be improved by perturbation theory.

Suppose we approximate Ψ_{rve} to the simple product form

$$\Psi_{rve}^0 = \psi_e^n(r_i, R)\phi_{rv}^n(R, \phi, \theta), \tag{2.133}$$

where $\psi_e^n(r_i, R)$ is an eigenfunction of $\mathcal{H}_{el}$ in (2.127) and ϕ_{rv}^n is approximately equal to ψ_{rv}^n in equation (2.126). In this approximation the electrons follow the nuclear motion *adiabatically*; the electronic states are not mixed and an electronic state is itself deformed progressively by the nuclear displacements, maintaining its integrity. The electronic wave function is dependent upon the nuclear coordinates but independent of the nuclear momenta. We can use the variational principle to arrive at the best choice for the function ϕ_{rv}^n in (2.133). We minimise the energy $\mathcal{E}_{rve}$ with respect to small changes in ϕ_{rv}^n where

$$\mathcal{E}_{rve} = \int\int \Psi_{rve}^{0*}\mathcal{H}\Psi_{rve}^0\, dr_i\, dr \Big/ \int\int \Psi_{rve}^{0*}\Psi_{rve}^0\, dr_i\, dr. \tag{2.134}$$

We can always choose ψ_e^n to be a real function, even if it represents one component of a doubly-degenerate state, so that we can put

$$\int \psi_e^n(\partial/\partial R)\psi_e^n\, dr_i = (1/2)(\partial/\partial R)\int \psi_e^n\psi_e^n\, dr_i = 0. \tag{2.135}$$

With this relation, substitution of (2.133) in (2.134) leads to the result that the best choice for ϕ_{rv}^n is one of the solutions of

$$\left[E_e^n(R) - \frac{\hbar^2}{2\mu}\int \psi_e^n\nabla_R^2\psi_e^n\, dr_i + \mathcal{H}_{nucl}\right]\phi_{rv}^n = \mathcal{E}_{rve}\phi_{rv}^n,$$

$$\text{i.e. } \left[E_e^n(R) + C_{n,n} + \mathcal{H}_{nucl}\right]\phi_{rv}^n = \mathcal{E}_{rve}\phi_{rv}^n. \tag{2.136}$$

Thus in this approximation the ϕ_{rv}^n constitute a complete set of rovibrational wave functions for each electronic state ψ_e^n and $[E_e^n(R) + C_{n,n} + \mathcal{H}_{nucl}(R)]$ is an effective potential function governing the motion of the nuclei. This choice for ϕ_{rv}^n is called the *Born adiabatic approximation* and amounts to neglect of the off-diagonal terms $C_{n',n}$ in equation (2.131) which mix different electronic states.

If $E_e^n(R)$ is a doubly-degenerate eigenvalue of $\mathcal{H}_{el}$ and ψ_e^{na}, ψ_e^{nb} are the two real, orthogonal components of this electronic state, it can be shown that $C_{na,nb}$ is zero for a diatomic molecule because there is only one, totally symmetric, vibrational mode. Thus the Born approximation is equally valid for degenerate and non-degenerate electronic states of a diatomic molecule.

In order to determine the form of $C_{n,n}$, we first expand ∇_R^2 from equation (2.120) and group together terms which are diagonal in the electronic states and those which are off-diagonal. We obtain

$$\nabla_R^2 = \frac{1}{R^2}\frac{\partial}{\partial R}\left(R^2\frac{\partial}{\partial R}\right) - \frac{1}{R^2}(\boldsymbol{J}^2 - 2J_zP_z + \boldsymbol{P}^2) + \frac{2}{R^2}(J_xP_x + J_yP_y). \quad (2.137)$$

Integrating $(-\hbar^2/2\mu)\nabla_R^2$ over electronic coordinates and using (2.135) we obtain

$$C_{n,n} = Q_{n,n} + P_{n,n} + \frac{\hbar^2}{2\mu R^2}\Omega_n^2 - \frac{\hbar^2}{\mu R^2}\Omega_n J_z \quad (2.138)$$

where

$$Q_{n,n} = -\frac{\hbar^2}{2\mu}\int \psi_e^{n*}(\partial^2\psi_e^n/\partial R^2)\,\mathrm{d}r_i, \quad (2.139)$$

and

$$P_{n,n} = \frac{\hbar^2}{2\mu R^2}\int \psi_e^{n*}\{(P_x^2 + P_y^2)\psi_e^n\}\,\mathrm{d}r_i. \quad (2.140)$$

Note that, because ψ_e^n is independent of the Euler angles θ and ϕ, there is no contribution from the term in $\boldsymbol{J}^2$. Substituting (2.138) into (2.136), we obtain the wave equation for the rotation–vibration wave functions in the Born adiabatic approximation:

$$\left\{-\frac{\hbar^2}{2\mu}\nabla_R^2 + E_e^n(R) + Q_{nn} + P_{nn} - \frac{\hbar^2}{2\mu R^2}(2\Omega_n J_z - \Omega_n^2) + V_{\text{nucl}}(R)\right\}\phi_{rv}^n = \mathcal{E}_{rv e}\phi_{rv}^n. \quad (2.141)$$

The third, fourth and fifth terms in braces in equation (2.141) represent small adiabatic corrections to the potential energy function. They all have a μ^{-1} reduced mass dependence, unlike $V_{\text{nucl}}(R)$, and so are the origin of the isotopic shifts in the electronic energy [5].

The corresponding rotation-vibration wave equation in the Born–Oppenheimer approximation, in which all coupling of electronic and nuclear motions is neglected, is

$$\left\{-\frac{\hbar^2}{2\mu}\nabla_R^2 + E_e^n(R) + V_{\text{nucl}}(R) - \mathcal{E}_{rve}\right\}\phi_{rv}^n = 0. \quad (2.142)$$

It is, therefore, still necessary to solve equation (2.127) for the electronic energies and wave functions.

2.8. Separation of the vibrational and rotational wave equations

We can separate the coordinates in ψ_{rv}^n by writing it as a product

$$\phi_{rv}^n = \chi^n(R)\mathrm{e}^{iM_J\phi}\Theta^n(\theta)\mathrm{e}^{ik\chi} \quad (2.143)$$

where M_J and k are constants taking integral or half-integral values. M_J will later be identified as the quantum number labelling the component of total angular momentum $\boldsymbol{J}$ along the space-fixed Z axis; it takes $2J+1$ values from $-J$ to $+J$. However, the quantum number k associated with the redundant Euler angle χ is restricted in the values that it can take. From equation (2.128) we have

$$J_z \Psi^n_{rve} = P_z \Psi^n_{rve} = \Omega_n \Psi^n_{rve} \tag{2.144}$$

or

$$J_z \phi^n_{rv} = \Omega_n \phi^n_{rv}. \tag{2.145}$$

Hence, by combining (2.143) and (2.145) we obtain the important result that $k = \Omega_n$. Now substitution in (2.141), and use of the standard methods of separating the variables, yields separate equations for $\chi^n(R)$ and $e^{iM_J\phi}\Theta^n(\theta)e^{ik\chi}$ which are

$$\left\{ -\frac{\hbar^2}{2\mu} \frac{1}{R^2} \frac{\partial}{\partial R}\left(R^2 \frac{\partial}{\partial R}\right) + E^n_e(R) + Q_{n,n}(R) + P_{n,n}(R) + V_{nucl}(R) \right.$$
$$\left. + \frac{\hbar^2}{2\mu R^2}\Omega_n^2 + E_{rot}(R) - E_{rve}(R) \right\}\chi^n(R) = 0, \tag{2.146}$$

$$\left\{ \frac{-\hbar^2}{2\mu R^2}\operatorname{cosec}\theta \frac{\partial}{\partial \theta}\left(\sin\theta \frac{\partial}{\partial \theta}\right) - \frac{\hbar^2}{2\mu R^2}\operatorname{cosec}^2\theta\left(\frac{\partial^2}{\partial \phi^2} + \frac{\partial^2}{\partial \chi^2} - 2\cos\theta \frac{\partial^2}{\partial \phi \partial \chi}\right) \right.$$
$$\left. - \frac{\hbar^2}{2\mu R^2}2i\Omega_n \frac{\partial}{\partial \chi} - E_{rot}(R) \right\} e^{iM_J\phi}\Theta^n(\theta)e^{i\Omega_n\chi} = 0. \tag{2.147}$$

Equation (2.146) governs the vibrational motion of the nuclei and (2.147) describes the rotational motion of the molecule-fixed axis system. We deal with the latter by separating off the variables ϕ and χ to yield an equation in θ which is

$$\left\{ -\frac{\hbar^2}{2\mu R^2}\left[\operatorname{cosec}\theta \frac{\partial}{\partial \theta}\left(\sin\theta \frac{\partial}{\partial \theta}\right) + \operatorname{cosec}^2\theta\left(-M_J^2 - \Omega_n^2 + 2\cos\theta M_J\Omega_n\right)\right] \right.$$
$$\left. + \frac{\hbar^2}{\mu R^2}\Omega_n^2 - E_{rot}(R) \right\}\Theta^n(\theta) = 0. \tag{2.148}$$

The eigenfunctions of (2.148) are given by

$$\Theta^n(\theta) = \left(\frac{2J+1}{2}\right)^{1/2}\left[\frac{(J+M_J)!(J-M_J)!}{(J+\Omega_n)!(J-\Omega_n)!}\right]^{1/2}$$
$$\times [\cos(\theta/2)]^{M_J+\Omega_n}[\sin(\theta/2)]^{M_J-\Omega_n}\mathcal{P}^{(M_J-\Omega_n, M_J+\Omega_n)}_{J-\Omega_n}(\cos\theta) \tag{2.149}$$

where $\mathcal{P}$ is a Jacobi polynomial; $\Theta^n(\theta)$ is normalised so that

$$\int_0^\pi \Theta^{n*}(\theta)\Theta^n(\theta)\sin\theta \, d\theta = 1. \tag{2.150}$$

The eigenvalues of (2.148) are given by

$$E_{rot}(R) = \frac{\hbar^2}{2\mu R^2}\left[J(J+1) - \Omega_n^2 \right]. \tag{2.151}$$

We can therefore substitute for $E_{\text{rot}}(R)$ in the vibrational wave equation (2.146) to give

$$\left\{\frac{\hbar^2}{2\mu}\frac{1}{R^2}\frac{\partial}{\partial R}\left(R^2\frac{\partial}{\partial R}\right) + E_{\text{rve}} - E_{\text{e}}^{\text{n}}(R) - Q_{\text{n,n}}(R) - P_{\text{n,n}}(R) - V_{\text{nucl}}(R)\right.$$
$$\left. - \frac{\hbar^2}{2\mu R^2}[J(J+1) - \Omega_{\text{n}}^2]\right\}\chi^{\text{n}}(R) = 0. \qquad (2.152)$$

2.9. The vibrational wave equation

Equation (2.152) is the wave equation of the vibrating rotator,

$$\frac{\hbar^2}{2\mu}\frac{1}{R^2}\frac{d}{dR}R^2\frac{d\chi^{\text{n}}(R)}{dR} + \left\{E_{\text{rve}} - V - \frac{\hbar^2}{2\mu R^2}J(J+1)\right\}\chi^{\text{n}}(R) = 0, \qquad (2.153)$$

in which the potential function V is given by

$$V = E_{\text{e}}^{\text{n}}(R) + Q_{\text{n,n}}(R) + P_{\text{n,n}}(R) + V_{\text{nucl}}(R) - \frac{\hbar^2}{2\mu R^2}\Omega_{\text{n}}^2 \qquad (2.154)$$

and $\chi^{\text{n}}(R)$ is normalised thus:

$$\int_0^\infty \chi^{\text{n}*}(R)\chi^{\text{n}}(R)R^2\,dR = 1. \qquad (2.155)$$

The main difficulty in solving (2.153) lies in the evaluation of the potential energy term (2.154). Even in the case of H_2, calculation of V from the electronic wavefunctions for different values of R is no easy matter. Usually, therefore, the vibrational wave equation is solved by inserting a restricted form of the potential; experimental data on the rovibrational levels are then expressed in terms of constants introduced semiempirically, as we shall show.

In a classic paper, Dunham [10] introduced a dimensionless vibrational variable ξ defined by

$$\xi = \frac{R - R_e}{R_e}, \qquad (2.156)$$

where R_e is the equilibrium nuclear separation, that is, the value of R when the potential energy is a minimum. With the substitution of this variable, the vibrational wave equation becomes

$$\frac{d^2\psi^{\text{n}}(\xi)}{d\xi^2} + \frac{2\mu R_e^2}{\hbar^2}\left\{E - V - \frac{\hbar^2}{2\mu R_e^2(1+\xi)^2}J(J+1)\right\}\psi^{\text{n}}(\xi) = 0, \qquad (2.157)$$

where $\psi^{\text{n}}(\xi) = R\chi^{\text{n}}(R)$. Alternatively one can introduce the mass-weighted normal coordinate Q, defined by

$$Q = \mu^{1/2}(R - R_e). \qquad (2.158)$$

The eigenfunctions of the resultant wave equation, $\psi^n(Q)$, are now normalised as

$$\int_0^\infty \psi^{n*}(Q)\psi^n(Q)\,dQ = 1 \tag{2.159}$$

which must be compared with (2.155). Neglecting, for the moment, the rotational term in (2.120), the vibrational Hamiltonian then becomes

$$\mathcal{H}_{\text{vib}} = -\frac{\hbar^2}{2}\frac{d^2}{dQ^2} + V \tag{2.160}$$

$$= \frac{1}{2}P_Q^2 + V \tag{2.161}$$

where P_Q is the momentum conjugate to Q,

$$P_Q = \frac{h}{2\pi i}\frac{d}{dQ}. \tag{2.162}$$

The simplest approximation for V is to assume that the vibration is harmonic, in which case the Hamiltonian becomes

$$\mathcal{H}_0 = (1/2)(P_Q^2 + \lambda Q^2) \tag{2.163}$$

where

$$\lambda = (\hbar\gamma)^2 \tag{2.164}$$

and

$$\gamma = \frac{2\pi\nu}{\hbar}. \tag{2.165}$$

This form of the harmonic oscillator equation is particularly convenient for solution by the methods of matrix mechanics, based on the commutation relationships:

$$[P_Q, Q] = -i\hbar. \tag{2.166}$$

The eigenvalues and eigenfunctions of the simple harmonic oscillator are well known. A detailed account of the solution of the wave equation in (2.157) is given by Pauling and Wilson [11]. The solution of equation (2.163) using creation and annihilation operators is described in the book by Bunker and Jensen [12]. The energy levels of the harmonic oscillator are given by

$$\begin{aligned}E_v &= (v + 1/2)h\nu \quad v = 0,\,1,\,2,\text{ etc.},\\ &= (v + 1/2)\hbar^2\gamma\end{aligned} \tag{2.167}$$

and the eigenfunctions are given by

$$\psi_v(Q) = \left[\left(\frac{\gamma}{\pi}\right)^{1/2}\frac{1}{2^v(v!)}\right]^{1/2} e^{-(1/2)\gamma Q^2} H_v(\gamma^{1/2}Q) \tag{2.168}$$

where the $H_v(\gamma^{1/2}Q)$ are Hermite polynomials of degree v in Q. The first four Hermite polynomials are

$$H_0(y) = 1, \tag{2.169}$$
$$H_1(y) = 2y, \tag{2.170}$$
$$H_2(y) = 4y^2 - 2, \tag{2.171}$$
$$H_3(y) = 8y^3 - 6y. \tag{2.172}$$

In general, $H_n(y)$ contains y to the powers $v, v-2, v-4, \ldots$, 1 or 0, that is, either all even powers or all odd powers as v is even or odd respectively.

The wavefunctions for the harmonic oscillator in terms of the Dunham coordinate ξ are

$$\psi_v(\xi) = \left\{ \frac{1}{R_e} \left(\frac{\alpha}{\pi} \right)^{1/2} \frac{1}{2^v(v)!} \right\}^{1/2} \exp\left(-\frac{1}{2}y^2\right) H_v(y), \tag{2.173}$$

where

$$y = \alpha^{1/2}\xi,$$
$$\alpha = \mu R_e^2 \gamma. \tag{2.174}$$

Although the vibrational motion of a diatomic molecule conforms quite closely to that of a harmonic oscillator, in practice the anharmonic deviations are quite significant and must be taken into account if vibrational energy levels are to be modelled accurately. A general form of the potential function V in equation (2.157) was proposed by Dunham [10]:

$$V = a_0 \xi^2 \{1 + a_1 \xi + a_2 \xi^2 + a_3 \xi^3 + \cdots\}, \tag{2.175}$$

where

$$a_0 = h\omega_e^2 / 4B_e$$
$$= (1/2)\hbar^2 \gamma^2 \mu R_e^2. \tag{2.176}$$

Since the rotational term in equation (2.157) can also be expanded as a power series in ξ, the complete perturbation to the harmonic oscillator Hamiltonian is

$$\mathcal{H}' = a_0 \xi^2 \{a_1 \xi + a_2 \xi^2 + a_3 \xi^3 + \cdots\} + \frac{\hbar^2}{2\mu R_e^2} J(J+1)\{1 + c_1 \xi + c_2 \xi^2 + c_3 \xi^3 + \cdots\}$$
$$= k_0 + k_1 \xi + k_2 \xi^2 + k_3 \xi^3 + \cdots, \tag{2.177}$$

where, in the rotational term,

$$c_n = (-1)^n (n+1) \tag{2.178}$$

from the binomial expansion.

The effects of $\mathcal{H}'$ may be treated using ordinary non-degenerate perturbation theory or, as in Dunham's original work, by means of the Wentzel–Kramers–Brillouin method

[13], which is described in chapter 6. The result is that the rovibrational energies are given by

$$E_{v,J} = \sum_{k\ell} Y_{k\ell}(v+1/2)^k J^\ell(J+1)^\ell \qquad (2.179)$$

$$= Y_{00} + Y_{10}(v+1/2) + Y_{20}(v+1/2)^2 + \cdots + Y_{01}J(J+1)$$
$$+ Y_{02}J^2(J+1)^2 + \cdots + Y_{11}(v+1/2)J(J+1) + \cdots. \qquad (2.180)$$

Formulae for the leading coefficients $Y_{k\ell}$ have been given explicitly by Dunham.

The generalised potential function (2.175) can be cumbersome to use and more restricted functions have often been employed. Of these the most important and satisfactory is the Morse potential [14],

$$V = D\left(1 - e^{-\beta(R-R_e)}\right)^2 \qquad (2.181)$$

in which D is the dissociation energy of the molecule and β is a constant. A graphical representation of the Morse potential is shown and discussed in chapter 6. Using the Morse potential in the vibrational wave equation, the rovibrational energies are given exactly by

$$\frac{E_{v,J}}{hc} = \omega_e(v+1/2) - x_e\omega_e(v+1/2)^2 + B_e J(J+1)$$
$$- D_e J^2(J+1)^2 - \alpha_e(v+1/2)J(J+1) \qquad (2.182)$$

in which the parameters, in cm^{-1}, are given by

$$\omega_e = \frac{\beta}{2\pi c}\sqrt{\frac{2D}{\mu}}, \quad x_e = \frac{hc\omega_e}{4D}, \quad B_e = \frac{\hbar}{4\pi\mu R_e^2 c}, \quad D_e = \frac{\hbar^3}{16\pi^3 \mu^3 \omega_e^2 R_e^6 c^3} = \frac{4B_e^3}{\omega_e^2}$$

$$\alpha_e = \frac{3\hbar^2 \omega_e}{4\mu R_e^2 D}\left(\frac{1}{aR_e} - \frac{1}{a^2 R_e^2}\right) = 6\sqrt{\frac{x_e B_e^3}{\omega_e}} - 6\frac{B_e^2}{\omega_e}. \qquad (2.183)$$

Comparing the results obtained from the Morse potential with those from the Dunham expansion, we see that the coefficients are related as follows,

$$Y_{10} = \omega_e, \quad Y_{20} = -x_e\omega_e, \quad Y_{01} = B_e, \quad Y_{02} = -D_e, \quad Y_{11} = -\alpha_e. \qquad (2.184)$$

These relationships ignore some small, higher-order corrections which arise from the Dunham expansion in equation (2.177). The five terms in (2.182) can be identified with the solutions obtained using more restricted potential functions. The first term has the same form as that obtained for a pure vibrator with a harmonic potential, the second term is obtained with a cubic term (anharmonic) in the potential, the third term is obtained in the treatment of the rigid rotator, the fourth term comes from centrifugal stretching of the rotating molecule, and the final term allows for change in the average moment of inertia on vibrational excitation.

In conclusion, we note that experimental data, usually from infrared or ultraviolet spectroscopy, are often expressed by giving values of the parameters presented in equation (2.182). The formula is able to model low-lying vibrational levels of a molecule in a closed shell state quite accurately.

2.10. Rotational Hamiltonian for space-quantised electron spin

All of the results derived so far are for the situation in which the electronic spin orientation is coupled to the molecular orientation. The essential steps were described in section 2.5 where we transformed both the spin and space coordinates from space-fixed to molecule-fixed axes. Now the situation often arises in which the electron spin is quantised in the space-fixed axis system, i.e. is not strongly coupled to the molecular axis. Evidently, therefore, we need only to omit the spin transformation in deriving a Hamiltonian suitable for this situation. Our previous equations must be modified, simply by replacing $\boldsymbol{P}$, the total electronic angular momentum, by $\boldsymbol{L}$, the orbital angular momentum, and also by replacing $\boldsymbol{J}$ by $\boldsymbol{N}$, the total angular momentum exclusive of spin; in Hund's case (b), $\boldsymbol{J}$ will be constructed by adding $\boldsymbol{N}$ and $\boldsymbol{S}$. Thus the operator form of the Hamiltonian for space-quantised spin is

$$\mathcal{H} = \frac{\hbar^2}{2\mu R^2}(\boldsymbol{N}-\boldsymbol{L})^2 - \frac{\hbar^2}{2\mu R^2}\frac{\partial}{\partial R}\left(R^2\frac{\partial}{\partial R}\right) + V_{\text{nucl}}(R) + \mathcal{H}_{\text{el}}. \tag{2.185}$$

This is, of course, similar to (2.120) for molecule-quantised electron spin; the expansion of $\mathcal{H}_{\text{el}}$ will, however, be somewhat different, as we shall see.

2.11. Non-adiabatic terms

It will be recalled that our use of the Born adiabatic approximation in section 2.6 enabled us to separate the nuclear and electronic parts of the total wave function. This separation led to wave equations for the rotational and vibrational motions of the nuclei. We now briefly reconsider this approximation, with the promise that we shall study it at greater length in chapters 6 and 7.

We could return to the exact equation (2.131) and examine the matrix elements of the $C_{n',n}$ terms in the same manner as we dealt with the diagonal $C_{n,n}$ terms. It is, however, easier to turn to the form of the exact Hamiltonian given in equations (2.113) and (2.120). The terms in this Hamiltonian which cause a breakdown of the adiabatic separation of nuclear and electronic motion are

$$-\frac{\hbar^2}{2\mu R^2}\frac{\partial}{\partial R}\left(R^2\frac{\partial}{\partial R}\right) - \frac{\hbar^2}{\mu R^2}(J_x P_x + J_y P_y). \tag{2.186}$$

If we define the shift operators by

$$J_+ = J_x + iJ_y, \quad J_- = J_x - iJ_y, \quad P_+ = P_x + iP_y, \quad P_- = P_x - iP_y \tag{2.187}$$

we can rewrite the second term in (2.186) as

$$-\frac{\hbar^2}{2\mu R^2}(J_+ P_- + J_- P_+). \tag{2.188}$$

This term has matrix elements off-diagonal in Ω, i.e. between different electronic states.

We shall show in chapter 7 that these matrix elements are of the form

$$\langle \eta, v, \Omega | \frac{-\hbar^2}{2\mu R^2} P_\mp | \eta', v', \Omega \pm 1 \rangle [(J \pm \Omega + 1)(J \mp \Omega)]^{1/2} \quad (2.189)$$

and they therefore remove the degeneracy of an electronic state with orbital angular momentum. The most important consequence of this is the observation of Λ-doubling, which is particularly significant in diatomic hydrides and other light molecules which rotate rapidly (i.e. they have a large B value). The matrix elements (2.189) thus represent a coupling of the rotational and electronic motions of the molecule.

It is important to note that the Hamiltonian (2.120) contains the terms which produce both the adiabatic and non-adiabatic effects. In chapter 7 we shall show how the total Hamiltonian can be reduced to an effective Hamiltonian which operates only in the rotational subspace of a single vibronic state, the non-adiabatic effects being treated by perturbation theory and incorporated into the molecular parameters which define the effective Hamiltonian. Almost for the first time in this book, this introduces an extremely important concept and tool, outlined in chapter 1, the effective Hamiltonian. Observed spectra are analysed in terms of an appropriate effective Hamiltonian, and this process leads to the determination of the values of what are best called 'molecular parameters'. An alternative terminology of 'molecular constants', often used, seems less appropriate. The quantitative interpretation of the molecular parameters is the link between experiment and electronic structure.

2.12. Effects of external electric and magnetic fields

The equations derived thus far take no account of the effects of applied electric or magnetic fields, or even of the fields created by the motion of the nuclei and electrons. We shall discuss these effects explicitly in the derivation of the electronic Hamiltonian, but for the moment we aim to correct our equations for the motion of the nuclei by appealing to classical mechanics. We here sketch the main points which are covered in detail by Landau and Lifshitz [15], among others.

In non-relativistic classical mechanics a mechanical system can be characterised by a function called the Lagrangian, $\mathscr{L}(q, \dot{q})$ where q denotes the coordinates, and the motion of the system is such that the action S, defined by

$$S = \int_{t_1}^{t_2} \mathscr{L}(q, \dot{q}, t) \, dt \quad (2.190)$$

is minimised. The Lagrangian for a system of n particles is given by

$$\mathscr{L} = \sum_{a=1}^{n} (1/2) m_a V_a^2 - U(r_1, r_2, \ldots, r_n) \quad (2.191)$$

where the two terms denote the kinetic and potential energies respectively.

The equations of motion are written

$$\frac{d}{dt}\left(\frac{\partial \mathcal{L}}{\partial \dot{q}_i}\right) - \frac{\partial \mathcal{L}}{\partial q_i} = 0, \qquad (2.192)$$

and are called Lagrange's equations.

Now for a single particle in an external field, the Lagrangian is given by

$$\mathcal{L} = (1/2)mV^2 - U(\mathbf{r}, t). \qquad (2.193)$$

The three equations of motion of the particle can be written by (2.192) as

$$m\dot{V} = -\frac{\partial U}{\partial \mathbf{r}}. \qquad (2.194)$$

Since the momentum P_i conjugate to the coordinate q_i is defined by

$$P_i = \frac{\partial \mathcal{L}}{\partial \dot{q}_i}, \qquad (2.195)$$

the total momentum of the particle is given by

$$\mathbf{P} = m\mathbf{V}. \qquad (2.196)$$

These equations are modified when we turn to relativistic mechanics. The Lagrangian for a free particle is now given by

$$\mathcal{L} = -mc^2\sqrt{1 - \frac{V^2}{c^2}} \qquad (2.197)$$

and the momentum of the particle is given by

$$\mathbf{P} = \frac{\partial \mathcal{L}}{\partial \mathbf{V}} = \frac{m\mathbf{V}}{\sqrt{1 - (V^2/c^2)}}. \qquad (2.198)$$

An electromagnetic field is described in relativistic theory by a four-vector $\mathbf{A}_i$, where the three space components $A_{1,2,3} = A_{x,y,z}$ are called the vector potential $\mathbf{A}$ and the fourth (time) component A_4 is equal to $i\phi$ where ϕ is called the scalar potential. The Lagrangian for a particle in an electromagnetic field is now given by

$$\mathcal{L} = -mc^2\sqrt{1 - (V^2/c^2)} + q\mathbf{A} \cdot \mathbf{V} - q\phi \qquad (2.199)$$

where q is the charge on the particle. By analogy with equation (2.198), the momentum $\mathbf{P}$ which is now a generalised conjugate momentum, is given by

$$\mathbf{P} = \frac{\partial \mathcal{L}}{\partial \mathbf{V}} = \frac{m\mathbf{V}}{\sqrt{1 - (V^2/c^2)}} + q\mathbf{A} = \boldsymbol{\pi} + q\mathbf{A} \qquad (2.200)$$

where $\boldsymbol{\pi}$ is called the mechanical momentum. If we adopt the accepted convention that e is a positive quantity, the charge on the electron is then $-e$, and the charge of nucleus α is $Z_\alpha e$. It is assumed that the results from classical mechanics may be taken over into quantum mechanics. Since our previous equations involved mechanical momentum only, we make the replacements

$$P_i \to \pi_i = P_i + eA_i \tag{2.201}$$
$$P_\alpha \to \pi_\alpha = P_\alpha - Z_\alpha e A_\alpha \tag{2.202}$$

for electrons and nuclei respectively, to take account of the effects of external fields. We do not specify the form of A_α at this stage; in general it will arise from the motion of other electrons and nuclei, and from the presence of an external magnetic field.

The nuclear kinetic energy, which was written initially in equation (2.35) as $P_R^2/2\mu$ and related to P_α should now evidently be rewritten as

$$\frac{1}{(M_1 + M_2)^2} \{-M_2(P_1 - Z_1 e A_1) + M_1(P_2 - Z_2 e A_2)\}^2. \tag{2.203}$$

We wish to find the corrected form of the Hamiltonian (2.120). On expansion (2.203) leads to

$$\mathcal{H}_{\text{nucl}} = \frac{\mu}{2} \left\{ \left(\frac{P_1}{M_1} + \frac{P_2}{M_2} \right)^2 + 2e \left(-\frac{P_1}{M_1} + \frac{P_2}{M_2} \right) \cdot \left(\frac{Z_1 A_1}{M_1} - \frac{Z_2 A_2}{M_2} \right) \right.$$
$$\left. + e^2 \left(\frac{Z_1 A_1}{M_1} - \frac{Z_2 A_2}{M_2} \right)^2 \right\} + V_{\text{nucl}}(R) \tag{2.204}$$

since $(P_\alpha \cdot A_\alpha) = (A_\alpha \cdot P_\alpha)$. The components of P_1 and P_2 in the original space-fixed coordinate system are given in equation (2.78). It is convenient to transform part of this expression to the molecule-fixed coordinate system, thus:

$$P_\alpha = -i\hbar \left\{ \frac{M_\alpha}{M} \left[i' \frac{\partial}{\partial X_0} + j' \frac{\partial}{\partial Y_0} + k' \frac{\partial}{\partial Z_0} \right] - \frac{M_\alpha}{M_N} \sum_i \left[i \frac{\partial}{\partial x_i} + j \frac{\partial}{\partial y_i} + k \frac{\partial}{\partial z_i} \right] \right.$$
$$\mp \left[i \frac{1}{R} \left(\csc\theta \sin\chi \frac{\partial}{\partial \phi} + \cos\chi \frac{\partial}{\partial \theta} - \cot\theta \sin\chi \frac{\partial}{\partial \chi} \right) \right.$$
$$\left. + j \frac{1}{R} \left(\csc\theta \cos\chi \frac{\partial}{\partial \phi} - \sin\chi \frac{\partial}{\partial \theta} - \cot\theta \cos\chi \frac{\partial}{\partial \chi} \right) + k \frac{\partial}{\partial R} \right] \right\}, \tag{2.205}$$
$$= -i\hbar \left\{ \frac{M_\alpha}{M} \left[i' \frac{\partial}{\partial X_0} + j' \frac{\partial}{\partial Y_0} + k' \frac{\partial}{\partial Z_0} \right] - \frac{M_\alpha}{M_N} \sum_i \left[i \frac{\partial}{\partial x_i} + j \frac{\partial}{\partial y_i} + k \frac{\partial}{\partial z_i} \right] \right.$$
$$\left. \mp \left[-\frac{\hbar}{R} (k \wedge J) - i\hbar k \frac{\partial}{\partial R} \right], \right. \tag{2.206}$$

where the upper sign refers to $\alpha = 1$ and the lower to $\alpha = 2$. The partial derivatives in (2.206) are performed with space-fixed electronic coordinates held constant. When these coordinates are transformed to the molecule-fixed axis system and the resultant expression is substituted in $\mathcal{H}_{\text{nucl}}$ we obtain

$$\mathcal{H}_{\text{nucl}} = \frac{\hbar^2}{2\mu R^2} (J - P)^2 - \frac{\hbar^2}{2\mu R^2} \frac{\partial}{\partial R} \left(R^2 \frac{\partial}{\partial R} \right) + e \left\{ -\frac{\hbar}{R} k \wedge (J - P) + i\hbar k \frac{\partial}{\partial R} \right\} \cdot$$
$$\left\{ \frac{Z_1 A_1}{M_1} - \frac{Z_2 A_2}{M_2} \right\} + \frac{\mu e^2}{2} \left(\frac{Z_1 A_1}{M_1} - \frac{Z_2 A_2}{M_2} \right)^2 + V_{\text{nucl}}(R). \tag{2.207}$$

It will, of course, be necessary to replace the conjugate momenta in $\mathcal{H}_{el}$ by the mechanical momenta; this matter is dealt with at length in the next chapter.

The corresponding corrected expression for a molecule in which the electron spins are spatially quantised is

$$\mathcal{H}_{nucl} = \frac{\hbar^2}{2\mu R^2}(N-L)^2 - \frac{\hbar^2}{2\mu R^2}\frac{\partial}{\partial R}\left(R^2 \frac{\partial}{\partial R}\right) + e\left\{-\frac{\hbar}{R}\boldsymbol{k} \wedge (N-L) + i\hbar \boldsymbol{k}\frac{\partial}{\partial R}\right\} \cdot \left\{\frac{Z_1 A_1}{M_1} - \frac{Z_2 A_2}{M_2}\right\} + \frac{\mu e^2}{2}\left(\frac{Z_1 A_1}{M_1} - \frac{Z_2 A_2}{M_2}\right)^2 + V_{nucl}(R). \qquad (2.208)$$

We shall see later how, with the appropriate expressions for the magnetic vector potentials, the effects of an external magnetic vector potentials, the effects of an external magnetic field can be introduced into the vibration–rotation Hamiltonian.

Appendix 2.1. Derivation of the momentum operator

We now give an explicit derivation of equation (2.14). The chain rule is:

$$\frac{\partial}{\partial X} = \frac{\partial A}{\partial X}\frac{\partial}{\partial A} + \frac{\partial B}{\partial X}\frac{\partial}{\partial B} + \cdots. \qquad (2.209)$$

Writing $\boldsymbol{P}_i$ in terms of $\boldsymbol{R}'_i$, $\boldsymbol{R}$ and $\boldsymbol{R}_O$ we have:

$$\boldsymbol{P}_i = -i\hbar \frac{\partial}{\partial \boldsymbol{R}_i} = -i\hbar \left\{\frac{\partial}{\partial X_i}\boldsymbol{i}' + \frac{\partial}{\partial Y_i}\boldsymbol{j}' + \frac{\partial}{\partial Z_i}\boldsymbol{k}'\right\}. \qquad (2.210)$$

If we consider only the X component along the $\boldsymbol{i}'$ axis, we have:

$$[\boldsymbol{P}_i]_X = -i\hbar\left[\frac{\partial}{\partial X_i}\right] = -i\hbar\left\{\frac{\partial X'_i}{\partial X_i}\frac{\partial}{\partial X'_i} + \frac{\partial Y'_i}{\partial X_i}\frac{\partial}{\partial Y'_i} + \frac{\partial Z'_i}{\partial X_i}\frac{\partial}{\partial Z'_i} \right.$$
$$+ \frac{\partial X}{\partial X_i}\frac{\partial}{\partial X} + \frac{\partial Y}{\partial X_i}\frac{\partial}{\partial Y} + \frac{\partial Z}{\partial X_i}\frac{\partial}{\partial Z}$$
$$\left. + \frac{\partial X_O}{\partial X_i}\frac{\partial}{\partial X_O} + \frac{\partial Y_O}{\partial X_i}\frac{\partial}{\partial Y_O} + \frac{\partial Z_O}{\partial X_i}\frac{\partial}{\partial Z_O}\right\}. \qquad (2.211)$$

Given the definition of $\boldsymbol{P}'_i$, we can write each of the above terms as the X component of the dot product of $\boldsymbol{P}'_i$, $\boldsymbol{P}_R$ and $\boldsymbol{P}_O$ with $\boldsymbol{R}'_i$, $\boldsymbol{R}$ and $\boldsymbol{R}_O$, i.e.

$$[\boldsymbol{P}_i]_X = \frac{\partial}{\partial X_i}\boldsymbol{R}'_i \cdot (\boldsymbol{P}'_i) + \frac{\partial}{\partial X_i}\boldsymbol{R} \cdot (\boldsymbol{P}_R) + \frac{\partial}{\partial X_i}\boldsymbol{R}_O \cdot (\boldsymbol{P}_O). \qquad (2.212)$$

Taking all three cartesian components gives us

$$\boldsymbol{P}_i = \frac{\partial}{\partial \boldsymbol{R}_i}\boldsymbol{R}'_i \cdot \boldsymbol{P}'_i + \frac{\partial}{\partial \boldsymbol{R}_i}\boldsymbol{R} \cdot \boldsymbol{P}_R + \frac{\partial}{\partial \boldsymbol{R}_i}\boldsymbol{R}_O \cdot \boldsymbol{P}_O. \qquad (2.213)$$

For n electrons this gives us

$$\boldsymbol{P}_i = \frac{\partial}{\partial \boldsymbol{R}_i}\boldsymbol{R}_O \cdot \boldsymbol{P}_O + \frac{\partial}{\partial \boldsymbol{R}_i}\boldsymbol{R} \cdot \boldsymbol{P}_R + \sum_{j=1}^{n}\frac{\partial}{\partial \boldsymbol{R}_i}\boldsymbol{R}'_j \cdot \boldsymbol{P}'_j \qquad (2.214)$$

which is equation (2.14).

References

[1] M. Born and R. Oppenheimer, *Ann. Phys.*, **84**, 4571 (1927).
[2] R. de L. Kronig, *Band Spectra and Molecular Structure*, Cambridge University Press, Cambridge, 1930.
[3] J.H. Van Vleck, *J. Chem. Phys.*, **4**, 327 (1936).
[4] W. Kolos and L. Wolniewicz, *Rev. Mod. Phys.*, **35**, 473 (1963).
[5] P.R. Bunker, *J. Mol. Spectrosc.*, **28**, 422 (1968).
[6] R.J. Pack and J.O. Hirschfelder, *J. Chem. Phys.*, **49**, 4009 (1968).
[7] R.N. Zare, *Angular Momentum*, John Wiley and Sons, New York, 1988.
[8] W. Pauli, *Z. Phys.*, **43**, 601 (1927).
[9] M. Born and K. Huang, *Dynamical Theory of the Crystal Lattice*, chapter 4, Clarendon Press, Oxford, 1956.
[10] J.L. Dunham, *Phys. Rev.*, **41**, 721 (1932).
[11] L. Pauling and E.B. Wilson, *Introduction to Quantum Mechanics*, McGraw-Hill, New York, 1935.
[12] P.R. Bunker and P. Jensen, *Molecular Symmetry and Spectroscopy*, 2nd edn., N.R.C.Research Press, Ottawa, 1998.
[13] M.S. Child, *Semiclassical Mechanics with Molecular Applications*, Clarendon Press, Oxford, 1991.
[14] P.M. Morse, *Phys. Rev.*, **34**, 57 (1929).
[15] L.D. Landau and E.M. Lifshitz, *Mechanics*, Pergamon Press, Oxford, 1960.

3   The electronic Hamiltonian

3.1. The Dirac equation

The analysis of molecular spectra requires the choice of an effective Hamiltonian, an appropriate basis set, and calculation of the eigenvalues and eigenvectors. The effective Hamiltonian will contain molecular parameters whose values are to be determined from the spectral analysis. The theory underlying these parameters requires detailed consideration of the fundamental electronic Hamiltonian, and the effects of applied magnetic or electrostatic fields. The additional complications arising from the presence of nuclear spins are often extremely important in high-resolution spectra, and we shall describe the theory underlying nuclear spin hyperfine interactions in chapter 4. The construction of effective Hamiltonians will then be described in chapter 7.

In this section we outline the steps which lead to a wave equation for the electron satisfying the requirements of the special theory of relativity. This equation was first proposed by Dirac, and investigation of its eigenvalues and eigenfunctions, particularly in the presence of an electromagnetic field, leads naturally to the property of electron spin and its associated magnetic moment. Our procedure is to start from classical mechanics, and then to convert the equations to quantum mechanical form; we obtain a relativistically-correct second-order wave equation known as the Klein–Gordon equation. Dirac's wave equation is linear in the momentum operator and is so constructed that its eigenvalues and eigenfunctions are also solutions of the Klein–Gordon equation. We shall show that the electron kinetic energy term introduced in the previous chapter emerges from the Dirac equation, together with additional terms previously represented by $\mathcal{H}(S)$ in the electronic Hamiltonian. Our discussion is by no means a derivation of the Dirac equation; it is only intended to explain some of the more important steps, and to lead the reader into the more complete descriptions provided by Dirac [1], Schiff [2], and Rose [3], among others. Somewhat more gentle introductions to the theory which we have found particularly valuable are the books by Moss [4] and Hannabuss [5].

The problem of units is discussed in the Preface to this book. In this and all of the following chapters, we use SI units. It is a fact, however, that almost all of the relevant original literature gives expressions in terms of cgs units. The relationships between cgs and SI units are straightforward and are summarised in General Appendix E. We therefore hope that the reader of chapters 8 to 11 will not find it too difficult to relate the theory and analysis of specific spectra described in this book to those presented in the original literature.

We start by considering the relativistic classical mechanics of a particle in free space. As we have already seen in chapter 2, the momentum $\boldsymbol{P}$ of the particle is given by

$$\boldsymbol{P} = \frac{m\boldsymbol{V}}{[1 - (V^2/c^2)]^{1/2}}, \tag{3.1}$$

and the relativistic Lagrangian is

$$\mathcal{L} = -mc^2[1 - (V^2/c^2)]^{1/2}. \tag{3.2}$$

Now the energy of the particle is given by the sum of the kinetic and potential energies, and is equal to

$$E = \boldsymbol{P} \cdot \boldsymbol{V} - \mathcal{L} \tag{3.3}$$

$$= \frac{mc^2}{[1 - (V^2/c^2)]^{1/2}}. \tag{3.4}$$

Note from this equation that when the particle is at rest ($V = 0$) the energy E is equal to mc^2, which is therefore known as the rest energy. Note also that for small velocities,

$$E \simeq mc^2 + (1/2)mV^2, \tag{3.5}$$

i.e. the rest energy plus the non-relativistic kinetic energy.

Now by combining equations (3.1) and (3.4) we obtain an important relationship between the energy and momentum of the particle, namely,

$$E^2 = c^2 P^2 + m^2 c^4. \tag{3.6}$$

We now pass from classical to quantum mechanics by means of the usual substitutions,

$$E \to i\hbar \frac{\partial}{\partial t}, \quad \boldsymbol{P} \to -i\hbar \nabla. \tag{3.7}$$

Equation (3.6) may then be rewritten as a wave equation,

$$\left(\nabla^2 - \frac{1}{c^2} \frac{\partial^2}{\partial t^2} - \frac{m^2 c^2}{\hbar^2} \right) \Psi = 0. \tag{3.8}$$

This important equation is known as the Klein–Gordon equation, and was proposed by various authors [6, 7, 8, 9] at much the same time. It is, however, an inconvenient equation to use, primarily because it involves a second-order differential operator with respect to time. Dirac therefore sought an equation linear in the momentum operator, whose solutions were also solutions of the Klein–Gordon equation. Dirac also required an equation which could more easily be generalised to take account of electromagnetic fields. The wave equation proposed by Dirac was [10]

$$(E - c\boldsymbol{\alpha} \cdot \boldsymbol{P} - \beta mc^2)\Psi = 0, \tag{3.9}$$

and we can compare this with the Klein–Gordon equation in order to learn about the quantities α and β, which make the eigenfunctions of (3.9) also eigenfunctions of (3.8). If we premultiply (3.9) by $(E + c\boldsymbol{\alpha} \cdot \boldsymbol{P} + \beta mc^2)$ we obtain the expression

$$\{E^2 - c^2[\alpha_x^2 P_x^2 + \alpha_y^2 P_y^2 + \alpha_z^2 P_z^2 + (\alpha_x \alpha_y + \alpha_y \alpha_x) P_x P_y \\ + (\alpha_y \alpha_z + \alpha_z \alpha_y) P_y P_z + (\alpha_z \alpha_x + \alpha_x \alpha_z) P_z P_x] \\ - mc^3[(\alpha_x \beta + \beta \alpha_x) P_x + (\alpha_y \beta + \beta \alpha_y) P_y \\ + (\alpha_z \beta + \beta \alpha_z) P_z] - m^2 c^4 \beta^2\} \Psi = 0. \tag{3.10}$$

Noting that the Klein–Gordon equation can be written in the form

$$\{E^2 - c^2(P_x^2 + P_y^2 + P_z^2) - m^2 c^4\} \Psi = 0, \tag{3.11}$$

we see that equations (3.10) and (3.11) agree with each other if

$$\alpha_x^2 = \alpha_y^2 = \alpha_z^2 = \beta^2 = 1, \tag{3.12}$$

$$\alpha_x \alpha_y + \alpha_y \alpha_x = \alpha_y \alpha_z + \alpha_z \alpha_y = \alpha_z \alpha_x + \alpha_x \alpha_z = 0, \tag{3.13}$$

$$\alpha_x \beta + \beta \alpha_x = \alpha_y \beta + \beta \alpha_y = \alpha_z \beta + \beta \alpha_z = 0. \tag{3.14}$$

Thus the four quantities α_x, α_y, α_z and β anticommute with each other, and their squares are unity. Investigation of these properties reveals that β and the components of $\boldsymbol{\alpha}$ can be represented by 4×4 matrices which are, explicitly

$$\beta = \begin{bmatrix} 1 & 0 & 0 & 0 \\ 0 & 1 & 0 & 0 \\ 0 & 0 & -1 & 0 \\ 0 & 0 & 0 & -1 \end{bmatrix} \quad \alpha_x = \begin{bmatrix} 0 & 0 & 0 & 1 \\ 0 & 0 & 1 & 0 \\ 0 & 1 & 0 & 0 \\ 1 & 0 & 0 & 0 \end{bmatrix} \quad \alpha_y = \begin{bmatrix} 0 & 0 & 0 & -i \\ 0 & 0 & i & 0 \\ 0 & -i & 0 & 0 \\ i & 0 & 0 & 0 \end{bmatrix} \tag{3.15}$$

$$\alpha_z = \begin{bmatrix} 0 & 0 & 1 & 0 \\ 0 & 0 & 0 & -1 \\ 1 & 0 & 0 & 0 \\ 0 & -1 & 0 & 0 \end{bmatrix}.$$

These matrices may be conveniently abbreviated by writing them in the form

$$\beta = \begin{bmatrix} 1 & 0 \\ 0 & -1 \end{bmatrix}, \quad \alpha = \begin{bmatrix} 0 & \sigma' \\ \sigma' & 0 \end{bmatrix}, \tag{3.16}$$

where each element is a 2×2 matrix. Note that the matrices α_x, α_y, α_z are regenerated from (3.16) if the components of σ' are the Pauli matrices introduced in the previous chapter.

Hence the Dirac Hamiltonian is given by

$$\mathcal{H} = c\boldsymbol{\alpha} \cdot \boldsymbol{P} + \beta mc^2, \tag{3.17}$$

and for an electron in the presence of an electromagnetic field this becomes

$$\mathcal{H} = c\boldsymbol{\alpha} \cdot (\boldsymbol{P} + e\boldsymbol{A}) + \beta mc^2 - e\phi \tag{3.18}$$

where, as in chapter 2, the charge on the electron is $-e$, and A and ϕ are the magnetic vector potential and scalar potential respectively.

Since α and β are represented by 4×4 matrices, the wave function Ψ must also be a four-component function and the Dirac wave equation (3.9) is actually equivalent to four simultaneous first-order partial differential equations which are linear and homogeneous in the four components of Ψ. According to the Pauli spin theory, introduced in the previous chapter, the spin of the electron requires the wave function to have only two components. We shall see in the next section that the wave equation (3.9) actually has two solutions corresponding to states of positive energy, and two corresponding to states of negative energy. The two solutions in each case correspond to the spin components.

3.2. Solutions of the Dirac equation in field-free space

The wave function $\Psi = \Psi(\boldsymbol{R}, t)$ may be written as a four-component spinor

$$\Psi(\boldsymbol{R}, t) = \begin{bmatrix} \psi_1(\boldsymbol{R}, t) \\ \psi_2(\boldsymbol{R}, t) \\ \psi_3(\boldsymbol{R}, t) \\ \psi_4(\boldsymbol{R}, t) \end{bmatrix}, \tag{3.19}$$

and solutions of the form

$$\psi_j(\boldsymbol{R}, t) = u_j e^{i(\boldsymbol{k}\cdot\boldsymbol{R}-\omega t)} \quad (j = 1, 2, 3, 4), \tag{3.20}$$

where the u_j are numbers, can be found. These functions are eigenfunctions of the operators $i\hbar(\partial/\partial t)$ and $-i\hbar\nabla$ with eigenvalues $\hbar\omega$ and $\hbar\boldsymbol{k}$ respectively. Substitution of (3.20) into the Dirac equation (3.9) yields four simultaneous equations in the u_j coefficients, which are

$$\begin{aligned}
(E - mc^2)u_1 - cP_z u_3 - c(P_x - iP_y)u_4 &= 0, \\
(E - mc^2)u_2 + cP_z u_4 - c(P_x + iP_y)u_3 &= 0, \\
(E + mc^2)u_3 - cP_z u_1 - c(P_x - iP_y)u_2 &= 0, \\
(E + mc^2)u_4 + cP_z u_2 - c(P_x + iP_y)u_1 &= 0.
\end{aligned} \tag{3.21}$$

The condition for non-trivial solutions is that the determinant of the coefficients is zero, yielding

$$(E^2 - m^2 c^4 - c^2 \boldsymbol{P}^2)^2 = 0. \tag{3.22}$$

Note that this relationship is in agreement with our previous equation (3.6) connecting the energy and momentum. Now equation (3.22) reveals the existence of positive and negative energy solutions, i.e.

$$E_+ = +(c^2 \boldsymbol{P}^2 + m^2 c^4)^{1/2}, \tag{3.23}$$

$$E_- = -(c^2 \boldsymbol{P}^2 + m^2 c^4)^{1/2}. \tag{3.24}$$

For the positive energy value we obtain two linearly independent solutions, which are

$$u_1 = 1, \quad u_2 = 0, \quad u_3 = \frac{cP_z}{E_+ + mc^2}, \quad u_4 = \frac{c(P_x + iP_y)}{E_+ + mc^2}, \qquad (3.25)$$

$$u_1 = 0, \quad u_2 = 1, \quad u_3 = \frac{c(P_x - iP_y)}{E_+ + mc^2}, \quad u_4 = -\frac{cP_z}{E_+ + mc^2}. \qquad (3.26)$$

Similarly, for the negative energy value, two new solutions are obtained,

$$u_1 = \frac{cP_z}{E_- - mc^2}, \quad u_2 = \frac{c(P_x + iP_y)}{E_- - mc^2}, \quad u_3 = 1, \quad u_4 = 0, \qquad (3.27)$$

$$u_1 = \frac{c(P_x - iP_y)}{E_- - mc^2}, \quad u_2 = -\frac{cP_z}{E_- - mc^2}, \quad u_3 = 0, \quad u_4 = 1. \qquad (3.28)$$

Hence, in general, three of the components are non-zero. In the non-relativistic limit, as $E_+ = E_- \approx mc^2$, the eigenfunctions corresponding to positive energy states approach $\psi_1(\mathbf{R}, t)$ and $\psi_2(\mathbf{R}, t)$ and, conversely, $\psi_3(\mathbf{R}, t)$ and $\psi_4(\mathbf{R}, t)$ become the solutions with negative energy. Now we shall only be seriously interested in the positive energy solutions, which correspond to electron wave functions. Nevertheless the existence of coupling terms between the positive and negative energy states (the latter are referred to as positron states) must be taken into account if the theory is to be relativistically correct. The positron was, in fact, discovered some five years later by Anderson [11]. Foldy and Wouthuysen [12] showed that the four-component Dirac equation can be reduced to a relativistically correct two-component wave equation for the electron by means of a unitary transformation. The importance of the Foldy–Wouthuysen transformation is that it can also be applied to the Dirac Hamiltonian for an electron in an electromagnetic field (equation (3.18)); we shall describe the method in detail in section 3.4.

3.3. Electron spin magnetic moment and angular momentum

We now show that the Dirac Hamiltonian (3.18) leads naturally to the result that the electron has an intrinsic magnetic moment arising from angular momentum of magnitude $(1/2)\hbar$, which we describe as spin angular momentum. The Dirac Hamiltonian for an electron (charge $-e$) in an electromagnetic field (3.18) is

$$\mathcal{H} = -e\phi + c\boldsymbol{\alpha} \cdot (\mathbf{P} + e\mathbf{A}) + \beta mc^2. \qquad (3.29)$$

Rearranging and squaring leads to the result,

$$\left\{\frac{\mathcal{H}}{c} + \frac{e\phi}{c}\right\}^2 = \{\boldsymbol{\alpha} \cdot (\mathbf{P} + e\mathbf{A}) + \beta mc\}^2 \qquad (3.30)$$

$$= [\boldsymbol{\alpha} \cdot (\mathbf{P} + e\mathbf{A})]^2 + m^2c^2. \qquad (3.31)$$

In passing from (3.30) to (3.31) we have made use of the anticommutation relations (3.14) and also of the fact that $\beta^2 = 1$. It is now convenient to express $\boldsymbol{\alpha}$ as $\rho\boldsymbol{\sigma}$ where $\boldsymbol{\sigma}$ is a three-component vector, each component of which is represented by a 4×4

matrix, called a Dirac spin matrix. To be explicit,

$$\rho = \begin{bmatrix} 0 & 0 & 1 & 0 \\ 0 & 0 & 0 & 1 \\ 1 & 0 & 0 & 0 \\ 0 & 1 & 0 & 0 \end{bmatrix} \quad \sigma_x = \begin{bmatrix} 0 & 1 & 0 & 0 \\ 1 & 0 & 0 & 0 \\ 0 & 0 & 0 & 1 \\ 0 & 0 & 1 & 0 \end{bmatrix}$$

(3.32)

$$\sigma_y = \begin{bmatrix} 0 & -i & 0 & 0 \\ i & 0 & 0 & 0 \\ 0 & 0 & 0 & -i \\ 0 & 0 & i & 0 \end{bmatrix} \quad \sigma_z = \begin{bmatrix} 1 & 0 & 0 & 0 \\ 0 & -1 & 0 & 0 \\ 0 & 0 & 1 & 0 \\ 0 & 0 & 0 & -1 \end{bmatrix};$$

ρ commutes with σ and with $[P + eA]$; also $\rho^2 = 1$, therefore

$$[\alpha \cdot (P + eA)]^2 = [\sigma \cdot (P + eA)][\sigma \cdot (P + eA)]. \quad (3.33)$$

The commutation properties of σ allow one to establish the vector identity,

$$(\sigma \cdot B)(\sigma \cdot C) = (B \cdot C) + i\sigma(B \wedge C). \quad (3.34)$$

We thus obtain the results

$$[\sigma \cdot (P + eA)][\sigma \cdot (P + eA)] = (P + eA)^2 + i\sigma \cdot (P + eA) \wedge (P + eA) \quad (3.35)$$
$$= (P + eA)^2 + ie\sigma \cdot [(A \wedge P) + (P \wedge A)] \quad (3.36)$$
$$= (P + eA)^2 + i\sigma \cdot [-i\hbar e\, \mathrm{curl}\, A] \quad (3.37)$$
$$= (P + eA)^2 + e\hbar(\sigma \cdot B). \quad (3.38)$$

We have here made use of Maxwell's result that curl A gives the magnetic field intensity B. Combining equations (3.31) and (3.38) we obtain the result,

$$\left\{ \frac{\mathcal{H}}{c} + \frac{e\phi}{c} \right\}^2 = (P + eA)^2 + e\hbar(\sigma \cdot B) + m^2c^2. \quad (3.39)$$

Now for an electron which is moving slowly (i.e. with small momentum) we may put $\mathcal{H}$ equal to $mc^2 + \mathcal{H}_1$, where $\mathcal{H}_1$ is small compared with mc^2. If we make this replacement in (3.39), neglect $\mathcal{H}_1^2$ and terms involving c^{-2}, and divide through by $2m$, we obtain a non-relativistic Hamiltonian,

$$\mathcal{H}_1 = \frac{1}{2m}(P + eA)^2 + \frac{\hbar e}{2m}(\sigma \cdot B) - e\phi. \quad (3.40)$$

This equation is the same as the classical Hamiltonian for a slowly moving electron, except for the middle term. This term represents an additional potential energy and may be interpreted as arising from the electron having a magnetic moment $-(e\hbar/2m)\sigma$. It was shown in equations (3.25) and (3.26) that the ψ_3 and ψ_4 components of the wave function are large in comparison with the ψ_1 and ψ_2 components for the positive energy solutions. Now the relations

$$\alpha = \begin{bmatrix} 0 & \sigma' \\ \sigma' & 0 \end{bmatrix}, \quad \sigma = \begin{bmatrix} \sigma' & 0 \\ 0 & \sigma' \end{bmatrix}, \quad (3.41)$$

show that σ operating on the four-component wave function is the same as σ' operating on the large components alone.

Equation (3.40) reveals the existence of the spin magnetic moment through its coupling with an external magnetic field. The electron spin carries no energy itself and can therefore only be observed through its coupling with the orbital motion of the electron. This coupling can be made apparent, either through demonstration of the conservation of total angular momentum, or through derivation of the spin–orbit coupling energy. We shall be dealing with spin–orbit coupling later in this chapter, so we here consider the first proposition.

We use the Dirac Hamiltonian in field-free space,

$$\mathcal{H} = c\boldsymbol{\alpha} \cdot \boldsymbol{P} + \beta mc^2, \qquad (3.42)$$

and see if the orbital angular momentum $\boldsymbol{L}$ commutes with the relativistic Hamiltonian $\mathcal{H}$, as it does with the classical Hamiltonian. We can deal with each component of $\boldsymbol{L}$ in turn, and consider first the L_z component.

We have

$$[\mathcal{H}, L_z] = \mathcal{H} L_z - L_z \mathcal{H} = c[\boldsymbol{\alpha} \cdot (\boldsymbol{P} L_z) - \boldsymbol{\alpha} \cdot (L_z \boldsymbol{P})] \qquad (3.43)$$

$$= c\rho \boldsymbol{\sigma} \cdot [\boldsymbol{P}, L_z] \qquad (3.44)$$

$$= c\rho \{\sigma_x [P_x, L_z] + \sigma_y [P_y, L_z] + \sigma_z [P_z, L_z]\}. \qquad (3.45)$$

Remembering that

$$\hbar L_z = x P_y - y P_x, \qquad (3.46)$$

and also noting the commutation relations between P_x, P_y, P_z we readily establish the results

$$[P_x, L_z] = -iP_y, \quad [P_y, L_z] = iP_x, \quad [P_z, L_z] = 0. \qquad (3.47)$$

Consequently (3.45) yields the result that $\mathcal{H}$ and L_z do not commute, but that

$$[\mathcal{H}, L_z] = -ic\rho \{\sigma_x P_y - \sigma_y P_x\}. \qquad (3.48)$$

Let us now consider the commutation of $\mathcal{H}$ and σ_z. We have

$$[\mathcal{H}, \sigma_z] = c\rho [(\boldsymbol{\sigma} \cdot \boldsymbol{P}), \sigma_z]$$

$$= c\rho \{[\sigma_x, \sigma_z] P_x + [\sigma_y, \sigma_z] P_y + [\sigma_z, \sigma_z] P_z\}. \qquad (3.49)$$

Noting now the commutation relations

$$[\sigma_x, \sigma_z] = -2i\sigma_y, \quad [\sigma_y, \sigma_z] = 2i\sigma_x, \quad [\sigma_z, \sigma_z] = 0, \qquad (3.50)$$

we find that $\mathcal{H}$ and σ_z do not commute, but that

$$[\mathcal{H}, \sigma_z] = 2ic\rho \{-\sigma_y P_x + \sigma_x P_y\}. \qquad (3.51)$$

Hence the relativistic Hamiltonian $\mathcal{H}$ does not commute with either L_z or σ_z separately, but by combining equations (3.48) and (3.51) we find that $\mathcal{H}$ does commute with

$L_z+(1/2)\sigma_z$, i.e.

$$[\mathcal{H}, L_z+(1/2)\sigma_z] = [\mathcal{H}, L_z]+(1/2)[\mathcal{H}, \sigma_z] \tag{3.52}$$
$$= -ic\rho\{\sigma_x P_y - \sigma_y P_x - \sigma_x P_y + \sigma_y P_x\} \tag{3.53}$$
$$= 0. \tag{3.54}$$

Now similar results are readily obtained for the x and y components of L and σ, so that although $\mathcal{H}$ does not commute separately with L and σ, it does commute with $L+(1/2)\sigma$,

$$[\mathcal{H}, L+(1/2)\sigma] = 0. \tag{3.55}$$

$L+(1/2)\sigma$ is therefore a constant of the motion and we conclude that, in addition to possessing orbital angular momentum L, the electron also possesses spin angular momentum $\hbar S$, where $S = (1/2)\sigma$. Moreover, if the spin magnetic moment μ_S is given by

$$\mu_S = -g_S\mu_B S, \tag{3.56}$$

where μ_B is the Bohr magneton, equal to $e\hbar/2m$, we see from (3.40) that the electron spin g value, g_S, is equal to 2. The corresponding result for the orbital magnetic moment is that $g_L = 1$. These predictions are in agreement with experiment. Thus the Dirac theory not only predicts the existence of electron spin angular momentum, but also accounts for what had previously been called the anomalous magnetic moment of the electron.

3.4. The Foldy–Wouthuysen transformation

In section 3.2 we pointed out that the Dirac Hamiltonian contains operators which connect states of positive and negative energy. What we now seek is a Hamiltonian which is relativistically correct but which operates on the two-component electron functions of positive energy only. We require that this Hamiltonian contain terms representing electromagnetic fields, and Foldy and Wouthuysen [12] showed, by a series of unitary transformations, that such a Hamiltonian can be derived. The Dirac Hamiltonian

$$\mathcal{H} = \beta mc^2 - e\phi + c\alpha \cdot (P + eA), \tag{3.57}$$

contains the even operator $\mathcal{E} = -e\phi$, which has vanishing matrix elements between electron and positron wave functions, and the odd operator $\mathcal{O} = c\alpha \cdot [P + eA]$ whose corresponding matrix elements are non-vanishing. Thus we can write (3.57) as

$$\mathcal{H} = \beta mc^2 + \mathcal{E} + \mathcal{O}, \tag{3.58}$$

and seek a unitary transformation which, to a certain level of accuracy, yields a Hamiltonian containing only even operators. The term βmc^2 is an even operator but is kept separate from the other even operator $(-e\phi)$ because they are of different orders of magnitude. In the Foldy–Wouthuysen transformation the odd operator $\mathcal{O}$ in the Dirac Hamiltonian is removed by a unitary transformation, as we shall see. Although other odd operators are introduced by the transformation, they are of higher order in α; the aim is to obtain a Hamiltonian correct to order $mc^2\alpha^4$.

Before proceeding we must state a few important properties of the operators $\mathcal{E}$ and $\mathcal{O}$. These are

(i) $\mathcal{E}$ commutes with β, but $\mathcal{O}$ anticommutes with β, i.e.

$$\mathcal{E}\beta = \beta\mathcal{E}, \quad \mathcal{O}\beta = -\beta\mathcal{O}, \quad \beta\mathcal{E}\beta = \mathcal{E}, \quad \beta\mathcal{O}\beta = -\mathcal{O} \quad \text{since } \beta^2 = 1; \quad (3.59)$$

(ii)

$$\mathcal{E} \times \mathcal{E} = \mathcal{E}, \quad \mathcal{E} \times \mathcal{O} = \mathcal{O}, \quad \mathcal{O} \times \mathcal{O} = \mathcal{E}. \quad (3.60)$$

Now the Schrödinger equation is written

$$i\hbar \frac{\partial \psi}{\partial t} = \mathcal{H}\psi, \quad (3.61)$$

and we consider the unitary transformation $\psi' = e^{i\bar{S}}\psi$ where $\bar{S}$ is a Hermitian operator. We find that

$$i\hbar \frac{\partial \psi'}{\partial t} = i\hbar \frac{\partial}{\partial t}(e^{i\bar{S}}\psi) \quad (3.62)$$

$$= i\hbar \left(\frac{\partial e^{i\bar{S}}}{\partial t}\right)\psi + i\hbar e^{i\bar{S}}\left(\frac{\partial \psi}{\partial t}\right) \quad (3.63)$$

$$= i\hbar \left(\frac{\partial e^{i\bar{S}}}{\partial t}\right)e^{-i\bar{S}}\psi' + e^{i\bar{S}}\mathcal{H}e^{-i\bar{S}}\psi'. \quad (3.64)$$

Hence ψ' is an eigenfunction satisfying the transformed Schrödinger equation

$$i\hbar \frac{\partial \psi'}{\partial t} = \mathcal{H}'\psi', \quad (3.65)$$

where

$$\mathcal{H}' = e^{i\bar{S}}\mathcal{H}e^{-i\bar{S}} + i\hbar \left(\frac{\partial e^{i\bar{S}}}{\partial t}\right)e^{-i\bar{S}}. \quad (3.66)$$

The transformed Hamiltonian $\mathcal{H}'$ can be written in a more useful form by expanding the exponentials in (3.66), i.e.

$$\mathcal{H}' = \left[1 + i\bar{S} + \frac{i^2}{2}\bar{S}^2 + \cdots\right]\mathcal{H}\left[1 - i\bar{S} + \frac{i^2}{2}\bar{S}^2 - \cdots\right]$$

$$+ i\hbar \left[i\frac{\partial \bar{S}}{\partial t} + \frac{i^2}{2}\bar{S}\frac{\partial \bar{S}}{\partial t} + \frac{i^2}{2}\frac{\partial \bar{S}}{\partial t}\bar{S} + \cdots\right]\left[1 - i\bar{S} + \frac{i^2}{2}\bar{S}^2 - \cdots\right] \quad (3.67)$$

$$= \mathcal{H} + i[\bar{S}, \mathcal{H}] + \frac{i^2}{2!}[\bar{S},[\bar{S}, \mathcal{H}]] + \frac{i^3}{3!}[\bar{S},[\bar{S},[\bar{S}, \mathcal{H}]]] + \cdots$$

$$- \hbar \frac{\partial \bar{S}}{\partial t} - \frac{i\hbar}{2}\left[\bar{S}, \frac{\partial \bar{S}}{\partial t}\right] - \frac{i^2}{2!}\frac{\hbar}{3}\left[\bar{S},\left[\bar{S}, \frac{\partial \bar{S}}{\partial t}\right]\right] - \frac{i^3}{3!}\frac{\hbar}{4}\left[\bar{S},\left[\bar{S},\left[\bar{S}, \frac{\partial \bar{S}}{\partial t}\right]\right]\right] - \cdots. \quad (3.68)$$

Now Foldy and Wouthuysen suggested that if $\mathcal{H}$ is the Dirac Hamiltonian (3.58), the operator $\bar{S}$ should be given by

$$\bar{S} = -\frac{i\beta\mathcal{O}}{2mc^2}. \tag{3.69}$$

The transformed Hamiltonian $\mathcal{H}'$ will then be given by (3.68) and we therefore evaluate the commutator brackets occurring in that expression. We find, for example,

$$[\bar{S}, \mathcal{H}] = \left[-\frac{i\beta\mathcal{O}}{2mc^2}, \beta mc^2\right] + \left[-\frac{i\beta\mathcal{O}}{2mc^2}, \mathcal{E}\right] + \left[-\frac{i\beta\mathcal{O}}{2mc^2}, \mathcal{O}\right] \tag{3.70}$$

$$= i\left\{\mathcal{O} - \frac{\beta}{2mc^2}[\mathcal{O}, \mathcal{E}] - \frac{\beta\mathcal{O}^2}{mc^2}\right\}, \tag{3.71}$$

$$[\bar{S}, [\bar{S}, \mathcal{H}]] = \frac{\beta\mathcal{O}^2}{mc^2} + \frac{1}{4m^2c^4}[\mathcal{O}, [\mathcal{O}, \mathcal{E}]] + \frac{\mathcal{O}^3}{m^2c^4}, \tag{3.72}$$

$$[\bar{S}, [\bar{S}, [\bar{S}, \mathcal{H}]]] = \frac{i\mathcal{O}^3}{m^2c^4} - \frac{i\beta}{8m^3c^6}[\mathcal{O}, [\mathcal{O}, [\mathcal{O}, \mathcal{E}]]] - \frac{i\beta\mathcal{O}^4}{m^3c^6}, \tag{3.73}$$

etc.
Similarly we find that

$$\frac{\partial \bar{S}}{\partial t} = -\frac{i\beta}{2mc^2}\frac{\partial \mathcal{O}}{\partial t}, \tag{3.74}$$

$$\left[\bar{S}, \frac{\partial \bar{S}}{\partial t}\right] = \frac{1}{4m^2c^4}\left[\mathcal{O}, \frac{\partial \mathcal{O}}{\partial t}\right], \tag{3.75}$$

$$\left[\bar{S}, \left[\bar{S}, \frac{\partial \bar{S}}{\partial t}\right]\right] = -\frac{i\beta}{8m^3c^6}\left[\mathcal{O}, \left[\mathcal{O}, \frac{\partial \mathcal{O}}{\partial t}\right]\right], \tag{3.76}$$

etc.
We substitute these results in equation (3.68) and obtain the transformed Hamiltonian

$$\mathcal{H}' = \beta mc^2 + \mathcal{E} + \frac{\beta\mathcal{O}^2}{2mc^2} - \frac{1}{8m^2c^4}\left\{\left[\mathcal{O}, [\mathcal{O}, \mathcal{E}] + i\hbar\frac{\partial \mathcal{O}}{\partial t}\right]\right\} - \frac{\beta\mathcal{O}^4}{8m^3c^6} + \cdots$$

$$+ \frac{\beta}{2mc^2}\left\{[\mathcal{O}, \mathcal{E}] + i\hbar\frac{\partial \mathcal{O}}{\partial t}\right\} - \frac{\mathcal{O}^3}{3m^2c^4} + \text{higher-order terms}. \tag{3.77}$$

Now the first row in the transformed Hamiltonian (3.77) consists of terms which contain even operators only, and the second row contains odd operators only. We have carried out the expansion up to terms including $1/c^2$, higher-order terms being neglected. However the significant feature of $\mathcal{H}'$ is that whereas it contains even terms of order c^2, c^0 and c^{-2}, the odd terms are of order c^{-1}; in the starting Hamiltonian (3.58) the odd term was of order c^{+1}. The transformation has therefore reduced the importance of the odd terms, and if we wish to proceed further and obtain a Hamiltonian which, to order c^{-2}, contains only even operators, we can repeat the transformation process. The new starting Hamiltonian is written as

$$\mathcal{H}' = \beta mc^2 + \mathcal{E}' + \mathcal{O}', \tag{3.78}$$

where

$$\mathcal{E}' = \mathcal{E} + \frac{\beta\mathcal{O}^2}{2mc^2} - \frac{1}{8m^2c^4}\left\{\left[\mathcal{O},[\mathcal{O},\mathcal{E}] + i\hbar\frac{\partial\mathcal{O}}{\partial t}\right]\right\} - \frac{\beta\mathcal{O}^4}{8m^3c^6} + O(mc^2\alpha^6), \quad (3.79)$$

and

$$\mathcal{O}' = \frac{\beta}{2mc^2}\left\{[\mathcal{O},\mathcal{E}] + i\hbar\frac{\partial\mathcal{O}}{\partial t}\right\} - \frac{\mathcal{O}^3}{3m^2c^4} + O(mc^2\alpha^5). \quad (3.80)$$

To order $1/c^2$, the new transformed Hamiltonian $\mathcal{H}''$ is given by

$$\mathcal{H}'' = \mathcal{H}' + i[\bar{S}',\mathcal{H}'], \quad (3.81)$$

where $\mathcal{H}'$ is given by (3.77) and $\bar{S}'$ is put equal to $-i\beta\mathcal{O}'/2mc^2$, by analogy with (3.69). We readily find that

$$i[\bar{S},\mathcal{H}'] = -\frac{\beta}{2mc^2}\left\{[\mathcal{O},\mathcal{E}] + i\hbar\frac{\partial\mathcal{O}}{\partial t}\right\} + \frac{\mathcal{O}^3}{3m^2c^4} + \text{higher–order terms} \quad (3.82)$$

and substitution in (3.81) yields the result

$$\mathcal{H}'' = \beta mc^2 + \mathcal{E} + \frac{\beta\mathcal{O}^2}{2mc^2} - \frac{1}{8m^2c^4}\left\{\left[\mathcal{O},[\mathcal{O},\mathcal{E}] + i\hbar\frac{\partial\mathcal{O}}{\partial t}\right]\right\} - \frac{\beta\mathcal{O}^4}{8m^3c^6}. \quad (3.83)$$

We have therefore achieved our objective in that equation (3.83), which is correct to order $1/c^2$, contains even operators only. It would, of course, be possible to proceed further with the Foldy–Wouthuysen transformation but there is little point in doing so, since the theory is inaccurate in other respects. For example, we have treated the electromagnetic field classically, instead of using quantum field theory. Furthermore, we shall ultimately be interested in many-electron diatomic molecules, for which it will be necessary to make a number of assumptions and approximations.

Since equation (3.83) contains even operators only, we may set β equal to 1, and substituting for $\mathcal{E}$ and $\mathcal{O}$ we obtain the Hamiltonian which is appropriate to positive energy states only,

$$\mathcal{H}'' = mc^2 - e\phi + \frac{1}{2m}(\boldsymbol{\alpha}\cdot\boldsymbol{\pi})^2 - \frac{1}{8m^2c^2}\left[(\boldsymbol{\alpha}\cdot\boldsymbol{\pi}),[(\boldsymbol{\alpha}\cdot\boldsymbol{\pi}),(-e\phi)] + i\hbar\frac{\partial}{\partial t}(\boldsymbol{\alpha}\cdot\boldsymbol{\pi})\right]$$
$$- \frac{1}{8m^3c^2}(\boldsymbol{\alpha}\cdot\boldsymbol{\pi})^4, \quad (3.84)$$

where $\boldsymbol{\pi} = \boldsymbol{P} + e\boldsymbol{A}$. We now expand (3.84) and examine the various terms which arise. We deal first with the third term in (3.84) and using equation (3.34), together with the fact that $\rho^2 = 1$, we obtain

$$(\boldsymbol{\alpha}\cdot\boldsymbol{\pi})^2 = \rho^2(\boldsymbol{\sigma}\cdot\boldsymbol{\pi})^2 \quad (3.85)$$
$$= \pi^2 + i\boldsymbol{\sigma}\cdot(\boldsymbol{\pi}\wedge\boldsymbol{\pi}). \quad (3.86)$$

Equation (3.86) can be further expanded using the previous results given in (3.35) to

(3.38) and we find that

$$\frac{1}{2m}(\boldsymbol{\alpha}\cdot\boldsymbol{\pi})^2 = \frac{1}{2m}\{\pi^2 + e\hbar(\boldsymbol{\sigma}\cdot\boldsymbol{B})\} \qquad (3.87)$$

$$= \frac{\pi^2}{2m} + g_S\mu_B \boldsymbol{S}\cdot\boldsymbol{B} \qquad (3.88)$$

We next consider the fourth term in the Hamiltonian (3.84). First we note the results

$$[(\boldsymbol{\alpha}\cdot\boldsymbol{\pi}),(-e\phi)] = -e\boldsymbol{\alpha}\cdot(\boldsymbol{\pi}\phi - \phi\boldsymbol{\pi}) \qquad (3.89)$$

$$= i\hbar e\boldsymbol{\alpha}\cdot(\nabla\phi), \qquad (3.90)$$

$$i\hbar\frac{\partial}{\partial t}(\boldsymbol{\alpha}\cdot\boldsymbol{\pi}) = i\hbar\boldsymbol{\alpha}\cdot\frac{\partial}{\partial t}(\boldsymbol{P} + e\boldsymbol{A}) \qquad (3.91)$$

$$= i\hbar e\boldsymbol{\alpha}\cdot\frac{\partial\boldsymbol{A}}{\partial t}. \qquad (3.92)$$

We then combine equations (3.90) and (3.92), obtaining the result

$$[(\boldsymbol{\alpha}\cdot\boldsymbol{\pi}),(-e\phi)] + i\hbar\frac{\partial}{\partial t}(\boldsymbol{\alpha}\cdot\boldsymbol{\pi}) = i\hbar e\boldsymbol{\alpha}\cdot\left\{(\nabla\phi) + \frac{\partial\boldsymbol{A}}{\partial t}\right\} \qquad (3.93)$$

$$= -i\hbar e\boldsymbol{\alpha}\cdot\boldsymbol{E}. \qquad (3.94)$$

In passing from (3.93) to (3.94) we have made use of the definition of the electric field intensity $\boldsymbol{E}$. We complete our expansion of the fourth term in (3.84) by using (3.94) and find that

$$[(\boldsymbol{\alpha}\cdot\boldsymbol{\pi}),(-i\hbar e\boldsymbol{\alpha}\cdot\boldsymbol{E})] = -i\hbar e[(\boldsymbol{\alpha}\cdot\boldsymbol{\pi}),(\boldsymbol{\alpha}\cdot\boldsymbol{E})] \qquad (3.95)$$

$$= -i\hbar e\{\boldsymbol{\pi}\cdot\boldsymbol{E} - \boldsymbol{E}\cdot\boldsymbol{\pi} + i\boldsymbol{\sigma}\cdot(\boldsymbol{\pi}\wedge\boldsymbol{E} - \boldsymbol{E}\wedge\boldsymbol{\pi})\} \qquad (3.96)$$

$$= -\hbar^2 e\nabla\cdot\boldsymbol{E} + 2\hbar e\boldsymbol{S}\cdot(\boldsymbol{\pi}\wedge\boldsymbol{E} - \boldsymbol{E}\wedge\boldsymbol{\pi}). \qquad (3.97)$$

Finally, the last term in (3.84) gives, on expansion,

$$-\frac{1}{8m^3c^2}(\boldsymbol{\alpha}\cdot\boldsymbol{\pi})^4 = -\frac{1}{8m^3c^2}\{\pi^2 + e\hbar(\boldsymbol{\sigma}\cdot\boldsymbol{B})\}^2 \qquad (3.98)$$

$$= -\frac{1}{8m^3c^2}\{\pi^4 + 2e\hbar(\boldsymbol{\sigma}\cdot\boldsymbol{B})\pi^2 + e^2\hbar^2(\boldsymbol{\sigma}\cdot\boldsymbol{B})^2\} \qquad (3.99)$$

$$= -\frac{\pi^4}{8m^3c^2} - \frac{g_S\mu_B}{m^2c^2}(\boldsymbol{S}\cdot\boldsymbol{B})\pi^2 + \cdots. \qquad (3.100)$$

The last term in equation (3.99) is of too high an order to be retained.

We can now collect the results given in equations (3.88), (3.97) and (3.100) and obtain the required Hamiltonian (we drop the primes on $\mathcal{H}''$),

$$\mathcal{H} = mc^2 - e\phi + \frac{\pi^2}{2m} + g_S\mu_B\boldsymbol{S}\cdot\boldsymbol{B} + \frac{e\hbar^2}{8m^2c^2}\nabla\cdot\boldsymbol{E} - \frac{e\hbar}{4m^2c^2}\boldsymbol{S}\cdot(\boldsymbol{\pi}\wedge\boldsymbol{E} - \boldsymbol{E}\wedge\boldsymbol{\pi})$$

$$-\frac{\pi^4}{8m^3c^2} - \frac{g_S\mu_B}{m^2c^2}(\boldsymbol{S}\cdot\boldsymbol{B})\pi^2. \qquad (3.101)$$

The first term is the rest mass energy and the second term represents the electric potential (which we shall examine in detail later). The third term is the kinetic energy of the electron, whilst the seventh term $(-\pi^4/8m^3c^2)$ is a relativistic correction to the kinetic energy. The fourth term $g_S\mu_B \mathbf{S}\cdot\mathbf{B}$ represents the interaction of the spin magnetic moment with the magnetic field $\mathbf{B}$; we met this term previously in equation (3.40). The fifth term $(e\hbar^2\nabla\cdot\mathbf{E}/8m^2c^2)$, which has no analogue in non-relativistic quantum mechanics, is called the Darwin term [13]. The sixth term represents the fact that a moving magnetic moment creates a perpendicular electric moment, which interacts with the electric field $\mathbf{E}$. Finally, the last term is a relativistic correction to the electron g factor.

In due course we shall expand the Hamiltonian (3.101) still further by replacing π by $(\mathbf{P} + e\mathbf{A})$ and proceed to examine the magnetic vector potential and scalar electric potential in detail.

3.5. The Foldy–Wouthuysen and Dirac representations for a free particle

One can gain some insight into the nature of the Dirac wave equation and the spin angular momentum of the electron by considering the Foldy–Wouthuysen transformation for a free particle. In the absence of electric and magnetic interactions, the Dirac Hamiltonian is

$$\mathcal{H} = \beta mc^2 + c\boldsymbol{\alpha}\cdot\mathbf{P}. \tag{3.102}$$

In field-free, four-dimensional space, the sequence of Foldy–Wouthuysen transformations can be summed to infinity and written in closed form,

$$\psi' = e^{i\bar{S}}\psi, \tag{3.103}$$

where the operator $\bar{S}$ is now given by

$$\bar{S} = -\frac{i\beta(\boldsymbol{\alpha}\cdot\mathbf{P})}{2P}f. \tag{3.104}$$

The f is a function of (P/mc), to be chosen later, which will remove odd operators from the transformed Hamiltonian, and $P = (\mathbf{P}^2)^{1/2}$. Now since $\bar{S}$ is not explicitly time-dependent, the transformed Hamiltonian is given from equation (3.66) by

$$\mathcal{H}' = e^{i\bar{S}}\mathcal{H}e^{-i\bar{S}}. \tag{3.105}$$

$\bar{S}$ and $\mathcal{H}$ anticommute, and by expanding the exponentials it may readily be proved that

$$\mathcal{H}e^{-i\bar{S}} = e^{i\bar{S}}\mathcal{H}, \tag{3.106}$$

from which

$$\mathcal{H}' = e^{2i\bar{S}}\mathcal{H}. \tag{3.107}$$

We now expand the exponential in equation (3.107) obtaining

$$e^{2i\bar{S}} = 1 + 2i\bar{S} + \frac{4i^2\bar{S}^2}{2!} + \frac{8i^3\bar{S}^3}{3!} + \frac{16i^4\bar{S}^4}{4!} + \frac{32i^5\bar{S}^5}{5!} + \cdots \quad (3.108)$$

$$= \left[1 - \frac{4\bar{S}^2}{2!} + \frac{16\bar{S}^4}{4!} - \cdots\right] + \left[2i\bar{S} + \frac{8i^3\bar{S}^3}{3!} + \frac{32i^5\bar{S}^5}{5!} + \cdots\right]. \quad (3.109)$$

Now consider the first series in (3.109) containing even powers of $\bar{S}$. From (3.104) we obtain

$$\bar{S}^2 = -\frac{[\beta(\boldsymbol{\alpha}\cdot\boldsymbol{P})]^2}{4P^2}f^2 = \frac{1}{4}f^2, \quad \text{since } [\beta(\boldsymbol{\alpha}\cdot\boldsymbol{P})]^2 = -P^2. \quad (3.110)$$

Similarly $\bar{S}^4 = (1/16)f^4$, etc., so that the first series may be written

$$\left[1 - \frac{4\bar{S}^2}{2!} + \frac{16\bar{S}^4}{4!} - \cdots\right] = \left[1 - \frac{f^2}{2!} + \frac{f^4}{4!} - \cdots\right] = \cos f. \quad (3.111)$$

Next consider the second series in (3.109) which contains odd powers of $\bar{S}$. From (3.104) it is clear that

$$2i\bar{S} = \frac{\beta(\boldsymbol{\alpha}\cdot\boldsymbol{P})}{P}f, \quad (3.112)$$

$$\frac{8i^3\bar{S}^3}{3!} = \frac{(2i\bar{S})(4\bar{S}^2)}{3!} = \frac{\beta(\boldsymbol{\alpha}\cdot\boldsymbol{P})}{P}\frac{f^3}{3!}, \quad (3.113)$$

etc.
The series may therefore be expressed as

$$\left[2i\bar{S} + \frac{8i^3\bar{S}^3}{3!} + \frac{32i^5\bar{S}^5}{5!} + \cdots\right] = \frac{\beta(\boldsymbol{\alpha}\cdot\boldsymbol{P})}{P}\left[f - \frac{f^3}{3!} + \frac{f^5}{5!} - \cdots\right] \quad (3.114)$$

$$= \beta(\boldsymbol{\alpha}\cdot\boldsymbol{P})P^{-1}\sin f. \quad (3.115)$$

Hence by using (3.111) and (3.115) we see that the transformation (3.107) can be written

$$\mathcal{H}' = [\cos f + \beta(\boldsymbol{\alpha}\cdot\boldsymbol{P})P^{-1}\sin f]\mathcal{H} \quad (3.116)$$

$$= [\cos f + \beta(\boldsymbol{\alpha}\cdot\boldsymbol{P})P^{-1}\sin f][\beta mc^2 + c\boldsymbol{\alpha}\cdot\boldsymbol{P}] \quad (3.117)$$

$$= \beta mc^2 \cos f - \beta^2 mc^2(\boldsymbol{\alpha}\cdot\boldsymbol{P})P^{-1}\sin f + c\boldsymbol{\alpha}\cdot\boldsymbol{P}\cos f$$
$$+ \beta c(\boldsymbol{\alpha}\cdot\boldsymbol{P})^2 P^{-1}\sin f \quad (3.118)$$

$$= \beta[mc^2\cos f + cP\sin f] + c(\boldsymbol{\alpha}\cdot\boldsymbol{P})P^{-1}[P\cos f - mc\sin f]. \quad (3.119)$$

Note that the minus sign appears in (3.118) because β anticommutes with the components of $\boldsymbol{\alpha}$, equation (3.14). We have now to make a choice for the function f and it is clear that our aim should be to eliminate the second term in (3.119) which contains the odd operators. The appropriate choice is

$$f = \tan^{-1}(P/mc), \quad (3.120)$$

in which case

$$\sin f = (P/mc)\cos f. \tag{3.121}$$

Substituting for $\sin f$ in (3.119) we obtain the result

$$\mathcal{H}' = \beta[mc^2 + (P^2/m)]\cos f. \tag{3.122}$$

We can eliminate $\cos f$ from this expression by squaring, using equation (3.121) to substitute for $\cos f$, and then taking the square root,

$$\mathcal{H}' = \beta\left[\left(mc^2 + \frac{P^2}{m}\right)^2 \cos^2 f\right]^{1/2} \tag{3.123}$$

$$= \beta\left[\left(m^2c^4 + \frac{P^4}{m^2} + 2c^2P^2\right)(m^2c^2/(P^2 + m^2c^2))\right]^{1/2} \tag{3.124}$$

$$= \beta[m^2c^4 + c^2P^2]^{1/2}. \tag{3.125}$$

Equation (3.125) is the required transformed Hamiltonian, and we see that in the representation in which β is diagonal, the Dirac equation decomposes into uncoupled equations for the upper and lower components of the wave function, i.e. for electron and positron wave functions. Setting β equal to $+1$ gives the positive energy (electron) states, whilst β equals -1 gives the negative energy (positron) states.

In performing the similarity transformation above, we have changed our representation of the particle from a Dirac representation to the so-called Foldy–Wouthuysen representation. This new representation provides a very simple link with the non-relativistic Schrödinger–Pauli representation. The latter, which is a two-component representation, just corresponds to the two upper components of the Foldy–Wouthuysen representation (3.125). It must be noted that under a similarity transformation such as (3.105), the operators which represent physical observables are also transformed. For example, the position observable whose operator is R in the Dirac representation is transformed to R'' in the Foldy–Wouthuysen representation where

$$R'' = e^{i\bar{S}} R e^{-i\bar{S}}$$
$$= R + i[\bar{S}, R] + (i^2/2)[\bar{S}, [\bar{S}, R]] + \cdots. \tag{3.126}$$

Appreciation of this fact leads to an understanding of some of the peculiar properties of the electron, including electron spin.

The explicit form of the position operator R'' in the Foldy–Wouthuysen representation can be obtained by substituting $\bar{S}$ from equations (3.104) and (3.120) into (3.126). After some manipulation, the final result is

$$R'' = R - \frac{i\hbar c \beta \alpha}{2E_+} + \frac{i\hbar c^3 \beta(\alpha \cdot P)P}{2E_+^2(E_+ + mc^2)} - \frac{\hbar c^2 \sigma \wedge P}{2E_+(E_+ + mc^2)} \tag{3.127}$$

where

$$E_+ = (m^2c^4 + c^2P^2)^{1/2}. \tag{3.128}$$

We must therefore ask what is the significance of the operator $\mathbf{R}''$ in the Foldy–Wouthuysen representation. This new observable, which is called the mean position, has an operator representative in the Dirac representation $\bar{\mathbf{R}}$ where

$$\bar{\mathbf{R}} = e^{-i\tilde{S}} \mathbf{R} e^{i\tilde{S}} = \mathbf{R} + \frac{i\hbar c\beta\boldsymbol{\alpha}}{2E_+} - \frac{i\hbar c^3 \beta(\boldsymbol{\alpha}\cdot\mathbf{P})\mathbf{P}}{2E_+^2(E_+ + mc^2)} - \frac{\hbar c^2 \boldsymbol{\sigma}\wedge\mathbf{P}}{2E_+(E_+ + mc^2)}. \quad (3.129)$$

The significance of (3.127) is that a wave function which corresponds to the particle being located at a definite point in the Dirac representation changes to a wave function in the Foldy–Wouthuysen representation which now corresponds to the particle being spread out over a finite region. Equation (3.129) shows that the converse is also true. These relationships are more easily appreciated if we consider the motion executed by a free electron. The instantaneous value of any one component of the electron's velocity is $+c$ or $-c$, but the periodic frequency of such motion is very high and the amplitude of the motion is very small. This motion, which Schrödinger [14] called the Zitterbewegung (literally, quivering motion), is superimposed on whatever slow motion the electron may have. Thus we can visualise the electron following a tight helical spiral on top of a comparatively lazy orbital motion through space. We now see that the mean position operator, $\bar{\mathbf{R}}$, corresponds to the mean position of the Zitterbewegung; the electron is actually executing a very rapid motion of very small amplitude about this position and so it appears to have a non-local distribution in the Foldy–Wouthuysen representation.

The electron moves in a similar way in the presence of an external electromagnetic field except that the Zitterbewegung is no longer exactly periodic. In the Foldy–Wouthuysen representation, therefore, the interaction between the electron and the field is similar to that arising from a spatially extended charge and current distribution. It is the spiralling motion of the electron that gives rise to its magnetic moment rather than any spinning motion about an axis through the particle, a concept which is entirely bogus. The effective spatial extension of the electron about its mean position is also responsible for the Darwin correction to the electrostatic energy that occurs in equation (3.101).

We saw in section 3.3 that neither the orbital ($\hbar\mathbf{L} = \mathbf{R}\wedge\mathbf{P}$) nor the spin $(1/2)\hbar\boldsymbol{\sigma}$ angular momenta commute with the Dirac Hamiltonian and are not therefore separate constants of motion, although their sum is. However, we can construct the mean orbital angular momentum and mean spin angular momentum operators in the same way as in equation (3.129). These operators are, respectively,

$$\mathbf{R}\wedge\mathbf{P} \quad \text{and} \quad (1/2)\hbar\boldsymbol{\sigma} \quad (3.130)$$

in the Foldy–Wouthuysen representation, and

$$\bar{\mathbf{R}}\wedge\mathbf{P} \quad \text{and} \quad (1/2)\hbar\left\{\boldsymbol{\sigma} - \frac{ic\beta(\boldsymbol{\alpha}\wedge\mathbf{P})}{E_+} - \frac{c^2\mathbf{P}\wedge(\boldsymbol{\sigma}\wedge\mathbf{P})}{E_+(E_+ + mc^2)}\right\} \quad (3.131)$$

in the Dirac representation. Both operators in (3.131) commute with the Dirac Hamiltonian and are separately constants of motion for the free particle. Note that it is the mean spin observable which is to be associated with the spin observable in the Schrödinger–Pauli representation, not $(1/2)\hbar\sigma$ itself.

3.6. Derivation of the many-electron Hamiltonian

In section 3.4 we derived the Hamiltonian for a single electron in the presence of external magnetic and electric fields. The starting point was the Dirac equation, which is invariant to a Lorentz transformation of the space and time coordinates. When we consider more than one electron, however, we encounter the problem that a wave equation which is properly Lorentz invariant is not available. In a later section we shall discuss the Breit equation for two electrons, which is approximately Lorentz invariant. For the moment we follow the usual course of assuming that the Dirac equation can be generalised to many electrons, the main justification for which is that the predictions of the resulting theory are in remarkable agreement with experiment. The effects of electron–electron and electron–nuclear interactions are incorporated into the theory by calculating the magnetic and electric potentials at electron i which arise from the motions and positions of all the other electrons and nuclei. In this section we derive a Hamiltonian which excludes effects that arise from external magnetic and electric fields; in the following section we shall include external fields, and in the next chapter we will examine the additional effects arising from nuclear spin. We use classical theory to calculate the magnetic and electric potentials, but Itoh [15] has shown that essentially the same Hamiltonian is obtained if quantum field theory is employed.

From classical magnetostatic theory, the magnetic vector potential at electron i is given by

$$A_i^e = \frac{1}{(4\pi\varepsilon_0 c^2)} \left\{ -\sum_{j \neq i} \frac{\mu_j \wedge R_{ji}}{R_{ji}^3} - \sum_{j \neq i} \frac{e_j V_j}{R_{ji}} \right\} \tag{3.132}$$

$$= \frac{g_s \hbar}{(8\pi\varepsilon_0 c^2)} \sum_{j \neq i} \frac{e_j}{m_j R_{ji}^3} (S_j \wedge R_{ji}) - \frac{1}{(4\pi\varepsilon_0 c^2)} \sum_{j \neq i} \frac{e_j}{m_j R_{ji}} \pi_j. \tag{3.133}$$

The first term arises from the spin magnetic moments of the other electrons, and the second term comes from the orbital motions. R_{ji} is equal to $R_j - R_i$, the vector which gives the position of electron j relative to electron i. Strictly speaking the second term in (3.133) is incorrect for several reasons, but we will return to this point later.

The electric potential at electron i is given by

$$\phi_i = \sum_\alpha \frac{Z_\alpha e}{4\pi\varepsilon_0 R_{\alpha i}} - \sum_{j \neq i} \frac{e_j}{4\pi\varepsilon_0 R_{ji}}, \tag{3.134}$$

where the first term describes the potential due to the nuclei (α) of charge $Z_\alpha e$, and the second term arises from the other electrons (j). Again we shall consider a more general form of the electrostatic potential in the next chapter, but (3.134) will suffice for our present needs.

In section 3.4 we carried out a Foldy–Wouthuysen transformation on the Dirac Hamiltonian and obtained the result (3.84), correct to order c^{-2},

$$\mathcal{H} = -e\phi + \frac{1}{2m}(\boldsymbol{\alpha}\cdot\boldsymbol{\pi})^2 - \frac{1}{8m^2c^2}\left[(\boldsymbol{\alpha}\cdot\boldsymbol{\pi}), [(\boldsymbol{\alpha}\cdot\boldsymbol{\pi}), (-e\phi)] + i\hbar\frac{\partial}{\partial t}(\boldsymbol{\alpha}\cdot\boldsymbol{\pi})\right]$$

$$- \frac{1}{8m^3c^2}(\boldsymbol{\alpha}\cdot\boldsymbol{\pi})^4, \tag{3.135}$$

where $\boldsymbol{\pi} = \boldsymbol{P} + e\boldsymbol{A}$. This Hamiltonian operates on the positive energy states only; in other words, it has a two-component representation. In addition, the constant energy term mc^2 has been subtracted.

We now expand the different terms in (3.135), retaining only terms which are linear in $\boldsymbol{A}$, and obtain the results,

$$\frac{1}{2m}(\boldsymbol{\alpha}\cdot\boldsymbol{\pi})^2 = \frac{1}{2m}\{\pi^2 + g_s e\hbar \boldsymbol{S}\cdot(\nabla\wedge\boldsymbol{A})\}, \tag{3.136}$$

$$\frac{1}{8m^3c^2}(\boldsymbol{\alpha}\cdot\boldsymbol{\pi})^4 = \frac{1}{8m^3c^2}\{\pi^2 + g_s e\hbar \boldsymbol{S}\cdot(\nabla\wedge\boldsymbol{A})\}^2 \tag{3.137}$$

$$= \frac{\pi^4}{8m^3c^2} + \frac{g_s\mu_B}{2m^2c^2}\boldsymbol{S}\cdot(\nabla\wedge\boldsymbol{A})\pi^2 + \cdots, \tag{3.138}$$

$$\frac{1}{8m^2c^2}\left[(\boldsymbol{\alpha}\cdot\boldsymbol{\pi}), [(\boldsymbol{\alpha}\cdot\boldsymbol{\pi}), (-e\phi)] + i\hbar\frac{\partial}{\partial t}(\boldsymbol{\alpha}\cdot\boldsymbol{\pi})\right]$$

$$= -\frac{e\hbar^2}{8m^2c^2}\nabla\cdot\boldsymbol{\varepsilon} - \frac{g_s\mu_B}{2mc^2}\boldsymbol{S}\cdot(\boldsymbol{\varepsilon}\wedge\boldsymbol{\pi}), \tag{3.139}$$

where $\boldsymbol{\varepsilon} = -\nabla\phi$ and is the electric field intensity at electron i due to the other electrons and nuclei. Equation (3.136) is obtained by making use of (3.35) to (3.37), and (3.139) follows from (3.90), (3.94) and (3.97), except that $\boldsymbol{\varepsilon} = -\nabla\phi$ where ϕ is given by (3.134). Our previous equations contained $\boldsymbol{E}$, the applied electric field intensity, rather than $\boldsymbol{\varepsilon}$. If we now substitute (3.136), (3.138) and (3.139) in (3.135) we obtain the Hamiltonian for electron i in the presence of other electrons:

$$\mathcal{H}_i = -e_i\phi_i + \frac{\pi_i^2}{2m_i} + g_s\mu_B \boldsymbol{S}_i\cdot\left(\nabla_i\wedge\boldsymbol{A}_i^e\right)$$

$$- \frac{\pi_i^4}{8m_i^3c^2} + \frac{e_i\hbar^2}{8m_i^2c^2}(\nabla_i\cdot\boldsymbol{\varepsilon}_i) + \frac{g_s\mu_B}{2m_ic^2}\boldsymbol{S}_i\cdot(\boldsymbol{\varepsilon}_i\wedge\boldsymbol{\pi}_i). \tag{3.140}$$

The procedure is now to expand the terms in (3.140), using (3.133) and (3.134) to evaluate $\boldsymbol{A}_i^e$ and ϕ_i and replacing $\boldsymbol{\pi}_i$ by $\boldsymbol{P}_i + e\boldsymbol{A}_i^e$. Clearly, many terms are obtained but we can simplify matters by neglecting all terms which are quadratic or higher order in $\boldsymbol{A}_i^e$ and by neglecting terms in c^{-3} or higher. We examine each of the six terms in (3.140) in turn. It is, of course, obvious that $e_i = e_j = e$ and $m_i = m_j = m$ in what follows but we retain the subscripts to keep track of the two identical particles.

(i) $-e_i\phi_i$

From equation (3.134) we obtain

$$-e_i\phi_i = \sum_{j\neq i}\frac{e_ie_j}{4\pi\varepsilon_0 R_{ji}} - \sum_{\alpha}\frac{Z_\alpha ee_i}{4\pi\varepsilon_0 R_{\alpha i}}. \tag{3.141}$$

These are just the familiar Coulomb potentials.

(ii) $\pi_i^2/2m_i$

Since $\boldsymbol{\pi}_i = \boldsymbol{P}_i + e_i \boldsymbol{A}_i^e$ we have

$$\frac{\pi_i^2}{2m_i} = \frac{1}{2m_i}\{P_i^2 + 2e_i \boldsymbol{A}_i^e \cdot \boldsymbol{P}_i + Order(A_i^e)^2\}. \tag{3.142}$$

Now from the first term of equation (3.133) we find

$$\frac{e_i}{m_i} \boldsymbol{A}_i^e \cdot \boldsymbol{P}_i = \frac{g_S \hbar e_i}{(8\pi\varepsilon_0 m_i c^2)} \sum_{j\neq i} \frac{e_j}{m_j R_{ji}^3} (\boldsymbol{S}_j \wedge \boldsymbol{R}_{ji}) \cdot \boldsymbol{P}_i \tag{3.143}$$

$$= \frac{g_S \hbar e_i}{(8\pi\varepsilon_0 m_i c^2)} \sum_{j\neq i} \frac{e_j}{m_j R_{ji}^3} \boldsymbol{S}_j \cdot (\boldsymbol{R}_{ji} \wedge \boldsymbol{P}_i). \tag{3.144}$$

Equation (3.144) represents the interaction of the spin of electron j with the orbital motion of electron i relative to electron j; it therefore makes a contribution to the 'spin–other-orbit' interaction.

We could now proceed to substitute the second term of (3.133) for A_i^e in (3.142) but, as mentioned earlier, the second term of (3.133) is incorrect. It has been found that the correct form of the interaction between the momentum $\boldsymbol{P}_i$ and this part of the vector potential is obtained only if the retardation of the electromagnetic potentials is included. We do not go into the details here but simply quote the resulting contribution to (3.142),

$$\frac{e_i}{m_i} \boldsymbol{A}_i^e \cdot \boldsymbol{P}_i = -\frac{e_i}{(8\pi\varepsilon_0 m_i c^2)} \sum_{j\neq i} \frac{e_j}{m_j} \left[\boldsymbol{P}_j \cdot \frac{1}{R_{ji}} \boldsymbol{P}_i + (\boldsymbol{P}_j \cdot \boldsymbol{R}_{ji}) \frac{1}{R_{ji}^3} (\boldsymbol{R}_{ji} \cdot \boldsymbol{P}_i) \right]. \tag{3.145}$$

This term represents the interaction between the orbital motions of the electrons and is therefore known as the 'orbit–orbit' interaction.

(iii) $g_S \mu_B \boldsymbol{S}_i \cdot (\nabla_i \wedge \boldsymbol{A}_i^e)$

This term represents the scalar interaction between the spin of electron i and the magnetic field created by the spin and orbital motions of the other electrons. Substituting the first term of (3.133) for A_i^e yields

$$g_S \mu_B \boldsymbol{S}_i \cdot (\nabla_i \wedge \boldsymbol{A}_i^e) = \frac{g_S^2 \hbar^2}{(16\pi\varepsilon_0 c^2)} \sum_{j\neq i} \frac{e_i e_j}{m_i m_j} \boldsymbol{S}_i \cdot \left\{ \frac{\nabla_i \wedge (\boldsymbol{S}_j \wedge \boldsymbol{R}_{ji})}{R_{ji}^3} \right\} \tag{3.146}$$

$$= \frac{g_S^2 \hbar^2}{(16\pi\varepsilon_0 c^2)} \sum_{j\neq i} \frac{e_i e_j}{m_i m_j} \boldsymbol{S}_i \cdot \left\{ \boldsymbol{S}_j \left(\nabla_i \cdot \frac{\boldsymbol{R}_{ji}}{R_{ji}^3} \right) - (\boldsymbol{S}_j \cdot \nabla_i) \frac{\boldsymbol{R}_{ji}}{R_{ji}^3} \right\}. \tag{3.147}$$

The first term in (3.147) involves

$$\left(\nabla_i \cdot \frac{\boldsymbol{R}_{ji}}{R_{ji}^3} \right) = -\nabla_i^2 \left(\frac{1}{R_{ji}} \right) = -4\pi \delta^{(3)}(\boldsymbol{R}_{ji}), \tag{3.148}$$

where $\delta^{(3)}(\boldsymbol{R}_{ji})$ is a Dirac delta function. Equation (3.148) is only strictly true for a non-relativistic situation. Difficulties arise because the Foldy–Wouthuysen transformation is non-local over a domain characterised by the Zitterbewegung (see section 3.4). The problem is particularly acute because it is just at the source of the potentials that A_i^e and ϕ_i cease to be small enough to be treated non-relativistically. However if we effect a cut-off of the singular potential at a distance from the source which corresponds to the Zitterbewegung ($\approx \hbar/mc$) it can be shown that the non-relativistic approximation is valid provided that the energy of the potential interaction is much less than mc^2 at the cut-off point, which is true for the cases that we deal with here.

The second term in (3.147) depends on

$$(S_j \cdot \nabla_i) \frac{R_{ji}}{R_{ji}^3} = S_j \frac{1}{R_{ji}^3} \nabla_i \cdot R_{ji} + S_j \cdot R_{ji} \nabla_i \frac{1}{R_{ji}^3} - S_j \frac{4\pi}{3} \delta^{(3)}(R_{ji}) \quad (3.149)$$

$$= -S_j \frac{1}{R_{ji}^3} + 3(S_j \cdot R_{ji}) \frac{R_{ji}}{R_{ji}^5} - S_j \frac{4\pi}{3} \delta^{(3)}(R_{ji}). \quad (3.150)$$

If we now combine (3.147) and (3.150) we obtain

$$g_S \mu_B S_i \cdot \left(\nabla_i \wedge A_i^e\right) = \frac{g_S^2 \hbar^2}{(16\pi\varepsilon_0 c^2)} \sum_{j \neq i} \frac{e_i e_j}{m_i m_j}$$

$$\times \left\{ \frac{S_i \cdot S_j}{R_{ji}^3} - \frac{3(S_i \cdot R_{ji})(S_j \cdot R_{ji})}{R_{ji}^5} - \frac{8\pi}{3} \delta^{(3)}(R_{ji}) S_i \cdot S_j \right\}.$$

$$(3.151)$$

This important term represents the 'spin–spin' interaction between pairs of electrons.

We must now consider the orbital contribution to the magnetic vector potential in (3.133). It can be shown that the magnetic field derived from this term in A_i^e is given accurately by the expression in (3.133) since the curl of the correction terms is zero. Thus we have

$$g_S \mu_B S_i \cdot \left(\nabla_i \wedge A_i^e\right) = -\frac{g_S e_i \hbar}{8\pi\varepsilon_0 m_i c^2} \sum_{j \neq i} \frac{e_j}{m_j} S_i \cdot \left(\nabla_i \wedge \frac{1}{R_{ji}} P_j\right) \quad (3.152)$$

$$= -\frac{g_S e_i \hbar}{8\pi\varepsilon_0 m_i c^2} \sum_{j \neq i} \frac{e_j}{m_j} S_i \cdot \left(\frac{R_{ji}}{R_{ji}^3} \wedge P_j\right). \quad (3.153)$$

This term is another contribution to the spin–other-orbit interaction.

(iv) $-\dfrac{\pi_i^4}{8m_i^3 c^2}$

As before, we substitute $P_i + eA_i^e$ for π_i; however only the leading term is acceptable on the basis of our order of magnitude criterion so that we have

$$-\frac{\pi_i^4}{8m_i^3 c^2} = -\frac{P_i^4}{8m_i^3 c^2} + \text{order}(1/c^4). \quad (3.154)$$

(v) $\dfrac{e\hbar^2}{8m_i^2 c^2}(\nabla_i \cdot \boldsymbol{\varepsilon}_i)$

This term is very similar to the Darwin term. It is evaluated using equation (3.134):

$$\dfrac{e_i \hbar^2}{8m_i^2 c^2}(\nabla_i \cdot \boldsymbol{\varepsilon}_i) = -\dfrac{e_i \hbar^2}{8m_i^2 c^2}(\nabla_i^2 \phi_i) \qquad (3.155)$$

$$= -\dfrac{e_i \hbar^2}{32\pi\varepsilon_0 m_i^2 c^2}\left[\nabla_i^2 \left\{\sum_\alpha \dfrac{Z_\alpha e}{R_{\alpha i}} - \sum_{j \neq i} \dfrac{e}{R_{ji}}\right\}\right] \qquad (3.156)$$

$$= \dfrac{e_i \hbar^2}{8\varepsilon_0 m_i^2 c^2}\left\{\sum_\alpha Z_\alpha e \delta^{(3)}(\boldsymbol{R}_{\alpha i}) - \sum_{j \neq i} e_j \delta^{(3)}(\boldsymbol{R}_{ji})\right\}. \qquad (3.157)$$

Once again we have introduced the Dirac delta functions in a non-relativistic approximation. The terms in (3.157) represent a correction to the Coulomb potential.

(vi) $\dfrac{g_S \mu_B}{2m_i c^2} \boldsymbol{S}_i \cdot (\boldsymbol{\varepsilon}_i \wedge \boldsymbol{\pi}_i)$

This term is potentially the most complicated of the six terms in (3.140) since $\boldsymbol{\pi}_i$ can be replaced by a sum of three terms (i.e. $\boldsymbol{P}_i + e\boldsymbol{A}_i^e$ with $\boldsymbol{A}_i^e$ given by (3.133)), whilst from (3.134), ϕ_i gives rise to two terms. Thus six separate contributions are possible. Once again, however, many of the possible terms are smaller than permissible. We therefore simply replace $\boldsymbol{\pi}_i$ by $\boldsymbol{P}_i$ and use (3.134) to evaluate $\boldsymbol{\varepsilon}_i$, obtaining

$$\boldsymbol{\varepsilon}_i = -\nabla_i \phi_i = -\sum_\alpha \dfrac{Z_\alpha e}{4\pi\varepsilon_0 R_{\alpha i}^3} \boldsymbol{R}_{\alpha i} + \sum_{j \neq i} \dfrac{e_j}{4\pi\varepsilon_0 R_{ji}^3} \boldsymbol{R}_{ji}. \qquad (3.158)$$

Thus we find that

$$\dfrac{g_S \mu_B}{2m_i c^2} \boldsymbol{S}_i \cdot (\boldsymbol{\varepsilon}_i \wedge \boldsymbol{\pi}_i) = -\dfrac{g_S e_i \hbar}{16\pi\varepsilon_0 m_i^2 c^2} \boldsymbol{S}_i \cdot \sum_\alpha \dfrac{Z_\alpha e}{R_{\alpha i}^3}(\boldsymbol{R}_{\alpha i} \wedge \boldsymbol{P}_i)$$

$$+ \dfrac{g_S e_i \hbar}{16\pi\varepsilon_0 m_i^2 c^2} \sum_{j \neq i} e_j \dfrac{\boldsymbol{S}_i}{R_{ji}^3} \cdot (\boldsymbol{R}_{ji} \wedge \boldsymbol{P}_i) + \text{order}(1/c^4). \qquad (3.159)$$

Equation (3.159) describes the spin–orbit coupling, the two terms involving the nuclear and electronic potentials respectively. It is interesting to note that these terms arise from the interaction between the electron spin magnetic moment and the effective magnetic field created by the passage of the electron through the electric field created by the other particles.

We are now in a position to present the total electronic Hamiltonian by summing over all possible electrons i. We must be very careful, however, not to count the various interactions twice. Thus on summing over i, we modify all terms which are symmetric in i and j by a factor of $1/2$. These terms are the electron–electron Coulomb interaction (3.141), the orbit–orbit interaction (3.145), the spin–spin interaction (3.151) and the spin–other-orbit interaction from (3.144) and (3.153).

This last term actually has the explicit form

$$\frac{g_S e_i \hbar}{8\pi\varepsilon_0 m_i c^2} \sum_{j\neq i} \frac{e_j}{m_j}\left\{\frac{1}{R_{ji}^3}\boldsymbol{S}_j\cdot(\boldsymbol{R}_{ji}\wedge\boldsymbol{P}_i) - \frac{1}{R_{ji}^3}\boldsymbol{S}_i\cdot(\boldsymbol{R}_{ji}\wedge\boldsymbol{P}_j)\right\}. \qquad (3.160)$$

It can be seen that this term is indeed symmetric to interchange of i and j. The second term in equation (3.157) is not exactly symmetrical with respect to interchange of i and j and therefore does not require a factor of $1/2$.

We have now completed our derivation of the electronic Hamiltonian when external fields and nuclear spin effects are absent. In summary, the Hamiltonian is as follows:

$$\mathcal{H}_{el} =$$

$$\frac{1}{2m_i}\sum_i \boldsymbol{P}_i^2 \qquad \text{:electron kinetic energy}$$

$$-\frac{1}{8m_i^3 c^2}\sum_i \boldsymbol{P}_i^4 \qquad \text{:relativistic correction}$$

$$+\sum_{i,j\neq i}\frac{e^2}{8\pi\varepsilon_0 R_{ji}} - \sum_{i,\alpha}\frac{Z_\alpha e^2}{4\pi\varepsilon_0 R_{\alpha i}} \qquad \text{:Coulomb energy}$$

$$+\frac{\hbar^2}{8\varepsilon_0 c^2}\left\{\sum_{i,\alpha}\frac{Z_\alpha e^2}{m^2}\delta^{(3)}(\boldsymbol{R}_{\alpha i}) - \sum_{i,j\neq i}\frac{e^2}{m^2}\delta^{(3)}(\boldsymbol{R}_{ji})\right\} \qquad \begin{array}{l}\text{:Darwin correction to}\\ \text{Coulomb energy}\end{array}$$

$$-\frac{g_S\hbar}{16\pi\varepsilon_0 c^2}\sum_i \frac{e}{m^2}\boldsymbol{S}_i\cdot\left\{\sum_\alpha \frac{Z_\alpha e}{R_{\alpha i}^3}(\boldsymbol{R}_{\alpha i}\wedge\boldsymbol{P}_i) - \sum_{j\neq i}\frac{e}{R_{ji}^3}(\boldsymbol{R}_{ji}\wedge\boldsymbol{P}_i)\right\} \qquad \begin{array}{l}\text{:spin–orbit}\\ \text{coupling}\end{array}$$

$$-\frac{g_S\hbar}{8\pi\varepsilon_0 c^2}\sum_{i,j\neq i}\frac{e^2}{m^2}\frac{1}{R_{ji}^3}\boldsymbol{S}_i\cdot(\boldsymbol{R}_{ji}\wedge\boldsymbol{P}_j) \qquad \text{:spin–other-orbit coupling}$$

$$-\frac{1}{16\pi\varepsilon_0 c^2}\sum_{i,j\neq i}\frac{e^2}{m^2}\left\{\boldsymbol{P}_i\cdot\frac{1}{R_{ji}}\boldsymbol{P}_j + (\boldsymbol{P}_i\cdot\boldsymbol{R}_{ji})\frac{1}{R_{ji}^3}(\boldsymbol{R}_{ji}\cdot\boldsymbol{P}_j)\right\} \qquad \text{:orbit–orbit coupling}$$

$$+\frac{g_S^2 \hbar^2}{32\pi\varepsilon_0 c^2}\sum_{i,j\neq i}\frac{e^2}{m^2}\left\{\frac{\boldsymbol{S}_i\cdot\boldsymbol{S}_j}{R_{ji}^3} - \frac{3(\boldsymbol{S}_i\cdot\boldsymbol{R}_{ji})(\boldsymbol{S}_j\cdot\boldsymbol{R}_{ji})}{R_{ji}^5} - \frac{8\pi}{3}\delta^{(3)}(\boldsymbol{R}_{ji})\boldsymbol{S}_i\cdot\boldsymbol{S}_j\right\}.$$

$$\text{:spin–spin coupling} \qquad (3.161)$$

3.7. Effects of applied static magnetic and electric fields

We now derive the additional terms in the Hamiltonian which arise from the application of an external magnetic field. The magnetic vector potential will be a sum of the contributions from the electrons (A_i^e) and from the additional term

$$A_i^B = (1/2)(\boldsymbol{B}\wedge\boldsymbol{R}_i), \qquad (3.162)$$

where $\boldsymbol{B}$ is the intensity of the applied magnetic field, assumed to be uniform (i.e. the same at all electrons). Hence we re-examine the terms in equation (3.140) which are linear in A_i, but this time we substitute A_i^B rather than A_i^e. We also carry out the summation over electrons at the same time.

With the substitution of (3.162), the contribution from the scalar product of A_i^B and P_i in the second term of equation (3.140) is given by

$$\frac{e}{m}\sum_i A_i^B \cdot P_i = \frac{e}{2m}\sum_i (B \wedge R_i) \cdot P_i \tag{3.163}$$

$$= \mu_B \hbar^{-1} B \cdot \sum_i (R_i \wedge P_i). \tag{3.164}$$

This term represents the interaction of the external magnetic field with the magnetic moment arising from the orbital motion of the electrons. Next we investigate the third term in (3.140) and find

$$\sum_i g_S \mu_B S_i \cdot (\nabla_i \wedge A_i^B) = g_S \mu_B \sum_i S_i \cdot B. \tag{3.165}$$

This term represents the coupling of the spin magnetic dipole moment with the applied field, and was derived previously for a single electron in equation (3.101). Turning to the fourth term in (3.140) we note that in the field free case we retained only the leading term (i.e. P_i^4). However, the term linear in A_i^B, although formally of order c^{-3}, has been found to contribute measurably to the Zeeman effect in some cases. The resulting term

$$-\sum_i \frac{\pi_i^4}{8m^3c^2} = -\frac{1}{8m^3c^2}\sum_i 4e(A_i^B \cdot P_i)P_i^2 \tag{3.166}$$

$$= -\frac{\mu_B}{2m^2c^2\hbar} B \cdot \sum_i (R_i \wedge P_i)P_i^2 \tag{3.167}$$

represents a correction to (3.164), arising from the relativistic increase in electron mass and consequent decrease in its orbital magnetic moment.

A similar relativistic correction to the spin magnetic moment which is again of order c^{-3} but nevertheless measurable comes from equations (3.135) and (3.138); if we replace π_i by P_i we find that

$$-\frac{g_S \mu_B}{2m^2c^2}\sum_i S_i \cdot (\nabla_i \wedge A_i^B)\pi_i^2 = -\frac{g_S \mu_B}{2m^2c^2}\sum_i (S_i \cdot B)P_i^2. \tag{3.168}$$

The fifth term in (3.140) does not contain the magnetic vector potential and we proceed to look at the last term. The expansion proceeds as outlined in (3.159) except that we now consider the A_i^B term in π_i. Thus

$$\frac{g_S \mu_B}{2mc^2}\sum_i S_i \cdot (\varepsilon_i \wedge \pi_i) = \frac{g_S \mu_B}{8\pi\varepsilon_0 mc^2}\sum_i S_i \cdot \left\{\sum_\alpha \frac{Z_\alpha e}{R_{\alpha i}^3} R_{\alpha i} - \sum_{j \neq i} \frac{e R_{ji}}{R_{ji}^3}\right\} \wedge e A_i^B \tag{3.169}$$

$$= \frac{g_S \mu_B e^2}{16\pi\varepsilon_0 mc^2} \sum_{i,\alpha} \frac{Z_\alpha}{R_{\alpha i}^3} \mathbf{S}_i \cdot [\mathbf{R}_{\alpha i} \wedge (\mathbf{B} \wedge \mathbf{R}_i)]$$

$$- \frac{g_S \mu_B e^2}{16\pi\varepsilon_0 mc^2} \sum_{i,j \neq i} \frac{1}{R_{ji}^3} \mathbf{S}_i \cdot [\mathbf{R}_{ji} \wedge (\mathbf{B} \wedge \mathbf{R}_i)]. \tag{3.170}$$

Equation (3.170), which is again of order c^{-3}, represents the influence of the external magnetic field on the spin–orbit coupling, and although we have now dealt with all the terms linear in A_i^B we might expect to find similar modifications to the spin–other-orbit and orbit–orbit interactions. In section 3.6 we neglected terms in $(A_i^e)^2$ since they were necessarily of order c^{-4}. This is not true, however, of terms containing $(A_i^B)^2$ or the cross-product $A_i^e \cdot A_i^B$. Inclusion of such terms in the expansion of π_i^2 yields

$$\sum_i \frac{\pi_i^2}{2m} \to \frac{e^2}{2m} \sum_i (A_i)^2$$

$$= \frac{g_S e^3 \hbar}{16\pi\varepsilon_0 m^2 c^2} \sum_{i,j \neq i} \frac{1}{R_{ji}^3} (\mathbf{S}_j \wedge \mathbf{R}_{ji}) \cdot (\mathbf{B} \wedge \mathbf{R}_i)$$

$$- \frac{e^3}{8\pi\varepsilon_0 m^2 c^2} \sum_{i,j \neq i} \frac{1}{R_{ji}} \mathbf{P}_j \cdot (\mathbf{B} \wedge \mathbf{R}_i) + \frac{e^2}{8m} \sum_i (\mathbf{B} \wedge \mathbf{R}_i)^2 + \cdots \tag{3.171}$$

$$= \frac{g_S e^3 \hbar}{16\pi\varepsilon_0 m^2 c^2} \sum_{i,j \neq i} \frac{1}{R_{ji}^3} \mathbf{S}_i \cdot [\mathbf{R}_{ij} \wedge (\mathbf{B} \wedge \mathbf{R}_j)]$$

$$- \frac{e^2}{8\pi\varepsilon_0 m^2 c^2} \sum_{i,j \neq i} \frac{1}{R_{ji}} (\mathbf{B} \wedge \mathbf{R}_i) \cdot \mathbf{P}_j + \frac{e^2}{8m} \sum_i (\mathbf{B} \wedge \mathbf{R}_i)^2 + \cdots . \tag{3.172}$$

The first term represents the modification to the spin–other-orbit coupling by the external field, the second term is the corresponding modification of the orbit–orbit coupling, and the third term represents the electronic contribution to the diamagnetism. As with the original orbit–orbit coupling term, the second term in (3.172) is incorrect since it does not take account of the retardation effects. The correct form of this term is actually

$$-\frac{e^2}{16\pi\varepsilon_0 m^2 c^2} \sum_{i,j \neq i} \left\{ \frac{1}{R_{ji}} (\mathbf{B} \wedge \mathbf{R}_i) \cdot \mathbf{P}_j + \frac{1}{R_{ji}^3} [\mathbf{R}_{ij} \cdot (\mathbf{B} \wedge \mathbf{R}_i)(\mathbf{R}_{ij} \cdot \mathbf{P}_j)] \right\}. \tag{3.173}$$

We now collect together the results of this section and summarise the Zeeman Hamiltonian as

$$\mathcal{H}_Z =$$

$$g_S \mu_B \sum_i \mathbf{S}_i \cdot \mathbf{B} \left(1 - \frac{\mathbf{P}_i^2}{2m^2 c^2} \right) \qquad \text{:interaction of } \mathbf{B} \text{ with spin magnetic moment}$$

$$+ \mu_B \hbar^{-1} \sum_i (\mathbf{R}_i \wedge \mathbf{P}_i) \cdot \mathbf{B} \left(1 - \frac{\mathbf{P}_i^2}{2m^2 c^2} \right) \qquad \text{:interaction of } \mathbf{B} \text{ with orbital magnetic moment}$$

$$-\frac{g_S\mu_B e^2}{16\pi\varepsilon_0 mc^2}\sum_{i,j\neq i}\frac{1}{R_{ji}^3}\boldsymbol{S}_i\cdot[\boldsymbol{R}_{ji}\wedge(\boldsymbol{B}\wedge\boldsymbol{R}_i)-2\boldsymbol{R}_{ij}\wedge(\boldsymbol{B}\wedge\boldsymbol{R}_j)]$$

:effect of $\boldsymbol{B}$ on spin–orbit and spin–other-orbit interactions

$$-\frac{e^2}{16\pi\varepsilon_0 m^2 c^2}\sum_{i,j\neq i}\left\{\frac{1}{R_{ji}}(\boldsymbol{B}\wedge\boldsymbol{R}_i)\cdot\boldsymbol{P}_j+\frac{1}{R_{ji}^3}\boldsymbol{R}_{ij}\cdot(\boldsymbol{B}\wedge\boldsymbol{R}_i)(\boldsymbol{R}_{ij}\cdot\boldsymbol{P}_j)\right\}$$

:effect of $\boldsymbol{B}$ on spin–orbit interaction

$$+\frac{e^2}{8m}\sum_i(\boldsymbol{B}\wedge\boldsymbol{R}_i)^2. \qquad \text{:electron diamagnetism}$$

(3.174)

Now we consider the results of applying a static uniform electric field $\boldsymbol{E}$. The electrostatic potential ϕ_i (see (3.134)) will now contain the extra term

$$\phi_i = -\boldsymbol{E}\cdot\boldsymbol{R}_i \qquad (3.175)$$

and hence the first, fifth and sixth terms of equation (3.140) will, in principle, be modified. Substitution of (3.175) in the first term of (3.140) yields directly the result

$$\sum_i -e\phi_i = e\sum_i \boldsymbol{E}\cdot\boldsymbol{R}_i, \qquad (3.176)$$

which describes the interaction of the electronic charge cloud with the applied field $\boldsymbol{E}$. The fifth term in (3.140) does not give a contribution for homogeneous $\boldsymbol{E}$ since

$$\nabla_i\cdot\boldsymbol{\varepsilon}_i = \nabla_i\cdot\nabla_i(\boldsymbol{E}\cdot\boldsymbol{R}_i) = \nabla_i\cdot\boldsymbol{E} = 0. \qquad (3.177)$$

Term six, however, yields the result

$$\frac{g_S\mu_B}{2mc^2}\sum_i \boldsymbol{S}_i\cdot(\boldsymbol{\varepsilon}_i\wedge\boldsymbol{\pi}_i) = \frac{g_S\mu_B}{2mc^2}\sum_i \boldsymbol{S}_i\cdot(\nabla_i(\boldsymbol{E}\cdot\boldsymbol{R}_i)\wedge\boldsymbol{\pi}_i) \qquad (3.178)$$

$$= \frac{g_S\mu_B}{2mc^2}\sum_i \boldsymbol{S}_i\cdot(\boldsymbol{E}\wedge\boldsymbol{\pi}_i). \qquad (3.179)$$

This term was encountered previously in the Hamiltonian for a single electron (3.101); $(\boldsymbol{E}\wedge\boldsymbol{\pi}_i)$ is essentially equivalent to a magnetic field, which interacts with the spin magnetic moment. The term is, however, usually negligible in laboratory experiments.

Summarising, the electric field, or Stark, Hamiltonian may be written

$$\mathcal{H}_E = e\sum_i \boldsymbol{E}\cdot\boldsymbol{R}_i + \frac{g_S\mu_B}{2mc^2}\sum_i \boldsymbol{S}_i\cdot(\boldsymbol{E}\wedge\boldsymbol{\pi}_i). \qquad (3.180)$$

3.8. Retarded electromagnetic interaction between electrons

3.8.1. Introduction

In the derivation of the many-electron Hamiltonian (section 3.6) the interactions between electrons were introduced in the expressions for the magnetic and electric potentials. In (3.133) the magnetic vector potential at electron i due to the motion of the other electrons (j) was given by

$$A_i^e = -\frac{g_S \mu_B}{4\pi\varepsilon_0 c^2} \sum_{j \neq i} (\boldsymbol{S}_j \wedge \boldsymbol{R}_{ji}) \frac{1}{R_{ji}^3} - \sum_{j \neq i} \frac{e}{4\pi\varepsilon_0 m c^2} \frac{1}{R_{ji}} \boldsymbol{\pi}_j, \qquad (3.181)$$

in which the first term arises from the spin motion of the other electrons and the second term comes from their orbital motion. The scalar electric potential arising from the electrons was given as

$$\phi_i = -\sum_{j \neq i} \frac{e_j}{4\pi\varepsilon_0 R_{ji}}. \qquad (3.182)$$

We commented that the second term in (3.133) is incorrect, and gave for the orbit–orbit interaction the correct form in (3.145), but without justification. We now examine the interaction between electrons more carefully, both to justify the earlier assumption and also to prepare the ground for our later discussion of the Breit equation.

The second term in (3.133) is incorrect for several reasons. First, as pointed out in section 3.6, R_{ji}^{-1} and $\boldsymbol{\pi}_j$ do not commute; this could, however, be remedied by taking a Hermitian average. Second, the first contribution to A_i^e in (3.133) satisfies the condition that div $A_i^e = 0$ whereas the second term does not. We shall discuss the significance of this later in this section. Third, and most important, we have not taken account of the fact that the potential interaction is actually retarded. If we consider the interaction of just two electrons, we must take account of the fact that the potential at electron 1 will depend upon the position and motion of electron 2 at some earlier time, since it takes a finite time for the effect of electron 2 to be felt by electron 1. Similarly, electron 1 is also not stationary. Although the interaction travels at the speed of light, we are seeking a Hamiltonian which is relativistically correct, and we must therefore take account of this retardation effect in the electromagnetic potential. Since we are concerned with the consequences of the relative motion of two electrons, we commence the discussion by examining aspects of the theory of special relativity that lead to the so-called Lorentz transformation. We have mentioned this transformation already in this chapter, and must now describe it in some detail.

3.8.2. Lorentz transformation

We consider two inertial frames of reference, S and S′, with origins O and O′ and axes Ox, Oy, Oz in S and O′x', O′y', O′z' in S′. (An inertial frame of reference is defined as a coordinate frame in which the laws of Newtonian mechanics hold; one of the consequences of the special theory of relativity is that any pair of such inertial frames can only move with a uniform velocity relative to each other.) Now an observer at the origin O will describe an event in his frame by values of x, y, z, t where t is the time measured by a clock at rest in S. Similarly an observer at O′ will describe the same event in terms of the corresponding values x', y', z', t' measured in S′.

We are interested in the relationship between observations in S and S′ when the two frames of reference are in uniform relative motion. For the sake of simplicity, we

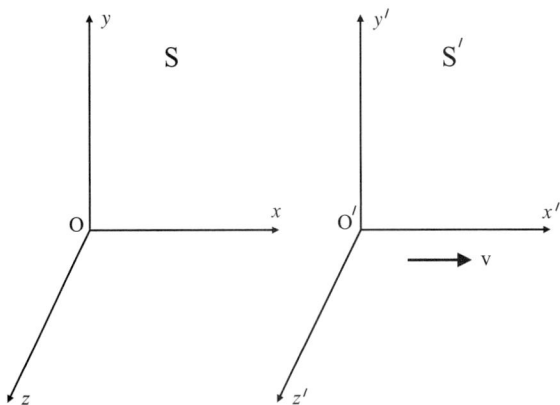

Figure 3.1. Relationship between the inertial frames S and S′.

choose the two frames so that they are coincident at $t = t' = 0$ and so that S′ is in uniform motion in the x direction with velocity v relative to S (see figure 3.1). The Lorentz transformation is a coordinate transformation which enables the interval between two events measured in the S frame to be related to the interval between the same two events measured in the S′ frame. For our choice of inertial frames, it has the explicit form:

$$x' = \gamma(x - vt),$$
$$y' = y,$$
$$z' = z,$$
$$t' = \gamma\left(t - \frac{vx}{c^2}\right), \quad (3.183)$$

where $\gamma = \{1 - (v^2/c^2)\}^{-1/2}$. The pair of events referred to by this transformation are at $(0, 0, 0, 0)$ and at (x, y, z, t) or at (x', y', z', t'), the second event depending on the frame of reference. The complete symmetry between space and 'time' coordinates can be seen by rewriting (3.183) as

$$x' = \gamma\{x - (v/c)ct\}, \quad (3.184)$$
$$ct' = \gamma\{ct - (v/c)x\}. \quad (3.185)$$

In other words, it is the distance, ct, travelled by light in a given time interval which fulfills the role of the fourth coordinate, rather than the time interval itself. The special theory of relativity requires that after a Lorentz transformation the new form of all laws of physics is the same as the old form. The Dirac equation, for example, is invariant under a Lorentz transformation.

3.8.3. Electromagnetic potentials due to a moving electron

We have already made implicit use of the fact that an electromagnetic field can be represented as a four-dimensional vector with components A_x, A_y, A_z and $i\phi$, where the three-dimensional vector potential A is the magnetic vector potential and ϕ is the scalar electric potential. Under a Lorentz transformation the components of this four vector transform in a manner similar to (3.183), i.e.

$$A_{x'} = \gamma\{A_x - (v/c^2)\phi\}, \quad A_{y'} = A_y, \quad A_{z'} = A_z, \quad \phi' = \gamma\{\phi - vA_x\}. \quad (3.186)$$

The inverse relations to these are

$$A_x = \gamma\{A_{x'} + (v/c^2)\phi'\}, \quad A_y = A_{y'}, \quad A_z = A_{z'}, \quad \phi = \gamma\{\phi' + vA_{x'}\}. \quad (3.187)$$

We now wish to derive the potentials due to an electron with charge $-e$ moving with uniform velocity v along the x axis of the inertial frame S introduced in the previous section, i.e. the electron is at rest in the S′ frame (see figure 3.1). Suppose that we wish to calculate the potentials at the field point (x, y, z) at time $t = 0$ in S. By (3.183) this corresponds to the space-time point (x', y', z', t') in S′ where

$$x' = \gamma x, \quad y' = y, \quad z' = z, \quad t' = \gamma vx/c^2. \quad (3.188)$$

Now the potentials at this point in S′ are easily derived since the electron is stationary in this frame:

$$\phi'(R', t') = -e/4\pi\varepsilon_0 R', \quad A'(R', t') = 0. \quad (3.189)$$

The expressions are, of course, time independent. If we transform these potentials to the S frame we obtain

$$\phi(R, 0) = \gamma\phi'(R', t') = -\frac{\gamma e}{4\pi\varepsilon_0 R'}, \quad (3.190)$$

$$A_x(R, 0) = \gamma\frac{v}{c^2}\phi'(R', t') = -\left(\frac{v}{c^2}\right)\frac{\gamma e}{4\pi\varepsilon_0 R'}. \quad (3.191)$$

Since $R' = [(x')^2 + (y')^2 + (z')^2]^{1/2}$, we have the final results

$$\phi(R, 0) = -\frac{\gamma e}{4\pi\varepsilon_0[\gamma^2 x^2 + y^2 + z^2]^{1/2}}, \quad (3.192)$$

$$A_x(R, 0) = -\frac{v\gamma e}{4\pi\varepsilon_0 c^2[\gamma^2 x^2 + y^2 + z^2]^{1/2}}. \quad (3.193)$$

Note that these expressions refer to a particular instant of time in the S frame. The potentials in this frame are time-dependent, since they follow the charge during its motion. At a general time t, for example,

$$\phi(R, t) = -\frac{\gamma e}{4\pi\varepsilon_0[\gamma^2(x - vt)^2 + y^2 + z^2]^{1/2}}. \quad (3.194)$$

It is important to realise that the Lorentz transformation describes accurately both the relativistic effects which are significant because of the high velocity of the electron and also the retardation effects which occur because of the finite time the field takes to reach the field point from the charge (which is a non-relativistic effect).

Now for an electron, $(v/c) \approx 10^{-2}$ so that we can expand the square roots in the denominators of (3.192) and (3.193) using the binomial theorem to give

$$\phi(\mathbf{R}, 0) = -\frac{e}{4\pi\varepsilon_0 R}\left[1 + \frac{v^2}{2c^2} - \frac{v^2 x^2}{2c^2 R^2} + \cdots\right], \qquad (3.195)$$

$$A_x(\mathbf{R}, 0) = \frac{v}{c^2}\phi(\mathbf{R}, 0). \qquad (3.196)$$

Since we require our final results to be accurate to $1/c^2$, we need only calculate A to this order in equation (3.196). However, ϕ and A as given in (3.195) and (3.196) are still not satisfactory for our purposes since they do not conform to a Coulomb gauge. We consider this in more detail in the next section.

If we use the potentials derived above in our molecular Hamiltonian, they are open to the further serious objection that they refer only to an electron moving with uniform velocity, a situation which is not very realistic in the context of the molecular problem. However, the theory of special relativity does not provide a means of describing the motion of a rapidly moving and accelerating particle exactly. An approximate treatment is possible, but since the effects of the non-uniform motion of an electron on its vector and scalar potentials give terms with higher powers of $1/c$ than we require in the final expansion of our Hamiltonian, we can ignore them.

3.8.4. Gauge invariance

From classical field theory we know that the force on a particle of mass m, charge e, and velocity v which is moving in an electromagnetic field is given by

$$\text{Force} = \frac{d\boldsymbol{\pi}}{dt} = e\left[-\nabla\phi - \frac{\partial \boldsymbol{A}}{\partial t}\right] + e\mathbf{v} \wedge (\nabla \wedge \boldsymbol{A}). \qquad (3.197)$$

The electric and magnetic field intensities, $\boldsymbol{E}$ and $\boldsymbol{B}$, are

$$\boldsymbol{E} = -\nabla\phi - \frac{\partial \boldsymbol{A}}{\partial t}, \qquad (3.198)$$

$$\boldsymbol{B} = \nabla \wedge \boldsymbol{A}. \qquad (3.199)$$

Now if we know the vector and scalar potentials A and ϕ, equations (3.198) and (3.199) show that the field intensities $\boldsymbol{E}$ and $\boldsymbol{B}$ are uniquely determined. The converse is not true, however, as we now show. Suppose that A and ϕ are transformed to A' and ϕ' according to

$$\boldsymbol{A}' = \boldsymbol{A} - \nabla f, \qquad (3.200)$$

$$\phi' = \phi + \frac{\partial f}{\partial t}. \qquad (3.201)$$

where f is any scalar function. Then from equation (3.198)

$$\boldsymbol{E}' = -\nabla\phi' - \frac{\partial \boldsymbol{A}'}{\partial t} \qquad (3.202)$$

$$= -\nabla\phi - \nabla\frac{\partial f}{\partial t} - \frac{\partial A}{\partial t} + \frac{\partial}{\partial t}\nabla f \qquad (3.203)$$
$$= E.$$

Similarly from equation (3.199),

$$B' = \nabla \wedge A'$$
$$= \nabla \wedge A - \nabla \wedge (\nabla f) \qquad (3.204)$$
$$= B.$$

Hence the transformation of A and ϕ to A' and ϕ' does not change the fields E and B and it therefore follows that A and ϕ are not uniquely determined. The transformation (3.200), (3.201) is known as a gauge transformation and, in general, we require expressions for B and E to be invariant to such a transformation, i.e. to be gauge-invariant. We therefore have some freedom in choosing A and ϕ and two particular choices are common. In the so-called Coulomb gauge we define A such that

$$\nabla \cdot A = 0, \qquad (3.205)$$

whilst in the Lorentz gauge the relationship to be satisfied is

$$\nabla \cdot A + \frac{1}{c^2}\frac{\partial \phi}{\partial t} = 0. \qquad (3.206)$$

These gauges are equivalent for time-independent problems; our previous expressions for A have satisfied (3.205) and we shall continue to use the Coulomb gauge.

Turning now to our previous expressions for A and ϕ we note that the result given in (3.193) is not satisfactory since $\nabla \cdot A \neq 0$. The recipe is therefore to make a gauge transformation

$$A \rightarrow A + \nabla f, \qquad (3.207)$$
$$\phi \rightarrow \phi - \frac{\partial f}{\partial t}, \qquad (3.208)$$

choosing the scalar function f so that $\nabla \cdot A = 0$. With the choice

$$f = \frac{e}{8\pi\varepsilon_0 c^2}\frac{\mathbf{v}\cdot\mathbf{R}}{R} \qquad (3.209)$$

we find that

$$\frac{\partial f}{\partial t} = \frac{e}{8\pi\varepsilon_0 c^2}\left\{-\frac{v^2}{R} + \frac{(\mathbf{v}\cdot\mathbf{R})^2}{R^3}\right\}, \qquad (3.210)$$

$$\nabla f = \frac{e}{8\pi\varepsilon_0 c^2}\left\{\frac{\mathbf{v}}{R} - \frac{(\mathbf{v}\cdot\mathbf{R})\mathbf{R}}{R^3}\right\}. \qquad (3.211)$$

(In deriving these equations it must be appreciated that $(\partial \mathbf{R}/\partial t) = -\mathbf{v}$.) By substituting (3.210) and (3.211) in (3.207) and (3.208) we obtain the results

$$A = -\frac{e}{8\pi\varepsilon_0 c^2}\left\{\frac{\mathbf{v}}{R} + \frac{(\mathbf{v}\cdot\mathbf{R})\mathbf{R}}{R^3}\right\} + \text{order}(v/c)^3, \qquad (3.212)$$

$$\phi = -\frac{e}{4\pi\varepsilon_0 R} + \text{order}(v/c)^4. \tag{3.213}$$

These expressions now correspond to a Coulomb gauge (3.205) to order $1/c^2$; note that the very simple result for ϕ is correct to order $1/c^2$.

3.8.5. Classical Lagrangian and Hamiltonian

We can rewrite equations (3.212) and (3.213) so that they represent the vector and scalar potentials at a point R_1 due to a moving electron at R_2:

$$A_1 = -\frac{e}{8\pi\varepsilon_0 c^2}\left\{\frac{\dot{R}_2}{R_{12}} + \frac{(\dot{R}_2 \cdot R_{12})R_{12}}{R_{12}^3}\right\} + \text{order}(1/c^3), \tag{3.214}$$

$$\phi_1 = -\frac{e}{4\pi\varepsilon_0 R_{12}} + \text{order}(1/c^4). \tag{3.215}$$

Now suppose that at the instant in time to which these expressions refer, we have another electron at R_1. The Lagrangian for this electron is

$$\mathcal{L} = -mc^2\gamma_1^{-1} + e\phi_1 - e(\dot{R}_1 \cdot A_1), \quad \text{where } \gamma_1 = [1 - (\dot{R}_1/c)^2]^{-1/2}. \tag{3.216}$$

If we substitute for A_1 and ϕ_1 using (3.214) and (3.215) we obtain the Lagrangian for electron 1 when the motion of electron 2 is regarded as known,

$$\mathcal{L} = -mc^2\gamma_1^{-1} - \frac{e^2}{4\pi\varepsilon_0 R_{12}} + \frac{e^2}{8\pi\varepsilon_0 c^2}\left\{\frac{(\dot{R}_1 \cdot \dot{R}_2)}{R_{12}} + \frac{(\dot{R}_1 \cdot R_{12})(R_{12} \cdot \dot{R}_2)}{R_{12}^3}\right\}. \tag{3.217}$$

If we now add the term $-mc^2\gamma_2^{-1}$ we obtain the expression

$$\mathcal{L} = -mc^2\gamma_1^{-1} - mc^2\gamma_2^{-1} - \frac{e^2}{4\pi\varepsilon_0 R_{12}} + \frac{e^2}{8\pi\varepsilon_0 c^2}\left\{\frac{(\dot{R}_1 \cdot \dot{R}_2)}{R_{12}} + \frac{(\dot{R}_1 \cdot R_{12})(R_{12} \cdot \dot{R}_2)}{R_{12}^3}\right\}, \tag{3.218}$$

which is symmetrical with respect to electrons 1 and 2; it is, in fact, the Lagrangian for the simultaneous motion of the two electrons. The interaction term is only correct to $1/c^2$; although γ_1 and γ_2 should only be expanded to this order, it is appropriate to retain the next higher-order term:

$$-mc^2\gamma_1^{-1} = -mc^2 + \frac{m\dot{R}_1^2}{2} - \frac{m\dot{R}_1^4}{8c^2} + \cdots. \tag{3.219}$$

Hence

$$\mathcal{L} = -2mc^2 + \sum_i \frac{m\dot{R}_1^2}{2} - \sum_i \frac{m\dot{R}_1^4}{8c^2} - \frac{e^2}{4\pi\varepsilon_0 R_{12}} + \frac{e^2}{8\pi\varepsilon_0 c^2}\left\{\frac{(\dot{R}_1 \cdot \dot{R}_2)}{R_{12}} + \frac{(\dot{R}_1 \cdot R_{12})(R_{12} \cdot \dot{R}_2)}{R_{12}^3}\right\}, \tag{3.220}$$

where $i = 1, 2$. The corresponding classical Hamiltonian is

$$\mathcal{H} = 2mc^2 + \sum_i \frac{P_i^2}{2m} - \sum_i \frac{P_i^4}{8m^3c^2} + \frac{e^2}{4\pi\varepsilon_0 R_{12}}$$
$$- \frac{e^2}{8\pi\varepsilon_0 m^2 c^2} \left\{ \frac{P_1 \cdot P_2}{R_{12}} + \frac{(P_1 \cdot R_{12})(R_{12} \cdot P_2)}{R_{12}^3} \right\}, \qquad (3.221)$$

which may be converted into a quantum mechanical Hamiltonian by making the usual replacement $P = -i\hbar\nabla$ and regarding all terms as operators. One must, however, then take care over the order of the operators in the interaction term, the correct term being

$$- \frac{e^2}{8\pi\varepsilon_0 m^2 c^2} \left\{ \pi_1 \cdot \frac{1}{R_{12}} \pi_2 + (\pi_1 \cdot R_{12}) \frac{1}{R_{12}^3} (R_{12} \cdot \pi_2) \right\} \qquad (3.222)$$

where we have replaced P by π to take account of any external fields. Our previous term given in equation (3.161) for the orbit–orbit interaction was of this form, and we shall see soon that the interaction term in the Breit equation is closely similar to (3.222).

3.9. The Breit Hamiltonian

3.9.1. Introduction

The most unsatisfactory features of our derivation of the molecular Hamiltonian from the Dirac equation stem from the fact that the Dirac equation is, of course, a single particle equation. Hence all of the inter-electron terms have been introduced by including the effects of other electrons in the magnetic vector and electric scalar potentials. A particularly objectionable aspect is the inclusion of electron spin terms in the magnetic vector potential A_i^e, with the use of classical field theory to derive the results. It is therefore of interest to examine an alternative development and in this section we introduce the Breit Hamiltonian [16] as the starting point. We eventually arrive at the same molecular Hamiltonian as before, but the derivation is more satisfactory, although fundamental difficulties are still present.

The Breit Hamiltonian for two electrons consists essentially of a Dirac Hamiltonian for each electron, with interaction terms. It may be written

$$\mathcal{H} = \beta_1 m_1 c^2 - e_1 \phi_1 + c\alpha_1 \cdot (P_1 + e_1 A_1) + \beta_2 m_2 c^2 - e_2 \phi_2 + c\alpha_2 \cdot (P_2 + e_2 A_2)$$
$$+ \frac{e_1 e_2}{4\pi\varepsilon_0 R_{12}} - \frac{e_1 e_2}{8\pi\varepsilon_0 R_{12}} \left\{ \alpha_1 \cdot \alpha_2 + \frac{(\alpha_1 \cdot R_{12})(\alpha_2 \cdot R_{12})}{R_{12}^2} \right\}. \qquad (3.223)$$

The first interaction term is, of course, the Coulomb interaction; the second interaction term has the same form as the classical expression for the retarded interaction of two particles, derived in section 3.8. However, the Breit Hamiltonian suffers from the defect that the interaction terms are not Lorentz invariant. Detailed investigations using quantum electrodynamics show that the interaction term is satisfactory, provided it is treated only to the first order in perturbation theory. The wave function corresponding to the Hamiltonian (3.223) has sixteen components, so that the operators in (3.223) are represented by matrices of dimension sixteen. Nevertheless we anticipate that it should be possible to reduce the wave equation to a non-relativistic form appropriate for two electrons in positive energy states; the associated wave function would then have four components, with two spin orientations for each electron.

3.9.2. Reduction of the Breit Hamiltonian to non-relativistic form

The first stage in deriving a molecular Hamiltonian is to reduce the Breit equation to non-relativistic form and Chraplyvy [17] has shown how this reduction can be performed by using an extension of the Foldy–Wouthuysen transformation. First let us remind ourselves of the most important features in the transformation of the Dirac Hamiltonian. The latter was written (see (3.57) and (3.58)) as

$$\mathcal{H} = \beta mc^2 - e\phi + c\boldsymbol{\alpha} \cdot (\boldsymbol{P} + e\boldsymbol{A}) \qquad (3.224)$$

$$= \beta mc^2 + \mathcal{E} + \mathcal{O}, \qquad (3.225)$$

where the even operator $\mathcal{E}(= -e\phi)$ has vanishing matrix elements between the upper (electron) and lower (positron) components of the four-component spinor, and the odd operator $\mathcal{O}(=c\boldsymbol{\alpha}\cdot(\boldsymbol{P}+e\boldsymbol{A}))$ has corresponding non-vanishing matrix elements. The aim of the Foldy–Wouthuysen transformation is to convert (3.225) into a Hamiltonian which, to order (c^{-2}), contains only even operators. A transformed time-independent Hamiltonian $\mathcal{H}'$ is related to the starting Hamiltonian $\mathcal{H}$ by

$$\mathcal{H}' = \mathcal{H} + \mathrm{i}[\bar{S}, \mathcal{H}] + \frac{\mathrm{i}^2}{2!}[\bar{S},[\bar{S},\mathcal{H}]] + \frac{\mathrm{i}^3}{3!}[\bar{S},[\bar{S},[\bar{S},\mathcal{H}]]] + \cdots \qquad (3.226)$$

and if $\bar{S}$ is given by

$$\bar{S} = -\frac{\mathrm{i}\beta\mathcal{O}}{2mc^2}, \qquad (3.227)$$

the transformed Hamiltonian $\mathcal{H}'$ contains even operators of order c^2, c^0, c^{-2}, etc., but odd operators of order c^{-1}, c^{-3}, etc. (in the initial Hamiltonian, the odd operator is of order c^{+1}). Repetition of the transformation converts $\mathcal{H}'$ into $\mathcal{H}''$ which to order c^{-2} now contains only even operators; thus each transformation reduces the importance of the odd operators and, after two transformations, we obtain a Hamiltonian which, to order c^{-2}, operates only on the upper (electron) components of the wave function.

Now the Breit Hamiltonian for two electrons in the presence of electromagnetic fields is, as we have seen,

$$\mathcal{H} = \beta_1 m_1 c^2 + \beta_2 m_2 c^2 - e_1\phi_1 - e_2\phi_2 + \frac{e_1 e_2}{4\pi\varepsilon_0 R_{12}} + c\boldsymbol{\alpha}_1 \cdot \boldsymbol{\pi}_1 + c\boldsymbol{\alpha}_2 \cdot \boldsymbol{\pi}_2$$

$$- \frac{e_1 e_2}{8\pi\varepsilon_0 R_{12}}\left\{\boldsymbol{\alpha}_1 \cdot \boldsymbol{\alpha}_2 + \frac{(\boldsymbol{\alpha}_1 \cdot \boldsymbol{R}_{12})(\boldsymbol{\alpha}_2 \cdot \boldsymbol{R}_{12})}{R_{12}^2}\right\}. \qquad (3.228)$$

The Breit Hamiltonian (3.228) can be written in a form analogous to (3.225), namely,

$$\mathcal{H} = \beta_1 m_1 c^2 + \beta_2 m_2 c^2 + \mathcal{EE} + \mathcal{OE} + \mathcal{EO} + \mathcal{OO}. \tag{3.229}$$

By comparison with (3.228) the even–even, odd–even, even–odd and odd–odd operators are

$$\mathcal{EE} = -e_1 \phi_1 - e_2 \phi_2 + \frac{e_1 e_2}{4\pi\varepsilon_0 R_{12}},$$

$$\mathcal{OE} = c\boldsymbol{\alpha}_1 \cdot \boldsymbol{\pi}_1,$$

$$\mathcal{EO} = c\boldsymbol{\alpha}_2 \cdot \boldsymbol{\pi}_2,$$

$$\mathcal{OO} = -\frac{e_1 e_2}{8\pi\varepsilon_0 R}\left\{\boldsymbol{\alpha}_1 \cdot \boldsymbol{\alpha}_2 + \frac{(\boldsymbol{\alpha}_1 \cdot \boldsymbol{R}_{12})(\boldsymbol{\alpha}_2 \cdot \boldsymbol{R}_{12})}{R_{12}^2}\right\}. \tag{3.230}$$

The Breit Hamiltonian operates on sixteen-component spinor functions which contain four types of function, designated ψ_{uU}, ψ_{uL}, $\psi_{\ell U}$, $\psi_{\ell L}$, which represent upper and lower components as previously defined, the small letters u, ℓ referring to the first particle (1) and the capital letters U, L referring to the second particle (2). Our aim is to find a transformation which gives a Hamiltonian operating only on the components ψ_{uU}; in other words, we seek a Hamiltonian which, to order c^{-2}, contains only terms which are overall even–even in character.

As a preliminary to this it is worthwhile noting a few properties of the matrix operators (3.230). As examples of the four types we may list,

$$\begin{aligned}
\mathcal{EE} &: \quad \text{I(unit)}, \beta_1, \beta_2, \\
\mathcal{EO} &: \quad \alpha_{2x}, \alpha_{2y}, \alpha_{2z}, \\
\mathcal{OE} &: \quad \alpha_{1x}, \alpha_{1y}, \alpha_{1z}, \\
\mathcal{OO} &: \quad \text{direct product of } \boldsymbol{\alpha}_1 \text{ and } \boldsymbol{\alpha}_2 \text{ matrices.}
\end{aligned} \tag{3.231}$$

The commutation relations which are important are

$$\begin{aligned}
[\beta_1, \mathcal{EE}] &= [\beta_1, \mathcal{EO}] = 0, & [\beta_1, \mathcal{OE}]_+ &= [\beta_1, \mathcal{OO}]_+ = 0, \\
[\beta_2, \mathcal{EE}] &= [\beta_2, \mathcal{OE}] = 0, & [\beta_2, \mathcal{OO}]_+ &= [\beta_2, \mathcal{EO}]_+ = 0,
\end{aligned} \tag{3.232}$$

where $[A, B]_+ \equiv AB + BA$ denotes the anticommutator of A and B.

In his first paper Chraplyvy used the transformation (3.226) with $i\bar{S}$ given by

$$i\bar{S} = \frac{\beta_1}{2m_1 c^2}\mathcal{OE} + \frac{\beta_2}{2m_2 c^2}\mathcal{EO} + \frac{\beta_1 m_1 - \beta_2 m_2}{2(m_1^2 - m_2^2)c^2}\mathcal{OO}. \tag{3.233}$$

These three terms remove the $\mathcal{OE}$, $\mathcal{EO}$ and $\mathcal{OO}$ terms respectively from (3.229) and repetition of the transformation ultimately yields a Hamiltonian which, with $\beta_1 = \beta_2 = 1$, operates only on the ψ_{uU} functions, as desired (to order c^{-2}). However, as (3.233) shows, this transformation is unacceptable if the two particles have equal masses (e.g. two electrons). It was realised subsequently that there exists a family of related transformations, of which (3.233) is just one member. The transformation using (3.233) in fact goes further than we require, in that it leads to complete separation of all four types of function. We would be satisfied with the more limited objective of separating

ψ_{uU} from the other components, and in his second paper Chraplyvy showed that a transformation which accomplishes this is

$$i\bar{S} = \frac{1}{4m_1c^2}(\beta_1 + \beta_1\beta_2)\mathcal{OE} + \frac{1}{4m_2c^2}(\beta_2 + \beta_1\beta_2)\mathcal{EO} + \frac{1}{4(m_1+m_2)c^2}(\beta_1+\beta_2)\mathcal{OO}. \tag{3.234}$$

If we take account of the fact that $\mathcal{OE}$ and $\mathcal{EO}$ commute, the resulting Hamiltonian to order c^{-2} is

$$\mathcal{H} = m_1c^2 + m_2c^2 + \mathcal{EE} + \frac{1}{2m_1c^2}(\mathcal{OE})^2 + \frac{1}{2m_2c^2}(\mathcal{EO})^2 + \frac{1}{8m_1^2c^4}[[\mathcal{OE}, \mathcal{EE}], \mathcal{OE}]$$

$$+ \frac{1}{8m_2^2c^4}[[\mathcal{EO}, \mathcal{EE}], \mathcal{EO}] + \frac{1}{4m_1m_2c^4}[[\mathcal{OE}, \mathcal{OO}]_+, \mathcal{EO}]_+ - \frac{1}{8m_1^3c^6}(\mathcal{OE})^4$$

$$- \frac{1}{8m_2^3c^6}(\mathcal{EO})^4 + \frac{1}{2(m_1+m_2)c^2}(\mathcal{OO})^2 + \cdots. \tag{3.235}$$

Noting that $\mathcal{E}\times\mathcal{E}=\mathcal{E}$, $\mathcal{E}\times\mathcal{O}=\mathcal{O}\times\mathcal{E}=\mathcal{O}$ and $\mathcal{O}\times\mathcal{O}=\mathcal{E}$ we see that each term is overall of the desired type, $\mathcal{EE}$. We can keep track of the two electrons by writing m_1 and m_2, although in this case the masses are, of course, equal.

It now remains to expand the operators in (3.235) using the definitions given in (3.230) but before we do so we must draw attention to a difficulty with (3.235). The final term, containing the operator $(\mathcal{OO})^2$ is not obtained if a more sophisticated treatment starting from the Bethe–Salpeter equation is used. The reader will recall our earlier comment that the interaction term in the Breit Hamiltonian is acceptable provided it is treated by first-order perturbation theory. Rather than launch into quantum electrodynamics at this stage, we shall proceed to develop (3.235) but will omit the $(\mathcal{OO})^2$ term without further comment.

Straightforward vector algebra yields the following results for the operators in (3.235):

$$(\mathcal{OE})^2 = c^2\{\pi_1^2 + e_1\hbar(\boldsymbol{\sigma}_1 \cdot \boldsymbol{B}_1)\}, \tag{3.236}$$

$$(\mathcal{EO})^2 = c^2\{\pi_2^2 + e_2\hbar(\boldsymbol{\sigma}_2 \cdot \boldsymbol{B}_2)\}, \tag{3.237}$$

$$(\mathcal{OE})^4 = c^4\{\pi_1^4 + 2e_1\hbar(\boldsymbol{\sigma}_1 \cdot \boldsymbol{B}_1)\pi_1^2 + e_1^2\hbar^2(\boldsymbol{\sigma}_1 \cdot \boldsymbol{B}_1)^2\}, \tag{3.238}$$

$$(\mathcal{EO})^4 = c^4\{\pi_2^4 + 2e_2\hbar(\boldsymbol{\sigma}_2 \cdot \boldsymbol{B}_2)\pi_2^2 + e_2^2\hbar^2(\boldsymbol{\sigma}_2 \cdot \boldsymbol{B}_2)^2\}, \tag{3.239}$$

$$[[\mathcal{OE}, \mathcal{EE}], \mathcal{OE}] = -\frac{e_1e_2c^2}{4\pi\varepsilon_0}\left\{4\pi\hbar^2\delta(\boldsymbol{R}_{12}) + \frac{2\hbar}{R_{12}^3}\boldsymbol{\sigma}_1 \cdot (\boldsymbol{R}_{12} \wedge \boldsymbol{\pi}_1)\right\}$$

$$+ e_1c^2\{2\hbar\boldsymbol{\sigma}_1 \cdot (\boldsymbol{E}_1 \wedge \boldsymbol{\pi}_1) + \hbar^2\nabla_1 \cdot \boldsymbol{E}_1\}, \tag{3.240}$$

$$[[\mathcal{EO}, \mathcal{EE}], \mathcal{EO}] = -\frac{e_1e_2c^2}{4\pi\varepsilon_0}\left\{4\pi\hbar^2\delta(\boldsymbol{R}_{12}) + \frac{2\hbar}{R_{12}^3}\boldsymbol{\sigma}_2 \cdot (\boldsymbol{R}_{12} \wedge \boldsymbol{\pi}_2)\right\}$$

$$+ e_2c^2\{2\hbar\boldsymbol{\sigma}_2 \cdot (\boldsymbol{E}_2 \wedge \boldsymbol{\pi}_2) + \hbar^2\nabla_2 \cdot \boldsymbol{E}_2\}, \tag{3.241}$$

$$[[\mathcal{OE}, \mathcal{OO}]_+, \mathcal{EO}]_+ = \frac{e_1 e_2 c^2}{4\pi\varepsilon_0} \left\{ \frac{\hbar^2}{R_{12}^3}(\boldsymbol{\sigma}_1 \cdot \boldsymbol{\sigma}_2) - \frac{8\pi\hbar^2}{3}\delta(\boldsymbol{R}_{12})(\boldsymbol{\sigma}_1 \cdot \boldsymbol{\sigma}_2) \right.$$

$$- \frac{3\hbar^2}{R_{12}^5}(\boldsymbol{\sigma}_1 \cdot \boldsymbol{R}_{12})(\boldsymbol{\sigma}_2 \cdot \boldsymbol{R}_{12}) + \frac{2\hbar}{R_{12}^3}\boldsymbol{\sigma}_1 \cdot (\boldsymbol{R}_{12} \wedge \boldsymbol{\pi}_2)$$

$$\left. - \frac{2\hbar}{R_{12}^3}\boldsymbol{\sigma}_2 \cdot (\boldsymbol{R}_{12} \wedge \boldsymbol{\pi}_1) - \frac{2}{R_{12}}(\boldsymbol{\pi}_1 \cdot \boldsymbol{\pi}_2) - \frac{2}{R_{12}^3}(\boldsymbol{\pi}_1 \cdot \boldsymbol{R}_{12})(\boldsymbol{R}_{12} \cdot \boldsymbol{\pi}_2) \right\}. \tag{3.242}$$

We now substitute these results in equation (3.235) to obtain the molecular Hamiltonian, which may be divided into terms which are additive for each electron, and terms which describe interaction between the electrons:

$$\mathcal{H} = \sum_i \mathcal{H}_i + \mathcal{H}_{ij}, \tag{3.243}$$

where

$$\mathcal{H}_i = \sum_{i=1}^{2} \left\{ m_i c^2 + \frac{1}{2m_i}\pi_i^2 - e_i\phi_i + \frac{e_i\hbar}{2m_i}(\boldsymbol{\sigma}_i \cdot \boldsymbol{B}_i) - \frac{1}{8m_i^3 c^2}\pi_i^4 \right.$$

$$\left. + \frac{e_i\hbar^2}{8m_i^2 c^2}\nabla_i \cdot \boldsymbol{E}_i - \frac{e_i\hbar}{8m_i^2 c^2}\boldsymbol{\sigma}_i \cdot (\boldsymbol{\pi}_i \wedge \boldsymbol{E}_i - \boldsymbol{E}_i \wedge \boldsymbol{\pi}_i) - \frac{e_i\hbar}{4m_i^3 c^2}(\boldsymbol{\sigma}_i \cdot \boldsymbol{B}_i)\pi_i^2 \right\}, \tag{3.244}$$

$$\mathcal{H}_{ij} = \frac{e_i e_j}{4\pi\varepsilon_0 R_{ij}} - \frac{e_i e_j}{8\pi\varepsilon_0 m_i m_j c^2}\left\{ \boldsymbol{\pi}_i \cdot \boldsymbol{\pi}_j \left(\frac{1}{R_{ij}}\right) + \frac{(\boldsymbol{\pi}_i \cdot \boldsymbol{R}_{ij})(\boldsymbol{R}_{ij} \cdot \boldsymbol{\pi}_j)}{R_{ij}^3} \right\}$$

$$+ \frac{e_i e_j \hbar}{8\pi\varepsilon_0 m_i m_j c^2}\frac{1}{R_{ij}^3}\{\boldsymbol{\sigma}_i \cdot (\boldsymbol{R}_{ij} \wedge \boldsymbol{\pi}_j) - \boldsymbol{\sigma}_j \cdot (\boldsymbol{R}_{ij} \wedge \boldsymbol{\pi}_i)\}$$

$$- \frac{e_i e_j \hbar}{16\pi\varepsilon_0 c^2}\frac{1}{R_{ij}^3}\left\{ \frac{1}{m_i^2}\boldsymbol{\sigma}_i \cdot (\boldsymbol{R}_{ij} \wedge \boldsymbol{\pi}_i) - \frac{1}{m_j^2}\boldsymbol{\sigma}_j \cdot (\boldsymbol{R}_{ij} \wedge \boldsymbol{\pi}_j) \right\}$$

$$- \frac{e_i e_j \hbar^2 \pi}{8\pi\varepsilon_0 c^2}\left\{ \frac{1}{m_i^2} + \frac{1}{m_j^2} \right\}\delta(\boldsymbol{R}_{ij}) + \frac{e_i e_j \hbar^2}{16\pi\varepsilon_0 m_i m_j c^2}$$

$$\times \left\{ \frac{1}{R_{ij}^3}(\boldsymbol{\sigma}_i \cdot \boldsymbol{\sigma}_j) - \frac{3}{R_{ij}^5}(\boldsymbol{\sigma}_i \cdot \boldsymbol{R}_{ij})(\boldsymbol{\sigma}_j \cdot \boldsymbol{R}_{ij}) - \frac{8\pi}{3}\delta(\boldsymbol{R}_{ij})\boldsymbol{\sigma}_i \cdot \boldsymbol{\sigma}_j \right\}. \tag{3.245}$$

The last stage is now to replace $\boldsymbol{\pi}_i$ by $\boldsymbol{P}_i + e\boldsymbol{A}_i$ and to use explicit expressions for the potentials $\boldsymbol{A}_i$ and ϕ_i. Our previous expressions for $\boldsymbol{A}_i$ and ϕ_i were given in equations (3.133) and (3.134). However, the inter-electron interactions have now been derived more naturally by starting with the Breit Hamiltonian and the vector and scalar potentials therefore contain only terms describing external fields or electrostatic interactions involving the nuclear charge. Hence we make the substitutions

$$\boldsymbol{A}_i = (1/2)(\boldsymbol{B}_i \wedge \boldsymbol{R}_i), \tag{3.246}$$

$$\phi_i = \sum_\alpha \frac{Z_\alpha e}{4\pi\varepsilon_0 R_{\alpha i}} - \boldsymbol{E}_i \cdot \boldsymbol{R}_i, \tag{3.247}$$

and further assume the magnetic and electric fields to be uniform, i.e. $\boldsymbol{B}_i = \boldsymbol{B}_j$ and $\boldsymbol{E}_i = \boldsymbol{E}_j$. This leads to a Hamiltonian which, on generalisation to many electrons, is identical to that derived from the Dirac equation, given in equations (3.161), (3.174) and (3.180). The same Hamiltonian has also been derived by Itoh [15] from quantum electrodynamics.

3.10. Electronic interactions in the nuclear Hamiltonian

In order to complete our derivation of the molecular Hamiltonian we must consider the nuclear Hamiltonian in more detail. A thorough relativistic treatment analogous to that for the electron is not possible within the limitations of quantum mechanics, since nuclei are not Dirac particles and they can have large anomalous magnetic moments. However, the use of quantum electrodynamics [18] shows that we can derive the correct Hamiltonian to order $1/c^2$ by taking the non-relativistic Hamiltonian:

$$\mathcal{H}_{\text{nucl}} = \sum_\alpha \frac{1}{2M_\alpha}\left(\boldsymbol{P}_\alpha - Z_\alpha e \boldsymbol{A}_\alpha^e\right)^2 + \frac{Z_1 Z_2 e^2}{4\pi\varepsilon_0 R}, \tag{3.248}$$

and including the effects of other particles on nucleus α in a vector potential $\boldsymbol{A}_\alpha^e$ provided we make the additional restriction that we retain only terms involving M_α^{-1}. Thus, for example, the relativistic correction to the nuclear kinetic energy, which by analogy with equation (3.161) is of order $1/M_\alpha^3 c^2$, is not included in (3.248). The explicit form that we use for $\boldsymbol{A}_\alpha^e$ is

$$\boldsymbol{A}_\alpha^e = \frac{g_S \mu_B}{4\pi\varepsilon_0 c^2} \sum_i \frac{1}{R_{i\alpha}^3}(\boldsymbol{S}_i \wedge \boldsymbol{R}_{i\alpha}) - \frac{e}{8\pi\varepsilon_0 m c^2}\sum_i \left\{\boldsymbol{P}_i \frac{1}{R_{i\alpha}} + (\boldsymbol{P}_i \cdot \boldsymbol{R}_{i\alpha})\frac{1}{R_{i\alpha}^3}\boldsymbol{R}_{i\alpha}\right\}, \tag{3.249}$$

where $\boldsymbol{R}_{i\alpha}$ is equal to $(\boldsymbol{R}_i - \boldsymbol{R}_\alpha)$. The first term represents the vector potential at nucleus α arising from the spin magnetic moments of the electrons, while the second represents the potential derived from the orbital motion of the electrons. We have attempted to include the latter in the correct relativistic form by taking an expression analogous to that required in the electronic Hamiltonian. There is a third possible contribution to $\boldsymbol{A}_\alpha^e$, namely, that arising from the orbital motion of the second nucleus relative to α; however, it produces terms of order $1/M_\alpha^2 c^2$ and so need not be considered.

Substitution of (3.249) into (3.248) leads in a straightforward manner to the result:

$$\mathcal{H}_{\text{nucl}} = \sum_\alpha \frac{1}{2M_\alpha}\boldsymbol{P}_\alpha^2 - \frac{g_S \mu_B e}{4\pi\varepsilon_0 c^2}\sum_i\sum_\alpha \frac{Z_\alpha}{M_\alpha}\frac{1}{R_{i\alpha}^3}\boldsymbol{S}_i\cdot(\boldsymbol{R}_{i\alpha}\wedge \boldsymbol{P}_\alpha)$$
$$+ \frac{e^2}{8\pi\varepsilon_0 m c^2}\sum_i\sum_\alpha \frac{Z_\alpha}{M_\alpha}\left\{\boldsymbol{P}_i \frac{1}{R_{i\alpha}}\cdot \boldsymbol{P}_\alpha + (\boldsymbol{P}_i\cdot\boldsymbol{R}_{i\alpha})\frac{1}{R_{i\alpha}^3}(\boldsymbol{R}_{i\alpha}\cdot\boldsymbol{P}_\alpha)\right\} + \frac{Z_1 Z_2 e^2}{4\pi\varepsilon_0 R}. \tag{3.250}$$

The second term in (3.250) yields the spin–rotation and spin–vibration interactions, whilst the third term leads to the orbit–rotation and orbit–vibration interactions. We will examine these terms in more detail in the next section.

A point which we discuss here before developing the molecular Hamiltonian further is the value of g_S. The electron magnetic moment was at one time described as being anomalous because the g value was found to be 2, rather than 1 as for the orbital g factor, g_L. It was one of the triumphs of the Dirac theory that it predicted a g value of exactly 2, although later more accurate measurements gave a value slightly larger, 2.002 319. This value, too, is now well understood, the correction being the result of effects of quantum electrodynamics. We will not go into the details, but note that experiment and theory are in agreement to at least eight decimal places.

3.11. Transformation of coordinates in the field-free total Hamiltonian

When we combine the results of sections 3.6 and 3.10, we can write down a total molecular Hamiltonian in field-free space:

$$\mathcal{H}_{\text{total}} = \mathcal{H}_{\text{el}} + \mathcal{H}_{\text{nucl}}$$

$$= \frac{1}{2m}\sum_i P_i^2 - \frac{1}{8m^3c^2}\sum_i P_i^4 + \sum_{i,j\neq i}\frac{e^2}{8\pi\varepsilon_0 R_{ji}} - \sum_{i,\alpha}\frac{Z_\alpha e^2}{4\pi\varepsilon_0 R_{\alpha i}}$$

$$+ \frac{e^2\hbar^2}{8\varepsilon_0 m^2 c^2}\sum_i\left\{\sum_\alpha Z_\alpha \delta^{(3)}(R_{\alpha i}) - \sum_{j\neq i}\delta^{(3)}(R_{ji})\right\}$$

$$- \frac{g_S\mu_B}{8\pi\varepsilon_0 mc}\sum_i S_i\cdot\left\{\sum_\alpha \frac{Z_\alpha e}{R_{\alpha i}^3}(R_{\alpha i}\wedge P_i) - \sum_{j\neq i}\frac{e}{R_{ji}^3}(R_{ji}\wedge P_i)\right\}$$

$$+ \frac{g_S\mu_B e}{4\pi\varepsilon_0 mc}\sum_{i,j\neq i}\frac{1}{R_{ji}^3}S_i\cdot(R_{ij}\wedge P_j)$$

$$- \frac{1}{4\pi\varepsilon_0}\left(\frac{e}{2mc}\right)^2\sum_{i,j\neq i}\left\{P_i\frac{1}{R_{ji}}\cdot P_j + (P_i\cdot R_{ij})\frac{1}{R_{ij}^3}(R_{ij}\cdot P_j)\right\}$$

$$+ \frac{g_S^2\mu_B^2}{8\pi\varepsilon_0 c^2}\sum_{i,j\neq i}\left\{\frac{1}{R_{ji}^3}(S_i\cdot S_j) - \frac{3}{R_{ji}^5}(S_i\cdot R_{ji})(S_j\cdot R_{ji}) - \frac{8\pi}{3}\delta^{(3)}(R_{ji})(S_i\cdot S_j)\right\}$$

$$+ \sum_\alpha \frac{P_\alpha^2}{2M_\alpha} - \frac{g_S\mu_B e}{4\pi\varepsilon_0 c^2}\sum_{i,\alpha}\frac{Z_\alpha}{M_\alpha}\frac{1}{R_{i\alpha}^3}S_i\cdot(R_{i\alpha}\wedge P_\alpha)$$

$$+ \frac{e^2}{8\pi\varepsilon_0 mc^2}\sum_{i,\alpha}\frac{Z_\alpha}{M_\alpha}\left\{P_i\frac{1}{R_{i\alpha}}P_\alpha + (P_i\cdot R_{i\alpha})\frac{1}{R_{i\alpha}^3}(R_{i\alpha}\cdot P_\alpha)\right\} + \frac{Z_1 Z_2 e^2}{4\pi\varepsilon_0 R}.$$

(3.251)

This represents the total field-free Hamiltonian for a diatomic molecule, to order $1/c^2$ in the purely electronic terms and to order $1/M_\alpha c^2$ in the nuclear terms, in a space-fixed axis system of arbitrary origin. We showed in chapter 2 that the solution of a Hamiltonian of this type was simplified if we first performed various coordinate transformations. In section 2.2 we re-expressed the Hamiltonian in terms of the coordinates R_O, the molecular centre of mass, R, the nuclear coordinate and R_i'', the electronic coordinates measured in a non-rotating frame with origin at the nuclear centre of mass and the momenta conjugate to these coordinates. Specifically we used

$$R_O = \frac{1}{M}\left\{m\sum_i R_i + \sum_\alpha M_\alpha R_\alpha\right\}, \quad (3.252)$$

$$R = R_2 - R_1, \quad (3.253)$$

$$R_i'' = R_i - \frac{1}{(M_1 + M_2)}\sum_\alpha M_\alpha R_\alpha, \quad (3.254)$$

from which one can readily show

$$P_i = \frac{m_i}{M}P_O + P_i'', \quad (3.255)$$

$$P_\alpha = \frac{M_\alpha}{M}P_O \mp P_R - \frac{M_\alpha}{(M_1 + M_2)}\sum_i P_i'', \quad (3.256)$$

where, as before, the upper sign in (3.256) refers to $\alpha = 1$ and the lower to $\alpha = 2$. This transformation enabled us to separate off the translational motion of the molecule completely, and also went some way towards separating electronic and nuclear motions. Then in section 2.5 we transformed the electronic spatial and spin coordinates to an axis system rotating with the nuclear framework but with origin still at the nuclear centre of mass. This removed awkward cross terms between nuclear and electronic coordinates in the Coulombic interaction term and gave as complete a separation of nuclear and electronic motions as possible. The explicit relationships between coordinates and momenta that we used were

$$R_i'' = \mathfrak{M} r_i, \quad (3.257)$$

$$P_i'' = \mathfrak{M} p_i, \quad (3.258)$$

where $\mathfrak{M}$ is a unitary matrix defined in equation (2.40) and each vector equation (3.257) and (3.258) represents three transformations, one for each component. Similarly for the spin operators

$$S_i = \mathfrak{U}^{-1} s_i \mathfrak{U}, \quad (3.259)$$

where $\mathfrak{U}$ is defined in (2.111) and s_i is defined by

$$s_i = s_x \mathbf{i} + s_y \mathbf{j} + s_z \mathbf{k}. \quad (3.260)$$

Because the unit vectors in the two coordinate systems are also related by the $\mathfrak{M}$ matrix, it is easy to show that

$$\frac{\partial}{\partial X_i''}\mathbf{i}' + \frac{\partial}{\partial Y_i''}\mathbf{j}' + \frac{\partial}{\partial Z_i''}\mathbf{k}' = \frac{\partial}{\partial x_i}\mathbf{i} + \frac{\partial}{\partial y_i}\mathbf{j} + \frac{\partial}{\partial z_i}\mathbf{k} \quad (3.261)$$

and
$$S_X \boldsymbol{i}' + S_Y \boldsymbol{j}' + S_Z \boldsymbol{k}' = s_x \boldsymbol{i} + s_y \boldsymbol{j} + s_z \boldsymbol{k}. \quad (3.262)$$

Thus the coordinate transformation can be realised simply by making the replacements
$$\boldsymbol{R}_i'' \to \boldsymbol{r}_i, \quad (3.263)$$
$$\boldsymbol{P}_i'' \to \boldsymbol{p}_i, \quad (3.264)$$
$$\boldsymbol{S}_i \to \boldsymbol{s}_i. \quad (3.265)$$

We now consider the overall effect of these coordinate transformations on the total Hamiltonian (3.251). We have already shown in chapter 2 that

$$\frac{1}{2m}\sum_i P_i^2 + \sum_\alpha \frac{1}{2M_\alpha} P_\alpha^2 = \frac{1}{2M} P_O^2 + \frac{1}{2m}\sum_i p_i^2 + \frac{1}{2(M_1+M_2)}\sum_{i,j} \boldsymbol{p}_i \cdot \boldsymbol{p}_j$$
$$- \frac{\hbar^2}{2\mu R^2}\frac{\partial}{\partial R}\left(R^2 \frac{\partial}{\partial R}\right) + \frac{\hbar^2}{2\mu R^2}(\boldsymbol{J}-\boldsymbol{P})^2. \quad (3.266)$$

When we substitute equations (3.255) and (3.256) in the remaining terms of (3.251) we find that we retain a number of awkward cross terms involving $\boldsymbol{P}_O$ whereas symmetry considerations suggest that translational terms should be completely separable for the field free case. The explanation seems to be that we have made a coordinate transformation to the centre of rest mass of all particles rather than to the centre of relativistic mass. Since the translational velocities of molecules are very much less than the speed of light, the contributions of these cross terms in $\boldsymbol{P}_O$ are expected to be very small and we ignore them in further discussion.

With this reservation, the Hamiltonian appropriate to a molecule with electronic spin quantised in the molecular frame of reference becomes

$$\mathcal{H}_{\text{total}} = \frac{1}{2M} P_O^2 + \frac{1}{2m}\sum_i p_i^2 + \frac{1}{2(M_1+M_2)}\sum_{i,j}\boldsymbol{p}_i \cdot \boldsymbol{p}_j - \frac{1}{8m^3 c^2}\sum_i p_i^4$$
$$+ \frac{1}{8\pi\varepsilon_0}\sum_{i,j\neq i}\frac{e^2}{r_{ij}} - \sum_{i,\alpha}\frac{Z_\alpha e^2}{4\pi\varepsilon_0 r_{\alpha,i}} + \frac{Z_1 Z_2 e^2}{4\pi\varepsilon_0 R} + \frac{e^2 \hbar^2}{8\varepsilon_0 m^2 c^2}\sum_i\left\{\sum_\alpha Z_\alpha \delta^{(3)}(\boldsymbol{r}_{\alpha i})\right.$$
$$\left. - \sum_{j\neq i}\delta^{(3)}(\boldsymbol{r}_{ji})\right\} - \frac{g_S \mu_B}{8\pi\varepsilon_0 mc}\sum_i \boldsymbol{s}_i \cdot \left\{\sum_\alpha \frac{Z_\alpha e}{r_{\alpha i}^3}(\boldsymbol{r}_{\alpha i}\wedge \boldsymbol{p}_i) - \sum_{j\neq i}\frac{e}{r_{ji}^3}(\boldsymbol{r}_{ji}\wedge \boldsymbol{p}_i)\right\}$$
$$+ \frac{g_S \mu_B e}{4\pi\varepsilon_0 mc^2}\sum_{i,j\neq i}\frac{1}{r_{ji}^3}\boldsymbol{s}_i \cdot (\boldsymbol{r}_{ij}\wedge \boldsymbol{p}_j)$$
$$- \left(\frac{e}{2mc}\right)^2 \frac{1}{4\pi\varepsilon_0}\sum_{i,j\neq i}\left\{\boldsymbol{p}_i \frac{1}{r_{ji}}\cdot \boldsymbol{p}_j + (\boldsymbol{p}_i \cdot \boldsymbol{r}_{ji})\frac{1}{r_{ji}^3}(\boldsymbol{r}_{ji}\cdot \boldsymbol{p}_j)\right\}$$
$$+ \frac{g_S^2 \mu_B^2}{8\pi\varepsilon_0 c^2}\sum_{i,j\neq i}\left\{\frac{\boldsymbol{s}_i \cdot \boldsymbol{s}_j}{r_{ji}^3} - \frac{3(\boldsymbol{s}_i \cdot \boldsymbol{r}_{ji})(\boldsymbol{s}_j \cdot \boldsymbol{r}_{ji})}{r_{ji}^5} - \frac{8\pi}{3}\delta^{(3)}(\boldsymbol{r}_{ji})\boldsymbol{s}_i \cdot \boldsymbol{s}_j\right\}$$

$$-\frac{\hbar^2}{2\mu R^2}\frac{\partial}{\partial R}\left(R^2\frac{\partial}{\partial R}\right) + \frac{\hbar^2}{2\mu R^2}(\boldsymbol{J}-\boldsymbol{P})^2 + \mathcal{H}_{\text{spin,nucl}} + \mathcal{H}_{\text{orb,nucl}}$$

$$+ \frac{g_S\mu_B e}{4\pi\varepsilon_0 c^2}\sum_{i,\alpha}\frac{Z_\alpha}{(M_1+M_2)}\frac{1}{r_{i\alpha}^3}\boldsymbol{s}_i\cdot\left(\boldsymbol{r}_{i\alpha}\wedge\sum_j\boldsymbol{p}_j\right)$$

$$- \frac{e^2}{8\pi\varepsilon_0 mc^2}\sum_{i,\alpha}\frac{Z_\alpha}{(M_1+M_2)}\left\{\boldsymbol{p}_i\frac{1}{r_{ij}}\cdot\sum_j\boldsymbol{p}_j + (\boldsymbol{p}_i\cdot\boldsymbol{r}_{ji})\frac{1}{r_{ji}^3}\left(\boldsymbol{r}_{ji}\cdot\sum_j\boldsymbol{p}_j\right)\right\}.$$

(3.267)

The last two terms in (3.267) represent mass polarisation corrections to the spin–orbit and orbit–orbit interactions respectively. From equations (2.73) and (3.250), the term $\mathcal{H}_{\text{spin,nucl}}$ is given by

$$\mathcal{H}_{\text{spin,nucl}} = -\frac{g_S\mu_B e}{4\pi\varepsilon_0 c^2}\sum_{i,\alpha}(\mp)\frac{Z_\alpha}{M_\alpha}\frac{1}{r_{i\alpha}^3}\boldsymbol{s}_i\cdot(\boldsymbol{r}_{i\alpha}\wedge\boldsymbol{P}_R)$$

$$= -\frac{g_S\mu_B e\hbar}{4\pi\varepsilon_0 c^2}\frac{1}{\mu R^2}\sum_i\boldsymbol{s}_i\cdot\left\{\frac{Z_1 M_2}{M_1+M_2}\frac{\boldsymbol{r}_{i1}}{r_{i1}^3} - \frac{Z_2 M_1}{M_1+M_2}\frac{\boldsymbol{r}_{i2}}{r_{i2}^3}\right\}\wedge\{\boldsymbol{R}\wedge(\boldsymbol{J}-\boldsymbol{P})\}$$

$$- \frac{g_S\mu_B e}{4\pi\varepsilon_0 c^2}\sum_i\boldsymbol{s}_i\cdot\left[\left\{\frac{Z_1}{M_1}\frac{\boldsymbol{r}_{i1}}{r_{i1}^3} - \frac{Z_2}{M_2}\frac{\boldsymbol{r}_{i2}}{r_{i2}^3}\right\}\wedge\boldsymbol{k}\right]i\hbar\frac{\partial}{\partial R}. \quad (3.268)$$

The first of these two terms describes the interaction of the magnetic moments due to electron spin and nuclear rotation and it is therefore called the spin–rotation interaction; the second term is the corresponding spin–vibration interaction. Similarly we have,

$$\mathcal{H}_{\text{orb,nucl}} = \frac{e^2}{8\pi\varepsilon_0 mc^2}\sum_{i,\alpha}(\mp)\frac{Z_\alpha}{M_\alpha}\left\{\boldsymbol{p}_i\frac{1}{r_{i\alpha}}\cdot\boldsymbol{P}_R + (\boldsymbol{p}_i\cdot\boldsymbol{r}_{i\alpha})\frac{1}{r_{i\alpha}^3}(\boldsymbol{r}_{i\alpha}\cdot\boldsymbol{P}_R)\right\}$$

$$= \frac{e^2\hbar}{8\pi\varepsilon_0 mc^2}\frac{1}{\mu R^2}\sum_i\boldsymbol{p}_i\cdot\left\{\frac{Z_1 M_2}{M_1+M_2}\left(\frac{1}{r_{i1}} + \boldsymbol{r}_{i1}\frac{1}{r_{i1}^3}\boldsymbol{r}_{i1}\right)\right.$$

$$\left. - \frac{Z_2 M_1}{M_1+M_2}\left(\frac{1}{r_{i2}} + \boldsymbol{r}_{i2}\frac{1}{r_{i2}^3}\boldsymbol{r}_{i2}\right)\right\}\cdot[\boldsymbol{R}\wedge(\boldsymbol{J}-\boldsymbol{P})]$$

$$+ \frac{e^2}{8\pi\varepsilon_0 mc^2}\sum_i\boldsymbol{p}_i\cdot\left\{\frac{Z_1}{M_1}\left(\frac{1}{r_{i1}} + \boldsymbol{r}_{i1}\frac{1}{r_{i1}^3}\boldsymbol{r}_{i1}\right) - \frac{Z_2}{M_2}\left(\frac{1}{r_{i2}} + \boldsymbol{r}_{i2}\frac{1}{r_{i2}^3}\boldsymbol{r}_{i2}\right)\right\}\cdot\boldsymbol{k}i\hbar\frac{\partial}{\partial R},$$

(3.269)

where the first term represents the orbit–rotation interaction and the second the orbit–vibration interaction.

Finally, in section 2.10 we considered the alternative transformation scheme in which the electron spin remained quantised in the space-fixed coordinate system. The Hamiltonian for this situation is easily derived from (3.267), (3.268) and (3.269) by making the substitutions

$$\boldsymbol{s}_i \to \boldsymbol{S}_i, \quad (3.270)$$

$$(\boldsymbol{J}-\boldsymbol{P}) \to (\boldsymbol{N}-\boldsymbol{L}). \quad (3.271)$$

Note that when we come to consider terms in this Hamiltonian such as $\boldsymbol{S}_i \cdot (\boldsymbol{r}_{ji} \wedge \boldsymbol{p}_i)$ which represents the scalar product of two vectors that are defined in different coordinate systems, we must necessarily transform one operator to the coordinate system of the other before we can evaluate these terms.

3.12. Transformation of coordinates for the Zeeman and Stark terms in the total Hamiltonian

We dealt with the effects of applied static fields on the electronic Hamiltonian in section 3.7. In this section we first give the relevant terms for the nuclear Zeeman and Stark Hamiltonians and then perform the same coordinate transformations that proved to be convenient for the field-free molecular Hamiltonian.

We introduced the field-free nuclear Hamiltonian in section 3.10. Again by analogy with the electronic Hamiltonian, we include the effects of external magnetic fields by replacing $\boldsymbol{P}_\alpha$ by $[\boldsymbol{P}_\alpha - Z_\alpha e \boldsymbol{A}_\alpha^B]$ in equation (III.248) and the effects of an external electric field by addition of the term $\sum_\alpha Z_\alpha e \phi_\alpha$; this treatment is only really justified if the nuclei behave as Dirac particles. The nuclear Zeeman Hamiltonian is thus:

$$\mathcal{H}_Z^N = -e \sum_\alpha \frac{Z_\alpha}{M_\alpha} \boldsymbol{A}_\alpha^B \cdot \boldsymbol{P}_\alpha + \frac{e^2}{2} \sum_\alpha \frac{Z_\alpha^2}{M_\alpha} (A_\alpha^B)^2 + \text{order } (1/c^3) \quad (3.272)$$

where we have chosen $\boldsymbol{A}_\alpha^B$ to conform to a Coulomb gauge, $(\nabla_\alpha \cdot \boldsymbol{A}_\alpha) = 0$. The nuclear Stark Hamiltonian is simply

$$\mathcal{H}_E^N = \sum_\alpha Z_\alpha e \phi_\alpha. \quad (3.273)$$

We could now choose explicit vector and scalar potential functions as we did in section 3.7. However, when we come to perform the various coordinate transformations outlined in the previous section, it is more convenient to treat the Zeeman and Stark Hamiltonians for the molecule as a whole. Accordingly we combine the expressions from section 3.7 with equations (3.272) and (3.273) to give

$$\mathcal{H}_Z = \frac{e}{m} \sum_i \boldsymbol{A}_i^B \cdot \boldsymbol{P}_i + \sum_i g_S \mu_B \boldsymbol{S}_i \cdot (\nabla_i \wedge \boldsymbol{A}_i^B) - \frac{e}{2m^3 c^2} \sum_i (\boldsymbol{A}_i^B \cdot \boldsymbol{P}_i) P_i^2$$
$$- \frac{g_S \mu_B}{2m^2 c^2} \sum_i \boldsymbol{S}_i \cdot (\nabla_i \wedge \boldsymbol{A}_i^B) P_i^2 + \frac{e^2}{2m} \sum_i (A_i^B)^2 - e \sum_\alpha \frac{Z_\alpha}{M_\alpha} \boldsymbol{A}_\alpha^B \cdot \boldsymbol{P}_\alpha$$
$$+ \frac{e^2}{2} \sum_\alpha \frac{Z_\alpha^2}{M_\alpha} (A_\alpha^B)^2, \quad (3.274)$$

and

$$\mathcal{H}_E = \sum_i -e\phi_i + \sum_\alpha Z_\alpha e \phi_\alpha - \frac{g_S \mu_B}{2mc^2} \sum_i \boldsymbol{S}_i \cdot \{(\nabla_i \phi_i) \wedge \boldsymbol{P}_i\}. \quad (3.275)$$

We deal with the Zeeman Hamiltonian first. We must first choose an origin for the vector potentials; this origin is completely arbitrary (indeed, this is the physical significance of gauge invariance). In our expansions above we have selected a Coulomb gauge so we adopt the obvious choice

$$A_i^B = (1/2)(\mathbf{B} \wedge \mathbf{R}_i), \quad A_\alpha^B = (1/2)(\mathbf{B} \wedge \mathbf{R}_\alpha), \tag{3.276}$$

where the potentials are referred to the arbitrary origin that we introduced in chapter 2. We now proceed to make the coordinate transformations

$$\mathbf{R}_i = \mathbf{r}_i + \mathbf{R}_O - \frac{m}{M} \sum_i \mathbf{r}_i, \tag{3.277}$$

$$\mathbf{R}_\alpha = \mp \frac{\mu}{M_\alpha} \mathbf{R} + \mathbf{R}_O - \frac{m}{M} \sum_i \mathbf{r}_i, \tag{3.278}$$

and the associated momentum transformations

$$\mathbf{P}_i = \frac{m}{M} \mathbf{P}_O + \mathbf{p}_i, \tag{3.279}$$

$$\mathbf{P}_\alpha = \frac{M_\alpha}{M} \mathbf{P}_O \mp \mathbf{P}_R - \frac{M_\alpha}{(M_1 + M_2)} \sum_i \mathbf{p}_i. \tag{3.280}$$

We deal with the terms in (3.274) in turn. The first term gives

$$\frac{e}{m} \sum_i A_i^B \cdot \mathbf{P}_i = \frac{e}{2m} \mathbf{B} \cdot \left\{ \frac{m}{M} \sum_i \left(\mathbf{r}_i - \frac{m}{M} \sum_j \mathbf{r}_j + \mathbf{R}_O \right) \wedge \mathbf{P}_O \right.$$
$$\left. + \sum_i \left(\mathbf{r}_i - \frac{m}{M} \sum_j \mathbf{r}_j + \mathbf{R}_O \right) \wedge \mathbf{p}_i \right\}. \tag{3.281}$$

The second and fourth terms in equation (3.274) lead directly to the interaction between the electron spins and the external field and its relativistic correction respectively:

$$\sum_i g_S \mu_B \mathbf{S}_i \cdot \left(\nabla_i \wedge A_i^B \right) - \sum_i \frac{g_S \mu_B}{2m^2 c^2} \mathbf{S}_i \cdot \left(\nabla_i \wedge A_i^B \right) \mathbf{P}_i^2$$
$$= \sum_i g_S \mu_B \mathbf{S}_i \cdot \mathbf{B} \left\{ 1 - \frac{(\mathbf{p}_i)^2}{2m^2 c^2} \right\} + \text{smaller terms.} \tag{3.282}$$

The third term in (3.274) gives the corresponding relativistic correction to the interaction between the field and the orbital motion of the electrons:

$$-\frac{e}{2m^3 c^2} \sum_i (A_i^B \cdot \mathbf{P}_i) \mathbf{P}_i^2 = -\frac{e}{4m^3 c^2} \mathbf{B} \cdot \sum_i (\mathbf{r}_i \wedge \mathbf{p}_i) p_i^2 + \text{smaller terms.} \tag{3.283}$$

The fifth term represents the electronic contribution to molecular diamagnetism

$$\frac{e^2}{2m} \sum_i (A_i^B)^2 = \frac{e^2}{8m} \sum_i \{ \mathbf{B}^2 (\mathbf{r}_i + \mathbf{R}_O)^2 - (\mathbf{B} \cdot (\mathbf{r}_i + \mathbf{R}_O))^2 \} + \text{smaller terms.} \tag{3.284}$$

Turning now to the sixth term in equation (3.274) we obtain the results,

$$-e \sum_\alpha \frac{Z_\alpha}{M_\alpha} A_\alpha^B \cdot P_\alpha$$

$$= -\frac{e}{2M} B \cdot \left\{ \mu \left(-\frac{Z_1}{M_1} + \frac{Z_2}{M_2} \right) R \wedge P_O - (Z_1 + Z_2) \left(\frac{m}{M} \sum_j r_j - R_O \right) \wedge P_O \right\}$$

$$-\frac{e}{2} B \cdot \left\{ \sum_\alpha \frac{Z_\alpha}{M_\alpha^2} \mu R \wedge P_R + \left(\frac{Z_1}{M_1} - \frac{Z_2}{M_2} \right) \left(\frac{m}{M} \sum_j r_j - R_O \right) \wedge P_R \right\}$$

$$-\frac{e}{2(M_1 + M_2)} B \cdot \left\{ \mu \left(\frac{Z_1}{M_1} - \frac{Z_2}{M_2} \right) R \wedge \sum_i p_i \right.$$

$$\left. + (Z_1 + Z_2) \left(\frac{m}{M} \sum_j r_j - R_O \right) \wedge \sum_i p_i \right\}. \quad (3.285)$$

Finally, the last term in (3.274) gives the nuclear contribution to the molecular diamagnetism:

$$\frac{e^2}{2} \sum_\alpha \frac{Z_\alpha^2}{M_\alpha} (A_\alpha^B)^2 = \frac{e^2}{4} \sum_\alpha \frac{Z_\alpha^2}{M_\alpha} \left\{ B^2 \left(\mp \frac{\mu}{M_\alpha} R + R_O \right)^2 \right.$$

$$\left. - \left(B \cdot \left(\mp \frac{\mu}{M_\alpha} R + R_O \right) \right)^2 \right\} + \text{smaller terms.} \quad (3.286)$$

We now collect together terms in P_O, P_R and p_i. We shall need the electric dipole moment operator which we define by

$$\mu_e = -e \left\{ \sum_i r_i + \mu \left(\frac{Z_1}{M_1} - \frac{Z_2}{M_2} \right) R \right\}, \quad (3.287)$$

and a residual charge number by

$$q = \sum_\alpha Z_\alpha - n, \quad (3.288)$$

where n is the number of electrons. Thus for a neutral molecule q is zero and we obtain from (3.281) and (3.285) the results given below:

$$\frac{e}{m} \sum_i A_i^B \cdot P_i + e \sum_\alpha \frac{1}{M_\alpha} A_\alpha^B \cdot P_\alpha$$

$$= -\frac{1}{2M} B \cdot (\mu_e \wedge P_O) - \frac{e\mu}{2} B \cdot \sum_\alpha \frac{Z_\alpha}{M_\alpha^2} (R \wedge P_R) - \frac{e}{2} \left(\frac{Z_1}{M_1} - \frac{Z_2}{M_2} \right)$$

$$\times B \cdot \left\{ \left(\frac{m}{M} \sum_j r_j - R_O \right) \wedge P_R \right\} + \frac{e}{2m} B \cdot \sum_i \left(r_i - \frac{m}{M} \sum_j r_j \right) \wedge p_i$$

$$+ \frac{e}{2m} B \cdot \left(R_O \wedge \sum_i p_i \right) - \frac{e}{2(M_1 + M_2)} B \cdot \sum_i \left\{ \mu \left(\frac{Z_1}{M_1} - \frac{Z_2}{M_2} \right) R \wedge p_i \right.$$

$$\left. + \sum_\alpha Z_\alpha \left(\frac{m}{M} \sum_j r_j - R_O \right) \wedge p_i \right\}. \quad (3.289)$$

The first term of (3.289) represents a translational 'Stark' effect. A molecule with a permanent dipole moment experiences a moving magnetic field as an electric field and hence shows an interaction; the term could equally well be interpreted as a Zeeman effect. The second term represents the nuclear rotation and vibration Zeeman interactions; we shall deal with this more fully below. The fourth term gives the interaction of the field with the orbital motion of the electrons and its small polarisation correction. The other terms are probably not important but are retained to preserve the gauge invariance of the Hamiltonian. For an ionic species ($q \neq 0$) we have the additional translational term

$$\frac{e}{2M} q \, \boldsymbol{B} \cdot \left\{ \left(\frac{m}{M} \sum_j \boldsymbol{r}_j - \boldsymbol{R}_O \right) \wedge \boldsymbol{P}_O \right\}. \quad (3.290)$$

We now simplify the term which describes the interaction between the external field and the rotational and vibrational motion of the nuclei. We have from equation (2.73) in chapter 2:

$$\boldsymbol{P}_R = -\frac{\hbar}{R} \boldsymbol{k} \wedge (\boldsymbol{J} - \boldsymbol{P}) - i\hbar \boldsymbol{k} \frac{\partial}{\partial R}. \quad (3.291)$$

We also need the result

$$\boldsymbol{R} \wedge [\boldsymbol{k} \wedge (\boldsymbol{J} - \boldsymbol{P})] = \boldsymbol{k}[\boldsymbol{R} \cdot (\boldsymbol{J} - \boldsymbol{P})] - (\boldsymbol{k} \cdot \boldsymbol{R})(\boldsymbol{J} - \boldsymbol{P}) = -R(\boldsymbol{J} - \boldsymbol{P}), \quad (3.292)$$

in which we have made use of the fact that the component of $(\boldsymbol{J} - \boldsymbol{P})$ along the molecule-fixed z axis is zero. Thus we have

$$-\frac{e\mu}{2} \boldsymbol{B} \cdot \sum_\alpha \frac{Z_\alpha}{M_\alpha^2} (\boldsymbol{R} \wedge \boldsymbol{P}_R) = -\frac{e\hbar\mu}{2} \sum_\alpha \frac{Z_\alpha}{M_\alpha^2} \boldsymbol{B} \cdot (\boldsymbol{J} - \boldsymbol{P}). \quad (3.293)$$

This term describes the rotational Zeeman effect, that is, the coupling between the external field and the magnetic moment of the rotating nuclei. We note that there is no corresponding vibrational contribution since $\boldsymbol{R} \wedge \boldsymbol{k}$ is zero. The physical reason for this lack is that it is not possible to generate vibrational angular momentum in a diatomic molecule because it possesses only one, non-degenerate, vibrational mode.

In summary we have the total Zeeman Hamiltonian for a neutral molecule:

$$\mathcal{H}_Z = g_S \mu_B \boldsymbol{B} \cdot \sum_i \boldsymbol{s}_i \left\{ 1 - \frac{p_i^2}{2m^2c^2} \right\} + \frac{e}{2m} \boldsymbol{B} \cdot \sum_i \left(\boldsymbol{r}_i - \frac{m}{M} \sum_j \boldsymbol{r}_j \right) \wedge \boldsymbol{p}_i$$

$$- \frac{e}{4m^3c^2} \boldsymbol{B} \cdot \sum_i (\boldsymbol{r}_i \wedge \boldsymbol{p}_i) p_i^2 - \frac{1}{2M} \boldsymbol{B} \cdot (\boldsymbol{\mu}_e \wedge \boldsymbol{P}_O) - \frac{e\hbar\mu}{2} \sum_\alpha \frac{Z_\alpha}{M_\alpha^2} \boldsymbol{B} \cdot (\boldsymbol{J} - \boldsymbol{P})$$

$$- \frac{e}{2} \left(\frac{Z_1}{M_1} - \frac{Z_2}{M_2} \right) \boldsymbol{B} \cdot \left\{ \left(\frac{m}{M} \sum_j \boldsymbol{r}_j - \boldsymbol{R}_O \right) \wedge \boldsymbol{P}_R \right\} + \frac{e}{2m} \boldsymbol{B} \cdot \left\{ \boldsymbol{R}_O \wedge \sum_i \boldsymbol{p}_i \right\}$$

$$- \frac{e}{2(M_1 + M_2)} \boldsymbol{B} \cdot \sum_i \left\{ \mu \left(\frac{Z_1}{M_1} - \frac{Z_2}{M_2} \right) \boldsymbol{R} \wedge \boldsymbol{p}_i \right.$$

$$\left. + \sum_\alpha Z_\alpha \left(\frac{m}{M} \sum_j \boldsymbol{r}_j - \boldsymbol{R}_O \right) \wedge \boldsymbol{p}_i \right\} + \mathcal{H}_{\text{diam}}, \quad (3.294)$$

where $\mathcal{H}_{\text{diam}}$ is the sum of (3.284) and (3.286).

The Stark Hamiltonian is more straightforward. We use the scalar potentials

$$\phi_i = -\boldsymbol{E} \cdot \boldsymbol{R}_i, \qquad \phi_\alpha = -\boldsymbol{E} \cdot \boldsymbol{R}_\alpha, \qquad (3.295)$$

where $\boldsymbol{E}$ is the applied electric field intensity. We substitute equations (3.277), (3.278) and (3.295) into (3.275) and obtain

$$\mathcal{H}_E = -\boldsymbol{E} \cdot \boldsymbol{\mu}_e - qe\boldsymbol{E} \cdot \left(\boldsymbol{R}_O - \frac{m}{M} \sum_j \boldsymbol{r}_j \right) + \frac{g_S \mu_B}{2mc} \sum_i \boldsymbol{s}_i \cdot (\boldsymbol{E} \wedge \boldsymbol{p}_i). \qquad (3.296)$$

The first term in (3.296) is the usual Stark interaction while the second term vanishes for a neutral molecule. We have met the third term before in section 3.4; one physical explanation of its occurence is that a moving spin magnetic moment creates an electric moment perpendicular to both its direction and its velocity which interacts with the applied electric field.

3.13. Conclusions

In this chapter our aim has been to derive a Hamiltonian for the electronic motion of a diatomic molecule, starting from first principles. Our 'first principles' have been the Dirac equation for a single particle, and the Breit equation for two interacting particles, but we have nevertheless met a number of limitations; a satisfactory derivation of the molecular Hamiltonian must use the methods of quantum electrodynamics. The derivation represented here is therefore a compromise, but a not too unhappy one, since the resulting Hamiltonian is undoubtedly accurate enough to provide a quantitative interpretation of most spectroscopic investigations. In any case we shall encounter further difficulties in the next chapter when we deal with nuclear spin effects, although the methods expounded in this chapter will again provide an acceptable final Hamiltonian.

The most important effect arising from quantum electrodynamics, which is quantitatively significant, is a correction to the electron spin g factor. Up to now the symbol g_S has been taken to represent the Dirac g factor for the electron and to have the value of 2 exactly. As we mentioned earlier, quantum electrodynamics shows the existence of radiative corrections to this value, and the best theoretical value of g_S, which agrees with experiment [19], is 2.002 32.

In conclusion we summarise the total Hamiltonian (excluding nuclear spin effects), written in a molecule-fixed rotating coordinate system with origin at the nuclear centre of mass, for a diatomic molecule with electron spin quantised in the molecular axis system. We number the terms sequentially, and then describe their physical significance. The Hamiltonian is as follows:

$$\mathcal{H}_{\text{total}} = \frac{1}{2M} \boldsymbol{P}_O^2 \qquad :(1)$$

$$+ \frac{1}{2m} \sum_i \boldsymbol{p}_i^2 \qquad :(2)$$

$$+\frac{1}{2(M_1+M_2)}\sum_{i,j}\bm{p}_i\cdot\bm{p}_j \qquad :(3)$$

$$-\frac{1}{8m^3c^2}\sum_i\bm{p}_i^4 \qquad :(4)$$

$$+\frac{\hbar^2}{2\mu R^2}(\bm{J}-\bm{P})^2 \qquad :(5)$$

$$-\frac{\hbar^2}{2\mu R^2}\frac{\partial}{\partial R}\left(R^2\frac{\partial}{\partial R}\right) \qquad :(6)$$

$$+\sum_{i,j\neq i}\frac{e^2}{8\pi\varepsilon_0 r_{ij}} \qquad :(7)$$

$$-\sum_{i,\alpha}\frac{Z_\alpha e^2}{4\pi\varepsilon_0 r_{\alpha i}} \qquad :(8)$$

$$+\frac{Z_1 Z_2 e^2}{4\pi\varepsilon_0 R} \qquad :(9)$$

$$+\frac{e^2\hbar^2}{8\varepsilon_0 m^2 c^2}\sum_i\left\{\sum_\alpha Z_\alpha\delta(\bm{r}_{\alpha i})-\sum_{j\neq i}\delta(\bm{r}_{ji})\right\} \qquad :(10)$$

$$-\frac{g_S\mu_B}{8\pi\varepsilon_0 mc^2}\sum_i \bm{s}_i\cdot\left\{\sum_\alpha\frac{Z_\alpha e}{r_{\alpha i}^3}(\bm{r}_{\alpha i}\wedge\bm{p}_i)-\sum_{j\neq i}\frac{e}{r_{ji}^3}(\bm{r}_{ji}\wedge\bm{p}_i)\right\} \qquad :(11)$$

$$+\frac{g_S\mu_B e}{4\pi\varepsilon_0 mc^2}\sum_{i,j\neq i}\frac{1}{r_{ji}^3}\bm{s}_i\cdot(\bm{r}_{ij}\wedge\bm{p}_i) \qquad :(12)$$

$$-\frac{1}{4\pi\varepsilon_0}\left(\frac{e}{2mc}\right)^2\sum_{i,j\neq i}\left\{\bm{p}_i\frac{1}{r_{ij}}\cdot\bm{p}_j+(\bm{p}_i\cdot\bm{r}_{ji})\frac{1}{r_{ji}^3}(\bm{r}_{ji}\cdot\bm{p}_j)\right\} \qquad :(13)$$

$$+\frac{g_S\mu_B e}{4\pi\varepsilon_0 c^2}\sum_{i,\alpha}\frac{Z_\alpha}{(M_1+M_2)}\frac{1}{r_{\alpha i}^3}\bm{s}_i\cdot\left(\bm{r}_{i\alpha}\wedge\sum_j\bm{p}_j\right) \qquad :(14)$$

$$-\frac{e^2}{8\pi\varepsilon_0 mc^2}\sum_{i,\alpha}\frac{Z_\alpha}{(M_1+M_2)}\left\{\bm{p}_i\frac{1}{r_{ji}}\cdot\sum_j\bm{p}_j+(\bm{p}_i\cdot\bm{r}_{ji})\frac{1}{r_{ji}^3}\left(\bm{r}_{ji}\cdot\sum_j\bm{p}_j\right)\right\} \qquad :(15)$$

$$+\frac{g_S^2\mu_B^2}{8\pi\varepsilon_0 c^2}\sum_{i,j\neq i}\left\{\frac{\bm{s}_i\cdot\bm{s}_j}{r_{ji}^3}-\frac{3(\bm{s}_i\cdot\bm{r}_{ji})(\bm{s}_j\cdot\bm{r}_{ji})}{r_{jl}^5}-\frac{8\pi}{3}\delta(\bm{r}_{ji})\bm{s}_i\cdot\bm{s}_j\right\} \qquad :(16)$$

$$-\frac{g_S\mu_B e\hbar}{4\pi\varepsilon_0 c^2}\frac{1}{\mu R^2}\sum_i\bm{s}_i\cdot\left\{\frac{Z_1 M_2}{(M_1+M_2)}\frac{\bm{r}_{i1}}{r_{i1}^3}-\frac{Z_2 M_1}{(M_1+M_2)}\frac{\bm{r}_{i2}}{r_{i2}^3}\right\}\wedge\{\bm{R}\wedge(\bm{J}-\bm{P})\} \qquad :(17)$$

$$-\frac{g_S\mu_B e}{4\pi\varepsilon_0 c^2}\sum_i\bm{s}_i\cdot\left\{\left[\frac{Z_1}{M_1}\frac{\bm{r}_{i1}}{r_{i1}^3}-\frac{Z_2}{M_2}\frac{\bm{r}_{i2}}{r_{i2}^3}\right]\wedge\bm{k}\right\}i\hbar\frac{\partial}{\partial R} \qquad :(18)$$

$$+ \frac{e^2\hbar}{8\pi\varepsilon_0 mc^2} \frac{1}{\mu R^2} \sum_i \mathbf{p}_i \cdot \left\{ \frac{Z_1 M_2}{(M_1+M_2)} \left(\frac{1}{r_{i1}} + \mathbf{r}_{i1} \frac{1}{r_{i1}^3} \mathbf{r}_{i1} \right) \right.$$
$$\left. - \frac{Z_2 M_1}{(M_1+M_2)} \left(\frac{1}{r_{i2}} + \mathbf{r}_{i2} \frac{1}{r_{i2}^3} \mathbf{r}_{i2} \right) \right\} \cdot \mathbf{R} \wedge (\mathbf{J} - \mathbf{P}) \qquad :(19)$$

$$+ \frac{e^2}{8\pi\varepsilon_0 mc^2} \sum_i \mathbf{p}_i \cdot \left\{ \frac{Z_1}{M_1} \left(\frac{1}{r_{i1}} + \mathbf{r}_{i1} \frac{1}{r_{i1}^3} \mathbf{r}_{i1} \right) - \frac{Z_2}{M_2} \left(\frac{1}{r_{i2}} + \mathbf{r}_{i2} \frac{1}{r_{i2}^3} \mathbf{r}_{i2} \right) \right\} \cdot ki\hbar \frac{\partial}{\partial \mathbf{R}}$$
$$\qquad :(20)$$

$$+ g_S \mu_B \mathbf{B} \cdot \sum_i \mathbf{s}_i \left\{ 1 - \frac{p_i^2}{2m^2 c^2} \right\} \qquad :(21)$$

$$+ \frac{e}{2m} \mathbf{B} \cdot \sum_i \left\{ \mathbf{r}_i - \frac{m}{M} \sum_j \mathbf{r}_j \right\} \wedge \mathbf{p}_i \qquad :(22)$$

$$- \frac{e}{4m^3 c^2} \mathbf{B} \cdot \sum_i (\mathbf{r}_i \wedge \mathbf{p}_i) p_i^2 \qquad :(23)$$

$$- \frac{1}{2M} \mathbf{B} \cdot (\boldsymbol{\mu}_e \wedge \mathbf{P}_O) \qquad :(24)$$

$$- \frac{e\hbar\mu}{2} \sum_\alpha \frac{Z_\alpha}{M_\alpha^2} \mathbf{B} \cdot (\mathbf{J} - \mathbf{P}) \qquad :(25)$$

$$- \frac{e}{2} \left(\frac{Z_1}{M_1} - \frac{Z_2}{M_2} \right) \mathbf{B} \cdot \left\{ \left(\frac{m}{M} \sum_j \mathbf{r}_j - \mathbf{R}_O \right) \wedge \mathbf{P}_R \right\} + \frac{e}{2m} \mathbf{B} \cdot \left\{ \mathbf{R}_O \wedge \sum_i \mathbf{p}_i \right\}$$

$$- \frac{e}{2(M_1+M_2)} \mathbf{B} \cdot \sum_i \left\{ \mu \left(\frac{Z_1}{M_1} - \frac{Z_2}{M_2} \right) \mathbf{R} \wedge \mathbf{p}_i \right.$$
$$\left. + \sum_\alpha Z_\alpha \left(\frac{m}{M} \sum_j \mathbf{r}_j - \mathbf{R}_O \right) \wedge \mathbf{p}_i \right\} \qquad :(26)$$

$$+ \frac{e}{2M} q \mathbf{B} \cdot \left\{ \left(\frac{m}{M} \sum_j \mathbf{r}_j - \mathbf{R}_O \right) \wedge \mathbf{P}_O \right\} \qquad :(27)$$

$$- \mathbf{E} \cdot \boldsymbol{\mu}_e \qquad :(28)$$

$$- qe\mathbf{E} \cdot \left(\mathbf{R}_O - \frac{m}{M} \sum_j \mathbf{r}_j \right) \qquad :(29)$$

$$+ \frac{g_S \mu_B}{2mc} \sum_i \mathbf{s}_i \cdot (\mathbf{E} \wedge \mathbf{p}_i) \qquad :(30)$$

$$+ \frac{e^2}{8m} \sum_i \{ B^2 (\mathbf{r}_i + \mathbf{R}_O)^2 - (\mathbf{B} \cdot (\mathbf{r}_i + \mathbf{R}_O))^2 \} \qquad :(31)$$

$$+ \frac{e^2}{4} \sum_\alpha \frac{Z_\alpha^2}{M_\alpha} \left\{ B^2 \left(\mp \frac{\mu}{M_\alpha} \mathbf{R} + \mathbf{R}_O \right)^2 - \left(\mathbf{B} \cdot \left(\mp \frac{\mu}{M_\alpha} \mathbf{R} + \mathbf{R}_O \right) \right)^2 \right\}. \qquad :(32)$$

$$(3.297)$$

As before, this Hamiltonian can be converted into a Hamiltonian for a diatomic molecule with space-quantised spin by making the replacements,

$$s_i \to S_i, \quad (J - P) \to (N - L). \tag{3.298}$$

The thirty-two terms in equation (3.297) have the following physical significance.

(1) Translational kinetic energy of the whole molecule.
(2) Kinetic energy of the electrons.
(3) Mass polarisation correction to the electron kinetic energy.
(4) Relativistic correction to the electron kinetic energy.
(5) Rotational kinetic energy of the nuclei.
(6) Vibrational kinetic energy.
(7) Electrostatic Coulomb potential energy between pairs of electrons.
(8) Electron–nuclear Coulomb potential energy.
(9) Electrostatic Coulomb potential energy between the nuclei.
(10) Darwin-type correction to the Coulomb potential.
(11) Spin–orbit coupling.
(12) Spin-other-orbit coupling.
(13) Orbit–orbit coupling.
(14) Mass polarisation correction to the spin–orbit coupling.
(15) Mass polarisation correction to the orbit–orbit coupling.
(16) Spin–spin interaction.
(17) Spin–rotation interaction.
(18) Spin–vibration interaction.
(19) Orbit–rotation interaction.
(20) Orbit–vibration interaction.
(21) Electron spin Zeeman interaction with relativistic correction.
(22) Orbital Zeeman interaction.
(23) Relativistic correction to the orbital Zeeman interaction.
(24) Translational 'Stark' effect.
(25) Rotational Zeeman interaction.
(26) Additional small Zeeman terms.
(27) Translational 'Stark' effect for a charged molecule.
(28) Stark effect.
(29) Electrostatic interaction for a charged molecule.
(30) Interaction of the spin-induced electric moment with an applied electric field.
(31) Electronic contribution to molecular diamagnetism.
(32) Nuclear contribution to molecular diamagnetism.

Appendix 3.1. Power series expansion of the transformed Hamiltonian

In section 3.4 the transformed Hamiltonian is expressed as a formal power series in $1/c$ so that the limitations imposed by neglect of specifically quantum electrodynamic effects can be assessed in a later section. From other considerations it is perhaps more useful to order the terms according to their magnitude, that is, according to the exponent n in the expression $(V/c)^n mc^2$, where V is a typical electron velocity. The terms in the original Dirac Hamiltonian can be classified by comparison with a simple model (e.g. the hydrogen atom), yielding

$$\beta mc^2 \approx (V/c)^0 mc^2$$
$$c\boldsymbol{\alpha} \cdot \boldsymbol{\pi} \approx (V/c)^1 mc^2$$
$$e\phi \approx (V/c)^2 mc^2 \qquad (3.299)$$
$$e\boldsymbol{\alpha} \cdot \boldsymbol{A} \approx (V/c)^3 mc^2.$$

The orders of magnitude of the terms in the transformed Hamiltonian are then easily calculated from their explicit form. This order of magnitude scheme is readily related to the power series expansion; thus $(V/c)^4 mc^2$ corresponds to the $1/c^2$ term, etc.

References

[1] P.A.M. Dirac, *The Principles of Quantum Mechanics*, Oxford University Press, 1958.
[2] L.I. Schiff, *Quantum Mechanics*, McGraw-Hill Book Company, Inc., 1949.
[3] M.E. Rose, *Relativistic Electron Theory*, John Wiley and Sons, Inc., 1961.
[4] R.E. Moss, *Advanced Molecular Quantum Mechanics*, Chapman and Hall, London, 1973.
[5] K. Hannabuss, *An Introduction to Quantum Theory*, Oxford University Press, Oxford, 1997.
[6] E. Schrödinger, *Ann. Physik*, **79**, 489 (1926).
[7] O. Klein, *Z. Physik*, **37**, 895 (1926).
[8] W. Gordon, *Z. Physik*, **40**, 117 (1926).
[9] V. Fock, *Z. Physik*, **38**, 242 (1926); **39**, 226 (1926).
[10] P.A.M. Dirac, *Proc. R. Soc. Lond.*, **A117**, 610 (1928); **A118**, 351 (1928).
[11] C.D. Anderson, *Phys. Rev.*, **43**, 491 (1933).
[12] L.L. Foldy and S.A. Wouthuysen, *Phys. Rev.*, **78**, 29 (1950); S. Tani, *Progr. Theoret. Phys.*, **6**, 267 (1957).
[13] C.G. Darwin, *Proc. R. Soc. Lond.*, **A118**, 654 (1928).
[14] E. Schrödinger, *Berlin Ber.*, 419 (1930); 63 (1931).
[15] T. Itoh, *Rev. Mod. Phys.*, **37**, 159 (1965).
[16] G. Breit, *Phys. Rev.*, **34**, 553 (1929); **36**, 383 (1930); **39**, 616 (1932).
[17] Z.V. Chraplyvy, *Phys. Rev.*, **91**, 388 (1953); **92**, 1310 (1953).
[18] H.A. Bethe and E.E. Salpeter, *Quantum Mechanics of One- and Two-Electron Atoms*, Springer-Verlag, 1957.
[19] S. Koenig, A.G. Pradell and P. Kusch, *Phys. Rev.*, **88**, 191 (1952).

4 Interactions arising from nuclear magnetic and electric moments

4.1. Nuclear spins and magnetic moments

In the course of developing a Hamiltonian for diatomic molecules, we have so far introduced and discussed two nuclear properties. We considered at length the nuclear kinetic energy in chapter 2, and in chapter 3 we took account of the nuclear charge in considering the potential energy arising from the electrostatic interaction between electrons and nuclei. With respect to the electrostatic interaction, however, we have implicitly treated the nucleus as an electric monopole, and this assumption is re-examined in section 4.4. First, however, we consider another important property of many nuclei, namely their spin and the important magnetic interactions within a molecule which arise from the property of nuclear spin. The possibility that a nucleus may have a spin and an associated magnetic moment was first postulated by Pauli [1], following the observation of unexpected structure in atomic spectra. The first quantitative theory of the interaction between a nuclear magnetic moment and the 'outer' electrons of an atom was provided by Fermi [2], Hargreaves [3], Breit and Doermann [4] and Fermi and Segrè [5]. In the case of diatomic molecules with closed shell electronic states, the magnetic interaction of the nuclear moment with the magnetic angular momentum vector, an $I \cdot J$ coupling, was treated by a number of authors [6, 7, 8]. The interaction between the nuclear electric quadrupole moment and the electronic charges, an interaction which has nothing to do with nuclear spins or magnetic moments, was treated by Bardeen and Townes [9]. The most important and pioneering theoretical study of the magnetic interactions between nuclei and electron spins in diatomic molecules was described by Frosch and Foley [10]. Their treatment, which was based upon the Dirac equation, is described in the next section.

The main relevant facts concerning nuclei are as follows.

(a) Many nuclei have spin angular momentum $\hbar I$, where I is dimensionless. The value of the spin I is defined as the maximum possible component of I in any given direction.
(b) The nuclear spin I is half-integral (strictly half-odd) if the mass number of the nucleus is odd, and integral (including zero) if the mass number is even.

(c) A nucleus with spin I has a resulting magnetic moment μ_I given by

$$\mu_I = \gamma_I \hbar I = g_I \mu_N I, \tag{4.1}$$

where γ_I and g_I are called the nuclear gyromagnetic ratio and nuclear g-factor respectively. μ_N is the nuclear magneton and is equal to $e\hbar/2M_p$, where M_p is the proton mass.

(d) A nucleus of spin I has $2I+1$ allowed orientations with respect to any chosen direction. These are distinguished by different values of the quantum number M_I which measures the component of I along an arbitrary space-fixed axis.

(e) A nucleus with spin I equal to 1 or greater possesses an electric quadrupole moment, i.e. the nuclear charge distribution departs from spherical symmetry.

General Appendix B lists the spins, magnetic moments and electric moments of all the known naturally occurring nuclei. The molecular spectroscopist is usually prepared to accept such data as part of the starting point in any investigation, and a detailed discussion of nuclear structure theory is certainly beyond the scope of this book and its authors. It is, however, of interest to note that many of the observations can be rationalised in terms of the shell theory of the nucleus. According to this theory the nucleons (neutrons and protons) are grouped in energy shells in a manner analogous to the grouping of electrons in atoms. There appears to be particular stability associated with 2, 8, 50, 82 and 126 protons or neutrons which can be rationalised in terms of nucleon shells 1s, 1p, 1d, 2s, etc. An important difference in the description of nuclear structure compared with atoms is that there are two types of particle to be put into the shells, protons and neutrons, whereas there is only one for atoms, namely, the electron. Thus the first filled-shell nucleus, with 2 protons and 2 neutrons, is the α-particle, ^{4}He; this is the nuclear analogue of the helium atom which contains two electrons. Both species are particularly stable in comparison with their neighbours. The observed nuclear spin is the resultant of the coupling of the nucleon angular momenta; in other words, the nuclear spin is the total angular momentum of the nucleus. If there is an even number of either neutrons or protons the angular momenta compensate each other with zero resultant. If, however, there is an odd number of either neutrons or protons, the nuclear spin results from the uncompensated angular momentum of a single nucleon. Finally, with an odd number of neutrons and an odd number of protons it is necessary to take account of both the spin and orbital angular momenta of the nucleons within the nucleus, and several resultants are possible and observed. The magnitude of the nuclear magnetic moment can be readily calculated if the total number of nucleons is odd, the calculation and formulae being very similar to the Landé formulae for atoms.

As an example, let us consider the deuteron for a moment. It is built up from one proton and one neutron, each of which has a spin of 1/2. When these two angular momenta couple through nuclear interactions, they give rise to a resultant angular momentum of 1 or 0. The former state, in which the angular momenta are parallel, is lower in energy and so constitutes the ground state. Thus we see that the deuteron has a nuclear spin of 1. There is also an excited state, with lower binding energy, in which the angular momenta are antiparallel ($I = 0$). However, the energy required to access this state is many orders of magnitude greater than that encountered in molecular physics. In fact, the interactions in this state are so weak that it is not even bound.

In conclusion it should be noted that detailed study of the magnetostatics of an assembly of nuclei and electrons shows that only 2^ℓ-pole nuclear magnetic moments, where ℓ is odd, are allowed. In contrast to this we shall show later that in the expansion of the electrostatic interaction, it is the 2^ℓ-pole electric moments with ℓ even which are non-vanishing. The most important nuclear magnetic moment is, of course, the magnetic dipole moment. The next highest moment ($\ell = 3$), which is non-vanishing if $I \geq 3/2$, is expected to be very small, but evidence for nuclear octupole magnetic moments in iodine, indium and gallium atoms has been obtained. In molecular spectroscopy, however, it appears at present that we may safely confine our attention to nuclear magnetic dipole moments.

4.2. Derivation of nuclear spin magnetic interactions through the magnetic vector potential

We now show how the many-electron Hamiltonian developed in the previous chapter may be extended to include magnetic interactions which arise from the presence of nuclear spin magnetic moments. Equation (3.140) represents the Hamiltonian for electron i in the presence of other electrons; we present it again here:

$$\mathcal{H}_i = -e_i\phi_i + \frac{\pi_i^2}{2m_i} + g_S\mu_B S_i \cdot (\nabla_i \wedge A_i^e) - \frac{\pi_i^4}{8m_i^3 c^2} + \frac{e_i \hbar^2}{8m_i^2 c^2}(\nabla_i \cdot \varepsilon_i)$$
$$+ \frac{g_S\mu_B}{2m_i c^2} S_i \cdot (\varepsilon_i \wedge \pi_i). \tag{4.2}$$

The effects of the other electrons were introduced through the magnetic vector potential A_i^e defined by equation (3.133), and also through the electrostatic scalar potential (3.134). Similarly the effects of external magnetic fields were included by adding an additional term A_i^B to the vector potential, given in (3.162). It therefore follows that the effects of nuclear magnetic moments may be incorporated by adding a further contribution A_i^N to the magnetic vector potential. This was indeed the approach introduced by Breit and Doermann [4]. Thus, the total magnetic vector potential A_i at electron i is given by

$$A_i = A_i^e + A_i^B + A_i^N \tag{4.3}$$

where, by (3.133) and (3.162),

$$A_i^e = \frac{g_S\mu_B}{4\pi\varepsilon_0 c^2} \sum_{j \neq i}(S_j \wedge R_{ji})R_{ji}^{-3} - \frac{e}{8\pi\varepsilon_0 mc^2} \sum_{j \neq i}\left\{P_j \cdot \frac{1}{R_{ji}} + (P_j \cdot R_{ji})\frac{1}{R_{ji}^3}R_{ji}\right\} \tag{4.4}$$

$$A_i^B = \frac{1}{2}(B \wedge R_i), \tag{4.5}$$

$$A_i^N = \frac{-\mu_N}{4\pi\varepsilon_0 c^2} \sum_\alpha g_\alpha (\boldsymbol{I}_\alpha \wedge \boldsymbol{R}_{\alpha i}) R_{\alpha i}^{-3}. \tag{4.6}$$

As before, the vector $\boldsymbol{R}_{\alpha i}$ gives the position of the nucleus α relative to electron i.

In writing equation (4.6), we have assumed that the nuclei can be treated as Dirac particles, that is, particles which are described by the Dirac equation and behave in the same way as electrons. This is a fairly desperate assumption because it suggests, for example, that all nuclei have a spin of $1/2$. This is clearly not correct; a wide range of values, integral and half-integral, is observed in practice. Furthermore, nuclei with integral spins are bosons and do not even obey Fermi–Dirac statistics. Despite this, if we proceed on the basis that the nuclei are Dirac particles but that most of them have anomalous spins, the resultant theory is not in disagreement with experiment. If the problem is treated by quantum electrodynamics, the approach can be shown to be justified provided that only terms of order (nuclear mass)$^{-1}$ are retained.

Let us therefore re-examine the lowest-order terms in (3.140) containing the magnetic vector potential, including the effects of nuclear spin and an applied magnetic field, but excluding A_i^e whose consequences were investigated fully in chapter 3. The important terms are

$$\sum_i \left\{ \frac{\pi_i^2}{2m} + g_S \mu_B \boldsymbol{S}_i \cdot (\nabla_i \wedge \boldsymbol{A}_i) \right\} = \frac{1}{2m} \sum_i (\boldsymbol{P}_i + e\boldsymbol{A}_i^B + e\boldsymbol{A}_i^N)^2$$
$$+ g_S \mu_B \sum_i \boldsymbol{S}_i \cdot (\nabla_i \wedge \boldsymbol{A}_i^B) + g_S \mu_B \sum_i \boldsymbol{S}_i \cdot (\nabla_i \wedge \boldsymbol{A}_i^N). \tag{4.7}$$

On expansion of the first term in (4.7) we obtain

$$\frac{1}{2m} \sum_i (\boldsymbol{P}_i + e\boldsymbol{A}_i^B + e\boldsymbol{A}_i^N)^2$$
$$= \frac{1}{2m} \sum_i \boldsymbol{P}_i^2 + \frac{e^2}{2m} \sum_i (\boldsymbol{A}_i^B)^2 + \frac{e^2}{2m} \sum_i (\boldsymbol{A}_i^N)^2 + \frac{e}{m} \sum_i \boldsymbol{A}_i^B \cdot \boldsymbol{P}_i$$
$$+ \frac{e}{m} \sum_i \boldsymbol{A}_i^N \cdot \boldsymbol{P}_i + \frac{e^2}{m} \sum_i \boldsymbol{A}_i^B \cdot \boldsymbol{A}_i^N. \tag{4.8}$$

Three of these terms have, of course, been examined already in the previous chapter. These are the first term, representing the electron kinetic energy, the second term representing the diamagnetic energy of the electrons, and the fourth term, which yields an expression for the interaction of the applied field $\boldsymbol{B}$ with the electron orbital magnetic moment, i.e. (3.164),

$$\frac{e}{m} \sum_i \boldsymbol{A}_i^B \cdot \boldsymbol{P}_i = \frac{e}{2m} \sum_i (\boldsymbol{B} \wedge \boldsymbol{R}_i) \cdot \boldsymbol{P}_i = \frac{e}{2m} \sum_i \boldsymbol{B} \cdot (\boldsymbol{R}_i \wedge \boldsymbol{P}_i). \tag{4.9}$$

The remaining terms in (4.8) are new, however, and we develop each of them in turn. The third term gives, on expansion,

$$\frac{e^2}{2m} \sum_i (\boldsymbol{A}_i^N)^2 = \frac{e^2 \mu_N^2}{2m(4\pi\varepsilon_0 c^2)^2} \sum_{i,\alpha,\alpha'} g_\alpha g_{\alpha'} \frac{(\boldsymbol{I}_\alpha \wedge \boldsymbol{R}_{\alpha i})}{R_{\alpha i}^3} \cdot \frac{(\boldsymbol{I}_{\alpha'} \wedge \boldsymbol{R}_{\alpha' i})}{R_{\alpha' i}^3}. \tag{4.10}$$

This expression represents a coupling between the nuclear spins which we consider again in due course. We merely note in passing that it is of order c^{-4} and thus strictly smaller than the acceptable magnitude of c^{-2} adopted in the previous chapter. The fifth term in (4.8) gives

$$\frac{e}{m}\sum_i A_i^N \cdot P_i = -\frac{e\mu_N}{4\pi m\varepsilon_0 c^2}\sum_{i,\alpha} g_\alpha R_{\alpha i}^{-3}(I_\alpha \wedge R_{\alpha i}) \cdot P_i$$
$$= -\frac{e\mu_N}{4\pi m\varepsilon_0 c^2}\sum_{i,\alpha} g_\alpha I_\alpha \cdot (R_{\alpha i} \wedge P_i)R_{\alpha i}^{-3}, \quad (4.11)$$

which describes interactions between the nuclear moments and the electron orbital magnetic moment. The sixth and last term in (4.8) yields

$$\frac{e^2}{m}\sum_i A_i^B \cdot A_i^N = -\frac{e^2\mu_N}{8\pi m\varepsilon_0 c^2}\sum_{i,\alpha}(B \wedge R_i) \cdot (I_\alpha \wedge R_{\alpha i})R_{\alpha i}^{-3}g_\alpha, \quad (4.12)$$

which represents coupling between the nuclear spin magnetic moments and the magnetic field arising from currents induced by the orbital precession of the electrons in the applied magnetic field B. We note that this term is formally of order c^{-3}.

So much for the first term on the right-hand side of in equation (4.7). The second term is a familiar one and yields the interaction between B and the electron spin magnetic moment. The third term is new, however, and may be developed in much the same manner as the corresponding term involving A_i^e in chapter 3 (see (3.146) to (3.151)). We have

$$g_S\mu_B \sum_i S_i \cdot \left(\nabla_i \wedge A_i^N\right)$$
$$= -\frac{g_S\mu_B\mu_N}{4\pi\varepsilon_0 c^2}\sum_{i,\alpha} g_\alpha S_i \cdot \left\{\nabla_i \wedge (I_\alpha \wedge R_{\alpha i})R_{\alpha i}^{-3}\right\}$$
$$= -\frac{g_S\mu_B\mu_N}{4\pi\varepsilon_0 c^2}\sum_{i,\alpha} g_\alpha S_i \cdot \left\{I_\alpha(\nabla_i \cdot R_{\alpha i})R_{\alpha i}^{-3} - (I_\alpha \cdot \nabla_i)R_{\alpha i}R_{\alpha i}^{-3}\right\}$$
$$= -\frac{g_S\mu_B\mu_N}{4\pi\varepsilon_0 c^2}\sum_{i,\alpha} g_\alpha S_i \cdot \left\{-4\pi I_\alpha \delta^{(3)}(R_{\alpha i}) + I_\alpha R_{\alpha i}^{-3}\right.$$
$$\left. - 3(I_\alpha \cdot R_{\alpha i})R_{\alpha i}R_{\alpha i}^{-5} + I_\alpha(4\pi/3)\delta^{(3)}(R_{\alpha i})\right\}$$
$$= \sum_{i,\alpha}\frac{g_S\mu_B g_\alpha \mu_N}{4\pi\varepsilon_0 c^2}(8\pi/3)\delta^{(3)}(R_{\alpha i})S_i \cdot I_\alpha$$
$$- \sum_{i,\alpha}\frac{g_S\mu_B g_\alpha \mu_N}{4\pi\varepsilon_0 c^2}\left\{\frac{S_i \cdot I_\alpha}{R_{ui}^3} - \frac{3(S_i \cdot R_{\alpha i})(I_\alpha \cdot R_{\alpha i})}{R_{\alpha i}^5}\right\}. \quad (4.13)$$

This is a very important result. The first term in the last line of (4.13) represents the so-called Fermi contact interaction between the electron and nuclear spin magnetic moments, and the second term is the electron–nuclear dipolar coupling, analogous to the electron–electron dipolar coupling derived previously in (3.151). The Fermi contact interaction occurs only when the electron and nucleus occupy the same position in Euclidean space, as required by the Dirac delta function $\delta^{(3)}(R_{\alpha i})$. This seemingly impossible requirement is a wholly relativistic effect; the two particles do not have to occupy the same position in four-dimensional space. The dipolar coupling term on the other hand is much more familiar; it corresponds to the classical dipole–dipole interaction.

It will be recalled that in section 3.10 we developed the nuclear Hamiltonian by calculating the magnetic vector potential A_α^e at nucleus α arising from the spin and orbital motion of the electrons. Clearly we should now also include the nuclear spin contribution to A_α, the complete magnetic vector potential being (3.249) plus the additional term from the other nucleus α',

$$A_\alpha^N = -\frac{\mu_N g_{\alpha'}}{4\pi\varepsilon_0 c^2}(\bm{I}_{\alpha'} \wedge \bm{R}_{\alpha\alpha'})R_{\alpha\alpha'}^{-3}, \tag{4.14}$$

where $\alpha' \neq \alpha$.

Hence, the nuclear kinetic energy is given by

$$\sum_\alpha \frac{1}{2M_\alpha}\{\bm{P}_\alpha - Z_\alpha e \bm{A}_\alpha^N\}^2 = \sum_\alpha \frac{1}{2M_\alpha}\bm{P}_\alpha^2 - \sum_\alpha \frac{Z_\alpha e}{M_\alpha}\bm{A}_\alpha^N \cdot \bm{P}_\alpha + \cdots$$

$$= \sum_\alpha \frac{1}{2M_\alpha}\bm{P}_\alpha^2 + \sum_\alpha \frac{Z_\alpha e}{M_\alpha 4\pi\varepsilon_0 c^2}g_{\alpha'}\mu_N R_{\alpha\alpha'}^{-3}\bm{I}_{\alpha'} \cdot (\bm{R}_{\alpha\alpha'} \wedge \bm{P}_\alpha).$$

$$\tag{4.15}$$

The second term in (4.15) above yields terms representing coupling between the rotational and vibrational motions of the nuclei and the nuclear spin magnetic moments. These terms will become more explicit when we later transform to new coordinates, analogous to those used in section 3.11.

There remain two other important magnetic interactions involving the nuclear spin magnetic moments, which cannot be derived from the present analysis, although their presence is reasonably self evident by analogy with corresponding electron spin terms which we have derived earlier. They are the nuclear Zeeman interaction,

$$\mathcal{H} = -\sum_\alpha g_\alpha \mu_N \bm{I}_\alpha \cdot \bm{B} \tag{4.16}$$

and the direct dipole–dipole interaction between the nuclear spin magnetic moments. The latter is analogous to the electron–electron (3.151) and electron–nuclear (4.13) dipolar couplings and may be written, for a diatomic molecule,

$$\mathcal{H} = \frac{g_1\mu_N g_2\mu_N}{4\pi\varepsilon_0 c^2}\left\{\frac{\bm{I}_1 \cdot \bm{I}_2}{R^3} - \frac{3(\bm{I}_1 \cdot \bm{R})(\bm{I}_2 \cdot \bm{R})}{R^5}\right\}, \tag{4.17}$$

where g_1 and g_2 are the nuclear g factors of nuclei 1 and 2 respectively. We shall show later how these terms may, in fact, be derived from the Breit equation.

In summary the complete nuclear spin Hamiltonian is given by

$$\mathcal{H}(\bm{I}_\alpha) = -\sum_\alpha g_\alpha \mu_N \bm{I}_\alpha \cdot \bm{B} \qquad \text{:nuclear Zeeman interaction}$$

$$+ \frac{g_1\mu_N g_2\mu_N}{4\pi\varepsilon_0 c^2}\left\{\frac{\bm{I}_1 \cdot \bm{I}_2}{R^3} - \frac{3(\bm{I}_1 \cdot \bm{R})(\bm{I}_2 \cdot \bm{R})}{R^5}\right\} \qquad \text{:direct nuclear–nuclear dipole coupling}$$

$$+ \sum_{i,\alpha} \frac{g_S \mu_B g_\alpha \mu_N}{4\pi\varepsilon_0 c^2} (8\pi/3) \delta^{(3)}(\boldsymbol{R}_{\alpha i}) \boldsymbol{S}_i \cdot \boldsymbol{I}_\alpha \quad \text{:electron–nuclear Fermi contact interaction}$$

$$- \sum_{i,\alpha} \frac{g_S \mu_B g_\alpha \mu_N}{4\pi\varepsilon_0 c^2} \left\{ \frac{\boldsymbol{S}_i \cdot \boldsymbol{I}_\alpha}{R_{\alpha i}^3} - \frac{3(\boldsymbol{S}_i \cdot \boldsymbol{R}_{\alpha i})(\boldsymbol{I}_\alpha \cdot \boldsymbol{R}_{\alpha i})}{R_{\alpha i}^5} \right\} \quad \text{:electron–nuclear dipole coupling}$$

$$- \frac{e \mu_N}{4\pi m \varepsilon_0 c^2} \sum_{i,\alpha} g_\alpha \boldsymbol{I}_\alpha \cdot (\boldsymbol{R}_{\alpha i} \wedge \boldsymbol{P}_i) R_{\alpha i}^{-3} \quad \text{:nuclear spin–electron orbital magnetic interaction}$$

$$- \frac{e^2 \mu_N}{8\pi m \varepsilon_0 c^2} \sum_{i,\alpha} g_\alpha (\boldsymbol{B} \wedge \boldsymbol{R}_i) \cdot (\boldsymbol{I}_\alpha \wedge \boldsymbol{R}_{\alpha i}) R_{\alpha i}^{-3} \quad \text{:interaction between nuclear moments and fields arising from diamagnetic electron currents}$$

$$+ \frac{e^2 \mu_N^2}{2m(4\pi\varepsilon_0 c^2)^2} \sum_{i,\alpha,\alpha'} g_\alpha g_{\alpha'} \frac{(\boldsymbol{I}_\alpha \wedge \boldsymbol{R}_{\alpha i})}{R_{\alpha i}^3} \cdot \frac{(\boldsymbol{I}_{\alpha'} \wedge \boldsymbol{R}_{\alpha' i})}{R_{\alpha' i}^3} \quad \text{:high-order nuclear–nuclear dipole coupling}$$

$$+ \sum_{\alpha \neq \alpha'} \frac{\mp Z_\alpha e}{4 M_\alpha \pi \varepsilon_0 c^2} g_{\alpha'} \mu_N R^{-3} \boldsymbol{I}_{\alpha'} \cdot (\boldsymbol{R} \wedge \boldsymbol{P}_\alpha) \quad \text{:nuclear spin–nuclear rotation and vibration coupling.} \qquad (4.18)$$

As elsewhere, the upper sign choice in the last contribution is for nucleus 1 and the lower for nucleus 2. The last three terms in equation (4.18) are, strictly speaking, too small to be completely reliable. This point is discussed in more detail in the next section.

The reader who is familiar with nuclear magnetic resonance spectroscopy will notice the absence in (4.18) of certain well-known terms, for example, the electron-coupled nuclear spin–spin interaction, and the correction to the nuclear Zeeman interaction known as the 'chemical shift'. The reason is that these are terms which belong to an *effective* Hamiltonian, obtained after integration over electronic spatial coordinates. We shall discuss the effective Hamiltonian in detail in chapter 7, but it might be worthwhile at this stage to indicate some of the developments so far as equation (4.18) is concerned. We shall find that the nuclear Zeeman interaction in molecules can be represented by a term in the effective Hamiltonian of the form $\boldsymbol{I} \cdot \boldsymbol{\sigma} \cdot \boldsymbol{B}$ where $\boldsymbol{\sigma}$ is a second-rank tensor, called the screening tensor (not to be confused with the σ matrices of chapter 3). The fifth term in (4.18) gives a first-order diamagnetic contribution to σ, whilst a cross-term between the electron orbital Zeeman interaction (4.9) and the sixth term in (4.18) gives, in second order, a paramagnetic contribution to the screening tensor. Nuclear spin–nuclear spin interaction terms in the effective Hamiltonian come from the second and seventh terms of (4.18) in the first order, the Fermi contact interaction alone in second-order, and various less important cross-terms in second order.

It should be noted that the interaction terms derived in this section are expressed in a space-fixed axis system of arbitrary origin. We shall later investigate the results of coordinate transformations analogous to those described in the previous two chapters. Our analysis is essentially the same as that presented by Frosch and Foley [10] in their determination of the electron spin–nuclear spin interaction terms.

4.3. Derivation of nuclear spin interactions from the Breit equation

It is possible to obtain the nuclear spin magnetic interaction terms by starting from the Breit equation. We recall that the Breit Hamiltonian describes the interaction of two electrons of spin 1/2, each of which may be separately represented by a Dirac Hamiltonian:

$$\mathcal{H} = \beta_1 m_1 c^2 - e_1\phi_1 + c\boldsymbol{\alpha}_1 \cdot (\boldsymbol{P}_1 + e_1 \boldsymbol{A}_1) + \beta_2 m_2 c^2 - e_2\phi_2 + c\boldsymbol{\alpha}_2 \cdot (\boldsymbol{P}_2 + e_2 \boldsymbol{A}_2)$$
$$+ \frac{e_1 e_2}{4\pi\varepsilon_0 R} - \frac{e_1 e_2}{8\pi\varepsilon_0 R}\left\{ \boldsymbol{\alpha}_1 \cdot \boldsymbol{\alpha}_2 + \frac{(\boldsymbol{\alpha}_1 \cdot \boldsymbol{R})(\boldsymbol{\alpha}_2 \cdot \boldsymbol{R})}{R^2} \right\}. \quad (4.19)$$

In chapter 3 we showed how the relativistic Breit Hamiltonian can be reduced to non-relativistic form by means of a Foldy–Wouthysen transformation. We obtained equations (3.244) and (3.245) which represent the non-relativistic Hamiltonian for two particles of masses m_i and m_j and electrostatic charges $-e_i$ and $-e_j$ and from this Hamiltonian we were able to derive the interelectronic interactions. We could, however, consider using (3.244) and (3.245) as the Hamiltonian for an electron of charge $-e_i = -e$ and mass $m_i = m$, and a nucleus of mass $m_j = M_\alpha$ and charge $-e_j = +Z_\alpha e$. As before, we make the assumption that the nucleus has spin 1/2, behaves like a Dirac particle and has an anomalous magnetic moment compared with that given by the Dirac theory. Consequently we may rewrite (3.245) by making the replacements

$$\frac{e_i \hbar}{m_i} = g_S \mu_B, \quad (4.20)$$

$$\frac{e_j \hbar}{m_j} = -g_\alpha \mu_N, \quad (4.21)$$

$$\boldsymbol{\sigma}_i = 2\boldsymbol{S}_i, \quad \boldsymbol{\sigma}_j = 2\boldsymbol{I}_\alpha. \quad (4.22)$$

Confining attention to those terms in (3.245) which contain $\boldsymbol{\sigma}_j$ and which are of order M_α^{-1} we obtain expressions for the electron–nuclear interactions which are

$$-\frac{e}{4\pi\varepsilon_0 mc^2} g_\alpha \mu_N \frac{1}{R_{\alpha i}^3} \boldsymbol{I}_\alpha \cdot (\boldsymbol{R}_{\alpha i} \wedge \boldsymbol{\pi}_i) - \frac{g_S \mu_B g_\alpha \mu_N}{4\pi\varepsilon_0 c^2}\left\{ \frac{\boldsymbol{S}_i \cdot \boldsymbol{I}_\alpha}{R_{\alpha i}^3} - \frac{3(\boldsymbol{S}_i \cdot \boldsymbol{R}_{\alpha i})(\boldsymbol{I}_\alpha \cdot \boldsymbol{R}_{\alpha i})}{R_{\alpha i}^5} \right\}$$
$$+ \frac{g_S \mu_B g_\alpha \mu_N}{4\pi\varepsilon_0 c^2}(8\pi/3)\delta^{(3)}(\boldsymbol{R}_{\alpha i})\boldsymbol{S}_i \cdot \boldsymbol{I}_\alpha. \quad (4.23)$$

(Other electron–nuclear interaction terms involving $\boldsymbol{\pi}_\alpha$ rather than $\boldsymbol{I}_\alpha$ arise from this treatment. However, these terms have all been dealt with in the previous chapter and we do not repeat them here.) The terms in (4.23) are the same as those obtained previously starting from the Dirac equation. Equation (3.244) will yield both the electron and nuclear Zeeman terms and a Breit equation for two nuclei, reduced to non-relativistic form, would yield the nuclear–nuclear interaction terms. Although many nuclei have spins other than 1/2, and even the proton with spin 1/2 has an anomalous magnetic moment which does not fit the simple Dirac theory, the approach outlined here is fully endorsed by quantum electrodynamics provided that only terms involving M_α^{-1} are retained (see equation (4.23)). The interested reader is referred to Bethe and Salpeter [11] for further details. In our present application we see that the expressions for both the direct dipole–dipole interaction between nuclear spin magnetic moments, and the nuclear spin–rotation interactions, involve M_α^{-2} and so are not completely reliable.

4.4. Nuclear electric quadrupole interactions

4.4.1. Spherical tensor form of the Hamiltonian operator

So far we have considered the magnetic interactions between the nuclear spin dipole moment and other magnetic dipole moments arising from electronic and nuclear motion in molecules. We now examine the electrostatic interactions which might be expected to exist between a nucleus, containing positively charged protons, and the surrounding negatively charged electrons (and also the other nucleus). We allow the nucleus to have a finite size and define the position vectors of the protons in the nucleus and the outer electrons relative to the centre of charge of the relevant nucleus. The classical electrostatic interactions involving the protons in the nucleus and the surrounding electrons are represented by a Hamiltonian containing an infinite series of Legendre polynomials,

$$\mathcal{H}_{el} = \frac{1}{4\pi\varepsilon_0} \sum_{i,p} \frac{e_i e_p}{|\boldsymbol{R}_i - \boldsymbol{R}_p|} = \frac{1}{4\pi\varepsilon_0} \sum_{i,p,\ell} e_i e_p \frac{R_p^\ell}{R_i^{\ell+1}} P_\ell(\cos\theta_{ip}). \tag{4.24}$$

The coordinates are illustrated in figure 4.1. e_p is the charge of the pth proton in the nucleus, having position vector $\boldsymbol{R}_p$, and e_i is the charge of the ith electron $(-e)$ with position vector $\boldsymbol{R}_i$ in the remainder of the atom or molecule. θ_{ip} is the angle between the vectors $\boldsymbol{R}_i$ and $\boldsymbol{R}_p$.

We now expand the Legendre polynomial $P_\ell(\cos\theta_{ip})$ using the spherical harmonic addition theorem,

$$P_\ell(\cos\theta_{ip}) = \frac{4\pi}{2\ell+1} \sum_q Y_{\ell q}(\theta_p, \phi_p)^* Y_{\ell q}(\theta_i, \phi_i), \tag{4.25}$$

where the polar angles θ_i, ϕ_i and θ_p, ϕ_p are defined in figure 4.1, and q takes all integer values from $+\ell$ to $-\ell$. If we introduce the definition

$$C_q^{(\ell)}(\theta, \phi) = \left(\frac{4\pi}{2\ell+1}\right)^{1/2} Y_{\ell q}(\theta, \phi), \tag{4.26}$$

then equation (4.25) may be rewritten in the form

$$P_\ell(\cos\theta_{ip}) = \sum_q C_q^{(\ell)}(\theta_p, \phi_p)^* C_q^{(\ell)}(\theta_i, \phi_i)$$

$$= \sum_q (-1)^q C_q^{(\ell)}(\theta_i, \phi_i) C_{-q}^{(\ell)}(\theta_p, \phi_p). \tag{4.27}$$

Hence the electrostatic interaction (4.24) may be expressed in the form

$$\mathcal{H}_{el} = \sum_{i,p,\ell} \frac{e_i e_p}{4\pi\varepsilon_0} \frac{R_p^\ell}{R_i^{\ell+1}} \sum_q (-1)^q C_q^{(\ell)}(\theta_i, \phi_i) C_{-q}^{(\ell)}(\theta_p, \phi_p). \tag{4.28}$$

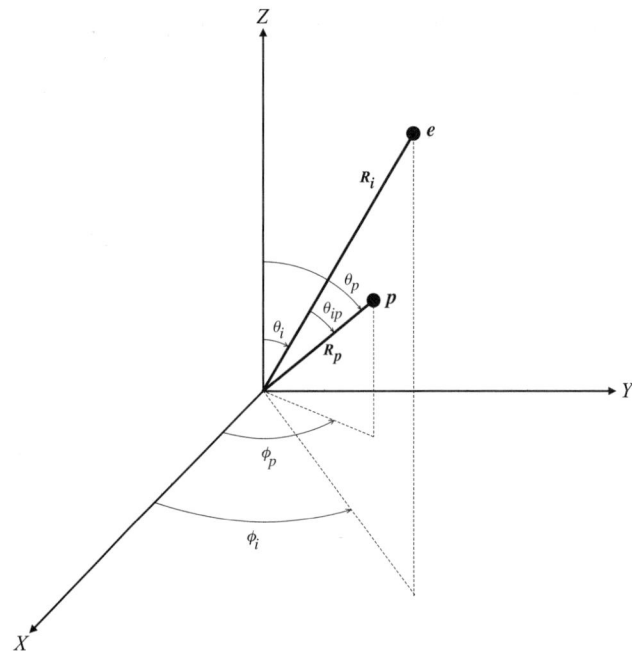

Figure 4.1. Electron and proton coordinates. X, Y, Z are space-fixed axes with origin at the nucleus.

The electrostatic quadrupole term in this sum is that for which $\ell = 2$, so that the quadrupole Hamiltonian is given by

$$\mathcal{H}_Q = \sum_{i,p,q} (-1)^q \frac{e_i e_p}{4\pi\varepsilon_0} \frac{R_p^2}{R_i^3} C_q^{(2)}(\theta_i, \phi_i) C_{-q}^{(2)}(\theta_p, \phi_p). \tag{4.29}$$

This may be written as the scalar product of two second-rank irreducible tensors,

$$\mathcal{H}_Q = -e \mathbf{T}^2(\nabla \boldsymbol{E}) \cdot \mathbf{T}^2(\boldsymbol{Q}), \tag{4.30}$$

where the first of these,

$$\mathbf{T}^2(\nabla \boldsymbol{E}) = -\frac{1}{4\pi\varepsilon_0} \sum_i \frac{e_i}{R_i^3} \boldsymbol{C}^2(\theta_i, \phi_i), \tag{4.31}$$

involves electron and proton coordinates *outside* the nucleus only, and defines what is known as the electric field gradient. The other second-rank tensor in (4.30),

$$e\mathbf{T}^2(\boldsymbol{Q}) = e \sum_p R_p^2 \boldsymbol{C}^2(\theta_p, \phi_p), \tag{4.32}$$

involves proton coordinates *inside* the nucleus only, and defines the quadrupole moment of the nucleus. Spherical tensors are dealt with in detail in the next chapter.

Some authors use equation (4.30) to represent the quadrupole interaction, as we will in this book, and others prefer to use the form

$$\mathcal{H}_Q = \mathbf{T}^2(\nabla^2 V) \cdot \mathbf{T}^2(\boldsymbol{Q}), \tag{4.33}$$

which differs only by a factor of e because $\nabla^2 V = -\nabla \boldsymbol{E}$.

4.4.2. Cartesian form of the Hamiltonian operator

We now return to the spherical harmonic form of the quadrupole Hamiltonian, equation (4.29), which is

$$\mathcal{H}_Q = \sum_{i,p,q} (-1)^q \frac{e_i e_p}{4\pi\varepsilon_0} \frac{R_p^2}{R_i^3} C_q^{(2)}(\theta_i, \phi_i) C_{-q}^{(2)}(\theta_p, \phi_p). \tag{4.34}$$

The dominant term is that with $q = 0$; it may be rewritten in the form

$$\mathcal{H}_Q = -\frac{e^2}{8\pi\varepsilon_0} \sum_{i,p} \frac{R_p^2}{R_i^3} (3\cos^2\theta_{ip} - 1), \tag{4.35}$$

provided that we consider only the electronic contributions to the electric field gradient.

Equation (4.35) can now expanded in the Cartesian coordinate system defined in figure 4.1. We replace $R_p R_i \cos\theta_{ip}$ by $\sum_j X_{pj} X_{ij}$ where the X_j represent X, Y, Z. Hence equation (4.35) becomes

$$\mathcal{H}_Q = -\frac{e^2}{8\pi\varepsilon_0} \sum_{i,p} \frac{1}{R_i^5} \left\{ 3 \sum_{j,k} X_{ij} X_{ik} X_{pj} X_{pk} - R_i^2 R_p^2 \right\}. \tag{4.36}$$

This result may be written in the form

$$\mathcal{H}_Q = \frac{1}{6} \sum_{j,k} Q_{jk} V_{jk}, \tag{4.37}$$

where

$$Q_{jk} = \sum_p e(3 X_{pj} X_{pk} - \delta_{jk} R_p^2), \tag{4.38}$$

$$V_{jk} = -\sum_i \frac{e}{4\pi\varepsilon_0 R_i^5} (3 X_{ij} X_{ik} - \delta_{jk} R_i^2). \tag{4.39}$$

These results follow since

$$\frac{1}{6} \sum_{j,k} (3 X_{ij} X_{ik} - \delta_{jk} R_i^2)(3 X_{pj} X_{pk} - \delta_{jk} R_p^2)$$

$$= \sum_{j,k} \left\{ \frac{3}{2} X_{ij} X_{ik} X_{pj} X_{pk} - \frac{1}{2} X_{ij} X_{ik} \delta_{jk} R_p^2 - \frac{1}{2} X_{pj} X_{pk} \delta_{jk} R_i^2 + \frac{1}{6} \delta_{jk} R_p^2 R_i^2 \right\}$$

$$= \frac{3}{2} \sum_{j,k} X_{ij} X_{ik} X_{pj} X_{pk} - \frac{1}{2} R_i^2 R_p^2, \tag{4.40}$$

because $\sum_{j,k} \delta_{jk} = 3$ and $\sum_j X_{ij} X_{ij} = R_i^2$. Thus we see that, as in equation (4.30), we have reduced the quadrupole interaction to the product of two operators, one of which depends only on the coordinates of protons in the nucleus and the other of which depends only on the coordinates of electrons and protons outside that nucleus.

The Q_{jk} are components of a second-rank cartesian tensor called the nuclear quadrupole tensor, and the V_{jk} are components of the electric field gradient tensor. Both $\mathbf{Q}$ and $\mathbf{V}$ are traceless, symmetric tensors of the second rank. The electric field gradient tensor components can be written in a more compact form by noting that

$$\frac{\partial}{\partial X_j} \frac{\partial}{\partial X_k} \left(\frac{1}{R} \right) = \frac{(3 X_j X_k - \delta_{jk} R^2)}{R^5}. \tag{4.41}$$

This relationship may be demonstrated by considering the two possible cases, $j = k$ and $j \ne k$. First we let $X_j = X_k = X$; then

$$\begin{aligned}
\frac{\partial^2}{\partial X^2} \left(\frac{1}{R} \right) &= \frac{\partial}{\partial X} \frac{\partial}{\partial X} (X^2 + Y^2 + Z^2)^{-1/2} \\
&= -\frac{\partial}{\partial X} X (X^2 + Y^2 + Z^2)^{-3/2} \\
&= -(X^2 + Y^2 + Z^2)^{-3/2} + 3 X^2 (X^2 + Y^2 + Z^2)^{-5/2} \\
&= \frac{1}{R^5} (3 X^2 - R^2).
\end{aligned} \tag{4.42}$$

Second, we let $X_j = X$, $X_k = Y$; then

$$\frac{\partial}{\partial X} \frac{\partial}{\partial Y} (X^2 + Y^2 + Z^2)^{-1/2} = -\frac{\partial}{\partial X} \{Y(X^2 + Y^2 + Z^2)^{-3/2}\} = \frac{3XY}{R^5}. \tag{4.43}$$

Clearly these relationships can be established for all other derivatives, thereby verifying equation (4.40). Hence, we can write the field gradient tensor components in the form

$$V_{jk} = -\frac{1}{4\pi\varepsilon_0} \sum_i e \frac{\partial}{\partial X_{ij}} \frac{\partial}{\partial X_{ik}} \left(\frac{1}{R_i} \right). \tag{4.44}$$

We note that the quantity V_{jk} is the second derivative of the electrostatic potential which is actually the *negative* of the electric field gradient.

4.4.3. Matrix elements of the quadrupole Hamiltonian

In both of the above treatments, spherical tensor and cartesian, we have factored the quadrupole interaction into the product of two terms, one of which operates only on functions of proton coordinates within the nucleus and the other only on functions of coordinates of electrons and protons outside the nucleus. We shall see in subsequent chapters that the spherical tensor form is rather more convenient for the calculation of matrix elements of $\mathcal{H}_Q$. However, we shall find this easier to appreciate once we have considered some of the theory of angular momentum in chapter 5 so we defer discussion until later.

Both equations (4.30) and (4.37) are rather inconvenient for our purposes since their explicit evaluation demands that we treat the nucleus as a many particle system. In fact, these forms would allow us to treat problems of greater complexity than those encountered in molecular spectroscopy. In general, we shall only be concerned with the nucleus in its ground state, and it is only necessary to characterise the nuclear eigenstates in terms of the total angular momentum $\boldsymbol{I}$ and its $2I+1$ components in a given direction, M_I. Consequently, the nuclear quadrupole moment tensor components Q_{jk} in equation (4.38) or $\sum_p eR_p^2 C^2(\theta_p, \phi_p)$ in equation (4.32) can be expressed in an equivalent operator form which is adequate for our purposes. Although this subject is more properly part of the next chapter, we shall continue the development for the cartesian coordinates form, equation (4.37), here since in molecular spectroscopy the electrostatic interaction is always expressed in terms of nuclear spin operators, even though the interaction does not, in fact, involve the nuclear spin directly. The interaction depends upon $\boldsymbol{I}$ only to the extent that the angular dependence of the nuclear state is related to the total nuclear angular momentum $\boldsymbol{I}$.

The matrix elements which arise in molecular spectroscopy are always diagonal in I, but may be off-diagonal in M_I. It can be shown that the operator

$$\sum_{j,k}\left\{3\frac{(\boldsymbol{I})_j(\boldsymbol{I})_k+(\boldsymbol{I})_k(\boldsymbol{I})_j}{2}-\delta_{jk}\boldsymbol{I}^2\right\} \tag{4.45}$$

has the same angular dependence with respect to nuclear orientation as $\sum_{j,k} Q_{jk}$. Hence Q_{jk} may be replaced by

$$Q_{jk}=C\left\{3\frac{(\boldsymbol{I})_j(\boldsymbol{I})_k+(\boldsymbol{I})_k(\boldsymbol{I})_j}{2}-\delta_{jk}(\boldsymbol{I})^2\right\}, \tag{4.46}$$

where the constant C can be expressed in terms of a scalar quantity Q conventionally called the nuclear quadrupole moment. Q is defined by

$$\begin{aligned}eQ &= \langle I, M_I=I|\sum_p e(3Z_p^2 - R_p^2)|I, M=I\rangle \\ &= C\langle I, I|3(\boldsymbol{I})_z^2 - \boldsymbol{I}^2|I, I\rangle \\ &= C\{3I^2 - I(I+1)\} \\ &= CI(2I-1).\end{aligned} \tag{4.47}$$

Hence the nuclear quadrupole tensor components are given by

$$Q_{jk}=\frac{eQ}{I(2I-1)}\left\{3\frac{(\boldsymbol{I})_j(\boldsymbol{I})_k+(\boldsymbol{I})_k(\boldsymbol{I})_j}{2}-\delta_{jk}\boldsymbol{I}^2\right\}, \tag{4.48}$$

and the computation of matrix elements in a nuclear spin basis set is straightforward. The procedure described in this section, in which the matrix elements of the electric quadrupole operator are replaced by those of a combination of nuclear spin operators, is based on the replacement theorem. This theorem is justified and described in more detail in section 5.5.3.

We will return to the quadrupole interaction in following chapters, but we now re-examine the general expansion of the electrostatic interaction and, in particular, the possibility of other nuclear electrostatic multipole moments. Because our multipole expansion is performed in a coordinate system with origin at the centre of charge of the protons p in the nucleus, the nuclear electric dipole moment is zero. However, this result arises only from our choice of origin and we now show that there are much more general restrictions on the expectation values of electric multipole moments. First we assume that all nuclear eigenstates have a definite parity, an assumption which is supported by all the available evidence. It can then be shown from parity considerations that no odd (ℓ odd) nuclear electrostatic multipole moment can exist. If all possible nuclear states have eigenfunctions ψ_n that are multiplied by $+1$ or -1 on inversion of coordinates, it follows that the product of a particular function ψ_n and its complex conjugate ψ_n^* is unaltered. In spherical polar coordinates, the 2^ℓ-pole moment tensor operator is given by

$$Q_q^{(\ell)} = \sum_p e R_p^\ell C_q^{(\ell)}(\theta_p, \phi_p) \tag{4.49}$$

(see equation (4.32)) and its expectation value is

$$\int \psi_n^* \sum_p e R_p^\ell C_q^{(\ell)}(\theta_p, \phi_p) \psi_n \, d\tau_n. \tag{4.50}$$

For this integral to be non-zero, the integrand must be invariant to the inversion operation. We have shown above that the product $\psi_n^* \psi_n$ is invariant under inversion, and the changes in the spherical polar coordinates under this operation are $\theta \to \pi - \theta$, $\phi \to \pi + \phi$ and $R \to R$. Hence inversion of $C_q^{(\ell)}(\theta, \phi)$ gives

$$C_q^{(\ell)}(\pi - \theta, \pi + \phi) = (-1)^\ell C_q^{(\ell)}(\theta, \phi), \tag{4.51}$$

i.e. a change of sign when ℓ is odd. Hence the integral (4.50) vanishes if ℓ is odd and therefore no odd nuclear electrostatic multipole moment can exist.

So far as even values of ℓ are concerned, it can be further shown that for a nuclear spin I, the 2^ℓ-pole electrostatic moment is zero if ℓ is greater than $2I$. The expectation value of the 2^ℓ multipole moment is given in equation (4.50). In this equation ψ_n and ψ_n^* are eigenfunctions corresponding to angular momentum I, and since R_p^ℓ has no angular dependence, $R_p^\ell \psi_n^*$ is also an eigenfunction with angular momentum I. However, we shall see in chapter 5 that $C_q^{(\ell)}(\theta_p, \phi_p)$ corresponds to angular momentum ℓ; the product of $C_q^{(\ell)}$ and ψ_n must therefore correspond to any angular momentum between $\ell + I$ and $\ell - I$. Now the integral (4.50) vanishes unless $R_p^\ell \psi_n^*$ and $C_q^{(\ell)} \psi_n$ correspond to eigenfunctions with the same angular momentum eigenvalues and I must therefore lie between $\ell + I$ and $|\ell - I|$, i.e. it is necessary that ℓ be less than or equal to $2I$. Hence, a nucleus with an electric quadrupole moment must have a nuclear spin I of at least 1; the next highest electrostatic multipole moment is the hexadecapole moment ($\ell = 4$) for nuclei of $I = 2$ or more. Interactions involving the nuclear electrostatic hexadecapole moment are too small to be detected at present.

4.5. Transformation of coordinates for the nuclear magnetic dipole and electric quadrupole terms

The nuclear spin magnetic dipole interactions are listed in equation (4.18) in a space-fixed coordinate system of arbitrary origin. The two forms of the electric quadrupole interaction are given in equations (4.30) and (4.37); each is expressed in a space-fixed coordinate system with origin at the centre of the nucleus concerned, this choice of origin being necessary to ensure the validity of the multipole expansion (4.25) and the vanishing of odd electric multipole moments. In chapters 2 and 3, we showed that the electronic and nuclear Hamiltonian operators could be simplified somewhat by a change of variables. For the sake of consistency, we now apply the same transformation of coordinates to the terms derived in this chapter. Since the method follows exactly the same lines as those laid down in sections 3.11 and 3.12, we do not repeat the details here.

We deal with the effect of the transformations on the nuclear quadrupole term first, because it is straightforward. Examination of equation (4.30) or (4.37) shows that the interactions depend only on the relative coordinates of the electrons (or protons) and the nuclei, R_i. Thus, when the quadrupole interactions are expressed in the molecule-fixed rotating frame, they have exactly the same form, except that R_i is replaced by r_i. We also note in passing that when the cartesian field gradient tensor V_{jk}, given in equation (4.44), is expressed in a coordinate system which is instantaneously coincident with the molecule-fixed axes, the axial symmetry of the molecule results in only the diagonal elements being non-zero. For a diatomic molecule with z as the internuclear axis, the component V_{zz} is given by

$$V_{zz} = -\sum_i \frac{e}{4\pi\varepsilon_0} \frac{\partial^2}{\partial z^2}\left(\frac{1}{r_i}\right). \qquad (4.52)$$

It is often written simply as q and called, somewhat erroneously, the field gradient. The quantity $(V_{xx} - V_{yy})/V_{zz}$ measures the deviation from axial symmetry and is called the asymmetry parameter, η.

We are now in a position to write down the Hamiltonian operator for all nuclear spin and quadrupole terms for a diatomic molecule; we allow for the possibility that both nuclei are involved and therefore sum over the nuclear index α. The terms are expressed in a molecule-fixed rotating coordinate system with origin at the nuclear centre of mass, except that we retain I_α as being quantised in a space-fixed axis system. We number the terms sequentially and then describe their physical significance.

We split the Hamiltonian into a sum of two parts, $\mathcal{H}(I_\alpha)$ and $\mathcal{H}_Q$, where

$$\mathcal{H}(I_\alpha) = \frac{g_1 g_2 \mu_N^2}{4\pi\varepsilon_0 c^2}\left\{\frac{I_1 \cdot I_2}{R^3} - \frac{3(I_1 \cdot R)(I_2 \cdot R)}{R^5}\right\} \qquad :(1)$$

$$+ \sum_{i,\alpha} \frac{g_S \mu_B g_\alpha \mu_N}{4\pi\varepsilon_0 c^2}(8\pi/3)\delta^{(3)}(r_{\alpha i})s_i \cdot I_\alpha \qquad :(2)$$

$$- \sum_{i,\alpha} \frac{g_S \mu_B g_\alpha \mu_N}{4\pi\varepsilon_0 c^2}\left\{\frac{s_i \cdot I_\alpha}{r_{\alpha i}^3} - \frac{3(s_i \cdot r_{\alpha i})(I_\alpha \cdot r_{\alpha i})}{r_{\alpha i}^5}\right\} \qquad :(3)$$

$$- \frac{e\mu_N}{4\pi\varepsilon_0 mc^2}\sum_{i,\alpha} g_\alpha I_\alpha \cdot \left\{\frac{1}{r_{\alpha i}^3} r_{\alpha i} \wedge p_i\right\} \qquad :(4)$$

$$+ \frac{e\mu_N \hbar}{4\pi\varepsilon_0 c^2 R^3}\left\{\frac{Z_1 g_2}{M_1} I_2 + \frac{Z_2 g_1}{M_2} I_1\right\} \cdot (J - P) \qquad :(5)$$

$$+ \sum_{\alpha \neq \alpha'} \frac{\mp Z_\alpha e}{(M_1 + M_2) 4\pi\varepsilon_0 c^2} \frac{g_{\alpha'} \mu_N}{R^3} \mathbf{I}_{\alpha'} \cdot \left(\mathbf{R} \wedge \sum_i \mathbf{p}_i \right) \qquad :(6)$$

$$- \sum_{\alpha = 1,2} g_\alpha \mu_N \mathbf{I}_\alpha \cdot \mathbf{B} \qquad :(7)$$

$$- \frac{e^2 \mu_N}{8\pi m \varepsilon_0 c^2} \sum_{i,\alpha} \frac{g_\alpha}{r_{\alpha i}^3} (\mathbf{B} \wedge \mathbf{r}_i) \cdot (\mathbf{I}_\alpha \wedge \mathbf{r}_{\alpha i}). \qquad :(8)$$

(4.53)

$$\mathcal{H}_Q = -\frac{e^2}{4\pi\varepsilon_0} \sum_\alpha \sum_{c,p} \sum_q (-1)^q \frac{r_{p\alpha}^2}{r_{c\alpha}^3} C_q^{(2)}(\theta_{c\alpha}, \phi_{c\alpha}) C_{-q}^{(2)}(\theta_{p\alpha}, \phi_{p\alpha}). \qquad (4.54)$$

The eight terms in equation (4.53) have the following physical significance.

(1) Nuclear spin–nuclear spin dipolar interaction.
(2) Fermi contact interaction between the electron and nuclear spins.
(3) Electron spin–nuclear spin dipolar interaction.
(4) Nuclear spin–electron orbital interaction.
(5) Nuclear spin–rotation interaction.
(6) Mass polarisation correction to the nuclear spin–electron orbital interaction.
(7) Nuclear Zeeman interaction.
(8) Interaction between nuclear magnetic moments and fields arising from diamagnetic electron currents.

The term in equation (4.54) represents the nuclear electric quadrupole interaction.

We note that the nuclear spin–vibration interaction, anticipated in equation (4.15), is actually identically zero. This is because the only vibrational mode for a diatomic molecule is that associated with bond stretching. Such motion does not generate any angular momentum and so does not produce a magnetic field with which the nuclear spin can interact.

References

[1] W. Pauli, *Naturwiss.*, **12**, 741 (1924).
[2] E. Fermi, *Z. Physik*, **60**, 320 (1930).
[3] J. Hargreaves, *Proc. R. Soc. Lond.*, **A124**, 568 (1929); **A127**, 141, 407 (1930).
[4] G. Breit and F.W. Doermann, *Phys. Rev.*, **36**, 1732 (1930).
[5] E. Fermi and E. Segrè, *Z. Physik*, **82**, 729 (1933).
[6] J.M.B. Kellogg, I.I. Rabi, N.F. Ramsey and J.R. Zacharias, *Phys. Rev.*, **57**, 677 (1940).
[7] H. Zieger and D.I. Bolef, *Phys. Rev.*, **85**, 788, 799 (1952).
[8] H.M. Foley, *Phys. Rev.*, **72**, 504 (1947); G.C. Wick, *Phys. Rev.*, **73**, 51 (1948).
[9] J. Bardeen and C.H. Townes, *Phys. Rev.*, **73**, 97 (1948).
[10] R.A. Frosch and H.M. Foley, *Phys. Rev.*, **88**, 1337 (1952).
[11] H.A. Bethe and E.E. Salpeter, *Quantum Mechanics of One- and Two-Electron Atoms*, Springer-Verlag, Berlin, 1957.

5 Angular momentum theory and spherical tensor algebra

5.1. Introduction

Much of the beauty of high-resolution molecular spectroscopy arises from the patterns formed by the fine and hyperfine structure associated with a given transition. All of this structure involves angular momentum in some sense or other and its interpretation depends heavily on the proper description of such motion. Angular momentum theory is very powerful and general. It applies equally to rotations in spin or vibrational coordinate space as to rotations in ordinary three-dimensional space.

All the laws of physics are easier to accept (and even to understand) when the underlying symmetry of the problem is appreciated. For example, classical Euclidean space is isotropic and a physical system is invariant to any rotation in this space. By this we mean that all the measurable properties of the system are unaffected by the rotation. An investigation of the behaviour of a quantum state under such rotations allows the properties of the state to be defined. These properties are most succinctly expressed as quantum numbers. Although quantum numbers are frequently used to label the eigenstates or eigenvalues of a molecule, they really carry information about the symmetry properties of the associated eigenfunctions.

In this chapter we give only a brief description of angular momentum theory, sufficient to introduce the various techniques required for the description of molecular energy levels and the transitions between them. In particular, we will present the ideas of spherical tensor algebra, a discipline which allows considerable simplification of the theoretical model when there are several different angular momenta involved. Our treatment is neither rigorous nor complete. Fortunately there are several excellent text books on this beguiling topic already in existence [1, 2, 3, 4, 5, 6, 7]. The reader who wishes to know more should turn to them. The book on angular momentum by Zare [4] is nowadays the textbook of choice for workers in the field of molecular physics. There is also a companion volume [6] which lists the typographical errors in the original book [4] and contains a large number of useful practice problems.

In section 5.2, we introduce the symmetry operations and irreducible representations of the full rotation group. This is followed in section 5.3 by the application of these ideas to the rotation of a rigid body in space; to a first-order approximation, we can think of a molecule as a rigid body. In section 5.4, we explain how to deal with the situation in which there are two or more independent angular momenta which are coupled together by a physical interaction. This situation arises very often in this book. The scene is then set for the introduction of spherical tensor operators in section 5.5. This provides the framework for a very powerful method for the description of angular momenta in molecules, a method which is capable of dealing with the most complex situations imaginable. Some special aspects of the application of spherical tensor methods to molecular problems and molecule-fixed coordinate systems are dealt with at this point. Finally, in an appendix at the end of this chapter, we collect together all the useful spherical tensor relationships for the application to molecular problems. This has something of the character of a recipe book and indeed can be used as such with a little experience.

5.2. Rotation operators

5.2.1. Introduction

The symmetry group with which we are concerned in the discussion of angular momentum is the full rotation group. This group is defined by the infinite set of rotation operators $R(\alpha \hat{n})$ which cause a rotation through an angle α about an axis pointing in the direction of the unit vector $\hat{n}$. A positive rotation is taken to be one which causes a (right-handed!) corkscrew to advance along the positive $\hat{n}$ direction. We need also to define the physical meaning of $R(\alpha \hat{n})$ and there are two possible conventions, both unfortunately with their adherents. The operator $R(\alpha \hat{n})$ can either be considered as being applied to the system under discussion (the *active* convention) or to the coordinate axes used to describe it (the *passive* convention). In this book we shall use the active convention, that is that $R(\alpha \hat{n})$ means a rotation of the physical system through an angle α about the axis $\hat{n}$, which is equivalent to a rotation of the coordinate system through $-\alpha$. This is the convention which is used in most text books on the subject (Rose [2], Brink and Satchler [3], Zare [4]) but not all (most notably, not by Edmonds [1]).

The full rotation group is infinite and its operators are continuous. In this context, it is useful to define an operator $J_{\hat{n}}$ which produces an infinitesimally small rotation

$$J_{\hat{n}} = i \lim_{\alpha \to 0} \frac{R(\alpha \hat{n}) - 1}{\alpha}. \tag{5.1}$$

For small α this equation can be rewritten to make $R(\alpha \hat{n})$ the subject:

$$R(\alpha \hat{n}) = 1 - i\alpha J_{\hat{n}} + O(\alpha^2) + \cdots. \tag{5.2}$$

Since successive rotations about a given axis are additive, we can form the *finite rotation operator* $R(\alpha \hat{n})$ from a series of m infinitesimal rotations, each through an angle α/m

as follows:

$$R(\alpha\hat{n}) = \lim_{m\to\infty} \left(1 - \frac{i\alpha}{m} J_{\hat{n}}\right)^m$$

$$= \left(1 - i\alpha J_{\hat{n}} - \frac{1}{2!}\alpha^2 J_{\hat{n}}^2 + \frac{1}{3!}i\alpha^3 J_{\hat{n}}^3 + \cdots\right)$$

$$= \exp(-i\alpha J_{\hat{n}}). \tag{5.3}$$

Let us consider a rotation about the laboratory-fixed Z axis. The effect of $R(\alpha\hat{Z})$ on a general function f is to transform it to a new function f', related to f by

$$f'(\phi) = R(\alpha\hat{Z})f(\phi) = f(\phi - \alpha) \tag{5.4}$$

as shown in figure 5.1. It then follows from equation (5.1) that

$$J_Z f(\phi) = i \lim_{\alpha\to 0}\left\{\frac{f(\phi-\alpha) - f(\phi)}{\alpha}\right\} = -i\frac{\partial}{\partial\alpha} f(\phi). \tag{5.5}$$

Thus the infinitesimal rotation operator J_Z is the same as the familiar angular momentum operator $\hbar J_Z$ (note that we use *dimensionless* angular momentum operators in this book).

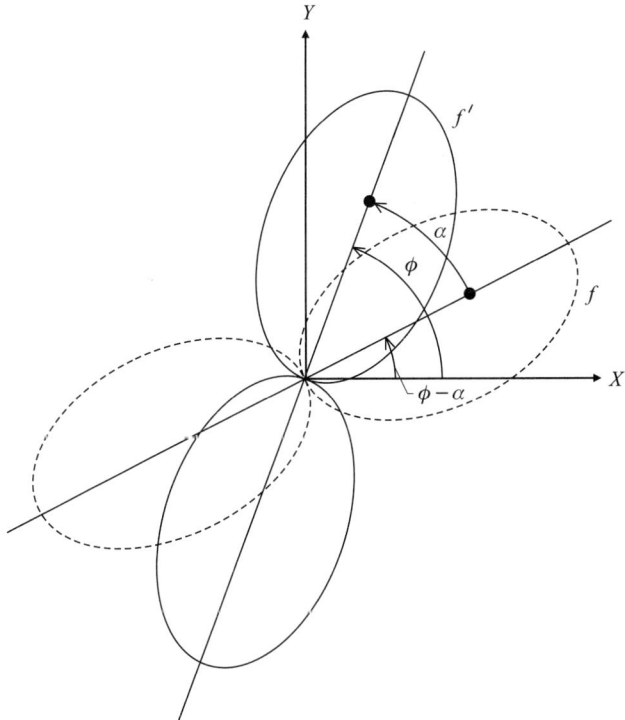

Figure 5.1. Rotation about the space-fixed Z axis, $R(\alpha\hat{Z})$, through an angle α of a function f to produce a new function f'. The value of the transformed function at a particular point with azimuthal angle ϕ is the same as that of the original function at azimuthal angle $(\phi - \alpha)$.

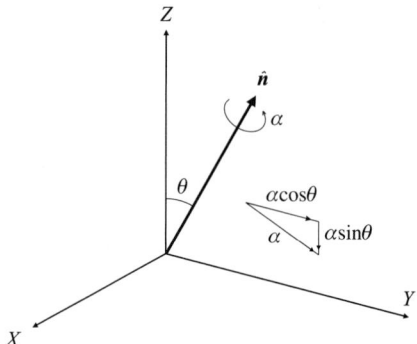

Figure 5.2. Rotation through α about $\hat{n}$, $R(\alpha\hat{n})$, in the space-fixed YZ plane.

5.2.2. Decomposition of rotational operators

Let us now consider a rotation $R(\alpha\hat{n})$ in which $\hat{n}$ is an axis in the YZ plane and α is very small (see figure 5.2). If θ is the angle between $\hat{n}$ and the Z axis, a rotation through α about $\hat{n}$ can be decomposed into a rotation through $\alpha\cos\theta$ about Z, followed by a rotation through $\alpha\sin\theta$ about Y (in either order to this level of approximation) so that

$$R(\alpha\hat{n}) = R([\alpha\sin\theta]\hat{Y})R([\alpha\cos\theta]\hat{Z}) + O(\alpha^2),$$
i.e. $1 - i\alpha J_{\hat{n}} = (1 - i\alpha\sin\theta J_Y)(1 - i\alpha\cos\theta J_Z) + O(\alpha^2),\quad (5.6)$
or $J_{\hat{n}} = J_Y\sin\theta + J_Z\cos\theta.$

A generalisation of this result allows us to express the infinitesimal rotation operator about any axis $\hat{n}$ in the form

$$J_{\hat{n}} = n_X J_X + n_Y J_Y + n_Z J_Z, \quad (5.7)$$

where (n_X, n_Y, n_Z) are the direction cosines of $\hat{n}$. Thus any infinitesimal rotation and hence any finite rotation can be expressed in terms of J_X, J_Y and J_Z. The transformation (5.7) demonstrates that $J_{\hat{n}}$ transforms as a vector.

5.2.3. Commutation relations

All rotations through a finite angle α form a single class of the full rotation group. Thus

$$SR(\alpha\hat{n})S^{-1} = R(\alpha S\hat{n}), \quad (5.8)$$

where S is a rotation which transforms the rotation axis $\hat{n}$ to a new one $\hat{n}' = S\hat{n}$. Thus for the situation in figure 5.2 with any α and θ, we have

$$R(\alpha\hat{n}) = R(-\theta\hat{X})R(\alpha\hat{Z})R(\theta\hat{X}), \quad (5.9)$$

or

$$R(\theta\hat{X})R(\alpha\hat{n}) = R(\alpha\hat{Z})R(\theta\hat{X}). \quad (5.10)$$

Expanding equation (5.10) as a power series in the infinitesimal rotations, we obtain

$$(1 - i\theta J_X + \cdots)(1 - i\alpha[J_Y \sin\theta + J_Z \cos\theta] + \cdots) = (1 - i\alpha J_Z + \cdots)(1 - i\theta J_X + \cdots). \tag{5.11}$$

If we now expand $\sin\theta$ and $\cos\theta$ as power series and equate coefficients of $\alpha\theta$, we obtain

$$-J_X J_Z - iJ_Y = -J_Z J_X \tag{5.12}$$

or

$$[J_Z, J_X] \equiv J_Z J_X - J_X J_Z = iJ_Y. \tag{5.13}$$

We have thus arrived at a familiar commutation relationship for components of $\boldsymbol{J}$ without any mention of quantum mechanics. It can therefore be appreciated that the properties of J_i follow simply from the geometric properties of rotations.

The more general form of the commutation relations is

$$[J_i, J_j] = i\varepsilon_{ijk} J_k, \tag{5.14}$$

where ε_{ijk} is known as the Levi–Civita symbol, or the unit antisymmetric tensor in three dimensions. It equals $+1$ if $ijk = XYZ$ or any permutation which preserves this cyclic order, it equals -1 for any permutation which disrupts this order, and it equals zero if any two of ijk are identical. Defining $\boldsymbol{J}^2 = J_X^2 + J_Y^2 + J_Z^2$, it is easy to show that $\boldsymbol{J}^2$ commutes with J_X, J_Y and J_Z. It is also convenient to define the *shift* or *ladder* operators $J_\pm = J_X \pm iJ_Y$. They also commute with $\boldsymbol{J}^2$. In addition,

$$[J_Z, J_\pm] = \pm J_\pm, \tag{5.15}$$
$$J_+ J_- = \boldsymbol{J}^2 - J_Z^2 + J_Z, \tag{5.16}$$
$$J_- J_+ = \boldsymbol{J}^2 - J_Z^2 - J_Z. \tag{5.17}$$

5.2.4. Representations of the rotation group

We next seek the irreducible representations of the full rotation group, formed by the infinite number of finite rotations $R(\alpha\hat{n})$. Because all such rotations can be expressed in terms of the infinitesimal rotation operators J_X, J_Y and J_Z (or equivalently J_+, J_- and J_Z), we start from these.

It is a well-known piece of bookwork which we do not reproduce here (see, for example, Mandl [8]) to show that, starting from the commutation relations, the simultaneous eigenfunctions of $\boldsymbol{J}^2$ and J_Z are:

$$\boldsymbol{J}^2 |j, m\rangle = j(j+1)|j, m\rangle, \tag{5.18}$$
$$J_Z |j, m\rangle = m|j, m\rangle, \tag{5.19}$$

where m can take any of the $(2j+1)$ possible values for a given j, $m = -j, -j+1, -j+2, \ldots, +j$.

The various eigenfunctions for a given j can be interconverted by use of the ladder operators J_+ and J_-. For example, J_+ has the effect of *raising* the value of m by one:

$$J_+|j,m\rangle = [j(j+1) - m(m+1)]^{1/2}|j, m+1\rangle, \tag{5.20}$$

whereas J_- lowers it by one:

$$J_-|j,m\rangle = [j(j+1) - m(m-1)]^{1/2}|j, m-1\rangle. \tag{5.21}$$

There is a phase convention implicit in these two equations, the so-called Condon and Shortley convention [9], which is universally adopted.

The $(2j+1)$ eigenfunctions $|j,m\rangle$ for a given value of j and for m values ranging from j in integer steps down to $-j$ are transformed among themselves and with no other functions by the operators J_Z and $J_\pm$, and hence by rotations in general. They thus form the basis for an irreducible representation of dimension $(2j+1)$ which must be an integer. From this we see that j can take the possible values $0, 1/2, 1, 3/2, 2, \ldots$.

5.2.5. Orbital angular momentum and spherical harmonics

Orbital angular momentum is associated with rotational motion in three-dimensional space. In terms of the operators representing the position r and linear momentum p of a particle, we have the important expression for the orbital angular momentum

$$\hbar L = r \wedge p. \tag{5.22}$$

From this we obtain the following results:

$$L_X = -i\left(Y\frac{\partial}{\partial Z} - Z\frac{\partial}{\partial Y}\right) = i\left(\sin\phi\frac{\partial}{\partial\theta} + \cot\theta\cos\phi\frac{\partial}{\partial\phi}\right), \tag{5.23}$$

$$L_Y = -i\left(Z\frac{\partial}{\partial X} - X\frac{\partial}{\partial Z}\right) = i\left(-\cos\phi\frac{\partial}{\partial\theta} + \cot\theta\sin\phi\frac{\partial}{\partial\phi}\right), \tag{5.24}$$

$$L_Z = -i\left(X\frac{\partial}{\partial Y} - Y\frac{\partial}{\partial X}\right) = -i\frac{\partial}{\partial\phi}, \tag{5.25}$$

$$L^2 = -\frac{1}{\sin\theta}\frac{\partial}{\partial\theta}\left(\sin\theta\frac{\partial}{\partial\theta}\right) - \frac{1}{\sin^2\theta}\frac{\partial^2}{\partial\phi^2}. \tag{5.26}$$

Here θ and ϕ are the spherical polar angles (only two angles are required to define the orientation of the vector r in space). Since these operators are the same as the infinitesimal rotation operators, all the results of the previous sections apply. The eigenfunctions of L^2 and L_Z are known as the *spherical harmonics*,

$$Y_{\ell m}(\theta, \phi) = \Theta_{\ell m}(\theta)\Phi_m(\phi), \tag{5.27}$$

where the $\Phi_m(\phi)$ obey the equation

$$L_Z\Phi_m(\phi) = m\Phi_m(\phi), \tag{5.28}$$

Table 5.1. *Explicit forms of the first few spherical harmonics $Y_{\ell m}(\theta, \phi)$*

ℓ	m	$Y_{\ell m}(\theta, \phi)$
0	0	$(1/4\pi)^{1/2}$
1	0	$(3/4\pi)^{1/2} \cos\theta$
1	± 1	$\mp(3/8\pi)^{1/2} \sin\theta \exp(\pm i\phi)$
2	0	$(5/16\pi)^{1/2} (3\cos^2\theta - 1)$
2	± 1	$\mp(15/8\pi)^{1/2} \cos\theta \sin\theta \exp(\pm i\phi)$
2	± 2	$(15/32\pi)^{1/2} \sin^2\theta \exp(\pm 2i\phi)$
3	0	$(7/16\pi)^{1/2} (5\cos^3\theta - 3\cos\theta)$
3	± 1	$\mp(21/64\pi)^{1/2} (5\cos^2\theta - 1) \sin\theta \exp(\pm i\phi)$
3	± 2	$(105/32\pi)^{1/2} \cos\theta \sin^2\theta \exp(\pm 2i\phi)$
3	± 3	$\mp(35/64\pi)^{1/2} \sin^3\theta \exp(\pm 3i\phi)$

with solutions

$$\Phi_m(\phi) = \sqrt{\frac{1}{2\pi}} \exp(im\phi). \tag{5.29}$$

For a single-valued solution, m must be an integer (and so therefore must ℓ). The functions $\Theta_{\ell n}(\theta)$ can be found by solving

$$\mathbf{L}^2 Y_{\ell n} = \ell(\ell+1) Y_{\ell n}, \tag{5.30}$$

which, after factoring out the ϕ dependence, becomes the associated Legendre equation. The functions $\Theta_{\ell n}$ are proportional to the associated Legendre functions $P_\ell^m(\cos\theta)$:

$$\Theta_{\ell n}(\theta) = (-1)^m \left[\frac{2\ell+1}{2} \frac{(\ell-m)!}{(\ell+m)!} \right]^{1/2} P_\ell^m(\cos\theta) \text{ for } m \geq 0$$
$$\text{or } (-1)^m \Theta_{\ell-m}(\theta) \text{ for } m < 0. \tag{5.31}$$

The spherical harmonics are normalised with respect to integration

$$\int\int Y_{\ell n}(\theta, \phi)^* Y_{\ell n}(\theta, \phi) \sin\theta \, d\theta \, d\phi = \delta_{\ell\ell'} \delta_{mm'}. \tag{5.32}$$

The explicit forms of the first few spherical harmonics are given in table 5.1. It is sometimes more convenient to use modified spherical harmonics $C_{\ell i}$, defined by

$$C_{\ell n}(\theta, \phi) = \sqrt{\frac{4\pi}{2\ell+1}} Y_{\ell n}(\theta, \phi), \tag{5.33}$$

which satisfy

$$C_{\ell n}(0, 0) = \delta_{m,0}. \tag{5.34}$$

5.3. Rotations of a rigid body

5.3.1. Introduction

It is easy to see that the orientation of a vector in space can be described by the two spherical polar angles, θ and ϕ. From a reference position parallel to the Z axis, the required orientation is obtained as follows:

(i) rotate the vector through an angle θ about the space-fixed Y axis,
(ii) rotate the vector through an angle ϕ about the space-fixed Z axis.

For a general body, however, a third angle is needed. The three angles are known as *Euler* angles and are usually labelled ϕ, θ and χ. ϕ and θ are used to define the orientation of some defined axis in the body itself (the so-called *figure* axis) in the same way as ϕ and θ respectively are used to define the orientation of a vector; χ then measures a rotation about the figure axis. In order to describe the final orientation of the body, let us attach a second coordinate system (x, y, z) to it according to some prescription; this is the *body-fixed* axis system. We start from a reference orientation with x, y, z coincident with the space-fixed axis system X, Y, Z, and then carry out the following three rotations:

(i) rotate (in a positive sense) through χ about the space-fixed Z axis,
(ii) rotate through θ about the space-fixed Y axis,
(iii) rotate through ϕ about the space-fixed Z axis.

The overall rotation is then defined by

$$R(\phi, \theta, \chi) = \exp(-i\phi J_Z) \exp(-i\theta J_Y) \exp(-i\chi J_Z). \tag{5.35}$$

We note that the three Euler angles have the following ranges:

$$0 \leq \phi \leq 2\pi; \quad 0 \leq \theta \leq \pi; \quad 0 \leq \chi \leq 2\pi. \tag{5.36}$$

The symbol ω is commonly used as a short-hand for the orientation (ϕ, θ, χ). The volume element for integration is

$$d\omega = \sin\theta \, d\phi \, d\theta \, d\chi, \tag{5.37}$$

and

$$\int d\omega = 8\pi^2. \tag{5.38}$$

The inverse of the rotation in equation (5.35) is

$$R^{-1}(\phi, \theta, \chi) = R(-\chi, -\theta, -\phi). \tag{5.39}$$

It can be shown that a sequence of rotations about axes fixed in space is equivalent to the same sequence of rotations but performed in the reverse order about axes which rotate with the body (provided that the body-fixed and space-fixed axes coincide initially).

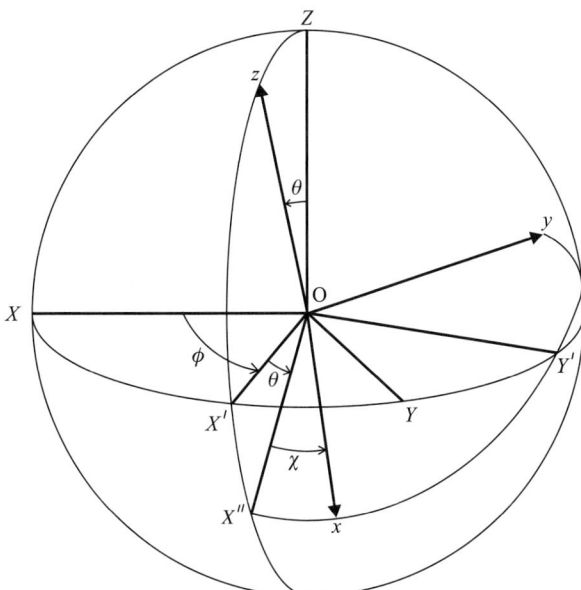

Figure 5.3. The Euler angles ϕ, θ, χ which define a general orientation of the body-fixed x, y, z axes relative to the space-fixed axes X, Y, Z. The line OY', which is the intersection of the XY and xy planes, is called the line of nodes. Note that: $x = X'''$, $Y' = Y''$, $y = Y'''$, $z = Z'' = Z'''$, $Z' = Z$, $\angle YOY' = \phi$, $\angle X'OX'' = \theta$ and $\angle Y'Oy = \chi$.

The seeds of this idea can be seen in equation (5.10). For our general rotation, this means

$$R_1 R_2 R_3 = R'_3 R'_2 R'_1 \qquad (5.40)$$

where we have written the space-fixed rotations on the left-hand side and the body-fixed rotations on the right-hand side. Consequently, the Euler angles can be defined equivalently by specifying the following sequence of rotations. As before, we start with a set of axes (x, y, z) fixed in the body and coincident with the space-fixed axes (X, Y, Z). We then perform the following sequence, which is shown in figure 5.3:

(i) rotate in a positive sense through ϕ about the body-fixed z axis (coincident at this stage with the space-fixed Z axis), bringing the xyz system into position $X'Y'Z'$,
(ii) rotate through θ about the y or Y' axis which brings the xyz system into position $X''Y''Z''$,
(iii) rotate through χ about the z or Z'' axis which brings xyz to its final position $X'''Y'''Z'''$.

This definition occurs in several textbooks, perhaps because it is easier to appreciate physically. However, it is not so convenient mathematically because it refers to rotations in different axis systems.

It can be seen from figure 5.3 that, while the polar coordinates of z in XYZ are (θ, ϕ), the polar coordinates of Z in xyz are $(\theta, \pi - \chi)$. In addition, the inverse rotation through ω^{-1} in equation (5.39) is

$$R^{-1}(\phi, \theta, \chi) = R(\pi - \chi, \theta, \pi - \phi). \tag{5.41}$$

We can also define infinitesimal rotation operators J_x, J_y and J_z for rotations about the body-fixed axes in accordance with equation (5.7). These commute with the usual space-fixed infinitesimal rotation operators J_X, J_Y and J_Z. In addition, because of equation (5.40), they obey *anomalous* commutation relationships with each other:

$$[J_x, J_y] = -iJ_z, \text{ etc.} \tag{5.42}$$

Now it is obvious that

$$J_x^2 + J_y^2 + J_z^2 = J_X^2 + J_Y^2 + J_Z^2, \tag{5.43}$$

so that both share the same eigenvalue $j(j+1)$. However, because J_z commutes with J_Z, it is possible to find eigenfunctions which are eigenfunctions of J_z with eigenvalue k (say) as well as being eigenfunctions of J_Z with eigenvalue m; m and k both take values from $-j$ to $+j$ in unit steps.

At first sight, the anomalous result in equation (5.42) is very surprising. It comes about because, although J_x, J_y and J_z commute with the space-fixed components J_X, J_Y and J_Z in equation (5.7), they do *not* commute with the direction cosines n_X, n_Y and n_Z [10]. Physically speaking, a body-fixed rotation is not strictly a rotation about an axis in the moving body but rather a rotation about a space-fixed axis which is instantaneously coincident with that axis. Van Vleck [10] is thus careful to describe these components as being *referred* to the body-fixed axes, not measured along them.

5.3.2. Rotation matrices

We recall that the square of the total angular momentum $\boldsymbol{J}^2$ commutes with all the components of $\boldsymbol{J}$ and hence with the rotation operator $R(\phi, \theta, \chi)$. Consequently, when a rotation operator is applied to an angular momentum eigenfunction $|j, m\rangle$, the result is also an eigenfunction of $\boldsymbol{J}^2$ with the same eigenvalue $j(j+1)$:

$$R(\phi, \theta, \chi)|j, m\rangle = \sum_{m'} |j, m'\rangle \mathcal{D}_{m', m}^{(j)}(\phi, \theta, \chi), \tag{5.44}$$

where the expansion coefficients $\mathcal{D}_{m', m}^{(j)}(\phi, \theta, \chi)$ are elements of a $(2j+1) \times (2j+1)$ matrix called a *Wigner rotation matrix*. It can be seen that

$$\mathcal{D}_{m'm}^{(j)}(\phi, \theta, \chi) = \langle j, m'|R(\phi, \theta, \chi)|j, m\rangle$$
$$= \langle j, m'|\exp(-i\phi J_Z)\exp(-i\theta J_Y)\exp(-i\chi J_Z)|j, m\rangle$$
$$= \exp(-i\phi m' - i\chi m)\langle j, m'|\exp(-i\theta J_Y)|j, m\rangle$$
$$\equiv \exp(-i\phi m' - i\chi m)d_{m'm}^{(j)}(\theta). \tag{5.45}$$

The quantity $d_{m'm}^{(j)}(\theta)$ is an element of a *reduced* rotation matrix.

The $\mathfrak{D}^{(j)}$ matrix forms an irreducible representation of the full rotation group. Because the basis vectors $|j, m\rangle$ are orthonormal and remain so on rotation, the matrices are unitary:

$$\mathfrak{D}^{-1} = \mathfrak{D}^{\dagger} \quad \text{or} \quad \sum_{m''} \{\mathfrak{D}^{(j)}_{m'm''}(\omega)\}^{\dagger} \mathfrak{D}^{(j)}_{m''m}(\omega) = \sum_{m''} \mathfrak{D}^{(j)}_{m''m'}(\omega)^* \mathfrak{D}^{(j)}_{m''m}(\omega) = \delta_{m'm}, \quad (5.46)$$

where the dagger (†) denotes the adjoint, that is the transpose complex conjugate, of the matrix. Explicitly we have

$$\mathfrak{D}^{(j)}_{mn}(\phi, \theta, \chi)^* = \mathfrak{D}^{(j)}_{nm}(-\chi, -\theta, -\phi), \quad (5.47)$$

where $(-\chi, -\theta, -\phi)$ is the rotation inverse to (ϕ, θ, χ). The orthogonality theorem of finite group theory has its analogue in the continuous rotation group (which is an infinite group); the sum over group elements becomes an integration over the rotation angles:

$$\int \mathfrak{D}^{(j)}_{mn}(\omega)^* \mathfrak{D}^{(j')}_{m'n'}(\omega)\, \mathrm{d}\omega = \frac{8\pi^2}{(2j+1)} \delta_{j'j} \delta_{m'm} \delta_{n'n}. \quad (5.48)$$

The elements of the reduced matrix are real and can be expressed explicitly as

$$d^{(j)}_{mn}(\theta) = \sum_t (-1)^t \frac{[(j+m)!(j-m)!(j+n)!(j-n)!]^{1/2}}{(j+m-t)!(j-n-t)!(t)!(t+n-m)!}$$
$$\times \left(\cos\frac{\theta}{2}\right)^{2j+m-n-2t} \left(\sin\frac{\theta}{2}\right)^{2t-m+n}, \quad (5.49)$$

where the sum is taken over all integral values of t which do not lead to negative factorials. The following symmetry properties of $d^{(j)}_{mn}(\theta)$ can be readily derived from the above expression:

$$\begin{aligned}
d^{(j)}_{mn}(\theta) &= d^{(j)}_{-n,-m}(\theta), \\
d^{(j)}_{mn}(\theta) &= (-1)^{m-n} d^{(j)}_{nm}(\theta), \\
d^{(j)}_{mn}(-\theta) &= d^{(j)}_{nm}(\theta), \\
d^{(j)}_{mn}(\pi+\theta) &= (-1)^{j+m} d^{(j)}_{-m,n}(\theta), \\
d^{(j)}_{mn}(\pi-\theta) &= (-1)^{j+m} d^{(j)}_{n,-m}(\theta).
\end{aligned} \quad (5.50)$$

In addition,

$$d^{(j)}_{mn}(0) = \delta_{mn}. \quad (5.51)$$

The rotation matrix elements reduce to spherical harmonics when j is an integer and m or n is zero:

$$\mathfrak{D}^{(j)}_{m0}(\phi, \theta, \chi) = C_{jm}(\theta, \phi)^* = \sqrt{\frac{2j+1}{4\pi}} Y_{jm}(\theta, \phi)^*. \quad (5.52)$$

5.3.3. Spin 1/2 systems

We recall from section 5.2.4 that, from the general theory of angular momentum, j can take half-integral (more strictly half-odd) values as well as integral ones. The particular case of $j = 1/2$ deserves special mention because of its importance in the discussion of electron or proton spin. For $j = 1/2$, there are two possible states $|1/2, 1/2\rangle$ and $|1/2, -1/2\rangle$ which are often denoted $|\alpha\rangle$ and $|\beta\rangle$ respectively. The spin operators which define these states are particularly simple. For example,

$$S_+|\beta\rangle = \left[\frac{1}{2}\left(\frac{1}{2}+1\right) - \left(-\frac{1}{2}\right)\left(-\frac{1}{2}+1\right)\right]^{1/2} |\alpha\rangle = |\alpha\rangle,$$

$$S_+|\alpha\rangle = 0. \tag{5.53}$$

From such considerations, we can write out the matrices which represent S_+, S_- and S_Z using $|\alpha\rangle$ and $|\beta\rangle$ as a basis set:

$$S_+ = \begin{bmatrix} 0 & 1 \\ 0 & 0 \end{bmatrix}, \quad S_- = \begin{bmatrix} 0 & 0 \\ 1 & 0 \end{bmatrix}, \quad S_Z = \begin{bmatrix} 1/2 & 0 \\ 0 & -1/2 \end{bmatrix}. \tag{5.54}$$

It is often convenient to use the Pauli spin operator $\sigma = 2S$, whence:

$$\sigma_X = S_+ + S_- = \begin{bmatrix} 0 & 1 \\ 1 & 0 \end{bmatrix}, \quad \sigma_Y = -\mathrm{i}(S_+ - S_-) = \begin{bmatrix} 0 & -\mathrm{i} \\ \mathrm{i} & 0 \end{bmatrix}, \quad \sigma_Z = 2S_Z = \begin{bmatrix} 1 & 0 \\ 0 & -1 \end{bmatrix}. \tag{5.55}$$

The components of σ *anticommute*, that is:

$$[\sigma_i, \sigma_j]_+ \equiv [\sigma_i \sigma_j + \sigma_j \sigma_i] = 2\delta_{ij} \begin{bmatrix} 1 & 0 \\ 0 & 1 \end{bmatrix}. \tag{5.56}$$

The three matrices $\sigma_X, \sigma_Y, \sigma_Z$ plus the unit 2×2 matrix $\mathbf{1}$ form a complete representation of a $j = 1/2$ system.

The rotation matrices take a particularly simple form for $j = 1/2$ because $J_Y = (1/2)\sigma_Y$ and $\sigma_Y^2 = \mathbf{1}$. Using these relationships, we can write

$$\begin{aligned}
d^{(1/2)}(\theta) &= \exp(-\mathrm{i}\theta S_Y) \\
&= \exp(-(1/2)\mathrm{i}\theta \sigma_Y) \\
&= \mathbf{1} - \mathrm{i}\sigma_Y(1/2)\theta - (1/2!)\sigma_Y^2(\theta/2)^2 + (\mathrm{i}/3!)\sigma_Y^3(\theta/2)^3 + (1/4!)\sigma_Y^4(\theta/2)^4 - \cdots \\
&= \mathbf{1}\cos(\theta/2) - \mathrm{i}\sigma_Y \sin(\theta/2) \\
&= \begin{bmatrix} \cos(\theta/2) & -\sin(\theta/2) \\ \sin(\theta/2) & \cos(\theta/2) \end{bmatrix}.
\end{aligned} \tag{5.57}$$

5.3.4. Symmetric top wave functions

The Hamiltonian which represents the rotational kinetic energy of an asymmetric top with three unequal moments of inertia is

$$\mathcal{H}_{\mathrm{rot}} = A P_x^2 + B P_y^2 + C P_z^2, \tag{5.58}$$

where $\boldsymbol{P}$ represents the rotational angular momentum of the body, x, y and z refer to the molecule-fixed principal inertial axes, and the rotational constants $A = \hbar^2/2I_{aa}$, etc., with $A \geq B \geq C$. Since the rotations about space-fixed axes commute with rotations about body-fixed axes, the operators $\boldsymbol{J}^2$ and J_Z both commute with $\mathcal{H}_{\text{rot}}$ and so J and M are good quantum numbers (we use upper case letters for the rotational angular momentum quantum numbers in this section). We therefore write the eigenfunctions of $\mathcal{H}_{\text{rot}}$ as $\phi_M^J(\omega) \equiv \phi_M^J(\phi, \theta, \chi)$.

Let us consider the behaviour of this function, with a particular value $\phi_M^J(\omega_1)$ at $(\phi_1, \theta_1, \chi_1)$, under a rotation of the body through ω_2. Under the rotation, the eigenfunction is transformed to a new function ϕ' where

$$\phi'(\omega_1) \equiv R(\omega_2)\phi_M^J(\omega_1) = \sum_K \phi_K^J(\omega_1) \mathfrak{D}_{KM}^{(J)}(\omega_2). \tag{5.59}$$

We know that the value of the transformed wave function in the new orientation is the same as that of the original wave function ϕ_M^J in the original orientation, i.e.

$$\phi'(\omega_2, \omega_1) = \phi_M^J(\omega_1) \quad \text{or} \quad \phi'(\omega_1) = \phi_M^J(\omega_2^{-1}, \omega_1). \tag{5.60}$$

The second equation states that the value of the new function at a particular coordinate position (ϕ, θ, χ) is the same as the value of the original function at the point (ϕ', θ', χ') which transforms into (ϕ, θ, χ) under the rotation through $(\phi_2, \theta_2, \chi_2)$ (see figure 5.1). From equations (5.59) and (5.60) we have

$$\phi_M^J(\omega_2^{-1}, \omega_1) = R(\omega_2)\phi_M^J(\omega_1)$$
$$= \sum_K \phi_K^J(\omega_1) \mathfrak{D}_{KM}^{(J)}(\omega_2). \tag{5.61}$$

If we start out with the principal inertial axes coincident with (X, Y, Z), that is, $\omega_1 = (0, 0, 0)$ and set $\omega_2^{-1} = (\phi, \theta, \chi) = \omega$, we can write

$$\phi_M^J(\omega) = \sum_K \phi_K^J(0) \mathfrak{D}_{KM}^{(J)}(-\omega)$$
$$= \sum_K \phi_K^J(0) \mathfrak{D}_{KM}^{(J)}(\omega)^*. \tag{5.62}$$

This is the general expression for the wave function of an asymmetric top molecule.

For a symmetric top, two of the three rotational constants are equal. Either $B = A$ and we have an oblate symmetric top:

$$\mathcal{H}_{\text{rot}} = B\boldsymbol{P}^2 - (B - C)P_z^2, \tag{5.63}$$

or $B = C$ and we have a prolate symmetric top:

$$\mathcal{H}_{\text{rot}} = B\boldsymbol{P}^2 + (A - B)P_z^2. \tag{5.64}$$

In either case, P_z also commutes with $\mathcal{H}_{\text{rot}}$ and the wave functions satisfy

$$P_z \phi_M^J(\omega) = -i\frac{\partial}{\partial \chi} \phi_M^J(\omega) = K\phi_M^J(\omega), \tag{5.65}$$

for some value of K. As a result of this condition, only one term in the general solution (5.62) survives, i.e.

$$\phi_M^J(\omega) = \phi_K^J(0) \exp(i\phi M) d_{MK}^{(J)}(\theta) \exp(i\chi K). \tag{5.66}$$

In other words, the wave function for a symmetric top molecule is given by

$$\phi_{MK}^J(\phi, \theta, \chi) = \mathscr{D}_{MK}^{(J)}(\phi, \theta, \chi)^* \times \text{constant}. \tag{5.67}$$

From equation (5.48), the constant is $[(2J+1)/8\pi^2]^{1/2}$ if the function is to be normalised to unity.

5.4. Addition of angular momenta

5.4.1. Introduction

Let us consider two angular momenta $\boldsymbol{J}_1$ and $\boldsymbol{J}_2$ which are independent, that is, act on different parts of a system, and have eigenfunctions $|j_1, m_1\rangle$ and $|j_2, m_2\rangle$ respectively. If the two parts of the system interact through some physical mechanism, the two angular momenta become *coupled* and it is meaningful to define a resultant angular momentum,

$$\boldsymbol{J} = \boldsymbol{J}_1 + \boldsymbol{J}_2. \tag{5.68}$$

It is easy to show that the components of $\boldsymbol{J}$ obey the standard commutation relationships (5.13). The commuting operators $\boldsymbol{J}$ and J_Z therefore have eigenfunctions $|j, m\rangle$. In this section we describe how these eigenfunctions are related to those of $\boldsymbol{J}_1^2$ and $\boldsymbol{J}_2^2$. We can describe the combined system by a simple product of the wavefunctions $|j_1, m_1\rangle |j_2, m_2\rangle$ which we write as $|j_1 m_1; j_2 m_2\rangle$; this is known as the *uncoupled* or *decoupled* representation. These products are still eigenfunctions of $\boldsymbol{J}_1^2$, J_{1Z}, $\boldsymbol{J}_2^2$ and J_{2Z}. They are also eigenfunctions of J_Z with eigenvalue $m = m_1 + m_2$:

$$J_Z |j_1, m_1; j_2, m_2\rangle = (J_{1Z} + J_{2Z}) |j_1, m_1; j_2, m_2\rangle = (m_1 + m_2) |j_1, m_1; j_2, m_2\rangle. \tag{5.69}$$

However, they are not eigenfunctions of $\boldsymbol{J}^2$ since

$$\begin{aligned}
\boldsymbol{J}^2 &= (\boldsymbol{J}_1 + \boldsymbol{J}_2) \cdot (\boldsymbol{J}_1 + \boldsymbol{J}_2) \\
&= \boldsymbol{J}_1^2 + \boldsymbol{J}_2^2 + 2\boldsymbol{J}_1 \cdot \boldsymbol{J}_2 \\
&= \boldsymbol{J}_1^2 + \boldsymbol{J}_2^2 + 2J_{1Z} J_{2Z} + J_{1+} J_{2-} + J_{1-} J_{2+}.
\end{aligned} \tag{5.70}$$

Remembering that the $\boldsymbol{J}_1$ operator acts only on the $|j_1, m_1\rangle$ part of the decoupled representation and $\boldsymbol{J}_2$ acts only on the $|j_2, m_2\rangle$ part, we see that $\boldsymbol{J}^2$ connects states with different values of m_1 and m_2 but with constant $(m_1 + m_2)$.

In order to determine the allowed values of the total angular momentum quantum number j, let us apply $\boldsymbol{J}^2$ in equation (5.70) to the uncoupled state $|j_1, j_1; j_2, j_2\rangle$, that is, with the maximum values for m_1 and m_2:

$$\begin{aligned}
\boldsymbol{J}^2 |j_1, j_1; j_2, j_2\rangle &= [j_1(j_1+1) + j_2(j_2+1) + 2j_1 j_2] |j_1, j_1; j_2, j_2\rangle \\
&= [(j_1 + j_2)(j_1 + j_2 + 1)] |j_1, j_1; j_2, j_2\rangle.
\end{aligned} \tag{5.71}$$

This particular product wavefunction is therefore an eigenfunction of $\mathbf{J}^2$ with eigenvalue $j = j_1 + j_2$:

$$|j = j_1 + j_2, m = j_1 + j_2\rangle = |j_1, m_1 = j_1; j_2, m_2 = j_2\rangle. \quad (5.72)$$

This is the largest possible value for j because $m_{\max} = j_1 + j_2$. We can then obtain other coupled eigenfunctions $|j_1, j_2, j, m\rangle$ by applying the lowering operator $J_- = J_{1-} + J_{2-}$:

$$J_-|j_1, j_2, j = j_1 + j_2, m = j_1 + j_2\rangle = (J_{1-} + J_{2-})|j_1, j_1; j_2, j_2\rangle. \quad (5.73)$$

The effects of the lowering operators on each side of this equation are given in equation (5.21). With these results we obtain

$$(2j)^{1/2}|j_1, j_2, j = j_1 + j_2, m = j_1 + j_2 - 1\rangle$$
$$= (2j_1)^{1/2}|j_1, j_1 - 1; j_2, j_2\rangle + (2j_2)^{1/2}|j_1, j_1; j_2, j_2 - 1\rangle. \quad (5.74)$$

States with the same value of j but lower values of m can be obtained in this way by successive operations of J_-. Equation (5.74) describes a state with $m = j_1 + j_2 - 1$. At the same time, there exists a second, independent state with the same value for m:

$$|j_1, j_2, j', m = j_1 + j_2 - 1\rangle$$
$$= (j)^{-1/2}\{(j_1)^{1/2}|j_1, j_1 - 1; j_2, j_2\rangle - (j_2)^{1/2}|j_1, j_1; j_2, j_2 - 1\rangle\}. \quad (5.75)$$

Operation on this state with $J_+ = J_{1+} + J_{2+}$, using equation (5.20) leads to zero. This means that m already possesses its maximum value so this second state has angular momentum $j' = j_1 + j_2 - 1$. This can be confirmed by operating on the state with $\mathbf{J}^2$. All lower values of m for this value of j' can then be obtained by successive application of the lowering operator $J_- = J_{1-} + J_{2-}$.

For the next lower value of m ($=j_1 + j_2 - 2$), there are three possible decoupled states. Two linear combinations of these states are associated with $j = j_1 + j_2$ and $j = j_1 + j_2 - 1$; the remaining orthogonal combination must therefore be the one with the maximum value of m for $j = j_1 + j_2 - 2$. This procedure is continued until either m_1 reaches $-j_1$, or m_2 reaches $-j_2$. Thereafter, no new values for j can be generated. Consequently, the smallest value for j is $(j_1 - j_2)$ or $(j_2 - j_1)$, whichever is the larger, i.e.,

$$j_{\min} = |j_1 - j_2|. \quad (5.76)$$

The relationship between the coupled states $|j, m\rangle$ and the uncoupled states $|j_1, m_1\rangle|j_2, m_2\rangle$ can be written in the general form:

$$|j, m\rangle = \sum_{m_1, m_2} |j_1, m_1\rangle|j_2, m_2\rangle\langle j_1, j_2, m_1, m_2|j, m\rangle. \quad (5.77)$$

This form follows directly from the closure relationship

$$|a\rangle = \sum_b |b\rangle\langle b|a\rangle. \quad (5.78)$$

The coefficients $\langle j_1, j_2, m_1, m_2|j, m\rangle$ are called the Clebsch–Gordan coefficients. As we have seen, they are zero unless $m_1 + m_2 = m$. The coefficients are defined by equation (5.77) only when j takes one of the values $j_1 + j_2, j_1 + j_2 - 1, \ldots, |j_1 - j_2|$. This condition can be expressed more symmetrically: none of the j values may exceed the sum of the other two and the sum of all three j values must be an integer. It is usually called the *triangle condition* and can be written as $\Delta(j_1, j_2, j)$, which is 1 if the triangle condition is satisfied and 0 otherwise. The Clebsch–Gordan coefficients are thus zero when the triangle condition is not satisfied. These two conditions for the Clebsch–Gordan coefficients to be non-zero reflect the fact that the angular momenta add vectorially (the triangle condition) and their components along an arbitrary Z axis add algebraically ($m_1 + m_2 = m$).

The Clebsch–Gordan coefficients for given values of j_1 and j_2 form a square matrix labelled by the j, m values one way and by m_1, m_2 the other. This matrix is always real and orthogonal, so that the inverse transformation to (5.77) is

$$|j_1, m_1\rangle |j_2, m_2\rangle = \sum_{j,m} |j, m\rangle \langle j_1, j_2, m_1, m_2 | j, m\rangle. \tag{5.79}$$

5.4.2. Wigner 3-j symbols

The Clebsch–Gordan coefficient is unsymmetrical between j_1 and j_2 on the one hand and j on the other. There is also an asymmetry between j_1 and j_2 which arises from the *order* of the coupling and leads to a different phase factor. A more symmetrical formulation of the coupling coefficients is possible, as shown by Wigner [11]. Consider first the coupling of two angular momenta, both with quantum number j, to give a scalar ($j = 0$) result. We have

$$|0, 0\rangle = \sum_m \langle j, j, m, -m | 0, 0 \rangle |j, m\rangle |j, -m\rangle. \tag{5.80}$$

It is relatively easy to show that all the possible coefficients on the right-hand side are equal in magnitude, specifically

$$\langle j, j, m, -m | 0, 0 \rangle = \frac{(-1)^{j-m}}{\sqrt{2j+1}}. \tag{5.81}$$

We next consider the coupling of three angular momenta to give a scalar. We first couple j_1 and j_2 together to give a resultant of magnitude j_3 and then couple this with j_3 to produce a scalar, thus:

$$|0, 0\rangle = \sum_{m_3 m'_3} \langle j_3, j_3, m_3, m'_3 | 0, 0\rangle |j_3, m_3\rangle \sum_{m_1 m_2} \langle j_1, j_2, m_1, m_2 | j_3, m'_3\rangle |j_1, m_1\rangle |j_2, m_2\rangle$$

$$= \delta_{m'_3, -m_3} \sum_{m_1 m_2 m_3} (-1)^{j_3 - m_3} (2j_3 + 1)^{-1/2}$$

$$\times \langle j_1, j_2, m_1, m_2 | j_3, -m_3\rangle |j_1, m_1\rangle |j_2, m_2\rangle |j_3, m_3\rangle. \tag{5.82}$$

This equation provides the basis for the *Wigner 3-j symbol*, defined as

$$\begin{pmatrix} j_1 & j_2 & j_3 \\ m_1 & m_2 & m_3 \end{pmatrix} \equiv (-1)^{j_1 - j_2 - m_3} (2j_3 + 1)^{-1/2} \langle j_1, j_2, m_1, m_2 | j_3, -m_3\rangle \tag{5.83}$$

where a further factor of $(-1)^{j_1-j_2-j_3}$ has been introduced to make the 3-j symbol as symmetrical as possible between the three angular momenta. The contraction to the scalar product in equation (5.82) thus becomes

$$|0,0\rangle = \sum_{m_1 m_2 m_3} \begin{pmatrix} j_1 & j_2 & j_3 \\ m_1 & m_2 & m_3 \end{pmatrix} |j_1, m_1\rangle |j_2, m_2\rangle |j_3, m_3\rangle \tag{5.84}$$

to within a phase factor; the equal status of each angular momentum is evident.

The 3-j symbol is invariant under an even or cyclic permutation in the order of its columns. An odd permutation of the columns multiplies the 3-j symbol by the phase factor $(-1)^{j_1+j_2+j_3}$, as does a change of sign of all the projection quantum numbers:

$$\begin{pmatrix} j_1 & j_2 & j_3 \\ m_1 & m_2 & m_3 \end{pmatrix} = (-1)^{j_1+j_2+j_3} \begin{pmatrix} j_2 & j_1 & j_3 \\ m_2 & m_1 & m_3 \end{pmatrix}$$

$$= (-1)^{j_1+j_2+j_3} \begin{pmatrix} j_1 & j_2 & j_3 \\ -m_1 & -m_2 & -m_3 \end{pmatrix}. \tag{5.85}$$

The 3-j symbol also has the advantage that the projection quantum numbers are shown below their associated vectors, consistent with their subordinate role. The orthogonality properties of the 3-j symbols are written as

$$\sum_{j_3 m_3} (2j_3+1) \begin{pmatrix} j_1 & j_2 & j_3 \\ m_1 & m_2 & m_3 \end{pmatrix} \begin{pmatrix} j_1 & j_2 & j_3 \\ m'_1 & m'_2 & m_3 \end{pmatrix} = \delta_{m_1 m'_1} \delta_{m_2 m'_2},$$

$$\sum_{m_1 m_2} \begin{pmatrix} j_1 & j_2 & j_3 \\ m_1 & m_2 & m_3 \end{pmatrix} \begin{pmatrix} j_1 & j_2 & j'_3 \\ m_1 & m_2 & m'_3 \end{pmatrix} = (2j_3+1)^{-1} \delta_{j_3 j'_3} \delta_{m_3 m'_3} \Delta(j_1, j_2, j_3). \tag{5.86}$$

The corresponding form of the coupled angular momentum functions given in equation (5.77) can thus be written:

$$|j, m\rangle = (-1)^{j_1-j_2+m} (2j+1)^{1/2} \sum_{m_1 m_2} \begin{pmatrix} j_1 & j_2 & j \\ m_1 & m_2 & -m \end{pmatrix} |j_1, m_1\rangle |j_2, m_2\rangle. \tag{5.87}$$

The 3-j symbols have explicit functional forms dependent on their arguments. We give the simplest of these formulae in General Appendix C at the end of this book. It should also be appreciated that expressions for particular symbols can be derived algebraically by computer software [12] in less time than it takes to look them up in a table.

5.4.3. Coupling of three or more angular momenta: Racah algebra, Wigner 6-j and 9-j symbols

The general procedure for coupling two angular momenta can be extended to describe the coupling of three or more angular momenta. The uncoupled basis for three angular momenta is $|j_1, m_1\rangle|j_2, m_2\rangle|j_3, m_3\rangle$ but there are several alternative coupling schemes for arriving at the coupled description. We can, for example, first couple j_1 and j_2 to give j_{12} and then couple j_{12} with j_3 to give the resultant j. We note that the resulting wave function $|((j_1, j_2)j_{12}, j_3)j, m\rangle$ for given values of j and m will be different for differing values of j_{12} whose value must therefore be specified. Alternatively, we can choose to couple j_1 with the resultant j_{23} of j_2 and j_3 to give $|(j_1,(j_2,j_3)j_{23})j,m\rangle$ and again the value of the intermediate quantum number j_{23} must be specified. It is also important to state the *order* of the coupling clearly; the coupling $(j_1,j_2)j_{12}$ differs from $(j_2,j_1)j_{12}$ by a phase factor.

Either of these two types of coupled function can be expressed in terms of the other using the closure relationship:

$$|((j_1,j_2)j_{12},j_3)j,m\rangle$$
$$= \sum_{j_{23}} |(j_1,(j_2,j_3)j_{23})j,m\rangle \langle (j_1,(j_2,j_3)j_{23})j,m|((j_1,j_2)j_{12},j_3)j,m\rangle. \quad (5.88)$$

Operating on both sides of this equation with $J_{\pm}$ shows that the transformation coefficients are independent of m. We can therefore rewrite this equation more simply (with an obvious change of notation) as

$$|((a,b)e,d)c\rangle = \sum_f |(a,(b,d)f)c\rangle \langle (a,(b,d)f)c|((a,b)e,d)c\rangle$$
$$\equiv \sum_f [(2e+1)(2f+1)]^{1/2} W(a,b,c,d;e,f)|(a,(b,d)f)c\rangle. \quad (5.89)$$

The quantity $W(a,b,c,d;e,f)$ defined by equation (5.89) is called the *Racah coefficient*; the normalisation factor $[(2e+1)(2f+1)]^{1/2}$ is chosen to simplify the symmetry properties. A closely related quantity is the *Wigner 6-j symbol*:

$$\begin{Bmatrix} a & b & e \\ d & c & f \end{Bmatrix} = (-1)^{a+b+c+d} W(a,b,c,d;e,f). \quad (5.90)$$

The 6-j symbol has a higher symmetry, being invariant with respect to interchange of any two columns and also under the interchange of the upper and lower arguments in each of any two columns.

It can be seen from equations (5.88) and (5.89) that the 6-j symbols can be expressed in terms of products of the 3-j symbols:

$$\begin{Bmatrix} a & b & e \\ d & c & f \end{Bmatrix} = \sum_{\alpha\beta\gamma\delta\varepsilon\phi} (-1)^{d+c+f+\delta+\gamma+\phi} \begin{pmatrix} a & b & e \\ \alpha & \beta & \varepsilon \end{pmatrix} \begin{pmatrix} a & c & f \\ \alpha & \gamma & -\phi \end{pmatrix}$$
$$\times \begin{pmatrix} d & b & f \\ -\delta & \beta & \phi \end{pmatrix} \begin{pmatrix} d & c & e \\ \delta & -\gamma & \varepsilon \end{pmatrix}. \quad (5.91)$$

Another useful formula is obtained from this expression, using the second of the equations (5.86):

$$\begin{pmatrix} a & b & e \\ \alpha & \beta & \varepsilon \end{pmatrix} \begin{Bmatrix} a & b & e \\ d & c & f \end{Bmatrix} = \sum_{\delta\gamma\phi} (-1)^{d+c+f+\delta+\gamma+\phi} \begin{pmatrix} a & c & f \\ \alpha & \gamma & -\phi \end{pmatrix}$$
$$\times \begin{pmatrix} d & b & f \\ -\delta & \beta & \phi \end{pmatrix} \begin{pmatrix} d & c & e \\ \delta & -\gamma & \varepsilon \end{pmatrix}. \quad (5.92)$$

Explicit formulae for the 6-j symbols can be obtained from equations (5.91) and (5.92);

a valuable collection is given in Edmonds' book [1] (see also General Appendix D). Once again, they can be generated analytically by Mathematica [12].

When there are four angular momenta to couple together, there are again several alternative coupling schemes and as before the basis functions in any two schemes are related. In this case, the relationship defines a *Wigner 9-j symbol*:

$$|((a,d)g,(b,e)h)i\rangle = \sum_{cf} [(2c+1)(2f+1)(2g+1)(2h+1)]^{1/2}$$
$$\times \begin{Bmatrix} a & b & c \\ d & e & f \\ g & h & i \end{Bmatrix} |((a,b)c,(d,e)f)i\rangle. \quad (5.93)$$

The 9-j symbol is multiplied by $(-1)^p$, where p is the sum of all 9 arguments, on exchanging any two rows or columns. It is unchanged by an even permutation of rows or columns, or by a reflection about a diagonal. When one argument of the 9-j symbol is zero, it reduces to a 6-j symbol:

$$[(2c+1)(2g+1)]^{1/2} \begin{Bmatrix} a & b & c \\ d & e & f \\ g & h & 0 \end{Bmatrix} = \delta_{cf}\delta_{gh}(-1)^{b+c+d+g} \begin{Bmatrix} a & b & c \\ e & d & g \end{Bmatrix}. \quad (5.94)$$

Note that the 9-j symbol, like the 6-j symbol, is independent of m and therefore makes no reference to the projection quantum numbers.

Wigner's n-j symbols may well look unfriendly and even intimidating on first acquaintance. To overcome this impression, it is helpful to remember that they are only coefficients in the linear expression of a wave function of a quantum system and simply take numerical values in any given application. Furthermore, what is often important in practice is the ability to recognise when they are zero from constraints such as the triangle rule because this leads to useful selection rules, as we shall see later in this chapter and elsewhere in this book.

5.4.4. Clebsch–Gordan series

It is possible to derive some general and useful properties of the rotation matrices from the relationship between the decoupled functions $|j_1,m_1\rangle|j_2,m_2\rangle$ and the coupled functions $|j_1,j_2,j_3,m_3\rangle$, equation (5.79). Operating with $R(\phi,\theta,\chi)$ on both sets of functions, we obtain:

$$\sum_{m'_1}\sum_{m'_2} \mathscr{D}^{(j_1)}_{m'_1 m_1}(\omega)\mathscr{D}^{(j_2)}_{m'_2 m_2}(\omega)|j_1,m'_1\rangle|j_2,m'_2\rangle$$
$$= \sum_{j_3}\sum_{m'_3} \langle j_1,m_1;j_2,m_2|j_1,j_2,j_3,m_3\rangle \mathscr{D}^{(j_3)}_{m'_3 m_3}(\omega)|j_1,j_2,j_3,m'_3\rangle. \quad (5.95)$$

If we multiply each side from the left by $\langle j_1, m_1; j_2, m_2 |$ and introduce 3-j symbols, we obtain:

$$\mathcal{D}^{(j_1)}_{m'_1 m_1}(\omega) \mathcal{D}^{(j_2)}_{m'_2 m_2}(\omega) = \sum_{j_3} (2j_3 + 1) \begin{pmatrix} j_1 & j_2 & j_3 \\ m'_1 & m'_2 & m'_3 \end{pmatrix} \begin{pmatrix} j_1 & j_2 & j_3 \\ m_1 & m_2 & m_3 \end{pmatrix} \mathcal{D}^{(j_3)}_{m'_3 m_3}(\omega)^*. \tag{5.96}$$

Similarly,

$$\mathcal{D}^{(j_3)}_{m'_3 m_3}(\omega)^* = \sum_{m_1 m'_1 m_2 m'_2} (2j_3 + 1) \begin{pmatrix} j_1 & j_2 & j_3 \\ m'_1 & m'_2 & m'_3 \end{pmatrix} \begin{pmatrix} j_1 & j_2 & j_3 \\ m_1 & m_2 & m_3 \end{pmatrix} \mathcal{D}^{(j_1)}_{m'_1 m_1}(\omega) \mathcal{D}^{(j_2)}_{m'_2 m_2}(\omega). \tag{5.97}$$

The expressions are particularly useful in the evaluation of integrals over products of rotational matrices, as we shall see. They are widely used in many branches of physics and chemistry from multipole expansions through to statistical mechanical averaging.

5.4.5. Integrals over products of rotation matrices

If we recall the definition of the rotation matrix $\mathcal{D}^{(j)}_{m'm}(\phi, \theta, \chi)$ from equation (5.45) and integrate over the volume element $d\omega = \sin\theta\, d\phi\, d\theta\, d\chi$, we obtain

$$\int \mathcal{D}^{(j)}_{m'm}(\phi, \theta, \chi) \, d\omega = \delta_{j0} \delta_{m'0} \delta_{m0}. \tag{5.98}$$

We can then use this result in the integration of equation (5.96) to give

$$\int \mathcal{D}^{(j_1)}_{m'_1 m_1}(\omega)^* \mathcal{D}^{(j_2)}_{m'_2 m_2}(\omega) \, d\omega = \frac{8\pi^2}{(2j_1 + 1)} \delta_{j_1 j_2} \delta_{m'_1 m'_2} \delta_{m_1 m_2}, \tag{5.99}$$

a normalisation equation which we have anticipated in equation (5.48). Using a similar approach, we can derive an expression for the integral over a product of three rotation matrix elements which arises in atomic theory and elsewhere. We multiply equation (5.96) by $\mathcal{D}^{(j_3)}_{m'_3 m_3}(\omega)$ and integrate over ω to obtain

$$\int \mathcal{D}^{(j_1)}_{m'_1 m_1}(\omega) \mathcal{D}^{(j_2)}_{m'_2 m_2}(\omega) \mathcal{D}^{(j_3)}_{m'_3 m_3}(\omega) \, d\omega = 8\pi^2 \begin{pmatrix} j_1 & j_2 & j_3 \\ m'_1 & m'_2 & m'_3 \end{pmatrix} \begin{pmatrix} j_1 & j_2 & j_3 \\ m_1 & m_2 & m_3 \end{pmatrix}. \tag{5.100}$$

The evaluation of this integral by conventional methods (even for the special case $m_1 = m_2 = m_3 = 0$ when the rotation matrices reduce to spherical harmonics) is extremely laborious. We shall use this result in the derivation of the Wigner–Eckart theorem and other angular momentum relationships later in this chapter.

5.5. Irreducible spherical tensor operators

5.5.1. Introduction

We recall from section 5.3.2 that the transformation of an angular momentum state $|j, m\rangle$ under a finite rotation of the axis system through Euler angles (ϕ, θ, χ) is described by

$$R(\phi, \theta, \chi)|j, m\rangle = |j, m\rangle' = \sum_{m'} |j, m'\rangle \mathfrak{D}^{(j)}_{m'm}(\phi, \theta, \chi). \tag{5.101}$$

This equation specifies the eigenstates of $J_{Z'}$ in the new coordinate system (X', Y', Z') in terms of the eigenstates of J_Z, in the original coordinate system (X, Y, Z). It tells us that the state $|j, m\rangle$ transforms only into states with the same value for j under rotations in three-dimensional space. It may seem a little wayward to drop the active convention (rotation of the physical system) in favour of the passive convention (rotation of the axis system) at this point. However, it is conventional to define cartesian tensors by the way they are affected by transformations of the coordinate system, and the same is true for spherical tensors. It is worthwhile remembering that a rotation of the coordinate system through Euler angles $\omega = (\phi, \theta, \chi)$ is equivalent to a rotation of the physical system through $\omega^{-1} = (-\chi, -\theta, -\phi)$. The behaviour of the spherical harmonic functions under rotation of the coordinate system,

$$R(\omega)C_{\ell n}(\theta, \phi) \equiv C_{\ell n}(\theta', \phi') = \sum_{m'} \mathfrak{D}^{(\ell)}_{m'm}(\omega) C_{\ell n}(\theta, \phi), \tag{5.102}$$

is a special example of the transformation rule stated above. Such transformation properties form the basis for the definition of a general spherical tensor,

$$T^k_p(\boldsymbol{T}) = \sum_{p'} T^k_{p'}(\boldsymbol{T}) \mathfrak{D}^{(k)}_{p'p}(\omega), \tag{5.103}$$

where ω stands for the Euler angles of the rotation which takes the old, unprimed to the new, primed axis system. Since the matrix $\mathfrak{D}^{(k)}(\omega)$ forms an irreducible representation of the rotation group of rank k, it follows directly that $T^k_p(\boldsymbol{T})$ is also an irreducible representation of the same rank.

The same ideas can be readily extended to cover quantum mechanical operators. The formulation is slightly different because, although a transformation S turns a wave function ψ into $S\psi$, it turns an operator T into $S\,T\,S^{-1}$. Thus an irreducible spherical tensor operator (usually abbreviated to spherical tensor operator) $T^k(\boldsymbol{T})$ of rank k is defined as an entity with $(2k + 1)$ components, $T^k_p(\boldsymbol{T})$, which transform under rotations as

$$R(\omega)T^k_p(\boldsymbol{T})R^{-1}(\omega) = \sum_{p'} T^k_{p'}(\boldsymbol{T}) \mathfrak{D}^{(k)}_{p'p}(\omega). \tag{5.104}$$

Each component is labelled by a different value for p which, as usual, runs from $-k$ to $+k$ in steps separated by unity. This definition can be re-cast in a more useful form as follows. We consider a small, finite rotation through an angle $\delta\chi$ about the space-fixed axis ξ. From equation (5.2), we can write

$$R(\delta\chi) = 1 - i\delta\chi J_\xi + \cdots$$
$$R^{-1}(\delta\chi) = 1 + i\delta\chi J_\xi + \cdots \tag{5.105}$$

We then substitute these expressions in equation (5.104) and equate coefficients linear in $\delta\chi$ to obtain

$$J_\xi T_p^k(\boldsymbol{T}) - T_p^k(\boldsymbol{T}) J_\xi = \sum_{p'} T_{p'}^k(\boldsymbol{T}) \langle k, p' | J_\xi | k, p \rangle. \tag{5.106}$$

From this, it is a short step to the alternative definition of a spherical tensor operator $T^k(\boldsymbol{T})$:

$$[J_Z, T_p^k(\boldsymbol{T})] = p T_p^k(\boldsymbol{T}),$$
$$[J_\pm, T_p^k(\boldsymbol{T})] = [k(k+1) - p(p\pm 1)]^{1/2} T_{p\pm 1}^k(\boldsymbol{T}). \tag{5.107}$$

5.5.2. Examples of spherical tensor operators

The most obvious example of a spherical tensor operator is the angular momentum itself (a spherical tensor of rank one):

$$T_1^1(\boldsymbol{J}) = -\frac{1}{\sqrt{2}}(J_X + iJ_Y) = -\frac{1}{\sqrt{2}} J_+,$$
$$T_0^1(\boldsymbol{J}) = J_Z, \tag{5.108}$$
$$T_{-1}^1(\boldsymbol{J}) = \frac{1}{\sqrt{2}}(J_X - iJ_Y) = \frac{1}{\sqrt{2}} J_-.$$

The difference between the definitions of the shift operators $J_\pm$ and the spherical tensor components $T_{\pm 1}^1(\boldsymbol{J})$ should be noted because it often causes confusion. Because $\boldsymbol{J}$ is a vector and because all vector operators transform in the same way under rotations, that is, according to equation (5.104) with $k = 1$, it follows that any cartesian vector $\boldsymbol{V}$ has spherical tensor components defined in the same way (see table 5.2). There is a one-to-one correspondence between the cartesian vector and the first-rank spherical tensor. Common examples of such quantities in molecular quantum mechanics are the position vector $\boldsymbol{r}$ and the electric dipole moment operator $\boldsymbol{\mu}_e$.

Just as angular momentum wave functions can be coupled together using the Clebsch–Gordan coefficients, so too can spherical tensors. Two spherical tensors $\boldsymbol{R}^{k_1}$ and $\boldsymbol{S}^{k_2}$ can be combined to form a tensor of rank K which takes all possible values from $(k_1 + k_2)$ to $(k_1 - k_2)$, assuming $k_1 \geq k_2$:

$$T_P^K(\boldsymbol{R}^{k_1}, \boldsymbol{S}^{k_2}) = \sum_{p_1 p_2} \langle k_1, k_2, p_1, p_2 | K, P \rangle T_{p_1}^{k_1}(\boldsymbol{R}) T_{p_2}^{k_2}(\boldsymbol{S}). \tag{5.109}$$

This result may be compared with equation (5.77). It often happens that $k_1 = k_2$, in which case a zeroth-rank tensor $K = 0$, as well as others, can be produced by their

Table 5.2. *The relationship between spherical tensors of ranks 1 and 2 and cartesian vectors V and second-rank tensors T*

	spherical tensors	cartesian vectors and tensors
vector	$T_0^1(V)$	V_Z
	$T_{\pm 1}^1(V)$	$\mp(1/\sqrt{2})(V_X \pm iV_Y)$
second-rank tensor	$T_0^0(T)$	$-(1/\sqrt{3})(T_{XX} + T_{YY} + T_{ZZ})$
	$T_0^1(T)$	$(i/\sqrt{2})(T_{XY} - T_{YX})$
	$T_{\pm 1}^1(T)$	$\mp(i/2)\{(T_{YZ} - T_{ZY}) \pm i(T_{ZX} - T_{XZ})\}$
	$T_0^2(T)$	$(1/\sqrt{6})\{2T_{ZZ} - T_{XX} - T_{YY}\}$
	$T_{\pm 1}^2(T)$	$\mp(1/2)\{(T_{XZ} + T_{ZX}) \pm i(T_{YZ} + T_{ZY})\}$
	$T_{\pm 2}^2(T)$	$(1/2)\{(T_{XX} - T_{YY}) \pm i(T_{XY} + T_{YX})\}$

interaction:

$$T_0^0(\boldsymbol{R}^k, \boldsymbol{S}^k) = \sum_p \langle k, k, p, -p | 0, 0 \rangle T_p^k(\boldsymbol{R}) T_{-p}^k(\boldsymbol{S})$$

$$= \sum_p (-1)^{k-p}(2k+1)^{-1/2} T_p^k(\boldsymbol{R}) T_{-p}^k(\boldsymbol{S}), \quad (5.110)$$

by use of equation (5.81). Such a resultant is invariant under rotations and so is a scalar quantity. However, in spherical tensor algebra, the *scalar product* is defined slightly differently, namely

$$T^k(\boldsymbol{R}) \cdot T^k(\boldsymbol{S}) = \sum_p (-1)^p T_p^k(\boldsymbol{R}) T_{-p}^k(\boldsymbol{S}). \quad (5.111)$$

This differs from equation (5.110) in both phase and normalisation factors. We have seen that, for $k = 1$, the spherical tensor corresponds to a cartesian vector; the spherical scalar product in this case is the same as the cartesian scalar product of two vectors:

$$T^1(\boldsymbol{A}) \cdot T^1(\boldsymbol{B}) = \sum_p (-1)^p T_p^1(\boldsymbol{A}) T_{-p}^1(\boldsymbol{B})$$

$$= A_X B_X + A_Y B_Y + A_Z B_Z$$

$$= \boldsymbol{A} \cdot \boldsymbol{B}. \quad (5.112)$$

Although there is a one-to-one correspondence between the first-rank cartesian and spherical tensors, the same is not true for second- and higher-rank cartesian tensors.

From equation (5.109) we see that two vectors $\boldsymbol{u}$ and $\boldsymbol{v}$ can be coupled together to give a scalar product $T^0(\boldsymbol{u}, \boldsymbol{v})$, a vector $T^1(\boldsymbol{u}, \boldsymbol{v})$ which is proportional to the vector product $\boldsymbol{u} \wedge \boldsymbol{v}$ and a second-rank spherical tensor $T^2(\boldsymbol{u}, \boldsymbol{v})$. The explicit expressions for their components can be obtained as follows:

$$T_0^0(\boldsymbol{u}, \boldsymbol{v}) = \sum_{mm'} \langle 1, 1, m, m' | 0, 0 \rangle u_m v_{m'} = \frac{1}{\sqrt{3}}(u_1 v_{-1} - u_0 v_0 + u_{-1} v_1)$$

$$= -\frac{1}{\sqrt{3}}(u_X v_X + u_Y v_Y + u_Z v_Z), \quad (5.113)$$

$$T_0^1(\boldsymbol{u}, \boldsymbol{v}) = \sum_{mm'} \langle 1, 1, m, m' \mid 1, 0 \rangle u_m v_{m'} = \frac{1}{\sqrt{2}}(u_1 v_{-1} + u_{-1} v_1)$$

$$= \frac{1}{\sqrt{2}}(u_X v_Y - u_Y v_X), \tag{5.114}$$

$$T_{\pm 1}^1(\boldsymbol{u}, \boldsymbol{v}) = \frac{1}{\sqrt{2}}[J_\pm, T_0^1] = \pm \frac{1}{\sqrt{2}}(u_{\pm 1} v_0 - u_0 v_{\pm 1})$$

$$= \frac{1}{2}[(u_X v_Z - u_Z v_X) \mp i(u_Y v_Z - u_Z v_Y)], \tag{5.115}$$

$$T_0^2(\boldsymbol{u}, \boldsymbol{v}) = \sum_{mm'} \langle 1, 1, m, m' \mid 2, 0 \rangle u_m v_{m'} = \frac{1}{\sqrt{6}}(u_1 v_{-1} + 2 u_0 v_0 + u_{-1} v_1)$$

$$= \frac{1}{\sqrt{6}}(2 u_Z v_Z - u_X v_X - u_Y v_Y), \tag{5.116}$$

$$T_{\pm 1}^2(\boldsymbol{u}, \boldsymbol{v}) = \frac{1}{\sqrt{6}}[J_\pm, T_0^2] = \frac{1}{\sqrt{2}}(u_{\pm 1} v_0 + u_0 v_{\pm 1})$$

$$= \mp \frac{1}{2}[u_X v_Z + u_Z v_X \pm i(u_Y v_Z + u_Z v_Y)], \tag{5.117}$$

$$T_{\pm 2}^2(\boldsymbol{u}, \boldsymbol{v}) = \frac{1}{2}[J_\pm, T_{\pm 1}^2] = u_{\pm 1} v_{\pm 1}$$

$$= \frac{1}{2}[u_X v_X - u_Y v_Y \pm i(u_X v_Y + u_Y v_X)]. \tag{5.118}$$

Since a second-rank cartesian tensor $T_{\alpha\beta}$ transforms in the same way as the set of products $u_\alpha v_\beta$, it can also be expressed in terms of a scalar (which is the trace $\sum_\alpha T_{\alpha\alpha}$), a vector (the three components of the antisymmetric tensor $(1/2)(T_{\alpha\beta} - T_{\beta\alpha})$), and a second-rank spherical tensor (the five components of the traceless, symmetric tensor, $(1/2)(T_{\alpha\beta} + T_{\beta\alpha}) - (1/3) \sum_\alpha T_{\alpha\alpha}$). The explicit irreducible spherical tensor components can be obtained from equations (5.114) to (5.118) simply by replacing $u_\alpha v_\beta$ by $T_{\alpha\beta}$. These results are collected in table 5.2. It often happens that these three spherical tensors with $k = 0$, 1 and 2 occur in real, physical situations. In any given situation, one or more of them may vanish; for example, all the components of T^1 are zero if the tensor is symmetric, $T_{\alpha\beta} = T_{\beta\alpha}$. A well-known example of a second-rank spherical tensor is the electric quadrupole moment. Its components are defined by

$$eQ_{\alpha\beta} = \sum_i q_i \alpha_i \beta_i, \tag{5.119}$$

where the sum is performed over all charges which make up the system, for example, a molecule or a nucleus. Expressed as the components of a second-rank spherical tensor, these become

$$eT_q^2(\boldsymbol{Q}) = \sum_i q_i r_i^2 C_q^2(\theta_i, \phi_i) \tag{5.120}$$

where (r_i, θ_i, ϕ_i) are the spherical polar coordinates of the point charge q_i.

5.5.3. Matrix elements of spherical tensor operators: the Wigner–Eckart theorem

We now consider how to evaluate the matrix elements of a spherical tensor operator, written as $\langle \eta, j, m | T_q^k(A) | \eta', j', m' \rangle$ where η and η' denote any further quantum numbers required to characterise the states (for example, vibrational quantum numbers v). If we now rotate the bra, operator and ket through Euler angles ω using equations (5.44) and (5.102), the result must be unaffected. Thus:

$$\langle \eta, j, m | T_q^k(A) | \eta', j', m' \rangle = \sum_{nn'p} \mathcal{D}_{nm}^{(j)}(\omega)^* \mathcal{D}_{pq}^{(k)}(\omega) \mathcal{D}_{n'm'}^{(j')}(\omega) \langle \eta, j, n | T_p^k(A) | \eta', j', n' \rangle. \tag{5.121}$$

We now integrate over all ω, making use of equation (5.100), and divide each side by $8\pi^2$. The result is

$$\langle \eta, j, m | T_q^k(A) | \eta', j', m' \rangle$$

$$= \sum_{nn'p} (-1)^{n-m} \begin{pmatrix} j & k & j' \\ -m & q & m' \end{pmatrix} \begin{pmatrix} j & k & j' \\ -n & p & n' \end{pmatrix} \langle \eta, j, n | T_p^k(A) | \eta', j', n' \rangle$$

$$= (-1)^{j-m} \begin{pmatrix} j & k & j' \\ -m & q & m' \end{pmatrix} \Bigg\{ \sum_{nn'p} (-1)^{j-n} \begin{pmatrix} j & k & j' \\ -n & p & n' \end{pmatrix}$$

$$\times \langle \eta, j, n | T_p^k(A) | \eta', j', n' \rangle \Bigg\}. \tag{5.122}$$

The quantity in braces in the second line of (5.122) is independent of the projection quantum numbers because it is a summation over all possible values of n, n' and p. Thus we have

$$\langle \eta, j, m | T_q^k(A) | \eta', j', m' \rangle = (-1)^{j-m} \begin{pmatrix} j & k & j' \\ -m & q & m' \end{pmatrix} \langle \eta, j \| T^k(A) \| \eta', j' \rangle. \tag{5.123}$$

This is the *Wigner–Eckart theorem*, a very important result which underpins most applications of angular momentum theory to quantum mechanics. It states that the required matrix element can be written as the product of a 3-j symbol and a phase factor, which expresses all the angular dependence, and the *reduced matrix element* $\langle \eta, j \| T^k(A) \| \eta', j' \rangle$ which is independent of component quantum numbers and hence of orientation. Thus one quantity is sufficient to determine all $(2j+1) \times (2k+1) \times (2j'+1)$ possible matrix elements $\langle \eta, j, m | T_q^k(A) | \eta', j', m' \rangle$. The phase factor arises because the bra $\langle \eta, j, m |$ transforms in the same way as the ket $(-1)^{j-m} | \eta, j, -m \rangle$. The definition of the reduced matrix element in equation (5.123), which is due to Edmonds [1] and also favoured by Zare [4], is the one we shall use throughout this book. The alternative definition, promoted by Brink and Satchler [3],

$$(J \| T^k(A) \| J') = [2J+1]^{-1/2} \langle J \| T^k(A) \| J' \rangle, \tag{5.124}$$

is less widely used in molecular quantum mechanics. The Wigner–Eckart theorem, when expressed in terms of Clebsch–Gordan coefficients, has the form

$$\langle \eta, j, m | T^k_q(A) | \eta', j', m' \rangle = (-1)^{2k} \langle j'km'q | jm \rangle [2j+1]^{-1/2} \langle \eta, j \| T^k(A) \| \eta', j' \rangle. \tag{5.125}$$

We next consider how to evaluate the reduced matrix element. Although it is defined in equation (5.123), this is not a useful relationship for its evaluation. The usual approach is to calculate the matrix element and then derive $\langle \eta, j \| T^k(A) \| \eta', j' \rangle$ from the result. The evaluation of the reduced matrix element for the total angular momentum $\boldsymbol{J}$ itself is a good example. From the Wigner–Eckart theorem, we write

$$\langle j, m | T^1_p(\boldsymbol{J}) | j', m' \rangle = (-1)^{j-m} \begin{pmatrix} j & 1 & j' \\ -m & p & m' \end{pmatrix} \langle j \| T^1(\boldsymbol{J}) \| j' \rangle. \tag{5.126}$$

Taking the $p = 0$ component we can write

$$\langle j, m | J_Z | j', m' \rangle = \delta_{jj'} \delta_{mm'} m. \tag{5.127}$$

The analytical expansion of the 3-j symbol is (see General Appendix C):

$$\begin{pmatrix} j & 1 & j \\ -m & 0 & m \end{pmatrix} = (-1)^{j-m} m [j(j+1)(2j+1)]^{-1/2}. \tag{5.128}$$

Hence we have the result

$$\langle j \| T^1(\boldsymbol{J}) \| j' \rangle = \delta_{jj'} [j(j+1)(2j+1)]^{1/2}. \tag{5.129}$$

We can also form a second rank tensor by coupling the angular momentum $\boldsymbol{J}$ with itself. In this case, the Wigner–Eckart theorem is expressed as;

$$\langle j, m | T^2_p(\boldsymbol{J}, \boldsymbol{J}) | j', m' \rangle = (-1)^{j-m} \begin{pmatrix} j & 2 & j' \\ -m & p & m' \end{pmatrix} \langle j \| T^2(\boldsymbol{J}, \boldsymbol{J}) \| j' \rangle. \tag{5.130}$$

If we again take the $p = 0$ component (from Table 5.2), we have

$$\langle j, m | (1/\sqrt{6})[3J_Z^2 - \boldsymbol{J}^2] | j', m' \rangle = \delta_{jj'} \delta_{mm'} (1/\sqrt{6}) \{3m^2 - j(j+1)\}. \tag{5.131}$$

Again we use the analytical expression for the 3-j symbol in (5.130) which is:

$$\begin{pmatrix} j & 2 & j \\ -m & 0 & m \end{pmatrix} = (-1)^{j-m} \frac{2[3m^2 - j(j+1)]}{[(2j-1)(2j)(2j+1)(2j+2)(2j+3)]^{1/2}}. \tag{5.132}$$

Combining the results given in the last three equations we obtain:

$$\langle j \| T^2(\boldsymbol{J}, \boldsymbol{J}) \| j' \rangle = \delta_{jj'} \frac{1}{2\sqrt{6}} [(2j-1)(2j)(2j+1)(2j+2)(2j+3)]^{1/2}. \tag{5.133}$$

Smith and Thornley [13] have derived a general expression for the reduced matrix elements of this type, namely,

$$\langle j \| T^k(\boldsymbol{J}, \ldots, \boldsymbol{J}) \| j' \rangle = \delta_{jj'} k! \left[\frac{(2j+k+1)!}{2^k (2k)! (2j-k)!} \right]^{1/2}. \tag{5.134}$$

Another immediate corollary of the Wigner–Eckart theorem is the *replacement theorem*, which allows one to write the matrix elements of one spherical tensor operator,

$T^k(S)$, in terms of another $T^k(R)$:

$$\langle \eta, j, m | T_p^k(S) | \eta', j', m' \rangle = \frac{\langle \eta, j \| T^k(S) \| \eta', j' \rangle}{\langle \eta, j \| T^k(R) \| \eta', j' \rangle} \langle \eta, j, m | T_p^k(R) | \eta', j', m' \rangle. \quad (5.135)$$

When the replacement theorem is used, $T_p^k(R)$ is often constructed from components of $\boldsymbol{J}$ (total angular momentum) or $\boldsymbol{I}$ (total nuclear spin). The first- and second-rank spherical tensor components of $\boldsymbol{J}$ (or $\boldsymbol{I}$) are given in equations (5.108) and equations (5.116) to (5.118) respectively. However, since $\boldsymbol{J}$ only connects states of the same j value, these operators can only be used as replacements for other angular momentum operators when the effects of mixing of states with $j \neq j'$ can be ignored. Similar remarks apply to the use of $\boldsymbol{I}$ as the replacement operator although the effects of mixing of states with $\boldsymbol{I}' \neq \boldsymbol{I}$ are not usually significant in molecular physics.

5.5.4. Matrix elements for composite systems

When states or operators can be characterised by a single angular momentum label, that is one j value and one m value, the evaluation of a matrix element such as $\langle j, m | T_p^k(j) | j', m' \rangle$ is simply a matter of coupling j' and k together and seeking the component which transforms like j under rotations, or equivalently, of coupling j, k and j' together to form a scalar. When states are characterised by several angular momenta, however, life becomes more complicated and we must use the more powerful techniques developed by Racah [14] to evaluate matrix elements; we have already introduced these methods in section 5.4.3.

It is a common occurrence that we wish to evaluate the reduced matrix element of an operator which acts on only one part of a coupled scheme. For example, the general formula for the reduced matrix element of an operator $T^k(A_1)$ which acts only on part 1 of a coupled scheme $\boldsymbol{j}_1 + \boldsymbol{j}_2 = \boldsymbol{j}$ is:

$$\langle j_1, j_2, j \| T^k(A_1) \| j_1', j_2', j' \rangle$$
$$= \delta_{j_2 j_2'} (-1)^{j_1' + j_2 + j' + k} [(2j+1)(2j'+1)]^{1/2} \begin{Bmatrix} j_1' & j' & j \\ j & j_1 & k \end{Bmatrix} \langle j_1 \| T^1(A_1) \| j_1' \rangle. \quad (5.136)$$

To derive this result, we express a typical matrix element in terms of the coupled representation on the one hand and in terms of the decoupled on the other:

$$\langle j_1, j_2, j, m | T_p^k(A_1) | j_1', j_2', j', m' \rangle$$

$$= (-1)^{j-m} \begin{pmatrix} j & k & j' \\ -m & p & m' \end{pmatrix} \langle j_1, j_2, j \| T^1(A_1) \| j_1', j_2', j' \rangle$$

$$= \sum_{m_1 m_2 m_1' m_2'} \langle j_1, j_2, m_1, m_2 | T_p^k(A_1) | j_1', j_2', m_1', m_2' \rangle$$

$$\times \langle j_1, j_2, m_1, m_2 | j, m \rangle \langle j_1', j_2', m_1', m_2' | j', m' \rangle$$

$$= \sum_{m_1 m_2 m_1'} \delta_{j_2 j_2'} \delta_{m_2 m_2'} \langle j_1 \| T^k(A_1) \| j_1' \rangle (-1)^{j_1 - m_1} \begin{pmatrix} j_1 & k & j_1' \\ -m_1 & p & m_1' \end{pmatrix}$$

$$\times \langle j_1, j_2, m_1, m_2 | j, m \rangle \langle j_1', j_2', m_1', m_2' | j, m \rangle. \quad (5.137)$$

The rest of the derivation is performed by expressing the Clebsch–Gordan coefficients in terms of 3-j symbols and making use of equation (5.92). The corresponding equation for a spherical tensor operator $T^k(A_2)$ which acts only on part 2 of the coupled scheme is

$$\langle j_1, j_2, j \| T^k(A_2) \| j_1', j_2', j' \rangle = \delta_{j_1 j_1'} (-1)^{j_1 + j_2' + j + k} [(2j+1)(2j'+1)]^{1/2}$$

$$\times \begin{Bmatrix} j_2' & j' & j \\ j & j_2 & k \end{Bmatrix} \langle j_2 \| T^k(A_2) \| j_2' \rangle. \tag{5.138}$$

Equations (5.136) and (5.138) are special cases of a more general result for the reduced matrix elements of a tensor operator $T^K(k_1, k_2)$ obtained by coupling together $T^{k_1}(A_1)$ and $T^{k_2}(A_2)$ which act on parts 1 and 2 respectively of the composite system. In this case, we require the reduced matrix element $\langle j_1, j_2, j \| T^K(k_1, k_2) \| j_1', j_2', j' \rangle$. We get it by considering the transformation from the coupling scheme $((j_1', j_2')j', (k_1, k_2)K)j$ to the coupling $((j_1', k_1)j_1, (j_2', k_2)j_2)j$. The result is

$$\langle j_1, j_2, j \| T^k(k_1, k_2) \| j_1', j_2', j' \rangle = [(2j+1)(2k+1)(2j'+1)]^{1/2} \begin{Bmatrix} j_1' & j_2' & j' \\ k_1 & k_2 & K \\ j_1 & j_2 & j \end{Bmatrix}$$

$$\times \langle j_1 \| T^{k_1}(A_1) \| j_1' \rangle \langle j_2 \| T^{k_2}(A_2) \| j_2' \rangle. \tag{5.139}$$

Equation (5.136) results from this equation when we set $T^{k_2}(A_2) = 1$ with $k_2 = 0$ and equation (5.138) follows when $T^{k_1}(A_1) = 1$ and $k_1 = 0$. Another important result arises when $k_1 = k_2 = k$ and $K = 0$. In this case T_0^0 is just the scalar product $T^k(A_1) \cdot T^k(A_2)$, apart from a phase and a normalisation factor, see equation (5.110). In this case, we can use equation (5.94) to replace the 9-j symbol by a 6-j symbol and so arrive at the important result

$$\langle j_1, j_2, j \| T^k(A_1) \cdot T^k(A_2) \| j_1', j_2', j' \rangle$$

$$= \delta_{j,j'} (2j+1)^{1/2} (-1)^{j_1' + j_2 + j} \begin{Bmatrix} j_1' & j_2' & j \\ j_2 & j_1 & k \end{Bmatrix} \langle j_1 \| T^k(A_1) \| j_1' \rangle \langle j_2 \| T^k(A_2) \| j_2' \rangle.$$

$$\tag{5.140}$$

Finally, we consider the composite tensor $T^k(A_1, B_1)$ which is the tensor product of $T^{k_1}(A_1)$ and $T^{k_2}(B_1)$, both of which act on part 1 only of the coupled scheme:

$$T_p^K(A_1, B_1) = (-1)^{k_1 - k_2 + p} (2K+1)^{1/2} \sum_{p_1 p_2} \begin{pmatrix} k_1 & k_2 & K \\ p_1 & p_2 & -p \end{pmatrix} T_{p_1}^{k_1}(A_1) T_{p_2}^{k_2}(B_1). \tag{5.141}$$

In this case, the reduced matrix element of $T^K(A_1, B_1)$ is related to those of $T^{k_1}(A_1)$ and $T^{k_2}(B_1)$ by

$$\langle \eta, j \| T^K(A_1, B_1) \| \eta', j' \rangle = (2K+1)^{1/2} (-1)^{K+j+j'} \sum_{\eta'' j''} \begin{Bmatrix} k_1 & k_2 & K \\ j' & j & j'' \end{Bmatrix}$$

$$\times \langle \eta, j \| T^{k_1}(A_1) \| \eta'', j'' \rangle \langle \eta'', j'' \| T^{k_2}(B_1) \| \eta', j' \rangle.$$

$$\tag{5.142}$$

5.5.5. Relationship between operators in space-fixed and molecule-fixed coordinate systems

In molecular quantum mechanics, we often find ourselves manipulating expressions so that one of a pair of interacting operators is expressed in laboratory-fixed coordinates while the other is expressed in molecule-fixed. A typical example is the Stark effect, where the molecular electric dipole moment is naturally described in the molecular framework, but the direction of an applied electric field is specified in space-fixed coordinates. We have seen already that if the molecule-fixed axes are obtained by rotation of the space-fixed axes through the Euler angles $(\phi, \theta, \chi) = \omega$, the spherical tensor operator in the laboratory-fixed system $T_p^k(A)$ can be expressed in terms of the molecule-fixed components by the standard transformation

$$T_p^k(A) = \sum_q \mathfrak{D}_{pq}^{(k)}(\omega)^* T_q^k(A), \tag{5.143}$$

where $\mathfrak{D}_{pq}^{(k)}(\omega)^*$ is the complex conjugate of the pq element of the kth rank rotation matrix $\mathfrak{D}^{(k)}(\omega)$. In this book, we use p to label space-fixed components and q to label molecule-fixed. The inverse of equation (5.143) is

$$T_q^k(A) = \sum_p \mathfrak{D}_{pq}^{(k)}(\omega) T_p^k(A) = \sum_p (-1)^{p-q} \mathfrak{D}_{-p,-q}^{(k)}(\omega)^* T_p^k(A). \tag{5.144}$$

We recall also that the $\mathfrak{D}_{pq}^{(k)}(\omega)^*$ are proportional to the eigenfunctions of a symmetric top rotational Hamiltonian (section 5.3.4). Realising that a diatomic molecule behaves as a symmetric top (albeit a rather special one), we write the rotational part of the wave function as

$$|J, \Omega, M\rangle = [(2J+1)/8\pi^2]^{1/2} \mathfrak{D}_{M\Omega}^{(J)}(\omega)^*. \tag{5.145}$$

From this, we can derive two other useful relationships for tensor operators which can be quantised in both molecule- and space-fixed coordinate systems:

$$\langle J, \Omega, M | \mathfrak{D}_{pq}^{(k)}(\omega)^* | J', \Omega', M' \rangle = (-1)^{M-\Omega} [(2J+1)(2J'+1)]^{1/2}$$
$$\times \begin{pmatrix} J & k & J' \\ -\Omega & q & \Omega' \end{pmatrix} \begin{pmatrix} J & k & J' \\ -M & p & M' \end{pmatrix}. \tag{5.146}$$

This result follows directly from equation (5.100) in conjunction with the relationship

$$\mathfrak{D}_{pq}^{(k)}(\omega)^* = (-)^{p-q} \mathfrak{D}_{-p,-q}^{(k)}(\omega). \tag{5.147}$$

Applying the Wigner–Eckart theorem to this equation, we obtain an expression for the reduced matrix element:

$$\langle J, \Omega, M \| \mathfrak{D}_{\cdot q}^{(k)}(\omega)^* \| J', \Omega', M' \rangle = (-1)^{J-\Omega} [(2J+1)(2J'+1)]^{1/2} \begin{pmatrix} J & k & J' \\ -\Omega & q & \Omega' \end{pmatrix}. \tag{5.148}$$

The dot in the first subscript of the rotation matrix shows that this matrix element is reduced as far as orientation in the space-fixed coordinate system is concerned, but not for the molecule-fixed axes.

We will make use of the results derived in this section many times elsewhere in this book.

5.5.6. Treatment of the anomalous commutation relationships of rotational angular momenta by spherical tensor methods

We have seen in section 5.5.1 that the definition of a spherical tensor depends on the components of the rotational angular momentum obeying standard commutation relationships:

$$[J_X, J_Y] = i J_Z, \text{ etc.} \tag{5.149}$$

For example, the components of the spherical tensor operator $T_p^1(\boldsymbol{J})$ obey the definition:

$$\begin{aligned} [J_Z, T_p^1(\boldsymbol{J})] &= p T_p^1(\boldsymbol{J}), \\ [J_\pm, T_p^1(\boldsymbol{J})] &= [(1)(2) - p(p \pm 1)]^{1/2} T_{p\pm 1}^1(\boldsymbol{J}), \end{aligned} \tag{5.150}$$

only if the normal commutation relationships hold. We have also seen in section 5.3.1 that, if we refer the angular momentum operator to a rotating, molecule-fixed axis system,

$$J_x = \lambda_{xX} J_X + \lambda_{xY} J_Y + \lambda_{xZ} J_Z, \text{ etc.} \tag{5.151}$$

(where $\lambda_{xX}, \lambda_{xY}$ and λ_{xZ} are the direction cosines), we find that such components obey anomalous commutation relationships:

$$[J_x, J_y] = -i J_z, \text{ etc.} \tag{5.152}$$

It is tempting to construct spherical tensors from $\boldsymbol{J}$ acting within the molecule-fixed coordinate system. From Table 5.2 the components would be expected to have the form:

$$\begin{aligned} T_{q=0}(\boldsymbol{J}) &= J_z, \\ T_{q=\pm 1}(\boldsymbol{J}) &= \mp (1/\sqrt{2})(J_x \pm i J_y). \end{aligned} \tag{5.153}$$

However, when we come to check whether such operators satisfy the general definition of spherical tensor operators, equation (5.107), we find that they do *not* because of the anomalous commutation relations, equation (5.42). It is important to appreciate that such difficulties arise only for the operators which represent rotational angular momenta in three-dimensional space. Thus it is only the angular momentum $\boldsymbol{J}$ which shows anomalous behaviour for a Hund's case (a) coupling scheme, and the angular momentum $\boldsymbol{N}$ for Hund's case (b). These are angular momenta associated with rotation of the molecule-fixed frame; the operators depend on the same angular coordinates as those which define the orientation of the coordinate system. For the internal angular momenta of a molecule, such as $\boldsymbol{L}$ or $\boldsymbol{S}$, the normal commutation relationships are obeyed by components in either space- or molecule-fixed coordinate systems. There is therefore no difficulty in constructing spherical tensors for such operators in any axis system. The essential difference in behaviour of these operators can be appreciated by writing the expressions for the molecule-fixed components such as:

$$L_x = \lambda_{xX} L_X + \lambda_{xY} L_Y + \lambda_{xZ} L_Z, \text{ etc.} \tag{5.154}$$

Each term on the right-hand side of this equation consists of the product of a direction cosine and an orbital angular momentum operator. Of these two factors, only the first depends on the Euler angles which define the instantaneous orientation of the molecule. In the corresponding equation (5.151) for J_x, however, both factors depend on the Euler angles.

Various methods have been developed for dealing with the anomalous commutation relationships in molecular quantum mechanics, chief among them being Van Vleck's reversed angular momentum method [10]. Most of these methods are rather complicated and require the introduction of an array of new symbols. Brown and Howard [15], however, have pointed out that it is quite possible to handle these difficulties within the standard framework of spherical tensor algebra. If matrix elements are evaluated directly in laboratory-fixed coordinates and components are referred to axes mounted on the molecule only when necessary, it is possible to avoid the anomalous commutation relationships completely. Only the standard equations given earlier in this chapter are used to derive the required results; it is just necessary to keep a cool head in the process!

Let us consider the scalar product of a rotation operator $\boldsymbol{J}$ and an internal angular momentum $\boldsymbol{P}$ which is quantised in the molecule-fixed axis system. Although it might appear sensible to evaluate $\boldsymbol{J} \cdot \boldsymbol{P}$ in a molecule-fixed axis system where both angular momenta operate, we shall instead expand the tensor product in a space-fixed axis system and then refer the components of $\boldsymbol{P}$ to the molecular axis system using a rotation matrix $\mathfrak{D}^{(1)}_{pq}(\omega)$:

$$\boldsymbol{J} \cdot \boldsymbol{P} = \sum_p (-1)^p \mathrm{T}^1_p(\boldsymbol{J}) \mathrm{T}^1_{-p}(\boldsymbol{P}), \tag{5.155}$$

and

$$\mathrm{T}^1_p(\boldsymbol{P}) = \sum_q \mathfrak{D}^{(1)}_{pq}(\omega)^* \mathrm{T}^1_q(\boldsymbol{P}). \tag{5.156}$$

Here, as usual, the suffices p and q refer to space- and molecule-fixed components respectively and ω stands for the three Euler angles (ϕ, θ, χ) which relate the two coordinate systems. From equations (5.155) and (5.156) we have

$$\boldsymbol{J} \cdot \boldsymbol{P} = \sum_{p,q} (-1)^p \mathrm{T}^1_p(\boldsymbol{J}) \mathfrak{D}^{(1)}_{-pq}(\omega)^* \mathrm{T}^1_q(\boldsymbol{P}). \tag{5.157}$$

The right-hand side of this equation appears to be non-Hermitian since $\mathrm{T}^1_p(\boldsymbol{J})$ does not commute with $\mathfrak{D}^{(1)}_{pq}(\omega)^*$ but the summation over p removes all non-commuting terms.

To take a specific example, let us consider $\boldsymbol{P} = \boldsymbol{S}$, the electron spin angular momentum for a diatomic molecule in a Hund's case (a) coupling scheme where the basis functions are simple products of orbital, rotational and spin functions. Using standard nomenclature, we have

$$\langle \eta, \Lambda; J, \Omega, M; S, \Sigma | \mathbf{J} \cdot \mathbf{S} | \eta', \Lambda'; J', \Omega', M'; S', \Sigma' \rangle$$
$$= \sum_{p,q} (-1)^p \langle J, \Omega, M | T_p^1(\mathbf{J}) \mathfrak{D}_{-pq}^{(1)}(\omega)^* | J', \Omega', M' \rangle \langle \eta, \Lambda, S, \Sigma | T_q^1(\mathbf{S}) | \eta', \Lambda', S', \Sigma' \rangle$$
$$= \sum_{p,q} (-1)^p \langle J, \Omega, M | T_p^1(\mathbf{J}) \mathfrak{D}_{-pq}^{(1)}(\omega)^* | J', \Omega', M' \rangle \langle \eta, \Lambda | \eta', \Lambda' \rangle \langle S, \Sigma | T_q^1(\mathbf{S}) | S', \Sigma' \rangle. \quad (5.158)$$

The spin term is straightforward to evaluate by the Wigner–Eckart theorem but the rotational term requires further consideration. Let us introduce the projection operator onto the complete set of rotational functions between the operators $T_p^1(\mathbf{J})$ and $\mathfrak{D}_{-pq}^{(1)}(\omega)^*$ (the closure relationship):

$$\sum_p (-1)^p \langle J, \Omega, M | T_p^1(\mathbf{J}) \mathfrak{D}_{-pq}^{(1)}(\omega)^* | J', \Omega', M' \rangle$$
$$= \sum_p \sum_{J''\Omega''M''} (-1)^p \langle J, \Omega, M | T_p^1(\mathbf{J}) | J'', \Omega'', M'' \rangle$$
$$\times \langle J'', \Omega'', M'' | \mathfrak{D}_{-pq}^{(1)}(\omega)^* | J', \Omega', M' \rangle. \quad (5.159)$$

For non-zero results, the first matrix element requires $J'' = J$, $M'' = M - p$ and $\Omega'' = \Omega$, see equation (5.129). Thus

$$\text{RHS} = \sum_p (-1)^p \langle J, \Omega, M | T_p^1(\mathbf{J}) | J, \Omega, M - p \rangle \langle J, \Omega, M - p | \mathfrak{D}_{-pq}^{(1)}(\omega)^* | J', \Omega', M' \rangle$$
$$= \sum_p (-1)^p (-1)^{J-M} \begin{pmatrix} J & 1 & J \\ -M & p & M-p \end{pmatrix}$$
$$\times [J(J+1)(2J+1)]^{1/2} (-1)^{J-M+p} (-1)^{J-\Omega} \begin{pmatrix} J & 1 & J' \\ -\Omega & q & \Omega' \end{pmatrix}$$
$$\times \begin{pmatrix} J & 1 & J' \\ -M+p & -p & M' \end{pmatrix} [(2J+1)(2J'+1)]^{1/2}, \quad (5.160)$$

by the use of equation (5.146). Using the orthogonality of the 3-j symbols,

$$\sum_p \begin{pmatrix} J & 1 & J \\ -M & p & M-p \end{pmatrix} \begin{pmatrix} J & 1 & J' \\ -M+p & -p & M' \end{pmatrix} = (2J+1)^{-1} \delta_{JJ'} \delta_{MM'}, \quad (5.161)$$

we can simplify equation (5.160) to give

$$\sum_p (-1)^p \langle J, \Omega, M | T_p^1(\mathbf{J}) \mathfrak{D}_{-pq}^{(1)}(\omega)^* | J', \Omega', M' \rangle$$
$$= (-1)^{J-\Omega} \begin{pmatrix} J & 1 & J' \\ -\Omega & q & \Omega' \end{pmatrix} [J(J+1)(2J+1)]^{1/2} \delta_{JJ'} \delta_{MM'}. \quad (5.162)$$

In this way, it can be seen that all the equipment required to evaluate matrix elements in molecular angular momenta coupling problems exists in standard spherical tensor theory, irrespective of complexity. It may appear rather laborious to evaluate the matrix elements of the rotational angular momentum operator in this way but it should be appreciated that the result in equation (5.162) needs only to be derived once; thereafter it is only the result which is required. There is thus some point in defining a symbol for the operator combination on the left hand side of equation (5.162), such as

$$J_q \equiv \sum_p (-1)^p T_p^1(\boldsymbol{J}) \mathfrak{D}_{-pq}^{(1)}(\omega)^*. \tag{5.163}$$

Although J_q behaves rather like a tensor operator in the molecule-fixed axis system, we must remember that *it is not a tensor operator* because it does not satisfy the conditions in (5.152). Elsewhere in this book, we shall often make use of this short hand. We expand a scalar product $\boldsymbol{J} \cdot \boldsymbol{P}$ as

$$\boldsymbol{J} \cdot \boldsymbol{P} = \sum_q J_q T_q^1(\boldsymbol{P}), \tag{5.164}$$

and then use equation (5.162) for the matrix elements of J_q.

Appendix 5.1. Summary of standard results from spherical tensor algebra

In our own work on both diatomic and polyatomic molecules, we have found it valuable to have a summary of the most important results from irreducible spherical tensor algebra, particularly those relating to the evaluation of matrix elements in various angular momentum coupling schemes. We now provide a summary of those results; detailed derivations are, of course, to be found in the main body of the text.

(i) *Angular momenta and tensor operators*
  $T_{p_1}^{k_1}(A_1)$ is a tensor of rank k_1 with components p_1 which operates on angular momentum $\boldsymbol{j}_1$.
  $T_{p_2}^{k_2}(A_2)$ is a tensor of rank k_2 with components p_2 which operates on angular momentum $\boldsymbol{j}_2$.
  $T_{p_3}^{k_3}(A_3)$ is a tensor of rank k_3 with components p_3 which operates on angular momentum $\boldsymbol{j}_3$.
  $\boldsymbol{j}_1, \boldsymbol{j}_2$ and $\boldsymbol{j}_3$ all commute with each other.

(ii) *Tensor product*
  The tensor product of $T_{p_1}^{k_1}(A_1)$ and $T_{p_2}^{k_2}(A_2)$ is defined by

$$T^{k_1}(A_1) \times T^{k_2}(A_2) = W_{p_{12}}^{k_{12}}(k_1, k_2)$$

$$= \sum_{p_1} T_{p_1}^{k_1}(A_1) T_{p_{12}-p_1}^{k_2}(A_2) \begin{pmatrix} k_1 & k_2 & k_{12} \\ p_1 & p_{12}-p_1 & -p_{12} \end{pmatrix}$$

$$\times (2k_{12}+1)^{1/2}(-1)^{-k_1+k_2-p_{12}}. \tag{5.165}$$

(iii) *Scalar product*

The scalar product is obtained from the tensor product by putting $k_{12} = p_{12} = 0$; it is only defined for $k_1 = k_2 = k$ (say). Then

$$W_0^0(k,k) = \sum_p T_p^k(A_1) T_{-p}^k(A_2) \begin{pmatrix} k & k & 0 \\ p & -p & 0 \end{pmatrix}$$

$$= \sum_p T_p^k(A_1) T_{-p}^k(A_2) (-1)^{k-p} (2k+1)^{-1/2}$$

$$= T^k(A_1) \cdot T^k(A_2) (-1)^k (2k+1)^{-1/2}, \quad (5.166)$$

where the scalar product is given by

$$T^k(A_1) \cdot T^k(A_2) = \sum_p (-1)^p T_p^k(A_1) T_{-p}^k(A_2). \quad (5.167)$$

Hence

$$T^k(A_1) \cdot T^k(A_2) = W_0^0(k,k) (-1)^k (2k+1)^{1/2}. \quad (5.168)$$

(iv) *Matrix elements of a tensor product*

If angular momenta j_1 and j_2 couple to form j_{12} and the tensor product is defined by $T^{k_1}(A_1) \times T^{k_2}(A_2) = W^{k_{12}}(k_1, k_2)$, then the reduced matrix elements of the tensor product are given by

$$\langle j_1, j_2, j_{12} \| W^{k_{12}} \| j_1', j_2', j_{12}' \rangle$$

$$= [(2j_{12}+1)(2j_{12}'+1)(2k_{12}+1)]^{1/2} \begin{Bmatrix} j_{12} & j_{12}' & k_{12} \\ j_1 & j_1' & k_1 \\ j_2 & j_2' & k_2 \end{Bmatrix}$$

$$\times \langle j_1 \| T^{k_1}(A_1) \| j_1' \rangle \langle j_2 \| T^{k_2}(A_2) \| j_2' \rangle. \quad (5.169)$$

Similarly if j_{12} is coupled with j_3 to form j, and

$$[T^{k_1}(A_1) \times T^{k_2}(A_2)] \times T^{k_3}(A_3) = X^k, \quad (5.170)$$

then the reduced matrix elements of X^k are as follows:

$$\langle j_1, j_2, j_{12}, j_3, j \| X^k \| j_1', j_2', j_{12}', j_3', j' \rangle$$

$$= [(2j+1)(2j'+1)(2k+1)]^{1/2} \begin{Bmatrix} j & j' & k \\ j_{12} & j_{12}' & k_{12} \\ j_3 & j_3' & k_3 \end{Bmatrix} \langle j_{12} \| W^{k_{12}} \| j_{12}' \rangle \langle j_3 \| T^{k_3}(A_3) \| j_3' \rangle$$

$$= [(2j_{12}+1)(2j_{12}'+1)(2k_{12}+1)(2j+1)(2j'+1)(2k+1)]^{1/2} \begin{Bmatrix} j & j' & k \\ j_{12} & j_{12}' & k_{12} \\ j_3 & j_3' & k_3 \end{Bmatrix}$$

$$\times \begin{Bmatrix} j_{12} & j_{12}' & k_{12} \\ j_1 & j_1' & k_1 \\ j_2 & j_2' & k_2 \end{Bmatrix} \langle j_1 \| T^{k_1}(A_1) \| j_1' \rangle \langle j_2 \| T^{k_2}(A_2) \| j_2' \rangle \langle j_3 \| T^{k_3}(A_3) \| j_3' \rangle. \quad (5.171)$$

(v) *Wigner–Eckart theorem*
Reduced matrix elements are defined by using the Wigner–Eckart theorem to evaluate the dependence on projection quantum numbers:

$$\langle j, m | T^k_p(A) | j', m' \rangle = (-1)^{j-m} \begin{pmatrix} j & k & j' \\ -m & p & m' \end{pmatrix} \langle j \| T^k(A) \| j' \rangle. \quad (5.172)$$

(vi) *Matrix elements of a scalar product*
Using the result in equation (5.169) with $k_1 = k_2 = k$ and $k_{12} = 0$, we obtain:

$$\langle j_1, j_2, j_{12}, m | T^k(A_1) \cdot T^k(A_2) | j'_1, j'_2, j'_{12}, m' \rangle$$

$$= (-1)^{j'_1 + j_{12} + j_2} \delta_{j_{12}, j'_{12}} \delta_{m,m'} \begin{Bmatrix} j'_2 & j'_1 & j_{12} \\ j_1 & j_2 & k \end{Bmatrix} \langle j_1 \| T^k(A_1) \| j'_1 \rangle \langle j_2 \| T^k(A_2) \| j'_2 \rangle. \quad (5.173)$$

(vii) *Matrix elements of a single operator in a coupled scheme*
For the matrix elements of $T^{k_1}(A_1)$ only we put $T^{k_2}(A_2) = 1$ and $k_2 = 0, k_{12} = k_1$ in equation (5.169); we then obtain

$$\langle j_1, j_2, j_{12} \| T^{k_1}(A_1) \| j'_1, j'_2, j'_{12} \rangle$$

$$= [(2j_{12} + 1)(2j'_{12} + 1)(2k_1 + 1)]^{1/2} \begin{Bmatrix} j_{12} & j'_{12} & k_1 \\ j_1 & j'_1 & k_1 \\ j_2 & j_2 & 0 \end{Bmatrix} \langle j_1 \| T^{k_1}(A_1) \| j'_1 \rangle \langle j_2 \| 1 \| j'_2 \rangle$$

$$= \delta_{j_2 j'_2} (-1)^{j'_{12} + j_1 + k_1 + j_2} [(2j_{12} + 1)(2j'_{12} + 1)]^{1/2} \begin{Bmatrix} j'_1 & j'_{12} & j_2 \\ j_{12} & j_1 & k_1 \end{Bmatrix} \langle j_1 \| T^{k_1}(A_1) \| j'_1 \rangle. \quad (5.174)$$

Similarly, for the matrix elements of $T^{k_2}(A_2)$ we have

$$\langle j_1, j_2, j_{12} \| T^{k_2}(A_2) \| j'_1, j'_2, j'_{12} \rangle = \delta_{j_1 j'_1} (-1)^{j_{12} + j_1 + k_2 + j'_2} [(2j_{12} + 1)(2j'_{12} + 1)]^{1/2}$$

$$\times \begin{Bmatrix} j'_2 & j'_{12} & j_1 \\ j_{12} & j_2 & k_2 \end{Bmatrix} \langle j_2 \| T^k(A_2) \| j'_2 \rangle. \quad (5.175)$$

(viii) *Tensor operators acting on the same inner part of a coupled system*
If the scalar product is formed from spherical tensor operators which both act on the same inner part of a coupled scheme, it is intuitively obvious that

$$\langle j_1, j_2, j_{12}, m_{12} | T^k(A_1) \cdot T^k(B_1) | j'_1, j'_2, j'_{12}, m'_{12} \rangle$$
$$= \delta_{j_{12} j'_{12}} \delta_{m_{12} m'_{12}} \delta_{j_2 j'_2} \langle j_1 | T^k(A_1) \cdot T^k(B_1) | j'_1 \rangle. \quad (5.176)$$

This result can be proved formally by application of the Wigner–Eckart theorem, equation (5.172), followed by equation (5.174).

For a tensor operator $W^{k_{12}}(k_1, k_2)$ which acts only on, say, part 1 of a coupled scheme, the reduced matrix element is given by

$$\langle j_1, j_2, j \| W^{k_{12}}(k_1, k_2) \| j_1', j_2', j' \rangle$$

$$= \delta_{jj'} \delta_{j_2 j_2'} (2k_{12}+1)^{1/2} (-1)^{k_{12}+j_1+j_1'} \sum_{j_1''} \begin{Bmatrix} k_1 & j_1 & j_1'' \\ j_1' & k_2 & k_{12} \end{Bmatrix}$$

$$\times \langle j_1 \| T^{k_1}(A_1) \| j_1'' \rangle \langle j_1'' \| T^{k_2}(B_1) \| j_1 \rangle. \tag{5.177}$$

Hence, for the scalar product of two operators which both act on the j_1 part of a coupled scheme, we have

$$\langle j_1, j_2, j_{12}, m_{12} | T^{k_1}(A_1) \cdot T^{k_2}(B_1) | j_1', j_2', j_{12}', m_{12}' \rangle$$

$$= \delta_{j_{12} j_{12}'} \delta_{m_{12} m_{12}'} \delta_{j_2 j_2'} \delta_{j_1 j_1'} \sum_{j_1''} (-1)^{j_1-j_1''} (2j_1+1)^{-1} \langle j_1 \| T^{k_1}(A_1) \| j_1'' \rangle \langle j_1'' \| T^{k_2}(B_1) \| j_1 \rangle.$$

$$\tag{5.178}$$

A similar result applies when two operators act on part 2 of the coupled scheme.

(ix) *Evaluation of reduced matrix elements*

The reduced matrix element of a first-rank tensor is given by

$$\langle j \| T^1(j) \| j' \rangle = \delta_{j,j'} [j(j+1)(2j+1)]^{1/2}. \tag{5.179}$$

The reduced matrix elements of second and third rank tensors are to be found elsewhere in this book (see equations (5.133) and (7.167)).

(x) *Rotational matrices*

Symmetric top eigenfunctions can be expressed in terms of rotational matrices as follows for a Hund's case (a) coupling scheme:

$$|J, \Omega, M\rangle = \sqrt{\frac{(2J+1)}{8\pi^2}} \mathcal{D}_{M\Omega}^{(J)}(\omega)^*. \tag{5.180}$$

The corresponding equation for Hund's case (b) is

$$|N, \Lambda, M\rangle = \sqrt{\frac{(2N+1)}{8\pi^2}} \mathcal{D}_{M\Lambda}^{(N)}(\omega)^*. \tag{5.181}$$

The transformations of a spherical tensor from space (p) to molecular (q) axes, and vice versa, are given by:

$$T_p^k(A) = \sum_q \mathcal{D}_{pq}^{(k)}(\omega)^* T_q^k(A). \tag{5.182}$$

$$T_q^k(A) = \sum_p \mathcal{D}_{pq}^{(k)}(\omega) T_p^k(A). \tag{5.183}$$

The matrix elements of rotational matrices are given by

$$\langle J, \Omega, M | \mathcal{D}_{pq}^{(k)}(\omega)^* | J', \Omega', M' \rangle = [(2J+1)(2J'+1)]^{1/2} (-1)^{M-\Omega}$$

$$\times \begin{pmatrix} J & k & J' \\ -M & p & M' \end{pmatrix} \begin{pmatrix} J & k & J' \\ -\Omega & q & \Omega' \end{pmatrix}. \tag{5.184}$$

From the Wigner–Eckart theorem we obtain

$$\langle J, \Omega, M|\mathfrak{D}^{(k)}_{pq}(\omega)^*|J', \Omega', M'\rangle = (-1)^{J-M} \begin{pmatrix} J & k & J' \\ -M & p & M' \end{pmatrix} \langle J, \Omega\|\mathfrak{D}^{(k)}_{\cdot q}(\omega)^*\|J', \Omega'\rangle. \quad (5.185)$$

Hence by combining equations (5.185) and (5.184) we obtain the result:

$$\langle J, \Omega\|\mathfrak{D}^{(k)}_{\cdot q}(\omega)^*\|J', \Omega'\rangle = (-1)^{J-\Omega} \begin{pmatrix} J & k & J' \\ -\Omega & q & \Omega' \end{pmatrix} [(2J+1)(2J'+1)]^{1/2}. \quad (5.186)$$

(xi) *Relationship between the matrix elements of a scalar product in coupled and decoupled representations*

The coupled and decoupled representations are related by

$$|j_1, j_2, j, m\rangle = (-1)^{j_2-j_1+m}(2j+1)^{1/2} \sum_{m_1 m_2} \begin{pmatrix} j_1 & j_2 & j \\ m_1 & m_2 & -m \end{pmatrix} |j_1, m_1\rangle |j_2, m_2\rangle. \quad (5.187)$$

Hence the matrix elements of a scalar product $T^k(A_1) \cdot T^k(A_2)$ in a decoupled basis set can be obtained from the corresponding expressions in the coupled basis set simply by making the replacement:

$$\delta_{mm'} \delta_{jj'} (-1)^{j_1+j_2+j} \begin{Bmatrix} j'_1 & j'_2 & j \\ j_2 & j_1 & k \end{Bmatrix}$$

$$\to \sum_p (-1)^p (-1)^{j_1-m_1} \begin{pmatrix} j_1 & k & j'_1 \\ -m_1 & p & m'_1 \end{pmatrix} (-1)^{j_2-m_2} \begin{pmatrix} j_2 & k & j'_2 \\ -m_2 & -p & m'_2 \end{pmatrix}. \quad (5.188)$$

References

[1] A.R. Edmonds, *Angular Momentum in Quantum Mechanics*, Princeton University Press, Princeton, 1960.
[2] M.E. Rose, *Elementary Theory of Angular Momentum*, John Wiley and Sons, Inc., New York, 1957.
[3] D.M. Brink and G.R. Satchler, *Angular Momentum*, Oxford University Press, Oxford, 1962.
[4] R.N. Zare, *Angular Momentum*, John Wiley and Sons, Inc., New York, 1988.
[5] B.L. Silver, *Irreducible Tensor Methods: An Introduction for Chemists*, Academic Press, New York, 1976.
[6] V.D. Kleinman, H. Park, R.J. Gordon and R.N. Zare, *Companion to Angular Momentum*, Wiley Interscience, New York, 1988.
[7] D.A. Varshalovich, A.N. Moskalev and V.K. Khersonskii, *Quantum Theory of Angular Momentum*, World Scientific Publications, Singapore and Teaneck, N.J., 1988.
[8] F. Mandl, *Quantum Mechanics*, J. Wiley and Sons, Chichester, 1992.
[9] E.U. Condon and G.H. Shortley, *The Theory of Atomic Spectra*, Cambridge University Press, Cambridge, 1963.

[10] J.H. Van Vleck, *Rev. Mod. Phys.*, **23**, 213 (1951).
[11] E.P. Wigner, *Group Theory and its Applications to the Quantum Mechanics of Atomic Spectra*, Academic Press, New York, 1959.
[12] *The Mathematica Book*, Cambridge University Press and Wolfram Media, 3rd Edn., 1998.
[13] D. Smith and J.H.M. Thornley, *Proc. Phys. Soc.*, **89**, 779 (1966).
[14] G. Racah, *Phys. Rev.*, **61**, 186 (1942); **61**, 438 (1942); **63**, 367 (1943).
[15] J.M. Brown and B.J. Howard, *Mol. Phys.*, **31**, 1517 (1976); **32**, 1197 (1976).

6 Electronic and vibrational states

6.1. Introduction

In chapter 3 we derived a Hamiltonian to describe the electronic motion in a diatomic molecule, starting from first principles. In our case, the first principles were the Dirac equation for a single particle, and the Breit equation for two interacting particles. We pointed out that even at this level our treatment was a compromise because it did not include quantum electrodynamics explicitly. Nevertheless we concluded the chapter with a rather complete and complicated Hamiltonian, and added yet more complications in chapter 4 with the inclusion of nuclear spin effects. In the next chapter, chapter 7, we will show how terms in the 'true' Hamiltonian may be reduced to 'effective' Hamiltonians designed to handle the particular cases which arise in spectroscopy. We will make extensive use of angular momentum theory, described in chapter 5, to describe the electronic and nuclear dynamics in diatomic molecules, and the interactions with applied magnetic and electric fields. The experimental study of these dynamical effects is dealt with at length in chapters 8 to 11. We will be classifying these studies according to molecular electronic states, and demonstrating how the high-resolution spectroscopic methods described probe the structural details of these electronic states. That, indeed, is one of the main purposes of spectroscopy.

Before we proceed to these details we must describe some aspects of the theory of the electronic and vibrational states of diatomic molecules. To this end we return to the 'master equation' displayed at the end of chapter 3, and develop the consequences of some of the terms contained therein. This is a huge subject, described in many textbooks, and at any level of detail which one might require. In this chapter we present what we consider to be the *minimum* required for a satisfactory understanding of what follows in later chapters. What is satisfactory is a subjective matter for the reader, and in many cases there are aspects to be explored in much greater depth than is to be found here. Some of these aspects are presented in later chapters, but here we deal with the essential fundamentals.

6.2. Atomic structure and atomic orbitals

6.2.1. The hydrogen atom

We start by considering the hydrogen atom, the simplest possible system, in which one electron interacts with a nucleus of unit positive charge. Only two terms are required from the master equation (3.161) in chapter 3, namely, those describing the kinetic energy of the electron and the electron–nuclear Coulomb potential energy. In the space-fixed axes system and SI units these terms are

$$\mathcal{H} = \frac{1}{2m}\boldsymbol{p}^2 - \frac{e^2}{4\pi\varepsilon_0 r}, \tag{6.1}$$

where m is the electron mass, r is the electron–nuclear distance, and e is the elementary charge. The linear momentum $\boldsymbol{p}$ is given by the quantum mechanical expression

$$\boldsymbol{p} = -i\hbar\left\{\frac{\partial}{\partial X}\boldsymbol{i} + \frac{\partial}{\partial Y}\boldsymbol{j} + \frac{\partial}{\partial Z}\boldsymbol{k}\right\}, \tag{6.2}$$

where $\boldsymbol{i}$, $\boldsymbol{j}$, $\boldsymbol{k}$ are the unit vectors in the X, Y, Z directions. Therefore $\boldsymbol{p}^2$ is given by

$$\boldsymbol{p}^2 = -\hbar^2\left\{\frac{\partial^2}{\partial X^2} + \frac{\partial^2}{\partial Y^2} + \frac{\partial^2}{\partial Z^2}\right\} \equiv -\hbar^2\nabla^2, \tag{6.3}$$

where ∇^2 is called the Laplacian operator. We may rewrite equation (6.1) by referring the coordinate system to one with an origin fixed at the centre of mass, so that we replace the electron mass m by the reduced mass of the system, μ. We thereby obtain the result

$$\mathcal{H} = -\frac{\hbar^2}{2\mu}\nabla^2 - \frac{e^2}{4\pi\varepsilon_0 r}. \tag{6.4}$$

The Schrödinger equation for the hydrogen atom may therefore be written

$$\mathcal{H}\psi = E\psi, \tag{6.5}$$

where the Hamiltonian operator $\mathcal{H}$ is given by (6.4). The task is to find the eigenfunctions ψ and eigenvalues E of (6.5).

It turns out that the solutions of (6.5) are much simpler if one transforms from cartesian to spherical polar coordinates, as defined in figure 6.1. The relationships between the two are

$$X = r\sin\theta\cos\phi, \quad Y = r\sin\theta\sin\phi, \quad Z = r\cos\theta, \tag{6.6}$$

so that, in spherical polar coordinates, the Laplacian is given by

$$\nabla^2 = \frac{1}{r^2}\frac{\partial}{\partial r}\left(r^2\frac{\partial}{\partial r}\right) + \frac{1}{r^2\sin\theta}\frac{\partial}{\partial\theta}\left(\sin\theta\frac{\partial}{\partial\theta}\right) + \frac{1}{r^2\sin^2\theta}\frac{\partial^2}{\partial\phi^2}. \tag{6.7}$$

The solutions of the Schrödinger equation in these spherical polar coordinates are described in many books. They can be factorised and have the following form:

$$\psi(r,\theta,\phi) = R_{nl}(r)\,Y_{lm}(\theta,\phi). \tag{6.8}$$

Table 6.1. *The first few* Θ_{lm} *wave functions for the hydrogen atom*

l	m	$\Theta_{lm}(\theta)$	l	m	$\Theta_{lm}(\theta)$
0	0	$1/\sqrt{2}$	2	0	$(\sqrt{10}/4)(3\cos^2\theta - 1)$
1	0	$(\sqrt{6}/2)\cos\theta$	2	± 1	$(\sqrt{15}/2)(\sin\theta\cos\theta)$
1	± 1	$(\sqrt{3}/2)\sin\theta$	2	± 2	$(\sqrt{15}/4)(\sin^2\theta)$

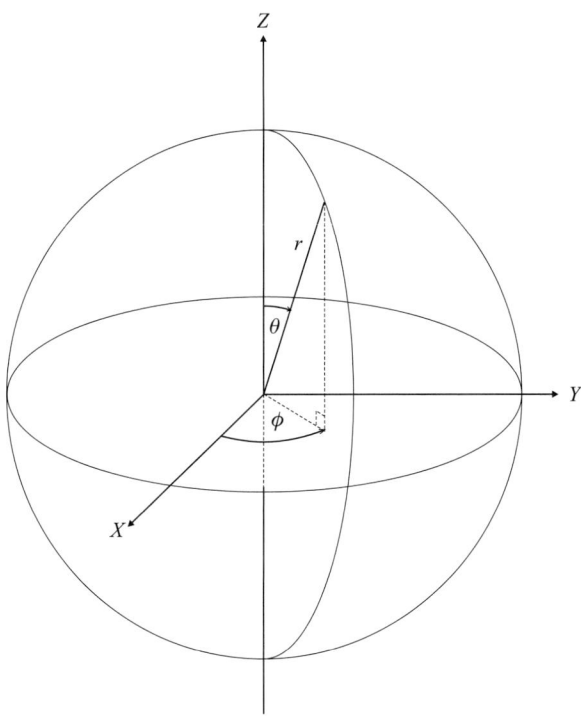

Figure 6.1. Relationship between cartesian (X, Y, Z) and spherical polar coordinates (r, θ, ϕ).

The $Y_{lm}(\theta, \phi)$ functions are angular functions known as spherical harmonics, which we have already met in chapter 5. They can be further factorised into θ- and ϕ-dependent parts:

$$Y_{lm}(\theta, \phi) = \frac{1}{2\pi} \Theta_{lm}(\theta) \exp(im\phi). \tag{6.9}$$

The $\Theta_{lm}(\theta)$ functions are related to the associated Legendre polynomials, and the first few are listed in table 6.1. The $R_{nl}(r)$ are the radial wave functions, known as associated Laguerre functions, the first few of which are listed in table 6.2. The quantities n, l and m in (6.8) are known as quantum numbers, and have the following allowed values:

$$\begin{aligned} n &= 1, 2, 3, \ldots, \infty \\ l &= 0, 1, 2, \ldots, (n-1) \\ m &= 0, \pm 1, \pm 2, \ldots, \pm l. \end{aligned} \tag{6.10}$$

Table 6.2. *The first few radial wave functions $R_{nl}(r)$ for the hydrogen atom (note that $\rho = Zr/a_0$ where Ze is the nuclear charge)*

n	l	$R_{nl}(r)$
1	0	$(Z/a_0)^{3/2} 2 \exp(-\rho)$
2	0	$\left(\dfrac{Z}{a_0}\right)^{3/2} \left(\dfrac{1}{\sqrt{2}}\right) \left(1 - \dfrac{\rho}{2}\right) \exp\left(-\dfrac{\rho}{2}\right)$
2	1	$\left(\dfrac{Z}{a_0}\right)^{3/2} \left(\dfrac{1}{2}\right) \left(\dfrac{1}{\sqrt{6}}\right) \rho \exp\left(-\dfrac{\rho}{2}\right)$

Here n is the principal quantum number, l is called the azimuthal quantum number, and m is the magnetic quantum number. We return to discuss these quantum numbers shortly, but note that the energies E obtained from the Schrödinger equation are given by

$$E = -\frac{\mu e^4}{32\pi^2 \varepsilon_0^2 \hbar^2 n^2}. \tag{6.11}$$

Alternatively, this formula can be rewritten as

$$E = -\frac{\hbar^2}{2\mu a_0^2 n^2}, \tag{6.12}$$

where a_0 is the radius of the Bohr orbit for $n = 1$,

$$a_0 = \frac{4\pi\varepsilon_0 \hbar^2}{\mu e^2}. \tag{6.13}$$

The important point to emerge from (6.11) is that the energies depend only upon the value of n, and not upon l and m.

The wave functions (6.8) are known as *atomic orbitals*; for $l = 0, 1, 2, 3$, etc., they are referred to as s, p, d, f, respectively, with the value of n as a prefix, i.e. $1s$, $2s$, $2p$, $3s$, $3p$, $3d$, etc., From the explicit forms of the wave functions we can calculate both the sizes and shapes of the atomic orbitals, important properties when we come to consider molecule formation and structure. It is instructive to examine the angular parts of the hydrogen atom functions (the spherical harmonics) in a polar plot but noting from (6.9) that these are complex functions, we prefer to describe the angular wave functions by real linear combinations of the complex functions, which are also acceptable solutions of the Schrödinger equation. This procedure may be illustrated by considering the $2p$ orbitals. From equations (6.8) and (6.9) the complex wave functions are

$$\psi_{2p,+1} = R_{21}(r) Y_{1,+1}(\theta, \phi) = R_{21}(r) \frac{1}{\sqrt{2\pi}} \Theta_{1,+1}(\theta) \exp(i\phi)$$

$$= R_{21}(r) \frac{1}{\sqrt{2\pi}} \frac{\sqrt{3}}{2} \sin\theta \exp(i\phi),$$

$$\psi_{2p,-1} = R_{21}(r) Y_{1,-1}(\theta, \phi) = R_{21}(r) \frac{1}{\sqrt{2\pi}} \Theta_{1,-1}(\theta) \exp(i\phi)$$

$$= R_{21}(r) \frac{1}{\sqrt{2\pi}} \frac{\sqrt{3}}{2} \sin\theta \exp(-i\phi). \tag{6.14}$$

Linear combinations of these functions may then be formed:

$$\psi_{2p_x} = \frac{1}{\sqrt{2}}[\psi_{2p,+1} + \psi_{2p,-1}] = \frac{\sqrt{3}}{2\sqrt{4\pi}} R_{21}(r) \sin\theta [\exp(i\phi) + \exp(-i\phi)]$$

$$= \sqrt{\frac{3}{4\pi}} R_{21}(r) \sin\theta \cos\phi,$$

$$\psi_{2p_y} = -\frac{i}{\sqrt{2}}[\psi_{2p,+1} - \psi_{2p,-1}] = \frac{i\sqrt{3}}{2\sqrt{4\pi}} R_{21}(r) \sin\theta [\exp(i\phi) - \exp(-i\phi)]$$

$$= \sqrt{\frac{3}{4\pi}} R_{21}(r) \sin\theta \sin\phi, \qquad (6.15)$$

In addition to these two functions, the third member of the set is already real, i.e.,

$$\psi_{2p_z} = \psi_{2p_0} = \sqrt{\frac{3}{4\pi}} R_{21}(r) \cos\theta. \qquad (6.16)$$

Similar treatments may be applied to the other wave functions; the ψ_{ns} wave functions are always real, as also are those members of the d, f, etc., sets which have $m = 0$. These real wave functions may now be plotted in the form of polar diagrams, as shown in figure 6.2. When we refer to the shape or spatial orientation of an atomic orbital, we are actually referring to a specific member of the set shown in figure 6.2. We could, of course, continue the process to include f orbitals, and higher.

The quantum number m, which distinguishes the three p orbitals, or the five degenerate d orbitals, is known as the magnetic quantum number. It actually gives the z component of the orbital angular momentum vector, $\hbar l_z$, that is

$$l_z = m_l. \qquad (6.17)$$

This is strictly an operator equation. For example, for a p state, with $l = 1$, the projections of the orbital angular momentum in the z direction, l_z, have the values $+1$, 0, -1 as shown in figure 6.3.

So far we have three quantum numbers. However, we know from the relativistic treatment of the electron due to Dirac, which we described in chapter 3, that there is a fourth quantum number, called electron spin. In the first instance the need for a fourth quantum number became evident from experiment; in the Dirac theory it is a consequence of introducing time as the fourth dimension. The spin angular momentum $\hbar s$ has the value $(1/2)\hbar$, so that the magnitude of the spin angular momentum is $[(1/2)(3/2)]^{1/2} \hbar$. It can be oriented in two possible directions, with the fourth quantum number m_s taking the values $+1/2$ or $-1/2$. Conventionally, these two orientations are described as α or β respectively.

6.2.2. Many-electron atoms

In an atom containing more than one electron, each electron can be considered as having its own set of values of n, l, m_l and m_s. An extremely important rule, known

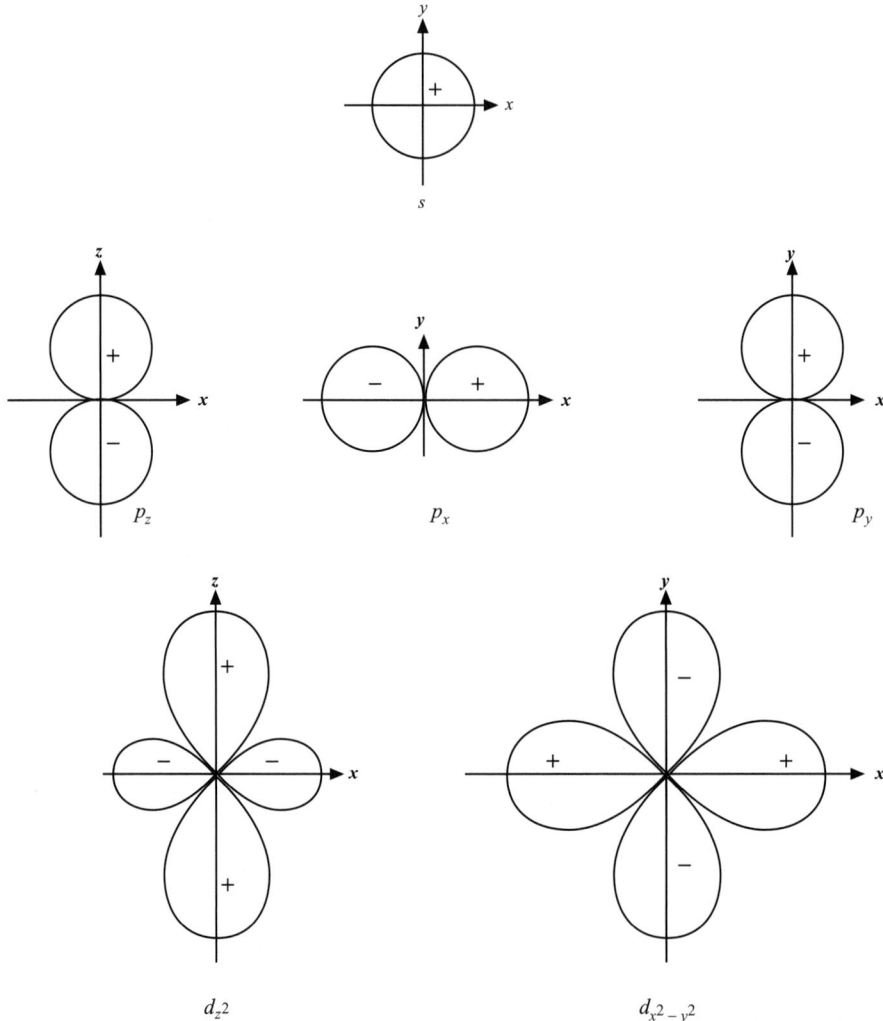

Figure 6.2. Polar diagrams for $1s$, $2p$ and two of the five $3d$ atomic orbitals, illustrating the angular parts of the wave functions.

as the Pauli exclusion principle, is that no two electrons in the same atom may have the same values for all four quantum numbers. An equivalent but more fundamental statement of the Pauli principle is that the *total* wave function for electrons, including spins, must be antisymmetric with respect to the interchange of any pair of electrons. It is instructive to consider a two-electron system, such as the helium atom in its lowest energy state where two electrons occupy the $1s$ atomic orbital. Knowing that each electron may have α or β spin, there are four possible combinations:

$$\alpha(1)\alpha(2), \quad \alpha(1)\beta(2), \quad \beta(1)\alpha(2), \quad \beta(1)\beta(2).$$

The first and fourth of these combinations have a definite symmetry when electrons 1

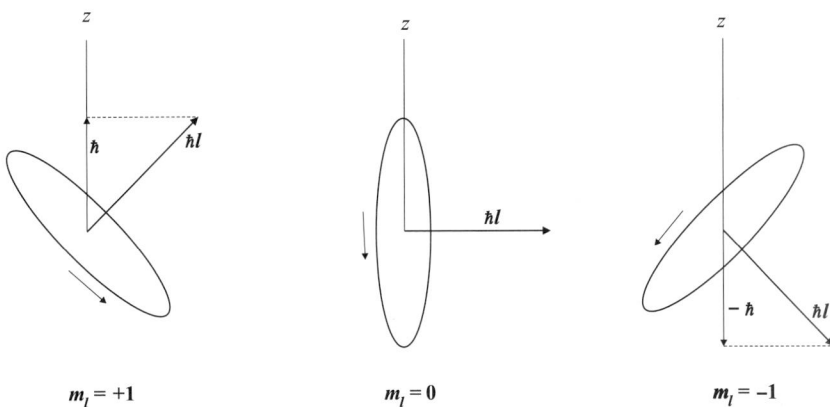

Figure 6.3. Space-quantised orientations for the angular momentum vector $\hbar \boldsymbol{l}$ of a state with $l = 1$. The length of $\hbar \boldsymbol{l}$ is $[l(l+1)\hbar^2]^{1/2}$.

and 2 are exchanged; they are unchanged, that is, *symmetric* with respect to this symmetry operation. The second and third combinations do not have a definite symmetry, but the two linear combinations of them do, i.e.

$$(1/\sqrt{2})[\alpha(1)\beta(2) + \beta(1)\alpha(2)], \quad (1/\sqrt{2})[\alpha(1)\beta(2) - \beta(1)\alpha(2)].$$

The first combination is symmetric, and the second is antisymmetric with respect to permutation. In the ground state of the helium atom both electrons occupy the same spatial orbital (1s), so that they must have the antisymmetric spin function for the total wave function to be antisymmetric; they therefore form a singlet spin state. In order to find a helium atom with a triplet spin state (so-called spins parallel), the spatial part of the wave function must be antisymmetric with respect to interchange.

Whilst we are discussing the Pauli principle, it is worthwhile to introduce a further method of expressing the antisymmetric nature of the electronic wave function, namely, the Slater determinant [1]. Since both electrons in the ground state of the helium atom occupy the same space orbital, the wave function may be written in the form

$$\psi(r_1, r_2) = \psi_{1s}(r_1)\psi_{1s}(r_2)(1/\sqrt{2})\{\alpha(1)\beta(2) - \alpha(2)\beta(1)\}$$
$$= \frac{1}{\sqrt{2}} \begin{vmatrix} \psi_{1s}(r_1)\alpha(1) & \psi_{1s}(r_1)\beta(1) \\ \psi_{1s}(r_2)\alpha(2) & \psi_{1s}(r_2)\beta(2) \end{vmatrix}. \quad (6.18)$$

This may be expressed in a more abbreviated form by introducing the *spin-orbital* $\psi_{1s}^\sigma(1)$, with $\sigma = \alpha$ or β, so that (6.18) can be written in the form

$$^1\psi(1,2) = \frac{1}{\sqrt{2}} \begin{vmatrix} \psi_{1s}^\alpha(1) & \psi_{1s}^\beta(1) \\ \psi_{1s}^\alpha(2) & \psi_{1s}^\beta(2) \end{vmatrix}. \quad (6.19)$$

The antisymmetric nature of the wave function when written as a Slater determinant is now apparent, since exchanging electrons 1 and 2 means that the first and second rows of the determinant are interchanged; under such an operation, the sign of a determinant changes. Obviously, if electrons 1 and 2 have exactly the same quantum numbers, interchange of the first and second rows of the determinant shows that $^1\psi(1,2)$ must be zero, as required by the Pauli principle. We will meet determinantal wave functions frequently elsewhere in this book.

In the hydrogen atom the $2s$ and $2p$ orbitals have the same energy, but in polyelectronic atoms the $2s$ electrons have a lower energy than the $2p$. In general the orbital energies for a given value of the principal quantum number n follow the general order

$$ns < np < nd < nf \ldots.$$

The reasons for this are described in many books on atomic structure, but are connected with the fact that as l increases for a given n value, the electrons are increasingly shielded from the full electrostatic interaction with the positively charged nucleus because of shielding due to the inner electrons. Figure 6.4 shows what is known as the Grotrian diagram for a sodium atom; it displays the energies of the atomic orbitals, largely determined from experimental atomic spectroscopy. Note that the shielding effects are sufficiently large for the $4s$ level to lie below the $3d$.

6.2.3. Russell–Saunders coupling

The electronic structure of an atom depends on the way in which the various orbital and spin angular momenta interact with each other. In the lighter atoms the electronic states can usually be described in terms of Russell–Saunders coupling. The individual electron spin angular momenta s, each with the value $1/2$, combine to give a resultant spin S. For two electrons, each with $s = 1/2$, the total spin S may be 1 or 0. Similarly, the individual orbital angular momenta of the electrons l combine to give a resultant orbital angular momentum L. For two p electrons, each with l equal to 1, L may take values of 2, 1 or 0. Now L and S couple together to give a total resultant angular momentum J; J is also quantised and takes values from $L + S$ to $|L - S|$. The vector coupling scheme is illustrated in figure 6.5. As we shall see later, similar vector coupling schemes are important in describing the electronic states of diatomic molecules.

We see that a single electron configuration can give rise to several different electronic states because of the coupling of angular momenta. These states are referred to as *terms* and are described by *term symbols*; the value of L may be 0, 1, 2, 3, etc., and this is denoted by the symbol S, P, D, F, respectively. If the value of the total spin is S, the corresponding spin multiplicity is $2S + 1$ and this is indicated as a pre-superscript. Finally the value of J is indicated as a post-subscript. Consequently for two p electrons we might appear to have the following possible electronic states, or term symbols:

$L = 2, S = 1$: $^3D_3, {}^3D_2, {}^3D_1$. $L = 2, S = 0$: $^1D_2$.

$L = 1, S = 1$: $^3P_2, {}^3P_1, {}^3P_0$. $L = 1, S = 0$: $^1P_1$.

$L = 0, S = 1$: $^3S_1$. $L = 0, S = 0$: $^1S_0$.

However, when two electrons are in equivalent orbitals (same value of n), the Pauli principle must be taken into account so that the permitted states are restricted to ^{1}D,

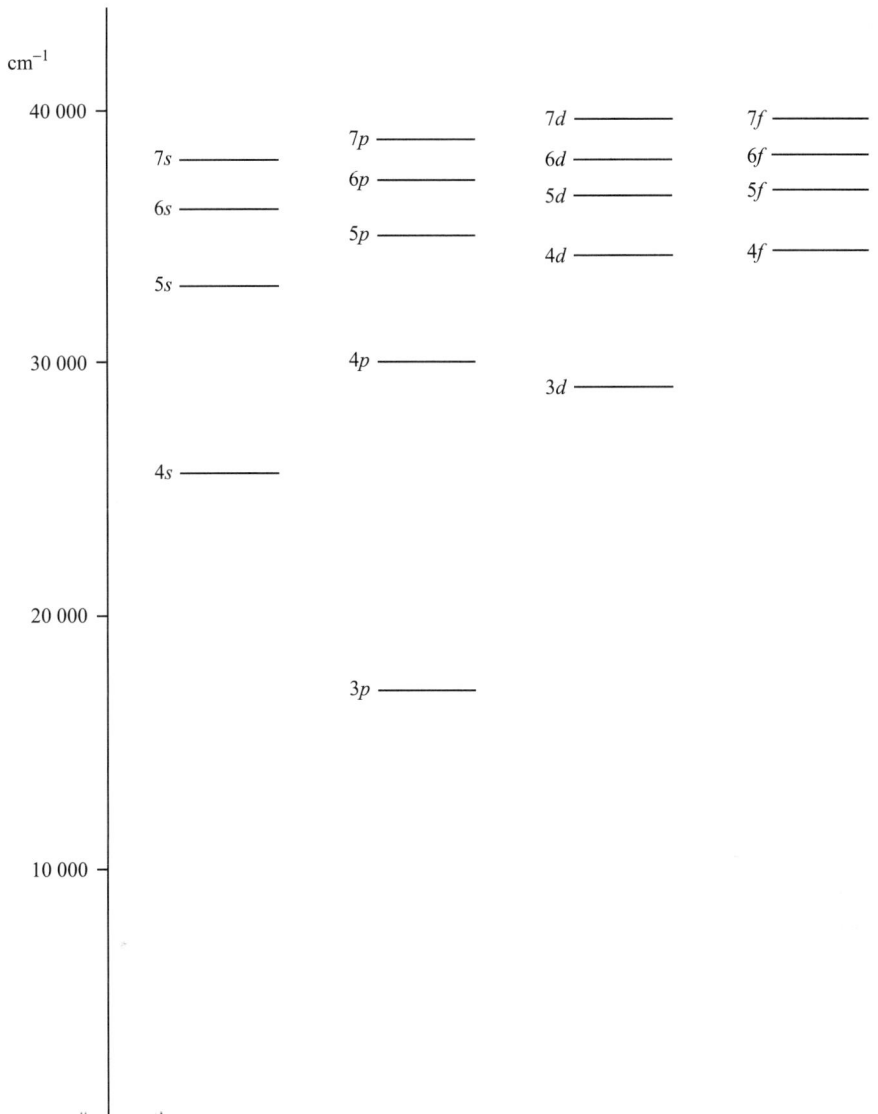

Figure 6.4. Grotrian diagram for the sodium atom, showing the energies of the atomic orbitals.

^{3}P and ^{1}S. Note that states with different values of L or S have different energies, partly because of the electron–electron interaction term in the electronic Hamiltonian. A further symmetry classification that should be mentioned is the parity of an atomic state which depends on the behaviour of the total wave function under space-fixed inversion. This is either even (g) or odd (u) and is determined by $\sum l_i$, summed over all the electrons in the atom.

The electron spin and orbital angular momenta give rise to intrinsic magnetic moments which interact with other; this interaction is known as the spin–orbit coupling

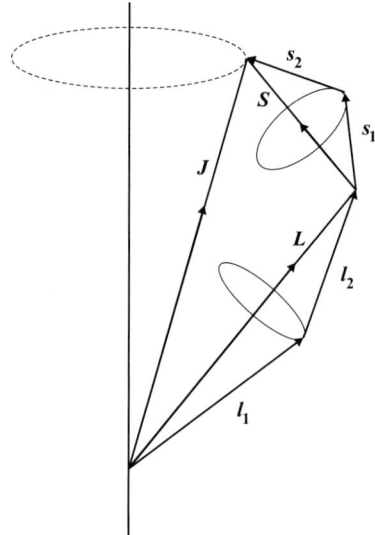

Figure 6.5. Russell–Saunders vector coupling scheme for two electrons.

and it can be represented by the simple expression

$$\mathcal{H}_{so} = \zeta \mathbf{L} \cdot \mathbf{S}, \tag{6.20}$$

where ζ is the spin–orbit coupling parameter. Since $\mathbf{J} = \mathbf{L} + \mathbf{S}$, it follows at once that $\mathbf{J}^2 = (\mathbf{L} + \mathbf{S})^2$. Consequently

$$\mathbf{J}^2 = \mathbf{L}^2 + 2\mathbf{L} \cdot \mathbf{S} + \mathbf{S}^2, \tag{6.21}$$

so that

$$\mathbf{L} \cdot \mathbf{S} = (1/2)\{\mathbf{J}^2 - \mathbf{L}^2 - \mathbf{S}^2\} = (1/2)\{J(J+1) - L(L+1) - S(S+1)\}. \tag{6.22}$$

The spin–orbit coupling therefore removes the degeneracy of the so-called 'fine-structure' states listed above. For example, simply by substituting the appropriate values of J, L and S, the three fine-structure components arising from the $L = 1$, $S = 1$ configuration are calculated to have the following first-order spin–orbit energies:

$$^3P_2 : E = \zeta, \quad ^3P_1 : E = -\zeta, \quad ^3P_0 : E = -2\zeta.$$

Values of the spin–orbit coupling constants for different atomic states have been determined from atomic spectra, or may be calculated from relativistic quantum mechanics. A more detailed examination of the matrix elements reveals that the spin–orbit coupling operator mixes states of different L and S values; spin–orbit matrices for many of the most common electron configurations have been given by Condon and Shortley [2].

Russell–Saunders, or LS, coupling is one limiting case which is usually a very good approximation for light atoms. However, for heavy atoms, jj coupling is often a better approximation. Considering again a two-electron system, for each electron l_i and s_i combine to give a resultant angular momentum j_i, and the individual values of j_i are then coupled to give a resultant total angular momentum $\mathbf{J}$. In detail for a system of, for example, one s and one p electron:

$$l_1 = 0, \; s_1 = 1/2: \; j_1 = 1/2; \qquad l_2 = 1, \; s_2 = 1/2: \; j_2 = 3/2, \; 1/2;$$
$$j_1 = 1/2, \; j_2 = 3/2: \; J = 2, \; 1; \qquad j_1 = 1/2, \; j_2 = 1/2: \; J = 1, \; 0.$$

In practice, for many atomic states, the coupling will be intermediate between LS and jj.

Our understanding of the electron configurations of the atoms leads ultimately to the periodic classification of the elements. As we have seen, a given electron configuration often leads to more than one term symbol, and additional criteria are required to establish which term will be the lowest in energy. Hund's rules, which address this problem are (i), the state with the highest spin multiplicity will be lowest in energy, (ii), if more than one term with the same spin multiplicitiy exists, that with the highest L value will be lowest in energy. In addition, for a term giving rise to more than one level, that with the lowest J value is lowest in energy for a less than half-full outer shell; conversely for a more than half-full shell (i.e. p^4, d^6, etc.) the level of highest J value will be lowest in energy.

We will see in due course that there are important correlation rules between atomic term symbols and molecular electronic states, rules that are important in understanding both the formation and dissociation of diatomic molecules. Elementary accounts of the theory of atomic structure are to be found in books by Softley [3] and Richards and Scott [4]. Among the more comprehensive descriptions of the quantum mechanical aspects, that by Pauling and Wilson [5] remains as good as any whilst group theoretical aspects are described by Judd [6].

6.2.4. Wave functions for the helium atom

In many places elsewhere in this book we describe the analysis of spectra, the definition and determination of molecular parameters from the spectra, and the relationships between these parameters and the wave functions for the molecules in question. Later in this chapter we will outline the principles and practice of calculating accurate wave functions for diatomic molecules. Before we can do that, however, we must discuss the calculation of atomic wave functions; the methods originally developed for atoms were subsequently extended to deal with molecules. This is not the book for an exhaustive discussion of these topics, and so many accounts exist elsewhere that such a discussion is not necessary. Nevertheless we must pay some attention to this topic because the interpretation of spectroscopic data in terms of molecular wave functions is one of the primary motivations for obtaining the data in the first place.

We have already dealt with the calculation of the wave functions of the hydrogen atom. We now proceed to consider many-electron atoms, first dealing with the simplest such example, the helium atom which possesses two electrons. The Hamiltonian for a helium-like atom with an infinitely heavy nucleus can be obtained by selecting the appropriate terms from the master equation in chapter 3. The Hamiltonian we use is

$$\mathcal{H} = -\frac{\hbar^2}{2m}\nabla_1^2 - \frac{Ze^2}{4\pi\varepsilon_0 r_1} - \frac{\hbar^2}{2m}\nabla_2^2 - \frac{Ze^2}{4\pi\varepsilon_0 r_2} + \frac{e^2}{4\pi\varepsilon_0 r_{12}}, \qquad (6.23)$$

where Ze is the nuclear charge, m is the electron mass, r_1 and r_2 are the electron–nucleus distances for the electrons (1) and (2) respectively, and r_{12} is the interelectronic distance. The Schrödinger equation is, as usual,

$$\mathcal{H}\psi(\mathbf{r}_1, \mathbf{r}_2) = E\psi(\mathbf{r}_1, \mathbf{r}_2), \quad (6.24)$$

and our aim is to find solutions to (6.24) which are transferable to more complex atoms and subsequently to molecules.

Almost all approaches to many-electron wave functions, for both atoms and molecules, involve their formulation as products of one-electron orbitals. If the interelectronic repulsion term in (6.23) is small compared with the other terms, the Hamiltonian is approximately separable into independent operators for each electron and the two-electron wave function, $\psi(1, 2)$, can be written as a simple product of one-electron functions,

$$\psi(1, 2) = \phi_1(1)\phi_1(2), \quad (6.25)$$

in which $\phi_i(j)$ denotes the ith spatial orbital containing the jth electron.

There are several different ways of proceeding from here, but following others we use the variation method. Multiplication of both sides of the Schrödinger equation by ψ^* and integration over all coordinates gives an expression for the energy,

$$W_0 = \frac{\int \psi^* \mathcal{H} \psi \, d\tau}{\int \psi^* \psi \, d\tau}. \quad (6.26)$$

Any approximate wave function ψ will give a calculated energy W_0 from (6.26) which is greater than the true energy E_0, and the objective is to find a wave function which minimises the value of W_0. Ultimately a trial wave function will contain one or more variable parameters, and we will look to find parameter values which minimise the calculated energy; however we first develop (6.26).

If we substitute (6.25) in (6.26) we obtain

$$E_0 \leq W_0 = \frac{\int_1 \int_2 \psi^*(1,2) \mathcal{H} \psi(1,2) \, dV_1 \, dV_2}{\int_1 \int_2 \psi^*(1,2) \psi(1,2) \, dV_1 \, dV_2}$$

$$= \frac{\int_1 \int_2 \phi_1^*(1)\phi_1^*(2)[-(\hbar^2/2m)\nabla_1^2 - (Ze^2/4\pi\varepsilon_0 r_1) - (\hbar^2/2m)\nabla_2^2 - (Ze^2/4\pi\varepsilon_0 r_2)}{\int_1 \phi_1^*(1)\phi_1(1) \, dV_1 \int_2 \phi_1^*(2)\phi_1(2) \, dV_2}.$$

$$\quad (6.27)$$

The spatial integral and volume elements in (6.27) are given by

$$\int_1 dV_1 = \int_0^{2\pi} \int_0^{\pi} \int_0^{\infty} r_1^2 \sin\theta_1 \, dr_1 \, d\theta_1 \, d\phi_1, \quad \int_2 dV_2 = \int_0^{2\pi} \int_0^{\pi} \int_0^{\infty} r_2^2 \sin\theta_2 \, dr_2 \, d\theta_2 \, d\phi_2,$$

$$\quad (6.28)$$

using spherical polar coordinates for each electron.

The problem now is to evaluate (6.27) for various choices of $\phi_1(i)$, and the most obvious first choice is the ground state 1s hydrogen-like orbitals which we calculated earlier in this chapter:

$$\phi_1(i) = \frac{1}{\sqrt{\pi}}\left(\frac{Z}{a_0}\right)^{3/2} \exp\left(-\frac{Zr_i}{a_0}\right). \tag{6.29}$$

For the hydrogen-like atom,

$$E_1 = -\frac{Z^2 e^2}{8 a_0 \pi \varepsilon_0}, \tag{6.30}$$

where the Schrödinger equation is

$$\left(-\frac{\hbar^2}{2m}\nabla_1^2 - \frac{Ze^2}{4\pi\varepsilon_0 r_1}\right)\phi_1(1) = E_1 \phi_1(1). \tag{6.31}$$

Substituting the orbitals (6.29) into (6.27) and making use of (6.31) we obtain the following result:

$$W_0 = E_1(1) + E_1(2) + \int_1 \int_2 \phi_1^*(1)\phi_1^*(2)\left(\frac{e^2}{4\pi\varepsilon_0 r_{12}}\right)\phi_1(1)\phi_1(2)\,dV_1\,dV_2$$

$$= 2E_1 + \frac{1}{\pi^2}\left(\frac{Z}{a_0}\right)^6 \int_1 \int_2 \exp\left[-\left(\frac{2Z}{a_0}\right)(r_1 + r_2)\right]\left(\frac{e^2}{4\pi\varepsilon_0 r_{12}}\right) dV_1\,dV_2. \tag{6.32}$$

The integral in (6.32) is evaluated in a number of standard texts, for example that by Kauzmann [7], and is found to be equal to $(5Ze^2/32\pi\varepsilon_0 a_0)$. Consequently (6.32) becomes

$$W_0 = 2E_1 + \frac{5}{8}Z\frac{e^2}{4\pi\varepsilon_0 a_0}$$

$$= \left(-\frac{Z^2 e^2}{a_0} + \frac{5}{8}Z\frac{e^2}{a_0}\right)\left(\frac{1}{4\pi\varepsilon_0}\right)$$

$$= \left(\frac{e^2}{4\pi\varepsilon_0 a_0}\right)\left(-Z^2 + \frac{5}{8}Z\right). \tag{6.33}$$

The quantity $(e^2/4\pi\varepsilon_0 a_0)$ is the atomic unit of energy, called the Hartree, and it has the value 27.2116 eV. If we put $Z = 2$ for the helium nucleus, (6.33) gives $W_0 = -74.832$ eV, compared with the experimental energy required to remove both electrons from a helium atom which is 79.0052 eV. This is a remarkably good result (94.7% of the true value) for such a simple orbital model, containing no variable parameters as yet.

The energy of the helium atom calculated above is the first-order energy, which differs from the true energy by an amount called the correlation energy; this is a measure of the tendency of the electrons to avoid each other. The simplest improvement to the trial wave function is to allow Z in (6.29) to be a variable parameter, which we call ζ (not to be confused with the spin-orbit coupling parameter in equation (6.20)); Z in the Hamiltonian (6.23) remains the same. The expression for the calculated energy, equation (6.27) and (6.32), now becomes the following:

$$W_0 = \frac{1}{\pi^2}\left(\frac{\zeta}{a_0}\right)^6 \int_1 \int_2 \exp\left[-\left(\frac{\zeta}{a_0}\right)(r_1+r_2)\right]\left[\left(-\frac{\hbar^2}{2m}\nabla_1^2 - \frac{\zeta e^2}{4\pi\varepsilon_0 r_1}\right)\right.$$
$$\left. + \left(-\frac{\hbar^2}{2m}\nabla_2^2 - \frac{\zeta e^2}{4\pi\varepsilon_0 r_2}\right) + \frac{e^2}{r_{12}} - \frac{(Z-\zeta)}{4\pi\varepsilon_0}\left(\frac{e^2}{r_1}+\frac{e^2}{r_2}\right)\right]$$
$$\times \exp\left[-\left(\frac{\zeta}{a_0}\right)(r_1+r_2)\right] dV_1\, dV_2. \qquad (6.34)$$

It is not difficult to show [8] that this equation reduces to

$$W_0 = \left[-\frac{\zeta^2 e^2}{a_0} + \frac{5}{8}\zeta\frac{e^2}{a_0} + 2(\zeta - Z)\zeta\frac{e^2}{a_0}\right]\left[\frac{1}{4\pi\varepsilon_0}\right] = \frac{e^2}{4\pi\varepsilon_0 a_0}\left[\zeta^2 + \zeta\left(\frac{5}{8} - 2Z\right)\right]. \qquad (6.35)$$

Minimising W_0 with respect to ζ gives $\zeta = Z - 5/16$, and putting $Z=2$ for helium we obtain $W_0 = -77.490$ eV, which is 98% of the true value, a considerable improvement over the previous value. In this case the value of ζ is $27/16 = 1.6875$, rather than 2, a measure of the shielding of the nuclear charge by the second electron. Further improvements to the calculated ionisation energy may be made by increasing the number of variable parameters in the trial wave function. Hylleraas [9] used a fourteen-parameter function and obtained an ionisation energy which was almost exactly equal to the true value.

6.2.5. Many-electron wave functions: the Hartree–Fock equation

In section 6.2.2 we introduced the Pauli principle and showed how, in the case of a two-electron system like the helium atom, the wave function could be written in the form of a determinant, called a Slater determinant (6.19). In a two-electron system the spin and spatial parts of the wave function can be separated, but for more than two electrons the general form of the wave function, written as a Slater determinant, is

$$\psi(1,2,3,\ldots,n) = \left(\frac{1}{n!}\right)^{1/2} \begin{vmatrix} \phi_1(1)\alpha(1) & \phi_1(1)\beta(1) & \phi_2(1)\alpha(1) & \ldots & \phi_{n/2}(1)\beta(1) \\ \phi_1(2)\beta(2) & \phi_1(2)\beta(2) & \cdot & \ldots & \phi_{n/2}(2)\beta(2) \\ \phi_1(3)\alpha(3) & \phi_2(2)\alpha(2) & \cdot & \ldots & \phi_{n/2}(3)\alpha(3) \\ \phi_1(4)\beta(4) & \phi_2(2)\beta(2) & \cdot & \ldots & \cdot \\ \phi_1(5)\alpha(5) & \phi_3(2)\alpha(2) & \cdot & \ldots & \cdot \\ \cdot & \cdot & \cdot & \ldots & \cdot \\ \phi_1(n)\beta(n) & \cdot & \cdot & \ldots & \phi_{n/2}(n)\beta(n) \end{vmatrix}, \qquad (6.36)$$

where $\phi_i(a)$ denotes the ith spatial orbital containing the ath electron; each orbital holds 2 electrons so that for n electrons there are $n/2$ orbitals for a closed shell configuration.

The complete Hamiltonian for the many-electron system is written

$$\mathcal{H} = \sum_a \mathcal{H}^0(a) + \sum_{a>b} \frac{e^2}{4\pi\varepsilon_0 r_{ab}}, \tag{6.37}$$

where $\mathcal{H}^0(a)$ is the ath one-electron hydrogen-like operator, called the *core* Hamiltonian, and defined by

$$\mathcal{H}^0(a) = -\frac{\hbar^2}{2m}\nabla_a^2 - \frac{Ze^2}{4\pi\varepsilon_0 r_a}. \tag{6.38}$$

Since the determinantal wave function (6.36) is normalised to unity and the spin functions are orthonormal, the variation function (6.26) takes a simple form, given by,

$$W_0 = \int_1 \int_2 \int_3 \ldots \int_n \psi^*(1,2,3,\ldots,n)\mathcal{H}\psi(1,2,3,\ldots,n)\,dV_1\,dV_2\ldots dV_n$$

$$= \sum_{i=1}^{n/2}\left[2\mathcal{H}_{ii}^0 + \sum_{j=1}^{n/2}(\langle ij|ij\rangle - \langle ij|ji\rangle)\right]. \tag{6.39}$$

Note that i and j run over the orbitals in this closed shell case. There are three different types of integral in equation (6.39), and they are defined in the manner shown below.

(i) *Core integrals*

$$\mathcal{H}_{ii}^0 = \int_a \phi_i^*(a)\mathcal{H}^0(a)\phi_i(a)\,dV_a = \int_a \phi_i^*(a)\left(-\frac{\hbar^2}{2m}\nabla_a^2 - \frac{Ze^2}{4\pi\varepsilon_0 r_a}\right)\phi_i(a)\,dV_a. \tag{6.40}$$

(ii) *Coulomb integrals*

$$\langle ij|ij\rangle = \int_a\int_b \phi_i^*(a)\phi_j^*(b)\left(\frac{e^2}{4\pi\varepsilon_0 r_{ab}}\right)\phi_i(a)\phi_j(b)\,dV_a\,dV_b$$

$$= \int_a \phi_i^*(a)\left[\int_b \phi_j^*(b)\left(\frac{e^2}{4\pi\varepsilon_0 r_{ab}}\right)\phi_j(b)\,dV_b\right]\phi_i(a)\,dV_a$$

$$= \int_a \phi_i^*(a)J_j(a)\phi_i(a)\,dV_a. \tag{6.41}$$

$J_j(a)$ is called the Coulomb operator.

(iii) *Exchange integrals*

$$\langle ij|ji\rangle = \int_a\int_b \phi_i^*(a)\phi_j^*(b)\left(\frac{e^2}{4\pi\varepsilon_0 r_{ab}}\right)\phi_j(a)\phi_i(b)\,dV_a\,dV_b$$

$$= \int_a \phi_i^*(a)\left[\int_b \phi_j^*(b)\left(\frac{e^2}{4\pi\varepsilon_0 r_{ab}}\right)\phi_i(b)\,dV_b\right]\phi_j(a)\,dV_a$$

$$= \int_a \phi_i^*(a)K_j(a)\phi_j(a)\,dV_a. \tag{6.42}$$

$K_j(a)$ is called the exchange operator; it exchanges electrons a and b in the i and j orbitals.

The problem to be solved is expressed by Parr [10] in the following way. For an electronic system containing an even number of electrons, how can we find the best single determinantal wave function of the form (6.36)? We calculate the energy using (6.39) for successive sets of orthonormal trial functions ϕ_i until the minimum energy is obtained. This process is known as the Hartree–Fock or self-consistent-field (SCF) method [11, 12].

The process in detail is as follows. We use what is known as Lagrange's method of undetermined multipliers, introducing constants ε_{ji} such that the quantity W', defined by

$$W' = W_0 - 2 \sum_{i,j}^{n/2} \varepsilon_{ji} S_{ij}, \tag{6.43}$$

is a minimum; S_{ij} is the overlap integral

$$S_{ij} = \int \phi_i^*(a) \phi_j(a) \, dV_a. \tag{6.44}$$

On expansion, equation (6.43) takes the form

$$W' = 2 \sum_i^{n/2} \mathcal{H}_{ii}^0 + \sum_{i,j}^{n/2} (2\langle ij|ij\rangle - \langle ij|ji\rangle) - 2 \sum_{i,j}^{n/2} \varepsilon_{ji} S_{ij}. \tag{6.45}$$

Minimisation is achieved by requiring that the energy is unchanged with respect to first-order changes in the wave functions, i.e.

$$\delta W' = 0 = 2 \sum_i \delta \mathcal{H}_{ii}^0 + \sum_{i,j} (2\delta\langle ij|ij\rangle - \delta\langle ij|ji\rangle - 2\varepsilon_{ji} \delta S_{ij}). \tag{6.46}$$

The variations in the four different types of integral must now be derived. For the core integral we have

$$\delta \mathcal{H}_{ii} = \delta \int_a \phi_i^*(a) \mathcal{H}^0(a) \phi_i(a) \, dV_a$$

$$= \int_a \delta\phi_i^*(a) \mathcal{H}^0(a) \phi_i(a) \, dV_a + \int_a \phi_i^*(a) \mathcal{H}^0(a) \delta\phi_i(a) \, dV_a$$

$$= 2 \int_a \delta\phi_i^*(a) \mathcal{H}^0(a) \phi_i(a) \, dV_a. \tag{6.47}$$

The one-electron Hamiltonian for the ath electron, $\mathcal{H}^0(a)$, is defined in equation (6.40); it is a Hermitian operator.

Next we look at the variation of the Coulomb integral:

$$\delta\langle ij|ij\rangle = 2 \int_a \int_b \delta\phi_i^*(a) \phi_j^*(b) \left(\frac{e^2}{4\pi\varepsilon_0 r_{ab}} \right) \phi_i(a) \phi_j(b) \, dV_a \, dV_b$$

$$+ 2 \int_a \int_b \phi_i^*(a) \delta\phi_j^*(b) \left(\frac{e^2}{4\pi\varepsilon_0 r_{ab}} \right) \phi_i(a) \phi_j(b) \, dV_a \, dV_b$$

$$= 2 \int_a \delta\phi_i^*(a) J_j(a) \phi_i(a) \, dV_a + 2 \int_b \delta\phi_j^*(b) J_i(b) \phi_j(b) \, dV_b. \tag{6.48}$$

Similarly the variation of the exchange integral is given by

$$\delta\langle ij|ji\rangle = 2\int_a \delta\phi_i^*(a)K_j(a)\phi_i(a)\,\mathrm{d}V_a + 2\int_b \delta\phi_j^*(b)K_i(b)\phi_j(b)\,\mathrm{d}V_b. \quad (6.49)$$

Finally, the variation in the overlap integral is given by

$$\delta S_{ij} = \delta\int \phi_i^*(a)\phi_j(a)\,\mathrm{d}V_a$$
$$= \int_a \delta\phi_i^*(a)\phi_j(a)\,\mathrm{d}V_a + \int_a \phi_i^*(a)\delta\phi_j(a)\,\mathrm{d}V_a. \quad (6.50)$$

We now substitute the expressions for the variations of the four different types of integral into equation (6.46) and obtain the following result:

$$0 = \sum_i 4\int \delta\phi_i^* \mathcal{H}^0 \phi_i\,\mathrm{d}V + \sum_i \sum_j \left(4\int \delta\phi_i^* J_j \phi_i\,\mathrm{d}V + 4\int \delta\phi_j^* J_i \phi_j\,\mathrm{d}V\right.$$
$$\left. -2\int \delta\phi_i^* K_j \phi_i\,\mathrm{d}V - 2\int \delta\phi_j^* K_i \phi_j\,\mathrm{d}V - 2\varepsilon_{ji}\int \delta\phi_i^* \phi_j\,\mathrm{d}V - 2\varepsilon_{ji}\int \delta\phi_j^* \phi_i\,\mathrm{d}V\right). \quad (6.51)$$

The index identifying the electrons is unnecessary and has been dropped, and if the order of summation over i and j is interchanged, (6.51) simplifies to

$$0 = \sum_i \int \delta\phi_i^* \left[\mathcal{H}^0 \phi_i + \sum_j (2J_j\phi_i - K_j\phi_i - \varepsilon_{ji}\phi_j)\right]\mathrm{d}V. \quad (6.52)$$

The magnitude of $\delta\phi_i$ in this equation is arbitrary, and therefore not zero; consequently the term in square brackets must be zero, so that we obtain the result

$$\left[\mathcal{H}^0 + \sum_j (2J_j - K_j)\right]\phi_i \equiv F\phi_i = \sum_j \phi_j \varepsilon_{ji}. \quad (6.53)$$

This is the *Hartree–Fock equation* for atoms, and F is the Hartree–Fock operator.

Equation (6.53) can be written in matrix form,

$$F\phi = \phi\varepsilon \quad (6.54)$$

where ε is not necessarily a diagonal matrix. A set ϕ must be chosen in order to compute the Hartree–Fock operator F, and then a new set ϕ is computed with (6.54). The new set is then used to recompute F and a new solution is again obtained. The process is repeated until the functions used to compute F are equal to the final solutions: at this point the self-consistent-field (SCF) limit has been reached. Numerical methods have been developed to solve for the atomic orbitals ϕ_i and eigenvalues ε_i. Normally only the radial functions are determined numerically, the angular parts being represented by spherical harmonics. It should be noted that a modified treatment is required for open shell systems, and the *unrestricted Hartree–Fock* method is applied to atoms which have unpaired electrons in different orbitals.

6.2.6. Atomic orbital basis set

We have not yet addressed the problem of choosing the set of atomic orbitals ϕ with which to solve the Hartree–Fock problem. This is a huge subject in itself but we should indicate one general method which has been much used. Let us return to the problem of the helium atom which we left in section 6.2.4 having used hydrogen-like orbitals but with a variational parameter ζ replacing the nuclear charge Z. The hydrogen-like radial functions, the first few of which were listed in table 6.2, can be represented in general normalised form by the expression

$$\eta_{nl}(r) = -\left[\left(\frac{2Z}{na}\right)^3 \frac{(n-l-1)!}{2n[(n+l)!]^3}\right]^{1/2} \left(\frac{2Zr}{na}\right)^l \exp\left(\frac{-Zr}{na}\right) L_{n+l}^{2l+1}\left(\frac{2Zr}{na}\right). \tag{6.55}$$

In this expression $l = 0, 1, 2, 3, \ldots$, $n = 1, 2, 3, \ldots$; $n \geq l+1$ and a is the Bohr radius for the one-electron atom or ion with reduced mass μ, i.e. $a = 4\pi\varepsilon_0\hbar^2/\mu e^2$. The L_{n+l}^{2l+1} are associated Laguerre functions. Slater-type orbitals (STOs) have been defined as the product of functions (6.55) with $l = n - 1$, and spherical harmonics:

$$S_{nlm}(r, \theta, \phi) = (2\zeta/a)^{n+1/2}[(2n)!]^{-1/2} r^{n-1} \exp(-\zeta r/a) Y_{lm}(\theta, \phi). \tag{6.56}$$

Equation (6.56) comes from (6.55) if $\zeta = Z/n$ but more generally ζ is defined by

$$\zeta = \frac{Z - S}{n}, \tag{6.57}$$

where S is the shielding factor, Z is the atomic number and n is now a variable not necessarily integral in value. If both ζ and n are treated as variables in the Slater orbital (6.56), the variation method used earlier gives an ionisation energy for the helium atom of -77.667 eV, now 98.3% of the true value.

Our main reason for introducing the Slater atomic orbitals, however, is that linear combinations of them have often been used to approximate the SCF Hartree–Fock numerical orbitals. If χ represents a set of analytical orbitals, such as the STOs of equation (6.56), then we may expand ϕ in the Slater determinant (6.36) in terms of χ,

$$\phi = \chi a, \tag{6.58}$$

representing a linear combination of atomic orbitals. We repeat the development beginning at equation (6.39), using (6.58) for the atomic orbitals ϕ_i; the variation is, however, applied to the coefficients in the a column matrix. If N is the order of the expansion basis, χ, then N equations similar to (6.53) are obtained:

$$\left\{\int \chi^{tr}\left[\mathcal{H}^0 + \sum_j(2J_j - K_j)\right]\chi \, dV\right\} a_i = \sum_j \varepsilon_{ji} S^\chi a_j. \tag{6.59}$$

Here S^χ is the overlap matrix in the χ basis. From (6.59) we may write

$$\left[\int \chi^{tr} F \chi \, dV\right] a_i = \sum_j S^\chi a_j \varepsilon_{ji}, \tag{6.60}$$

so that

$$F^\chi a_i = \sum_j S^\chi a_j \varepsilon_{ji}. \qquad (6.61)$$

If ε_{ji} is restricted to $\delta_{ij}\varepsilon_{ji}$ we obtain

$$F^\chi a_i = \sum_j S^\chi a_j \varepsilon_{ji}\delta_{ij} = S^\chi a_i \varepsilon_{ii}, \qquad (6.62)$$

from which

$$(F^\chi - S^\chi \varepsilon_{ii})a_i = 0. \qquad (6.63)$$

If the determinant of the coefficients of a_i is set equal to zero, the secular determinant is obtained,

$$\mathrm{Det}(F^\chi - S^\chi \varepsilon_{ii}) = 0, \qquad (6.64)$$

where the number of roots equals the order of the χ basis.

The problem is now solved again by an iterative process, which starts with a choice of the χ set and the expansion (6.58). The Hartree–Fock operator F and the matrix representation F^χ are calculated, (6.64) is solved for the orbital energies, and these are used to compute a new set of coefficients in (6.63). If these are different from the starting set, the cycle is repeated until the self-consistent-field limit is reached. The total electronic energy is obtained by adding the SCF energy to the core energy for the lowest occupied $n/2$ levels:

$$W_0 = \sum_{i=1}^{n/2} (\mathcal{H}_{ii} + \varepsilon_{ii}). \qquad (6.65)$$

Many calculations for atoms have led to the development of a number of recipes for deciding the best values of ζ and n. A further important issue is the size of the basis set. A *minimal basis set* of STOs for an atom would include one function for each SCF occupied orbital with different n and l quantum numbers in equation (6.56); for the chlorine atom, therefore, the minimal basis set would include $1s$, $2s$, $2p$, $3s$ and $3p$ functions, each with an optimised Slater orbital exponent ζ. A higher order of approximation would be to double the number of STOs (the *double zeta basis set*), with orbital exponents optimised; ultimately the Hartree–Fock limit is reached, as it has been for all atoms from He to Xe [13].

Among the alternative basis functions which have been used in SCF calculations, mention should be made of Gaussian orbitals, which by analogy with the Slater orbitals of (6.56) may be written

$$G_{nlm}(r, \theta, \phi) = Nr^n \exp(-\alpha r^2) Y_{lm}(\theta, \phi), \qquad (6.66)$$

where N is the normalisation constant, α is the Gaussian orbital exponent and n is the analogue of n in the Slater-type orbital. Alternatively the Gaussian orbitals may be expressed in cartesian form as

$$G_{nlm}(x, y, z) = N' x^n y^l z^m \exp(-\alpha r^2). \qquad (6.67)$$

Compared with the STOs, there are both advantages and disadvantages to Gaussian orbitals. The main disadvantage is that larger basis sets are required for Gaussians; on the other hand several of the integrals which arise in molecular calculations are easier to compute.

Perhaps the most important point to emphasise in conclusion is that the Hartree–Fock atomic orbitals are represented by a *single* determinant.

6.2.7. Configuration interaction

The difference between the true non-relativistic electron energy and the best single determinant Hartree–Fock energy is called the correlation error; it arises from the true instantaneous repulsion ($1/r_{ij}$) energy being repaced by a self-consistent average. To improve matters beyond this point the effects of mixing of the ground state Hartree–Fock determinant with additional determinants representing alternative electron configurations in the atom are included. The resulting many-electron wave function is written

$$\Psi(1, 2, \ldots, n) = \sum_I C_I \psi_I, \qquad (6.68)$$

where the C_I are variation coefficients and ψ_I is the Ith determinantal wave function of the type (6.36). The method is called, not surprisingly, *configuration interaction* (CI). The coefficients in (6.68) are determined by using the variation method already outlined, which leads to the secular equation

$$(\mathcal{H} - E\mathbf{S})\mathbf{C} = 0, \quad \text{Det}(\mathcal{H} - E\mathbf{S}) = 0. \qquad (6.69)$$

$\mathbf{C}$ is a column matrix whose elements are the C_I in (6.68) and E are the eigenvalues. The other matrix elements are given by

$$\mathcal{H}_{JI} = \int_1 \int_2 \cdots \int_n \psi_J^* \mathcal{H} \psi_I \, dV_1 \, dV_2 \ldots dV_n,$$

$$S_{JI} = \int_1 \int_2 \cdots \int_n \psi_J^* \psi_I \, dV_1 \, dV_2 \ldots dV_n. \qquad (6.70)$$

The question always arises as to how many configurations should be included. In the helium atom the radial correlation is improved by expanding in s orbitals, for example,

$$\Psi(1, 2) = C_1 \Psi(1s^2) + C_2 \Psi(1s, 2s) + C_3 \Psi(2s^2) + \cdots, \qquad (6.71)$$

where only states of the same symmetry (^{1}S) are included. Extensive calculations have been performed for many different atoms, giving accurate descriptions of both the ground and excited states, including open shell systems. Configuration interaction calculations are important for molecular wave functions also, as we shall see later in this chapter.

6.3. Molecular orbital theory

In chapter 2 we discussed at length the separation of nuclear and electronic coordinates in the solution of the Schrödinger equation. We described the Born–Oppenheimer approximation which allows us to solve the Schrödinger equation for the motion of the electrons in the electrostatic field produced by fixed nuclear charges. There are certain situations, particularly with polyatomic molecules, when the separation of nuclear and electronic motions cannot be made satisfactorily, but with most diatomic molecules the Born-Oppenheimer separation is acceptable. The discussion of molecular electronic wave functions presented in this chapter is therefore based upon the Born–Oppenheimer approximation.

There are a number of different approaches to the description of molecular electronic states. In this section we describe molecular orbital theory, which has been by far the most significant and popular approach to both the qualitative and quantitative description of molecular electronic structure. In subsequent sections we will describe the theory of the correlation of molecular states to the Russell–Saunders states of the separated atoms; we will also discuss what is known as the united atom approach to the description of molecular electronic states, an approach which is confined to diatomic molecules.

The most common and important approach to the description of the wave functions for a molecule is the linear combination of atomic orbitals ($l.c.a.o.$) method. As the name implies, this method involves consideration of the interaction of the atomic orbitals on the separated atoms to form molecular orbitals which accomodate the available electrons. The method takes a relatively simple form in diatomic molecules, as figure 6.6 illustrates for a homonuclear molecule. The $1s$ atomic orbitals on the two atoms a and b can interact in-phase to form a bonding molecular orbital, or out-of-phase to form an antibonding orbital. The bonding orbital is called $\sigma_g 1s$, and the antibonding orbital is $\sigma_u^* 1s$; in the H$_2$ molecule, the two electrons are accomodated in the bonding molecular orbital, forming a stable system. Similar considerations apply to the interaction of the next highest atomic orbitals, the $2s$ orbitals, and the Li$_2$ molecule, with six electrons, has the electron configuration $(\sigma_g 1s)^2 (\sigma_u^* 1s)^2 (\sigma_g 2s)^2$.

The $2p$ atomic orbitals fall into one of two categories. If z is the internuclear axis, the $2p_z$ atomic orbitals interact to form a bonding $\sigma_g 2p$ and an antibonding $\sigma_u^* 2p$ molecular orbital. On the other hand, the $2p_x$ and $2p_y$ atomic orbitals, whose axes are perpendicular to the internuclear axis, interact to form degenerate bonding $\pi_u 2p$ and antibonding $\pi_g^* 2p$ orbitals. A molecular orbital energy level diagram for homonuclear diatomic molecules is shown in figure 6.7; by progressively filling the molecular orbitals with electrons, having due regard to the Pauli principle, the sequence of electron configurations for the lighter diatomic molecules can be established, as shown in table 6.3. We will, in due course, deal with the more complex situations which arise when atomic d orbitals are involved.

We have included in table 6.3 the electronic state nomenclature, the definition of which we now describe. First, we deal with the resultant orbital angular momentum about the internuclear axis, which is denoted Λ, and is equal to the sum of the individual

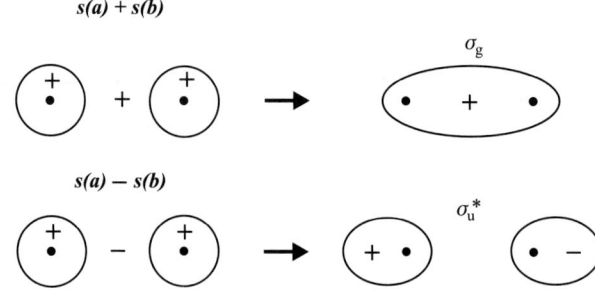

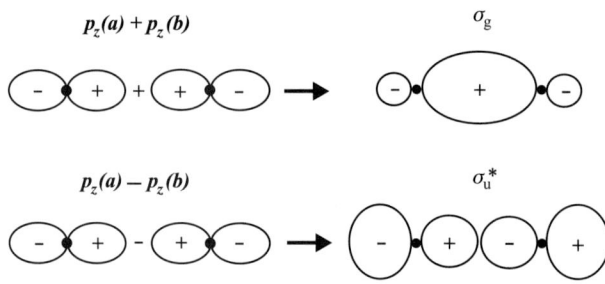

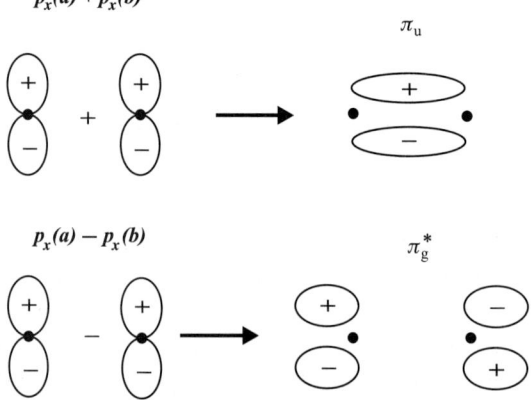

Figure 6.6. The symmetries of the linear combinations of atomic orbitals giving the lowest energy bonding and antibonding molecular orbitals.

electron orbital angular momenta, λ_i ; that is

$$\Lambda = \sum_i \lambda_i. \qquad (6.72)$$

The vectors lie along the internuclear axis, so that we have a simple algebraic addition.

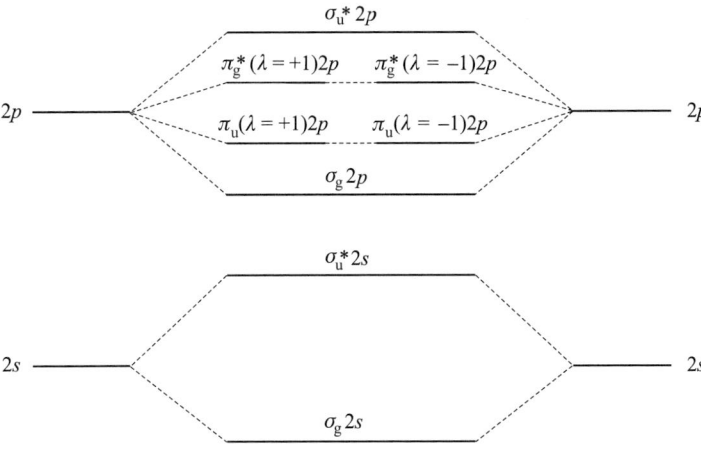

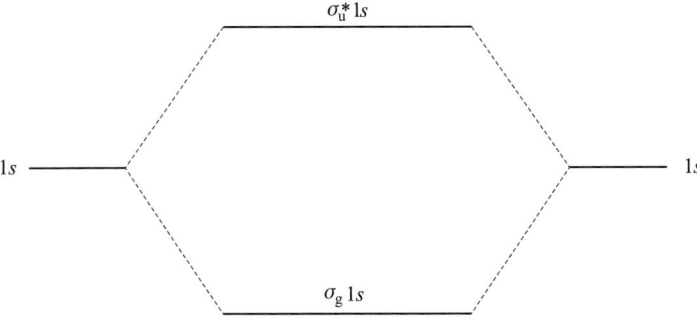

Figure 6.7. Molecular orbital energy level diagram for the homonuclear diatomic molecules of the first row of the periodic table.

For a single π electron, $\Lambda = \lambda = \pm 1$, and the electronic state is called a Π state. For two π electrons, the resultant Λ is equal to ± 2 or 0, each of these states arising in two different ways. For $|\Lambda| = 2$ we have a Δ state, which is two-fold degenerate. For $\Lambda = 0$ the electronic state is a Σ state; although there are two such states arising from the vectorial addition, they are not degenerate but split into a Σ^+ and a Σ^- state. We will define the difference between these states in due course. If $\Lambda = \pm 3$, which can arise when d orbitals are involved, the state is a Φ state, and so on. The resultant spin is equal to the sum of the individual spins, that is,

$$S = \sum_i s_i. \tag{6.73}$$

The resulting spin multiplicity, $2S + 1$, is indicated in the electronic state symbol as a

Table 6.3. *Ground state electron configurations and states for homonuclear diatomic molecules in the first row of the periodic table*

Molecule	Electron configuration	State
H_2^+	$(\sigma_g 1s)^1$	$^2\Sigma_g^+$
H_2	$(\sigma_g 1s)^2$	$^1\Sigma_g^+$
He_2^+	$(\sigma_g 1s)^2(\sigma_u^* 1s)^1$	$^2\Sigma_u^+$
He_2	$(\sigma_g 1s)^2(\sigma_u^* 1s)^2$	$^1\Sigma_g^+$
Li_2^+	$(\sigma_g 1s)^2(\sigma_u^* 1s)^2(\sigma_g 2s)^1$	$^2\Sigma_g^+$
Li_2	$(\sigma_g 1s)^2(\sigma_u^* 1s)^2(\sigma_g 2s)^2$	$^1\Sigma_g^+$
Be_2	$(\sigma_g 1s)^2(\sigma_u^* 1s)^2(\sigma_g 2s)^2(\sigma_u^* 2s)^2$	$^1\Sigma_g^+$
B_2	$(\sigma_g 1s)^2(\sigma_u^* 1s)^2(\sigma_g 2s)^2(\sigma_u^* 2s)^2(\pi_u 2p)^2$	$^3\Sigma_g^-$
C_2	$(\sigma_g 1s)^2(\sigma_u^* 1s)^2(\sigma_g 2s)^2(\sigma_u^* 2s)^2(\pi_u 2p)^3(\sigma_g 2p)^1$	$^3\Pi_u$
N_2^+	$(\sigma_g 1s)^2(\sigma_u^* 1s)^2(\sigma_g 2s)^2(\sigma_u^* 2s)^2(\pi_u 2p)^4(\sigma_g 2p)^1$	$^2\Sigma_g^+$
N_2	$(\sigma_g 1s)^2(\sigma_u^* 1s)^2(\sigma_g 2s)^2(\sigma_u^* 2s)^2(\pi_u 2p)^4(\sigma_g 2p)^2$	$^1\Sigma_g^+$
O_2^+	$(\sigma_g 1s)^2(\sigma_u^* 1s)^2(\sigma_g 2s)^2(\sigma_u^* 2s)^2(\pi_u 2p)^4(\sigma_g 2p)^2(\pi_g^* 2p)^1$	$^2\Pi_g$
O_2	$(\sigma_g 1s)^2(\sigma_u^* 1s)^2(\sigma_g 2s)^2(\sigma_u^* 2s)^2(\pi_u 2p)^4(\sigma_g 2p)^2(\pi_g^* 2p)^2$	$^3\Sigma_g^-$
F_2	$(\sigma_g 1s)^2(\sigma_u^* 1s)^2(\sigma_g 2s)^2(\sigma_u^* 2s)^2(\pi_u 2p)^4(\sigma_g 2p)^2(\pi_g^* 2p)^4$	$^1\Sigma_g^+$
Ne_2	$(\sigma_g 1s)^2(\sigma_u^* 1s)^2(\sigma_g 2s)^2(\sigma_u^* 2s)^2(\pi_u 2p)^4(\sigma_g 2p)^2(\pi_g^* 2p)^4(\sigma_u^* 2p)^2$	$^1\Sigma_g^+$

prefix superscript. Third, the subscripts g and u, which only arise for the homonuclear systems, are readily decided; only an odd number of u electrons can give rise to an overall u state. The ground electronic state is labelled the X state, and successive excited states of the same spin multiplicity but increasing energy are labelled A, B, C, etc. The lowest energy excited state of different spin multiplicity (higher or lower) is the a state, with higher energy states being labelled b, c, d, etc.

We now return to the question of the distinction between Σ^+ and Σ^- states, which requires a more detailed consideration of the electronic wave function. In a diatomic molecule any plane containing the internuclear axis is a plane of symmetry, and the electronic eigenfunction of a non-degenerate Σ state either remains unchanged when reflected through such a plane, in which case it is a Σ^+ state, or changes sign, in which case it is a Σ^- state. To illustrate this and other aspects, let us consider the O_2 molecule which, according to table 6.3, has its two outermost electrons in a degenerate $\pi_g^* 2p$ molecular orbital. Since each electron can occupy the π orbital with $\lambda = +1$ or $\lambda = -1$, there are four possible arrangments:

$$\phi_{+1}(1)\phi_{+1}(2): \Lambda = +2; \quad \phi_{+1}(1)\phi_{-1}(2): \Lambda = 0;$$
$$\phi_{-1}(1)\phi_{+1}(2): \Lambda = 0; \quad \phi_{-1}(1)\phi_{-1}(2): \Lambda = -2. \quad (6.74)$$

The permutation operator P_{12} which interchanges the electrons leaves the $\Lambda = \pm 2$ functions unchanged; they are symmetric and form the degenerate components of a Δ_g state which, since it must be combined with the antisymmetric spin function

$(1/\sqrt{2})\{\alpha(1)\beta(2) - \beta(1)\alpha(2)\}$ by the Pauli principle, is actually a $^1\Delta_g$ state, the lowest excited electronic state of the O$_2$ molecule. The $\Lambda = 0$ functions above are converted into each other by the permutation operator P_{12}, but the symmetric and antisymmetric combinations have definite symmetry under P_{12}:

$$P_{12}\{\phi_{+1}(1)\phi_{-1}(2) \pm \phi_{-1}(1)\phi_{+1}(2)\} = \pm\{\phi_{+1}(1)\phi_{-1}(2) \pm \phi_{-1}(1)\phi_{+1}(2)\}. \quad (6.75)$$

The symmetric combination corresponds to a Σ^+ state, and since it must be combined with the antisymmetric spin function, we have a $^1\Sigma_g^+$ state, the highest energy state arising from the ground electron configuration. The antisymmetric combination in (6.75) is a Σ^- state, and combined with the symmetric triplet spin functions forms the ground state, $^3\Sigma_g^-$. We discuss permutation and inversion symmetry in much more detail later in this chapter.

We mentioned at the beginning of this section that there is a related approach to the electronic structure of diatomic molecules, namely, the united atom approach. We consider the motion of an electron about a core which consists of two nuclei, at a fixed distance from each other, and of the remaining electrons. For this system, which we may regard as a two-centre system, the field of force acting on the outer electron may be regarded as axially symmetric about the internuclear axis. If we disregard electron spin, the stationary states of this atom-like system can be characterised by three quantum numbers. Of these, only one, the component of orbital angular momentum of the electron about the internuclear axis (λ), is precisely defined for all separations of the two nuclei. The possible values of λ are 0, ± 1, ± 2, ± 3, etc., and the electron is described as a σ, π, δ, ϕ, etc., electron respectively. Two other quantum numbers, which are only strict for the true united atom with internuclear distance $R = 0$, are n and l. These quantum numbers have the same significance as we described in the previous section on atomic wave functions, but become less significant as R increases. A correlation diagram connecting the molecular orbitals and the united atom descriptions for homonuclear molecules is shown in figure 6.8. The united atom orbitals are shown on the extreme left, and the separated atom molecular orbitals are on the extreme right. This diagram, taken from Herzberg [14], shows the molecular orbitals for a large range of R values, and indicates the positions of certain particular molecules. Note that the notation distinguishes between united atom and molecular orbitals; the symbols denoting the n, l values are listed first in the united atom description, but last in the molecular orbital case.

Heteronuclear diatomic molecules are naturally somewhat more complicated than the homonuclear; comprehensive comparisons with homonuclear molecules were given by Mulliken [15]. The atomic orbital coefficients in the molecular orbitals of heteronuclear diatomic molecules are no longer determined by symmetry alone, and the electrons in the molecular orbitals may be shared equally between atoms, or may be almost localised on one atom. The molecular orbitals can still be classified as σ or π, but in the absence of a centre-of-symmetry the g/u classification naturally disappears. Some heteronuclear molecules contain atoms which are sufficiently similar that the molecular orbitals resemble those shown in figure 6.7. In many other cases, however, the atoms are very different. This is particularly the case for hydride systems, like the HCl molecule,

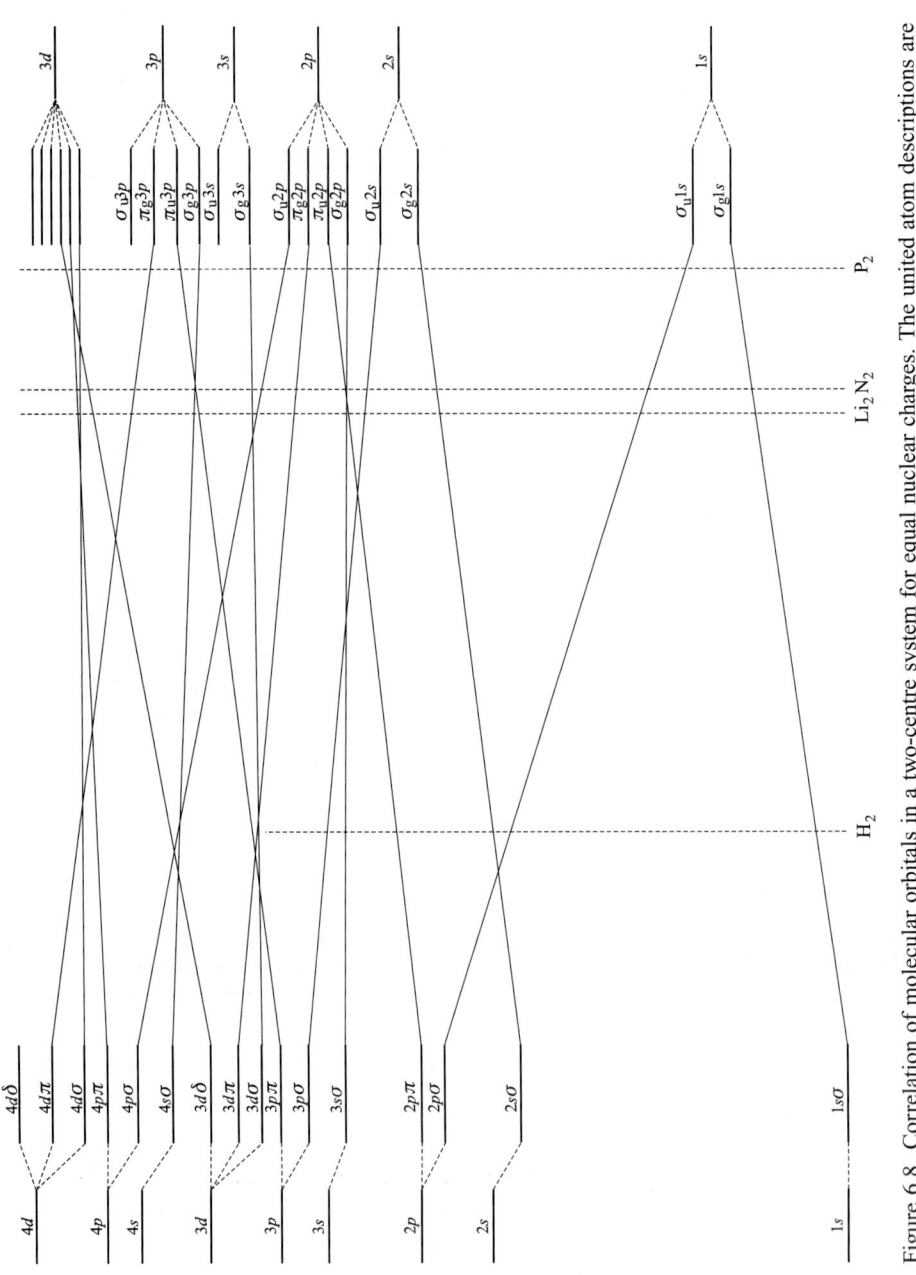

Figure 6.8. Correlation of molecular orbitals in a two-centre system for equal nuclear charges. The united atom descriptions are shown on the left, with the molecular orbital descriptions on the right.

where the hydrogen 1s electron has an energy similar to that of a chlorine $3p_z$ electron, and the main σ-type molecular orbital will be formed between the two corresponding orbitals; even so, the electron distribution is likely to be strongly polarised in favour of the chlorine atom. Electrons in the $3p_x$ and $3p_y$ chlorine orbitals remain essentially lone pair electrons.

A correlation diagram relating the molecular orbitals to the united atom orbitals can be drawn for heteronuclear systems, as shown in figure 6.9; it is similar to that for homonuclear systems. The combination of like atomic orbitals on atoms a and b will, in each case, give rise to two complementary molecular orbitals, as shown on the right-hand side of figure 6.9.

6.4. Correlation of molecular and atomic electronic states

We have described the orbital approaches to the electron configurations of diatomic molecules, both the molecular orbital and the united atom models. We now turn to the question of what types of molecular states result from given states of the separate atoms. If Russell–Saunders coupling is valid for the separate atoms, the correlation rules, due to Wigner and Witmer [16] provide a valid and complete summary of the molecular states. This information is extremely important for an understanding of both the formation and dissociation of diatomic molecules.

If two atoms with orbital and spin angular momentum values L_1, S_1 and L_2, S_2 are brought up to each other, an electric field is created in the direction of a line joining the nuclei; this produces space quantisation of $\boldsymbol{L}_1$ and $\boldsymbol{L}_2$ with respect to this direction, with components M_1 and M_2. The resultant angular momentum about the internuclear axis is therefore $M_1 + M_2$, and the molecule formed has values of the quantum number Λ given by $M_1 + M_2$. Clearly, consideration of all of the possible values of M_1 and M_2 will give the possible values of Λ. This is best appreciated by working through a particular example, so we consider the interaction of two atoms, both in P states (electron spin will be considered later).

The problem is most readily understood with the aid of a vector coupling diagram, such as that presented in figure 6.10. Each atom has $L_1 = L_2 = 1$, and the three allowed orientations of the vectors $\boldsymbol{L}_1$ and $\boldsymbol{L}_2$ with respect to the direction of the internuclear axis are shown. As can be seen from the figure, there is a total of nine relative orientations of the two vectors; combining $M_1 = +1, 0, -1$ with $M_2 = +1, 0, -1$ gives nine values of $M_1 + M_2$, with two different ways of obtaining total components of $+1$, and -1, and three ways of obtaining 0, as shown. We therefore obtain a Δ state, two Π states and three Σ states. The Δ and Π states retain their two-fold degeneracy, but the two similar Σ states split into Σ^+ and Σ^- states of different energies. Table 6.4 is an abbreviated version of a similar table in Herzberg's book [14], giving the molecular electronic states resulting from given states of the separated unlike atoms.

The possible spin multiplicities are readily determined because the resultant spin quantum number S is given by

$$S = (S_1 + S_2), (S_1 + S_2 - 1), (S_1 + S_2 - 2), \ldots, |S_1 - S_2|, \qquad (6.76)$$

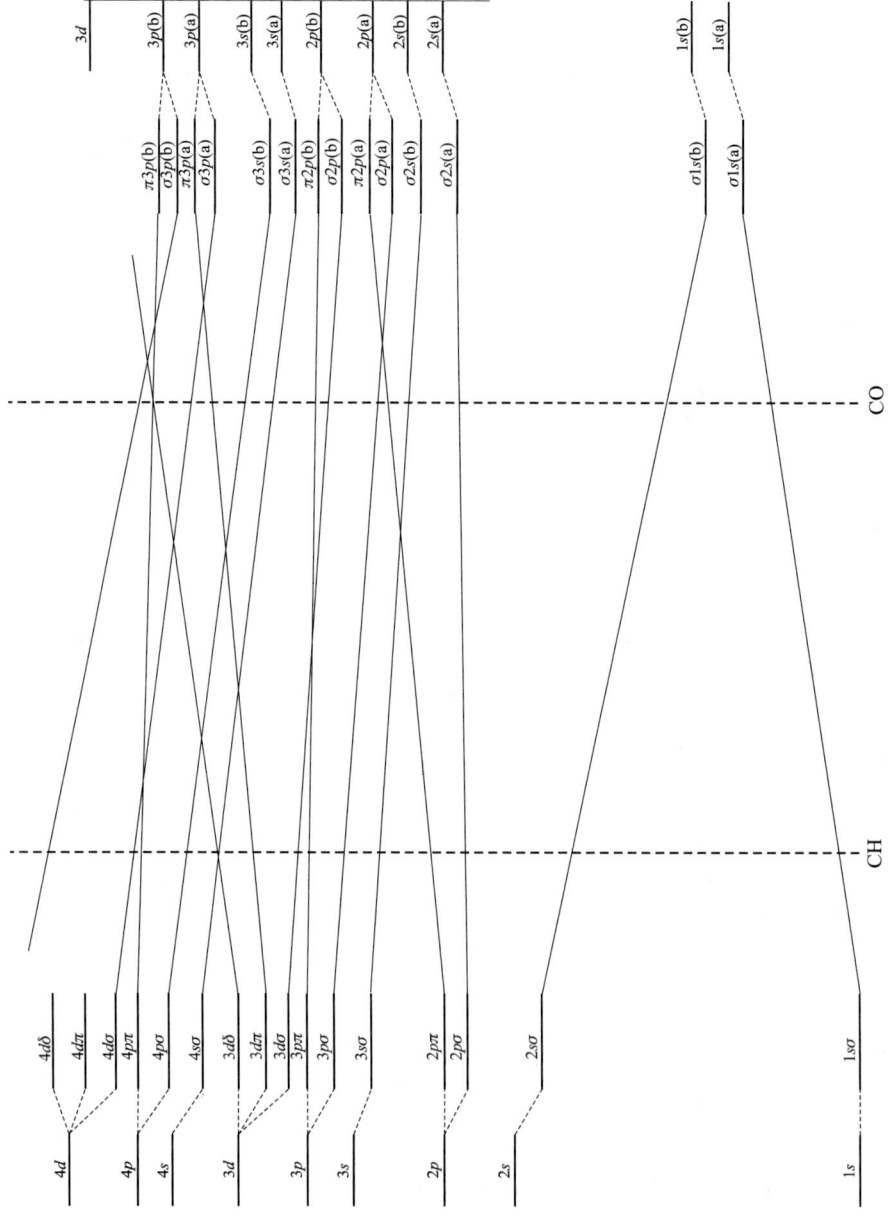

Figure 6.9. Correlation of molecular orbitals and united atom orbitals in a two-centre system for unequal nuclear charges.

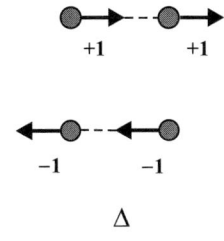
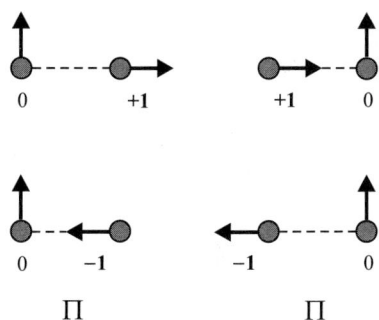
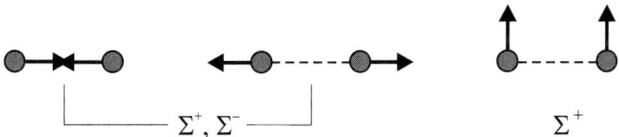

Figure 6.10. Vector diagram showing the molecular states arising from the interaction of two atoms, both in P states. The arrows show the orientations of the orbital angular momentum vectors L_1 and L_2 with respect to the direction of the internuclear axis, and the numbers show the values of M_1 and M_2.

and each of the orbital states listed in table 6.4 can be combined with each of the spin states. For example, if both atoms are in quartet states ($S = 3/2$), the resulting molecular states can be septet ($S = 3$), quintet ($S = 2$), triplet ($S = 1$) and singlet ($S = 0$).

Similar rules hold for the combination of two like atoms, the main difference being that the resulting molecular states must be either even or odd. Again an abbreviated list is presented in table 6.5; more comprehensive lists are available in the literature. Note that it is not necessary to specify the parities of the separated atoms, since we are dealing with combinations of like atoms.

Similar correlation rules exist for atoms which follow jj coupling, rather than Russell–Saunders coupling, but such cases are relatively rare and will not be discussed here.

Table 6.4. *Molecular electronic states resulting from given states of the separated unlike atoms*

States of the separated atoms	Resulting molecular states
$S_g + S_g$ or $S_u + S_u$	Σ^+
$S_g + S_u$	Σ^-
$S_g + P_g$ or $S_u + P_u$	Σ^-, Π
$S_g + P_u$ or $S_u + P_g$	Σ^+, Π
$S_g + D_g$ or $S_u + D_u$	Σ^+, Π, Δ
$S_g + D_u$ or $S_u + D_g$	Σ^-, Π, Δ
$S_g + F_g$ or $S_u + F_u$	$\Sigma^-, \Pi, \Delta, \Phi$
$S_g + F_u$ or $S_u + F_g$	$\Sigma^+, \Pi, \Delta, \Phi$
$P_g + P_g$ or $P_u + P_u$	$\Sigma^+(2), \Sigma^-, \Pi(2), \Delta$
$P_g + P_u$	$\Sigma^+, \Sigma^-(2), \Pi(2), \Delta$

Table 6.5. *Molecular electronic states resulting from identical states of the separated like atoms*

States of the separated atoms	Molecular states
$^1S + {}^1S$	$^1\Sigma_g^+$
$^2S + {}^2S$	$^1\Sigma_g^+, {}^3\Sigma_u^+$
$^3S + {}^3S$	$^1\Sigma_g^+, {}^3\Sigma_u^+, {}^5\Sigma_g^+$
$^4S + {}^4S$	$^1\Sigma_g^+, {}^3\Sigma_u^+, {}^5\Sigma_g^+, {}^7\Sigma_u^+$
$^1P + {}^1P$	$^1\Sigma_g^+(2), {}^1\Sigma_u^-, {}^1\Pi_g, {}^1\Pi_u, {}^1\Delta_g$
$^2P + {}^2P$	$^1\Sigma_g^+(2), {}^1\Sigma_u^-, {}^1\Pi_g, {}^1\Pi_u, {}^1\Delta_g, {}^3\Sigma_u^+(2), {}^3\Sigma_g^-, {}^3\Pi_g, {}^3\Pi_u, {}^3\Delta_u$

6.5. Calculation of molecular electronic wave functions and energies

6.5.1. Introduction

Many of the principles and techniques for calculations on atoms, described in section 6.2 of this chapter, can be applied to molecules. In atoms the electronic wave function was written as a determinant of one-electron atomic orbitals which contain the electrons; these atomic orbitals could be represented by a range of different analytical expressions. We showed how the Hartree–Fock self-consistent-field methods could be applied to calculate the single determinantal best energy, and how configuration interaction calculations of the mixing of different determinantal wave functions could be performed to calculate the correlation energy. We will now see that these technques can be applied to the calculation of molecular wave functions, the atomic orbitals of section 6.2 being replaced by one-electron molecular orbitals, constructed as linear combinations of atomic orbitals (*l.c.a.o.* method).

Before discussing the problems of many-electron wave functions for molecules, it is instructive to consider the special cases of the H_2^+ and H_2 molecules, containing one and two electrons respectively. The electronic wave function for H_2^+ can actually be calculated exactly within the Born–Oppenheimer approximation, not analytically but by using series expansion methods; an excellent description of the calculation has been given by Teller and Sahlin [17], and we give a summary of the method in appendix 6.1. We will, however, use the H_2^+ and H_2 molecules to illustrate the *l.c.a.o.* method in the next two subsections.

6.5.2. Electronic wave function for the H_2^+ molecular ion

The hydrogen molecular ion has long been a test bed for quantum theoretical methods, and continues so to be. We now present first the simplest quantitative treatment, leading to calculations of the electronic energy as a function of the internuclear distance, R.

The Hamiltonian for the H_2^+ molecule may be written in the form

$$\mathcal{H} = -\frac{\hbar^2}{2m}\nabla^2 - \frac{e^2}{4\pi\varepsilon_0 r_a} - \frac{e^2}{4\pi\varepsilon_0 r_b} + \frac{e^2}{4\pi\varepsilon_0 R}, \quad (6.77)$$

where r_a and r_b are the distances from the nuclei a and b to the electron. The molecular orbitals (ϕ) for H_2^+ must reflect the symmetry of the molecule, and as we have seen in our earlier qualitative descriptions, they may be written as linear combinations of atomic orbitals (χ) located on each atom:

$$\phi_\pm = \frac{1}{[2 \pm 2S_{ab}]^{1/2}}[\chi_a \pm \chi_b]. \quad (6.78)$$

S_{ab} is the overlap integral defined by

$$S_{ab} = \int \chi_a^* \chi_b \, dV. \quad (6.79)$$

In order to determine the ground state energy of the molecule at a fixed value of R we use the Variation Theorem, that is,

$$E_0 \leq W = \int \phi^* \mathcal{H} \phi \, dV. \quad (6.80)$$

Substituting for the Hamiltonian, equation (6.77), and the molecular orbitals (6.78), equation (6.80) becomes

$$W_\pm = \frac{\int (\chi_a \pm \chi_b)^*[-(\hbar^2/2m)\nabla^2 - e^2/4\pi\varepsilon_0 r_a - e^2/4\pi\varepsilon_0 r_b + e^2/4\pi\varepsilon_0 R](\chi_a \pm \chi_b) \, dV}{2(1 \pm S_{ab})}$$

$$= \frac{\mathcal{H}_{aa} + \mathcal{H}_{bb} \pm 2\mathcal{H}_{ab}}{2(1 \pm S_{ab})}. \quad (6.81)$$

In order to progress we must now choose analytical forms for the atomic orbitals χ_a and χ_b, and the most obvious choice is the hydrogen atom $1s$ orbitals given earlier. All of the integrals required to evaluate (6.81) are to be found in the literature [18] and we may proceed as follows. For $\mathcal{H}_{aa}$ we have

$$\mathcal{H}_{aa} = \int \chi_a^* \left(-\frac{\hbar^2}{2m}\nabla^2 - \frac{e^2}{4\pi\varepsilon_0 r_a}\right)\chi_a\, dV - e^2 \int \frac{\chi_a^*\chi_a}{4\pi\varepsilon_0 r_b}\, dV + \frac{e^2}{4\pi\varepsilon_0 R}\int \chi_a^*\chi_a\, dV$$

$$= \left(\frac{1}{4\pi\varepsilon_0}\right)\left\{-\frac{e^2}{2a_0} - e^2\int \frac{\chi_a^*\chi_a}{r_b}\, dV + \frac{e^2}{R}\right\}. \tag{6.82}$$

The first matrix element in (6.82) arises in the treatment of the hydrogen atom, and the second can be evaluated by transforming to elliptical coordinates [18], giving

$$\int \frac{\chi_a^*\chi_a}{r_b}\, dV = -\left(\frac{1}{a_0} + \frac{1}{R}\right)\exp\left(-\frac{2R}{a_0}\right) + \frac{1}{R}. \tag{6.83}$$

Consequently

$$\mathcal{H}_{aa} = \left(\frac{1}{4\pi\varepsilon_0}\right)\left\{-\frac{e^2}{2a_0} + e^2\left(\frac{1}{a_0} + \frac{1}{R}\right)\exp\left(-\frac{2R}{a_0}\right)\right\} = \mathcal{H}_{bb}. \tag{6.84}$$

The remaining integrals in (6.81) can be evaluated by similar methods [18] and lead to the final result for the energies:

$$W_{\pm} = \frac{-(e^2/2a_0) + (e^2/a_0)[1 + (1/r)]\exp(-2r) \pm (e^2/a_0)\{[(r^2/3) + r + 1][-(1/2) + (1/r)] - (r+1)\}\exp(-r)}{(4\pi\ \varepsilon_0)\{1 \pm [(r^2/3) + r + 1]\exp(-r)\}}. \tag{6.85}$$

In this expression $r = R/a_0$. Equation (6.85) gives a dissociation energy of 1.76 eV and a minimum in the ground state potential energy at $R = 1.32$ Å, compared with the experimental values of 2.79 eV and 1.058 Å. For such a simple model the results are not too bad.

6.5.3. Electronic wave function for the H₂ molecule

In the H$_2$ molecule the lowest energy electron configuration is obtained by placing both electrons in the $\sigma_g 1s$ molecular orbital with their spins paired so as to satisfy the Pauli exclusion principle. The Slater determinant for this arrangement is

$$\psi(1,2) = \left(\frac{1}{\sqrt{2}}\right)\begin{vmatrix} \phi_{\sigma_g 1s}(1)\alpha(1) & \phi_{\sigma_g 1s}(1)\beta(1) \\ \phi_{\sigma_g 1s}(2)\alpha(2) & \phi_{\sigma_g 1s}(2)\beta(2) \end{vmatrix}$$

$$= \left(\frac{1}{\sqrt{2}}\right)[\phi_{\sigma_g 1s}(1)\phi_{\sigma_g 1s}(2)][\alpha(1)\beta(2) - \alpha(2)\beta(1)]. \tag{6.86}$$

If we now represent the $\sigma_g 1s$ molecular orbital as a linear combination of the 1s atomic orbitals on the two atoms, we can expand (6.86) as follows:

$$\psi(1,2) = \frac{1}{2\sqrt{2}}[\chi_{1s_a}(1) + \chi_{1s_b}(1)][\chi_{1s_a}(2) + \chi_{1s_b}(2)][\alpha(1)\beta(2) - \beta(1)\alpha(2)]$$

$$= \frac{1}{2\sqrt{2}}[\chi_{1s_a}(1)\chi_{1s_a}(2) + \chi_{1s_a}(1)\chi_{1s_b}(2) + \chi_{1s_b}(1)\chi_{1s_a}(2)$$

$$+ \chi_{1s_b}(1)\chi_{1s_b}(2)][\alpha(1)\beta(2) - \beta(1)\alpha(2)]. \tag{6.87}$$

Note that, in this two-electron system, the spatial and spin parts of the wave function can be separated.

The electronic Hamiltonian for H_2, including the nuclear repulsion term is

$$\mathcal{H} = -\frac{\hbar^2}{2m}\nabla_1^2 - \frac{\hbar^2}{2m}\nabla_2^2 - \left(\frac{1}{4\pi\varepsilon_0}\right)\left\{\frac{e^2}{r_{1a}} + \frac{e^2}{r_{1b}} + \frac{e^2}{r_{2a}} + \frac{e^2}{r_{2b}} - \frac{e^2}{r_{12}} - \frac{e^2}{R}\right\}. \quad (6.88)$$

This Hamiltonian can be combined with the wave function given in (6.87) and, using the hydrogen atom 1s atomic orbitals once more, Coulson [19] obtained a calculated dissociation energy of 2.681 eV and an equilibrium internuclear separation of $R = 0.850$ Å, compared with the experimental values of 4.75 eV and 0.740 Å.

A large number of variation calculations have been performed for H_2, some of the most accurate employing elliptical coordinates with the nuclei as foci to describe electronic positions relative to the nuclei. The electron coordinates may be written

$$\xi = (r_a + r_b)/R, \quad \eta = (r_a - r_b)/R, \quad (6.89)$$

together with an angle ϕ measured about the internuclear axis. A trial variation function for hydrogen is then written in the form

$$\Phi(1,2) = \sum_{p,q,r,s,t} A_{pqrst}\left(\xi_1^p \eta_1^q \xi_2^r \eta_2^s + \xi_1^r \eta_1^s \xi_2^p \eta_2^q\right)\exp[-\alpha(\xi_1 + \xi_2)]r_{12}^t. \quad (6.90)$$

One can use as many terms as one might wish with this function. In 1933 James and Coolidge [20] used 13 terms and obtained an energy only 0.02 eV less than the full value of 4.75 eV, whilst Kolos and Roothaan [21] extended the basis set to 50 terms and obtained a calculated energy in exact agreement with experiment.

A modification of (6.90) which replaces the difficult r_{12} term by an appropriate expansion is

$$\Phi(1,2) = \sum_{p,q,r,s,m} B_{pqrsm}\left(\xi_1^p \eta_1^q \xi_2^r \eta_2^s + \xi_1^r \eta_1^s \xi_2^p \eta_2^q\right)\exp[-\alpha(\xi_1 + \xi_2)]\cos[m(\phi_1 - \phi_2)], \quad (6.91)$$

and using 40 such terms Kolos and Roothaan [21] obtained a binding energy only 0.01 eV less than the true value. A slightly different and more general form is

$$\Phi(1,2) = \sum_{i,j} C_{ij}[\chi_a^i(1)\chi_b^j(2) + \chi_b^j(1)\chi_a^i(2)], \quad (6.92)$$

with, for example, on nucleus a,

$$\chi_a(i) = \exp(-\delta_a \xi_i - \alpha_a \eta_i)\xi_i^{n_a}\eta_i^{m_a}\exp(i\nu_a\phi_i)\left[(\xi_i^2 - 1)(1 - \eta_i^2)\right]^{1/2|\nu_a|}. \quad (6.93)$$

Harris [22] and Davidson [23] used such a function in variation calculations for both the ground and excited states of H_2, with excellent results.

So much for complicated analytical expressions which are useful for H_2 but less likely to be valuable for more complex molecules. We return to the use of atomic orbitals, the first and most famous example being the Heitler–London treatment [24] using $1s$ orbitals on each atom. This valence bond Heitler–London wave function may be written

$$\Phi_{vb}(1,2) = \frac{1}{[2+2S^2]^{1/2}}[1s_a(1)1s_b(2) + 1s_b(1)1s_a(2)], \qquad (6.94)$$

where S is, as usual, the overlap integral. Alternatively we may use the *l.c.a.o.* molecular orbital wave function, placing both electrons in the bonding orbital:

$$\Phi_{mo}(1,2) = \frac{1}{2(1+S)}[1s_a(1) + 1s_b(1)][1s_a(2) + 1s_b(2)]. \qquad (6.95)$$

Using the hydrogen $1s$ orbitals with optimised ζ parameters shows that there is not much to choose between these two approaches; neither is particularly good. It is, however, possible to obtain a better description, even confining attention to the $1s_a$ and $1s_b$ atomic orbitals. First, following Parr [10], we note that the valence bond function (6.94) represents a covalent structure, with the electrons shared between the two atoms. Alternatively, a function which places both electrons on the same atom,

$$1s_a(1)1s_a(2) + 1s_b(1)1s_b(2), \qquad (6.96)$$

represents an ionic structure. It seems likely that a mixture of the covalent and ionic structures would be a better representation, and Weinbaum [25] confirmed this with a function

$$\Psi = c_1[1s_a(1)1s_b(2) + 1s_b(1)1s_a(2)] + c_2[1s_a(1)1s_a(2) + 1s_b(1)1s_b(2)], \qquad (6.97)$$

finding the best energy with $c_1/c_2 = 3.9$.

An apparently different approach is to use a molecular orbital description which includes configuration interaction between the ground electron configuration (6.95) and the doubly excited configuration in which both electrons occupy the lowest antibonding orbital,

$$\Phi'_{mo} = \frac{1}{2(1-S)}[1s_a(1) - 1s_b(1)][1s_a(2) - 1s_b(2)]. \qquad (6.98)$$

A mixture of (6.95) and (6.98) actually gives an identical result to that obtained with the function (6.97). This is, in fact, an example of a general result that the valence bond method, including ionic terms, is equivalent to the molecular orbital method including configuration interaction.

In order to obtain increased accuracy it is necessary to include atomic orbitals higher than $1s$ and such a configuration interaction calculation by McLean, Weiss and Yoshimine [26] which employed atomic $1s$, $2s$ and $2p$ atomic orbitals gave a binding energy of 4.55 eV. Later calculations, particularly by Kolos and Wolniewicz [27], produced energies for dissociation and ionisation of H_2 which were within 10^{-4} eV or less of the experimental values.

Many other calculations on H_2 in its electronic ground state have been described, and they continue to be reported. Calculations of excited electronic states of H_2 are also important, particularly in the light of experimental studies described elsewhere in this book, notably in chapters 8 and 11. We should, therefore, say something about these calculations.

In equation (6.98) we wrote the wave function for an excited state of H_2 in which both electrons occupy the $\bar{\sigma}_u 1s$ molecular orbital; this symmetric spatial function must be combined with the antisymmetric spin function

$$(1/\sqrt{2})[\alpha(1)\beta(2) - \beta(1)\alpha(2)], \tag{6.99}$$

to give an excited $^1\Sigma_g^+$ state. Lower energy excited states, however, will arise from promotion of a *single* electron from the $\sigma_g 1s$ bonding to the $\bar{\sigma}_u 1s$ antibonding orbital. Both singlet and triplet states are possible; the singlet state has the following wave function:

$$\Phi_S = (1/\sqrt{2})[\sigma_g 1s(1)\bar{\sigma}_u 1s(2) + \sigma_g 1s(2)\bar{\sigma}_u 1s(1)](1/\sqrt{2})[\alpha(1)\beta(2) - \alpha(2)\beta(1)]. \tag{6.100}$$

The triplet spin state is associated with the antisymmetric spatial function:

$$\Phi_t = (1/\sqrt{2})[\sigma_g 1s(1)\bar{\sigma}_u 1s(2) - \sigma_g 1s(2)\bar{\sigma}_u 1s(1)](1/\sqrt{2})[\alpha(1)\beta(2) + \alpha(2)\beta(1)],$$
$$\alpha(1)\alpha(2),$$
$$\beta(1)\beta(2). \tag{6.101}$$

This triplet state is actually the $b\,^3\Sigma_u^+$ state; it is repulsive at all internuclear distances and, like the ground state, correlates with $H(1s) + H(1s)$ at the dissociation limit. Indeed it is responsible for the continuum ultraviolet and visible light produced by a hydrogen discharge lamp. The singlet state (6.100) is the $B\,^1\Sigma_u^+$ state, a high-lying state which nevertheless possesses a potential minimum and is therefore stable. The even higher energy state with both electrons in the lowest antibonding orbital, represented by (6.98), actually has two distinct minima in its potential energy curve, which have been identified spectroscopically. These two virtually distinct states are known as the $E\,^1\Sigma_g^+$ and $F\,^1\Sigma_g^+$ states. There are therefore four states, namely, the ground state and three excited states (one being a triplet state) which can be constructed from the $1s$ atomic orbitals. A potential energy diagram showing these states is presented in figure 6.11. Many more excited states of H_2 exist, involving the $2s$ and $2p$ atomic orbitals. The singly excited states involve one electron remaining in the $\sigma_g 1s$ molecular orbital, whilst the second electron occupies a higher energy orbital; in effect one has an H_2^+ core, with a second electron orbiting at long range. Highly excited states of this type are known as molecular Rydberg states; several of them have been studied by high resolution spectroscopy and are discussed at length in chapters 8 and 11.

It is hoped that we have now explored the theory of the pre-eminent H_2 molecule in sufficient detail for us to approach many-electron molecules with increased insight. We emphasise again that all of the preceding discussion has presumed the validity of the Born–Oppenheimer separation. Life is not always so kind, and we return to this aspect later in this chapter.

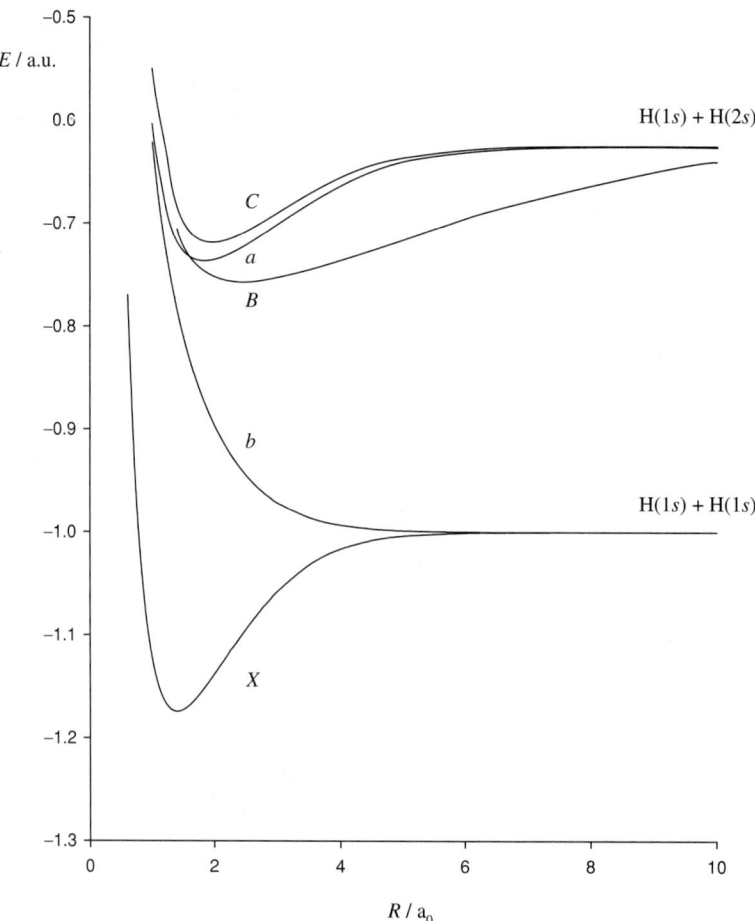

Figure 6.11. Potential energy curves for the ground state and first few excited electronic states of H_2. In the order of increasing energy the states are $X\,^1\Sigma_g^+$, $b\,^3\Sigma_u^+$, $B\,^1\Sigma_u^+$, $a\,^3\Sigma_g^+$ and $C\,^1\Pi_u$.

6.5.4. Many-electron molecular wave functions

(i) Introduction

Throughout this book we refer to the interpretation and understanding of molecular parameters determined from spectroscopic analysis. The first essential objective in analysing a spectrum is to establish the most suitable form of an *effective Hamiltonian* which represents the important intramolecular interactions present in a specific case; procedures for deriving an effective Hamiltonian are described in chapter 7. Using an effective Hamiltonian, one proceeds to calculate the appropriate energy levels and the frequencies and intensities of the transitions between them which give rise to the observed spectrum. Algebraic expressions for the transition frequencies will contain quantities which can be called the molecular constants or, as we prefer, the molecular parameters. The precise meaning of these molecular parameters is defined in the process of deriving the effective Hamiltonian. The values of the parameters are related directly to the electronic wave function, so that the ultimate value of the parameters lies in the information they provide about the electronic structure of the molecule under investigation. The links between the molecular parameters and the electronic wave function are not simple, however. There are, in general, two different approaches to understanding these links, and both are important. We may describe one approach as *semi-empirical*; the electronic structure is described in terms of a somewhat idealised model which, in some sense, provides a physical picture which we can visualise. A successful model is one which provides a semi-quantitative understanding of the values of the molecular parameters. In addition, it will often be predictive and rationalise trends to some level of satisfaction.

The other approach to the electronic wave function and its links with the molecular parameters is the *ab initio* calculation; we often use this term or mode of description elsewhere in this book. An *ab initio* calculation is one which uses only the fundamental laws of physics and the values of the fundamental constants. It is easy to define, but efforts to achieve success for molecules with many electrons, even diatomic molecules, have occupied the attention of many for the past seventy years. The electronic wave function produced by an *ab initio* calculation is often too complicated algebraically to provide something which can reasonably be described as a physical picture. In this situation the value of a molecular parameter, determined by experiment, is best regarded as a *benchmark* which can be used to test the accuracy of an *ab initio* calculation.

The development of *ab initio* methods, which has come to be known as quantum chemistry, is one of the outstanding cumulative intellectual and technical achievements of the past fifty years. Fifty years ago, at the time of the publication of Herzberg's classic book [14], quantum mechanics had been applied to the calculation of the wave functions of the very simplest molecules, but for most systems the problems were considered intractable. At the turn of the millenium we have come a long way, but many difficult problems remain. Progress has been closely related to the development of the digital computer, and the technical achievements in the use of computers to solve problems in quantum mechanics have been impressive. There has always, however, been an accompanying and continual need for intellectual advances to make use of the technology. This section describes the nature of the problems to be solved, and some of the methods which have been developed to tackle them.

(ii) The electronic Hamiltonian

Within the Born–Oppenheimer approximation [28], the electron distribution is supposed to depend only on the instantaneous positions of the nuclei and not on their velocities. The Schrödinger equation for the electrons in the field of the fixed nuclei, which we have met before, and was derived in chapter 2, is

$$\mathcal{H}^{\text{elec}} \Psi^{\text{elec}}(r, R) = E^{\text{elec}}(R) \Psi^{\text{elec}}(r, R). \quad (6.102)$$

The electronic wave function $\Psi^{\text{elec}}(r, R)$ depends on both the electronic coordinates,

r, and nuclear coordinates, R, whilst the electronic Hamiltonian corresponds to the motion of the electrons in the field of the fixed nuclei. It is the sum of two terms,

$$\mathcal{H}^{\text{elec}} = T^{\text{elec}} + V. \tag{6.103}$$

The electronic kinetic energy is given by

$$T^{\text{elec}} = -\left(\frac{\hbar^2}{2m}\right) \sum_i^{\text{electrons}} \left(\frac{\partial^2}{\partial x_i^2} + \frac{\partial^2}{\partial y_i^2} + \frac{\partial^2}{\partial z_i^2}\right). \tag{6.104}$$

The Coulomb potential energy, V, is the sum of three different types of term, as follows:

$$V = -\sum_i^{\text{electrons}} \sum_s^{\text{nuclei}} \frac{e^2 Z_s}{4\pi\varepsilon_0 r_{is}} + \sum_{i<j}^{\text{electrons}} \frac{e^2}{4\pi\varepsilon_0 r_{ij}} + \sum_{s<t}^{\text{nuclei}} \frac{e^2 Z_s Z_t}{4\pi\varepsilon_0 R_{st}}. \tag{6.105}$$

The third term does not involve the electron coordinates and is a constant for any fixed configuration of the nuclei; this sum reduces to a single term for a diatomic system.

A summary of the theory has been provided by Pople [29]. The convention in this field is to use atomic units, such that $\hbar = e = m = 1$. In this system, length is measured as a multiple of the Bohr radius a_0 and energy as a multiple of twice the ionisation energy of the H atom in its ground state (an amount of energy known as the Hartree). Thus the Rydberg constant R_H and the Bohr magneton μ_B each have a value of $1/2$ in this scheme. As we discussed earlier for atoms, the task is to find a wave function $\Psi^{\text{elec}}(r, R)$ which minimises the energy, calculated as the expectation value of the full Hamiltonian $\mathcal{H}$. In the foundations of the theory due to Hartree, Fock and Slater, the n electrons in a closed shell molecule are assigned to a set of $n/2$ molecular orbitals $\psi_i (i = 1, \ldots, n/2)$ and the total wave function is written as a single configuration Slater determinant, which we introduced for atoms in (6.36), and which we repeat for a molecule here:

$$\Psi = \frac{1}{(n!)^{1/2}}$$

$$\times \begin{vmatrix} \psi_1(1)\alpha(1) & \psi_1(1)\beta(1) & \psi_2(1)\alpha(1) & \psi_2(1)\beta(1) & \cdots & \psi_{n/2}(1)\alpha(1) & \psi_{n/2}(1)\beta(1) \\ \psi_1(2)\alpha(2) & \psi_1(2)\beta(2) & \psi_2(2)\alpha(2) & \psi_2(2)\beta(2) & \cdots & \psi_{n/2}(2)\alpha(2) & \psi_{n/2}(2)\beta(2) \\ \cdot & \cdot & \cdot & \cdot & \cdots & \cdot & \cdot \\ \cdot & \cdot & \cdot & \cdot & \cdots & \cdot & \cdot \\ \cdot & \cdot & \cdot & \cdot & \cdots & \cdot & \cdot \\ \psi_1(n)\alpha(n) & \psi_1(n)\beta(n) & \psi_2(n)\alpha(n) & \psi_2(n)\beta(n) & \cdots & \psi_{n/2}(n)\alpha(n) & \psi_{n/2}(n)\beta(n) \end{vmatrix}$$

(6.106)

An important development was the introduction by Roothaan [12] and Hall [30] of the expansion of the molecular orbitals as linear combinations of three-dimensional one-electron functions, χ_μ, where $\mu = 1, 2, \ldots, N$, $N > n$,

$$\psi_i = \sum_{\mu=1}^{N} c_{\mu i} \chi_\mu. \tag{6.107}$$

The set $\{\chi_\mu\}$ is referred to as the basis set, the basis functions being normally chosen to be centred at the nuclei and to depend only on the positive charge of the nucleus. The chosen functions may be atomic orbitals of the component atoms, but are not necessarily so. Variation of the total energy with respect to the coefficients $c_{\mu i}$ leads to a set of algebraic equations which can be written in matrix form:

$$\mathbf{FC} = \mathbf{SCE}. \tag{6.108}$$

We now examine the elements of the matrices in (6.108). First, $\mathbf{F}$ is called the Fock matrix, with elements:

$$F_{\mu\nu} = \mathcal{H}_{\mu\nu} + \sum_{\lambda\sigma} P_{\lambda\sigma}[(\mu\nu|\lambda\sigma) - (1/2)(\mu\lambda|\nu\sigma)]. \tag{6.109}$$

In this expression, $\mathcal{H}_{\mu\nu}$ is a matrix representing the energy of a single electron in the field of the bare nuclei. Its elements are given by, for example,

$$\mathcal{H}_{\mu\nu}(1) = \int \chi_\mu^*(1) \left\{ -\frac{1}{2}\left(\frac{\partial^2}{\partial x_1^2} + \frac{\partial^2}{\partial y_1^2} + \frac{\partial^2}{\partial z_1^2}\right) - \sum_{s=1,2}\frac{Z_s}{r_{1s}} \right\} \chi_\nu(1) \, d\tau. \tag{6.110}$$

$P_{\lambda\sigma}$ are the elements of a one-electron density matrix,

$$P_{\lambda\sigma} = 2\sum_i^n c_{\lambda i} c_{\sigma i}, \tag{6.111}$$

where the sum is over the occupied orbitals only. Finally in (6.109), we have, for electrons 1 and 2, the repulsion integrals,

$$(\mu\nu|\lambda\sigma) = \int\int \chi_\mu(1)\chi_\nu(1)\frac{1}{r_{12}}\chi_\lambda(2)\chi_\sigma(2) \, d\tau_1 \, d\tau_2. \tag{6.112}$$

Returning to the matrix equation (6.108), the elements of the $N \times N$ overlap matrix $\mathbf{S}$ are given by

$$S_{\mu\nu} = \int \chi_\mu(1)\chi_\nu(1) \, d\tau_1. \tag{6.113}$$

Finally,

$$E_{lj} = \varepsilon_l \delta_{lj}, \tag{6.114}$$

where the eigenvalues ε_i are the one-electron Fock energies.

The Roothaan–Hall equations are nonlinear because the Fock matrix $F_{\mu\nu}$ depends upon the orbital coefficients $c_{\mu i}$ through the density matrix expression (6.111). Solution therefore involves an iterative process, as we discussed previously for atomic systems, and the technique is therefore called *self-consistent-field* (SCF) theory.

The equations require to be modified for open-shell systems, in which some orbitals are doubly occupied and some singly (this is called *spin-restricted Hartree–Fock* theory). A further extension to the theory involves electrons of α and β spin being assigned to different molecular orbitals, ψ^α and ψ^β, so that there are two sets of coefficients $c_{\mu i}^\alpha$ and $c_{\mu i}^\beta$. The corresponding Roothaan-type equations are described as *unrestricted Hartree–Fock* [31].

(iii) Atomic basis functions

Evaluation of the integrals that arise in the calculations was for some time a primary problem in the field. A most important development in this context was the introduction of Gaussian-type basis functions by Boys [32], who showed that all of the integrals in SCF theory could be evaluated analytically if the radial parts of the basis functions were of the form $P(x, y, z) \exp(-r^2)$. The first ten functions are listed by Hehre, Radom, Schleyer and Pople [33] and we repeat them here:

$$g_s(\alpha, r) = \left(\frac{2\alpha}{\pi}\right)^{3/4} \exp(-\alpha r^2), \tag{6.115}$$

$$g_x(\alpha, r) = N_1 x \exp(-\alpha r^2), \quad g_y(\alpha, r) = N_1 y \exp(-\alpha r^2),$$
$$g_z(\alpha, r) = N_1 z \exp(-\alpha r^2) \quad \text{where } N_1 = \left(\frac{128\alpha^5}{\pi^3}\right)^{1/4}, \tag{6.116}$$

$$g_{xx}(\alpha, r) = N_2 x^2 \exp(-\alpha r^2), \quad g_{yy}(\alpha, r) = N_2 y^2 \exp(-\alpha r^2),$$
$$g_{zz}(\alpha, r) = N_2 z^2 \exp(-\alpha r^2), \quad g_{xy}(\alpha, r) = N_2 xy \exp(-\alpha r^2),$$
$$g_{xz}(\alpha, r) = N_2 xz \exp(-\alpha r^2), \quad g_{yz}(\alpha, r) = N_2 yz \exp(-\alpha r^2),$$
$$\text{where } N_2 = \left(\frac{2048\alpha^7}{\pi^3}\right)^{1/4}. \tag{6.117}$$

The function (6.115) has the angular symmetry of an s orbital and the functions (6.116) have the angular symmetry of p orbitals. Three of the second-order functions, g_{xy}, g_{xz}, g_{yz}, have the angular symmetry of atomic d orbitals, whilst linear combinations of the other three give the remaining d orbitals and a further s orbital:

$$g_{3zz-rr} = (1/2)(2g_{zz} - g_{xx} - g_{yy}), \quad g_{xx} - g_{yy} = (3/4)^{1/2}(g_{xx} - g_{yy}),$$
$$g_{rr} = (5)^{-1/2}(g_{xx} + g_{yy} + g_{zz}). \tag{6.118}$$

Higher-order gaussians may be combined to produce f-type functions. In many cases linear combinations of gaussians have been used as basis functions; such combinations are known as *contracted* gaussians.

Many different variants in the choice of basis functions have been used in electronic structure calculations, and computer programs developed for general use; some of these variants are described by Pople [29]

(iv) Multiple-determinant wave functions: configuration interaction

Our earlier discussion of electronic wave functions for many-electron atoms drew attention to the main inadequacy of the Hartree–Fock single determinant treatment; it does not take account of the correlation between the motions of electrons with opposite spins. In molecules this can even lead to qualitative deficiencies in the description of electronic structure, such as the failure to describe dissociation correctly. For example, the correct wave function for the singlet state of the hydrogen molecule at large internuclear separation,

$$(1/2)[1s_a(1)1s_b(2) + 1s_b(1)1s_a(2)][\alpha(1)\beta(2) - \beta(1)\alpha(2)], \tag{6.119}$$

cannot be expressed in terms of a single determinant.

In (6.106) we gave a single-determinantal wave function for a system of n electrons; if there are N functions ψ_μ, we can define $2N$ so-called spin-orbital basis functions of the type $\psi_\mu(\alpha)$ and $\psi_\mu(\beta)$ which may be linearly combined into $2N$ spin-orbitals χ_i. Solution of the Hartree–Fock problem yields the single-determinant wave function, written in abbreviated form,

$$\Psi_0 = (n!)^{-1/2}|\chi_1\chi_2 \cdots \chi_n|. \tag{6.120}$$

The spin-orbitals appearing in (6.120) are a subset of the total set determined in the variational calculation; the unoccupied or virtual spin-orbitals are denoted χ_a where $a = n+1, n+2, \ldots, 2N$.

The convention is to denote occupied spin-orbitals by the subscripts $i, j, k, \ldots$ and virtual or unoccupied spin-orbitals by subscripts $a, b, c, \ldots$. Further determinantal wave functions can now be written down by replacing one or more of the occupied spin-orbitals χ_i, χ_j, which appear in the Hartree–Fock function Ψ_0 above (6.120), by virtual spin-orbitals $\chi_a, \chi_b, \ldots$. These determinants may be classified as *single-substitution* functions, Ψ_i^a, if χ_i is replaced by χ_a, *double-substitution* functions, Ψ_{ij}^{ab}, in which χ_i is replaced by χ_a and χ_j by χ_b, and so on. The general substitution determinant may be written $\Psi_{ijk\ldots}^{abc\ldots}$ where $i < j < k \ldots$ and $a < b < c \ldots$, a series which can continue until all occupied spin-orbitals are replaced by virtual spin-orbitals. A general multi-determinant wave function can then be written as

$$\Psi = a_0\Psi_0 + \sum_{ia} a_i^a \Psi_i^a + \sum_{ijab} a_{ij}^{ab} \Psi_{ij}^{ab} + \cdots. \tag{6.121}$$

This series could be continued to include all possible substituted determinants, in which case it would correspond to a full configuration interaction wave function,

$$\Psi = a_0\Psi_0 + \sum_{s>0} a_s \Psi_s, \tag{6.122}$$

the summation being over all substituted determinants. The unknown coefficients a_s are determined by the usual variation method, leading to the result

$$\sum_s (\mathcal{H}_{st} - E_i\delta_{st})a_{si} = 0, \tag{6.123}$$

where $t = 0, 1, 2, \ldots$. $\mathcal{H}_{st}$ is now a configurational matrix element

$$\mathcal{H}_{st} = \int \cdots \int \Psi_s \mathcal{H} \Psi_t \, d\tau_1 \, d\tau_2 \ldots d\tau_n. \tag{6.124}$$

Equations (6.123) are similar to the Roothaan–Hall equations used to obtain the Hartree–Fock energy. A full configuration interaction treatment is feasible only for the simplest molecular systems, and therefore much effort has been expended on establishing the best ways to achieve the optimum limited configuration interaction. One method of choosing the configurations that has been widely employed is known as CASSCF (complete active space SCF); a helpful review of this and other methods, particularly for open shell systems, has been provided by Bally and Borden [34]. It should also be pointed out that some of the most acute computational problems do not arise for diatomic molecules.

Three other approaches towards the problem of incorporating electron correlation should be mentioned. The first is *Moeller–Plesset perturbation theory*, a method first introduced in 1934 by Moeller and Plesset [35]. Suppose that a perturbed Hamiltonian is defined by

$$\mathcal{H}(\lambda) = F_0 + \lambda\{\mathcal{H} - F_0\}, \tag{6.125}$$

where F_0 is the Fock Hamiltonian for which the single determinants in (6.122) are exact eigenfunctions. If $\lambda = 0$ then Ψ_0 is the eigenfunction, but if $\lambda = 1$ the exact full configuration interaction wave function in (6.122) is obtained. The computed energy is expanded in powers of λ,

$$E(\lambda) = E_0 + \lambda E_1 + \lambda^2 E_2 + \lambda^3 E_3 + \cdots, \tag{6.126}$$

the series cut off at some point, and λ then set equal to 1. This perturbation method is denoted by MPn if it is terminated at order n. The MP1 energy is therefore $E_0 + E_1$, identical to the Hartree–Fock value. MP2 and MP3 incorporate the effects of double substitutions, as described above, MP4 includes all singles and doubles, with some triples and quadruple substitutions also [36, 37].

The second general approach to correlation theory, also based on perturbation theory, is the *coupled-cluster* method, which can be thought of as an infinite-order perturbation method. The coupled-cluster wave function Ψ_{CC} is expressed as a power series,

$$\Psi_{CC} = \exp(T)\Psi_0 = \left(1 + T + \frac{1}{2!}T^2 + \frac{1}{3!}T^3 + \cdots\right)\Psi_0. \tag{6.127}$$

The exponential operator T creates excitations from Ψ_0 according to $T = T_1 + T_2 + T_3 + \cdots$, where the subscript indicates the excitation level (single, double, triple, etc.). This excitation level can be truncated. If excitations up to T_N (where N is the number of electrons) were included, Ψ_{CC} would become equivalent to the full configuration interaction wave function. One does not normally approach this limit, but higher excitations are included at lower levels of coupled-cluster calculations, so that convergence towards the full CI limit is faster than for MP calculations.

The third and final approach to the electron correlation problem included briefly here is *density functional theory* (DFT), a review of which has been given by Kohn in his Nobel lecture [38]. The Hohenberg–Kohn theorem [39] states that there is a one-to-one mapping between the potential $V(r)$ in which the electrons in a molecule move, the associated electron density $\rho(r)$, and the ground state wave function Ψ_0. A consequence of this is that given the density $\rho(r)$, the potential and wave function Ψ_0 are functionals of that density. An additional theorem provided by Kohn and Sham [40] states that it is possible to construct an auxiliary reference system of non-interacting electrons that has exactly the same electron density $\rho(r)$ as that of the real system of interacting electrons. There must therefore be an associated potential $V_n(r)$ in which the non-interacting electrons move, with a wave function Ψ_n, which gives the same electron density as that in the real system. Ψ_n can be formulated as a Slater determinant; the molecular orbitals in the determinant may be described as linear combinations of atomic orbitals. In both DFT and Hartree–Fock calculations one iteratively solves for a set of orbitals so that the associated density minimises the energy. There are, however, important differences between the two methods in their respective treatments of electron exchange and correlation [41].

Many of the objectives that exercise *ab initio* theorists do not exist for diatomic molecules. These objectives for polyatomic systems include obvious aspects like molecular geometry, chemical reaction mechanisms of formation and destruction, thermochemistry, transition states, etc. In diatomic molecules the main thrust of the spectroscopic studies is to provide very detailed and accurate information about the electronic wave function. In the past this information has related primarily to electronic ground states, but an increasing amount of complementary information about excited states is now being obtained. Furthermore the development of techniques that have the sensitivity required to probe much heavier molecules means that problems arising from the close proximity of several electronic states are often encountered. Molecules in which electronic and vibrational energy spacings are similar in magnitude are also being studied, and formidable problems of spectroscopic assignment arise. The interplay of experiment and *ab initio* calculations which have sufficient accuracy is likely to become increasingly important in the future.

6.6. Corrections to Born–Oppenheimer calculations for H_2^+ and H_2

It is a fortunate fact that, for the vast majority of molecules and their electronic states, the Born–Oppenheimer approximation is so good as to be adopted implicitly without further consideration. If it were not so, almost all of the guiding rules and principles which we take for granted would be invalid. In polyatomic molecules there are certain special cases involving electronic degeneracy when the Born–Oppenheimer approximation collapses. Naturally these special cases have attracted a great deal of interest; there are many papers which discuss the Jahn–Teller effect in polyatomic molecules of high symmetry, or the Renner–Teller effect in linear molecules. In diatomic molecules breakdown of the Born–Oppenheimer approximation is, in most cases, much less spectacular and its effects on the energy levels much smaller. Nevertheless, because the Born–Oppenheimer approximation is so central to an understanding of the energy levels and spectra even of diatomic molecules, it is important to understand the range and nature of its validity.

Much of the detailed attention paid to calculations beyond the Born–Oppenheimer limit has applied to one- and two-electron molecules; the literature up to 1980 was reviewed by Bishop and Cheung [42], and for the hydrogen molecular ion more recent summaries have been given by Carrington and Kennedy [43], Carrington, McNab and

Montgomerie [44] and Leach and Moss [45]. We will make use of the analyses given in these reviews. Not surprisingly, the most accurate treatments have been for one-electron molecules, the principal example being H_2^+ and its isotopomers. Spectacular breakdown of the Born–Oppenheimer approximation in diatomic molecules is observed for the HD^+ ion, which we will describe in some detail.

As we have mentioned before, the complete non-relativistic Hamiltonian for a system of point charges interacting electrostatically and moving through field-free space can be written in the form

$$\mathcal{H} = \sum_i \frac{-\hbar^2 \nabla_i^2}{2m_i} + \sum_i \sum_{j>i} \frac{z_i z_j e^2}{4\pi\varepsilon_0 r_{ij}}. \tag{6.128}$$

The motion of the centre-of-mass of the particles is separated out by applying the transformation

$$\begin{pmatrix} r_g \\ R \\ R_{cm} \end{pmatrix} = \begin{pmatrix} -\frac{1}{2} & -\frac{1}{2} & 1 \\ -1 & 1 & 0 \\ \frac{m_1}{M} & \frac{m_2}{M} & \frac{m_e}{M} \end{pmatrix} \begin{pmatrix} r_1 \\ r_2 \\ r_e \end{pmatrix}, \tag{6.129}$$

where $M = m_1 + m_2 + m_e$. The coordinate systems are illustrated in figure 6.12; r_1, r_2 and r_e are the position vectors of the three particles relative to an arbitrary space-fixed origin O. The new basis vectors are the internuclear vector ($R = r_2 - r_1$), the position vector R_{cm} of the centre-of-mass of the system, CM, relative to the space-fixed origin, and the position vector of the electron r_g relative to the geometric centre of the nuclei G. Applying the transformation to the Hamiltonian (6.128), converting to atomic units, and removing the translational term describing the motion of the centre-of-mass we

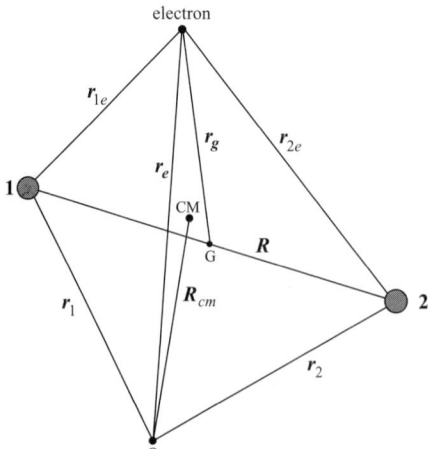

Figure 6.12. Coordinate system for the hydrogen molecular ion. O is the arbitrary space-fixed origin, CM is the centre-of-mass of the system, and G is the geometric centre of the nuclei. This diagram, although similar to figure 2.1, differs in some important aspects.

obtain the new Hamiltonian

$$\mathcal{H} = -\frac{\nabla_g^2}{2} - \frac{\nabla_R^2}{2\mu} - \frac{\nabla_g^2}{8\mu} - \frac{\nabla_g \cdot \nabla_R}{2\mu_a} + \frac{1}{R} - \frac{1}{r_{1e}} - \frac{1}{r_{2e}}. \quad (6.130)$$

Here μ is given by $1/\mu = (1/m_1) + (1/m_2)$ and $1/\mu_a = (1/m_1) - (1/m_2)$; consequently $1/\mu_a$ is zero for the homonuclear case.

Equation (6.130), which we will call the complete non-relativistic Hamiltonian, contains terms which couple the electronic and nuclear motions, making it impossible to obtain exact eigenfunctions and eigenvalues. This is where the Born–Oppenheimer approximation enters, in a method suggested by Born and Huang [46]. We choose to expand the complete molecular wave function as the series

$$\psi_{mol}(\mathbf{R}, \mathbf{r}_g) = \sum_t F_t(\mathbf{R}) \phi_t(R, \mathbf{r}_g), \quad (6.131)$$

where the $\phi_t(R, \mathbf{r}_g)$ are the exact solutions of the electronic Born–Oppenheimer equation for the hydrogen molecular ion

$$\left(-\frac{\nabla_g^2}{2} - \frac{1}{r_{1e}} - \frac{1}{r_{2e}} + \frac{1}{R} \right) \phi_t(R, \mathbf{r}_g) = E_t(R) \phi_t(R, \mathbf{r}_g). \quad (6.132)$$

The Born–Oppenheimer electronic Hamiltonian was given previously in (6.77) and a method of obtaining exact solutions is described in appendix 6.1. If the Born expansion (6.131) is substituted into the complete non-relativistic Schrödinger equation, using the Hamiltonian (6.130), we obtain a set of coupled differential equations for the functions $F_t(\mathbf{R})$, which are

$$\mathcal{H} \sum_t F_t(\mathbf{R}) \phi_t(R, \mathbf{r}_g) = E \sum_t F_t(\mathbf{R}) \phi_t(R, \mathbf{r}_g). \quad (6.133)$$

We now premultiply by $\phi_s^*(R, \mathbf{r}_g)$ and integrate over the electronic coordinates $\mathbf{r}_g$ to obtain

$$E_s(R) F_s(\mathbf{R}) + \int \phi_s^*(R, \mathbf{r}_g) \left[-\frac{\nabla_R^2}{2\mu} - \frac{\nabla_g^2}{8\mu} - \frac{\nabla_g \cdot \nabla_R}{2\mu_a} \right] F_s(\mathbf{R}) \phi_s(R, \mathbf{r}_g) \, d\mathbf{r}_g$$

$$+ \sum_{t \neq s} \int \phi_s^*(R, \mathbf{r}_g) \left[-\frac{\nabla_R^2}{2\mu} - \frac{\nabla_g^2}{8\mu} - \frac{\nabla_g \cdot \nabla_R}{2\mu_a} \right] F_t(\mathbf{R}) \phi_t(R, \mathbf{r}_g) \, d\mathbf{r}_g = E F_s(\mathbf{R}).$$

$$(6.134)$$

Fortunately we are able to simplify this expression somewhat by using symmetry arguments. The functions $\phi_s^*(R, \mathbf{r}_g)$ must be either symmetric or antisymmetric with respect to exchange of nuclei, and electron inversion through the geometric centre of the nuclei. The operator ∇_R is antisymmetric with respect to nuclear permutation, while ∇_g is antisymmetric under electron inversion. Consequently by symmetry we have the results

$$\int \phi_s^*(R, r_g) \nabla_R \phi_s(R, r_g) \, dr_g = 0,$$

$$\int \phi_s^*(R, r_g) \nabla_g \phi_s(R, r_g) \, dr_g = 0, \quad (6.135)$$

$$\int \phi_s^*(R, r_g) \nabla_g \cdot \nabla_R \phi_s(R, r_g) \, dr_g = 0.$$

Consequently equation (6.134) is simplified to the following:

$$\left\{ E_s(R) - \frac{\nabla_R^2}{2\mu} - \int \phi_s^*(R, r_g) \left[\frac{\nabla_g^2}{8\mu} + \frac{\nabla_R^2}{2\mu} \right] \phi_s(R, r_g) \, dr_g \right\} F_s(R)$$

$$+ \sum_{t \neq s} \left\{ \int \phi_s^*(R, r_g) \left[-\frac{\nabla_g^2}{8\mu} - \frac{\nabla_R^2}{2\mu} - \frac{\nabla_g \cdot \nabla_R}{2\mu_a} \right] \phi_t(R, r_g) \, dr_g \right.$$

$$\left. + \int \phi_s^*(R, r_g) \left[-\frac{\nabla_R}{\mu} - \frac{\nabla_g}{2\mu_a} \right] \phi_t(R, r_g) \, dr_g \cdot \nabla_R \right\} F_t(R) = E F_s(R). \quad (6.136)$$

If all terms in (6.136) coupling the electronic and nuclear motions are neglected, we obtain the *Born–Oppenheimer* equation for nuclear motion:

$$\{E_s(R) - (\nabla_R^2 / 2\mu)\} F_s^{BO}(R) = E^{BO} F_s^{BO}(R). \quad (6.137)$$

The next level of approximation is to retain the terms in (6.136) which couple the electronic and nuclear motion, but which are diagonal in the electronic state. This is known as the *adiabatic approximation*, and at this level, equation (6.136) becomes

$$\left\{ E_s(R) - \int \phi_s^*(R, r_g) \frac{\nabla_g^2}{8\mu} \phi_s(R, r_g) \, dr_g - \int \phi_s^*(R, r_g) \frac{\nabla_R^2}{2\mu} \phi_s(R, r_g) \, dr_g - \frac{\nabla_R^2}{2\mu} \right\} F_s^{AD}(R)$$

$$= E^{AD} F_s^{AD}(R). \quad (6.138)$$

The result of the approximation is that the nuclear motion is now governed by an effective potential

$$E_s(R) - \int \phi_s^*(R, r_g) \frac{\nabla_g^2}{8\mu} \phi_s(R, r_g) \, dr_g - \int \phi_s^*(R, r_g) \frac{\nabla_R^2}{2\mu} \phi_s(R, r_g) \, dr_g, \quad (6.139)$$

which is obtained by averaging the complete Hamiltonian (6.130) over the Born–Oppenheimer electronic wave function $\phi_s(R, r_g)$. This effective potential is now isotope dependent because of the presence of μ, whereas the Born–Oppenheimer potential $E_s(R)$ is isotope independent. The electronic and nuclear motions are still effectively separated, but the inclusion of the diagonal corrections increases the accuracy. Provided there are no close-lying electronic states which can be coupled non-adiabatically to the state of interest, the adiabatic approximation gives an accurate potential. This is true for H_2^+, but it is not true for HD^+ where, as we will see, the $\nabla_g \cdot \nabla_R / 2\mu_a$ term in equation (6.136) leads to strong coupling of the ground and first excited electronic states, particularly for levels close to the dissociation limit.

Table 6.6. *Born–Oppenheimer potential energies (with respect to dissociation at zero) and adiabatic corrections (all in cm^{-1}) for the H_2^+ molecular ion*

$R(a_0)$	B–O energy	∇_R^2 correction	∇_g^2 correction
1.0	10 581.681	21.738	58.138
2.0	−22 525.607	20.893	35.983
3.0	−17 023.081	21.914	28.382
4.0	−10 114.463	24.178	26.185
5.0	−5359.635	26.551	26.387
6.0	−2626.902	28.233	27.414
7.0	−1227.742	29.135	28.406
8.0	−564.135	29.548	29.083
9.0	−262.372	29.728	29.472
10.0	−127.016	29.806	29.677

The adiabatic corrections for H_2^+ were calculated by Bishop and Wetmore [47] and an abbreviated selection of their results is given in table 6.6; the third and fourth columns of this table give the corrections due to the third and second terms respectively in (6.139). Notice that the corrections are remarkably insensitive to the value of the internuclear distance R.

Finally we come to the terms in equation (6.136) which result in the mixing of different electronic states, the so-called *non-adiabatic* terms. Of these terms, the most important, which only arises in the heteronuclear molecule, HD$^+$, is that involving the operator product $\nabla_g \cdot \nabla_R/2\mu_a$. The non-adiabatic couplings in H_2^+ were treated using variational methods by Bishop and Cheung [48] and Bishop and Solunac [49], but only the lowest vibrational levels of the ground state could be treated adequately. A different approach, which was designed particularly for the HD$^+$ ion, was developed by Carrington and Kennedy [50]. The problem with HD$^+$ is that the two dissociation limits, H$^+$ + D and H + D$^+$ are not degenerate, but differ in energy by 29.84 cm^{-1}; this difference can be attributed to the dissimilar electron reduced masses (and hence ionisation energies) of the hydrogen and deuterium atoms. The ground $1s\sigma$ and excited $2p\sigma$ electronic states cannot be associated uniquely with either of the two dissociation limits, and it is not possible to draw potential energy curves. Carrington and Kennedy [50] formed suitably asymmetric electronic wave functions for the two electronic states by considering normalised linear combinations of the $1s\sigma_g (\phi_1)$ and $2p\sigma_u(\phi_2)$ adiabatic states,

$$\Phi_{1s\sigma} = a\phi_1 + b\phi_2, \quad \Phi_{2p\sigma} = b\phi_1 - a\phi_2. \quad (6.140)$$

We note that the g and u labels are not strict symmetry labels for HD$^+$. The adiabatic states are mixed by the operator which couples nuclear and electronic motion in the Hamiltonian ($-\nabla_g \cdot \nabla_R/2\mu_a$), and the electronic Hamiltonian matrix for the two states is decomposed according to

$$\mathcal{H}(R) = \begin{pmatrix} E^{AD}_{1s\sigma}(R) & H(R) \\ H(R) & E^{AD}_{2p\sigma}(R) \end{pmatrix} + \begin{pmatrix} 0 & A(R) \\ -A(R) & 0 \end{pmatrix}, \quad (6.141)$$

where

$$H(R) = \frac{1}{2}\left\{\int \phi^*_{1s\sigma}\mathcal{H}\phi_{2p\sigma}\,d\mathbf{r}_g + \int \phi^*_{2p\sigma}\mathcal{H}\phi_{1s\sigma}\,d\mathbf{r}_g\right\},$$

$$A(R) = \frac{1}{2}\left\{\int \phi^*_{1s\sigma}\mathcal{H}\phi_{2p\sigma}\,d\mathbf{r}_g - \int \phi^*_{2p\sigma}\mathcal{H}\phi_{1s\sigma}\,d\mathbf{r}_g\right\}.$$

(6.142)

$\mathcal{H}$ is the complete non-relativistic Hamiltonian given in (6.130). The Hermitian matrix in (6.141) is diagonalised to obtain the coefficients in the wave functions (6.140) and effective potential curves for the coupled states using these wave functions as a function of R.

Other methods of treating the HD$^+$ ion have been developed by Moss and Sadler [51] and relativistic [52] and radiative [53] corrections have also been calculated. These are small, but still significant in comparison with the accuracy of the experimental data. We should also note that the other one-electron molecule to have been studied theoretically [54] is HeH^{2+}; although the ground state is predicted to be repulsive, some excited states are calculated to have potential minima. No spectroscopic studies of this molecular ion have been described.

Adiabatic corrections for H$_2$ were first calculated by Kolos and Wolniewicz [27], and much later confirmed by Bishop and Cheung [55]. The best potential curves for H$_2$ and D$_2$, incorporating both adiabatic and relativistic corrections, have been tabulated by Bishop and Shih [56]. Bishop and Cheung [55] have also carried out non-adiabatic calculations for H$_2$, the energy of the lowest level being lowered by 0.42 cm^{-1} compared with the adiabatic value. A small number of calculations for excited states have also been reported.

Formidable problems arise for many-electron molecules, and the non-adiabatic effects will, in general, be smaller for molecules with heavier atoms.

6.7. Coupling of electronic and rotational motion: Hund's coupling cases

6.7.1. Introduction

We have already seen in chapter 5 the importance of angular momenta in diatomic molecules. We now consider the various ways in which these angular momenta can be coupled in diatomic molecules, giving rise to Hund's coupling cases [57]. As we will see many times elsewhere in this book, Hund's coupling cases are idealised situations which help us to understand the pattern of rotational levels and the resulting spectra. They are also central to the theory underlying the quantitative analysis of spectra and the consequent definition and determination of molecular parameters.

The angular momenta which are involved are as follows:

L – the electronic orbital angular momentum,
S – the electronic spin angular momentum,
J – the total angular momentum,
N – the total angular momentum excluding electron spin, so that $N = J - S$,
R – the rotational angular momentum of the nuclei, so that $R = N - L$.

In addition there will often be nuclear spin angular momentum (*I*), which is coupled to the electronic orbital and spin angular momenta; these coupling cases are described in section 6.7.8.

6.7.2. Hund's coupling case (a)

In Hund's case (a), illustrated in the vector diagram shown in figure 6.13, the orbital angular momentum *L* is strongly coupled to the internuclear axis by electrostatic forces; the electron spin angular momentum in turn is strongly coupled to *L* through spin–orbit coupling. The axial components of *L* and *S* are well defined and are denoted Λ and Σ; their sum is denoted Ω, i.e. $\Omega = \Lambda + \Sigma$. The angular momentum of the rotating nuclei is *R*; this is coupled to a vector Ω pointing along the axis to form the resulting total angular momentum *J*. The precession of *L* and *S* about the internuclear axis is presumed to be much faster than the nutation of Ω and *R* about *J*.

The precessions of *L* and *S* about the internuclear axis have two equal and opposite senses, so that the projections also have magnitudes $\pm \Lambda, \pm \Sigma$. Consequently the total

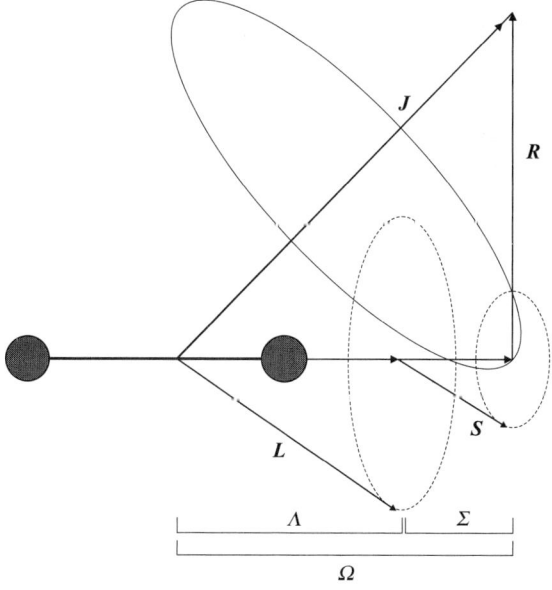

Figure 6.13. Vector coupling diagram for Hund's case (a).

Table 6.7. *Hund's coupling cases*

Coupling case	Good quantum numbers	Requirements
(a)	$\eta, \Lambda, S, \Sigma, J, \Omega$	$A\Lambda \gg BJ$
(b)	η, Λ, N, S, J	$A\Lambda \ll BJ$
(c)	$\eta, (J_a), \Omega, J$	$A\Lambda \gg \Delta E_{el}$
(d)	η, L, R, N, S, J	$BJ \gg \Delta E_{el}$
(e)	η, J_a, R, J	$A\Lambda \gg BJ \gg \Delta E_{el}$

projection of electronic angular momentum also has two senses, $\pm\Omega$. The two-fold degeneracy in Λ is called Λ-doubling, and that in Ω is called Ω-doubling; we will see elsewhere how the molecular rotation can remove this degeneracy.

The various Hund's coupling cases can be defined more rigorously in terms of their good quantum numbers (see table 6.7). Thus case (a) wave functions may be written in ket notation as $|\eta, \Lambda; S, \Sigma; J, \Omega, M_J\rangle$. The symbol η here denotes all other quantum numbers not expressed explicitly, for example, electronic and vibrational. M_J is the component of $\boldsymbol{J}$ in a space-fixed Z direction, and is important when we discuss the effects of external magnetic or electric fields. It should be appreciated that Hund's case (a) is a *decoupled* basis set, that is, $\boldsymbol{L}$ and $\boldsymbol{S}$ are decoupled along the internuclear axis. The operator describing the rotational energy in Hund's case (a) notation is given by

$$\mathcal{H}_{\text{rot}} = B\boldsymbol{R}^2 = B[\boldsymbol{J} - \boldsymbol{L} - \boldsymbol{S}]^2. \tag{6.143}$$

B is called the rotational constant, and will be discussed in detail later; the operator form of the rotational energy in (6.143) has profound consequences, as we shall see.

Case (a) is a good representation whenever $A\Lambda$ is much greater than BJ, A being the spin–orbit coupling constant. In a good case (a) system there are $2S+1$ fine-structure states, characterised by their Ω values, with spin-orbit energies $A\Lambda\Sigma$. Each fine-structure state has a pattern of rotational levels, with relative energies $BJ(J+1)$, the lowest rotational level having $J=\Omega$. As J increases, case (a) becomes less appropriate; we will see why in due course.

6.7.3. Hund's coupling case (b)

When $\Lambda=0$ and $S\neq 0$ the spin vector $\boldsymbol{S}$ is no longer coupled to the internuclear axis because spin–orbit coupling vanishes in this case. Consequently Ω is not defined. Even in some very light molecules with $\Lambda\neq 0$, the coupling of the spin to the internuclear axis is so weak that case (a) coupling does not apply. Hund's case (b) may then be appropriate, and the corresponding vector coupling diagram is shown in figure 6.14. Once again, $\boldsymbol{L}$ precesses very rapidly about the internuclear axis with a well-defined component, Λ. Λ is coupled to $\boldsymbol{R}$ to form $\boldsymbol{N}$; $\boldsymbol{N}$ is then coupled with $\boldsymbol{S}$ to form the total angular momentum $\boldsymbol{J}$. Basis functions in this coupling scheme are written $|\eta, \Lambda; N, S, J, M_J\rangle$

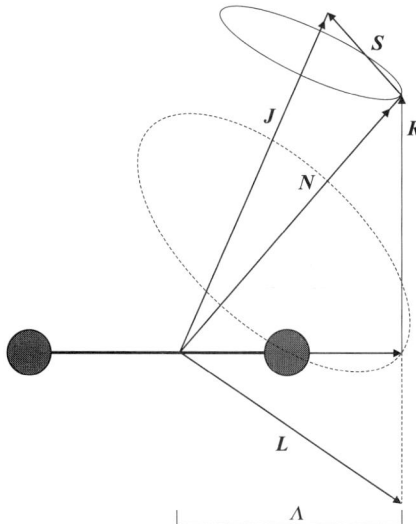

Figure 6.14. Vector coupling diagram for Hund's case (b).

(see table 6.7). The projection of N on the internuclear axis is Λ. In the older literature, the symbol K is used rather than N.

The operator which describes the rotational kinetic energy in Hund's case (b) notation is

$$\mathcal{H}_{\text{rot}} = B\boldsymbol{R}^2 = B[\boldsymbol{N} - \boldsymbol{L}]^2, \qquad (6.144)$$

so that the rotational levels have energies $BN(N+1)$, with the lowest N level having $N = \Lambda$. If $S = 1/2$, each N level for $N \geq 1$ is then split into a doublet by the spin–rotation interaction, represented in the effective Hamiltonian by a term $\gamma \boldsymbol{N} \cdot \boldsymbol{S}$; the resulting levels are characterised by values of the total angular momentum J. The series which has $J = N + 1/2$ is called the $F_1(J)$ series, whilst that with $J = N - 1/2$ is the $F_2(J)$ series. If $S \geq 1$ there is still a spin–rotation splitting but the pattern of levels is more complicated, as we will see elsewhere.

The operator form for the rotational kinetic energy, given in (6.144), can be expanded to give

$$\mathcal{H}_{\text{rot}} = B[\boldsymbol{N} - \boldsymbol{L}]^2 = B[\boldsymbol{N}^2 + \boldsymbol{L}^2 - 2\boldsymbol{N} \cdot \boldsymbol{L}] = B[N(N+1) + \boldsymbol{L}^2 - 2\boldsymbol{N} \cdot \boldsymbol{L}]. \quad (6.145)$$

The second and third terms in (6.145) have no effect *within* a case (b) state, but do have non-zero matrix elements between different electronic states.

Cases (a) and (b) are the most widely observed for diatomic molecules; most molecules, indeed, conform to coupling which is intermediate between cases (a) and (b). We will discuss the nature and origin of this intermediate coupling in due course, but first deal with three other limiting coupling cases which can be important in certain specific situations.

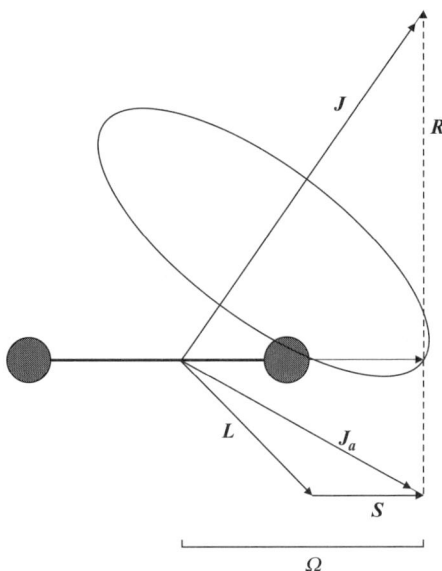

Figure 6.15. Vector coupling diagram for Hund's case (c).

6.7.4. Hund's coupling case (c)

In some cases the coupling between L and S may be stronger than the interaction with the internuclear axis; this situation can arise in heavy molecules. In this case the projections Λ and Σ are not defined; instead L and S first couple to form a resultant J_a, which then precesses rapidly about the internuclear axis with a component Ω. The nuclear rotational angular momentum now adds vectorially to Ω to form the total angular momentum J. The vector diagram illustrating case (c) coupling is given in figure 6.15. Case (c) basis functions are specified in the form $|\eta, J_a, J, \Omega, M_J\rangle$; different electronic states are characterised by different values of Ω. The component of total angular momentum along the internuclear axis is also Ω. If $\Omega = 0$ the states are non-degenerate, but for $\Omega \neq 0$ we have states which would be degenerate for a non-rotating molecule. This degeneracy is removed by rotation, and is called Ω-doubling. The rotational levels have relative energies $BJ(J+1)$, just as in case (a).

A particularly clear example of Hund's case (c) coupling has been observed for the HeAr$^+$ ion in its near-dissociation vibration–rotation levels [58]; it also occurs for the I$_2$ molecule [59].

6.7.5. Hund's coupling case (d)

In Hund's case (d) the coupling between L and the nuclear rotation R is much stronger than that between L and the internuclear axis. As shown in figure 6.16, the result of the coupling between L and R is N, which can be further coupled with S in suitable open

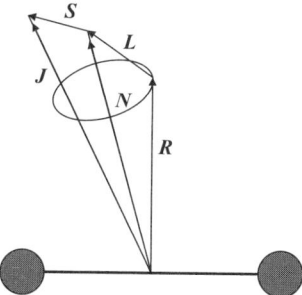

Figure 6.16. Vector coupling diagram for Hund's case (d).

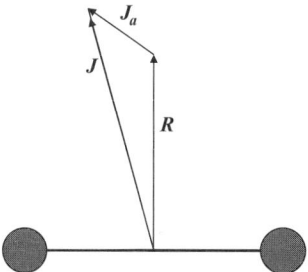

Figure 6.17. Vector coupling diagram for Hund's case (e).

shell systems. The value of N is given by

$$N = (R+L), (R+L-1), (R+L-2), \ldots, |R-L|, \qquad (6.146)$$

so that there are $2L+1$ different N values for each R, except when $R < L$. The rotational energy in case (d) is given by

$$E_{\text{rot}} = BR(R+1), \qquad (6.147)$$

each level being then further split into $2L+1$ components. Hund's case (d) coupling is an appropriate description for the electronic states of many Rydberg molecules, where the Rydberg electron interacts only very weakly with the molecular core. We will discuss the transition from case (b) to case (d) coupling in section 6.7.7.

6.7.6. Hund's coupling case (e)

Although not identified by Hund, a fifth vector coupling case is possible, which we will call Hund's case (e); it is described in the vector coupling diagram shown in figure 6.17. L and S are strongly coupled to each other, to form a resultant J_a. However, the interaction of L and S with the internuclear axis is very weak; their resultant J_a is combined with the nuclear rotation R to form the total angular momentum J. Different J components of the rotational levels arise, since for given values of J_a and R,

$$J = R + J_a, R + J_a - 1, R + J_a - 2, \ldots, |R - J_a|. \qquad (6.148)$$

Notice that in both case (d) and case (e) there is no molecular projection quantum number. An example of case (e) coupling, probably the first, has been observed [60] for vibration–rotation levels of the HeKr$^+$ ion which lie very close to the dissociation limit. The Kr$^+$ atomic ion has $L=1$ and $S=1/2$, so that J_a is 3/2 or 1/2, and the spin–orbit interaction is strong. When a very weak bond is formed with a He atom, J_a remains a good quantum number, at least for the most weakly bound levels, but there are nevertheless series of rotation levels, with rotational energy $BR(R+1)$. The details are described in chapter 10, where we show that case (e) coupling is identified, both by the observed pattern of the rotational levels, and by the measured Zeeman effects and effective g factors for individual rotational levels.

6.7.7. Intermediate coupling

As Lefebvre-Brion and Field [61] point out, the only coupling cases for which the electronic and nuclear motions can be separated are cases (a) and (c); consequently only in these cases can potential curves be defined unambiguously and accurately. However, as we have already pointed out, Hund's coupling cases are idealised descriptions and for most molecules the actual coupling corresponds to an intermediate situation. Moreover, the best description of the vector coupling often changes as the molecular rotation increases. In this section we consider the nature of the intermediate coupling schemes in more detail; some of these will appear elsewhere in this book in connection with the observed spectra of specific molecules.

The general procedure for analysing a spectrum, particularly one which involves the fine and hyperfine structure of rotational levels, is to choose a convenient Hund's case basis set for defining the effective Hamiltonian, and for calculating its matrix elements and eigenvalues. Whilst, in a sense, the choice of basis may be irrelevant, it always makes both physical and analytical sense to choose, if possible, the basis which most nearly diagonalises the effective Hamiltonian. Amongst other things, this points to the most appropriate quantum numbers to use to label the eigenstates. It is, however, always possible to express the basis functions of one Hund's coupling case in terms of those of another. For example, case (b) basis functions can be expressed as linear combinations of case (a) functions, by using Wigner 3-j symbols:

$$|\eta, \Lambda; N, S, J\rangle = \sum_{\Sigma=-S}^{+S} (-1)^{J-S+\Lambda}(2N+1)^{1/2} \begin{pmatrix} J & S & N \\ \Omega & -\Sigma & -\Lambda \end{pmatrix} |\eta, \Lambda; S, \Sigma; J, \Omega\rangle. \tag{6.149}$$

In many instances in this book we shall discuss the transition from one Hund's coupling case to another, in connection with the spectra of specific molecules. We now summarise a particularly common example, the transition from case (a) to case (b). This is probably the most frequently encountered example; it is discussed in considerable detail in chapter 9 so we will confine ourselves to a brief outline here. Figure 6.18, which reappears in chapter 9 in our description of the NO molecule, shows the correlation between case (a) levels on the left-hand side, and case (b) on the right-hand side.

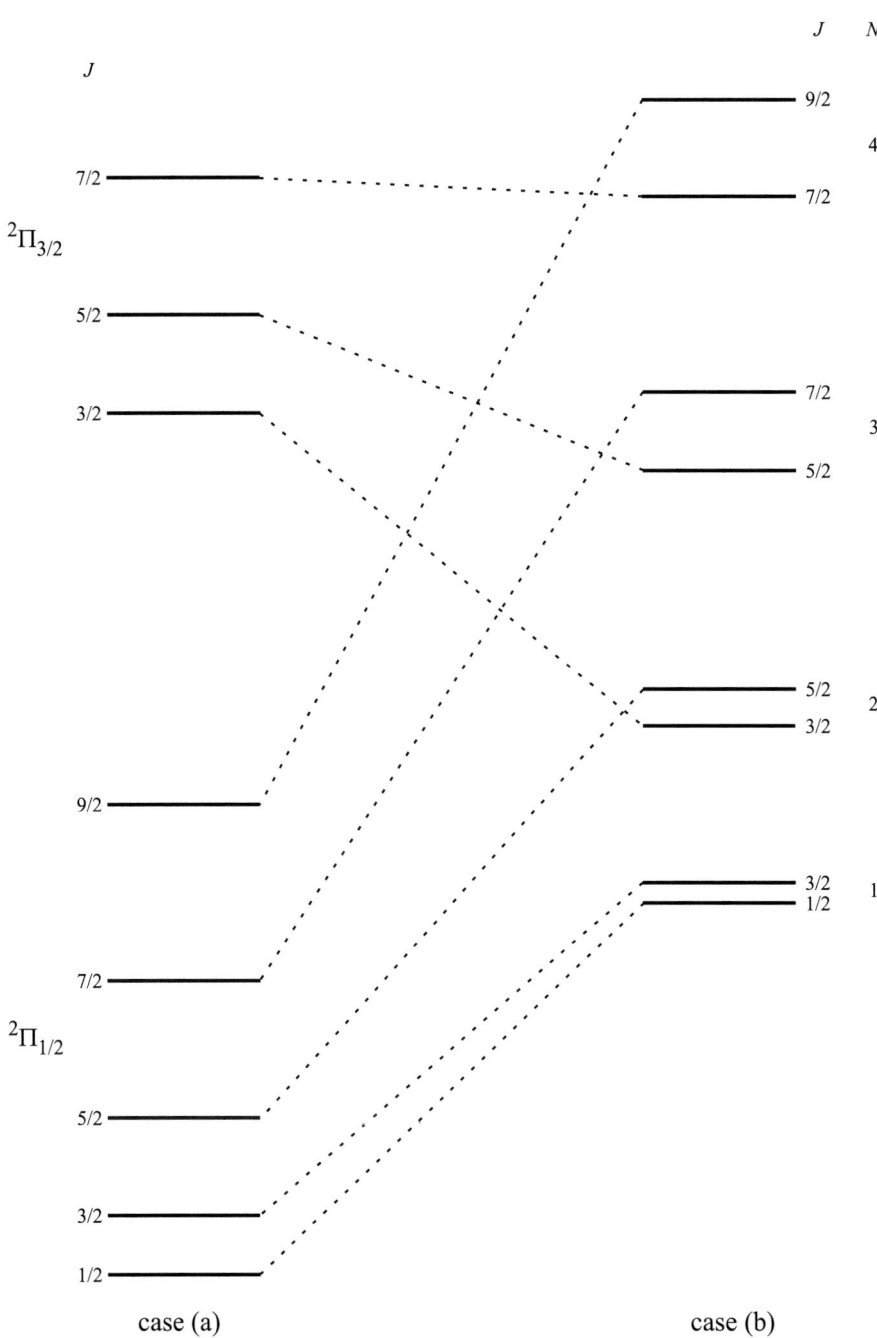

Figure 6.18. Transition from case (a) to case (b) for the rotational levels of a regular $^2\Pi$ state.

For NO the $^2\Pi_{1/2}$ fine structure component lies lower than the $^2\Pi_{3/2}$; this is called a 'regular' $^2\Pi$ state. Each level in figure 6.18 is actually doubly-degenerate, because of the Λ-doubling, which is indeed resolved in the rotational spectrum of NO. The transition from case (a) to case (b) coupling, which is also called spin-decoupling, can be understood by considering the rotational kinetic energy term in the effective Hamiltonian, equation (6.143), which may be expanded in a molecule-fixed coordinate system (q), in spherical tensor form, as follows:

$$B\{T^1(\boldsymbol{J}) - T^1(\boldsymbol{L}) - T^1(\boldsymbol{S})\}^2$$
$$= B\left\{J(J+1) + S(S+1) + 2\sum_q (-1)^q [T_q^1(\boldsymbol{L})T_{-q}^1(\boldsymbol{S}) + (1/2)T_q^1(\boldsymbol{L})T_{-q}^1(\boldsymbol{L})]\right\}$$
$$- 2B\sum_q \{T_q^1(\boldsymbol{J})T_q^1(\boldsymbol{S}) + T_q^1(\boldsymbol{J})T_q^1(\boldsymbol{L})\}. \tag{6.150}$$

The important term in this equation so far as the transition towards case (b) coupling is concerned is the penultimate term, which in a case (a) basis has matrix elements given by

$$\langle \eta, \Lambda; S, \Sigma; J, \Omega | -2B\sum_q T_q^1(\boldsymbol{J})T_q^1(\boldsymbol{S}) | \eta, \Lambda; S, \Sigma'; J, \Omega' \rangle$$
$$= -2B\sum_q (-1)^{J+S-\Omega-\Sigma} \begin{pmatrix} J & 1 & J \\ -\Omega & q & \Omega' \end{pmatrix} \begin{pmatrix} S & 1 & S \\ -\Sigma & q & \Sigma' \end{pmatrix}$$
$$\times \{J(J+1)(2J+1)S(S+1)(2S+1)\}^{1/2}. \tag{6.151}$$

The $q = \pm 1$ terms link the fine structure states; the importance of this mixing is maximised when the separation between the fine-structure states is small (i.e. small spin–orbit splitting), and when the rotational constant B is large. Consequently molecules like ClO, BrO and IO are good case (a) $^2\Pi$ systems, whereas CH obeys case (b) coupling, even in its lowest rotational levels. The OH radical approximates more closely to case (a) in its lowest rotational levels, but goes over to case (b) as the rotational quantum number increases. As we mentioned above, the transition from case (a) to case (b) is discussed in detail in chapter 9; it is not, of course, confined to $^2\Pi$ molecules.

6.7.8. Nuclear spin coupling cases

In chapter 4 we described the theory of magnetic and electric nuclear hyperfine interactions in diatomic molecules, and we will encounter many examples of observed hyperfine structure elsewhere in this book. In the analysis of such structure we will choose an appropriate form of the effective hyperfine Hamiltonian, as well as a basis set which describes the coupling of the nuclear spin angular momentum with other angular momenta in a molecule. The best coupling scheme in any particular molecule is usually fairly self-evident, and in this subsection we outline possible schemes associated with case (a) or case (b) coupling of the electron spin, orbital, and rotational angular momenta. The first person to present these coupling schemes in a systematic manner seems to have been Dunn [62].

In case (a) coupling two main possibilities arise. The first, which is expected to arise very rarely, if at all, implies that the magnetic interaction of the nuclear spin magnetic moment with the electronic orbital and spin moments is sufficiently strong to force the nuclear spin to be quantised in the molecular axis system. The basis kets may be expressed in the form $|\eta, \Lambda; S, \Sigma, \Lambda, \Omega; \Omega, I_z, \Omega'; \Omega', N, J\rangle$; this scheme is known as case (a_α) and we will not discuss it further. The most common situation, described as case (a_β), is a straightforward extension of Hund's case (a) in which the nuclear spin angular momentum I is coupled to J to form a grand total angular momentum, F. The basis kets are therefore expressed in the form $|\eta, \Lambda; S, \Sigma, \Lambda, \Omega; J, I, F\rangle$ and we will encounter many examples in which the matrix elements of the hyperfine Hamiltonian are calculated in this basis.

When the coupling of the angular momenta, exclusive of nuclear spin, is described by Hund's case (b), three ways of including the nuclear spin may be considered. In the first, known as case ($b_{\beta N}$), the nuclear spin I is coupled to N, forming an intermediate F_1 which is then coupled with S to form F. The corresponding basis kets take the form $|\eta, \Lambda; N, \Lambda, I, F_1; F_1, S, F\rangle$, but they are unlikely to be used because the coupling of S to N is invariably much stronger than that between I and N.

In the second scheme to be considered, labelled case ($b_{\beta S}$), S and I are coupled to form an intermediate G, which is then coupled with N to form F. This scheme is appropriate when the hyperfine interaction between S and I is strong compared with spin–rotation coupling, and we will meet it elsewhere, most notably in the H_2^+ ion. The basis kets are written in the form $|\eta, \Lambda; S, I, G; G, N, F\rangle$ and the hyperfine matrix elements is this basis are calculated in later chapters. The most natural extension of Hund's case (b), known as case ($b_{\beta J}$), is that in which J, the resultant of N and S coupling, is coupled with I to form F. The corresponding basis kets are $|\eta, \Lambda; N, S, J; J, I, F\rangle$ and we will often meet matrix elements calculated in this basis.

6.8. Rotations and vibrations of the diatomic molecule

6.8.1. The rigid rotor

In chapter 2 we showed how the wave equation of a vibrating rotator was derived through a series of coordinate transformations. We discussed the solutions of this wave equation in section 2.8, and the particular problem of representing the potential in which the nuclei move. We outlined the relatively simple solutions obtained for a harmonic oscillator, the corrections which are introduced to take account of anharmonicity, and derived an expression for the rovibrational energies. Our treatment was relatively brief, so we now return to this subject in rather more detail.

It is instructive to start with the simplest possible model of a rotating diatomic molecule, the so-called dumbbell model, as illustrated in figure 6.19. The two atoms, of masses m_1 and m_2, are regarded as point-like, and are fastened a distance R apart

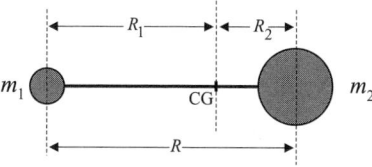

Figure 6.19. The rigid dumbbell model of a diatomic molecule.

to the ends of a rigid weightless rod. In classical mechanics the energy of rotation E_{rot} of this body is given by

$$E_{\text{rot}} = (1/2)I\omega^2, \tag{6.152}$$

where ω is the angular velocity of the rotation (about the centre of mass), related to the frequency of rotation ν_{rot} by $\omega = 2\pi\nu_{\text{rot}}$. I is the moment of inertia, defined by

$$I = m_1 R_1^2 + m_2 R_2^2, \tag{6.153}$$

where R_1 and R_2, the distances of the masses m_1 and m_2 from the centre-of-gravity (CG), are given by

$$R_1 = \frac{m_2}{m_1 + m_2} R, \quad R_2 = \frac{m_1}{m_1 + m_2} R. \tag{6.154}$$

Substitution for R_1 and R_2 in equation (6.153) gives an important result for the moment of inertia,

$$I = \frac{m_1 m_2}{m_1 + m_2} R^2 = \mu R^2. \tag{6.155}$$

Here μ is called the reduced mass of the molecule, and the above analysis shows that this simple system is actually equivalent to the rotation of a single point of mass μ at a fixed distance R from the axis of rotation; such a system is called a rigid rotor (or rotator).

This result leads us naturally into a quantum mechanical description of the system because, for the wave motion of a single point of mass μ, the Schrödinger equation is

$$\frac{\partial^2 \psi}{\partial x^2} + \frac{\partial^2 \psi}{\partial y^2} + \frac{\partial^2 \psi}{\partial z^2} + \frac{2\mu}{\hbar^2} E\psi = 0. \tag{6.156}$$

The standard transformation of the Laplacian from cartesian coordinates to polar coordinates leads to the following Hamiltonian, derived previously in chapter 2:

$$\mathcal{H} = -\frac{\hbar^2}{2\mu R^2} \frac{\partial}{\partial R}\left(R^2 \frac{\partial}{\partial R}\right) - \frac{\hbar^2}{2\mu R^2 \sin^2\theta}\left[\sin\theta \frac{\partial}{\partial \theta}\left(\sin\theta \frac{\partial}{\partial \theta}\right) + \frac{\partial^2}{\partial \phi^2}\right]. \tag{6.157}$$

Two points should be made about this expression. First it has exactly the same form as that introduced in our discussion of the hydrogen atom, equation (6.7). Second, for the *rigid* rotor the first term in (6.157) disappears, and the Schrödinger equation becomes

$$-\frac{\hbar^2}{2\mu R^2}\left[\frac{1}{\sin\theta}\frac{\partial}{\partial \theta}\left(\sin\theta \frac{\partial}{\partial \theta}\right) + \frac{1}{\sin^2\theta}\frac{\partial^2}{\partial \phi^2}\right]\psi(\theta, \phi) = E\psi(\theta, \phi), \tag{6.158}$$

which may be rewritten in the form

$$\frac{\hbar^2}{2\mu R^2} J^2 \psi(\theta, \phi) = E\psi(\theta, \phi), \tag{6.159}$$

where $\boldsymbol{J}$ is the (dimensionless) rotational angular momentum. We have already shown that μR^2 is the moment of inertia, I, for the two-body system, and the angular eigenfunctions $\psi(\theta, \phi)$ are spherical harmonics. We have, therefore, obtained the solutions of the rotational Schrödinger equation for a linear molecule,

$$\frac{\hbar^2 J^2}{2I} Y_{JM}(\theta,\phi) = \frac{\hbar^2}{2I} J(J+1) Y_{JM}(\theta,\phi). \qquad (6.160)$$

The rotational constant, B, in frequency units (Hz), is given by

$$B = \frac{1}{h}\frac{\hbar^2}{2I}. \qquad (6.161)$$

Substituting this expression, the rotational term value (in Hz) is given by

$$F(J) = BJ(J+1), \qquad (6.162)$$

In practice values of B are also often quoted in cm^{-1}. For the simple rigid rotor the rotational quantum number J takes integral values, $J = 0, 1, 2$, etc. The rotational energy levels therefore have energies $0, 2B, 6B, 12B$, etc. Elsewhere in this book we will describe the theory of electric dipole transition probabilities and will show that for a diatomic molecule possessing a permanent electric dipole moment, transitions between the rotational levels obey the simple selection rule $\Delta J = \pm 1$. The rotational spectrum of the simple rigid rotor therefore consists of a series of equidistant absorption lines with frequencies $2B, 4B, 6B$, etc.

6.8.2. The harmonic oscillator

The next stage in our development of the internal dynamics of a diatomic molecule is to recognise that the molecule is not rigid. The atoms with point masses m_1 and m_2 are at distances R_1 and R_2 from the centre-of-mass at any given *instant*, but actually move with respect to each other, with displacements from each equilibrium position given by

$$q_1 = R_1 - R_{1e}, \quad q_2 = R_2 - R_{2e}. \qquad (6.163)$$

The simplest definition of the vibrational coordinate, q, for a diatomic molecule is, in fact,

$$q = R - R_e, \qquad (6.164)$$

where R and R_e are the instantaneous and equilibrium bond lengths respectively. The condition for the centre-of-mass requires that

$$m_1 R_1 = m_2 R_2 \quad \text{or} \quad m_1 q_1 = m_2 q_2, \qquad (6.165)$$

and the total displacement $(q_1 + q_2)$ is the vibrational coordinate q. It follows that

$$q_1 = \frac{m_2 q}{(m_1 + m_2)}, \quad q_2 = \frac{m_1 q}{(m_1 + m_2)}. \qquad (6.166)$$

The kinetic energy is given by

$$k.e. = (1/2) m_1 v_1^2 + (1/2) m_2 v_2^2, \qquad (6.167)$$

and since the velocities are given by

$$v_1 = dR_1/dt = dq_1/dt, \quad v_2 = dR_2/dt = dq_2/dt, \tag{6.168}$$

we find that the kinetic energy is given by

$$k.e. = \frac{1}{2}\left[\frac{m_1 m_2}{(m_1 + m_2)}\right]\left(\frac{dq}{dt}\right)^2 = \frac{1}{2}\mu\left(\frac{dq}{dt}\right)^2 = \frac{1}{2}\frac{p^2}{\mu}. \tag{6.169}$$

The final result in (6.169) follows from the fact that the momentum p is equal to $\mu(dq/dt)$.

Equation (6.169) gives the kinetic energy, but we have still to obtain a classical expression for the potential energy. A harmonic oscillator can be defined as a mass point m which is acted upon by a force F proportional to the distance x from the equilibrium position and directed towards the equilibrium position. Since force equals mass times acceleration, we have

$$F = m\frac{d^2 x}{dt^2} = -kx, \tag{6.170}$$

where the proportionality constant k is called the force constant. The solution to the differential equation (6.170) is

$$x = x_0 \sin(2\pi \nu_{osc} t + \varphi), \tag{6.171}$$

where the vibrational frequency ν_{osc} is given by

$$\nu_{osc} = \frac{1}{2\pi}\sqrt{\frac{k}{m}}, \tag{6.172}$$

and x_0 is the amplitude of the vibration. The force F is the negative derivative of the potential energy V; it therefore follows that

$$V = (1/2)kx^2, \tag{6.173}$$

since

$$-\frac{dV}{dx} = F = -kx. \tag{6.174}$$

In terms of the vibrational coordinate, we have by analogy with (6.173),

$$V = \left(\frac{1}{2}\right)kq^2, \tag{6.175}$$

so that the total vibrational energy is given by

$$W_{vib} = \frac{p^2}{2\mu} + \frac{kq^2}{2}. \tag{6.176}$$

By referring the motion to the centre of mass, the two-body problem has been reduced to a one-body problem of the vibrational motion of a particle of mass μ against a fixed point, under the restraining influence of a spring of length R with a force constant k.

We can now convert the problem from one in classical mechanics to one in quantum mechanics by making the changes,

$$q \to q, \quad p \to p = -i\hbar(\partial/\partial q), \tag{6.177}$$

so that the Schrödinger equation for the harmonic oscillator becomes

$$-\frac{\hbar^2}{2\mu}\frac{d^2\Psi_v}{dq^2} + \frac{1}{2}kq^2\Psi_v = E_v\Psi_v. \tag{6.178}$$

Equation (6.178) is an ordinary second-order differential equation, the solution of which is discussed in detail by Pauling and Wilson [5]. The approach to its solution may be divided into two stages; first we express Ψ_v in the trial form

$$\Psi_v = \exp(-\alpha q^2/2)H(q), \tag{6.179}$$

where $\alpha^2 = k\mu/\hbar^2$. This ensures that the wave function shows the correct behaviour as $q \to \infty$, that is, it goes exponentially to zero. We are then left with a standard differential equation for the function $H(q)$:

$$\frac{d^2H}{d\xi^2} - 2\xi\frac{dH}{d\xi} + (\lambda/\alpha - 1)H = 0, \tag{6.180}$$

where $\lambda = 2\mu E_v/\hbar^2$ and ξ is a dimensionless coordinate equal to q/R_e. The solutions of this differential equation are a set of special polynomial functions, known as the Hermite polynomials $H_v(\xi)$, which were described in chapter 2 and are discussed again below.

Let us first consider the energy levels, E_v, which arise from (6.178). These are

$$E_v = (v + 1/2)h\nu, \tag{6.181}$$

where v is the vibrational quantum number which takes integral values 0, 1, 2, . . . , and ν is the harmonic vibrational frequency in Hz,

$$\nu = \frac{1}{2\pi}\left(\frac{k}{\mu}\right)^{1/2}, \tag{6.182}$$

(compare equation (6.172)). In fact, the unit used most often in vibrational spectroscopy is the wavenumber unit, cm^{-1}, designated by the symbol ω_e which is equal to ν/c. It is important to note, from (6.181), that for $v = 0$, the lowest vibrational level, there remains a zero-point energy of $(1/2)hc\omega_e$. We now return to the eigenfunctions, given earlier in (6.179). In normalised form these may be written

$$\Psi_v(q) = \left(\frac{v!\pi^{1/2}}{2^v}\right)^{1/2} H_v(y)\exp(-y^2/2), \tag{6.183}$$

where y is a dimensionless coordinate defined by

$$y = 2\pi\left(\frac{\mu\nu}{h}\right)^{1/2} q. \tag{6.184}$$

The first few Hermite polynomials are as follows:

$$H_0(y) = 1$$
$$H_1(y) = 2y$$
$$H_2(y) = 4y^2 - 2$$
$$H_3(y) = 8y^3 - 12y \qquad (6.185)$$
$$H_4(y) = 16y^4 - 48y^2 + 12$$
$$H_5(y) = 32y^5 - 160y^3 + 120y.$$

The wave functions for $v = 0$ to 4 are plotted in figure 6.20; the point where the function crosses through zero is called a *node*, and we note that the wave function for level v has v nodes. The probability density distribution for each vibrational level is shown in figure 6.21, and the difference between quantum and classical behaviour is a notable feature of this diagram. For example, in the $v = 0$ level the probability is a maximum at $y = 0$, whereas for a classical harmonic oscillator it would be a minimum at $y = 0$, with maxima at the classical turning points. Furthermore the probability density is small but finite outside the classical region, a phenomenon known as *quantum mechanical tunnelling*.

A harmonic oscillator potential with vibrational levels is illustrated in figure 6.22. The levels are equally spaced and, as we show later, the main vibrational transitions occur between adjacent levels, with $\Delta v = \pm 1$. The $v = 1 \leftarrow 0$ transition is called the *fundamental* and occurs at the harmonic frequency v. However transitions which obey the selection rule $\Delta v = \pm 2$, known as *overtone transitions*, also have finite intensity and are often observed.

6.8.3. The anharmonic oscillator

Equation (6.173) shows that for the harmonic oscillator the potential energy curve is represented by a parabola, as shown in figure 6.22; the potential energy, and therefore the restoring force, increases indefinitely with increasing distance from the equilibrium position. This cannot, however, be a correct representation for a real molecule because when the atoms are an infinite distance apart, the attractive force between them must be zero, and the potential energy then has a constant value. As a first approximation to the true potential energy we might add a cubic term to the quadratic term of equation (6.173), representing the potential energy by the equation

$$V(R) = f(R - R_e)^2 - g(R - R_e)^3, \qquad (6.186)$$

with the constant g being much smaller than f. This combination would give a better representation around the minimum and, of course, one could continue to add higher terms to (6.186), or adopt an alternative functional form for the potential. A particularly important analytical potential is that due to Morse [63], written in the form

$$V(R) = D_e \{1 - \exp[-\beta(R - R_e)]\}^2. \qquad (6.187)$$

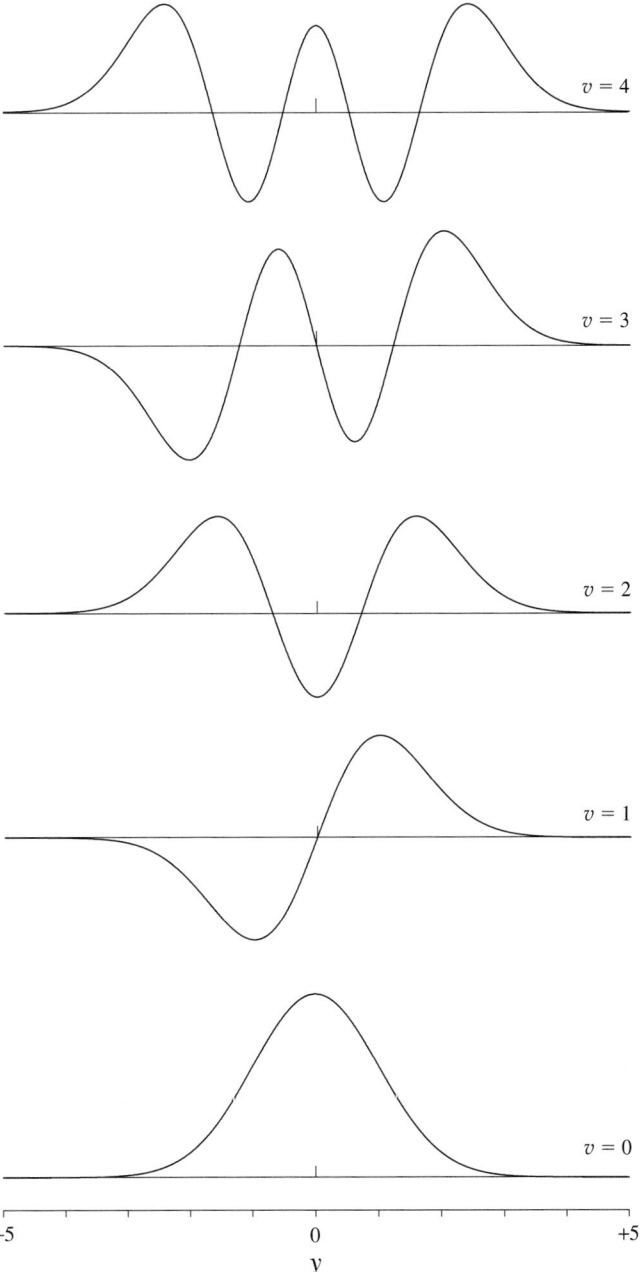

Figure 6.20. Form of the wave functions for the first few vibrational levels of a harmonic oscillator. These are plotted as a function of y, defined in the text.

Figure 6.23 shows the potential for the ground state of H_2 plotted from this equation; β is a constant, and D_e is the dissociation energy, defined in the figure. A mass point which moves under the influence of a potential such as that shown in figure 6.23 is called an *anharmonic oscillator*. Apart from its simplicity, an advantage of the Morse

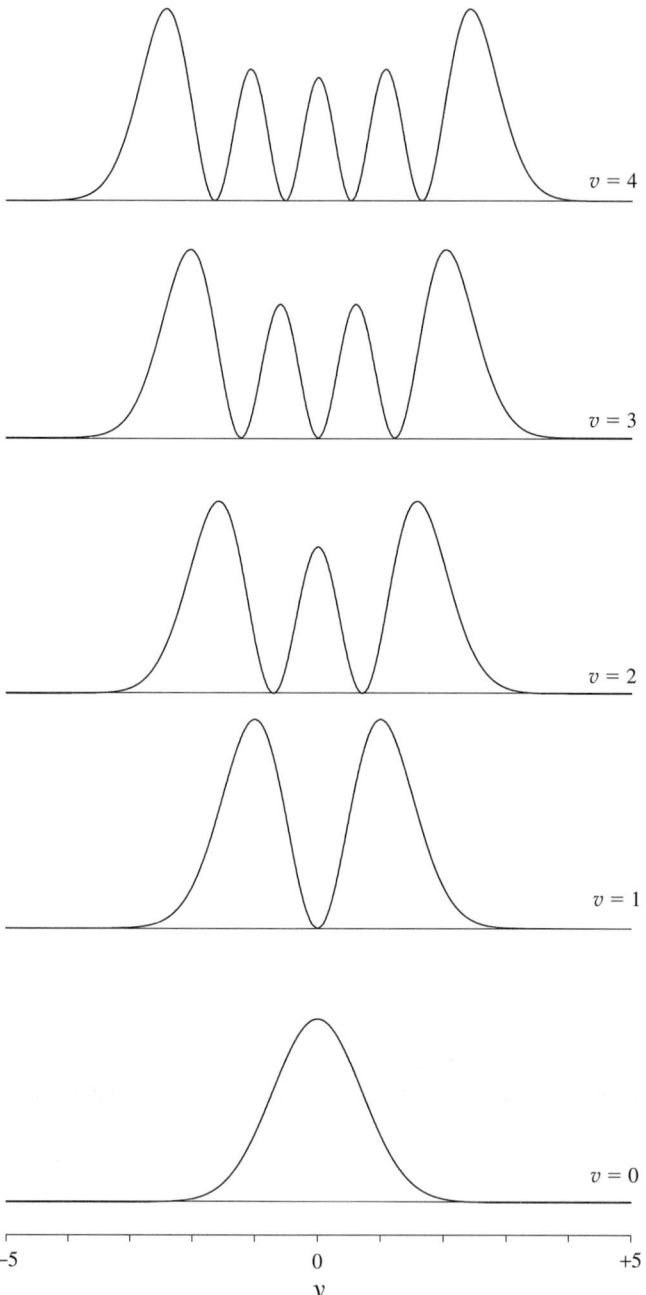

Figure 6.21. Probability density distribution for the $v = 0$ to 4 vibrational levels of the harmonic oscillator.

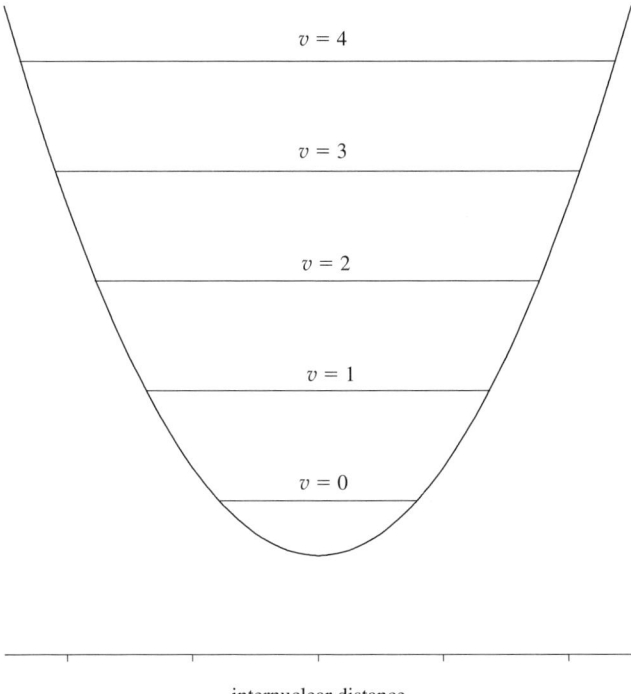

Figure 6.22. Potential energy curve for a harmonic oscillator, and the first few vibrational levels.

potential is that it enables the wave equation to be solved rigorously. In the general case it is found that the term values for the vibrational levels of the anharmonic oscillator are given by

$$G(v) = \omega_e(v + 1/2) - \omega_e x_e(v + 1/2)^2 + \omega_e y_e(v + 1/2)^3 + \cdots, \qquad (6.188)$$

although for the Morse potential the series in (6.188) does not extend higher than the quadratic term. As a consequence of the quadratic and higher terms in (6.188) the vibrational spacings decrease as v increases, with the levels essentially converging to the dissociation asymptote.

The 'true' potential energy curve can be determined from experiment if sufficient data about the vibrational levels are obtained; we will describe how this is accomplished later in this chapter. As we mentioned earlier, improvements over the Morse potential have been described, a particularly important one being due to Hulburt and Hirschfelder [64]:

$$V(x) = D_e[(1 - \exp(-\beta x))^2 + c\beta^3 x^3 \exp(-2\beta x)(1 + b\beta x)]. \qquad (6.189)$$

Here $x = R - R_e$ and b and c are constants which depend upon both vibrational and rotational constants, as we will describe later. This form of the potential works very well for a large number of molecules and electronic states.

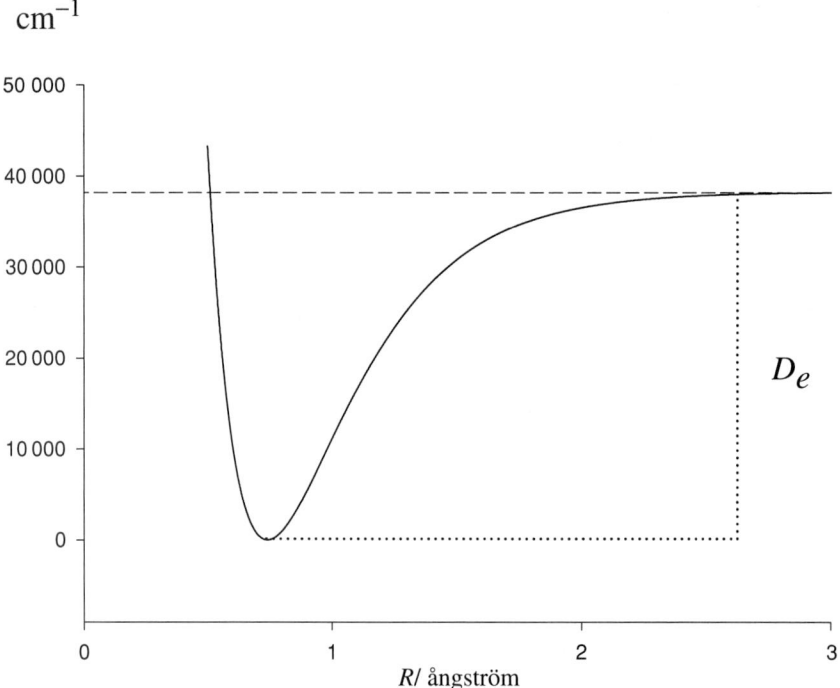

Figure 6.23. Potential curve for an anharmonic oscillator, the potential being represented by the Morse function, equation (6.187).

6.8.4. The non-rigid rotor

So far we have used the models of the rigid rotor, and the harmonic or anharmonic oscillator to describe the internal dynamics of the diatomic molecule. Since the period for rotational motion is of the order 10^{-11} s, and that for vibrational motion is 10^{-14} s, the Born–Oppenheimer type separation of the two different types of motion is justified. It is not rigorous, however, and a better model for describing the rotations of the molecule is that of the non-rigid rotor. As a result of centrifugal forces when the molecule rotates, the internuclear distance and hence the moment of inertia will increase with increasing rotation. In the rotating molecule the internuclear distance, R_c, is determined by the requirement that the centrifugal force F_c is balanced by the restoring force $k(R_c - R_e)$. If ω is the angular velocity, the angular momentum P is given by

$$P = I\omega = \mu R_c^2 \omega. \tag{6.190}$$

The centrifugal force F_c is given by

$$F_c = \mu \omega^2 R_c = \frac{P^2}{\mu R_c^3}, \tag{6.191}$$

so that since $F_c = k(R_c - R_e)$, we have

$$R_c - R_e = \frac{P^2}{\mu R_c^3 k} \approx \frac{P^2}{\mu R_e^3 k}. \tag{6.192}$$

The total rotational energy is now the sum of the kinetic and potential energies, given by

$$E = \frac{P^2}{2I_c} + \frac{1}{2}k(R_c - R_e)^2 = \frac{P^2}{2\mu R_c^2} + \frac{1}{2}k(R_c - R_e)^2. \tag{6.193}$$

If we substitute for R_c using equation (6.192), expand the denominator in the first term in (6.193) and neglect cubic and higher powers of $R_c - R_e$ we obtain

$$E = \frac{P^2}{2\mu R_e^2} - \frac{P^4}{2\mu^2 R_e^6 k} + \cdots. \tag{6.194}$$

We have already seen that, in quantum mechanics, the eigenvalue of $\boldsymbol{J}^2$ is $J(J+1)$ so that from (6.194) we obtain, for the non-rigid rotor,

$$E = \frac{\hbar^2}{2\mu R_e^2} J(J+1) - \frac{\hbar^4}{2\mu^2 R_e^6 k} J^2(J+1)^2 + \cdots. \tag{6.195}$$

The rotational term values are therefore given by the power series

$$F(J) = BJ(J+1) - DJ^2(J+1)^2 + HJ^3(J+1)^3 + \cdots. \tag{6.196}$$

The coefficients B, D, H, etc., are determined from an analysis of the experimental spectrum; it is rarely necessary to go beyond the cubic term, except when very high J values are involved. The parameters D, H, etc., are known as the centrifugal distortion corrections to the rotational kinetic energy.

As we will see later, there is an alternative formulation of the rotational (and vibrational) energies due to Dunham [65], which is often used.

6.8.5. The vibrating rotor

If there was no interaction between vibration and rotation, the energy levels would be given by the simple sum of the expression giving the vibrational levels for the anharmonic oscillator, equation (6.188), and that describing the rotational levels of the rigid rotor, equation (6.162). There is an interaction, however; during a vibration the moment of inertia of the molecule changes, and therefore so also does the rotational constant. We may therefore use a mean value of B_v for the rotational constant of the vibrational level considered, i.e.

$$B_v = \frac{h}{8\pi^2 c \mu} \left[\overline{\frac{1}{R^2}} \right], \tag{6.197}$$

where $[\overline{1/R^2}]$ is the value of $1/R^2$ averaged over the vibrational motion [66]. In general B_v will be smaller than B_e because of the anharmonicity. To a first approximation the rotational constant B_v is given by

$$B_v = B_e - \alpha_e(v + 1/2) + \gamma_e(v + 1/2)^2 + \cdots, \qquad (6.198)$$

and similarly a mean rotational constant D_v for the vibrational level v is given by

$$D_v = D_e + \beta_e(v + 1/2) + \cdots. \qquad (6.199)$$

Consequently for the rotational levels in a given vibrational level, the term values are given by

$$F_v(J) = B_v J(J+1) - D_v J^2(J+1)^2 + H_v J^3(J+1)^3 + \cdots. \qquad (6.200)$$

The full expression for the term values of the vibrating rotor is therefore

$$T = G(v) + F_v(J) = \omega_e(v + 1/2) - \omega_e x_e(v + 1/2)^2 + \omega_e y_e(v + 1/2)^3$$
$$+ \cdots + B_v J(J+1) - D_v J^2(J+1)^2 + H_v J^3(J+1)^3 + \cdots. \qquad (6.201)$$

Relationships connecting α_e, β_e, γ_e with ω_e, $\omega_e x_e$ and B_e have been given by Pekeris [67] and Dunham [65]. In a very detailed study of vibration–rotation interactions, Dunham [65] has shown that the term values for a vibrating rotor should actually be expressed as a double power series, given by

$$T = \sum_{kl} Y_{kl} \left(v + \frac{1}{2}\right)^k [J(J+1)]^l. \qquad (6.202)$$

The coefficients in (6.202) are very closely related to the standard spectroscopic parameters according to the following:

$$Y_{10} \approx \omega_e, \quad Y_{20} \approx \omega_e x_e, \quad Y_{01} \approx B_e, \quad Y_{02} \approx D_e, \quad Y_{11} \approx \alpha_e, \quad \cdots. \qquad (6.203)$$

The approximate equalities arise because there are higher order corrections, of the order $(B_e/\omega_e)^2$, in the Dunham treatment. For very light molecules, like H_2 and some hydrides, the precise interpretation of the spectroscopic parameters obtained from experiment has to be considered carefully in the light of the Dunham theory. For most molecules, however, it is satisfactory to use the parameters given above in spectroscopic analyses.

In this book, which is concerned predominantly with rotational transitions and their fine and hyperfine structure, we will have only a peripheral interest in the details of vibrational structure. Similarly we will not usually be concerned directly with electronic transitions, except in double resonance studies. Nevertheless it is important to see the broader picture, in order to understand better the detailed structure.

6.9. Inversion symmetry of rotational levels

6.9.1. The space-fixed inversion operator

The inversion operator E^* is defined as the operator which transforms a function $f(X_i, Y_i, Z_i)$ into a new function which has the same value as $f(-X_i, -Y_i, -Z_i)$

where (X_i, Y_i, Z_i) are the coordinates of a point measured in an arbitrary space-fixed axis system:

$$E^* f(X_i, Y_i, Z_i) = f'(X_i, Y_i, Z_i)$$
$$= f(-X_i, -Y_i, -Z_i). \quad (6.204)$$

Clearly, if this operator is applied twice to a wave function, the system reverts to its original configuration:

$$E^* E^* \psi(X_i, Y_i, Z_i) = \psi(X_i, Y_i, Z_i). \quad (6.205)$$

The following behaviour is consistent with this result:

$$E^* \psi(X_i, Y_i, Z_i) = \pm \psi(X_i, Y_i, Z_i). \quad (6.206)$$

If a quantum system transforms according to the upper sign, the state has a *positive parity* and, if according to the lower sign, a *negative parity*.

It is important to distinguish between the space-fixed inversion operator E^* defined here and the molecule-fixed inversion operator, denoted i. The latter defines the g,u character of functions of molecule-fixed coordinates in appropriate systems (i.e. those with a centre of symmetry) but says nothing about the overall parity of the state. It is therefore a less powerful operator than E^*.

6.9.2. The effect of space-fixed inversion on the Euler angles and on molecule-fixed coordinates

In chapter 2 we introduced the molecule-fixed axis system (x, y, z) which rotates in space with the molecule. This axis system is related to the non-rotating (but translating) axis system (X, Y, Z) by the three Euler angles (ϕ, θ, χ). The coordinates of a point i in the molecule-fixed axis system are related to its coordinates expressed in the space-fixed axis system by the transformation

$$\begin{bmatrix} x_i \\ y_i \\ z_i \end{bmatrix} = \begin{bmatrix} \cos\phi \cos\theta \cos\chi - \sin\phi \sin\chi & \sin\phi \cos\theta \cos\chi + \cos\phi \sin\chi & -\sin\theta \cos\chi \\ -\sin\phi \cos\chi - \cos\phi \cos\theta \sin\chi & \cos\phi \cos\chi - \sin\phi \cos\theta \sin\chi & \sin\theta \sin\chi \\ \cos\phi \sin\theta & \sin\phi \sin\theta & \cos\theta \end{bmatrix} \begin{bmatrix} X_i \\ Y_i \\ Z_i \end{bmatrix}.$$
$$(6.207)$$

Thus if we know how ϕ, θ, and χ transform under E^*, we can easily determine how x_i, y_i and z_i transform.

A particular problem arises for linear molecules in that it is not possible to define the third Euler angle χ uniquely. The first two Euler angles (which correspond to the two spherical polar angles) bring the space-fixed Z axis into coincidence with the molecule-fixed z axis. The third angle χ is a rotation about the z axis and, since there are no off-axis nuclei to define this rotation, we have to adopt an arbitrary convention about the way in which χ transforms under E^*. The convention we choose is that the molecule-fixed y axis is unchanged in direction after space-fixed inversion, as a result of which the x coordinate of a point is unaltered after inversion:

$$E^* x = x. \tag{6.208}$$

We note that the molecule-fixed axis system must be placed back on the inverted molecule so that it is still right-handed. Since the first two Euler angles are defined to be the spherical polar angles, it is fairly easy to see that they transform as follows under E^*:

$$E^* \phi = (\pi + \phi), \quad E^* \theta = (\pi - \theta). \tag{6.209}$$

The transformation properties of χ can be best appreciated from figure 6.24, from which we obtain:

$$E^* \chi = (\pi - \chi). \tag{6.210}$$

Using these results we have

$$E^* \sin \phi = -\sin \phi, \quad E^* \sin \theta = \sin \theta, \quad E^* \sin \chi = \sin \chi,$$
$$E^* \cos \phi = -\cos \phi, \quad E^* \cos \theta = -\cos \theta, \quad E^* \cos \chi = -\cos \chi. \tag{6.211}$$

Substituting these results in equation (6.207), we obtain the transformation properties of the molecule-fixed coordinates of a point i under space-fixed inversion:

$$E^* x_i = x_i, \quad E^* y_i = -y_i, \quad E^* z_i = z_i. \tag{6.212}$$

We note the interesting result that the transformation properties of functions of coordinates which are defined in the molecule-fixed axis system (that is, electronic or vibrational) are the same under E^* as they are under σ_v in the xz plane in our convention. Thus we can use E^* to determine the reflection symmetry of electronic or vibrational states of diatomic molecules.

In conclusion, we have determined that a function of the Euler angles transforms under E^* as

$$E^* f(\phi, \theta, \chi) = f(\pi + \phi, \pi - \theta, \pi - \chi) \tag{6.213}$$

and a function of molecule-fixed coordinates transforms as

$$E^* f(x_i, y_i, z_i) = f(x_i, -y_i, z_i). \tag{6.214}$$

In the next section we use these results to show how a total wave function (and hence state) transforms under E^*.

6.9.3. The transformation of general Hund's case (a) and case (b) functions under space-fixed inversion

In Hund's case (a) coupling scheme, the electron spin angular momentum is quantised along the internuclear axis. The symmetry properties of such a function can be obtained

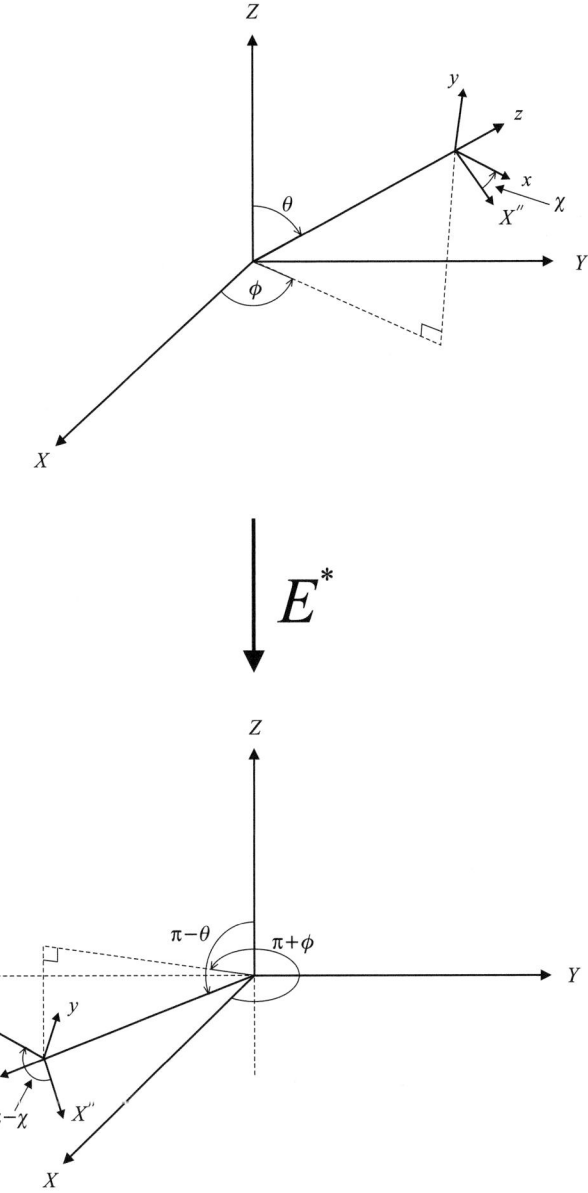

Figure 6.24. The effect of the space-fixed inversion operator E^* on the molecule-fixed coordinate system (x, y, z). The molecule-fixed coordinate system is always taken to be right-handed. After the inversion of the electronic and nuclear coordinates in laboratory-fixed space, the (x, y, z) coordinate system is fixed back onto the molecule so that the z axis points from nucleus 1 to nucleus 2 and the y axis is arbitrarily chosen to point in the same direction as before the inversion. As a result, the new values of the Euler angles (ϕ', θ', χ') are related to the original values (ϕ, θ, χ) by $\phi' = \pi + \phi$, $\theta' = \pi - \theta$, and $\chi' = \pi - \chi$.

from a consideration of the product representation

$$|\eta, \Lambda; v; S, \Sigma; J, \Omega, M\rangle = |\eta, \Lambda\rangle |v\rangle |S, \Sigma\rangle |J, \Omega, M\rangle. \tag{6.215}$$

We deal with each of the factors of the right-hand side of (6.215) in turn.

(i) Electronic orbital function $|\eta, \Lambda\rangle$

There is not a generally accepted form for the electronic basis function $|\eta, \Lambda\rangle$. The only common factor is a term $\exp(i\Lambda\phi_e)$ where ϕ_e is an electronic azimuthal angle, which is required so that

$$\hbar L_z |\eta, \Lambda\rangle = -i\hbar \frac{\partial}{\partial \phi_e} |\eta, \Lambda\rangle = \hbar \Lambda |\eta, \Lambda\rangle. \tag{6.216}$$

In this book, we adopt a form in which the function is expressed as a linear combination of spherical harmonics. This form is particularly appropriate for systems with near-spherical symmetry (such as Rydberg states or molecules which conform to Van Vleck's pure precession hypothesis [68, 69]) and is also consistent with the spirit of spherical tensors, which have the same transformation properties under rotations as spherical harmonics. The functional form of the ket $|\eta, \Lambda\rangle$ is written

$$|\eta, \Lambda\rangle = \sum_L F_L(\rho_e) Y_{L,\Lambda}(\theta_e, \phi_e), \tag{6.217}$$

where $(\rho_e, \theta_e, \phi_e)$ can be regarded as spherical polar coordinates of an electron measured in the molecule-fixed axis system. This interpretation is consistent with a many-electron wave function also.

For a general function $f(\rho_e, \theta_e, \phi_e)$ we know that

$$E^* f(\rho_e, \theta_e, \phi_e) = f(\rho_e, \theta_e, 2\pi - \phi_e) \tag{6.218}$$

from equation (6.214). Thus

$$E^* |\eta, \Lambda\rangle = \sum_L F_L(\rho_e) Y_{L,\Lambda}(\theta_e, 2\pi - \phi_e)$$

$$= \sum_L F_L(\rho_e) Y_{L,\Lambda}(\theta_e, \phi_e)^*$$

$$= (-1)^\Lambda \sum_L F_L(\rho_e) Y_{L,-\Lambda}(\theta_e, \phi_e)$$

$$\equiv (-1)^\Lambda |\eta, -\Lambda\rangle. \tag{6.219}$$

In addition, we must treat $\Sigma^\pm$ states as a special case. There are two possible functions depending on their behaviour under σ_v; the Σ^- combination requires at the very least a two-electron wave function. Let us take as a model wave function one constructed with two electrons in π orbitals:

$$\psi_{\text{el}}(\Sigma^s) = \frac{1}{\sqrt{2}} \{|\pi_{+1}\pi_{-1}\rangle + (-1)^s |\pi_{-1}\pi_{+1}\rangle\}, \tag{6.220}$$

where electron 1 is in the first orbital and electron 2 in the second. Using the result in equation (6.219) we have

$$E^*|\Sigma^s\rangle = \frac{1}{\sqrt{2}}\{(-1)^{1-1}|\pi_{-1}\pi_{+1}\rangle + (-1)^s(-1)^{-1+1}|\pi_{+1}\pi_{-1}\rangle\}$$
$$= (-1)^s \frac{1}{\sqrt{2}}\{|\pi_{+1}\pi_{-1}\rangle + (-1)^s|\pi_{-1}\pi_{+1}\rangle\}$$
$$= (-1)^s|\Sigma^s\rangle. \tag{6.221}$$

Equations (6.221) and (6.219) can be combined to give

$$E^*|\eta, \Lambda^s\rangle = (-1)^{\Lambda+s}|\eta, -\Lambda^s\rangle, \tag{6.222}$$

where s is even for Σ^+ (and higher Λ) states and odd for Σ^-.

(ii) Vibrational wave function $|v\rangle$

The vibrational coordinate is simply R, the magnitude of the separation between the two nuclei ($0 \leq R \leq \infty$). We see that this coordinate is unaffected by the operation E^* and so therefore is any function of R:

$$E^*\psi_{\text{vib}}(R) = \psi_{\text{vib}}(R). \tag{6.223}$$

Thus for a diatomic molecule, the vibrational factor is always symmetric with respect to E^* and so does not play any part in the symmetry classification scheme.

(iii) Electron spin function $|S, \Sigma\rangle$

In a case (a) basis set, the electron spin angular momentum is quantised along the linear axis, the quantum number Σ labelling the allowed components along this axis. Because we have chosen this axis of quantisation, the wave function is an implicit function of the three Euler angles and so is affected by the space-fixed inversion operator E^*. An electron spin wave function which is quantised in an arbitrary space-fixed axis system, $|S, M_S\rangle$, is not affected by E^*, however. This is because E^* operates on functions of coordinates in ordinary three-dimensional space, not on functions in spin space. The analogous operator to E^* in spin space is the time reversal operator.

Thus to see how $|S, \Sigma\rangle$ transforms under E^*, we must make its Euler angle dependence explicit:

$$|S, \Sigma\rangle = \sum_{M_S} |S, M_S\rangle \mathcal{D}^{(S)}_{M_S, \Sigma}(\phi, \theta, \chi). \tag{6.224}$$

Therefore

$$E^*|S, \Sigma\rangle = \sum_{M_S} |S, M_S\rangle \mathcal{D}^{(S)}_{M_S, \Sigma}(\pi + \phi, \pi - \theta, \pi - \chi)$$
$$= \sum_{M_S} |S, M_S\rangle \exp(-\mathrm{i}M_S\pi)\exp(-\mathrm{i}M_S\phi)d^{(S)}_{M_S, \Sigma}(\pi - \theta)\exp(-\mathrm{i}\Sigma\pi)\exp(\mathrm{i}\Sigma\chi)$$

$$= \sum_{M_S} |S, M_S\rangle (-1)^{-M_S-\Sigma} \exp(-iM_S\phi)(-1)^{S+M_S} d^{(S)}_{M_S,-\Sigma}(\theta) \exp(-i\Sigma\chi)$$
$$= (-1)^{S-\Sigma}|S, -\Sigma\rangle. \qquad (6.225)$$

We have used the relationship

$$d^{(J)}_{M,N}(\pi - \theta) = (-1)^{J+M} d^{(J)}_{M,-N}(\theta) \equiv (-1)^{3J-M} d^{(J)}_{M,-N}(\theta) \qquad (6.226)$$

from chapter 5 to obtain our result.

(iv) Rotational function $|J, \Omega, M\rangle$

With our choice of phases, the symmetric top wave function is related to the corresponding rotation matrix element by

$$|J, \Omega, M\rangle = [(2J+1)/8\pi^2]^{1/2} \mathcal{D}^{(J)}_{M,\Omega}(\phi, \theta, \chi)^*. \qquad (6.227)$$

Hence

$$E^*|J, \Omega, M\rangle = [(2J+1)/8\pi^2]^{1/2} \mathcal{D}^{(J)}_{M,\Omega}(\pi+\phi, \pi-\theta, \pi-\chi)^*$$
$$= [(2J+1)/8\pi^2]^{1/2} (-1)^{M+\Omega}(-1)^{3J-M} \mathcal{D}^{(J)}_{M,-\Omega}(\phi, \theta, \chi)^*$$
$$= (-1)^{3J+\Omega}|J, -\Omega, M\rangle \equiv (-1)^{J-\Omega}|J, -\Omega, M\rangle. \qquad (6.228)$$

We have now dealt with the effect of E^* on each of the factors in the case (a) function, (6.215). The results, which are quoted in equations (6.222), (6.225) and (6.228), can be combined to give the overall effect:

$$E^*|\eta, \Lambda^s; v; S, \Sigma; J, \Omega, M\rangle = (-1)^{\Lambda+s}(-1)^{S-\Sigma}(-1)^{J-\Omega}$$
$$\times |\eta, -\Lambda^s; v; S, -\Sigma; J, -\Omega, M\rangle. \qquad (6.229)$$

For a linear (diatomic) molecule we have the restriction

$$\Omega = \Lambda + \Sigma. \qquad (6.230)$$

Thus we have

$$E^*|\eta, \Lambda^s; v; S, \Sigma; J, \Omega, M\rangle = (-1)^p |\eta, -\Lambda^s; v; S, -\Sigma; J, -\Omega, M\rangle \qquad (6.231)$$

where

$$p = J - S + s. \qquad (6.232)$$

It often happens that it is more convenient to set up the problem in a Hund's case (b) basis set, $|\eta, \Lambda^s; v; N, \Lambda, S, J, M\rangle$. Using the same conventions as above, the transformation which corresponds to that given in equation (6.231) is

$$E^*|\eta, \Lambda^s; v; N, \Lambda, S, J, M\rangle = (-1)^{N+s}|\eta, -\Lambda^s; v; N, -\Lambda, S, J, M\rangle. \qquad (6.233)$$

6.9.4. Parity combinations of basis functions

It can be seen from equations (6.231) and (6.233) that the simple case (a) or case (b) basis functions are not eigenfunctions of E^*. In the language of group theory, they are not irreducible representations of the space-fixed inversion group and the states which they represent do not have a definite parity. The appropriate combinations which do have a definite parity are easily projected out (we are never concerned with a degeneracy higher than two for diatomic molecules). For the case (a) functions, the combinations of positive and negative parity are given by

$$|\eta, \Lambda^s; J, M; +\rangle = \frac{1}{\sqrt{2}} \{|\eta, \Lambda^s; S, \Sigma; J, \Omega, M\rangle + (-1)^p |\eta, -\Lambda^s; S, -\Sigma; J, -\Omega, M\rangle\} \quad (6.234)$$

$$|\eta, \Lambda^s; J, M; -\rangle = \frac{1}{\sqrt{2}} \{|\eta, \Lambda^s; S, \Sigma; J, \Omega, M\rangle - (-1)^p |\eta, -\Lambda^s; S, -\Sigma; J, -\Omega, M\rangle\}.$$

If there is a perturbation which lifts the two-fold degeneracy without destroying parity, the two functions in (6.234) are eigenfunctions of the system in the limit of a small perturbation.

The $(-1)^{J-S+s}$ phase factor in equation (6.234) causes the parity labels to alternate as J increases, that is, the lower of the near-degenerate pair for a given J might be $+$ and the upper $-$, the designation swaps over for the next J value and then back again and so on. To avoid this alternation, an alternative parity labelling system, the so-called e/f convention has been introduced [70]. For integral J values, levels with parities $(-1)^J$ or $(-1)^{J+1}$ are designated as e or f respectively; for half-integral J values, levels with parities $(-1)^{J-1/2}$ or $(-1)^{J+1/2}$ are designated e or f respectively. Using this convention, all the lower components of parity doublets have the same label, say e, and all the upper components have the opposite label, say f. For the situation where $\Lambda = 0$, a Σ^+ state has rotational levels of even $(+)$ or odd $(-)$ parity for N even or odd respectively; for Σ^- states this rule is reversed. Thus all the levels of a $^1\Sigma^+$ state are e levels and those of a $^1\Sigma^-$ state are f.

6.10. Permutation symmetry of rotational levels

6.10.1. The nuclear permutation operator for a homonuclear diatomic molecule

A second useful symmetry operation exists for homonuclear diatomic molecules, namely the permutation of two identical nuclei, P_{12}. In the same way that E^* has two possible eigenfunctions ± 1 in equation (6.206), so there are two possible ways in which the molecular wave function can transform under P_{12}:

$$P_{12} \psi_{\text{tot}} = \pm \psi_{\text{tot}}. \quad (6.235)$$

The upper sign choice (*symmetric* behaviour) corresponds to the interchange of identical nuclei with integral spins; such particles are called *bosons*. The lower sign choice (*antisymmetric* behaviour) is the transformation followed by *fermions*, with half-integral spin. This symmetry property is a generalisation of the Pauli exclusion principle as applied to the permutation of pairs of electrons. Electrons, with $S = 1/2$, are fermions and therefore show antisymmetric behaviour.

Let us suppose that nuclei 1 and 2 have coordinates (X_1, Y_1, Z_1) and (X_2, Y_2, Z_2) in the translating but not rotating coordinate system. Since the origin of the axis system is at the nuclear centre of mass of the homonuclear molecule, we have

$$X_2 = -X_1, \quad Y_2 = -Y_1, \quad Z_2 = -Z_1. \tag{6.236}$$

Now

$$P_{12}(X_1, Y_1, Z_1; X_2, Y_2, Z_2) = (X_2, Y_2, Z_2; X_1, Y_1, Z_1)$$
$$= (-X_1, -Y_1, -Z_1; -X_2, -Y_2, -Z_2). \tag{6.237}$$

Thus, so far as the nuclei are concerned, P_{12} produces the same effect as E^* and, since the position of the molecule-fixed axes depends on the location of the nuclei only, P_{12} has the same effect on a function of the Euler angles as does E^*:

$$P_{12} f(\phi, \theta, \chi) = f(\pi + \phi, \pi - \theta, \pi - \chi). \tag{6.238}$$

Now, the permutation of the nuclei P_{12} has no effect on the positions of the electrons measured in the space-fixed coordinate system. Thus for an electron at (X_i, Y_i, Z_i) we have

$$P_{12}(X_i, Y_i, Z_i) = (X_i, Y_i, Z_i). \tag{6.239}$$

P_{12} will, however, affect the molecule-fixed coordinates of the electrons through its effect on the orientation of this axis system. Substitution of equations (6.238) and (6.239) into the transformation equation (6.207) leads directly to the result

$$P_{12} f(x_i, y_i, z_i) = f(-x_i, y_i, -z_i). \tag{6.240}$$

By applying E^* and P_{12} in succession, it is easy to show that

$$P_{12} E^* (x_i, y_i, z_i) = E^* P_{12}(x_i, y_i, z_i) = (-x_i, -y_i, -z_i). \tag{6.241}$$

Thus we have the interesting result that $P_{12} E^* = E^* P_{12}$ corresponds to inversion i in the molecule-fixed axis system.

6.10.2. The transformation of general Hund's case (a) and case (b) functions under nuclear permutation P_{12}

As in section 6.9.3, we start out with the case (a) functions and consider the effect of P_{12} on each of its three factors in turn. For a homonuclear diatomic molecule, we write the electronic orbital wave function as a linear combination of spherical harmonics:

$$|\eta, \Lambda_t\rangle = \sum_L{}' F_L(\rho_e) Y_{L,\Lambda}(\theta_e, \phi_e) \tag{6.242}$$

where the subscript t identifies the state as either g or u. In the summation on the right-hand side, L is even for g wave functions and odd for u wave functions. Thus, using equation (6.240), we have

$$\begin{aligned}
P_{12}|\eta, \Lambda_t\rangle &= \sum_L{}' F_L(\rho_e) Y_{L,\Lambda}(\pi - \theta_e, \pi - \phi_e) \\
&= \sum_L{}' F_L(\rho_e)(-1)^L Y_{L,\Lambda}(\theta_e, \phi_e)^* \\
&= \sum_L{}' F_L(\rho_e)(-1)^{L+\Lambda} Y_{L,-\Lambda}(\theta_e, \phi_e) \\
&= (-1)^{t+\Lambda}|\eta, -\Lambda_t\rangle,
\end{aligned} \quad (6.243)$$

where the exponent t is even for g states or odd for u states. If the electronic state is a Σ^s state we must include an additional factor of $(-1)^s$ in equation (6.243) to distinguish between Σ_g^+ and Σ_g^- states.

In the same way that we dealt with the transformation of the vibrational, case (a) spin and rotational wave functions in section 6.9.3, it is easy to show that

$$P_{12}|v\rangle = |v\rangle, \quad (6.244)$$

$$P_{12}|S, \Sigma\rangle = (-1)^{S-\Sigma}|S, \Sigma\rangle, \quad (6.245)$$

$$P_{12}|J, \Omega, M\rangle = (-1)^{J-\Omega}|J, -\Omega, M\rangle. \quad (6.246)$$

The corresponding transformation for a Hund's case (b) function can be derived in a similar fashion:

$$P_{12}|\eta, \Lambda_t^s; v; N, \Lambda, S, J, M\rangle = (-1)^s(-1)^t(-1)^N |\eta, -\Lambda_t^s; v; N, -\Lambda, S, J, M\rangle. \quad (6.247)$$

When considering the effect of P_{12} on a Hund's case (a) or case (b) wave function, we must also take the nuclear spin wave function into account. The nuclear spin is usually very weakly coupled to the other angular momenta and so can be described by a separate factor $|\psi_{\text{ns}}\rangle$. We shall discuss the detailed form of this function shortly but for the moment, we simply need to recognise that there are two types of functions which can arise from the coupling scheme

$$\boldsymbol{I}_1 + \boldsymbol{I}_2 = \boldsymbol{I}_T. \quad (6.248)$$

These are *ortho* functions which are symmetric with respect to P_{12} and *para* functions which are antisymmetric. The ortho combinations always have the greater weight, that is, the greater number of independent nuclear spin functions. Thus we have

$$P_{12}|\psi_{\text{ns}}\rangle = \pm 1|\psi_{\text{ns}}\rangle = (-1)^{I_T - I_1 - I_2}|I_1, I_2, I_T\rangle. \quad (6.249)$$

It will not come as a great surprise to learn that there is a connection between the way in which the wave functions of homonuclear diatomic molecules transform under E^* and under P_{12}. We recall from equation (6.234) that the positive and negative parity combinations of the case (a) wave functions are given by

$$|\eta, \Lambda_t^s; J, M; \pm\rangle = \frac{1}{\sqrt{2}}\{|\eta, \Lambda; S, \Sigma; J, \Omega, M\rangle \pm (-1)^p |\eta, -\Lambda; S, -\Sigma; J, -\Omega, M\rangle\}. \tag{6.250}$$

Using equations (6.243) to (6.246) we have

$$P_{12}\left[\frac{1}{\sqrt{2}}\{|\Lambda_t; S, \Sigma; J, \Omega, M\rangle \pm (-1)^p |-\Lambda_t; S, -\Sigma; J, -\Omega, M\rangle\}\right]$$

$$= \pm(-1)^t\left[\frac{1}{\sqrt{2}}\{|\Lambda_t; S, \Sigma; J, \Omega, M\rangle \pm (-1)^p |-\Lambda_t; S, -\Sigma; J, -\Omega, M\rangle\}\right]. \tag{6.251}$$

Thus for a Λ_g state, positive parity states are symmetric and negative parity states are antisymmetric with respect to P_{12} and vice versa for a Λ_u state.

6.10.3. Nuclear statistical weights

The requirement set by the exclusion principle, given in equation (6.235), restricts the nuclear spin states which can combine with a rotational level of a given parity. They can be either symmetric or antisymmetric with respect to P_{12}. The number of symmetric (*ortho*) states is always greater than the number of antisymmetric (*para*) states. For equivalent nuclei with spin I_1,

$$\text{number of symmetric states} = (I_1 + 1)(2I_1 + 1),$$
$$\text{number of antisymmetric states} = I_1(2I_1 + 1). \tag{6.252}$$

Consequently the rotational levels of a given parity with which they combine will have different weights (the so-called *nuclear statistical weights*). The consequences of this are readily observable in spectroscopy, as we will see elsewhere.

Let us consider some specific examples whose spectra occur in this book. A very simple case is O_2 in its excited $^1\Delta_g$ state. The predominant nucleus, ^{16}O, has $I = 0$ and so is a boson. I_T is also zero and there is only one nuclear spin function ($I_T = 0$, $M_{I_T} = 0$) which is symmetric with respect to P_{12}. Thus for each value of J only one Λ-doublet component is allowed by the exclusion principle, namely, that which has positive parity; all the rotational levels of O_2 in its $^1\Delta_g$ state therefore have positive parity. The other Λ-doublet component is missing.

A slightly more complicated example is that of H_2 in its $X\,^1\Sigma_g^+$ state. The hydrogen nucleus is, of course, the proton with $I = 1/2$. In this case there are four possible nuclear spin wave functions, which are

$$\alpha(1)\alpha(2), \quad \beta(1)\beta(2), \quad (1/\sqrt{2})(\alpha(1)\beta(2) + \beta(1)\alpha(2)), \quad (1/\sqrt{2})(\alpha(1)\beta(2) - \beta(1)\alpha(2)). \tag{6.253}$$

The first three of these functions are symmetric with respect to P_{12} and constitute the three components of a spin triplet with $I_T = 1$; the fourth spin function is antisymmetric with respect to P_{12} and represents a spin singlet. Recalling that the total wave function consists of factors for electron orbital, electron spin, vibrational, rotational and nuclear spin motion, we have

$$P_{12}\psi_{\text{tot}} = \psi(^1\Sigma_g^+)\psi_{\text{vib}}(-1)^J\psi_{\text{rot}}(-1)^{I_T-1}\psi_{\text{ns}} = -\psi_{\text{tot}}. \quad (6.254)$$

The second part of this equation follows because the proton is a fermion. Thus for the lowest rotational level with $J = 0$, the nuclear wave function must be the fourth in equation (6.253) with $I_T = 0$. In the second rotational level, on the other hand, ψ_{ns} must be symmetric and so corresponds to the triplet spin function. In general, therefore, rotational levels with even J are associated with the singlet nuclear spin state and are called *para*-H$_2$ (with the lower nuclear statistical weight of one). Rotational levels with odd J have triplet nuclear spin states and are called *ortho*-H$_2$ (with the higher nuclear statistical weight of three).

Finally, we consider the N$_2$ molecule in its $A\ ^3\Sigma_u^+$ state. In the predominant isotopomer, both nuclei are ^{14}N with $I = 1$ (bosons); the total wave function must therefore satisfy

$$P_{12}\psi_{\text{tot}} = \psi_{\text{tot}}. \quad (6.255)$$

The coupling scheme for N$_2$ in this state is Hund's case (b). Therefore we have

$$P_{12}\psi_{\text{tot}} = (-1)\psi(^3\Sigma_u^+)\psi_{\text{vib}}(-1)^N\psi_{\text{rot}}(-1)^{I_T}\psi_{\text{ns}} \quad (6.256)$$

where N labels the rotational levels. Hence, for levels of even N, we require that

$$P_{12}\psi_{\text{ns}} = -\psi_{\text{ns}}, \quad (6.257)$$

whilst for odd N the opposite result is required, i.e.

$$P_{12}\psi_{\text{ns}} = \psi_{\text{ns}}. \quad (6.258)$$

We must therefore examine the possible nuclear spin states and classify them according to (6.257) or (6.258). Since each ^{14}N nucleus has spin $I = 1$, and three spatial orientations with $M_I = +1, 0, -1$, there are nine basis spin functions, which are

$$a_1b_1,\ a_1b_0,\ a_1b_{-1},\ a_0b_1,\ a_0b_0,\ a_0b_{-1},\ a_{-1}b_1,\ a_{-1}b_0,\ a_{-1}b_{-1}, \quad (6.259)$$

where a and b refer to nuclei 1 and 2 and the subscripts are the individual values of M_I. It is necessary to form linear combinations of these basis functions which have a definite permutation symmetry. The vector addition of the two nuclear spins gives a total spin I_T of 2, 1 or 0, and it is a straightforward exercise to form the appropriate linear combinations; they are as follows:

$$\begin{aligned}
I_T = 2\,\text{(symmetric):}\quad & M_I = 2;\ a_1b_1 \\
& M_I = 1;\ (1/\sqrt{2})(a_1b_0 + a_0b_1) \\
& M_I = 0;\ (1/\sqrt{6})(2a_0b_0 - a_1b_{-1} - a_{-1}b_1) \\
& M_I = -1;\ (1/\sqrt{2})(a_{-1}b_0 + a_0b_{-1}) \\
& M_I = -2;\ a_{-1}b_{-1} \quad (6.260) \\
I_T = 1\,\text{(antisymmetric):}\quad & M_I = 1;\ (1/\sqrt{2})(a_1b_0 - a_0b_1) \\
& M_I = 0;\ (1/\sqrt{2})(a_1b_{-1} - a_{-1}b_1) \\
& M_I = -1;\ (1/\sqrt{2})(a_{-1}b_0 - a_0b_{-1}) \quad (6.261)
\end{aligned}$$

$$I_T = 0 \text{ (symmetric)}: \quad M_I = 0; \quad (1/\sqrt{2})(a_1 b_{-1} + a_{-1} b_1). \tag{6.262}$$

Clearly the symmetric $I_T = 2$ and 0 functions satisfy (6.258) and are therefore associated with rotational levels having odd N. Conversely, the antisymmetric $I_T = 1$ functions satisfy (6.257) and must be associated with even N levels. These nuclear spin statistical results are very important, because they determine the nature of the nuclear hyperfine structure; if the different nuclear spin states remain degenerate, the spin statistics then help to determine rotational level populations and hence spectroscopic intensities.

One final symmetry aspect for homonuclear diatomic molecules to be mentioned here is that g/u states can, under some circumstances, be mixed by the nuclear spin part of the molecular Hamiltonian. This mixing, which is explained by Bunker and Jensen [71], has some interesting spectroscopic consequences, particularly in the H_2^+ molecular ion, which are described elsewhere in this book.

6.11. Theory of transition probabilities

6.11.1. Time-dependent perturbation theory

Spectroscopy is concerned with the observation of transitions between stationary states of a system, with the accompanying absorption or emission of electromagnetic radiation. In this section we consider the theory of transition probabilities, using time-dependent perturbation theory, and the selection rules for transitions, particularly those relevant for rotational spectroscopy.

We will consider a two-level system with states ψ_a and ψ_b which have energies E_a and E_b. The system is acted upon by a time-dependent perturbation,

$$\mathcal{H}'(t) = V f(t), \tag{6.263}$$

where V is a time-independent operator and $f(t)$ is a fluctuating factor which measures the strength of V at different times. We will be more specific about the form of $f(t)$ in due course. We suppose that the Hamiltonian matrix takes the form

$$\begin{vmatrix} E_a & V_{ab} f(t) \\ V_{ba} f(t) & E_b \end{vmatrix}, \tag{6.264}$$

where the off-diagonal elements $V_{ab} = V_{ba}^*$ induce transitions between the levels.

We look for a solution of the time-dependent Schrödinger equation

$$i\hbar \frac{\partial \psi}{\partial t} = \mathcal{H} \psi, \tag{6.265}$$

which has the following form:

$$\psi = C_a(t) \psi_a \exp(-i E_a t / \hbar) + C_b(t) \psi_b \exp(-i E_b t / \hbar). \tag{6.266}$$

If we substitute (6.266) into (6.265) and (6.264) we find that the coefficients satisfy the equations

$$i\hbar \frac{\partial C_a}{\partial t} = f(t)\exp(i[E_a - E_b]t/\hbar)V_{ab}C_b, \qquad (6.267)$$

$$i\hbar \frac{\partial C_b}{\partial t} = f(t)\exp(i[E_b - E_a]t/\hbar)V_{ba}C_a. \qquad (6.268)$$

We now suppose that at time $t=0$ the system starts in ψ_a; in other words $C_a(0) = 1$, $C_b(0) = 0$. Integration of equation (6.268) with $C_a = 1$ gives the first-order correction to C_b, and after a time T we find

$$C_b(T) = -\frac{i}{\hbar}V_{ba}\int_0^T f(t)\exp[i(E_b - E_a)t/\hbar]\,dt. \qquad (6.269)$$

The probability that the system has made a transition to state b is $|C_b(T)|^2$, or

$$P(a,b) = \frac{1}{\hbar^2}|V_{ab}|^2 \int_0^T dt_1 \int_0^T dt_2\, f(t_1)f(t_2)\exp(i(E_b - E_a)(t_1 - t_2)/\hbar). \qquad (6.270)$$

We may now simplify the time integral in (6.270) by changing to the new variables $t_1 = (t + \tau)$, $t_2 = t$ so that (6.270) becomes

$$P(a,b) = \frac{1}{\hbar^2}|V_{ab}|^2 \int_0^T dt \int_{-t}^{T-t} f(t+\tau)f(t)\exp(i(E_b - E_a)\tau/\hbar)\,d\tau. \qquad (6.271)$$

We now consider the specific case when $f(t)$ is a periodic function, given by

$$f(t) = 2\cos\omega t. \qquad (6.272)$$

We substitute for $f(t)$ in equation (6.269) and integrate directly, obtaining the result

$$C_b(t) = \frac{V_{ba}}{\hbar}\left\{\frac{\exp(i[\omega_{ba} - \omega]t) - 1}{(\omega_{ba} - \omega)} + \frac{\exp(i[\omega_{ba} + \omega]t) - 1}{(\omega_{ba} + \omega)}\right\}, \qquad (6.273)$$

where $\hbar\omega_{ba} = E_b - E_a$. We consider the case that $E_b > E_a$; the first term in (6.273) is very large at resonance and the second may be neglected, so that the transition probability is given by

$$P_{ab} = |C_b(t)|^2 = \frac{2\pi}{\hbar^2}|V_{ab}|^2 \left\{\frac{\sin^2[(1/2)(\omega_{ba} - \omega)T]}{2\pi[(1/2)(\omega_{ba} - \omega)]^2}\right\}. \qquad (6.274)$$

The expression in curly brackets is plotted as a function of the frequency in figure 6.25; the condition $\omega = \omega_{ba}$ is called the resonance condition. This expression represents an idealised situation because in practice one does not use a strictly monochromatic perturbing field, and one is not concerned with transitions to a perfectly defined state.

One of the most important quantities in equation (6.274) is $|V_{ba}|^2$ because it contains all of the information concerning the detailed nature of the interaction between the

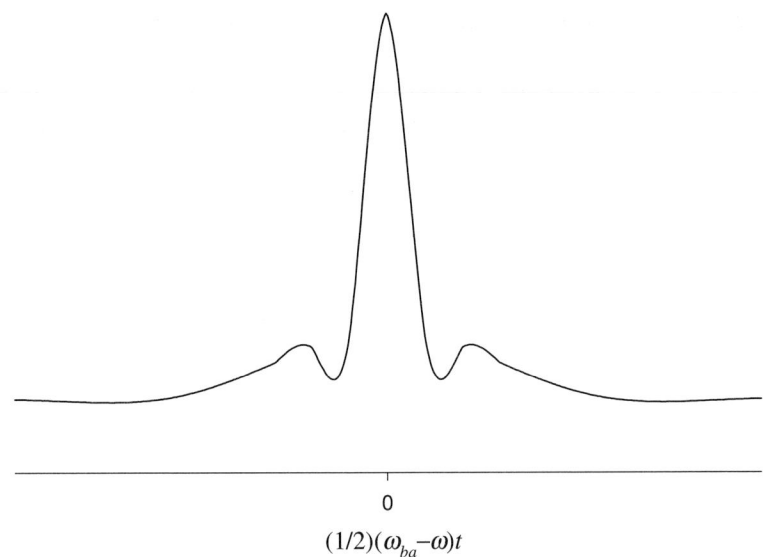

Figure 6.25. Dependence of the transition probability P_{ab} from $|a\rangle$ to $|b\rangle$ on the frequency of the driving radiation.

electromagnetic radiation and the appropriate properties of the molecule under investigation. We will return to a detailed discussion of this aspect a little later.

6.11.2. The Einstein transition probabilities

As we have seen, a transition from one state to another will be accompanied by the absorption or emission of radiation of frequency ν_{ba}, where

$$\nu_{ba} = \frac{E_b - E_a}{h}. \tag{6.275}$$

We now assume that a system in the lower energy state a is immersed in a bath of radiation which has a density $\rho(\nu_{ba})$; the radiation density $\rho(\nu)$ is defined such that the energy of radiation between frequencies ν and $\nu + d\nu$ in unit volume is $\rho(\nu)\,d\nu$. The probability that the system will absorb a quantum of radiation and undergo a transition to the upper state b in unit time is given by

$$B_{ab}\rho(\nu_{ba}), \tag{6.276}$$

where B_{ab} is called the *Einstein B coefficient of absorption* [68]. When it comes to emission, however, it is necessary to postulate that the probability is the sum of two parts, one of which is independent of the radiation density and the other proportional to it. These are referred to as spontaneous and induced processes respectively. We assume, therefore, that the probability that the system in the upper state b will undergo a transition to the lower state a with the emission of radiation is given by

$$A_{ba} + B_{ba}\rho(\nu_{ba}). \tag{6.277}$$

A_{ba} is the *Einstein coefficient of spontaneous emission* and B_{ba} is the *Einstein coefficient of induced emission*.

We now consider an assembly of systems identical to that described above which are in equilibrium with radiation at a temperature T. The density of radiation is given by Planck's radiation law as

$$\rho(\nu) = \frac{8\pi h \nu^3}{c^3} \frac{1}{\exp(h\nu/kT) - 1}, \qquad (6.278)$$

where k is the Boltzmann constant. Now let the initial number of systems in state b be N_b, and that in state a be N_a. The number of systems undergoing transitions from the lower state a to the upper state b is then

$$N_a B_{ab} \rho(\nu_{ba}), \qquad (6.279)$$

whilst the number undergoing the reverse transition is

$$N_b \{A_{ba} + B_{ba} \rho(\nu_{ba})\}. \qquad (6.280)$$

At equilibrium these two numbers are equal,

$$N_a B_{ab} \rho(\nu_{ba}) = N_b \{A_{ba} + B_{ba} \rho(\nu_{ba})\}, \qquad (6.281)$$

so that

$$\frac{N_a}{N_b} = \frac{A_{ba} + B_{ba} \rho(\nu_{ba})}{B_{ab} \rho(\nu_{ba})}. \qquad (6.282)$$

However, the Boltzmann distribution law states that at thermal equilibrium the ratio of the numbers in the lower and upper states is given by

$$\frac{N_a}{N_b} = \exp[-(E_a - E_b)/kT] = \exp(h\nu_{ba}/kT). \qquad (6.283)$$

Consequently from equations (6.282) and (6.283) combined we obtain an expression for the radiation density at the transition frequency:

$$\rho(\nu_{ba}) = \frac{A_{ba}}{B_{ab} \exp(h\nu_{ba}/kT) - B_{ba}}. \qquad (6.284)$$

This is identical with equation (6.278) if the three Einstein coefficients are related as follows:

$$B_{ab} = B_{ba}, \qquad (6.285)$$

$$A_{ba} = \frac{8\pi h \nu_{ba}^3}{c^3} B_{ba}. \qquad (6.286)$$

Notice the presence of the ν^3 factor in (6.286); whilst spontaneous emission is a characteristically fast phenomenon for electronic transitions occurring in the visible and ultraviolet regions of the spectrum, it is a slow process in the microwave region. This is a very important difference, as we will appreciate in many other places in this book.

Let us now consider a number of different situations and their experimental consequences with the aid of figure 6.26. This diagram illustrates four different situations concerning the relative populations of the two non-degenerate states. In case (i) the

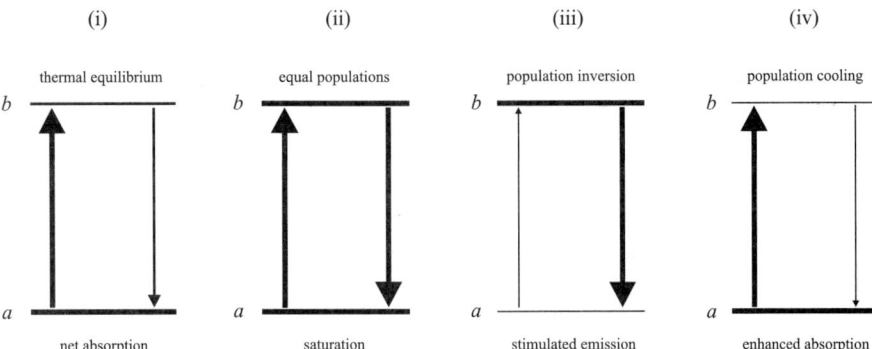

Figure 6.26. Population cases in a two-level system and the consequences of stimulated transitions.

populations are determined by thermal equilibrium conditions and obey the Boltzmann distribution law in which the relative populations are given by the simple expression

$$\frac{N_b}{N_a} = \exp[-(E_b - E_a)/kT], \qquad (6.287)$$

where E_b and E_a are the energies of the states and T is the absolute temperature of the bulk gas. This is by far the most common case, and would apply in the majority of the studies described in chapters 9 and 10.

Case (ii), where the populations are equal, is known as the saturation condition, or infinite temperature case. It can be achieved in many experiments if sufficient electromagnetic radiation power is available, and in some double resonance experiments it is actually an aim. Case (iii) represents a population inversion, sometimes referred to as a negative temperature, whilst case (iv) is described as a population cooling, in the sense that it corresponds to an abnormally low temperature, even though in other respects the temperature may be considered to be normal.

We now make use of our knowledge of the Einstein A and B coefficients in the light of the population cases outlined above. If N_a is the number of molecules in the lower (a) state, then the rate of absorption of energy from the radiation field is

$$I_{ab} = N_a B_{ab} \rho(\nu_{ba}) h \nu_{ab}. \qquad (6.288)$$

The power returned by stimulated emission is

$$I_{ba} = N_b B_{ba} \rho(\nu_{ba}) h \nu_{ab}, \qquad (6.289)$$

so that the net change is

$$\Delta I = I_{ab} - I_{ba} = (N_a - N_b) B \rho(\nu) h \nu. \qquad (6.290)$$

We may now examine the four population cases in the light of equation (6.290). In case (i) when the populations are in accord with the Boltzmann distribution law, ΔI is positive and we observe a net absorption of power. Substituting the Boltzmann distribution into (6.290) we obtain for the absorption intensity

$$I_{abs} = \Delta I = N_a[1 - \exp(-h\nu/kT)]B\rho(\nu)h\nu. \tag{6.291}$$

Since at microwave frequencies and normal temperatures, $h\nu \ll kT$, we can expand the exponential as a convergent power series to yield

$$\begin{aligned}I_{abs} &= \frac{N_a(h\nu)^2}{kT}B\rho(\nu)\left[1 - \frac{1}{2}\frac{h\nu}{kT} + \cdots\right] \\ &\simeq \frac{N_a(h\nu)^2}{kT}B\rho(\nu).\end{aligned} \tag{6.292}$$

Provided a Boltzmann distribution is maintained the absorption intensity is proportional to the square of the frequency. It is important, however, that there be collisions in the gas which maintain thermal equilibrium; if there is insufficient thermal relaxation, the population difference is reduced by absorption of radiation. Consequently case (ii) is approached and ultimately, from (6.290), there is no net observable absorption of radiation.

Population case (iii), where we have an inversion ($N_b > N_a$), is a particularly interesting and important one. Equation (6.290) tells us that ΔI is negative and that therefore stimulated emission of radiation will be observed. The difficulty is that in order to obtain a population inversion and maintain it, the effects of molecular collisions must be overcome; the radiation-induced transitions themselves lead towards equalisation of the populations. This problem is overcome in molecular beam experiments by presenting the radiation with a continuous supply of fresh molecules in state b, using electric or magnetic state selection to reject molecules in state a. The result of the experiment is essentially an amplification of the incident microwave radiation and the first successful device was termed a *MASER* by its inventors. In experiments involving infrared, visible or ultraviolet radiation, population inversions are usually produced by optical pumping and collisional processes, and the Light Amplification produced by the Stimulated Emission of Radiation gives rise to the *LASER* device.

Finally, case (iv) leads to the observation of enhanced stimulated absorption; again molecular beam techniques which select the lower (a) state can be used. It is interesting to note that all four population cases are observed in different interstellar gas clouds.

6.11.3. Einstein transition probabilities for electric dipole transitions

As promised at the end of section 6.11.1 we now consider a specific example of the time-dependent perturbation introduced in equations (6.263) and (6.272). The radiation density for electromagnetic radiation with unit dielectric constant and magnetic permeability is given by [5, 72]

$$\rho(\nu) = \frac{\varepsilon_0}{2}\overline{E^2(\nu)}, \tag{6.293}$$

where $\overline{E^2(\nu)}$ is the time-averaged value of the square of the electric field strength corresponding to the radiation. For plane wave radiation, we can write

$$E(\nu) = E_X(\nu),$$
$$E_Y(\nu) = E_Z(\nu) = 0. \tag{6.294}$$

$E_X(\nu)$ represents the component of the electric field in the X direction, and we may describe the time variation, by analogy with (6.272), by writing

$$E_X(\nu) = 2E_X^0(\nu)\cos 2\pi \nu t, \tag{6.295}$$

Since the average value of $\cos^2 2\pi \nu t$ is $1/2$, we find that

$$\rho(\nu) = \varepsilon_0 \left[E_X^0(\nu)\right]^2. \tag{6.296}$$

The perturbation of a molecular system of electrically charged particles (electrons and nuclei) by an oscillating electric field applied parallel to the X axis is represented by the operator

$$\mathcal{H}' = -E_X \sum_j q_j X_j, \tag{6.297}$$

q_j and X_j being the charge and X coordinate of the jth particle in the system. When the sum is taken over all particles of the system, the quantity

$$\sum_j q_j X_j = \mu_X, \tag{6.298}$$

is called the component of the electric dipole moment (μ) of the system along the X axis. The analysis now follows the same lines as those described in section 6.11.1; if $C_b(t)$ is the coefficient of the upper state, initially with zero value, its time evolution is given by

$$\begin{aligned}\frac{\partial C_b(t)}{\partial t} &= -\frac{2\pi i}{h}\int \psi_b^{0*}\mathcal{H}'\psi_a^0\,d\tau \\ &= \frac{2\pi i}{h}\int \psi_b^{0*}\exp[(2\pi i/h)E_b t]E_X^0(\nu)[\exp(2\pi i\nu t) \\ &\quad + \exp(-2\pi i\nu t)]\sum_j q_j X_j \psi_a^0 \exp[(-2\pi i/h)E_a t]\,d\tau.\end{aligned} \tag{6.299}$$

We now use the expression $\mu_X(b,a)$ to represent the integral

$$\mu_X(b,a) = \int \psi_b^{0*}\sum_j q_j X_j \psi_a^0 \,d\tau$$

$$= \int \psi_b^{0*}\mu_X\psi_a^0\,d\tau. \tag{6.300}$$

Integration of (6.299) and retention, as before, of the resonance term leads to the result

$$C_b^*(t)C_b(t) = 4[\mu_X(b,a)]^2\left[E_X^0(\nu)\right]^2\frac{\sin^2\{(\pi/h)(E_b - E_a - h\nu)t\}}{(E_b - E_a - h\nu)^2}. \tag{6.301}$$

This expression includes only the terms due to a single frequency, but we can integrate over the range of frequencies concerned. Since the integrand in (6.299) makes a significant contribution only when ν is near the resonance value ν_{ba} we obtain the final result

$$C_b^*(t)C_b(t) = 4[\mu_X(b,a)]^2 [E_X^0(\nu_{ba})]^2 \int \frac{\sin^2\{(\pi/h)(E_b - E_a - h\nu)t\}}{(E_b - E_a - h\nu)^2} d\nu$$

$$= \frac{4\pi^2}{h^2}[\mu_X(b,a)]^2 [E_X^0(\nu_{ba})]^2 t. \quad (6.302)$$

Equation (6.302) tells us that the probability of an electric dipole transition per unit time from state a to state b under the influence of electromagnetic radiation polarised in the X direction is

$$\frac{4\pi^2}{\varepsilon_0 h^2}[\mu_X(b,a)]^2 \rho(\nu_{ba}), \quad (6.303)$$

where we have used (6.296) for the radiation density. Similar expressions would be obtained for the components of the oscillating electric fields in the Y and Z directions, so that we obtain a result for the Einstein B_{ab} coefficient for absorption, which is

$$B_{ab} = \frac{4\pi^3}{3\varepsilon_0 h^2}\{[\mu_X(b,a)]^2 + [\mu_Y(b,a)]^2 + [\mu_Z(b,a)]^2\}. \quad (6.304)$$

One can repeat the above exercise for the time evolution of $C_a(t)$, yielding an expression for the complementary Einstein coefficient B_{ba}; this is found to be identical to B_{ab} above, as required in equation (6.285).

The important result of the analysis above is that the calculation of spectroscopic line intensities and the determination of selection rules is reduced to an assessment of the electric moment integrals given in (6.300). We now consider the main examples in more detail.

6.11.4. Rotational transition probabilities

We recall from our discussion (section 6.8.1) of the rigid rotor model of a diatomic molecule that the Schrödinger equation is

$$-\frac{\hbar^2}{2\mu R^2}\left[\frac{1}{\sin\theta}\frac{\partial}{\partial \theta}\left(\sin\theta \frac{\partial}{\partial \theta}\right) + \frac{1}{\sin^2\theta}\frac{\partial^2}{\partial \phi^2}\right]\psi(\theta,\phi) = E\psi(\theta,\phi), \quad (6.305)$$

which may be rewritten in the more compact form

$$\frac{\hbar^2}{2\mu R^2}\boldsymbol{J}^2 \psi(\theta,\phi) = E\psi(\theta,\phi), \quad (6.306)$$

where $\boldsymbol{J}^2$ is the rotational angular momentum. The wave functions can be written,

$$\psi(\theta,\phi) = K_{lm} P_{lm}(\cos\theta)\exp(im\phi), \quad (6.307)$$

where the $P_{lm}(\cos\theta)$ are the associated Legendre polynomials and the K_{lm} are normalisation constants. The rotational energy levels, $E(J)$, are given by

$$E(J) = \frac{\hbar^2}{2\mu R^2} J(J+1), \qquad (6.308)$$

where, in this simple example, J takes integral values 0, 1, 2, etc., and μ is the reduced mass.

The components of the dipole moment operator in polar coordinates are

$$\begin{aligned} \mu_X &= \mu_e \sin\theta \cos\phi, \\ \mu_Y &= \mu_e \sin\theta \sin\phi, \\ \mu_Z &= \mu_e \cos\theta. \end{aligned} \qquad (6.309)$$

The integrals which determine the transition probabilities between states l, m and l', m' of a molecule in a non-degenerate state are, from equation (6.300),

$$\mu_X(lm, l'm') = \mu_e K_{lm} K_{l'm'} \int \sin\theta P_{lm}(\cos\theta) P_{l'm'}(\cos\theta) \sin\theta \, d\theta$$

$$\times \int \cos\phi \exp[i(m'-m)\phi] d\phi,$$

$$\mu_Y(lm, l'm') = \mu_e K_{lm} K_{l'm'} \int \sin\theta P_{lm}(\cos\theta) P_{l'm'}(\cos\theta) \sin\theta \, d\theta$$

$$\times \int \sin\phi \exp[i(m'-m)\phi] d\phi, \qquad (6.310)$$

$$\mu_Z(lm, l'm') = \mu_e K_{lm} K_{l'm'} \int \cos\theta P_{lm}(\cos\theta) P_{l'm'}(\cos\theta) \sin\theta \, d\theta$$

$$\times \int \exp[i(m'-m)\phi] d\phi.$$

We now make use of some relations; first we have the familiar

$$\cos\phi = (1/2)[\exp(i\phi) + \exp(-i\phi)], \quad \sin\phi = (1/2i)[\exp(i\phi) - \exp(-i\phi)]. \qquad (6.311)$$

Next we make use of three recurrence relationships involving the associated Legendre polynomials:

$$(2l+1)\sin\theta P_{lm}(\cos\theta) = P_{l+1,m+1}(\cos\theta) - P_{l-1,m+1}(\cos\theta). \qquad (6.312)$$

$$(2l+1)\cos\theta P_{lm}(\cos\theta) = (l-m+1)P_{l+1,m}(\cos\theta) + (l+m)P_{l-1,m}(\cos\theta). \qquad (6.313)$$

$$(2l+1)\sin\theta P_{lm}(\cos\theta) = (l+m)(l+m-1)P_{l-1,m-1}(\cos\theta)$$
$$- (l-m+1)(l-m+2)P_{l+1,m-1}(\cos\theta). \qquad (6.314)$$

One can now show that

$$\sin\theta P_{lm}(\cos\theta) = [P_{l+1,m+1}(\cos\theta) - P_{l-1,m+1}(\cos\theta)]/(2l+1)$$
$$= [(l+m-1)(l+m)P_{l-1,m-1}(\cos\theta)$$
$$- (l-m+2)(l-m+1)P_{l+1,m-1}(\cos\theta)]/(2l+1). \qquad (6.315)$$

$$\cos\theta P_{lm}(\cos\theta) = [(l-m+1)P_{l+1,m}(\cos\theta) + (l+m)P_{l-1,m}(\cos\theta)]/(2l+1). \qquad (6.316)$$

From these results we obtain the selection rules

$$\mu_X(lm, l'm') = \mu_Y(lm, l'm') = 0 \text{ unless } m = m' \pm 1 \text{ and } l = l' \pm 1,$$
$$\mu_Z(lm, l'm') = 0 \text{ unless } m = m' \text{ and } l = l' \pm 1. \quad (6.317)$$

The important result, therefore, is the selection rule $\Delta l = \pm 1$, or in terms of the conventional rotational quantum number J, $\Delta J = \pm 1$.

In most of the examples described in this book, the rotational angular momentum is coupled to other angular momenta within the molecule, and the selection rules for transitions are more complicated than for the simplest example described above. Spherical tensor methods, however, offer a powerful way of determining selection rules and transition intensities. Let us consider, as an example, rotational transitions in a good case (a) molecule. The perturbation due to the oscillating electric component of the electromagnetic radiation, interacting with the permanent electric dipole moment of the molecule, is represented by the operator

$$\mathcal{H}'(t) = -\mathbf{T}^1(\boldsymbol{E}_t) \cdot \mathbf{T}^1(\boldsymbol{\mu}_e)$$
$$= -\mathbf{T}^1_{p=0}(\boldsymbol{E}_t)\mathbf{T}^1_{p=0}(\boldsymbol{\mu}_e). \quad (6.318)$$

We have taken the direction of the electric field to define the space-fixed $p=0$ direction in the expansion of the scalar product. The electric dipole moment of the molecule, $\boldsymbol{\mu}_e$, is, however, quantised in the molecule-fixed axis system (q); we therefore rotate the space-fixed component of $\boldsymbol{\mu}_e$ into the molecular axis system using a rotation matrix, so that the perturbation (6.318) becomes

$$\mathcal{H}'(t) = -\mathbf{T}^1_{p=0}(\boldsymbol{E}_t) \sum_q \mathcal{D}^{(1)}_{0q}(\omega)^* \mathbf{T}^1_q(\boldsymbol{\mu}_e)$$
$$= -E_0(t)\mathcal{D}^{(1)}_{00}(\omega)^* \mathbf{T}^1_0(\boldsymbol{\mu}_e). \quad (6.319)$$

In the second line of (6.319) we have used the fact that the permanent electric dipole moment of the molecule lies along the internuclear axis ($q=0$). The matrix elements of (6.319) in a case (a) basis are

$$\langle J, \Omega, M_J| - E_0(t)\mathbf{T}^1_0(\boldsymbol{\mu}_e)\mathcal{D}^{(1)}_{00}(\omega)^*|J', \Omega', M'_J\rangle$$
$$= -E_0(t)\mu_0(-1)^{J-M_J}\begin{pmatrix} J & 1 & J' \\ -M_J & 0 & M_J \end{pmatrix}\langle J, \Omega\|\mathcal{D}^{(1)}_{\cdot 0}(\omega)^*\|J', \Omega'\rangle,$$
$$= -E_0(t)\mu_0(-1)^{J-M_J}\begin{pmatrix} J & 1 & J' \\ -M_J & 0 & M_J \end{pmatrix}(-1)^{J-\Omega}\begin{pmatrix} J & 1 & J' \\ -\Omega & 0 & \Omega' \end{pmatrix}$$
$$\times \{(2J+1)(2J'+1)\}^{1/2}. \quad (6.320)$$

This expression gives the selection rules immediately; the 3-j symbols are non-zero if $J' = J, J \pm 1$, and we note the additional selection rule $\Delta \Omega = 0$. In the event that $\Omega = 0$, the second 3-j symbol in (6.320) is only non-zero for $J' = J \pm 1$.

We now repeat the exercise for a case (b) open shell molecule like the CN radical which has a $^2\Sigma^+$ ground state. We again transform the perturbation Hamiltonian into the molecule-fixed axis system, and find the following matrix element in a case (b) basis:

$$\langle \eta, \Lambda; N, \Lambda, S, J, M_J | - E_0(t) \mathrm{T}_0^1(\boldsymbol{\mu}_e) \mathscr{D}_{00}^{(1)}(\omega)^* | \eta, \Lambda'; N', \Lambda', S, J', M_J' \rangle$$

$$= -E_0(t)\mu_0(-1)^{J-M_J} \begin{pmatrix} J & 1 & J' \\ -M_J & 0 & M_J \end{pmatrix} \langle \eta, \Lambda; N, \Lambda, S, J \| \mathscr{D}_{\cdot 0}^{(1)}(\omega)^* \| \eta, \Lambda'; N', \Lambda', S, J' \rangle$$

$$= -E_0(t)\mu_0(-1)^{J-M_J} \begin{pmatrix} J & 1 & J' \\ -M_J & 0 & M_J \end{pmatrix}$$

$$\times (-1)^{J'+N+1+S}\{(2J'+1)(2J+1)\}^{1/2} \begin{Bmatrix} J & N & S \\ N' & J' & 1 \end{Bmatrix} \langle N, \Lambda \| \mathscr{D}_{\cdot 0}^{(1)}(\omega)^* \| N', \Lambda' \rangle$$

$$= E_0(t)\mu_0(-1)^{J+J'+N+S-M_J} \begin{pmatrix} J & 1 & J' \\ -M_J & 0 & M_J \end{pmatrix} \{(2J'+1)(2J+1)\}^{1/2}$$

$$\times \begin{Bmatrix} J & N & S \\ N' & J' & 1 \end{Bmatrix} (-1)^{N-\Lambda} \begin{pmatrix} N & 1 & N' \\ -\Lambda & 0 & \Lambda \end{pmatrix} \{(2N+1)(2N'+1)\}^{1/2}. \quad (6.321)$$

For an electronic Σ state we have $\Lambda = 0$; the last 3-j symbol in (6.321) then requires that $N' = N \pm 1$ in order to be non-zero, which gives the most important selection rule. We see also that there is a selection rule for J, namely $\Delta J = 0, \pm 1$. We will meet examples of other coupling cases elsewhere in this book.

We should not leave this discussion of the intensity of rotational transitions without some mention of the parity selection rule. Electric dipole transitions involve the interaction between the oscillating electric field and the oscillating electric dipole moment of the molecule. The latter is represented in quantum mechanics by the transition moment $\mu_X(b,a)$ given in equation (6.300). For this transition moment to be non-zero, the integrand $\psi_b^{0*} \mu_X \psi_a^0$ must be totally symmetric with respect to all appropriate symmetry operations, which includes the space-fixed inversion operator E^*. Now the electric dipole moment operator,

$$\mu_X = \sum_i q_i X_i, \quad (6.322)$$

changes sign under E^* because of the result

$$E^*(X_i) = -X_i. \quad (6.323)$$

Therefore the product $\psi_b^{0*} \psi_a^0$ must also be antisymmetric with respect to E^* if the transition moment is to be non-zero. Put another way, electric dipole transitions must obey the parity selection rule $+ \leftrightarrow -$, that is, they only connect states of opposite parity. This selection rule is general for *all* electric dipole transitions.

6.11.5. Vibrational transition probabilities

Vibrational transitions play only a minor role in the spectroscopy described in this book, although they do occur in some of the double resonance experiments described in chapter 11. It does seem important, however, to understand the factors which govern their intensity, at least in a qualitative manner.

In section 6.8.2 we described and solved the Schrödinger equation for a harmonic oscillator, equation (6.178). The potential energy was expressed in terms of a vibrational coordinate q which was equal to $R - R_e$, R_e being the equilibrium bond length. The dependence of the electric dipole moment on the internuclear distance may be expressed as a Taylor series,

$$\mu_e = \mu_0 + \left(\frac{\partial \mu_e}{\partial R}\right)_0 q + \frac{1}{2!}\left(\frac{\partial^2 \mu_e}{\partial R^2}\right)_0 q^2 + \frac{1}{3!}\left(\frac{\partial^3 \mu_e}{\partial R^3}\right)_0 q^3 + \cdots, \qquad (6.324)$$

where μ_0 is the permanent electric dipole moment at equilibrium with $q = 0$ and $(\partial \mu_e/\partial R)_0$ represents the change in the dipole moment with distance evaluated at the equilibrium internuclear distance. Since the displacement of the nuclei for a vibration in a diatomic molecule is only a few percent of the interatomic distance, the series (6.324) converges quite rapidly.

The electric dipole transition moment for a harmonic oscillator may now be written

$$\mu_e(v, v') = \int \Psi_v^*(q) \left[\mu_0 + \left(\frac{d\mu_e}{dR}\right)_0 q + \frac{1}{2!}\left(\frac{d^2\mu_e}{dR^2}\right)_0 q^2 + \frac{1}{3!}\left(\frac{d^3\mu_e}{dR^3}\right)_0 q^3 + \cdots \right] \Psi_{v'}(q)\, dq$$

$$= \mu_0 + \left(\frac{d\mu_e}{dR}\right)_0 \langle v|q|v \pm 1\rangle + \frac{1}{2}\left(\frac{d^2\mu_e}{dR^2}\right)_0 \langle v|q^2|v, v \pm 2\rangle + \cdots. \qquad (6.325)$$

The dominant term for vibrational transitions is, of course, the second, which gives the primary selection rule for a harmonic oscillator of $\Delta v = \pm 1$. The overtone transitions $\Delta v = \pm 2, \pm 3$, etc., are very much weaker because of the rapid convergence of (6.325).

As we have described before, a more accurate potential function takes account of the vibrational anharmonicity. For a Morse potential the vibrational Hamiltonian becomes

$$\mathcal{H} = -\frac{\hbar^2}{2\mu}\frac{d^2}{dq^2} + D_e[1 - \exp(-\beta q)]^2 \qquad (6.326)$$

and the matrix representation of this Hamiltonian in a harmonic oscillator basis can be readily computed. Using the wave functions which, as linear combinations of harmonic oscillator functions diagonalise (6.326), the terms in (6.325) can be re-examined. The effect of increased anharmonicity is to enhance the intensity of the overtone transitions.

6.11.6. Electronic transition probabilities

Electronic transitions in diatomic molecules are determined by matrix elements of the dipole operator,

$$\mu_e(f, i) = \int \psi_f^* \mu_e(f, i) \psi_i \, d\tau, \qquad (6.327)$$

where the f and i subscripts indicate the final and initial states, respectively. Let us consider the situation when a plane-polarised oscillating electric field in the radiation is incident at an angle θ to the molecular dipole axis. The component of the electric dipole moment $\boldsymbol{\mu}_e$ projected along the field axis is given by

$$\mu_Z = \mu_e \cos\theta. \tag{6.328}$$

Since $\boldsymbol{\mu}_e$ arises from a sum over nuclear and electronic charges, we may write

$$\boldsymbol{\mu}_e = e\left(\sum_\alpha Z_\alpha \boldsymbol{r}_\alpha - \sum_i \boldsymbol{r}_i\right), \tag{6.329}$$

where e is the elementary charge, Z_α is the atomic number of the αth nucleus, $\boldsymbol{r}_\alpha$ is the centre-of-mass position of the αth nucleus, and $\boldsymbol{r}_i$ is the centre-of-mass position of the ith electron in the molecule, both measured in the molecule-fixed axis system. We note that, because of the electronic contribution, there are instantaneous non-zero components of $\boldsymbol{\mu}_e$ perpendicular to the internuclear axis as well as parallel to it.

Now the total wave functions for the initial and final states can be written in the usual form as products of electronic, vibrational and rotational functions:

$$\psi_i = \psi'_e(R_\alpha, r_i)\psi'_v(R_\alpha)Y_{J'M'}, \quad \psi_f = \psi_e(R_\alpha, r_i)\psi_v(R_\alpha)Y_{JM}. \tag{6.330}$$

Substituting (6.330), (6.329) and (6.328) in (6.327) we obtain

$$\mu_e(f,i) = \langle J, M | \cos\theta | J', M'\rangle \int\int_{i\ \alpha} \psi_e^* \psi_v^* e\left(\sum_\alpha Z_\alpha \boldsymbol{r}_\alpha - \sum_i \boldsymbol{r}_i\right)\psi'_e \psi'_v \, dV_i \, dV_\alpha,$$
$$\tag{6.331}$$

where the integrations are over electronic (i) and nuclear (α) coordinates. The matrix elements of $\cos\theta$ lead to selection rules $\Delta J = 0, \pm 1$, $\Delta M = 0, \pm 1$ for the rotational part of the transition. For the remaining integrals in (6.331) we have

$$\text{INT} = e\int\int_{i\ \alpha} \psi_e^* \psi_v^* \left(\sum_\alpha Z_\alpha \boldsymbol{r}_\alpha - \sum_i \boldsymbol{r}_i\right)\psi'_e \psi'_v \, dV_i \, dV_\alpha,$$

$$= e\int\int_{\alpha\ i} \psi_e^* \psi_v^* \sum_\alpha Z_\alpha \boldsymbol{r}_\alpha \psi'_e \psi'_v \, dV_i \, dV_\alpha - e\int\int_{\alpha\ i} \psi_e^* \psi_v^* \sum_i \boldsymbol{r}_i \psi'_e \psi'_v \, dV_i \, dV_\alpha. \tag{6.332}$$

We now make use of the Born–Oppenheimer approximation which allows us to separate the electronic and vibrational wave functions, yielding

$$\text{INT} = e\int_i \psi_e^*(r_i, R_\alpha)\psi'_e(r_i, R_\alpha)\, dV_i \int_\alpha \psi_v^*(R_\alpha)\left(\sum_\alpha Z_\alpha \boldsymbol{r}_\alpha\right)\psi'_v(R_\alpha)\, dV_\alpha$$

$$- e\int_\alpha \psi_v^*(R_\alpha)\psi'_v(R_\alpha)\, dV_\alpha \int_i \psi_e^*(r_i, R_\alpha)\sum_i \boldsymbol{r}_i \psi'_e(r_i, R_\alpha)\, dV_i. \tag{6.333}$$

The first term in this equation is zero because the two electronic states have orthogonal eigenfunctions. The final expression for the matrix element of the electric dipole moment operator is therefore

$$\mu_e(f,i) = \langle J, M | \cos\theta | J', M' \rangle \left[\int_\alpha \psi_v^*(R_\alpha) \psi_v'(R_\alpha) \, dV_\alpha \right]$$

$$\times \left[-e \int_i \psi_e^*(r_i, R_\alpha) \sum_i r_i \psi_e'(r_i, R_\alpha) \, dV_i \right], \quad (6.334)$$

the product of a rotational, vibrational and electronic part. The vibrational integral in (6.334) describes the overlap of the vibrational functions in the ground and excited electronic states; its square is referred to as the *Franck–Condon factor*.

Notice a very important feature of equation (6.334). Electronic transitions do not depend for their intensity on the presence of a permanent electric dipole moment in the molecule, so that they exist for both homonuclear and heteronuclear diatomic molecules. This is in contrast to rotational and vibrational transitions which have electric dipole intensity only in heteronuclear molecules (apart from one extraordinary exception for the H_2^+ molecule, described in chapter 10.)

6.11.7. Magnetic dipole transition probabilities

The oscillating magnetic component of electromagnetic radiation can also interact with a magnetic moment in a molecule, and spectroscopic transitions known as magnetic dipole transitions then occur. There are several possible sources of a magnetic moment in a molecule, the most important occurring in electronic open shell systems. Electron spin gives rise to a magnetic moment

$$\mu_S = -g_S \mu_B S, \quad (6.335)$$

the origin of which was discussed extensively in chapter 3, where the presence of electron spin was shown to arise naturally in the Dirac theory of the electron; $\hbar S$ is the spin angular momentum, with a value $(1/2)\hbar$, g_S is the free electron g-value with a value of 2.0023, and μ_B is the electron Bohr magneton, with a value $9.274\,015\,4 \times 10^{-24}$ J T^{-1}. The interaction of the electron spin magnetic moment with an applied static magnetic field is the basis of magnetic resonance spectroscopy, which is discussed extensively in chapter 9. However, interaction of μ_S with an oscillating magnetic field can give rise to magnetic dipole transitions; we will look at an important example shortly.

Open shell molecules may also have a magnetic moment arising from electron orbital angular momentum,

$$\mu_L = -g_L \mu_B L, \quad (6.336)$$

which may be either the sole source of an electronic magnetic moment, or one in addition to an electron spin magnetic moment. In this case g_L has the value 1.0000. In many cases, as we shall see elsewhere, L and S are strongly coupled to each other, and to other angular momenta in the molecule. There are, however, examples of molecules where the orbital magnetic moment is the main source of both a static magnetic moment, and also magnetic dipole transitions.

We will return to consider the magnitudes of the interaction of a magnetic field with the electronic spin and orbital magnetic moments, but first we consider other sources of magnetic moment. The most important of these is nuclear spin angular momentum, conferring a magnetic moment given by

$$\mu_I = g_N \mu_N \boldsymbol{I}, \qquad (6.337)$$

where $\boldsymbol{I}$ is the nuclear spin angular momentum, which takes either integral or half-integral (strictly speaking, half-odd) values, and μ_N is the nuclear magneton with a value $5.050\,786\,6 \times 10^{-27}$ J T^{-1}. Chapter 4 was devoted to a detailed description of interactions arising from nuclear spin magnetic moments, but we did not discuss magnetic dipole transition probabilities. Although these are a factor of $(2 \times 10^3)^2$ smaller than electron spin magnetic dipole probabilities, and a further five orders-of-magnitude smaller than electric dipole probabilities, they are nevertheless of extreme importance. Combined with static Zeeman interactions, they form the basis of nuclear magnetic resonance in condensed phases, and the molecular beam magnetic resonance studies which are described extensively in chapter 8. The most important magnetic nucleus is the proton, with a spin $I = 1/2$ and a nuclear g-factor, g_N, of 2.792 847 in units of the nuclear magneton. Many other nuclei have magnetic moments, and appear in this book.

Finally in this parade of magnetic moments we have the rotational magnetic moment,

$$\mu_R(J) = g_R \mu_N \boldsymbol{J}. \qquad (6.338)$$

Two protons rotating about their common centre-of-mass with an angular momentum $\hbar$ will possess a magnetic moment of one nuclear magneton [73]. In the hydrogen molecule the presence of the negatively charged electrons reduces this value to 0.8829 nuclear magnetons [74]; the value of g_R is, indeed, the difference between a contribution from the nuclei (g_R^N) and a contribution from the electrons (g_R^e).

As we described above, magnetic dipole transition probabilities in closed shell systems are many orders-of-magnitude smaller than electric dipole probabilities. In an experiment where a spectroscopic transition is detected *directly* by measuring the absorption of electromagnetic radiation, this difference is very significant. In many other experiments, however, an *indirect* detection method which involves energy level population transfer is used. This is the case in most molecular beam magnetic or electric resonance studies, and in most double resonance investigations. Under these circumstances, optimum detection sensitivity is usually achieved by approaching saturation of the transition concerned, and this simply requires that adequate radiation power be available.

One other important difference between electric and magnetic dipole transition probabilities involves the inversion symmetry of all spatial coordinates (i.e. parity). A magnetic dipole moment is an axial vector that does not change sign under inversion, unlike an electric dipole moment. Consequently magnetic dipole transitions occur only between states of the same parity.

We shall encounter many examples of magnetic dipole spectra elsewhere in this book but note briefly here a few examples which again illustrate the importance of the Wigner–Eckart theorem in determining the selection rules. Rotational transitions in the metastable $^1\Delta_g$ state of O_2 provide an important example for an open shell system which does not possess an electric dipole moment [75]. The $^1\Delta_g$ state arises from the presence of the two highest energy electrons in degenerate π-molecular orbitals; if these orbitals are denoted π_{+1} and π_{-1} the wave functions for the $^1\Delta_g$ state may be written

$$\psi_{+2} = \frac{1}{\sqrt{2}} \pi_{+1}(1)\pi_{+1}(2)\{\alpha(1)\beta(2) - \beta(1)\alpha(2)\}, \quad \Lambda = +2$$

$$\psi_{-2} = \frac{1}{\sqrt{2}} \pi_{-1}(1)\pi_{-1}(2)\{\alpha(1)\beta(2) - \beta(1)\alpha(2)\}, \quad \Lambda = -2. \quad (6.339)$$

These two functions do not have definite parity, but the symmetric and antisymmetric combinations of them do, that is,

$$\psi_s = \frac{1}{2}\{\pi_{+1}(1)\pi_{+1}(2) + \pi_{-1}(1)\pi_{-1}(2)\}\{\alpha(1)\beta(2) - \beta(1)\alpha(2)\},$$

$$\psi_a = \frac{1}{2}\{\pi_{+1}(1)\pi_{+1}(2) - \pi_{-1}(1)\pi_{-1}(2)\}\{\alpha(1)\beta(2) - \beta(1)\alpha(2)\}. \quad (6.340)$$

These two functions are the components of a Λ-doublet but in the homonuclear molecule $^{16}O^{16}O$ symmetry requirements dictate that each rotational level can be associated with only one Λ-doublet component, and all of the rotation levels in $^1\Delta_g$ O_2 have positive parity. (We have already discussed this result in section 6.10.3). Transitions between them are obviously magnetic dipole transitions, and we can calculate their relative intensities by considering the matrix elements of the perturbation:

$$\mathcal{H}_B(t) = -\boldsymbol{\mu}_L \cdot \boldsymbol{B}(t). \quad (6.341)$$

We again expand in the space-fixed direction, choosing $p=0$ to be the direction of the oscillating magnetic field, and then rotate the orbital angular momentum into the molecule-fixed system, retaining only the $q=0$ component:

$$\mathcal{H}_B(t) = g_L \mu_B B_{p=0}(t) T^1_{p=0}(\boldsymbol{L})$$

$$= g_L \mu_B B_{p=0}(t) \mathcal{D}^{(1)}_{00}(\omega)^* T^1_{q=0}(\boldsymbol{L}) \quad (6.342)$$

$$= g_L \mu_B \Lambda B_{p=0}(t) \mathcal{D}^{(1)}_{00}(\omega)^*.$$

The matrix element we now require, using the unsymmetrised states (6.339), is

$$\langle \eta, \Lambda; J, \Lambda, M_J | \mathcal{H}_B(t) | \eta, \Lambda; J', \Lambda, M_J \rangle$$

$$= g_L \mu_B \Lambda B_0(t) \langle \eta, \Lambda; J, \Lambda, M_J | \mathcal{D}^{(1)}_{00}(\omega)^* | \eta, \Lambda; J', \Lambda, M_J \rangle$$

$$= g_L \mu_B \Lambda B_0(t) (-1)^{J-M_J} \begin{pmatrix} J & 1 & J' \\ -M_J & 0 & M_J \end{pmatrix} (-1)^{J-\Lambda}$$

$$\times \begin{pmatrix} J & 1 & J' \\ -\Lambda & 0 & \Lambda \end{pmatrix} \{(2J+1)(2J'+1)\}^{1/2}. \quad (6.343)$$

In terms of the symmetrised combinations (6.340) the matrix element between the symmetric and antisymmetric combination is zero. In other words, the magnetic dipole $\Delta J = \pm 1$ rotational transitions occur between states of the same parity, as we expect.

For O_2 in its $^3\Sigma_g^-$ ground state the rotational transitions are again magnetic dipole allowed, but in this case arise through interaction of the oscillating magnetic field with the electron spin magnetic moment,

$$\mathcal{H}_B(t) = g_S \mu_B \mathbf{T}^1(\boldsymbol{B}(t)) \cdot \mathbf{T}^1(\boldsymbol{S}). \tag{6.344}$$

The details of the rotational spectrum are described in chapter 10; only odd N rotational levels exist in the homonuclear $^{16}O^{16}O$ molecule.

Finally in this brief review of magnetic dipole transitions we look at the proton magnetic resonance spectrum of H_2 in its $J = 1$ rotational level; in this level the two proton spins are combined to form a total spin $I = 1$ (this is *ortho*-H_2). An applied static magnetic field defines the space-fixed $p = 0$ direction, the effective Hamiltonian being

$$\mathcal{H}_Z = -g_H \mu_N \mathrm{T}_0^1(\boldsymbol{B}) \mathrm{T}_0^1(\boldsymbol{I}). \tag{6.345}$$

As we show in chapter 8, the effective applied magnetic field is actually modified by a screening factor, but we ignore this for the moment. The Zeeman effect arising from (6.345) may be derived using the Wigner–Eckart theorem:

$$\langle \eta, I, M_I | \mathcal{H}_Z | \eta, I, M_I \rangle = -g_H \mu_N B_0 (-1)^{I-M_I} \begin{pmatrix} I & 1 & I \\ -M_I & 0 & M_I \end{pmatrix} \{I(I+1)(2I+1)\}^{1/2}. \tag{6.346}$$

This would imply a very simple linear Zeeman effect but, as we show in chapter 8, additional terms describing the nuclear spin–rotation interaction and the spin–spin interaction make the system much more interesting. The nuclear spin transitions are induced by an oscillating magnetic field applied perpendicular to the static magnetic field, the perturbation being represented, for example, by the term

$$\langle I, M_I | -g_N \mu_N B_{-1}(t) \mathrm{T}_1^1(\boldsymbol{I}) | I, M_I' \rangle = -g_N \mu_N B_{-1}(t) (-1)^{I-M_I}$$
$$\times \begin{pmatrix} I & 1 & I \\ -M_I & 1 & M_I - 1 \end{pmatrix} \{I(I+1)(2I+1)\}^{1/2}. \tag{6.347}$$

The selection rule given by the 3-j symbol is $\Delta M_I = +1$; there will, of course, be a similar term for $p = -1$, giving the selection rule $\Delta M_I = -1$. Magnetic dipole transitions arising from coupling of the rotational magnetic moment with the oscillating magnetic field are also possible.

Apart from electric and magnetic dipole transitions, time-dependent interactions involving higher-pole electric and magnetic moments can, in principle, occur. However, no examples of such transitions appear in this book; they are of academic rather than practical interest.

6.12. Line widths and spectroscopic resolution

6.12.1. Natural line width

Line broadening processes are important in themselves, in the information they can provide concerning the important physics of a molecular gas system. They are also important to the experimental spectroscopist in determining the spectroscopic resolution and hence the amount and accuracy of the information which can be derived. We here review briefly the principal effects which determine spectroscopic line widths in different experiments.

Many of the processes which determine line widths can be removed by appropriately designed experiments, but it is almost impossible to avoid so-called natural line broadening. This arises from the spontaneous emission process (governed by the Einstein A coefficient) described in the previous section. Spontaneous emission terminates the lifetime of the upper state involved in a transition, and the Heisenberg uncertainty principle states that the lifetime of the state (Δt) and uncertainty in its energy (ΔE) are related by the expression

$$\Delta t \cdot \Delta E \approx \hbar. \tag{6.348}$$

The corresponding spread in frequency, $\Delta \nu$, is

$$\Delta \nu = \frac{\Delta E}{h} \approx \frac{1}{2\pi \Delta t}. \tag{6.349}$$

We must therefore determine Δt and this is straightforward since it is simply equal to the inverse of the Einstein A coefficient,

$$\Delta t = \frac{1}{A} = \frac{3hc^3}{16\pi^3 \nu^3 |\mu|^2}. \tag{6.350}$$

Here ν and μ, the frequency and transition dipole moment, refer to a specific transition, but a given level can usually decay by spontaneous emission to a number of different levels. However if we take the frequency to be 10 GHz, and the dipole matrix element to be 3×10^{-30} C m, we obtain a natural line width from (6.350) of 10^{-9} Hz. In the microwave region, therefore, this contribution is negligible, but in the near ultraviolet the natural line width of an excited electronic state is of the order 1 MHz, unless the state is metastable.

6.12.2. Transit time broadening

A source of line broadening which is fairly common in molecular beam studies, and often dominant in ion beam studies is transit time broadening. This arises when the interaction time between the electromagnetic radiation field and the molecule is limited by the time the molecule spends in the radiation field. The transit time t is equal to d/v, where v is the molecular velocity and d is the length of the radiation field. A typical molecular velocity is 10^3 m s^{-1}, so that a 1 m radiation field would give a line width of the order 1 kHz. A more severe limitation arises when a laser beam crosses a molecular beam; the interaction zone can then be very short, being determined by the laser beam waist, and the transit time broadening correspondingly large. In ion beam experiments which employ ions accelerated to kilovolt potentials, transit times are of the order 1 μs, which often determines a limiting spectroscopic resolution of 1 MHz.

6.12.3. Doppler broadening

The Doppler effect plays a major role in spectroscopic resolution, in both beam and non-beam experiments. The sonic form of the Doppler shift, when a moving vehicle emits a sound heard by the stationary observer, is familiar to everyone. The electromagnetic radiation equivalent can be expressed in a very simple form. If a molecular source is moving with a velocity v relative to a receiver, and is emitting radiation of frequency ν, the observed frequency f is given by

$$f = \frac{1}{\{1/\nu + v/\nu c\}} = \frac{\nu}{\{1 + v/c\}}. \tag{6.351}$$

The observed frequency f may be smaller or larger than the emitted frequency, depending upon the direction of the moving source relative to the observer. The fractional Doppler shift in wavelength, z, is given by

$$z = \frac{\lambda - \lambda_0}{\lambda_0} = \frac{v}{c}, \tag{6.352}$$

provided v $\ll$ c; if v is sufficiently high, as with ion beam experiments involving light ions, a relativistic theory is necessary, but this is a straightforward extension. In rotational spectroscopy there is a preference for measuring frequencies rather than wavelengths. For a source moving relative to an observer, the Doppler shift in frequency, $\Delta \nu = f - \nu$, is given by

$$\Delta \nu = \nu\{-z/(1+z)\} \cong -v\nu/c. \tag{6.353}$$

We note the important result that the Doppler shift is directly proportional to the frequency.

In the microwave and radiofrequency regions of the spectrum the Doppler contribution to the line width is usually negligible because of the relatively low frequency of the electromagnetic radiation. In the infrared, visible and ultraviolet regions, however, Doppler broadening becomes increasingly important. The Doppler effect can contribute to line widths in several different ways, depending on the experimental technique. In molecular beam experiments the direction of propagation of the electromagnetic radiation is usually arranged to be perpendicular to the trajectory direction of the molecular beam. In a perfect perpendicular experiment the Doppler effect would be zero. However, even in a well-collimated molecular beam the molecular trajectories are not all perfectly parallel to the beam direction, and there is also a spread of molecular velocities within the beam. A good summary of these factors has been provided by Demtröder [76]; despite the reservations above, the Doppler effect is drastically minimised and very high resolution is obtained, particularly for electronic spectroscopy in the visible and near-ultraviolet regions of the spectrum. In a few experiments, particularly with those employing laser beams, the electromagnetic radiation is propagated parallel to and coincident with the molecular beam direction. There is then a maximised Doppler shift, often for both parallel and antiparallel radiation, and a corresponding Doppler line width.

In studies of bulk gas samples, as in conventional microwave absorption experiments, one must take account of the fact that one is studying an assembly of molecules moving in different directions at different velocities, and suffering frequent collisions which change both the velocity and direction. It may be shown that for a gas at thermal equilibrium, the Doppler full line width $\Delta\nu$ at half-height is given by

$$\Delta\nu = \frac{2\nu}{c}\left(\frac{2N_A kT \ln 2}{M}\right)^{1/2} = 7.15 \times 10^{-7}(T/M)^{1/2}\nu, \quad (6.354)$$

where N_A is Avogadro's number and M is the relative molecular mass of the molecule. For $T = 300$ and $M = 30$, at a microwave frequency of 10 GHz the Doppler width is about 23 kHz, which is small compared with other contributions to the line width. In the ultraviolet, however, the line width from bulk gas samples is usually limited by the Doppler effect, and is typically of the order of 0.1 cm^{-1} ($\sim$3000 MHz).

6.12.4. Collision broadening

The most important source of line broadening in microwave studies of bulk gas samples is collisional or pressure broadening, the theory of which was first developed by Van Vleck and Weisskopf [77]. They developed the line shape function

$$S(\nu, \nu_0) = \frac{\nu}{\pi\nu_0}\left[\frac{\Delta\nu}{(\nu-\nu_0)^2 + (\Delta\nu)^2} + \frac{\Delta\nu}{(\nu+\nu_0)^2 + (\Delta\nu)^2}\right], \quad (6.355)$$

where $\Delta\nu = 1/(2\pi\tau)$, τ being the average time between collisions and ν_0 being the resonant frequency. $\Delta\nu$ is here the half-width at half peak height. The mechanisms which operate during molecular collisions and their relationship to the intermolecular forces have been considered by Anderson [78]. In standard microwave experiments, for which $\Delta\nu \ll \nu$ and $\nu \approx \nu_0$, the second term in (6.355) can be neglected, leaving a Lorentzian line shape function

$$S(\nu, \nu_0) = \frac{1}{\pi}\left[\frac{\Delta\nu}{(\nu-\nu_0)^2 + (\Delta\nu)^2}\right]. \quad (6.356)$$

$\Delta\nu$ then becomes the half-width at half-peak height. As the gas pressure is increased, however, the second term becomes significant and the line shape departs from the simple Lorentzian.

When the populations of the two levels involved in a transition are determined by thermal equilibrium conditions, the Van Vleck–Weisskopf line shape function may be combined with the transition probability function to yield an equation giving the line absorption coefficient, which is

$$\gamma = \frac{8\pi^2 N_v f_a}{3ckT} |\mu_{ab}|^2 \nu^2 \frac{\Delta\nu}{(\nu-\nu_0)^2 + (\Delta\nu)^2}. \tag{6.357}$$

N_v is the number of molecules per unit volume and f_a is the fraction in the lower state a involved in the transition.

We see that the factors which determine line widths in microwave experiments can, to a considerable extent, be controlled. Molecular beam methods generally give the highest resolution, particularly if long transit times in the radiation field can be arranged. Double resonance experiments involving the observation of microwave transitions are subject to the same factors as those for simpler experiments, unless a radiative decay process limits the lifetime of an excited electronic state being studied.

The usual situation in molecular spectroscopy is that there is a single dominant contribution to the spectral line broadening. It sometimes happens that two contributions are of similar magnitude. For example, rotational transitions of light molecules occur in the far-infrared region of the spectrum. For sample pressures of around 1 torr, pressure broadening is similar in magnitude to Doppler broadening at these frequencies. The resultant lineshape, which is a convolution of a Lorentzian and a Gaussian line shape, is known as a Voigt profile. It cannot be described in analytic form and must be constructed numerically in simulation.

6.13. Relationships between potential functions and the vibration–rotation levels

6.13.1. Introduction

Earlier in this chapter we discussed the *ab initio* calculation of diatomic molecule potential functions; these may be expressed as the sum of the potential for a rotationless molecule, and a rotational energy term,

$$V_J(R) = V_0(R) + J(J+1)\hbar^2/2\mu R^2. \tag{6.358}$$

One of the problems we address in this section is that of calculating the vibration–rotation energies and wave functions, given a theoretical potential function for the molecule. We also tackle the reciprocal problem; given measurements of a set of vibration–rotation energies, how best is the potential function determined? The first problem has already been solved for the simple case when an analytical potential function is assumed, for example, the Morse potential, for which the vibrational energies and wave functions follow in a straightforward manner. We are now concerned with the more general and realistic problem of a theoretical potential which cannot be represented by a simple analytical expression. Among many books which deal with this topic we have found those by Child [79, 80] and Lefebvre-Brion and Field [61] to be particularly helpful.

6.13.2. The JWKB semiclassical method

The Schrödinger equation can be written in the form

$$\hbar^2 \frac{d^2\psi}{dR^2} + p^2(R)\psi = 0, \tag{6.359}$$

where $p(R)$ is the classical momentum

$$p(R) = \{2\mu[E - V_J(R)]\}^{1/2}. \tag{6.360}$$

Given the potential energy curve $V_J(R)$ it is possible, through an iterative procedure which we describe shortly, to locate the vibrational levels using the semiclassical quantisation condition

$$\int_{R_1}^{R_2} [p(R)/\hbar]\,dR = (v + 1/2)\pi = \frac{\sqrt{2\mu}}{\hbar} \int_{R_1}^{R_2} [E - V_J(R)]^{1/2}\,dR. \tag{6.361}$$

R_1 and R_2 are the lower and upper classical turning points at which the total energy E is equal to the potential energy $V_J(R)$. The integral in equation (6.361) is known as the Bohr–Sommerfeld integral and the semiclassical vibrational eigenfunctions are known as JWKB wave functions (after **Jeffreys** [81], **Wentzel** [82], **Kramers** [83] and **Brillouin** [84]). In order to see the relationship between (6.361) and the Schrödinger equation (6.359) we search for solutions to (6.359) of the form

$$\psi(R) = \exp[iS(R)/\hbar]. \tag{6.362}$$

The first- and second-derivatives which we need are

$$\frac{d\psi}{dR} = \frac{i\psi(R)S'(R)}{\hbar}, \tag{6.363}$$

$$\frac{d^2\psi}{dR^2} = i\frac{d\psi(R)}{dR}\frac{S'(R)}{\hbar} + \frac{i\psi(R)S''(R)}{\hbar}$$

$$= -\psi(R)\left\{\frac{(S'(R))^2}{\hbar^2} - \frac{iS''(R)}{\hbar}\right\}. \tag{6.364}$$

$S'(R)$ and $S''(R)$ are the first- and second-derivatives of $S(R)$ with respect to R. The one-dimensional Schrödinger equation,

$$-\frac{\hbar^2}{2\mu}\frac{d^2\psi(R)}{dR^2} = [E - V(R)]\psi(R), \tag{6.365}$$

therefore becomes

$$-\frac{\hbar^2}{2\mu}\left\{-\psi(R)\left[\frac{(S'(R))^2}{\hbar^2} - \frac{iS''(R)}{\hbar}\right]\right\} = [E - V(R)]\psi(R). \tag{6.366}$$

From this equation we obtain the result

$$(S')^2 - i\hbar S'' = 2\mu[E - V(R)]. \tag{6.367}$$

We now expand the phase function $S(R)$ in powers of $\hbar$:

$$S = S_0 + \hbar S_1 + \frac{\hbar^2}{2} S_2 + \cdots . \tag{6.368}$$

We substitute in (6.367), equate terms with like powers of $\hbar$, and obtain the result:

$$[-(S_0')^2 + p^2(R)] + \hbar[-2S_0'S_1' + iS_0''] + \hbar^2[-2S_0'S_2' - (S_1')^2 + iS_1''] + \cdots = 0. \tag{6.369}$$

The term independent of $\hbar$ gives the result

$$(S_0')^2 = 2\mu[E - V(R)] = p^2(R). \tag{6.370}$$

If we now integrate (6.370) we obtain the result

$$S_0 = \pm \int p(R)\, dR. \tag{6.371}$$

We now return to equations (6.367) and (6.368), and extract the terms which are linear in $\hbar$, obtaining the results

$$S' = S_0' + \hbar S_1', \tag{6.372}$$

$$(S')^2 = (S_0')^2 + 2\hbar S_0' S_1' + \hbar^2 (S_1')^2. \tag{6.373}$$

It follows that for terms linear in $\hbar$,

$$2S_0' S_1' - iS_0'' = 0. \tag{6.374}$$

Now from equation (6.371)

$$S_0' = p(R), \tag{6.375}$$

so that

$$2p(R) S_1' - ip'(R) = 0,$$

$$S_1' = \frac{ip'(R)}{2p(R)}. \tag{6.376}$$

The solution to equation (6.376) is

$$S_1 = i \ln[p(R)]^{1/2}. \tag{6.377}$$

Hence the solution (6.362) to the radial Schrödinger equation is a linear combination of the functions

$$\begin{aligned}
\psi_\pm(R) &= \exp[iS(R)/\hbar] \\
&= \exp[iS_0/\hbar + iS_1 + \cdots] \\
&= \exp\left\{ i/\hbar \left\{ \pm \int p(R)\, dR \right\} + i\{i \ln[p(R)]^{1/2}\} + \cdots \right\} \\
&= \exp\left\{ i/\hbar \left\{ \pm \int p(R)\, dR + i\hbar \ln[p(R)]^{1/2} \right\} + \cdots \right\}.
\end{aligned} \tag{6.378}$$

Taking the first term in the exponent, we obtain the first-order JWKB functions, appropriately normalised:

$$\psi_\pm(R) = \frac{1}{[p(R)]^{1/2}} \exp\left\{\pm \frac{i}{\hbar} \int p(R)\,dR\right\}. \tag{6.379}$$

The validity of equation (6.379) is restricted to classically accessible regions, where $p(R)$ is real. It breaks down catastrophically at any classical turning points where $p(R)$ is zero but methods are available to correct JWKB solutions around such singularities. Proper real combinations of $\psi_\pm(R)$ in the region $R_1 < R < R_2$ may be expressed as [79, 80]

$$\begin{aligned}\psi(R) &= \frac{A}{[p(R)]^{1/2}} \sin\left[\frac{1}{\hbar}\int_{R_1}^{R} p(R)\,dR + \frac{\pi}{4}\right] \\ &= \frac{A'}{[p(R)]^{1/2}} \sin\left[\frac{1}{\hbar}\int_{R}^{R_2} p(R)\,dR + \frac{\pi}{4}\right].\end{aligned} \tag{6.380}$$

The first and second forms of this wave function ensure the proper connections around $R = R_1$ and $R = R_2$ respectively. Consistency between equations (6.380) requires that

$$\frac{1}{\hbar}\int_{R_1}^{R_2} p(R)\,dR = (v + 1/2)\pi \tag{6.381}$$

which is the Bohr–Sommerfeld quantisation condition, and the coefficients A and A' are related by

$$A' = (-1)^v A. \tag{6.382}$$

The numerical evaluation of the Bohr–Sommerfeld integral in the equation,

$$(v + 1/2)\pi = \frac{\sqrt{2\mu}}{\hbar}\int_{R_1}^{R_2} [E - V_J(R)]^{1/2}\,dR \tag{6.383}$$

is carried out by a process known as Gaussian quadrature. We define the function

$$F(R) = \left(\frac{\sqrt{2\mu}}{\hbar}\right)\left\{\frac{E - V_J(R)}{(R_2 - R)(R - R_1)}\right\}^{1/2}, \tag{6.384}$$

so that

$$\{E - V_J(R)\}^{1/2} = (\hbar/\sqrt{2\mu})\{(R_2 - R)(R - R_1)\}^{1/2} F(R). \tag{6.385}$$

The JWKB integral may therefore be rewritten in the form

$$v + 1/2 = \int_{R_1}^{R_2} \{(R_2 - R)(R - R_1)\}^{1/2} F(R)\,dR. \tag{6.386}$$

The reason for the substitution is that the right-hand side of (6.386) can be rewritten by making use of a standard expansion, given by Abramowitz and Segun [85], which is

$$v + 1/2 = \left(\frac{R_2 - R_1}{2}\right)^2 \sum_{i=1}^{n} w_i F(R_i), \tag{6.387}$$

where

$$\begin{aligned}
R_i &= \frac{R_2 + R_1}{2} + \frac{R_2 - R_1}{2} x_i, \\
x_i &= \cos(i\pi/n + 1), \\
w_i &= (\pi/n + 1)(1 - x_i^2).
\end{aligned} \tag{6.388}$$

The calculation of the vibrational energies from a given potential $V(R)$ therefore requires the solution of (6.387) for integral values of v, R_2 and R_1 being the outer and inner turning points at energy E. The number of terms n in the summation in equation (6.387) is taken to achieve adequate convergence.

A numerical method for solving the Schrödinger radial equation, which makes use of JWKB wave functions, was described by Cooley [86]. This method is readily available as a computer program called LEVEL from Le Roy [87]; a helpful summary of the method is given by Cashion [88].

6.13.3. Inversion of experimental data to calculate the potential function (RKR)

The most important method of obtaining the potential function from the experimental vibrational and rotational spectroscopic term values is the RKR method (from **R**ydberg [89], **K**lein [90], and **R**ees [91]). This semi-classical procedure exploits the dependence of the quantisation condition on phase integrals which involve the underlying potential function explicitly. These phase integrals can be subjected to an Abelian transformation which provides the classical turning points at any energy. The resultant energy-dependent turning points can then be used to define the required potential function. Thus we see that the RKR potential is not generated as an analytical function. If it is required in this form, a suitable $V(R)$ function must be least-squares fitted to the turning points.

The required form of the potential function $V_J(R)$ for a diatomic molecule has already been given in equation (6.358). It can be seen that it is a function of both R and J. Let us consider the integral

$$X(E, J) = \int_{R_1(E)}^{R_2(E)} [E - V_J(R)] \, dR, \tag{6.389}$$

where R_1 and R_2 are the inner and outer classical turning points at energy E. This function has the convenient properties that

$$\frac{\partial X}{\partial E} = \int_{R_1(E)}^{R_2(E)} 1 \, dR = R_2(E) - R_1(E), \tag{6.390}$$

and

$$\begin{aligned}\frac{\partial X}{\partial J} &= \frac{(2J+1)\hbar^2}{2\mu} \int_{R_1(E)}^{R_2(E)} \frac{1}{R^2} \, dR \\ &= -(2J+1)\frac{\hbar^2}{2\mu}\left[\frac{1}{R_2(E)} - \frac{1}{R_1(E)}\right].\end{aligned} \tag{6.391}$$

Hence, for $J=0$,

$$\frac{\partial X}{\partial J} = \frac{\hbar^2}{2\mu}\left[\frac{1}{R_1(E)} - \frac{1}{R_2(E)}\right]. \tag{6.392}$$

Thus, a knowledge of the derivatives of X with respect to E and J will allow us to determine the turning points at energy E.

The experimental input into the RKR calculation comes from the vibrational energy levels $G(v)$ and the rotational constants for each level $B(v)$. If it is possible to define the integral $X(E, J)$ in terms of $G(v)$ and $B(v)$, we can use equations (6.390) and (6.392) to determine $V_0(R)$. The link is provided by the semi-classical quantisation condition, which has a form very similar to that of equation (6.389):

$$(v+1/2)\pi = \frac{\sqrt{2\mu}}{\hbar} \int_{R_1(E)}^{R_2(E)} [E - V_J(R)]^{1/2} \, dR. \tag{6.393}$$

The Abelian transformation is carried out at this point. After a certain amount of work (see, for example, Zare [92] or Miller [93]), the variable R in the integral (6.383) can be replaced by v, which we recall is continuous in the semi-classical world:

$$X(E, J) = 2\pi \int_{v(0)=v_{\min}}^{v(E)} [E - E(v, J)]^{1/2} \, dv. \tag{6.394}$$

By taking the partial derivatives of X with respect to E and J at $J = 0$, we obtain

$$\frac{\partial X}{\partial E} = \pi \int_{v_{\min}}^{v(E)} [E - G(v)]^{-1/2} \, dv, \tag{6.395}$$

$$\frac{\partial X}{\partial J} = \pi \int_{v_{\min}}^{v(E)} [E - G(v)]^{-1/2} \frac{\partial E}{\partial J} \, dv. \tag{6.396}$$

Now we know that

$$\frac{\partial E}{\partial J} = (2J+1)B(v) - (4J^3 + 6J^2 + 2J)D(v) + \cdots . \quad (6.397)$$

so that, for $J = 0$,

$$\frac{\partial X}{\partial J} = \pi \int_{v_{min}}^{v(E)} B(v)[E - G(v)]^{-1/2} \, dv. \quad (6.398)$$

Equations (6.395) and (6.398) thus provide us with the quantities which we need to determine the turning points $R_1(E)$ and $R_2(E)$ in equations (6.390) and (6.392) from a knowledge of $G(v)$ and $B(v)$. It is obvious from their form that RKR inversion can only be performed up to the energy which corresponds to the highest observed vibrational level. Furthermore, although the integrands in these equations become infinite at the upper limits of integration, the resultant singularity is integrable by various methods [94].

In the quantum mechanical derivation of the vibrational levels, the value of v_{min} at the bottom of the potential function is $-1/2$, that is,

$$G(-1/2) = 0. \quad (6.399)$$

However, this is not so in the semi-classical treatment (see, for example, Dunham [65]) for which the zero of energy is defined by

$$E_{v0} = Y_{00} + G(v), \quad (6.400)$$

where

$$Y_{00} = \frac{B_e}{4} + \frac{\alpha_e \omega_e}{12 B_e} + \left(\frac{\alpha_e \omega_e}{12 B_e}\right)^2 \frac{1}{B_e} - \frac{\omega_e x_e}{4}. \quad (6.401)$$

This quantity is the leading term in the Dunham expansion which is obtained by higher-order quantisation than that given in equation (6.358). In the semi-classical treatment therefore,

$$v_{min} \cong -\frac{1}{2} - \frac{Y_{00}}{\omega_e}. \quad (6.402)$$

Finally, we note that this discussion of the RKR method has been given in energy units (J). The equations must be divided throughout by hc if the equivalent expressions in wavenumber units, for example, are required.

6.14. Long-range near-dissociation interactions

Most spectroscopic studies involve the lowest energy vibration–rotation levels, and the determination of the values of the molecular parameters at or near the equilibrium position. This is equally true of most theoretical studies; indeed there are many published accurate *ab initio* calculations of equilibrium properties which do not even extrapolate with the correct analytical form to the dissociation asymptote. Calculations which give an accurate description of equilibrium properties can be wildly inaccurate for the long-range part of the potential. The reverse is not true, however. In order to calculate the near-dissociation vibration–rotation levels accurately, it is necessary to obtain an accurate description of the whole potential, not just the region close to the minimum. It should also be recognised that the long-range part of the potential is relevant to important aspects of the molecular physics, particularly, in the case of diatomic molecules, the reactive and non-reactive scattering of atoms.

We may define long-range as being the region where the electron clouds on the separate atoms no longer overlap, so that chemical valence forces, attractive or repulsive, are no longer significant. It is then usual to express the long-range part of the potential near the dissociation limit as a sum of inverse (integer) power terms in the interatomic separation R,

$$V(R) = D_e - \sum_n \frac{C_n}{R^n}, \qquad (6.403)$$

where D_e is the dissociation energy. The n value of the lowest-order term in this series expansion is determined by the nature of the two atoms to which the molecular state dissociates adiabatically [95]. We summarise the main rules here, but will return to some of them in more detail in due course.

$n = 1$: This occurs when both atoms are charged.

$n = 2$: This very unusual case can occur if one atom is charged and the other is in an electronic state with a permanent electric dipole moment [96].

$n = 3$: This could occur if both atoms are uncharged and in electronic states with permanent dipole moments. More commonly, it occurs in the interaction between two identical uncharged atoms in electronic states whose total angular momenta differ by one (i.e. $\Delta L = 1$). This interaction is a first-order dipole resonance [97] without a classical analogue.

$n = 4$: This is an important and common case which arises in the interaction between a charged and a neutral atom; it is often called the charge-induced dipole interaction. The coefficient C_4 in (6.403) is equal to $(1/8\pi\varepsilon_0)Z^2 e^2 \alpha$ where Ze is the charge on the ion and α is the polarisability of the neutral atom. An important example, to which we will return, is the H_2^+ long-range potential of the molecular ion. The $n = 4$ case can also arise in the interaction between an atom with a permanent electric dipole moment and a non-S-state atom with a permanent quadrupole moment.

$n = 5$: This is the first-order quadrupole-quadrupole interaction involving pairs of non-S-state uncharged atoms; theoretical values of C_5 are available for a wide range of systems. C_5 coefficients may be expressed [98] as the product of an angular factor, and the product of the expectation values [99] for the squares of the electron radii in the unfilled valence shells on the interacting atoms, $[\langle r_A^2 \rangle \langle r_B^2 \rangle]$.

$n = 6$: This arises for the London induced-dipole induced-dipole interaction, to which all interacting species are subject.

Higher values of n can also arise, but for obvious reasons will seldom be significant in determining the long-range potential.

If the relative energies and vibrational numbers of several levels lying close to the dissociation asymptote are known, it is possible to derive information about the dissociation energy and the long-range terms in the potential. The traditional method of extrapolation to the dissociation limit has been the Birge–Sponer plot [100], but a more recent theory and method has been described by LeRoy and Bernstein [101]. Their starting point is again the first-order JWKB quantum condition for the eigenvalues of a potential $V(R)$:

$$v + 1/2 = \frac{(2\mu)^{1/2}}{\pi\hbar} \int_{R_1}^{R_2} [E(v) - V(R)]^{1/2} \, dR, \qquad (6.404)$$

where $E(v)$ is the energy of level v and R_1 and R_2 are its inner and outer classical turning points. The allowed eigenvalues correspond to integer values of v, but it is convenient to treat v as a continuous variable.

We differentiate (6.404) with respect to $E(v)$ and obtain

$$\frac{dv}{dE(v)} = \frac{(2\mu)^{1/2}}{2\pi\hbar} \int_{R_1}^{R_2} [E(v) - V(R)]^{-1/2} \, dR. \qquad (6.405)$$

Now the integral will be nearly unchanged if the exact potential $V(R)$ is replaced by an approximate function which is accurate near the outer turning point $R_2(v)$; in other words the approximation

$$\frac{dv}{dE(v)} \simeq \frac{(2\mu)^{1/2}}{2\pi\hbar} \int_{0}^{R_2} \left[E(v) - D_e + \frac{C_n}{R^n} \right]^{-1/2} dR \qquad (6.406)$$

becomes increasingly valid as $E(v)$ approaches D_e, where $R_2(v)$ is given by

$$E(v) = V(R_2) \simeq D_e - \frac{C_n}{[R_2(v)]^n}. \qquad (6.407)$$

We now make the substitution

$$y = R_2(v)/R, \qquad (6.408)$$

so that (6.406) then becomes

$$\left(\frac{dv}{dE} \right) = \frac{(2\mu)^{1/2}}{2\pi\hbar} C_n^{1/n} (D_e - E(v))^{-(n+2)/2n} \int_{1}^{\infty} y^{-2}(y^n - 1)^{-1/2} \, dy. \qquad (6.409)$$

After evaluation of the integral this can be rearranged [102] to give the result

$$\left(\frac{dv}{dE} \right) = K_n (D_e - E(v))^{(n+2)/2n}, \qquad (6.410)$$

where

$$K_n = \frac{2\pi\hbar n}{(2\mu)^{1/2}} C_n^{-1/n} \frac{\Gamma(1 + 1/n)}{\Gamma(1/2 + 1/n)}. \qquad (6.411)$$

The Γ functions are tabulated by Abramowitz and Segun [85]. An alternative and more useful version of (6.410) is

$$E(v) = D_e - X_n(v_D - v)^{2n/(n-2)}, \tag{6.412}$$

where v_D is the non-integral effective value of v at the dissociation limit and

$$X_n = [(n-2)K_n/2n]^{2n/(n-2)}. \tag{6.413}$$

As a final version, equations (6.410) and (6.412) can be rearranged to give the result

$$\left(\frac{dE}{dv}\right) = \left(\frac{2n}{n-2}\right) X_n (v_D - v)^{(n+2)/(n-2)}. \tag{6.414}$$

In the application of (6.414) to experimental results, (dE/dv) can be satisfactorily approximated by $E(v+1) - E(v)$. LeRoy and Bernstein [102] have shown how very well this approach works for I_2 in its excited $B\,^3\Pi(0_u^+)$ state.

Extensions of the above method to the near-dissociation behaviour of the rotational constants have been described by LeRoy [103, 104, 105] and Stwalley [106]. The rotational constant B_v may be expressed in terms of the expectation value of R^{-2}, and expressing this expectation value semi-classically yields the result

$$B_v = \left(\frac{\hbar}{4\pi\mu c}\right) \frac{\int_{R_1}^{R_2} R^{-2}[E(v) - V(R)]^{-1/2}\,dR}{\int_{R_1}^{R_2} [E(v) - V(R)]^{-1/2}\,dR}. \tag{6.415}$$

We now substitute (6.403) into (6.415) and again replace the variable of integration by $y \equiv R_2(v)/R$, obtaining

$$B_v = \left(\frac{\hbar}{4\pi\mu c}\right) \left[\frac{D-E(v)}{C_n}\right]^{2/n} \frac{\int_1^{R_2/R_1} [y^n - 1]^{-1/2}\,dy}{\int_1^{R_2/R_1} y^{-2}[y^n - 1]^{-1/2}\,dy}. \tag{6.416}$$

The last approximation is allowing $R_1(v) \to 0$ so that $R_2(v)/R_1(v) \to \infty$; provided $n \geq 2$, equation (6.416) becomes

$$\begin{aligned} B_v &= \left(\frac{\hbar}{4\pi c\mu}\right) \left[\frac{\Gamma(1+1/n)\Gamma(1/2-1/n)}{\Gamma(1/2+1/n)\Gamma(1-1/n)}\right] \left[\frac{D-E(v)}{C_n}\right]^{2/n} \\ &= P_n[D - E(v)]^{2/n}. \end{aligned} \tag{6.417}$$

An alternative way of expressing B_v and the other rotational constants, valid in the region close to dissociation, is as follows [107]:

$$\begin{aligned} B_v &= X_1(n)(v_D - v)^{2n/(n-2)-2}, \\ D_v &= -X_2(n)(v_D - v)^{2n/(n-2)-4}, \\ H_v &= X_3(n)(v_D - v)^{2n/(n-2)-6}, \end{aligned} \tag{6.418}$$

etc.,

where

$$X_i(n) = \frac{\bar{X}_i(n)}{[(C_n)^2 \mu^n]^{1/(n-2)}}, \quad (6.419)$$

with $\bar{X}_i(n)$ being a known numerical constant [104].

Theoretical calculations of the long-range coefficients, C_n, have been described by many authors. The simplest cases, where exact solutions are possible, involve the interaction of a proton with a hydrogen atom. When the hydrogen atom is in its ground electronic state, the long-range interaction has been shown by Coulson [108] to take the form

$$\begin{aligned} E_{ab} = & -\frac{e^2}{4\pi\varepsilon_0 a_0} \left[\frac{9/4}{(R_{ab}/a_0)^4} + \frac{0}{(R_{ab}/a_0)^5} + \frac{15/2}{(R_{ab}/a_0)^6} \right. \\ & \left. + \frac{213/4}{(R_{ab}/a_0)^7} + \frac{7755/64}{(R_{ab}/a_0)^8} + \cdots \right]. \end{aligned} \quad (6.420)$$

Similar expressions have been derived by Coulson and Gillam [109] and Krogdahl [110] for the interaction of a proton with an electronically excited hydrogen atom; the interaction of a proton with a helium atom has also been studied by Krogdahl [111]. In order to calculate the long-range interaction coefficients for more complex atoms a knowledge of their static dipole polarisabilities is required; a useful list compiled from *ab initio* calculations was given by Teachout and Pack [112].

6.15. Predissociation

A vibration–rotation level of a diatomic molecule which lies above the lowest dissociation limit may be quasibound and able to undergo spontaneous dissociation into the separate atoms. This process is known as *predissociation*, and two different cases may be distinguished for diatomic molecules, as we will see shortly. Predissociation does not normally play an important role in rotational spectroscopy but merits a brief discussion here for the sake of completeness.

The two cases which arise in diatomic molecules are rotational predissociation and electronic predissociation; the latter case applies only to excited electronic states. We deal first with rotational predissociation, with can arise for either ground or excited states. The potential energy curve shown for a Morse oscillator in section 6.8 is for a rotationless ($J = 0$) molecule. For a rotating molecule, however, we must add a centrifugal term to the potential,

$$E_{\text{rot}} = \frac{\hbar^2}{2\mu R^2} J(J+1), \quad (6.421)$$

the effects of which become increasingly important as J increases. This is illustrated in figure 6.27, which shows the Morse potential of figure 6.23 with $J = 0$ and $J = 10$ (say). The curve for $J = 10$ exhibits an energy maximum, known as a centrifugal barrier, the height of which becomes progressively larger as J increases. As the figure shows, it is possible to have vibration–rotation levels which lie above the dissociation limit,

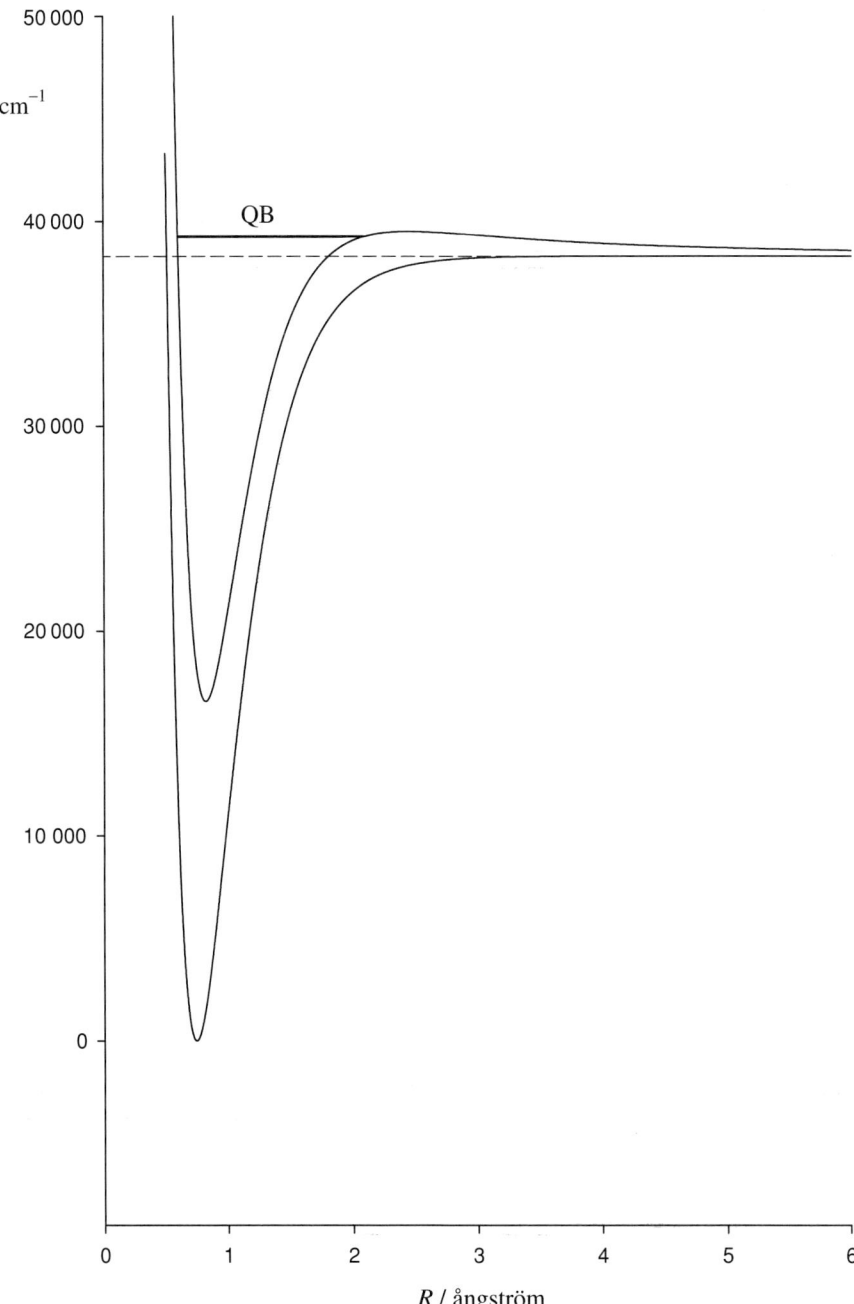

Figure 6.27. Morse potential for $J = 0$ and $J = 10$, showing the presence of the centrifugal barrier in the $J = 10$ case, and a quasibound level, QB.

but below the centrifugal barrier maximum. Such levels are metastable, because they can predissociate by tunnelling through the centrifugal barrier. The predissociation lifetimes cover an almost infinite range, and at the short lifetime end of the range they often determine spectroscopic linewidths. At the long lifetime end, quasibound vibration–rotation levels are much like normal bound levels, and there is no reason why rotational transitions between such levels should not be observed. We know of no examples of such spectra, but vibration–rotation spectra involving only rotationally quasibound levels have been observed in the HD^+ ion [113].

The second type of predissociation observed for diatomic molecules is known as electronic predissociation; the principles are illustrated in figure 6.28. A vibrational level v of a bound state E_1 lies *below* the dissociation asymptote of that state, but above the dissociation asymptote of a second state E_2. This second state, E_2, is a repulsive state which crosses the bound state E_1 as shown. The two states are mixed, and the level v can predissociate via the unbound state. It is not, in fact, necessary for the potential curves of the two states to actually cross. It is, however, necessary that they be mixed and there are a number of different interaction terms which can be responsible for the mixing. We do not go into the details here because electronic predissociation, though an important phenomenon in electronic spectroscopy, seldom plays a role in rotational spectroscopy. Since it involves excited electronic states it could certainly be involved in some double resonance cases.

The primary observable result of predissociation is, of course, line broadening. A transition involving a predissociating level has a width at half-height Γ given

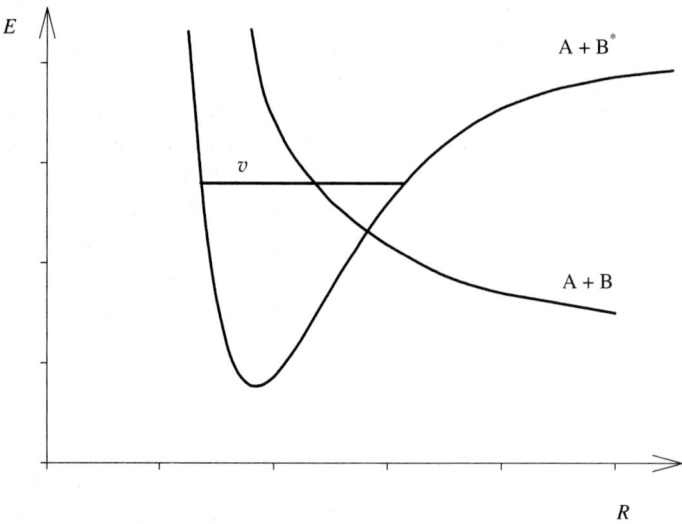

Figure 6.28. Electronic predissociation. The vibrational level v belongs to an electronic state which dissociates into a ground state atom A and an excited atom B*. The potential curve for this state crosses that of a second repulsive state which dissociates into ground state atoms A and B. Coupling between the two electronic states leads to predissociation of the level v into ground state atoms.

by

$$\Gamma(\text{cm}^{-1}) = (2\pi c\tau)^{-1} = 5.3 \times 10^{-12}/\tau(s), \tag{6.422}$$

where τ is the predissociation lifetime. In electronic spectroscopy there are dramatic examples of rotation band contours which suddenly break off when predissociation becomes exceptionally fast.

Appendix 6.1. Calculation of the Born–Oppenheimer potential for the H_2^+ ion

(i) Hamiltonian in elliptical coordinates

We shall show how the Born–Oppenheimer potential energy for the H_2^+ ion can be calculated exactly using series expansion methods, even though an exact analytical solution cannot be obtained. Figure 6.29 shows the coordinate system used for an electron moving in the field of two clamped nuclei. In atomic units the Hamiltonian is

$$\mathcal{H} = -\frac{1}{2}\nabla^2 - \frac{Z_1}{r_1} - \frac{Z_2}{r_2}, \tag{6.423}$$

where r_1 and r_2 are the distances from the electron to the nuclei, which have charges Z_1 and Z_2. We now reformulate the Hamiltonian in elliptical coordinates because, as we shall see, the Schrödinger equation is then separable into three equations, each of which involves only a single coordinate.

We deal first with the potential energy, described by the second and third terms in (6.423). If R is the distance between the two nuclei, we see from figure 6.29 that

$$r_1^2 = [z + (R/2)]^2 + y^2, \quad r_2^2 = [z - (R/2)]^2 + y^2, \tag{6.424}$$

so that

$$r_1^2 - r_2^2 = 2zR. \tag{6.425}$$

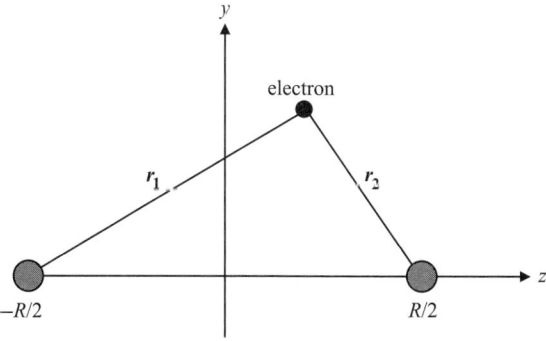

Figure 6.29. Coordinate system for the description of the H_2^+ ion, showing the two-dimensional transformation from cartesian to elliptical coordinates.

Electronic and vibrational states

We now transform to new elliptical coordinates λ and μ such that

$$\lambda = (r_1 + r_2)/R \quad \text{with } 1 \leq \lambda < \infty,$$
$$\mu = (r_1 - r_2)/R \quad \text{with } -1 \leq \mu \leq 1. \tag{6.426}$$

Hence we have

$$r_1 = (R/2)(\lambda + \mu), \quad r_2 = (R/2)(\lambda - \mu). \tag{6.427}$$

We can now rewrite the potential energy in terms of the new coordinates, as follows:

$$V = -\frac{2Z_1}{(R/2)(\lambda + \mu)} - \frac{2Z_2}{(R/2)(\lambda - \mu)} \tag{6.428}$$

$$= -\frac{4}{R}\left\{\frac{Z_1}{(\lambda + \mu)} + \frac{Z_2}{(\lambda - \mu)}\right\}$$

$$= -\frac{4}{R(\lambda^2 - \mu^2)}\{Z_1(\lambda - \mu) + Z_2(\lambda + \mu)\}. \tag{6.429}$$

Putting $Z_1 = Z_2 = 1$ for H_2^+, we obtain the final result,

$$V = -\frac{8\lambda}{R(\lambda^2 - \mu^2)}. \tag{6.430}$$

We now turn to the transformation of the Laplacian, which is rather more complicated. In cartesian coordinates we have

$$\nabla^2 = \frac{\partial^2}{\partial x^2} + \frac{\partial^2}{\partial y^2} + \frac{\partial^2}{\partial z^2}. \tag{6.431}$$

The transformation of the cartesian coordinates to elliptical coordinates is given by

$$x = (R/2)\cos\omega[(\lambda^2 - 1)(1 - \mu^2)]^{1/2},$$
$$y = (R/2)\sin\omega[(\lambda^2 - 1)(1 - \mu^2)]^{1/2}, \tag{6.432}$$
$$z = (R/2)\mu\lambda.$$

The lines of constant λ are ellipses and the lines of constant μ are hyperbolas; ω is the angle between the plane passing through all three particles and some fixed plane which passes through the two protons. By a process of straightforward but tedious partial differentiation we can obtain the following results:

$$\frac{\partial \lambda}{\partial x} = \frac{2\cos\omega\lambda[(\lambda^2 - 1)(1 - \mu^2)]^{1/2}}{R(\lambda^2 - \mu^2)}, \quad \frac{\partial \lambda}{\partial y} = \frac{2\sin\omega\lambda[(\lambda^2 - 1)(1 - \mu^2)]^{1/2}}{R(\lambda^2 - \mu^2)},$$

$$\frac{\partial \lambda}{\partial z} = \frac{2\mu(\lambda^2 - 1)}{R(\lambda^2 - \mu^2)},$$

$$\frac{\partial \mu}{\partial x} = -\frac{2\cos\omega\mu[(\lambda^2 - 1)(1 - \mu^2)]^{1/2}}{R(\lambda^2 - \mu^2)}, \quad \frac{\partial \mu}{\partial y} = -\frac{2\sin\omega\mu[(\lambda^2 - 1)(1 - \mu^2)]^{1/2}}{R(\lambda^2 - \mu^2)},$$

$$\frac{\partial \mu}{\partial z} = \frac{2\lambda(1 - \mu^2)}{R(\lambda^2 - \mu^2)},$$

$$\frac{\partial \omega}{\partial x} = -\frac{2\sin\omega}{R[(\lambda^2-1)(1-\mu^2)]^{1/2}}, \quad \frac{\partial \omega}{\partial y} = \frac{2\cos\omega}{R[(\lambda^2-1)(1-\mu^2)]^{1/2}},$$

$$\frac{\partial \omega}{\partial z} = 0. \tag{6.433}$$

We now make use of these results in evaluating the following expression:

$$\frac{\partial}{\partial x} = \frac{\partial \lambda}{\partial x} \cdot \frac{\partial}{\partial \lambda} + \frac{\partial \mu}{\partial x} \cdot \frac{\partial}{\partial \mu} + \frac{\partial \omega}{\partial x} \cdot \frac{\partial}{\partial \omega}, \tag{6.434}$$

from which

$$\frac{\partial}{\partial x} = \frac{2\cos\omega\lambda[(\lambda^2-1)(1-\mu^2)]^{1/2}}{R(\lambda^2-\mu^2)} \cdot \frac{\partial}{\partial \lambda} - \frac{2\cos\omega\mu[(\lambda^2-1)(1-\mu^2)]^{1/2}}{R(\lambda^2-\mu^2)} \cdot \frac{\partial}{\partial \mu}$$

$$- \frac{2\sin\omega}{R[(\lambda^2-1)(1-\mu^2)]^{1/2}} \cdot \frac{\partial}{\partial \omega}. \tag{6.435}$$

Since we require the expression for $\partial^2/\partial x^2$, we square (6.435) and evaluate the resulting partial differentials with respect to λ, μ, and ω; the process is straightforward but leads to a long and messy combination of terms.

The whole process is repeated for the partial differentials $\partial/\partial y$, $\partial/\partial z$ and their squares; on addition of the squares, much cancellation of terms occurs and we obtain the relatively simple result:

$$\nabla^2 = \frac{4}{R^2(\lambda^2-\mu^2)} \left\{ \frac{\partial}{\partial \lambda}(\lambda^2-1)\frac{\partial}{\partial \lambda} + \frac{\partial}{\partial \mu}(1-\mu^2)\frac{\partial}{\partial \mu} + \frac{(\lambda^2-\mu^2)}{(\lambda^2-1)(1-\mu^2)}\frac{\partial^2}{\partial \omega^2} \right\}. \tag{6.436}$$

(ii) Separation of the Schrödinger equation

Combining equations (6.436) and (6.430) we obtain the Schrödinger equation for the H_2^+ molecular ion in elliptical coordinates:

$$\frac{4}{R^2(\lambda^2-\mu^2)} \left\{ \frac{\partial}{\partial \lambda}(\lambda^2-1)\frac{\partial \Psi}{\partial \lambda} + \frac{\partial}{\partial \mu}(1-\mu^2)\frac{\partial \Psi}{\partial \mu} + \frac{(\lambda^2-\mu^2)}{(\lambda^2-1)(1-\mu^2)}\frac{\partial^2 \Psi}{\partial \omega^2} \right\}$$

$$+ \left\{ \frac{8\lambda}{R(\lambda^2-\mu^2)} + E \right\} \Psi = 0, \tag{6.437}$$

which can be rewritten in the more transparent form:

$$\frac{\partial}{\partial \lambda}(\lambda^2-1)\frac{\partial \Psi}{\partial \lambda} + \frac{\partial}{\partial \mu}(1-\mu^2)\frac{\partial \Psi}{\partial \mu} + \frac{(\lambda^2-\mu^2)}{(\lambda^2-1)(1-\mu^2)}\frac{\partial^2 \Psi}{\partial \omega^2}$$

$$+ \left\{ \frac{1}{4}R^2E(\lambda^2-\mu^2) + 2R\lambda \right\} \Psi = 0. \tag{6.438}$$

This equation has been given by a number of authors, including Bates, Ledsham and Stewart [114], Hunter, Gray and Pritchard [115], Barber and Hasse [116], and Teller and Sahlin [17].

It can now be seen at once that the substitution

$$\Psi(R, \lambda, \mu, \omega) = L(R, \lambda) M(R, \mu) N(\omega) \qquad (6.439)$$

yields three separated differential equations, which are

$$\left\{ \frac{\partial^2}{\partial \omega^2} + m^2 \right\} N = 0, \qquad (6.440)$$

$$\left\{ \frac{\partial}{\partial \lambda} (\lambda^2 - 1) \frac{\partial}{\partial \lambda} + A_1 - \frac{m^2}{(\lambda^2 - 1)} + 2R\lambda - p^2 \lambda^2 \right\} L = 0, \qquad (6.441)$$

$$\left\{ \frac{\partial}{\partial \mu} (1 - \mu^2) \frac{\partial}{\partial \mu} - A_2 - \frac{m^2}{(1 - \mu^2)} + p^2 \mu^2 \right\} M = 0, \qquad (6.442)$$

where

$$p^2 = -\frac{1}{4} E R^2. \qquad (6.443)$$

Equation (6.440) can be solved directly to yield

$$N(\omega) = \frac{1}{\sqrt{2\pi}} \exp(im\omega), \qquad (6.444)$$

where $m = 0, \pm 1, \pm 2$, etc. We now seek solutions to equations (6.441) and (6.442) which satisfy the condition $A_1 = A_2$, necessary if the separated equations are to be equivalent to the unseparated equation.

(iii) Solution of the λ equation

The λ equation can be written in the form

$$(\lambda^2 - 1) \frac{d^2 L}{d\lambda^2} + 2\lambda \frac{dL}{d\lambda} + \left[A - \frac{m^2}{(\lambda^2 - 1)} + 2R\lambda - p^2 \lambda^2 \right] L = 0. \qquad (6.445)$$

We now show that solutions may be obtained using the Jaffé expansion [117]:

$$L(\lambda) = (\lambda^2 - 1)^{m/2} (\lambda + 1)^\sigma \exp(-p\lambda) \sum_{n=0}^{\infty} b_n \left[\frac{\lambda - 1}{\lambda + 1} \right]^n, \qquad (6.446)$$

where

$$\sigma = (R/p) - m - 1 \quad \text{and} \quad p = (1/2) R |E|^{1/2}. \qquad (6.447)$$

We make use of a series of successive substitutions.
(a) Put $L = (\lambda^2 - 1)^{m/2} L_1(\lambda)$. Then

$$\frac{dL}{d\lambda} = (m/2)(\lambda^2 - 1)^{m/2 - 1} 2\lambda L_1 + (\lambda^2 - 1)^{m/2} \frac{dL_1}{d\lambda}$$

$$= m\lambda (\lambda^2 - 1)^{m/2 - 1} L_1 + (\lambda^2 - 1)^{m/2} \frac{dL_1}{d\lambda}, \qquad (6.448)$$

$$\frac{d^2 L}{d\lambda^2} = m(\lambda^2 - 1)^{m/2-1} L_1 + m\lambda\left(\frac{m}{2} - 1\right)(\lambda^2 - 1)^{m/2-2} 2\lambda L_1$$
$$+ 2m\lambda(\lambda^2 - 1)^{m/2-1}\frac{dL_1}{d\lambda} + (\lambda^2 - 1)^{m/2}\frac{d^2 L_1}{d\lambda^2}. \tag{6.449}$$

Substituting in equation (6.445) we obtain

$$\left\{(\lambda^2 - 1)\frac{d^2}{d\lambda^2} + 2\lambda(m+1)\frac{d}{d\lambda} + m(m+1) + A + 2R\lambda - p^2\lambda^2\right\} L_1 = 0. \tag{6.450}$$

(b) We now make a second substitution, $L_1(\lambda) = \exp(-p\lambda) L_2(\lambda)$.
We obtain

$$\frac{dL_1}{d\lambda} = \exp(-p\lambda)\left\{-pL_2 + \frac{dL_2}{d\lambda}\right\}, \tag{6.451}$$

$$\frac{d^2 L_1}{d\lambda^2} = \exp(-p\lambda)\left\{p^2 L_2 - 2p\frac{dL_2}{d\lambda} + \frac{d^2 L_2}{d\lambda^2}\right\}. \tag{6.452}$$

Substituting these results in (6.450) we obtain

$$\left\{(\lambda^2 - 1)\frac{d^2}{d\lambda^2} + 2[(m+1)\lambda - p(\lambda^2 - 1)]\frac{d}{d\lambda} + A + m(m+1)\right.$$
$$\left. - p^2 + [2R - 2p(m+1)]\lambda\right\} L_2(\lambda) = 0. \tag{6.453}$$

(c) We next make a change of variable.

Put $\quad \lambda = \dfrac{1+\psi}{1-\psi} \quad$ from which $\psi = \dfrac{\lambda - 1}{\lambda + 1}$. $\tag{6.454}$

Now $\quad \dfrac{d\psi}{d\lambda} = \dfrac{d}{d\lambda}\left(\dfrac{\lambda - 1}{\lambda + 1}\right) = \dfrac{2}{(\lambda + 1)^2}$. $\tag{6.455}$

Hence $\quad \dfrac{d}{d\lambda} = \dfrac{2}{(\lambda + 1)^2}\dfrac{d}{d\psi}$, $\tag{6.456}$

$$\frac{d^2}{d\lambda^2} = -\frac{4}{(\lambda + 1)^3}\frac{d}{d\psi} + \frac{4}{(\lambda + 1)^4}\frac{d^2}{d\psi^2}. \tag{6.457}$$

We now substitute for λ, $d/d\lambda$ and $d^2/d\lambda^2$ in equation (6.453) and obtain the following equation in which the variable λ has been replaced by the variable ψ:

$$\psi(1-\psi)^2\frac{d^2 L_2}{d\psi^2} + \{(m+1)(1-\psi) + \psi(1-\psi)(m-1) - 4p\psi\}\frac{dL_2}{d\psi}$$
$$+ \left\{A + m(m+1) - p^2 + 2p\sigma\frac{1+\psi}{1-\psi}\right\} L_2 = 0. \tag{6.458}$$

(d) The final term in equation (6.458) becomes singular at $\psi = 1$ and may be removed by the substitution

$$L_2(\psi) = 4(1-\psi)^{-\sigma} L_3(\psi). \tag{6.459}$$

One then finds that

$$\frac{dL_2}{d\psi} = 4\sigma(1-\psi)^{-(\sigma+1)}L_3(\psi) + 4(1-\psi)^{-\sigma}\frac{dL_3}{d\psi}, \qquad (6.460)$$

$$\frac{d^2L_2}{d\psi^2} = 4(1-\psi)^{-\sigma}\left\{\sigma(\sigma+1)(1-\psi)^{-2}L_3(\psi)\right.$$
$$\left. + 2\sigma(1-\psi)^{-1}\frac{dL_3}{d\psi} + \frac{d^2L_3}{d\psi^2}\right\}. \qquad (6.461)$$

Substituting (6.459), (6.460) and (6.461) in (6.458) we obtain

$$\psi(1-\psi)^2\frac{d^2L_3}{d\psi^2} + \{\psi^2(1-m-2\sigma) + 2\psi(\sigma-2p-1) + (m+1)\}\frac{dL_3}{d\psi}$$
$$+ \{A + 2p\sigma - p^2 + (\sigma+m)(m+1) + \sigma(\sigma+m)\psi\}L_3 = 0. \qquad (6.462)$$

(e) We assume a power series solution to equation (6.462) and make the substitution

$$L_3(\psi) = \sum_{n=0}^{\infty} b_n \psi_n = b_0 + b_1\psi + b_2\psi^2 + b_3\psi^3 + \cdots + b_{n-1}\psi^{n-1} + b_n\psi^n + \cdots. \qquad (6.463)$$

It then follows that

$$\frac{dL_3}{d\psi} = \sum_{j=0}^{n-1} b_{j+1}(j+1)\psi^j, \qquad (6.464)$$

$$\frac{d^2L_3}{d\psi^2} = \sum_{j=0}^{n-2} (j+1)(j+2)b_{j+2}\psi^j. \qquad (6.465)$$

Substituting (6.463), (6.464) and (6.465) in equation (6.462) yields the result

$$\psi(1-\psi)^2 \sum_{j=0}^{n-2}(j+1)(j+2)b_{j+2}\psi^j + \{\psi^2(1-m-2\sigma)$$
$$+ 2\psi(\sigma-2p-1) + (m+1)\}\sum_{j=0}^{n-1} b_{j+1}(j+1)\psi^j + \{A+2p\sigma - p^2$$
$$+ (\sigma+m)(m+1) + \sigma(\sigma+m)\psi\}\sum_{j=0}^{n} b_j\psi^j = 0, \qquad (6.466)$$

which can be rearranged in order of descending powers of ψ as

$$\sum_{j=0}^{n-2}(j+1)(j+2)b_{j+2}\psi^{j+3} - 2\sum_{j=0}^{n-2}(j+1)(j+2)b_{j+2}\psi^{j+2}$$
$$+ (1-m-2\sigma)\sum_{j=0}^{n-1}(j+1)b_{j+1}\psi^{j+2} + \sum_{j=0}^{n-2}(j+1)(j+2)b_{j+2}\psi^{j+1}$$
$$+ 2(\sigma-2p-1)\sum_{j=0}^{n-1}(j+1)b_{j+1}\psi^{j+1} + \sigma(\sigma+m)\sum_{j=0}^{n} b_j\psi^{j+1}$$

$$+ (m+1) \sum_{j=0}^{n-1} (j+1) b_{j+1} \psi^j + [A + 2p\sigma - p^2$$

$$+ (\sigma + m)(m+1)] \sum_{j=0}^{n} b_j \psi^j = 0. \tag{6.467}$$

We now put $j = n-3, n-2, n-1, n$ and compare coefficients of ψ^n: after some rearrangement we obtain the recurrence relationship

$$b_{n+1}\{(n+1)(n+m+1)\} + b_n\{A - p^2 + 2p\sigma + (\sigma+m)(m+1) + 2n(\sigma - 2p) - 2n^2\}$$
$$+ b_{n-1}\{(n-1-\sigma)(n-1-\sigma-m)\} = 0. \tag{6.468}$$

This result holds for all values of n except $n = 0$; we then require $b_{-1} = 0$. In this special case

$$b_1(m+1) + b_0\{A - p^2 + 2p\sigma + (\sigma + m)(m+1)\} = 0. \tag{6.469}$$

This completes the solution of the λ equation. Combining all the replacements made above we see that the complete solution is of the form

$$L(\lambda) = (\lambda^2 - 1)^{m/2} (\lambda + 1)^\sigma \exp(-p\lambda) \sum_{n=0}^{\infty} b_n \left[\frac{\lambda - 1}{\lambda + 1} \right]^2. \tag{6.470}$$

This is known as the Jaffé series [117]. The important recurrence relations (6.468) can be represented by the matrix eigenvalue equation [118]

$$\boldsymbol{B} \cdot \boldsymbol{b} = -A\boldsymbol{b}, \tag{6.471}$$

where $\boldsymbol{b}$ is the column vector $\{b_i(R)\}$ and the matrix elements of $\boldsymbol{B}$ are

$$\begin{aligned}
B_{j,j-1} &= (j-1-\sigma)(j-1-\sigma-m) \quad j \geq 1 \\
B_{j,j} &= 2j(\sigma - 2p - j) + m(m+\sigma+1) + \sigma(1+2p) - p^2 \quad j \geq 0 \\
B_{j,j+1} &= (j+1)(j+1+m) \quad j \geq 0 \\
B_{j,k} &= 0 \quad |j-k| > 1.
\end{aligned} \tag{6.472}$$

(iv) Solution of the μ equation

The μ equation obtained earlier, (6.442), was

$$\left\{ \frac{d}{d\mu}(1-\mu^2)\frac{d}{d\mu} - A - \frac{m^2}{(1-\mu^2)} + p^2\mu^2 \right\} M(\mu) = 0, \tag{6.473}$$

which may be rewritten in the form

$$(\mu^2 - 1)\frac{d^2 M}{d\mu^2} + 2\mu \frac{dM}{d\mu} + \left[A + \frac{m^2}{(1-\mu^2)} - p^2\mu^2 \right] M = 0. \tag{6.474}$$

We now outline two different methods for dealing with the μ equation. In the first method we make a first substitution

$$M(\mu) = (1 - \mu^2)^{m/2} M_1(\mu). \tag{6.475}$$

It then follows that

$$\frac{dM}{d\mu} = \left(\frac{m}{2}\right)(1 - \mu^2)^{m/2-1}(-2\mu)M_1 + (1 - \mu^2)^{m/2}\frac{dM_1}{d\mu}, \tag{6.476}$$

$$\frac{d^2M}{d\mu^2} = 2m\mu^2\left(\frac{m}{2} - 1\right)(1 - \mu^2)^{m/2-2}M_1 - m(1 - \mu^2)^{m/2-1}M_1$$

$$- 2m\mu(1 - \mu^2)^{m/2-1}\frac{dM_1}{d\mu} + (1 - \mu^2)^{m/2}\frac{d^2M_1}{d\mu^2}. \tag{6.477}$$

Substituting for M, $dM/d\mu$ and $d^2M/d\mu^2$ in equation (6.474) we obtain

$$(\mu^2 - 1)\frac{d^2M_1}{d\mu^2} + 2\mu(m+1)\frac{dM_1}{d\mu} + [A + m(m+1) - p^2\mu^2]M_1 = 0. \tag{6.478}$$

We now make the series substitution

$$M_1(\mu) = \sum_n a_n \mu^n, \tag{6.479}$$

from which we find

$$\frac{dM_1}{d\mu} = \sum_{j=0}^{n-1} a_{j+1}(j+1)\mu^j, \tag{6.480}$$

$$\frac{d^2M_1}{d\mu^2} = \sum_{j=0}^{n-2} a_{j+2}(j+2)(j+1)a_{j+2}\mu^j. \tag{6.481}$$

Substitution of (6.479), (6.480) and (6.481) in (6.478) gives the result

$$\sum_{j=0}^{n-2}(j+1)(j+2)a_{j+2}\mu^{j+2} - p^2\sum_{j=0}^{n}a_j\mu^{j+2} + 2(m+1)\sum_{j=0}^{n-1}(j+1)a_{j+1}\mu^{j+1}$$

$$- \sum_{j=0}^{n-2}(j+1)(j+2)a_{j+2}\mu^j + \{A + m(m+1)\}\sum_{j=0}^{n}a_j\mu^j = 0. \tag{6.482}$$

Comparing coefficients of μ^n gives the recursion relationship

$$n(n-1)a_n - p^2 a_{n-2} + 2n(m+1)a_n$$
$$- (n+1)(n+2)a_{n+2} + \{A + m(m+1)\}a_n = 0. \tag{6.483}$$

This relationship holds for all a_n with n odd, or all a_n with n even; a_{-2} must be zero for even solutions, and a_{-1} must be zero for odd solutions. Hence

$$2a_2 = a_0[A + m(m+1)], \tag{6.484}$$

$$6a_3 = a_1[A + (m+1)(m+2)]. \tag{6.485}$$

The second method of dealing with the μ equation is used by Wind [119], but was probably first employed by Hylleraas [120]. Returning to equation (6.474) we expand $M(\mu)$ in terms of Legendre functions:

$$M(\mu) = \sum_{\ell=0}^{\infty} c_\ell P_\ell(\mu). \qquad (6.486)$$

There is a standard recursion relationship involving the differentials of Legendre functions, which is

$$(1-\mu^2)\frac{d^2 P_\ell(\mu)}{d\mu^2} - 2\mu\frac{dP_\ell(\mu)}{d\mu} + \ell(\ell+1)P_\ell(\mu) = 0. \qquad (6.487)$$

When (6.486) is used to replace $M(\mu)$ and its derivatives in equation (6.474), the recursion relationship (6.487) leads to a further recurrence relationship for the coefficients c_ℓ which is

$$\frac{\ell(\ell-1)}{(2\ell-3)(2\ell-1)}p^2 c_{\ell-2}$$
$$+ \left\{-A - \ell(\ell+1) + \left[\frac{(\ell+1)^2}{(2\ell+3)(2\ell+1)} + \frac{\ell^2}{(2\ell+1)(2\ell-1)}\right]p^2\right\} c_\ell$$
$$+ \frac{(\ell+2)(\ell+1)}{(2\ell+5)(2\ell+3)}p^2 c_{\ell+2} = 0. \qquad (6.488)$$

(v) Calculation of the potential energy curve

For the $1s\sigma_g$ ground state of H_2^+ the value of m is zero, and ℓ in (6.488) must be even. One of the problems in dealing with the series solutions of the λ and μ equations is that one does not know where to truncate; Wind [119] chose to truncate the $M(\mu)$ series at ten terms, the criterion being that addition of further terms does not affect the final results. The recurrence relation (6.488) is a set of homogeneous linear equations, in which the coefficient matrix determinant must be zero. We choose a value of R, an input value of p, and find the value of A for which the determinant is zero. The value of the determinant is extremely sensitive to the value of A, and changes sign as it passes through the required null value. This exercise is repeated for the λ equation; acceptable solutions to the problem require that the A value be the same for both the λ and μ equations. When this condition is satisfied, the corresponding value of p gives the energy E for the chosen value of R since, we recall from (6.443), $p^2 = -(1/4)ER^2$. The potential energy curve is then obtained by repeating the calculations for an appropriate series of R values; curves for the ground state and first excited state are shown in figure 6.30. From the numerical viewpoint the calculations are trivial for a modern desk top or portable personal computer.

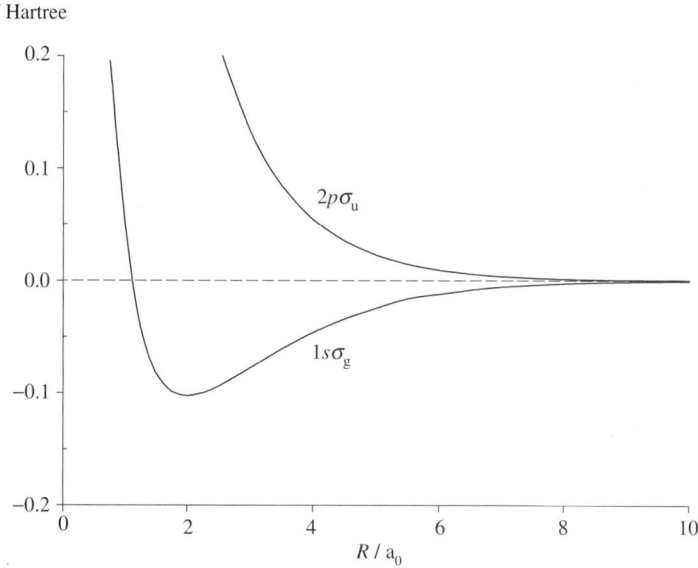

Figure 6.30. Potential energy curves for the ground and first excited electronic states of the H_2^+ molecular ion, calculated within the Born–Oppenheimer approximation.

References

[1] J.C. Slater, *Quantum Theory of Atomic Structure*, Vol. II, McGraw-Hill Book Company, New York, 1960.
[2] E.U. Condon and G.H. Shortley, *The Theory of Atomic Spectra*, Cambridge University Press, 1963.
[3] T.P. Softley, *Atomic Spectra*, Oxford University Press, 1994.
[4] W.G. Richards and P.R. Scott, *Energy Levels in Atoms and Molecules*, Oxford University Press, 1994.
[5] L. Pauling and E.B. Wilson, *Introduction to Quantum Mechanics*, McGraw-Hill Book Company, 1935.
[6] B.R. Judd, *Operator Techniques in Atomic Spectroscopy*, McGraw-Hill Book Company, 1963.
[7] W. Kauzmann, *Quantum Chemistry*, Academic Press Inc., New York, 1957, p. 287.
[8] W.H. Flygare, *Molecular Structure and Dynamics*, Prentice-Hall, Inc., New Jersey, 1978, p. 268.
[9] E.A. Hylleraas, *Z. Physik.*, **48**, 469 (1928); **54**, 347 (1929); **65**, 209 (1930).
[10] R.G. Parr, *The Quantum Theory of Molecular Electronic Structure*, W.A. Benjamin, Inc., New York, 1963.
[11] D.R. Hartree, *The Calculation of Atomic Structures*, Wiley, New York, 1957.
[12] C.C.J. Roothaan, *Rev. Mod. Phys.*, **23**, 69 (1951).
[13] E. Clementi and C. Roetti, *At. Data Nucl. Data Tables*, **14**, 177 (1974).
[14] G. Herzberg, *Spectra of Diatomic Molecules*, D. Van Nostrand Company, Inc., Princeton, New Jersey, 1950.

[15] R.S. Mulliken, *Rev. Mod. Phys.*, **2**, 60 (1930); **2**, 506 (1930); **3**, 90 (1931); **4**, 1 (1932).
[16] E. Wigner and E.E. Witmer, *Z. Physik*, **51**, 859 (1928).
[17] E. Teller and H.L. Sahlin, *Physical Chemistry, An Advanced Treatise*, ed. H. Eyring, Vol. 5, Academic Press, New York, 1970.
[18] H. Eyring, J. Walter and G.F. Kimball, *Quantum Chemistry*, John Wiley and Sons, Inc., New York, 1944.
[19] C.A. Coulson, *Trans. Faraday Soc.*, **33**, 1479 (1937).
[20] H.M. James and A.S. Coolidge, *J. Chem. Phys.*, **1**, 825 (1933).
[21] W. Kolos and C.C.J. Roothaan, *Rev. Mod. Phys.*, **32**, 219 (1960).
[22] F.E. Harris, *J. Chem. Phys.*, **27**, 812 (1957).
[23] E.R. Davidson, *J. Chem. Phys.*, **35**, 1189 (1961).
[24] W. Heitler and F. London, *Z. Physik*, **44**, 455 (1927).
[25] S. Weinbaum, *J. Chem. Phys.*, **1**, 593 (1933).
[26] A.D. McLean, A. Weiss and M. Yoshimine, *Rev. Mod. Phys.*, **32**, 211 (1960).
[27] W. Kolos and L. Wolniewicz, *J. Chem. Phys.*, **41**, 3663 (1964); **43**, 2429 (1965); **49**, 404 (1968); **51**, 1417 (1969); **45**, 509 (1966); **48**, 3672 (1968).
[28] M. Born and R. Oppenheimer, *Ann. Physik*, **84**, 457 (1927).
[29] J.A. Pople, *Rev. Mod. Phys.*, **71**, 1267 (1998).
[30] G.G. Hall, *Proc. R. Soc. Lond.*, **A205**, 541 (1951).
[31] J.A. Pople and R.K. Nesbet, *J. Chem. Phys.*, **22**, 571 (1954).
[32] S.F. Boys, *Proc. R. Soc. Lond.*, **A200**, 542 (1950).
[33] W.J. Hehre, L. Radom, P.v.R. Schleyer and J.A. Pople, *Ab initio Molecular Orbital Theory*, John Wiley and Sons, Inc., New York, 1986.
[34] T. Bally and W.T. Borden, *Revs. Comp. Chem.*, **13**, 1 (1999).
[35] C. Moeller and M.S. Plesset, *Phys. Rev.*, **46**, 618 (1934).
[36] R. Krishnan, M.J. Frisch and J.A. Pople, *J. Chem. Phys.*, **72**, 4244 (1980).
[37] R. Krishnan and J.A. Pople, *Int. J. Quantum Chem.*, **14**, 91 (1978).
[38] W. Kohn, *Rev. Mod. Phys.*, **71**, 1253 (1999).
[39] P. Hohenberg and W. Kohn, *Phys. Rev.*, **136**, B864 (1964).
[40] W. Kohn and L.J. Sham, *Phys. Rev.*, **140**, A1133 (1965).
[41] A.D. Becke, *Phys. Rev.*, **38**, A3098 (1988).
[42] D.M. Bishop and L.M. Cheung, *Adv. Quant. Chem.*, **12**, 1 (1980).
[43] A. Carrington and R.A. Kennedy, *Gas Phase Ion Chemistry*, ed. M.T. Bowers, vol. 3., p. 393, 1984.
[44] A. Carrington, I.R. McNab and C.A. Montgomerie, *J. Phys. B. At. Mol. Opt. Phys.*, **22**, 3551 (1989).
[45] C.A. Leach and R.E. Moss, *Ann. Rev. Phys. Chem.*, **46**, 55 (1995).
[46] M. Born and K. Huang, *Dynamical Theory of Crystal Lattices*, Oxford University Press, London, 1954.
[47] D.M. Bishop and R.W. Wetmore, *Mol. Phys.*, **26**, 145 (1973).
[48] D.M. Bishop and L.M. Cheung, *Phys. Rev.*, **A16**, 640 (1977).
[49] D.M. Bishop and S.A. Solunac, *Phys. Rev. Lett.*, **55**, 1986 (1985).
[50] A. Carrington and R.A. Kennedy, *Mol. Phys.*, **56**, 935 (1985).
[51] R.E. Moss and I.A. Sadler, *Mol. Phys.*, **61**, 905 (1987); **64**, 165 (1988); **66**, 591 (1989).

[52] M.H. Howells and R.A. Kennedy, *J. Chem. Soc. Faraday Trans.*, **86**, 3495 (1990).
[53] R. Bukowski, B. Jeziorski, R. Moszynski and W. Kolos, *Int. J. Quant. Chem.*, **42**, 287 (1992).
[54] D.R. Bates and T.R. Carson, *Proc. R. Soc. Lond.*, **A234**, 207 (1956).
[55] D.M. Bishop and L.M. Cheung, *Phys. Rev.*, **A18**, 1846 (1978).
[56] D.M. Bishop and S. Shih, *J. Chem. Phys.*, **61**, 162 (1976); **67**, 4313 (1977).
[57] F. Hund, *Handbuch der Physik*, **24**, 561 (1933).
[58] A. Carrington, J.M. Hutson, M.M. Law, C.A. Leach, A.J. Marr, A.M. Shaw and M.R. Viant, *J. Chem. Phys.*, **102**, 2379 (1995).
[59] R.S. Mulliken, *J. Chem. Phys.*, **55**, 288 (1971).
[60] A. Carrington, J.M. Hutson, M.M. Law, C.H. Pyne, A.M. Shaw and S.M. Taylor, *J. Chem. Phys.*, **105**, 8602 (1996).
[61] H. Lefebvre-Brion and R.W. Field, *Perturbations in the Spectra of Diatomic Molecules*, Academic Press Inc., Orlando, 1986.
[62] T.M. Dunn, *Molecular Spectroscopy: Modern Research*, eds. K.N. Rao and C.W. Mathews, Academic Press, New York, 1972.
[63] P.M. Morse, *Phys. Rev.*, **34**, 57 (1929).
[64] H.M. Hulburt and J.O. Hirschfelder, *J. Chem. Phys.*, **9**, 61 (1941).
[65] J.L. Dunham, *Phys. Rev.*, **41**, 721 (1932).
[66] E.A. Hylleraas, *Z. Physik*, **96**, 643 (1935); **96**, 599 (1936).
[67] C.L. Pekeris, *Phys. Rev.*, **45**, 98 (1934).
[68] J.H. Van Vleck, *Phys. Rev.*, **33**, 467 (1929).
[69] R.S. Mulliken and A. Christy, *Phys. Rev.*, **38**, 87 (1931).
[70] J.M. Brown, J.T. Hougen, K.-P. Huber, J.W.C. Johns, I. Kopp, H. Lefebvre-Brion, A.J. Merer, D.A. Ramsay, J. Rostas and R.N. Zare, *J. Mol. Spectrosc.*, **55**, 500 (1975).
[71] P.R. Bunker and Per Jensen, *Molecular Symmetry and Spectroscopy*, NRC Research Press, Ottawa, 1998.
[72] A. Corney, *Atomic and Laser Spectroscopy*, Clarendon Press, Oxford, 1977.
[73] N.F. Ramsey, *Molecular Beams*, Oxford University Press, London, 1956.
[74] N.F. Ramsey, *Phys. Rev.*, **58**, 226 (1940).
[75] C.A. Arrington, A.M. Falick and R.J. Meyers, *J. Chem. Phys.*, **55**, 909 (1971).
[76] W. Demtröder, *Atomic and Molecular Beam Methods*, vol. 2, ed. G. Scoles, Oxford University Press, 1992.
[77] J.H. Van Vleck and V.F. Weisskopf, *Revs. Mod. Phys.*, **17**, 227 (1945).
[78] P.W. Anderson, *Phys. Rev.*, **76**, 647, 471A (1949).
[79] M.S. Child, *Semiclassical Mechanics with Molecular Applications*, Clarendon Press, Oxford, 1991.
[80] M.S. Child, *Molecular Collision Theory*, Appendix C, Academic Press, London, 1974.
[81] H. Jeffreys, *Proc. London Math. Soc.*, **23**, 428 (1925).
[82] G. Wentzel, *Z. Phys.*, **38**, 518 (1926).
[83] H.A. Kramers, *Z. Phys.*, **39**, 828 (1926).
[84] L. Brillouin, *CR Acad. Sci., Paris*, **183**, 24 (1926).
[85] M. Abramowitz and I.A. Segun, *Handbook of Mathematical Functions*, Dover Publications, Inc, New York, 1964, p. 889.

[86] J.W. Cooley, *Math. Computations*, **15**, 363 (1961).
[87] R.J. Le Roy, *Guelph–Waterloo Centre for Graduate Work in Chemistry, University of Waterloo, Ontario N2L 3G1, Canada.*
[88] J.K. Cashion, *J. Chem. Phys.*, **39**, 1872 (1963).
[89] R. Rydberg, *Ann. Physik*, **73**, 376 (1931).
[90] O. Klein, *Z. Physik*, **76**, 226 (1932).
[91] A.L.G. Rees, *Proc. Phys. Soc.*, **A59**, 998 (1947).
[92] R.N. Zare, *J. Chem. Phys.*, **40**, 1934 (1964).
[93] W.H. Miller, *J. Chem. Phys.*, **54**, 4174 (1971).
[94] A.W. Mantz, J.K.G. Watson, K.N. Rao, D.L. Albritton, A.L. Schmeltekopf and R.N. Zare, *J. Mol. Spectrosc.*, **39**, 180 (1971).
[95] J.O. Hirschfelder, C.F. Curtiss and R.B. Bird, *Molecular Theory of Gases and Liquids*, John Wiley and Sons, Inc., New York, 1954.
[96] J.O. Hirschfelder and W.J. Meath, *Adv. Chem. Phys.*, **12**, 3 (1967).
[97] G.W. King and J.H. Van Vleck, *Phys. Rev.*, **55**, 1165 (1939).
[98] J.K. Knipp, *Phys. Rev.*, **53**, 734 (1938).
[99] C.F. Fischer, *Can. J. Phys.*, **46**, 2336 (1968).
[100] R.T. Birge and H. Sponer, *Phys. Rev.*, **28**, 259 (1926).
[101] R.J. LeRoy and R.B. Bernstein, *J. Chem. Phys.*, **52**, 3869 (1970).
[102] R.J. LeRoy and R.B. Bernstein, *J. Mol. Spectrosc.*, **37**, 109 (1971).
[103] R.J. LeRoy, *Semiclassical Methods in Molecular Scattering and Spectroscopy*, ed. M.S. Child, NATO ASI Series C, Ch. 3, Reidel, Dordrecht (1980).
[104] R.J. LeRoy, *Can. J. Phys.*, **50**, 953 (1972).
[105] R.J. LeRoy, *J. Chem. Phys.*, **73**, 6003 (1980).
[106] W.C. Stwalley, *Contemp. Phys.*, **1**, 65 (1978).
[107] J.W. Tromp and R.J. LeRoy, *Can. J. Phys.*, **60**, 26 (1982).
[108] C.A. Coulson, *Proc. R. Soc. Edinburgh*, **A61**, 21 (1941).
[109] C.A. Coulson and C.M. Gillam, *Proc. R. Soc. Edinburgh*, **A62**, 360 (1948).
[110] M.K. Krogdahl, *Astrophys. J.*, **100**, 311 (1944).
[111] M.K. Krogdahl, *Astrophys. J.*, **100**, 333 (1944).
[112] R.R. Teachout and R.T. Pack, *Atomic Data*, **3**, 195 (1971).
[113] A. Carrington, C.A. Leach, A.J. Marr, R.E. Moss, C.H. Pyne, M.R. Viant, Y.D. West, R.A. Kennedy and I.R. McNab, *Chem. Phys.*, **166**, 145 (1992).
[114] D.R. Bates, K. Ledsham and A.L. Stewart, *Phil. Trans. R. Soc. Lond.*, **A246**, 215 (1954).
[115] G. Hunter, B.F. Gray and H.O. Pritchard, *J. Chem. Phys.*, **45**, 3806 (1966).
[116] W.G. Barber and H.R. Hasse, *Proc. Phil. Soc.*, **31**, 564 (1935).
[117] G. Jaffé, *Z. Physik.*, **87**, 535 (1934).
[118] G. Hunter and H.O. Pritchard, *J. Chem. Phys.*, **46**, 2146 (1967).
[119] H. Wind, *J. Chem Phys.*, **42**, 2371 (1965).
[120] E.A. Hylleraas, *Z. Physik* **71**, 739 (1931).

7 Derivation of the effective Hamiltonian

7.1. Introduction

The Born–Oppenheimer approximation [1] is an important linch pin in the description of molecular energy levels. It reveals the difference between electronic and nuclear motions in a molecule, as a result of which we expect the separation between different electronic states to be much larger than that between vibrational levels within an electronic state. An extension of these ideas shows that the separation between vibrational levels is correspondingly larger than the separation between the rotational levels of a molecule. We thus have a hierarchy of energy levels which reveals itself in the electronic, vibrational and rotational structure of molecular spectra. This gradation in the magnitude of the different types of quanta also provides the inspiration for an energy operator known as the effective Hamiltonian.

In this chapter we introduce and derive the effective Hamiltonian for a diatomic molecule. The effective Hamiltonian operates only within the levels (rotational, spin and hyperfine) of a single vibrational level of the particular electronic state of interest. It is derived from the full Hamiltonian described in the previous chapters by absorbing the effects of off-diagonal matrix elements, which link the vibronic level of interest to other vibrational and electronic states, by a perturbation procedure. It has the same eigenvalues as the full Hamiltonian, at least to within some prescribed accuracy.

The motivation for constructing the effective Hamiltonian is one of economy and perhaps even of feasibility. It reproduces the eigenstates of the vibronic state of interest but with a much smaller representation than that of the full Hamiltonian. The effective Hamiltonian provides a natural resting point in the journey from experiment to theory. It permits data to be fitted in an unprejudiced fashion, the parameters being determined by statistical criteria only. These parameters in turn can be interpreted in terms of various theoretical models for the electronic states, and provide a point of comparison for *ab initio* calculations. A soundly based effective Hamiltonian makes allowance for all possible admixtures of electronic states; the relative importance of the perturbations by these different states is determined by a detailed comparison of the parameter values with theoretical predictions. In this way, the task of data fitting is clearly separated from that of theoretical interpretation.

It took several decades for the effective Hamiltonian to evolve to its modern form. It will come as no surprise to learn that Van Vleck played an important part in this development; for example, he was the first to describe the form of the operator for a polyatomic molecule with quantised orbital angular momentum [2]. The present formulation owes much to the derivation of the effective spin Hamiltonian by Pryce [3] and Griffith [4]. Miller published a pivotal paper in 1969 [5] in which he built on these ideas to show how a general effective Hamiltonian for a diatomic molecule can be constructed. He has applied his approach in a number of specific situations, for example, to the description of N_2 in its $A\ ^3\Sigma_u^+$ state [6], described in chapter 8. In this book, we follow the treatment of Brown, Colbourn, Watson and Wayne [7], except that we incorporate spherical tensor methods where advantageous. It is a strange fact that the standard form of the effective Hamiltonian for a polyatomic molecule [2] was established many years before that for a diatomic molecule [7].

7.2. Derivation of the effective Hamiltonian by degenerate perturbation theory: general principles

We commence with some groundwork which will enable us to define the problem and state our ultimate goal. Although the method we present in this section is completely general for systems with discrete eigenstates, consideration of a particular example is often a helpful way of appreciating the principles involved. With this in mind, we shall exemplify points in this section by outlining how the method may be applied to derive an effective Hamiltonian which operates entirely within a given vibronic state. The actual details of how the method can be applied to this as well as the other cases of interest are given later in this chapter.

It is always possible to divide the total Hamiltonian $\mathcal{H}$ into a major part $\mathcal{H}^0$ (the zeroth-order Hamiltonian) and a perturbation $\lambda\mathcal{H}'$:

$$\mathcal{H} = \mathcal{H}^0 + \lambda\mathcal{H}', \tag{7.1}$$

where λ is a dimensionless parameter in the range $0 \leq \lambda \leq 1$. Thus, for example, in the derivation of an effective Hamiltonian operating in a particular vibronic state, $\mathcal{H}^0$ might be comprised of the non-relativistic electronic kinetic energy operators, the vibrational kinetic energy and the electrostatic potential terms; $\lambda\mathcal{H}'$ would then be composed of all the remaining terms in the total Hamiltonian. It is now helpful to introduce two different sorts of quantum number. The first sort, which we denote in general by η, label the distinct eigenstates of the zeroth-order Hamiltonian; each of these eigenstates is g_η-fold degenerate. In our particular example, η would represent the quantum numbers required to specify a particular vibronic state; the energy separation between these states is typically between 10^3 and 10^4 cm^{-1}. Following a well-established convention, the lowest of these vibronic states (the ground state) is labelled $|0\rangle$. The second sort of quantum number, denoted in general by i, identifies the individual substates in a particular state η. In our example, i would represent the rotational, electron spin and nuclear spin quantum numbers. Substates distinguished by different i are, of course, degenerate in the solution of $\mathcal{H}^0$. However, there are perturbing terms in the Hamiltonian which connect different vibronic states. These terms lift the degeneracy of some or all of the g_η states by amounts which are typically around 1 cm^{-1}. Although these terms in $\lambda\mathcal{H}'$ are usually small compared with the separation of different vibronic states, their effects can dominate high-resolution microwave or radiofrequency spectra.

Next, we show how these two types of quantum number are used to label the eigenstates involved. The zeroth-order eigenstates and eigenvalues are defined by

$$\mathcal{H}^0 |0, i\rangle^0 = E_0^0 |0, i\rangle^0, \qquad (7.2)$$

$$\vdots$$

$$\mathcal{H}^0 |\eta, i\rangle^0 = E_\eta^0 |\eta, i\rangle^0, \qquad (7.3)$$

$$\vdots$$

We note that each of these equations actually represents g_η equations, corresponding to the g_η distinct values of i in each state, η. These g_η functions $|\eta, i\rangle^0$ are, for convenience, chosen to be orthonormal to each other, that is,

$$^0\langle \eta, i \mid \eta, j\rangle^0 = \delta_{ij}. \qquad (7.4)$$

In addition, the functions $|\eta, i\rangle^0$ define what is called a g_η-fold subspace of total Hilbert space. Furthermore, the eigenfunctions of two states with different values of η must be orthogonal since $\mathcal{H}^0$ is a Hermitian operator. Once again we normalise these functions to unity so that (7.4) may be generalised to

$$^0\langle \eta, i \mid \eta', j\rangle^0 = \delta_{\eta\eta'}\delta_{ij}. \qquad (7.5)$$

Thus the members of the orthonormal set $|\eta, i\rangle^0$ are all orthogonal to those of the orthonormal set $|\eta', j\rangle^0$ for $\eta' \neq \eta$; correspondingly the subspaces defined by these two sets are said to be orthogonal to each other. For example, all the subspaces defined by $|\eta, i\rangle^0$ with $\eta \neq 0$ (i.e. excited vibronic states) are orthogonal to the subspace of the ground vibronic state. The sum total of all these subspaces with $\eta \neq 0$ (itself a subspace of total Hilbert space) is called the subspace complementary to the subspace of the ground vibronic state.

Besides these zeroth-order eigenfunctions, we shall also be interested in the eigenfunctions of the total Hamiltonian $\mathcal{H}$. In what follows, we shall assume that the effects of the perturbation $\lambda\mathcal{H}'$ are small enough that we can correlate the exact g_η eigenstates with their parent zero-order eigenstate η; thus we can usefully retain the quantum numbers η to define the exact eigenstates. (This is convenient but not essential). Thus after including the effects of the perturbation $\lambda\mathcal{H}'$, we have

$$\mathcal{H}|0, k\rangle = \left(E_0^0 + E_{0k}\right)|0, k\rangle, \qquad (7.6)$$

$$\vdots$$

$$\mathcal{H}|\eta, k\rangle = \left(E_\eta^0 + E_{\eta k}\right)|\eta, k\rangle, \qquad (7.7)$$

where the functions $|\eta, k\rangle$ are orthogonal and normalised to unity. Here we distinguish the g_η substates of a given η state by the quantum number k. There are, of course, g_η distinct values of k but we use it here rather than i to emphasise, as we shall see shortly, that there is not a one-to-one correspondence between the g_η exact eigenstates $|\eta, k\rangle$ and the g_η zeroth-order states $|\eta, i\rangle^0$ from which they arise. Since the sum total of the zeroth-order eigenfunctions $|\eta, i\rangle^0$ form a complete basis set (spanning total Hilbert space), we can expand each exact $|0, k\rangle$ eigenfunction in terms of this basis set:

$$|0, k\rangle = \sum_{\eta,i} c_{\eta i}^k |\eta, i\rangle^0. \tag{7.8}$$

The coefficients $c_{\eta i}^k$ can, in principle, be determined by solving the equations (7.6), (7.7), etc. Finally, we note that the states $|0, k\rangle$ may all be non-degenerate.

Now in microwave or radiofrequency spectroscopic investigations we are generally interested in groups of levels characterised by different values of one or two quantum numbers of the i type, but with the vibronic quantum number η remaining unchanged. For example, in rotational spectroscopy we measure primarily the energy dependence on rotational quantum number within a particular vibronic state, usually the ground state. It would be extremely inconvenient to have to work with the complete Hamiltonian $\mathcal{H}$ and to try to solve (7.6), (7.7), etc., completely; even with present day computers this would be a formidable task. Instead we seek to define an effective Hamiltonian $\mathcal{H}_{\text{eff}}(\eta)$ which contains the terms of primary interest, gives the correct eigenvalues $(E_\eta^0 + E_{\eta k})$, but which operates solely within the zeroth-order subspace of the vibronic state $|\eta\rangle^0$. The effects of terms off-diagonal in η will be contained in constants or parameters appearing in $\mathcal{H}_{\text{eff}}$. In our example, we are interested in deriving an effective Hamiltonian which operates solely within the i-dimensional subspace of the ground vibronic state. This effective Hamiltonian, $\mathcal{H}_{\text{eff}}(0)$, must therefore satisfy an eigenvalue equation of the form

$$\mathcal{H}_{\text{eff}}(0) \sum_i d_{0i}^k |0, i\rangle^0 = \left(E_0^0 + E_{0k}\right) \sum_i d_{0i}^k |0, i\rangle^0. \tag{7.9}$$

Each eigenfunction of $\mathcal{H}_{\text{eff}}(0)$ is some linear combination of the zeroth-order orthonormal set $|0, i\rangle^0$, as yet unspecified. Hence the problem we have to solve is to derive a general expression for $\mathcal{H}_{\text{eff}}(0)$ in terms of $\mathcal{H}^0$ and $\mathcal{H}'$.

Following Bloch [8], we shall tackle the problem through the use of certain projection operators, which we now define. The operator P_0 is defined by

$$P_0 = \sum_i |0, i\rangle^0 {}^0\langle 0, i|, \tag{7.10}$$

and projects any function onto the subspace spanned by the zeroth-order eigenfunctions with $\eta = 0$; for example,

$$\begin{aligned} P_0 |0, k\rangle &= \sum_i |0, i\rangle^0 {}^0\langle 0, i | 0, k\rangle, \\ &= \sum_i |0, i\rangle^0 {}^0\langle 0, i| \sum_{\eta,j} c_{\eta j}^k |\eta, j\rangle^0, \\ &= \sum_i c_{0i}^k |0, i\rangle^0. \end{aligned} \tag{7.11}$$

We note that the set of g_0 functions formed by projecting the eigenfunctions $|0, k\rangle$ onto the zeroth-order subspace of the ground vibronic state are, in general, neither orthogonal to each other nor normalised to unity. Also, as mentioned earlier, there is not necessarily a one-to-one correspondence between the quantum numbers k and i.

The operator Q_0 which is complementary to P_0 is given by

$$Q_0 = 1 - P_0 = \sum_i \sum_{\eta \neq 0} |\eta, i\rangle^0 \; {}^0\langle \eta, i|. \tag{7.12}$$

The effect of Q_0 on a particular function is to give the function's projection in the subspace complementary to the zeroth-order subspace of the ground vibronic state.

We also define the operator U, whose effect is essentially the opposite to that of P_0, i.e.

$$U \sum_i c_{0i}^k |0, i\rangle^0 = |0, k\rangle. \tag{7.13}$$

It is not difficult to show that the operator U must be given by

$$U = \sum_k \sum_j b_{0j}^k |0, k\rangle \; {}^0\langle 0, j|, \tag{7.14}$$

where the b coefficients are related to the c coefficients defined in (7.8) by the equation

$$\sum_i b_{0i}^k c_{0i}^{k'} = \delta_{kk'}. \tag{7.15}$$

We can readily prove the definition (7.14) of the operator U by expanding (7.13):

$$U \sum_i c_{0i}^k |0, i\rangle^0 = \sum_{j,k'} b_{0j}^{k'} |0, k'\rangle \; {}^0\langle 0, j| \sum_i c_{0i}^k |0, i\rangle^0 \tag{7.16}$$

$$= \sum_{j,k'} b_{0j}^{k'} |0, k'\rangle c_{0j}^k \tag{7.17}$$

$$= |0, k\rangle. \tag{7.18}$$

Equation (7.17) follows from the previous equation because of the orthogonality properties of the basis set, given by equation (7.4).

In our subsequent discussion we shall also require the results

$$U P_0 = U, \tag{7.19}$$

and

$$P_0 U = P_0. \tag{7.20}$$

Equation (7.19) follows readily from the definitions of P_0 and U. To verify (7.20) we consider the effect of the operator $P_0 U$ on an exact eigenfunction $|0, k\rangle$:

$$P_0 U |0, k\rangle = P_0 \sum_{i,k'} b_{0i}^{k'} |0, k'\rangle \; {}^0\langle 0, i | 0, k\rangle$$

$$= \sum_{i,j,k'} b_{0i}^{k'} c_{0j}^{k'} |0, j\rangle^0 \; {}^0\langle 0, i| \sum_{\eta, \ell} c_{\eta \ell}^k |\eta, \ell\rangle^0$$

$$= \sum_{i,j,k'} b_{0i}^{k'} c_{0j}^{k'} c_{0i}^{k} |0, j\rangle^0$$

$$= \sum_j c_{0j}^k |0, j\rangle^0$$

$$= P_0 |0, k\rangle. \tag{7.21}$$

Consideration of the requirements of the effective Hamiltonian $\mathcal{H}_{\text{eff}}(0)$, equation (7.9), suggests that we might, with advantage, investigate the properties of the operator product $P_0 \lambda \mathcal{H}' U$, since U essentially projects the exact eigenfunctions $|0, k\rangle$ out of the zeroth-order subspace spanned by $|0, i\rangle^0$, $\lambda \mathcal{H}'$ operates on the exact eigenfunction according to (7.6) and (7.7), and P_0 projects back into the basis set within which we wish $\mathcal{H}_{\text{eff}}(0)$ to operate. Accordingly we operate with $\lambda P_0 \mathcal{H}' U$ on the zeroth-order basis functions with the following results:

$$\lambda P_0 \mathcal{H}' U \sum_i c_{0i}^k |0, i\rangle^0$$

$$= \lambda \sum_j |0, j\rangle^0 \,^0\langle 0, j | (\mathcal{H} - \mathcal{H}^0) U \sum_i c_{0i}^k |0, i\rangle^0 \tag{7.22}$$

$$= \lambda \sum_j |0, j\rangle^0 \left\{ {}^0\langle 0, j | \mathcal{H} U \sum_i c_{0i}^k |0, i\rangle^0 - {}^0\langle 0, j | \mathcal{H}^0 U \sum_i c_{0i}^k |0, i\rangle^0 \right\}$$

$$= \lambda \sum_j |0, j\rangle^0 \left\{ {}^0\langle 0, j | \mathcal{H} |0, k\rangle - {}^0\langle 0, j | \mathcal{H}^0 \sum_\ell b_{0\ell}^{k'} |0, k'\rangle \, {}^0\langle 0, \ell | \sum_i c_{0i}^k |0, i\rangle^0 \right\}$$

$$= \lambda \sum_j |0, j\rangle^0 \left\{ {}^0\langle 0, j | 0, k\rangle (E_0^0 + E_{0k}) - {}^0\langle 0, j | \mathcal{H}^0 \sum_\ell b_{0\ell}^{k'} |0, k\rangle c_{0\ell}^k \right\}$$

$$= \lambda \sum_j |0, j\rangle^0 \left\{ {}^0\langle 0, j | \sum_{\eta, \ell} c_{\eta\ell}^k |\eta, \ell\rangle^0 (E_0^0 + E_{0k}) - {}^0\langle 0, j | \mathcal{H}^0 \sum_{\eta, i} c_{\eta i}^k |\eta, i\rangle^0 \right\}$$

$$= \lambda \sum_j |0, j\rangle^0 \{ c_{0j}^k (E_0^0 + E_{0k}) - c_{0j}^k E_0^0 \}$$

$$= \lambda E_{0k} \sum_i c_{0i}^k |0, i\rangle^0. \tag{7.23}$$

We have now obtained the desired result because equation (7.23) is of the same form as (7.9) provided that we set $d_{0i}^k = c_{0i}^k$; in other words, the eigenfunctions of $P_0 \mathcal{H}' U$ are just the projections of the exact eigenfunctions $|0, k\rangle$ onto the zeroth-order subspace of the ground vibronic state. Comparison of equations (7.23) and (7.9) yields the relationship

$$\mathcal{H}_{\text{eff}}(0) = \mathcal{H}^0 + \lambda P_0 \mathcal{H}' U. \tag{7.24}$$

The final, but fairly lengthy, stage of this analysis is to seek a recurrence relation for U which allows us to express U in terms of P_0, Q_0 and $\mathcal{H}'$. First we note from (7.6)

that
$$\left(\mathcal{H} - E_0^0\right)|0, k\rangle = E_{0k}|0, k\rangle. \tag{7.25}$$

Multiplying (7.25) on the left by P_0 yields
$$\lambda P_0 \mathcal{H}'|0, k\rangle = E_{0k} \sum_i c_{0i}^k |0, i\rangle^0. \tag{7.26}$$

The left-hand side of (7.26) follows because
$$P_0\left(\mathcal{H} - E_0^0\right) = P_0\left(\mathcal{H}^0 + \lambda \mathcal{H}' - E_0^0\right) = P_0 \lambda \mathcal{H}'. \tag{7.27}$$

We now multiply (7.26) from the left by U to give
$$\lambda U P_0 \mathcal{H}'|0, k\rangle = E_{0k}|0, k\rangle. \tag{7.28}$$

Noting that $UP_0 = U$ (equation (7.19)) we rewrite (7.28) in the form
$$\lambda U \mathcal{H}'|0, k\rangle = E_{0k}|0, k\rangle, \tag{7.29}$$
$$= \left(\mathcal{H} - E_0^0\right)|0, k\rangle. \tag{7.30}$$

We therefore obtain the equation for the exact eigenfunctions $|0, k\rangle$,
$$\left(\mathcal{H} - E_0^0 - \lambda U \mathcal{H}'\right)|0, k\rangle = 0. \tag{7.31}$$

Multiplication on the right by $\sum_j b_{0j}^k{}^0\langle 0, j|$ and summation of the resultant expression over k gives:
$$\left(\mathcal{H} - E_0^0 - \lambda U \mathcal{H}'\right) \sum_{j,k} b_{0j}^k |0, k\rangle{}^0\langle 0, j| = \left(\mathcal{H} - E_0^0 - \lambda U \mathcal{H}'\right) U = 0. \tag{7.32}$$

Equation (7.32) can be rewritten as
$$\left(E_0^0 - \mathcal{H}^0\right) U = \lambda(\mathcal{H}' U - U \mathcal{H}' U) \tag{7.33}$$

Now since $Q_0 = 1 - P_0$, we can write
$$U = P_0 U + Q_0 U. \tag{7.34}$$

If we substitute this expression for U in the left-hand side of (7.33), and note that $(E_0^0 - \mathcal{H}^0) P_0$ is identically zero, we have
$$\left(E_0^0 - \mathcal{H}^0\right) Q_0 U = \lambda(\mathcal{H}' U - U \mathcal{H}' U). \tag{7.35}$$

Now in the subspace complementary to that which is spanned by our zeroth-order basis set $|0, i\rangle^0$, the operator $(E_0^0 - \mathcal{H}^0)$ has a well defined inverse, which we denote by a^{-1}. Thus if we multiply (7.35) by $Q_0 a^{-1}$ we have
$$Q_0 Q_0 U = \lambda \frac{Q_0}{a}(\mathcal{H}' U - U \mathcal{H}' U), \tag{7.36}$$

i.e.
$$Q_0 U = \lambda \frac{Q_0}{a}(\mathcal{H}'U - U\mathcal{H}'U). \tag{7.37}$$

Substituting (7.20) and (7.37) in (7.34), we finally obtain the recursion equation in U which we are seeking:
$$U = P_0 + \lambda \frac{Q_0}{a}(\mathcal{H}'U - U\mathcal{H}'U). \tag{7.38}$$

Now U can be expanded as a power series in the perturbation $\lambda \mathcal{H}'$; if we write U as a sum of zeroth-order, first-order, second-order, etc., contributions,
$$U = U^{(0)} + \lambda U^{(1)} + \lambda^2 U^{(2)} + \cdots, \tag{7.39}$$

we can then rewrite (7.38) in the form
$$\begin{aligned}
&U^{(0)} + \lambda U^{(1)} + \lambda^2 U^{(2)} + \cdots \\
&= P_0 + \lambda \frac{Q_0}{a}\{\mathcal{H}'U^{(0)} + \lambda \mathcal{H}'U^{(1)} + \lambda^2 \mathcal{H}'U^{(2)} + \cdots\} \\
&\quad - \lambda \frac{Q_0}{a}\{U^{(0)}\mathcal{H}'U^{(0)} + \lambda U^{(1)}\mathcal{H}'U^{(0)} + \lambda^2 U^{(2)}\mathcal{H}'U^{(0)} + \cdots\} \\
&\quad - \lambda \frac{Q_0}{a}\{\lambda U^{(0)}\mathcal{H}'U^{(1)} + \lambda^2 U^{(1)}\mathcal{H}'U^{(1)} + \lambda^3 U^{(2)}\mathcal{H}'U^{(1)} + \cdots\} \\
&\quad - \lambda \frac{Q_0}{a}\{\lambda^2 U^{(0)}\mathcal{H}'U^{(2)} + \lambda^3 U^{(1)}\mathcal{H}'U^{(2)} + \lambda^4 U^{(2)}\mathcal{H}'U^{(2)} + \cdots\} + \cdots.
\end{aligned} \tag{7.40}$$

Equating like powers of λ in equation (7.40) we obtain the results
$$U^{(0)} = P_0,$$
$$U^{(1)} = \frac{Q_0}{a}\{\mathcal{H}'U^{(0)} - U^{(0)}\mathcal{H}'U^{(0)}\},$$
$$U^{(2)} = \frac{Q_0}{a}\{\mathcal{H}'U^{(1)} - U^{(1)}\mathcal{H}'U^{(0)} - U^{(0)}\mathcal{H}'U^{(1)}\},$$
$$\ldots,$$
$$\ldots,$$
$$U^{(n)} = \frac{Q_0}{a}\left\{\mathcal{H}'U^{(n-1)} - \sum_{p=0}^{n-1} U^{(p)}\mathcal{H}'U^{(n-p-1)}\right\}. \tag{7.41}$$

Noting that $U^{(0)} = P_0$ and that $(Q_0/a)P_0$ vanishes since Q_0 and P_0 are projection operators onto orthogonal subspaces, we see that we can remove the term $p = 0$ from the summation in the last line of (7.41). In other words, the general expression for $U^{(n)}$ is
$$U^{(n)} = \frac{Q_0}{a}\left\{\mathcal{H}'U^{(n-1)} - \sum_{p=1}^{n-1} U^{(p)}\mathcal{H}'U^{(n-p-1)}\right\}. \tag{7.42}$$

Using these results in equation (7.24) we are finally able to obtain an expression for the desired effective Hamiltonian, which is,

$$\mathcal{H}_{\text{eff}}(0) = \mathcal{H}^0 + \lambda P_0 \mathcal{H}' U$$
$$= \mathcal{H}^0 + \lambda P_0 \mathcal{H}' U^{(0)} + \lambda^2 P_0 \mathcal{H}' U^{(1)} + \lambda^3 P_0 \mathcal{H}' U^{(2)} + \lambda^4 P_0 \mathcal{H}' U^{(3)} + \cdots$$
$$= \mathcal{H}^0 + \lambda P_0 \mathcal{H}' P_0 + \lambda^2 P_0 \mathcal{H}' \frac{Q_0}{a} \mathcal{H}' P_0$$
$$+ \lambda^3 \left\{ P_0 \mathcal{H}' \frac{Q_0}{a} \mathcal{H}' \frac{Q_0}{a} \mathcal{H}' P_0 - P_0 \mathcal{H}' \frac{Q_0}{a^2} \mathcal{H}' P_0 \mathcal{H}' P_0 \right\}$$
$$+ \lambda^4 \left\{ P_0 \mathcal{H}' \frac{Q_0}{a} \mathcal{H}' \frac{Q_0}{a} \mathcal{H}' \frac{Q_0}{a} \mathcal{H}' P_0 - P_0 \mathcal{H}' \frac{Q_0}{a} \mathcal{H}' \frac{Q_0}{a^2} \mathcal{H}' P_0 \mathcal{H}' P_0 \right.$$
$$- P_0 \mathcal{H}' \frac{Q_0}{a^2} \mathcal{H}' \frac{Q_0}{a} \mathcal{H}' P_0 \mathcal{H}' P_0 + P_0 \mathcal{H}' \frac{Q_0}{a^3} \mathcal{H}' P_0 \mathcal{H}' P_0 \mathcal{H}' P_0$$
$$\left. - P_0 \mathcal{H}' \frac{Q_0}{a^2} \mathcal{H}' P_0 \mathcal{H}' \frac{Q_0}{a} \mathcal{H}' P_0 \right\} + \cdots. \tag{7.43}$$

This expression is correct to fourth-order and higher-order terms can be obtained if required from the general expression for $U^{(n)}$. To remind the reader at this point, we recall that from equation (7.10),

$$P_0 = \sum_i |0, i\rangle^0 \, {}^0\langle 0, i|$$

and

$$(Q_0/a^r) = \sum_{\eta \neq 0} \sum_i |\eta, i\rangle^0 \, {}^0\langle \eta, i|/(E_0^0 - E_\eta^0)^r. \tag{7.44}$$

We note that the operator $\mathcal{H}_{\text{eff}}(0)$ is constrained to act within the manifold of states $|0\rangle$ because each term on the right-hand side is sandwiched between a pair of projection operators, P_0.

The various contributions to the effective Hamiltonian operator on the right-hand side of equation (7.43) can be represented diagrammatically in a way which is very helpful in practice. These diagrams are shown in figure 7.1. For each diagram, the position at the bottom refers to the chosen zeroth-order manifold $|0\rangle^0$ and the position at the top refers to the states in the complementary subspace. Each line joining these two positions stands for a connecting matrix element.

We shall return to the derivation of the individual terms of the effective Hamiltonian listed in (7.43) for some particular cases later in this chapter. To conclude this section, we now consider some of the general properties of the operator $\mathcal{H}_{\text{eff}}(0)$ and its eigenfunctions. Each of the terms listed in (7.43) is composed of products of the three operators P_0, Q_0 and $\mathcal{H}'$. Although each of these operators is individually Hermitian (i.e. the operator is self-adjoint, $P_0^\dagger = P_0$), a product of any of them is not necessarily Hermitian. In fact, a term is only Hermitian when it is palindromic, that is, when it reads the same forwards as backwards. Inspection of equation (7.43) reveals that $\mathcal{H}_{\text{eff}}(0)$ is Hermitian up to and including terms in λ^2 but that there are non-Hermitian terms in the λ^3 and higher-order contributions. The nature of these non-Hermitian properties can

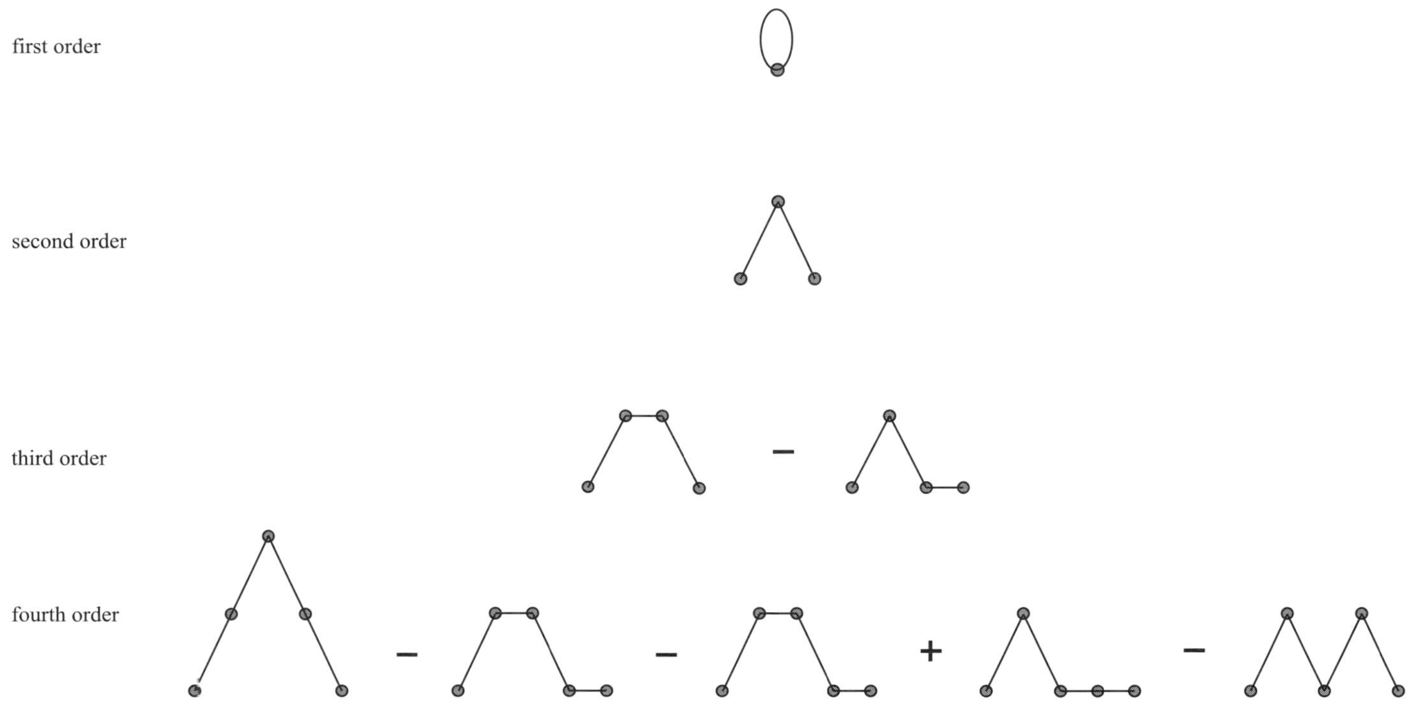

first order

second order

third order

fourth order

Figure 7.1. Diagrammatic representation of the various contributions to the effective Hamiltonian given in equation (7.43) (see text).

be understood by referring back to eigenvalue equations for $\mathcal{H}_{\text{eff}}(0)$, equations (7.9) and (7.23). Because we have constructed $\mathcal{H}_{\text{eff}}(0)$ to have eigenvalues identical with those of the total Hamiltonian $\mathcal{H}$ (which is, of course, Hermitian), $\mathcal{H}_{\text{eff}}(0)$ will have real eigenvalues, even though it is non-Hermitian. However, we recall that the eigenfunctions of $\mathcal{H}_{\text{eff}}(0)$ are $\sum_i c_{0i}^k |0, i\rangle^0$, the projections of the actual eigenfunctions of $\mathcal{H}$ onto a zeroth-order subspace. These eigenfunctions are not necessarily orthogonal or normalised with respect to each other and it is in this way that the non-Hermitian properties of the operator $\mathcal{H}_{\text{eff}}(0)$ manifest themselves.

Since we shall usually be more interested in the eigenvalues than the eigenfunctions of $\mathcal{H}_{\text{eff}}(0)$, the possible non-Hermitian properties need not concern us too deeply. However, there may be certain situations, for example in the use of computer programmes to extract the eigenvalues of $\mathcal{H}_{\text{eff}}(0)$, where we must take account of the non-orthogonal eigenfunctions. In these situations, it is easiest to use a symmetrised form of the non-Hermitian terms in $\mathcal{H}_{\text{eff}}(0)$; in the third order, this would be

$$-\frac{1}{2}\lambda^3 \left\{ P_0 \mathcal{H}' \frac{Q_0}{a^2} \mathcal{H}' P_0 \mathcal{H}' P_0 + P_0 \mathcal{H}' P_0 \mathcal{H}' \frac{Q_0}{a^2} \mathcal{H}' P_0 \right\}. \quad (7.45)$$

Soliverez [9] has shown that errors introduced by this modification of equation (7.43) only appear in fifth and higher orders. Since in practice it is rarely worthwhile to use perturbation theory beyond third order, these discrepancies need not worry us.

7.3. The Van Vleck and contact transformations

We now describe two other methods of deriving an effective Hamiltonian, both of which are widely used. Although we shall not go into details, the mathematical development will show that the two methods are exactly equivalent and, in addition, that they are very nearly equivalent to the method based on projection operators given in the previous section. The equivalence of the three methods is not really very surprising since they are all solutions of the problem by perturbation theory, differing only in the mathematical techniques employed.

Again we start by partitioning the total Hamiltonian,

$$\mathcal{H} = \mathcal{H}^0 + \mathcal{H}', \quad (7.46)$$

where all terms off-diagonal in the quantum number η are included in $\mathcal{H}'$. We note that since we choose our zeroth-order Hamiltonian to be independent of operators involving the quantum numbers i (see previous section), our zeroth-order eigenfunctions $|\eta, i\rangle^0$ can be completely factorised into the product

$$|\eta, i\rangle^0 = |\eta\rangle^0 |i\rangle^0, \quad (7.47)$$

where $|\eta\rangle^0$ are the eigenfunctions of $\mathcal{H}^0$,

$$\mathcal{H}^0 |\eta\rangle^0 = E_n^0 |\eta\rangle^0. \quad (7.48)$$

We first imagine that we have constructed a matrix representation of $\mathcal{H}$ in the zeroth-order basis set (7.47), and we then subject this matrix to a similarity transformation to give a transformed matrix $\tilde{\mathcal{H}}$ where

$$\tilde{\mathcal{H}} = T^{-1}\mathcal{H}T. \tag{7.49}$$

Since the matrix T is unitary ($T^\dagger = T^{-1}$), $\tilde{\mathcal{H}}$ will have the same eigenvalues as $\mathcal{H}$ but different eigenfunctions. We then set

$$T = e^{i\lambda S_1} e^{i\lambda^2 S_2} e^{i\lambda^3 S_3} \ldots, \tag{7.50}$$

where S_n is Hermitian and $S_1, S_2, S_3, \ldots$, are chosen in the way outlined below to remove the effects of matrix elements off-diagonal in η to order $\lambda, \lambda^2, \lambda^3, \ldots$. Each of the successive transformations produced by the development of T (7.50) is known as a *contact transformation* [10, 11], while the first ($e^{-i\lambda S_1}\mathcal{H}e^{i\lambda S_1}$) is what is usually referred to as the *Van Vleck transformation* [12, 13].

We now expand $\tilde{\mathcal{H}}$ as

$$\tilde{\mathcal{H}} = \tilde{\mathcal{H}}_0 + \lambda\tilde{\mathcal{H}}_1 + \lambda^2\tilde{\mathcal{H}}_2 + \ldots, \tag{7.51}$$

and equate the right-hand side with

$$\ldots e^{-i\lambda^2 S_2} e^{-i\lambda S_1}(\mathcal{H}^0 + \lambda\mathcal{H}')e^{i\lambda S_1} e^{i\lambda^2 S_2} \ldots. \tag{7.52}$$

Comparing coefficients of λ, we can obtain the results

$$\lambda^0 : \quad \tilde{\mathcal{H}}_0 = \mathcal{H}^0. \tag{7.53}$$

$$\lambda^1 : \quad \tilde{\mathcal{H}}_1 = \mathcal{H}' + i[\mathcal{H}^0, S_1]. \tag{7.54}$$

$$\lambda^2 : \quad \tilde{\mathcal{H}}_2 = i[\mathcal{H}^0, S_2] + i[\mathcal{H}', S_1] + \frac{1}{2}[[S_1, \mathcal{H}^0], S_1]. \tag{7.55}$$

$$\lambda^3 : \quad \tilde{\mathcal{H}}_3 = i[\mathcal{H}^0, S_3] + i[\mathcal{H}', S_2] - \frac{i}{6}[[[\mathcal{H}^0, S_1], S_1], S_1]$$

$$+ \frac{1}{2}[[S_1, \mathcal{H}'], S_1] + [S_2, [\mathcal{H}^0, S_1]]. \tag{7.56}$$

Substituting (7.54) into (7.55) and (7.56), we have

$$\tilde{\mathcal{H}}_2 = i[\mathcal{H}^0, S_2] + \frac{i}{2}[\mathcal{H}', S_1] + \frac{i}{2}[\tilde{\mathcal{H}}_1, S_1]. \tag{7.57}$$

$$\tilde{\mathcal{H}}_3 = i[\mathcal{H}^0, S_3] + i[\tilde{\mathcal{H}}_1, S_2] - \frac{1}{3}[[\mathcal{H}', S_1], S_1] - \frac{1}{6}[[\tilde{\mathcal{H}}_1, S_1], S_1]. \tag{7.58}$$

We now show how the matrix elements of S_1 can be selected to remove the effects on $\mathcal{H}$ of matrix elements off-diagonal in η. Since it can easily be verified that $^0\langle\eta, i|[\mathcal{H}^0, S_1]|\eta, j\rangle^0$ is always zero, irrespective of the form of S_1, we see from (7.54) that the diagonal matrix elements of $\tilde{\mathcal{H}}_1$ are just the diagonal matrix elements of $\mathcal{H}'$:

$$^0\langle\eta, i|\tilde{\mathcal{H}}_1|\eta, j\rangle^0 = {}^0\langle\eta, i|\mathcal{H}'|\eta, j\rangle^0. \tag{7.59}$$

Without any loss in generality therefore, we can set the matrix elements of S_1 diagonal in η equal to zero:

$$^0\langle\eta, i|S_1|\eta, j\rangle^0 = 0 \quad \text{for all } \eta. \tag{7.60}$$

In addition, we require the matrix elements of $\tilde{\mathcal{H}}_1$ off-diagonal in η to be zero, that is, from (7.54):

$$^0\langle\eta, i|\tilde{\mathcal{H}}_1|\eta', j\rangle^0 = {}^0\langle\eta, i|\mathcal{H}'|\eta', j\rangle^0 + i\;{}^0\langle\eta, i|S_1|\eta', j\rangle^0 (E_\eta^0 - E_{\eta'}^0) = 0 \quad \text{for } \eta' \neq \eta. \tag{7.61}$$

Thus

$$i\;{}^0\langle\eta, i|S_1|\eta', j\rangle^0 = -\frac{{}^0\langle\eta, i|\mathcal{H}'|\eta', j\rangle^0}{(E_\eta^0 - E_{\eta'}^0)}. \tag{7.62}$$

The matrix elements of S_2 are chosen in a similar fashion so that ${}^0\langle\eta, i|\tilde{\mathcal{H}}_2|\eta', j\rangle^0$ is zero, and so on.

We now consider the diagonal matrix elements of $\tilde{\mathcal{H}}_2$ and $\tilde{\mathcal{H}}_3$ using the results (7.60) and (7.62). It is a simple matter to establish that for $\tilde{\mathcal{H}}_2$, equation (7.57), only the second term on the right-hand side has non-zero diagonal matrix elements,

$$^0\langle\eta, i|\tilde{\mathcal{H}}_2|\eta, j\rangle^0 = \frac{i}{2}\;{}^0\langle\eta, i|[\mathcal{H}', S_1]|\eta, i\rangle^0, \tag{7.63}$$

while for $\tilde{\mathcal{H}}_3$, equation (7.58), only the third and fourth terms have non-zero diagonal matrix elements, irrespective of the forms of S_2 and S_3:

$$^0\langle\eta, i|\tilde{\mathcal{H}}_3|\eta, j\rangle^0 = -\frac{1}{3}\;{}^0\langle\eta, i|[[\mathcal{H}', S_1], S_1]|\eta, j\rangle^0 - \frac{1}{6}\;{}^0\langle\eta, i|[[\tilde{\mathcal{H}}_1, S_1], S_1]|\eta, j\rangle^0. \tag{7.64}$$

Thus if we are only interested in deriving an effective Hamiltonian up to order λ^3, we need not concern ourselves with the explicit forms of S_2 and S_3. Furthermore, for the particular situation where $\mathcal{H}'$ only has matrix elements which are off-diagonal in η (i.e. all diagonal matrix elements of $\mathcal{H}'$ are zero), the second term in (7.64) also vanishes and we have

$$^0\langle\eta, i|\tilde{\mathcal{H}}_3|\eta, j\rangle^0 = -\frac{1}{3}\;{}^0\langle\eta, i|[[\mathcal{H}', S_1], S_1]|\eta, j\rangle^0. \tag{7.65}$$

Thus for the general case, we can separate the matrix elements of $\mathcal{H}'$ into diagonal and off-diagonal categories and use (7.65) to evaluate contributions to $\tilde{\mathcal{H}}_3$ involving off-diagonal matrix elements of $\mathcal{H}'$ and to use, from (7.59) and (7.64),

$$^0\langle\eta, i|\tilde{\mathcal{H}}_3|\eta, j\rangle^0 = -\frac{1}{2}\;{}^0\langle\eta, i|[[\mathcal{H}', S_1], S_1]|\eta, j\rangle^0, \tag{7.66}$$

for contributions from diagonal elements.

In vibration–rotation theory, the λ, λ^2 and λ^3 contributions to the contact-transformed Hamiltonian are commonly evaluated directly from the relationships (7.59), (7.63), (7.65) and (7.66). This is because the particularly simple commutation relationships which exist between the normal coordinate operator Q, its conjugate momentum P and the simple harmonic oscillator Hamiltonian, which is commonly taken as $\mathcal{H}^0$ in equation (7.46), make the evaluation of the commutator brackets quite straightforward. However, equations (7.59), (7.63), (7.65) and (7.66) are not very suitable for the derivation of an effective Hamiltonian that includes the effects of matrix elements of $\mathcal{H}'$ off-diagonal in electronic state, since the commutation relations between $\mathcal{H}^0$ and the operators involved are generally much more complicated. In this situation, it is easier to develop the various contributions in terms of matrix element products using equation (7.62). After some straightforward algebra, we obtain the alternative but completely equivalent results,

$$^0\langle \eta, i|\tilde{\mathcal{H}}_0|\eta, j\rangle^0 = E_\eta^0. \tag{7.67}$$

$$^0\langle \eta, i|\tilde{\mathcal{H}}_1|\eta, j\rangle^0 = {}^0\langle \eta, i|\mathcal{H}'|\eta, j\rangle^0. \tag{7.68}$$

$$^0\langle \eta, i|\tilde{\mathcal{H}}_2|\eta, j\rangle^0 = \sum_{\eta'\neq\eta}\sum_k \frac{{}^0\langle \eta, i|\mathcal{H}'|\eta', k\rangle^0\ {}^0\langle \eta', k|\mathcal{H}'|\eta, j\rangle^0}{(E_\eta^0 - E_{\eta'}^0)}. \tag{7.69}$$

$$^0\langle \eta, i|\tilde{\mathcal{H}}_3|\eta, j\rangle^0 = \sum_{\eta',\eta''\neq\eta}\sum_{k,\ell} \frac{{}^0\langle \eta, i|\mathcal{H}'|\eta', k\rangle^0\ {}^0\langle \eta', k|\mathcal{H}'|\eta'', \ell\rangle^0\ {}^0\langle \eta'', \ell|\mathcal{H}'|\eta, j\rangle^0}{(E_\eta^0 - E_{\eta'}^0)(E_\eta^0 - E_{\eta''}^0)}$$

$$- \frac{1}{2}\sum_{\eta'\neq\eta}\sum_{k,\ell}\left\{\frac{{}^0\langle \eta, i|\mathcal{H}'|\eta, k\rangle^0\ {}^0\langle \eta, k|\mathcal{H}'|\eta', \ell\rangle^0\ {}^0\langle \eta', \ell|\mathcal{H}'|\eta, j\rangle^0}{(E_\eta^0 - E_{\eta'}^0)^2}\right.$$

$$\left. + \frac{{}^0\langle \eta, i|\mathcal{H}'|\eta', k\rangle^0\ {}^0\langle \eta', k|\mathcal{H}'|\eta, \ell\rangle^0\ {}^0\langle \eta, \ell|\mathcal{H}'|\eta, j\rangle^0}{(E_\eta^0 - E_{\eta'}^0)^2}\right\}. \tag{7.70}$$

If we compare these equations with the projection operator expansion given in equation (7.43), we find that the expressions are identical up to and including the λ^2 contribution but that the λ^3 term derived here corresponds not to the λ^3 term in the expansion (7.43) but to its symmetrised (Hermitian) form discussed at the end of section 7.2. Since the discrepancies that arise from these two different forms are of order λ^5 or higher, the effective Hamiltonians derived by the two methods are identical to order λ^3. In the literature the Van Vleck transformation is normally implemented by use of equations (7.67) to (7.70) although the λ^3 contribution (7.70) has often been ignored.

In conclusion, we note that thus far we have derived matrix elements of the transformed Hamiltonian $\tilde{\mathcal{H}}$ for a given block in the complete matrix labelled by a particular value of η rather than an effective Hamiltonian operating only within the subspace of the state η. It is an easy matter to cast our results in the form of an effective Hamiltonian for any particular case since the matrix elements involved in either the commutator bracket formulation (contact transformation) or the explicit matrix element formulation (Van Vleck transformation) can always be factorised into a product of a matrix element of operators involved in $\mathcal{H}^0$ associated with the quantum number η and a matrix element of operators that act only within the subspace levels of a given η state, associated with the quantum number i. This follows because the basis set can be factorised as in equation (7.47). The matrix element involving the η quantum number can then either be evaluated or included as a parameter to be determined experimentally, while the matrix element involving the i quantum number can be restored to operator form with the restriction that it only operates on the functions $|i\rangle^0$ which span a particular state $|\eta\rangle^0$. We shall see later in this chapter how this process is carried out in detail.

7.4. Effective Hamiltonian for a diatomic molecule in a given electronic state

7.4.1. Introduction

It is well known from the Born–Oppenheimer separation [1] that the pattern of energy levels for a typical diatomic molecule consists first of widely separated electronic states ($\Delta E_{\text{elec}} \approx 20\,000$ cm^{-1}). Each of these states then supports a set of more closely spaced vibrational levels ($\Delta E_{\text{vib}} \approx 1000$ cm^{-1}). Each of these vibrational levels in turn is spanned by closely spaced rotational levels ($\Delta E_{\text{rot}} \approx 1$ cm^{-1}) and, in the case of open shell molecules, by fine and hyperfine states ($\Delta E_{\text{fs}} \approx 100$ cm^{-1} and $\Delta E_{\text{hfs}} \approx 0.01$ cm^{-1}). The objective is to construct an effective Hamiltonian which is capable of describing the detailed energy levels of the molecule in a single vibrational level of a particular electronic state. It is usual to derive this Hamiltonian in two stages because of the different nature of the electronic and nuclear coordinates. In the first step, which we describe in the present section, we derive a Hamiltonian which acts on all the vibrational states of a single electronic state. The operators thus remain explicitly dependent on the vibrational coordinate R (the internuclear separation). In the second step, described in section 7.55, we remove the effects of terms in this intermediate Hamiltonian which couple different vibrational levels. The result is an effective Hamiltonian for each vibronic state.

In this section, we shall use the degenerate perturbation theory approach to derive the form of the effective Hamiltonian for a diatomic molecule in a given electronic state. Exactly the same result can be obtained by use of the Van Vleck or contact transformations [12, 13]. The general expression for the operator up to fourth order in perturbation theory is given in equation (7.43). Fourth order can be considered as the practical limit to this type of approach. Indeed, even its implementation is very laborious and has only been used to investigate the form of certain special terms in the effective Hamiltonian. We shall consider some of these terms later in this chapter. For the moment we confine our attention to first- and second-order effects only.

To simplify matters, we shall start out by considering the effective Hamiltonian for a molecule in the absence of external fields. The additional effects of magnetic or electric fields will be dealt with later. We have derived the fundamental Hamiltonian for a diatomic molecule in chapters 2 to 4. For the present purposes it can be written:

$$\mathcal{H} = \mathcal{H}_{\text{elec}} + \frac{1}{2\mu}\boldsymbol{P}_R^2 + hcB(R)(\boldsymbol{N} - \boldsymbol{L})^2 + \frac{1}{2M}\boldsymbol{P}^2$$
$$+ \mathcal{H}_{\text{SO}}^{(e)} + \mathcal{H}_{\text{SO}}^{(n)} + \mathcal{H}_{\text{SS}}^{(\text{scal})} + \mathcal{H}_{\text{SS}}^{(\text{tens})} + \mathcal{H}_{\text{hfs}}. \quad (7.71)$$

The first term consists of both the kinetic energy of the electrons and the complete Coulomb energy, the second is the vibrational kinetic energy, the third is the rotational kinetic energy, the fourth is the 'mass polarisation' energy involving the total linear momentum of the electrons $\boldsymbol{P} = \sum_i \boldsymbol{p}_i$, the fifth and sixth are the parts of the electronic spin–orbit and spin–other-orbit operators which contain the electronic and nuclear momenta respectively, the seventh and eight terms are the scalar and tensor parts of the electron spin–spin interaction, and the last term describes the nuclear hyperfine effects (both magnetic and electric). In equation (7.71), μ and M are the reduced and total masses of the nuclei respectively, the rotational variable $B(R)$ is $\hbar^2/2hc\mu R^2$ (in cm^{-1}), and $\hbar \boldsymbol{L}$ is the electron orbital angular momentum about the nuclear centre of mass. $\boldsymbol{N}$ is equal to $\boldsymbol{J} - \boldsymbol{S}$, where $\hbar \boldsymbol{J}$ is the total angular momentum apart from nuclear spin, and $\hbar \boldsymbol{S}$ is the electron spin angular momentum; as elsewhere in this book, the angular momenta are dimensionless.

The explicit forms of terms five to eight in equation (7.71) are as follows:

$$\mathcal{H}_{so}^{(e)} = \frac{\hbar e^2 g_S}{16\pi\varepsilon_0 m_e^2 c^2} \sum_i \boldsymbol{s}_i \cdot \left\{ \sum_\alpha Z_\alpha r_{i\alpha}^{-3}(\boldsymbol{r}_i - \boldsymbol{r}_\alpha) \wedge \boldsymbol{p}_i - \sum_{j \neq i} r_{ij}^{-3}(\boldsymbol{r}_i - \boldsymbol{r}_j) \wedge (\boldsymbol{p}_i - 2\boldsymbol{p}_j) \right\}. \quad (7.72)$$

$$\mathcal{H}_{so}^{(n)} = -\frac{\hbar e^2 g_S}{8\pi\varepsilon_0 m_e c^2} \sum_i \boldsymbol{s}_i \cdot \left\{ \sum_\alpha Z_\alpha M_\alpha^{-1} r_{i\alpha}^{-3}(\boldsymbol{r}_i - \boldsymbol{r}_\alpha) \wedge \boldsymbol{P}_\alpha \right\}. \quad (7.73)$$

$$\mathcal{H}_{ss}^{(scal)} = -\frac{\hbar^2 e^2 g_S^2}{6\varepsilon_0 m_e^2 c^2} \sum_i \sum_{j>i} (\boldsymbol{s}_i \cdot \boldsymbol{s}_j)\delta(\boldsymbol{r}_{ij}). \quad (7.74)$$

$$\mathcal{H}_{ss}^{(tens)} = \frac{\hbar^2 e^2 g_S^2}{16\pi\varepsilon_0 m_e^2 c^2} \sum_i \sum_{j>i} [\boldsymbol{s}_i \cdot \boldsymbol{s}_j - 3(\boldsymbol{s}_i \cdot \boldsymbol{r}_{ij})(\boldsymbol{s}_j \cdot \boldsymbol{r}_{ij})r_{ij}^{-2}][r_{ij}^{-3} - (4\pi/3)\delta(\boldsymbol{r}_{ij})]. \quad (7.75)$$

The subscripts i, j refer to electrons, α refers to nuclei, and ε_0 is the permittivity of free space. These terms were listed at the end of chapter 3, where their derivation was described.

We now proceed to remove the effects of the terms in equation (7.71) which couple different electronic states. For a molecule in which the spin–orbit coupling matrix elements are small compared with the electronic intervals, a convenient starting point is the set of Hund's case (a) electronic state vectors $|\eta, \Lambda, (S), \Sigma; R\rangle$ which are obtained by solving the electronic Schrödinger equation,

$$\mathcal{H}_{elec}|\eta, \Lambda, (S), \Sigma; R\rangle = |\eta, \Lambda, (S), \Sigma; R\rangle V_\eta(R), \quad (7.76)$$

as a function of the internuclear separation R. Here η is a state symbol (such as $X^2\Pi$) specifying the electronic state in question, which has two-fold degenerate orbital components $\Lambda = \pm|\Lambda|$ if $\Lambda \neq 0$. Although $\mathcal{H}_{elec}$ is a purely orbital operator, the electronic permutation symmetry of the orbital eigenvectors determines the electron spin quantum number S through the Pauli principle [14]. The $(2S+1)$ different Σ components of S are degenerate with each other at the level of approximation defined by equation (7.76).

The notation for the state vector in this equation is redundant since S is contained in η but it is useful to emphasise that S has a definite value (unlike L for a diatomic molecule).

We shall use equation (7.76) as the basis for a perturbation treatment of the full Hamiltonian $\mathcal{H}$. Thus $\mathcal{H}_{\text{elec}}$ is the zeroth-order Hamiltonian, see equation (7.1). Consequently the kets $|\eta, \Lambda, (S), \Sigma; R\rangle$ are the zeroth-order eigenstates and $V_\eta(R)$ are the zeroth-order eigenvalues, see equation (7.2). The perturbation process has the effect of removing terms which give matrix elements off-diagonal in η without altering the eigenvalues of $\mathcal{H}$ (at least, to a desired level of accuracy). In performing these calculations, one should remember that the electronic matrix elements of $\mathcal{H}$ are operators in the space of the vibrational, rotational and spin degrees of freedom. From equation (7.43), we can write the matrix elements of the transformed Hamiltonian $\mathcal{H}'$ to the second order as follows:

$$\langle \eta, \Lambda, (S), \Sigma; R | \mathcal{H}' | \eta', \Lambda', (S'), \Sigma'; R \rangle$$

$$= \delta_{\eta\eta'} \delta_{SS'} \left\{ \langle \eta, \Lambda, (S), \Sigma; R | \mathcal{H} | \eta, \Lambda', (S), \Sigma'; R \rangle \right.$$

$$+ \sum_{\eta'' \neq \eta, \Lambda'', \Sigma''} \langle \eta, \Lambda, (S), \Sigma; R | \mathcal{H} | \eta'', \Lambda'', (S''), \Sigma''; R \rangle$$

$$\left. \times \langle \eta'', \Lambda'', (S''), \Sigma''; R | \mathcal{H} | \eta, \Lambda', (S), \Sigma'; R \rangle / [V_\eta(R) - V_{\eta''}(R)] \right\}. \quad (7.77)$$

The condition for the validity of this approach to the perturbation treatment is

$$|\langle \eta, \Lambda, (S), \Sigma; R | \mathcal{H} | \eta'', \Lambda'', (S), \Sigma''; R \rangle| \ll |V_\eta(R) - V_{\eta''}(R)| \quad (7.78)$$

for all $\Lambda, \Sigma, \Lambda'', \Sigma''$ with $\eta \neq \eta''$. If this relationship is not satisfied, a different zeroth-order coupling case must be adopted. Instead of working in terms of matrix elements, it is usually more illuminating to employ an effective Hamiltonian operator $\mathcal{H}_\eta$ for the electronic state η. This is an operator in the space of the Λ and Σ components of η as well as the space of the vibrational and rotational degrees of freedom and is an equivalent operator defined so that its matrix elements are equal to those of $\mathcal{H}'$:

$$\langle \Lambda, S, \Sigma | \mathcal{H}_\eta | \Lambda', S, \Sigma' \rangle = \langle \eta, \Lambda, (S), \Sigma; R | \mathcal{H}' | \eta, \Lambda', (S), \Sigma'; R \rangle. \quad (7.79)$$

Thus, what we call $\mathcal{H}_\eta$ here is the same as the operator $\mathcal{H}_{\text{eff}}(0)$ in equation (7.43) which acts only within the manifold of states $|0\rangle$.

In the present treatment, we retain essentially all the diagonal matrix elements of $\mathcal{H}$; these are the first-order contributions to the effective electronic Hamiltonian. There are many possible off-diagonal matrix elements but we shall consider only those due to the terms in $\mathcal{H}_{\text{rot}}$ and $\mathcal{H}_{\text{so}}^{\text{e}}$ here since these are the largest and provide readily observable effects. The appropriate part of the rotational Hamiltonian is $-2hcB(R)(N_x L_x + N_y L_y)$. The matrix elements of this operator are comparatively sparse because they are subject to the selection rules $\Delta \Lambda = \pm 1$, $\Delta S = 0$ and $\Delta \Sigma = 0$. The spin–orbit coupling term, on the other hand, has a much more extensive set of matrix elements allowed by the selection rules $\Delta\Omega = 0$, $\Delta S = 0, \pm 1$ and $\Delta\Sigma = -\Delta\Lambda = 0, \mp 1$. In order to see how the effective Hamiltonian is derived, we now deal with some of the contributions explicitly.

7.4.2. The rotational Hamiltonian

The simplest example of the way in which the various terms arise in the effective electronic Hamiltonian involves the rotational kinetic energy operator, $\mathcal{H}_{\text{rot}}$:

$$\begin{aligned}\mathcal{H}_{\text{rot}} &= hcB(R)\boldsymbol{R}^2 \\ &= hcB(R)(\boldsymbol{N}-\boldsymbol{L})^2,\end{aligned} \tag{7.80}$$

where $\hbar\boldsymbol{R}$ is the end-over-end rotational angular momentum of the nuclei, and

$$B(R) = \hbar^2/(2\mu R^2 hc). \tag{7.81}$$

The operators $B(R)$ and $\boldsymbol{N}$ act only within each electronic state while the orbital angular momentum $\boldsymbol{L}$ acts both within and between such states.

The first-order contribution to the effective Hamiltonian involves the diagonal matrix element of the operator in equation (7.80):

$$\begin{aligned}\mathcal{H}_{\text{eff}}^{(1)}/hc &= |0\rangle\langle 0|B(R)(\boldsymbol{N}-\boldsymbol{L})^2|0\rangle\langle 0| \\ &= B^{(1)}(R)(\boldsymbol{N}^2 - L_z^2),\end{aligned} \tag{7.82}$$

where

$$B^{(1)}(R) = \langle 0, \Lambda|B(R)|0, \Lambda\rangle, \tag{7.83}$$

because the z component of $(\boldsymbol{N}-\boldsymbol{L})$, which is the only component of $\boldsymbol{L}$ which gives a diagonal matrix element, vanishes. Note that when we use the term *diagonal* in this context, we mean diagonal within the zeroth-order states of $|0\rangle$. The second-order contribution arises from the admixture of other electronic states $|\eta\rangle$ into the state in question $|0\rangle$ through the operator components L_x and L_y in the form $-hcB(R)(N_+L_- + N_-L_+)$:

$$\begin{aligned}&\mathcal{H}_{\text{eff}}^{(2)}/hc \\ &= hc|0\rangle \sum_{\eta\neq 0} \frac{\langle 0|-B(R)(N_+L_- + N_-L_+)|\eta\rangle\langle\eta|-B(R)(N_+L_- + N_-L_+)|0\rangle\langle 0|}{[V_0(R) - V_\eta(R)]} \\ &= hc \sum_{\eta\neq 0, \Lambda'} \frac{\langle 0, \Lambda|B(R)L_\mp|\eta, \Lambda'\rangle\langle\eta, \Lambda'|B(R)L_\pm|0, \Lambda\rangle}{[V_0(R) - V_\eta(R)]} (N_\pm N_\mp) \\ &= B^{(2)}(R)(N_x^2 + N_y^2) \\ &= B^{(2)}(R)(\boldsymbol{N}^2 - N_z^2).\end{aligned} \tag{7.84}$$

The second-order contribution to $B(R)$ is thus given by

$$B^{(2)}(R) = hc \sum_{\eta \neq 0} \frac{\langle 0, \Lambda | B(R) L_\mp | \eta, \Lambda' \rangle \langle \eta, \Lambda' | B(R) L_\pm | 0, \Lambda \rangle}{[V_0(R) - V_\eta(R)]}. \tag{7.85}$$

Hence, to second order, we can write the effective rotational Hamiltonian as

$$\mathcal{H}_{\text{rot}} = B(R)(N^2 - N_z^2), \tag{7.86}$$

where

$$B(R) = B^{(1)}(R) + B^{(2)}(R). \tag{7.87}$$

Since the matrix element of N_z^2 is constant for a given electronic state ($= \Lambda^2$), we see from equation (7.87) that the second-order contribution has exactly the same operator form as the first-order contribution. The effects of the matrix elements off-diagonal in electronic state are therefore absorbed in the Hamiltonian to give an operator of exactly the same form as in the full Hamiltonian, but with a slightly modified coefficient $B(R)$. Physically, the first-order contribution $B^{(1)}(R)$ describes the rotational kinetic energy from the motion of the bare nuclei and the second-order contribution $B^{(2)}(R)$ adds in the contribution to the kinetic energy from the electrons which rotate in laboratory space with the nuclei. (The electrons follow the motion of the nuclei very closely because of the Coulombic attraction between them.) We see also from its definition in (7.81) that the first-order contribution to $B(R)$ is proportional to μ^{-1} whereas $B^{(2)}(R)$ is proportional to μ^{-2}, provided that $[V_0(R) - V_\eta(R)]$ is large compared with the off-diagonal matrix elements. The ratio of the two contributions will therefore vary between different isotopic forms of the molecule and can thus, in principle, be separated. This difference in reduced mass dependence must be taken into account when modelling the effects of isotopic substitution on rotational energy levels.

We see from the way in which the effective rotational Hamiltonian is constructed that it is naturally expressed in terms of the angular momentum operator N. In the scientific literature, however, it is frequently written in terms of the vector R (which represents the rotational angular momentum of the nuclei) rather than N. While $R = N - L$ occurs in the fundamental Hamiltonian (7.71), its use in the effective Hamiltonian is not satisfactory because R has matrix elements (due to L) which connect different electronic states and so is not block diagonal in the electronic states. In practice, authors who claim to be using R in their formulations usually ignore any matrix elements which they find inconvenient such as those of L_x and L_y. We shall return to this point in more detail later in this chapter.

7.4.3. Hougen's isomorphic Hamiltonian

There is a particular difficulty in the formulation of the full Hamiltonian for a linear molecule (and hence for a diatomic molecule) which was first identified by Hougen [15]. The source of this difficulty is the fact that only two rotational coordinates are required to define the orientation of a diatomic molecule in laboratory space, the Euler angles ϕ and θ. It is customary to define the molecule-fixed z axis to point along the diatomic molecule axis from nucleus 1 to nucleus 2. Without any loss of generality, we can take the third Euler angle χ to be zero so that the transformation of the coordinates R_i of a particle i in the space-fixed system to the coordinates of the same particle in the rotating system is given by

$$\begin{bmatrix} x'_i \\ y'_i \\ z'_i \end{bmatrix} = \begin{bmatrix} \cos\theta\cos\phi & \cos\theta\sin\phi & -\sin\theta \\ -\sin\phi & \cos\phi & 0 \\ \sin\theta\cos\phi & \sin\theta\sin\phi & \cos\theta \end{bmatrix} \begin{bmatrix} X_i \\ Y_i \\ Z_i \end{bmatrix}. \tag{7.88}$$

Using standard methods to transform differential operators, Hougen attempted to achieve as great a separation of electronic, vibrational, rotational and spin coordinates as possible. The resultant operator representing the rotational kinetic energy has the form

$$\mathcal{H}_{\text{rot}} = \frac{\hbar^2}{2\mu R^2}\left[(\mathcal{J}_{x'} - W_{x'})^2 + \frac{1}{\sin\theta}(\mathcal{J}_{y'} - W_{y'})\sin\theta(\mathcal{J}_{y'} - W_{y'})\right]. \tag{7.89}$$

The angular momentum operator $\mathcal{J}$ which emerges from this treatment has the form:

$$i\hbar\mathcal{J}_{x'} + j\hbar\mathcal{J}_{y'} + k\hbar\mathcal{J}_{z'} = i(\cot\theta\hbar W_{z'} - \operatorname{cosec}\theta p_\phi) + j p_\theta + k\hbar W_{z'}, \tag{7.90}$$

where $\hbar W$ represents all the angular momenta apart from that due to the rotation of the nuclei:

$$W = L + S, \tag{7.91}$$

and

$$p_\theta = -i\hbar(\partial/\partial\theta), \tag{7.92}$$
$$p_\phi = -i\hbar(\partial/\partial\phi). \tag{7.93}$$

The operators $\mathcal{J}_{x'}$, $\mathcal{J}_{y'}$ and $\mathcal{J}_{z'}$ are the instantaneous components of the total angular momentum in the molecule-fixed axis system, including nuclear, electron orbital and spin motion, as computed by an observer fixed in space (what Van Vleck calls the components *referred* to the molecule-fixed axis system). They are not, however, what one might naively expect from a knowledge of elementary angular momentum theory. Because of the absence of the third Euler angle in the rotational variables for a diatomic molecule, the components referred to the rotating axis system do not obey the usual commutation relations for angular momenta components (even though the space-fixed components $\mathcal{J}_X$, $\mathcal{J}_Y$ and $\mathcal{J}_Z$ do behave normally). Instead, the commutation relations are more complicated:

$$[\mathcal{J}_{x'}, \mathcal{J}_{y'}] = -i\cot\theta\mathcal{J}_{x'} - iW_z, \tag{7.94}$$
$$[\mathcal{J}_{y'}, \mathcal{J}_{z'}] = [\mathcal{J}_{z'}, \mathcal{J}_{x'}] = 0, \tag{7.95}$$
$$[\mathcal{J}_{x'}, W_{y'}] = i\cot\theta\sum_k \varepsilon_{z'jk} W_k, \tag{7.96}$$
$$[\mathcal{J}_{y'}, W_j] = 0, \tag{7.97}$$
$$[\mathcal{J}_{z'}, W_j] = i\sum_k \varepsilon_{z'jk} W_k, \tag{7.98}$$

where $i, j, k = x', y'$ or z' and ε_{ijk} is equal to $+1$ if $ijk = x'y'z'$ or any cyclic permutation, equals -1 if any pair of these $x'y'z'$ are interchanged, and equals zero for any other combination. The components of W (or L and S separately) in the molecule-fixed axis system obey the normal commutation relations:

$$[W_i, W_j] = i \sum_k \varepsilon_{ijk} W_k. \tag{7.99}$$

Thus we see that the operator $\mathcal{J}$ is not strictly an angular momentum operator in the quantum mechanical sense, which is why we have assigned it a different symbol. More importantly for the present purposes, we cannot use the armoury of angular momentum theory and spherical tensor methods to construct representations of the molecular Hamiltonian. In addition, the rotational kinetic energy operator, equation (7.89), takes a more complicated form than it has for a nonlinear molecule where there are three Euler angles (rotational coordinates).

This lack of a third Euler angle, which might at the outset have seemed to be a simplification, therefore turns out to be something of a nuisance. In order to avoid this problem, Hougen devised a construct which permitted the re-introduction of the Euler angle χ in the description of the rotational motion of a linear molecule; he called the resulting operator the *isomorphic Hamiltonian* [15]. Since this Hamiltonian has one more degree of freedom than the true Hamiltonian, it has additional eigenstates which are not eigenstates of the true Hamiltonian. However, it is simple to use a restricted set of basis states to diagonalise the isomorphic Hamiltonian so that these extra eigenstates are excluded.

To obtain the isomorphic Hamiltonian for a diatomic molecule, χ is introduced as an independent variable and the coordinates of the particles which make up the molecule are measured in an axis system (x, y, z) whose orientation is described by the Euler angles (ϕ, θ, χ) in the (X, Y, Z) axis system. We recall that we chose χ to be zero in constructing the true Hamiltonian. The (x, y, z) axes are therefore obtained by rotation of the (x', y', z') axis system about the $z'(=z)$ axis through the angle χ. As a result, we have

$$(J_x - W_x) = \cos\chi(J_{x'} - W_{x'}) + \sin\chi(J_{y'} - W_{y'}), \tag{7.100}$$

$$(J_y - W_y) = -\sin\chi(J_{x'} - W_{x'}) + \cos\chi(J_{y'} - W_{y'}), \tag{7.101}$$

$$(J_z - W_z) = (J_{z'} - W_{z'}) = 0. \tag{7.102}$$

The components of J referred to the new coordinate system then have completely standard definitions:

$$\hbar J_x = \cos\chi\left(\cot\theta p_\chi - \operatorname{cosec}\theta p_\phi\right) + \sin\theta p_\theta, \tag{7.103}$$

$$\hbar J_y = -\sin\chi\left(\cot\theta p_\chi - \operatorname{cosec}\theta p_\phi\right) + \cos\chi p_\theta, \tag{7.104}$$

$$\hbar J_z = p_\chi. \tag{7.105}$$

These components are now independent of the orbital and spin variables and so commute with L and S. They also obey the commutation relations for a general angular momentum, provided only that the anomalous sign of i is used.

In the isomorphic Hamiltonian, the rotational kinetic energy operator has the simpler (and familiar) form

$$\mathcal{H}_{\text{rot}}^{\text{iso}} = \frac{\hbar^2}{2\mu R^2}[(J_x - W_x)^2 + (J_y - W_y)^2]$$
$$= hcB(\boldsymbol{J} - \boldsymbol{W})^2. \tag{7.106}$$

The second line of this equation follows from (7.102) above. We note that the awkward $\sin\theta$ factors in (7.89) have now disappeared. As Hougen points out, the eigenfunctions of the true Hamiltonian involve one less variable and so one less quantum number than the eigenfunctions of the artificial Hamiltonian and consequently the two operators cannot be completely isomorphic. However, a simple restriction on the extra quantum number in the artificial problem identifies that part of the full artifical Hamiltonian which *is* isomorphic with the true operator. Since the isomorphic Hamiltonian commutes with $(J_z - W_z)$, the two operators have a set of simultaneous eigenfunctions. Equation (7.102) states that only those eigenfunctions of the isomorphic Hamiltonian which have an eigenvalue of zero for $(J_z - W_z)$ are eigenfunctions of the true Hamiltonian.

Hougen's construct thus provides considerable simplification of the theory of the general linear molecule and has been used, often unknowingly, by countless authors. We use it with gratitude throughout this book.

7.4.4. Fine structure terms: spin–orbit, spin–spin and spin–rotation operators

In section 7.4.2 we dealt with perhaps the simplest contribution to the effective Hamiltonian for a particular electronic state, that of the rotational kinetic energy. We now turn our attention to contributions which are only slightly more complicated, the so-called fine structure interactions involving the electron spin angular momentum $\boldsymbol{S}$. Obviously for $\boldsymbol{S}$ to be non-zero, the molecule must be in an open shell electronic state with a general multiplicity $(2S+1)$.

The terms in the fundamental Hamiltonian given in equation (7.71) which are involved are $\mathcal{H}_{\text{rot}}$, $\mathcal{H}_{\text{so}}^{(e)}$, $\mathcal{H}_{\text{so}}^{(n)}$ and $\mathcal{H}_{\text{ss}}^{(\text{tensor})}$. The first-order contributions can be written down directly from equation (7.43) as

$$\mathcal{H}_{\text{eff}}^{(1)} = hcP_0\{A^{(1)}(R)\text{T}^1_{q=0}(\boldsymbol{L})\text{T}^1_{q=0}(\boldsymbol{S}) + B^{(1)}(R)\text{T}^1(\boldsymbol{J}-\boldsymbol{S})\ \text{T}^1(\boldsymbol{J}-\boldsymbol{S})$$
$$+ \gamma^{(1)}(R)\text{T}^1(\boldsymbol{J}-\boldsymbol{S})\cdot\text{T}^1(\boldsymbol{S}) + (2\sqrt{6}/3)\lambda^{(1)}(R)\text{T}^2_{q=0}(\boldsymbol{S},\boldsymbol{S})\}P_0 \tag{7.107}$$

with the linear molecule restriction

$$\text{T}^1_{q=0}(\boldsymbol{J}-\boldsymbol{L}-\boldsymbol{S}) = 0. \tag{7.108}$$

We recall that the component q is referred to the molecule-fixed axis system. The first-order contribution to the spin–orbit interaction involves $\mathcal{H}_{so}^{(e)}$,

$$A^{(1)}(R) = \frac{1}{hc\Lambda\Sigma}\langle \eta, \Lambda(S), \Sigma; R|\mathcal{H}_{so}^{(e)}|\eta, \Lambda(S), \Sigma; R\rangle, \tag{7.109}$$

where the quantum number product in the denominator is just the expectation value of $T^1_{q=0}(L)T^1_{q=0}(S)$. The first-order contribution to the spin–rotation interaction on the other hand involves $\mathcal{H}_{so}^{(n)}$:

$$\gamma^{(1)}(R) = -\frac{g_S e^2 B(R)}{4\pi\varepsilon_0 m_e c^2 S(S+1)}$$
$$\times \langle \eta, \Lambda(S), \Sigma; R|\sum_{i,\alpha}(S\cdot s_i)Z_\alpha r_{i\alpha}^{-3}(z_i - z_\alpha)z_\alpha|\eta, \Lambda(S), \Sigma; R\rangle. \tag{7.110}$$

Here, as elsewhere, the subscripts i and α stand for electrons and nuclei respectively. The factor $(S\cdot s_i)/S(S+1)$ is used to project the contribution from each open shell electron i onto the total spin angular momentum S. We remind ourselves that the effective Hamiltonian is constructed to operate within an electronic state with a given multiplicity $(2S+1)$.

The first-order contribution to the spin–spin coupling term involves just the expectation value of $\mathcal{H}_{ss}^{(tensor)}$:

$$\lambda^{(1)}(R) = \frac{1}{hc}\frac{\langle \eta, \Lambda(S), \Sigma; R|\mathcal{H}_{ss}^{(tensor)}|\eta, \Lambda(S), \Sigma; R\rangle}{(2/3)\{3\Sigma^2 - S(S+1)\}}. \tag{7.111}$$

Once again, the quantum numbers in brackets in the denominator are there simply to produce the correct matrix element, so that equation (7.79) is satisfied. Since the spin operators act on the spin coordinates only, the diagonal matrix element of the operator in $\mathcal{H}_{eff}^{(1)}$, equation (7.107), is

$$\langle S, \Sigma|T^2_{q=0}(S,S)|S, \Sigma\rangle = \langle S, \Sigma|(1/\sqrt{6})(3S_z^2 - S^2)|S, \Sigma\rangle$$
$$= (1/\sqrt{6})[3\Sigma^2 - S(S+1)]. \tag{7.112}$$

The last contribution in equation (7.107) is the first-order rotational kinetic energy, which we have already dealt with in section 7.4.2.

We now turn our attention to the second-order contributions. In order to see how these are derived, let us consider in particular the contributions of the spin–orbit interaction, $\mathcal{H}_{so}^{(e)}$. Before we can use second-order perturbation theory to evaluate these contributions, we need to write down the general matrix elements of this operator. We can do this easily if we write the expression in equation (7.72) in the simplified form

$$\mathcal{H}_{so}^{(e)} = \sum_i a_i(R)T^1(l_i)\cdot T^1(s_i), \tag{7.113}$$

where the summation is over all open shell electrons and $\hbar l_i$ represents the orbital angular momentum of an individual electron i. It requires some work to show that this form of the spin–orbit operator, known as the microscopic form, is almost exactly equivalent to the full expression in equation (7.72); it neglects only the spin–other-orbit interactions between unpaired electrons [16]. However, all we need to accept at this stage is that the matrix elements of (7.113) obey exactly the same selection rules as those of (7.72). We can then evaluate the matrix elements of (7.113) in a Hund's case (a) representation by standard spherical tensor methods:

$$\langle \eta, \Lambda; S, \Sigma | \mathcal{H}_{\mathrm{so}}^{(e)} | \eta', \Lambda'; S', \Sigma' \rangle$$
$$= \sum_q (-1)^q \sum_i \langle \eta, \Lambda | a_i T^1_{-q}(l_i) | \eta', \Lambda' \rangle \langle S, \Sigma | T^1_q(s_i) | S', \Sigma' \rangle. \quad (7.114)$$

By the Wigner–Eckart theorem,

$$\langle S, \Sigma | T^1_q(s_i) | S', \Sigma' \rangle = (-1)^{S-\Sigma} \begin{pmatrix} S & 1 & S' \\ -\Sigma & q & \Sigma' \end{pmatrix} \langle S \| T^1(s_i) \| S' \rangle. \quad (7.115)$$

Hence

$$\langle \eta, \Lambda; S, \Sigma | \mathcal{H}_{\mathrm{so}}^{(e)} | \eta', \Lambda'; S', \Sigma' \rangle$$
$$= \sum_q (-1)^q (-1)^{S-\Sigma} \begin{pmatrix} S & 1 & S' \\ -\Sigma & q & \Sigma' \end{pmatrix} \sum_i \langle \eta, \Lambda | a_i T^1_{-q}(l_i) | \eta', \Lambda' \rangle \langle S \| T^1(s_i) \| S' \rangle.$$
(7.116)

The selection rules of $\mathcal{H}_{\mathrm{so}}^{(e)}$ for Hund's case (a) quantum numbers can be seen immediately from this equation: $\Delta S = 0, \pm 1$ (by the triangle rule), $\Delta \Lambda = 0, \pm 1$, $\Delta \Sigma = 0, \mp 1$ so that $\Delta \Omega = 0$ where $\Omega = \Lambda + \Sigma$.

The second-order contribution of the spin–orbit coupling can now be written down using the expression in equation (7.43), namely:

$$\mathcal{H}^{(2)} = P_0 \Biggl\{ \sum_{\eta' \Lambda'} \sum_{S' \Sigma'} (V_{\eta \Lambda}(R) - V_{\eta' \Lambda'}(R))^{-1} \Biggl[\sum_q (-1)^q (-1)^{S-\Sigma} \begin{pmatrix} S & 1 & S' \\ -\Sigma & q & \Sigma' \end{pmatrix}$$
$$\times \sum_i \langle S \| T^1(s_i) \| S' \rangle \langle \eta, \Lambda | T^1_{-q}(a_i l_i) | \eta', \Lambda' \rangle \Biggr] \Biggl[\sum_{q'} (-1)^{q'} (-1)^{S'-\Sigma'}$$
$$\times \begin{pmatrix} S' & 1 & S'' \\ -\Sigma' & q' & \Sigma'' \end{pmatrix} \sum_j \langle S' \| T^1(s_j) \| S \rangle \langle \eta', \Lambda' | T^1_{-q'}(a_j l_j) | \eta, \Lambda'' \rangle \Biggr] \Biggr\} P_0.$$
(7.117)

Only the unpaired electrons contribute to the sums over i and j because only these electrons contribute to the total spin. The operator that represents this contribution in the effective Hamiltonian is designed to act only within the manifold of the state of interest, $|\eta, S\rangle$. In other words, it should not make any explicit reference to the quantum numbers S' and Σ' which must therefore be suppressed. With this in mind, the pair of 3-j symbols in equation (7.117) which both include S' and Σ', can be re-expressed by making use of the relationship

$$\begin{pmatrix} j_1 & \ell_2 & \ell_3 \\ m_1 & \mu_2 & -\mu_3 \end{pmatrix} \begin{pmatrix} \ell_1 & j_2 & \ell_3 \\ -\mu_1 & m_2 & \mu_3 \end{pmatrix} = (-1)^p \sum_{j_3, m_3} (2j_3 + 1) \begin{pmatrix} j_1 & j_2 & j_3 \\ m_1 & m_2 & m_3 \end{pmatrix}$$
$$\times \begin{pmatrix} \ell_1 & \ell_2 & j_3 \\ \mu_1 & -\mu_2 & m_3 \end{pmatrix} \begin{Bmatrix} j_1 & j_2 & j_3 \\ \ell_1 & \ell_2 & \ell_3 \end{Bmatrix}, \quad (7.118)$$

where $p = \ell_1 + \ell_2 + \ell_3 + \mu_1 + \mu_2 + \mu_3$. Anticipating from equation (7.107) an operator which is diagonal in the quantum numbers Λ and Σ, we set $\Lambda'' = \Lambda$ and $\Sigma'' = \Sigma$ to obtain the general second-order contribution

$$\mathcal{H}^{(2)} = P_0 \Bigg\{ \sum_{\eta',\Lambda'} \sum_{S',\Sigma'} (V_{\eta,\Lambda}(R) - V_{\eta',\Lambda'}(R))^{-1} (-1)^{S-\Sigma} \sum_k (2k+1) \begin{pmatrix} S & k & S \\ -\Sigma & 0 & \Sigma \end{pmatrix}$$

$$\times (-1)^{2S'} \begin{Bmatrix} S & 1 & S' \\ 1 & S & k \end{Bmatrix} \sum_{i,j} \langle S \| T^1(s_i) \| S' \rangle \langle S' \| T^1(s_j) \| S \rangle \sum_q \begin{pmatrix} 1 & k & 1 \\ -q & 0 & q \end{pmatrix}$$

$$\times \langle \eta, \Lambda | T^1_{-q}(a_i l_i) | \eta', \Lambda' \rangle \langle \eta', \Lambda' | T^1_q(a_j l_j) | \eta, \Lambda \rangle \Bigg\} P_0. \qquad (7.119)$$

The rank k can take values 0, 1 and 2 by the triangle rule. Of these, the scalar term with $k=0$ has no Σ dependence and hence does not affect the relative positions of the ro-vibrational energy levels. It just makes a small contribution to the electronic energy of the state $|\eta, \Lambda\rangle$. The first-rank term produces a second-order contribution to the spin–orbit interaction because it is directly proportional to the quantum number Σ from the 3-j symbol in the first line of (7.119). The contribution to the spin–orbit parameter $A(R)$ which arises in this way is given (in cm^{-1}) by

$$A^{(2)}(R) = \frac{1}{hc\Lambda[S(S+1)(2S+1)]^{1/2}} \sum_{\eta',\Lambda'} \sum_{S'} (V_{\eta,\Lambda}(R) - V_{\eta',\Lambda'}(R))^{-1} 3 (-1)^{2S'}$$

$$\times \begin{Bmatrix} S & 1 & S' \\ 1 & S & 1 \end{Bmatrix} \sum_{i,j} \langle S \| T^1(s_i) \| S' \rangle \langle S' \| T^1(s_j) \| S \rangle \sum_q \begin{pmatrix} 1 & 1 & 1 \\ -q & 0 & q \end{pmatrix}$$

$$\times \langle \eta, \Lambda | T^1_{-q}(a_i l_i) | \eta', \Lambda' \rangle \langle \eta', \Lambda' | T^1_q(a_j l_j) | \eta, \Lambda \rangle. \qquad (7.120)$$

If $\Lambda = 0$, we know that this contribution must vanish because there is no direct spin–orbit coupling for a Σ state. This expected result can be proved by replacing q by $-q$ in equation (7.119). The effect of this replacement is to multiply the whole expression by $(-1)^k$. Thus the summation over q causes the $k=1$ term to vanish for Σ states.

The third contribution from equation (7.119), that with $k=2$, can be written as

$$\mathcal{H}^{(2)}_{ss} = (-1)^{S-\Sigma} \begin{pmatrix} S & 2 & S \\ -\Sigma & 0 & \Sigma \end{pmatrix} P_0 \Bigg\{ \sum_{\eta',\Lambda'} \sum_{S'} (V_{\eta,\Lambda}(R) - V_{\eta',\Lambda'}(R))^{-1} 5 \begin{Bmatrix} S & 1 & S' \\ 1 & S & 2 \end{Bmatrix}$$

$$\times \sum_{i,j} \langle S \| T^1(s_i) \| S' \rangle \langle S' \| T^1(s_j) \| S \rangle \sum_q \begin{pmatrix} 1 & 2 & 1 \\ -q & 0 & q \end{pmatrix} \langle \eta, \Lambda | T^1_{-q}(a_i l_i) | \eta', \Lambda' \rangle$$

$$\times \langle \eta', \Lambda' | T^1_q(a_j l_j) | \eta, \Lambda \rangle \Bigg\} P_0. \qquad (7.121)$$

This result now has exactly the same form as the first-order contribution to the spin–spin coupling term in equation (7.107) with the term in braces being equated with $(2/3)\sqrt{6}\lambda^{(2)} \langle S \| T^2(S, S) \| S \rangle$ to ensure this result.

In a similar manner, there is a second-order contribution to the spin–rotation parameter which arises from the cross-term between the spin–orbit coupling and the rotational kinetic energy. We do not go through the details, but merely quote the result:

$$\gamma^{(2)}(R) = -2 \sum_{\eta'}{}' \frac{M_1 + M_2}{[\{V_\eta(R) - V_{\eta'}(R)\} \langle S, \Sigma | S_- | S, \Sigma + 1 \rangle]}, \quad (7.122)$$

where

$$\begin{aligned} M_1 &= \langle \eta, \Lambda(S), \Sigma; R | B(R) L_- | \eta', \Lambda + 1(S), \Sigma; R \rangle \\ &\quad \times \langle \eta', \Lambda + 1(S), \Sigma; R | \mathcal{H}_{so}^{(e)} | \eta, \Lambda(S), \Sigma + 1; R \rangle, \\ M_2 &= \langle \eta, \Lambda(S), \Sigma; R | \mathcal{H}_{so}^{(e)} | \eta', \Lambda - 1(S), \Sigma + 1; R \rangle \\ &\quad \times \langle \eta', \Lambda - 1(S), \Sigma + 1; R | B(R) L_- | \eta, \Lambda(S), \Sigma + 1; R \rangle. \quad (7.123) \end{aligned}$$

In summary, we can write down the effective Hamiltonian which represents the fine-structure terms for a diatomic molecule in a $^{2S+1}\Lambda$ state as:

$$\begin{aligned} \mathcal{H}_{\text{eff}}(R) &= hcP_0 \big\{ V_\eta^{(\text{ad})}(R) + V_\eta^{(sp)}(R) + A(R) \mathbf{T}_{q=0}^1(\boldsymbol{L}) \mathbf{T}_{q=0}^1(\boldsymbol{S}) + B(R) \mathbf{T}^1(\boldsymbol{J} - \boldsymbol{S}) \cdot \mathbf{T}^1(\boldsymbol{J} - \boldsymbol{S}) \\ &\quad + \gamma(R) \mathbf{T}^1(\boldsymbol{J} - \boldsymbol{S}) \cdot \mathbf{T}^1(\boldsymbol{S}) + (2/3)\sqrt{6} \lambda(R) \mathbf{T}_{q=0}^2(\boldsymbol{S}, \boldsymbol{S}) \big\} P_0, \quad (7.124) \end{aligned}$$

where

$$A(R) = A^{(1)}(R) + A^{(2)}(R), \quad (7.125)$$

$$B(R) = B^{(1)}(R) + B^{(2)}(R), \quad (7.126)$$

$$\gamma(R) = \gamma^{(1)}(R) + \gamma^{(2)}(R), \quad (7.127)$$

$$\lambda(R) = \lambda^{(1)}(R) + \lambda^{(2)}(R). \quad (7.128)$$

In other words, each of the parameters is the sum of a first-order and a second-order contribution. We have met equation (7.126) for the effective rotational constant operator before, in an earlier section, where we pointed out that the second-order contribution $B^{(2)}$ is very much smaller than $B^{(1)}$ and that these two contributions have a different reduced mass dependence. It is important to realise that this is not generally true. Indeed, except for molecules with very light atoms such as H_2, the second-order contribution to the spin–rotation parameter is usually very much larger in magnitude than the first-order contribution. The same is also often true for the spin–spin coupling parameter λ. The reduced mass dependences of the two contributions to the spin–rotation parameter γ are different from each other and quite complicated. However, Brown and Watson [17] were able to show the rather remarkable result that when one takes the first- and second-order contributions together as in equation (7.127), the reduced mass dependence of the resultant parameter $\gamma(R)$ is simply μ^{-1}.

It will be noticed that two contributions to the vibrational potential energy, of the general form $V(R)$ and independent of the rotational and spin quantum numbers, have also been included in equation (7.124). These are corrections to $V_\eta(R)$, the zeroth-order contribution to the electronic energy defined in equation (7.76). The first term, $V_\eta^{(\text{ad})}(R)$, is the adiabatic contribution to the electronic energy, which we have discussed in section 2.7. It describes the first-order effect of the nuclear kinetic energy within the electronic state $|\eta, \Lambda\rangle$:

$$hcV_\eta^{(ad)}(R)$$
$$= \langle \eta, \Lambda(S), \Sigma; R | \left\{ \frac{1}{2\mu} P_R^2 + hcB(R)(\boldsymbol{L}^2 - 2L_z^2) + \frac{1}{2M} \boldsymbol{P}^2 \right\} | \eta, \Lambda(S), \Sigma; R \rangle. \tag{7.129}$$

This is a relatively small contribution when compared with $V_\eta(R)$ but it is nevertheless important when considering the isotopic dependence of the electronic energy. The second term, $V_\eta^{(sp)}(R)$, describes the contribution from spin–spin coupling (in first order) and spin–orbit coupling (in second order) to the electronic energy:

$$hcV_\eta^{(sp)}(R) = \langle \eta, \Lambda(S), \Sigma; R | \mathcal{H}_{ss}^{(s)} | \eta, \Lambda(S), \Sigma; R \rangle + \frac{1}{2S+1} \sum_\Sigma \sum_{\eta', \Lambda', \Sigma'}$$
$$\times \frac{|\langle \eta', \Lambda'(S'), \Sigma'; R | \mathcal{H}_{so}^{(e)} | \eta, \Lambda(S), \Sigma; R \rangle|^2}{[V_\eta(R) - V_{\eta'}(R)]}. \tag{7.130}$$

The second of these contributions is just the $k=0$ contribution to the energy in equation (7.119). Neither of the contributions shows any dependence on the spin quantum number Σ.

7.4.5. Λ-doubling terms for a Π electronic state

The two component states of orbital degeneracy in a diatomic molecule have opposite parity. As we described in chapter 6, parity is the symmetry label associated with the behaviour of a wave function under the space-fixed inversion operator E^*:

$$E^* f(X, Y, Z) = f(-X, -Y, -Z). \tag{7.131}$$

Under circumstances where this operation leaves the physical system invariant, the quantum states have either positive or negative parity,

$$E^* \psi(X, Y, Z) = \pm \psi(X, Y, Z), \tag{7.132}$$

that is, they transform with the upper or lower sign choice in the above equation.

The two-fold degeneracy of these orbital levels is exact for the non-rotating molecule (strictly we have to consider spin–orbit levels: those with $\Omega \geq 1/2$ are degenerate for a non-rotating molecule). However, when the molecule rotates, the degeneracy is lifted in a manner which increases as J increases. The precise J dependence depends on the Ω value for Hund's case (a) states or on the Λ value for case (b). The phenomenon is known as lamda-doubling. All lamda-doubling effects originate from the admixture of the rotational levels of the degenerate electronic state with the corresponding levels of a Σ electronic state. Since each of these Σ state levels is non-degenerate, each has a particular parity, either positive or negative. Each therefore interacts with only *one* of the two degenerate components, namely, that with the same parity. This component is therefore shifted either up or down, away from its partner with the opposite parity.

In this way, a splitting of the Λ-doublets arises. Λ-doubling can be described very succinctly by the effective Hamiltonian method.

For reasons which will become clear later, Λ-doubling effects are largest for molecules in Π electronic states. The electronic orbital part of the wave function for such a state can be represented by the pair of functions $|\Lambda = +1\rangle$ and $|\Lambda = -1\rangle$, which correspond to the angular momentum vector precessing in clockwise and anticlockwise directions about the internuclear axis. We shall instead take sum and difference combinations of these functions, $(1/\sqrt{2})\{|\Lambda = +1\rangle \pm |\Lambda = -1\rangle\}$, since they reflect the parity properties of the two Λ-doublets more closely (although they are still not true representations of states with a definite parity because we have not included the rotational and electron spin parts of the wave function). In the effective Hamiltonian, the Λ-doubling effects arise from the perturbations by the matrix elements of the spin–orbit and rotational Hamiltonians, $\mathcal{H}_{so}$ and $\mathcal{H}_{rot}$, with $\Delta \Lambda = \pm 1$. In second-order perturbation theory, these operators are applied twice giving rise to two types of term in the effective Hamiltonian, those with $\Delta \Lambda = 0$ overall and those with $\Delta \Lambda = \pm 2$. The former type makes equal contributions of the same sign to both Λ-doublets; Λ-doubling is not resolved by this type of term. On the other hand, the terms with $\Delta \Lambda = \pm 2$ make equal contributions of opposite sign to each Λ-doublet. These are then the terms which are responsible for Λ-doubling effects.

As mentioned above, the terms which are responsible for the coupling of the $^{2S+1}\Pi$ state to the $^{2S+1}\Sigma$ states are the spin–orbit coupling and the rotational electronic Coriolis term. Thus in the second-order perturbation expression in equation (7.43), the perturbation term is

$$\mathcal{H}' = \mathcal{H}_{so} + \mathcal{H}_{rot}, \tag{7.133}$$

where, for the present purposes, we only need consider the terms with $\Delta \Lambda = \pm 1$:

$$\mathcal{H}_{so} = \sum_{q=\pm 1}{}' (-1)^q T_q^1(a_i l_i) T_{-q}^1(s_i), \tag{7.134}$$

$$\mathcal{H}_{rot} = -2hc \sum_{q=\pm 1}{}' N_q T_q^1(B\boldsymbol{L}). \tag{7.135}$$

The latter expression comes from equation (V.157). We recall that $\mathcal{H}_{so}$ connects states with $\Delta S = 0, \pm 1$, whereas $\mathcal{H}_{rot}$ couples in states of the same multiplicity only, $\Delta S = 0$.

In order to construct the Λ-doubling terms, we substitute the perturbation operator $\mathcal{H}'$ in the second-order expressions and look for terms which link the $|\eta, \Lambda = +1\rangle$ basis state with $|\eta, \Lambda = -1\rangle$. Because $\mathcal{H}'$ contains two independent operators, we finish up with three independent contributions to the effective Hamiltonian which are as follows:

(i) the squared term $\mathcal{H}_{so} \times \mathcal{H}_{so}$ which gives rise to an operator of the form $T_q^1(\boldsymbol{S})T_q^1(\boldsymbol{S})$,
(ii) the cross term $2 \times \mathcal{H}_{so} \times \mathcal{H}_{rot}$ which gives rise to an operator of the form $T_q^1(\boldsymbol{S})T_q^1(\boldsymbol{N})$,
(iii) the squared term $\mathcal{H}_{rot} \times \mathcal{H}_{rot}$ which gives rise to an operator of the form $T_q^1(\boldsymbol{N})T_q^1(\boldsymbol{N})$.

In all of these expressions, the spherical tensor component q can take the values ± 1 only.

The operator terms in the effective Λ-doubling Hamiltonian can all be expressed more concisely by combining the spherical tensors from equation (7.135),

$$T^2_{\pm 2}(\mathbf{A}, \mathbf{B}) = T^1_{\pm 1}(\mathbf{A}) T^1_{\pm 1}(\mathbf{B}). \tag{7.136}$$

In this way, we arrive at the following effective Hamiltonian for Λ-doubling for a molecule in a $^{2S+1}\Pi$ state:

$$\mathcal{H}_{LD}(R) = \sum_{q=\pm 1}{}' \exp(-2iq\phi)\left[-o(R)T^2_{2q}(\mathbf{S}, \mathbf{S}) + p(R)T^2_{2q}(\mathbf{S}, \mathbf{N}) - q(R)T^2_{2q}(\mathbf{N}, \mathbf{N})\right]. \tag{7.137}$$

The coordinate ϕ is the electron orbital azimuthal angle and the presence of the exponential factor $\exp(-2iq\phi)$ ensures that only the matrix elements which connect the component $|\eta, \Lambda = +1\rangle$ with $|\eta, \Lambda = -1\rangle$ are non-zero. The electron orbital basis functions used in this book imply a choice of phase factor which leads to

$$\langle \eta, \Lambda = \pm 1 | \exp(\pm 2i\phi) | \eta, \Lambda = \mp 1 \rangle = -1. \tag{7.138}$$

We see that for a molecule in a $^{2S+1}\Pi$ state, there are in general three Λ-doubling parameters. However, for a singlet state $(S=0)$, only q is non-zero and for a doublet state, p and q are non-zero. To second order in perturbation theory, these parameters are given by the following expressions (in cm^{-1}):

$$o(R) = o^{(1)}(R) + o^{(2)}(R), \tag{7.139}$$

where

$$o^{(1)}(R) = -\left(\frac{6}{5}\right)^{1/2} g_S^2 \mu_B^2 (\mu_0/4\pi hc)[S(S+1)(2S+1)]^{-1} \begin{Bmatrix} S & 1 & S \\ 1 & S & 2 \end{Bmatrix}^{-1}$$

$$\times \sum_{i, j > i} \langle S \| T^2(\mathbf{s}_i, \mathbf{s}_j) \| S \rangle \langle \eta, \Lambda = 1 | C_2^2(\theta, \phi) r_{ij}^{-3} | \eta, \Lambda = -1 \rangle, \tag{7.140}$$

$$o^{(2)}(R) = -(hc)^{-1} [S(S+1)(2S+1)]^{-1} \begin{Bmatrix} S & 1 & S \\ 1 & S & 2 \end{Bmatrix}^{-1}$$

$$\times \sum_{\eta', \Lambda'} \sum_{S'} (V_{\eta, \Pi}(R) - V_{\eta, \Sigma}(R))^{-1} (-1)^s \begin{Bmatrix} S & 1 & S' \\ 1 & S & 2 \end{Bmatrix} \sum_{i,j} \langle S \| T^1(\mathbf{s}_i) \| S' \rangle \langle S' \| T^1(\mathbf{s}_j) \| S \rangle$$

$$\times \langle \eta, \Lambda = 1 | T^1_1(a_i l_i) | \eta', \Lambda' = 0^s \rangle \langle \eta', \Lambda' = 0^s | T^1_{-1}(a_j l_j) | \eta =, \Lambda = 1 \rangle, \tag{7.141}$$

$$p(R) = -4[S(S+1)(2S+1)]^{-1/2} \sum_{\eta', \Lambda'} (V_{\eta, \Pi}(R) - V_{\eta, \Sigma}(R))^{-1} (-1)^s$$

$$\times \langle \eta, \Lambda = 1 | B(R) T^1_1(\mathbf{L}) | \eta', \Lambda = 0^s \rangle \sum_i \langle \eta', \Lambda = 0^s | a_i T^1_{-1}(l_i) | \eta, \Lambda = 1 \rangle$$

$$\times \langle S \| T^1(\mathbf{s}_i) \| S \rangle, \tag{7.142}$$

$$q(R) = 4hc \sum_{\eta',\Lambda'} (V_{\eta,\Pi}(R) - V_{\eta,\Sigma}(R))^{-1}(-1)^s |\langle \eta, \Lambda = 1|B(R)T_1^1(\boldsymbol{L})|\eta', \Lambda = 0^s\rangle|^2.$$

(7.143)

These expressions apply for a state with total spin angular momentum S. The exponent s is even for Σ^+ states and odd for Σ^- states. The factorisation of the spin and orbital matrix elements in these expressions has been achieved by making a single configuration approximation. If configuration interaction is not negligible, more general expressions, given in a paper on the effective Hamiltonian by Brown, Colbourn, Watson and Wayne [7] are used. The Λ-doubling parameters p and q were originally introduced by Mulliken and Christy [18].

The expressions in equations (7.139) to (7.143) are exact to second order in perturbation theory. There are also higher-order terms of the same operator form as given in (7.137) but such contributions are much smaller as long as the interaction terms are small compared with the separation of the Π and Σ states; this is usually the case. It is important to appreciate that the form of the Λ-doubling operator is the same even when these higher order effects are included. This is a real advantage of the effective Hamiltonian approach. The correct form of the Hamiltonian can be established by a limited perturbation treatment. Thus, no approximation is made in fitting the parameters of this Hamiltonian to experimental data. The limitations, such as they are, arise only when the parameters so determined are compared with theoretical expectations.

We note that the perturbation terms in the full Hamiltonian which are responsible for the Λ-doubling effects are the same as those which give rise to the spin–spin and spin–rotation terms as discussed in the previous section, namely $\mathcal{H}_{\text{rot}}$ and $\mathcal{H}_{\text{so}}$. The only difference is that we select the effective operators with $\Delta \Lambda = 0$ for the fine-structure terms and those with $\Delta \Lambda = \pm 2$ for the Λ-doubling terms. Because of this, the Λ-doubling effects manifest themselves in the second order of perturbation theory, that is, they depend on the admixture of Σ electronic states, whereas the fine-structure terms have both first- and second-order contributions, as given in equations (7.125) to (7.128). This is an important point to appreciate. In addition to terms in the effective Hamiltonian which have the same operator form as in the full Hamiltonian, there are others which have a completely different form. The common origin of the fine-structure and Λ-doubling effects in the effective Hamiltonian suggests that the two sets of parameters are related. These relationships can have a very simple form in particular circumstances. We shall discuss them after we have included the effects of vibrational averaging (over R) in the effective Hamiltonian.

7.4.6. Nuclear hyperfine terms

When a molecule contains an atom with a nuclear spin $I \geq 1/2$, its energy levels acquire an additional $(2I + 1)$ degeneracy. This degeneracy is lifted in practice by magnetic and electric interactions which are called nuclear hyperfine interactions; they have been described in detail in chapter 4. The magnitude of these interactions is usually relatively small, only rarely larger than 0.03 cm^{-1} or 1 GHz. The size of nuclear hyperfine effects is therefore very much less than electron spin fine structure effects, which in turn are smaller than the electronic and vibrational contributions to the energy. Consequently, it is usually only necessary to include the first-order effects of the nuclear hyperfine terms in the effective Hamiltonian.

We recall from chapter 4 that there are many individual types of interaction which involve the nuclear magnetic dipole and electric quadrupole moments. Let us take just three of these to exemplify how the effective Hamiltonian is constructed. They are as follows:

(i) the Fermi contact interaction

$$\mathcal{H}_F = (2/3) \sum_\alpha \sum_i g_S \mu_B g_\alpha \mu_N \mu_0 \delta(\boldsymbol{r}_{\alpha i}) \boldsymbol{s}_i \cdot \boldsymbol{I}_\alpha, \qquad (7.144)$$

(ii) the dipole–dipole coupling term,

$$\mathcal{H}_{\text{dip}} = -\sum_\alpha \sum_i g_S \mu_B g_\alpha \mu_N \frac{\mu_0}{4\pi} \left\{ \frac{\boldsymbol{s}_i \cdot \boldsymbol{I}_\alpha}{r_{\alpha i}^3} - \frac{3(\boldsymbol{s}_i \cdot \boldsymbol{r}_{\alpha i})(\boldsymbol{I}_\alpha \cdot \boldsymbol{r}_{\alpha i})}{r_{\alpha i}^5} \right\}, \qquad (7.145)$$

(iii) the electric quadrupole coupling term,

$$\mathcal{H}_Q = -e \sum_\alpha \mathrm{T}^2(\nabla \boldsymbol{E}_\alpha) \cdot \mathrm{T}^2(\boldsymbol{Q}_\alpha). \qquad (7.146)$$

In these expressions the index i runs over electrons and α runs over nuclei. The Fermi contact term describes the magnetic interaction between the electron spin and nuclear spin magnetic moments when there is electron spin density at the nucleus. This condition is imposed by the presence of the Dirac delta function $\delta(\boldsymbol{r}_{\alpha i})$ in the expression. The dipole–dipole coupling term describes the classical interaction between the magnetic dipole moments associated with the electron and nuclear spins. It depends on the relative orientations of the two moments described in equation (7.145) and falls off as the inverse cube of the separations of the two dipoles. The cartesian form of the dipole–dipole interaction to some extent masks the simplicity of this term. Using the results of spherical tensor algebra from the previous chapter, we can bring this into the open as

$$\mathcal{H}_{\text{dip}} = -(10)^{1/2} \sum_\alpha \sum_i g_S \mu_B g_\alpha \mu_N (\mu_0/4\pi) \mathrm{T}^1(\boldsymbol{I}_\alpha) \cdot \mathrm{T}^1(\boldsymbol{S}_i, \boldsymbol{C}_i). \qquad (7.147)$$

In other words, it is a first-rank scalar product involving the nuclear spin angular momentum with an operator defined by

$$\mathrm{T}_q^1(\boldsymbol{S}_i, \boldsymbol{C}_i) = -(-1)^q \sum_{q_1, q_2} \mathrm{T}_{q_1}^1(\boldsymbol{s}_i) C_{q_2}^2(\theta_{\alpha i}, \phi_{\alpha i}) r_{\alpha i}^{-3} (3)^{1/2} \begin{pmatrix} 1 & 2 & 1 \\ q_1 & q_2 & -q \end{pmatrix}, \qquad (7.148)$$

where $(r_{\alpha i}, \theta_{\alpha i}, \phi_{\alpha i})$ are the spherical polar coordinates of electron i relative to nucleus α. Finally, the quadrupole term describes the interaction between the electric quadrupole moment of nucleus α and the electric field gradient $\nabla \boldsymbol{E}_\alpha$ at the nucleus; the latter is made up of contributions from all nearby charges, both electrons and nuclei. This electric field gradient also falls off as $r_{\alpha i}^{-3}$.

The first-order contribution of these hyperfine interactions to the effective electronic Hamiltonian involves the diagonal matrix elements of the individual operator terms over the electronic wave function, see equation (7.43). As before, we factorise out those terms which involve the electronic spin and spatial coordinates. For example, for the Fermi contact term we need to evaluate matrix elements of the type:

$$\langle \eta, \Lambda(S) | \delta(\boldsymbol{r}_{\alpha i}) \boldsymbol{s}_i | \eta, \Lambda(S) \rangle. \tag{7.149}$$

The nuclear hyperfine operators therefore have essentially the same form in the effective Hamiltonian as they do in the full Hamiltonian, certainly as far as the nuclear spin terms are concerned. Throughout our derivation, we have assumed that the electronic state $|\eta, \Lambda\rangle$ which is to be described by our effective Hamiltonian has a well-defined spin angular momentum S. It is therefore desirable to write the effective Hamiltonian in terms of the associated operator $\boldsymbol{S}$ rather than the individual spin angular momenta $\boldsymbol{s}_i$. We introduce the projection operators Φ_i^s for each electron i,

$$\boldsymbol{s}_i = \Phi_i^s \boldsymbol{S}. \tag{7.150}$$

The operator can be evaluated by taking the scalar product of each side with $\boldsymbol{S}$ and rearranging to give

$$\Phi_i^s = \boldsymbol{s}_i \cdot \boldsymbol{S} / \{S(S+1)\}. \tag{7.151}$$

For a molecule in a Hund's case (a) state where the component of $\boldsymbol{S}$ along the z axis is well defined as Σ, we can write

$$\Phi_i^s = s_{iz}/\Sigma. \tag{7.152}$$

Using these results, we can write

$$\mathcal{H}_F(R) = \sum_\alpha b_F^\alpha \boldsymbol{I}_\alpha \cdot \boldsymbol{S}, \tag{7.153}$$

where

$$b_F^\alpha = \sum_i \frac{2 g_S \mu_B g_\alpha \mu_N \mu_0}{3} \langle \eta, \Lambda(S) | \Phi_i^s \delta(\boldsymbol{r}_{\alpha i}) | \eta, \Lambda(S) \rangle. \tag{7.154}$$

In a similar fashion, the dipole–dipole coupling term becomes

$$\mathcal{H}_{\mathrm{dip}}(R) = \sum_\alpha (\sqrt{6}/3) c^\alpha \mathrm{T}_{q=0}^2(\boldsymbol{I}_\alpha, \boldsymbol{S}), \tag{7.155}$$

where

$$c^\alpha = \sum_i \frac{3 g_S \mu_B g_\alpha \mu_N \mu_0}{8\pi} \langle \eta, \Lambda(S) | \Phi_i^s (3\cos^2 \theta_{i\alpha} - 1) r_{i\alpha}^{-3} | \eta, \Lambda(S) \rangle, \tag{7.156}$$

and the component $q = 0$ refers as always to the molecule-fixed z axis. We remember that the parameters b_F and c^α in equations (7.154) and (7.156) are still at this stage functions of the molecular bond length R, albeit weak ones.

Since the electric quadrupole interaction does not involve the electron spin operators, its form remains the same as in equation (7.146) with the operator $\mathrm{T}^2(\nabla \boldsymbol{E})$

evaluated over the electronic wave function:

$$\langle \eta, \Lambda(S) | T^2_{q=0}(\nabla E) | \eta, \Lambda(S) \rangle. \tag{7.157}$$

The effective operator is the product of one operator on three-dimensional spatial coordinates and another which acts on nuclear spin space. This distinction can be brought out even more clearly by making use of the operator replacement theorem in section 5.5.3 to give

$$\mathcal{H}_Q(R) = \sum_\alpha \frac{eQ_\alpha q_0^\alpha}{4I_\alpha(2I_\alpha - 1)} \sqrt{6} T^2_{q=0}(I_\alpha, I_\alpha). \tag{7.158}$$

The expectation value over the electric field gradient is represented by the parameter q_0^α:

$$q_0^\alpha = -2\langle \eta, \Lambda | T^2_{q=0}(\nabla E) | \eta, \Lambda \rangle. \tag{7.159}$$

This parameter is also R-dependent at this stage of the development. We note the curiosity that, because of the way in which it is defined, the parameter actually represents the *negative* of the electric field gradient.

In equations (7.155) and (7.158) we have taken the diagonal $q = 0$ component of the second-rank spherical tensors $T^2(I_\alpha, S)$ and $T^2(I_\alpha, I_\alpha)$. In general, these interactions and others like them will have off-diagonal terms also, with $q = \pm 1$ and ± 2. The $q = \pm 2$ components are particularly interesting because, for a molecule in a Π electronic state, they connect the $|\Lambda = +1\rangle$ and $|\Lambda = -1\rangle$ components directly. They therefore make additional hyperfine contributions to the Λ-doubling of molecules in Π electronic states. As a result, the nuclear hyperfine splitting of one component of a Λ-doublet is different from that of the other component. The two contributions are:

$$\mathcal{H}_{\text{dip}}(R) = \sum_\alpha {\sum_{q=\pm 1}}' \exp(-2iq\phi) d^\alpha T^2_{2q}(I_\alpha, S), \tag{7.160}$$

$$\mathcal{H}_Q(R) = -\sum_\alpha {\sum_{q=\pm 1}}' \exp(-2iq\phi) \frac{eQ_\alpha q_2^\alpha}{4I_\alpha(2I_\alpha - 1)} T^2_{2q}(I_\alpha, I_\alpha), \tag{7.161}$$

where

$$d^\alpha = \sum_i \frac{3g_S \mu_B g_\alpha \mu_N \mu_0}{8\pi} \langle \eta, \Lambda = \pm 1 | \Phi_i^s \sin^2 \theta_\alpha r_{i\alpha}^{-3} | \eta, \Lambda = \pm 1 \rangle, \tag{7.162}$$

and

$$q_2^\alpha = -2\sqrt{6} \langle \eta, \Lambda = \pm 1 | T^2_{\pm 2}(\nabla_\alpha E) | \eta, \Lambda = \mp 1 \rangle. \tag{7.163}$$

The other off-diagonal terms, with $q = \pm 1$, mix different electronic states with $\Delta \Lambda = \pm 1$. In the great majority of cases this mixing is too small to be significant, even with the most accurate measurements. However when the states involved happen to lie close together, or when the mixing term is unusually large, these off-diagonal effects produce additional higher-order terms in the effective Hamiltonian. We will not discuss these effects any further.

7.4.7. Higher-order fine structure terms

Thus far, we have investigated the various contributions to the effective Hamiltonian for a diatomic molecule in a particular electronic state which arise from the spin–orbit and rotational kinetic energy terms treated up to second order in degenerate perturbation theory. Higher-order effects of such mixing will also contribute and we now consider some of their characteristics.

If we refer back to the explicit forms for third- and fourth-order contributions, given in equation (7.43) and depicted in figure 7.1, we see that the number of possible pathways between the various electronic states involved increases rapidly with the order of perturbation theory. This makes the mathematical treatment of such terms considerably more arduous and the work involved in going beyond fourth order renders the exercise futile. Fortunately, in the great majority of cases, these higher order contributions can be safely neglected. There are two reasons for this. First, the contributions get progressively smaller as the order of perturbation theory increases and second, there are angular momentum constraints on the effects of such terms.

The first of these reasons can be easily appreciated by reference to equation (7.43) again. It can be seen that the nth order terms have the general form $\langle i|\mathcal{H}'|j\rangle^n/(E_0 - E_i)^{n-1}$, that is, the successive orders of terms in the perturbation expansion are decreased by the factor $\langle i|\mathcal{H}'|j\rangle/(E_0 - E_i)$. If the perturbation treatment is well constructed from a convergence point of view, this factor will be considerably smaller than unity; for light molecules, it is typically on the order of 10^{-2}. Consequently, the higher-order effects become rapidly smaller and, for all but the most precise measurements, are negligible compared with the experimental uncertainty. The second consideration is more subtle but in the end more powerful because it depends on symmetry properties of the molecular system. It is related to the rank of the electron spin operator in the effective Hamiltonian. Though not always desirable, it is always possible to collect the various spin terms together and to express a typical term in the effective Hamiltonian as the scalar product

$$\mathbf{T}^k(\mathbf{X}) \cdot \mathbf{T}^k(\mathbf{S}), \tag{7.164}$$

where the operator $\mathbf{X}$ stands for the non-spin terms such as position $\mathbf{r}$, angular momentum $\mathbf{N}$ or orbital angular momentum $\mathbf{L}$. For a molecule in a state with a well-defined spin angular momentum S, simple application of the Wigner–Eckart theorem to the spin term leads to a 3-j symbol with upper row arguments S, k and S. We recall that such a symbol is subject to the triangle rule $\Delta(S, k, S)$. The matrix elements of the effective operator (7.164) within the electronic state of interest will therefore be rigorously zero unless $2S \geq k$. We have already met examples of this constraint in section 7.4.4. The spin–rotation term is first rank in electron spin and so appears in states of doublet or higher multiplicity. The spin–spin fine structure term is second rank and so only occurs in states of triplet or higher multiplicity. When the perturbation term involves the electron spin through spin–orbit coupling, the higher-order terms are of higher rank in the operator $\mathbf{S}$ and so their effects do not show up in states of lower multiplicity.

Table 7.1. *Some higher-order contributions to the effective Hamiltonian*

Perturbation form	Parameter	Order of perturbation	Rank of spin operator
$(N \cdot S)(L \cdot S)(L \cdot S)$	γ_S	3	3
$(L_z \cdot S_z)(L \cdot S)(L \cdot S)$	η	3	3
$(L \cdot S)(L \cdot S)(L \cdot S)(L \cdot S)$	Θ	4	4

In order to appreciate this point more clearly, we confine our attention to the contributions to $\mathcal{H}_{\text{eff}}$ produced by perturbations from the spin–orbit coupling $\mathcal{H}_{\text{so}}$ and the electronic Coriolis mixing $\mathcal{H}_{\text{rot}}$. If we represent an off-diagonal matrix element of the former by $(L \cdot S)$ and the latter by $(N \cdot L)$, we can describe some examples of these higher order terms, as shown in table 7.1. The third-rank terms appear only in states of quartet or higher multiplicity and the fourth-rank terms in states of quintet (or higher) multiplicity. With the important exception of transition metal compounds, the vast majority of electronic states encountered in practice have triplet multiplicity or lower.

Let us consider the derivation of the higher-order spin–rotation term γ_S in more detail. The contribution of such a term to the energy levels of molecules in quartet states was first suggested by Hougen [19] and developed in detail by Brown and Milton [20]. The terms arise in the third order of perturbation theory, the general form of which is given in equation (7.43) and illustrated in figure 7.1. Two examples of the way in which these perturbations affect the levels of a $^4\Sigma$ state are shown in figure 7.2. In the first example, the admixture of a $^4\Pi$ state, the normalisation terms have to be written in a symmetrised form to ensure that the resulting operator is Hermitian. The second example of a $^2\Sigma$ state shows that states of different multiplicity can also contribute to these higher order effects. Using the same approach as that described for the fine structure terms in section 7.4.4 but with rather more algebra, we can obtain the following expression for the third-order spin–rotation effects,

$$\mathcal{H}_{\text{sr}}^{(3)} = C^{(3)}(R)\mathrm{T}^3(L^2, N) \cdot \mathrm{T}^3(S, S, S), \tag{7.165}$$

where

$$C^{(3)}(R) = -(10)^{-1/2} \langle \Lambda | \mathrm{T}_0^2(L^2) | \Lambda \rangle^{-1} \langle S \| \mathrm{T}^3(S,S,S) \| S \rangle^{-1} (35)$$
$$\times \sum_{\eta', \Lambda', \Sigma', q} \left(E_\eta^0(R) - E_{\eta'}^0(R) \right)^{-2} (-1)^{q+S-\Sigma+S'-\Sigma} |\langle \eta, \Lambda, S \| \sum_i \mathrm{T}^1(s_i) \mathrm{T}_q^1(a_i l_i) \| \eta', \Lambda', S' \rangle|^2$$
$$\times \begin{pmatrix} 1 & 2 & 3 \\ q & -q \pm 1 & \mp 1 \end{pmatrix} \begin{Bmatrix} 1 & S & S' \\ S & 2 & 3 \end{Bmatrix}$$
$$\times \begin{pmatrix} 1 & 1 & 2 \\ -q & \pm 1 & q \mp 1 \end{pmatrix} [S(S+1)(2S+1)]^{1/2} \begin{Bmatrix} 1 & S' & S \\ S & 1 & 2 \end{Bmatrix}$$
$$\times [B'(R) - B(R)], \tag{7.166}$$

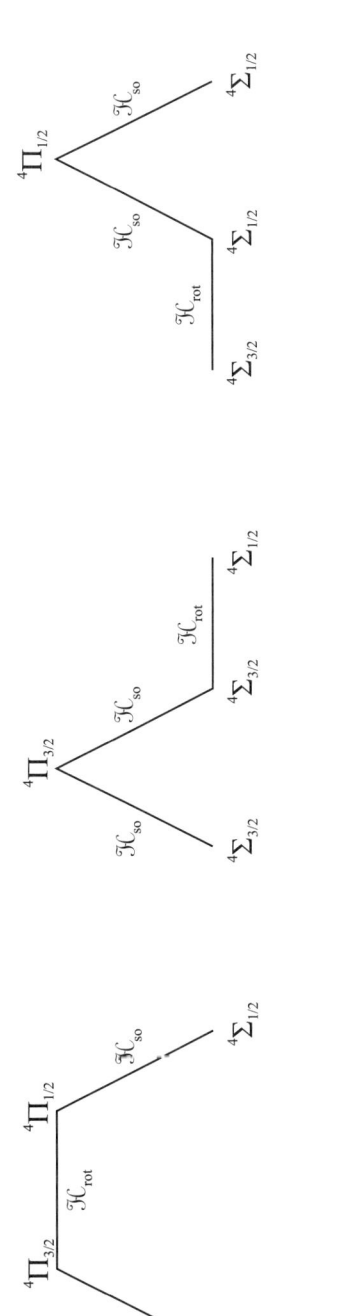

Figure 7.2. Diagrammatic representation of some third-order contributions to the matrix elements $\langle \Omega = 1/2 | \mathcal{H}_{\text{eff}} | \Omega = 3/2 \rangle$ in a $^4\Sigma$ state. The three top diagrams illustrate the admixture of $^4\Pi$ states, whilst the lower diagram shows the admixture of a $^2\Sigma$ state.

and

$$\langle S \| T^3(\boldsymbol{S}, \boldsymbol{S}, \boldsymbol{S}) \| S \rangle = (1/4)(10)^{-1/2}[(2S-2)(2S-1)(2S)(2S+1)$$
$$\times (2S+2)(2S+3)(2S+4)]^{1/2}. \tag{7.167}$$

The parameter γ_S, first introduced by Brown and Milton [20], is related to $C^{(3)}(R)$ in equation (7.165) by

$$\gamma_S(R) = (6)^{1/2}(10)^{-1} C^{(3)}(R) \langle \eta, \Lambda | T_0^2(\boldsymbol{L}^2) | \eta, \Lambda \rangle. \tag{7.168}$$

This expression for the higher-order spin–rotation interaction probably seems at best daunting and at worse cumbersome. However the great merit of the formulation in equation (7.165) is that it is very straightforward to work out its matrix elements, as we shall see later.

The experimental evidence for such a contribution to the spin–rotation interaction in the effective Hamiltonian was somewhat elusive in the early days although there are now well documented cases of its involvement, for example for CH in its $^4\Sigma^-$ state [21]. Equation (7.166) suggests one reason why this parameter is not as important in practice as might be expected. The last factor on the right-hand side of (7.166) is just the difference of the rotational constant operators for the upper and lower states. This causes a considerable degree of cancellation in a typical situation because the B value is not expected to vary markedly between the electronic states.

We shall not go through the details of the derivations of the other terms in table 7.1. Suffice it to say that they are similar to that described for the parameter $\gamma_S(R)$. We shall write down their operator form once we have dealt with the next stage of the development of the effective Hamiltonian, that of vibrational averaging.

7.5. Effective Hamiltonian for a single vibrational level

7.5.1. Vibrational averaging and centrifugal distortion corrections

Once we have removed the terms which couple different electronic states (at least to a certain level of accuracy), we can deal with the motion in the other degrees of freedom of the molecule for each electronic state separately. The next step in the process is to consider the vibrational degree of freedom which is usually responsible for the largest energy separations within each electronic state. If we perform a suitable transformation to uncouple the different vibrational states, we obtain an effective Hamiltonian for each vibronic state. Once again, we adopt a perturbation approach.

In this case, the zeroth-order Hamiltonian is chosen to represent the vibrational energy of the anharmonic oscillator:

$$\{P_R^2/2\mu + V_\eta(R)\} | \eta, v \rangle = | \eta, v \rangle hc \{T_{\eta e} + G_{\eta v}^{(0)}\}. \tag{7.169}$$

Here $|\eta, v\rangle$ is the radial eigenvector with vibrational quantum number v in the electronic state $|\eta, \Lambda\rangle$. It is assumed that the potential curve $V_\eta(R)$ has a minimum value $V_\eta(R_{\eta e})$ equal to $hc T_{\eta e}$ at the equilibrium bond length $R_{\eta e}$, i.e. this defines the energy origin of the electronic state in question. The zeroth-order vibrational term value relative to this minimum is then $G_{\eta v}^{(0)}$ where

$$G_{\eta v}^{(0)} = \left(v + \tfrac{1}{2}\right)\omega_{\eta e} - \left(v + \tfrac{1}{2}\right)^2 \omega_{\eta e} x_e + \left(v + \tfrac{1}{2}\right)^3 \omega_{\eta e} y_e + \cdots. \quad (7.170)$$

We recall that the vibrational potential energy function $V_\eta(R)$ is just the zeroth-order eigenvalue in the description of the electronic motion given in equation (7.76). Although this is not the most accurate choice we could make at this level (it is possible, for example, to add the adiabatic correction in equation (7.129)), it has the great advantage that it is isotopically independent. For heavier molecules, the spin correction term $V_\eta^{(\text{sp})}(R)$ in equation (7.130) could be included without spoiling the isotopic independence but for lighter molecules we prefer to develop the Hamiltonian with a power series dependence on the spin operators. This is achieved by avoiding spin terms in the zeroth-order vibrational energies.

We use the operator on the left-hand side of equation (7.169) as the zeroth-order vibrational Hamiltonian. The remaining terms in the effective electronic Hamiltonian, given for example in equations (7.124) and (7.137), are treated as perturbations. In a similar vein to the electronic problem, we consider only first- and second-order corrections as given in equations (7.68) and (7.69) to produce an effective Hamiltonian $\mathcal{H}_{\eta v}$ which is confined to act within a single vibronic state $|\eta, v\rangle$ only. Once again, the condition for the validity of this approximation is that the perturbation matrix elements should be small compared with the vibrational intervals. It will therefore tend to fail for loosely bound states with low vibrational frequencies.

The first-order perturbation contribution is obtained by replacing each operator function $X(R)$ by its anharmonic expectation value,

$$X_{\eta v} = \langle \eta, v | X_\eta(R) | \eta, v \rangle = X_{\eta e} + (B_{\eta e}/\omega_{\eta e})(X''_{\eta e} - 3a_{1\eta} X'_{\eta e})(v + 1/2) + \cdots, \quad (7.171)$$

where X corresponds to any one of $V^{(\text{ad})}$, $V^{(\text{sp})}$, B, A, γ, λ, o, p, q or the nuclear hyperfine parameters. The expansion in powers of $(v + 1/2)$ can be obtained by a semi-classical perturbation treatment [22]. The subscripts η and e refer to the value at $R_{\eta e}$ and the derivatives are defined by

$$X'_{\eta e} = (\mathrm{d}X_\eta(R)/\mathrm{d}\xi)_{R_{\eta e}}, \quad X''_{\eta e} = (\mathrm{d}^2 X_\eta(R)/\mathrm{d}\xi^2)_{R_{\eta e}}, \quad (7.172)$$

where $\xi = (R - R_{\eta e})/R_{\eta e}$, the dimensionless vibrational coordinate. The coefficient $a_{1\eta}$ in equation (7.171) is the leading anharmonic potential coefficient in Dunham's power series expansion of $V_\eta(R)$ [23]:

$$V_\eta(R) = \left(hc\omega_{\eta e}^2/4B_{\eta e}\right)\xi^2 (1 + a_{1\eta}\xi + a_{2\eta}\xi^2 + a_{3\eta}\xi^3 + \cdots). \quad (7.173)$$

This form of $V_\eta(R)$ has the advantage of generality and converges well for the lower part of the potential function. It is clear from the form of equation (7.171) that if $X_\eta(R)$ has only a weak dependence on R, the corresponding parameter in the effective Hamiltonian will not depend strongly on the vibrational quantum number either. For example, the isotropic magnetic hyperfine interaction samples the electronic wave function in the immediate region around the nucleus. In consequence, it is virtually unaware of the electrons around the other nucleus in a diatomic molecule and consequently the hyperfine parameter is hardly affected by vibrational excitation.

We now turn to the second-order perturbation terms; they are rather numerous since all possible combinations of squared and cross terms can contribute. In practice, we can often be selective and take only those terms which are significant into consideration. For example, the second-order terms produce only minor modifications of the fine structure patterns and are often difficult to separate from the effects of the first-order terms. Probably the most easily observable of the second-order terms are the centrifugal distortion effects associated with the off-diagonal matrix elements of $B_\eta(R)$. There are two reasons for this. First, $B_\eta(R)$ has a fairly strong dependence on r:

$$B_\eta(R) = (\hbar^2/2hc\mu R^2) = (\hbar^2/2hc\mu R_{ne}^2)(1 - 2\xi + 3\xi^2 - 4\xi^3 + \cdots), \quad (7.174)$$

and second, levels with relatively high values of the rotational quantum number N (or J, where appropriate) can often be observed in practice.

Let us first consider the contribution of the rotational kinetic energy,

$$\mathcal{H}_{\text{rot}} = B(R)N^2, \quad (7.175)$$

with itself. In second order of perturbation theory, this produces an operator

$$P_0 \mathcal{H}' \frac{Q_0}{a} \mathcal{H}' P_0 = \sum_{v'}{}' \sum_{J'} |v;J\rangle \frac{\langle v;J|B(R)N^2|v';J'\rangle\langle v';J'|B(R)N^2|v;J\rangle}{[G_v^{(0)} - G_{v'}^{(0)}]} \langle v;J|, \quad (7.176)$$

where the restriction to the particular electronic state η is implied. The expression on the right-hand side can be separated into vibrational and rotational factors:

$$P_0 \mathcal{H}' \frac{Q_0}{a} \mathcal{H}' P_0 = \left\{ \sum_{v'}{}' |v\rangle \frac{\langle v|B(R)|v'\rangle\langle v'|B(R)|v\rangle}{[G_v^{(0)} - G_{v'}^{(0)}]} \langle v| \right\}$$
$$\times \left\{ \sum_{J'} |J\rangle\langle J|N^2|J'\rangle\langle J'|N^2|J\rangle\langle J| \right\}. \quad (7.177)$$

The rotational factor is further reduced to the matrix element of the operator product $(N^2)(N^2)$ by the closure rule. The contribution can then be expressed in terms of an operator of the form

$$\mathcal{H}_{\text{cd}} = -D_{\eta v}(N^2)(N^2) \quad (7.178)$$

which, in the spirit of the effective Hamiltonian, is understood to act only on spin-rotational levels of the vibrational level v in the electronic state $|\eta, \Lambda\rangle$.

The centrifugal distortion constant $D_{\eta v}$ is the result of the second-order vibrational contributions:

$$D_{\eta v} = -\sum_{v'}{}' \frac{\langle \eta, v|B_\eta(R)|\eta, v'\rangle\langle \eta, v'|B_\eta(R)|\eta, v\rangle}{[G_{\eta v}^{(0)} - G_{\eta v'}^{(0)}]}$$

$$= \frac{4B_{\eta e}^3}{\omega_{\eta e}^2} + \left(\frac{6B_{\eta e}^4}{\omega_{\eta e}^3}\right)(19 + 18a_{1\eta} + 9a_{1\eta}^2 - 8a_{2\eta})(v + 1/2) + \cdots. \quad (7.179)$$

As with the first-order contributions, we see that each parameter in $\mathcal{H}_{\text{eff}}$ has a weak vibrational dependence which can in general be modelled as

$$P_v = P_e + \alpha_P(v + 1/2) + \beta_P(v + 1/2)^2 + \cdots. \qquad (7.180)$$

Because we have chosen the potential function $V_\eta(R)$ to be independent of the reduced mass in the zeroth-order Hamiltonian, the isotopic dependences of the various terms in (7.179) are quite explicit. The leading term D_{ve} is proportional to μ^{-2} and the vibrational dependent term α_{Dv} is proportional to $\mu^{-5/2}$.

We have seen that the dependence of $B_\eta(R)$ on the vibrational coordinate ξ causes a mixing of the vibrational level of interest with neighbouring levels. This mixing results in centrifugal distortion corrections to all the various parameters $X_\eta(R)$ in the perturbation Hamiltonian $\mathcal{H}'$ when combined in a cross term. The operator has the same form as in the original term, for example, $(2/3)\sqrt{6}T^2_{q=0}(\boldsymbol{S}, \boldsymbol{S})$ for the spin–spin dipolar term, multiplied by $\boldsymbol{N}^2$. The coefficient which qualifies this term has the general form

$$X_{\eta Dv} = \sideset{}{'}\sum_{v'} \{\langle \eta, v|B_\eta(R)|\eta, v'\rangle\langle \eta, v'|X_\eta(R)|\eta, v\rangle + \langle \eta, v|X_\eta(R)|\eta, v'\rangle$$
$$\times \langle \eta, v'|B_\eta(R)|\eta, v\rangle\}/[G^{(0)}_{\eta v} - G^{(0)}_{\eta v'}]. \qquad (7.181)$$

Following the perturbation treatment of Watson [22], the matrix elements can be evaluated to give

$$X_{\eta Dv} = (4B^2_{\eta e}/\omega^2_{\eta e})X'_{\eta e} + (2B^3_{\eta e}/\omega^3_{\eta e})\{2X'''_{\eta e} - 3(1 + 3a_{1\eta})X''_{\eta e}$$
$$+ 3(8 + 9a_{1\eta} + 9a^2_{1\eta} - 8a_{2\eta})X'_{\eta e}\}(v + 1/2) + \cdots. \qquad (7.182)$$

We see from this equation that the centrifugal distortion of interactions which depend only weakly on R will be very small.

7.5.2. The form of the effective Hamiltonian

At this point, it is helpful to collect all the various terms in the zero-field effective Hamiltonian together. We remind ourselves that the effective Hamiltonian is an operator which is confined to act only on the rotational, electron spin and nuclear spin states spanning a given vibrational level of a single electronic state $|\eta, \Lambda\rangle$:

$$\mathcal{H}_{\eta v}/hc = T_{\eta e} + G_{\eta v} + \mathcal{H}_{\text{rot}} + \mathcal{H}_{\text{cd}} + \mathcal{H}_{\text{so}} + \mathcal{H}_{\text{ss}} + \mathcal{H}_{\text{sr}} + \mathcal{H}_{\Lambda d} + \mathcal{H}_{\text{hfs}} + \mathcal{H}_Q + \mathcal{H}_{\text{fs}}. \qquad (7.183)$$

In this expression, $T_{\eta e}$ is the energy origin of the electronic state $|\eta, \Lambda\rangle$ and $G_{\eta v}$ has absorbed the adiabatic and spin contributions to the vibrational potential energy (vibrationally averaged):

$$G_{\eta v} = G^{(0)}_{\eta v} + V^{(\text{ad})}_{\eta v} + V^{(\text{spin})}_{\eta v}. \qquad (7.184)$$

The other contributions are as follows:

rotational kinetic energy

$$\mathcal{H}_{\text{rot}} = B_{\eta v} \mathbf{N}^2 = \{\langle \eta, v | B_\eta(R) | \eta, v \rangle + V^{(\text{ad})}_{\eta D v} + V^{(\text{spin})}_{\eta D v}\} \mathbf{N}^2 \quad (7.185)$$

centrifugal distortion

$$\mathcal{H}_{\text{cd}} = -D_{\eta v}(\mathbf{N}^2)^2 + H_{\eta v}(\mathbf{N}^2)^3 + \cdots \quad (7.186)$$

spin–orbit coupling

$$\mathcal{H}_{\text{so}} = (1/2)\big[(A_{\eta v} + A_{\eta D v}\mathbf{N}^2),\, \mathrm{T}^1_{q=0}(\mathbf{L})\mathrm{T}^1_{q=0}(\mathbf{S})\big]_+ \quad (7.187)$$

spin–spin coupling

$$\mathcal{H}_{\text{ss}} = (\sqrt{6}/3)\big[(\lambda_{\eta v} + \lambda_{\eta D v}\mathbf{N}^2),\, \mathrm{T}^2_{q=0}(\mathbf{S},\mathbf{S})\big]_+ \quad (7.188)$$

spin–rotation interaction

$$\mathcal{H}_{\text{sr}} = (1/2)\big[(\gamma_{\eta v} + \gamma_{\eta D v}\mathbf{N}^2),\, \mathrm{T}^1(\mathbf{N}) \cdot \mathrm{T}^1(\mathbf{S})\big]_+ \quad (7.189)$$

Λ-doubling

$$\mathcal{H}_{\Lambda d} = \sum_{q=\pm 1}{}' \exp(-2iq\phi)(1/2)\Big\{-\big[(o_{\eta v} + o_{\eta D v}\mathbf{N}^2),\, \mathrm{T}^2_{2q}(\mathbf{S},\mathbf{S})\big]_+$$
$$+ \big[(p_{\eta v} + p_{\eta D v}\mathbf{N}^2),\, \mathrm{T}^2_{2q}(\mathbf{N},\mathbf{S})\big]_+ - \big[(q_{\eta v} + q_{\eta D v}\mathbf{N}^2),\, \mathrm{T}^2_{2q}(\mathbf{N},\mathbf{N})\big]_+\Big\} \quad (7.190)$$

magnetic hyperfine interactions

$$\mathcal{H}_{\text{hfs}} = \sum_\alpha \Big\{ a^\alpha_{\eta v} \mathrm{T}^1_{q=0}(\mathbf{I}_\alpha)\mathrm{T}^1_{q=0}(\mathbf{L}) + b^\alpha_{F\eta v} \mathrm{T}^1(\mathbf{I}_\alpha)\cdot\mathrm{T}^1(\mathbf{S}) + (\sqrt{6}/3)c^\alpha_{\eta v}\mathrm{T}^2_{q=0}(\mathbf{I}_\alpha,\mathbf{S})$$
$$+ \sum_{q=\pm 1}{}' \exp(-2iq\phi) d^\alpha_{\eta v}\mathrm{T}^2_{2q}(\mathbf{I}_\alpha,\mathbf{S}) \Big\} \quad (7.191)$$

electric quadrupole interaction

$$\mathcal{H}_Q = \sum_\alpha \frac{eQ_\alpha}{4I_\alpha(2I_\alpha - 1)} \Big\{ \sqrt{6}q^\alpha_0 \mathrm{T}^2_{q=0}(\mathbf{I}_\alpha,\mathbf{I}_\alpha) + \sum_{q=\pm 1}{}' \exp(-2iq\phi) q^\alpha_2 \mathrm{T}^2_{2q}(\mathbf{I}_\alpha,\mathbf{I}_\alpha) \Big\}$$
$$(7.192)$$

higher-order fine structure

$$\mathcal{H}_{fs} = C^{(3)}_{\eta v}\mathrm{T}^3(\mathbf{L}^2,\mathbf{N}) \cdot \mathrm{T}^3(\mathbf{S},\mathbf{S},\mathbf{S}) + (\sqrt{10}/5)\eta_{\eta v}\mathrm{T}^1_{q=0}(\mathbf{L})\mathrm{T}^3_{q=0}(\mathbf{S},\mathbf{S},\mathbf{S})$$
$$+ (\sqrt{70}/6)\Theta_{\eta v}\mathrm{T}^4_{q=0}(\mathbf{S},\mathbf{S},\mathbf{S},\mathbf{S}). \quad (7.193)$$

In these equations, the square brackets [,]$_+$ indicate anticommutators which are introduced to ensure that the operator is Hermitian. Note that the expression for $B_{\eta v}$ in equation (7.185) is slightly modified from equation (7.171) by the inclusion of the small correction terms $V^{(\text{ad})}_{\eta D v}$ and $V^{(\text{spin})}_{\eta D v}$. These centrifugal distortion corrections show up only in the electronic contributions to the isotope effects in $B_{\eta v}$ where they are responsible for the adiabatic change in bond length on isotopic substitution. The last contribution to the effective Hamiltonian, in equation (7.193), describes the higher-order fine structure effects. The first of these, a spin–rotation interaction, has already been described in some detail in section 7.4.7. The third term, involving $\Theta_{\eta v}$, is a spin–spin interaction first described by Brown and Milton [20] while the second term, involving $\eta_{\eta v}$, is a higher-order spin–orbit interaction which was introduced by Brown, Milton, Watson, Zare, Albritton, Horani and Rostas [24]. From the arguments presented in section 7.4.7, the $C_{\eta v}^{(3)}$ and $\eta_{\eta v}$ terms only occur in states of quartet or higher multiplicity while the $\Theta_{\eta v}$ term requires the state to be of at least quintet multiplicity. We remind ourselves also that the Λ-doubling terms in equations (7.232), (7.191) and (7.192) are formulated for Π electronic states only.

The parameters in the effective Hamiltonian $\mathcal{H}_{\eta v}$ may not all be determinable from spectroscopic data. When indeterminacies occur, they can often be resolved by utilising data from different isotopic forms. The parameters $X_{\eta v}$ and $X_{\eta D v}$ in the equations above do not themselves have simple isotopic ratios but the coefficients of the powers of $(v + 1/2)$ in the expansions (7.168) and (7.182) do. The isotopic ratios for $X_e, X'_e, X''_e, \ldots$, are the same as for $X(R)$ itself and the ratio B_e/ω_e is proportional to $\mu^{-1/2}$. Thus the isotopic ratio for any of the coefficients in $X_{\eta v}$ and $X_{\eta D v}$ is readily determined.

7.5.3. The N^2 formulation of the effective Hamiltonian

We have been careful in our formulation of the effective Hamiltonian to represent the square of the rotational angular momentum by the operator N^2, as in equation (7.185), for example. There are many papers in the literature, however, where the authors use the operator $R^2 = (N - L)^2$ instead. Their justification for doing this is that the expression for the rotational kinetic energy in the full Hamiltonian is

$$\mathcal{H}_{\text{rot}} = B(R)R^2 \qquad (7.194)$$

and so this operator might be expected to give a more faithful representation of the rotational motion. The difficulty with the R^2 operator is that, as we have seen in section 7.4.2, it involves the orbital angular momentum operator L which in turn has matrix elements which are off-diagonal in the electronic state as well as diagonal. In section 7.4.2, we worked through the implications of this mixing to show that the *effective* rotational Hamiltonian for a given electronic state has the form

$$\mathcal{H}_{\text{rot}}(\text{eff}) = B_\eta(R)N^2 \qquad (7.195)$$

where the effects of the off-diagonal elements have been absorbed into the parameter $B_\eta(R)$. The same remarks apply to all second-order terms in the effective Hamiltonian which involve the mixing of electronic states through the rotational kinetic energy operator, such as the spin–rotation interaction or Λ-doubling. For all of these operators in the effective Hamiltonian the rotational angular momentum is represented by N rather than R. Therefore if the effective Hamiltonian is developed in the way described in this

Table 7.2. *The interconversion of parameters in the N^2 and R^2 Hamiltonians. The symbol X represents any molecular parameter other than G, B or D*

$$G_{\eta v}(N^2) = G_{\eta v}(R^2) - \Lambda^2 B_{\eta v}(R^2) - \Lambda^4 D_{\eta v}(R^2) - \Lambda^6 H_{\eta v}(R^2) - \cdots$$
$$B_{\eta v}(N^2) = B_{\eta v}(R^2) + 2\Lambda^2 D_{\eta v}(R^2) + 3\Lambda^4 H_{\eta v}(R^2) + \cdots$$
$$D_{\eta v}(N^2) = D_{\eta v}(R^2) + 3\Lambda^2 H_{\eta v}(R^2) + \cdots$$
$$X_{\eta v}(N^2) = X_{\eta v}(R^2) - \Lambda^2 X_{\eta v D}(R^2) + \Lambda^4 X_{\eta v H}(R^2) + \cdots$$
$$X_{\eta v D}(N^2) = X_{\eta v D}(R^2) - 2\Lambda^2 X_{\eta v H}(R^2) + \cdots$$
$$X_{\eta v H}(N^2) = X_{\eta v H}(R^2) + \cdots$$

book, there are strong grounds for using the N^2 formulation. The use of R^2 is not justified and indeed causes some difficulties in practice. Those parts of its matrix elements which depend on L or L^2 are not well defined because the diatomic molecule wave functions are not eigenfunctions of L^2; the quantity $\langle \eta, \Lambda | L^2 | \eta, \Lambda \rangle$ must be evaluated by *ab initio* methods. Workers who claim to be using the R^2 formulation get round this difficulty by dropping the L^2 term altogether, which can be very deceptive. Another tricky aspect of using R^2 for the formulation of centrifugal distortion terms is that

$$\langle J | R^4 | J \rangle \neq \{\langle J | R^2 | J \rangle\}^2. \tag{7.196}$$

Failure to recognise this leads to the introduction of extra terms in the Hamiltonian. To summarise, a Hamiltonian formulated in terms of N^2 has the highly desirable characteristic that each term is the product of a determinable parameter which governs the magnitude of the interaction and an angular momentum operator whose matrix elements are fully defined. The difficulties associated with the effect of the L^2 terms are confined to the interpretation of parameters in the effective Hamiltonian and their comparison with *ab initio* calculations.

When comparing one's results with those of other workers, it is important to check the detailed form of the Hamiltonian which has been used. If it is an R^2 Hamiltonian, the parameters will have different definitions from those in the N^2 Hamiltonian and so will have different values. The relationship between the two sets of parameters is easy to work out provided that the R^2 formulation has suppressed the off-diagonal matrix elements of L in the manner described above, that is, R is defined as

$$R = N_x \mathbf{i} + N_y \mathbf{j} + (N_z - L_z) \mathbf{k}. \tag{7.197}$$

The relationships between the two sets of parameters for an electronic state $|\eta, \Lambda\rangle$ are given in table 7.2.

7.5.4. The isotopic dependence of parameters in the effective Hamiltonian

To a good approximation, all the parameters in the effective Hamiltonian presented in section 7.5.2 have a well-defined and easily constructed dependence on the reduced mass μ of the diatomic molecule. However, as we have seen in the case of the rotational constant B_v in equation (7.85), there are some higher-order corrections which tend to spoil the simple isotopic dependence. In the particular case of B_v, these corrections arise from non-adiabatic mixing of different electronic states, but there are also higher-order terms arising in the vibrational averaging process which have a similar effect. This more detailed isotopic dependence has been investigated by several authors [25, 26] culminating in a definitive paper by Watson [27].

Watson's treatment applies to vibration–rotation effects in diatomic molecules in singlet electronic states; it has not yet been extended to include spin–dependent phenomena. It is based on the Dunham expansion [23] which we met earlier in the previous chapter,

$$E(v, J) = hc \sum_{k\ell} Y_{k\ell}(v + 1/2)^k [J(J+1) - \Lambda^2]^\ell, \tag{7.198}$$

where v and J are, as usual, the vibrational and rotational quantum numbers. Values for the Dunham coefficients $Y_{k\ell}$ can be determined from experimental data; the point at which the expansion in equation (7.198) is truncated depends on the range of v and J values available and on the precision of the data. For different isotopic forms of the same molecule, which have essentially the same potential energy function $V_\eta(R)$ in equation (7.169), the $Y_{k\ell}$ are related approximately by

$$Y_{k\ell} \cong \mu^{-(k+2\ell)/2} U_{k\ell}, \tag{7.199}$$

where $U_{k\ell}$ is isotopically invariant. This corresponds to the level of treatment in the previous section. It is, however, not adequate for the most accurate of modern measurements which require corrections to be added to equation (7.199).

These corrections are of two types. The first arises from quantum effects in the calculation of vibration–rotation energy levels and was originally discussed by Dunham himself [23]. He used semiclassical methods, discussed in section 6.13.2, to derive his results. Use of the first-order quantisation condition

$$(2\mu)^{1/2} \int \left[E_\eta^0(v, J) - V_{\eta J}(R) \right]^{1/2} dR = h(v + 1/2) \tag{7.200}$$

gives the familiar formula in equation (7.198) with the isotopic dependence described in equation (7.199). The corrections to this result arise from the use of the second-order quantisation condition,

$$(2\mu)^{1/2} \int \left[E_\eta^0(v, J) - V_{\eta J}(R) \right]^{1/2} dR - \{\hbar^2/32(2\mu)^{1/2}\}$$
$$\times \int \{V'_{\eta J}(R)\}^2 \left[E_\eta^0(v, J) - V_{\eta J}(R) \right]^{-5/2} dR = h(v + 1/2), \tag{7.201}$$

where

$$V'_{\eta J}(R) = \frac{dV_{\eta J}(R)}{dR}. \tag{7.202}$$

Specifically, the corrections arise from the second-order integral in this condition. The same result can, of course, be derived in a purely quantum calculation using perturbation theory. We shall not go into details but the way in which these higher-order corrections arise can be appreciated from the following discussion.

The anharmonic corrections to the vibrational energy levels can be derived from the solutions of the Schrödinger equation with a potential of the form

$$V(q) = (1/2)kq^2 + gq^3 + hq^4 + \cdots \tag{7.203}$$

as discussed in section 6.8.3; in this equation, the vibrational coordinate q is equal to $(R_e - R)$. The effects of the anharmonic terms in this potential can be treated by perturbation theory using the harmonic oscillator eigenfunctions $|v\rangle$ as the zeroth-order basis set. The non-zero matrix elements in this treatment are of the form

$$\langle v+1|q|v\rangle = A(v+1/2)^{1/2} \tag{7.204}$$

where

$$A = (\hbar/4\pi\mu\nu)^{1/2}. \tag{7.205}$$

The quartic term hq^4 makes a contribution in first order of perturbation theory of the form

$$\begin{aligned}\langle v|hq^4|v\rangle &= hA^4\{(2v+1)^2 + 2v^2 + 2v + 2\} \\ &= hA^4\{6(v+1/2)^2 + 3/2\}.\end{aligned} \tag{7.206}$$

Use of equation (7.205) shows that A^4 is proportional to μ^{-1}. We thus see that there is a contribution $6hA^4$ to the Dunham coefficient Y_{20} (or $\omega_e x_e$) since it is the coefficient of $(v+1/2)^2$ in the expansion (7.198); this term has the expected reduced mass dependence. However, there is also a small contribution $3hA^4/2$ to the coefficient Y_{00} with the wrong reduced mass dependence (μ^{-1}). This is an example of the higher-order corrections revealed by Dunham's calculation.

The other way in which these corrections can arise is from non-adiabatic correction terms, that is, through the breakdown of the Born–Oppenheimer separation. They were first identified by Van Vleck [28]. We have seen how these contributions can be given explicit form in the derivation of the effective electronic Hamiltonian described in section 7.4. Some of the reduced mass dependencies are easy to track because they enter explicitly into the calculation as shown, for example, in the derivation of the second order contribution $B^{(2)}$ to the rotational constant in section 7.4.2. Other effects are more subtle. For example, one might naively believe that the electron orbital angular momentum $\boldsymbol{L}$ is independent of isotopic substitution. However, this is not so because the angular momentum is measured relative to the *nuclear* centre of mass in our formulation. Since the geometrical position of the nuclear centre of mass changes on isotopic substitution, so too does the orbital angular momentum. The details are all laid out clearly in Watson's paper [27]. Generally speaking, the Dunham corrections are rather small compared with those from the Born–Oppenheimer breakdown.

Collecting all these various contributions together, Watson showed that a more accurate description of the isotopic mass dependence is given by

$$Y_{k\ell} = \mu_C^{-(k+2\ell)/2} U_{k\ell} \left\{ 1 + \frac{m_e}{M_1} \Delta_{k\ell}^{(1)} + \frac{m_e}{M_2} \Delta_{k\ell}^{(2)} + O\left(\frac{m_e^2}{M_i^2}\right) \right\}, \qquad (7.207)$$

where m_e is the electron mass, M_1 and M_2 are here the *atomic* masses of atoms 1 and 2, and μ_C is a charge-modified reduced mass defined by

$$\mu_C = \frac{M_1 M_2}{(M_1 + M_2 - Cm_e)} \qquad (7.208)$$

in which C is the charge number of the molecule. The denominator of μ_C is thus the total mass of the molecule or ion; for a neutral species it is the reduced atomic mass (because to a good approximation the electrons follow the nuclear motion closely). The dimensionless quantities $\Delta_{k\ell}^{(i)} (i = 1, 2)$ are isotopically invariant and are expected to be of the order unity for a well isolated electronic state. The even higher order corrections $O(m_e/M_i)^2$ in equation (7.207) are usually too small to be characterised experimentally.

The $U_{k\ell}$ are isotopically invariant parameters. Our discussion of the RKR potential in section 6.13.3 showed that the potential $V_\eta(R)$ can be determined from a knowledge of G_v and B_v. Since the parameters $U_{k\ell}$ with $\ell \geq 2$ can be calculated from $V_\eta(R)$, it follows that they are exactly determined by the values of $U_{k\ell}$ with $\ell = 0$ and 1. The simplest case of U_{02} is a familiar one. The relationship in question,

$$Y_{02} \simeq -4Y_{01}^3/Y_{01}^2, \qquad (7.209)$$

more usually written as

$$D_e \simeq 4B_e^3/\omega_e^2, \qquad (7.210)$$

is not exact because the correction terms on each side do not balance (see equation (7.179)). However the corresponding relationship between the $U_{k\ell}$ parameters,

$$U_{02} = -4U_{01}^3/U_{01}^2, \qquad (7.211)$$

is exact because the correction terms have been removed. This and similar relationships provide convenient constraints when fitting data.

7.6. Effective Zeeman Hamiltonian

Many informative experiments can be carried out on diatomic molecules in the presence of an external magnetic field, irrespective of whether they are in closed or open shell states. An external magnetic field destroys the isotropy of free space, as a consequence of which the degeneracy of the different orientations of the molecule is lifted. In other words, the states of different orientation have different energies in the magnetic field. These splittings are observable and provide a measurement of the magnetic dipole moment of the molecule in the particular spin–rotational level involved. The magnetic moment depends on the electronic structure of the molecule; measurement of this quantity therefore provides information about the structure.

The effect of a magnetic field on a diatomic molecule has been described in detail in section 3.7. An extra term is added to the magnetic vector potential of each electron i of the form

$$A_i^B = (1/2)(\boldsymbol{B} \wedge \boldsymbol{R}_i) \tag{7.212}$$

where $\boldsymbol{B}$ is the uniform applied flux density. When this additional contribution is substituted into the electronic Hamiltonian and the coordinates are transformed to the molecule-fixed axis system, the various Zeeman terms given in chapter 3 are obtained. There are several of them, with the largest arising from the interactions with the magnetic moments produced by the orbital and spin motion of the electron. Most of the terms are linear in $\boldsymbol{B}$ and represent the interaction of the flux density with the magnetic dipole moment of the molecule:

$$\mathcal{H}_Z = -\boldsymbol{\mu}_m \cdot \boldsymbol{B}. \tag{7.213}$$

There are also some smaller terms which are quadratic in $\boldsymbol{B}$ and have the general form

$$\mathcal{H}_{\text{susc}} = -\boldsymbol{B} \cdot \boldsymbol{\chi} \cdot \boldsymbol{B}, \tag{7.214}$$

where $\boldsymbol{\chi}$ is the magnetic susceptibility tensor. It is more convenient to express this interaction using spherical tensor notation [29]:

$$\mathcal{H}_{\text{susc}} = -\frac{1}{2} \sum_{k=0,2}{}' T^k(\boldsymbol{\chi}) \cdot T^k(\boldsymbol{B}, \boldsymbol{B}). \tag{7.215}$$

The scalar contribution, with $k = 0$, is constant for all levels of a molecule and so cannot be measured in practice. The anisotropic $k = 2$ contribution, on the other hand, is readily determinable.

We see that there are many contributions to the Zeeman Hamiltonian, all of which must be taken into account in the construction of the effective Hamiltonian. In order to focus our attention, let us consider just one of the dipolar terms, namely, the interaction involving the orbital motion of the electrons:

$$\mathcal{H}'_Z = \frac{e}{2m} \sum_i \boldsymbol{B} \cdot \left\{ \boldsymbol{r}_i - \frac{m}{M} \sum_j \boldsymbol{r}_j \right\} \wedge \boldsymbol{p}_i. \tag{7.216}$$

We can write this more simply as

$$\mathcal{H}'_Z = g_L \mu_B \boldsymbol{B} \cdot \boldsymbol{L}, \tag{7.217}$$

where $\mu_B = e\hbar/2m$ is the Bohr magneton and $\hbar \boldsymbol{L}$ is the orbital angular momentum of the electrons measured relative to the origin of the molecule-fixed axis system (x, y, z). The parameter g_L is the orbital g-factor. Since it has a value of 1.0, one might query the need for its introduction. We shall see later that the value will deviate from unity when it is absorbed into the effective Hamiltonian. The direction of the uniform flux density $\boldsymbol{B}$ in equation (7.217) serves to define the orientation of the laboratory-fixed coordinate system. By convention, $\boldsymbol{B}$ is taken to point along the Z

axis so that B_X and B_Y are zero. The interaction therefore involves L_Z which can be written in terms of the molecule-fixed components by use of the direction cosines,

$$L_Z = -\sin\theta \cos\chi\, L_x + \sin\theta \sin\chi\, L_y + \cos\theta\, L_z, \tag{7.218}$$

or in spherical tensor notation by the use of rotation matrices

$$T^1_{p=0}(\boldsymbol{L}) = \sum_q \mathcal{D}^{(1)}_{0q}(\omega)^* T^1_q(\boldsymbol{L}). \tag{7.219}$$

Here, ω represents the Euler angles (ϕ, θ, χ). Thus when we come to work out the effects of the orbital Zeeman interaction, we see that it has off-diagonal matrix elements which link electronic states with $\Delta\Lambda = 0, \pm 1$, as well as purely diagonal elements. It is clearly desirable to remove the effect of these matrix elements by a suitable perturbative transformation to achieve an effective Zeeman Hamiltonian which acts only within the spin–rotational levels of a given electronic state $|\eta, \Lambda, v\rangle$, in the same way as the zero-field effective Hamiltonian in equation (7.183).

In order to appreciate how the effective Zeeman Hamiltonian is derived, let us consider the mixing effects of the orbital magnetic interaction $\mathcal{H}'_Z$ in equation (7.217) along with the rotational Hamiltonian $\mathcal{H}_{\text{rot}}$, equation (7.80), and the spin–orbit coupling term $\mathcal{H}_{\text{so}}$ in equation (7.72). These are absorbed into the effective Hamiltonian in a procedure which will by now be quite familiar and involves degenerate perturbation theory. The first-order contribution of $\mathcal{H}'_Z$ simply introduces this leading term into the Zeeman Hamiltonian. The second-order contributions involve both squared and cross terms. It can be seen that the cross term of $\mathcal{H}'_Z$ with $\mathcal{H}_{\text{rot}}$ or $\mathcal{H}_{\text{so}}$ produces a term which is linear in the magnetic flux density $\boldsymbol{B}$ and will therefore have the same form as the first order contribution. The cross term between $\mathcal{H}'_Z$ and $\mathcal{H}_{\text{so}}$ gives an operator of the following form:

$$P_0 \mathcal{H}' \frac{Q_0}{a} \mathcal{H}' P_0 = P_0 \left\{ \left[\sum_{q=\pm 1}{}' \sum_{\eta'\Lambda'} \sum_{S'\Sigma'} \sum_{J'\Omega'M'} \frac{ME_1 \times ME_2 + \text{transpose}}{V_\eta(R) - V_{\eta'}(R)} \right] \right\} P_0 B_Z, \tag{7.220}$$

where

$$ME_1 = \langle \eta, \Lambda; S, \Sigma; J, \Omega, M | (-1)^q A T^1_q(\boldsymbol{L}) T^1_{-q}(\boldsymbol{S}) | \eta', \Lambda'; S', \Sigma'; J', \Omega', M' \rangle,$$

$$ME_2 = \langle \eta', \Lambda'; S', \Sigma'; J', \Omega', M' | g_L \mu_B \mathcal{D}^{(1)}_{0,-q}(\omega)^* T^1_{-q}(\boldsymbol{L}) | \eta, \Lambda; S, \Sigma; J, \Omega, M \rangle.$$

This expression can be factorised into orbital, spin and rotational parts and then tidied up to produce an effective operator of the form

$$g_l \mu_B B_Z \sum_{q=\pm 1}{}' \mathcal{D}^{(1)}_{0q}(\omega)^* T^1_q(\boldsymbol{S}). \tag{7.221}$$

The dimensionless parameter g_l is defined by

$$g_l = -g_L \sum_{q=\pm 1}{}' \sum_{\eta'\Lambda'} (V_{\eta\Lambda}(R) - V_{\eta'\Lambda'}(R))^{-1} \langle \eta, \Lambda | A T^1_q(\boldsymbol{L}) | \eta', \Lambda' \rangle$$

$$\times \langle \eta', \Lambda' | T^1_{-q}(\boldsymbol{L}) | \eta, \Lambda \rangle. \tag{7.222}$$

The restricted summation in equation (7.221) means that, for this contribution, the Zeeman energy depends on the x and y components of $\mathbf{S}$ only (i.e. it has cylindrical symmetry). It is therefore referred to as the anisotropic correction to the electron spin Zeeman term, which has the isotropic form $g_S \mu_B \mathbf{B} \cdot \mathbf{S}$.

The cross term between $\mathcal{H}'_Z$ and $\mathcal{H}_{\mathrm{rot}}$ can be treated in exactly the same way. The result is a second-order contribution to the effective Zeeman Hamiltonian of the form

$$g_r^e \mu_B B_Z T^1_{p=0}(\mathbf{N}) \tag{7.223}$$

where

$$g_r^e = 2 g_L \sum_{q=\pm 1}{}' \sum_{\eta' \Lambda'} (V_{\eta \Lambda}(R) - V_{\eta' \Lambda'}(R))^{-1} \langle \eta, \Lambda | T^1_q(\mathbf{L}) | \eta', \Lambda' \rangle$$

$$\times \langle \eta', \Lambda' | B(R) T^1_{-q}(\mathbf{L}) | \eta, \Lambda \rangle. \tag{7.224}$$

This term describes the electronic contribution to the rotational g-factor. The contribution (7.223) represents the interaction between the applied magnetic field B_Z and the magnetic moment produced by the electrons in the molecule as it rotates in laboratory space. It has an operator form identical to that of the first-order nuclear orbital (i.e. rotational) Zeeman interaction

$$-g_r^N \mu_B B_Z T^1_{p=0}(\mathbf{N}) \tag{7.225}$$

where

$$g_r^N = \frac{m \left(Z_1 M_2^2 + Z_2 M_1^2 \right)}{M_1 M_2 (M_1 + M_2)}. \tag{7.226}$$

These terms are usually combined into a single operator which describes the magnetic interaction between the external field and the rotation of the molecule as a whole, both electrons and nuclei,

$$\mathcal{H}_{Z,\mathrm{rot}} = -g_r \mu_B B_Z T^1_{p=0}(\mathbf{N}), \tag{7.227}$$

with

$$g_r = g_r^N - g_r^e. \tag{7.228}$$

In rotational motion the electrons tend to follow the nuclei quite closely because of the strong Coulombic interaction. As a consequence, there is a fair degree of cancellation implied by equation (7.228).

The second-order contribution of the orbital Zeeman term $\mathcal{H}'_Z$ with itself produces a term in the effective Hamiltonian which is quadratic in $\mathbf{B}$. It therefore has the same form as the diamagnetic susceptibility contribution to the energy; it provides the paramagnetic or 'high-frequency' contribution to the susceptibility of the molecular system. The resultant term in the effective Zeeman Hamiltonian is

$$\mathcal{H}_{Z,\mathrm{susc}} = \mathcal{L} T^2_{q=0}(\mathbf{B}, \mathbf{B}). \tag{7.229}$$

The parameter $\mathscr{L}$ is the sum of first- and second-order contributions, $\mathscr{L}^{(1)}$ and $\mathscr{L}^{(2)}$, where

$$\mathscr{L}^{(1)} = -(e^2/8m) \sum_i \langle \eta, \Lambda | T_0^2(\mathbf{r}_i, \mathbf{r}_i) | \eta, \Lambda \rangle \qquad (7.230)$$

and

$$\mathscr{L}^{(2)} = g_L^2 \mu_B^2 (5)^{1/2} \sum_{\eta' \Lambda' q} (V_{\eta\Lambda}(R) - V_{\eta'\Lambda'}(R))^{-1} (-1)^q \begin{pmatrix} 1 & 2 & 1 \\ -q & 0 & q \end{pmatrix}$$
$$\times |\langle \eta, \Lambda | T_q^1(\mathbf{L}) | \eta', \Lambda' \rangle|^2. \qquad (7.231)$$

The form of the complete Zeeman effective Hamiltonian for a diatomic molecule in a given vibrational level of an open shell state has been given by Brown, Kaise, Kerr and Milton [30]. It is the sum of the following terms:

electron spin (parameter g_S)

$$g_S \mu_B B_Z T_{p=0}^1(\mathbf{S})$$

electron orbital motion (parameter g'_L, corrected for relativistic, diamagnetic [29] and non-adiabatic [31] effects)

$$g'_L \mu_B B_Z T_{p=0}^1(\mathbf{L})$$

rotational motion (parameter g_r)

$$-g_r \mu_B B_Z T_{p=0}^1(\mathbf{N})$$

anisotropic correction to the electron spin interaction (parameter g_l)

$$g_l \mu_B B_Z \sum_{q=\pm 1}{}' \mathscr{D}_{0q}^{(1)}(\omega)^* T_q^1(\mathbf{S})$$

nuclear spin interaction (parameter g_N^α, where α labels the nuclei)

$$-\sum_\alpha g_N^\alpha \mu_N B_Z T_{p=0}^1(\mathbf{I}^\alpha)$$

anisotropic susceptibility (parameter $\mathscr{L}$)

$$\mathscr{L} T_{q=0}^2(\mathbf{B}, \mathbf{B})$$

parity-dependent contributions for a Π state (parameters g'_l and $g_r^{\prime e}$)

$$\mu_B B_Z \sum_{q=\pm 1}{}' \exp(-2iq\phi) \Big\{ g'_l \mathscr{D}_{0,-q}^{(1)}(\omega)^* T_q^1(\mathbf{S})$$
$$-g_r^{c'} \sum_p (-1)^p \mathscr{D}_{-p,-q}^{(1)}(\omega)^* T_p^1(\mathbf{N}) \mathscr{D}_{0,-q}^{(1)}(\omega)^* \Big\}. \qquad (7.232)$$

It should be appreciated that the Zeeman interactions are usually dominated by the first two terms in (7.232), the electron spin and orbital terms. The other terms are typically between two and four orders of magnitude smaller. For a molecule in a closed shell $^1\Sigma$ state, only the rotational Zeeman term, the nuclear spin contribution and the susceptibility term survive.

A recent paper [32] has suggested that the primary g-factors, g_S and g'_L, should be defined as *negative* quantities so that they reveal the alignment of the magnetic dipole moment relative to the angular momentum. If this convention is adopted, the signs of the two contributions to the effective Zeeman Hamiltonian, given above, must be reversed.

7.7. Indeterminacies: rotational contact transformations

If the values of the parameters for the vibronic state $|\eta, v\rangle$ are known *a priori*, the spin–rotational energy levels can be calculated in a straightforward manner from the effective Hamiltonian, equation (7.183), by computing the matrix elements in a suitable basis set and diagonalising the sub-matrices numerically for each value of the quantum number J (or F if hyperfine effects are included). In practice, however, $\mathcal{H}_{\eta v}$ is usually used as an empirical Hamiltonian with parameters to be determined by optimisation of the agreement between computed and observed spectra. In this inversion procedure, the possibility of indeterminacy arises, because different terms in equation (7.183) can make indistinguishable contributions to the eigenvalues, in much the same way as $B^{(1)}(R)$ and $B^{(2)}(R)$, for example, make indistinguishable contributions to the effective rotational Hamiltonian, equation (7.86). These are theoretical indeterminacies, caused by the non-uniqueness of the solution of these equations, rather than experimental indeterminacies due to shortage of data. The latter type of indeterminacy may also be present, of course, and often is.

The solution to this problem which is usually adopted in practice is to constrain particular parameters to preset values, usually zero, in a least-squares fit of the Hamiltonian to the data. This has the effect of modifying the values obtained for the other parameters in the Hamiltonian. In order to interpret the results in this situation, it is necessary to subject the effective Hamiltonian $\mathcal{H}_{\eta v}$ to yet another transformation to bring it into the form which is actually used in the empirical fit. Before we do this, however, let us consider a particular example to help us understand the nature of these indeterminacies. There is a well-known indeterminacy between the parameters $\gamma_{\eta v}$ and $A_{\eta D v}$ in the Hamiltonian for a molecule in a $^2\Pi$ state, first pointed out by Veseth [33]. If we confine ourselves to the simple Hamiltonian

$$\mathcal{H} = \mathcal{H}_{\text{so}} + \mathcal{H}_{\text{rot}} + \mathcal{H}_{\text{sr}} \qquad (7.233)$$

where the three terms are given by equations (7.187), (7.185) and (7.189), the representation for a given J value is the following 2×2 matrix.

	$\|J, 3/2\rangle$	$\|J, 1/2\rangle$
$\|J, 3/2\rangle$	$(1/2)\{A + A_D z\} + Bz$	$-(B - \gamma/2)\sqrt{z}$
$\|J, 1/2\rangle$	$-(B - \gamma/2)\sqrt{z}$	$-(1/2)\{A + A_D(z+2)\} + B(z+2) - \gamma$

In this matrix

$$z = (J - 1/2)(J + 3/2) = (J + 1/2)^2 - 1. \qquad (7.234)$$

The eigenvalues of the matrix are

$$E_\pm = B(z+1) - \gamma/2 - A_D/2 \pm (1/2)[\{(A - 2B + \gamma) + A_D(z+1)\}^2 + 4(B - \gamma/2)^2 z]^{1/2} \tag{7.235}$$

The terms inside the square root part of this expression may be rearranged in increasing powers of z, to give

$$[\ ]^{1/2} = [\{(A - 2B + \gamma) + A_D\}^2 + \{2A_D(A - 2B + \gamma) + 4(B - \gamma/2)^2\}z + A_D^2 z^2]^{1/2}. \tag{7.236}$$

If we make measurements of the energy levels for different J values, the coefficients of the different powers of z are determinable. Thus the following combinations can be determined:

$$(1/2)(\mathcal{H}_{11} + \mathcal{H}_{22}) = B(z+1) - \gamma/2 - A_D/2, \tag{7.237}$$

and, inside the square root,

$$\text{coefficient of } z^0 = \{A - 2B + \gamma + A_D\}^2, \tag{7.238}$$

$$\text{coefficient of } z^1 = \{2A_D(A - 2B + \gamma) + 4(B - \gamma/2)^2\}, \tag{7.239}$$

$$\text{coefficient of } z^2 = A_D^2. \tag{7.240}$$

The combination in equation (7.237) allows the value for B to be determined from the z dependence. With this value, it is possible to determine $(A + \gamma + A_D)$ from equation (7.238). Since it is preferable to produce definite values for the major parameters, let us say that this coefficient gives a value for A. The coefficient of z^1, in equation (7.239), on the other hand, gives us a linear combination of A_D and γ; this is the equation which defines the nature of the indeterminacy between A_D and γ. If the coefficient of z^2 in equation (7.240) can be determined (it is very small in practice), it would appear that γ and A_D can be separated. In other words, the values of all four parameters in the Hamiltonian can be obtained. However, at this level of approximation, we have also to include the next two terms in the centrifugal expansion of $\mathcal{H}_{so}$ and $\mathcal{H}_{sr}$ (A_H and γ_D) which contribute to the coefficient of z^2 as well. Thus only three of the four parameters in equation (7.235) can be determined. We choose these to be B, A and the coefficient of z^1 in equation (7.239). In practice, we might choose to constrain γ to zero in the fit and then determine an effective value for A_D, denoted by $\tilde{A}_D$. Substitution in equation (7.239) gives

$$\{2\tilde{A}_D(A - 2B) + 4B^2\} = \{2A_D(A - 2B + \gamma) + 4(B - \gamma/2)^2\}$$
$$\simeq 2A_D(A - 2B) + 4B^2 - 4B\gamma, \tag{7.241}$$

provided the magnitude of γ is much smaller than B. Hence we see that the effective parameter is related to the true value for A_D and γ by

$$\tilde{A}_D = A_D - 2B\gamma/(A - 2B). \tag{7.242}$$

Conversely, if we choose to constrain A_D to zero, we determine a value for the effective spin–rotation parameter $\tilde{\gamma}$ where

$$\tilde{\gamma} = \gamma - (A - 2B)A_D/2B. \tag{7.243}$$

Brown and Watson [17] have shown that it is possible to break the indeterminacy implicit in equation (7.239) by combining data for different isotopic forms. This separation makes use of the fact that all three parameters on the right-hand side of equation (7.242) are, to a good approximation, proportional to μ^{-1} where μ is the reduced mass of the diatomic molecule. Thus the first term on the right-hand side is proportional to μ^{-1} while the second is proportional to μ^{-2}. This allows the true values for γ and A_D to be determined. Similar arguments apply to equation (7.243). The resolution of this method depends directly on the fractional change in μ on isotopic substitution.

Brown, Colbourn, Watson and Wayne [7] have shown that this type of indeterminacy occurs for any $^{2S+1}\Lambda$ state which conforms to Hund's case (a) coupling. This result can be established most easily by applying a contact transformation to the effective Hamiltonian. Let us divide $\mathcal{H}_{\eta v}$ into a principal part $\mathcal{H}_{\eta v}^{(0)}$ and a remainder $\mathcal{H}_{\eta v}^{(1)}$:

$$\mathcal{H}_{\eta v} = \mathcal{H}_{\eta v}^{(0)} + \mathcal{H}_{\eta v}^{(1)} \tag{7.244}$$

with

$$\mathcal{H}_{\eta v}^{(0)} = T_{\eta e} + G_{\eta v} + B_{\eta v} N^2 + A_{\eta v} L_z S_z + (2/3)\lambda_{\eta v}(3S_z^2 - S^2). \tag{7.245}$$

Following the procedure described in section 7.3, we apply a similarity transformation to the Hamiltonian to produce a transformed Hamiltonian,

$$\tilde{\mathcal{H}}_{\eta v} = e^{iF} \mathcal{H}_{\eta v} e^{-iF}, \tag{7.246}$$

where F is the transformation operator for a particular vibronic state $|\eta, v\rangle$. Following equations (7.53), (7.54) and (7.55), we expand the exponential operators to give

$$\tilde{\mathcal{H}}_{\eta v} = \mathcal{H}_{\eta v}^{(0)} + \{\mathcal{H}_{\eta v}^{(1)} + i[F, \mathcal{H}_{\eta v}^{(0)}]\} + \{i[F, \mathcal{H}_{\eta v}^{(1)}] - (1/2)[F, [F, \mathcal{H}_{\eta v}^{(0)}]]\} + \cdots. \tag{7.247}$$

Here, terms within the same set of braces are assumed to have the same order of magnitude. As long as F is chosen to be small, the expansion in equation (7.247) will converge rapidly and can be truncated after the first set of braces. Apart from this truncation, the eigenvalues of $\tilde{\mathcal{H}}_{\eta v}$ and $\mathcal{H}_{\eta v}$ are equal. The operator F is chosen suitably to remove terms in $\mathcal{H}_{\eta v}^{(1)}$ and so to eliminate redundancies. At the same time, the coefficients of other terms in $\mathcal{H}_{\eta v}^{(1)}$ may be modified. It is these modified parameters whose values are determined empirically.

The operator F must be taken to be Hermitian (to ensure that the transformation (7.246) is unitary), totally symmetric and of odd degree in the angular momenta. The last requirement follows from the fact that the non-vanishing commutation relationships between angular momentum components, say J_α, reduce the power of the operators by one. The result of the transformation is therefore of even power in J which is required if the term in $\tilde{\mathcal{H}}_{\eta v}$ is to be symmetric with respect to time reversal. To illustrate the procedure, let us consider the operator chosen by Brown and Watson [17]:

$$F_1 = s_1(\boldsymbol{J} \wedge \boldsymbol{S})_z L_z = s_1(J_x S_y - J_y S_x) L_z. \tag{7.248}$$

When this operator is substituted into equation (7.246), it corresponds to a small rotation in rotational–spin–orbital space. The parameter s_1 governs the magnitude of this rotation. It is a variable parameter which can be chosen to eliminate terms from the transformed Hamiltonian. Using this form for F_1 and the well known commutation relations between the molecule-fixed components of $\boldsymbol{J}$, $\boldsymbol{S}$ and $\boldsymbol{L}$, it is easy to show that

$$i[F_1, H_{\eta v}^{(0)}] = s_1\{-A_{\eta v}\Lambda^2\boldsymbol{S}^2 + (A_{\eta v}\Lambda^2 - 4\lambda_{\eta v}\boldsymbol{S}^2)L_zS_z + (A_{\eta v} + 4\lambda_{\eta v})\Lambda^2 S_z^2 \\ + 4\lambda_{\eta v}L_zS_z^3 - (A_{\eta v} - 2B_{\eta v})\Lambda^2(\boldsymbol{N}\cdot\boldsymbol{S}) - B_{\eta v}(\boldsymbol{N}^2 L_zS_z + L_zS_z\boldsymbol{N}^2) \\ - 2\lambda_{\eta v}[(\boldsymbol{N}\cdot\boldsymbol{S})L_zS_z + L_zS_z(\boldsymbol{N}\cdot\boldsymbol{S})]\}. \tag{7.249}$$

The fifth term on the right-hand side has the operator form $\Lambda^2(\boldsymbol{N}\cdot\boldsymbol{S})$. Hence if s_1 is chosen as

$$s_1 = \gamma_{\eta v}/(A - 2B_{\eta v})\Lambda^2, \tag{7.250}$$

we can eliminate the term in $(\boldsymbol{N}\cdot\boldsymbol{S})$ from $\tilde{\mathcal{H}}_{\eta v}$. This choice of the transformation also modifies the coefficients of the surviving terms in $\tilde{\mathcal{H}}_{\eta v}$ as follows:

$$\tilde{A}_{\eta v} = A_{\eta v} + \gamma_{\eta v}(A_{\eta v}\Lambda^2 - 4\lambda_{\eta v}\boldsymbol{S}^2)/(A_{\eta v} - 2B_{\eta v})\Lambda^2, \tag{7.251}$$
$$\tilde{A}_{\eta Dv} = A_{\eta Dv} - 2B_{\eta v}\gamma_{\eta v}/(A_{\eta v} - 2B_{\eta v})\Lambda^2. \tag{7.252}$$

Equation (7.252) is a generalisation of (7.242). Alternatively we can choose

$$s_1 = A_{\eta Dv}/2B_{\eta v} \tag{7.253}$$

to eliminate the $A_{\eta Dv}$ term in $\tilde{\mathcal{H}}_{\eta v}$. In this case, the parameters $\tilde{A}_{\eta v}$ and $\tilde{\gamma}_{\eta v}$ are modified to:

$$\tilde{A}_{\eta v} = A_{\eta v} + A_{\eta Dv}(A_{\eta v}\Lambda^2 - 4\lambda_{\eta v}\boldsymbol{S}^2)/2B_{\eta v}, \tag{7.254}$$
$$\tilde{\gamma}_{\eta v} = \gamma_{\eta v} - A_{\eta Dv}(A_{\eta v} - 2B_{\eta v})\Lambda^2/2B_{\eta v}. \tag{7.255}$$

As well as modifying the coefficients of existing terms in $\mathcal{H}_{\eta v}$, the transformation also introduces new terms in $L_zS_z^3$ and $\{(\boldsymbol{N}\cdot\boldsymbol{S})L_zS_z + L_zS_z(\boldsymbol{N}\cdot\boldsymbol{S})\}$. For a doublet state, these terms do not arise because $\lambda_{\eta v}$ must equal zero; the elimination therefore modifies the coefficients in the Hamiltonian without altering its form. This neat property does not hold for states of higher multiplicity. However, we note that the additional contribution in $L_zS_z^3$ has essentially the same form as the higher-order spin–orbit interaction $\eta_{\eta v}$ in section 7.4.7 and so will modify that parameter if it is included. Alternatively, we could choose the form of F_1 so as to eliminate it from the transformed Hamiltonian.

Before we leave this topic, it is well to remember that we have confined our attention to the leading contribution in the transformed Hamiltonian in (7.247). If the transformation operator is not small, we may have to include higher-order effects from the second and subsequent pairs of braces in this equation. This corresponds to retention of the terms in γ^2 in equation (7.241). Furthermore, the commutator $[F, \mathcal{H}_{\eta v}^{(1)}]$ in the second brace has the potential to modify the parameters and forms of all the terms in $\mathcal{H}_{\eta v}^{(1)}$. If a particular choice is made for s_1, as say in equations (7.250) or (7.253), parameters determined in the empirical fit will also be modified. This modification must be taken into account if the parameters so obtained are to be used to provide structural information on the molecule. For example, some of the parameters in the Zeeman Hamiltonian, equation (7.232), are affected by the transformation given in equation (7.248). The details are presented elsewhere [34, 35].

7.8. Estimates and interpretation of parameters in the effective Hamiltonian

7.8.1. Introduction

As we explained earlier in this chapter, one of the great merits of the effective Hamiltonian is that it allows the two tasks of fitting experimental data and interpreting the resultant parameters to be separated. In this section we discuss the latter aspect and explain how the quantities obtained from a fit of experimental data can be interpreted in terms of the geometric and electronic structure of the molecule concerned. We have seen how the process of averaging the parameters over the vibrational motion of the molecule leads to additional terms which describe the vibrational dependence of the parameters. We shall assume in what follows that all such vibrational averaging effects have been properly taken into account and that we are left to deal with the equilibrium value of the parameter, P_e, in equation (7.180).

7.8.2. Rotational constant

It is well known, and shown earlier in this book, that the rotational constant of a diatomic molecule is very simply related to its bond length (R) and reduced mass (μ),

$$B_e(\text{cm}^{-1}) = \hbar^2/(2\mu R_e^2 hc). \tag{7.256}$$

The dominant contribution to the rotational constant is from the nuclear masses. There is also a small contribution from the electrons in the molecule which is of the order $(m_e/m_N)B_e = \kappa^4 B_e \approx 10^{-4} B_e$, where m_e and m_N are the masses of an electron and a typical nucleus respectively and $\kappa = (m_e/m_N)^{1/4}$ is the Born–Oppenheimer expansion parameter discussed in chapter 6. Since the electrons interact strongly with the nuclei, they tend to follow their motion closely. The bulk of the electronic contribution to the moment of inertia can therefore be taken into account by the use of atomic rather than nuclear masses in the reduced mass μ in equation (7.256). However, as we have seen in section 7.4.2, equation (7.256) gives only the first-order contribution to the rotational constant. There is also a small but not insignificant second-order contribution which describes the slippage of the electronic contribution to the moment of inertia, that is, the extent to which the electrons do not follow the nuclear motion. Although this second-order contribution is large compared with the measurement errors of high resolution spectroscopy, it only reveals itself as small inconsistencies which arise when the equilibrium rotational constants of different isotopomers are compared with each other according to elementary theory (because it is proportional to μ^{-2} rather than

μ^{-1}). In effect, the equilibrium bond lengths of different isotopomers are found to be slightly different. As we have discussed elsewhere, Watson [25] has shown that the equilibrium bond length R_e obtained by the use of equation (7.256) can be related to the Born–Oppenheimer equilibrium bond length R_e^{BO} (the bond length at the minimum of the potential energy curve) by the equation

$$R_e = R_e^{BO}\left[1 + m_e\left\{\frac{d_1}{M_1} + \frac{d_2}{M_2}\right\}\right], \tag{7.257}$$

where M_1, M_2 are the masses of the atoms 1 and 2 and d_1, d_2 are dimensionless, isotopically invariant parameters. These parameters can be determined either by fitting the experimentally determined values of R_e to equation (7.257) or from theoretical estimates described by Watson. They give information on the non-adiabatic mixing effects, that is, the breakdown of the Born–Oppenheimer approximation.

7.8.3. Spin–orbit coupling constant, A

The spin–orbit coupling constant A is the coefficient of the operator term $L_z S_z$ in the effective Hamiltonian. It arises from the operator $\mathcal{H}_{so}^{(e)}$ in equation (7.71); we have seen in section 7.4.4 how the matrix elements of this operator lead to first-, second- and higher-order contributions in the effective Hamiltonian,

$$A = A^{(1)} + A^{(2)} + \cdots. \tag{7.258}$$

If accurate electronic wave functions are available, $A^{(1)}$ and $A^{(2)}$ can be estimated from equations (7.109) and (7.120) respectively. All nearby electronic states which contribute by spin–orbit mixing to $A^{(2)}$ must be included if the result is to be reliable.

A simpler but less accurate approach to the estimation of A is often worthwhile because it provides greater physical insight. Often this estimate is just the expectation value of the microscopic operator

$$\mathcal{H}_{so}^{(e)} = \sum_i a_i \mathbf{T}^1(l_i) \cdot \mathbf{T}^1(s_i), \tag{7.259}$$

as described in equation (7.113), using a molecular orbital approximation to the wavefunction. The summation in equation (7.259) is restricted to open shell electrons. To see why this is so, consider the spin–orbit coupling for a molecule in a $^1\Sigma^+$ state arising from the closed shell configuration π^4. As we described in section 6.5.4, the four-electron wave function for this state can be expressed as a Slater determinant

$$|\psi\rangle = (1/2\sqrt{6})|\pi_{+1}^\alpha \pi_{+1}^\beta \pi_{-1}^\alpha \pi_{-1}^\beta|. \tag{7.260}$$

The expression on the right-hand side is a convenient shorthand for a 4×4 determinant, listing the diagonal elements of the full determinant, which is

$$|\pi_{+1}^\alpha \pi_{+1}^\beta \pi_{-1}^\alpha \pi_{-1}^\beta| \equiv \begin{vmatrix} \pi_{+1}^\alpha(1) & \pi_{+1}^\beta(1) & \pi_{-1}^\alpha(1) & \pi_{-1}^\beta(1) \\ \pi_{+1}^\alpha(2) & \pi_{+1}^\beta(2) & \pi_{-1}^\alpha(2) & \pi_{-1}^\beta(2) \\ \pi_{+1}^\alpha(3) & \pi_{+1}^\beta(3) & \pi_{-1}^\alpha(3) & \pi_{-1}^\beta(3) \\ \pi_{+1}^\alpha(4) & \pi_{+1}^\beta(4) & \pi_{-1}^\alpha(4) & \pi_{-1}^\beta(4) \end{vmatrix} \quad (7.261)$$

The contribution to the spin–orbit energy is therefore given by

$$E_{\text{so}} = \langle \psi | \mathcal{H}_{\text{so}}^{(e)} | \psi \rangle = (1/2)a_1 - (1/2)a_2 - (1/2)a_3 + (1/2)a_4$$
$$= a[(1/2) - (1/2) - (1/2) + (1/2)] = 0, \quad (7.262)$$

since the spin–orbit coupling constant for each electron in the *same* molecular orbital is the same.

Let us now consider a molecule which does show a spin–orbit splitting, namely, OH in its $^2\Pi$ ground state. The orbital configuration for this state is $\sigma^2 \pi^3$ and confining our attention to the open shell orbital, the spin–orbital wave functions and energies for the two spin components are as follows:

$$\begin{aligned} ^2\Pi_{3/2} \quad &(1/\sqrt{6})|\pi_{+1}^\alpha \pi_{+1}^\beta \pi_{-1}^\alpha| \quad E_{\text{so}} = a[(1/2) - (1/2) - (1/2)] = -a/2, \\ ^2\Pi_{1/2} \quad &(1/\sqrt{6})|\pi_{+1}^\alpha \pi_{+1}^\beta \pi_{-1}^\beta| \quad E_{\text{so}} = a[(1/2) - (1/2) + (1/2)] = a/2. \end{aligned} \quad (7.263)$$

Thus the spin–orbit splitting is

$$\Delta E_{\text{so}} = E(\Omega = 3/2) - E(\Omega = 1/2) = -a. \quad (7.264)$$

When compared with the diagonal element of the spin–orbit operator in the effective Hamiltonian, $A\Lambda\Sigma$ in a Hund's case (a) representation, we see that

$$A = -a \quad (7.265)$$

(strictly this is the value for $A^{(1)}$ in equation (7.125)). A common approximation for the molecular orbital is to represent it by a linear combination of suitable atomic orbitals. In the case of a π orbital, these atomic orbitals are p orbitals, one on each atom of the diatomic molecule:

$$|\pi\rangle = c_1|p\pi_1\rangle + c_2|p\pi_2\rangle. \quad (7.266)$$

With this approximation, the microscopic spin–orbit coupling parameter is

$$a_\pi = \langle \pi | \mathcal{H}_{\text{so}} | \pi \rangle = c_1^2 \zeta_1(p) + c_2^2 \zeta_2(p), \quad (7.267)$$

where $\zeta_i(p)$ is the atomic spin–orbit coupling parameter for an electron in the appropriate p orbital. In other words, it is possible to relate the molecular parameter to the corresponding atomic ones. This result was first exploited by Dixon and Kroto [36]. In the case of OH, the LCAO approximation suggests that the π molecular orbital is satisfactorily described by a $2p$ orbital on the O atom. Using the value for the atomic spin–orbit coupling parameter for O of 151 cm^{-1}, we obtain a first-order estimate of the spin–orbit coupling constant of OH in its $X\,^2\Pi$ state of -151 cm^{-1}, which compares reasonably well with the experimental value of -139 cm^{-1}. The second-order contribution to the effective spin–orbit coupling parameter comes from the admixture of the first excited electronic state of OH, the $A\,^2\Sigma^+$ state, which arises from the $\sigma^1\pi^4$ configuration. Using the LCAO approximation, the matrix element of $\mathcal{H}_{so}$ between these two states is

$$\langle ^2\Sigma_{1/2}|\mathcal{H}_{so}|^2\Pi_{1/2}\rangle = (a/2)\langle\lambda=0|l_-|\lambda=1\rangle \simeq (a/2)[l(l+1)]^{1/2} = (a/\sqrt{2}). \tag{7.268}$$

In writing this equation, we have made use of Van Vleck's *pure precession* hypothesis [12], in which the molecular orbital $|\lambda\rangle$ is approximated by an atomic orbital with well-defined values for the quantum numbers n, l and λ. Such an orbital implies a spherically symmetric potential and its use is most appropriate when the electronic distribution is nearly spherical. Examples of this situation occur quite often in the description of Rydberg states. It is also appropriate for hydrides like OH where the molecule is essentially an oxygen atom with a small pimple, the hydrogen atom, on its side. Accepting the pure precession hypothesis allows the matrix elements of the orbital operators to be evaluated since

$$l_\pm|n,l,\lambda\rangle = [l(l+1)-\lambda(\lambda\pm 1)]^{1/2}|n,l,\lambda\pm 1\rangle. \tag{7.269}$$

The second-order effect of the mixing between the $^2\Pi$ and $^2\Sigma^+$ states is

$$\Delta E_{so}^{(2)} = a^2/2\lfloor E(^2\Sigma) - E(^2\Pi)\rfloor = 0.35\,\text{cm}^{-1} \tag{7.270}$$

for an energy separation of 32 600 cm^{-1}. This interaction pushes the $^2\Pi_{1/2}$ component down and so reduces the spin–orbit splitting, that is, $A^{(2)} = 0.35$ cm^{-1}. This is a move in the right direction but not nearly large enough to improve significantly the agreement with the experimental value.

For a non-hydride molecule such as NO, we must use equation (7.267) to estimate the spin–orbit coupling parameter. NO also has a $^2\Pi$ ground state but it arises from a π^1 configuration. Estimates from the magnetic hyperfine interactions suggest that $c_N^2 = 0.734$ and $c_O^2 = 0.470$ [37]. From this we calculate a value for $A^{(1)}$ of 124.8 cm^{-1} using $\zeta_O = 151$ cm^{-1} and $\zeta_N = 73.3$ cm^{-1}. The agreement with the experimentally determined value of 123.3 cm^{-1} is probably fortuitously good.

Finally let us consider one more example, the more complicated case of FeH in its low-lying $a^6\Delta$ state. The lowest spin component is $^6\Delta_{9/2}$ which is reasonably well approximated by the wave function

$$|^6\Delta_{9/2}\rangle = 1/(7!)^{1/2}|\delta^\alpha_{+2}\delta^\beta_{+2}\delta^\alpha_{-2}\pi^\alpha_{+1}\pi^\alpha_{-1}\sigma^\alpha\bar{\sigma}^\alpha|, \tag{7.271}$$

where the first six molecular orbitals are essentially non-bonding $3d$ orbitals on the Fe atom and the $\bar{\sigma}$ orbital is an antibonding combination of the $1s$ orbital on the H atom and an sp hybrid ($n=4$) on the Fe atom. (There is also a σ^2 contribution to the electron configuration which has been omitted from equation (7.271) for the sake of simplicity. It is the bonding combination of the two atomic orbitals and is primarily responsible for holding the atoms together.) The spin–orbit energy for this wave function using the microscopic Hamiltonian is

$$E_{so} = a(1-1-1+1/2-1/2+0+0) = -a. \tag{7.272}$$

The wave function for the next spin component of the $^6\Delta$ state can be easily generated from (7.271) by application of the total spin lowering operator S_-, given by

$$S_- = \sum_{i=1}^{7} s_{i-}. \tag{7.273}$$

We obtain

$$|^6\Delta_{7/2}\rangle = (1/\sqrt{7!})(1/\sqrt{5}) \times \{|\delta^\alpha_{+2}\delta^\beta_{+2}\delta^\beta_{-2}\pi^\alpha_{+1}\pi^\alpha_{-1}\sigma^\alpha\bar{\sigma}^\alpha| + |\delta^\alpha_{+2}\delta^\beta_{+2}\delta^\alpha_{-2}\pi^\beta_{+1}\pi^\alpha_{-1}\sigma^\alpha\bar{\sigma}^\alpha|$$
$$+ |\delta^\alpha_{+2}\delta^\beta_{+2}\delta^\alpha_{-2}\pi^\alpha_{+1}\pi^\beta_{-1}\sigma^\alpha\bar{\sigma}^\alpha| + |\delta^\alpha_{+2}\delta^\beta_{+2}\delta^\alpha_{-2}\pi^\alpha_{+1}\pi^\alpha_{-1}\sigma^\beta\bar{\sigma}^\alpha|$$
$$+ |\delta^\alpha_{+2}\delta^\beta_{+2}\delta^\alpha_{-2}\pi^\alpha_{+1}\pi^\alpha_{-1}\sigma^\alpha\bar{\sigma}^\beta|\}. \tag{7.274}$$

For this wave function

$$E_{\text{so}} = (a/5)(1 - 2 + 0 - 1 - 1) = -(3a/5). \tag{7.275}$$

The separation between adjacent spin components is therefore $-2a/5$ which equates with $2A^{(1)}$ from the effective Hamiltonian. Hence $A^{(1)} = -a/5$ or -83.4 cm^{-1}, using $\zeta_{\text{Fe}} = 417$ cm^{-1}. The value obtained from experiment is -77.3 cm^{-1} although in practice it is difficult to model the spin–rotation levels of FeH with an effective Hamiltonian because of large spin–orbit perturbations [38]. For molecules like FeH, one would expect second- and higher-order contributions to A to be significant.

7.8.4. Spin–spin and spin–rotation parameters, λ and γ

A theoretical estimate of the spin–spin parameter in the effective Hamiltonian can be made by the use of equations (7.111) and (7.121) using either *ab initio* or simpler, less accurate wave functions. This was done for the case of O_2 in its $^3\Sigma_g^-$ ground state in an influential paper by Kayama and Baird [39]. They demonstrated that, in the case of a π^2 configuration (which occurs quite commonly in practice), there is significant spin–orbit mixing between the lowest $^3\Sigma^-$ state and the metastable $^1\Sigma^+$ state which arises from the same electron configuration. This has the effect of lowering the $\Omega = 0$ component of the $^3\Sigma^-$ state and so makes a positive contribution to the spin–spin parameter. For molecules with heavy atoms and large spin–orbit interactions, this becomes the dominant contribution to λ. In their paper Kayama and Baird calculate a value of 0.822 cm^{-1} for $\lambda^{(1)}$ and 0.86 cm^{-1} for $\lambda^{(2)}$ for O_2, giving a total value for λ of 1.682 cm^{-1} (to be compared with the experimental value of 1.985 cm^{-1}). A later calculation by Wayne and Colbourne [40] gave $\lambda^{(1)} = 0.750$ cm^{-1}.

The way in which the second-order contribution arises can be appreciated from the following. The spin–orbital wave functions for the $^3\Sigma_0^-$ and $^1\Sigma_0^+$ states are approximated by

$$|^3\Sigma_0^-\rangle = (1/\sqrt{2})\{|\pi^\alpha_{+1}\pi^\beta_{-1}| + |\pi^\beta_{+1}\pi^\alpha_{-1}|\},$$
$$|^1\Sigma_0^+\rangle = (1/\sqrt{2})\{|\pi^\alpha_{+1}\pi^\beta_{-1}| - |\pi^\beta_{+1}\pi^\alpha_{-1}|\}. \tag{7.276}$$

Hence the spin–orbit matrix element is given by

$$\langle ^3\Sigma_0^- | \mathcal{H}_{so} | ^1\Sigma_0^+ \rangle = (a/2)\{(1/2 + 1/2) - (-1/2 - 1/2)\} = a. \tag{7.277}$$

In addition equation (7.267) tells us that for a homonuclear diatomic molecule, $a = \zeta$. Consequently

$$\lambda^{(2)} = -(1/2)\langle ^3\Sigma_0^- | \mathcal{H}_{so} | ^1\Sigma_0^+ \rangle^2 / \{E(^3\Sigma_0) - E(^1\Sigma_0)\}. \tag{7.278}$$

Using $\zeta_O = 151$ cm^{-1} and a value of $-13\,120$ cm^{-1} for the energy denominator, we obtain a value of 0.864 cm^{-1} for $\lambda^{(2)}$.

A very similar treatment can be applied to the spin–rotation parameter γ. Once again there is a first- and second-order contribution with the second-order contribution dominating in all but the lightest molecules. Full expressions for $\gamma^{(1)}$ and $\gamma^{(2)}$ are given in equations (7.110) and (7.122).

Semi-empirical estimates of the second-order contribution to γ provide a useful means to interpret experimental data. Reverting to the case of OH in its $X\,^2\Pi$ state, we need the matrix elements

$$\langle ^2\Pi_{1/2} | \mathcal{H}_{so} | ^2\Sigma_{1/2}^+ \rangle = a/\sqrt{2}, \tag{7.279}$$

$$\langle ^2\Sigma_{1/2} | -BL_- | ^2\Pi_{3/2} \rangle = (1/5!)\langle || \sigma^\alpha \pi_{+1}^\alpha \pi_{+1}^\beta \pi_{-1}^\alpha \pi_{-1}^\beta || - BL_- || \sigma^\alpha \sigma^\beta \pi_{+1}^\alpha \pi_{+1}^\beta \pi_{-1}^\alpha || \rangle$$
$$= \langle \lambda = -1 | Bl_- | \lambda = 0 \rangle \cong \sqrt{2}B. \tag{7.280}$$

The first result has been obtained earlier; in obtaining the second result, we have to take account of the sign change which results from the permutation of the order of the orbitals (required to set up the one-electron result). We have also used the pure precession hypothesis to evaluate the orbital matrix element. Using $a = 151$ cm^{-1}, $B = 18.535$ cm^{-1} and $\{E(^2\Pi) - E(^2\Sigma^+)\} = -32\,600$ cm^{-1}, we obtain a value of -0.172 cm^{-1} for $\gamma^{(2)}$ for OH in the $X\,^2\Pi$ state, somewhat larger in magnitude than the experimental value of -0.119 cm^{-1}. More accurate calculations using *ab initio* wave functions suggest that it is mainly the estimate of the off-diagonal element of $\mathcal{H}_{so}$ in equation (7.279) which is too large [41]. The *ab initio* value for this matrix element corresponds to a value for a of 111.7 cm^{-1} which gives much better agreement with the experimental value.

Before we leave OH, it is instructive to use the same semi-empirical model to estimate the value of $\gamma^{(2)}$ for the $A\,^2\Sigma^+$ state. In this case, we find that the combined effect of spin–orbit and rotational coupling with the $X\,^2\Pi$ state gives a value for $\gamma^{(2)}$ which has the opposite sign (because of the change in sign of the energy denominator) and which is twice as big, that is, 0.344 cm^{-1}. The latter difference occurs because the $|\Lambda = 0\rangle$ orbital of the $^2\Sigma^+$ state interacts with both $|\Lambda = +1\rangle$ and $|\Lambda = -1\rangle$ components of the $^2\Pi$ state. The experimental value of γ for OH in the $A\,^2\Sigma^+$ state is 0.201 cm^{-1}.

By this stage, it will be apparent that even the sign of the spin–rotation parameter contains valuable information on the electronic structure of the molecule. To appreciate this point fully, let us consider a $^3\Sigma$ state which arises from a π^2 electron configuration. The effective spin–rotation interaction arises from second-order mixing with $^3\Pi$ states. Let us assume that there is only one such state which lies higher in energy so that the energy denominator $\{E(^3\Sigma) - E(^3\Pi)\}$ in equation (7.122) is negative. If the $^3\Pi$ state is obtained by promotion of an electron from an inner σ orbital ($\sigma^2\pi^2 \to \sigma^1\pi^3$), the matrix elements involved in the calculation of $\gamma^{(2)}$ are of the form

$$\langle^3\Pi_1| - BL_+|^3\Sigma_0\rangle = \langle ||\sigma^\beta\pi^\alpha_{+1}\pi^\beta_{+1}\pi^\alpha_{-1}|| - BL_+||\sigma^\alpha\sigma^\beta\pi^\beta_{+1}\pi^\alpha_{-1}||\rangle,$$
$$\langle^3\Pi_1|\mathcal{H}_{so}|^3\Sigma_1\rangle = \langle ||\sigma^\beta\pi^\alpha_{+1}\pi^\beta_{+1}\pi^\alpha_{-1}||\mathcal{H}_{so}||\sigma^\alpha\sigma^\beta\pi^\alpha_{+1}\pi^\alpha_{-1}||\rangle. \quad (7.281)$$

Permutation of the order of the molecular orbitals in the Slater determinant to reduce them to one-electron integrals shows that both these matrix elements are positive and so their product is also positive. Consequently, $\gamma^{(2)}$ is negative for this particular electronic excitation. By contrast, when we investigate mixing by a $^3\Pi$ state obtained by promotion of a π electron to an outer σ orbital ($\pi^2 \to \pi\sigma$), the corresponding matrix elements are:

$$\langle^3\Pi_1| - BL_+|^3\Sigma_0\rangle = \langle ||\pi^\alpha_{+1}\sigma^\beta|| - BL_+||\pi^\alpha_{+1}\pi^\beta_{-1}||\rangle,$$
$$\langle^3\Pi_1|\mathcal{H}_{so}|^3\Sigma_1\rangle = \langle ||\pi^\alpha_{+1}\sigma^\beta||\mathcal{H}_{so}||\pi^\alpha_{+1}\pi^\alpha_{-1}||\rangle. \quad (7.282)$$

The rotational kinetic energy matrix element is now negative while that of the spin–orbit coupling remains positive. In this case, therefore, $\gamma^{(2)}$ is positive.

7.8.5. Λ-doubling parameters

As we have seen in section 7.4.5, the Λ-doubling parameters in the effective Hamiltonian, like the spin–spin and spin–rotation parameters, arise from the mixing of electronic states by the spin–orbit interaction and rotational coupling. However, Λ-doubling effects must involve a non-degenerate Σ electronic state at some stage whereas the spin–spin and spin–rotation couplings do not have to do so. For a molecule in a Π state, Λ-doubling arises through second order mixing with a remote Σ state. For a molecule in a Δ state, on the other hand, Λ-doubling arises from fourth-order mixing with a remote Σ state through an intermediate Π state. For this reason, Λ-doubling effects are smaller for molecules in Δ states than for those in Π states. The other important feature of the Λ-doubling parameters is that, with the exception of the parameter o, they do not have a first-order contribution in the effective Hamiltonian. The leading contribution is the second-order one described in section 7.4.5 for molecules in Π electronic states.

For OH in its $X^2\Pi$ state, there are two Λ-doubling parameters, p and q. Using the explicit expressions given in equations (7.142) and (7.143), these parameters can be calculated very accurately using good quality *ab initio* wave functions. A simple estimate can be made using the pure precession values for the relevant matrix elements which we have derived earlier. This approximation assumes that the Λ-doubling effects arise wholly from the perturbations with the $A^2\Sigma^+$ state. For this approximation,

$$p = -4B\zeta/\{E(\Pi) - E(\Sigma^+)\} = 0.3434 \text{ cm}^{-1}, \quad cf.\text{expt}, 0.2503 \text{ cm}^{-1}, \quad (7.283)$$
$$q = 4B^2/\{E(\Pi) - E(\Sigma^+)\} = -0.0422 \text{ cm}^{-1}, \quad cf.\text{expt}, -0.0387 \text{ cm}^{-1}. \quad (7.284)$$

Once again, we see that the values of these two parameters are reasonably well reproduced by this very simple calculation.

Later on in this book, we discuss the properties of the CH radical in its $X^2\Pi$ state in some detail. The electronic structure of this radical is rather more complicated than that of OH. Despite this, a simple pure precession calculation of the Λ-doubling parameters reproduces the experimental values (particularly q) reasonably well (see section 10.6.3).

7.8.6. Magnetic hyperfine interactions

Nuclear hyperfine interactions may be small in magnitude but it would be wrong to dismiss them as insignificant. Indeed, they provide highly specific and accurate information on the electronic wave function. This is because the interactions fall off very rapidly with distance from the nucleus (typically as the inverse cube of the separation of the electron from the nucleus although the Fermi contact interaction provides information of the electron density only at the nucleus). In addition, the magnetic hyperfine interactions depend on the open shell electrons only. As with the other interactions, the effective Hamiltonian provides explicit formulae for the calculation of the hyperfine parameters (see, for example, equations (7.154), (7.156) and (7.162)). Because of the small size of these interactions, it is usually only the first-order contribution which is significant. It is therefore relatively easy to calculate their values using *ab initio* wave functions.

If we confine attention to molecules in a $^{2S+1}\Pi$ electronic state, there are four magnetic hyperfine parameters, a, b_F, c and d. The first of these describes the strength of the nuclear-spin/electron–orbital interaction and gives information on the spatial distribution of the unpaired electrons. The other three parameters give information on the electron spin distribution within the molecule. Though often similar, these two distribution functions are not identical.

A complete characterisation of the magnetic hyperfine parameters has been achieved for several free radicals in $^2\Pi$ electronic states. One such example is the CF radical which is isoelectronic with NO. The values of the four hyperfine parameters for the ^{19}F nucleus ($I = 1/2$) are given in table 7.3 together with the expectation values of the corresponding distribution functions over the electronic wave function. It can be seen that the hyperfine parameter a is related to the orbital distribution $\langle r^{-3}\rangle_l$ where r is the distance between the unpaired electron and the F nucleus, the Fermi contact parameter b_F is a measure of the spin density at the nucleus $|\Psi(0)|^2$, whilst c and d are associated with the angular spin distributions $\langle(3\cos^2\theta - 1)/r^3\rangle_s$ and $\langle\sin^2\theta/r^3\rangle_s$ respectively, where θ is the polar angle between the vector r and the internuclear axis. The value for $\langle r^{-3}\rangle_l$ can be compared with the corresponding value for the fluorine atom in its ground ^2P state (4.96×10^{31} m^{-3} [42]), suggesting that the square of the atomic orbital coefficient c_1 in equation (7.266) is 0.192. The corresponding spin average can be calculated from the sum $(d + c/3)$; the value obtained is 9.075×10^{30} m^{-3}, close to but different from that for $\langle r^{-3}\rangle_l$. The angular distribution of the

Table 7.3. ^{19}F hyperfine structure parameters [43] and expectation values of distribution functions over the electronic wave function of CF in its $X\,^2\Pi$ state

hyperfine parameter	experimental value	function	value /m^{-3}
a	705.94 (14)	$\langle r^{-3}\rangle_l$	9.502×10^{30}
b_F	151.19 (49)	$|\Psi(0)|^2$	2.429×10^{30}
c	−351.6 (14)	$\langle(3\cos^2\theta - 1)/r^3\rangle_s$	-3.151×10^{30}
d	792.195 (98)	$\langle \sin^2\theta/r^3\rangle_s$	7.100×10^{30}

spin density is well modelled by a p_π orbital. For such an orbital,

$$\langle \pi | \sin^2\theta | \pi \rangle = 4/5, \tag{7.285}$$

$$\langle \pi | (3\cos^2\theta - 1) | \pi \rangle = -2/5. \tag{7.286}$$

In this approximation, the ratios $\langle r^{-3}\rangle_s : \langle(3\cos^2\theta - 1)/r^3\rangle_s : \langle\sin^2\theta/r^3\rangle_s$ are $1 : -0.4 : 0.8$, to be compared with the actual values for CF from table 7.3 of $1 : -0.35 : 0.78$.

A striking example of the specific information carried in the hyperfine parameters is provided by a study of the molecule CuO in its so-called $A'^2\Sigma^-$ state [44]. This molecule shows a large magnetic hyperfine splitting from the ^{63}Cu nucleus ($I = 3/2$), so large in fact that it can be resolved in the Doppler-limited optical spectrum associated with the $A'^2\Sigma^- - X^2\Pi$ transition. The hyperfine structure can be interpreted in terms of a large Fermi contact interaction and a fit of the data to the effective Hamiltonian for a $^2\Sigma^s$ state gives a value of -0.052 (22) cm^{-1} for the parameter b_F. We note that this value is both large and *negative*. The sign, in particular, cannot be explained by the simplistic definition of the Fermi contact parameter,

$$b_F = (2g_S\mu_B g_N\mu_N\mu_0/3)|\Psi(0)|^2, \tag{7.287}$$

where all the terms on the right-hand side are positive. The proper explanation provides insight into the nature of the electronic wave function. The likely configuration for the A' state is $\ldots 1\delta^4 9\sigma^2 4\pi^2 10\sigma$ where the 1δ orbital is essentially localised on the Cu atom, the 9σ and 4π orbitals are predominantly O $2p$ atomic orbitals and the 10σ orbital is a $4s$ orbital localised on the metal atom. The states which arise from this configuration are $^2\Sigma^+, ^2\Sigma^-, ^2\Delta_r$ and $^4\Sigma^-$. The spin–orbital wave functions for the $^2\Sigma^+$ and $^2\Sigma^-$ states are [45]

$$|^2\Sigma^+\rangle = (2)^{-1/2}(6)^{-1/2}\{|4\pi_{+1}^\beta 4\pi_{-1}^\alpha 10\sigma^\alpha| - |4\pi_{+1}^\alpha 4\pi_{-1}^\beta 10\sigma^\alpha|\}, \tag{7.288}$$

$$|^2\Sigma^-\rangle = (6)^{-1}\{2|4\pi_{+1}^\alpha 4\pi_{-1}^\alpha 10\sigma^\beta| - |4\pi_{+1}^\beta 4\pi_{-1}^\alpha 10\sigma^\alpha| - |4\pi_{+1}^\alpha 4\pi_{-1}^\beta 10\sigma^\alpha|\}. \tag{7.289}$$

Using the full form of the Fermi contact hyperfine interaction given in equation (7.144), which differs from equation (7.287) in that it contains an explicit summation over the open shell electrons, we have that

$$\langle^2\Sigma^+|\delta(\boldsymbol{r})s_z|^2\Sigma^+\rangle = (1/2)|\Psi_{10\sigma}(0)|^2, \tag{7.290}$$

$$\langle^2\Sigma^-|\delta(\boldsymbol{r})s_z|^2\Sigma^-\rangle = -(1/6)|\Psi_{10\sigma}(0)|^2. \tag{7.291}$$

where $\delta(r)$ and s_z act on the 10σ electron only. A value for $|\Psi_{10\sigma}(0)|^2$ can be obtained from the optical spectrum of the Cu atom which, in its ground state, has one unpaired electron in a $4s$ orbital which produces a Fermi contact parameter of 0.0967 cm^{-1}. Using this value, we can make the following estimates,

$$b_F(^2\Sigma^+) = 0.0967 \text{ cm}^{-1}, \tag{7.292}$$
$$b_F(^2\Sigma^-) = -0.0322 \text{ cm}^{-1}. \tag{7.293}$$

The experimental value for b_F clearly supports the assignment of the A' state as $^2\Sigma^-$. The value for the $^2\Sigma^-$ state is negative because the contribution of the β spin in the $\sigma(4s)$ orbital in equation (7.289) outweighs that for the α spins.

We see that the analysis of the hyperfine structure in this case provides a simple and direct way of distinguishing between Σ^+ and Σ^- states. It depends on the precise form of the electron spin part of the total molecular wave function which is permitted by the Pauli exclusion principle. It has the advantage that it does not require a knowledge of the parities of the individual states. This contrasts with the traditional way of making the Σ^+/Σ^- assignment which is based on a consideration of the orbital part of the wave function.

7.8.7. Electric quadrupole hyperfine interaction

The form of the nuclear electric quadrupole interaction in the effective Hamiltonian for a diatomic molecule is given in equations (7.158) and (7.161), with the latter applying only to molecules in Π electronic states. The two parameters which can be determined from a fit of the experimental data are $eq_0 Q$ and $eq_2 Q$ respectively. Since the electric quadrupole moment eQ is known for most nuclei, an experimental observation gives information on q_0 (and perhaps q_2), the electric field gradient at the nucleus. This quantity depends on the electronic structure of the molecule according to the expression

$$\begin{aligned} q_0 &= -2\langle \eta, \Lambda | T^2_{q=0}(\nabla E) | \eta, \Lambda \rangle \\ &= \langle \eta, \Lambda | \partial^2 V / \partial z^2 | \eta, \Lambda \rangle, \end{aligned} \tag{7.294}$$

and is strictly the negative of the electric field gradient.

Let us consider a nucleus with $I \geq 1$ in atom 1 of a diatomic molecule. We shall use a local coordinate system (x, y, z) with its origin at the nucleus and z lying along the molecular bond. An unscreened electric charge q at a distance r from the nucleus gives rise to an electrostatic potential

$$V = q/4\pi\varepsilon_0 r \tag{7.295}$$

at the nucleus. The negative of the electric field gradient at the nucleus is therefore

$$q_0 \equiv \frac{\partial^2 V}{\partial z^2} = \frac{q}{4\pi\varepsilon_0} \frac{\partial^2}{\partial z^2} [x^2 + y^2 + z^2]^{-1/2} = q \frac{(3\cos^2\theta - 1)}{4\pi\varepsilon_0 r^3}, \tag{7.296}$$

where θ is the polar angle between r and the z axis. If, as is most likely, the charged particle is an electron, its contribution to the electric field gradient at the nucleus may be obtained by averaging the quantity in equation (7.296) over the molecular orbital ϕ_i of the electron,

$$q_{0,i} = -\frac{e}{4\pi\varepsilon_0} \langle \phi_i | (3\cos^2\theta_i - 1)/r_i^3 | \phi_i \rangle. \quad (7.297)$$

The total electric field gradient at the nucleus due to all the electrons in the molecule is therefore

$$q_0 = -\frac{e}{4\pi\varepsilon_0} \sum_i n_i \langle \phi_i | (3\cos^2\theta_i - 1)/r_i^3 | \phi_i \rangle \quad (7.298)$$

where n_i (=0, 1 or 2) is the number of electrons in the ith orbital.

This discussion shows that the most accurate interpretation of the electric quadrupole coupling constant is obtained by evaluating

$$q_0 = -\frac{e}{4\pi\varepsilon_0} \langle \eta, \Lambda | \sum_i \frac{(3\cos^2\theta_i - 1)}{r^3} | \eta, \Lambda \rangle + \frac{Z_{\alpha'}e}{4\pi\varepsilon_0} \langle \eta, \Lambda | \frac{2}{R^3} | \eta, \Lambda \rangle, \quad (7.299)$$

where $|\eta, \Lambda\rangle$ is the *ab initio* electronic wave function and i is the sum over *all* electrons. The second term on the right-hand side is the contribution to the electric field gradient from the charge on the other nucleus ($\alpha' = 2$) in the diatomic molecule. Because of the R^{-3} dependence, this contribution is small. It is not, however, negligible in the most accurate calculations; for example, it makes a large (81%) contribution to the quadrupole coupling constant for ^{17}OH in its ground $^2\Pi$ state because the electronic contribution is accidentally very small in this case [46]. In addition, equation (7.299) shows that, unlike the magnetic hyperfine interactions, the electric quadrupole interaction depends on all the electrons in the molecule, which makes its calculation more arduous.

Short of the *ab initio* calculations, there are several semi-empirical approaches to the calculation and interpretation of electric quadrupole coupling constants. These were developed originally by Townes and Dailey [47, 48] and are well documented in the book by Gordy and Cook [49]. They are based on the linear combination of atomic orbitals approximation for molecular orbitals, mentioned earlier in equation (7.266) and described in more detail in chapter 6:

$$|\eta, \Lambda\rangle \simeq c_1 |n_1, l_1, \lambda_1\rangle + c_2 |n_2, l_2, \lambda_2\rangle. \quad (7.300)$$

This allows the local electron distribution to be modelled by atomic orbitals. The symmetry of these orbitals leads to a number of useful simplifications. First, the spherically symmetric charge distribution of s orbitals produce zero average field at the central nucleus and so do not contribute to q_0. The same remark applies to electrons in the inner closed shells of an atom, if we ignore the polarisation effects of electrons in valence shells. Furthermore, electrons in orbitals with $l = 2$ (d orbitals) or greater produce a much smaller electric field gradient at the nucleus than those in $l = 1$ (p orbitals). Finally, the contributions to q_0 of the electrons and nuclear charge of the *other* atom approximately cancel and so can be neglected. The conclusion of these simplifications is that the electric field gradient at the nucleus can be calculated by restricting attention to the p electrons in the valence shells of the atom in question, since these electrons make the predominant contribution to q_0. The resultant expression for the electric field gradient at nucleus 1 is thus

$$q_0 = \sum_i n_i c_i^2 q_i \qquad (7.301)$$

where q_i is defined by equation (7.297) and the summation is over valence electrons in p orbitals only. This model produces reasonably reliable values for q_0 but, more importantly, it allows trends in a series of molecules to be rationalised.

In addition to the main axial component $eq_0 Q$ of the electric quadrupole interaction, there is also a perpendicular component $eq_2 Q$ for molecules in Π electronic states. Recalling that

$$T_{\pm 2}^2(\nabla \boldsymbol{E}) = -\frac{1}{2\sqrt{6}} \left[\frac{\partial^2 V}{\partial x^2} - \frac{\partial^2 V}{\partial y^2} \right], \qquad (7.302)$$

we see that this component measures the asymmetry of the electric field gradient perpendicular to the z axis. The particular component of the electric field gradient is given by

$$\begin{aligned} q_2 &= -2\sqrt{6} \langle \eta, \Lambda = \pm 1 | T_{\pm 2}^2(\nabla \boldsymbol{E}) | \eta, \Lambda = \mp 1 \rangle \\ &= \frac{3e}{4\pi\varepsilon_0} \langle \eta, \Lambda = \pm 1 | \sum_i (\sin^2 \theta_i / r_i^3) | \eta, \Lambda = \pm 1 \rangle. \end{aligned} \qquad (7.303)$$

Because this term is a measure of the deviation from cylindrical symmetry, there is no contribution from the charge on the other nucleus nor, to a good approximation, from the electrons in closed shell orbitals. The magnetic hyperfine parameter d also depends on the deviation of the electronic distribution from cylindrical symmetry, in this case on the non-cylindrical distribution of electron spin for a molecule in a Π electronic state. The two parameters are therefore related since they give independent estimates of the expectation values of $\langle \sin^2 \theta / r^3 \rangle$. For example, the following values for the two parameters have been measured for ^{35}ClO in its $^2\Pi$ state [50]:

$$d = 173.030(20)\,\text{MHz}, \quad eq_2 Q = -116.0(56)\,\text{MHz}.$$

These correspond to expectation values

$$\langle \sin^2 \theta / r^3 \rangle_S = 14.88 \times 10^{30}\,\text{m}^{-3}, \quad \langle \sin^2 \theta / r^3 \rangle_l = -14.06 \times 10^{30}\,\text{m}^{-3},$$

which are similar in magnitude. The negative sign for the quadrupole coupling constant arises because the ground state of ClO is derived from a π^3 configuration. The electric field gradient at the nucleus is therefore caused by a hole in the π orbital, that is, by an effective distribution of positive charge rather than negative.

Appendix 7.1. Molecular parameters or constants

We present below a summary of the molecular parameters which appear in effective Hamiltonians. Often the appropriate vibrational level is indicated by the subscript v and, more rarely, other vibronic quantum numbers, particularly electronic, are indicated by the subscript η. These molecular parameters are combined with operator expressions, given in the equations referenced below.

A or A_v – fine structure or molecular spin–orbit coupling constant for vibration level v. Equation (7.187).

A_D or A_{Dv} – centrifugal distortion correction to the spin–orbit coupling constant. Equation (7.187).

B or B_v – rotational constant for vibrational level v. The subscript v is often omitted when only the ground vibrational level is involved. Equation (7.185).

D or D_v – first centrifugal distortion constant. Equation (7.186).

H or H_v – second centrifugal distortion constant. Equation (7.186).

λ_v – electron spin–spin constant in vibrational level v. Equation (7.188).

λ_{Dv} – centrifugal distortion correction to the spin–spin constant. Equation (7.188).

γ_v – electron spin–rotation constant in vibrational level v. Equation (7.189).

γ_{Dv} – centrifugal distortion correction to the electron spin–rotation constant. Equation (7.189).

$C_v^{(3)}$, η_v, Θ_v – higher-order fine structure constants. Equation (7.193).

o_v, p_v, q_v – Λ-doubling parameters for Π electronic states. Equation (7.190).

o_{Dv}, p_{Dv}, q_{Dv} – centrifugal distortion corrections to the Λ-doubling constants. Equation (7.190).

a or a_v or a_v^{α} – electron orbital magnetic hyperfine constant. Equation (7.191).

b_F or b_{Fv} or b_{Fv}^{α} – Fermi contact hyperfine constant. Equation (7.191).

t_0, t_{0v}, or c_v – axial component of the electron–nuclear dipolar interaction.

c_I – nuclear spin–rotation constant.

d or t_2 – nuclear spin–electron spin dipolar interaction constant for molecules in Π electronic states.

q_0 – axial component of (the negative of) the electric field gradient. Equation (7.192).

$eq_0 Q$ – axial nuclear quadrupole coupling constant. Equation (7.192).

q_2 – perpendicular component of the electric field gradient. Equation (7.303).

g_S – electron spin g-factor with value 2 in the Dirac theory, and 2.002 319 with quantum electrodynamical correction. Equation (7.232). See also reference [32].

g_L – electron orbital g-factor with value 1. In some instances admixture of other electronic states is indicated by the alternative symbol g'_L with a value slightly different from 1. Equation (7.232). See also reference [32].

g_r – rotational g-factor. Equation (7.232).

g_l – anisotropic correction to the electron spin Zeeman interaction. Equation (7.232).

g_N or g_N^{α} – nuclear g-factor for nucleus α. Equation (7.232).

ξ – anisotropic magnetic susceptibility constant. Equation (7.232).

$\sigma(J)$ – nuclear screening constant.

References

[1] M. Born and R. Oppenheimer, *Ann. Phys.*, **84**, 4571 (1927).
[2] J.H. Van Vleck, *Rev. Mod. Phys.*, **23**, 213 (1951).
[3] M.H.L. Pryce, *Proc. Phys. Soc.*, **A63**, 25 (1950).
[4] J.S. Griffith, *Mol. Phys.*, **3**, 79 (1960).
[5] T.A. Miller, *Mol. Phys.*, **16**, 105 (1969).
[6] D. De Santis, A. Lurio, T.A. Miller and R.S. Freund, *J. Chem. Phys.*, **58**, 4625 (1972).
[7] J.M. Brown, E.A. Colbourn, J.K.G. Watson and F.D. Wayne, *J. Mol. Spectrosc.*, **74**, 294 (1979).
[8] C. Bloch, *Nucl. Phys.*, **6**, 329 (1958).
[9] C.E. Soliverez, *J. Phys. C.*, **2**, 2161 (1969).
[10] H. Lefebvre-Brion and R.W. Field, *Perturbations in the Spectra of Diatomic Molecules*, Academic Press, Inc., Orlando, 1986.
[11] M.R. Aliev and J.K.G. Watson, *Molecular Spectroscopy: Modern Research*, ed. K.N. Rao, **3**, 2, 1985.
[12] J.H. Van Vleck, *Phys. Rev.*, **33**, 467 (1929).
[13] E.C. Kemble, *The Fundamental Principles of Quantum Mechanics*, Dover, New York, 1958.
[14] E.P. Wigner, *Group Theory*, Academic Press, New York, 1959.
[15] J.T. Hougen, *J. Chem. Phys.*, **36**, 519 (1962).
[16] L. Veseth, *Theor. Chim. Acta*, **18**, 368 (1970).
[17] J.M. Brown and J.K.G. Watson, *J. Mol. Spectrosc.*, **65**, 65 (1977).
[18] R.S. Mulliken and A. Christy, *Phys. Rev.*, **38**, 87 (1931).
[19] J.T. Hougen, *Can. J. Phys.*, **40**, 598 (1962).
[20] J.M. Brown and D.J. Milton, *Mol. Phys.*, **31**, 409 (1976).
[21] T. Nelis, J.M. Brown and K.M. Evenson, *J. Chem. Phys.*, **92**, 4067 (1990).
[22] J.K.G. Watson, *J. Mol. Spectrosc.*, **74**, 319 (1979).
[23] J.L. Dunham, *Phys. Rev.*, **41**, 721 (1932).
[24] J.M. Brown, D.J. Milton, J.K.G. Watson, R.N. Zare, D.L. Albritton, M. Horani and J. Rostas, *J. Mol. Spectrosc.*, **89**, 139 (1981).
[25] J.K.G. Watson, *J. Mol. Spectrosc.*, **45**, 99 (1973).
[26] P.R. Bunker, *J. Mol. Spectrosc.*, **68**, 367 (1977).
[27] J.K.G. Watson, *J. Mol. Spectrosc.*, **80**, 411 (1980).
[28] J.H. Van Vleck, *J. Chem. Phys.*, **4**, 327 (1936).
[29] T.A. Miller, *J. Chem. Phys.*, **54**, 330 (1971).
[30] J.M. Brown, M. Kaise, C.L.M. Kerr and D.J. Milton, *Mol. Phys.*, **36**, 553 (1978).
[31] J.M. Brown and H. Uehara, *Mol. Phys.*, **24**, 1169 (1972).
[32] J.M. Brown, R.J. Buenker, A. Carrington, C. Di Lauro, R.N. Dixon, R.W. Field, J.T. Hougen, W. Hüttner, K. Kuchitsu, M. Mehring, A.J. Merer, T.A. Miller, M. Quack, D.A. Ramsay, L. Veseth and R.N. Zare, *Mol. Phys.*, **98**, 1597 (2000).
[33] L. Veseth, *J. Mol. Spectrosc.*, **38**, 228 (1971).
[34] F. Tamassia, J.M. Brown and K.M. Evenson, *J. Chem. Phys.*, **110**, 7273 (1999).
[35] F. Tamassia, J.M. Brown and J.K.G. Watson, *Mol. Phys.*, **100**, 3485 (2002).
[36] R.N. Dixon and H.W. Kroto, *Trans. Faraday Soc.*, **59**, 1484 (1963).
[37] J.M. Brown, C.R. Byfleet, B.J. Howard and D.K. Russell, *Mol. Phys.*, **23**, 457 (1972).

[38] C. Wilson, H.M. Cook and J.M. Brown, *J. Chem. Phys.*, **115**, 5943 (2001).
[39] K. Kayama and J.C. Baird, *J. Chem. Phys.*, **46**, 2604 (1967).
[40] F.D. Wayne and E.A. Colbourne, *Mol. Phys.*, **34**, 1141 (1977).
[41] R.K. Hinkley, J.A. Hall, T.E.H. Walker and W.G. Richards, *J. Phys. B.*, **5**, 204 (1972).
[42] J.S.M. Harvey, *Proc. R. Soc. Lond.*, **A**285, 581 (1965).
[43] J.M. Brown, J.E. Schubert, R.J. Saykally and K.M. Evenson, *J. Mol. Spectrosc.*, **120**, 421 (1986).
[44] O. Appelbad, I. Renhorn, M. Dulick, M.R. Purnell and J.M. Brown, *Physica Scripta*, **28**, 539 (1983).
[45] J. Raftery, P.R. Scott and W.G. Richards, *J. Phys.*, **B5**, 1293 (1972).
[46] A. Carrington and N.J.D. Lucas, *Proc. R. Soc. Lond.*, **A**314, 567 (1970).
[47] C.H. Townes and B.P. Dailey, *J. Chem. Phys.*, **17**, 782 (1949).
[48] C.H. Townes and B.P. Dailey, *J. Chem. Phys.*, **23**, 118 (1955).
[49] W. Gordy and R.L. Cook, *Microwave Molecular Spectroscopy*, Interscience, New York, 1970.
[50] E.A. Cohen, H.M. Pickett and M. Geller, *J. Mol. Spectrosc.*, **106**, 4301 (1984).

8 Molecular beam magnetic and electric resonance

8.1. Introduction

We have seen that the fine and hyperfine structure of vibration–rotation levels arises almost entirely from interactions involving electron spin, nuclear spin, and the rotational motion of the nuclei, with or without the additional presence of applied magnetic or electric fields. In this chapter we concentrate on the study of direct transitions between rotational, fine or hyperfine energy levels. Such transitions occur mainly in those regions of the spectrum which extend from the radiofrequency, through the microwave and millimetre wave, to the far infrared region. They are therefore transitions that involve very low energy photons, with the absorption or emission of very small amounts of energy. Specialised techniques have been developed to carry out spectroscopic studies in this frequency range. In particular one often does not attempt to detect the low energy photons directly, but to make use of indirect detection methods which rely on the energy level population transfer resulting from spectroscopic transitions. We shall describe a number of the indirect methods which have been employed.

Radiofrequency spectroscopy, in particular, is frequently combined with molecular beam techniques, or other methods using gas pressures which are low enough to remove the effects of molecular collisions. There are two main reasons for this. First, experiments which depend upon population transfer can only be successful if collisional relaxation or equilibration is absent. Second, radiofrequency spectroscopy usually provides a spectroscopic resolution which is intrinsically very high because Doppler broadening is negligible; in such a desirable situation, collisional broadening is obviously to be avoided. The very high resolution afforded by radiofrequency methods means that intramolecular interactions which are very small can be studied in great detail and with great accuracy. Some of these interactions, such as those between nuclear spin magnetic moments, or between a nuclear spin and an applied magnetic field, are of profound importance and application, as we shall see.

Many of the classic studies we will describe were carried out more than sixty years ago, when the available radiofrequency and high-vacuum technology was primitive compared with that enjoyed now, at the beginning of the new millennium. We will describe the original experiments faithfully, whilst also giving an outline of how such studies might be performed today. At the same time we will also attempt to achieve consistency by sometimes redeveloping the theory used to analyse the spectra, employing the modern methods and notation of angular momentum theory used elsewhere in this book. The experimental results and spectral analyses of fundamental molecular species such as H_2 have never needed repetition, such was the quality of the original work. These studies are part of the structural foundations of our subject; they were well built indeed.

Molecular beam techniques are now very well established, for the study of both spectroscopy and fundamental reactive and non-reactive scattering dynamics. They have been described in numerous review articles and books, so in this chapter we content ourselves with details sufficient to understand the beautiful experiments which have been performed, but insufficient for those readers with ambitions to enter the field experimentally. For those requiring more details, the 1988 and 1992 two-volume collection of specialist articles edited by Scoles [1] constitutes an excellent summary, with many references to the original literature.

8.2. Molecular beam magnetic resonance of closed shell molecules

8.2.1. H_2, D_2 and HD in their $X\,^1\Sigma^+$ ground states

(a) Principles of molecular beam magnetic resonance

(i) MOLECULAR BEAM FORMATION

Molecules injected into a collision-free environment travel indefinitely in straight lines. Consequently transfer from a high-pressure region in which molecular trajectories are constantly changing because of collisions, through a small orifice into a very low-pressure region, results in a molecular spray. By means of a secondary orifice, suitably placed, we may select a small proportion of the molecules whose trajectories are essentially in the same direction, forming a beam. The collimation efficiency depends upon the size and nature of the orifices, the gas pressure before the first orifice, and the distances between the orifices. Very high pumping speeds in both the low- and high-pressure regions are also important because scattering with background gas molecules obviously destroys the beam collimation. In many cases two or more differentially-pumped vacuum chambers are employed, each separated by a small orifice which increases the definition of the molecular beam at the price of reduced beam flux.

A wide variety of beam orifices and background pressures have been used, but in essence there are just two types of molecular beam which can be produced. If the primary gas pressure (called the stagnation pressure) before the first orifice is substantially less than one bar, and the orifice, be it a circular hole or a slit, is relatively large, the resulting beam is called an 'effusive' beam. The studies on hydrogen which we are to describe shortly employed an effusive beam, and we give details of the pressures and orifice dimensions used in due course. A more recent development, which is used in almost all contemporary molecular beam studies, is that of the 'nozzle' or 'supersonic' beam. Generation of such a beam requires a much higher stagnation pressure (up to 100 bar or more, although a few bars is more normal), and a much smaller orifice. The price for this is a much higher gas throughput, so that extremely fast pumping is required. In many cases the molecules of interest are seeded in an inert gas carrier (often helium) which is in large excess. The primary advantage of nozzle sources is the beam flux, which is increased by several orders-of-magnitude over that characteristic of effusive sources. Other advantages include preferential directionality in the beam propagation, and cooling processes for the beam molecules which result in much higher populations of the lowest rotational and vibrational levels. Atoms and molecules can also stick together in nozzle beams, so that unusual molecular species can be formed and studied. In some experiments involving refractory materials the nozzle is operated at high temperatures; in other cases nozzles incorporating electrical discharges lead to the production of beams containing free radicals, or electronically excited molecules. Nozzles can be operated with continuous flows, or used in a pulsed mode which drastically reduces the gas load on the pumping system. We will encounter and study more of the special properties of nozzle beams elsewhere in this chapter, but for the moment we confine ourselves to the less remarkable attributes of effusive beams. The rather general remarks made in this subsection will acquire more substance when we examine the details of the particular experiments described.

(ii) MOLECULAR BEAM DETECTION

Almost all spectroscopic studies of molecules in molecular beams require efficient detection and measurement of the beam flux; spectroscopic transitions are invariably detected through consequent changes in the beam flux. In certain special cases very high detection efficiencies can be obtained using surface ionisation detectors. A hot surface of a material with a high work function will ionise species having low ionisation potentials. The resulting ion current can then be amplified by means of an electron multiplier, and monitored, with or without ion mass selection. This method was used to detect beams of alkali metal atoms, and more recently diatomic alkali metal molecules, with almost unit efficiency. Closely related are the surface ionisation detectors which are used to detect beams of species excited to metastable electronic states by monitoring the Auger electrons produced. In many cases spectroscopic detection involving laser-induced fluorescence or photoionisation has been successful, and bolometer detectors have been used to detect vibrationally-hot molecules. By far the most common detection method, however, is that of electron bombardment to produce molecular ions which can be detected with essentially unit efficiency. The overall sensitivity of this detection method depends primarily upon the ionisation efficiency which is achieved. Consequently much attention has been paid to the design of suitable electron guns. These usually involve electron emission from a hot filament, combined with ion optics and mass selection schemes to discriminate against ions produced from background gas; some detectors are operated in special ultrahigh vacuum chambers to reduce background interference. The highest ionisation efficiencies which have been achieved are close to 10^{-3}, but efficiencies in the range 10^{-4} to 10^{-5} are more common.

The studies of hydrogen described later in this section all involved a Pirani gauge detector. The detector works by directing the beam into a cavity where pressure changes resulting from beam intensity changes are measured by monitoring the thermal conduction of the gas. Pirani detectors are slow and relatively insensitive and seldom, if ever, used now. Nevertheless at the time they were instrumental in enabling the very fine fundamental studies of hydrogen to be performed successfully.

(iii) STATE SELECTION AND POPULATION TRANSFER

Molecular beam resonance studies depend upon the preparation of a spatially-aligned or partially-aligned beam; the flux of the aligned beam is measured, the alignment is reduced or even destroyed by population transfer resulting from spectroscopic transitions, and the resulting change in the beam flux is recorded. The alignment, or state selection, can be achieved by passage of the beam through suitable magnetic or electric fields. Electric fields are effective only when the molecule under investigation possesses a permanent electric dipole moment; this is obviously not the case for molecular hydrogen, so that in this section we deal only with magnetic alignment. The necessary magnetic moment of the molecule arises primarily from the presence of non-zero nuclear spins. We will describe examples of electric field alignment later in this chapter.

The first experiments using magnetic alignment, or state selection, were described by Rabi, Millman, Kusch and Zacharias [2]. Their studies were of beams of LiCl, LiF, NaF and Li$_2$, with the purpose of determining the magnetic moments of the ^{6}Li, ^{7}Li and ^{19}F nuclei. Figure 8.1 is essentially that shown in their paper and serves to explain both the principles and the techniques used. The beam source (O), the collimating slit (S) and the detector (D) are all on the same central axis, and only molecules which have passed through S can reach the detector. The solid line shows the trajectory of a molecule which has a magnetic moment μ_z, but which, in the absence of any applied magnetic field, would fail to pass through the slit S and therefore not reach the detector. However, a magnetic field gradient (d$\boldsymbol{B}$/dz)$_A$ is produced by magnet A, which bends the trajectory of the molecule; provided μ_z has the correct sign, the molecule will pass through the slit S. This molecule is clearly not heading towards the detector, but

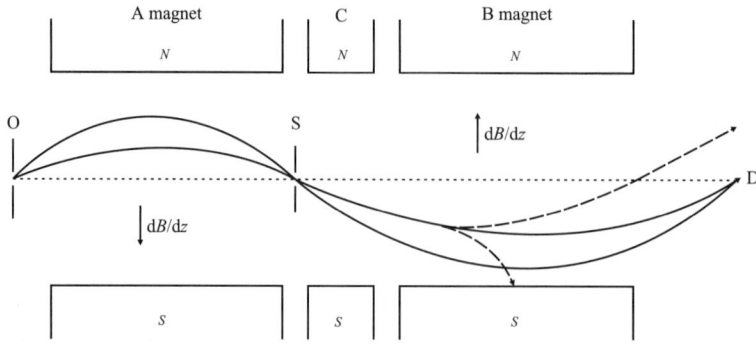

Figure 8.1. Principles of magnetic state selection and molecular beam magnetic resonance.

if it passes through a second magnetic field gradient $(\mathrm{d}\boldsymbol{B}/\mathrm{d}z)_\mathrm{B}$, which is of the same magnitude as that produced in A but points in the opposite direction, the trajectory of the molecule will again be changed, bringing it to the detector. The combination of the A and B fields thus acts as a state selector for a molecule with a given magnetic moment and velocity vector. Both the magnitude and the sign of μ_z are important; a molecule with the same velocity vector but μ_z of opposite *sign* will not reach the detector. In terms of the projection quantum numbers M_I, a non-thermal population distribution of states with M_I opposite in sign is produced. States with $M_I = 0$ are, of course, unaffected by the A or B fields.

Now consider the addition of a homogeneous C field, of magnitude B, situated symmetrically between the A and B fields. This does not affect the trajectories of any molecules, but does remove the degeneracy of levels with different M_I values. The further addition of an oscillating magnetic field of frequency ν, applied perpendicular to the static field C, will induce transitions between states differing in M_I by ± 1 if the resonance condition,

$$h\nu = g_N \mu_N B, \tag{8.1}$$

is satisfied. The molecule which reached the detector in figure 8.1 now suffers a change in magnetic moment after passing through the slit S; its trajectory in the B field therefore changes, as shown by the dashed line in figure 8.1, and it fails to reach the detector. Consequently the occurrence of a resonant transition in the C field is detected as a decrease in the beam flux reaching the detector; this is known as the 'flop-out' mode of detection. In practice, of course, one is dealing with a molecular ensemble with a range of molecular velocity vectors, and M_I values ranging from $+I$ to $-I$. Moreover the simple resonance condition (8.1) involves only the nuclear spin and its resulting magnetic moment, whereas in general the coupling of the nuclear spin to the rotational angular momentum must be considered in detail. In the experiments on ^{7}LiCl described above, the resonant frequency was found to be 5.610 MHz in a C field of 3400 G; the spectral resolution was too low to reveal the effects of molecular rotation, but good enough to yield a value for the ^{7}Li nuclear magnetic moment. Values of the ^{6}Li and ^{19}F nuclear magnetic moments were also obtained.

Many variants of the experiment described in figure 8.1 have been performed, and we shall encounter some of them later in this chapter. Perhaps the most important variant is that it is often possible to arrange the state selection so that resonant transitions result in an *increase* in the detected beam flux, ideally against a very low off-resonance background. This is known as the 'flop-in' mode of detection, and it can be very sensitive. Again we shall meet examples of this later.

(b) Details of the apparatus used

The pioneering experiments on H_2, HD and D_2 were performed by Kellogg, Rabi, Ramsey and Zacharias in 1939 [3]. Their principal objectives were to determine the magnetic moments of the proton and deuteron, and the electric quadrupole moment of the deuteron. The sorry events of 1939 to 1945 brought the work to an untimely halt, but the experiments were resumed in 1950 with substantially improved apparatus, which we now describe.

Figure 8.2 shows a simplified block diagram of the apparatus used by Kolsky, Phipps, Ramsey and Silsbee [4]. As described in the earlier work and in section 8.2.1. **a**(iii), a beam of molecules was spread by the inhomogeneous magnetic field A and refocused onto the detector by the second inhomogeneous magnetic field B; the A and B fields were each 39.2 cm in length. The homogeneous C field of length 150 cm was symmetrically located between the A and B fields, with a centrally placed collimating slit of height 0.8 cm and width 0.0015 cm. The overall beam length from source to detector was 269 cm. Radiofrequency transitions were induced in the homogeneous C field by copper strip lines, the orientation providing a radio frequency magnetic field perpendicular to the static C field. Under correct operating conditions the beam flux reaching the detector was the same whether or not the A and B fields were switched on. Resonant transitions induced in the C field region, however, resulted in partial reorientation of the magnetic moment, so that the number of molecules reaching the detector was decreased; the mode of detection was therefore the 'flop-out' mode. It is possible to scan either the magnetic C field, or the radiation frequency; both modes of operation have indeed been employed. Off-resonance beam fluxes were about 10^{13} molecules per cm^2 per second, the main chamber pressure with the H_2 beam on being about 10^{-6} torr. Differential pumping was provided by separating the source and main chambers with a separating chamber. Resonance line half-widths of about 13 kHz were obtained using a single rf field inside the static C field, but later work used two separated oscillating fields [5, 6], with linewidths reduced to 0.7 kHz. We will illustrate some of the resonance lines observed after discussing the theory of the energy levels and the transitions between them.

(c) Effective Hamiltonian, energy levels and spectroscopic transitions

The effective Hamiltonian for H_2 in its $^1\Sigma_g^+$ ground state may be written as the sum of five terms:

$$\mathcal{H} = \mathcal{H}_{IZ} + \mathcal{H}_{JZ} + \mathcal{H}_{IJ} + \mathcal{H}_{dip} + \mathcal{H}_{diam}. \tag{8.2}$$

These represent the nuclear spin Zeeman interaction, the rotational Zeeman interaction, the nuclear spin–rotation interaction, the nuclear spin–nuclear spin dipolar interaction, and the diamagnetic interactions. Using irreducible tensor methods we examine the matrix elements of each of these five terms in turn, working first in the decoupled basis set $|\eta; J, M_J; I, M_I\rangle$, where η specifies all other electronic and vibrational quantum numbers; this is the basis which is most appropriate for high magnetic field studies. In due course we will also calculate the matrix elements and energy levels in a $|\eta; J, I, F, M_F\rangle$ coupled basis which is appropriate for low field investigations. Most of the experimental studies involved *ortho*-H_2 in its lowest rotational level, $J = 1$. If the proton nuclear spins are denoted I_1 and I_2, each with value $1/2$, *ortho*-H_2 has total nuclear spin I equal to 1. *Para*-H_2 has a total nuclear spin I equal to 0.

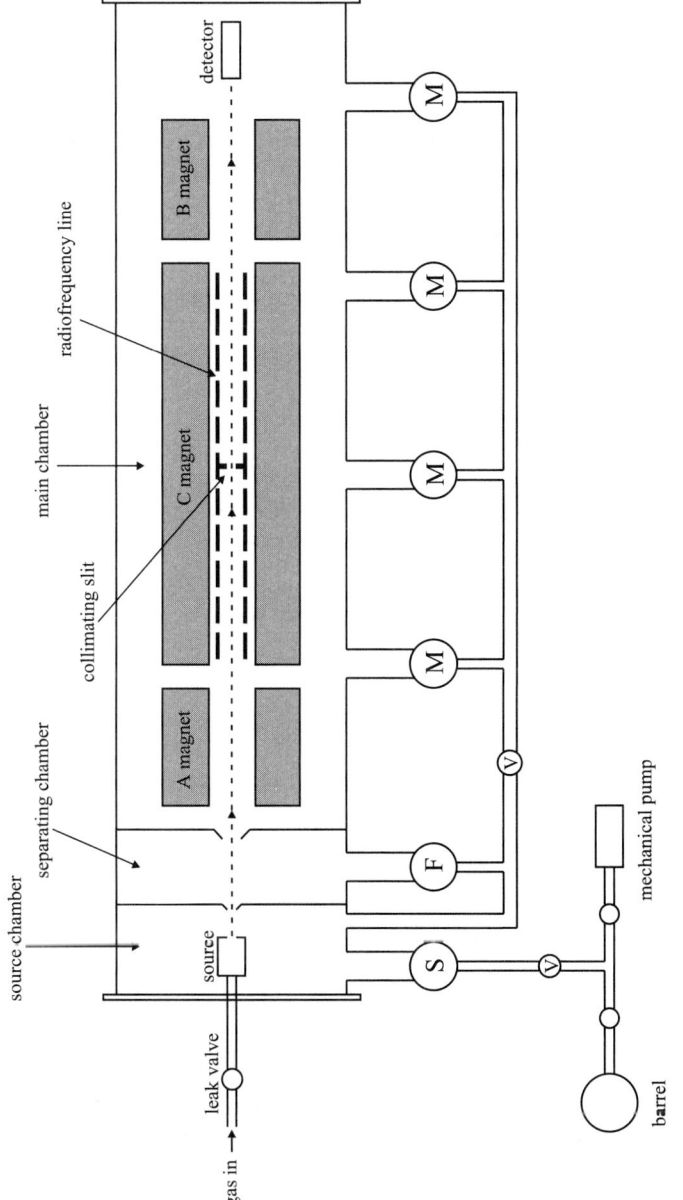

Figure 8.2. Simplified block diagram of the apparatus used to study the magnetic resonance spectrum of H_2. The molecular beam source was cooled to liquid nitrogen temperature. M denotes fractionating pumps, whilst S and F are diffusion pumps. The appropriate molecular trajectories are not shown in this diagram; the reader is referred back to figure 8.1.

From the treatment described in chapter 4 the nuclear spin Zeeman interaction may be written as the sum of two terms, one for each nucleus,

$$\mathcal{H}_{IZ} = -g_{N_1}\mu_N\{1 - \sigma_1(J)\}\mathbf{T}^1(I_1) \cdot \mathbf{T}^1(B) - g_{N_2}\mu_N\{1 - \sigma_2(J)\}\mathbf{T}^1(I_2) \cdot \mathbf{T}^1(B). \quad (8.3)$$

This expression is not quite the same as that shown previously in that the magnetic field intensity B is modified by a screening factor, $\sigma_i(J)$, for each nucleus, the nature and origin of which will be discussed in due course. Since we are dealing with two equivalent protons, (8.3) may be simplified to a single term with $I = 1$,

$$\mathcal{H}_{IZ} = -g_H\mu_N \mathbf{T}^1(I) \cdot \mathbf{T}^1(B)\{1 - \sigma_H(J)\}, \quad (8.4)$$

Here we have made the replacements $g_{N_1} = g_{N_2} = g_H$; $\sigma_1 = \sigma_2 = \sigma_H$. Notice that the screening factor is dependent on the J value. Equation (8.4) may be further simplified by taking the space-fixed $p = 0$ component of the scalar product to be defined by the direction of the magnetic field, so that

$$\mathcal{H}_{IZ} = -g_H\mu_N \mathbf{T}^1_0(I)\mathbf{T}^1_0(B)\{1 - \sigma_H(J)\}. \quad (8.5)$$

The matrix elements of (8.5) are diagonal in M_I and would be independent of M_J were it not for the $\sigma_H(J)$ term.

The rotational Zeeman term is equally simple,

$$\mathcal{H}_{JZ} = -\{1 - \sigma_H(J)\}g_r\mu_N \mathbf{T}^1(J) \cdot \mathbf{T}^1(B), \quad (8.6)$$

and for a magnetic field in the Z ($p=0$) direction, the matrix elements (shown in table 8.1), are diagonal in M_J and independent of M_I; g_r is the rotational g factor.

The third term in equation (8.2), the nuclear spin–rotation term, may be written

$$\mathcal{H}_{JI} = c_I \mathbf{T}^1(J) \cdot \mathbf{T}^1(I), \quad (8.7)$$

and application of the Wigner–Eckart theorem yields the matrix elements,

$$\langle \eta; J, M_J; I, M_I | c_I \mathbf{T}^1(J) \cdot \mathbf{T}^1(I) | \eta; J, M'_J; I, M'_I \rangle$$
$$= \langle J, M_J; I, M_I | c_I \sum_p (-1)^p \mathbf{T}^1_p(J)\mathbf{T}^1_{-p}(I) | J, M'_J; I, M'_I \rangle$$
$$= c_I \sum_p (-1)^p (-1)^{J-M_J}(-1)^{I-M_I} \begin{pmatrix} J & 1 & J \\ -M_J & p & M'_J \end{pmatrix} \begin{pmatrix} I & 1 & I \\ -M_I & -p & M'_I \end{pmatrix}$$
$$\times \{J(J+1)(2J+1)I(I+1)(2I+1)\}^{1/2}. \quad (8.8)$$

Again the explicit evaluation of these matrix elements is given with table 8.1.

The fourth term in (8.2) describes the dipole–dipole interaction between the nuclear spin magnetic moments. For two equivalent protons this term takes the form,

$$\mathcal{H}_{\text{dip}} = g_H^2 \mu_N^2 (\mu_0/4\pi) \left\{ \frac{I_1 \cdot I_2}{R^3} - \frac{3(I_1 \cdot R)(I_2 \cdot R)}{R^5} \right\}, \quad (8.9)$$

where g_H is the proton g factor. This term may be written as the scalar product of two

Table 8.1. Energy matrix for H_2 in the $J=1$ level in the decoupled representation

		$M_J=+1$			$M_J=0$			$M_J=-1$		
		$M_I=+1$	$M_I=0$	$M_I=-1$	$M_I=+1$	$M_I=0$	$M_I=-1$	$M_I=+1$	$M_I=0$	$M_I=-1$
$M_J=+1$	$M_I=+1$	m_{11}	0	0	0	0	0	0	0	0
	$M_I=0$	0	m_{22}	0	m_{24}	0	0	0	0	0
	$M_I=-1$	0	0	m_{33}	0	m_{35}	0	m_{37}	0	0
$M_J=0$	$M_I=+1$	0	m_{42}	0	m_{44}	0	0	0	0	0
	$M_I=0$	0	0	m_{53}	0	m_{55}	0	m_{57}	0	0
	$M_I=-1$	0	0	0	0	0	m_{66}	0	m_{68}	0
$M_J=-1$	$M_I=+1$	0	0	m_{73}	0	m_{75}	0	m_{77}	0	0
	$M_I=0$	0	0	0	0	0	m_{86}	0	m_{88}	0
	$M_I=-1$	0	0	0	0	0	0	0	0	m_{99}

irreducible second-rank tensors,

$$\mathcal{H}_{\text{dip}} = -g_H^2 \mu_N^2 (\mu_0/4\pi) \sqrt{6} T^2(\boldsymbol{C}) \cdot T^2(\boldsymbol{I}_1, \boldsymbol{I}_2), \tag{8.10}$$

where

$$T_p^2(\boldsymbol{I}_1, \boldsymbol{I}_2) = (-1)^p \sqrt{5} \sum_{p_1, p_2} \begin{pmatrix} 1 & 1 & 2 \\ p_1 & p_2 & -p \end{pmatrix} T_{p_1}^1(\boldsymbol{I}_1) T_{p_2}^1(\boldsymbol{I}_2), \tag{8.11}$$

and

$$T_q^2(\boldsymbol{C}) = \langle \eta | C_q^2(\theta, \phi) R^{-3} | \eta \rangle. \tag{8.12}$$

The equivalence of (8.9) and (8.10) is shown in appendix 8.1. Application of the Wigner–Eckart theorem using (8.10) yields the result

$$\langle \eta; J, M_J; I, M_I | \mathcal{H}_{\text{dip}} | \eta; J, M'_J; I', M'_I \rangle$$

$$= -\sqrt{6} g_H^2 \mu_N^2 (\mu_0/4\pi) \sum_p (-1)^p (-1)^{J-M_J} \begin{pmatrix} J & 2 & J \\ -M_J & p & M'_J \end{pmatrix} \langle \eta; J \| T^2(\boldsymbol{C}) \| \eta; J \rangle$$

$$\times (-1)^{I-M_I} \begin{pmatrix} I & 2 & I' \\ -M_I & -p & M'_I \end{pmatrix} \langle I \| T^2(\boldsymbol{I}_1, \boldsymbol{I}_2) \| I' \rangle. \tag{8.13}$$

The reduced matrix element involving the nuclear spin is given by

$$\langle I \| T^2(\boldsymbol{I}_1, \boldsymbol{I}_2) \| I' \rangle = \sqrt{5} \begin{Bmatrix} I_1 & I_1 & 1 \\ I_2 & I_2 & 1 \\ I & I' & 2 \end{Bmatrix} \{(2I+1)(2I'+1) I_1(I_1+1)(2I_1+1)$$

$$\times I_2(I_2+1)(2I_2+1)\}^{1/2}. \tag{8.14}$$

For $I' = I = 1$, $I_1 = I_2 = 1/2$, the reduced matrix element in (8.14) has the value $\sqrt{5}/2$ and for $J = I = 1$ the explicit values of the matrix elements are again given with table 8.1.

The fifth term in (8.2) describes the diamagnetic interaction of the molecule with the external magnetic field. Ramsey [7] showed that this interaction could be represented by the term,

$$\mathcal{H}_{\text{diam}} = -\frac{5f}{(2J-1)(2J+3)} \{3(T^1(\boldsymbol{J}) \cdot T^1(\boldsymbol{B}))^2 - J^2 B^2\} - g, \tag{8.15}$$

and, again selecting the $p = 0$ component of the scalar product, we obtain matrix elements which are diagonal in M_J:

$$\langle \eta; J, M_J; I, M_I | \mathcal{H}_{\text{diam}} | \eta; J, M_J; I, M_I \rangle$$

$$= -\frac{5f B_Z^2}{(2J-1)(2J+3)} \{3M_J^2 - J(J+1)\} - g. \tag{8.16}$$

The constants f and g are functions of the magnetic susceptibility of the molecule when the magnetic field is aligned perpendicular or parallel to the internuclear axis. They are very small in magnitude and were only included in the analysis of the high field spectra. We will discuss these quantities at length in section 8.2.2. They are closely related to the shielding parameter $\sigma_H(\boldsymbol{J})$, as we shall see.

We can now summarise the results of our calculations of the matrix elements of $\mathcal{H}$ in table 8.1. The matrix elements are as follows:

$$m_{11} = -a\{1 - \sigma_H(\boldsymbol{J})\} - b\{1 - \sigma_H(\boldsymbol{J})\} + c_I + d/2 - f/3 - g$$
$$m_{22} = -b\{1 - \sigma_H(\boldsymbol{J})\} - d - f/3 - g$$
$$m_{33} = a\{1 - \sigma_H(\boldsymbol{J})\} - b\{1 - \sigma_H(\boldsymbol{J})\} - c_I + d/2 - f/3 - g$$
$$m_{44} = -a\{1 - \sigma_H(\boldsymbol{J})\} - d + 2f/3 - g$$
$$m_{55} = 2d + 2f/3 - g$$
$$m_{66} = a\{1 - \sigma_H(\boldsymbol{J})\} - d + 2f/3 - g$$
$$m_{77} = -a\{1 - \sigma_H(\boldsymbol{J})\} + b\{1 - \sigma_H(\boldsymbol{J})\} - c_I + d/2 - f/3 - g$$
$$m_{88} = b\{1 - \sigma_H(\boldsymbol{J})\} - d - f/3 - g$$
$$m_{99} = a\{1 - \sigma_H(\boldsymbol{J})\} + b\{1 - \sigma_H(\boldsymbol{J})\} + c_I + d/2 - f/3 - g$$
$$m_{24} = m_{42} = m_{68} = m_{86} = c_I + 3d/2$$
$$m_{35} = m_{53} = m_{57} = m_{75} = c_I - 3d/2$$
$$m_{37} = m_{73} = 3d$$

Note that in the above expressions we conform with Ramsey [7] by making the replacements,

$$g_H \mu_N B_Z \equiv a, \quad g_r \mu_N B_Z \equiv b,$$

and our dipolar constant is defined by

$$d = \frac{1}{\sqrt{30}} g_H^2 \mu_N^2 (\mu_0/4\pi) \langle J \| \mathbf{T}^2(\boldsymbol{C}) \| J \rangle. \tag{8.17}$$

We note that this matrix factorises into two 1×1 matrices, two 2×2 and one 3×3; we will calculate the eigenvalues in due course.

As stated earlier, the decoupled representation used above is most appropriate for the high field measurements of the magnetic resonance spectrum. Before examining the field dependence of the energy levels and transition probabilities, we recalculate the matrix elements of $\mathcal{H}$ using the coupled basis, $|\eta; J, I, F, M_F\rangle$. We note that for $J = 1$ and $I = 1$, we can have $F = 2$ (with $M_F = \pm 2, \pm 1, 0$), $F = 1$ (with $M_F = \pm 1, 0$), and $F = 0$ ($M_F = 0$).

First, the nuclear Zeeman term, using equation (8.5):

$$\langle \eta; J, I, F, M_F | - g_H \mu_N \mathbf{T}_0^1(\boldsymbol{I}) \mathbf{T}_0^1(\boldsymbol{B}) \{1 - \sigma_H(\boldsymbol{J})\} | \eta; J, I, F', M_F \rangle$$
$$= -g_H \mu_N B_Z \{1 - \sigma_H(\boldsymbol{J})\} (-1)^{F - M_F} \begin{pmatrix} F & 1 & F' \\ -M_F & 0 & M_F \end{pmatrix} (-1)^{F+J+1+I}$$
$$\times \{(2F+1)(2F'+1)\}^{1/2} \begin{Bmatrix} I & F' & J \\ F & I & 1 \end{Bmatrix} \{I(I+1)(2I+1)\}^{1/2}. \tag{8.18}$$

Second, the rotational Zeeman term (8.6):

$$\langle \eta; J, I, F, M_F| - g_J \mu_N \mathrm{T}_0^1(\boldsymbol{J})\mathrm{T}_0^1(\boldsymbol{B})\{1 - \sigma_J(\boldsymbol{J})\}|\eta; J, I, F', M_F\rangle$$

$$= -g_J \mu_N B_Z \{1 - \sigma_{\mathrm{H}}(\boldsymbol{J})\}(-1)^{F-M_F} \begin{pmatrix} F & 1 & F' \\ -M_F & 0 & M_F \end{pmatrix} (-1)^{F'+J+1+I}$$

$$\times \{(2F+1)(2F'+1)\}^{1/2} \begin{Bmatrix} J & F' & I \\ F & J & 1 \end{Bmatrix} \{J(J+1)(2J+1)\}^{1/2}. \quad (8.19)$$

Third, the nuclear spin–rotation term (8.7):

$$\langle \eta; J, I, F, M_F| c_I \mathrm{T}^1(\boldsymbol{J}) \cdot \mathrm{T}^1(\boldsymbol{I})|\eta; J, I, F', M_F\rangle$$

$$= c_I(-1)^{J+F+I} \delta_{FF'} \begin{Bmatrix} I & J & F \\ J & I & 1 \end{Bmatrix} \{J(J+1)(2J+1)I(I+1)(2I+1)\}^{1/2}. \quad (8.20)$$

Fourth, the nuclear spin dipolar interaction for terms diagonal in J and I (8.10):

$$\langle \eta; J, I, F, M_F| -\sqrt{6}g_{\mathrm{H}}^2 \mu_N^2 (\mu_0/4\pi)\mathrm{T}^2(\boldsymbol{C}) \cdot \mathrm{T}^2(\boldsymbol{I}_1, \boldsymbol{I}_2)|\eta; J, I, F', M_F'\rangle$$

$$= -\delta_{FF'} \delta_{M_F M_F'} \sqrt{6} g_{\mathrm{H}}^2 \mu_N^2 (\mu_0/4\pi)(-1)^{J+F+I} \begin{Bmatrix} I & J & F \\ J & I & 2 \end{Bmatrix}$$

$$\times \langle \eta; J\|\mathrm{T}^2(\boldsymbol{C})\|\eta; J\rangle \langle I\|\mathrm{T}^2(\boldsymbol{I}_1, \boldsymbol{I}_2)\|I\rangle$$

$$= -\delta_{FF'} \delta_{M_F M_F'} \sqrt{30} g_{\mathrm{H}}^2 \mu_N^2 (\mu_0/4\pi)(-1)^{J+F+I} \begin{Bmatrix} I & J & F \\ J & I & 2 \end{Bmatrix} \begin{Bmatrix} I_1 & I_1 & 1 \\ I_2 & I_2 & 1 \\ I & I & 2 \end{Bmatrix}$$

$$\times (2I+1)\{I_1(I_1+1)(2I_1+1)I_2(I_2+1)(2I_2+1)\}^{1/2} \langle \eta; J\|\mathrm{T}^2(\boldsymbol{C})\|\eta; J\rangle. \quad (8.21)$$

The diamagnetic terms were not calculated by Ramsey [7] for the coupled representation, which was used only for analysis of the low field spectra.

The structure of the 9×9 energy matrix in the coupled representation is shown in table 8.2, with the explicit matrix elements given below.

$$m_{11} = -a\{1 - \sigma_{\mathrm{H}}(\boldsymbol{J})\} - b\{1 - \sigma_{\mathrm{H}}(\boldsymbol{J})\} + c_I - d/2$$
$$m_{22} = -(a/2)\{1 - \sigma_{\mathrm{H}}(\boldsymbol{J})\} - (b/2)\{1 - \sigma_{\mathrm{H}}(\boldsymbol{J})\} + c_I - d/2$$
$$m_{33} = c_I - d/2$$
$$m_{44} = (a/2)\{1 - \sigma_{\mathrm{H}}(\boldsymbol{J})\} + (b/2)\{1 - \sigma_{\mathrm{H}}(\boldsymbol{J})\} + c_I - d/2$$
$$m_{55} = a\{1 - \sigma_{\mathrm{H}}(\boldsymbol{J})\} + b\{1 - \sigma_{\mathrm{H}}(\boldsymbol{J})\} + c_I - d/2$$
$$m_{66} = -(a/2)\{1 - \sigma_{\mathrm{H}}(\boldsymbol{J})\} - (b/2)\{1 - \sigma_{\mathrm{H}}(\boldsymbol{J})\} - c_I + 5d/2$$
$$m_{77} = -c_I + 5d/2$$
$$m_{88} = (a/2)\{1 - \sigma_{\mathrm{H}}(\boldsymbol{J})\} + (b/2)\{1 - \sigma_{\mathrm{H}}(\boldsymbol{J})\} - c_I + 5d/2$$
$$m_{99} = -2c_I - 5d$$
$$m_{26} = m_{48} = m_{62} = m_{84} = (1/2)(a - b)$$
$$m_{37} = m_{73} = (1/\sqrt{3})(a - b)$$
$$m_{79} = m_{97} = (\sqrt{2}/\sqrt{3})(a - b)$$

The energy matrix in the coupled representation again factorises into two 1×1 matrices (for $M_F = \pm 2$), two 2×2 matrices (for $M_F = \pm 1$) and one 3×3 (for $M_F = 0$).

Table 8.2. Energy matrix for H_2 in the $J = 1$ level in the coupled representation

		$F=2$					$F=1$			$F=0$
		$M_F=+2$	$M_F=+1$	$M_F=0$	$M_F=-1$	$M_F=-2$	$M_F=+1$	$M_F=0$	$M_F=-1$	$M_F=0$
$F=2$	$M_F=+2$	m_{11}	0	0	0	0	0	0	0	0
	$M_F=+1$	0	m_{22}	0	0	0	m_{26}	0	0	0
	$M_F=0$	0	0	m_{33}	0	0	0	m_{37}	0	0
	$M_F=-1$	0	0	0	m_{44}	0	0	0	m_{48}	0
	$M_F=-2$	0	0	0	0	m_{55}	0	0	0	0
$F=1$	$M_F=+1$	0	m_{62}	0	0	0	m_{66}	0	0	0
	$M_F=0$	0	0	m_{73}	0	0	0	m_{77}	0	m_{79}
	$M_F=-1$	0	0	0	m_{84}	0	0	0	m_{88}	0
$F=0$	$M_F=0$	0	0	0	0	0	0	m_{97}	0	m_{99}

We are now in a position to calculate the energies of the nine levels as a function of magnetic field strength, given values of the molecular constants involved. In the original analysis of the spectra, perturbation theory was employed to derive energy expressions for the mixed states. We will adopt the easier contemporary method of employing the values of the constants derived from the original analyses, and using a computer to diagonalise the energy matrices. Our purpose is to show how the experimentally observed spectra are related to the calculated energy levels.

The following values of the constants (in kHz) were determined:

$$a = 4.258, \quad b = 0.6717, \quad c_I = -113.904, \quad d = 57.671.$$

Using these values we diagonalise the 9×9 matrix in the coupled representation, and plot the energies as functions of magnetic field strength. The results are shown in figure 8.3, which is essentially identical with Figure 1 in Ramsey's paper [7]. Each level is labelled with quantum numbers in the decoupled and coupled representations, but as the field strength is increased the decoupled quantum numbers M_I and M_J become increasingly appropriate. In separate papers two different types of transition

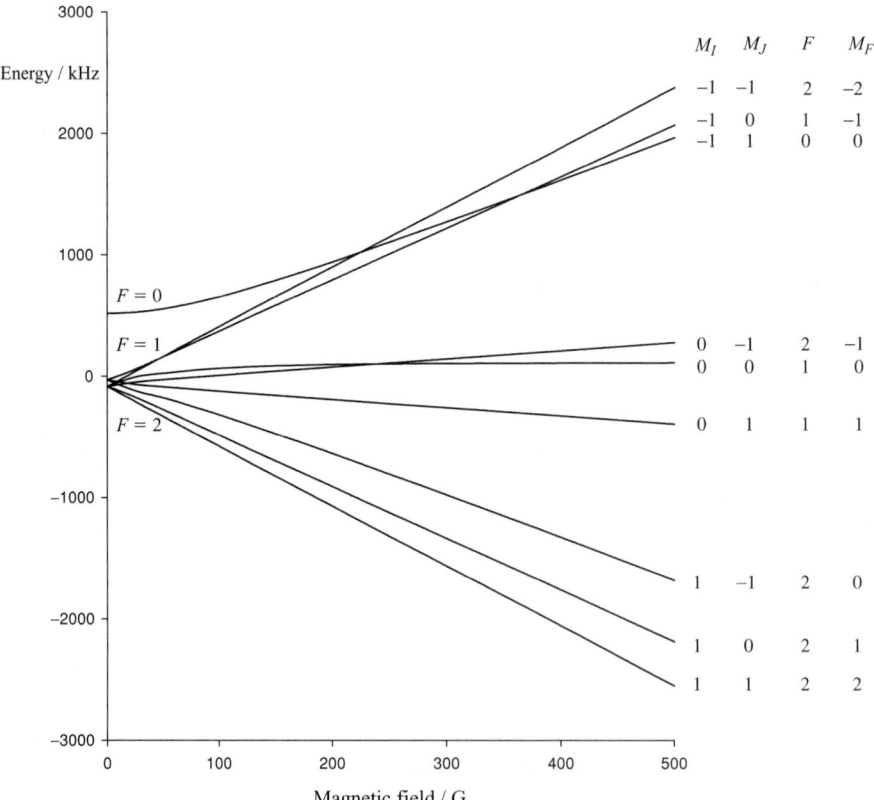

Figure 8.3. Energy levels of H_2 in its $J = 1$ level as a function of magnetic field strength. The levels are labelled with both their decoupled (M_I, M_J) and coupled (F, M_F) quantum numbers.

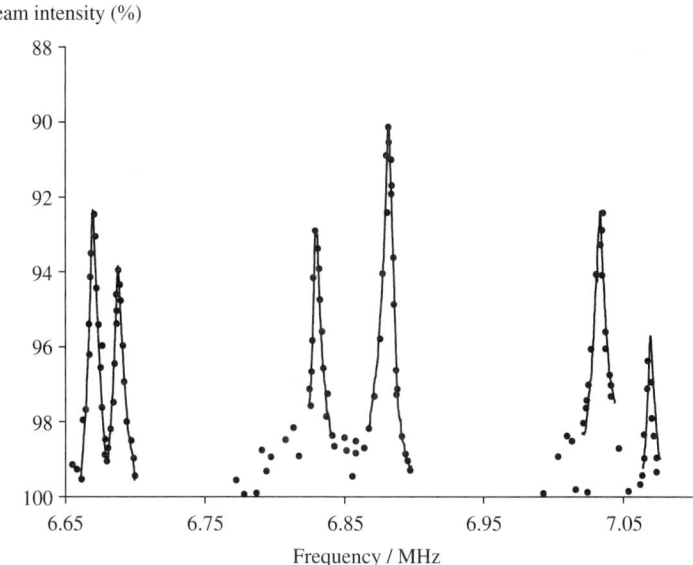

Figure 8.4. Six-line radiofrequency spectrum of *ortho*-H$_2$ ($J = 1$) observed in a magnetic field of 1611 G, obeying the selection rules $\Delta M_I = \pm 1$, $\Delta M_J = 0$, observed by Kolsky, Phipps, Ramsey and Silsbee [4]. The original spectrum was recorded manually point by point as shown.

were described. Kolsky, Phipps, Ramsey and Silsbee [4] made careful measurements of the six $\Delta M_I = \pm 1$, $\Delta M_J = 0$ transitions in a magnetic field of 1611 G. Their spectrum was recorded manually point by point; their envelopes of the six lines are shown in figure 8.4. Note that the strongest resonance corresponds to a 10% reduction in the detected beam intensity, leading to an excellent signal-to-noise ratio. The original paper also shows the resolution enhancement obtained using two separated rf fields. The six-line pattern spans the frequency range 6.6 to 7.1 MHz. Harrick and Ramsey [8] subsequently studied the six complementary transitions which obey the selection rules $\Delta M_J = \pm 1$, $\Delta M_I = 0$ at higher magnetic fields ranging from 3 to 7 kG. From these measurements they were able to obtain an accurate value for the rotational magnetic moment of H$_2$ in its $J = 1$ level; they were also able to determine the dependence of the diamagnetic susceptibility on molecular orientation. Additional studies [4] at very low magnetic fields in the range 1 to 5 G confirmed the intermediate and high field analyses. We examine the appropriate theory of the molecular parameters in section 8.2.2.

We turn now to the corresponding studies of the isotopic species D$_2$ and HD. The deuterium nucleus has spin I_D equal to 1, so that the two equivalent deuterium nuclei in D$_2$ have their spins coupled to give total nuclear spin I equal to 2, 1 or 0. The states with I equal to 2 or 0 correspond to *ortho*-D$_2$, whilst that with I equal to 1 is known as *para*-D$_2$. The molecular beam magnetic resonance studies have been performed on *para*-D$_2$, in the $J = 1$ rotational level. Formally, therefore, the effective Hamiltonian is the same as that described above for experimental studies of *ortho*-H$_2$, also in the $J = 1$ rotational level. There is one extremely important difference, however, in that the deuterium nucleus possesses an electric quadrupole moment, so that the quadrupole interaction must be included in the analysis. There should therefore be an additional term in the effective Hamiltonian describing this interaction; however, even though the quadrupole interaction does not involve nuclear spin directly, it may be represented by a term in the effective Hamiltonian which has exactly the same dependence on the angular momentum quantum numbers as the dipole–dipole interaction. We now prove this perhaps surprising result.

The quadrupole interaction between the electric field gradient and the nuclear quadrupole moment is represented by a sum of terms, one for each nucleus k, as follows:

$$\mathcal{H}_Q = -e \sum_k \mathbf{T}^2(\nabla \mathbf{E}_k) \cdot \mathbf{T}^2(\mathbf{Q}_k). \tag{8.22}$$

The nuclear spins I_1 and I_2 may be coupled to form the total nuclear spin I, which in turn is coupled to J to form F. In this coupled basis set the matrix elements of $\mathcal{H}_Q$ are given by

$$\langle \eta, \Lambda; J, I_1, I_2, I, F | \mathcal{H}_Q | \eta, \Lambda'; J', I_1, I_2, I', F' \rangle$$
$$= -e\delta_{FF'} \sum_k (-1)^{J'+I+F} \begin{Bmatrix} I & J & F \\ J' & I' & 2 \end{Bmatrix} \langle \eta, J, \Lambda \| \mathbf{T}^2(\nabla \mathbf{E}_k) \| \eta, J', \Lambda' \rangle$$
$$\times \langle I_1, I_2, I \| \mathbf{T}^2(\mathbf{Q}_k) \| I_1, I_2, I' \rangle. \tag{8.23}$$

The first reduced matrix element in (8.23) may be evaluated by noting that, in the molecule-fixed axis system q, the following relationship holds:

$$\langle \eta, J, \Lambda, \| \sum_q \mathcal{D}^{(2)}_{\cdot q}(\omega)^* \mathbf{T}^2_q(\nabla \mathbf{E}_k) \| \eta, J', \Lambda' \rangle$$
$$= \sum_q (-1)^{J-\Lambda} \{(2J+1)(2J'+1)\}^{1/2} \begin{pmatrix} J & 2 & J' \\ -\Lambda & q & \Lambda' \end{pmatrix}$$
$$\times \langle \eta, \Lambda | \mathbf{T}^2_q(\nabla \mathbf{E}_k) | \eta, \Lambda' \rangle, \tag{8.24}$$

where $\mathcal{D}^{(2)}_{\cdot q}(\omega)^*$ is the second-rank rotation matrix. The second reduced matrix element in equation (8.23) is obtained from the results

$$\langle I_1, I_2, I \| \mathbf{T}^2(\mathbf{Q}_1) \| I_1, I_2, I' \rangle = (-1)^{I_1+I_2+I'+2} \{(2I+1)(2I'+1)\}^{1/2}$$
$$\times \begin{Bmatrix} I_1 & I & I_2 \\ I' & I_1 & 2 \end{Bmatrix} \langle I_1 \| \mathbf{T}^2(\mathbf{Q}_1) \| I_1 \rangle \tag{8.25}$$

for $k = 1$, and

$$\langle I_1, I_2, I \| \mathbf{T}^2(\mathbf{Q}_2) \| I_1, I_2, I' \rangle = (-1)^{I_1+I_2+I+2} \{(2I+1)(2I'+1)\}^{1/2}$$
$$\times \begin{Bmatrix} I_2 & I & I_1 \\ I' & I_2 & 2 \end{Bmatrix} \langle I_2 \| \mathbf{T}^2(\mathbf{Q}_2) \| I_2 \rangle \tag{8.26}$$

for $k = 2$. We now make use of two important definitions. First, the nuclear quadrupole moment for a nucleus with spin I_1, Q_1, is defined by the equation

$$\langle I_1\|T^2(\boldsymbol{Q}_1)\|I_1\rangle = \left(\frac{Q_1}{2}\right)\begin{pmatrix} I_1 & 2 & I_1 \\ -I_1 & 0 & I_1 \end{pmatrix}^{-1} \tag{8.27}$$

with an identical result for Q_2. Second, the negative of the electric field gradient at the nucleus, q_0, is defined by

$$\langle \eta, \Lambda | T_0^2(\nabla \boldsymbol{E}) | \eta, \Lambda \rangle = -(q_0/2). \tag{8.28}$$

Consequently for matrix elements diagonal in Λ, equation (8.23) gives the result:

$$\langle \eta, \Lambda; J, I_1, I_2, I, F | \mathcal{H}_Q | \eta, \Lambda; J', I_1, I_2, I', F \rangle$$
$$= \frac{eQq_0}{4}(-1)^{J'+I+F}\begin{Bmatrix} I & J & F \\ J' & I' & 2 \end{Bmatrix}(-1)^{J-\Lambda}\begin{pmatrix} J & 2 & J' \\ -\Lambda & 0 & \Lambda \end{pmatrix}\{(2J+1)(2J'+1)\}^{1/2}$$
$$\times [(-1)^I + (-1)^{I'}](-1)^{I_1+I_2}\{(2I+1)(2I'+1)\}^{1/2}$$
$$\times \begin{Bmatrix} I_2 & I & I_1 \\ I' & I_2 & 2 \end{Bmatrix}\begin{pmatrix} I_1 & 2 & I_1 \\ -I_1 & 0 & I_1 \end{pmatrix}^{-1}. \tag{8.29}$$

We now show that the nuclear spin dipolar interaction has matrix elements of exactly the same form. We take the dipolar Hamiltonian to have the form given previously in equation (8.10) and find that its matrix elements are given by

$$\langle \eta, \Lambda; J, I_1, I_2, I, F | \mathcal{H}_{\text{dip}} | \eta, \Lambda'; J', I_1, I_2, I', F \rangle$$
$$= -g_{I_1}g_{I_2}\mu_N^2(\mu_0/4\pi)\sqrt{6}(-1)^{J'+I+F}\begin{Bmatrix} I & J & F \\ J' & I' & 2 \end{Bmatrix}\langle \eta, J, \Lambda\|T^2(\boldsymbol{C})\|\eta, J', \Lambda'\rangle$$
$$\times \langle I_1, I_2, I\|T^2(\boldsymbol{I}_1, \boldsymbol{I}_2)\|I_1, I_2, I'\rangle. \tag{8.30}$$

The first reduced matrix element in (8.30) is readily evaluated:

$$\langle \eta, J, \Lambda\|T^2(\boldsymbol{C})\|\eta, J', \Lambda'\rangle = \langle \eta, J, \Lambda\|\sum_q \mathcal{D}_{\cdot q}^{(2)}(\omega)^* T_q^2(\boldsymbol{C})\|\eta, J', \Lambda'\rangle$$
$$= \sum_q (-1)^{J-\Lambda}\{(2J+1)(2J'+1)\}^{1/2}\begin{pmatrix} J & 2 & J' \\ -\Lambda & q & \Lambda' \end{pmatrix}\langle \eta, \Lambda | C_q^2(\theta, \phi) R^{-3} | \eta, \Lambda'\rangle.$$
$$\tag{8.31}$$

We have made use of equation (8.12) in deriving (8.31). The second reduced matrix element in (8.30) has also been evaluated previously in (8.14):

$$\langle I_1, I_2, I\|T^2(\boldsymbol{I}_1, \boldsymbol{I}_2)\|I_1, I_2, I'\rangle = \sqrt{5}\begin{Bmatrix} I_1 & I_1 & 1 \\ I_2 & I_2 & 1 \\ I' & I & 2 \end{Bmatrix}\{(2I+1)(2I'+1)I_1(I_1+1)$$
$$\times (2I_1+1)I_2(I_2+1)(2I_2+1)\}^{1/2}. \tag{8.32}$$

By combining (8.31) and (8.32), equation (8.30) becomes, for $q = 0$,

$$\langle \eta, \Lambda; J, I_1, I_2, I, F | \mathcal{H}_{\text{dip}} | \eta, \Lambda; J', I_1, I_2, I', F \rangle$$

$$= -g_{I_1} g_{I_2} \mu_N^2 (\mu_0/4\pi) \sqrt{6} (-1)^{J'+I+F} \begin{Bmatrix} I & J & F \\ J' & I' & 2 \end{Bmatrix} (-1)^{J-\Lambda}$$

$$\times \{(2J+1)(2J'+1)\}^{1/2} \begin{pmatrix} J & 2 & J' \\ -\Lambda & 0 & \Lambda \end{pmatrix} \langle \eta, |C_0^2(\theta, \phi) R^{-3}| \eta \rangle \sqrt{5} \begin{Bmatrix} I_1 & I_1 & 1 \\ I_2 & I_2 & 1 \\ I' & I & 2 \end{Bmatrix}$$

$$\times \{(2I+1)(2I'+1) I_1(I_1+1)(2I_1+1) I_2(I_2+1)(2I_2+1)\}^{1/2}. \tag{8.33}$$

We are now in a position to compare the results for the quadrupole and dipolar interactions, as given by (8.29) and (8.33). For the matrix elements diagonal in I (which can take the values 0, 1 or 2), the dependence on the quantum numbers J and Λ is clearly the same. The dependence on the value of I is best determined by evaluating the appropriate 3-j, 6-j and 9-j symbols using computer programs. One finds that for $I = 1$ and $I = 2$ the *ratios* of the dipolar and quadrupole energies are the same; for $I = 0$ the contributions are zero for both interactions because the triangle rule $\Delta(I, I, 2)$ is not satisfied. For the off-diagonal matrix elements connecting the $I = 0$ and 2 states, the ratio of the two contributions is different from that found for the diagonal elements; it would therefore be possible to separate the contributions for *ortho*-D_2, provided the off-diagonal elements were sufficiently large. The above conclusions hold for any homonuclear diatomic molecule in which the relevant nucleus has a quadrupole moment.

The effective Hamiltonian matrix for D_2 with $J = 1$ and $I = 1$ is exactly the same as that previously described for H_2, but the values of the four main constants are (in kHz)

$$a = 0.6536, \quad b = 0.3368, \quad c_I = -8.773, \quad d = 25.237.$$

These are all considerably smaller than the corresponding constants in H_2. The calculated energy levels in magnetic fields ranging from 0 to 1000 G are shown in figure 8.5, whilst a section of the predicted Zeeman pattern from 1800 to 2200 G is shown in figure 8.6. The six transitions indicated are those obeying the selection rules $\Delta M_I = \pm 1$, $\Delta M_J = 0$, and the spectrum obtained in the early work [9] of Kellogg, Rabi, Ramsey and Zacharias is shown in figure 8.7. Apart from the six transitions mentioned above, this spectrum also shows a very strong central line arising from *ortho*-D_2 ($I = 2$ and 0) in the $J = 0$ level. The six short dotted lines shown in figure 8.7 indicate the predicted positions of the six lines if the quadrupole interaction was zero; the experimentally observed pattern is therefore dominated by the quadrupole interaction. In later work this spectrum was recorded at much higher resolution [10], with consequent refinements in the values of the molecular constants.

Finally in this section we turn to the heteronuclear species HD, which was studied in both the 1940 phase [9] and the post-war higher-resolution phase [11]. The three coupled angular momenta $\mathbf{J}$, $\mathbf{I}_D$ and $\mathbf{I}_H$ have space-fixed projections, which are appropriate for strong field studies, of M_J, M_D and M_H. The experimental studies involved HD again in its $J = 1$ level, so that in the decoupled representation there are 18 possible levels ($M_J = 0, \pm 1$; $M_D = 0, \pm 1$; $M_H = \pm 1/2$). Consequently there are three types of transition involving these energy levels which occur in clusters at different

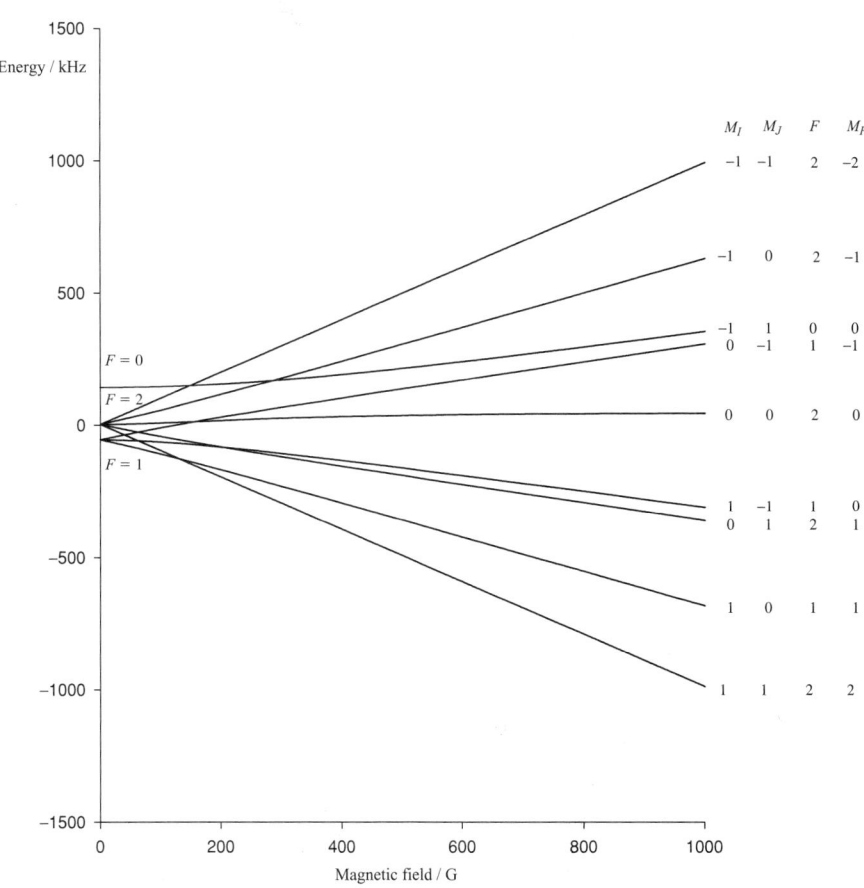

Figure 8.5. Calculated Zeeman levels for D_2 ($J = 1$, $I = 1$) in the range 0 to 1000 G.

radiofrequency regions. There are twelve transitions for which M_D changes by ± 1 (with the other projection quantum numbers unchanged), nine transitions for which M_H changes by ± 1, and twelve transitions for which M_J changes by ± 1. Figure 8.8 shows the *proton* magnetic resonance spectrum, recorded at a frequency of 15.75 MHz. Eight of the expected nine lines are clearly resolved, but the remaining line in this group is predicted to occur at the same position as the deep central minimum, which is itself due to molecules in the $J = 0$ level. We will not repeat the theoretical analysis of the HD spectrum, which follows closely the routes taken for the homonuclear species described above. The most important new feature which arises in the analysis of the HD spectrum is that the proton–deuteron dipolar interaction can now be separated from the deuterium quadrupole interaction. Indeed the primary purpose of the work on HD and D_2, indicated in the title of the paper, was the determination of the deuteron nuclear quadrupole moment.

The rotational magnetic resonance spectrum ($\Delta M_J = \pm 1$, $\Delta M_H = \Delta M_D = 0$) was the subject of a separate paper by Ramsey [12].

The magnitudes of the various molecular constants determined from these experiments were examined in depth, particularly in a series of papers by Ramsey. We now

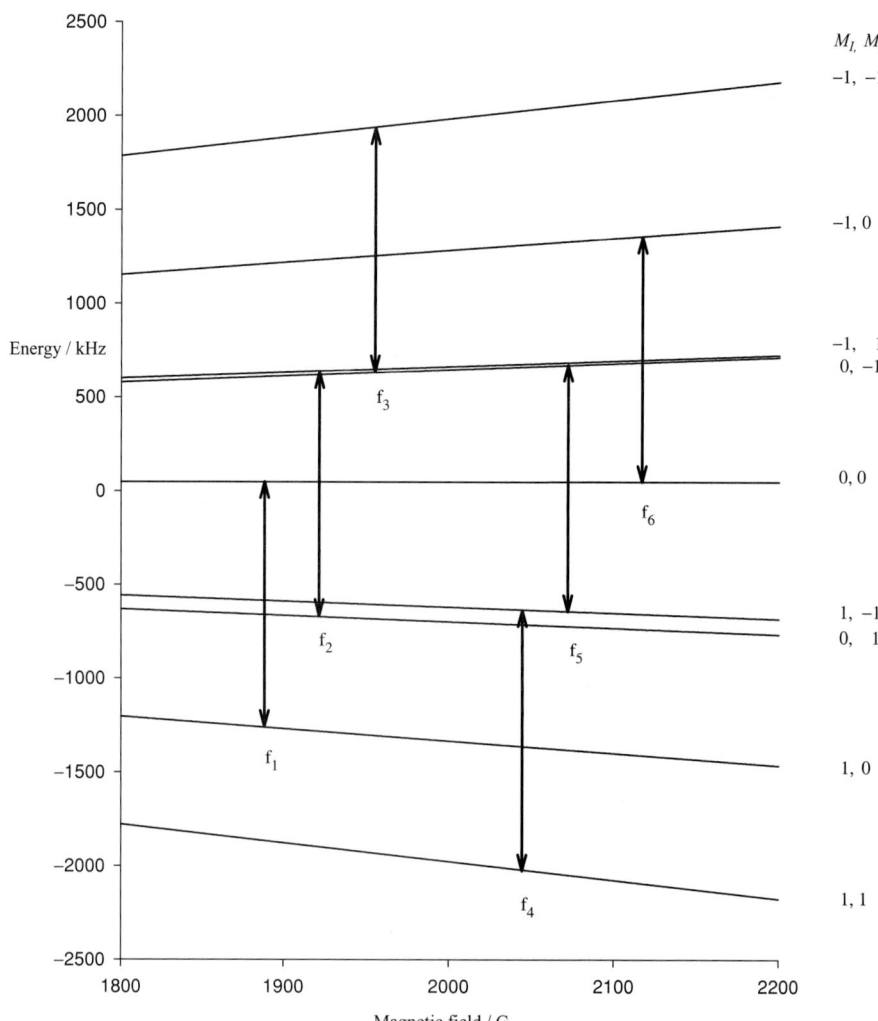

Figure 8.6. Calculated Zeeman levels for D_2 ($J = 1, I = 1$) in the magnetic field range 1800 to 2200 G. The six predicted $\Delta M_I = \pm 1$, $\Delta M_J = 0$ transitions are shown for a frequency of 1.3 MHz; these predictions are to be compared with the experimental observations presented in figure 8.7.

proceed to discuss these quantitative interpretations, which form part of the foundations of nuclear magnetic resonance.

8.2.2. Theory of Zeeman interactions in $^1\Sigma^+$ states

(a) Introduction

We have described the principles and experimental techniques involved in the molecular beam magnetic resonance studies of H_2 and its deuterium isotopes. We have shown

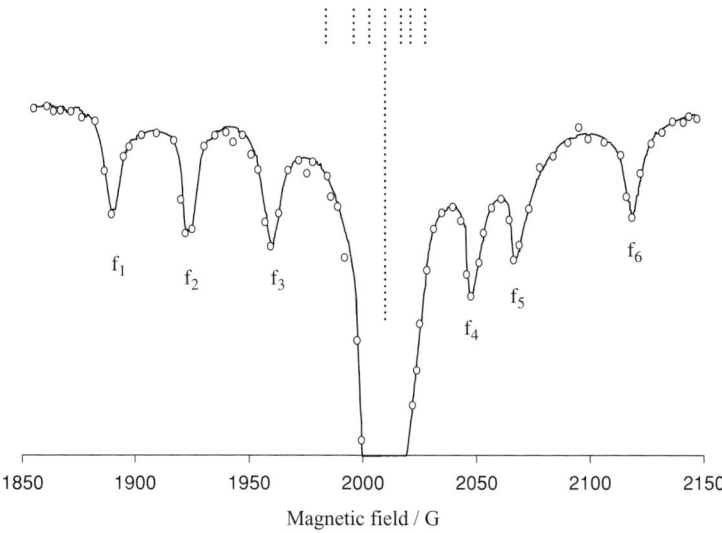

Figure 8.7. Radiofrequency magnetic resonance spectrum of D_2 ($J=1$, $I=1$) observed [9] at a frequency of 1.3 MHz. This spectrum is to be compared with the predictions presented in figure 8.6.

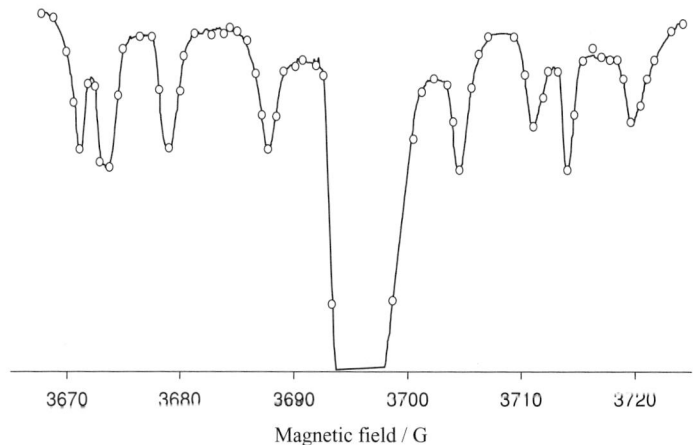

Figure 8.8. The *proton* magnetic resonance spectrum of HD observed at a frequency of 15.75 MHz [9]. This spectrum satisfies the selection rules $\Delta M_{\rm H} = \pm 1$, $\Delta M_{\rm D} = \Delta M_J = 0$.

how the spectra have been analysed in terms of appropriate effective Hamiltonians, and the values of various molecular parameters determined. It is now time to describe the theory underlying these parameters, and the information they provide about electronic structure. Whilst much of the theory was originally developed in order to understand the experimental studies of H_2, it is, of course, generally applicable to other, more complex molecules. As we shall see, there is a series of subtle relationships between magnetic susceptibility, magnetic shielding of nuclei, rotational magnetic moments, and molecular electric quadrupole moments. A thorough description of these relationships could, in itself, constitute the contents of a large book. We must therefore content ourselves here with an outline discussion, which might prove a useful introduction to the original literature.

First we remind ourselves of some elementary definitions. An external magnetic field $\boldsymbol{H}$ is associated with a magnetic flux density $\boldsymbol{B}$ through the simple equation

$$\boldsymbol{B} = \mu_0 \boldsymbol{H}, \qquad (8.34)$$

where μ_0 is the permittivity of free space, with the value $4\pi \times 10^{-7}$ H m^{-1}. We are usually more interested in the flux density through molecular matter, and in this case there is an important modification to (8.34) which we write

$$\boldsymbol{B} = \mu_0 (1 + \chi) \boldsymbol{H}. \qquad (8.35)$$

The quantity χ is described as the magnetic susceptibility of the molecular material; if χ is negative the material is described as diamagnetic, and if χ is positive, the material is paramagnetic. Paramagnetism is a characteristic of atoms or molecules with open shell electronic structures, and we encounter many examples of such systems elsewhere in this book. In this section, however, our attention is confined to closed shell systems. Even so, we will see that an overall diamagnetic susceptibility consists of diamagnetic and paramagnetic parts, although in this case the paramagnetic part is not directly associated with an open shell electronic structure. We should be careful to distinguish between the bulk magnetic susceptibility and that for a single molecule. In this book we will invariably be dealing with single molecules. We will use the symbol χ to represent the single molecule susceptibility, and its components, rather than the symbol ξ which appears in Ramsey's papers and book [12]. The symbol χ has been used by most later authors.

We commence our discussion by considering the diamagnetic susceptibility of an atom, following the classical theory presented by Pople, Schneider and Bernstein [13] and by Lamb [14].

(b) Semi-classical theory of diamagnetism for a spherically symmetric atom

An electron in an atom will circulate about the direction of an applied magnetic field of strength $\boldsymbol{B}$, and thereby give rise to a magnetic moment. Referring to figure 8.9, we consider an electron at a distance r from the origin (which we could take as the position

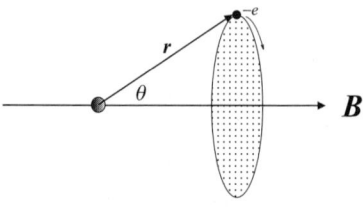

Figure 8.9. Electron rotation about an applied magnetic field of strength $\boldsymbol{B}$.

of the nucleus), the vector r making an angle θ with the direction of the applied flux density B, as shown. The axis of the electron rotation is parallel to B, and the result of this rotation is to create a current of magnitude $-e^2 B/4\pi m$. The electron traces out a circular area (the dotted area in figure 8.9) of $\pi r^2 \sin^2\theta$, and gives rise to a magnetic dipole moment of magnitude equal to the product of the current and the area, that is,

$$-\frac{e^2 B r^2 \sin^2\theta}{4m}. \tag{8.36}$$

The direction of the induced magnetic moment opposes the direction of the applied field, giving rise to diamagnetism. Evidently also, the effective magnetic field at the nucleus will be reduced; the nucleus is 'screened'. In practice we must now calculate the probability distribution of all electrons (i), giving a total moment,

$$-\frac{e^2 B}{6m}\sum_i r_i^2 = \chi_{diam} B. \tag{8.37}$$

This simple treatment applies to an atom in a spherically symmetric state, but does not usually apply to molecules, where the induced magnetic moment will, in general, depend upon the direction of the applied field with respect to the molecular orientation. In other words, the magnetic susceptibility for a molecule is usually a tensor quantity which is anisotropic. The simple treatment does apply to a diatomic molecule (or other linear molecule in a Σ state) when the external magnetic field is applied along the internuclear axis, which we define as the molecular z axis. The electron cloud is then free to rotate about the axis and the magnetic susceptibility component along the axis is given by

$$\chi_{diam} = -\frac{e^2}{4m}\sum_i \left(x_i^2 + y_i^2\right). \tag{8.38}$$

If the secondary induced magnetic field at a nucleus is denoted B', its relationship with the applied field B is given by the expression

$$B' = -\sigma B \tag{8.39}$$

where σ is a second-rank tensor, called the screening tensor. We now describe expressions for the components of this tensor, developing the theory first presented by Ramsey [15].

(c) Diamagnetism in a closed shell diatomic molecule

(i) SEMI-CLASSICAL THEORY

The fundamental expressions which describe the interaction of an external magnetic field with the electrons and nuclei within a molecule were developed from the Dirac and Breit equations in chapters 3 and 4. In this section we develop the theory again, making use of the approach described by Flygare [107]. We start with the classical description of the interaction of a free particle of mass m and charge q with an electromagnetic field, the Hamiltonian for which is

$$H = \frac{1}{2m}[(p_x - qA_x)^2 + (p_y - qA_y)^2 + (p_z - qA_z)^2] + q\phi; \quad (8.40)$$

p_x, p_y and p_z are the components of the linear momentum of the particle, in a cartesian coordinate system which we need not define further at this stage. The electromagnetic field is described by the vector potential A and scalar potential ϕ. Using the operator form for the linear momentum gives us the semi-classical expression,

$$\mathcal{H} = \frac{1}{2m}\left[\left(-i\hbar\frac{\partial}{\partial x} - qA_x\right)^2 + \left(-i\hbar\frac{\partial}{\partial y} - qA_y\right)^2 \right.$$
$$\left. + \left(-i\hbar\frac{\partial}{\partial z} - qA_z\right)^2\right] + q\phi \quad (8.41)$$
$$= \frac{1}{2m}\left[-\hbar^2\left(\frac{\partial^2}{\partial x^2} + \frac{\partial^2}{\partial y^2} + \frac{\partial^2}{\partial z^2}\right) + q^2(A_x^2 + A_y^2 + A_z^2)\right]$$
$$+ \frac{i\hbar q}{2m}\left(\frac{\partial}{\partial x}A_x + A_x\frac{\partial}{\partial x} + \frac{\partial}{\partial y}A_y + A_y\frac{\partial}{\partial y} + \frac{\partial}{\partial z}A_z + A_z\frac{\partial}{\partial z}\right) + q\phi$$
$$= \frac{1}{2m}(-\hbar^2\nabla^2 + i\hbar q\nabla\cdot A + i\hbar qA\cdot\nabla + q^2A^2) + q\phi. \quad (8.42)$$

Choosing the Coulomb gauge and setting ϕ to zero, (8.42) reduces to

$$\mathcal{H} = \frac{1}{2m}(-\hbar^2\nabla^2 + 2i\hbar qA\cdot\nabla + q^2A^2). \quad (8.43)$$

If we now take the particle to be an electron with charge $-e$ and potential energy V, the semi-classical Hamiltonian becomes

$$\mathcal{H} = -\frac{\hbar^2}{2m}\nabla^2 + V - \frac{i\hbar e}{m}A\cdot\nabla + \frac{e^2}{2m}A^2. \quad (8.44)$$

We now consider, following Flygare, the model system of a particle in a ring, with a magnetic field defining the z axis perpendicular to the plane of the ring (see figure 8.10). This model system has obvious similarities to real molecular systems

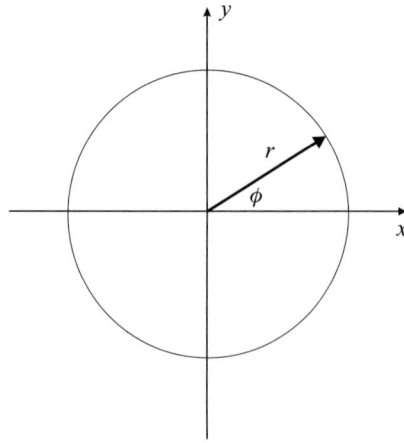

Figure 8.10. Coordinate system for the particle in a ring.

possessing axial symmetry. The magnetic vector potential is given, in the general case, by the expression

$$A = -\frac{1}{2} r \wedge B \tag{8.45}$$

so that the vector components in our case are

$$A_x = -\frac{y}{2} B_z, \quad A_y = \frac{x}{2} B_z, \quad A_z = 0. \tag{8.46}$$

Since we know that

$$A \cdot \nabla = A_x \frac{\partial}{\partial x} + A_y \frac{\partial}{\partial y} + A_z \frac{\partial}{\partial z} = -\frac{y}{2} B_z \frac{\partial}{\partial x} + \frac{x}{2} B_z \frac{\partial}{\partial y}, \tag{8.47}$$

we see that the third term in equation (8.44) becomes

$$-\frac{i\hbar e}{m} A \cdot \nabla = -\frac{i\hbar e B_z}{2m} \left(x \frac{\partial}{\partial y} - y \frac{\partial}{\partial x} \right). \tag{8.48}$$

The last term in equation (8.44) contains A^2 which, using (8.46), is readily expanded:

$$A^2 = A_x^2 + A_y^2 + A_z^2 = \frac{1}{4} B_z^2 (x^2 + y^2). \tag{8.49}$$

Consequently equation (8.44) becomes

$$\mathcal{H} = -\frac{\hbar^2}{2m} \nabla^2 + V - \frac{i\hbar e B_z}{2m} \left(x \frac{\partial}{\partial y} - y \frac{\partial}{\partial x} \right) + \frac{e^2 B_z^2}{8m} (x^2 + y^2). \tag{8.50}$$

The last step is to substitute the electron Bohr magneton for $e\hbar/2m$, and to make the replacement

$$L_z = -i \left(x \frac{\partial}{\partial y} - y \frac{\partial}{\partial x} \right). \tag{8.51}$$

We then obtain the Hamiltonian for an electron in a magnetic field applied in the z direction, which is:

$$\mathcal{H} = -\frac{\hbar^2}{2m} \nabla^2 + V + \mu_B B_z L_z + \frac{e^2 B_z^2}{8m} (x^2 + y^2). \tag{8.52}$$

The more general form of the magnetic field part of this Hamiltonian, when the direction of the magnetic field is not specified is

$$\mathcal{H} = \mu_B B \cdot L + \frac{e^2}{8m} B \cdot (r^2 \mathbf{1} - rr) \cdot B, \tag{8.53}$$

where $\mathbf{1}$ is the unit matrix.

The one-electron Hamiltonian (8.53) may be generalised for a many-electron system, to become

$$\mathcal{H} = \mu_B B \cdot L + \frac{e^2}{8m} B \cdot \sum_i (r_i^2 \mathbf{1} - r_i r_i) \cdot B, \tag{8.54}$$

where the sum i is over all the electrons in the molecule, and $\boldsymbol{L}$ is the total electronic orbital angular momentum, given by

$$\hbar \boldsymbol{L} = \sum_i \boldsymbol{r}_i \wedge \boldsymbol{p}_i. \tag{8.55}$$

We wish to add the effects of molecular rotation before proceeding further, in order to establish important results concerning the dependence of the magnetic susceptibility on the rotational state, and its relationship to the rotational magnetic moment of a molecule.

(ii) MOLECULAR ROTATION

Much of the following exposition was already presented in chapter 2, but it is fundamental and can bear repetition. The coordinate system employed to describe the motion of the particles in a molecule, both electrons and nuclei, is illustrated in figure 8.11. O is an arbitrary laboratory-fixed origin and c.m. is the centre-of-mass of the many-particle system; $\boldsymbol{R}_O$ is the vector from O to the centre-of-mass. The position of each particle i (electron or nucleus) is defined by the vectors $\boldsymbol{R}_i$ and $\boldsymbol{r}_i$ from the origin O and the centre-of-mass respectively.

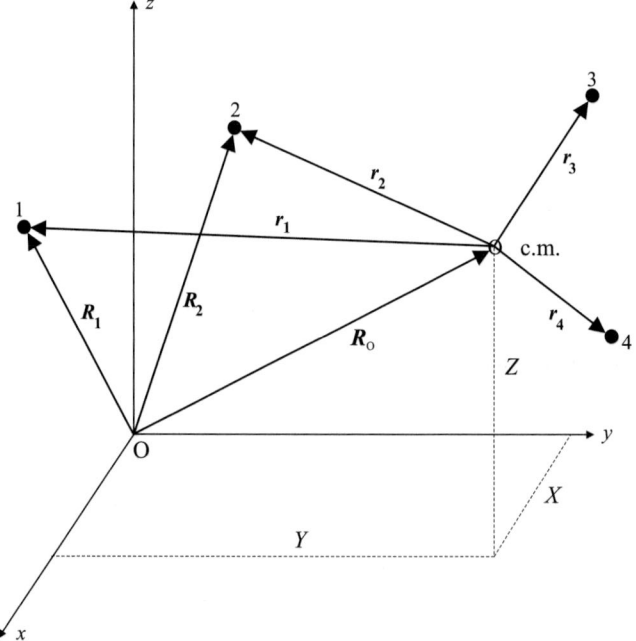

Figure 8.11. Cartesian coordinate system for describing the position vectors of the particles (electrons and nuclei) in a molecule. O(X, Y, Z) is the laboratory-fixed frame of arbitrary origin, and c.m. is the centre-of-mass in the molecule-fixed frame. For the purposes of illustration four particles are indicated, but for most molecular systems there will be many more than four.

First we need to define several quantities which enter into our discussion of the classical mechanics of the system. The symbols m_k, V_k, P_k, $\dot{P}_k$ denote the mass, velocity, linear momentum and time rate of change of the linear momentum of the kth particle in the laboratory-fixed frame. We remember that the momentum is defined by

$$P_k = m_k V_k. \tag{8.56}$$

We remember also a number of other basic results. The angular momentum in the laboratory frame for particle k is given by

$$\hbar J_k = R_k \wedge P_k, \tag{8.57}$$

the total mass M of the system of n particles is

$$M = \sum_{k=1}^{n} m_k, \tag{8.58}$$

the vector R_O from the laboratory origin O to the centre-of-mass origin (c.m.) is given by

$$R_O = \frac{1}{M} \sum_k m_k R_k, \tag{8.59}$$

and, finally, the linear momentum of the centre-of-mass is given by

$$P_O = M \frac{d}{dt} R_O = \sum_k m_k \frac{dR_k}{dt} = \sum_k m_k V_k. \tag{8.60}$$

We now proceed to develop equation (8.57) for the total angular momentum in the laboratory frame:

$$\hbar J_O = \sum_k R_k \wedge P_k = \sum_k m_k R_k \wedge V_k. \tag{8.61}$$

From figure 8.11 we see that

$$R_k = R_O + r_k, \tag{8.62}$$

from which it follows that

$$\frac{d}{dt}(R_k) = V_k = V_O + v_k, \tag{8.63}$$

where V_O is the velocity of the centre-of-mass, $\dot{R}_O$, and v_k is the velocity of particle k measured relative to the centre-of-mass coordinate system.

We can now substitute (8.62) and (8.63) into (8.61) to obtain a new expression for the total angular momentum in the laboratory frame:

$$\hbar J_O = \sum_k m_k R_k \wedge V_k = \sum_k m_k (R_O + r_k) \wedge (V_O + v_k)$$
$$= R_O \wedge V_O \sum_k m_k + R_O \wedge \sum_k m_k v_k - V_O \wedge \sum_k m_k r_k + \sum_k m_k r_k \wedge v_k. \tag{8.64}$$

The second and third terms vanish (because they both contain $\sum_k m_k \, r_k$), so that (8.64) becomes

$$\hbar J_O = M R_O \wedge V_O + \sum_k m_k r_k \wedge v_k = M R_O \wedge V_O + \sum_k \hbar J_k = M R_O \wedge V_O + \hbar J. \tag{8.65}$$

This is an important result. It tells us that the total angular momentum determined in the laboratory frame is the sum of a centre-of-mass contribution and the total angular momentum $\hbar J$ measured *in* the centre-of-mass frame. In a rigid rotor the velocity vector of the kth particle in the centre-of-mass frame is related to the angular velocity, $\boldsymbol{\omega}$, of the rotating particle by

$$v_k = \omega \wedge r_k, \tag{8.66}$$

where $\boldsymbol{\omega}$ has its origin at the centre-of-mass and the same value for all particles. If we now substitute (8.66) for J in (8.65) we obtain the result

$$\hbar J = \sum_k m_k (r_k \wedge v_k) = \sum_k m_k [r_k \wedge (\omega \wedge r_k)]. \tag{8.67}$$

Using the standard result for a triple vector product, equation (8.67) becomes

$$\hbar J = \sum_k m_k [\omega(r_k \cdot r_k) - r_k (r_k \cdot \omega)] = \sum_k m_k \left(r_k^2 \mathbf{1} - r_k r_k \right) \cdot \omega = \mathbf{I} \cdot \omega. \tag{8.68}$$

This result requires some additional explanation. The expression $(r^2 \mathbf{1} - rr)$ will occur frequently in the following analysis, summed for both electrons and nuclei. The symbol $\mathbf{1}$ is the unit dyadic and is represented by a unit matrix. The product rr, which is not to be confused with either a scalar product or a vector product, is also a dyadic. Explicitly the above expression is evaluated in the following manner:

$$r^2 \mathbf{1} - rr = (x^2 + y^2 + z^2) \begin{pmatrix} 1 & 0 & 0 \\ 0 & 1 & 0 \\ 0 & 0 & 1 \end{pmatrix} - \begin{pmatrix} x \\ y \\ z \end{pmatrix} \begin{pmatrix} x & y & z \end{pmatrix}$$

$$= \begin{pmatrix} x^2+y^2+z^2 & 0 & 0 \\ 0 & x^2+y^2+z^2 & 0 \\ 0 & 0 & x^2+y^2+z^2 \end{pmatrix} - \begin{pmatrix} xx & xy & xz \\ yx & yy & yz \\ zx & zy & zz \end{pmatrix}$$

$$= \begin{pmatrix} y^2+z^2 & -xy & -xz \\ -xy & x^2+z^2 & -yz \\ -xz & -yz & x^2+y^2 \end{pmatrix}. \tag{8.69}$$

As is indicated by the result, the matrix representation is usually symmetric, and if the x, y, z axes are the symmetry axes of the molecule, the off-diagonal elements are also zero. In all of the examples given in the following pages, the result is

$$r^2 \mathbf{1} - rr = \begin{pmatrix} y^2+z^2 & 0 & 0 \\ 0 & x^2+z^2 & 0 \\ 0 & 0 & x^2+y^2 \end{pmatrix}, \tag{8.70}$$

when x, y, z are the molecular symmetry axes. For symmetric top or linear molecules the result is even simpler, as we shall see.

The total kinetic energy, T, of the system is given by

$$T = \frac{1}{2} \sum_k m_k \mathbf{V}_k \cdot \mathbf{V}_k, \tag{8.71}$$

and by using (8.63), this may be expanded as follows:

$$T = \frac{1}{2} \sum_k m_k \mathbf{V}_k \cdot \mathbf{V}_k = \frac{1}{2} \sum_k m_k (\mathbf{V}_O + \mathbf{v}_k) \cdot (\mathbf{V}_O + \mathbf{v}_k)$$

$$= \frac{1}{2} V_O^2 \sum_k m_k + \frac{1}{2} \sum_k m_k v_k^2 + \mathbf{V}_O \cdot \sum_k m_k \mathbf{v}_k. \tag{8.72}$$

The last term in (8.72) is zero because

$$\sum_k m_k \mathbf{v}_k = \frac{\mathrm{d}}{\mathrm{d}t} \sum_k m_k \mathbf{r}_k = 0. \tag{8.73}$$

Consequently the total kinetic energy is the sum of the kinetic energy of the total mass at the centre-of-mass, and the internal kinetic energy in the centre-of-mass frame, i.e.

$$T = \frac{1}{2} M V_O^2 + \frac{1}{2} \sum_k m_k v_k^2. \tag{8.74}$$

For the rigid system of particles we may use (8.66) and (8.65) to reformulate the expression for the centre-of-mass kinetic energy as

$$T_{\mathrm{cm}} = \frac{1}{2} \sum_k m_k \mathbf{v}_k \cdot \mathbf{v}_k = \frac{1}{2} \sum_k m_k \mathbf{v}_k \cdot (\boldsymbol{\omega} \wedge \mathbf{r}_k) = \frac{1}{2} \boldsymbol{\omega} \cdot \sum_k m_k (\mathbf{r}_k \wedge \mathbf{v}_k) = \frac{\hbar}{2} \boldsymbol{\omega} \cdot \mathbf{J}. \tag{8.75}$$

This may be rewritten in the form

$$T_{\mathrm{cm}} = \frac{\hbar}{2} \boldsymbol{\omega} \cdot \mathbf{J} = \frac{1}{2} \boldsymbol{\omega} \cdot \mathbf{I} \cdot \boldsymbol{\omega} = \frac{1}{2} (\omega_x, \omega_y, \omega_z) \begin{pmatrix} I_{xx} & I_{xy} & I_{xz} \\ I_{yx} & I_{yy} & I_{yz} \\ I_{zx} & I_{zy} & I_{zz} \end{pmatrix} \begin{pmatrix} \omega_x \\ \omega_y \\ \omega_z \end{pmatrix}. \tag{8.76}$$

$\mathbf{I}$ is the moment of inertia tensor; if the x, y, z axes are chosen to be the principal inertial axes of the molecule (a, b, c), $\mathbf{I}$ is then diagonal with principal components I_{aa}, I_{bb}, I_{cc}. For a linear molecule (including diatomics), $I_{aa} = 0$ and $I_{bb} = I_{cc}$. In the inertial axis system equation (8.76) becomes simply

$$T_{\mathrm{cm}} = \frac{1}{2} \left(\omega_a^2 I_{aa} + \omega_b^2 I_{bb} + \omega_c^2 I_{cc} \right) = \frac{1}{2} \omega^2 I_{bb} \ (diatomic). \tag{8.77}$$

As a result of the preceding analyses, we can express the velocity of the kth particle in a molecule with respect to a laboratory-fixed framework in the following terms:

$$\mathbf{V}_k = \dot{\mathbf{R}}_O + \boldsymbol{\omega} \wedge \mathbf{r}_k + \mathbf{v}_k. \tag{8.78}$$

The kinetic energy of the system of particles, still with respect to the laboratory-fixed system, is

$$T = \frac{1}{2} \sum_k m_k \mathbf{V}_k \cdot \mathbf{V}_k$$

$$= \frac{M}{2} \dot{\mathbf{R}}_O^2 + \frac{1}{2} \sum_k m_k (\boldsymbol{\omega} \wedge \mathbf{r}_k) \cdot (\boldsymbol{\omega} \wedge \mathbf{r}_k) + \frac{1}{2} \sum_k m_k v_k^2$$

$$+ \dot{\mathbf{R}}_O \cdot \left(\boldsymbol{\omega} \wedge \left(\sum_k m_k \mathbf{r}_k \right) \right) + \dot{\mathbf{R}}_O \cdot \left(\sum_k m_k \mathbf{v}_k \right) + \boldsymbol{\omega} \cdot \left(\sum_k m_k \mathbf{r}_k \wedge \mathbf{v}_k \right)$$

$$= \frac{M}{2} \dot{\mathbf{R}}_O^2 + \frac{1}{2} \sum_k m_k (\boldsymbol{\omega} \wedge \mathbf{r}_k) \cdot (\boldsymbol{\omega} \wedge \mathbf{r}_k) + \frac{1}{2} \sum_k m_k v_k^2 + \boldsymbol{\omega} \cdot \left(\sum_k m_k \mathbf{r}_k \wedge \mathbf{v}_k \right). \tag{8.79}$$

The simplification from the second to the third lines is a consequence of the result given in (8.73). It is now time to separate the third and fourth terms in (8.79) into sums relating to electrons (i) and nuclei (α) respectively,

$$\frac{1}{2} \sum_k m_k v_k^2 = \frac{1}{2} \sum_i m_i v_i^2 + \frac{1}{2} \sum_\alpha m_\alpha v_\alpha^2, \tag{8.80}$$

$$\boldsymbol{\omega} \cdot \left(\sum_k m_k \mathbf{r}_k \wedge \mathbf{v}_k \right) = \boldsymbol{\omega} \cdot m \left(\sum_i \mathbf{r}_i \wedge \mathbf{v}_i \right) + \boldsymbol{\omega} \cdot \left(\sum_\alpha m_\alpha \mathbf{r}_\alpha \wedge \mathbf{v}_\alpha \right). \tag{8.81}$$

Here, m and m_α are the electron and nuclear masses. We now define the displacement vector of the α th nucleus, $\mathbf{s}_\alpha$, relative to its equilibrium position in the rotating coordinate system, $\mathbf{a}_\alpha$, by $\mathbf{s}_\alpha = \mathbf{r}_\alpha - \mathbf{a}_\alpha$. Then for small displacements,

$$\boldsymbol{\omega} \cdot \left(\sum_\alpha m_\alpha \mathbf{r}_\alpha \wedge \mathbf{v}_\alpha \right) \simeq \boldsymbol{\omega} \cdot \left(\sum_\alpha m_\alpha \mathbf{s}_\alpha \wedge \mathbf{v}_\alpha \right). \tag{8.82}$$

We can now bring together the results presented in equations (8.82), (8.81), (8.80), (8.79), and drop the translational term, to obtain an expression for the kinetic energy of the whole system, divided into separate contributions. The result is

$$T = T_{\text{rot}} + T_{\text{el}} + T_{\text{nucl}} + T_{\text{rot-el}} + T_{\text{rot-nucl}} \tag{8.83}$$

where

$$T_{\text{rot}} = \frac{1}{2}\sum_{k} m_k(\boldsymbol{\omega} \wedge \boldsymbol{r}_k) \cdot (\boldsymbol{\omega} \wedge \boldsymbol{r}_k) \quad : \textit{classical rotational kinetic energy} \quad (8.84)$$

$$T_{\text{el}} = \frac{m}{2}\sum_{i} v_i^2 \quad : \textit{electronic kinetic energy} \quad (8.85)$$

$$T_{\text{nucl}} = \frac{1}{2}\sum_{\alpha} m_\alpha v_\alpha^2 \quad : \textit{nuclear kinetic energy} \quad (8.86)$$

$$T_{\text{rot-el}} = \boldsymbol{\omega} \cdot m\left(\sum_{i} \boldsymbol{r}_i \wedge \boldsymbol{v}_i\right) \quad : \textit{rotational electronic coupling} \quad (8.87)$$

$$T_{\text{rot-nucl}} = \boldsymbol{\omega} \cdot \left(\sum_{\alpha} m_\alpha \boldsymbol{s}_\alpha \wedge \boldsymbol{v}_\alpha\right) \quad : \textit{rotational vibrational coupling.} \quad (8.88)$$

The final term, (8.88), is zero for a rigid molecule.

The potential energy of an isolated molecule can be represented as the sum of three terms,

$$V = V_{\text{el-el}} + V_{\text{nucl-nucl}} + V_{\text{el-nucl}}, \quad (8.89)$$

where

$$V_{\text{el-el}} = \sum_{i>j} \frac{e^2}{4\pi\varepsilon_0 r_{ij}} \quad : \textit{electron–electron potential energy} \quad (8.90)$$

$$V_{\text{nucl-nucl}} = \sum_{\alpha>\beta} \frac{Z_\alpha Z_\beta e^2}{4\pi\varepsilon_0 R_{\alpha\beta}} \quad : \textit{nuclear–nuclear potential energy} \quad (8.91)$$

$$V_{\text{el-nucl}} = -\sum_{i,\alpha} \frac{Z_\alpha e^2}{4\pi\varepsilon_0 r_{i\alpha}} \quad : \textit{electron–nuclear potential energy.} \quad (8.92)$$

Before considering the effects of an external magnetic field, it is desirable to reformulate the expression for the total kinetic energy (8.83) in terms of the total angular momentum $\boldsymbol{J}$ and the total electronic orbital angular momentum $\boldsymbol{L}$. Equation (8.83) can be rewritten for a rigid molecule ($v_\alpha = 0$) in the form

$$T = \frac{m}{2}\sum_{i} \boldsymbol{v}_i \cdot \boldsymbol{v}_i + \frac{1}{2}\boldsymbol{\omega} \cdot \boldsymbol{I}_{\text{nucl}} \cdot \boldsymbol{\omega} + \frac{1}{2}\boldsymbol{\omega} \cdot \boldsymbol{I}_{\text{el}} \cdot \boldsymbol{\omega} + m\boldsymbol{\omega} \cdot \sum_{i} \boldsymbol{r}_i \wedge \boldsymbol{v}_i \quad (8.93)$$

where the rotational term (8.84) has been separated into pure nuclear and electronic terms defined by

$$\boldsymbol{I}_{\text{nucl}} = \sum_{\alpha} M_\alpha \left(r_\alpha^2 \boldsymbol{1} - \boldsymbol{r}_\alpha \boldsymbol{r}_\alpha\right), \quad (8.94)$$

$$\boldsymbol{I}_{\text{el}} = m\sum_{i} \left(r_i^2 \boldsymbol{1} - \boldsymbol{r}_i \boldsymbol{r}_i\right). \quad (8.95)$$

We can now introduce the total angular momentum $\boldsymbol{J}$ by noting that it is obtained by taking the first derivative of the kinetic energy with respect to the angular velocity, i.e.

$$\hbar \bm{J} = \frac{\partial T}{\partial \bm{\omega}} = \bm{\omega} \cdot \bm{I}_{\text{nucl}} + \bm{\omega} \cdot \bm{I}_{\text{el}} + m \sum_i \bm{r}_i \wedge \bm{v}_i. \tag{8.96}$$

The linear momentum of the ith electron is also obtained by taking the first derivative of T with respect to the velocity,

$$\bm{p}_i = \frac{\partial T}{\partial \bm{v}_i} = m\bm{v}_i + m\bm{\omega} \wedge \bm{r}_i, \tag{8.97}$$

from which we also obtain

$$\bm{v}_i = \frac{\bm{p}_i}{m} - (\bm{\omega} \wedge \bm{r}_i). \tag{8.98}$$

Substituting (8.96), (8.97) and (8.98) into (8.93) gives

$$\begin{aligned} T &= \frac{\hbar}{2} \bm{\omega} \cdot \bm{J} + \frac{1}{2} \sum_i \bm{p}_i \cdot \bm{v}_i \\ &= \frac{\hbar}{2} \bm{\omega} \cdot \bm{J} + \frac{1}{2m} \sum_i p_i^2 - \frac{1}{2} \sum_i \bm{p}_i \cdot (\bm{\omega} \wedge \bm{r}_i) \\ &= \frac{\hbar}{2} \bm{\omega} \cdot \bm{J} + \frac{1}{2m} \sum_i p_i^2 - \frac{1}{2} \bm{\omega} \cdot \sum_i \bm{r}_i \wedge \bm{p}_i \\ &= \frac{\hbar}{2} \bm{\omega} \cdot \bm{J} + \frac{1}{2m} \sum_i p_i^2 - \frac{1}{2} \bm{\omega} \cdot \hbar \bm{L} \\ &= \frac{\hbar}{2} \bm{\omega} \cdot (\bm{J} - \bm{L}) + \frac{1}{2m} \sum_i p_i^2. \end{aligned} \tag{8.99}$$

If we now substitute (8.97) into (8.99) and compare the resulting expression for T with that given in (8.93) we obtain the important results

$$\hbar(\bm{J} - \bm{L}) = \bm{\omega} \cdot \bm{I}_{\text{nucl}}, \quad \bm{\omega} = \hbar(\bm{J} - \bm{L}) \cdot \bm{I}_{\text{nucl}}^{-1}. \tag{8.100}$$

$\bm{I}_{\text{nucl}}^{-1}$ is the inverse principal inertial tensor.

Finally we substitute for $\bm{\omega}$ in equation (8.99) and obtain the result

$$\begin{aligned} T &= \frac{\hbar^2}{2} (\bm{J} - \bm{L}) \cdot \bm{I}_{\text{nucl}}^{-1} \cdot (\bm{J} - \bm{L}) \\ &= \frac{\hbar^2}{2} \bm{J} \cdot \bm{I}_{\text{nucl}}^{-1} \cdot \bm{J} - \hbar^2 \bm{J} \cdot \bm{I}_{\text{nucl}}^{-1} \cdot \bm{L} + \frac{\hbar^2}{2} \bm{L} \cdot \bm{I}_{\text{nucl}}^{-1} \cdot \bm{L}. \end{aligned} \tag{8.101}$$

This is an important result. The first term leads to the rotational eigenvalues, whilst the second term describes the rotational–electronic coupling and, as we shall see, contributes to the rotational magnetic moment and the spin–rotation interaction. The third term is small and can be neglected for $^1\Sigma$ states where $\Lambda = 0$. We have omitted the electron kinetic energy term from (8.101) because it is part of the zeroth-order Hamiltonian which determines the electronic eigenvalues and eigenfunctions.

We will now proceed to combine our knowledge of the rigid body rotation with the Zeeman interactions discussed in part (*i*).

(iii) INTERACTIONS WITH AN APPLIED MAGNETIC FIELD

The effects of an applied magnetic field on the electrons in a non-vibrating molecule were described in equation (8.54), which we repeat here:

$$\mathcal{H} = \mu_B \, \boldsymbol{B} \cdot \boldsymbol{L} + \frac{e^2}{8m} \boldsymbol{B} \cdot \sum_i \left(r_i^2 \mathbf{1} - \boldsymbol{r}_i \boldsymbol{r}_i \right) \cdot \boldsymbol{B}. \quad (8.102)$$

This expression was derived rigorously for a diatomic molecule in section 3.7. The two analogous contributions for the field interaction with the nuclei are

$$-\frac{e}{2} \boldsymbol{B} \cdot \sum_\alpha Z_\alpha \boldsymbol{r}_\alpha \wedge \boldsymbol{v}_\alpha = -\frac{e}{2} \boldsymbol{B} \cdot \sum_\alpha Z_\alpha \boldsymbol{r}_\alpha \wedge \boldsymbol{\omega}_\alpha \wedge \boldsymbol{r}_\alpha$$

$$= -\boldsymbol{B} \cdot \left[\frac{e}{2} \sum_\alpha Z_\alpha \left(r_\alpha^2 \mathbf{1} - \boldsymbol{r}_\alpha \boldsymbol{r}_\alpha \right) \right] \cdot \boldsymbol{\omega}, \quad (8.103)$$

and

$$\frac{e^2}{8} \boldsymbol{B} \cdot \sum_\alpha \frac{Z_\alpha^2}{M_\alpha} \left(r_\alpha^2 \mathbf{1} - \boldsymbol{r}_\alpha \boldsymbol{r}_\alpha \right) \cdot \boldsymbol{B}, \quad (8.104)$$

respectively. They were derived rigorously in section 3.12. If these nuclear terms are added to the electronic Zeeman terms (8.102) and the rotational kinetic energy term (8.101) we obtain the total perturbation Hamiltonian, as follows:

$$\mathcal{H}' = \frac{\hbar^2}{2} \boldsymbol{J} \cdot \boldsymbol{I}_{\text{nucl}}^{-1} \cdot \boldsymbol{J} + \frac{e^2}{8m} \boldsymbol{B} \cdot \left[\sum_i \left(r_i^2 \mathbf{1} - \boldsymbol{r}_i \boldsymbol{r}_i \right) + m \sum_\alpha \frac{Z_\alpha^2}{M_\alpha} \left(r_\alpha^2 \mathbf{1} - \boldsymbol{r}_\alpha \boldsymbol{r}_\alpha \right) \right] \cdot \boldsymbol{B}$$

$$- \frac{e\hbar}{2} \sum_\alpha Z_\alpha \boldsymbol{B} \cdot \left(r_\alpha^2 \mathbf{1} - \boldsymbol{r}_\alpha \boldsymbol{r}_\alpha \right) \cdot \boldsymbol{I}_{\text{nucl}}^{-1} \cdot \boldsymbol{J} + \frac{\hbar^2}{2} \boldsymbol{L} \cdot \boldsymbol{I}_{\text{nucl}}^{-1} \cdot \boldsymbol{L} - \hbar \, \boldsymbol{L} \cdot \bar{\boldsymbol{\omega}}, \quad (8.105)$$

where

$$\bar{\boldsymbol{\omega}} = \hbar \, \boldsymbol{I}_{\text{nucl}}^{-1} \cdot \boldsymbol{J} - \frac{e}{2m} \boldsymbol{B} - \frac{e}{2} \boldsymbol{I}_{\text{nucl}}^{-1} \cdot \sum_\alpha Z_\alpha \left(r_\alpha^2 \mathbf{1} - \boldsymbol{r}_\alpha \boldsymbol{r}_\alpha \right) \cdot \boldsymbol{B}. \quad (8.106)$$

The magnetic field induced frequency $\nu = \bar{\omega}/2\pi$ is called the Larmor frequency; it is one-half the cyclotron frequency for an electron.

The next stage in the analysis is to examine the effects of (8.105) on the zeroth-order electronic eigenvalues and eigenfunctions.

(iv) MAGNETIC FIELD PERTURBATIONS OF THE ELECTRONIC GROUND STATE

The zeroth-order electronic states are eigenfunctions of the Hamiltonian

$$\mathcal{H}_{\text{el}} = T_{\text{el}} + V_{\text{el-el}} + V_{\text{el-nucl}} \quad (8.107)$$

which we derived previously in (8.85), (8.90) and (8.92); for a fixed nuclear configuration we obtained the result

$$\mathcal{H}_{el} = \frac{m}{2}\sum_i v_i^2 + \sum_{i>j}\frac{e^2}{4\pi\varepsilon_0 r_{ij}} - \sum_{i,\alpha}\frac{Z_\alpha e^2}{4\pi\varepsilon_0 r_{i\alpha}}. \tag{8.108}$$

The electronic eigenfunctions of this Hamiltonian are $\psi_0, \ldots, \psi_k$ where ψ_0 is the ground state, which we presume does not possess any electronic angular momentum, and ψ_k are the excited states, which may possess electronic angular momentum.

We calculate the effects of the Hamiltonian (8.105) on these zeroth-order states using perturbation theory. This is exactly the same procedure as that which we used to construct the effective Hamiltonian in chapter 7. Our objective here is to formulate the terms in the effective Hamiltonian which describe the nuclear spin–rotation interaction and the susceptibility and chemical shift terms in the Zeeman Hamiltonian. We deal with them in much more detail at this point so that we can interpret the measurements on closed shell molecules by molecular beam magnetic resonance. The first-order corrections of the perturbation Hamiltonian are readily calculated to be

$$E^{(1)} = \frac{\hbar^2}{2}\boldsymbol{J}\cdot\boldsymbol{I}_{\text{nucl}}^{-1}\cdot\boldsymbol{J} + \frac{1}{2}\boldsymbol{B}\cdot\left[\langle 0|\frac{e^2}{4m}\sum_i(r_i^2\boldsymbol{1} - \boldsymbol{r}_i\boldsymbol{r}_i)|0\rangle + \right.$$

$$\left.\frac{e^2}{4}\sum_\alpha\frac{Z_\alpha^2}{M_\alpha}(r_\alpha^2\boldsymbol{1} - \boldsymbol{r}_\alpha\boldsymbol{r}_\alpha)\right]\cdot\boldsymbol{B} - \hbar\boldsymbol{B}\cdot\left[\frac{e}{2}\sum_\alpha Z_\alpha(r_\alpha^2\boldsymbol{1} - \boldsymbol{r}_\alpha\boldsymbol{r}_\alpha)\cdot\boldsymbol{I}_{\text{nucl}}^{-1}\cdot\boldsymbol{J}\right]. \tag{8.109}$$

The second-order terms are more complicated. They all involve the matrix elements of $\boldsymbol{L}\cdot\tilde{\boldsymbol{\omega}}$, which itself consists of three terms, as we see from (8.106). If we label these three terms t_1, t_2 and t_3, the second order corrections have the form

$$E^{(2)} = \hbar^2\sum_{n>0}\frac{\langle 0|\boldsymbol{L}\cdot\tilde{\boldsymbol{\omega}}|n\rangle\langle n|\boldsymbol{L}\cdot\tilde{\boldsymbol{\omega}}|0\rangle}{E_0 - E_n}$$

$$= \sum_{n>0}\frac{\langle 0|t_1 + t_2 + t_3|n\rangle\langle n|t_1 + t_2 + t_3|0\rangle}{E_0 - E_n}. \tag{8.110}$$

The nine terms arising from (8.110), are reduced by commutation properties to six, given below; each contains the product of matrix elements of $\boldsymbol{L}$ which we denote by the symbol $\boldsymbol{A}$:

$$\boldsymbol{A} = \hbar^2\sum_{n>0}\frac{\langle 0|\boldsymbol{L}|n\rangle\langle n|\boldsymbol{L}|0\rangle}{E_0 - E_n}. \tag{8.111}$$

$$t_1 t_1 = \hbar^2 \mathbf{J} \cdot \mathbf{I}_{\text{nucl}}^{-1} \cdot \mathbf{A} \cdot \mathbf{I}_{\text{nucl}}^{-1} \cdot \mathbf{J},$$

$$t_2 t_2 = \frac{e^2}{4m^2} \mathbf{B} \cdot \mathbf{A} \cdot \mathbf{B},$$

$$t_3 t_3 = \frac{e^2}{4} \mathbf{B} \cdot \mathbf{I}_{\text{nucl}}^{-1} \cdot \sum_\alpha Z_\alpha \left(r_\alpha^2 \mathbf{1} - \mathbf{r}_\alpha \mathbf{r}_\alpha\right) \cdot \mathbf{A} \cdot \sum_\alpha Z_\alpha \left(r_\alpha^2 \mathbf{1} - \mathbf{r}_\alpha \mathbf{r}_\alpha\right) \cdot \mathbf{I}_{\text{nucl}}^{-1} \cdot \mathbf{B},$$

$$t_1 t_2 + t_2 t_1 = -\frac{e\hbar}{m} \mathbf{J} \cdot \mathbf{A} \cdot \mathbf{I}_{\text{nucl}}^{-1} \cdot \mathbf{B}, \tag{8.112}$$

$$t_1 t_3 + t_3 t_1 = -e\hbar \, \mathbf{J} \cdot \mathbf{I}_{\text{nucl}}^{-1} \cdot \mathbf{A} \cdot \mathbf{I}_{\text{nucl}}^{-1} \cdot \mathbf{B} \cdot \sum_\alpha Z_\alpha \left(r_\alpha^2 \mathbf{1} - \mathbf{r}_\alpha \mathbf{r}_\alpha\right),$$

$$t_2 t_3 + t_3 t_2 = \frac{e^2}{2m} \mathbf{B} \cdot \mathbf{I}_{\text{nucl}}^{-1} \cdot \mathbf{A} \cdot \sum_\alpha Z_\alpha \left(r_\alpha^2 \mathbf{1} - \mathbf{r}_\alpha \mathbf{r}_\alpha\right) \cdot \mathbf{B}.$$

The first terms of (8.109) and (8.112) together constitute the rigid rotor Hamiltonian, as follows:

$$\mathcal{H}_{\text{rigid-rot}} = \frac{\hbar^2}{2} \mathbf{J} \cdot \mathbf{I}_{\text{nucl}}^{-1} \cdot \mathbf{J} + \hbar^2 \mathbf{J} \cdot \mathbf{I}_{\text{nucl}}^{-1} \cdot \mathbf{A} \cdot \mathbf{I}_{\text{nucl}}^{-1} \cdot \mathbf{J}$$

$$= \frac{\hbar^2}{2} \mathbf{J} \cdot \left[\mathbf{I}_{\text{nucl}}^{-1} \cdot \left(1 + 2\mathbf{A} \cdot \mathbf{I}_{\text{nucl}}^{-1}\right)\right] \cdot \mathbf{J}$$

$$= \frac{\hbar^2}{2} \mathbf{J} \cdot \mathbf{I}_{\text{eff}}^{-1} \cdot \mathbf{J}. \tag{8.113}$$

$\mathbf{I}_{\text{eff}}^{-1}$ is the effective inverse moment of inertia tensor that would be measured experimentally. The remaining first- and second-order terms coming from (8.109) and (8.112), combined with (8.113) give the rotational plus Zeeman Hamiltonian correct to the second order; it may be written

$$\mathcal{H} = \frac{\hbar^2}{2} \mathbf{J} \cdot \mathbf{I}_{\text{eff}}^{-1} \cdot \mathbf{J} - \mathbf{B} \cdot \left[\frac{e\hbar}{2} \sum_\alpha Z_\alpha \left(r_\alpha^2 \mathbf{1} - \mathbf{r}_\alpha \mathbf{r}_\alpha\right) \cdot \mathbf{I}_{\text{eff}}^{-1}\right] \cdot \mathbf{J} - \mathbf{B} \cdot \left(\frac{e\hbar}{m} \mathbf{A} \cdot \mathbf{I}_{\text{nucl}}^{-1}\right) \cdot \mathbf{J}$$

$$- \frac{1}{2} \mathbf{B} \cdot \boldsymbol{\chi}^d \cdot \mathbf{B} - \frac{1}{2} \mathbf{B} \cdot \boldsymbol{\chi}^p \cdot \mathbf{B} + \frac{1}{2} \mathbf{B} \cdot \boldsymbol{\gamma} \cdot \mathbf{B}. \tag{8.114}$$

This is the important result towards which we have been working throughout this section, particularly the terms involving χ^d and χ^p. These terms expanded are:

$$\chi^d = -\frac{e^2}{4m} \langle 0| \sum_i \left(r_i^2 \mathbf{1} - \mathbf{r}_i \mathbf{r}_i\right) |0\rangle \tag{8.115}$$

$$\chi^p = -\frac{e^2}{2m^2} \mathbf{A} = -\frac{e^2 \hbar^2}{2m^2} \sum_{n>0} \frac{\langle 0|\mathbf{L}|n\rangle \langle n|\mathbf{L}|0\rangle}{E_0 - E_n}, \tag{8.116}$$

where χ^d is the diamagnetic susceptibility. As equation (8.115) shows, it involves the electron distribution in the ground electronic state only. Similarly, χ^p is the paramagnetic susceptibility. Equation (8.116) shows that it arises from mixing of excited states with the ground state through the electronic orbital angular momentum; it is often referred to as the 'high-frequency paramagnetism' or 'high-temperature paramagnetism'.

The total magnetic susceptibility χ is the sum of the diamagnetic and paramagnetic parts, i.e.

$$\chi = \chi^d + \chi^p. \tag{8.117}$$

We shall return to the magnetic susceptibility later, to examine the simplifications which occur for a diatomic molecule, and to consider the form of this interaction used by Ramsey [52] in effective Hamiltonians for the analysis of magnetic resonance spectra.

Returning to equation (8.114) we find that the tensor γ is given by

$$\gamma = \frac{e^2}{2} \mathbf{I}_{\text{nucl}}^{-1} \cdot \sum_\alpha Z_\alpha \left(r_\alpha^2 \mathbf{1} - \mathbf{r}_\alpha \mathbf{r}_\alpha \right) \cdot \mathbf{A} \cdot \sum_\alpha Z_\alpha \left(r_\alpha^2 \mathbf{1} - \mathbf{r}_\alpha \mathbf{r}_\alpha \right) \cdot \mathbf{I}_{\text{nucl}}^{-1}$$

$$+ \frac{e^2}{m} \mathbf{I}_{\text{nucl}}^{-1} \cdot \mathbf{A} \cdot \sum_\alpha Z_\alpha \left(r_\alpha^2 \mathbf{1} - \mathbf{r}_\alpha \mathbf{r}_\alpha \right) + \frac{e^2}{4} \sum_\alpha \frac{Z_\alpha^2}{M_\alpha} \left(r_\alpha^2 \mathbf{1} - \mathbf{r}_\alpha \mathbf{r}_\alpha \right). \tag{8.118}$$

Each of the terms in this expression is smaller than the magnetic susceptibility terms by the electron to proton mass ratio, and may therefore be neglected. With this simplification equation (8.114) can be written in the more compact form

$$\mathcal{H} = \frac{\hbar^2}{2} \mathbf{J} \cdot \mathbf{I}_{\text{eff}}^{-1} \cdot \mathbf{J} - \mu_N \mathbf{B} \cdot \mathbf{g}_\text{r} \cdot \mathbf{J} - \frac{1}{2} \mathbf{B} \cdot \chi \cdot \mathbf{B}. \tag{8.119}$$

Here, μ_N is the nuclear magneton, equal to $e\hbar/2M_p$, where M_p is the proton mass. The first and third terms have already been discussed. The second term contains the molecular rotational magnetic moment $\mathbf{g}_\text{r}$ tensor, and equation (8.114) shows that it consists of the sum of a nuclear and electronic contribution, i.e.

$$\mathbf{g}_\text{r} = \mathbf{g}_{\text{nucl}} + \mathbf{g}_{\text{el}}. \tag{8.120}$$

From equation (8.114) we see that the nuclear contribution is given by

$$\mathbf{g}_{\text{nucl}} = M_p \sum_\alpha Z_\alpha \left(r_\alpha^2 \mathbf{1} - \mathbf{r}_\alpha \mathbf{r}_\alpha \right) \cdot \mathbf{I}_{\text{eff}}^{-1}, \tag{8.121}$$

whilst the electronic contribution is

$$\mathbf{g}_{\text{el}} = \frac{2M_p \hbar^2}{m} \mathbf{I}_{\text{eff}}^{-1} \cdot \sum_{n>0} \frac{\langle 0|\mathbf{L}|n\rangle \langle n|\mathbf{L}|0\rangle}{E_0 - E_n}. \tag{8.122}$$

The nuclear contribution to the $\mathbf{g}_\text{r}$ tensor is positive, but the electronic contribution is negative because $E_0 < E_n$ when the molecule is in its ground electronic state.

Up to this stage our discussion applies to any polyatomic molecule. We now look at the simplifications which occur for a diatomic system. Dealing first with the diamagnetic part of the susceptibility, given in (8.115), we note from (8.70) that if z lies along the internuclear axis, the x and y components of the susceptibility are equivalent, but different from the parallel component. To be specific,

$$\chi^d_{xx} = \chi^d_{yy} = -\frac{e^2}{4m}\langle 0|\sum_i (x_i^2 + z_i^2)|0\rangle \equiv \chi_\sigma \equiv \chi_{\text{perp}}, \tag{8.123}$$

$$\chi^d_{zz} = -\frac{e^2}{4m}\langle 0|\sum_i (x_i^2 + y_i^2)|0\rangle \equiv \chi_\pi \equiv \chi_{\text{par}}. \tag{8.124}$$

The subscripts σ and π are used by Ramsey [52] and others, but we use the more explicit subscripts *parallel* and *perpendicular*.

The paramagnetic part of the susceptibility tensor, given in equation (8.116), has a zero component in the z direction because it depends upon the mixing brought about by the x and y components of the orbital angular momentum **L**. Hence

$$\chi^p_{xx} = \chi^p_{yy} = -\frac{e^2\hbar^2}{2m^2}\sum_{n>0}\frac{\langle 0|L_x|n\rangle\langle n|L_x|0\rangle}{E_0 - E_n} = \frac{3}{2}\chi^{\text{HF}}, \tag{8.125}$$

$$\chi^p_{zz} = 0 = \chi^p_{\text{par}}. \tag{8.126}$$

The net result for the total magnetic susceptibility is

$$\chi_{\text{perp}} = \chi^d_{\text{perp}} + \frac{3}{2}\chi^{\text{HF}}, \tag{8.127}$$

$$\chi_{\text{par}} = \chi^d_{\text{par}}. \tag{8.128}$$

Somewhat similar conclusions apply to the rotational magnetic moment **g** tensor for a diatomic molecule. The component of the moment of inertia tensor along the internuclear axis is zero, and the two perpendicular components are, of course, equal. Consequently the rotational magnetic moment Zeeman interaction can be represented by the simple term

$$\mathcal{H} = -g_J\mu_N \mathbf{B}\cdot\mathbf{J}. \tag{8.129}$$

The rotational magnetic moment of a diatomic molecule is defined by

$$\boldsymbol{\mu}_J = g_J\mu_N \mathbf{J}, \tag{8.130}$$

and Wick [16] and Ramsey [17] showed that there is a relationship between the high-frequency paramagnetism and the rotational magnetic moment, expressed as follows:

$$\chi^{\text{HF}} = \frac{e^2 R^2}{12m}\left\{\frac{2Z_1 Z_2}{(Z_1 + Z_2)} + 2(Z_1 + Z_2)\frac{(D^2 - d^2)}{R^2} - \frac{2\mu_J\mu'}{M_p J \mu_N}\right\}. \tag{8.131}$$

The various quantities in this expression are:

Z_1, Z_2: nuclear charges,
R: internuclear distance,
D: distance between the centre-of-mass and the centroid of the nuclear charge distribution,

d: distance between the centre-of-mass and the centroid of the electronic charge distribution,
μ': reduced mass of the molecule,
M_p: proton mass.

For a homonuclear molecule, $D = d$, so that the second term in (8.131) vanishes. Corrections to (8.131) for molecular vibration and centrifugal stretching have been given by Ramsey [18]. The above result means that if the rotational magnetic moment μ_J is measured, the high-frequency part of the diamagnetic susceptibility can be determined.

(v) REPRESENTATION OF THE DIAMAGNETIC SUSCEPTIBILITY IN THE EFFECTIVE HAMILTONIAN

Before leaving this aspect of the subject we must see how the term used by Ramsey [52] to describe the magnetic susceptibility in the effective Hamiltonian (8.15) arises. In what follows, the direction of the magnetic field defines the $p = 0$ space-fixed direction, but the components of the magnetic susceptibility tensor are defined in the molecule-fixed axis system (q). Note that $\boldsymbol{B}$ and $\mu_0 \boldsymbol{H}$ are equivalent in a vacuum.

We can write the diamagnetic Zeeman term as the scalar product of two second-rank tensors:

$$\mathcal{H}_{\text{diam}} = -\frac{1}{2} T^2(\chi) \cdot T^2(\boldsymbol{B}, \boldsymbol{B}) \tag{8.132}$$

where the space-fixed components of the new second-rank tensor $T^2(\boldsymbol{B}, \boldsymbol{B})$ are defined by

$$T^2_p(\boldsymbol{B}, \boldsymbol{B}) = (-1)^p \sqrt{5} \sum_{p_1, p_2} \begin{pmatrix} 1 & 1 & 2 \\ p_1 & p_2 & -p \end{pmatrix} T^1_{p_1}(\boldsymbol{B}) T^1_{p_2}(\boldsymbol{B}). \tag{8.133}$$

Since the direction of the magnetic field defines the $p = p_1 = p_2 = 0$ direction, (8.133) reduces to

$$T^2_0(\boldsymbol{B}, \boldsymbol{B}) = (2/3)^{1/2} B_Z^2. \tag{8.134}$$

Consequently (8.132) may be rewritten in the space-fixed axis system

$$\mathcal{H}_{\text{diam}} = -\frac{1}{2} T^2_0(\chi) T^2_0(\boldsymbol{B}, \boldsymbol{B}) = -\frac{1}{\sqrt{6}} B_Z^2 T^2_0(\chi). \tag{8.135}$$

The components of the magnetic susceptibility tensor are defined in the molecule-fixed axes system so that we use the transformation

$$T^2_{p=0}(\chi) = \sum_q \mathcal{D}^{(2)}_{0q}(\omega)^* T^2_q(\chi). \tag{8.136}$$

The matrix elements of (8.132) are now readily calculated:

$$\langle \eta, \Lambda; J, M_J | \mathcal{H}_{\text{diam}} | \eta, \Lambda'; J', M_J \rangle$$
$$= -\frac{1}{\sqrt{6}} B_Z^2 (-1)^{J-M_J} \begin{pmatrix} J & 2 & J' \\ -M_J & 0 & M_J \end{pmatrix} \sum_q (-1)^{J-\Lambda} \begin{pmatrix} J & 2 & J' \\ -\Lambda & q & \Lambda' \end{pmatrix}$$
$$\times \{(2J'+1)(2J+1)\}^{1/2} \langle T_q^2(\chi) \rangle. \tag{8.137}$$

Since we are concerned only with the $^1\Sigma$ ground state we take the $q = 0$ component of (8.137) and obtain the result

$$\langle \eta, \Lambda; J, M_J | \mathcal{H}_{\text{diam}} | \eta, \Lambda; J, M_J \rangle = -\frac{1}{\sqrt{6}} B_Z^2 \frac{\{3M_J^2 - J(J+1)\}}{(2J+3)(2J-1)} \langle T_0^2(\chi) \rangle_\eta. \tag{8.138}$$

Now the spherical component $T_0^2(\chi)$ is related to the cartesian components by

$$T_0^2(\chi) = \frac{1}{\sqrt{6}} \{2\chi_{zz} - \chi_{xx} - \chi_{yy}\} = \sqrt{\frac{2}{3}} (\chi_{\text{par}} - \chi_{\text{perp}}). \tag{8.139}$$

Substituting in (8.138)

$$\langle \eta, \Lambda; J, M_J | \mathcal{H}_{\text{diam}} | \eta, \Lambda; J, M_J \rangle = -B_Z^2 \frac{\{3M_J^2 - J(J+1)\}}{3(2J-1)(2J+3)} (\chi_{\text{par}} - \chi_{\text{perp}}). \tag{8.140}$$

This is the result used by Ramsey [52] for the Zeeman interaction involving the diamagnetic part of the susceptibility.

(vi) MOLECULAR QUADRUPOLE MOMENTS

From equations (8.123), (8.124) (8.127) and (8.128), together with the knowledge that, because of symmetry, $\langle x_i^2 \rangle$ and $\langle y_i^2 \rangle$ are equivalent, we can derive the following important result:

$$\chi_{\text{perp}} - \chi_{\text{par}} = -\frac{e^2}{8m} \sum_i (3z_i^2 - r_i^2) + \frac{3}{2} \chi^{\text{HF}}. \tag{8.141}$$

Now the quadrupole moment of the electron distribution in a molecule, Q_{el}, is defined by

$$Q_{\text{el}} = -e \left\langle \sum_i (3z_i^2 - r_i^2) \right\rangle = \frac{8m}{e} \left\{ \chi_{\text{perp}} - \chi_{\text{par}} - \frac{3}{2} \chi^{\text{HF}} \right\}. \tag{8.142}$$

Since χ^{HF} can be determined from the rotational magnetic moment, and the anisotropy of the susceptibility from the Zeeman effect, we are able to determine the molecular electronic quadrupole moment, as was first shown by Ramsey [19]. The total electric quadrupole moment of a molecule is the sum of the electronic contribution Q_{el} and a nuclear contribution Q_{nucl}; the latter contribution is given by

$$Q_{\text{nucl}} = e \sum_\alpha Z_\alpha r_\alpha^2, \tag{8.143}$$

where, for a diatomic molecule, the summation is over the two nuclei. The application of these results to H_2 and its deuterium isotopes will be described later.

(d) Nuclear spin effects: magnetic shielding and nuclear spin–rotation interaction

The rotational and Zeeman perturbation Hamiltonian ($\mathcal{H}'$) to the electronic eigenstates was given in equation (8.105). It did not, however, contain terms which describe the interaction effects arising from nuclear spin. These are of primary importance in molecular beam magnetic resonance studies, so we must now extend our treatment and, in particular, demonstrate the origin of the terms in the effective Hamiltonian already employed to analyse the spectra. Again the treatment will apply to any molecule, but we shall subsequently restrict attention to diatomic systems.

The coordinates used to describe the position vectors of the electrons and nuclei are shown in figure 8.12. The internal magnetic field intensity at the kth nucleus, $\boldsymbol{B}^k$, arises from the sum of electronic, $\boldsymbol{B}_{\text{el}}^k$, and nuclear, $\boldsymbol{B}_{\text{nucl}}^k$, terms given by the following expression:

$$\boldsymbol{B}^k = \boldsymbol{B}_{\text{el}}^k + \boldsymbol{B}_{\text{nucl}}^k = -e \sum_i \frac{\boldsymbol{r}_{ik} \wedge \boldsymbol{v}_{ik}}{r_{ik}^3} + e \sum_\alpha Z_\alpha \frac{\boldsymbol{R}_{\alpha k} \wedge \boldsymbol{v}_{\alpha k}}{R_{\alpha k}^3}. \qquad (8.144)$$

The vectors $\boldsymbol{r}_{ik}$ and $\boldsymbol{R}_{\alpha k}$ give the position of electron i or nucleus α relative to the kth nucleus (figure 8.12), and the sums are over all electrons i in the molecule, and all *other* nuclei α.

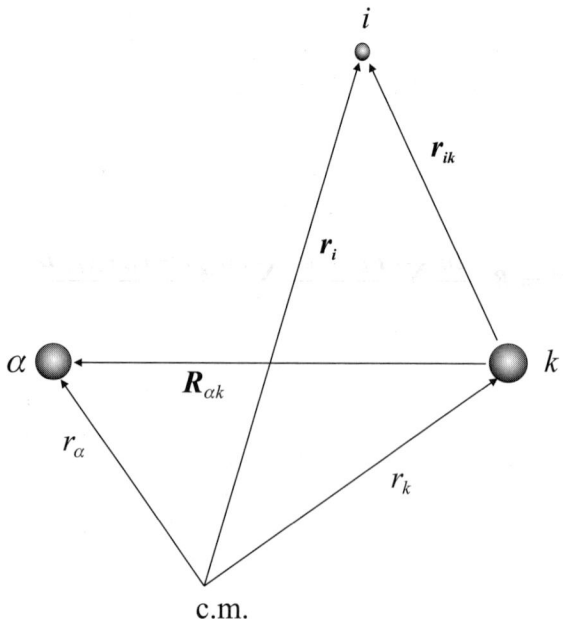

Figure 8.12. Coordinate system defining the vector positions of the electrons (i) and nuclei (k or α), relative to the centre-of-mass.

The nuclear term in (8.144) involves only nuclear coordinates in a rigid molecule, so that we may substitute for $v_{\alpha k}$ as follows:

$$\begin{aligned}
\boldsymbol{B}_{\text{nucl}}^k &= e \sum_\alpha Z_\alpha \frac{\boldsymbol{R}_{\alpha k} \wedge (\boldsymbol{\omega} \wedge \boldsymbol{R}_{\alpha k})}{R_{\alpha k}^3} = e \sum_\alpha \frac{Z_\alpha}{R_{\alpha k}^3} \left(R_{\alpha k}^2 \mathbf{1} - \boldsymbol{R}_{\alpha k} \boldsymbol{R}_{\alpha k} \right) \cdot \boldsymbol{\omega} \\
&= e\hbar \sum_\alpha \frac{Z_\alpha}{R_{\alpha k}^3} \left(R_{\alpha k}^2 \mathbf{1} - \boldsymbol{R}_{\alpha k} \boldsymbol{R}_{\alpha k} \right) \cdot \boldsymbol{I}_{\text{nucl}}^{-1} \cdot (\boldsymbol{J} - \boldsymbol{L}).
\end{aligned} \quad (8.145)$$

We have here substituted for $\boldsymbol{\omega}$ using the result given in equation (8.100).

The next stage is to add the effects of an applied magnetic field. The velocity of an electron is modified according to the equation

$$v_{ik} = \frac{\boldsymbol{p}_{ik}}{m} + \frac{e}{m} \boldsymbol{A}_{ik}, \quad (8.146)$$

where the vector potential arising from the uniform magnetic field is given by

$$\boldsymbol{A}_{ik} = \frac{1}{2} \boldsymbol{B} \wedge \boldsymbol{r}_{ik}. \quad (8.147)$$

The origin is thus chosen to be at the kth nucleus. Using equation (8.146) and defining the orbital angular momentum of the ith electron with respect to the kth nucleus as origin by

$$\hbar \boldsymbol{l}_{ik} = \boldsymbol{r}_{ik} \wedge \boldsymbol{p}_{ik}, \quad (8.148)$$

we can rewrite the first term in (8.144) in the following manner:

$$\begin{aligned}
\boldsymbol{B}_{\text{el}}^k &= -e \sum_i \frac{\boldsymbol{r}_{ik} \wedge \boldsymbol{v}_{ik}}{r_{ik}^3} = -e \sum_i \frac{\boldsymbol{r}_{ik} \wedge [\boldsymbol{p}_{ik}/m]}{r_{ik}^3} - \frac{e^2}{2m} \sum_i (\boldsymbol{r}_{ik} \wedge \boldsymbol{B} \wedge \boldsymbol{r}_{ik}) \\
&= -\frac{e\hbar}{m} \sum_i \frac{\boldsymbol{l}_{ik}}{r_{ik}^3} - \frac{e^2}{2m} \sum_i \frac{(r_{ik}^2 \mathbf{1} - \boldsymbol{r}_{ik} \boldsymbol{r}_{ik}) \cdot \boldsymbol{B}}{r_{ik}^3}.
\end{aligned} \quad (8.149)$$

If we now combine (8.145), (8.149) with the applied magnetic field $\boldsymbol{B}$, we obtain an expression for the total field intensity at the nucleus k, which is

$$\begin{aligned}
\boldsymbol{B}^k &= \boldsymbol{B} - \frac{e\hbar}{m} \sum_i \frac{\boldsymbol{l}_{ik}}{r_{ik}^3} - \frac{e^2}{2m} \sum_i \frac{(r_{ik}^2 \mathbf{1} - \boldsymbol{r}_{ik} \boldsymbol{r}_{ik}) \cdot \boldsymbol{B}}{r_{ik}^3} \\
&\quad + e\hbar \sum_\alpha \frac{Z_\alpha}{R_{\alpha k}^3} \left(R_{\alpha k}^2 \mathbf{1} - \boldsymbol{R}_{\alpha k} \boldsymbol{R}_{\alpha k} \right) \cdot \boldsymbol{I}_{\text{nucl}}^{-1} \cdot (\boldsymbol{J} - \boldsymbol{L}).
\end{aligned} \quad (8.150)$$

We are now able to write an expression describing the interaction between the magnetic dipole moment of nucleus k and the total field at the nucleus. It is as follows:

$$\begin{aligned}
\mathcal{H}_{\text{nucl}} &= -\boldsymbol{\mu}_k \cdot \boldsymbol{B}^k = -\gamma_k \boldsymbol{I}_k \cdot \boldsymbol{B}^k = -\mu_N g_k \boldsymbol{I}_k \cdot \boldsymbol{B}^k \\
&= -\gamma_k \boldsymbol{I}_k \cdot \boldsymbol{B} + \frac{e\hbar}{m} \gamma_k \boldsymbol{I}_k \cdot \sum_i \frac{\boldsymbol{l}_{ik}}{r_{ik}^3} + \frac{e^2}{2m} \gamma_k \boldsymbol{I}_k \cdot \sum_i \frac{(r_{ik}^2 \mathbf{1} - \boldsymbol{r}_{ik} \boldsymbol{r}_{ik}) \cdot \boldsymbol{B}}{r_{ik}^3} \\
&\quad - e\hbar \gamma_k \boldsymbol{I}_k \cdot \sum_\alpha \frac{Z_\alpha}{R_{\alpha k}^3} \left(R_{\alpha k}^2 \mathbf{1} - \boldsymbol{R}_{\alpha k} \boldsymbol{R}_{\alpha k} \right) \cdot \boldsymbol{I}_{\text{nucl}}^{-1} \cdot (\boldsymbol{J} - \boldsymbol{L}).
\end{aligned} \quad (8.151)$$

γ_k is the magnetogyric ratio of the kth nucleus, and is equal to $g_k \mu_N$. We may now add (8.151) to (8.105) to obtain the complete perturbation Hamiltonian:

$$\mathcal{H} = \mathcal{H}' + \mathcal{H}_{\text{nucl}} = \frac{\hbar^2}{2} \boldsymbol{J} \cdot \boldsymbol{I}_{\text{nucl}}^{-1} \cdot \boldsymbol{J} + \frac{e^2}{8m} \boldsymbol{B} \cdot \left[\sum_i (r_i^2 \boldsymbol{1} - \boldsymbol{r}_i \boldsymbol{r}_i) \right.$$

$$\left. + m \sum_\alpha \frac{Z_\alpha^2}{M_\alpha} (r_\alpha^2 \boldsymbol{1} - \boldsymbol{r}_\alpha \boldsymbol{r}_\alpha) \right] \cdot \boldsymbol{B} - \frac{e\hbar}{2} \sum_\alpha Z_\alpha \boldsymbol{B} \cdot (r_\alpha^2 \boldsymbol{1} - \boldsymbol{r}_\alpha \boldsymbol{r}_\alpha) \cdot \boldsymbol{I}_{\text{nucl}}^{-1} \cdot \boldsymbol{J}$$

$$+ \frac{\hbar^2}{2} \boldsymbol{L} \cdot \boldsymbol{I}_{\text{nucl}}^{-1} \cdot \boldsymbol{L} - \hbar \boldsymbol{L}_k \cdot \bar{\omega} - \gamma_k \boldsymbol{I}_k \cdot \boldsymbol{B} + \frac{e\hbar}{m} \gamma_k \boldsymbol{I}_k \cdot \sum_i \frac{\boldsymbol{l}_{ik}}{r_{ik}^3}$$

$$+ \frac{e^2}{2m} \gamma_k \boldsymbol{I}_k \cdot \sum_i \frac{(r_{ik}^2 \boldsymbol{1} - \boldsymbol{r}_{ik} \boldsymbol{r}_{ik}) \cdot \boldsymbol{B}}{r_{ik}^3}$$

$$- e\hbar \gamma_k \boldsymbol{I}_k \cdot \sum_\alpha \frac{Z_\alpha}{R_{\alpha k}^3} (R_{\alpha k}^2 \boldsymbol{1} - \boldsymbol{R}_{\alpha k} \boldsymbol{R}_{\alpha k}) \cdot \boldsymbol{I}_{\text{nucl}}^{-1} \cdot (\boldsymbol{J} - \boldsymbol{L}), \quad (8.152)$$

where $\bar{\omega}$ is defined in equation (8.106) and $\hbar \boldsymbol{L}_k$ is the total electronic angular momentum with the origin at the kth nucleus.

We now follow exactly the same procedure as was used previously to calculate the first- and second-order perturbations to the zeroth-order electronic states, given in (8.114). We calculate the additional terms which arise from interactions involving the kth nuclear spin; the first-order corrections are

$$E^{(1)} = -\gamma_k \boldsymbol{I}_k \cdot \boldsymbol{B} + \frac{e^2}{2m} \gamma_k \boldsymbol{I}_k \cdot \langle 0| \sum_i \frac{(r_{ik}^2 \boldsymbol{1} - \boldsymbol{r}_{ik} \boldsymbol{r}_{ik})}{r_{ik}^3} |0\rangle \cdot \boldsymbol{B}$$

$$- e\hbar \gamma_k \boldsymbol{I}_k \cdot \sum_\alpha \frac{Z_\alpha}{R_{\alpha k}^3} (R_{\alpha k}^2 \boldsymbol{1} - \boldsymbol{R}_{\alpha k} \boldsymbol{R}_{\alpha k}) \boldsymbol{I}_{\text{nucl}}^{-1} \cdot \boldsymbol{J}. \quad (8.153)$$

The second-order corrections are

$$E^{(2)} = \frac{e^2 \hbar^2}{2m^2} \gamma_k \boldsymbol{I}_k \cdot \left\{ \sum_{n>0} \frac{\langle 0| \sum_i \boldsymbol{l}_{ik}/r_{ik}^3 |n\rangle \langle n| \boldsymbol{L}_k |0\rangle + \text{c.c.}}{E_0 - E_n} \right\} \cdot \boldsymbol{B}$$

$$- 2e\hbar \gamma_k \boldsymbol{I}_k \cdot \sum_\alpha \frac{Z_\alpha}{R_{\alpha k}^3} (R_{\alpha k}^2 \boldsymbol{1} - \boldsymbol{R}_{\alpha k} \boldsymbol{R}_{\alpha k}) \cdot \boldsymbol{I}_{\text{nucl}}^{-1} \boldsymbol{A} \cdot \boldsymbol{I}_{\text{nucl}}^{-1} \cdot \boldsymbol{J}$$

$$- e\hbar^3 \gamma_k \boldsymbol{I}_k \left\{ \sum_{n>0} \frac{\langle 0| \sum_i \boldsymbol{l}_{ik}/r_{ik}^3 |n\rangle \langle n| \boldsymbol{L}_k |0\rangle + \text{c.c.}}{E_0 - E_n} \right\} \cdot \boldsymbol{I}_{\text{nucl}}^{-1} \cdot \boldsymbol{J} \ldots, \quad (8.154)$$

where c.c. means the complex conjugate of the preceding term. We now combine the first- and second-order terms, change $\boldsymbol{I}_{\text{nucl}}^{-1}$ to $\boldsymbol{I}_{\text{eff}}^{-1}$ as described in equation (8.113), and obtain the important result:

$$\mathcal{H} = -\gamma_k \boldsymbol{I}_k \cdot (\boldsymbol{1} - \boldsymbol{\sigma}) \cdot \boldsymbol{B} + \boldsymbol{I}_k \cdot \boldsymbol{c}_I \cdot \boldsymbol{J}. \quad (8.155)$$

We have finally reached our goal. The first term in (8.155) describes the nuclear Zeeman interaction, and was introduced in equation (8.4). σ is called the 'shielding', 'screening', or 'chemical shift' tensor at the kth nucleus; we shall examine it in detail shortly. c_I is the nuclear spin–rotation tensor for the kth nucleus, first introduced in equation (8.7).

The magnetic shielding tensor σ is the sum of a diamagnetic part σ^d and a paramagnetic part σ^p,

$$\sigma = \sigma^d + \sigma^p. \tag{8.156}$$

As we can readily see, the diamagnetic part comes from the above first-order correction (8.153), so that σ^d for the kth nucleus is given by

$$\sigma^d = \frac{e^2}{2m} \langle 0| \sum_i \frac{\left(r_{ik}^2 \mathbf{1} - \mathbf{r}_{ik}\mathbf{r}_{ik}\right)}{r_{ik}^3} |0\rangle. \tag{8.157}$$

It is, of course, very closely related to the diamagnetic part of the magnetic suceptibility tensor, given previously in (8.115), and depends only upon the ground state electron distribution. The paramagnetic part of the shielding tensor for the kth nucleus, σ^p, arises from the first term of the second-order correction, given in equation (8.154):

$$\sigma^p = \frac{e^2\hbar^2}{2m^2} \sum_{n>0} \frac{\langle 0|\sum_i \mathbf{l}_{ik}/r_{ik}^3|n\rangle \langle n|\mathbf{L}_k|0\rangle + c.c.}{E_0 - E_n}. \tag{8.158}$$

Again this is closely related to the paramagnetic part of the susceptibility, given in equation (8.116), involving mixing of the ground electronic state with excited states, through the orbital angular momentum operator. The diamagnetic term is positive, thereby decreasing the effective field at the nucleus, whereas the paramagnetic term is negative and increases the effective field. The observed screening is, of course, a net balance of these two opposing effects.

We now turn to the nuclear spin–rotation tensor c_I for the kth nucleus in equation (8.155). This contains a purely nuclear contribution, arising from the third term of the first-order expression (8.153), and a purely electronic contribution coming from the third term of the second-order expression (8.154). To be specific:

$$c_I = c_{\text{nucl}} + c_{\text{el}}, \tag{8.159}$$

where

$$c_{\text{nucl}} = -e\gamma_k \hbar \sum_\alpha \frac{Z_\alpha}{R_{\alpha k}^3} \left(R_{\alpha k}^2 \mathbf{1} - \mathbf{R}_{\alpha k}\mathbf{R}_{\alpha k}\right) \cdot \mathbf{I}_{\text{eff}}^{-1}, \tag{8.160}$$

$$c_{\text{el}} = -\frac{2e}{m}\gamma_k \hbar^3 \left\{ \sum_{n>0} \frac{\langle 0|\sum_i \mathbf{l}_{ik}/r_{ik}^3|n\rangle \langle n|\mathbf{L}_k|0\rangle}{E_0 - E_n} \right\} \cdot \mathbf{I}_{\text{eff}}^{-1}. \tag{8.161}$$

The nuclear term is negative, and the electronic term is positive.

If x, y and z are the principal axes of the molecule, the shielding and nuclear spin–rotation tensors for the kth nucleus are diagonal. The xx components, for example, are readily obtained from the general expressions and are found to be

$$\sigma_{xx} = \sigma_{xx}^d + \sigma_{xx}^p = \frac{e^2}{2m}\langle 0|\sum_i \frac{(y_{ik}^2 + z_{ik}^2)}{r_{ik}^3}|0\rangle + \frac{e^2\hbar^2}{2m^2}$$
$$\times \sum_{n>0}\left[\frac{\langle 0|\sum_i (l_{ik})_x/r_{ik}^3|n\rangle\langle n|(L_k)_x|0\rangle + \langle 0|(L_k)_x|n\rangle\langle n|\sum_i (l_{ik})_x/r_{ik}^3|0\rangle}{E_0 - E_n}\right],$$
(8.162)

$$(c_I)_{xx} = (c_I)_{xx(\text{nucl})} + (c_I)_{xx(\text{el})} = -\frac{\hbar e g_k \mu_N}{I_{xx}}\sum_\alpha \frac{Z_\alpha}{R_{\alpha k}^3}\left(R_{\alpha k}^2 - x_{\alpha k}^2\right) - \frac{\hbar^3 e g_k \mu_N}{m I_{xx}}$$
$$\times \sum_{n>0}\left[\frac{\langle 0|\sum_i (l_{ik})_x/r_{ik}^3|n\rangle\langle n|(L_k)_x|0\rangle + \langle 0|(L_k)_x|n\rangle\langle n|\sum_i (l_{ik})_x/r_{ik}^3|0\rangle}{E_0 - E_n}\right],$$
(8.163)

The yy and zz components are obtained by cyclic permutations.

Earlier in this chapter when dealing with the nuclear Zeeman interaction we calculated the behaviour of the nuclear spin levels of H_2 ignoring the effects of nuclear shielding. We now return to this question in more detail. We have shown that the Zeeman interaction for a nucleus of spin I should be written in the form

$$\mathcal{H} = -\mu_N g_N \mathbf{I} \cdot (\mathbf{1} - \boldsymbol{\sigma}) \cdot \mathbf{B} = -\mu_N g_N \mathbf{I} \cdot \mathbf{B} + \mu_N g_N \mathbf{I} \cdot \boldsymbol{\sigma} \cdot \mathbf{B}, \quad (8.164)$$

where $\boldsymbol{\sigma}$ is the second-rank shielding tensor. The matrix elements of the first term are straightforward, but those of the shielding term are not. There are several ways of handling the problem, but we use the same method as that described above for the susceptibility tensor, choosing to define a new second-rank tensor, called the spin-field tensor [20], so that we make the following replacement:

$$\mu_N g_N \mathbf{I} \cdot \boldsymbol{\sigma} \cdot \mathbf{B} = \mu_N g_N \mathrm{T}^2(\boldsymbol{\sigma}) \cdot \mathrm{T}^2(\mathbf{I}, \mathbf{B}). \quad (8.165)$$

We now have the scalar product of two second-rank spherical tensors. The space-fixed components of the new tensor are defined by

$$\mathrm{T}_p^2(\mathbf{I}, \mathbf{B}) = (-1)^p \sqrt{5} \sum_{p_1, p_2} \begin{pmatrix} 1 & 1 & 2 \\ p_1 & p_2 & -p \end{pmatrix} \mathrm{T}_{p_1}^1(\mathbf{I})\mathrm{T}_{p_2}^1(\mathbf{B}). \quad (8.166)$$

The direction of the external magnetic field defines the $p_2 = 0$ direction, and since we are interested in the strong field behaviour, where M_I is a good quantum number, we also take $p_1 = 0$. Consequently $p = 0$ and the matrix elements we seek (in the space-fixed axis system) are of the form

$$\langle \eta, \Lambda; J, M_J; I, M_I | \mu_N g_N T_0^2(\sigma) T_0^2(\boldsymbol{I}, \boldsymbol{B}) | \eta, \Lambda'; J', M_J'; I, M_I \rangle$$

$$= \sqrt{\frac{2}{3}} \mu_N g_N B_Z M_I \langle \eta, \Lambda; J, M_J | \sum_q \mathcal{D}_{0q}^{(2)}(\omega)^* T_q^2(\sigma) | \eta, \Lambda'; J', M_J' \rangle$$

$$= \sqrt{\frac{2}{3}} \mu_N g_N B_Z M_I (-1)^{J-M_J} \begin{pmatrix} J & 2 & J' \\ -M_J & 0 & M_J \end{pmatrix} \sum_q (-1)^{J-\Lambda} \begin{pmatrix} J & 2 & J' \\ -\Lambda & q & \Lambda' \end{pmatrix}$$

$$\times \{(2J+1)(2J'+1)\}^{1/2} \langle T_q^2(\sigma) \rangle. \tag{8.167}$$

We are concerned with the $^1\Sigma$ ground state only, so that $\Lambda' = \Lambda = 0$, $q = 0$, and the important matrix elements are those diagonal in J. Equation (8.167) therefore reduces to

$$\langle \eta, \Lambda; J, M_J; I, M_I | \mu_N g_N T_0^2(\sigma) T_0^2(\boldsymbol{I}, \boldsymbol{B}) | \eta, \Lambda; J, M_J; I, M_I \rangle$$

$$= \sqrt{\frac{2}{3}} \mu_N g_N B_Z M_I \frac{\{3M_J^2 - J(J+1)\}}{(2J+3)(2J-1)} T_0^2(\sigma)$$

$$= \sqrt{\frac{2}{3}} \mu_N g_N B_Z M_I \frac{\{3M_J^2 - J(J+1)\}}{(2J+3)(2J-1)} \frac{1}{\sqrt{6}} \{2\sigma_{zz} - \sigma_{xx} - \sigma_{yy}\}$$

$$= B_Z M_I \mu_N g_N \frac{2\{3M_J^2 - J(J+1)\}}{3(2J+3)(2J-1)} (\sigma_{\text{par}} - \sigma_{\text{perp}}). \tag{8.168}$$

This is the result first obtained by Ramsay [21]. We note finally that an important quantity, measured in isotropic liquid phase n.m.r. studies, is the average shielding factor σ_{av}. This is given by the trace of the shielding tensor σ, i.e.

$$\sigma_{\text{av}} = \frac{1}{3}(\sigma_{xx} + \sigma_{yy} + \sigma_{zz}) = \frac{1}{3}\sigma_{\text{par}} + \frac{2}{3}\sigma_{\text{perp}}. \tag{8.169}$$

The nuclear spin–rotation interaction becomes very simple for a diatomic molecule. The principal components of the tensor $\boldsymbol{c}_I$ for a polyatomic molecule were described in equation (8.163); this expression reveals that for a diatomic system the axial component $(c_I)_{zz}$ is zero and, of course, the two perpendicular components are equal. The nuclear spin–rotation interaction for a diatomic molecule is therefore described by a single parameter c_I. The appropriate term in the effective Hamiltonian, first presented in equation (8.7), is

$$\mathcal{H}_{JI} = c_I \, \mathrm{T}^1(\boldsymbol{J}) \cdot \mathrm{T}^1(\boldsymbol{I}). \tag{8.170}$$

The major interaction between the nuclear spin magnetic moments in H_2 and D_2 is the dipolar interaction, equation (8.10). We should at least mention the existence of an electron-coupled scalar interaction; this is very small compared with the dipolar interaction, and plays a very minor role in the gas phase measurements. In liquids, however, the dipolar interaction averages to zero, and the scalar coupling becomes the important observable interaction between nuclear spins. The power and range of applications of high-resolution n.m.r. in liquids depends ultimately upon the scalar shielding and spin–spin interactions.

Table 8.3. *Measured values of selected molecular parameters for H_2 and D_2. All of the symbols are defined in the text except for B', which is the rotational magnetic flux density at the nucleus*

Parameter	Units	H_2	D_2
c_I	kHz	−113.904	−8.788
d	kHz	57.671	25.237
σ_{av}	—	2.62×10^{-5}	2.63×10^{-5}
σ^{HF}	—	-0.59×10^{-5}	—
$\chi_{perp} - \chi_{par}$	J G^{-2} mol^{-1}	-9.15×10^{-38}	-8.75×10^{-38}
χ^{HF}	J G^{-2} mol^{-1}	1.719×10^{-38}	1.622×10^{-38}
f/B^2	Hz G^{-2}	-27.6×10^{-6}	-26.2×10^{-6}
$-\langle Q_{el}\rangle/e$	m^2	0.333×10^{-20}	0.318×10^{-20}
$\langle r^2 \rangle$	m^2	0.7258×10^{-20}	—
$\langle x^2 \rangle = \langle y^2 \rangle$	m^2	0.2144×10^{-20}	—
$\langle z^2 \rangle$	m^2	0.2969×10^{-20}	—
$eq_0 Q(D)$	kHz	—	224.992
B'	G	26.73	13.44

(e) Values of the molecular constants

From the numerous studies of the magnetic resonance spectra of H_2 and D_2, described in the preceding pages, values of the molecular parameters have been obtained. We summarise the results for the main parameters in table 8.3. Many *ab initio* calculations of these parameters have been described.

8.2.3. Na_2 in the $X\,^1\Sigma_g^+$ ground state: optical state selection and detection

The magnetic resonance experiments described thus far depend upon magnetic field gradients (A and B fields) to produce quantum state selection before spectroscopic transitions are induced, and to reselect afterwards. In an important and far-reaching experiment, Rosner, Holt and Gaily [22] showed that in favourable cases the A and B magnetic fields can be replaced by laser optical pumping. Their studies were carried out on a supersonic beam of Na_2 molecules produced from an oven operated at 620 °C. The optical spectrum of Na_2 arising from the $B\,^1\Pi_u \leftarrow X\,^1\Sigma_g^+$ transition was well known and understood, and Rosner, Holt and Gaily noted that the B, $v'=6, J=27 \leftarrow X, v''=0, J''=28$ rovibronic component is coincident with the 476.5 nm line from an Ar$^+$ laser. They therefore crossed the molecular beam with the laser beam in the region normally occupied by the A field, exciting the optical transition. The absorption probability is M-dependent so that the component sub-states of the $v''=0, J''=28$ rotational level are unequally populated after the molecule has traversed the A crossing; so far as this particular rotational level is concerned, the molecules in the beam are spatially aligned. In the B region, the molecular beam is again crossed by the same laser beam, which excites the same vibration–rotation level, producing fluorescence from the upper rovibronic state which is detected. A radiofrequency field in the C region redistributes the molecules among the various sublevels of the lower level; those sublevels with high absorption probabilities are repopulated and hence the fluorescence at the B region increases. Amplitude modulation of the radiofrequency power results in modulation of the fluorescence intensity, which could be detected with high sensitivity.

This experiment is, in some ways, a double resonance experiment and should perhaps be described in chapter 11. We include it here because optical detection with molecular beams has become a general and powerful technique, which we shall meet again in this chapter. Optical pumping and state selection is also a general approach. In the particular case of the Na_2 study it depended initially upon a fortuitous coincidence but, as Rosner, Holt and Gaily realised in advance, the subsequent development and ready availability of tunable lasers would transform this approach. Ten years later Van Esbroeck, McLean, Gaily, Holt and Rosner [23] used a visible dye laser to provide greater versatility, sensitivity and resolution. We will come to their experiments after first describing the initial studies.

The radiofrequency resonances observed arise from transitions between the nuclear hyperfine components of the $J'' = 28$ rotational level. The ^{23}Na nuclei have spin $I_1 = I_2 = 3/2$, so that the total nuclear spin I can take the values 3, 2, 1 and 0. A rotational level with J even combines with $I = 2$ and 0, so that for $I = 2$, the total angular momentum F takes all integral values from 26 to 30; when $I = 0$ we have $F = J = 28$.

The most important terms in the effective hyperfine Hamiltonian are those which describe the nuclear quadrupole and nuclear spin–rotation interactions:

$$\mathcal{H}_{\text{eff}} = -e \sum_{k=1,2} \mathbf{T}^2(\nabla \mathbf{E}_k) \cdot \mathbf{T}^2(\mathbf{Q}_k) + c_I \mathbf{T}^1(\mathbf{J}) \cdot \mathbf{T}^1(\mathbf{I}). \tag{8.171}$$

Additional terms involving the scalar and tensor interactions between the two nuclear spins were found to be too small to be significant in the first study, but we will meet them later. We encountered the quadrupole term in (8.171) in our earlier discussion of the D_2 molecule, and obtained the following results for the matrix elements in the coupled representation:

$$\langle \eta, \Lambda; J, I_1, I_2, I, F | -e \sum_{k=1,2} \mathbf{T}^2(\nabla \mathbf{E}_k) \cdot \mathbf{T}^2(\mathbf{Q}_k) | \eta, \Lambda; J', I_1, I_2, I', F \rangle$$

$$= \frac{eQq_0}{4}(-1)^{J'+I+F} \begin{Bmatrix} I & J & F \\ J' & I' & 2 \end{Bmatrix} (-1)^{J-\Lambda} \begin{pmatrix} J & 2 & J' \\ -\Lambda & 0 & \Lambda \end{pmatrix} \{(2J+1)(2J'+1)\}^{1/2}$$

$$\times [(-1)^I + (-1)^{I'}](-1)^{I_1+I_2}\{(2I+1)(2I'+1)\}^{1/2} \begin{Bmatrix} I_2 & I & I_1 \\ I' & I_2 & 2 \end{Bmatrix} \begin{pmatrix} I_1 & 2 & I_1 \\ -I_1 & 0 & I_1 \end{pmatrix}^{-1}.$$

$$\tag{8.172}$$

The nuclear spin–rotation term has relatively simple matrix elements, as follows:

$$\langle \eta, \Lambda; J, I, F, M_F | c_I \mathbf{T}^1(\mathbf{J}) \cdot \mathbf{T}^1(\mathbf{I}) | \eta, \Lambda; J, I, F, M_F \rangle$$
$$= c_I (-1)^{J+F+I} \begin{Bmatrix} I & J & F \\ J & I & 1 \end{Bmatrix} \{J(J+1)(2J+1)I(I+1)(2I+1)\}^{1/2}. \quad (8.173)$$

Upon evaluation for $J = 28$, $I = 2$ and 0 one obtains very simple results. The diagonal matrix elements of the quadrupole interaction in the $|\eta, \Lambda; J, I, F\rangle$ basis set are either zero (for $I = 0$) or very nearly constant (for $I = 2$). The only really significant matrix element of the quadrupole interaction is that which mixes the $I = 2$ and 0 states with $F = 28$; the resulting symmetric and antisymmetric combinations have quadrupole energies $\pm 0.25\, eq_0 Q$. The nuclear spin–rotation interaction has non-zero diagonal elements, but zero off-diagonal elements. Consequently the hyperfine level structure can be summarised in a simple diagram, as shown in figure 8.13, with the observed transitions

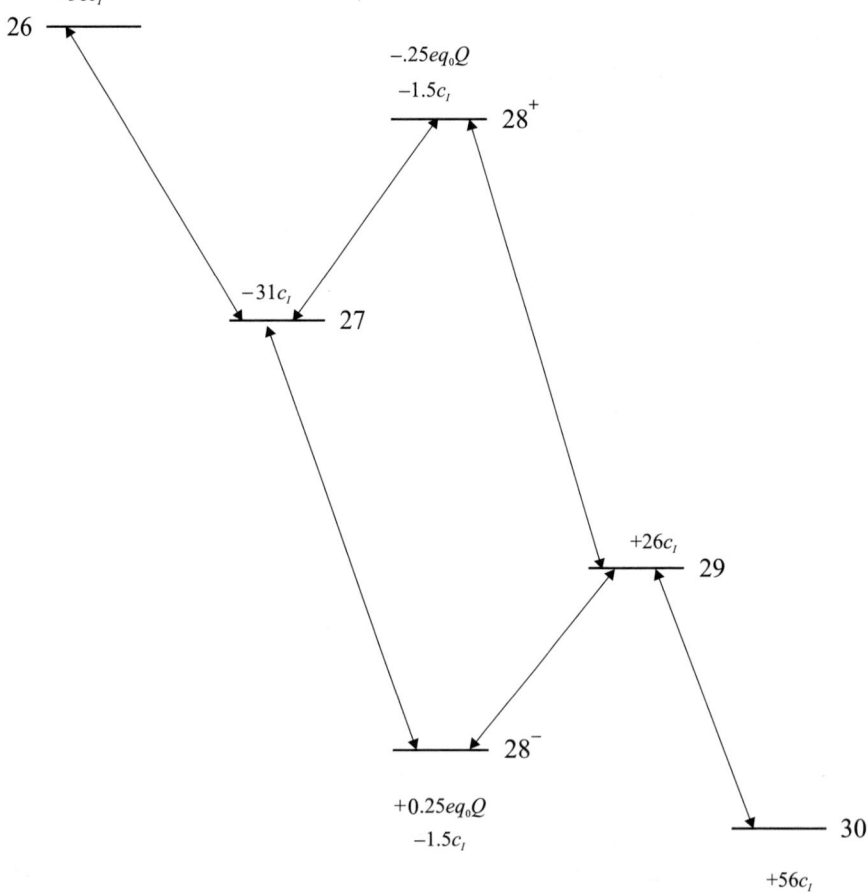

Figure 8.13. Hyperfine level structure of the $J = 28$ rotational level in the $v = 0$, $X\,^1\Sigma_g^+$ state of Na$_2$. The quadrupole coupling constant, $eq_0 Q$, is found to have a negative value [22]. The levels are labelled with their F values.

also indicated. The values of the constants were determined as $eq_0Q = -463.7$ kHz, $c_I = -0.201$ kHz. The negative sign of the quadrupole coupling constant is, of course, the reason for the relative energies of the $F = 28$ symmetric and antisymmetric states.

We come now to the second study, described ten years later [23]. The main development was the employment of a tunable dye laser to pump the $A\,^1\Sigma_u^+ \leftarrow X\,^1\Sigma_g^+$ transition. Rotational levels in the ground state with $J = 1$ to 29, in the $v = 0$ vibrational level, were pumped by the laser and radiofrequency hyperfine transitions studied. The range of J levels studied meant that the effective Hamiltonian required the addition of terms describing the dipolar and scalar interactions between the ^{23}Na nuclear spins. These terms were given earlier in our discussion of the D$_2$ molecule, and the complete effective Hamiltonian is:

$$\mathcal{H}_{\text{eff}} = -e \sum_{k=1,2} T^2(\nabla E_k) \cdot T^2(Q_k) + c_I T^1(J) \cdot T^1(I)$$
$$- \sqrt{6}\, g_N^2 \mu_N^2 (\mu_0/4\pi) T^2(C) \cdot T^2(I_1, I_2) + \delta T^1(I_1) \cdot T^1(I_2). \quad (8.174)$$

The matrix elements of the first two terms have already been derived, and from equation (8.21) we obtain the matrix elements of the dipolar term, which are

$$\langle \eta; J, \Lambda, I, F| - \sqrt{6} g_{\text{Na}}^2 \mu_N^2 (\mu_0/4\pi) T^2(C) \cdot T^2(I_1, I_2) |\eta; J', \Lambda, I', F'\rangle$$

$$= -\delta_{FF'}\sqrt{30}\, g_{\text{Na}}^2 \mu_N^2 (\mu_0/4\pi)(-1)^{J'+F+I} \begin{Bmatrix} I' & J' & F \\ J & I & 2 \end{Bmatrix} \begin{Bmatrix} I_1 & I_1 & 1 \\ I_2 & I_2 & 1 \\ I & I' & 2 \end{Bmatrix}$$

$$\times \{(2I+1)(2I'+1)I_1(I_1+1)(2I_1+1)I_2(I_2+1)(2I_2+1)\}^{1/2}$$
$$\times \langle \eta, J, \Lambda = 0 \| T^2(C) \| \eta, J', \Lambda = 0\rangle$$

$$= -\delta_{FF'}\sqrt{30}\, g_{\text{Na}}^2 \mu_N^2 (\mu_0/4\pi)(-1)^{J'+F+I} \begin{Bmatrix} I' & J' & F \\ J & I & 2 \end{Bmatrix} \begin{Bmatrix} I_1 & I_1 & 1 \\ I_2 & I_2 & 1 \\ I & I' & 2 \end{Bmatrix}$$

$$\times \{(2I+1)(2I'+1)\}^{1/2}\{I_1(I_1+1)(2I_1+1)I_2(I_2+1)(2I_2+1)\}^{1/2}(-1)^J$$

$$\times \begin{pmatrix} J & 2 & J' \\ 0 & 0 & 0 \end{pmatrix} \{(2J+1)(2J'+1)\}^{1/2} \langle C_0^2(\theta, \phi)\rangle_\eta. \quad (8.175)$$

The constant d listed by Van Esbroeck, McLean, Gaily, Holt and Rosner [23] is defined by

$$d = g_{\text{Na}}^2 \mu_N^2 (\mu_0/4\pi) \langle C_0^2(\theta, \phi) R^{-3}\rangle_\eta = g_{\text{Na}}^2 \mu_N^2 (\mu_0/4\pi) \langle 1/R^3\rangle_\eta, \quad (8.176)$$

where R is the internuclear distance in the vibronic state η. We shall confine attention to matrix elements which are diagonal in J.

The matrix elements of the scalar interaction between the nuclear spins in a homonuclear molecule were not given in our earlier discussion, but they are obtained very simply by noting that

$$I_1 \cdot I_2 = \frac{1}{2}\{I^2 - I_1^2 - I_2^2\} = \frac{1}{2}\{I(I+1) - I_1(I_1+1) - I_2(I_2+1)\}. \quad (8.177)$$

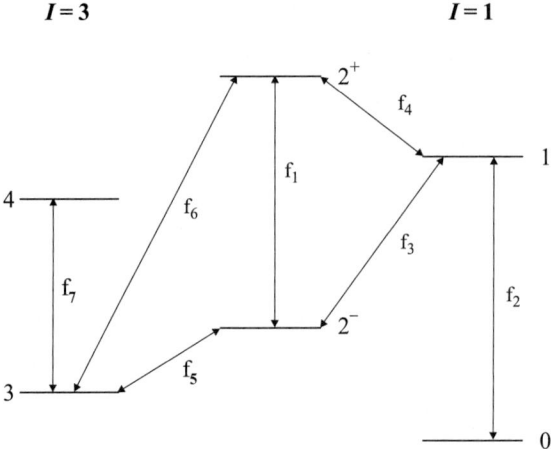

Figure 8.14. Hyperfine levels for the $J = 1$ rotational level of Na_2 in its ground electronic state, and the observed transitions [23]. Each level is labelled by its F value.

The lowest level studied experimentally was $J = 1$, for which seven hyperfine transitions were observed and the molecular constants determined [23]. For an odd J value the allowed values of I are 1 and 3, so that the following hyperfine levels exist:

$$J = 1, I = 3, F = 4, 3, 2: \quad J = 1, I = 1, F = 2, 1, 0.$$

Four of the states are not mixed and their energies are listed below. As we have seen earlier, the quadrupole interaction mixes the I states differing in value by 2 so that for $F = 2$ (which remains a good quantum number), we diagonalise a 2×2 matrix to obtain the final energies, also listed below. The resulting states are very close to being symmetric and antisymmetric combinations, denoted 2^+ and 2^-. We may then calculate the energies of the six hyperfine levels, using the values of the molecular constants determined from the experiments.

We now list the results, which are summarised in the hyperfine level diagram shown in figure 8.14, where the observed transitions are also indicated. The constants determined from the experiments [23] were (in kHz):

$$eq_0 Q = -458.98 - 0.007\,28\, J(J+1), \quad c_I = 0.2429, \quad d = 0.3026, \quad \delta = 1.0667.$$

The diagonal and off-diagonal matrix elements are (in kHz), for $J = 1$:

$$\langle I = 3, F = 4 | \mathcal{H}_{\text{eff}} | I = 3, F = 4 \rangle$$
$$= -eq_0 Q(1/10) + d(9/10) + 3c_I + \delta(9/4) = 49.2991$$
$$\langle I = 3, F = 3 | \mathcal{H}_{\text{eff}} | I = 3, F = 3 \rangle$$
$$= +eq_0 Q(3/10) - d(27/10) - c_I + \delta(9/4) = -136.3583$$
$$\langle I = 3, F = 2 | \mathcal{H}_{\text{eff}} | I = 3, F = 2 \rangle$$
$$= -eq_0 Q(6/25) + d(54/25) - 4c_I + \delta(9/4) = 112.2409$$

$$\langle I = 1, F = 2|\mathcal{H}_{\text{eff}}|I = 1, F = 2\rangle$$
$$= +eq_0 Q(1/25) + d(17/50) + c_I - \delta(11/4) = -20.9474$$
$$\langle I = 1, F = 1|\mathcal{H}_{\text{eff}}|I = 1, F = 1\rangle$$
$$= -eq_0 Q(1/5) - d(17/10) - c_I - \delta(11/4) = 88.1083$$
$$\langle I = 1, F = 0|\mathcal{H}_{\text{eff}}|I = 1, F = 0\rangle$$
$$= +eq_0 Q(2/5) + d(17/5) - 2c_I - \delta(11/4) = -185.9884$$
$$\langle I = 3, F = 2|\mathcal{H}_{\text{eff}}|I = 1, F = 2\rangle$$
$$= -eq_0 Q(3\sqrt{14}/50) - d(9\sqrt{14}/25) = 102.6365. \tag{8.178}$$

The energies in kHz of the mixed states (with $F = 2$) are calculated to be:

$$2^+ : 167.995 \quad 2^- : -76.701.$$

The calculated frequencies of the seven transitions shown in figure 8.14 agree exactly with those listed by Van Esbroeck, McLean, Gaily, Holt and Rosner [23] which are, in turn, in very good agreement with the experimentally measured frequencies (in kHz):

$$\textit{experiment}: \quad f_1 = 244.752, \quad f_2 = 274.214, \quad f_3 = 164.896, \quad f_4 = 79.904,$$
$$f_5 = 59.684, \quad f_6 = 304.340, \quad f_7 = 185.684.$$
$$\textit{calculated}: \quad f_1 = 244.697, \quad f_2 = 274.097, \quad f_3 = 164.810, \quad f_4 = 79.887,$$
$$f_5 = 59.657, \quad f_6 = 304.354, \quad f_7 = 185.659.$$

It should be noted that there are only five independent frequencies. Transitions in twelve other rotational levels were also measured, and again the calculated frequencies using the above constants were in excellent agreement with experiment.

Quantitative interpretation of the magnetic hyperfine parameters is a far from trivial task because, as we discussed in chapter 7, they involve the calculation of reduced matrix elements requiring electronic wavefunctions for both the ground and excited electronic states. The nuclear spin–rotation constant, c_I, for example, contains both first- and second-order parts which are opposite in sign. Even the dipolar tensor constant d, contains higher order contributions, although its value is dominated by the classical through-space contribution. The scalar spin–spin constant δ, usually known as the electron-coupled spin–spin interaction constant, arises entirely from admixture of higher electronic states with the ground state. There is, therefore, no simple semi-empirical interpretation of the constants which is worth presenting.

There have been many other spectroscopic studies of the Na_2 molecule, particularly of its visible and ultraviolet electronic spectra.

8.2.4. Other $^1\Sigma^+$ molecules

Many other diatomic molecules with $^1\Sigma^+$ ground states have been studied by molecular beam magnetic resonance. Where magnetic nuclei are present, magnetic focusing is based upon the nuclear Zeeman effects. This is the case with $^{15}N_2$ for which the

Table 8.4. *Molecular beam magnetic resonance studies of alkali halide molecules*

LiF [26, 27]	NaF [27, 28]	KF [29]	CsF [29]	RbF [30]
LiCl [26, 27]	NaCl [27]	KCl [27]	CsCl [29]	RbCl [30]
LiBr [26, 27]	NaBr [27]	KBr [31]		
LiI [26, 27]	NaI [27]			

rotational magnetic moment and the nuclear spin–rotation constant have been determined [24]. For a molecule like CO, where magnetic nuclei are not normally present, observation of the molecular beam magnetic resonance spectrum makes use of the rotational magnetic moment [25]. The alkali halide molecules have also been studied extensively by molecular beam magnetic and electric resonance. In many cases both the alkali metal and halogen nuclei have magnetic dipole and electric quadrupole moments. We shall describe in detail some of the electric resonance studies later in this chapter, but here summarise the magnetic resonance studies, with references, in table 8.4.

8.3. Molecular beam magnetic resonance of electronically excited molecules

8.3.1. H_2 in the $c\ ^3\Pi_u$ state

(a) Introduction

In chapter 6 we described the theory of molecular electronic states, particularly as it applies to diatomic molecules. We introduced the united atom nomenclature for describing the orbitals, and pointed out that this was particularly useful for tightly bound molecules with small internuclear distances, like H_2. We also discussed the more conventional nomenclature for describing electronic states, which is based upon the assumption that the component of electronic orbital angular momentum along the direction of the internuclear axis is conserved, i.e. is a *good* quantum number. The latter description is therefore appropriate for molecules in electronic states which conform to Hund's case (a) or case (b) coupling.

In H_2 the ground state electron configuration is described, in the united atom nomenclature, as $(1s\sigma)^2$ with the electron spins paired; alternatively the ground electronic state is described as $X\ ^1\Sigma_g^+$. In the molecular orbital description the ground state electron configuration is $(\sigma_g 1s)^2$. The first excited electronic state in the united atom description is $(1s\sigma)^1(2p\sigma)^1$, whilst in the molecular orbital description it is $(\sigma_g 1s)^1(\sigma_u 1s)^1$. In either case the electron spins may be paired, giving a $^1\Sigma_u^+$ state, or they may be unpaired to give a $^3\Sigma_u^+$ state. The excited singlet state is bound, but the triplet state, which correlates at the limit of infinite nuclear separation with two H $(1s)$ atoms, is repulsive. Singly excited electronic states, arising from promotion of one electron only, constitute two large families of either singlet or triplet spin states.

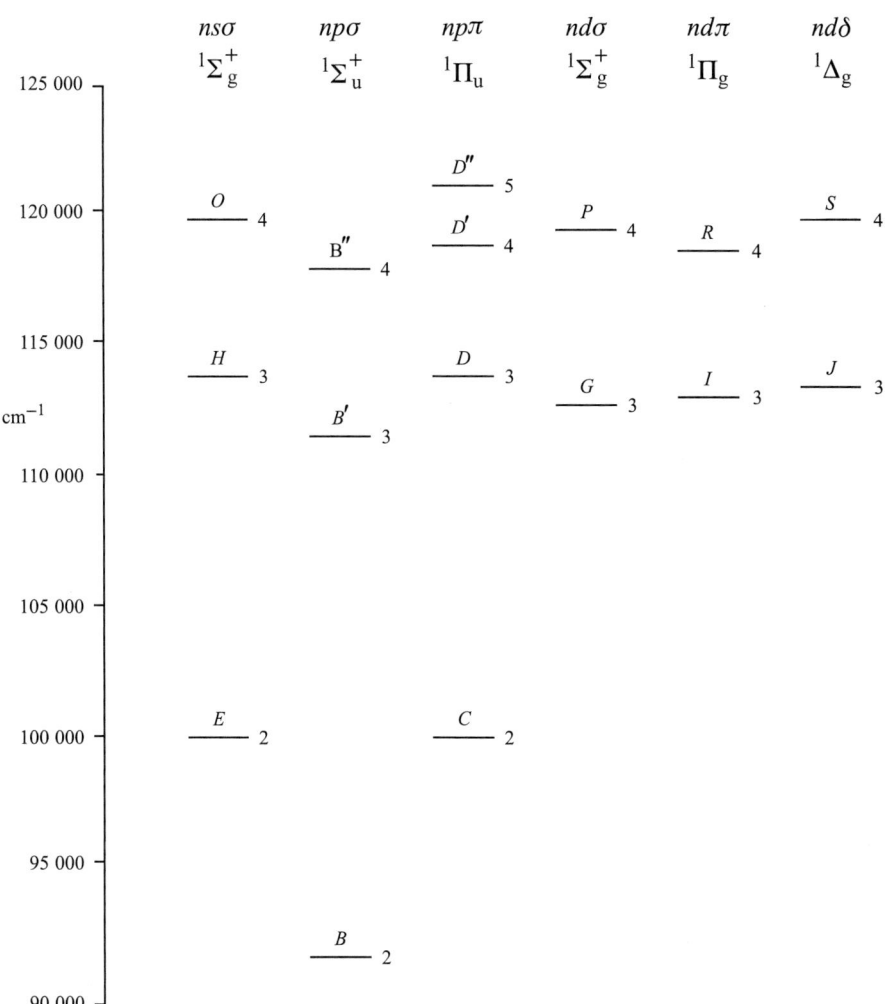

Figure 8.15. The excited singlet states of H_2 arising from promotion of a single electron. The zero of the energy scale relates to the ground electronic state, which is not shown in the figure. The orbital description at the top of each column is in the united atom designation. The number beside each level is the value of the principal quantum number n.

The relative energies of the excited singlet states are shown in figure 8.15, whilst the corresponding triplet states are shown in figure 8.16. These diagrams were originally constructed by Richardson [32], are shown by Herzberg [33] in his classic book, and have simply been updated by us from the data given by Herzberg and Huber [34]. It should be noted that many other excited states, which arise from promotion of both electrons, also exist. These are not shown in figures 8.15 and 8.16.

In this chapter we are concerned with the excited triplet states. The $c\,^3\Pi_u$ state is the subject of this section, whilst the d and $k\,^3\Pi_u$ states will appear later when we discuss microwave/optical double resonance studies. These triplet states are metastable,

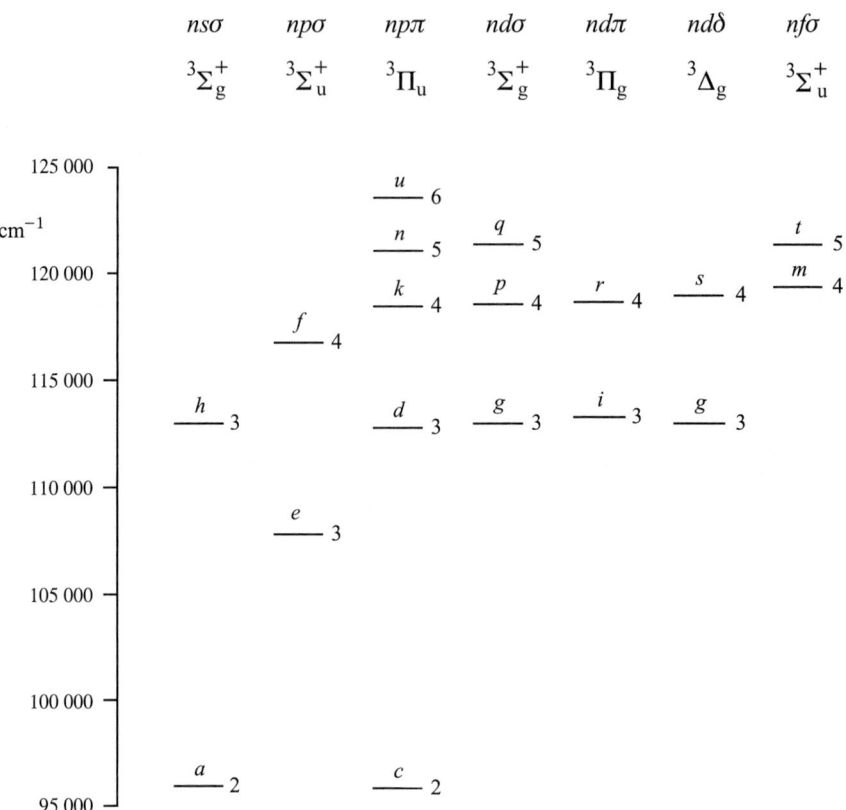

Figure 8.16. The excited triplet states of H_2 which arise from promotion of a single electron. The energy scale relates to the ground singlet state; the first excited triplet state $2p\sigma_u\,^3\Sigma_u^+$ is wholly repulsive and is not shown in the figure. The orbital description at the top of each column is in the united atom designation. The number beside each level is the value of the principal quantum number n.

at least in some of their vibrational levels, whereas most of the other excited states, singlets and triplets, decay rapidly through radiative transitions to lower states.

(b) **Experimental studies**

Molecular beam magnetic resonance studies of H_2 in its $c\,^3\Pi_u$ state were first described in two classic papers by Lichten [35]. In earlier work [36] Lichten had obtained strong but not conclusive evidence that the c state is metastable in its $v = 0$ level (higher vibrational levels are almost certainly not metastable), with a lifetime of approximately 10^{-3} s. He therefore designed and built an unusually small magnetic resonance beam machine in which the overall distance from source to detector was about 20 cm (compared with the 269 cm beam length employed in the earlier studies of ground state H_2). The H_2 beam emerging from the source was subjected to electron beam bombardment, and the electronically excited H_2 molecules were detected, with high sensitivity, by means of a nickel secondary electron emission detector; the apparatus was operated in the conventional 'flop-out' mode. In his first paper Lichten reported the observation of radiofrequency low field Zeeman transitions for the $N=2$ levels of *para*-H_2 but in his second paper he described the measurement of Zeeman components of fine-structure transitions in the $N=2$ level which occur at much higher frequencies in the microwave region.

We first describe the theory of the Zeeman effect and determination of the effective g-factors for both *para*- and *ortho*-H_2, and then proceed to discuss the hyperfine splitting in *ortho*-H_2.

(c) Effective Zeeman Hamiltonian in a case (b) basis

For H_2 the rotational constant is very large while the spin–orbit coupling effects are very small; the $c\,^3\Pi_u$ state of H_2 therefore conforms very closely to Hund's case (b). The nuclear rotational angular momentum $\boldsymbol{R}$ is coupled to the electronic orbital angular momentum $\boldsymbol{L}$, to form the total orbital angular momentum $\boldsymbol{N}$. $\boldsymbol{N}$ is now coupled to the electron spin $\boldsymbol{S}$ to form $\boldsymbol{J}$, the total angular momentum for *para*-H_2. In *ortho*-H_2 the total nuclear spin vector $\boldsymbol{I}$ is further coupled to $\boldsymbol{J}$ to form $\boldsymbol{F}$, the grand total angular momentum. This scheme is summarised as follows:

$$\boldsymbol{R}+\boldsymbol{L} = \boldsymbol{N}: \quad N = 1, 2, 3, \ldots$$
$$\boldsymbol{N}+\boldsymbol{S} = \boldsymbol{J}: \quad J = N-1, N, N+1$$
$$\boldsymbol{J}+\boldsymbol{I} = \boldsymbol{F}: \quad F = J-1, J, J+1 \quad \text{(for *ortho* levels)}.$$

The first level to be studied in detail by Lichten [35] was the $N=2$ level of both *para*-H_2 and *ortho*-H_2. He measured a series of fixed-frequency magnetic resonance transitions, determining effective g-values and proving the identification of the $c\,^3\Pi_u$ state in the process. An effective Zeeman Hamiltonian may be written, in the space-fixed axis system,

$$\mathcal{H}_{\text{eff}} = g_S \mu_B B_Z T^1_{p=0}(\boldsymbol{S}) + g_L \mu_B B_Z T^1_{p=0}(\boldsymbol{L}) - g_I \mu_N B_Z T^1_{p=0}(\boldsymbol{I}), \quad (8.179)$$

where $p=0$ is the direction of the applied magnetic field. This equation does not contain a term describing the rotational Zeeman interaction, and the third term in (8.179) applies only to *ortho*-H_2.

The matrix elements of the three terms in (8.179) in a case (b) basis are as follows:

$$\langle \eta, \Lambda; N, S, J, M_J | g_S \mu_B B_Z T^1_0(\boldsymbol{S}) | \eta, \Lambda; N, S, J', M_J \rangle$$
$$= g_S \mu_B B_Z (-1)^{J-M_J} \begin{pmatrix} J & 1 & J' \\ -M_J & 0 & M_J \end{pmatrix} (-1)^{J+N+1+S} \begin{Bmatrix} J & S & N \\ S & J' & 1 \end{Bmatrix}$$
$$\times \{(2J'+1)(2J+1)S(S+1)(2S+1)\}^{1/2}. \quad (8.180)$$

Table 8.5. *Calculated and observed g-factors for H_2 in the $c\,^3\Pi_u$ state*

N	J	g_J (para) first-order	F	g_F (ortho) first-order	g_J (para) exact	g_F (ortho) exact	g observed
1	0	—	1	−0.003 04	—	−0.168	−0.170
	1	1.251 15					
			0	—			
			1	0.624 05		0.8298	0.830
			2	0.624 05		0.648	0.651
	2	1.251 15					
			1	1.878 24		1.837	1.85
			2	1.042 11		1.019	1.026
			3	0.833 09		0.833	0.833
2	2	0.4726					0.4734
	3	0.778 54					0.7794

$$\langle \eta, \Lambda; N, S, J, M_J | g_S \mu_B B_Z T_0^1(L) | \eta, \Lambda; N', S, J', M_J \rangle$$

$$= g_L \mu_B B_Z (-1)^{J-M_J} \begin{pmatrix} J & 1 & J' \\ -M_J & 0 & M_J \end{pmatrix} \langle \eta, \Lambda; N, S, J \| T^1(L) \| \eta, \Lambda; N', S, J' \rangle$$

$$= g_L \mu_B B_Z \begin{pmatrix} J & 1 & J' \\ -M_J & 0 & M_J \end{pmatrix} (-1)^{J-M_J+J'+N+1+S} \{(2J'+1)(2J+1)\}^{1/2}$$

$$\times \begin{Bmatrix} J & N & S \\ N' & J' & 1 \end{Bmatrix} \langle N, \Lambda \| \mathfrak{D}_{\cdot 0}^{(1)}(\omega)^* \| N', \Lambda \rangle \langle \eta, \Lambda | T_0^1(L) | \eta, \Lambda \rangle$$

$$= g_L \mu_B B_Z \begin{pmatrix} J & 1 & J' \\ -M_J & 0 & M_J \end{pmatrix} (-1)^{J-M_J+J'+N+1+S} \{(2J'+1)(2J+1)\}^{1/2}$$

$$\times \begin{Bmatrix} J & N & S \\ N' & J' & 1 \end{Bmatrix} (-1)^{N-\Lambda} \begin{pmatrix} N & 1 & N' \\ -\Lambda & 0 & \Lambda \end{pmatrix} \{(2N+1)(2N'+1)\}^{1/2} \Lambda. \quad (8.181)$$

We have used the first-rank rotation matrix $\mathfrak{D}_{\cdot 0}^{(1)}(\omega)^*$ to transform into the molecule-fixed ($q=0$) axis system. Before considering the third term in (8.179), which arises for *ortho*-H_2, we calculate the diagonal elements from (8.180) and (8.181) in order to obtain the first-order effective g-factors for *para*-H_2. One obtains

$$g(N, J) = g_S \frac{\{J(J+1)+S(S+1)-N(N+1)\}}{2J(J+1)}$$
$$+ g_L \Lambda^2 \frac{\{J(J+1)+N(N+1)-S(S+1)\}}{2J(J+1)N(N+1)}, \quad (8.182)$$

which is the result obtained by Lichten [35] by a somewhat different route. If one now substitutes the values $g_L = \Lambda = S = 1$, and $g_S = 2.0023$, the calculated values of $g(N, J)$ for *para*-H_2 are as shown in table 8.5.

We shall return to the results for *para*-H$_2$ in due course, but first repeat the Zeeman calculations for *ortho*-H$_2$, including the nuclear spin Zeeman term in equation (8.179). The required matrix elements are as follows:

$$\langle \eta, \Lambda; N, S, J, I, F, M_F | g_S \mu_B B_Z T^1_{p=0}(\boldsymbol{S}) | \eta, \Lambda; N, S, J', I, F', M_F \rangle$$

$$= g_S \mu_B B_Z (-1)^{F-M_F} \begin{pmatrix} F & 1 & F' \\ -M_F & 0 & M_F \end{pmatrix} \langle N, S, J, I, F \| T^1(\boldsymbol{S}) \| N, S, J', I, F' \rangle$$

$$= g_S \mu_B B_Z (-1)^{F-M_F} \begin{pmatrix} F & 1 & F' \\ -M_F & 0 & M_F \end{pmatrix} (-1)^{F'+J+1+I} \{(2F'+1)(2F+1)\}^{1/2}$$

$$\times \begin{Bmatrix} F & J & I \\ J' & F' & 1 \end{Bmatrix} \langle N, S, J \| T^1(\boldsymbol{S}) \| N, S, J' \rangle$$

$$= g_S \mu_B B_Z (-1)^{F-M_F} \begin{pmatrix} F & 1 & F' \\ -M_F & 0 & M_F \end{pmatrix} (-1)^{F'+J+1+I} \{(2F'+1)(2F+1)\}^{1/2}$$

$$\times \begin{Bmatrix} F & J & I \\ J' & F' & 1 \end{Bmatrix} (-1)^{J+N+1+S} \{(2J'+1)(2J+1)\}^{1/2} \begin{Bmatrix} J & S & N \\ S & J' & 1 \end{Bmatrix}$$

$$\times \{S(S+1)(2S+1)\}^{1/2}. \tag{8.183}$$

$$\langle \eta, \Lambda; N, S, J, I, F, M_F | g_L \mu_B B_Z T^1_{p=0}(\boldsymbol{L}) | \eta, \Lambda; N', S, J', I, F', M_F \rangle$$

$$= g_L \mu_B B_Z (-1)^{F-M_F} \begin{pmatrix} F & 1 & F' \\ -M_F & 0 & M_F \end{pmatrix} \langle N, S, J, I, F \| T^1(\boldsymbol{L}) \| N', S, J', I, F' \rangle$$

$$= g_L \mu_B B_Z (-1)^{F-M_F} \begin{pmatrix} F & 1 & F' \\ -M_F & 0 & M_F \end{pmatrix} (-1)^{F'+J+1+I} \{(2F'+1)(2F+1)\}^{1/2}$$

$$\times \begin{Bmatrix} F & J & I \\ J' & F' & 1 \end{Bmatrix} \langle N, S, J \| T^1(\boldsymbol{L}) \| N', S, J' \rangle$$

$$= g_L \mu_B B_Z (-1)^{F-M_F} \begin{pmatrix} F & 1 & F' \\ -M_F & 0 & M_F \end{pmatrix} (-1)^{F'+J+1+I} \{(2F'+1)(2F+1)\}^{1/2}$$

$$\times \begin{Bmatrix} F & J & I \\ J' & F' & 1 \end{Bmatrix} (-1)^{J'+N+1+S} \{(2J'+1)(2J+1)\}^{1/2} \begin{Bmatrix} J & N & S \\ N' & J' & 1 \end{Bmatrix}$$

$$\times (-1)^{N-\Lambda} \begin{pmatrix} N & 1 & N' \\ -\Lambda & 0 & \Lambda \end{pmatrix} \{(2N+1)(2N'+1)\}^{1/2} \Lambda. \tag{8.184}$$

We have made use of the results given in equation (8.181) in order to complete the last line of (8.184). Finally for the nuclear Zeeman term we have

$$\langle \eta, \Lambda; N, S, J, I, F, M_F | -g_I \mu_N B_Z T^1_{p=0}(\boldsymbol{I}) | \eta, \Lambda; N, S, J, I, F', M_F \rangle$$

$$= -g_I \mu_N B_Z (-1)^{F-M_F} \begin{pmatrix} F & 1 & F' \\ -M_F & 0 & M_F \end{pmatrix} \langle J, I, F \| T^1(\boldsymbol{I}) \| J, I, F' \rangle$$

$$= -g_I \mu_N B_Z (-1)^{F-M_F} \begin{pmatrix} F & 1 & F' \\ -M_F & 0 & M_F \end{pmatrix} (-1)^{F+J+1+I} \{(2F'+1)(2F+1)\}^{1/2}$$

$$\times \begin{Bmatrix} F & I & J \\ I & F' & 1 \end{Bmatrix} \{I(I+1)(2I+1)\}^{1/2}. \tag{8.185}$$

We are now in a position to calculate the first-order g-factors for the hyperfine levels in *ortho*-H_2, making the same numerical substitutions as before, but additionally with $I = 1$. The result is

$$g(N, J, F) = g_S\{J(J + 1) + F(F + 1) - I(I + 1)\}\{J(J + 1) + S(S + 1)$$
$$- N(N + 1)\}/\{4F(F + 1)J(J + 1)\} + g_L \Lambda^2\{J(J + 1) + F(F + 1)$$
$$- I(I + 1)\}\{J(J + 1) + N(N + 1) - S(S + 1)\}/\{4F(F + 1)$$
$$\times J(J + 1)N(N + 1)\} - g_I(\mu_N/\mu_B)\{F(F + 1) + I(I + 1)$$
$$- J(J + 1)\}/2F(F + 1). \tag{8.186}$$

This result agrees with that given by Lichten in a later paper [37], and also by Jette and Cahill [38]. In his first paper [35] Lichten included an experimental recording of his early g-factor determinations; his spectrum is shown in figure 8.17. The experimental results for both *para-* and *ortho-*H_2 are listed in table 8.5, where they are compared with the first-order values calculated from equations (8.182) and (8.186) respectively. The 'exact' theoretical values listed in the Table are obtained after a detailed consideration of the complete zero-field effective Hamiltonian, which is presented in the next subsection.

(d) Effective zero-field Hamiltonian for para-$H_2(^3\Pi_u)$ in a case (b) basis

There are four important types of interaction that must be represented in the zero-field effective Hamiltonian for H_2 in its $c\,^3\Pi_u$ state. They can be summarised as follows:

$$\mathcal{H}_{eff} = \mathcal{H}_{sr} + \mathcal{H}_{dip} + \mathcal{H}_{so} + \mathcal{H}_{hfs}. \tag{8.187}$$

The first three terms are present for both *para-* and *ortho-*H_2; they represent the electron spin-rotation, electron spin–spin dipolar and spin–orbit interactions respectively. The fourth term in (8.187) represents the magnetic hyperfine interactions, which we will come to a little later. We deal first, however, with the terms that do not involve nuclear spin interactions.

(i) SPIN–ROTATION INTERACTION

The electron spin–rotation interaction is similar to the nuclear spin–rotation interaction we met earlier, and may be written as a simple scalar interaction,

$$\mathcal{H}_{sr} = \gamma T^1(N) \cdot T^1(S). \tag{8.188}$$

The matrix elements of (8.188) are diagonal in a case (b) basis, and are given by

$$\langle \eta, \Lambda; N, S, J, M_J | \gamma T^1(N) \cdot T^1(S) | \eta, \Lambda; N, S, J, M_J \rangle$$
$$= \gamma(-1)^{N+J+S} \begin{Bmatrix} S & N & J \\ N & S & 1 \end{Bmatrix} \langle N\|T^1(N)\|N\rangle\langle S\|T^1(S)\|S\rangle$$
$$= \gamma(-1)^{N+J+S} \begin{Bmatrix} S & N & J \\ N & S & 1 \end{Bmatrix} \{N(N + 1)(2N + 1)S(S + 1)(2S + 1)\}^{1/2}. \tag{8.189}$$

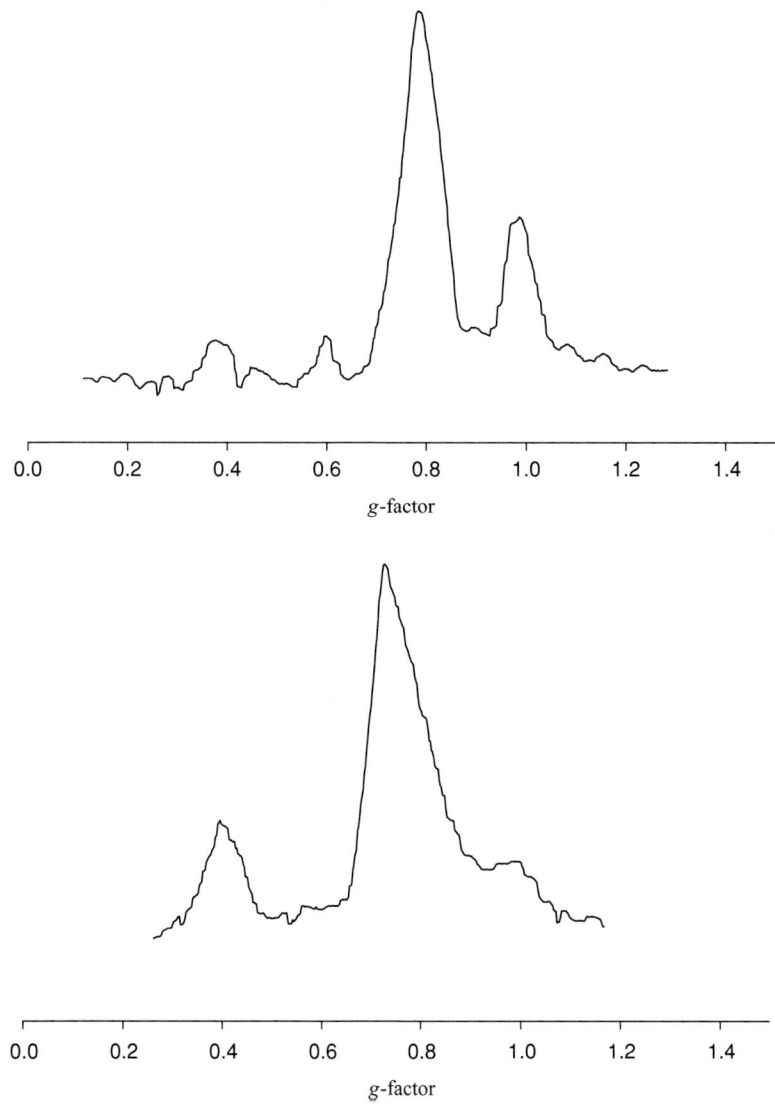

Figure 8.17. Low-frequency Zeeman transitions for the $c\,^3\Pi_u$ state of H_2, showing the g-factors [35]. The top spectrum was obtained from natural H_2 (75% *ortho*, 25% *para*), and the bottom spectrum was obtained from *para*-H_2. The spectra were obtained by scanning the frequency.

Evaluation of the 6-j symbol gives the results for the three spin components of the rotational level N:

$$J = N+1: \quad E = \gamma N$$
$$J = N: \quad E = -\gamma \qquad (8.190)$$
$$J = N-1: \quad E = -\gamma(N+1).$$

(ii) ELECTRON SPIN–SPIN DIPOLAR INTERACTION

The dipolar interaction between the electron spin magnetic moments is also similar to the interaction between two nuclear spins. It may therefore be represented by a classical dipolar interaction term, but with the electron spin magnetic moments represented by their quantum mechanical operators:

$$\mathcal{H}_{\text{dip}} = \frac{g_S^2 \mu_B^2 \mu_0}{4\pi} \left\{ \frac{\mathbf{S}_1 \cdot \mathbf{S}_2}{r_{12}^3} - \frac{3(\mathbf{S}_1 \cdot \mathbf{r}_{12})(\mathbf{S}_2 \cdot \mathbf{r}_{12})}{r_{12}^5} \right\}. \tag{8.191}$$

Here r_{12} is the distance between the unpaired electrons, which must be obtained from a suitable electronic wave function describing the electron distribution in the molecule. In terms of irreducible tensor operators (8.191) may be written

$$\mathcal{H}_{\text{dip}} = -\frac{g_S^2 \mu_B^2 \mu_0}{4\pi} \sqrt{6} \mathrm{T}^2(\mathbf{C}) \cdot \mathrm{T}^2(\mathbf{S}_1, \mathbf{S}_2), \tag{8.192}$$

where

$$\mathrm{T}_p^2(\mathbf{S}_1, \mathbf{S}_2) = (-1)^p \sqrt{5} \sum_{p_1, p_2} \begin{pmatrix} 1 & 1 & 2 \\ p_1 & p_2 & -p \end{pmatrix} \mathrm{T}_{p_1}^1(\mathbf{S}_1) \mathrm{T}_{p_2}^1(\mathbf{S}_2), \tag{8.193}$$

and

$$\mathrm{T}_q^2(\mathbf{C}) = \langle \eta, \Lambda | C_q^2(\theta_{12}, \phi_{12}) r_{12}^{-3} | \eta, \Lambda' \rangle. \tag{8.194}$$

This form of the electron spin–spin dipolar Hamiltonian is discussed in appendix 8.3. The diagonal ($q=0$) component of (8.192) may be written

$$-g_S^2 \mu_B^2 \frac{\mu_0}{4\pi} \langle \eta, \Lambda | C_0^2(\theta_{12}, \phi_{12}) r_{12}^{-3} | \eta, \Lambda \rangle \sqrt{6} \mathrm{T}_0^2(\mathbf{S}_1, \mathbf{S}_2) \equiv \frac{2}{3} \lambda \sqrt{6} \mathrm{T}_0^2(\mathbf{S}, \mathbf{S}), \tag{8.195}$$

and by making use of the relationship (8.193) one finds that

$$\sqrt{6} \mathrm{T}_0^2(\mathbf{S}, \mathbf{S}) = 3S_z^2 - \mathbf{S}^2. \tag{8.196}$$

The combination of (8.195) and (8.196) gives the cartesian form of the spin–spin interaction which is often encountered in the literature. However, we shall in due course discuss the $q \neq 0$ components of (8.194) which, in the case of H_2 in its $c\,^3\Pi_u$ state, are not negligible.

The matrix elements of the spin–spin Hamiltonian are as follows:

$$\langle \eta, \Lambda; N, S, J, M_J | -(\mu_0/4\pi)g_S^2\mu_B^2\sqrt{6}\mathbf{T}^2(\mathbf{S}_1, \mathbf{S}_2) \cdot \mathbf{T}^2(\mathbf{C}) | \eta', \Lambda'; N', S, J, M_J \rangle$$

$$= \sqrt{6}(-1)^{J+N'+S} \begin{Bmatrix} S & N' & J \\ N & S & 2 \end{Bmatrix} \langle \eta, N, \Lambda | (\mu_0/4\pi)g_S^2\mu_B^2 \mathbf{T}^2(\mathbf{C}) | \eta', N', \Lambda' \rangle$$

$$\times \langle S \| \mathbf{T}^2(\mathbf{S}_1, \mathbf{S}_2) \| S \rangle$$

$$= \sqrt{6}(-1)^{J+N'+S} \begin{Bmatrix} S & N' & J \\ N & S & 2 \end{Bmatrix} \sum_q \langle \eta, \Lambda | (\mu_0/4\pi)g_S^2\mu_B^2 C_q^2(\theta_{12}, \phi_{12})(r_{12}^{-3}) | \eta', \Lambda' \rangle$$

$$\times \langle N, \Lambda \| \mathcal{D}_{\cdot q}^{(2)}(\omega_{12})^* \| N', \Lambda' \rangle \sqrt{5} \begin{Bmatrix} S_1 & S_1 & 1 \\ S_2 & S_2 & 1 \\ S & S & 2 \end{Bmatrix} (2S+1)\{S_1(S_1+1)$$

$$\times (2S_1+1)S_2(S_2+1)(2S_2+1)\}^{1/2}. \qquad (8.197)$$

The reduced matrix element of $\mathbf{T}^2(\mathbf{S}_1, \mathbf{S}_2)$ was previously evaluated (for proton nuclear spins) in (8.14). Confining attention, for the moment, to matrix elements diagonal in the electronic state, we take the $q=0$ component of (8.197). Hence we have the result

$$\langle \eta, \Lambda; N, S, J, M_J | -(\mu_0/4\pi)g_S^2\mu_B^2\sqrt{6}\mathbf{T}^2(\mathbf{S}_1, \mathbf{S}_2) \cdot \mathbf{T}^2(\mathbf{C}) | \eta', \Lambda'; N', S, J, M_J \rangle$$

$$= -\sqrt{6}(-1)^{J+N'+S} \begin{Bmatrix} S & N' & J \\ N & S & 2 \end{Bmatrix} \langle \eta, \Lambda | (\mu_0/4\pi)g_S^2\mu_B^2 C_0^2(\theta_{12}, \phi_{12})(r_{12}^{-3}) | \eta, \Lambda \rangle$$

$$\times \langle N, \Lambda \| \mathcal{D}_{\cdot 0}^{(2)}(\omega_{12})^* \| N', \Lambda \rangle (\sqrt{5}/2)$$

$$= -\frac{\sqrt{30}}{2}(-1)^{J+N'+S} \begin{Bmatrix} S & N' & J \\ N & S & 2 \end{Bmatrix} \langle \eta, \Lambda | (\mu_0/4\pi)g_S^2\mu_B^2 C_0^2(\theta_{12}, \phi_{12})(r_{12}^{-3}) | \eta, \Lambda \rangle$$

$$\times (-1)^{N-\Lambda} \begin{pmatrix} N & 2 & N' \\ -\Lambda & 0 & \Lambda \end{pmatrix} \{(2N+1)(2N'+1)\}^{1/2}$$

$$= \frac{2}{3}\lambda_0\sqrt{30}(-1)^{J+N'+S} \begin{Bmatrix} S & N' & J \\ N & S & 2 \end{Bmatrix} (-1)^{N-\Lambda} \begin{pmatrix} N & 2 & N' \\ -\Lambda & 0 & \Lambda \end{pmatrix}$$

$$\times \{(2N+1)(2N'+1)\}^{1/2}. \qquad (8.198)$$

In the last line of equation (8.198) we have introduced the parameter λ_0, which is defined by the identity

$$\lambda_0 \equiv -(3/4)\langle(\mu_0/4\pi)g_S^2\mu_B^2 C_0^2(\theta_{12}, \phi_{12})r_{12}^{-3}\rangle_\eta. \qquad (8.199)$$

Since the rotational levels of different N values are very widely spaced in H_2, we need consider only the matrix elements diagonal in N so that substituting $\Lambda = S = 1$ we derive the following expression for the spin–spin energies:

$$E(N, J)$$
$$= \lambda_0 \frac{\{3[N(N+1) - J(J+1) + 2][N(N+1) - J(J+1) + 1] - 8N(N+1)\}}{(2N-1)(2N)(2N+1)(2N+2)(2N+3)}.$$

$$(8.200)$$

For the three spin components of a given N level we now obtain the results:

$$J = N+1: \quad E(N,J) = \frac{4}{3}\lambda_0 \frac{\{3-N(N+1)\}}{(2N+2)(2N+3)}.$$

$$J = N: \quad E(N,J) = -\frac{4}{3}\lambda_0 \frac{\{3-N(N+1)\}}{2N(N+1)}. \qquad (8.201)$$

$$J = N-1: \quad E(N,J) = \frac{4}{3}\lambda_0 \frac{\{3-N(N+1)\}}{2N(2N-1)}.$$

Up to this point we have taken Λ to have the value $+1$, ignoring the fact that the true wave functions are Λ-doublets. The two components can be described in terms of basis functions which have either $+$ or $-$ parity:

$$|N, \Lambda, S, J; \pm\rangle = \frac{1}{\sqrt{2}}\{|+1; N, +1, S, J, M_J\rangle \pm (-1)^N |-1; N, -1, S, J, M_J\rangle\}. \qquad (8.202)$$

There are, therefore *additional* matrix elements of the spin–spin interaction (8.197) involved with $\Delta\Lambda = \pm 2$, as was first pointed out by Fontana [39]; these are as follows:

$$\langle \eta, \Lambda; N, S, J, M_J | (\mu_0/4\pi) g_S^2 \mu_B^2 \sqrt{6} \mathbf{T}^2(\mathbf{S}_1, \mathbf{S}_2) \cdot \mathbf{T}^2(\mathbf{C}) | \eta, \Lambda'; N', S, J, M_J \rangle$$

$$= \sqrt{6}(-1)^{J+N'+S} \begin{Bmatrix} S & N' & J \\ N & S & 2 \end{Bmatrix} \langle \eta, N, \Lambda \| (\mu_0/4\pi) g_S^2 \mu_B^2 \mathbf{T}^2(\mathbf{C}) \| \eta, N', \Lambda' \rangle$$

$$\times \langle S \| \mathbf{T}^2(\mathbf{S}_1, \mathbf{S}_2) \| S \rangle$$

$$= \sqrt{6}(-1)^{J+N'+S} \begin{Bmatrix} S & N' & J \\ N & S & 2 \end{Bmatrix} \sum_{q=\pm 2} \langle \eta, \Lambda | (\mu_0/4\pi) g_S^2 \mu_B^2 C_q^2(\theta_{12}, \phi_{12})(r_{12}^{-3}) | \eta, \Lambda' \rangle$$

$$\times \langle N, \Lambda \| \mathfrak{D}_{\cdot q}^{(2)}(\omega_{12})^* \| N', \Lambda' \rangle \sqrt{5} \begin{Bmatrix} S_1 & S_1 & 1 \\ S_2 & S_2 & 1 \\ S & S & 2 \end{Bmatrix} (2S+1)$$

$$\times \{S_1(S_1+1)(2S_1+1)S_2(S_2+1)(2S_2+1)\}^{1/2}$$

$$= \frac{\sqrt{30}}{2}(-1)^{J+N'+S} \begin{Bmatrix} S & N' & J \\ N & S & 2 \end{Bmatrix} \sum_{q=\pm 2} \langle \eta, \Lambda | (\mu_0/4\pi) g_S^2 \mu_B^2 C_q^2(\theta_{12}, \phi_{12})(r_{12}^{-3}) | \eta, \Lambda' \rangle$$

$$\times \langle N, \Lambda \| \mathfrak{D}_{\cdot q}^{(2)}(\omega_{12})^* \| N', \Lambda' \rangle$$

$$= \frac{\sqrt{30}}{2}\lambda_2 (-1)^{J+N'+S} \begin{Bmatrix} S & N' & J \\ N & S & 2 \end{Bmatrix} \sum_{q=\pm 2} (-1)^{N-\Lambda} \begin{pmatrix} N & 2 & N' \\ -\Lambda & q & \Lambda' \end{pmatrix}$$

$$\times \{(2N+1)(2N'+1)\}^{1/2}. \qquad (8.203)$$

Non-vanishing matrix elements also arise from the $q = \pm 1$ terms, but they involve the mixing of other electronic states with the $^3\Pi_u$ state. These matrix elements, and others that are off-diagonal in N, are listed by Chiu [40]. In an obvious notation, specified below, we have introduced the parameter λ_2. We are now in a position to examine the matrix elements (diagonal in N) of (8.203) for the states $|N, \Lambda, S, J; +\rangle$ and $|N, \Lambda, S, J; -\rangle$ given in (8.202). The main contributions to the diagonal elements involve λ_0 and come from (8.198) but there are additional contributions which involve λ_2, namely,

$$\langle N, \Lambda, S, J; +|\mathcal{H}_{\text{dip}}|N, \Lambda, S, J; +\rangle$$
$$= \frac{1}{2}(-1)^N\{\langle \Lambda = +1|\mathcal{H}_{\text{dip}}|\Lambda = -1\rangle + \langle \Lambda = -1|\mathcal{H}_{\text{dip}}|\Lambda = +1\rangle\}$$
$$= (-1)^N \frac{\sqrt{30}}{4}\lambda_2(-1)^{J+N+S}\begin{Bmatrix} S & N & J \\ N & S & 2 \end{Bmatrix}(2N+1)\left[(-1)^{N-1}\begin{pmatrix} N & 2 & N \\ -1 & 2 & -1 \end{pmatrix}\right.$$
$$\left. + (-1)^{N+1}\begin{pmatrix} N & 2 & N \\ 1 & -2 & 1 \end{pmatrix}\right]$$
$$= (-1)^N \frac{\sqrt{30}}{4}\lambda_2(-1)^{J+N+S}\begin{Bmatrix} S & N & J \\ N & S & 2 \end{Bmatrix}\left[\frac{6N(N+1)(2N+1)}{(2N+3)(2N-1)}\right]^{1/2}. \quad (8.204)$$

The other matrix element between states of negative parity is

$$\langle N, \Lambda, S, J; -|\mathcal{H}_{\text{dip}}|N, \Lambda, S, J; -\rangle = -\langle N, \Lambda, S, J; +|\mathcal{H}_{\text{dip}}|N, \Lambda, S, J; +\rangle. \quad (8.205)$$

Because parity is a good quantum label in the absence of electric fields, all matrix elements between states of $+$ and $-$ parity are zero.

Consequently we see that the λ_2 term makes equal and opposite contributions to the spin-spin energies of the Λ-doublet components. These contributions to the levels of $+$ and $-$ parity are:

$$J = N+1: \quad E(+) = (-1)^N \frac{\sqrt{6}}{4}\lambda_2 \frac{N}{(2N+3)}, \quad E(-) = -(-1)^N \frac{\sqrt{6}}{4}\lambda_2 \frac{N}{(2N+3)},$$

$$J = N: \quad E(+) = -(-1)^N \frac{\sqrt{6}}{4}\lambda_2, \quad E(-) = (-1)^N \frac{\sqrt{6}}{4}\lambda_2, \quad (8.206)$$

$$J = N-1: \quad E(+) = (-1)^N \frac{\sqrt{6}}{4}\lambda_2 \frac{(N+1)}{(2N-1)}, \quad E(-) = -(-1)^N \frac{\sqrt{6}}{4}\lambda_2 \frac{(N+1)}{(2N-1)}.$$

These results agree with those derived by Chin [40] with the following identity.

$$\lambda_2 \equiv \langle \Lambda = \pm 1|(\mu_0/4\pi)g_S^2\mu_B^2 C_{\pm 2}^2(\theta_{12}, \phi_{12})r_{12}^{-3}|\Lambda = \mp 1\rangle_\eta \quad (8.207)$$

We are now almost in a position to calculate the zero-field spin–rotation and spin–spin dipolar energies of the N, J levels, but we have first to discuss the parity restrictions on the levels for *para-* and *ortho-*H_2(see also chapter 6).

The overall symmetry of a given level must be antisymmetric with respect to the permutation P_{12} of the two H nuclei to satisfy the Pauli principle. The $+$ and $-$ parity combinations defined by equation (8.202) are antisymmetric and symmetric respectively with respect to P_{12} (because the electronic wavefunction has u character, see equation (8.251)). Since the *ortho* and *para* nuclear spin states are symmetric and antisymmetric respectively, we see that the $+$ parity states combine with the *ortho*

Table 8.6. *Zero-field spin–spin, spin–rotation and spin–orbit energies of the lower rotational levels of* $c\,^3\Pi_u\,H_2$

para-H$_2$			ortho-H$_2$		
N	J	energy	N	J	energy
1	0	$2\lambda_0/3 + \sqrt{6}\lambda_2/2 - 2\gamma - A$	1	0	$2\lambda_0/3 - \sqrt{6}\lambda_2/2 - 2\gamma - A$
	1	$-\lambda_0/3 - \sqrt{6}\lambda_2/4 - \gamma - A/2$		1	$-\lambda_0/3 + \sqrt{6}\lambda_2/4 - \gamma - A/2$
	2	$\lambda_0/15 + \sqrt{6}\lambda_2/20 + \gamma + A/2$		2	$\lambda_0/15 - \sqrt{6}\lambda_2/20 + \gamma + A/2$
2	1	$-\lambda_0/3 - \sqrt{6}\lambda_2/4 - 3\gamma - A/2$	2	1	$-\lambda_0/3 + \sqrt{6}\lambda_2/4 - 3\gamma - A/2$
	2	$\lambda_0/3 + \sqrt{6}\lambda_2/4 - \gamma - A/6$		2	$\lambda_0/3 - \sqrt{6}\lambda_2/4 - \gamma - A/6$
	3	$-2\lambda_0/21 - \sqrt{6}\lambda_2/14 + 2\gamma + A/3$		3	$-2\lambda_0/21 + \sqrt{6}\lambda_2/14 + 2\gamma + A/3$
3	2	$-2\lambda_0/5 + \sqrt{6}\lambda_2/5 - 4\gamma - A/3$	3	2	$-2\lambda_0/5 - \sqrt{6}\lambda_2/5 - 4\gamma - A/3$
	3	$\lambda_0/2 - \sqrt{6}\lambda_2/4 - \gamma - A/12$		3	$\lambda_0/2 + \sqrt{6}\lambda_2/4 - \gamma - A/12$
	4	$-\lambda_0/6 + \sqrt{6}\lambda_2/12 + 3\gamma + A/4$		4	$-\lambda_0/6 - \sqrt{6}\lambda_2/12 + 3\gamma + A/4$
4	3	$-17\lambda_0/42 - 5\sqrt{6}\lambda_2/28 - 5\gamma - A/4$	4	3	$-17\lambda_0/42 + 5\sqrt{6}\lambda_2/28 - 5\gamma - A/4$
	4	$17\lambda_0/30 + \sqrt{6}\lambda_2/4 - \gamma - A/20$		4	$17\lambda_0/30 - \sqrt{6}\lambda_2/4 - \gamma - A/20$
	5	$-34\lambda_0/165 - \sqrt{6}\lambda_2/11 + 4\gamma + A/5$		5	$-34\lambda_0/165 + \sqrt{6}\lambda_2/11 + 4\gamma + A/5$
5	4	$-2\lambda_0/5 + \sqrt{6}\lambda_2/6 - 6\gamma - A/5$	5	4	$-2\lambda_0/5 - \sqrt{6}\lambda_2/6 - 6\gamma - A/5$
	5	$9\lambda_0/10 - \sqrt{6}\lambda_2/4 - \gamma - A/30$		5	$9\lambda_0/10 + \sqrt{6}\lambda_2/4 - \gamma - A/30$
	6	$-3\lambda_0/13 + 5\sqrt{6}\lambda_2/52 + 5\gamma + A/6$		6	$-3\lambda_0/13 - 5\sqrt{6}\lambda_2/52 + 5\gamma + A/6$
6	5	$-13\lambda_0/33 - 7\sqrt{6}\lambda_2/44 - 7\gamma - A/6$	6	5	$13\lambda_0/33 + 7\sqrt{6}\lambda_2/44 - 7\gamma - A/6$
	6	$13\lambda_0/21 + \sqrt{6}\lambda_2/4 - \gamma - A/42$		6	$13\lambda_0/21 - \sqrt{6}\lambda_2/4 - \gamma - A/42$
	7	$-26\lambda_0/105 - \sqrt{6}\lambda_2/10 + 6\gamma + A/7$		7	$-26\lambda_0/105 + \sqrt{6}\lambda_2/10 + 6\gamma + A/7$

nuclear spin states and the $-$ parity states with *para*:

$$\text{para-H}_2: \quad \text{combines with} |N, \Lambda, S, J; -\rangle$$
$$\text{ortho-H}_2: \quad \text{combines with} |N, \Lambda, S, J; +\rangle.$$

Note, for completeness, that the vibrational and electronic spin parts of the total wave function are both unaffected by P_{12}, that is, they are symmetric. We now have all the information required to derive expressions for the zero-field spin–spin and spin–rotation energies of the N, J levels for both *para*- and *ortho*-H$_2$, excluding nuclear magnetic hyperfine interaction for *ortho*-H$_2$ which we will come to in due course. These are given in table 8.6.

(iii) SPIN–ORBIT INTERACTIONS

The effects of spin–orbit coupling in the H$_2$ case (b) $^3\Pi$ state have been discussed in detail by Fontana [39] and Chiu [40, 41]. Chiu [41] starts by writing the full spin–orbit Hamiltonian for the two-electron, two-nucleus system as a sum of one-electron terms,

$$\mathcal{H}_{so} = a\mathbf{T}^1(\mathbf{l}_1) \cdot \mathbf{T}^1(\mathbf{s}_1) + b\mathbf{T}^1(\mathbf{l}_2) \cdot \mathbf{T}^1(\mathbf{s}_2) + c\mathbf{T}^1(\mathbf{l}_1) \cdot \mathbf{T}^1(\mathbf{s}_2) + d\mathbf{T}^1(\mathbf{l}_2) \cdot \mathbf{T}^1(\mathbf{s}_1). \quad (8.208)$$

The first two terms represent spin–orbit coupling, whilst the second two are normally described as spin–other-orbit terms. Following Fontana [39], Chiu pointed out that for matrix elements diagonal in the total electron spin $S\,(S = S_1 + S_2)$, (8.208) is contracted to the sum of two terms,

$$\mathcal{H}'_{so} = a'\,\mathrm{T}^1(l_1) \cdot \mathrm{T}^1(S) + b'\,\mathrm{T}^1(l_2) \cdot \mathrm{T}^1(S). \tag{8.209}$$

Both terms now contain spin–orbit and spin–other-orbit contributions. The constants a' and b' are defined (in atomic units) as follows:

$$a' = \frac{1}{4}\left[\frac{Z_a}{r_{1a}^3} + \frac{Z_b}{r_{1b}^3} - \frac{3}{r_{12}^3}\right], \quad b' = \frac{1}{4}\left[\frac{Z_b}{r_{2b}^3} + \frac{Z_a}{r_{2a}^3} - \frac{3}{r_{12}^3}\right]. \tag{8.210}$$

Z_a and Z_b are the nuclear charges of nuclei a and b, and 1 and 2 refer to the two electrons.

Equation (8.209) may be rewritten in the form

$$\mathcal{H}'_{so} = \frac{1}{2}(a'+b')(\mathrm{T}^1(l_1) + \mathrm{T}^1(l_2)) \cdot \mathrm{T}^1(S) + \frac{1}{2}(a'-b')(\mathrm{T}^1(l_1) - \mathrm{T}^1(l_2)) \cdot \mathrm{T}^1(S)$$

$$= A\,\mathrm{T}^1(L) \cdot \mathrm{T}^1(S) + \frac{1}{2}(a'-b')(\mathrm{T}^1(l_1) - \mathrm{T}^1(l_2)) \cdot \mathrm{T}^1(S). \tag{8.211}$$

The first term in (8.211) has the familiar form of the spin–orbit coupling operator, and we are interested in its matrix elements which are diagonal in the case (b) basis set. We first deal with the coupling of N and S to form J:

$$\langle \eta, \Lambda; N, S, J | A\,\mathrm{T}^1(L) \cdot \mathrm{T}^1(S) | \eta, \Lambda; N, S, J \rangle$$

$$= (-1)^{N+J+S} \begin{Bmatrix} S & N & J \\ N & S & 1 \end{Bmatrix} \langle \eta, \Lambda; N \| \mathrm{T}^1(L) \| \eta, \Lambda; N \rangle \langle S \| \mathrm{T}^1(S) \| S \rangle. \tag{8.212}$$

The first reduced matrix element in (8.212) is now expanded by using a first-rank rotation matrix to transform from the space- to the molecule-fixed axis system, and retaining only the wholly diagonal matrix elements:

$$\langle \eta, \Lambda; N, \Lambda \| \mathrm{T}^1(L) \| \eta, \Lambda; N, \Lambda \rangle = \langle N, \Lambda \| \mathcal{D}^{(1)}_{\cdot 0}(\omega)^* \| N, \Lambda \rangle \langle \eta, \Lambda | \mathrm{T}^1_0(L) | \eta, \Lambda \rangle$$

$$= (-1)^{N-\Lambda} \begin{pmatrix} N & 1 & N \\ -\Lambda & 0 & \Lambda \end{pmatrix} (2N+1)\Lambda. \tag{8.213}$$

Combining the results in (8.212) and (8.213) we obtain

$$\langle \eta, \Lambda; N, \Lambda, S, J | \mathcal{H}'_{so} | \eta, \Lambda; N, \Lambda, S, J \rangle$$

$$= A(-1)^{N+S+J} \begin{Bmatrix} S & N & J \\ N & S & 1 \end{Bmatrix} (-1)^{N-\Lambda} \begin{pmatrix} N & 1 & N \\ -\Lambda & 0 & \Lambda \end{pmatrix}$$

$$\times (2N+1)\Lambda\{S(S+1)(2S+1)\}^{1/2}$$

$$= A\frac{\{J(J+1) - N(N+1) - S(S+1)\}}{2N(N+1)}. \tag{8.214}$$

The final result in (8.214) is obtained by putting $\Lambda^2 = 1$. The constant A is half the sum of a spin–orbit coupling constant A_1 and a spin–other-orbit coupling constant A_2,

Table 8.7. *Energy matrix for $M_J = 0$ or ± 1 for the $N = 2$ level of para-H_2 in the $c\,^3\Pi_u$ state*

	$J = 1$	$J = 2$	$J = 3$
$J = 1$	$-\lambda^*/3 - A/2$ $+\mu_B B_z M_J\{-g_S/2 + g_L/4\}$	$\mu_B B_z\{4 - M_J^2\}^{1/2}$ $\times\{-g_S(\sqrt{3/2}\sqrt{5}) + g_L/4\sqrt{3}\}$	0
$J = 2$	$\mu_B B_z\{4 - M_J^2\}^{1/2}$ $\times\{-g_S(\sqrt{3/2}\sqrt{5}) + g_L/4\sqrt{3}\}$	$\lambda^*/3 - A/6$ $+\mu_B B_z M_J\{g_S/6 + 5g_L/36\}$	$\mu_B B_z\{9 - M_J^2\}^{1/2}$ $\times\{-g_S(\sqrt{2/3}\sqrt{5})$ $+\sqrt{2}g_L/18\}$
$J = 3$	0	$\mu_B B_z\{9 - M_J^2\}^{1/2}$ $\times\{-g_S(\sqrt{2/3}\sqrt{5}) + \sqrt{2}g_L/18\}$	$-2\lambda^*/21 + A/3$ $+\mu_B B_z M_J\{g_S/3 + g_L/9\}$

defined by

$$A_1 = \frac{1}{4}\left\langle \left(\frac{Z_a}{r_{1a}^3} + \frac{Z_b}{r_{1b}^3}\right) T_0^1(l_1) + \left(\frac{Z_a}{r_{2a}^3} + \frac{Z_b}{r_{2b}^3}\right) T_0^1(l_2) \right\rangle_\eta, \quad (8.215)$$

$$A_2 = \frac{3}{4}\left\langle \left[\frac{(\mathbf{r}_1 - \mathbf{r}_2) \wedge (\mathbf{p}_2 - \mathbf{p}_1)}{r_{12}^3}\right]_0 \right\rangle_\eta. \quad (8.216)$$

Both of these expressions are defined in the molecule-fixed ($q = 0$) coordinate system, and are expectation values over the electronic wave function for the vibronic state η. The contributions of (8.214) are included in the expressions for the first-order energies of the rotational levels given in table 8.6. There are, of course, many non-zero matrix elements in the case (b) basis, all of which are listed by Chiu [40].

Lichten [35] studied the magnetic resonance spectrum of the *para*-H_2, $N = 2$ level, and was able to determine the zero-field spin–spin and spin–orbit parameters; we will describe how this was done below. Before we come to that we note, from table 8.6, that in $N = 2$ it is not possible to separate λ_0 and λ_2. Measurements of the relative energies of the J spin components in $N = 2$ give values of $\lambda_0 + \sqrt{6}\lambda_2$, and the spin–orbit constant A; the spin–rotation constant γ is too small to be determined. In figure 8.18 we show a diagram of the lower rotational levels for both *para*- and *ortho*-H_2 in its $c\,^3\Pi_u$ state, which illustrates the difference between the two forms of H_2. This diagram does not show any details of the nuclear hyperfine splitting, which we will come to in due course.

We are now in a position to examine the details of the Zeeman effect in the *para*-H_2, $N = 2$ level, and thereby to understand Lichten's magnetic resonance studies. For each M_J component we may set up an energy matrix, using equations (8.180) and (8.181) which describe the Zeeman interactions, and equations (8.201), (8.206) and (8.214) which give the zero-field energies. Since $M_J = \pm 3$ components exist only for $J = 3$, diagonalisation in this case is not required. For $M_J = \pm 2$ the $J = 2$ and 3 states are involved. For $M_J = 0$ and ± 1, however, the matrices involve all three fine-structure states and take the form shown below in table 8.7. Note that λ^* is equal to $\lambda_0 + 3\sqrt{6}\lambda_2/4$ and the spin–rotation terms have been omitted. The diagonal Zeeman matrix elements are

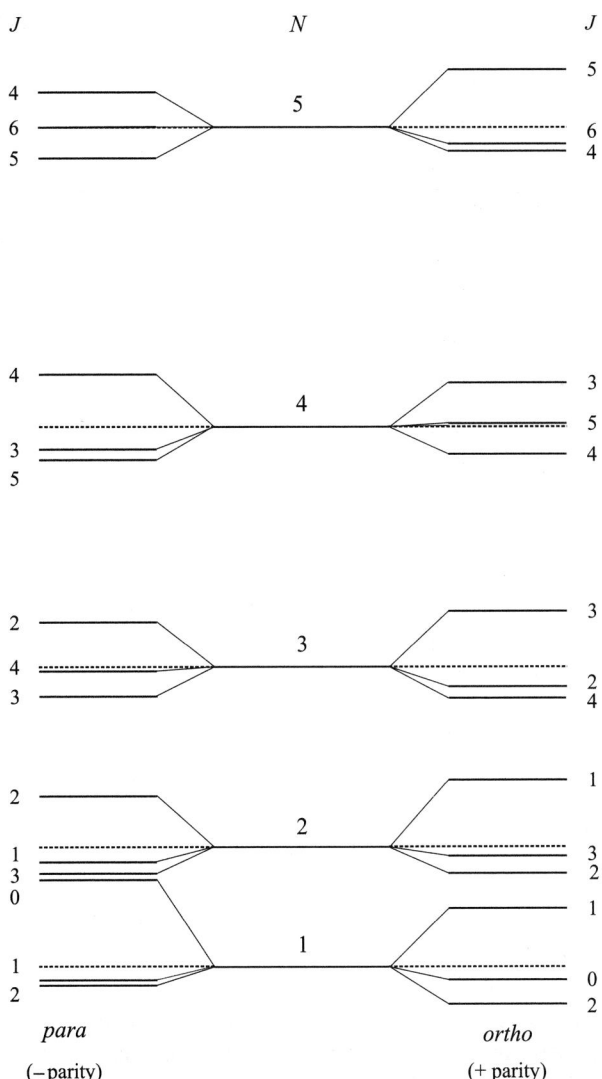

Figure 8.18. The lower rotational levels of the $c\,^3\Pi_u$ state of H_2, illustrating the difference between *para*- and *ortho*-H_2. The spin–spin splittings are drawn to scale, but the separations between N levels are actually very much larger than shown [39].

calculated from equation (8.182) and the off-diagonal Zeeman elements from equations (8.180) and (8.181). Matrix elements off-diagonal in N are ignored because these levels are widely separated. Each matrix is diagonalised for a given value of M_J, and values of the zero-field parameters. The experiments were performed at a fixed magnetic field sufficiently small that the Zeeman effect was essentially first order, as shown in figure 8.19. The $\Delta M_J = -1$ components of the $J = 2 \leftarrow 3$ transition occur at a lower frequency than the $\Delta M_J = +1$, and the off-diagonal Zeeman matrix elements result in separation of the five Zeeman components for each set. This is shown in the

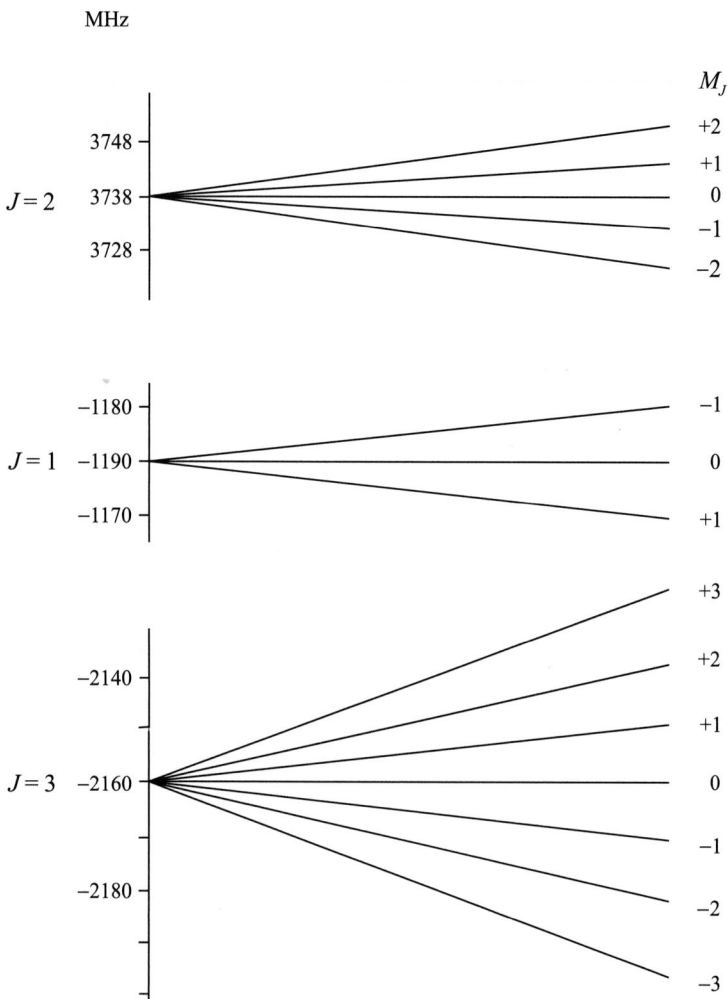

Figure 8.19. First-order Zeeman splitting of the $N = 2$, $J = 1, 2, 3$ fine-structure states for the $c\,^3\Pi_u$ state of H_2.

experimental spectrum obtained by Lichten, illustrated in figure 8.20 (*top*). Similarly the $J = 2 \leftarrow 1$ transition is split into two sets of three Zeeman components (figure 8.20 (*bottom*)). From these spectra the zero-field constants that determine the separation of the J levels are obtained (in MHz):

$$\lambda^* = 9303.08, \quad A = -3822.14.$$

As we shall see in due course, later studies of the *ortho* species enabled the two contributions to λ^* to be separated. We will then also compare the experimental values of the constants with those obtained from *ab initio* calculations. Two final points concerning these studies of *para*-H_2 are worth making. First, the spectra shown in figure 8.20 include $\Delta M_J = 0$ transitions, particularly at the centre where their line shape is anomalous, the reasons for which are discussed by Lichten [35]. Second, it

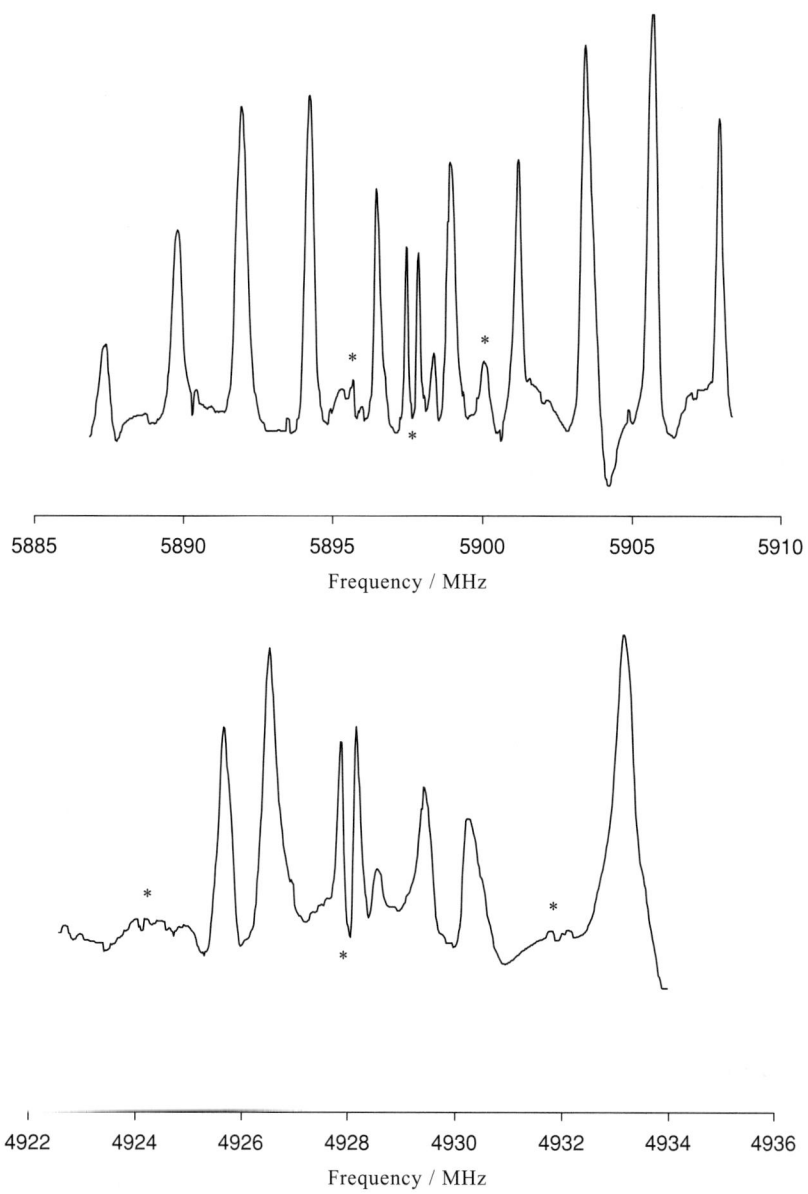

Figure 8.20. Magnetic resonance spectrum of H_2 in its $c\,^3\Pi_u$, $v = 0$, $N = 2$ state. The *top* spectrum arises from the $J = 3 \leftarrow 2$ transition, and the *bottom* from $J = 2 \leftarrow 1$. Lines marked * arise from $\Delta M_J = \pm 0$ transitions, the remainder obeying the $\Delta M_J = \pm 1$ selection rule [35].

is instructive to calculate the Zeeman energies in high magnetic fields, up to 10 kG. The effect of a strong magnetic field is to decouple the electron spin S from the molecular framework, so that ultimately the Zeeman levels separate into three groups characterised by the three allowed values of $M_S = 0, +1$ and -1. This decoupling corresponds to the well-known Paschen–Back effect in atomic spectroscopy; we will meet further examples later in this chapter.

(e) Effective zero-field Hamiltonian for ortho-H$_2$ ($^3\Pi_u$) in a case (b) basis

Experimental studies of *ortho*-H$_2$ in the $N = 1$ level of the $c\,^3\Pi_u$ state were described by Brooks, Lichten and Reno [42]. All of the interactions already described for *para*-H$_2$ are present, but in addition there is extensive magnetic hyperfine interaction. The theory of this has been developed by Jette and Cahill [38], whose analysis forms the basis of our treatment.

There are three separate contributions to the total magnetic hyperfine interaction, namely, the Fermi contact term, the orbital hyperfine term, and the electron spin–nuclear spin dipolar term:

$$\mathcal{H}_{\text{hfs}} = \mathcal{H}_{\text{F}} + \mathcal{H}_{\text{IL}} + \mathcal{H}_{\text{dip}}. \tag{8.217}$$

We will deal with each of these in turn, calculating the matrix elements in a coupled hyperfine basis, $\boldsymbol{F} = \boldsymbol{J} + \boldsymbol{I}$.

(i) FERMI CONTACT INTERACTION

The Fermi contact hyperfine interaction is written in the form

$$\mathcal{H}_{\text{F}} = b_{\text{F}} \mathbf{T}^1(\boldsymbol{S}) \cdot \mathbf{T}^1(\boldsymbol{I}), \tag{8.218}$$

where the Fermi contact constant b_{F} is defined by

$$b_{\text{F}} = \frac{2\mu_0}{3} g_S g_N \mu_B \mu_N \int \psi_{\text{el}}^* \delta(\boldsymbol{r}) \psi_{\text{el}} \, d\boldsymbol{r}, \tag{8.219}$$

where $\delta(\boldsymbol{r})$ is the Dirac delta function, and the integral represents the density of the electronic wave function at the nucleus. The matrix elements of (8.218) in the hyperfine-coupled case (b) basis set are derived as follows:

$$\langle \eta, \Lambda; N, S, J, I, F, M_F | b_{\text{F}} \mathbf{T}^1(\boldsymbol{S}) \cdot \mathbf{T}^1(\boldsymbol{I}) | \eta, \Lambda; N', S, J', I, F', M_F' \rangle$$

$$= b_{\text{F}} \delta_{M_F M_F'} \delta_{FF'} (-1)^{J'+F+I} \begin{Bmatrix} I & J' & F \\ J & I & 1 \end{Bmatrix} \langle N, S, J \| \mathbf{T}^1(\boldsymbol{S}) \| N', S, J' \rangle \langle I \| \mathbf{T}^1(\boldsymbol{I}) \| I \rangle$$

$$= b_{\text{F}} \delta_{M_F M_F'} \delta_{FF'} \delta_{NN'} (-1)^{J'+F+I} \begin{Bmatrix} I & J' & F \\ J & I & 1 \end{Bmatrix} \begin{Bmatrix} J & S & N \\ S & J' & 1 \end{Bmatrix} (-1)^{J+N+1+S}$$

$$\times \{(2J'+1)(2J+1)\}^{1/2} \{S(S+1)(2S+1)I(I+1)(2I+1)\}^{1/2}. \tag{8.220}$$

The matrix elements off-diagonal in J by 1 are important, because the hyperfine interaction is comparable with the electron spin splitting.

(ii) ORBITAL HYPERFINE INTERACTION

The orbital hyperfine interaction is written in the form

$$\mathcal{H}_{\text{IL}} = a\, \mathbf{T}^1(\boldsymbol{L}) \cdot \mathbf{T}^1(\boldsymbol{I}), \tag{8.221}$$

where the constant a is defined by

$$a = 2g_N\mu_B\mu_N(\mu_0/4\pi) \sum_j \langle \eta, \Lambda | 1/r_{jN}^3 | \eta, \Lambda \rangle; \qquad (8.222)$$

r_{jN} measures the position of electron j with respect to nucleus N; its expectation value can be calculated from the electronic wave function.

The matrix elements of (8.221) in the coupled–hyperfine case (b) basis are calculated in a similar manner to that described for the Fermi contact interaction. First we deal with the coupling of the N, S, J, I, F angular momenta:

$$\langle \eta, \Lambda; N, S, J, I, F, M_F | a\mathrm{T}^1(\boldsymbol{L}) \cdot \mathrm{T}^1(\boldsymbol{I}) | \eta, \Lambda'; N', S, J', I, F', M_F' \rangle$$

$$= a\delta_{M_F M_F'}\delta_{FF'}(-1)^{J'+F+I} \begin{Bmatrix} I & J' & F \\ J & I & 1 \end{Bmatrix} \langle \eta, \Lambda; N, S, J \| \mathrm{T}^1(\boldsymbol{L}) \| \eta, \Lambda'; N', S, J' \rangle$$

$$\times \langle I \| \mathrm{T}^1(\boldsymbol{I}) \| I \rangle$$

$$= a\delta_{M_F M_F'}\delta_{FF'}(-1)^{J'+F+I} \begin{Bmatrix} I & J' & F \\ J & I & 1 \end{Bmatrix} (-1)^{J'+N+1+S}\{(2J'+1)(2J+1)\}^{1/2}$$

$$\times \begin{Bmatrix} J & N & S \\ N' & J' & 1 \end{Bmatrix} \langle \eta, \Lambda; N, \Lambda \| \mathrm{T}^1(\boldsymbol{L}) \| \eta, \Lambda'; N', \Lambda' \rangle \{I(I+1)(2I+1)\}^{1/2}.$$

$$(8.223)$$

We have already met the matrix element of $\mathrm{T}^1(\boldsymbol{L})$ in equation (8.213) and, if we again restrict ourselves to the important matrix elements diagonal in N and Λ, we can make use of our earlier results to obtain the required matrix elements of the orbital hyperfine interaction:

$$\langle \eta, \Lambda; N, S, J, I, F, M_F | aT^1(\boldsymbol{L}) \cdot T^1(\boldsymbol{I}) | \eta, \Lambda; N, S, J', I, F, M_F \rangle$$

$$= a(-1)^{J'+F+I} \begin{Bmatrix} I & J' & F \\ J & I & 1 \end{Bmatrix} (-1)^{J'+N+1+S}\{(2J'+1)(2J+1)\}^{1/2} \begin{Bmatrix} J & N & S \\ N & J' & 1 \end{Bmatrix}$$

$$\times \langle \eta, \Lambda; N, \Lambda \| T^1(\boldsymbol{L}) \| \eta, \Lambda; N, \Lambda \rangle \{I(I+1)(2I+1)\}^{1/2}$$

$$= a(-1)^{J'+F+I} \begin{Bmatrix} I & J' & F \\ J & I & 1 \end{Bmatrix} (-1)^{J'+N+1+S}\{(2J'+1)(2J+1)\}^{1/2} \begin{Bmatrix} J & N & S \\ N & J' & 1 \end{Bmatrix}$$

$$\times (-1)^{N-\Lambda} \begin{pmatrix} N & 1 & N \\ -\Lambda & 0 & \Lambda \end{pmatrix} (2N+1)\Lambda\{I(I+1)(2I+1)\}^{1/2}. \qquad (8.224)$$

As we mentioned above for the Fermi contact interaction, it is important to retain the matrix elements off-diagonal in J because their effects are significant.

(iii) ELECTRON SPIN–NUCLEAR SPIN DIPOLAR INTERACTION

The dipolar hyperfine interaction is a through-space interaction of the electron and nuclear spin magnetic moments. As such, it is similar to the nuclear spin–nuclear spin dipolar interaction discussed earlier in connection with the H_2 molecule in its ground electronic state. We shall meet the dipolar hyperfine interaction in many examples described later, so at the risk of seeming somewhat pedantic and repetitive, we here describe the details of the derivation of the electron–nuclear dipolar Hamiltonian in irreducible tensor form.

The classical expression for the interaction energy of two magnetic moments μ_S and μ_N is given by

$$E = \frac{\mu_0}{4\pi}\left\{\frac{\mu_S \cdot \mu_N}{r^3} - \frac{3(\mu_S \cdot r)(\mu_N \cdot r)}{r^5}\right\} \tag{8.225}$$

where r is the radius vector from μ_S to μ_N and r is the distance between the two moments. The quantum mechanical version of (8.225) is obtained simply by substituting

$$\mu_S = -g_S\mu_B S, \quad \mu_N = g_N\mu_N I \tag{8.226}$$

yielding the dipolar interaction Hamiltonian

$$\mathcal{H}_{\text{dip}} = -g_S\mu_B g_N\mu_N \frac{\mu_0}{4\pi}\left\{\frac{I \cdot S}{r^3} - \frac{3(I \cdot r)(S \cdot r)}{r^5}\right\}. \tag{8.227}$$

Note that this is similar to the Hamiltonian describing the dipolar interaction of two nuclear magnetic moments, presented in equation (8.9), but is *opposite* in sign. Using the results derived in the first part of appendix 8.1, we see that the dipolar Hamiltonian (8.227) may be written as a cartesian tensorial operator:

$$\mathcal{H}_{\text{dip}} = [S_x, S_y, S_z]\begin{bmatrix} T_{xx} & T_{xy} & T_{xz} \\ T_{yx} & T_{yy} & T_{yz} \\ T_{zx} & T_{zy} & T_{zz} \end{bmatrix}\begin{bmatrix} I_x \\ I_y \\ I_z \end{bmatrix}. \tag{8.228}$$

We will discuss the details of the cartesian components of the second-rank tensor T in due course.

We now wish to reformulate the problem using spherical rather than cartesian tensors. It is by no means obvious that an equivalent form of the operator (8.227) is

$$\mathcal{H}_{\text{dip}} = -\sqrt{10}g_S\mu_B g_N\mu_N(\mu_0/4\pi)T^1(I) \cdot T^1(S, C^2), \tag{8.229}$$

where the spherical components of the new tensor $T^1(S, C^2)$ are defined by

$$T_q^1(S, C^2) = -\sqrt{3}\sum_{q_1 q_2}(-1)^q T_{q_1}^1(S) T_{q_2}^2(C)\begin{pmatrix} 1 & 2 & 1 \\ q_1 & q_2 & -q \end{pmatrix} \tag{8.230}$$

where

$$T_{q_2}^2(C) = \langle \eta, \Lambda|C_{q_2}^2(\theta, \phi)(r^{-3})|\eta, \Lambda\rangle. \tag{8.231}$$

The components $C_{q_2}^2(\theta, \phi)$ are spherical harmonics, with the angles θ and ϕ defined in figure 8.52, shown in appendix 8.1. Equation (8.229) is similar to (8.10), except that we have chosen to couple the vectors differently because of the basis set used in the present problem. Clearly the components of the cartesian tensor T are related to those of the spherical tensor $T^2(C)$; these relationships are derived in appendix 8.2.

We now examine the matrix elements of (8.229) in our case (b) basis. The nuclear spin dependence is readily evaluated:

$$\langle \eta, \Lambda; N, S, J, I, F, M_F | \mathcal{H}_{\text{dip}} | \eta, \Lambda'; N', S, J', I, F', M_F' \rangle$$

$$= \sqrt{30} g_S \mu_B g_N \mu_N (\mu_0/4\pi) \delta_{M_F M_F'} \delta_{FF'} (-1)^{J'+F+I+1} \begin{Bmatrix} I & J' & F \\ J & I & 1 \end{Bmatrix}$$

$$\times \langle \eta, \Lambda; N, S, J \| T^1(S, C^2) \| \eta, \Lambda'; N', S, J' \rangle \langle I \| T^1(I) \| I \rangle$$

$$= \sqrt{30} g_S \mu_B g_N \mu_N (\mu_0/4\pi) \delta_{M_F M_F'} \delta_{FF'} (-1)^{J'+F+I+1} \begin{Bmatrix} I & J' & F \\ J & I & 1 \end{Bmatrix}$$

$$\times \{I(I+1)(2I+1)\}^{1/2} \langle \eta, \Lambda; N, S, J \| T^1(S, C^2) \| \eta, \Lambda'; N', S, J' \rangle. \quad (8.232)$$

The complications lie in the remaining matrix elements of $T^1(S, C^2)$. First we note a result which we have used previously which deals with the coupling of N and S:

$$\langle \eta, \Lambda; N, \Lambda, S, J \| T^1(S, C^2) \| \eta', \Lambda'; N', \Lambda', S, J' \rangle$$

$$= \{(3)(2J+1)(2J'+1)\}^{1/2} \begin{Bmatrix} J & J' & 1 \\ N & N' & 2 \\ S & S & 1 \end{Bmatrix} \langle \eta, \Lambda; N, \Lambda \| T^2(C) \| \eta', \Lambda'; N', \Lambda' \rangle$$

$$\times \{S(S+1)(2S+1)\}^{1/2}. \quad (8.233)$$

In order to proceed further, and to evaluate the remaining matrix element in (8.233), we must now take note of the fact that Λ is a *signed* quantity and therefore use the parity-conserving combinations introduced earlier:

$$|\eta; N, \Lambda, +\rangle = \frac{1}{\sqrt{2}} \{|+1\rangle + (-1)^N |-1\rangle\},$$

$$|\eta; N, \Lambda, -\rangle = \frac{1}{\sqrt{2}} \{|+1\rangle - (-1)^N |-1\rangle\}. \quad (8.234)$$

With these functions as bases, and with the neglect of matrix elements off-diagonal in N, we have the following results:

$$\langle \eta; N, \Lambda, + \| T^2(C) \| \eta; N, \Lambda, + \rangle = (-1)^{N-\Lambda} (2N+1) \left\{ \begin{pmatrix} N & 2 & N \\ -1 & 0 & 1 \end{pmatrix} \langle T_0^2(C) \rangle_\eta \right.$$

$$\left. + (-1)^N \begin{pmatrix} N & 2 & N \\ -1 & 2 & -1 \end{pmatrix} \langle T_2^2(C) \rangle_\eta \right\}, \quad (8.235)$$

$$\langle \eta; N, \Lambda, - \| T^2(C) \| \eta; N, \Lambda, - \rangle = (-1)^{N-\Lambda} (2N+1) \left\{ \begin{pmatrix} N & 2 & N \\ -1 & 0 & 1 \end{pmatrix} \langle T_0^2(C) \rangle_\eta \right.$$

$$\left. - (-1)^N \begin{pmatrix} N & 2 & N \\ -1 & 2 & -1 \end{pmatrix} \langle T_2^2(C) \rangle_\eta \right\}. \quad (8.236)$$

We can relate the components of $T^2(C)$ in the above equations to the constants c and

d given by Jette and Cahill [38] by noting the following identities:

$$\begin{aligned}
\langle T_0^2(\boldsymbol{C})\rangle &= g_S\mu_B g_N\mu_N(\mu_0/4\pi)\left(\frac{4\pi}{5}\right)^{1/2}\langle r^{-3}Y_{20}(\theta,\phi)\rangle \\
&= \frac{1}{2}g_S\mu_B g_N\mu_N(\mu_0/4\pi)\left\langle\frac{(3\cos^2\theta-1)}{r^3}\right\rangle = \frac{1}{3}c, \\
\langle T_2^2(\boldsymbol{C})\rangle &= g_S\mu_B g_N\mu_N(\mu_0/4\pi)\left(\frac{4\pi}{5}\right)^{1/2}\langle r^{-3}Y_{22}(\theta,\phi)\rangle \\
&= \frac{3}{2\sqrt{6}}g_S\mu_B g_N\mu_N(\mu_0/4\pi)\left\langle\frac{\sin^2\theta}{r^3}\right\rangle = -\frac{d}{\sqrt{6}}.
\end{aligned} \qquad (8.237)$$

Our results agree with those of Jette and Cahill if $g_S = 2$.

We are now in a position to apply our results to the fine and hyperfine structure of the $N = 1$ level of *ortho*-H$_2$, which was studied experimentally by Brooks, Lichten and Reno [42]. There are seven basis states to consider:

$$N = 1: \quad J = 2, F = 3, 2, 1; \quad J = 1, F = 2, 1, 0; \quad J = 0, F = 1.$$

For the zero-field problem F remains a good quantum number, but J is not because of the hyperfine mixing. The spin–spin, spin–orbit and spin–rotation energies have already been listed in table 8.6. The complete zero-field energy matrix, including the hyperfine terms, is as follows.

		$J=2$			$J=1$			$J=0$
		$F=3$	$F=2$	$F=1$	$F=2$	$F=1$	$F=0$	$F=1$
$J=2$	$F=3$	m_{11}	0	0	0	0	0	0
	$F=2$	0	m_{22}	0	m_{24}	0	0	0
	$F=1$	0	0	m_{33}	0	m_{35}	0	0
$J=1$	$F=2$	0	m_{42}	0	m_{44}	0	0	0
	$F=1$	0	0	m_{53}	0	m_{55}	0	m_{57}
	$F=0$	0	0	0	0	0	m_{66}	0
$J=0$	$F=1$	0	0	0	0	m_{75}	0	m_{77}

$$\begin{aligned}
m_{11} &= \lambda_0/15 - \sqrt{6}\lambda_2/20 + \gamma + A/2 + b_F + a/2 + x/5 \\
m_{22} &= \lambda_0/15 - \sqrt{6}\lambda_2/20 + \gamma + A/2 - b_F/2 - a/4 - x/10 \\
m_{33} &= \lambda_0/15 - \sqrt{6}\lambda_2/20 + \gamma + A/2 - 3b_F/2 - 3a/4 - 3x/10 \\
m_{44} &= -\lambda_0/3 + \sqrt{6}\lambda_2/4 - \gamma - A/2 + b_F/2 + a/4 - x/2 \\
m_{55} &= -\lambda_0/3 + \sqrt{6}\lambda_2/4 - \gamma - A/2 - b_F/2 - a/4 + x/2 \\
m_{66} &= -\lambda_0/3 + \sqrt{6}\lambda_2/4 - \gamma - A/2 - b_F - a/2 + x
\end{aligned}$$

$$m_{77} = 2\lambda_0/3 - \sqrt{6}\lambda_2/2 - 2\gamma - A$$
$$m_{24} = m_{42} = (\sqrt{3})b_F/2 - a(\sqrt{3})/4 - 3x/(\sqrt{75})$$
$$m_{35} = m_{53} = (\sqrt{15})b_F/6 - a(\sqrt{15})/12 - x/(\sqrt{15})$$
$$m_{57} = m_{75} = 2b_F/(\sqrt{3}) - a/(\sqrt{3}) + x/(\sqrt{3})$$
(8.238)

In these matrix elements $x = (c/3) - d$.

Given values of the constants appearing in the expressions for the matrix elements listed in (8.238), we can calculate the energies of the hyperfine levels for $N = 1$ and hence calculate the transition frequencies. This we shall now do, but note that this is the easy direction in which to proceed; Brooks, Lichten and Reno [42] had the more difficult problem of determining the constants from the experimental data. Their values [42] are as follows (MHz):

$$A = -3717.120, \quad \lambda_0 - 3\sqrt{6}\lambda_2/4 = -7171.866, \quad b_F = 450.479,$$
$$a = 26.6, \quad c - 3d = 104.177.$$

The calculated energy levels and the observed transitions are illustrated in figure 8.21; the agreement between experiment and theory was found to be excellent for $N = 1$. It was also clear that transitions involving two different vibrational levels (probably $v = 0$ and 1) were being observed, and that it was possible to separate them.

Comparing results for the $N = 2$ levels of para-H_2 and the $N = 1$ levels of ortho-H_2 it is clear that, in principle, λ_0 and λ_2 can be separated; the values (in MHz) thus obtained

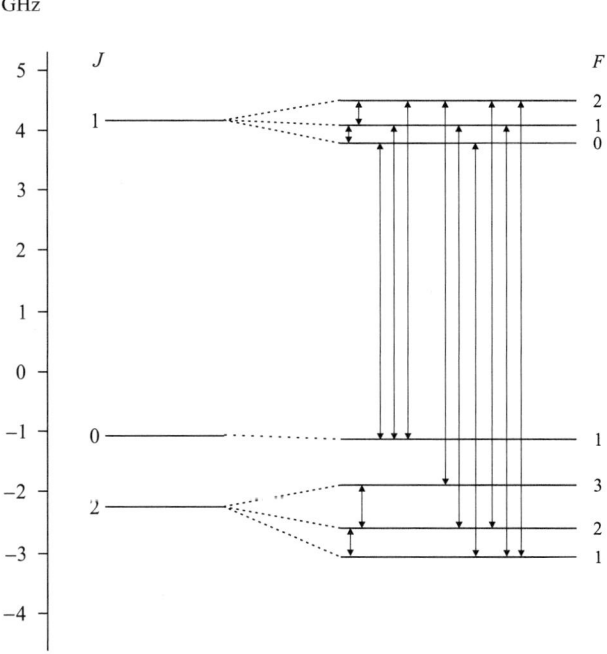

Figure 8.21. Fine and hyperfine structure of the $N = 1$ level of ortho-H_2, and the observed transitions.

are $\lambda_0 = 1420.806$, $\lambda_2 = 4483.911$. There is, however, a problem with this procedure in that one would hope to obtain the same value of the spin–orbit coupling constant, A, from both rotational levels, but the actual values differed by about 2.5%. Brooks, Lichten and Reno [42] pointed out that there is nothing in the experimental data to identify the vibrational level involved in the spectra, and it may not be the same level in the $N = 1$ and $N = 2$ studies, which were made several years apart. Relevant questions concerning the relative predissociation rates of different vibrational levels of the excited electronic state are complicated. There will, however, be a vibrational dependence of the parameters determined from the magnetic resonance spectra. An equally plausible explanation, since H_2 is such a light molecule, is that the difference between the two A values is caused by centrifugal distortion effects.

Many theoretical calculations of the molecular constants have been described, using a variety of models for the electronic structure. We do not intend to go into the details here, except to note that one of the simplest models of the excited electronic states of H_2 describes them as an H_2^+ core coupled with a Rydberg electron. In this connection it is significant to note that the Fermi contact interaction constant b_F for the $c\ ^3\Pi_u$ state of H_2 is very close to the value determined for H_2^+ itself, described in chapter 11.

Figure 8.16 shows that the next excited $^3\Pi_u$ state of H_2 is the d state. This has been studied by means of some elegant double resonance studies, which are described in detail in chapter 11.

8.3.2. N_2 in the $A\ ^3\Sigma_u^+$ state

(a) Introduction

Molecular nitrogen, N_2, is one of the most extensively studied diatomic molecules and optical spectroscopy has provided a wealth of information about its ground and excited electronic states. Molecular beam magnetic resonance studies of N_2 in its ground state have yielded information about ^{14}N nuclear spin dipolar and quadrupole interactions. Similar studies of N_2 in its electronically excited $A\ ^3\Sigma_u^+$ state were described in two very extensive papers by Freund, Miller, De Santis and Lurio [43] (paper I) and De Santis, Lurio, Miller and Freund [44] (paper II). We will describe their results and analysis in detail, but first note in passing that, strictly speaking, the lowest excited triplet state should be labelled the a state; the label A has been used by all concerned in the past, so we will continue to do so.

The electron configurations of the ground ($X\ ^1\Sigma_g^+$) and lowest excited ($A\ ^3\Sigma_u^+$) states may be written in terms of molecular orbital theory as

$$X\ ^1\Sigma_g^+ : (1s\ \sigma_g)^2(1s\ \sigma_u^*)^2(2s\ \sigma_g)^2(2s\ \sigma_u^*)^2(2p\ \pi_u)^4(2p\sigma_g)^2$$
$$A\ ^3\Sigma_u^+ : (1s\ \sigma_g)^2(1s\ \sigma_u^*)^2(2s\ \sigma_g)^2(2s\ \sigma_u^*)^2(2p\ \pi_u)^3(2p\sigma_g)^2(2p\ \pi_g^*)^1.$$

The A state lies 50 203.6 cm^{-1} above the ground state. The next excited state is a $B\ ^3\Pi_g$ state and the $B - A$ electronic band system, which is known as the first positive system, was studied by a number of workers [45, 46]. Benesch, Vanderslice, Tilford and Wilkinson [47] re-analysed the available data to produce the best set of vibrational constants for the A state, which are (in cm^{-1})

$$\omega_e = 1460.518, \quad \omega_e x_e = 13.8313, \quad \omega_e y_e = 5.999 \times 10^{-3}, \quad \omega_e z_e = -1.853 \times 10^{-3}.$$

Information about the rotational and fine-structure constants was also obtained.

The $A\,^3\Sigma_u^+$ state has a radiative lifetime of 2 s, and is therefore long-lived on the time scale of the molecular beam experiments, where it was produced by electron impact on a beam of N$_2$ diluted with Ar, cooled to liquid nitrogen temperature. Calculations of the Franck–Condon factors show that direct excitation of the A state is expected to populate many vibrational levels, and a major feature of the magnetic resonance studies was that spectra involving the first thirteen vibrational levels ($v = 0$ to 12) were obtained. The accurate and extensive determination of the vibrational constants, given above, was therefore of considerable importance.

(b) Experimental studies and results

The experimental studies were performed using a conventional molecular beam magnetic resonance machine with dipole A and B fields separated by a homogeneous C field containing a radiofrequency 'hairpin' device to induce transitions in the frequency range 0 to 150 MHz. The electronically excited N$_2$ molecules were detected by means of an Auger detector coupled to an electron multiplier; the distance from the electron impact region to the final beam detector was 50 cm. In paper I [43] pure magnetic resonance transitions ($\Delta F = 0$) were reported, but in paper II [44] direct transitions between nuclear hyperfine levels ($\Delta F = \pm 1$) were described.

In order to understand the observed spectra it is first necessary to consider the importance of nuclear spin in determining the nature of the allowed rotational levels. Each ^{14}N nucleus has spin $\boldsymbol{I} = 1$, and in the homonuclear N$_2$ system the individual spins are coupled to form a total nuclear spin I_T of 2, 1 or 0. The most appropriate basis system for $^3\Sigma_u^+$ N$_2$ is Hund's case (b) coupling:

$$\boldsymbol{N} + \boldsymbol{S} = \boldsymbol{J}, \quad \boldsymbol{J} + \boldsymbol{I}_T = \boldsymbol{F}.$$

Consideration of symmetry with respect to nuclear interchange shows, however, that there are restrictions on the allowed combinations of N and I_T. Even values of I_T (0 or 2) correspond to *ortho*-N$_2$, and can only combine with *odd* values of N. The *odd* value of I_T (1), which corresponds to *para*-N$_2$, can only combine with *even* values of N. This association is the reverse of that which occurs in the ground state of N$_2$ because of the u character of the excited state. The situation for the first four rotational levels ($N = 0$ to 3) is summarised schematically in figure 8.22; this pattern, of course, applies for all vibrational levels.

In an applied magnetic field each F level shown in figure 8.22 splits into $2F + 1$ components, each characterised by a different value of M_F. The first magnetic resonance paper I [43] described transitions obeying the selection rules $\Delta F = 0$, $\Delta M_F = \pm 1$, whilst paper II [44] dealt with $\Delta F = \pm 1$, $\Delta M_F = 0, \pm 1$ transitions.

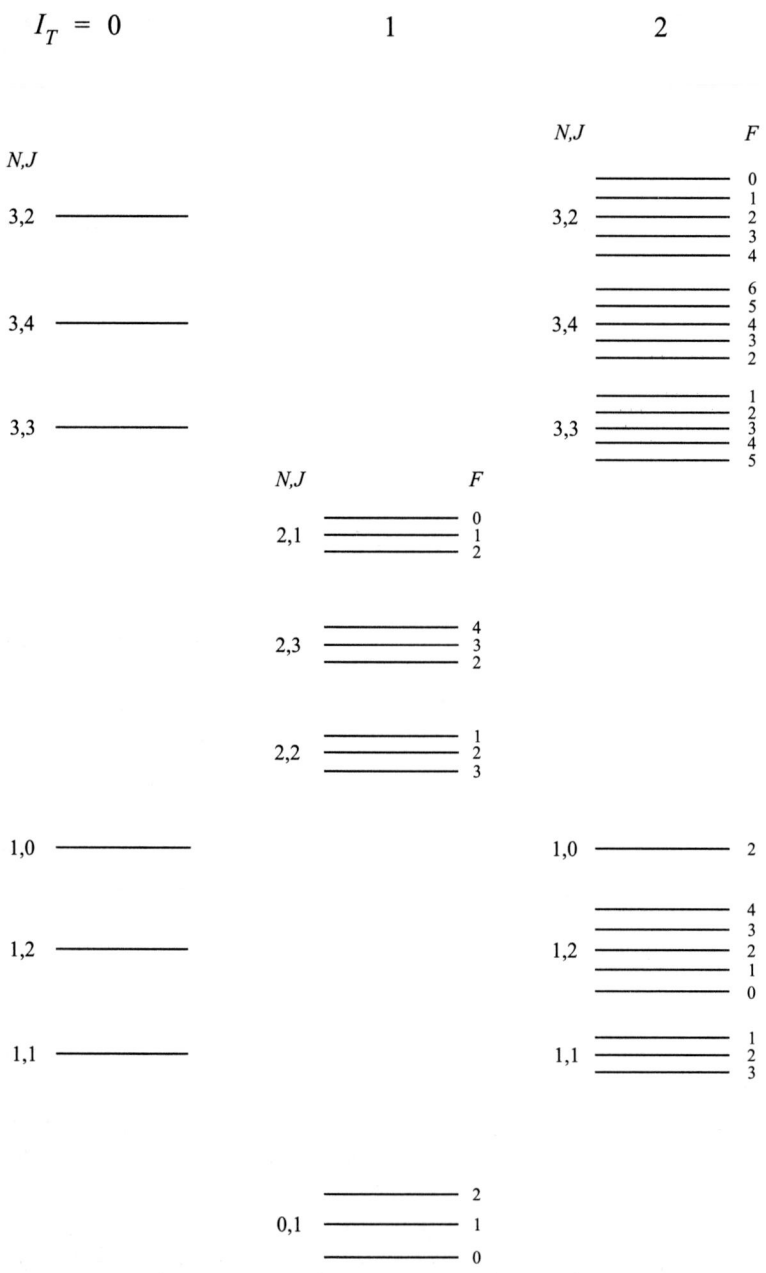

Figure 8.22. Schematic energy level diagram for the first four rotational levels of N_2 in its $A\,^3\Sigma_u^+$ state, showing the nuclear hyperfine states which are allowed to combine with each N level. Relative vertical spacings are not drawn to scale [43].

It is clearly time to consider the effects of an applied magnetic field in more detail.

(c) Zeeman effect

The simplest possible description of the Zeeman interaction is a single-term effective Hamiltonian describing the magnetic interaction between the applied field and the electron spin magnetic moment,

$$\mathcal{H}_Z = g_S \mu_B T_0^1(\boldsymbol{B}) T_0^1(\boldsymbol{S}). \tag{8.239}$$

The direction of the magnetic field defines the space-fixed $p = 0$ (or Z) direction. Equation (8.239) represents a very simplified version, in that it neglects the nuclear and rotational Zeeman effects, as well as the second-order effects of spin–orbit coupling, none of which are negligible. Nevertheless (8.239) will allow us to derive theoretical values for the first-order effective g-factors, for comparison with the experimental spectra [43]. The required matrix elements of (8.239) in a case (b) hyperfine-coupled basis are as follows:

$$\langle \eta, N, S, J, I_T, F, M_F | g_S \mu_B T_0^1(\boldsymbol{B}) T_0^1(\boldsymbol{S}) | \eta, N, S, J', I_T, F', M_F \rangle$$

$$= g_S \mu_B B_Z (-1)^{F-M} \begin{pmatrix} F & 1 & F' \\ -M & 0 & M \end{pmatrix} \langle \eta, N, S, J, I_T, F \| T^1(\boldsymbol{S}) \| \eta, N, S, J', I_T, F' \rangle$$

$$= g_S \mu_B B_Z (-1)^{F-M} \begin{pmatrix} F & 1 & F' \\ -M & 0 & M \end{pmatrix} (-1)^{F'+J+1+I} \{(2F'+1)(2F+1)\}^{1/2}$$

$$\times \begin{Bmatrix} F & J & I_T \\ J' & F' & 1 \end{Bmatrix} \langle \eta, N, S, J \| T^1(\boldsymbol{S}) \| \eta, N, S, J' \rangle$$

$$= g_S \mu_B B_Z (-1)^{F-M} \begin{pmatrix} F & 1 & F' \\ -M & 0 & M \end{pmatrix} (-1)^{F'+J+1+I} \{(2F'+1)(2F+1)\}^{1/2}$$

$$\times \begin{Bmatrix} F & J & I_T \\ J' & F' & 1 \end{Bmatrix} (-1)^{J+N+1+S} \{(2J'+1)(2J+1)\}^{1/2} \begin{Bmatrix} J & S & N \\ S & J' & 1 \end{Bmatrix}$$

$$\times \{S(S+1)(2S+1)\}^{1/2}. \tag{8.240}$$

Equation (8.240) shows that neither J nor F remains a good quantum number in the presence of a magnetic field, and the Zeeman effect will become highly nonlinear as the Zeeman energy becomes comparable with the hyperfine and spin coupling energy. However, we are interested in the first-order effective g-factors that will be appropriate at low magnetic fields and we therefore confine our attention at present to the diagonal elements of (8.240). Expanding the 3-j and 6-j symbols, and putting $S = 1$ and $g_S = 2$, we obtain the following result:

$$g_F(N, J, I_T, F) = \frac{\{F(F+1) + J(J+1) - I_T(I_T+1)\}\{J(J+1) - N(N+1) + 2\}}{2F(F+1)J(J+1)}. \tag{8.241}$$

The theoretical first-order g-factors calculated from this equation for the lowest

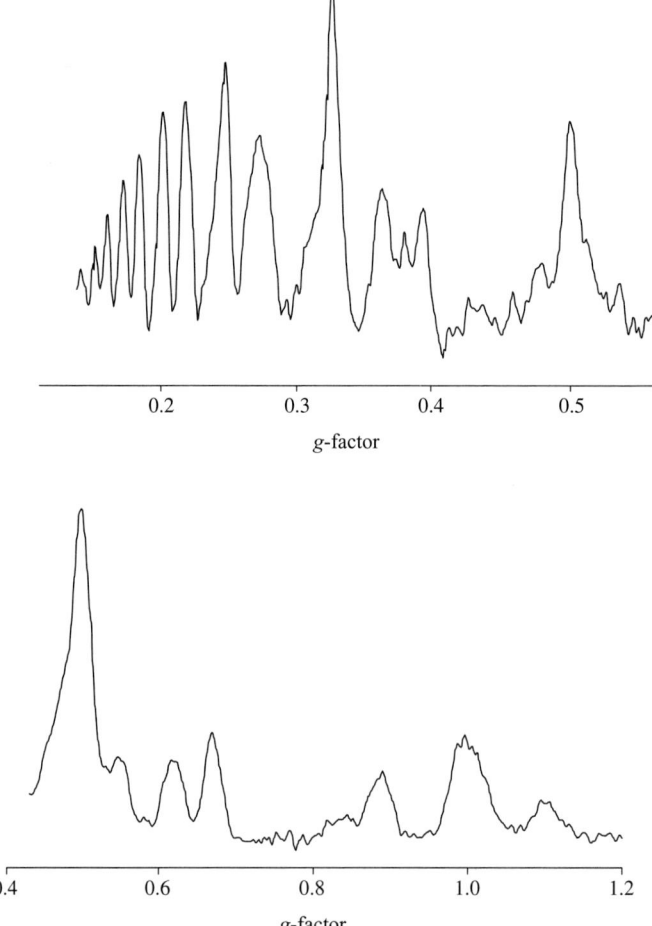

Figure 8.23. Low magnetic field spectrum of N_2 $A\,^3\Sigma_u^+$, showing effective g-factors. The upper trace was obtained at 3.55 G and the lower trace at 1.90 G [43].

rotational levels are listed in table 8.8. These predictions can be compared with the experimental spectra described in paper I and shown in figure 8.23. It is clear that resonances for a large number of N, J, F levels can be unambiguously identified. The observed sequence of resonances with g-factors less than 0.2 arises from levels with increasing values of N. The results constituted the first definitive identification of the $\Delta F = 0$ magnetic resonance spectrum. The authors were able to go much further by examining magnetic resonance spectra at intermediate field strengths, and thereby to make approximate determinations of the hyperfine constants. However, in paper II they described new measurements in which the line width was reduced to approximately 10 kHz. They were able to study direct hyperfine transitions ($\Delta F = \pm 1$) which provide much more accurate information for an analysis of the hyperfine interactions. We now consider the appropriate theory needed to understand the additional experimental results and their analysis.

Table 8.8. *Calculated first-order g-factors for the lowest N, J, F levels of N_2 $A\,^3\Sigma_u^+$*

N	J	I_T	g_J	F	g_F
0	1	1	—	2	1.000
			—	1	1.000
			—	0	—
1	1	0	1.000	1	—
1	2	0	1.000	2	—
1	0	0	—	0	—
1	1	2	—	3	0.333
			—	2	0.167
			—	1	−0.500
1	2	2	—	4	0.500
			—	3	0.500
			—	2	0.500
			—	1	0.500
			—	0	—
1	0	2	—	2	—
2	2	1	—	3	0.222
			—	2	0.278
			—	1	0.500
2	3	1	—	4	0.500
			—	3	0.611
			—	2	0.889
2	1	1	—	2	−0.500
			—	1	−0.500
			—	0	—
3	3	0	0.167	3	—
3	4	0	0.500	4	—
3	2	0	−0.667	2	—
3	3	2	—	5	0.100
			—	4	0.108
			—	3	0.125
			—	2	0.167
			—	1	0.333
3	4	2	—	6	0.333
			—	5	0.367
			—	4	0.425
			—	3	0.541
			—	2	0.833
3	2	2	—	4	−0.333
			—	3	−0.333
			—	2	−0.333
			—	1	−0.333
			—	0	—

(d) Electron and nuclear spin interactions

(i) EFFECTIVE HAMILTONIAN

In a detailed consideration of the full Hamiltonian, De Santis, Lurio, Miller and Freund [44] in paper II show that the required effective Hamiltonian for a given vibrational level v can be written as the sum of a part describing the rotational motion with electron spin interactions, and a part describing the magnetic and electric hyperfine interactions. The first part may be written:

$$\mathcal{H}_{\text{eff}} = \mathcal{H}_{\text{rot}} + \mathcal{H}_{\text{ss}} + \mathcal{H}_{\text{sr}}. \tag{8.242}$$

The rotational Hamiltonian, including the effects of centrifugal distortion, is given by

$$\mathcal{H}_{\text{rot}} = B_v \mathbf{N}^2 - D_v \mathbf{N}^4. \tag{8.243}$$

The electron spin–spin interaction term was written in paper II as

$$\mathcal{H}_{\text{ss}} = 2\lambda_v \, \mathbf{T}^2(\mathbf{S}, \mathbf{S}) \cdot \mathbf{T}^2(\mathbf{n}, \mathbf{n}) + \lambda_D \left[\mathbf{T}^2(\mathbf{S}, \mathbf{S}) \cdot \mathbf{T}^2(\mathbf{n}, \mathbf{n}), \mathbf{N}^2 \right]_+ \tag{8.244}$$

where $\mathbf{n}$ is the unit vector along the molecule-fixed z axis.

The second term in (8.244) describes the centrifugal correction to the spin–spin interaction, where the subscript $+$ denotes the anticommutator; we shall, however, omit this term from further consideration since its effects were very small. It is more convenient to take the spin–spin dipolar interaction term to have the form

$$\mathcal{H}_{ss} = -g_S^2 \mu_B^2 \, (\mu_0/4\pi) \sqrt{6} \, \mathbf{T}^2(\mathbf{C}) \cdot \mathbf{T}^2(\mathbf{S}_1, \mathbf{S}_2) \tag{8.245}$$

in order to remain consistent with our earlier treatment of H$_2$ in its excited triplet state (see also appendix 8.3.)

Finally in this part of the effective Hamiltonian, the spin–rotation interaction takes the simple form

$$\mathcal{H}_{\text{sr}} = \gamma_v \, \mathbf{T}^1(\mathbf{N}) \cdot \mathbf{T}^1(\mathbf{S}). \tag{8.246}$$

The magnetic hyperfine interaction is represented by the sum of two terms, $\mathcal{H}_F$ representing the Fermi contact interaction, and $\mathcal{H}_{\text{dip}}$ representing the electron spin-nuclear spin dipolar interaction. They are written as follows:

$$\mathcal{H}_F = b_F(v) \, \mathbf{T}^1(\mathbf{S}) \cdot \mathbf{T}^1(\mathbf{I}_T). \tag{8.247}$$

$$\mathcal{H}_{\text{dip}} = -\sqrt{10} \, g_S \mu_B g_N \mu_N (\mu_0/4\pi) \mathbf{T}^1(\mathbf{I}_T) \cdot \mathbf{T}^1(\mathbf{S}, \mathbf{C}^2). \tag{8.248}$$

Note that in both equations we take the total spin I_T; one would obtain the same results for the matrix elements using individual nuclear spins. The form of the dipolar interaction is the same as that used in our treatment of $^3\Pi_u$ H$_2$, given previously in equation (8.229).

Three more terms involving nuclear spin remain to be considered. The most important of these is the nuclear electric quadrupole interaction, which we take to have the same form as in our earlier treatment of the D$_2$ molecule, equation (8.22),

$$\mathcal{H}_Q = -e \sum_{k=1,2} \mathrm{T}^2(\nabla \boldsymbol{E}_k) \cdot \mathrm{T}^2(\boldsymbol{Q}_k). \tag{8.249}$$

As we shall see, each of these two terms, one for each nucleus, describes a second-rank scalar interaction between the electric field gradient at each nucleus and the nuclear quadrupole moment. De Santis, Lurio, Miller and Freund [44] included two other terms which involve the nuclear spins. One is the direct dipolar coupling of the ^{14}N nuclear magnetic moments, an interaction which we discussed earlier in connection with the magnetic resonance spectrum of D$_2$; its matrix elements were given in equation (8.33). The other is the nuclear spin–rotation interaction, also discussed in connection with H$_2$ and its deuterium isotopes. It is represented by the term

$$\mathcal{H}_{\mathrm{nsr}} = c_I(v) \, \mathrm{T}^1(\boldsymbol{N}) \cdot \mathrm{T}^1(\boldsymbol{I}_T). \tag{8.250}$$

Both of these interactions are small in relation to the resolution obtained in the studies of N$_2$ in its $^3\Sigma_u^+$ state. The effect of the nuclear dipolar interaction was included as an estimated correction, but was not actually determined in the spectroscopic analysis.

In summary, the effective Hamiltonian is the sum of seven terms,

$$\mathcal{H}_{\mathrm{eff}} = \mathcal{H}_{\mathrm{rot}} + \mathcal{H}_{\mathrm{ss}} + \mathcal{H}_{\mathrm{sr}} + \mathcal{H}_{\mathrm{F}} + \mathcal{H}_{\mathrm{dip}} + \mathcal{H}_{\mathrm{Q}} + \mathcal{H}_{\mathrm{nsr}}, \tag{8.251}$$

given explicitly by equations (8.243), (8.245), (8.246), (8.247), (8.248), (8.249) and (8.250). The spectral analysis would be fairly complex even if it was confined to a single vibrational level. In fact De Santis, Lurio, Miller and Freund [44] were able to disentangle the spectra of no fewer that thirteen different vibrational levels ($v = 0$ to 12), and to determine the values of the molecular constants for each of them. It was a spectroscopic *tour-de-force*. In order to give a flavour of the spectral complexity, we show a small section of the radiofrequency spectrum in figure 8.24, where resonances from eight vibrational levels ($v = 0$ to 7) are assigned; they involve two different hyperfine transitions. It is, however, easier to discuss the observed spectra after we have dealt with the spectroscopic theory which underpins the analysis. We therefore proceed to calculate the matrix elements and energy levels resulting from the effective Hamiltonian (8.251).

(ii) MATRIX ELEMENTS OF THE EFFECTIVE HAMILTONIAN

We continue to use the case (b) hyperfine-coupled basis set used earlier. The matrix elements of the first three terms in the effective Hamiltonian (8.251) do not involve the nuclear spins and are therefore independent of, and diagonal in, the quantum number F. The required matrix elements are now tabulated.

For the rotational Hamiltonian we have the simple result:

$$\langle \eta, N, S, J | \mathcal{H}_{\mathrm{rot}} | \eta, N, S, J \rangle = B_v N(N+1) - D_v N^2(N+1)^2 \tag{8.252}$$

The matrix elements of the electron spin-spin interaction were discussed earlier in connection with H$_2$ in its $^3\Pi_u$ state and we can make use of equation (8.197) to obtain

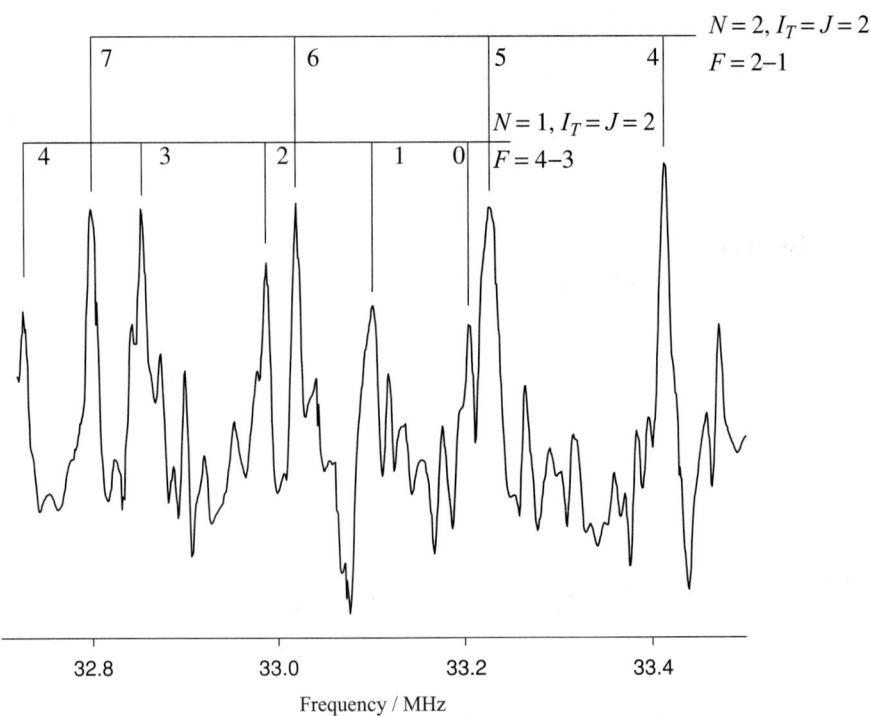

Figure 8.24. Section of the radiofrequency spectrum of N_2 in its $A\,^3\Sigma_u^+$ state. The observed resonances involve two different hyperfine transitions in two different rotational levels, and arise from eight different vibrational levels, $v = 0$ to 7, as shown. The transitions are all $\Delta M_F = 0$ and were recorded at a magnetic field of 150 mG [44].

the result

$$\langle \eta, \Lambda; N, S, J, M_J | - g_S^2 \mu_B^2 (\mu_0/4\pi)\sqrt{6}\mathbf{T}^2(\mathbf{S}_1, \mathbf{S}_2) \cdot \mathbf{T}^2(\mathbf{C}) | \eta', \Lambda'; N', S, J, M_J \rangle$$

$$= -\sqrt{6}(-1)^{J+N'+S} \begin{Bmatrix} S & N' & J \\ N & S & 2 \end{Bmatrix} \langle \eta, \Lambda; N, \Lambda \| (\mu_0/4\pi) g_S^2 \mu_B^2 \, \mathbf{T}^2(\mathbf{C}) \| \eta', \Lambda'; N', \Lambda' \rangle$$

$$\times \langle S \| \mathbf{T}^2(\mathbf{S}_1, \mathbf{S}_2) \| S \rangle$$

$$= -\sqrt{6}(-1)^{J+N'+S} \begin{Bmatrix} S & N' & J \\ N & S & 2 \end{Bmatrix} \sum_q \langle \eta, \Lambda | (\mu_0/4\pi) g_S^2 \mu_B^2 C_q^2(\theta_{12}, \phi_{12})(r_{12}^{-3}) | \eta', \Lambda' \rangle$$

$$\times \langle N, \Lambda \| \mathcal{D}_{\cdot q}^{(2)}(\omega_{12})^* \| N', \Lambda' \rangle \sqrt{5} \begin{Bmatrix} S_1 & S_1 & 1 \\ S_2 & S_2 & 1 \\ S & S & 2 \end{Bmatrix} (2S+1)$$

$$\times \{S_1(S_1+1)(2S_1+1)S_2(S_2+1)(2S_2+1)\}^{1/2}. \qquad (8.253)$$

For a $^3\Sigma$ state, Λ has the value 0, and we can choose to neglect the admixture of other electronic states by putting $q = 0$ in equation (8.253). If we also insert the appropriate values of the electron spin quantum numbers in (8.253) we obtain the result:

$$\langle \eta, \Lambda; N, S, J, M_J | - g_S^2 \mu_B^2 (\mu_0/4\pi)\sqrt{6} T^2(S_1, S_2) \cdot T^2(C) | \eta', \Lambda; N', S, J, M_J \rangle$$
$$= \frac{2}{3}\sqrt{30}\lambda_v (-1)^{J+N'+S} \begin{Bmatrix} S & N' & J \\ N & S & 2 \end{Bmatrix} (-1)^N \begin{pmatrix} N & 2 & N' \\ 0 & 0 & 0 \end{pmatrix} \{(2N+1)(2N'+1)\}^{1/2}.$$
(8.254)

As before, the parameter λ_v is defined by the identity

$$\lambda_v \equiv -(3/4)\langle g_S^2 \mu_B^2 (\mu_0/4\pi) C_0^2(\theta_{12}, \phi_{12}) r_{12}^{-3} \rangle_\eta.$$
(8.255)

If matrix elements off-diagonal in N are neglected, equation (8.254) yields the following simple results for the energies of the three spin components of a given N level in a case (b) $^3\Sigma$ state:

$$J = N+1: \quad E = -\frac{2}{3}\lambda_v \frac{N}{(2N+3)}$$
$$J = N: \quad E = \frac{2}{3}\lambda_v \qquad \qquad (8.256)$$
$$J = N-1: \quad E = -\frac{2}{3}\lambda_v \frac{(N+1)}{(2N-1)}.$$

However, equation (8.254) shows that there are matrix elements off-diagonal in N (i.e. N mixes with $N \pm 2$), so that (8.256) is of limited value. We discuss this aspect further in due course.

The matrix elements of the spin–rotation interaction in a case (b) basis were also presented in our discussion of *para*-H_2 in its $c\,^3\Pi_u$ state. They are diagonal in N, and given by equation (8.189) as:

$$\langle \eta, \Lambda; N, S, J, M_J | \gamma_v T^1(N) \cdot T^1(S) | \eta, \Lambda; N, S, J, M_J \rangle$$
$$= \gamma_v (-1)^{N+J+S} \begin{Bmatrix} S & N & J \\ N & S & 1 \end{Bmatrix} \langle N \| T^1(N) \| N \rangle \langle S \| T^1(S) \| S \rangle$$
$$= \gamma_v (-1)^{N+J+S} \begin{Bmatrix} S & N & J \\ N & S & 1 \end{Bmatrix} \{N(N+1)(2N+1)S(S+1)(2S+1)\}^{1/2}. \quad (8.257)$$

We come now to the magnetic hyperfine interaction terms. Again we make use of the results derived earlier for *ortho*-H_2 in the $c\,^3\Pi_u$ state. For the Fermi contact interaction, from equation (8.220):

$$\langle \eta, \Lambda; N, S, J, I_T, F, M_F | b_F T^1(S) \cdot T^1(I_T) | \eta, \Lambda; N, S, J', I_T, F', M'_F \rangle$$
$$= b_F \delta_{M_F M'_F} \delta_{FF'} (-1)^{J'+F+I_T} \begin{Bmatrix} I_T & J' & F \\ J & I_T & 1 \end{Bmatrix} \langle N, S, J \| T^1(S) \| N, S, J' \rangle \langle I_T \| T^1(I_T) \| I_T \rangle$$
$$= b_F \delta_{M_F M'_F} \delta_{FF'} (-1)^{J'+F+I_T} \begin{Bmatrix} I_T & J' & F \\ J & I_T & 1 \end{Bmatrix} \begin{Bmatrix} J & S & N \\ S & J' & 1 \end{Bmatrix} (-1)^{J+N+1+S}$$
$$\times \{(2J'+1)(2J+1)\}^{1/2} \{S(S+1)(2S+1)I_T(I_T+1)(2I_T+1)\}^{1/2}. \quad (8.258)$$

For the electron spin–nuclear spin dipolar interaction we use equations (8.232), (8.233) and (8.248) so that we obtain:

$$\langle \eta, \Lambda; N, S, J, I_T, F, M_F | \mathcal{H}_{\text{dip}} | \eta, \Lambda'; N', S, J', I_T, F', M_F' \rangle$$
$$= \sqrt{10} g_S \mu_B g_N \mu_N (\mu_0/4\pi) \delta_{M_F M_F'} \delta_{FF'} (-1)^{J'+F+I_T+1} \begin{Bmatrix} I_T & J' & F \\ J & I_T & 1 \end{Bmatrix}$$
$$\times \langle \eta, \Lambda; N, S, J \| \mathbf{T}^1(\mathbf{S}, \mathbf{C}^2) \| \eta, \Lambda'; N', S, J' \rangle \langle I_T \| \mathbf{T}^1(\mathbf{I_T}) \| I_T \rangle$$
$$= \sqrt{10} g_S \mu_B g_N \mu_N (\mu_0/4\pi) \delta_{M_F M_F'} \delta_{FF'} (-1)^{J'+F+I_T+1} \begin{Bmatrix} I_T & J' & F \\ J & I_T & 1 \end{Bmatrix}$$
$$\times \{I_T(I_T+1)(2I_T+1)\}^{1/2} \langle \eta, \Lambda; N, S, J \| \mathbf{T}^1(\mathbf{S}, \mathbf{C}^2) \| \eta, \Lambda'; N', S, J' \rangle$$
$$= \sqrt{30} g_S \mu_B g_N \mu_N (\mu_0/4\pi) \delta_{M_F M_F'} \delta_{FF'} (-1)^{J'+F+I_T+1} \begin{Bmatrix} I_T & J' & F \\ J & I_T & 1 \end{Bmatrix}$$
$$\times \{I_T(I_T+1)(2I_T+1)\}^{1/2} \{(2J+1)(2J'+1)\}^{1/2} \{S(S+1)(2S+1)\}^{1/2}$$
$$\times \begin{Bmatrix} J & J' & 1 \\ N & N' & 2 \\ S & S & 1 \end{Bmatrix} \langle \eta, \Lambda; N, \Lambda \| T^2(\mathbf{C}) \| \eta', \Lambda'; N', \Lambda' \rangle. \quad (8.259)$$

Further development is simpler than it was before because we do not have the problem of Λ-doubling in a Σ state. The remaining reduced matrix element in (8.259) may be simplified by confining attention to the $^3\Sigma_u^+$ state (i.e. putting $\Lambda = 0$ and ignoring matrix elements off-diagonal in Λ). With this simplification we use the result:

$$\langle \eta, N, \Lambda = 0 \| T^2(\mathbf{C}) \| \eta, N', \Lambda = 0 \rangle$$
$$= (-1)^N \{(2N+1)(2N'+1)\}^{1/2} \begin{pmatrix} N & 2 & N' \\ 0 & 0 & 0 \end{pmatrix} \langle T_0^2(\mathbf{C}) \rangle_\eta. \quad (8.260)$$

In this equation we note that

$$\langle T_0^2(\mathbf{C}) \rangle_\eta = \langle C_0^2(\theta, \phi) r^{-3} \rangle_\eta, \quad (8.261)$$

and for the axial component of the dipolar hyperfine interaction we use the notation

$$t_v = g_S \mu_B g_N \mu_N (\mu_0/4\pi) \langle \eta, \Lambda; v | C_0^2(\theta, \phi) r^{-3} | \eta, \Lambda; v \rangle; \quad (8.262)$$

v denotes any specific vibrational level in the present problem.

We now come to the nuclear electric quadrupole interaction, and draw upon the results derived earlier for D_2 in its ground state. The present problem is marginally more complicated because of presence of electron spin. We note first that if the quadrupole interaction is given by (8.249), i.e.

$$\mathcal{H}_Q = -e \sum_{k=1,2} T^2(\nabla E_k) \cdot T^2(\mathbf{Q}_k), \quad (8.263)$$

the following steps take note of the couplings of $\boldsymbol{I}_T$ with $\boldsymbol{J}$, and $\boldsymbol{N}$ with $\boldsymbol{S}$ to form $\boldsymbol{J}$:

$$\langle \eta, \Lambda; N, S, J, I_1, I_2, I_T, F | \mathcal{H}_Q | \eta, \Lambda'; N', S, J', I_1, I_2, I_T', F \rangle$$

$$= -e \sum_k (-1)^{J'+I_T+F} \begin{Bmatrix} J & I_T & F \\ I_T' & J' & 2 \end{Bmatrix} \langle \eta, \Lambda; N, S, J \| T^2(\nabla \boldsymbol{E}_k) \| \eta', \Lambda'; N', S, J' \rangle$$

$$\times \langle I_1, I_2, I_T \| T^2(\boldsymbol{Q}_k) \| I_1, I_2, I_T' \rangle$$

$$= -e \sum_k (-1)^{J'+I_T+F} \begin{Bmatrix} J & I_T & F \\ I_T' & J' & 2 \end{Bmatrix} (-1)^{J'+N+2+S}$$

$$\times \{(2J'+1)(2J+1)\}^{1/2} \begin{Bmatrix} J & N & S \\ N' & J' & 2 \end{Bmatrix} \langle \eta, \Lambda; N, \Lambda \| T^2(\nabla \boldsymbol{E}_k) \| \eta', \Lambda'; N', \Lambda' \rangle$$

$$\times \langle I_1, I_2, I_T \| T^2(\boldsymbol{Q}_k) \| I_1, I_2, I_T' \rangle. \tag{8.264}$$

The nuclear spin reduced matrix element, summed over k, was given earlier in (8.25) to (8.27). The result is

$$\langle I_1, I_2, I_T \| \sum_{k=1,2} T^2(\boldsymbol{Q}_k) \| I_1, I_2, I_T' \rangle$$

$$= (-1)^{I_1+I_2+2} \{(2I_T+1)(2I_T'+1)\}^{1/2} \left[(-1)^{I_T'} \begin{Bmatrix} I_T & I_1 & I_2 \\ I_1 & I_T' & 2 \end{Bmatrix} \langle I_1 \| T^2(\boldsymbol{Q}_1) \| I_1 \rangle \right.$$

$$\left. + (-1)^{I_T} \begin{Bmatrix} I_T & I_2 & I_1 \\ I_2 & I_T' & 2 \end{Bmatrix} \langle I_2 \| T^2(\boldsymbol{Q}_2) \| I_2 \rangle \right]. \tag{8.265}$$

This equation may be simplified by making use of the equivalence of the nuclei ($I_1 = I_2 = I_N$), by neglecting matrix elements off-diagonal in I_T, and by noting the definition of the nuclear quadrupole moment Q_N, which is given by,

$$\langle I_N \| T^2(\boldsymbol{Q}_N) \| I_N \rangle = \left(\frac{Q_N}{2} \right) \begin{pmatrix} I_N & 2 & I_N \\ -I_N & 0 & I_N \end{pmatrix}^{-1}. \tag{8.266}$$

Equation (8.265) then becomes

$$\langle I_1, I_2, I_T \| \sum_{k=1,2} T^2(\boldsymbol{Q}_k) \| I_1, I_2, I_T \rangle$$

$$= (-1)^{2I_N+I_T} Q_N (2I_T+1) \begin{Bmatrix} I_N & I_T & I_N \\ I_T & I_N & 2 \end{Bmatrix} \begin{pmatrix} I_N & 2 & I_N \\ -I_N & 0 & I_N \end{pmatrix}^{-1}. \tag{8.267}$$

The other reduced matrix element in equation (8.264) is evaluated in the following manner; first we note that

$$\langle \eta, \Lambda; N, \Lambda \| T^2(\nabla \boldsymbol{E}) \| \eta, \Lambda'; N', \Lambda' \rangle$$

$$= \langle \eta, \Lambda; N, \Lambda \| \sum_q \mathcal{D}^{(2)}_{\cdot q}(\omega)^* T^2_q(\nabla \boldsymbol{E}) \| \eta', \Lambda'; N', \Lambda' \rangle$$

$$= \sum_q (-1)^{N-\Lambda} \{(2N+1)(2N'+1)\}^{1/2} \begin{pmatrix} N & 2 & N' \\ -\Lambda & q & \Lambda' \end{pmatrix} \langle \eta, \Lambda | T^2_q(\nabla \boldsymbol{E}) | \eta', \Lambda' \rangle. \tag{8.268}$$

Now we note that $\Lambda = 0$ for a Σ state, and neglect the mixing of excited electronic states by putting $q = 0$. We also note the definition of the electric field gradient at the nucleus, q_0, given in equation (8.28). With these simplifications the result is

$$\langle \eta, \Lambda; N, \Lambda \| T^2(\nabla \boldsymbol{E}) \| \eta, \Lambda; N', \Lambda \rangle$$
$$= -(-1)^N \{(2N+1)(2N'+1)\}^{1/2} \begin{pmatrix} N & 2 & N' \\ 0 & 0 & 0 \end{pmatrix} (q_0/2). \quad (8.269)$$

Gathering together the above results, retaining only matrix elements off-diagonal in J, and putting $I_1 = I_2 = 1$, we obtain the final result for the quadrupole interaction:

$$\langle \eta, \Lambda = 0; N, J, I_T, F | \mathcal{H}_Q | \eta, \Lambda = 0; N', J', I_T, F \rangle$$
$$= \left(\frac{eq_0 Q}{2} \right)(-1)^{F+J'+N} \{30(2J'+1)(2J+1)\}^{1/2}(2I_T+1)\{(2N+1)(2N'+1)\}^{1/2}$$
$$\times \begin{Bmatrix} J & I_T & F \\ I_T & J' & 2 \end{Bmatrix} \begin{Bmatrix} J & N & 1 \\ N' & J' & 2 \end{Bmatrix} \begin{Bmatrix} I_T & 1 & 1 \\ 1 & I_T & 2 \end{Bmatrix} \begin{pmatrix} N & 2 & N' \\ 0 & 0 & 0 \end{pmatrix}. \quad (8.270)$$

The final term in the effective Hamiltonian describes the nuclear spin–rotation interaction and its matrix elements are relatively straightforward:

$$\langle \eta, \Lambda; N, S, J, I_T, F | c_I(v) T^1(\boldsymbol{N}) \cdot T^1(\boldsymbol{I}_T) | \eta, \Lambda; N, S, J', I_T, F \rangle$$
$$= c_I(v)(-1)^{J'+F+I_T} \begin{Bmatrix} I_T & J' & F \\ J & I_T & 1 \end{Bmatrix} \langle \eta, \Lambda; N, S, J \| T^1(\boldsymbol{N}) \| \eta, \Lambda; N, S, J' \rangle$$
$$\times \langle I_T \| T^1(\boldsymbol{I}_T) \| I_T \rangle$$
$$= c_I(v)(-1)^{J'+F+I_T} \begin{Bmatrix} I_T & J' & F \\ J & I_T & 1 \end{Bmatrix} (-1)^{J'+N+1+S} \{(2J'+1)(2J+1)\}^{1/2}$$
$$\times \begin{Bmatrix} J & N & S \\ N & J' & 1 \end{Bmatrix} \{N(N+1)(2N+1)I_T(I_T+1)(2I_T+1)\}^{1/2}. \quad (8.271)$$

(iii) CALCULATION OF ENERGY LEVELS AND TRANSITION FREQUENCIES

We now come to the problem of relating the fairly complex theoretical expressions given in the previous subsection to the experimental transition frequencies. A selection of the experimental data is presented in table 8.9. Measurements were made in four different rotational levels as shown in the table, which should be read in conjunction with figure 8.22. In every case the measurements were made for levels in which $J = I_T$, and in most cases involved vibrational levels from $v = 0$ to 12. They were performed in a very small magnetic field (150 mG), so that the $\Delta M_F = 0$ components observed were essentially superimposed.

In essence there are five molecular constants to be determined for each vibrational level; they are summarised as follows, where we present our notation as well as that

Table 8.9. *Selection of hyperfine transition frequencies involving the lower rotational levels of N_2 in its $A\,^3\Sigma_u^+$ state [43, 44]. Transition frequencies for $v = 0$ and 10 are shown*

I_T	N	J	F	$\Delta F = \pm 1$	f_0(MHz)	f_{10}(MHz)
2	3	4	6, 5, 4, 3, 2			
		3	5, 4, 3, 2, 1			
		2	4, 3, 2, 1, 0	$4 \leftrightarrow 3$	40.851	39.118
				$3 \leftrightarrow 2$	31.720	—
				$2 \leftrightarrow 1$	—	21.097
2	1	2	4, 3, 2, 1, 0	$4 \leftrightarrow 3$	34.449	31.855
				$3 \leftrightarrow 2$	25.022	23.350
		1	3, 2, 1			
		0	2			
1	2	3	4, 3, 2			
		2	3, 2, 1			
		1	2, 1, 0	$2 \leftrightarrow 1$	33.358	31.390
1	0	1	2, 1, 0	$2 \leftrightarrow 1$	20.846	18.532

originally used by De Santis, Lurio, Miller and Freund [44].

electron spin–spin:	λ_v	$(4/3)\lambda_v$
Fermi contact:	$b_{F,v}$	α_v
dipolar hyperfine:	t_v	β_v
quadrupole constant:	$eq_0 Q_v$	Q_v
electron spin–rotation:	γ_v	γ_v
nuclear spin–rotation:	$c_{I,v}$	p_v

The electron spin–spin constant λ_v is not determined directly through the radiofrequency spectrum, but nevertheless plays a very important role, as does also the rotational constant B_v. In both cases the required values were taken from the optical electronic spectrum.

It is instructive to consider in quantitative detail the analysis of a particular hyperfine transition; a simple example would seem to be the $F = 2 \leftrightarrow 1$ transition in the level $I_T = 1$, $N = 0$, $J = 1$, which is observed at a frequency of 20.846 MHz for the $v = 0$ level. The expressions for the matrix elements of the magnetic and electric hyperfine terms, (8.258), (8.259) and (8.270), show that for $N = 0$ only the Fermi contact interaction is non-zero and the energies of the hyperfine levels are

$$F = 2:\ E = b_F,\quad F = 1:\ E = -b_F.$$

The transition frequency would be equal to $2b_F$; De Santis, Lurio, Miller and Freund [44] give $b_F = 13.145$ MHz for the $v = 0$ level, so that the predicted transition frequency would be 26.290 MHz.

The observed frequency is 20.846 MHz, so what is the origin of the large discrepancy? The answer is that the electron spin–spin interaction mixes levels N with $N \pm 2$, and the dipolar and quadrupole interactions then come into play. The spin-spin mixing may be represented by the following 2×2 matrix.

	$\|I_T = 1, N = 0, J = 1\rangle$	$\|I_T = 1, N = 2, J = 1\rangle$
$\langle I_T = 1, N = 0, J = 1\|$	0	$4\lambda_0/3\sqrt{2}$
$\langle I_T = 1, N = 2, J = 1\|$	$4\lambda_0/3\sqrt{2}$	$6B_0 - 2\lambda_0/3$

The values of the constants used were $\lambda_0 = -1.768$, $B_0 = 1.4366$ cm^{-1} and if these are substituted into the above matrix we obtain the following results for the eigenvalues (in cm^{-1}) and eigenvectors:

$$E_1 = -0.1617: \quad \Psi_1 = 0.9917|N=0, J=1\rangle + 0.1282|N=2, J=1\rangle$$
$$E_2 = 9.6653: \quad \Psi_2 = 0.1282|N=0, J=1\rangle - 0.9917|N=2, J=1\rangle. \tag{8.272}$$

The admixture is, in this example, relatively small but it has important consequences for the hyperfine splitting in the state Ψ_1; we are now interested in the expectation value,

$$\langle \Psi_1|\mathcal{H}_F + \mathcal{H}_{dip} + \mathcal{H}_Q|\Psi_1\rangle = 0.9835 \langle N=0, J=1|\mathcal{H}_F + \mathcal{H}_{dip} + \mathcal{H}_Q|N=0, J=1\rangle$$
$$+ 0.2543 \langle N=0, J=1|\mathcal{H}_F + \mathcal{H}_{dip} + \mathcal{H}_Q|N=2, J=1\rangle$$
$$+ 0.0164 \langle N=2, J=1|\mathcal{H}_F + \mathcal{H}_{dip} + \mathcal{H}_Q|N=2, J=1\rangle.$$
$$(8.273)$$

Although only the Fermi contact interaction contributes to the first of these three matrix elements, all three hyperfine interactions contribute to the other two, and they are dependent on the value of the quantum number F. Expansion of the relevant equations gives the following results, all for $I_T = 1$,
For $F = 2$:

$$\langle N=0, J=1|\mathcal{H}_F + \mathcal{H}_{dip} + \mathcal{H}_Q|N=0, J=1\rangle = b_F,$$
$$\langle N=0, J=1|\mathcal{H}_F + \mathcal{H}_{dip} + \mathcal{H}_Q|N=2, J=1\rangle = t/\sqrt{2}, \tag{8.274}$$
$$\langle N=2, J=1|\mathcal{H}_F + \mathcal{H}_{dip} + \mathcal{H}_Q|N=2, J=1\rangle = -b_F/2 + t + eq_0 Q(\sqrt{3}/8\sqrt{35}).$$

For $F = 1$:

$$\langle N=0, J=1|\mathcal{H}_F + \mathcal{H}_{dip} + \mathcal{H}_Q|N=0, J=1\rangle = -b_F,$$
$$\langle N=0, J=1|\mathcal{H}_F + \mathcal{H}_{dip} + \mathcal{H}_Q|N=2, J=1\rangle = -t/\sqrt{2}, \tag{8.275}$$
$$\langle N=2, J=1|\mathcal{H}_F + \mathcal{H}_{dip} + \mathcal{H}_Q|N=2, J=1\rangle = b_F/2 - t - eq_0 Q(\sqrt{15}/8\sqrt{7}).$$

We now combine these results with (8.273) to obtain the total hyperfine energies of the two states, as follows:

$$F = 2: \quad E = (0.9835)(b_F) + (0.2543)(0.7071t) + (0.0164)$$
$$\times (-0.5b_F + t + 0.0366 eq_0Q)$$
$$= 0.9753 b_F + 0.1962 t + 0.0006 eq_0 Q$$
$$F = 1: \quad E = (0.9835)(-b_F) - (0.2543)(0.7071t) + (0.0164)$$
$$\times (0.5b_F - t - 0.1830 eq_0 Q)$$
$$= -0.9753 b_F - 0.1962 t - 0.0030 eq_0 Q. \quad (8.276)$$

De Santis, Lurio, Miller and Freund [44] give the following values of the constants for N_2 in the $v = 0$ level:

$$b_F = 13.145 \text{ MHz}, \quad t = -12.660 \text{ MHz}, \quad eq_0 Q = -2.518 \text{ MHz}.$$

Substituting these values in (8.276) we obtain the following values for the hyperfine energies, and the frequency of the hyperfine transition (all in MHz):

$$F = 2: \quad E = 10.335; \quad F = 1: E = -10.329; \quad \text{freq}(F = 2 \leftrightarrow F = 1) = 20.664;$$
$$\text{freq (exp)} = 20.846.$$

The agreement between experiment and theory is now much better than before, the discrepancy having been reduced from 5.444 to 0.182 MHz, but it is still poor compared with the experimental accuracy which is quoted as ±0.01 MHz. However, our theory is still approximate because the electron spin–spin interaction mixes $N = 2$ with $N = 4$, which introduces more hyperfine matrix elements off-diagonal in both N and J. The nuclear spin–rotation term, equation (8.271), does not contribute to the first-order energy of the $N = 0$ level, and makes a negligible second-order contribution. We will not pursue this analysis any further, our aim having been to illustrate the complexity of the fitting process; moreover this was achieved for 13 different vibrational levels.

One of the most interesting and important results of the study was to show how the molecular constants change as the vibrational quantum number v increases. This behaviour is presented in table 8.10. The electron spin–spin and rotational constant values came, initially, from the analysis of the optical electronic spectrum [47], although the values of the spin–spin constants for different vibrational levels were refined by the analysis of the radiofrequency spectrum. The nuclear hyperfine parameters are obtained solely from the magnetic resonance experiments. We will discuss the significance of these constants in the following subsection.

(iv) INTERPRETATION OF THE MOLECULAR PARAMETERS

The data presented in table 8.10 show that the vibrational dependence of the magnetic hyperfine constants (from $v = 0$ to 12) is quite small, whereas the quadrupole coupling constant exhibits a strong percentage variation. Direct calculation of the Fermi contact

Table 8.10. *Vibrational dependence of the molecular parameters of N_2 in the $A\,^3\Sigma_u^+$ state*

v	B_v (cm^{-1})	λ_v (cm^{-1})	b_F (MHz)	t (MHz)	eq_0Q (MHz)
0	1.4366	−1.326	13.145	−12.660	−2.518
1	1.4274	−1.320	13.094	−12.644	−2.446
2	1.4091	−1.314	13.024	−12.626	−2.358
3	1.3905	−1.307	12.949	−12.606	−2.290
4	1.3718	−1.300	12.854	−12.589	−2.190
5	1.3529	−1.293	12.763	−12.564	−2.107
6	1.3339	−1.285	12.657	−12.540	−2.023
7	1.3147	−1.276	12.542	−12.513	−1.945
8	1.2952	−1.268	12.425	−12.476	−1.864
9	1.2757	−1.259	12.295	−12.441	−1.771
10	1.2559	−1.249	12.162	−12.405	−1.672
11	1.2360	−1.239	12.028	−12.358	−1.610
12	1.2159	−1.229	11.886	−12.308	−1.545

interaction is difficult because the $(2p\pi_u)^3(2p\pi_g^*)^1$ electron configuration for the A state would predict a zero value; the non-zero value of b_F must arise from configuration interaction with other excited states having unfilled $s\sigma_g$ orbitals. The same problem arises for the N atom in its ^4S ground state, where the unpaired electrons occupy $2p$ atomic orbitals. Indeed, it is pointed out [44] that the magnitudes of the Fermi contact and dipolar interactions are similar in both the A state of the N_2 molecule and in the ground state of the N atom. It is concluded [44] that the best simple picture of N_2 in its $A\,^3\Sigma_u^+$ state is of two weakly-interacting N atoms.

The strong vibrational dependence of the quadrupole coupling constant eq_0Q is consistent with this very simple picture. The electric field gradient q_0 must be zero at the nucleus for the isolated N atom. The approach of the N_2 molecule towards the dissociation limit, with its increasing internuclear distance as v increases, must therefore result in decreasing values of the electric field gradient.

The calculation of the electron spin-spin parameter λ, including its vibrational dependence, is even more complicated. We have derived the irreducible tensor form of this interaction (in appendix 8.3) by starting with the classical form of the dipolar interaction. This is the origin of the first-order contribution to λ. However, we know from the detailed discussion of the effective Hamiltonian in chapter 7 that the effects of spin–orbit mixing of other electronic states lead to a contribution to λ (the so-called second-order contribution). This second-order contribution is often the dominant contribution to λ, as we shall see later in our discussion of heavier molecules. Only in H_2 and a few other very light molecules is the value of the λ constant determined essentially by the direct dipolar coupling. Consequently an accurate theoretical calculation of λ requires very good wave functions, both for the vibronic state under examination, and for other states coupled to it by the spin–orbit interaction. This is a demanding requirement, and a challenge for *ab initio* theory.

8.4. Molecular beam electric resonance of closed shell molecules

8.4.1. Principles of electric resonance methods

Molecular beam electric resonance has much in common with the magnetic resonance method described earlier; it uses the same molecular beam production methods, and often the same detection methods. Above all, it depends upon molecular state selection but through the use of electric rather than magnetic fields. Electric resonance methods are therefore restricted to molecules which possess a permanent electric dipole moment. On the other hand, electric resonance techniques could be regarded as more general than magnetic resonance; the latter is usually concerned with transitions between nuclear spin states in a weak magnetic field, which occur in the low radiofrequency region of the spectrum. In contrast electric resonance methods have been used to study different kinds of transitions, including rotational transitions occurring in the microwave and millimetre wave regions. They have been applied to a large range of molecules, including molecular complexes which are weakly bound by van der Waals or hydrogen-bonding forces. In this section we confine our attention to diatomic molecules in closed shell electronic states, usually the ground states. In the following section we deal with open shell systems, which raise new problems of spectroscopic analysis.

As the name suggests, electric resonance experiments make use of electric fields to achieve molecular state selection. Figure 8.25 shows a schematic diagram of a molecular beam electric resonance instrument, which we will discuss in more detail when we describe experiments on the CsF molecule. In contrast to the magnetic resonance apparatus discussed earlier, the A, B and C fields in figure 8.25 are all electric fields. In

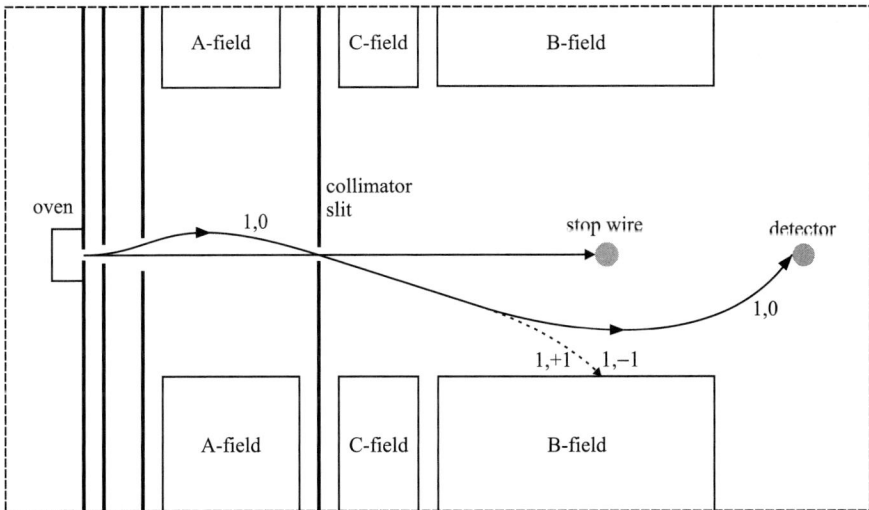

Figure 8.25. Schematic diagram of a molecular beam electric resonance spectrometer, employing dipole A and B fields [48]. The trajectories of levels characterised by J, M quantum numbers are shown. Transitions occurring in the C field region which obey the selection rule $\Delta M = \pm 1$ result in a reduction in the number of molecules reaching the detector; this is therefore an example of *flop-out* detection.

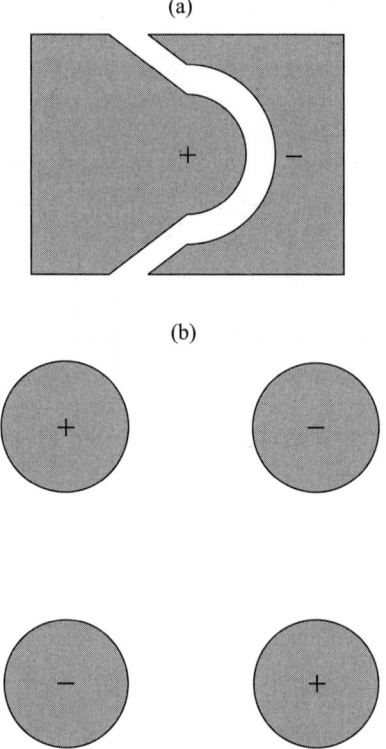

Figure 8.26. Geometry of the electrodes for (a) a dipole electric field, (b) a quadrupole electric field.

the early work the A field was a dipole field but in most subsequent work a quadrupole field was used; we will meet examples later. The analysing B field may be either a quadrupole or a dipole field. Strictly speaking, the electrodes are arranged to produce dipole or quadrupole electrostatic potentials. It is, however, common practice to refer to such arrangements as fields. Figure 8.26 shows typical examples of the geometry of electrostatic lenses for producing such fields; the quadrupole lens arrangement produces maximum field at the electrodes, with zero field at the centre, and is essentially a focussing device. The dipole field simply produces a deflection of the molecular trajectory perpendicular to the transmission axis. The molecular trajectories shown in figure 8.25 refer to molecules in one or other of two energy levels which are mixed by an electric field to show a second-order Stark effect. The dipole fields are arranged so that molecules in one Stark level, indicated by the continuous line, are focussed on to the detector; molecules in other quantum states are removed from the beam. If, however, radiofrequency or microwave transitions are induced in the homogeneous C field region, the number of focussed molecules decreases, so that the detector current decreases; this mode of operation is naturally known as 'flop-out' detection. In reality the Stark behaviour is usually quite complicated and successful spectroscopic detection requires full appreciation of the subtleties, as we shall see.

8.4.2. CsF in the $X\,^1\Sigma^+$ ground state

(a) Stark effect

The first successful electric resonance experiment was reported by Hughes [48] who studied the CsF molecule, an appropriate beam being produced from a hot oven. He used both A and B electric dipole fields, separated by a homogeneous electric C field combined with a radiofrequency electric field at right angles to the static field. In order to understand both the deflection and state selection in the dipole fields, as well as the electric resonance spectrum, we first consider the details of the Stark effect.

An applied electric field (E) interacts with the electric dipole moment (μ_e) of a polar diatomic molecule, which lies along the direction of the internuclear axis. The applied field defines the space-fixed $p=0$ direction, or Z direction, whilst the molecule-fixed $q=0$ direction corresponds to the internuclear axis. Transformation from one axis system to the other is accomplished by means of a first-rank rotation matrix, so that the interaction may be represented by the effective Hamiltonian as follows:

$$\begin{aligned}\mathcal{H}' &= -\mathbf{T}^1(\boldsymbol{\mu}_e)\cdot\mathbf{T}^1(\boldsymbol{E}) = -\mathbf{T}^1_{p=0}(\boldsymbol{\mu}_e)\mathbf{T}^1_{p=0}(\boldsymbol{E})\\ &= -E_Z\sum_q \mathcal{D}^{(1)}_{0q}(\omega)^*\,\mathbf{T}^1_q(\boldsymbol{\mu}_e)\\ &= -\mu_0 \mathcal{D}^{(1)}_{00}(\omega)^* E_Z.\end{aligned} \quad (8.277)$$

μ_0 here is the permanent electric dipole moment lying along the molecular z axis. Note the potential confusion with the permeability of free space which has the same symbol. For this reason we replace μ_0 by $(\varepsilon_0 c^2)^{-1}$ in this section.

We choose to work in the simple basis set $|\eta,\Lambda;J,M_J\rangle$ where Λ is the component of electronic angular momentum along the internuclear axis; η represents all relevant unspecified quantum numbers, including the vibrational quantum number. We use J, which is appropriate for $^1\Sigma$ molecules, rather than N; there is no distinction between J and N for molecules in singlet states. The matrix elements of (8.277) in this basis are given by

$$\langle\eta,\Lambda;J,M_J|\mathcal{H}'|\eta,\Lambda;J',M_J\rangle = -\mu_0 E_Z\{(2J+1)(2J'+1)\}^{1/2}(-1)^{M-\Lambda}$$
$$\times\begin{pmatrix}J & 1 & J'\\ -M_J & 0 & M_J\end{pmatrix}\begin{pmatrix}J & 1 & J'\\ -\Lambda & 0 & \Lambda\end{pmatrix}. \quad (8.278)$$

Now for CsF in its $^1\Sigma^+$ ground state the value of Λ is zero; the second 3-j symbol in (8.278) is then non-zero only if $1+J+J'$ is even, so that $J'=J\pm1$ is a requirement. In other words, there can be no first-order Stark effect in this case. Equation (8.278) tells us that each rotational level J is mixed by the electric field with the adjacent rotational levels $J\pm1$, and the Stark behaviour may therefore be represented by the following 3×3 truncated matrix.

	$\|J-1\rangle$	$\|J\rangle$	$\|J+1\rangle$
$\langle J-1\|$	$BJ(J-1)$	$-\mu_0 E_Z \left\{ \dfrac{J^2 - M^2}{(2J-1)(2J+1)} \right\}^{1/2}$	0
$\langle J\|$	$-\mu_0 E_Z \left\{ \dfrac{J^2 - M^2}{(2J-1)(2J+1)} \right\}^{1/2}$	$BJ(J+1)$	$-\mu_0 E_Z \left\{ \dfrac{(J+1)^2 - M^2}{(2J+1)(2J+3)} \right\}^{1/2}$
$\langle J+1\|$	0	$-\mu_0 E_Z \left\{ \dfrac{(J+1)^2 - M^2}{(2J+1)(2J+3)} \right\}^{1/2}$	$B(J+1)(J+2)$

B is the rotational constant, and M remains a good quantum number. The above 3×3 matrix is, of course, something of an approximation since, in reality, the Stark matrix is infinite and the accuracy with which the electric field mixing is calculated depends upon the number of rotational states included in the calculation.

The energies of the levels in an electric field can be calculated by numerical diagonalisation of the above matrix for different values of the electric field and the J, M quantum numbers. However, perturbation theory has also often been used and we may readily derive an expression for the second-order Stark energy using the above matrix elements. The result is as follows:

$$\Delta E^{(2)} = -\frac{\mu_0^2 E_Z^2}{2B} \left\{ \frac{(J+1)^2 - M^2}{(J+1)(2J+1)(2J+3)} \right\} + \frac{\mu_0^2 E_Z^2}{2B} \left\{ \frac{J^2 - M^2}{(2J-1)(2J+1)J} \right\}$$
$$= \frac{\mu_0^2 E_Z^2}{2B} \left\{ \frac{J(J+1) - 3M^2}{J(J+1)(2J-1)(2J+3)} \right\}. \qquad (8.279)$$

This is a very well-known and often quoted result, first presented by Kronig [49]. For $J = 0, M = 0$, we have the special case

$$\Delta E^{(2)} = -\frac{\mu_0^2 E_Z^2}{6B}. \qquad (8.280)$$

Diagonalisation of the Stark matrices enables us to plot the Stark energies, given values of B and μ_0, and the results are shown in figure 8.27 for the first three rotational levels, $J = 0, 1$ and 2. The parameter λ is defined by $\lambda^2 = \mu_0^2 E_Z^2 / B$. In figure 8.28 we show plots of the effective electric moment of the molecule in the different J, M states listed in figure 8.27. With the aid of both diagrams, we are able to understand the principles of electric state selection, and the electric resonance transitions.

Figure 8.28 shows that the $J = 1$ components exhibit strong dependence of the effective electric moment, $-dW/dE_z$, on λ; in particular, the $J = 1, M = 0$ component has a negative electric moment at $\lambda = 2$ which becomes strongly positive at $\lambda = 6$. It is therefore a particularly good candidate for state selection. The apparatus designed by Hughes is shown schematically in figure 8.25, and the trajectory of molecules in $J = 1, M = 0$ is also indicated. The A and B fields are both electric dipole fields, with the electric field and field gradient pointing in the same direction. However the A field corresponds to $\lambda \approx 6.6$, whereas for the B field $\lambda \approx 2$. Consequently the sign of the deflection changes between the A and B fields, which are designed to focus molecules in $J = 1, M = 0$ on to the detector, as shown. Molecules in higher rotational levels

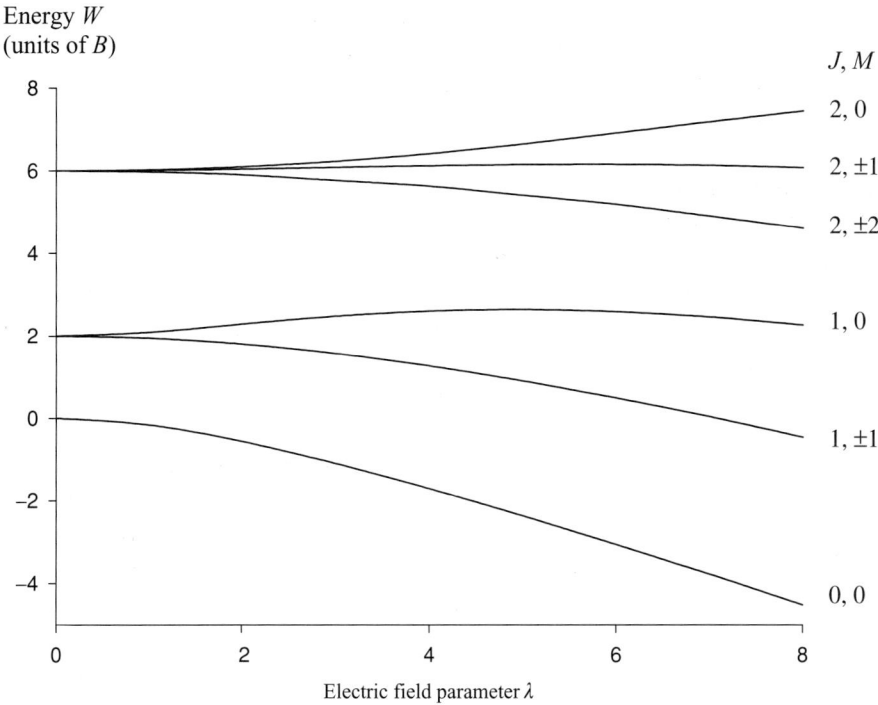

Figure 8.27. Second-order Stark energies for the first three rotational levels of a heteronuclear diatomic molecule in a $^1\Sigma$ state [48]. The parameter λ is defined by $\lambda^2 = \mu_0^2 E_Z^2/B$. Note that the states with $M = \pm 1$ or ± 2 remain rigorously degenerate, irrespective of field strength.

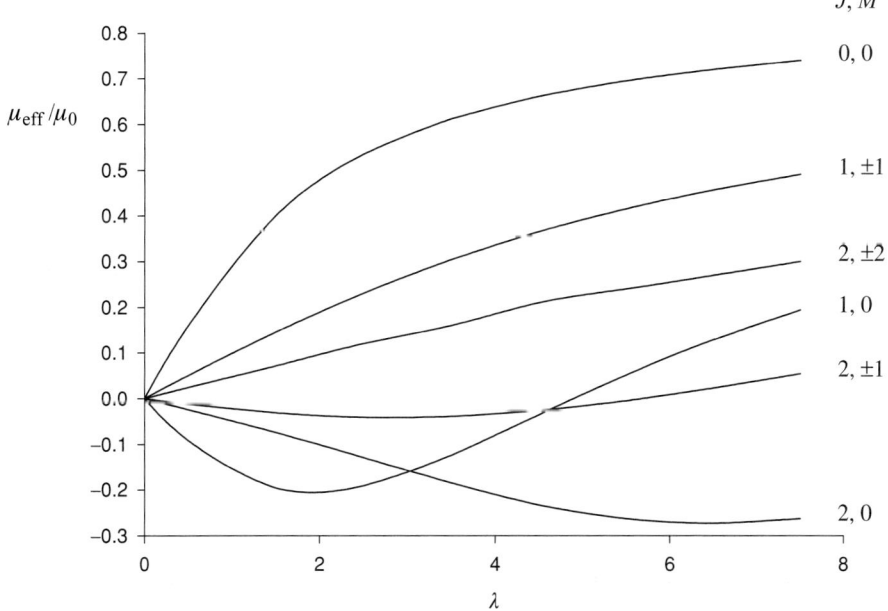

Figure 8.28. Effective electric moments (μ_{eff}) for different J, M states. These are calculated from the result $\mu_{\text{eff}}/\mu_0 = -\partial W/\partial E$ where W is the energy and E is the electric field strength.

have smaller Stark effects, as indicated by equation (8.279), and are blocked by the stop wire. In the homogeneous C field region electric resonance transitions of the type $J = 1, M = 0 \leftrightarrow J = 1, M = \pm 1$ are induced by radiofrequency radiation in the range 5 to 300 MHz. Population transfer from $J = 1, M = 0$ to the $M = \pm 1$ components results in a decrease in the detected beam current ('flop-out'). It will be clear from this description that optimum sensitivity in the experiment requires very careful adjustment of the deflection fields.

Hughes [48] was able to detect transitions involving $J = 2$, as well as the $J = 1$ described above, and from his results could determine both the electric dipole moment of CsF (7.3 ± 0.5 D) and an effective rotational constant ($B = 0.147$ cm^{-1}). The apparatus was subsequently improved by Trischka [50], particularly with respect to the quality of the homogeneous C field, and much higher resolution was obtained. Figure 8.29 shows resonance lines arising from molecules in $v = 0$ to 4, for which nuclear hyperfine interactions are not resolved. A higher-resolution spectrum is shown in figure 8.30 where nine components arising from molecules in $v = 0$ are observed, with additional lines from $v = 1$ and 2 also being recorded. Almost twenty years later the spectrum was recorded again, with even higher accuracy, by English and Zorn [51]. Their apparatus was similar to the earlier instrument used by Hughes and by Trischka, except that the A dipole field was replaced by a quadrupole field (see figure 8.26). Most recent instruments are of this type, and we shall describe the details later in this chapter.

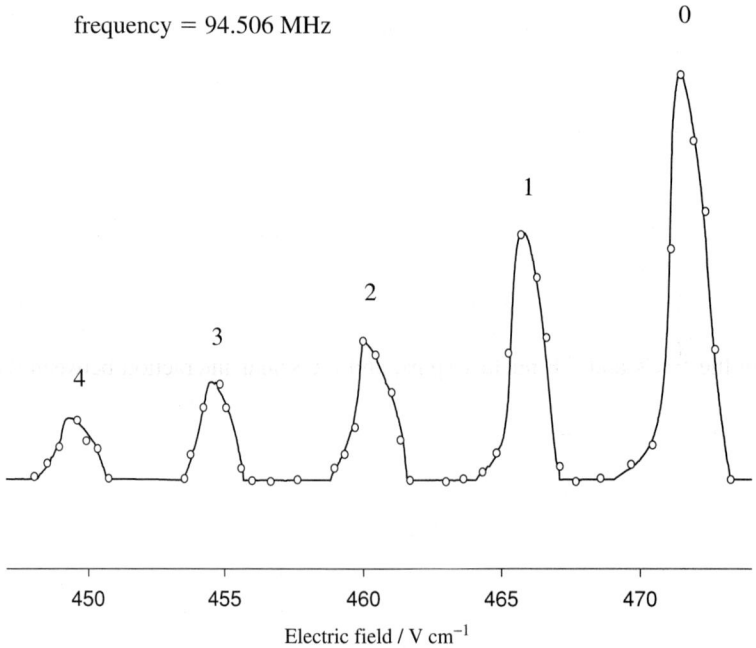

Figure 8.29. Electric resonance spectrum of CsF in 'strong' fields, showing resonance from molecules in five different vibrational levels ($v = 0$ to 4). The hyperfine structure resulting from nuclear–molecular interactions is not resolved [50].

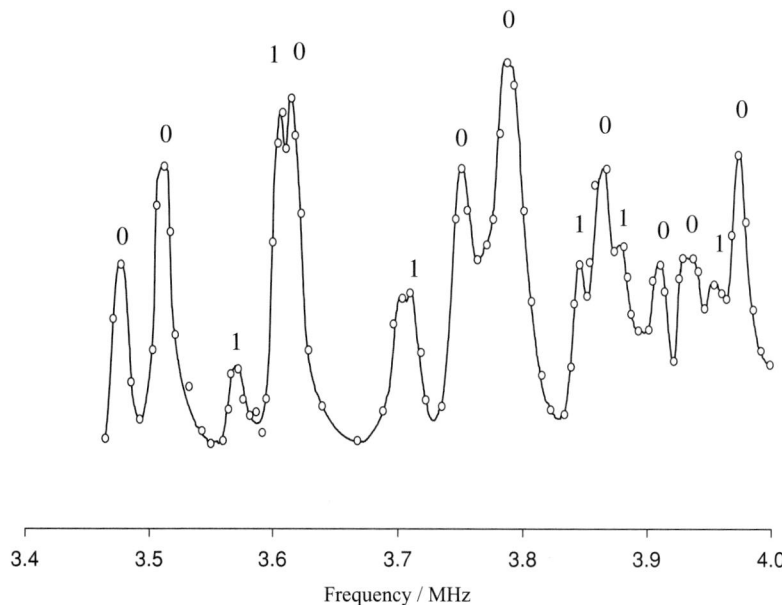

Figure 8.30. 'Strong' field electric resonance spectrum of CsF showing the hyperfine structure produced by nuclear–molecular interactions. Nine of the components arise from molecules in $v = 0$, as indicated, six from $v = 1$, and further components from $v = 2$. The analysis of this spectrum [50] is described in the text.

Our theoretical analysis of the zero field spectrum is directed towards a quantitative understanding of the most recent results.

The predominant isotope of cesium is ^{133}Cs which has a nuclear spin I_1 of 7/2; its quadrupole moment and g-factor will be denoted by Q_1 and g_1. The ^{19}F nucleus has spin I_2 of 1/2 (and therefore no quadrupole moment) and a nuclear g-factor denoted g_2. The nuclear hyperfine Hamiltonian used by English and Zorn [51] was the sum of five terms representing the ^{133}Cs quadrupole interaction, the ^{133}Cs nuclear spin–rotation interaction, the ^{19}F nuclear spin–rotation interaction, the dipolar (tensorial) interaction between the ^{133}Cs and ^{19}F nuclear spins, and the scalar interaction between the two nuclear spins. Consistent with the conventions in use at the time, this Hamiltonian was written in the following form:

$$\mathcal{H}_{\text{hf}} = -\frac{eq_1 Q_1}{2I_1(2I_1 - 1)(2J - 1)(2J + 3)}\left[3(\boldsymbol{I}_1 \cdot \boldsymbol{J})^2 + \frac{3}{2}(\boldsymbol{I}_1 \cdot \boldsymbol{J}) - I_1(I_1 + 1)J(J + 1)\right]$$
$$+ c_1 \boldsymbol{I}_1 \cdot \boldsymbol{J} + c_2 \boldsymbol{I}_2 \cdot \boldsymbol{J} + \frac{g_1 g_2 \mu_N^2}{R^3(2J + 3)(2J - 1)}\left(\frac{1}{4\pi\varepsilon_0 c^2}\right)[3(\boldsymbol{I}_1 \cdot \boldsymbol{J})(\boldsymbol{I}_2 \cdot \boldsymbol{J})$$
$$+ 3(\boldsymbol{I}_2 \cdot \boldsymbol{J})(\boldsymbol{I}_1 \cdot \boldsymbol{J}) - 2\boldsymbol{I}_1 \cdot \boldsymbol{I}_2 J(J + 1)] + c_4 \boldsymbol{I}_1 \cdot \boldsymbol{I}_2. \tag{8.281}$$

The nuclear spin subscripts 1 and 2 refer to the ^{133}Cs and ^{19}F nuclei respectively. In the spirit of this book, we rewrite the Hamiltonian in terms of irreducible tensors as follows:

$$\mathcal{H}_{hf} = -eT^2(\nabla E_1) \cdot T^2(Q_1) + c_1 T^1(I_1) \cdot T^1(J) + c_2 T^1(I_2) \cdot T^1(J) \\ + \sqrt{10} g_1 g_2 \mu_N^2 (1/4\pi\varepsilon_0 c^2) T^1(C, I_1) \cdot T^1(I_2) + c_4 T^1(I_1) \cdot T^1(I_2). \quad (8.282)$$

The values of the constants appearing in (8.282) are, in principle, vibrationally-dependent. Although we shall evaluate the matrix elements of (8.282) without initially specifying the value of J, the experimental studies were for the $J = 1$ level.

It is possible to work in a basis set in which the nuclear spins are coupled, both to each other and also to J; the analysis of the spectrum by Trischka [50] was accomplished using a coupled basis set for weak field experiments, and a fully decoupled basis for the strong field experiments. We will evaluate the matrix elements in both bases.

(b) 'Weak' field coupled basis

The coupled basis functions are taken in the form $|\eta, \Lambda; J, I_1, F_1, I_2, F, M_F\rangle$. For completeness, we have included Λ, although for a $^1\Sigma$ state Λ has the value 0. η represents all other relevant quantum numbers, but particularly v in the current problem, and we use the coupling scheme

$$F_1 = J + I_1, \quad F = F_1 + I_2. \quad (8.283)$$

The matrix elements of the four terms in equation (8.282) are now evaluated. First, the quadrupole interaction follows along lines similar to those outlined for D_2 in equations (8.22) to (8.29):

$$\langle \eta, \Lambda; J, I_1, F_1, I_2, F, M_F | -eT^2(\nabla E_1) \cdot T^2(Q_1) | \eta, \Lambda; J', I_1, F_1', I_2, F', M_F' \rangle$$

$$= -e\delta_{FF'}\delta_{M_F M_F'}\delta_{F_1 F_1'}(-1)^{J'+F_1+I_1} \begin{Bmatrix} I_1 & J' & F_1 \\ J & I_1 & 2 \end{Bmatrix} \langle \eta, \Lambda, J \| T^2(\nabla E_1) \| \eta, \Lambda', J' \rangle$$

$$\times \langle I_1 \| T^2(Q_1) \| I_1 \rangle$$

$$= -e\delta_{FF'}\delta_{M_F M_F'}\delta_{F_1 F_1'}(-1)^{J'+F_1+I_1} \begin{Bmatrix} I_1 & J' & F_1 \\ J & I_1 & 2 \end{Bmatrix}$$

$$\times \langle \eta, \Lambda, J | \sum_q \mathfrak{D}_{\cdot q}^{(2)}(\omega)^* T_q^2(\nabla E_1) | \eta, \Lambda', J' \rangle \begin{pmatrix} Q_1 \\ 2 \end{pmatrix} \begin{pmatrix} I_1 & 2 & I_1 \\ -I_1 & 0 & I_1 \end{pmatrix}^{-1}. \quad (8.284)$$

Confining attention to the molecule-fixed $q = 0$ component, and making use of the definition (8.28) of the electric field gradient q_1 at nucleus 1, we obtain the final result

$$\langle \eta, \Lambda; J, I_1, F_1, I_2, F, M_F | -eT^2(\nabla E_1) \cdot T^2(Q_1) | \eta, \Lambda; J', I_1, F_1, I_2, F, M_F \rangle$$

$$= \frac{eq_1 Q_1}{4}(-1)^{J'+F_1+I_1} \begin{Bmatrix} I_1 & J' & F_1 \\ J & I_1 & 2 \end{Bmatrix} \begin{pmatrix} I_1 & 2 & I_1 \\ -I_1 & 0 & I_1 \end{pmatrix}^{-1}$$

$$\times (-1)^{J-\Lambda} \{(2J+1)(2J'+1)\}^{1/2} \begin{pmatrix} J & 2 & J' \\ -\Lambda & 0 & \Lambda \end{pmatrix}. \quad (8.285)$$

For the $^1\Sigma^+$ ground state of CsF, $\Lambda = 0$ in this equation. This result should now be compared with the term used by English and Zorn [51] and Ramsey [52] to represent the quadrupole interaction, which is

$$\mathcal{H}_Q = -\frac{eqQ}{2I_1(2I_1-1)(2J-1)(2J+3)}\left\{3(\boldsymbol{I}_1\cdot\boldsymbol{J})^2 + \frac{3}{2}(\boldsymbol{I}_1\cdot\boldsymbol{J}) - I_1(I_1+1)J(J+1)\right\}. \tag{8.286}$$

It is convenient to expand this equation, making use of the fact that $\boldsymbol{F}_1 = \boldsymbol{J} + \boldsymbol{I}_1$, and it is easy to show that (8.286) becomes

$$\mathcal{H}_Q = -\frac{eqQ}{2I_1(2I_1-1)(2J-1)(2J+3)}\left\{\frac{3}{4}\{I_1^2 + J^2 - F^2\}[\{I_1^2 + J^2 - F^2\} - 1]\right. $$
$$\left. - I_1(I_1+1)J(J+1)\right\}. \tag{8.287}$$

We now compare this with the analytic expansion of (8.285), confining attention to the diagonal elements. We have

$$\langle \eta, \Lambda; J, I_1, F_1, I_2, F, M_F | -e\mathbf{T}^2(\nabla\boldsymbol{E}_1)\cdot\mathbf{T}^2(\boldsymbol{Q}_1)|\eta, \Lambda; J, I_1, F_1, I_2, F, M_F\rangle$$

$$= \frac{eq_1Q_1}{4}(-1)^{J'+F_1+I_1}\begin{Bmatrix} I_1 & J' & F_1 \\ J & I_1 & 2 \end{Bmatrix}\begin{pmatrix} I_1 & 2 & I_1 \\ -I_1 & 0 & I_1 \end{pmatrix}^{-1}(-1)^{J-\Lambda}\{(2J+1)(2J'+1)\}^{1/2}\begin{pmatrix} J & 2 & J \\ -\Lambda & 0 & \Lambda \end{pmatrix}$$

$$= \frac{eq_1Q_1}{4}\frac{2\{3X(X-1) - 4I_1(I_1+1)J(J+1)\}}{\{(2I_1-1)(2I_1)(2I_1+1)(2I_1+2)(2I_1+3)(2J-1)2J(2J+1)(2J+2)(2J+3)\}^{1/2}}$$

$$\times \left\{\frac{(2I_1+3)(2I_1+2)(2I_1+1)}{2I_1(2I_1-1)}\right\}^{1/2}(-1)^{J}(2J+1)(-1)^{J+1}$$

$$\times 2\left\{\frac{1}{(2J+3)(2J+2)(2J+1)2J(2J-1)}\right\}^{1/2}, \tag{8.288}$$

where $X = I_1(I_1+1) + J(J+1) - F_1(F_1+1) = I_1^2 + J^2 - F_1^2$. After some algebraic manipulation, (8.288) becomes

$$\langle \eta, \Lambda; J, I_1, F_1, I_2, F, M_F | -e\mathbf{T}^2(\nabla\boldsymbol{E}_1)\cdot\mathbf{T}^2(\boldsymbol{Q}_1)|\eta, \Lambda; J, I_1, F_1, I_2, F, M_F\rangle$$

$$= -\frac{eq_1Q_1}{2I_1(2I_1-1)(2J-1)(2J+3)}\left\{\frac{3}{4}\{I_1^2 + J^2 - F_1^2\}[\{I_1^2 + J^2 - F_1^2\} - 1]\right.$$
$$\left. - I_1(I_1+1)J(J+1)\right\}. \tag{8.289}$$

Comparing our (8.289) with (8.287) we see that they are in agreement. Full details of the definitions used here, and throughout this book, are given in appendix 8.4.

Second, the nuclear spin–rotation terms are as follows:

$$\langle \eta, \Lambda; J, I_1, F_1, I_2, F, M_F | c_1(\eta) \mathbf{T}^1(\boldsymbol{J}) \cdot \mathbf{T}^1(\boldsymbol{I}_1) | \eta, \Lambda; J', I_1, F_1', I_2, F', M_F \rangle$$

$$= c_1(\eta) \delta_{FF'} \delta_{F_1 F_1'} (-1)^{J'+F_1+I_1} \begin{Bmatrix} I_1 & J' & F_1 \\ J & I_1 & 1 \end{Bmatrix} \langle J \| T^1(\boldsymbol{J}) \| J' \rangle \langle I_1 \| T^1(\boldsymbol{I}_1) \| I_1 \rangle$$

$$= c_1(\eta) \delta_{FF'} \delta_{F_1 F_1'} \delta_{JJ'} (-1)^{J+F_1+I_1} \begin{Bmatrix} I_1 & J & F_1 \\ J & I_1 & 1 \end{Bmatrix}$$

$$\times \{J(J+1)(2J+1) I_1(I_1+1)(2I_1+1)\}^{1/2}. \tag{8.290}$$

$$\langle \eta, \Lambda; J, I_1, F_1, I_2, F, M_F | c_2(\eta) \mathbf{T}^1(\boldsymbol{J}) \cdot \mathbf{T}^1(\boldsymbol{I}_2) | \eta, \Lambda; J', I_1, F_1', I_2, F', M_F \rangle$$

$$= c_2(\eta) \delta_{FF'} (-1)^{F_1'+F+I_2} \begin{Bmatrix} I_2 & F_1' & F \\ F_1 & I_2 & 1 \end{Bmatrix} \langle J, I_1, F_1 \| T^1(\boldsymbol{J}) \| J', I_1, F_1' \rangle \langle I_2 \| T^1(\boldsymbol{I}_2) \| I_2 \rangle$$

$$= c_2(\eta) \delta_{FF'} \delta_{JJ'} (-1)^{F_1'+F+I_2} \begin{Bmatrix} I_2 & F_1' & F \\ F_1 & I_2 & 1 \end{Bmatrix} (-1)^{F_1'+J+I_1+1} \{(2F_1'+1)(2F_1+1)\}^{1/2}$$

$$\times \begin{Bmatrix} F_1 & J & I_1 \\ J & F_1' & 1 \end{Bmatrix} \{J(J+1)(2J+1) I_2(I_2+1)(2I_2+1)\}^{1/2}. \tag{8.291}$$

These results are the same as those presented in Trischka's analysis.

Third, we deal with the dipolar coupling of the nuclear spins, which is evaluated below; note the form of the dipolar Hamiltonian (equation (8.282)), which is the one appropriate for the particular angular momentum coupling scheme used.

$$\langle \eta, \Lambda; J, I_1, F_1, I_2, F, M_F | \mathcal{H}_{\text{dip}} | \eta, \Lambda'; J', I_1, F_1', I_2, F, M_F \rangle$$

$$= \sqrt{10} g_1 g_2 \mu_N^2 (1/4\pi\varepsilon_0 c^2) (-1)^{F_1'+F+I_2} \begin{Bmatrix} I_2 & F_1' & F \\ F_1 & I_2 & 1 \end{Bmatrix}$$

$$\times \langle \eta, \Lambda; J, I_1, F_1 \| T^1(\boldsymbol{C}^2, \boldsymbol{I}_1) \| \eta, \Lambda'; J', I_1, F_1' \rangle \langle I_2 \| T^1(\boldsymbol{I}_2) \| I_2 \rangle$$

$$= \sqrt{10} g_1 g_2 \mu_N^2 (1/4\pi\varepsilon_0 c^2) (-1)^{F_1'+F+I_2} \begin{Bmatrix} I_2 & F_1' & F \\ F_1 & I_2 & 1 \end{Bmatrix} \{(3)(2F_1+1)(2F_1'+1)\}^{1/2}$$

$$\times \begin{Bmatrix} J & J' & 2 \\ I_1 & I_1 & 1 \\ F_1 & F_1' & 1 \end{Bmatrix} \langle \eta, \Lambda; J \| T^2(\boldsymbol{C}) \| \eta, \Lambda'; J' \rangle \{I_2(I_2+1)(2I_2+1) I_1(I_1+1)$$

$$\times (2I_1+1)\}^{1/2}. \tag{8.292}$$

We restrict attention to matrix elements diagonal in Λ (with value 0 for the $^1\Sigma^+$ ground state of CsF), so that the final result is

$$\langle \eta, \Lambda; J, I_1, F_1, I_2, F, M_F | \mathcal{H}_{\text{dip}} | \eta, \Lambda; J', I_1, F_1', I_2, F, M_F \rangle$$

$$= \sqrt{10} g_1 g_2 \mu_N^2 (1/4\pi\varepsilon_0 c^2) \langle \eta | C_0^2(\theta, \phi) R^{-3} | \eta \rangle (-1)^{F_1'+F+I_2} \begin{Bmatrix} I_2 & F_1' & F \\ F_1 & I_2 & 1 \end{Bmatrix}$$

$$\times \{(3)(2F_1+1)(2F_1'+1)\}^{1/2} \begin{Bmatrix} J & J' & 2 \\ I_1 & I_1 & 1 \\ F_1 & F_1' & 1 \end{Bmatrix} (-1)^J \begin{pmatrix} J & 2 & J' \\ 0 & 0 & 0 \end{pmatrix}$$

$$\times \{(2J+1)(2J'+1)I_2(I_2+1)(2I_2+1)I_1(I_1+1)(2I_1+1)\}^{1/2}. \quad (8.293)$$

We have used equation (8.31) in deriving the above in equation (8.293). The distribution operator in the vibronic matrix element is simply R^{-3}, since $\theta=0$ in this problem. Hence our final result is

$$\langle \eta, \Lambda; J, I_1, F_1, I_2, F, M_F | \mathcal{H}_{\text{dip}} | \eta, \Lambda; J', I_1, F_1', I_2, F, M_F \rangle$$
$$= \sqrt{30} g_1 g_2 \mu_N^2 (1/4\pi\varepsilon_0 c^2) \langle R^{-3} \rangle_\eta (-1)^{F_1'+F+I_2} \begin{Bmatrix} I_2 & F_1' & F \\ F_1 & I_2 & 1 \end{Bmatrix} \{(2F_1+1)(2F_1'+1)\}^{1/2}$$

$$\times \begin{Bmatrix} J & J' & 2 \\ I_1 & I_1 & 1 \\ F_1 & F_1' & 1 \end{Bmatrix} (-1)^J \begin{pmatrix} J & 2 & J' \\ 0 & 0 & 0 \end{pmatrix} \{(2J+1)(2J'+1)I_2(I_2+1)$$

$$\times (2I_2+1)I_1(I_1+1)(2I_1+1)\}^{1/2}. \quad (8.294)$$

On reduction to analytical form, our result agrees with that of English and Zorn [51].

The final term in equation (8.282) represents the scalar interaction between the two nuclear spins; its matrix elements are as follows:

$$\langle \eta, \Lambda; J, I_1, F_1, I_2, F, M_F | c_4 \mathbf{T}^1(\mathbf{I}_1) \cdot \mathbf{T}^1(\mathbf{I}_2) | \eta, \Lambda'; J', I_1, F_1', I_2, F, M_F \rangle$$
$$= c_4 (-1)^{F_1'+F+I_2} \begin{Bmatrix} I_2 & F_1' & F \\ F_1 & I_2 & 1 \end{Bmatrix} \langle \eta, \Lambda; J, I_1, F_1 \| \mathbf{T}^1(\mathbf{I}_1) \| \eta, \Lambda'; J', I_1, F_1' \rangle$$
$$\times \{I_2(I_2+1)(2I_2+1)\}^{1/2}$$
$$= c_4 (-1)^{F_1'+F+I_2} \begin{Bmatrix} I_2 & F_1' & F \\ F_1 & I_2 & 1 \end{Bmatrix} (-1)^{F_1+J+1+I_1} \begin{Bmatrix} F_1 & I_1 & J \\ I_1 & F_1' & 1 \end{Bmatrix}$$
$$\times \{(2F_1'+1)(2F_1+1)I_1(I_1+1)(2I_1+1)I_2(I_2+1)(2I_2+1)\}^{1/2}. \quad (8.295)$$

One final result is required to analyse the 'weak' field spectrum of CsF. We derived an expression for the matrix elements of the electric field perturbation earlier, without the inclusion of nuclear spin effects, in equation (8.278). We now repeat this derivation using the basis set employed above. Taking the direction of the electric field to define the $p=0$ (Z) direction, the results are as follows:

$$\langle \eta, \Lambda; J, I_1, F_1, I_2, F, M_F | - \mathbf{T}^1_{p=0}(\boldsymbol{\mu}_e) \mathbf{T}^1_{p=0}(\mathbf{E}) | \eta, \Lambda; J', I_1, F_1', I_2, F', M_F \rangle$$
$$= -E_Z (-1)^{F-M_F} \begin{pmatrix} F & 1 & F' \\ -M_F & 0 & M_F \end{pmatrix}$$
$$\times \langle \eta, \Lambda; J, I_1, F_1, I_2, F \| \mathbf{T}^1(\boldsymbol{\mu}_e) \| \eta, \Lambda; J', I_1, F_1', I_2, F' \rangle$$

$$= -E_Z(-1)^{F-M_F} \begin{pmatrix} F & 1 & F' \\ -M_F & 0 & M_F \end{pmatrix} (-1)^{F'+F_1+1+I_2} \{(2F'+1)(2F+1)\}^{1/2}$$

$$\times \begin{Bmatrix} F & F_1 & I_2 \\ F'_1 & F' & 1 \end{Bmatrix} \langle \eta, \Lambda; J, I_1, F_1 \| T^1(\mu_e) \| \eta, \Lambda; J', I_1, F'_1 \rangle$$

$$= -E_Z(-1)^{F-M_F} \begin{pmatrix} F & 1 & F' \\ -M_F & 0 & M_F \end{pmatrix} (-1)^{F'+F_1+1+I_2} \{(2F'+1)(2F+1)\}^{1/2}$$

$$\times \begin{Bmatrix} F & F_1 & I_2 \\ F'_1 & F' & 1 \end{Bmatrix} (-1)^{F'_1+J+1+I_1} \{(2F'_1+1)(2F_1+1)\}^{1/2} \begin{Bmatrix} F_1 & J & I_1 \\ J' & F'_1 & 1 \end{Bmatrix}$$

$$\times \langle \eta, \Lambda; J \| \mathfrak{D}^{(1)}_{00}(\omega)^* T^1_{q=0}(\mu_e) \| \eta, \Lambda; J' \rangle$$

$$= -\mu_0 E_Z(-1)^{F-M_F} \begin{pmatrix} F & 1 & F' \\ -M_F & 0 & M_F \end{pmatrix} (-1)^{F'+F_1+1+I_2} \{(2F'+1)(2F+1)\}^{1/2}$$

$$\times \begin{Bmatrix} F & F_1 & I_2 \\ F'_1 & F' & 1 \end{Bmatrix} (-1)^{F'_1+J+1+I_1} \{(2F'_1+1)(2F_1+1)\}^{1/2} \begin{Bmatrix} F_1 & J & 1 \\ J' & F'_1 & I_1 \end{Bmatrix}$$

$$\times \{(2J'+1)(2J+1)\}^{1/2}(-1)^J \begin{pmatrix} J & 1 & J' \\ 0 & 0 & 0 \end{pmatrix}. \tag{8.296}$$

Equation (8.296) shows that as the magnitude of the applied electric field increases, our coupled basis set will become less and less meaningful because of the mixing of states with different values for F, F_1 and J.

It is now instructive to make use of the 'weak' field results to obtain an energy level pattern for the $J = 1$ rotational level. The complete zero-field matrix is as follows.

		$F_1 = 9/2$		$F_1 = 7/2$		$F_1 = 5/2$	
		$F = 5$	$F = 4$	$F = 4$	$F = 3$	$F = 3$	$F = 2$
$F_1 = 9/2$	$F = 5$	m_{11}	0	0	0	0	0
	$F = 4$	0	m_{22}	m_{23}	0	0	0
$F_1 = 7/2$	$F = 4$	0	m_{32}	m_{33}	0	0	0
	$F = 3$	0	0	0	m_{44}	m_{45}	0
$F_1 = 5/2$	$F = 3$	0	0	0	m_{54}	m_{55}	0
	$F = 2$	0	0	0	0	0	m_{66}

The matrix elements are:

$$m_{11} = -eq_1 Q_1(1/20) + c_1(7/2) + c_2(1/2) + c_3(7/10) + c_4(7/4)$$
$$m_{22} = -eq_1 Q_1(1/20) + c_1(7/2) - c_2(11/18) - c_3(77/90) - c_4(77/36)$$
$$m_{33} = +eq_1 Q_1(1/7) - c_1 + c_2(1/9) - c_3(13/9) + c_4(59/36)$$

$$m_{44} = +eq_1Q_1(1/7) - c_1 - c_2(1/7) + c_3(13/7) - c_4(59/28)$$
$$m_{55} = -eq_1Q_1(3/28) - c_1(9/2) - c_2(5/14) + c_3(9/14) + c_4(15/28)$$
$$m_{66} = -eq_1Q_1(3/28) - c_1(9/2) + c_2(1/2) - c_3(9/10) - c_4(9/4)$$
$$m_{23} = m_{32} = -c_2(\sqrt{35}/9) - c_3(23\sqrt{7}/18\sqrt{5}) + c_4(\sqrt{35}/9)$$
$$m_{45} = m_{54} = -c_2(3\sqrt{3}/7) + c_3(15\sqrt{3}/14) + c_4(3\sqrt{3}/7).$$
(8.297)

The dipolar constant c_3 is defined by

$$c_3 = g_1 g_2 \mu_N^2 (1/4\pi\varepsilon_0 c^2) \langle R^{-3} \rangle_\eta. \quad (8.298)$$

The off-diagonal elements have been included for the sake of completeness but their effects are very small. The values of the constants for $v = 0$ obtained from the analysis of the spectrum [51] were (in kHz):

$$eq_1Q_1 = 1237.0, \quad c_1 = 0.70, \quad c_2 = 15.1, \quad c_3 = 0.92, \quad c_4 = 0.61.$$

Insertion of these values of the constants into the diagonal elements listed in (8.297) enables us to construct a hyperfine energy level diagram for the $J = 1$ rotational level, as shown in figure 8.31. We also show the eight transitions observed by English and Zorn [51]; the diagram is constructed for zero electric field in the C field region, but was actually recorded in a very weak field. The presence of an electric field, which mixes

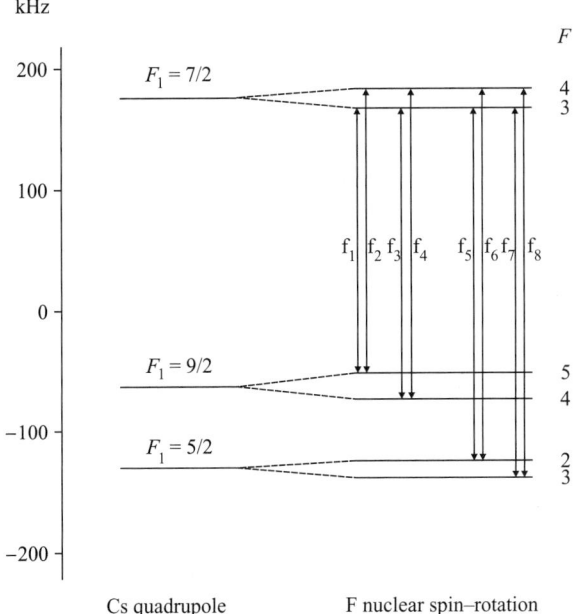

Figure 8.31. Nuclear hyperfine splitting of the $J = 1$ rotational level of CsF in its $X\,^1\Sigma^+$ state. The major splitting is due to the ^{133}Cs quadrupole interaction, and the smaller doublet splitting arises from the ^{19}F nuclear spin–rotation coupling. The diagram is drawn for zero applied electric field, and the strongest electric dipole transitions are indicated.

Table 8.11. *Hyperfine transition frequencies in the $J = 1$, $v = 0$ state of $^{133}\text{Cs}^{19}\text{F}$ (in kHz)*

	F_1, F	F_1', F'	Observed	Calculated
f_1	(7/2, 3)	(9/2, 5)	224.60	224.30
f_2	(7/2, 4)	(9/2, 5)	228.05	227.98
f_3	(7/2, 3)	(9/2, 4)	245.80	245.36
f_4	(7/2, 4)	(9/2, 4)	249.05	249.04
f_5	(7/2, 3)	(5/2, 2)	304.90	304.50
f_6	(7/2, 4)	(5/2, 2)	308.10	308.18
f_7	(7/2, 3)	(5/2, 3)	314.40	314.58
f_8	(7/2, 4)	(5/2, 3)	317.60	318.26

different J levels, is necessary in order to transfer electric dipole intensity into the transitions shown; in zero electric field the $\Delta F = 0, \pm 1$ transitions would have magnetic dipole intensity only, and the $\Delta F = 2$ transitions would be totally forbidden. The measured transition frequencies are presented in table 8.11, and the frequencies calculated from the above analysis are also shown. The agreement between the two is excellent.

We shall discuss the information about the electronic structure of CsF obtained from the measured constants later, but first consider the earlier measurements of Trischka made in 'strong' electric fields, and the assignment of the spectrum shown in figure 8.30.

(c) 'Strong' field decoupled basis

We now re-examine the problem when a moderately strong electric field is applied in the C field region; spectra were recorded by Trischka under these conditions, as shown in figure 8.30. The matrix elements of the electric field perturbation in the coupled basis set were given in equation (8.296), but it would make little sense to use this basis set for the calculation. Only M_F remains a good quantum number, and the Stark effect operates through the mixing of $J = 1$ with $J = 0$ and 2, as was shown in equation (8.278). A suitable nuclear spin decoupled basis set would be $|\eta, \Lambda; J, M_J, I_1, M_{I_1}, I_2, M_{I_2}\rangle$ and even for $J = 1$, the number of levels given by $(2J + 1)(2I_1 + 1)(2I_2 + 1)$ is equal to 48. Trischka [50] did not give details of the analysis, but since our aim is to understand the details of the 'strong' field spectrum shown in figure 8.30, we will present a full account of the theory.

The required matrix elements in the decoupled representation are now calculated; all of the scalar products which occur are expanded in the space-fixed coordinate system.

(i) QUADRUPOLE INTERACTION

$$\langle \eta, \Lambda; J, M_J, I_1, M_{I_1}, I_2, M_{I_2}| \sum_p (-1)^p (-e) \text{T}_p^2(\nabla E) \text{T}_{-p}^2(\boldsymbol{Q}_1) |\eta, \Lambda'; J', M_J', I_1', M_{I_1}', I_2', M_{I_2}'\rangle$$

$$= -e\delta_{I_2 I_2'} \delta_{M_{I_2} M_{I_2}'} \sum_p (-1)^p \langle \eta, \Lambda; J, M_J | T_p^2(\nabla E) | \eta, \Lambda'; J', M_J' \rangle \langle I_1, M_{I_1} | T_{-p}^2(\boldsymbol{Q}_1) | I_1', M_{I_1}' \rangle$$

$$= -e\delta_{I_1 I_1'} \sum_p (-1)^p (-1)^{J-M_J} \begin{pmatrix} J & 2 & J' \\ -M_J & p & M_J' \end{pmatrix} \langle \eta, \Lambda; J \| T^2(\nabla E) \| \eta, \Lambda'; J' \rangle$$

$$\times (-1)^{I_1-M_{I_1}} \begin{pmatrix} I_1 & 2 & I_1 \\ -M_{I_1} & -p & M_{I_1}' \end{pmatrix} \langle I_1 \| T^2(\boldsymbol{Q}_1) \| I_1 \rangle$$

$$= \frac{eq_0 Q_1}{4} \sum_p (-1)^p (-1)^{J-M_J} \begin{pmatrix} J & 2 & J' \\ -M_J & p & M_J' \end{pmatrix} (-1)^{J-\Lambda} \{(2J+1)(2J'+1)\}^{1/2}$$

$$\times \begin{pmatrix} J & 2 & J' \\ -\Lambda & 0 & \Lambda \end{pmatrix} (-1)^{I_1-M_{I_1}} \begin{pmatrix} I_1 & 2 & I_1 \\ -M_{I_1} & -p & M_{I_1}' \end{pmatrix} \begin{pmatrix} I_1 & 2 & I_1 \\ -I_1 & 0 & I_1 \end{pmatrix}^{-1}. \quad (8.299)$$

(ii) ^{133}Cs NUCLEAR SPIN–ROTATION INTERACTION

$$\langle \eta, \Lambda; J, M_J, I_1, M_{I_1}, I_2, M_{I_2} | \sum_p (-1)^p c_1 T_p^1(\boldsymbol{J}) T_{-p}^1(\boldsymbol{I}_1) | \eta, \Lambda'; J', M_J', I_1', M_{I_1}', I_2', M_{I_2}' \rangle$$

$$= c_1 \delta_{\Lambda, \Lambda'} \delta_{I_2 I_2'} \delta_{M_{I_2} M_{I_2}'} \sum_p (-1)^p \langle \eta, \Lambda; J, M_J | T_p^1(\boldsymbol{J}) | \eta, \Lambda; J', M_J' \rangle$$

$$\times \langle I_1, M_{I_1} | T_{-p}^1(\boldsymbol{I}_1) | I_1, M_{I_1}' \rangle$$

$$= c_1 \sum_p (-1)^p (-1)^{J-M_J} \begin{pmatrix} J & 1 & J \\ -M_J & p & M_J' \end{pmatrix} \langle J \| T^1(\boldsymbol{J}) \| J \rangle (-1)^{I_1 - M_{I_1}}$$

$$\times \begin{pmatrix} I_1 & 1 & I_1 \\ -M_{I_1} & -p & M_{I_1}' \end{pmatrix} \langle I_1 \| T^1(\boldsymbol{I}_1) \| I_1 \rangle$$

$$= c_1 \sum_p (-1)^p (-1)^{J-M_J} (-1)^{I_1-M_{I_1}} \{J(J+1)(2J+1) I_1(I_1+1)(2I_1+1)\}^{1/2}$$

$$\times \begin{pmatrix} J & 1 & J \\ -M_J & p & M_J' \end{pmatrix} \begin{pmatrix} I_1 & 1 & I_1 \\ -M_{I_1} & -p & M_{I_1}' \end{pmatrix}. \quad (8.300)$$

(iii) ^{19}F NUCLEAR SPIN–ROTATION INTERACTION

$$\langle \eta, \Lambda; J, M_J, I_1, M_{I_1}, I_2, M_{I_2} | \sum_p (-1)^p c_2 T_p^1(\boldsymbol{J}) T_{-p}^1(\boldsymbol{I}_2) | \eta, \Lambda'; J', M_J', I_1', M_{I_1}', I_2', M_{I_2}' \rangle$$

$$= c_2 \delta_{\Lambda, \Lambda'} \delta_{I_1 I_1'} \delta_{M_{I_1} M_{I_1}'} \sum_p (-1)^p \langle \eta, \Lambda; J, M_J | T_p^1(\boldsymbol{J}) | \eta, \Lambda; J', M_J' \rangle$$

$$\times \langle I_2, M_{I_2} | T_{-p}^1(\boldsymbol{I}_2) | I_2', M_{I_2}' \rangle$$

$$= c_2 \delta_{JJ'} \sum_p (-1)^p (-1)^{J-M_J} \begin{pmatrix} J & 1 & J \\ -M_J & p & M'_J \end{pmatrix} \langle J\|T^1(J)\|J\rangle (-1)^{I_2-M_{I_2}}$$

$$\times \begin{pmatrix} I_2 & 1 & I_2 \\ -M_{I_2} & -p & M'_{I_2} \end{pmatrix} \langle I_2\|T^1(I_2)\|I_2\rangle$$

$$= c_2 \sum_p (-1)^p (-1)^{J-M_J} (-1)^{I_2-M_{I_2}} \{J(J+1)(2J+1)I_2(I_2+1)(2I_2+1)\}^{1/2}$$

$$\times \begin{pmatrix} J & 1 & J \\ -M_J & p & M'_J \end{pmatrix} \begin{pmatrix} I_2 & 1 & I_2 \\ -M_{I_2} & -p & M'_{I_2} \end{pmatrix}. \tag{8.301}$$

(iv) NUCLEAR SPIN DIPOLAR INTERACTION

$$\langle \eta, \Lambda; J, M_J, I_1, M_{I_1}, I_2, M_{I_2} | \mathcal{H}_{\text{dip}} | \eta, \Lambda'; J', M'_J, I'_1, M'_{I_1}, I_2, M'_{I_2} \rangle$$

$$= \sqrt{10} g_1 g_2 \mu_N^2 (1/4\pi\varepsilon_0 c^2) \sum_p (-1)^{p+I_2-M_{I_2}} \begin{pmatrix} I_2 & 1 & I_2 \\ -M_{I_2} & -p & M'_{I_2} \end{pmatrix}$$

$$\times \{I_2(I_2+1)(2I_2+1)\}^{1/2} \langle \eta, \Lambda; J, M_J, I_1, M_{I_1} | T_p^1(C^2, I_1) | \eta, \Lambda'; J', M'_J, I'_1, M'_{I_1} \rangle$$

$$= -\sqrt{30} g_1 g_2 \mu_N^2 (1/4\pi\varepsilon_0 c^2) \sum_p (-1)^{p+I_2-M_{I_2}} \begin{pmatrix} I_2 & 1 & I_2 \\ -M_{I_2} & -p & M'_{I_2} \end{pmatrix}$$

$$\times \{I_2(I_2+1)(2I_2+1)\}^{1/2}$$

$$\times \langle \eta, \Lambda; J, M_J, I_1, M_{I_1} | \sum_{p_1, p_2} (-1)^p T_{p_1}^2(C) T_{p_2}^1(I_1) | \eta, \Lambda'; J', M'_J, I'_1, M'_{I_1} \rangle$$

$$\times \begin{pmatrix} 1 & 2 & 1 \\ p_1 & p_2 & -p \end{pmatrix}$$

$$= -\sqrt{30} g_1 g_2 \mu_N^2 (1/4\pi\varepsilon_0 c^2) \sum_p (-1)^{p+I_2-M_{I_2}} \begin{pmatrix} I_2 & 1 & I_2 \\ -M_{I_2} & -p & M'_{I_2} \end{pmatrix}$$

$$\times \{I_2(I_2+1)(2I_2+1)\}^{1/2} \{I_1(I_1+1)(2I_1+1)\}^{1/2}$$

$$\times \sum_{p_1 p_2} (-1)^p \langle \eta, \Lambda; J, M_J | T_{p_1}^2(C) | \eta, \Lambda'; J', M'_J \rangle$$

$$\times (-1)^{I_1-M_{I_1}} \begin{pmatrix} I_1 & 1 & I_1 \\ -M_{I_1} & p_2 & M'_{I_1} \end{pmatrix} \begin{pmatrix} 1 & 2 & 1 \\ p_1 & p_2 & -p \end{pmatrix}$$

$$= -\sqrt{30} g_1 g_2 \mu_N^2 (1/4\pi\varepsilon_0 c^2) \sum_p (-1)^{I_2-M_{I_2}+I_1-M_{I_1}} \begin{pmatrix} I_2 & 1 & I_2 \\ -M_{I_2} & -p & M'_{I_2} \end{pmatrix}$$

$$\times \{I_1(I_1+1)(2I_1+1)I_2(I_2+1)(2I_2+1)\}^{1/2}$$

$$\times \sum_{p_1 p_2} \langle \eta, \Lambda; J\|T^2(C)\|\eta, \Lambda'; J'\rangle (-1)^{J-M_J}$$

$$\times \begin{pmatrix} J & 1 & J' \\ -M_J & p_1 & M'_J \end{pmatrix} \begin{pmatrix} I_1 & 1 & I_1 \\ -M_{I_1} & p_2 & M'_{I_1} \end{pmatrix} \begin{pmatrix} 1 & 2 & 1 \\ p_1 & p_2 & -p \end{pmatrix}. \tag{8.302}$$

The reduced matrix element of $T^2(C)$ can be treated in the usual manner but we shall not pursue this rather messy analysis any further, particularly since the dipolar interaction in CsF is very small!

(v) ELECTRIC FIELD INTERACTION

The applied electric field interacts with the electric dipole moment of the molecule; nuclear spin is not involved in the decoupled basis set, so that we need only the results of our earlier analysis, given in equation (8.278), i.e.

$$\langle \eta, \Lambda; J, M_J, I_1, M_{I_1}, I_2, M_{I_2} | - T^1_{p=0}(\boldsymbol{\mu}_e) T^1_{p=0}(\boldsymbol{E}) | \eta, \Lambda; J', M_J, I'_1, M'_{I_1}, I'_2, M'_{I_2} \rangle$$
$$= -\delta_{I_1 I'_1} \delta_{M_{I_1} M'_{I_1}} \delta_{I_2 I'_2} \delta_{M_{I_2} M'_{I_2}} \mu_0 E_Z \{(2J+1)(2J'+1)\}^{1/2}$$
$$\times (-1)^{M-\Lambda} \begin{pmatrix} J & 1 & J' \\ -M_J & 0 & M_J \end{pmatrix} \begin{pmatrix} J & 1 & J' \\ -\Lambda & 0 & \Lambda \end{pmatrix}. \tag{8.303}$$

As we have already seen, the applied electric field operates by mixing different rotational levels, and within the $J = 1$ level the result is a splitting between the $M_J = 0$ component, and the $M_J = \pm 1$ components, the latter remaining degenerate (see figure 8.27). For a dipole moment μ_0 of 7.882 98 D and an electric field of 93.45 V cm^{-1} (the value used by Trischka) we calculate the Stark separation to be 3742 kHz, the $M_J = \pm 1$ components being the lower in energy.

We now consider the electric quadrupole interaction, using equation (8.299), with the index p being equal to 0, ± 1 and ± 2 in turn. Neglecting for the moment the effects of the ^{19}F nuclear spin, we have 24 states to consider, i.e. $(2J + 1)(2I_1 + 1)$, each characterised by values of M_J and M_{I_1}. The $p = 0$ matrix elements are diagonal in the decoupled basis set, but the $p = \pm 2$ elements connect the $M_J = \pm 1$ components and remove their degeneracy; the results are shown in figure 8.32, which is essentially drawn to scale. The $p = \pm 1$ terms in equation (8.299) are off-diagonal elements which mix $M_J = 0$ with $M_J = \pm 1$ components, producing much smaller shifts or splittings. We should not forget that each of the 24 states listed in figure 8.32 has a further two-fold degeneracy because of the ^{19}F nuclear spin (i.e. $I_2 = 1/2$, $M_{I_2} = \pm 1/2$).

The radiofrequency electric field is applied perpendicular to the static C field so that the selection rules for the transitions are $\Delta M_J = \pm 1$, $\Delta M_{I_1} = 0$. As we show in figure 8.32, there are eight such transitions, which essentially accounts for the experimental spectrum shown in figure 8.30, except that the latter exhibits *nine* components for the $v = 0$ level. The additional complexity arises because of other terms in the hyperfine Hamiltonian, but we will not pursue the details of the analysis. Measurements were made by Graff and Runolfsson [53] using higher electric fields, and with improvements in resolution they were able to measure further splittings of the resonance lines. They also carried out measurements using an applied magnetic field and were able to measure components of both the screening and susceptibility tensors, and also the rotational magnetic moment. The molecular constants determined from the 'strong' field and 'zero' field measurements were in very good agreement. The most

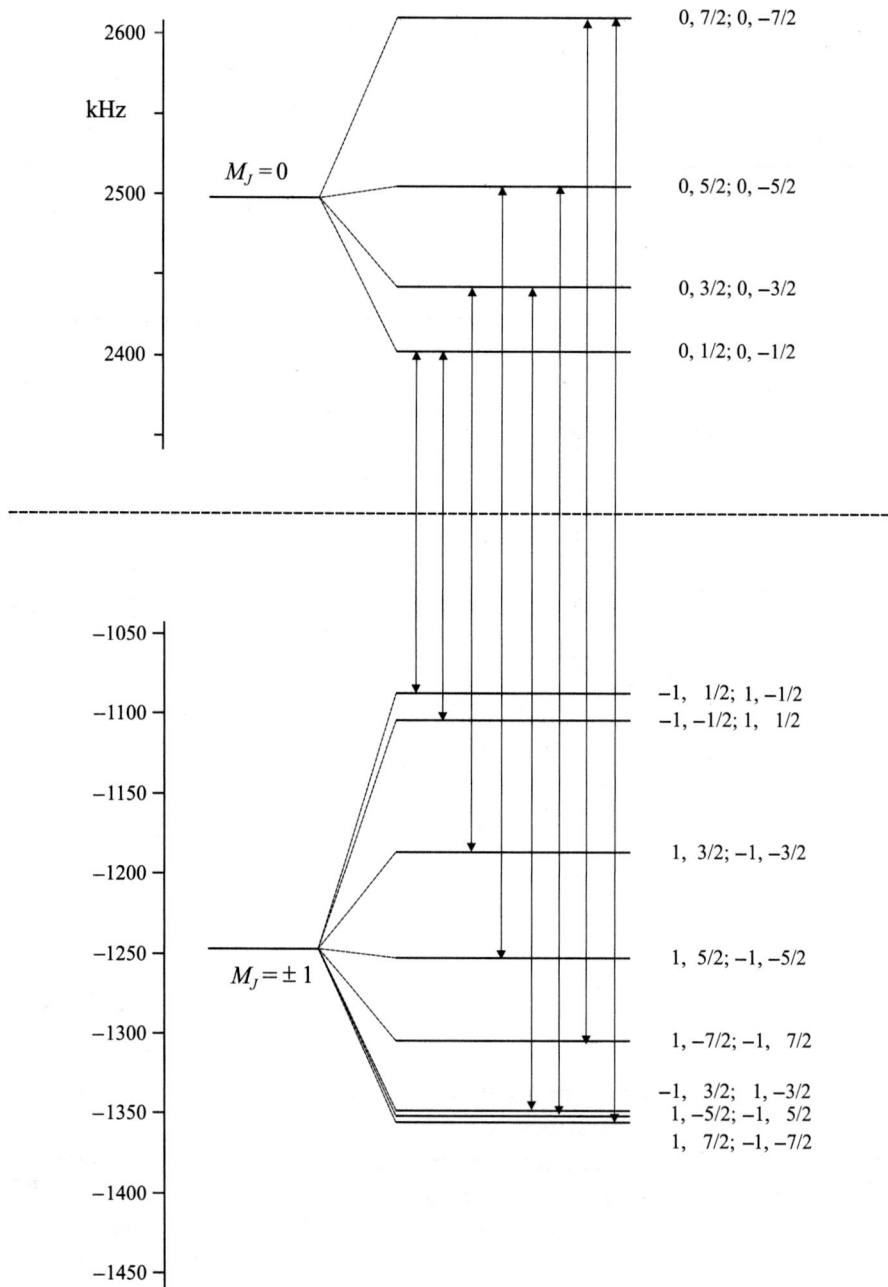

Figure 8.32. 'Strong' field levels for the $J = 1$ level in CsF. An electric field produces the Stark splitting between the $M_J = 0$ and ± 1 components, and ^{133}Cs quadrupole interaction splits each Stark component into four levels. Each level is two-fold degenerate; the two components are labelled by their values of M_J and M_{I_1}. The eight electric dipole transitions which become allowed in the electrostatic C field are shown.

Table 8.12. *Vibrational dependence of the molecular parameters for CsF in its $X^1\Sigma^+$ state*

	B_v (cm^{-1})	μ_v (Debye)	eqQ (kHz)	c_1 (kHz)	c_2 (kHz)	c_3 (kHz)	c_4 (kHz)
$v=0$	0.183 782	7.8783	1237.0	0.70	15.1	0.92	0.61
$v=1$	0.182 610	7.9506	1223.0	0.68	15.0	0.90	0.62
$v=2$	0.181 442	8.0247	1209.0	0.66	14.7	0.89	—

complete sets of data relate to the $v=0$ level, but information for higher vibrational levels was also obtained. Finally, some sophisticated triple resonance experiments were performed by Zorn, Stephenson, Dickinson and English [54] who were able to measure the low frequency $\Delta F_1 = 0$, $\Delta F = \pm 1$ hyperfine transitions in essentially zero electric field.

In a particularly novel extension of the studies of CsF, Freund, Fisk, Herschbach and Klemperer [55] have used electric resonance to probe the internal state distribution of CsF formed by the crossed beam reaction of Cs and SF$_6$. Transitions involving $J = 1$ to 4 and $v = 0$ to 4 were observed, and details of the kinematics thereby unravelled. A block diagram of the electric resonance spectrometer is shown in figure 8.33.

(d) Interpretation of the molecular constants

The vibrational dependence of the molecular constants is summarised by English and Zorn [51] for $v = 0, 1, 2$ and the results are listed in table 8.12. The electric dipole moment of the CsF molecule is large (over 7 D) but, according to Hughes [48], is not as large as one would expect for a purely ionic molecule. The decrease in the electric quadrupole constant as v increases is attributed by English and Zorn [51] to an increasing asymmetry of the internuclear potential. The measured c_3 constant is actually the sum of two separate contributions

$$c_3 = (c_3)_{\text{dir}} + (c_3)_{\text{ec}}. \tag{8.304}$$

The direct contribution $(c_3)_{\text{dir}}$ arises from the through-space dipolar coupling of the nuclear magnetic moments and, as expected, it decreases as the internuclear distance increases with increasing v, because of its R^{-3} dependence. The second contribution $(c_3)_{\text{ec}}$ is the axial component of the tensorial electron-coupled spin–spin interaction; the scalar part of this interaction is given by the value of c_4. In the $v=0$ level the direct contribution is estimated by English and Zorn [51] to be 1.15 kHz, and the electron-coupled part is -0.23 kHz.

Several other alkali fluorides have been studied by molecular beam electric resonance. The analyses are similar to that described above and similar sets of molecular constants have been determined; the molecules studied include ^{85}Rb^{19}F and ^{87}Rb^{19}F [56], ^{39}K^{19}F [57], ^{23}Na^{19}F [58, 59] and ^{7}Li^{19}F [60, 61].

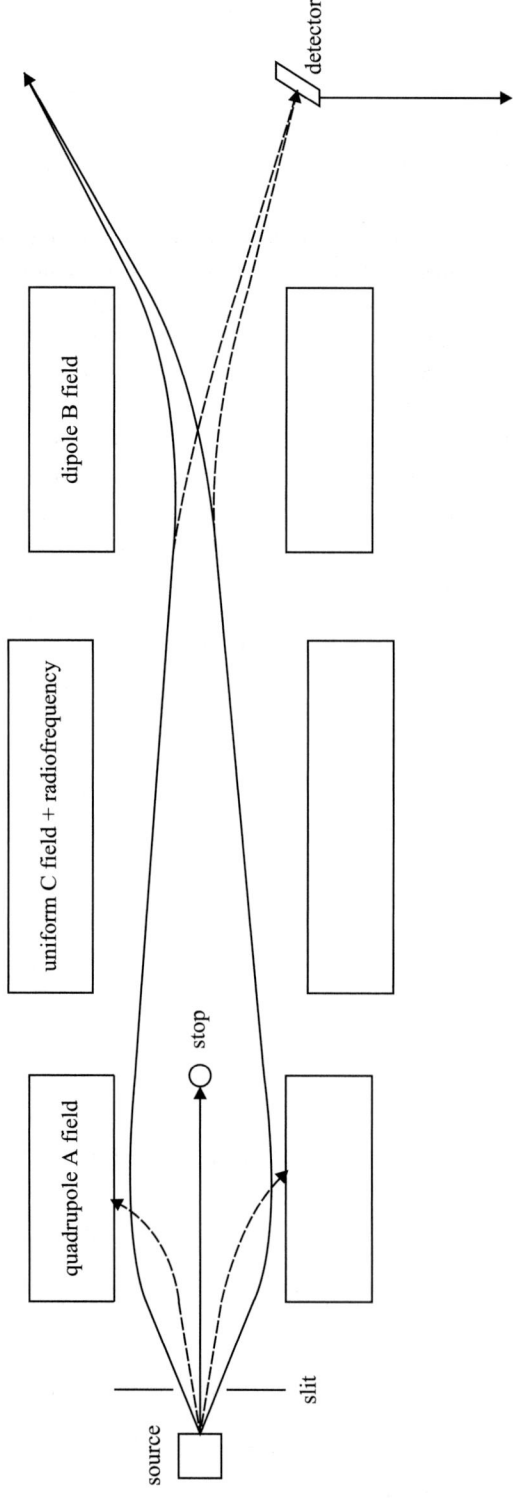

Figure 8.33. Block diagram of the molecular beam electric resonance instrument used by Freund, Fisk, Herschbach and Klemperer [55]. Transitions induced in the C field region which obey the selection rule $\Delta M = \pm 1$ result in an increase in the number of molecules reaching the detector; this is the *flop-in* mode of detection.

8.4.3. LiBr in the $X\,^1\Sigma^+$ ground state

(a) Introduction

We have chosen to describe the electric resonance studies of CsF in some detail because, apart from being the pioneering experiments in this field, they are also typical of many others on closed shell molecules. Our second chosen example is LiBr, studied first by Hilborn, Gallagher and Ramsey [62] and subsequently by Cecchi and Ramsey [63]. The first of these papers describes a conventional electric resonance study, complicated by the occurrence of two natural isotopic forms, namely, $^7\text{Li}^{79}\text{Br}$ and $^7\text{Li}^{81}\text{Br}$, whose relative abundances are essentially equal. Both the Li and Br nuclei possess magnetic dipole and electric quadrupole moments. In the later work Cecchi and Ramsey [63] chose to replace the electric C field with a strong dc magnetic field, so that the analysis of their spectra required consideration of the details of the Zeeman effect. Having noted in our discussion of the CsF studies that a weak electric C field is necessary in order to transfer electric dipole intensity from the $\Delta J = \pm 1$ rotational transitions into the $\Delta J = 0$ transitions studied in the electric resonance measurements, one might ask how Cecchi and Ramsey could dispense with the electric C field. The answer is that the molecules in the rapidly moving beam pass through a static magnetic field and so experience an effective electric field which is sufficient to provide the required electric dipole transition moment. We will say more about this aspect in due course, noting here that the magnetic fields used were close to 5 kG. Armed with the knowledge of the zero-field constants obtained by Hilborn, Gallagher and Ramsey [62], Cecchi and Ramsey [63] were able to investigate the magnetic field effects. These studies of LiBr were, in fact preceded by investigations of LiCl [64] where full details of the apparatus are to be found.

The spectra of LiBr were again for molecules in the $J = 1$ rotational level, and quadrupole lenses were operated to focus molecules in the $M_J = 0$ state. $\Delta M_J = \pm 1$ resonances were therefore detected using the 'flop-out' mode. We shall consider the details of the 'weak' field investigation first, subsequently describing the 'strong' magnetic field studies.

(b) Effective Hamiltonian, matrix elements and energy levels in the 'weak' field case

The effective Hamiltonian used is an extension of that previously described in equations (8.284) to (8.296) for CsF, but for the case of two quadrupolar nuclei. It therefore takes the following form:

$$\mathcal{H} = B\boldsymbol{J}^2 - e\mathrm{T}^2(\nabla \boldsymbol{E}_1)\cdot \mathrm{T}^2(\boldsymbol{Q}_1) - e\mathrm{T}^2(\nabla \boldsymbol{E}_2)\cdot \mathrm{T}^2(\boldsymbol{Q}_2) + c_1\mathrm{T}^1(\boldsymbol{I}_1)\cdot \mathrm{T}^1(\boldsymbol{J})$$
$$+ c_2\mathrm{T}^1(\boldsymbol{I}_2)\cdot \mathrm{T}^1(\boldsymbol{J}) + \sqrt{10}g_1 g_2 \mu_N^2 (1/4\pi\varepsilon_0 c^2)\mathrm{T}^1(\boldsymbol{C}^2, \boldsymbol{I}_1)\cdot \mathrm{T}^1(\boldsymbol{I}_2)$$
$$+ c_4\mathrm{T}^1(\boldsymbol{I}_1)\cdot \mathrm{T}^1(\boldsymbol{I}_2) - \mathrm{T}^1(\boldsymbol{\mu}_e)\cdot \mathrm{T}^1(\boldsymbol{E}). \tag{8.305}$$

The subscripts 1 and 2 refer to the Br and Li nuclei respectively. This Hamiltonian differs from that used previously for CsF, equation (8.282), only through the additional quadrupole term and the explicit addition of a Stark effect term. Although the weak electric fields (a few V cm^{-1}) used in this work were employed mainly to transfer electric dipole intensity into the resonance transitions, the resulting Stark shifts were measurable because of the extremely small linewidths obtained (about 300 Hz).

Following Hilborn, Gallagher and Ramsey [62] we use the same nuclear spin coupled basis set as we used for the 'weak' field CsF analysis, i.e. $|\eta, \Lambda; J, I_1, F_1, I_2, F, M_F\rangle$. All of the required matrix elements were calculated earlier except for those of the quadrupole interaction involving nucleus 2 (Li), which are,

$$\langle \eta, \Lambda; J, I_1, F_1, I_2, F, M_F | -e\mathrm{T}^2(\nabla E_2) \cdot \mathrm{T}^2(Q_2) | \eta, \Lambda'; J', I_1, F_1', I_2, F', M_F' \rangle$$

$$= -e\delta_{M_F M_F'}\delta_{F F'}(-1)^{F_1'+F+I_2} \begin{Bmatrix} I_2 & F_1' & F \\ F_1 & I_2 & 2 \end{Bmatrix}$$

$$\times \langle \eta, \Lambda; J, I_1, F_1 \| \mathrm{T}^2(\nabla E_2) \| \eta, \Lambda'; J', I_1, F_1' \rangle \langle I_2 \| \mathrm{T}^2(Q_2) \| I_2 \rangle$$

$$= \delta_{M_F M_F'}\delta_{F F'}(-1)^{F_1'+F+I_2} \begin{Bmatrix} I_2 & F_1' & F \\ F_1 & I_2 & 2 \end{Bmatrix} (-1)^{F_1'+J+2+I_1} \{(2F_1'+1)(2F_1+1)\}^{1/2}$$

$$\times \begin{Bmatrix} F_1 & J & I_1 \\ J' & F_1' & 2 \end{Bmatrix} (-1)^J \{(2J+1)(2J'+1)\}^{1/2} \begin{pmatrix} J & 2 & J' \\ 0 & 0 & 0 \end{pmatrix}$$

$$\times \left(\frac{q_2}{2}\right)\left(\frac{eQ_2}{2}\right) \begin{pmatrix} I_2 & 2 & I_2 \\ -I_2 & 0 & I_2 \end{pmatrix}^{-1}. \tag{8.306}$$

Note a possible source of confusion here. We have put $\Lambda = 0$ and taken only the axial component of the field gradient; the subscript 2 on q_2 and Q_2 refers to nucleus 2, and not to non-axial components of the second-rank tensors.

Both nuclei have spin $I = 3/2$ so that the possible values of the quantum numbers for $J = 1$ are

$$F_1 = 5/2, F = 4, 3, 2, 1: \quad F_1 = 3/2, F = 3, 2, 1, 0: \quad F_1 = 1/2, F = 2, 1.$$

The energies of the hyperfine components of the $J = 1$ rotational level may now be calculated in terms of the constants appearing in the above equations. The quadrupole coupling constant for ^{79}Br is large enough that matrix elements of the quadrupole interaction coupling $J = 1$ and 3 must be included in the analysis (there is no coupling with $J = 0$ and 2 for small electric fields because of the parity selection rule). Using the constants given by Hilborn, Gallagher and Ramsey we can construct the hyperfine energy level diagram for the $J = 1$ rotational level as shown in figure 8.34. A portion of the electric resonance spectrum is shown in figure 8.35, and we have related the observed resonance lines to the transitions indicated in figure 8.34. A total of 16 lines was measured for each of the two main isotopic species, involving transitions

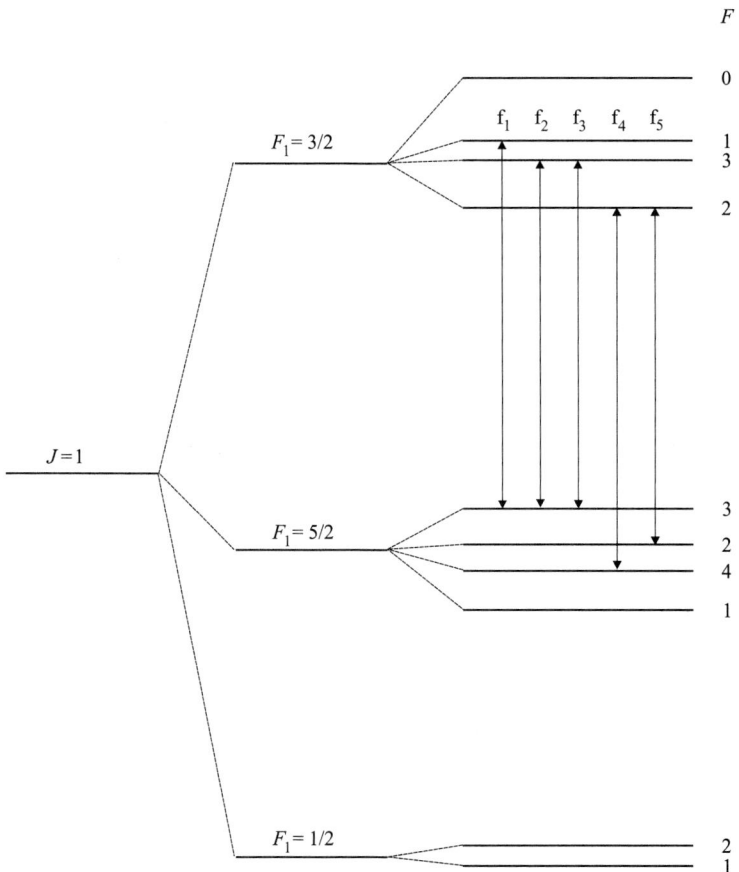

Figure 8.34. Nuclear quadrupole hyperfine structure of the $J=1$ level of $^7\text{Li}^{79}\text{Br}$ ($v=0$) in zero field. The largest splitting is due to the quadrupole interaction involving the ^{79}Br nucleus, and the smaller splitting arises from the ^6Li nucleus. The transitions indicated are resonsible for the resonances shown in the experimental spectrum presented in figure 8.35.

between the $F_1 = 3/2$ and $5/2$ components, and obeying the additional selection rules $\Delta F = 0, \pm 1$. Subsequent studies involving the ^6Li isotope enabled the comprehensive data set presented in table 8.13 to be compiled.

(c) 'Strong' magnetic field spectrum

The effective Hamiltonian used by Cecchi and Ramsey [63] to analyse the strong magnetic field spectrum was the sum of four terms, describing the molecular rotation, nuclear spin interactions, Stark interactions and Zeeman interactions. Specifically the Hamiltonian is the following,

$$\mathcal{H} = \mathcal{H}_{\text{rot}} + \mathcal{H}_{\text{hyp}} + \mathcal{H}_E + \mathcal{H}_Z. \tag{8.307}$$

Table 8.13. *Molecular parameters for the four isotopic forms of LiBr, determined from the zero- and strong-magnetic field measurements*

Parameter		^{6}Li^{79}Br	^{6}Li^{81}Br	^{7}Li^{79}Br	^{7}Li^{81}Br
$(eq_0Q)_{Br}$	(kHz)	38 463.48	32 128.23	38 368.104	32 050.860
$(eq_0Q)_{Li}$	(kHz)	4.42	3.11	211.04	211.03
c_{Br} (c_1)	(kHz)	9.069	9.993	7.8816	8.4740
c_{Li} (c_2)	(kHz)	0.052	−0.015	0.859	0.815
$\frac{g_1 g_2 \mu_N^2}{4\pi\varepsilon_0 c^2}\langle R^{-3}\rangle$	(kHz)	0.370	0.294	1.0710	1.1789
c_4	(kHz)	0.039	0.079	0.0604	0.0711
μ_e	(D)	7.268	7.268	7.265	7.265
B	(MHz)	19 194.592	19 161.026	16 650.179	16 616.622

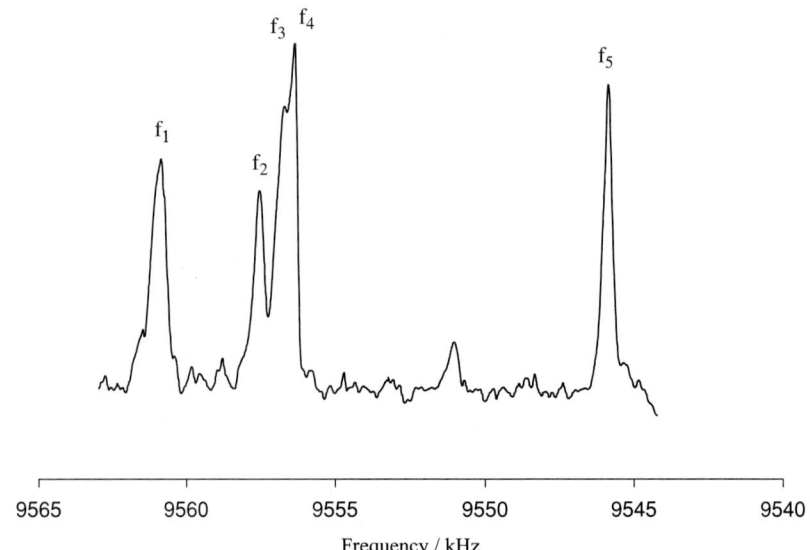

Figure 8.35. Section of the electric resonance hyperfine spectrum for ^{7}Li^{79}Br in the $v = 0$, $J = 1$ rotational level [62]. The effective 'motional' electric field was 2.33 V cm^{-1}. The resonances are labelled to permit comparison with the transitions indicated in figure 8.34. Resonances f_2 and f_3 are Stark components of the same transition.

The first three terms were discussed in the previous section; the new term is, of course, the Zeeman interaction:

$$\mathcal{H}_Z = -g_1\mu_N\{\mathbf{T}^1(\mathbf{I}_1)\cdot\mathbf{T}^1(\mathbf{B}) - T^2(\boldsymbol{\sigma}_1)\cdot T^2(\mathbf{I}_1,\mathbf{B})\} - g_2\mu_N\{\mathbf{T}^1(\mathbf{I}_2)\cdot\mathbf{T}^1(\mathbf{B})$$
$$- T^2(\boldsymbol{\sigma}_2)\cdot T^2(\mathbf{I}_2,\mathbf{B})\} - g_J\mu_N\mathbf{T}^1(\mathbf{J})\cdot\mathbf{T}^1(\mathbf{B}) - (1/2)\mathbf{T}^2(\boldsymbol{\chi})\cdot\mathbf{T}^2(\mathbf{B},\mathbf{B}). \quad (8.308)$$

We met these Zeeman terms earlier in this chapter, notably when describing the

Table 8.14. *Alkali halides studied by electric resonance methods*

LiF [60, 61]	NaF [58, 59]	KF [57]	CsF [48, 50, 51, 54]	RbF [56, 65, 66]
LiCl [67, 68, 64]	NaCl [69, 70, 71]	KCl [72, 69]	CsCl [73, 74, 69]	RbCl [69]
LiBr [75, 76, 62, 63]	NaBr [69]	KBr [77, 78]		
LiI [79, 80]	NaI [69, 81]			

magnetic resonance spectrum of H_2. The last term in (8.308) describing the diamagnetic susceptibility was, however, expressed differently in equation (8.15).

We shall not go through all the details of the analysis of the spectrum, but will give some idea of the complexity of the problem. The largest terms in the effective Hamiltonian are the nuclear spin Zeeman terms. Cecchi and Ramsey [63] made measurements at three different magnetic fields; at their highest field (5.262 kG) it is a simple matter to calculate the nuclear spin Zeeman energies and the results are shown in figure 8.36. At this level of theory there are 16 Zeeman components, as shown, but addition of the rotational Zeeman term splits each of the levels shown in figure 8.36 into a further triplet corresponding to $M_J = 0, \pm 1$. The final energies of the resulting 48 levels depend upon the quadrupole and nuclear spin–rotation interactions for both nuclei, as well as screening and diamagnetic effects; the quantum mechanics is straightforward but the numerical analysis must have been tricky. And all of this was accomplished for all four isotopic species!

The origin of the electric dipole intensity for the $\Delta M_J = \pm 1$ transitions studied merits further consideration. If the static magnetic field is 5 kG, the motional electric field has a magnitude of approximately 3 V cm^{-1} and is perpendicular to the applied magnetic field. This electric field mixes a state $|J, M_J\rangle$ with the states $|J \pm 1, M_J \pm 1\rangle$ and in order to obtain non-zero electric dipole transition moments for the transitions $|J, M_J\rangle \leftrightarrow |J, M_J \pm 1\rangle$, the oscillating electric field must be applied *parallel* to the static magnetic field.

The results for the four isotopic LiBr species are shown in table 8.13, assembled from both the zero-field and strong field studies.

(d) Summary of electric resonance studies of alkali halides

The alkali halide molecules have been studied comprehensively by molecular beam electric resonance methods. Table 8.14 presents a summary with references. In most cases the electric quadrupole coupling constants have been determined, and usually also the nuclear spin-rotation constants.

8.4.4. Alkaline earth and group IV oxides

The diatomic oxides of the alkaline earth and group IV elements are refractory materials for which electric resonance, preceded by a high-temperature source, is an ideal

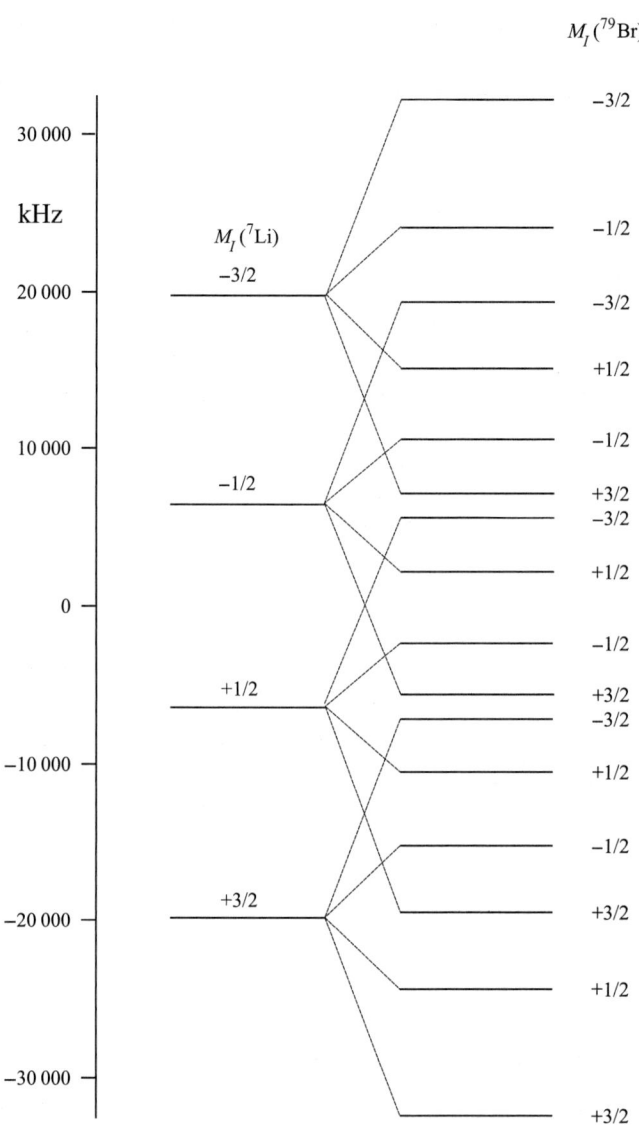

Figure 8.36. Energy level diagram showing the nuclear spin Zeeman energies for the ^{7}Li and ^{79}Br nuclei in LiBr. The nuclear g-factor for ^{7}Li (3.256) is larger than that for ^{79}Br (2.106). Each level shown is split into a further triplet by the rotational Zeeman interaction which removes the M_J threefold-degeneracy for $J = 1$.

technique. The molecules which have been investigated all have $^1\Sigma^+$ ground states and the experiments have mainly been directed towards determining their electric dipole moments. These molecules are somewhat simpler than the alkali halides discussed in the previous section, in that the predominant isotope of oxygen, ^{16}O, has zero nuclear spin. We will therefore not go into details; the reader who has mastered our descriptions of the analyses of the electric resonance spectra of CsF and LiBr will have no difficulty in understanding these oxide spectra. We confine ourselves to an abbreviated summary of the literature, noting that the molecules studied include BaO [82], SrO [83], SiO [84, 85] and GeO [84, 85].

8.4.5. HF in the $X\,^1\Sigma^+$ ground state

(a) Introduction

The hydrogen halides have been studied extensively by both magnetic and electric resonance methods. To be specific, electric resonance measurements have been described for HF [86, 87, 88, 89], HCl [90, 89] and HBr [91], whilst magnetic resonance studies were described for HCl [92] and HF [93]. All of these molecules have $^1\Sigma^+$ ground states, and rotational transitions will occur at very high frequencies. Hyperfine transitions occur in the radiofrequency region, and we first describe the conventional electric resonance studies of HCl and HF by Weiss [86], Kaiser [90] and Muenter and Klemperer [87]. Later in this section we will describe the extensive studies of de Leeuw and Dymanus [89] on the HF and HCl molecules who concentrated on the magnetic and electric properties by making measurements in relatively large magnetic and electric fields.

The experiments described by Weiss [86] and by Muenter and Klemperer [87] on HF and DF were performed on molecules in the $J=1$ rotational level and used quadrupole A and B fields, focusing $M=0$ molecules onto the detector. The homogeneous electric C field was large enough (up to approximately 3000 V cm^{-1}) to separate the $|M|=\pm 1$ and $M=0$ levels by about 2 MHz and electric dipole transitions between them were detected. Each M level is split by nuclear spin interactions, such that the $\Delta M = \pm 1$ transitions in $J=1$ showed nine transitions, as illustrated in figure 8.37. We now see how this spectrum arises, developing the general theory first, and then using the methods of Weiss [86] to apply the theory to the HF molecule.

(b) Effective Hamiltonian and matrix elements

HF is similar to CsF, discussed earlier, except that both nuclei have spins I of 1/2 and therefore no quadrupole moments. The effective Hamiltonian for HF employed by Weiss [86] and by Muenter and Klemperer [87] was the same as that used for CsF in equation (8.281), without the quadrupole term, but with the addition of an electric field term. In the case of DF, described later, the quadrupole term is present. In irreducible tensor form, therefore, our Hamiltonian follows equation (8.282), and is written:

$$\mathcal{H}_{\text{eff}} = -\mathbf{T}^1(\boldsymbol{\mu}_e)\cdot\mathbf{T}^1(\boldsymbol{E}) + c_H\mathbf{T}^1(\boldsymbol{I}_H)\cdot\mathbf{T}^1(\boldsymbol{J}) + c_F\mathbf{T}^1(\boldsymbol{I}_F)\cdot\mathbf{T}^1(\boldsymbol{J})$$
$$+\sqrt{10}g_H g_F \mu_N^2(1/4\pi\varepsilon_0 c^2)\mathbf{T}^1(\boldsymbol{C},\boldsymbol{I}_H)\cdot\mathbf{T}^1(\boldsymbol{I}_F) + c_4\mathbf{T}^1(\boldsymbol{I}_H)\cdot\mathbf{T}^1(\boldsymbol{I}_F), \quad (8.309)$$

where the subscripts H and F refer to the proton and fluorine nuclei respectively. Since the Stark energy is much greater than the intramolecular interactions, we use the nuclear spin decoupled representation; we can ignore the very small effects of matrix elements

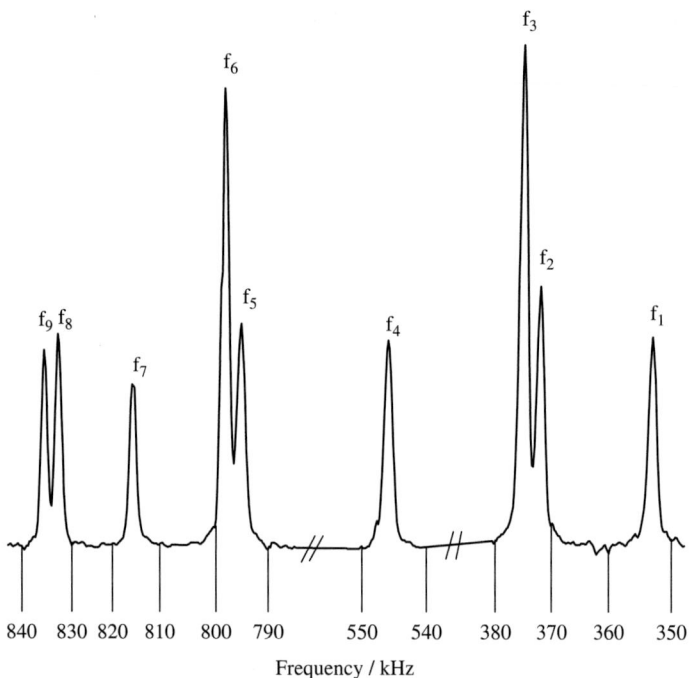

Figure 8.37. The $J = 1$, $\Delta M = \pm 1$ electric resonance spectrum of HF shown by Muenter and Klemperer [87], recorded in an electric field of 1500 V cm^{-1}.

off-diagonal in J. We summarise the results previously obtained for CsF (the 'strong field' case):

(i) STARK INTERACTION

$$\langle \eta, \Lambda = 0; J, M_J, I_H, M_H, I_F, M_F | - \mathrm{T}^1_{p=0}(\boldsymbol{\mu}_e)\mathrm{T}^1_{p=0}(\boldsymbol{E}) | \eta, \Lambda = 0; J', M_J, I_H, M_H, I_F, M_F \rangle$$
$$= -\mu_0 E_Z \{(2J+1)(2J'+1)\}^{1/2}(-1)^{M_J} \begin{pmatrix} J & 1 & J' \\ -M_J & 0 & M_J \end{pmatrix} \begin{pmatrix} J & 1 & J' \\ 0 & 0 & 0 \end{pmatrix}. \tag{8.310}$$

In our discussion of the Stark effect for CsF, we pointed out that (8.310) vanishes unless $1 + J' + J$ is *even* in the 3-j symbol with zero arguments in the lower row; therefore $J' = J \pm 1$ of necessity, and the Stark effect is second order. We showed that the second-order Stark energy could be obtained from second-order perturbation theory, to give the well-known expression (8.279) which we repeat again:

$$\Delta E^{(2)} = \frac{\mu_0^2 E_Z^2}{2B} \left\{ \frac{J(J+1) - 3M^2}{J(J+1)(2J-1)(2J+3)} \right\}. \tag{8.311}$$

We shall, in fact, diagonalise the matrix involving the $J = 0, 1, 2$ and 3 rotational levels to obtain results for $J=1$. The Stark effect matrix is readily found to be the following.

	$J=0$	$J=1$	$J=2$	$J=3$
$J=0$	0	$-\mu_0 E_Z \times \left\{\dfrac{1-M^2}{3}\right\}^{1/2}$	0	0
$J=1$	$-\mu_0 E_Z \times \left\{\dfrac{1-M^2}{3}\right\}^{1/2}$	$2B$	$-\mu_0 E_Z \times \left\{\dfrac{4-M^2}{15}\right\}^{1/2}$	0
$J=2$	0	$-\mu_0 E_Z \times \left\{\dfrac{4-M^2}{15}\right\}^{1/2}$	$6B$	$-\mu_0 E_Z \left\{\dfrac{9-M^2}{70}\right\}^{1/2}$
$J=3$	0	0	$-\mu_0 E_Z \times \left\{\dfrac{9-M^2}{70}\right\}^{1/2}$	$12B$

For a dipole moment of 1.826 526 D and an electric field of 1500 V cm^{-1} the Stark energies of the $M=0$ and $|M|=1$ levels are calculated to be 308.31 and -154.16 kHz respectively. These shifts are very small because of the large separation between the rotational levels.

(ii) NUCLEAR SPIN–ROTATION TERMS

For the proton we have:

$$\langle \eta, \Lambda; J, M_J; I_H, M_H; I_F, M_F | \sum_p (-1)^p c_H T^1_p(\boldsymbol{J}) T^1_{-p}(\boldsymbol{I}_H) | \eta, \Lambda; J, M'_J; I_H, M'_H; I_F, M_F \rangle$$

$$= c_H \sum_p (-1)^p (-1)^{J-M_J} (-1)^{I_H - M_H} \{ J(J+1)(2J+1) I_H(I_H+1)(2I_H+1) \}^{1/2}$$

$$\times \begin{pmatrix} J & 1 & J \\ -M_J & p & M'_J \end{pmatrix} \begin{pmatrix} I_H & 1 & I_H \\ -M_H & -p & M'_H \end{pmatrix}. \tag{8.312}$$

For the fluorine we have the similar result:

$$\langle \eta, \Lambda; J, M_J; I_H, M_H; I_F, M_F | \sum_p (-1)^p c_F T^1_p(\boldsymbol{J}) T^1_{-p}(\boldsymbol{I}_F) | \eta, \Lambda; J, M'_J; I_H, M_H; I_F, M'_F \rangle$$

$$= c_F \sum_p (-1)^p (-1)^{J-M_J} (-1)^{I_F - M_F} \{ J(J+1)(2J+1) I_F(I_F+1)(2I_F+1) \}^{1/2}$$

$$\times \begin{pmatrix} J & 1 & J \\ -M_J & p & M'_J \end{pmatrix} \begin{pmatrix} I_F & 1 & I_F \\ -M_F & -p & M'_F \end{pmatrix}. \tag{8.313}$$

(iii) NUCLEAR SPIN DIPOLAR INTERACTION

From equations (8.302) and (8.309),

$$\langle \eta, \Lambda; J, M_J; I_H, M_H; I_F, M_F | \mathcal{H}_{\text{dip}} | \eta, \Lambda'; J', M_J'; I_H, M_H'; I_F, M_F' \rangle$$
$$= \sqrt{10} g_H g_F \mu_N^2 (1/4\pi\varepsilon_0 c^2) \sum_p (-1)^p (-1)^{I_F - M_F} \begin{pmatrix} I_F & 1 & I_F \\ -M_F & -p & M_F' \end{pmatrix}$$
$$\times \{I_F(I_F + 1)(2I_F + 1)\}^{1/2}$$
$$\times \langle \eta, \Lambda; J, M_J; I_H, M_H | T_p^1(\boldsymbol{C}^2, \boldsymbol{I}_H) | \eta, \Lambda'; J', M_J'; I_H, M_H' \rangle. \quad (8.314)$$

The remaining matrix element in (8.314) is now expanded:

$$\langle \eta, \Lambda; J, M_J; I_H, M_H | T_p^1(\boldsymbol{C}^2, \boldsymbol{I}_H) | \eta, \Lambda'; J', M_J'; I_H, M_H' \rangle$$
$$= -\langle \eta, \Lambda; J, M_J; I_H, M_H | \sqrt{3} \sum_{p_1 p_2} (-1)^p T_{p_1}^2(\boldsymbol{C}) T_{p_2}^1(\boldsymbol{I}_H) | \eta, \Lambda'; J', M_J'; I_H, M_H' \rangle$$
$$\times \begin{pmatrix} 1 & 2 & 1 \\ p_1 & p_2 & -p \end{pmatrix}$$
$$= -(-1)^p \sqrt{3} \sum_{p_1 p_2} \begin{pmatrix} 1 & 2 & 1 \\ p_1 & p_2 & -p \end{pmatrix} (-1)^{I_H - M_H} \begin{pmatrix} I_H & 1 & I_H \\ -M_H & p_2 & M_H' \end{pmatrix}$$
$$\times \{I_H(I_H + 1)(2I_H + 1)\}^{1/2} \langle \eta, \Lambda; J, M_J | T_{p_1}^2(\boldsymbol{C}) | \eta, \Lambda'; J', M_J' \rangle$$
$$= -(-1)^p \sqrt{3} \sum_{p_1 p_2} (-1)^{I_H - M_H + J - M_J} \begin{pmatrix} 1 & 2 & 1 \\ p_1 & p_2 & -p \end{pmatrix} \begin{pmatrix} I_H & 1 & I_H \\ -M_H & p_2 & M_H' \end{pmatrix}$$
$$\times \begin{pmatrix} J & 1 & J' \\ -M_J & p_1 & M_J' \end{pmatrix} \{I_H(I_H + 1)(2I_H + 1)\}^{1/2} \langle \eta, \Lambda \| T^2(\boldsymbol{C}) \| \eta, \Lambda', J' \rangle. \quad (8.315)$$

The final matrix element in (8.315) is evaluated in the following manner:

$$\langle \eta, \Lambda; J \| T^2(\boldsymbol{C}) \| \eta, \Lambda'; J \rangle = \sum_q \langle \eta, \Lambda; J \| \mathcal{D}_{\cdot q}^{(2)}(\omega)^* T_q^2(\boldsymbol{C}) \| \eta, \Lambda'; J' \rangle$$
$$= \sum_q (-1)^{J - \Lambda} \begin{pmatrix} J & 2 & J' \\ -\Lambda & q & \Lambda' \end{pmatrix} \{(2J + 1)(2J' + 1)\}^{1/2}$$
$$\times \langle \eta, \Lambda | T_q^2(\boldsymbol{C}) | \eta, \Lambda' \rangle. \quad (8.316)$$

We have used the second-rank rotation matrix to transform into the molecule-fixed axis system, q, and have now almost reached our conclusion because we are only concerned with the $q = 0$ component, and are dealing with point magnetic dipoles for the two nuclei. Consequently

$$\langle \eta, \Lambda | T_0^2(\boldsymbol{C}) | \eta, \Lambda \rangle = \langle C_0^2(\theta, \phi)(R^{-3}) \rangle_\eta = \langle R^{-3} \rangle_\eta. \quad (8.317)$$

The final result is therefore obtained by combining the results in equations (8.317), (8.316), (8.315) and (8.314). In the most general case it is complicated but for the present problem we can make a number of simplifications, as we shall see. We define the dipolar constant t_0 to be

$$t_0 = g_F g_H \mu_N^2 (1/4\pi\varepsilon_0 c^2) \langle R^{-3} \rangle_\eta. \tag{8.318}$$

(iv) ISOTROPIC NUCLEAR SPIN–SPIN INTERACTION

The last term in the effective Hamiltonian, equation (8.309), is the isotropic or scalar nuclear spin-spin interaction, which was included by Muenter and Klemperer [87] but not by Weiss [86]. Its matrix elements are readily calculated in our basis set, as follows:

$$\langle \eta, \Lambda; J, M_J; I_H, M_H; I_F, M_F | c_4 \mathbf{T}^1(\mathbf{I}_H) \cdot \mathbf{T}^1(\mathbf{I}_F) | \eta, \Lambda; J, M_J; I_H, M'_H; I_F, M'_F \rangle$$

$$= c_4 \sum_p (-1)^p \langle I_H, M_H | T_p^1(\mathbf{I}_H) | I_H, M'_H \rangle \langle I_F, M_F | T_{-p}^1(\mathbf{I}_F) | I_F, M'_F \rangle$$

$$= c_4 \sum_p (-1)^p (-1)^{I_H - M_H} \begin{pmatrix} I_H & 1 & I_H \\ -M_H & p & M'_H \end{pmatrix} (-1)^{I_F - M_F} \begin{pmatrix} I_F & 1 & I_F \\ -M_F & -p & M'_F \end{pmatrix}$$

$$\times \{I_H(I_H+1)(2I_H+1)I_F(I_F+1)(2I_F+1)\}^{1/2}. \tag{8.319}$$

(c) Calculation of the energy levels and electric resonance spectrum

We now apply these general results to the specific problem of HF. Apart from the Stark effect, we shall otherwise ignore the very small matrix elements which are off-diagonal in J, and evaluate the terms for $J = 1$, $M_J = 0, \pm 1$, $I_H = I_F = 1/2$. As Weiss pointed out, the total magnetic component $M_Z = M_J + M_F + M_H$ is a good quantum number and may be used to set up a decoupled representation in which states of different M_Z value are diagonalised separately. For $J = 1$ there are twelve primitive basis states, which we write below in the form $|M_F, M_H, M_J\rangle$.

The appropriate basis states chosen to block diagonalise the matrix of the effective Hamiltonian are as follows:

$$
\begin{aligned}
|M_Z| = 2: \quad & |1/2, 1/2, 1\rangle, |-1/2, -1/2, -1\rangle \\
|M_Z| = 1: \quad & |1/2, 1/2, 0\rangle, |1/2, -1/2, 1\rangle, |-1/2, 1/2, 1\rangle \\
& |-1/2, -1/2, 0\rangle, |1/2, 1/2, -1\rangle, |-1/2, 1/2, -1\rangle \\
M_Z = 0: \quad & \psi_\pm(0) = (1/\sqrt{2})\{|1/2, -1/2, 0\rangle \pm |-1/2, 1/2, 0\rangle\} \\
& \psi_\pm(1) = (1/\sqrt{2})\{|1/2, 1/2, -1\rangle \pm |-1/2, -1/2, 1\rangle\}.
\end{aligned} \tag{8.320}
$$

For $|M_Z| = 2$ and 1 there is a two-fold degeneracy in each case which is not removed.

Calculation of the matrix elements using our earlier general results is straightforward. For $|M_Z| = 2$, where there is only one state, the energy is given by

$$E_1 = W(1, 1) + (1/2)\{c_F + c_H + t_0/5\}, \tag{8.321}$$

where $W(J, |M_J|)$ is the Stark energy. For $|M_Z| = 1$ the 3×3 matrix to be diagonalised is the following.

	$\|1/2, 1/2, 0\rangle$	$\|1/2, -1/2, 1\rangle$	$\|-1/2, 1/2, 1\rangle$
$\|1/2, 1/2, 0\rangle$	$W(1, 0) - t_0/5$	$(1/\sqrt{2})\{c_H + 3t_0/10\}$	$(1/\sqrt{2})\{c_F + 3t_0/10\}$
$\|1/2, -1/2, 1\rangle$	$(1/\sqrt{2}) \times \{c_H + 3t_0/10\}$	$W(1, 1) + (1/2) \times \{c_F - c_H - t_0/5\}$	$-t_0/10$
$\|-1/2, 1/2, 1\rangle$	$(1/\sqrt{2}) \times \{c_F + 3t_0/10\}$	$-t_0/10$	$W(1, 1) + (1/2) \times \{-c_F + c_H - t_0/5\}$

For $M_Z = 0$ we have the following 4×4 matrix.

	$\psi_+(0)$	$\psi_-(0)$	$\psi_+(1)$	$\psi_-(1)$
$\psi_+(0)$	$W(1, 0) + 2t_0/5$	0	$(1/\sqrt{2}) \times \{c_F + c_H - 3t_0/5\}$	0
$\psi_-(0)$	0	$W(1, 0)$	0	$(1/\sqrt{2})\{c_H - c_F\}$
$\psi_+(1)$	$(1/\sqrt{2}) \times \{c_F + c_H - 3t_0/5\}$	0	$W(1, 1) + (1/2) \times \{-c_F - c_H + 7t_0/5\}$	0
$\psi_-(1)$	0	$(1/\sqrt{2}) \times \{c_H - c_F\}$	0	$W(1, 1) + (1/2) \times \{-c_F - c_H - t_0\}$

These matrices do not include the scalar spin–spin terms, which may be calculated if required from equation (8.319).

The energy levels and transition frequencies can be obtained by diagonalising the above matrices, given suitable values of the molecular constants. The most accurate values were obtained by Muenter and Klemperer [87], and are as follows:

$$\mu_e(v = 0) = 1.826\,526\,\text{D}, \quad c_F = 307.637\,\text{kHz}, \quad c_H = -71.128\,\text{kHz},$$
$$t_0 = 143.375\,\text{kHz}, \quad c_4 = 0.529\,\text{kHz}.$$

In an electric field of 1475 V cm^{-1} the Stark energies are $+298.119$ kHz for $M_J = 0$, and are -149.064 kHz for $M_J = \pm 1$. Using these values and the molecular constants given above, we may construct the hyperfine energy level diagram shown in figure 8.38. We label the levels with the basis state labels used in the above

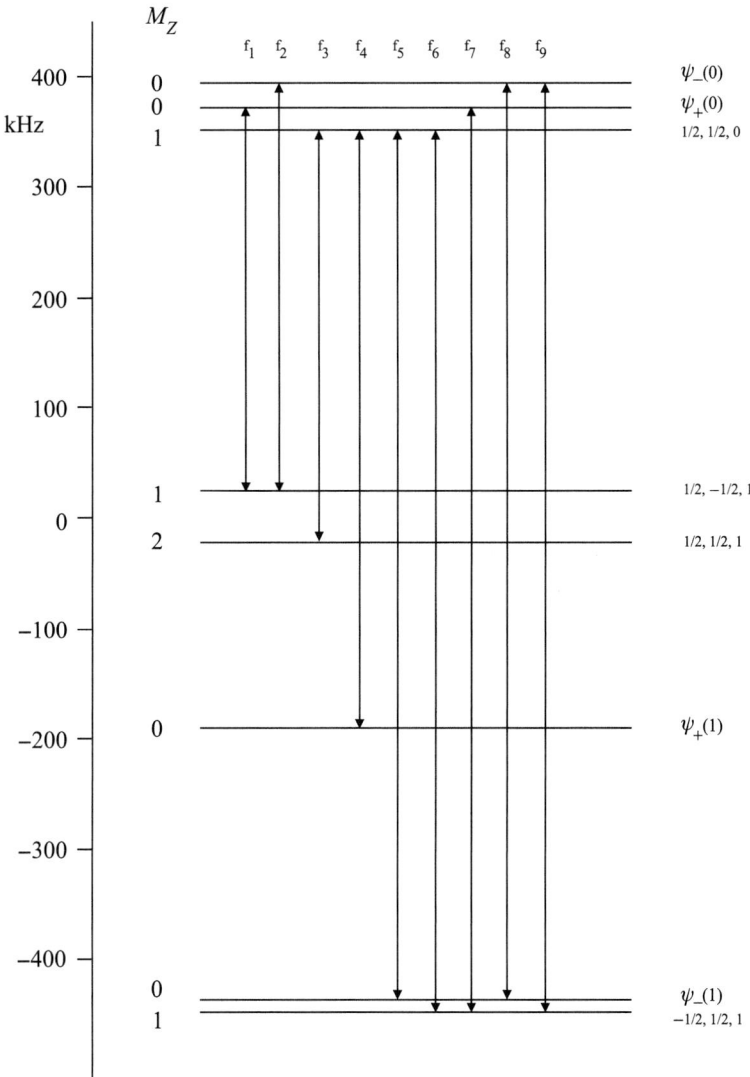

Figure 8.38. Hyperfine energy levels and calculated structure for the $\Delta M_Z = \pm 1$ transitions in the $J = 1$ level of HF, assuming an electric field of 1475 V cm^{-1} and the values of the constants given in the text. The predicted structure may be compared with that observed, as shown in figure 8.37. The quantum numbers on the right hand side are the values for M_F, M_H and M_J (and are only approximate). The total quantum number $M_Z = M_J + M_H + M_F$ is a good one, however.

matrices, although most of these are now only approximate labels because of the state mixing.

M_Z remains a rigorous quantum number and the allowed electric dipole transitions in the electric resonance experiment should satisfy the selection rule $\Delta M_Z = \pm 1$. In fact only seven of the nine observed transitions satisfy this rule, the remaining two apparently involving $\Delta M_Z = 0$. In addition the observed transitions should obey the rule $\Delta M_J = \pm 1$, which they all do, but it is important to note that M_J is no longer a good quantum number. With these rules in mind, the observed transitions can be identified, as shown in figures 8.37 and 8.38. As in many other cases described in this book, we have taken the easy route of using the values of the molecular constants to calculate the energy levels and transition frequencies; the reverse route is not quite so easy.

Although we will not go into the details of the spectroscopic analysis, we should note that Muenter and Klemperer [87] also examined the electric resonance spectrum of DF and obtained the following values of the molecular constants:

$$\mu_e(v = 0) = 1.818\,805 \text{ D}, \quad c_F = 158.356 \text{ kHz}, \quad c_D = -5.755 \text{ kHz},$$
$$t_0 = 4.434 \text{ kHz}, \quad eq_0 Q = 354.238 \text{ kHz}.$$

The difference in electric dipole moment (0.007 72 D) between the proton and deuteron species is discussed by Muenter and Klemperer [87] and attributed to the difference in zero-point amplitude averaged over the same dipole moment function. If the difference is purely vibrational in origin, the dipole moment of the vibrationless molecule is calculated to be 1.7965 D, which compares with a theoretical value of 1.942 D obtained from Hartree–Fock calculations by Huo [94]. The nuclear spin–rotation constants of non-rigid diatomic molecules have been discussed theoretically by Hindermann and Cornwell [95].

The simplest molecular constant to understand is the nuclear spin dipolar interaction constant, t_0, which is found to be, within experimental error, that calculated from the classical interaction of two magnetic moments, i.e. $g_F g_H \mu_N^2 (1/4\pi\varepsilon_0 c^2)\langle R^{-3}\rangle_{v=0}$. On the other hand, calculation of scalar electron-coupled spin–spin interaction constants is notoriously difficult, requiring a molecular electronic wave function of the highest quality. The best available calculation for HF quoted by Muenter and Klemperer is one due to O'Reilly [96].

(d) Electric resonance spectrum in the presence of a strong magnetic field

The magnetic resonance spectrum of HF was studied some nine years earlier than the electric resonance spectrum by Baker, Nelson, Leavitt and Ramsey [93]; in this case the transitions studied were magnetic dipole, corresponding to reorientation of the proton and fluorine nuclear spins. Values of the nuclear spin–rotation and dipolar constants were essentially confirmed by the later electric resonance measurements. We now describe measurements of the electric resonance spectrum in the additional presence of a strong magnetic field, carried out by de Leeuw and Dymanus [89].

The effective Hamiltonian for HF in a strong electric field was given in equation (8.309). The simplest extension to include the effects of an additional strong magnetic field would involve the inclusion of the terms

$$\mathcal{H}_Z = -g_J \mu_N \mathbf{T}^1(\boldsymbol{B}) \cdot \mathbf{T}^1(\boldsymbol{J}) - g_H \mu_N \mathbf{T}^1(\boldsymbol{B}) \cdot \mathbf{T}^1(\boldsymbol{I}_H) - g_F \mu_N \mathbf{T}^1(\boldsymbol{B}) \cdot \mathbf{T}^1(\boldsymbol{I}_F). \quad (8.322)$$

We have used this type of Zeeman Hamiltonian in our earlier discussions of magnetic resonance spectra, but it is an oversimplification because all three interactions should more correctly be represented by second-rank tensors with components described in the molecular axis system. Similarly, there should be an additional term describing the magnetic susceptibility tensor. We will come to these important matters in due course, because their elucidation was the objective of de Leeuw and Dymanus [89]. It is, however, instructive to examine the behaviour of the energy levels, and the transitions between them, when the Zeeman interaction is represented by the simple expression (8.322), with the magnetic field applied in the same direction ($p = 0$) as the electric field. The Stark matrix was described earlier and we will denote the eigenvalues for the Stark interaction in $J = 1$ by $W(J, |M_J|)$, observing that the Stark degeneracy of the $M_J = \pm 1$ states is not removed by the Stark interaction.

For $M_Z = \pm 2$ we have the following simple results:

$$M_Z = +2: \quad |1/2, 1/2, 1\rangle \quad E_2 = W(1,1) + (1/2)\{c_F + c_H + t_0/5\}$$
$$- (g_F + g_H + 2g_J)\mu_N(B_Z/2),$$
$$M_Z = -2: \quad |-1/2, -1/2, -1\rangle \quad E_{-2} = W(1,1) + (1/2)\{c_F + c_H + t_0/5\}$$
$$+ (g_F + g_H + 2g_J)\mu_N(B_Z/2).$$

Again we express the basis states in the form $|M_F, M_H, M_J\rangle$.

For the $M_Z = \pm 1$ states we now have two separate 3×3 matrices. For $M_Z = +1$ we have the following.

	$\|1/2, 1/2, 0\rangle$	$\|1/2, -1/2, 1\rangle$	$\|-1/2, 1/2, 1\rangle$
$\|1/2, 1/2, 0\rangle$	$W(1,0) - t_0/5$ $-\dfrac{(g_F + g_H)\mu_N B_Z}{2}$	$\dfrac{\{c_H + 3t_0/10\}}{\sqrt{2}}$	$\dfrac{\{c_F + 3t_0/10\}}{\sqrt{2}}$
$\|1/2, -1/2, 1\rangle$	$\dfrac{\{c_H + 3t_0/10\}}{\sqrt{2}}$	$W(1,1)$ $+\dfrac{\{c_F - c_H - t_0/5\}}{2}$ $-\dfrac{(g_F - g_H + 2g_J)\mu_N B_Z}{2}$	$-t_0/10$
$\|-1/2, 1/2, 1\rangle$	$\dfrac{\{c_F + 3t_0/10\}}{\sqrt{2}}$	$-t_0/10$	$W(1,1)$ $+\dfrac{\{-c_F + c_H - t_0/5\}}{2}$ $+\dfrac{(g_F - g_H - 2g_J)\mu_N B_Z}{2}$

For $M_Z = -1$ the matrix is as follows.

	$\|-1/2, -1/2, 0\rangle$	$\|1/2, -1/2, -1\rangle$	$\|-1/2, 1/2, -1\rangle$
$\|-1/2, -1/2, 0\rangle$	$W(1,0) - t_0/5$ $+\dfrac{(g_F + g_H)\mu_N B_Z}{2}$	$\dfrac{\{c_H + 3t_0/10\}}{\sqrt{2}}$	$\dfrac{\{c_F + 3t_0/10\}}{\sqrt{2}}$
$\|1/2, -1/2, -1\rangle$	$\dfrac{\{c_H + 3t_0/10\}}{\sqrt{2}}$	$W(1,1)$ $+\dfrac{\{c_F - c_H - t_0/5\}}{2}$ $-\dfrac{(g_F - g_H - 2g_J)\mu_N B_Z}{2}$	$-t_0/10$
$\|-1/2, 1/2, -1\rangle$	$\dfrac{\{c_F + 3t_0/10\}}{\sqrt{2}}$	$-t_0/10$	$W(1,1)$ $+\dfrac{\{-c_F + c_H - t_0/5\}}{2}$ $+\dfrac{(g_F - g_H - 2g_J)\mu_N B_Z}{2}$

Finally, for $M_Z = 0$, the 4×4 matrix to be diagonalised is as follows.

	$\psi_+(0)$	$\psi_-(0)$	$\psi_+(1)$	$\psi_-(1)$
$\psi_+(0)$	$W(1,0) + 2t_0/5$	$-\dfrac{(g_F - g_H)\mu_N B_Z}{2}$	$\dfrac{\{c_F + c_H - 3t_0/5\}}{\sqrt{2}}$	0
$\psi_-(0)$	$-\dfrac{(g_F - g_H)\mu_N B_Z}{2}$	$W(1,0)$	0	$\dfrac{\{c_H - c_F\}}{\sqrt{2}}$
$\psi_+(1)$	$\dfrac{\{c_F + c_H - 3t_0/5\}}{\sqrt{2}}$	0	$W(1,1)$ $+\dfrac{\{-c_F - c_H + 7t_0/5\}}{2}$	$-\dfrac{(g_F + g_H - 2g_J)\mu_N B_Z}{2}$
$\psi_-(1)$	0	$\dfrac{\{c_H - c_F\}}{\sqrt{2}}$	$-\dfrac{(g_F + g_H - 2g_J)\mu_N B_Z}{2}$	$W(1,1)$ $+\dfrac{\{-c_F - c_H - t_0\}}{2}$

De Leeuw and Dymanus [89] show the behaviour of the energy levels in an electric field of 2952 V cm^{-1} and a range of magnetic field values. Inserting the values of the constants into the above matrices and performing diagonalisation for different values of the magnetic field we can reproduce this behaviour. Figure 8.39 shows the energy levels at magnetic fields from 0 to 8 kG. At the higher fields the fully decoupled representation is appropriate and the levels can be labelled according to their magnetic projection quantum numbers, as shown in figure 8.39. Five electric resonance transitions with $\Delta M_J = \pm 1$ were measured by de Leeuw and Dymanus [89], and these are indicated in figure 8.39. De Leeuw and Dymanus did not illustrate the spectra obtained, and they did

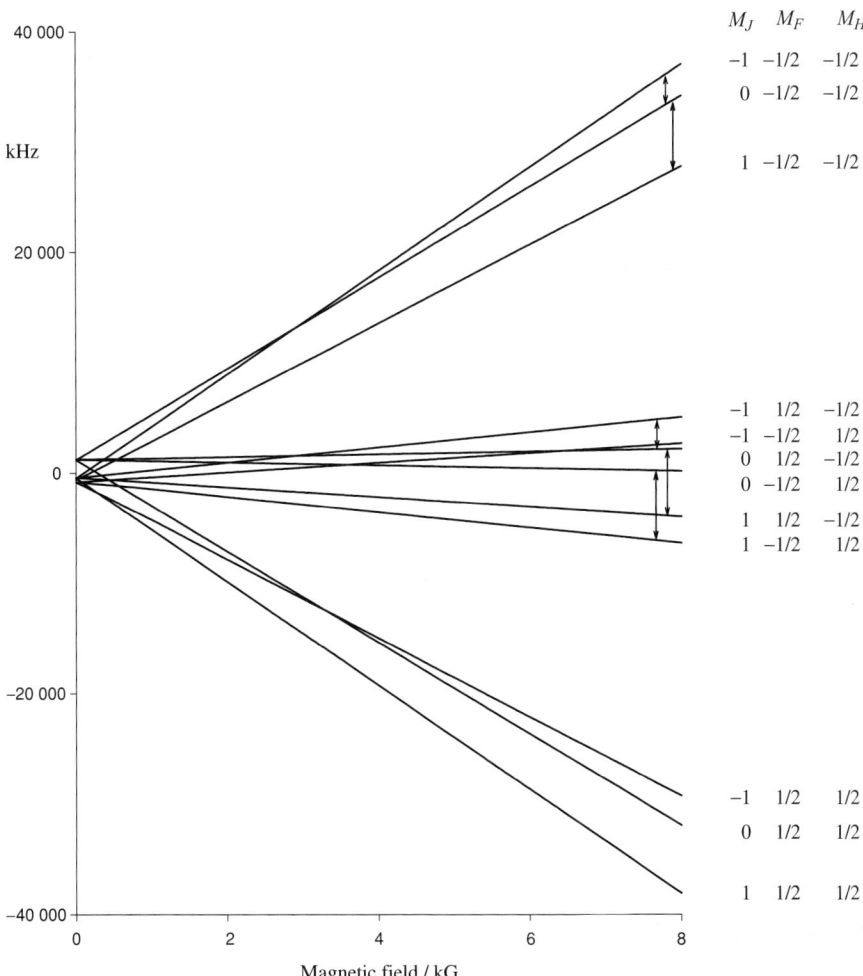

Figure 8.39. Behaviour of the energy levels of HF in the $J = 1$ level in magnetic fields from 0 to 8 kG. The electric field was 2952 V cm^{-1}. The vertical arrows indicate the five transitions measured by de Leeuw and Dymanus [89]. For the zero field energy level pattern, see figure 8.38.

not provide any quantitative details of the experimental results. It would also be possible to use our theoretical results to calculate the proton and fluorine nuclear magnetic resonance spectra (putting the electric field equal to zero) but unfortunately the original paper, although containing examples of the spectra, gives insufficient experimental data to make this exercise worthwhile.

The Zeeman Hamiltonian given in equation (8.322) is sufficient to provide a semi-quantitative description of the magnetic effects but, as was described in our discussion of the magnetic resonance spectrum of H$_2$, it is an approximate form. The local magnetic field experienced by the H and F nuclei is not quite the same as the applied laboratory field because of shielding effects due to the surrounding electrons. In addition the rotational Zeeman interaction should be described not by the single constant

g_J, but by a rotational magnetic moment tensor. Consequently in sufficiently accurate and detailed work, equation (8.322) should be replaced by the more precise Zeeman Hamiltonian,

$$\mathcal{H}_Z = -\mu_N T^2(\boldsymbol{G}) \cdot T^2(\boldsymbol{J}, \boldsymbol{B}) - \mu_N g_H \{T^1(\boldsymbol{I}_H) \cdot T^1(\boldsymbol{B}) - T^2(\boldsymbol{\sigma}_H) \cdot T^2(\boldsymbol{I}_H, \boldsymbol{B})\}$$
$$- \mu_N g_F \{T^1(\boldsymbol{I}_F) \cdot T^1(\boldsymbol{B}) - T^2(\boldsymbol{\sigma}_F) \cdot T^2(\boldsymbol{I}_F, \boldsymbol{B})\}. \quad (8.323)$$

Here we have introduced three second-rank tensors, the rotational magnetic moment tensor $T^2(\boldsymbol{G})$ and the proton and fluorine screening tensors, $T^2(\boldsymbol{\sigma}_H)$ and $T^2(\boldsymbol{\sigma}_F)$. In addition to the rotational magnetic moment and chemical shift (shielding) tensors, a further term which should be included in the Zeeman Hamiltonian describes the diamagnetic susceptibility. This term takes the form

$$\mathcal{H}_{\text{diam}} = -(1/2) T^2(\boldsymbol{\chi}) \cdot T^2(\boldsymbol{B}, \boldsymbol{B}), \quad (8.324)$$

the diamagnetic susceptibility being described by the second-rank tensor $T^2(\boldsymbol{\chi})$.

The principal components (defined in the molecular axis system) of the shielding and susceptibility tensors can be determined from solid state studies which have sufficient accuracy and resolution. In the case of HF in the gas phase, with $J = 1$, the spin–rotation and dipolar constants were determined accurately from the earlier electric resonance studies, so that de Leeuw and Dymanus [89] were able to use their Zeeman studies to measure the anisotropy of the screening and susceptibility tensors, with the following results:

$$\chi(parallel) - \chi(perpendicular) = 0.132 \text{ kHz kG}^{-2},$$
$$\sigma_F(parallel) - \sigma_F(perpendicular) = 108 \text{ p.p.m.}, \quad (8.325)$$
$$\sigma_H(parallel) - \sigma_H(perpendicular) = 24 \text{ p.p.m.}$$

'*parallel*' and '*perpendicular*' are, of course, defined with respect to the direction of the internuclear axis, and 'p.p.m.' stands for 'parts per million'. The shielding anisotropy for the ^{19}F nucleus is particularly large; the resulting reduction in the effective magnetic field at the nucleus results in significant shifts in the resonance frequencies.

It now remains to see how *ab initio* calculations of the molecular parameters agree with experiment. De Leeuw and Dymanus [89] presented a detailed comparison of experiment and theory, which is summarised in table 8.15. Details of the calculations may be obtained by consulting the references listed; as may be seen, the agreement is generally very good.

8.4.6. HCl in the $X^1\Sigma^+$ ground state

An important property of molecules is the behaviour of the electric dipole moment function near the equilibrium configuration, and the changes which occur on vibrational excitation. Electric resonance studies of the HCl molecule in its electronic ground state, carried out by Kaiser [90] are important in this respect, and also in showing, through the ^{35}Cl quadrupole interaction, how the electric field gradient changes on vibrational excitation.

Table 8.15. *Comparison of experimental and theoretical parameters for the HF molecule at its equilibrium internuclear separation ($R_e = 1.7328$ a.u.). Superscripts d or p stand for diamagnetic or paramagnetic respectively*

Parameter	Units	Obs. [87]	[97]	[98]	[99]	[100]
$(\sigma_{ave})_F^d$	p.p.m.	—	481.6	482.2	482.2	482.1
$(\sigma_{ave})_H^d$	p.p.m.	108.9	108.4	108.5	108.5	108.5
$(\sigma_{par} - \sigma_{perp})_F^d$	p.p.m.	−1	—	—	—	—
$(\sigma_{par} - \sigma_{perp})_H^d$	p.p.m.	−96	—	—	—	—
χ_{ave}^p	10^{-5} J T^{-2} mol^{-1}	0.5579	0.64	—	—	—
$(\chi_{par} - \chi_{perp})^d$	10^{-5} J T^{-2} mol^{-1}	1.342	—	1.222	1.224	—
μ_e	D	1.7965	—	1.941	1.934	1.941
Q_{mol}	10^{-40} C m^2	7.37	—	7.84	7.77	7.34
μ_J/J	μ_N	0.74104	0.738	—	—	—

Kaiser's studies employed a conventional spectrometer with A and B electric quadrupole fields, and by passing the HCl gas through a microwave discharge situated prior to the molecular beam source, populations in the ratios 21 : 3 : 1 for the $v = 0$, 1 and 2 vibrational levels were obtained. An effusion source was operated at 170 K and line widths close to 1 kHz were obtained; similar studies of DCl were described, except that in this case the gas was preheated to 1440 K to produce increased vibrational excitation. Kaiser was able to observe spectra of H^{35}Cl in $J = 1$, $v = 0$, 1, 2 and $J = 2$, $v = 0$. He also studied D^{35}Cl in $v = 0$, $J = 1$, 2, 3 and $v = 1$, $J = 1$.

Kaiser did not show examples of the spectra obtained, and did not list experimental frequencies. We cannot, therefore, reconstruct his spectra, but note that the spectroscopic analysis follows that outlined earlier for CsF closely. ^{35}Cl has a nuclear spin of 3/2 and a large electric quadrupole moment whilst the proton is, of course, like ^{19}F in having a spin of 1/2. D^{35}Cl is rather more complicated in that the ^{2}D nucleus also has a quadrupole moment. Table 8.16 summarises the results obtained for H^{35}Cl, whilst Table 8.17 presents the data for D^{35}Cl. The data for H^{35}Cl illustrate the vibrational dependence of the molecular parameters, whilst the data for D^{35}Cl involve their rotational dependence. We now describe Kaiser's analysis of these dependences.

The ultimate objective of the analysis is to express the vibrational and rotational dependence of the various molecular parameters as power series in their derivatives with respect to the internuclear distance. In chapter 6 we discussed various analytic representations of the potential energy in terms of the internuclear distance. Perhaps the most general representation is that due to Dunham [101] who assumed a power series for the potential of the form

$$V(R) = hca_0\xi^2\{1 + a_1\xi + a_2\xi^2 + a_3\xi^3 + \cdots\}, \qquad (8.326)$$

where

$$\xi = (R - R_e)/R_e. \qquad (8.327)$$

Table 8.16. *Molecular parameters for $H^{35}Cl$ determined from the electric resonance spectra. All parameters are in kHz except for the electric dipole moment, $\mu(v)$, which is in Debye units (D). The rotational and centrifugal distortion constants were obtained by Rank, Rao and Wiggins [102]. S_{12} is the spin–spin interaction constant, equal to $[g_1 g_2 \mu_N^2 (\mu_0/4\pi)/(2J+3)(2J-1)]\langle R^{-3}\rangle_{vJ}$*

	$v=0, J=1$	$v=1, J=1$	$v=2, J=1$	$v=0, J=2$
$eq_0 Q_H$	0	0	0	0
$eq_0 Q_{Cl}$	−67 618.93	−69 272.89	−70 908.1	−67 638.53
c_H	−41.80	−41.09	−39.94	−41.68
c_{Cl}	53.851	58.597	63.68	53.887
S_{12}	1.117	1.081	1.047	0.266
$\mu(v)$	1.1085	1.1390	1.1685	1.1085
B_v	3.129 909 8 × 10^8	3.038 765 1 × 10^8	2.948 358 8 × 10^8	3.129 909 8 × 10^8
D_v	15 837.47	15 636.28	15 459.16	15 837.47

Table 8.17. *Molecular parameters for $D^{35}Cl$ determined from the electric resonance spectra. All parameters are in kHz except for the electric dipole moment, $\mu(v)$, which is in Debye units (D). The rotational and centrifugal distortion constants were obtained by Rank, Eastman, Rao and Wiggins [103]*

	$v=0, J=1$	$v=0, J=2$	$v=0, J=3$	$v=1, J=1$
$eq_0 Q_D$	187.36	187.02	187.36	184.8
$eq_0 Q_{Cl}$	−67 393.38	−67 403.32	−67 418.01	−68 583.1
c_D	−3.295	−3.308	−3.295	−3.25
c_{Cl}	27.426	27.430	27.426	29.121
S_{12}	0.172	0.041	0.019	0.168
$\mu(v)$	1.1033	1.1033	1.1033	1.1256
B_v	1.616 559 4 × 10^8	1.616 559 4 × 10^8	1.616 559 4 × 10^8	1.582 849 2 × 10^8
D_v	4198.5	4198.5	4198.5	4155.5

Dunham then showed that for this potential the vibration–rotation energies could be expressed by

$$E_{v,J} = \sum_{kl} Y_{k,l}(v+1/2)^k J^l(J+1)^l \qquad (8.328)$$

where the k and l are positive integral summation indices. The Y_{kl}s of equation (8.328) can be related directly to the a_n coefficients of equation (8.326). Kaiser [90] and de Leeuw and Dymanus [89] used these results to show that the expectation value of any operator O can be expressed as a power series in B_e/ω_e and, up to quadratic terms in

Table 8.18. *Derivatives (with respect to R) of the ^{35}Cl electric field gradient in $H^{35}Cl$*

Coefficient	Experiment	Theory
q_e (C m^{-3})	39.291×10^{11}	38.59×10^{11}
$(dq/dR)_e$ (C m^{-3} Å^{-1})	45.10×10^{11}	45.0×10^{11}
$(d^2q/dR^2)_e$ (C m^{-3} Å^{-2})	-79.7×10^{11}	-73.1×10^{11}

this function, obtained the result

$$\langle O \rangle_{v,J} = O_e + (B_e/\omega_e)^2 \left\{ O'_e R_e \left[-\frac{15}{4}a_3 + \frac{23}{4}a_1 a_2 - \frac{21}{8}a_1^3 \right] + O''_e R_e^2 \left[-\frac{3}{4}a_2 + \frac{7}{8}a_1^2 \right] \right.$$
$$\left. -\frac{7}{24}a_1 O'''_e R_e^3 + \frac{1}{16} O''''_e R_e^4 \right\} + (B_e/\omega_e)(v+1/2)\{O'_e R_e(-3a_1) + O''_e R_e^2\}$$
$$+ (B_e/\omega_e)^2 (v+1/2)^2 \left\{ O'_e R_e \left[-15 a_3 + 39 a_1 a_2 - \frac{45}{2} a_1^3 \right] \right.$$
$$\left. + O''_e R_e^2 \left[-3a_2 + \frac{15}{2}a_1^2 \right] - \frac{15}{6} a_1 O'''_e R_e^3 + \frac{1}{4} O''''_e R_e^4 \right\}$$
$$+ 4(B_e/\omega_e)^2 J(J+1) O'_e R_e. \qquad (8.329)$$

The primes denote successive derivatives of the operator O_e with respect to R at the equilibrium internuclear separation R_e. The values of the coefficients a_n, as well as B_e and ω_e are known from analysis of the vibration–rotation spectrum [102].

Kaiser [90] pointed out that using only equation (8.329) to determine the derivatives of any chosen operator is not possible, an observation proved by Trischka and Salwen [104]. It is necessary to observe both centrifugal distortion and vibrational variation of an expectation value in order to separate first and second derivatives. We will not go through the details of this problem here, but present some of the results achieved. Kaiser found that the chlorine quadrupole constants for $v = 0$, 1 and 2 could be fitted to a second-order power series in $(v + 1/2)$ adjusted to $J = 0$:

$$eq_0 Q_{Cl} = a + b(v+1/2) + c(v+1/2)^2, \qquad (8.330)$$

with $a = -66775.1$ kHz, $b = -1672.8$ kHz, $c = 9.4$ kHz. By combining this vibrational information with the observed centrifugal distortion, the first and second derivatives of the quadrupole coupling constant with respect to R evaluated at R_e were obtained. The quadrupole moment for the ^{35}Cl nucleus is known, so that the derivatives of the electric field gradient q can be obtained. These are listed in table 8.18, and compared with Hartree–Fock calculations of Huo [105]; the agreement is remarkably good.

We now turn to the spin–rotation constants, noting first that the proton constant was determined in earlier molecular beam magnetic resonance studies by Code, Khosla, Ozier, Ramsey and Yi [106]. First there is a relativistic acceleration correction c_H (acc) to the spin–rotation constant,

$$c_H(\text{obs}) = c'_H + c_H(\text{acc}), \qquad (8.331)$$

which is given by [95],

$$c_H(\text{acc}) = [hm_{Cl}/2m_H(m_H + m_{Cl})][2(v+1/2)B_e\omega_e - 4J(J+1)B_e^2]. \quad (8.332)$$

The corrected value of the spin–rotation constant was then fitted to the expression

$$c'_H = g_H B_{v,J}[a + b(v+1/2)]. \quad (8.333)$$

Using the values of c'_H observed in the $v = 0$ and 1 states of HCl, the values of a and b were determined to be -0.712 and -0.0114 kHz·cm respectively. Kaiser tested these results by calculating the expected values of the spin–rotation constants for DCl. He defined a quantity $(v + 1/2)_{\text{eff}}$ for DCl by the relationship

$$(v+1/2)_{\text{eff}} = (\mu_{HCl}/\mu_{DCl})^{1/2}(v+1/2), \quad (8.334)$$

where μ is the reduced molecular mass and using equations (8.332) and (8.333) above obtained the reasonably satisfactory results,

$v = 0$: c_D (observed) $= -3.295$ kHz, c_D(calculated) $= -3.299$
$v = 1$: c_D (observed) $= -3.015$ kHz, c_D (calculated) $= -3.246$.

A similar analysis was carried out for the ^{35}Cl spin–rotation constants, except that an additional vibrational term was added:

$$c_{Cl} = B_{v,J}[a + b(v+1/2) + c(v+1/2)^2]. \quad (8.335)$$

The constants were determined to be

$$a = 4.874 \text{ kHz·cm}, \quad b = 0.552, \quad c = 0.036.$$

As before, these constants were used to calculate the values of c_{Cl} expected for DCl, with the results,

$v = 0$: c_{Cl} (observed) $= 27.426$ kHz, c_{Cl} (calculated) $= 27.37$
$v = 1$: c_{Cl} (observed) $= 29.12$ kHz, c_{Cl} (calculated) $= 29.08$.

The agreement is again satisfactory.

The molecular physics underlying the nuclear spin–rotation interaction has been discussed by Flygare [107]. In the general case of a polyatomic molecule the spin–rotation interaction is represented by a second-rank tensor; in a molecule fixed coordinate system x, y, z, the diagonal component in the x direction may be written as the sum of a nuclear part (k labelling the nucleus under consideration) and an electronic part:

$$\mathcal{H}_{\text{nsr}} = \sum_k \mathbf{T}^2(\mathbf{M}^k) \cdot \mathbf{T}^2(\mathbf{I}_k, \mathbf{J}) \quad (8.336)$$

where

$$M^k_{xx} = \frac{e\hbar\, g_k \mu_N}{I_{xx}} \frac{\mu_0}{4\pi} \sum_\alpha \frac{Z_\alpha}{R^3_{\alpha k}} (R^2_{\alpha k} - x^2_{\alpha k}) + \frac{e\hbar^3 g_k \mu_N}{2m I_{xx}} \frac{\mu_0}{4\pi}$$
$$\times \sum_{n>0} \left[\frac{\langle 0| \sum_i (l_{ik})_x / r^3_{ik} |n\rangle \langle n|(L_k)_x|0\rangle + \langle 0|(L_k)_x|n\rangle \langle n| \sum_i (l_{ik})_x / r^3_{ik}|0\rangle}{E_0 - E_n} \right]$$
(8.337)

and cyclic permutations for the yy and zz components. I_{xx} in this expression is a component of the molecular inertial tensor. The tensor components are expressed in energy units (J). There is a positive nuclear contribution, and a negative electronic term involving excited electronic states which are coupled in by orbital angular momentum. The ground state electronic distribution of electrons does not contribute to the nuclear spin–rotation interaction. In a linear molecule, where the internuclear axis is labelled (a), only the perpendicular component $M_{bb} = M_{cc}$ is non-zero; this is the quantity we have labelled c_k.

Perhaps the most interesting aspect of Kaiser's work on the HCl molecule is the vibrational dependence of the molecular electric dipole moment. In particular the aim was to obtain a complete description of the dipole moment operator $M(R)$ as a function of the internuclear distance R. It has been shown by Trischka and Salwen [104] that the dipole moment function can be obtained unambiguously if all of the elements in a row or a column of the dipole moment matrix in the vibrational wave function basis are known. The diagonal elements of this matrix correspond to the dipole moment in specific vibrational states, and can be determined from the Stark effect measurements of the molecular beam electric resonance spectrum. More specifically, the *difference* in the dipole moment of two different vibrational states is obtained from the high-resolution studies. The off-diagonal matrix elements can be obtained from absolute intensity measurements of vibrational bands, except that, as we shall see below, there remains an ambiguity in the sign of these off-diagonal elements. Absolute intensity measurements have been made [108] for the 1-0, 2-0, 3-0 bands of HCl and DCl, and for the 2-1 and 3-2 bands of HCl. There is more than one way of defining the dipole moment matrix, but since the measurements are usually restricted to the lower vibrational levels, the necessary truncation procedure is best handled by means of what is known as the 'wave function' approximation. If the complete set of vibrational wavefunctions is denoted $\Psi_v(R)$, we may write the expansion

$$M(R)\Psi_v(R) = \sum_{v'} \Psi_{v'}(R) \langle \Psi_{v'}|M(R)|\Psi_v\rangle. \quad (8.338)$$

Making the arbitrary choice of $v = 0$, and dividing both sides of (8.338) by $\Psi_0(R)$, we obtain

$$M(R) = \sum_{v'} [\Psi_{v'}(R)/\Psi_0(R)]\langle \Psi_{v'}(R)|M(R)|\Psi_0(R)\rangle. \quad (8.339)$$

Clearly, if the matrix elements in this equation, and the vibrational wave functions, are known, the dipole moment function $M(R)$ can be obtained. The square of the matrix

Table 8.19. *Measured relative dipole moments of $H^{35}Cl$ and $D^{35}Cl$. The zero error values were used as reference values for those that follow in the table*

Molecule	Vibrational level (v)	Rotational level (J)	$\mu_e(v)$(Debye)
HCl	0	1	1.108 47 ± 0.0
HCl	0	2	1.108 48 ± 0.000 25
HCl	1	1	1.138 88 ± 0.000 26
HCl	2	1	1.168 35 ± 0.0006
DCl	0	1	1.103 12 ± 0.000 08
DCl	0	1	1.103 12 ± 0.0
DCl	0	2	1.103 18 ± 0.000 35
DCl	1	1	1.125 15 ± 0.000 25

Table 8.20. *Absolute intensity measurements [108] of $M^{v,0}$ in HCl and DCl*

Matrix element (Debye)	HCl	DCl
$|M^{1,0}|$	$(6.7 \pm 0.14) \times 10^{-2}$	$(5.6 \pm 0.25) \times 10^{-2}$
$|M^{2,0}|$	$(7.02 \pm 0.14) \times 10^{-3}$	$(5.0 \pm 0.2) \times 10^{-3}$
$|M^{3,0}|$	$(5.15 \pm 0.2) \times 10^{-4}$	$(3.08 \pm 0.12) \times 10^{-4}$

element for $v' \neq 0$ can be obtained from the absolute intensity of the $v' - 0$ band,

$$I_{v,v'} = (8\pi^3 v_{v,v'}/3hc)(N_v - N_{v'})|\langle \Psi_{v'}|M(R)|\Psi_v \rangle|^2, \tag{8.340}$$

where N_v is the population of level v and $v_{v,v'}$ is the frequency of the $v' - v$ transition. Clearly the *sign* of the dipole matrix element is not determined from these intensity measurements.

The vibrational wave functions can be calculated from an *RKR* representation of the potential well, which is based on the rotational constants and vibrational levels of the molecule concerned. This leaves the problem of determining the signs of the matrix elements. From equation (8.339) we may write

$$M^{a,a} - M^{0,0} = \sum_{v' \neq 0} C_{v'aa} M^{v',0}, \tag{8.341}$$

where

$$M^{v,v'} = \int \Psi_v(R) M(R) \Psi_{v'}(R) \, d\tau,$$

$$C_{vij} = \int (\Psi_v \Psi_i \Psi_j / \Psi_0) \, d\tau. \tag{8.342}$$

The C coefficients are obtained from the vibrational wave functions calculated as described above; the measured relative dipole moments are listed in table 8.19 and the absolute infrared intensity measurements [108] are given in table 8.20.

Table 8.21. *Distance derivatives of the dipole moment operator for HCl and DCl*

Derivative	HCl	DCl
M_e (D)	$+1.0933 \pm 0.0005$	$+1.0922 \pm 0.0005$
$(dM/dR)_e$ (D Å^{-1})	$+0.925 \pm 0.02$	$+0.935 \pm 0.025$
$(d^2M/dR^2)_e$ (D Å^{-2})	$+0.16 \pm 0.11$	$+0.14 \pm 0.13$
$(d^3M/dR^3)_e$ (D Å^{-3})	-3.83 ± 0.90	-3.81 ± 1.1
$(d^4M/dR^4)_e$ (D Å^{-4})	-9.3 ± 4.5	-7.6 ± 4.8

Because of the limited data set it is necessary to truncate the series (8.341); consequently the difference $(M^{1,1} - M^{0,0})$ is taken as

$$M^{1,1} - M^{0,0} = C_{111}M^{1,0} + C_{211}M^{2,0}. \qquad (8.343)$$

If the values of the M matrix elements from table 8.20 are combined with the calculated C coefficients, the result obtained is

$$M^{1,1} - M^{0,0} = \pm 0.0394 \pm 0.0106 \text{D}. \qquad (8.344)$$

Kaiser [90] showed how the sign ambiguity above can be resolved, and the absolute signs of the M matrix elements determined; the interested reader is referred to his paper. By combining the M values with the theoretical vibrational wave functions, he was able to use equation (8.339) to derive values of the dipole moment function at a range of values of $R-R_e$. He then fitted the dipole moment function to a sixth-order polynomial in $R-R_e$, from which he was able to calculate the distance derivatives given in table 8.21.

The relative dipole moments of the three observed vibrational levels of HCl can be fitted to a second-order equation in $(v + 1/2)$, with the results

$$\mu_e = \mu_e^0 + \mu_e^I(v + 1/2) + \mu_e^{II}(v + 1/2)^2, \qquad (8.345)$$

with

$$\begin{aligned} \mu_e^0 &= +1.09290 \pm 0.00055 \text{D}, \\ \mu_e^I &= +0.0314 \pm 0.0014 \text{D}, \\ \mu_e^{II} &= -0.0047 \pm 0.00055 \text{D}. \end{aligned} \qquad (8.346)$$

A plot of the dipole moment of HCl versus $(v + 1/2)$ is thus close to linear; so also is a similar plot for DCl, but the plots are not coincident, as they should be if the Born–Oppenheimer approximation holds. It therefore appears that there is a noticeable violation of the Born–Oppenheimer approximation in these molecules, which is due to vibrational–electronic interaction [109]. A similar effect has been observed for the HD molecule [110, 111]. One concludes that the customary practice of inferring the vibrational behaviour from isotopically related molecules must be treated with caution when dealing with light molecules, like hydrides.

Table 8.22. *Comparison of experimental and theoretical parameters for the $H^{35}Cl$ molecule at its equilibrium internuclear separation ($R_e = 2.4085$ a.u.). Superscripts d or p stand for diamagnetic or paramagnetic respectively*

Parameter	Units	Observed	[98]	[99]
$(\sigma_{ave})^d_{Cl}$	p.p.m.	942	1150.3	1150.3
$(\sigma_{ave})^d_H$	p.p.m.	—	141.8	141.9
$(\sigma_{par} - \sigma_{perp})^d_{Cl}$	p.p.m.	−4	—	—
$(\sigma_{par} - \sigma_{perp})^d_H$	p.p.m.	−148	—	—
χ^p_{ave}	10^{-5} J T^{-2} mol^{-1}	2.46	—	—
$(\chi_{par} - \chi_{perp})^d$	10^{-5} J T^{-2} mol^{-1}	3.49	—	—
μ_e	D	1.0933	1.196	1.215
Q_{mol}	10^{-40} C m^2	11.77	12.67	12.47
μ_J/J	μ_N	0.459 35	—	—
eq_0Q_{Cl}	MHz	−66.80	−66.4	−66.5

Finally we note that the electric resonance spectrum of H^{35}Cl was also studied in the presence of a strong magnetic field by de Leeuw and Dymanus [89]; the results are summarised in table 8.22, which may be compared with the earlier results tabulated for HF.

8.5. Molecular beam electric resonance of open shell molecules

8.5.1. Introduction

The years from 1960 to 1975 represented a golden era in the radiofrequency and microwave spectroscopy of open shell diatomic molecules. Molecular beam electric resonance was one of the most important experimental approaches, but microwave, far-infrared and magnetic resonance studies of bulk gaseous samples were equally important and our understanding of these open shell species is derived from a combination of different experimental approaches. In this book we have chosen to organise our descriptions according to the experimental techniques employed, but as with any such scheme, we run the risk, which we wish to avoid, of not connecting the results from different types of experiment in a coherent manner. As we shall see, the OH radical is the example *par excellence* which illustrates the pitfalls of an approach which is technique-oriented, rather than molecule-oriented.

Many of the open shell species which have been studied have $^2\Pi$ ground states and therefore share certain common features in their spectroscopic study. In this section we choose to describe, in some detail, the LiO, NO and OH molecules, all of which have $^2\Pi$ ground states. The complexity increases, however, partly for reasons intrinsic to the species, and partly because of the increased range of techniques which have been employed. Later in this section we come to the $^3\Pi$ excited electronic state of CO, which introduces new complexities in its spectroscopy, some of which we shall meet again in our descriptions of the magnetic resonance spectroscopy of open shell species (chapter 9).

8.5.2. LiO in the $X\,^2\Pi$ ground state

(a) Introduction

One of the first applications of molecular beam electric resonance to an open-shell molecule was described in a classic paper by Freund, Herbst, Mariella and Klemperer [112]. The molecule concerned was LiO, and at the time of their work there was no other gas phase study of this species; unpublished calculations by Wahl [113] were of considerable help. Other theoretical calculations were being performed concurrently by Yoshimine [114], although Yoshimine and the experimentalists appeared to be unaware of their mutual interest in the LiO molecule. The experimental work was therefore performed against the background of essentially no previous information, which makes it the more remarkable. Beams of LiO were produced by heating Li_2O to 1800 K, and the resonance experiments were carried out using a spectrometer similar to that described in figure 8.33, operated in the 'flop-in' mode with a surface ionisation detector. Attention was focused on the naturally occurring isotopic form $^7Li^{16}O$, because of interest in the ^7Li hyperfine structure ($I = 3/2$).

In terms of single electron configurations, the ground and first excited electronic states of LiO may be written

$$X^2\Pi:\quad (1\sigma)^2(2\sigma)^2(3\sigma)^2(4\sigma)^2(1\pi)^3$$
$$A^2\Sigma^+:\quad (1\sigma)^2(2\sigma)^2(3\sigma)^2(4\sigma)(1\pi)^4. \qquad (8.347)$$

The electric resonance experiments were carried out on LiO in its ground $^2\Pi$ state, but we shall see that the nearby $^2\Sigma^+$ state, lying only 2330 cm^{-1} above the ground state, has very important effects on the ground state spectrum. Figure 8.40 gives an approximate idea of the relevant energy levels, which are not drawn to scale. The lowest fine-structure component of the $X^2\Pi$ state is the $^2\Pi_{3/2}$, with the $^2\Pi_{1/2}$ component lying 112 cm^{-1} higher in energy. Each rotational level is split primarily into Λ-doublet components, as shown. Mixing of the ground state with the $A^2\Sigma^+$ excited state is very important, so we include the first few rotational levels of the A state in figure 8.40, using case (a) nomenclature (for both states). Figure 8.41 shows the nuclear hyperfine splitting in the $J = 5/2$ rotational level, and the types of transitions studied in the electric resonance experiments. Resonances were observed involving the first three vibrational levels of the ground electronic state, but we shall concentrate on the $v = 0$ spectrum, which exhibited the best resolution. The nuclear spin I of ^7Li is 3/2, and both magnetic and electric quadrupole interactions were observed and analysed.

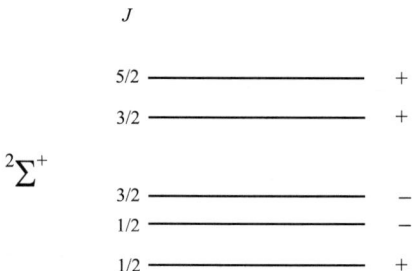
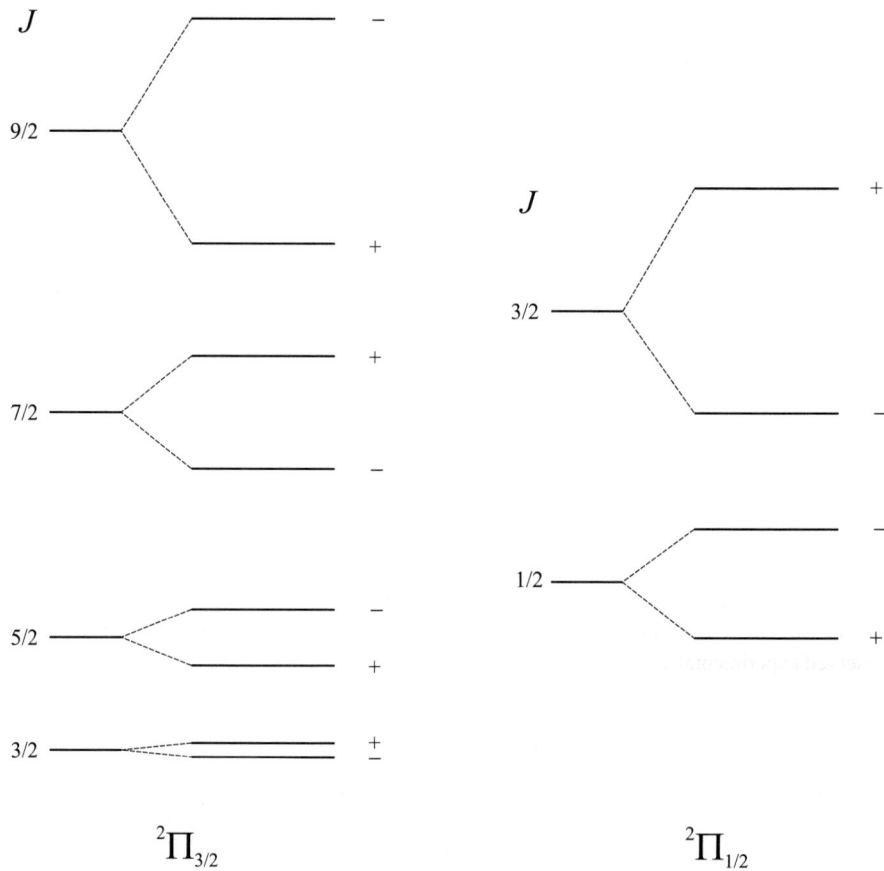

Figure 8.40. Lower rotational levels of the $X\,^2\Pi$ and $A\,^2\Sigma^+$ states of LiO, showing the Λ-doublet splitting. The diagram is not drawn to scale, but does demonstrate the increasing Λ-doublet splitting as J increases, and the much larger splitting in the $^2\Pi_{1/2}$ state. The Λ-doublet splitting in the $^2\Pi_{3/2}$ state increases from about 10 MHz in $J=3/2$ to 223 MHz in $J=9/2$. In the $^2\Pi_{1/2}$ state, which lies 112 cm^{-1} above $^2\Pi_{3/2}$, it is about 12 300 MHz in $J=3/2$. The $^2\Sigma^+$ state lies 2330 cm^{-1} above the ground state.

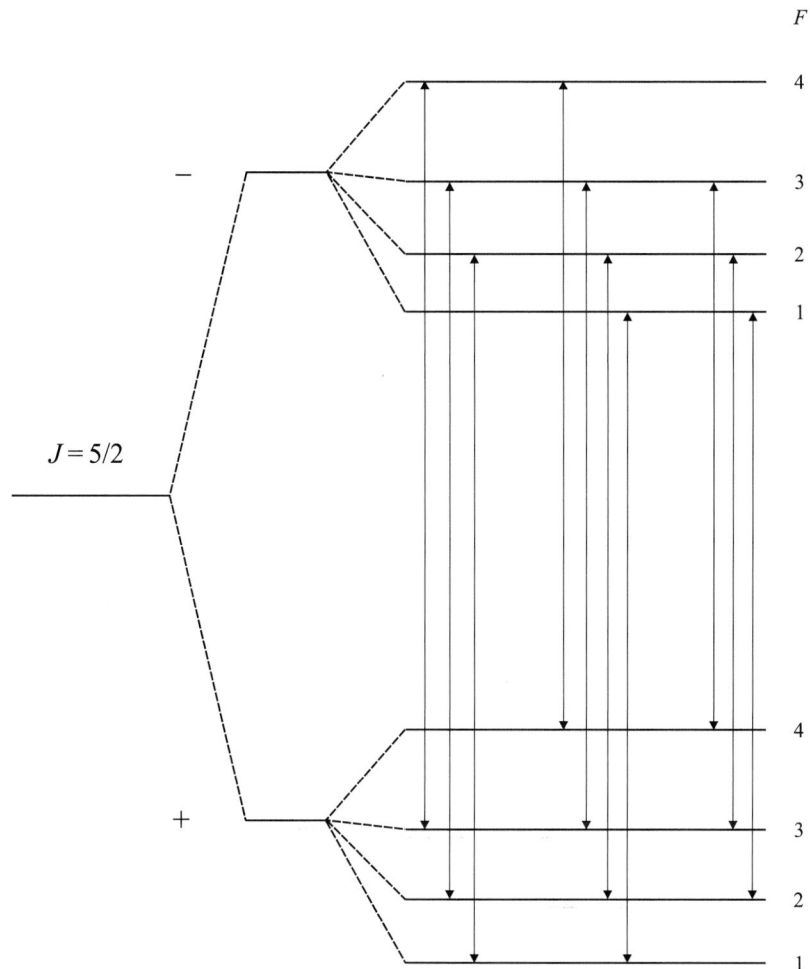

Figure 8.41. ^{7}Li nuclear hyperfine splitting of the $J = 5/2$ Λ-doublet levels of LiO, and the types of transitions detected in the electric resonance experiments [112]. The electric dipole transitions obey the selection rules $\Delta J = 0$, $\Delta F = 0, \pm 1$ and all ten transitions shown were observed experimentally.

(b) Effective Hamiltonian and basis set

The effective Hamiltonian may be regarded as the sum of four terms,

$$\mathcal{H}_{\text{eff}} = \mathcal{H}_{\text{so}} + \mathcal{H}_{\text{rot}} + \mathcal{H}_{\text{hfs}} + \mathcal{H}_{\text{Q}}, \quad (8.348)$$

representing the spin–orbit coupling, rigid body rotation, magnetic hyperfine and electric quadrupole interactions respectively. These four terms are written explicitly as:

$$\mathcal{H}_{\text{so}} = A\,T^1(\boldsymbol{L}) \cdot T^1(\boldsymbol{S}), \quad (8.349)$$

$$\mathcal{H}_{\text{rot}} = B\,T^1(\boldsymbol{R}) \cdot T^1(\boldsymbol{R}) = B\,T^1(\boldsymbol{J} - \boldsymbol{L} - \boldsymbol{S}) \cdot T^1(\boldsymbol{J} - \boldsymbol{L} - \boldsymbol{S}), \quad (8.350)$$

$$\mathcal{H}_{\mathrm{hfs}} = a\,\mathrm{T}^1(\boldsymbol{L})\cdot\mathrm{T}^1(\boldsymbol{I}) + b_{\mathrm{F}}\mathrm{T}^1(\boldsymbol{S})\cdot\mathrm{T}^1(\boldsymbol{I})$$
$$- \sqrt{10}g_S\mu_B g_N\mu_N(\mu_0/4\pi)\,\mathrm{T}^1(\boldsymbol{S},\boldsymbol{C}^2)\cdot\mathrm{T}^1(\boldsymbol{I}), \tag{8.351}$$
$$\mathcal{H}_Q = -e\,\mathrm{T}^2(\nabla\boldsymbol{E})\cdot\mathrm{T}^2(\boldsymbol{Q}). \tag{8.352}$$

We have met most of these terms earlier in this chapter, except for that describing the rigid body rotation $\boldsymbol{R}^2$; the expansion in equation (8.350) anticipates the basis set to be employed. Note that we have chosen to express the rotational kinetic energy in terms of the bare nuclear angular momentum $\boldsymbol{R}$ here rather than N as recommended in chapter 7. This is because we wish to include the mixture of the $^2\Sigma^+$ state of LiO explicitly in our description. The effective Hamiltonian does not contain terms describing centrifugal distortion and later in this book we will describe studies of other $^2\Pi$ radicals, notably NO, OH and CH, where the spectroscopic analysis is considerably more detailed than that described here for LiO. What follows for LiO may be regarded as a fairly gentle introduction to the theory of $^2\Pi$ states, with much more to come later.

The eigenvalue problem is set up using a case (a) basis set for both the $^2\Pi$ and $^2\Sigma^+$ states of LiO. The quantum numbers specified are therefore Λ, the component of electronic orbital angular momentum along the internuclear axis, Σ, the component of electron spin S along the internuclear axis, and Ω, the sum of Λ and Σ. In addition we specify the total angular momentum J, exclusive of nuclear spin I, and ultimately F, the total angular momentum including nuclear spin. We start by noting that there are six 'primitive' basis states to be considered, which are as follows:

$$\begin{aligned}
^2\Pi_{\Omega=+3/2}:\ &|\eta,\Lambda=+1;\ S,\Sigma=+1/2;\ J,M_J,\Omega=+3/2\rangle,\\
^2\Pi_{\Omega=-3/2}:\ &|\eta,\Lambda=-1;\ S,\Sigma=-1/2;\ J,M_J,\Omega=-3/2\rangle,\\
^2\Pi_{\Omega=+1/2}:\ &|\eta,\Lambda=+1;\ S,\Sigma=-1/2;\ J,M_J,\Omega=+1/2\rangle,\\
^2\Pi_{\Omega=-1/2}:\ &|\eta,\Lambda=-1;\ S,\Sigma=+1/2;\ J,M_J,\Omega=-1/2\rangle,\\
^2\Sigma_{\Omega=+1/2}:\ &|\eta,\Lambda=0;\ S,\Sigma=+1/2;\ J,M_J,\Omega=+1/2\rangle,\\
^2\Sigma_{\Omega=-1/2}:\ &|\eta,\Lambda=0;\ S,\Sigma=-1/2;\ J,M_J,\Omega=-1/2\rangle.
\end{aligned} \tag{8.353}$$

We shall calculate the matrix elements of the effective Hamiltonian within the basis of these six primitive states in due course. These states do not, however, have definite parities. Since parity is conserved (except in the presence of an applied electric field), we construct a basis set of six functions, three of each parity type, so that for a given J level we are left with the diagonalisation of 3×3 matrices, rather than 6×6. More importantly, we are aiming to understand the electric dipole radiofrequency and microwave spectra, and know that transitions must occur between states of *opposite* parity.

We follow the conventions described by Brown, Kaise, Kerr and Milton [115] in order to form parity-conserved functions, as discussed in detail in, section 6.9. Parity is related to the behaviour of a state or function under the space-fixed inversion operator E^*:

$$E^* f(X, Y, Z) = f(-X, -Y, -Z), \tag{8.354}$$

where X, Y, Z are space-fixed coordinates. So far as the transformation of the molecule-fixed coordinates x, y, z is concerned, the convention adopted is that the molecule-fixed y axis is unchanged in direction by the application of E^* to the molecular system. This implies that the Euler angles (ϕ, θ, χ) relating the molecule- and space-fixed axes transform under E^* in the following way:

$$E^* f(\phi, \theta, \chi) = f(\pi + \phi, \pi - \theta, \pi - \chi), \tag{8.355}$$

which is equivalent to the transformation

$$E^* f(x, y, z) = f(x, -y, z). \tag{8.356}$$

Noting from (8.353) that each primitive basis function in case (a) is of the form

$$|\eta, \Lambda; S, \Sigma; J, M_J, \Omega\rangle = |\eta, \Lambda\rangle |S, \Sigma\rangle |J, M_J, \Omega\rangle, \tag{8.357}$$

we may apply E^* to each of the three factors on the right-hand side of (8.357) in turn. The results are

$$\begin{aligned} E^*|\eta, \Lambda\rangle &= (-1)^{\Lambda+s}|\eta, -\Lambda\rangle, \\ E^*|S, \Sigma\rangle &= (-1)^{S-\Sigma}|S, -\Sigma\rangle, \\ E^*|J, M_J, \Omega\rangle &= (-1)^{J-\Omega}|J, M_J, -\Omega\rangle. \end{aligned} \tag{8.358}$$

The exponent s in the first of these results refers to Σ states only, and is even for a Σ^+ state or odd for a Σ^- state; the current problem involves an excited Σ^+ state, so we will omit the exponent s in what follows. The net result of the above operations is therefore

$$E^*|\eta, \Lambda; S, \Sigma; J, M_J, \Omega\rangle = (-1)^{J-S}|\eta, -\Lambda; S, -\Sigma; J, M_J, -\Omega\rangle. \tag{8.359}$$

Using (8.359) we are now able to construct linear combinations of the primitive functions (8.353) which have definite parity. These linear combinations are as follows:

$$+ \text{parity}: \quad \frac{1}{\sqrt{2}}\{|\eta, \Lambda; S, \Sigma; J, M_J, \Omega\rangle + (-1)^{J-S}|\eta, -\Lambda; S, -\Sigma; J, M_J, -\Omega\rangle\},$$

$$- \text{parity}: \quad \frac{1}{\sqrt{2}}\{|\eta, \Lambda; S, \Sigma; J, M_J, \Omega\rangle - (-1)^{J-S}|\eta, -\Lambda; S, -\Sigma; J, M_J, -\Omega\rangle\}. \tag{8.360}$$

Our procedure now is to calculate the matrix elements of the effective Hamiltonian in the primitive basis set, and then to reformulate these elements in the parity-conserved set (8.360).

(c) Matrix elements without nuclear spin in the primitive basis set

Ultimately we will use a basis set in which the nuclear spin I is coupled to J to form F, but first we tackle the nuclear spin-free problem, dealing with the first two terms in

(8.348). Note that in the expressions which follow, Λ, Σ, and Ω are *signed* quantities. All of the matrix elements are strictly diagonal in F and M_F (or in J and M_J for the nuclear spin-free problem), and we ignore any matrix elements off-diagonal in S and I. The symbol η denotes vibronic state quantum numbers not otherwise specified in the basis states.

It is customary to take the spin–orbit and rigid body rotation terms together, for reasons which will become immediately apparent. On expansion of the rigid body term (8.350) we obtain

$$\begin{aligned}\mathcal{H}_{so}+\mathcal{H}_{rot}=\mathcal{H}_{rso}&=A\mathbf{T}^1(\boldsymbol{L})\cdot\mathbf{T}^1(\boldsymbol{S})+B\,\{\mathbf{T}^1(\boldsymbol{J})\cdot\mathbf{T}^1(\boldsymbol{J})+\mathbf{T}^1(\boldsymbol{S})\cdot\mathbf{T}^1(\boldsymbol{S})\\ &\quad+\mathbf{T}^1(\boldsymbol{L})\cdot\mathbf{T}^1(\boldsymbol{L})-2\mathbf{T}^1(\boldsymbol{J})\cdot\mathbf{T}^1(\boldsymbol{S})-2\mathbf{T}^1(\boldsymbol{J})\cdot\mathbf{T}^1(\boldsymbol{L})+2\mathbf{T}^1(\boldsymbol{L})\cdot\mathbf{T}^1(\boldsymbol{S})\}\\ &=(A+2B)\,\mathbf{T}^1(\boldsymbol{L})\cdot\mathbf{T}^1(\boldsymbol{S})+B\,\{\mathbf{T}^1(\boldsymbol{J})\cdot\mathbf{T}^1(\boldsymbol{J})+\mathbf{T}^1(\boldsymbol{S})\cdot\mathbf{T}^1(\boldsymbol{S})\\ &\quad+\mathbf{T}^1(\boldsymbol{L})\cdot\mathbf{T}^1(\boldsymbol{L})-2\mathbf{T}^1(\boldsymbol{J})\cdot\mathbf{T}^1(\boldsymbol{S})-2\mathbf{T}^1(\boldsymbol{J})\cdot\mathbf{T}^1(\boldsymbol{L})\}.\end{aligned} \quad (8.361)$$

The first two terms in the purely rotational part of (8.361) are wholly diagonal in our basis set and may be replaced by their respective eigenvalues. The remaining scalar products are expanded in the molecule-fixed coordinate system, q, and in the sum over q we separate the $q=0$ terms from those with $q=\pm 1$ (denoted by a superscript prime). We also take note of the anomalous commutation rules for the components of $\boldsymbol{J}$. Equation (8.361) becomes

$$\begin{aligned}\mathcal{H}_{rso}&=\mathcal{H}^{(0)}+\mathcal{H}'\quad\text{where}\\ \mathcal{H}^{(0)}&=(A+2B)\mathrm{T}^1_0(\boldsymbol{L})\mathrm{T}^1_0(\boldsymbol{S})+B\,\{J(J+1)+S(S+1)\\ &\quad+\mathrm{T}^1_0(\boldsymbol{L})\mathrm{T}^1_0(\boldsymbol{L})-2\mathrm{T}^1_0(\boldsymbol{J})\mathrm{T}^1_0(\boldsymbol{S})-2\mathrm{T}^1_0(\boldsymbol{J})\mathrm{T}^1_0(\boldsymbol{L})\}\\ \mathcal{H}'&=\sum_{q=\pm 1}{}'\{(-1)^q(A+2B)\mathrm{T}^1_q(\boldsymbol{L})\mathrm{T}^1_{-q}(\boldsymbol{S})+(-1)^q\,B\mathrm{T}^1_q(\boldsymbol{L})\mathrm{T}^1_{-q}(\boldsymbol{L})\\ &\quad-2B\mathrm{T}^1_q(\boldsymbol{J})\mathrm{T}^1_q(\boldsymbol{S})-2B\mathrm{T}^1_q(\boldsymbol{J})\mathrm{T}^1_q(\boldsymbol{L})\}.\end{aligned} \quad (8.362)$$

We are now ready to examine the matrix elements. For $\mathcal{H}^{(0)}$, which contains terms with $q=0$, the matrix elements are wholly diagonal:

$$\begin{aligned}\langle\eta,\Lambda;S,\Sigma;J,\Omega|\mathcal{H}^{(0)}|\eta',\Lambda';S,\Sigma';J',\Omega'\rangle\\ =\delta_{\eta\eta'}\delta_{\Lambda\Lambda'}\delta_{\Sigma\Sigma'}\delta_{\Omega\Omega'}\{(A+2B)\Lambda\Sigma+B[J(J+1)+S(S+1)+\Lambda^2-2\Omega^2]\}.\end{aligned} \quad (8.363)$$

The four terms in $\mathcal{H}'$, however, require closer attention. The second term, which in molecule-fixed cartesian coordinates may be expressed as $L_x^2+L_y^2$, affects all levels equally and is therefore usually omitted. The third term, which does not involve the orbital angular momentum, is known as the spin uncoupling term. Its matrix elements are readily obtained:

$$\langle \eta, \Lambda; S, \Sigma; J, \Omega | -2B \sum_{q=\pm 1}{}' T_q^1(J) T_q^1(S) | \eta', \Lambda'; S, \Sigma'; J', \Omega' \rangle$$

$$= -\delta_{\eta\eta'} \delta_{\Lambda\Lambda'} \delta_{JJ'} 2B \sum_{q=\pm 1}{}' \langle J, \Omega | T_q^1(J) | J, \Omega' \rangle \langle S, \Sigma | T_q^1(S) | S, \Sigma' \rangle$$

$$= -2B \sum_{q=\pm 1}{}' (-1)^{J+S-\Omega-\Sigma} \begin{pmatrix} J & 1 & J \\ -\Omega & q & \Omega' \end{pmatrix} \begin{pmatrix} S & 1 & S \\ -\Sigma & q & \Sigma' \end{pmatrix}$$

$$\times \{J(J+1)(2J+1)S(S+1)(2S+1)\}^{1/2}. \tag{8.364}$$

The two remaining terms contained within $\mathcal{H}'$ in equation (8.362) require special attention because they mix the Π and Σ electronic states and thereby give rise to Λ-doubling. We extract these two terms from $\mathcal{H}'$ to define a Hamiltonian which causes Λ-doubling:

$$\mathcal{H}'' = \sum_{q=\pm 1}{}' \{(-1)^q (A+2B) T_q^1(L) T_{-q}^1(S) - 2B T_q^1(J) T_q^1(L)\}. \tag{8.365}$$

The matrix elements of the two terms in (8.365) can be partially evaluated as follows:

$$\langle \eta, \Lambda; S, \Sigma; J, \Omega | \sum_{q=\pm 1} (-1)^q (A+2B) T_q^1(L) T_{-q}^1(S) | \eta', \Lambda'; S, \Sigma'; J, \Omega' \rangle$$

$$= -2\delta_{\Omega\Omega'} \sum_{q=\pm 1} \langle \eta, \Lambda | (A/2 + B) T_q^1(L) | \eta', \Lambda' \rangle \langle S, \Sigma | T_{-q}^1(S) | S, \Sigma' \rangle$$

$$= -2\delta_{\Omega\Omega'} \sum_{q=\pm 1} \langle \eta, \Lambda | (A/2 + B) T_q^1(L) | \eta', \Lambda' \rangle (-1)^{S-\Sigma}$$

$$\times \begin{pmatrix} S & 1 & S \\ -\Sigma & -q & \Sigma' \end{pmatrix} \{S(S+1)(2S+1)\}^{1/2}. \tag{8.366}$$

$$\langle \eta, \Lambda; S, \Sigma; J, \Omega | -\sum_{q=\pm 1} 2B T_q^1(J) T_q^1(L) | \eta', \Lambda'; S, \Sigma'; J, \Omega' \rangle$$

$$= -2\delta_{\Sigma\Sigma'} \sum_{q=\pm 1} \langle \eta, \Lambda | B T_q^1(L) | \eta', \Lambda' \rangle (-1)^{J-\Omega}$$

$$\times \begin{pmatrix} J & 1 & J \\ -\Omega & q & \Omega' \end{pmatrix} \{J(J+1)(2J+1)\}^{1/2}. \tag{8.367}$$

In both of these equations there remain matrix elements of the off-diagonal components of L; these elements couple the Π and Σ electronic states and they cannot be expanded further at this stage. As we shall see, they constitute molecular parameters to be determined from the experiments.

(d) Matrix elements without nuclear spin in the parity-conserved basis set

We now make use of the results derived above, using the parity-conserved basis functions defined in equations (8.360). For states of *positive* parity we have the following three functions:

$$|^2\Pi_{3/2}^{(+)}\rangle = \frac{1}{\sqrt{2}}\{|\eta, \Lambda = 1; S, \Sigma = 1/2; J, \Omega = 3/2\rangle$$
$$+ (-1)^{J-1/2}|\eta, \Lambda = -1; S, \Sigma = -1/2; J, \Omega = -3/2\rangle\},$$

$$|^2\Pi_{1/2}^{(+)}\rangle = \frac{1}{\sqrt{2}}\{|\eta, \Lambda = 1; S, \Sigma = -1/2; J, \Omega = 1/2\rangle$$
$$+ (-1)^{J-1/2}|\eta, \Lambda = -1; S, \Sigma = 1/2; J, \Omega = -1/2\rangle\},$$

$$|^2\Sigma_{1/2}^{(+)}\rangle = \frac{1}{\sqrt{2}}\{|\eta, \Lambda = 0; S, \Sigma = 1/2; J, \Omega = 1/2\rangle$$
$$+ (-1)^{J-1/2}|\eta, \Lambda = 0; S, \Sigma = -1/2; J, \Omega = -1/2\rangle\}. \quad (8.368)$$

It is now a straightforward matter to construct the 3×3 matrix for the positive parity states; the results are summarised below.

	$^2\Pi_{3/2}^{(+)}$	$^2\Pi_{1/2}^{(+)}$	$^2\Sigma_{1/2}^{(+)}$
$^2\Pi_{3/2}^{(+)}$	$A/2 + B_\Pi\{J(J+1) - 7/4\}$	$-B_\Pi\{(J+3/2) \times (J-1/2)\}^{1/2}$	$-M_1\{(J-1/2) \times (J+3/2)\}^{1/2}$
$^2\Pi_{1/2}^{(+)}$	$-B_\Pi\{(J+3/2)(J-1/2)\}^{1/2}$	$-A/2 + B_\Pi\{J(J+1)+1/4\}$	$M_2 - M_1(-1)^{J-1/2} \times (J+1/2)$
$^2\Sigma_{1/2}^{(+)}$	$-M_1\{(J-1/2)(J+3/2)\}^{1/2}$	$M_2 - M_1(-1)^{J-1/2} \times (J+1/2)$	$\Delta E + B_\Sigma\{J(J+1) +1/4\} - B_\Sigma(-1)^{J-1/2} \times (J+1/2)$

In these matrix elements ΔE is the separation between the Π and Σ electronic states, and the electronic matrix elements M_1 and M_2 are defined by

$$M_1 = \langle \Pi | BL_+ | \Sigma \rangle, \quad M_2 = \langle \Pi | (A/2 + B)L_+ | \Sigma \rangle.$$

The basis states of *negative* parity are

$$|^2\Pi_{3/2}^{(-)}\rangle = \frac{1}{\sqrt{2}}\{|\eta, \Lambda = 1; S, \Sigma = 1/2; J, \Omega = 3/2\rangle$$
$$- (-1)^{J-1/2}|\eta, \Lambda = -1; S, \Sigma = -1/2; J, \Omega = -3/2\rangle\},$$

$$|^2\Pi_{1/2}^{(-)}\rangle = \frac{1}{\sqrt{2}}\{|\eta, \Lambda = 1; S, \Sigma = -1/2; J, \Omega = 1/2\rangle$$
$$- (-1)^{J-1/2}|\eta, \Lambda = -1; S, \Sigma = 1/2; J, \Omega = -1/2\rangle\},$$

$$|^2\Sigma_{1/2}^{(-)}\rangle = \frac{1}{\sqrt{2}}\{|\eta, \Lambda = 0; S, \Sigma = 1/2; J, \Omega = 1/2\rangle$$
$$- (-1)^{J-1/2}|\eta, \Lambda = 0; S, \Sigma = -1/2; J, \Omega = -1/2\rangle\}. \quad (8.369)$$

The 3 × 3 matrix for the *negative* parity states differs from that for the *positive* parity states only in the signs of certain terms.

	$^2\Pi_{3/2}^{(-)}$	$^2\Pi_{1/2}^{(-)}$	$^2\Sigma_{1/2}^{(-)}$
$^2\Pi_{3/2}^{(-)}$	$A/2 + B_\Pi\{J(J+1) - 7/4\}$	$-B_\Pi\{(J+3/2)(J-1/2)\}^{1/2}$	$-M_1\{(J-1/2) \times (J+3/2)\}^{1/2}$
$^2\Pi_{1/2}^{(-)}$	$-B_\Pi\{(J+3/2)(J-1/2)\}^{1/2}$	$-A/2 + B_\Pi\{J(J+1) + 1/4\}$	$M_2 + M_1(-1)^{J-1/2} \times (J+1/2)$
$^2\Sigma_{1/2}^{(-)}$	$-M_1\{(J-1/2)(J+3/2)\}^{1/2}$	$M_2 + M_1(-1)^{J-1/2}(J+1/2)$	$\Delta E + B_\Sigma\{J(J+1) + 1/4\} + B_\Sigma(-1)^{J-1/2} \times (J+1/2)$

These results are the same as those obtained by Freund, Herbst, Mariella and Klemperer [112] except for the J-dependent phase factors in our matrices. These arise because of our specific definitions of the parity-conserved basis function and are necessary if the energies of the Λ-doublet components are to alternate with J. If we know the values of the five molecular constants appearing in these matrices, we can calculate the energies of the levels, of both parity types, for each value of J. In practice, of course, it was the task of the experimental spectroscopists to solve the reverse problem of determining the molecular parameters from the observed transition frequencies.

The Λ-doubling arises from mixing of the $^2\Pi$ components with the $^2\Sigma^+$ state. Freund, Herbst, Mariella and Klemperer [112] do not give quite enough information for us to reproduce their theoretical data. They give the following values of the molecular constants (in cm^{-1}):

$$A = -112.0, \quad B_\Pi = 1.222, \quad B_\Sigma = 1.35, \quad M_1 M_2 = -139.3, \quad \Delta E = 2800.$$

Using these numbers we find that the value $M_1 = 1.485$ cm^{-1} reproduces their observed Λ-doubling frequencies up to $J = 25/2$ extremely well. They point out, however, that the molecular parameters are strongly correlated; we shall return to discuss their values later in this section. We will also discuss the relationship of this relatively simple example of Λ-doubling (because mixing with only one excited electronic state is considered) to the more complicated cases encountered for other molecules.

(e) Matrix elements of the nuclear hyperfine terms in the primitive basis set

The magnetic hyperfine interaction terms were given in equation (8.351) and the electric quadrupole interaction in equation (8.352). We extend the basis functions by inclusion of the ^{7}Li nuclear spin I, coupled to J to form F; the value of I is $3/2$. We deal with each term in turn, first deriving expressions for the matrix elements in the primitive basis set (8.353), and then extending these results to the parity-conserved basis. All matrix elements are diagonal in F, and any elements off-diagonal in S and I can of course be ignored.

For the orbital hyperfine term the calculation proceeds as follows:

$$\langle \eta, \Lambda; S, \Sigma; J, \Omega; I, F | a\mathbf{T}^1(\mathbf{L}) \cdot \mathbf{T}^1(\mathbf{I}) | \eta', \Lambda'; S, \Sigma'; J', \Omega'; I, F \rangle$$

$$= (-1)^{J'+F+I} \begin{Bmatrix} I & J' & F \\ J & I & 1 \end{Bmatrix} \langle \eta, \Lambda; S, \Sigma; J, \Omega \| a\mathbf{T}^1(\mathbf{L}) \| \eta', \Lambda'; S, \Sigma'; J', \Omega' \rangle$$

$$\times \langle I \| \mathbf{T}^1(\mathbf{I}) \| I \rangle$$

$$= (-1)^{J'+F+I} \begin{Bmatrix} I & J' & F \\ J & I & 1 \end{Bmatrix} \{I(I+1)(2I+1)\}^{1/2}$$

$$\times \langle \eta, \Lambda; S, \Sigma; J, \Omega \| a\mathbf{T}^1(\mathbf{L}) \| \eta', \Lambda'; S, \Sigma'; J', \Omega' \rangle. \tag{8.370}$$

In order to evaluate the reduced matrix element of $\mathbf{T}^1(\mathbf{L})$, we first rotate into the molecule-fixed axis system, q, using a first rank rotational matrix, so that

$$\langle \eta, \Lambda; S, \Sigma; J, \Omega \| a\mathbf{T}^1(\mathbf{L}) \| \eta', \Lambda'; S, \Sigma'; J', \Omega' \rangle$$

$$= \langle \eta, \Lambda; S, \Sigma; J, \Omega \| a \sum_q \mathcal{D}^{(1)}_{\cdot q}(\omega)^* \mathbf{T}^1_q(\mathbf{L}) \| \eta', \Lambda'; S, \Sigma'; J', \Omega' \rangle$$

$$= \delta_{\Sigma\Sigma'} \sum_q (-1)^{J-\Omega} \begin{pmatrix} J & 1 & J' \\ -\Omega & q & \Omega' \end{pmatrix} \{(2J+1)(2J'+1)\}^{1/2}$$

$$\times \langle \eta, \Lambda | a\mathbf{T}^1_q(\mathbf{L}) | \eta', \Lambda' \rangle. \tag{8.371}$$

We may therefore combine equations (8.371) and (8.370), and separate the diagonal ($q = 0$) from the off-diagonal elements, with the following results.
For $q = 0$ we have

$$\langle \eta, \Lambda; S, \Sigma; J, \Omega; I, F | a\mathbf{T}^1(\mathbf{L}) \cdot \mathbf{T}^1(\mathbf{I}) | \eta', \Lambda'; S, \Sigma'; J', \Omega'; I, F \rangle$$

$$= (-1)^{J'+F+I} \begin{Bmatrix} I & J' & F \\ J & I & 1 \end{Bmatrix} \langle \eta, \Lambda; S, \Sigma; J, \Omega \| a\mathbf{T}^1(\mathbf{L}) \| \eta', \Lambda'; S, \Sigma'; J', \Omega' \rangle$$

$$\times \langle I \| \mathbf{T}^1(\mathbf{I}) \| I \rangle$$

$$= (-1)^{J'+F+I} \begin{Bmatrix} I & J' & F \\ J & I & 1 \end{Bmatrix} \{I(I+1)(2I+1)\}^{1/2}$$

$$\times \langle \eta, \Lambda; S, \Sigma; J, \Omega \| a\mathcal{D}^{(1)}_{\cdot 0}(\omega)^* \mathbf{T}^1_0(\mathbf{L}) \| \eta, \Lambda; S, \Sigma; J', \Omega \rangle$$

$$= a_\eta \Lambda (-1)^{J'+F+I} \begin{Bmatrix} I & J' & F \\ J & I & 1 \end{Bmatrix} \{I(I+1)(2I+1)\}^{1/2} (-1)^{J-\Omega}$$

$$\times \begin{pmatrix} J & 1 & J' \\ -\Omega & 0 & -\Omega \end{pmatrix} \{(2J+1)(2J'+1)\}^{1/2}. \tag{8.372}$$

For $q = \pm 1$ we have

$$\langle \eta, \Lambda; S, \Sigma; J, \Omega; I, F | a\mathbf{T}^1(\mathbf{L}) \cdot \mathbf{T}^1(\mathbf{I}) \eta', \Lambda'; S, \Sigma'; J', \Omega'; I, F \rangle$$

$$= (-1)^{J'+F+I} \begin{Bmatrix} I & J' & F \\ J & I & 1 \end{Bmatrix}$$

$$\times \langle \eta, \Lambda; S, \Sigma; J, \Omega \| aT^1(\boldsymbol{L}) \| \eta', \Lambda'; S, \Sigma'; J', \Omega' \rangle \langle I \| T^1(\boldsymbol{I}) \| I \rangle$$

$$= (-1)^{J'+F+I} \begin{Bmatrix} I & J' & F \\ J & I & 1 \end{Bmatrix} \{I(I+1)(2I+1)\}^{1/2}$$

$$\times \langle \eta, \Lambda; S, \Sigma; J, \Omega \| a \sum_{q=\pm 1} \mathscr{D}^{(1)}_{\cdot q}(\omega)^* T^1_q(\boldsymbol{L}) \| \eta', \Lambda'; S, \Sigma'; J', \Omega' \rangle$$

$$= (-1)^{J'+F+I} \begin{Bmatrix} I & J' & F \\ J & I & 1 \end{Bmatrix} \{I(I+1)(2I+1)\}^{1/2} \sum_{q=\pm 1} (-1)^{J-\Omega}$$

$$\times \begin{pmatrix} J & 1 & J' \\ -\Omega & q & \Omega' \end{pmatrix} \{(2J+1)(2J'+1)\}^{1/2} \langle \eta, \Lambda | aT^1_q(\boldsymbol{L}) | \eta', \Lambda' \rangle. \quad (8.373)$$

Note that the constant a can be separated out in (8.372) and has a well defined value, a_η, whereas in (8.373) it is part of a matrix element which cannot be further reduced.

The next term in the magnetic hyperfine Hamiltonian (8.351) describes the Fermi contact interaction and the calculation of its matrix elements proceeds in a manner similar to that just described for the orbital hyperfine term, as follows:

$$\langle \eta, \Lambda; S, \Sigma; J, \Omega; I, F | b_F T^1(\boldsymbol{S}) \cdot T^1(\boldsymbol{I}) | \eta', \Lambda'; S, \Sigma'; J', \Omega'; I, F \rangle$$

$$= (-1)^{J'+F+I} \begin{Bmatrix} I & J' & F \\ J & I & 1 \end{Bmatrix}$$

$$\times \langle \eta, \Lambda; S, \Sigma; J, \Omega \| b_F T^1(\boldsymbol{S}) \| \eta', \Lambda'; S, \Sigma'; J', \Omega' \rangle \langle I \| T^1(\boldsymbol{I}) \| I \rangle$$

$$= (-1)^{J'+F+I} \begin{Bmatrix} I & J' & F \\ J & I & 1 \end{Bmatrix} \{I(I+1)(2I+1)\}^{1/2}$$

$$\times \langle \eta, \Lambda; S, \Sigma; J, \Omega \| b_F \sum_q D^{(1)}_{\cdot q}(\omega)^* T^1_q(\boldsymbol{S}) \| \eta, \Lambda; S, \Sigma'; J', \Omega' \rangle$$

$$= b_{F(\eta)} (-1)^{J'+F+I} \begin{Bmatrix} I & J' & F \\ J & I & 1 \end{Bmatrix} \{I(I+1)(2I+1)\}^{1/2}$$

$$\times \sum_q (-1)^{J-\Omega} \begin{pmatrix} J & 1 & J' \\ -\Omega & q & \Omega' \end{pmatrix} \{(2J+1)(2J'+1)\}^{1/2} (-1)^{S-\Sigma}$$

$$\times \begin{pmatrix} S & 1 & S \\ \Sigma & q & \Sigma' \end{pmatrix} \{S(S+1)(2S+1)\}^{1/2} \quad (8.374)$$

We have neglected admixture with excited electronic states, so that the Fermi contact constant, b_F, is defined by

$$b_F = (2\mu_0/3) g_S \mu_B g_N \mu_N \delta(r)_\eta. \quad (8.375)$$

The matrix elements of the third term in equation (8.351) are somewhat more complicated. Dealing first with the nuclear spin coupling, we have, as usual,

$$\langle \eta, \Lambda; S, \Sigma; J, \Omega; I, F | -\sqrt{10} g_S \mu_B g_N \mu_N (\mu_0/4\pi) T^1(\boldsymbol{S}, \boldsymbol{C}^2) \cdot T^1(\boldsymbol{I}) | \eta', \Lambda'; S, \Sigma'; J', \Omega'; I, F \rangle$$

$$= (-1)^{J'+F+I} \begin{Bmatrix} I & J' & F \\ J & I & 1 \end{Bmatrix}$$

$$\times \langle \eta, \Lambda; S, \Sigma; J, \Omega \| -\sqrt{10} g_S \mu_B g_N \mu_N (\mu_0/4\pi) \mathbf{T}^1(\mathbf{S}, \mathbf{C}^2) \| \eta', \Lambda'; S, \Sigma'; J', \Omega' \rangle$$
$$\times \langle I \| \mathbf{T}^1(\mathbf{I}) \| I \rangle$$
$$= -\sqrt{10} g_S \mu_B g_N \mu_N (\mu_0/4\pi)(-1)^{J'+F+I} \begin{Bmatrix} I & J' & F \\ J & I & 1 \end{Bmatrix}$$
$$\times \{I(I+1)(2I+1)\}^{1/2} \langle \eta, \Lambda; S, \Sigma; J, \Omega \| \mathbf{T}^1(\mathbf{S}, \mathbf{C}^2) \| \eta', \Lambda'; S, \Sigma'; J', \Omega' \rangle$$
$$= -\sqrt{10} g_S \mu_B g_N \mu_N (\mu_0/4\pi)(-1)^{J'+F+I} \begin{Bmatrix} I & J' & F \\ J & I & 1 \end{Bmatrix} \{I(I+1)(2I+1)\}^{1/2}$$
$$\times \sum_q \langle \eta, \Lambda; S, \Sigma; J, \Omega \| \mathcal{D}^{(1)}_q(\omega)^* \mathbf{T}^1_q(\mathbf{S}, \mathbf{C}^2) \| \eta', \Lambda'; S, \Sigma'; J', \Omega' \rangle. \quad (8.376)$$

The remaining reduced matrix element in the last line is developed further by noting the construction of the first-rank tensor,

$$\mathbf{T}^1_q(\mathbf{S}, \mathbf{C}^2) = -\sum_{q_1, q_2} (-1)^q \mathbf{T}^1_{q_1}(\mathbf{S}) \mathbf{C}^2_{q_2}(\theta, \phi)(\sqrt{3}) \begin{pmatrix} 1 & 2 & 1 \\ q_1 & q_2 & -q \end{pmatrix} (r^{-3}), \quad (8.377)$$

in which the spherical harmonic is defined by

$$C^k_q(\theta, \phi) = \left(\frac{4\pi}{2k+1}\right)^{1/2} Y_{kq}(\theta, \phi). \quad (8.378)$$

With these definitions the matrix element in (8.376) can be expanded:

$$\sum_q \langle \eta, \Lambda; S, \Sigma; J, \Omega | \mathcal{D}^{(1)}_q(\omega)^* \mathbf{T}^1_q(\mathbf{S}, \mathbf{C}^2) | \eta', \Lambda'; S, \Sigma'; J', \Omega' \rangle$$
$$= -\sum_q (-1)^{J-\Omega} \begin{pmatrix} J & 1 & J' \\ -\Omega & q & \Omega' \end{pmatrix} \{(3)(2J+1)(2J'+1)\}^{1/2}$$
$$\times \sum_{q_1, q_2} (-1)^q \begin{pmatrix} 1 & 2 & 1 \\ q_1 & q_2 & -q \end{pmatrix} (-1)^{S-\Sigma} \begin{pmatrix} S & 1 & S \\ -\Sigma & q_1 & \Sigma' \end{pmatrix}$$
$$\times \{S(S+1)(2S+1)\}^{1/2} \langle \eta, \Lambda | C^2_{q_2}(\theta, \phi)(r^{-3}) | \eta', \Lambda' \rangle. \quad (8.379)$$

This is an important result because, as we shall see when we deal with the parity-conserved basis functions, matrix elements with $\Delta \Lambda = 0, \pm 1, \pm 2$ are significant. The complete expression for the matrix elements of the dipolar interaction is obtained by combining (8.376) with (8.379) to yield:

$$\langle \eta, \Lambda; S, \Sigma; J, \Omega; I, F | -\sqrt{10} g_S \mu_B g_N \mu_N (\mu_0/4\pi) \mathbf{T}^1(\mathbf{S}, \mathbf{C}^2) \cdot \mathbf{T}^1(\mathbf{I}) | \eta', \Lambda'; S, \Sigma'; J', \Omega'; I, F \rangle$$

$$= \sqrt{30} g_S \mu_B g_N \mu_N (\mu_0/4\pi)(-1)^{J'+F+I} \begin{Bmatrix} I & J' & F \\ J & I & 1 \end{Bmatrix} \{I(I+1)(2I+1)\}^{1/2}$$

$$\times \sum_q (-1)^{J-\Omega+q} \begin{pmatrix} J & 1 & J' \\ -\Omega & q & \Omega' \end{pmatrix} \{(2J+1)(2J'+1)\}^{1/2}$$

$$\times \sum_{q_1,q_2} \begin{pmatrix} 1 & 2 & 1 \\ q_1 & q_2 & -q \end{pmatrix} (-1)^{S-\Sigma} \begin{pmatrix} S & 1 & S \\ -\Sigma & q_1 & \Sigma' \end{pmatrix}$$

$$\times \{S(S+1)(2S+1)\}^{1/2} \langle \eta, \Lambda | C^2_{q_2}(\theta,\phi)(r^{-3}) | \eta', \Lambda' \rangle. \tag{8.380}$$

Finally in this subsection we come to the electric quadrupole interaction, which is discussed at some length in appendix 8.4. From equation (8.23) we have

$$\langle \eta, \Lambda; S, \Sigma; J, \Omega; I, F | - e\mathrm{T}^2(\nabla E) \cdot \mathrm{T}^2(Q) | \eta', \Lambda'; S, \Sigma'; J', \Omega'; I, F \rangle$$

$$= (-1)^{J'+I+F} \begin{Bmatrix} I & J & F \\ J' & I & 2 \end{Bmatrix} \langle \eta, \Lambda; S, \Sigma; J, \Omega \| \mathrm{T}^2(\nabla E) \| \eta', \Lambda'; S, \Sigma'; J', \Omega' \rangle$$

$$\times \langle I \| -e\mathrm{T}^2(Q) \| I \rangle. \tag{8.381}$$

The nuclear spin part of this result is evaluated in (8.27) and the electronic part is developed by rotation into the molecule-fixed axis system through the use of a second-rank rotation matrix, so that (8.381) becomes

$$\langle \eta, \Lambda; S, \Sigma; J, \Omega; I, F | - e\mathrm{T}^2(\nabla E) \cdot \mathrm{T}^2(Q) | \eta', \Lambda'; S, \Sigma'; J', \Omega'; I, F \rangle$$

$$= (-1)^{J'+I+F} \begin{Bmatrix} I & J & F \\ J' & I & 2 \end{Bmatrix} \left(\frac{-eQ}{2}\right) \begin{pmatrix} I & 2 & I \\ -I & 0 & I \end{pmatrix}^{-1}$$

$$\times \langle \eta, \Lambda; S, \Sigma; J, \Omega \| \sum_q \mathcal{D}^{(2)}_{\cdot q}(\omega)^* \mathrm{T}^2_q(\nabla E) \| \eta', \Lambda'; S, \Sigma'; J', \Omega' \rangle$$

$$= \delta_{\Sigma\Sigma'} (-1)^{J'+I+F} \begin{Bmatrix} I & J & F \\ J' & I & 2 \end{Bmatrix} \left(\frac{-eQ}{2}\right) \begin{pmatrix} I & 2 & I \\ -I & 0 & I \end{pmatrix}^{-1} \sum_q (-1)^{J-\Omega}$$

$$\times \begin{pmatrix} J & 2 & J' \\ -\Omega & q & \Omega' \end{pmatrix} \{(2J+1)(2J'+1)\}^{1/2} \langle \eta, \Lambda | \mathrm{T}^2_q(\nabla E) | \eta', \Lambda' \rangle. \tag{8.382}$$

The remaining matrix element in (8.382) leads to definitions of the $q = 0, \pm 2$ components of the electric field gradient tensor (actually, its negative), which are

$$\frac{1}{2} q_0 = -\langle \Lambda | \mathrm{T}^2_0(\nabla E) | \Lambda \rangle = \frac{1}{2} \langle \Lambda | \sum_i \frac{e_i}{4\pi\varepsilon_0} (3\cos^2\theta_i)/r_i^3 | \Lambda \rangle,$$

$$\frac{1}{2\sqrt{6}} q_2 = -\langle \Lambda = \pm 1 | \mathrm{T}^2_{\pm 2}(\nabla E) | \Lambda' = \mp 1 \rangle = \frac{1}{2\sqrt{6}} \langle \Lambda | \sum_i \frac{e_i}{4\pi\varepsilon_0} \sin^2\theta_i / r_i^3 | \Lambda \rangle. \tag{8.383}$$

These definitions are consistent with those of Gallagher and Johnson [116]. This completes our calculation of the nuclear hyperfine terms in the primitive basis set.

(f) Matrix elements of the nuclear hyperfine terms in the parity-conserved basis set

The simplest approach to the 'book-keeping' problem of calculating all the nuclear spin-dependent matrix elements is first to evaluate all the terms in the primitive basis set, leaving only J, I and F as variables. We therefore construct the following 6×6 matrix, using the functions listed in (8.353). The rotational levels are widely spaced compared with the hyperfine terms, so that we also confine attention to matrix elements diagonal in J. The required 6×6 matrix is as follows.

	$^2\Pi_{\Omega=+3/2}$	$^2\Pi_{\Omega=-3/2}$	$^2\Pi_{\Omega=+1/2}$	$^2\Pi_{\Omega=-1/2}$	$^2\Sigma_{\Omega=+1/2}$	$^2\Sigma_{\Omega=-1/2}$
$^2\Pi_{\Omega=+3/2}$	m_{11}	0	m_{13}	m_{14}	m_{15}	0
$^2\Pi_{\Omega=-3/2}$	0	m_{22}	m_{23}	m_{24}	0	m_{26}
$^2\Pi_{\Omega=+1/2}$	m_{31}	m_{32}	m_{33}	m_{34}	m_{35}	m_{36}
$^2\Pi_{\Omega=-1/2}$	m_{41}	m_{42}	m_{43}	m_{44}	m_{45}	m_{46}
$^2\Sigma_{\Omega=+1/2}$	m_{51}	0	m_{53}	m_{54}	m_{55}	m_{56}
$^2\Sigma_{\Omega=-1/2}$	0	m_{62}	m_{63}	m_{64}	m_{65}	m_{66}

With the definitions $X = J(J+1) + I(I+1) - F(F+1)$,
$$Y = \{3X(X-1) - 4I(I+1)J(J+1)\}/I(2I-1),$$

the matrix elements are as follows:

$$m_{11} = -a_\Pi\{3X/4J(J+1)\} - b_F\{X/2J(J+1)\}\{3/4\} + t_0\{3X/8J(J+1)\}$$
$$+ (eq_0Q/2)\{Y/(2J-1)2J(2J+2)(2J+3)\}\{27/4 - J(J+1)\},$$

$$m_{22} = -a_\Pi\{3X/4J(J+1)\} - b_F\{X/2J(J+1)\}\{3/4\} + t_0\{3X/8J(J+1)\}$$
$$+ (eq_0Q/2)\{Y/(2J-1)2J(2J+2)(2J+3)\}\{27/4 - J(J+1)\},$$

$$m_{33} = -a_\Pi\{3X/4J(J+1)\} + b_F\{X/2J(J+1)\}\{1/4\} - t_0\{X/8J(J+1)\}$$
$$+ (eq_0Q/2)\{Y/(2J-1)2J(2J+2)(2J+3)\}\{3/4 - J(J+1)\},$$

$$m_{44} = -a_\Pi\{3X/4J(J+1)\} + b_F\{X/2J(J+1)\}\{1/4\} - t_0\{X/8J(J+1)\}$$
$$+ (eq_0Q/2)\{Y/(2J-1)2J(2J+2)(2J+3)\}\{3/4 - J(J+1)\},$$

$$m_{55} = -b_{F(\Sigma)}\{X/2J(J+1)\}\{1/4\} + t_{0(\Sigma)}\{X/8J(J+1)\}$$
$$+ (eq_{0\Sigma}Q/2)\{Y/(2J-1)2J(2J+2)(2J+3)\}\{3/4 - J(J+1)\},$$

$$m_{66} = -b_{F(\Sigma)}\{X/2J(J+1)\}\{1/4\} + t_{0(\Sigma)}\{X/8J(J+1)\}$$
$$+ (eq_{0\Sigma}Q/2)\{Y/(2J-1)2J(2J+2)(2J+3)\}\{3/4 - J(J+1)\}.$$

(8.384)

$$m_{13} = m_{31} = -b_F\{X/4J(J+1)\}\{(J+3/2)(J-1/2)\}^{1/2}$$
$$- t_0\{X/8J(J+1)\}\{(J+3/2)(J-1/2)\}^{1/2}$$

$$m_{14} = m_{41} = eq_2Q\{Y/16(2J-1)J(J+1)(2J+3)\}\{J+1/2\}\{(J+3/2)(J-1/2)\}^{1/2}$$

$$m_{15} = m_{51} = -\langle\Pi|aL_+|\Sigma\rangle\{X/4J(J+1)\}\{(J-1/2)(J+3/2)\}^{1/2}$$
$$+ [t_{+1} + t_{-1}]\{3X/8J(J+1)\}\{(J+3/2)(J-1/2)\}^{1/2}$$

$$\begin{aligned}
m_{23} = m_{32} &= eq_2 Q\{Y/16(2J-1)J(J+1)(2J+3)\}\{J+1/2\}\{(J+3/2)(J-1/2)\}^{1/2}\\
m_{24} = m_{42} &= -b_F\{X/4J(J+1)\}\{(J+3/2)(J-1/2)\}^{1/2}\\
&\quad - t_0\{X/8J(J+1)\}\{(J+3/2)(J-1/2)\}^{1/2}\\
m_{26} = m_{62} &= -\langle\Pi|aL_+|\Sigma\rangle\{X/4J(J+1)\}\{(J-1/2)(J+3/2)\}^{1/2}\\
&\quad + [t_{+1}+t_{-1}]\{3X/8J(J+1)\}\{(J+3/2)(J-1/2)\}^{1/2}\\
m_{34} = m_{43} &= [t_{+2}-t_{-2}]\{3X/8J(J+1)\}\{J+1/2\}\\
m_{35} = m_{53} &= t_{+1}\{3X/8J(J+1)\}\\
m_{36} = m_{63} &= -\langle\Pi|aL_+|\Sigma\rangle\{X/4J(J+1)\}\{J+1/2\}\\
&\quad + [t_{+1}+t_{-1}]\{3X/8J(J+1)\}\{J+1/2\}\\
m_{45} = m_{54} &= -\langle\Pi|aL_+|\Sigma\rangle\{X/4J(J+1)\}\{J+1/2\}\\
&\quad + [t_{+1}+t_{-1}]\{3X/8J(J+1)\}\{J+1/2\}\\
m_{46} = m_{64} &= -t_{-1}\{3X/8J(J+1)\}\\
m_{56} = m_{65} &= -b_F\{X/4J(J+1)\}\{J+1/2\} - t_{0\Sigma}\{X/8J(J+1)\}\{J+1/2\}.
\end{aligned}$$
(8.385)

The dipolar hyperfine constants are defined by

$$\begin{aligned}
t_0 &= g_S\mu_B g_N\mu_N(\mu_0/4\pi)\frac{1}{2}\left\langle\frac{(3\cos^2\theta-1)}{r^3}\right\rangle_\Pi,\\
t_{+1} &= t_{-1} = g_S\mu_B g_N\mu_N(\mu_0/4\pi)\left\langle\frac{\cos\theta\sin\theta}{r^3}\right\rangle_{\Pi-\Sigma},\\
t_{+2} &= t_{-2} = g_S\mu_B g_N\mu_N(\mu_0/4\pi)\left\langle\frac{\sin^2\theta}{r^3}\right\rangle_\Pi.
\end{aligned}$$
(8.386)

There are also non-vanishing matrix elements of the quadrupole interaction ($q = \pm 1$) connecting the Π and Σ states, but these were found to be insignificant by Freund, Herbst, Mariella and Klemperer [112].

It is now a simple matter to use the above results for the primitive basis functions to generate matrix elements for the parity-conserved basis. For the *positive*-parity states the hyperfine matrix is as follows.

	$^2\Pi^{(+)}_{3/2}$	$^2\Pi^{(+)}_{1/2}$	$^2\Sigma^{(+)}_{1/2}$
$^2\Pi^{(+)}_{3/2}$	$\frac{1}{2}(m_{11}+m_{22})$	$\frac{1}{2}(-1)^{J-1/2}(m_{14}+m_{23})$	$\frac{1}{2}(m_{15}+m_{26})$
$^2\Pi^{(+)}_{1/2}$	$\frac{1}{2}(-1)^{J-1/2} \times (m_{41}+m_{32})$	$\frac{1}{2}(m_{33}+m_{44})$	$\frac{1}{2}\{m_{35}+(-1)^{J-1/2} \times (m_{36}+m_{45})+m_{46}\}$
$^2\Sigma^{(+)}_{1/2}$	$\frac{1}{2}(m_{51}+m_{62})$	$\frac{1}{2}\{m_{53}+(-1)^{J-1/2} \times (m_{63}+m_{54})+m_{64}\}$	$\frac{1}{2}(m_{55}+m_{66})$

For the *negative*-parity states it is as follows.

	$^2\Pi_{3/2}^{(-)}$	$^2\Pi_{1/2}^{(-)}$	$^2\Sigma_{1/2}^{(-)}$
$^2\Pi_{3/2}^{(-)}$	$\frac{1}{2}(m_{11} + m_{22})$	$-\frac{1}{2}(-1)^{J-1/2}(m_{14} + m_{23})$	$\frac{1}{2}(m_{15} + m_{26})$
$^2\Pi_{1/2}^{(-)}$	$-\frac{1}{2}(-1)^{J-1/2}$ $\times (m_{41} + m_{32})$	$\frac{1}{2}(m_{33} + m_{44})$	$\frac{1}{2}\{m_{35} - (-1)^{J-1/2}$ $\times (m_{36} + m_{45}) + m_{46}\}$
$^2\Sigma_{1/2}^{(-)}$	$\frac{1}{2}(m_{51} + m_{62})$	$\frac{1}{2}\{m_{53} - (-1)^{J-1/2}$ $\times (m_{63} + m_{54}) + m_{64}\}$	$\frac{1}{2}(m_{55} + m_{66})$

These results agree with those given by Freund, Herbst, Mariella and Klemperer [112] in their equations (10), (11) and (12).

One final piece of information obtained from the electric resonance spectrum of LiO was the electric dipole moment, which was determined to be 6.84 D.

(g) Values of the molecular hyperfine constants

Freund, Herbst, Mariella and Klemperer [112] expressed their magnetic hyperfine constants in the form originally given by Frosch and Foley [117]. As discussed elsewhere in this book, particularly in chapters 9, 10 and 11, we prefer to separate the different physical interactions, particularly the Fermi contact and dipolar interactions, in our effective Hamiltonian. This separation is usually made by other authors even when the effective Hamiltonian is expressed in terms of Frosch and Foley constants, because it is the natural route if the molecular physics of a problem is to be understood. Nevertheless since so many authors, particularly of the earlier papers, use the magnetic hyperfine theory presented by Frosch and Foley, we present in appendix 8.5 a detailed comparison of their effective Hamiltonian with that adopted in this book. The merit of the Frosch and Foley parameters is that they form the linear combination of parameters which is best determined (i.e. with least correlation) for a molecule which conforms to Hund's case (a) coupling. The values of the constants determined experimentally from the ^{7}LiO spectrum were therefore, in our notation (in MHz):

$$a_\Pi = 6.12, \quad b_F = -14.54, \quad t_0 = 15.4, \quad t_2 = 1.94.$$

The simplest description of the LiO molecule would be in terms of an ionic complex, Li$^+$O$^-$, so that the orbital and dipolar hyperfine constants would be those of the O$^-$ atomic ion. Given a suitable wave function for an atomic $2p$ orbital located on the oxygen atom, it is then a straightforward matter to calculate the dipolar constants above, with the following results (in MHz):

$$a_\Pi = 2\mu_B g_N \mu_N (\mu_0/4\pi) \left\langle \frac{1}{r^3} \right\rangle = 5.4,$$

$$t_0 = g_S \mu_B g_N \mu_N (\mu_0/4\pi) \frac{1}{2} \left\langle \frac{(3\cos^2\theta - 1)}{r^3} \right\rangle = 13.2, \quad (8.387)$$

$$t_2 = g_S \mu_B g_N \mu_N (\mu_0/4\pi) \left\langle \frac{\sin^2\theta}{r^3} \right\rangle = 1.0.$$

These agree rather well with the experimental values listed above, suggesting that the ionic model is a good one. On the other hand, the negative Fermi contact constant can only arise through polarisation of the electron spins in a covalent bond between the two atoms. The electric dipole moment also seems to be inconsistent with a purely ionic model, yet the quadrupole coupling constant $eq_0 Q$ is very close to that of the ionic molecule LiF.

The pure microwave rotational spectrum of LiO was measured seventeen years later by Yamada, Fujitake and Hirota [118], and will be discussed in chapter 10. Their conclusions were generally in agreement with those based on the earlier electric resonance spectrum, except for a reassignment of some of the transitions.

(h) The Λ-doubling constants and frequencies

Λ-doubling is a characteristic feature of electronic states possessing orbital degeneracy, which has been discussed extensively by many authors, particularly van Vleck [119], Mulliken and Christy [120], and Dousmanis, Sanders and Townes [121]. We shall discuss the details at some length in our descriptions of the spectra of NO and, particularly, OH. In the case of LiO described above, the important matrix elements are those defined earlier as M_1 and M_2, i.e.

$$M_1 = \langle \Pi | BL_+ | \Sigma \rangle, \quad M_2 = \langle \Pi | (A/2 + B) | \Sigma \rangle. \quad (8.388)$$

For LiO it is sufficient to include mixing of the $^2\Pi$ ground state with a single $^2\Sigma^+$ excited state which happens to be particularly low-lying in energy. Ab initio calculations of the matrix elements by Cooper and Richards [122] gave results in satisfactory agreement with experiment. More generally the calculation of Λ-doubling frequencies involves the use of perturbation theory, and summations over more than one excited electronic state, as we described in chapter 7. Further details of this are presented later in this chapter, and also in chapter 9, where we describe the magnetic resonance studies of the OH radical.

A particularly simple approach to handling the matrix elements in (8.388) involves what is known as van Vleck's model of pure precession [119], see section 7.8. Applied to LiO the model assumes L to be a good quantum number, with $L = 1$ since the ground state of O^- is ^2P. As a result of this assumption,

$$\langle \Pi | L_+ | \Sigma \rangle = \langle L = 1, \Lambda = 1 | L_+ | L = 1, \Lambda = 0 \rangle = \sqrt{2}. \quad (8.389)$$

Furthermore, consistent with the pure precession model, one can factorise the M_1 and M_2 matrix elements, with the results,

$$\langle \Pi | (A/2) L_+ | \Sigma \rangle = (A/2) \langle \Pi | L_+ | \Sigma \rangle = A/\sqrt{2},$$
$$\langle \Pi | B L_+ | \Sigma \rangle = (1/2)(B_\Pi + B_\Sigma) \langle \Pi | L_+ | \Sigma \rangle = (1/\sqrt{2})(B_\Pi + B_\Sigma). \quad (8.390)$$

Using their experimental results and the relationships above, Freund, Herbst, Mariella and Klemperer [112] obtained the value 1.407 for the matrix element of L_+, which is very close to $\sqrt{2}$.

8.5.3. NO in the $X\,^2\Pi$ ground state

(a) Introduction

Nitric oxide, NO, is a chemically stable molecule and not surprisingly has been studied extensively by a range of techniques. Its microwave and far-infrared laser magnetic resonance spectra are discussed in chapter 9. These involve an understanding of both the zero-field levels and also the interactions with an external magnetic field. The pure microwave and millimetre wave spectra are described in chapter 10, but they provide information, which we will use, relevant to the radiofrequency electric resonance spectrum described in this section.

The first description of the radiofrequency electric resonance spectrum of $^{14}N^{16}O$ was provided by Neumann [123], and his work was followed by an extremely comprehensive study, which included the isotopic species $^{15}N^{16}O$, described by Meerts and Dymanus [124]. Later refinements of the theory were provided by Meerts [125] and Kristiansen [126]. The analysis of the Λ-doubling and hyperfine electric resonance spectrum depended on measurements of rotational level spacings obtained from microwave and millimetre wave spectra recorded by Burrus and Gordy [127], Gallagher and Johnson [116], and Favero, Mirri and Gordy [128]. Far-infrared transitions which probed some of the higher energy rotational levels were studied by Hall and Dowling [129], and direct transitions between the fine structure states were observed by Brown, Cole and Honey [130]. In this section we concentrate on the results obtained by Meerts and Dymanus [124].

The ground state molecular orbital configuration of NO is

$$X\,^2\Pi : (1\sigma)^2 (2\sigma)^2 (3\sigma)^2 (4\sigma)^2 (5\sigma)^2 (1\pi)^4 (2\pi)^1. \quad (8.391)$$

Apart from a lower-lying quartet state, the first excited electronic state is the $A\,^2\Sigma^+$ state, formed by promotion of an electron into the 6σ orbital, and lying 43 966 cm^{-1} above the ground state. Lying within the next 25 000 cm^{-1} there are many other excited doublet states; there is no excited state lying very close to the ground state, as was the case described for LiO, so we can expect the theory of the Λ-doubling to be somewhat more complicated. The ground state spin–orbit coupling constant A is

+123.1394 cm^{-1}, so that in this case the lowest fine-structure component is the $^2\Pi_{1/2}$, unlike that of LiO. In figure 8.42 we indicate schematically the lower rotational levels for both fine-structure components; this diagram is not to scale, the Λ-doublet splittings being exaggerated for the sake of clarity. The rotational levels observed by Meerts and Dymanus [124] are specified in the figure caption, the transitions measured being Λ-doublet transitions within a rotational level, further split by hyperfine interaction involving the ^{14}N nucleus ($I = 1$) or the ^{15}N nucleus ($I = 1/2$) in the case of artificially enriched ^{15}N^{16}O. Figure 8.43 shows the Λ-doublet and ^{14}N hyperfine splittings, as well as the observed transitions, for the $J = 13/2$ level of the $^2\Pi_{3/2}$ state. Similar sets of transitions were observed for other rotational levels. The molecular beam apparatus, described by de Leeuw and Dymanus [113], contained quadrupole A and B fields; a homogeneous electric C field is not required because the Λ-doublet transitions have intrinsic electric dipole intensity. Beam detection was accomplished by electron impact ionisation, and electron multiplier detection which was preceded by a quadrupole mass filter. Broadening effects due to the earth's magnetic field were minimised by suitable screening, and line widths of less than 10 kHz were obtained for most transitions. Meerts and Dymanus [124] provided a comprehensive list of transition frequencies for J values up to 7/2 in the $\Omega = 1/2$ state and 17/2 in the $\Omega = 3/2$ state.

(b) Theory of the Λ-doubling

The analysis of the spectrum was accomplished using a case (a) basis although the increasing tendency towards case (b) coupling as J increases must be taken into account. We can, therefore, make use of most of the matrix elements derived in our earlier discussion of the LiO spectrum. The first-order rotational energies were shown to be given by equation (8.363):

$$\langle \eta, \Lambda; S, \Sigma; J, \Omega | \mathcal{H}^{(0)} | \eta', \Lambda'; S, \Sigma'; J', \Omega' \rangle$$
$$= \delta_{\eta\eta'}\delta_{\Lambda\Lambda'}\delta_{\Sigma\Sigma'}\delta_{\Omega\Omega'}\{(A + 2B)\Lambda\Sigma + B$$
$$\times [J(J+1) + S(S+1) + \Lambda^2 - 2\Omega^2]\}, \tag{8.392}$$

and the rotational mixing of the fine structure states (rotational distortion) was given by

$$\langle \eta, \Lambda; S, \Sigma; J, \Omega | -2B \sum_{q=\pm 1} T_q^1(\boldsymbol{J}) T_q^1(\boldsymbol{S}) | \eta', \Lambda'; S, \Sigma'; J', \Omega' \rangle$$

$$= -2B\delta_{\eta\eta'}\delta_{\Lambda\Lambda'}\delta_{JJ'} \sum_{q=\pm 1} (-1)^{J-\Omega+S-\Sigma} \begin{pmatrix} J & 1 & J \\ -\Omega & q & \Omega' \end{pmatrix} \begin{pmatrix} S & 1 & S \\ -\Sigma & q & \Sigma' \end{pmatrix}$$
$$\times \{J(J+1)(2J+1)S(S+1)(2S+1)\}^{1/2}. \tag{8.393}$$

This is the term which causes the transition from case (a) to case (b). The terms of interest in the present subsection are those which remain, given in equation (8.365),

$$\mathcal{H}'' = \sum_{q=\pm 1} \{(-1)^q (A + 2B) T_q^1(\boldsymbol{L}) T_{-q}^1(\boldsymbol{S}) - 2B T_q^1(\boldsymbol{J}) T_q^1(\boldsymbol{L})\}. \tag{8.394}$$

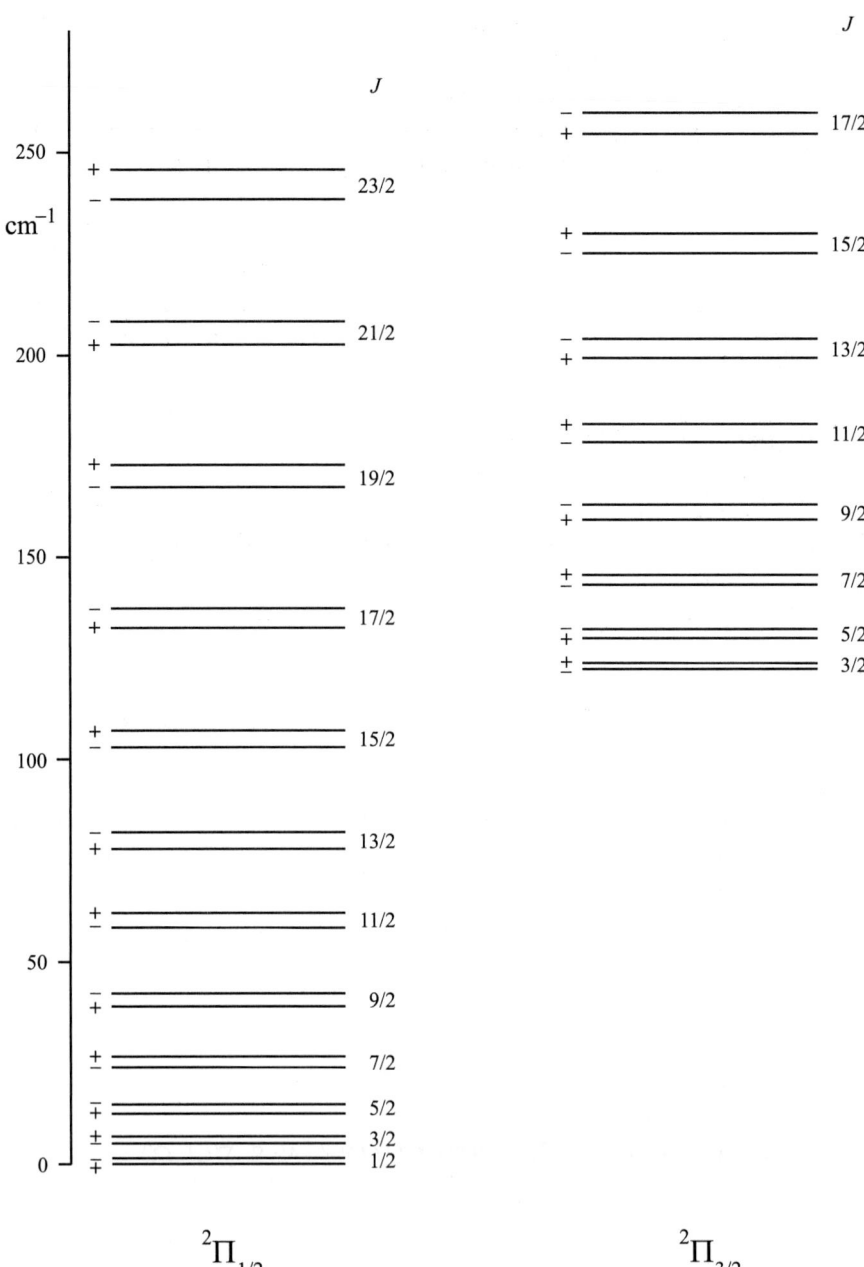

Figure 8.42. Schematic diagram of the lower rotational levels of NO, showing the Λ-doublet splittings exaggerated for the sake of clarity, but not drawn correctly to scale. The levels of ^{14}N^{16}O studied by Meerts and Dymanus [124] were $^2\Pi_{1/2}$, $J = 1/2$ to $7/2$, and $^2\Pi_{3/2}$, $J = 3/2$ to $17/2$.

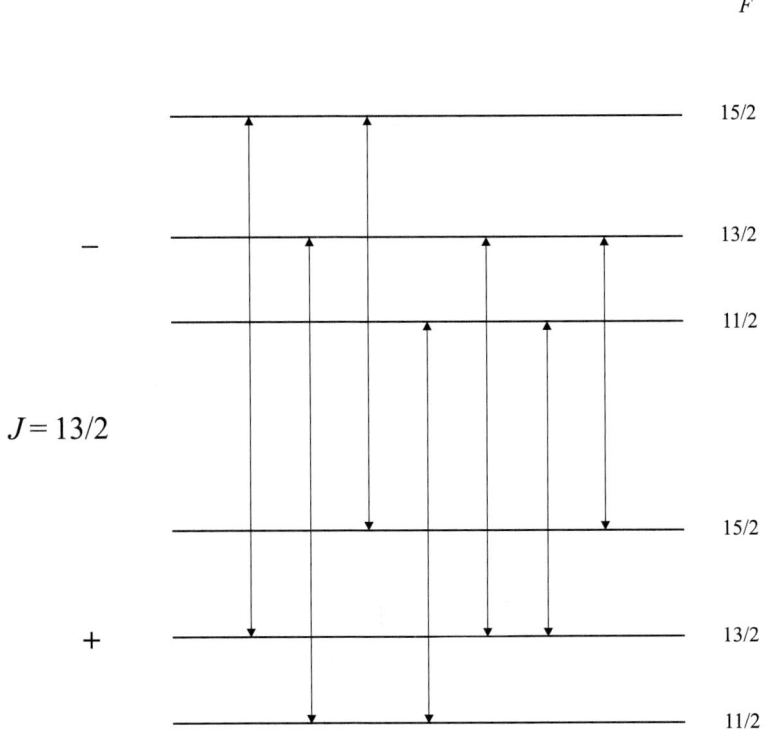

Figure 8.43. Λ-doubling and ^{14}N hyperfine splitting for the $^2\Pi_{3/2}$, $J = 13/2$ rotational level of NO, and the transitions studied by Meerts and Dymanus [124].

The matrix elements of the two terms in (8.394) were obtained as follows:

$$\langle \eta, \Lambda; S, \Sigma; J, \Omega | \sum_{q=\pm 1} (-1)^q (A + 2B) T_q^1(L) T_{-q}^1(S) | \eta', \Lambda'; S, \Sigma'; J, \Omega' \rangle$$

$$= -2\delta_{\Omega\Omega'} \sum_{q=\pm 1} \langle \eta, \Lambda | (A/2 + B) T_q^1(L) | \eta', \Lambda' \rangle (-1)^{S-\Sigma}$$

$$\times \begin{pmatrix} S & 1 & S \\ -\Sigma & -q & \Sigma' \end{pmatrix} \{S(S+1)(2S+1)\}^{1/2}, \quad (8.395)$$

$$\langle \eta, \Lambda; S, \Sigma; J, \Omega | - \sum_{q=\pm 1} 2B T_q^1(J) T_q^1(L) | \eta', \Lambda'; S, \Sigma'; J, \Omega' \rangle$$

$$= -2\delta_{\Sigma\Sigma'} \sum_{q=\pm 1} \langle \eta, \Lambda | B T_q^1(L) | \eta', \Lambda' \rangle (-1)^{J-\Omega} \begin{pmatrix} J & 1 & J \\ -\Omega & q & \Omega' \end{pmatrix}$$

$$\times \{J(J+1)(2J+1)\}^{1/2}. \quad (8.396)$$

In the case of LiO we considered mixing with only a single excited vibronic state η', namely, the low-lying $^2\Sigma^+$ state. We also ignored the fact that there should be a sum over all of the vibrational levels of the excited state. In general there should be a sum over all vibrational levels of all relevant excited electronic states, which leads to the definition of Λ-doubling constants for $^2\Pi$ states, given by Mulliken and Christy [120] following earlier work by Van Vleck [119]. These constants (q and p) were defined as follows:

$$q = C_1 - C_2 \quad \text{where}$$
$$C_1 = 8\sum_{\eta'} |\langle \eta, \Lambda |BL_x|\eta', \Lambda'\rangle|^2/(E_\eta - E_{\eta'})$$

where η' represents all v levels of all $^2\Sigma^+$ states, (8.397)

$$C_2 = 8\sum_{\eta'} |\langle \eta, \Lambda |BL_x|\eta', \Lambda'\rangle|^2/(E_\eta - E_{\eta'})$$

where η' represents all v levels of all $^2\Sigma^-$ states.

$$p = a_1 - a_2 \quad \text{where}$$
$$a_1 = 8\sum_{\eta'} |\langle \eta, \Lambda |AL_x|\eta', \Lambda'\rangle\langle\eta', \Lambda'|BL_x|\eta, \Lambda\rangle|/(E_\eta - E_{\eta'})$$

for all v levels of all $^2\Sigma^+$ states, (8.398)

$$a_2 = 8\sum_{\eta'} |\langle \eta, \Lambda |AL_x|\eta', \Lambda'\rangle\langle\eta', \Lambda'|BL_x|\eta, \Lambda\rangle|/(E_\eta - E_{\eta'})$$

for all v levels of all $^2\Sigma^-$ states.

The Λ-doubling constants may then be written in spherical tensor form as

$$q = 4\sum_{\eta'} (-1)^s |\langle \eta, \Lambda = 1|BT_1^1(L)|\eta', \Lambda' = 0^s\rangle|^2/(E_\eta - E_{\eta'}),$$
$$p = -4\sum_{\eta'} (-1)^s \langle \eta, \Lambda = 1|AT_1^1(L)|\eta', \Lambda' = 0^s\rangle \qquad (8.399)$$
$$\times \langle \eta', \Lambda' = 0^s|BT_{-1}^1(L)|\eta, \Lambda = 1\rangle/(E_\eta - E_{\eta'}).$$

In some other papers they are expressed in terms of the shift operators. The superscript s on the zero distinguishes between Σ^+ and Σ^- states; the phase factor $(-1)^s$ is $+1$ for the former and -1 for the latter.

Our objective is to replace the Hamiltonian terms in (8.394) by an effective Hamiltonian which *operates only* within the $^2\Pi$ state, but which contains the molecular parameters p and q describing the admixture of excited $^2\Sigma$ states. As we discussed in chapter 7, a suitable effective Hamiltonian for case (a), given by Brown and Merer [132] is

$$\mathcal{H}_{LD} = \frac{1}{2}q_v(J_+^2 + J_-^2) - \frac{1}{2}(p_v + 2q_v)(J_+S_+ + J_-S_-) + \frac{1}{2}(o_v + p_v + q_v)(S_+^2 + S_-^2).$$
(8.400)

This Hamiltonian has been generalised so that it applies to Π states of any multiplicity. It may be readily rewritten in terms of spherical tensor operators as follows:

$$\mathcal{H}_{LD} = \sum_{q=\pm 1} \{q_v T_{2q}^2(J, J) - (p_v + 2q_v)T_{2q}^2(J, S) + (o_v + p_v + q_v)T_{2q}^2(S, S)\}.$$
(8.401)

The third term in (8.401) has only zero diagonal matrix elements for doublet spin states and will therefore be disregarded. The matrix elements of (8.401) may now be calculated for the parity-conserved basis functions of the $^2\Pi$ fine-structure components (8.368). It is important to note that all matrix elements must satisfy the requirement $\Delta \Lambda = \pm 2$. The matrix elements of the first term in (8.401) are calculated by making use of the Wigner–Eckart theorem, and noting the evaluation of the reduced matrix element,

$$\langle J \| T^2(\boldsymbol{J}, \boldsymbol{J}) \| J \rangle = \frac{1}{2\sqrt{6}} \{(2J+3)(2J+2)(2J+1)(2J)(2J-1)\}^{1/2}, \quad (8.402)$$

as described in appendix 8.3. The matrix elements of the second term in (8.401) are obtained by using the result

$$T_q^2(\boldsymbol{J}, \boldsymbol{S}) = (-1)^q \sum_{q_1, q_2} \sqrt{5}\, T_{q_1}^1(\boldsymbol{J})\, T_{q_2}^1(\boldsymbol{S}) \begin{pmatrix} 1 & 1 & 2 \\ q_1 & q_2 & -q \end{pmatrix}. \quad (8.403)$$

We also take account of the anomalous commutation rules for the components of $\boldsymbol{J}$ in the molecule-fixed coordinate system by making the replacement

$$T_q^1(\boldsymbol{J}) \rightarrow (-1)^q T_{-q}^1(\boldsymbol{J}), \quad (8.404)$$

which should be regarded as no more than a successful recipe (see chapter 5)!

There is a further term which should be included in the effective Hamiltonian, derived in chapter 7, describing the electron spin–nuclear rotation interaction. This may be written in the form

$$\mathcal{H}_{sr} = \gamma \{T^1(\boldsymbol{J}) - T^1(\boldsymbol{S})\} \cdot T^1(\boldsymbol{S}) = \gamma T^1(\boldsymbol{J}) \cdot T^1(\boldsymbol{S}) - \gamma S(S+1). \quad (8.405)$$

The $q = \pm 1$ components of the first term on the right-hand side of this equation represents a correction to the so-called rotational distortion term, whose matrix elements were given in equations (8.363) and (8.364). In the light of this term we should replace $-2B$ by $(-2B + \gamma)$ in the off-diagonal elements (8.364), add nothing to the energy of the $^2\Pi_{3/2}$ component, and subtract γ from the energy of the $^2\Pi_{1/2}$ component.

The 2×2 Λ-doubling matrix for *positive*-parity components is as follows.

	$^2\Pi_{3/2}^{(+)}$	$^2\Pi_{1/2}^{(+)}$
$^2\Pi_{3/2}^{(+)}$	0	$(-1)^{J-1/2}\left[\dfrac{q_v}{2}\right]\left[J+\dfrac{1}{2}\right]\left[\left(J+\dfrac{1}{2}\right)^2 - 1\right]^{1/2}$
$^2\Pi_{1/2}^{(+)}$	$(-1)^{J-1/2}\left[\dfrac{q_v}{2}\right]\left[J+\dfrac{1}{2}\right]\left[\left(J+\dfrac{1}{2}\right)^2 - 1\right]^{1/2}$	$-(-1)^{J-1/2}\dfrac{1}{2}[p_v + 2q_v]\left[J+\dfrac{1}{2}\right]$

For the *negative*-parity states it is as follows.

	$^2\Pi_{3/2}^{(-)}$	$^2\Pi_{1/2}^{(-)}$
$^2\Pi_{3/2}^{(-)}$	0	$-(-1)^{J-1/2}\left[\dfrac{q_v}{2}\right]\left[J+\dfrac{1}{2}\right]\left[\left(J+\dfrac{1}{2}\right)^2 - 1\right]^{1/2}$
$^2\Pi_{1/2}^{(-)}$	$-(-1)^{J-1/2}\left[\dfrac{q_v}{2}\right]\left[J+\dfrac{1}{2}\right]\left[\left(J+\dfrac{1}{2}\right)^2 - 1\right]^{1/2}$	$(-1)^{J-1/2}\dfrac{1}{2}[p_v + 2q_v]\left[J+\dfrac{1}{2}\right]$

Comparison of these matrices shows at once that the Λ-doublet splitting in the $^2\Pi_{1/2}$ fine-structure state is determined primarily by the diagonal elements, and will therefore be much larger than in the $^2\Pi_{3/2}$ component which is only affected in second order. It will also increase linearly with J in the $^2\Pi_{1/2}$ component.

Complete matrices for the parity-conserved $^2\Pi$ fine-structure states (exclusive of nuclear spin terms) may now be constructed by combining the Λ-doubling matrices given above with the spin–orbit and rigid body rotation matrices given in our discussion of the LiO spectrum. The matrix representation is block diagonal for each value of J and each parity. The results for the positive and negative parity states are as follows.

	$^2\Pi_{3/2}^{(+)}$	$^2\Pi_{1/2}^{(+)}$	$^2\Pi_{3/2}^{(-)}$	$^2\Pi_{1/2}^{(-)}$
$^2\Pi_{3/2}^{(+)}$	m_{11}	m_{12}	0	0
$^2\Pi_{1/2}^{(+)}$	m_{21}	m_{22}	0	0
$^2\Pi_{3/2}^{(-)}$	0	0	m_{33}	m_{34}
$^2\Pi_{1/2}^{(-)}$	0	0	m_{43}	m_{44}

The matrix elements are given by

$$m_{11} = \frac{A_v}{2} + B_v\{J(J+1) - \tfrac{7}{4}\}$$
$$m_{22} = -\frac{A_v}{2} + B_v\{J(J+1) + \tfrac{1}{4}\} - \gamma_v - (-1)^{J-1/2}\tfrac{1}{2}[p_v + 2q_v][J + \tfrac{1}{2}]$$
$$m_{12} = -\{(J+\tfrac{3}{2})(J-\tfrac{1}{2})\}^{1/2}\{(B_v - \tfrac{\gamma_v}{2}) - (-1)^{J-1/2}[\tfrac{q_v}{2}][J + \tfrac{1}{2}]\}$$
$$m_{21} = m_{12}$$
$$m_{33} = \frac{A_v}{2} + B_v\{J(J+1) - \tfrac{7}{4}\} \tag{8.406}$$
$$m_{44} = -\frac{A_v}{2} + B_v\{J(J+1) + \tfrac{1}{4}\} - \gamma_v + (-1)^{J-1/2}\tfrac{1}{2}[p_v + 2q_v][J + \tfrac{1}{2}]$$
$$m_{34} = -\{(J+\tfrac{3}{2})(J-\tfrac{1}{2})\}^{1/2}\{(B_v - \tfrac{\gamma_v}{2}) + (-1)^{J-1/2}[\tfrac{q_v}{2}][J + \tfrac{1}{2}]\}$$
$$m_{43} = m_{34}.$$

At this stage of the theoretical analysis there are no matrix elements off-diagonal J; this relatively simple situation will change when we consider nuclear spin magnetic and electric interactions.

(c) Nuclear hyperfine interactions

If there were no nuclear hyperfine interactions to be considered, we could now insert the values of the molecular constants which have been determined into the above matrices, and compare the calculated Λ-doublet splittings with those measured by Meerts and

Dymanus [124]. However the Λ-doubling transitions show extensive ^{14}N hyperfine splittings, which we must also take into account.

All of the magnetic and electric hyperfine matrix elements were derived in our discussion of the LiO spectrum. We now use the symbol m to denote nuclear spin-free terms, as listed above in equations (8.406), and denote the hyperfine terms, previously given in equations (8.384) and (8.385), by the symbol hf. Since the ^{14}N nucleus has spin $I = 1$, each J level is split into three hyperfine levels, characterised by $F = J$, $J \pm 1$, except for the $J = 1/2$ level which has only two hyperfine components, with $F = 1/2$ and $3/2$. Consequently if we neglect matrix elements off-diagonal in J for the moment, each characteristic set of J, F levels is described by a 4×4 matrix, or two 2×2 matrices, as follows.

	$^2\Pi^{(+)}_{3/2}$	$^2\Pi^{(+)}_{1/2}$	$^2\Pi^{(-)}_{3/2}$	$^2\Pi^{(-)}_{1/2}$
$^2\Pi^{(+)}_{3/2}$	$m_{11} + hf_{11}$	$m_{12} + hf_{12}$	0	0
$^2\Pi^{(+)}_{1/2}$	$m_{21} + hf_{21}$	$m_{22} + hf_{22}$	0	0
$^2\Pi^{(-)}_{3/2}$	0	0	$m_{33} + hf_{33}$	$m_{34} + hf_{34}$
$^2\Pi^{(-)}_{1/2}$	0	0	$m_{43} + hf_{43}$	$m_{44} + hf_{44}$

Note that the total parity is still conserved. Using equations (8.384) and (8.385) to obtain explicit expressions for the hf$_{ij}$ terms, we could derive analytic expressions for the total matrix elements, and ultimately for the energies of the hyperfine components. In the past this has usually been achieved using perturbation theory, which has led to the definition of a large number of molecular constants. Meerts and Dymanus [124] listed no fewer than 22 adjustable parameters, but Kristiansen [126] showed subsequently that a better fit to the observed spectrum could be obtained with 16 parameters, provided matrix elements of the hyperfine and quadrupole terms off-diagonal in J were taken into account. In principle this leads to a matrix of dimension 6×6 for each parity-conserved set of basis functions, but in practice it is sufficient to include only the mixing of each J with the $J \pm 1$ functions. Even so, the problem becomes quite complex. Consider, for example, the energy levels illustrated in figure 8.43, for which $J = 13/2$. For $F = 15/2$ and $11/2$ we have 4×4 matrices for each parity-conserved state, whilst for $F = 13/2$ we have 6×6 matrices.

Most recently in theoretical approaches to the analysis of NO spectra, Varberg, Stroh and Evenson [133] have used their tunable far-infrared measurements, described in chapter 10, to refine, yet again, the values of the constants. Using results from the Λ-doublet, rotational and fine-structure spectra of ^{14}NO ($v = 0$) which have been recorded, the following values of the constants (in MHz) are listed [133]:

$$B = 50848.13072, \quad A = 3691813.855, \quad \gamma = -193.9879, \quad p = 350.405,$$
$$q = 2.822100, \quad a = 84.20378, \quad b_F = 22.3792, \quad t_0 = -19.6273,$$
$$t_2 = 75.06479, \quad eq_0Q = -1.85671, \quad eq_2Q = 23.3115.$$

In addition several centrifugal distortion constants were listed; these become significant as J values increase and, as we shall see, are very important in the case of the OH radical. We will defer our discussion of centrifugal distortion until we deal with OH.

Let us first consider the simplest approach to the quantitative assignment of the radiofrequency spectrum for $J = 13/2$, the seven observed transitions being illustrated in figure 8.43. We neglect all matrix elements which are off-diagonal in J, which should be quite a good approximation since the different rotational (J) levels are widely separated in energy compared with the size of the nuclear hyperfine interaction terms. For each F level, with values $11/2$, $13/2$ and $15/2$, we have a pair of 2×2 matrices, one for each parity component as shown previously. The specific expressions for the hyperfine matrix elements are obtained from equations (8.372), (8.374), (8.380) and (8.382), using the parity-conserved basis functions. The nuclear spin-free terms m_{ij} are, of course, independent of the F value. The results are given below, the composite hyperfine constants being given by:

$$hf_1 = 2a + b_F + 2t_0 = 2h_{3/2}$$
$$hf_2 = 2a - b_F - 2t_0 = 2h_{1/2} \qquad (8.407)$$
$$bt = b_F - t_0.$$

For + parity levels ($J = 13/2$)

$m_{11} = A/2 + 47B$
$m_{22} = -A/2 + 49B - \gamma - 3.5(p + 2q)$
$m_{23} = -(B - \gamma/2)(4\sqrt{3}) + q(14\sqrt{3}) = m_{32}$

$F = 11/2$
$hf_{11} = -hf_1(3/26) - eq_0 Q(7/52)$
$hf_{22} = -hf_2(1/26) - eq_0 Q(2/13) + t_2(21/26)$
$hf_{12} = +eq_2 Q(7\sqrt{3}/156) - bt(4\sqrt{3}/13) = hf_{21}$

$F = 13/2$ $\qquad (8.408)$
$hf_{11} = -hf_1(1/65) + eq_0 Q(14/65)$
$hf_{22} = -hf_2(1/195) + eq_0 Q(16/65) + t_2(7/65)$
$hf_{12} = -eq_2 Q(14\sqrt{3}/195) - bt(8\sqrt{3}/195) = hf_{21}$

$F = 15/2$
$hf_{11} = +hf_1(1/10) - eq_0 Q(7/80)$
$hf_{22} = +hf_2(1/30) - eq_0 Q(1/10) - t_2(7/10)$
$hf_{12} = +eq_2 Q(7\sqrt{3}/240) + bt(4\sqrt{3}/15) = hf_{21}.$

For − parity levels ($J = 13/2$)

$m_{33} = A/2 + 47B$
$m_{44} = -A/2 + 49B - \gamma + 3.5(p + 2q)$
$m_{34} = -(B - \gamma/2)(4\sqrt{3}) - q(14\sqrt{3}) = m_{43}$

$F = 11/2$
$$\text{hf}_{33} = -hf_1(3/26) - eq_0Q(7/52)$$
$$\text{hf}_{44} = -hf_2(1/26) - eq_0Q(2/13) - t_2(21/26)$$
$$\text{hf}_{34} = -eq_2Q(7\sqrt{3}/156) - bt(4\sqrt{3}/13) = \text{hf}_{43}$$
$F = 13/2$ \hfill (8.409)
$$\text{hf}_{33} = -hf_1(1/65) + eq_0Q(14/65)$$
$$\text{hf}_{44} = -hf_2(1/195) + eq_0Q(16/65) - t_2(7/65)$$
$$\text{hf}_{34} = +eq_2Q(14\sqrt{3}/195) - bt(8\sqrt{3}/195) = \text{hf}_{43}$$
$F = 15/2$
$$\text{hf}_{33} = +hf_1(1/10) - eq_0Q(7/80)$$
$$\text{hf}_{44} = +hf_2(1/30) - eq_0Q(1/10) + t_2(7/10)$$
$$\text{hf}_{34} = -eq_2Q(7\sqrt{3}/240) + bt(4\sqrt{3}/15) = \text{hf}_{43}.$$

We are now in a position to compare experiment and theory. From the large data set provided by Meerts and Dymanus [124] we have selected the two sets of hyperfine components which involve the lowest rotational levels of the two fine structure states, and one involving the relatively high $J = 13/2$ level of the $\Omega = 3/2$ state. The experimental accuracy was quoted as 1 kHz. The simplest theoretical interpretation involves the 11 molecular parameters given earlier and the neglect of $\Delta J = \pm 1$ matrix elements; the agreement between experiment and theory is remarkably good, with a mean least squares difference (δ) of 0.025 MHz for $J = 13/2$. For the $J = 3/2$ transitions with $\Omega = 1/2$ and $3/2$ it is 0.028 and 0.008 MHz respectively. Columns 8 and 9 in table 8.23 show that inclusion of the $\Delta J = \pm 1$ matrix elements has very little effect on the results for $J = 13/2$ ($\delta = 0.026$ MHz), or for those with $J = 3/2, \Omega = 3/2$ ($\delta = 0.009$ MHz), but produces a marked improvement for the $J = 3/2, \Omega = 1/2$ results ($\delta = 0.007$ MHz).

An obvious improvement to the theory, which should be most significant for the $J = 13/2$ levels, would be the inclusion of centrifugal distortion terms. There is, however, another interaction which might be significant, namely, the nuclear spin–rotation term. This is represented by an additional term in the effective Hamiltonian,

$$\mathcal{H}_{nsr} = c_I \mathbf{T}^1(\mathbf{I}) \cdot \mathbf{T}^1(\mathbf{J} - \mathbf{S}) = c_I \{\mathbf{T}^1(\mathbf{I}) \cdot \mathbf{T}^1(\mathbf{J}) - \mathbf{T}^1(\mathbf{I}) \cdot \mathbf{T}^1(\mathbf{S})\}. \tag{8.410}$$

The second part of this term may be regarded as a small correction to the Fermi contact interaction; the matrix elements of the first part are diagonal in Ω, J and F, with matrix elements given by

$$\langle \eta, \Lambda; S, \Sigma; J, \Omega; I, F | c_I \mathbf{T}^1(\mathbf{J}) \cdot \mathbf{T}^1(\mathbf{I}) | \eta, \Lambda; S, \Sigma; J', \Omega; I, F \rangle$$
$$= \delta_{JJ'} c_I (-1)^{J+F+I} \begin{Bmatrix} I & J & F \\ J & I & 1 \end{Bmatrix} \{J(J+1)(2J+1)I(I+1)(2I+1)\}^{1/2}. \tag{8.411}$$

Table 8.23. *Comparison of experimental and calculated frequencies (in MHz) for the hyperfine components of three Λ-doublet transitions in NO. The final two columns (∗) denote the 12 parameter results obtained with the inclusion of the nuclear spin–rotation term*

Ω	J	$F(-)$	$F(+)$	ν_{exp}	ν_{calc} ($\Delta J=0$)	ν_{diff}	ν_{calc} (MHz) ($\Delta J=0, \pm 1$)	ν_{diff}	$\nu_{calc}(*)$	$\nu_{diff}(*)$
3/2	13/2	11/2	11/2	49.405	49.387	0.018	49.387	0.018	49.387	0.018
		13/2	13/2	48.578	48.553	0.025	48.553	0.025	48.553	0.025
		15/2	15/2	51.260	51.260	0.000	51.261	−0.001	51.261	−0.001
		15/2	13/2	63.640	63.539	0.101	63.534	0.106	63.622	0.018
		13/2	15/2	36.196	36.275	−0.079	36.280	−0.084	36.187	0.009
		13/2	11/2	59.742	59.645	0.097	59.642	0.100	59.722	0.020
		11/2	13/2	38.243	38.296	−0.053	38.299	−0.056	38.218	0.025
δ						0.025		0.026		0.007
1/2	3/2	1/2	1/2	560.854	560.820	0.034	560.851	0.003	560.851	0.003
		3/2	3/2	651.543	651.438	0.105	651.544	−0.001	651.544	−0.001
		5/2	5/2	801.196	801.147	0.049	801.193	0.003	801.193	0.003
		3/2	5/2	758.911	758.855	0.056	758.879	0.032	758.910	0.001
		5/2	3/2	693.828	693.731	0.097	693.858	−0.030	693.827	0.001
		1/2	3/2	624.649	624.554	0.095	624.630	0.019	624.649	0.000
		3/2	1/2	587.747	587.704	0.043	587.765	−0.018	587.746	0.001
δ						0.028		0.007		0.001
3/2	3/2	3/2	3/2	0.612	0.609	0.003	0.609	0.003	0.609	0.003
		5/2	5/2	1.029	1.030	−0.001	1.030	−0.001	1.030	−0.001
		3/2	5/2	74.931	74.904	0.027	74.901	0.030	74.932	−0.001
		5/2	3/2	73.286	73.265	0.021	73.262	0.024	73.293	−0.007
		1/2	3/2	46.464	46.443	0.021	46.438	0.026	46.456	0.008
δ						0.008		0.009		0.002

From this equation the diagonal corrections to the energies for $J = 13/2$, $F = 11/2$, 13/2 and 15/2 are found to be $-c_I(15/2)$, $-c_I$, and $c_I(13/2)$, respectively. The value of c_I given by Varberg, Stroh and Evenson [133] is 0.012 347 MHz, and inclusion of this term yields the results given in the last two columns of table 8.23. The agreement between experiment and theory for the low J levels, with a total of 12 molecular parameters, now approaches the experimental accuracy.

Further improvement in the agreement between experiment and theory, particularly for the higher J levels, would be obtained by the inclusion of centrifugal distortion terms. We will, however, defer discussion of these terms until we come to describe the results for the OH radical, where centrifugal distortion is much more important. Our reduction in the number of parameters required for a satisfactory quantitative interpretation of the radiofrequency electric resonance spectrum arises because of a number of factors. At the time of the measurements reported by Meerts and Dymanus [124], with the subsequent refinements by Meerts [125] and Kristiansen [126], the numerical analysis of such a spectrum depended heavily on the use of perturbation theory. Whilst perturbation theory still has a role to play, the development of computer programs which enable exact diagonalisation of large matrices, as well as least-squares refinement of molecular parameters, has greatly simplified the analytical process. Moreover in the case of the NO molecule, we now have accurate data from other spectroscopic studies which greatly assists the quantitative analysis. For example, the spin–orbit (A) and rotational (B) constants are not well determined from the Λ-doubling hyperfine spectrum described in this section. The spin–orbit interaction has been measured directly through observation of the $^2\Pi_{3/2} \leftarrow {}^2\Pi_{1/2}$ magnetic dipole spectrum in the far-infrared [130, 134] whilst the best determinations of the rotational and centrifugal distortion constants come from pure microwave [127, 128], millimetre wave [135] and far-infrared [136, 129] measurements. Infrared/microwave double resonance studies [137, 138] will be described in chapter 11. Magnetic resonance studies in both the microwave [139] and far-infrared [133] regions have provided a wealth of detail concerning the magnetic interactions in NO; these aspects are discussed at length in chapter 10. We should add that in many of the papers cited above, the spectra of the isotopic form ^{15}NO are also discussed.

(d) Interpretation of the molecular constants

As we have discussed many times elsewhere, the magnetic and electric nuclear hyperfine constants provide information about the electronic structure of the molecule. The latter was outlined and summarised briefly in (8.391), where the unpaired electron is placed in a π-type molecular orbital, which may be regarded as a linear combination of the N and O atomic $2p$ orbitals. Dousmanis [140] was among to first to show the relationships between the dipolar hyperfine constants and the electronic wave function. The orbital hyperfine constant, a, which in NO is found to have the value 84.20378 MHz, is given by the expression

$$a = 2\mu_B g_N \mu_N (\mu_0/4\pi) \left\langle \frac{1}{r^3} \right\rangle_{\text{av}} \tag{8.412}$$

where r is the electron–nuclear distance. Substitution of the values of the constants in the above definition leads to the value $\langle r^{-3} \rangle_{\text{av}} = 15.0 \times 10^{30}$ m^{-3}. Dousmanis calculates that for a Hartree–Fock $2p$ atomic orbital on the N atom, the theoretical value would be 22.5×10^{30} m^{-3}. There will also be a contribution from electron density in the O $2p$ atomic orbital but this will be much smaller because of the larger value of r; Dousmanis estimates a value of 0.49×10^{30} m^{-3}. The result of this simple analysis is that the molecular orbital containing the unpaired electron has a density of 0.65 on the N atom and 0.35 on the O atom. A check on this conclusion can be made by looking at the electron spin–nuclear spin dipolar constants, t_0 and t_2, which are given by the expressions

$$t_0 = g_S \mu_B g_N \mu_N (\mu_0/4\pi) \frac{1}{2} \left\langle \frac{(3\cos^2\theta - 1)}{r^3} \right\rangle,$$

$$t_2 = g_S \mu_B g_N \mu_N (\mu_0/4\pi) \left\langle \frac{\sin^2\theta}{r^3} \right\rangle.$$
(8.413)

For a $2p\pi$ electron ($m_l = \pm 1$) we can use the results

$$(3\cos^2\theta - 1)_{av} = -\frac{2}{5}, \quad (\sin^2\theta)_{av} = \frac{4}{5}.$$
(8.414)

Comparison with the value of the orbital hyperfine constant a therefore suggests that t_0 and t_2 should have the values -16.8 and 67.4 MHz, which are in quite good agreement with the experimental values -19.6273 and 75.06479 MHz, considering the simplicity of the model.

The Fermi contact constant b_F, which has the value 22.3792 MHz, gives a value for the wave function at the nucleus, $\psi^2(0)$, of 0.85×10^{30} m^{-3}. Although small, this value indicates that there is approximately 2.5% s character in the electronic wave function, which must arise from configuration interaction with appropriate higher energy electronic states.

Finally we come to the quadrupole coupling constant, eq_0Q, which Dousmanis interprets in terms of valence bond theory; the electric field gradient arises from all charged particles in the molecule and so its evaluation involves consideration of the total electronic wave function, not simply the molecular orbital containing the unpaired electron. Dousmanis advances plausible auguments to show that the observed value is consistent with the magnetic hyperfine results, but a semi-empirical treatment of this kind is always open to debate. As we have stated elsewhere for other molecules, the determined magnetic and electric hyperfine constants should probably best be regarded as benchmarks for testing the accuracy of *ab initio* calculations. This also applies to the Λ-doubling constants, the quantitative interpretation of which requires good wave functions for excited electronic states, as well as the ground state.

8.5.4. OH in the $X\,^2\Pi$ ground state

(a) Introduction

The hydroxyl radical, OH, occupies an extremely important position in spectroscopy, in free radical laboratory chemistry, and in atmospheric, cometary and interstellar chemistry. Its ultraviolet electronic spectrum has been described in many papers published over the past seventy years. It was the first short lived gaseous free radical to be studied by microwave spectroscopy, described in a classic paper by Dousmanis, Sanders and Townes [121] in 1955. The details of this work are presented in chapter 10. It was the first free radical to be studied by microwave magnetic resonance, in pioneering work by Radford [141]; the microwave and far-infrared laser magnetic resonance studies are described in detail in chapter 9. It was also the first free radical to be studied by molecular beam spectroscopy, using both electric resonance and maser detection methods; this work is the subject of the present section. OH will also appear again in chapter 11, as a very early example of the use of microwave/optical double resonance methods.

The ground state of OH is $^2\Pi$, arising from the molecular orbital configuration

$$X^2\Pi : (1s)^2(2s\sigma)^2(2p\sigma)^2(2p\pi)^3. \tag{8.415}$$

The lowest fine structure component is the $^2\Pi_{3/2}$ state, and the lower rotational levels for both OH and OD are shown schematically in figure 8.44. Each rotational level is split into a Λ-doublet, indicated in figure 8.44, but not at all to scale. Moreover, each Λ-doublet level in OH is further split into a doublet by magnetic hyperfine interaction with the proton. This is illustrated in figure 8.45 for the lowest rotational level, $J = 3/2$, and the observed electric dipole transitions are also shown; each rotational level possesses a similar Λ-doublet and hyperfine structure.

The high-resolution spectroscopy of OH has been perhaps the most important test bed for the development of the theory of the molecular energy levels, both in zero field and in the presence of applied magnetic fields. In this section, we concentrate on the Λ-doubling and hyperfine structure, as probed by the molecular beam studies. In chapter 9 we discuss the complex theory of the Zeeman effect, and in chapter 10 deal with rotational transitions. Our discussion therefore follows a pattern similar to that adopted for the NO molecule.

(b) Molecular beam electric resonance and beam maser studies

The first successful application of molecular beam electric resonance to the study of a short-lived free radical was achieved by Meerts and Dymanus [142] in their study of OH. They were also able to report spectra of OD, SH and SD. Their electric resonance instrument was conventional except for a specially designed free radical source, in which OH radicals were produced by mixing H atoms, formed from a microwave discharge in H_2, with NO_2 (or H_2S in the case of SH radicals). In table 8.24 we present a complete Λ-doublet data set for OH, including the sets determined by Meerts and Dymanus, with $J = 3/2$ to $11/2$ for the $^2\Pi_{3/2}$ state, and $1/2$ to $9/2$ for the $^2\Pi_{1/2}$ state. Notice that, for the lowest rotational level ($J = 3/2$ in $^2\Pi_{3/2}$), the accuracy of the data is higher. These transitions were observed by ter Meulen and Dymanus [143], not by electric resonance methods, but by beam maser spectroscopy, with the intention of providing particularly accurate data for astronomical purposes. This is the moment for a small diversion into the world of beam maser spectroscopy. It has been applied to a large number of polyatomic molecules, but apparently OH is the only diatomic molecule to be studied by this method.

The principles of a beam maser spectrometer are illustrated in figure 8.46. Beam maser spectroscopy depends upon the ability to create artificially a large population difference (and inversion) between two energy levels. As we know from our discussion of electric resonance, a quadrupole electric field achieves just this for levels which are

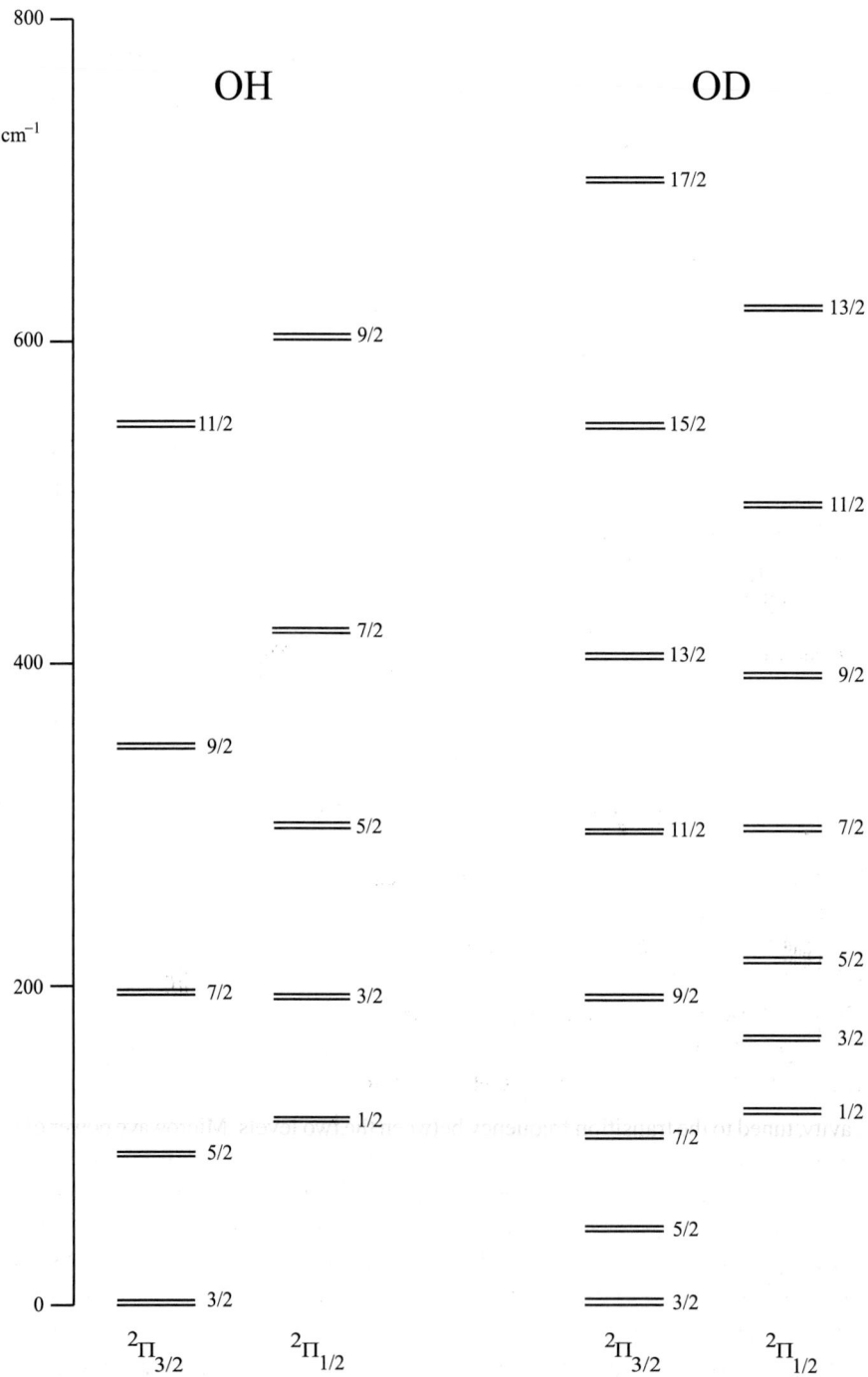

Figure 8.44. Lower rotational levels of OH and OD, labelled with their respective J values. The Λ-doubling is suggested only and the splittings are not at all drawn to scale.

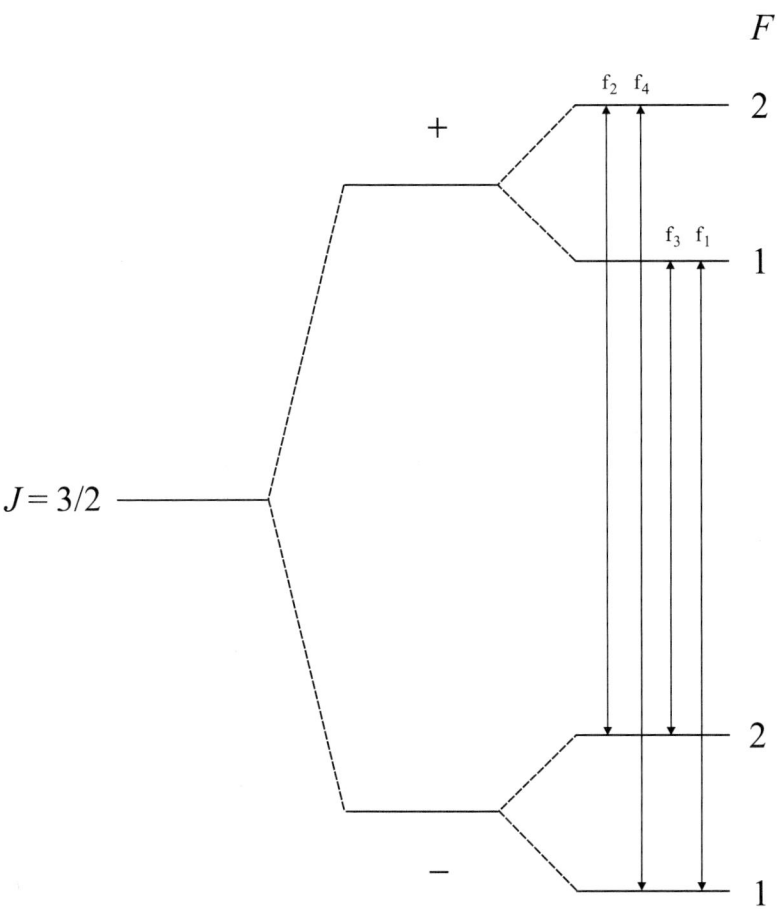

Figure 8.45. Λ-doublet and proton hyperfine splitting for the lowest rotational level in OH ($J = 3/2$, $^2\Pi_{3/2}$), and the allowed electric dipole transitions.

coupled through the Stark effect. In a beam maser spectrometer, therefore, the C and B fields which normally follow the A quadrupole field are replaced by a microwave cavity, tuned to the transition frequency between the two levels. Microwave power of the correct frequency is fed to the cavity; because of the population inversion, the induced transitions result in enhanced emission, which is detected. An advantage of the beam maser technique is that the molecular beam does not need to be as well collimated as in a beam deflection experiment, so that larger beam fluxes may be used. The disadvantage of the maser technique is that the range of frequency scanning is very small for a high Q resonance cavity; the transition frequency therefore needs to be known fairly accurately in advance. Ter Meulen and Dymanus took care to reduce residual earth magnetic field effects, and obtained line widths of 2.5 kHz, determined by the residence time of the beam molecules inside the microwave cavity. Their transition frequencies, given in table 8.24, are accurate to 0.1 kHz.

Table 8.24. Λ-doublet hyperfine transition frequencies recorded for the OH radical

Ω	J	F(e)[a]	F(f)[a]	ν_{obs} (MHz)	Reference
3/2	3/2	1	1	1665.401 84	[143]
		2	2	1667.359 03	[143]
		2	1	1612.231 01	[143]
		1	2	1720.529 98	[143]
	5/2	2	2	6030.747	[142]
		3	3	6035.092	[142]
		2	3	6016.746	[142]
		3	2	6049.084	[144]
	7/2	3	3	13 434.62	[145]
		4	4	13 441.36	[145]
	9/2	4	4	23 817.6153	[146]
		5	5	23 826.6211	[146]
		4	5	23 838.46	[145]
		5	4	23 805.13	[145]
	11/2	5	5	36 983.47	[145]
		6	6	36 994.43	[145]
1/2	1/2	1	1	4750.656	[144]
		0	1	4660.242	[144]
		1	0	4765.562	[144]
	3/2	1	1	7761.747	[147]
		2	2	7820.125	[147]
		2	1	7749.909	[147]
		1	2	7831.962	[147]
	5/2	2	2	8135.870	[142]
		3	3	8189.587	[142]
		2	3	8118.051	[142]
		3	2	8207.402	[142]
	7/2	3	3	5473.045	[142]
		4	4	5523.438	[142]
		4	3	5449.436	[142]
		3	4	5547.042	[142]
	9/2	4	4	164.7960	[142]
		5	5	117.1495	[142]
		4	5	192.9957	[142]
		5	4	88.9504	[142]

[a]: the e and f levels have parities of $(-1)^{J-1/2}$ and $-(-1)^{J-1/2}$ respectively.

(c) Theory and analysis of the Λ-doubling hyperfine transitions

A comprehensive fit of all of the high resolution data for OH has been performed first by Brown, Kaise, Kerr and Milton [115], and subsequently by Brown, Kerr, Wayne,

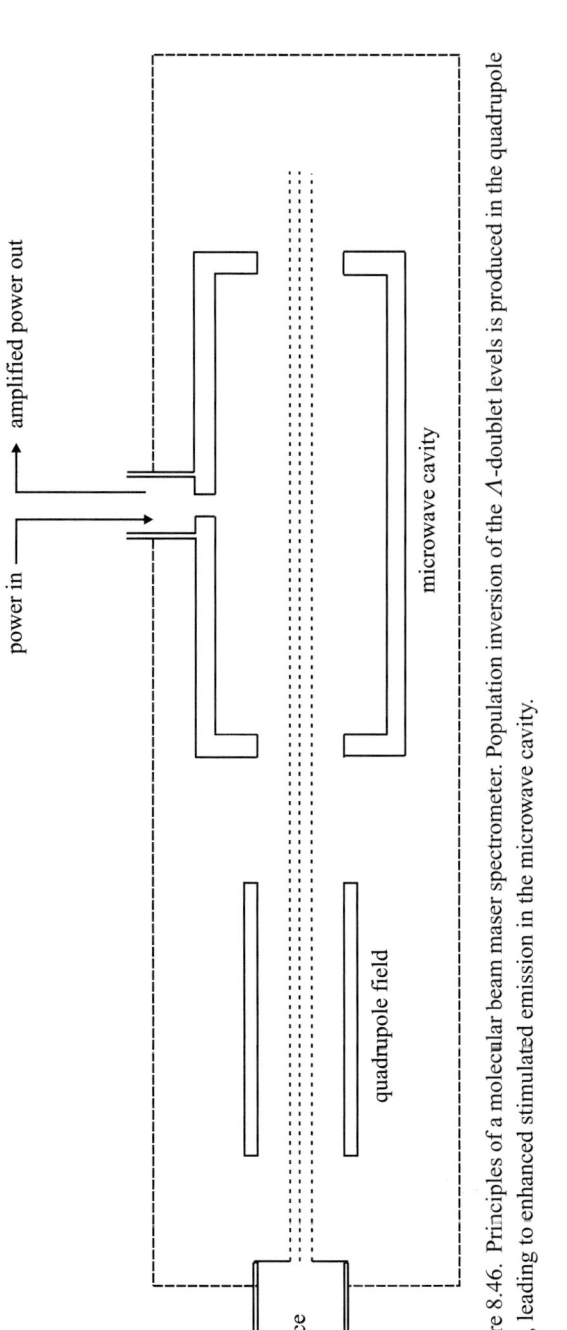

Figure 8.46. Principles of a molecular beam maser spectrometer. Population inversion of the Λ-doublet levels is produced in the quadrupole field, leading to enhanced stimulated emission in the microwave cavity.

Evenson and Radford [148]. We may start our discussion by using their values for the same set of 10 parameters which were introduced earlier for analysing the spectrum of NO. (Note that there is no quadrupole interaction in this case, which reduces the number of parameters from 12 to 10). For OH these molecular parameters (in MHz) are as follows:

$$B = 555\,660.97, \quad A = -4\,168\,639.13, \quad \gamma = -3574.88, \quad p = 7053.098\,46,$$
$$q = -1159.991\,65, \quad a = 86.1116, \quad b_F = -73.2537, \quad t_0 = 43.547,$$
$$t_2 = 37.7892, \quad c_I = -0.099\,71.$$

These values differ slightly from those given by Meerts [149]. Since the $J = 3/2$ level of the $^2\Pi_{3/2}$ state is particularly important, we commence our analysis with this level. We may anticipate that the mixing of different J levels should be even less important than it was in NO, since the rotational levels are much more widely spaced. Centrifugal distortion effects should also be small for the low J levels.

We therefore return to the pair of 2×2 matrices described immediately before (8.406), and evaluate first the terms which do not involve the proton nuclear spin. For the (+) parity states with $J = 3/2$, the matrix elements are

$$m_{11} = A/2 + 2B,$$
$$m_{22} = -A/2 + 4B - \gamma + (p + 2q), \qquad (8.416)$$
$$m_{12} = -(\sqrt{3})(B - \gamma/2) - q(\sqrt{3}) = m_{21},$$

and for the (−) parity states they are

$$m_{33} = A/2 + 2B,$$
$$m_{44} = -A/2 + 4B - \gamma - (p + 2q), \qquad (8.417)$$
$$m_{34} = -(\sqrt{3})(B - \gamma/2) + q(\sqrt{3}) = m_{43}.$$

Since $I = 1/2$ for the proton, the allowed values of F (for $J = 3/2$) are 2 and 1. The magnetic hyperfine terms are as follows:

$F = 2$	$F = 1$	
$hf_{11} = +hf_1(3/20)$	$hf_{11} = -hf_1(1/4)$	
$hf_{22} = +hf_2(1/20) + t_2(3/10)$	$hf_{22} = -hf_2(1/12) - t_2(1/2)$	
$hf_{12} = (b_F - t_0)(\sqrt{3}/10) = hf_{21}$	$hf_{12} = -(b_F - t_0)(1/\sqrt{12}) = hf_{21}$	(8.418)
$hf_{33} = +hf_1(3/20)$	$hf_{33} = -hf_1(1/4)$	
$hf_{44} = +hf_2(1/20) - t_2(3/10)$	$hf_{44} = -hf_2(1/12) + t_2(1/2)$	
$hf_{34} = (b_F - t_0)(\sqrt{3}/10) = hf_{43}$	$hf_{34} = -(b_F - t_0)(1/\sqrt{12}) = hf_{43}.$	

In these expressions the composite constants are as defined previously for NO, i.e.

$$hf_1 = 2a + b_F + 2t_0, \quad hf_2 = 2a - b_F - 2t_0. \qquad (8.419)$$

Substitution of the molecular constants listed above and diagonalisation of the 2×2 matrices gives the energies and transition frequencies for $J = 3/2$ in both the $^2\Pi_{3/2}$

Table 8.25. *Comparison of measured and calculated transition frequencies (in MHz) for the hyperfine components of Λ-doublet transitions in OH. The origins of the experimental results are given in table 8.24*

Ω	J	$F'(e) - F''(f)$	ν_{exp}	$\nu_{calc}(1)$	$\Delta\nu(1)$	$\nu_{calc}(2)$	$\Delta\nu(2)$
3/2	3/2	1-1	1665.402	1664.666	0.736	1665.414	−0.01
		2-2	1667.359	1666.616	0.743	1667.364	−0.00
		1-2	1612.231	1611.546	0.685	1612.195	0.036
		2-1	1720.530	1719.736	0.794	1720.583	−0.05
		δ			0.740		0.027
1/2	3/2	1-1	7761.747	7763.776	−2.029	7761.713	0.034
		2-2	7820.125	7822.289	−2.164	7820.195	−0.07
		1-2	7749.909	7751.983	−2.074	7749.950	−0.04
		2-1	7831.962	7834.081	−2.119	7831.958	0.004
		δ			−2.097		0.037
3/2	9/2	4-4	23 817.615	23 965.577	−147.9	23 817.165	0.450
		5-5	23 826.621	23 976.614	−150.0	23 826.199	0.422
		5-4	23 838.46	23 986.963	−148.5	23 838.549	−0.08
		4-5	23 805.13	23 953.228	−148.1	23 804.815	0.315
		δ			−148.6		0.319
1/2	9/2	4-4	164.796	331.492	−166.7	163.689	1.107
		5-5	117.150	283.273	−166.1	115.467	1.683
		5-4	192.996	359.443	−166.4	191.639	1.357
		4-5	88.950	255.322	−166.4	87.517	1.433
		δ			−166.4		1.395

and $^2\Pi_{1/2}$ states. The labelling of the four transitions for each $J = 3/2$ level is given in figure 8.45, which is correct for the $^2\Pi_{3/2}$ component. In the $^2\Pi_{1/2}$ component the energies of the $F = 2$ and 1 hyperfine levels are reversed.

The results of these, the simplest possible calculations ($\nu_{calc}(1)$), are presented in table 8.25 and compared with experiment. The agreement is remarkably good, and the discrepancy is systematic, as we see from the table. If the radiofrequency measurements involving $J = 3/2$ were the *only* experimental results available, it would be possible to adjust the values of the parameters, particularly the Λ-doubling constants p and q, to improve the agreement between experiment and theory. However, we have also included in table 8.25 the results for a higher J level, $J = 9/2$ in both fine structure states, and note that the difference between experiment and theory is very much larger. It is clear, therefore, that the effects of centrifugal distortion must be taken into account, the theory of which is described in the following subsection. Brown, Kaise, Kerr and Milton [115] derived values for the centrifugal distortion parameters, and we include three of their parameters, namely, the centrifugal corrections p_D, q_D and D. The values of these additional constants (in MHz) are

$$p_D = -1.350\,962, \quad q_D = 0.442\,032, \quad D = 57.1785,$$

and the new calculated transition frequencies, $\nu_{\text{calc}}(2)$, are also given in table 8.25. The agreement with experiment is now good for both $J = 3/2$ and $9/2$. Centrifugal distortion corrections for other parameters, in particular the hyperfine constants, have also been derived and further improve the agreement between experiment and theory. Increasing the number of molecular parameters, however, increases also the correlation between parameters so that the process is sometimes of uncertain value.

Finally we note that, as anticipated, the matrix elements off-diagonal in J have very little effect on the calculated transition frequencies.

(d) Centrifugal distortion of the Λ-doubling

As a molecule rotates faster, the bond between the two atoms stretches; this centrifugal effect increases the moment of inertia so that there is a reduction in the rotational constant which increases with increasing J. This effect can be described by the addition of an extra term to the effective Hamiltonian for the rotational kinetic energy:

$$\mathcal{H}_{\text{eff}} = BN^2 - DN^2N^2 \tag{8.420}$$

where $N = J - S$. All molecular properties which depend on the bond length R will show similar centrifugal effects as the molecule rotates. This includes the Λ-doubling contributions in the OH radical which occur because excited $^2\Sigma$ electronic states are mixed into the $^2\Pi$ ground state by the rotational motion of the molecule. There is also a somewhat weaker dependence of the electronic wave function on the internuclear separation.

These effects can be described by an additional pair of terms in the effective Hamiltonian:

$$\mathcal{H}_{\text{LDcd}} = (1/2)q_D(J_+^2 + J_-^2)(J-S)^2 - (1/2)(p_D + 2q_D)(J_+S_+ + J_-S_-)(J-S)^2. \tag{8.421}$$

The form of this correction can be appreciated by comparison with the expression for the Λ-doubling terms themselves, equation (8.400). There is, however, a problem with this form for the Hamiltonian operator because the two operator factors, such as $(J_+^2 + J_-^2)$ and $(J-S)^2$ do not commute with each other. The Hamiltonian (8.421) is therefore not Hermitian and so has complex eigenvalues. The operator can be made to have Hermitian form by taking the so-called Hermitian average,

$$\mathcal{H}_{\text{LDcd}} = (1/4)q_D[(J_+^2 + J_-^2), (J-S)^2]_+ \\ - (1/4)(p_D + 2q_D)[(J_+S_+ + J_-S_-), (J-S)^2]_+, \tag{8.422}$$

where the symbol $[A, B]_+$ is the *anti-commutator bracket*,

$$[A, B]_+ = AB + BA. \tag{8.423}$$

The evaluation of the matrix elements of these centrifugal distortion corrections appears rather daunting. However, they can be derived quite simply by matrix multiplication. We have already constructed the matrix representations of the two operators involved for a molecule in a $^2\Pi$ state. Since the operator in equation (8.422) consists of products of these two operators, its matrix elements are given simply by the products of the matrices concerned:

$$[AB]_{ij} = \sum_k [A]_{ik}[B]_{kj}. \tag{8.424}$$

In this way we can obtain the matrix representation for the centrifugal distortion of the rotational kinetic energy, in equation (8.421), as

$$\begin{aligned} hcd_{11} &= -D(z^2 + z), \\ hcd_{22} &= -D[(z+2)^2 + z], \\ hcd_{12} &= 2Dz^{1/2}(z+1) = hcd_{21}, \end{aligned} \tag{8.425}$$

where z is an abbreviation for $[(J+1/2)^2 - 1]$. For the $J = 3/2$ levels of OH described in the previous section, $z = 3$. The corresponding corrections to the Λ-doubling terms can first be written for a general set of levels J, analogous to those given in section 8.5.3. For the positive parity levels, the 2×2 Λ-doubling matrix is as follows.

	$^2\Pi_{3/2}^{(+)}$	$^2\Pi_{1/2}^{(+)}$
$^2\Pi_{3/2}^{(+)}$	$-(-1)^p(1/2)q_D z(J+1/2)$	$+(-1)^p(1/2)q_D z^{1/2}(J+1/2)^3$ $+(-1)^p(1/4)(p_D + 2q_D)(J+1/2)z^{1/2}$
$^2\Pi_{1/2}^{(+)}$	$+(-1)^p(1/2)q_D z^{1/2}(J+1/2)^3$ $+(-1)^p(1/4)(p_D + 2q_D)(J+1/2)z^{1/2}$	$-(-1)^p(1/2)q_D z(J+1/2)$ $-(-1)^p(1/2)(p_D + 2q_D)(J+1/2)(z+2)$

For negative parity levels the matrix is as follows.

	$^2\Pi_{3/2}^{(-)}$	$^2\Pi_{1/2}^{(-)}$
$^2\Pi_{3/2}^{(-)}$	$+(-1)^p(1/2)q_D z(J+1/2)$	$-(-1)^p(1/2)q_D z^{1/2}(J+1/2)^3$ $-(-1)^p(1/4)(p_D + 2q_D)(J+1/2)z^{1/2}$
$^2\Pi_{1/2}^{(-)}$	$-(-1)^p(1/2)q_D z^{1/2}(J+1/2)^3$ $-(-1)^p(1/4)(p_D + 2q_D)(J+1/2)z^{1/2}$	$+(-1)^p(1/2)q_D z(J+1/2)$ $+(-1)^p(1/2)(p_D + 2q_D)(J+1/2)(z+2)$

In these matrices the exponent p stands for $J - S$ or $J - 1/2$ in this case. For the levels with $J = 3/2$, the correction terms which have to be added to the 2×2 matrix representations are

$$\begin{aligned} mcd_{11} &= -12D \pm 3q_D, \\ mcd_{22} &= -28D \pm 3q_D \pm 5(p_D + 2q_D), \\ mcd_{12} &= 8\sqrt{3}D \mp 4\sqrt{3}q_D \mp (1/2)\sqrt{3}(p_D + 2q_D) = mcd_{21}, \end{aligned} \tag{8.426}$$

for positive and negative parity levels respectively. These matrix elements were used to calculate the transition frequencies for OH listed in the penultimate column of

Table 8.26. Λ-doubling and magnetic hyperfine parameters (in MHz) determined by Meerts and Dymanus [142] for the OH, OD, SH and SD radicals in the $v = 0$ level of the $X\,^2\Pi$ state

Parameter	OH	OD	SH	SD
p	7068.068	3765.112	8999.6	4667.2
q	−1165.22	−329.38	−283.8	−76.0
a	86.014	13.296	32.579	5.02
b_F	−74.040	−11.285	−52.627	−7.99
t_0	44.040	6.783	10.813	1.66
t_2	37.763	5.847	—	—

table 8.25. The theory can be further refined by the inclusion of centrifugal distortion corrections to the magnetic hyperfine constants.

(e) Comparison of OH, OD, SH and SD

In addition to the extensive data on the Λ-doublet transitions of OH already described, Meerts and Dymanus [142] also measured similar spectra of the species OD, SH and SD. The SH and SD radicals were produced by reacting H atoms with either H_2S or D_2S in the molecular beam source. In all cases hyperfine components of the Λ-doublet transitions were measured for a number of rotational levels in both fine structure states. The theoretical analysis of the spectra was similar to that already described for OH, with the addition of deuterium quadrupole interactions in OD and SD. In table 8.26 we list the Λ-doubling and nuclear hyperfine constants determined for the four species. Note that the parameters listed above for OH differ slightly from those given subsequently by Brown, Kaise, Kerr and Milton [115].

Meerts and Dymanus [142] compared their measured hyperfine constants with those calculated from *ab initio* wave functions for both OH and SH. There are a series of direct relations between the four hyperfine constants and the following average quantities calculated from an electronic wave function:

$$\begin{aligned}
a &= 2\mu_B g_N \mu_N (\mu_0/4\pi)\langle 1/r^3 \rangle, \\
b_F &= (2/3) g_S \mu_B g_N \mu_N \mu_0 \langle \psi^2(0) \rangle, \\
t_0 &= (1/2) g_S \mu_B g_N \mu_N (\mu_0/4\pi)\langle (3\cos^2\theta - 1)/r^3 \rangle, \\
t_2 &= g_S \mu_B g_N \mu_N (\mu_0/4\pi)\langle \sin^2\theta / r^3 \rangle,
\end{aligned} \quad (8.427)$$

where θ and r, whose average values must be calculated from the electronic wave function, have been defined in a number of places elsewhere in this book. Meerts and Dymanus [142] listed a number of calculations of the electronic averages listed in (8.427) and their results are summarised for OH and SH in table 8.27.

The calculated values listed in table 8.27 were obtained by a variety of methods, the details of which are given by Meerts and Dymanus [142]. Notice in particular that

Table 8.27. *Experimental and calculated molecular constants of OH and SH (in units of 10^{30} m^{-3})*

		$\langle 1/r^3 \rangle$	$\langle (3\cos^2\theta - 1)/r^3 \rangle$	$\langle \sin^2\theta/r^3 \rangle$	$\langle \psi^2(0) \rangle$
OH	observed	1.093	1.117	0.480	−0.1115
	calc [150]	1.015	1.037	0.331	−0.103
	calc [151] a	1.064	1.165		−0.111
	calc [151] b	1.014	1.018		−0.094
	calc [152] a				−0.128
	calc [152] b				−0.115
SH	observed	0.413	0.274	0.231	−0.0795
	calc [151] a	0.379	0.276		−0.035
	calc [151] b	0.306	0.098		−0.009
	calc [152] a				−0.047
	calc [152] b				−0.054

the Fermi contact interaction has a negative sign, which arises in the theory because of interaction between different electronic configurations. The agreement between experiment and theory is generally good.

A further interesting feature arises in the comparison of proton and deuteron hyperfine constants for the same radical. One normally expects these constants to be in the ratio of the nuclear gyromagnetic ratios, so that deuteron constants are expected to be a factor 6.514 39 smaller than those for the corresponding proton. We may calculate this ratio for the four constants of OH (OD) and SH (SD) with the following results:

OH(OD): a: 6.469, b_F: 6.561, t_0: 6.493, t_2: 6.459.

SH(SD): a: 6.490, b_F: 6.587, t_0: 6.514.

In most cases these differ significantly from the theoretical value of 6.514; an explanation is not given in the original papers, but the differences may reflect subtle vibrational averaging effects or breakdown of the Born–Oppenheimer approximation.

(f) Electric dipole moment of OH

Meerts and Dymanus [142, 153] extended their studies of the OH and SH radicals by examining the Stark effect and determining the electric dipole moments, but an even more extensive study of the Stark effect for OH and OD in several different vibrational levels was described by Peterson, Fraser and Klemperer [154]. The effect of an applied electric field on the hyperfine components of the Λ-doublets for the $J = 3/2$ level of the $^2\Pi_{3/2}$ state is illustrated in figure 8.47. Measurements were made of the $M_F = 2$, $\Delta M_F = 0$ transition in a calibrated electric field of approximately 700 V cm^{-1} and the Stark shift from the zero-field line position measured. The observations were made on resonances from $v = 0$, 1 and 2 for OH, and $v = 0$ and 1 for OD.

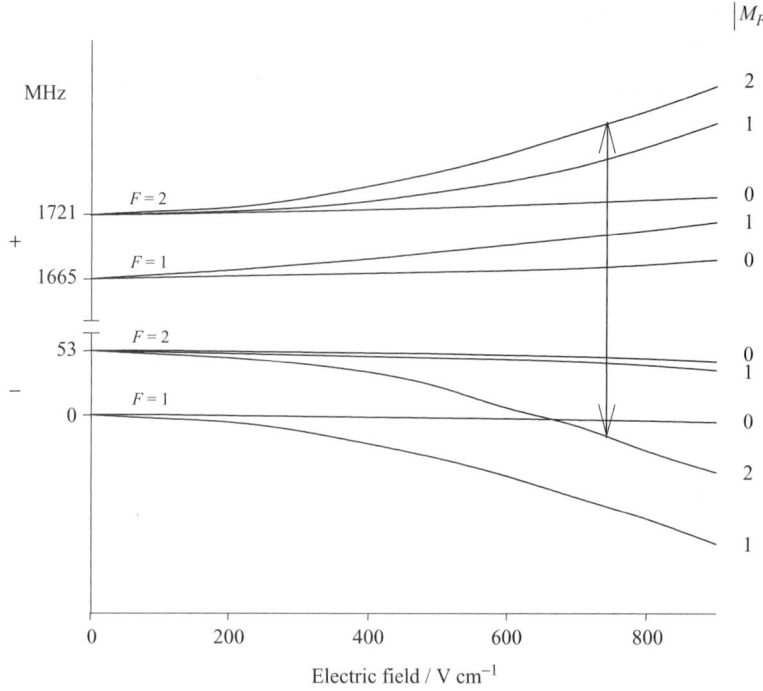

Figure 8.47. Stark effect for the hyperfine and Λ-doublet components of the $J = 3/2$ level of the $^2\Pi_{3/2}$ state of the OH radical [154].

The interaction between the molecular electric dipole moment and an applied magnetic field is represented by the simple operator

$$\mathcal{H}_E = -\mathbf{T}^1(\boldsymbol{\mu}_e) \cdot \mathbf{T}^1(\boldsymbol{E}), \tag{8.428}$$

and, as usual, we may take the direction of the electric field to define the $p = 0$ space-fixed axis. Since the electric dipole-moment lies along the molecule-fixed $q = 0$ axis, the matrix elements may be calculated for primitive basis functions, using a first-rank rotational matrix, as follows:

$$\langle J, \Omega, I, F, M_F | -\mathbf{T}^1_{p=0}(\boldsymbol{\mu}_e) \mathbf{T}^1_{p=0}(\boldsymbol{E}) | J', \Omega', I, F', M_F \rangle$$

$$= -\mu_0 E_0 (-1)^{F-M_F} \begin{pmatrix} F & 1 & F' \\ -M_F & 0 & M_F \end{pmatrix} (-1)^{F'+J+1+I}$$

$$\times \{(2F'+1)(2F+1)\}^{1/2} \begin{Bmatrix} J' & F' & I \\ F & J & 1 \end{Bmatrix} \langle J, \Omega \| \mathfrak{D}^{(1)}_{\cdot 0}(\omega)^* \| J', \Omega' \rangle$$

$$= -\mu_0 E_0 (-1)^{F-M_F} \begin{pmatrix} F & 1 & F' \\ -M_F & 0 & M_F \end{pmatrix} (-1)^{F'+J+1+I} \{(2F'+1)(2F+1)\}^{1/2}$$

$$\times \begin{Bmatrix} J' & F' & I \\ F & J & 1 \end{Bmatrix} (-1)^{J-\Omega} \begin{pmatrix} J & 1 & J' \\ -\Omega & 0 & \Omega \end{pmatrix} \{(2J+1)(2J'+1)\}^{1/2}. \tag{8.429}$$

Table 8.28. *Electric dipole moment of OH and OD in several different vibrational, rotational and fine-structure states (from Peterson, Fraser and Klemperer [154])*

	Ω	J	$|\Omega|_{\text{eff}}$	v	μ_e (D)
OH	1/2	1/2	0.5	0	1.6549
	1/2	9/2	0.6516	0	1.665 72
	3/2	3/2	1.4697	0	1.655 20
	3/2	3/2	1.4714	1	1.662 57
	3/2	3/2	1.4732	2	1.6648
OD	1/2	1/2	0.5	0	1.653 12
	3/2	3/2	1.4887	0	1.652 83
	3/2	3/2	1.4893	1	1.6550

Peterson, Fraser and Klemperer [154] chose to neglect matrix elements off-diagonal in F and J, since the experiments were performed in relatively weak electric fields. Under these circumstances equation (8.429) reduces to

$$\langle J, \Omega, I, F, M_F | - T^1_{p=0}(\boldsymbol{\mu}_e) T^1_{p=0}(\boldsymbol{E}) | J, \Omega, I, F, M_F \rangle$$
$$= -\frac{\mu_0 E_0 M_F \Omega [J(J+1) + F(F+1) - I(I+1)]}{2F(F+1)J(J+1)}. \quad (8.430)$$

This is, in essence, the result needed to construct figure 8.47. There is more to be done, however, because it is necessary to use (8.430) to derive matrix elements for the parity-conserved functions, and then to take note of the rotational distortion which mixes the fine-structure states. This mixing can be represented by an effective Ω value, which is designated $|\Omega|_{\text{eff}}$ in table 8.28, where the results of the Stark experiments are listed.

The data presented in table 8.28 show a number of interesting features. There is a significant isotopic dependence of the dipole moment for the same quantum state. Peterson, Fraser and Klemperer [154] draw attention to two possible reasons for the difference between OH and OD. A difference of 0.001 D for OH and OD in the $v=0$ state was calculated by Werner, Rosmus and Reinsch [155] as a result of vibrational averaging of their dipole moment function. The remaining difference is probably due to breakdown of the Born–Oppenheimer approximation, and is similar in both magnitude and sign to a difference between HCl and DCl observed by Kaiser [156]. The vibrational dependence of the dipole moment of OH is also determined accurately, providing further important information concerning vibrational averaging of the dipole moment function. Peterson, Fraser and Klemperer [154] point out that such information is needed if spectroscopic measurements are to be accurate in determining OH concentrations in the earth's atmosphere [157] and in interstellar space [158, 159].

(g) Concluding remarks

In this section we have described in considerable detail just one aspect of the spectroscopy of OH, namely, the measurement of Λ-doubling frequencies and their nuclear hyperfine structure. This has led us to develop the theory of the fine and hyperfine levels in zero field as well as a brief discussion of the Stark effect. We should note at this point, however, that OH was the first transient gas phase free radical to be studied by pure microwave spectroscopy [121]. We will describe these experiments in chapter 10. We note also that magnetic resonance investigations using microwave or far-infrared laser frequencies have also provided much of the most important and accurate information; these studies are described in chapter 9, where we are also able to compare OH with the equally important radical, CH, a species which, until very recently, had not been detected and studied by either electric resonance techniques or pure microwave spectroscopy.

8.5.5. CO in the $a\,^3\Pi$ state

(a) Introduction and experimental results

The electronically excited $a\,^3\Pi$ state of CO might well be the most thoroughly studied excited electronic state of all. Before reviewing briefly the many spectroscopic studies that have been described, we summarise the molecular orbital description of the CO molecule, and describe the ground and excited state electron configurations. The molecular orbitals for both homonuclear and heteronuclear diatomic molecules were described, as a function of internuclear distance, in chapter 6, and in figure 6.9 we indicated the position of the CO molecule on a molecular orbital energy level diagram. The ground state and two of the excited electronic state configurations may be written as follows:

$$\begin{aligned} X\,^1\Sigma^+ : \quad & (1\sigma)^2(2\sigma)^2(3\sigma)^2(4\sigma)^2(5\sigma)^2(1\pi)^4 \\ a\,^3\Pi : \quad & (1\sigma)^2(2\sigma)^2(3\sigma)^2(4\sigma)^2(5\sigma)^1(1\pi)^4(2\pi)^1 \\ a'\,^3\Sigma^+ : \quad & (1\sigma)^2(2\sigma)^2(3\sigma)^2(4\sigma)^2(5\sigma)^2(1\pi)^3(2\pi)^1. \end{aligned} \quad (8.431)$$

We have included the $^3\Sigma^+$ state because, as we shall see, it plays an important role in certain aspects of the spectroscopy of the $a\,^3\Pi$ state. The $a\,^3\Pi$ state lies about 49 000 cm^{-1} above the ground state, and is relatively metastable with respect to radiative decay, having a lifetime of about 60 ms.

The $a\,^3\Pi$ state of CO was first identified through its ultraviolet emission spectrum to the ground state, producing what are now known as the Cameron bands [160, 161, 162]. Its radiofrequency spectrum was then described by Klemperer and his colleagues in a classic series of molecular beam electric resonance experiments. Its microwave rotational spectrum was measured by Saykally, Dixon, Anderson, Szanto and Woods [163], and the far-infrared laser magnetic resonance spectrum was recorded by Saykally, Evenson, Comben and Brown [164]. In the infrared region both electronic and vibrational spectra involving the $a\,^3\Pi$ state have been studied. In this section we shall describe in some detail the molecular beam radiofrequency studies, but will leave an in-depth discussion of the theoretical analysis until we come to describe the laser magnetic resonance spectrum. Saykally, Evenson, Comben and Brown [164] were in the advantageous position of having all of the different experimental measurements available and were therefore able to produce a unified theory which made maximum use of all the information. We have already described the principles of the molecular beam electric resonance method and confine ourselves here to those aspects which were particular to the initial discovery of the radiofrequency spectrum by Freund and Klemperer [165]. Their molecular beam electric resonance apparatus was similar to that shown in figure 8.33. They formed an effusive beam of CO from a liquid nitrogen cooled source, which was then bombarded with an electron beam of 18 eV energy, exciting a small proportion to the $a\,^3\Pi$ state. The beam was detected with an Auger detector, formed by depositing sodium on a metal surface. The A field was an electric quadrupole of length 8 cm, whilst the B field was an electric dipole of length 10 cm. The homogeneous electric C field, with radiofrequency components both parallel and perpendicular to the static field, had an effective central length of 9 cm. Resonance absorption lines were modulated either by on-off modulation of the radiofrequency power, or by modulation of the static electric field.

The lowest rotational levels of the $a\,^3\Pi$ state of CO are fairly well described by case (a) wave functions $|\eta, J, \Omega, M_J\rangle$, where $\Omega = \Lambda + \Sigma$ is the component of total electronic angular momentum along the internuclear axis, taking the values 0, 1 and 2. The lower fine-structure states and the associated rotational levels of $a\,^3\Pi$ CO are illustrated in figure 8.48; the Λ-doubling in the $\Omega = 0$ state is large and relatively independent of the rotational quantum number J. The Λ-doubling in the $\Omega = 1$ state is much smaller, and in the $\Omega = 2$ state even smaller still; these doublet splittings are exaggerated in the figure. The transitions detected by Freund and Klemperer [165] were between the Λ-doublets of a particular J level in both the $\Omega = 1$ and 2 fine-structure states. They obtained line widths of 150 kHz or more, but in later work Stern, Gammon, Lesk, Freund and Klemperer [166] were able to reduce this to 7 kHz, revealing additional structure in the resonances (see also figure 8.49). In order to understand both the electric state focusing (and defocusing) and the observed transitions it is necessary to consider the Stark effect in some detail.

(b) Stark effect

The correct zero-field basis functions for the Λ-doublet components of particular J and Ω states were given by Freed [167] as

$$- \text{combination}, \text{parity} - (-1)^J: \quad \frac{1}{\sqrt{2}}\{|+\Omega, \pm M_J\rangle - |-\Omega, \pm M_J\rangle\}.$$
$$+ \text{combination}, \text{parity}\,(-1)^J: \quad \frac{1}{\sqrt{2}}\{|+\Omega, \pm M_J\rangle + |-\Omega, \pm M_J\rangle\}.$$
(8.432)

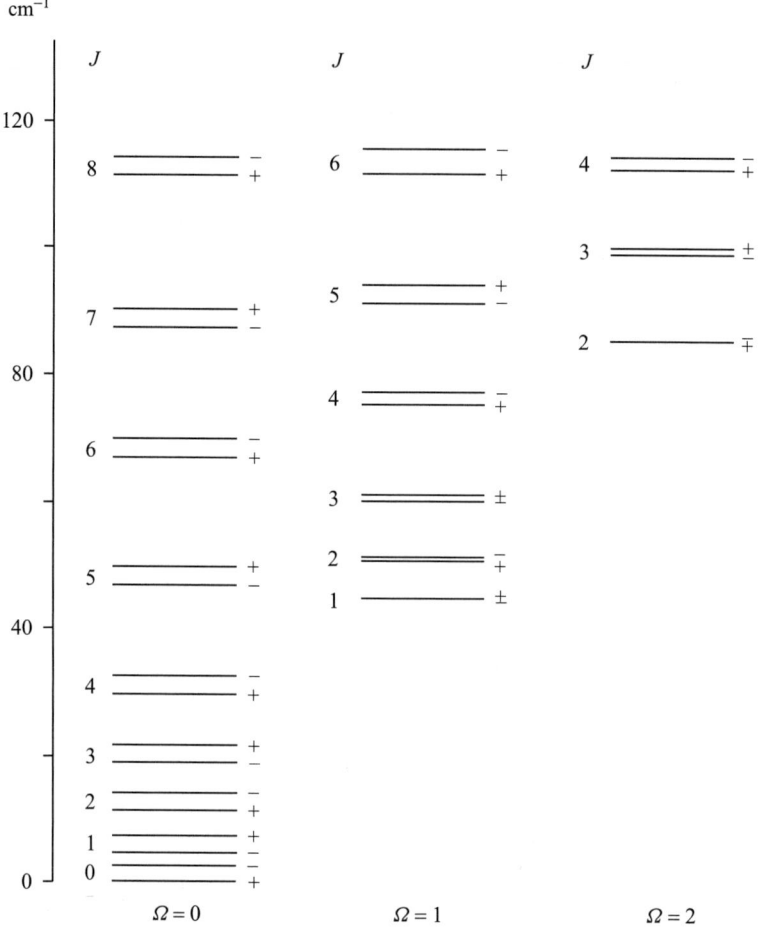

Figure 8.48. Lower rotational levels and fine structure states of CO in its $a\,^3\Pi$ state.

The matrix elements of an applied electric field, diagonal in M_J and Ω, are

$$\langle J, \Omega, M_J | - \mathbf{T}^1(\boldsymbol{\mu}_e) \cdot \mathbf{T}^1(\mathbf{E}) | J', \Omega, M_J \rangle$$

$$= -\mu_0 E_0 (-1)^{J-M_J} \begin{pmatrix} J & 1 & J' \\ -M_J & 0 & M_J \end{pmatrix} (-1)^{J-\Omega} \begin{pmatrix} J & 1 & J' \\ -\Omega & 0 & \Omega \end{pmatrix} \{(2J+1)(2J'+1)\}^{1/2}.$$

(8.433)

Using (8.432) and (8.433) the Stark energies for $J = 2$, $\Omega = 2$ can be readily calculated and the results are presented in figure 8.50; the initial splitting of the Λ-doublets was determined from the electric resonance study to be 7.351 MHz for the $v = 0$ level. In small electric fields the parities of the states are essentially preserved, and transitions between the Λ-doublets have their full electric dipole intensities. At higher electric fields, however, the opposite parity states are mixed and the electric dipole intensity decreases. It follows that so far as the intensities of the electric resonance transitions are concerned, low electric fields are desirable. On the other hand, Stern, Gammon,

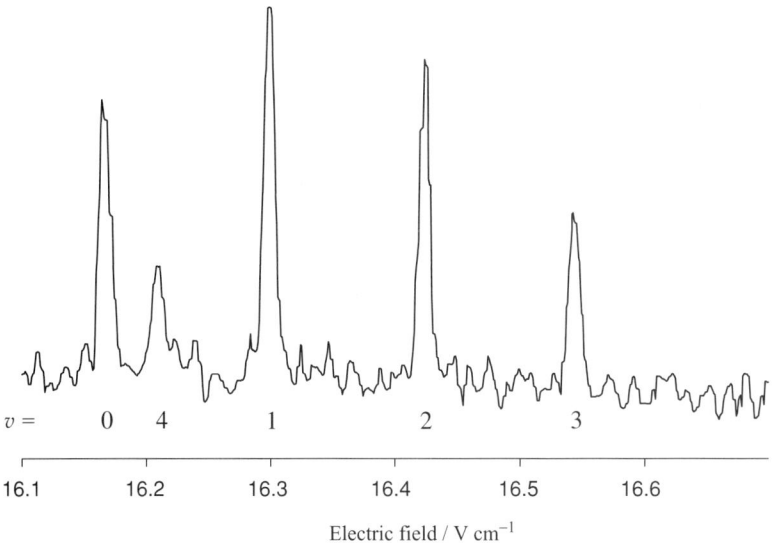

Figure 8.49. Molecular beam electric resonance spectrum of the $M_J = J$, $\Delta M_J = \Delta J = 0$ transition across the Λ-doublets in the states $|v, \Omega, J\rangle = |0 \text{ to } 4, 2, 2\rangle$ of $a\, ^3\Pi$ CO. The radiofrequency was 16.515 MHz and the spectrum was recorded [166] by sweeping the electric C field. The unexpected position of the $v = 4$ resonance is due to a perturbation (see text).

Lesk, Freund and Klemperer point out that efficient state selection requires electric fields sufficiently high that the levels studied in their experiments have linear Stark effects in the A and B deflecting fields; this corresponds to the right hand side of figure 8.50. The quadrupole A field focusses molecules with $M\Omega < 0$ onto the beam axis, whilst the dipole B field horizontally deflects resonant molecules with $M\Omega > 0$ onto a geometrically shielded off-axis detector; the experiments were therefore set up for the 'flop-in' detection mode. In terms of the high-field levels shown in figure 8.50, the top-half of the diagram survive passage through the A field; radiofrequency transitions in the C field region result in *re-population* of levels lying in the bottom half of the diagram, which are then deflected onto the detector by the B field.

Studies were made of Λ-doublet transitions in levels with $|\Omega| = 1$ and 2, J values from 1 to 7, and v from 0 to 4. Figure 8.49 shows an example of a high-resolution spectrum, in which the vibrational dependence of the Λ-doubling is clearly resolved; the reasons for the anomalous position of the $v = 4$ resonance will be explained in due course. The Λ-doublet splitting ranges from 6.529 MHz in the $v = 3$, $|\Omega| = 2$, $J = 2$ level, to 1150.934 MHz in the $v = 0$, $|\Omega| = 1$, $J = 2$ level. The electric dipole moments in a number of different v, $|\Omega|$, J levels were also determined; they range from 1.375 to 1.378 D. We return later to a more quantitative discussion of the analysis of the results, particularly the Λ-doublet splittings.

A particularly fascinating aspect of the radiofrequency spectrum shown in figure 8.49 is the anomalous position of the $v = 4$ resonance, to which we referred above. Gammon, Stern and Klemperer [168] made a careful study of this effect, which

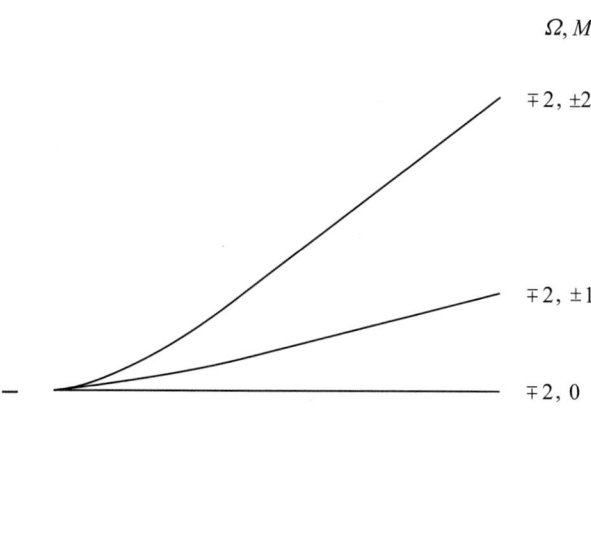
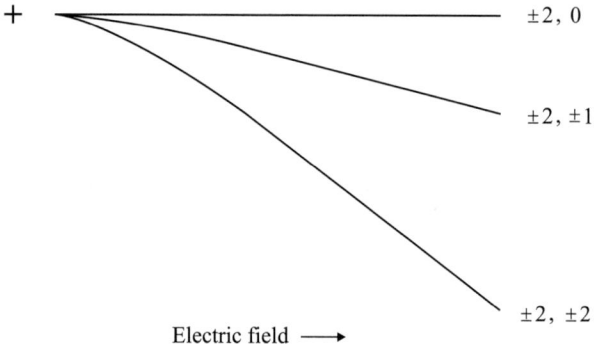

Figure 8.50. Stark energies of the Λ-doublet levels for $|\Omega| = 2$, $J = 2$. On the left-hand side, in zero field, the wave functions are the parity-conserved combinations given in equation (8.432). On the right-hand side, in strong field, the wave functions are the simple combinations shown, with parity not conserved.

is attributed to a perturbation between the $v = 4$ level of the $a\,^3\Pi$ state and the $v = 0$ level of the $a'\,^3\Sigma^+$ state; the electron configurations of both states were given in (8.431). The relevant sections of the potential energy curves for the two electronic states are shown in figure 8.51, where the near-coincidence of the interacting vibrational levels is also shown.

(c) Theoretical analysis

The theory of the Λ-doubling in the $^3\Pi$ state, and the perturbations with the $^3\Sigma^+$ state, is quite involved. We give here an outline of the problem, to which we will return in chapter 10. The problem can be represented in the form of a 5×5 matrix, constructed

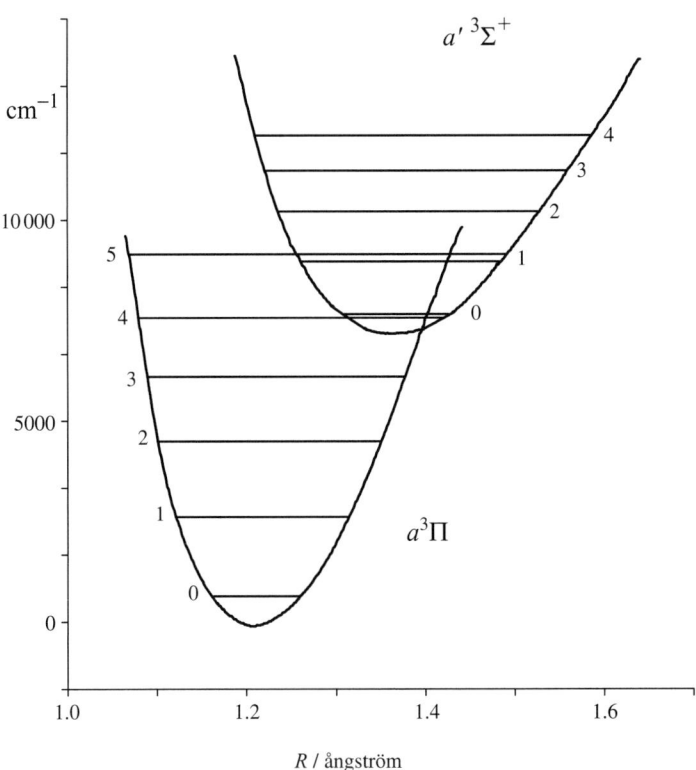

Figure 8.51. Potential energy curves for the $a\,^3\Pi$ and $a'\,^3\Sigma^+$ excited electronic states of CO, showing the near-coincidence of the $v=4$ and $v=0$ vibrational levels of the two respective electronic states [168].

in a case (a) basis, involving the three fine-structure components of the $^3\Pi$ state, and the two fine-structure components of the $^3\Sigma^+$ state. The matrix elements have been presented [166, 167, 168] and the results may be summarised in the following matrix. The structure of this matrix is readily understood; the top left 3×3 block applies to the $^3\Pi$ state with its three fine-structure components, whilst the bottom right 2×2 block represents the $^3\Sigma^+$ state. The remaining off-diagonal elements describe the perturbations between the two electronic states.

	$\lvert^3\Pi_2\rangle^\pm$	$\lvert^3\Pi_1\rangle^\pm$	$\lvert^3\Pi_0\rangle^\pm$	$\lvert^3\Sigma_1^s\rangle^\pm$	$\lvert^3\Sigma_0^s\rangle^\pm$
$^\pm\langle^3\Pi_2\rvert$	m_{11}	m_{12}	m_{13}	m_{14}	m_{15}
$^\pm\langle^3\Pi_1\rvert$	m_{21}	m_{22}	m_{23}	m_{24}	m_{25}
$^\pm\langle^3\Pi_0\rvert$	m_{31}	m_{32}	m_{33}	m_{34}	m_{35}
$^\pm\langle^3\Sigma_1^s\rvert$	m_{41}	m_{42}	m_{43}	m_{44}	m_{45}
$^\pm\langle^3\Sigma_0^s\rvert$	m_{51}	m_{52}	m_{53}	m_{54}	m_{55}

Since parity is conserved in the absence of external electric fields, one such matrix exists for each of the two opposite parity states. The matrix elements involving the $^3\Pi$ state only involve the following interactions:

m_{11}, m_{22}, m_{33}: vibronic energy + spin-orbit coupling + rotation
+ diagonal rotation–electronic coupling + spin–rotation
+ spin–spin dipolar coupling

m_{12}, m_{13}, m_{23}: rotation + off-diagonal rotation–electronic coupling
+ spin–orbit mixing.

The matrix elements involving the $^3\Sigma^+$ state only describe the following interactions:

m_{44}, m_{55}: vibronic energy + rotation + spin–spin dipolar coupling + spin–rotation.
m_{45}: rotation + spin–rotation.

Finally we come to the matrix elements which mix the two electronic states and thus give rise to the observed perturbations:

m_{14}: rotation + off-diagonal rotation–electronic coupling
m_{24}, m_{35}: off-diagonal spin–orbit and rotation–electronic coupling
m_{25}, m_{34}: rotation + off-diagonal rotation–electronic coupling.

Full details of the matrix elements are given by Gammon, Stern and Klemperer [168] who obtained a satisfactory quantitative interpretation of their experimental results; the electronic state interaction affects both the Λ-doubling intervals and the electric dipole moment.

This is not the end of the CO $a\,^3\Pi$ radiofrequency story because Gammon, Stern, Lesk, Wicke and Klemperer [169] studied the spectrum of ^{13}CO and observed the hyperfine structure arising from ^{13}C which has a nuclear spin $I = 1/2$. They were able to obtain values of the parameters describing the Fermi contact interaction, the electron–nuclear spin dipolar and orbital hyperfine interactions.

Appendix 8.1. Nuclear spin dipolar interaction

We show here the equivalence of the two forms of the nuclear spin dipolar interaction, equations (8.9) and (8.10). The most familiar representation of this interaction is equation (8.9),

$$\mathcal{H}_{dip} = g_1 g_2 \mu_N^2 (\mu_0/4\pi) \left\{ \frac{\mathbf{I}_1 \cdot \mathbf{I}_2}{R^3} - \frac{3(\mathbf{I}_1 \cdot \mathbf{R})(\mathbf{I}_2 \cdot \mathbf{R})}{R^5} \right\}. \quad (8.434)$$

We now expand the scalar products in this expression using the cartesian coordinate system shown in figure 8.52. We obtain

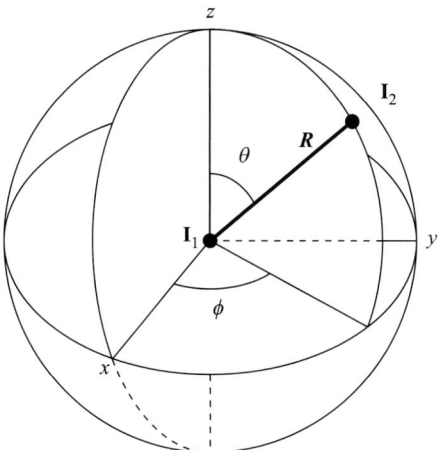

Figure 8.52. Axis system and polar coordinates for the dipolar coupling between two nuclei.

$$\begin{aligned} \boldsymbol{I}_1 \cdot \boldsymbol{I}_2 &= I_{1x}I_{2x} + I_{1y}I_{2y} + I_{1z}I_{2z}, \\ \boldsymbol{I}_1 \cdot \boldsymbol{R} &= I_{1x}x + I_{1y}y + I_{1z}z, \\ \boldsymbol{I}_2 \cdot \boldsymbol{R} &= I_{2x}x + I_{2y}y + I_{2z}z, \end{aligned} \qquad (8.435)$$

where $\boldsymbol{R} = x\boldsymbol{i} + y\boldsymbol{j} + z\boldsymbol{k}$. Substitution of these results in the above expression for the dipolar interaction yields the result:

$$\begin{aligned} &\mathcal{H}_{\text{dip}}/g_1 g_2 \mu_N^2 (\mu_0/4\pi) \\ &= [I_{1x}, I_{1y}, I_{1z}] \begin{bmatrix} (R^2 - 3x^2)/R^5 & -3xy/R^5 & -3xz/R^5 \\ -3xy/R^5 & (R^2 - 3y^2)/R^5 & -3yz/R^5 \\ -3xz/R^5 & -3yz/R^5 & (R^2 - 3z^2)/R^5 \end{bmatrix} \begin{bmatrix} I_{2x} \\ I_{2y} \\ I_{2z} \end{bmatrix}, \\ &= \boldsymbol{I}_1 \cdot \boldsymbol{D} \cdot \boldsymbol{I}_2. \end{aligned} \qquad (8.436)$$

$\boldsymbol{D}$ is the dipolar coupling (cartesian) tensor, which although not necessarily diagonal in the axis system shown in figure 8.52, can be made so by choosing the z axis to lie along the internuclear vector $\boldsymbol{R}$.

Now we examine the irreducible spherical tensorial form (8.10) and expand the scalar product in the molecule-fixed coordinate system:

$$\begin{aligned} T^2(\boldsymbol{C}) \cdot T^2(\boldsymbol{I}_1, \boldsymbol{I}_2) &= \sum_q (-1)^q T_q^2(\boldsymbol{C}) T_{-q}^2(\boldsymbol{I}_1, \boldsymbol{I}_2), \\ &= T_0^2(\boldsymbol{C})T_0^2(\boldsymbol{I}_1, \boldsymbol{I}_2) - T_1^2(\boldsymbol{C})T_{-1}^2(\boldsymbol{I}_1, \boldsymbol{I}_2) - T_{-1}^2(\boldsymbol{C})T_1^2(\boldsymbol{I}_1, \boldsymbol{I}_2) \\ &\quad + T_2^2(\boldsymbol{C})T_{-2}^2(\boldsymbol{I}_1, \boldsymbol{I}_2) + T_{-2}^2(\boldsymbol{C})T_2^2(\boldsymbol{I}_1, \boldsymbol{I}_2). \end{aligned} \qquad (8.437)$$

Using the expansion (8.11) we treat each nuclear spin term in (8.437) in turn; for example,

$$T_0^2(I_1, I_2) = \sqrt{5}\begin{pmatrix}1 & 1 & 2\\ 1 & -1 & 0\end{pmatrix}T_1^1(I_1)T_{-1}^1(I_2) + \sqrt{5}\begin{pmatrix}1 & 1 & 2\\ 0 & 0 & 0\end{pmatrix}T_0^1(I_1)T_0^1(I_2)$$
$$+ \sqrt{5}\begin{pmatrix}1 & 1 & 2\\ -1 & 1 & 0\end{pmatrix}T_{-1}^1(I_1)T_1^1(I_2). \quad (8.438)$$

Noting the relationships between the spherical tensor and cartesian components,

$$T_0^1(I) = I_z, \quad T_{\pm 1}^1(I) = \mp \frac{1}{\sqrt{2}}\{I_x \pm iI_y\}, \quad (8.439)$$

we find that

$$T_0^2(I_1, I_2) = -\frac{1}{\sqrt{6}}(I_{1x}I_{2x} + I_{1y}I_{2y}) + \frac{\sqrt{2}}{\sqrt{3}}I_{1z}I_{2z}. \quad (8.440)$$

The other four components of $T^2(I_1, I_2)$ may be similarly written in cartesian form:

$$\begin{aligned}T_{-1}^2(I_1, I_2) &= \frac{1}{2}\{I_{1x}I_{2z} - iI_{1y}I_{2z} + I_{1z}I_{2x} - iI_{1z}I_{2y}\},\\ T_1^2(I_1, I_2) &= -\frac{1}{2}\{I_{1x}I_{2z} + iI_{1y}I_{2z} + I_{1z}I_{2x} + iI_{1z}I_{2y}\},\\ T_{-2}^2(I_1, I_2) &= \frac{1}{2}\{I_{1x}I_{2x} - iI_{1x}I_{2y} - iI_{1y}I_{2x} - I_{1y}I_{2y}\},\\ T_2^2(I_1, I_2) &= \frac{1}{2}\{I_{1x}I_{2x} + iI_{1x}I_{2y} + iI_{1y}I_{2x} - I_{1y}I_{2y}\}.\end{aligned} \quad (8.441)$$

Noting now the cartesian forms of the spherical tensor components of $T^2(C)$, we may write,

$$\begin{aligned}T_0^2(C) &= C_0^2(\theta, \phi)(R^{-3}) = \frac{1}{2}(R^{-5})(3z^2 - R^2),\\ T_{\pm 1}^2(C) &= C_{\pm 1}^2(\theta, \phi)(R^{-3}) = \mp \frac{\sqrt{3}}{\sqrt{2}}z(x \pm iy)(R^{-5}),\\ T_{\pm 2}^2(C) &= C_{\pm 2}^2(\theta, \phi)(R^{-3}) = \frac{\sqrt{3}}{2\sqrt{2}}(x \pm iy)^2(R^{-5}).\end{aligned} \quad (8.442)$$

If we now combine (8.442) and (8.441) with equation (8.437) we obtain the desired result:

$$-\sqrt{6}T^2(C) \cdot T^2(I_1, I_2)$$
$$= [I_{1x}, I_{1y}, I_{1z}]\begin{bmatrix}(R^2 - 3x^2)/R^5 & -3xy/R^5 & -3xz/R^5\\ -3xy/R^5 & (R^2 - 3y^2)/R^5 & -3yz/R^5\\ -3xz/R^5 & -3yz/R^5 & (R^2 - 3z^2)/R^5\end{bmatrix}\begin{bmatrix}I_{2x}\\ I_{2y}\\ I_{2z}\end{bmatrix}, \quad (8.443)$$

which is clearly equivalent to (8.436).

In order to represent the dipolar coupling of two electron spins, it is merely necessary to change I to S, replace g_1 and g_2 by g_S, and μ_N by μ_B. Corresponding replacements can be made to represent the dipolar coupling of an electron spin with a nuclear spin, but in this case the overall *sign* of the effective operator must be reversed. We now summarise the three cases so that, hopefully, there should be no confusion:

(i) nuclear–nuclear dipolar interaction

$$\mathcal{H}_{\text{dip}} = -\sqrt{6} g_1 g_2 \mu_N^2 (\mu_0/4\pi) \mathbf{T}^2(\mathbf{C}) \cdot \mathbf{T}^2(\mathbf{I}_1, \mathbf{I}_2) \qquad (8.444)$$

(ii) electron–electron dipolar interaction

$$\mathcal{H}_{\text{dip}} = -\sqrt{6} g_S^2 \mu_B^2 (\mu_0/4\pi) \mathbf{T}^2(\mathbf{C}) \cdot \mathbf{T}^2(\mathbf{S}_1, \mathbf{S}_2) \qquad (8.445)$$

(iii) electron–nuclear dipolar interaction

$$\mathcal{H}_{\text{dip}} = +\sqrt{6} g_S g_N \mu_B \mu_N (\mu_0/4\pi) \mathbf{T}^2(\mathbf{C}) \cdot \mathbf{T}^2(\mathbf{S}, \mathbf{I}). \qquad (8.446)$$

In appendix 8.2 we shall describe an alternative way of expressing the dipolar interaction between two spin magnetic moments.

Appendix 8.2. Relationship between the cartesian and spherical tensor forms of the electron spin–nuclear spin dipolar interaction

We first make use of the results derived for the dipolar coupling of two proton spins in appendix 8.1. If we replace $\mathbf{I}_1$ by $\mathbf{I}$, $\mathbf{I}_2$ by $\mathbf{S}$ and note the overall change in sign, we see that the conventional dipolar Hamiltonian,

$$\mathcal{H}_{\text{dip}} = -g_S \mu_B g_N \mu_N (\mu_0/4\pi) \left\{ \frac{\mathbf{I} \cdot \mathbf{S}}{r^3} - \frac{3(\mathbf{I} \cdot \mathbf{r})(\mathbf{S} \cdot \mathbf{r})}{r^5} \right\}, \qquad (8.447)$$

may be written in cartesian tensor form

$$\mathcal{H}_{\text{dip}}/g_S g_N \mu_B \mu_N (\mu_0/4\pi)$$

$$= [I_x, I_y, I_z] \begin{bmatrix} (3x^2-r^2)/r^5 & 3xy/r^5 & 3xz/r^5 \\ 3xy/r^5 & (3y^2-r^2)/r^5 & 3yz/r^5 \\ 3xz/r^5 & 3yz/r^5 & (3z^2-r^2)/r^5 \end{bmatrix} \begin{bmatrix} S_x \\ S_y \\ S_z \end{bmatrix}, \qquad (8.448)$$

where $\mathbf{r}$ is the position of the electron relative to the nucleus. We now expand the spherical tensor form and show that it is equivalent to (8.448). We remind ourselves of the results presented in equations (8.229) to (8.231). We stated that the dipolar Hamiltonian in irreducible tensor form can be written as the scalar product

$$\mathcal{H}_{\text{dip}} = -\sqrt{10} g_S \mu_B g_N \mu_N (\mu_0/4\pi) \mathbf{T}^1(\mathbf{I}) \cdot \mathbf{T}^1(\mathbf{S}, \mathbf{C}^2), \qquad (8.449)$$

where the spherical components of the new tensor $\mathbf{T}^1(\mathbf{S}, \mathbf{C}^2)$ are defined by

$$\mathbf{T}^1_q(\mathbf{S}, \mathbf{C}^2) = -\sqrt{3} \sum_{q_1 q_2} (-1)^q \mathbf{T}^1_{q_1}(\mathbf{S}) \mathbf{T}^2_{q_2}(\mathbf{C}) \begin{pmatrix} 1 & 2 & 1 \\ q_1 & q_2 & -q \end{pmatrix} \qquad (8.450)$$

with

$$T^2_{q_2}(C) = C^2_{q_2}(\theta, \phi)(r^{-3}). \tag{8.451}$$

We now expand the scalar product in (8.449):

$$\begin{aligned}
T^1(I) \cdot T^1(S, C^2) &= \sum_q (-1)^q T^1_q(I) T^1_{-q}(S, C^2) \\
&= T^1_0(I) T^1_0(S, C^2) - T^1_1(I) T^1_{-1}(S, C^2) - T^1_{-1}(I) T^1_1(S, C^2).
\end{aligned} \tag{8.452}$$

We note the relationships between the spherical and cartesian components of the spin vector I in the molecule-fixed coordinate system:

$$T^1_0(I) = I_z, \quad T^1_1(I) = -\frac{1}{\sqrt{2}}\{I_x + iI_y\}, \quad T^1_{-1}(I) = \frac{1}{\sqrt{2}}\{I_x - iI_y\}, \tag{8.453}$$

with similar relationships for the components of S. We are now in a position to evaluate the components of $T^1(S, C^2)$ using equation (8.450). We obtain the following results:

$$\begin{aligned}
T^1_0(S, C^2) = -\sqrt{3} \Bigg\{ & T^1_0(S) T^2_0(C) \begin{pmatrix} 1 & 2 & 1 \\ 0 & 0 & 0 \end{pmatrix} + T^1_1(S) T^2_{-1}(C) \begin{pmatrix} 1 & 2 & 1 \\ 1 & -1 & 0 \end{pmatrix} \\
& + T^1_{-1}(S) T^2_1(C) \begin{pmatrix} 1 & 2 & 1 \\ -1 & 1 & 0 \end{pmatrix} \Bigg\},
\end{aligned} \tag{8.454}$$

$$\begin{aligned}
T^1_{-1}(S, C^2) = \sqrt{3} \Bigg\{ & T^1_{-1}(S) T^2_0(C) \begin{pmatrix} 1 & 2 & 1 \\ -1 & 0 & 1 \end{pmatrix} + T^1_0(S) T^2_{-1}(C) \begin{pmatrix} 1 & 2 & 1 \\ 0 & -1 & 1 \end{pmatrix} \\
& + T^1_1(S) T^2_{-2}(C) \begin{pmatrix} 1 & 2 & 1 \\ 1 & -2 & 1 \end{pmatrix} \Bigg\},
\end{aligned} \tag{8.455}$$

$$\begin{aligned}
T^1_1(S, C^2) = \sqrt{3} \Bigg\{ & T^1_1(S) T^2_0(C) \begin{pmatrix} 1 & 2 & 1 \\ 1 & 0 & -1 \end{pmatrix} + T^1_0(S) T^2_1(C) \begin{pmatrix} 1 & 2 & 1 \\ 0 & 1 & -1 \end{pmatrix} \\
& + T^1_{-1}(S) T^2_2(C) \begin{pmatrix} 1 & 2 & 1 \\ -1 & 2 & -1 \end{pmatrix} \Bigg\}.
\end{aligned} \tag{8.456}$$

We can rewrite all of the terms in the above three equations in terms of cartesian components using the relationships for the components of S, and the following additional relationships for the components of $T^2(C)$, given earlier in appendix 8.1:

$$\begin{aligned}
T^2_0(C) &= C^2_0(\theta, \phi)(r^{-3}) = \frac{1}{2}(r^{-5})(3z^2 - r^2), \\
T^2_{\pm 1}(C) &= C^2_{\pm 1}(\theta, \phi)(r^{-3}) = \mp \frac{\sqrt{3}}{\sqrt{2}} z(x \pm iy)(r^{-5}), \\
T^2_{\pm 2}(C) &= C^2_{\pm 2}(\theta, \phi)(r^{-3}) = \frac{\sqrt{3}}{2\sqrt{2}}(x \pm iy)^2(r^{-5}).
\end{aligned} \tag{8.457}$$

It is now a matter of straightforward but tedious algebra to obtain the required result:

$$\mathcal{H}_{\text{dip}}/g_S g_N \mu_B \mu_N (\mu_0/4\pi)$$
$$= [I_x, I_y, I_z] \begin{bmatrix} (3x^2 - r^2)/r^5 & 3xy/r^5 & 3xz/r^5 \\ 3xy/r^5 & (3y^2 - r^2)/r^5 & 3yz/r^5 \\ 3xz/r^5 & 3yz/r^5 & (3z^2 - r^2)/r^5 \end{bmatrix} \begin{bmatrix} S_x \\ S_y \\ S_z \end{bmatrix}. \quad (8.458)$$

Comparison of (8.458), (8.448) and (8.449) shows that we have achieved our aim.

The same result can be established more succinctly by the following piece of spherical tensor algebra:

$$-\sqrt{10}\, \mathrm{T}^1(\mathbf{I}) \cdot \mathrm{T}^1(\mathbf{S}, \mathbf{C}^2)$$
$$= -\sqrt{10} \sum_q (-1)^q \mathrm{T}_q^1(\mathbf{I}) \mathrm{T}_{-q}^1(\mathbf{S}, \mathbf{C}^2)$$
$$= \sum_q (-1)^q \mathrm{T}_q^1(\mathbf{I}) \sqrt{30} \sum_{q_1 q_2} (-1)^q \mathrm{T}_{q_1}^1(\mathbf{S}) \mathrm{T}_{q_2}^2(\mathbf{C}^2) \begin{pmatrix} 1 & 2 & 1 \\ q_1 & q_2 & q \end{pmatrix}$$
$$= \sqrt{6} \sum_{q_2} (-1)^{q_2} \mathrm{T}_{q_2}^2(\mathbf{C}^2) \sqrt{5} \sum_{q q_1} (-1)^{-q_2} \begin{pmatrix} 1 & 1 & 2 \\ q & q_1 & q_2 \end{pmatrix} \mathrm{T}_q^1(\mathbf{I}) \mathrm{T}_{q_1}^1(\mathbf{S})$$
$$= \sqrt{6} \sum_{q_2} (-1)^{q_2} \mathrm{T}_{q_2}^2(\mathbf{C}^2) \mathrm{T}_{-q_2}^2(\mathbf{I}, \mathbf{S})$$
$$= \sqrt{6}\, \mathrm{T}^2(\mathbf{C}^2) \cdot \mathrm{T}^2(\mathbf{I}, \mathbf{S}). \quad (8.459)$$

This equivalence corresponds to a re-coupling of the various angular momenta involved.

It follows that the nuclear–nuclear and electron–electron dipolar interactions may also be represented in irreducible tensor form by operator expressions equivalent to (8.449), with the appropriate replacements, and with an overall positive sign. Again to avoid confusion we list the appropriate operators.

(i) Nuclear–nuclear dipolar interaction

$$\mathcal{H}_{\text{dip}} = \sqrt{10} g_1 g_2 \mu_N^2 (\mu_0/4\pi) \mathrm{T}^1(\mathbf{I}_1) \cdot \mathrm{T}^1(\mathbf{I}_2, \mathbf{C}^2). \quad (8.460)$$

(ii) Electron–electron dipolar interaction

$$\mathcal{H}_{\text{dip}} = \sqrt{10} g_S^2 \mu_B^2 (\mu_0/4\pi) \mathrm{T}^1(\mathbf{S}_1) \cdot \mathrm{T}^1(\mathbf{S}_2, \mathbf{C}^2). \quad (8.461)$$

(iii) Electron–nuclear dipolar interaction

$$\mathcal{H}_{\text{dip}} = -\sqrt{10} g_S \mu_B g_N \mu_N (\mu_0/4\pi) \mathrm{T}^1(\mathbf{I}) \cdot \mathrm{T}^1(\mathbf{S}, \mathbf{C}^2). \quad (8.462)$$

Appendix 8.3. Electron spin–electron spin dipolar interaction

In this book and in the scientific literature, one frequently encounters the dipolar interaction between two electron spin magnetic moments; it always occurs in the description of electronic states of triplet or higher spin multiplicity. Several different operator representations for this interaction have been used and the purpose of this appendix is to summarise these, and to show the relationships between them.

At least *four* different forms of the spin–spin operator are encountered, as follows.

(i) *Classical form converted into quantum mechanical operators*

$$\mathcal{H}_{\text{dip}}^{(i)} = g_S^2 \mu_B^2 (\mu_0/4\pi) \left\{ \frac{S_1 \cdot S_2}{r^3} - \frac{3(S_1 \cdot r)(S_2 \cdot r)}{r^5} \right\}; \tag{8.463}$$

r is the vector from electron 1 to electron 2, r being the magnitude of the vector. In practice an appropriate average of r must be calculated from the electronic wave function.

(ii) *First irreducible tensor form*

$$\mathcal{H}_{\text{dip}}^{(ii)} = -g_S^2 \mu_B^2 (\mu_0/4\pi) \sqrt{6} T^2(S_1, S_2) \cdot T^2(C), \tag{8.464}$$

where C stands for $C^2(\theta, \phi)r^{-3}$. S_1 and S_2 refer to the individual electron spins. The second-rank tensors occurring in (8.464) are defined below. This is the form we use most often in this book. Note that the *relative* signs of (8.463) and (8.464) are important, as we shall show. Equation (8.464) is the most general form of the dipolar interaction, since it allows for matrix elements off-diagonal in S as well as Σ. An alternative but completely equivalent way of writing equation (8.464) is

$$\mathcal{H}_{\text{dip}}^{(ii)} = -g_S^2 \mu_B^2 (\mu_0/4\pi)(3/r^5) T^2(S_1, S_2) \cdot T^2(r, r). \tag{8.465}$$

(iii) *Second irreducible tensor form*

$$\mathcal{H}_{\text{dip}}^{(iii)} = 2\lambda T^2(S, S) \cdot T^2(n, n). \tag{8.466}$$

S is the total electron spin; n is a unit vector along the internuclear axis. The matrix elements of (8.466) are diagonal in S. The advantage of this particular formulation is that it is explicitly scalar in form.

(iv) *Cartesian form*

$$\mathcal{H}_{\text{dip}}^{(iv)} = \frac{2}{3}\lambda' \{3S_z^2 - S^2\}, \tag{8.467}$$

where z lies along the internuclear axis. We have chosen to distinguish between λ in equation (8.466) and λ' in equation (8.467) but we derive the relationship between them below. Equation (8.467) is the oldest form of the electron spin dipolar interaction, and is the one most frequently used. It is a restrictive form in that its matrix elements are diagonal in both S and Σ.

Equivalence of forms (i) and (ii)

Equation (8.463) is the standard form of the dipolar interaction and we have already expanded this form in terms of its cartesian components in appendix 8.1 for two nuclear spins. In the present case, therefore, the equivalent expression for two electron spins is

$$\mathcal{H}_{\text{dip}}^{(i)}/g_S^2 \mu_B^2(\mu_0/4\pi)$$

$$= [S_{1x}, S_{1y}, S_{1z}] \begin{bmatrix} (r^2-3x^2)/r^5 & -3xy/r^5 & -3xz/r^5 \\ -3xy/r^5 & (r^2-3y^2)/r^5 & -3yz/r^5 \\ -3xz/r^5 & -3yz/r^5 & (r^2-3z^2)/r^5 \end{bmatrix} \begin{bmatrix} S_{2x} \\ S_{2y} \\ S_{2z} \end{bmatrix}. \quad (8.468)$$

We now expand the second form,

$$\mathcal{H}_{\text{dip}}^{(ii)} = -g_S^2\mu_B^2(\mu_0/4\pi)\sqrt{6}\,\text{T}^2(\boldsymbol{S}_1, \boldsymbol{S}_2) \cdot \text{T}^2(\boldsymbol{C}), \quad (8.469)$$

as follows:

$$-\sqrt{6}\,\text{T}^2(\boldsymbol{S}_1, \boldsymbol{S}_2) \cdot \text{T}^2(\boldsymbol{C})$$
$$= -\sqrt{6}\sum_q(-1)^q \text{T}_q^2(\boldsymbol{S}_1,\boldsymbol{S}_2)\text{T}_{-q}^2(\boldsymbol{C})$$
$$= -\sqrt{6}\,\{\text{T}_2^2(\boldsymbol{S}_1,\boldsymbol{S}_2)\text{T}_{-2}^2(\boldsymbol{C})+\text{T}_{-2}^2(\boldsymbol{S}_1,\boldsymbol{S}_2)\text{T}_2^2(\boldsymbol{C})-\text{T}_1^2(\boldsymbol{S}_1,\boldsymbol{S}_2)\text{T}_{-1}^2(\boldsymbol{C})$$
$$-\text{T}_{-1}^2(\boldsymbol{S}_1,\boldsymbol{S}_2)\text{T}_1^2(\boldsymbol{C})+\text{T}_0^2(\boldsymbol{S}_1,\boldsymbol{S}_2)\text{T}_0^2(\boldsymbol{C})\}. \quad (8.470)$$

We wish to express each of these terms in terms of cartesian components. First we note that the components of the second-rank spin tensor are defined by

$$\text{T}_q^2(\boldsymbol{S}_1, \boldsymbol{S}_2) = (-1)^q\sqrt{5}\sum_{q_1,q_2}\begin{pmatrix} 1 & 1 & 2 \\ q_1 & q_2 & -q \end{pmatrix}\text{T}_{q_1}^1(\boldsymbol{S}_1)\text{T}_{q_2}^1(\boldsymbol{S}_2), \quad (8.471)$$

so that, for example,

$$\text{T}_0^2(\boldsymbol{S}_1, \boldsymbol{S}_2) = \sqrt{5}\left\{\begin{pmatrix} 1 & 1 & 2 \\ 1 & -1 & 0 \end{pmatrix}\text{T}_1^1(\boldsymbol{S}_1)\text{T}_{-1}^1(\boldsymbol{S}_2)+\begin{pmatrix} 1 & 1 & 2 \\ -1 & 1 & 0 \end{pmatrix}\text{T}_{-1}^1(\boldsymbol{S}_1)\text{T}_1^1(\boldsymbol{S}_2)\right.$$
$$\left.+\begin{pmatrix} 1 & 1 & 2 \\ 0 & 0 & 0 \end{pmatrix}\text{T}_0^1(\boldsymbol{S}_1)\text{T}_0^1(\boldsymbol{S}_2)\right\}$$
$$= \frac{1}{\sqrt{6}}\{\text{T}_1^1(\boldsymbol{S}_1)\text{T}_{-1}^1(\boldsymbol{S}_2)+\text{T}_{-1}^1(\boldsymbol{S}_1)\text{T}_1^1(\boldsymbol{S}_2)+2\text{T}_0^1(\boldsymbol{S}_1)\text{T}_0^1(\boldsymbol{S}_2)\}. \quad (8.472)$$

The first-rank spherical spin tensor components in (8.472) may now be rewritten in cartesian form using the definitions

$$\text{T}_0^1(\boldsymbol{S}_1)=S_{1z}, \quad \text{T}_1^1(\boldsymbol{S}_1)=-\frac{1}{\sqrt{2}}\{S_{1x}+iS_{1y}\}, \quad \text{T}_{-1}^1(\boldsymbol{S}_1)=\frac{1}{\sqrt{2}}\{S_{1x}-iS_{1y}\}, \quad (8.473)$$

with similar relationships for the components of $\text{T}^1(\boldsymbol{S}_2)$.

Returning to equation (8.470) we now seek to rewrite the components of the second-rank spherical tensor $\text{T}^2(\boldsymbol{C})$ in cartesian form. We make use of the following results:

$$T_0^2(C) = C_0^2(\theta, \phi)(r^{-3}) = \left(\frac{4\pi}{5}\right)^{1/2} Y_{2,0}(\theta, \phi)(r^{-3}) = \frac{1}{2}\{2z^2 - x^2 - y^2\}(r^{-5}),$$

$$T_{\pm 1}^2(C) = C_{\pm 1}^2(\theta, \phi)(r^{-3}) = \left(\frac{4\pi}{5}\right)^{1/2} Y_{2,\pm 1}(\theta, \phi)(r^{-3}) = \mp\sqrt{\frac{3}{2}}z\{x \pm iy\}(r^{-5}),$$
(8.474)

$$T_{\pm 2}^2(C) = C_{\pm 2}^2(\theta, \phi)(r^{-3}) = \left(\frac{4\pi}{5}\right)^{1/2} Y_{2,\pm 2}(\theta, \phi)(r^{-3}) = \frac{\sqrt{3}}{2\sqrt{2}}\{x \pm iy\}^2(r^{-5}).$$

The $Y_{\ell m}(\theta, \phi)$ are spherical harmonics which are rewritten in terms of harmonic polynomials. We now substitute (8.472), (8.473) and (8.474) into (8.470) and after some tedious but straightforward algebra succeed in regenerating equation (8.468).

Relationship between forms (ii) and (iii)

We can determine the relationship between $T^2(S_1, S_2)$ and $T^2(S, S)$ by comparing their reduced matrix elements. The first of these is straightforward (see equation 5.139):

$$\langle S_1, S_2, S \| T^2(S_1, S_2) \| S_1, S_2, S \rangle$$

$$= \sqrt{5} \begin{Bmatrix} S_1 & S_1 & 1 \\ S_2 & S_2 & 1 \\ S & S & 2 \end{Bmatrix} \{(2S+1)(2S+1)S_1(S_1+1)(2S_1+1)S_2(S_2+1)(2S_2+1)\}^{1/2}$$

$$= \frac{\sqrt{5}}{2} \text{ for } S_1 = S_2 = 1/2, S = 1. \tag{8.475}$$

Evaluation of the reduced matrix element $\langle S \| T^2(S, S) \| S \rangle$ is slightly more complicated. First we use the Wigner–Eckart theorem to obtain

$$\langle S, \Sigma | T_0^2(S, S) | S, \Sigma \rangle = (-1)^{S-\Sigma} \begin{pmatrix} S & 2 & S \\ \Sigma & 0 & -\Sigma \end{pmatrix} \langle S \| T^2(S, S) \| S \rangle, \tag{8.476}$$

where Σ is the axial component ($q = 0$) of S. The problem reduces to that of determining the left-hand side of (8.476) and to do this we make use of equation (8.472), replacing the individual electron spins by the total spin S. Using also the definitions (8.473) we obtain the result

$$T_0^2(S, S) = \frac{1}{\sqrt{6}}\{3S_z^2 - S^2\}. \tag{8.477}$$

We may now evaluate the left-hand side of (8.476) taking, without loss of generality, Σ to have its maximum value S. Consequently

$$\langle S, \Sigma = S | T_0^2(S, S) | S, \Sigma = S \rangle = \frac{1}{\sqrt{6}}\{3\Sigma^2 - S(S+1)\}$$

$$= \frac{1}{\sqrt{6}}\{3S^2 - S(S+1)\} = \frac{1}{\sqrt{6}}S(2S-1). \tag{8.478}$$

Substituting for the left-hand side of (8.476) we obtain

$$\langle S\|T^2(\boldsymbol{S},\boldsymbol{S})\|S\rangle = \frac{1}{\sqrt{6}}S(2S-1)\begin{pmatrix} S & 2 & S \\ S & 0 & -S \end{pmatrix}^{-1}$$

$$= \frac{1}{2\sqrt{6}}\{(2S+3)(2S+2)(2S+1)(2S)(2S-1)\}^{1/2}$$

$$= \sqrt{5} \text{ for } S=1. \qquad (8.479)$$

Comparison of (8.479) with (8.475) gives the result, for $S=1$,

$$T^2(\boldsymbol{S}_1,\boldsymbol{S}_2) = \frac{1}{2}T^2(\boldsymbol{S},\boldsymbol{S}). \qquad (8.480)$$

We can now compare the $q=0$ components of (8.464) and (8.466). For (8.464) we find that

$$g_S^2\mu_B^2(\mu_0/4\pi)\sqrt{6}T_0^2(\boldsymbol{S}_1,\boldsymbol{S}_2)T_0^2(\boldsymbol{C}) = g_S^2\mu_B^2(\mu_0/4\pi)\frac{\sqrt{6}}{2}T_0^2(\boldsymbol{S},\boldsymbol{S})T_0^2(\boldsymbol{C})$$

$$= (\mu_0/4\pi)\sqrt{\frac{3}{2}}g_S^2\mu_B^2 T_0^2(\boldsymbol{S},\boldsymbol{S})\langle C_0^2(\theta,\phi)(r^{-3})\rangle. \qquad (8.481)$$

On the other hand, the $q=0$ component of (8.466) gives the result

$$2\lambda T_0^2(\boldsymbol{S},\boldsymbol{S})T_0^2(\boldsymbol{n},\boldsymbol{n}) = \frac{2\sqrt{6}}{3}\lambda T_0^2(\boldsymbol{S},\boldsymbol{S}), \qquad (8.482)$$

where $T_0^2(\boldsymbol{n},\boldsymbol{n})$ has been replaced by its average value of $2/\sqrt{6}$. The result of this analysis is that $\mathcal{H}_{\text{dip}}^{(ii)}$ and $\mathcal{H}_{\text{dip}}^{(iii)}$ are equivalent to each other provided that

$$-g_S^2\mu_B^2(\mu_0/4\pi)\langle C_0^2(\theta,\phi)(r^{-3})\rangle = \frac{4}{3}\lambda. \qquad (8.483)$$

We recall that (r,θ,ϕ) are the spherical polar coordinates of electron 2 relative to electron 1.

Relationship between forms (iii) and (iv)

Combining equation (8.477) with the right-hand side of (8.482) gives the result

$$2\sqrt{\frac{2}{3}}\lambda T_0^2(\boldsymbol{S},\boldsymbol{S}) = 2\sqrt{\frac{2}{3}}\lambda \frac{1}{\sqrt{6}}\{3S_z^2 - \boldsymbol{S}^2\} = \frac{2}{3}\lambda\{3S_z^2 - \boldsymbol{S}^2\}, \qquad (8.484)$$

so that, in form (iv), $\lambda' = \lambda$. All of the required relationships have now been established.

Appendix 8.4. Matrix elements of the quadrupole Hamiltonian

The nuclear electric quadrupole interaction was described in detail in chapter 4, where we showed that it could be represented by the scalar product of two second-rank tensors:

$$\mathcal{H}_Q = -e\mathrm{T}^2(\nabla E) \cdot \mathrm{T}^2(Q). \tag{8.485}$$

The first second-rank tensor in this expression is given by

$$\mathrm{T}^2(\nabla E) = -\sum_i \frac{e_i}{4\pi\varepsilon_0 R_i^3} C^2(\theta_i, \phi_i), \tag{8.486}$$

and involves electron and proton coordinates outside the quadrupolar nucleus. The second tensor in (8.485) is defined by

$$e\mathrm{T}^2(Q) = e\sum_p R_p^2 C^2(\theta_p, \phi_p), \tag{8.487}$$

and involves proton coordinates inside the quadrupolar nucleus.

The matrix elements of the quadrupole interaction are calculated in various places, for different coupling cases, in the main text. Here we shall carry out the calculation in a case (a) coupled representation, which will enable us to define the nuclear quadrupole moment, the electric field gradient, and the quadrupole coupling constant.

We work in the basis set $|\eta, J, \Omega, I, F, M_F\rangle$ where $\mathbf{F} = \mathbf{J} + \mathbf{I}$ and Ω is the component of electronic angular momentum along the internuclear axis. We shall ignore any possibility of Ω-degeneracy; η refers to any other unspecified quantum numbers. Using the standard results for the matrix elements of the scalar product of two irreducible tensor operators, we obtain

$$\langle \eta, J, \Omega, I, F, M_F| - e\mathrm{T}^2(\nabla E) \cdot \mathrm{T}^2(Q)|\eta', J', \Omega', I', F', M'_F\rangle$$
$$= \delta_{FF'}\delta_{MM'}(-1)^{J'+I+F} \begin{Bmatrix} J & I & F \\ I & J' & 2 \end{Bmatrix}$$
$$\times \langle \eta, J, \Omega\|\mathrm{T}^2(\nabla E)\|\eta', J', \Omega'\rangle\langle I\| - e\mathrm{T}^2(Q)\|I\rangle, \tag{8.488}$$

where we have neglected matrix elements involving excited nuclear states. The problem therefore reduces to that of evaluating the reduced matrix elements in equation (8.488).

For the purely nuclear term we have to evaluate the reduced matrix element of $\mathrm{T}^2(Q)$ in (8.491), and to achieve this we make use of the Wigner–Eckart theorem:

$$\langle I, M_I|\mathrm{T}^2_{p=0}(Q)|I, M_I\rangle = (-1)^{I-M_I} \begin{pmatrix} I & 2 & I \\ -M_I & 0 & M_I \end{pmatrix} \langle I\|\mathrm{T}^2(Q)\|I\rangle, \tag{8.489}$$

We note also that the nuclear quadrupole moment is defined by the following relationship:

$$eQ = \langle I, M_I = I | \sum_p e_p (3Z^2 - R_p^2) | I, M_I = I \rangle$$

$$= \langle I, M_I = I | 2 \sum_p e_p R_p^2 C_0^2(\theta_p, \phi_p) | I, M_I = I \rangle. \quad (8.490)$$

Combining (8.489) and (8.490) gives the results

$$\langle I \| T^2(Q) \| I \rangle = \langle I, M_I = I | \sum_p e_p R_p^2 C_0^2(\theta_p, \phi_p) | I, M_I = I \rangle$$

$$\times (-1)^{M_I - I} \begin{pmatrix} I & 2 & I \\ -M_I & 0 & M_I \end{pmatrix}^{-1}$$

$$= \frac{1}{2} eQ \begin{pmatrix} I & 2 & I \\ -M_I & 0 & M_I \end{pmatrix}^{-1}$$

$$= \frac{1}{2} eQ \begin{pmatrix} I & 2 & I \\ -I & 0 & I \end{pmatrix}^{-1}. \quad (8.491)$$

We now return to equation (8.488) and consider the reduced matrix element of $T^2(\nabla E)$. Without loss of generality we may compute the Z component of $T^2(\nabla E)$, i.e.

$$T_0^2(\nabla E) = -\sum_i \frac{e_i}{4\pi\varepsilon_0 R_i^3} C_0^2(\theta_i, \phi_i). \quad (8.492)$$

Since we shall be interested in the electric field gradient with respect to a molecule-fixed coordinate system, we need to transform (8.492) from space-fixed to molecule-fixed axes; the relationships between the two are illustrated in figure 8.53. Denoting molecule-fixed axes with primes, and space-fixed axes without primes, the spherical harmonic addition theorem gives the result:

$$C_0^2(\theta_i, \phi_i) = P_2(\cos\theta_i) = \frac{4\pi}{5} \sum_{q'} Y_{2q'}(\Theta_i, \Phi_i)^* Y_{2q'}(\Theta, \Phi)$$

$$= \sum_{q'} C_{q'}(\Theta_i, \Phi_i)^* C_{q'}(\Theta, \Phi)$$

$$= \sum_{q'} (-1)^{q'} C_{q'}(\Theta, \Phi) C_{-q'}(\Theta_i, \Phi_i). \quad (8.493)$$

Hence we obtain the result

$$T_0^2(\nabla E) = -\sum_{q'} (-1)^{q'} C_{q'}^2(\Theta, \Phi) \sum_i \frac{e_i}{4\pi\varepsilon_0 r_i^3} C_{-q'}^2(\Theta_i, \Phi_i). \quad (8.494)$$

Note that the $C_{q'}^2(\Theta, \Phi)$ term simply relates the two coordinate systems, whilst the $C_{-q'}^2(\Theta_i, \Phi_i)$ term describes the positions of the electrons relative to the molecule-fixed axes.

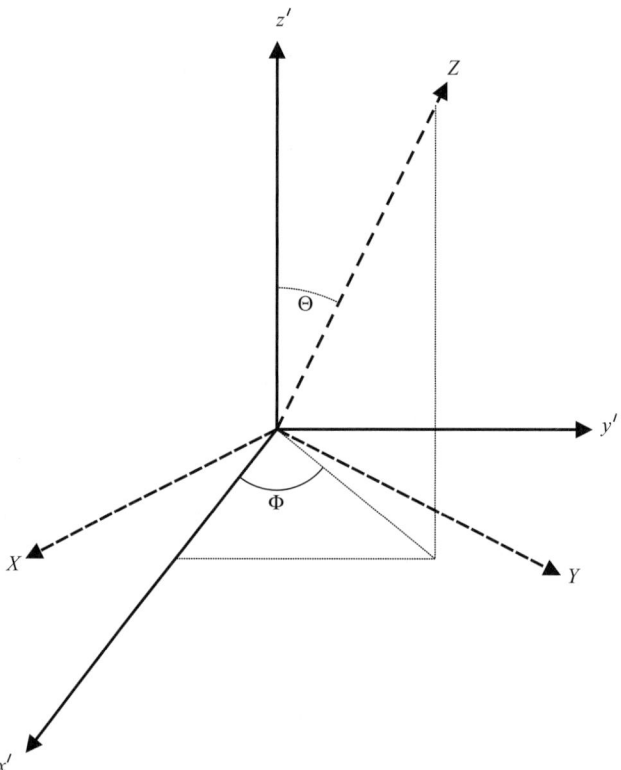

Figure 8.53. Relationship between space-fixed (unprimed) and molecule-fixed (primed) axes for the description of the electric quadrupole interaction.

By making use of (8.494), we can now factorise the general matrix element of $T_0^2(\nabla E)$ as follows:

$$\langle \eta, J, \Omega, M | T_0^2(\nabla E) | \eta', J', \Omega', M' \rangle$$
$$= \sum_{q'} (-1)^{q'} \langle J, \Omega, M | C_{q'}^2(\Theta, \Phi) | J', \Omega', M' \rangle \langle \eta | \sum_i \frac{e_i}{4\pi\varepsilon_0 r_i^3} C_{-q'}^2(\Theta_i, \Phi_i) | \eta' \rangle. \quad (8.495)$$

The value of the right-hand matrix element in (8.495) gives the electric field gradient and we will calculate the $q' = 0$ component. Then $\eta' = \eta$ and we obtain

$$\langle \eta | \sum_i \frac{e_i}{4\pi\varepsilon_0 r_i^3} C_0^2(\Theta_i, \Phi_i) | \eta \rangle = \langle \eta | (4\pi/5)^{1/2} \sum_i \frac{e_i}{4\pi\varepsilon_0 r_i^3} Y_{20}(\Theta_i, \Phi_i) | \eta \rangle$$
$$= \langle \eta | (4\pi/5)^{1/2} \sum_i \frac{e_i}{4\pi\varepsilon_0 r_i^3} \frac{1}{2}(5/4\pi)^{1/2} \left(\frac{3z_i^2 - r_i^2}{r_i^2} \right) | \eta \rangle$$

$$= \langle\eta|\frac{1}{2}\sum_i e_i \frac{(3z_i^2 - r_i^2)}{4\pi\varepsilon_0 r_i^5}|\eta\rangle$$

$$= \langle\eta|\frac{1}{2}\sum_i e_i \frac{1}{4\pi\varepsilon_0}\frac{\partial^2}{\partial z^2}\left(\frac{1}{r_i}\right)|\eta\rangle$$

$$= \frac{1}{2}\left\langle\frac{\partial^2 V}{\partial z^2}\right\rangle_\eta$$

$$= q_{zz}/2, \tag{8.496}$$

where q_{zz}, or q_0, is called the electric field gradient (as we have mentioned before, it is actually the negative of the electric field gradient).

The analysis is not yet complete because we have to consider the left-hand term in equation (8.495) which relates the space-fixed and molecule-fixed coordinate systems. Although we have already selected the $q' = 0$ component, we will retain q' as a variable in order to facilitate later discussion. First we note that Θ and Φ can be expressed in terms of the Euler angles θ and χ as follows:

$$\Theta = \theta, \quad \Phi = \pi - \chi. \tag{8.497}$$

The bra and ket functions in equation (8.495) may also be expressed in terms of rotational matrices:

$$|J', \Omega', M'\rangle = \left(\frac{2J'+1}{8\pi^2}\right)^{1/2} \mathcal{D}_{M'\Omega'}^{J'}(\phi, \theta, \chi)^*,$$
$$\langle J, \Omega, M| = \left(\frac{2J+1}{8\pi^2}\right)^{1/2} \mathcal{D}_{M\Omega}^{J}(\phi, \theta, \chi). \tag{8.498}$$

Consequently the required matrix element in (8.495) may be expressed as the triple integral

$$\langle J, \Omega, M|C_{q'}^2(\Theta, \Phi)|J', \Omega', M'\rangle$$
$$= \{(2J'+1)(2J+1)\}^{1/2}\frac{1}{8\pi^2}$$
$$\times \int\int\int \mathcal{D}_{M\Omega}^{J}(\phi, \theta, \chi) C_{q'}^2(\theta, \pi - \chi) \mathcal{D}_{M'\Omega'}^{J'}(\phi, \theta, \chi)^* \, d\phi \sin\theta \, d\theta \, d\chi. \tag{8.499}$$

Now we note the relationships

$$C_{q'}^2(\theta, \pi - \chi) = (4\pi/5)^{1/2} Y_{2q'}(\theta, \pi - \chi)$$
$$= (4\pi/5)^{1/2}(5/4\pi)^{1/2}\mathcal{D}_{0q'}^{(2)}(\phi, \theta, \pi - \chi)$$
$$= \mathcal{D}_{0-q'}^{(2)}(\phi, \theta, \chi)(-1)^{q'}. \tag{8.500}$$

We incorporate these results into equation (8.499) in the following manner:

$$\langle J, \Omega, M | C_{q'}^2(\Theta, \Phi) | J', \Omega', M' \rangle$$
$$= (-1)^{q'} \{(2J'+1)(2J+1)\}^{1/2} \frac{1}{8\pi^2}$$
$$\times \int \int \int \mathcal{D}_{M\Omega}^J(\phi, \theta, \chi) \mathcal{D}_{0-q'}^{(2)}(\phi, \theta, \chi) \mathcal{D}_{M'\Omega'}^{J'}(\phi, \theta, \chi)^* \, d\phi \sin\theta \, d\theta \, d\chi$$
$$= (-1)^{q'} \{(2J'+1)(2J+1)\}^{1/2} \frac{1}{8\pi^2} (-1)^{M'-\Omega'}$$
$$\times \int \int \int \mathcal{D}_{M\Omega}^J(\phi, \theta, \chi) \mathcal{D}_{0-q'}^{(2)}(\phi, \theta, \chi) \mathcal{D}_{-M',-\Omega'}^{J'}(\phi, \theta, \chi) \, d\phi \sin\theta \, d\theta \, d\chi.$$
(8.501)

We now make use of a standard result for integrating over the product of three rotational matrices (see chapter 5, equation (5.100)),

$$\frac{1}{8\pi^2} \int \int \int \mathcal{D}_{m_1' m_1}^{j_1}(\phi, \theta, \chi) \mathcal{D}_{m_2' m_2}^{j_2}(\phi, \theta, \chi) \mathcal{D}_{m_3' m_3}^{j_3}(\phi, \theta, \chi) \, d\phi \sin\theta \, d\theta \, d\chi$$
$$= \begin{pmatrix} j_1 & j_2 & j_3 \\ m_1' & m_2' & m_3' \end{pmatrix} \begin{pmatrix} j_1 & j_2 & j_3 \\ m_1 & m_2 & m_3 \end{pmatrix},$$
(8.502)

so that equation (8.501) becomes

$$\langle J, \Omega, M | C_{q'}^2(\Theta, \Phi) | J', \Omega', M' \rangle = (-1)^{q'} \{(2J'+1)(2J+1)\}^{1/2} (-1)^{M'-\Omega'}$$
$$\times \begin{pmatrix} J & 2 & J' \\ M & 0 & -M' \end{pmatrix} \begin{pmatrix} J & 2 & J' \\ \Omega & -q' & -\Omega' \end{pmatrix}. \quad (8.503)$$

We can remove the M dependence by making use of the Wigner–Eckart theorem, so that

$$\langle \eta, J, \Omega \| T^2(\nabla E) \| \eta', J', \Omega' \rangle = (-1)^{M-J} \langle \eta, J, \Omega, M | T_{q'}^2(\nabla E) | \eta, J', \Omega', M' \rangle$$
$$\times \begin{pmatrix} J & 2 & J' \\ -M & q' & M' \end{pmatrix}^{-1}. \quad (8.504)$$

We now substitute equations (8.496) and (8.503) into (8.504). Putting $q' = 0$, and noting that the matrix elements must then be diagonal in Ω, we obtain

$$\langle \eta, J, \Omega \| T^2(\nabla E) \| \eta', J', \Omega' \rangle = (-1)^{J-\Omega} \{(2J+1)(2J'+1)\}^{1/2} \begin{pmatrix} J & 2 & J' \\ -\Omega & 0 & \Omega \end{pmatrix} \frac{q_0}{2}.$$
(8.505)

The final result is obtained by combining (8.488), (8.491) and (8.505):

$$\langle \eta, J, \Omega, I, F, M_F | -eT^2(\nabla E) \cdot T^2(Q) | \eta, J', \Omega, I, F, M_F \rangle$$
$$= \frac{eq_0 Q}{4} (-1)^{J'+I+F} (-1)^{J-\Omega} \{(2J+1)(2J'+1)\}^{1/2}$$
$$\times \begin{Bmatrix} I & J & F \\ J' & I & 2 \end{Bmatrix} \begin{pmatrix} J & 2 & J' \\ -\Omega & 0 & \Omega \end{pmatrix} \begin{pmatrix} I & 2 & I \\ -I & 0 & I \end{pmatrix}^{-1}. \quad (8.506)$$

Appendix 8.5. Magnetic hyperfine Hamiltonian and hyperfine constants

In a classic paper, Frosch and Foley [117] derived an effective Hamiltonian to describe the magnetic nuclear hyperfine interactions of a diatomic molecule in an open shell electronic state. Their Hamiltonian was expressed in the following form:

$$\mathcal{H}_{\text{hfs}}(FF) = a\mathbf{I} \cdot \mathbf{L} + b\mathbf{I} \cdot \mathbf{S} + cI_z S_z + (1/2)d\{e^{2i\varphi}I^-S^- + e^{-2i\varphi}I^+S^+\}, \quad (8.507)$$

where z is the molecule-fixed internuclear axis. The term involving the parameter d only produces first-order effects for molecules in Π electronic states. The first three terms can, of course, be written in terms of spherical tensors, as follows:

$$\mathcal{H}_{\text{hfs}}(FF) = a\mathbf{T}^1(\mathbf{I}) \cdot \mathbf{T}^1(\mathbf{L}) + b\mathbf{T}^1(\mathbf{I}) \cdot \mathbf{T}^1(\mathbf{S}) + c\mathbf{T}^1_{q=0}(\mathbf{I})\mathbf{T}^1_{q=0}(\mathbf{S}), \quad (8.508)$$

where the direction of the internuclear axis is denoted by $q = 0$. The constants in these equations were defined by Frosch and Foley [117] as follows:

$$a = 2\mu_B g_N \mu_N (\mu_0/4\pi)\langle \eta, \Lambda | 1/r^3 | \eta, \Lambda \rangle,$$

$$b = g_S \mu_B g_N \mu_N (\mu_0/4\pi) \left\{ \frac{8\pi}{3}\langle \eta, \Lambda | \delta(r) | \eta, \Lambda \rangle - \frac{1}{2}\langle \eta, \Lambda | (3\cos^2\theta - 1)/r^3 | \eta, \Lambda \rangle \right\},$$

$$c = g_S \mu_B g_N \mu_N (\mu_0/4\pi) \frac{3}{2}\langle \eta, \Lambda | (3\cos^2\theta - 1)/r^3 | \eta, \Lambda \rangle. \quad (8.509)$$

The constant b therefore contains contributions from two quite different magnetic interactions, the Fermi contact and the electron-nuclear dipolar interactions. Interpretation of the magnitudes of these constants in terms of electronic structure theory always involves the separate assessment of these different effects, so that we prefer to use an effective Hamiltonian which separates them at the outset. Consequently the effective magnetic hyperfine Hamiltonian used throughout this book is

$$\mathcal{H}_{\text{hfs}} = a\mathbf{T}^1(\mathbf{I}) \cdot \mathbf{T}^1(\mathbf{L}) + b_F \mathbf{T}^1(\mathbf{I}) \cdot \mathbf{T}^1(\mathbf{S})$$
$$- \sqrt{10} g_S \mu_B g_N \mu_N (\mu_0/4\pi) \mathbf{T}^1(\mathbf{I}) \cdot \mathbf{T}^1(\mathbf{S}, \mathbf{C}^2). \quad (8.510)$$

The orbital hyperfine constant a is the same in both (8.509) and (8.510), but the second term in (8.510) describes *only* the Fermi contact interaction, the constant b_F being equal to the first part of b in (8.509). It is clear from a comparison of the two equations that

$$b_F = b + (1/3)c. \quad (8.511)$$

The third term in equation (8.510) describes *only* the electron–nuclear spin dipolar interaction, with the first-rank tensor $\mathbf{T}^1(\mathbf{S}, \mathbf{C}^2)$ being constructed so that

$$\mathbf{T}^1_q(\mathbf{S}, \mathbf{C}^2) = -\sum_{q_1, q_2} (-1)^q \mathbf{T}^1_{q_1}(\mathbf{S}) \mathbf{T}^2_{q_2}(\mathbf{C}) \begin{pmatrix} 1 & 2 & 1 \\ q_1 & q_2 & -q \end{pmatrix}. \quad (8.512)$$

As we discussed in appendix 8.2, all of the spatial and angular dependence of the dipolar interaction is contained within the components of the second-rank tensor in

(8.512), defined by

$$T_q^2(\mathbf{C}) = C_q^2(\theta, \phi)(r^{-3}) = \left(\frac{4\pi}{5}\right)^{1/2} Y_{2q}(\theta, \phi)\left(\frac{1}{r^3}\right), \tag{8.513}$$

where $Y_{2q}(\theta, \phi)$ is a spherical harmonic. The relationship with the Frosch and Foley constants b and c becomes clear when the $q = 0$ component of the dipolar interaction is considered, since its matrix elements must be diagonal in Λ and the expectation value of (8.513) is given by

$$\langle T_0^2(\mathbf{C}) \rangle = \frac{1}{2}\left\langle \frac{(3\cos^2\theta - 1)}{r^3} \right\rangle. \tag{8.514}$$

We use the symbol t_0, defined by

$$t_0 = \frac{1}{2} g_S \mu_B g_N \mu_N (\mu_0/4\pi) \langle \eta, \Lambda | (3\cos^2\theta - 1)/r^3 | \eta, \Lambda \rangle, \tag{8.515}$$

which is now clearly related to the Frosch and Foley c constant by

$$t_0 = \frac{c}{3}. \tag{8.516}$$

In many cases in this book, particularly when Λ-doubling is involved, we also consider the $q \neq 0$ components of the dipolar interaction, which follow naturally from the irreducible tensor formalism. The matrix elements of such components are off-diagonal in Λ, and their angular dependence is described by the spherical harmonics $Y_{2,\pm1}(\theta, \phi)$ or $Y_{2,\pm2}(\theta, \phi)$. In such cases, we shall use the constants $t_{\pm1}$ and $t_{\pm2}$, defined by analogy with t_0 in equation (8.515). To be specific, the parameter d of Frosch and Foley is related to our t_2 by the expression

$$t_2 = g_S \mu_B g_N \mu_N (\mu_0/4\pi) \langle \eta, \Lambda | \sin^2\theta/r^3 | \eta, \Lambda \rangle = (2/3)d. \tag{8.517}$$

References

[1] *Atomic and Molecular Beam Methods*, Vol.1 (1988), Vol.2 (1992), ed. G. Scoles, Oxford University Press Inc., New York.
[2] I.I. Rabi, S. Millman, P. Kusch and J.R. Zacharias, *Phys. Rev.*, **55**, 526 (1939).
[3] J.M.B. Kellogg, I.I. Rabi, N.F. Ramsey, Jr. and J.R. Zacharias, *Phys. Rev.*, **56**, 728 (1939).
[4] H.G. Kolsky, T.E. Phipps, Jr., N.F. Ramsey and H.B. Silsbee, *Phys. Rev.*, **87**, 395 (1952).
[5] N.F. Ramsey, *Phys. Rev.*, **76**, 996 (1949); **78**, 695 (1950).
[6] H.G. Kolsky, T.E. Phipps, N.F. Ramsey and H.B. Silsbee, *Phys. Rev.*, **79**, 883 (1950).
[7] N.F. Ramsey, *Phys. Rev.*, **85**, 60 (1952).
[8] N.J. Harrick and N.F. Ramsey, *Phys. Rev.*, **88**, 228 (1952).
[9] J.M.B. Kellogg, I.I. Rabi, N.F. Ramsey and J.R. Zacharias, *Phys. Rev.*, **57**, 677 (1940).
[10] N.J. Harrick, R.G. Barnes, P.J. Bray and N.F. Ramsey, *Phys. Rev.*, **90**, 260 (1953).
[11] R.F. Code and N.F. Ramsey, *Phys. Rev.*, **A4**, 1945 (1971).
[12] N.F. Ramsey, *Phys. Rev.*, **58**, 226 (1940).

[13] J.A. Pople, W.G. Schneider and H.J. Bernstein, *High-Resolution Nuclear Magnetic Resonance*, McGraw-Hill Book Company, Inc., New York, 1959.
[14] W.E. Lamb, Jr., *Phys. Rev.*, **60**, 817 (1941).
[15] N.F. Ramsey, *Phys. Rev.*, **78**, 699 (1950); *Physica*, **17**, 303 (1951); *Phys. Rev.*, **86**, 243 (1952).
[16] G.C. Wick, *Z. Physik*, **85**, 25 (1933).
[17] N.F. Ramsey, *Phys. Rev.*, **58**, 226 (1940).
[18] N.F. Ramsey, *Phys. Rev.*, **87**, 1075 (1952).
[19] N.F. Ramsey, *Phys. Rev.*, **78**, 221 (1950).
[20] S.A. Smith, W.E. Palke and J.T. Gerig, *Concepts in Magnetic Resonance*, **4**, 107 (1992).
[21] N.F. Ramsey, *Phys. Rev.*, **83**, 540 (1951).
[22] S.D. Rosner, R.A. Holt and T.D. Gaily, *Phys. Rev. Lett.*, **35**, 785 (1975).
[23] P.E. Van Esbroeck, R.A. McLean, T.D. Gaily, R.A. Holt and S.D. Rosner, *Phys. Rev.*, **A32**, 2595 (1985).
[24] S.I. Chan, M.R. Baker and N.F. Ramsey, *Phys. Rev.*, **A136**, 1224 (1964).
[25] I. Ozier, P.N. Yi, A. Khoshla and N.F. Ramsey, *J. Chem. Phys.*, **46**, 1530 (1967).
[26] P. Kusch, *Phys. Rev.*, **75**, 887 (1949); **76**, 138 (1949).
[27] R.A. Logan, R.E. Cote and P. Kusch, *Phys. Rev.*, **86**, 280 (1952).
[28] H.J. Zeiger and D.I. Boleff, *Phys. Rev.*, **85**, 788 (1952).
[29] F. Mehran, R.A. Brooks and N.F. Ramsey, *Phys. Rev.*, **141**, 93 (1966).
[30] D.I. Boleff and H. J. Zeiger, *Phys. Rev.*, **85**, 799 (1952).
[31] R.E. Cote and P. Kusch, *Phys. Rev.*, **90**, 103 (1953).
[32] O.W. Richardson, *Molecular Hydrogen and its Spectrum*, Yale University Press, 1934.
[33] G. Herzberg, *Spectra of Diatomic Molecules*, D. Van Nostrand Company, Inc., Princeton, 1950.
[34] G. Herzberg and K.P. Huber, *Constants of Diatomic Molecules*, Van Nostrand Reinhold Company, New York, 1979.
[35] W. Lichten, *Phys. Rev.*, **120**, 848 (1960); *Phys. Rev.*, **126**, 1020 (1962).
[36] W. Lichten, *J. Chem. Phys.*, **26**, 306 (1957).
[37] W. Lichten, *Phys. Rev.*, **A3**, 594 (1971).
[38] A.N. Jette and P. Cahill, *Phys. Rev.*, **160**, 35 (1967).
[39] P.R. Fontana, *Phys. Rev.*, **125**, 220 (1962).
[40] L.-Y. Chow Chiu, *J. Chem. Phys.*, **40**, 2276 (1964).
[41] L.-Y. Chow Chiu, *Phys. Rev.*, **137**, A384 (1965).
[42] P.R. Brooks, W. Lichten and R. Reno, *Phys. Rev.*, **A4**, 2217 (1971).
[43] R.S. Freund, T.A. Miller, D. De Santis and A. Lurio, *J. Chem. Phys.*, **53**, 2290 (1970).
[44] D. De Santis, A. Lurio, T.A. Miller and R.S. Freund, *J. Chem. Phys.*, **58**, 4625 (1973).
[45] S.M. Naude, *Proc. R. Soc. Lond.*, **A136**, 114 (1932).
[46] P.K. Carroll, *Proc. R. Irish Acad.*, **A54**, 369 (1952).
[47] W. Benesch, J.T. Vanderslice, S.G. Tilford and P.G. Wilkinson, *Astrophys. J.*, **142**, 1227 (1965).
[48] H.K. Hughes, *Phys. Rev.*, **72**, 614 (1947).
[49] R. de L. Kronig, *Proc. Nat. Acad. Wash.*, **12**, 608 (1926).
[50] J.W. Trischka, *Phys. Rev.*, **74**, 718 (1948).
[51] T.C. English and J.C. Zorn, *J. Chem. Phys.*, **47**, 3896 (1967).

[52] N.F. Ramsey, *Molecular Beams*, Oxford University Press, Oxford, 1956.
[53] G. Graff and O. Runolfsson, *Z. Physik*, **187**, 140 (1965).
[54] J.C. Zorn, D.A. Stephenson, J.T. Dickinson and T.C. English, *J. Chem. Phys.*, **47**, 3904 (1967).
[55] S.M. Freund, G.A. Fisk, D.R. Herschbach and W. Klemperer, *J. Chem. Phys.*, **54**, 2510 (1971).
[56] J.C. Zorn, T.C. English, J.T. Dickinson and D.A. Stephenson, *J. Chem. Phys.*, **45**, 3731 (1966).
[57] G. Graff and O. Runolfsson, *Z. Physik*, **176**, 90 (1963).
[58] R. Van Wachem and A. Dymanus, *J. Chem. Phys.*, **46**, 3749 (1967).
[59] C.D. Hollowell, A.J. Herbert and K. Street, Jr., *J. Chem. Phys.*, **41**, 3540 (1964).
[60] G. Graff and G. Werth, *Z. Physik*, **183**, 223 (1965).
[61] L. Wharton, L.P. Gold and W. Klemperer, *Phys. Rev.*, **133**, B270 (1964).
[62] R.C. Hilborn, T.F. Gallagher and N.F. Ramsey, *J. Chem. Phys.*, **56**, 855 (1972).
[63] J.L. Cecchi and N.F. Ramsey, *J. Chem. Phys.*, **60**, 53 (1974).
[64] T.F. Gallagher, R.C. Hilborn and N.F. Ramsey, *J. Chem. Phys.*, **56**, 5972 (1972).
[65] P.A. Bonczyk and V.W. Hughes, *Phys. Rev.*, **161**, 15 (1967).
[66] V. Hughes and L. Grabner, *Phys. Rev.*, **79**, 314 (1950).
[67] D.T.F. Marple and J.W. Trischka, *Phys. Rev.*, **103**, 597.
[68] R.L. Matcha, *J. Chem. Phys.*, **47**, 4595 (1967).
[69] A.J. Hebert, F.J. Lovas, C.A. Melendres, C.D. Hollowell, T.L. Story and K. Street, *J. Chem. Phys.*, **48**, 2824 (1968).
[70] J.W. Cederberg and C.E. Miller, *J. Chem. Phys.*, **50**, 3547 (1969).
[71] F.H. de Leeuw, R. van Wachem and A. Dymanus, *J. Chem. Phys.*, **53**, 981 (1970).
[72] R. van Wachem and A. Dymanus, *J. Chem. Phys.*, **46**, 3749 (1967).
[73] R.G. Luce and J.W. Trischka, *Phys. Rev.*, **82**, 323 (1951); **83**, 851 (1951); *J. Chem. Phys.*, **21**, 105 (1953).
[74] J.W. Trischka, *J. Chem. Phys.*, **25**, 784 (1956).
[75] A.J. Hebert and K. Street, Jr., *Phys. Rev.*, **178**, 205 (1969).
[76] A.J. Hebert, F.W. Breivogel, Jr. and K. Street, Jr., *J. Chem. Phys.*, **41**, 2368 (1964).
[77] R. van Wachem, F.H. de Leeuw and A. Dymanus, *J. Chem. Phys.*, **47**, 2256 (1967).
[78] F.H. de Leeuw, R. van Wachem and A. Dymanus, *J. Chem. Phys.*, **50**, 1393 (1969).
[79] F.W. Breivogel, Jr., A.J. Hebert and K. Street, Jr., *J. Chem. Phys.*, **42**, 1555 (1965).
[80] A.R. Jacobson and N.F. Ramsey, *J. Chem. Phys.*, **65**, 1211 (1976).
[81] R.C. Miller and J.C. Zorn, *J. Chem. Phys.*, **50**, 3748 (1969).
[82] L. Wharton, M. Kaufman and W. Klemperer, *J. Chem. Phys.*, **37**, 621 (1962); **39**, 240 (1963).
[83] M. Kaufman, L. Wharton and W. Klemperer, *J. Chem. Phys.*, **43**, 943 (1965).
[84] J.W. Raymonda, J.S. Muenter and W. Klemperer, *J. Chem. Phys.*, **52**, 3458 (1970).
[85] R.E. Davis and J.S. Muenter, *J. Chem. Phys.*, **61**, 2940 (1974).
[86] R. Weiss, *Phys. Rev.*, **131**, 659 (1963).
[87] J.S. Muenter and W. Klemperer, *J. Chem. Phys.*, **52**, 6033 (1970).
[88] J.S. Muenter, *J. Chem. Phys.*, **56**, 5409 (1972).
[89] F.H. de Leeuw and A. Dymanus, *J. Mol. Spectrosc.*, **48**, 427 (1973).
[90] E.W. Kaiser, *J. Chem. Phys.*, **53**, 1686 (1970).

[91] O.B. Dabbousi, W.L. Meerts, F.H. de Leeuw and A. Dymanus, *Chem. Phys.*, **2**, 473 (1973).
[92] R.F. Code, A. Khosla, I. Ozier, N.F. Ramsey and P.N. Yi, *J. Chem. Phys.*, **49**, 1895 (1968).
[93] M.R. Baker, H.M. Nelson, J.A. Leavitt and N.F. Ramsey, *Phys. Rev.*, **121**, 807 (1961).
[94] W.M. Huo, unpublished work (1970).
[95] P.K. Hindermann and C.D. Cornwell, *J. Chem. Phys.*, **48**, 4148 (1968).
[96] D.E. O'Reilly, *J. Chem. Phys.*, **36**, 274 (1962).
[97] R.M. Stevens and W.N. Lipscomb, *J. Chem. Phys.*, **41**, 184 (1964).
[98] P.E. Cade and W.M. Huo, *J. Chem. Phys.*, **47**, 614, 648 (1967).
[99] A.D. McLean and M. Yoshimine, *J. Chem. Phys.*, **47**, 3256 (1967).
[100] C.F. Bender and E.R. Davidson, *Phys. Rev.*, **183**, 23 (1969).
[101] J.L. Dunham, *Phys. Rev.*, **41**, 713, 721 (1932); **49**, 797 (1936); **34**, 446 (1929).
[102] D.H. Rank, B.S. Rao and T.A. Wiggins, *J. Mol. Spectrosc.*, **17**, 122 (1965).
[103] D.H. Rank, D.P. Eastman, B.S. Rao and T.A. Wiggins, *J. Opt. Soc. Am.*, **52**, 1 (1962).
[104] J. Trischka and H. Salwen, *J. Chem. Phys.*, **31**, 218 (1959).
[105] W. Huo, unpublished results (1973).
[106] R.F. Code, A. Khosla, I. Ozier, N.F. Ramsey and P.N. Yi, *J. Chem. Phys.*, **48**, 4148 (1968).
[107] W.H. Flygare, *Molecular Structure and Dynamics*, Prentice-Hall, Inc., New Jersey, 1978.
[108] W.S. Benedict, R. Herman, G.E. Moore and S. Silverman, *J. Chem. Phys.*, **26**, 1671 (1957).
[109] C. Schlier, *Fortschr. Physik*, **9**, 455 (1961).
[110] S.M. Blinder, *J. Chem. Phys.*, **35**, 974 (1961).
[111] W. Kolos and L. Wolniewicz, *J. Chem. Phys.*, **45**, 944 (1966).
[112] S.M. Freund, E. Herbst, R.P. Mariella and W. Klemperer, *J. Chem. Phys.*, **56**, 1467 (1972).
[113] A.C. Wahl, unpublished work (1972).
[114] M. Yoshimine, *J. Chem. Phys.*, **57**, 1108 (1972).
[115] J.M. Brown, M. Kaise, C.M.L. Kerr and D.J. Milton, *Mol. Phys.*, **36**, 553 (1978).
[116] J.J. Gallagher and C.M. Johnson, *Phys. Rev.*, **103**, 1727 (1956).
[117] R.A. Frosch and H.M. Foley, *Phys. Rev.*, **88**, 1337 (1952).
[118] C. Yamada, M. Fujitake and E. Hirota, *J. Chem. Phys.*, **91**, 137 (1989).
[119] J.H. van Vleck, *Phys. Rev.*, **33**, 467 (1929).
[120] R. S. Mulliken and A. Christy, *Phys. Rev.*, **38**, 87 (1931).
[121] G.C. Dousmanis, T.M. Sanders and C.H. Townes, *Phys. Rev.*, **100**, 1735 (1955).
[122] D.L. Cooper and W.G. Richards, *J. Chem. Phys.*, **73**, 3515 (1980).
[123] R.M. Neumann, *Astrophys. J.*, **161**, 779 (1970).
[124] W.L. Meerts and A. Dymanus, *J. Mol. Spectrosc.*, **44**, 320 (1972).
[125] W.L. Meerts, *Chem. Phys.*, **14**, 421 (1976).
[126] P. Kristiansen, *J. Mol. Spectrosc.*, **66**, 177 (1977).
[127] C.A. Burrus and W. Gordy, *Phys. Rev.*, **92**, 1437 (1953).
[128] P.G. Favero, A.M. Mirri and W. Gordy, *Phys. Rev.*, **114**, 1534 (1959).
[129] R.T. Hall and J.M. Dowling, *J. Chem. Phys.*, **45**, 1899 (1966).
[130] J.M. Brown, A.R.H. Cole and F.R. Honey, *Mol. Phys.*, **23**, 287 (1972).
[131] F.H. de Leeuw and A. Dymanus, *J. Mol. Spectrosc.*, **48**, 427 (1973).
[132] J.M. Brown and A.J. Merer, *J. Mol. Spectrosc.*, **74**, 488 (1979).
[133] T.D. Varberg, F. Stroh and K.M. Evenson, *J. Mol. Spectrosc.*, **196**, 5 (1999).
[134] A.H. Salek, G. Winnewisser and K.M.T. Yamada, *Mol. Phys.*, **76**, 1443 (1992).

[135] F.C. van den Heuvel, W.L. Meerts and A. Dymanus, *J. Mol. Spectrosc.*, **84**, 162 (1980).
[136] A.H. Salek, K.M.T. Yamada and G. Winnewisser, *Mol. Phys.*, **72**, 1135 (1991).
[137] R.M. Dale, J.W.C. Johns, A.R.W. McKellar and M. Riggin, *J. Mol. Spectrosc.*, **67**, 440 (1977).
[138] R.S. Lowe, A.R.W. McKellar, P. Veillette and W.L. Meerts, *J. Mol. Spectrosc.*, **88**, 372 (1981).
[139] R.L. Brown and H.E. Radford, *Phys. Rev.*, **147**, 6 (1966).
[140] G.C. Dousmanis, *Phys. Rev.*, **97**, 967 (1955).
[141] H.E. Radford, *Phys. Rev.*, **122**, 114 (1961); *Phys. Rev.*, **126**, 1035 (1962).
[142] W.L. Meerts and A. Dymanus, *Can. J. Phys.*, **53**, 2123 (1975).
[143] J.J. ter Meulen and A. Dymanus, *Astrophys. J.*, **172**, L21 (1972).
[144] H.E. Radford, *Rev. Sci. Instr.*, **39**, 1687 (1968).
[145] R.L. Poynter and R.A. Beaudet, *Phys. Rev. Lett.*, **21**, 305 (1968).
[146] J.J. ter Meulen, unpublished (1970).
[147] J.A. Ball, D.F. Dickinson, C.A. Gottlieb and H.E. Radford, *Astron. J.*, **75**, 762 (1970).
[148] J.M. Brown, C.M.L. Kerr, F.D. Wayne, K.M. Evenson and H.E. Radford., *J. Mol. Spectrosc.*, **86**, 544 (1981).
[149] W.L. Meerts, *Chem. Phys. Lett.*, **46**, 24 (1977).
[150] K. Kayama, *J. Chem. Phys.*, **39**, 1507 (1963).
[151] Y. Kotake, M. Ono and K. Kuwata, *Bull. Chem. Soc. Japan*, **44**, 2056 (1971).
[152] G.L. Bendazzoli, F. Bernadi and P. Palmieri, *Mol. Phys.*, **23**, 193 (1972).
[153] W.L. Meerts and A. Dymanus, *Astrophys. J.*, **180**, L93 (1973); *Chem. Phys. Lett.*, **23**, 45 (1973).
[154] K.I. Peterson, G.T. Fraser and W. Klemperer, *Can. J. Phys.*, **62**, 1502 (1984).
[155] H.-J. Werner, P. Rosmus and E.-A. Reinsch, *J. Chem. Phys.*, **79**, 905 (1983).
[156] E.W. Kaiser, *J. Chem. Phys.*, **53**, 1686 (1970).
[157] B.A. Thrush, *Acc. Chem. Res.*, **14**, 116 (1981).
[158] D.M. Rank, C.H. Townes and W.J. Welch, *Science*, **174**, 1083 (1971).
[159] R.M. Crutcher and W.D. Watson, *Astrophys. J.*, **203**, L123 (1976).
[160] A. Budó, *Z. Physik.*, **96**, 219 (1935); **98**, 437 (1936).
[161] L. Gerö, G. Herzberg and R. Schmid, *Phys. Rev.*, **52**, 467 (1937).
[162] R.T. Birge, *Phys. Rev.*, **28**, 1157 (1926); B.S. Beer, *Z. Physik*, **107**, 73 (1937); L. Gerö, *Z. Physik*, **109**, 204 (1938).
[163] R.J. Saykally, T.A. Dixon, T.G. Anderson, P.G. Szanto and R.C. Woods, *J. Chem. Phys.*, **87**, 6423 (1987).
[164] R.J. Saykally, K.M. Evenson, E.R. Comben and J.M. Brown, *Mol. Phys.*, **58**, 735 (1986).
[165] R.S. Freund and W. Klemperer, *J. Chem. Phys.*, **43**, 2422 (1965).
[166] R.C. Stern, R.H. Gammon, M.E. Lesk, R.S. Freund and W. Klemperer, *J. Chem. Phys.*, **52**, 3467 (1970).
[167] K. Freed, *J. Chem. Phys.*, **45**, 4214 (1966).
[168] R.H. Gammon, R.C. Stern and W. Klemperer, *J. Chem. Phys.*, **54**, 2151 (1971).
[169] R.H. Gammon, R.C. Stern, M.E. Lesk, B.G. Wicke and W. Klemperer, *J. Chem. Phys.*, **54**, 2136 (1971).

9 Microwave and far-infrared magnetic resonance

9.1. Introduction

Microwave magnetic resonance, which has often been called 'gas phase electron resonance', and far-infrared laser magnetic resonance have been extremely important techniques in the study of free radicals. These techniques depend upon the presence of a large magnetic moment for the species under investigation, because both rely on the ability to tune the energy levels with a magnetic field into resonance with fixed frequency radiation. Although historically the first study of the rotational levels and their fine structure involved fairly conventional swept-frequency microwave spectroscopy, applied to the OH radical [1], subsequent development of the subject depended initially on the success of magnetic resonance methods. Pure microwave spectroscopy of gaseous free radicals has now become almost routine, and many examples will be described in chapter 10. We have, after some deliberation, decided to present the exciting subject of magnetic resonance with due regard to its historical development. Nevertheless, this and the two following chapters should be taken together for a balanced view. Laser magnetic resonance techniques are still widely used, but microwave magnetic resonance has now been largely superseded by swept-frequency methods, without the presence of an external magnetic field. Zeeman effects can, of course, still be investigated, but they are not an essential part of the detection method.

9.2. Experimental methods

9.2.1. Microwave magnetic resonance

Electron spin resonance (e.s.r.) spectroscopy, applied to free radicals in condensed phases, is a long established technique with several commercially available spectrometers. The gas phase applications we will describe have little in common with condensed phase studies, and are much more a part of rotational spectroscopy. However, the experimental methods used for condensed phase studies can be applied to the study of gases with rather little change, so it is appropriate first to describe a typical microwave magnetic resonance spectrometer, as illustrated schematically in figure 9.1.

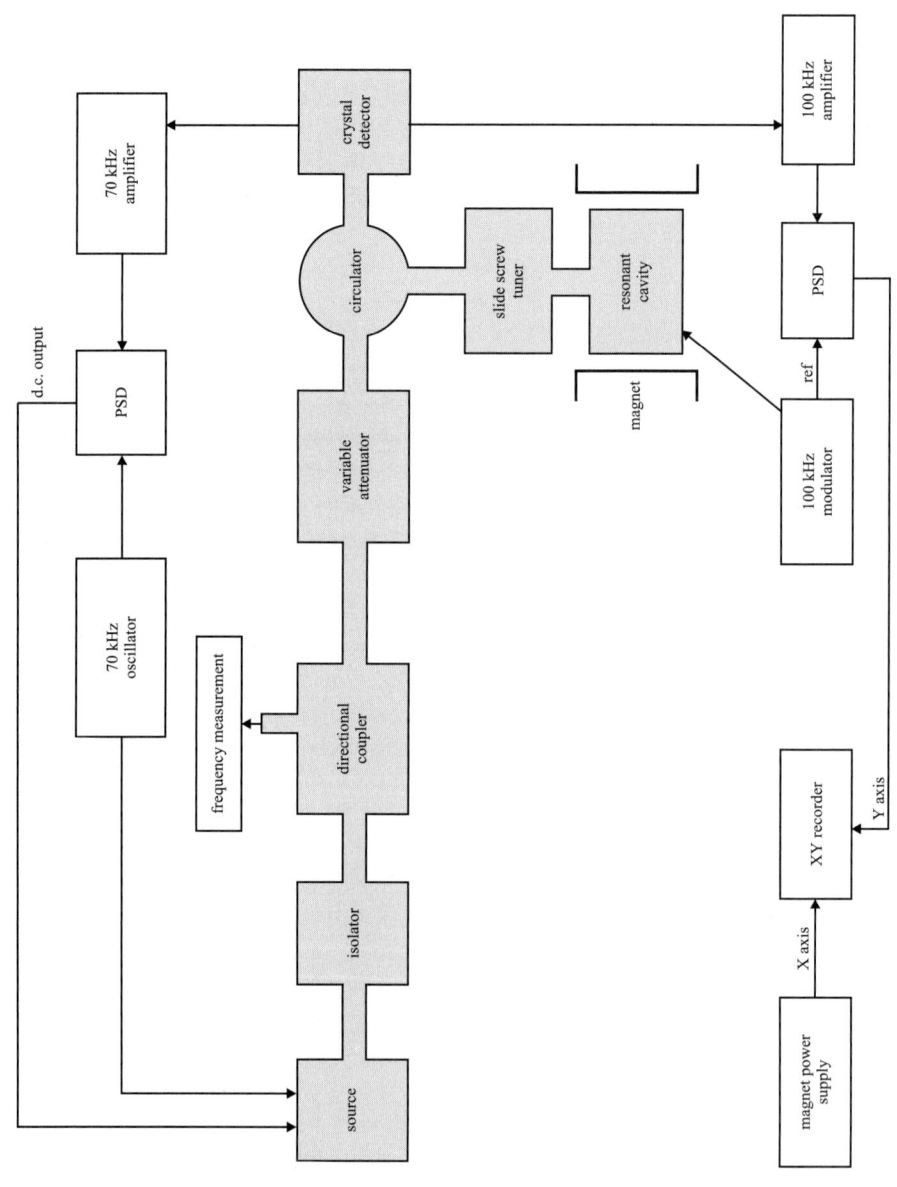

Figure 9.1. Block diagram of a microwave magnetic resonance spectrometer.

Most e.s.r. spectrometers operate in the so-called X-band microwave frequency range; 9.5 GHz is a typical operating frequency, and it is convenient because the magnetic fields required for resonance are in the range accessible to conventional electromagnets. Spectrometers operating up to 40 GHz have been used but they are less suitable for gas phase studies, mainly because of the smaller size of the resonant cavity.

The operating frequency of the spectrometer depicted in figure 9.1 is determined by the natural resonance frequency of the cavity, which in turn depends upon its shape, dimensions and fundamental microwave mode. These important matters are discussed in some detail below. The microwave radiation is generated by a suitable source, usually a klystron or backward-wave oscillator, and its frequency is locked to the resonant frequency of the cavity by means of a feed-back control system. The radiation is propagated through waveguide to a three-port circulator via an isolator, to prevent back-reflection, a directional coupler to provide a fraction for accurate frequency measurement, and a variable attenuator to allow for control of the power level. Microwaves enter the first port of the circulator and are coupled to the resonant cavity through a slide-screw tuner and a variable iris. In the initial setting up of the instrument, all of the incident power is stored in the cavity; any reflected power emerges through the third port where it is converted to an electrical current by means of a crystal detector. Spectroscopic resonance lines are sought by scanning the magnetic field, and when the resonance condition is satisfied, power in the cavity is absorbed by the sample, the impedance of the cavity arm is changed, and the reflected power level to the detector changes. It is usual to modulate the resonant absorption frequency by superimposing a small oscillating magnetic or electric field on the static magnetic field; this converts the detector current into an a.c. signal, so that a.c. detection circuitry may be used.

Everything described thus far would be common to most e.s.r. spectrometers employed for condensed phase studies; the part which is different and somewhat specific for gas phase investigations is the resonant cavity. In designing a cavity one must decide whether the cavity is to be an integral part of a vacuum system, or will simply hold within it a quartz cell containing, usually, a flowing gas sample. One must also decide whether the spectroscopic objective is to detect electric or magnetic dipole transitions, and therefore whether the relevant microwave field should be perpendicular to or parallel with the static magnetic field. We must therefore investigate certain key aspects of the theory of microwave cavities, and show how these determine critical aspects of their design and operation. Further details may be found in specialist books on microwave theory and design; that written by Seeger [2] is an excellent introduction.

Conventional condensed phase e.s.r. studies usually employ a rectangular cavity, but most gas phase studies have been made using cylindrical cavities operating in the so-called TE_{011} mode. In a TE_{mnp} mode, TE stand for 'transverse electric', the integer m represents the number of E field maxima in a 180° angle measured in a plane perpendicular to the axis of the cylinder, n represents the number of E field maxima between the centre and the wall, and p is the number of E field maxima along the axis

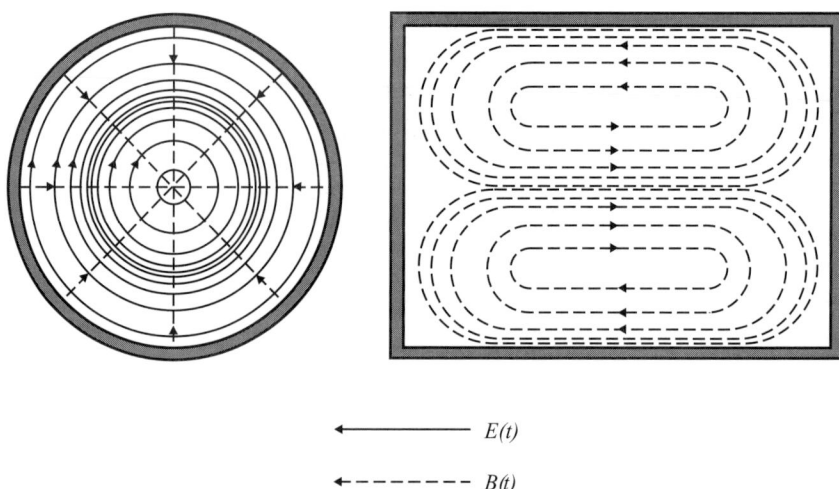

Figure 9.2. Microwave electric and magnetic fields in a cylindrical TE$_{011}$ cavity.

of the cylinder. The resonant wavelength of a cylindrical cavity, λ_0, is given by

$$\lambda_0 = \frac{2}{\{(2\chi_{mn}/D)^2 + (p/L)^2\}^{1/2}} \quad (9.1)$$

where D is the diameter of the cavity, L is its length and χ_{mn} is a constant related to a Bessel function. The distributions of the microwave electric and magnetic fields in a TE$_{011}$ mode cavity are illustrated in figure 9.2. The oscillating electric field is parallel to the cylinder walls, with zero amplitude at the centre of the cylinder, and maximum amplitude halfway between the centre and the walls. The oscillating magnetic field is parallel to the axis of the cylinder, except at the ends; it has maximum amplitude at the centre and at the cavity walls. Hence for magnetic dipole transitions the gas sample should ideally be at the centre of the cavity, which should itself be oriented with the external static magnetic field perpendicular to the cylinder axis to drive $\Delta M = \pm 1$ transitions. Conversely, for electric dipole transitions the external magnetic field should be parallel to the cylinder axis, and hence perpendicular to the microwave electric field. The microwave fields cause induced wall currents, and since in the TE$_{011}$ mode these do not flow across the junctions between the cylinder and the end walls, the latter can be electrically insulated from the cylinder body. Microwave radiation is coupled through a small coupling iris located at the midpoint of the cylinder walls. The final general point to make is that although any resonant cavity has an optimum resonant frequency, it also has a bandwidth $\Delta \nu$ which is typically about 1 MHz. The performance of a cavity is usually described by the value of its Q factor, defined by $Q = \nu_0/\Delta \nu$. At a resonant frequency of 10 GHz and a bandwidth of 1 MHz, the Q factor is 10 000, which is a satisfactory figure of merit and provides high sensitivity.

In order to see how these principles are put into practice we now describe a cavity designed by Carrington, Levy and Miller [3] specifically for gas phase studies of free radicals; it is illustrated in figure 9.3. It operates in the cylindrical TE$_{011}$ mode and

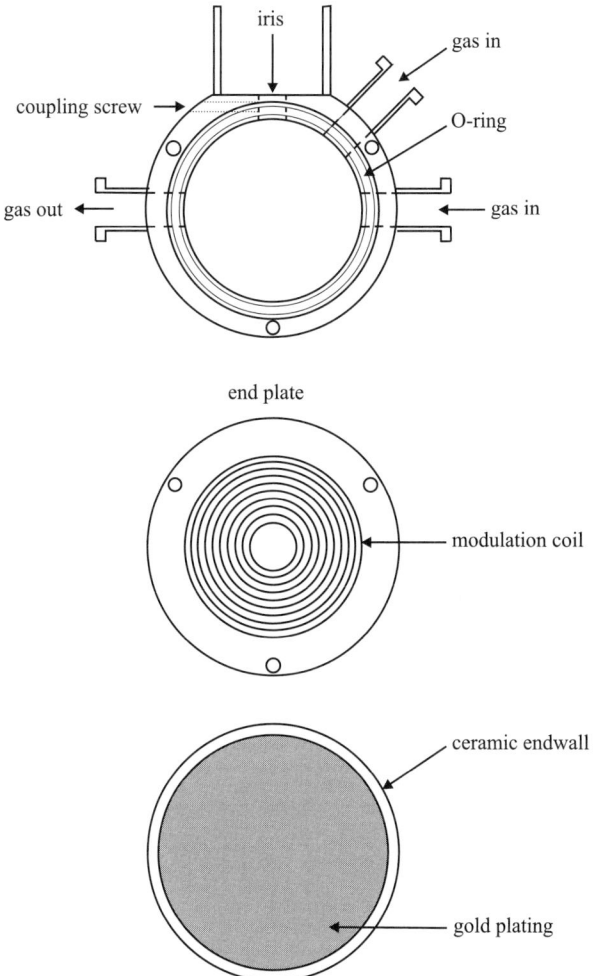

Figure 9.3. Schematic diagram of a dual purpose cylindrical TE$_{011}$ cavity used for gaseous free radical studies [3].

is an integral part of a gas flow system, in that the two inlet ports may be used to allow the mixing of two different gas flows inside the cavity. The cavity is oriented so that the static magnetic field is parallel to the cylinder axis, and it may be used with either electric or magnetic field modulation. For electric field modulation the end plates are of solid gold-plated brass and are electrically insulated from the cylindrical body; they may therefore be used directly as Stark electrodes, with the modulating field also parallel to the cylinder axis. Alternatively for magnetic field modulation the two end walls consist of ceramic plates with modulation coils mounted outside. The insides of the ceramic plates are gold plated, the gold thickness being small enough to allow penetration of the 100 kHz modulating field, but large enough to preserve the high Q of the cavity. Coupling to the microwave radiation occurs through an iris located at the top of the cavity, with a metal-tipped Teflon screw provided to optimise the coupling.

Several different methods have been used successfully to generate a detectable concentration of short-lived free radicals inside the resonant cavity. The simplest method is a microwave discharge in the flowing gas, located upstream of the resonant cavity; discharges in water vapour, for example, yield readily detectable concentrations of OH radicals. Shorter lived species have been produced by atom abstraction reactions inside the cavity, for example, by mixing fluorine atoms, produced by a microwave discharge in CF_4, with a suitable secondary gas. Reaction of F atoms with OCS, for example, produces detectable concentrations of the SF radical [4].

9.2.2. Far-infrared laser magnetic resonance

The extension to higher frequencies of magnetic resonance methods for studying open shell molecules was clearly desirable for several reasons. Firstly, light molecules, particularly hydrides like OH and CH, have even their lowest rotational transitions in the far-infrared region. Secondly, increased frequency would be accompanied by increased sensitivity, so that free radicals which could not be detected in the microwave region (CH being a notable example) might then become accessible in the far-infrared. There is, however, a tricky spectral region lying between the microwave/millimetre region, and the far-infrared. Resonant cavities become so small, even in the millimetre wave region, as to be impractical for gas phase studies, and the transition to optical methods has to be made as one approaches the far-infrared region. Far-infrared spectroscopy was, in any case, the Cinderella region of the electromagnetic spectrum. However, all of the apparent difficulties were overcome in the brilliant development of far-infrared laser magnetic resonance, using optically-pumped far-infrared lasers as the radiation sources. Microwave magnetic resonance, the precursor technique, has now become largely redundant, but far-infrared (FIR) laser magnetic resonance continues to compete successfully with other methods. Laser magnetic resonance methods have also been extended into the mid-infrared, but since one is then involved in the study of vibrational transitions, this aspect is not discussed in this book.

Figure 9.4 shows a diagram of a far-infrared laser magnetic resonance spectrometer designed by Evenson [5] and constructed at the National Bureau of Standards Laboratory (now called NIST) in Boulder, Colorado. The FIR radiation is generated by pumping a transition in an appropriate gas with high-powered radiation from a mid-infrared carbon dioxide laser. Well over 1000 FIR laser lines have been tabulated and many of these have sufficient power for spectroscopic use. The FIR laser cavity extends through the pole gap of a large electromagnet, with the lasing medium and the sample region separated from each other by a polypropylene window, set at the Brewster angle. Apart from acting as the vacuum seal between the two gaseous regions, the Brewster window also serves to restrict the polarisation of the FIR radiation, which can be oriented either parallel or perpendicular to the magnetic field. The laser cavity end mirrors are located 91 cm apart; the pump laser power enters the FIR cavity through a zinc selenide window, and a small amount of FIR radiation is coupled out of the cavity for detection. The detector employed is a liquid helium-cooled photoconductor

Experimental methods | 585

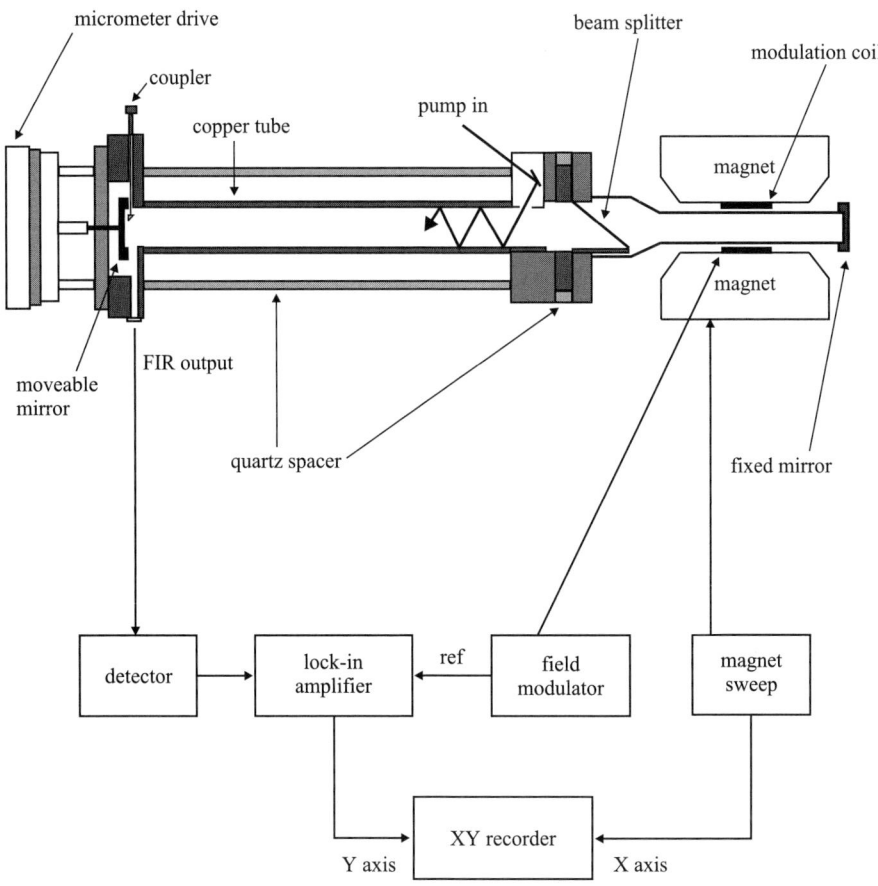

Figure 9.4. Schematic diagram of a far-infrared laser magnetic resonance spectrometer, designed by Evenson [5] and constructed at N.B.S. Boulder.

whose response is sufficiently rapid for magnetic field modulation at 40 kHz to be used. This is generated by modulation coils located inside the static magnetic field and also serves to define the detected sample volume. The laser cavity is further defined by a cylindrical copper tube of 10 cm internal diameter, which reflects and refocuses the CO_2 laser beam. Quartz spacers are used to provide high thermal stability to the length of the laser cavity, and the beam splitter is rotatable about the laser axis so that the polarisation can easily be altered.

Since the laser magnetic resonance experiment relies on a chance near-coincidence between a laser line and a molecular transition frequency, and the range over which spectroscopic transitions can be magnetically tuned is often quite small, it is desirable to have a large number of FIR laser lines available. This is now seldom a major problem, and table 9.1 lists a restricted sample of FIR laser lines that have been used for magnetic resonance studies. It is, of course, necessary to be able to measure the FIR frequency accurately and this is accomplished in Evenson's laboratory by measuring the beat

Table 9.1. *A selection of discharge and CO_2-pumped far-infrared laser lines available for magnetic resonance studies*

laser gas	CO_2 laser line	wavelength (μm)	wavenumber (cm^{-1})	frequency(GHz)
HCOOH	9R(40)	742.57	13.466 810	403.7215
CH_2CHCl	10P(38)	601.90	16.614 267	498.0791
CH_3OH	9P(16)	570.6	17.526 375	525.4275
CH_3OH	9P(36)	392.1	25.505 732	764.6426
DCOOD	10R(12)	380.6	26.276 695	787.7555
$^{13}CD_3OD$	10P(16)	333.3	30.006 721	899.5717
CH_3OD	9R(4)	332.1	30.108 749	902.6302
N_2H_4	9P(12)	331.7	30.150 752	903.8894
N_2H_4	9P(12)	331.3	30.184 446	904.8995
CH_3NH_2	9R(4)	314.8	31.761 473	952.1850
N_2H_4	9P(20)	311.1	32.146 619	963.7314
HCOOH	9R(4)	302.3	33.082 116	991.7769
N_2H_4	10R(12)	301.3	33.192 223	995.0778
CH_3OD	9R(8)	294.8	33.920 039	1016.8972
CD_3OH	10P(18)	287.3	34.805 896	1043.4545
CH_3OH	9R(10)	232.9	42.929 682	1286.9995
N_2H_4	10P(24)	192.9	51.838 395	1554.0760
CH_3OH	10R(10)	191.6	52.186 726	1564.5187
CH_2F_2	9R(22)	166.7	59.996 898	1798.6470
CH_2F_2	9R(20)	166.6	60.013 319	1799.1393
CH_3OH	9P(16)	164.6	60.753 203	1821.3352
CH_3OH	10R(38)	163.0	61.337 076	1838.8393
CH_3NH_2	9P(24)	147.84	67.639 101	2027.7526
CH_3OH	10R(34)	129.55	77.191 077	2314.1113
CH_3OH	9P(36)	118.8	84.150 936	2522.7816
H_2O[a]	—	118.6	84.323 402	2527.9313
CH_3OH	9P(36)	110.7	90.321 535	2707.7493
CH_3OD	9P(30)	103.125	96.970 843	2907.0889
CH_3OH	9R(10)	96.52	103.603 749	3105.9368
$^{13}CH_3OH$	9P(22)	85.3	117.209 553	3513.854
CH_3OH	9R(8)	77.41	129.190 603	3873.0051
CH_3OH	9P(34)	70.51	141.821 742	4251.6740

[a] The far-infrared lasing transition is excited by an electrical discharge.

frequency between the laser line and a FIR frequency generated as the difference of two known CO_2 frequencies on a metal–insulator–metal (MIM) diode.

The initial FIR laser magnetic resonance studies [6] were performed using a sample cell which was external to the laser cavity. All experiments now use an intracavity arrangement as shown in figure 9.4 which is estimated to be 10^3 times more sensitive than the extracavity arrangement. It is the very high sensitivity which continues to make

FIR laser magnetic resonance such an important technique for the study of short-lived transient molecular species. The flow methods used to generate free radicals are very much the same for both microwave and FIR laser magnetic resonance, although one notable difference is that FIR methods have been used to study molecular ions, as we shall discuss later. We now describe in detail a number of microwave and FIR studies of free radicals in differing electronic states, to see the nature of the observed spectra, the modes of analysis and the determination of molecular constants. We present these in order of increasing complexity, rather than historical precedent. The first microwave magnetic resonance studies were, in fact, described by Radford [7] for the OH radical; we will come to these important investigations in due course.

9.3. $^1\Delta$ states

A number of species which have $^3\Sigma$ ground states also have low lying $^1\Delta$ states which arise from the same electron configuration (see chapter 6); these include O_2, SO, SeO and NF. The $^1\Delta$ states are, in general, long-lived and readily studied by magnetic resonance methods. Since the Zeeman effect for a molecule in a $^1\Delta$ state is particularly simple, this seems a good system with which to introduce the principles of the magnetic resonance experiments, and we first describe the experiments of Carrington, Levy and Miller [8] on SO in its $^1\Delta$ state.

9.3.1. SO in the $a\,^1\Delta$ state

(a) Introduction

$^1\Delta$ SO was produced by reacting the products of a microwave discharge in O_2 with OCS, using the resonant cavity described in figure 9.3. A microwave discharge in O_2 produces $^1\Delta\,O_2$, and reaction of O atoms with OCS produces SO in its $^3\Sigma$ ground state, so it seems probable that $^1\Delta$ SO is produced by an energy transfer process. Using electric field modulation, a strong spectrum was obtained as shown in figure 9.5 (*bottom*). This spectrum, which arises from $^1\Delta$ SO in its lowest rotational level ($J = 2$) consists of four separated lines; note that with this mode of modulation, a second-derivative absorption line shape is obtained. The resonant microwave frequency was close to 10 GHz, and the lines span the magnetic field region from 9.6 to 10.1 kG. We must explain the appearance of the four-line pattern, and describe the information which is obtained from its analysis. In subsequent work Uehara [9] observed a six-line pattern from the $J = 3$ level, and Brown and Uehara [10] measured an eight-line pattern from $J = 4$.

There are no nuclear spin magnetic moments in the predominant isotopic form of SO (i.e. $^{32}S^{16}O$), so that hyperfine interactions are absent. The orbital angular momentum vector $\boldsymbol{L}$ is coupled to the rotational angular momentum vector $\boldsymbol{R}$ to form the total angular momentum $\boldsymbol{J}$. For a Δ state the projection of the value of $\boldsymbol{L}$ on the internuclear

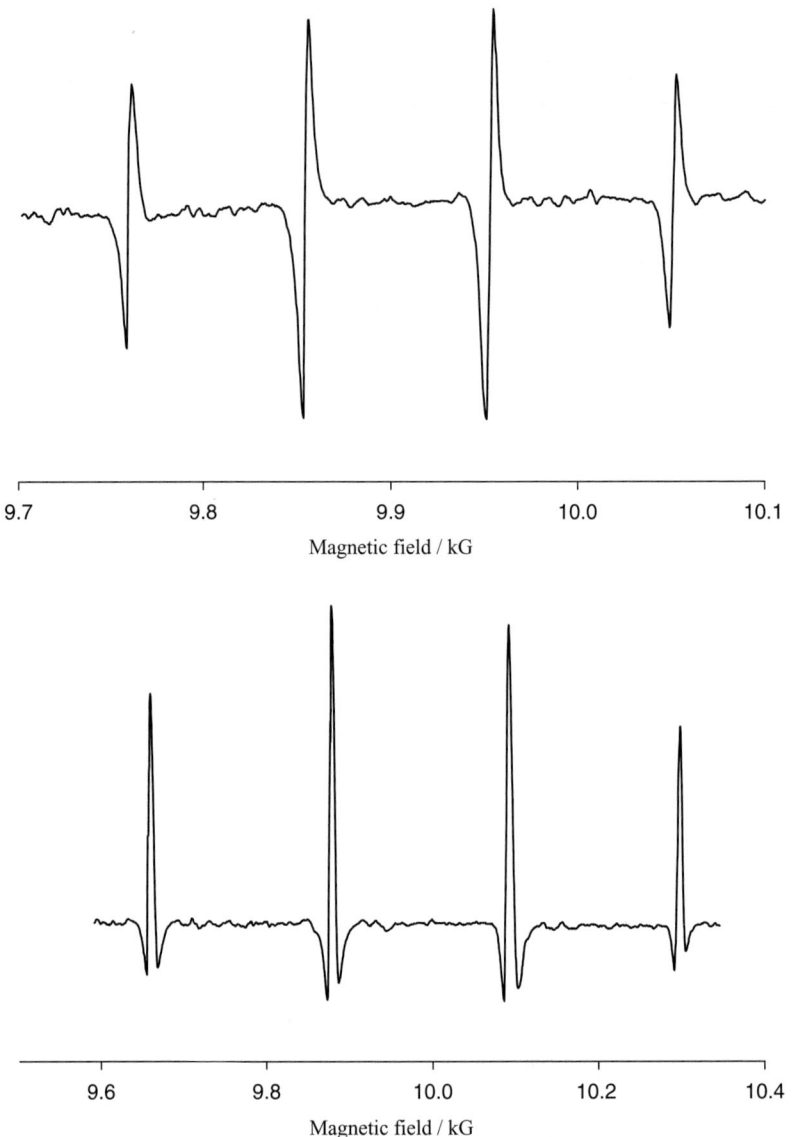

Figure 9.5. Observed microwave magnetic resonance spectra of $^1\Delta$ O_2 (*top*) and $^1\Delta$ SO (*bottom*). The microwave frequency was close to 10 GHz in both cases. The top spectrum is obtained by magnetic field modulation, the bottom by electric field modulation.

axis is $\Lambda = \pm 2$. The two components of the Λ-doublet have opposite parities, and in a $^1\Delta$ state they are essentially degenerate; this is certainly true at the level of the spectroscopic resolution common to conventional microwave experiments. Nevertheless the existence of the Λ-doublets is extremely important in determining the character of the magnetic resonance transitions, as we shall see. For the lowest rotational level ($R = 0$), the value of J is 2.

(b) Effective Hamiltonian, matrix elements and assignment

In a theoretical analysis of the Zeeman effect for a diatomic molecule in a singlet state, Brown and Uehara [10] show that the effective Hamiltonian may be written,

$$\mathcal{H}_Z^{\text{eff}} = g'_L \mu_B T_0^1(\boldsymbol{B})T_0^1(\boldsymbol{L}) - g_r \mu_B T_0^1(\boldsymbol{B})T_0^1(\boldsymbol{J}-\boldsymbol{L}), \tag{9.2}$$

where, as usual, we have taken the space-fixed $p = 0$ direction to be defined by the direction of the magnetic field. The g-factors in (9.2) are defined as follows:

$$\begin{aligned} g'_L &= g_L + \Delta g_L, \\ g_r &= g_r^N - g_r^e. \end{aligned} \tag{9.3}$$

In these equations g_L is the orbital g-factor corrected for quantum electrodynamic, relativistic and diamagnetic effects [11]. Δg_L is a small correction to the orbital g factor arising from non-adiabatic mixing of excited electronic states, whilst g_r^N and g_r^e are the nuclear and electronic contributions to the rotational g-factor, g_r.

The matrix elements of the first term in (9.2) are given by:

$$\langle \eta, L, \Lambda; J, \Lambda, M_J | g'_L \mu_B T_0^1(\boldsymbol{B})T_0^1(\boldsymbol{L}) | \eta', L', \Lambda'; J', \Lambda', M'_J \rangle$$
$$= \langle \eta, L, \Lambda; J, \Lambda, M_J | g'_L \mu_B B_Z \sum_q \mathcal{D}_{0q}^{(1)}(\omega)^* T_q^1(\boldsymbol{L}) | \eta', L', \Lambda'; J', \Lambda', M' \rangle$$
$$= g'_L \mu_B B_Z \delta_{M_J M'_J} \sum_q \langle \eta, \Lambda | T_q^1(\boldsymbol{L}) | \eta', \Lambda' \rangle \{(2J+1)(2J'+1)\}^{1/2} (-1)^{M_J - \Lambda}$$
$$\times \begin{pmatrix} J & 1 & J' \\ -M_J & 0 & M_J \end{pmatrix} \begin{pmatrix} J & 1 & J' \\ -\Lambda & q & \Lambda' \end{pmatrix}. \tag{9.4}$$

In the second line of (9.4) we have transformed from space to molecule-fixed axes (q). If we now retain only terms diagonal in the $^1\Delta$ state by putting $q = 0$, we obtain the diagonal and off-diagonal Zeeman matrix elements, as follows:

$$\langle J, M_J | \mathcal{H}_Z^{\text{eff}} | J, M_J \rangle = g'_L \mu_B B_Z \frac{M_J \Lambda^2}{J(J+1)} = 4 g'_L \mu_B B_Z \frac{M_J}{J(J+1)}, \tag{9.5}$$

$$\langle J, M_J | \mathcal{H}_Z^{\text{eff}} | J+1, M_J \rangle$$
$$= g'_L \mu_B B_Z \frac{\Lambda \{(J+M_J+1)(J-M_J+1)\}^{1/2} \{(J+\Lambda+1)(J-\Lambda+1)\}^{1/2}}{(J+1)\{(2J+1)(2J+3)\}^{1/2}}. \tag{9.6}$$

Equation (9.5) shows that the first-order effective g-factors for the $J = 2, 3, 4$ rotational levels are $2/3$, $1/3$ and $1/5$ respectively. Equation (9.6) shows that the second-order effect of the magnetic field is to mix adjacent rotational levels, so that J is no longer a perfectly good quantum number. This is because the isotropy of field-free space is replaced by the cylindrical symmetry of the applied magnetic field. In consequence, the four transitions observed at a fixed frequency for the $J = 2$ level occur at different magnetic fields. This second-order Zeeman splitting is exhibited in the experimental spectra shown in figure 9.5, and illustrated in the Zeeman energy levels shown on the right-hand side of figure 9.6. The Zeeman mixing depends upon the square of

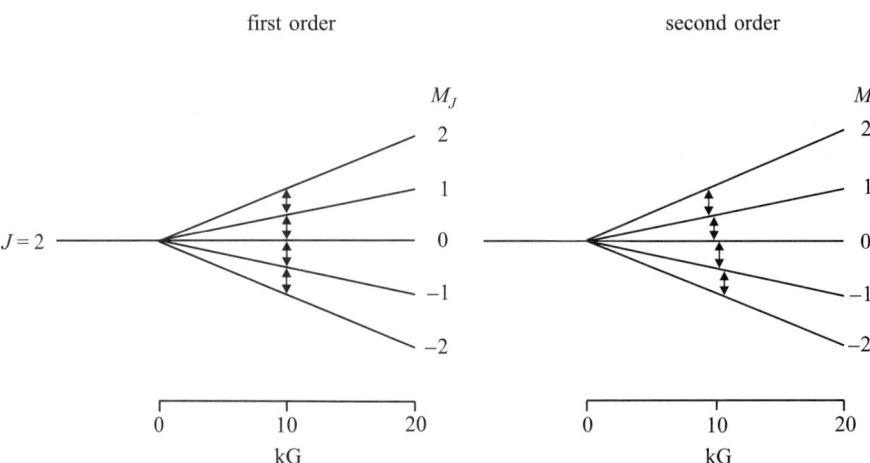

Figure 9.6. Zeeman splitting of the $J = 2$ rotational level of a $^1\Delta$ state, and the observed $\Delta M_J = \pm 1$ transitions.

the magnetic field strength, and is inversely proportional to the spacing between the rotational levels being mixed.

The second term in equation (9.2) may be rewritten as the sum of two terms,

$$-g_r \mu_B T_0^1(\boldsymbol{B}) T_0^1(\boldsymbol{J}) + g_r \mu_B T_0^1(\boldsymbol{B}) T_0^1(\boldsymbol{L}), \qquad (9.7)$$

and we note that the second of these has the same operator form as the first term in (9.2), so that it merely provides a small correction to our earlier results; g_L' is replaced by $g_L' + g_r$. The matrix elements of the first term in (9.7) are obtained directly by using the Wigner–Eckart theorem in the space-fixed axis system; they are diagonal in both J and M_J:

$$\langle J, \Lambda, M_J | - g_r \mu_B T_0^1(\boldsymbol{B}) T_0^1(\boldsymbol{J}) | J, \Lambda, M_J \rangle = -g_r \mu_B B_Z M_J. \qquad (9.8)$$

In order to assign the Zeeman patterns for the three lowest rotational levels quantitatively, one must determine the spacings between the rotational levels, and the values of g_L' and g_r. In the simplest model which neglects centrifugal distortion, the rotation spacings are simply $B_0 J(J+1)$; this approximation was used by Brown and Uehara [10], who used the rotational constant $B_0 = 21295$ MHz obtained by Saito [12] from pure microwave rotational spectroscopy (see later in the next chapter). The values of the g-factors were found to be $g_L' = 0.99982$, $g_r = -(1.35) \times 10^{-4}$. Note that because of the off-diagonal matrix elements (9.6), the Zeeman matrices (one for each value of M_J) are actually infinite in size and must be truncated at some point to achieve the desired level of accuracy. In subsequent work Miller [14] observed the spectrum of $^1\Delta\,^{33}$SO in natural abundance; ^{33}S has a nuclear spin of 3/2 and from the hyperfine structure Miller was able to determine the magnetic hyperfine constant a (see below for the definition of this constant).

Similar analyses have been carried out for O_2 in its $J = 2$ and 3 rotational levels of the $^1\Delta$ state [13, 14]. In a particularly careful and accurate study Miller [14] showed that apart from determining the rotational constant and the rotational and orbital g-factors, it was also possible to determine the anisotropy of the diamagnetic susceptibility, and from that the electric quadrupole moment of O_2 in its $^1\Delta$ state. FIR laser magnetic resonance studies of $^1\Delta$ $^{16}O^{18}O$ have also been described [15]; the transitions studied were magnetic dipole rotational transitions up to $J = 9 \leftarrow 8$.

9.3.2. NF in the a $^1\Delta$ state

(a) Nuclear magnetic and electric hyperfine interactions

Perhaps the most interesting spectrum of a molecule in a $^1\Delta$ state, illustrated in figure 9.7, is that of the NF radical, obtained by Curran, MacDonald, Stone and Thrush [16] by reacting NF_2 with hydrogen atoms inside a cavity similar to that shown in figure 9.3. Because of the presence of ^{14}N and ^{19}F nuclei, with spins of 1 and 1/2 respectively, a beautiful resolved hyperfine pattern was obtained. In order to understand this spectrum additional terms describing the nuclear Zeeman, magnetic hyperfine and nitrogen quadrupole interactions must be added to the effective Hamiltonian (9.2). The total Hamiltonian therefore becomes

$$\mathcal{H}_{\text{total}} = \mathcal{H}_Z^{\text{eff}} - g_N \mu_N T^1(\boldsymbol{B}) \cdot T^1(\boldsymbol{I}_N) - g_F \mu_N T^1(\boldsymbol{B}) \cdot T^1(\boldsymbol{I}_F)$$
$$+ a_N T^1(\boldsymbol{L}) \cdot T^1(\boldsymbol{I}_N) + a_F T^1(\boldsymbol{L}) \cdot T^1(\boldsymbol{I}_F) + \mathcal{H}_{Q_N}. \tag{9.9}$$

The basis set will be that used earlier with the addition of nuclear spin-decoupled

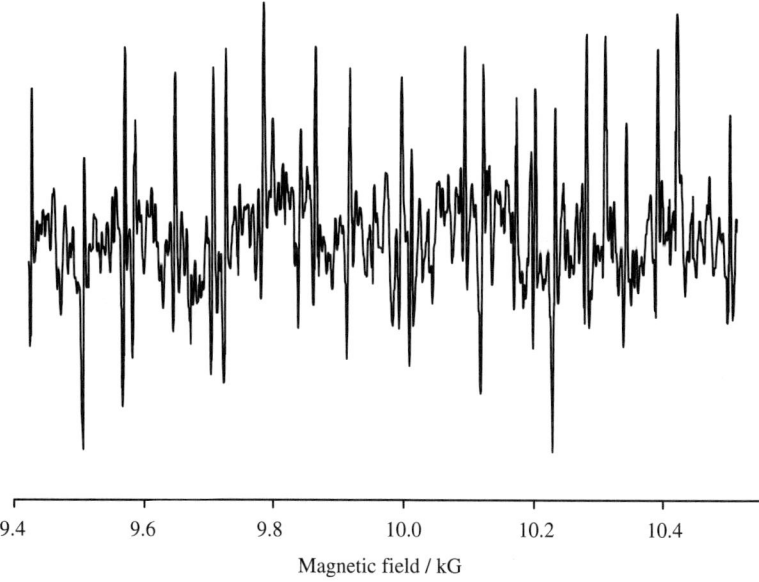

Figure 9.7. Microwave magnetic resonance spectrum of $^1\Delta$ NF ($J = 2$) recorded at a frequency of 9320 MHz [16].

functions, i.e. $|\eta, \Lambda; J, M_J; I_N, M_N; I_F, M_F\rangle$; the N and F subscripts refer, of course, to the two nuclei involved, not as elsewhere to the projections of N and F.

The nuclear electric quadrupole interaction in equation (9.9) was introduced in chapter 4; it describes the sum of all the electric interactions between the protons in the nucleus and the electrons in the molecule, averaged over the positions of the latter:

$$\mathcal{H}_{Q_N} = -\sum_p \sum_i (e^2/4\pi\varepsilon_0) r_p^2 \boldsymbol{C}^2(\theta_p, \phi_p) \cdot \boldsymbol{C}^2(\theta_i, \phi_i) r_i^{-3} = -e\mathrm{T}^2(\boldsymbol{Q}) \cdot \mathrm{T}^2(\nabla \boldsymbol{E}). \tag{9.10}$$

The scalar product of the two second rank tensors in equation (9.10) is expanded in the space-fixed axis system, so that

$$\langle \eta, \Lambda; J, M_J; I_N, M_N; I_F, M_F | - e\sum_p (-1)^p \mathrm{T}_p^2(\boldsymbol{Q}_N) \mathrm{T}_{-p}^2(\nabla \boldsymbol{E}) | \eta', \Lambda'; J', M_J'; I_N, M_N'; I_F, M_F' \rangle$$

$$= -\delta_{M_F, M_F'} e \sum_p (-1)^p \langle I_N, M_N | \mathrm{T}_p^2(\boldsymbol{Q}_N) | I_N, M_N' \rangle$$

$$\times \langle \eta, \Lambda; J, M_J | \mathrm{T}_{-p}^2(\nabla \boldsymbol{E}) | \eta', \Lambda'; J', M_J' \rangle. \tag{9.11}$$

The matrix elements are diagonal in, and independent of M_F, which we will therefore omit from the subsequent development of (9.11); we have also omitted terms involving excited nuclear states (with $I_N' \neq I_N$)!

We deal first with the nuclear part of equation (9.11), using the Wigner–Eckart theorem to obtain

$$\langle I_N, M_N | \mathrm{T}_p^2(\boldsymbol{Q}_N) | I_N, M_N' \rangle = (-1)^{I_N - M_N} \begin{pmatrix} I_N & 2 & I_N \\ -M_N & p & M_N' \end{pmatrix} \langle I_N \| \mathrm{T}^2(\boldsymbol{Q}_N) \| I_N \rangle. \tag{9.12}$$

The reduced matrix element in (9.12) is evaluated by noting the definition of the nuclear quadrupole moment Q_N, and using the Wigner–Eckart theorem as follows:

$$Q_N/2 = \langle I_N, M_N = I_N | \mathrm{T}_0^2(\boldsymbol{Q}_N) | I_N, M_N = I_N \rangle = \begin{pmatrix} I_N & 2 & I_N \\ -I_N & 0 & I_N \end{pmatrix} \langle I_N \| \mathrm{T}^2(\boldsymbol{Q}_N) \| I_N \rangle. \tag{9.13}$$

Expanding the 3-j symbol in (9.13) we obtain the required result,

$$\langle I_N \| \mathrm{T}^2(\boldsymbol{Q}_N) \| I_N \rangle = \frac{Q_N}{2} \left\{ \frac{(I_N + 1)(2I_N + 1)(2I_N + 3)}{I_N(2I_N - 1)} \right\}^{1/2}. \tag{9.14}$$

We will only be concerned with matrix elements diagonal in M_N, so that (9.12) gives the result

$$\langle I_N, M_N | \mathrm{T}_0^2(\boldsymbol{Q}_N) | I_N, M_N \rangle = Q_N \frac{\{3M_N^2 - I_N(I_N + 1)\}}{2I_N(2I_N - 1)}. \tag{9.15}$$

We now turn to the electronic part of the quadrupole interaction in equation (9.11). Since this part of the interaction is most sensibly described in a molecule-fixed frame, we make the necessary transformation

$$\langle \eta, \Lambda; J, M_J | T^2_{-p}(\nabla E) | \eta', \Lambda'; J', M'_J \rangle$$
$$= \langle \eta, \Lambda; J, M_J | \sum_q \mathcal{D}^{(2)}_{-pq}(\omega)^* T^2_q(\nabla E) | \eta', \Lambda'; J', M'_J \rangle$$
$$= \sum_q \langle \eta, \Lambda | T^2_q(\nabla E) | \eta', \Lambda' \rangle \langle J, \Lambda, M_J | \mathcal{D}^{(2)}_{-pq}(\omega)^* | J', \Lambda', M'_J \rangle. \quad (9.16)$$

We have already restricted our attention to the $p = 0$ components, and we now further limit ourselves to the $q = 0$ components of (9.16) since the $q = \pm 1, \pm 2$ terms involve mixing of excited electronic states. The $q = 0$ term leads to the definition of the (negative of the) electric field gradient q_N at the ^{14}N nucleus, i.e.

$$\langle \eta, \Lambda | T^2_0(\nabla E) | \eta, \Lambda \rangle = -q_N/2. \quad (9.17)$$

Consequently we obtain the required diagonal matrix element,

$$\langle \eta, \Lambda; J, M_J | T^2_{-p}(\nabla E) | \eta, \Lambda; J, M_J \rangle$$
$$= -\frac{q_N}{2}(2J+1)(-1)^{M_J-\Lambda} \begin{pmatrix} J & 2 & J \\ -M_J & 0 & M_J \end{pmatrix} \begin{pmatrix} J & 2 & J \\ -\Lambda & 0 & \Lambda \end{pmatrix}. \quad (9.18)$$

Setting $J = \Lambda = 2$ in (9.18) and $I_N = 1$ in (9.15) we obtain the diagonal quadrupole coupling contributions to the energies,

$$E_{Q_N}(M_J, M_N, M_F) = eq_N Q_N \{3M_J^2 - 6\}\{3M_N^2 - 2\}/84, \quad (9.19)$$

which is the result obtained by Curran, MacDonald, Stone and Thrush [16].

We deal next with the ^{14}N magnetic hyperfine interaction, and again restrict attention to the diagonal (first-order) contribution:

$$\langle \eta, \Lambda; J, M_J; I_N, M_N | a_N T^1(L) \cdot T^1(I_N) | \eta, \Lambda; J, M_J; I_N, M_N \rangle$$
$$= a_N \langle \eta, \Lambda; J, M_J | T^1_0(L) | \eta, \Lambda; J, M_J \rangle \langle I_N, M_N | T^1_0(I_N) | I_N, M_N \rangle$$
$$= a_N \langle \eta, \Lambda; J, M_J | \mathcal{D}^{(1)}_{00}(\omega)^* T^1_{q=0}(L) | \eta, \Lambda; J, M_J \rangle M_N$$
$$= a_N \langle \eta, \Lambda | T^1_{q=0}(L) | \eta, \Lambda \rangle (2J+1)(-1)^{M_J-\Lambda} \begin{pmatrix} J & 1 & J \\ -M_J & 0 & M_J \end{pmatrix} \begin{pmatrix} J & 1 & J \\ -\Lambda & 0 & \Lambda \end{pmatrix} M_N$$
$$= \frac{2}{3} a_N M_J M_N. \quad (9.20)$$

The fourth line above follows from equation (9.4) and the last line from putting $J = \Lambda = 2$. A similar analysis shows the diagonal contribution from the fluorine magnetic hyperfine interaction to be $(2/3)a_F M_J M_F$. Finally the nuclear Zeeman energies are simply $M_N g_N \mu_N B_Z$ and $M_F g_F \mu_N B_Z$. In summary, the total first-order Zeeman energies for $J = 2$ are:

$$E_{\text{first-order}}(M_J, M_N, M_F)$$
$$= \left(\frac{2}{3}g'_L - \frac{1}{3}g_r\right)\mu_B B_Z M_J - g_F \mu_N B_Z M_F - g_N \mu_N B_Z M_N$$
$$+ \frac{2}{3}a_F M_J M_F + \frac{2}{3}a_N M_J M_N + (eq_N Q_N/84)(3M_N^2 - 2)(3M_J^2 - 6). \quad (9.21)$$

The rotational mixing of the $J = 2$ and 3 levels by the electronic Zeeman effect, (9.6), may be calculated by second-order perturbation theory, noting that the separation between the two levels is $6B_0$, where B_0 is the rotational constant in the $v = 0$ level; the result is

$$E_{\text{second-order}}(M_J, M_N, M_F) = -\lfloor 2(g_L + g_r)^2 \mu_B^2 B_Z^2 \rfloor (9 - M_J^2)[1/189 B_0]. \quad (9.22)$$

Analysis of the microwave magnetic resonance spectrum recorded at a frequency of 9320 MHz gave the following values of the constants:

$$g_L = 1.0000, \quad g_r = -1 \times 10^{-4}, \quad a_F = 758.06 \text{ MHz},$$
$$a_N = 109.92 \text{ MHz}, \quad eq_N Q_N = 4.1 \text{ MHz}.$$

The authors do not mention the values of the nuclear g-factors, but we may take them to be $g_F = +2.62887$ and $g_N = +0.40376$ nuclear Bohr magnetons. Consequently it is now a simple matter to calculate the energies of the 30 levels for a range of magnetic fields between 9400 and 10 600 G; the magnetic resonance transitions are those which obey the selection rules $\Delta M_J = \pm 1$, $\Delta M_N = \Delta M_F = 0$ and their frequencies may also be calculated.

Our purpose in this analysis is to show how the observed spectrum arises, but we again point out that this is a simple matter if one knows the values of the molecular constants involved. In practice one has the much more difficult but interesting task of determining the constants from the observed Zeeman pattern. We have described this particular system in some detail because it illustrates a number of the features which we shall encounter later. It is not often that one observes a fully resolved hyperfine pattern from two different nuclei, with both magnetic and electric interactions present.

(b) Parity doubling and Stark effect in $^1\Delta$ states

We must now say more about the nature of the resonance transitions, and also describe additional measurements of the Stark effect which enable the electric dipole moment of the molecule to be determined. In both SO and NF the transitions detected are actually electric-dipole allowed, so perhaps the spectrum ought not to be described as a magnetic resonance spectrum.

O_2 and NF are isoelectronic with $^3\Sigma^-$ ground states which arise from the two highest energy electrons occupying two degenerate π-molecular orbitals; the excited $^1\Delta$ states have both electrons in the same orbital. If the π-molecular orbitals are denoted

π_{+1} and π_{-1} the wave functions for the $^1\Delta$ state may be written as

$$\psi_{+2} = \frac{1}{\sqrt{2}}\pi_{+1}(1)\pi_{+1}(2)\{\alpha(1)\beta(2) - \beta(1)\alpha(2)\}, \quad \Lambda = +2,$$

$$\psi_{-2} = \frac{1}{\sqrt{2}}\pi_{-1}(1)\pi_{-1}(2)\{\alpha(1)\beta(2) - \beta(1)\alpha(2)\}, \quad \Lambda = -2.$$
(9.23)

These two functions do not have definite parities but the symmetric and antisymmetric combinations of them do; we use these combinations to calculate both the Stark effect and the electric dipole transition probabilities.

The two parity combinations represent the components of a Λ-doublet, and they are degenerate in the non-rotating molecule. Rotational-electronic coupling removes the degeneracy, giving rise to Λ-doubling. This doubling can be very large for a light molecule in a $^2\Pi$ state, like OH, but it is very small for a $^1\Delta$ state; in the three cases described in this section it is negligible compared with the spectroscopic linewidth. However, the Λ-doublets have opposite parities, and the total parity alternates with the rotational quantum number J. If the molecule possesses an electric dipole moment, the magnetic resonance transitions within a single rotational level, of the type described above, occur between the Λ-doublet components of *opposite* parity. The O_2 molecule is different because, being a homonuclear molecule with zero-spin nuclei, only one of the Λ-doublet components exists for each rotational level, namely that with positive parity. Furthermore the homonuclear molecule has no electric dipole moment, so the magnetic resonance transitions are necessarily magnetic dipole allowed only, and occur between states of the same overall parity.

Since the Λ-doublet components of a given rotational level are of opposite parity, a static electric field will split them apart, producing a splitting in the magnetic resonance spectrum which is known as a Stark splitting. The Stark effect occurs through interaction of the applied electric field (E_0) with the molecular electric dipole moment (μ_e); if the static electric field is applied parallel to the static magnetic field (i.e. in the $p=0$ direction), the coupling may be represented by the perturbation

$$\mathcal{H}_E = -\mu_e E_0 \mathcal{D}^{(1)}_{00}(\omega)^*$$
(9.24)

since the dipole moment is oriented in the $q=0$ direction. The matrix elements of $\mathcal{H}_E$ are calculated as follows:

$$\langle \eta, J, \Lambda, M_J | \mathcal{H}_E | \eta, J, \Lambda, M_J \rangle$$
$$= -\mu_e E_0 (2J+1)(-1)^{M_J - \Lambda} \begin{pmatrix} J & 1 & J \\ -M_J & 0 & M_J \end{pmatrix} \begin{pmatrix} J & 1 & J \\ -\Lambda & 0 & \Lambda \end{pmatrix}.$$
(9.25)

The value of this matrix element for $J=2$ is $-\mu_e E_0 M_J \Lambda /6$ where $\Lambda = +2$ or -2. We actually require the matrix of $\mathcal{H}_E$ for the states ϕ_s and ϕ_a, the symmetric and antisymmetric combinations of ψ_{+2} and ψ_{-2} in equation (9.23). One readily finds that

$$\langle \phi_s | \mathcal{H}_E | \phi_s \rangle = \langle \phi_a | \mathcal{H}_E | \phi_a \rangle = 0, \quad \langle \phi_s | \mathcal{H}_E | \phi_a \rangle = \langle \phi_a | \mathcal{H}_E | \phi_s \rangle = -\frac{1}{3}\mu_e E_0 M_J$$
(9.26)

so that the Stark energies of the two states mixed by the electric field are $\pm(1/3)\mu_e E_0 M_J$. The electric field splits each line into a doublet, from which the electric dipole moment μ_e can be determined. This method of determining electric dipole moments from microwave magnetic resonance spectra was first developed by Carrington, Levy and Miller [17], using the resonance cavity illustrated in figure 9.3. They determined the dipole moments of a number of diatomic free radicals, including that of $^1\Delta$ SO which was found to be 1.47 D. The dipole moment of $^1\Delta$ SeO was subsequently determined by Byfleet, Carrington and Russell [18] to be 2.01 D, whilst $^1\Delta$ NF was found [16] to have a dipole moment of 0.37 ± 0.06 D.

A simple extension of the Stark analysis given above enables one to derive an expression for the intensities of the electric dipole transitions. The oscillating microwave electric field is applied perpendicular to the static magnetic field, so that the Zeeman levels experience a time-dependent perturbation, represented by the operator

$$\mathcal{H}_E(t) = \mu_e E_{p=\pm 1}(t) \mathfrak{D}^{(1)}_{-p0}(\omega)^*. \tag{9.27}$$

For $p = +1$ the matrix elements for the unsymmetrised states (9.23) are given by

$$\langle J, \Lambda, M_J | \mathcal{H}_E(t) | J, \Lambda, M'_J \rangle$$
$$= \mu_e E(t)(2J+1)(-1)^{M_J-\Lambda} \begin{pmatrix} J & 1 & J \\ -M_J & -1 & M'_J \end{pmatrix} \begin{pmatrix} J & 1 & J \\ -\Lambda & 0 & \Lambda \end{pmatrix} \tag{9.28}$$

in which the first 3-j symbol shows that $p = +1$ component leads to $\Delta M_J = +1$ Zeeman components, whilst the $p = -1$ component gives $\Delta M_J = -1$. The relative intensities of the Zeeman components for $J = 2$ are given by the squares of the first 3-j symbol in (9.28) and if we put $J = 2$ we obtain the result

$$|\langle J, \Lambda, M_J | \mathcal{H}_E(t) | J, \Lambda, M_J + 1 \rangle|^2 \approx \mu_e^2 E^2(t)(2 - M_J)(3 + M_J). \tag{9.29}$$

The relative intensities of the four Zeeman components in the spectrum of $^1\Delta$ SO in the $J = 2$ level, shown in figure 9.5, are therefore predicted to be $2:3:3:2$, matching the observations very well.

Curran *et al.* [16] conclude their analysis of the $^1\Delta$ NF spectrum by comparing the magnitudes of the magnetic hyperfine, quadrupole and dipole moment parameters with predictions from self-consistent field calculations due to O'Hare and Wahl [19]. The agreement is satisfactory, and one supposes that contemporary calculations would be in even better agreement with experiment. One of the main purposes of the measurements and spectral analysis described in this section is to provide accurate benchmarks for theoretical calculations, and also physical insight into the nature of the molecular bonding.

9.4. $^2\Pi$ states

9.4.1. Introduction

It is probably true that the majority of free radicals which have been studied by microwave magnetic resonance have $^2\Pi$ ground states. In part this might be because

$^2\Pi$ states that conform closely to case (a) coupling have readily predictable magnetic properties, which eases the spectroscopic search problem. In this section we distinguish studies of good Hund's case (a) systems from those where the coupling is case (b) or, more usually, intermediate between cases (a) and (b). Case (a) coupling arises when the spin–orbit coupling is large compared with the separation of rotational levels; when this is not the case, the tendency is towards case (b) but in many cases it is a matter of personal choice as to which basis to use, as we shall see. We begin, therefore, by looking more closely at the correlation of rotational levels in the two coupling cases. This is an abbreviated summary of the much more detailed analysis presented in chapter 6.

The essential aspects are presented in figure 9.8. On the left-hand side we present a typical pattern of rotational levels for case (a); the axial component of total electronic angular momentum $|\Omega|$ is a good quantum number, and takes values 3/2 or 1/2, corresponding to the 'fine-structure' states. These states are split by spin–orbit coupling, their energy ordering being determined by the sign of the spin–orbit coupling constant; figure 9.8 would be appropriate for the NO molecule, whereas in ClO the order of the $^2\Pi_{3/2}$ and $^2\Pi_{1/2}$ states is inverted. Note that N is not a good quantum number. On the right hand side we see the corresponding pattern of levels for case (b); Ω is no longer a conserved quantity, but the rotational quantum number N is now meaningful, and each N level exhibits a spin-doubling. The quantum number J remains good in both situations. It is important to note that, throughout figure 9.8, each level retains an extra two-fold degeneracy, or near-degeneracy, irrespective of whether the coupling is close to case (a) or case (b). In case (b) this is Λ-doubling. In case (a) this should be described as Ω-doubling, since Λ and Σ (the axial components of $\boldsymbol{L}$ and $\boldsymbol{S}$) are not separately conserved. This doubling is not shown in figure 9.8.

9.4.2. ClO in the $X^2\Pi$ ground state

(a) Introduction

The simplest spectra to understand and analyse are those obtained from molecules which exhibit good case (a) coupling, and for which the Λ-doublets are degenerate to within the spectroscopic resolution. Typical examples are ClO [20] and BrO [21], the spectra of which are shown in figure 9.9. In both cases the Stark modulation cavity described in figure 9.3 was used; ClO was obtained by flowing a mixture of chlorine and oxygen through a microwave discharge located upstream of the cavity, whilst BrO was generated inside the cavity by the reaction of oxygen atoms with bromine. Both radicals have $^2\Pi_{3/2}$ ground states and the spectra shown in figure 9.9 arise from molecules in the lowest rotational level, with $J = 3/2$. Our discussion of these spectra is similar to that of the $^1\Delta$ states, but with additional features arising from the presence of electron spin. In both cases there are two isotopes present in natural abundance: ^{35}Cl and ^{37}Cl for ClO, and ^{79}Br and ^{81}Br for BrO. All of these isotopes have nuclear spin $I = 3/2$, and large nuclear quadrupole moments.

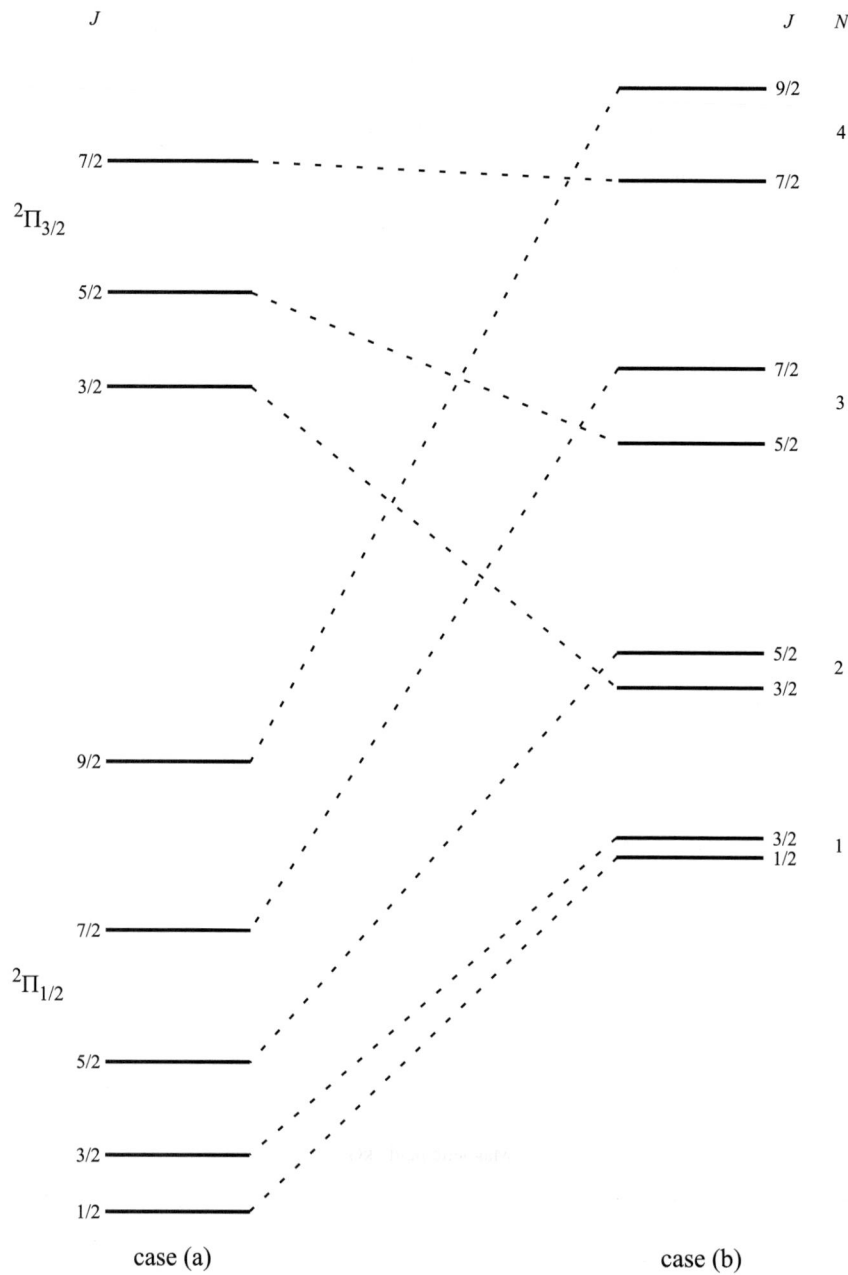

Figure 9.8. Correlation between the Hund's case (a) and case (b) rotational levels of a $^2\Pi$ state. The diagram is not drawn to scale; for a good case (a) molecule the spin–orbit splitting is very much larger than the rotational level spacing.

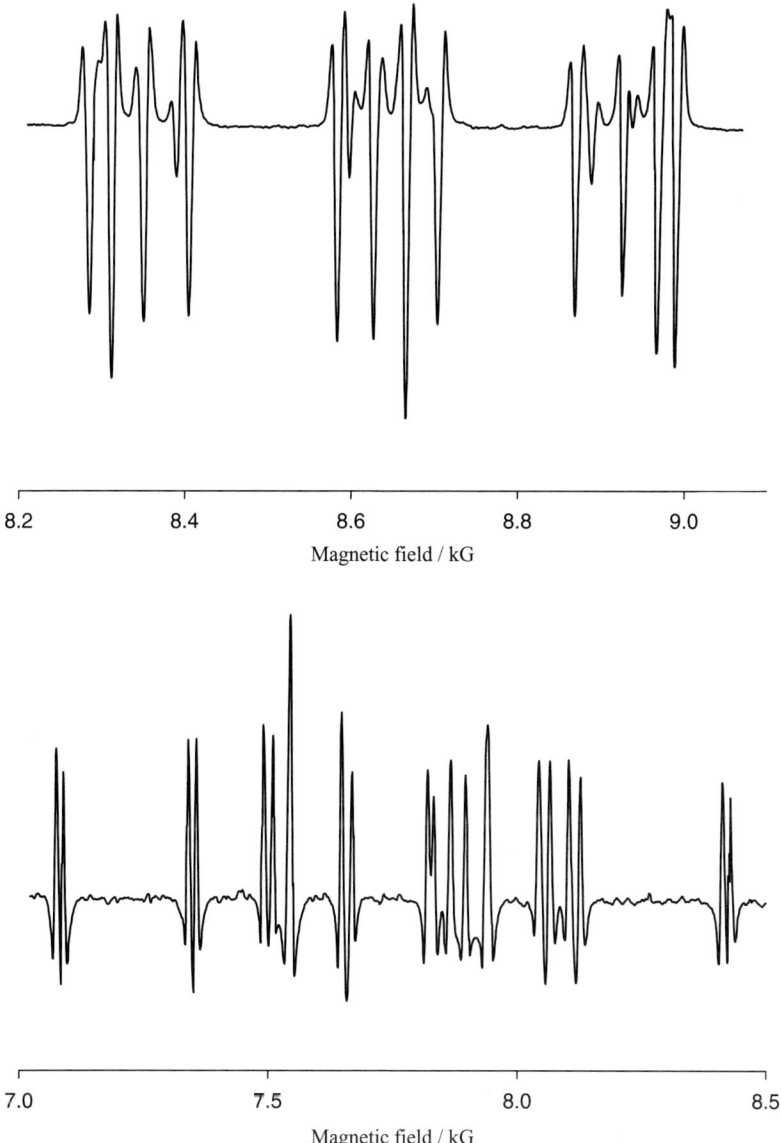

Figure 9.9. Microwave magnetic resonance spectra of (*top*) ClO [20] and (*bottom*) BrO [21], both in their $J = 3/2$, $^2\Pi_{3/2}$ states.

(b) Effective Hamiltonian

The effective Hamiltonian which was used to analyse these spectra is of the same form for all four isotopic species and may be written as the sum of four parts,

$$\mathcal{H}_{\text{eff}} = \mathcal{H}_{\text{rso}} + \mathcal{H}_{\text{hf}} + \mathcal{H}_{\text{Q}} + \mathcal{H}_{\text{Z}} \quad (9.30)$$

representing the rigid body rotation and spin–orbit coupling, the magnetic hyperfine interactions, the electric quadrupole interaction and the Zeeman interactions. This type of effective Hamiltonian has already been discussed in chapter 8.

The first term in (9.30) may be written in the form

$$\mathcal{H}_{\text{rso}} = B_0 \{T^1(J) - T^1(L) - T^1(S)\}^2 + A T^1(L) \cdot T^1(S) \tag{9.31}$$

where B_0 is the rotational constant in the lowest vibrational level and A is the spin–orbit coupling constant. Expanded in the molecule-fixed coordinate system this may be written

$$\mathcal{H}_{\text{rso}} = A \sum_q (-1)^q T_q^1(L) T_{-q}^1(S) + B_0 \Big\{ J(J+1) + S(S+1)$$
$$+ 2 \sum_q (-1)^q \{T_q^1(L) T_{-q}^1(S) + (1/2) T_q^1(L) T_{-q}^1(L)\} \Big\}$$
$$- 2 B_0 \sum_q \{T_q^1(J) T_q^1(S) + T_q^1(J) T_q^1(L)\}. \tag{9.32}$$

Note the separation of the terms into those involving J, with anomalous commution rules in the molecule-fixed system, and those which do not involve J. Because both ClO and BrO are good case (a) molecules we choose to work in the case (a) basis set $|\eta, \Lambda; S, \Sigma; J, \Omega, M_J\rangle$; we will subsequently add nuclear spin to the basis set. If we separate the $q = 0$ components of (9.32) from the $q = \pm 1$ it is readily apparent that the matrix elements within the case (a) basis have three different types of contribution,

$$\langle \eta, \Lambda; S, \Sigma; J, \Omega | \mathcal{H}_{\text{rso}} | \eta', \Lambda'; S, \Sigma'; J, \Omega' \rangle$$
$$= \delta_{\eta\eta'} \delta_{\Lambda\Lambda'} \delta_{\Omega\Omega'} \delta_{\Sigma\Sigma'} [B_0\{J(J+1) + S(S+1) - 2\Omega\Sigma - \Lambda^2\} + A\Lambda\Sigma]$$
$$- \delta_{\eta\eta'} \delta_{\Lambda\Lambda'} 2 B_0 \langle J, \Omega, \Sigma | \sum_{q=\pm 1} T_q^1(J) T_q^1(S) | J, \Omega', \Sigma' \rangle$$
$$- \langle \eta, \Lambda; J, \Omega, \Sigma | \mathcal{H}'_{\text{rso}} | \eta', \Lambda'; J, \Omega', \Sigma' \rangle, \tag{9.33}$$

where

$$\mathcal{H}'_{\text{rso}} = (A + 2 B_0) \sum_{q=\pm 1} (-1)^q T_q^1(L) T_{-q}^1(S)$$
$$- 2 B_0 \sum_{q=\pm 1} T_q^1(J) T_q^1(L) + B_0 \sum_{q=\pm 1} (-1)^q T_q^1(L) T_{-q}^1(L). \tag{9.34}$$

Equation (9.34) contains three separate contributions; the third of these involving components of L only contributes a constant amount to each level and may be omitted. The first two terms have non-vanishing matrix elements only between levels differing in Λ by ± 1 and so mix different electronic states, leading to the effect of Λ-doubling. Since we do not observe Λ-doubling in the ClO and BrO spectra, we neglect these terms in this analysis. They will, however, become important in other species discussed later. Consequently we retain only the first two terms of equation (9.33).

We turn now to the magnetic hyperfine Hamiltonian in (9.30) which may be written as the sum of three terms representing the orbital, Fermi contact and dipolar hyperfine interactions:

$$\mathcal{H}_{hf} = \mathcal{H}_{IL} + \mathcal{H}_F + \mathcal{H}_{dip}$$
$$= a\mathrm{T}^1(\boldsymbol{I}) \cdot \mathrm{T}^1(\boldsymbol{L}) + b_F \mathrm{T}^1(\boldsymbol{I}) \cdot \mathrm{T}^1(\boldsymbol{S}) - \sqrt{10} g_S \mu_B g_N \mu_N (\mu_0/4\pi) \mathrm{T}^1(\boldsymbol{I}) \cdot \mathrm{T}^1(\boldsymbol{S}, \boldsymbol{C}^2). \quad (9.35)$$

The inclusion of nuclear spin in the basis set may now be accomplished using either the coupled representation $|\eta, \Lambda; S, \Sigma; J, \Omega, I, F, M_F\rangle$, or the decoupled representation $|\eta, \Lambda; S, \Sigma; J, \Omega, M_J; I, M_I\rangle$; we choose to use the former because M_F is the only rigorously good quantum number in the presence of a magnetic field. Note that, at this stage, we include Λ in our basis functions but since the effective Hamiltonian will operate exclusively within the ground vibronic state, all terms which would mix excited states with the ground state can be excluded. This approximation is valid for ClO and BrO, but later in this chapter the theory will be developed to take account of excited state mixing.

The third term in (9.30), the electric quadrupole interaction, is again represented by the scalar product,

$$\mathcal{H}_Q = -e\mathrm{T}^2(\boldsymbol{Q}) \cdot \mathrm{T}^2(\nabla \boldsymbol{E}). \quad (9.36)$$

Finally the interactions with the external magnetic field $\boldsymbol{B}$ are summarised with the Zeeman term:

$$\mathcal{H}_Z = g_L \mu_B \mathrm{T}^1(\boldsymbol{B}) \cdot \mathrm{T}^1(\boldsymbol{L}) + g_S \mu_B \mathrm{T}^1(\boldsymbol{B}) \cdot \mathrm{T}^1(\boldsymbol{S})$$
$$- g_N \mu_N \mathrm{T}^1(\boldsymbol{B}) \cdot \mathrm{T}^1(\boldsymbol{I}) - g_r \mu_B \mathrm{T}^1(\boldsymbol{B}) \cdot \{\mathrm{T}^1(\boldsymbol{J}) - \mathrm{T}^1(\boldsymbol{L}) - \mathrm{T}^1(\boldsymbol{S})\}. \quad (9.37)$$

Consequently the total effective Hamiltonian, operating within the subspace of the ground vibronic state is the sum of equations (9.33), (9.35), (9.36) and (9.37).

(c) Matrix elements in a case (a) basis

We now calculate the matrix elements of each of the four main terms in $\mathcal{H}_{eff}$ in turn; for simplicity we will omit the primes when the matrix elements are diagonal in those quantum numbers.

(i) RIGID BODY + SPIN−ORBIT COUPLING

$$\langle \eta, \Lambda; S, \Sigma; J, \Omega, M_J | \mathcal{H}_{rso} | \eta, \Lambda; S, \Sigma'; J, \Omega', M_J \rangle$$
$$= \delta_{\Omega\Omega'} \delta_{\Sigma\Sigma'} [B_0\{J(J+1) + S(S+1) - 2\Omega\Sigma - \Lambda^2\} + A\Lambda\Sigma]$$
$$- 2B_0 \sum_{q=\pm 1} \langle J, \Omega | \mathrm{T}^1_q(\boldsymbol{J}) | J, \Omega' \rangle \langle S, \Sigma | \mathrm{T}^1_q(\boldsymbol{S}) | S, \Sigma' \rangle$$
$$= \delta_{\Omega\Omega'} \delta_{\Sigma\Sigma'} [B_0\{J(J+1) + S(S+1) - 2\Omega\Sigma - \Lambda^2\} + A\Lambda\Sigma]$$
$$- 2B_0 \sum_{q=\pm 1} (-1)^{J+S-\Omega-\Sigma} \begin{pmatrix} J & 1 & J \\ -\Omega & q & \Omega' \end{pmatrix} \begin{pmatrix} S & 1 & S \\ -\Sigma & q & \Sigma' \end{pmatrix}$$
$$\times \{J(J+1)(2J+1)S(S+1)(2S+1)\}^{1/2}. \quad (9.38)$$

(ii) MAGNETIC HYPERFINE COUPLING

For the magnetic hyperfine interaction we deal with the three terms separately, expanding each in the space-fixed axis system, as follows:

$$\mathcal{H}_{\text{IL}} = (2/r^3)\mu_B g_N \mu_N (\mu_0/4\pi) \sum_{q,p} (-1)^p T_p^1(\mathbf{I}) \mathcal{D}^{(1)}_{-pq}(\omega)^* T_q^1(\mathbf{L}), \tag{9.39}$$

$$\mathcal{H}_F = (2/3) g_S \mu_B g_N \mu_N \mu_0 \delta(r) \sum_{q,p} (-1)^p T_p^1(\mathbf{I}) \mathcal{D}^{(1)}_{-pq}(\omega)^* T_q^1(\mathbf{S}), \tag{9.40}$$

$$\mathcal{H}_{\text{dip}} = -\sqrt{10} g_S \mu_B g_N \mu_N (\mu_0/4\pi) \sum_{q,p} (-1)^p T_p^1(\mathbf{I}) \mathcal{D}^{(1)}_{-pq}(\omega)^* T_q^1(\mathbf{S},\mathbf{C}^2). \tag{9.41}$$

In equation (9.41) the tensor $T^1(\mathbf{S},\mathbf{C}^2)$ is constructed according to the equation

$$T_q^1(\mathbf{S},\mathbf{C}^2) = -\sum_{q_1,q_2} (-1)^q (3)^{1/2} T_{q_1}^1(\mathbf{S}) C_{q_2}^2(\theta,\phi) \begin{pmatrix} 1 & 2 & 1 \\ q_1 & q_2 & -q \end{pmatrix} (r^{-3}) \tag{9.42}$$

with the spherical harmonic defined by

$$C_q^k(\theta,\phi) = \left(\frac{4\pi}{2k+1}\right)^{1/2} Y_{kq}(\theta,\phi). \tag{9.43}$$

Note that there is more than one way of representing the dipolar interaction in irreducible tensors, as discussed in appendix 8.2. Equation (9.41) can be compared with the coupling choice used in our analysis of the H_2^+ spectrum described in chapter 11. It is partly a matter of choice, but more especially a matter of the basis set used in the analysis.

We now examine the matrix elements of each of these three terms in turn.

$$\langle \eta, \Lambda; S, \Sigma; J, \Omega, I, F, M_F | \mathcal{H}_{\text{IL}} | \eta, \Lambda'; S, \Sigma; J', \Omega', I, F, M_F \rangle$$
$$= a \sum_q (-1)^{J'+I+F+J-\Omega} \{I(I+1)(2I+1)(2J+1)(2J'+1)\}^{1/2}$$
$$\times \begin{Bmatrix} J' & I & F \\ I & J & 1 \end{Bmatrix} \begin{pmatrix} J & 1 & J' \\ -\Omega & q & \Omega' \end{pmatrix} \langle \eta, \Lambda | T_q^1(\mathbf{L}) | \eta, \Lambda' \rangle \tag{9.44}$$

where $a = 2\mu_B g_N \mu_N (\mu_0/4\pi) \langle r^{-3} \rangle$. We note that this interaction could mix excited states $|\eta', \Lambda'\rangle$ with the ground state, but we neglect these effects by including only the $q = 0$ terms. Consequently the simplified version of (9.44) which we use is:

$$\langle \eta, \Lambda; S, \Sigma; J, \Omega, I, F, M_F | \mathcal{H}_{\text{IL}} | \eta, \Lambda; S, \Sigma; J', \Omega, I, F, M_F \rangle$$
$$= a\Lambda (-1)^{J'+I+F+J-\Omega} \{I(I+1)(2I+1)(2J+1)(2J'+1)\}^{1/2}$$
$$\times \begin{Bmatrix} J' & I & F \\ I & J & 1 \end{Bmatrix} \begin{pmatrix} J & 1 & J' \\ -\Omega & 0 & \Omega' \end{pmatrix}, \tag{9.45}$$

where $J' = J, J \pm 1$.

The next term in the magnetic hyperfine Hamiltonian (9.35) is the Fermi contact interaction (9.40), whose matrix elements are:

$$\langle \eta, \Lambda; S, \Sigma; J, \Omega, I, F, M_F | \mathcal{H}_F | \eta', \Lambda'; S, \Sigma'; J', \Omega', I, F, M_F \rangle$$
$$= \left(\frac{2}{3}\right) g_S \mu_B g_N \mu_N \mu_0 \sum_q (-1)^{I+J'+F+S-\Sigma+J-\Omega}$$
$$\times \{I(I+1)(2I+1)(2J+1)(2J'+1)S(S+1)(2S+1)\}^{1/2}$$
$$\times \begin{Bmatrix} J' & I & F \\ I & J & 1 \end{Bmatrix} \begin{pmatrix} J & 1 & J' \\ -\Omega & q & \Omega' \end{pmatrix} \begin{pmatrix} S & 1 & S \\ -\Sigma & q & \Sigma' \end{pmatrix} \langle \eta, \Lambda \| \delta(\mathbf{r}) \| \eta', \Lambda' \rangle. \quad (9.46)$$

If the Fermi contact interaction constant b_F is defined to be $g_S \mu_B g_N \mu_N (2\mu_0/3) \delta(\mathbf{r})_\eta$, and we again neglect the admixture of excited states, equation (9.46) becomes

$$\langle \eta, \Lambda; S, \Sigma; J, \Omega, I, F, M_F | \mathcal{H}_F | \eta, \Lambda; S, \Sigma'; J', \Omega', I, F, M_F \rangle$$
$$= b_F \sum_q (-1)^{I+J'+F+S-\Sigma+J-\Omega}$$
$$\times \{I(I+1)(2I+1)(2J+1)(2J'+1)S(S+1)(2S+1)\}^{1/2}$$
$$\times \begin{Bmatrix} J' & I & F \\ I & J & 1 \end{Bmatrix} \begin{pmatrix} J & 1 & J' \\ -\Omega & q & \Omega' \end{pmatrix} \begin{pmatrix} S & 1 & S \\ -\Sigma & q & \Sigma' \end{pmatrix}. \quad (9.47)$$

Finally we deal with the magnetic dipolar interaction (9.41) whose general matrix elements are given by

$$\langle \eta, \Lambda; S, \Sigma; J, \Omega, I, F, M_F | \mathcal{H}_{\text{dip}} | \eta', \Lambda'; S, \Sigma'; J', \Omega', I, F, M_F \rangle$$
$$= g_S \mu_B g_N \mu_N (\mu_0/4\pi) \sum_q (-1)^{J'+I+F+q+J-\Omega}$$
$$\times \{I(I+1)(2I+1)(2J+1)(2J'+1)S(S+1)(2S+1)\}^{1/2}$$
$$\times \sqrt{30} \begin{Bmatrix} J' & I & F \\ I & J & 1 \end{Bmatrix} \begin{pmatrix} J & 1 & J' \\ -\Omega & q & \Omega' \end{pmatrix} \sum_{q_1 q_2} \begin{pmatrix} 1 & 2 & 1 \\ q_1 & q_2 & -q \end{pmatrix} (-1)^{S-\Sigma}$$
$$\times \begin{pmatrix} S & 1 & S \\ -\Sigma & q_1 & \Sigma' \end{pmatrix} \langle \eta, \Lambda | C_{q_2}^2(\theta, \phi)(r^{-3}) | \eta', \Lambda' \rangle. \quad (9.48)$$

Again we confine attention to the ground vibronic state by putting $q_2 = 0$ (so that $q_1 = q$):

$$\langle \eta, \Lambda; S, \Sigma; J, \Omega, I, F, M_F | \mathcal{H}_{\text{dip}} | \eta, \Lambda; S, \Sigma'; J', \Omega', I, F, M_F \rangle$$
$$= (\sqrt{30}/2) g_S \mu_B g_N \mu_N (\mu_0/4\pi) \sum_q (-1)^{I+J'+F+S+q-\Sigma+J-\Omega}$$
$$\times \{I(I+1)(2I+1)(2J+1)(2J'+1)S(S+1)(2S+1)\}^{1/2} \begin{Bmatrix} J' & I & F \\ I & J & 1 \end{Bmatrix}$$
$$\times \begin{pmatrix} J & 1 & J' \\ -\Omega & q & \Omega' \end{pmatrix} \begin{pmatrix} 1 & 2 & 1 \\ q & 0 & -q \end{pmatrix} \begin{pmatrix} S & 1 & S \\ -\Sigma & q & \Sigma' \end{pmatrix} \left\langle \frac{3\cos^2\theta - 1}{r^3} \right\rangle_\eta. \quad (9.49)$$

If we combine equations (9.45), (9.47) and (9.49) and expand the 3j- and 6-j symbols we obtain the following simple results for those matrix elements of the magnetic hyperfine interaction which are diagonal in all of the case (a) quantum numbers except for J:

$$\langle \eta, J|\mathcal{H}_{hf}|\eta, J\rangle = \{a\Lambda + (b_F + (2/3)c)\Sigma\}\Omega \frac{\{F(F+1) - J(J+1) - I(I+1)\}}{2J(J+1)},$$

(9.50)

$$\langle \eta, J|\mathcal{H}_{hf}|\eta, J-1\rangle$$
$$= -\{a\Lambda + (b_F + (2/3)c)\Sigma\}$$
$$\times \frac{(J^2 - \Omega^2)^{1/2}\{(F - I + J)(F + I + J + 1)(J + I - F)(F - J + I + 1)\}^{1/2}}{2J(4J^2 - 1)^{1/2}}.$$

(9.51)

In addition there are matrix elements off-diagonal in Ω and Σ. The constant c arises from the dipolar term, which is discussed in more detail later.

The main conclusion from these results is that the observed hyperfine splitting is determined primarily by a linear combination of the hyperfine constants corresponding to the three separate interactions. The spectrum depends upon the *axial* component of the total magnetic hyperfine interaction, which we designate $h_{3/2} (= a + (1/2)(b_F + 2c/3))$, and in a good case (a) system it is not usually possible to separate the individual contributions from the microwave magnetic resonance spectrum alone. The solution to the problem lies in the combination of these studies with pure rotational spectroscopy, as we shall see later in this chapter.

(iii) NUCLEAR ELECTRIC QUADRUPOLE COUPLING

The electric quadrupole interaction is handled in exactly the same way as the magnetic hyperfine interactions, by expanding the scalar product (9.36) first in the space-fixed axis system, and then transforming the electronic part of the interaction into the molecule-fixed system. One obtains the result:

$$\langle \eta, \Lambda; S, \Sigma; J, \Omega, I, F, M_F|\mathcal{H}_Q|\eta', \Lambda'; S, \Sigma; J', \Omega', I, F, M_F\rangle$$
$$= -\frac{1}{2}eQ\sum_q (-1)^{J'+I+F+J-\Omega}\{(2J+1)(2J'+1)\}^{1/2}$$
$$\times \begin{Bmatrix} J' & I & F \\ I & J & 2 \end{Bmatrix} \begin{pmatrix} J & 2 & J' \\ -\Omega & q & \Omega' \end{pmatrix} \begin{pmatrix} I & 2 & I \\ -I & 0 & I \end{pmatrix}^{-1} \langle \eta, \Lambda|T_q^2(\nabla E)|\eta', \Lambda'\rangle. \quad (9.52)$$

Confining attention to the $q = 0$ component, we obtain the results

$$\langle \eta, \Lambda; S, \Sigma; J, \Omega, I, F, M_F|\mathcal{H}_Q|\eta, \Lambda; S, \Sigma; J', \Omega', I, F, M_F\rangle$$
$$= -\frac{eQ}{2}(-1)^{J'+I+F+J-\Omega}\{(2J+1)(2J'+1)\}^{1/2}$$
$$\times \begin{Bmatrix} J' & I & F \\ I & J & 2 \end{Bmatrix} \begin{pmatrix} J & 2 & J' \\ -\Omega & 0 & \Omega' \end{pmatrix} \begin{pmatrix} I & 2 & I \\ -I & 0 & I \end{pmatrix}^{-1} \langle \eta, \Lambda|T_0^2(\nabla E)|\eta, \Lambda\rangle$$

$$= \frac{eq_0Q}{4}(-1)^{J'+I+F+J-\Omega}\{(2J+1)(2J'+1)\}^{1/2}$$

$$\times \begin{Bmatrix} J' & I & F \\ I & J & 2 \end{Bmatrix} \begin{pmatrix} J & 2 & J' \\ -\Omega & 0 & \Omega \end{pmatrix} \begin{pmatrix} I & 2 & I \\ -I & 0 & I \end{pmatrix}^{-1}, \quad (9.53)$$

where q_0 is the negative of the electric field gradient, and eq_0Q is the quadrupole coupling constant. The 3-j symbol indicates that matrix elements with $J' = J, J \pm 1$ and $J \pm 2$ are non-zero, but the diagonal elements are, of course, the most significant.

(iv) EXTERNAL MAGNETIC FIELD INTERACTIONS

The Zeeman Hamiltonian (9.37) may be rewritten in the form

$$\mathcal{H}_Z = [g_L\mu_B + g_r\mu_B]\mathbf{T}^1(\boldsymbol{B}) \cdot \mathbf{T}^1(\boldsymbol{L}) + [g_S\mu_B + g_r\mu_B]\mathbf{T}^1(\boldsymbol{B}) \cdot \mathbf{T}^1(\boldsymbol{S})$$
$$- g_N\mu_N\mathbf{T}^1(\boldsymbol{B}) \cdot \mathbf{T}^1(\boldsymbol{I}) - g_r\mu_B\mathbf{T}^1(\boldsymbol{B}) \cdot \mathbf{T}^1(\boldsymbol{J}), \quad (9.54)$$

and if the space-fixed Z ($p=0$) direction is defined by the direction of the magnetic field, the Zeeman Hamiltonian may be expressed in space-fixed coordinates as

$$\mathcal{H}_Z = B_Z\{[g_L\mu_B + g_r\mu_B]\mathrm{T}_0^1(\boldsymbol{L}) + [g_S\mu_B + g_r\mu_B]\mathrm{T}_0^1(\boldsymbol{S}) - g_N\mu_N\mathrm{T}_0^1(\boldsymbol{I}) - g_r\mu_B\mathrm{T}_0^1(\boldsymbol{J})\}. \quad (9.55)$$

The first two terms may now be treated by rotation from the space- to the molecule-fixed axis system, as follows.

(a) Orbital Zeeman interaction

The first term in equation (9.55) has matrix elements

$$\langle \eta, \Lambda; S, \Sigma; J, \Omega, I, F, M_F | B_Z[g_L\mu_B + g_r\mu_B]\sum_q \mathcal{D}_{0q}^{(1)}(\omega)^* \mathrm{T}_q^1(\boldsymbol{L}) | \eta, \Lambda'; S, \Sigma; J', \Omega', I, F', M_F\rangle$$

$$= B_Z\mu_B[g_L + g_r]\sum_q (-1)^{F-M_F+F'+J+I+1+J-\Omega} \begin{Bmatrix} J & F & I \\ F' & J' & 1 \end{Bmatrix}$$

$$\times \begin{pmatrix} F & 1 & F' \\ -M_F & 0 & M_F \end{pmatrix}\{(2F+1)(2F'+1)(2J+1)(2J'+1)\}^{1/2}$$

$$\times \begin{pmatrix} J & 1 & J' \\ -\Omega & q & \Omega' \end{pmatrix} \langle \eta, \Lambda | \mathrm{T}_q^1(\boldsymbol{L}) | \eta', \Lambda'\rangle. \quad (9.56)$$

The $q = \pm 1$ components of $T_q^1(\boldsymbol{L})$ will be neglected because they involve the mixing of excited states with the ground vibronic state, giving rise to temperature-independent paramagnetism. With this simplification the matrix elements of the orbital Zeeman interaction operating within the ground vibronic state are

$$\langle \eta, \Lambda; S, \Sigma; J, \Omega, I, F, M_F | B_Z[g_L\mu_B + g_r\mu_B]\sum_q \mathcal{D}_{0q}^{(1)}(\omega)^* \mathrm{T}_q^1(\boldsymbol{L}) | \eta, \Lambda; S, \Sigma; J', \Omega, I, F', M_F\rangle$$

$$= B_Z \mu_B \Lambda [g_L + g_r](-1)^{F-M_F+F'+J+I+1+J-\Omega} \begin{Bmatrix} J & F & I \\ F' & J' & 1 \end{Bmatrix} \begin{pmatrix} F & 1 & F' \\ -M_F & 0 & M_F \end{pmatrix}$$

$$\times \{(2F+1)(2F'+1)(2J+1)(2J'+1)\}^{1/2} \begin{pmatrix} J & 1 & J' \\ -\Omega & 0 & \Omega \end{pmatrix}. \qquad (9.57)$$

(b) Electron spin Zeeman interaction

The second term in equation (9.55) has matrix elements

$$\langle \eta, \Lambda; S, \Sigma; J, \Omega, I, F, M_F | B_Z [g_S \mu_B + g_r \mu_B] \sum_q \mathcal{D}^{(1)}_{0q}(\omega)^* T^1_q(S) | \eta, \Lambda; S, \Sigma; J', \Omega', I, F', M_F \rangle$$

$$= B_Z \mu_B [g_S + g_r] \sum_q (-1)^{F-M_F+J+I+F'+1+J-\Omega+S-\Sigma} \begin{Bmatrix} J & F & I \\ F' & J' & 1 \end{Bmatrix}$$

$$\times \begin{pmatrix} F & 1 & F' \\ -M_F & 0 & M_F \end{pmatrix} \{(2F+1)(2F'+1)(2J+1)(2J'+1)S(S+1)(2S+1)\}^{1/2}$$

$$\times \begin{pmatrix} J & 1 & J' \\ -\Omega & q & \Omega' \end{pmatrix} \begin{pmatrix} S & 1 & S \\ -\Sigma & q & \Sigma' \end{pmatrix}. \qquad (9.58)$$

It operates solely within the ground vibronic state.

(c) Nuclear spin Zeeman interaction

We treat the nuclear Zeeman interaction by remaining in the space-fixed axis system, and obtain

$$\langle \eta, \Lambda; S, \Sigma; J, \Omega, I, F, M_F | -g_N \mu_N B_Z T^1_{p=0}(I) | \eta, \Lambda; S, \Sigma; J', \Omega, I, F', M_F \rangle$$

$$= -g_N \mu_N B_Z (-1)^{F-M_F+J+I+F+1} \{I(I+1)(2I+1)(2F+1)(2F'+1)\}^{1/2}$$

$$\times \begin{Bmatrix} I & F & J \\ F' & I & 1 \end{Bmatrix} \begin{pmatrix} F & 1 & F' \\ -M_F & 0 & M_F \end{pmatrix}. \qquad (9.59)$$

(d) Rotational Zeeman interaction

The final term in equation (9.55) is the rotational Zeeman interaction whose matrix elements are again obtained by remaining in the space-fixed axis system:

$$\langle \eta, \Lambda; S, \Sigma; J, \Omega, I, F, M_F | -g_r \mu_B B_Z T^1_{p=0}(J) | \eta, \Lambda; S, \Sigma; J', \Omega, I, F', M_F \rangle$$

$$= -\delta_{JJ'} g_r \mu_B B_Z (-1)^{F-M_F+J+I+F'+1} \{J(J+1)(2J+1)(2F+1)(2F'+1)\}^{1/2}$$

$$\times \begin{Bmatrix} J & F & I \\ F' & J' & 1 \end{Bmatrix} \begin{pmatrix} F & 1 & F' \\ -M_F & 0 & M_F \end{pmatrix}. \qquad (9.60)$$

(d) Analysis of the ClO spectrum

Confining our attention to the more abundant ^{35}ClO spectrum shown in figure 9.9, we note the main features which can be understood immediately, at least qualitatively. The spectrum consists of three quartets, centred at an effective g-value close to 0.8, which is the value expected for a case (a) $^2\Pi_{3/2}$ state in its lowest rotational level, $J = 3/2$. (The

$^2\Pi_{1/2}$ fine-structure state has a g value very close to zero and is therefore unobservable by magnetic resonance.) The separation of the three groups, each of which arises from different $\Delta M_J = \pm 1$ components, is due to the second-order Zeeman effect, which comes from the mixing of the $J = 3/2$ and $5/2$ levels. This is analogous to the $2:3:3:2$ quartet splitting observed in the $^1\Delta$ spectra discussed earlier. The quartet splitting within each Zeeman component in the ClO spectrum is due primarily to the magnetic hyperfine interaction; the asymmetry in the quartet hyperfine spacings arises from the electric quadrupole interaction. Each of these aspects will now be discussed in more detail.

Even though matrix elements which mix excited vibronic states with the ground state have been neglected above, the analysis is still complicated because of matrix elements off-diagonal in J, Ω and F. Carrington, Dyer and Levy [20] carried out the analysis of their spectrum for the $J = 3/2$ level with a basis set including both the $|\Omega| = 3/2$ and $1/2$ states, with J values up to $7/2$. This involved the diagonalisation of 26×26 matrices for each M value, but ultimately it was possible to determine the values of four molecular constants for ^{35}ClO, as follows:

$$A = -282 \pm 9 \text{ cm}^{-1}, \quad B_0 = 0.622 \pm 0.001 \text{ cm}^{-1},$$
$$h_{3/2} = 111 \pm 2 \text{ MHz}, \quad eq_0 Q = -88 \pm 6 \text{ MHz}.$$

The constant $h_{3/2}$ was defined earlier and is defined again below. The corresponding constants for the less-abundant ^{37}ClO species were also determined. The only previous spectroscopic studies of ClO were described in one of the earliest flash-photolysis investigations by Porter [22], and a more detailed study of its near-ultraviolet spectrum was made by Durie and Ramsay [23]. They were unable to prove that the ground state is $^2\Pi_{3/2}$ but the determination of the sign of A through the magnetic resonance studies proved that it is so. The value of the rotational constant B_0 from the magnetic resonance spectrum also pointed to an error in the earlier analysis of the electronic spectrum, which was subsequently corrected [24] through a recording of the electronic spectrum at higher resolution.

The axial component of the magnetic hyperfine interaction for the $^2\Pi_{3/2}$ component is designated $h_{3/2}$; in terms of the original Frosch and Foley constants [25] $h_{3/2}$ is equal to $a + (1/2)(b + c)$, and in terms of our preferred hyperfine constants it is $a + (1/2)(b_F + 2t)$, the latter constants describing the orbital, Fermi contact and dipolar hyperfine interactions separately. Specifically, our constants are given by,

$$a = 2\mu_B g_N \mu_N (\mu_0/4\pi) \left\langle \frac{1}{r^3} \right\rangle_\eta,$$
$$b_F = \frac{2}{3} g_S \mu_B g_N \mu_N \mu_0 \Psi_\eta^2(0), \quad (9.61)$$
$$t = g_S \mu_B g_N \mu_N (\mu_0/4\pi) \left\langle \frac{3\cos^2\theta - 1}{r^3} \right\rangle_\eta.$$

The analysis of the $J = 3/2$ magnetic resonance spectrum gave only the value of $h_{3/2}$; as we shall see later, the microwave rotational spectrum enables the separate contributions to $h_{3/2}$ to be determined. It is possible to rationalise the values of the hyperfine constants in terms of a simple model of the electronic structure [20]; ultimately, however, it is best to regard these constants as benchmarks against which to test *ab initio* calculations of the electronic wave function. Similar conclusions may be applied to the determination of the quadrupole coupling constant. In subsequent work it proved possible to study the Stark splittings in the ClO spectrum [26] and to determine the electric dipole moment to be 1.26 D.

(e) Other $^2\Pi$ case (a) molecules

(i) MICROWAVE MAGNETIC RESONANCE

The microwave magnetic resonance spectrum of BrO has been observed [21] and its analysis is similar to that of ClO, although it is not so immediately apparent because the two bromine isotopes are present in essentially equal natural abundance, and the separation of the second-order Zeeman splitting and the nuclear hyperfine effects is not so clear. Nevertheless the analysis is straightforward, but there is one important new feature which was clarified by Brown, Byfleet, Howard and Russell [27] through additional studies of the $J = 5/2$ rotational level, and a more accurate theoretical treatment. If the matrix elements of the spin–orbit coupling operator are examined more closely, they are found to mix excited states with the ground vibronic state through $\Delta\Lambda = \pm 1$ and $\Delta\Sigma = \mp 1$ terms, so that although Ω is preserved as a good quantum number, Λ and Σ are not. This behaviour marks the departure from case (a) coupling towards case (c), so that the distinction between orbital and spin contributions to the magnetic moment is partially lost. The Zeeman Hamiltonian (9.54) is modified by the inclusion of extra terms to become

$$\mathcal{H}_Z = \{g_L + g_r + \Delta g_L + \Delta g_\Omega\}\mu_B T^1(\boldsymbol{B}) \cdot T^1(\boldsymbol{L}) + \{g_S + g_r + \Delta g_\Omega\}\mu_B T^1(\boldsymbol{B}) \cdot T^1(\boldsymbol{S}) \\ - g_N \mu_N T^1(\boldsymbol{B}) \cdot T^1(\boldsymbol{I}) - g_r \mu_B T^1(\boldsymbol{B}) \cdot T^1(\boldsymbol{J}). \tag{9.62}$$

The effects of Δg_L are negligible, but Δg_Ω and g_r are significant and their inclusion in the Zeeman analysis gives values for A which are more consistent with determinations from other spectroscopic investigations, and with theoretical expectations.

A further refinement in the analysis of the case (a) spectra was described by Carrington and Howard [28] in their study of the CF radical. We have already pointed out that in ClO, for example, it is only possible to determine the total axial magnetic hyperfine constant $h_{3/2}$ from studies of the $J = 3/2$ level alone. In CF, however, the rotational constant B_0 is relatively large and the spin–orbit constant A is relatively small. This means that the ratio B_0/A is considerably larger for CF (0.0183) than for ClO (0.0022), so that the rotational mixing of the $^2\Pi_{3/2}$ and $^2\Pi_{1/2}$ states (equation (9.38)) is more important. This fact, taken with additional measurements of the $J = 5/2$ level in the $^2\Pi_{3/2}$ state, enables both $h_{3/2}$ and b to be determined uniquely, though with less accuracy than one would wish.

In subsequent work the corresponding spectrum of IO was observed and analysed [21]; the dipole moments of BrO and IO [29] were also determined from Stark splittings

Table 9.2. *Experimental parameters for diatomic molecules in good case (a)* $^2\Pi$ *states*

Molecule	J	B_0/cm^{-1}	A/cm^{-1}	$h_{3/2}$/MHz	eq_0Q/MHz	μ_e/D
^{35}ClO	3/2	0.622	-282	111	-88	1.26
^{37}ClO	3/2	0.611	-282	93	-69	
^{79}BrO	3/2, 5/2	0.4281	-980	504.5	649.8	1.61
^{81}BrO	3/2, 5/2	0.4263	-980	543.9	542.7	
^{127}IO	3/2, 5/2	0.3385	-2330	582.1	-1907.0	2.45
S^{19}F	3/2	0.5527	-387	428.4	—	0.87
Se^{19}F	3/2, 5/2	0.3624	-1790	325.6	—	1.52
^{14}NS	3/2	0.7722	(223.03)	57.0	-2.86	1.86
C^{19}F	3/2, 5/2	1.4083	(77.11)	662.9	—	0.65

in the magnetic resonance spectra. Table 9.2 summarises the magnetic resonance results that have been obtained for a number of essentially good case (a) diatomic radicals in their $^2\Pi$ ground vibronic states. Some of these molecules will be discussed again in chapter 10, and more complete determinations of the hyperfine parameters presented.

(ii) FAR-INFRARED LASER MAGNETIC RESONANCE OF HF$^+$, HCl$^+$ AND HBr$^+$

Another milestone in the history of magnetic resonance spectroscopy was passed with the detection of the rotational spectra of molecular ions. This was first achieved by Saykally and Evenson [30] in their observation of rotational transitions in the cation HBr$^+$. This molecule is isoelectronic with SeH and has an inverted $^2\Pi$ ground state. The ion was generated in a d.c. electric discharge through a low pressure gas which consisted mostly of helium with about 1% of HBr. The FIR laser was of the familiar carbon dioxide pumped design. The section of the laser cavity filled by the sample volume was surrounded by a liquid nitrogen-cooled solenoid, capable of generating magnetic field strengths of up to 0.5 T. In this configuration only perpendicular $\Delta M = \pm 1$ transitions can be detected. In this way Saykally and Evenson were able to detect LMR spectra arising from the two lowest rotational transitions in the $^2\Pi_{3/2}$ component, $J = 5/2 \leftarrow 3/2$ and $J = 7/2 \leftarrow 5/2$. Part of the spectrum arising from the second transition is shown in figure 9.10. The figure shows the structure associated with a single Zeeman (M_J) component. Transitions for H^{79}Br$^+$ and H^{81}Br$^+$, which are present in almost equal natural abundance, were observed. For each isotopomer, there is a widely spaced quartet structure due to the $I = 3/2$ spin of the Br nucleus, and a small splitting due to Λ-doubling. The proton hyperfine structure was not resolved in this experiment.

Similar experiments were later carried out in Berkeley by Ray, Lubic and Saykally [31] who detected the FIR LMR spectrum of HCl$^+$, and by Hovde, Schaefer, Strahan, Ferrari, Ray, Lubic and Saykally [32] who studied HF$^+$. These experiments differed from those of Saykally and Evenson [30] in that a large conventional electromagnet was used to provide the variable magnetic field, and the electric discharge was struck

Table 9.3. *Spin–orbit and rotational constants (in cm^{-1}) for the hydrogen halide cations in their $X\,^2\Pi$ ground states*

	HF$^+$	HCl$^+$	HBr$^+$
A	-291.4	-643.4	-2651.6
B	17.1143	9.7880	7.9538
B/A	-0.0587	-0.0152	-0.0030

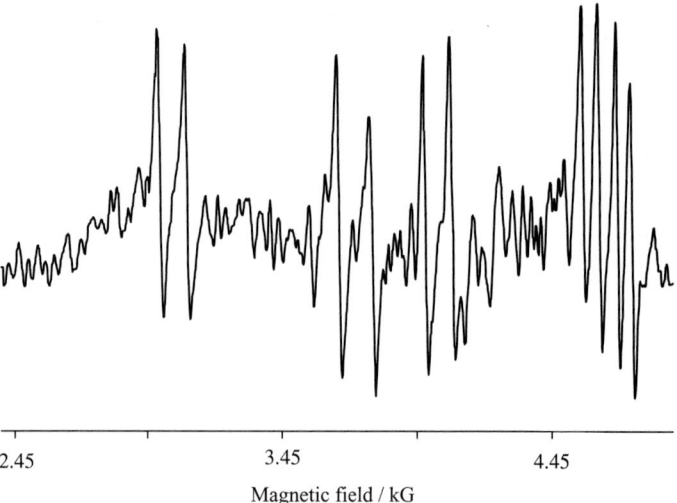

Figure 9.10. Laser magnetic resonance spectrum of HBr$^+$, associated with the $J = 7/2 - 5/2$ transition and recorded using the 180.7 μm line of CD$_3$OH, with σ polarisation ($\Delta M_J = \pm 1$) [30].

over a much shorter path length, within the pole gap of the electromagnet and parallel to the field. This enabled magnetic fields up to 2 T to be used and so provided a much larger tuning range. The three molecules are obviously related and have similar characteristics. They all have inverted $^2\Pi$ ground states with large spin–orbit coupling constants and large rotational constants (see table 9.3). It can be seen from the B/A ratios that they conform reasonably well to Hund's case (a) coupling, although HF$^+$ is beginning to drift towards case (b).

The transitions which have been detected are all pure rotational transitions in the lower fine-structure component, $^2\Pi_{3/2}$. The detection of the corresponding transitions in the upper $^2\Pi_{1/2}$ component is difficult to achieve by LMR experiments. The energy levels lie a long way above the $J = 3/2$ level of the $^2\Pi_{3/2}$ component, so that their population is fairly small. In addition, the transitions are barely tunable by a magnetic field because the molecules have very small magnetic moments in the $^2\Pi_{1/2}$ state. A nice example of the LMR spectra obtained is shown in figure 9.11, involving part of

$^2\Pi$ states

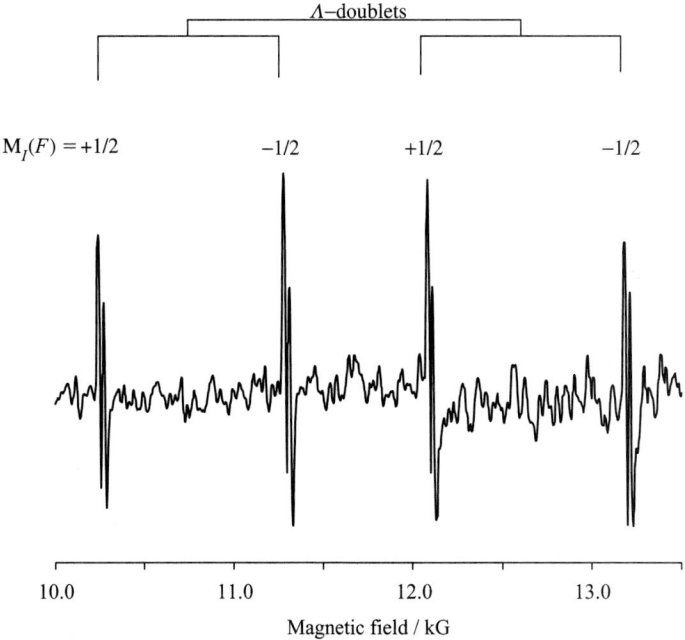

Figure 9.11. A portion of the FIR LMR spectrum of HF$^+$, recorded with a CH$_2$F$_2$ laser line at 122.5 μm [32]. This spectrum arises from the $M_J = -1/2 \leftarrow -3/2$ Zeeman component of the $J = 5/2 \leftarrow 3/2$ rotational transition in the $^2\Pi_{3/2}$ state. The small doubling of each resonance arises from the proton hyperfine interaction.

the $J = 5/2 \leftarrow 3/2$ transition in HF$^+$. The pattern shows doublings from Λ-doubling, and hyperfine interaction with ^{19}F and ^{1}H. Full analyses of the spectra of HCl$^+$ and HBr$^+$ were published later [33, 34].

The LMR spectra of this class of molecules provide accurate measurements of some rotational intervals and some hyperfine splittings. Independent measurements of the molecules in the $^2\Pi_{1/2}$ component are needed to provide a complete determination of the parameters in the effective Hamiltonian, such as the magnetic hyperfine parameters.

The experiments described demonstrate that it is possible to detect the spectra of molecular cations in the presence of a large magnetic field, despite the expected spatial displacement. Several other cations have since been detected by FIR LMR, both atoms and molecules.

(iii) MICROWAVE MAGNETIC RESONANCE SPECTRUM OF NO

Historically the first open shell molecule to be studied by magnetic resonance methods was nitric oxide, NO. The $^2\Pi_{1/2}$ fine-structure component is lower in energy than the $^2\Pi_{3/2}$ component by 123 cm^{-1} and is only weakly magnetic. However, the $^2\Pi_{3/2}$ component is substantially populated at room temperature and its microwave magnetic resonance spectrum is readily recorded. Spectra of the lowest rotational level,

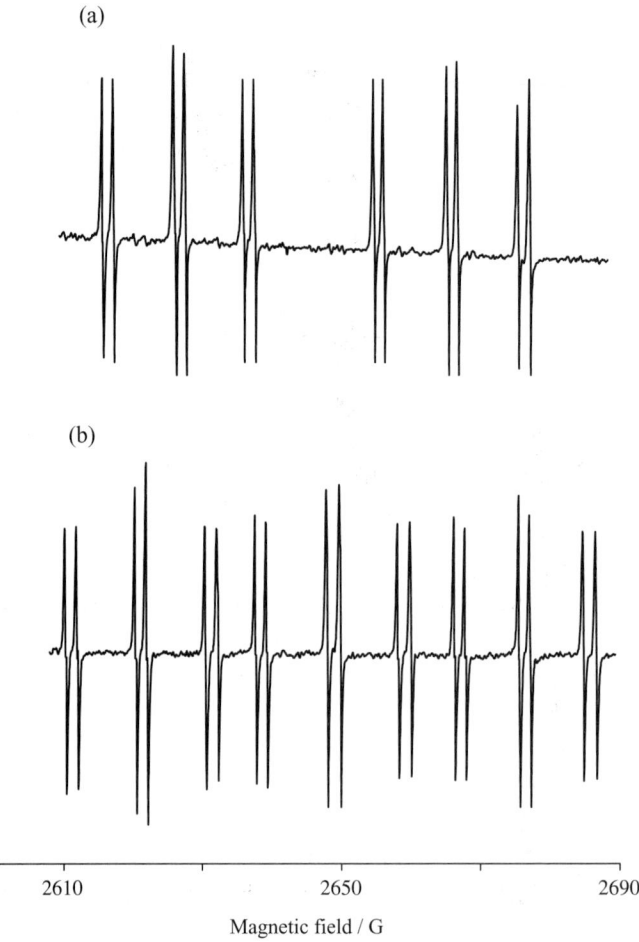

Figure 9.12. Microwave magnetic resonance spectra of NO in the $J = 3/2$ level of the $^2\Pi_{3/2}$ component, recorded by Brown and Radford [37]. Part (a) shows the ^{15}N^{16}O spectrum, with a very small Λ-doublet splitting, a larger second-order Zeeman splitting of the three $\Delta M_J = \pm 1$ components, and a doublet splitting from the ^{15}N nucleus, which has $I = 1/2$. Part (b) shows the ^{14}N^{16}O spectrum, which is similar to that shown in (a), except that there is now a triplet splitting from the ^{14}N nucleus, which has $I = 1$. The microwave frequency was 2879.9 MHz.

$J = 3/2$, in the $^2\Pi_{3/2}$ state were first obtained by Beringer, Rawson and Henry [35], and a theoretical analysis was provided by Lin and Mizushima [36]. The nicest spectra, however, were obtained later by Brown and Radford [37], and examples are shown in figure 9.12. These spectra are for the $J = 3/2$ rotational level, and exhibit nuclear hyperfine structure (doublet splitting for ^{15}N^{16}O, triplet splitting for ^{14}N^{16}O), second-order Zeeman splitting of the three $\Delta M_J = \pm 1$ components, and a very small doublet splitting arising from Λ-doubling. We will not go into the details of the quantitative analysis here because we described a very thorough electric resonance study of the Λ-doubling spectrum in chapter 8, and will return to NO in chapter 10 to discuss the far-infrared field-free rotational spectrum. Both types of study related to both fine-structure states, so that a more accurate separation of the individual magnetic hyperfine constants was obtained. Brown and Radford [37] were able to determine the rotational g-factors, and to use earlier field-free microwave studies to interpret the hyperfine spectrum, including the nitrogen electric quadrupole interaction for ^{14}N^{16}O.

9.4.3. OH in the $X\,^2\Pi$ ground state

We now turn to radicals with $^2\Pi_{3/2}$ ground states which do not conform to the simple case (a) model described above. Foremost among these are OH and CH, both of which occupy extremely important places in chemistry, molecular physics, and astrophysics. OH, in particular, is probably the free radical that has been most extensively studied by high-resolution spectroscopy; the story is complex, but ultimately very well understood. OH was the first gas-phase free radical to be detected by conventional microwave rotational spectroscopy, and we described its electric resonance Λ-doubling spectrum in detail in chapter 8. It has also been investigated by both microwave and far-infrared laser magnetic resonance, as we shall describe in due course. Following that we will describe pioneering laser magnetic resonance experiments on the CH radical.

(a) Microwave magnetic resonance

(i) INTRODUCTION

The ratio B_0/A_0 for the OH radical in its ground $^2\Pi$ state is -0.1333, large compared with, for example, the ratio for ClO which is -0.0022. Consequently its coupling scheme shows a considerable departure from the Hund's case (a) limit towards case (b), particularly as J increases. Its rotational energy levels are more complicated than those of ClO and are best described as showing intermediate coupling behaviour. It is still possible to think of the pattern as two sets of rotational levels, one associated with the lower spin component and the other with the upper. However, for OH these two sets of levels are not well separated from each other but are extensively intermingled, as shown in figure 9.13. We shall see that terms in the rotational kinetic energy operator mix levels with the same J in the two spin components quite heavily. As a result, the quantum number Ω which we have used so far to label the spin components ceases to be a good one. In this situation, the lower level for a given J is referred to as the F_1 spin component and the upper one as F_2. For OH, with a negative spin–orbit coupling constant ($A_0 = -139$ cm^{-1}), the F_1 component correlates with $^2\Pi_{3/2}$ in the case (a) limit and the F_2 component with $^2\Pi_{1/2}$. This correlation is quite reliable at low J but becomes increasingly less so as J increases. We note a curiosity that, although there are always two spin components for levels with J greater than or equal to 3/2, for $J = 1/2$ there is only one spin component which always has pure $\Omega = 1/2$ character. Nearer the case (b) limit, there can be some uncertainty whether to associate the $J = 1/2$

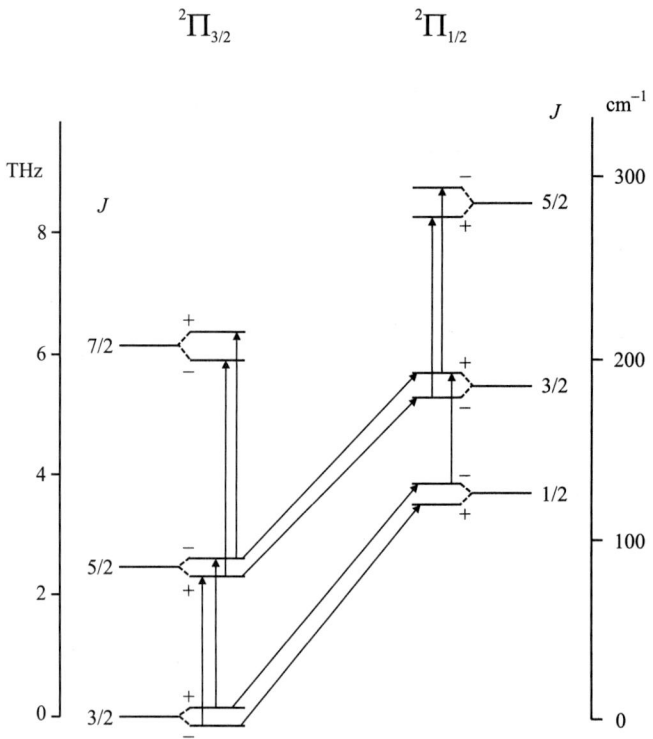

Figure 9.13. Lowest rotational levels of the OH radical and the far-infrared transitions observed. The Λ-doublet splittings are exaggerated for the sake of clarity.

level with the F_1 set or the F_2 set. Fortunately, in the case of OH, the situation is quite clear. The $J = 1/2$ level is associated with the upper set of levels and it is appropriate to label it as F_2. In addition to showing intermediate coupling behaviour, OH has a very small moment of inertia and hence a large rotational constant. As a result, its energy levels show two additional effects compared with ClO, centrifugal distortion and Λ-type doubling; we discussed the latter extensively in chapter 8 when describing the molecular beam electric resonance spectrum.

(ii) EXPERIMENTAL STUDIES

The OH radical has been studied extensively by both microwave and far-infrared laser magnetic resonance. In the microwave region, the pioneering work of Radford [7] effectively launched the field of gas phase electron resonance, which in this book we are calling microwave magnetic resonance. Radford first studied OH in the $J = 3/2, 5/2$ and $7/2$ levels of the $^2\Pi_{3/2}$ component, and also OD in the $J = 3/2$ level. Figure 9.14 shows the behaviour of the $J = 3/2$ Λ-doublet and hyperfine levels in a magnetic field, together with the magnetic resonance electric dipole transitions. We also show a stick diagram of the observed spectrum. Radford did not show an experimental recording of any of his OH F_1 ($^2\Pi_{3/2}$) spectra, but to compensate we show, in figure 9.15, the

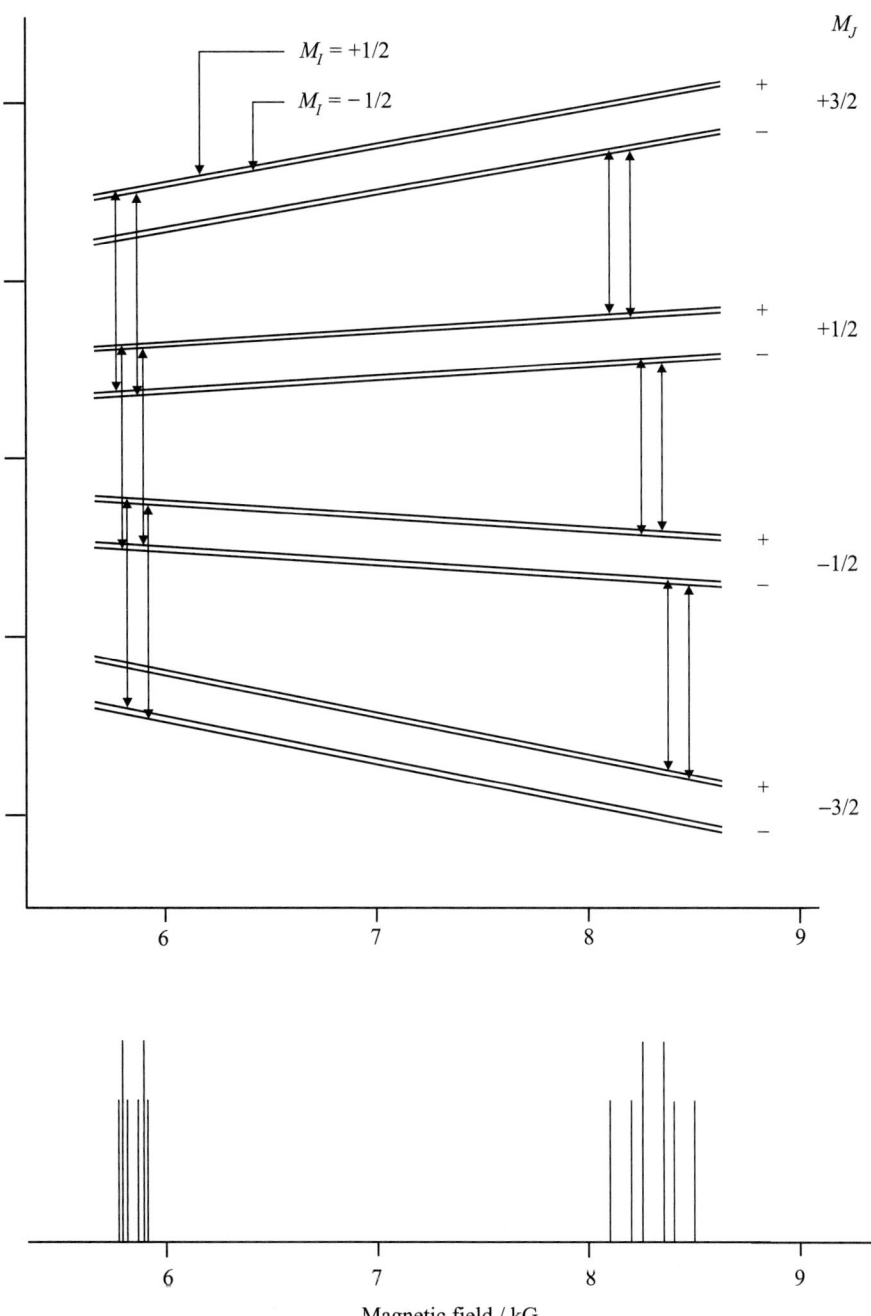

Figure 9.14. *Upper*: Zeeman behaviour of the Λ-doublet and proton hyperfine levels of OH in the $J = 3/2$ $F_1(^2\Pi_{3/2})$ rotational level, and the electric dipole transitions. *Lower*: stick diagram of the magnetic resonance spectrum obtained by Radford at a frequency of 9263 MHz [7].

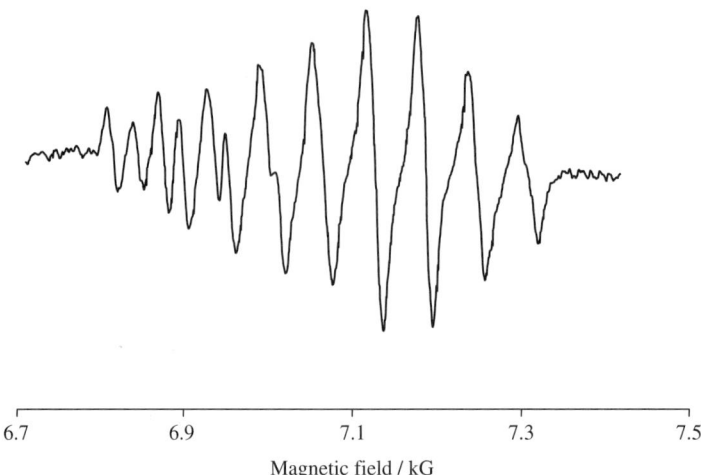

Figure 9.15. Microwave Λ-doublet magnetic resonance spectrum of OH in the $J = 9/2$ F_1 ($^2\Pi_{3/2}$) rotational level [66]. The resonant microwave frequency was 26254 MHz. Eighteen transitions are expected but only 13 lines are resolved.

spectrum obtained by Brown, Kaise, Kerr and Milton [66] arising from the $J = 9/2$ F_1 ($^2\Pi_{3/2}$) level. Radford realised that, because of the intermediate coupling behaviour in OH, the levels of the $^2\Pi_{1/2}$ component also show strong magnetic effects. He then went on to detect the microwave magnetic resonance spectrum of OH in the $J = 3/2$ and 5/3 levels of this upper component (F_2); his experimental spectrum for $J = 3/2$ is shown in figure 9.16. The six lines labelled in figure 9.16 arise from the following transitions.

f_1 : $M_J = -3/2$, $M_I = -1/2$ (+parity) $\leftrightarrow$ $M_J = -1/2$, $M_I = -1/2$ (−parity)
f_2 : $M_J = -1/2$, $M_I = -1/2$ (+parity) $\leftrightarrow$ $M_J = +1/2$, $M_I = -1/2$ (−parity)
f_3 : $M_J = +1/2$, $M_I = +1/2$ (+parity) $\leftrightarrow$ $M_J = +3/2$, $M_I = +1/2$ (−parity)
f_4 : $M_J = -1/2$, $M_I = +1/2$ (+parity) $\leftrightarrow$ $M_J = +1/2$, $M_I = +1/2$ (−parity)
f_5 : $M_J = +1/2$, $M_I = -1/2$ (+parity) $\leftrightarrow$ $M_J = +3/2$, $M_I = -1/2$ (−parity)
f_6 : $M_J = -3/2$, $M_I = +1/2$ (+parity) $\leftrightarrow$ $M_J = -1/2$, $M_I = +1/2$ (−parity)

Radford's analysis of the magnetic resonance spectrum was ground-breaking by his extension of the theory of the Zeeman effect to the general intermediate coupling case. In addition to Radford's observations, Brown, Kaise, Kerr and Milton [66] recorded microwave magnetic resonance spectra at higher resonant frequencies and were able to observe, as we have seen, the $J = 9/2$ and also the $J = 11/2$ levels in the F_1 ($^2\Pi_{3/2}$) spin component. The frequencies of the zero magnetic field transitions involved in the microwave magnetic resonance studies had all been measured previously and much more accurately in molecular beam work by Meerts [38], as we described in chapter 8. Brown, Kaise, Kerr and Milton [66] therefore used the magnetic resonance measurements to provide a full determination of the magnetic properties of the OH radical. In their paper they gave the complete Zeeman Hamiltonian for a molecule

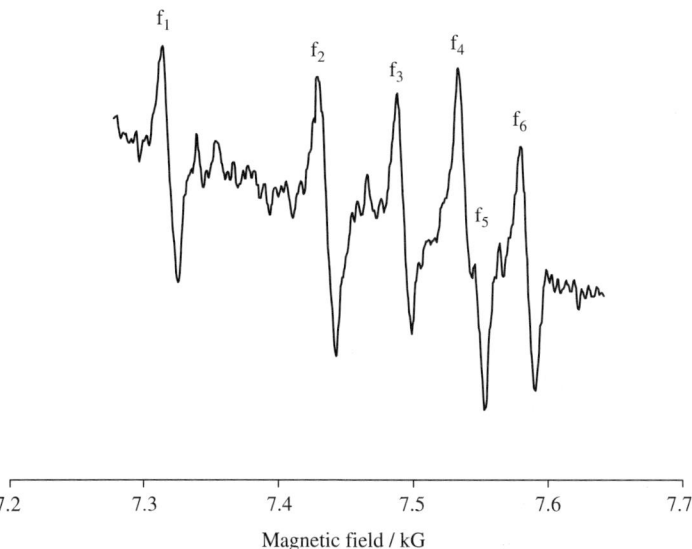

Figure 9.16. Recording of the Λ-doublet microwave magnetic resonance spectrum of OH in the $J = 3/2$ $F_2(^2\Pi_{1/2})$ rotational level [7]. The assignment of the lines is given in the text. The resonant microwave frequency was 9200 MHz.

in a $^{2S+1}\Pi$ electronic state and showed that it involves six determinable parameters (g-factors) in addition to the nuclear spin g-factor for the proton. Carrington and Lucas [39], in earlier work, extended our knowledge of the OH radical by studying ^{17}OH and determining the ^{17}O magnetic and electric hyperfine constants.

(iii) THEORY OF THE Λ-DOUBLING

As we discussed earlier in chapter 8, the Λ-doubling in a molecule such as OH arises from the $\Delta\Lambda = \pm 1$ terms in the spin–orbit and rotational Hamiltonians which mix the $^2\Pi$ ground state with excited $^2\Sigma$ states. These terms also mix the $^2\Pi$ state, $\Lambda = \pm 1$, with $^2\Delta$ states, for which $\Lambda = \pm 2$, but such mixing does not contribute to the Λ-doubling. In the case of OH, like LiO, the situation is particularly simple because the mixing is predominantly with a single excited state, the $A\,^2\Sigma^+$ state, which lies some 32 680 cm^{-1} above the ground state. It is therefore possible to analyse the observed Λ-doublet splittings explicitly in terms of the $X\,^2\Pi$ and $A\,^2\Sigma^+$ states. However, in general there will be varying contributions to the Λ-doubling from a host of $^2\Sigma$ states, both $^2\Sigma^+$ and $^2\Sigma^-$ in character. In order to cope with such a situation, the effects of the terms in the full Hamiltonian which are off-diagonal in the quantum number Λ are described by perturbation theory to produce an effective Hamiltonain. This effective Hamiltonian operates only within the manifold of levels of a particular electronic state, in this case the ground $^2\Pi$ state, but still reproduces the molecular eigenstates to the required accuracy. In principle, this procedure can be carried on indefinitely to any order of perturbation theory, as we showed in chapter 7. In practice, it becomes very hard work beyond third order.

The effects of the off-diagonal terms when folded-in by perturbation theory are of two types. They can either produce operators of the same form as those which already exist in the Hamiltonian constructed from the $\Delta\Lambda = 0$ matrix elements (the zeroth-order Hamiltonian), or they can have a completely novel form. A good example of the former type is the second-order contribution to the rotational constant which arises from admixture of excited Σ and Δ states,

$$B = B^{(1)} + B^{(2)}. \tag{9.63}$$

The first-order part describes the nuclear and the second-order the electronic contribution to the molecular moment of inertia. The Λ-doubling terms, on the other hand, have no counterpart in the zeroth-order Hamiltonian. For a $^2\Pi$ state, the operator form is

$$\mathcal{H}_{LD} = -(1/2)q(J_+^2 e^{-2i\phi} + J_-^2 e^{2i\phi}) + (1/2)(p+2q)(J_+ S_+ e^{-2i\phi} + J_- S_- e^{2i\phi}), \tag{9.64}$$

where p and q are the Λ-doubling parameters. (For a Π state of triplet or higher multiplicity, there is a third parameter, o. However, the terms which involve this parameter have no effect on $^2\Pi$ or $^1\Pi$ states.) The operators $J_\pm$ are shift operators which act on the rotational basis functions with the selection rules $\Delta\Omega = \mp 1$, and $S_\pm$ are the corresponding operators which act on the spin basis functions with $\Delta\Sigma = \pm 1$ selection rules. In addition, the coordinate ϕ in equation (9.64) represents an electronic azimuthal angle, associated with the orbital angular momentum about the z axis. The exponential terms which involve ϕ ensure that $\mathcal{H}_{LD}$ only has matrix elements which connect states whose Λ values differ by ± 2. Since a Π electronic state can be represented by a pair of basis functions $|\Lambda = +1\rangle$ and $|\Lambda = -1\rangle$, which are linked directly by $\mathcal{H}_{LD}$, we see that the operator has a first-order effect for a $^{2S+1}\Pi$ electronic state. The full form of the effective Hamiltonian for Λ-doubling was first given by Brown, Kopp, Malmberg and Rydh [63] although the seeds of the idea were sown in just about the very first treatment of this topic, by Mulliken and Christy [40] in 1931.

The Λ-doubling operator can be written in spherical tensor notation:

$$\mathcal{H}_{LD} = \sum_{q=\pm 1} \exp(-2iq\phi)\{-q\mathrm{T}_{2q}^2(\boldsymbol{J},\boldsymbol{J}) + (p+2q)\mathrm{T}_{2q}^2(\boldsymbol{J},\boldsymbol{S})\}. \tag{9.65}$$

In a Hund's case (a) basis set the matrix elements of (9.65) are as follows:

$$\langle \eta, \Lambda; S, \Sigma; J, \Omega, M_J | \mathcal{H}_{LD} | \eta, \Lambda'; S, \Sigma'; J, \Omega', M_J \rangle$$
$$= \sum_{q=\pm 1} \delta_{\Lambda',\Lambda\mp 2} \left\{ \delta_{\Sigma,\Sigma'} (1/2\sqrt{6}) q (-1)^{J-\Omega} \begin{pmatrix} J & 2 & J \\ -\Omega & -2q & \Omega' \end{pmatrix} [(2J-1)(2J) \right.$$
$$\times (2J+1)(2J+2)(2J+3)]^{1/2} + (p+2q)(-1)^{J-\Omega+S-\Sigma} \begin{pmatrix} J & 1 & J \\ -\Omega & -q & \Omega' \end{pmatrix}$$
$$\left. \times \begin{pmatrix} S & 1 & S \\ -\Sigma & q & \Sigma' \end{pmatrix} [J(J+1)(2J+1)S(S+1)(2S+1)]^{1/2} \right\}. \tag{9.66}$$

We have used the result

$$\langle \Lambda = \pm 1 | \exp(\pm 2i\phi) | \Lambda = \mp 1 \rangle = -1, \quad (9.67)$$

which implies a particular phase choice for the orbital function [66]. The role which these terms play in the matrix and their ultimate contribution to the energy levels can best be appreciated by re-casting the matrix elements in parity-conserving combinations of case (a) functions,

$$|\Lambda; J, |\Omega|; \pm\rangle$$
$$= \frac{1}{\sqrt{2}}\{|\Lambda = 1; S, \Sigma; J, \Omega, M_J\rangle \pm (-1)^{J-S}|\Lambda = -1; S, -\Sigma; J, -\Omega, M_J\rangle\}. \quad (9.68)$$

With appropriately chosen basis functions [66], the upper sign choice is a function with positive parity and the lower sign choice has negative parity. In this basis, the matrix elements of $\mathcal{H}_{LD}$ are:

$$\langle \Lambda; J, |\Omega|; \pm | \mathcal{H}_{LD} | \Lambda; J, |\Omega'|; \pm \rangle$$
$$= \pm(-1)^{J-S} \left\{ \delta_{\Sigma,\Sigma'}(1/2\sqrt{6})q(-1)^{J+\Omega} \begin{pmatrix} J & 2 & J \\ \Omega & -2 & \Omega' \end{pmatrix} [(2J-1)(2J)(2J+1)\right.$$
$$\times (2J+2)(2J+3)]^{1/2} + (p+2q)(-1)^{J+\Omega+S+\Sigma} \begin{pmatrix} J & 1 & J \\ \Omega & -1 & \Omega' \end{pmatrix} \begin{pmatrix} S & 1 & S \\ \Sigma & 1 & \Sigma' \end{pmatrix}$$
$$\left. \times [J(J+1)(2J+1)S(S+1)(2S+1)]^{1/2} \right\}. \quad (9.69)$$

These matrix elements contribute with one sign for the positive parity matrix and with the opposite sign for the negative parity matrix. Note that, because $\mathcal{H}_{LD}$ commutes with the laboratory-fixed inversion operator E*, which defines parity, matrix elements between states of opposite parity are rigorously zero.

The matrix representation of the Λ-doubling terms for a molecule in a $^2\Pi$ state has been given in chapter 8. For completeness and continuity, we repeat it here. There is a 2×2 matrix for the states with a given J value, as follows.

	$^2\Pi_{3/2}$	$^2\Pi_{1/2}$
$^2\Pi_{3/2}$	0	$\pm(-1)^{J-1/2}(q/2)(J+1/2)[(J+1/2)^2-1]^{1/2}$
$^2\Pi_{1/2}$	$\pm(-1)^{J-1/2}(q/2)(J+1/2)$ $\times [(J+1/2)^2-1]^{1/2}$	$\mp(-1)^{J-1/2}(1/2)(p+2q)(J+1/2)$

The upper and lower signs refer to states of positive and negative parity respectively.

As mentioned above, the small mass of OH leads to a pronounced centrifugal distortion of its rotational and spin energy levels. The description of these effects has already been given earlier in section 8.5.4(d), where the explicit matrix elements for a molecule in a $^2\Pi$ state were given.

(iv) ZEEMAN EFFECT

Perhaps the most important interaction in the microwave magnetic resonance experiment is that between the molecular magnetic dipole moment and the applied magnetic field, represented by the Zeeman Hamiltonian. We have already introduced this operator in our discussion of the spectra of ClO and BrO, equation (9.37). However, this form is tailored to the description of the energy levels of a molecule in a $^2\Pi_{3/2}$ spin component which shows good case (a) behaviour and a negligible Λ-doubling. For a more complicated situation, such as that for OH, other terms have to be added to this operator. We now give the complete form for the Zeeman Hamiltonian, foreshadowed by the work of Radford [7] and put into effective operator form by Brown and Uehara [10] and by Brown, Kaise, Kerr and Milton [66], as follows:

$$
\begin{aligned}
\mathcal{H}_Z = {} & g'_L \mu_B B_Z T^1_{p=0}(\boldsymbol{L}) & \text{(i)} \\
& + g_S \mu_B B_Z T^1_{p=0}(\boldsymbol{S}) & \text{(ii)} \\
& - g_r \mu_B B_Z T^1_{p=0}(\boldsymbol{J} - \boldsymbol{L} - \boldsymbol{S}) & \text{(iii)} \\
& + g_l \mu_B B_Z \sum_{q=\pm 1} \mathcal{D}^{(1)}_{0q}(\omega)^* T^1_q(\boldsymbol{S}) & \text{(iv)} \\
& - g_N \mu_N B_Z T^1_{p=0}(\boldsymbol{I}) & \text{(v)} \\
& + g'_l \mu_B B_Z \sum_{q=\pm 1} \exp(-2iq\phi) \mathcal{D}^{(1)}_{0,-q}(\omega)^* T^1_q(\boldsymbol{S}) & \text{(vi)} \\
& - g^{e'}_r \mu_B B_Z \sum_{q=\pm 1} \sum_p \exp(-2iq\phi)(-1)^p \mathcal{D}^{(1)}_{-p,-q}(\omega)^* T^1_p(\boldsymbol{J}-\boldsymbol{S}) \mathcal{D}^{(1)}_{0,-q}(\omega)^*. & \text{(vii)}
\end{aligned}
$$
(9.70)

The seven terms in this expression represent the following interactions:

(i) the orbital Zeeman effect. The orbital g-factor, g'_L, is so called (rather than simply g_L) because it contains relativistic, diamagnetic and non-adiabatic contributions;
(ii) the electron spin isotropic contribution;
(iii) the rotational magnetic moment contribution to the Zeeman effect;
(iv) the anisotropic contribution to the electron spin Zeeman effect, which takes account of the cylindrical symmetry shown by a diatomic molecule;
(v) the nuclear spin Zeeman effect;
(vi) and (vii) are two terms which are parity-dependent and effectively mark the reduction of the electronic distribution from cylindrical to a lower symmetry.

There are seven g-factors in equation (9.70). The two major ones are g'_L and g_S which are expected to have values very near to 1.0 and 2.002. In a real molecule, however, the values will differ slightly but significantly from these values. In practice, their values are determined along with those of all the others from measurements of the molecule in a known magnetic field. Only the nuclear spin g factor, g_N, is effectively a known quantity. Its apparent value will be subject to chemical shifts in any given molecular environment, of course, but such effects are much too small to be observable in a gas phase magnetic resonance experiment. Ignoring the nuclear spin

g-factor, therefore, we see that there are six determinable parameters, or g-factors, in equation (9.70) which are required to describe the Zeeman interaction for a molecule in a $^{2S+1}\Pi$ state fully.

The matrix elements of the Zeeman Hamiltonian given in equation (9.70) are evaluated most appropriately in an I-decoupled basis set because the nuclear spin is decoupled in almost all situations in quite modest magnetic fields. The matrix elements of the simpler Zeeman Hamiltonian have been given earlier in equations (9.56) to (9.60). For the sake of completeness, we give the matrix elements of the full Zeeman Hamiltonian here:

$$\langle \Lambda; S, \Sigma; J, \Omega, M_J; I, M_I | \mathcal{H}_Z | \Lambda'; S, \Sigma'; J', \Omega', M_J; I, M_I \rangle$$

$$= \mu_B B_Z \delta_{\Lambda, \Lambda'} \sum_q (-1)^{J-M_J+J-\Omega} [(2J+1)(2J'+1)]^{1/2} \begin{pmatrix} J & 1 & J' \\ -M_J & 0 & M_J \end{pmatrix}$$

$$\times \begin{pmatrix} J & 1 & J' \\ -\Omega_J & q & \Omega' \end{pmatrix} \left\{ (g'_L + g_r)\Lambda\delta_{\Sigma,\Sigma'} + (g_S + g_r + g_l)(-1)^{S-\Sigma} \right.$$

$$\times [S(S+1)(2S+1)]^{1/2} \begin{pmatrix} S & 1 & S \\ -\Sigma & q & \Sigma' \end{pmatrix} - g_l \Sigma \delta_{\Sigma,\Sigma'} \right\}$$

$$- (g_r \mu_B B_Z M_J + g_N \mu_N B_Z M_I) \delta_{J,J'} \delta_{\Sigma,\Sigma'} \delta_{\Lambda,\Lambda'}$$

$$- \mu_B B_Z \sum_{q=\pm 1} \delta_{\Lambda,\Lambda\mp 2} (-1)^{J-M_J} [(2J+1)(2J'+1)]^{1/2} \begin{pmatrix} J & 1 & J' \\ -M_J & 0 & M_J \end{pmatrix}$$

$$\times \left[(g_l^{e'} - g_r^{e'})(-1)^{S-\Sigma} \begin{pmatrix} S & 1 & S \\ -\Sigma & q & \Sigma' \end{pmatrix} [S(S+1)(2S+1)]^{1/2} (-1)^{J-\Omega} \right.$$

$$\times \begin{pmatrix} J & 1 & J' \\ -\Omega & -q & \Omega' \end{pmatrix} - g_r^{e'} \delta_{\Sigma,\Sigma'} (-1)^{J-\Omega} \sum_{\Omega''} (1/2) \left\{ (-1)^{J-\Omega''} \begin{pmatrix} J & 1 & J \\ -\Omega & -q & \Omega'' \end{pmatrix} \right.$$

$$\times \begin{pmatrix} J & 1 & J' \\ -\Omega'' & -q & \Omega' \end{pmatrix} [J(J+1)(2J+1)]^{1/2} + (-1)^{J'-\Omega''} \begin{pmatrix} J & 1 & J' \\ -\Omega & -q & \Omega'' \end{pmatrix}$$

$$\times \begin{pmatrix} J' & 1 & J' \\ -\Omega'' & -q & \Omega' \end{pmatrix} [J'(J'+1)(2J'+1)]^{1/2} \right\} \right]. \tag{9.71}$$

A complete determination of the Zeeman parameters for OH has been made by Brown, Kaise, Kerr and Milton [66]. They first re-fitted Meerts' data on the Λ-doubling frequencies [38] to provide accurate zero-field frequencies. They then combined their measurements with those made earlier by Radford [7] to determine the g-factors. The results of their fits are given in table 9.4. The values given in table 9.4 form a representative set and demonstrate some general points. First, the broad features of the Zeeman interaction are described by two major parameters, g'_L and g_S, which are typically three orders-of-magnitude larger than the other four g-factors. Second, these two parameters have values close to those expected for a free electron (1.0 and 2.002 respectively) so the Zeeman effect can be predicted quite reliably before any experimental observations are attempted. Finally, it can be seen that the two major g-factors differ significantly from the expected free electron values, and the other four g-parameters have significant non-zero values.

Table 9.4. g-Factors for OH in the $X\,^2\Pi$ state [66]. The calculations are made by assuming that the $X\,^2\Pi$ and $A\,^2\Sigma^+$ states are in pure precession

Parameter	Experimental value	Calculated value
g'_L	1.001 07 (15)	1.000 93
g_S	2.001 52 (36)	2.002 06
$10^3\, g_l$	4.00 (56)	4.29
$10^3\, g_r$	−0.633 (19)	−0.55
$10^3\, g'_l$	6.386 (30)	8.58
$10^3\, g_r^{e'}$	2.0446 (23)	2.19

All of the experimentally determined quantities carry information on the electronic structure of the OH radical. A very simple picture of this structure which actually works very well is to describe the properties in terms of the wave functions for two electronic states only, the $X\,^2\Pi$ and $A\,^2\Sigma^+$ states, with electronic configurations

$$\cdots (2p\sigma)^2(2p\pi)^3 \quad X\,^2\Pi$$
$$\cdots (2p\sigma)^1(2p\pi)^4 \quad A\,^2\Sigma^+.$$

We can then make an even more drastic approximation and represent the molecular orbitals in these configurations by pure $2p$ atomic orbitals on the O atom. This approximation was called pure precession by Van Vleck [41]; in this approximation the electrons in these outermost orbitals are in a spherically symmetric environment and they have a well defined value of the orbital angular momentum quantum number l (unity for a p orbital). In the pure precession approximation, we can derive very simple expressions for the g-factors [66]. The values for OH predicted on the basis of this very simple model are given in table 9.4. The fact that they agree reasonably well with the experimental numbers suggests that the theoretical model is essentially correct.

(b) Far-infrared laser magnetic resonance spectrum of OH

The OH molecule was one of the first molecules to be studied by the fledgling technique of far-infrared laser magnetic resonance (FIR LMR) in the early 1970s [42]. At that stage there were very few FIR laser lines available; for all of them, the population inversion was created in the gain medium by electric discharge, through H_2O, D_2O or HCN vapour. It is a remarkable coincidence that many of the water discharge lines are very close in frequency to OH transitions in the $v=0$ level, particularly when one considers that OH is a light molecule with a very open rotational spectrum. It was for this reason that the water discharge laser was familiarly referred to as the 'OH laser' by workers in the field. The various FIR LMR studies of OH in the $v=0$ level of its $X\,^2\Pi$ state were collected together in a paper by Brown, Kerr, Wayne, Evenson and Radford [43]. The energy level diagram for OH in this state, showing the transitions studied, has already been illustrated in figure 9.13. Of the seven laser lines used in this study, five are lines from the water discharge laser. The transitions are of two types: pure rotational transitions within a given spin component, and fine-structure transitions between the two spin components. The latter are usually weaker than the former; they are actually electric dipole forbidden in the case (a) limit.

The dominant source of line broadening in a spectral transition in the far-infrared is the Doppler effect. For a light molecule such as OH at room temperature, the Doppler linewidth at 100 cm^{-1} (corresponding to a wavelength of 100 µm) is 0.000 30 cm^{-1} or 9.0 MHz. This is comparable with, and sometimes larger than, the proton hyperfine splittings in OH. For this reason, the proton hyperfine splittings cannot always be observed on resonances in the LMR spectrum. In such a situation, the intra-cavity aspect of the FIR LMR apparatus comes into its own because it provides a natural environment to produce saturation or Lamb dips on the absorption signals. Lamb dips arise when radiation of a given frequency passes through the same region of the molecular sample simultaneously from opposite directions. The linewidth of a Lamb dip is usually dictated by lifetime limiting effects of the levels involved; in the case of OH this is pressure broadening. The increase in resolution obtainable with Lamb dips is demonstrated in figure 9.17, which shows part of the LMR spectrum of OH recorded with the 84.3 µm line of the water discharge laser. The broader background is the residual Doppler profile (or rather, its first derivative) on which are superimposed the two Lamb dips; these correspond to the proton hyperfine doublet, also as first derivatives, but with opposite phases.

Measurements of the FIR LMR spectrum of OH are less accurate than those of the field-free Λ-doubling spectrum in the microwave region, partly because they are made at much higher frequencies (with correspondingly larger linewidths), and partly

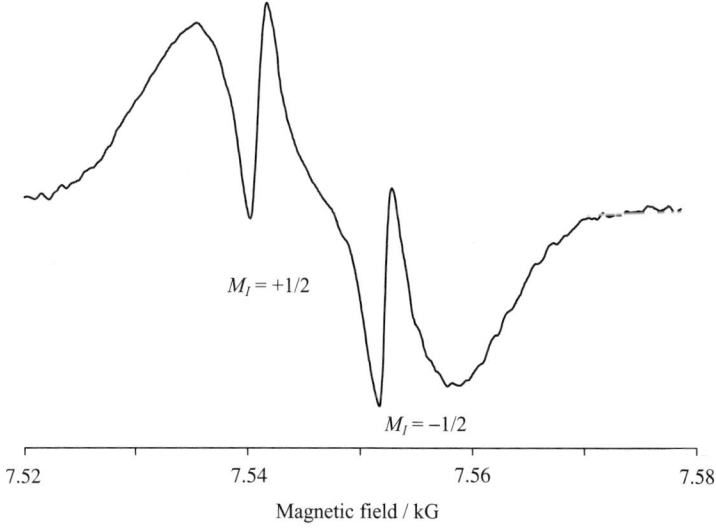

Figure 9.17. A line in the 84 µm LMR spectrum of OH; this line is one Zeeman component of the $J = 7/2 \leftrightarrow 5/2\ F_1(^2\Pi_{3/2})$ rotational transition. The Lamb dips correspond to the proton hyperfine splitting which is otherwise hidden in the overall Doppler profile [43].

because they involve magnetic field rather than frequency measurements. Brown, Kerr, Wayne, Evenson and Radford [43] therefore used the data from the FIR LMR spectrum to determine the values of the major parameters A_0, γ_0, B_0 and D_0, constraining the Λ-doubling, hyperfine and Zeeman parameters to the values obtained by earlier workers. The values obtained were a considerable improvement on the best available at the time which had been determined from the $A\,^2\Sigma^+ - X\,^2\Pi$ electronic transition in the ultraviolet region. Since then they have been superseded by observations of the rotational and fine structure spectrum of OH by field-free tunable FIR techniques [44], which are described in detail in chapter 10.

The dominant isotopic form of the radical is, of course, $^{16}\text{O}^1\text{H}$. Following its study by FIR LMR, observations were also made of the spectra of other less-abundant isotopomers, including $^{16}\text{O}^2\text{D}$ [45] and $^{18}\text{O}^1\text{H}$ [46] in natural abundance, and artificially enriched $^{17}\text{O}^1\text{H}$ [47]. Much of the motivation for these studies was the detection of OH in remote regions such as the interstellar gas clouds, or the middle regions of the earth's atmosphere.

9.4.4. Far-infrared laser magnetic resonance of CH in the $X\,^2\Pi$ ground state

(a) Introduction

The CH radical is an important molecule in both chemistry and physics. It has long been familiar to optical astronomers through its interstellar absorption spectrum. The fact that the spectrum is so simple, consisting of a single line, the $R_2(1)$ line in the $A\,^2\Delta - X\,^2\Pi$ transition, was the first indication that not only do the interstellar gas clouds contain molecular species but also that their temperature is very low. Apart from its astronomical significance, the CH radical is an enthusiastic participant in most combustion processes. A proper knowledge of the way in which it is involved requires a full determination of its physical properties and the development of experimental methods for its detection at low concentrations. These objectives have fuelled a large amount of research on the spectroscopy of CH, not all of it successful. It was one of the early triumphs of FIR LMR to provide a laboratory detection of a rotational transition in CH [48]; this followed several unsuccessful attempts to detect its microwave Λ-doubling spectrum by either magnetic resonance or field-free techniques. Equally, it was a major coup for the radioastronomers to be able to detect and measure the 3.3 GHz Λ-doubling transition in the interstellar medium [49, 50], long before detection in the laboratory. Such has been the progress made in subsequent years that laboratory studies of CH are now almost routine.

(b) Hund's case (b) behaviour in $^2\Pi$ states

From a spectroscopic point of view, CH is quite similar to OH. It has a $^2\Pi$ ground state and a large rotational constant. In consequence, Λ-doubling and centrifugal distortion effects are important in the description of its energy levels. However, because carbon has a lower nuclear charge than oxygen, the spin–orbit coupling effects are much smaller; $A = 28.1$ cm^{-1} for CH compared with -139.1 cm^{-1} for OH. Note also the opposite sign, which means that the $^2\Pi_{1/2}$ fine-structure component is now the lower in energy. The coupling scheme for the electron spin in CH is therefore no longer intermediate between case (a) and case (b), as for OH, but right over at the case (b) limit. Its pattern of energy levels differs considerably from that of OH, as we shall see.

The behaviour of the spin and rotational levels of a molecule in a $^2\Pi$ state near the case (b) limit can be appreciated by restricting our attention to the two major contributions to the energy, the rotational kinetic energy

$$\mathcal{H}_{\text{rot}} = B(\boldsymbol{J} - \boldsymbol{S})^2, \tag{9.72}$$

and the spin–orbit coupling,

$$\mathcal{H}_{\text{so}} = A L_z S_z. \tag{9.73}$$

Despite our objective of describing case (b) behaviour, it is easier and more familiar to formulate the problem in a case (a) basis set, $|\Lambda; S, \Sigma; J, \Omega, M\rangle$. We have already calculated the required matrix in chapter 8 in connection with our discussion of LiO, with the following result.

	$|^2\Pi_{3/2}\rangle$	$|^2\Pi_{1/2}\rangle$
$|^2\Pi_{3/2}\rangle$	$\dfrac{A}{2} + B\left[\left(J + \dfrac{1}{2}\right)^2 - 1\right]$	$-B\left[\left(J + \dfrac{1}{2}\right)^2 - 1\right]^{1/2}$
$|^2\Pi_{1/2}\rangle$	$-B\left[\left(J + \dfrac{1}{2}\right)^2 - 1\right]^{1/2}$	$-\dfrac{A}{2} + B\left[\left(J + \dfrac{1}{2}\right)^2 + 1\right]$

(9.74)

The eigenvalues of the matrix are

$$E_\pm = B\left(J + \frac{1}{2}\right)^2 \pm \frac{1}{2}\left[(A - 2B)^2 + 4B^2\left\{\left(J + \frac{1}{2}\right)^2 - 1\right\}\right]^{1/2}. \tag{9.75}$$

This result is completely general, applying for all relative magnitudes of A and B.

In the case (a) limit, the separation of the two spin components, $|A - 2B|$, is very large compared with the off-diagonal mixing term, $-B[(J + 1/2)^2 - 1]^{1/2}$, and we get two separate sets of energy levels, as we have seen, for example, in the case of ClO. In this limit, we can expand the square root as a power series in $(J + 1/2)^2$ to obtain

$$E_\pm \approx \pm\frac{1}{2}(A - 2B) + B_{\text{eff}}\left(J + \frac{1}{2}\right)^2, \tag{9.76}$$

where

$$B_{\text{eff}} = \left[1 \pm \frac{B}{(A - 2B)}\right]. \tag{9.77}$$

We have assumed that $(A - 2B)$ is a positive quantity in these equations; if it is negative, the role of the $\pm$ signs on the right-hand side is reversed, that is, the upper sign gives the lower level and vice versa. Equation (9.77) tells us that the effective B value is slightly larger than the true value in the upper spin component and slightly smaller in the lower spin component. Physically, the Hund's case (a) coupling scheme corresponds to the situation where L and S are decoupled from each other and projected instead onto the internuclear axis of the diatomic molecule. This scheme is therefore really a *decoupled* scheme, which explains why it is easier to evaluate matrix elements using such a basis set.

The off-diagonal term in the matrix (9.74), arising from the rotational Hamiltonian, mixes the levels of the two spin components. It is sometimes called the spin-uncoupling term because it has the effect of decoupling the electron spin angular momentum from the internuclear axis. However, this is something of a misnomer because, as the effect of the off-diagonal term increases, the electron spin becomes progressively more strongly coupled to the angular momentum N, the resultant of the rotational and electron orbital angular momenta,

$$J = N + S, \tag{9.78}$$

until, in the case (b) limit, it is completely coupled with no quantisation along the internuclear axis. The consequence of this is that, as we move away from the case (a) limit, the projection quantum numbers Σ and Ω which we have used to label our basis functions, become progressively less meaningful. The two spin components can no longer be labelled as $\Omega = 1/2$ and $3/2$ and some other scheme is needed. By convention [51], the lower level for a given J value in equation (9.76) is labelled as F_1 and the upper as F_2. Since this labelling scheme is based simply on energy ordering, it can be used for any degree of coupling. It can also be extended to states of higher multiplicity quite easily: the three levels of a given J in a triplet state are labelled F_1, F_2 and F_3 in order of increasing energy, and so on. It can be seen that, near the case (a) limit where Ω is well-defined, the F_1 component is $^2\Pi_{1/2}$ and the F_2 component is $^2\Pi_{3/2}$ for positive A values, and vice versa for negative A.

In the case (b) limit the spin–orbit coupling constant A is zero and the eigenvalues in equation (9.75) become

$$E_\pm = B(J + 1/2)^2 \pm B(J + 1/2). \tag{9.79}$$

If we make the substitution $J = N + 1/2$ for the lower (F_1) component and $J = N - 1/2$ for the upper (F_2) spin component, we obtain in each case

$$E_\pm = BN(N + 1), \tag{9.80}$$

which is the case (b) expression for the rotational energy levels with $N = 1, 2, 3, \ldots$, etc. To this order of approximation, there is no fine structure splitting.

We have, elsewhere, shown a correlation diagram for typical case (a) and case (b) rotational levels. A rather more general correlation can be better appreciated with the help of the diagram in figure 9.18, where the eigenvalues of equation (9.75) are plotted against A. In the usual way for a 2×2 matrix, the pair of eigenvalues F_1 and

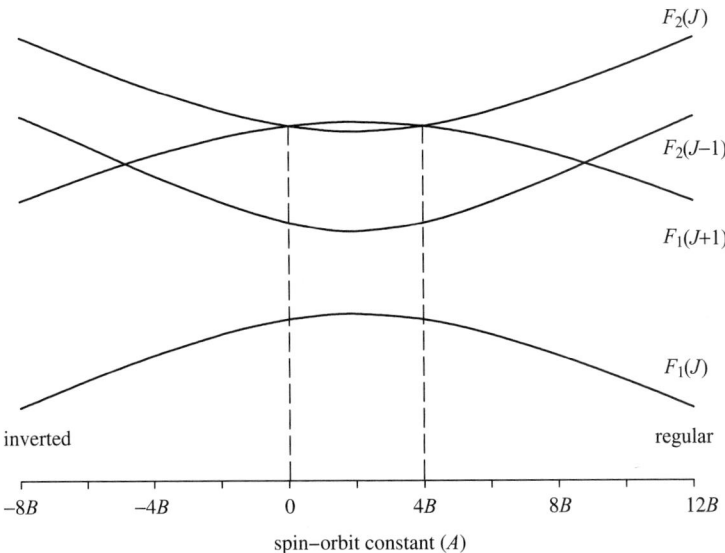

Figure 9.18. Plot of some eigenvalues for a molecule in a $^2\Pi$ state against the spin–orbit coupling constant A.

F_2 for a given J value are closest when $A = 2B$, and repel each other further apart as the difference of the two diagonal elements increases. The diagram is therefore symmetrical about $A = 2B$. We see that at the points $A = 0$ and $A = 4B$, the pair of levels $F_1(J)$ and $F_2(J-1)$ are degenerate for any particular value of J. The first of these coincidences is just the case (b) limit which we have discussed above. The diagram shows that, for $A = 4B$ also, the rotational levels show exact case (b) behaviour with zero fine structure splitting as in equation (9.80). Between these two case (b) crossing points, the F_1 component for a given N level lies *above* the F_2 component. For the special case of $A = 2B$, the splitting is

$$F_1(N) - F_2(N) = B\{(2N+1) - [(N+1)^2 - 1]^{1/2} - [N^2 - 1]^{1/2}\}$$
$$\to 0 \text{ for larger } N \text{ values.} \qquad (9.81)$$

Outside these crossing points, that is for $A < 0$ or $A > 4B$, the order of the two spin components for a given N level is reversed, with F_2 now lying above F_1. Of course, the further the value of A is from these case (b) crossing points, the less meaningful is the quantum number N.

It may seem slightly strange that the Hund's case (b) pattern of rotational energy levels seen for $A = 0$ recurs when $A = 4B$. It is certainly true that these two situations cannot be distinguished from a consideration of the eigenvalues alone. However, they are easily distinguished in practice because the corresponding eigenfunctions are different:

$$|+\rangle = \cos\beta|3/2\rangle + \sin\beta|1/2\rangle,$$
$$|-\rangle = -\sin\beta|3/2\rangle + \cos\beta|1/2\rangle, \qquad (9.82)$$

where

$$\sin 2\beta = -\frac{2Bz^{1/2}}{[(A-2B)^2 + 4B^2z]^{1/2}}, \quad \cos 2\beta = \frac{(A-2B)}{[(A-2B)^2 + 4B^2z]^{1/2}}. \quad (9.83)$$

In these equations z is shorthand for $[(J+1/2)^2 - 1]$. Thus, for $A = 0$,

$$\sin 2\beta = -z^{1/2}/(J+1/2), \quad \cos 2\beta = -1/(J+1/2). \quad (9.84)$$

For $A = 4B$, on the other hand, the result is the same for $\sin 2\beta$ but has the opposite sign for $\cos 2\beta$. Thus molecular observables which depend on the phase of the eigenfunctions, such as transition intensities or magnetic dipole moments, will differ in the two cases.

Before leaving this general discussion of $^2\Pi$ states, we must consider the special case of the $J = 1/2$ level. Since this level does not exist for the $^2\Pi_{3/2}$ spin component, the 2×2 matrix description (9.74) is not appropriate. The $J = 1/2$ level remains a pure $^2\Pi_{1/2}$ state for all possible values of A and B and its eigenvalue is given by

$$E(J = 1/2) = -A/2 + B[(J+1/2)^2 + 1] = -A/2 + 2B. \quad (9.85)$$

Because of the relationship (9.78) we know that the $J = 1/2$ level is always associated with the level $N = 1$ in the case (b) regime and it is tempting to say that it must always be F_2 also (because $J = N - 1/2$). However, this is not the case because there are not two spin components for $J = 1/2$ and the distinction between F_1 and F_2 on the basis of energy ordering is not applicable. However, the eigenvalue in equation (9.85) tells us that it is appropriate to associate the $J = 1/2$ level with the lower spin component when A is positive (*regular* $^2\Pi$ state) and with the upper spin component when A is negative (*inverted* $^2\Pi$ state). More explicitly,

$$J = 1/2 \text{ is } F_1 \quad \text{for } A > 2B, \quad \text{or } F_2 \quad \text{for } A < 2B. \quad (9.86)$$

(c) Experimental studies

As we mentioned in the introduction to this section, it was known forty years ago from optical spectroscopy that CH is a component of interstellar gas clouds and the search was on for a spectroscopic detection of the radical at higher resolution so that the Λ-doubling in the lowest rotational level ($J = 1/2$) could be measured or predicted accurately. This would enable the detection of CH by radio-astronomers and so allow the distribution of CH in these remote sources to be mapped out. The race was won by Evenson, Radford and Moran [48] using the then new technique of far-infrared LMR in the Boulder laboratories of the NBS (now known as NIST). They realised that there was a good near-coincidence between the water discharge laser line at 118.6 μm (84.249 cm^{-1}) and the $N = 3 \leftarrow 2$, $J = 7/2 \leftarrow 5/2$ transition of CH in the F_1 spin component. They formed the CH radical in an oxyacetylene flame and were rewarded with the novel spectrum shown in figure 9.19. The strong lines in this spectrum are all attributable to OH which, we recall, was detected on several different water discharge lines by LMR, including the 118.6 μm line. The new, additional, lines conformed to the pattern expected for CH.

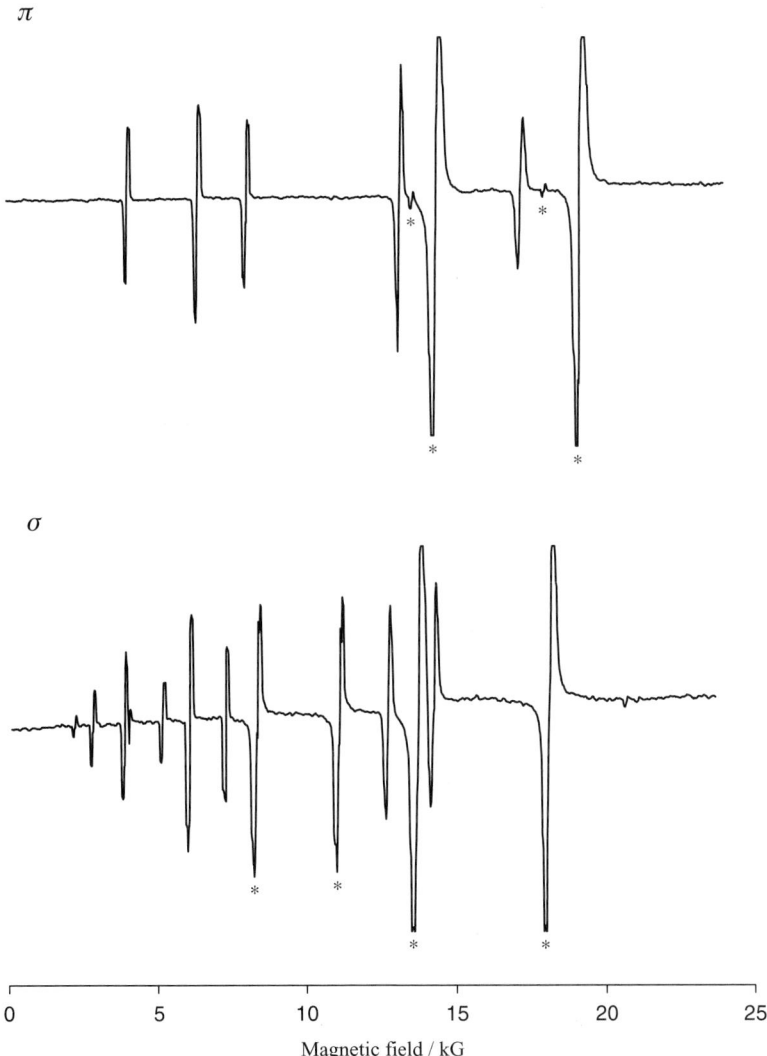

Figure 9.19. The 118.6 μm magnetic resonance spectrum of a low-pressure oxyacetylene flame [48]. The lines marked * are due to OH, with the remaining lines due to CH. The spectra are recorded for two different polarisations (see text).

We remember that there were very few far-infrared laser lines available at the time, all of them discharge laser lines. Unfortunately, although there is a spectacular series of coincidences between the lines of the water discharge laser and the spectrum of the OH radical, the same is not true for CH. The only good coincidence is the 118.6 μm line used by Evenson, Radford and Moran [48]. Progress on the far-infrared spectrum of CH had therefore to await the development of new far-infrared laser lines, optically pumped by an infrared carbon dioxide laser. With these laser lines, two further studies of CH in the ground $^2\Pi$ state were made in the Boulder laboratories, by Hougen,

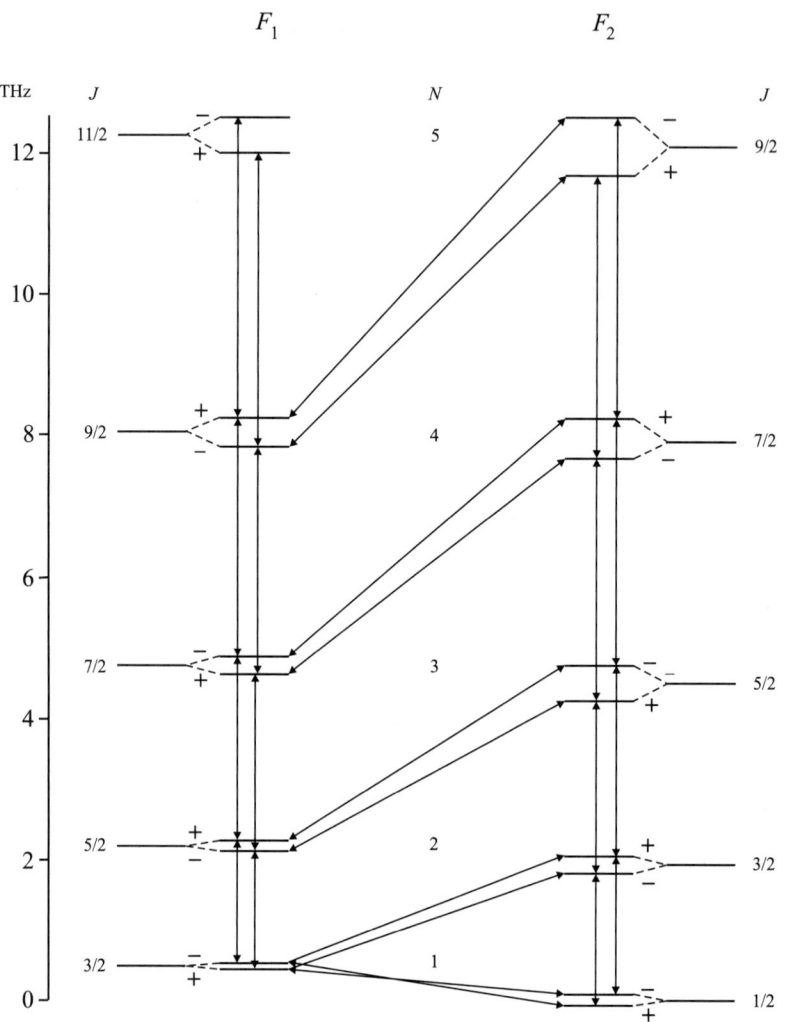

Figure 9.20. Lower rotational levels of CH in the $v=0$ level of the $X\,^2\Pi$ ground state. The size of the Λ-doublet splitting is exaggerated for the sake of clarity. The observed FIR LMR transitions are shown [53]. Other transitions were later observed using tunable far-infrared radiation, and are described in chapter 10.

Mucha, Jennings and Evenson [52] and by Brown and Evenson [53]. An important contribution [52] was the discovery of a better method for the generation of CH, the reaction between F atoms and CH_4. The reaction was easier to control than the oxyacetylene flame, produced the radical at lower temperatures, and was less damaging to the apparatus!

In figure 9.20 we give the energy level diagram for CH in the zero-point vibrational level of it ground $^2\Pi$ state, including the transitions reported by Brown and Evenson [53]. For CH, by a happy accident of nature, A is very close to $2B$, and the energy level scheme shows the expected case (b) behaviour. The quantum number N provides a

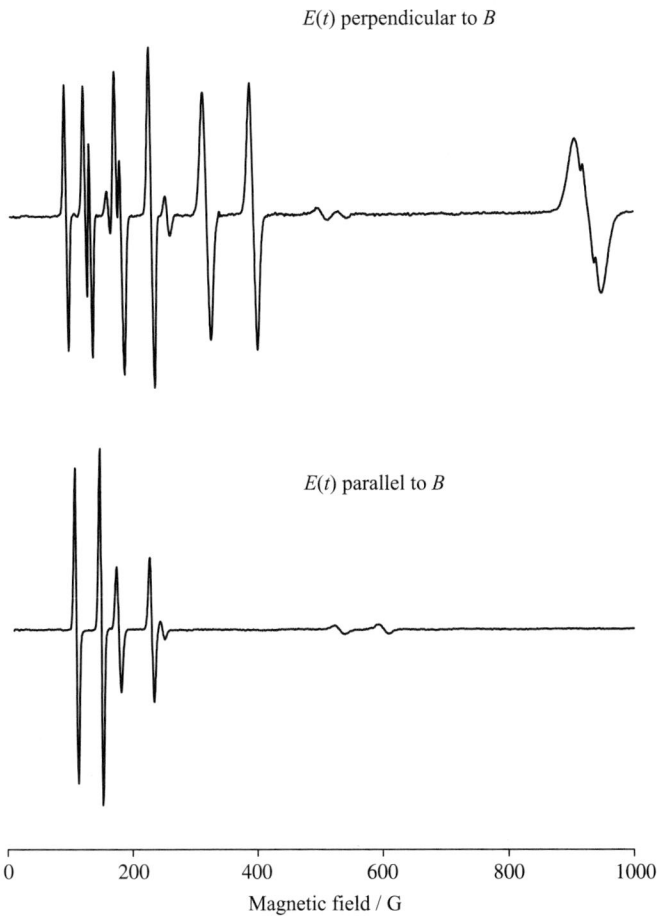

Figure 9.21. FIR LMR spectrum of CH [53]. The spectra were recorded using a laser wavelength of 124.4 μm. The lasing medium was CH_2DOH, pumped by the 10P(34) infrared line of carbon dioxide. The rotational transition involved is $N = 3 \leftarrow 2$, $J = 5/2 \leftarrow 5/2$, $- \leftarrow +$.

better label than J; levels with the same N value are very close together whereas those with the same J values are some way apart. We see also, as expected from figure 9.18, that for the two components with the same N value, F_1 lies above F_2. In addition, it is far from clear whether the level $J = 1/2$ should be associated with the F_1 or the F_2 state. Because A is very slightly smaller than $2B$, we have placed it in the F_2 stack but really it could equally well be associated with either. We have shown the very first LMR spectrum of CH in figure 9.19, but in figure 9.21, taken from Brown and Evenson [53], we show a later spectrum obtained through clean production of CH, with a superb signal-to-noise ratio, and Lamb dips on the highest field resonance in the perpendicular spectrum, revealing the proton hyperfine structure. All of the resonances shown in this spectrum arise from the Zeeman M-components and proton hyperfine structure for one Λ-doubling transition; the other Λ-doubling transitions occur at much higher magnetic fields because the splitting is quite large.

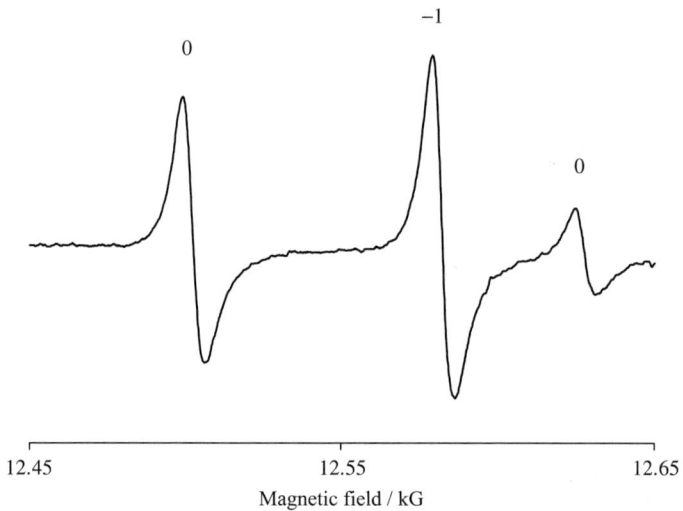

Figure 9.22. Proton hyperfine pattern observed for the LMR transition F_1, $N' = 1$, $J' = 3/2 \leftarrow F_2$, $N'' = 1$, $J'' = 1/2$, recorded using a laser line at 554.4 μm from CH_2F_2. The numbers refer to the value of the quantum number M_F in the lower state ($M_J = -1/2$ combined with $M_I = +1/2$ or $-1/2$). The spectrum is recorded in perpendicular polarisation ($\Delta M_F = \pm 1$).

The LMR transitions which involve the $J = 1/2$ level all show a peculiar proton hyperfine structure; a typical example from the 554.4 μm spectrum is shown in figure 9.22. The normal hyperfine pattern consists of two lines of equal intensity, one associated with the nuclear spin state $M_I = +1/2$ and the other with $M_I = -1/2$. In other words, the pattern reflects the fact that the nuclear spin is decoupled from the other angular momenta in the magnetic fields used to produce the resonance. For the transitions which involve the $J = 1/2$ level, three hyperfine components are seen, usually two stronger but unequal and one weaker. The explanation for these observations lies in the peculiar nature of the $J = 1/2$ level as discussed in the previous section. This level has pure $^2\Pi_{1/2}$ character even in the Hund's case (b) limit to which CH conforms. In the $^2\Pi_{1/2}$ component, the electron orbital and spin contributions to the magnetic moment cancel almost exactly and the magnetic moment of CH in this level is very small (7×10^{-4} μ_B). In consequence, the two nuclear spin states, labelled by the total angular momentum quantum $F = 0$ and 1, are not mixed strongly even by a large applied magnetic field and the nuclear spin remains coupled to the rotational angular momentum. For all other levels of CH, the nuclear spin is decoupled by modest magnetic fields. The LMR transition giving the spectrum shown in figure 9.22 is therefore between two states with very different coupling character. Nuclear spin 'forbidden' transitions (forbidden, that is, on the basis of the I-decoupled selection rule $\Delta M_I = 0$) are therefore seen, consistent with the overall selection rule $\Delta M_F = 0$ or ± 1, depending on the polarisation.

The outcome of the analysis of the FIR LMR spectrum of CH in its $X\,^2\Pi$ state was an accurate determination of several parameters in the effective Hamiltonian.

The two primary parameters for a $^2\Pi$ state were determined to be $A_0 = 28.147$ cm^{-1} and $B_0 = 14.192$ cm^{-1} so that $A/B = 1.983$. The two Λ-doubling parameters $p = 0.0335$ cm^{-1} and $q = 0.0387$ cm^{-1} have very similar magnitudes, again reflecting case (b) behaviour (for case (a), p is much larger than q in magnitude). The full set of four hyperfine parameters was determined for the first time, carrying valuable information on the electronic wave function for CH. We will return to the CH radical in chapter 10, describing the recent field-free tunable far-infrared studies, which provide even more accurate values of the molecular parameters. We shall also describe the interpretation of these parameters in some detail in terms of *ab initio* calculations of the electronic wave function.

Following their work on the LMR spectrum of CH, Brown and Evenson also studied the far-infrared LMR spectrum of CD [54]. The energy level diagram for CD in the $v = 0$ level of the $X\,^2\Pi$ state is shown in figure 9.23. Like that of CH, it conforms very closely to Hund's case (b) coupling. Indeed it is an almost perfect example of exact case (b) behaviour because in this case A is almost exactly $4B$ (A for CD is very similar to A for CH, being an electronic property of the molecule, whereas B for CD is half that for CH). The F_1, F_2 spin–rotation splittings for a given N level are therefore extremely small. Analysis of the LMR spectrum once again yielded a rich crop of molecular parameters which made interesting comparison with the corresponding parameters for CH. These are discussed in chapter 10.

9.5. $^2\Sigma$ states

9.5.1. Introduction

Free radicals in $^2\Sigma$ ground or excited states have been studied extensively by a range of spectroscopic methods, but not in general by magnetic resonance techniques. We shall discuss the microwave spectroscopy of $^2\Sigma$ free radicals and ions in detail in chapter 10. It is, however, important to understand why magnetic resonance methods are usually unsuitable. In many ways $^2\Sigma$ diatomic molecules may be regarded as prototypes for nonlinear polyatomic molecules. In addition, some of the interactions which are important in $^2\Sigma$ states also occur in $^3\Sigma$ states, for which magnetic resonance methods have been important, as we shall see in due course. We therefore develop the theory of the rotational, fine and hyperfine levels of $^2\Sigma$ states, including magnetic field effects, with the promise that the results will become important when we discuss the pure microwave spectra of species like CN, SiN and CO$^+$ in chapter 10. The theory will also become relevant in understanding microwave/optical double resonance results, described in chapter 11.

9.5.2. CN in the $X\,^2\Sigma^+$ ground state

We choose to describe the CN radical in its $X\,^2\Sigma^+$ ground state since it illustrates most of the interactions which are likely to be encountered. It conforms to Hund's case (b)

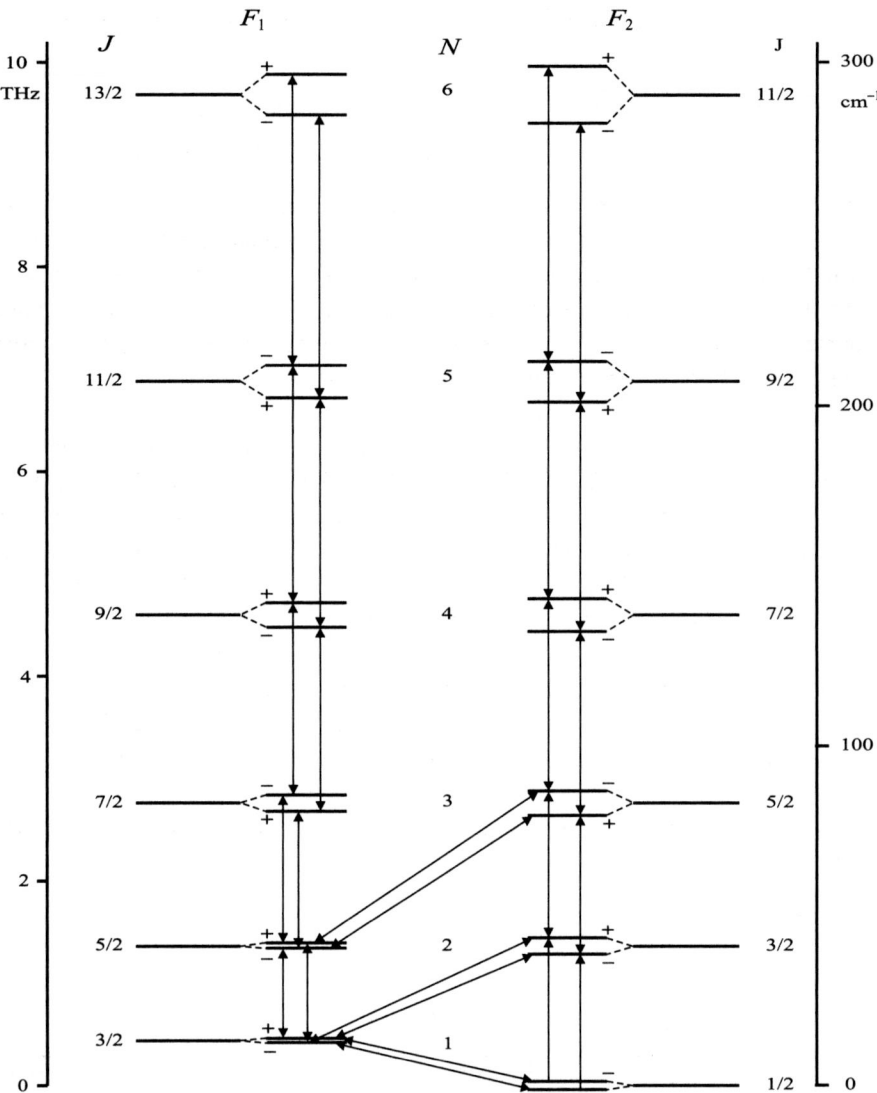

Figure 9.23. Lower rotational levels of the CD radical in the $v = 0$ level of the $X\,^2\Pi$ ground state, and the observed FIR LMR transitions [54]. Note the unexpected inversion of the parity-doublet in the $J = 3/2$, F_1 level; this is another symptom of the transition from case (a) to case (b) coupling.

coupling (see chapter 6), where the rotational angular momentum N is coupled to the electron spin S to form the total angular momentum J. Nuclear spin interactions can then be treated in either the hyperfine-coupled representation, $J + I = F$, or the hyperfine decoupled representation. We shall examine both, the coupled representation being convenient for zero-field spectroscopy, whilst the decoupled representation is more useful for analysing Zeeman effects.

The effective Hamiltonian for CN $^2\Sigma^+$ in zero field may therefore be expressed as the sum of five terms:

$$\mathcal{H}_{\text{eff}} = \{B_v - D_v N^2\} N^2 + \gamma_v \mathbf{T}^1(\mathbf{N}) \cdot \mathbf{T}^1(\mathbf{S}) + b_F \mathbf{T}^1(\mathbf{I}) \cdot \mathbf{T}^1(\mathbf{S})$$
$$- \sqrt{10} g_S \mu_B g_N \mu_N (\mu_0 / 4\pi) \mathbf{T}^1(\mathbf{S}, \mathbf{C}^2) \cdot \mathbf{T}^1(\mathbf{I}) - e \mathbf{T}^2(\mathbf{Q}) \cdot \mathbf{T}^2(\nabla \mathbf{E}). \quad (9.87)$$

This effective Hamiltonian is very similar to that used in the analysis of the radiofrequency spectrum of H_2^+, described in chapter 11, but with the addition of the quadrupole term. We have also added the term representing the rotational kinetic energy because we will wish to derive expressions for the frequencies of rotational transitions in due course. A further very important comment should be added concerning the electron spin–rotation term in the effective Hamiltonian. As we discussed in chapter 7, the first-order contribution to γ_v originates directly from the magnetic interaction between the electron spin magnetic moment and the magnetic moment arising from rotation of the nuclei plus electrons. In H_2^+ this is indeed the physical origin of the interaction. In almost all other molecules, however, the first-order contribution is outweighed by a second-order contribution arising from spin–orbit mixing of excited electronic states with the ground state $^2\Sigma$ state, which generates some orbital angular momentum. We showed in chapter 7 that this effect can be accurately represented by the term shown in equation (9.87).

The case (b) basis functions are of the form $|\eta, \Lambda; N, S, J, I, F, M_F\rangle$ in the hyperfine-coupled basis set and the matrix elements of the effective Hamiltonian are given below. We include Λ in the basis set because although $\Lambda = 0$ for a Σ state, there are terms in the effective Hamiltonian which can mix the ground state with excited states having $\Lambda \neq 0$. As before, the absence of primes on the right-hand side denotes that the matrix elements are diagonal in the relevant quantum numbers.

(i) ROTATION

$$\langle \eta, \Lambda, N, S, J, I, F, M_F | \{B_v - D_v N^2\} N^2 | \eta, \Lambda, N, S, J, I, F, M_F \rangle$$
$$= \{B_v - D_v N(N+1)\} N(N+1). \quad (9.88)$$

Note that a centrifugal distortion term has been included; the resolution and accuracy of pure microwave spectroscopy warrants this inclusion, even for the lowest rotational levels.

(ii) SPIN–ROTATION INTERACTION

The spin–rotation interaction is diagonal in our basis set, and independent of quantum numbers involving nuclear spin:

$$\langle \eta, \Lambda, N, S, J, I, F, M_F | \gamma_v \mathbf{T}^1(\mathbf{N}) \cdot \mathbf{T}^1(\mathbf{S}) | \eta, \Lambda, N, S, J, I, F, M_F \rangle$$
$$= \gamma_v (-1)^{N+J+S} \begin{Bmatrix} S & N & J \\ N & S & 1 \end{Bmatrix} \{S(S+1)(2S+1)N(N+1)(2N+1)\}^{1/2}. \quad (9.89)$$

(iii) FERMI CONTACT INTERACTION

There are matrix elements off-diagonal in J which will become significant if the spin–rotation interaction is sufficiently small; the interaction is, however, largely diagonal.

$$\langle \eta, \Lambda, N, S, J, I, F, M_F | b_F \, T^1(S) \cdot T^1(I) | \eta, \Lambda, N, S, J', I, F, M_F \rangle$$

$$= b_F (-1)^{J'+F+I} \begin{Bmatrix} I & J' & F \\ J & I & 1 \end{Bmatrix} \{I(I+1)(2I+1)\}^{1/2} \langle N, S, J \| T^1(S) \| N, S, J \rangle$$

$$= b_F (-1)^{J'+F+I} \begin{Bmatrix} I & J' & F \\ J & I & 1 \end{Bmatrix} (-1)^{J'+N+1+S} \{(2J+1)(2J'+1)\}^{1/2} \begin{Bmatrix} S & J' & N \\ J & S & 1 \end{Bmatrix}$$

$$\times \{I(I+1)(2I+1)S(S+1)(2S+1)\}^{1/2}. \tag{9.90}$$

(iv) DIPOLAR HYPERFINE INTERACTION

The analysis is similar to that presented for ClO, but is more complicated; from equation (9.87):

$$\langle \eta, \Lambda, N, S, J, I, F, M_F | \mathcal{H}_{\text{dip}} | \eta', \Lambda', N', S, J', I, F, M_F \rangle$$

$$= -\sqrt{10} g_S \mu_B g_N \mu_N (\mu_0/4\pi)(-1)^{J'+F+I}$$

$$\times \begin{Bmatrix} I & J' & F \\ J & I & 1 \end{Bmatrix} \{I(I+1)(2I+1)\}^{1/2} \langle \eta, \Lambda, N, S, J \| T^1(S, C^2) \| \eta', \Lambda', N', S, J' \rangle$$

$$= -\sqrt{10} g_S \mu_B g_N \mu_N (\mu_0/4\pi)(-1)^{J'+F+I} \begin{Bmatrix} I & J' & F \\ J & I & 1 \end{Bmatrix} \{I(I+1)(2I+1) 3(2J+1)$$

$$\times (2J'+1)S(S+1)(2S+1)\}^{1/2} \begin{Bmatrix} J & J' & 1 \\ N & N' & 2 \\ S & S & 1 \end{Bmatrix} \langle \eta, \Lambda, N \| C^2(\theta, \phi)(r^{-3}) \| \eta', \Lambda', N \rangle$$

$$= -\sqrt{30} g_S \mu_B g_N \mu_N (\mu_0/4\pi)(-1)^{J'+F+I} \begin{Bmatrix} I & J' & F \\ J & I & 1 \end{Bmatrix} \begin{Bmatrix} J & J' & 1 \\ N & N' & 2 \\ S & S & 1 \end{Bmatrix}$$

$$\times \sum_q (-1)^{N-\Lambda} \begin{pmatrix} N & 2 & N' \\ -\Lambda & q & \Lambda' \end{pmatrix} \{I(I+1)(2I+1)(2J+1)(2J'+1)S(S+1)(2S+1)$$

$$\times (2N+1)(2N'+1)\}^{1/2} \langle \eta, \Lambda | C_q^2(\theta, \phi)(r^{-3}) | \eta', \Lambda' \rangle. \tag{9.91}$$

Fortunately the terms arising from $q \neq 0$ which involve excited electronic states are probably always going to be negligible, and in the molecular species which have been studied so far, matrix elements off-diagonal in N are also negligible. Consequently it will usually be sufficiently accurate to consider only matrix elements off-diagonal in J, so we will use the result:

$$\langle \eta, \Lambda, N, S, J, I, F, M_F | \mathcal{H}_{\text{dip}} | \eta, \Lambda, N, S, J', I, F, M_F \rangle$$

$$= -\sqrt{30} t (-1)^{J'+F+I+N} \begin{Bmatrix} I & J' & F \\ J & I & 1 \end{Bmatrix} \begin{Bmatrix} J & J' & 1 \\ N & N & 2 \\ S & S & 1 \end{Bmatrix} \begin{pmatrix} N & 2 & N \\ 0 & 0 & 0 \end{pmatrix}$$

$$\times \{I(I+1)(2I+1)(2J+1)(2J'+1)S(S+1)(2S+1)\}^{1/2} (2N+1), \tag{9.92}$$

where t, the axial component of the dipolar interaction, is defined in equation (9.61). 9-j symbols present something of a headache if one seeks a neat algebraic expression of the result; with the availability of computer programs for their evaluation, we leave (9.92) as it is until we need to evaluate it further in the next section.

(v) NUCLEAR ELECTRIC QUADRUPOLE INTERACTION

Our treatment of the quadrupole interaction is similar to that described previously:

$$\langle \eta, \Lambda; N, S, J, I, F, M_F | -e\mathrm{T}^2(\boldsymbol{Q}) \cdot \mathrm{T}^2(\nabla \boldsymbol{E}) | \eta', \Lambda'; N', S, J', I, F, M_F \rangle$$

$$= -e(-1)^{J'+F+I} \begin{Bmatrix} I & J' & F \\ J & I & 2 \end{Bmatrix} \langle I \| \mathrm{T}^2(\boldsymbol{Q}) \| I \rangle \langle \eta, \Lambda; N, S, J \| \mathrm{T}^2(\nabla \boldsymbol{E}) \| \eta', \Lambda'; N', S, J' \rangle$$

$$= -e(-1)^{J'+F+I} \begin{Bmatrix} I & J' & F \\ J & I & 2 \end{Bmatrix} \langle I \| \mathrm{T}^2(\boldsymbol{Q}) \| I \rangle (-1)^{J'+N+S} \{(2J'+1)(2J+1)\}^{1/2}$$

$$\times \begin{Bmatrix} N' & J' & S \\ J & N & 2 \end{Bmatrix} \langle \eta, \Lambda; N \| \mathrm{T}^2(\nabla \boldsymbol{E}) \| \eta', \Lambda'; N' \rangle$$

$$= -e(-1)^{J'+F+I} \begin{Bmatrix} I & J' & F \\ J & I & 2 \end{Bmatrix} \langle I \| \mathrm{T}^2(\boldsymbol{Q}) \| I \rangle (-1)^{J'+N+S} \{(2J'+1)(2J+1)\}^{1/2}$$

$$\times \begin{Bmatrix} N' & J' & S \\ J & N & 2 \end{Bmatrix} \sum_q (-1)^{N-\Lambda} \begin{pmatrix} N & 2 & N' \\ -\Lambda & q & \Lambda' \end{pmatrix} \{(2N+1)(2N'+1)\}^{1/2}$$

$$\times \langle \eta, \Lambda | \mathrm{T}_q^2(\nabla \boldsymbol{E}) | \eta', \Lambda' \rangle. \tag{9.93}$$

As was the case with the dipolar interaction, the quadrupole interaction has matrix elements which connect the ground state with excited states; we neglect these by putting $q = 0$. The reduced matrix element of $\mathrm{T}^2(\boldsymbol{Q})$ was evaluated in (9.14), so with these replacements, including the neglect of matrix elements off-diagonal in N, we obtain the simplified result:

$$\langle \eta, \Lambda; N, S, J, I, F, M_F | -e\mathrm{T}^2(\boldsymbol{Q}) \cdot \mathrm{T}^2(\nabla \boldsymbol{E}) | \eta, \Lambda; N, S, J', I, F, M_F \rangle$$

$$= (-1)^{J'+F+I} \begin{Bmatrix} I & J' & F \\ J & I & 2 \end{Bmatrix} \left(\frac{eQ}{2} \right) \{(I+1)(2I+1)(2I+3)/I(2I-1)\}^{1/2}$$

$$\times (-1)^{J'+N+S} \{(2J'+1)(2J+1)\}^{1/2} \begin{Bmatrix} N & J' & S \\ J & N & 2 \end{Bmatrix}$$

$$\times (-1)^N \begin{pmatrix} N & 2 & N \\ 0 & 0 & 0 \end{pmatrix} (2N+1) \left(\frac{q_0}{2} \right). \tag{9.94}$$

We shall make use of this result when we discuss pure microwave rotational transitions in chapter 10.

(vi) ZEEMAN INTERACTIONS

One of the purposes of this section is to show the limitations of magnetic resonance methods in studying $^2\Sigma$ states. We do not need to include nuclear spin interactions, and we consider only the interaction of the electron spin S with the magnetic field applied in the Z ($p=0$) direction. The required matrix elements are therefore

$$\langle \eta, \Lambda, N, S, J, M_J | g_S \mu_B B_Z T_0^1(S) | \eta, \Lambda, N, S, J', M_J \rangle$$
$$= g_S \mu_B B_Z (-1)^{J-M_J} \begin{pmatrix} J & 1 & J' \\ -M_J & 0 & M_J \end{pmatrix} (-1)^{J+N+1+S} \begin{Bmatrix} S & J' & N \\ J & S & 1 \end{Bmatrix}$$
$$\times \{S(S+1)(2S+1)(2J'+1)(2J+1)\}^{1/2}. \qquad (9.95)$$

Let us now consider the $N=1$ and 2 rotational levels; the spin–rotation interaction produces a doublet splitting of each level according to equation (9.89), the energies being given by

$$E(J) = \frac{\gamma_v}{2} \{J(J+1) - N(N+1) - S(S+1)\}. \qquad (9.96)$$

Substituting the values for N, S and J, the spin–rotation energies of the levels are

$$N = 1: \quad E(J = 3/2) = \frac{\gamma_v}{2}; \quad E(J = 1/2) = -\gamma_v$$
$$N = 2: \quad E(J = 5/2) = \gamma_v; \quad E(J = 3/2) = -\frac{3\gamma_v}{2}. \qquad (9.97)$$

The rotational transition $N=2 \leftarrow 1$ is electric dipole allowed, and if the oscillating electric field is applied in the p direction, the relative intensities of the spin components are obtained by evaluating the expression

$$|\langle N, S, J, M_J | -\mu_e T_p^1(E) | N', S, J', M_J' \rangle|^2$$
$$= \delta_{N',N\pm1} \mu_e^2 E_p^2 \begin{pmatrix} J & 1 & J' \\ -M_J & p & M_J' \end{pmatrix}^2 (2J+1)(2J'+1)$$
$$\times \begin{Bmatrix} N' & J' & S \\ J & N & 1 \end{Bmatrix}^2 (1/2)(2N+1\pm1). \qquad (9.98)$$

This expression tells us that for a rotational transition $\Delta N = \pm 1$, the selection rule for J is $\Delta J = 0, \pm 1$, with the $\Delta J = \Delta N$ components being the strongest. The zero-field swept-frequency rotational transitions are shown in figure 9.24; if there is no hyperfine structure, the spectrum will consist of a main doublet, with a weaker $\Delta J = 0$ component.

In a magnetic resonance experiment designed to record the rotational spectrum, the *fixed* radiation frequency would be close to the resonant zero-field frequency, with the mismatch being tuned with a swept magnetic field. Equation (9.95) enables us to calculate the behaviour of the levels as a function of magnetic field strength; we show the results for both the $N = 1$ and 2 rotational levels in figure 9.25, appropriate for the CN radical where $\gamma_{v=0}$ has the value 217.5 MHz [55].

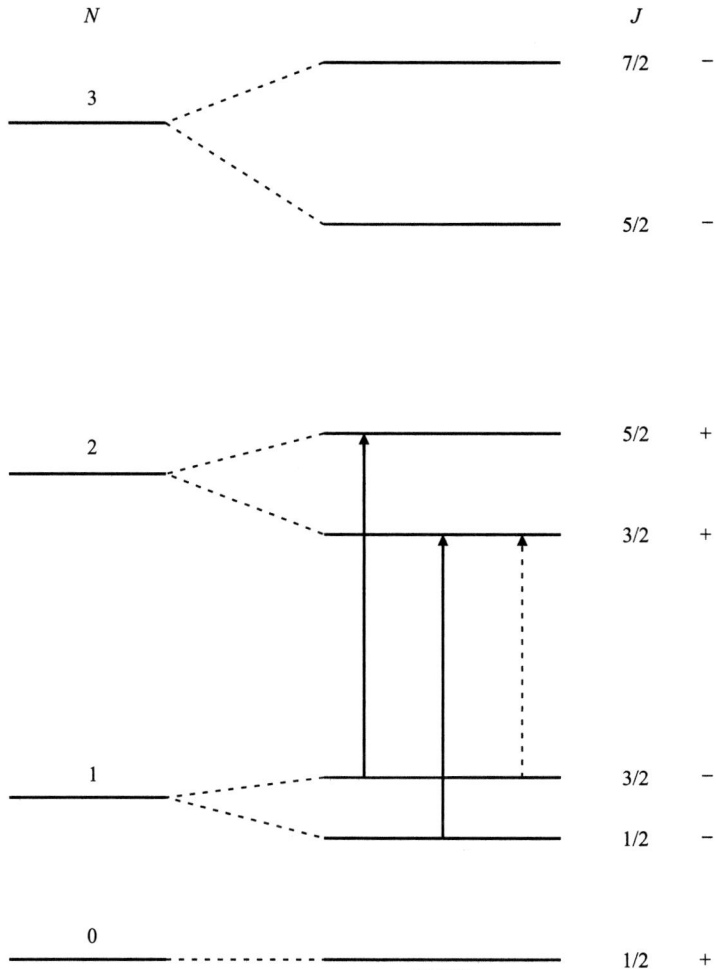

Figure 9.24. Lower rotational levels of a $^2\Sigma^+$ molecule, like the CN radical, showing the spin-doubling and the allowed electric dipole transitions between $N = 2$ and 1.

The Zeeman diagram can be divided into three different regions. At low field the splitting of the M_J levels is linear in the magnetic field (region 1). As the field increases the energies become nonlinear (region 2) until eventually at sufficiently high field (region 3) they become linear again. Region 1 corresponds to the situation when the Zeeman energies are small compared with the spin–rotation energies; region 3 corresponds to the opposite situation, in which the electron spin is decoupled from the molecular frame, and quantised by the external magnetic field. This decoupling is known as the Paschen–Back effect. Region 2 is, of course, the intermediate region. Magnetic resonance experiments would usually be performed with the oscillating electric vector perpendicular to the static field, so that $\Delta M_J = \pm 1$ transitions are tuned into resonance with the fixed frequency radiation in region 1. In this region they retain their electric dipole intensity. Conversely, in region 3 the $\Delta M_J = \pm 1$ electric dipole transition

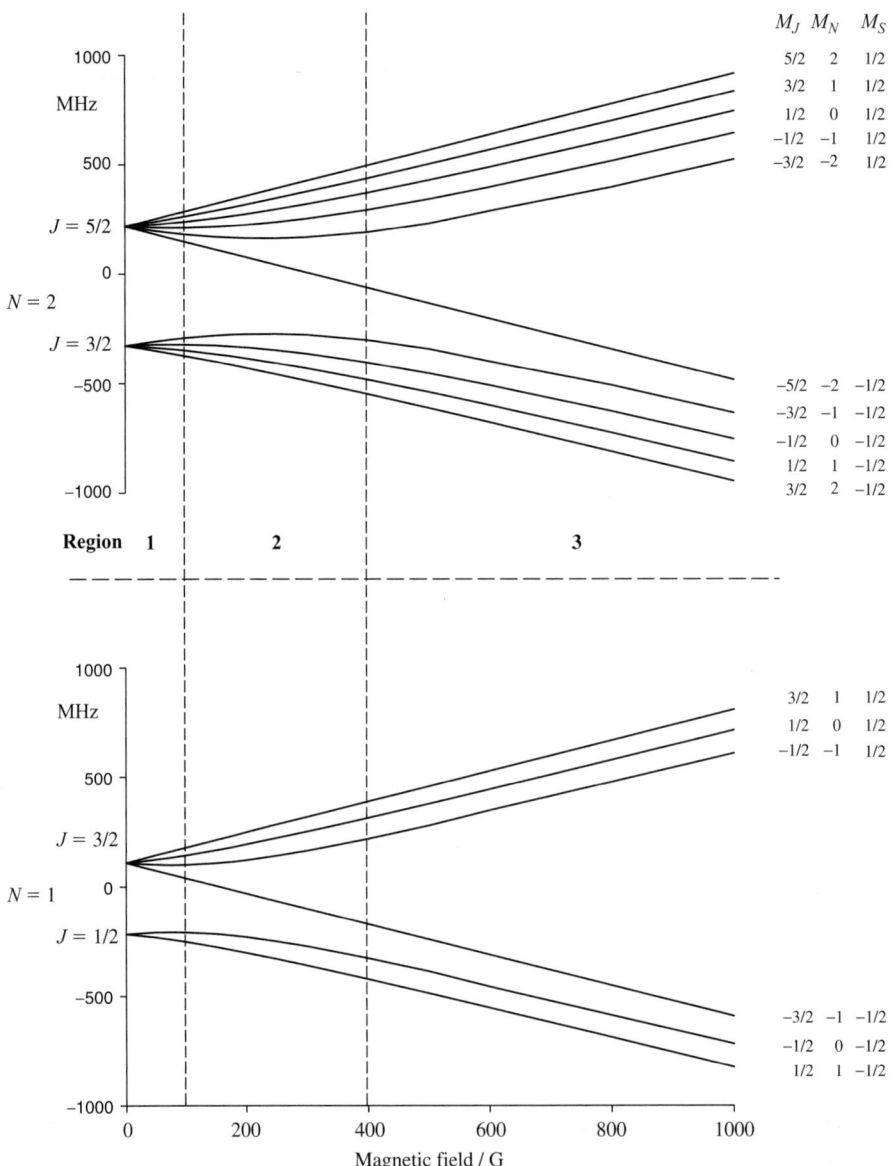

Figure 9.25. Zeeman splitting of the $N = 1$ and 2 rotational levels in the CN radical. In region 1 the rotational transition is electric dipole allowed and magnetically tunable. In region 3 the magnetically-tunable transitions are magnetic dipole electron spin transitions; the electric dipole transitions are not magnetically tunable. Region 2 is intermediate between these limiting cases.

energies become independent of the magnetic field strength; they obey $\Delta M_S = 0$ and have lost their magnetic tunability. There are other transitions whose frequencies are strongly field-dependent, but they are very weak. In the high-field region 3, the Zeeman levels are better described in terms of their decoupled projection quantum numbers, M_N

and M_S, and the magnetically tunable transitions have the selection rules $\Delta M_S = \pm 1$, $\Delta M_N = 0$. These conclusions, which come from electric dipole intensity calculations based upon equations (9.95) and (9.98) have a straightforward physical interpretation. Electric dipole transitions involve a reorientation of the molecular electric dipole moment, which lies along the internuclear axis. Their magnetic tunability comes from the electron spin magnetic moment; if this is also coupled to the molecular axis, all is well. But in region 2 we see the onset of the process through which the spin is decoupled from the molecular axis by the increasing magnetic field strength (as the matrix elements off-diagonal in J become larger). In region 3 the electron spin is wholly spaced-quantised with respect to the magnetic field direction. Electron spin transitions with $\Delta N = 0$ are *not* associated with reorientation of the molecular electric dipole moment, so they can only possess magnetic dipole intensity. Magnetic tunability of electric dipole transitions therefore depends upon the strength of the spin coupling to the molecular frame, and in $^2\Sigma$ molecules only the spin–rotation interaction provides the necessary coupling; it is generally rather weak. In $^2\Pi$ molecules, like ClO, the first order spin–orbit coupling is immensely strong compared with the Zeeman interaction, at all feasible magnetic field strengths. As we shall see, $^3\Sigma$ molecules are intermediate, in that the electron spin–spin interaction is often strong enough to withstand the decoupling by an applied magnetic field until much higher field strengths.

These, then, are the reasons why magnetic resonance methods, microwave or far-infrared laser, have had limited success with $^2\Sigma$ diatomic radicals. Similar considerations apply to nonlinear polyatomic radicals in doublet states; success in far-infrared laser magnetic resonance depends upon the magnitude of the spin–rotation coupling, and the size of the energy mismatch between the transition frequency and the laser frequency, since the mismatch has to be magnetically tuned. This becomes less of a limitation as more laser frequencies become available, except that one then needs to know in advance which laser frequency to choose. It becomes part of the search problem!

9.6. $^3\Sigma$ states

9.6.1. SO in the $X^3\Sigma^-$ ground state

Several molecules with $^3\Sigma$ ground states have been studied by both microwave and far-infrared laser magnetic resonance; they include O_2, SO and SeO. In O_2 the observed transitions are necessarily magnetic dipole, and they are frequently used to calibrate the sensitivity of a FIR laser magnetic resonance spectrometer. The other species have electric dipole transitions, and we shall illustrate the situation by describing the studies of SO carried out by Carrington, Levy and Miller [56]. SO was also one of the first free radicals to be studied by pure microwave methods, which we will describe in chapter 10. The analysis of the magnetic resonance spectrum actually made use of the parameters determined earlier by pure microwave studies. SO is an easy radical to study experimentally since it is relatively unreactive and has a lifetime of several seconds under most conditions. We commence by considering the dominant isotopic species ^{32}S^{16}O where hyperfine interactions are not involved. We then discuss the spectrum of ^{33}S^{16}O which exhibits ^{33}S hyperfine structure, and finally NH which, as in the case of NF in its $^1\Delta$ state described earlier, has hyperfine structure arising from both nuclei.

The microwave magnetic resonance spectrum of SO was recorded at a resonance frequency of 8762 MHz and involved Zeeman components of rotational transitions between the $N = 1$ and 2 rotational levels. We first describe the effective Hamiltonian and its matrix elements, and then the observed spectrum and its analysis.

The effective Hamiltonian used to analyse the spectrum was written in the form

$$\mathcal{H}_{\text{eff}} = B_0 N^2 + \frac{2}{3}\lambda(3S_z^2 - S^2) + \gamma N \cdot S, \tag{9.99}$$

where z is the internuclear axis. Each of the three molecular parameters in (9.99) is the sum of first-order and second-order terms,

$$B_0 = B_0^{(1)} + B_0^{(2)}, \quad \lambda = \lambda^{(1)} + \lambda^{(2)}, \quad \gamma = \gamma^{(1)} + \gamma^{(2)}, \tag{9.100}$$

where, as we discussed in chapter 7, the second-order terms arise from spin–orbit coupling which mixes the $^3\Sigma$ ground state with excited states for which $\Lambda = 0, \pm 1$ and $\Delta S = 0, \pm 1$. The extra term in equation (9.99) is the second one, the first-order part of which represents the dipole–dipole interaction between the spins of the two unpaired electrons. This form may seem strange, although it is in conventional use, but we can readily understand its expression in irreducible tensor form as follows. By analogy with the dipole-dipole coupling of two nuclear spins (see appendix 8.1), we would expect the dipolar coupling of two electron spins to be represented by the term

$$\mathcal{H}_{\text{dip}} = -g_S^2 \mu_B^2 (\mu_0/4\pi)\sqrt{6}T^2(C) \cdot T^2(S_1, S_2), \tag{9.101}$$

where

$$T_p^2(S_1, S_2) = (-1)^p \sqrt{5} \sum_{p_1, p_2} \begin{pmatrix} 1 & 1 & 2 \\ p_1 & p_2 & -p \end{pmatrix} T_{p_1}^1(S_1) T_{p_2}^1(S_2), \tag{9.102}$$

and

$$T_q^2(C) = \langle \eta | C_q^2(\theta_{12}, \phi_{12}) r_{12}^{-3} | \eta \rangle. \tag{9.103}$$

The diagonal ($q = 0$) component of (9.101) may be written

$$-g_S^2 \mu_B^2 (\mu_0/4\pi) C_0^2(\theta_{12}, \phi_{12}) r_{12}^{-3} \sqrt{6} T_0^2(S_1, S_2) = \frac{2}{3}\lambda \sqrt{6} T_0^2(S, S), \tag{9.104}$$

where S is the total spin. By making use of the relationship (9.102) one finds that one can make the replacement

$$\sqrt{6} T_0^2(S, S) = 3S_z^2 - S^2. \tag{9.105}$$

We shall, however, represent the electron spin dipolar interaction by (9.101) and then relate our results to (9.99). Consequently the effective Hamiltonian can now be written in irreducible tensor form as

$$\mathcal{H}_{\text{eff}} = B_0 \mathbf{N}^2 - g_S^2 \mu_B^2 (\mu_0/4\pi) \sqrt{6} \mathbf{T}^2(\mathbf{C}) \cdot \mathbf{T}^2(\mathbf{S}_1, \mathbf{S}_2) + \gamma \mathbf{T}^1(\mathbf{N}) \cdot \mathbf{T}^1(\mathbf{S}). \quad (9.106)$$

It is defined to operate within the ground vibronic state by (9.104), but the second-order part of the parameter λ, (9.100), arises from spin–orbit mixing of excited states with Λ equal to 0, ± 1 and $\Delta S = 0$, ± 1. In molecules containing an atom beyond the first row of the periodic table, the second-order contribution to λ (i.e. $\lambda^{(2)}$) is far larger than the first-order part, $\lambda^{(1)}$.

Equation (9.106) is a field-free Hamiltonian, to which must be added the Zeeman terms; the complete effective Hamiltonian for the problem is therefore

$$\mathcal{H}_{\text{eff}} = B_0 \mathbf{N}^2 - g_S^2 \mu_B^2 (\mu_0/4\pi) \sqrt{6} \mathbf{T}^2(\mathbf{C}) \cdot \mathbf{T}^2(\mathbf{S}_1, \mathbf{S}_2) + \gamma \mathbf{T}^1(\mathbf{N}) \cdot \mathbf{T}^1(\mathbf{S})$$
$$+ \mu_B \mathbf{T}^1(\mathbf{B}) \cdot g_S \mathbf{T}^1(\mathbf{S}) + g_l^e \mu_B \sum_p (-1)^p T_p^1(\mathbf{B}) \sum_{q=\pm 1} \mathfrak{D}_{-pq}^{(1)}(\omega)^* T_q^1(\mathbf{S})$$
$$- g_r \mathbf{T}^1(\mathbf{B}) \cdot \mathbf{T}^1(\mathbf{N}). \quad (9.107)$$

The anisotropic corrections to the electron spin g-factor appear in (9.107) because of the admixture of excited electronic states with the $^3\Sigma$ ground state.

The matrix elements of the effective Hamiltonian (9.107) may now be calculated in a case (b) basis set.

(i) ROTATIONAL ENERGY

$$\langle \eta, \Lambda; N, S, J, M_J | B_0 \mathbf{N}^2 | \eta, \Lambda; N, S, J, M_J \rangle = B_0 N(N+1). \quad (9.108)$$

(ii) SPIN–SPIN INTERACTION

Using the form (9.101) for the electron spin–spin interaction we have

$$\langle \eta, \Lambda; N, S, J, M_J | - g_S^2 \mu_B^2 (\mu_0/4\pi) \sqrt{6} \mathbf{T}^2(\mathbf{S}_1, \mathbf{S}_2) \cdot \mathbf{T}^2(\mathbf{C}) | \eta', \Lambda; N', S, J, M_J \rangle$$

$$= -\sqrt{6} (-1)^{J+N'+S} \begin{Bmatrix} S & N' & J \\ N & S & 2 \end{Bmatrix}$$

$$\times \langle \eta, N, \Lambda \| g_S^2 \mu_B^2 (\mu_0/4\pi) \mathbf{T}^2(\mathbf{C}) \| \eta', N', \Lambda' \rangle \langle S \| \mathbf{T}^2(\mathbf{S}_1, \mathbf{S}_2) \| S \rangle$$

$$= -\sqrt{6} (-1)^{J+N'+S} \begin{Bmatrix} S & N' & J \\ N & S & 2 \end{Bmatrix} \sum_q \langle \eta, \Lambda | g_S^2 \mu_B^2 (\mu_0/4\pi) C_q^2 (\theta_{12}, \phi_{12}) (r_{12}^{-3}) | \eta', \Lambda' \rangle$$

$$\times \langle N, \Lambda \| \mathfrak{D}_{\cdot q}^{(2)}(\omega)^* \| N', \Lambda' \rangle \sqrt{5} \begin{Bmatrix} S_1 & S_1 & 1 \\ S_2 & S_2 & 1 \\ S & S & 2 \end{Bmatrix}$$

$$\times (2S+1) \{ S_1(S_1+1)(2S_1+1) S_2(S_2+1)(2S_2+1) \}^{1/2}. \quad (9.109)$$

The reduced matrix element of $\mathbf{T}^2(\mathbf{S}_1, \mathbf{S}_2)$ was previously evaluated in appendix 8.3, and to be consistent with the form (9.99) of the spin–spin interaction we take the $q = 0$

component of (9.109). Hence we have the result

$$\langle \eta, \Lambda; N, S, J, M_J | - g_S^2 \mu_B^2 (\mu_0/4\pi) \sqrt{6} \mathbf{T}^2(\mathbf{S}_1, \mathbf{S}_2) \cdot \mathbf{T}^2(\mathbf{C}) | \eta', \Lambda'; N', S, J, M_J \rangle$$

$$= -\sqrt{6}(-1)^{J+N'+S} \begin{Bmatrix} S & N' & J \\ N & S & 2 \end{Bmatrix} \langle \eta, \Lambda | g_S^2 \mu_B^2 (\mu_0/4\pi) C_0^2(\theta_{12}, \phi_{12}) (r_{12}^{-3}) | \eta, \Lambda \rangle$$

$$\times \langle N, \Lambda | \mathfrak{D}_{.0}^{(2)}(\omega)^* | N', \Lambda \rangle (\sqrt{5}/2)$$

$$= -\frac{\sqrt{30}}{2}(-1)^{J+N'+S} \begin{Bmatrix} S & N' & J \\ N & S & 2 \end{Bmatrix} \langle \eta, \Lambda | g_S^2 \mu_B^2 (\mu_0/4\pi) C_0^2(\theta_{12}, \phi_{12}) (r_{12}^{-3}) | \eta, \Lambda \rangle$$

$$\times (-1)^N \begin{pmatrix} N & 2 & N' \\ 0 & 0 & 0 \end{pmatrix} \{(2N+1)(2N'+1)\}^{1/2}$$

$$= \lambda \frac{2\sqrt{30}}{3}(-1)^{J+N'+S} \begin{Bmatrix} S & N' & J \\ N & S & 2 \end{Bmatrix} (-1)^N \begin{pmatrix} N & 2 & N' \\ 0 & 0 & 0 \end{pmatrix} \{(2N+1)(2N'+1)\}^{1/2}.$$

(9.110)

In the last line of equation (9.110) we have introduced the parameter λ, which we first met in equation (9.99). If we now consider only the matrix elements diagonal in N we can calculate the spin–spin energies for the three spin components to be

$$J = N+1: \quad E = -\frac{2\lambda N}{3(2N+3)},$$

$$J = N: \quad E = \frac{2}{3}\lambda, \qquad (9.111)$$

$$J = N-1: \quad E = -\frac{2\lambda(N+1)}{3(2N-1)}.$$

These are the first-order spin–spin energies but we see from (9.110) that there are also matrix elements off-diagonal in N, which cannot, in general, be neglected. The 3-j symbol in (9.110) tells us that $N' = N \pm 1$ matrix elements are zero, but that $N' = N \pm 2$ elements are non-zero. Specifically we find that

$$\langle \eta, \Lambda; N, S, J, M_J | - g_S^2 \mu_B^2 (\mu_0/4\pi) \sqrt{6} \mathbf{T}^2(\mathbf{S}_1, \mathbf{S}_2) \cdot \mathbf{T}^2(\mathbf{C}) | \eta, \Lambda; N \pm 2, S, J, M_J \rangle$$

$$= \lambda \frac{2\sqrt{30}}{3}(-1)^{J+N+S} \begin{Bmatrix} S & N \pm 2 & J \\ N & S & 2 \end{Bmatrix} (-1)^N \begin{pmatrix} N & 2 & N \pm 2 \\ 0 & 0 & 0 \end{pmatrix}$$

$$\times \{(2N+1)(2[N \pm 2]+1)\}^{1/2}. \qquad (9.112)$$

Evaluation of this expression for different members of the spin triplet shows that there are just two non-zero elements which must, of course, be diagonal in J:

$$\langle \eta, \Lambda; N, S, J, M_J | - g_S^2 \mu_B^2 (\mu_0/4\pi) \sqrt{6} \mathbf{T}^2(\mathbf{S}_1, \mathbf{S}_2) \cdot \mathbf{T}^2(\mathbf{C}) | \eta, \Lambda; N+2, S, J, M_J \rangle$$

$$= 2\lambda \left\{ \frac{(N+1)(N+2)}{(2N+3)^2} \right\}^{1/2} \quad \text{for } J = N+1, \qquad (9.113)$$

$$\langle \eta, \Lambda; N, S, J, M_J | - g_S^2 \mu_B^2 (\mu_0/4\pi)\sqrt{6} T^2(S_1, S_2) \cdot T^2(C)|\eta, \Lambda; N-2, S, J, M_J\rangle$$

$$= 2\lambda \left\{ \frac{N(N-1)}{(2N-1)^2} \right\}^{1/2} \quad \text{for } J = N-1, \tag{9.114}$$

(iii) SPIN–ROTATION INTERACTION

The matrix elements of the spin–rotation interaction are diagonal in a case (b) basis:

$$\langle \eta, \Lambda; N, S, J, M_J | \gamma T^1(N) \cdot T^1(S)|\eta, \Lambda; N, S, J, M_J\rangle$$

$$= \gamma (-1)^{N+J+S} \begin{Bmatrix} S & N & J \\ N & S & 1 \end{Bmatrix} \langle N\|T^1(N)\|N\rangle \langle S\|T^1(S)\|S\rangle$$

$$= \gamma (-1)^{N+J+S} \begin{Bmatrix} S & N & J \\ N & S & 1 \end{Bmatrix} \{N(N+1)(2N+1)S(S+1)(2S+1)\}^{1/2}. \tag{9.115}$$

Evaluation of the 6-j symbol gives the results for the three spin components of rotational level N:

$$\begin{aligned} J &= N+1: & E &= \gamma N \\ J &= N: & E &= -\gamma \\ J &= N-1: & E &= -\gamma(N+1). \end{aligned} \tag{9.116}$$

(iv) ELECTRON SPIN ZEEMAN INTERACTION

As usual, the direction of the applied magnetic field is taken to define the Z ($p=0$) direction:

$$\langle \eta, \Lambda; N, S, J, M_J | g_S \mu_B T_0^1(\boldsymbol{B}) T_0^1(\boldsymbol{S})|\eta, \Lambda; N, S, J', M_J\rangle$$

$$= g_S \mu_B B_Z (-1)^{J-M} \begin{pmatrix} J & 1 & J' \\ -M_J & 0 & M_J \end{pmatrix} (-1)^{J+N+1+S} \begin{Bmatrix} S & J' & N \\ J & S & 1 \end{Bmatrix}$$

$$\times \{(2J'+1)(2J+1)S(S+1)(2S+1)\}^{1/2}. \tag{9.117}$$

Expansion of the Wigner symbols in (9.117) gives the results:

$$\langle \eta, \Lambda; N, S, J, M_J | g_S \mu_B T_0^1(\boldsymbol{B}) T_0^1(\boldsymbol{S})|\eta, \Lambda; N, S, J, M_J\rangle$$

$$= g_S \mu_B B_Z M_J/(N+1) \quad \text{for } J = N+1,$$
$$\quad g_S \mu_B B_Z M_J/N(N+1) \quad \text{for } J = N,$$
$$\quad -g_S \mu_B B_Z M_J/N \quad \text{for } J = N-1.$$

$$\langle \eta, \Lambda; N, S, J, M_J | g_S \mu_B T_0^1(\boldsymbol{B}) T_0^1(\boldsymbol{S})|\eta, \Lambda; N, S, J-1, M_J\rangle$$

$$= g_S \mu_B B_Z \{N[(N+1)^2 - M_J^2]/(N+1)^2(2N+1)\}^{1/2} \quad \text{for } J = N+1,$$
$$\quad g_S \mu_B B_Z \{(N^2 - M_J^2)(N+1)/N^2(2N+1)\}^{1/2} \quad \text{for } J = N. \tag{9.118}$$

(v) ANISOTROPIC SPIN ZEEMAN INTERACTION

The matrix elements of the anisotropic spin Zeeman interaction are:

$$\langle \eta, \Lambda; N, S, J, M_J | g_l^e \mu_B T_0^1(\boldsymbol{B}) \sum_{q=\pm 1} \mathcal{D}_{0q}^{(1)}(\omega)^* T_q^1(\boldsymbol{S}) | \eta, \Lambda; N, S, J, M_J \rangle$$

$$= 2g_l^e \mu_B B_Z M_J / (2N+3) \quad \text{for } J = N+1,$$
$$ 0 \quad \text{for } J = N,$$
$$-2g_l^e \mu_B B_Z M_J / (2N-1) \quad \text{for } J = N-1.$$

$$\langle \eta, \Lambda; N, S, J, M_J | g_l^e \mu_B T_0^1(\boldsymbol{B}) \sum_{q=\pm 1} \mathcal{D}_{0q}^{(1)}(\omega)^* T_q^1(\boldsymbol{S}) | \eta, \Lambda; N, S, J-1, M_J \rangle$$

$$= g_l^e \mu_B B_Z \{ N[(N+1)^2 - M_J^2]/[4(N+1)^2 - 1](2N+3) \}^{1/2} \quad \text{for } J = N+1,$$
$$g_l^e \mu_B B_Z \{ (N+1)(N^2 - M_J^2)/(4N^2 - 1)(2N-1) \}^{1/2} \quad \text{for } J = N,$$
$$0 \quad \text{for } J = N-1.$$

(9.119)

(vi) ROTATIONAL ZEEMAN INTERACTION

By replacing N by $(\boldsymbol{J}-\boldsymbol{S})$, we see that the rotational Zeeman interaction has straightforward diagonal matrix elements only:

$$\langle \eta, \Lambda; N, S, J, M_J | -g_r \mu_B T_0^1(\boldsymbol{B}) T_0^1(\boldsymbol{J}) | \eta, \Lambda; N, S, J, M_J \rangle$$
$$= -g_r \mu_B B_Z (-1)^{J-M} \begin{pmatrix} J & 1 & J \\ -M_J & 0 & M_J \end{pmatrix} \{ J(J+1)(2J+1) \}^{1/2}$$
$$= -g_r \mu_B B_Z M_J \quad (9.120)$$

together with a correction $+g_r$ to the electron spin g factor in equation (9.117).

We now use the results we have obtained to calculate the energy levels in a magnetic field, determine the field values for the allowed electric dipole transitions, and compare the results with the experimental spectrum [56]. It is already clear that in the case (b) basis set we shall have to take note of the extensive mixing of different rotational levels by the $\Delta N = \pm 2$ off-diagonal matrix elements of the spin–spin interaction. In SO the spin–spin parameter λ is comparable with the rotational constant B_0, and, as we shall see, in heavier molecules like SeO, λ is so much larger than B_0, because of spin–orbit coupling, that a case (a) basis is more appropriate.

In their analysis of the SO microwave magnetic resonance spectrum, Carrington, Levy and Miller [56] set up and diagonalised a series of matrices of size up to 22×22, one for each M_J value, which alone remains a good quantum number. They included rotational levels from $N = 0$ to 7, but to illustrate the nature of the problem we will set up the 9×9 matrix for $N = 0$ to 3, for $M_J = 1$; in practice one includes levels of N value high enough to avoid truncation errors. The individual matrix elements for our

$^3\Sigma$ states | 647

Table 9.5. *Energy matrix for SO $^3\Sigma$ in a magnetic field, including $N = 0$ to 3, for $M_J = 1$; note that $N = 1$, $J = 0$ is not included because $M_J = 1$ does not exist for $J = 0$*

		$N=0$	$N=1$		$N=2$			$N=3$		
		$J=1$	$J=2$	$J=1$	$J=3$	$J=2$	$J=1$	$J=4$	$J=3$	$J=2$
$N=0$	$J=1$	m_{11}	0	0	0	0	m_{16}	0	0	0
$N=1$	$J=2$	0	m_{22}	m_{23}	0	0	0	0	0	m_{29}
	$J=1$	0	m_{32}	m_{33}	0	0	0	0	0	0
$N=2$	$J=3$	0	0	0	m_{44}	m_{45}	0	0	0	0
	$J=2$	0	0	0	m_{54}	m_{55}	m_{56}	0	0	0
	$J=1$	m_{61}	0	0	0	m_{65}	m_{66}	0	0	0
$N=3$	$J=4$	0	0	0	0	0	0	m_{77}	m_{78}	0
	$J=3$	0	0	0	0	0	0	m_{87}	m_{88}	m_{89}
	$J=2$	0	m_{92}	0	0	0	0	0	m_{98}	m_{99}

example are as follows:

$$m_{11} = +g'_S \mu_B B_Z + g^e_l \mu_B B_Z (2/3) - g_r \mu_B B_Z$$
$$m_{22} = 2B_0 - 2\lambda/15 + \gamma + g'_S \mu_B B_Z (1/2) + g^e_l \mu_B B_Z (2/5) - g_r \mu_B B_Z$$
$$m_{33} = 2B_0 + 2\lambda/3 - \gamma + g'_S \mu_B B_Z (1/2) - g_r \mu_B B_Z$$
$$m_{44} = 6B_0 - 4\lambda/21 + 2\gamma + g'_S \mu_B B_Z (1/3) + g^e_l \mu_B B_Z (2/7) - g_r \mu_B B_Z$$
$$m_{55} = 6B_0 + 2\lambda/3 - \gamma + g'_S \mu_B B_Z (1/6) - g_r \mu_B B_Z$$
$$m_{66} = 6B_0 - 2\lambda/3 - 3\gamma - g'_S \mu_B B_Z (1/2) - g^e_l \mu_B B_Z (2/3) - g_r \mu_B B_Z$$
$$m_{77} = 12B_0 - 2\lambda/9 + 3\gamma + g'_S \mu_B B_Z (1/4) + g^e_l \mu_B B_Z (2/9) - g_r \mu_B B_Z$$
$$m_{88} = 12B_0 + 2\lambda/3 - \gamma + g'_S \mu_B B_Z (1/12) - g_r \mu_B B_Z$$
$$m_{99} = 12B_0 - 8\lambda/15 - 4\gamma - g'_S \mu_B B_Z (1/3) - g^e_l \mu_B B_Z (2/5) - g_r \mu_B B_Z$$
$$m_{16} = m_{61} = 2\lambda (2/9)^{1/2}$$
$$m_{23} = m_{32} = g'_S \mu_B B_Z \{(4 - M_J^2)/12\}^{1/2} + g^e_l \mu_B B_Z \{(4 - M_J^2)/75\}^{1/2}$$
$$m_{29} = m_{92} = 2\lambda (6/25)^{1/2}$$
$$m_{45} = m_{54} = g'_S \mu_B B_Z \{(18 - 2M_J^2)/45\}^{1/2} + g^e_l \mu_B B_Z \{(18 - 2M_J^2)/245\}^{1/2}$$
$$m_{56} = m_{65} = g'_S \mu_R B_Z \{(12 - 3M_J^2)/20\}^{1/2} + g^e_l \mu_B B_Z \{(12 - 3M_J^2)/45\}^{1/2}$$
$$m_{78} = m_{87} = g'_S \mu_B B_Z \{(48 - 3M_J^2)/112\}^{1/2} + g^e_l \mu_B B_Z \{(48 - 3M_J^2)/567\}^{1/2}$$
$$m_{89} = m_{98} = g'_S \mu_B B_Z \{(36 - 4M_J^2)/63\}^{1/2} + g^e_l \mu_B B_Z \{(36 - 4M_J^2)/175\}^{1/2}$$

with $g'_S = g_S + g_r$. The matrix shown in table 9.5 factorises into a 4 × 4 and a 5 × 5 because there are no matrix elements connecting odd and even N states but, especially in the presence of a magnetic field, many of the states are heavily mixed. Figure 9.26

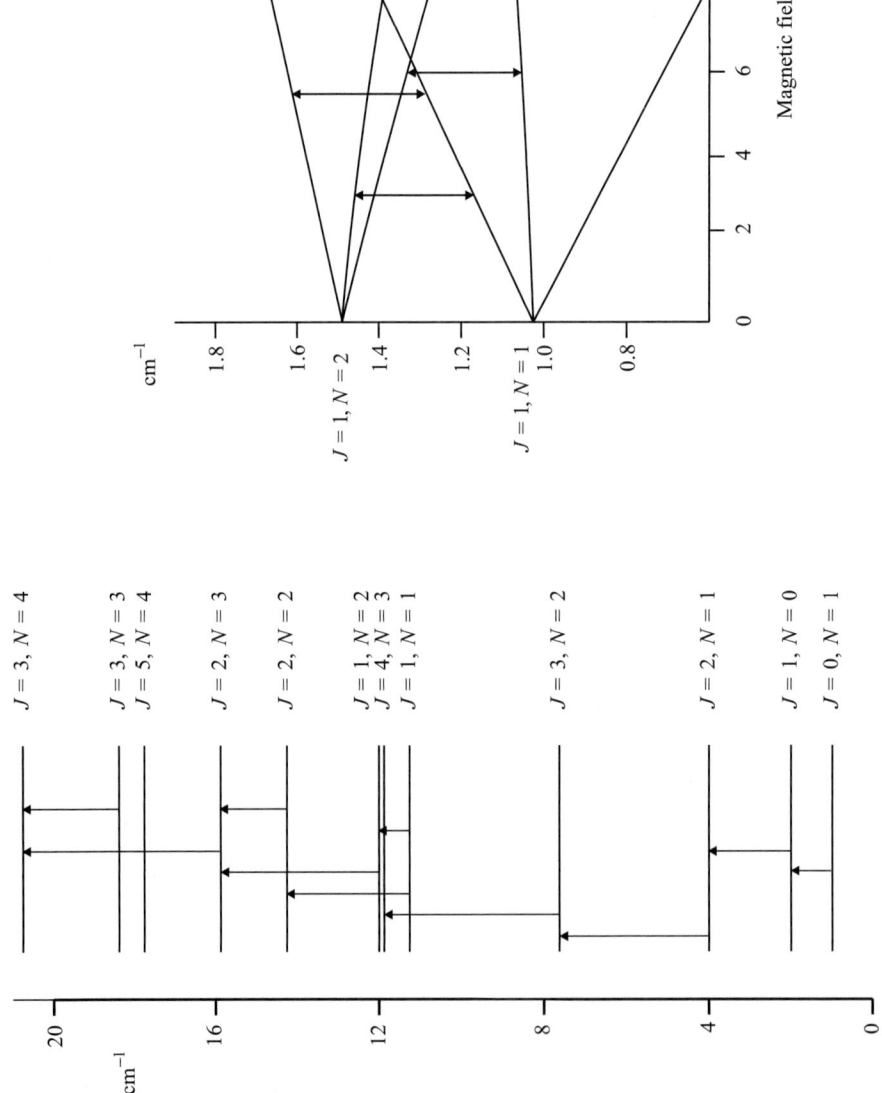

Figure 9.26. *Left*: lower rotational levels of SO $^3\Sigma$ in zero magnetic field, and the transitions observed (see chapter 10). *Right*: microwave magnetic resonance transitions observed in SO at 8762 MHz [56].

shows the relative energies of some of the lowest field-free rotational levels and indicates the transitions which have been detected by pure microwave spectroscopy; we shall describe these studies in the next chapter. Figure 9.26 also shows the Zeeman behaviour of two of the rotational levels of SO and the microwave magnetic resonance transitions which were observed. It would be possible to determine most of the important parameters in the effective Hamiltonian (9.107) from the magnetic resonance spectrum alone. However the values of B_0, λ and γ had already been determined accurately from the field-free microwave spectrum; these values were used in the analysis of the magnetic resonance spectrum, so that the Zeeman parameters could then be determined.

The case (b) representation is satisfactory for $^3\Sigma$ SO and also for O_2, but its usefulness depends upon the ratio λ/B_0. In $^3\Sigma$ SO this ratio is 7.350 but in $^3\Sigma$ SeO it is 185.408 and a case (a) description is much more appropriate [57]; figure 9.27 shows a correlation diagram for the correspondence of rotational levels in case (b) and case (a) for a $^3\Sigma$ state.

9.6.2. SeO in the $X\,^3\Sigma^-$ ground state

The spectrum of SeO was obtained [57] by reacting oxygen atoms with carbonyl selenide inside the Stark cavity described earlier. Figure 9.27 shows how case (a) levels for a $^3\Sigma$ state can be arranged in two separate stacks. One consists of widely spaced single levels, whilst the other consists of pairs of levels which are nearly degenerate for low J values; the latter are reminiscent of Λ-doublets in a case (a) $^2\Pi$ state. We will now see how this diagram arises by examining the matrix elements of the effective Hamiltonian in a case (a) basis.

The essential feature of a case (a) basis set for a $^3\Sigma$ molecule is that the electron spin S is strongly coupled to the rotational angular momentum N to form the total angular momentum J. The component of S along the internuclear axis, Σ, is therefore a conserved quantum number, but N is not and the basis functions are of the form $|\eta, \Lambda; J, S, \Sigma, M_J\rangle$. In this basis the effective Hamiltonian can be conveniently written in the form

$$\mathcal{H}_{\text{eff}} = B_0 T^1(J) \cdot T^1(J) + (\gamma - 2B_0) T^1(J) \cdot T^1(S) + (2\sqrt{6}/3)\lambda T_0^2(S) + (B_0 - \gamma_0)S^2. \tag{9.121}$$

The rotational kinetic energy, which in case (b) was $B_0 N^2$, now becomes $B_0(J - S)^2$, giving rise to the first term in (9.121), and also the B_0 part of the second term. The spin–spin interaction term in (9.121) is used in the form described earlier in equation (9.105), involving the molecule-fixed $q = 0$ component of the second rank tensor $T^2(S)$.

The matrix elements of the first three terms in (9.121) are as follows:

$$\langle \eta, \Lambda; S, \Sigma; J, M_J | B_0 T^1(J) \cdot T^1(J) | \eta, \Lambda; S, \Sigma; J, M_J \rangle = B_0 J(J+1). \tag{9.122}$$

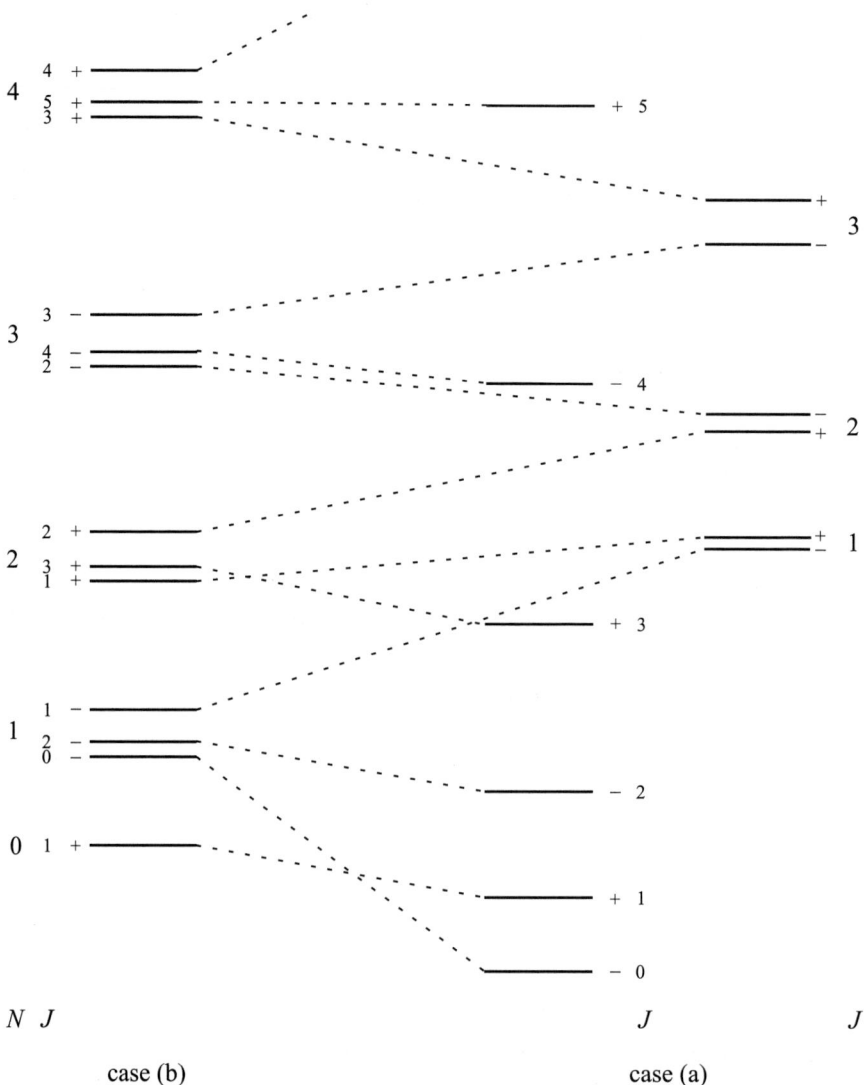

Figure 9.27. Correlation of case (b) and case (a) rotational levels for a $^3\Sigma^+$ state [51]. For a $^3\Sigma^-$ state the parities are reversed.

$$\langle \eta, \Lambda; S, \Sigma; J, M_J | (\gamma - 2B_0)\mathbf{T}^1(\mathbf{J}) \cdot \mathbf{T}^1(\mathbf{S}) | \eta, \Lambda; S, \Sigma'; J, M_J \rangle$$

$$= (\gamma - 2B_0) \sum_q (-1)^{J+S-2\Sigma} \{J(J+1)(2J+1)S(S+1)(2S+1)\}^{1/2}$$

$$\times \begin{pmatrix} J & 1 & J \\ -\Sigma & q & \Sigma' \end{pmatrix} \begin{pmatrix} S & 1 & S \\ -\Sigma & q & \Sigma' \end{pmatrix}. \tag{9.123}$$

$$\langle \eta, \Lambda; S, \Sigma; J, M_J | (2\sqrt{6}/3)\lambda \mathbf{T}_0^2(\mathbf{S}) | \eta, \Lambda; S, \Sigma; J, M_J \rangle$$

$$= (2\sqrt{6}/3)\lambda(-1)^{S-\Sigma} \begin{pmatrix} S & 2 & S \\ -\Sigma & 0 & \Sigma \end{pmatrix} \langle S \| \mathbf{T}^2(\mathbf{S}) \| S \rangle. \tag{9.124}$$

The reduced matrix element in (9.124) is evaluated by decomposing the second-rank tensor into its constituent first-rank tensors; one finds that

$$\langle S \| T^2(\boldsymbol{S}) \| S \rangle = (1/\sqrt{6}) S(2S-1) \begin{pmatrix} S & 2 & S \\ -S & 0 & S \end{pmatrix}^{-1}. \tag{9.125}$$

Evaluating the 3-j symbol in (9.125) and substituting in (9.124) we obtain the result:

$$\langle \eta, \Lambda; S, \Sigma; J, M_J | (2\sqrt{6}/3) \lambda T_0^2(\boldsymbol{S}) | \eta, \Lambda; S, \Sigma; J, M_J \rangle$$

$$= \frac{2}{3} \lambda (-1)^{S-\Sigma} \begin{pmatrix} S & 2 & S \\ -\Sigma & 0 & \Sigma \end{pmatrix} S(2S-1) \begin{pmatrix} S & 2 & S \\ -S & 0 & S \end{pmatrix}^{-1}$$

$$= \frac{2}{3} \lambda \{ 3\Sigma^2 - S(S+1) \}. \tag{9.126}$$

This equation largely explains the pattern of levels shown in the case (a) limit in figure 9.27. For $\Sigma = 0$ we have one series of non-degenerate rotational levels, as shown. For $\Sigma = \pm 1$ we have a second rotational series of doublets, with origin some 2λ higher in energy than the $\Sigma = 0$ series. The splitting of the doublet levels in the $\Sigma = \pm 1$ series is due to the $q = \pm 1$ terms in equation (9.123), but in order to understand the situation more clearly, it is necessary to consider the parity symmetry of the levels. The case (a) basis functions do not have definite parity, but the following combinations are more appropriate:

$$\psi_J^{\pm}(M_J) = \frac{1}{\sqrt{2}} \{ |\eta, \Lambda; S, \Sigma; J, M_J \rangle \pm (-1)^{J-S+s} |\eta, \Lambda; S, -\Sigma; J, M_J \rangle \}_{|\Sigma|=1}, \tag{9.127}$$

$$\psi_J^0(M_J) = |\eta, \Lambda; S, \Sigma; J, M_J \rangle_{\Sigma=0}. \tag{9.128}$$

Applying the inversion operator E^* we find:

$$E^* \psi_J^{\pm}(M_J) = \pm \psi_J^{\pm}(M_J),$$
$$E^* \psi_J^0(M_J) = (-1)^{J-S+s} \psi_J^0(M_J). \tag{9.129}$$

The new basis functions (9.127) and (9.128) are now of definite parity; equation (9.123) is again used to calculate the matrix elements, this time with the basis functions of definite parity. The off-diagonal matrix elements of (9.123) produce the doublet splittings shown in the $\Sigma = \pm 1$ stack in figure 9.27, and the levels may be labelled with their parities, as shown.

Two further aspects need to be considered in order to understand the magnetic resonance spectrum, namely, the effects of an applied magnetic field, and the electric dipole transition probabilities. The effective Hamiltonian describing the interactions with an applied magnetic field, expressed in the molecule-fixed axis system q, is:

$$\mathcal{H}_Z = g_S' \mu_B T_0^1(\boldsymbol{B}) T_0^1(\boldsymbol{S}) + \sum_{q=\pm 1} (-1)^q g_l^e \mu_B T_q^1(\boldsymbol{B}) T_{-q}^1(\boldsymbol{S}) - g_r \mu_B T^1(\boldsymbol{B}) \cdot T^1(\boldsymbol{J}). \tag{9.130}$$

In this expression $g_S' = g_S + g_r$ as shown in equation (9.107). The matrix elements of

(9.130) within the case (a) basis are as follows:
for $q = 0$,

$$\langle \eta, \Lambda; S, \Sigma; J, M_J | \mathcal{H}_Z | \eta, \Lambda; S, \Sigma; J', M_J \rangle$$

$$= \mu_B B_Z (-1)^{J-M_J} [(2J'+1)(2J+1)]^{1/2} \begin{pmatrix} J & 1 & J' \\ -M_J & 0 & M_J \end{pmatrix}$$

$$\times \left\{ g'_S (-1)^{S-\Sigma} [S(S+1)(2S+1)]^{1/2} (-1)^{J-\Sigma} \begin{pmatrix} J & 1 & J' \\ -\Sigma & 0 & \Sigma \end{pmatrix} \begin{pmatrix} S & 1 & S \\ -\Sigma & 0 & \Sigma \end{pmatrix} \right.$$

$$\left. - g_r \delta_{JJ'} [J(J+1)(2J+1)]^{1/2} \right\}, \tag{9.131}$$

for $q = \pm 1$,

$$\langle \eta, \Lambda; S, \Sigma; J, M_J | \mathcal{H}_Z | \eta, \Lambda; S, \Sigma'; J', M_J \rangle$$

$$= g'_S \mu_B B_Z (-1)^{J-M_J+J-\Sigma} [(2J'+1)(2J+1)]^{1/2} \begin{pmatrix} J & 1 & J' \\ -M_J & 0 & M_J \end{pmatrix}$$

$$\times \sum_q (-1)^{S-\Sigma} [S(S+1)(2S+1)]^{1/2} \begin{pmatrix} J & 1 & J' \\ -\Sigma & q & \Sigma' \end{pmatrix} \begin{pmatrix} S & 1 & S \\ -\Sigma & q & \Sigma' \end{pmatrix}. \tag{9.132}$$

The magnetic resonance spectrum recorded by Carrington, Currie, Levy and Miller [57] involved the $J = 1$ level, the lowest level of the $\Sigma = \pm 1$ stack shown in figure 9.27. The net results of (9.131) and (9.132) are Zeeman splittings of the $M_J = +1$, 0 and -1 levels, each M_J component possessing a further doublet splitting of the opposite parity components caused by the off-diagonal matrix elements of (9.123). The observed spectrum consisted of electric dipole transitions between opposite parity states, each obeying the normal selection rule $\Delta M_J = \pm 1$. Additional transitions were observed in the presence of an electric field because of the consequent mixing of opposite parity states.

9.6.3. NH in the $X\,^3\Sigma^-$ ground state

Nuclear hyperfine structure was observed and analysed in both the naturally abundant ^{33}SO and ^{77}SeO, but to illustrate the principles we turn to another $^3\Sigma$ system, the imine radical (NH) studied by Wayne and Radford [58] using far-infrared laser magnetic resonance. The spectra of ^{14}NH, ^{15}NH and ^{14}ND show hyperfine structure from all the nuclei involved. The related radicals PH and PD, both of which show rich hyperfine structure arising from ^{31}P and ^{1}H or ^{2}D, have been studied by Ohashi, Kawaguchi and Hirota [59] and by Davies and his coworkers [60], again using far-infrared laser magnetic resonance.

The NH radical was produced by the reaction of fluorine atoms with ammonia. In its $^3\Sigma^-$ ground state NH is well described by a case (b) basis set, and the transitions observed by Wayne and Radford were rotational transitions involving, in different

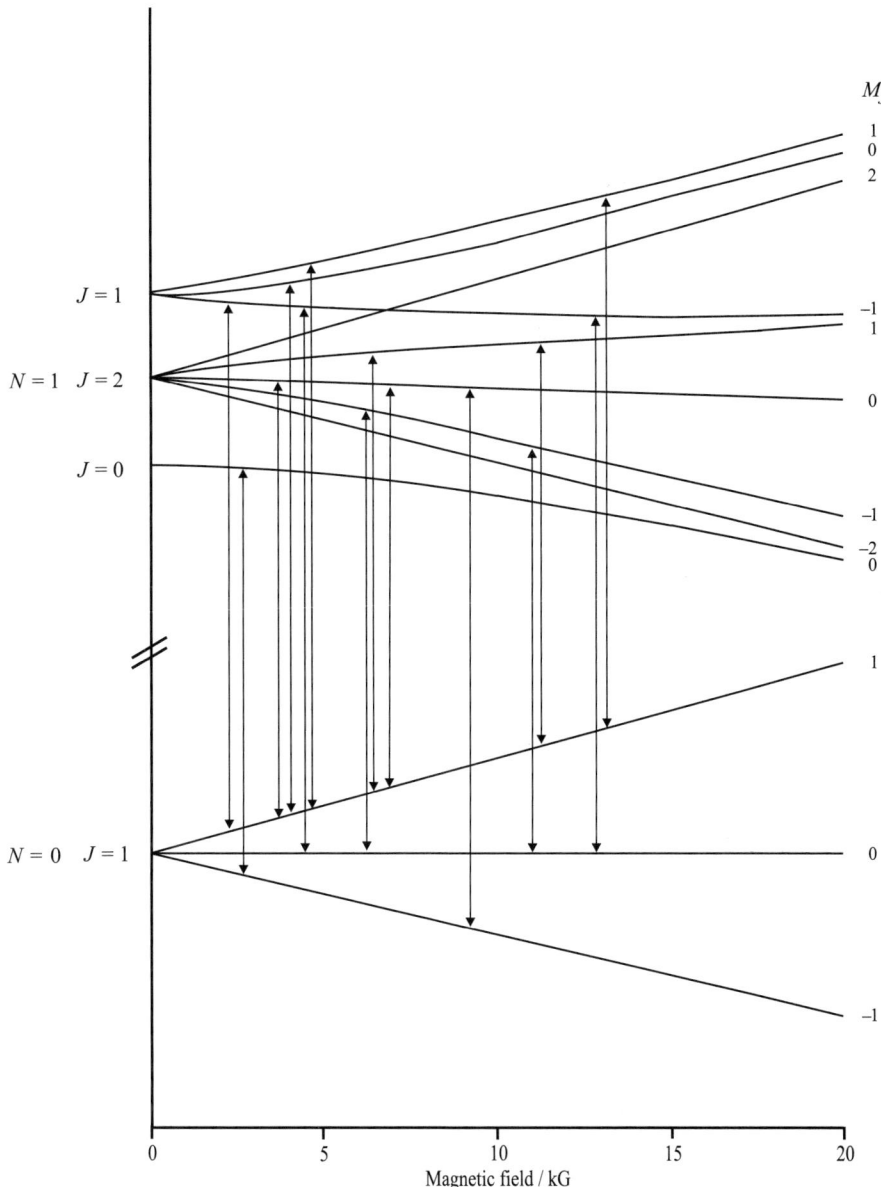

Figure 9.28. Zeeman levels for $N=0$ and 1 of NH $^3\Sigma^-$ ($v=0$) and the observed far-infrared laser magnetic resonance transitions [58]. These were recorded using four different FIR lines at 31.7615, 32.1466, 33.0822 and 33.1922 cm^{-1}.

isotopic species, the $N=0$, 1, 2 and 3 rotational levels in both $v=0$ and 1. Figure 9.28 shows a Zeeman diagram for the $N=0$ and 1 rotational levels of NH in its $v=0$ level; the nuclear hyperfine splitting is not shown. The four far-infrared laser lines used are listed in the figure caption, and the transitions detected are shown in the figure. All but one of the transitions recorded showed fully resolved hyperfine structure from

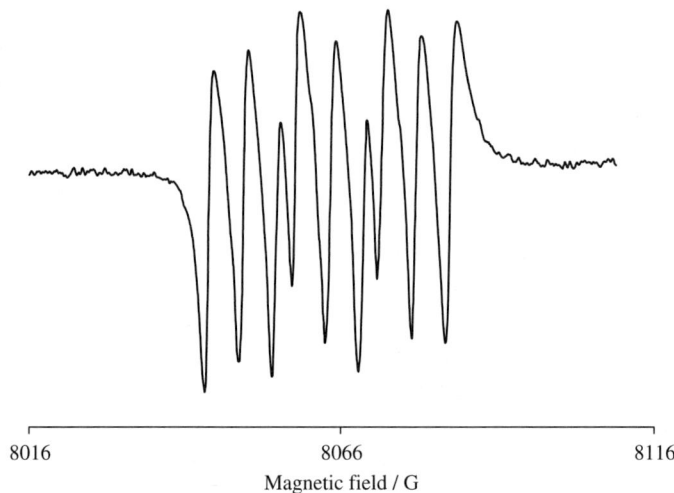

Figure 9.29. The laser magnetic resonance transition $N = 1, J = 1, M = 0 \rightarrow N = 2, J = 1, M = 1$ for ND $^3\Sigma^-$, showing fully resolved hyperfine splitting from both nuclei, each with spin 1 [58]. The laser frequency was 991.7778 GHz.

both ^{14}N and ^{1}H. Figure 9.29 shows a recording of the $N = 1, J = 1, M = 0 \rightarrow N = 2, J = 1, M = 1$ Zeeman component of ND; this beautiful pattern is an example of a fully resolved nine-line hyperfine structure arising from ^{14}N and ^{2}D, both with nuclear spin $I = 1$.

The effective Hamiltonian used by Wayne and Radford was essentially that described earlier for $^3\Sigma$ SO, with the addition of a centrifugal distortion term:

$$\mathcal{H}_{\text{eff}} = BN^2 + \frac{2\sqrt{6}}{3}\lambda T_0^2(\boldsymbol{S}, \boldsymbol{S}) + \gamma T^1(\boldsymbol{N}) \cdot T^1(\boldsymbol{S}) - DN^4. \tag{9.133}$$

The nuclear hyperfine and magnetic field interaction terms must be added to this effective Hamiltonian; they are similar to those introduced earlier in this section. For the hyperfine interactions:

$$\mathcal{H}_{\text{hfs}} = \sum_k \left\{ b_{F_k} T^1(\boldsymbol{I}_k) \cdot T^1(\boldsymbol{S}) - t_k \sqrt{10}\, T^1(\boldsymbol{I}_k) \cdot T^1(\boldsymbol{S}, \boldsymbol{C}^2(\omega)) \right\}$$

$$+ eq_0 Q \sqrt{\frac{3}{2}} \frac{1}{2I(2I-1)} T_{q=0}^2(\boldsymbol{I}, \boldsymbol{I}). \tag{9.134}$$

The sum over k represents the terms for both nuclei, whereas the quadrupole term exists only for the ^{14}N nucleus. Equation (9.134) also recognises implicitly that only terms diagonal in the ground vibronic state will be included. For the magnetic field interactions the effective Hamiltonian used by Wayne and Radford was

$$\mathcal{H}_Z = \mu_B(g_S + g_r) T^1(\boldsymbol{B}) \cdot T^1(\boldsymbol{S}) - \mu_B g_r T^1(\boldsymbol{B}) \cdot T^1(\boldsymbol{N})$$

$$+ \mu_B g_l^e T_0^1(\boldsymbol{B}) \sum_{q=\pm 1} \mathfrak{D}_{0q}^{(1)}(\omega)^* T_q^1(\boldsymbol{S}) + \sum_k g_{N_k} \mu_N T^1(\boldsymbol{B}) \cdot T^1(\boldsymbol{I}_k). \tag{9.135}$$

The presence of two nuclear spins means that there is considerable choice in the selection of basis functions; the reader who wishes to practice virtuosity in irreducible tensor algebra is invited to calculate the matrix elements in the different coupled representations that are possible! In fact the sensible choice, particularly when a strong magnetic field is to be applied, is the nuclear spin-decoupled basis set $|\eta, \Lambda; N, S, J, M_J; I_N, M_N; I_H, M_H\rangle$. Again note the possible source of confusion here; M_N is the space-fixed component of the nitrogen nuclear spin I_N, not the space-fixed component of N. This nuclear spin-decoupled basis set was the one chosen by Wayne and Radford in their analysis of the NH spectrum.

Wayne and Radford analysed the spectra of NH for $v = 0$ and 1, with ^{14}N, ^{15}N, ^{1}H and ^{2}D nuclei, and they obtained valuable new information, particularly concerning the nuclear hyperfine constants. *Ab initio* calculations of the dipolar constants were found to agree well with experiment, but for the Fermi contact constants the agreement was poor. This is not surprising because, in the restricted SCF approximation, the unpaired electrons in the molecule occupy a π orbital which is located almost entirely on the nitrogen atom. Calculation of the dipolar hyperfine constants is therefore straightforward, but the Fermi contact interaction arises from configuration interaction involving excited electronic states, which is more difficult to calculate accurately.

Before leaving $^3\Sigma$ molecules we should mention studies by Gruebele, Müller and Saykally [61] in which they used FIR LMR to measure rotational transitions in the OH$^+$ and OD$^+$ ions, which are isoelectronic with NH and also conform to Hund's case (b) coupling.

9.7. $^3\Pi$ states

9.7.1. CO in the $a\,^3\Pi$ state

The most important and comprehensive laser magnetic resonance study of a $^3\Pi$ state is that of the CO excited a state which, as we described in chapter 8, has also been studied by radiofrequency spectroscopy (using molecular beam electric resonance), pure microwave rotational spectroscopy and mid-infrared laser magnetic resonance. In this section we present a detailed description, both of the FIR laser magnetic resonance studies, and also the theoretical analysis which applies to all of the experimental studies. It is always important, but not always possible, to bring together measurements made in different regions of the electromagnetic spectrum under a unified theoretical umbrella.

The apparatus used by Saykally, Evenson, Comben and Brown [62] was similar to that shown in figure 9.4, but with one important difference. Attempts to detect radicals or ions in a gas discharge were thwarted by the transverse magnetic field used in figure 9.4. The transverse field was therefore replaced by an axial field, produced by a solenoid magnet of 7.6 cm diameter and 33 cm length, cooled by liquid nitrogen. This enabled a d.c. glow discharge to be maintained inside the sample region between a cylindrical copper anode and a water-cooled copper cathode located in a sidearm outside the laser cavity. A gas mixture of helium containing 10% CO at a total pressure

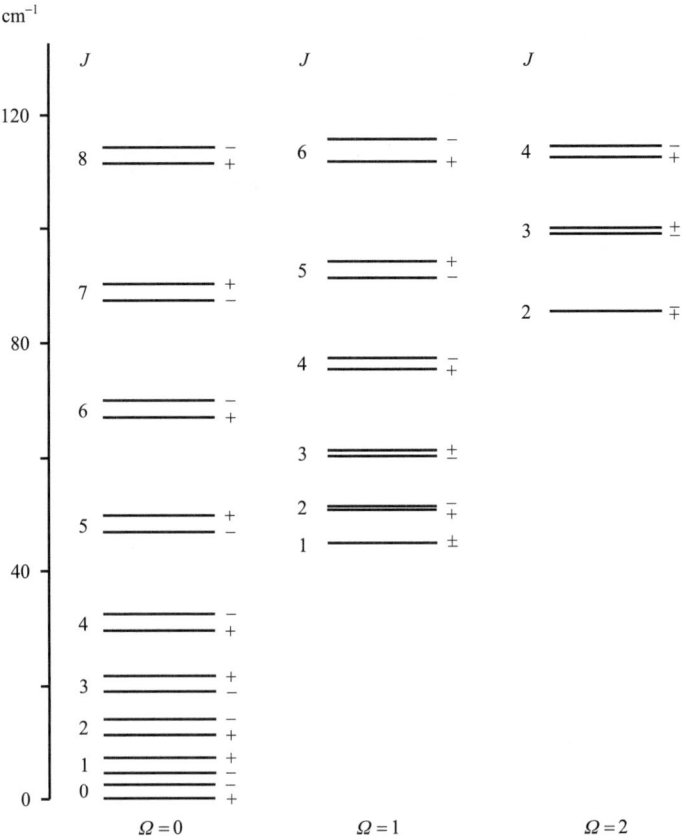

Figure 9.30. Lower rotational levels and fine-structure splitting for the $a\,^3\Pi$ state of CO. This diagram is a repeat of figure 8.49.

of 1 torr was used, and the discharge resulted in sufficient excitation of the $a\,^3\Pi$ state for strong FIR laser magnetic resonance transitions to be observed. The fine-structure components of the $a\,^3\Pi$ state with the lower rotational levels are shown in figure 9.30. The transitions studied by Saykally, Evenson, Comben and Brown [62] were rotational transitions within the $^3\Pi_2$ and $^3\Pi_0$ states and examples of the spectra obtained are shown in figures 9.31 and 9.32.

The effective Hamiltonian used by Saykally, Evenson, Comben and Brown [62] contains terms which we have already met in this chapter, and which we will therefore deal with fairly briefly, with appropriate references to the details given elsewhere, particularly in this section. The theory has been developed in a number of papers, particularly by Brown, Kopp, Malmberg and Rydh [63], Brown and Merer [64], who dealt with Π states of triplet and higher spin multiplicity, and Steimle and Brown [65] who specifically addressed the theory of the Λ-doubling of CO in the $^3\Pi$ state. The theory of the Zeeman interactions follows closely that developed to analyse the magnetic resonance spectra of OH by Brown, Kaise, Kerr and Milton [66]. All of these

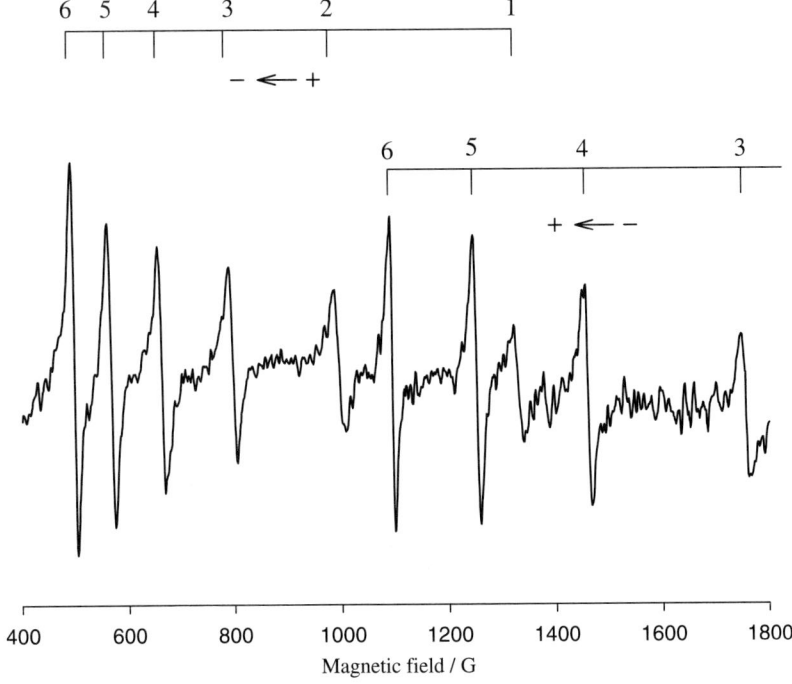

Figure 9.31. FIR laser magnetic resonance spectrum of CO in the $a\,^3\Pi$ state, observed using the 393.6 μm line from formic acid [62]. This spectrum arises from the $J = 7 \leftarrow 6$ rotational transition in the $\Omega = 2$ fine-structure state, and the transitions obey the selection rule $\Delta M_J = +1$. The lower M_J states are indicated in the diagram.

treatments are related to a more general exposition concerning the effective Hamiltonian for diatomic molecules given by Brown, Colbourn, Watson and Wayne [67].

The effective Hamiltonian may be summarised as follows:

$$\mathcal{H}_{\text{eff}} = \mathcal{H}_{\text{so}} + \mathcal{H}_{\text{rot}} + \mathcal{H}_{\text{ss}} + \mathcal{H}_{\text{sr}} + \mathcal{H}_{\text{LD}} + \mathcal{H}_Z. \tag{9.136}$$

The matrix elements are calculated in a case (a) basis, and for the spin–orbit coupling we have the simple result

$$\langle \eta, \Lambda; S, \Sigma; J, \Omega | A\, T_0^1(\boldsymbol{L}) T_0^1(\boldsymbol{S}) | \eta, \Lambda; S, \Sigma; J, \Omega \rangle = A\Lambda\Sigma, \tag{9.137}$$

where only the axial ($q = 0$) component was included. For the rotational Hamiltonian, $B\,\boldsymbol{N}^2$, we note that, in case (a), $\boldsymbol{N} = \boldsymbol{J} - \boldsymbol{S}$. We use the result previously derived, equation (9.33), in our treatment of the ClO spectrum:

$$\langle \eta, \Lambda; S, \Sigma; J, \Omega | \mathcal{H}_{\text{rot}} | \eta, \Lambda; S, \Sigma'; J, \Omega' \rangle$$
$$= B\{\delta_{\Sigma\Sigma'}\delta_{\Omega\Omega'}[J(J+1) + S(S+1) - 2\Omega\Sigma]\}$$
$$- 2B \sum_{q=\pm 1} (-1)^{J-\Omega+S-\Sigma} \begin{pmatrix} J & 1 & J \\ -\Omega & q & \Omega' \end{pmatrix} \begin{pmatrix} S & 1 & S \\ -\Sigma & q & \Sigma' \end{pmatrix}$$
$$\times [J(J+1)(2J+1)S(S+1)(2S+1)]^{1/2}. \tag{9.138}$$

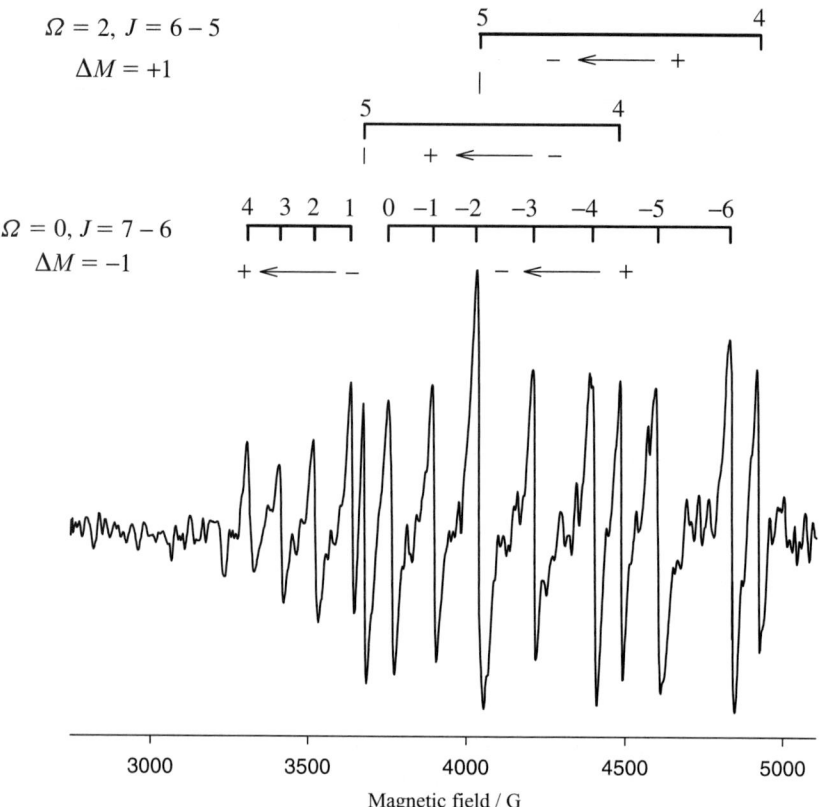

Figure 9.32. FIR laser magnetic resonance spectrum of CO in its $a\,^3\Pi$ state, recorded with the 458.5 μm laser line [62].

The electron spin–spin interaction Hamiltonian and its case (a) matrix elements were previously derived, equation (9.124), for the analysis of the magnetic resonance spectrum of $^3\Sigma$ SeO:

$$\langle \eta, \Lambda; S, \Sigma; J, \Omega | (2/3)\lambda\{3S_z^2 - \boldsymbol{S}^2\}|\eta, \Lambda; S, \Sigma; J, \Omega\rangle = (2/3)\lambda\{3\Sigma^2 - S(S+1)\}. \tag{9.139}$$

The spin–rotation interaction is given by

$$\mathcal{H}_{sr} = \gamma \mathbf{T}^1(\boldsymbol{N}) \cdot \mathbf{T}^1(\boldsymbol{S}) = \gamma\{\mathbf{T}^1(\boldsymbol{J}) - \mathbf{T}^1(\boldsymbol{S})\} \cdot \mathbf{T}^1(\boldsymbol{S}), \tag{9.140}$$

and its matrix elements are similar to those given in (9.138):

$$\langle \eta, \Lambda; S, \Sigma; J, \Omega | \gamma \mathbf{T}^1(\boldsymbol{N}) \cdot \mathbf{T}^1(\boldsymbol{S}) | \eta, \Lambda; S, \Sigma'; J, \Omega'\rangle$$

$$= \gamma \delta_{\Sigma\Sigma'}\delta_{\Omega\Omega'}[\Omega\Sigma - S(S+1)] + \gamma \sum_{q=\pm 1}(-1)^{J-\Omega+S-\Sigma}\begin{pmatrix}J & 1 & J\\ -\Omega & q & \Omega'\end{pmatrix}$$

$$\times \begin{pmatrix}S & 1 & S\\ -\Sigma & q & \Sigma'\end{pmatrix}[J(J+1)(2J+1)S(S+1)(2S+1)]^{1/2}. \tag{9.141}$$

The remaining terms in (9.136) are rather more complicated in that they include the effects of mixing other excited electronic states with the $^3\Pi$ state. Both terms have already been discussed extensively in connection with the OH magnetic resonance spectra. It was shown there that the Λ-doubling Hamiltonian could be written in the form

$$\mathcal{H}_{LD} = \sum_{q=\pm 1} \exp(-2iq\phi)\left[-q\mathrm{T}^2_{2q}(\boldsymbol{J}, \boldsymbol{J}) + (p+2q)\mathrm{T}^2_{2q}(\boldsymbol{J}, \boldsymbol{S}) - (o+p+q)\mathrm{T}^2_{2q}(\boldsymbol{S}, \boldsymbol{S})\right], \quad (9.142)$$

or alternatively,

$$\mathcal{H}_{LD} = (1/2)(o+p+q)(S_+^2 + S_-^2) \\ - (1/2)(p+2q)(J_+S_+ + J_-S_-) + (1/2)q(J_+^2 + J_-^2), \quad (9.143)$$

with the assumption that this operator connects states with $\Lambda = +1$ and -1 only. The matrix elements have been evaluated by Brown and Merer [64], using (9.143), as follows:

$$\langle \Lambda = \mp 1, \Sigma \pm 2, J, \Omega | \mathcal{H}_{LD} | \Lambda = \pm 1, \Sigma, J, \Omega \rangle \\ = (1/2)(o+p+q)\{[S(S+1) - \Sigma(\Sigma \pm 1)][S(S+1) - (\Sigma \pm 1)(\Sigma \pm 2)]\}^{1/2}. \quad (9.144)$$

$$\langle \Lambda = \mp 1, \Sigma \pm 1, J, \Omega \mp 1 | \mathcal{H}_{LD} | \Lambda = \pm 1, \Sigma, J, \Omega \rangle \\ = -(1/2)(p+2q)\{[S(S+1) - \Sigma(\Sigma \pm 1)][J(J+1) - \Omega(\Omega \mp 1)]\}^{1/2}. \quad (9.145)$$

$$\langle \Lambda = \mp 1, \Sigma, J, \Omega \mp 2 | \mathcal{H}_{LD} | \Lambda = \pm 1, \Sigma, J, \Omega \rangle \\ = (1/2)q\{[J(J+1) - \Omega(\Omega \mp 1)][J(J+1) - (\Omega \mp 1)(\Omega \mp 2)]\}^{1/2}. \quad (9.146)$$

In these equations Λ, Σ and Ω are to be taken as *signed* quantities, and we see that the effect of the Λ-doubling operator is to mix $\Lambda = +1$ and $\Lambda = -1$ components of the Π state.

Before examining the effect of an applied magnetic field it is instructive and hopefully helpful to look at the matrix of the above five terms. For each of the three fine-structure components with a given J value there are two parity states, labelled e and f. According to the now accepted convention [68] for integral J values, levels with parity $+(-1)^J$ are called e levels and levels with parity $-(-1)^J$ are called f levels. The matrix is as follows.

	$|^3\Pi_2\rangle e$	$|^3\Pi_1\rangle e$	$|^3\Pi_0\rangle e$	$|^3\Pi_2\rangle f$	$|^3\Pi_1\rangle f$	$|^3\Pi_0\rangle f$
$\langle ^3\Pi_2|e$	m_{11}	m_{12}	m_{13}	0	0	0
$\langle ^3\Pi_1|e$	m_{21}	m_{22}	m_{23}	0	0	0
$\langle ^3\Pi_0|e$	m_{31}	m_{32}	m_{33}	0	0	0
$\langle ^3\Pi_2|f$	0	0	0	m_{44}	m_{45}	m_{46}
$\langle ^3\Pi_1|f$	0	0	0	m_{54}	m_{55}	m_{56}
$\langle ^3\Pi_0|f$	0	0	0	m_{64}	m_{65}	m_{66}

We now look at the matrix elements, in e and f pairs for each fine-structure component.

e states

$m_{11} = A + (B/3)[J(J+1) - 2] + 2\lambda/3,$
$m_{22} = (B/3)[J(J+1) + 2] - 2\gamma$
$\quad - (1/2)qJ(J+1),$
$m_{33} = -A + (B/3)[J(J+1) + 2]$
$\quad + 2\lambda/3 - 2\gamma - (o + p + q),$
$m_{12} = -\{2[J(J+1) - 2]\}^{1/2}(B - \gamma/2),$
$m_{21} = m_{12},$
$m_{23} = -[2J(J+1)]^{1/2}[B - (\gamma/2)$
$\quad - (1/2)(p + 2q)],$
$m_{32} = m_{23},$
$m_{13} = -\{J(J+1)[J(J+1) - 2]\}^{1/2}(q/2),$

f states

$m_{44} = m_{11},$
$m_{55} = (B/3)[J(J+1) + 2]$
$\quad - 2\gamma + (1/2)qJ(J+1),$
$m_{66} = -A + (B/3)[J(J+1) + 2]$
$\quad + 2\lambda/3 - 2\gamma + (o + p + q),$
$m_{45} = m_{12},$
$m_{54} = m_{45},$
$m_{56} = -[2J(J+1)]^{1/2}[B - (\gamma/2)$
$\quad + (1/2)(p + 2q)],$
$m_{65} = m_{56},$
$m_{46} = \{J(J+1)[J(J+1) - 2]\}^{1/2}(q/2).$

The effects of centrifugal distortion are omitted from these matrix elements, but their inclusion is necessary for a final fit to experiment. The structure of the matrix accounts for some of the observations described previously. The Λ-doubling in the $^3\Pi_0$ state is determined primarily by the first-order contribution (see m_{33} and m_{66}) so that it is both relatively large, and essentially independent of J. For all of the fine-structure components it is a matter of diagonalising the matrix and determining the values of the Λ-doubling parameters by fitting the calculated frequencies to experiment. In the most recent analysis [65], which included constants determined in the earlier molecular beam work (see chapter 8), agreement between experiment and theory for nine Λ-doubling frequencies was good to a few kHz.

Quantitative assignment of the FIR laser magnetic resonance spectrum requires detailed attention to the form of the Zeeman Hamiltonian, $\mathcal{H}_Z$, in equation (9.136). This was taken from work described earlier in this section [66] aimed at providing a comprehensive description of the Zeeman interactions in the OH radical. In the case of $^3\Pi$ CO a slightly simpler Hamiltonian was adopted:

$$\mathcal{H}_Z = g_S \mu_B B_Z T^1_{p=0}(\boldsymbol{S}) + g_L \mu_B B_Z T^1_{p=0}(\boldsymbol{L}) - g_r \mu_B B_Z T^1_{p=0}(\boldsymbol{J} - \boldsymbol{S})$$
$$+ g_l^e \mu_B B_Z \sum_{q=\pm 1} \mathcal{D}^{(1)}_{0q}(\omega)^* T^1_q(\boldsymbol{S}). \quad (9.147)$$

The Zeeman Hamiltonian for OH contained two further parameters, as well as modifications to the electron spin and orbital g-factors. The principal parameter determined from the laser magnetic resonance data was g_r, the rotational g-factor. Agreement between experimentally measured resonant magnetic fields (from figures 9.30 and 9.31) and the calculated values was excellent.

We discuss the pure microwave measurements for $^3\Pi$ CO in the next chapter. As in the case of OH, it is the combination of results from different types of spectroscopic study that provides the complete picture. It would, for example, have been much more difficult to analyse the magnetic resonance spectra without the detailed zero-field radiofrequency measurements of the Λ-doubling frequencies.

9.8. $^4\Sigma$ states

9.8.1. CH in the $a\ ^4\Sigma^-$ state

We described the laser magnetic resonance spectrum of CH in its ground $^2\Pi$ state in section 9.4.4. The lowest quartet excited state of CH lies only 0.7 eV above the ground state and the reaction between fluorine atoms and methane was found by Nelis, Brown and Evenson [69] to produce the CH radical in both electronic states. An example of the FIR laser magnetic resonance spectrum of the excited state is shown in figure 9.34, and the appropriate energy level diagram with the observed transitions is presented in figure 9.33. The observed spectrum arises from rotational transitions in the $v = 0$ level, split by both fine and hyperfine interactions which we now consider in more detail.

The effective Hamiltonian will be evaluated in a case (b) basis and, as one would expect, it is similar to that previously developed for $^3\Sigma$ states, but with some important additions. The effective Hamiltonian used [69] was as follows:

$$\mathcal{H}_{\text{eff}} = \mathcal{H}_{\text{rot}} + \mathcal{H}_{\text{ss}} + \mathcal{H}_{\text{sr}} + \mathcal{H}_{\text{hfs}} + \mathcal{H}_{\text{Z}}. \tag{9.148}$$

We now compare the various terms in (9.148) with those used previously for the analysis of $^3\Sigma$ spectra. The rotational term is given by

$$\mathcal{H}_{\text{rot}} = B_0 N^2 - D_0 N^4 + H_0 N^6, \tag{9.149}$$

which is the same as the first term of (9.99) but with the addition of two centrifugal distortion terms. These are required because CH is a much lighter molecule than SO. The spin–spin interaction is given by

$$\mathcal{H}_{\text{ss}} = \frac{2}{3}\lambda\sqrt{6}T^2_{q=0}(\boldsymbol{S},\boldsymbol{S}) + \frac{1}{3}\lambda_D\sqrt{6}\left[T^2_{q=0}(\boldsymbol{S},\boldsymbol{S}), N^2\right]_+. \tag{9.150}$$

We have made explicit the fact that this term is evaluated in the molecule-fixed axis system (with $q = 0$). The first term was introduced previously in equation (9.104) but the second term is less familiar; it represents a centrifugal distortion correction to the spin–spin interaction but was not, in fact, included in the analysis of the CH spectrum so we shall not discuss it further here.

The spin–rotation interaction term given by Nelis, Brown and Evenson [69] was written in the form

$$\mathcal{H}_{\text{sr}} = \gamma T^1(\boldsymbol{N}) \cdot T^1(\boldsymbol{S}) + \gamma_D(T^1(\boldsymbol{N}) \cdot T^1(\boldsymbol{S}))N^2 + CT^3(\boldsymbol{L}^2, \boldsymbol{N}) \cdot T^3(\boldsymbol{S},\boldsymbol{S},\boldsymbol{S}). \tag{9.151}$$

Again the first term is familiar (see (9.99)), whilst the second term represents a centrifugal distortion correction to the spin–rotation interaction and was neglected

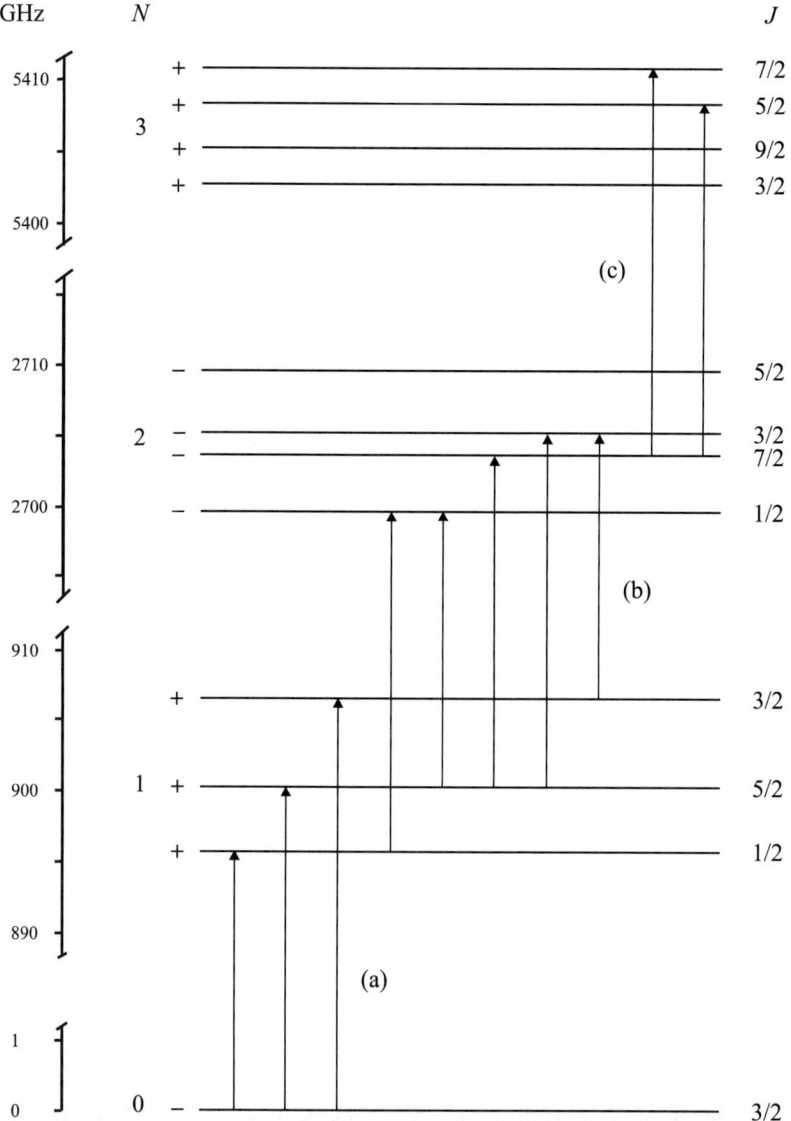

Figure 9.33. Energy level diagram showing the transitions involving the four lowest rotational levels of CH in the $v = 0$ level of the $^4\Sigma^-$ state [69]. The laser wavelengths used were (a) 333 µm, (b) 167 µm, (c) 111 µm, full details of which are given in table 9.1.

in the quantitative analysis of the CH spectrum. The third term is new (the first time in this chapter we have encountered third-rank tensors), and represents a third-order spin–orbit coupling effect. We will return to its origin and matrix elements below.

The magnetic hyperfine terms are now familiar, representing the Fermi contact interaction and axial dipolar interaction:

$$\mathcal{H}_{\text{hfs}} = b_F \mathbf{T}^1(\mathbf{I}) \cdot \mathbf{T}^1(\mathbf{S}) + \frac{1}{3} c \sqrt{6} \mathbf{T}^2_{q=0}(\mathbf{I}, \mathbf{S}). \tag{9.152}$$

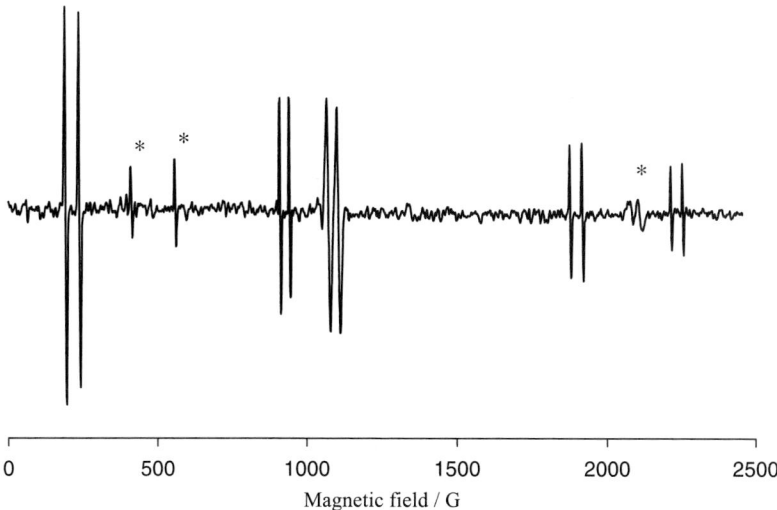

Figure 9.34. Laser magnetic resonance spectrum of CH in its $a\,^4\Sigma^-$ state recorded in parallel polarisation ($\Delta M_J = 0$) with the 166.6 μm laser line of CH_2F_2. The rotational transition is $N = 2 \leftarrow 1$, and the quintet fine structure may be understood by reference to the energy level diagram in figure 9.33. The lines marked with an asterisk arise from an impurity species; the doublet splittings of the CH lines are due to proton hyperfine interaction [69].

The magnetic field interaction terms are also familiar:

$$\mathcal{H}_Z = g_S \mu_B B_Z T^1_{p=0}(\boldsymbol{S}) - g_r \mu_B B_Z T^1_{p=0}(\boldsymbol{N}). \quad (9.153)$$

They appeared as the first two terms in the analysis of the Zeeman interaction for $^3\Sigma$ NH (see equation (9.135)).

The only new term to be considered, therefore, is the third-order spin–orbit coupling term, whose matrix elements in a case (b) basis are evaluated as follows:

$$\langle \eta, \Lambda, N, S, J, M_J | C T^3(\boldsymbol{L}^2, \boldsymbol{N}) \cdot T^3(\boldsymbol{S}, \boldsymbol{S}, \boldsymbol{S}) | \eta', \Lambda'; N', S', J', M'_J \rangle$$
$$= \delta_{JJ'} \delta_{M_J M'_J} C (-1)^{N'+S+J'} \begin{Bmatrix} S & N & J \\ N' & S & 3 \end{Bmatrix}$$
$$\times \langle \eta, N, \Lambda \| T^3(\boldsymbol{L}^2, \boldsymbol{N}) \| \eta', N', \Lambda' \rangle \langle S \| T^3(\boldsymbol{S}, \boldsymbol{S}, \boldsymbol{S}) \| S' \rangle. \quad (9.154)$$

The reduced matrix element of $T^3(\boldsymbol{S}, \boldsymbol{S}, \boldsymbol{S})$, which is diagonal in S, is calculated in appendix 9.1 and is found to be given by

$$\langle S \| T^3(\boldsymbol{S}, \boldsymbol{S}, \boldsymbol{S}) \| S \rangle$$
$$= (1/4\sqrt{10})\{(2S-2)(2S-1)(2S)(2S+1)(2S+2)(2S+3)(2S+4)\}^{1/2}. \quad (9.155)$$

The reduced matrix element of $T^3(\boldsymbol{L}^2, \boldsymbol{N})$ is calculated by noting the following results:

$$\begin{aligned}
T_p^3(L^2, N) &= (-1)^{1-p}\sqrt{7}\sum_{p_1,p_2}\begin{pmatrix} 2 & 1 & 3 \\ p_1 & p_2 & -p \end{pmatrix} T_{p_1}^2(L^2) T_{p_2}^1(N) \\
&= (-1)^{1-p}\sqrt{7}\sum_{p_1,p_2}\begin{pmatrix} 2 & 1 & 3 \\ p_1 & p_2 & -p \end{pmatrix}\sum_q \mathcal{D}_{p_1,q}^{(2)}(\omega)^* T_q^2(L^2) T_{p_2}^1(N) \\
&= T_p^3(\mathcal{D}_{\cdot 0}^{(2)}(\omega)^*, N) T_0^2(L^2).
\end{aligned} \qquad (9.156)$$

We have confined attention to the $q = 0$ component in the third line of (9.156). We therefore require the following reduced matrix element:

$$\begin{aligned}
\langle N, \Lambda &\| T^3(\mathcal{D}_{\cdot 0}^{(2)}(\omega)^*, N) \| N', \Lambda' \rangle \\
&= \sqrt{7}(-1)^{3+N+N'}\sum_{N'',\Lambda''}\begin{Bmatrix} 2 & 1 & 3 \\ N' & N & N'' \end{Bmatrix} \\
&\quad \times \langle N, \Lambda \| \mathcal{D}_{\cdot 0}^{(2)}(\omega)^* \| N'', \Lambda'' \rangle \langle N'', \Lambda'' \| T^1(N) \| N', \Lambda' \rangle \\
&= \sqrt{7}(-1)^{3+N+N'}\begin{Bmatrix} 2 & 1 & 3 \\ N' & N & N \end{Bmatrix}(-1)^{N-\Lambda}\begin{pmatrix} N & 2 & N' \\ -\Lambda & 0 & \Lambda \end{pmatrix} \\
&\quad \times \{(2N+1)(2N'+1)N'(N'+1)(2N'+1)\}^{1/2}.
\end{aligned} \qquad (9.157)$$

Combining (9.154), (9.155), (9.156) and (9.157) gives the required final result, which is:

$$\begin{aligned}
\langle \eta, \Lambda; N, S, J, M_J | &CT^3(L^2, N) \cdot T^3(S, S, S) | \eta, \Lambda; N', S, J, M_J \rangle \\
&= \gamma_S(\sqrt{70}/4\sqrt{6})(-1)^{N'+S+J+1}\begin{Bmatrix} S & N & J \\ N' & S & 3 \end{Bmatrix}(-1)^{N+N'}\begin{Bmatrix} 2 & 1 & 3 \\ N' & N & N' \end{Bmatrix}(-1)^{N-\Lambda} \\
&\quad \times \begin{pmatrix} N & 2 & N' \\ -\Lambda & 0 & \Lambda \end{pmatrix}\{(2N+1)(2N'+1)\}^{1/2}\{N'(N'+1)(2N'+1) \\
&\quad \times (2S-2)(2S-1)(2S)(2S+1)(2S+2)(2S+3)(2S+4)\}^{1/2}.
\end{aligned} \qquad (9.158)$$

In this equation we have made use of the result

$$C\langle \eta, \Lambda | T_{q=0}^2(L^2) | \eta, \Lambda \rangle = (10/\sqrt{6})\gamma_S. \qquad (9.159)$$

Calculation of the energy levels as a function of the molecular parameters, and subsequent assignment of the laser magnetic resonance spectra, was a complicated exercise accomplished by an extensive process of trial and error. Moreover there were no other spectroscopic data available to assist in the analysis. Nevertheless Nelis, Brown and Evenson [69] were able to determine the following molecular constants for CH in the $v = 0$ level in its $^4\Sigma^-$ state:

$B_0 = 451\,138.434$ MHz, $D_0 = 44.427$ MHz, $\lambda = 2785.83$ MHz, $\gamma = -1.74$ MHz, $\gamma_S = 0.154$ MHz, $b_F = 106.56$ MHz, $c = 56.6$ MHz, $g_r = -0.000\,164$.

Notice that not all of the constants appearing in the effective Hamiltonian were actually determined in the fit; this is often the case in high-resolution spectroscopy, and it can cause some confusion for the reader! It is interesting to compare the above values of the constants with those obtained, either from *ab initio* calculations, or experimentally for CH in its $^2\Pi$ ground state, as described earlier in this section. The rotational constants determined for the two electronic states are very similar, but show that the vibrationally averaged bond length r_0 is actually slightly smaller for the excited quartet state. The Fermi contact constant b_F is much larger for the excited quartet state because it contains a direct contribution from spin density in a σ-type orbital possessing appreciable H 1s character. In the $^2\Pi$ ground state, where the unpaired electron occupies a π-type molecular orbital, the contact interaction arises only indirectly from spin-polarisation effects due to configuration interaction with excited electronic states. The so-called spin–spin constant λ arises partly from spin–orbit mixing of other electronic states possessing orbital angular momentum. The contribution from the ground state is, however, estimated to be 313 MHz and the observed λ value of 2786 MHz is therefore thought to arise predominantly from the first-order dipolar coupling of the electron spins. The spin–rotation constant γ has a very small value (-1.74 MHz) for the quartet state, compared with -771 MHz for the ground state; this suggests that spin–orbit mixing with excited $^4\Pi$ states is very weak.

9.9. $^4\Delta$, $^3\Phi$, $^2\Delta$ and $^6\Sigma^+$ states

9.9.1. Introduction

In recent years a number of diatomic first-row transition metal hydrides have been studied by FIR laser magnetic resonance. We will review the results, which include some of the most beautiful laser magnetic resonance spectra yet observed, without going into the details as deeply as we have earlier in this chapter.

It is helpful to remind ourselves of the ground state electron configurations (outside the KLM shells) of the first row transition metal elements, which are as follows.

Sc	Ti	V	Cr	Mn	Fe	Co	Ni	Cu	Zn
$3d^14s^2$	$3d^24s^2$	$3d^34s^2$	$3d^54s^1$	$3d^54s^2$	$3d^64s^2$	$3d^74s^2$	$3d^84s^2$	$3d^{10}4s^1$	$3d^{10}4s^2$
$^2D_{1/2}$	$^3F_2$	$^4F_{3/2}$	$^7S_3$	$^6S_{5/2}$	$^5D_4$	$^4F_{9/2}$	$^3F_4$	$^2S_{1/2}$	$^1S_0$

Cr and Cu are anomalous in this series in not possessing filled 4s shells in the ground state; in both cases the state with a $4s^2$ configuration is a low-lying excited state. In the axial field produced by bonding to a hydrogen atom, the five 3d orbitals of the first row transition elements split into one $3d\sigma$, two $3d\pi$ and two $3d\delta$ orbitals. Four diatomic hydrides have been studied so far by FIR laser magnetic resonance. Their ground state symmetries and multiplicities have been established, and we *may* choose to write their ground state electron configurations as follows.

$$\text{CrH:} \ (3d\sigma)^2(3d\pi)^2(3d\delta)^1(4s)^2 \quad X^6\Sigma^+$$
$$\text{FeH:} \ (3d\sigma)^2(3d\pi)^2(3d\delta)^3(4s)^2 \quad X^4\Delta_{7/2}$$
$$\text{CoH:} \ (3d\sigma)^2(3d\pi)^3(3d\delta)^3(4s)^2 \quad X^3\Phi_4$$
$$\text{NiH:} \ (3d\sigma)^2(3d\pi)^4(3d\delta)^3(4s)^2 \quad X^2\Delta_{5/2}$$

It should be noted, however, that the above single electron configurations are, in some cases, likely to be poor descriptions. This is because different electron configurations and spin states for a particular molecule are often very close in energy, so that extensive configurational mixing occurs. We have also not committed ourselves to the role of the H $1s$ orbital, which must be involved in a σ-bond, perhaps primarily with the metal $4s$ orbital.

We now describe briefly the FIR laser magnetic resonance studies of the four molecules listed above. FeH has also been studied by mid-infrared laser magnetic resonance, and NiH by microwave/optical double resonance; these investigations are discussed elsewhere.

9.9.2. CrH in the $X\,{}^6\Sigma^+$ ground state

The FIR laser magnetic resonance spectrum of CrH was obtained by Corkery, Brown, Beaton and Evenson [70] by flowing helium over crystals of $Cr(CO)_6$ and reacting the carbonyl with atomic hydrogen produced in a suitable discharge. A beautiful but nevertheless low-resolution scan is shown in figure 9.35; under higher resolution many of the lines shown are split into doublets due to proton hyperfine interaction. The lower rotational levels of CrH are illustrated in figure 9.36, and the spectrum shown in figure 9.35 arises from the $N = 3 \leftarrow 2$ rotation transition. Each rotational level is split by the spin–rotation interaction into a number of components (six for $N \geq 3$), characterised by different J quantum numbers.

Since each J level is split into $2J + 1$ components by the applied magnetic field, the rotational transition is split into many Zeeman components and the assignment

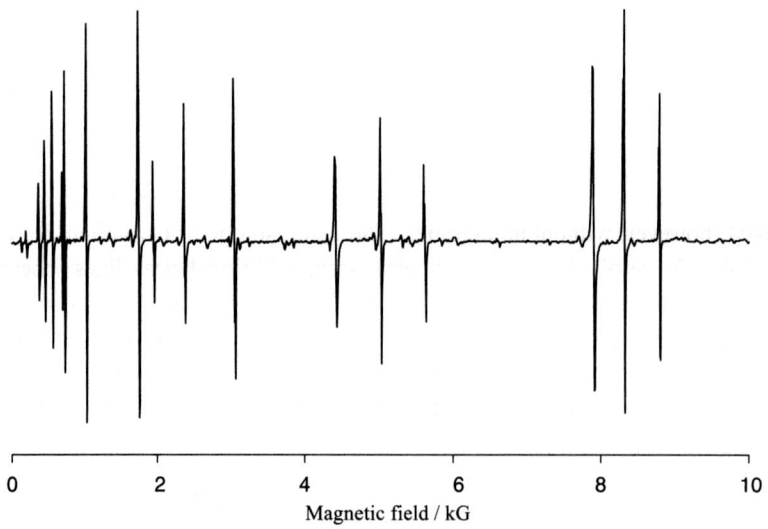

Figure 9.35. Laser magnetic resonance spectrum of CrH ($X\,{}^6\Sigma^+$) arising from the $N = 3 \leftarrow 2$ rotational transition. The laser frequency was 1100.8067 GHz and the spectrum was recorded in π polarisation [70].

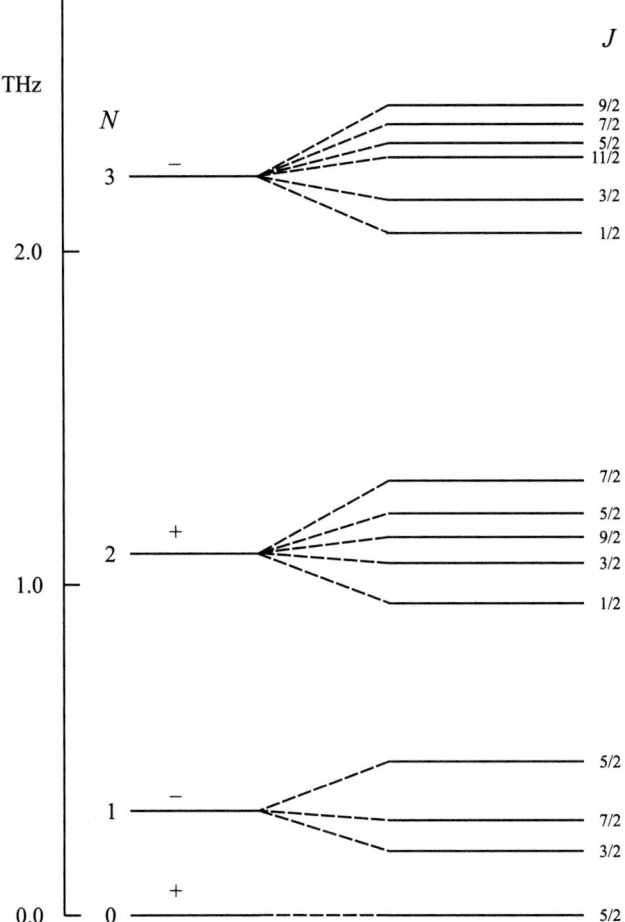

Figure 9.36. Lower rotational levels of CrH in the $X\,^6\Sigma^+$ state. The spin–rotation splittings are exaggerated for the sake of clarity.

of the spectrum shown in figure 9.35 is by no means readily apparent. It has been achieved, however, without ambiguities. The main isotope of chromium, ^{52}Cr, has a natural abundance of 83.8% and no nuclear spin. Additional weak lines appearing in the spectrum are due to ^{53}Cr, with a natural abundance of 9.6% and a spin of 3/2.

The effective Hamiltonian used to analyse the spectrum was that previously described for the $^4\Sigma$ state of CH, equation (9.148), but with the addition of two extra terms. The first is a fourth-order spin–orbit coupling term, described by Brown and Milton [71]:

$$\mathcal{H}_{so}^{(4)} = (1/12)\theta\left\{35S_z^4 - 30\boldsymbol{S}^2 S_z^2 + 25S_z^2 - 6\boldsymbol{S}^2 + 3\boldsymbol{S}^4\right\} = (\sqrt{70}/6)\theta \mathrm{T}_{q=0}^4(\boldsymbol{S}). \tag{9.160}$$

This term does not arise for states of quartet or lower multiplicity, but makes a significant contribution in the present case. It arises from fourth-order perturbation effects of spin–orbit coupling in the effective Hamiltonian (see chapter 7). Its matrix elements in a case (b) basis are given by

$$\langle \eta, \Lambda; N, S, J, M_J | \mathcal{H}_{so}^{(4)} | \eta, \Lambda'; N', S, J', M'_J \rangle$$
$$= \delta_{JJ'} \delta_{M_J M'_J} (\theta/24)(-1)^{N'+S+J} \begin{Bmatrix} S & N & J \\ N' & S & 4 \end{Bmatrix} (-1)^{N-\Lambda} \begin{pmatrix} N & 4 & N' \\ -\Lambda & 0 & \Lambda' \end{pmatrix}$$
$$\times \{(2N+1)(2N'+1)\}^{1/2} \{(2S+5)!/(2S-4)!\}^{1/2}. \tag{9.161}$$

The second additional term is an anisotropic correction to the electron spin Zeeman interaction, which in a molecule-fixed axis system is given by

$$\mathcal{H}_Z = g_l^e \mu_B (B_x S_x + B_y S_y). \tag{9.162}$$

Corkery, Brown, Beaton and Evenson [70] measured almost 500 resonances in CrH. The analysis proved complicated because the Zeeman effect was found to be highly nonlinear, a consequence of the decoupling of the electron spin from the molecular framework. This decoupling also means that magnetically tunable transitions become weaker in higher magnetic fields, or that strong transitions become less tunable. Nevertheless rotational transitions involving N up to 5 in the $v=0$ level were measured and analysed, providing the following values of the molecular parameters (in cm^{-1}):

$$B_0 = 6.131\,7456, \quad D_0 = 3.4951 \times 10^{-4}, \quad H_0 = 1.59 \times 10^{-8},$$
$$\gamma = 5.033\,23 \times 10^{-2}, \quad \gamma_D = 3.451 \times 10^{-6}, \quad \lambda = 0.232\,8341,$$
$$\lambda_D = 9.831 \times 10^{-6}, \quad \theta = -2.317 \times 10^{-3}.$$

In addition, three g-factors (unitless) and two proton hyperfine constants (MHz) were determined:

$$g_S = 2.001\,663, \quad g_r = -1.280 \times 10^{-3}, \quad g_l^e = -4.201 \times 10^{-3},$$
$$b_F = -34.80, \quad c = 41.81.$$

The net effect of the five unpaired electrons in the 3d orbitals is to produce a spherically symmetric charge distribution centred on the chromium atom. Using the internuclear distance R_0 calculated from B_0 above, the dipolar constant c is calculated to be 51.2 MHz which compares quite well with the measured value of 41.81 MHz. The Fermi contact constant b_F is found to be negative and small, consistent with a spin polarisation mechanism. The determined value of g_l^e is in remarkably good agreement with the value calculated from a relationship due to Curl [72], $g_l^e = -\gamma/2B = -0.414 \times 10^{-2}$.

Finally we note that it was possible to determine the value of the fourth-order spin–spin splitting constant θ; the value of this parameter depends upon the spin–orbit mixing of excited electronic states with the ground state.

9.9.3. FeH in the $X\,^4\Delta$ ground state

A rotational transition, $J = 11/2 \leftarrow 9/2$, in the ground $^4\Delta_{7/2}$ state of FeH has been reported by Beaton, Evenson, Nelis and Brown [73] using FIR laser magnetic resonance. Their spectrum shows three $\Delta M_J = 0$ (π polarisation) Zeeman components, two more which involve $M_J = 3/2$ and $1/2$ occurring outside the range of the electromagnet. The predominant isotope of iron is ^{56}Fe, with a natural abundance of 92%, but weak lines from ^{54}Fe (6%) and ^{57}Fe are also observed. A marked doublet splitting in the spectrum is due to Λ-doubling, which is unusually large for a molecule in a Δ electronic state. A detailed analysis of the spectrum is still awaited, but the observed g-value confirms the electronic state to be $^4\Delta_{7/2}$.

9.9.4. CoH in the $X\,^3\Phi$ ground state

The FIR laser magnetic resonance spectrum of CoH was obtained by Beaton, Evenson and Brown [74] using the reaction between hydrogen atoms and cobalt carbonyl vapour. A very small part of the spectrum is presented in figure 9.37, showing the exquisite octet hyperfine structure of the ^{59}Co nucleus, which has a nuclear spin of 7/2 and is in 100% natural abundance. The octets illustrated are two Zeeman components of the $J = 6 \leftarrow 5$ rotational transition in the $^3\Phi_4$ fine structure state, with $v = 0$. High-resolution recordings are presented in figure 9.38, showing two cobalt hyperfine

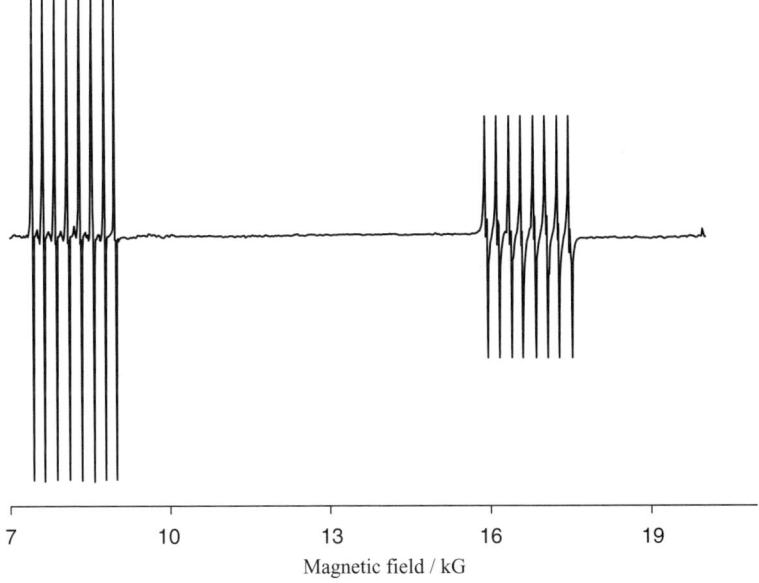

Figure 9.37. Section of the FIR laser magnetic resonance spectrum of CoH ($^3\Phi_4$) arising from the $J = 6 \leftarrow 5$ rotation transition, at low resolution, showing the ^{59}Co hyperfine structure [74].

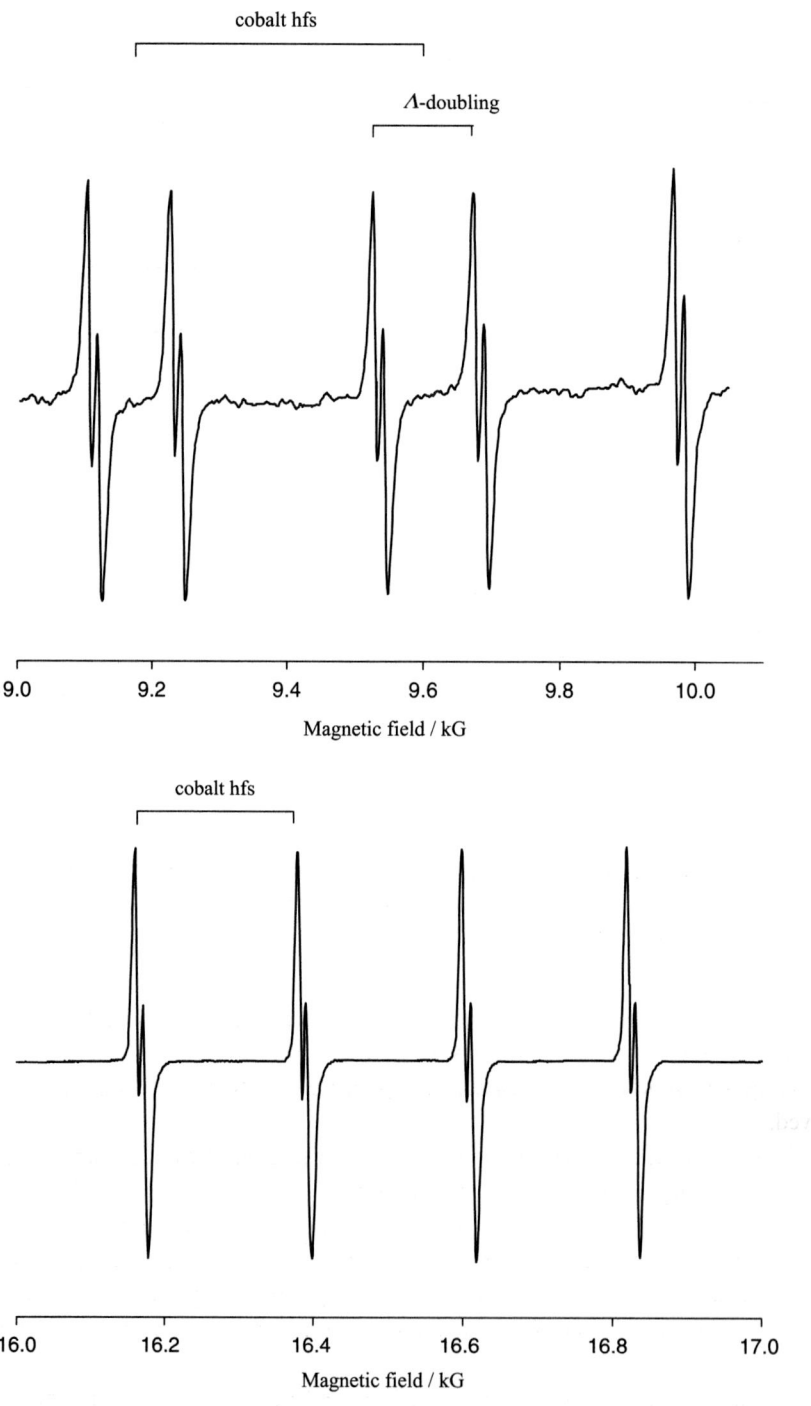

Figure 9.38. *Top*: section of the FIR LMR spectrum of CoH recorded with the 138.3 μm laser line, arising from the $J = 5 \leftarrow 4$ transition in the $\Omega = 3$ spin component. The very small splitting is due to proton hyperfine coupling [74]. *Bottom*: $J = 5 \leftarrow 4$ transition in the $\Omega = 4$ spin component (138.3 μm laser line).

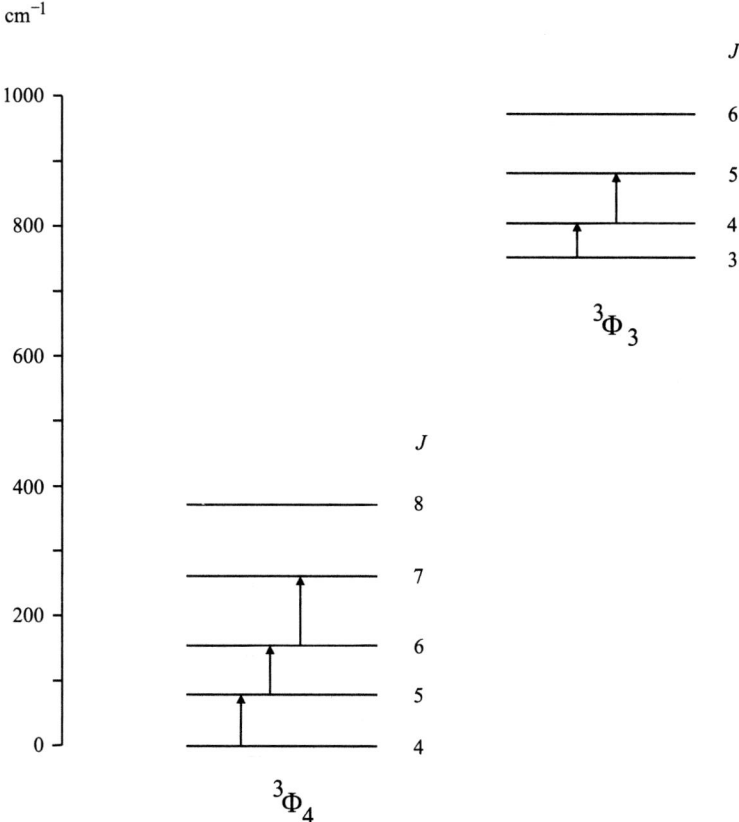

Figure 9.39. Fine-structure components, rotational levels, and observed rotational transitions for CoH in its $v = 0$ X $^3\Phi_4$ and $^3\Phi_3$ fine-structure components. The remaining $\Omega = 2$ spin component lies at about 1500 cm^{-1} above the lowest one.

lines, split into doublets from Λ-doubling, with a further very small doublet splitting arising from proton hyperfine interaction. Figure 9.39 shows an energy level diagram including the two lowest fine-structure components, and the rotational transitions observed.

This spectrum is a spectroscopist's dream because apart from its beauty, its assignment is straightforward. The Zeeman effect is essentially linear, and the g-factor identifies both the fine-structure state ($^3\Phi_4$ or $^3\Phi_3$ in different cases) and the J values of the levels involved, the latter being confirmed by the number of second-order Zeeman components. The hyperfine and Λ-doublet splittings are readily apparent. Many resonances arising from transitions within the two lowest fine-structure states were observed and assigned.

The effective Hamiltonian used to analyse the CoH spectrum was similar to that described elsewhere. It may be summarised in the form

$$\mathcal{H}_{\text{eff}} = \mathcal{H}_{\text{rot}} + \mathcal{H}_{\text{so}} + \mathcal{H}_{\text{sr}} + \mathcal{H}_{\text{ss}} + \mathcal{H}_{\ell d} + \mathcal{H}_{\text{hfs}} + \mathcal{H}_{Z}. \tag{9.163}$$

The rotational term, including centrifugal distortion, is

$$\mathcal{H}_{\text{rot}} = B_v N^2 - D_v N^4. \tag{9.164}$$

For the spin–orbit coupling we again confine attention to the molecule-fixed axial component:

$$\mathcal{H}_{\text{so}} = A_v L_z S_z. \tag{9.165}$$

The spin–rotation interaction, including the correction for centrifugal distortion is

$$\mathcal{H}_{\text{sr}} = \gamma_v \boldsymbol{N} \cdot \boldsymbol{S} + \gamma_{vD}(\boldsymbol{N} \cdot \boldsymbol{S})N^2, \tag{9.166}$$

and the spin–spin interaction is, as before,

$$\mathcal{H}_{\text{ss}} = (2/3)\lambda_v \left(3S_z^2 - \boldsymbol{S}^2\right). \tag{9.167}$$

Finally we come to the Λ-doubling which was modelled by adding the term

$$\mathcal{H}_{\ell d} = (1/2) q_{\Phi v}(N_+^6 + N_-^6). \tag{9.168}$$

The addition of this term is more than intuition; the Λ-doubling in a Π state arises from second-order mixing with a Σ state, Λ-doubling in a Δ state involves fourth-order mixing with, successively, a Π and a Σ state, so that in a Φ state the Λ-doubling must involve sixth-order spin–orbit coupling to a Δ, Π and Σ state in turn.

The Zeeman interaction and nuclear hyperfine terms have to be added to this effective Hamiltonian. The Zeeman terms are, as in previous examples,

$$\mathcal{H}_Z = (g_L + g_r)\mu_B B_Z L_Z + g_S \mu_B B_Z S_Z - g_r \mu_B B_Z N_Z - g_N \mu_N B_Z I_Z \\ + g_l \mu_B (B_X S_X + B_Y S_Y). \tag{9.169}$$

All of these terms are written in the space-fixed axis system, with Z being the direction of the applied magnetic field, except for the last (the Zeeman anisotropy term), which is expressed in the molecule-fixed system. It is, of course, necessary to transform from space- to molecule-fixed axes in order to evaluate the matrix elements in a case (a) basis.

The magnetic hyperfine interaction was expressed in the molecule-fixed axis system by Beaton, Evenson and Brown [74] in the Frosch and Foley [25] form:

$$\mathcal{H}_{\text{hfs}} = a I_z L_z + b \boldsymbol{I} \cdot \boldsymbol{S} + c I_z S_z. \tag{9.170}$$

Finally, the ^{59}Co nucleus has an electric quadrupole moment so that the quadrupole interaction may be written in the normal way for calculation in a case (a) basis as

$$\mathcal{H}_Q = -e\text{T}^2(\boldsymbol{Q}) \cdot \text{T}^2(\nabla \boldsymbol{E}). \tag{9.171}$$

Even if we omit the nuclear hyperfine and Zeeman terms, the zero-field problem which must be solved is fairly complicated, in that the three fine-structure components are mixed by one or more of the terms shown in (9.163). The matrix of the effective Hamiltonian for a given value of J has the following form.

	$^3\Phi_4$	$^3\Phi_3$	$^3\Phi_2$
$^3\Phi_4$	m_{11}	m_{12}	m_{13}
$^3\Phi_3$	m_{21}	m_{22}	m_{23}
$^3\Phi_2$	m_{31}	m_{32}	m_{33}

The full details of the matrix elements are as follows.

$$m_{11} = 3A + 2\gamma + (2/3)\lambda + B[J(J+1) - 6] - D\{[J(J+1) - 6]^2 \\ + 2[J(J+1) - 12]\} + \gamma_D J(J+1)$$

$$m_{22} = -2\gamma - (4/3)\lambda + B[J(J+1) + 2] - D\{[J(J+1) + 2]^2 + 4[J(J+1) - 9]\} \\ - \gamma_D[4J(J+1) - 14] \pm (1/2)q_\Phi[J(J+1)][J(J+1) - 2][J(J+1) - 6]$$

$$m_{33} = -3A - 4\gamma + (2/3)\lambda + B[J(J+1) + 6] - D\{[J(J+1) + 6]^2 \\ + 2[J(J+1) - 6]\} - \gamma_D[5J(J+1) + 18]$$

$$m_{12} = -[2J(J+1) - 24]^{1/2}\{[B - (1/2)\gamma] - D[2J(J+1) - 4] \\ - (1/2)\gamma_D[J(J+1) - 2]\}$$

$$m_{13} = -\{[J(J+1) - 12][J(J+1) - 6]\}^{1/2}\{2D + \gamma_D \\ \mp (1/2)q_\Phi J(J+1)[J(J+1) - 2]\}$$

$$m_{23} = -\{2[J(J+1) - 6]\}^{1/2}\{[B - (1/2)\gamma] - D[2J(J+1) + 8] \\ - (1/2)\gamma_D[J(J+1) + 10]\}.$$

The upper and lower signs refer to e and f levels respectively. In the presence of a magnetic field, J is no longer a good quantum number so that a series of 3×3 matrices similar to that above are mixed by the Zeeman terms to form an infinite matrix. This is truncated at some suitable point, and diagonalised as a function of the magnetic field strength. The result is a complex system of Zeeman components and transitions; not surprisingly, 511 FIR laser magnetic resonance lines were observed. Quantitative assignment is not difficult with appropriate computer routines and Beaton, Evenson and Brown [74] were able to determine the values of the parameters appearing in the effective Hamiltonian. Their results (in cm^{-1}) were as follows:

$$B_0 = 7.313\,713, \quad D_0 = 5.4545 \times 10^{-4}, \quad A_0 = -221.5, \quad \gamma_0 = -1.2134,$$
$$\gamma_D = 0.004\,339, \quad q_\Phi = 5.560 \times 10^{-7}.$$

In addition, the g-factors (unitless) and hyperfine constants (in MHz) were determined:

$$g_S = 1.942\,42, \quad g_L = 1.025\,775, \quad g_r = -0.020\,689, \quad a = 621.01,$$
$$(b+c) = -320.08, \quad b = 136.2, \quad eq_0Q = -92.5.$$

The spin–spin constant λ does not appear in the above list; it was not determinable in the analysis, primarily because resonances from the third spin component ($^3\Phi_2$) were not observed.

The parameter values listed above may be regarded as benchmarks for *ab initio* calculations, although some qualitative remarks may be made. The Fermi constant b_F has the value $(b + c/3) = -15.9$ MHz; as in the case of CrH discussed previously, this small negative value indicates that the unpaired electrons occupy orbitals which are essentially $3d$ orbitals. Spin density at the proton nucleus can arise only through configuration interaction with other excited electronic states. The orbital hyperfine parameter a can be used to determine the expectation value of the operator $\sum_i r_i^{-3}$, where the summation is over the electrons which are responsible for the orbital angular momentum. From the observed value of a this expectation value is 3.322×10^{31} m^{-3}, compared with the theoretical value of 4.528×10^{31} for an electron in a pure $3d$ orbital on cobalt [75]. One concludes that the $3d\pi$ and $3d\delta$ molecular orbitals in CoH are similar to $3d$ atomic orbitals, but slightly more diffuse. The interpretation of other parameters, for example, the Λ-doubling and spin-rotation constants, requires knowledge of nearby excited electronic states.

9.9.5. NiH in the $X\,^2\Delta$ ground state

The NiH radical has been studied quite thoroughly, by FIR laser magnetic resonance as described here [76], but also by microwave/optical double resonance and by mid-infrared laser magnetic resonance, discussed elsewhere.

In the far-infrared experiments the NiH radicals were again generated by reaction of the hydrogen atoms with the metal carbonyl vapour. Four rotational transitions of NiH in its $X\,^2\Delta$ state, two in each of the two fine-structure states, were observed. These transitions are identified in figure 9.40; a total of 327 magnetic resonances was observed, complexity arising from the presence of Λ-doubling as well as the removal of the $2J + 1$ degeneracy of each J level by the applied magnetic field. A further complication is that nickel contains the following isotopes, with their natural abundances:

$$^{58}\text{Ni}(67.9\%), \quad ^{60}\text{Ni}(26.2\%), \quad ^{61}\text{Ni}(1.2\%), \quad ^{62}\text{Ni}(3.7\%), \quad ^{64}\text{Ni}(1.1\%)$$

The sensitivity of the FIR laser magnetic resonance experiments was high enough to permit observations of all of these isotopes, including quartet hyperfine structure from ^{61}Ni.

The zero-field effective Hamiltonian used to analyse the spectrum was similar to those described earlier for other transition metal hydrides, except for a much more complex set of terms to describe the Λ-doubling. It was expressed in a form suitable for calculations in a case (a) basis as follows:

$$\begin{aligned}
\mathcal{H}_{\text{eff}} = {}& B\mathbf{N}^2 - D\mathbf{N}^4 + AL_zS_z + (\gamma + \gamma_D\mathbf{N}^2)\mathbf{N}\cdot\mathbf{S} + (1/2)q_\Delta(J_+^4 + J_-^4) \\
& - (1/2)(p_\Delta + 4q_\Delta)(J_+^3 S_+ + J_-^3 S_-) + (1/2)(o_\Delta + 3p_\Delta + 6q_\Delta)(J_+^2 S_+^2 + J_-^2 S_-^2) \\
& - (1/2)(n_\Delta + 2o_\Delta + 3p_\Delta + 4q_\Delta)(J_+ S_+^3 + J_- S_-^3) \\
& + (1/2)(m_\Delta + n_\Delta + o_\Delta + p_\Delta + q_\Delta)(S_+^4 + S_-^4). \quad\quad (9.172)
\end{aligned}$$

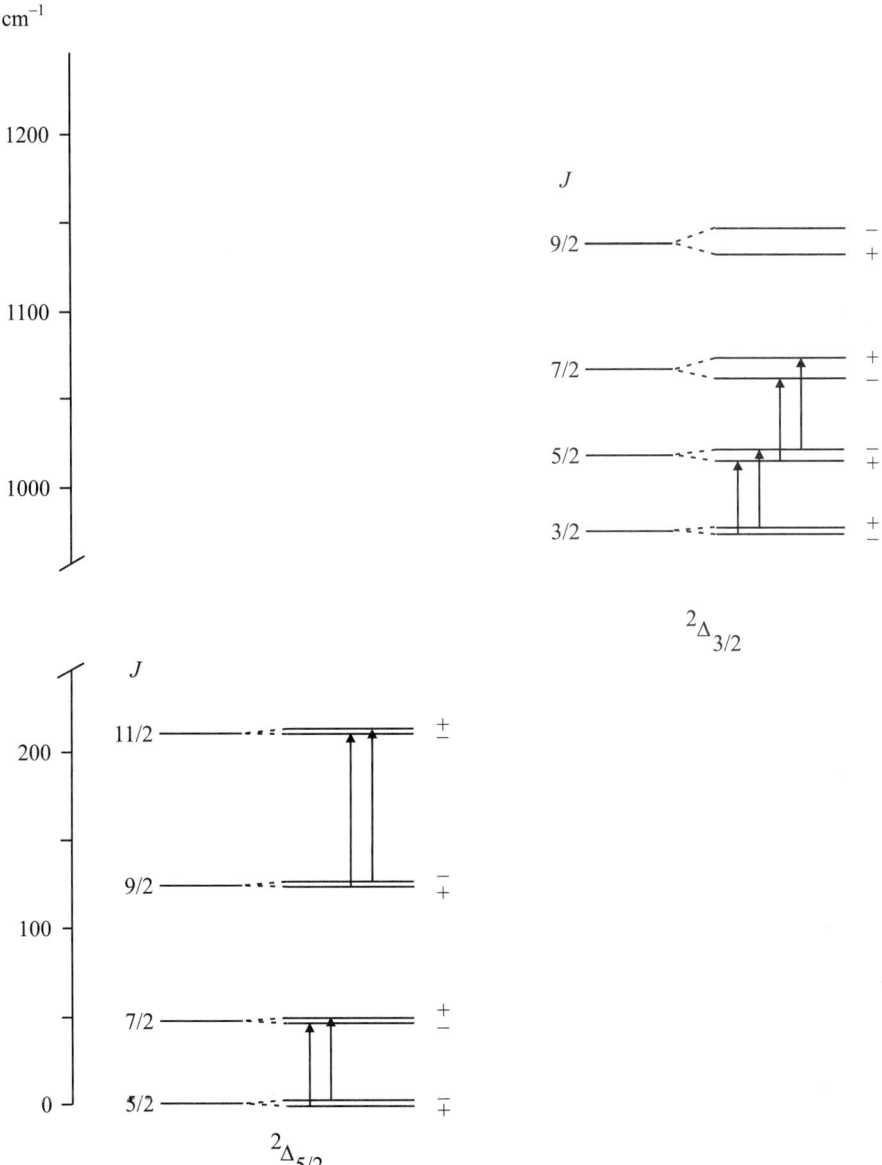

Figure 9.40. Lower rotational levels in the two fine-structure components of the $X\,^2\Delta$ ground state of NiH, and the observed transitions [76]. The Λ-doubling in both states has been exaggerated for the sake of clarity, and is not to scale. In particular, the Λ-doubling in the $^2\Delta_{5/2}$ state is actually very small.

In the analysis of the CoH spectrum only one sixth-order term was included in the effective Hamiltonian to describe the Λ-doubling, although a complete representation would be similar to that shown in equation (9.172), with even more terms. The reasons for the differences between (9.172) and (9.168) are twofold; the Λ-doubling in NiH is much larger than in CoH, and much more experimental data were obtained for NiH. The matrix elements of (9.172) have been calculated by Brown, Cheung and Merer [77] and the results may be summarised in the following matrix for a given J value.

	$^2\Delta_{5/2}$	$^2\Delta_{3/2}$
$^2\Delta_{5/2}$	m_{11}	m_{12}
$^2\Delta_{3/2}$	m_{21}	m_{22}

The matrix elements for a given value of J are as follows

$$m_{11} = A + \gamma/2 + \gamma_D + B[J(J+1) - 7/4] - D[J(J+1) + 1/4][J(J+1) - 11/4]$$

$$m_{22} = -A - 3\gamma/2 - \gamma_D[2J(J+1) + 3/2] + B[J(J+1) + 9/4] - D[J(J+1) + 1/4][J(J+1) + 21/4] \mp (1/2)(p_\Delta + 4q_\Delta)(J+1/2)[J(J+1) - 3/4]$$

$$m_{12} = -[J(J+1) - 15/4]^{1/2}\{B - \gamma/2 - (\gamma_D/2)[J(J+1) + 5/4] - 2D[J(J+1) + 1/4] \mp (q_\Delta/2)(J+1/2)[J(J+1) - 3/4]\}$$

$$m_{21} = m_{12}$$

In these expressions, the upper and lower sign choices refer to e and f levels respectively, of opposite parity.

Terms describing the Zeeman and magnetic hyperfine interactions must be added to (9.172). The complete Zeeman Hamiltonian for a molecule in a $^{2S+1}\Delta$ state is given by Nelis, Beaton, Evenson and Brown [76] as

$$\mathcal{H}_Z = (g_L + g_r)\mu_B B_Z L_Z + g_S\mu_B B_Z S_Z - g_r\mu_B B_Z N_Z + (1/2)g_l^e\mu_B(B_+S_- + B_-S_+)$$
$$- (1/2)g_{rD}\mu_B(B_+J_-J_+J_- + B_-J_+J_-J_+) + (1/2)g_{lD}\mu_B(B_+S_-J_+J_- + B_-S_+J_-J_+) - (1/2)g_{rS}\mu_B(B_+J_-S_+S_- + B_-J_+S_-S_+)$$
$$+ (1/2)g_{lS}\mu_B(B_+S_-S_+S_- + B_-S_+S_-S_+) + (1/2)g'_{rD}\mu_B(B_+J_+^3 + B_-J_-^3)$$
$$+ (1/2)g'_{lD}\mu_B(B_+J_+^2S_+ + B_-J_-^2S_-) + (1/2)g'_{rS}\mu_B(B_+J_+S_+^2 + B_-J_-S_-^2)$$
$$+ (1/2)g'_{lS}\mu_B(B_+S_+^3 + B_-S_-^3). \tag{9.173}$$

This expression is written in the space-fixed axis system, but transformations to the molecule-fixed system are necessary to evaluate the matrix elements in a case (a) basis. We have met the first four terms previously, but the remaining eight terms are new. The first four are parity-independent while the remainder are parity-dependent and implicitly couple basis states with $\Delta\Lambda = +4$ or -4, so that they give diagonal contributions in a Δ state. The last two terms in (9.173) have zero diagonal terms for a doublet spin state, and the matrix elements of the terms involving g_{lS} are indistinguishable from those involving g_l^e. Similarly matrix elements involving g_{rS} are indistinguishable from those involving g_r. Consequently there are 8 independent g factors for a molecule in a $^2\Delta$ state, rather than 12 as equation (9.173) implies. Nelis, Beaton, Evenson and Brown [76] were able to determine six of the possible eight g-factors.

Finally we add the magnetic hyperfine Hamiltonian which is required to analyse the proton hyperfine structure. It is given, in the molecule-fixed axis system, as [76]

$$\mathcal{H}_{\text{hfs}} = aI_zL_z + b\mathbf{I}\cdot\mathbf{S} + cI_zS_z - (1/2)d_\Delta(J_+^2 I_+S_+ + J_-^2 I_-S_-), \tag{9.174}$$

where the first three terms are the familiar Frosch and Foley [25] terms whilst the fourth term represents a parity-dependent hyperfine interaction for a molecule in a Δ state [78]. We will discuss this term further in chapter 11 on double resonance spectroscopy.

In analysing the FIR laser magnetic resonance spectrum, Nelis, Beaton, Evenson and Brown [76] fitted data for the different isotopic forms of NiH simultaneously, using appropriate scaling factors for the different parameters [67]. They encountered one interesting problem in this procedure, in that it proved necessary to take account of non-adiabatic effects as described by Watson [79] when scaling the rotational constant B_e, the spin–rotation constant γ and the spin–orbit coupling constant A_e. The scaling parameter used for B_e and γ was

$$\mu^{-1}\left[1 + \frac{m_e}{m_{\text{Ni}}}\Delta_{01}(\text{Ni}) + \frac{m_e}{m_{\text{H}}}\Delta_{01}(\text{H})\right], \tag{9.175}$$

where μ is the reduced mass of the diatomic molecule, m_e is the mass of the electron and m_{Ni} and m_{H} are the masses of the Ni and H atoms, respectively. For A_e a mass-dependence of similar form was used except that the factor outside the bracket was μ^0, in which case the Born–Oppenheimer correction factor was referred to as $\Delta_{0A}(\text{Ni})$.

A remarkably large proportion of the constants in the effective Hamiltonian were determined from the analysis of the FIR laser magnetic resonance spectrum, supplemented by data from other spectroscopic regions which allowed equilibrium values of some of the constants to be determined. The following rotational and vibrational constants are listed by Nelis, Beaton, Evenson and Brown [76] (in cm^{-1}):

$$B_0 = 7.753\,016, \quad \alpha_e = 0.256\,403, \quad D_0 = 5.5439 \times 10^{-4}, \quad \beta_e = -2.73 \times 10^{-5},$$
$$A_0 = -491.4849, \quad \gamma_0 = 1.3370, \quad \gamma_D = 6.781 \times 10^{-3}.$$

The Λ-doubling and hyperfine constants are listed in MHz:

$$p_\Delta + 4q_\Delta = 188.614, \quad (p_\Delta + 4q_\Delta)_D = 0.108\,11, \quad q_\Delta = -0.4647,$$
$$b(\text{H}) = -81.23, \quad d_\Delta(\text{H}) = 0.757, \quad h_{5/2}(\text{H}) = 40.0, \quad h_{3/2}(\text{H}) = 51.1,$$
$$h_{5/2}(^{61}\text{Ni}) = 441.2, \quad eq_0Q(^{61}\text{Ni}) = 42.3.$$

In the values listed above, the h constants are the axial components of the total hyperfine interaction, given by

$$h_{3/2} = 2a - (1/2)(b+c), \quad h_{5/2} = 2a + (1/2)(b+c). \tag{9.176}$$

Finally, the six determined g factors have the values:

$$g_L = 1.039\,759, \quad g_S = 1.837\,08, \quad g_r = -4.019 \times 10^{-2}, \quad g_l^e = 0.8351,$$
$$g_{lD} = 1.58 \times 10^{-3}, \quad g'_{lD} = 1.425 \times 10^{-3}.$$

One now asks how these parameters can be related to the energy levels and electronic structure of the radical. Knowledge of the excited electronic states is at present too sparse for a full quantitative assessment, but the values of the Λ-doubling parameters can be sensibly rationalised. So far as the magnetic hyperfine constants are concerned, the a parameter gives an expectation value $\langle 1/r^3 \rangle = 2.89 \times 10^{29}$ m^{-3}, where r is the separation of the open shell electron from the proton in NiH. The determined value of B_e gives $r_e^{-3} = 3.145 \times 10^{29}$ m^{-3}, which is satisfactorily close to the value determined from a above, and confirms that the distribution of the open shell electron is very well described by an orbital centred on the nickel atom.

The g-factors also point to the problem of excited state mixing, the values of g_L and g_S in particular being too far from the free electron values for comfort. The single state effective Hamiltonian, derived by perturbation theory, may be inadequate in molecules where there are several close-lying electronic states which are appreciably mixed.

Appendix 9.1. Evaluation of the reduced matrix element of T^3 (S, S, S)

In our discussion of the FIR laser magnetic resonance spectrum of CH in its $a\,^4\Sigma^-$ state we encountered the reduced matrix element of $T^3(S, S, S)$. The result was presented in equation (9.155), which we now derive. First we note that, by the Wigner–Eckart theorem, the following result applies in the molecule-fixed coordinate system with $q = 0$:

$$\langle S, \Sigma | T_0^3(S, S, S) | S, \Sigma \rangle = (-1)^{S-\Sigma} \begin{pmatrix} S & 3 & S \\ -\Sigma & 0 & \Sigma \end{pmatrix} \langle S \| T^3(S, S, S) \| S \rangle. \quad (9.177)$$

The 3-j symbol in (9.177) may be evaluated using a recursion relation, such as that given by Edmonds [80] in his equation (3.7.13). One then finds that

$$\begin{aligned}
&\langle S, \Sigma | T_0^3(S, S, S) | S, \Sigma \rangle \\
&= (-1)^{S-\Sigma}(-1)^{S-\Sigma-1} \\
&\quad \times \frac{4\Sigma\{3S(S+1) - 5\Sigma^2 - 1\}}{\{(2S+4)(2S+3)(2S+2)(2S+1)(2S)(2S-1)(2S-2)\}^{1/2}} \\
&\quad \times \langle S \| T^3(S, S, S) \| S \rangle. \quad (9.178)
\end{aligned}$$

The problem therefore reduces to that of evaluating the left-hand side of equation (9.178). We first decompose the third-rank tensor by making use of the result

$$T_p^3(S, S, S) = (-1)^{p-1}\sqrt{7} \sum_{p_1, p_2} \begin{pmatrix} 1 & 2 & 3 \\ p_1 & p_2 & -p \end{pmatrix} T_{p_1}^1(S) T_{p_2}^2(S, S) \quad (9.179)$$

where, since $p = 0$ in our case, $p_2 = 0$ or ± 1. We now make use of the further decomposition result,

$$T_q^2(S, S) = (-1)^q \sqrt{5} \sum_{q_1, q_2} \begin{pmatrix} 1 & 1 & 2 \\ q_1 & q_2 & -q \end{pmatrix} T_{q_1}^1(S) T_{q_2}^1(S), \quad (9.180)$$

and we are back on the more familiar ground of first-rank tensors. We remind ourselves of the relationships between the spherical tensor and cartesian components,

$$T_0^1(\boldsymbol{S}) = S_z, \quad T_1^1(\boldsymbol{S}) = -\frac{1}{\sqrt{2}}(S_x + iS_y), \quad T_{-1}^1(\boldsymbol{S}) = \frac{1}{\sqrt{2}}(S_x - iS_y), \quad (9.181)$$

which we will need in evaluating (9.180). It will also be necessary to take account of the commutation relations between the components of $\boldsymbol{S}$, which are:

$$S_x S_y - S_y S_x = iS_z, \quad S_y S_z - S_z S_y = iS_x, \quad S_z S_x - S_x S_z = iS_y. \quad (9.182)$$

The three required components of $T_q^2(\boldsymbol{S}, \boldsymbol{S})$ are now evaluated using (9.180) with $q = 0$ and ± 1. The results are as follows:

$$T_0^2(\boldsymbol{S},\boldsymbol{S}) = \sqrt{5}\left\{ \begin{pmatrix} 1 & 1 & 2 \\ 1 & -1 & 0 \end{pmatrix} T_1^1(\boldsymbol{S})T_{-1}^1(\boldsymbol{S}) \right.$$
$$+ \begin{pmatrix} 1 & 1 & 2 \\ -1 & 1 & 0 \end{pmatrix} T_{-1}^1(\boldsymbol{S})T_1^1(\boldsymbol{S}) + \begin{pmatrix} 1 & 1 & 2 \\ 0 & 0 & 0 \end{pmatrix} T_0^1(\boldsymbol{S})T_0^1(\boldsymbol{S}) \left. \right\}$$
$$= \frac{1}{\sqrt{6}}\{3S_z^2 - \boldsymbol{S}^2\}. \quad (9.183)$$

$$T_1^2(\boldsymbol{S},\boldsymbol{S}) = -\sqrt{5}\left\{ \begin{pmatrix} 1 & 1 & 2 \\ 1 & 0 & -1 \end{pmatrix} T_{-1}^1(\boldsymbol{S})T_0^1(\boldsymbol{S}) + \begin{pmatrix} 1 & 1 & 2 \\ 0 & 1 & -1 \end{pmatrix} T_0^1(\boldsymbol{S})T_{-1}^1(\boldsymbol{S}) \right\}$$
$$= -\frac{1}{2}\{S_x S_z + iS_y S_z + S_z S_x + iS_z S_y\}. \quad (9.184)$$

$$T_{-1}^2(\boldsymbol{S},\boldsymbol{S}) = -\sqrt{5}\left\{ \begin{pmatrix} 1 & 1 & 2 \\ 1 & 0 & -1 \end{pmatrix} T_1^1(\boldsymbol{S})T_0^1(\boldsymbol{S}) + \begin{pmatrix} 1 & 1 & 2 \\ 0 & 1 & -1 \end{pmatrix} T_0^1(\boldsymbol{S})T_1^1(\boldsymbol{S}) \right\}$$
$$= \frac{1}{2}\{S_x S_z - iS_y S_z + S_z S_x - iS_z S_y\}. \quad (9.185)$$

These results are now to be inserted into equation (9.179), which takes the explicit form

$$T_0^3(\boldsymbol{S},\boldsymbol{S},\boldsymbol{S}) = -\sqrt{7}\left\{ \begin{pmatrix} 1 & 2 & 3 \\ 1 & -1 & 0 \end{pmatrix} T_1^1(\boldsymbol{S})T_{-1}^2(\boldsymbol{S},\boldsymbol{S}) + \begin{pmatrix} 1 & 2 & 3 \\ 1 & 1 & 0 \end{pmatrix} T_{-1}^1(\boldsymbol{S})T_1^2(\boldsymbol{S},\boldsymbol{S}) \right.$$
$$+ \begin{pmatrix} 1 & 2 & 3 \\ 0 & 0 & 0 \end{pmatrix} T_0^1(\boldsymbol{S})T_0^2(\boldsymbol{S},\boldsymbol{S}) \left. \right\}. \quad (9.186)$$

The required result is now obtained by substituting the first- and second-rank tensor components from (9.181) which yields

$$T_0^3(\boldsymbol{S},\boldsymbol{S},\boldsymbol{S}) = -\frac{1}{\sqrt{10}} S_z \{3\boldsymbol{S}^2 - 5S_z^2 - 1\}. \quad (9.187)$$

The left-hand side of (9.178) is therefore

$$\langle S, \Sigma | T_0^3(\boldsymbol{S},\boldsymbol{S},\boldsymbol{S}) | S, \Sigma \rangle = -\frac{1}{\sqrt{10}} \Sigma \{3S(S+1) - 5\Sigma^2 - 1\} \quad (9.188)$$

and we obtain the final result:

$$\langle S \| T^3(\boldsymbol{S}, \boldsymbol{S}, \boldsymbol{S}) \| S \rangle$$
$$= (1/4\sqrt{10})\{(2S+4)(2S+3)(2S+2)(2S+1)(2S)(2S-1)(2S-2)\}^{1/2}. \quad (9.189)$$

The derivation above is not the only way to obtain the required result, but it is straightforward, if somewhat tedious. The reduced matrix element of the fourth-rank spin tensor, $T^4(\boldsymbol{S}, \boldsymbol{S}, \boldsymbol{S}, \boldsymbol{S})$, which can arise in the analysis of higher spin states, is obtained by further use of the recursion relationship given by Edmonds [80]. See also the general expression given in equation (5.134) of chapter 5.

References

[1] G.C. Dousmanis, T.M. Sanders and C.H. Townes, *Phys. Rev.*, **100**, 1735 (1955).
[2] J.A. Seeger, *Microwave Theory, Components and Devices*, Prentice-Hall, New Jersey, 1986.
[3] A. Carrington, D.H. Levy and T.A. Miller, *Rev. Sci. Instr.*, **38**, 1183 (1967).
[4] A. Carrington, G.N. Currie, D.H. Levy and T.A. Miller, *J. Chem. Phys.*, **50**, 2726 (1968).
[5] K.M. Evenson, *Faraday Disc.*, **71**, 7 (1981).
[6] K.M. Evenson, H.P. Broida, J.S. Wells, R.J. Mahler and M. Mizushima, *Phys. Rev. Lett.*, **21**, 1038 (1968).
[7] H.E. Radford, *Phys. Rev.*, **122**, 114 (1961); **126**, 1035 (1962).
[8] A. Carrington, D.H. Levy and T.A. Miller, *Proc. R. Soc. Lond.*, **A293**, 108 (1966); *Trans. Faraday Soc.*, **62**, 2994 (1966).
[9] H. Uehara, *Mol. Phys.*, **21**, 407 (1971).
[10] J.M. Brown and H. Uehara, *Mol. Phys.*, **24**, 1169 (1972).
[11] A. Carrington, D.H. Levy and T.A. Miller, *Adv. Chem. Phys.*, **18**, 149 (1970).
[12] S. Saito, *J. Chem. Phys.*, **53**, 2544 (1970).
[13] C.A. Arrington, A.M. Falick and R.J. Myers, *J. Chem. Phys.*, **55**, 909 (1971).
[14] T.A. Miller, *J. Chem. Phys.*, **54**, 330 (1971); **54**, 1658 (1971).
[15] A. Scalabrin, R.J. Saykally, K.M. Evenson, H.E. Radford and M. Mizushima, *J. Mol. Spectrosc.*, **89**, 344 (1981).
[16] A.H. Curran, R.G. MacDonald, A.J. Stone and B.A. Thrush, *Proc. R. Soc. Lond.*, **A332**, 355 (1973).
[17] A. Carrington, D.H. Levy and T.A. Miller, *J. Chem. Phys.*, **47**, 3801 (1967).
[18] C.R. Byfleet, A. Carrington and D.K. Russell, *Mol. Phys.*, **20**, 271 (1971).
[19] P.A.G. O'Hare and A.C. Wahl, *J. Chem. Phys.*, **54**, 4563 (1971).
[20] A. Carrington, P.N. Dyer and D.H. Levy, *J. Chem. Phys.*, **47**, 1756 (1967).
[21] A. Carrington, P.N. Dyer and D.H. Levy, *J. Chem. Phys.*, **52**, 309 (1970).
[22] G. Porter, *Disc. Faraday Soc.*, **9**, 60 (1950).
[23] R.A. Durie and D.A. Ramsay, *Can. J. Phys.*, **36**, 35 (1958).
[24] J.A. Coxon and D.A. Ramsay, *Can. J. Phys.*, **54**, 1034 (1976).
[25] R.A. Frosch and H.M. Foley, *Phys. Rev.*, **88**, 1337 (1952).
[26] A. Carrington, D.H. Levy and T.A. Miller, *J. Chem. Phys.*, **47**, 3801 (1967).
[27] J.M. Brown, C.R. Byfleet, B.J. Howard and D.K. Russell, *Mol. Phys.*, **23**, 457 (1972).
[28] A. Carrington and B.J. Howard, *Mol. Phys.*, **18**, 225 (1970).

[29] C.R. Byfleet, A. Carrington and D.K. Russell, *Mol. Phys.*, **20**, 271 (1971).
[30] R.J. Saykally and K.M. Evenson, *Phys. Rev. Lett.*, **43**, 515 (1970).
[31] D. Ray, K.G. Lubic and R.J. Saykally, *Mol. Phys.*, **46**, 217 (1982).
[32] D.C. Hovde, E. Schaefer, S.E. Strahan, C.A. Ferrari, D. Ray, K.G. Lubic and R.J. Saykally, *Mol. Phys.*, **52**, 2451 (1984).
[33] K.G. Lubic, D. Ray, D.C. Hovde, L. Veseth and R.J. Saykally, *J. Mol. Spectrosc.*, **134**, 1 (1989).
[34] K.G. Lubic, D. Ray, D.C. Holvde, L. Veseth and R.J. Saykally, *J. Mol. Spectrosc.*, **134**, 21 (1989).
[35] R. Beringer, E.B. Rawson and A.F. Henry, *Phys. Rev.*, **94**, 343 (1954).
[36] C.C. Lin and M. Mizushima, *Phys. Rev.*, **100**, 1726 (1955).
[37] R.L. Brown and H.E. Radford, *Phys. Rev.*, **147**, 6 (1966).
[38] W.L. Meerts, *Chem. Phys. Lett.*, **46**, 24 (1977).
[39] A. Carrington and N.J.D. Lucas, *Proc. R. Soc. Lond.*, **A314**, 567 (1970).
[40] R.S. Mulliken and A. Christy, *Phys. Rev.*, **38**, 87 (1931).
[41] J.H. Van Vleck, *Phys. Rev.*, **33**, 467 (1929).
[42] K.M. Evenson, J.S. Wells and H.E. Radford, *Phys. Rev. Lett.*, **25**, 199 (1970).
[43] J.M. Brown, C.M.L. Kerr, F.D. Wayne, K.M. Evenson and H.E. Radford, *J. Mol. Spectrosc.*, **86**, 544 (1981).
[44] T.D. Varberg and K.M. Evenson, *J. Mol. Spectrosc.*, **157**, 55 (1993).
[45] J.M. Brown, J.E. Schubert, J.S. Geiger and D.R. Smith, *J. Mol. Spectrosc.*, **114**, 185 (1985).
[46] E.R. Comben, J.M. Brown, T.C. Steimle, K.R. Leopold and K.M. Evenson, *Astrophys. J.*, **305**, 513 (1986).
[47] K.R. Leopold, K.M. Evenson, E.R. Comben and J.M. Brown, *J. Mol. Spectrosc.*, **122**, 440 (1987).
[48] K.M. Evenson, H.E. Radford and M.M. Moran, *App. Phys. Lett.*, **18**, 426 (1971).
[49] O.E.H. Rydbeck, J. Elldér and W.M. Irvine, *Nature*, **246**, 466 (1974).
[50] B.E. Turner and B. Zuckerman, *Astrophys. J. Lett.*, **187**, L59 (1974).
[51] G. Herzberg, *Spectra of Diatomic Molecules*, Van Nostrand, Princeton, N.J., 1950.
[52] J.T. Hougen, J.A. Mucha, D.A. Jennings and K.M. Evenson, *J. Mol. Spectrosc.*, **72**, 463 (1978).
[53] J.M. Brown and K.M. Evenson, *J. Mol. Spectrosc.*, **98**, 392 (1983).
[54] J.M. Brown and K.M. Evenson, *J. Mol. Spectrosc.*, **136**, 68 (1989).
[55] T.A. Dixon and R.C. Woods, *J. Chem. Phys.*, **67**, 3956 (1977).
[56] A. Carrington, D.H. Levy and T.A. Miller, *Proc. R. Soc. Lond.*, **A298**, 340 (1967).
[57] A. Carrington, G.N. Currie, D.H. Levy and T.A. Miller, *Mol. Phys.*, **17**, 535 (1969).
[58] F.D. Wayne and H.E. Radford, *Mol. Phys.*, **32**, 1407 (1976).
[59] N. Ohashi, K. Kawaguchi and E. Hirota, *J. Mol. Spectrosc.*, **103**, 337 (1984).
[60] P.B. Davies, D.K. Russell and B.A. Thrush, *Chem. Phys. Lett.*, **36**, 280 (1975); P.B. Davies, D.K. Russell, D.R. Smith and B.A. Thrush, *Can. J. Phys.*, **57**, 522 (1979).
[61] M.H.W. Gruebele, R.P. Müller and R.J. Saykally, *J. Chem. Phys.*, **84**, 2489 (1986).
[62] R.J. Saykally, K.M. Evenson, E.R. Comben and J.M. Brown, *Mol. Phys.*, **58**, 735 (1986).
[63] J.M. Brown, I. Kopp, C. Malmberg and B. Rydh, *Physica Scripta*, **17**, 55 (1978).
[64] J.M. Brown and A.J. Merer, *J. Mol. Spectrosc.*, **74**, 488 (1979).
[65] T.C. Steimle and J.M. Brown, *J. Chem. Phys.*, **82**, 1681 (1985).

[66] J.M. Brown, M. Kaise, C.M.L. Kerr and D.J. Milton, *Mol. Phys.*, **36**, 553 (1978).
[67] J.M. Brown, E.A. Colbourn, J.K.G. Watson and F.D. Wayne, *J. Mol. Spectrosc.*, **74**, 294 (1979).
[68] J.M. Brown, J.T. Hougen, K.-P. Huber, J.W.C. Johns, I. Kopp, H. Lefebvre-Brion, A.J. Merer, D.A. Ramsay, J. Rostas and R.N. Zare, *J. Mol. Spectrosc.*, **55**, 500 (1975).
[69] T. Nelis, J.M. Brown and K.M. Evenson, *J. Chem. Phys.*, **92**, 4067 (1990).
[70] S.M. Corkery, J.M. Brown, S.P. Beaton and K.M. Evenson, *J. Mol. Spectrosc.*, **149**, 257 (1991).
[71] J.M. Brown and D.J. Milton, *Mol. Phys.*, **31**, 409 (1976).
[72] R.F. Curl, *Mol. Phys.*, **9**, 585 (1965).
[73] S.P. Beaton, K.M. Evenson, T. Nelis and J.M. Brown, *J. Chem. Phys.*, **89**, 4446 (1988).
[74] S.P. Beaton, K.M. Evenson and J.M. Brown, *J. Mol. Spectrosc.*, **164**, 395 (1994).
[75] W. Weltner, Jr., *Magnetic Atoms and Molecules*, pp. 345–348, Van-Nostrand-Reinhold, New York, 1983.
[76] T. Nelis, S.P. Beaton, K.M. Evenson and J.M. Brown, *J. Mol. Spectrosc.*, **148**, 462 (1991).
[77] J.M. Brown, A.S.-C. Cheung and A.J. Merer, *J. Mol. Spectrosc.*, **124**, 464.
[78] T.C. Steimle, D.F. Nachman, J.E. Shirley, D.A. Fletcher and J.M. Brown, *Mol. Phys.*, **69**, 923 (1990).
[79] J.K.G. Watson, *J. Mol. Spectrosc.*, **80**, 411 (1980).
[80] A.R. Edmonds, *Angular Momentum in Quantum Mechanics*, Princeton University Press, 1960.

10 Pure rotational spectroscopy

10.1. Introduction and experimental methods

10.1.1. Simple absorption spectrograph

The first direct observations of pure rotational transitions induced by microwave or millimetre-wave radiation date from about 1945. Research during the Second World War, particularly on radiation sources in these wavelength regions, led to rapid developments in this branch of molecular spectroscopy. However, the observation of rotational structure in electronic and vibrational spectra, which had been routine for at least twenty years previously, meant that much was already understood about molecular rotations. Indeed most of the theory was developed in the years immediately following the birth of modern quantum mechanics. We have already described much of this theory in earlier chapters, and we will draw upon the results obtained in this chapter. We have also described molecular beam resonance and maser techniques for studying rotational spectra, but in this chapter we describe somewhat more conventional experiments in which the absorption of microwave radiation by molecules in the bulk gas phase is detected directly. As we will see, some ingenious techniques have been developed in more recent years to enable the study of short-lived transient species, neutral or charged free radicals. Molecular radio astronomy is also, in part, a branch of microwave absorption or emission spectroscopy, and we shall deal with that in subsection 10.1.6. We will also describe, in the next chapter, double resonance experiments in which microwave radiation is combined simultaneously with a second source of radiation of either similar or quite different wavelength.

The experimental studies described in this chapter cover a very wide frequency or wavelength range. We have already used the 'term' microwave in the previous paragraph but we should perhaps define our terms more precisely. There is no generally accepted convention, so we present ours in figure 10.1, where we define the boundaries between the radiofrequency, microwave, millimetre wave, sub-millimetre wave and far-infrared regions of the electromagnetic spectrum. Many experiments will cross these articially defined boundaries, embracing different regions.

In order to focus our attention we now outline the basic essential features of a microwave absorption spectrometer, and then examine the constituent parts in more

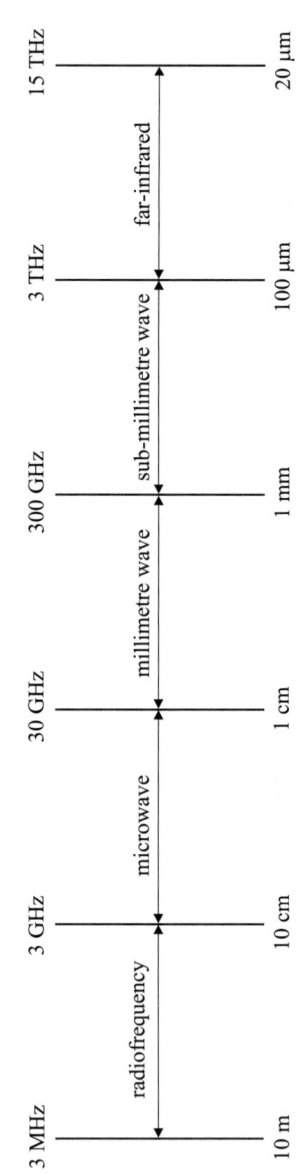

Figure 10.1. A convention for describing regions of the electromagnetic spectrum from the radiofrequency to the far-infrared.

Introduction and experimental methods | 685

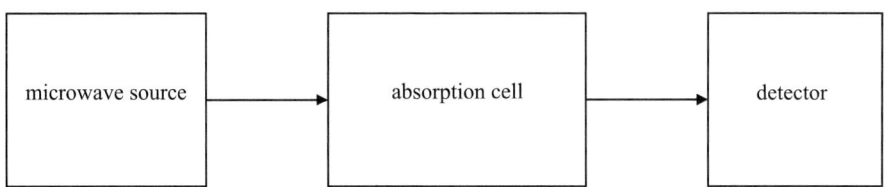

Figure 10.2. The three main elements of a simple microwave absorption spectrometer.

detail in subsequent subsections. Figure 10.2 shows a simple block diagram of such a spectrometer. There are three main elements. The first is a tunable source of radiation which might be anywhere in the (nominal) range 1 to 1000 GHz. 1 GHz corresponds to a wavelength of about 30 cm, whilst 1000 GHz (or 1 Terahertz, THz) represents a wavelength of about 0.3 mm. The range actually extends from the radiofrequency, through the microwave and millimetre wave, to the sub-millimetre region. The second main element is an absorption cell containing the gas under investigation, typically in the pressure range 0.01 to 0.1 Torr. As we shall see, many different types of absorption cell have been employed, particularly to optimise the study of short-lived transient species like free radicals or molecular ions. The third and final main element is a detector whose purpose is to convert incident microwave power to an output voltage, which can be measured. Absorption of microwave power in the absorption cell results in a decrease in the detector voltage, which is measured as a function of the source frequency. Such a simple spectrometer system, employing d.c. detection, has been successful for studying stable molecules with large electric dipole moments, but it is extremely insensitive.

During the past fifty years extensive effort in many laboratories has led to enhancements of the simple system, turning microwave spectroscopy into an extremely sensitive and versatile tool. We now review some of these developments. We shall also describe the essential features of a radio telescope because almost thirty diatomic molecular species, many of which would be transient species in the laboratory, have been detected in interstellar gas clouds. Molecular radio astronomy is closely linked with and complementary to laboratory microwave spectroscopy. Or, if you wish, you can reverse the emphasis of that last statement!

10.1.2. Microwave radiation sources

The workhorse of microwave spectroscopy until recent times has been the reflex klystron, which is now being replaced by synthesisers at the lower frequency end of the region, and backward-wave oscillators (BWO) at the high-frequency end. Both the klystron and the BWO make use of the fact that accelerating electrons emit radiation, and the principles of their design are illustrated in figures 10.3 and 10.4. In both cases a cathode is heated electrically to produce a continuous beam of electrons through thermionic emission. In the klystron the electrons are accelerated towards the positively charged cavity, passing through holes in the cavity wall, and are then repelled by the

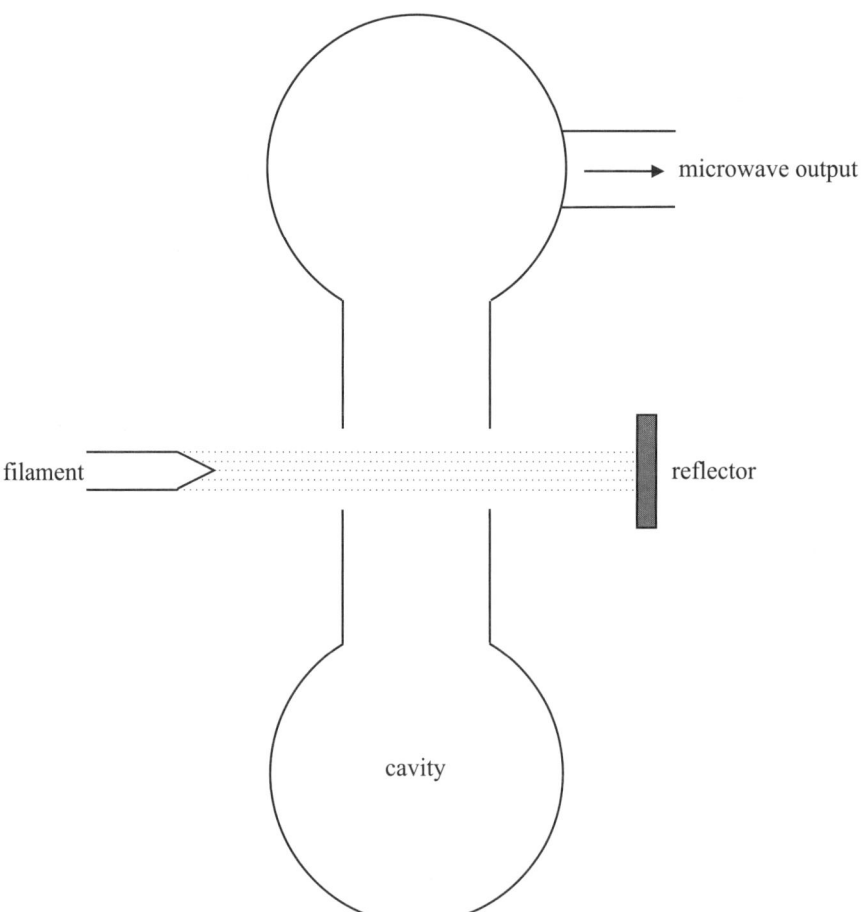

Figure 10.3. Principles of the reflex klystron.

negatively charged reflector which turns them back towards the cathode. The cavity is a metallic box which stores microwave energy of a particular wavelength, depending on the cavity dimensions. Resonant microwave cavities were discussed at some length in chapter 9. As the electrons pass through the cavity they are subjected to the oscillating microwave electric field supported by the cavity, which either increases or decreases their velocities. This velocity modulation leads to bunching of the electrons, and if the microwave phase is such as to reduce the velocity of an electron bunch, microwave energy is received by the cavity. A small portion of this microwave radiation is tapped off through an appropriate window. The microwave radiation frequency depends primarily on the cavity dimensions which can be changed manually. Further limited tuning of the microwave frequency is achieved by changing the reflector voltage. Consequently narrow band frequency tuning can be obtained electronically, but broad band tuning can be achieved only by stepwise mechanical adjustment of the cavity dimensions. Power outputs of up to 100 mW are readily obtained, and klystrons which operate at frequencies up to several hundred GHz are commercially available.

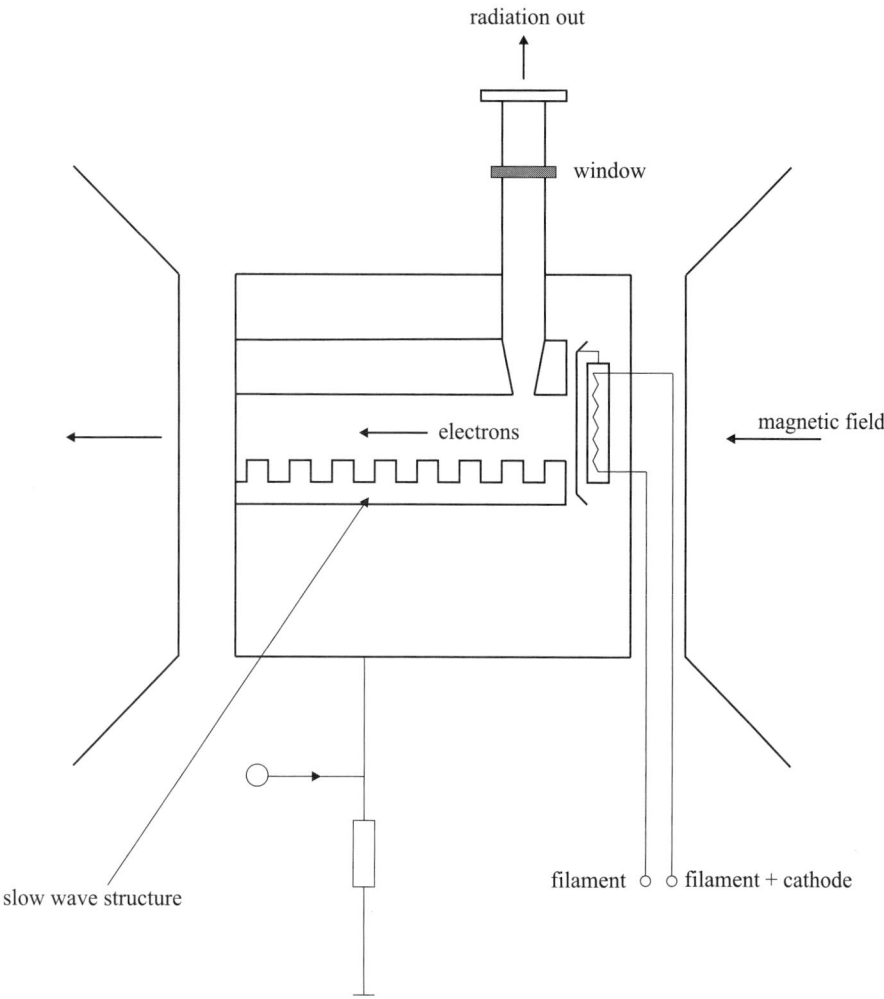

Figure 10.4. Principles of the backward-wave oscillator.

The backward-wave oscillator also uses an electron beam which is collimated by the grid and anode, accelerated by the positively charged accelerator, focussed by a permanent magnetic field and collected at the positively charged collector plate. The focussed beam passes through a wire helix, which is a microwave transmission line equal in length to several wavelengths of the longest required output wavelength. Under operating conditions random noise in the electron beam induces voltages in the helix which velocity modulate the beam and produce bunches of electrons which move towards the collector. The electric fields produced by the electron bunches appear outside the helix, and at the correct accelerating potential the electric fields are synchronous with the electron bunches, generating a backward moving wave in the helix. This wave further bunches the electron beam, which amplifies the backward wave, thereby maximising the bunching. The microwave signal generated on the helix is coupled out of the tube via a d.c. blocking capacitor. The oscillation frequency is determined by the potentials of the accelerating electrode, the helix, and the collector.

The great advantage of the backward-wave oscillator is that it is electronically tunable over its entire range of oscillation, unlike the klystron which has only limited electronic tuning. In both cases the frequency of the microwave source is usually locked to a harmonic of a stable reference oscillator. Recent advances in the design of backward-wave oscillators have enabled them to be used at ever higher frequencies, thereby opening up the millimetre to sub-millimetre region of the spectrum more effectively than before. We shall return to a discussion of very high frequency backward-wave oscillators later in this chapter.

Within the frequency range up to 200 GHz the klystron can be replaced (at a price) by synthesisers, coupled with solid state microwave amplifiers and passive or active frequency multipliers. These devices have very high frequency stability, are easily modulated either in frequency or power, and are readily compatible with computer control of all their main functions. The klystron is gradually becoming redundant, but has an honoured place in the development of microwave spectroscopy.

10.1.3. Modulation spectrometers

(a) Stark modulation

The simple spectrometer system illustrated in block diagrammatic form in figure 10.2 would be rather insensitive, but there are many refinements which greatly improve the situation. Perhaps the most important of these is signal modulation and in this section we consider a number of different modulation schemes which have been used to great effect. Modulation has several objectives, but one of them is to convert the output detector signal from d.c. to a.c.; a.c. amplification and detection techniques, including phase-sensitive detection, can then be used.

The detector used to convert incident microwave power into an output voltage is often a crystal detector consisting of a fine metal whisker in point contact with a semiconductor. The contact resistance is greater in one direction than in the other, and the small contact capacitance means that the crystal acts as a fast rectifier which is sensitive to microwave radiation. Incident microwave power on the crystal causes a voltage drop so that a current flows. At very low incident microwave power levels the rectified current is proportional to the power and, since this is proportional to the square of the voltage drop across the crystal, the detector is known as a square-law detector. At higher incident powers the rectified current is proportional to the first power of the voltage and the detector is then a linear detector.

The sensitivity of a detector depends primarily on its noise power output. Modulation of the microwave power incident on the crystal results in an alternating output current. The efficiency with which a small input signal is converted to a change in the output current increases with increasing current in the square-law region up to a maximum constant value in the linear region. It is therefore common practice to bias the crystal to ensure that it is operating in the linear region.

Introduction and experimental methods | 689

The noise output power (P) depends on the thermal noise (which depends upon the temperature, T) and the crystal conversion noise which is proportional to the square of the crystal current (I):

$$P = \left(kT + \frac{CI^2}{\nu}\right)\Delta\nu. \qquad (10.1)$$

Here C is a constant for a particular crystal (typically 10^{-7} ohms s^{-1}), ν is the output (modulation) frequency, and $\Delta\nu$ is the bandwidth of the detector system. The noise will be minimised by working at a high enough modulation frequency ν to render the second term in (10.1) negligible, and by reducing the detector bandwidth. It is also common practice to cool the detector to reduce the thermal noise. The usual method of modulating the incident power is to modulate the microwave absorption by perturbing the energy levels with a suitable oscillating electric or magnetic field. In this section we describe the use of an oscillating electric field, which is called Stark modulation. Later on we will describe magnetic field modulation, which has proved to be particularly important in the study of transient molecular species with open shell electronic structures.

Let us consider the simple two-level system illustrated in figure 10.5 where, in the absence of any external field, transitions will be induced between levels E_1 and E_2 by radiation at the correct frequency ν_0, which we take to be a microwave frequency. A single line spectrum would be obtained by slowly scanning the microwave frequency,

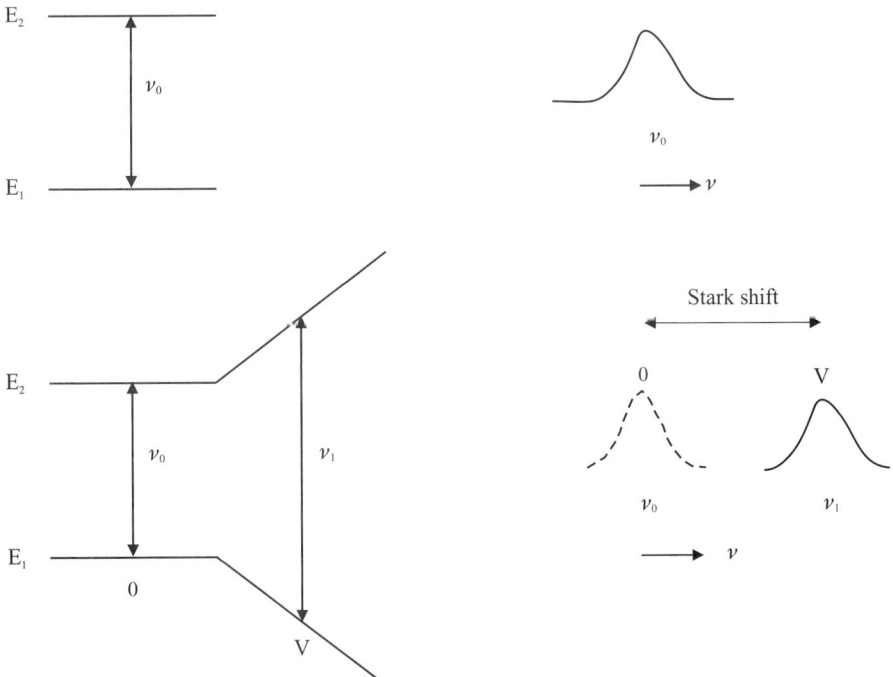

Figure 10.5. Stark effect and line shift for a two-level system.

and absorption would be detected as a decrease in the microwave power incident at the crystal detector. This simple method, which is illustrated in figure 10.2, is called crystal video detection. If now a static electric field is applied, and the molecule under investigation possesses an electric dipole moment, the resulting Stark effect will shift the resonant frequency to a new value ν_1. If we use a square-wave oscillating electric field, which is zero based, we will observe the original unshifted absorption line, plus the Stark-shifted line.

Now consider the microwave power level at the crystal detector as the microwave frequency ν is swept through the same range as in figure 10.5. When the square-wave voltage is in the first half of its cycle, at 0 volts, the absorption centred at ν_0 is obtained. However, in the second half of its cycle, at V volts, the Stark component at ν_1 is recorded. The detector output will be an alternating current and amplification with an amplifier tuned to the modulation frequency will yield a signal output as shown in figure 10.6. For the unshifted line ν_0 the peaks of the output correspond to the bottom of the square wave, i.e. at 0 volts. The signal is therefore 180° out-of-phase with the square-wave modulation. Conversely, the ν_1 component, produced at the top of the square-wave (V volts), is in-phase with the modulation. In practice the outputs are fed to a phase-sensitive detector which produces a d.c. output proportional to the cosine of the phase difference between the square-wave voltage and the output signal. Since the phase difference is either 180° or 0°, the d.c. output is proportional to -1 or $+1$; we obtain the recording shown in figure 10.6. The Stark component, often referred to as a Stark lobe, is inverted with respect to the unshifted line.

We have achieved the aim of providing the detector with a modulated power level, and if the modulation frequency is high enough, say 100 kHz, the crystal conversion noise is reduced to a level well below the thermal noise. In the example shown the Stark voltage is large enough to separate the unshifted and shifted lines, but at a smaller voltage the two lines will overlap, and the resultant will look like a first-derivative absorption line, with positive and negative lobes. In the more complicated examples often encountered, the mix of shifted and unshifted lines can be quite difficult to disentangle. Note that it is important that the Stark voltage be zero-based; if it is not, but ranges from $-V$ to $+V$ volts, only the Stark component will be observed and the microwave power at the detector is then not modulated. Sine-wave Stark modulation can be used but the amplitude must be chosen carefully because absorption will occur at all frequencies between ν_0 and ν_1; this can result in severe line broadening. It is again important that the modulation be zero-based.

We now consider how Stark modulation has been achieved in practice. Perhaps the earliest application was that of Hughes and Wilson [1]. A single flat metal plate is located in the centre of a rectangular waveguide cell, but insulated from it (see figure 10.7(a)). Since the Stark field is almost entirely parallel to the microwave electric field in this arrangement, observed transitions will obey the selection rule $\Delta M = 0$. An alternative arrangement is to cut the waveguide into two halves, which are held in place by insulators, and to apply the Stark voltage between the two halves (figure 10.7(b)). A third type of Stark cell contains two parallel plates, located inside a cell through which the microwave radiation is transmitted. Such a cell is illustrated in figure 10.8;

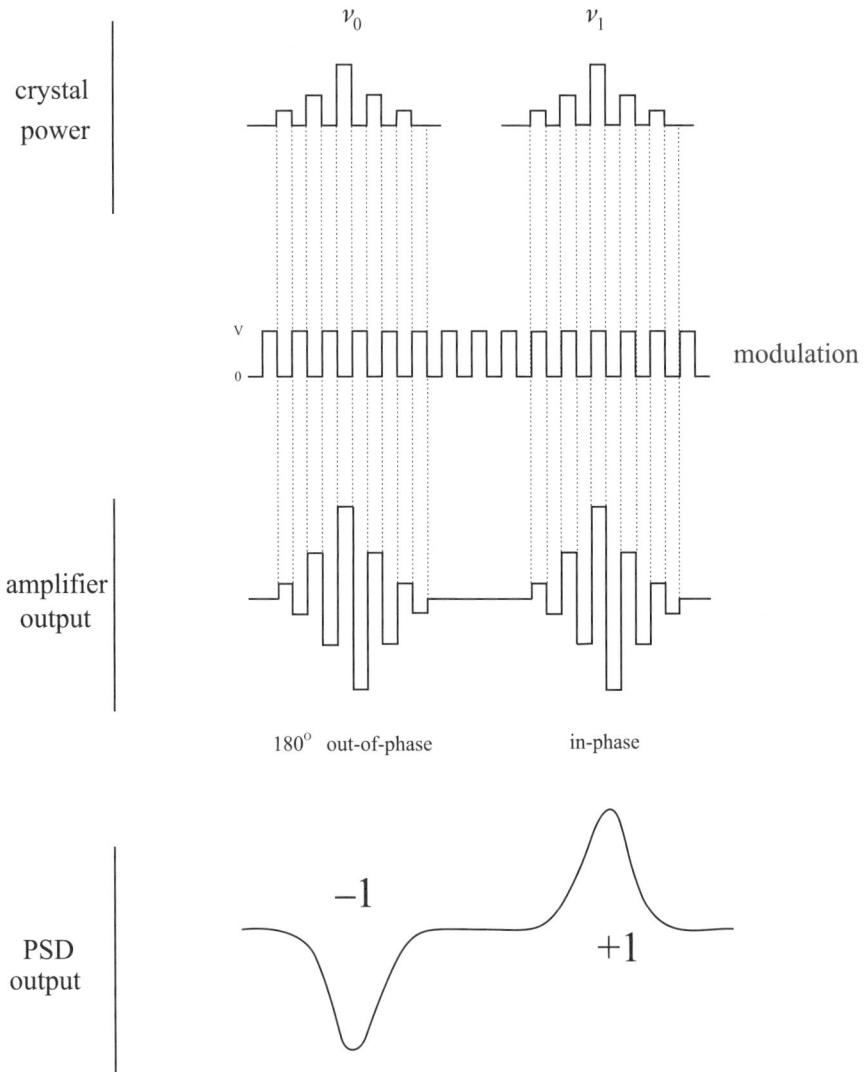

Figure 10.6. The effects of Stark modulation and phase-sensitive detection.

the microwave radiation is propagated and collected by a pair of horns, and transmitted by means of a pair of parallel metal plates, which also function as Stark electrodes. The cell illustrated is a design due to Johnson [2] but many others have used this type of cell successfully up to frequencies of at least 130 GHz [3].

Whichever method is used for modulation, it also possible to measure the electric dipole moment of the molecule under investigation. This can be achieved either by measuring the separation between the unshifted and Stark-shifted lines as a function of the modulation amplitude, or by direct measurement of the Stark splitting produced by a static electric field.

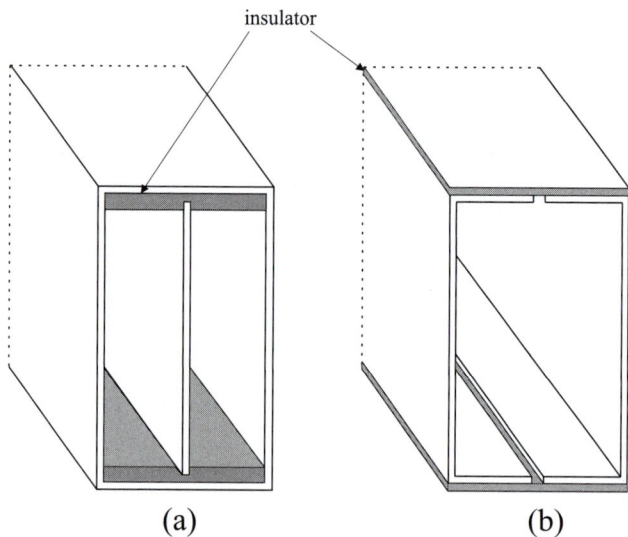

Figure 10.7. Methods for introducing Stark modulation into a rectangular waveguide cell: (a) Stark septum, (b) split waveguide.

(b) Zeeman modulation

If the molecule under investigation is a free radical with an open shell electronic structure, the energy levels are sensitive to an applied magnetic field, as we have seen elsewhere. Under these circumstance an oscillating magnetic field can be used for modulation, in much the same way as it is used in microwave magnetic resonance (see chapter 9). The first and most important example of this approach was the detection of the Λ-doublet spectrum of the OH radical by Dousmanis, Sanders and Townes [4]. They used a fairly simple arrangement which is illustrated in figure 10.9. The cylindrical absorption cell was made of brass, 150 cm in length and 3 cm diameter, coupled at the ends to rectangular waveguide through tapered sections. Inside the brass tube was a cylindrical quartz tube forming part of the gas system, whilst outside was a wire-wound solenoid to produce an axial modulating magnetic field. Rectangular slots were cut into the brass tube to allow for penetration of the modulating field, and a small d.c. bias magnetic field was also applied. Water vapour was pumped continuously through a radiofrequency discharge, situated immediately upstream of the microwave cell, and the OH radicals produced were estimated to have an average lifetime of 300 ms (being regenerated downstream in the gas flow). Absorption lines were observed over the range 7.7 to 37 GHz. We shall discuss the nature and assignment of the spectrum later in this chapter.

(c) Source modulation

There are often reasons why the gas sample cell needs to be free of Stark plates, and in such cases the free-space cell has been used, sometimes with Zeeman modulation coils situated outside the cell. However, when Zeeman modulation is not appropriate

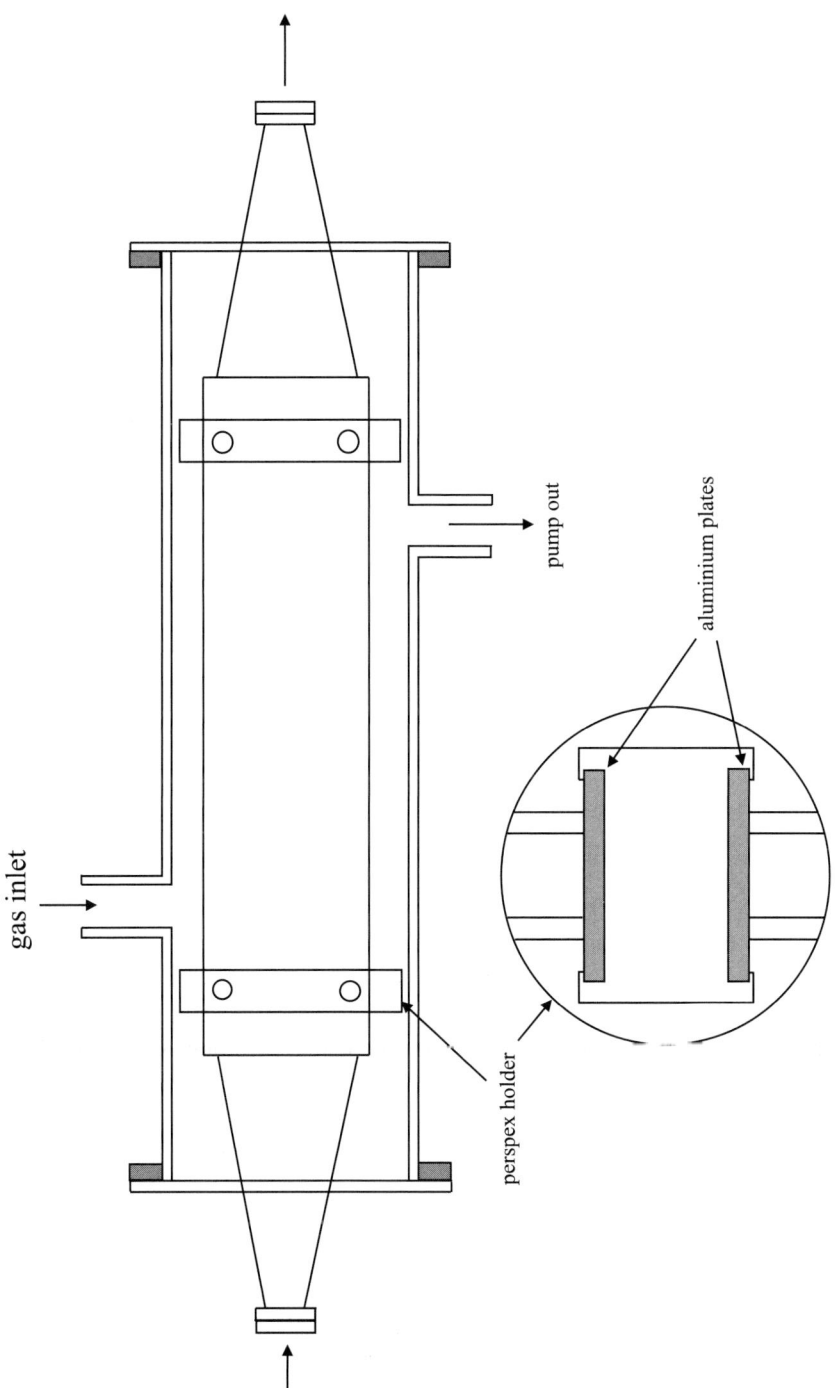

Figure 10.8. Parallel plate cell; the aluminium plates function both as the waveguide walls and as the Stark plates [2].

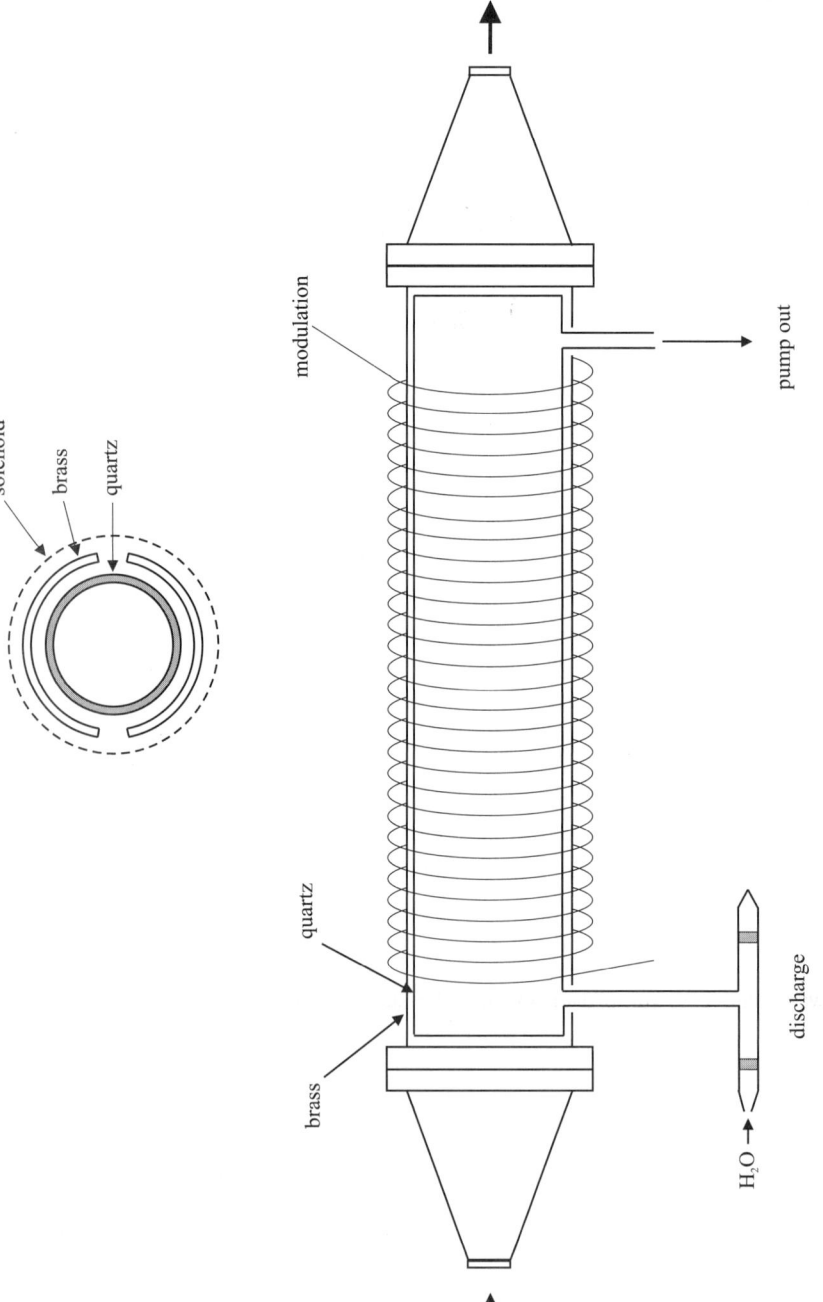

Figure 10.9. Cylindrical cell used by Dousmanis, Sanders and Townes [4] to study the Λ-doublet microwave spectrum of the OH radical.

(e.g. for a closed shell molecule), direct frequency modulation of the microwave source frequency has often been employed. This achieves some of the advantages of molecular modulation, including the removal of the $1/\nu$ crystal noise, but not all. One particular problem is that when scanning the microwave frequency there are bound to be fluctuations in the power level incident on the detector crystal. These arise either from changes in the power level of the source itself, or unavoidable changes in the transmission efficiency of the microwave line and sample cell. If source modulation is used, these power fluctuations result in a.c. voltage fluctuations from the crystal detector. These are detected by the phase-sensitive detector and a sloping or erratic baseline is the consequence.

Sometimes one has to live with this problem, and some of the finest work, particularly on transient molecular species, has been achieved using source modulation. This includes some of the earliest work on neutral free radicals, by Kewley, Sastry, Winnewisser and Gordy [5], who studied the SO and CS species. With the benefit of hindsight, we know these radicals to be very long-lived. We shall illustrate the details of source modulation by describing two other experiments. The first is the classic work of Woods [6] and Dixon and Woods [7] who obtained the first microwave spectrum of a molecular ion, namely CO^+. The second is an example of a high-temperature microwave cell for the study of refractory materials.

The objective of Woods [6] was to design and operate a microwave absorption cell which also supported a d.c. electrical discharge in the sample gas, with the aim of producing a detectable concentration of molecular ions. The principal features of the cell are illustrated in figure 10.10. The main part of the apparatus was a cylindrical Pyrex tube, of length 305 cm and diameter 15 cm, which served as the discharge tube and vacuum envelope. This glass tube was extended by stainless steel sections, 30 cm long at one end and 7.5 cm long at the other, which contained all the vacuum gauges, pumping ports, gas inlets and electrical feedthroughs. The complete tube was sealed, vacuum tight, at each end by Teflon lenses designed to collimate the microwave radiation. Just inside the glass tube, at the ends, were hollow cylinders of copper, 10 cm long and small enough to slide inside the glass. These copper cylinders were the electrodes for the d.c. discharge, but Woods actually found that the discharge at the cathode end always tended to jump over and use the stainless steel section as a larger cathode. In normal operation this was accepted as a feature of the tube which had some advantages. At the anode end the stainless steel section was floated electrically and the copper cylinder carried all the current, corresponding to a maximum power of 0.5 kW dissipated under operating conditions. The discharge voltage used was in the range 500 to 800 V. It was found to be advantageous to cool the discharge tube with a continuous flow of liquid nitrogen through coils wrapped around the tube.

The frequency source used for the study of CO^+ was a klystron operating close to 120 GHz. The microwave radiation was launched through a microwave horn onto the Teflon lens at the source end of the discharge tube. The radiation emerging from the cell was focused by the second Teflon lens into a complementary horn for transmission to a diode detector. The mounting and precision adjustment of the horns was extremely important, and the whole assembly was mounted on an optical bench. The klystron

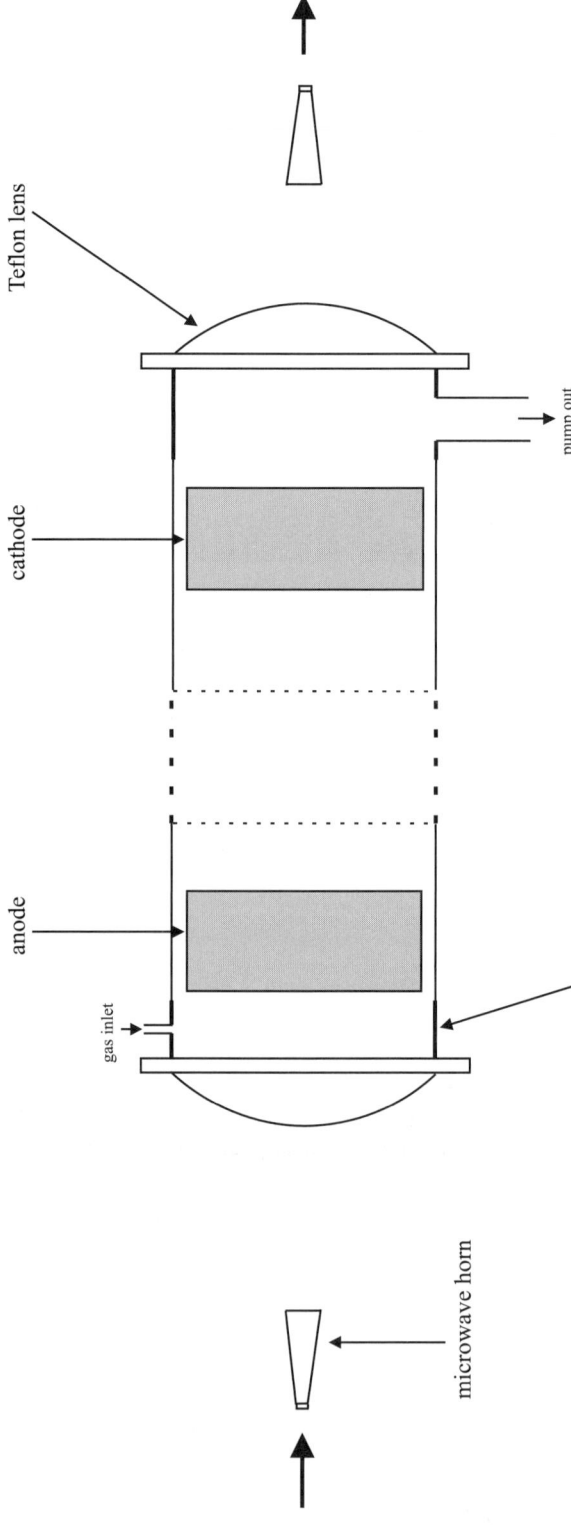

Figure 10.10. Microwave cell/discharge tube built by Woods [6] for the study of transient molecules, including molecular ions.

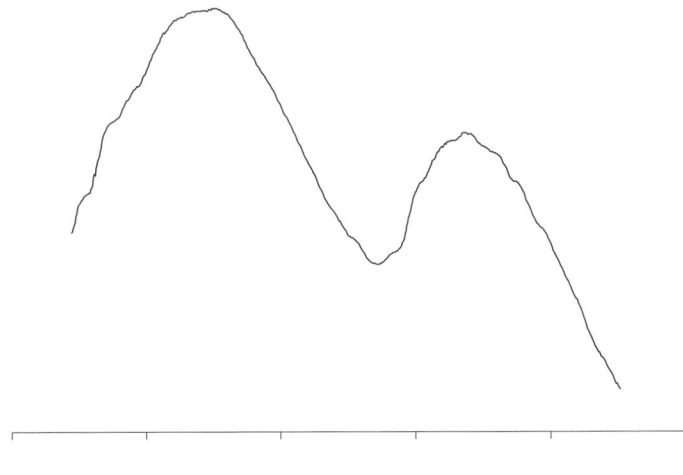

5 MHz scan width

Figure 10.11. Rotational transition of the CO$^+$ ion at a frequency of 118 101.99 MHz, observed by Dixon and Woods [7]. The overall scan width is 5 MHz, the line was obtained by averaging 200 scans, and the centre of the resonance is the central dip. Note the confusing baseline drift.

frequency was stabilised by phase-locking it to a harmonic of a stable lower frequency source. Source frequency modulation was accomplished with a 30 kHz square-wave voltage, centred at zero voltage.

In his inital experiments, Woods [6] observed very strong lines from the OH radical when water vapour was the discharge medium. Dixon and Woods [7] subsequently observed lines from the CO$^+$ ion using a helium/carbon monoxide mixture; previous work on the electronic spectrum [8] had provided approximate values for the requisite microwave frequencies, and two resonances were observed, corresponding to the lowest rotational transition in the ground vibrational level, split by spin–rotation interaction. The assignment of the spectrum is described later in this chapter. One of the observed resonances in shown in figure 10.11, the central dip being the resonance. This spectrum required one hour of signal averaging, and it demonstrates the problem of the sloping baseline characteristic of source modulation experiments. Perhaps the most surprising observation, however, was that observation conditions could be found under which noise from the gas discharge was negligible. This was contrary to both expectation and the conventional wisdom at the time.

The second type of free space cell we describe is one of a number designed to operate at a high temperature so that species of low volatility can be studied. We outline briefly a cell designed by Yamada, Fujitake and Hirota [9] to study the rotational spectrum of NaO, formed by the reaction of sodium vapour with N_2O at a pressure of 100 m Torr. This cell was operated at a temperature of 350 °C; other workers have built cells which operate at much higher temperatures.

The main body of the cell, shown in figure 10.12, was a cylindrical Pyrex tube of diameter 10 cm and length 2 m, sealed at the ends with Teflon lenses designed to

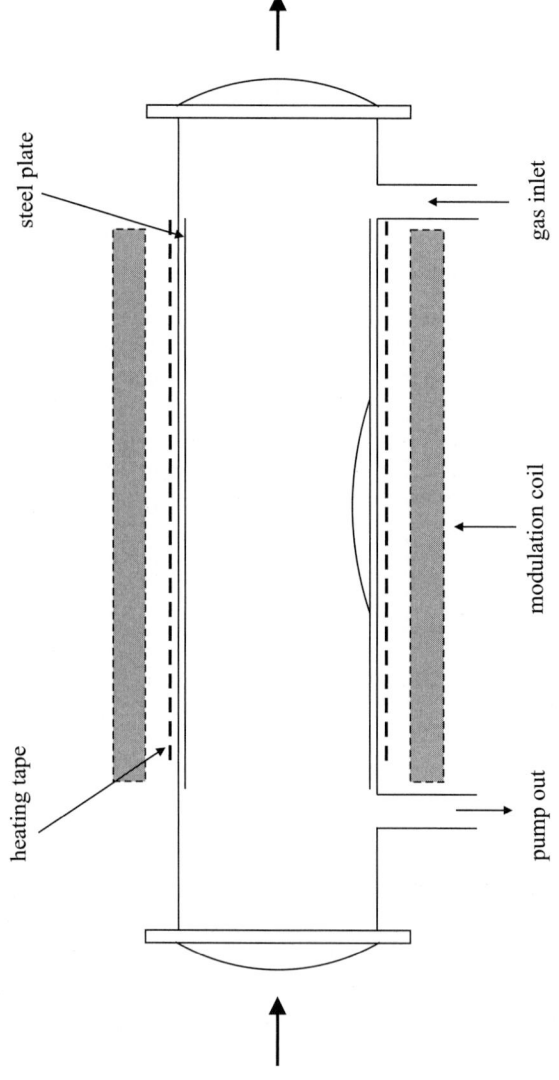

Figure 10.12. Free space microwave absorption cell used by Yamada, Fujitake and Hirota [9] for the study of NaO at a temperature of 350 °C.

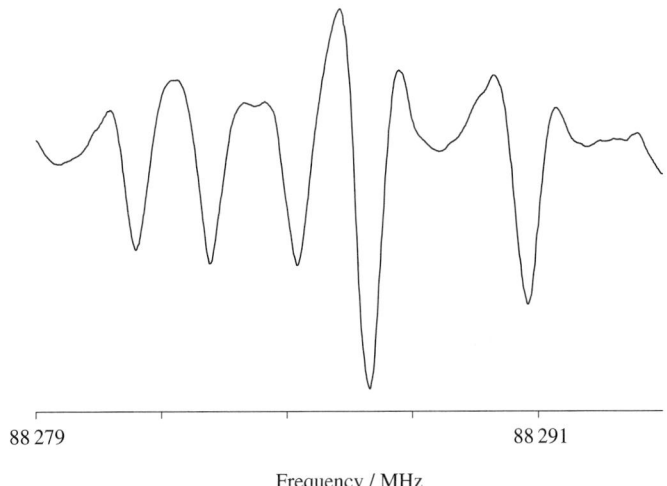

Figure 10.13. Microwave spectrum of NaO, arising from a $J = 7/2 \leftarrow 5/2$ rotational transition in the $v = 0$ level of the $^2\Pi_{3/2}$ ground state [9].

collimate the microwave beam. Inside the cell was a stainless steel plate supporting sodium metal. The cell was heated by means of heating tape wound around the outside; outside this tape was a solenoid which was used either for cancelling the earth's magnetic field or for studying Zeeman effects. The frequency region was covered up to 400 GHz using klystrons and Schottky barrier multipliers, and source frequency modulation was employed for a.c. detection. An observed spectrum is shown in figure 10.13. The NaO molecule was determined to have a $^2\Pi_{3/2}$ ground state, and the spectrum shown in the figure arises from the $J = 5/2$ to $7/2$ rotational transition in the $v = 0$ level, split by Λ-doubling and hyperfine interaction from the ^{23}Na nucleus. Again the details are described later in this chapter. Baseline drift is not so apparent in this spectrum, mainly because the signal-to-noise ratio is very high.

(d) Velocity modulation

We conclude this section on modulation spectrometers by describing a particularly novel and important method, known as velocity modulation, which was originally developed by Gudeman, Begemann, Pfaff and Saykally [10]. It applies specifically to ionic species, and has been used primarily to study the infrared vibration–rotation spectra of molecular ions. If we need an excuse to include it in this book, however, it is provided by Matsushima, Oka and Takagi [11] who used the method to study the far-infrared rotational spectrum of the HeH$^+$ ion.

A block diagram of the spectrometer system designed by Gudeman, Begemann, Pfaff and Saykally [10] is shown in figure 10.14. As shown, this system employed a mid-infrared laser as the radiation source, and was designed to measure the vibration–rotation spectrum of the HCO$^+$ ion. The essential principle is very simple. The ions are produced by a square-wave a.c. discharge; during the first half of the square-wave

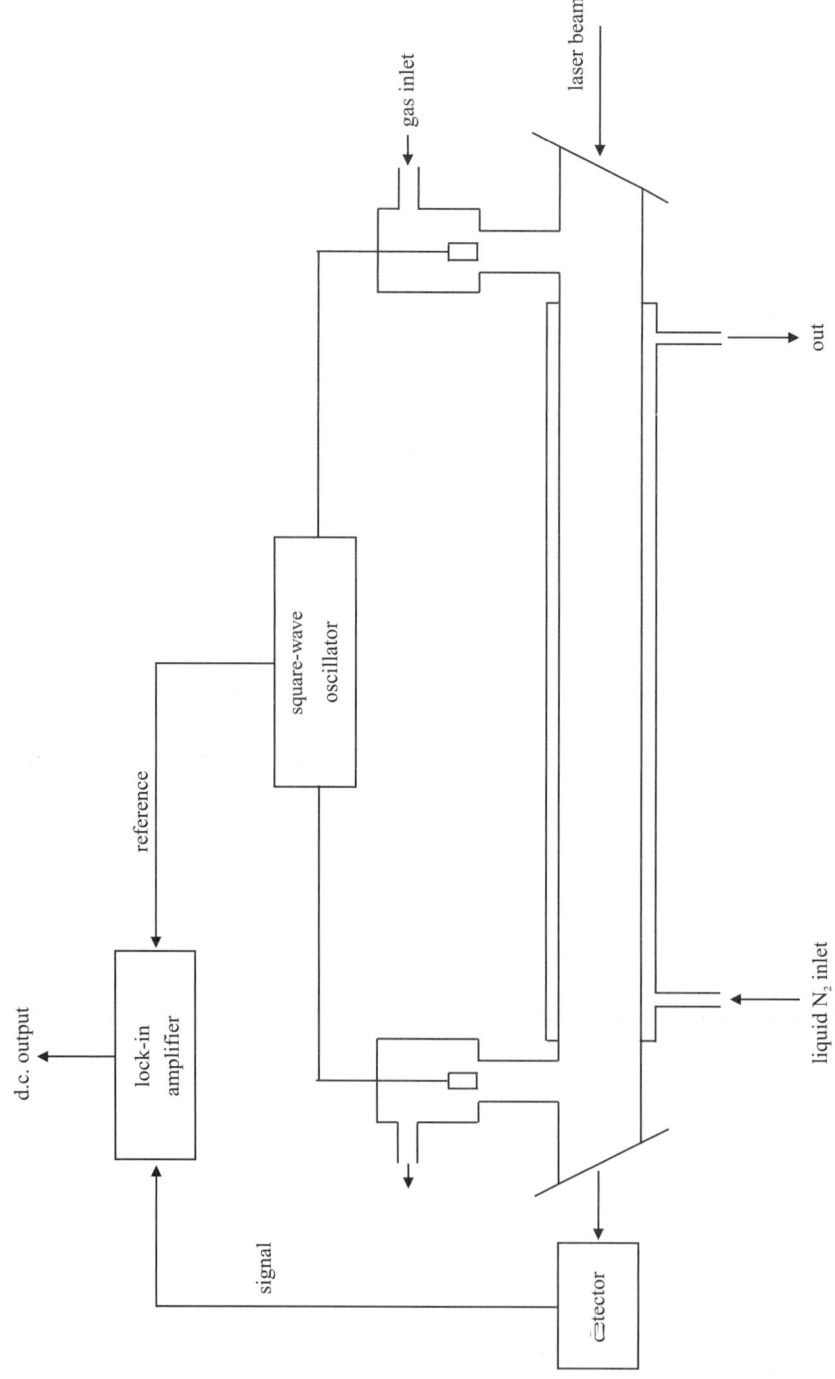

Figure 10.14. Velocity modulation spectrometer introduced by Gudeman, Begemann, Pfaff and Saykally [10] to study the infrared vibration–rotation spectra of molecular ions.

cycle the positive ions migrate towards the negative electrode and any resonance with
the laser beam passing through the cell will be Doppler-shifted in one direction. In the
second half of the square-wave cycle the electrode polarities are reversed, the migration
direction of the ions is reversed, and the Doppler shift is also reversed. Consequently the
a.c. discharge modulates the direction of the ion velocity vector, and hence the Doppler
shift; there is a resulting frequency modulation of the resonance, and conventional a.c.
detection techniques can be used.

There are at least two huge advantages with the technique. The first is that only
molecular *ions* are affected, so that there is total detection discrimination in favour of
ions. The second is that positive and negative ions present in the discharge will move in
opposite directions, with resulting Doppler shifts of opposite sign. There is therefore a
180° phase change in the detected signals from cations and anions, leading to a relative
inversion of the resonances from these species. If both positive and negative ion spectra
are detected, they are clearly separated [12]. A beautiful example was the detection of
OH^+ and OH^- ions in the same discharge.

The simplified block diagram of the apparatus used by Gudeman, Begemann, Pfaff
and Saykally, shown in figure 10.14, indicates an infrared laser beam as the radiation
source. The authors point out, however, that there is no reason why the method should
not be applied in lower frequency regions of the spectrum, including the microwave.
The far-infrared studies of the HeH^+ ion by Matsushima, Oka and Takagi [11] were performed using a laser system in which two carbon dioxide lasers at different frequencies
are mixed together and also with a microwave frequency from a tunable synthesiser.
Coarse tuning is achieved by choosing different pairs of carbon dioxide laser frequencies, and fine tuning by scanning the microwave synthesiser. They used a sinusoidal
a.c. discharge at a frequency of 1.2 kHz, with a peak-to-peak voltage of 5 kV. We shall
describe the analysis and interpretation of the rotational spectrum of HeH^+, including
its isotopic variants, later in this chapter.

10.1.4. Superheterodyne detection

As we have seen, modulation of the energy levels involved in a transition results in a
signal at the detector oscillating at the modulation frequency, with all the advantages
of discrimination and removal of the $1/\nu$ noise. It is, however, not always possible to
arrange the apparatus to modulate the energy levels, and this is where superheterodyne
detection becomes important. The essential principle is to mix the primary microwave
frequency ν_1 at the crystal detector with a secondary frequency ν_2 from what is called a
'local oscillator'. One then detects at the difference frequency, $\nu_1 - \nu_2$. Typical values
of the difference frequency are 30 or 60 MHz, and at this frequency the crystal noise
term is negligible. It is necessary to stabilise both the primary and the local oscillator
frequency, and the difference between them. Signal detection and amplification is, of
course, at the beat frequency. When the requirements of the experiment permit, it is
possible to combine superheterodyne detection with molecular modulation, in which
case the bandwidth of the detection system can be made very small.

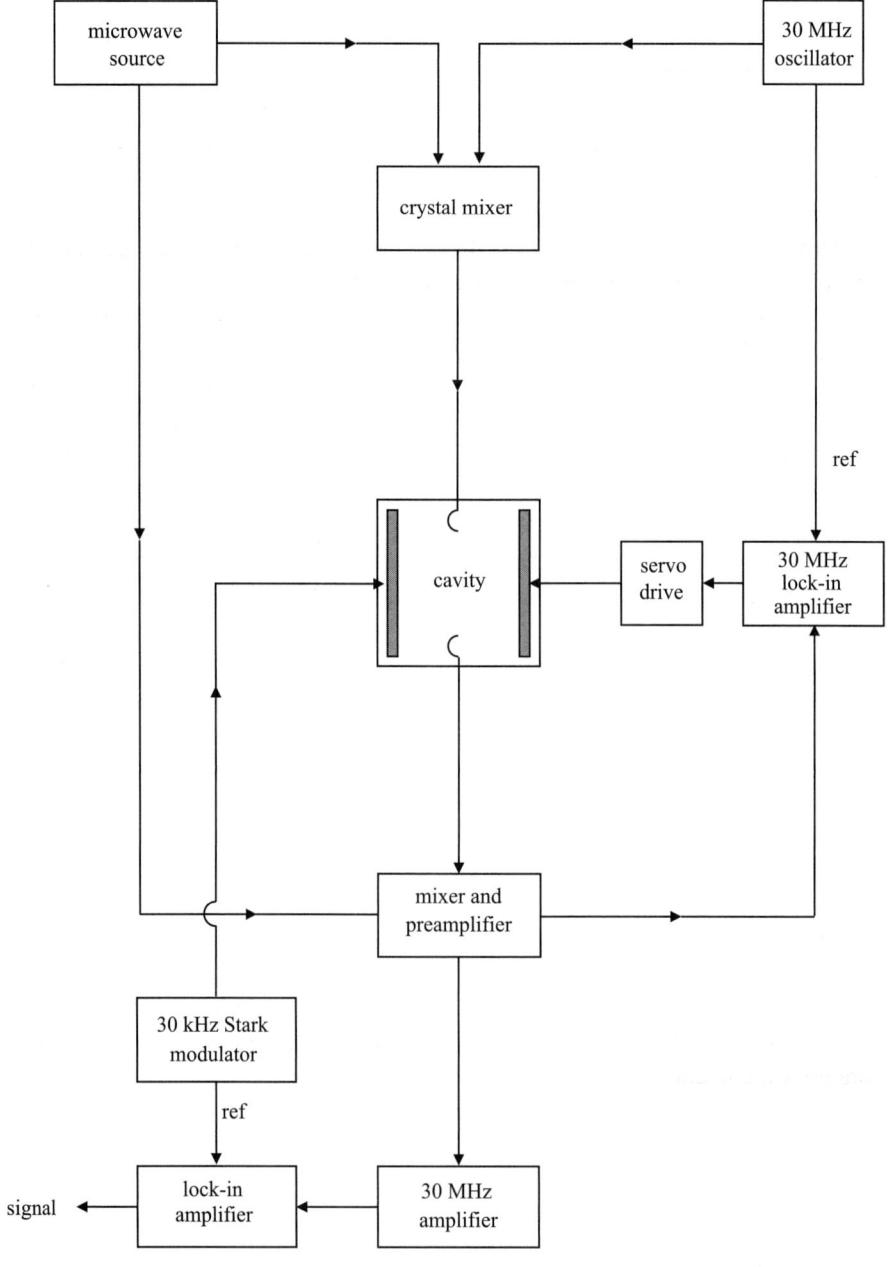

Figure 10.15. Block diagram of the tunable-cavity spectrometer built by Radford [13], and used by him to study the microwave spectra of OH and OD.

An unusual example of a microwave spectrometer which uses superheterodyne detection and molecular modulation is a tunable-cavity spectrometer designed and built by Radford [13]. Microwave cavities were described in chapter 9, where they form the heart of a microwave magnetic resonance spectrometer. Compared with non-resonant microwave cells, such as we have described earlier, a resonant cavity has a small volume which should offer filling factor advantages for the study of short-lived free radicals. On the other hand a resonant cavity operates at a specific frequency, and molecular energy levels are usually tuned by some other method, such as an external magnetic field. Radford [13] designed a cylindrical cavity whose resonant frequency could be tuned by changing its length. It was then necessary to design a system so that the microwave source frequency tracked the resonant frequency of the cavity as the latter was changed. The details of Radford's spectrometer system are illustrated in figure 10.15. The cavity is designed to be mechanically tunable over the range 3.5 to 7 GHz. The output of the microwave source is frequency modulated at 30 MHz, and the cavity tuned to one of the 30 MHz sidebands; as a transmission cavity it passes only the sideband frequency chosen. This sideband is mixed with the original microwave source and the resulting 30 MHz intermediate frequency signal is amplified and fed to a lock-in detector where it is compared with a reference signal from the 30 MHz oscillator. The d.c. output of the lock-in detector is then passed to a servo amplifier which mechanically adjusts the resonant frequency of the cavity. Frequency scanning is thus accomplished by sweeping the microwave source, and relying on the servo system to keep the cavity in tune. In addition a 30 kHz modulating Stark voltage is applied across the cavity, as shown. The final molecular resonance signal is a 30 kHz signal, detected with a lock-in amplifier, with the noise advantages of 30 MHz superheterodyne detection. Radford succeeded in observing the microwave Λ-doublet spectra of OH and OD, but attempts to observe the spectrum of CH were unsuccessful. CH is a very short-lived species; it was observed eventually by far-infrared laser magnetic resonance, as we described in chapter 9, and microwave transitions have been observed by microwave/optical double resonance, as we shall describe in chapter 11. The pure rotational spectrum of CH, however, has only been observed very recently.

Superherodyne spectrometers are now not common in laboratory microwave experiments, but superheterodyne detection plays a major role in radio astronomy, as we shall see later. The reasons are obvious; one cannot modulate the energy levels of extra-terrestrial molecules, and a radio telescope collects radiant energy at all frequencies simultaneously. One does not have a primary monochromatic source of radiation, as in laboratory experiments.

10.1.5. Fourier transform spectrometer

(a) Introduction

A quite different approach to radiofrequency, microwave and infrared spectroscopy is that known as Fourier transform (FT) spectroscopy. As we shall see, this method of recording the spectra of transient molecular species is particularly appropriate in combination with the use of pulsed gas nozzles. For this reason it has proved to be a powerful technique for the study of weakly bound dimer complexes formed in supersonic gas expansions. It has, however, also been used for the study of diatomic molecules, both stable and transient, and is likely to have increasing value in this field. It is also closely related to techniques used in radio astronomy. Similar Fourier transform methods are of primary importance in nuclear magnetic resonance spectroscopy of condensed phases.

In the field of microwave rotational spectroscopy pioneering developments were described by Flygare and his colleagues; an important paper which discusses in considerable detail the major instrumental factors and the underlying theory is that due to Balle and Flygare [14]. We will outline the essential features of the Flygare spectrometer, using the description given by Legon [15] as our guide, and then explore different aspects of the instrument and their underlying principles in more depth.

(b) Pulsed-nozzle Fourier transform spectrometer

A schematic diagram of the spectrometer is shown in figure 10.16; its successful operation depends critically upon the ability to achieve accurate timing for a sequence of several events. First, a short pulse of gas is produced from a pulsed-nozzle source, the gas travelling in a direction perpendicular to the axis of an evacuated Fabry–Perot cavity, described later. This gas pulse lasts for about 1 ms, and the expansion in the cavity is in an essentially collision-free environment.

Second, after an appropriate time interval to allow the gas pulse to reach an optimum position between the cavity mirrors, a 1 μs pulse of monochromatic microwave radiation is introduced into the cavity, which is itself tuned to the correct matching resonant frequency. The pulse carries with it a band of frequencies $\Delta \nu \approx 1$ MHz, centred at the resonant frequency ν of the cavity. The cavity has a bandwidth of approximately 1 MHz, so that the microwave radiation density is high. If the molecular species under investigation has one or more resonant frequencies within this bandwidth, an appreciable macroscopic polarisation is induced, corresponding classically to a phase-coherent oscillation of the molecular electric dipole moments. The microwave pulse must arrive at the correct time interval after the gas pulse.

An essential requirement is that the characteristic time, T_2, for the decay of the macroscopic polarisation must be much longer than the time taken for the polarising radiation pulse to dissipate. This requirement is readily satisfied; the pin-diode S_2 is held closed until the pulsed radiation has dissipated, and is then opened to capture the coherent radiation emitted by the polarised gas, due to one or more rotational transitions producing spontaneous emission. If all is well, the emission is detected against a near-zero radiation background.

Microwave emission at a frequency ν_m is detected by means of a double superheterodyne system. The radiation at frequency ν_m is mixed with a phase-coherent signal at $\nu - 30$ MHz to give an intermediate frequency $|\nu - \nu_m| \pm 30$ MHz. After amplification this signal is mixed in a second stage with further phase-coherent radiation at 30 MHz, to yield the final signal. This consists essentially of an exponentially damped oscillating electric field $E(t)$ of frequency $\Delta = |\nu - \nu_m|$; after suitable averaging of a chosen number of pulses, a corresponding intensity versus frequency spectrum is obtained. Conversion of data obtained in the time domain to data in the frequency domain is achieved by fast Fourier transformation, using a computer. The principles behind

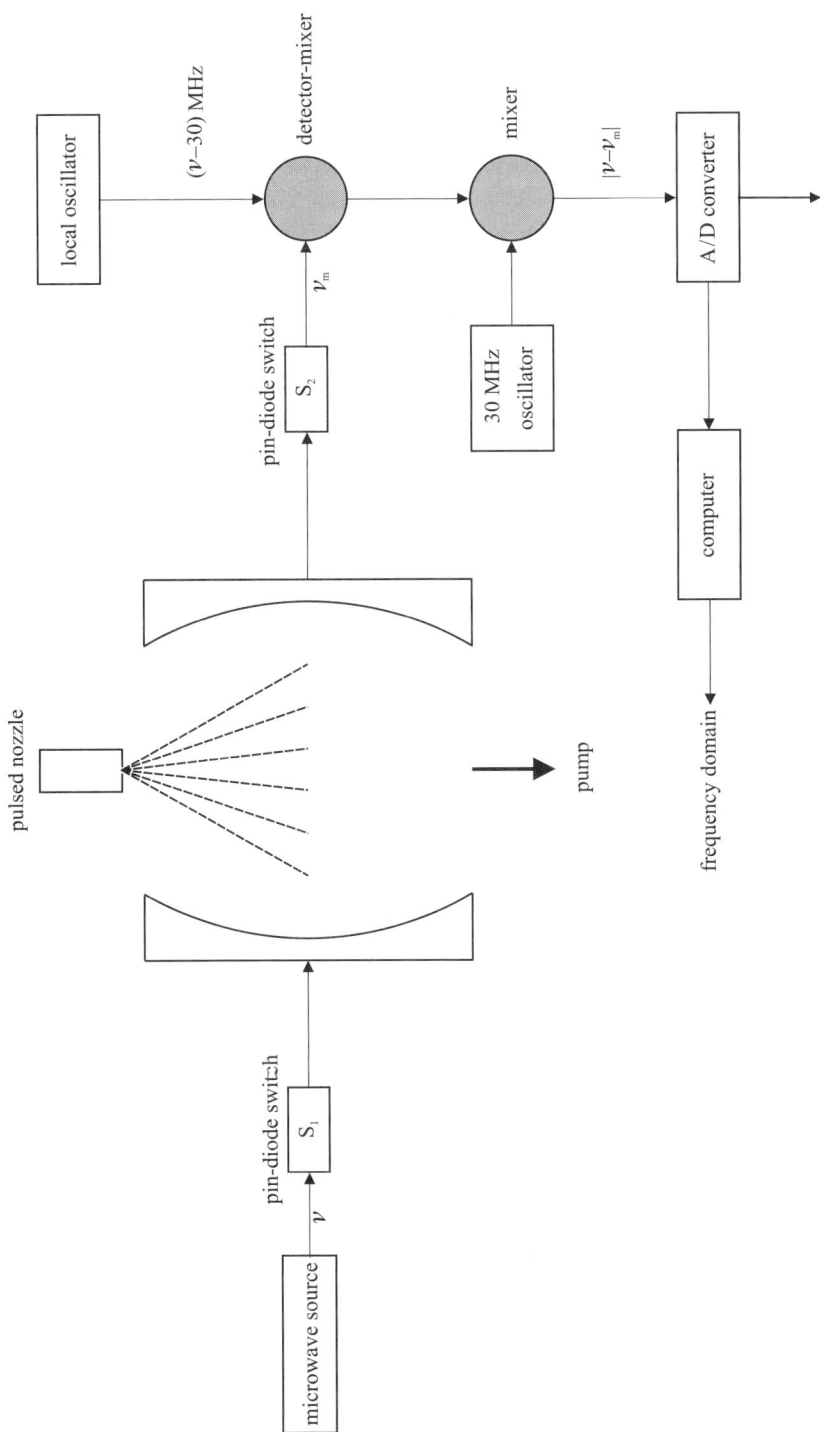

Figure 10.16. Schematic block diagram of a pulsed-nozzle Fourier transform microwave spectrometer [15].

this transformation are described in the next subsection; we deal with this fundamental aspect, and then return to the details of the spectrometer.

(c) Fourier transformation from the time domain to the frequency domain

The conversion of an oscillating electric field $E(t)$, the so-called time domain spectrum, into a frequency domain spectrum is known as a Fourier transformation. A simple but neat description of this transformation is given by Hollas [16]. The oscillating electric field arising from a molecular emission line following the radiation pulse is converted into an oscillating voltage $f(t)$ with a frequency ν, which we may write

$$f(t) = A \cos 2\pi\nu t, \qquad (10.2)$$

where A is the amplitude of $f(t)$. Figure 10.17 shows a very simple radiofrequency case where $\nu = 100$ MHz; the time domain spectrum shown in figure 10.17(a) is transformed into a frequency domain spectrum, shown in figure 10.17(b) with a single line at a frequency of 100 MHz. This transformation is a specific example of a more general result: if a square integrable function $F(x)$ exists, then its Fourier transform, $f(k)$, is

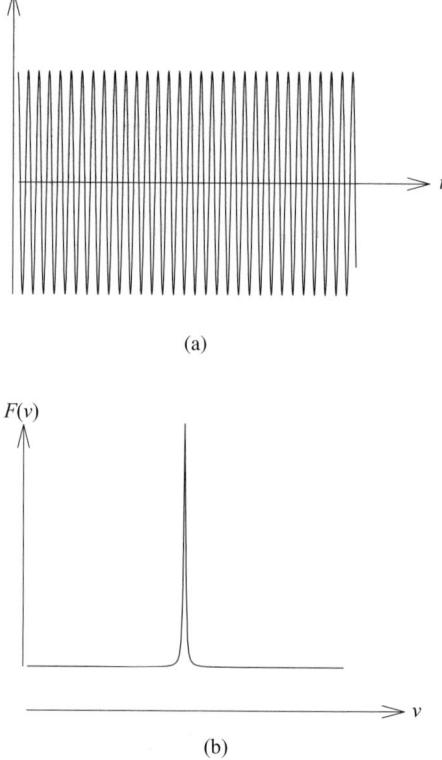

Figure 10.17. (a) Time domain spectrum, (b) frequency domain spectrum for radiation of a single frequency.

defined by

$$f(k) = \frac{1}{2\pi} \int_{-\infty}^{+\infty} F(x) \exp(ikx) \, dx. \tag{10.3}$$

The reverse transformation is

$$F(x) = \int_{-\infty}^{+\infty} f(k) \exp(-ikx) \, dk, \tag{10.4}$$

and $F(x)$ and $f(k)$ are called Fourier transform pairs. In the radiofrequency example given above the time domain spectrum can be expressed as

$$f(t) = \int_{-\infty}^{+\infty} F(\nu) \exp(i2\pi\nu t) \, d\nu, \tag{10.5}$$

where $F(\nu)$ is the frequency domain spectrum we seek. Since the exponential function in (10.5) can be written in the form

$$\exp(i2\pi\nu t) = \cos 2\pi\nu t + i \sin 2\pi\nu t, \tag{10.6}$$

equation (10.5) can be rewritten:

$$f(t) = \int_{-\infty}^{+\infty} F(\nu)(\cos 2\pi\nu t + i \sin 2\pi\nu t) \, d\nu. \tag{10.7}$$

Neglecting the imaginary part of (10.7) we see that $f(t)$ is a sum of cosine waves, as shown in figure 10.17(a). In like manner, the Fourier transform can be re-expressed as

$$F(\nu) = \int_{-\infty}^{+\infty} f(t)(\cos 2\pi\nu t - i \sin 2\pi\nu t) \, dt, \tag{10.8}$$

and again the imaginary part of (10.8) can be neglected.

This is the simplest possible case. In real examples there could be two or more molecular resonances within the bandwidth of the cavity. Suppose, in fact, that there are three resonances at frequencies ν, 0.7ν and 0.5ν, with relative intensities 10, 15 and 7. In this case $f(t)$ would be given by

$$f(t) = A(\cos 2\pi\nu t + (15/10) \cos 2\pi(0.7)\nu t + (7/10) \cos 2\pi(0.5)\nu t), \tag{10.9}$$

and the time domain spectrum would be as shown in figure 10.18(a). The corresponding frequency domain spectrum is shown in figure 10.18(b). The complex time domain spectrum can be broken down into its component waves of characteristic frequency and amplitude, and the corresponding frequency domain spectrum reconstructed, as shown in figure 10.18(b).

In the microwave region, where a typical cavity bandwidth is 1 MHz, it is unlikely that more than one rotational resonance will occur within the bandwidth, and it is then

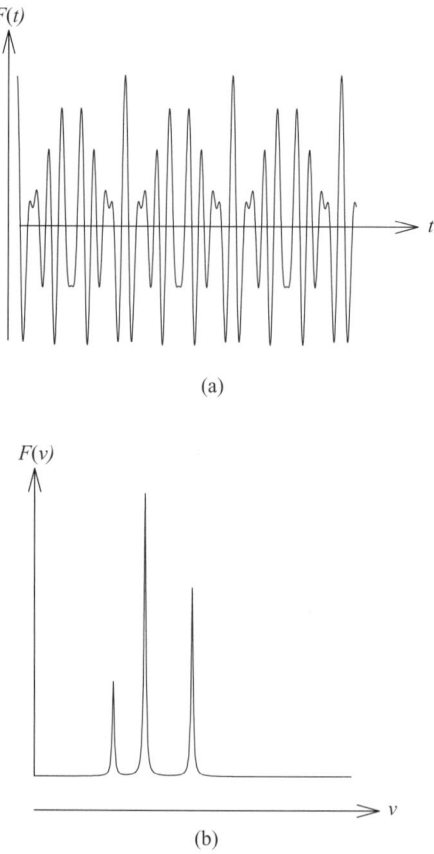

Figure 10.18. (a) Time domain spectrum for radiation of three different frequencies, ν, $0.7\,\nu$ and $0.5\,\nu$, with relative intensities 10, 15 and 7. (b) Frequency domain spectrum.

necessary to tune the resonant frequency of the Fabry–Perot cell. It should be clear that the technique needs detectors with extremely fast response times. A further requirement is a computer which can accomplish the necessary mathematical processing quickly enough in real time. We shall return to these matters later. We also note that a resonance line will have a finite width, often determined by the radiative lifetime of the upper state involved in the transition.

(d) Fabry–Perot cavities

The most important and unique part of a Fourier transform microwave spectrometer is the Fabry–Perot cavity. A fairly complete description of the principles has been given by Balle and Flygare [14] and we here summarise the main aspects, with the aid of figure 10.19. We use the cavity built by Balle and Flygare as a typical example. It is formed by two parallel, spherical concave mirrors made from solid aluminium. The mirrors are 36 cm in diameter, have a radius of curvature (R) of 84 cm, and are situated

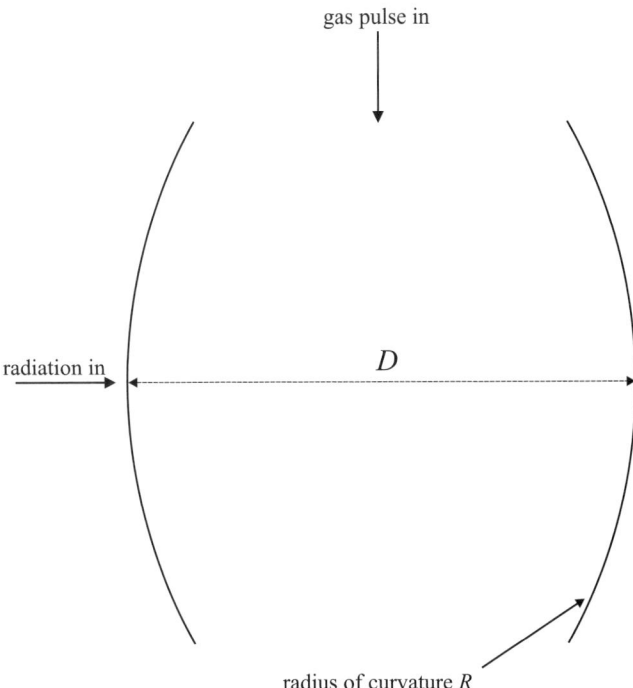

Figure 10.19. Geometry of a microwave Fabry–Perot cavity.

typically 50 to 70 cm apart (D). The fundamental resonance frequency depends upon the mirror separation, and the mirrors are mounted so that the separation can be accurately adjusted. The cavity is designed to operate in the frequency range 8 to 18 GHz. The microwave radiation is coupled into the cavity through an iris hole situated in the centre of one of the mirrors. For the designed frequency range the hole is 1 cm in diameter, and at the iris position the wall thickness of the mirror is ground down to about 0.5 mm.

The cavity supports an infinite number of standing wave patterns, called modes, which are designated TEM_{npq}, where n, p and q are integers. The microwave electric field distribution can be calculated for any mode, from a formula given by Balle and Flygare [14]. However, the dominant modes are those of the type TEM_{00q} and for these the resonant frequencies are

$$\nu_c = \left(\frac{c}{2D}\right)\left[q + \pi^{-1}\cos^{-1}\left(1 - \frac{D}{R}\right)\right]. \tag{10.10}$$

The term $c/2D$ gives the frequency separation between consecutive modes and is approximately 300 MHz for a mirror separation D of 50 cm. The shape of the cavity resonance is Lorentzian, and its full width at half height ($\Delta\nu_c$) is typically 1 MHz at a frequency ν_c of 10 GHz. The quality factor, Q, which is equal to $\nu_c/\Delta\nu_c$, is approximately 10^4, which is high.

The electric dipole interaction of the standing wave electric field in a Fabry–Perot cavity with a two-level system has been treated theoretically by Campbell, Buxton,

Balle, Keenan and Flygare [17], and by Campbell, Buxton, Balle and Flygare [18]. They found that the maximum polarisation of the gas in the cavity can, in a typical example, be achieved within a radiation pulse length of 1 μs. This means that a transition whose resonant frequency lies anywhere within the 1 MHz cavity bandwidth will be polarised. The macroscopic polarisation of the gas decays relatively slowly, with a typical relaxation time of 100 μs. The resulting electric field which is coupled out of the cavity is a decaying function of time, and is subjected to very fast Fourier transformation. The resulting frequency domain spectrum usually exhibits a doublet splitting, arising from the Doppler effect [17]. In addition to the original papers referenced, a clear summary of the most important aspects has been given by Legon [19].

(e) Signal detection

It is customary to use signal averaging to obtain a sufficient signal-to-noise ratio and this requires precise timing of the sequence of events. In order to eliminate any coherent content in repeated signals that are not molecular in origin, the pin-diode switches S_1 and S_2 can be pulsed at twice the rate of the solenoid valve which controls the gas pulses. Each cycle consists of a gas pulse, a microwave pulse synchronised to interact with the gas inside the cavity, and a second microwave pulse timed to enter the cavity after the gas has been evacuated. The digitised signal from the first radiation pulse is stored in a computer, whilst that arising from the second radiation pulse is subtracted from the first. With the fast pumping achieved with a large diffusion pump, the gas pulse can be repeated at a rate up to 10 Hz. It is convenient to display the accumulated time domain signal arising from successive cycles, and then to carry out the Fourier transformation needed to display the frequency domain spectrum.

Campbell, Read and Shea [20] were able to measure Stark effects by placing a pair of stainless steel parallel plates between the Fabry–Perot mirrors. Zeeman effects were observed by Campbell and Read [21] by placing both the gas nozzle and the Fabry–Perot cavity inside a superconducting solenoid. We may conclude that the pulsed FT microwave technique has high sensitivity and considerable versatility. The high sensitivity arises in part from the fact that by recording in the time domain, all frequencies in the spectrum are recorded at the same time. This is known as the multiplex or Fellgett advantage. Other pulsed Fourier transform microwave spectrometers have been built [22, 23] and this now seems to be a well established technique. Perhaps the only disadvantages are that the upper limit to the microwave frequency is about 40 GHz, partly because of the decreasing size of the Fabry–Perot cavity at higher frequencies, and partly because of the need for very fast detectors.

(f) Fourier transform far-infrared interferometry

Fourier transform techniques are used throughout the whole spectroscopic region, particularly in the infrared and visible. As we pass from the microwave region to the far-infrared, Fourier transform methods are still used, but based now on interferometry rather than pulsed methods. Perhaps this region of the spectrum will, in

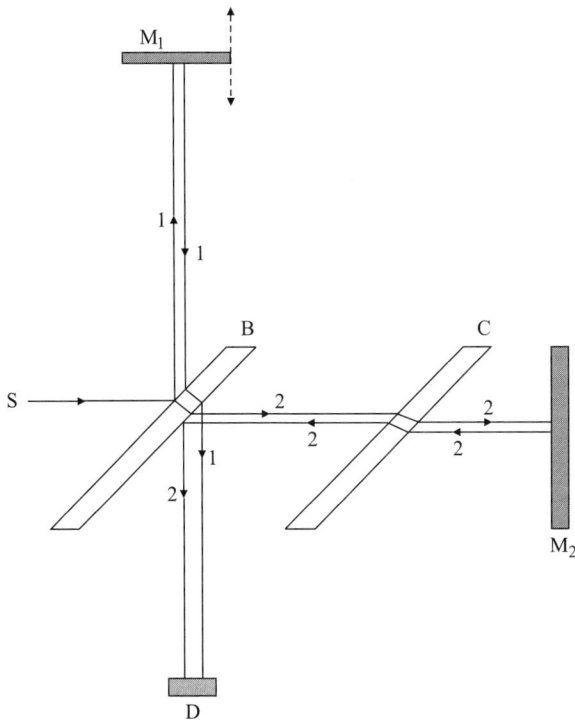

Figure 10.20. Principles of a Michelson interferometer [16].

the future, be dominated by the terahertz backward-wave oscillators, but interferometric methods are still important and we now describe some of the principles and methods.

We start by considering the original experiment of Michelson which is appropriate for the visible region of the spectrum. The essential idea is to split a beam of light into two equal parts which travel different path lengths before being recombined and detected; interference between the two parts can then be arranged to be either constructive or destructive, depending upon their respective transit times. Various ways of arranging this have been described but one of the simplest in illustrated in figure 10.20. We consider first a beam of *monochromatic* radiation from a source S entering the apparatus and striking a beam splitter B; half of the light beam (ray 2) passes through the beam splitter, but the other half (ray 1) is reflected through ninety degrees. Ray 1 is reflected by mirror M_1 back through B to a detector D. The position of the mirror M_1 is adjustable so that the total path length of ray 1 can be changed, the effects of which we will see in a moment. Meanwhile ray 2 proceeds to a second mirror, M_2, where it is reflected back to the backside of the beam splitter B; this side of B is also reflective so that ray 2 is reflected through ninety degrees to the detector D. The only other addition to the instrument is a compensating plate C, made of the same material as B, which is there to ensure that *both* rays have passed through this material twice before reaching the detector.

Although the path length for both rays from source to detector could be made identical, in general ray 2 will have travelled a distance δ further than ray 1. This distance is called a retardation, and it can be adjusted by moving the position of mirror M_1. If the wavelength of the radiation is λ, then if $\delta = 0, \lambda, 2\lambda, \ldots$, etc., the two rays interfere *constructively*, that is to say they add. On the other hand if $\delta = \lambda/2, 3\lambda/2, 5\lambda/2, \ldots$, etc., the two rays interfere *destructively* and cancel each other out; no signal is received from the detector. So this simple arrangement provides a method of determining the wavelength of the radiation; mirror M_1 is the wavelength- (or frequency-) determining element which any spectrometer system needs.

The detector output when the rays are constructively combining is a cosine wave which can be recorded and digitised. As we have seen, this time domain spectrum can then be subjected to Fourier transformation to provide a frequency domain spectrum (strictly speaking, a wavelength is transformed to its inverse or wavenumber). If instead of a source of monochromatic radiation we have the emission or absorption spectrum of a molecule against a continuous background radiation source, the length domain spectrum is more complex than a simple cosine function, but it can still be Fourier-transformed to yield the wavenumber domain spectrum we require. It will be clear that a practical and sensitive instrument requires very accurate and uniform movement of the mirror M_1; moreover the resolution depends upon making the displacement δ of M_1 as large as possible, perhaps as large as 1 m in a high-resolution instrument. The sensitivity of an interferometer is high because it is not necessary to reduce the source and detector apertures of the radiation to achieve high resolution as it is in a conventional grating spectrometer; this is known as the Jacqinot advantage. The simultaneous detection of a broad band of radiation is a further advantage for sensitivity, known as the multiplex or Fellgett advantage.

Careful and thorough analyses of the significant factors have been given by Strong and Vanasse [24, 25], particularly in relation to the application of interferometric methods in the far-infrared region of the spectrum. These considerations led them to design a lamellar grating far-infrared interferometer, covering a wavelength range from greater than 4 mm to less than 15 μm. This, in wavenumber units, is 2.5 to 67 cm^{-1}, or in frequency units from 75 to over 2000 GHz. Beam splitters, which are radiation amplitude-dividers are not efficient devices for producing coherent radiation in this wavelength range; Strong and Vanasse therefore turned to a special type of grating, which is a wavefront divider. Their grating is illustrated schematically in figure 10.21. It is constructed of two interleaved sets of glass plates which can be moved with respect to each other, either to produce a groove depth (h) of almost one inch, or at the other extreme, to form an almost uninterrupted plane mirror. One surface is moved with respect to the other at a rate v to achieve the desired modulation. The grating has a spacing (a) of 0.5 in, with a total of 24 grooves, and is combined with Czerny–Turner optics, described in detail in the original paper [25]. The instrument is effective over a wavelength range from 4 mm to 15 μm, and was used to obtain the first extensive rotational spectrum of CO, as we will describe later in this chapter.

Other authors have described Fourier transform spectrometers for operation in the millimeter and far-infrared regions. Plummer, Winnewisser, Winnewisser, Hahn

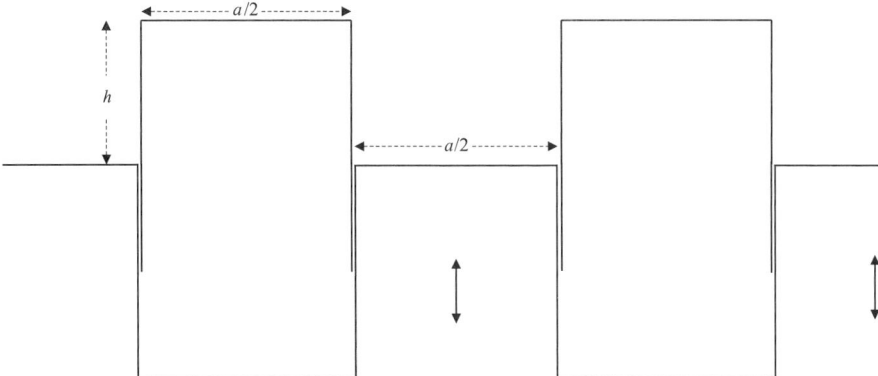

Figure 10.21. Profile of the lamellar grating designed by Strong and Vanesse [25] for a far-infrared Fourier transform spectrometer.

and Reinartz [26] have modified a Bruker IFS 120 HR infrared instrument, replacing the normal sample compartment with a custom-built cell and using a liquid helium cooled silicon bolometer as the detector. Schwarz, Guarnieri, Grabow and Doose [27] have built a novel Fourier transform instrument which operates in the millimeter-wave region, up to 100 GHz.

10.1.6. Radio telescopes and radio astronomy

(a) Introduction

The existence of molecular species in interstellar space has been known for almost seventy years. The first observations involved the electronic spectra, seen in absorption in the near-ultraviolet, of the CN, CH [28] and CH$^+$ [29] species. Radiofrequency lines due to hydrogen atoms in emission [30] and absorption [31], and from the recombination of H$^+$ ions with electrons were also known. However, molecular radio astronomy started with the observation of the OH radical by Weinreb, Barrett, Meeks and Henry [32] in 1963; in due course, this was followed by the discovery of CO [33] In the subsequent years over 110 molecules have been observed in a variety of astronomical sources, including some in galaxies other than our own. Nearly a third of these are diatomic molecules, with both closed and open shell electronic ground states, and some were observed by astronomers prior to being detected in the laboratory.

In this section, we shall restrict ourselves to those aspects of radio astronomy which are relevant to the study of the rotational spectra of diatomic molecules. We will not deal with the study of continuum sources, with cosmology, or with the detailed structure, dynamics and chemistry of interstellar clouds. These are important parts of astrophysics, covered in many research articles, reviews and books [34, 35, 36]. We will describe the main features of the dishes which collect radiation (i.e. the telescope), the detectors and signal processing equipment, and the analysis of the spectra. Many of the microwave spectra of diatomic molecules are now used as important probes to study the physics of interstellar clouds, but again we leave descriptions of these matters to the experts in these fields.

(b) Telescope dishes

The first element of a radio telescope is the receiving dish, always the most visually impressive part of the instrument. For the study of spectral lines the ideal dish must satisfy a number of requirements.

(a) The dish must collect incoming radiation and focus it on to the detector with maximum efficiency.
(b) The angular resolution of the dish should be as high as possible. The resolution depends on the physical size of the dish and the wavelength; it is not an adjustable parameter. Some molecular line sources are highly localised, and good angular resolution is necessary to locate and study them.
(c) The dish should be mounted such that its detection direction can be aligned in two independent orthogonal directions.
(d) The surface accuracy of the dish must be sufficiently high that desired signals are not degraded. Line measurements at frequencies of 300 GHz are now common; the corresponding wavelength is 1 mm, which imposes very severe requirements on the consistency of the dish surface.
(e) The dish and its mounting must be mechanically stable, so that distortions are not produced by weather fluctuations or gravitational forces. Telescope dishes are frequently contained within a protective housing, called a radome, to minimise problems arising from temperature, pressure and wind fluctuations.
(f) The earth's atmosphere results in absorption and diffraction of incoming radiation in the microwave region. In particular water vapour and ozone are strong absorbers in certain specific ranges of the microwave spectrum. Consequently radio telescopes are usually located at positions of high altitude in dry regions of the world. Mauna Kea in Hawaii, Pico Veleta in Spain, Mount Graham in Arizona and Effelsberg in the Alps are among the most important sites. Even in these optimum locations there are frequency regions of the microwave spectrum which are almost opaque. Another important consideration regarding the geographical location of a telescope is that different regions of the sky come into view from different locations. For example, it is not possible to see the galactic centre from the Northern hemisphere whereas it is from the Southern hemisphere.

Figure 10.22 illustrates the principles of the so-called altitude–azimuth drive which allows for adjustment of the parabolic reflector dish in two orthogonal directions. The azimuthal axis is vertical, whilst the elevation axis is horizontal. This figure describes the simplest arrangement, with only single focussing of the incoming radiation, the detector being placed at the focus of the dish. The orientation and alignment of the dish is computer controlled, and must be related to equatorial coordinates, which in turn are related to celestial coordinates. Because the earth is constantly changing its orientation, the transformation from equatorial to celestial cordinates is also changing.

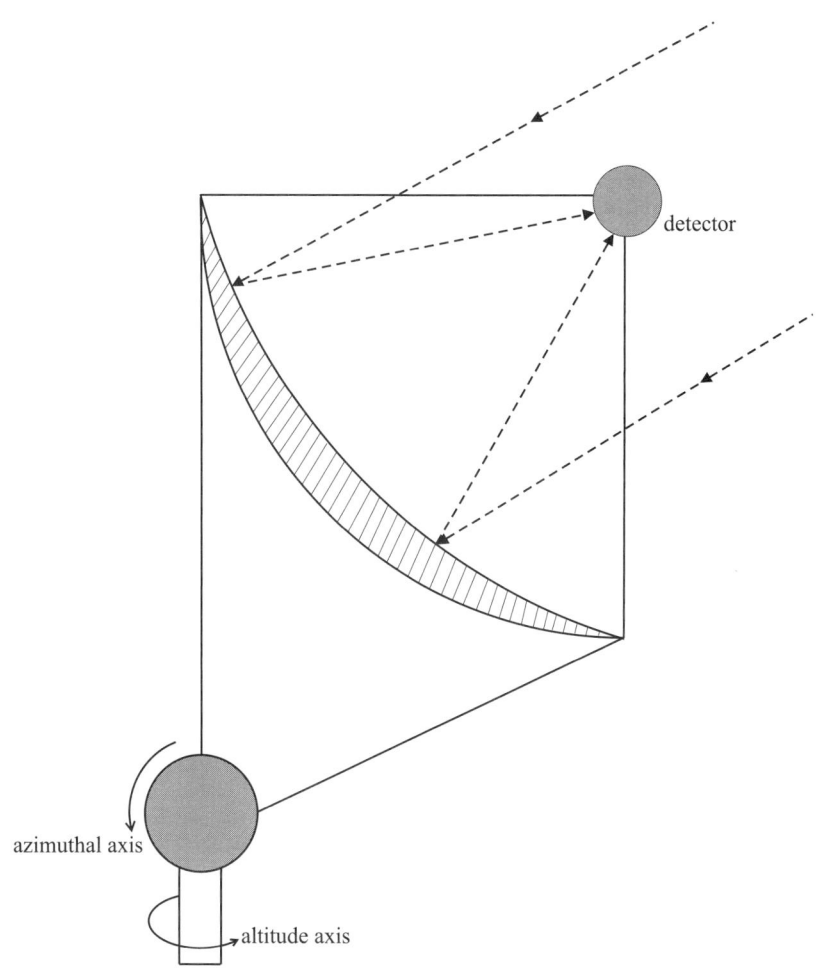

Figure 10.22. Principles of the altitude–azimuth drive for orientation of the telescope dish.

Consequently if a particular interstellar source is being studied for any length of time, the dish orientation must be continuously and precisely adjusted. In practice an overall pointing accuracy of 10 arcsec is commonly achieved.

In practice more sophisticated double focussing arrangements are usually used, several of which are shown in figure 10.23. The Cassegrain, Gregory and Nasmyth systems are in common use, and the offset Cassegrain system has the advantage that the secondary reflector does not block the field of view of the primary dish.

(c) Receiver/detector systems

The receiving antenna, placed at the focus of the dish, is a device for transforming an electromagnetic wave in free space into a guided wave; we have already encountered examples of such devices in some of the laboratory spectrometers described earlier

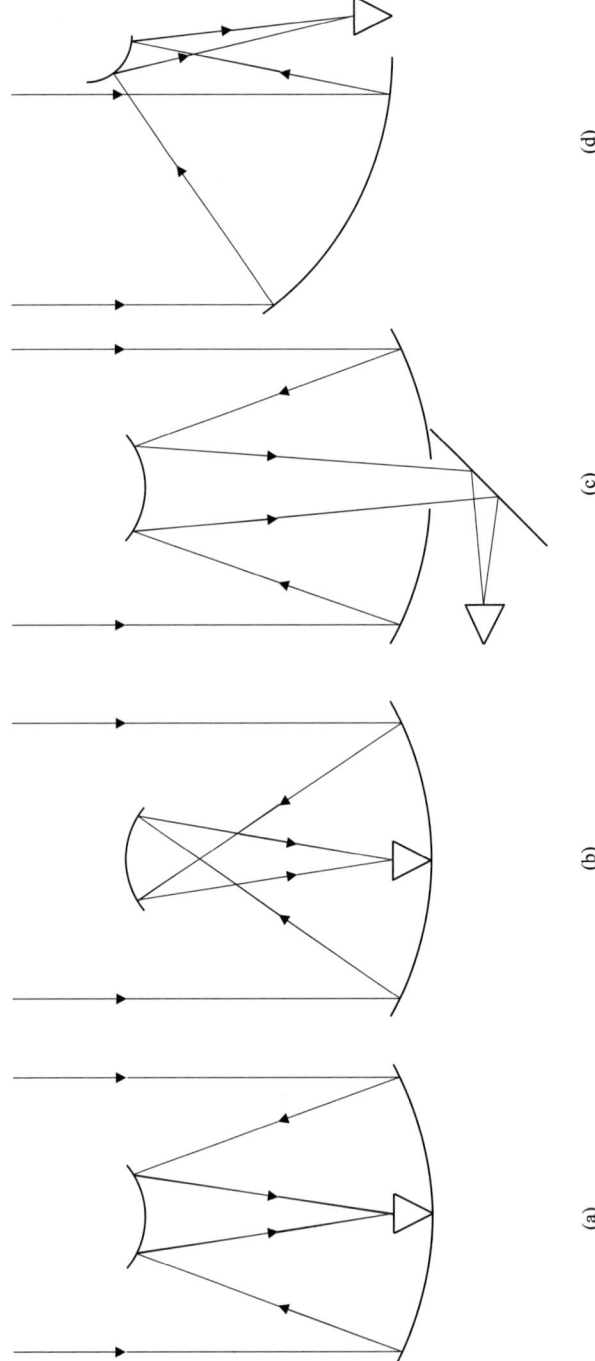

Figure 10.23. Geometry of different focusing systems and detector positions with parabolic radio telescope reflector dishes. (a) Cassegrain, (b) Gregory, (c) Nasmyth, (d) offset Cassegrain [36].

in this chapter. In early telescopes simple dipole antennae were used but these have now been replaced by microwave horns. Most parabolic dishes use circular horns, with rings forming a periodic structure, of depth $\lambda/4$, so that the microwave electric field in the aperture is oriented in the direction of propagation. Although microwave horns are relatively broad-banded devices, it is desirable to choose a horn feed which is most appropriate to the wavelength of the incoming radiation to be studied.

Following the receiver system we find a detector system whose ultimate purpose is to convert the incident microwave radiation into a measured voltage. The main elements of a typical radio telescope for studying frequency-discrete absorption or emission lines are illustrated in figure 10.24. The incoming radiation, which is generally broad band, is first amplified before being mixed with a local oscillator frequency; this is the element which makes the telescope frequency sensitive, and results in superheterodyne detection at an intermediate frequency ν_{IF}, which might be anywhere from a few MHz to more than 1 GHz. The intermediate frequency ν_{IF} is, of course, the difference between the source frequency ν_S and the local oscillator frequency ν_{LO}. In most cases the mixer itself is either a cooled field effect transistor amplifier, or a superconducting mixer. The first stage of the front end amplifier is often cooled to 3 to 5 K, both to

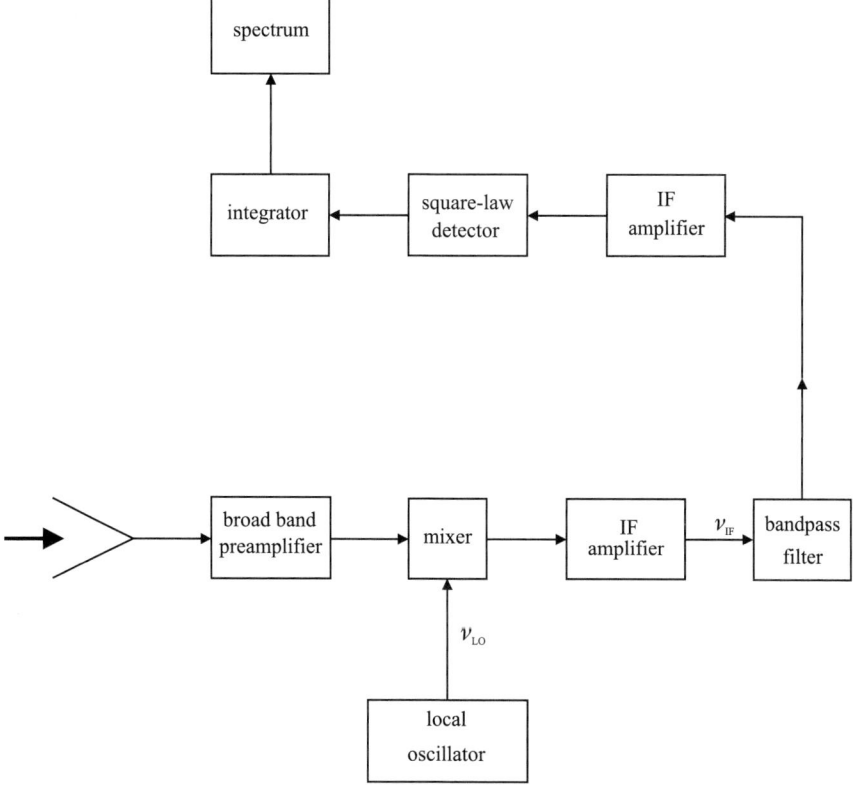

Figure 10.24. The principal elements of a radio telescope employing superheterodyne detection for recording frequency-discrete line emissions or absorptions [34].

reduce noise, and to allow the mixer to operate. In some instruments even sections of the microwave horn are cooled. In earlier work masers or parametric amplifiers were included, but these have been almost entirely superseded by HEMT (High Electron Mobility Transistors) transistor amplifiers, up to 100 GHz. At higher frequencies SIS (superconductor–isolator–superconductor) devices are used.

Following the frequency mixer, the intermediate frequency is amplified and filtered before being detected with a square-law detector. Signal-to-noise enhancement is then often achieved through the use of some kind of signal integrator before final display and recording. A frequency-resolved spectrum of the incoming radiation is obtained by scanning the local oscillator frequency, ν_{LO}. Alternatively, and probably preferably, the local oscillator frequency is fixed and the spectrum obtained by using a chain of IF preamplifiers and narrow band amplifiers, as shown in figure 10.25. This system is known as a multichannel superheterodyne receiver, and its intrinsic frequency resolution is determined by the IF step size, and hence the number of following amplifiers and detectors; up to 512 contiguous filters have been used!

Not all radio telescopes are earth bound, and we should mention a far-infrared telescope designed by Storey, Watson and Townes [37] which is designed to be flown aboard an aircraft at altitudes over 40 000 feet. A block diagram of the instrument is shown in figure 10.26. It comprises two separate chambers, the second of which houses the important optics and detector and is evacuated. The infrared radiation from an astronomical source is collected with a 91 cm telescope which has an angular resolution of one arc minute. The radiation is processed with two Fabry–Perot filters arranged in tandem. The first is a scanning instrument, whilst the second is a fixed wavelength filter with a bandwidth of about 5 μm. The detector is a liquid helium cooled gallium-doped germanium photoconductor. The first chamber contains a gas cell which can be filled with water vapour or ammonia to provide wavelength calibration.

This spectrometer has been used to study both CO and OH in extra-terrestial environments, and the results are discussed later in this chapter.

(d) Nature of discrete line spectra

Molecular line spectra may be observed either in absorption or emission. We now remind ourselves of the fundamental rules governing these processes which were described in chapter 6. Consider a pair of levels with energies E_2 and E_1, with N_2 and N_1 being the density of molecules in each state in a defined sample volume (see figure 10.27). There are three types of transition which can occur between the two levels. In the presence of a radiation field of energy density U, which may be the ubiquitous isotropic black body radiation at 2.7 K, or continuum radiation from another emitting source, stimulated absorption and emission will occur with equal probability (i.e. $B_{21}U = B_{12}U$). In addition, however, spontaneous emission from the upper to the lower state will also occur (A_{21} in figure 10.27), provided the upper level is populated. Einstein showed that the A and B coefficients are related in the following way. If our defined system is at thermal equilibrium, the number of absorbed and emitted photons

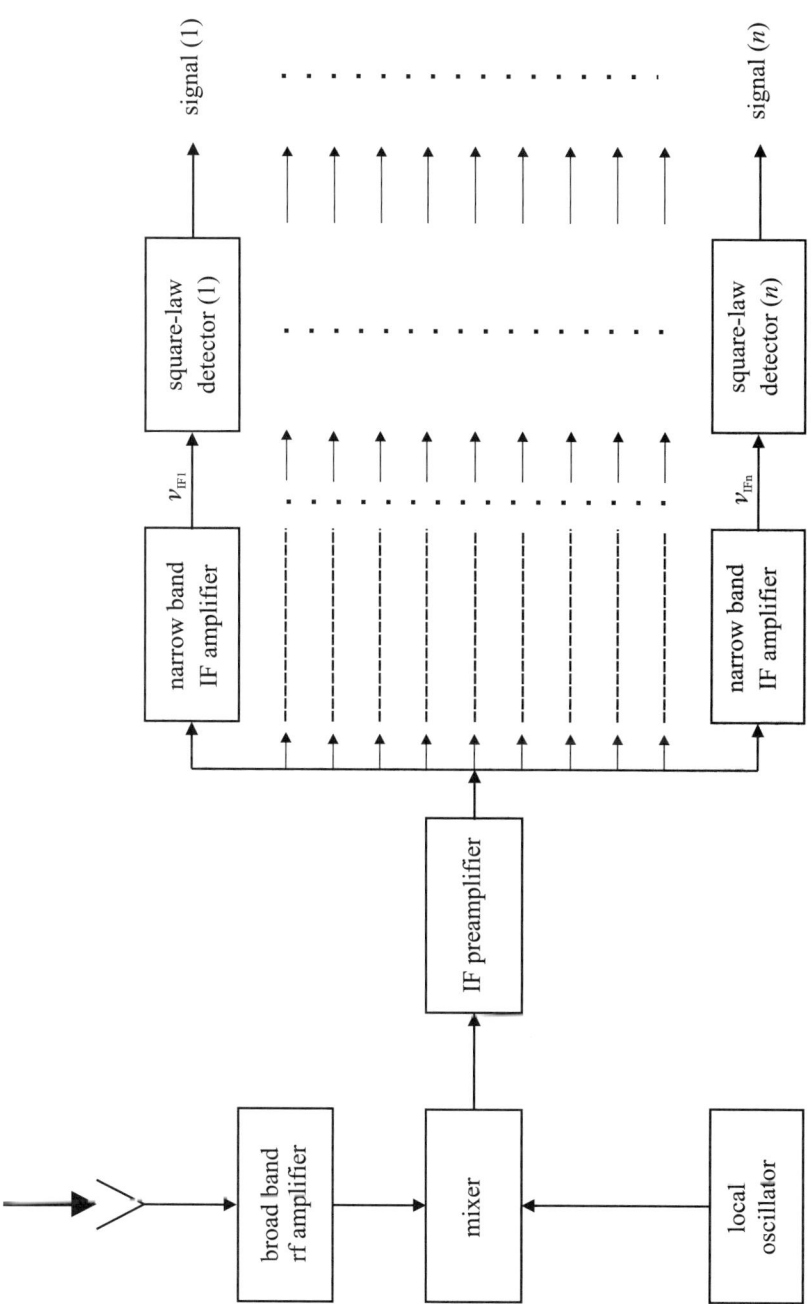

Figure 10.25. Multichannel superheterodyne receiver [34].

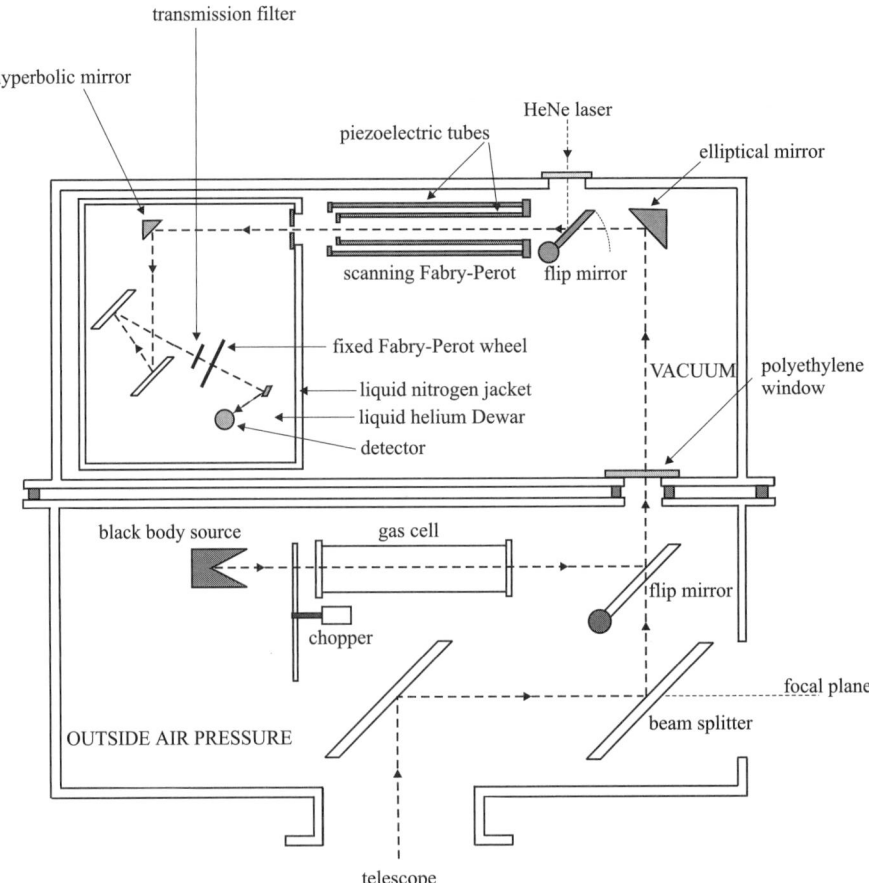

Figure 10.26. Block diagram of a far-infrared spectrometer designed for use aboard an aircraft flying at 41 000 feet [37].

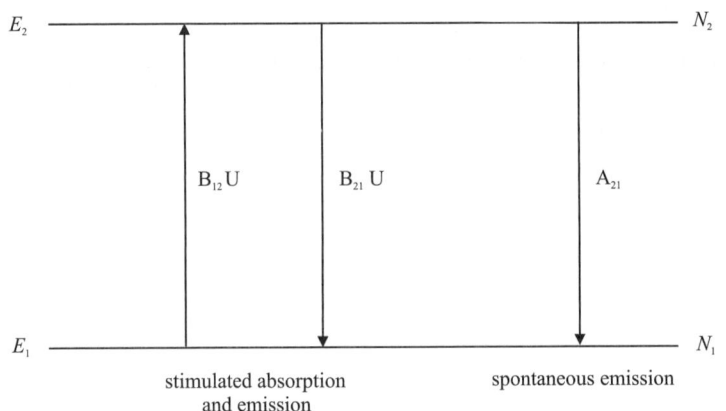

Figure 10.27. Spontaneous and stimulated transitions in a simple two-level system.

must be equal, so that

$$N_2 A_{21} + N_2 B_{21} U = N_1 B_{12} U. \tag{10.11}$$

If we assume a Boltzmann distribution of molecules between the two levels we may write

$$\frac{N_2}{N_1} = \frac{g_2}{g_1} \exp\left(-\frac{h\nu}{kT}\right), \tag{10.12}$$

where T is the temperature of the system in K, g_1 and g_2 are the statistical weights of the two states, and $\nu = (E_2 - E_1)/h$. From equations (10.11) and (10.12) we obtain the result

$$U = \frac{A_{21}}{(N_1/N_2)B_{12} - B_{21}} = \frac{A_{21}}{(g_1/g_2)\exp(h\nu/kT)B_{12} - B_{21}}. \tag{10.13}$$

Now for a system at thermal equilibrium, Planck's law for black body radiation tells us that the brightness distribution is given by

$$B_\nu(T) = \frac{2h\nu^3}{c^2} \frac{1}{\exp(h\nu/kT) - 1}, \tag{10.14}$$

so that the radiation density is given by

$$U = \frac{4\pi}{c} B_\nu(T) = \frac{8\pi h\nu^3}{c^3} \frac{1}{\exp(h\nu/kT) - 1}. \tag{10.15}$$

Comparing equations (10.13) and (10.15) we obtain the important results

$$g_1 B_{12} = g_2 B_{21}, \tag{10.16}$$

$$A_{21} = \frac{8\pi h\nu^3}{c^3} B_{21}. \tag{10.17}$$

If the effective temperature of our defined system is less than the universal radiation background temperature of 2.7 K, transitions between the two levels can be observed in absorption. This is the case with interstellar formaldehyde. Alternatively absorption can be observed against the continuum radiation from a nearby bright source. Spontaneous emission will always occur provided the upper of the two levels is populated, and can be observed if the populations are different. There are, in addition, examples of the exceptional situation in which $N_2 > N_1$; the result of this population inversion is that stimulated emission dominates, and maser emission is observed. Interstellar OH and SiO provide diatomic examples of this unusual situation, as also does interstellar H_2O; we shall describe the results for OH later in this chapter. Departures from local thermodynamic equilibrium are very common, and the concept of temperature in interstellar gas clouds is not simple; this is a major part of astrophysics which is, however, beyond the scope of this book.

Molecular line spectra in the astrophysical literature are often exhibited as a function of radial velocity of the source (in km s^{-1}), rather than as a function of frequency. If the rest frequency for a spectroscopic line is known accurately from laboratory measurements, the astronomical frequency can be used to determine the radial velocity of

Table 10.1. *Diatomic molecules observed in interstellar and circumstellar sources up to January 2000*

Molecule	State	Transition type	Location	Reference
CH	$A^2\Delta - X^2\Pi$	electronic	many	[28]
CN	$B^2\Sigma^+ - X^2\Sigma^+$	electronic	many	[28]
	$X^2\Sigma^+$	rotational	IRC+10216	[39]
CH$^+$	$A^1\Pi - X^1\Sigma^+$	electronic	many	[29]
OH	$X^2\Pi$	Λ-doublet	many	[32, 38]
CO	$X^1\Sigma^+$	rotational	IRC+10216, many	[33, 39]
SiO	$X^1\Sigma^+$	rotational	Sgr B2	[40]
CS	$X^1\Sigma^+$	rotational	IRC+10216	[39]
SO	$X^3\Sigma^-$	rotational	Orion A, Sgr B2	[41]
SiS	$X^1\Sigma^+$	rotational	IRC+10216	[42]
NS	$X^2\Pi_{1/2}$	rotational	Sgr B2	[43]
H$_2$	$X^1\Sigma^+$	vibrational	Orion Nebula	[44]
C$_2$	$A^1\Pi_u - X^1\Sigma_g^+$	electronic	Cygnus OB2	[45]
NO	$X^2\Pi_{1/2}$	rotational	Sgr B2	[46]
HCl	$X^1\Sigma^+$	rotational	OMC-1	[47]
PN	$X^1\Sigma^+$	rotational	Ori(KL), Sgr B2	[48]
NaCl	$X^1\Sigma^+$	rotational	IRC+10216	[49]
AlCl	$X^1\Sigma^+$	rotational	IRC+10216	[49]
KCl	$X^1\Sigma^+$	rotational	IRC+10216	[49]
SiC	$X^3\Pi$	rotational	IRC+10216	[50]
CP	$X^2\Sigma^+$	rotational	IRC+10216	[51]
NH	$A^3\Pi - X^3\Sigma^-$	electronic	zPER, HD 2778	[52]
CO$^+$	$X^2\Sigma^+$	rotational	M17SW, NGC7027	[53]
SiN	$X^2\Sigma^+$	rotational	IRC+10216	[54]
HF	$X^1\Sigma^+$	rotational	Sgr B2	[55]
SO$^+$	$X^2\Pi$	rotational	DR21-OH, Sgr B2	[56]

the source, through calculation of the Doppler shift. The conversion requires a calculation of the velocity of the receiving system, which contains contributions from the earth's rotation and also the motion of the centre of the earth relative to the barycentre of the solar system. Knowledge of the source velocity is obviously important in astronomy. In some cases multiple splitting of a spectroscopic line is observed in astronomical sources. Turbulence causes consequent differing radial velocities within the source, which leads to groups of molecules with differing Doppler shifts.

The number of molecules, diatomic and polyatomic, detected in interstellar space, continues to increase, and now totals well over one hundred. At the beginning of the third millenium 26 diatomic species have been observed; these are listed in table 10.1. Details of some of the spectroscopic transitions observed and their analyses are described later in this chapter.

10.1.7. Terahertz (far-infrared) spectrometers

A major development during the past few years has been the development in Russia of backward-wave oscillators which operate in the terahertz (THz) frequency range, and hence bridge the gap between the millimetre-wave and far-infrared regions of the spectrum. In consequence rotational transitions of very light molecules like the hydrides become accessible. Several groups are currently involved in terahertz or near-terahertz spectroscopy and we summarise the main aspects. As we will see, successful experiments using tunable far-infrared radiation generated without the use of high-frequency backward-wave oscillators have also been described; we come to those in due course.

Winnewisser and his colleagues have described several versions of what they call the Cologne terahertz spectrometer, and we shall consider two of these versions. The first, described by Winnewisser and his colleagues in 1994 [57, 58], is illustrated schematically in figure 10.28. The most important element is a high-frequency backward-wave oscillator (BWO) which produces radiation up to a frequency of 2 THz, and, depending on the frequency, can produce output powers up to 50 mW or more. A unique feature of the high-frequency BWO is the presence of a three-dimensional periodic array of square posts; many thousands of such posts form a slow wave structure, in which the electrons are alternately accelerated and decelerated, producing high-frequency radiation. A small portion of this radiation is mixed with a high harmonic of a synthesiser, and a phase-lock loop (PLL) system stabilises the BWO frequency; scanning and modulating the synthesiser frequency results in scanning and modulation (at typically 20 kHz) of the primary BWO frequency. The mixer-multiplier consists of a planar Schottky diode, mounted at the end of a grooved millimeter waveguide; the diode is placed at the focal point of a semiparabolic mirror illuminated with the submillimeter radiation. The major portion of the radiation from the BWO passes through a free space absorption cell and is detected with a helium-cooled InSb hot electron bolometer. The final 20 kHz signal is fed to a lock-in amplifier. This spectrometer has been combined with a pulsed supersonic jet to study van der Waals complexes [59].

The second Cologne spectrometer [60] we describe is illustrated schematically in figure 10.29. The tunable far-infrared radiation is generated by mixing the far-infrared radiation from a fixed-frequency laser (typically a carbon dioxide laser-pumped methylene fluoride laser oscillating at 1626.6 GHz), with a tunable BWO oscillating typically over the range 280 to 380 GHz. The BWO is locked to a harmonic of a microwave synthesiser, which controls the scanning and the modulation.

Other groups have built tunable far-infrared spectrometers which do not involve high-frequency backward-wave oscillators. Verhoeve, Zwart, Versluis, Drabbels, ter Meulen, Meerts, Dymanus and McLay [61] have described a system in which fixed frequency far-infrared radiation is mixed with tunable microwave radiation in Schottky barrier diodes. This instrument has been operated up to 2.7 THz, and used to study OD and N_2H^+. A similar system, combined with a continuous supersonic jet, has been described by Cohen, Busarow, Laughlin, Blake, Havenith, Lee and Saykally [62]. This instrument was used to study rare gas/water clusters.

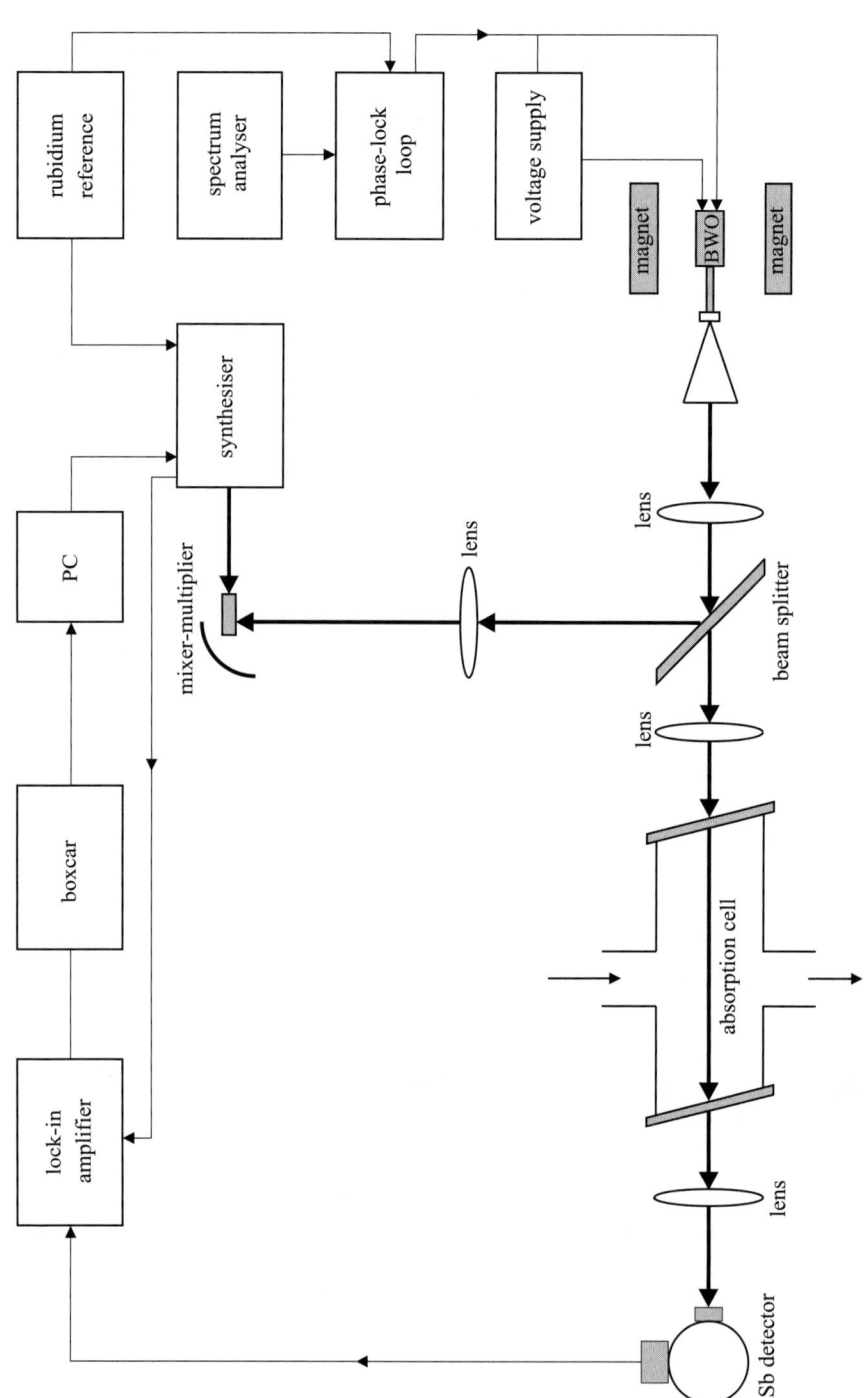

Figure 10.28. Schematic diagram of the first Cologne terahertz spectrometer [58].

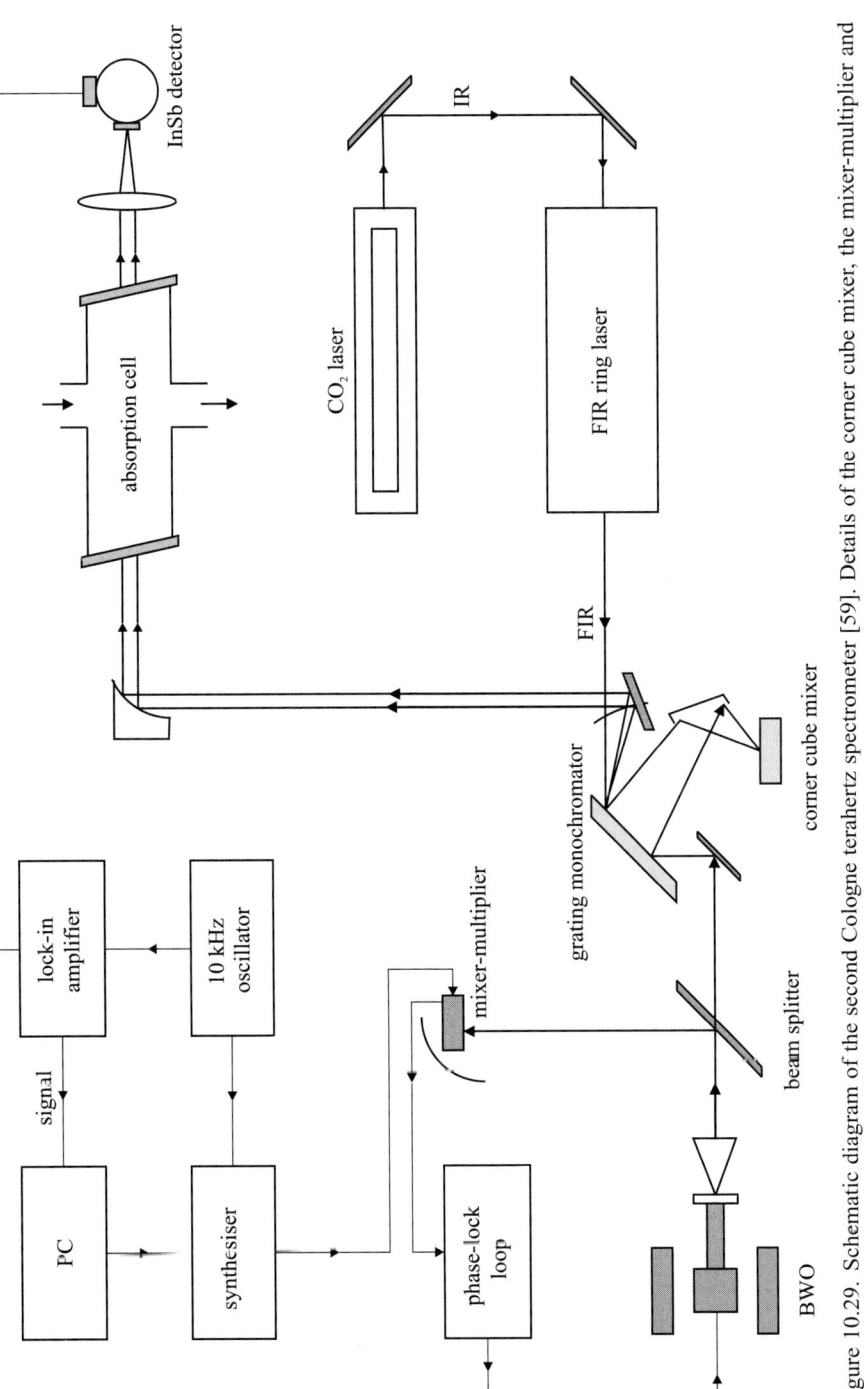

Figure 10.29. Schematic diagram of the second Cologne terahertz spectrometer [59]. Details of the corner cube mixer, the mixer-multiplier and the far-infrared laser are not included.

One of the most important examples of terahertz spectroscopy is the recent detection of the lowest rotational transition of the CH radical by Amano [63] and we shall also describe his spectrometer in some detail. Strictly speaking Amano's work, which is in the 500 to 600 GHz region, does not reach 1 THz. However, CH is a very important species and we make no apology for describing Amano's work in some detail. As we have described earlier, backward-wave oscillators have the advantage of being frequency-tunable over a wide range, but their intrinsic frequency and output power stabilities are not high. The BWO is therefore locked to the harmonic of a Gunn oscillator whose fundamental range is 80 to 110 GHz; the Gunn oscillator is itself phase-locked to the harmonic of a frequency synthesiser (I) operating over the range 12 to 18 GHz. Efficient mixing of the frequency pairs is achieved with GeAs harmonic mixers designed by Winnewisser and his colleagues. The beat frequency (IF frequency) between the BWO and the Gunn oscillator was chosen to be 345 MHz, generated from a second frequency synthesiser (II). Synthesiser (II) was frequency modulated at 10 kHz, so that this modulation was carried through to the BWO (i.e. source modulation). Any absorption resonance line was therefore frequency-modulated at 10 kHz; secondary modulation at 11 Hz was applied to the discharge used to produce free radicals in a free-space absorption cell. The power transmitted through the cell was detected with an indium-antimonide detector, and the resonance signal first demodulated with a lock-in amplifier at 10 kHz, and subsequently demodulated with a second lock-in amplifier referenced at 11 Hz. The complete system is summarised in figure 10.30; it should be added that the double modulation system employed by Amano [63], which largely overcomes baseline drift problems, was introduced earlier by Amano and Hirota [64].

Amano succeeded in observing the lowest rotational transition of the CH radical, generated by means of a glow discharge in mixture of CH_4 and helium. The observation of this spectrum is a major achievement in microwave spectroscopy; we show a part of the spectrum in figure 10.31, exhibiting two proton hyperfine lines, but defer a detailed discussion of the energy levels and spectrum until later in this chapter.

Although there seems little doubt that the development of the high-frequency BWO will have a major impact on the development of spectroscopy in this region, it is not the only route into this hitherto unexplored area. As we have already described, other methods of generating tunable far-infrared radiation have been developed. Evenson, Jennings and Petersen [65] and Evenson, Jennings and Vanek [66] have described two different far-infrared sources, both of which depend primarily on mixing the radiation from two mid-infrared carbon dioxide lasers. One of these instruments is illustrated schematically in figure 10.32. Infrared radiation from two carbon dioxide lasers is first passed through 90 MHz opto-acoustic modulators and then focussed onto a tungsten or cobalt MIM diode along with microwave radiation (ν_m) from a 2 to 20 GHz synthesiser. The latter adds tunable sidebands to the carbon dioxide difference frequency. The opto-acoustic modulators (AOM) serve to isolate the lasers from the diode, and to provide an additional 180 MHz of frequency range. The far-infrared radiation of a few tenths of a microwatt has a frequency

$$\nu_{FIR} = (\nu_I - \nu_{II}) \pm \nu_m. \qquad (10.18)$$

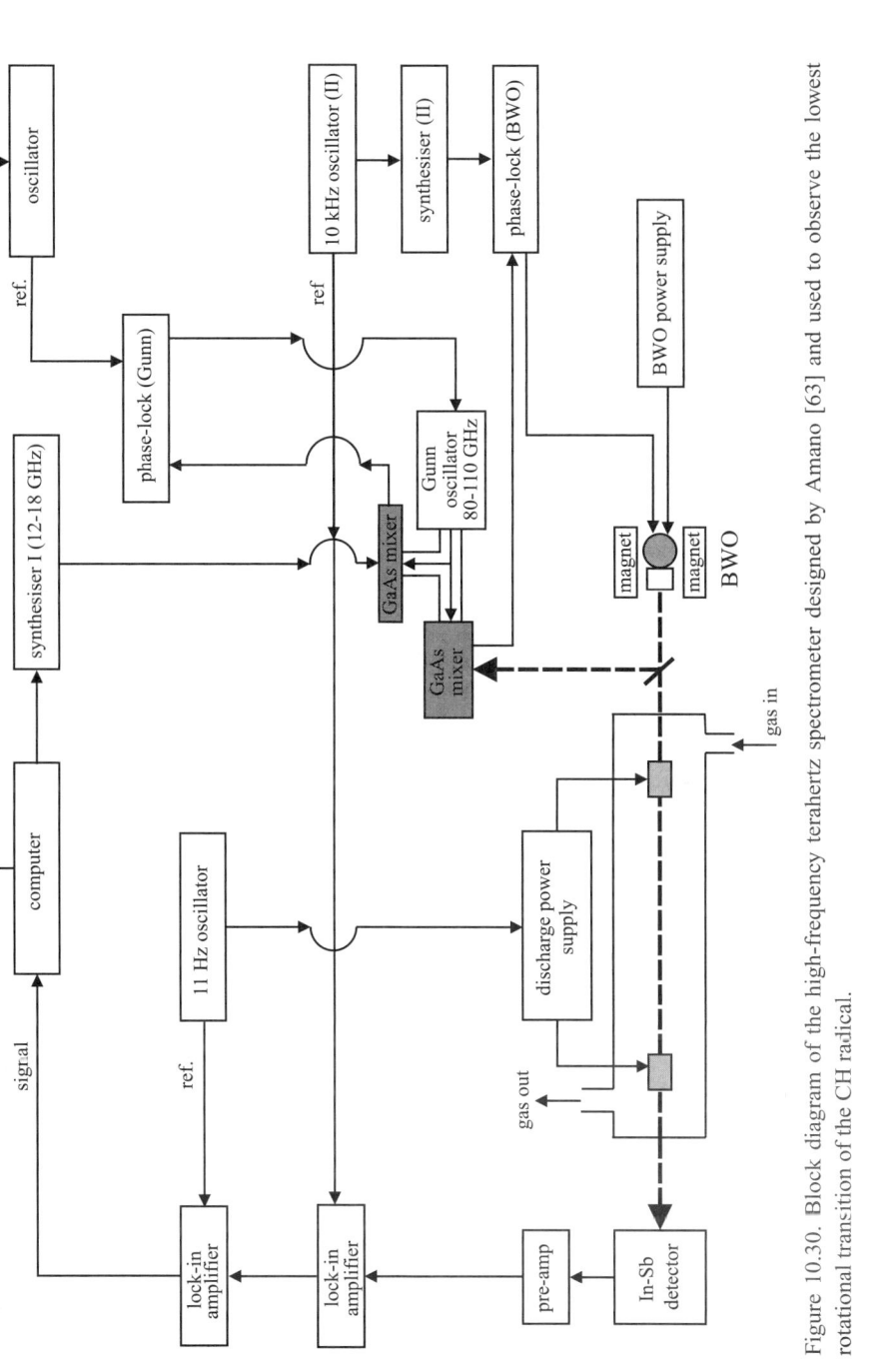

Figure 10.30. Block diagram of the high-frequency terahertz spectrometer designed by Amano [63] and used to observe the lowest rotational transition of the CH radical.

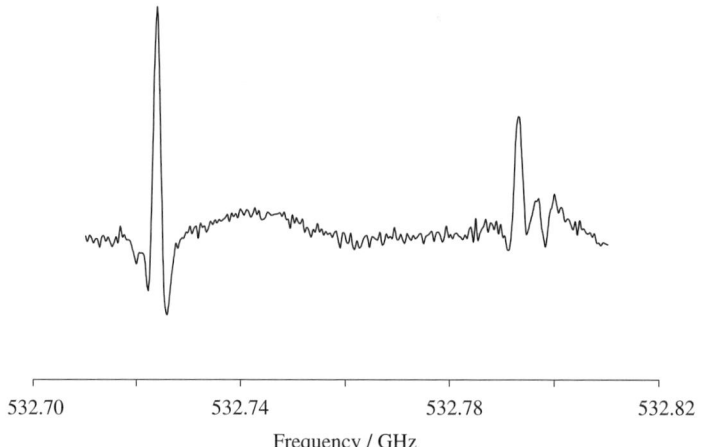

Figure 10.31. Part of the $J = 3/2 \leftarrow 1/2$ transition in CH recorded by Amano [63], with proton hyperfine splitting. The assignment of the spectrum is discussed later in this chapter.

It passes through the sample absorption cell made of Pyrex with polyethylene windows and is detected with a liquid helium cooled bolometer. One of the lasers is frequency-modulated at 1 kHz and the detector output is processed with a lock-in amplifier, as shown. Far-infrared rotational spectra of CO, HCl and HF have been recorded [67], and as an example of the excellent sensitivity achieved, we refer the reader to the spectrum of the OH radical [68] shown later in this chapter. Evenson's spectrometer operates over a wide range of the far-infrared region up to 9 THz, with excellent frequency stability.

It will be clear from this subsection that much skillful and imaginative instrument design, by a number of different groups, has been directed towards the development of far-infrared spectroscopy. Quite apart from the developments in laboratory spectroscopy, the impact on astronomy in this region of the spectrum is of major importance. A high power tunable far-infrared source can serve as the local oscillator for the detection of far-infrared interstellar radiation. We can anticipate exciting developments in this field.

10.1.8. Ion beam techniques

In recent years Carrington and his colleagues have developed ion beam techniques for studying the spectra of molecular ions, particularly in vibration–rotation levels lying close to the dissociation limit. Some of the spectra observed were vibrational spectra, obtained using mid-infrared laser sources; we will not deal with these results. In other cases, however, the spectra have been obtained in the microwave region, and they include the observation of rotational transitions. In reality, the conventional classification of spectra as being rotational, vibrational or electronic becomes increasingly unsatisfactory for levels close to dissociation, and microwave spectroscopy can involve all three types of spectroscopic transition. In some ways, chapters 8 or 11 might have been

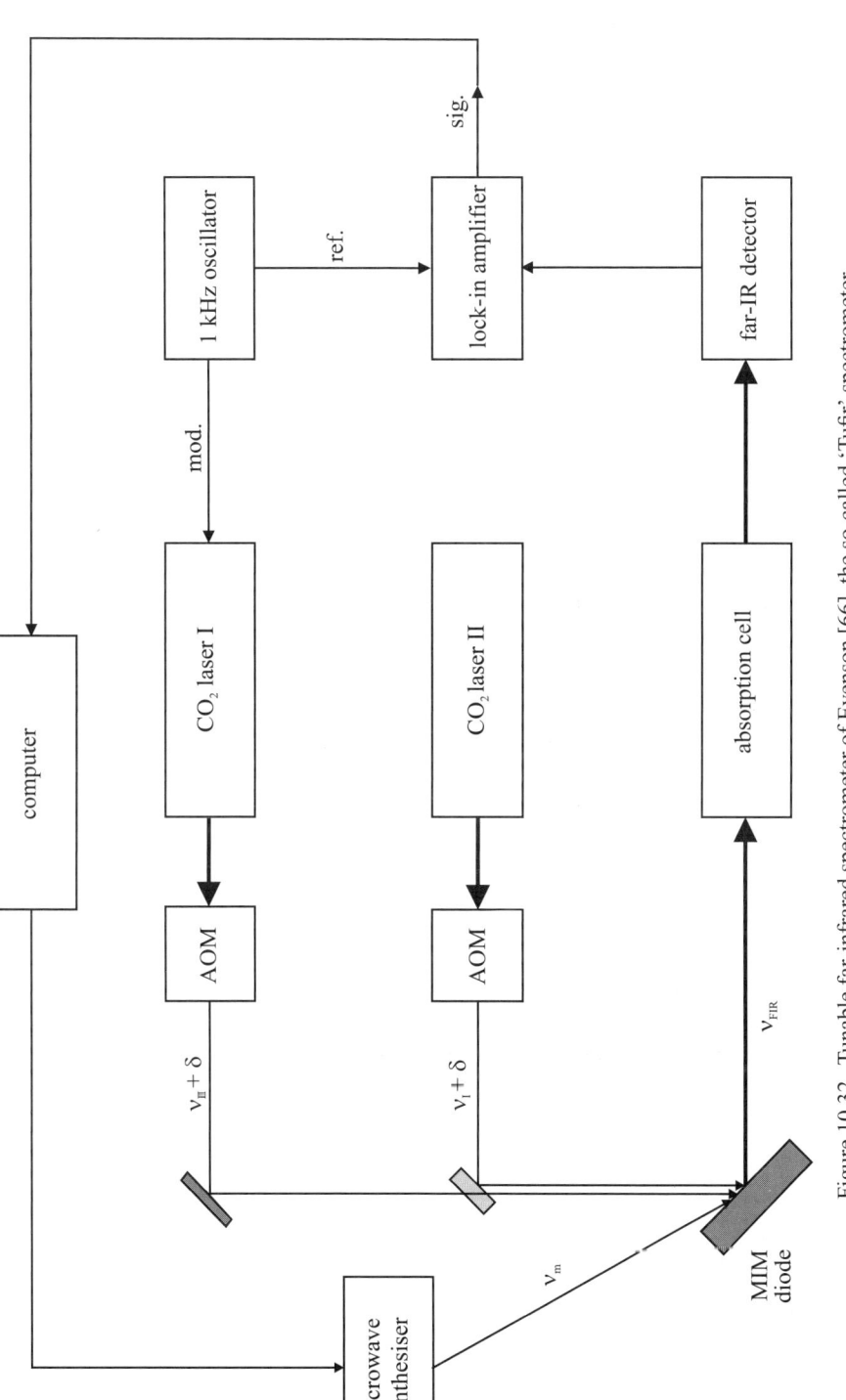

Figure 10.32. Tunable far-infrared spectrometer of Evenson [66], the so-called 'Tufir' spectrometer.

considered as more appropriate homes for descriptions of this work, and indeed, the ion beam double resonance studies which have been performed will be described in chapter 11. On the other hand, the single resonance experiments do not depend upon the specific state selection characteristic of conventional molecular beam magnetic or electric resonance, although some state selection does enhance the sensitivity, as we shall see. Pure microwave rotational transitions are involved, at least in part, so these studies do belong in this chapter!

There is another classification difficulty. In this chapter we have arranged our discussion of open shell molecules on the basis of Hund's case (a) or case (b) coupling. However the near-dissociation spectra of the molecular ions studied are much better understood in terms of Hund's case (c) coupling, even tending towards Hund's case (e). Consequently we group the results together in section 10.7, for case (c) molecules.

A number of different ion beam instruments have been developed, and an excellent review has been provided by Cox, Critchley, McNab and Smith [69] but figure 10.33 illustrates the instrument used by Carrington, Shaw and Taylor [70] for the study of microwave spectra involving near-dissociation vibration–rotation levels in rare gas diatomic ions. The instrument is essentially a commercially available tandem mass spectrometer, designed and manufactured by Vacuum Generators Ltd. The molecular ions are produced by electron bombardment of either a slowly flowing gas or gas mixture, or a supersonic neutral molecular beam. The ions are accelerated out of the source to potentials up to 10 kV, forming a tightly focussed ion beam. This beam is directed through a magnetic sector, which acts as a mass analyser, and ions of the desired charge-to-mass ratio can be selected by fixing the magnetic field at an appropriate value. The ion beam passes through an intermediate region, to which we

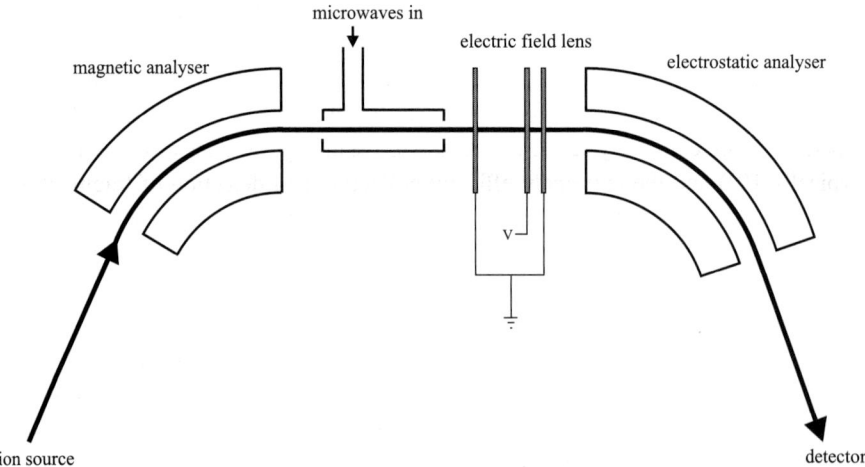

Figure 10.33. Tandem mass spectrometer system employing electric field dissociation, designed to enable the study of microwave spectra of molecular ions involving vibration–rotation levels lying close to the dissociation limit.

return shortly, and fragment ions produced in this region are separated from the parent ions by means of an electrostatic analyser (ESA), which is a kinetic energy analyser. After deflection through the ESA the fragment ions are detected by means of a high-gain electron multiplier. As we will see, spectra are detected by monitoring changes in the fragment ion intensity; no attempt is made to detect the absorption or emission of radiation directly. In this respect, the ion beam experiments are similar to the magnetic and electric resonance experiments described in chapter 8.

Two important elements of the overall experiment are to be found in the intermediate region between the magnetic and electric sectors. First, the molecular ion beam passes through a section of waveguide where it is exposed to radiation in the frequency range 6 to 170 GHz. Vibration–rotation levels lying close to the dissocation limit are populated by the ion bombardment and formation processes occurring naturally in the ion source. No other means of populating these high-lying levels is employed. Spectroscopic transitions induced in the microwave field result in population transfer between vibration–rotation levels, and this population transfer results, as we shall see, in fragment ion intensity changes. These changes are detected by the electron multiplier, as a function of the radiation frequency.

The second important element in the intermediate region is an electric field lens, which follows the waveguide cell. Figure 10.33 shows the form of this lens; a strong electric field is established between the centre plate which is held at a high positive or negative potential and a closely spaced earth plate. The first earth plate is situated further upstream, so that the electric field between it and the central plate is relatively small. Ions in weakly bound levels are dissociated in the lens to form fragments having characteristic kinetic energies which depend upon the potential within the lens at which fragmentation occurs. The lens therefore acts as a partial state selector; only weakly bound levels are affected at all, and there is some discrimination between these levels, all of which lie less than 10 cm^{-1} below the dissociation limit.

As already stated, spectra are detected by monitoring changes in the electric field-induced fragment ion intensity as a function of the microwave frequency. These changes are, in practice, measured by amplitude modulation of the microwave power, and detection of the resonant fragment ion current by means of a lock-in amplifier. The experiments are extremely sensitive because of the high gain of the electron multiplier (typically 10^6) and the extremely efficient collection and detection of fragment ions (close to 100%). Resonant spectra involving the detection of 10 to 100 ions s^{-1} are routinely recorded; the parent ion beams have total flux intensities of about 10^9 to 10^{10} ions s^{-1} in most cases, of which only a very small proportion occupy the high-lying energy levels of interest.

There are several other important aspects of the experiment which should be mentioned. The waveguide cell is surrounded by a solenoid coil which can produce a magnetic field parallel to the ion beam direction; the magnitude of this field (up to 50 G) is often sufficient to produce observable Zeeman splittings which greatly assist spectroscopic assignment, as we will see. It is also possible to expose the molecular ion beam to two different microwave frequencies; this so-called double resonance technique enables two different microwave transitions to be connected, if they share a common energy level. We will see the power of these techniques later in this chapter. A further aspect is that the microwave radiation propagates in directions both parallel and antiparallel to the direction of the ion beam. The ions are moving rapidly at the high beam potentials used, so that there are always Doppler shifts of the resonances, which can be accurately calculated.

The techniques described have been used to study the fundamental H_2^+ ion, and its important isotopomers D_2^+ and HD^+. Our description of these particular experiments is postponed to chapter 11. Later in this chapter we will describe microwave experiments on the rare gas ions $HeAr^+$, $HeKr^+$ and Ne_2^+, for which rotational transitions, among others, have been studied.

10.2. $^1\Sigma^+$ states

The most extensively studied group of diatomic molecules are those with $^1\Sigma^+$ ground states. These do not exhibit the interesting and informative complications characteristic of open shell systems, but, as we have seen earlier, the presence of nuclei with non-zero spins or electric quadrupole moments can give rise to hyperfine structure. Electric resonance experiments have been an important source of information, but this technique depends upon efficient electrostatic focussing, and is therefore confined to the study of the lowest rotational levels. An important aspect of conventional microwave absorption spectroscopy is the information it can provide about an extended sequence of rotational levels and, in some cases, the additional effects arising from vibrational excitation. In this section, we describe examples of increasing complexity. We start with the simplest cases of molecules with two nuclei of zero spin, and work towards the most complicated cases where both nuclei have spin magnetic moments and electric quadrupole moments. In particular we will deal with cases where a sequence of rotational levels has been studied; the recent development of very high frequency sources has made possible the study of higher rotational transitions in heavy molecules, and lower rotational transitions in light molecules like hydrides. We also note that electric dipole rotational transitions require the presence of a permanent electric dipole moment, and therefore occur only for heteronuclear molecules.

10.2.1. CO in the $X\,^1\Sigma^+$ ground state

Carbon monoxide may be bad for your health, but it has helped to provide a living for generations of spectroscopists, working in all regions of the electromagnetic spectrum. The reasons are clear. It is a stable, readily available gas, with a ground state dissociation energy of more than 11 eV. Consequently the ground state potential supports a large number of vibrational levels, which are accessible through vibrational or electronic spectroscopy. It has been a test bed over many years for the development of new experimental and theoretical techniques. It exists in six different naturally occurring isotopic variants, namely $^{12}C^{16}O$, $^{13}C^{16}O$, $^{12}C^{17}O$, $^{13}C^{17}O$, $^{12}C^{18}O$ and $^{13}C^{18}O$. When required, isotopic enrichment is relatively straightforward, and the existence of these different species has provided the opportunity to examine isotopic and mass dependencies in molecular spectroscopy. The absence of nuclear spin effects in the dominant isotopic form, $^{12}C^{16}O$, means that spectroscopic features exhibiting important variations with vibrational or rotational excitation are more readily amenable to accurate study. In this section we concentrate on the rotational spectroscopy of CO, but with some reference to the associated vibration–rotation spectroscopy. Following the initial discovery in 1970 of CO in interstellar space by Wilson, Jefferts and Penzias [33], its widespread distribution in space has made it almost as important to astronomers as it is to ground based spectroscopists. In our following discussion references to CO will apply to the dominant isotopic species, unless otherwise indicated.

The first direct measurement of the $J = 1 \leftarrow 0$ rotational transition in the $v = 0$ level of the electronic ground state was due to Gilliam, Johnson and Gordy [71]. This was followed by more accurate measurements, due first to Gordy and Cowan [72], and then to Rosenblum, Nethercot and Townes [73]. The most recent measurements of the lowest rotational transition, and many higher transitions, are those by Varberg and Evenson [74] using the tunable far-infrared spectrometer described in the previous section, and by Winnewisser, Belov, Klaus and Schieder [75] and Belov, Lewen, Klaus and Winnewisser [76] using the terahertz spectrometer, also described earlier. Almost fifty years separate the first and the most recent studies, and during this period other measurements in the far-infrared using interferometric methods were described by Loewenstein [77]. Figure 10.34 illustrates the progress over a forty year period; figure 10.34(a) shows a sequence of rotational transitions for CO observed with a grating interferometer [75]; the line width is 0.4 cm^{-1}. Figure 11.34(b) shows the line arising from the $J = 4 \leftarrow 3$ transition, with a Lamb dip, obtained using the Cologne terahertz spectrometer [75]. The line width is 40 kHz. The most recent results are summarised in table 11.2, which is taken from reference [75].

The calculated values in the table were obtained by fitting the results to the conventional formula for the rotational energies

$$E_0(J) = B_0 J(J+1) - D_0 J^2 (J+1)^2 + H_0 J^3 (J+1)^3, \quad (10.19)$$

and the values of the constants obtained were (in MHz),

$$B_0 = 57\,635.968\,019\,(28), \quad D_0 = 0.183\,504\,89\,(16), \quad H_0 = 1.716\,8\,(10) \cdot 10^{-7},$$

where the values in parenthesis are 1σ standard errors.

It should be noted that the rotational spectroscopy of CO confined to a single vibrational level, usually the ground $v = 0$ level, provides only a limited amount of information about molecular structure. In the field of vibration–rotation spectroscopy, however, CO has been studied extensively and particular attention paid to the variation of the rotational and centrifugal distortion constants with vibrational quantum number. Vibrational transitions involving v up to 37 have been studied with high accuracy [78, 79, 80], and the measurements extended to other isotopic species [81] to test the conventional isotopic relationships. CO is, however, an extremely important and widespread molecule in the interstellar medium. CO distribution maps are now commonplace and with the advent of far-infrared telescopes, it is also an important

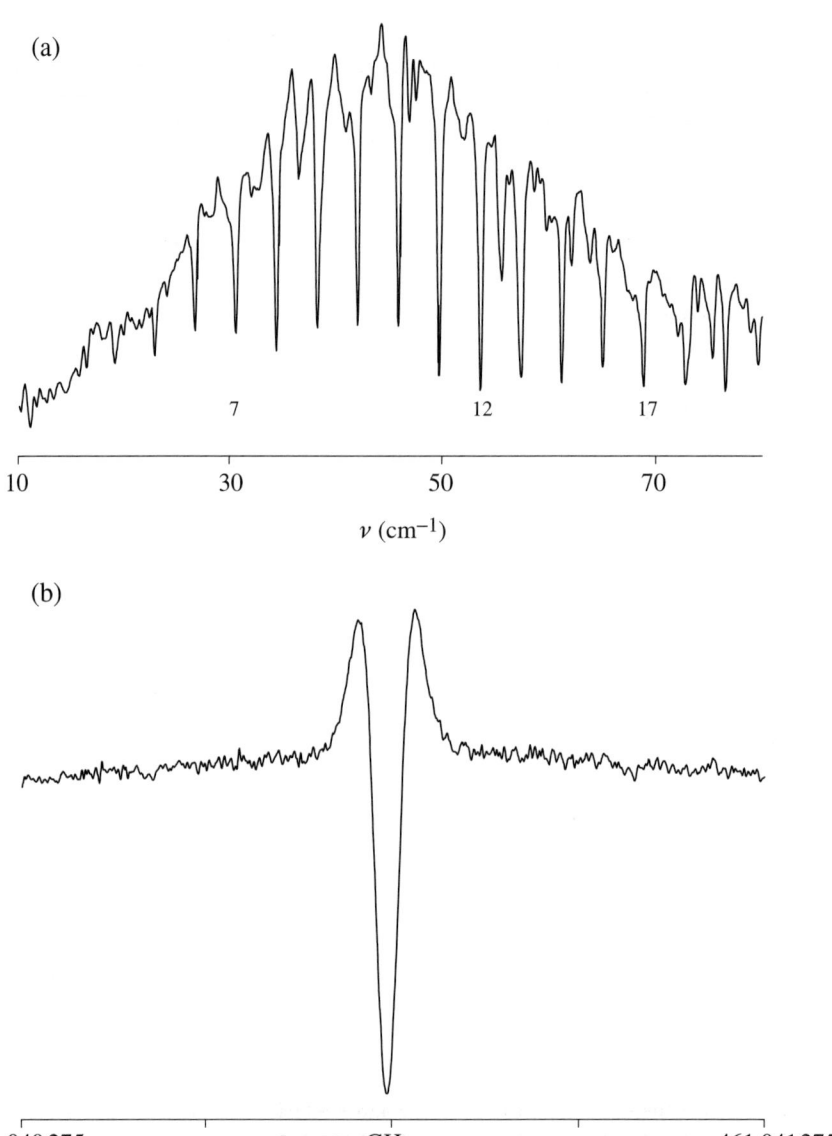

Figure 10.34. (a) Sequence of CO rotational transitions recorded in 1960 by means of a grating interferometer [77]. The lower J values are indicated for three of the lines. (b) Lamb dip spectrum of the $J = 4 \leftarrow 3$ rotational transition, recorded in 1997 [75].

temperature and density probe. Rotational transitions involving J up to 34 have been detected in the shocked gas associated with Orion [170] and it has been estimated that 20% of the cosmic abundance of carbon exists in the form of CO. The very simplicity of the rotational spectrum of CO, which makes it one of the less interesting molecules in the laboratory, makes it a valuable probe of both chemical and physical conditions in interstellar environments.

Table 10.2. *Observed and calculated rotational transition frequencies of $^{12}C\,^{16}O$*

J'	J''	ν_{exp} (MHz)	$\Delta\nu_{exp}$ (kHz)	ν_{calc} (MHz)	σ_{calc} (kHz)	O–C (kHz)
1	0	115 271.2018	0.5	115 271.202 02	0.06	−0.22
2	1	230 538.0000	0.5	230 537.999 96	0.11	0.04
3	2	345 795.9899	0.5	345 795.989 85	0.16	0.05
4	3	461 040.7682	0.5	461 040.767 98	0.21	0.22
5	4	576 267.9305	0.5	576 267.931 01	0.25	−0.51
6	5	691 473.0763	0.5	691 473.076 09	0.29	0.21
7	6	806 651.806	5	806 651.8010	0.3	5
8	7	921 799.700	5	921 799.7042	0.4	−4
9	8	1 036 912.393	5	1 036 912.3852	0.5	8
10	9	1 151 985.452	11	1 151 985.4442	0.6	8
11	10	1 267 014.486	5	1 267 014.4828	0.8	3
12	11	1 381 995.105	13	1 381 995.1036	1.0	1
13	12	1 496 922.909	12	1 496 922.9108	1.2	−2
14	13	1 611 793.518	11	1 611 793.5099	1.5	8
15	14			1 726 602.508	2	
16	15	1 841 345.506	11	1 841 345.514	2	−8
17	16	1 956 018.139	11	1 956 018.139	3	0
18	17	2 070 615.993	14	2 070 615.995	3	−2
19	18	2 185 134.680	13	2 185 134.698	4	−18
20	19	2 299 569.842	10	2 299 569.863	5	−21
21	20	2 413 917.113	11	2 413 917.112	6	1
22	21	2 528 172.060	11	2 528 172.065	6	−5
23	22			2 642 330.347	7	
24	23	2 756 387.584	17	2 756 387.586	8	−2
25	24	2 870 339.407	13	2 870 339.411	10	−4
26	25	2 984 181.455	14	2 984 181.455	11	0
27	26	3 097 909.361	17	3 097 909.354	12	7
28	27			3 211 518.748	14	
29	28			3 325 005.279	15	
30	29	3 438 364.611	10	3 438 364.594	17	17
31	30	3 551 592.361	10	3 551 592.342	19	19
32	31			3 664 684.177	20	
33	32	3 777 635.728	16	3 777 635.755	22	−26
34	33	3 890 442.717	13	3 890 442.738	25	−21
35	34			4 003 100.791	27	
36	35	4 115 605.585	22	4 115 605.584	29	1
37	36			4 227 952.790	32	
38	37	4 340 138.112	43	4 340 138.088	35	24

10.2.2. HeH$^+$ in the $X\,^1\Sigma^+$ ground state

The HeH$^+$ molecule is the simplest cationic molecule and, possessing only two electrons, it is one of the most fundamental species for molecular structure studies. It might also be important in astronomy since hydrogen and helium are the most abundant elements in the universe. It has been studied extensively in the laboratory, the first observation of its fundamental vibration–rotation band being made by Tolliver, Kyrala and Wing [82] using a novel ion beam method. Vibration–rotation transitions involving the levels lying near to or above the dissociation limit, for several different isotopomers, were observed by Carrington, Kennedy, Softley, Fournier and Richard [83]. Bernath and Amano [84] and Crofton, Altman, Haese and Oka [85] used similar tunable infrared sources to observe vibration–rotation transitions in the fundamental and hot bands of all possible isotopic species, i.e. ^{4}HeH$^+$, ^{3}HeH$^+$, ^{4}HeD$^+$ and ^{3}HeD$^+$. The first observations of rotational transitions involving the low J levels were made by Matsushima, Oka and Takagi [11]. We now describe their results, and also their theoretical analysis which combined the mid-infrared and far-infrared measurements.

Matsushima, Oka and Takagi [11] generated the HeH$^+$ ions using an a.c. discharge in a 100 : 1 mixture of helium and hydrogen, employing the velocity modulation method of detection [10]. Their tunable far-infrared source was similar to that of Evenson, Jennings and Petersen [65], in which the fixed difference-frequency of two carbon dioxide lasers was mixed with tunable microwave radiation. Their recording of the $J = 1 \leftarrow 0$ rotation transition in the $v = 0$ level of ^{4}HeH$^+$ is shown in figure 10.35; they were able to observe similar low J transitions in all four isotopomers, covering a frequency range from 2.0 to 4.8 THz. For each vibrational level the vibration-rotation

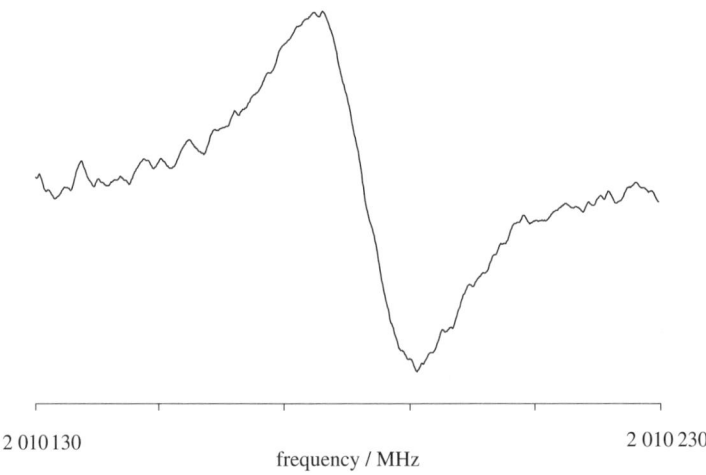

Figure 10.35. Observed $J = 1 \leftarrow 0$ rotational transition of ^{4}HeH$^+$ in the $v = 0$ level [65]. This line was obtained using six scans, each lasting 1 min, with a PSD time constant of 0.3 s.

Table 10.3. *Molecular parameters (in cm^{-1}) of HeH$^+$ for all four isotopomers*

	^{4}HeH$^+$	^{4}HeD$^+$	^{3}HeH$^+$	^{3}HeD$^+$
$T_1 - T_0$	2910.957 42(55)	2310.485 8(18)	2995.048 49(70)	2423.424 4(33)
B_0	33.558 670 00(98)	20.349 486 2(11)	35.720 188 3(16)	22.540 741 8(34)
$D_0 \times 10^2$	1.621 774(14)	0.586 906 4(41)	1.841 163(31)	0.722 276(27)
$H_0 \times 10^6$	5.931 1(33)	1.266	7.202(24)	1.807
$L_0 \times 10^9$	-2.832	-0.345	-3.67	—
B_1	30.839 982(47)	19.083 895(84)	32.728 20(12)	21.061 35(57)
$D_1 \times 10^2$	1.586 062(88)	0.575 680(67)	1.798 33(43)	0.706 9(18)
$H_1 \times 10^6$	5.528 9(68)	1.187	6.538(44)	1.677
$L_1 \times 10^9$	-3.206	-0.355	-4.17	—

term values were fitted to the usual formula

$$E_v(J) = T_v + B_v J(J+1) - D_v[J(J+1)]^2 + H_v[J(J+1)]^3 + L_v[J(J+1)]^4. \tag{10.20}$$

The final results for $v = 0$ and 1 are summarised in table 10.3, using data from rotational and vibration–rotation transitions for all four isotopomers.

The results given in table 10.3 constitute a good data set with which to test theoretical relationships [86, 87] relating to breakdown of the Born–Oppenheimer approximation. As we have described elsewhere, the vibration–rotation term values may be expressed as power series using the Dunham parameters Y_{kl}:

$$E_{kl} = \sum_{kl} Y_{kl}(v + 1/2)^k [J(J+1)]^l. \tag{10.21}$$

The subscript k gives the vibrational dependence and l gives the rotational dependence. The parameters Y_{kl} are isotopically variant, but Watson [87] has shown that they may be written in terms of isotopically invariant parameters according to the expression

$$Y_{kl} = \mu_c^{-(k/2+l)} U_{kl} \left[1 + \frac{m_e \Delta_{kl}^A}{M_A} + \frac{m_e \Delta_{kl}^B}{M_B} \right]. \tag{10.22}$$

U_{kl}, Δ_{kl}^A and Δ_{kl}^B are the invariant parameters, and μ_c is given by

$$\mu_c = \frac{M_A M_B}{(M_A + M_B - C m_e)}, \tag{10.23}$$

where M_A, M_B and m_e are the masses of the atoms A and B and the electron e, and C is the charge number of the ion, which is 1 for HeH$^+$. The terms involving the Δ parameters in (10.22) are very small compared with unity, but they represent the breakdown of the Born–Oppenheimer approximation. The results of a least-squares analysis yielding the isotopically invariant parameters are presented in table 10.4. A more detailed discussion of the treatment of Born–Oppenheimer breakdown is given in the subsection dealing with the rotational spectrum of LiH.

Table 10.4. *Isotopically invariant parameters for* HeH^+. *The unit of U_{kl} is* $cm^{-1}(amu)^{(k/2+l)}$, *but the Δ parameters are dimensionless*

U_{10}	2888.95(14)	U_{03}	$3.2070(91) \times 10^{-6}$
U_{20}	$-124.233(93)$	U_{13}	$-1.70(12) \times 10^{-7}$
U_{30}	0.142(19)	U_{23}	$-3.52(59) \times 10^{-8}$
U_{01}	28.115 425(95)	U_{04}	$-1.237(39) \times 10^{-9}$
U_{11}	$-1.967\,71(23)$	Δ_{10}^{He}	1.048(18)
U_{21}	$6.26(16) \times 10^{-3}$	Δ_{10}^{H}	$-0.613(20)$
U_{31}	$-2.658(32) \times 10^{-3}$	Δ_{01}^{He}	0.7736(18)
U_{02}	$-1.065\,58(13) \times 10^{-2}$	Δ_{01}^{H}	$-0.1459(19)$
U_{12}	$2.999(26) \times 10^{-4}$	Δ_{02}^{H}	2.346(91)
U_{22}	$-3.84(11) \times 10^{-5}$	Δ_{02}^{He}	0.0(fixed)

The calculation of these molecular parameters would provide a significant challenge to *ab initio* calculations which, so far as we know, have not yet been performed at the highest level.

The lowest rotational transition of HeH^+ observed at a frequency of 2010.1839 GHz should be a powerful probe for the detection in astronomical objects; the detection of HeH^+ has not yet been reported, but the progress in far-infrared astronomy might well lead to its observation in the future.

10.2.3. CuCl and CuBr in their $X\,^1\Sigma^+$ ground states

Earlier in this chapter we described experiments on species formed by high-temperature vapourisation, and we also described the pulsed Fourier transform microwave methods pioneered by Flygare and his colleagues. In more recent years the study of refractory materials has been accomplished by a combination of laser ablation and Fourier transform spectroscopy. Perhaps the first workers to develop this powerful combination of techniques were Suenram, Lovas, Fraser and Matsumura [88] who used it to study the rotational spectra of YO, LaO, ZrO and HfO; the first two molecules in this set have $^2\Sigma^+$ ground states, whilst the last two have $^1\Sigma^+$ ground states. A simple schematic diagram of the apparatus used is shown in figure 10.36, and a similar instrument was subsequently used by Low, Varberg, Connelly, Auty, Howard and Brown [89] to study the CuCl and CuBr molecules. We will use this work to illustrate the method, the spectra being more complex (and therefore more interesting) because of the presence of two nuclei with spins and quadrupole moments.

The gaseous sample was produced by using argon as a carrier gas, passing over CuCl or CuBr powder exposed to pulses from an ArF excimer laser, and injected through a pulsed nozzle into the Fabry–Perot cavity. In contrast to the earlier work on the rare earth oxides mentioned above, the nozzle expansion was injected *along* the axis of the microwave cavity, rather than with the perpendicular orientation illustrated

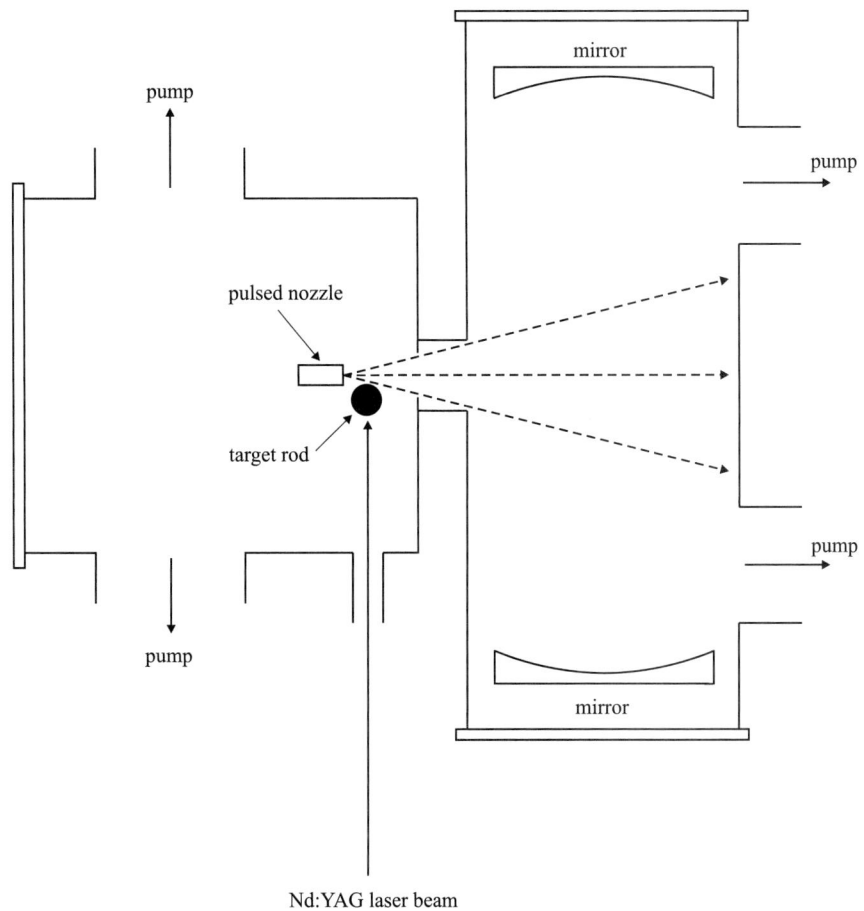

Figure 10.36. Schematic diagram of the laser vapourisation source and pulsed Fourier transform microwave spectrometer developed to study rare earth oxides [88].

in figure 10.36. Synchronous pulses of microwave energy led to line widths of 6 to 7 kHz, and Doppler-shifted components (parallel and antiparallel); these are shown in figure 10.37 for both CuCl and CuBr. The bandwidth of the Fabry–Perot cavity at a microwave frequency of 10 GHz was 1 MHz, and spectra were obtained by averaging the results of typically 1000 pulses. Earlier microwave work using a conventional high temperature absorption cell by Tiemann and Hoeft [90] and Hoeft and Nair [91] had provided good values for the rotational transition frequencies, which acted as a guide for the much higher resolution Fourier transform studies.

The principal isotope of copper is ^{63}Cu which has a nuclear spin I of 3/2; both chlorine and bromine have naturally occurring isotopes (^{35}Cl, ^{37}Cl, ^{79}Br, ^{81}Br) with spins of 3/2, so that extensive quadrupole hyperfine splitting is to be expected. The spectrum of CuCl shown in figure 10.37 arises from the $J = 1 \leftarrow 0$ rotational transition; the $J = 0$ level does not have a quadrupole splitting but if the ^{35}Cl and ^{63}Cu nuclear

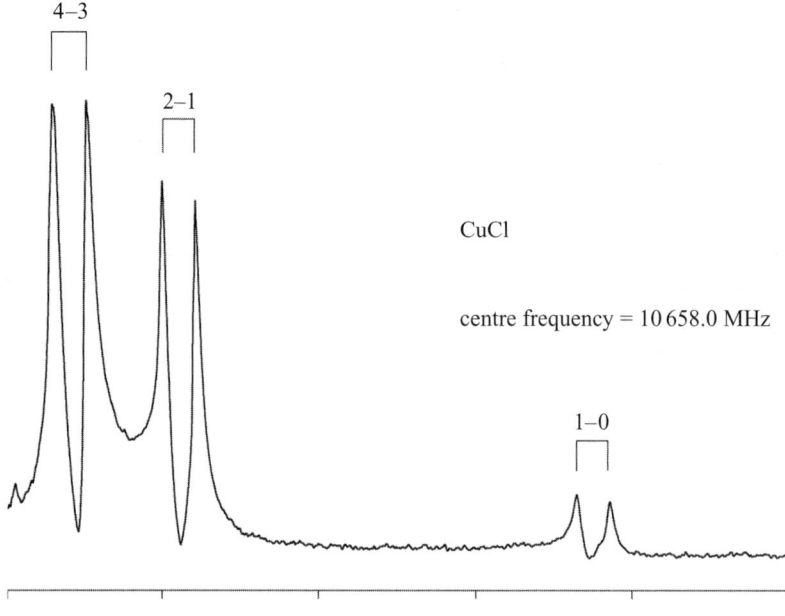
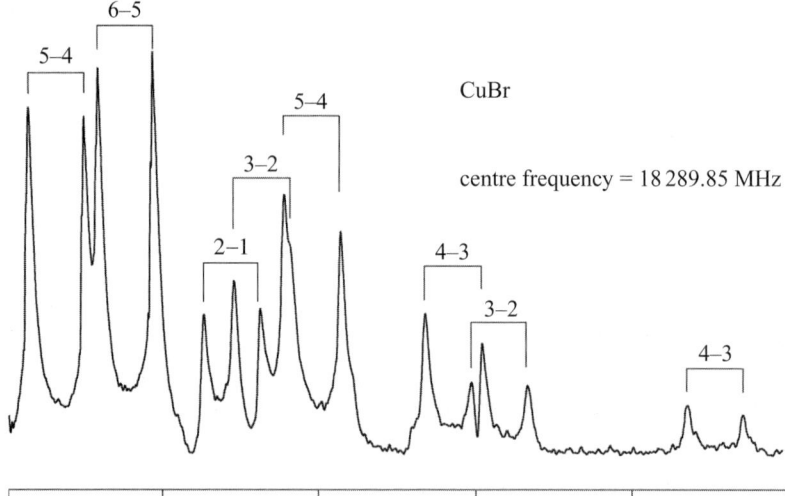

Figure 10.37. *Top*: $J = 1 \leftarrow 0$ rotational transition of $^{63}\text{Cu}^{35}\text{Cl}$ showing quadrupole hyperfine splitting; the transitions are labelled with their appropriate F values. Each transition shows a Doppler splitting arising from parallel and antiparallel orientations within the Fabry–Perot cavity [89]. *Bottom*: $J = 3 \leftarrow 2$ rotational transition of $^{63}\text{Cu}^{79}\text{Br}$ showing similar quadrupole and Doppler splittings [89]. In both spectra the scan range was ± 1 MHz about the centre frequency.

spins are denoted I_1 and I_2, we may anticipate a coupling scheme,

$$J + I_1 = F_1, \quad F_1 + I_2 = F. \tag{10.24}$$

There is a weak coupling between I_1 and I_2 for $J = 0$. For $J = 1$, F_1 takes the values 5/2, 3/2 and 1/2; the possible values of F are then 4, 3, 2, 1 and 0. The strongest transitions are those for which $\Delta F = \Delta J = \pm 1$, as shown in figure 10.37. The pattern observed for CuBr is more complicated because the rotational transition is $J = 3 \leftarrow 2$, so that both rotational levels involved have a more extensive quadrupole splitting.

The spectrum follows rules similar to those described in chapter 8 for the molecular beam magnetic resonance study of ^{7}Li^{79}Br, which also has a $^1\Sigma^+$ ground state and two nuclear spins of 3/2. In that case the transitions studied were nuclear spin transitions within the $J = 1$ level, but the effective Hamiltonian is similar in the two cases. The Hamiltonian used by Low, Varberg, Connelly, Auty, Howard and Brown [89] contained rotational, quadrupolar and nuclear spin–rotation terms,

$$\mathcal{H} = \mathcal{H}_{\text{rot}} + \mathcal{H}_Q + \mathcal{H}_{\text{nsr}}, \tag{10.25}$$

where

$$\mathcal{H}_{\text{rot}} = B_v \boldsymbol{J}^2 - D_v \boldsymbol{J}^4 + H_v \boldsymbol{J}^6, \tag{10.26}$$

$$\mathcal{H}_Q = -e\mathrm{T}^2(\nabla \boldsymbol{E}_1) \cdot \mathrm{T}^2(\boldsymbol{Q}_1) - e\mathrm{T}^2(\nabla \boldsymbol{E}_2) \cdot \mathrm{T}^2(\boldsymbol{Q}_2), \tag{10.27}$$

$$\mathcal{H}_{\text{nsr}} = c_1 \mathrm{T}^1(\boldsymbol{I}_1) \cdot \mathrm{T}^1(\boldsymbol{J}) + c_2 \mathrm{T}^1(\boldsymbol{I}_2) \cdot \mathrm{T}^1(\boldsymbol{J}). \tag{10.28}$$

The matrix elements of the effective Hamiltonian in the basis $|\eta, \Lambda; J, I_1, F_1, I_2, F, M_F\rangle$ were given in chapter 8 and the calculation of the energies and transition frequencies is relatively straightforward. The analysis was carried out for all four isotopomers of each molecule, and the constants determined.

The experimental techniques described in this section, combining laser ablation with a Fourier transform pulsed microwave spectrometer, have proved to be of great value in studying refractory materials which would otherwise pose considerable problems. The techniques have been discussed by Walker and Gerry [92] and illustrated in their study of the lowest rotational transition in the MgS molecule. Figure 10.38 shows their arrangement for injection of the pulsed gas into the Fabry–Perot cavity, making use of the 'parallel' arrangement described earlier for CuCl and CuBr.

10.2.4. SO, NF and NCl in their $b\,^1\Sigma^+$ states

In chapter 6 we described the low-lying electronic states of the O_2 molecule. In terms of molecular orbital theory the lowest energy electron configuration may be written

$$(\sigma_g 1s)^2 (\sigma_u^* 1s)^2 (\sigma_g 2s)^2 (\sigma_u^* 2s)^2 (\pi_u 2p)^4 (\sigma_g 2p)^2 (\pi_g^* 2p)^2,$$

and three electronic states are possible; these are as follows:

$$\begin{aligned} X\,^3\Sigma_g^- &: T_e = 0, \\ a\,^1\Delta_g &: T_e = 7918.1\ \text{cm}^{-1}, \\ b\,^1\Sigma_g^+ &: T_e = 13\,195.1\ \text{cm}^{-1}. \end{aligned} \tag{10.29}$$

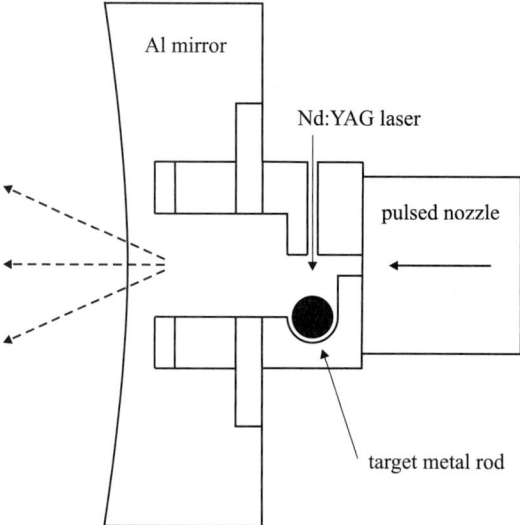

Figure 10.38. Laser ablation pulsed nozzle source, for gas injection parallel to the axis of the Fabry–Perot cavity, described by Walker and Gerry [92].

The NF radical is isoelectronic with O_2 and the three low-lying electronic states in this case have relative energies of 0, 12 003.6 and 18 877.05 cm^{-1}. The rotational spectra of all three states have been studied, and we will describe the $^3\Sigma$ and $^1\Delta$ state spectra later in this chapter. Similarly, the SO and NCl radicals are isoelectronic, with the outermost pair of electrons occupying a $(\pi 3p)$ molecular orbital, which again gives rise to the three electronic states described above. All three states for both SO and NCl have been studied by pure rotational spectroscopy, and the open shell states have also been investigated by magnetic resonance methods. In this section we concentrate on the rotational spectra of the $b\,^1\Sigma^+$ states.

We deal first with SO, which is the most extensively studied and which, in its predominant naturally occurring isotopomer, does not exhibit nuclear spin effects. The $b\,^1\Sigma^+$ state of SO lies 10 509.97 cm^{-1} above the ground state and its pure rotational spectrum seems to have been detected and measured first by Yamamoto [93] who used a d.c.-glow discharge in a mixture of H_2S, O_2 and He to produce electronically excited SO in vibrational levels up to $v = 8$. His studies covered a frequency range from 83 to 462 GHz. Subsequent work up to 900 GHz by Bogey, Civis, Delcroix, Demuynck, Krupnov, Quiguer, Tretyakov and Walters [94] extended the measurements to $v = 11$, and further studies by Klaus, Belov and Winnewisser [95], up to 1070 GHz using the Cologne terahertz spectrometer, provided data for high rotational levels, up to $J = 25$. These authors also included all of the previously published data to perform an accurate quantitative analysis, using the Dunham expansion for the vibration–rotation energies,

$$E_{vJ} = \sum_{kl} \mu^{-k/2-l} U_{kl}(v+1/2)^k[J(J+1)]^l, \tag{10.30}$$

where μ is the appropriate reduced mass. To a certain level of approximation, the U_{kl}

coefficients are isotopically invariant. The analysis actually included data for three different isotopomers, $^{32}S^{16}O$, $^{34}S^{16}O$ and $^{32}S^{18}O$, and, as we discussed earlier in chapter 7, Watson [96] and Tiemann [97] have shown that deviations from the Born–Oppenheimer approximation can be taken into account by replacing the U_{kl} parameters by the expression

$$U_{kl}\left\{1 + \frac{m_e}{M_1}\Delta_{kl}^{(1)} + \frac{m_e}{M_2}\Delta_{kl}^{(2)}\right\}. \tag{10.31}$$

Here M_1 and M_2 are the masses of atoms 1 and 2, m_e is the electron mass, and $\Delta_{kl}^{(1)}$ and $\Delta_{kl}^{(2)}$ are dimensionless, isotopically invariant corrections to the Born–Oppenheimer approximation. Accurate values of fourteen U_{kl} coefficients were listed by Klaus, Belov and Winnewisser [95]. Later in this chapter we will list the equilibrium bond lengths determined for SO in the three lowest-lying electronic states.

Kobayashi and Saito have studied the sub-millimetre rotational spectra of NF [98] and NCl [99] in their $b\,^1\Sigma^+$ states. The measurements for NF were confined to the $v = 0$ and 1 vibrational levels, and for the lowest rotational transitions, electric quadrupole interaction for the ^{14}N nucleus was observed. Similarly the ^{35}Cl quadrupole interaction was observed for the lowest rotational transitions of NCl. We will compare the data for the three low-lying electronic states in NF and NCl when we discuss their $^1\Delta$ spectra later in this chapter.

10.2.5. Hydrides (LiH, NaH, KH, CuH, AlH, AgH) in their $X\,^1\Sigma^+$ ground states

In the last subsection we encountered an example of Born–Oppenheimer breakdown in the SO molecule; one might expect such breakdown to be most noticeable in the lightest molecules, and indeed it is an important aspect in the quantitative analysis of the rotational spectrum of LiH. The $J = 1 \leftarrow 0$ rotational transition of ^{6}LiH, ^{7}LiH, ^{6}LiD and ^{7}LiD, and the $J = 2 \leftarrow 1$ transition in the deuterium isotopomers have been measured by Plummer, Herbst and De Lucia [100], following much earlier work by Pearson and Gordy [101] The original Dunham expansion for the vibration–rotation energies was given in the form

$$E(v, J) = h \sum_{kl} Y_{kl}(v + 1/2)^k J^l(J + 1)^l. \tag{10.32}$$

From this expression one can calculate that, within the vibrational state v, the $J + 1 \leftarrow J$ rotational transition frequencies are given by

$$\nu(v; J+1 \leftarrow J) = 2Y_{01}(J+1) + 2Y_{11}(v+1/2)(J+1) + 2Y_{21}(v+1/2)^2(J+1)$$
$$+ 4Y_{02}(J+1)^3 + 4Y_{12}(v+1/2)(J+1)^3 + \cdots. \tag{10.33}$$

In order to determine uniquely the first five Y coefficients in (10.33) one would need at least five rotational transition frequencies for a given isotopomer, but for a light molecule like LiH these occur at extremely high frequencies. Instead, measurements

Table 10.5. *Dunham coefficients determined for four isotopomers of LiH [100]. The parameters are given in MHz, except for the Δ parameters which are dimensionless*

	^{6}LiH	^{7}LiH	^{6}LiD	^{7}LiD
B_e^{BO}	230 161.654	225 449.805	131 671.220	126 959.371
Y_{11}	−6695.380	−6490.835	−2897.090	−2742.982
Y_{21}	63.733	61.151	20.859	19.392
Y_{02}	−26.832	−25.745	−8.782	−8.164
Y_{12}	0.476	0.452	0.118	0.108
Y_{01}	229 965.070	225 257.584	131 614.252	126 904.633
Δ_{01}^H	−1.549 76	−1.549 76	−1.549 76	−1.549 76
Δ_{01}^{Li}	−0.115 60	−0.115 60	−0.115 60	−0.115 60

were made for four different isotopomers; within the Born–Oppenheimer approximation, the relationship

$$Y'_{kl} = Y_{kl}[\mu/\mu']^{(k/2+l)} \qquad (10.34)$$

can be used, in which μ is the reduced isotopic mass. This expression is not exact within the Dunham expansion, but it is satisfactory except for the coefficient Y_{01}. The other four coefficients determined by Plummer, Herbst and De Lucia [100] are listed for the four isotopomers in table 10.5.

The remaining coefficients listed in table 10.5 arise from an analysis originally given by Herman and Asgharian [102] and Watson [103]. Dunham showed that Y_{01} may be expressed in terms of the Born–Oppenheimer equilibrium constant B_e^{BO} and an additive correction term ΔY_{01},

$$Y_{01} = B_e^{BO} + \Delta Y_{01}. \qquad (10.35)$$

The equilibrium bond lengths, R_e, which are given by

$$R_e = (h/8\pi^2 \mu Y_{01})^{1/2}, \qquad (10.36)$$

differ from one isotopomer to another. Watson showed, however, that they are related to the isotopically-invariant bond lengths R_e^{BO} through the expression

$$R_e = R_e^{BO}\left[1 + \frac{m_e d_a}{M_a} + \frac{m_e d_b}{M_b}\right], \qquad (10.37)$$

where m_e is the electron mass and M_a and M_b are the masses of the atoms a and b. The quantities d_a and d_b are isotopically invariant and are given by

$$d_a = d_a^{ad} - \frac{\mu \Delta Y_{01}}{2m_e B_e^{BO}} - \frac{(\mu g_J)_b}{2m_p},$$

$$d_b = d_b^{ad} - \frac{\mu \Delta Y_{01}}{2m_e B_e^{BO}} - \frac{(\mu g_J)_a}{2m_p}. \qquad (10.38)$$

The quantities $(\mu g_J)_a$ and $(\mu g_J)_b$ are the values of μg_J calculated in coordinate systems fixed at the nuclei, and the centre-of-mass rotational g_J factor is given by

$$g_J = \frac{(\mu g_J)_b}{M_a} + \frac{(\mu g_J)_a}{M_b}. \tag{10.39}$$

The remaining terms in equations (10.38) are the adiabatic corrections to the bond length, and are given by Watson [103] in terms of the Born–Oppenheimer force constant f as

$$d_i^{\text{ad}} = -\frac{\left[d/dR\langle\eta|P_i^2|\eta\rangle\right]_{R_e^{\text{BO}}}}{2m_e f R_e^{\text{BO}}}, \tag{10.40}$$

where i equals a or b and P_i is the nuclear momentum. From a combination of the above expressions we obtain the following result for Y_{01}:

$$Y_{01} = B_e^{\text{BO}}\left[1 + \frac{m_e \Delta_{01}^a}{M_a} + \frac{m_e \Delta_{01}^b}{M_b}\right]. \tag{10.41}$$

In this expression,

$$\Delta_{01}^a = -2d_a, \tag{10.42}$$

$$\Delta_{01}^b = -2d_b, \tag{10.43}$$

and we can use equations (10.38) to substitute for d_a and d_b.

In table 10.5 we include the values of Y_{01}, Δ_{01}^{H} and Δ_{01}^{Li}; Y_{01} is the equilibrium rotational constant. The Δ parameters are a measure of the departure from Born–Oppenheimer behaviour, and the values for LiH are the largest known for a closed shell diatomic molecule. We should note here that in molecular beam electric resonance studies described by Freeman, Jacobson, Johnson and Ramsey [104], accurate values of the rotational g-factor and nuclear spin-rotation coupling constants for ^{7}LiH and ^{7}LiD were determined. The lowest rotational transitions in NaH and KH have been studied by Okabayashi and Tanimoto [105] and the electric quadrupole constants for ^{23}Na and ^{39}K determined. Okabayashi and Tanimoto have also measured the lowest rotational transition in CuH [106] and in AgH and AgD [107]. Goto and Saito [108] have made similar studies of AlH at frequencies near 387 GHz; in all of these investigations of hydrides, attention has been focussed on departures from Born–Oppenheimer behaviour. Finally, Goto, Namiki and Saito [109] have measured the lowest rotational transition in ZnH and ZnD; in this case, however, the ground electronic state is $^2\Sigma^+$. Nuclear hyperfine splittings and spin–rotation splittings were determined, but since our next section is devoted to the details of rotational transitions in $^2\Sigma$ states, we will not elaborate further at this point.

10.3. $^2\Sigma$ states

10.3.1. CO$^+$ in the $X\,^2\Sigma^+$ ground state

One of the most important microwave rotational studies of a molecule in a $^2\Sigma$ state was that of Dixon and Woods [7] on the CO$^+$ ion, the first pure rotational study of a molecular ion. We described the details of their experiment earlier in this chapter, and showed their recording of the lowest rotational transition, occurring at 118 101.99 MHz. The data recorded for the predominant isotopomer, $^{12}C^{16}O^+$, enabled Erickson, Snell, Loren, Mundy and Plambeck [110] to establish the presence of the ion in interstellar clouds. The lowest rotational transitions in the isotopomers $^{12}C^{18}O^+$ and $^{13}C^{16}O^+$ were studied subsequently by Piltch, Szanto, Anderson, Gudeman, Dixon and Woods [111]. Of these the most interesting is $^{13}C^{16}O^+$ because it exhibits a large hyperfine splitting from the ^{13}C nucleus, which has a spin of $1/2$. Indeed, to anticipate the theory and analysis presented shortly, the most suitable angular momentum coupling scheme turned out to be unusual:

$$\boldsymbol{I} + \boldsymbol{S} = \boldsymbol{G}: \quad G = 1, 0$$
$$\boldsymbol{G} + \boldsymbol{N} = \boldsymbol{F}: \quad N = 0, G = F = 1, 0$$
$$N = 1, G = 1, F = 2, 1, 0$$
$$N = 1, G = 0, F = 1. \tag{10.44}$$

This is because the hyperfine (Fermi contact) interaction is larger than the spin–rotation coupling. The lowest rotational transition, $N = 1 \leftarrow 0$, was found to be split by a combination of hyperfine interaction and electron spin-rotation interaction into four observed lines. The frequencies of these lines were as follows:

$f_1 = 112\,902.610$ MHz, $\quad f_2 = 112\,753.426$, $\quad f_3 = 112\,694.956$, $\quad f_4 = 112\,468.609$.

The energy level diagram and assignment of these transitions is shown in figure 10.39; we now show how this analysis was accomplished.

One form of the effective Hamiltonian for the CN radical in its $^2\Sigma^+$ ground state was given in chapter 9. Omitting the quadrupole term, which is not required for $^{13}CO^+$, our starting point might be

$$\mathcal{H}_{\text{eff}} = \{B - D\boldsymbol{N}^2\}\boldsymbol{N}^2 + \gamma \mathbf{T}^1(\boldsymbol{N}) \cdot \mathbf{T}^1(\boldsymbol{S}) + b_{\text{F}} \mathbf{T}^1(\boldsymbol{I}) \cdot \mathbf{T}^1(\boldsymbol{S})$$
$$- \sqrt{10} g_S \mu_B g_N \mu_N (\mu_0/4\pi) \mathbf{T}^1(\boldsymbol{S}, \boldsymbol{C}^2) \cdot \mathbf{T}^1(\boldsymbol{I}). \tag{10.45}$$

However the fourth term, representing the dipolar hyperfine interaction, is unsuitable for the basis set chosen. Instead we use a different form for this interaction, discussed in chapter 8, and particularly appendix 8.1, which recognises the strong coupling of $\boldsymbol{S}$ and $\boldsymbol{I}$. This form is

$$\mathcal{H}_{\text{dip}} = \sqrt{6} g_S \mu_B g_N \mu_N (\mu_0/4\pi) \mathbf{T}^2(\boldsymbol{C}) \cdot \mathbf{T}^2(\boldsymbol{S}, \boldsymbol{I}). \tag{10.46}$$

We now calculate the matrix elements of the effective Hamiltonian in the basis set appropriate for CO^+. For the first term, the rotational energy, we have the following result:

$$\langle \eta, \Lambda; I, S, G, N, F | \{B - D\boldsymbol{N}^2\}\boldsymbol{N}^2 | \eta, \Lambda; I, S, G, N, F \rangle$$
$$= \{B - DN(N+1)\}N(N+1), \tag{10.47}$$

The remaining terms are more complicated. For the electron spin–rotation

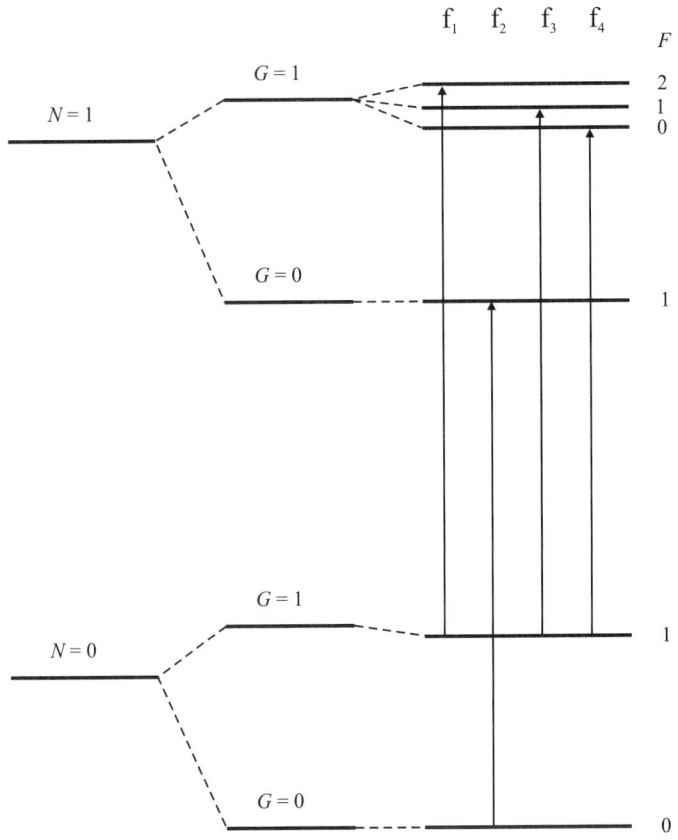

Figure 10.39. ^{13}C hyperfine and electron spin–rotation splitting of the $N=0$ and 1 rotational levels of ^{13}CO$^+$, and the observed transitions [111]. The large splitting is mainly due to the ^{13}C Fermi contact interaction. The smaller splittings are due to the spin–rotation interaction and the dipolar hyperfine coupling.

interaction

$$\langle \eta, \Lambda; I, S, G, N, F | \gamma \, \mathbf{T}^1(\mathbf{S}) \cdot \mathbf{T}^1(\mathbf{N}) | \eta, \Lambda; I, S, G', N', F \rangle$$

$$= \gamma (-1)^{G'+F+N} \begin{Bmatrix} N' & G' & F \\ G & N & 1 \end{Bmatrix} \langle I, S, G \| \mathbf{T}^1(\mathbf{S}) \| I, S, G' \rangle \langle N \| \mathbf{T}^1(\mathbf{N}) \| N' \rangle$$

$$= \gamma \delta_{NN'} (-1)^{G'+F+N} \begin{Bmatrix} N' & G' & F \\ G & N & 1 \end{Bmatrix} (-1)^{G+I+1+S} \{(2G'+1)(2G+1)\}^{1/2}$$

$$\times \begin{Bmatrix} G & S & I \\ S & G' & 1 \end{Bmatrix} \{S(S+1)(2S+1)N(N+1)(2N+1)\}^{1/2}. \quad (10.48)$$

The Fermi contact interaction is independent of N so that the matrix elements are simply

$$\langle \eta, \Lambda; I, S, G; G, N, F | b_\text{F} \mathbf{T}^1(\mathbf{I}) \cdot \mathbf{T}^1(\mathbf{S}) | \eta, \Lambda; I, S, G; G, N, F \rangle$$

$$= b_\text{F}(-1)^{I+G+S} \begin{Bmatrix} S & I & G \\ I & S & 1 \end{Bmatrix} \{I(I+1)(2I+1)S(S+1)(2S+1)\}^{1/2}. \quad (10.49)$$

The dipolar hyperfine matrix elements are the most complicated:

$$\langle \eta, \Lambda; I, S, G; G, N, F | \sqrt{6} g_S \mu_B g_N \mu_N (\mu_0/4\pi) \mathbf{T}^2(\mathbf{S}, \mathbf{I}) \cdot \mathbf{T}^2(\mathbf{C}) | \eta, \Lambda; I, S, G; G', N', F \rangle$$

$$= \sqrt{6} g_S \mu_B g_N \mu_N (\mu_0/4\pi)(-1)^{G'+F+N} \begin{Bmatrix} N' & G' & F \\ G & N & 1 \end{Bmatrix}$$

$$\times \langle I, S, G \| \mathbf{T}^2(\mathbf{S}, \mathbf{I}) \| I, S, G' \rangle \langle \eta, \Lambda; N, \Lambda \| \mathbf{T}^2(C) \| \eta, \Lambda; N', \Lambda \rangle$$

$$= \sqrt{6} g_S \mu_B g_N \mu_N (\mu_0/4\pi)(-1)^{G'+F+N} \begin{Bmatrix} N' & G' & F \\ G & N & 1 \end{Bmatrix}$$

$$\times \sqrt{5} \{(2G'+1)(2G+1)I(I+1)(2I+1)S(S+1)(2S+1)\}^{1/2}$$

$$\times \begin{Bmatrix} G & G' & 2 \\ I & I & 1 \\ S & S & 1 \end{Bmatrix} (-1)^N \{(2N+1)(2N'+1)\}^{1/2} \begin{pmatrix} N & 2 & N' \\ 0 & 0 & 0 \end{pmatrix}$$

$$\times \langle \eta, \Lambda | C_0^2(\theta, \phi) r^{-3} | \eta, \Lambda \rangle. \quad (10.50)$$

The evaluation of the reduced matrix elements in the second line of this equation is described in chapter 11 in connection with the H_2^+ ion. In the last line of (10.50) we have substituted $\Lambda = 0$ for the $^2\Sigma^+$ ground state of CO^+. The final matrix element in (10.50) involves the molecular parameter t, or $c/3$ where c is the Frosch and Foley parameter.

The matrix elements off-diagonal in N are negligible, and with this approximation the matrix of the effective Hamiltonian for each value of N factors in the following manner.

	$G'=1,$ $F'=N+1$	$G'=1,$ $F'=N$	$G'=0,$ $F'=N$	$G'=1,$ $F'=N-1$
$G=1,$ $F=N+1$	$\dfrac{\gamma N}{2} + \dfrac{b_\text{F}}{4} - t\dfrac{N}{4N+6}$	0	0	0
$G=1,$ $F=N$	0	$-\dfrac{\gamma}{2} + \dfrac{b_\text{F}}{4} + \dfrac{t}{2}$	$\dfrac{\gamma}{2}\sqrt{N(N+1)}$	0
$G=0,$ $F=N$	0	$\dfrac{\gamma}{2}\sqrt{N(N+1)}$	$-\dfrac{3b_\text{F}}{4}$	0
$G=1,$ $F=N-1$	0	0	0	$-\dfrac{\gamma(N+1)}{2} + \dfrac{b_\text{F}}{4} - \dfrac{t(N+1)}{(4N-2)}$

(10.51)

From their analyses of the rotational spectra of the three isotopomers, Piltch, Szanto, Anderson, Gudeman, Dixon and Woods [111] determined the values of the

Table 10.6. *Molecular constants (in MHz) determined for three isotopomers of CO^+ in the $X\,^2\Sigma^+$ ground state*

Species	B_0	γ	b_F	t	b	c
$^{12}C^{16}O^+$	58 983.06	273.1	—	—	—	—
$^{13}C^{16}O^+$	56 388.96	260.4	1506	48.2	1458	144.6
$^{12}C^{18}O^+$	56 174.61	259.8	—	—	—	—

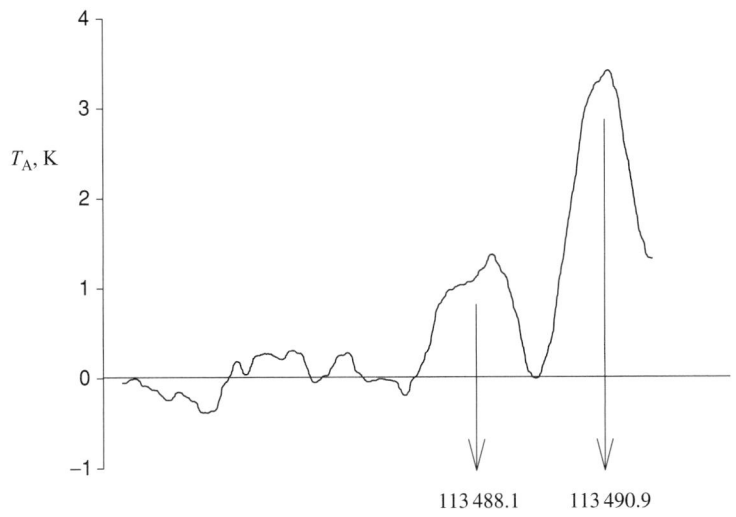

Figure 10.40. $N = 1 \to 0$ rotational emission lines of CN observed from the Orion nebula [114]. The two lines correspond to the two transitions marked with asterisks in figure 10.41.

molecular constants presented in table 10.6. The Fermi contact constant is not well determined in this experiment, but the value obtained nevertheless agreed well with an earlier value obtained from an ion beam study of the electronic spectrum by Carrington, Milverton and Sarre [112]. Comparison of the isotopic variation of the spin rotation constant for the three isotopomers suggests a degree of Born–Oppenheimer breakdown in the CO^+ molecular ion.

10.3.2. CN in the $X\,^2\Sigma^+$ ground state

The microwave rotational spectrum of the CN radical has been elusive in the laboratory, and the first observations were made by astronomers in 1970 [113], with more extensive studies four years later [114, 115]. Figure 10.40 shows two emission lines observed from the Orion nebula, the initial assignment being based upon constants obtained from the electronic spectrum. Three years later the first observations of the

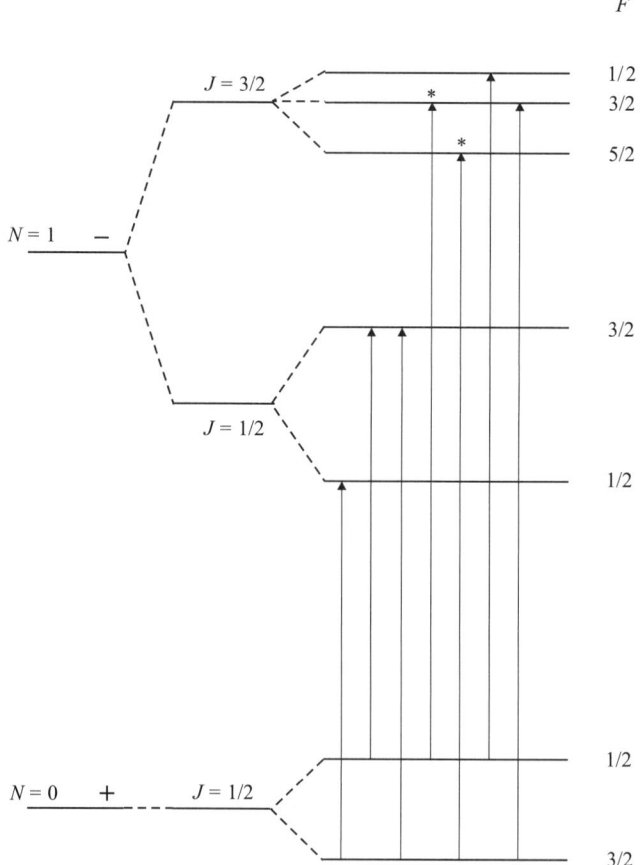

Figure 10.41. Spin–rotation and hyperfine splitting of the $N = 0$ and 1 rotational levels of CN in its $X\,^2\Sigma^+$ ground state, and the observed transitions. The asterisks denote the transitions responsible for the interstellar lines shown in figure 10.40.

laboratory microwave rotational spectrum were described by Dixon and Woods [116] using the apparatus described earlier for studies of the CO^+ ion, but with a discharge in nitrogen/cyanogen mixtures. The $N = 1 \leftarrow 0$ rotational transitions were observed in absorption; they exhibited extensive hyperfine structure from ^{14}N, and lines from both the $v = 0$ and 1 levels were recorded. In figure 10.41 we show an energy level diagram giving the assignment of these transitions, and now briefly examine the theory which lies behind the assignment.

The CN radical in its $^2\Sigma^+$ ground state shows fine and hyperfine structure of the rotational levels which is more conventional than that of CO^+, in that the largest interaction is the electron spin–rotation coupling. J is once more a good quantum number, and the effective Hamiltonian is that given in equation (10.45), with the addition of the nuclear electric quadrupole term given in chapter 9. The matrix elements in the conventional hyperfine-coupled case (b) basis set were derived in detail in chapter 9,

Table 10.7. *Molecular constants (in MHz) determined [116] for $X^2\Sigma^+$ CN in the $v = 0$ and 1 levels; nuclear hyperfine and nuclear spin–rotation constants refer to ^{14}N*

	$v = 0$	$v = 1$	equilibrium values
B_v	56 693.096	56 170.738	$B_e = 56\,954.106$
γ_v	217.488	215.072	$\alpha_e = 520.880$
b_F	-13.860	-12.996	$\gamma_e(vib) = 0.740$
t	20.107	20.168	$R_e(\text{Å}) = 1.171\,807$
eq_0Q	-1.287	-1.271	
c_I	0.009	0.009	

so we simply summarise the results here, as follows.

$$\langle \eta, \Lambda; N, S, J; I, F | \{B_v - D_v \mathbf{N}^2\} \mathbf{N}^2 | \eta, \Lambda; N, S, J; I, F \rangle$$
$$= \{B_v - D_v N(N+1)\} N(N+1). \tag{10.52}$$

$$\langle \eta, \Lambda; N, S, J; I, F | \gamma_v \mathbf{T}^1(\mathbf{N}) \cdot \mathbf{T}^1(\mathbf{S}) | \eta, \Lambda; N, S, J; I, F \rangle$$
$$= \gamma_v (-1)^{N+J+S} \begin{Bmatrix} S & N & J \\ N & S & 1 \end{Bmatrix} \{S(S+1)(2S+1)N(N+1)(2N+1)\}^{1/2}. \tag{10.53}$$

$$\langle \eta, \Lambda; N, S, J; I, F | b_F \mathbf{T}^1(\mathbf{S}) \cdot \mathbf{T}^1(\mathbf{I}) | \eta, \Lambda; N, S, J'; I, F \rangle$$
$$= b_F (-1)^{J'+F+I} \begin{Bmatrix} I & J' & F \\ J & I & 1 \end{Bmatrix} (-1)^{N+J+S+1} \begin{Bmatrix} S & J' & N \\ J & S & 1 \end{Bmatrix}$$
$$\times \{(2J+1)(2J'+1)I(I+1)(2I+1)S(S+1)(2S+1)\}^{1/2}. \tag{10.54}$$

$$\langle \eta, \Lambda; N, S, J; I, F | - \sqrt{10} g_S \mu_B g_N \mu_N (\mu_0/4\pi) \mathbf{T}^1(\mathbf{S}, \mathbf{C}^2) \cdot \mathbf{T}^1(\mathbf{I}) | \eta, \Lambda; N, S, J'; I, F \rangle$$
$$= -\sqrt{30}\, t (-1)^{J'+F+I+N} \begin{Bmatrix} I & J' & F \\ J & I & 1 \end{Bmatrix} \begin{Bmatrix} J & J' & 1 \\ N & N & 2 \\ S & S & 1 \end{Bmatrix} \begin{pmatrix} N & 2 & N \\ 0 & 0 & 0 \end{pmatrix}$$
$$\times \{I(I+1)(2I+1)(2J+1)(2J'+1)S(S+1)(2S+1)\}^{1/2}(2N+1). \tag{10.55}$$

$$\langle \eta, \Lambda; N, S, J; I, F | - e\mathbf{T}^2(\mathbf{Q}) \cdot \mathbf{T}^2(\nabla \mathbf{E}) | \eta, \Lambda; N, S, J'; I, F \rangle$$
$$- (-1)^{J'+F+I} \begin{Bmatrix} I & J' & F \\ J & I & 2 \end{Bmatrix} \left(\frac{eQ}{2}\right) \{(I+1)(2I+1)(2I+3)/I(2I-1)\}^{1/2}$$
$$\times (-1)^{J'+N+S} \{(2J'+1)(2J+1)\}^{1/2}$$
$$\times \begin{Bmatrix} N & J' & S \\ J & N & 2 \end{Bmatrix} (-1)^N \begin{pmatrix} N & 2 & N \\ 0 & 0 & 0 \end{pmatrix} (2N+1) \left(\frac{q_0}{2}\right). \tag{10.56}$$

In the matrix elements of the dipolar and quadrupolar interactions, equations (10.55) and (10.56), we have, as usual, neglected terms off-diagonal in N and Λ.

The spin–rotation and nuclear hyperfine structure of the $N = 0$ and 1 rotational levels is shown in figure 10.41, together with the observed transitions. The spectra of CN radicals in both $v = 0$ and $v = 1$ were observed, and the molecular constants determined are listed in table 10.7. The final column of table 10.7 shows the equilibrium values of certain parameters and also the internuclear distance. The final row gives a value for the nuclear spin–rotation constant which was necessary to achieve an accurate fit between experiment and theory. The laboratory frequencies, of course, provide accurate rest frequencies which are important in astrophysical studies of the velocity distribution of CN radicals in interstellar clouds.

Among other $^2\Sigma$ molecules which have been investigated we should mention SiN, investigated in the laboratory by Saito, Endo and Hirota [117], and in space by Turner [54].

10.4. $^3\Sigma$ states

10.4.1. Introduction

We have already described some of the important aspects of the high-resolution spectroscopy of molecules in $^3\Sigma$ states. In chapter 8 we described the molecular beam magnetic resonance spectrum of N_2 in its excited $A\,^3\Sigma_u^+$ state, concentrating on the nuclear hyperfine levels arising from the presence of two identical ^{14}N nuclei (each with spin 1) and the magnetic dipole transitions between them. All of the transitions described occurred *within* a single rotational level. We also described the theory of the Zeeman effect. In chapter 9 we turned our attention to the microwave and far-infrared magnetic resonance, where transitions both *within* and *between* different rotational levels are observed. We discussed the difference between case (b) and case (a) descriptions of $^3\Sigma$ states, and again explored the theory of the Zeeman effect, which is, of course, central to an understanding of magnetic resonance spectra. We now deal with the field-free rotational spectra, which are in general simpler. First we review briefly the important aspects of the theory of the rotational levels, drawing from our earlier discussions, before reviewing a selection of typical studies.

If, for the moment, we ignore magnetic field and nuclear hyperfine effects, the important terms in the effective Hamiltonian for a $^3\Sigma$ state are the following:

$$\mathcal{H}_{\text{eff}} = \mathcal{H}_{\text{rot}} + \mathcal{H}_{ss} + \mathcal{H}_{sr}. \tag{10.57}$$

The rotational term is given by

$$\mathcal{H}_{\text{rot}} = B_v \mathbf{N}^2 - D_v \mathbf{N}^4, \tag{10.58}$$

the so-called spin–spin interaction term is given by

$$\mathcal{H}_{ss} = -g_S^2 \mu_B^2 (\mu_0/4\pi)\sqrt{6}\mathbf{T}^2(\mathbf{C}) \cdot \mathbf{T}^2(\mathbf{S}_1, \mathbf{S}_2), \tag{10.59}$$

and the spin–rotation term is given by

$$\mathcal{H}_{sr} = \gamma_v \mathbf{T}^1(\mathbf{N}) \cdot \mathbf{T}^1(\mathbf{S}). \tag{10.60}$$

In a case (b) basis with $\Lambda = 0$ the matrix elements of these three terms are as follows:

$$\langle \eta, \Lambda; N, \Lambda; N, S, J, M_J | \mathcal{H}_{\text{rot}} | \eta, \Lambda; N, \Lambda; N, S, J, M_J \rangle$$
$$= B_v N(N+1) - D_v N^2 (N+1)^2. \tag{10.61}$$

$$\langle \eta, \Lambda; N, \Lambda; N, S, J, M_J | \mathcal{H}_{ss} | \eta, \Lambda; N, \Lambda; N, S, J, M_J \rangle$$
$$= \lambda \frac{2\sqrt{30}}{3} (-1)^{J+N+S} \begin{Bmatrix} S & N & J \\ N & S & 2 \end{Bmatrix} (-1)^N \begin{pmatrix} N & 2 & N \\ 0 & 0 & 0 \end{pmatrix} \{2N+1\}. \quad (10.62)$$

$$\langle \eta, \Lambda; N, \Lambda; N, S, J, M_J | \mathcal{H}_{ss} | \eta, \Lambda; N \pm 2, \Lambda; N, S, J, M_J \rangle$$
$$= \lambda \frac{2\sqrt{30}}{3} (-1)^{J+N+S} \begin{Bmatrix} S & N \pm 2 & J \\ N & S & 2 \end{Bmatrix} (-1)^N \begin{pmatrix} N & 2 & N \pm 2 \\ 0 & 0 & 0 \end{pmatrix}$$
$$\times \{(2N+1)(2[N \pm 2]+1)\}^{1/2}. \quad (10.63)$$

In these expressions, S is taken as 1 and the parameter λ is defined by

$$\lambda = -\frac{3}{4} \langle g_S^2 \mu_B^2 (\mu_0/4\pi) C_0^2 (\theta_{12}, \phi_{12}) r_{12}^{-3} \rangle_\eta, \quad (10.64)$$

when the dipole–dipole coupling of the electron spin magnetic moments is indeed the major contribution. However, we pointed out earlier, and showed in chapter 7, that the parameter λ is actually the sum of a first-order part, $\lambda^{(1)}$, representing the dipolar coupling, and a second-order part, $\lambda^{(2)}$, arising from spin–orbit mixing. In heavier molecules, the second-order part usually dominates; we return to this point later. We note from (10.64) that only the axial ($q = 0$) term has been retained.

Finally the spin–rotation term has simple matrix elements, which are diagonal in the case (b) basis, and which are given by

$$\langle \eta, \Lambda; N, \Lambda; N, S, J, M_J | \mathcal{H}_{sr} | \eta, \Lambda; N, \Lambda; N, S, J, M_J \rangle$$
$$= \gamma_v (-1)^{N+J+S} \begin{Bmatrix} S & N & J \\ N & S & 1 \end{Bmatrix} \{N(N+1)(2N+1)S(S+1)(2S+1)\}^{1/2}. \quad (10.65)$$

Since $\boldsymbol{J} = \boldsymbol{N} + \boldsymbol{S}$, each rotational level N is split into a spin triplet by the spin–spin term in the effective Hamiltonian and from equation (10.62) we see that the spin triplet energies are

$$J = N+1: \; E = -2\lambda \frac{N}{3(2N+3)}, \quad J = N: \; E = \frac{2}{3}\lambda,$$
$$J = N-1: \; E = -2\lambda \frac{(N+1)}{3(2N-1)}. \quad (10.66)$$

These results hold only if the off-diagonal elements (10.63) are small. If, however, the spin-orbit coupling is strong, and the rotational constant is small, the off-diagonal elements (10.63) become very important. N is then no longer a good quantum number, and a case (a) basis becomes more appropriate.

In a case (a) basis the component of the electron spin along the internuclear axis, Σ, is a specified quantum number. The matrix elements of the spin–spin and spin–rotation terms in the effective Hamiltonian were given in chapter 9 in our discussion of the spectrum of SeO. The rotational and spin–rotation Hamiltonians in the case (a) basis are

$$\mathcal{H}_{\text{rot}} + \mathcal{H}_{\text{sr}} = B\{T^1(\boldsymbol{J}) - T^1(\boldsymbol{S})\}^2 + \gamma\{T^1(\boldsymbol{J}) - T^1(\boldsymbol{S})\} \cdot T^1(\boldsymbol{S})$$
$$= B\boldsymbol{J}^2 + (\gamma - 2B)T^1(\boldsymbol{J}) \cdot T^1(\boldsymbol{S}) + (B - \gamma)\boldsymbol{S}^2, \quad (10.67)$$

and the matrix elements (neglecting the term in S^2) are given by

$$\langle \eta; S, \Sigma; J, M_J | \mathcal{H}_{\text{rot}} + \mathcal{H}_{\text{sr}} | \eta; S, \Sigma'; J, M_J \rangle$$
$$= \delta_{\Sigma\Sigma'} B J(J+1) + (\gamma - 2B) \sum_q (-1)^{J+S-2\Sigma}$$
$$\times \{J(J+1)(2J+1)S(S+1)(2S+1)\}^{1/2}$$
$$\times \begin{pmatrix} J & 1 & J \\ -\Sigma & q & \Sigma' \end{pmatrix} \begin{pmatrix} S & 1 & S \\ -\Sigma & q & \Sigma' \end{pmatrix} + (B-\gamma)S(S+1). \quad (10.68)$$

The spin–spin interaction term was written in the form

$$\mathcal{H}_{\text{ss}} = (2\sqrt{6}/3)\lambda T_0^2(\boldsymbol{S}, \boldsymbol{S}), \quad (10.69)$$

and the equivalence of this form to that used earlier in equation (10.59) was demonstrated in Appendix 8.3. The matrix elements in a case (a) basis are given by

$$\langle \eta; S, \Sigma; J, M_J | (2\sqrt{6}/3)\lambda T_0^2(\boldsymbol{S}, \boldsymbol{S}) | \eta; S, \Sigma; J, M_J \rangle$$
$$= \frac{2}{3}\lambda(-1)^{S-\Sigma} \begin{pmatrix} S & 2 & S \\ -\Sigma & 0 & \Sigma \end{pmatrix} S(2S-1) \begin{pmatrix} S & 2 & S \\ -S & 0 & S \end{pmatrix}^{-1}$$
$$= \frac{2}{3}\lambda\{3\Sigma^2 - S(S+1)\}. \quad (10.70)$$

The matrix elements in case (b) and case (a) bases enable us to construct the correlation diagram shown in figure 10.42, which is a repeat of that presented in chapter 9. With the matrix elements given above, and the correlation diagram, we are in a position to describe the experimental studies which have been made on molecules in $^3\Sigma$ states.

10.4.2. O$_2$ in its $X\,^3\Sigma_g^-$ ground state

There can be no question that the most important species with a $^3\Sigma$ ground state is molecular oxygen and, not surprisingly, it was one of the first molecules to be studied in detail when microwave and millimetre-wave techniques were first developed. It was also one of the first molecules to be studied by microwave magnetic resonance, notably by Beringer and Castle [118]. In this section we concentrate on the field-free rotational spectrum, but note at the outset that this is an atypical system; O$_2$ is a homonuclear diatomic molecule in its predominant isotopomer, ^{16}O^{16}O, and as such does not possess an electric dipole moment. Spectroscopic transitions must necessarily be magnetic dipole only.

Because of the symmetry of the homonuclear diatomic molecule, every alternate rotational level is missing; those that exist have N *odd* and positive parity, as shown for the first three rotational levels in figure 10.43. The magnetic dipole transitions arise from coupling of the electron spin magnetic moment with the oscillating magnetic field, represented by the interaction term

$$\mathcal{H}'(t) = g_S \mu_B \boldsymbol{T}^1(\boldsymbol{B}(t)) \cdot \boldsymbol{T}^1(\boldsymbol{S}). \quad (10.71)$$

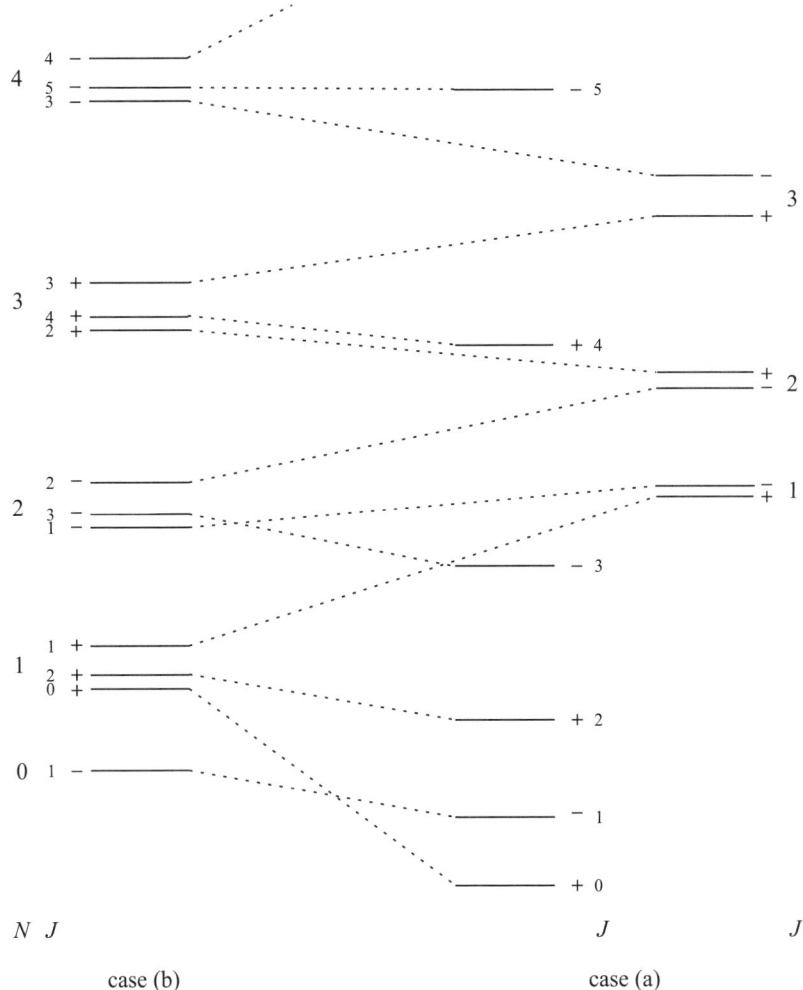

Figure 10.42. Correlation diagram for case (b) and case (a) rotational levels of a $^3\Sigma^-$ state.

If, for convenience, we take the direction of the oscillating field to define the space-fixed $p=0$ direction, the relative transition probabilities in the case (b) basis will be given by

$$\text{T.P.} \approx \left|\langle N, S, J, M_J | g_S \mu_B \text{T}_0^1(\boldsymbol{B}(t))\text{T}_0^1(\boldsymbol{S}) | N', S, J', M_J \rangle\right|^2$$

$$= \delta_{NN'} g_S^2 \mu_B^2 B_Z^2 \begin{pmatrix} J & 1 & J' \\ -M_J & 0 & M_J \end{pmatrix}^2 \{(2J'+1)(2J+1)\}$$

$$\times \begin{Bmatrix} J & S & N \\ S & J' & 1 \end{Bmatrix}^2 S(S+1)(2S+1). \tag{10.72}$$

This equation gives the selection rules $\Delta J = \pm 1$, $\Delta N = 0$. Six of the transitions shown

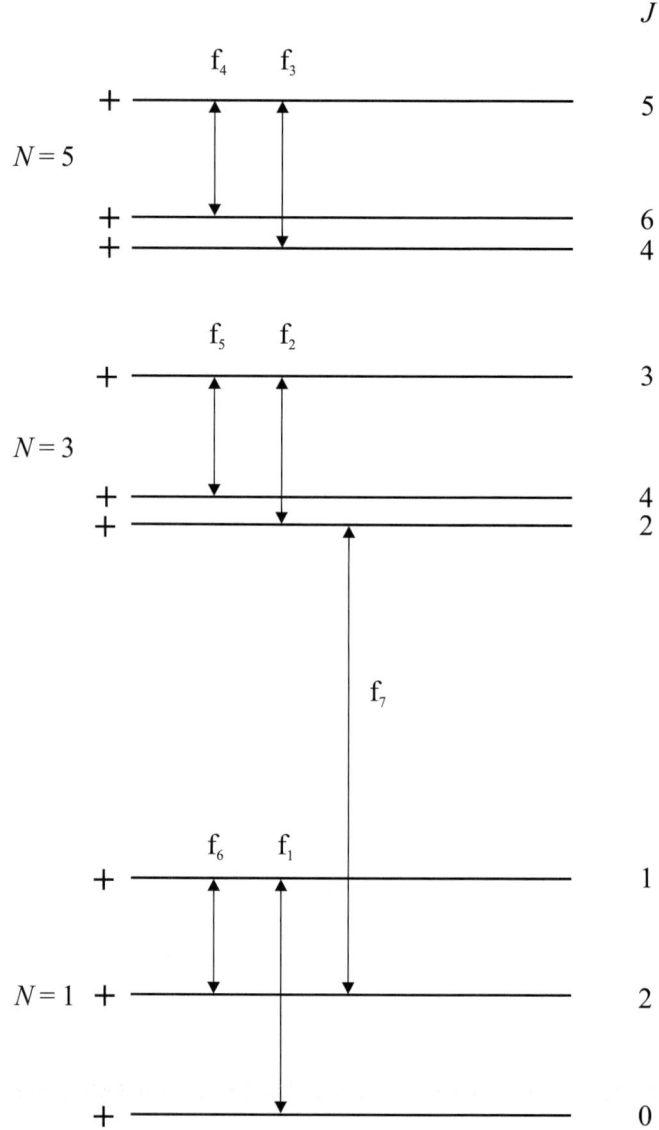

Figure 10.43. Three lowest rotational levels of O_2 $X\,^3\Sigma_g^-$ and the observed transitions.

in figure 10.43 obey these selection rules. However, it is important to note that the spin–spin interaction term has off-diagonal matrix elements, with $\Delta N = \pm 2$ (see equation (10.63)). Consequently transitions which obey the *apparent* selection rule $\Delta N = \pm 2$, such as f_7 in figure 10.43, acquire magnetic dipole intensity and are readily observable.

Many authors have contributed to our knowledge of the fine-structure and rotational spectrum of O_2 and the results were summarised by Amano and Hirota [119]. Table 10.8 lists the transitions which have been measured, and also gives the calculated frequencies [119] which we discuss shortly; calculated frequencies by Welch

Table 10.8. *Fine-structure and rotational transitions for $^{16}O_2\ X\ ^3\Sigma_g^-$ in the $v = 0$ level*

lower state		upper state		observed	obs. − calc.	
N	J	N'	J'	(MHz)	(MHz)	reference
1	0	1	1	118 750.343	0.012	[123]
1	2	1	1	56 264.778	−0.004	[123, 124]
1	2	3	2	424 763.12	−0.013	[125]
3	2	3	3	62 486.255	0.000	[123, 126]
3	4	3	3	58 446.600	−0.006	[123]
5	4	5	5	60 306.044	−0.015	[126]
5	6	5	5	59 590.978	−0.004	[126]
7	6	7	7	59 164.215	0.007	[126]
7	8	7	7	60 434.776	−0.003	[126]
9	8	9	9	58 323.885	0.004	[126]
9	10	9	9	61 150.570	0.004	[126]
11	12	11	11	61 800.169	0.005	[124]
13	12	13	13	56 968.180	−0.034	[124]
13	14	13	13	62 411.223	−0.006	[126]
13	13	15	15	2 496 283	3.7	[121]
15	14	15	15	56 363.393	−0.005	[124]
17	16	17	17	55 783.819	0.012	[124]
17	18	17	17	63 568.520	−0.013	[126]
19	18	19	19	55 221.372	0.008	[124]
19	20	19	19	64 127.777	−0.001	[124]
21	20	21	21	54 671.145	0.001	[124]
21	21	23	23	3 865 810	−1.0	[121]
25	26	25	25	65 764.744	0.007	[124]

and Mizushima [120] gave closely similar results. Table 10.8 includes two far-infrared transitions involving higher N levels, measured by Evenson and Mizushima [121] using FIR LMR methods, which were discussed in more detail in chapter 9 with other magnetic resonance measurements.

Other isotopomers of oxygen have been studied, notably $^{16}O^{17}O$ by Miller and Townes [122], $^{18}O_2$ by Steinbach and Gordy [125] and $^{16}O^{18}O$ by Amano and Hirota [119]. Measurements of the predominant isotopomer in its $v = 1$ level have also been described by Amano and Hirota [119]. Values of the molecular constants obtained from these studies are listed in table 10.9. In the case of $^{16}O^{17}O$ magnetic hyperfine constants for ^{17}O, which has a spin of 5/2, have been determined by Miller and Townes [122]; we will come to these studies shortly, but first examine how the transition frequencies listed in table 10.8 depend upon the values of the molecular parameters listed in table 10.9.

The result of the spin-spin interaction is that whilst J remains a good quantum number, N does not. Hence for each value of J we may formulate a 3×3 matrix

Table 10.9. *Molecular parameters (in MHz) determined for different isotopomers of oxygen*

parameter	$^{16}O^{16}O\ (v=0)$	$^{16}O^{16}O\ (v=1)$	$^{16}O^{17}O$	$^{16}O^{18}O$	$^{18}O^{18}O$
B	43 099.795	42 626.9	43 102	40 708.0	38 313.721
D	0.145	0.149	—	0.1294	—
λ	59 501.471	59 646.3	59 501.6	59 499.17	59 496.708
γ	−252.5872	−253.23	−252.72	−238.518	−224.438

which incorporates the $\Delta N = 0$ and $\Delta N = \pm 2$ matrix elements. The general form of this matrix, expressed as a function of J, is given below. Our results agree with those presented by Miller and Townes [122].

	$N = J+1$	$N = J$	$N = J-1$
$N=J+1$	$B(J+1)(J+2)$ $-\frac{2}{3}\lambda\frac{(J+2)}{(2J+1)} - \gamma(J+2)$	0	$2\lambda\left\{\frac{J(J+1)}{(2J+1)^2}\right\}^{1/2}$
$N=J$	0	$BJ(J+1)$ $+\frac{2}{3}\lambda - \gamma$	0
$N=J-1$	$2\lambda\left\{\frac{J(J+1)}{(2J+1)^2}\right\}^{1/2}$	0	$BJ(J-1)$ $-\frac{2}{3}\lambda\frac{(J-1)}{(2J+1)} + \gamma(J-1)$

In the case of O_2, however, the situation is particularly simple. For *odd* values of J, there are only the $N = J$ states to be considered; they are pure states. For *even* values of J, only the $N = J+1$ and $N = J-1$ states exist. The fine-structure separations for a given value of N are therefore given by the following simple equations:

$$\begin{aligned}
\nu_+(N) &= E(J=N) - E(J=N+1) \\
&= \lambda - \gamma(N+1) - (B-\gamma/2)(2N+3) + \{\lambda^2 - 2\lambda(B-\gamma/2) \\
&\quad + (B-\gamma/2)^2(2N+3)^2\}^{1/2} \\
\nu_-(N) &= E(J=N) - E(J=N-1) \\
&= \lambda + \gamma N + (B-\gamma/2)(2N-1) - \{\lambda^2 - 2\lambda(B-\gamma/2) \\
&\quad + (B-\gamma/2)^2(2N-1)^2\}^{1/2}.
\end{aligned} \quad (10.73)$$

It is a relatively straightforward exercise to use these equations in order to determine the values of the molecular parameters.

Miller and Townes [122] studied the magnetic hyperfine structure arising from the ^{17}O nucleus in the $^{16}O^{17}O$ isotopomer and were able to obtain information about the magnetic hyperfine parameters. We do not go into the details here, but will discuss nuclear hyperfine structure when we describe the rotational spectra of NCl and PF. We have already, in chapter 9, described the nitrogen and proton hyperfine structure observed in the FIR LMR spectrum of NH. We should also mention here the studies of the Zeeman effect in O_2 described by Beringer and Castle [127] and by Henry [128]; again, we discussed at length the Zeeman effect in $^3\Sigma$ states in connection with SO and NH.

10.4.3. SO, S_2 and NiO in their $X\,^3\Sigma^-$ ground states

The energies of the three spin states for a given N value are as follows:

$$\begin{aligned}
E(J = N + 1) &= BN(N+1) + B(2N+3) - \lambda - \gamma/2 \\
&\quad - \{\lambda^2 - 2\lambda(B - \gamma/2) + (2N+3)^2(B - \gamma/2)^2\}^{1/2} \\
E(J = N) &= BN(N+1) \\
E(J = N - 1) &= BN(N+1) - B(2N-1) - \lambda - \gamma/2 \\
&\quad + \{\lambda^2 - 2\lambda(B - \gamma/2) + (2N-1)^2(B - \gamma/2)^2\}^{1/2}.
\end{aligned} \qquad (10.74)$$

These expressions are essentially the same as those originally derived by Schlapp [129]; they also agree with the formulae of Winnewisser, Sastry, Cook and Gordy [130], except that there is a typographical error in the latter's expression for $E(J = N - 1)$. The sharp-eyed reader will note that the above energies can be derived from the interaction matrix given in the previous section provided $(2\lambda/3 - \gamma)$ is subtracted from the diagonal elements. It should also be noted that for the $N = 1$, $J = 0$ level, the sign of the square root factor in the third expression in equation (10.74) should be reversed. This particular eigenvalue is given by the diagonal element in the top left-hand corner of the matrix.

Before describing the experimental studies of SO and S_2 it is instructive to use the results and the experimental parameters determined to construct rotational level diagrams, as shown in figure 10.44. In the case of SO, experimental data have been obtained by Winnewisser, Sastry, Cook and Gordy [130], by Powell and Lide [131], by Amano, Hirota and Morino [132], by Clark and De Lucia [133], and at sub-millimetre wavelengths by Klaus, Saleck, Belov, Winnewisser, Hirahara, Hayashi, Kagi and Kawaguchi [134]. As figure 10.44 shows, a large number of transitions had been measured for $^{32}S^{16}O$ in the $v = 0$ level, even before the very high frequency measurements were described [134] and the main molecular parameters determined to high accuracy. These are listed in table 10.10, together with values determined for the first excited vibrational level, $v = 1$. The very high sensitivity of the sub-millimetre wave measurements enabled transitions in $^{32}S^{17}O$ (in natural abundance, 0.038 %) to be measured [134] and these exhibit hyperfine splitting from the ^{17}O nucleus, which has a spin of 5/2; an example is shown in figure 10.45. The pattern is fairly complicated for this $J = 4 \leftarrow 3$ transition because with $I = 5/2$, F takes six possible values from 13/2 to 3/2 in the upper level and 11/2 to 1/2 in the lower level. Analysis is, however, straightforward and the magnetic hyperfine and electric quadrupole parameters were determined.

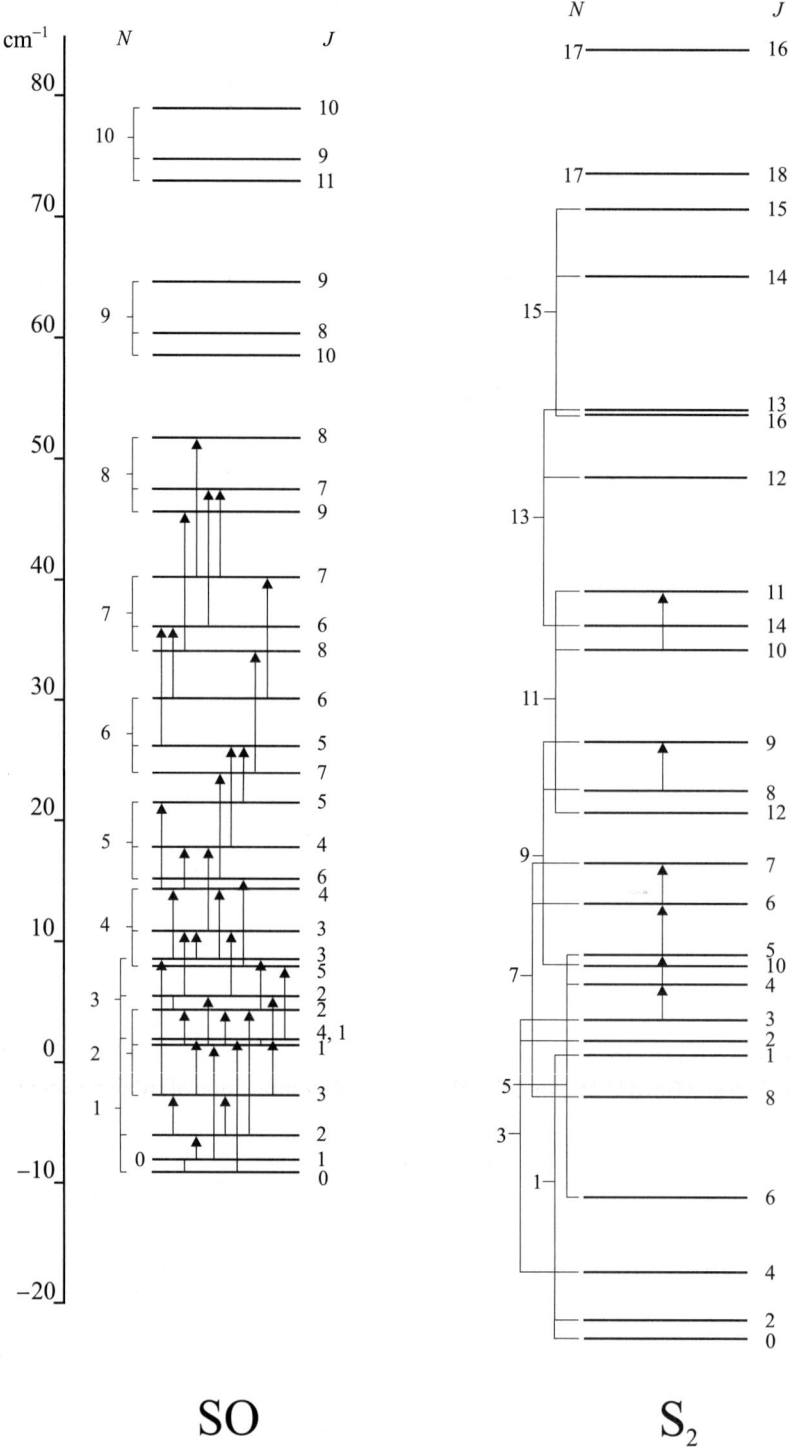

Figure 10.44. Lower rotational levels and observed transitions for SO and S$_2$ in the $v = 0$ level of their respective $X\,^3\Sigma^-$ and $X\,^3\Sigma_g^-$ ground states.

Table 10.10. *Molecular parameters (in MHz) determined for $^{32}S^{16}O$ ($v=0$ and 1) and $^{32}S_2$ in their electronic $^3\Sigma^-$ ground states*

parameter	SO ($X^3\Sigma^-$, $v=0$)	SO ($X^3\Sigma^-$, $v=1$)	S_2 ($X^3\Sigma_g^-$, $v=0$)
B	21 523.561	21 351.58	8831.8676
D	0.03399	0.034	0.0059
λ	158 254.387	159 204.7	35 3040
γ	−168.342	−171.5	−200.9

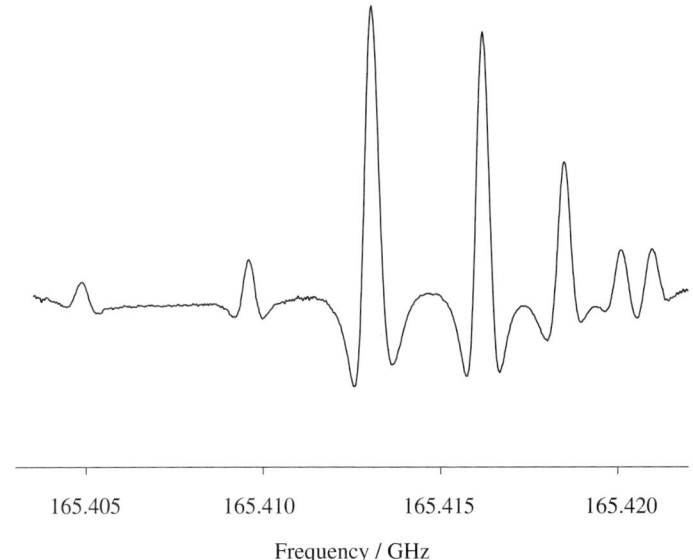

Figure 10.45. ^{17}O hyperfine structure of the $N=4 \leftarrow 3$, $J=4 \leftarrow 3$ rotational transition in $^{32}S^{17}O$ in its $X^3\Sigma^-$ state [134].

Rather less information about S_2 (in its predominant isotopomer $^{32}S_2$) is available but Pickett and Boyd [135], studying the thermal decomposition of sulphur vapour, were able to measure the six transitions shown in figure 10.44. In the predominant isotopomer, S_2 is a homonuclear system, so that only odd N levels exist. The molecular constants obtained by Pickett and Boyd [135] for the $v=0$ level are also listed in table 10.10.

The spin–orbit coupling constant for S_2 is, of course, larger than that for SO, so that the lower rotational levels of S_2 show a tendency towards case (a) coupling, as was described for SeO in chapter 9. A further example of case (a) coupling in a $^3\Sigma^-$ state has been provided by NiO, studied by Namiki and Saito [136]. NiO was produced in the gas phase by d.c.-sputtering of NiO powder, placed inside a nickel cathode with helium as a carrier gas. The predominant isotopes of nickel are ^{58}Ni (68.3%) and ^{60}Ni (26.1%), neither of which possesses a nuclear spin, and rotational progressions were observed for both isotopomers. The appropriate case (a) rotational level diagram and observed transitions are shown in figure 10.46. As we discussed for the SeO radical in chapter 9,

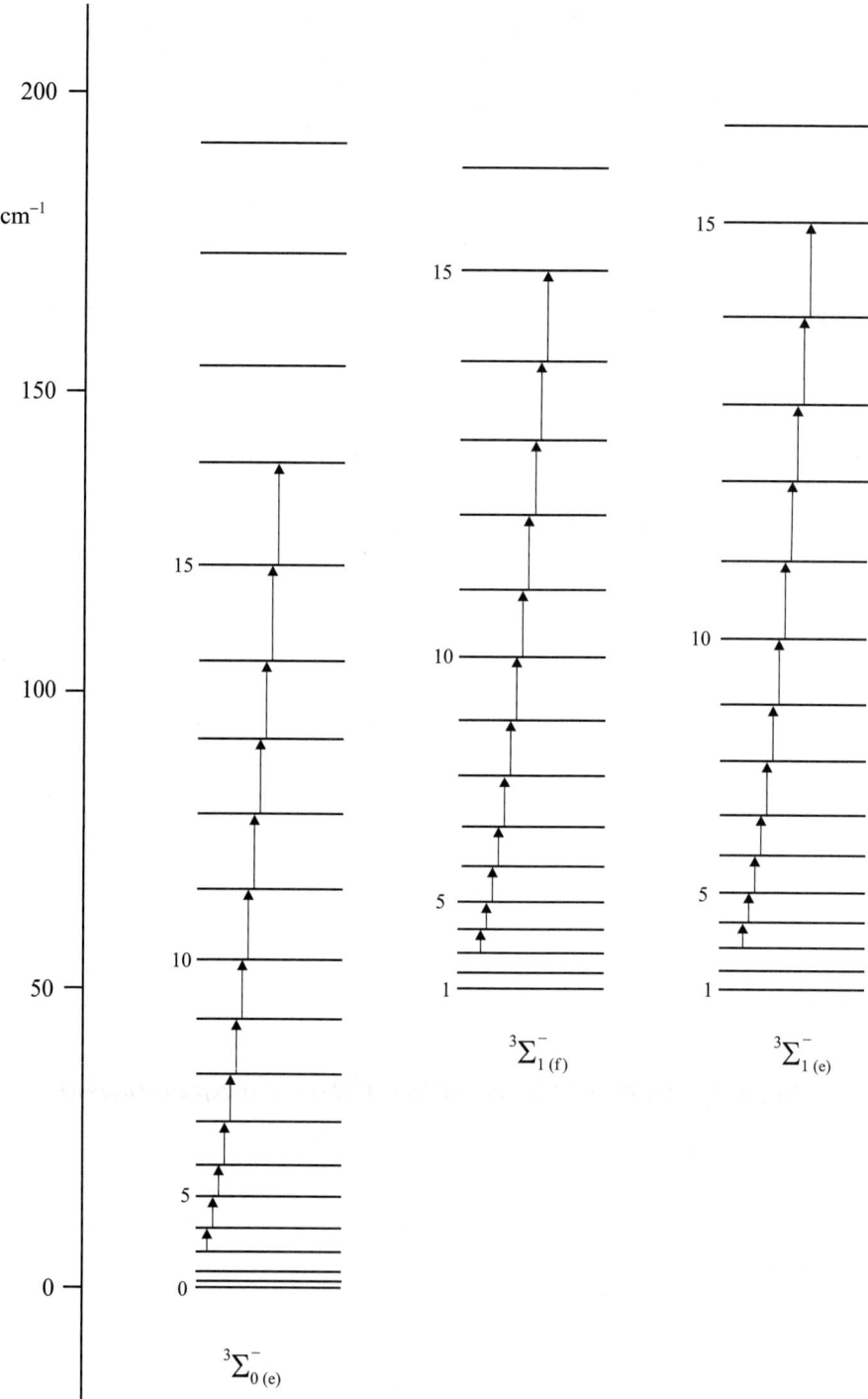

Figure 10.46. Case (a) rotational energy level diagram for NiO in its $X\,^3\Sigma^-$ state, and the transitions observed by Namiki and Saito [136].

the essential features of a case (a) $^3\Sigma$ state are that the electron spin $\mathbf{S}$ is strongly coupled to the rotational angular momentum $\mathbf{N}$ to form the total angular momentum $\mathbf{J}$. The component of $\mathbf{S}$ along the internuclear axis, Σ, is a good quantum number, but N is not and the appropriate basis functions are of the form $|\eta, \Lambda; J, S, \Sigma, M_J\rangle$. The effective rotational Hamiltonian for $v = 0$ is

$$\mathcal{H}_{\text{eff}} = B_0 \mathbf{T}^1(\mathbf{J}) \cdot \mathbf{T}^1(\mathbf{J}) + (\gamma_0 - 2B_0)\mathbf{T}^1(\mathbf{J}) \cdot \mathbf{T}^1(\mathbf{J})$$
$$+ (2\sqrt{6}/3)\lambda_0 \mathbf{T}_0^2(\mathbf{S}) + (B_0 - \gamma_0)\mathbf{S}^2, \qquad (10.75)$$

and Namiki and Saito [136] were able to determine the values of the constants B_0, γ_0 and λ_0.

10.4.4. PF, NCl, NBr and NI in their $X\ ^3\Sigma^-$ ground states

The PF radical is isoelectronic with SO and also has a $^3\Sigma^-$ ground state. Its rotational spectrum has been studied by Saito, Endo and Hirota [137] using a free-space microwave cell containing a glow discharge in mixtures of PH_3 and CF_4. The spectrum extends over a frequency range from 44 to 171 GHz. The lower rotational levels for the $v = 0$ level are shown in the left-hand side of figure 10.47, with the observed transitions indicated. The hyperfine splitting and structure for two of the rotational levels is shown on the right-hand side of the figure. Both ^{31}P and ^{19}F have nuclear spins of 1/2; the phosphorus nucleus has the larger hyperfine splitting so that if I_1 and I_2 denote the phosphorus and fluorine nuclear spins, an appropriate Hund's case (b) coupling scheme is

$$\mathbf{J} = \mathbf{N} + \mathbf{S}, \quad \mathbf{F}_1 = \mathbf{J} + \mathbf{I}_1, \quad \mathbf{F} = \mathbf{F}_1 + \mathbf{I}_2,$$

with the corresponding basis functions being expressed in the form $|N, S, J; I_1, F_1; I_2, F\rangle$. The effective Hamiltonian without nuclear spin interactions is the same as that given earlier for SO. The magnetic hyperfine Hamiltonian involves the Fermi contact and dipolar terms for both nuclei and, following earlier analyses, may be written in the form

$$\mathcal{H}_{\text{hfs}} = b_{F(1)}\mathbf{T}^1(\mathbf{S}) \cdot \mathbf{T}^1(\mathbf{I}_1) + b_{F(2)}\mathbf{T}^1(\mathbf{S}) \cdot \mathbf{T}^1(\mathbf{I}_2) - \sqrt{10} g_S \mu_B \mu_N (\mu_0/4\pi)$$
$$\times \{g_{N(1)}\mathbf{T}^1(\mathbf{I}_1) \cdot \mathbf{T}^1(\mathbf{S}, \mathbf{C}_{(1)}^2) + g_{N(2)}\mathbf{T}^1(\mathbf{I}_2) \cdot \mathbf{T}^1(\mathbf{S}, \mathbf{C}_{(2)}^2)\}. \qquad (10.76)$$

As before, the subscripts (1) and (2) refer to the ^{31}P and ^{19}F nuclei respectively.

The matrix elements of these four terms in the case (b) hyperfine-coupled basis set are as follows.

For the phosphorus Fermi contact interaction we have

$$\langle \eta, \Lambda; N, S, J, I_1, F_1, I_2, F | b_{F(1)}\mathbf{T}^1(\mathbf{S}) \cdot \mathbf{T}^1(\mathbf{I}_1) | \eta, \Lambda; N, S, J', I_1, F_1', I_2, F \rangle$$
$$= b_{F(1)} \delta_{F_1 F_1'} \langle \eta, \Lambda; N, S, J, I_1, F_1 | b_{F(1)}\mathbf{T}^1(\mathbf{S}) \cdot \mathbf{T}^1(\mathbf{I}_1) | \eta, \Lambda; N, S, J', I_1, F_1' \rangle$$
$$= b_{F(1)} \delta_{F_1 F_1'} (-1)^{J'+F_1+I_1} \begin{Bmatrix} I_1 & J' & F_1 \\ J & I_1 & 1 \end{Bmatrix} \langle N, S, J \| \mathbf{T}^1(\mathbf{S}) \| N, S, J' \rangle \langle I_1 \| \mathbf{T}^1(\mathbf{I}_1) \| I_1 \rangle$$

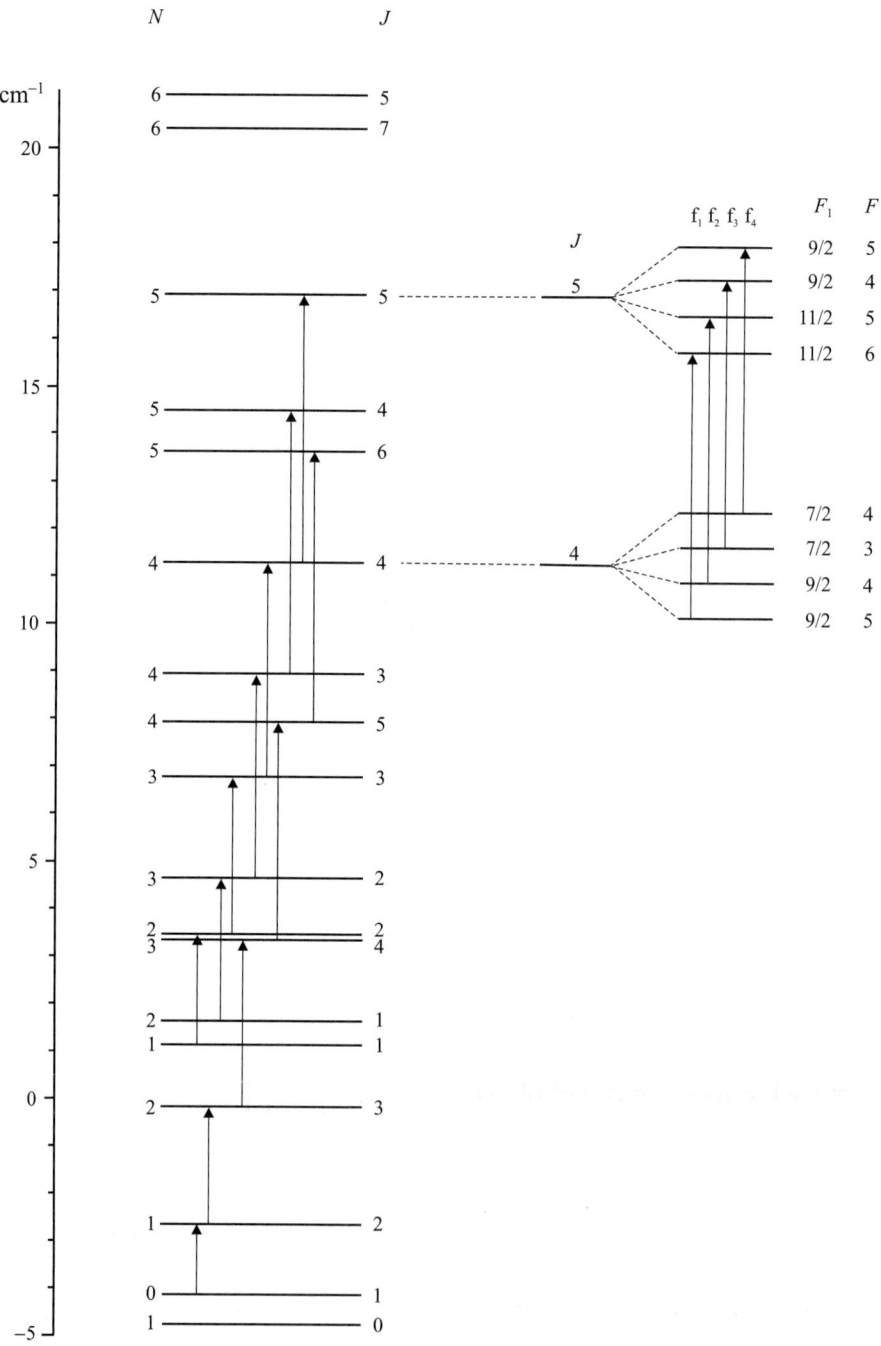

Figure 10.47. *Left*: lower rotation levels of the PF radical in its $X\,^3\Sigma^-$, $v=0$ state (to scale), and the transitions observed [137]. *Right*: ^{31}P and ^{19}F hyperfine splitting and observed transitions for the $N, J = 5,5 \leftarrow 4,4$ rotational transition (not to scale).

$$= b_{F(1)} \delta_{F_1 F_1'} (-1)^{J'+F_1+I_1} \begin{Bmatrix} I_1 & J' & F_1 \\ J & I_1 & 1 \end{Bmatrix} (-1)^{J+N+1+S} \{(2J'+1)(2J+1)\}^{1/2}$$

$$\times \begin{Bmatrix} J & S & N \\ S & J' & 1 \end{Bmatrix} \{S(S+1)(2S+1)I_1(I_1+1)(2I_1+1)\}^{1/2}. \quad (10.77)$$

For the fluorine Fermi contact interaction:

$$\langle \eta, \Lambda; N, S, J, I_1, F_1, I_2, F | b_{F(2)} \mathbf{T}^1(\mathbf{S}) \cdot \mathbf{T}^1(\mathbf{I}_2) | \eta, \Lambda; N, S, J', I_1, F_1', I_2, F \rangle$$

$$= b_{F(2)} (-1)^{F_1'+F+I_2} \begin{Bmatrix} I_2 & F_1' & F \\ F_1 & I_2 & 1 \end{Bmatrix} \langle J, I_1, F_1 \| \mathbf{T}^1(\mathbf{S}) \| J', I_1, F_1' \rangle \langle I_2 \| \mathbf{T}^1(\mathbf{I}_2) \| I_2 \rangle$$

$$= b_{F(2)} (-1)^{F_1'+F+I_2} \begin{Bmatrix} I_2 & F_1' & F \\ F_1 & I_2 & 1 \end{Bmatrix} (-1)^{F_1'+J+1+I_1} \{(2F_1'+1)(2F_1+1)\}^{1/2}$$

$$\times \begin{Bmatrix} F_1 & J & I_1 \\ J' & F_1' & 1 \end{Bmatrix} \langle N, S, J \| \mathbf{T}^1(\mathbf{S}) \| N', S, J' \rangle \langle I_2 \| \mathbf{T}^1(\mathbf{I}_2) \| I_2 \rangle$$

$$= b_{F(2)} (-1)^{F_1'+F+I_2} \begin{Bmatrix} I_2 & F_1' & F \\ F_1 & I_2 & 1 \end{Bmatrix} (-1)^{F_1'+J+1+I_1} \{(2F_1'+1)(2F_1+1)\}^{1/2}$$

$$\times \begin{Bmatrix} F_1 & J & I_1 \\ J' & F_1' & 1 \end{Bmatrix} (-1)^{J+N+1+S} \{(2J'+1)(2J+1)\}^{1/2} \begin{Bmatrix} J & S & N \\ S & J' & 1 \end{Bmatrix}$$

$$\times \{S(S+1)(2S+1)I_2(I_2+1)(2I_2+1)\}^{1/2}. \quad (10.78)$$

The electron–nuclear dipolar interaction (see equation (10.76)) for the phosphorus nucleus is as follows:

$$\langle \eta, \Lambda; N, \Lambda; S, J, I_1, F_1, I_2, F | \mathcal{H}_{\text{dip}(1)} | \eta, \Lambda'; N', \Lambda'; S, J', I_1, F_1', I_2, F \rangle$$

$$= -\sqrt{10} g_S \mu_B g_{N(1)} \mu_N (\mu_0/4\pi) \delta_{F_1 F_1'}$$

$$\times \langle \eta, \Lambda; N, S, J, I_1, F_1 | \mathbf{T}^1(\mathbf{S}, \mathbf{C}^2_{(1)}) \cdot \mathbf{T}^1(\mathbf{I}_1) | \eta, \Lambda'; N', S, J', I_1, F_1' \rangle$$

$$= -\sqrt{10} g_S \mu_B g_{N(1)} \mu_N (\mu_0/4\pi) (-1)^{J'+F_1+I_1} \begin{Bmatrix} I_1 & J' & F_1 \\ J & I_1 & 1 \end{Bmatrix}$$

$$\times \langle \eta, \Lambda; N, S, J \| \mathbf{T}^1(\mathbf{S}, \mathbf{C}^2_{(1)}) \| \eta, \Lambda'; N', S, J_1' \rangle \langle I_1 \| \mathbf{T}^1(\mathbf{I}_1) \| I_1 \rangle$$

$$= -\sqrt{10} g_S \mu_B g_{N(1)} \mu_N (\mu_0/4\pi) (-1)^{J'+F_1+I_1} \begin{Bmatrix} I_1 & J' & F_1 \\ J & I_1 & 1 \end{Bmatrix}$$

$$\times \{(3)(2J+1)(2J'+1)\}^{1/2} \begin{Bmatrix} J & J' & 1 \\ N & N' & 2 \\ S & S & 1 \end{Bmatrix} \langle \eta, \Lambda; N, \Lambda \| \mathbf{T}^2(\mathbf{C}_{(1)}) \| \eta', \Lambda'; N', \Lambda' \rangle$$

$$\times \{S(S+1)(2S+1)I_1(I_1+1)(2I_1+1)\}^{1/2}. \quad (10.79)$$

In the last line of (10.79) we have made use of a result derived earlier in chapter 8. As with our discussion of N_2 in its $^3\Sigma_u^+$ state, we put $\Lambda = \Lambda' = 0$, so that the final result is

$$\langle \eta, \Lambda; N, S, J, I_1, F_1, I_2, F | \mathcal{H}_{\text{dip}(1)} | \eta, \Lambda; N', S, J', I_1, F'_1, I_2, F \rangle$$
$$= -\sqrt{10} g_S \mu_B g_{N(1)} \mu_N (\mu_0/4\pi)(-1)^{J'+F_1+I_1}$$

$$\times \begin{Bmatrix} I_1 & J' & F_1 \\ J & I_1 & 1 \end{Bmatrix} \{(3)(2J+1)(2J'+1)\}^{1/2} \begin{Bmatrix} J & J' & 1 \\ N & N' & 2 \\ S & S & 1 \end{Bmatrix} (-1)^N \begin{pmatrix} N & 2 & N' \\ 0 & 0 & 0 \end{pmatrix}$$

$$\times \{(2N+1)(2N'+1)S(S+1)(2S+1)I_1(I_1+1)(2I_1+1)\}^{1/2} \langle T_0^2(\boldsymbol{C}_{(1)}) \rangle_\eta. \quad (10.80)$$

We shall simplify this result further in due course.

Finally, the matrix elements of the fluorine dipolar hyperfine interaction (equation (10.76)) are given by

$$\langle \eta, \Lambda; N, S, J, I_1, F_1, I_2, F | \mathcal{H}_{\text{dip}(2)} | \eta, \Lambda'; N', S, J', I_1, F'_1, I_2, F \rangle$$
$$= -\sqrt{10} g_S \mu_B g_{N(2)} \mu_N (\mu_0/4\pi)(-1)^{F'_1+F+I_2} \begin{Bmatrix} I_2 & F'_1 & F \\ F_1 & I_2 & 1 \end{Bmatrix} (-1)^{F'_1+J+1+I_1}$$

$$\times \{(2F'_1+1)(2F_1+1)\}^{1/2} \begin{Bmatrix} F_1 & J & I_1 \\ J' & F'_1 & 1 \end{Bmatrix}$$

$$\times \langle \eta, \Lambda; N, S, J \| T^1(\boldsymbol{S}, \boldsymbol{C}^2_{(2)}) \| \eta, \Lambda'; N', S, J' \rangle \langle I_2 \| T^1(\boldsymbol{I}_2) \| I_2 \rangle$$

$$= -\sqrt{10} g_S \mu_B g_{N(2)} \mu_N (\mu_0/4\pi)(-1)^{F'_1+F+I_2} \begin{Bmatrix} I_2 & F'_1 & F \\ F_1 & I_2 & 1 \end{Bmatrix} (-1)^{F'_1+J+1+I_1}$$

$$\times \{(2F'_1+1)(2F_1+1)\}^{1/2} \begin{Bmatrix} F_1 & J & I_1 \\ J' & F'_1 & 1 \end{Bmatrix} (\sqrt{3})\{(2J+1)(2J'+1)\}^{1/2}$$

$$\times \begin{Bmatrix} J & J' & 1 \\ N & N' & 2 \\ S & S & 1 \end{Bmatrix} \langle \eta, \Lambda; N, \Lambda \| T^2(\boldsymbol{C}_{(2)}) \| \eta', \Lambda'; N', \Lambda' \rangle$$

$$\times \{S(S+1)(2S+1)I_2(I_2+1)(2I_2+1)\}^{1/2}. \quad (10.81)$$

Again by setting $\Lambda = \Lambda' = 0$, we obtain the result

$$\langle \eta, \Lambda; N, S, J, I_1, F_1, I_2, F | \mathcal{H}_{\text{dip}(2)} | \eta, \Lambda; N', S, J', I_1, F'_1, I_2, F \rangle$$
$$= -\sqrt{10} g_S \mu_B g_{N(2)} \mu_N (\mu_0/4\pi)(-1)^{F'_1+F+I_2} \begin{Bmatrix} I_2 & F'_1 & F \\ F_1 & I_2 & 1 \end{Bmatrix} (-1)^{F'_1+J+1+I_1}$$

$$\times \{(2F'_1+1)(2F_1+1)\}^{1/2} \begin{Bmatrix} F_1 & J & I_1 \\ J' & F'_1 & 1 \end{Bmatrix}$$

$$\times \{(3)(2J+1)(2J'+1)\}^{1/2} \begin{Bmatrix} J & J' & 1 \\ N & N' & 2 \\ S & S & 1 \end{Bmatrix} (-1)^N \begin{pmatrix} N & 2 & N' \\ 0 & 0 & 0 \end{pmatrix} \{(2N+1)$$

$$\times (2N'+1)S(S+1)(2S+1)I_2(I_2+1)(2I_2+1)\}^{1/2} \langle T_0^2(\boldsymbol{C}_{(2)}) \rangle_\eta. \quad (10.82)$$

Finally for the dipolar matrix elements we recall that

$$g_S\mu_B g_N\mu_N(\mu_0/4\pi)\langle T_0^2(C)\rangle_\eta = t_0 \equiv (c/3). \tag{10.83}$$

The above expressions look complicated, especially for the dipolar interaction, but this is largely because they include all possible diagonal and off-diagonal elements. If we put $S = 1$, $I_1 = I_2 = 1/2$, we obtain the following simpler results for the four required matrix elements:

$$\langle \eta, \Lambda; N, S, J, I_1, F_1, I_2, F|b_{F(1)}T^1(S)\cdot T^1(I_1)|\eta, \Lambda; N, S, J', I_1, F_1, I_2, F\rangle$$
$$= b_{F(1)}(-1)^{J'+F_1+I_1+J+N}3\{(2J+1)(2J'+1)\}^{1/2}$$
$$\times \begin{Bmatrix} 1/2 & J' & F_1 \\ J & 1/2 & 1 \end{Bmatrix}\begin{Bmatrix} J' & 1 & N \\ 1 & J & 1 \end{Bmatrix}. \tag{10.84}$$

$$\langle \eta, \Lambda; N, S, J, I_1, F_1, I_2, F|b_{F(2)}T^1(S)\cdot T^1(I_2)|\eta, \Lambda; N, S, J', I_1, F_1', I_2, F\rangle$$
$$= b_{F(2)}(-1)^{F+N+1}3\{(2F_1+1)(2F_1'+1)(2J+1)(2J'+1)\}^{1/2}$$
$$\times \begin{Bmatrix} 1/2 & F_1' & F \\ F_1 & 1/2 & 1 \end{Bmatrix}\begin{Bmatrix} F_1' & J' & 1/2 \\ J & F_1 & 1 \end{Bmatrix}\begin{Bmatrix} J' & 1 & N \\ 1 & J & 1 \end{Bmatrix}. \tag{10.85}$$

$$\langle \eta, \Lambda; N, S, J, I_1, F_1, I_2, F|\mathcal{H}_{\text{dip}(1)}|\eta, \Lambda; N', S, J', I_1, F_1, I_2, F\rangle$$
$$= -t_{0(1)}3\sqrt{30}(-1)^{J'+F_1+I_1+N}\{(2J+1)(2J'+1)(2N+1)(2N'+1)\}^{1/2}$$
$$\times \begin{Bmatrix} 1/2 & J' & F_1 \\ J & 1/2 & 1 \end{Bmatrix}\begin{Bmatrix} J & J' & 1 \\ N & N' & 2 \\ 1 & 1 & 1 \end{Bmatrix}\begin{pmatrix} N & 2 & N' \\ 0 & 0 & 0 \end{pmatrix}. \tag{10.86}$$

$$\langle \eta, \Lambda; N, S, J, I_1, F_1, I_2, F|\mathcal{H}_{\text{dip}(2)}|\eta, \Lambda; N', S, J', I_1, F_1', I_2, F\rangle$$
$$= -t_{0(2)}3\sqrt{30}(-1)^{F-J+1+N}\{(2F_1+1)(2F_1'+1)(2J+1)(2J'+1)$$
$$\times (2N+1)(2N'+1)\}^{1/2}$$
$$\times \begin{Bmatrix} 1/2 & F_1' & F \\ F_1 & 1/2 & 1 \end{Bmatrix}\begin{Bmatrix} F_1' & J' & 1/2 \\ J & F_1 & 1 \end{Bmatrix}\begin{Bmatrix} J & J' & 1 \\ N & N' & 2 \\ 1 & 1 & 1 \end{Bmatrix}\begin{pmatrix} N & 2 & N' \\ 0 & 0 & 0 \end{pmatrix}. \tag{10.87}$$

Using these results we are in a position to derive expressions for the diagonal hyperfine energies of the eight levels shown on the right-hand side of figure 10.47, from which the values of the constants can be obtained. However, the diagonal energies are probably not sufficiently accurate, particularly for the pairs of hyperfine levels in a given rotational level which have the same F value. Equations (10.82), (10.81), (10.80) and (10.79) show that such levels are mixed by the hyperfine interaction. In order to obtain the values of the parameters, as well as the nuclear spin-free parameters, it is necessary to carry out the analysis for all twelve observed rotational transitions, including the hyperfine levels. Saito, Endo and Hirota [137] achieved this, with a final standard deviation in the fit of only 20 kHz. The values of the hyperfine constants are presented

Table 10.11. *Nuclear hyperfine coupling constants (in MHz) for PF and related free radicals [37]*

radical	c	t_0	b_F	spin density	s character	reference
^{31}P						
PF($^3\Sigma^-$)	−502.645	−167.548	116.809	91.3%	0.88%	[137]
PD($^3\Sigma^-$)	−476.6	−158.9	129.8	86.6	0.98	[139]
PO($^2\Pi$)	−415.3	−138.4	89.1	66.8	0.67	[140]
^{19}F						
CF($^2\Pi$)	−352.7	−117.6	151.6	15.3	0.29	[141]
SiF($^2\Pi$)	−175	−58	69	6.8	0.13	[142]
PF($^3\Sigma^-$)	−240.29	−80.10	89.433	9.1	0.17	[137]
SF($^2\Pi$)	−317	−106	104.3	11	0.20	[143]

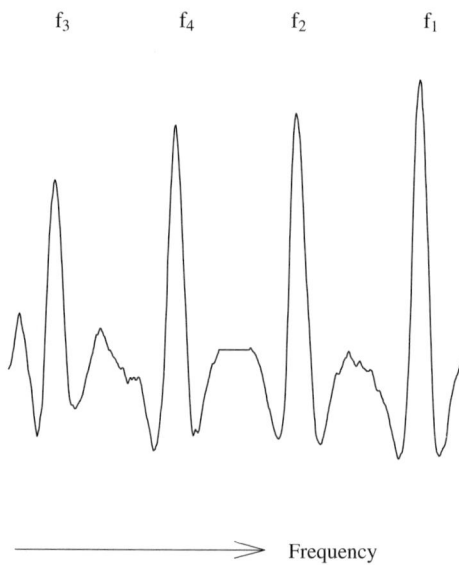

Figure 10.48. Observed ^{31}P and ^{19}F hyperfine lines [137] for the transitions shown on the right-hand side of figure 10.47 (PF in its $X\,^3\Sigma^-$, $v = 0$ state; rotational transition $N, J = 5, 5 \leftarrow 4, 4$). The observed frequencies are as follows:

$$f_1 = 169\,200.5\,\text{MHz}, \quad f_2 = 169\,196.5, \quad f_3 = 169\,187.3, \quad f_4 = 169\,191.3.$$

The F quantum number assignments are given in figure 10.47.

in table 10.11, where they are compared with other phosporus and fluorine-containing free radicals. Also given in this Table are the calculated spin densities [137] for the two atoms in PF; the unpaired electrons are largely localised on the phosphorus atom, in an orbital which is predominantly $3p_\pi$ in character. Part of the observed spectrum is shown in figure 10.48.

$^3\Sigma$ states | 769

We have discussed the origin of the so-called 'spin–spin' interaction constant λ in chapter 7.

We now turn to studies of the NCl radical, which also has a $^3\Sigma^-$ ground state, described by Yamada, Endo and Hirota [138]. In many ways NCl is similar to the other radicals described in this section, particularly PF, but is more complicated because the ^{14}N nucleus has a spin of 1, and ^{35}Cl has a spin of 3/2. Apart from the more extensive magnetic hyperfine structure, both nuclei also have electric quadrupole moments, thereby adding to the complexity.

The NCl radical conforms well to Hund's case (b), as may be seen from figure 10.49 which shows the rotational levels up to $N = 4$, with their associated triplet spin splitting. The figure also shows the eleven rotational transitions studied [138]. Each of these transitions exhibited a complex nuclear hyperfine splitting, and an example is shown in

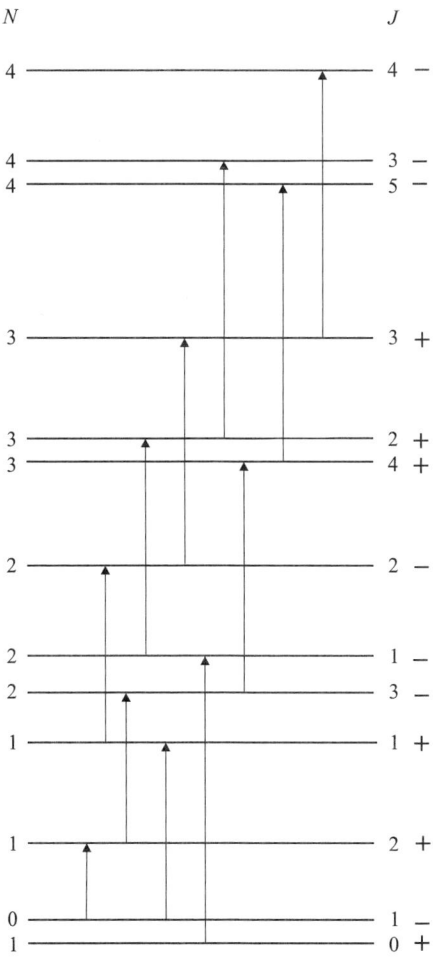

Figure 10.49. The observed rotation and fine-structure transitions observed [138] for NCl in its $X\,^3\Sigma^-$ ground state.

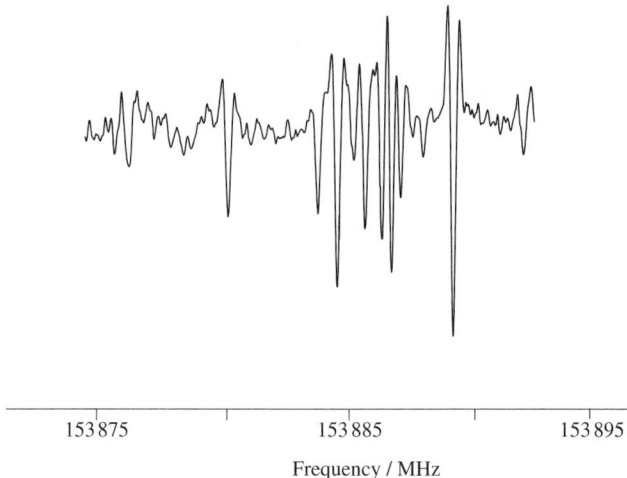

Figure 10.50. The $N=4, J=3 \leftarrow N=3, J=2$ transition in NCl, with ^{14}N and ^{35}Cl hyperfine structure [138].

figure 10.50. For the predominant isotopomer, ^{14}N^{35}Cl, almost 140 lines were measured over the frequency range 41 to 155 GHz; all were assigned with a standard error in the fit of only 29 kHz. We will discuss the molecular parameters determined in due course, but first consider the complexities arising from the presence of two nuclei with each having an electric quadrupole moment.

The theory of the magnetic hyperfine interactions in NCl is essentially the same as that already described for the PF radical in the previous section, except that the nuclear spins I are 1 for ^{14}N and 3/2 for ^{35}Cl. The form of the effective Hamiltonian for the quadrupole interaction and its matrix elements for two different quadrupolar nuclei was described in some detail in chapter 8 when we discussed the electric resonance spectra of CsF and LiBr. We now use the same case (b) hyperfine-coupled basis set as was used for PF. The quadrupole Hamiltonian for the two nuclei can be written as the sum of two independent terms as follows:

$$\mathcal{H}_Q = -e\mathrm{T}^2(\nabla \boldsymbol{E}_1) \cdot \mathrm{T}^2(\boldsymbol{Q}_1) - e\mathrm{T}^2(\nabla \boldsymbol{E}_2) \cdot \mathrm{T}^2(\boldsymbol{Q}_2), \tag{10.88}$$

where the subscripts 1 and 2 refer to the ^{14}N and ^{35}Cl nuclei respectively.

We deal first with the ^{14}N quadrupole interaction, using the first term in (10.88), and making use of the results described in chapter 8 for the 'weak' field coupled basis in CsF. The nuclear spin coupling scheme is

$$\boldsymbol{F}_1 = \boldsymbol{J} + \boldsymbol{I}_1, \quad \boldsymbol{F} = \boldsymbol{F}_1 + \boldsymbol{I}_2, \tag{10.89}$$

and the required matrix elements are obtained through the following sequence:

$$\langle \eta, \Lambda; N, \Lambda, S, J, I_1, F_1, I_2, F | -e\mathrm{T}^2(\nabla \boldsymbol{E}_1) \cdot \mathrm{T}^2(\boldsymbol{Q}_1) | \eta, \Lambda'; N', \Lambda', S, J', I_1, F_1', I_2, F' \rangle$$
$$= \delta_{FF'} \delta_{F_1 F_1'} \langle \eta, \Lambda; N, \Lambda, S, J, I_1, F_1 | -e\mathrm{T}^2(\nabla \boldsymbol{E}_1) \cdot \mathrm{T}^2(\boldsymbol{Q}_1) | \eta, \Lambda'; N', \Lambda', S, J', I_1, F_1 \rangle$$

$$= \delta_{FF'}\delta_{F_1 F_1'}(-e)(-1)^{J'+F_1+I_1} \begin{Bmatrix} I_1 & J' & F_1 \\ J & I_1 & 2 \end{Bmatrix}$$
$$\times \langle \eta, \Lambda; N, \Lambda, S, J \| T^2(\nabla E_1) \| \eta, \Lambda'; N', \Lambda', S, J' \rangle \langle I_1 \| T^2(Q_1) \| I_1 \rangle$$
$$= \delta_{FF'}\delta_{F_1 F_1'}(-e)(-1)^{J'+F_1+I_1} \begin{Bmatrix} I_1 & J' & F_1 \\ J & I_1 & 2 \end{Bmatrix} (-1)^{J'+N+S}\{(2J'+1)(2J+1)\}^{1/2}$$
$$\times \begin{Bmatrix} J & N & S \\ N' & J' & 2 \end{Bmatrix} \langle \eta, \Lambda; N, \Lambda \| T^2(\nabla E_1) \| \eta, \Lambda'; N', \Lambda' \rangle \left(\frac{Q_1}{2} \right) \begin{pmatrix} I_1 & 2 & I_1 \\ -I_1 & 0 & I_1 \end{pmatrix}^{-1}$$
$$= \delta_{FF'}\delta_{F_1 F_1'}(-e)(-1)^{J'+F_1+I_1} \begin{Bmatrix} I_1 & J' & F_1 \\ J & I_1 & 2 \end{Bmatrix} (-1)^{J'+N+S}\{(2J'+1)(2J+1)\}^{1/2}$$
$$\times \begin{Bmatrix} J & N & S \\ N' & J' & 2 \end{Bmatrix} \langle \eta, \Lambda; N, \Lambda | \sum_q \mathcal{D}_{\cdot q}^{(2)}(\omega)^* T_q^2(\nabla E_1) | \eta, \Lambda'; N', \Lambda' \rangle$$
$$\times \left(\frac{Q_1}{2} \right) \begin{pmatrix} I_1 & 2 & I_1 \\ -I_1 & 0 & I_1 \end{pmatrix}^{-1}$$
$$= \delta_{FF'}\delta_{F_1 F_1'}(-e)(-1)^{J'+F_1+I_1} \begin{Bmatrix} I_1 & J' & F_1 \\ J & I_1 & 2 \end{Bmatrix} (-1)^{J'+N+S}\{(2J'+1)(2J+1)\}^{1/2}$$
$$\times \begin{Bmatrix} J & N & S \\ N' & J' & 2 \end{Bmatrix} \sum_q (-1)^{N-\Lambda}\{(2N'+1)(2N+1)\}^{1/2}$$
$$\times \begin{pmatrix} N & 2 & N' \\ -\Lambda & q & \Lambda' \end{pmatrix} \langle \eta, \Lambda | T_q^2(\nabla E_1) | \eta, \Lambda' \rangle \left(\frac{Q_1}{2} \right) \begin{pmatrix} I_1 & 2 & I_1 \\ -I_1 & 0 & I_1 \end{pmatrix}^{-1}. \quad (10.90)$$

Everything is general so far but, as in previous cases, we confine attention to the $q = 0$ component (the molecule-fixed axial component) and make use of the definition of the electric field gradient at nucleus 1,

$$\langle \eta, \Lambda | T_0^2(\nabla E_1) | \eta, \Lambda \rangle = -q_1/2. \quad (10.91)$$

We also put $\Lambda = 0$ since we are dealing with a Σ state and thereby obtain the final result:

$$\langle \eta, \Lambda; N, \Lambda, S, J, I_1, F_1, I_2, F | -eT^2(\nabla E_1) \cdot T^2(Q_1) | \eta, \Lambda'; N', \Lambda', S, J', I_1, F_1, I_2, F \rangle$$
$$= (-1)^{2J'+F_1+I_1+S} \left(\frac{eq_1 Q_1}{4} \right) [(2J'+1)(2J+1)(2N'+1)(2N+1)]^{1/2}$$
$$\times \begin{Bmatrix} J & N & S \\ N' & J' & 2 \end{Bmatrix} \begin{Bmatrix} I_1 & J' & F_1 \\ J & I_1 & 2 \end{Bmatrix} \begin{pmatrix} N & 2 & N' \\ 0 & 0 & 0 \end{pmatrix} \begin{pmatrix} I_1 & 2 & I_1 \\ -I_1 & 0 & I_1 \end{pmatrix}^{-1}. \quad (10.92)$$

Note that for the ^{14}N nucleus in NCl, both I_1 and S are equal to 1, so that (10.92) becomes rather simpler.

We turn now to the quadrupole interaction for nucleus 2, the ^{35}Cl nucleus. Again, the reduction is as follows:

$$\langle \eta, \Lambda; N, \Lambda, S, J, I_1, F_1, I_2, F| -e\mathbf{T}^2(\nabla \mathbf{E}_2) \cdot \mathbf{T}^2(\mathbf{Q}_2) |\eta, \Lambda'; N', \Lambda', S, J', I_1, F_1', I_2, F'\rangle$$

$$= \delta_{FF'}(-e)(-1)^{F_1'+F+I_2} \begin{Bmatrix} I_2 & F_1' & F \\ F_1 & I_2 & 2 \end{Bmatrix}$$

$$\times \langle \eta, \Lambda; N, \Lambda, S, J, I_1, F_1 | \mathbf{T}^2(\nabla \mathbf{E}_2) |\eta, \Lambda'; N', \Lambda', S, J', I_1, F_1'\rangle \langle I_2 \| \mathbf{T}^2(\mathbf{Q}_2) \| I_2 \rangle$$

$$= \delta_{FF'}(-e)(-1)^{F_1'+F+I_2} \begin{Bmatrix} I_2 & F_1' & F \\ F_1 & I_2 & 2 \end{Bmatrix} (-1)^{F_1'+J+I_1} \{(2F_1'+1)(2F_1+1)\}^{1/2}$$

$$\times \begin{Bmatrix} F_1 & J & I_1 \\ J' & F_1' & 2 \end{Bmatrix} \langle \eta, \Lambda; N, \Lambda, S, J \| \mathbf{T}^2(\nabla \mathbf{E}_2) \| \eta, \Lambda'; N', \Lambda', S, J'\rangle$$

$$\times \left(\frac{Q_2}{2}\right)\begin{pmatrix} I_2 & 2 & I_2 \\ -I_2 & 0 & I_2 \end{pmatrix}^{-1}$$

$$= \delta_{FF'}(-e)(-1)^{F_1'+F+I_2} \begin{Bmatrix} I_2 & F_1' & F \\ F_1 & I_2 & 2 \end{Bmatrix} (-1)^{F_1'+J+I_1} \{(2F_1'+1)(2F_1+1)\}^{1/2}$$

$$\times \begin{Bmatrix} F_1 & J & I_1 \\ J' & F_1' & 2 \end{Bmatrix} (-1)^{J'+N+S} \{(2J'+1)(2J+1)\}^{1/2} \begin{Bmatrix} N' & J' & S \\ J' & N & 2 \end{Bmatrix}$$

$$\times \langle \eta, \Lambda; N, \Lambda \| \mathbf{T}^2(\nabla \mathbf{E}_2) \| \eta, \Lambda'; N', \Lambda'\rangle \left(\frac{Q_2}{2}\right)\begin{pmatrix} I_2 & 2 & I_2 \\ -I_2 & 0 & I_2 \end{pmatrix}^{-1}$$

$$= \delta_{FF'}(-e)(-1)^{F_1'+F+I_2} \begin{Bmatrix} I_2 & F_1' & F \\ F_1 & I_2 & 2 \end{Bmatrix} (-1)^{F_1'+J+I_1} \{(2F_1'+1)(2F_1+1)\}^{1/2}$$

$$\times \begin{Bmatrix} F_1 & J & I_1 \\ J' & F_1' & 2 \end{Bmatrix} (-1)^{J'+N+S} \{(2J'+1)(2J+1)\}^{1/2}$$

$$\times \begin{Bmatrix} N' & J' & S \\ J & N & 2 \end{Bmatrix} \langle \eta, \Lambda; N, \Lambda | \sum_q \mathcal{D}^{(2)}_{\cdot q}(\omega)^* \mathbf{T}^2_q(\nabla \mathbf{E}_2) |\eta, \Lambda'; N', \Lambda'\rangle$$

$$\times \left(\frac{Q_2}{2}\right)\begin{pmatrix} I_2 & 2 & I_2 \\ -I_2 & 0 & I_2 \end{pmatrix}^{-1}$$

$$= \delta_{FF'}(-1)^{F_1'+F+I_2} \begin{Bmatrix} I_2 & F_1' & F \\ F_1 & I_2 & 2 \end{Bmatrix} (-1)^{F_1'+J+I_1} \{(2F_1'+1)(2F_1+1)\}^{1/2}$$

$$\times \begin{Bmatrix} F_1 & J & I_1 \\ J' & F_1' & 2 \end{Bmatrix} (-1)^{J'+N+S} \{(2J'+1)(2J+1)\}^{1/2}$$

$$\times \begin{Bmatrix} N' & J' & S \\ J & N & 2 \end{Bmatrix} (-1)^N \{(2N+1)(2N'+1)\}^{1/2}$$

$$\times \begin{pmatrix} N & 2 & N' \\ 0 & 0 & 0 \end{pmatrix}\left(\frac{eq_2 Q_2}{4}\right)\begin{pmatrix} I_2 & 2 & I_2 \\ -I_2 & 0 & I_2 \end{pmatrix}^{-1}. \quad (10.93)$$

Again we have substituted $\Lambda = 0$ and $q = 0$ to obtain this result. Rearranging the terms in (10.93) and noting that since N is an integer, $(-1)^{2N} = +1$ we obtain the final result

$$\langle \eta, \Lambda; N, \Lambda, S, J, I_1, F_1, I_2, F | - e\mathbf{T}^2(\nabla \mathbf{E}_2) \cdot \mathbf{T}^2(\mathbf{Q}_2) | \eta, \Lambda'; N', \Lambda', S, J', I_1, F_1', I_2, F \rangle$$
$$= (-1)^{2F_1' + F + J + I_2 + I_1 + J' + S} \left(\frac{eq_2 Q_2}{4} \right)$$
$$\times \{(2F_1' + 1)(2F_1 + 1)(2J' + 1)(2J + 1)(2N + 1)(2N' + 1)\}^{1/2}$$
$$\times \begin{Bmatrix} F_1 & J & I_1 \\ J' & F_1' & 2 \end{Bmatrix} \begin{Bmatrix} N' & J' & S \\ J & N & 2 \end{Bmatrix} \begin{pmatrix} N & 2 & N' \\ 0 & 0 & 0 \end{pmatrix} \begin{Bmatrix} I_2 & F_1' & F \\ F_1 & I_2 & 2 \end{Bmatrix} \begin{pmatrix} I_2 & 2 & I_2 \\ -I_2 & 0 & I_2 \end{pmatrix}^{-1}.$$
(10.94)

This is the result given by Yamada, Endo and Hirota [138] in their equation (8), except that they have a typographical error in which F and F' are written instead of F_1 and F_1' in the square root expression.

We now have all of the required matrix elements but in order to understand fully the numerical complexity of the problem, consider the manifold of states which arise for $N = 4$. The parentage of the final hyperfine states is shown in figure 10.51. F is always a good quantum number in the absence of external fields, so that a matrix of the total effective Hamiltonian may be set up for each F value. The parity of each N level is also a good quantum number, but alternates with the N value; a complete F matrix will therefore exist for each parity, and will contain basis states representing alternate N levels. Note that the electric quadrupole interaction can mix each rotational level characterised by a value of N with those characterised by $N \pm 2$. The analytical problem is therefore to calculate the energies of the hyperfine levels for a given set of molecular parameters, and to converge towards the observed spectrum, usually by means of a least-squares routine.

It is, of course, also necessary to calculate the relative intensities of the hyperfine components of each rotational transition in order to assign the spectrum. As we have seen elsewhere, the perturbation due to the interaction of the microwave electric field $E(t)$ with the molecular electric dipole moment may be represented by the effective Hamiltonian

$$\mathcal{H}'(t) = -\mathbf{T}_0^1(\boldsymbol{\mu}_e)\mathbf{T}_0^1(\mathbf{E}(t)),$$
(10.95)

where we have chosen the direction of the oscillating electric field to define the space-fixed $p = 0$ direction. The relative intensity of each transition is calculated from the squared transition dipole matrix element summed over all directions in space and also all M_F values. In the basis set used that matrix element is as follows:

$$\langle \eta, \Lambda; N, \Lambda, S, J, I_1, F_1, I_2, F, M_F | -\mathbf{T}_0^1(\boldsymbol{\mu}_e)\mathbf{T}_0^1(\mathbf{E}) | \eta, \Lambda'; N', \Lambda', S, J', I_1, F_1', I_2, F', M_F \rangle$$
$$= E_0 (-1)^{F - M_F} \begin{pmatrix} F & 1 & F' \\ -M_F & 0 & M_F \end{pmatrix}$$

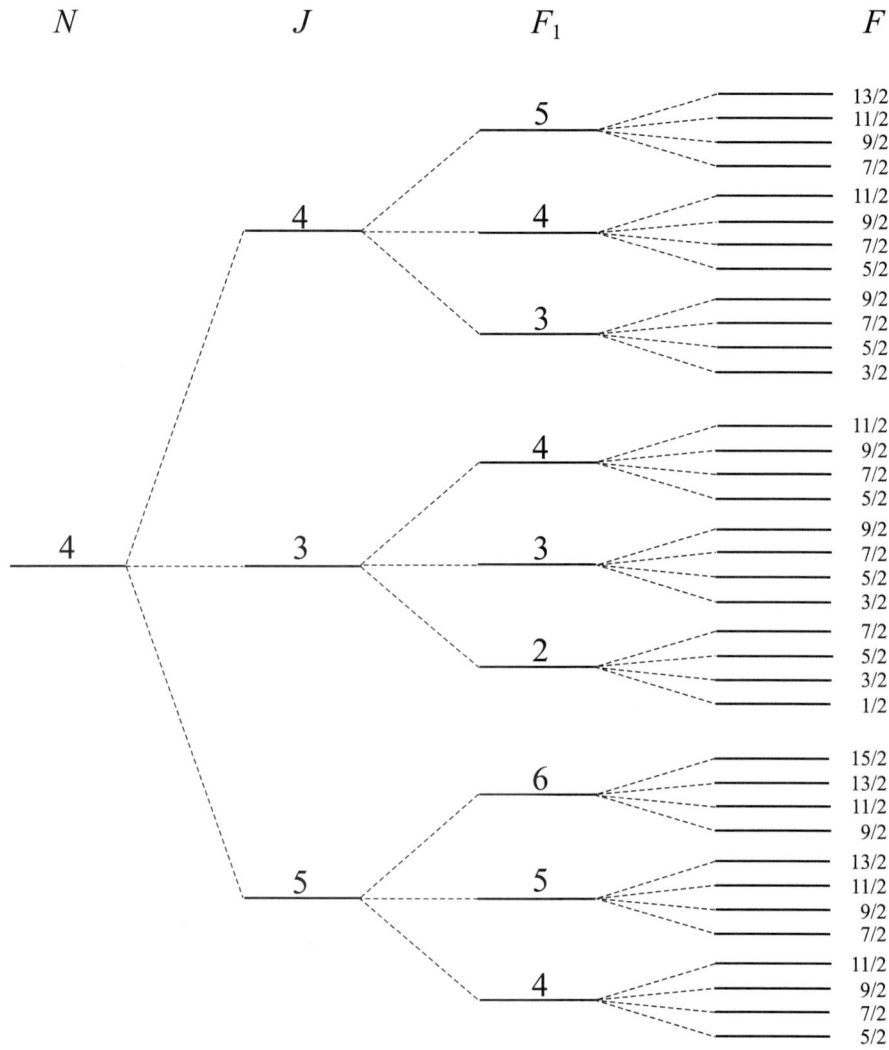

Figure 10.51. Hyperfine levels in the $N=4$ rotational level of ^{14}N^{35}Cl, with hyperfine splitting from both nuclei. This figure shows the parentage of the hyperfine levels only, and is not meant to represent either the relative or absolute energies of the hyperfine levels.

$$\times \langle \eta, \Lambda; N, \Lambda, S, J, I_1, F_1, I_2, F| - \mathrm{T}_0^1(\boldsymbol{\mu}_e)|\eta, \Lambda'; N', \Lambda', S, J', I_1, F_1', I_2, F'\rangle$$

$$= E_0(-1)^{F-M_F} \begin{pmatrix} F & 1 & F' \\ -M_F & 0 & M_F \end{pmatrix} (-1)^{F'+F_1+1+I_2}\{(2F'+1)(2F+1)\}^{1/2}$$

$$\times \begin{Bmatrix} F & F_1 & I_2 \\ F_1' & F' & 1 \end{Bmatrix} \langle \eta, \Lambda; N, \Lambda, S, J, I_1, F_1\| - \mathrm{T}_0^1(\boldsymbol{\mu}_e)\|\eta, \Lambda'; N', \Lambda', S, J', I_1, F_1'\rangle$$

$$= E_0(-1)^{F-M_F} \begin{pmatrix} F & 1 & F' \\ -M_F & 0 & M_F \end{pmatrix} (-1)^{F'+F_1+1+I_2}\{(2F'+1)(2F+1)\}^{1/2}$$

$$\times \begin{Bmatrix} F & F_1 & I_2 \\ F_1' & F' & 1 \end{Bmatrix} (-1)^{F_1'+J+1+I_1} \{(2F_1'+1)(2F_1+1)\}^{1/2}$$

$$\times \begin{Bmatrix} F_1 & J & I_1 \\ J' & F_1' & 1 \end{Bmatrix} \langle \eta, \Lambda; N, \Lambda, S, J \| - T_0^1(\boldsymbol{\mu}_e) \| \eta, \Lambda'; N', \Lambda', S, J' \rangle$$

$$= E_0 (-1)^{F-M_F} \begin{pmatrix} F & 1 & F' \\ -M_F & 0 & M_F \end{pmatrix} (-1)^{F'+F_1+1+I_2} \{(2F'+1)(2F+1)\}^{1/2}$$

$$\times \begin{Bmatrix} F & F_1 & I_2 \\ F_1' & F' & 1 \end{Bmatrix} (-1)^{F_1'+J+1+I_1} \{(2F_1'+1)(2F_1+1)\}^{1/2}$$

$$\times \begin{Bmatrix} F_1 & J & I_1 \\ J' & F_1' & 1 \end{Bmatrix} (-1)^{J'+N+1+S} \{(2J'+1)(2J+1)\}^{1/2} \begin{Bmatrix} J & N & S \\ N' & J' & 1 \end{Bmatrix}$$

$$\times \langle \eta, \Lambda; N, \Lambda \| - T_0^1(\boldsymbol{\mu}_e) \| \eta, \Lambda'; N', \Lambda' \rangle. \tag{10.96}$$

Our final step is to transform the space-fixed molecular dipole moment $p = 0$ component into the molecule-fixed component with $q = 0$ by means of a rotation matrix. The final line of (10.96) therefore becomes

$$\langle \eta, \Lambda; N, \Lambda, S, J, I_1, F_1, I_2, F, M_F | -T_0^1(\boldsymbol{\mu}_e) T_0^1(\boldsymbol{E}) | \eta, \Lambda'; N', \Lambda', S, J', I_1, F_1', I_2, F', M_F \rangle$$

$$= E_0 (-1)^{F-M_F+F'+F_1+F_1'+J+I_1+I_2+J'+N+S}$$

$$\times \{(2F'+1)(2F+1)(2F_1'+1)(2F_1+1)(2J'+1)(2J+1)\}^{1/2}$$

$$\times \begin{pmatrix} F & 1 & F' \\ -M_F & 0 & M_F \end{pmatrix} \begin{Bmatrix} F & F_1 & I_2 \\ F_1' & F' & 1 \end{Bmatrix} \begin{Bmatrix} F_1 & J & I_1 \\ J' & F_1' & 1 \end{Bmatrix} \begin{Bmatrix} J & N & S \\ N' & J' & 1 \end{Bmatrix}$$

$$\times \langle \eta, \Lambda; N, \Lambda | \mathcal{D}_{.0}^{(1)}(\omega)^* T_{q=0}^1(\boldsymbol{\mu}_e) | \eta, \Lambda'; N', \Lambda' \rangle$$

$$= E_Z \mu_0 (-1)^{F-M_F+F'+F_1+F_1'+J+I_1+I_2+J'+S+N} \{(2F'+1)(2F+1)$$

$$\times (2F_1'+1)(2F_1+1)(2J'+1)(2J+1)(2N'+1)(2N+1)\}^{1/2}$$

$$\times \begin{pmatrix} F & 1 & F' \\ -M_F & 0 & M_F \end{pmatrix} \begin{Bmatrix} F & F_1 & I_2 \\ F_1' & F' & 1 \end{Bmatrix} \begin{Bmatrix} F_1 & J & I_1 \\ J' & F_1' & 1 \end{Bmatrix}$$

$$\times \begin{Bmatrix} J & N & S \\ N' & J' & 1 \end{Bmatrix} \begin{pmatrix} N & 1 & N' \\ 0 & 0 & 0 \end{pmatrix}. \tag{10.97}$$

Again this result agrees with that given by Yamada, Endo and Hirota [138]; final results for the relative transition intensities require prior diagonalisation of the effective Hamiltonian matrix.

We will not pursue the analysis in detail any further. It does illustrate the power of spherical tensor methods, and one can only shudder at the possibility of developing the theory in a cartesian coordinate system, with direction cosines. We list the final values of the molecular parameters for $^{14}\text{N}^{35}\text{Cl}$ in table 10.12. The values of the hyperfine constants may be interpreted semi-empirically in the following way. The outmost pair of electrons occupy a $3p_\pi$ molecular orbital and the Fermi contact constants, given in table 10.12, may be compared with the atomic values [144] of 1811 and 5723 MHz for the nitrogen and chlorine atoms respectively; one concludes that the s electron character

Table 10.12. *Molecular constants determined for* $^{14}N^{35}Cl$ *(in MHz)*

Parameter	Value (MHz)
B	19 383.4655
D	0.047 95
λ	56 390.850
λ_D	−0.2568
γ	−208.6306
^{35}Cl coupling constants	^{14}N coupling constants
b_F 3.519	b_F 22.958
$c = 3t_0$ −57.764	$c = 3t_0$ −63.159
$eq_0 Q$ −63.13	$eq_0 Q$ 1.842
c_I 0.0152	

in the $3p_\pi$ molecular orbital is only 1.27% and 0.06% for the two atoms in the NCl molecule. In the case of the dipolar interaction one may assume that the angular part, $3\cos^2\theta - 1$, can be evaluated with a hydrogenic p_π wave function, so that

$$g_S \mu_B g_N \mu_N (\mu_0/4\pi)\langle 1/r^3 \rangle = -(5/3)c. \tag{10.98}$$

The measured c constants give the left-hand side of (10.98) to be 105.265 and 96.273 MHz for the nitrogen and chlorine atoms respectively. The atomic values are 138.8 and 439.0 MHz, so that the spin densities on the nitrogen and chlorine atoms in NCl are calculated to be 75.8% and 21.9% respectively. It is reassuring that these two values add up very nearly to 100 %.

Sakamaki, Okabayashi and Tanimoto have studied the spectra of NBr [145] and NI [146] in their $X\,^3\Sigma^-$ states. NBr is similar to NCl in having two bromine isotopes, ^{79}Br and ^{81}Br, with spins of 3/2; they are present is almost equal natural abundance. In the case of NI, the ^{127}I nucleus, in 100% natural abundance, has a spin of 5/2. Figure 10.52 shows the hyperfine patterns obtained for a sub-millimetre rotational transition in each molecule; in both cases there is resolved magnetic hyperfine splitting from the ^{14}N and halogen nuclei, with additional quadrupole splitting from the halogens. Analysis of the hyperfine structure leads to the conclusion that, in both molecules, the unpaired electron spin density on the nitrogen is close to 75%, with approximately 25% on the halogen.

10.5. $^1\Delta$ states

10.5.1. O_2 in its $a\,^1\Delta_g$ state

We discussed the Zeeman effect for $^1\Delta$ states in chapter 9, particularly with reference to SO where it provides a particularly simple example with which to illustrate the

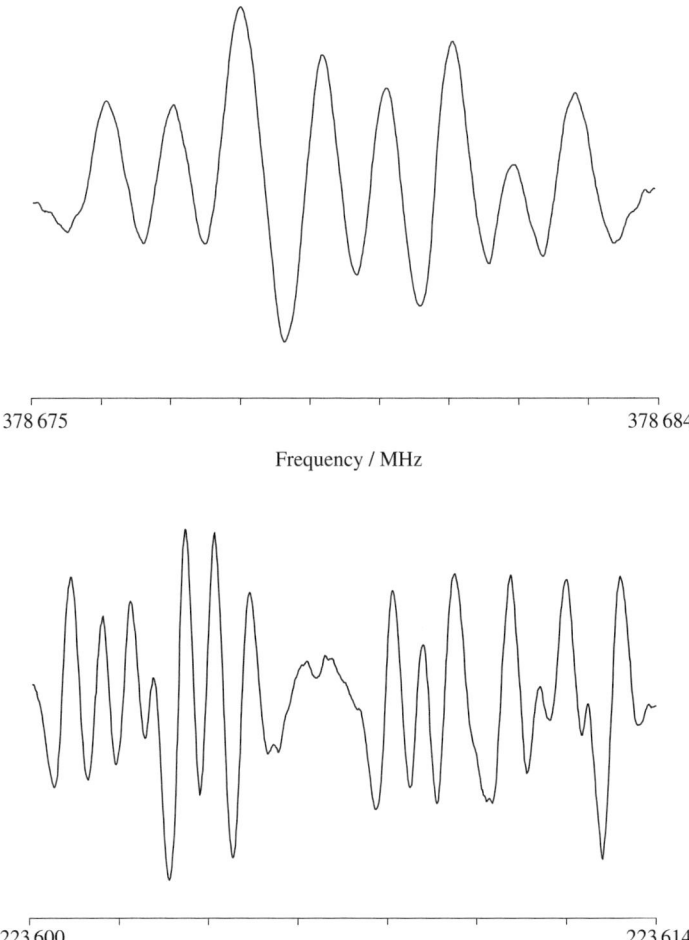

Figure 10.52. *Top*: the $N = 14 \leftarrow 13$, $J = 15 \leftarrow 14$ rotational transition in N^{81}Br, showing hyperfine splitting from both nuclei [145]. *Bottom*: the $N = 10 \leftarrow 9$, $J = 11 \leftarrow 10$ rotational transition in NI, showing hyperfine splitting from both nuclei [146].

principles of microwave magnetic resonance [147]. The most important example of a $^1\Delta$ state, however, is that of O$_2$. The ground state electronic configuration of O$_2$ may be written

$$(1s\sigma_g)^2(1s\sigma_u)^2(2s\sigma_u)^2(2s\sigma_g)^2(2p\sigma_g)^2(2p\pi_u)^4(2p\pi_g)^2$$

but we have, in addition, to specify the spatial and spin coordinates of the two electrons in the half filled $2p\pi_g$ orbital. If we denote the degenerate π orbitals as π_+ and π_-, we may distinguish three different electronic configurations, which are as follows.

$$\pi_+(1)\pi_-(2): \alpha(1)\alpha(2), (1/\sqrt{2})\{\alpha(1)\beta(2)+\beta(1)\alpha(2)\}, \beta(1)\beta(2) \quad X^3\Sigma_g^-$$

$$\pi_+(1)\pi_-(2): (1/\sqrt{2})\{\alpha(1)\beta(2) - \beta(1)\alpha(2)\} \quad b^1\Sigma_g^+$$

$$\pi_+(1)\pi_+(2), \pi_-(1)\pi_-(2): (1/\sqrt{2})\{\alpha(1)\beta(2) - \beta(1)\alpha(2)\} \quad a^1\Delta_g$$

The $a\ ^1\Delta_g$ state lies 7918.1 cm^{-1} above the ground state and is highly metastable, with a radiative lifetime of about 45 min. It is formed in the laboratory when O$_2$ is passed through a microwave discharge, or in the stratosphere and mesosphere by the photodecomposition of O$_3$.

The microwave magnetic resonance spectrum of $^1\Delta$ O$_2$ was first studied by Arrington, Falick and Myers [148], and the first observations of rotational transitions were made by Scalabrin, Saykally, Evenson, Radford and Mizushima [149] using far-infrared laser magnetic resonance. Pure microwave rotational transitions were first observed by Cazzoli, Degli Esposti and Favero [150], and the complete sequence of $\Delta J = +1$ transitions, from the lowest level with $J = 2$ up to $J(\text{lower}) = 9$ was measured by Hillig, Chiu, Read and Cohen [151]. The rotational spectrum of the homonuclear species ^{16}O^{16}O is necessarily a magnetic dipole spectrum, and it is particularly simple. Although $\Lambda = \pm 2$ degeneracy might be expected, symmetry requirements dictate that each rotational level can be associated with only one Λ component; all of the rotational levels in $^1\Delta_g$ O$_2$ have positive parity. Consequently the excited state mixing which, in a heteronuclear molecule, would lead to Λ-doublet splitting here produces only a small alternating shift in the energies of the rotational levels. Moreover there are no nuclear hyperfine effects in the predominant isotopomer.

The rotational Hamiltonian, and its expansion in the molecule-fixed axis system may be written

$$\begin{aligned}\mathcal{H}_{\text{rot}} &= B\mathbf{T}^1(\mathbf{R}) \cdot \mathbf{T}^1(\mathbf{R}) = B\mathbf{T}^1(\mathbf{J} - \mathbf{L}) \cdot \mathbf{T}^1(\mathbf{J} - \mathbf{L}) \\ &= B\{\mathbf{T}^1(\mathbf{J}) \cdot \mathbf{T}^1(\mathbf{J}) - 2\mathbf{T}^1(\mathbf{J}) \cdot \mathbf{T}^1(\mathbf{L}) + \mathbf{T}^1(\mathbf{L}) \cdot \mathbf{T}^1(\mathbf{L})\} \\ &= B\bigg\{J(J+1) - 2\mathrm{T}_0^1(\mathbf{J})\mathrm{T}_0^1(\mathbf{L}) - 2\sum_{q=\pm 1}\mathrm{T}_q^1(\mathbf{J})\mathrm{T}_q^1(\mathbf{L}) \\ &\quad + \mathrm{T}_0^1(\mathbf{L})\mathrm{T}_0^1(\mathbf{L}) - \sum_{q=\pm 1}\mathrm{T}_q^1(\mathbf{L})\mathrm{T}_{-q}^1(\mathbf{L})\bigg\}.\end{aligned} \quad (10.99)$$

Using the basis representation $|\eta, \Lambda; J, \Lambda\rangle$ we see that the first, second and fourth terms in (10.99) are diagonal, whereas the third and fifth terms have matrix elements off-diagonal in Λ, and therefore mix other electronic states with the $^1\Delta_g$ state. If we include the first rotational distortion term it is easy to show that, from (10.99), the rotational energies are given by

$$E(J) = B\{J(J+1) - 4\} - D\{J(J+1) - 4\}^2 \\ + (-1)^J(1/2)q_\Delta(J-1)J(J+1)(J+2) \quad (10.100)$$

where we have used the value $\Lambda = 2$ and consider states of positive parity only. The constant q_Δ produces the small shift in the energies, referred to above, and in a heteronuclear system would produce a very small Λ-doublet splitting of each rotational level. The values of the constants obtained [151] from the pure microwave far-infrared studies were

$$B_0 = 42\,504.5203\,\text{MHz}, \quad D_0 = 152.957\,\text{kHz}, \quad q_\Delta = -5.4\,\text{Hz}.$$

These values are very close to, but not identical with, the values obtained from the

Table 10.13. *Comparison of equilibrium bond lengths (in Å) of SO isotopomers in the electronic states* $X\,^3\Sigma^-$, $a\,^1\Delta$ *and* $b\,^1\Sigma^+$

Bond length	$X\,^3\Sigma^-$	$a\,^1\Delta$	$b\,^1\Sigma^+$
$R_e(\text{B–O})$	1.480 987 69	1.488 707 89	1.500 001 15
$R_e\,(^{32}\text{S}^{16}\text{O})$	1.481 082 02	1.488 805 95	1.500 109 47
$R_e\,(^{34}\text{S}^{16}\text{O})$	1.481 080 54	1.488 804 41	1.500 107 71
$R_e\,(^{32}\text{S}^{18}\text{O})$	1.481 074 31	1.488 797 93	1.500 100 75

laser magnetic resonance spectrum. In particular the value of q_Δ must be regarded as uncertain, even as regards its sign. If it is negative, it implies that rotational mixing with $^1\Sigma^+$ states outweighs that with $^1\Sigma^-$ states.

As we pointed out in chapter 9, the magnetic resonance methods enabled Miller [152] to determine the magnetic parameters, including the anisotropy of the magnetic susceptibility, from which he was able to calculate the electric quadrupole moment of the oxygen molecule in its $^1\Delta_g$ state. Magnetic resonance studies by Arrington, Falick and Myers [148] on the isotopomer $^{16}\text{O}^{17}\text{O}$ have provided information about the ^{17}O magnetic hyperfine constants.

10.5.2. SO and NCl in their $a\,^1\Delta$ states

The microwave magnetic resonance spectrum [147] of SO in its excited $^1\Delta$ state was described in chapter 9; subsequently the lowest rotational transition was observed through pure microwave spectroscopy by Saito [153] at a frequency of 127 770.47 MHz. Although in this heteronuclear case each rotational level possesses a two-fold Λ-degeneracy, the splitting is far too small to be resolved. The Λ-degeneracy does, however, lead to a first-order Stark effect, enabling Saito to determine the electric dipole moment to be 1.336 D, in good agreement with an earlier magnetic resonance value. More recently sub-millimetre studies by Klaus, Belov and Winnewisser [95] using frequencies up to 1063 GHz have provided measurements of rotational transitions up to $J = 25 \leftarrow 24$ for the predominant isotopomer, with additional transitions for the $^{34}\text{S}^{16}\text{O}$ and $^{32}\text{S}^{18}\text{O}$ species. The abundant data for SO provided particularly by the high-frequency studies have enabled Klaus, Belov and Winnewisser to determine accurate values of the equilibrium bond lengths for different isotopomers of all three low-lying electronic states. The results are given in table 10.13.

The pure microwave rotational spectrum of NCl in its $a\,^1\Delta$ state has been studied by Kobayashi, Goto, Yamamoto and Saito [154] at frequencies up to 400 GHz. Extensive hyperfine structure from both ^{14}N and ^{35}Cl was observed and the following unusual coupling scheme was found to be the most appropriate:

$$G = I_{\text{Cl}} + I_{\text{N}} = I_1 + I_2$$
$$F = J + G.$$
(10.101)

The effective Hamiltonian contained terms for both nuclei describing the orbital hyperfine interaction, the electric quadrupole interaction and the nuclear spin–rotation interaction:

$$\mathcal{H}_{\text{eff}} = a_{\text{Cl}} \mathbf{T}^1(\mathbf{I}_1) \cdot \mathbf{T}^1(\mathbf{L}) + a_{\text{N}} \mathbf{T}^1(\mathbf{I}_2) \cdot \mathbf{T}^1(\mathbf{L}) - e \mathbf{T}^2(\mathbf{Q}_1) \cdot \mathbf{T}^2(\nabla \mathbf{E}_1) - e \mathbf{T}^2(\mathbf{Q}_2) \cdot \mathbf{T}^2(\nabla \mathbf{E}_2)$$
$$+ c_{\text{Cl}} \mathbf{T}^1(\mathbf{I}_1) \cdot \mathbf{T}^1(\mathbf{J} - \mathbf{L}) + c_{\text{N}} \mathbf{T}^1(\mathbf{I}_2) \cdot \mathbf{T}^1(\mathbf{J} - \mathbf{L}). \tag{10.102}$$

The matrix elements of each term have been calculated in various different parts of this book. First, for the orbital hyperfine terms:

$$\langle \eta, \Lambda; I_1, I_2, G, J, \Lambda, F, M_F | a_{\text{Cl}} \mathbf{T}^1(\mathbf{I}_1) \cdot \mathbf{T}^1(\mathbf{L}) | \eta, \Lambda; I_1, I_2, G', J', \Lambda, F, M_F \rangle$$
$$= a_{\text{Cl}} (-1)^{G'+F+J} \begin{Bmatrix} J' & G' & F \\ G & J & 1 \end{Bmatrix} \langle I_1, I_2, G \| \mathbf{T}^1(\mathbf{I}_1) \| I_1, I_2, G' \rangle \langle J, \Lambda \| \mathbf{T}^1(\mathbf{L}) \| J', \Lambda \rangle$$
$$= a_{\text{Cl}} (-1)^{G'+F+J} \begin{Bmatrix} J' & G' & F \\ G & J & 1 \end{Bmatrix} (-1)^{G'+I_1+I_2+1} \{(2G'+1)(2G+1)\}^{1/2}$$
$$\times \begin{Bmatrix} G & I_1 & I_2 \\ I_1 & G' & 1 \end{Bmatrix} \{I_1(I_1+1)(2I_1+1)\}^{1/2} \langle J, \Lambda \| \mathbf{T}^1(\mathbf{L}) \| J', \Lambda \rangle$$
$$= a_{\text{Cl}} (-1)^{G'+F+J} \begin{Bmatrix} J' & G' & F \\ G & J & 1 \end{Bmatrix} (-1)^{G'+I_1+I_2+1} \{(2G'+1)(2G+1)\}^{1/2}$$
$$\times \begin{Bmatrix} G & I_1 & I_2 \\ I_1 & G' & 1 \end{Bmatrix} \{I_1(I_1+1)(2I_1+1)\}^{1/2}$$
$$\times (-1)^{J-\Lambda} \begin{pmatrix} J & 1 & J' \\ -\Lambda & 0 & \Lambda \end{pmatrix} \{(2J+1)(2J'+1)\}^{1/2} \Lambda, \tag{10.103}$$

$$\langle \eta, \Lambda; I_1, I_2, G; G, J, \Lambda, F, M_F | a_{\text{N}} \mathbf{T}^1(\mathbf{I}_2) \cdot \mathbf{T}^1(\mathbf{L}) | \eta, \Lambda; I_1, I_2, G'; G', J', \Lambda, F, M_F \rangle$$
$$= a_{\text{N}} (-1)^{G'+F+J} \begin{Bmatrix} J' & G' & F \\ G & J & 1 \end{Bmatrix} (-1)^{G+I_1+I_2+1} \{(2G'+1)(2G+1)\}^{1/2}$$
$$\times \begin{Bmatrix} G & I_2 & I_1 \\ I_2 & G' & 1 \end{Bmatrix} \{I_2(I_2+1)(2I_2+1)\}^{1/2}$$
$$\times (-1)^{J-\Lambda} \begin{pmatrix} J & 1 & J' \\ -\Lambda & 0 & \Lambda \end{pmatrix} \{(2J+1)(2J'+1)\}^{1/2} \Lambda. \tag{10.104}$$

Next, the electric quadrupole matrix elements for two equivalent nuclei were given in chapter 8 when we discussed the magnetic resonance spectrum of D_2. When the nuclei are not identical the results, for Λ diagonal, are

$$\langle \eta, \Lambda; J, I_1, I_2, G, F | -e \sum_{k=1,2} \mathbf{T}^2(\nabla \mathbf{E}_k) \cdot \mathbf{T}^2(\mathbf{Q}_k) | \eta, \Lambda; J', I_1, I_2, G', F \rangle$$
$$= -e \sum_{k=1,2} (-1)^{J'+I+F} \begin{Bmatrix} G & J & F \\ J' & G' & 2 \end{Bmatrix} \langle \eta, J, \Lambda \| \mathbf{T}^2(\nabla \mathbf{E}_k) \| \eta, J', \Lambda \rangle$$
$$\times \langle I_1, I_2, G \| \mathbf{T}^2(\mathbf{Q}_k) \| I_1, I_2, G' \rangle$$

$$= \sum_{k=1,2} (-1)^{J'+I+F} \begin{Bmatrix} G & J & F \\ J' & G' & 2 \end{Bmatrix} (-1)^{J-\Lambda} \{(2J+1)(2J'+1)\}^{1/2}$$

$$\times \begin{pmatrix} J & 2 & J' \\ -\Lambda & 0 & \Lambda \end{pmatrix} (eq_0(k)/2) \langle I_1, I_2, G \| T^2(\boldsymbol{Q}_k) \| I_1, I_2, G' \rangle. \quad (10.105)$$

For the remaining reduced matrix elements involving the nuclear spins in (10.105) we have

$$\langle I_1, I_2, G \| T^2(\boldsymbol{Q}_1) \| I_1, I_2, G' \rangle = (-1)^{I_1+I_2+G'} \{(2G+1)(2G'+1)\}^{1/2} \begin{Bmatrix} I_1 & G & I_2 \\ G' & I_1 & 2 \end{Bmatrix}$$

$$\times \langle I_1 \| T^2(\boldsymbol{Q}_1) \| I_1 \rangle, \quad (10.106)$$

$$\langle I_1, I_2, G \| T^2(\boldsymbol{Q}_2) \| I_1, I_2, G' \rangle = (-1)^{I_1+I_2+G'} \{(2G+1)(2G'+1)\}^{1/2} \begin{Bmatrix} I_2 & G & I_1 \\ G' & I_2 & 2 \end{Bmatrix}$$

$$\times \langle I_2 \| T^2(\boldsymbol{Q}_2) \| I_2 \rangle. \quad (10.107)$$

We complete the calculation by making use of the definition of the nuclear quadrupole moment of nucleus k:

$$\langle I_k \| T^2(\boldsymbol{Q}_k) \| I_k \rangle = \left(\frac{Q_k}{2} \right) \begin{pmatrix} I_k & 2 & I_k \\ -I_k & 0 & I_k \end{pmatrix}^{-1}. \quad (10.108)$$

Finally we have the nuclear spin–rotation terms:

$$\langle \eta, \Lambda; J, I_1, I_2, G, F | c_1 T^1(\boldsymbol{J}) \cdot T^1(\boldsymbol{I}_1) | \eta, \Lambda; J, I_1, I_2, G', F \rangle$$

$$= c_1(-1)^{J'+F+G} \begin{Bmatrix} G' & J' & F \\ J & G & 1 \end{Bmatrix} \langle J \| T^1(\boldsymbol{J}) \| J \rangle \langle I_1, I_2, G \| T^1(\boldsymbol{I}_1) \| I_1, I_2, G' \rangle$$

$$= c_1(-1)^{J'+F+G} \begin{Bmatrix} G' & J' & F \\ J & G & 1 \end{Bmatrix} \{J(J+1)(2J+1)\}^{1/2} (-1)^{G'+I_1+I_2+1}$$

$$\times \{(2G'+1)(2G+1)\}^{1/2} \begin{Bmatrix} G & I_1 & I_2 \\ I_1 & G' & 1 \end{Bmatrix} \{I_1(I_1+1)(2I_1+1)\}^{1/2}, \quad (10.109)$$

$$\langle \eta, \Lambda; J, I_1, I_2, G, F | c_2 T^1(\boldsymbol{J}) \cdot T^1(\boldsymbol{I}_2) | \eta, \Lambda; J, I_1, I_2, G', F \rangle$$

$$= c_2(-1)^{J'+F+G} \begin{Bmatrix} G' & J' & F \\ J & G & 1 \end{Bmatrix} \langle J \| T^1(\boldsymbol{J}) \| J \rangle \langle I_1, I_2, G \| T^1(\boldsymbol{I}_2) \| I_1, I_2, G' \rangle$$

$$= c_2(-1)^{J'+F+G} \begin{Bmatrix} G' & J' & F \\ J & G & 1 \end{Bmatrix} \{J(J+1)(2J+1)\}^{1/2} (-1)^{G'+I_1+I_2+1}$$

$$\times \{(2G'+1)(2G+1)\}^{1/2} \begin{Bmatrix} G & I_2 & I_1 \\ I_2 & G' & 1 \end{Bmatrix} \{I_2(I_2+1)(2I_2+1)\}^{1/2}. \quad (10.110)$$

Kobayashi, Goto, Yamamoto and Saito [154] were able to determine five of the six molecular parameters listed above, only the nuclear spin–rotation constant for the ^{14}N nucleus being too small to be significant. The molecular parameters and conclusions about the spin density distribution for both the $a\,^1\Delta$ and $X\,^3\Sigma^-$ states are listed in table 10.14.

Table 10.14. *Molecular parameters (in MHz) and derived spin distributions for NCl in the a $^1\Delta$ and X $^3\Sigma^-$ states*

Molecular parameter	NCl a $^1\Delta$ ($v=0$)	NCl X $^3\Sigma^-$
B	20 196.298	19 383.4655
D	0.045 238 1	0.047 95
a(Cl)	102.515	—
a(N)	91.43	—
$eq_0 Q$(Cl)	−52.54	−63.13
$eq_0 Q$(N)	1.72	1.842
c_{Cl}	0.0196	0.0152
N atom spin density	0.659	0.758
Cl atom spin density	0.234	0.219
Ionic character	0.29	0.21

10.6. $^2\Pi$ states

10.6.1. NO in the X $^2\Pi$ ground state

The most extensively studied studied open shell molecule is undoubtedly nitric oxide, NO. It is a stable gas which comes in cylinders, and in possessing electronic orbital and spin angular momentum, together with the nuclear spin ($I = 1$) of ^{14}N, it exhibits most of the features we expect to encounter in high-resolution spectroscopy. NO has already been discussed extensively in chapter 8 in connection with its Λ-doubling electric resonance spectrum, and its Zeeman properties were reviewed at length in chapter 9. The microwave spectroscopy of NO in its excited electronic states will feature in chapter 11, but in this section we describe studies of its rotational spectrum in the $^2\Pi$ ground electronic state, and bring together all of the detailed information derived from different spectroscopic studies. We remind ourselves that NO has two fine-structure states, $^2\Pi_{1/2}$ being the ground state with $^2\Pi_{3/2}$ lying 123 cm^{-1} higher in energy. Rotational levels in both states are appreciably populated at room temperature.

Figure 10.53, which is a repeat of figure 8.42, shows the lowest rotational levels of NO. For the predominant isotopic species almost all possible $\Delta J = \pm 1$ rotational transitions up to $J = 53/2$ in both fine structure states have been observed (see [155] for a summary), all showing Λ-doublet splitting and many exhibiting resolved hyperfine structure. In many rotational levels the $\Delta J = 0$, Λ-doublet transitions have been observed, particularly by Meerts and Dymanus [156], as we described in chapter 8. Figure 10.53 is only approximately drawn to scale, the Λ-doublet splittings being exaggerated. Although most studies have been concerned with the predominant naturally occurring isotopic species, ^{14}N^{16}O, other isotopic variants have also been extensively investigated and figure 10.54 shows a state-of-the-art (for 2000) recording of the $J = 17/2 \leftarrow 15/2$, $^2\Pi_{3/2}$ rotational transition of ^{14}N^{18}O, published by Klisch, Belov,

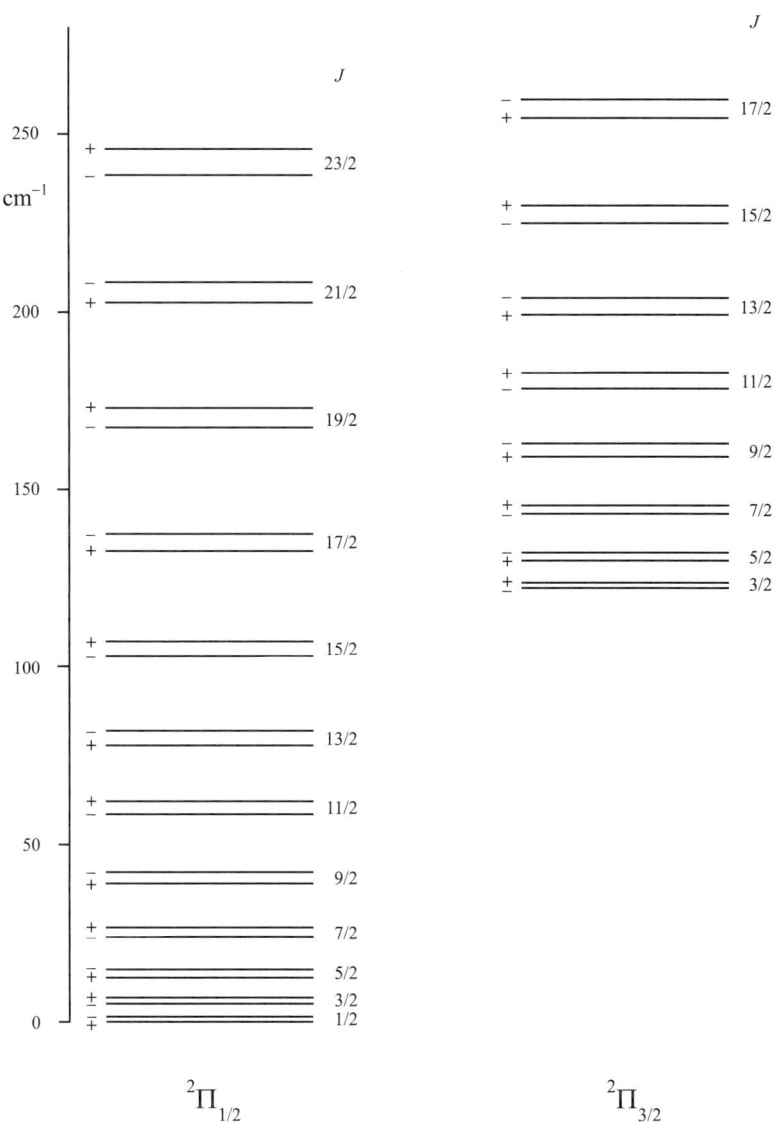

Figure 10.53. Lowest rotational levels in the fine-structure states of NO.

Schieder, Winnewisser and Herbst [157]. Hyperfine structure from the ^{14}N nucleus, with $I = 1$, is a characteristic feature of most high resolution spectra of NO; the detailed structure of the levels involved in the spectrum is shown in figure 10.55, and the observed spectrum is illustrated in figure 10.54. The quantitative aspects will be discussed in due course.

In the lower rotational levels NO exhibits good Hund's case (a) coupling, but as we progress to higher rotational levels, the coupling tends increasingly to Hund's case (b). Therein lie some of the complications in understanding the quantitative aspects of the spectra.

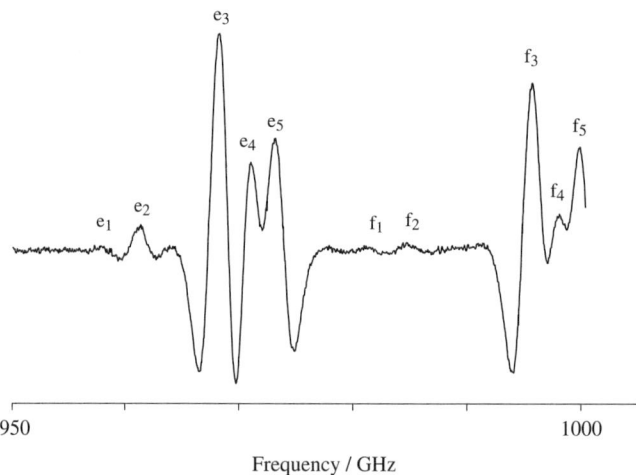

Figure 10.54. Recording of the $J = 17/2 \leftarrow 15/2$, $^2\Pi_{3/2}$ transition of $^{14}N^{18}O$ [157].

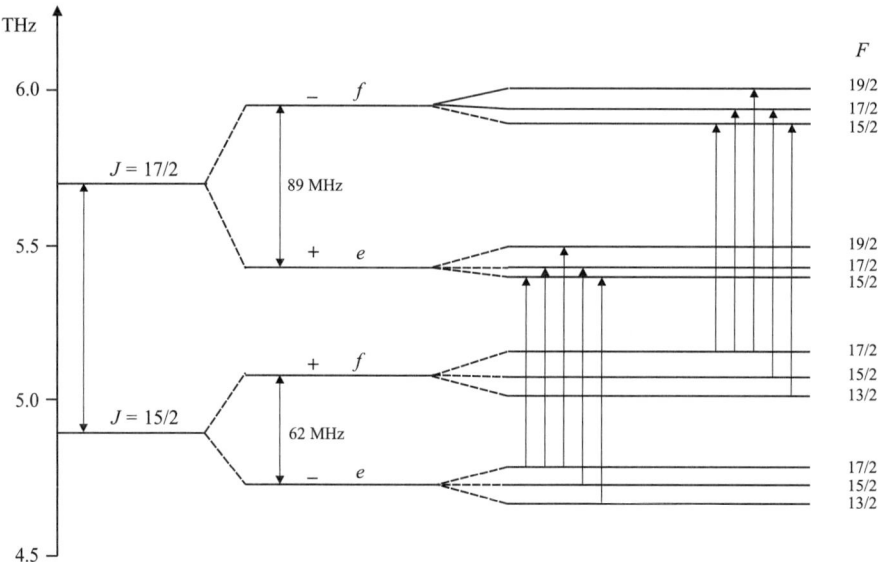

Figure 10.55. Rotational level structure and components observed for the $J = 17/2 \leftarrow 15/2$, $^2\Pi_{3/2}$ transition of $N^{18}O$, shown in figure 10.54. The diagram is not to scale: approximate values for the rotational and Λ-doublet splittings are shown.

We discussed the transition from Hund's case (a) to case (b) in chapter 9 and we present again figure 9.8, this time as figure 10.56. Although it is not shown in the diagram each rotational level in the case (a) set has a two-fold degeneracy which is usually called Λ-degeneracy but which might be called Ω-degeneracy because, in case (a), Ω is a good quantum number. On the other hand, each spin-doublet component J in the case (b) set also has a two-fold degeneracy which can now be correctly called Λ-degeneracy, because Λ is a good quantum number but Ω is not. We presented a

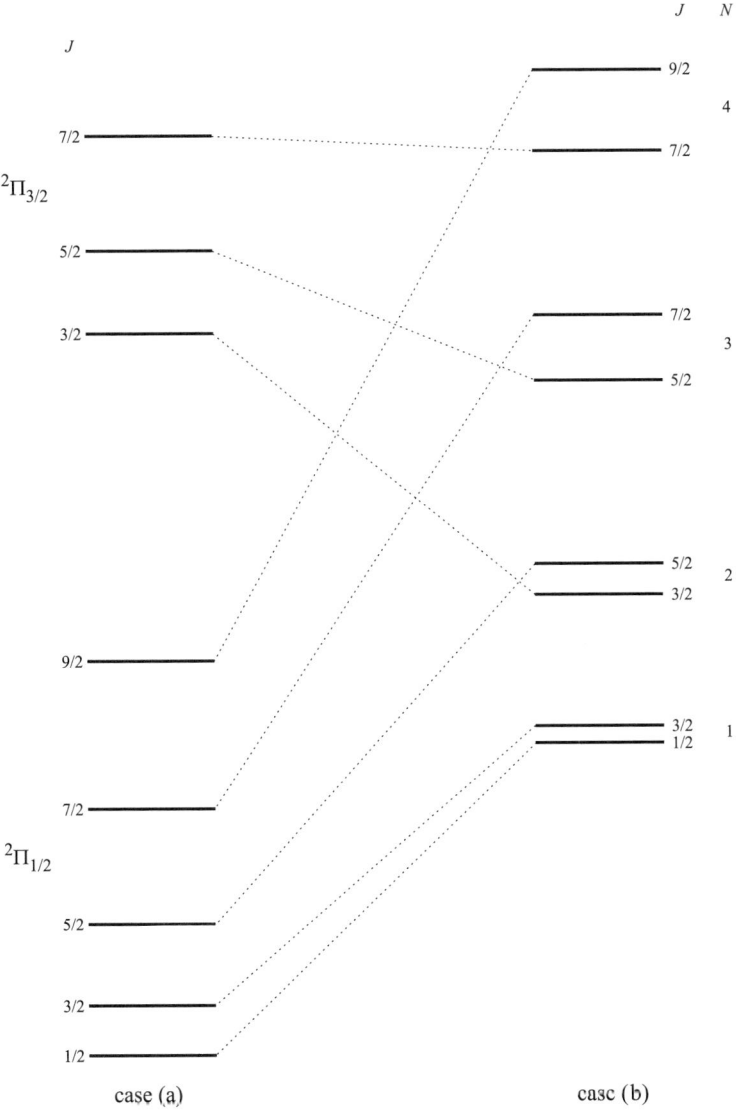

Figure 10.56. The transition from Hund's case (a) to case (b) in a $^2\Pi$ state.

thorough discussion of these important matters when discussing the electric resonance spectra of LiO and NO in chapter 8, so a brief summary will suffice here. Leaving aside nuclear spin interactions, which do not directly affect the situation, we may formulate what we called *primitive* case (a) basis functions as follows:

$$\begin{aligned}
&^2\Pi_{\Omega=+3/2} : |\eta, \Lambda = +1; S, \Sigma = +1/2; J, M_J, \Omega = +3/2\rangle, \\
&^2\Pi_{\Omega=-3/2} : |\eta, \Lambda = -1; S, \Sigma = -1/2; J, M_J, \Omega = -3/2\rangle, \\
&^2\Pi_{\Omega=+1/2} : |\eta, \Lambda = +1; S, \Sigma = -1/2; J, M_J, \Omega = +1/2\rangle, \\
&^2\Pi_{\Omega=-1/2} : |\eta, \Lambda = -1; S, \Sigma = +1/2; J, M_J, \Omega = -1/2\rangle.
\end{aligned} \quad (10.111)$$

In our earlier discussion of LiO in chapter 8 we also included an excited $^2\Sigma^+$ state but that was to facilitate our treatment of the Λ-doubling; in LiO there is one very low-lying $^2\Sigma^+$ excited state whose effects dominate the Λ-doubling, but that is an unusual situation. In NO the Λ-doubling arises from mixing with more than one excited state. However, our present concern is the transition from case (a) to case (b), so the four basis states in (10.111) will suffice for our purposes.

We now recall that the spin–orbit coupling and rigid body rotation terms in the effective Hamiltonian are

$$\mathcal{H}_{so} + \mathcal{H}_{rot} = \mathcal{H}_{rso} = A\mathbf{T}^1(\mathbf{L}) \cdot \mathbf{T}^1(\mathbf{S}) + B\{\mathbf{T}^1(\mathbf{J}) \cdot \mathbf{T}^1(\mathbf{J}) + \mathbf{T}^1(\mathbf{S}) \cdot \mathbf{T}^1(\mathbf{S})$$
$$+ \mathbf{T}^1(\mathbf{L}) \cdot \mathbf{T}^1(\mathbf{L}) - 2\mathbf{T}^1(\mathbf{J}) \cdot \mathbf{T}^1(\mathbf{S}) - 2\mathbf{T}^1(\mathbf{J}) \cdot \mathbf{T}^1(\mathbf{L}) - 2\mathbf{T}^1(\mathbf{L}) \cdot \mathbf{T}^1(\mathbf{S})\}$$
$$= (A + 2B)\mathbf{T}^1(\mathbf{L}) \cdot \mathbf{T}^1(\mathbf{S}) + B\{\mathbf{T}^1(\mathbf{J}) \cdot \mathbf{T}^1(\mathbf{J}) + \mathbf{T}^1(\mathbf{S}) \cdot \mathbf{T}^1(\mathbf{S})$$
$$+ \mathbf{T}^1(\mathbf{L}) \cdot \mathbf{T}^1(\mathbf{L}) - 2\mathbf{T}^1(\mathbf{J}) \cdot \mathbf{T}^1(\mathbf{S}) - 2\mathbf{T}^1(\mathbf{J}) \cdot \mathbf{T}^1(\mathbf{L})\}. \quad (10.112)$$

The first two terms in the purely rotational part of (10.112) are wholly diagonal in our basis set and may be replaced by their respective eigenvalues. The remaining scalar products are expanded in the molecule-fixed coordinate system, q, and we separate the $q = 0$ terms from $q = \pm 1$ so that (10.112) becomes

$$\mathcal{H}_{rso} = \mathcal{H}^{(0)} + \mathcal{H}' \quad \text{where}$$
$$\mathcal{H}^{(0)} = (A + 2B)\mathrm{T}_0^1(L)\mathrm{T}_0^1(S) + B\left\{J(J+1) + S(S+1) + \mathrm{T}_0^1(L)\mathrm{T}_0^1(L)\right.$$
$$\left. - 2\mathrm{T}_0^1(J)\mathrm{T}_0^1(S) - 2\mathrm{T}_0^1(J)\mathrm{T}_0^1(L)\right\}$$
$$\mathcal{H}' = \sum_{q=\pm 1}\{(-1)^q(A+2B)\mathrm{T}_q^1(L)\mathrm{T}_{-q}^1(S) + (-1)^q B\mathrm{T}_q^1(L)\mathrm{T}_{-q}^1(L)$$
$$- 2B\mathrm{T}_q^1(J)\mathrm{T}_q^1(S) - 2B\mathrm{T}_q^1(J)\mathrm{T}_q^1(L)\}. \quad (10.113)$$

We are now ready to examine the matrix elements. For $\mathcal{H}^{(0)}$, obtained with $q = 0$, the matrix elements are wholly diagonal:

$$\langle \eta, \Lambda; S, \Sigma; J, \Omega | \mathcal{H}^{(0)} | \eta', \Lambda'; S, \Sigma'; J', \Omega' \rangle$$
$$= \delta_{\eta\eta'}\delta_{\Lambda\Lambda'}\delta_{\Sigma\Sigma'}\delta_{\Omega\Omega'}\{(A+2B)\Lambda\Sigma + B[J(J+1) + S(S+1) + \Lambda^2 - 2\Omega^2]\}. \quad (10.114)$$

The four terms in $\mathcal{H}'$, however, require closer attention. The second term, which in molecule-fixed cartesian coordinates may be expressed as $L_x^2 + L_y^2$, affects all levels equally and is therefore usually omitted. The third term, which does not involve the orbital angular momentum, is known as the rotational distortion term. Its matrix elements are readily obtained:

$$\langle \eta, \Lambda; S, \Sigma; J, \Omega | - 2B\sum_{q=\pm 1}\mathrm{T}_q^1(J)\mathrm{T}_q^1(S) | \eta', \Lambda'; S, \Sigma'; J', \Omega' \rangle$$
$$= -\delta_{\eta\eta'}\delta_{\Lambda\Lambda'}\delta_{JJ'}2B\sum_{q=\pm 1}\langle J, \Omega | \mathrm{T}_q^1(J) | J, \Omega' \rangle \langle S, \Sigma | \mathrm{T}_q^1(S) | S, \Sigma' \rangle$$
$$= -2B\sum_{q=\pm 1}(-1)^{J+S-\Omega-\Sigma}\begin{pmatrix} J & 1 & J \\ -\Omega & q & \Omega' \end{pmatrix}\begin{pmatrix} S & 1 & S \\ -\Sigma & q & \Sigma' \end{pmatrix}\{J(J+1)(2J+1)$$
$$\times S(S+1)(2S+1)\}^{1/2}. \quad (10.115)$$

These are the matrix elements which concern us here. The two remaining terms contained within $\mathcal{H}'$ in equation (10.113) give rise to Λ-doubling, which we are not dealing with here.

Now, we recall, we must construct *parity-conserved* basis functions from the *primitive* functions, and these are

$$+parity : \frac{1}{\sqrt{2}}\{|\eta, \Lambda; S, \Sigma; J, M_J, \Omega\rangle + (-1)^{J-S}|\eta, -\Lambda; S, -\Sigma; J, M_J, -\Omega\rangle\},$$

$$-parity : \frac{1}{\sqrt{2}}\{|\eta, \Lambda; S, \Sigma; J, M_J, \Omega\rangle - (-1)^{J-S}|\eta, -\Lambda; S, -\Sigma; J, M_J, -\Omega\rangle\}.$$

(10.116)

We therefore have the following 2×2 matrices, which are identical for both parity states because we are ignoring Λ-doubling for the moment.

	$^2\Pi^{(\pm)}_{3/2}$	$^2\Pi^{(\pm)}_{1/2}$
$^2\Pi^{(\pm)}_{3/2}$	$A/2 + B\{J(J+1) - 7/4\}$	$-B\{(J+3/2)(J-1/2)\}^{1/2}$
$^2\Pi^{(\pm)}_{1/2}$	$-B\{(J+3/2)(J-1/2)\}^{1/2}$	$-A/2 + B\{J(J+1) + 1/4\}$

The off-diagonal elements mix the two fine-structure states; clearly the mixing increases with increasing B and J, and is also greater the smaller the spin–orbit coupling constant A. This mixing, called rotational distortion, is what gives rise to Hund's case (b) coupling, and in particular the increasing tendency towards case (b) in the higher rotational levels. We see also that, at this level, the two-fold Λ-degeneracy is preserved despite the rotational distortion.

The Λ-doubling and ^{14}N magnetic and electric hyperfine parameters were discussed at length in chapter 8. We have nothing significant to add here, except to note that the centrifugal distortion corrections are much more accurately defined by the studies of higher rotational levels.

Klisch, Belov, Schieder, Winnewisser and Herbst [157] have combined all of the data for NO to produce a current best set of molecular constants for three isotopomers, presented in table 10.15. The data used, apart from their own terahertz studies, included the Λ-doubling of Meerts and Dymanus [156, 158], the sub-millimetre transitions of ^{15}N^{16}O and ^{14}N^{18}O, and Fourier transform data from Salek, Winnewisser and Yamada [159]. These last authors were able to study the magnetic dipole transitions between the two fine-structure states. The values of the spin–orbit constant A for the less common isotopomers come from Amiot, Bacis and Guelachvili [160].

The information concerning the electronic structure of NO which can be deduced from the values of the molecular parameters was discussed in chapter 8, and we refer the interested reader back to that discussion. Finally we note a number of recent studies of NO in interstellar clouds, observed through emission involving the lowest rotational transition in the $^2\Pi_{1/2}$ state [161, 162, 163, 164].

Table 10.15. *Molecular parameters for NO and its isotopomers in the $X\,^2\Pi$ ground state*

Parameter	$^{14}\text{N}^{16}\text{O}$	$^{15}\text{N}^{16}\text{O}$	$^{14}\text{N}^{18}\text{O}$	Unit
A	3.692 064 2(28)	3.691 683	3.691 713	THz
A_D	7.133 6(29)	5.113(11)	4.9451(28)	MHz
B	50 847.801 6(15)	49 050.534 3(82)	48 211.777 2(22)	MHz
D	164.099 0(82)	152.715(56)	147.608(53)	kHz
H	682(41)	—	594(37)	mHz
p	350.440 80(16)	337.959 91(41)	332.207(33)	MHz
d_D	305(64)	170(32)	—	Hz
q	2.804 324(84)	2.641 46(16)	2.518(20)	MHz
q_D	43.71(49)	38.3(23)	76(14)	Hz
a	84.215 3(11)	−118.143 0(30)	84.225(31)	MHz
b	41.904 9(75)	−59.025(15)	42.05(34)	MHz
c	−58.795 0(75)	82.726(13)	−58.92(34)	MHz
d	112.597 38(29)	−157.947 84(70)	112.576(21)	MHz
d_D	−457(46)	−452(37)	—	Hz
c_I	12.45(10)	−16.24(20)	−9.3(22)	kHz
c'_I	1.24(55)	−1.34(70)	—	kHz
eq_0Q	−1.857 49(74)	—	−1.839(37)	MHz
eq_2Q	23.049(18)	—	23.30(67)	MHz

10.6.2. OH in the $X\,^2\Pi$ ground state

We have already discussed the high-resolution spectroscopy of the OH radical at some length. It occupies a special place in the history of the subject, being the first short-lived free radical to be detected and studied in the laboratory by microwave spectroscopy. The details of the experiment by Dousmanis, Sanders and Townes [4] were described in section 10.1. It was also the first interstellar molecule to be detected by radio-astronomy. In chapter 8 we described the molecular beam electric resonance studies of Λ-doubling transitions in the lowest rotational levels, and in chapter 9 we gave a comprehensive discussion of the microwave and far-infrared magnetic resonance spectra of OH. Our quantitative analysis of the magnetic resonance spectra made use of the results of pure field-free microwave studies of the rotational transitions, which we now describe.

The initial studies of Dousmanis, Sanders and Townes [4] were focussed on the hyperfine components of the Λ-doubling transitions in the $J = 7/2, 9/2$ and $11/2$ rotational levels of the ground $^2\Pi_{3/2}$ state, and the $J = 3/2$ and $5/2$ levels of the excited $^2\Pi_{1/2}$ state. Figure 10.57 illustrates the Λ-doublet and proton hyperfine splitting of the

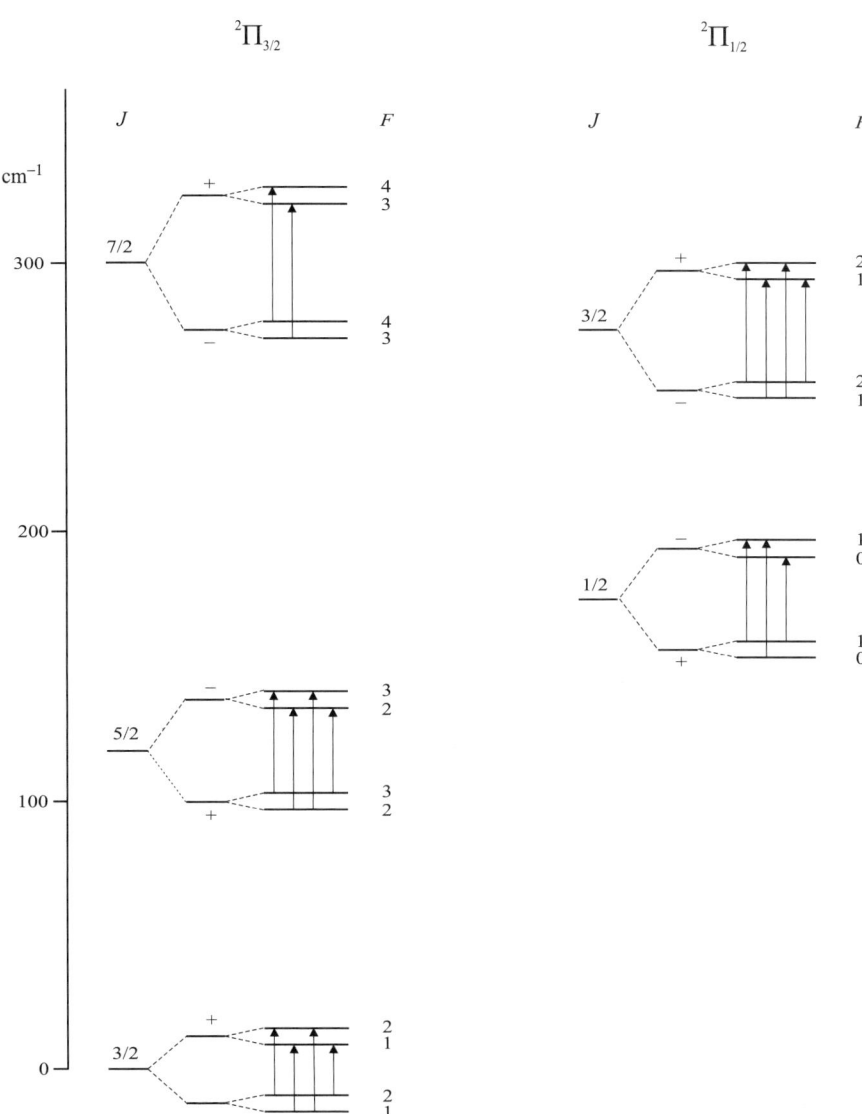

Figure 10.57. Λ-doublet and proton hyperfine splittings of the lower rotational levels of the OH radical, and the observed $\Delta J = 0$, Λ-doublet transitions. The diagram is drawn for the sake of clarity, and is not to scale.

$J - 3/2$, $5/2$ and $7/2$ levels of both fine-structure states. Also shown are the Λ-doublet transitions observed, first by Dousmanis, Sanders and Townes [4], and subsequently by ter Meulen and Dymanus [165] and Meerts and Dymanus [166]. The later studies [166] used molecular beam electric resonance methods which were described in chapter 8, and the most accurate laboratory measurements of transitions within the lowest rotational level were those of ter Meulen and Dymanus [165] using a beam maser spectrometer, also described in chapter 8. In the years following these field-free experiments, attention

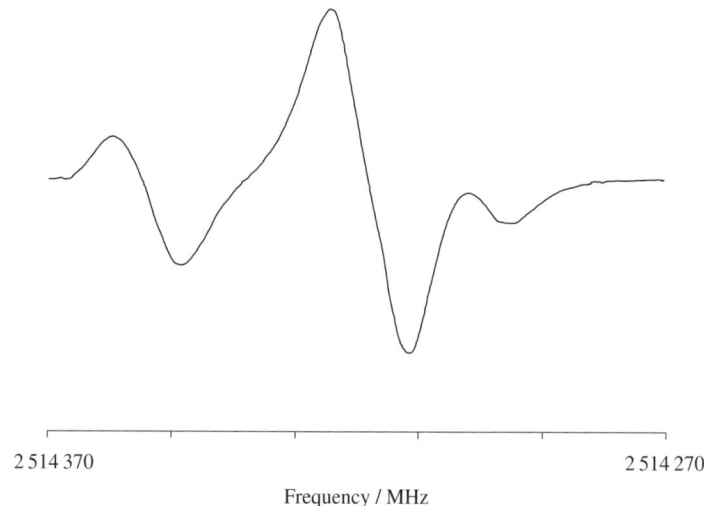

Figure 10.58. Observed spectrum for $J = 5/2 \leftarrow 3/2$ in the $^2\Pi_{3/2}$ ground state of the OH radical [68]. The three components are due to proton hyperfine interaction, as discussed in the text. See also figure 10.59 for an energy level diagram, with transitions, but without hyperfine splitting.

was concentrated on magnetic resonance studies, described at length in chapter 9. In recent years, however, the development of tunable far-infrared sources has enabled field-free rotational transitions to be studied. Figure 10.58 shows part of the spectrum obtained by Brown, Zink, Jennings, Evenson, Hinz and Nolt [68] whose experiments were described in the first section of this chapter. Similar results were also obtained independently by Farhoomand, Blake and Pickett [167] using a somewhat similar far-infrared spectrometer [168]. Figure 10.59 shows, again, the energy level diagram for the lowest rotational levels, this time without nuclear hyperfine splitting but indicating the $\Delta J = \pm 1$ rotational transitions observed.

As we have already described, the OH radical has played an important role in radio and far-infrared astronomy. It was the first molecule to be observed by radioastronomy, the hyperfine components of the Λ-doublet transition in the lowest rotational level being observed by Weinreb, Barrett, Meeks and Henry [32] in 1963 in a cloud near to the supernova remnant Cassiopeia A. Initially the two strongest hyperfine lines were observed in absorption, but later observations showed that for an interstellar cloud within a few parsecs of the sun, in the direction of Cassiopeia A, the highest frequency component appears in emission, whilst the other three components are seen in absorption. This observation shows a large departure from thermal equilibrium, the upper level being overpopulated and amplifying the continuum background. Further measurements showed the emission to be polarised, time-variable, and to come from very compact sources. This was the first observation of an interstellar maser. Many subsequent investigations have been described and the population inversion is probably due to infrared pumping involving a nearby star. Other interstellar masers have also been observed.

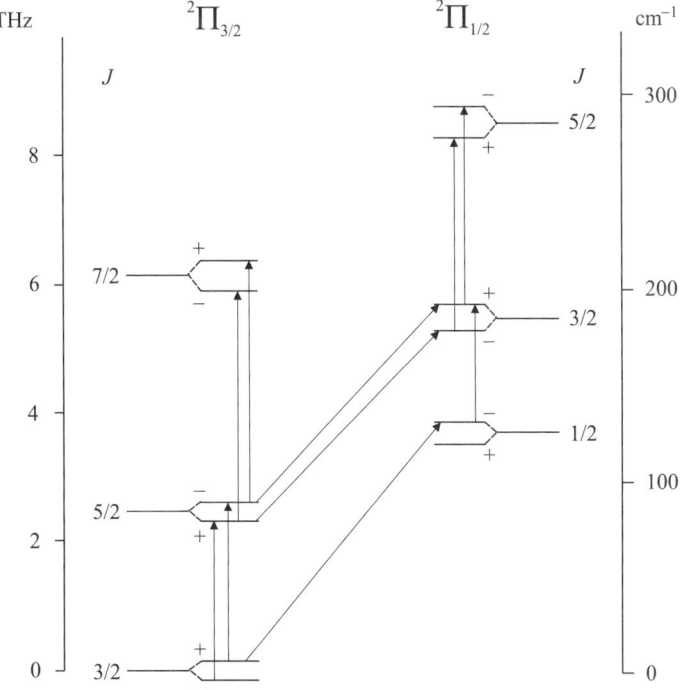

Figure 10.59. Lowest rotational levels of the OH radical and the far-infrared transitions observed [68]. The Λ-doublet splittings are exaggerated for the sake of clarity.

It should not be thought that OH always exhibits the unusual behaviour described above; the recent developments in tunable far-infrared sources have had an important impact in astronomy, so that interstellar rotational transitions can now be observed. We described an airborne far-infrared telescope in the first part of this chapter, and figure 10.60 shows two examples of interstellar OH rotational transitions, observed by Watson, Genzel, Townes and Storey [170].

All of the high quality data for the OH radical has been combined by Brown, Zink, Jennings, Evenson, Hinz and Nolt [68], building on an earlier analysis of the laser magnetic resonance spectrum [171] and more recent work by Varberg and Evenson [172], to produce a current best set of field-free molecular constants for the ^{16}OH radical. These are presented in table 10.16. The constants are defined by the following effective Hamiltonian which has been described extensively elsewhere in this book:

$$\begin{aligned}\mathcal{H} =\ & A \mathrm{T}^1_{q=0}(\boldsymbol{L})\, \mathrm{T}^1_{q=0}(\boldsymbol{S}) + B\boldsymbol{N}^2 - D(\boldsymbol{N}^2)^2 + H(\boldsymbol{N}^2)^3 + \gamma \mathrm{T}^1(\boldsymbol{J}-\boldsymbol{S}) \cdot \mathrm{T}^1(\boldsymbol{S}) \\ & + \gamma_D \{\mathrm{T}^1(\boldsymbol{J}-\boldsymbol{S}) \cdot \mathrm{T}^1(\boldsymbol{S})\} \boldsymbol{N}^2 + \sum_{q=\pm 1} \exp(-2 i q \phi) [-q \mathrm{T}^2_{2q}(\boldsymbol{J},\boldsymbol{J}) \\ & + (p+2q) \mathrm{T}^2_{2q}(\boldsymbol{J},\boldsymbol{S})] + \sum_{q=\pm 1} \exp(-2 i q \phi) \{-q_D (1/2) [\mathrm{T}^2_{2q}(\boldsymbol{J},\boldsymbol{J}) \boldsymbol{N}^2 + \boldsymbol{N}^2 \mathrm{T}^2_{2q}(\boldsymbol{J},\boldsymbol{J})]\end{aligned}$$

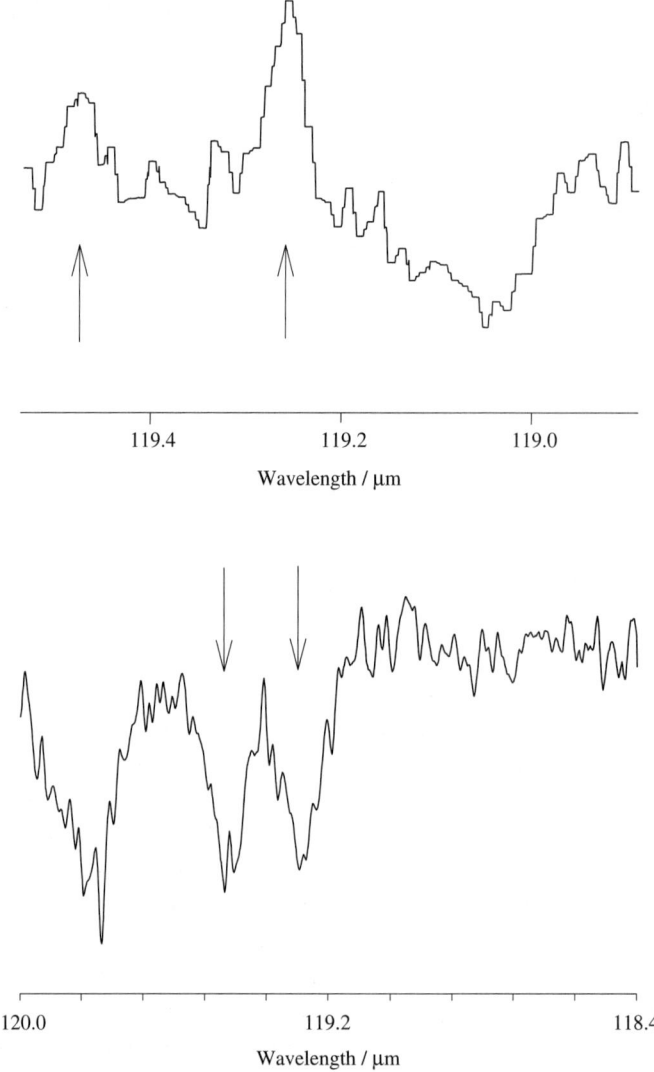

Figure 10.60. Far-infrared transitions of interstellar OH observed using the airborne telsecope described earlier [37]. The transitions are $J = 5/2 \to 3/2$ in the $^2\Pi_{3/2}$ state [169, 170]. The upper spectrum is observed in emission in Orion, the lower spectrum in absorption in Sagitarius B2. In both cases the shorter wavelength line close to 119.4 μm corresponds to the laboratory spectrum shown in figure 10.58, but without resolved hyperfine structure.

$$+ (p_D + 2q_D)(1/2)[T^2_{2q}(\boldsymbol{J},\boldsymbol{S})N^2 + N^2 T^2_{2q}(\boldsymbol{J},\boldsymbol{S})]\}$$
$$+ \sum_{q=\pm 1} \exp(-2iq\phi)\{-q_H(1/2)[T^2_{2q}(\boldsymbol{J},\boldsymbol{J})(N^2)^2 + (N^2)^2 T^2_{2q}(\boldsymbol{J},\boldsymbol{J})]$$
$$+ (p_H + 2q_H)(1/2)[T^2_{2q}(\boldsymbol{J},\boldsymbol{S})(N^2)^2 + (N^2)^2 T^2_{2q}(\boldsymbol{J},\boldsymbol{S})]\}$$
$$+ a\, T^1_{q=0}(\boldsymbol{I})T^1_{q=0}(\boldsymbol{L}) + b_F T^1(\boldsymbol{I}) \cdot T^1(\boldsymbol{S}) + \sqrt{(2/3)}\, c T^2_{q=0}(\boldsymbol{I},\boldsymbol{S})$$

Table 10.16. *Molecular parameters for OH in the v = 0 level of the X $^2\Pi$ ground state (in MHz)*

$A = -4\,168\,639.13(78)$			
$B = 555\,660.97(11)$	$D = 57.178\,5(86)$	$H = 0.4236 \times 10^{-2}$	
$\gamma = -3574.88(49)$	$\gamma_D = 0.7315$		
$q = -1159.991\,650$	$p = 7053.098\,46$		
$q_D = 0.442\,032\,0$	$p_D = -1.550\,962$		
$q_H = -0.8237 \times 10^{-4}$	$p_H = 0.1647 \times 10^{-3}$		
$a = 86.1116$	$b_F = -73.2537$	$c = 130.641$	$d = 56.6838$
$d_D = -0.022\,76$	$c_I = -0.099\,71$	$c'_I = 0.643 \times 10^{-2}$	

$$+ d \sum_{q=\pm 1} \exp(-2iq\phi)\, T^2_{2q}(\boldsymbol{I}, \boldsymbol{S}) + d_D \sum_{q=\pm 1} \exp(-2iq\phi)(1/2)$$
$$\times \left[T^2_{2q}(\boldsymbol{I}, \boldsymbol{S}) N^2 + N^2 T^2_{2q}(\boldsymbol{I}, \boldsymbol{S}) \right] + c_I\, T^1(\boldsymbol{I}) \cdot T^1(\boldsymbol{J} - \boldsymbol{S}) + c'_I$$
$$\times \sum_{q=\pm 1} \exp(-2iq\phi)(1/2) \left[T^2_{2q}(\boldsymbol{I}, \boldsymbol{J} - \boldsymbol{S}) + T^2_{2q}(\boldsymbol{J} - \boldsymbol{S}, \boldsymbol{I}) \right]. \quad (10.117)$$

This effective Hamiltonian is written for calculation of the matrix elements in a Hund's case (a) basis, in the molecule-fixed coordinate system. We have described the evaluation of the matrix elements elsewhere. The first line describes the spin–orbit coupling, nuclear rotation and spin–rotation interactions. The next three lines are the Λ-doubling terms and their centrifugal distortion corrections. The fifth line represents the magnetic hyperfine interactions, and the sixth line describes the centrifugal distortion correction to the hyperfine d constant. The final line describes higher-order corrections to the hyperfine parameters. Equation (10.117) is a complicated effective Hamiltonian, and arises because OH is a light molecule, showing considerable centrifugal distortion and non-adiabatic effects arising from the admixture of other vibrational and electronic states. The additional terms which arise from Zeeman interactions in an applied magnetic field are described in detail in chapter 9.

Studies of the OH radical have featured prominently in the progress of high-resolution microwave and far-infrared spectroscopy. Although we have concentrated on the predominant naturally occurring isotopic species, its isotopic relations have also been studied extensively. With the benefit of hindsight, OH is not a difficult free radical to form and study; since it is usually formed in the presence of water vapour, it seems likely that it regenerates itself. We might add that, in our own experiences over thirty years ago with microwave magnetic resonance, we found it almost impossible to not observe OH, even in systems where both hydrogen and oxygen were supposed to be absent! On many occasions we were excited at the discovery of a new free radical spectrum, only to discover that we had been fooled by OH once again. The CH radical is an altogether more elusive species, as we describe in the following sub-section.

10.6.3. CH in the $X^2\Pi$ ground state

(a) Observation and assignment of Λ-doubling and rotational transitions

The CH radical is the simplest hydrocarbon and its rotational or Λ-doublet spectrum has been sought by many. The first detection of rotational transitions was a triumph for far-infrared laser magnetic resonance; the experiments carried out by Evenson, Radford and Moran [173] were described in detail in chapter 9. The Λ-doublet transition in the lowest rotational level was first observed through radioastronomy by Rydbeck, Elldér, Irvine, Sume and Hjalmarson [174]. It was almost a further ten years before laboratory observations of the field-free spectrum were reported.

First we will find it helpful to have an overall view of the rotational levels. Although we could insist on calling the fine structure states $^2\Pi_{1/2}$ and $^2\Pi_{3/2}$, with $^2\Pi_{1/2}$ being the lower in the CH radical (a so-called 'regular' state), this is misleading because Ω is not a good quantum number and a case (b) description is much more appropriate. The spin–orbit coupling constant A is 27.95 cm^{-1} for CH, and the rotational constant B_0 is 14.190 cm^{-1}. Expressions for the rotational energies were given originally by Hill and Van Vleck [175] as follows:

$$F_1(J) = B_v[(J+1/2)^2 - \Lambda^2 - (1/2)\{4(J+1/2)^2 + Y(Y-4)\Lambda^2\}^{1/2}],$$
$$F_2(J) = B_v[(J+1/2)^2 - \Lambda^2 + (1/2)\{4(J+1/2)^2 + Y(Y-4)\Lambda^2\}^{1/2}]. \quad (10.118)$$

Here $Y = A/B_v$; $F_1(J)$ is the term series that forms levels with $J = N + 1/2$, whilst $F_2(J)$ forms levels with $J = N - 1/2$. A correction $B_v\Lambda^2$ needs to be added to convert to the form of the N^2 Hamiltonian. Centrifugal distortion terms can be added to (10.118) if necessary. The correlation with case (a) levels is shown in figure 10.56. The lower rotational levels of CH, without Λ-doublet or nuclear hyperfine splittings, are shown, approximately to scale, in figure 10.61. Each level shown actually exhibits Λ-doublet splitting and, apart from the far-infrared laser magnetic resonance studies described in chapter 9, the first successful experiments involved the detection of transitions between Λ-doublet components of a particular rotational level. In fact the first observations of such transitions for the lowest rotational level with $J = 1/2$ were made by astronomers [176, 177], and another nine years passed before the first laboratory observations. These were made by Brazier and Brown [178] using a novel microwave/optical double resonance method, which employed the reaction of F atoms with CH$_4$. Their initial experiments involved the $N = 2$, 3 and 4 rotational levels; the detailed Λ-doublet and proton hyperfine splitting for the $N = 3$, $J = 5/2$ (F_2) rotational level is shown in figure 10.62, as well as the observed transitions. The experimental spectrum corresponding to this energy level diagram is shown in figure 10.63. The Λ-doubling transition in the lowest rotational level ($J = 1/2$) was observed subsequently by Steimle, Woodward and Brown [179] for both ^{12}CH and ^{13}CH, the latter showing hyperfine splitting from both the proton and the ^{13}C nucleus. Figure 10.64 shows the energy level diagram for the lowest rotational level of ^{13}CH, with the Λ-doublet and nuclear hyperfine splittings. The transitions shown in the figure are correlated with the experimental spectra in figure 10.65. Λ-doubling transitions involving higher

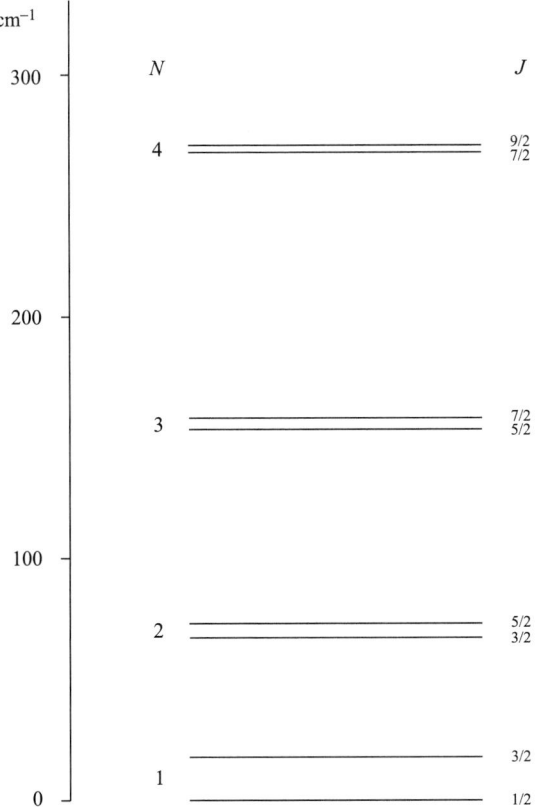

Figure 10.61. Lower rotational levels of the CH radical, with Hund's case (b) labels. Each level actually possesses Λ-doublet and nuclear hyperfine splitting, which is not shown in this diagram. Note that the spin–rotation splitting decreases with N, in accordance with equation (9.81).

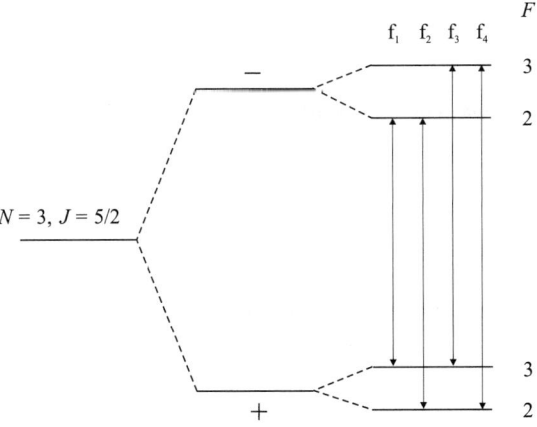

Figure 10.62. Details of the Λ-doubling and proton hyperfine splitting for the $N = 3$, $J = 5/2$ (F_2) level of CH, and the observed transitions.

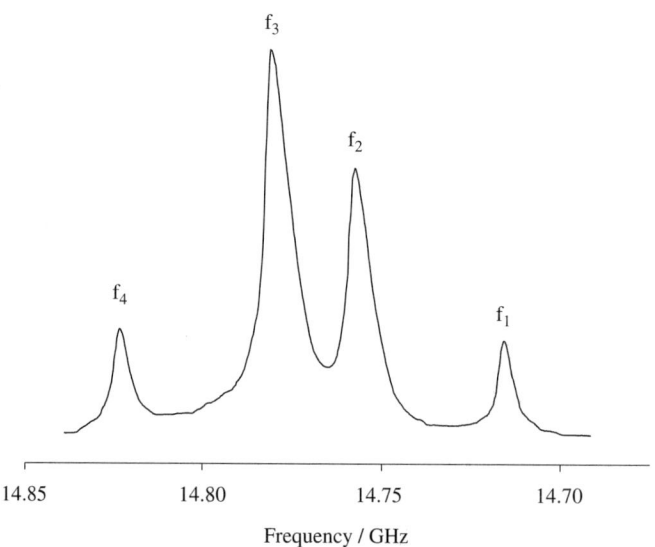

Figure 10.63. Microwave spectrum of CH in the $N = 3$, $J = 5/2$ (F_2) level observed by Brazier and Brown [178]. The corresponding energy level diagram is shown in figure 10.62.

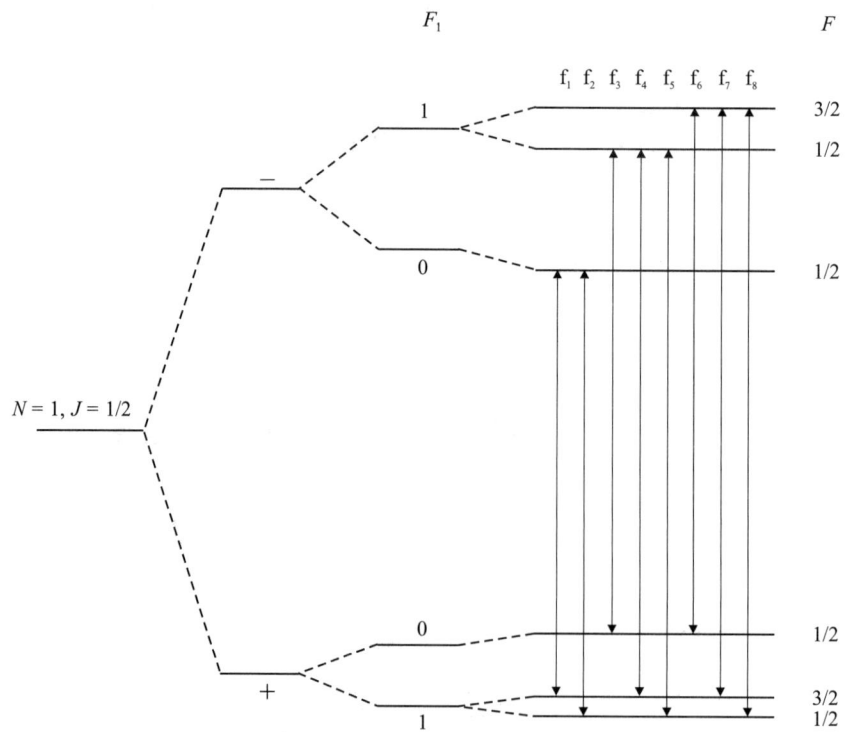

Figure 10.64. The Λ-doubling, proton and ^{13}C hyperfine splitting for the lowest rotational level of ^{13}CH, and the observed transitions [179].

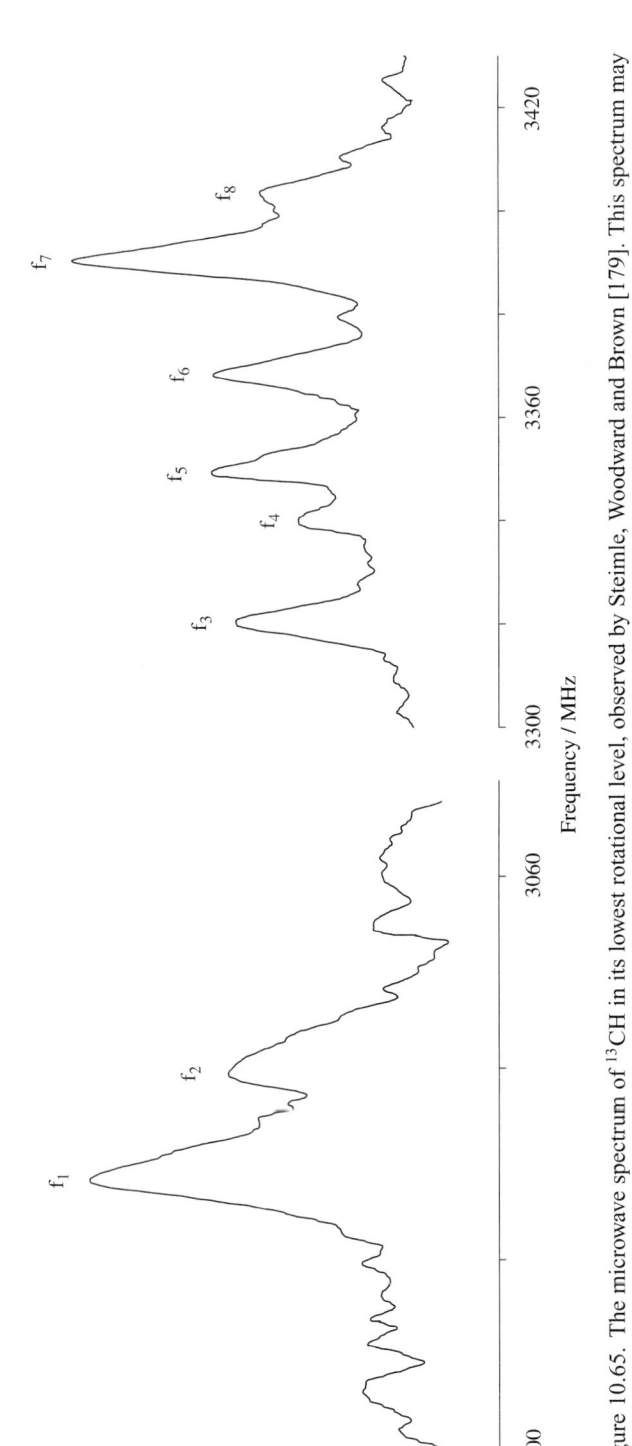

Figure 10.65. The microwave spectrum of ^{13}CH in its lowest rotational level, observed by Steimle, Woodward and Brown [179]. This spectrum may be correlated with the energy level diagram shown in figure 10.64.

Table 10.17. *Experimental and calculated values (in MHz) for the Λ-doubling frequencies in ^{12}CH*

F_i	J	$F' \leftarrow F''$	frequency (MHz)	obs. − calc. (kHz)	Ref.
F_1	3/2	$1^- \leftarrow 1^+$	724.789(7)	−2.5	[182]
		$2^- \leftarrow 2^+$	701.677(7)	0.9	[112]
F_1	5/2	$2^+ \leftarrow 2^-$	4847.84(10)	76.8	[108]
		$3^+ \leftarrow 3^-$	4870.12(10)	59.1	[108]
F_1	7/2	$3^- \leftarrow 4^+$	11 301.22(20)	29.3	[108]
		$3^- \leftarrow 3^+$	11 287.05(15)	74.6	[108]
		$4^- \leftarrow 4^+$	11 265.21(15)	231.0	[108]
		$4^- \leftarrow 3^+$	11 250.79(50)	25.9	[108]
F_1	13/2	$6^+ \leftarrow 6^-$	43 872.591(30)	−9.6	[110]
		$7^+ \leftarrow 7^-$	43 851.026(30)	−10.0	[110]
F_1	15/2	$7^- \leftarrow 7^+$	59 008.076(20)	6.6	[110]
		$8^- \leftarrow 8^+$	59 986.633(20)	1.4	[110]
F_1	17/2	$8^+ \leftarrow 8^-$	76 168.632(50)	−0.8	[110]
		$9^+ \leftarrow 9^-$	76 147.336(30)	1.6	[110]
F_2	1/2	$0^- \leftarrow 0^+$	3263.794(3)	−0.1	[104]
		$1^- \leftarrow 1^+$	3335.481(3)	0.2	[104]
		$1^- \leftarrow 0^+$	3349.193(3)	−0.1	[104]
F_2	3/2	$1^+ \leftarrow 2^-$	7274.78(15)	−280.0	[111]
		$1^+ \leftarrow 1^-$	7325.15(15)	−60.1	[111]
		$2^+ \leftarrow 2^-$	7348.28(15)	−141.9	[111]
		$2^+ \leftarrow 1^-$	7398.38(15)	−193.1	[111]
F_2	5/2	$2^- \leftarrow 3^+$	14 713.78(15)	91.6	[108]
		$2^- \leftarrow 2^+$	14 756.81(15)	129.0	[108]
		$3^- \leftarrow 3^+$	14 778.97(20)	−3.5	[108]
		$3^- \leftarrow 2^+$	14 821.88(15)	−85.6	[108]
F_2	7/2	$3^+ \leftarrow 4^-$	24 381.57(40)	246.1	[108]
		$3^+ \leftarrow 3^-$	24 420.65(10)	3.8	[108]
		$4^+ \leftarrow 4^-$	24 442.56(10)	−16.5	[108]
		$4^+ \leftarrow 3^-$	24 482.10(20)	202.0	[108]
F_2	11/2	$5^+ \leftarrow 5^-$	50 299.750(20)	1.8	[110]
		$6^+ \leftarrow 6^-$	50 321.276(20)	−2.9	[110]
F_2	13/2	$6^- \leftarrow 6^+$	66 400.098(30)	−0.8	[110]
		$7^- \leftarrow 7^+$	66 421.466(30)	1.6	[110]

rotational levels have also been observed by Bogey, Demuynck and Destombes [180] using a conventional microwave absorption technique, which has the required sensitivity at higher frequencies. A complete data set [181] of the Λ-doubling frequencies in the predominant isotopomer, ^{12}CH, with proton hyperfine structure, is presented in table 10.17. We deal with the theoretical analysis, which yields the calculated values in table 10.17 in due course. First, however, we describe the measurements of rotational transitions.

We have already described the experiments on CH of Amano [63], and showed part of his recording of the lowest rotational transition in figure 10.31. Figure 10.66 now presents his complete spectrum, whilst figure 10.67 gives the corresponding energy level diagram and the respective transitions. This diagram is, however, a magnified portion of a more comprehensive diagram which was first presented in chapter 9 as part of our discussion of the far-infrared laser magnetic resonance studies [183, 184] by Brown and Evenson, and is now shown again as figure 10.68. Their measurements provided the first accurate values of the important molecular parameters; zero-field measurements are the ultimate aim, because of the obvious reason that they do not require an accurate theory of the Zeeman effect. Subsequent observations were indeed made in the absence of magnetic fields. Very recently Davidson, Evenson and Brown [185] have greatly extended the measurement of far-infrared field-free rotational transitions and have provided the best set of molecular parameters for CH in the $v = 0$ level. In this connection it should also be noted that rotational transitions involving higher rotational levels have been observed by astronomers [186].

(b) Theoretical analysis and determination of molecular parameters

The CH radical conforms well to Hund's case (b) coupling, which we discussed at some length in connection with the c $^3\Pi_u$ state of H_2 in chapter 8. The emphasis there was on the Zeeman effect, which was examined again in chapter 9. We therefore repeat here only enough of the essential theory to define the molecular parameters whose values have been obtained from a combination of all of the experimental studies.

First we recall the details of the angular momentum coupling scheme for a $^2\Pi$ molecule described as Hund's case (b). It is as follows:

$$\begin{aligned}
&\boldsymbol{R} + \boldsymbol{L} = \boldsymbol{N}: \quad N = 1, 2, 3, \ldots, \\
&\boldsymbol{N} + \boldsymbol{S} = \boldsymbol{J}: \quad J = N - 1/2, N + 1/2, \\
&\boldsymbol{J} + \boldsymbol{I} = \boldsymbol{F}.
\end{aligned} \quad (10.119)$$

We have chosen to use the hyperfine-coupled representation, where for ^{12}CH, F is equal to $J \pm 1/2$. An appropriate basis set is therefore $|\eta, \Lambda; N, \Lambda; S, J, I, F\rangle$, with M_F also important when discussing Zeeman effects. As usual the effective zero-field Hamiltonian will be, at the least, a sum of terms representing the spin–orbit coupling, rigid body rotation, electron spin–rotation coupling and nuclear hyperfine interactions, i.e.

$$\mathcal{H}_{\text{eff}} = \mathcal{H}_{\text{so}} + \mathcal{H}_{\text{rot}} + \mathcal{H}_{\text{sr}} + \mathcal{H}_{\text{hfs}}. \quad (10.120)$$

The magnetic hyperfine interactions were discussed in chapter 8, where we followed rather closely the analysis of Jette and Cahill [187]; we will come to these a little later, but first consider briefly the nuclear spin-free terms.

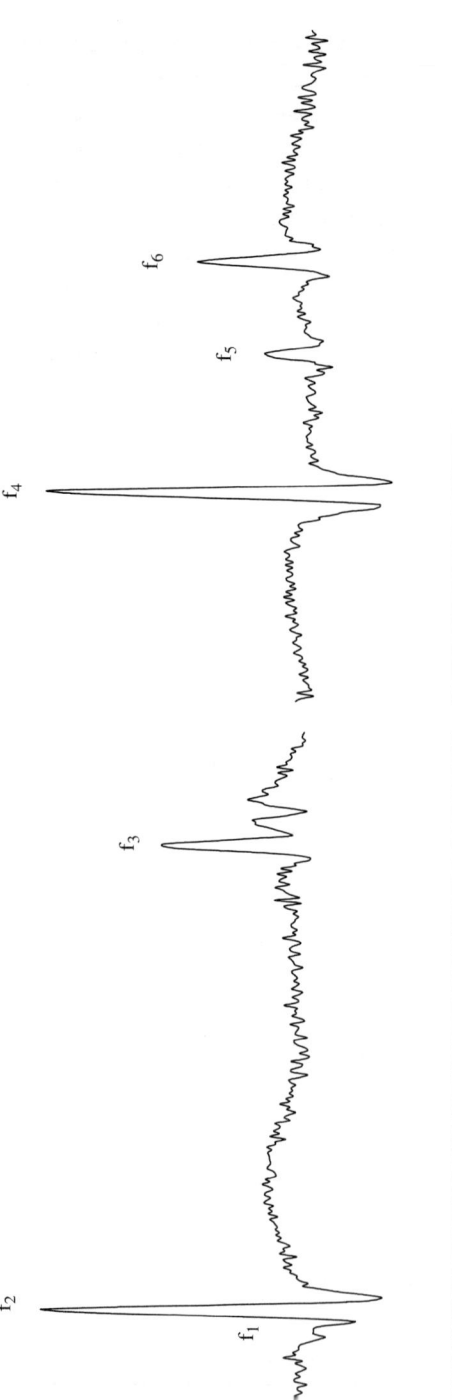

Figure 10.66. High-frequency, millimetre wave spectrum of CH arising from the lowest rotational transition [63]. The assignment of the lines is given in figure 10.67.

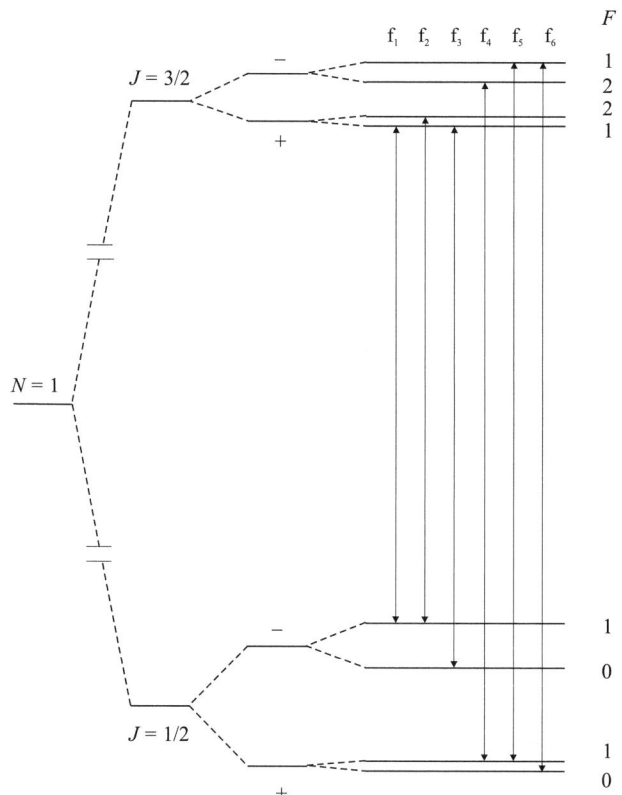

Figure 10.67. Energy level diagram and observed transitions for the lowest rotational transition of CH. This diagram should be compared with the experimental spectrum [63] shown in figure 10.66.

The spin–orbit coupling term is straightforward, with case (b) matrix elements:

$$\langle \eta, \Lambda; N, \Lambda; S, J| A T^1(\boldsymbol{L}) \cdot T^1(\boldsymbol{S}) |\eta, \Lambda; N', \Lambda; S, J\rangle$$
$$= A(-1)^{N'+S+J} \begin{Bmatrix} S & N' & J \\ N & S & 1 \end{Bmatrix} (-1)^{N-\Lambda} \begin{Bmatrix} N & 1 & N' \\ -\Lambda & 0 & \Lambda \end{Bmatrix} \{(2N+1)(2N'+1)\}^{1/2}$$
$$\times \Lambda \{S(S+1)(2S+1)\}^{1/2}$$
$$= A \frac{J(J+1) - N(N+1) - S(S+1)}{2N(N+1)}. \tag{10.121}$$

We have inserted the value $\Lambda = 1$ and $N' = N$ to obtain the last line of (10.121), and off-diagonal elements have been neglected in this equation; they are, however, involved in the Λ-doubling.

The rigid body rotation term in the effective Hamiltonian is

$$\mathcal{H}_{\text{rot}} = B\boldsymbol{N}^2 - D\boldsymbol{N}^4 + \cdots,$$
$$= B(\boldsymbol{J} - \boldsymbol{S})^2 - D(\boldsymbol{J} - \boldsymbol{S})^4 + \cdots. \tag{10.122}$$

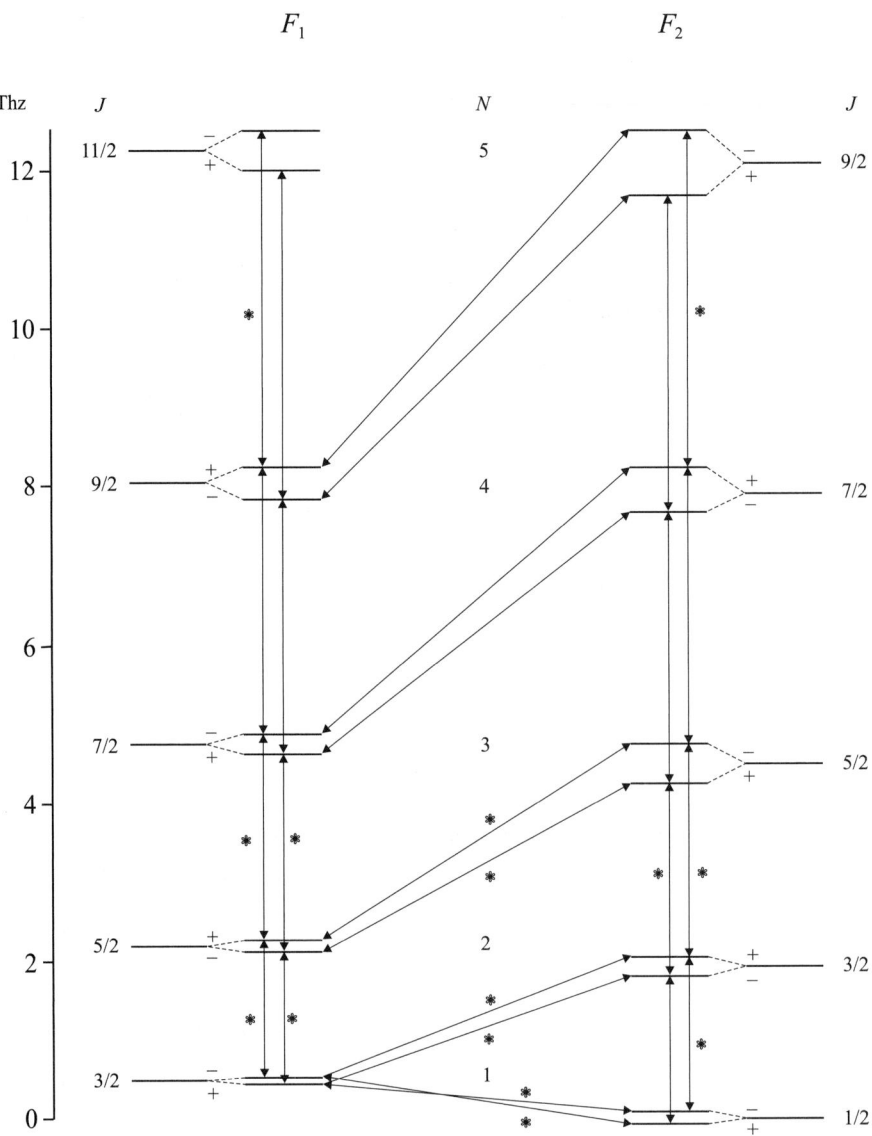

Figure 10.68. The low-energy rotational levels of CH, with the size of the Λ-doubling exaggerated for clarity. Transitions marked with an asterisk have been observed by far-infrared laser magnetic resonance. In addition many $\Delta J = 0$, Λ-doublet transitions have been observed field-free, as listed in table 10.17.

Again there are off-diagonal terms which, combined with the off-diagonal spin–orbit coupling terms, mix excited electronic states with the ground state and thereby give rise to Λ-doubling. These effects were described in our discussion of the NO spectrum in chapter 8, and the Λ-doubling constants, p and q, were defined in equation (8.399).

The spin–rotation interaction is, happily, very straightforward, with only diagonal matrix elements in a case (b) basis:

$$\langle \eta, \Lambda; N, \Lambda; S, J | \gamma \mathbf{T}^1(\mathbf{N}) \cdot \mathbf{T}^1(\mathbf{S}) | \eta, \Lambda; N, \Lambda; S, J \rangle$$

$$= \gamma(-1)^{N+J+S} \begin{Bmatrix} S & N & J \\ N & S & 1 \end{Bmatrix} \{N(N+1)(2N+1)S(S+1)(2S+1)\}^{1/2}$$

$$= \gamma\{J(J+1) - S(S+1) - N(N+1)\}. \quad (10.123)$$

We come now to the magnetic hyperfine interaction, which involves the sum of three terms representing the Fermi contact, orbital and dipolar interactions:

$$\mathcal{H}_{hfs} = \mathcal{H}_F + \mathcal{H}_{IL} + \mathcal{H}_{dip}. \quad (10.124)$$

The matrix elements were derived in a case (b) basis for each of these terms. For the Fermi contact interaction (8.220) we retain matrix elements off-diagonal in J which can be significant:

$$\langle \eta, \Lambda; N, \Lambda; S, J, I, F | b_F \mathbf{T}^1(\mathbf{S}) \cdot \mathbf{T}^1(\mathbf{I}) | \eta, \Lambda; N, \Lambda; S, J', I, F \rangle$$

$$= b_F (-1)^{J'+F+I} \begin{Bmatrix} I & J' & F \\ J & I & 1 \end{Bmatrix} \begin{Bmatrix} J & S & N \\ S & J' & 1 \end{Bmatrix} (-1)^{J+N+S+1}$$

$$\times \{(2J'+1)(2J+1)S(S+1)(2S+1)I(I+1)(2I+1)\}^{1/2}. \quad (10.125)$$

The Fermi contact interaction constant is defined, as usual, by

$$b_F = \frac{2\mu_0}{3} g_S g_N \mu_B \mu_N \int \psi_{el}^* \delta(\mathbf{r}) \psi_{el} \, d\mathbf{r}. \quad (10.126)$$

The case (b) matrix elements of the orbital hyperfine interaction were shown to be:

$$\langle \eta, \Lambda; N, \Lambda, S, J, I, F | a\mathbf{T}^1(\mathbf{L}) \cdot \mathbf{T}^1(\mathbf{I}) | \eta, \Lambda; N', \Lambda, S, J', I, F \rangle$$

$$= a(-1)^{J'+F+I} \begin{Bmatrix} I & J' & F \\ J & I & 1 \end{Bmatrix} (-1)^{J'+N+S+1} \{(2J'+1)(2J+1)\}^{1/2} \begin{Bmatrix} J & N & S \\ N' & J' & 1 \end{Bmatrix}$$

$$\times (-1)^{N-\Lambda} \begin{pmatrix} N & 1 & N' \\ -\Lambda & 0 & \Lambda \end{pmatrix} \{(2N+1)(2N'+1)\}^{1/2} \Lambda \{I(I+1)(2I+1)\}^{1/2}.$$

$$(10.127)$$

For a Π state we put $\Lambda = 1$, and the orbital hyperfine constant, a, is defined, as usual, by

$$a = 2g_N \mu_B \mu_N (\mu_0/4\pi) \sum_j \langle 1/r_{jN}^3 \rangle, \quad (10.128)$$

where r_{jN} measures the position of electron j with respect to nucleus N.

Finally the electron spin-nuclear spin dipolar interaction, which is more complicated, was given, initially, by equations (8.232) and (8.233):

$$\langle \eta, \Lambda; N, \Lambda, S, J, I, F | -\sqrt{10} g_S \mu_B g_N \mu_N (\mu_0/4\pi) \mathbf{T}^1(\mathbf{S}, \mathbf{C}^2) \cdot \mathbf{T}^1(\mathbf{I}) | \eta, \Lambda'; N', \Lambda', S, J', I, F \rangle$$

$$= -\sqrt{30} g_S \mu_B g_N \mu_N (\mu_0/4\pi) (-1)^{J'+F+I} \begin{Bmatrix} I & J' & F \\ J & I & 1 \end{Bmatrix} \{I(I+1)(2I+1)\}^{1/2}$$

$$\times \{(2J+1)(2J'+1)S(S+1)(2S+1)\}^{1/2} \begin{Bmatrix} J & J' & 1 \\ N & N' & 2 \\ S & S & 1 \end{Bmatrix}$$

$$\times \langle \eta, \Lambda; N, \Lambda \| T^2(C) \| \eta', \Lambda'; N', \Lambda' \rangle. \tag{10.129}$$

The complications arise in evaluating the reduced matrix element in (10.129) for which it is necessary to take account of the fact that Λ is a *signed* quantity. Consequently we must use the symmetrised combinations

$$\begin{aligned} \psi_+ &= (1/\sqrt{2})\{|+1\rangle + |-1\rangle\}, \\ \psi_- &= (1/\sqrt{2})\{|+1\rangle - |-1\rangle\}. \end{aligned} \tag{10.130}$$

With these functions as bases, and neglecting the very small matrix elements off-diagonal in N, we have

$$\langle \eta, \psi_+ \| T^2(C) \| \eta, \psi_+ \rangle = (-1)^{N-1}(2N+1) \Bigg\{ \begin{pmatrix} N & 2 & N \\ -1 & 0 & 1 \end{pmatrix} \langle T_0^2(C) \rangle_\eta$$

$$+ \begin{pmatrix} N & 2 & N \\ -1 & 2 & -1 \end{pmatrix} \langle T_2^2(C) \rangle_\eta \Bigg\}, \tag{10.131}$$

$$\langle \eta, \psi_- \| T^2(C) \| \eta, \psi_- \rangle = (-1)^{N-1}(2N+1) \Bigg\{ \begin{pmatrix} N & 2 & N \\ -1 & 0 & 1 \end{pmatrix} \langle T_0^2(C) \rangle_\eta$$

$$- \begin{pmatrix} N & 2 & N \\ -1 & 2 & -1 \end{pmatrix} \langle T_2^2(C) \rangle_\eta \Bigg\}. \tag{10.132}$$

The components of $T^2(C)$ in the above equations can be related to the constants c and d given by Jette and Cahill [187] by noting the following identities:

$$\begin{aligned} \langle T_0^2(C) \rangle &= g_S \mu_B g_N \mu_N (\mu_0/4\pi) \left\langle r^{-3} \left(\frac{4\pi}{5}\right)^{1/2} Y_{20}(\theta, \phi) \right\rangle \\ &= \frac{1}{2} g_S \mu_B g_N \mu_N (\mu_0/4\pi) \left\langle \frac{(3\cos^2\theta - 1)}{r^3} \right\rangle = \frac{1}{6} g_S c, \\ \langle T_2^2(C) \rangle &= g_S \mu_B g_N \mu_N (\mu_0/4\pi) \left\langle r^{-3} \left(\frac{4\pi}{5}\right)^{1/2} Y_{22}(\theta, \phi) \right\rangle \\ &= -\frac{3}{2\sqrt{6}} g_S \mu_B g_N \mu_N (\mu_0/4\pi) \left\langle \frac{\sin^2\theta}{r^3} \right\rangle = -\frac{1}{2\sqrt{6}} g_S d. \end{aligned} \tag{10.133}$$

Our results agree with those of Jette and Cahill if g_S is set equal to 2.

We may now list the values of the molecular parameters defined above for the $v = 0$ level of CH in its $X^2\Pi$ ground state. They are given in table 10.18. The magnetic parameters for ^{12}CH were listed in chapter 9. The corresponding set of field-free molecular parameters for ^{13}CH were determined by Steimle, Woodward and Brown [179]; the rotational, spin–rotation and Λ-doubling constants differ slightly from those

Table 10.18. *Molecular parameters (in MHz) for CH in the $v = 0$ level of the $X\,^2\Pi$ ground state, determined from a combination of the far-infrared laser magnetic resonance, field-free microwave measurements [181], and field-free far-infrared measurements [185]*

$A = 843\,818.508\,(55)$	$B = 425\,476.222\,(45)$	$D = 43.785\,7(28)$	$10^2 H = 0.3173$
$\gamma = -772.106(25)$	$\gamma_D = 0.258\,4(27)$		
$p = 1003.995\,8(57)$	$p_D = -0.273\,66(66)$	$10^4 p_H = 0.34(12)$	
$q = 1159.683\,2(28)$	$q_D = -0.457\,49(11)$	$10^4 q_H = 0.964\,(12)$	
$a = 54.006\,(80)$	$b_F = -57.777\,(68)$	$c = 56.52(25)$	$d = 43.513\,(11)$
$d_D = -0.015\,7(12)$			

of ^{12}CH, but the most significant new information concerns the ^{13}C magnetic hyperfine constants. The values determined were

$$h_{1/2} = 240.2 \text{ MHz}, \quad d = 274.9 \text{ MHz} \quad \text{where } h_{1/2} = a - (1/2)(b + c).$$

(c) Interpretation of the molecular parameters (a case study)

(i) ELECTRONIC STRUCTURE CALCULATIONS

The interpretation of the molecular parameters of CH has been discussed by Brazier and Brown [178], and *ab initio* calculations have been carried out by Lie, Hinze and Liu [188]. We will describe the theory in rather more detail than usual, partly because the theorists provided considerably more detail about their calculations than is often the case, and partly because high-resolution spectroscopic studies have been carried out for two excited electronic states, as well as for the ground state.

In terms of molecular orbital theory, the electronic ground state of CH has the following electron configuration:

$$X\,^2\Pi : 1\sigma^2\,2\sigma^2\,3\sigma^2\,1\pi^1.$$

The 3σ and 1π molecular orbitals are composed primarily of carbon $2p\sigma$ and $2p\pi$ atomic orbitals respectively, but the 3σ molecular orbital will have a small but significant admixture of the hydrogen $1s$ atomic orbital. The lowest excited electronic states arise from a $3\sigma \to 1\pi$ electronic excitation, and four different electronic states are possible:

$$1\sigma^2\,2\sigma^2\,3\sigma^1\,1\pi^2: \quad \begin{array}{ll} a\,^4\Sigma^-, & T_e = 5844 \text{ cm}^{-1}, \\ A\,^2\Delta, & 23189.8, \\ B\,^2\Sigma^-, & 26044, \\ C\,^2\Sigma^+, & 31801.5. \end{array} \quad (10.134)$$

The electronic excitation energies are known from electronic spectroscopy, or photoelectron spectroscopy in the case of the $a\,^4\Sigma^-$ state. For the purposes of our subsequent

Table 10.19. *Spin-orbital single configuration representations for the ground and lowest excited electronic states of CH. Only the spatial and spin coordinates of the three highest-energy electrons are specified*

$X\,^2\Pi$	$2p\sigma^2\,2p\pi^1$	$^2\Pi_{+3/2}$	$\left\|\sigma^\alpha \sigma^\beta\,\pi^\alpha_{+1}\right\|$
		$^2\Pi_{-3/2}$	$\left\|\sigma^\alpha\,\sigma^\beta\,\pi^\beta_{-1}\right\|$
		$^2\Pi_{+1/2}$	$\left\|\sigma^\alpha\,\sigma^\beta\,\pi^\beta_{+1}\right\|$
		$^2\Pi_{-1/2}$	$\left\|\sigma^\alpha\,\sigma^\beta\,\pi^\alpha_{-1}\right\|$
$a\,^4\Sigma^-$	$2p\sigma^1\,2p\pi^2$	$^4\Sigma^-_{+3/2}$	$\left\|\sigma^\alpha \pi^\alpha_{+1}\,\pi^\alpha_{-1}\right\|$
		$^4\Sigma^-_{-3/2}$	$\left\|\sigma^\beta \pi^\beta_{+1}\,\pi^\beta_{-1}\right\|$
		$^4\Sigma^-_{+1/2}$	$(1/\sqrt{3})\{\left\|\sigma^\alpha \pi^\alpha_{+1}\,\pi^\beta_{-1}\right\| + \left\|\sigma^\alpha \pi^\beta_{+1}\,\pi^\alpha_{-1}\right\| + \left\|\sigma^\beta \pi^\alpha_{+1}\,\pi^\alpha_{-1}\right\|\}$
		$^4\Sigma^-_{-1/2}$	$(1/\sqrt{3})\{\left\|\sigma^\beta \pi^\beta_{+1}\,\pi^\alpha_{-1}\right\| + \left\|\sigma^\beta \pi^\alpha_{+1}\,\pi^\beta_{-1}\right\| + \left\|\sigma^\alpha \pi^\beta_{+1}\,\pi^\beta_{-1}\right\|\}$
$A\,^2\Delta$	$2p\sigma^1\,2p\pi^2$	$^2\Delta_{+5/2}$	$\left\|\sigma^\alpha\,\pi^\alpha_{+1}\,\pi^\beta_{+1}\right\|$
		$^2\Delta_{-5/2}$	$\left\|\sigma^\beta\,\pi^\alpha_{-1}\,\pi^\beta_{-1}\right\|$
		$^2\Delta_{+3/2}$	$\left\|\sigma^\beta\,\pi^\alpha_{+1}\,\pi^\beta_{+1}\right\|$
		$^2\Delta_{-3/2}$	$\left\|\sigma^\alpha\,\pi^\alpha_{-1}\,\pi^\beta_{-1}\right\|$
$B\,^2\Sigma^-$	$2p\sigma^1\,2p\pi^2$	$^2\Sigma^-_{+1/2}$	$(1/\sqrt{6})\{2\left\|\sigma^\beta\,\pi^\alpha_{+1}\,\pi^\alpha_{-1}\right\| - \left\|\sigma^\alpha\,\pi^\beta_{+1}\,\pi^\alpha_{-1}\right\| - \left\|\sigma^\alpha\,\pi^\alpha_{+1}\,\pi^\beta_{-1}\right\|\}$
		$^2\Sigma^-_{-1/2}$	$(1/\sqrt{6})\{2\left\|\sigma^\alpha\,\pi^\beta_{+1}\,\pi^\beta_{-1}\right\| - \left\|\sigma^\beta\,\pi^\alpha_{+1}\,\pi^\beta_{-1}\right\| - \left\|\sigma^\beta\,\pi^\beta_{+1}\,\pi^\alpha_{-1}\right\|\}$
$C\,^2\Sigma^+$	$2p\sigma^1\,2p\pi^2$	$^2\Sigma^+_{+1/2}$	$(1/\sqrt{2})\{\left\|\sigma^\alpha\,\pi^\alpha_{+1}\,\pi^\beta_{-1}\right\| - \left\|\sigma^\alpha\,\pi^\beta_{+1}\,\pi^\alpha_{-1}\right\|\}$
		$^2\Sigma^+_{-1/2}$	$(1/\sqrt{2})\{\left\|\sigma^\beta\,\pi^\alpha_{+1}\,\pi^\beta_{-1}\right\| - \left\|\sigma^\beta\,\pi^\beta_{+1}\,\pi^\alpha_{-1}\right\|\}$

discussion it might be helpful to write down the spin-orbital single configurations for the ground and lowest excited states, expressed as determinantal wave functions. These are given in table 10.19, where π_{+1} and π_{-1} are the orbitals with $\Lambda = +1$ and -1. The spin state, α or β, with $\Sigma = +1/2$ or $-1/2$, is denoted by the superscript. Note that the configurations are written in a case (a) representation, with Ω also specified for the fine-structure components, although we know that case (a) is, in fact, a poor representation for the CH molecule. Lie, Hinze and Liu [188] carried out a series of configuration interaction calculations and their final results for the potential curves of the five lowest electronic states are summarised, both in table 10.20 and in figure 10.69. The method used involved the prior calculation of self-consistent field basis functions for the single configurations given in table 10.19, followed by configuration interaction calculations involving an extended number of excited configurations, not just those given in table 10.19.

Lie, Hinze and Liu [189] have used their calculated wave functions to compute expectation values of the magnetic hyperfine parameters, and we will come to their calculations shortly.

Table 10.20. *Potential curves (in Hartrees) for the five lowest electronic states of CH, calculated by Lie, Hinze and Liu [188] using an extended configuration interaction method*

$R(a_0)$	$X\,^2\Pi$	$a\,^4\Sigma^-$	$A\,^2\Delta$	$B\,^2\Sigma^-$	$C\,^2\Sigma^+$
1.00	−37.641 03	−37.634 99	−37.542 32	−37.507 26	−37.496 20
1.30	−38.150 61	−38.142 12	−38.052 55	−38.021 21	−38.007 96
1.60	−38.341 21	−38.326 61	−38.239 33	−38.213 46	−38.196 64
1.90	−38.401 94	−38.381 64	−38.296 75	−38.277 74	−38.255 43
2.00	−38.408 29	−38.385 93	−38.301 93	−38.285 52	−38.261 07
2.05	−38.409 81	−38.386 45	−38.302 92	−38.287 89	−38.262 30
2.10	−38.410 41	−38.386 08	−38.303 04	−38.289 46	−38.262 68
2.15	−38.410 25	−38.384 97	−38.302 44	−38.290 37	−38.262 35
2.20	−38.409 43	−38.383 23	−38.301 24	−38.290 75	−38.261 42
2.30				−38.290 27	
2.40	−38.401 56	−38.371 83	−38.292 28	−38.288 75	−38.253 54
2.70	−38.383 38	−38.348 39	−38.273 74	−38.281 92	−38.236 37
3.00	−38.362 52	−38.325 30	−38.257 07	−38.277 28	−38.223 42
3.25				−38.276 38	−38.219 42
3.50	−38.331 24	−38.299 11	−38.241 75	−38.276 59	−38.220 07
4.00	−38.308 70	−38.288 28	−38.236 32	−38.278 58	−38.225 96
5.00	−38.287 86	−38.283 47	−38.233 93	−38.281 44	−38.231 81
6.00	−38.282 87	−38.282 59	−38.233 49	−38.282 16	−38.233 42
8.00	−38.281 60	−38.282 26	−38.233 24	−38.282 23	−38.233 80
11.00	−38.281 59	−38.282 24	−38.233 27	−38.282 22	−38.233 93
15.00	−38.281 59	−38.282 23	−38.233 27	−38.282 22	−38.233 92
20.00	−38.281 59	−38.282 23	−38.233 27	−38.282 22	−38.233 92

(ii) Λ-DOUBLING CONSTANTS

We first look at the form of the calculations for the Λ-doubling parameters, p and q, which are a measure of the mixing of the $^2\Pi$ state with the $^2\Sigma^+$ and $^2\Sigma^-$ states brought about by the combined effects of spin–orbit coupling and Coriolis effects. Specifically we showed in chapter 8 the results [190] first obtained by Mulliken and Christy:

$$p = -2\sum_n \frac{\left\{\langle^2\Pi_{1/2}|\mathcal{H}_{so}|n\,^2\Sigma^s_{1/2}\rangle\langle n\,^2\Sigma^s_{1/2}|-BL_+|^2\Pi_{-1/2}\rangle + \langle^2\Pi_{1/2}|-BL_+|n\,^2\Sigma^s_{-1/2}\rangle\langle n\,^2\Sigma^s_{-1/2}|\mathcal{H}_{so}|^2\Pi_{-1/2}\rangle\right\}}{E_\Pi - E_\Sigma},$$

(10.135)

$$q = 2\sum_n \frac{\langle^2\Pi_{3/2}|-BL_+|n\,^2\Sigma^s_{1/2}\rangle\langle n\,^2\Sigma^s_{1/2}|-BL_+|^2\Pi_{-1/2}\rangle}{E_\Pi - E_\Sigma}.$$

(10.136)

Although these expressions involve sums over all excited $^2\Sigma^s$ states, Brazier and Brown [181] chose to investigate the results of the simplest possible calculation, which involves

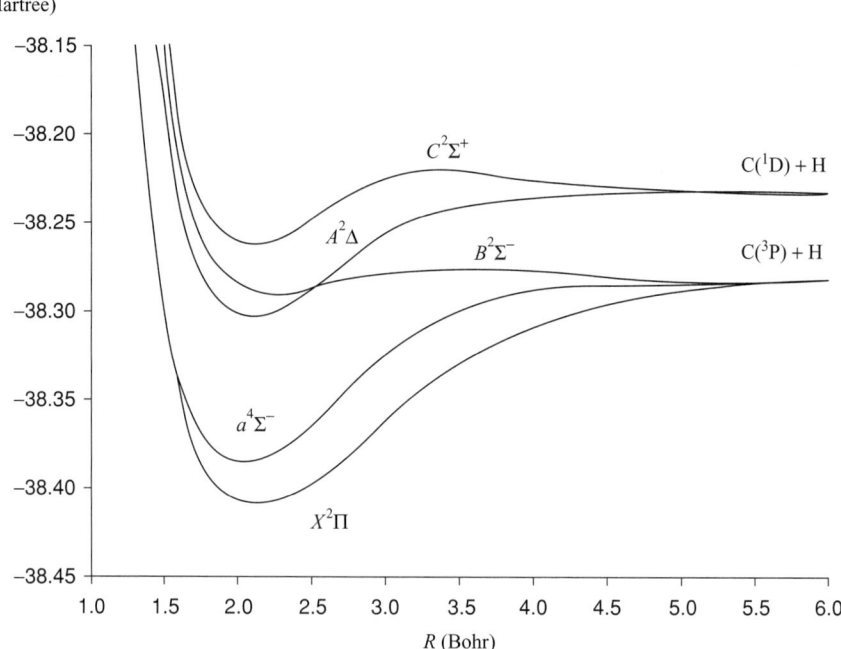

Figure 10.69. Potential energy curves for the five lowest energy electronic states of CH, calculated by Lie, Hinze and Liu [188] using an extended configuration interaction method.

only the B and $C\,^2\Sigma$ states described above. In order to evaluate the matrix elements in the above equations it is also necessary to make some further simplifying assumptions. The spin–orbit coupling is written as the sum of one-electron operators. One may also assume that the 3σ and 1π molecular orbitals can be replaced by the carbon $2p_z$ and $2p_x$, $2p_y$ atomic orbitals respectively [190]; this is called the pure precession approximation. If this is done, the required matrix elements of the spin–orbit coupling and rotational Coriolis mixing are as follows, where ζ is the $2p$ atomic spin–orbit coupling parameter for C.

Spin–orbit coupling:

$$\langle X\,^2\Pi_{1/2}|\mathcal{H}_{so}|^2\Sigma^+_{-1/2}\rangle = \langle ^2\Sigma^+_{-1/2}|\mathcal{H}_{so}|X\,^2\Pi_{-1/2}\rangle = (1/2\sqrt{2})\zeta\,\langle\sigma|l_+|\pi\rangle \cong (1/2)\zeta.$$

$$\langle X\,^2\Pi_{1/2}|\mathcal{H}_{so}|^2\Sigma^-_{-1/2}\rangle = \langle ^2\Sigma^-_{-1/2}|\mathcal{H}_{so}|X\,^2\Pi_{-1/2}\rangle = (1/2\sqrt{6})\zeta\,\langle\sigma|l_+|\pi\rangle \cong (1/2\sqrt{3})\zeta.$$

$$\langle ^2\Delta_{3/2}|\mathcal{H}_{so}|X\,^2\Pi_{3/2}\rangle = \langle ^2\Delta_{-3/2}|\mathcal{H}_{so}|X\,^2\Pi_{-3/2}\rangle = (1/2)\zeta\,\langle\pi_+|l_+|\sigma\rangle \cong (1/\sqrt{2})\zeta.$$

(10.137)

Rotational Coriolis mixing:

$$\langle X\,^2\Pi_{3/2}|BL_+|^2\Sigma^+_{1/2}\rangle = \langle ^2\Sigma^+_{-1/2}|BL_+|X\,^2\Pi_{-3/2}\rangle = -(1/\sqrt{2})\langle\sigma|Bl_+|\pi_-\rangle \cong -B.$$

$$\langle ^2\Sigma^+_{1/2}|BL_+|X\,^2\Pi_{-1/2}\rangle = \langle X\,^2\Pi_{1/2}|BL_+|^2\Sigma^+_{-1/2}\rangle = -(1/\sqrt{2})\,\langle\pi_+|Bl_+|\sigma\rangle \cong -B.$$

$$\langle X\,^2\Pi_{3/2}|\,BL_+|^2\Sigma^-_{-1/2}\rangle = \langle ^2\Sigma^-_{-1/2}|BL_+|X\,^2\Pi_{-3/2}\rangle = (\sqrt{3}/\sqrt{2})\,\langle\sigma|\,Bl_+\,|\pi_-\rangle \cong \sqrt{3}\,B.$$
$$\langle ^2\Sigma^-_{1/2}|BL_+|X\,^2\Pi_{-1/2}\rangle = \langle X\,^2\Pi_{1/2}|BL_+|^2\Sigma^-_{-1/2}\rangle = -(\sqrt{3}/\sqrt{2})\,\langle \pi_+|\,Bl_+\,|\sigma\rangle \cong -\sqrt{3}\,B.$$
$$\langle ^2\Delta_{3/2}|\,BL_+\,|X\,^2\Pi_{1/2}\rangle = \langle X\,^2\Pi_{-1/2}|\,BL_+\,|^2\Delta_{-3/2}\rangle = -\langle \pi_+|Bl_+|\sigma\rangle \cong -\sqrt{2}B.$$
$$\langle ^2\Delta_{5/2}|\,BL_+\,|X\,^2\Pi_{3/2}\rangle = \langle X\,^2\Pi_{-3/2}|\,BL_+\,|^2\Delta_{-5/2}\rangle = \langle\sigma|\,Bl_+\,|\pi_-\rangle \cong \sqrt{2}B.$$
(10.138)

We may now ignore the sums over excited states n in (10.135) and (10.136) and consider only the effects of the B and $C\,^2\Sigma$ states, and the $A\,^2\Delta$ state. If we take $B = B_0 = 14.19$ cm^{-1} for the $X\,^2\Pi$ state, use the value $\zeta = 27.5$ cm^{-1} for a carbon $2p$ orbital, and substitute the appropriate electronic excitation energies, we obtain the following results for p and q:

$$\begin{aligned}p &= 4\{-(1/2)\zeta B/(E(\Pi)-E(\Sigma^+))-(1/2)\zeta B/(E(\Pi)-E(\Sigma^-))\}\\ &= 1654\text{ MHz}\quad(\exp = 1004\text{ MHz}),\\ q &= 2\{B^2/(E(\Pi)-E(\Sigma^+))-3B^2/(E(\Pi)-E(\Sigma^-))\}\\ &= 1030\text{ MHz}\quad(\exp = 1160\text{ MHz}).\end{aligned}$$
(10.139)

Note of course that the mixing with the $^2\Delta$ state does not contribute to the Λ-doubling parameters. Considering the approximations made, the agreement between experiment and theory is very reasonable.

(iii) SPIN–ROTATION CONSTANT

The contribution of excited electronic state mixing to the electron spin–rotation constant γ is given by a second-order expression similar to those for p and q. It involves the mixing of both $^2\Sigma$ and $^2\Delta$ states with the ground state and is as follows:

$$\gamma = 2\sum_n \frac{\langle ^2\Pi_{3/2}|-BL_+|n\,^2\Sigma_{1/2}\rangle\langle n\,^2\Sigma_{1/2}|\mathcal{H}_{so}|^2\Pi_{1/2}\rangle}{E_\Pi - E_\Sigma}$$
$$+2\sum_n \frac{\langle ^2\Pi_{3/2}|\mathcal{H}_{so}|n\,^2\Delta_{3/2}\rangle\langle n\,^2\Delta_{3/2}|-Rl_{+1}|^2\Pi_{1/2}\rangle}{E_\Pi - E_\Delta}. \quad (10.140)$$

Once again using the pure precession approximation and the parameter values given above we obtain

$$\gamma \cong -B\zeta\left\{-\frac{1}{E(\Pi)-E(\Sigma^+)}+\frac{1}{E(\Pi)-E(\Sigma^-)}-\frac{2}{E(\Pi)-E(\Delta)}\right\}$$
$$\cong -920\text{ MHz}\quad(\exp = -771\text{ MHz}). \quad (10.141)$$

It is interesting to note that the contributions to γ from the $^2\Sigma^+$ and $^2\Sigma^-$ states almost cancel, so that it is the $^2\Delta$ state mixing which is primarily responsible for the value of γ. Again the agreement between experiment and theory is reasonable.

(iv) MAGNETIC HYPERFINE PARAMETERS

The magnetic hyperfine parameters have been defined in a number of different places in this book. Levy and Hinze [191] have used the single configuration Hartree–Fock *ab initio* wave function calculated by Lie, Hinze and Liu [188] to calculate the Frosch and Foley [192] magnetic hyperfine constants. The results, compared with experiment, are as follows:

$$a(\text{calc}) = 58.5 \pm 4.5 \, \text{MHz}, \quad a(\text{exp}) = 54.3$$
$$b(\text{calc}) = -72 \pm 10, \quad b(\text{exp}) = -76.7$$
$$c(\text{calc}) = 57 \pm 4, \quad c(\text{exp}) = 57.2.$$

The agreement is excellent.

Other molecular properties for CH have been calculated by Lie, Hinze and Liu [189], not all of which have been measured experimentally. An important exception is the electric dipole moment in the ground state which has been determined experimentally to be 1.46 ± 0.06 D, and is calculated to be 1.450 D, in the sense C^-H^+.

As we mentioned at the beginning of this section, two of the low-lying excited electronic states of CH have also been studied experimentally by high-resolution spectroscopy. The energy of the $a\ ^4\Sigma^-$ state was found to be 0.742 eV, measured by Kasdan, Herbst and Lineberger [193] using laser photoelectron spectrometry of CH^-. As we have described in chapter 9, the far-infrared laser magnetic resonance spectrum of CH in this state was detected and studied by Nelis, Brown and Evenson [194]. A full description of the analysis was presented in chapter 9. In addition the microwave/optical double resonance technique which enabled Brazier and Brown [178] to detect the rotational spectrum of CH in its ground state also yielded a rotational spectrum for the $A\ ^2\Delta$ state. We defer a discussion of the experimental details and analysis of the spectrum until chapter 11, but will understand the natural curiosity of the reader who wishes to advance to that description now!

The pursuit of the CH radical in spectroscopic laboratories has led to important advances in experimental techniques, a fairly complete description of its electronic structure in the ground state, and the establishment of firm foundations for astrophysical and astrochemical studies.

10.6.4. CF, SiF, GeF in their $X\ ^2\Pi$ ground states

The CF, SiF and GeF radicals have $^2\Pi$ ground states and are similar to NO in that the $^2\Pi_{1/2}$ fine-structure component is the lower in energy. The rotational energy level diagram shown for NO is therefore appropriate for these three species except that as the spin–orbit coupling increases, the $^2\Pi_{1/2}$–$^2\Pi_{3/2}$ separation also increases. The fine-structure splitting is 77.12 cm^{-1} for CF, compared with 119.82 cm^{-1} for NO. It increases to 161.88 cm^{-1} in SiF, and 934.33 cm^{-1} in GeF. CF in its lowest rotational levels approximates reasonably well to Hund's case (a) coupling, and case (a) becomes an increasingly good description as the fine-structure splitting increases for SiF and GeF.

Table 10.21. *Molecular constants (in MHz except for dipole moment) for the CF, SiF and GeF radicals in their* $X\ ^2\Pi$ *states (hyperfine constants are for* ^{19}F)

Parameter	CF	SiF	^{74}GeF
B_0	42 196.663	17 350.2752	10 924.499
D_0	0.1993	0.03188	0.013 225
p_0	255.95	−87.67	−656.792
q_0	0.69	−1.26	—
$a+(b+c)/2$	664.07	288.26	—
$a-(b+c)/2$	747.58	336.4	309.6
b	269.2	127	—
d	792.17	359.0	296.5
a	705.82	312.35	—
c	−352.7	−175	—
μ_e (D)	0.645	—	—

The first microwave spectrum of CF was observed by Carrington and Howard [195] using magnetic resonance methods. The $J = 11/2 \leftarrow 9/2$ rotational transition in the $^2\Pi_{3/2}$ state was observed by Saykally, Lubic, Scarabrin and Evenson [196] using far-infrared laser magnetic resonance, and high-frequency high-J rotational transitions were observed by Van den Heuvel, Meerts and Dymanus [197]. Pure rotational transitions involving the low rotational levels in both fine-structure components were studied by Saito, Endo, Takami and Hirota [198], using a glow discharge in CF$_4$, and these measurements provided the most complete and accurate data for the ^{19}F hyperfine constants. The frequency range involved was 124 to 300 GHz.

The pure rotational spectrum of SiF, produced by a dc glow discharge in SiF$_4$/SiH$_4$ mixtures, was studied by Tanimoto, Saito, Endo and Hirota [199] over the frequency range 86 to 192 GHz. Rotational transitions involving low rotational levels in both fine-structure components were measured. Similar studies by the same authors [200] of the GeF radical were described three years later; in this case, however, only rotational transitions in the lower fine-structure component, $^2\Pi_{1/2}$, were observed, the population of the higher component being too small for detection.

In all three radicals the main information obtained concerned the ^{19}F hyperfine interaction. The data are summarised in table 10.21. The electron spin density on the fluorine atom in CF is estimated to be 17.8%, and in SiF it is found to be close to 8%. It is probably even smaller in GeF but the data are limited because only the $^2\Pi_{1/2}$ component could be observed.

10.6.5. Other free radicals with $^2\Pi$ ground states

As we have indicated elsewhere, free radicals with $^2\Pi$ ground states are numerous and represent the largest class of molecules studied by microwave and far-infrared techniques, particularly magnetic resonance methods. We described the microwave magnetic resonance spectrum [201] of the ClO radical at length in chapter 9, and now complete this story by discussing the pure rotational spectrum observed and analysed by Amano, Saito, Hirota, Morino, Johnson and Powell [202]. ClO is a good case (a) molecule, with the $^2\Pi_{3/2}$ component being the ground state, and the $^2\Pi_{1/2}$ component being almost 300 cm^{-1} higher in energy. The magnetic resonance studies were confined to the $^2\Pi_{3/2}$ component because, in case (a) coupling, the $^2\Pi_{1/2}$ component is almost non-magnetic. A consequence of this restriction was that the magnetic resonance studies could provide only a composite chlorine magnetic hyperfine constant, $h_{3/2}$, the total axial component in the $^2\Pi_{3/2}$ state. It is also the case that, although the rotational levels in the $^2\Pi_{3/2}$ component possess two-fold Λ-degeneracy, the Λ-doublet splitting is very small and was not resolved in the magnetic resonance experiments. This is not true of the $^2\Pi_{1/2}$ component, where the Λ-doubling is much larger. A summary of the lowest rotational levels for both fine-structure components is shown in figure 10.70.

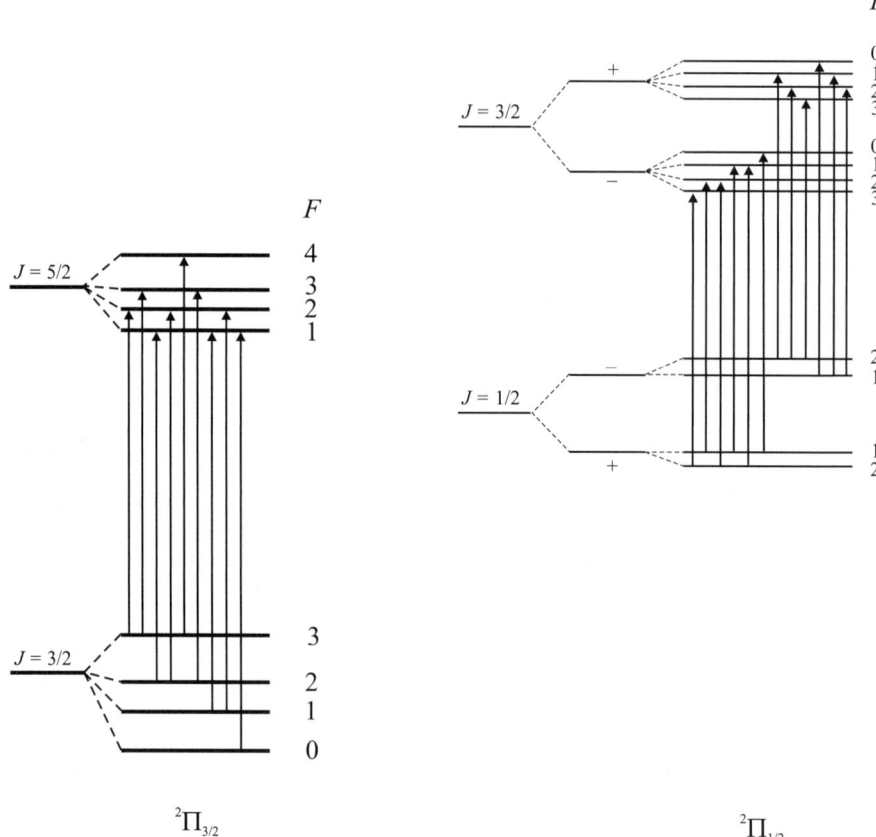

Figure 10.70. Lowest rotational levels and pure microwave transitions observed [202] for the ClO molecule in its $^2\Pi$ ground state. The diagram is not drawn to scale, the fine-structure splitting being very much larger than the hyperfine or Λ-doublet splittings.

In this diagram we have included the chlorine hyperfine splitting, which arises from both magnetic and electric quadrupole interactions. Chlorine possesses two important naturally occurring isotopes, ^{35}Cl and ^{37}Cl, which have similar magnetic and electric moments; both isotopomers are observed in the experiments.

The axial component of the total magnetic hyperfine interaction, $h_{3/2}$, in the $^2\Pi_{3/2}$ component is equal to $a + (b+c)/2$, where a, b and c are the Frosch and Foley [192] constants. In the $^2\Pi_{1/2}$ component the axial hyperfine constant, $h_{1/2}$, is equal to $a - (b+c)/2$. Since both fine-structure components can usually be studied by pure microwave experiments, a partial separation of the magnetic hyperfine constants can be achieved. The electric quadrupole coupling constant, eq_0Q, is obtained for both isotopes.

In table 10.22 we have summarised the data obtained for a number of different $^2\Pi$ molecules, from a combination of magnetic resonance studies involving the Zeeman components of a single rotational level, and pure microwave to far-infrared studies of transitions between rotational levels. We remind ourselves of the definitions of the Frosch and Foley parameters:

$$\begin{aligned} a &= 2\mu_B g_N \mu_N (\mu_0/4\pi) \langle 1/r^3 \rangle, \\ b &= g_S \mu_B g_N \mu_N (\mu_0/4\pi) \{(8\pi/3)\,\psi^2(0) - \langle (3\cos^2\theta - 1)/2r^3 \rangle\}, \\ c &= g_S \mu_B g_N \mu_N (\mu_0/4\pi)(3/2)\langle (3\cos^2\theta - 1)/r^3 \rangle, \\ d &= (3/2) g_S \mu_B g_N \mu_N (\mu_0/4\pi) \langle \sin^2\theta / r^3 \rangle. \end{aligned} \quad (10.142)$$

Note also that $b_F = b + (c/3)$.

10.7. Case (c) doublet state molecules

10.7.1. Studies of the HeAr$^+$ ion

(a) Experimental observations

We now come to studies which are different from anything described earlier. Carrington, Leach, Marr, Shaw, Viant, Hutson and Law [211] have carried out microwave experiments on the HeAr$^+$ ion, with the objective of investigating the energy levels lying very close to the dissociation limit, approximately within 0 to 8 cm^{-1}. Almost all of the spectroscopic studies described so far in this book have involved the ground vibrational level of the ground electronic state, and it is natural to ask if levels close to dissociation follow the same rules and classification. Supplementary but important questions concern the details of the long-range part of the potential, and the relationship between these high-lying yet bound levels to those parts of the overall potential which are important in considerations of chemical reactivity. The experimental techniques developed are applicable to molecular cations; they make use of mass spectrometry, and the very high sensitivity associated with detection of the ions themselves. Whilst it is possible to conceive of similar experiments on neutral molecules, none have yet been reported.

Table 10.22. *Summary of molecular parameters (all in MHz except for A which is in cm^{-1}) obtained for free radicals in their $X\,^2\Pi$ ground states, with $v = 0$. Results for OH, CH and NO have been given earlier. Nuclear interaction constants refer to the nuclei specified explicitly*

Param.	^{35}ClO	Si^{35}Cl	C^{35}Cl	^{14}NS	^{65}CuO	^{107}AgO	S^{19}F	^{7}LiO	^{23}NaO
A	−282	207.130	134.959	229.94	−279.02	−269.3	−387.0	−111.67	−107.15
B_0	18 602.9	7652.305	20 797.2	23 156.0	13 252.874	9035.497	16 576.9	36 091.42	12 662.68
p	674.22	138.260	401.696	397.32	409.275	−175.667	3.409	6284.11	2650.112
q	—	0.20	1.176	—	2.5886	−1.0399	—	−56.58	18.687
$h_{3/2}$	112.60	38.17	66.43	56.35	−3.62	69.75	428.44	4.018	−16.129
$h_{1/2}$	160.07	49.17	93.96	67.45	431.7	−108.25	535.74	12.338	32.371
a	136.24	43.67	80.199	61.90	214.1	−19.25	482.09	8.178	8.121
b	48.17	9.0	11.97		−507.8	206	209.7	−22.50	−52.4
c	−95.6	−21	−39.5		72.5	−28	−317.0	14.18	3.9
d	173.03	46.40	82.212	87.03	134.5	—	589.76	4.036	7.03
b_F	16.3	2.0	−1.2		−483.6	197	104.0	−17.773	−51.10
eq_0Q	−87.02	−23.13	−34.26	−2.62	−16.9	—	—	0.463	−6.81
μ (D)	1.239	—	—	1.81	—	—	—	—	—
Ref.	[202, 201]	[203]	[204]	[205]	[206]	[207]	[208]	[209]	[210]

HeAr⁺ ions were made by electron impact on He/Ar gas mixtures; it is likely that the primary process occurring is electronic excitation of He atoms to a metastable electronic state and a subsequent chemi-ionisation reaction with Ar atoms. This is an exothermic process and the HeAr⁺ ions are formed with extensive internal excitation. Once formed, the ions are accelerated out of the ion source to form a focussed beam; there is little or no opportunity for collisional relaxation. As we described earlier in this chapter (see figure 10.33), the levels close to dissociation are monitored by electric field dissociation, and the fragment ions (Ar⁺) are detected. Prior to the electric field dissociation lens, the molecular ion beam passes through a microwave radiation field. The transitions induced in this field result in population transfer, and the Ar⁺ ions are analysed according to their kinetic energies by means of an electrostatic analyser, and detected with an electron multiplier. Population transfer results in changes in the fragment Ar⁺ current; these changes are detected by amplitude modulation of the microwave power, the modulated Ar⁺ current being detected with a lock-in amplifier. Electric field dissociation operates effectively on levels which are bound by only a few cm⁻¹ at most, and these levels are populated by the molecular ion formation process. Most of the ions in the beam (which typically had a total flux of 3×10^9 ions s⁻¹) are not affected by the electric field lens, and play no part in the experiment. Microwave frequencies ranging from 6 to 170 GHz were used.

Potential curves for the three lowest electronic states of HeAr⁺ are shown in figure 10.71. These curves are labelled with case (a) labels although, as we shall see, Hund's case (c) proved to be a better basis for both qualitative and quantitative descriptions of the results. Although almost all of the levels studied were within 8 cm⁻¹ of the

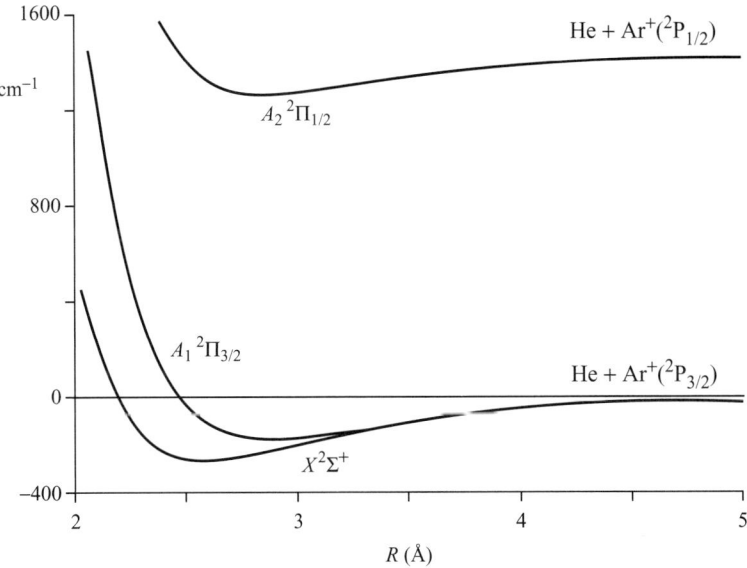

Figure 10.71. Potential curves for the three lowest electronic states of the HeAr⁺ ion, labelled in a Hund's case (a) notation [211].

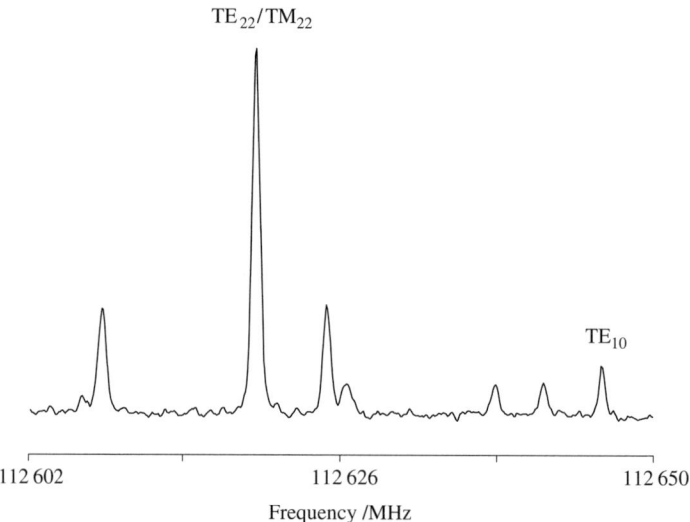

Figure 10.72. Multimode and associated parallel Doppler-shifted pattern observed for a HeAr$^+$ resonance with a rest frequency of 112 586.7 MHz, recorded in WR-28 waveguide [211]. Two of the modes are indicated in the figure, but all were identified unambiguously.

lowest dissociation limit shown in figure 10.71, the higher electronic state correlating with the first excited dissociation limit plays a role in the interpretation, as we will see. Previous work by Dabrowski, Herzberg and Yoshino [212] on the electronic emission spectrum from the higher energy $B\ ^2\Sigma^+$ state to the X and A_1 ($^2\Pi_{3/2}$) states showed that the latter states have dissociation energies of about 262 and 154 cm^{-1} respectively. An *ab initio* potential function due to Siska [213] indicated that the X state has 7 bound vibrational levels, and the A_1 state has 6.

Sixty-eight absorption lines were detected, and a typical example is shown in figure 10.72. The resonance corresponds to parallel alignment of the ion beam and propagation direction of the microwave radiation, and is Doppler-shifted to higher source frequencies. Antiparallel resonances are also observed, but are not shown in the figure. The structure apparent in figure 10.72, which is fairly typical of most of the observed resonances, arises because the microwave frequency is much higher than the optimum transmission frequency of the waveguide used. A consequence of this is that the microwave radiation is propagated in several different modes; each has its own characteristic phase velocity and hence Doppler shift. Fortunately these shifts can be easily and accurately calculated so there is no problem in calculating the rest frequency of the resonance.

The overall spectrum showed no recognisable pattern and without further experimental evidence, assignment would have been difficult, if not impossible. Fortunately two other diagnostic studies of each resonance line could be made. It was possible to carry out double resonance experiments, using two different microwave frequencies simultaneously. Suppose that two resonance lines with frequencies f_1 and f_2 have been observed. If they share a common energy level, a modulation signal on f_2 may be

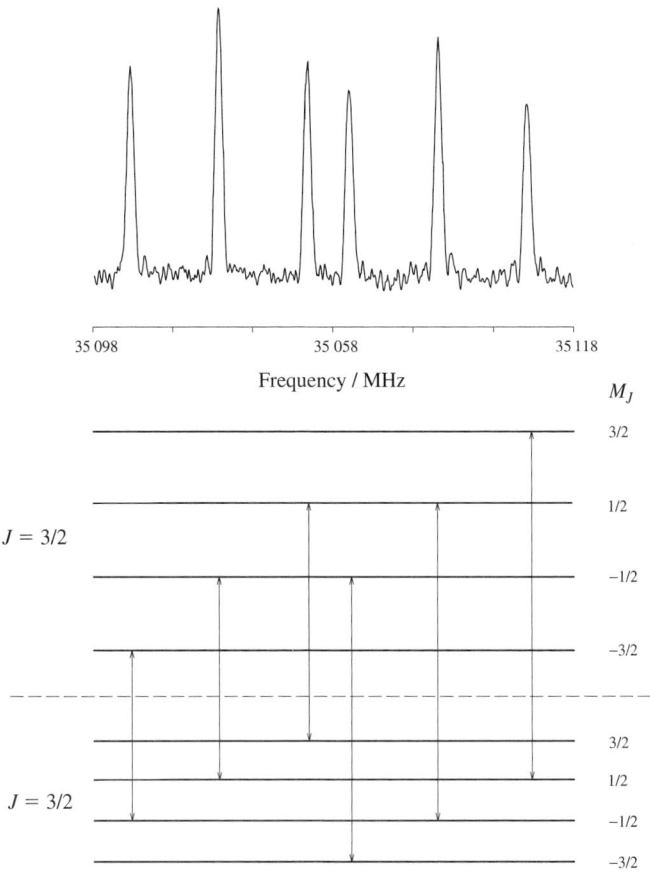

Figure 10.73. Observed Zeeman pattern and theoretical reconstruction for a $J = 3/2 \to 3/2$ transition in HeAr$^+$, with a rest frequency of 35 092.7 MHz [211]. The magnetic field was 4.85 G, using the TE$_{10}$ mode with parallel ion beam and microwave propagation, but perpendicular microwave electric field and static magnetic field ($\Delta M_J = \pm 1$).

detected by modulating f$_1$ on resonance, but scanning through the range around f$_2$. Double resonance was also used to make possible the detection of transitions involving levels very close to dissociation, by transferring population from a lower, more populated level.

The second diagnostic study made use of the Zeeman effect, observed when a small magnetic field was applied parallel to the ion beam direction. For almost every line in the spectrum a Zeeman splitting could be observed; figure 10.73 illustrates a particularly simple example, where the six-line Zeeman pattern shows conclusively that the resonance must arise from a $J = 3/2 \to 3/2$ transition. Effective g factors are obtained for both levels, providing valuable additional information which we shall discuss in detail later. It is important to know the microwave mode involved in a particular resonance because this determines the nature of the Zeeman components. The transitions are electric-dipole allowed, and if the Zeeman field is parallel to the

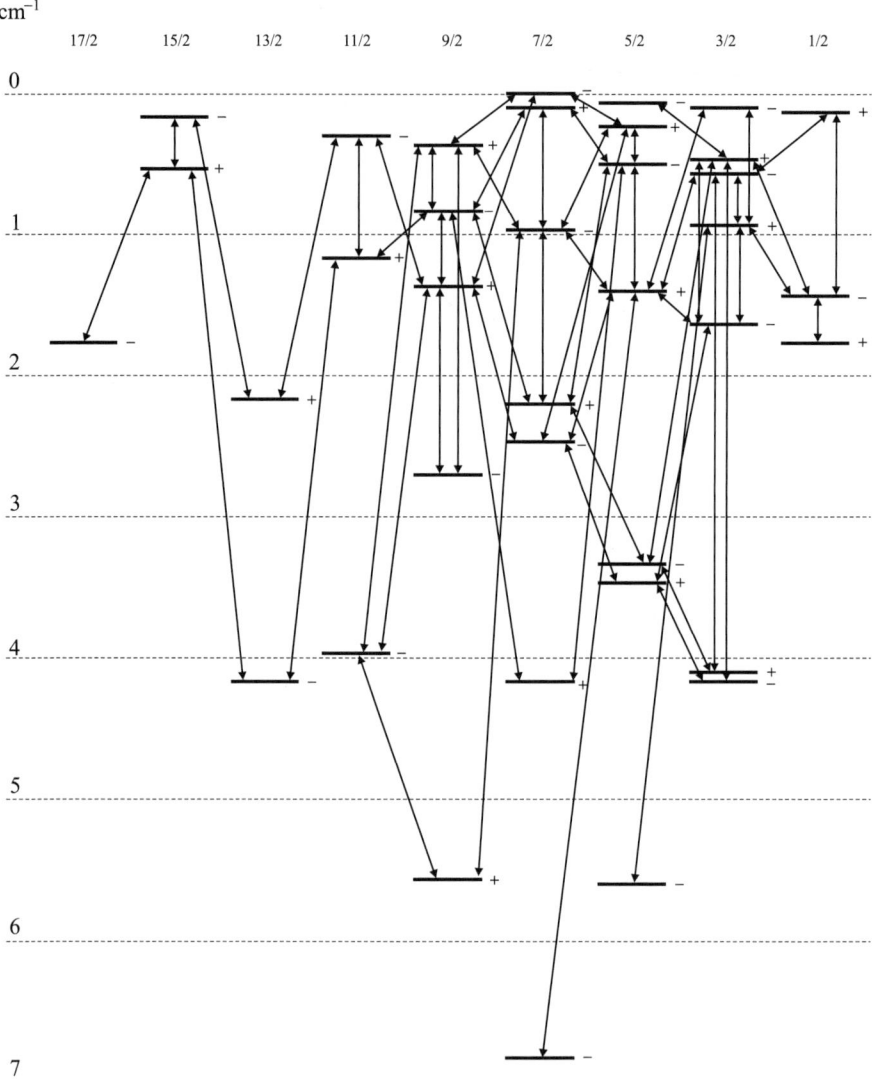

Figure 10.74. Experimentally determined energy level diagram for HeAr$^+$, arranged according to their J values and parities. The transitions observed are also shown. The diagram covers a range from the dissociation limit to about 7 cm^{-1} below.

microwave electric field, as was the case with most transitions studied (but not that shown in figure 10.73), the selection rule for the Zeeman components is $\Delta M_J = 0$.

Finally, there are numerical relationships between groups of transitions which share common levels; these relationships correspond to combination differences in other branches of spectroscopy. Through a combination of these relationships, double resonance studies and Zeeman effect measurements, it was possible to establish the energy level diagram shown in figure 10.74. Each level is characterised by its parity and J value; the observed transitions are also shown in figure 10.74. The important task remaining was to assign other quantum numbers to the levels, describing their electronic and vibrational states. We will come to this aspect shortly, but first we examine the angular momentum theory for case (c) in order to understand both the progressions of rotational levels, and also their Zeeman characteristics.

(b) Angular momentum theory for Hund's case (c) and the Zeeman behaviour

(i) VECTOR COUPLING AND QUANTUM NUMBERS

The three electronic states correlating with the dissociation limits involving Ar^+ in its $^2P_{3/2}$ and $^2P_{1/2}$ spin–orbit states are described in figure 10.71 by the case (a) labels $X\,^2\Sigma^+$, $A_1\,^2\Pi_{3/2}$ and $A_2\,^2\Pi_{1/2}$. However, even at the bottom of the potential well Λ is not a good quantum number, and it becomes worse as we approach dissociation. It is therefore much more satisfactory to use case (c) labels; the X and A_1 states arise from the $J_a = 3/2$ state of Ar^+, with the projection $|\Omega| = 1/2$ and $3/2$, respectively, and the A_2 state arises from the $J_a = 1/2$ state of Ar^+, with $|\Omega| = 1/2$. The case (c) coupling scheme was described in chapter 6, and the appropriate vector coupling diagram is repeated in figure 10.75. J_a and the rotational angular momentum vector R are coupled to form the total angular momentum J; Ω is therefore the projection of both J and J_a along the internuclear axis. In fact, as we shall see, even the case (c) labels are approximate near the dissociation limit, and we will make use of case (e), which is also illustrated in figure 10.75. We return to this aspect later.

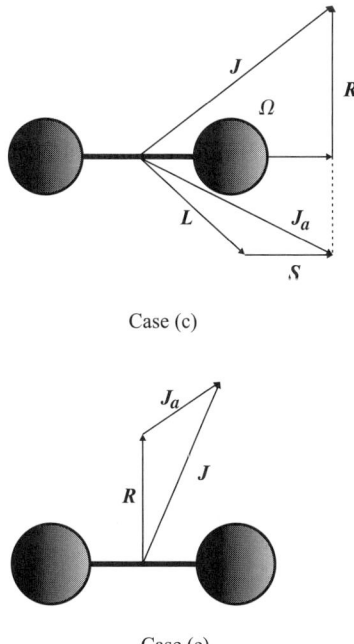

Figure 10.75. Vector coupling diagrams for Hund's case (c) and case (e).

The vibration–rotation levels will be described in terms of the total angular momentum J, the total parity ε, the vibrational quantum number $v(\Omega)$, and $|\Omega|$. In order to describe the total parity of a level we need to work with symmetric and antisymmetric combinations of the $\pm\Omega$ states, so that our basis functions take the form

$$|v(\Omega), J, |\Omega|, \varepsilon\rangle = \frac{1}{\sqrt{2}}\{|v(\Omega), J, +\Omega\rangle \pm |v(\Omega), J, -\Omega\rangle\}. \quad (10.143)$$

The symmetric combination corresponds to e levels and the antisymmetric to f levels [214], the e/f parity differing from the total parity by a factor $(-1)^{J-1/2}$.

(ii) ROTATIONAL HAMILTONIAN AND ROTATIONAL LEVELS IN CASE (c)

The rotational levels in case (c) coupling have been discussed in detail by Veseth [215]; we will follow the usual methods described in this book. The rigid-body rotational Hamiltonian may be written

$$\begin{aligned}\mathcal{H}_{\text{rot}} &= (\hbar^2/2\mu R^2)\boldsymbol{R}^2 = (\hbar^2/2\mu R^2)(\boldsymbol{J}-\boldsymbol{J}_a)^2 \\ &= hcB(R)\bigl(\boldsymbol{J}^2 + \boldsymbol{J}_a^2 - 2\,\mathbf{T}^1(\boldsymbol{J}_a)\cdot\mathbf{T}^1(\boldsymbol{J})\bigr).\end{aligned} \quad (10.144)$$

If we expand the scalar product in the molecule-fixed coordinate system, the diagonal elements ($q=0$) are readily seen to be

$$\langle v(\Omega), J_a; J, \Omega|\mathcal{H}_{\text{rot}}|v(\Omega), J_a; J, \Omega\rangle = B_v\{J(J+1)+J_a(J_a+1)-2\Omega^2\}. \quad (10.145)$$

The constant term $B_v J_a(J_a+1)$ is absorbed into the term value $T_v(\Omega)$ and is not considered to be part of the rotational Hamiltonian. There are matrix elements off-diagonal in v but these are small and are taken into account as centrifugal distortion terms. The result of (10.145) is that the first-order rotational energies of the two electronic states (for $J_a = 3/2$) are given by

$$\begin{aligned}|\Omega| &= 1/2: \quad E(v, J) = B_v[J(J+1)-1/2], \\ |\Omega| &= 3/2: \quad E(v, J) = B_v[J(J+1)-9/2].\end{aligned} \quad (10.146)$$

The off-diagonal matrix elements are as follows:

$$\langle v(\Omega), J_a; J, \Omega| - 2B(R)\sum_{q=\pm 1}\mathrm{T}^1_q(\boldsymbol{J}_a)\mathrm{T}^1_q(\boldsymbol{J})|v(\Omega), J_a; J, \Omega'\rangle$$

$$= -2B_v\sum_{q=\pm 1}\langle J_a, \Omega|\mathrm{T}^1_q(\boldsymbol{J}_a)|J_a, \Omega\rangle\langle J, \Omega|\mathrm{T}^1_q(\boldsymbol{J})|J, \Omega'\rangle$$

$$= -2B_v\sum_{q=\pm 1}(-1)^{J_a-\Omega}\begin{pmatrix}J_a & 1 & J_a \\ -\Omega & q & \Omega'\end{pmatrix}\langle J_a\|\mathrm{T}^1(\boldsymbol{J}_a)\|J_a\rangle(-1)^{J-\Omega}$$

$$\times \begin{pmatrix}J & 1 & J \\ -\Omega & q & \Omega'\end{pmatrix}\langle J\|\mathrm{T}^1(\boldsymbol{J})\|J\rangle$$

$$= -2B_v\sum_{q=\pm 1}(-1)^{J_a-\Omega}\begin{pmatrix}J_a & 1 & J_a \\ -\Omega & q & \Omega'\end{pmatrix}(-1)^{J-\Omega}\begin{pmatrix}J & 1 & J \\ -\Omega & q & \Omega'\end{pmatrix}$$

$$\times \{J_a(J_a+1)(2J_a+1)J(J+1)(2J+1)\}^{1/2}. \quad (10.147)$$

We have made use of the recipe, introduced in chapters 5 and 8, to handle the matrix elements of the total angular momentum in the molecule-fixed axis system. It is important to remember that Ω is a signed quantity. It is now worthwhile expanding the 3-j symbols in (10.147); for $q = +1$ we have

$$\langle v(\Omega), J_a; J, \Omega| - 2B_v T_1^1(\boldsymbol{J}_a) \, T_1^1(\boldsymbol{J})|v(\Omega), J_a; J, \Omega - 1\rangle$$
$$= -2B_v \{(J_a + \Omega)(J_a - \Omega + 1)(J + \Omega)(J - \Omega + 1)\}^{1/2}, \quad (10.148)$$

and for $q = -1$ we have

$$\langle v(\Omega), J_a; J, \Omega| - 2B_v T_{-1}^1(\boldsymbol{J}_a) T_{-1}^1(\boldsymbol{J})|v(\Omega), J_a; J, \Omega + 1\rangle$$
$$= -2B_v\{(J_a - \Omega)(J_a + \Omega + 1)(J - \Omega)(J + \Omega + 1)\}^{1/2}. \quad (10.149)$$

These matrix elements are of two different types. The matrix elements between $\Omega = +1/2$ and $-1/2$ connect states which are otherwise degenerate, making first-order contributions of opposite sign to the energy for the even (e) and odd (f) combinations of $\Omega = \pm 1/2$. These contributions to e and f states are represented in the effective Hamiltonian by diagonal terms which are $\mp P_v(J + 1/2)$ for $|\Omega| = 1/2$ states, and if J_a is a good quantum number, P_v has the value $2B_v$. This is known as the pure precession limit.

The matrix elements connecting $|\Omega| = 1/2$ and $3/2$ states are more complicated, because these states have different sets of vibrational wave functions, and there is no simple expression for the vibrational matrix elements for these highly anharmonic potential functions. These matrix elements are therefore treated as phenomenological spectroscopic parameters, $Q_{v,v'}$, where v and v' refer to the $|\Omega| = 1/2$ and $3/2$ states respectively. The addition of centrifugal distortion constants further complicates the analysis [211].

(iii) ZEEMAN EFFECT IN HUND'S CASE (c)

We have already shown the importance of the Zeeman effect, both in identifying the J quantum numbers involved in each line, and in providing effective g-factors for the levels. These g-factors serve as additional labels for each level, and provide information concerning the best angular momentum coupling scheme. We now develop the theory of the Zeeman effect in Hund's case (c).

The effective Hamiltonian describing the interaction between an applied magnetic field and the magnetic moments due to electron spin and orbital motion is, as we have seen elsewhere,

$$\mathcal{H}_Z = g_S \mu_B \, T^1(\boldsymbol{B}) \cdot T^1(\boldsymbol{S}) + g_L \mu_B \, T^1(\boldsymbol{B}) \cdot T^1(\boldsymbol{L}). \quad (10.150)$$

In a space-fixed coordinate system we define the direction of the magnetic field to be the $p = 0$ direction. However the spin $\boldsymbol{S}$ and orbital $\boldsymbol{L}$ angular momenta are quantised in the molecule-fixed axis system. Consequently equation (10.150) is expanded as follows:

$$\mathcal{H}_Z = g_S \mu_B \, T_0^1(\boldsymbol{B}) \sum_q \mathcal{D}_{0q}^{(1)}(\omega)^* T_q^1(\boldsymbol{S}) + g_L \mu_B \, T_0^1(\boldsymbol{B}) \sum_q \mathcal{D}_{0q}^{(1)}(\omega)^* T_q^1(\boldsymbol{L}). \quad (10.151)$$

For the electron spin term the matrix elements in a case (c) basis are

$$\langle v(\Omega), J_a; J, \Omega, M_J | g_S \mu_B T_0^1(\boldsymbol{B}) \sum_q \mathcal{D}_{0q}^{(1)}(\omega)^* T_q^1(\boldsymbol{S}) | v'(\Omega'), J_a; J', \Omega', M_J \rangle$$

$$= g_S \mu_B B_Z \sum_q \langle J, \Omega, M_J | \mathcal{D}_{0q}^{(1)}(\omega)^* | J', \Omega', M_J \rangle \langle J_a, \Omega | T_q^1(\boldsymbol{S}) | J_a, \Omega' \rangle$$

$$= g_S \mu_B B_Z \sum_q (-1)^{J-M_J} (-1)^{J-\Omega} \begin{pmatrix} J & 1 & J' \\ -M_J & 0 & M_J \end{pmatrix} \begin{pmatrix} J & 1 & J' \\ -\Omega & q & \Omega' \end{pmatrix}$$

$$\times [(2J+1)(2J'+1)]^{1/2} \langle J_a, \Omega | T_q^1(\boldsymbol{S}) | J_a, \Omega' \rangle. \tag{10.152}$$

The remaining matrix element in (10.152) is evaluated in the following manner:

$$\langle J_a, \Omega | T_q^1(\boldsymbol{S}) | J_a, \Omega' \rangle = \langle L, S, J_a, \Omega | T_q^1(\boldsymbol{S}) | L, S, J_a, \Omega' \rangle$$

$$= (-1)^{J_a - \Omega} \begin{pmatrix} J_a & 1 & J_a \\ -\Omega & q & \Omega' \end{pmatrix} \langle L, S, J_a \| T(\boldsymbol{S}) \| L, S, J_a \rangle$$

$$= (-1)^{J_a - \Omega} \begin{pmatrix} J_a & 1 & J_a \\ -\Omega & q & \Omega' \end{pmatrix} (-1)^{J_a + L + S + 1} \begin{Bmatrix} J_a & S & L \\ S & J_a & 1 \end{Bmatrix}$$

$$\times \{(2J_a + 1)(2J_a' + 1) S(S+1)(2S+1)\}^{1/2}. \tag{10.153}$$

If we combine equations (10.152) and (10.153), substitute the values $L = 1$, $S = 1/2$ and neglect the small matrix elements off-diagonal in J we obtain the results

$$\langle v(\Omega), J_a; J, \Omega, M_J | \mathcal{H}_Z(\text{spin}) | v(\Omega), J_a; J, \Omega, M_J \rangle$$

$$= \frac{g_S \mu_B B_Z M_J \Omega^2 [J_a(J_a+1) - 5/4] \delta'_{vv'}}{2J(J+1) J_a(J_a+1)}. \tag{10.154}$$

$$\langle v(\Omega), J_a; J, \Omega, M_J | \mathcal{H}_Z(\text{spin}) | v(\Omega \pm 1), J_a; J, \Omega \pm 1, M_J \rangle$$

$$= \frac{5 g_S \mu_B B_Z M_J [(J \mp \Omega)(J \pm \Omega + 1)(J_a \mp \Omega)(J_a \pm \Omega + 1)]^{1/2} \langle v(\Omega) | v'(\Omega \pm 1) \rangle}{8J(J+1) J_a(J_a+1)}. \tag{10.155}$$

If the above analysis is repeated for the orbital term in (10.151) we find an almost identical result, except that g_S is replaced by $2g_L$ (because $S = 1/2$ and $L = 1$). Consequently for pure case (c) states we obtain the following simple results for the energies E_Z,

$$E_Z = g_{\text{eff}} \mu_B B_Z M_J, \tag{10.156}$$

where the effective g-factors are given by:

$$|\Omega| = \frac{1}{2}, \; e \text{ states}: \quad g_{\text{eff}} = \frac{+(g_S + 2g_L)(4J+3)}{12J(J+1)},$$

$$|\Omega| = \frac{1}{2}, \; f \text{ states}: \quad g_{\text{eff}} = \frac{-(g_S + 2g_L)(4J+1)}{12J(J+1)}, \tag{10.157}$$

$$|\Omega| = \frac{3}{2}, \; e, f \text{ states}: \quad g_{\text{eff}} = \frac{+3(g_S + 2g_L)}{4J(J+1)}.$$

Many of the observed levels have measured g-factors which are closer to the pure case (c) values than to any alternative pure coupling case. However there is extensive rotational–electronic coupling which, in many instances, mixes the case (c) states; case (e) is then a better limiting basis, as we shall see in due course. First we investigate the electric dipole transition probabilities for the Zeeman components, so that we can understand the pattern of lines illustrated in figure 10.73.

(iv) ELECTRIC DIPOLE TRANSITION PROBABILITIES

The perturbation due to the microwave electric field $\boldsymbol{E}(t)$ is given by the operator

$$\mathcal{H}_{E(t)} = -\mathrm{T}^1(\boldsymbol{\mu}_e) \cdot \mathrm{T}^1[\boldsymbol{E}(t)], \tag{10.158}$$

where $\boldsymbol{\mu}_e$ is the electric dipole moment. The components of $\boldsymbol{\mu}_e$ are most naturally expressed in the molecule-fixed coordinate system, whilst those of $\boldsymbol{E}(t)$ are described in the space-fixed system. Consequently we expand (10.158) in the form

$$\mathcal{H}_E = -\sum_p (-1)^p \sum_q \mathcal{D}_{pq}^{(1)}(\omega)^* \mathrm{T}_q^1(\boldsymbol{\mu}_e) \mathrm{T}_{-p}^1[\boldsymbol{E}(t)], \tag{10.159}$$

For a molecular ion the electric dipole moment, measured relative to the molecular centre of mass, is determined primarily by the separation of charge and mass [216], and since it lies along the internuclear axis, we are interested in the $q = 0$ component of (10.159); consequently

$$\mathcal{H}_E = -\sum_p (-1)^p \mathcal{D}_{p0}^{(1)}(\omega)^* \mathrm{T}_0^1(\boldsymbol{\mu}_e) \mathrm{T}_{-p}^1[\boldsymbol{E}(t)]. \tag{10.160}$$

The matrix elements are therefore given by

$$\langle J, \Omega, M_J | \mathcal{H}_E | J', \Omega', M_J' \rangle$$
$$= -\mu_0 \sum_p (-1)^p E_{-p}(t) \langle J, \Omega, M_J | \mathcal{D}_{p0}^{(1)}(\omega)^* | J', \Omega, M_J' \rangle$$
$$= -\mu_0 \sum_p (-1)^p E_{-p}(t) (-1)^{J-M_J} (-1)^{J-\Omega}$$
$$\times \begin{pmatrix} J & 1 & J' \\ -M_J & p & M_J' \end{pmatrix} \begin{pmatrix} J & 1 & J' \\ -\Omega & 0 & \Omega \end{pmatrix} \{(2J+1)(2J'+1)\}^{1/2}, \tag{10.161}$$

and the transition intensities are proportional to the square of this matrix element.

In the microwave ion beam experiments described in this section, it is important to identify the microwave mode corresponding to the resonance line studied in a magnetic field. For a TM mode the microwave electric field along the central axis of the waveguide is parallel to the static magnetic field. We then put $p = 0$ in equation (10.161) so that the Zeeman components obey the selection rule $\Delta M_J = 0$. Alternatively in a TE mode the microwave electric field is perpendicular to the static magnetic field and the selection rule is $\Delta M_J = \pm 1$. This is the case for the Zeeman pattern shown in figure 10.73; each $J = 3/2$ level splits into four M_J components and the six allowed transitions should, according to (10.161), have relative intensities 3:4:3:3:4:3, in good agreement with experiment.

We have, we hope, provided enough detail about the Zeeman effect to show how almost every microwave resonance could be assigned, so far as the J values were concerned. A final remark should be made concerning the parity labels. These depend upon the identification of a $J = 1/2 \leftarrow 1/2$ transition, and the measured g-factors for the two $J = 1/2$ levels which identify their e/f, and hence total parities. The parities of all other levels then follow because all transitions are electric-dipole allowed, between states of opposite parity. As we have mentioned earlier, the combination of numerical relationships between the resonance frequencies, double resonance studies, and Zeeman effect measurements enabled the pattern of levels lying within 8 cm^{-1} of the dissociation limit to be established. The highest level, $J = 7/2$ (−parity), in figure 10.74, was thought to lie within 20 MHz (<0.001 cm^{-1}) of the dissociation limit.

The next stage was to convert the level pattern shown in figure 10.74 to a more conventional pattern, including electronic and vibrational quantum numbers. This required an accurate theory, the details of which we now outline.

(c) Theoretical interpretation of the spectrum

(i) INTRODUCTION

As we have seen many times throughout this book, the conventional route to the assignment and quantitative analysis of a molecular spectrum is the correct definition of an effective Hamiltonian, the calculation of its eigenvalues and eigenvectors, and the determination of the values of the molecular parameters appearing in the effective Hamiltonian. For HeAr$^+$ in its near dissociation levels, this process is of limited value because of very strong rotational–electronic coupling. In other words, it is not possible to find an effective Hamiltonian which is convergent. Consequently we bypass the effective Hamiltonian and proceed directly to a quantitative quantum mechanical theory.

(ii) COUPLED-CHANNEL THEORY

Carrington, Leach, Marr, Shaw, Viant, Hutson and Law [211] used a coupled-channel theory, described by Hutson [217], in order to solve the Schrödinger equation, written in the form

$$\left[-\frac{\hbar^2}{2\mu} R^{-1} \frac{d^2}{dR^2} R + \frac{\hbar^2 \boldsymbol{R}^2}{2\mu R^2} + V(R, \boldsymbol{r}_a) + \mathcal{H}_{so}(R) - E \right] \boldsymbol{\Psi}_\alpha^{\varepsilon J M_J} = 0. \quad (10.162)$$

R is the internuclear distance and μ is the reduced mass, so that the first term represents the vibrational motion of the nuclei. $\boldsymbol{R}$ is the angular momentum operator for rotation of the nuclear framework. The interaction potential for the He...Ar$^+$ system, $V(R, \boldsymbol{r}_a)$, is a function of the internuclear distance R and the electron coordinates $\boldsymbol{r}_a$; we will discuss the details in due course. The problem was set up in a Hund's case (e) basis (see figure 10.75) and the only rigorously good quantum numbers are the parity (ε), the total angular momentum J, and its space-fixed projection M_J. We shall use the index α to denote all other quantum numbers.

Case (e) basis functions are formed by combining atomic wave functions, denoted $\psi^{\text{atom}}_{LSJ_aM_a}$, with spherical harmonics $Y_{RM_R}(\omega)$ for the rotation of the nuclear framework. M_a and M_R are the projections of $\boldsymbol{J_a}$ and $\boldsymbol{R}$ onto the space-fixed Z axis. Consequently we may write for the case (e) basis functions

$$\psi^{JM_J}_{LSJ_aR} = \sum_{M_aM_R} (-1)^{J_a-R+M_J} (2J+1)^{1/2} \begin{pmatrix} J_a & R & J \\ M_a & M_R & -M_J \end{pmatrix} \psi^{\text{atom}}_{LSJ_aM_a} Y_{RM_R}(\omega). \tag{10.163}$$

The basis functions have parity $\varepsilon = (-1)^{R+1}$, which remains a good quantum number; the $+1$ in the exponent arises because the atom wave function is primarily $3p$ in character. For $J \geq 3/2$ there are three case (e) basis functions of each parity, as follows.

$$\begin{aligned}
R = 0, (\varepsilon = -1), \quad & J_a = 3/2, \quad J = 3/2 \\
& J_a = 1/2, \quad J = 1/2 \\
R = 1, (\varepsilon = +1), \quad & J_a = 3/2, \quad J = 5/2, 3/2, 1/2 \\
& J_a = 1/2, \quad J = 3/2, 1/2 \\
R = 2, (\varepsilon = -1), \quad & J_a = 3/2, \quad J = 7/2, 5/2, 3/2, 1/2 \\
& J_a = 1/2, \quad J = 5/2, 3/2 \\
R = 3, (\varepsilon = +1), \quad & J_a = 3/2, \quad J = 9/2, 7/2, 5/2, 3/2 \\
& J_a = 1/2, \quad J = 7/2, 5/2
\end{aligned}$$

The total wave function for the HeAr$^+$ molecular ion is now expanded as a product of the case (e) functions given by (10.163) and radial channel functions $\chi^{\varepsilon J}_{\alpha;LSJ_aR}(R)$,

$$\Psi^{\varepsilon JM_J}_\alpha = R^{-1} \sum_{LSJ_aR} \psi^{JM_J}_{LSJ_aR} \chi^{\varepsilon J}_{\alpha;LSJ_aR}(R). \tag{10.164}$$

The expansion (10.164) is truncated to include only the three electronic states that correlate with the $^2P_{3/2}$ and $^2P_{1/2}$ states of the Ar$^+$ ion shown in figure 10.71.

If we substitute the right-hand side of (10.164) into the total Schrödinger equation (10.162) and project onto a single case (e) basis function $\psi^{JM_J}_{LSJ_aR'}$, we obtain a set of coupled differential equations for the channel wave functions $\chi^{\varepsilon J}_{\alpha;LSJ_aR}(R)$. As we have seen above, for each J and ε there are three coupled equations corresponding to the allowed values of J_a and R. These equations may be written in matrix notation,

$$\frac{\hbar^2}{2\mu}\frac{d^2\chi}{dR^2} = [W(R) - E\mathbf{I}]\chi(R). \tag{10.165}$$

$\chi(R)$ is a column vector with three elements $\chi^{\varepsilon J}_{\alpha;LSJ_aR}(R)$ for the different allowed values of J_a and R. $\mathbf{I}$ is the 3×3 unit matrix, and $W(R)$ is a 3×3 coupling matrix whose elements we shall now investigate.

(iii) COUPLING MATRIX

We see from the starting Schrödinger equation (10.162) that the coupling matrix elements have the form

$$\langle L, S, J_a, R, J | W(R) | L, S, J'_a, R', J \rangle$$
$$= \langle L, S, J_a, R, J | \frac{\hbar^2 \boldsymbol{R}^2}{2\mu} + V(R, \theta_a) + \mathcal{H}_{\text{so}} | L, S, J'_a, R', J \rangle. \quad (10.166)$$

The only change is that $V(R, r_a)$ has been replaced by $V(R, \theta_a)$, where θ_a is the angle describing the p-electron hole of the Ar^+ ion, relative to the internuclear vector. Let us suppose we are seeking the coupling matrix for $J = 3/2$, with positive parity. We see from the results presented earlier that the three relevant states are

$$R = 1, J_a = 3/2, \quad R = 1, J_a = 1/2, \quad R = 3, J_a = 3/2.$$

First, the operator $\boldsymbol{R}^2$ is diagonal in the case (e) basis set:

$$\langle L, S, J_a, R, J | \boldsymbol{R}^2 | L, S, J'_a, R', J \rangle = R(R+1)\delta_{J_a J'_a}\delta_{RR'}. \quad (10.167)$$

Second, it is a good approximation for an ion like HeAr^+ to assume that the spin–orbit coupling operator is the same as that for the free Ar^+ ion, $\zeta \, \mathbf{L} \cdot \mathbf{S}$, where ζ is the atomic spin–orbit coupling constant. If the basis functions are confined to those arising from the $^2P_{3/2}$ and $^2P_{1/2}$ states, the spin–orbit operator is also diagonal in a case (e) basis set:

$$\langle L, S, J_a, R, J | \mathcal{H}_{\text{so}} | L, S, J'_a, R', J \rangle$$
$$= \langle L, S, J_a, R, J | \zeta \mathbf{L} \cdot \mathbf{S} | L, S, J'_a, R', J \rangle$$
$$= \frac{\zeta}{2}[J_a(J_a + 1) - L(L+1) - S(S+1)]\delta_{J_a J'_a}\delta_{RR'}. \quad (10.168)$$

The matrix elements of the interaction potential are more complicated. It is common practice to express the potential as a Legendre polynomial:

$$V(R, \theta_a) = V_0 + V_2 P_2(\cos\theta_a). \quad (10.169)$$

The matrix elements of the interaction potential in a case (e) basis are then given by

$$\langle L, S, J_a, R, J | V(R, \theta_a) | L, S, J'_a, R', J \rangle$$
$$= \sum_{\lambda = 0,2} V_\lambda(R)(-1)^{J-S+\lambda}(2L+1)[(2R+1)(2R'+1)(2J_a+1)(2J'_a+1)]^{1/2}$$
$$\times \begin{pmatrix} L & \lambda & L \\ 0 & 0 & 0 \end{pmatrix} \begin{pmatrix} R & \lambda & R' \\ 0 & 0 & 0 \end{pmatrix} \begin{Bmatrix} J_a & L & S \\ L & J'_a & \lambda \end{Bmatrix} \begin{Bmatrix} J_a & R & J \\ R' & J'_a & \lambda \end{Bmatrix}. \quad (10.170)$$

It is now a straightforward process to calculate the diagonal and off-diagonal matrix elements. The interaction matrix for the three $J = 3/2$ positive parity states described above is as follows.

	$\|R=1, J_a=3/2\rangle$	$\|R=1, J_a=1/2\rangle$	$\|R=3, J_a=3/2\rangle$
$\langle R=1, J_a=3/2\|$	$\dfrac{h^2}{\mu} + \dfrac{\zeta}{2} + V_0 - \dfrac{4V_2}{25}$	$-\dfrac{V_2}{5\sqrt{5}}$	$-\dfrac{3V_2}{25}$
$\langle R=1, J_a=1/2\|$	$-\dfrac{V_2}{5\sqrt{5}}$	$\dfrac{h^2}{\mu} - \zeta + V_0$	$\dfrac{3V_2}{5\sqrt{5}}$
$\langle R=3, J_a=3/2\|$	$-\dfrac{3V_2}{25}$	$\dfrac{3V_2}{5\sqrt{5}}$	$\dfrac{6h^2}{\mu} + \dfrac{\zeta}{2} + V_0 + \dfrac{4V_2}{25}$

We will come to the form of the interaction potential employed in the next section; given the appropriate form, the coupled equations (10.165) were solved by a numerical method due to Johnson [218] and described by Hutson [217]. For most states the coupled equations were solved for internuclear distances up to 25 Å, and the energies of the levels converged to ±2 MHz. A computer program for interactive nonlinear least squares fitting of the parameters to the observed transition frequencies has been described by Law and Hutson [219].

(iv) INTERACTION POTENTIAL

The functional form used for the interaction potential was as follows:

$$V_0(R) = A_0(1-a)\exp[-\beta_0(R-R_e)] + A_0 a \exp[-(\beta_0/2)(R-R_e)]$$
$$- \sum_{n=4,6,8,10} [C_n]_0 D_n(R)/R^n, \qquad (10.171)$$

$$V_2(R) = A_2 \exp[-\beta_2(R-R_e)] - [C_6]_2 D_6(R)/R^6.$$

The functions $D_n(R)$ in these expressions are Tang–Toennies damping functions [220], that prevent the inverse power terms from dominating at short range; they are given by

$$D_n(R) = 1 - \exp(-\beta R) \sum_{m=0}^{n} \frac{(\beta R)^m}{m!}, \qquad (10.172)$$

with β set to β_0 or β_2 for $V_0(R)$ or $V_2(R)$ respectively. Equation (10.171) contains eleven constants whose values are required. The expression for $V_0(R)$ contains Morse-type terms for the short and intermediate ranges, with four long-range terms. At long range the most important interaction is the charge-induced dipole term, $[C_4]_0$, which arises from the interaction between the charge on Ar$^+$ and the dipole moment that it induces on the He atom. It is determined by the dipole polarisability of the He atom, and is already known very accurately. The coefficient $[C_6]_0$ has contributions from both dispersion and the quadrupole polarisability of the He atom and has been calculated by Siska [213]. A_0 in the expression for $V_0(R)$ was replaced by the well depth. We note also that the anisotropy of the C_6 term, $[C_6]_2$, was included in the fit. The final aim was both to assign the microwave spectrum, and to match the values of the constants determined from the ultraviolet emission spectrum. The most novel aspect of the work was undoubtedly the accuracy with which the long-range part of the interaction potential could be determined. In due course we present the final near-dissociation energy level diagram, and a sample of the quantitative agreement between experiment and theory. Before that, however, we examine the theory of the Zeeman effect in case (e).

(v) ZEEMAN EFFECT IN CASE (e)

We recall from equation (10.150) that if the direction of the external magnetic field is defined to be $p = 0$, the primary Zeeman Hamiltonian is

$$\mathcal{H}_Z = \mu_B B_0 \left[g_S T_0^1(\boldsymbol{S}) + g_L T_0^1(\boldsymbol{L}) \right]. \tag{10.173}$$

In contrast to Hund's case (c), there are no molecule-fixed projection quantum numbers in case (e), so rotational matrices are not involved this time. The matrix elements of the electron spin term are:

$$\langle L, S, J_a, R, J, M_J | T_0^1(\boldsymbol{S}) | L, S, J_a', R, J, M_J \rangle$$

$$= (-1)^{J-M_J} \begin{pmatrix} J & 1 & J \\ -M_J & 0 & M_J \end{pmatrix} \langle J_a, R, J \| T^1(\boldsymbol{S}) \| J_a', R, J \rangle$$

$$= (-1)^{J-M_J} \begin{pmatrix} J & 1 & J \\ -M_J & 0 & M_J \end{pmatrix} (-1)^{J+J_a+R+1} (2J+1) \begin{Bmatrix} J & J_a & R \\ J_a' & J & 1 \end{Bmatrix}$$

$$\times \langle L, S, J_a \| T^1(\boldsymbol{S}) \| L, S, J_a' \rangle$$

$$= (-1)^{J-M_J} \begin{pmatrix} J & 1 & J \\ -M_J & 0 & M_J \end{pmatrix} (-1)^{J+J_a+R+1} (2J+1) \begin{Bmatrix} J & J_a & R \\ J_a' & J & 1 \end{Bmatrix}$$

$$\times (-1)^{J_a+L+1+S} \{(2J_a+1)(2J_a'+1)\}^{1/2} \begin{Bmatrix} J_a & S & L \\ S & J_a' & 1 \end{Bmatrix}$$

$$\times \{S(S+1)(2S+1)\}^{1/2}. \tag{10.174}$$

We have neglected the very small terms off-diagonal in J (because B_Z is so small). In a similar manner, the matrix elements of the orbital term are

$$\langle L, S, J_a, R, J, M_J | T_0^1(\boldsymbol{L}) | L, S, J_a', R, J, M_J \rangle$$

$$= (-1)^{J-M_J} \begin{pmatrix} J & 1 & J \\ -M_J & 0 & M_J \end{pmatrix} (-1)^{J+J_a+R+1} (2J+1) \begin{Bmatrix} J & J_a & R \\ J_a' & J & 1 \end{Bmatrix}$$

$$\times (-1)^{J_a'+L+1+S} \{(2J_a+1)(2J_a'+1)\}^{1/2} \begin{Bmatrix} J_a & L & S \\ L & J_a' & 1 \end{Bmatrix}$$

$$\times \{L(L+1)(2L+1)\}^{1/2}. \tag{10.175}$$

Pure case (e) effective g-factors are readily calculated from these expressions, and effective g-factors for the final wave functions, expressed as linear combinations of case (e) functions, are also easy to calculate.

(vi) RESULTS

Apart from a quantitative assignment of the microwave transitions, the theoretical analysis outlined enables vibrational and electronic quantum numbers to be assigned to the energy levels, as well as the J and parity labels previously known. As a result the energy level diagram shown earlier in figure 10.74 can be rearranged as a more conventional rovibronic diagram, shown in figure 10.76. The columns are labelled with case (c) electronic quantum numbers, and we have shown a selection of the observed rotational transitions ($\Delta J = \pm 1$) as continuous vertical lines. Also shown as continuous vertical lines are Ω-doubling transitions. In addition, many of the energy

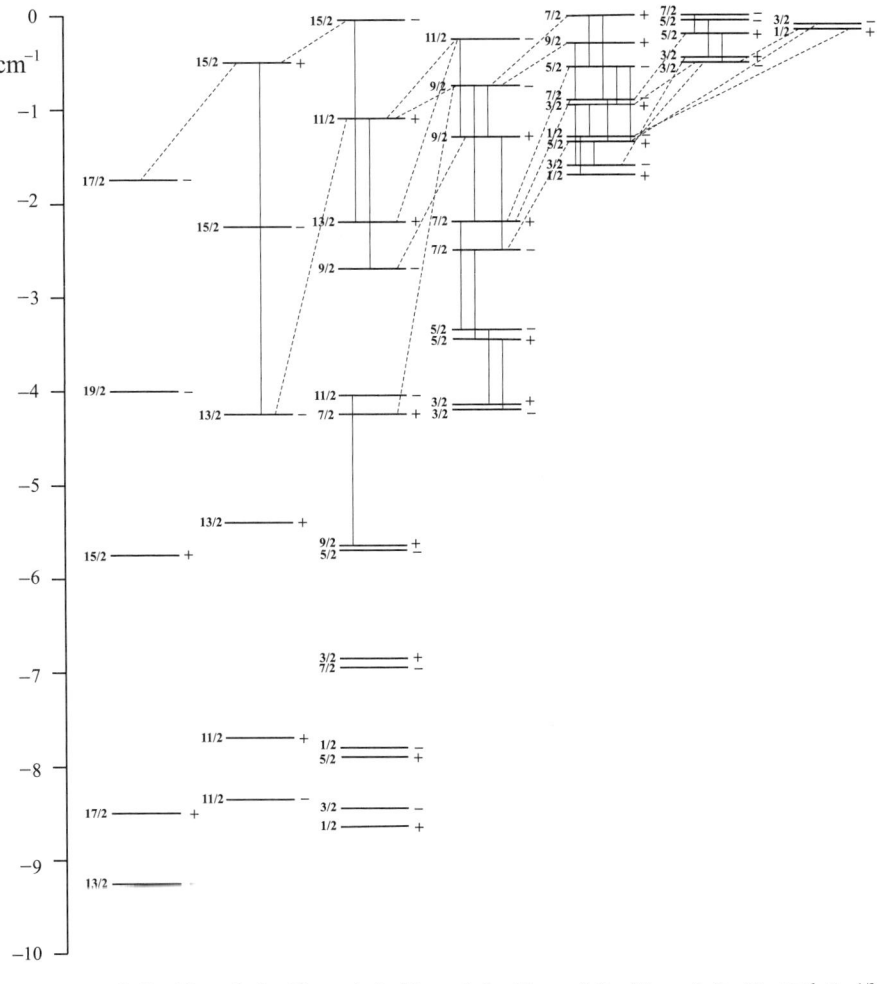

Figure 10.76. Energy level diagram for HeAr$^+$ rearranged according to vibronic quantum numbers in a case (c) basis. A few vibration–rotation levels predicted by theory but not observed experimentally are also included for the sake of completeness [211].

Table 10.23. *Comparison of observed and calculated transition frequencies (in MHz) for 36 transitions involving 37 identified energy levels*

Upper state				Lower state				Observed	Calculated	Difference
v	Ω	J	ε	v	Ω	J	ε	MHz	MHz	MHz
6	1/2	1/2	+	5	1/2	1/2	−	36 599.3	36 601.6	−2.3
6	1/2	3/2	−	5	1/2	5/2	+	38 312.6	38 311.6	1.0
5	1/2	1/2	−	5	1/2	1/2	+	11 795.7	11 797.1	−1.4
5	1/2	3/2	+	5	1/2	1/2	−	11 795.2	11 798.9	−3.7
4	3/2	5/2	−	4	3/2	3/2	+	11 919.9	11 918.3	1.6
4	3/2	3/2	+	3	3/2	5/2	−	86 573.9	86 551.5	22.4
4	3/2	3/2	+	5	1/2	3/2	−	35 092.7	35 094.9	−2.2
4	3/2	3/2	−	5	1/2	3/2	+	12 548.6	12 548.8	−0.2
4	3/2	3/2	−	3	3/2	3/2	+	110 075.9	110 054.3	21.6
5	1/2	5/2	−	5	1/2	3/2	+	12 682.4	12 686.3	−3.9
5	1/2	3/2	+	5	1/2	3/2	−	21 200.4	21 202.7	−2.3
5	1/2	3/2	+	4	1/2	5/2	−	142 637.8	142 666.0	−28.2
3	3/2	5/2	+	3	3/2	3/2	−	22 520.3	22 519.0	1.3
4	3/2	7/2	−	4	3/2	5/2	+	6 636.0	6 637.0	−1.0
4	3/2	5/2	+	3	3/2	7/2	−	66 961.2	66 943.4	17.8
4	3/2	5/2	+	5	1/2	5/2	−	8 122.6	8 115.9	6.7
5	1/2	5/2	−	5	1/2	5/2	+	25 693.3	25 696.3	−3.0
5	1/2	7/2	−	5	1/2	5/2	+	13 542.0	13 547.7	−5.7
5	1/2	5/2	+	4	1/2	7/2	−	167 524.7	167 570.8	−46.1
3	3/2	7/2	−	3	3/2	5/2	+	29 568.3	29 563.6	4.7
5	1/2	7/2	+	4	1/2	9/2	−	78 251.9	78 244.8	7.1
5	1/2	9/2	+	5	1/2	7/2	−	18 019.6	18 029.2	−9.6
3	3/2	9/2	−	4	1/2	7/2	+	104 397.4	104 427.9	−30.5
3	3/2	9/2	−	3	3/2	7/2	+	42 663.1	42 662.6	0.5
5	1/2	9/2	+	4	1/2	9/2	−	71 891.7	71 896.1	−4.4
3	3/2	9/2	−	3	3/2	9/2	+	16 712.4	16 728.9	−16.5
3	3/2	9/2	+	4	1/2	9/2	−	41 221.5	41 215.0	6.5
5	1/2	7/2	−	4	1/2	9/2	+	142 454.5	142 487.8	−33.3
4	1/2	11/2	+	4	1/2	9/2	−	47 583.2	47 599.2	−16.0
3	3/2	11/2	−	4	1/2	11/2	+	26 241.4	26 205.6	35.8
3	3/2	9/2	+	4	1/2	11/2	−	81 915.8	81 921.5	−5.7
3	3/2	11/2	−	4	1/2	13/2	+	58 934.2	58 890.3	43.9
4	1/2	15/2	−	4	1/2	13/2	+	61 500.6	61 546.5	−45.9
4	1/2	15/2	−	2	3/2	15/2	+	9 807.0	9 808.9	−1.9
4	1/2	11/2	+	2	3/2	13/2	−	94 102.5	94 096.8	5.7
2	3/2	15/2	+	3	1/2	17/2	−	38 463.7	38 621.5	−157.8

Table 10.24. *Comparison of observed and calculated g-factors for the near-dissociation energy levels in HeAr$^+$*

v	Ω	J	ε	observed g	calculated g	difference
5	1/2	1/2	−	−1.357	−1.333	−0.024
5	1/2	3/2	+	−0.150	−0.150	0.000
5	1/2	3/2	−	0.915	0.892	0.023
5	1/2	5/2	−	0.216	0.210	0.006
5	1/2	5/2	+	0.576	0.566	0.010
5	1/2	7/2	+	0.252	0.251	0.001
5	1/2	7/2	−	0.405	0.399	0.006
5	1/2	9/2	+	0.276	0.275	0.001
4	3/2	3/2	+	0.388	0.378	0.010
4	3/2	3/2	−	0.812	0.787	0.025
4	3/2	5/2	+	0.465	0.455	0.010
4	3/2	7/2	−	0.539	0.530	0.009
4	1/2	5/2	−	−0.207	−0.211	0.004
4	1/2	11/2	+	0.094	0.097	−0.003
4	1/2	13/2	+	0.216	0.216	0.000
4	1/2	15/2	−	0.152	0.154	−0.002
3	3/2	5/2	−	0.156	0.158	−0.002
3	3/2	5/2	+	0.323	0.320	0.003
3	3/2	7/2	+	−0.046	−0.046	0.000
3	3/2	7/2	−	0.198	0.196	0.002
3	3/2	9/2	−	−0.132	−0.131	−0.001
3	3/2	9/2	+	0.188	0.186	0.002
3	3/2	11/2	−	0.262	0.262	0.000
2	3/2	15/2	+	−0.126	−0.129	0.003

levels were located by vibronic transitions, and a selection of these are indicated by the sloping dashed lines in figure 10.76. It is tempting to categorise these transitions as being either vibrational or electronic, but one should remember that the case (c) electronic quantum numbers are only approximate.

Demonstrations of the quality of the fit of the theoretical model to experiment are presented in tables 10.23 and 10.24. Table 10.23 presents observed and calculated transition frequencies for 36 different transitions, which specify the relative energies of the 37 observed levels; the agreement is generally excellent. Table 10.24 gives a similar comparison of the observed and calculated g-factors; in a few cases only the *difference* between the g-factors for the pair of levels involved in a transition could be determined experimentally. We have omitted these cases from the table, but the results may be found in the original paper. The results presented show excellent agreement between experiment and theory.

There are several features of the study of HeAr$^+$ described here which are unusual. Microwave spectroscopy is normally associated with rotational transitions, which are, indeed, observed. It is not usually associated with vibrational and electronic transitions, which are also observed because the near-dissociation levels are so closely spaced. Another important feature is that the conventional approach of spectral analysis using an effective Hamiltonian is of limited value, and the comparison of experiment and theory described above bypasses the effective Hamiltonian. The scattering-type theory employed does not involve the Born–Oppenheimer separation. Another novelty is the observation of case (e) coupling, but we have not elaborated on this aspect here because our next example illustrates case (e) coupling even more clearly.

10.7.2. Studies of the HeKr$^+$ ion

Studies of the near-dissociation microwave spectrum of HeKr$^+$ similar to those just described for HeAr$^+$ have been carried out by Carrington, Pyne, Shaw, Taylor, Hutson and Law [221]. There were, however, some important differences. Twenty-five transitions were observed for He^{84}Kr$^+$ and eleven for He^{86}Kr$^+$; all of the energy levels involved lie within 2.5 cm^{-1} of the dissociation limit. Electric field dissociation for this relatively heavy ion does not affect levels with dissociation energies greater than 1.5 cm^{-1}. The pattern of energy levels and the selection rules governing the transitions were very well understood in terms of case (e) coupling as originally described, not by Hund, but by Mulliken [222]. In this coupling case there are no projection quantum numbers, and we are concerned only with the rotational quantum number R, the total angular momentum J, and the parity ε. The observed transitions all satisfied the selection rules $\Delta J = 0, \pm 1, \Delta R = \pm 1$. Figure 10.77 shows the pattern of energy levels involving the $v = 4$ level of the X state. Since the vector coupling scheme in case (e) is

$$\boldsymbol{J} = \boldsymbol{J}_a + \boldsymbol{R}, \qquad (10.176)$$

the lowest rotational levels, with their parities in brackets, are as follows:

$R = 0$ (−): $J = 3/2$,
$R = 1$ (+): $J = 5/2, 3/2, 1/2$,
$R = 2$ (−): $J = 7/2, 5/2, 3/2, 1/2$,
$R = 3$ (+): $J = 9/2, 7/2, 5/2, 3/2$,

etc. The observed and final calculated transition frequencies are listed in table 10.25, and a selection of g-factors is shown in table 10.26, where we give both the pure case (e), and final calculated values for comparison. The existence of this very simple coupling scheme was unambiguously established, seemingly for the first time. As with HeAr$^+$, the long-range part of the interaction potential was defined accurately and in considerable detail. It should be added that a few transitions involving levels lying very close to the second dissociation limit were also identified and accurately assigned.

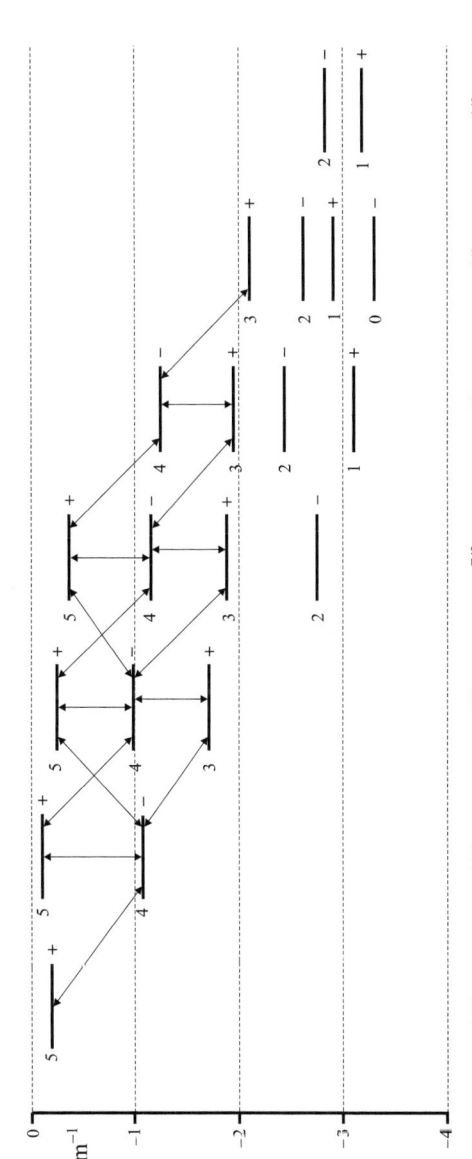

Figure 10.77. Energy level diagram for $v = 4$ of the X state at the lowest dissociation limit for He^{84}Kr$^+$, and the observed transitions [221]. The level stacks are arranged according to their J values, and individual levels are labelled with their parities and R values in case (e). The lowest levels shown were not observed experimentally, but their positions were calculated.

Table 10.25. *Comparison of observed and calculated transition frequencies (in MHz) for He ^{84}Kr$^+$ in the v = 4 level of the X state*

upper			lower			observed MHz	calculated MHz	difference MHz
J	R	ε	J	R	ε			
7/2	4	−	7/2	3	+	20 263.7	20 244.8	18.9
7/2	5	+	9/2	4	−	20 756.8	20 726.4	30.4
5/2	4	−	5/2	3	+	21 133.3	21 173.6	−40.3
9/2	5	+	9/2	4	−	23 108.4	23 088.5	19.9
9/2	4	−	7/2	3	+	24 404.6	24 430.3	−25.7
7/2	4	−	5/2	3	+	24 503.6	24 522.5	−18.9
7/2	5	+	7/2	4	−	24 896.7	24 911.8	−15.1
11/2	4	−	9/2	3	+	25 119.3	25 111.4	7.9
5/2	4	−	3/2	3	+	25 350.5	25 330.5	20.0
9/2	5	+	11/2	4	−	26 379.1	26 367.0	12.1
11/2	5	+	9/2	4	−	26 634.7	26 655.3	−20.6
9/2	5	+	7/2	4	−	27 248.6	27 273.9	−25.3
13/2	5	+	11/2	4	−	27 583.5	27 583.7	−0.24
7/2	5	+	5/2	4	−	28 267.9	28 260.8	7.1
9/2	4	−	9/2	3	+	28 389.4	28 389.9	−0.5
11/2	5	+	11/2	4	−	29 905.1	29 933.8	−28.7

Table 10.26. *Observed and calculated g-factors for a selection of near-dissociation levels of He^{84}Kr$^+$ in the v = 4 level of the X state. The calculated values are obtained from the detailed theory described above, but pure case (e) values are also listed*

J	R	ε	Obs. g	Calc. g	Diff.	case(e) g
7/2	4	−	−0.012	−0.010	−0.002	−0.021
5/2	4	−	−0.584	−0.564	−0.020	−0.572
9/2	5	+	−0.027	−0.027	0.000	−0.040
9/2	4	−	0.228	0.225	0.003	0.229
11/2	4	−	0.370	0.360	0.010	0.364
11/2	5	+	0.178	0.175	0.003	0.177
13/2	5	+	0.316	0.306	0.010	0.308

10.8. Higher spin/orbital states

10.8.1. CO in the $a\,^3\Pi$ state

Compared with laser magnetic resonance techniques, pure rotational spectroscopy has advanced to more complex spin and orbital systems more slowly, probably because it

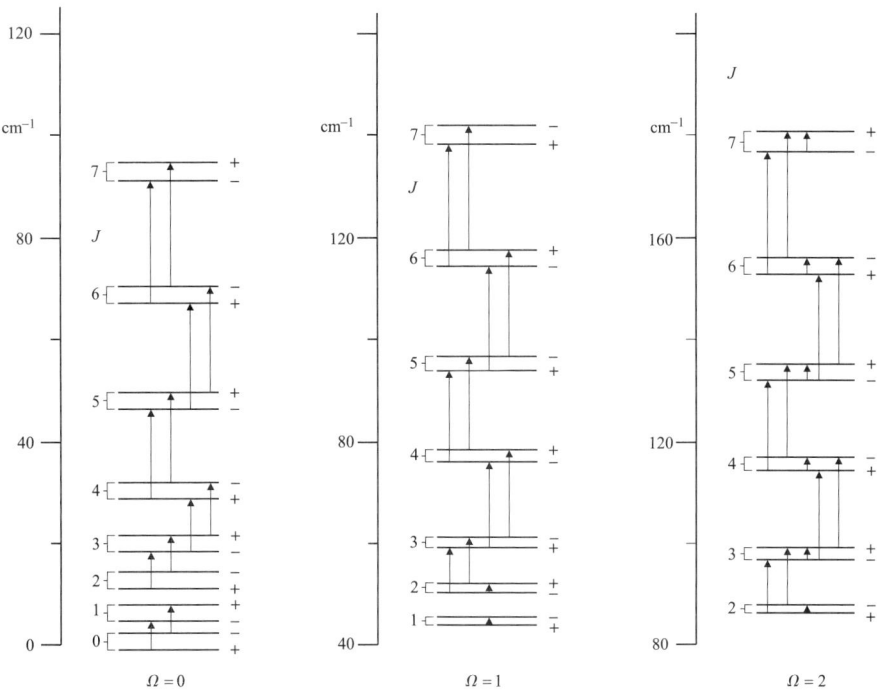

Figure 10.78. Lower rotational levels of the three fine-structure components of the $a\,^3\Pi$ state of CO, and the observed rotational and Λ-doublet transitions. The Λ-doublet splittings are exaggerated for the sake of clarity.

does not have the same intrinsic sensitivity. However there are a number of important studies to be described, particularly those of the CO molecule in its $a\,^3\Pi$ state. As we have already mentioned elsewhere, this is probably the most thoroughly studied excited molecular electronic state. We described the radiofrequency Λ-doublet spectrum, investigated by molecular beam electric resonance, in chapter 8. In chapter 9 we discussed the far-infrared laser magnetic resonance studies of rotational transitions. In this section we turn our attention to pure microwave and millimeter wave measurements Figure 10.78 shows the lowest rotational levels of the three fine structure components, $^3\Pi_0$, $^3\Pi_1$ and $^3\Pi_2$. This diagram also shows the Λ-doublet and some of the rotational transitions which have been studied. $J = 1 \leftarrow 0$ rotational transitions within the $^3\Pi_0$ component were measured by Saykally, Dixon, Anderson, Szanto and Woods [223] in the microwave range 86 to 93 GHz. Subsequent studies by Carballo, Warner, Gudeman and Woods [224] involved radiofrequency Λ-doublet transitions within the $^3\Pi_1$ and $^3\Pi_2$ components, and millimeter wave $\Delta J = \pm 1$ rotational transitions, with frequencies almost up to 500 GHz, in all three fine-structure states. Both papers described measurements for the first five vibrational levels ($v = 0$ to 4) of the $a\,^3\Pi$ state, in the three isotopomers $^{12}C^{16}O$, $^{13}C^{16}O$ and $^{12}C^{18}O$. With the further development of submillimetre frequency sources the measurements have been extended by Yamamoto and Saito [225] and by Wada and Kanamori [226] up to $J = 9$ and $v = 5$.

Table 10.27. *Radiofrequency Λ-doublet transition frequencies (in MHz) for $v = 0$ to 4 in the $a\,^3\Pi$ state of CO. The transitions are labelled with the appropriate J_Ω values*

transition	$v = 0$	$v = 1$	$v = 2$	$v = 3$	$v = 4$
2_2–2_2	7.351	7.078	6.808	6.529	7.108
3_2–3_2	34.064	32.847	31.632	30.390	33.427
4_2–4_2	92.907	89.723	86.548	83.322	92.891
5_2–5_2	194.080	187.743	181.427	175.043	198.360
6_2–6_2	343.486	332.843	322.245	311.630	359.910
7_2–7_2	542.591	526.678	510.847	495.180	584.408
8_2–8_2	789.256	767.360	745.630	724.448	876.111
1_1–1_1	394.065	385.141	374.911	360.830	321.148
2_1–2_1	1150.934	1125.295	1095.790	1054.964	—

We have already described the theory of the Λ-doubling in chapters 8 and 9 and will not repeat the analysis here, except to say that it arises predominantly through rotational mixing of the $^3\Pi$ state with a $^3\Sigma^+$ state. As we discussed in chapter 8, however, a particularly strong interaction occurs in the $v = 4$ level of the $^3\Pi$ state because of its near-degeneracy with the $v = 0$ level of the $a'\,^3\Sigma^+$ state. Table 10.27 shows the consequences of this perturbation on the Λ-doubling frequencies; the measurements were made by Carballo, Warner, Gudeman and Woods [224]. For the vibrational levels $v = 0$ to 3 there is a relatively smooth variation, but a discontinuity occurs for $v = 4$. This discontinuity was illustrated in the electric resonance studies described in chapter 8. A detailed theory of the perturbation, using matrix elements calculated by Freed [227], was presented by Field, Tilford, Howard and Simmons [228], and later refined by Saykally, Dixon, Anderson, Szanto and Woods [223]. The latter authors also presented details of the ^{13}C hyperfine interactions in ^{13}CO.

10.8.2. SiC in the $X\,^3\Pi$ ground state

Attempts to identify the SiC radical through a spectrum, in either the laboratory or in space, have been made for many years. Only as recently as 1988 was an electronic spectrum involving two excited states identified [229], but in the following year a remarkable paper by Cernicharo, Gottlieb, Guélin, Thaddeus and Vrtilek [50] described the identification of SiC through its millimetre wave rotational spectrum, in both the laboratory and in space. Millimetre wave lines associated with the circumstellar shell of the object IRC + 10216 were thought to be possible candidates, and laboratory studies using a glow discharge in mixture of SiH_4, CO and C_2H_2 confirmed these conclusions. Subsequently Mollaaghababa, Gottlieb, Vrtilek and Thaddeus [230] described laboratory studies of both ^{13}C-enriched and vibrationally excited SiC, and Mollaaghababa, Gottlieb and Thaddeus [231] described the hyperfine structure in the rotational spectrum

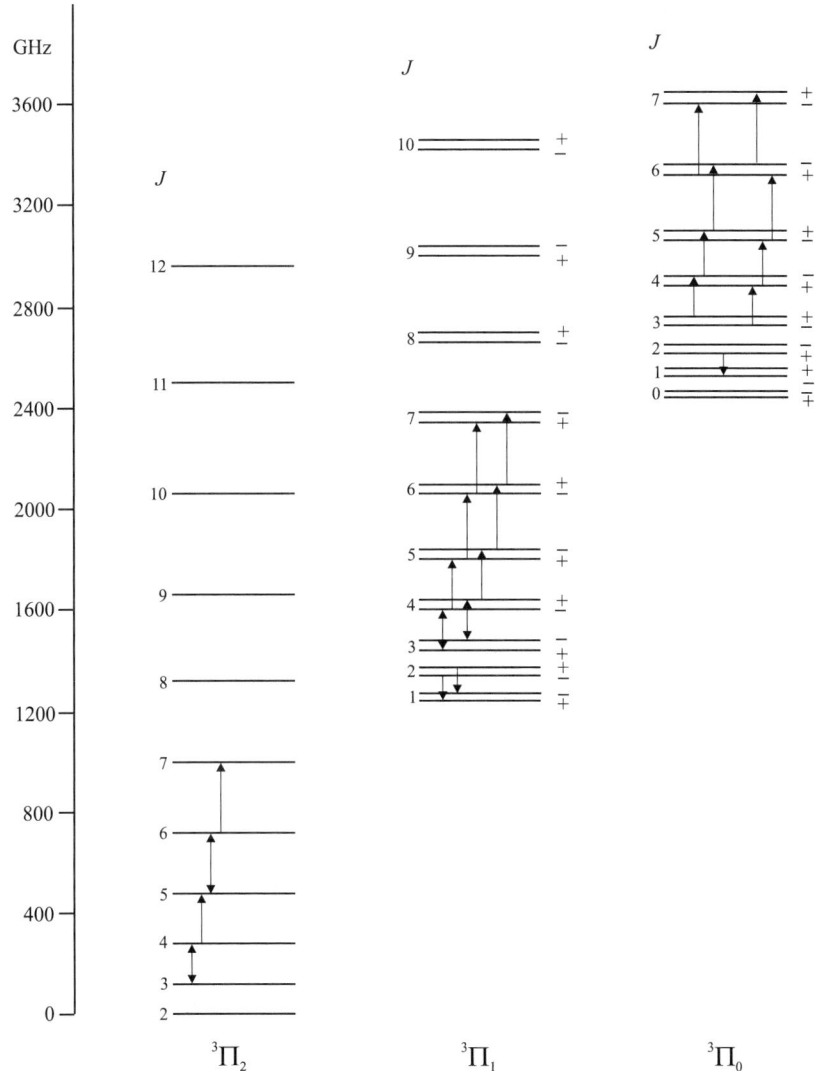

Figure 10.79. Fine structure components and lower rotational levels in the electronic and vibrational ground state of the SiC radical. Transitions detected in the laboratory are indicated by up-arrows, whilst those detected in space are shown as down-arrows [50].

of ^{29}SiC. Bogey, Demuynck and Destombes [232] have also observed high frequency transitions involving higher rotational levels.

The ground electronic state of SiC is $^3\Pi$; the three fine-structure states and lower rotational levels are illustrated in figure 10.79. Also shown are the transitions detected in the laboratory and in space. The laboratory frequencies range from 149 to 474 GHz; figure 10.80 shows examples of laboratory spectra, whilst figure 10.81 illustrates lines observed from IRC-10216. We will not discuss the astrophysical aspects of these observations, but confine ourselves to the spectroscopic and structural implications.

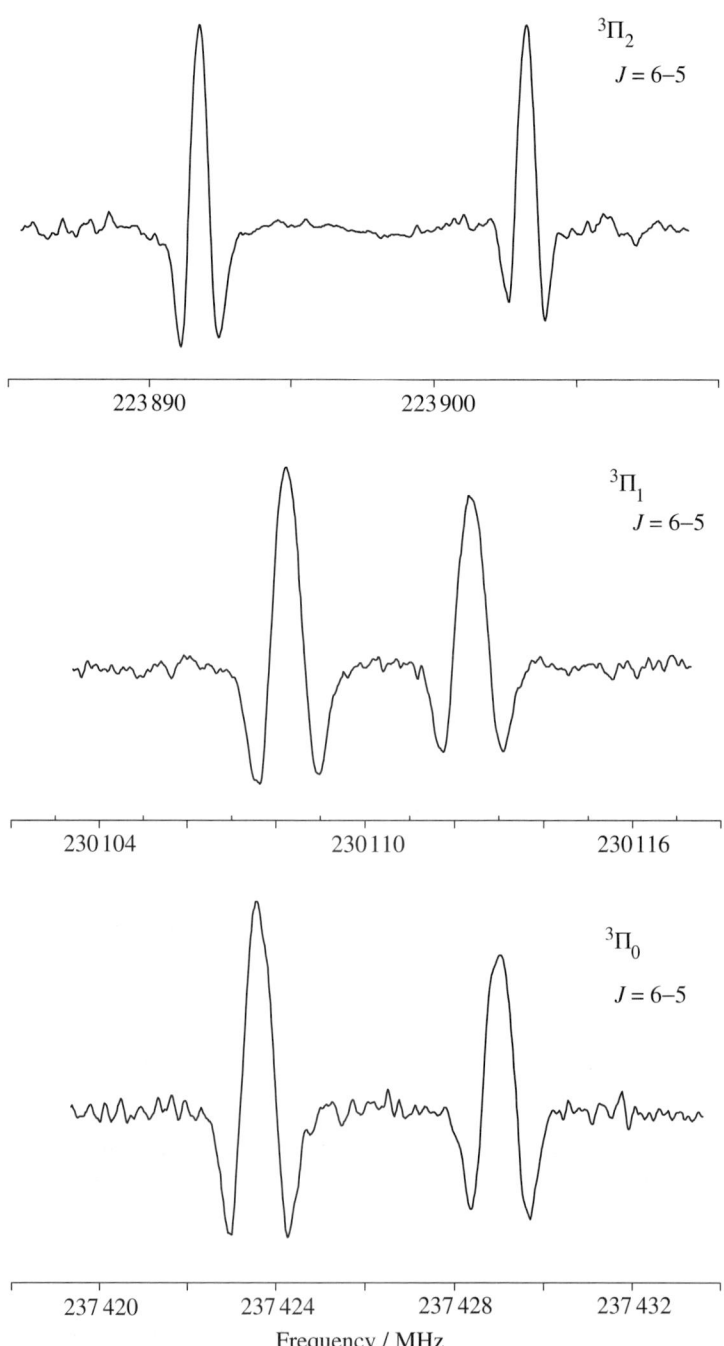

Figure 10.80. Rotational transitions observed for Si^{13}C in the laboratory, exhibiting ^{13}C hyperfine structure [201].

Higher spin/orbital states | 839

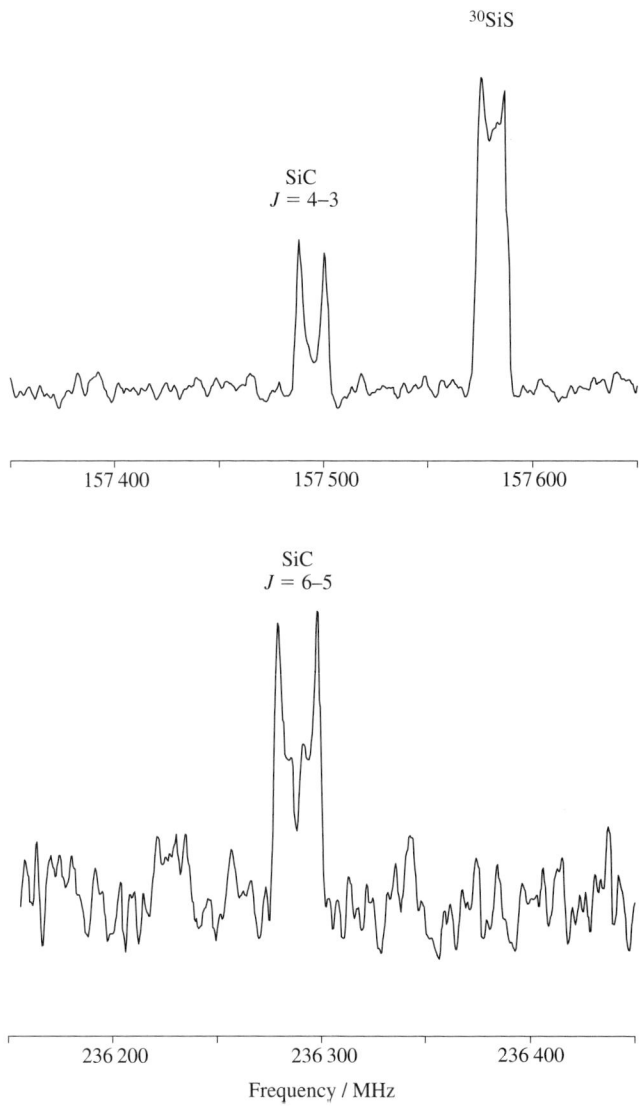

Figure 10.81. Emission lines of SiC observed from IRC-10216; the doublet structure arises from dynamical properties of the emitting source, not from intrinsic energy level structure. The transitions occur within the lowest-lying $^3\Pi_2$ fine-structure state, and each line consists of an unresolved Λ-doublet [50].

The effective Hamiltonian for case (a) $^3\Pi$ states was described by Brown and Merer [233]. The effective rotational Hamiltonian (excluding the Λ-doubling terms) was given, in a molecule-fixed axis system, as

$$\mathcal{H}_{\text{rot}} = A\text{T}_0^1(\boldsymbol{L})\text{T}_0^1(\boldsymbol{S}) + B\{\text{T}^1(\boldsymbol{J}-\boldsymbol{S})\}^2 - D\{\text{T}^1(\boldsymbol{J}-\boldsymbol{S})\}^4 \\ + (2/3)\lambda\{3\text{T}_0^1(\boldsymbol{S})\text{T}_0^1(\boldsymbol{S}) - \boldsymbol{S}^2\} + \gamma\text{T}^1(\boldsymbol{J}-\boldsymbol{S})\cdot\text{T}^1(\boldsymbol{S}), \quad (10.177)$$

and the Λ-doubling Hamiltonian as

$$\mathcal{H}_{\ell d} = (1/2)(o_v + p_v + q_v)(S_+^2 + S_-^2) - (1/2)(p_v + 2q_v)(J_+S_+ + J_-S_-)$$
$$+ (1/2)q_v(J_+^2 + J_-^2). \quad (10.178)$$

The matrix elements are calculated using symmetrised case (a) basis functions,

$$|J, \Omega, \pm\rangle = (1/\sqrt{2})\{|\Lambda = +1, S, \Sigma, J, \Omega\rangle \pm |\Lambda = -1, S, -\Sigma, J, -\Omega\rangle\}, \quad (10.179)$$

which have parities $\pm(-1)^{J-S}$ respectively. The matrix elements of (10.177) and (10.178) using the symmetrised basis functions (10.179) are, for a $^3\Pi$ state, given below. The matrix is diagonal in J, and $x = J(J+1)$. The parameter B^* is equal to $B - \gamma/2$, and the upper and lower signs refer to e and f parity levels respectively.

	$^3\Pi_0$	$^3\Pi_1$	$^3\Pi_2$
$^3\Pi_0$	$-A + B(x+2) + (2\lambda/3)$ $- 2\gamma - D(x^2 + 6x + 4)$ $\mp (o + p + q)$	$-\sqrt{2x}[B^* \mp (p/2 + q)$ $- 2D(x+2)]$	$-\sqrt{[x(x-2)]}(2D \pm q/2)$
$^3\Pi_1$		$-(4\lambda/3) + B(x+2)$ $- 2\gamma - D(x^2 + 8x)$ $\mp (qx/2)$	$-\sqrt{[2(x-2)]}(B^* - 2Dx)$
$^3\Pi_2$			$A + B(x-2) + (2\lambda/3)$ $- D(x^2 - 2x)$

The matrix contains eight parameters; their values (in MHz) for SiC were determined [50] as follows:

$$A = -1\,248\,200, \quad B = 20\,297.582, \quad D = 0.040\,51, \quad \lambda = -1159,$$
$$\gamma = 186, \quad o = 26\,705, \quad p = 132, q = -1.185.$$

The ^{13}C and ^{29}Si nuclei both have spin I equal to 1/2 (but with magnetic moments of opposite sign), and for each nucleus the hyperfine Hamiltonian may be written in the (improved) Frosch and Foley [192] form:

$$\mathcal{H}_{hf} = aI_zS_z + b_F\mathbf{I}\cdot\mathbf{S} + (c/3)(3I_zS_z - \mathbf{I}\cdot\mathbf{S}) - (d/2)(S_+I_+ + S_-I_-). \quad (10.180)$$

The parameters for the two nuclei (in MHz) were determined as follows [231]:

$$^{13}\text{C}: a = 84.1, \quad b_F = 138.3, \quad c = 54.4, \quad d = 63.6.$$
$$^{29}\text{Si}: a = -137.3, \quad b_F = 28.8, \quad c = -63.2, \quad d = -68.5.$$

The ground state electron configuration of SiC is $(5\sigma)^2 (6\sigma)^2 (7\sigma)^1 (2\pi)^3$ so that the hyperfine constants will reflect the electron distribution in both σ and π molecular orbitals. The hyperfine constants a and d are determined mainly by the unpaired π electron, the Fermi contact constant b_F depends mainly on the σ electron, whilst c is sensitive to both σ and π electrons. In fact the Fermi contact constant also depends, as usual, upon configurational mixing with excited electronic states, which makes a simple interpretation difficult. Not surprisingly, perhaps, the electron densities on the two atoms are actually rather similar; this is the main conclusion from a semi-quantitative consideration of the hyperfine constants, and it is supported by *ab initio* calculations.

Silicon carbide is thought to be an important component of the dust shells surrounding carbon-rich stars; it is likely, therefore, that astrophysical studies of the SiC radical will be significant in the future.

10.8.3. FeC in the $X\,^3\Delta$ ground state

There is a clear interest in the possibility of detecting the FeC molecule in interstellar clouds or in circumstellar shells, although such detection has not yet been achieved. A laboratory millimetre wave spectrum has, however, been observed by Allen, Pesch and Ziurys [234] and it confirms the ground electronic state as $X\,^3\Delta$. A $^3\Delta$ state has three fine-structure components, corresponding to Ω values of 3, 2 and 1, with spin–orbit energies of $2A$, 0 and $-2A$, where A is the spin–orbit coupling constant. An electronic spectrum observed by Balfour, Cao, Prasad and Qian [235] leads to a value for A of -124.2 cm^{-1}, so that the lowest energy fine-structure component is $^3\Delta_3$. The fine-structure components, their lower rotational levels, and the rotational transitions observed [234] are illustrated in figure 10.82. Figure 10.83 shows recordings of two rotational transitions in ^{56}Fe^{12}C; both nuclear spins are zero so that hyperfine splitting does not arise. The signal-to-noise ratio is excellent but, even so, transitions involving the higher fine-structure component, $^3\Delta_1$ were not observed, the reasons for which are not clear.

Although the rotational levels are doubly-degenerate through Ω-doubling, the splitting is expected to be extremely small in a $^3\Delta$ state, and was not observed. The effective Hamiltonian used to analyse the spectrum was very simple [236]:

$$\mathcal{H}_{\text{eff}} = AL_zS_z + B\{J(J+1) - \Omega^2 + S(S+1) - \Sigma^2\} + (2/3)\lambda(3S_z^2 - S^2).$$

(10.181)

The rotational constants in equation (10.181) were determined [234] and the values (in MHz) found to be as follows:

$$B(\Omega = 3) = 20\,075.3976, \quad B(\Omega = 2) = 20\,171.9625.$$

The spin–spin constant λ appears not to have been determined, however, and even the value of A is rather uncertain; more work is required. It remains to be seen whether FeC will be observed in interstellar or circumstellar atmospheres.

10.8.4. VO and NbO in their $X\,^4\Sigma^-$ ground states

Microwave spectra of VO and NbO arising from the lowest rotational transition in the ground state have been observed and analysed by Suenram, Fraser, Lovas and

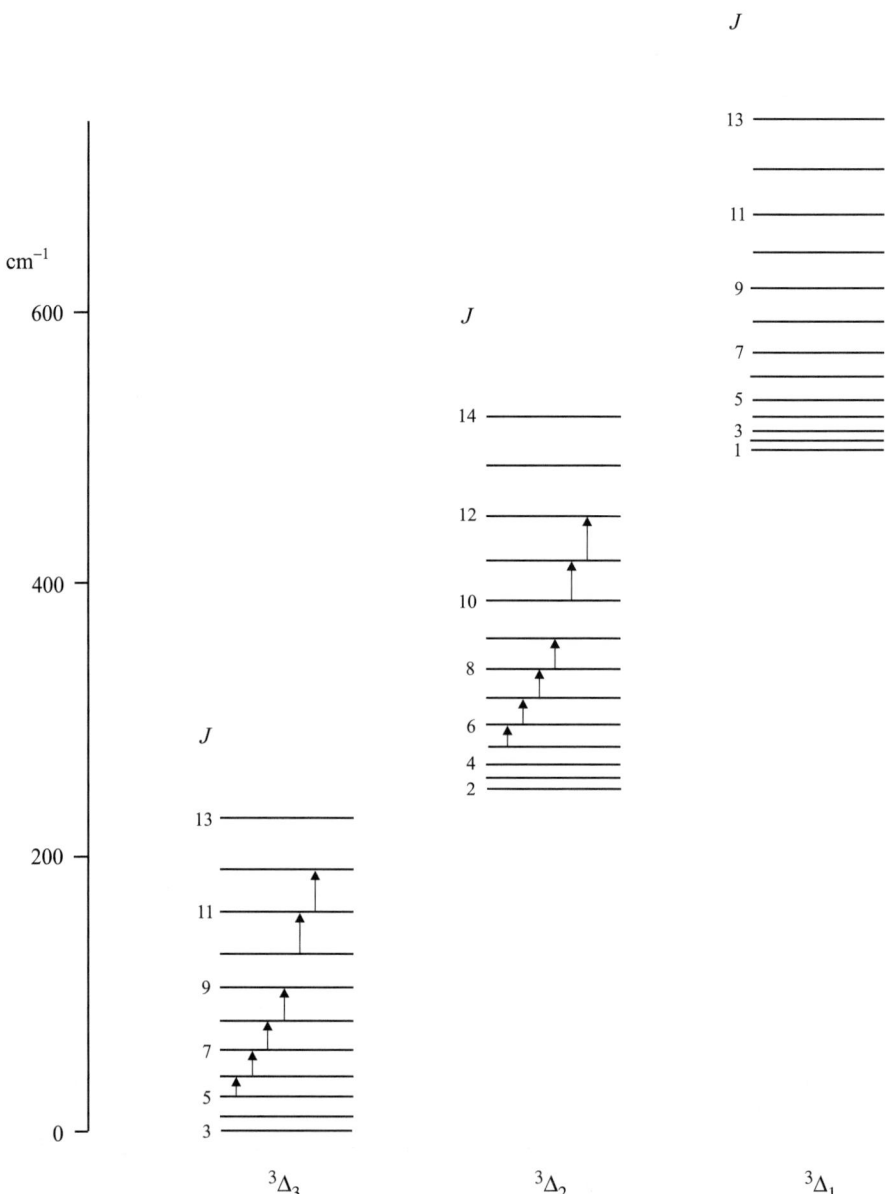

Figure 10.82. Fine-structure components, lower rotational levels, and rotational transitions observed [234] for FeC.

Gillies [237]. They used the technique of laser ablation to form nozzle beams of the molecules, injected into a Fourier transform microwave spectrometer. In both cases the ground electronic state is $^4\Sigma^-$, an example of which we met in chapter 9, the excited a state of the CH radical. However in that example the coupling was close to case (b), whereas VO and NbO provide examples of a case (a) $^4\Sigma^-$ state. The $\Sigma = \pm 1/2$ and $\pm 3/2$ states are split by second-order spin–orbit coupling; the splitting is

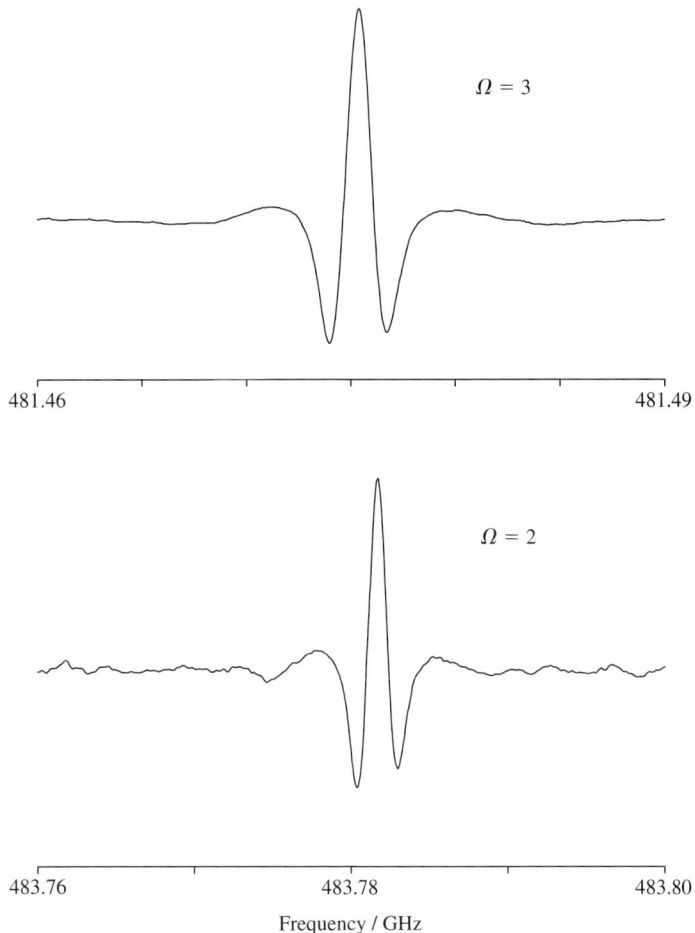

Figure 10.83. Recordings of the $J = 12 \leftarrow 11$ rotational transition in the $^3\Delta_3$ and $^3\Delta_2$ fine-structure states of FeC [234].

≈ 62 cm^{-1} in NbO with the $^4\Sigma^-_{1/2}$ state being the lower in energy. The lowest rotational transition is therefore $J = 3/2 \leftarrow 1/2$ and this is split by hyperfine interaction from the ^{93}Nb nucleus which has a spin of 9/2 and is present in 100% natural abundance. The $J = 1/2$ level has hyperfine components with F values of 5 and 4, whilst the $J = 3/2$ level has components with F values 6, 5, 4 and 3. An energy level diagram showing the splitting and the observed transitions for NbO is shown in figure 10.84; the situation for VO is similar, with the spin of ^{51}V being 7/2.

The case (a) effective rotational Hamiltonian for NbO in its ground state may be written

$$\mathcal{H}_{\text{eff}} = B(\boldsymbol{J} - \boldsymbol{S})^2 + (2/3)\lambda(3S_z^2 - \boldsymbol{S}^2) \\ + \gamma(\boldsymbol{J} - \boldsymbol{S}) \cdot \boldsymbol{S} + C\mathbf{T}^3(\boldsymbol{L}^2, \boldsymbol{N}) \cdot \mathbf{T}^3(\boldsymbol{S}, \boldsymbol{S}, \boldsymbol{S}). \quad (10.182)$$

Values of B, λ and γ were determined from an analysis of the $X\,^4\Sigma^- - C\,^4\Sigma^-$ electronic band system by Cheval, Féménias, Merer and Sassenberg [238], and the centrifugal distortion corrections were also determined. The matrix of the rotational Hamiltonian for a $^4\Sigma^-$ case (a) state, using properly symmetrised basis functions, was given [238] as follows.

	$||\Sigma| = 3/2\rangle$	$||\Sigma| = 1/2\rangle$
$||\Sigma| = 3/2\rangle$	$2\lambda + BX - 3\gamma/2 + 2\lambda_D$ $- DX(X+3) - 3\gamma_D X$	$-\sqrt{3X}\{(B - \gamma/2) - 2D[X + 2 \mp (J + 1/2)]\}$ $+ (\sqrt{3X}/2)\gamma_D[X + 7 \mp (2J + 1)] + \sqrt{3X}\gamma_S$
$||\Sigma| = 1/2\rangle$	symmetric	$-2\lambda + B(X+4) - 7\gamma/2 - 2\lambda_D(X+4)$ $- D[(X+4)^2 + 7X + 4] - \gamma_D(7X + 16)$ $\mp 2[(B - \gamma/2) - 2\lambda_D - 2D(X+4)$ $- (\gamma_D/2)(X+11) + 3\gamma_S/2](J + 1/2)$

In this matrix $X = (J - 1/2)(J + 3/2)$ and the upper and lower signs correspond to the e and f levels.

In addition, the ^{93}Nb magnetic hyperfine coupling was large enough to be readily observed, even in the electronic spectrum. It was fitted to the normal Frosch and Foley Hamiltonian for a Σ state,

$$\mathcal{H}_{\text{hfs}} = (b + c)I_z S_z + (b/2)(I_+ S_- + I_- S_+), \tag{10.183}$$

and the constants b and c determined to be 0.0549 and –0.0020 cm^{-1} respectively. As we described above, the $\Sigma = \pm 1/2$ and $\pm 3/2$ states are split by 4λ, arising essentially entirely from the second-order effects of spin–orbit coupling. As figure 10.84 shows, the fine, hyperfine and rotational splittings are similar in magnitude. Solid state esr studies of NbO [239] and the later gas phase studies are consistent with each other. Additional information was provided by measurements of the Stark effect [237] which gave the electric dipole moment of NbO as 3.498 D.

The electronic structure of NbO was discussed by Brom, Durham and Weltner [239] in terms of the molecular orbital diagram shown in figure 10.85. Niobium belongs to the $4d$ transition metal group and, as shown in the diagram, the three unpaired electrons in the electronic ground state occupy $4d\delta$ and $5s\sigma$ orbitals which are essentially non-bonding. The large isotropic hyperfine interaction in NbO is consistent with a $5s$ spin density of about 40%. At the same time the quadrupole interaction is expected to be too small to be observed because the electric field gradient at the nucleus is zero or very small for s or d electrons respectively.

Similar aspects arise for VO, except that vanadium belongs to the $3d$ transition element group. The electronic structure of VO is, however, similar to that of NbO and a large isotropic ^{53}V hyperfine interaction arises from a $4s$ spin density of around 30%.

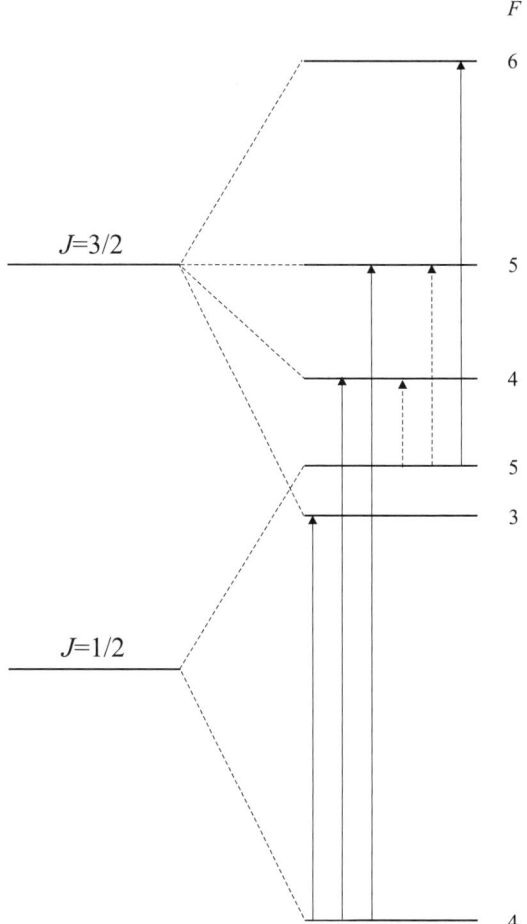

Figure 10.84. Hyperfine energy level diagram and observed transitions (solid lines) for the $J = 3/2 \leftarrow 1/2$ rotational transition of ^{93}NbO in the $^4\Sigma^-_{1/2}$ component [237].

10.8.5. FeF and FeCl in their $X\,^6\Delta$ ground states

A millimeter wave spectrum of FeCl was obtained by Tanimoto, Saito and Okabayashi [240] using a glow discharge in helium/aluminium trichloride mixtures in a free-space millimeter wave cell; the iron atoms originated from the stainless steel electrodes. Analysis of the electronic emission spectrum of FeCl by Delaval, Dufour and Schamps [241] had already established the ground electronic state to be $^6\Delta$. Subsequently a more extensive rotational spectrum of FeCl was described by Allen, Li and Ziurys [242]. A similar millimetre wave rotational spectrum of FeF, also with a $^6\Delta$ ground state, was studied by Allen and Ziurys [243, 244]. In a Hund's case (a) coupling scheme there are six fine structure components with different spin–orbit energies, with Ω values $9/2, 7/2, 5/2, 3/2, 1/2$ and $-1/2$, and since the sign of the spin–orbit constant had been determined to be negative, the $^6\Delta_{9/2}$ component must be the lowest in energy.

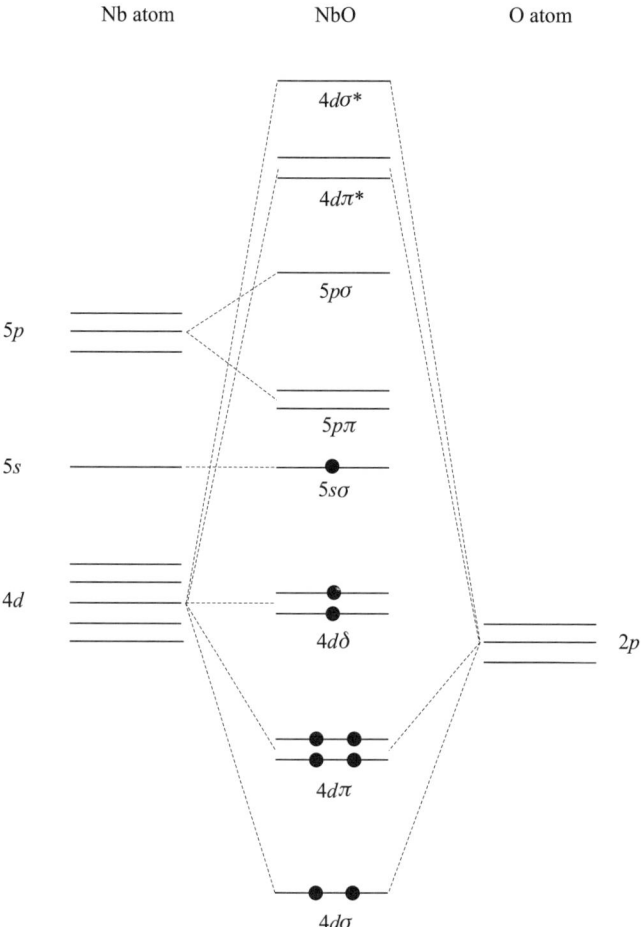

Figure 10.85. Schematic molecular orbital diagram for the valence electrons of NbO, forming the $X\,^4\Sigma^-$ ground state [239].

Figure 10.86 shows an energy level diagram illustrating the fine-structure components and the lower rotational levels; this diagram is actually drawn for FeF, but the diagram for FeCl is similar.

Tanimoto, Saito and Okabayashi [240] observed a number of transitions in the range 241 to 380 GHz, which they assigned as rotational transitions involving high J values in both the $^6\Delta_{9/2}$ and $^6\Delta_{7/2}$ fine structure states. The rotational constants in these two spin components were determined to be 4926.1478 and 4938.5177 MHz; although chlorine hyperfine and quadrupole structure might be expected, the splitting becomes very small for the high J values observed (47/2 to 77/2) and was not resolved. Subsequently, however, Allen, Li and Ziurys [242] observed rotational transitions over a very much larger range of J values and were able to observe ^{35}Cl hyperfine splitting in the $\Omega = 9/2$ and 7/2 components, and Λ-doublet splitting in smaller Ω levels. Similarly, rotational transitions in FeF were observed for all six fine-structure components [244], many

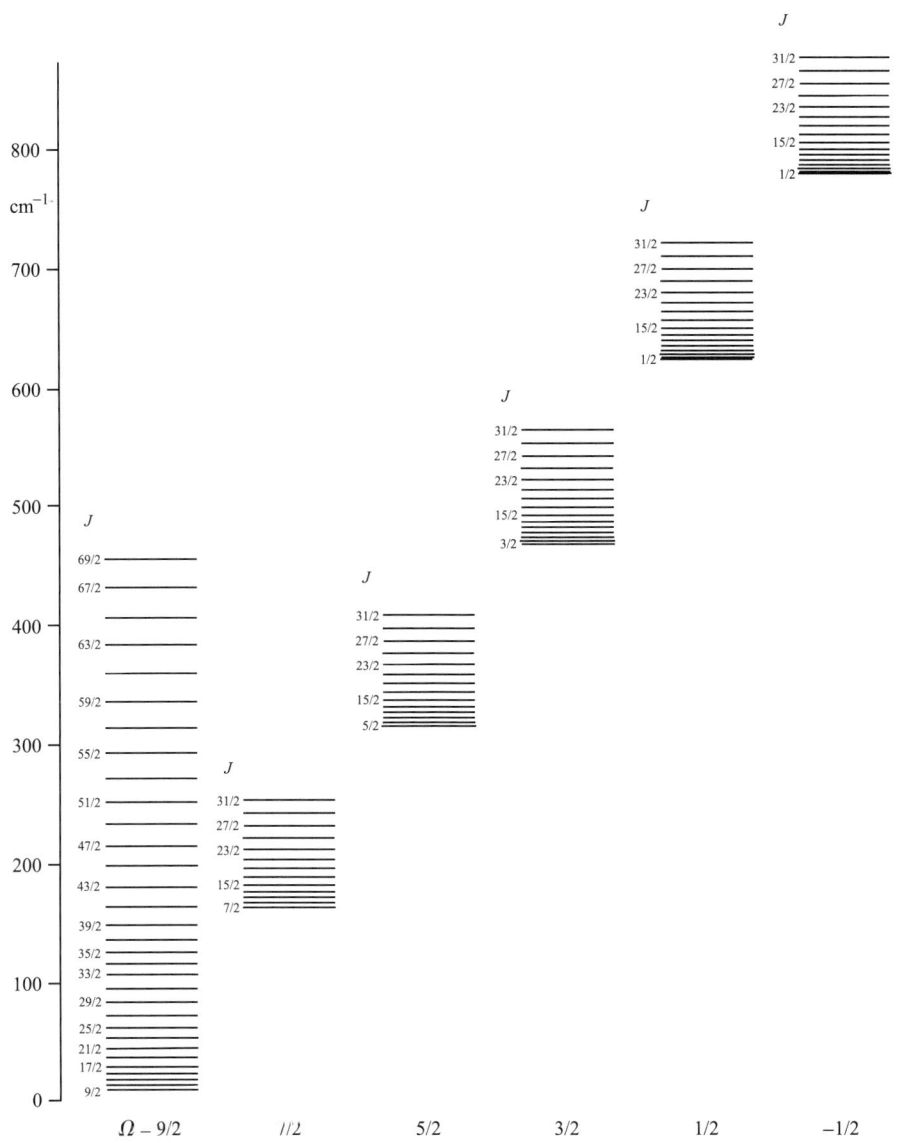

Figure 10.86. Fine-structure components and lower rotational levels for FeF in the $X\,^6\Delta$ ground state.

with ^{19}F hyperfine structure. Figure 10.87 shows recordings of rotational transitions in FeCl; in the high J transition the hyperfine splitting is too small to be resolved, but in the lower J transition the ^{35}Cl structure is readily observed. Figure 10.88 shows a recording of a rotational transition in FeF with the ^{19}F doublet splitting well resolved.

For FeF and FeCl the effective Hamiltonian used was similar, and was written in the form:

$$\mathcal{H}_{\text{eff}} = \mathcal{H}_{\text{rot}} + \mathcal{H}_{\text{cd}} + \mathcal{H}_{\text{so}} + \mathcal{H}_{\text{socd}} + \mathcal{H}_{\text{ss}} + \mathcal{H}_{\text{sscd}} + \mathcal{H}_{\ell d} + \mathcal{H}_{\text{hfs}}. \quad (10.184)$$

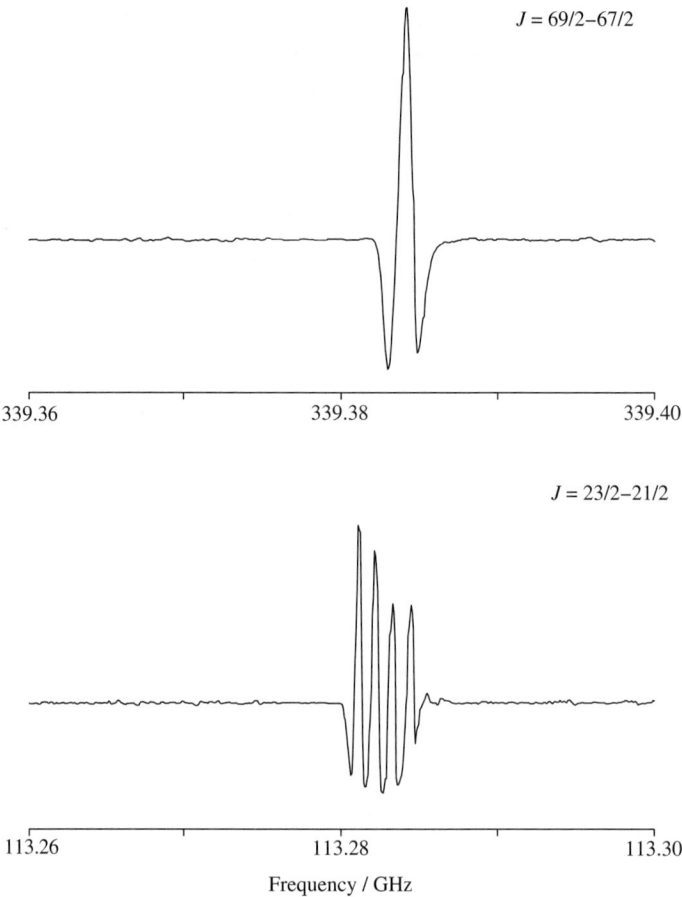

Figure 10.87. Rotational transitions in FeCl in its $X\,^6\Delta_{9/2}$ state. The lower J transition shows a quartet hyperfine splitting from ^{35}Cl, whereas the hyperfine splitting is too small to be resolved in the higher J transition [242].

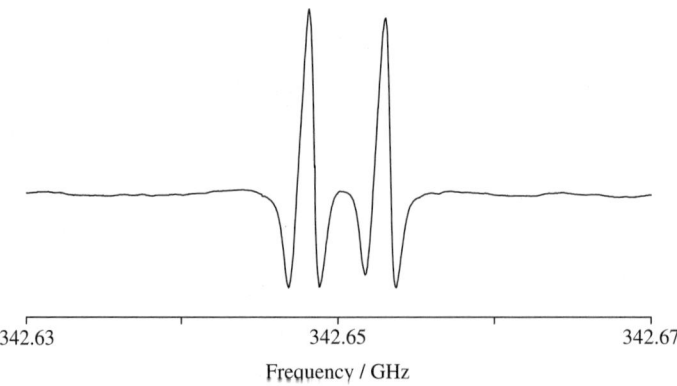

Figure 10.88. Observed $J = 31/2–29/2$ rotational transition for FeF in its $X\,^6\Delta_{9/2}$ state, showing ^{19}F doublet hyperfine structure [243].

We follow the original work in using the cartesian forms of the operators. The rotational Hamiltonian with its centrifugal distortion is given by

$$\mathcal{H}_{\text{rot}} + \mathcal{H}_{\text{cd}} = B(\boldsymbol{J} - \boldsymbol{S})^2 - D(\boldsymbol{J} - \boldsymbol{S})^4. \tag{10.185}$$

The spin–orbit terms, including centrifugal distortion, are written in the form

$$\mathcal{H}_{\text{so}} + \mathcal{H}_{\text{socd}} = AL_z S_z + (1/2)A_D[(\boldsymbol{J} - \boldsymbol{S})^2 L_z S_z + L_z S_z (\boldsymbol{J} - \boldsymbol{S})^2] \\ + \eta' L_z S_z [S_z^2 - (3\boldsymbol{S}^2 - 1)/5], \tag{10.186}$$

where the third term, containing η', describes a coupling between the spin–orbit and spin–spin interactions. The spin–spin terms, again including centrifugal distortion, are taken to be

$$\mathcal{H}_{\text{ss}} + \mathcal{H}_{\text{sscd}} = (2/3)\lambda\left(3S_z^2 - \boldsymbol{S}^2\right) + (1/3)\lambda_D\left[(\boldsymbol{J} - \boldsymbol{S})^2\left(3S_z^2 - \boldsymbol{S}^2\right) + \left(3S_z^2 - \boldsymbol{S}^2\right)\right. \\ \left. \times (\boldsymbol{J} - \boldsymbol{S})^2\right] + (1/12)\theta\left(35S_z^4 - 30\boldsymbol{S}^2 S_z^2 + 25S_z^2 - 6\boldsymbol{S}^2 + 3\boldsymbol{S}^4\right). \tag{10.187}$$

The Λ-doubling terms are,

$$\mathcal{H}_{\ell d} = (1/2)\tilde{m}_\Delta(S_+^4 + S_-^4) - (1/2)\tilde{n}_\Delta(S_+^3 J_+ + S_-^3 J_-) + (1/2)\tilde{o}_\Delta(S_+^2 J_+^2 + S_-^2 J_-^2) \\ - (1/2)\tilde{p}_\Delta(S_+ J_+^3 + S_- J_-^3) + (1/2)\tilde{q}_\Delta(J_+^4 + J_-^4), \tag{10.188}$$

where the five constants are defined by Brown, Cheung and Merer [236]. Finally, the hyperfine terms, including the quadrupole interaction for ^{35}Cl in FeCl, take the form

$$\mathcal{H}_{\text{hf}} = aL_z I_z + b\boldsymbol{I} \cdot \boldsymbol{S} + cI_z S_z - (1/2)d_\Delta(J_+^2 I_+ S_+ + J_-^2 S_- I_-) \\ + eq_0 Q\left(3I_z^2 - \boldsymbol{I}^2\right)/4I(2I - 1). \tag{10.189}$$

The values of the molecular parameters (in MHz) determined for FeCl in its $X\,^6\Delta$ state were as follows [242]:

$$A = -2\,274\,691.0, \quad A_D = -0.494\,30, \quad B = 4955.904\,45, \quad D = 0.003\,283\,38,$$
$$\eta' = 1147.7, \quad \lambda = 16\,376.7, \quad \lambda_D = 0.024\,168, \quad \theta_D = -0.001\,414,$$
$$\tilde{m}_\Delta = -455.57, \quad \tilde{n}_\Delta = 4.2196, \quad \tilde{o}_\Delta = -0.0122, \quad \tilde{p}_\Delta = 0.000\,168, \quad \tilde{q}_\Delta \cong 0,$$
$$a = 1.65, \quad b = 7.3, \quad c = 4.8, \quad eq_0 Q = -12.3.$$

The corresponding constants for the predominant isotopomer, ^{56}FeF ($v = 0$), are [244]

$$A = -2\,342\,900, \quad A_D = -1.759, \quad B = 11\,197.5884, \quad D = 0.014\,380\,1, \quad \eta' = 331.1,$$
$$\lambda = 2000, \quad \lambda_D = 0.134\,83, \quad \theta_D = -0.002\,893,$$
$$\tilde{m}_\Delta = -175.60, \quad \tilde{n}_\Delta = 2.5307, \quad \tilde{o}_\Delta = -0.0194, \quad \tilde{p}_\Delta = 0.000\,186, \quad \tilde{q}_\Delta \cong 0,$$
$$a = -0.45, \quad b = 74.5, \quad c = 51.7.$$

Constants for the $v = 1$ and 2 levels were also obtained.

A qualitative molecular orbital diagram involving the valence electrons in FeF was presented by Allen and Ziurys [244] and is shown in figure 10.89. The main covalent

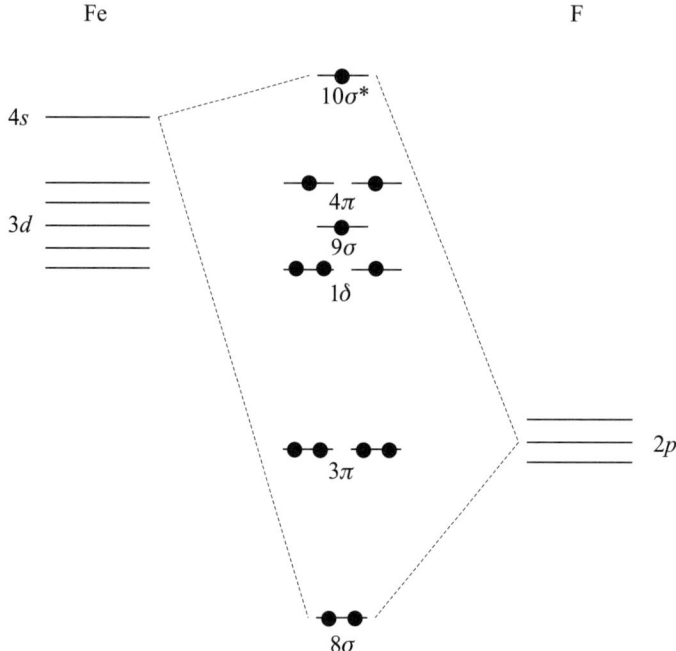

Figure 10.89. Qualitative molecular orbital diagram for FeF, involving the fluorine $2p$ and iron $4s$, $3d$ atomic orbitals [244]. The energies of the iron non-bonding orbitals (1δ, 9σ, 4π) are very close to that of the $10\sigma^*$ antibonding orbital.

interaction involves the $4s$ orbital on the Fe atom and the $2p\sigma$ orbital on the F atom; the Fe $3d$ orbitals are essentially non-bonding, as shown. The electron configuration of FeCl is similar, and in both molecules the five unpaired electrons are largely localised on the iron atom; the larger magnetic hyperfine constants b and c for FeF are due, in part, to the much larger nuclear magnetic moment of ^{19}F as compared with ^{35}Cl. Although FeF is more covalent than FeCl, both molecules seem to be largely ionic in their structure.

10.8.6. CrF, CrCl and MnO in their $X\,^6\Sigma^+$ ground states

A millimetre wave rotational spectrum of the CrF radical in the region 270 to 460 GHz has been described by Okabayashi and Tanimoto [245], and the related spectrum of CrCl has been reported by Oike, Okabayashi and Tanimoto [246, 247]. These molecules have $^6\Sigma^+$ ground states, and in both cases the spectra were assigned as rotational transitions involving high N values. The molecule MnO also has a $^6\Sigma^+$ ground state and its millimetre wave rotational spectrum has been described by Namiki and Saito [248].

In these molecules the spin-spin interaction and rotational constants are similar in magnitude and an analysis of the spectrum may be carried out using either a case (a) or a case (b) basis. Many of the observed spectra involved relatively high rotational

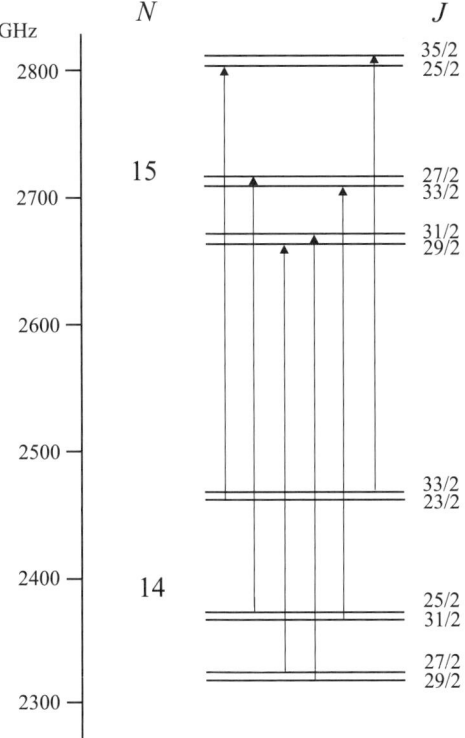

Figure 10.90. $N = 14$ and 15 rotational levels of CrF in its $X\,^6\Sigma^+$ state, with the spin splitting and spin component transitions observed. This diagram is approximately to scale.

levels, where case (b) becomes somewhat more appropriate, and figure 10.90 shows a typical pair of adjacent rotational levels in CrF, with their spin splittings and the observed transitions. This diagram is drawn approximately to scale; its purpose is mainly to explain the pattern of six spin components observed for each rotational transition. However the grouping of the spin components into three pairs shows the incipient case (a) behaviour, with $\Sigma = \pm 1/2, \pm 3/2, \pm 5/2$ in increasing order of energy. In CrF rotational levels with N values from 11 to 20 were observed, whilst in CrCl N ranged from 19 to 27; nuclear hyperfine structure was not observed in either case. The N values involved in the MnO spectrum were somewhat smaller ($N = 6$ to 14) and hyperfine structure from the ^{55}Mn nucleus, which has a spin I of 5/2, was observed. Figure 10.91 shows a recording of one electron spin component of the $N = 12 \leftarrow 11$ rotational transition, with six ^{55}Mn hyperfine lines cleanly resolved. They are labelled according to the F value of the lower hyperfine component. We will discuss the analysis of this hyperfine pattern in due course.

The effective Hamiltonian for all three molecules, excluding nuclear hyperfine interactions for the moment, may be written in a case (b) basis as

$$\mathcal{H}_{\text{eff}} = \mathcal{H}_{\text{rot}} + \mathcal{H}_{\text{ss}} + \mathcal{H}_{\text{sr}}, \tag{10.190}$$

Table 10.28. *Molecular parameters (in MHz) for $^6\Sigma^+$ molecules*

Parameter	$^{52}\mathrm{Cr}^{35}\mathrm{Cl}$	$^{52}\mathrm{Cr}^{19}\mathrm{F}$	$^{55}\mathrm{MnO}$
B	5009.346 95	11 369.614 33	15 025.814 87
D	3.528 92	15.054 44	0.0215 548 5
λ	7992.5	16162.4	17 198.00
λ_D	−0.013 214	−0.015 486	−0.069 646
γ	65.534	408.502	−70.7886
γ_D	−0.0839	−0.000 754	−0.001 058 0
θ	−2.11	−5.15	−14.67

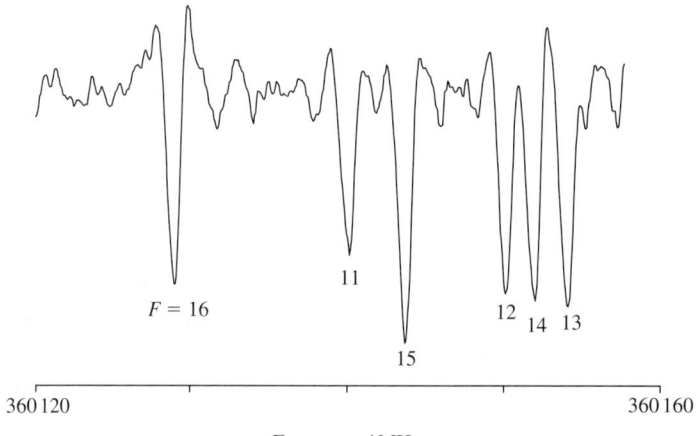

Figure 10.91. The $N = 12 \leftarrow 11$, $J = 29/2 \leftarrow 27/2$ rotational/spin transition of MnO in its $X\,^6\Sigma^+$ state, showing the ^{55}Mn hyperfine structure [248].

where, as usual,

$$\mathcal{H}_{\mathrm{rot}} = BN^2 - DN^4, \tag{10.191}$$

$$\mathcal{H}_{\mathrm{ss}} = (2/3)\lambda\left(3S_z^2 - S^2\right) + (1/3)\lambda_D\left[\left(3S_z^2 - S^2\right), N^2\right]_+, \tag{10.192}$$

$$\mathcal{H}_{\mathrm{sr}} = \gamma N \cdot S + \gamma_D(N \cdot S)N^2 + \gamma_H(N \cdot S)N^4. \tag{10.193}$$

The energy level diagram shown in figure 10.90 was drawn using the first-order contributions of these three terms. In addition, the higher-order spin–orbit term given in equation (10.187) was included in the original analysis:

$$\mathcal{H}_{\mathrm{so}}^{(4)} = (1/12)\theta\left(35S_z^4 - 30S^2S_z^2 + 25S_z^2 - 6S^2 + 3S^4\right)$$
$$= (\sqrt{70}/6)\theta T_{q=0}^4(S, S, S, S). \tag{10.194}$$

The required matrix elements are given elsewhere in this book, and were listed by Ram, Jarman and Bernath [249] for a case (a) basis. In table 10.28 we list the values of the constants, excluding hyperfine parameters, given for all three molecules.

The most interesting hyperfine interaction in the three molecules is that of ^{55}Mn in the MnO molecule, illustrated in figure 10.91. The Frosch and Foley [192] hyperfine constants and the electric quadrupole coupling constant are found to have the following values (in MHz):

$$b_\text{F} = 479.861, \quad c = -48.199, \quad eq_0Q = -25.65.$$

In terms of the molecular orbital scheme presented (for FeF) in figure 10.89, the ground electronic state of MnO may be approximated by the single configuration,

$$X\,^6\Sigma^+ : (\text{core})(9\sigma)^1(4\pi)^2(1\delta)^2,$$

where the five electrons specified have parallel spins. Calculations [250] suggest that the MnO bond is best described by a mainly ionic structure $\text{Mn}^+(3d^54s^1)\text{O}^-(2p^5)$; since the Fermi contact coupling constant b_F reflects the s orbital character of the orbitals occupied by the unpaired electrons, only the 9σ orbital need be considered. It is given by a combination of the $3d\sigma$ and $4s$ orbitals of the Mn atom,

$$|9\sigma\rangle = c_1|4s(\text{Mn}^+)\rangle - c_2|3d\sigma(\text{Mn}^+)\rangle. \tag{10.195}$$

Comparison of the observed value of b_F with the known contact interaction constant of the Mn$^+$ ion gives a value for c_1^2 of 0.573. In other words the $3d\sigma$ and $4s$ hybridisation is a nearly perfect one-to-one mixture. The dipolar hyperfine constant c depends upon a sum of contributions from the $3d$ unpaired electrons in the 9σ, 4π and 1δ orbitals; c is calculated to be -54.0 MHz, assuming the 4π and 1δ orbitals to be pure $3d$ atomic orbitals, a value which agrees well with the measured value of -48.199 MHz. As we have commented elsewhere, the quadrupole coupling constant involves all of the electrons, and is not readily amenable to a simple semi-empirical treatment.

CrF and CrCl also appear to be essentially ionic, with little delocalisation of the unpaired electrons on to the halogen nuclei, so that their hyperfine interaction is not observed. Similar studies have been described by Tanimoto, Sakamaki and Okabayashi [251] for NiF and by Yamazaki, Okabayashi and Tanimoto [252] for NiCl, both of which have $^2\Pi_{3/2}$ ground states. In both cases rotational transitions falling in the frequency range 200 to 400 GHz were observed; small hyperfine splittings from ^{19}F were observed for NiF, but no hyperfine splitting was seen for NiCl. The electronic structures were taken to be essentially ionic in both molecules. In the case of NiF, rotational transitions in the $A\,^2\Sigma^+$ state were also observed, this state lying sufficiently low in energy to be populated thermally.

10.8.7. FeO in the $X\,^5\Delta$ ground state

A rotational spectrum of FeO in its $X\,^5\Delta$ ground state was described by Endo, Saito and Hirota [253]. Transitions involving relatively low J values were observed for three fine-structure states, with $\Omega = 4$, 3 and 2. Subsequently a much more comprehensive rotational spectrum was measured by Kröckertskothen, Knöckel and Tiemann [254] using double resonance methods; we will discuss these results in the next chapter. More recently Allen, Ziurys and Brown [255] returned to the pure rotational spectrum and observed rotational transitions in all five fine-structure states. We will describe their studies in this section, but defer the detailed discussion of the theory and analysis until the next chapter; this is because they combined their results with the more accurate double resonance studies which used a molecular beam source. Allen, Ziurys and Brown [255] produced the FeO molecule by reacting iron vapour at 1400 °C with N_2O, and measured the rotational spectrum over the frequency range 93 to 376 GHz. The rotational levels of the $X\ ^5\Delta$ state are best described in a case (a) basis; there are five fine-structure components, with $\Omega = 0$ to 4, which are summarised in figure 10.92, together with their lower rotational levels and observed rotational transitions. Each rotational level is actually doubly degenerate because of the Λ-doubling; the splitting was too small to be resolved in the $\Omega = 4$ spectrum, but was readily observed for lower Ω values. Typical spectra are shown in figure 10.93; the $J = 11 \leftarrow 10$ ($\Omega = 4$) transition for the ^{56}FeO and ^{54}FeO isotopomers, with natural abundances of 91.7 and 5.9% respectively, does not exhibit resolved Λ-doublet splitting, but in the lower $\Omega = 0$ and 1 states the splitting is observed.

The effective Hamiltonian and analysis of the spectra is described in chapter 11 when we discuss the microwave/optical double resonance spectrum.

10.8.8. TiCl in the $X\ ^4\Phi$ ground state

The sub-millimetre wave spectrum of TiCl has been observed recently by Maeda, Hirao, Bernath and Amano [256], the frequency range being 407 to 604 GHz; the TiCl molecules were produced by means of a dc glow discharge in a flowing mixture of $TiCl_4$ and argon. The sub-millimetre wave radiation was produced by backward-wave oscillators, and a double modulation scheme, involving modulation of both the sub-millimetre frequency and the dc discharge power, was employed. Figure 10.94 shows the four fine-structure components of the $^4\Phi$ state and their lower rotational levels; the observed spectrum, however, involved transitions between high rotational levels ($J = 85/2$ to $127/2$) in all four fine-structure states. Spectra of both $Ti^{35}Cl$ and $Ti^{37}Cl$ in the $v = 0$ and 1 levels of the $^4\Phi$ ground state were observed, but nuclear hyperfine structure was not resolved.

The effective Hamiltonian used was

$$\mathcal{H}_{\text{eff}} = \mathcal{H}_{\text{rot}} + \mathcal{H}_{\text{so}} + \mathcal{H}_{\text{sr}} + \mathcal{H}_{\text{ss}} + \mathcal{H}_{\text{so}}^{(3)}, \quad (10.196)$$

where
$$\mathcal{H}_{\text{rot}} = BN^2 - DN^4 + HN^6, \quad (10.197)$$

$$\mathcal{H}_{\text{so}} = (1/2)[A + A_D N^2 + A_H N^4, L_z S_z]_+, \quad (10.198)$$

$$\mathcal{H}_{\text{sr}} = (\gamma + \gamma_D N^2 + \gamma_H N^4)(N \cdot S), \quad (10.199)$$

$$\mathcal{H}_{\text{ss}} = (1/3)\left[\lambda + \lambda_D N^2 + \lambda_H N^4, 3S_z^2 - S^2\right]_+, \quad (10.200)$$

$$\mathcal{H}_{\text{so}}^{(3)} = (1/2)\left[\eta' + \eta'_D N^2 + \eta'_H N^4, L_z S_z \left(S_z^2 - \{3S^2 - 1\}/5\right)\right]_+. \quad (10.201)$$

For calculation of the matrix elements in case (a) basis, $N = J - S$; note also that $[x, y]_+$ stands for the anticommutator $xy + yx$.

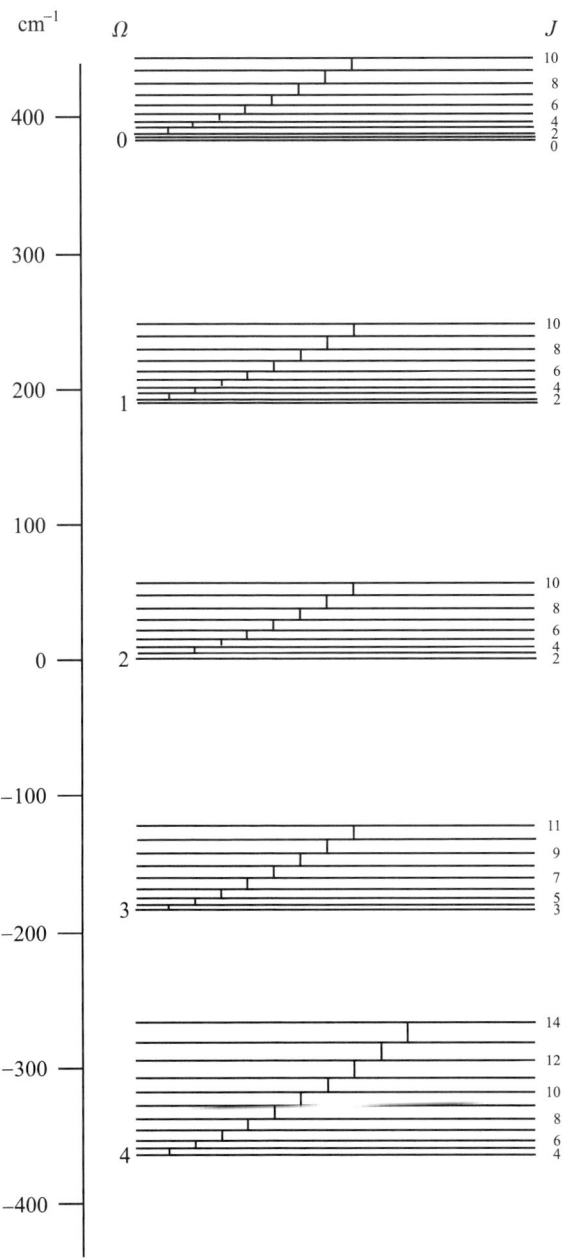

Figure 10.92. Fine-structure components, lower rotational levels, and observed rotational transitions for FeO in its $X\,^5\Delta$ state. Λ-doublet splitting is not resolved.

The spin–rotation constants in equation (10.199) could not be determined, but most of the other constants were listed. The results indicate that the electron configuration $3d^2 4s$ from $Ti^+(^4F)$ dominates in the ground state of TiCl. At this stage it is difficult to estimate reliably the degree of electron delocalisation, but the structure seems to be mainly ionic.

856 Pure rotational spectroscopy

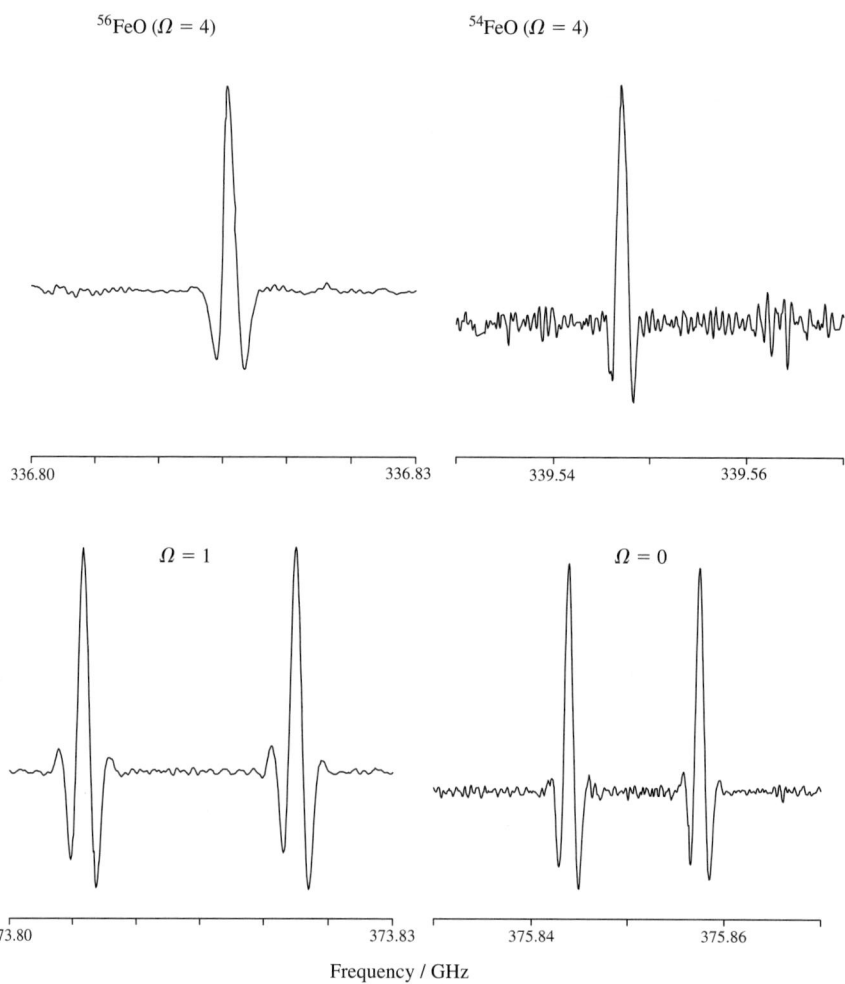

Figure 10.93. *Top*: spectra showing the $J = 11 \leftarrow 10$ rotational transition in the $\Omega = 4$ fine-structure state for ^{56}FeO and ^{54}FeO. *Bottom*: spectra showing the $J = 12 \leftarrow 11$ rotational transition in the $\Omega = 0$ and 1 fine-structure states of ^{56}FeO. The Λ-doublet splitting is clearly resolved [255].

It will be clear that pure rotational spectra of more complex orbital and spin states, most of which arise in molecules containing transition metal atoms, are still relatively sparse. This will almost certainly change as experimental techniques develop further; a further stimulus is the growing recognition of the importance of these molecules in interstellar and circumstellar space.

10.9. Observation of a pure rotational transition in the H_2^+ molecular ion

Our final discussion in this chapter concerns a very recent and remarkable observation of an electric dipole rotational transition in the H_2^+ molecular ion [257]. Since this

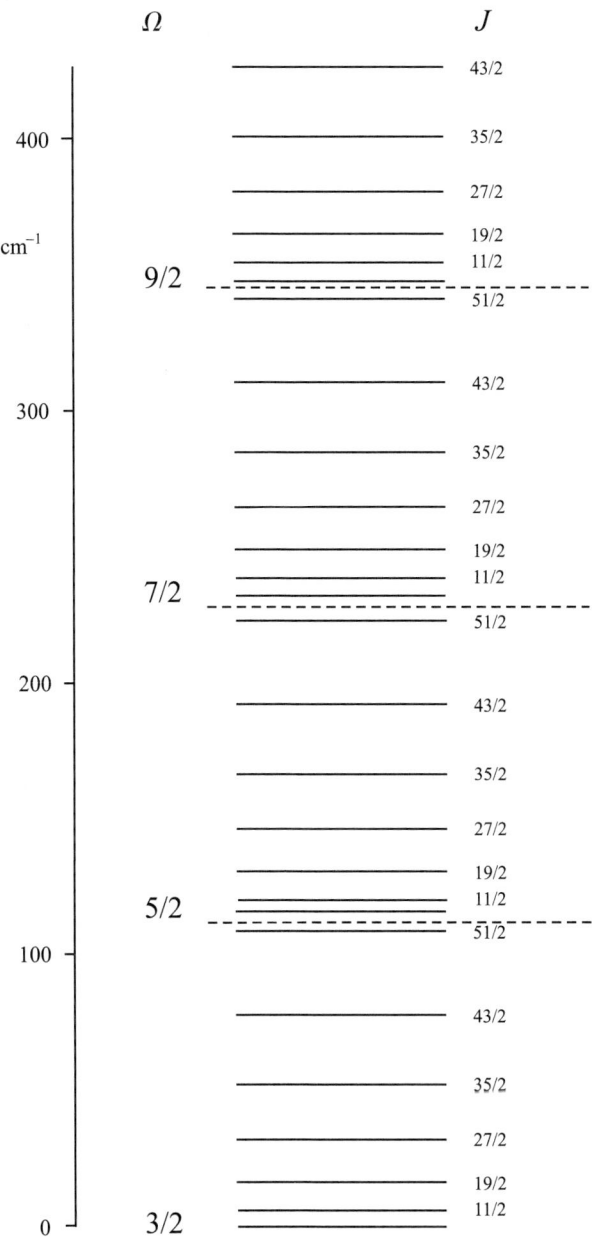

Figure 10.94. Fine-structure components and lower rotational levels for the $X\,^4\Phi$ state of TiCl.

homonuclear ion would not be expected to possess an electric dipole moment, and since adjacent rotational levels belong to *ortho* and *para* proton spin forms, a $\Delta J = \pm 1$ rotational transition should be very strongly electric dipole forbidden. How, then, can such a transition occur? In order to answer this question we first review earlier work on the electric dipole allowed *electronic* spectrum.

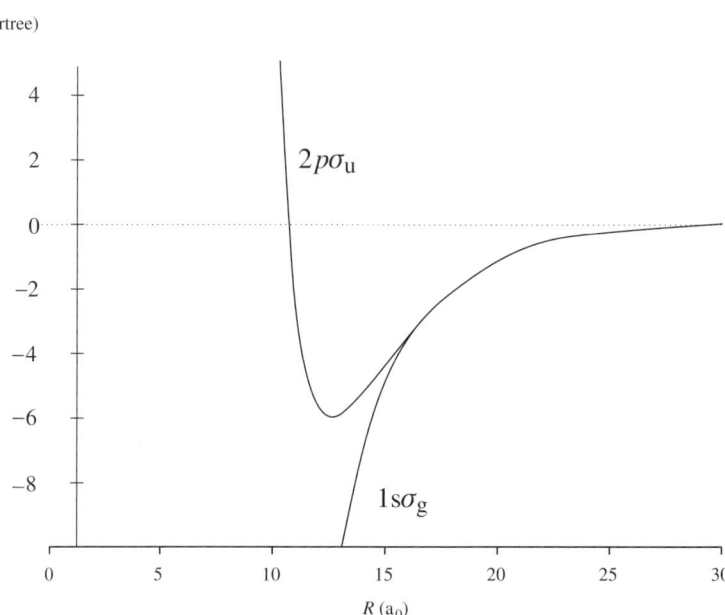

Figure 10.95. Potential energy curves for the $2p\sigma_u$ and $1s\sigma_g$ electronic states of the H_2^+ molecular ion in the near-dissociation region.

In the united atom nomenclature the electronic ground state of H_2^+ may be described as $1s\sigma_g$ and the first excited state as $2p\sigma_u$; these two states become degenerate at the dissociation limit. The ground state is strongly bound with a dissociation energy D_0 of 2.6507 eV and supports twenty vibrational levels. The excited state is repulsive at most internuclear distances, but possesses a long-range minimum because of the attractive charge-induced dipole interaction [258]. The dissociation energy D_e of this state is 13.346 cm^{-1} and it is calculated to support just one vibrational level, with three rotational components [259]. The potential curves in the near-dissociation region are shown in figure 10.95; the highest vibration–rotation levels in the $1s\sigma_g$ state are $v, N =$ 18,3, 19,0 and 19,1. Microwave and millimetre wave *electronic* transitions between these levels and those of the $2p\sigma_u$ excited state have been observed by Carrington, McNab and Montgomerie [260] and by Carrington, Leach and Viant [261]. Similar near-dissociation electronic transitions in D_2^+ have been observed [262].

The most unexpected feature of the observed spectra for H_2^+ was the proton hyperfine structure. An energy level diagram showing the nuclear hyperfine and spin–rotation energy level splittings expected to be observed for the $2p\sigma_u(v = 0, N = 2) \leftarrow 1s\sigma_g(v = 18, N = 3)$ transition is illustrated in figure 10.96. For *ortho*-H_2^+ with total nuclear spin $I = 1$, an appropriate coupling scheme is

$$\boldsymbol{S} + \boldsymbol{I} = \boldsymbol{G},$$
$$\boldsymbol{G} + \boldsymbol{N} = \boldsymbol{F}. \qquad (10.202)$$

The quantum number G takes values 3/2 and 1/2 and the splitting of the levels arises almost entirely from the large Fermi contact hyperfine interaction, which is 1067 MHz. The spin–rotation and dipolar hyperfine splittings are expected to be very small in comparison, and because the electron spin is only very weakly coupled to the nuclear framework in these high-lying energy levels, the hyperfine splitting was expected to be very similar in the two electronic states. Consequently the experimental hyperfine splitting was expected to be very small, probably unresolvable, and it was therefore a considerable surprise when splittings of 6 to 8 MHz were observed experimentally [261]. These splittings correspond to the separation of the $G = 1/2$ and $3/2$ levels, other transitions shown in figure 10.96 being too close to be resolved.

Clearly the proton hyperfine splittings are unexpectedly different in the two electronic states and the reason for this was described by Moss [263]; a detailed discussion of the group theoretical aspects is provided by Bunker and Jensen [264]. In brief, the Fermi contact interaction mixes the g and u electronic state levels which have the same values of N and G. Returning to the energy level diagram shown in figure 10.96, the $2p\sigma_u$ (0,2) level is mixed with both bound and continuum $N = 2$ levels of the $1s\sigma_g$ ground state; the quantum numbers in parenthesis are v and N respectively. However, only the $G = 1/2$ level is affected because $G = 3/2$ is not allowed for $N = 2$ levels of the ground state, since $I = 0$ for $para$-H_2^+. Similarly, (18,3) of the ground state couples with $N = 3$ continuum levels of the excited state, but again only $G = 1/2$ levels are involved; furthermore, there are no bound excited state $N = 3$ levels. The net result of the mixing, which is called symmetry-breaking, is that the $G = 1/2$ levels are pushed to lower energy compared with the $G = 3/2$, as shown in figure 10.96, the effect being greatest for (0,2). The result is to produce a readily observable hyperfine splitting in the electronic spectrum, an example for the (0,2)–(19,1) vibronic component being shown in figure 10.97. Moss [263] was able to calculate the magnitude of the hyperfine symmetry-breaking effect, obtaining results in excellent agreement with experiment.

The inversion operation i which leads to the g/u classification of the electronic states is not a true symmetry operation because it does not commute with the Fermi contact hyperfine Hamiltonian. The operator i acts within the molecule-fixed axis system on electron orbital and vibrational coordinates only. It does not affect electron or nuclear spin coordinates and therefore cannot be used to classify the total wave function of the molecule. Since g and u are not exact labels, it was realised by Bunker and Moss [265] that electric dipole pure rotational transitions were possible in H_2^+, the g/u symmetry breaking (and simultaneous *ortho–para* mixing) being relatively large for levels very close to the dissociation asymptote. The electric dipole transition moment for the 19,1 ← 19,0 rotational transition in the ground electronic state was calculated to be $-0.166\,10$ ea_0 or -0.4222 D; this transition has been observed [257] using the ion beam techniques combined with electric field dissociation described earlier in this chapter, and the spectrum is shown in figure 10.97. The transition frequency agrees very well with that obtained from *ab initio* calculations [266].

Symmetry-breaking arising from the presence of nuclear spin interactions, even in closed shell molecules, was discussed by Herzberg [267] and observed for the I_2 molecule by Pique, Hartmann, Churassy and Bacis [268]. As Bunker and Moss [265] point out, it will be most important for homonuclear open shell molecules with large

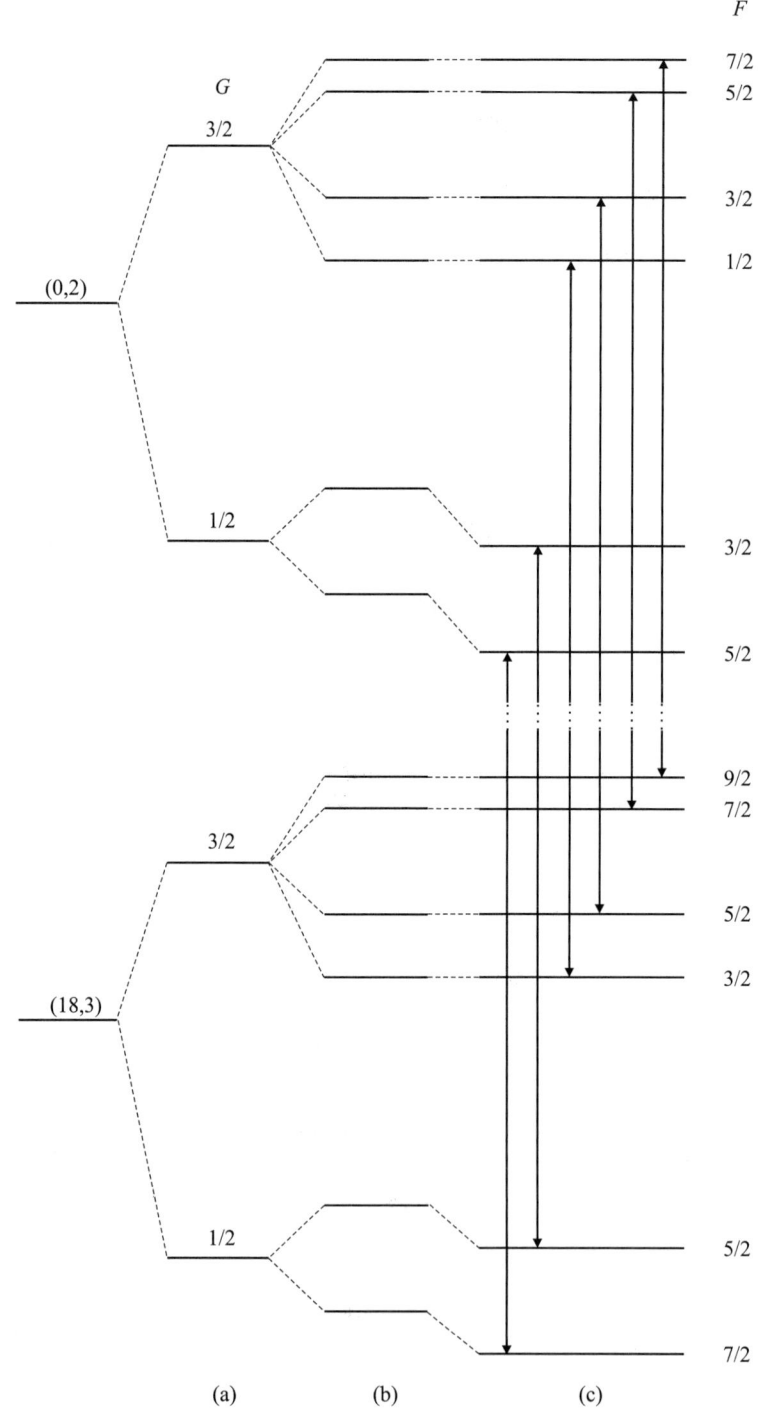

Figure 10.96. H_2^+ (0,2)–(18,3) hyperfine, spin–rotation and symmetry-breaking energy level diagram, showing the six $\Delta F = \Delta N$ transitions. (a) denotes the Fermi contact splitting, (b) is the spin–rotation splitting and (c) shows the effect of symmetry breaking.

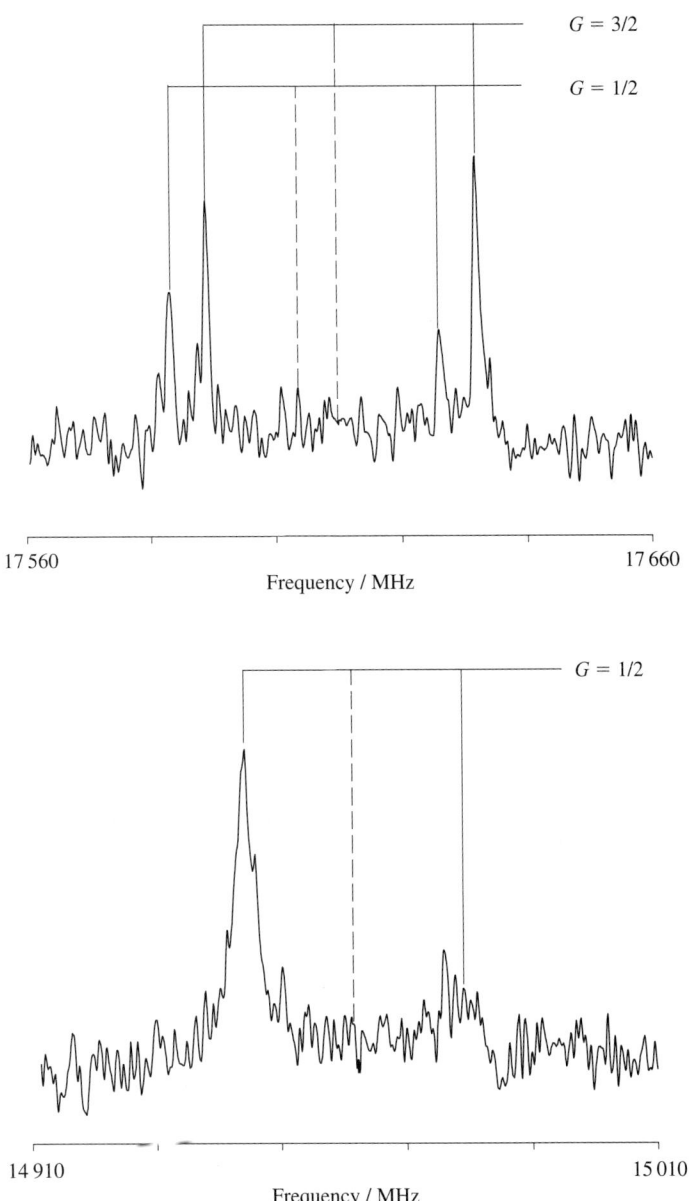

Figure 10.97. *Top:* $2p\sigma_u$ (0,2) ← $1s\sigma_g$ (19,1) electronic transition in H_2^+ showing the proton hyperfine splitting, the parallel and antiparallel Doppler-shifted components, and the position of the rest frequency. This spectrum was obtained after one scan. *Bottom:* $1s\sigma_g$ (19,1) ← $1s\sigma_g$ (19,0) rotational transition in H_2^+, with Doppler shifted components and the position of the rest frequency. This spectrum was obtained after 999 scans [257].

nuclear hyperfine interactions in levels close to a dissociation asymptote where g and u electronic states become degenerate.

References

[1] R.H. Hughes and E.B. Wilson, *Phys. Rev.*, **71**, 562 (1947).
[2] D.R. Johnson, F.X. Powell and W.H. Kirchhoff, *J. Mol. Spectrosc.*, **39**, 136 (1971).
[3] S. Saito and T. Amano, *J. Mol. Spectrosc.*, **34**, 383 (1970).
[4] G.C. Dousmanis, T.M. Sanders and C.H. Townes, *Phys. Rev.*, **100**, 1735 (1955).
[5] R. Kewley, K.V.L.N. Sastry, M. Winnewisser and W. Gordy, *J. Chem. Phys.*, **39**, 2856 (1963).
[6] R.C. Woods, *Rev. Sci. Instr.*, **44**, 282 (1973).
[7] T.A. Dixon and R.C. Woods, *Phys. Rev. Lett.*, **34**, 61 (1975).
[8] K.N. Rao, *Astrophys. J.*, **111**, 50 (1950).
[9] C. Yamada, M. Fujitake and E. Hirota, *J. Chem. Phys.*, **90**, 3033 (1989).
[10] C.S. Gudeman, M.H. Begemann, J. Pfaff and R.J. Saykally, *Phys. Rev. Lett.*, **50**, 727 (1983).
[11] F. Matsushima, T. Oka and K. Takagi, *Phys. Rev. Lett.*, **78**, 1664 (1997).
[12] M.H.W. Gruebele, R.P. Muller and R.J. Saykally, *J. Chem. Phys.*, **84**, 2489 (1986).
[13] H.E. Radford, *Rev. Sci. Instr.*, **39**, 1687 (1968).
[14] T.J. Balle and W.H. Flygare, *Rev. Sci. Instr.*, **52**, 33 (1981).
[15] A.C. Legon, *Ann. Rev. Phys. Chem.*, **34**, 275 (1983).
[16] J.M. Hollas, *Modern Spectroscopy (Third Edition)*, John Wiley and Sons, 1996.
[17] E.J. Campbell, L.W. Buxton, T.J. Balle, M.R. Keenan and W.H. Flygare, *J. Chem. Phys.*, **74**, 829 (1981).
[18] E.J. Campbell, L.W. Buxton, T.J. Balle and W.H. Flygare, *J. Chem. Phys.*, **74**, 813 (1981).
[19] A.C. Legon, *Atomic and Molecular Beam Methods Vol. 2* (ed. G. Scholes), p. 289, Oxford University Press, 1992.
[20] E.J. Campbell, W.G. Read and J.A. Shea, *Chem. Phys. Lett.*, **94**, 69 (1983).
[21] E.J. Campbell and W.G. Read, *J. Chem. Phys.*, **78**, 6490 (1983).
[22] R.D. Suenram, F.J. Lovas, G.T. Fraser, J.Z. Gillies, C.W. Gillies and M. Onda, *J. Mol. Spectrosc.*, **137**, 127 (1989).
[23] F.J. Lovas and R.D. Suenram, *J. Chem. Phys.*, **87**, 2010 (1987).
[24] J. Strong and G.A. Vanesse, *J. Opt. Soc. Amer.*, **49**, 844 (1959).
[25] J. Strong and G.A. Vanesse, *J. Opt. Soc. Amer.*, **50**, 113 (1960).
[26] G.M. Plummer, G. Winnewisser, M. Winnewisser, J. Hahn and K. Reinartz, *J. Mol. Spectrosc.*, **126**, 255 (1987).
[27] R. Schwarz, A. Guarnieri, J.-U. Grabow and J. Doose, *Rev. Sci. Instr.*, **63**, 4108 (1992).
[28] P. Swings and L. Rosenfeld, *Astrophys. J.*, **86**, 483 (1937).
[29] A.E. Douglas and G. Herzberg, *Astrophys. J.*, **94**, 381 (1941).
[30] H.I. Ewen and E.M. Purcell, *Nature*, **168**, 356 (1951).
[31] J.P. Hagen and E.F. McClain, *Astrophys. J.*, **120**, 368 (1954).
[32] S.A. Weinreb, A.H. Barrett, M.L. Meeks and J.C. Henry, *Nature*, **200**, 829 (1963).
[33] R.W. Wilson, K.B. Jefferts and A.A. Penzias, *Astrophys. J.*, **161**, L43 (1970).
[34] J.D. Krauss, *Radio Astronomy*, McGraw-Hill Book Company, 1966.

[35] B.F. Burke and F. Graham-Smith, *An Introduction to Radio Astronomy*, Cambridge University Press, 1997.
[36] K. Rohlfs and T.L. Wilson, *Tools of Radio Astronomy (Third Edition)*, Springer, 1999.
[37] J.W.V. Storey, D.M. Watson and C.H. Townes, *Int. J. IR and MM waves*, **1**, 15 (1980).
[38] J.M. Moran, A.H. Barrett, A.E. Rogers, B.F. Burke, B. Zuckerman, H. Penfield and M.L. Meeks, *Astrophys. J.*, **148**, L69 (1967).
[39] R.W. Wilson, P.M. Solomon, A.A. Penzias and K.B. Jefferts, *Astrophys. J.*, **169**, L35 (1971).
[40] R.W. Wilson, A.A. Penzias, K.B. Jefferts, M. Kutner and P. Thaddeus, *Astrophys. J.*, **167**, L97 (1971).
[41] C.A. Gottlieb and J.A. Ball, *Astrophys. J.*, **184**, L59 (1973).
[42] M. Morris, W. Gilmore, P. Palmer, B.E. Turner and B. Zuckerman, *Astrophys. J.*, **199**, L47 (1975).
[43] C.A. Gottlieb, J.A. Ball, E.W. Gottlieb, C.J. Lada and H. Penfield, *Astrophys. J.*, **200**, L147 (1975).
[44] T.N. Gautier, U. Fink, R.R. Treffers and H.P. Larson, *Astrophys. J.*, **207**, L129 (1976).
[45] S.P. Souza and B.L. Lutz, *Astrophys. J.*, **216**, L49 (1977).
[46] H.S. Liszt and B.E. Turner, *Astrophys. J.*, **224**, L73 (1978).
[47] G.A. Blake, J. Keene and T.G. Phillips, *Astrophys. J.*, **295**, 501 (1985).
[48] B.E. Turner and J. Bally, *Astrophys. J.*, **321**, L75 (1987).
[49] J. Cernicharo and M. Guélin, *Astron. Astrophys.*, **183**, L10 (1987).
[50] J. Cernicharo, C.A. Gottlieb, M.Guélin, P. Thaddeus and J.M. Vrtilek, *Astrophys. J.*, **341**, L25 (1989).
[51] M. Guélin, J. Cernicharo, G. Paubert and B.E. Turner, *Astron. Astrophys.*, **230**, L9 (1990).
[52] D.M. Meyer and K.C. Roth, *Astrophys. J.*, **376**, L49 (1991).
[53] W.B. Latter, C.K. Walker and P.R. Maloney, *Astrophys. J.*, **419**, L97 (1993).
[54] B.E. Turner, *Astrophys. J.*, **388**, L35 (1992).
[55] D.A. Neufeld, J. Zmuidzinas, P. Schilke and T.G. Phillips, *Astrophys. J.*, **488**, L141 (1997).
[56] R.C. Woods, *Phil. Trans. R. Soc. Lond.*, **A324**, 141 (1988).
[57] G. Winnewisser, *Vib. Spectrosc.*, **8**, 241 (1995).
[58] G. Winnewisser, A.F. Krupnov, M.Yu. Tretyakov, M. Liedtke, F. Lewen, A.H. Salek, R. Schieder, A.P. Shkaev and S.V. Volokhov, *J. Mol. Spectrosc.*, **165**, 294 (1994).
[59] F. Lewen, R. Gendriesch, I. Pak, D.G. Paveliev, M. Hepp, R. Schieder and G. Winnewisser, *Rev. Sci. Instr.*, **69**, 32 (1998).
[60] F. Lewen, E. Michael, R. Gendriesch, J. Stutzki and G. Winnewisser, *J. Mol. Spectrosc.*, **183**, 207 (1997).
[61] P. Verhoeve, E. Zwart, M. Versluis, M. Drabbels, J.J. ter Meulen, W.L. Meerts, A. Dymanus and D.B. McLay, *Rev. Sci. Instr.*, **61**, 1612 (1990).
[62] R.C. Cohen, K.L. Busarow, K.B. Laughlin, G.A. Blake, M. Havenith, Y.T. Lee and R.J. Saykally, *J. Chem. Phys.*, **89**, 4494 (1988).
[63] T. Amano, *Astrophys. J.*, **531**, L161 (2000).
[64] T. Amano and E. Hirota, *J. Mol. Spectrosc.*, **53**, 346 (1974).
[65] K.M. Evenson, D.A. Jennings and F.R. Petersen, *Appl. Phys. Lett.*, **44**, 576 (1984).
[66] K.M. Evenson, D.A. Jennings and M.D. Vanek, *Frontiers of Laser Spectroscopy* (eds. Alves, Brown and Hollas), Kluwer Academic Publishers, p. 43, 1987.

[67] I.G. Nolt, J.V. Radostitz, G. DiLonardo, K.M. Evenson, D.A. Jennings, K.R. Leopold, M.D. Vanek, L.R. Zink, A. Hinz and K.V. Chance, *J. Mol. Spectrosc.*, **125**, 274 (1987).
[68] J.M. Brown, L.R. Zink, D.A. Jennings, K.M. Evenson, A. Hinz and I.G. Nolt, *Astrophys. J.*, **307**, 410 (1986).
[69] S.G. Cox, A.D.J. Critchley, I.R. McNab and F.E. Smith, *Meas. Sci. Technol.*, **10**, R101 (1999).
[70] A. Carrington, A.M. Shaw and S.M. Taylor, *J. Chem. Soc. Faraday Trans.*, **91**, 3725 (1995).
[71] O.R. Gilliam, C.M. Johnson and W. Gordy, *Phys. Rev.*, **78**, 140 (1950).
[72] W. Gordy and M. Cowan, *Bull. Amer. Phys. Soc. Ser. II*, **2**, 212 (1957).
[73] B. Rosenblum, A.H. Nethercot, Jr., and C.H. Townes, *Phys. Rev.*, **109**, 400 (1958).
[74] T.D. Varberg and K.M. Evenson, *Astrophys. J.*, **385**, 763 (1992).
[75] G. Winnewisser, S.P. Belov, Th. Klaus and R. Schieder, *J. Mol. Spectrosc.*, **184**, 468 (1997).
[76] S.P. Belov, F. Lewen, Th. Klaus and G. Winnewisser, *J. Mol. Spectrosc.*, **174**, 606 (1995).
[77] E.V. Loewenstein, *J. Opt. Soc. Amer.*, **50**, 1163 (1960).
[78] A.W. Mantz, E.R. Nichols, B.D. Alpert and K.N. Rao, *J. Mol. Spectrosc.*, **35**, 325 (1970).
[79] J.T. Yardley, *J. Mol. Spectrosc.*, **35**, 314 (1970).
[80] W.B. Roh and K.N. Rao, *J. Mol. Spectrosc.*, **49**, 317 (1974).
[81] D.-W. Chen, K.N. Rao and R.S. McDowell, *J. Mol. Spectrosc.*, **61**, 71 (1976).
[82] D.E. Tolliver, G.A. Kyrala and W.H. Wing, *Phys. Rev. Lett.*, **43**, 1719 (1979).
[83] A. Carrington, R.A. Kennedy, T.P. Softley, P.G. Fournier and E.G. Richard, *Chem. Phys.*, **81**, 251 (1983).
[84] P. Bernath and T. Amano, *Phys. Rev. Lett.*, **48**, 20 (1982).
[85] M.W. Crofton, R.S. Altman, N.N. Haese and T. Oka, *J. Chem. Phys.*, **91**, 5882 (1989).
[86] P.R. Bunker, *J. Mol. Spectrosc.*, **68**, 367 (1977).
[87] J.K.G. Watson, *J. Mol. Spectrosc.*, **80**, 411 (1980).
[88] R.D. Suenram, F.J. Lovas, G.T. Fraser and K. Matsumura, *J. Chem. Phys.*, **92**, 4724 (1990).
[89] R.J. Low, T.D. Varberg, J.P. Connelly, A.R. Auty, B.J. Howard and J.M. Brown, *J. Mol. Spectrosc.*, **161**, 499 (1993).
[90] E. Tiemann and J. Hoeft, *Z. Naturforsch.*, **A32**, 1477 (1977).
[91] J. Hoeft and K.P.R. Nair, *Z. Naturforsch.*, **A34**, 1290 (1979).
[92] K.A. Walker and M.C.L. Gerry, *J. Mol. Spectrosc.*, **182**, 178 (1997).
[93] S. Yamamoto, *Chem. Phys. Lett.*, **212**, 113 (1993).
[94] M. Bogey, S. Civis, B. Delcroix, C. Demuynck, A.F. Krupnov, J. Quiguer, M. Yu Tretyakov and A. Walters, *J. Mol. Spectrosc.*, **182**, 85 (1997).
[95] Th. Klaus, S.P. Belov and G. Winnewisser, *J. Mol. Spectrosc.*, **186**, 416 (1997).
[96] J.K.G. Watson, *J. Mol. Spectrosc.*, **80**, 411 (1980).
[97] E. Tiemann, *J. Mol. Spectrosc.*, **91**, 60 (1982).
[98] K. Kobayashi and S. Saito, *J. Chem. Phys.*, **108**, 6606 (1998).
[99] K. Kobayashi and S. Saito, *J. Phys. Chem. A*, **101**, 1068 (1997).
[100] G.M. Plummer, E. Herbst and F.C. De Lucia, *J. Chem. Phys.*, **81**, 4893 (1984).
[101] E.F. Pearson and W. Gordy, *Phys. Rev.*, **177**, 59 (1969).
[102] R.M. Herman and A. Asgharian, *J. Mol. Spectrosc.*, **19**, 305 (1966).
[103] J.K.G. Watson, *J. Mol. Spectrosc.*, **45**, 99 (1973).
[104] R.R. Freeman, A.R. Jacobson, D.W. Johnson and N.F. Ramsey, *J. Chem. Phys.*, **63**, 2597 (1975).

[105] T. Okabayashi and M. Tanimoto, *Astrophys. J.*, **543**, 275 (2000).
[106] T. Okabayashi and M. Tanimoto, *Astrophys. J.*, **487**, 463 (1997).
[107] T. Okabayashi and M. Tanimoto, *J. Mol. Spectrosc.*, **204**, 159 (2000).
[108] M. Goto and S. Saito, *Astrophys. J.*, **452**, L147 (1995).
[109] M. Goto, K. Namiki and S. Saito, *J. Mol. Spectrosc.*, **173**, 585 (1995).
[110] N.R. Erickson, R.L. Snell, R.B. Loren, L. Mundy and R.L. Plambeck, *Astrophys. J.*, **245**, L83 (1981).
[111] N.D. Piltch, P.G. Szanto, T.G. Anderson, C.S. Gudeman, T.A. Dixon and R.C. Woods, *J. Chem. Phys.*, **76**, 3385 (1982).
[112] A. Carrington, D.R.J. Milverton and P.J. Sarre, *Mol. Phys.*, **35**, 1505 (1978).
[113] K.B. Jefferts, A.A. Penzias and R.W. Wilson, *Astrophys. J.*, **161**, L87 (1970).
[114] A.A. Penzias, R.W. Wilson and K.B. Jefferts, *Phys. Rev. Lett.*, **32**, 701 (1974).
[115] B.E. Turner and R.H. Gammon, *Astrophys. J.*, **198**, 71 (1975).
[116] T.A. Dixon and R.C. Woods, *J. Chem. Phys.*, **67**, 3956 (1977).
[117] S. Saito, Y. Endo and E. Hirota, *J. Chem. Phys.*, **78**, 6447 (1983).
[118] R. Beringer and J.G. Castle, *Phys. Rev.*, **78**, 581 (1950).
[119] T. Amano and E. Hirota, *J. Mol. Spectrosc.*, **53**, 346 (1974).
[120] W.M. Welch and M. Mizushima, *Phys. Rev.*, **A5**, 2692 (1972).
[121] K.M. Evenson and M. Mizushima, *Phys. Rev.*, **A6**, 2197 (1972).
[122] S.L. Miller and C.H. Townes, *Phys. Rev.*, **90**, 537 (1953).
[123] J.S. McKnight and W. Gordy, *Phys. Rev. Lett.*, **21**, 1787 (1968).
[124] B.G. West and M. Mizushima, *Phys. Rev.*, **143**, 31 (1966).
[125] W. Steinbach and W. Gordy, *Phys. Rev.*, **A8**, 1753 (1973).
[126] R.W. Zimmerer and M. Mizushima, *Phys. Rev.*, **121**, 152 (1961).
[127] R. Beringer and J.G. Castle, *Phys. Rev.*, **81**, 82 (1951).
[128] A.F. Henry, *Phys. Rev.*, **80**, 396 (1950).
[129] R. Schlapp, *Phys. Rev.*, **51**, 342 (1937).
[130] M. Winnewisser, K.V.L.N. Sastry, R.L. Cook and W. Gordy, *J. Chem. Phys.*, **41**, 1687 (1964).
[131] F.X. Powell and D.R. Lide, *J. Chem. Phys.*, **41**, 1413 (1964).
[132] T. Amano, E. Hirota and Y. Morino, *J. Phys. Soc. Japan*, **22**, 399 (1967).
[133] W.W. Clark and F.C. De Lucia, *J. Mol. Spectrosc.*, **60**, 332 (1976).
[134] Th. Klaus, A.H. Saleck, S.P. Belov, G. Winnewisser, Y. Hirahara, M. Hayashi, E. Kagi and K. Kawaguchi, *J. Mol. Spectrosc.*, **180**, 197 (1996).
[135] H.M. Pickett and T.L. Boyd, *J. Mol. Spectrosc.*, **75**, 53 (1979).
[136] K. Namiki and S. Saito, *Chem. Phys. Lett.*, **252**, 343 (1996).
[137] S. Saito, Y. Endo and E. Hirota, *J. Chem. Phys.*, **82**, 2947 (1985).
[138] C. Yamada, Y. Endo and E. Hirota, *J. Chem. Phys.*, **79**, 4159 (1983).
[139] N. Ohashi, K. Kawaguchi and E. Hirota, *J. Mol. Spectrosc.*, **103**, 337 (1984).
[140] K. Kawaguchi, S. Saito and E. Hirota, *J. Chem. Phys.*, **79**, 629 (1983).
[141] S. Saito, Y. Endo, M. Takami and E. Hirota, *J. Chem. Phys.*, **78**, 116 (1983).
[142] M. Tanimoto, S. Saito, Y. Endo and E. Hirota, *J. Mol. Spectrosc.*, **100**, 205 (1983).
[143] Y. Endo, S. Saito and E. Hirota, *J. Mol. Spectrosc.*, **92**, 443 (1982).
[144] J.R. Morton and K.F. Preston, *J. Mag. Res.*, **30**, 577 (1978).
[145] T. Sakamaki, T. Okabayashi and M. Tanimoto, *J. Chem. Phys.*, **109**, 7169 (1998).

[146] T. Sakamaki, T. Okabayashi and M. Tanimoto, *J. Chem. Phys.*, **111**, 6345 (1999).
[147] A. Carrington, D.H. Levy and T.A. Miller, *Proc. R. Soc. Lond.*, **A293**, 108 (1966).
[148] C.A. Arrington, A.M. Falick and R.J. Myers, *J. Chem. Phys.*, **55**, 909 (1971).
[149] A. Scalabrin, R.J. Saykally, K.M. Evenson, H.E. Radford and M. Mizushima, *J. Mol. Spectrosc.*, **89**, 344 (1981).
[150] G. Cazzoli, C. Degli Esposti and P.G. Favero, *Chem. Phys. Lett.*, **100**, 99 (1983).
[151] K.W. Hillig, C.C.W. Chiu, W.G. Read and E.A. Cohen, *J. Mol. Spectrosc.*, **109**, 205 (1985).
[152] T.A. Miller, *J. Chem. Phys.*, **54**, 330 (1971).
[153] S. Saito, *J. Chem. Phys.*, **53**, 2544 (1970).
[154] K. Kobayashi, M. Goto, S. Yamamoto and S. Saito, *J. Chem. Phys.*, **104**, 8865 (1996).
[155] A.H. Saleck, K.M.T. Yamada and G. Winnewisser, *Mol. Phys.*, **72**, 1135 (1991).
[156] W.L. Meerts and A. Dymanus, *J. Mol. Spectrosc.*, **44**, 320 (1972).
[157] E. Klisch, S.P. Belov, R. Schieder, G. Winnewisser and E. Herbst, *Mol. Phys.*, **97**, 65 (1999).
[158] W.L. Meerts, *Chem. Phys.*, **14**, 421 (1976).
[159] A.H. Salek, G. Winnewisser and K.M.T. Yamada, *Mol. Phys.*, **76**, 1443 (1992).
[160] C. Amiot, R. Bacis and G. Guelachvili, *Can. J. Phys.*, **56**, 251 (1978).
[161] D. McGonagle, W.M. Irvine, Y.C. Minh and L.M. Ziurys, *Astrophys. J.*, **359**, 121 (1990).
[162] M. Gerin, Y. Viala, F. Pauzat and Y. Ellinger, *Astron. Astrophys.*, **266**, 463 (1992).
[163] L.M. Ziurys, D. McGonagle, Y.C. Minh and W.M. Irvine, *Astrophys. J.*, **373**, 535 (1991).
[164] M. Gerin, Y. Viala and F. Casoli, *Astron. Astrophys.*, **268**, 212 (1993).
[165] J.J. ter Meulen and A. Dymanus, *Astrophys. J.*, **172**, L21 (1972).
[166] W.L. Meerts and A. Dymanus, *Can. J. Phys.*, **53**, 2123 (1975).
[167] J. Farhoomand, G.A. Blake and H.M. Pickett, *Astrophys. J.*, **291**, L19 (1985).
[168] J. Farhoomand, G.A. Blake, M.A. Frerking and H.M. Pickett, *J. App. Phys.*, **57**, 1763 (1985).
[169] J.W.V. Storey, D.M. Watson and C.H. Townes, *Astrophys. J.*, **244**, L27 (1981).
[170] D.M. Watson, R. Genzel, C.H. Townes and J.W.V. Storey, *Astrophys. J.*, **298**, 316 (1985).
[171] J.M. Brown, C.M.L. Kerr, F.D. Wayne, K.M. Evenson and H.E. Radford, *J. Mol. Spectrosc.*, **86**, 544 (1981).
[172] T.D. Varberg and K.M. Evenson, *J. Mol. Spectrosc.*, **157**, 551 (1993).
[173] K.M. Evenson, H.E. Radford and M.M. Moran, *App. Phys. Lett.*, **18**, 426 (1971).
[174] O.E.H. Rydbeck, J. Elldér, W.M. Irvine, A. Sume and A. Hjalmarson, *Astron. Astrophys.*, **34**, 479 (1974).
[175] E.L. Hill and J.H. Van Vleck, *Phys. Rev.*, **32**, 250 (1923).
[176] O.E.H. Rydbeck, J. Elldér and W.M. Irvine, *Nature*, **246**, 466 (1973).
[177] B.E. Turner and B. Zuckerman, *Astrophys. J.*, **187**, L59 (1974).
[178] C.R. Brazier and J.M. Brown, *J. Chem. Phys.*, **78**, 1608 (1983).
[179] T.C. Steimle, D.R. Woodward and J.M. Brown, *Astrophys. J.*, **294**, L59 (1985).
[180] M. Bogey, C. Demuynck and J.L. Destombes, *Chem. Phys. Lett.*, **100**, 105 (1983).
[181] C.R. Brazier and J.M. Brown, *Can. J. Phys.*, **62**, 1563 (1984).
[182] L.M. Ziurys and B.E. Turner, *Astrophys. J.*, **292**, L25 (1985).
[183] J.M. Brown and K.M. Evenson, *J. Mol. Spectrosc.*, **98**, 392 (1983).

[184] J.M. Brown and K.M. Evenson, *Astrophys. J.*, **268**, L51 (1983).
[185] S.A. Davidson, K.M. Evenson and J.M. Brown, *Astrophys. J.*, **546**, 330 (2001).
[186] G.J. Stacey, J.B. Lugten and R. Genzel, *Astrophys. J.*, **313**, 859 (1987).
[187] A.N. Jette and P. Cahill, *Phys. Rev.*, **160**, 35 (1967).
[188] G.C. Lie, J. Hinze and B. Liu, *J. Chem. Phys.*, **59**, 1872 (1973).
[189] G.C. Lie, J. Hinze and B. Liu, *J. Chem. Phys.*, **59**, 1887 (1973).
[190] R.S. Mulliken and A. Christy, *Phys. Rev.*, **38**, 87 (1931).
[191] D.H. Levy and J. Hinze, *Astrophys. J.*, **200**, 236 (1975).
[192] R.A. Frosch and H.M. Foley, *Phys. Rev.*, **88**, 1337 (1952).
[193] A. Kasdan, E. Herbst and W.C. Lineberger, *Chem. Phys. Lett.*, **31**, 78 (1975).
[194] T. Nelis, J.M. Brown and K.M. Evenson, *J. Chem. Phys.*, **92**, 4067 (1990).
[195] A. Carrington and B.J. Howard, *Mol. Phys.*, **18**, 225 (1970).
[196] R.J. Saykally, K.G. Lubic, A. Scalabrin and K.M. Evenson, *J. Chem. Phys.*, **77**, 58 (1982).
[197] F.C. Van den Heuvel, W.L. Meerts and A. Dymanus, *Chem. Phys. Lett.*, **88**, 59 (1982).
[198] S. Saito, Y. Endo, M. Takami and E. Hirota, *J. Chem. Phys.*, **78**, 116 (1983).
[199] M. Tanimoto, S. Saito, Y. Endo and E. Hirota, *J. Mol. Spectrosc.*, **100**, 205 (1983).
[200] M. Tanimoto, S. Saito, Y. Endo and E. Hirota, *J. Mol. Spectrosc.*, **116**, 499 (1986).
[201] A. Carrington, P.N. Dyer and D.H. Levy, *J. Chem. Phys.*, **47**, 1756 (1967).
[202] T. Amano, S. Saito, E. Hirota, Y. Morino, D.R. Johnson and F.X. Powell, *J. Mol. Spectrosc.*, **30**, 275 (1969).
[203] M. Tanimoto, S. Saito, Y. Endo and E. Hirota, *J. Mol. Spectrosc.*, **103**, 330 (1984).
[204] Y. Endo, S. Saito and E. Hirota, *J. Mol. Spectrosc.*, **94**, 199 (1982).
[205] T. Amano, S. Saito, E. Hirota and Y. Morino, *J. Mol. Spectrosc.*, **32**, 97 (1969).
[206] T.C. Steimle, K. Namiki and S. Saito, *J. Chem. Phys.*, **107**, 6109 (1997).
[207] T.C. Steimle, M. Tanimoto, K. Namiki and S. Saito, *J. Chem. Phys.*, **108**, 7616 (1998).
[208] Y. Endo, S. Saito and E. Hirota, *J. Mol. Spectrosc.*, **92**, 443 (1982).
[209] C. Yamada, M. Fujitake and E. Hirota, *J. Chem. Phys.*, **91**, 137 (1989).
[210] C. Yamada, M. Fujitake and E. Hirota, *J. Chem. Phys.*, **90**, 3033 (1989).
[211] A. Carrington, C.A. Leach, A.J. Marr, A.M. Shaw, M.R. Viant, J.M. Hutson and M.M. Law, *J. Chem. Phys.*, **102**, 2379 (1995).
[212] I. Dabrowski, G. Herzberg and K. Yoshino, *J. Mol. Spectrosc.*, **89**, 491 (1981).
[213] P.E. Siska, *J. Chem. Phys.*, **85**, 7497 (1986).
[214] J.M. Brown, J.T. Hougen, K.P. Huber, J.W.C. Johns, I. Kopp, H. Lefebvre Brion, A.J. Merer, D.A. Ramsay, J. Rostas and R.N. Zare, *J. Mol. Spectrosc.*, **55**, 500 (1975).
[215] L. Veseth, *J. Phys. B At. Mol. Phys.*, **6**, 1473 (1973).
[216] P.R. Bunker, *Chem. Phys. Lett.*, **27**, 332 (1974).
[217] J.M. Hutson, *Comp. Phys. Comm.*, **84**, 1 (1994).
[218] B.R. Johnson, *J. Chem. Phys.*, **69**, 4678 (1978).
[219] M.M. Law and J.M. Hutson, *Comp. Phys. Comm.*, **102**, 252 (1997).
[220] K.T. Tang and J.P. Toennies, *J. Chem. Phys.*, **80**, 3726 (1984).
[221] A. Carrington, C.H. Pyne, A.M. Shaw, S.M. Taylor, J.M. Hutson and M.M. Law, *J. Chem. Phys.*, **105**, 8602 (1996).
[222] R.S. Mulliken, *Rev. Mod. Phys.*, **2**, 60 (1930).
[223] R.J. Saykally, T.A. Dixon, T.G. Anderson, P.G. Szanto and R.C. Woods, *J. Chem. Phys.*, **87**, 6423 (1987).

[224] N. Carballo, H.E. Warner, C.S. Gudeman and R.C. Woods, *J. Chem. Phys.*, **88**, 7273 (1988).
[225] S. Yamamoto and S. Saito, *J. Chem. Phys.*, **89**, 1936 (1988).
[226] A. Wada and H. Kanamori, *J. Mol. Spectrosc.*, **200**, 196 (2000).
[227] K.F. Freed, *J. Chem. Phys.*, **45**, 4214 (1966).
[228] R.W. Field, S.G. Tilford, R.A. Howard and J.D. Simmons, *J. Mol. Spectrosc.*, **44**, 347 (1972).
[229] P.F. Bernath, S.A. Rogers, L.C. O'Brien, C.R. Brazier and A.D. McLean, *Phys. Rev. Lett.*, **60**, 197 (1988).
[230] R. Mollaaghababa, C.A. Gottlieb, J.M. Vrtilek and P. Thaddeus, *Astrophys. J.*, **352**, L21 (1990).
[231] R. Mollaaghababa, C.A. Gottlieb and P. Thaddeus, *J. Chem. Phys.*, **98**, 968 (1993).
[232] M. Bogey, C. Demuynck and J.L. Destombes, *Astron. Astrophys.*, **232**, L19 (1990).
[233] J.M. Brown and A.J. Merer, *J. Mol. Spectrosc.*, **74**, 488 (1979).
[234] M.D. Allen, T.C. Pesch and L.M. Ziurys, *Astrophys. J.*, **472**, L57 (1996).
[235] W.J. Balfour, J. Cao, C.V.V. Prasad and C.X.W. Qian, *J. Chem. Phys.*, **103**, 4046 (1995).
[236] J.M. Brown, A.S.-C. Cheung and A.J. Merer, *J. Mol. Spectrosc.*, **124**, 464 (1987).
[237] R.D. Suenram, G.T. Fraser, F.J. Lovas and C.W. Gillies, *J. Mol. Spectrosc.*, **148**, 114 (1991).
[238] G. Cheval, J.-L. Féménias, A.J. Merer and U. Sassenberg, *J. Mol. Spectrosc.*, **131**, 113 (1988).
[239] J.M. Brom, C.H. Durham and W. Weltner, *J. Chem. Phys.*, **61**, 970 (1974).
[240] M. Tanimoto, S. Saito and T. Okabayashi, *Chem. Phys. Lett.*, **242**, 153 (1995).
[241] J.M. Delaval, C. Dufour and J. Schamps, *J. Phys. B. Atom. Mol. Phys.*, **13**, 4757 (1980).
[242] M.D. Allen, B.Z. Li and L.M. Ziurys, *Chem. Phys. Lett.*, **270**, 517 (1997).
[243] M.D. Allen and L.M. Ziurys, *Astrophys. J.*, **479**, 1237 (1996).
[244] M.D. Allen and L.M. Ziurys, *J. Chem. Phys.*, **106**, 3494 (1997).
[245] T. Okabayashi and M. Tanimoto, *J. Chem. Phys.*, **105**, 7421 (1996).
[246] T. Oike, T. Okabayashi and M. Tanimoto, *Astrophys. J.*, **445**, L68 (1995).
[247] T. Oike, T. Okabayashi and M. Tanimoto, *J. Chem. Phys.*, **109**, 3501 (1998).
[248] K. Namiki and S. Saito, *J. Chem. Phys.*, **107**, 8848 (1997).
[249] R.S. Ram, C.N. Jarman and P.F. Bernath, *J. Mol. Spectrosc.*, **161**, 445 (1993).
[250] B. Pinchemel and J. Schamps, *Chem. Phys.*, **18**, 481 (1976).
[251] M. Tanimoto, T. Sakamaki and T. Okabayashi, *J. Mol. Spectrosc.*, **207**, 66 (2001).
[252] E. Yamazaki, T. Okabayashi and M. Tanimoto, *Astrophys. J.*, **551**, L199 (2001).
[253] Y. Endo, S. Saito and E. Hirota, *Astrophys. J.*, **278**, L131 (1984).
[254] T. Kröckertskothen, H. Knöckel and E. Tiemann, *Mol. Phys.*, **62**, 1031 (1987).
[255] M.D. Allen, L.M. Ziurys and J.M. Brown, *Chem. Phys. Lett.*, **257**, 130 (1996).
[256] A. Maeda, T. Hirao, P.F. Bernath and T. Amano, *J. Mol. Spectrosc.*, **210**, 250 (2001).
[257] A.D.J. Critchley, A.N. Hughes and I.R. McNab, *Phys. Rev. Lett.*, **86**, 1725 (2001).
[258] C.A. Coulson, *Proc. R. Soc. Edinburgh*, **A61**, 20 (1941).
[259] J.M. Peek, *J. Chem. Phys.*, **50**, 4595 (1969).
[260] A. Carrington, I.R. McNab and C.A. Montgomerie, *Chem. Phys. Lett.*, **160**, 237 (1989).
[261] A. Carrington, C.A. Leach and M.R. Viant, *Chem. Phys. Lett.*, **206**, 77 (1993).
[262] A. Carrington, I.R. McNab, C.A. Montgomerie and R.A. Kennedy, *Mol. Phys.*, **67**, 711 (1989).

[263] R.E. Moss, *Chem. Phys. Lett.*, **206**, 83 (1993).
[264] P.R. Bunker and Per Jensen, *Molecular Symmetry and Spectroscopy*, NRC Research Press, Ottawa, 1998.
[265] P.R. Bunker and R.E. Moss, *Chem. Phys. Lett.*, **316**, 266 (2000).
[266] R.E. Moss, *Mol. Phys.*, **80**, 1541 (1993).
[267] G. Herzberg, *Spectra of Diatomic Molecules*, D.Van Nostrand Company, Inc., Princeton, p. 139, 1950.
[268] J.P. Pique, F. Hartmann, S. Churassy and R. Bacis, *J. Phys. (Paris)*, **47**, 1909 (1986).

11 Double resonance spectroscopy

11.1. Introduction

Double resonance spectroscopy involves the simultaneous use of two spectroscopic radiation sources, often of quite different wavelengths. Figure 11.1(a) illustrates the simplest example of many possible variations. High-frequency electronic excitation (f_1) is combined with microwave or radiofrequency radiation (f_2); the objective is usually to observe and measure the lower frequency spectrum by making use of the sensitivity advantages provided by the higher frequency radiation. Detection of the fluorescence intensity from the intermediate state E_2 provides a monitor of the population of the state. The lower frequency transition f_2 changes the population of E_2, and hence changes the fluorescence intensity. Many of the experiments to be described in this chapter depend upon this simple scheme. Such experiments have been extremely valuable, particularly in the study of short-lived species such as neutral free radicals, molecular ions, or metastable excited electronic states. Their success usually depends on prior knowledge and study of the high-frequency spectrum, as we shall see. In other cases, however, the two radiation sources may be of similar wavelengths; microwave/microwave double resonance, for example, has proved to be a powerful method for confirming otherwise uncertain spectroscopic assignments.

As is often the case, the initial experiments were developed by atomic spectroscopists. Figure 11.1(b) illustrates an example from atomic physics, described by Brossel and Bitter [1]. Mercury atoms are excited by a mercury lamp from the ^{1}S ground state to the ^{3}P$_1$ excited state, in the presence of a small applied magnetic field. The ^{3}P$_1$ state exhibits a Zeeman splitting, as shown in the figure, each Zeeman sublevel being characterised by its M quantum number, which can take the values $+1, 0$ and -1. The incident light is polarised, and induces a $\Delta M = 0$ (π polarisation) electronic excitation. An oscillating magnetic field applied perpendicular to the static field induces $\Delta M = \pm 1$ transitions, which for a static field of 10 G are radiofrequency transitions. The result of these transitions is a change in the polarisation of the fluorescence, which can be easily detected. The experiment was also applied to more complex atomic systems, such as potassium atoms where nuclear hyperfine splittings were resolved [2]. This type of double resonance experiment has also been applied to molecular systems, as we shall see.

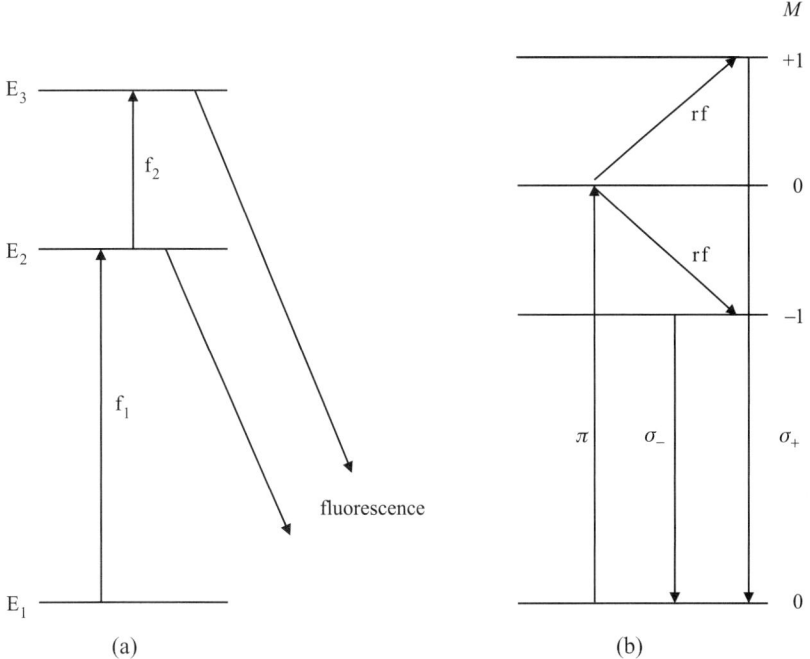

Figure 11.1. (a) Principles of the microwave/optical double resonance method. (b) Change of polarisation of fluorescent light resulting from $\Delta M = \pm 1$ radiofrequency transitions.

Before commencing our description of true double resonance studies of molecular systems, we describe an extremely important precursor study of the CN radical, which pointed the way to many later experiments.

11.2. Radiofrequency and microwave studies of CN in its excited electronic states

The CN radical can be produced by the reaction of nitrogen atoms with almost any organic compound, a process which has been extensively studied by electronic emission spectroscopy. The electronic structure of CN in its ground state and two relevant excited states may be written in the following simple molecular orbital form.

$$(1\sigma)^2(2\sigma)^2(3\sigma)^2(4\sigma)^2(1\pi)^4(5\sigma)^1: \quad X\,^2\Sigma^+(T_e = 0)$$
$$(1\sigma)^2(2\sigma)^2(3\sigma)^2(4\sigma)^2(1\pi)^3(5\sigma)^2: \quad A\,^2\Pi\,(T_e = 9241.7\text{ cm}^{-1})$$
$$(1\sigma)^2(2\sigma)^2(3\sigma)^2(4\sigma)^1(1\pi)^4(5\sigma)^2: \quad B\,^2\Sigma^+(T_e = 25751.8\text{ cm}^{-1})$$

The chemical reaction produces CN in excited vibrational levels of the A state, and energy transfer from the $v = 10$ level of the A state to the $v = 0$ level of the B state leads to strong fluorescent emission from the B state. The possibility of detecting microwave transitions in these excited states was investigated by Radford and Broida [3] and first realised experimentally by Evenson, Dunn and Broida [4]. We will discuss the nature

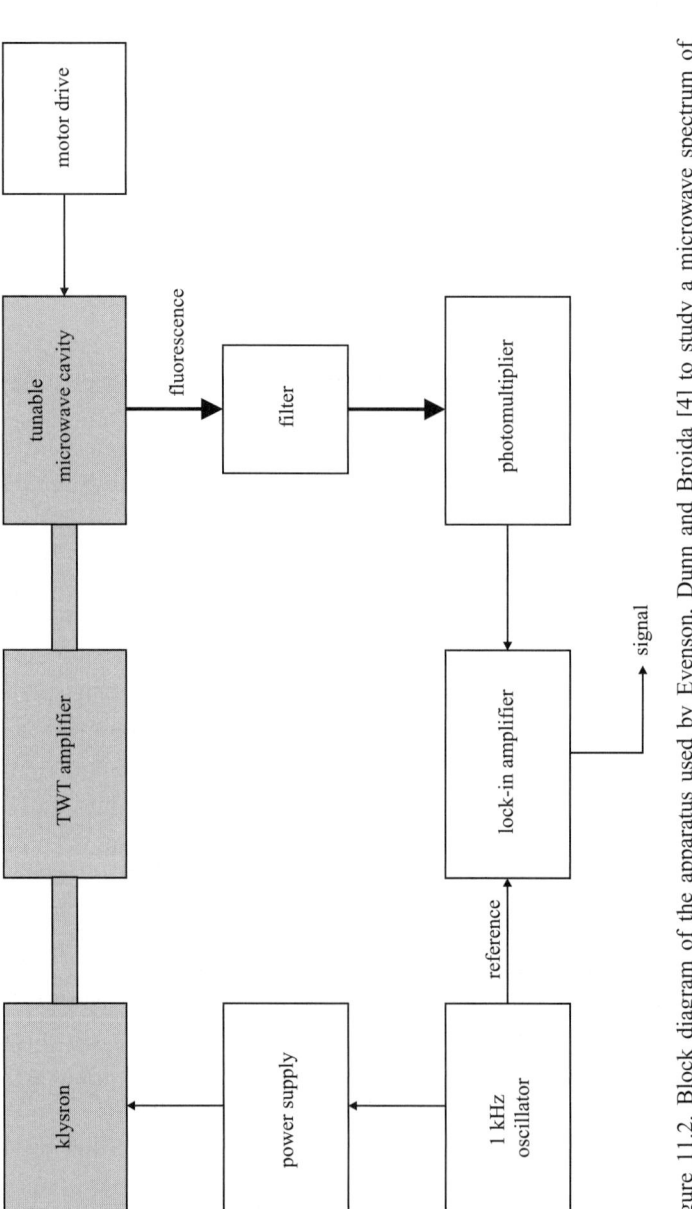

Figure 11.2. Block diagram of the apparatus used by Evenson, Dunn and Broida [4] to study a microwave spectrum of electronically excited CN radicals.

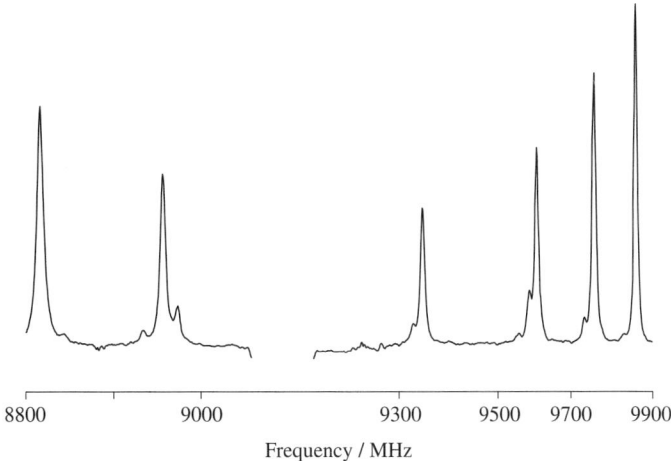

Figure 11.3. Microwave spectrum of electronically excited CN radicals [4].

of the transitions in due course, but first we describe the details of their experiment with the aid of figure 11.2. The heart of the apparatus was a frequency-tunable TE_{011} cavity, operating over the microwave range 8750 to 10 375 MHz. Active nitrogen produced in a 2450 MHz discharge cavity was mixed with methylene chloride vapour inside the resonant cavity, and the $B\,^2\Sigma^+ \to X\,^2\Sigma^+$ fluorescence detected with a photomultiplier, preceded by an appropriate filter. The microwave power, obtained from a reflex klystron, was amplitude modulated at 1 kHz, and the modulated fluorescence signal detected with a lock-in amplifier. The microwave spectrum obtained is shown in figure 11.3.

The transitions detected are indicated in the energy level diagram presented in figure 11.4. As we have already mentioned, the $v = 10$ level of the $A\,^2\Pi$ state lies close in energy to the $v = 0$ level of the $B\,^2\Sigma^+$ state, and the transitions detected by Evenson, Dunn and Broida [4] are the type (i) transitions shown in figure 11.4. They are, in fact, electronic transitions. In the absence of any interaction between the two electronic states we would expect each rotational level of the $^2\Sigma^+$ state to show a spin–rotation doublet splitting, and each rotational level of the $^2\Pi$ state to show a Λ-doublet splitting. In this instance, however, one spin component of the $^2\Sigma^+$ state interacts strongly with one Λ-doublet component of the $^2\Pi_{3/2}$ state, yielding two perturbed states which are labelled $\Sigma(p)$ and $\Pi(p)$ in figure 11.4. The other two states do not interact and therefore remain unperturbed; they are labelled $\Sigma(u)$ and $\Pi(u)$. The diagram refers to the $N = 4$ rotational level of the $^2\Sigma^+$ state ($v = 0$) and the $J = 7/2$ rotational level of the $^2\Pi_{3/2}$ state ($v = 10$). Hyperfine interaction with the $I = 1$ nuclear spin of ^{14}N produces the set of levels labelled by their F and J quantum numbers. The selection rule $\Delta F = \pm 1$ or 0 gives the possibility of 13 hyperfine components, all of which were observed and measured. For both types of electronic state we have already developed the rotational and hyperfine theory, including the effective Hamiltonian matrix elements. We will discuss the values of the molecular parameters obtained from the analysis in due course.

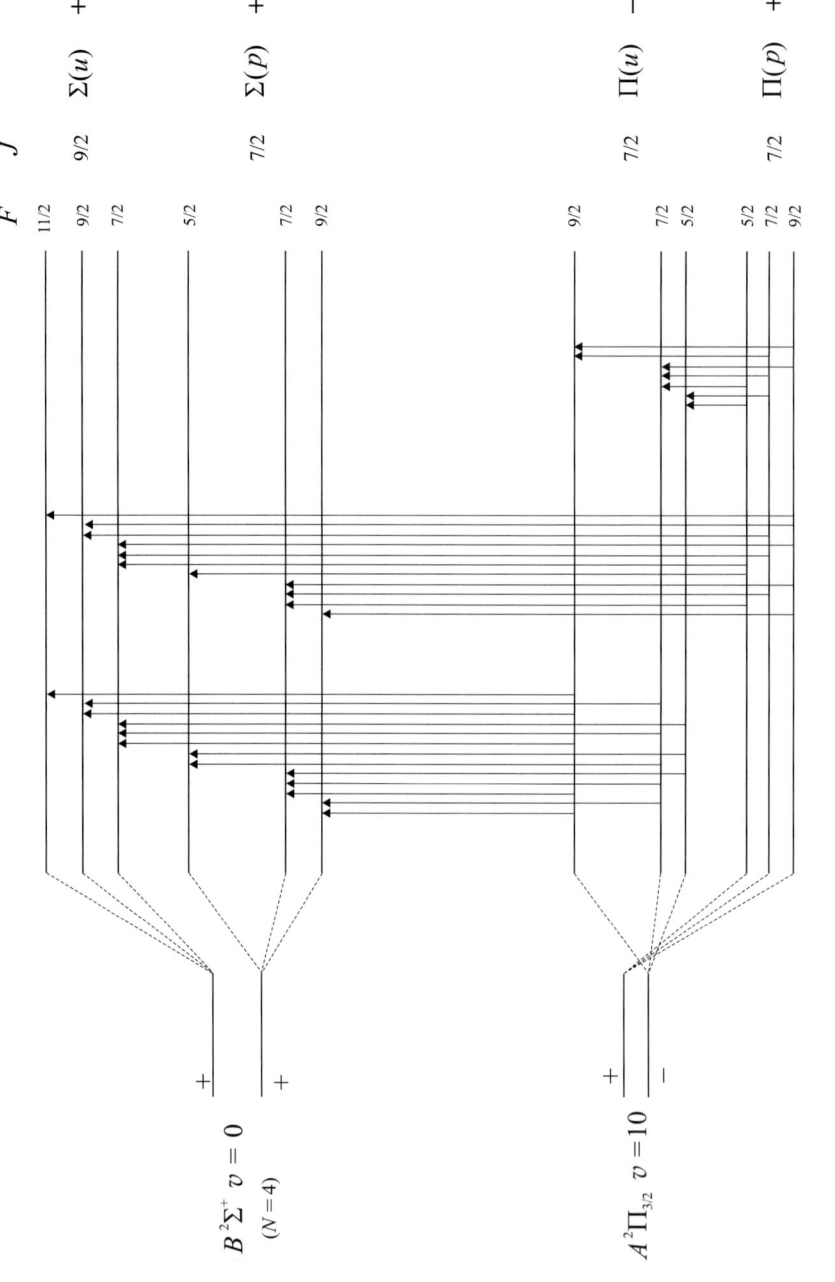

Figure 11.4. Energy levels and microwave transitions [4] involving the $A\,^2\Pi_{3/2}\,(v=10)$ and $B\,^2\Sigma^+\,(v=0)$ excited electronic states of CN.

Transitions of type (ii) were detected subsequently by Evenson [5]. He used a rectangular cavity operating in the TE_{108} mode, with a large optical window which enabled him to observe the optical emission at various locations along the axis of the cavity. In this way he was able to distinguish between electric and magnetic dipole microwave transitions. The type (i) transitions shown in figure 11.4 are conventional electronic, electric-dipole transitions, whereas the type (ii) transitions, whilst still electronic, occur between states of the same parity and are magnetic-dipole allowed only. This experiment demonstrates an important aspect of double resonance experiments, namely, that microwave magnetic-dipole transitions can be detected with essentially the same sensitivity as electric-dipole transitions, provided sufficient microwave power is available.

Finally, transitions of type (iii) in figure 11.4 were studied by Evenson [6] and by Pratt and Broida [7]. These are hyperfine components of an electric-dipole, Λ-doublet transition within the $^2\Pi_{3/2}$ state. They are therefore conventional, except for the enhanced splitting between the Λ-doublet states caused by the perturbation from the close-lying $^2\Sigma^+$ state. They are still detected through changes in fluorescent emission from the $B\ ^2\Sigma^+$ state.

The theory of the ^{14}N hyperfine interaction in the $B\ ^2\Sigma^+$ state of CN has been considered by Radford [8]; the effective Hamiltonian is the same as that used in chapter 10 for the ground state of CN. Using Frosch and Foley constants [9] it is written, in cartesian form,

$$\mathcal{H}_{\text{hfs}} = B\mathbf{N}^2 + \gamma\mathbf{S}\cdot\mathbf{N} + b\mathbf{I}\cdot\mathbf{S} + cI_zS_z + eq_0Q\frac{\{3I_z^2 - \mathbf{I}^2\}}{4I(2I-1)}. \quad (11.1)$$

In the alternative spherical tensor form, it is written

$$\mathcal{H}_{\text{hfs}} = B\mathbf{N}^2 + \gamma T^1(\mathbf{S})\cdot T^1(\mathbf{N}) + b_F T^1(\mathbf{I})\cdot T^1(\mathbf{S})$$
$$- \sqrt{10}\, g_S\mu_B g_N\mu_N(\mu_0/4\pi)T^1(\mathbf{S},\mathbf{C}^2)\cdot T^1(\mathbf{I}) - eT^2(\mathbf{Q})\cdot T^2(\nabla\mathbf{E}). \quad (11.2)$$

The matrix elements of (11.2) in a case (b) representation for CN in its ground state were derived in chapter 10; Radford [8] obtained the following values of the constants (in MHz):

$$b = 467, \quad c = 60, \quad b_F = b + (c/3) = 487, \quad t_0 = c/3 = 20, \quad eq_0Q = -5.$$

As we have shown in Appendix 8.5, and elsewhere, t_0 is the axial component of the dipolar interaction obtained from the fourth term in equation (11.2). The large value of the Fermi contact constant is consistent with a model in which the unpaired electron occupies a σ-type molecular orbital which has 45% N atom s character. Radford produced convincing arguments to show that the model is also consistent with the small dipolar hyperfine constant, and also the electric quadrupole coupling constant.

This pioneering study of the CN radical was not a true double resonance experiment because it did not need a separate radiation source to produce the initial electronic excitation. Nevertheless it has many of the other elements common to double resonance, as we shall see, and was a landmark study in the field.

11.3. Early radiofrequency or microwave/optical double resonance studies

11.3.1. Radiofrequency/optical double resonance of CS in its excited $A\,^1\Pi$ state

The first radiofrequency/optical double resonance studies of molecules were published almost simultaneously. Observations of OH and OD were described by German and Zare [10] late in 1969, and will be discussed in detail in the next subsection. A few months later studies of the CS molecule in its excited $A\,^1\Pi$ were reported by Silvers, Bergeman and Klemperer [11], with more detailed results described later by Field and Bergeman [12]. We now describe these investigations, which are in some ways simpler than those of OH because of the absence of electron and nuclear spin effects in the CS $^1\Pi$ state.

The electronic ground state of CS is the same as that of CO, that is, an $X\,^1\Sigma^+$ state. The four lowest excited electronic states are all triplet states, but close by in energy there is an $A\,^1\Pi$ state, lying 38 794 cm^{-1} above the ground state. This state can be populated by allowed electronic excitation from the ground state, and is therefore

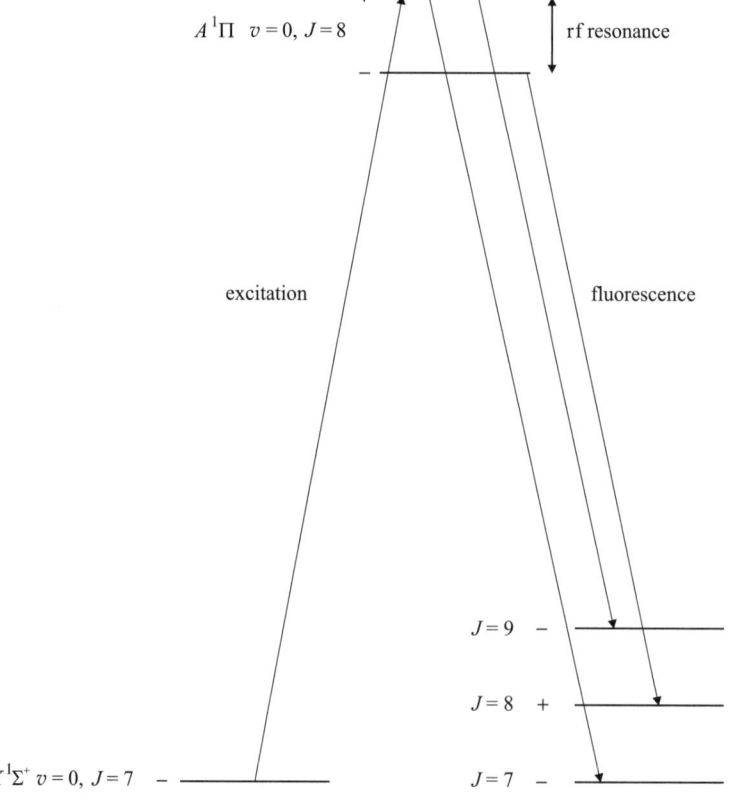

Figure 11.5. Ground and excited state levels involved in the radiofrequency/optical double resonance study of CS in its $A\,^1\Pi$ state [12].

a potential candidate for radiofrequency or microwave double resonance studies. The energy levels involved are illustrated in figure 11.5. The $J = 7$ level of the $X\,^1\Sigma^+$ state, which has $(-)$ parity, is excited to the upper $(+)$ parity Λ-doublet component of the $J = 8$ level of the $A\,^1\Pi$ state; an intense atomic manganese emission line at 2576 Å was used for the excitation. Fluorescence was observed in a direction perpendicular to the exciting light, and was not dispersed. Radiofrequency transitions between the Λ-doublet parity components resulted in changes in the fluorescence intensity, which were detected. The radiofrequency field was amplitude- modulated, and the modulated component of the fluorescence detected with a lock-in amplifier. The heart of the apparatus used is illustrated in figure 11.6, and an example of the resonances obtained is shown in figure 11.7(a); the signal-to-noise ratio is not large, but it is sufficient.

An important extension of the measurements was the application of a static electric field, applied between the Stark plates, and the consequent observation of Stark components in the double resonance lines, from which the electric dipole moment could be determined. An example of the spectra obtained is shown in figure 11.7(b). In their later work Field and Bergeman [12] replaced the manganese atomic emission line with white light from a CS molecular discharge lamp and they were then able to study the Λ-doublet transitions in rotational levels $J = 1$ to 9 of the $A\,^1\Pi$ state. Their

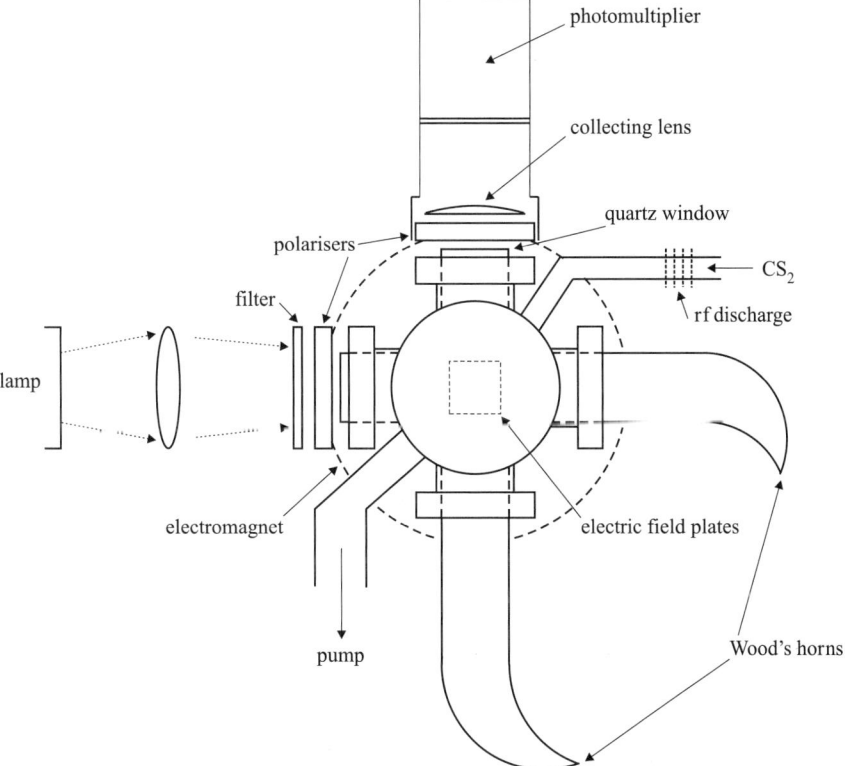

Figure 11.6. Schematic diagram of the apparatus used for the radiofrequency/optical double resonance study of CS in its $A\,^1\Pi$ state [11].

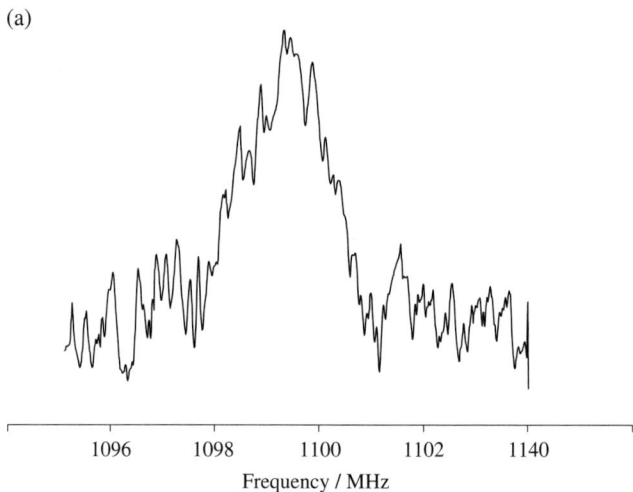
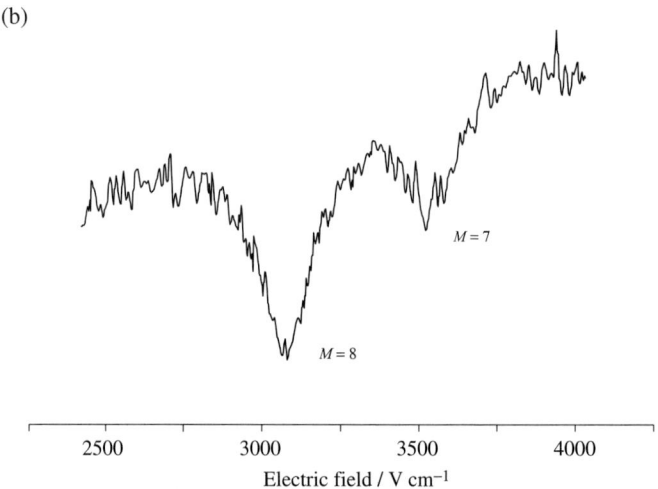

Figure 11.7. (a) Radiofrequency/optical double resonance line observed for the Λ-doublet transition in the $J = 8$ level of the $A\,^1\Pi$ state of CS. The Stark field was zero [11]. (b) Radiofrequency/optical double resonance with a Stark field, showing the $|M|=8$ and 7 resonances for $J = 8$. The frequency was fixed at 1124.4 MHz while the electric field was swept.

experimental (and theoretical) results, for both the Λ-doubling and dipole moments, are presented in table 11.1.

To understand the Λ-doubling in the $A\,^1\Pi\ v = 0$ state of CS it is necessary to consider the spin–orbit interaction with other excited electronic states; a very thorough analysis was presented by Field and Bergeman [12]. The CS molecule possesses 22 electrons, of which the first 16 form an inner core which need not concern us further. Of the remaining six electrons, two occupy a σ orbital (which we call 7σ) and the remaining four occupy a π orbital, which we call 2π. The lowest vacant orbital to be considered is a 3π orbital. Consequently the electron configurations for the ground

Table 11.1. Λ-doubling (MHz) and electric dipole moments (D) determined for CS in the $A\,^1\Pi$ ($v=0$) state. The calculated values are discussed in the text

	Λ-doubling		Electric dipole moment	
J	exp. (MHz)	calc. (MHz)	exp. (D)	calc. (D)
1	13.6	13.5		
2	42.1	42.08		
3	89.5	89.45	0.6703	0.6701
4	161.8	162.14	0.6699	0.6705
5	271.33	271.35	0.6713	0.6710
6	436.38	436.33	0.6718	0.6718
7	691.29	691.18	0.6733	0.6732
8	1099.37	1099.39	0.6752	0.6753
9	1787.19	1787.14	0.6786	0.6785

and lowest excited electronic states may be written as follows.

(inner core)$^{16}(7\sigma)^2(2\pi)^4(3\pi)^0$: $X\,^1\Sigma^+$

(inner core)$^{16}(7\sigma)^2(2\pi)^3(3\pi)^1$: $a'\,^3\Sigma^+$, $e\,^3\Sigma^-$, $d\,^3\Delta$, plus singlet states

(inner core)$^{16}(7\sigma)^1(2\pi)^4(3\pi)^1$: $A\,^1\Pi$, $a\,^3\Pi$

The triplet states have energies similar to that of the $A\,^1\Pi$ state, and there is thought to be an additional $k\,^3\Pi$ state lying nearby. More significantly, the $v=0$ level of the $A\,^1\Pi$ state is particularly close in energy to the $v=10$ level of the $a'\,^3\Sigma^+$ state, and the interaction between these two levels is thought to dominate the Λ-doubling. In a case (a) basis we may specify the value of $\Omega = |\Lambda + \Sigma|$ which is 1 for the $^1\Pi$ state, and 0 or 1 for the $a'\,^3\Sigma^+$ state. The spin–orbit mixing is diagonal in Ω so that the direct interaction does not involve the $a'\,^3\Sigma_0^+$ component; the interaction matrix given by Field and Bergeman is presented below. In this matrix the superscripts $(\pm)$ denote parity, and $\bar{B} = B - DJ(J+1)$; γ is the spin–rotation constant and λ is the spin–spin constant for the $^3\Sigma^+$ state. The other constants are self-explanatory; their values for the two electronic states, determined from the experimental results, are given below the energy matrix. Note that the matrix has dimension 3×3 for even J, positive parity states or odd J, negative parity states. Otherwise it is a 2×2 matrix because the appropriate level for the $^3\Sigma_0^{+(-)}$ component does not exist.

	$^1\Pi^{(\pm)}$	$^3\Sigma_1^{+(\pm)}$	$^3\Sigma_0^{+(\pm)}$
$^1\Pi^{(\pm)}$	$T_\Pi + (\bar{B}_\Pi \pm (-1)^J q/2)$ $\times J(J+1)$	$A_{\Pi\Sigma}$	$(1/2)(1 \pm (-1)^J)\gamma_{\Pi\Sigma}$ $\times [J(J+1)]^{1/2}$
$^3\Sigma_1^{+(\pm)}$	$A_{\Pi\Sigma}$	$T_\Sigma - \bar{B}_\Sigma + 2\lambda/3$ $- \gamma + \bar{B}_\Sigma J(J+1)$	$(1 \pm (-1)^J)(-\bar{B}_\Sigma + \gamma/2)$ $\times [J(J+1)]^{1/2}$
$^3\Sigma_0^{+(\pm)}$	$(1/2)(1 \pm (-1)^J)\gamma_{\Pi\Sigma}$ $\times [J(J+1)]^{1/2}$	$(1 \pm (-1)^J)(-\bar{B}_\Sigma + \gamma/2)$ $\times [J(J+1)]^{1/2}$	$T_\Sigma + \bar{B}_\Sigma - 4\lambda/3 - 2\gamma$ $+ (\bar{B}_\Sigma - \Delta B_\Sigma)J(J+1)$

$T_\Pi(v = 0) = 38\,797.656 \text{ cm}^{-1}$ $T_\Sigma(v = 10) = 38\,853.20 \text{ cm}^{-1}$

$B_\Pi(v = 0) = 0.773\,65 \text{ cm}^{-1}$ $B_\Sigma(v = 10) = 0.5868 \text{ cm}^{-1}$

$D_\Pi(v = 0) = -1.3 \times 10^{-6} \text{ cm}^{-1}$ $D_\Sigma(v = 10) = 2.1 \times 10^{-6} \text{ cm}^{-1}$

$q(v = 0) = 0.68 \text{ MHz}$ $\Delta B_\Sigma(v = 10) = -80 \text{ MHz}$

$\lambda(v = 10) = -1.28 \text{ cm}^{-1}$

$\gamma(v = 10) = 151 \text{ MHz}$

$A_{\Pi\Sigma} = 5.509 \text{ cm}^{-1}$ $\gamma_{\Pi\Sigma} = 298 \text{ MHz}$

Field and Bergeman [12] chose to diagonalise the above matrix, thereby generating a modified $^1\Pi$ wave function with $^3\Sigma^+$ admixture. Using the modified function as a basis, they then calculated the admixture of the other nearby triplet states (in appropriate vibrational levels) using perturbation theory. The calculated values of the Λ-doubling parameters and effective electric dipole moments obtained are listed in table 11.1.

We may anticipate that perturbations between electronic states are likely to be important for excited electronic states, simply because of the close proximity of different excited states. In contrast, the lower vibrational levels of ground electronic states are usually well separated from other electronic states.

11.3.2. Radiofrequency/optical double resonance of OH in its excited $A\,^2\Sigma^+$ state

As we have already mentioned, the first published observations of radiofrequency/optical double resonance in a molecular system were those of German and Zare [10] and German, Bergeman, Weinstock and Zare [13] on electronically excited OH and OD radicals. The radicals were generated by reaction of H or D atoms with NO_2 in a flow system. The relevant energy levels for OH are illustrated in figure 11.8. For OH the $v = 0$, $N = 2$, $J = 3/2$ level of the $A\,^2\Sigma^+$ state was excited from the $^2\Pi_{3/2}$ ground state by means of light from an electrodeless microwave discharge in Zn vapour at 3072.06 Å. The $v = 0$, $N = 1$, $J = 3/2$ level of OD in the $A\,^2\Sigma^+$ state was excited by the 3071.60 Å emission line from a discharge in Ba vapour. The sample volume was contained within a small, variable magnetic field, which removed the spatial degeneracy of the hyperfine levels, and the incident light was polarised perpendicular to the magnetic field. Detected fluorescence from the excited electronic state was polarised parallel ($\Delta M = 0$) or perpendicular ($\Delta M = \pm 1$) to the applied magnetic field, and a radiofrequency field applied perpendicular to the static field induced $\Delta M = \pm 1$ transitions, thereby changing the relative intensities of the parallel and perpendicular polarised fluorescence. Radiofrequency resonances in the range 1 to 8 MHz were observed by sweeping the static magnetic field, an example of the spectra obtained being shown in figure 11.9. The two resonances resolved at the higher frequency (8 MHz) have effective g-factors of 0.492 and 0.301. Note that the spectrum observed is actually a magnetic resonance spectrum.

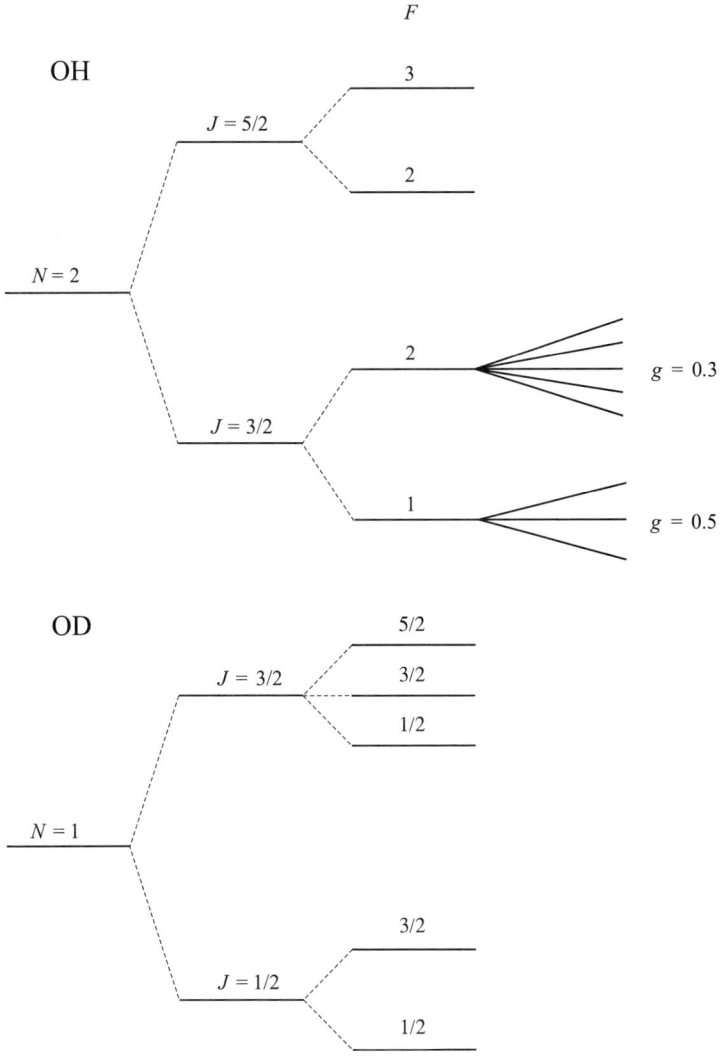

Figure 11.8. Hyperfine levels in OH and OD in the excited $A\,^2\Sigma^+$ state studied by radiofrequency/optical double resonance. The splittings are not to scale.

In a Hund's case (b) coupling scheme, the Zeeman interaction is described by the following matrix element:

$$\langle \eta, \Lambda; N, S, J, I, F, M_F | g_S \mu_B B_Z T_0^1(\boldsymbol{S}) | \eta, \Lambda; N, S, J', I, F', M_F \rangle$$

$$= g_S \mu_B B_Z (-1)^{F-M_F} \begin{pmatrix} F & 1 & F' \\ -M_F & 0 & M_F \end{pmatrix}$$

$$\times \langle \eta, \Lambda; N, S, J, I, F \| T^1(\boldsymbol{S}) \| \eta, \Lambda; N, S, J', I, F' \rangle$$

$$= g_S \mu_B B_Z (-1)^{F-M_F} \begin{pmatrix} F & 1 & F' \\ -M_F & 0 & M_F \end{pmatrix} (-1)^{F'+J+1+I} \{(2F'+1)(2F+1)\}^{1/2}$$

$$\times \begin{Bmatrix} F & J & I \\ J' & F' & 1 \end{Bmatrix} \langle \eta, \Lambda; N, S, J \| T^1(\boldsymbol{S}) \| \eta, \Lambda; N, S, J' \rangle$$

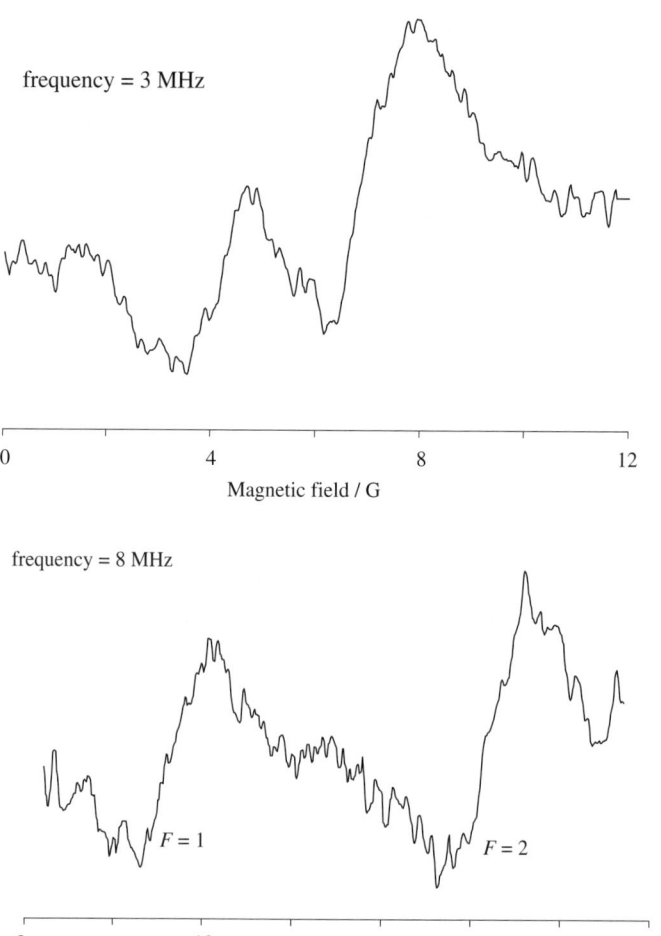

Figure 11.9. Radiofrequency magnetic resonance transitions observed by double resonance at two different frequencies for OH in the $A\,^2\Sigma^+$ state [10].

$$= g_S \mu_B B_Z (-1)^{F-M_F} \begin{pmatrix} F & 1 & F' \\ -M_F & 0 & M_F \end{pmatrix} (-1)^{F'+J+1+I} \{(2F'+1)(2F+1)\}^{1/2}$$

$$\times \begin{Bmatrix} F & J & I \\ J' & F' & 1 \end{Bmatrix} (-1)^{J+N+1+S} \{(2J'+1)(2J+1)\}^{1/2}$$

$$\times \begin{Bmatrix} J & S & N \\ S & J' & 1 \end{Bmatrix} \{S(S+1)(2S+1)\}^{1/2}. \tag{11.3}$$

Although there are off-diagonal elements, at the very weak magnetic fields used in the OH studies, only the diagonal elements in (11.3) are important. The Zeeman interaction may therefore be described by means of an effective g-factor, g_F, for each F level, where, from (11.3), the effective g-factors are given by

$$g_F = g_S(-1)^{F+2J+I+N+S}(2F+1)(2J+1)\{S(S+1)(2S+1)\}^{1/2}$$
$$\times \{F(F+1)(2F+1)\}^{-1/2} \begin{Bmatrix} F & J & I \\ J & F & 1 \end{Bmatrix} \begin{Bmatrix} J & S & N \\ S & J & 1 \end{Bmatrix}. \quad (11.4)$$

For OH, with $S = I = 1/2$, and $g_S = 2$, this becomes

$$g_F = \sqrt{6}(-1)^{F+2J+N+1}(2F+1)(2J+1)\{F(F+1)(2F+1)\}^{-1/2}$$
$$\times \begin{Bmatrix} F & J & I \\ J & F & 1 \end{Bmatrix} \begin{Bmatrix} J & S & N \\ S & J & 1 \end{Bmatrix}. \quad (11.5)$$

For the $N = 2, J = 3/2, F = 2$ and 1 levels, (11.5) yields effective g-factors of 0.3 and 0.5, which are very close to the experimental values given above. This result confirms the correctness of the assignments and that case (b) coupling is observed, but it does not provide much further information. Experiments at higher magnetic fields would result in second-order Zeeman splittings, from which information about the proton hyperfine coupling could be obtained. In the case of OD some hyperfine data were obtained from measurements at the highest magnetic fields available [13], but this was not true for OH.

11.3.3. Microwave/optical double resonance of BaO in its ground $X\,^1\Sigma^+$ and excited $A\,^1\Sigma^+$ states

The first observation of a rotational transition in a diatomic molecule in a short-lived, excited electronic state, namely BaO, was reported by Field, Bradford, Broida and Harris [14]. This work followed earlier double resonance studies by Field, Bradford, Harris and Broida [15] in which rotational transitions in the *ground* state of BaO were detected. A more comprehensive investigation of both ground and excited state rotational transitions in BaO was described by Field, English, Tanaka, Harris and Jennings [16]; we now review this work.

The elementary principles of double resonance experiments were described at the beginning of this chapter, and summarised in figure 11.1. We now explore these principles in more detail [16], emphasising that double resonance experiments involving electronic excitation can and have been used to study rotational transitions in either the ground state or the excited state. The two possibilities are summarised in figure 11.10. Field, English, Tanaka, Harris and Jennings [16] make a distinction between weak (linear) optical absorption, and strong (nonlinear) absorption. In the linear case, photoluminescence is linearly proportional to the optical pump power, and a ground state microwave transition between levels 1 and 2 is detectable if microwave transitions alter the small Boltzmann population difference between the levels. An excited state rotational transition, as shown in figure 11.10, is detectable if the light emitted from one of the two excited state rotational levels is preferentially detected, either through sufficient optical resolution or through a polarisation change. Strong nonlinear pumping is

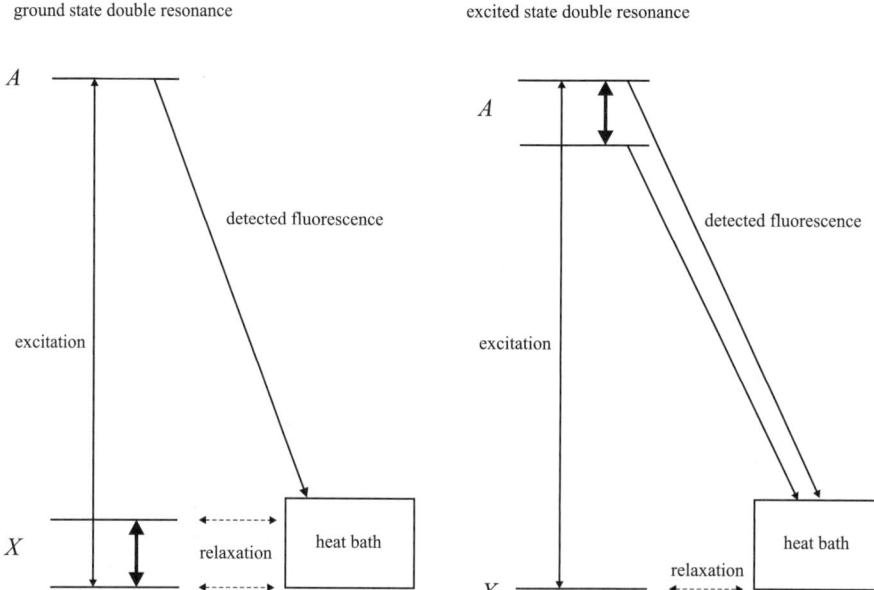

Figure 11.10. Principles of microwave/optical double resonance, permitting the observation of rotational transitions in either the ground or excited electronic state [16]. The ground state levels are in thermal equilibrium with the heat bath, and it is assumed that when the molecule in the excited state spontaneously emits a photon, it enters the heat bath rather than returning to the optically-depleted ground state level.

readily achieved through optical pumping with a tunable laser; it causes significant departures from equilibrium, and is then more sensitive. The BaO studies were performed with a dye laser tuned to a specific rovibronic transition; this is the best experiment, but depends upon prior knowledge and assignment of the electronic spectrum. Field, English, Tanaka, Harris and Jennings [16] gave a careful kinetic analysis of the relevant level population dynamics, which govern the sensitivity of a microwave/optical double resonance experiment. An even more thorough development of the theory of double resonance, using density matrix methods, was presented subsequently by Takami [17].

BaO was produced by allowing barium vapour in an argon carrier to interact with traces of oxygen. Microwave power was introduced by means of a simple horn situated close to the optical interaction zone. By tuning the dye laser frequency to coincide exclusively with a succession of rovibronic transitions, it was possible to observe and measure four rotational transitions in the $X\,^1\Sigma^+$ ground state, involving both $v = 0$ and $v = 1$, and thirteen rotational transitions in the $A\,^1\Sigma^+$ excited electronic state. From these measurements accurate values of the rotational constants were obtained, particularly for the excited state.

The studies of BaO were important pioneering experiments showing the power of microwave/optical double resonance methods. We shall describe a number of significant applications of these methods later in this chapter.

11.4. Microwave/optical magnetic resonance studies of electronically excited H_2

11.4.1. Introduction

In chapter 8 we described the elegant studies of Lichten [18] on the electronically excited $c\,^3\Pi_u$ state of H_2. Lichten's experiments involved electronic excitation of a beam of H_2 molecules by collision with an electron beam, but they were not double resonance experiments. Rather, they were classic molecular beam magnetic resonance studies of the type described extensively in chapter 8. In this section we discuss later experiments on H_2, again electronically excited by collision with electrons, but involving microwave/optical double resonance studies. Before we describe these experiments, however, we summarise the relevant excited states of H_2, repeating to some extent our discussion in chapter 8.

The ground state electron configuration of H_2 is described in the united atom nomenclature as $(1s\sigma)^2$ with the electron spins paired, or in the molecular orbital description as $(\sigma_g 1s)^2$. In either description we have the $X\,^1\Sigma_g^+$ state. In the first excited electronic state the united atom description, corresponding to excitation of one electron, is $(1s\sigma)^1(2p\sigma)^1$. In the molecular orbital description it is $(\sigma_g 1s)^1(\sigma_u 1s)^1$; in either case the electron spins may be paired to give a $^1\Sigma_u^+$ state, or unpaired to give a $^3\Sigma_u^+$ state. In this case the excited singlet state is bound, but the triplet state is unbound. Singly excited states corresponding to the promotion of one electron give rise to a family of excited singlet states, and a corresponding family of excited triplet states. The relative energies of these states are illustrated in figure 11.11, which is based on work by Richardson [19], summarised by Herzberg [20]. This figure is a combination of two such figures, given earlier in chapter 8.

11.4.2. H_2 in the $G\,^1\Sigma_g^+$ state

Four of the electronic states shown in figure 11.11 have been studied by high-resolution methods, namely, the $G\,^1\Sigma_g^+$, the $d\,^3\Pi_u$, the $k\,^3\Pi_u$, and the $c\,^3\Pi_u$ states [18] which was discussed in considerable detail in chapter 8. In the cases of the triplet states, both *ortho* and *para*-H_2 have been investigated. We now describe the experiments on the first three states listed above, starting with the G state. Freund and Miller [21] have described a technique which they call MOMRIE, standing for Microwave-Optical Magnetic Resonance Induced by Electrons. The technique is beautiful and elegant, even if the acronym is not, and is based on earlier experiments conducted on atoms. A sample of ground state H_2 molecules is located in a static magnetic field and subjected to bombardment by a beam of electrons moving parallel to the field. The electron bombardment results in excitation to excited electronic states, with unequal population of the Zeeman sublevels. This results in polarised visible fluorescence when the molecules in the excited states decay to lower energy excited states. Simultaneous driving of magnetic resonance transitions with a microwave field inside an

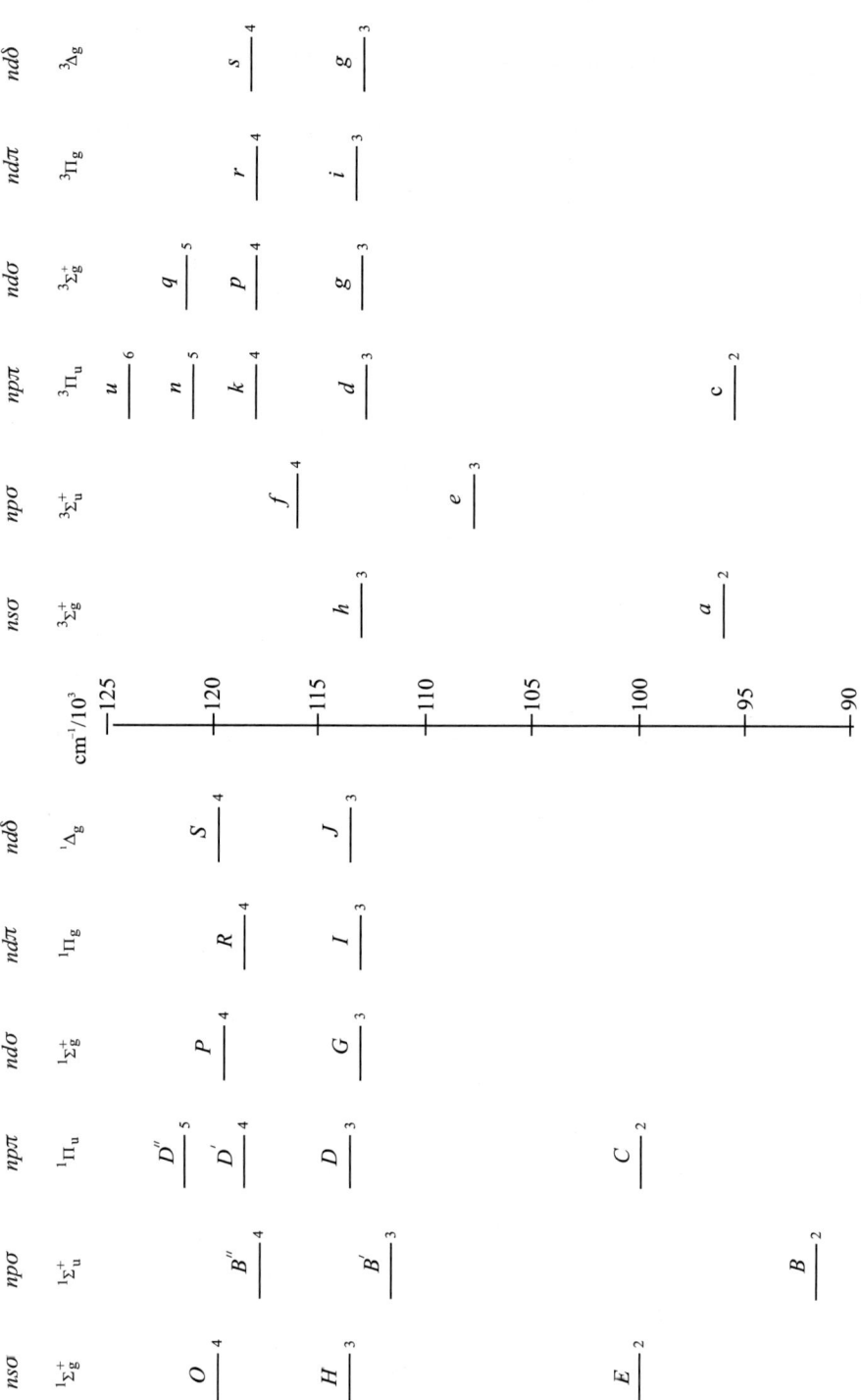

Figure 11.11. The excited singlet and triplet states of H₂ corresponding to the promotion of a single electron.

X-band cavity changes the fluorescence polarisation, which can be detected. The main features of the apparatus are illustrated in figure 11.12. One of the advantages of excitation with an electron beam, rather than with conventional monochromatic or white light sources, is that transitions between electronic states of different spin multiplicity are allowed; consequently both singlet and triplet excited states can be populated.

Figure 11.13 shows the energy levels involved and the transitions studied for H_2 in the $N = 1$ rotational level of the $G\,^1\Sigma_g^+$ state. The experiments were performed using a fixed microwave frequency, typically 9204 MHz, and the resonances detected by scanning the magnetic field; amplitude modulation of the microwave power at 100 kHz and lock-in amplifier detection were employed. Polarising filters were used to detect the fluorescence, so that changes in polarisation could be observed.

Figure 11.13 shows that, for the $N = 1$ level, two $\Delta M_N = \pm 1$ transitions are expected and the resonance line observed by Freund and Miller [21] exhibits a partially resolved doublet splitting, arising because the Zeeman effect is slightly nonlinear. At first sight it is surprising that a Zeeman splitting is observed at all because a pure singlet sigma state would be essentially non-magnetic. In fact Freund and Miller observed an effective g factor of 0.89077 for the $N = 1$ rotational level of the $G\,^1\Sigma_g^+$ state. The energy level diagram shown in figure 11.11 shows that the $I\,^1\Pi_g$ and $J\,^1\Delta_g$ states lie close in energy to the G state, and can be mixed with it by the rotational operator, as we outline below. It is this mixing which creates the magnetic moment in the G state measured by Freund and Miller.

The case (b) eigenkets for the three electronic states listed above and their parities may be written as follows.

$$^1\Sigma_g^+:\ |\eta, \Lambda = 0; N, \Lambda = 0\rangle \qquad (-1)^N$$

$$^1\Pi_g:\ |^1\Pi_g^+\rangle = (1/\sqrt{2})\{|\eta, \Lambda = 1; N, \Lambda = 1\rangle + |\eta, \Lambda = -1; N, \Lambda = -1\rangle\}\quad (-1)^N$$

$$|^1\Pi_g^-\rangle = (1/\sqrt{2})\{|\eta, \Lambda = 1; N, \Lambda = 1\rangle - |\eta, \Lambda = -1; N, \Lambda = -1\rangle\}\quad (-1)^{N+1}$$

$$^1\Delta_g:\ |^1\Delta_g^+\rangle = (1/\sqrt{2})\{|\eta, \Lambda = 2; N, \Lambda = 2\rangle + |\eta, \Lambda = -2; N, \Lambda = -2\rangle\}\quad (-1)^N$$

$$|^1\Delta_g^-\rangle = (1/\sqrt{2})\{|\eta, \Lambda = 2; N, \Lambda = 2\rangle - |\eta, \Lambda = -2; N, \Lambda = -2\rangle\}\quad (-1)^{N+1}$$

(11.6)

We have met these Λ-doublet functions several times already; the e and f combinations have definite parities, which are preserved in the absence of electric fields. For a given value of N, the three states of $(+)$ symmetry can be mixed together, as can also the two states of negative symmetry $(-)$. This restriction, however, applies only in zero field. An applied magnetic field mixes states of adjacent N value with opposite e/f character; however, the total parity, determined by inversion of all particles in the centre of mass must remain a good quantum number. The eigenvalue of the inversion operator, E^*, is listed for each of the five functions given above, and even in a magnetic field all coupled states must have the same eigenvalue of E^*. The overall situation with respect to symmetry and the mixing of the states is summarised in figure 11.14; the

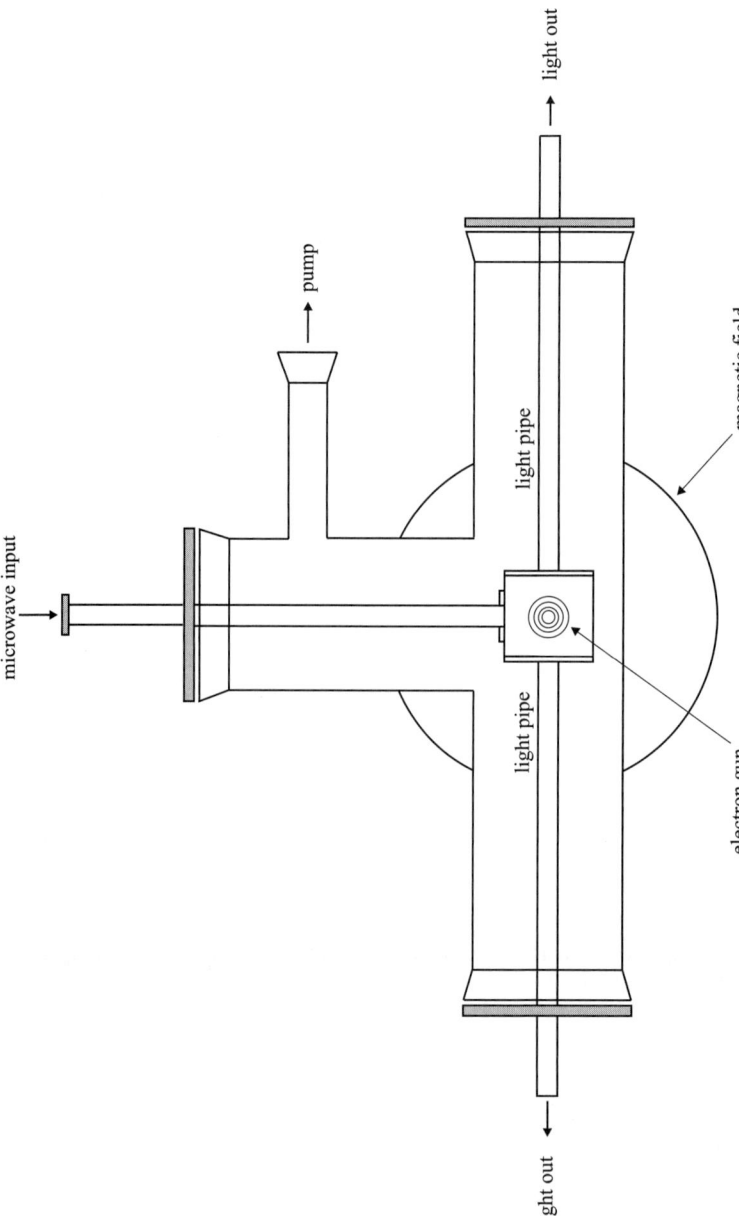

Figure 11.12. Cavity system used by Freund and Miller [21] in their magnetic resonance study of H_2 in its $G\,^1\Sigma_g^+$ state, excited from the ground state by electron bombardment.

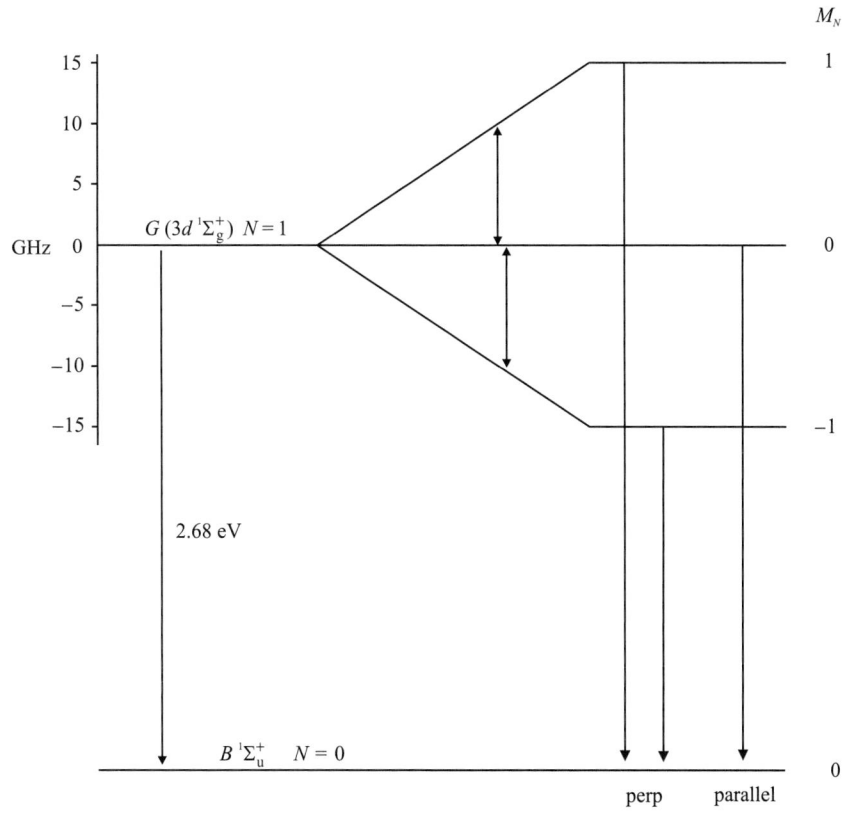

Figure 11.13. Energy levels of H$_2$ in the $N = 1$ level of the $G\,^1\Sigma_g^+$ state in an applied magnetic field, and the observed magnetic resonance transitions [21].

experiments relate to the $N = 1$ level, which is antisymmetric with respect to inversion (see chapter 6 for more details).

In a case (b) basis the field-free rotational Hamiltonian may be written, in the molecule fixed axis system, in the form

$$\mathcal{H}_{\text{rot}} = A_{\eta\Lambda} + B(\boldsymbol{N} - \boldsymbol{L})^2$$
$$= A_{\eta\Lambda} + BN^2 + BL^2 - 2B\sum_q T_q^1(\boldsymbol{N})T_q^1(\boldsymbol{L}). \quad (11.7)$$

$A_{\eta\Lambda}$ is the energy of a given rotationless electronic state characterised by particular values of η and Λ, and the expansion of the scalar product takes note of the anomalous

	e/f character	N = 1 parity	N = 2 parity
$^1\Sigma_g^+$	e	(−)	(+)
$^1\Pi_g^+$	e	(−)	(+)
$^1\Delta_g^+$	e	(−)	(+)
$^1\Pi_g^-$	f	(+)	(−)
$^1\Delta_g^-$	f	(+)	(−)

Figure 11.14. The e/f character and parities of the $N = 1$ and 2 rotational levels of the three electronic states, $G\ ^1\Sigma_g^+$, $I\ ^1\Pi_g$ and $J\ ^1\Delta_g$.

commutation rules for the molecule-fixed components of N. We now calculate the matrix elements of (11.7) for the primitive basis functions $|\eta, L, \Lambda; N, \Lambda\rangle$. For the diagonal elements we obtain the result

$$\langle \eta, L, \Lambda; N, \Lambda|\mathcal{H}_{\text{rot}}|\eta, L, \Lambda; N, \Lambda\rangle = A_{\eta\Lambda} + B_{\eta\Lambda}\{N(N+1) + L(L+1) - 2\Lambda^2\},$$
(11.8)

and for the off-diagonal elements we obtain

$$\langle \eta, L, \Lambda; N, \Lambda| - 2B \sum_{q=\pm 1} T_q^1(N)T_q^1(L)|\eta', L, \Lambda'; N, \Lambda'\rangle$$

$$= -2B_{\eta\Lambda,\eta'\Lambda'} \sum_q (-1)^{N-\Lambda} \begin{pmatrix} N & 1 & N \\ -\Lambda & q & \Lambda' \end{pmatrix} \{N(N+1)(2N+1)\}^{1/2} (-1)^{L-\Lambda}$$

$$\times \begin{pmatrix} L & 1 & L \\ -\Lambda & q & \Lambda' \end{pmatrix} \{L(L+1)(2L+1)\}^{1/2}.$$
(11.9)

Expansion of the 3-j symbols in (11.9) gives the results

$$\langle \eta, L, \Lambda; N, \Lambda| - 2B \sum_{q=\pm 1} T_q^1(N)T_q^1(L)|\eta', L, \Lambda \pm 1; N, \Lambda \pm 1\rangle$$

$$= -B_{\eta\Lambda,\eta'\Lambda'}\{(N \mp \Lambda)(N \pm \Lambda + 1)\}^{1/2}\{(L \mp \Lambda)(L \pm \Lambda + 1)\}^{1/2}. \quad (11.10)$$

An important point to note is that B in the field-free Hamiltonian (11.7) is the rotational moment *operator*, and can have different values in (11.8) and (11.9). From these general expressions we can calculate the matrix elements involving the five primitive functions as shown below. In order to evaluate these matrix elements it is necessary to specify a value of L; Freund and Miller pointed out that $L = 2$ for a d electron. This is the pure precession value, but in order to take account of any deviation from this limit, Freund and Miller multiplied the off-diagonal elements by a factor $(1 - \Delta)$.

	$\|\eta_\Sigma, \Lambda=0, N\rangle$	$\|\eta_\Pi, \Lambda=1, N\rangle$	$\|\eta_\Pi, \Lambda=-1, N\rangle$	$\|\eta_\Delta, \Lambda=2, N\rangle$	$\|\eta_\Delta, \Lambda=-2, N\rangle$
$\langle\eta_\Sigma, \Lambda=0, N\|$	m_{11}	m_{12}	m_{13}	0	0
$\langle\eta_\Pi, \Lambda=1, N\|$	m_{21}	m_{22}	0	m_{24}	0
$\langle\eta_\Pi, \Lambda=-1, N\|$	m_{31}	0	m_{33}	0	m_{35}
$\langle\eta_\Delta, \Lambda=2, N\|$	0	m_{42}	0	m_{44}	0
$\langle\eta_\Delta, \Lambda=-2, N\|$	0	0	m_{53}	0	m_{55}

The individual matrix elements are

$$m_{11} = A_\Sigma + B_\Sigma\{N(N+1) + 6\},$$
$$m_{22} = m_{33} = A_\Pi + B_\Pi\{N(N+1) + 4\},$$
$$m_{44} = m_{55} = A_\Delta + B_\Delta\{N(N+1) + 2\},$$
$$m_{12} = m_{21} = m_{13} = m_{31} = -B_{\Sigma\Pi}\{6N(N+1)\}^{1/2},$$
$$m_{24} = m_{42} = m_{35} = m_{53} = -2B_{\Pi\Delta}\{N(N+1) - 2\}^{1/2},$$
$$m_{14} = m_{41} = m_{15} = m_{51} = m_{23} = m_{32} = m_{25} = m_{52} = m_{34} = m_{43} = m_{45} = m_{54} = 0.$$
(11.11)

From the matrix elements derived with the primitive functions, we use the linear combinations given above and obtain the elements of a 3 × 3 matrix for the (+) parity states, and a 2 × 2 matrix for the (−) parity states. These matrices (using the elements defined above) are as follows.

	$^1\Sigma_g^+$	$^1\Pi_g^+$	$^1\Delta_g^+$	$^1\Pi_g^-$	$^1\Delta_g^-$
$^1\Sigma_g^+$	m_{11}	$\sqrt{2}m_{12}$	0	0	0
$^1\Pi_g^+$	$\sqrt{2}m_{12}$	m_{22}	m_{24}	0	0
$^1\Delta_g^+$	0	m_{24}	m_{44}	0	0
$^1\Pi_g^-$	0	0	0	m_{22}	m_{24}
$^1\Delta_g^-$	0	0	0	m_{24}	m_{44}

A further symmetry aspect of the problem concerns the nuclear spin state. The individual proton spins may be combined to form a total nuclear spin I_T of 1 (*ortho*) or 0 (*para*); since the $N = 1$ states are, from figure 11.14, antisymmetric with respect to inversion in the centre-of-mass, they must be combined with the symmetric nuclear spin functions, which are those corresponding to $I_T = 1$. This suggests that proton hyperfine structure might be observable but, as Freund and Miller [21] point out, only the mixing of some electronic orbital angular momentum into the singlet sigma state can produce significant hyperfine coupling. This is expected to produce splittings which are very much smaller than the experimental linewidth, and are therefore unobservable. The experimental linewidths are, in fact, quite large, being determined by the relatively short radiative lifetime of the $G\,^1\Sigma_g^+$ state (about 21 ns).

Freund and Miller included centrifugal distortion corrections in their theory. The numerical solution of the above matrices requires values for the various A and B parameters involved and these were derived from earlier studies of the electronic spectrum [22]. Once the zero-field energy and wave function of the $G\,^1\Sigma_g^+\,N=1$ odd-parity level has been determined, there remains only the Zeeman effect to be evaluated. The magnetic field interacts with the orbital angular momentum generated by the rotational mixing, and defining the $p=0$ direction to be the magnetic field direction, as usual, the perturbation due to the applied field is represented by the effective Hamiltonian

$$\mathcal{H}_Z = g_L \mu_B B_Z T_0^1(\boldsymbol{L}) = g_L \mu_B B_Z \sum_q \mathcal{D}_{0q}^{(1)}(\omega)^* T_q^1(\boldsymbol{L}). \tag{11.12}$$

In the case (b) basis used above the matrix elements of this perturbation are given by

$$\langle \eta, L, \Lambda; N, \Lambda, M_N | \mathcal{H}_Z | \eta, L, \Lambda'; N', \Lambda', M_N' \rangle$$
$$= g_L \mu_B B_Z \delta_{M_N, M_N'} \sum_q \langle \eta, L, \Lambda | T_q^1(\boldsymbol{L}) | \eta, L, \Lambda' \rangle \langle N, \Lambda, M_N | D_{0q}^{(1)}(\omega)^* | N', \Lambda', M_N' \rangle$$
$$= g_L \mu_B B_Z \delta_{M_N, M_N'} \sum_q \langle \eta, L, \Lambda | T_q^1(\boldsymbol{L}) | \eta, L, \Lambda' \rangle \{(2N+1)(2N'+1)\}^{1/2}$$
$$\times (-1)^{M_N - \Lambda} \begin{pmatrix} N & 1 & N' \\ -M_N & 0 & M_N \end{pmatrix} \begin{pmatrix} N & 1 & N' \\ -\Lambda & q & \Lambda' \end{pmatrix}. \tag{11.13}$$

Again, Freund and Miller [21] included a multiplying term $(1-\Delta)$ to include deviations of L from its free precession value in the off-diagonal elements $(q=\pm 1)$. Now note that (11.13) is applied to a wave function which represents a mixture of the three electronic states, and note also that there are matrix elements both diagonal and off-diagonal in N. The off-diagonal elements result in a slightly nonlinear Zeeman behaviour, which is why the two transitions indicated in figure 11.13 occur at slightly different magnetic fields for a fixed microwave frequency. Freund and Miller calculated an effective g factor of 1.074, compared with the experimental value given earlier of 0.890 77.

The main results obtained from this study were the g-factor and the radiative lifetime for the G state. However, the experimental techniques developed proved to be very powerful in the study of other electronic states of H_2, as we will see shortly.

11.4.3. H_2 in the $d\,^3\Pi_u$ state

Studies of H_2 in its $d\,^3\Pi_u$ state were described by Miller and Freund [23] and Freund and Miller [24] for *para*-H_2, and by Miller and Freund [25] for *ortho*-H_2. Similar but less extensive studies were made by Jost, Marechal and Lombardi [26]. The experimental method used [23] was essentially that described earlier, except that a specific light filter was used to isolate the fluorescence arising from radiative decay of the $d\,^3\Pi_u$ to the $a\,^3\Sigma_g^+$ state. Resonances involving the $N=1$ level with $v=0$ to 3 were observed for both *ortho* and *para* species, but the *para* spectrum was the simpler because of the

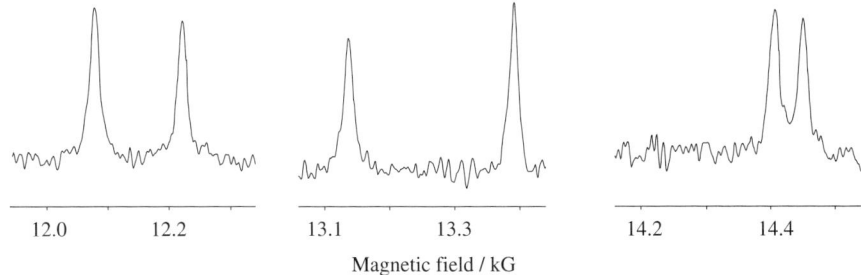

Figure 11.15. MOMRIE spectrum observed for *para*-H$_2$ in the $d\,^3\Pi_u$, $v = 1$, $N = 1$ state. The total recording time for each section was 32 minutes [24].

absence of nuclear hyperfine splitting. An example of the observed spectra for *para*-H$_2$ in the $N = 1$, $v = 1$ level is shown in figure 11.15, and the origin of the six resonances is given in the energy level diagram shown in figure 11.16. The energy levels on the left-hand side of the diagram are labelled in a coupled representation

$$N + S = J \tag{11.14}$$

and since $N = S = 1$, we see that $J = 2$, 1 or 0. In a magnetic field the $(2J + 1)$ spatial degeneracy of each J level is removed, and a total of nine Zeeman sublevels, characterised by their M_J values, is observed. On the right-hand side the levels are labelled in a spin-decoupled representation, N, S, M_N. The measurements were made as a function of magnetic field strength at a microwave frequency of 9202.8 MHz; all of the transitions obey a $\Delta M_N = \pm 1$ selection rule.

The molecular beam magnetic resonance spectrum of H$_2$ in its $c\,^3\Pi_u$ state was discussed in great detail in chapter 8, and much of the theory described there applies directly to the $d\,^3\Pi_u$ state discussed in this section. We summarise briefly the most important results here; a major difference between the two studies is that the MOMRIE experiments used strong magnetic fields, in the range 12 to 15 kG, whereas the molecular beam experiments used comparatively weak fields.

The zero-field effective Hamiltonian for *para*-H$_2$ in a $^3\Pi_u$ state is written as the sum of four terms,

$$\mathcal{H}_{\text{eff}} = \mathcal{H}_{\text{rot}} + \mathcal{H}_{\text{sr}} + \mathcal{H}_{\text{ss}} + \mathcal{H}_{\text{so}}, \tag{11.15}$$

representing the rotational energy, spin–rotation, spin–spin and spin–orbit interactions. In a case (b) notation these interaction terms take the following forms:

$$\mathcal{H}_{\text{rot}} = B(\mathbf{N} - \mathbf{L})^2, \tag{11.16}$$

$$\mathcal{H}_{\text{sr}} = \gamma \mathbf{T}^1(\mathbf{N}) \cdot \mathbf{T}^1(\mathbf{S}), \tag{11.17}$$

$$\mathcal{H}_{\text{ss}} = -g_S^2 \mu_B^2 (\mu_0/4\pi) \sqrt{6}\, \mathbf{T}^2(\mathbf{C}) \cdot \mathbf{T}^2(\mathbf{S}_1, \mathbf{S}_2), \tag{11.18}$$

$$\mathcal{H}_{\text{so}} = a' \mathbf{T}^1(\mathbf{l}_1) \cdot \mathbf{T}(\mathbf{S}) + b' \mathbf{T}^1(\mathbf{l}_2) \cdot \mathbf{T}^1(\mathbf{S})$$

$$= A\mathbf{T}^1(\mathbf{L}) \cdot \mathbf{T}^1(\mathbf{S}) + \frac{1}{2}(a' - b')(\mathbf{T}^1(\mathbf{l}_1) - \mathbf{T}^1(\mathbf{l}_2)) \cdot \mathbf{T}^1(\mathbf{S}). \tag{11.19}$$

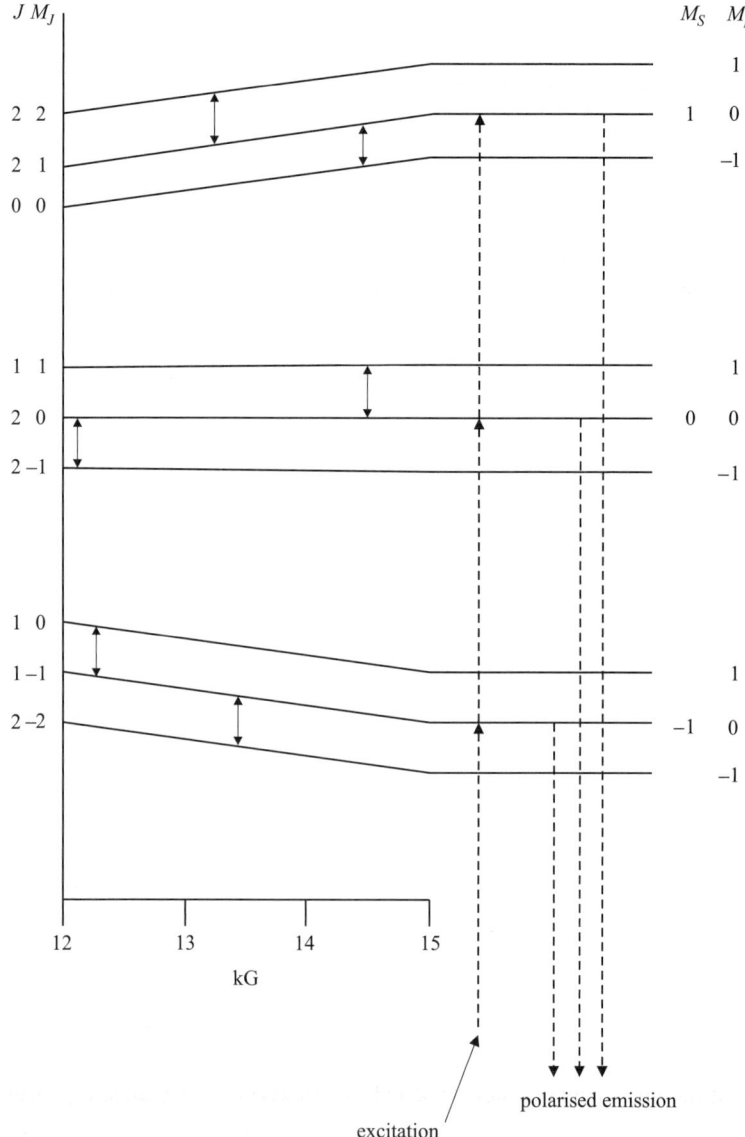

Figure 11.16. Zeeman energy level diagram for the $N = 1$ level of *para*-H_2 in the $^3\Pi_u$, $v = 1$ state, and the transitions responsible for the spectrum shown in figure 11.15.

This particular form of the spin–orbit interaction, due to Fontana [27] and Chiu [28], was discussed in chapter 8. It is also important to remember that the wave functions are Λ-doublets, so that parity-conserved combinations should be employed:

$$|\eta, \Lambda; N, S, J; \pm\rangle = \frac{1}{\sqrt{2}}\{|\eta, \Lambda = 1; N, S, J\rangle \pm (-1)^N |\eta, \Lambda = -1; N, S, J\rangle\}. \quad (11.20)$$

This is particularly important in the calculation of the spin–spin matrix elements, as we showed in chapter 8. There we introduced two spin–spin constants, defined as follows:

$$4\lambda_0/3 = -\langle g_S^2 \mu_B^2 (\mu_0/4\pi) C_0^2(\theta_{12}, \phi_{12}) r_{12}^{-3}\rangle_\eta,$$
$$\lambda_2 = \langle g_S^2 \mu_B^2 (\mu_0/4\pi) C_{\pm 2}^2(\theta_{12}, \phi_{12}) r_{12}^{-3}\rangle_\eta. \tag{11.21}$$

These constants are labelled B_0 and B_2 by Freund and Miller but, with respect, we will use the more common labels given in equation (11.21).

The Zeeman Hamiltonian for the $d\,^3\Pi_u$ state of *para*-H$_2$ takes the now familiar form

$$\mathcal{H}_Z = g_S \mu_B \mathbf{T}^1(\boldsymbol{B}) \cdot \mathbf{T}^1(\boldsymbol{S}) + g_L \mu_B \mathbf{T}^1(\boldsymbol{B}) \cdot \mathbf{T}^1(\boldsymbol{L}) - g_r \mu_B (\mathbf{T}^1(\boldsymbol{N}) - \mathbf{T}^1(\boldsymbol{L})) \cdot \mathbf{T}^1(\boldsymbol{B}). \tag{11.22}$$

In addition Freund and Miller [24] included the diamagnetic interaction between the magnetic susceptibility and the applied magnetic field; we will not describe the details here.

Freund and Miller [24] determined the values of a number of parameters, some composite, for the $N = 1$ level of *para*-H$_2$, $^3\Pi_u$, in its $v = 0$ to 3 levels. In particular they determined the value of a parameter A' containing contributions from both the spin–orbit and spin–rotation terms, and components of the spin–spin interaction tensor. They were also able to identify the effects of breakdown in the Born–Oppenheimer approximation, a subject explored theoretically in more depth by Miller [29].

The MOMRIE spectrum of *ortho*-H$_2$ $^3\Pi_u$, also in the $N = 1$, $v = 0$ to 3 levels, was also investigated by Miller and Freund [25]. This spectrum is considerably more complicated than that of *para*-H$_2$ because of the presence of proton hyperfine structure arising from the total nuclear spin, $I_T = 1$. An experimental spectrum is shown in figure 11.17. There are, in fact, eighteen hyperfine components but they are not all resolved in the spectrum shown; some of the components were located by line shape simulations, individual lines being assumed to have the same Lorentzian shapes and widths. The energy level diagram shown in figure 11.18 gives the origin of the components. At the high magnetic fields used (12 to 14 kG) the decoupled representation in which M_N, M_S and M_I are specified is the most appropriate, and the observed transitions all satisfy the selection rules $\Delta M_N = \pm 1$, $\Delta M_S = \Delta M_I = 0$. The labels for the coupled representation would be appropriate at low magnetic fields, and they are included in figure 11.18 for purposes of comparison.

The main additions to the theory for *ortho*-H$_2$ are, of course, the magnetic hyperfine terms. These were discussed with particular reference to the $c\,^3\Pi_u$ state of *ortho*-H$_2$ in chapter 8, the work of Jette and Cahill [30] being particularly important. The magnetic hyperfine Hamiltonian is usually written as the sum of three terms,

$$\mathcal{H}_{\text{hfs}} = b_F \mathbf{T}^1(\boldsymbol{S}) \cdot \mathbf{T}^1(\boldsymbol{I}) + a \mathbf{T}^1(\boldsymbol{L}) \cdot \mathbf{T}^1(\boldsymbol{I})$$
$$- \sqrt{10}\, g_S \mu_B g_N \mu_N (\mu_0/4\pi) \mathbf{T}^1(\boldsymbol{S}, \boldsymbol{C}^2) \cdot \mathbf{T}^1(\boldsymbol{I}), \tag{11.23}$$

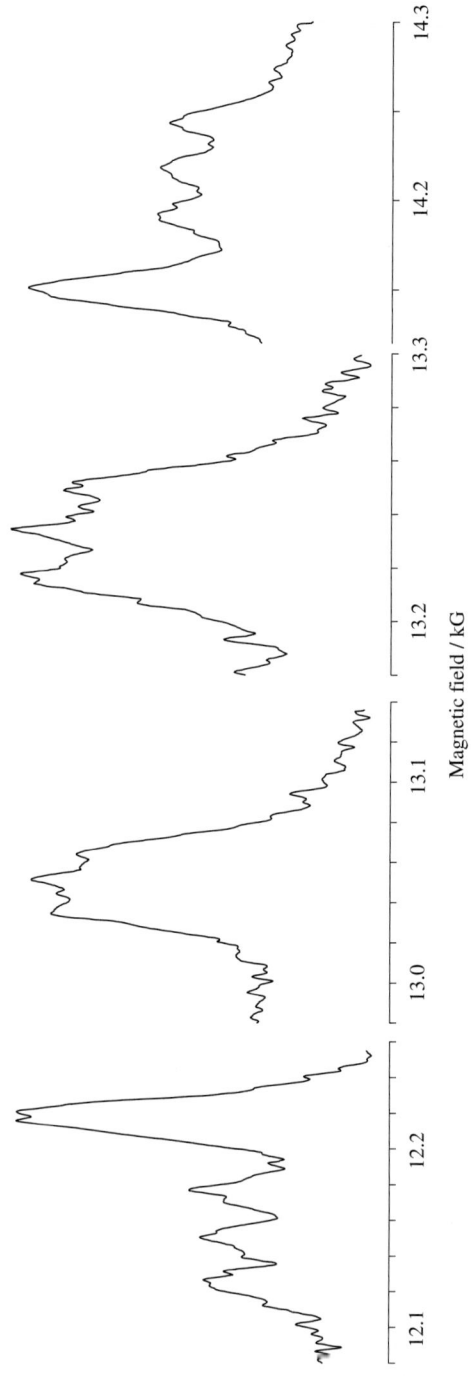

Figure 11.17. MOMRIE spectrum of *ortho*-H$_2$ in the $d\,^3\Pi_u$ $v = 0$, $N = 1$ level [25].

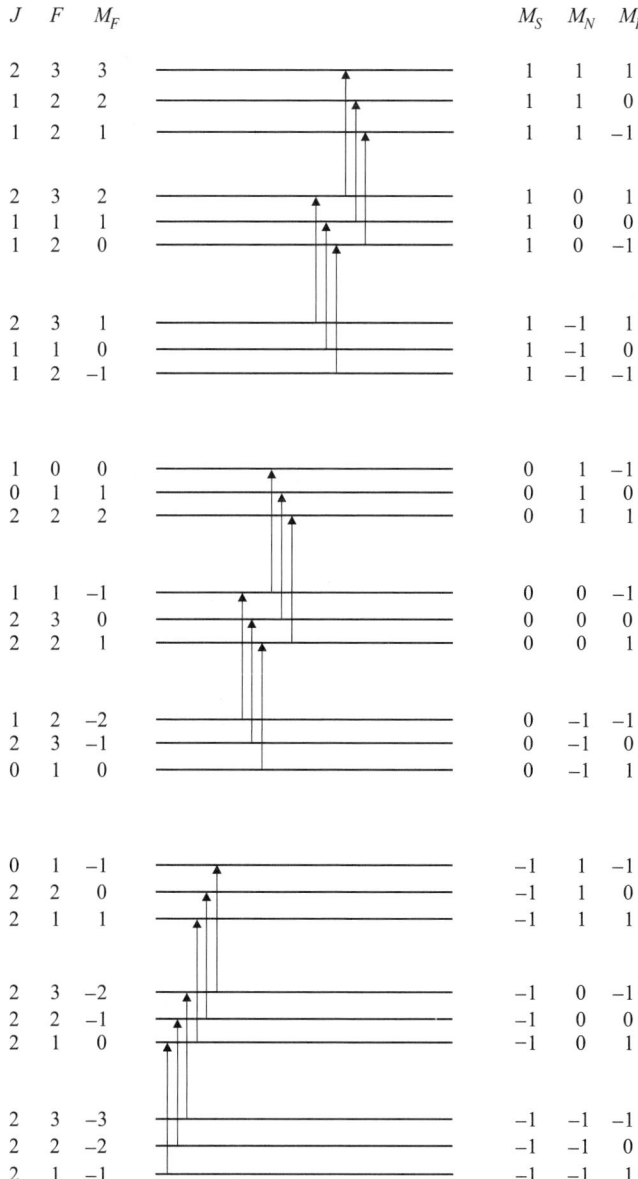

Figure 11.18. Energy levels for the $N = 1$ level of *ortho*-H_2 in its $d\,^3\Pi_u\,v = 0$ state at a magnetic field of about 13 kG, and the eighteen transitions which give rise to the spectrum shown in figure 11.17. This diagram is not drawn to scale.

representing the Fermi contact, electron orbital, and spin–dipolar interactions. The matrix elements in a case (b) basis were presented in chapter 8; the Fermi contact constant b_F and orbital hyperfine constant, a, are straightforward and defined by

$$b_\text{F} = \frac{2\mu_0}{3} g_S \mu_B g_N \mu_N \int \psi_\text{el}^* \delta(\boldsymbol{r}) \psi_\text{el}\, \text{d}r, \quad a = 2g_N \mu_B \mu_N (\mu_0/4\pi) \sum_j \langle 1/r_{jN}^3 \rangle. \quad (11.24)$$

Here, we recall, that the integral determining b_F represents the density of the electronic wave function at the nucleus, whilst a depends on the position of electron j with respect to the nucleus N. The dipolar hyperfine constants are more complicated; the matrix elements of the dipolar term are given by

$$\langle \eta,\Lambda;N,S,J,I,F,M_F | -\sqrt{10} g_S \mu_B g_N \mu_N (\mu_0/4\pi) \mathbf{T}^1(\mathbf{S},\mathbf{C}^2) \cdot \mathbf{T}^1(\mathbf{I}) | \eta,\Lambda';N',S,J',I,F',M'_F \rangle$$

$$= -\sqrt{10}\, g_S \mu_B g_N \mu_N (\mu_0/4\pi) \delta_{FF'} \delta_{M_F M'_F} (-1)^{J'+F+I} \begin{Bmatrix} I & J' & F \\ J & I & 1 \end{Bmatrix}$$

$$\times \{I(I+1)(2I+1)\}^{1/2} \langle \eta,\Lambda;N,S,J \| \mathbf{T}^1(\mathbf{S},\mathbf{C}^2) \| \eta,\Lambda';N',S,J' \rangle$$

$$= \sqrt{30} g_S \mu_B g_N \mu_N (\mu_0/4\pi) \delta_{FF'} \delta_{M_F M'_F} (-1)^{J'+F+I} \begin{Bmatrix} I & J' & F \\ J & I & 1 \end{Bmatrix}$$

$$\times \{I(I+1)(2I+1)\}^{1/2} \{(2J+1)(2J'+1)\}^{1/2} \begin{Bmatrix} J & J' & 1 \\ N & N' & 2 \\ S & S & 1 \end{Bmatrix}$$

$$\times \{S(S+1)(2S+1)\}^{1/2} \langle \eta,\Lambda;N,\Lambda \| \mathbf{T}^2(\mathbf{C}) \| \eta,\Lambda';N',\Lambda' \rangle. \tag{11.25}$$

In order to evaluate the remaining reduced matrix element in (11.25) it is necessary to use properly symmetrised combinations of the $\Lambda = \pm 1$ states, as we described in our discussion of the $c\,^3\Pi_u$ state of H_2 in chapter 8. When this is done we arrive at the definition [30] of two dipolar constants, c and d in Frosch and Foley's notation [9], given by

$$c = \frac{3}{2} g_S \mu_B g_N \mu_N (\mu_0/4\pi) \left\langle \frac{(3\cos^2\theta - 1)}{r^3} \right\rangle,$$

$$d = \frac{3}{2} g_S \mu_B g_N \mu_N (\mu_0/4\pi) \left\langle \frac{\sin^2\theta}{r^3} \right\rangle. \tag{11.26}$$

In their first full paper Miller and Freund [25] gave values of the hyperfine parameters b_F, a and the dipolar parameter $c - 3d$, but found that the high field MOMRIE studies are not satisfactory in providing an accurate value of the Fermi contact constant, b_F. In later studies Freund and Miller [31] were able to observe MOMRIE transitions at much lower magnetic fields (1600 to 2000 G); they measured four transitions in ortho-H_2 and also two in para-H_2. In figure 11.19 we show the Zeeman behaviour in the range 1500 to 2000 G of the hyperfine levels and the transitions involved, noting that the coupled representation $|J, I, F, M_F\rangle$ is now the more appropriate. The transitions observed, which have a $\Delta M_S = \pm 1$ selection rule, lose their intensity at high magnetic fields and become unobservable; their measurement is, however, important in obtaining a more accurate derivation of the hyperfine parameters. In figure 11.19 we have included only the M_F Zeeman components observed experimentally, but have also shown their correlation with the $|J, I, F, M_F\rangle$ zero-field levels. The most important result of these studies is that Freund and Miller [31] were able to determine accurate

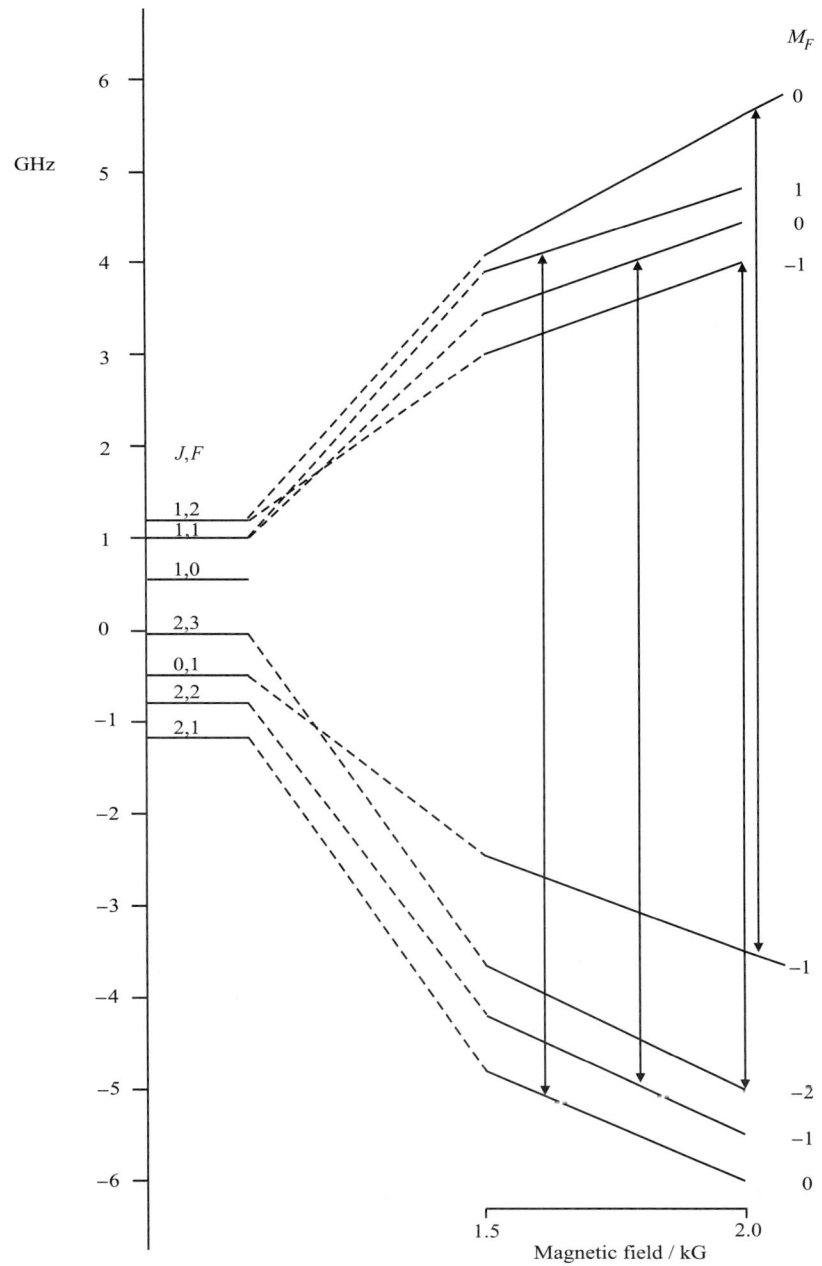

Figure 11.19. Zeeman behaviour of the hyperfine levels for H_2 in the $N = 1$ level of the $d\ ^3\Pi_u$ state in the range 1500 to 2100 G, and the transitions observed by Freund and Miller [31].

values of b_F for the four vibrational levels $v = 0$ to 3:

$v = 0$: $b_F = 459.4$ MHz, $v = 1$: 448.0, $v = 2$: 437.9, $v = 3$: 428.3.

Table 11.2. *Values of the hyperfine and fine structure constants (MHz) in the $np\pi\ ^3\Pi_u$ states at R_e*

np	b_F	$(c-3d)$	a	A	$(\lambda_0/3 - \sqrt{6}\lambda_2/4)$
$2p\ (c)$	456.3	104	26.6	-3717	975.9
$3p\ (d)$	465.3	74.4	6.9	-839	233.6
$4p\ (k)$	—	67.1	2.5	-314	91.9
$\infty p\ (H_2^+)$	446.1	66.2	0	0	0

They can be fitted to a quadratic equation in $(v + 1/2)$, with the result

$$b_F(v) = 465.3 - 12.5(v + 1/2) + 0.45(v + 1/2)^2.$$

Comparisons can then be made with the values of b_F obtained for the $c\ ^3\Pi_u$ state [32] of H_2, the value obtained by Jefferts [33] for the H_2^+ core itself, and theoretical calculations by Lombardi [34]. Finally, measurements of D_2 were made by Miller and Freund [35] and the resulting isotope effects interpreted in terms of perturbations between the $d\ ^3\Pi_u$ state and $^3\Sigma_u^+$ states.

11.4.4. H_2 in the $k\ ^3\Pi_u$ state

Figure 11.11 shows that the next highest energy state in the $np\pi\ ^3\Pi_u$ Rydberg series is the k state, for which $n = 4$. *Para*-H_2 in this state has been studied by Freund, Miller and Zegarski [36], and *ortho*-H_2 by Miller, Freund and Zegarski [37]. The measurements were again concerned with the $N = 1$ rotational level, with $v = 0$ for the *para* species, and $v = 0$ to 3 for the *ortho* species. A recording of the $v = 1$, $N = 1$ *ortho* spectrum is shown in figure 11.20; again the spectrum contains eighteen hyperfine components and although they are not all resolved, line shape simulations were used to obtain all of the resonant field values.

The analysis of the k state spectrum followed along the same lines as outlined already for the d state, and the most interesting result of the study, taken with the earlier studies of the d and c states was a comparison of the molecular parameters for the Rydberg states and with that of H_2^+ itself. This comparison is presented in table 11.2. The values given were obtained by extrapolation to the equilibrium internuclear distance R_e.

The Fermi contact constant for the k state was not determined because the low field transitions could not be observed. The overall behaviour of the parameters can be understood in terms of the $1s\sigma_g$ and $np\pi$ orbitals. The Fermi contact constant, which is proportional to the unpaired electron density at the nucleus, is determined almost entirely by the $1s\sigma_g$ electron. The nearly constant value of b_F is consistent with the $1s\sigma_g$ electron being essentially unaffected by the $np\pi_u$ electron. The orbital

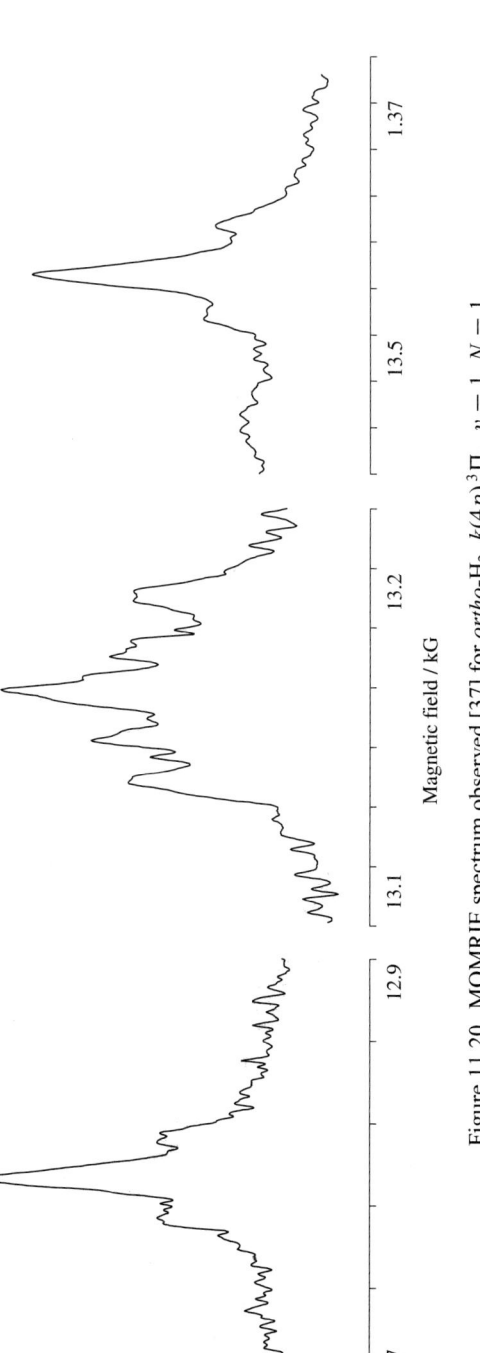

Figure 11.20. MOMRIE spectrum observed [37] for *ortho*-H$_2$, $k(4p)\,^3\Pi_u$, $v = 1$, $N = 1$.

interaction constant, however, depends on the $np\pi_u$ electron, and its inverse third-power dependence on the electron–nucleus distance r. Note that r increases as n increases and, of course, its inverse goes to zero when the electron is ionised to leave the H_2^+ core.

The interpretation of the dipolar constants c and d is somewhat more complicated, as we showed earlier. They are defined by equations (11.26) and the analysis of the spectrum yields the value of $c - 3d$; c and d are not separable experimentally. The observed behaviour of $c - 3d$, summarised in table 11.2, is consistent with a rapid decrease in the contribution of the $np\pi_u$ orbital to both c and d as n increases. The $1s\sigma_g$ contribution to c is relatively constant as n increases, and the limiting value of $c - 3d$, reached at H_2^+, represents this contribution.

Finally, the spin–orbit and spin–spin constants are expected to approach zero as we approach the limiting case of H_2^+ itself.

There are, of course, many other Rydberg states of H_2 which could be studied, at least in principle. Perhaps enough has already been done, however, to show the trends to be expected as the ionisation limit of H_2 is approached.

11.5. Radiofrequency or microwave/optical double resonance of alkaline earth molecules

11.5.1. Introduction

The alkaline earth elements are, in order of increasing atomic weight,

Be, Mg, Ca, Sr, Ba, Ra,

forming group IIA of the Periodic Table. Their basic electron configurations are as follows.

Be : $(1s)^2(2s)^2$
Mg: $(1s)^2(2s)^2(2p)^6(3s)^2$
Ca : $(1s)^2(2s)^2(2p)^6(3s)^2(3p)^6(4s)^2$
Sr : $(1s)^2(2s)^2(2p)^6(3s)^2(3p)^6(3d)^{10}(4s)^2(4p)^6(5s)^2$
Ba : $(1s)^2(2s)^2(2p)^6(3s)^2(3p)^6(3d)^{10}(4s)^2(4p)^6(4d)^{10}(5s)^2(5p)^6(6s)^2$
Ra : $(1s)^2(2s)^2(2p)^6(3s)^2(3p)^6(3d)^{10}(4s)^2(4p)^6(4d)^{10}(4f)^{14}(5s)^2(5p)^6$
$(5d)^{10}(6s)^2(6p)^6(7s)^2$

As can be seen, in each case the outermost pair of electrons occupies an s orbital, so that the molecules are normally divalent. A number of alkaline earth diatomic molecules have been studied by optical spectroscopy and double resonance methods. We have already described the very early microwave investigations of BaO in both its ground $X\,^1\Sigma^+$ and excited $A\,^1\Sigma^+$ states, and now proceed to discuss other molecules in this series. We describe these in their chronological order.

11.5.2. SrF, CaF and CaCl in their $X\,^2\Sigma^+$ ground states

Microwave/optical double resonance studies have been described for SrF by Domaille, Steimle and Harris [38], for CaCl by Domaille, Steimle and Harris [39], and for CaF by

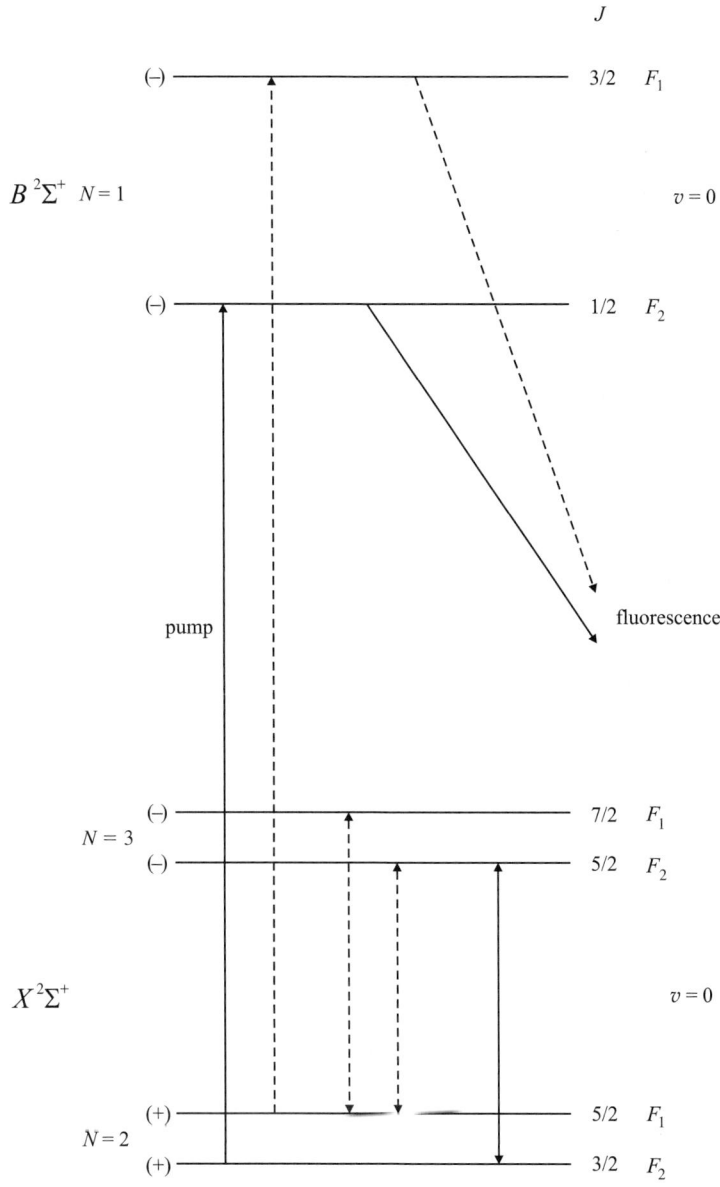

Figure 11.21. Energy level diagram and microwave/optical double resonance transitions for the $X\,^2\Sigma^+$ state in SrF and similar systems [38].

Nakagawa, Domaille, Steimle and Harris [40]. In all three cases rotational transitions within the $X\,^2\Sigma^+$ ground state were studied, through optical excitation and fluorescence involving the excited $B\,^2\Sigma^+$ state for CaCl and SrF, and the $A\,^2\Pi$ state for CaF. An energy level diagram for SrF is shown in figure 11.21, which does not include ^{19}F hyperfine splittings. The SrF radicals were produced by mixing strontium vapour and SF_6 in a flow cell, and optical excitation was accomplished with a single mode dye laser. The fluorescence was monitored in the direction perpendicular to the laser beam. As shown in figure 11.21, the laser preferentially excited the $P_2(2)$ spin-doublet component, and the detected microwave double resonance then involved the F_2 components of a rotational transition. Alternatively the laser could be adjusted to excite the $P_1(2)$ spin component, in which case the double resonance transitions shown by the dashed arrows could be detected. It will be appreciated that an experiment of this type is only really feasible if the electronic band system involved has been previously measured and analysed, which was the case here [41].

Six rotational transitions in the $v'' = 0$ level and three in the $v'' = 1$ level were observed, the microwave frequency spanning the range 29 to 60 GHz. Although ^{19}F hyperfine splittings were expected, the particular transitions necessary to determine the interaction constants were not observable. The experimental results were therefore fitted to the usual effective Hamiltonian,

$$\mathcal{H}_{\text{eff}} = [B_v - D_v N^2] N^2 + \gamma_v \, \mathbf{N} \cdot \mathbf{S} \qquad (11.27)$$

and accurate rotational and spin–rotation constants obtained for the electronic ground state. These were combined with the optical data obtained from the tunable laser study [41] to provide the following constants (in cm^{-1}) for both electronic states.

	$X\,^2\Sigma^+$:	$B\,^2\Sigma^+$:
B_0:	0.249 760	0.248 617
B_1:	0.248 214	0.247 060
B_e:	0.250 533	0.249 396
α_e:	1.546×10^{-3}	1.557×10^{-3}
D_e:	2.49×10^{-7}	2.52×10^{-7}
γ_0:	0.002 49	$-0.135\,37$
γ_1:	0.002 48	$-0.135\,28$

The most interesting aspect of these constants is the much larger spin–rotation constant observed for the upper state, which undoubtedly arises because of second-order mixing with other nearby electronic states.

Somewhat similar results were obtained for CaF [40], the main difference being that the excited electronic state involved in the optical pumping was the $A\,^2\Pi$ state. Four rotational transitions of CaF in the $v'' = 0$ level of the $X\,^2\Sigma^+$ state were observed, and three in the $v'' = 1$ level, spanning the frequency range from 41 to 61 GHz. Again, ^{19}F hyperfine splitting was too small, compared with the line width, to be observed. We note that these interactions are expected to be small given the predominantly ionic structure of the molecule, Ca$^+$F$^-$. In the case of CaCl [38] improved ground state rotational constants were obtained, but chlorine hyperfine structure was not observed.

An important development in microwave/optical double resonance, called microwave/optical polarisation spectroscopy, was described by Ernst and Törring [42]. The principles of this technique are illustrated in figure 11.22. A linearly polarised probe beam from a tunable laser is sent through the gas sample and a nearly crossed linear polariser, before its final detection. Polarised microwave radiation resonant with a rotational transition in the gas sample is introduced *via* a microwave horn as shown, and resonant absorption results in a partial change in polarisation of

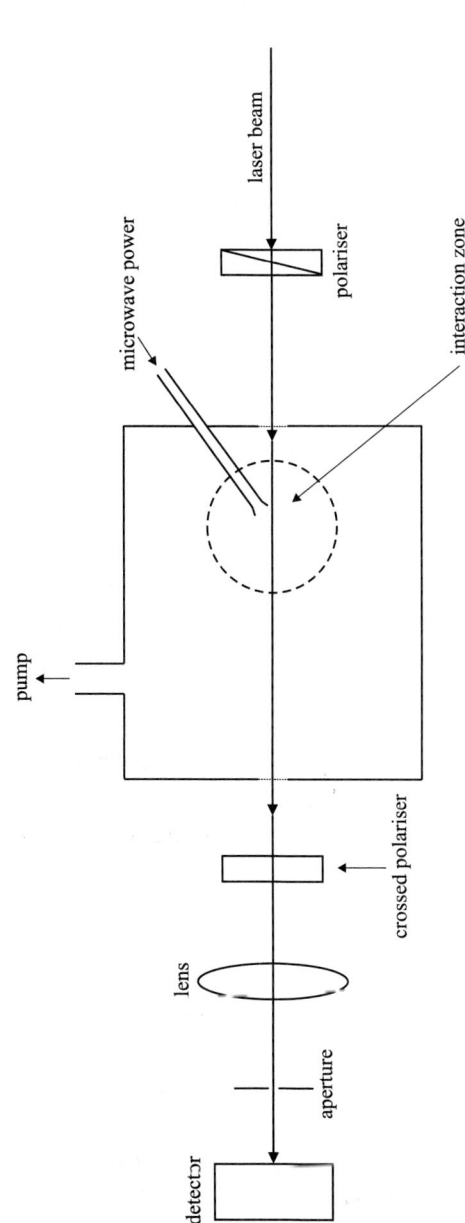

Figure 11.22. Experimental arrangement for microwave/optical polarisation spectroscopy [42].

the laser radiation; there is a consequent change in the laser beam intensity at the detector. The technique was tested using BaO as the resonant molecule, and the results compared with those obtained by Field, Bradford, Harris and Broida [15] described earlier in this chapter. Two important differences were noted, arising from the fact that both the laser and microwave powers necessary were very much lower in the polarisation experiment. The sensitivity of the polarisation method was much higher, and the double resonance line width was much smaller. Subsequent experiments by Ernst [43] on the $X\,^2\Sigma^+$ state of SrF demonstrated the resolution enhancement obtainable with the new technique. The fluorine hyperfine splitting of the lowest rotational transition, $N = 1 \leftarrow 0$, illustrated in the energy level diagram shown in figure 11.23, was readily resolved, with line widths down to 600 kHz obtained. An important aspect of the polarisation technique is that it does not depend upon fluorescence monitoring.

Remarkable enhancement of the spectroscopic resolution was obtained by Ernst and Kindt [44] in studies of a rotational transition of CaCl in its ground state, where line widths as small as 15 to 20 kHz enabled the chlorine hyperfine structure to be resolved. Their experimental arrangement is illustrated in figure 11.24; they used a pump/probe technique (PPMODR) similar to that described in chapter 8 which was used by Rosner, Holt and Gaily [45] to study radiofrequency hyperfine transitions in the Na_2 molecule. The molecular beam is exposed to a perpendicular laser beam in two regions of the apparatus, as shown, with exposure to microwave radiation in the intermediate (C) region. Ernst and Kindt were the first to use this method to study rotational transitions, a method which has since been employed by many others, as we describe later in this chapter. The high resolution is, in part, a consequence of the fact that low laser powers can be used to monitor the fluorescence. The use of molecular beams also removes problems associated with collisional broadening.

11.6. Radiofrequency or microwave/optical double resonance of transition metal molecules

11.6.1. Introduction

The first row of transition elements, with their lowest energy electron configurations, are as follows.

Sc: $(1s)^2(2s)^2(2p)^6(3s)^2(3p)^6(3d)^1(4s)^2$
Ti: $(1s)^2(2s)^2(2p)^6(3s)^2(3p)^6(3d)^2(4s)^2$
V: $(1s)^2(2s)^2(2p)^6(3s)^2(3p)^6(3d)^3(4s)^2$
Cr: $(1s)^2(2s)^2(2p)^6(3s)^2(3p)^6(3d)^5(4s)^1$
Mn: $(1s)^2(2s)^2(2p)^6(3s)^2(3p)^6(3d)^5(4s)^2$
Fe: $(1s)^2(2s)^2(2p)^6(3s)^2(3p)^6(3d)^6(4s)^2$
Co: $(1s)^2(2s)^2(2p)^6(3s)^2(3p)^6(3d)^7(4s)^2$
Ni: $(1s)^2(2s)^2(2p)^6(3s)^2(3p)^6(3d)^8(4s)^2$
Cu: $(1s)^2(2s)^2(2p)^6(3s)^2(3p)^6(3d)^{10}(4s)^1$
Zn: $(1s)^2(2s)^2(2p)^6(3s)^2(3p)^6(3d)^{10}(4s)^2$

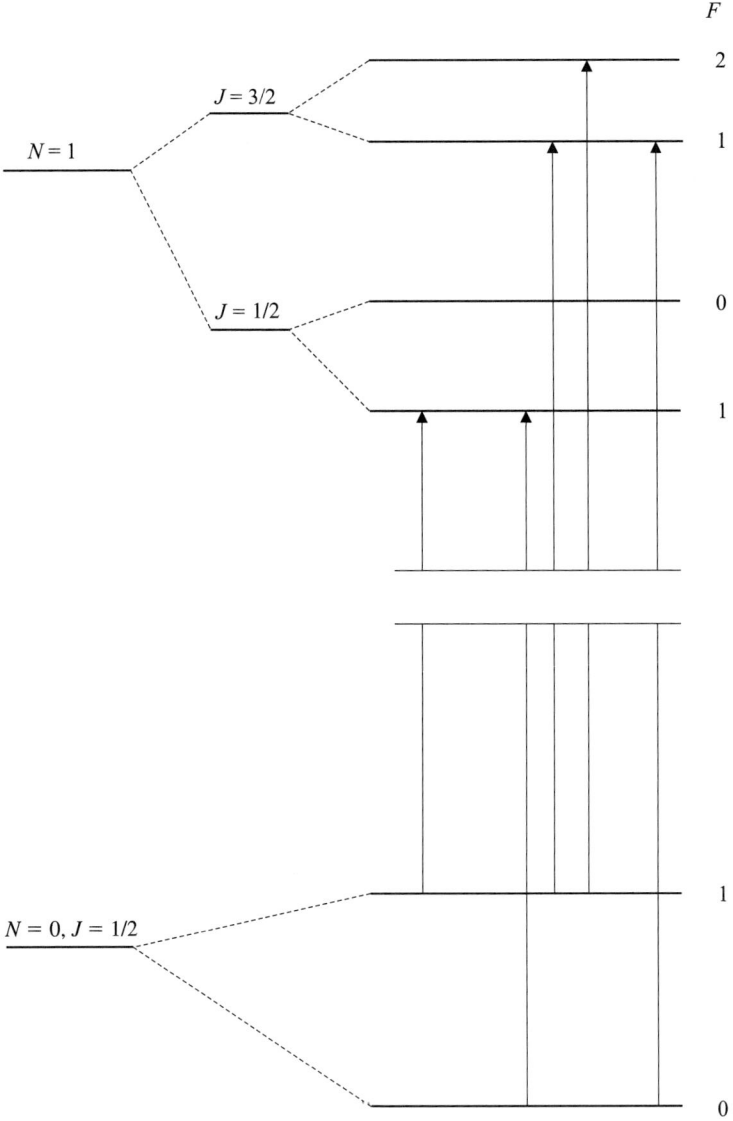

Figure 11.23. Energy level diagram and observed transitions for SrF in its $X^2\Sigma^+$ state, showing the ^{19}F hyperfine splitting [43], observed because of the higher resolution obtained with the microwave/optical polarisation method. This diagram may be compared with figure 11.21, appropriate for the earlier lower resolution studies employing conventional fluorescence detection.

As we have seen elsewhere in this book, the techniques of high-resolution rotational spectroscopy are beginning to be applied to transition metal molecules, and this is particularly true of double resonance methods. We now review some of the work which has been described, again in a mainly chronological order.

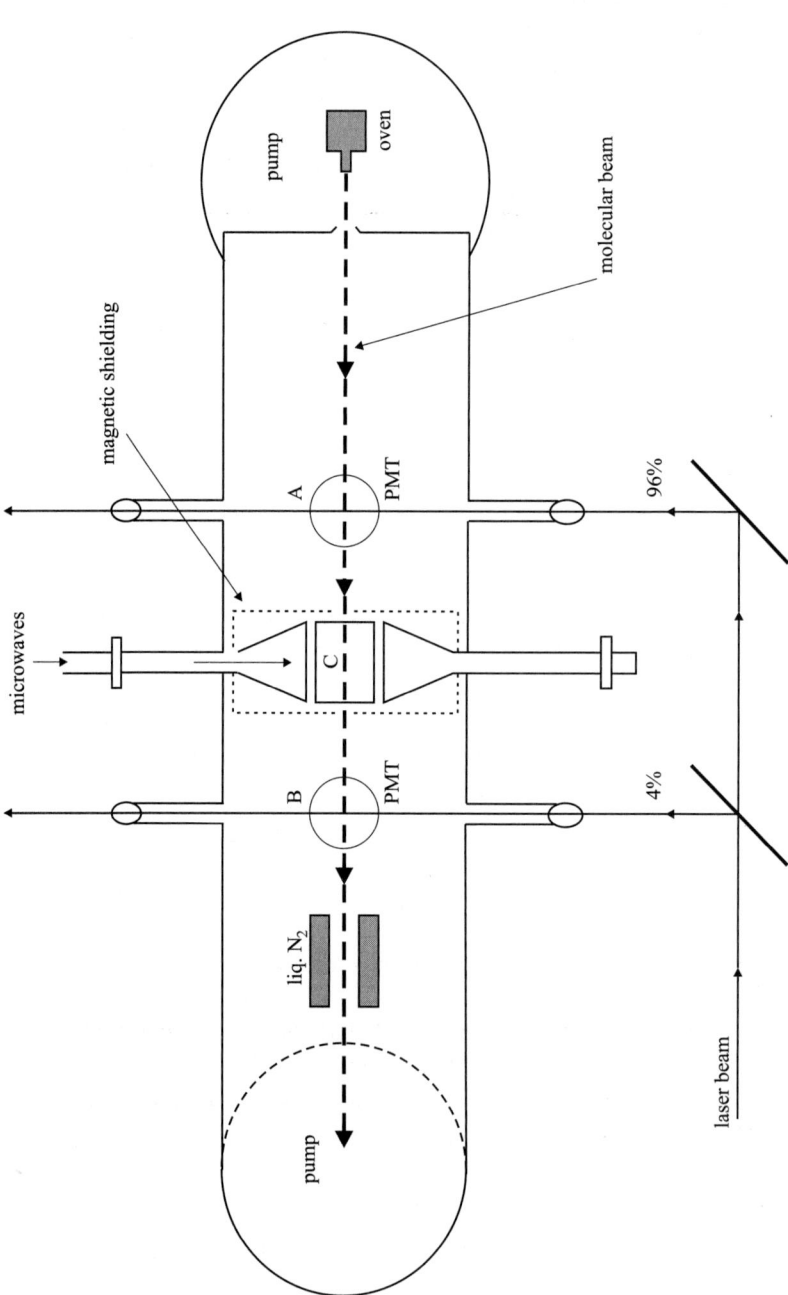

Figure 11.24. Experimental arrangement used by Ernst and Kindt [44] in their pump/probe microwave/optical double resonance study of a rotational transition (18.2 GHz) in the ground state of CaCl. The photomultiplier tubes which monitor fluorescence are situated on the axis perpendicular to both the laser beam and the molecular beam. The C region, where the molecular beam is exposed to microwave radiation, is magnetically shielded to minimise stray Zeeman effects. The microwave power was amplitude modulated at 160 Hz and the modulated fluorescence detected by photomultiplier B.

11.6.2. FeO in the $X\,^5\Delta$ ground state

The inherent sensitivity of double resonance methods has led to studies of transition metal and rare earth compounds which would be difficult with more conventional methods. In chapter 10 we described the study of rotational transitions of the FeO molecule by Endo, Saito and Hirota [46] using a conventional free-space absorption cell, and by Allen, Ziurys and Brown [47] who used a high-temperature oven to study the reaction between iron vapour and nitrous oxide. We also mentioned, however, that a comprehensive study had been made by Kröckertskothen, Knöckel and Tiemann [48] using a double resonance technique which we now describe in detail. Allen, Ziurys and Brown [47] combined the data from different experiments to produce the best analysis.

The principles of the double resonance experiment are illustrated in figure 11.25. A beam of FeO is produced by an oven/discharge combination [49]. A cw, single-frequency dye laser beam is split into two beams of equal intensity; one serves as a pump beam and intersects the FeO molecular beam at right angles just after a skimmer. The second laser beam intersects the molecular beam, again at right angles, some 30 cm downstream, and acts as the probe beam. The laser frequency is chosen to pump a specific rovibronic transition, and the fluorescence from the excited state is monitored. Between the two laser beams the molecular beam is subjected to microwave radiation, again propagated perpendicular to the molecular beam; when the resonance condition for a rotational transition is satisfied, the population of the upper rotational level is enhanced, leading to an increase in fluorescence intensity, which is detected. The microwave power is amplitude modulated, and the synchronous modulated component of the fluorescence is processed with a lock-in amplifier. As with most double resonance experiments, a detailed knowledge of the electronic spectrum is a prerequisite for success. This had been provided earlier by Taylor, Cheung and Merer [50].

A $^5\Delta$ state has five fine-structure components, corresponding to $\Omega = 4, 3, 2, 1, 0$. The spin–orbit constant A has the value -94.948 cm^{-1}, so that the $\Omega = 4$ component has the lowest energy, as shown in figure 11.26, which is a repeat of the energy level diagram given in chapter 10. Each of these fine-structure states has a two-fold Λ-degeneracy, and the Λ-doublet splitting was observed in the $\Omega = 0$ and 1 states in the pure rotational studies. The spin–orbit, spin–spin and spin–rotation effective Hamiltonian for a $^5\Delta$ state, expanded in the molecule-fixed axis system and assuming case (a) coupling, is as follows [50]:

$$\mathcal{H}_{so} + \mathcal{H}_{ss} + \mathcal{H}_{sr} = A T_0^1(\boldsymbol{L}) T_0^1(\boldsymbol{S}) + (2\sqrt{6}/3)\lambda T_{q=0}^2(\boldsymbol{S},\boldsymbol{S})$$
$$+ (\sqrt{10}/5)\eta T_{q=0}^1(\boldsymbol{L}) T_{q=0}^3(\boldsymbol{S},\boldsymbol{S},\boldsymbol{S}) + (\sqrt{70}/6)\theta T_{q=0}^4(\boldsymbol{S},\boldsymbol{S},\boldsymbol{S},\boldsymbol{S})$$
$$+ \gamma T^1(\boldsymbol{J}-\boldsymbol{S}) \cdot T^1(\boldsymbol{S}). \tag{11.28}$$

The most important terms are the first two, the two constants having the values $A = -94.948$ cm^{-1}, $\lambda = 0.9326$ cm^{-1}; figure 11.26 is based upon these two terms.

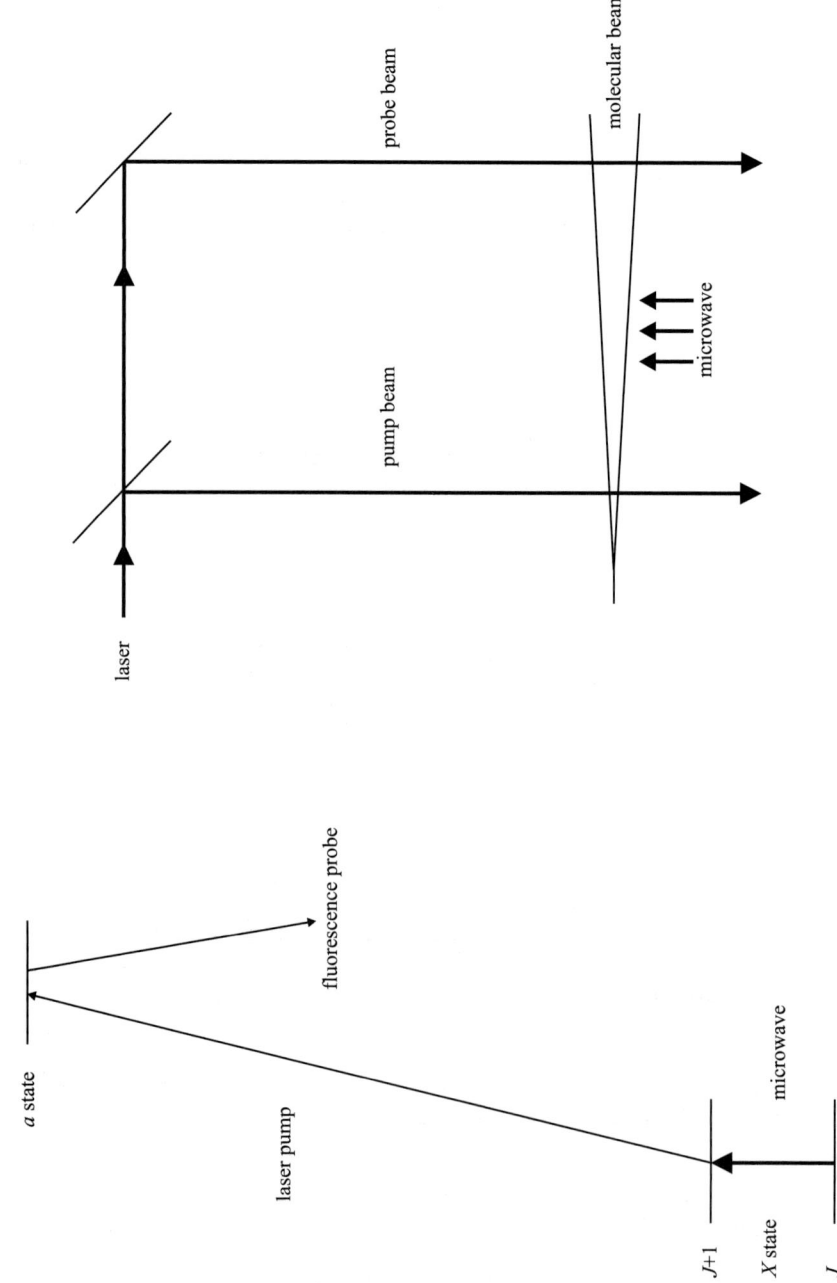

Figure 11.25. Principles of the microwave/optical double resonance experiment for studying rotational transitions in the ground state of FeO, and the experimental arrangement [49].

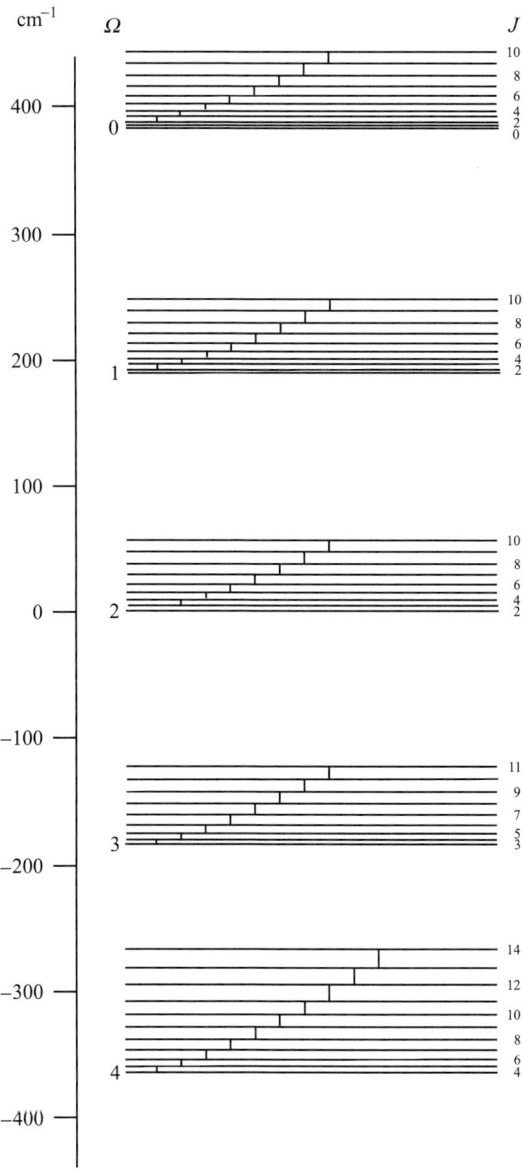

Figure 11.26. Fine structure splitting in the $X\,^5\Delta$ state of FeO, with the rotational levels and transitions studied [48]. This diagram is drawn to scale.

There is, then, a sequence of rotational levels for each fine-structure component, as illustrated in figure 11.26. The rotational Hamiltonian takes the conventional form

$$\mathcal{H}_{\text{rot}} = B[\mathrm{T}^1(\boldsymbol{J}) - \mathrm{T}^1(\boldsymbol{S})]^2 - D[\mathrm{T}^1(\boldsymbol{J} - \boldsymbol{S}) \cdot \mathrm{T}^1(\boldsymbol{J} - \boldsymbol{S})]^2. \qquad (11.29)$$

Within the $\Omega = 4$ fine-structure component, ten rotational transitions were observed in the double resonance studies, with J values ranging from 4 to 14, as shown

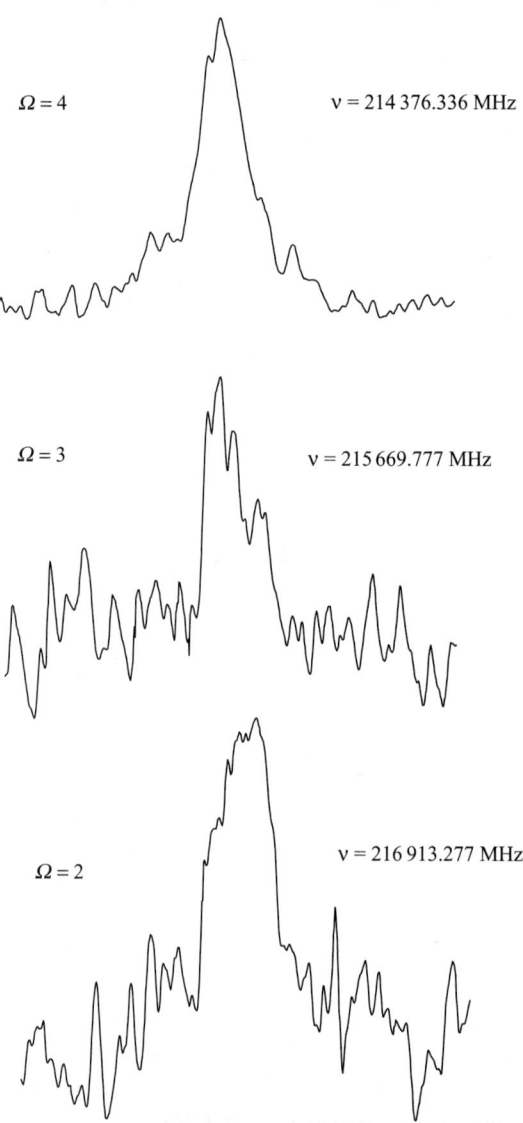

Figure 11.27. Microwave/optical double resonance lines observed [48] for the $J = 7 \leftarrow 6$ transition in the three lowest fine structure components of $X\,^5\Delta$ FeO.

in the figure. Three examples of the observed microwave resonances, all for $J = 7 \leftarrow 6$ transitions are shown in figure 11.27; the intensities decrease as Ω decreases, presumably because of decreasing thermal populations of the higher levels. Nevertheless eight rotational transitions within the $\Omega = 3$ fine-structure state were observed, and seven within the $\Omega = 2$ component. The observed transitions all occur in the range 123 to 429 GHz. From these data, combined with the earlier pure microwave data [46, 47] a set of the best constants for FeO in its ground state was derived. The electronic ground state of FeO is probably $(3d\delta)^3(4s\sigma)^1(3d\pi)^2$, but unfortunately there is no nuclear hyperfine structure to provide more detailed information about the nature of the highest occupied molecular orbitals because I is zero for both ^{56}Fe and ^{16}O, the dominant isotopes for each element.

The values of the molecular parameters (in MHz) obtained from a combination of the pure microwave and double resonance studies were as follows:

$$A = -2\,846\,420, \quad A_D = -2.403\,407, \quad B = 15\,493.632\,55, \quad D = 0.021\,628\,4,$$
$$\eta = 1079.08, \quad \lambda = 27\,928, \quad \lambda_D = -0.176\,394, \quad \theta \simeq 0, \quad \tilde{m}_\Delta = -397.55,$$
$$\tilde{n}_\Delta = 2.406, \quad \tilde{o}_\Delta = -0.021\,35.$$

The last three constants above are Λ-doubling parameters for a molecular Δ state.

11.6.3. CuF in the $b\,^3\Pi$ excited state

The low-lying electronic states of CuF are predominantly ionic in character; they have been studied both experimentally through their electronic spectra [51], and theoretically through *ab initio* calculations [52]. The ground $^1\Sigma^+$ state wave function arises mainly from the electron configuration $Cu^+(3d^{10})F^-(2p^6)$ and the low-lying excited electronic states result from promotion of an electron on the Cu^+ atom from the $3d$ to the $4s$ orbital. A manifold of singlet and triplet Σ, Π and Δ states arises, lying relatively close in energy. The $b\,^3\Pi$–$X\,^1\Sigma^+$ system has been studied by laser-induced fluorescence [53, 54], leading to the possibility of microwave double resonance studies of the excited $b\,^3\Pi$ state. These have been performed by Steimle, Brazier and Brown [55], and figure 11.28 shows the energy level diagram relevant to the study of the lowest rotational transition, $J = 1 - 0$, in the $^3\Pi_{0^+}$ fine-structure component. The triplet splitting of the $J = 1$ level arises from hyperfine interaction with the ^{63}Cu nucleus, which has spin $I = 3/2$, and the observed microwave spectrum exhibiting the splitting is shown in figure 11.29. Hyperfine splitting from the ^{19}F nucleus was too small to be resolved in this case, but was observed in other transitions. Apart from the transition described in figure 11.28 which was observed at 23 GHz, the $J = 2 - 1$ rotational transition in the $^3\Pi_0$ state was observed at 46 GHz, and an Ω-doubling transition in the $J = 7$ level of the $^3\Pi_1$ component was also measured at 16 GHz.

The analysis of the spectrum was accomplished using a case (a) basis, with the addition of two nuclear spins, I_1 and I_2, for ^{63}Cu and ^{19}F respectively. The basis functions therefore take the form $|\eta, \Lambda; S, \Sigma; J, \Omega, I_1, F_1, I_2, F, M_F\rangle$, and leaving aside nuclear spin interactions, the theory follows closely the same path as that already described for $^3\Pi$ CO in chapters 9 and 10. The effective Hamiltonian is the sum of terms representing the spin–orbit, rotational, spin–rotation, spin–spin and centrifugal distortion contributions and is written [56]:

$$\mathcal{H}_{\text{eff}} = A\mathbf{T}_0^1(\mathbf{L})\mathbf{T}_0^1(\mathbf{S}) + B(\mathbf{J} - \mathbf{S})^2 + \gamma(\mathbf{T}^1(\mathbf{J}) - \mathbf{T}^1(\mathbf{S})) \cdot \mathbf{T}^1(\mathbf{S})$$
$$+ (2\sqrt{6}/3)\lambda \mathbf{T}_0^2(\mathbf{S}, \mathbf{S}) + (1/2)(o + p + q)(S_+^2 + S_-^2)$$
$$- (1/2)(p + 2q)(J_+ S_+ + J_- S_-) + (1/2)q(J_+^2 + J_-^2). \quad (11.30)$$

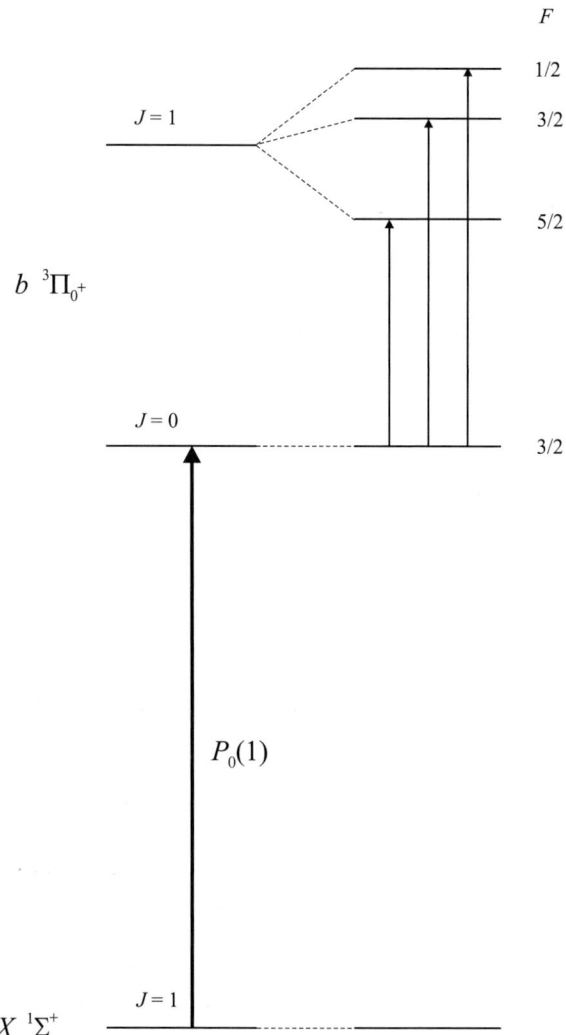

Figure 11.28. Energy level diagram and observed transitions for the microwave/optical double resonance spectrum of CuF. Optical excitation was accomplished using a rhodamine cw dye laser, pumped by an argon ion laser [53].

The matrix elements of these terms in the chosen case (a) basis were given originally in irreducible tensor form by Brown, Kopp, Malmberg and Rydh [57] and Brown and Merer [58], and are summarised in chapter 9. All of the molecular constants appearing in (11.30) were determined from the double resonance measurements; we discuss them in due course.

The magnetic hyperfine terms for the ^{63}Cu and ^{19}F nuclei may be expressed in molecule-fixed cartesian form as follows:

$$\mathcal{H}_{\text{hfs}} = \mathcal{H}_{\text{hfs},1} + \mathcal{H}_{\text{hfs},2}, \tag{11.31}$$

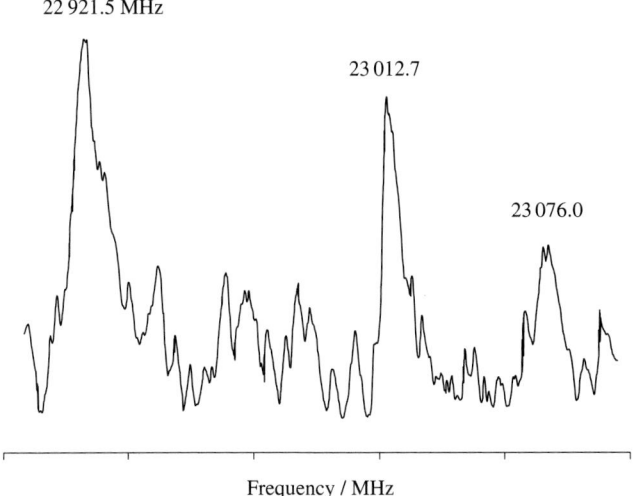

Figure 11.29. Experimental recording of the $J = 1 - 0$ rotation transition in CuF, showing hyperfine splitting from the ^{63}Cu nucleus [53].

where

$$\mathcal{H}_{\text{hfs},1} = a_1 I_{1z} L_z + b_{F,1} \mathbf{I}_1 \cdot \mathbf{S} + (1/3)c_1(3I_{1z}S_z - \mathbf{I}_1 \cdot \mathbf{S}) - (1/2)d_1(S_+ I_{1+} + S_- I_{1-}), \tag{11.32}$$

$$\mathcal{H}_{\text{hfs},2} = a_2 I_{2z} L_z + b_{F,2} \mathbf{I}_2 \cdot \mathbf{S} + (1/3)c_2(3I_{2z}S_z - \mathbf{I}_2 \cdot \mathbf{S}) - (1/2)d_2(S_+ I_{2+} + S_- I_{2-}). \tag{11.33}$$

Alternatively, in irreducible tensor form, we may write

$$\mathcal{H}_{\text{hfs},1} = a_1 T_0^1(\mathbf{L}) T_0^1(\mathbf{I}_1) + b_{F,1} T^1(\mathbf{I}_1) \cdot T^1(\mathbf{S})$$
$$- \sqrt{10}\, g_S \mu_B g_{N,1} \mu_N (\mu_0/4\pi) T^1(\mathbf{S}, \mathbf{C}^2) \cdot T^1(\mathbf{I}_1), \tag{11.34}$$

$$\mathcal{H}_{\text{hfs},2} = a_2 T_0^1(\mathbf{L}) T_0^1(\mathbf{I}_2) + b_{F,2} T^1(\mathbf{I}_2) \cdot T^1(\mathbf{S})$$
$$- \sqrt{10}\, g_S \mu_B g_{N,2} \mu_N (\mu_0/4\pi) T^1(\mathbf{S}, \mathbf{C}^2) \cdot T^1(\mathbf{I}_2). \tag{11.35}$$

In either case there are eight magnetic hyperfine constants, four for each nucleus, whose values will be discussed in due course.

Finally there is the electric quadrupole interaction for the ^{63}Cu nucleus, with matrix elements given by:

$$\langle \eta, \Lambda; S, \Sigma; J, \Omega, I_1, F, M_F | - e T^2(\nabla \mathbf{E}) \cdot T^2(\mathbf{Q}) | \eta, \Lambda'; S, \Sigma; J', \Omega', I_1, F, M_F \rangle$$
$$= -(-1)^{J'+I+F} \frac{eQ}{2} \{(2J+1)(2J'+1)\}^{1/2} \begin{Bmatrix} J' & I & F \\ I & J & 2 \end{Bmatrix} \begin{pmatrix} I & 2 & I \\ -I & 0 & I \end{pmatrix}^{-1}$$
$$\times \sum_q (-1)^{J-\Omega} \begin{pmatrix} J & 2 & J' \\ -\Omega & q & \Omega' \end{pmatrix} \langle \eta, \Lambda | T_q^2(\nabla \mathbf{E}) | \eta, \Lambda' \rangle. \tag{11.36}$$

For a Π state, two quadrupole parameters arise from (11.36), which are

$$eQq_0 = -eQ\langle \eta, \Lambda = \pm 1 | 2T_0^2(\nabla \boldsymbol{E}) | \eta, \Lambda = \pm 1 \rangle,$$
$$eQq_2 = -eQ\langle \eta, \Lambda = \pm 1 | 2\sqrt{6}T_{\pm 2}^2(\nabla \boldsymbol{E}) | \eta, \Lambda = \mp 1 \rangle. \quad (11.37)$$

The analysis described by Steimle, Brazier and Brown [55] used double resonance and fluorescence data obtained for the $b\,^3\Pi$ state and also, from microwave studies [59], the $X\,^1\Sigma^+$ ground state. The following values (in cm^{-1}) of the molecular constants for the $b\,^3\Pi$ state ($v=0$) were determined:

$A = -412.846$, $\quad \lambda = -18.799$, $\quad B = 0.374\,763$, $\quad D = 0.5102 \times 10^{-6}$, $\quad \gamma = 0.320$,

$o = 320$, $\quad p = -0.600$, $\quad q = 0.317 \times 10^{-3}$, $\quad a(\text{Cu}) = 0.0255$, $\quad b_\text{F}(\text{Cu}) = 0.1304$,

$c(\text{Cu}) = 0.0064$, $\quad d(\text{Cu}) = -0.0555$, $\quad eq_0Q(\text{Cu}) = -0.29 \times 10^{-3}$,

$eq_2Q(\text{Cu}) = 0.0161$, $\quad d(\text{F}) = 0.00927$.

At the beginning of this section we noted the fact that the $b\,^3\Pi$ state is one of a manifold of close-lying excited states. The spin–orbit interaction mixes singlet and triplet states and *ab initio* calculations result in the following wave functions for the fine structure components of the $b\,^3\Pi$ state:

$$\left| \Psi(b^3\Pi_2) \right\rangle = 1.0 |^3\Pi \rangle$$
$$\left| \Psi(b^3\Pi_1) \right\rangle = -0.18 |^3\Sigma^+\rangle + 0.95 |^3\Pi\rangle + 0.09 |^3\Delta\rangle - 0.25 |^1\Pi\rangle$$
$$\left| \Psi(b^3\Pi_{0^+}) \right\rangle = -0.45 |^1\Sigma^+\rangle + 0.89 |^3\Pi\rangle$$
$$\left| \Psi(b^3\Pi_{0^-}) \right\rangle = -0.24 |^3\Sigma^+\rangle + 0.97 |^3\Pi\rangle. \quad (11.38)$$

We see that the $^3\Pi_{0^+}$ component is heavily mixed with the $^1\Sigma^+$ state, and calculations of the Λ-doubling constants p and q arising solely from this mixing give values which are in good agreement with experiment. Similarly the mixing of the $^3\Pi$ state with the $^3\Sigma^+$ state is calculated to produce a spin-rotation constant γ of 0.286 cm^{-1}, which is quite close to the measured value of 0.320.

The nuclear hyperfine constants provide the best information about the electronic wave function. As we have seen elsewhere, the four magnetic hyperfine constants (in cm^{-1}) are related to the electron distribution by the expressions:

$$a = (\mu_0/4\pi hc) \sum_i 2\mu_B g_N \mu_N \langle 1/r_i^3 \rangle,$$

$$b = (\mu_0/hc) \sum_i (2/3) g_S \mu_B g_N \mu_N \langle \delta(\boldsymbol{r}_i) \rangle,$$

$$c = (\mu_0/4\pi hc) \sum_i (3/2) g_S \mu_B g_N \mu_N \langle (3\cos^2\theta_i - 1)/r_i^3 \rangle,$$

$$d = (\mu_0/4\pi hc) \sum_i (3/2) g_S \mu_B g_N \mu_N \langle (\sin^2\theta_i)/r_i^3 \rangle. \quad (11.39)$$

The summation is over the unpaired electrons which have spherical polar coordinates r_i and θ_i, and μ_0 is the permeability of free space. The angular averages over a $3d$

orbital needed for the evaluation of the above expressions are

$$\langle 3d\pi|(3\cos^2\theta - 1)|3d\pi\rangle = 2/7, \quad \langle 3d\pi|\sin^2\theta|3d\pi\rangle = 4/7. \tag{11.40}$$

Combined with a value of r^{-3} for a $3d$ electron [60], these angular averages give the following theoretical values for the ^{63}Cu magnetic hyperfine constants (in cm^{-1}):

$$a = 0.035 \quad b_F = 0.098, \quad c = 0.0075, \quad d = 0.015.$$

Considering the extreme simplicity of the model, these constants agree tolerably well with those measured, except for the d constant which is the wrong sign. As we have said many times elsewhere, the measured constants should be treated as benchmarks for the accuracy of *ab initio* calculations. At least the starting point of this discussion, that CuF in the excited $b\,^3\Pi$ state is essentially ionic, seems to be correct.

11.6.4. CuO in the $X\,^2\Pi$ ground state

A microwave/optical double resonance spectrum of CuO was first observed by Gerry, Merer, Sassenberg and Steimle [61] who studied the two lowest rotational transitions in the $X\,^2\Pi_{1/2}$ ground state. An energy level diagram for the lowest rotational transition, $J = 3/2 \leftarrow 1/2$, is shown in figure 11.30. Λ-doubling separates the e and f parity states, which are then further split by ^{63}Cu hyperfine interaction. In subsequent work, Steimle, Chang and Nachman [62] were able to observe the direct radiofrequency Λ-doubling transitions within the $J = 3/2$ level. The experiments relied upon dye laser pumping of appropriate rovibronic components of the $A\,^2\Sigma^- - X\,^2\Pi$ band system, and monitoring the visible fluorescence from the excited state. The CuO vapour was produced by heating powdered copper metal to 1500 °C, and reacting the gaseous Cu with N$_2$O in an argon carrier gas stream.

The observed transitions ranged in frequency from 778 to 66 550 MHz. The molecule conforms very well to Hund's case (a) coupling in the low rotational levels studied, and the effective Hamiltonian is therefore of the same form as that previously used in chapter 9 for the ClO radical, including the presence of one nucleus of spin 3/2 involved in both magnetic and electric quadrupole interactions. Unlike ClO, however, the Λ-doubling is resolved in CuO. The nuclear spin-free effective Hamiltonian for $v = 0$, representing the rigid body rotation and spin–orbit coupling is

$$\mathcal{H}_{\text{eff}} = B_0\{\mathrm{T}^1(\boldsymbol{J}) - \mathrm{T}^1(\boldsymbol{S})\}^2 + A\mathrm{T}^1(\boldsymbol{L})\cdot\mathrm{T}^1(\boldsymbol{S}) \tag{11.41}$$

whilst the nuclear hyperfine terms are, as usual,

$$\begin{aligned}\mathcal{H}_{\text{eff}} &= \mathcal{H}_{\text{hfs}} + \mathcal{H}_Q \\ &= a\mathrm{T}^1(\boldsymbol{I})\cdot\mathrm{T}^1(\boldsymbol{L}) + b_F\mathrm{T}^1(\boldsymbol{I})\cdot\mathrm{T}^1(\boldsymbol{S}) \\ &\quad - \sqrt{10}\,g_S\mu_B g_N\mu_N(\mu_0/4\pi)\mathrm{T}^1(\boldsymbol{I})\cdot\mathrm{T}^1(\boldsymbol{S},\boldsymbol{C}^2) - e\mathrm{T}^2(\boldsymbol{Q})\cdot\mathrm{T}^2(\nabla\boldsymbol{E}).\end{aligned} \tag{11.42}$$

The values of the molecular constants obtained from the analysis of both types of

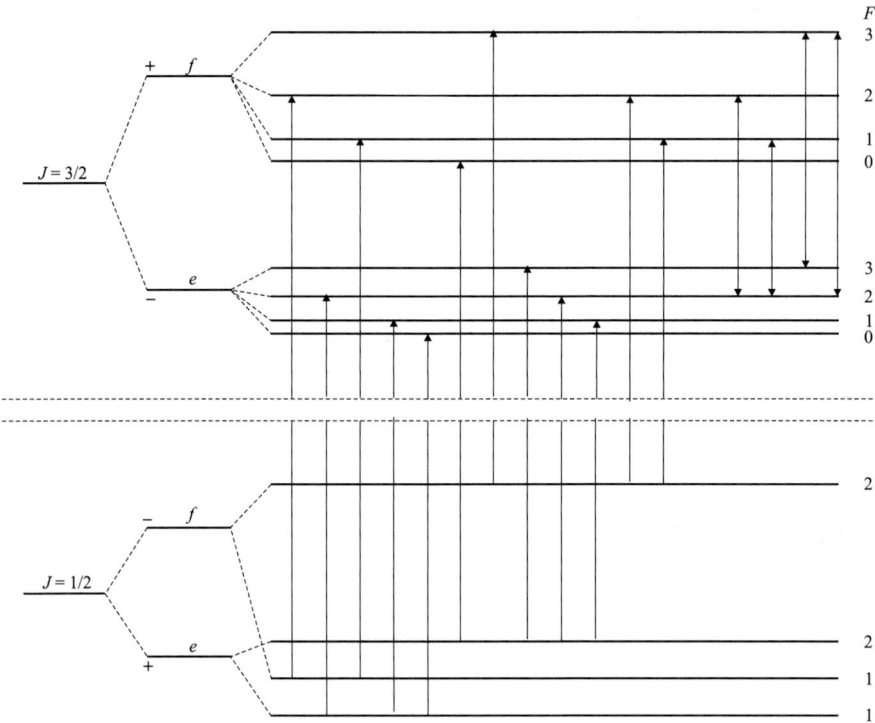

Figure 11.30. Energy level diagram and observed microwave transitions for CuO in the $X\,^2\Pi_{1/2}$ state [61]. The four $\Delta J = 0$ Λ-doublet transitions in the $J = 3/2$ level are shown on the right-hand side.

microwave spectra were as follows (in MHz):

$$A = -8\,278\,319.5 (= -276.137 \text{ cm}^{-1}), \quad B = 132\,53.2, \quad p = 413.2,$$
$$a = 218.2, \quad b_F = -429.5, \quad d = 139.8, \quad eq_0Q = -3.8.$$

Note that, as in other cases, complete separation of the Fermi contact and dipolar hyperfine terms cannot be achieved because only one of the fine-structure states, the $^2\Pi_{1/2}$, could be studied thoroughly.

High level *ab initio* calculations have been described for CuO [63, 64]. The ground state electronic wave function is fairly complicated, but the contributing configuration which gives rise to the ^{63}Cu hyperfine interaction may be written.

$$\text{Cu}(3d^9 4s 4p)\,\text{O}(2p^4): (8\sigma)^1\,(1\delta)^4\,(3\pi)^4(9\sigma)^2\,(4\pi)^3\,(10\sigma)^1.$$

The 8σ, 3π and 1δ orbitals are essentially the copper $3d$ atomic orbitals, and the 10σ is the copper $4s$ orbital. The 9σ molecular orbital is a $2p\sigma(\text{O}) + 4p\sigma(\text{Cu})$ bonding molecular orbital, whilst the 4π orbital is a mixture of $p\pi$ atomic orbitals on oxygen and copper. The electronic wavefunction for the $^2\Pi$ state resulting from this configuration is given by a linear combination of three-electron Slater determinants:

$$\Psi(^2\Pi) = (1/\sqrt{6}) \begin{Bmatrix} 2|8\sigma(1)\alpha(1)4\pi(2)\alpha(2)10\sigma(3)\beta(3)| \\ -|8\sigma(1)\alpha(1)4\pi(2)\beta(2)10\sigma(3)\alpha(3)| \\ -|8\sigma(1)\beta(1)4\pi(2)\alpha(2)10\sigma(3)\alpha(3)| \end{Bmatrix}. \quad (11.43)$$

In principle one can use this wave function to calculate the magnetic and electric hyperfine parameters. In practice, it is interesting to follow the arguments of Gerry, Merer, Sassenberg and Steimle [61] as they attempt to find a semi-empirical description of the bonding in CuO which also gives a reasonable quantitative interpretation of the molecular constants. It is, evidently, not easy to find a satisfactory compromise between the physically visual semi-empirical model, and the full blown *ab initio* calculations.

11.6.5. ScO in the $X\,^2\Sigma^+$ ground state

Scandium is the first member of the 3d transition metal elements and the ScO molecule has been studied through its electronic spectrum and microwave/optical double resonance spectrum. The ground state of ScO is $X\,^2\Sigma^+$ and the $A\,^2\Pi$–$X\,^2\Sigma^+$ emission band system has been recorded and analysed [65, 66], paving the way for more recent double resonance studies by Childs and Steimle [67]. Their investigations involved transitions within the electronic ground state, and employed the molecular beam pump/probe system, the principles of which were described in figure 11.25. The nature of the double resonance transitions studied is described in figure 11.31. There is a very large Fermi contact interaction between the electron spin and the $I = 7/2$ nuclear spin of ^{45}Sc which is present in 100% natural abundance. In the resulting coupling scheme $\mathbf{G} = \mathbf{I} + \mathbf{S}$, the quantum number G takes values of 4 or 3, and for each rotational level N, the total angular momentum F takes values from $|N - G|$ to $N + G$, as shown in figure 11.31 which is for $N = 40$. The observed double resonance transitions were either of the type $\Delta G = \pm 1$, occurring at around 8 GHz, or of the type $\Delta G = 0$, occurring in the radiofrequency region below 100 MHz. An example of an observed radiofrequency resonance, involving $v = 2$, $N = 30$, $G = 4$ is shown in figure 11.32. All of the observed resonances, microwave or radiofrequency, were magnetic-dipole allowed transitions. They involved the $v = 0, 1$ and 2 vibrational levels, with N values ranging from 10 to 75.

The resonance spectra were analysed in terms of the conventional effective Hamiltonian, written in terms of the Frosch and Foley hyperfine constants [9]:

$$\mathcal{H}_{\text{eff}} = \gamma \mathbf{N} \cdot \mathbf{S} + b\mathbf{S} \cdot \mathbf{I} + cI_z S_z + c_I \mathbf{N} \cdot \mathbf{I} + \frac{eq_0 Q}{4I(2I-1)}(3I_z^2 - I^2) + \mathcal{H}_{\text{cdsr}} + \mathcal{H}_{\text{cdhfs}}. \quad (11.44)$$

The centrifugal distortion terms in (11.44) were incorporated rather simply by making the replacements

$$b \to b + b_D N(N+1), \quad c \to c + c_D N(N+1),$$
$$\gamma \to \gamma + \gamma_D N(N+1) + \gamma_H N^2(N+1)^2. \quad (11.45)$$

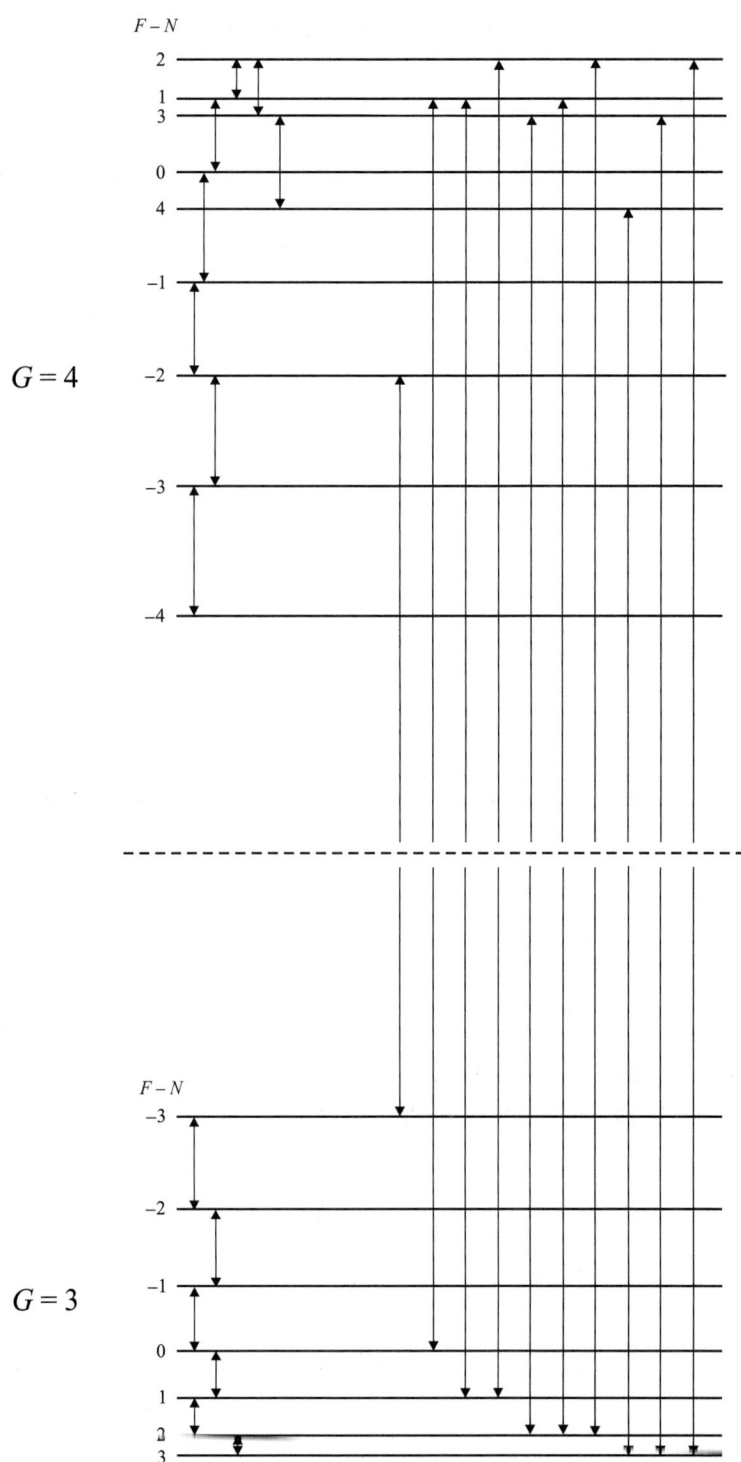

Figure 11.31. Energy level diagram for the $N = 40$ level of ScO in its $X\,^2\Sigma^+$ ground state, and the observed radiofrequency transitions [67].

Table 11.3. *Spectroscopic parameters (in MHz) for the $v = 0$, 1 and 2 levels of the $X\,^2\Sigma^+$ state of ScO, determined by radiofrequency/optical double resonance*

Parameter	$v = 0$	$v = 1$	$v = 2$
γ	3.217 5	4.434 4	5.702 9
b	1922.534	1923.848	1925.124
c	74.416	74.656	74.884
eq_0Q	72.240	71.663	71.177
c_I	0.021 81	0.022 08	0.022 38
γ_D	2.323×10^{-4}	2.432×10^{-4}	2.538×10^{-4}
γ_H	1.0×10^{-9}	0.79×10^{-9}	0.62×10^{-9}
b_D	4.615×10^{-4}	4.542×10^{-4}	4.49×10^{-4}
c_D	1.47×10^{-4}	1.49×10^{-4}	1.54×10^{-4}

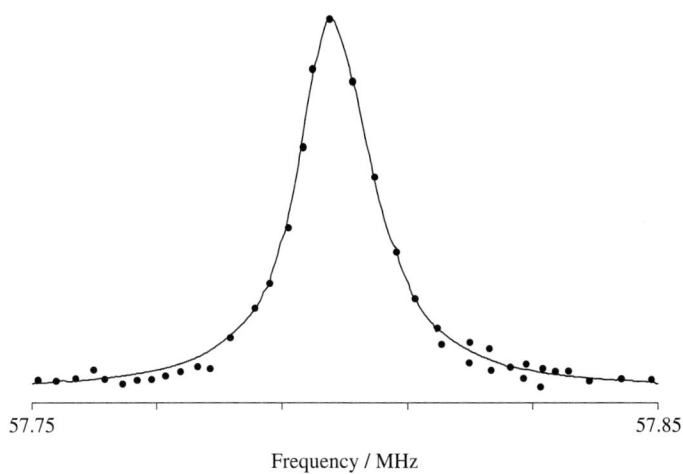

Figure 11.32. Observed radiofrequency/optical double resonance line for ScO [67]. This line arises from the $F = 26$–27 transition in the $v = 2$, $N = 30$, $G = 4$ level of the $X\,^2\Sigma^+$ ground state.

The final results of the analysis, which gave an excellent fit of observed and calculated transition frequencies, are presented in table 11.3.

Two features of the results presented in the table are of particular interest. The first is the very large Fermi contact interaction constant, given by $b_F = b + (c/3)$, of 1.947 GHz. The calculated value for an unpaired electron in a $4s$ atomic orbital of Sc is 2.823 GHz, suggesting that the orbital of the unpaired electron in ScO has 69% $4s$ character, or an atomic orbital coefficient of 0.83. *Ab initio* calculations by Bauschlicher and Langhoff [68] indicate that the primary electron configuration for the ground state of ScO is

$$(\text{core})(3\pi)^4(7\sigma)^2(8\sigma)^1$$

where the molecular orbitals can be represented approximately as

$$3\pi \approx 3d\pi(Sc) + 2p\pi(O), \quad 7\sigma \approx 3d\sigma(Sc) + 2p\sigma(O), \quad 8\sigma \approx 4sp(Sc).$$

The unpaired electron occupies the 8σ molecular orbital which is essentially an sp hybridised orbital, leading to the large Fermi contact interaction observed.

The other particularly interesting feature of the data in table 11.3 is the very large vibrational dependence of the spin–rotation constant γ. There is no doubt that this dependence arises because the spin–rotation interaction is dominated by the second-order mixing of excited electronic states with the ground state; specifically we may write

$$\gamma^{(2)} = 2 \sum_{\eta',v'} \frac{\langle X\,^2\Sigma_{1/2}^+ | B(r) L_{\mp} | \eta', \Lambda' \rangle \langle \eta', \Lambda' | \varsigma_i l_i^{\pm} | X\,^2\Sigma_{1/2}^+ \rangle}{E(\eta', \Lambda') - E(X\,^2\Sigma^+)}. \qquad (11.46)$$

The sum in (11.46) is taken over all vibronic $^2\Pi_{1/2}$ states, and the strong vibrational dependence of γ is thought to be due to the competing effects of at least two different excited $^2\Pi_{1/2}$ states. It is a problem to test the accuracy of *ab initio* calculations; there is certainly no simple semi-empirical explanation.

The interpretation of the ScO spectrum raises again the problem that even in this relatively simple three-valence electron problem there is a manifold of closely-spaced excited electronic states. Interactions between these states, and with the ground state, determine the values of some of the molecular constants, so that a theory which is satisfactory for understanding ground state properties must also be good for the excited states.

11.6.6. TiO in the $X\,^3\Delta$ ground state and TiN in the $X\,^2\Sigma^+$ ground state

Pure rotational transitions in the ground states of TiO and TiN have been described by Namiki, Saito, Scott Robinson and Steimle [69], following earlier experiments by Steimle, Shirley, Jung, Russon and Scurlock [70], which used microwave/optical double resonance molecular beam methods. The ground state of TiO is $X\,^3\Delta$, with three fine-structure components $^3\Delta_1$, $^3\Delta_2$ and $^3\Delta_3$. These have spin–orbit energies equal to $A\Lambda\Sigma$, and since $A = 50.651$ cm^{-1}, the lowest state is the $^3\Delta_1$. The three fine structure states are shown schematically in figure 11.33, with the lower rotational levels and observed rotational transitions indicated. As can be seen, transitions within all three fine-structure states were observed, covering a frequency range from 63 to 448 GHz. Each rotational level actually has a two-fold Λ-doubling (not shown in the figure) which could not be resolved in the conventional rotational spectrum. It could, however, be observed in the double resonance studies of the lowest rotational levels because of the optical selection rules; individual parity components could be selectively laser-pumped in the $B\,^3\Pi_0 \leftarrow X\,^3\Delta$ band system.

The effective Hamiltonian used to analyse the spectrum was, in cartesian form,

$$\mathcal{H}_{\text{eff}} = AL_z S_z + (A_D/2)[L_z S_z, \mathbf{N}^2]_+ + B\mathbf{N}^2 - D\mathbf{N}^4 + H\mathbf{N}^6 + (2\lambda/3)(3S_z^2 - \mathbf{S}^2)$$
$$+ (\lambda_D/3)\big[(3S_z^2 - \mathbf{S}^2), \mathbf{N}^2\big]_+ + (1/2)o_\Delta(S_+^2 J_+^2 + S_-^2 J_-^2). \qquad (11.47)$$

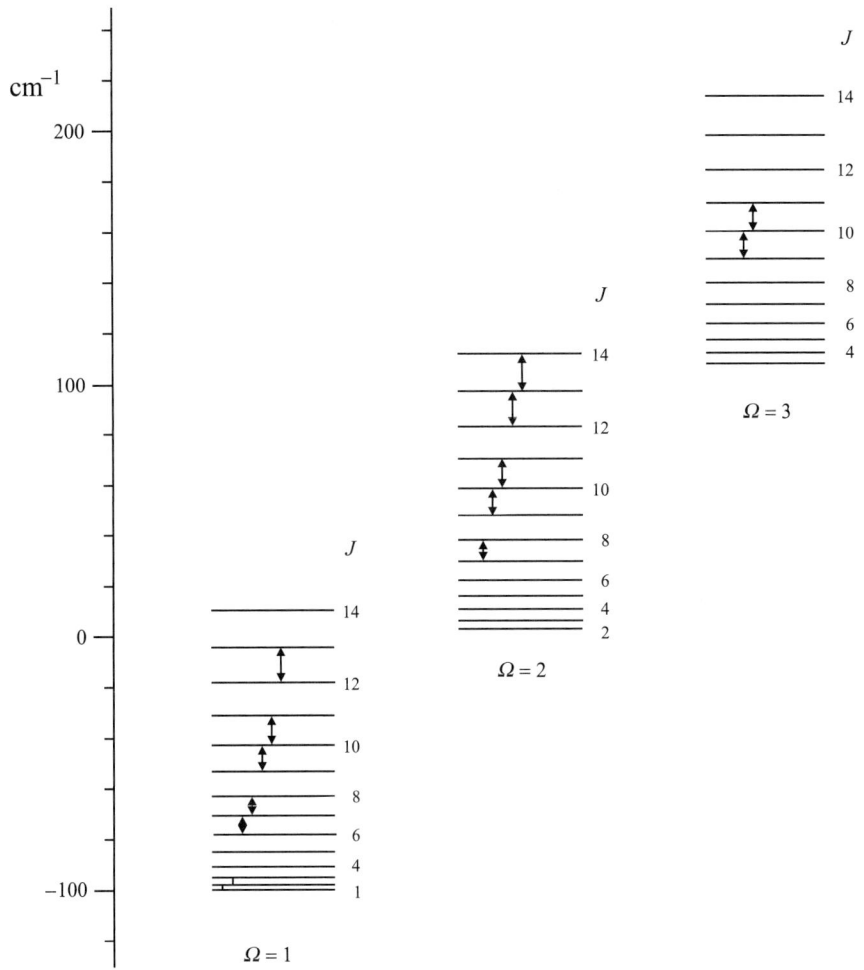

Figure 11.33. Energy level diagram and observed microwave transitions for TiO in its $^3\Delta$ state.

The last term in this effective Hamiltonian represents the Λ-doubling. The following values of the constants (in MHz) were determined for the $v = 0$ level:

$$A = 1\,518\,477.2, \quad A_D = -0.794\,22, \quad B = 16\,003.408\,14, \quad D = 0.018\,060\,6,$$
$$H = -1.53 \times 10^{-7}, \quad \lambda = 52\,380.0, \quad \lambda_D = 0.014\,94, \quad o_\Delta = 0.000\,73.$$

The dominant electronic configuration for TiO in its $X\,^3\Delta$ ground state may be written (core) $(9\sigma)^1 (1\delta)^1$ where the 9σ orbital is essentially the $4s$ orbital of the Ti^{2+} ion and the 1δ orbital is essentially a $3d$ orbital of Ti^{2+}. It is also necessary to provide wave functions for the low-lying excited electronic states of TiO because both the spin–spin constant λ and the Λ-doubling constant o_Δ for the ground state depend upon mixing of excited states produced by a combination of the rotational and spin–orbit interactions. Namiki, Saito, Robinson and Steimle [69] give an acceptable rationalisation, but the *ab initio* calculations one might wish to see have not yet been performed.

TiN has an $X\,^2\Sigma^+$ ground state and rotational transitions in the $v = 0$ level have been measured and analysed [69, 70]; pure millimetre wave and microwave/optical double resonance methods were used, over a frequency range from 37 to 446 GHz. ^{14}N hyperfine structure was observed for the two lowest rotational transitions, and the spectrum analysed using the conventional effective Hamiltonian, again expressed in cartesian form:

$$\mathcal{H}_{\text{eff}} = BN^2 - DN^4 + HN^6 + \gamma \mathbf{N} \cdot \mathbf{S} + b_F \mathbf{I} \cdot \mathbf{S} + (c/3)(3I_z S_z - \mathbf{I} \cdot \mathbf{S}) \\ + c_I \mathbf{I} \cdot \mathbf{N} + eq_0 Q \frac{3I_z^2 - \mathbf{I}^2}{4I(2I-1)}. \qquad (11.48)$$

TiN is different from ScO discussed earlier because the ^{14}N Fermi contact interaction is quite small; in fact the rotational levels and observed transitions for TiN are similar to those of the CN radical, described in chapter 10. The following molecular constants (in MHz) were determined for $X\,^2\Sigma^+$ TiN in the $v = 0$ level:

$$B = 18\,589.350\,81, \quad D = 0.026\,252, \quad H = -1.72 \times 10^{-7}, \quad \gamma = -52.2050,$$
$$\gamma_D = -0.000\,44, \quad b_F = 18.4936, \quad c = 0.1661, \quad c_I = 0.013\,70,$$
$$eq_0 Q = -1.5148.$$

The implications of these constants for the electronic structure of TiN were not seriously discussed.

11.6.7. CrN and MoN in their $X\,^4\Sigma^-$ ground states

Both CrN and MoN have $^4\Sigma^-$ ground states, and rotational transitions have been observed by Namiki and Steimle [71] using the molecular beam pump/probe microwave/optical double resonance method. Energy level diagrams for both molecules are shown in figure 11.34, together with the observed rotational transitions. We described the far-infrared laser magnetic resonance spectrum of CH in its $^4\Sigma^-$ excited state in chapter 9. In that case the most appropriate coupling scheme was Hund's case (b), whereas for CrN and MoN the lower rotational levels conform to Hund's case (a); there are two fine-structure components with $|\Omega| = 3/2$ or $1/2$ as shown in figure 11.34. Each rotational level possesses an additional doublet splitting due to Ω-doubling, particularly large for the $\Omega = 1/2$ state, as shown in the figure. Also resolved in both molecules was ^{14}N hyperfine structure; examples of the observed double resonance spectra are shown in figure 11.35. The microwave frequency ranged from 43 to 79 GHz for CrN, and from 14 to 76 GHz for MoN, and in both cases the optical pumping transition was $A\,^4\Pi \leftarrow X\,^4\Sigma^-$. We should add that figure 11.34 applies to the ^{52}CrN and ^{98}MoN isotopomers, where the metal nucleus has $I = 0$.

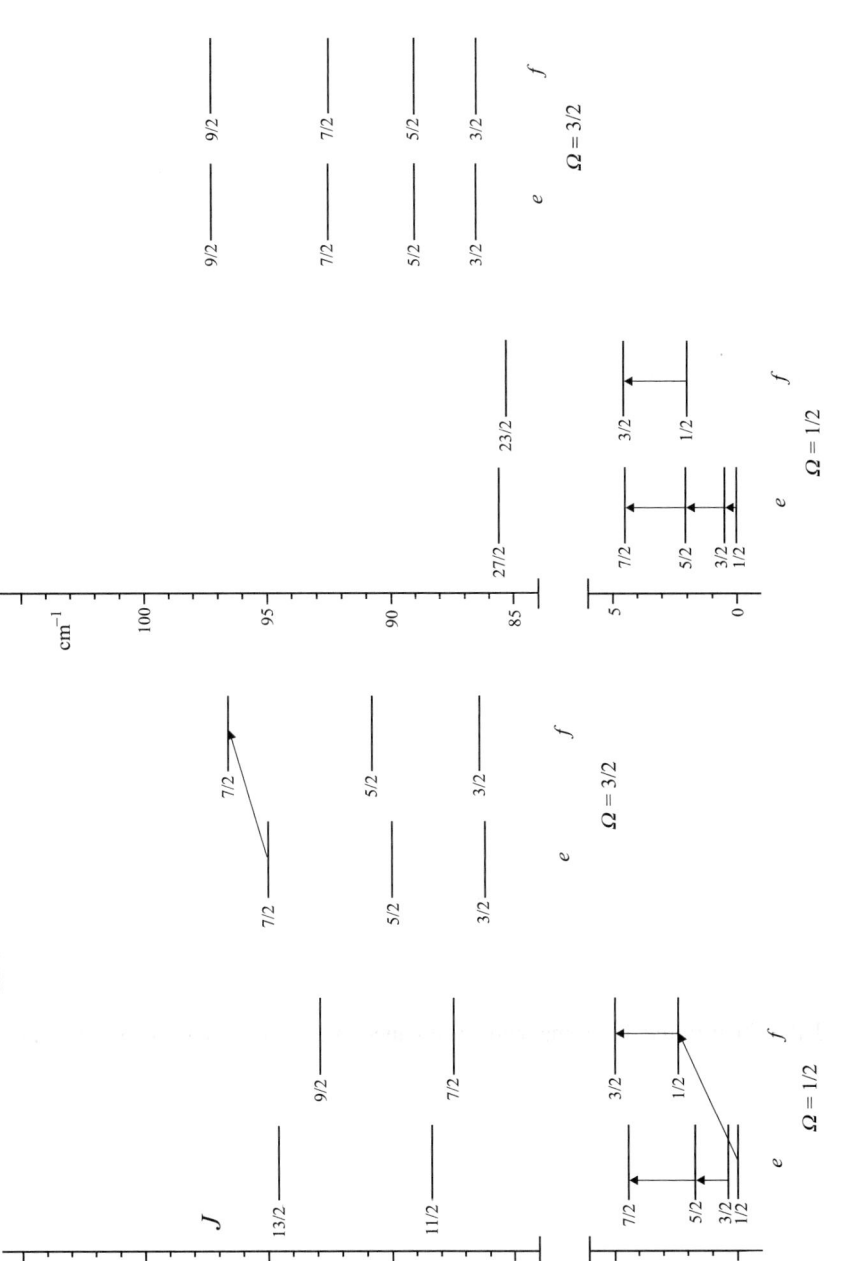

Figure 11.34. Energy level diagram for the $X\,^4\Sigma^-$ states of CrN and MoN, and the observed microwave transitions [71].

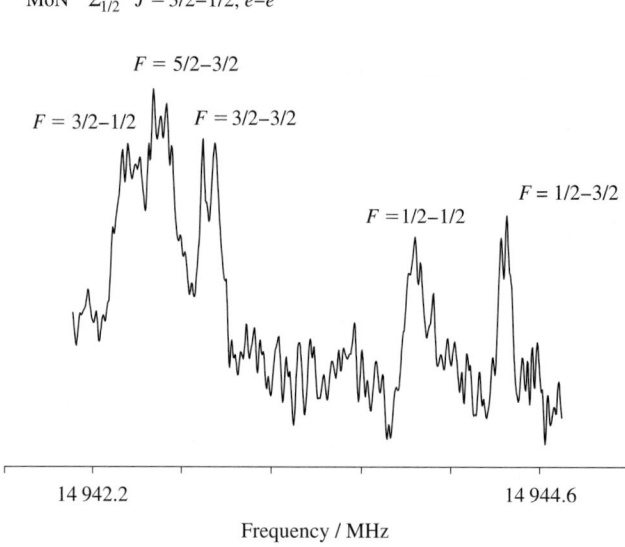

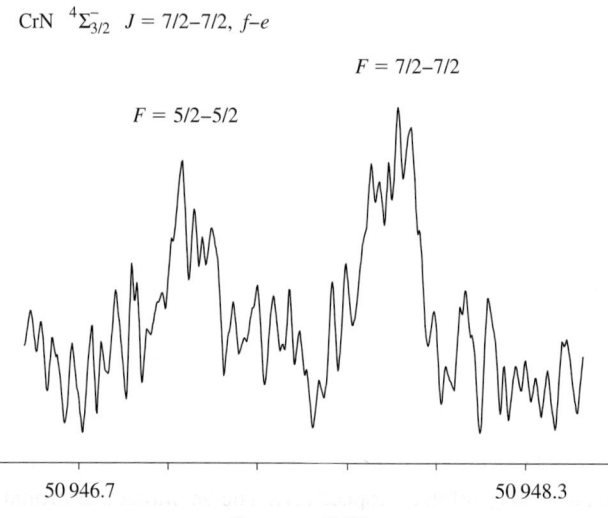

Figure 11.35. *Top*: microwave/optical double resonance transitions [71] in MoN ($X\,^4\Sigma^-_{1/2}$). *Bottom*: microwave/optical double resonance transitions [71] in CrN ($X\,^4\Sigma^-_{3/2}$).

The effective Hamiltonian used by Namiki and Steimle [71], written in spherical tensor form, was as follows:

$$\mathcal{H}_{\text{eff}} = B[\mathrm{T}^1(\boldsymbol{J}) \quad \mathrm{T}^1(\boldsymbol{S})]^2 + \lambda(2\sqrt{6}/3)T_0^2(\boldsymbol{S},\boldsymbol{S}) + \gamma[\mathrm{T}^1(\boldsymbol{J}) - \mathrm{T}^1(\boldsymbol{S})] \cdot \mathrm{T}^1(\boldsymbol{S})$$
$$+ 10\gamma_S \frac{\mathrm{T}^3(\boldsymbol{L}^2, \boldsymbol{J}) \cdot \mathrm{T}^3(\boldsymbol{S}, \boldsymbol{S}, \boldsymbol{S})}{\sqrt{6}\langle \Lambda|T_0^2(\boldsymbol{L}^2)|\Lambda\rangle} + b_{\mathrm{F}}\mathrm{T}^1(\boldsymbol{I}) \cdot \mathrm{T}^1(\boldsymbol{S})$$

$$-\sqrt{10}\,g_S\mu_B g_N\mu_N(\mu_0/4\pi)\mathbf{T}^1(\mathbf{I})\cdot\mathbf{T}^1(\mathbf{S},\mathbf{C}^2) - e\mathbf{T}^2(\mathbf{Q})\cdot\mathbf{T}^2(\nabla\mathbf{E})$$
$$+5\sqrt{14}\,b_S\frac{\mathbf{T}^1(\mathbf{I})\cdot\mathbf{T}^1\{\mathbf{T}^2(\mathbf{L}^2),\mathbf{T}^3(\mathbf{S},\mathbf{S},\mathbf{S})\}}{3\langle\Lambda|T_0^2(\mathbf{L}^2)|\Lambda\rangle}. \tag{11.49}$$

The fourth and last terms in (11.49) are spin–orbit distortion corrections to the spin-rotation and Fermi contact interactions. The hyperfine and quadrupole terms in this Hamiltonian refer to the ^{14}N nucleus.

The matrix elements of the effective Hamiltonian are calculated for case (a) basis functions of definite parity $(-1)^{J+S+s}$:

$$\Psi = (1/\sqrt{2})\{|\eta,\Lambda;S,\Sigma;J,+\Omega,I,F,M_F\rangle \pm |\eta,-\Lambda;S,-\Sigma;J,-\Omega,I,F,M_F\rangle\}. \tag{11.50}$$

Case (a) is the most appropriate basis because the spin–spin parameter λ for both molecules is large, as a result of strong spin–orbit coupling. The values of the molecular parameters (in MHz) for CrN are:

$$B = 18\,702.954,\quad D = 0.015,\quad \lambda = 78\,281.32,\quad \gamma = 209.521,\quad \gamma_S = -0.224,$$
$$b_F = -0.285,\quad b_S = -0.014,\quad c = 4.34,\quad eq_0Q = -2.079.$$

For the MoN the corresponding parameters are:

$$B = 15\,419.568,\quad D = 0.0148,\quad \lambda = 643\,474,\quad \gamma = 334.987,$$
$$b_F = 0.377,\quad c = 4.182,\quad eq_0Q = -2.310.$$

In addition to the above parameters, determined from the microwave data, the metal Fermi contact interaction constants in ^{97}MoN and ^{53}CrN were determined from the optical spectra [72] to be −508 and −179.8 MHz respectively. The element molybdenum belongs to the 4d transition metal row, which we will discuss further in due course. For the CrN molecule in its ground state, *ab initio* calculations [73] and molecular orbital considerations [74] suggest that the dominant electron configuration may be written

$$(\text{core})(8\sigma)^2(3\pi)^4(9\sigma)^1(1\delta)^2.$$

The 8σ and 3π molecular orbitals result from coupling the N $2p_0$ and $2p_{\pm1}$ orbitals with the Cr $3d_0$ and $3d_{\pm1}$ orbitals respectively. The 9σ molecular orbital is a $4p_0 + 4s$ hybrid pointing away from the N atom, whilst the 1δ is essentially a $3d_{\pm2}$ atomic-like orbital. The ^{53}Cr Fermi contact interaction is, of course, dependent on the s-orbital spin density. The spin–spin and spin–rotation constants are determined by the second-order mixing of low-lying electronic states with the ground state; there are, however, a large number of possible excited states, so that an unambiguous quantitative interpretation is difficult without recourse to *ab initio* calculations.

11.6.8. NiH in the $X\,^2\Delta$ ground state

The far-infrared laser magnetic resonance spectrum of NiH has been studied by Nelis, Beaton, Evenson and Brown [75] and was discussed in detail in chapter 9. The electronic ground state is $X\,^2\Delta$, with the $^2\Delta_{5/2}$ fine-structure component lying about 1000 cm^{-1} below the $^2\Delta_{3/2}$ state. Rotational transitions in both fine-structure states were studied in the magnetic resonance spectrum, and some information about the Λ-doubling and magnetic hyperfine parameters was obtained. However, more accurate information for the upper spin component was obtained by Steimle, Nachman, Shirley, Fletcher and Brown [76] using microwave/optical double resonance. Specifically they were able to make direct measurements of Λ-doubling transitions, split by proton hyperfine interaction. An energy level diagram for the lowest rotational levels of the $^2\Delta_{3/2}$ component, and the transitions measured, is shown in figure 11.36. These transitions fall in the range 1.137 to 22.953 GHz.

The laser-induced fluorescence spectrum arising from the $B\,^2\Delta_{3/2}\,(v'=0) - X\,^2\Delta_{3/2}\,(v''=0)$ band system was used, individual rovibronic transitions being pumped with a cw dye laser system operating over the range 658 to 659 nm. The lowest frequency Λ-doubling microwave transition, with hyperfine splitting, is shown in figure 11.37. Transitions from the naturally occurring isotopomers ^{58}NiH and ^{60}NiH were used in the final analysis.

In our discussion of the far-infrared laser magnetic resonance spectrum of NiH in chapter 9, a fairly general effective Hamiltonian was presented. This Hamiltonian included terms which would produce Λ-doubling in a Δ state, an unusual situation because one requires electronic orbital angular momentum operators to connect $\Lambda = +2$ and $\Lambda = -2$ components [77]. The effective Hamiltonian used to analyse the microwave/optical double resonance spectrum of NiH was as follows:

$$\mathcal{H}_{\text{eff}} = AL_z S_z + B\mathbf{N}^2 - D\mathbf{N}^2\mathbf{N}^2 + \gamma \mathbf{N}\cdot\mathbf{S} + (1/2)q_\Delta(J_-^4 + J_+^4)$$
$$- (1/2)(p_\Delta + 4q_\Delta)(S_+ J_+^3 + S_- J_-^3) + aI_z L_z + b_F \mathbf{I}\cdot\mathbf{S}$$
$$+ (1/3)c(3I_z S_z - \mathbf{I}\cdot\mathbf{S}) - (1/2)d_\Delta(J_+^2 I_+ S_+ + J_-^2 I_- S_-). \quad (11.51)$$

The last four terms in this expression describe the magnetic hyperfine interaction; three of them are familiar, but the final term is new to us and is a Λ-doubling magnetic hyperfine term [77]. Its matrix elements are found to be the following:

$$\langle \Lambda, S, \Sigma, J, \Omega, I, F | -(1/2)d_\Delta(J_+^2 I_+ S_+ + J_-^2 I_- S_-)|\Lambda', S, \Sigma', J', \Omega', I, F\rangle$$

$$= d_\Delta(-1)^{J'+I+F}\begin{Bmatrix} J & I & F \\ I & J' & 1 \end{Bmatrix}[I(I+1)(2I+1)(2J+1)(2J'+1)]^{1/2}$$

$$\times (1/2\sqrt{6})\sum_{q=\pm 1}\delta_{\Lambda\Lambda'\mp 4}(-1)^{S-\Sigma}\begin{pmatrix} S & 1 & S \\ -\Sigma & q & \Sigma' \end{pmatrix}[S(S+1)(2S+1)]^{1/2}$$

$$\times \left\{(-1)^{J-\Omega}\begin{pmatrix} J & 1 & J' \\ -\Omega & -q & \Omega'' \end{pmatrix}(-1)^{J'-\Omega''}\begin{pmatrix} J' & 2 & J' \\ -\Omega'' & -2q & \Omega' \end{pmatrix}\right.$$

$$\times [(2J'-1)(2J')(2J'+1)(2J'+2)(2J'+3)]^{1/2}$$

$$+ (-1)^{J-\Omega}\begin{pmatrix} J & 2 & J \\ -\Omega & -2q & \Omega''' \end{pmatrix}(-1)^{J-\Omega'''}\begin{pmatrix} J & 1 & J' \\ -\Omega''' & -q & \Omega' \end{pmatrix}$$

$$\left.\times [(2J-1)(2J)(2J+1)(2J+2)(2J+3)]^{1/2}\right\}. \quad (11.52)$$

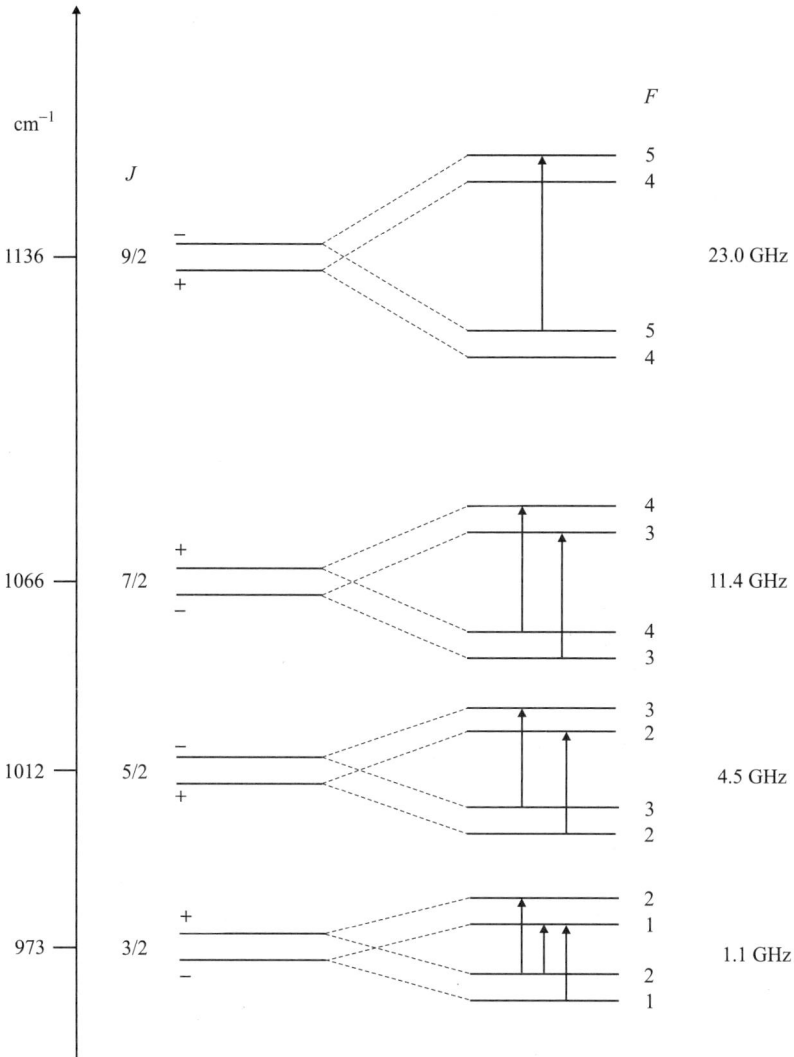

Figure 11.36. The four lowest rotational levels of the $X\,^2\Delta_{3/2}$ spin component of NiH, and the observed microwave transitions [76]. The energies are given relative to the lowest rotational level of the $X\,^2\Delta_{5/2}$ component (with $J = 5/2$). The Λ-doublet and proton hyperfine splittings are exaggerated for the sake of clarity.

This result shows that the operator connects states with $\Delta\Lambda = \pm 4$, $\Delta\Sigma = \pm 1$ and $\Delta\Omega = \pm 3$. The diagonal matrix element from (11.52) is

$$(^2\Delta; S=1/2, \Sigma=1/2; J, \Omega=3/2, I, F, M_F, \pm |\mathcal{H}_{\text{eff}}(d_\Delta)|^2\Delta; S=1/2, \Sigma = 1/2; J, \Omega=3/2, I, F, M_F, \pm)$$
$$= \mp(-1)^{J-1/2}\left(\frac{1}{2}\right)d_\Delta \times \frac{[F(F+1) - J(J+1) - I(I+1)]}{2J(J+1)}$$
$$\times [J(J+1) - 3/4](J+1/2). \tag{11.53}$$

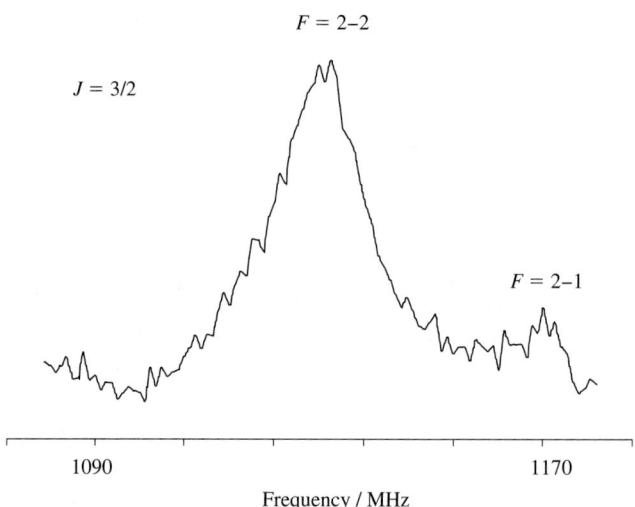

Figure 11.37. The Λ-doublet spectrum arising from the $J = 3/2$ level of the $X\,^2\Delta_{3/2}$ state of ^{58}NiH. The recording time was 20 min [76].

The upper and lower signs refer to states of positive and negative parity respectively, and the very small matrix elements off-diagonal in J have been neglected.

All of the other required matrix elements have been given elsewhere [77]; some of the constants in equation (11.51) were determined in the laser magnetic resonance study, and the remaining constants were obtained from the double resonance study. The final parameters for the $v = 0$ level were (in MHz) as follows:

$$A = -14\,768\,100, \quad B = 232\,109.3, \quad D = 15.156, \gamma = 89\,110, \quad q_\Delta = -0.570,$$
$$(p_\Delta + 4q_\Delta) = 188.638, \quad h_{3/2} = 2a - (1/2)(b_F + 2c/3) = 50.8, \quad d_\Delta = 0.767.$$

Note that only the total diagonal hyperfine constant, $h_{3/2}$, could be determined, as is usually the case for good case (a) systems when only one fine-structure state is studied. The final fit to the experimental Λ-doublet frequencies, including hyperfine interaction, had a standard deviation of 1 MHz. The values of the constants were interpreted in terms of the interaction of Ni($3d^94s$) with H($1s$), the electronic wave function being represented in terms of atomic Ni $3d$ orbitals.

11.6.9. 4d transition metal molecules: YF in the $X\,^1\Sigma^+$ ground state, YO and YS in their $X\,^2\Sigma^+$ ground states

The second row of transition metal elements, the $4d$ series, consists of the following members, with their lowest energy electron configurations.

$$\text{Y:} \quad (1s)^2(2s)^2(2p)^6(3s)^2(3p)^6(3d)^{10}(4s)^2(4p)^6(4d)^1(5s)^2$$
$$\text{Zr:} \quad (1s)^2(2s)^2(2p)^6(3s)^2(3p)^6(3d)^{10}(4s)^2(4p)^6(4d)^2(5s)^2$$

Nb: $(1s)^2(2s)^2(2p)^6(3s)^2(3p)^6(3d)^{10}(4s)^2(4p)^6(4d)^4(5s)^1$
Mo: $(1s)^2(2s)^2(2p)^6(3s)^2(3p)^6(3d)^{10}(4s)^2(4p)^6(4d)^5(5s)^1$
Tc: $(1s)^2(2s)^2(2p)^6(3s)^2(3p)^6(3d)^{10}(4s)^2(4p)^6(4d)^5(5s)^2$
Ru: $(1s)^2(2s)^2(2p)^6(3s)^2(3p)^6(3d)^{10}(4s)^2(4p)^6(4d)^7(5s)^1$
Rh: $(1s)^2(2s)^2(2p)^6(3s)^2(3p)^6(3d)^{10}(4s)^2(4p)^6(4d)^8(5s)^1$
Pd: $(1s)^2(2s)^2(2p)^6(3s)^2(3p)^6(3d)^{10}(4s)^2(4p)^6(4d)^{10}$

We have already discussed the microwave/optical double resonance spectrum of the MoN molecule in its $X^4\Sigma^-$ ground state. We now deal with studies of YF, YO and YS, yttrium being the first member of the $4d$ transition elements analogous to scandium in the $3d$ series.

We have described a number of molecular beam pump/probe microwave/optical double resonance studies carried out by Steimle and his collaborators, and we now discuss earlier experiments on YF and YO carried out by Fletcher, Jung, Scurlock and Steimle [78]. This paper described the details of their molecular beam method. A schematic block diagram of their apparatus is shown in figure 11.38. It is similar in some ways to those described earlier by Kröckertskothen, Knöckel and Tiemann [48], which was used to study the FeO molecule, Rosner, Holt and Gaily [79], and Childs and Goodman [80]. The heart of the apparatus is the molecular beam system, which made use of laser ablation of a rotating rod of the metal of interest (e.g. yttrium) produced by a pulsed Nd:YAG laser timed to occur as a pulse of gas was released from the valve. The gas pulse was mainly argon, containing a small amount of a suitable reactant, namely N_2O for the production of YO, or SF_6 for YF. The molecular beam was skimmed before entering the main chamber where it was interrogated with suitable electromagnetic radiation. Following the conventions of molecular beam magnetic and electric resonance, described in chapter 8, three regions may be designated. In region A, the molecular beam intersects a cw dye laser beam at right angles; the frequency of the laser is chosen to drive a known selected rovibronic transition in the molecule of interest. A portion of the same laser beam again intersects the molecular beam at right angles in region B; typically 96% of the total laser power is used in region A and 4% in region B. The purpose of the 'probe' laser beam in region B is to stimulate fluorescence, which is collected and focussed into a cooled photomultiplier tube. Between the two laser regions is the C region, where microwave or millimetre wave radiation is propagated through a suitable horn to intersect the molecular beam, again in a perpendicular direction. The result of the optical pumping in region A is to deplete the population of the ground state rotational level involved. The microwave radiation, when in resonance with a rotational transition, repopulates the ground state rotational level; this leads to an increase in the fluorescence intensity, which is detected. The microwave frequency is scanned stepwise, and the rotational spectrum recorded; a typical resonance line for YO is shown in figure 11.39. The line width is 45 kHz, determined by the transit time of the molecular beam through the microwave region. The apparatus provides a particularly powerful way of studying refractory materials, and a number of examples have been described in the preceding sections.

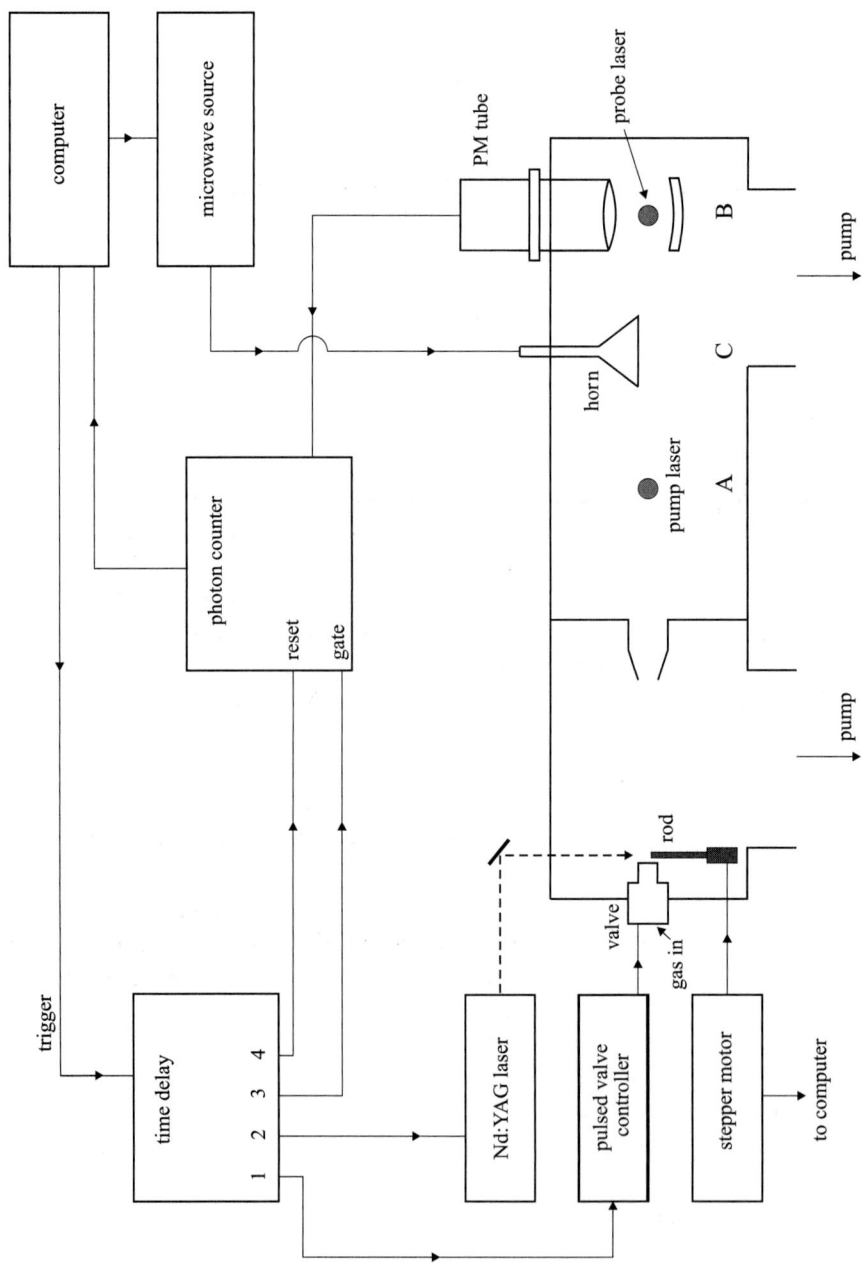

Figure 11.38. Block diagram of the molecular beam pump/probe microwave/optical double resonance spectrometer, developed by Fletcher, Jung, Scurlock and Steimle [78].

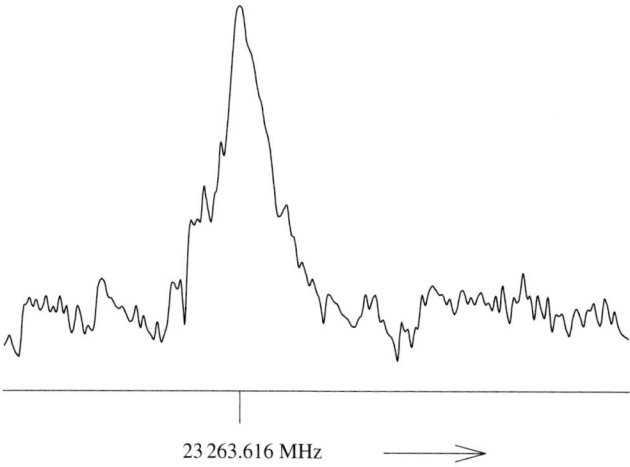

Figure 11.39. A hyperfine component of the $N = 1 \leftarrow 0$ rotational transition of YO in its $X\,^2\Sigma^+$ ground state [78].

The rotational spectrum of YF is particularly simple because although the ^{89}Y isotope, present in 100% natural abundance, has spin 1/2, the YF ground state is $^1\Sigma^+$ and any interactions involving the nuclear spins are extremely small. The $J = 1 \leftarrow 0, 2 \leftarrow 1$ and $3 \leftarrow 2$ rotational transitions were measured at 17 367.202, 34 734.210 and 52 100.882 MHz, from which the rotational constants $B_0 = 8683.6156$, $D_0 = 0.007\,521$ MHz were determined.

The YO molecule in its $X\,^2\Sigma^+$ ground state was the subject of earlier radiofrequency/optical double resonance studies by Childs, Poulsen and Steimle [81], who made observations for $v = 0$ to 4 and N values up to 91. The spin–rotation and nuclear hyperfine splitting for each rotational level takes a simple form; the largest interaction is the ^{89}Y Fermi contact interaction, so that the case (b) coupling scheme most appropriate is

$$S + I = G, \quad G + N = F.$$

This leads to the simple spin structure of each rotational level shown in figure 11.40 and the three magnetic dipole transitions indicated (all at frequencies less than 800 MHz). This triplet pattern was measured for many rotational levels, and analysed using the customary effective Hamiltonian,

$$\mathcal{H}_{\text{eff}} = \gamma \boldsymbol{S} \cdot \boldsymbol{N} + b \boldsymbol{I} \cdot \boldsymbol{S} + c I_z S_z + c_I \boldsymbol{I} \cdot \boldsymbol{N}. \tag{11.54}$$

Here b and c are the Frosch and Foley magnetic hyperfine constants for ^{89}Y; for the $v = 0$ level the values of the constants (in MHz) were determined to be

$$\gamma = -9.2254, \quad b = -762.976, \quad c = -28.236, \quad b_F = 772.388, \quad c_I = -0.002\,57.$$

The primary molecular orbital electron configuration for the $X\,^2\Sigma^+$ state of YO is

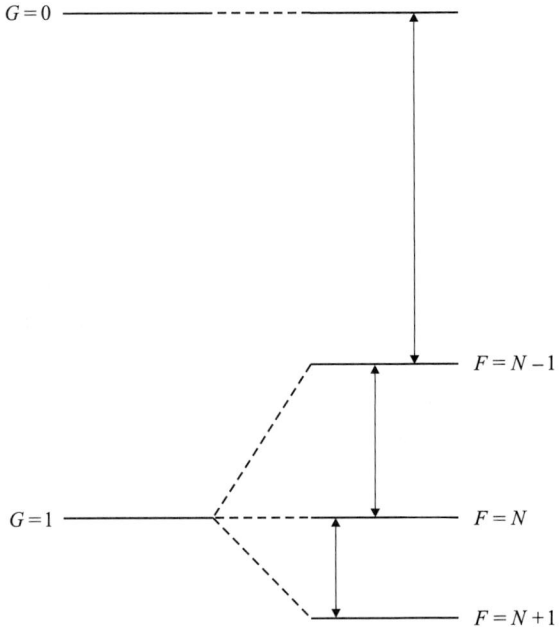

Figure 11.40. Hyperfine and spin–rotation splitting of a typical rotational level in the $X\,^2\Sigma^+$ state of YO, and the magnetic dipole transitions observed by radiofrequency/optical double resonance.

thought to be

$$(\text{core})(11\sigma)^2(5\pi)^4(12\sigma)^1$$

where the 12σ orbital is a non-bonding combination of the $5s$ and $5p$ atomic orbitals of yttrium. The large Fermi contact interaction observed is consistent with the unpaired electron occupying this orbital.

The YS molecule has been studied by Azuma and Childs [82], again using the molecular beam laser/radiofrequency double resonance technique. The electronic spectrum arising from the $B\,^2\Sigma^+ \to X\,^2\Sigma^+$ transition was studied through fluorescence arising from visible dye laser excitation and hyperfine structure from ^{89}Y with a spin of 1/2 was observed. Direct hyperfine transitions were then detected by radiofrequency double resonance and, as with YO, low rotational levels in the ground state were best described in terms of the hyperfine coupled representation $|\eta, \Lambda; S, I, G; G, N, F\rangle$. Rotational levels of the excited state, and higher rotational levels of the ground state were described in terms of the more common representation $|\eta, \Lambda; N, S, J; J, I, F\rangle$.

Examples of the electronic and radiofrequency transitions are shown in the energy level diagram presented in figure 11.41. The electronic transitions are shown as broken lines, whilst the radiofrequency double resonance transitions are denoted by continuous lines. Three double resonance transitions are possible for each ground state rotational level, and transitions involving N values from 2 to 48 were observed, over the frequency

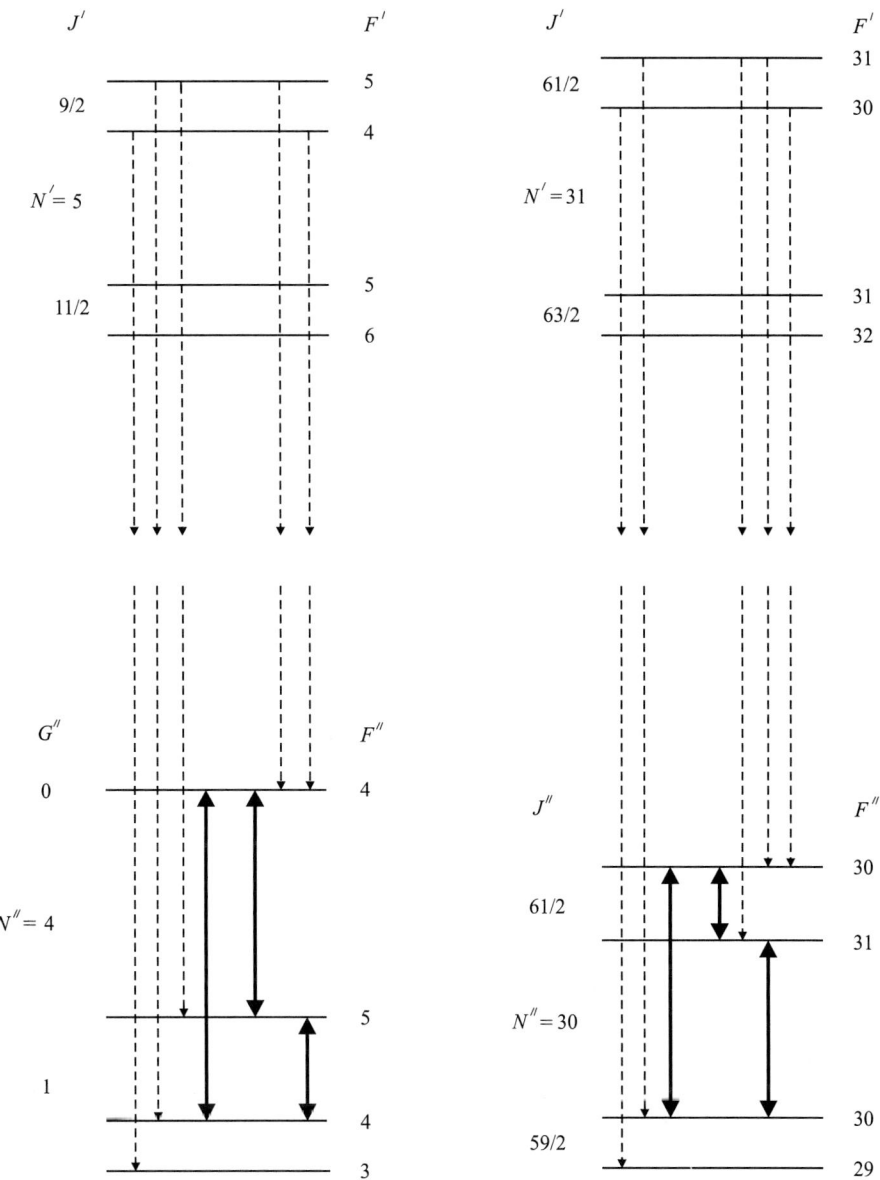

Figure 11.41. Energy level diagrams and transitions for two different R-branch transitions in the YS molecule. The broken lines denote laser-induced fluorescence transitions, whilst the continuous lines indicate radiofrequency double resonance transitions observed in rotational levels of the ground electronic state [82].

range 700 to 2200 MHz. Using the standard Frosch and Foley hyperfine Hamiltonian,

$$\mathcal{H}_{\text{hfs}} = \gamma \mathbf{S} \cdot \mathbf{N} + b \mathbf{I} \cdot \mathbf{S} + c I_z S_z + c_I \mathbf{I} \cdot \mathbf{N}, \tag{11.55}$$

Azuma and Childs [82] determined the following values of the constants (in MHz) for the $v=0$ level of the $X\,^2\Sigma^+$ ground state:

$$\gamma = 42.2382, \quad b = -653.251, \quad c = -42.684, \quad c_I = -0.0046.$$

The Fermi contact constant b_F is therefore found to be -667 MHz, a large value which reflects participation of the Y $5s$ atomic orbital in the unpaired electron wave function (note that the g-factor for [89]Y is negative).

It is possible from the optical measurements to use the ground state constants in order to determine values of the spin-rotation constant γ and hyperfine constant b for the excited B state of the YS molecule; the values obtained are $\gamma = -4620$ MHz and $b = -78$ MHz. These are quite different from the ground state values, but are not discussed by the original authors.

11.7. Microwave/optical double resonance of rare earth molecules

11.7.1. Radiofrequency/optical double resonance of YbF in its $X\,^2\Sigma^+$ ground state

The first radiofrequency/optical double resonance study of a simple lanthanide molecule, YbF, was described by Sauer, Wang and Hinds [83]. The Yb atom in its ground state electron configuration is [Xe] $(4f)^{14}(6s)^2$ and although it is normally trivalent in its stable compounds, the YbF radical has been studied through its electronic spectrum for many years and the ground electronic state is known to be $X\,^2\Sigma^+$. The $A\,^2\Pi_{1/2} - X\,^2\Sigma^+$ laser induced fluorescence spectrum was studied by Sauer, Wang and Hinds [84] and this enabled the same authors to set up a radiofrequency/optical double resonance experiment, very similar to that described earlier for FeO. An effusive beam of YbF was produced by the reaction of YbF$_3$ with Al in an oven maintained at 1200 °C, and crossed twice with a cw dye laser beam at 552 nm. Yb has six isotopes with significant natural abundance, but the most preominant is [174]Yb with 31.8% abundance (spin zero) and the studies described concern this isotope.

It is perhaps a pleasant surprise that the $X\,^2\Sigma^+$ state of YbF conforms to a simple case (b) coupling scheme, so that each rotational level N is split by the spin–rotation interaction into states characterised by $J = N \pm 1/2$; each of these states is further split into a doublet by the fluorine hyperfine interaction, giving final states $|N, S, J, I, F\rangle$ as shown in figure 11.42. The effective Hamiltonian is therefore written in the familiar form (see, for example, our discussion of the CN radical in chapter 9)

$$\mathcal{H}_{\text{eff}} = \{B - DN^2\}N^2 + \gamma T^1(N) \cdot T^1(S) + \gamma_D[T^1(N) \cdot T^1(S)]N^2 + b_F T^1(I) \cdot T^1(S)$$
$$- \sqrt{10} g_S \mu_B g_N \mu_N (\mu_0/4\pi) T^1(S, C^2) \cdot T^1(I) + c_I T^1(N) \cdot T^1(I). \quad (11.56)$$

The matrix elements of this effective Hamiltonian were given in chapter 9, and will

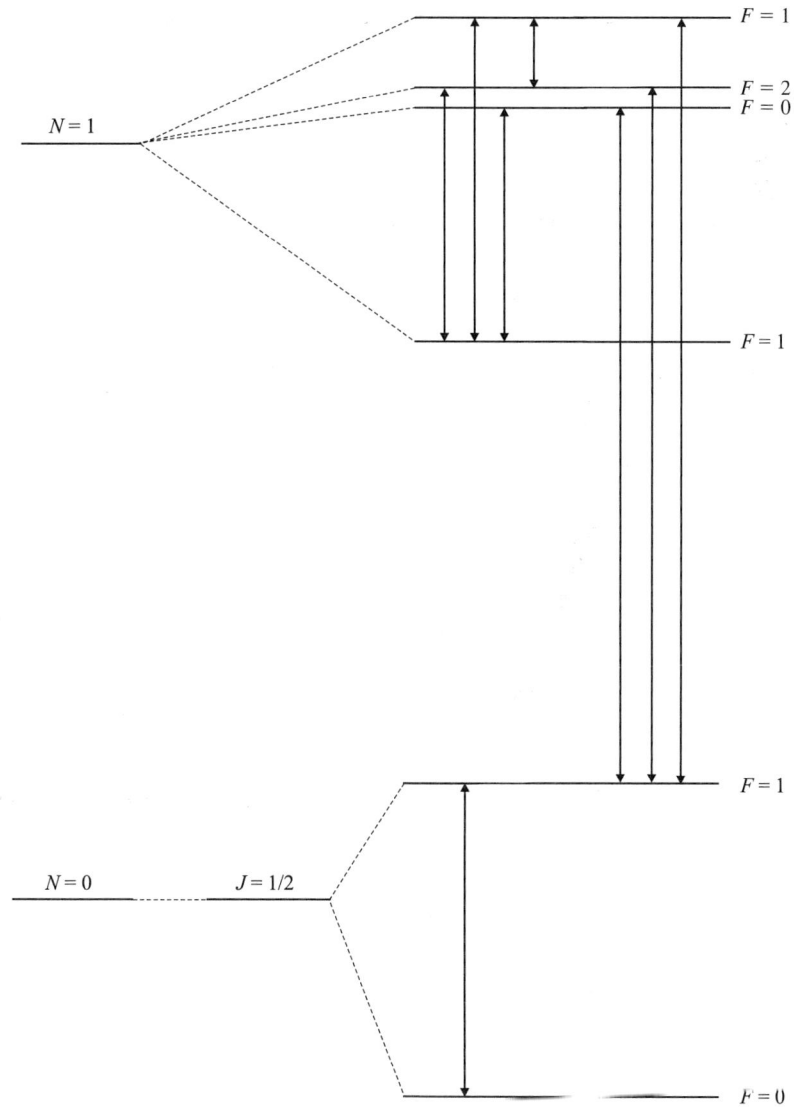

Figure 11.42. Lower rotational levels of YbF $X^2\Sigma^+$, and the radiofrequency transitions studied [83].

not be repeated here, except for the Fermi contact term, which is:

$$\langle \eta, \Lambda; N, S, J, I, F | b_F T^1(\boldsymbol{S}) \cdot T^1(\boldsymbol{I}) | \eta, \Lambda; N, S, J', I, F \rangle$$
$$= b_F (-1)^{J'+F+I} \begin{Bmatrix} I & J' & F \\ J & I & 1 \end{Bmatrix} (-1)^{N+J+S+1} \begin{Bmatrix} S & J' & N \\ J & S & 1 \end{Bmatrix}$$
$$\times \{(2J'+1)(2J+1)I(I+1)(2I+1)S(S+1)(2S+1)\}^{1/2}. \quad (11.57)$$

For the $N = 1$ rotational level, the spin-rotation interaction produces a splitting into levels which would normally be characterised as $J = 1/2$ and $3/2$. However the spin–rotation constant γ is found to have an abnormally small value, about -20 MHz, compared with the Fermi contact constant b_F which is close to 143 MHz. Consequently, as equation (11.57) shows, the $J = 3/2$ and $1/2$ levels are heavily mixed, so that J is a very poor quantum number.

Except for $N = 0$, which has no spin–rotation interaction, Sauer, Wang and Hinds [83] measured the four radiofrequency transitions indicated in figure 11.42 for $N = 1$ to 10 in the $v = 0$ level, and also obtained more limited data for $v = 1$. They also measured the lowest rotational transition at around 14.5 GHz, with hyperfine splitting, as indicated in figure 11.42. An unusual feature observed by Sauer, Wang and Hinds [83] was a very strong dependence of the spin–rotation constant on the rotational quantum number N, which led them to express γ by the power series expansion

$$\gamma = \gamma + \gamma_D N(N+1) + \gamma_H [N(N+1)]^2 + \cdots. \tag{11.58}$$

From their laser-induced fluorescence study they were able to identify transitions with N values up to 75, and to determine the following values of the constants in (11.58):

$$\gamma = -13.424 \text{ MHz}, \quad \gamma_D = 0.003\,982 \text{ MHz}, \quad \gamma_H = -0.025 \text{ kHz}.$$

The centrifugal distortion term involving γ_D is remarkably large, so much so that at $N = 60$, the order of the spin components actually changes sign. This is probably a unique observation.

An additional observation made by Sauer, Wang and Hinds [83] was the Stark effect which enabled them to determine the electric dipole moment to be 3.91 D. This value taken with the relatively small fluorine hyperfine interaction suggests that YbF has a largely ionic structure.

11.7.2. Radiofrequency/optical double resonance of LaO in its $X\,^2\Sigma^+$ and $B\,^2\Sigma^+$ states

The ground electronic state of $^{139}\text{La}^{16}\text{O}$ is $X\,^2\Sigma^+$ and its electronic spectrum involving the excited $B\,^2\Sigma^+$ has been studied by Doppler-free laser-induced fluorescence by Bacis, Collomb and Bessis [85] and by Bernard and Sibai [86]. Both states have therefore been well characterised and the system is ideal for radiofrequency/optical double resonance, as described by Childs, Goodman, Goodman and Young [87]. They used a collimated molecular beam, with the laser pump/probe technique described elsewhere in this chapter.

The ^{139}La nucleus has a spin of $7/2$; the Fermi contact hyperfine interaction is large in the B state and very large in the X state. Consequently the $B \to X$ fluorescence spectrum exhibits a distinctive hyperfine structure, an example of which

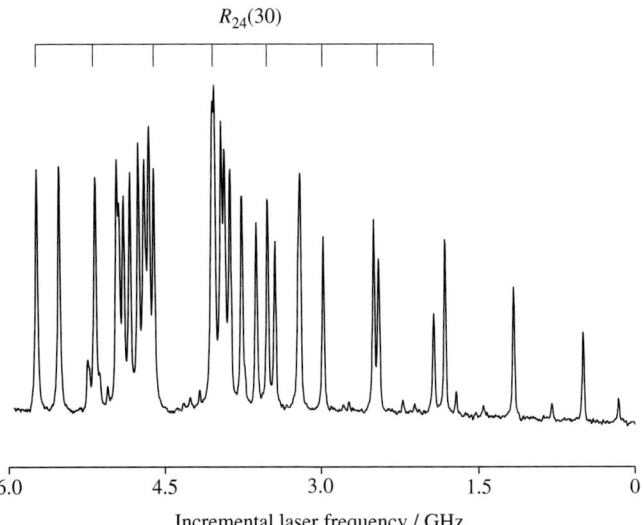

Figure 11.43. Section of the Doppler-free laser-induced fluorescence spectrum of LaO, arising from the $B\,^2\Sigma^+(v=0) \to X\,^2\Sigma^+\,(v=1)$ electronic transition [87]. Four different rotational components are present, one of which is marked, with the lower state N value (30) being given in the brackets. The region of the electronic spectrum scanned is 5866.75 to 5866.80 Å.

is shown in figure 11.43. Figure 11.44 shows the energy level diagram appropriate for the spectrum exhibited in figure 11.43. The labelling of the hyperfine levels requires further discussion, however. Because the hyperfine interaction is so large in the ground state, the best coupling scheme is one in which the electron (S) and nuclear (I) spin angular momenta are coupled to form a resultant G, which is then coupled with the rotational angular momentum N to form the grand total angular momentum F:

$$S + I = G,$$
$$G + N = F.$$

Since $I = 7/2$ and $S = 1/2$, G can take the values 4 and 3, as shown, and F takes values from $|N - G|$ to $N + G$. In the excited electronic state, B, we adopt the more conventional coupling scheme:

$$N + S = J,$$
$$J + I = F.$$

The hyperfine components of the R_{23} and R_{24} rotational branches are shown in figure 11.44.

Radiofrequency transitions between hyperfine components in the ground electronic state were induced in the beam and detected through strong increases in the

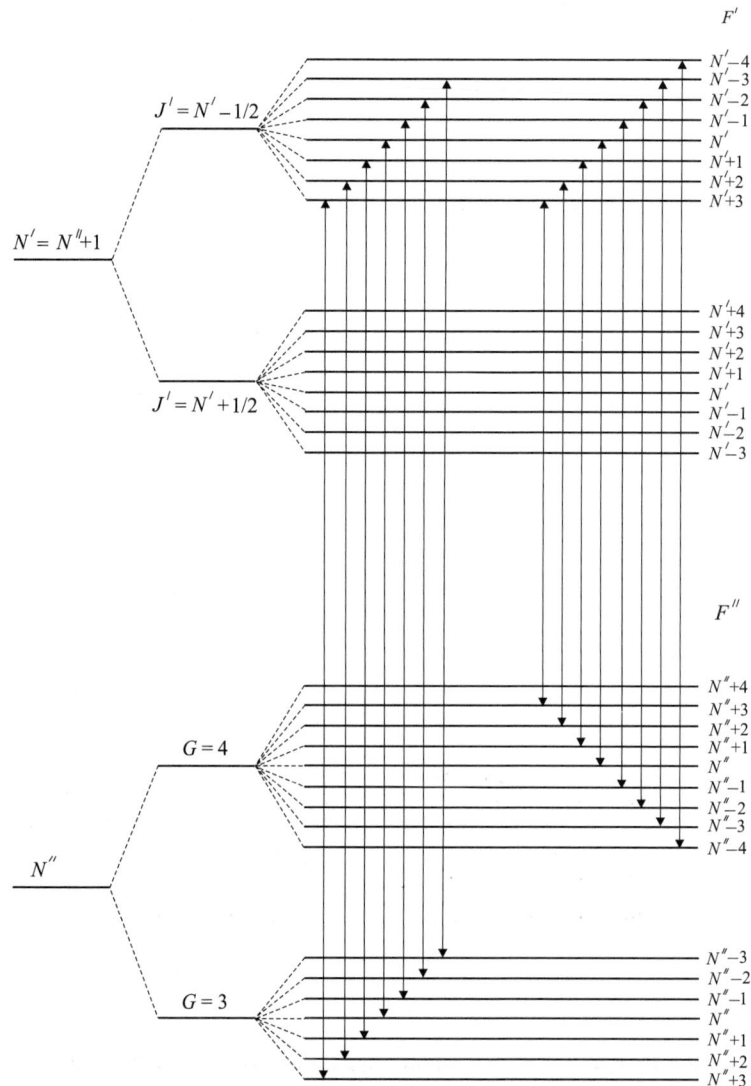

Figure 11.44. Energy level diagram [87], including ^{139}La nuclear hyperfine splitting, and transitions corresponding to the spectrum shown in figure 11.43.

fluorescence intensity. Many such transitions were measured over the frequency range 25 to 13 800 MHz, enabling very accurate values of the ground state molecular constants to be obtained. Double resonance transitions involving excited electronic state levels were not observed, but the much improved accuracy in the ground state spacings was used to provide more accurate values for the excited state constants. The measurements involved ground state vibrational levels up to $v = 1$ and excited levels $v = 0$ and 1. The effective Hamiltonian used for both states was, as

Table 11.4. *Molecular constants (in MHz) for the $X\,^2\Sigma^+$ and $B\,^2\Sigma^+$ electronic states of LaO, determined [87] from the laser/radiofrequency studies*

parameter	$X\,^2\Sigma^+$ state	$B\,^2\Sigma^+$ state
γ	66.0065	-7574.2
b	3630.63	586.7
c	94.416	199.2
b_F	3662.10	653.1
eq_0Q	-84.419	-193.4
c_I	0.014444	—

usual,

$$\mathcal{H}_{\text{eff}} = \gamma \boldsymbol{S} \cdot \boldsymbol{N} + b \boldsymbol{S} \cdot \boldsymbol{I} + c S_z I_z + c_I \boldsymbol{I} \cdot \boldsymbol{N} + eq_0 Q \frac{\{3I_z^2 - I(I+1)\}}{4I(2I-1)}. \quad (11.59)$$

This form of the Hamiltonian, using the Frosch and Foley constants, is less useful than the alternative form, written in terms of spherical tensor operators. This is particularly true when the basis functions for the two electronic states are different. For the ground state we use the functions $|\eta, \Lambda; S, I, G; N, G, F\rangle$ and for the excited state $|\eta, \Lambda; N, S, J; J, I, F\rangle$. As we have seen in chapters 9 and 10, the appropriate effective Hamiltonian when the excited state basis functions are used is

$$\mathcal{H}_{\text{eff}} = \{B_v - D_v \boldsymbol{N}^2\}\boldsymbol{N}^2 + \gamma_v \mathrm{T}^1(\boldsymbol{N}) \cdot \mathrm{T}^1(\boldsymbol{S}) + b_F \mathrm{T}^1(\boldsymbol{S}) \cdot \mathrm{T}^1(\boldsymbol{I})$$
$$- \sqrt{10} g_S \mu_B g_N \mu_N (\mu_0/4\pi) \mathrm{T}^1(\boldsymbol{S}, \boldsymbol{C}^2) \cdot \mathrm{T}^1(\boldsymbol{I}) - e \mathrm{T}^2(\boldsymbol{Q}) \cdot \mathrm{T}^2(\nabla \boldsymbol{E}). \quad (11.60)$$

For the ground state, however, the dipolar hyperfine term is written in the alternative form

$$\mathcal{H}_{\text{dip}} = \sqrt{6} g_S \mu_B g_N \mu_N (\mu_0/4\pi) \mathrm{T}^2(\boldsymbol{C}) \cdot \mathrm{T}^2(\boldsymbol{I}, \boldsymbol{S}), \quad (11.61)$$

which takes account of the fact that $\boldsymbol{I}$ and $\boldsymbol{S}$ are coupled to form $\boldsymbol{G}$ in the basis functions. The required matrix elements for both forms of the dipolar interaction are given in several places in this book.

The molecular constants determined for both electronic states are given in table 11.4. There are some very large differences in the values of some of the parameters for the two electronic states, but the authors [87] do not provide simple reasons for these differences, noting that there are a large number of low-lying states which can interact with each other. There is much scope for *ab initio* calculations in due course.

11.8. Double resonance spectroscopy of molecular ion beams

11.8.1. Radiofrequency and microwave/infrared double resonance of HD$^+$ in the $X\,^2\Sigma^+$ ground state

Double resonance spectra observed in a molecular ion beam were described by Carrington, McNab and Montgomerie [88] for the HD$^+$ ion. The ion is now known to possess 22 bound vibrational levels, and the double resonance studies were preceded by single photon infrared studies in which vibration–rotation transitions of the ion were tuned into resonance with a carbon dioxide infrared laser line using Doppler tuning of the ion beam. These resonances were detected by electric field dissociation of the near-dissociation upper state, producing H$^+$ or D$^+$ fragment ions which were preferentially detected. The principles of the experiment are illustrated in figure 11.45. A mass-selected HD$^+$ ion beam is accelerated to a potential which brings it into resonance with the infrared laser beam, propagated antiparallel to the direction of the ion beam. The ion beam then passes through a radiofrequency cell of length 36 cm, and subsequently through an electric field lens.

Electron impact ionisation of HD leads to population of all of the bound vibrational levels of the ground state, with a maximum population at $v = 3$. The populations gradually fall off with increasing v value, but the initial populations are preserved in the collision-free ion beam environment. The infrared beam excites transitions from the $v = 17$ to the $v = 21$ vibrational level, the penultimate bound level, which undergoes dissociation in the electric field lens; fragment protons are separated from the molecular beam with a magnetic sector and detected with an electron multiplier. An example of the single photon infrared spectrum is shown in figure 11.46 for the transition $v = 17$, $N = 1 \rightarrow v = 21$, $N = 2$. It shows extensive nuclear hyperfine structure from both the proton and the deuteron in the molecule. The radiofrequency field drives nuclear hyperfine transitions within the 17,1 level, and produces changes in the fragment H$^+$ current which are detected. The radiofrequency cell was based upon a design due to Rosner, Gaily and Holt [89] and consisted of a central copper rod supported inside a copper cylinder; it could be operated satisfactorily up to 1000 MHz, and the radiofrequency spectrum was observed over the region 600 to 800 MHz. Different double resonance hyperfine transitions are studied by driving different hyperfine components of the 21,2 ← 17,1 vibration–rotation transition, as shown in the figure.

An energy level diagram showing the nuclear hyperfine splitting of the 17,1 level is shown in figure 11.47. The most suitable coupling scheme is

$$S + I_H = G_1, \quad G_1 + I_D = G_2, \quad G_2 + N = F,$$

where I_H and I_D are the proton and deuteron nuclear spins (with values 1/2 and

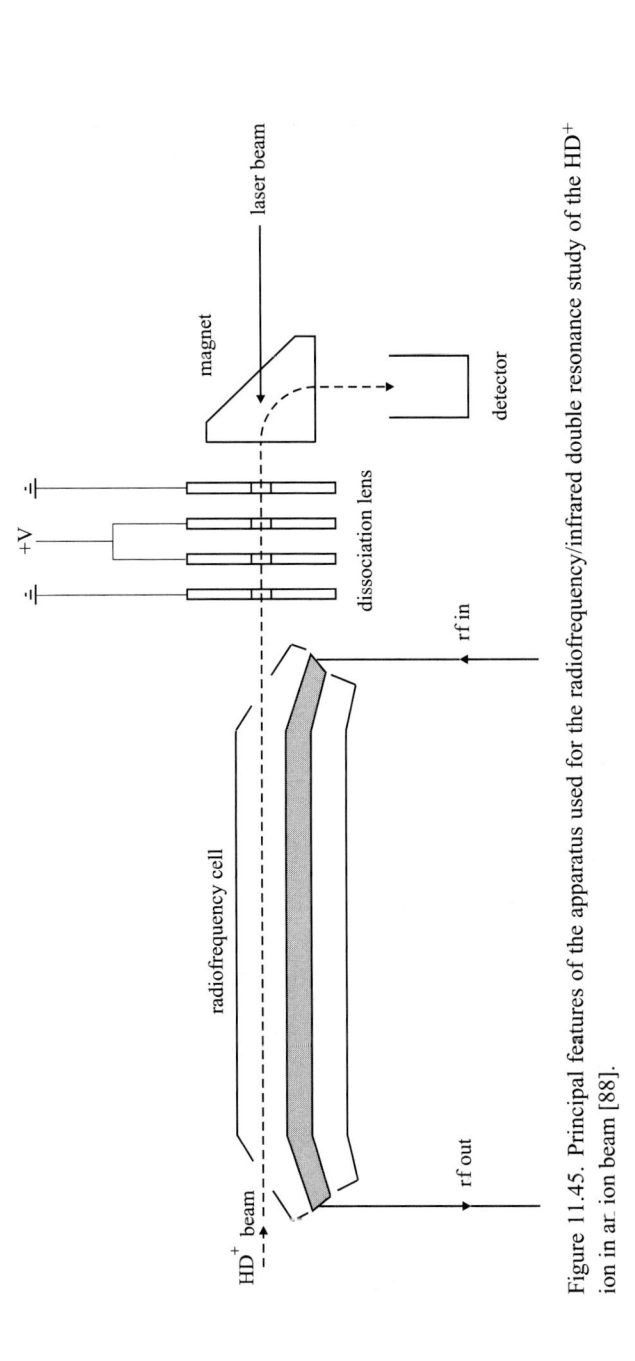

Figure 11.45. Principal features of the apparatus used for the radiofrequency/infrared double resonance study of the HD^+ ion in ar ion beam [88].

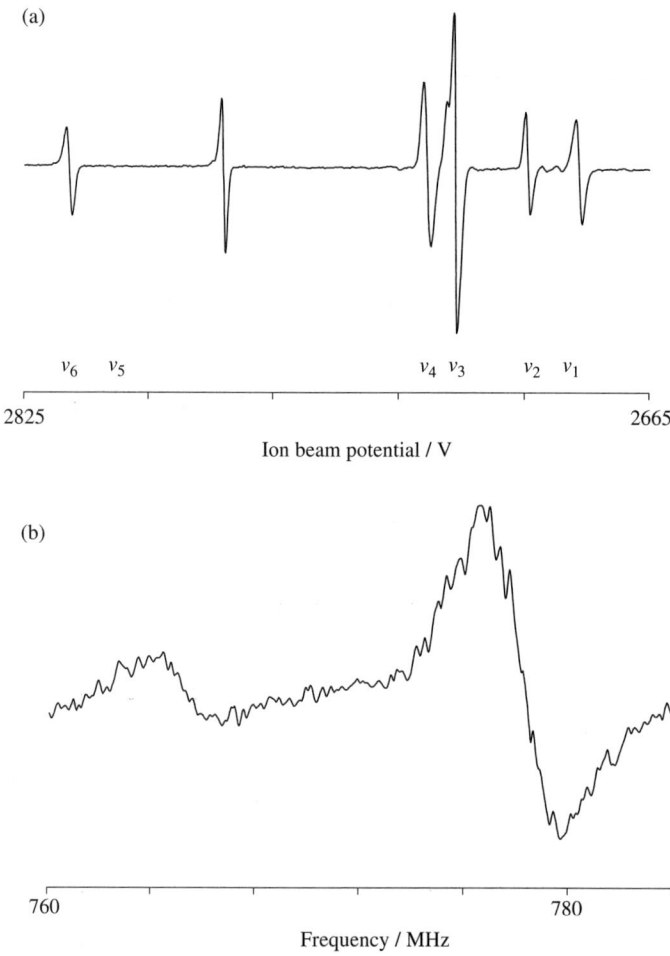

Figure 11.46. (a) Recording of the 21,2 ← 17,1 vibration–rotation transition of the HD$^+$ ion, obtained by Doppler tuning the ion beam into resonance with a carbon dioxide infrared laser beam [88]. (b) A radiofrequency/infrared double resonance spectrum obtained by pumping the ν_3 line shown in the infrared spectrum (see text for the assignment).

1 respectively). The possible values of the quantum numbers are as follows.

$S + I_H = G_1: G_1 =$		1			0
$G_1 + I_D = G_2: G_2 =$	2	1	0		1
$G_2 + N = F: F =$	$N+2$	$N+1$	N		$N+1$
	$N+1$	N			N
	N	$\|N-1\|$			$\|N-1\|$
	$\|N-1\|$				
	$\|N-2\|$				

For $N \geq 2$ there are 12 possible states; for $N = 1$ there are 11. As figure 11.47 shows,

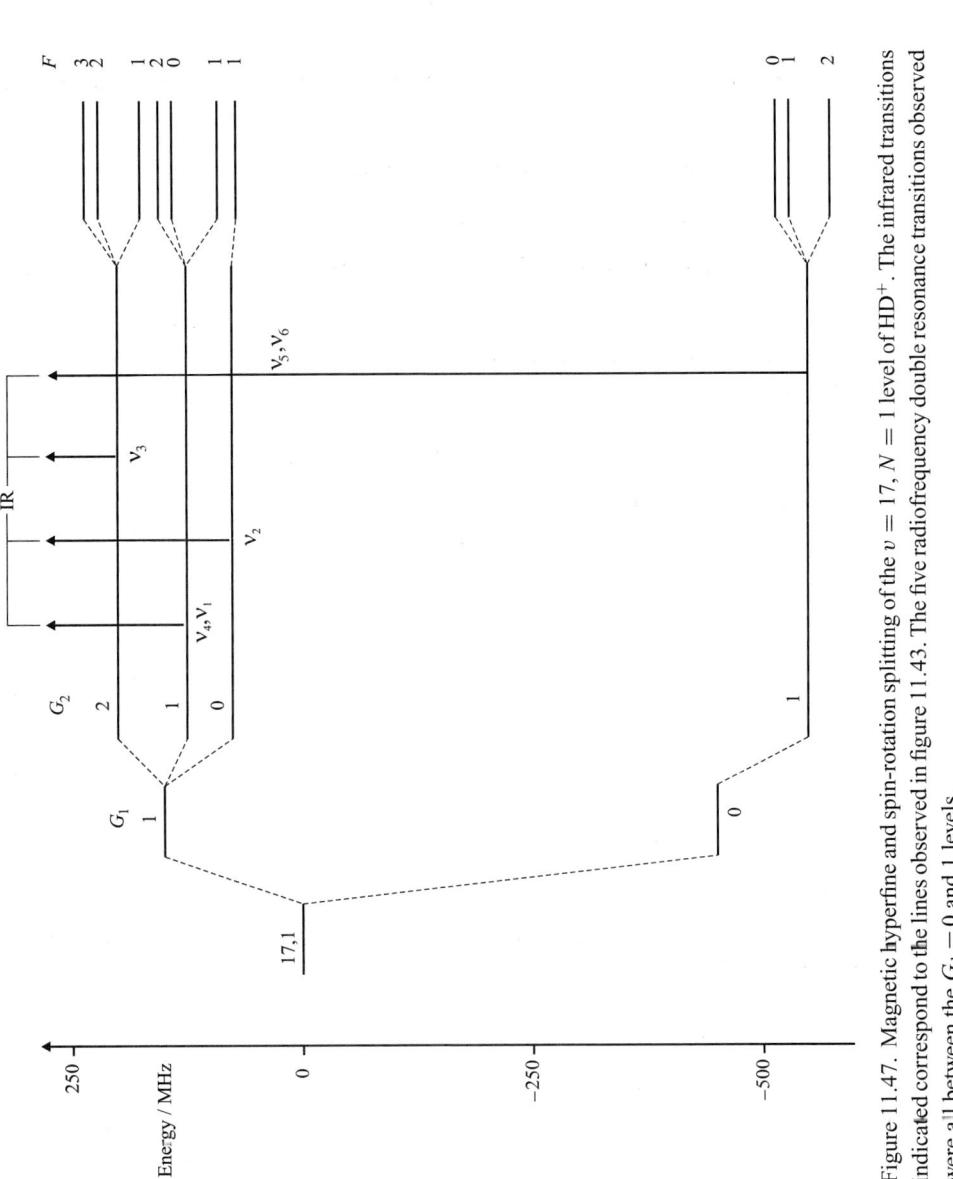

Figure 11.47. Magnetic hyperfine and spin-rotation splitting of the $v = 17$, $N = 1$ level of HD$^+$. The infrared transitions indicated correspond to the lines observed in figure 11.43. The five radiofrequency double resonance transitions observed were all between the $G_1 = 0$ and 1 levels.

the main splitting is due to the proton hyperfine interaction; each G_1 level is then split by the deuteron hyperfine interaction, and finally the spin–rotation splitting is added to give levels characterised by their F values. The lines in the infrared spectrum shown in figure 11.46 arise from the transitions (to 21, 2) indicated in figure 11.47.

Figure 11.47 is based upon the final analysis, to which we will return. The effective Hamiltonian for the magnetic hyperfine and spin–rotation interactions may be written as usual, except that we adopt the alternative formulation of the dipolar interactions which is more appropriate for our coupling scheme:

$$\mathcal{H}_{\text{eff}} = b_F(\text{H})\mathbf{T}^1(\mathbf{S}) \cdot \mathbf{T}^1(\mathbf{I}_H) + b_F(\text{D})\mathbf{T}^1(\mathbf{S}) \cdot \mathbf{T}^1(\mathbf{I}_D)$$
$$+ \sqrt{6} g_S \mu_B g_N(\text{H}) \mu_N (\mu_0/4\pi)\mathbf{T}^2(\mathbf{S}, \mathbf{I}_H) \cdot \mathbf{T}^2(\mathbf{C}_H)$$
$$+ \sqrt{6} g_S \mu_B g_N(\text{D}) \mu_N (\mu_0/4\pi)\mathbf{T}^2(\mathbf{S}, \mathbf{I}_D) \cdot \mathbf{T}^2(\mathbf{C}_D) + \gamma \mathbf{T}^1(\mathbf{N}) \cdot \mathbf{T}^1(\mathbf{S}). \quad (11.62)$$

The matrix elements in the case (b) nuclear spin coupled representation given above, $|\eta, \Lambda; S, I_H, G_1, I_D, G_2, N, F\rangle$, are now presented [90].

For the Fermi contact terms,

$$\langle \eta, \Lambda; S, I_H, G_1, I_D, G_2, N, F | b_F(\text{H})\mathbf{T}^1(\mathbf{S}) \cdot \mathbf{T}^1(\mathbf{I}_H) | \eta, \Lambda; S, I_H, G_1, I_D, G_2, N, F \rangle$$
$$= b_F(\text{H})(-1)^{S+G_1+I_H} \begin{Bmatrix} I_H & S & G_1 \\ S & I_H & 1 \end{Bmatrix} [S(S+1)(2S+1)I_H(I_H+1)(2I_H+1)]^{1/2}. \quad (11.63)$$

$$\langle \eta, \Lambda; S, I_H, G_1, I_D, G_2, N, F | b_F(\text{D})\mathbf{T}^1(\mathbf{S}) \cdot \mathbf{T}^1(\mathbf{I}_D) | \eta, \Lambda; S, I_H, G_1', I_D, G_2, N, F \rangle$$
$$= b_F(\text{D})(-1)^{G_1'+G_2+I_D} \begin{Bmatrix} I_D & G_1' & G_2 \\ G_1 & I_D & 1 \end{Bmatrix} \langle G_1 \| \mathbf{T}^1(\mathbf{S}) \| G_1' \rangle [I_D(I_D+1)(2I_D+1)]^{1/2}$$
$$= b_F(\text{D})(-1)^{G_1'+G_2+I_D} \begin{Bmatrix} I_D & G_1' & G_2 \\ G_1 & I_D & 1 \end{Bmatrix} (-1)^{G_1'+S+1+I_H}[(2G_1'+1)(2G_1+1)]^{1/2}$$
$$\times \begin{Bmatrix} G_1 & S & I_H \\ S & G_1' & 1 \end{Bmatrix} [S(S+1)(2S+1)I_D(I_D+1)(2I_D+1)]^{1/2}. \quad (11.64)$$

We calculate the dipolar terms giving more of the detail; for the proton (equation (11.62)) we have

$$\langle \eta, \Lambda; S, I_H, G_1, I_D, G_2, N, F | \mathcal{H}_{\text{dip(H)}} | \eta', \Lambda'; S, I_H, G_1', I_D, G_2', N', F \rangle$$
$$= \sqrt{6}\, g_S \mu_B g_N(\text{H}) \mu_N (\mu_0/4\pi)(-1)^{G_2'+F+N} \begin{Bmatrix} N' & G_2' & F \\ G_2 & N & 2 \end{Bmatrix}$$
$$\times \langle G_1, I_D, G_2 \| \mathbf{T}^2(\mathbf{S}, \mathbf{I}_H) \| G_1', I_D, G_2' \rangle \langle \eta, N, \Lambda \| \mathbf{T}^2(\mathbf{C}_H) \| \eta', N', \Lambda' \rangle$$
$$= \sqrt{6}\, g_S \mu_B g_N(\text{H}) \mu_N (\mu_0/4\pi)(-1)^{G_2'+F+N} \begin{Bmatrix} N' & G_2' & F \\ G_2 & N & 2 \end{Bmatrix}$$
$$\times (-1)^{G_2'+G_1+I_D} \{(2G_2'+1)(2G_2+1)\}^{1/2} \begin{Bmatrix} G_1' & G_2' & I_D \\ G_2 & G_1 & 2 \end{Bmatrix}$$
$$\times \langle G_1 \| \mathbf{T}^2(\mathbf{S}, \mathbf{I}_H) \| G_1' \rangle \langle \eta, N, \Lambda \| \mathbf{T}^2(\mathbf{C}_H) \| \eta', N', \Lambda' \rangle$$

$$= \sqrt{30}(-1)^{G_2'+F+N} \begin{Bmatrix} N' & G_2' & F \\ G_2 & N & 2 \end{Bmatrix} (-1)^{G_2'+G_1+I_D} \{(2G_2'+1)(2G_2+1)\}^{1/2}$$

$$\times \begin{Bmatrix} G_1' & G_2' & I_D \\ G_2 & G_1 & 2 \end{Bmatrix} \{(2G_1'+1)(2G_1+1)\}^{1/2} \{I_H(I_H+1)(2I_H+1)$$

$$\times S(S+1)(2S+1)\}^{1/2} \begin{Bmatrix} G_1 & G_1' & 2 \\ I_H & I_H & 1 \\ S & S & 1 \end{Bmatrix} (-1)^N \{(2N'+1)(2N+1)\}^{1/2}$$

$$\times \begin{pmatrix} N & 2 & N' \\ 0 & 0 & 0 \end{pmatrix} \langle \eta | g_S \mu_B g_N(\mathrm{H}) \mu_N (\mu_0/4\pi) C_0^2(\theta_\mathrm{H}, \phi_\mathrm{H}) r_\mathrm{H}^{-3} | \eta \rangle. \quad (11.65)$$

Here we have restricted attention to matrix elements diagonal in η and Λ, and also put $\Lambda = 0$.

Likewise for the deuteron dipolar interaction (see equation (11.62)):

$$\langle \eta, \Lambda; S, I_\mathrm{H}, G_1, I_\mathrm{D}, G_2, N, F | \mathcal{H}_{\mathrm{dip}(\mathrm{D})} | \eta', \Lambda'; S, I_\mathrm{H}, G_1', I_\mathrm{D}, G_2', N', F \rangle$$

$$= \sqrt{6} g_S \mu_B g_N(\mathrm{D}) \mu_N (\mu_0/4\pi) (-1)^{G_2'+F+N} \begin{Bmatrix} N' & G_2' & F \\ G_2 & N & 2 \end{Bmatrix}$$

$$\times \langle G_1, I_\mathrm{D}, G_2 \| T^2(\mathbf{S}, \mathbf{I}_\mathrm{D}) \| G_1', I_\mathrm{D}, G_2' \rangle \langle \eta, N, \Lambda \| T^2(\mathbf{C}_\mathrm{D}) \| \eta', N', \Lambda' \rangle$$

$$= \sqrt{30}(-1)^{G_2'+F+N} \begin{Bmatrix} N' & G_2' & F \\ G_2 & N & 2 \end{Bmatrix} \{(2G_2'+1)(2G_2+1)(2G_1'+1)(2G_1+1)\}^{1/2}$$

$$\times \begin{Bmatrix} G_2 & G_2' & 2 \\ G_1 & G_1' & 1 \\ I_\mathrm{D} & I_\mathrm{D} & 1 \end{Bmatrix} (-1)^{G_1'+S+1+I_\mathrm{H}} \{S(S+1)(2S+1) I_\mathrm{D}(I_\mathrm{D}+1)(2I_\mathrm{D}+1)\}^{1/2}$$

$$\times \begin{Bmatrix} S & G_1' & I_\mathrm{H} \\ G_1 & S & 1 \end{Bmatrix} (-1)^N \{(2N'+1)(2N+1)\}^{1/2} \begin{pmatrix} N & 2 & N' \\ 0 & 0 & 0 \end{pmatrix}$$

$$\times \langle \eta | g_S \mu_B g_N(\mathrm{D}) \mu_N (\mu_0/4\pi) C_0^2(\theta_\mathrm{D}, \phi_\mathrm{D}) r_\mathrm{D}^{-3} | \eta \rangle. \quad (11.66)$$

Although matrix elements off-diagonal in N do exist, they are negligible in this case.

Finally we have the spin–rotation interaction:

$$\langle \eta, \Lambda; S, I_\mathrm{H}, G_1, I_\mathrm{D}, G_2, N, F | \gamma T^1(\mathbf{S}) \cdot T^1(\mathbf{N}) | \eta', \Lambda'; S, I_\mathrm{H}, G_1', I_\mathrm{D}, G_2', N, F \rangle$$

$$= \gamma(-1)^{2G_2'+G_1+G_1'+F+N+I_\mathrm{D}+I_\mathrm{H}+S} \{N(N+1)(2N+1)(2G_2'+1)(2G_2+1)$$

$$\times (2G_1'+1)(2G_1+1) S(S+1)(2S+1)\}^{1/2} \begin{Bmatrix} N & G_2' & F \\ G_2 & N & 1 \end{Bmatrix}$$

$$\times \begin{Bmatrix} G_2 & G_1 & I_\mathrm{D} \\ G_1' & G_2' & 1 \end{Bmatrix} \begin{Bmatrix} G_1 & S & I_\mathrm{H} \\ S & G_1' & 1 \end{Bmatrix}. \quad (11.67)$$

The only rigorous quantum number is F, although N may also be taken as good since the separation of different N levels is very large compared with the dipolar hyperfine interaction. Hence the problem of deriving expressions for the energies of the hyperfine levels in 17,1 is tackled by setting up the five matrices for $F = N + 2$

to $F = N - 2$. These matrices are given explicitly by Carrington and Kennedy [90] and will not be repeated here. Ultimately the energies, and hence double resonance transition frequencies, depend upon the values of five constants for the $v = 17$ levels. These constants, and their values (in MHz), are as follows:

$$b_F(H) = 713.07, \quad b_F(D) = 111.36, \quad \gamma = 8.10, \quad t_0(H) = 9.75, \, t_0(D) = 1.50.$$

Note, as a reminder, the definition of the axial dipolar constants:

$$t_0(H) = \left\langle g_S \mu_B g_N(H) \mu_N (\mu_0/4\pi) C_0^2(\theta_H, \phi_H) r_H^{-3} \right\rangle_\eta,$$
$$t_0(D) = \left\langle g_S \mu_B g_N(D) \mu_N (\mu_0/4\pi) C_0^2(\theta_D, \phi_D) r_D^{-3} \right\rangle_\eta. \qquad (11.68)$$

Five radiofrequency lines were observed by Carrington, McNab and Montgomerie [88], each being the superposition of several very closely components; the hyperfine constants listed above were obtained from a combination of the infrared and radiofrequency data. It can be seen that the Fermi contact constants are by far the largest, and within the Born–Oppenheimer approximation one would expect the ratio of the proton to deuteron constant (6.4033) to be equal to the ratio of the nuclear g factors (6.5144). In fact these ratios are not in particularly good agreement and immediately suggest a breakdown of the Born–Oppenheimer approximation. We defer discussion of this until later because, as we shall see, the effect becomes progressively larger as the vibration–rotation levels studied approach the lowest dissociation limit.

We turn now to a microwave/infrared study of the HD^+ ion described by Carrington, McNab, Montgomerie and Brown [91], a block diagram of their apparatus being shown in figure 11.48. Again a carbon dioxide infrared laser line was tuned into resonance with a vibration–rotation transition by means of the Doppler effect, the transition this time being $v = 22, N = 1 \leftarrow v = 17, N = 2$. The observed infrared spectrum is shown in figure 11.49(a), the lines being identified in the energy level diagram shown in figure 11.50. In this case it turned out that the deuterium hyperfine splitting was much *larger* than the proton, so that the coupling scheme used was

$$S + I_D = G_1, \quad G_1 + I_H = G_2.$$

G_1 can therefore take the values $3/2$ or $1/2$; G_2 is 2 or 1 for $G_1 = 3/2$, and 1 or 0 for $G_1 = 1/2$. The vibration-rotation level 22,1 is actually the highest bound level in the HD^+ ground state, and Carrington, McNab, Montgomerie and Brown [91] were able to detect the rotational transition $22,1 \leftarrow 22,0$ by microwave/infrared double resonance. A particularly interesting feature of this experiment is that the electric field dissociation process can discriminate between the 22,0 and 22,1 vibration–rotation levels. The infrared pumping enhances the population of 22,1 and, with a suitable choice of potential applied to the field lens, together with adjustment of the magnetic field to select protons originating from this level, the double resonance is observed as a *decrease* in the proton fragment current (figure 11.49(b)). Alternatively one can choose to monitor the protons originating from field dissociation of 22,0, in which case the double resonance is observed as an *increase* in the proton fragment current (figure 11.49(c)). This

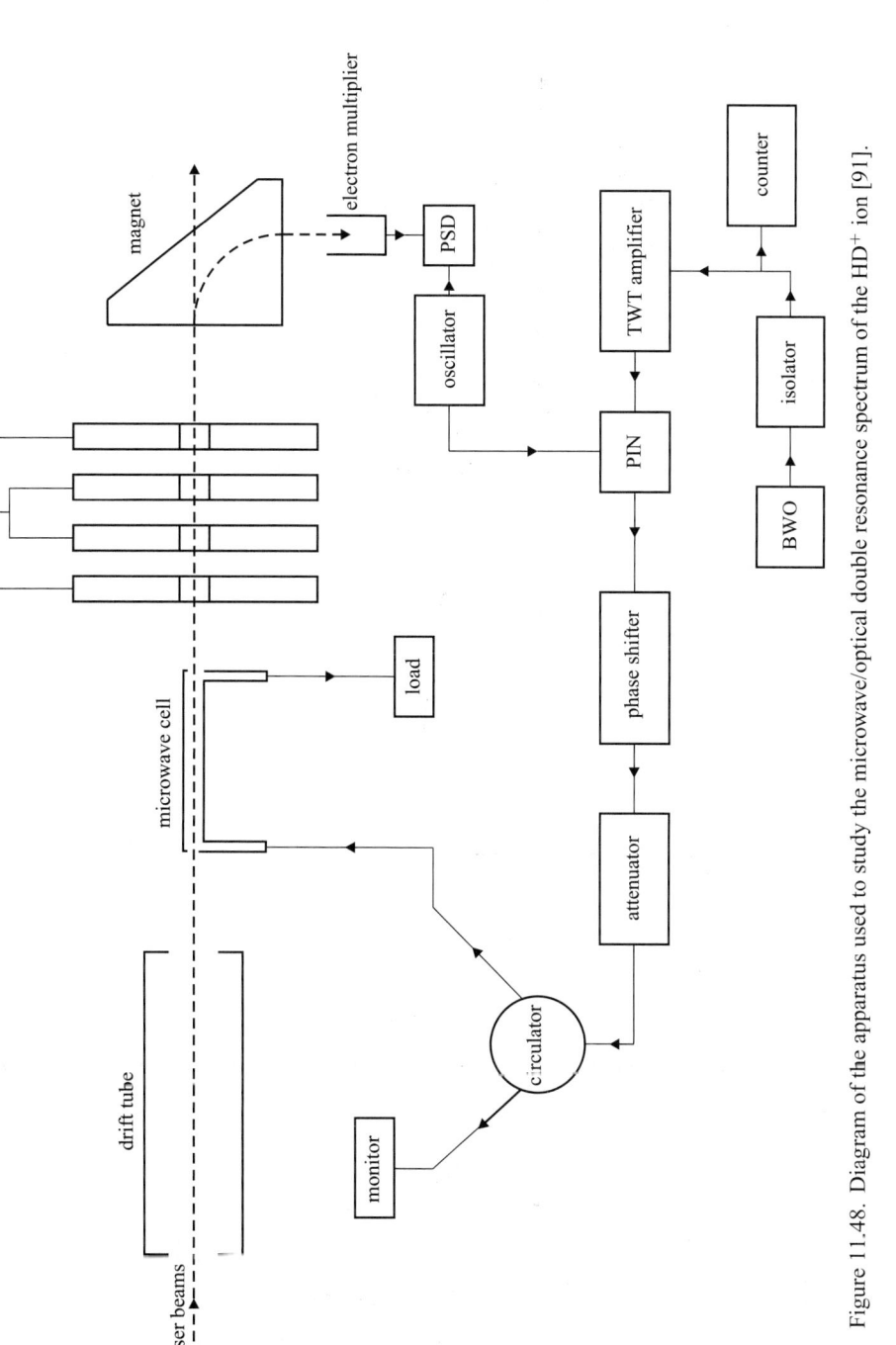

Figure 11.48. Diagram of the apparatus used to study the microwave/optical double resonance spectrum of the HD$^+$ ion [91].

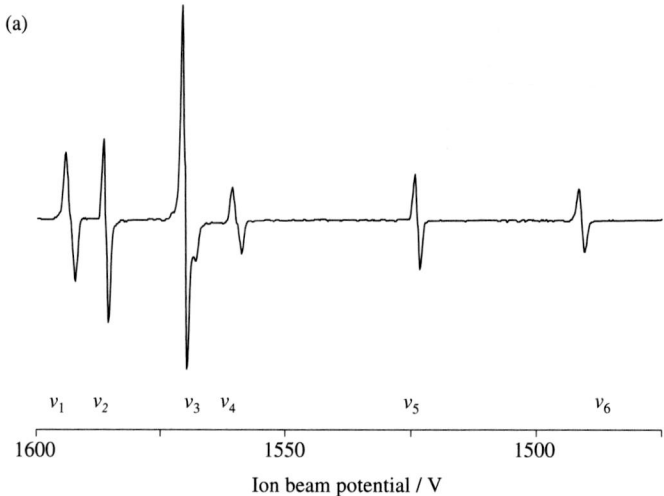

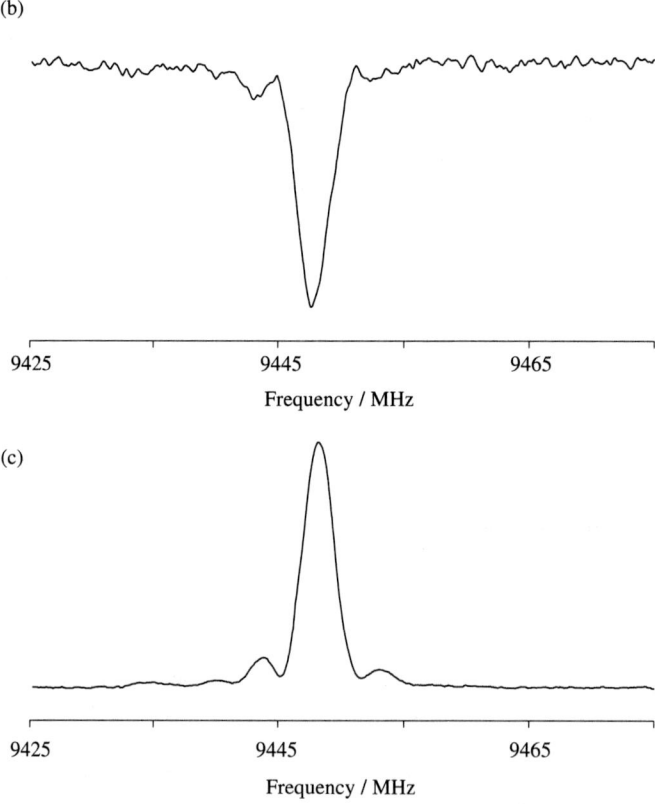

Figure 11.49. (a) Recording of the 22,1 ← 17,2 vibration–rotation transition in HD$^+$ obtained by Doppler tuning the ion beam into resonance with a carbon dioxide infrared laser line. (b) Microwave-infrared double resonance line arising from the 22,1 ← 22,0 rotational transition in HD$^+$, recording by monitoring H$^+$ fragments from the electric field dissociation of 22,1 [91]. (c) Microwave-infrared double resonance line arising from the 22,1 ← 22,0 rotational transition in HD$^+$, recording by monitoring H$^+$ fragments from the electric field dissociation of 22,0.

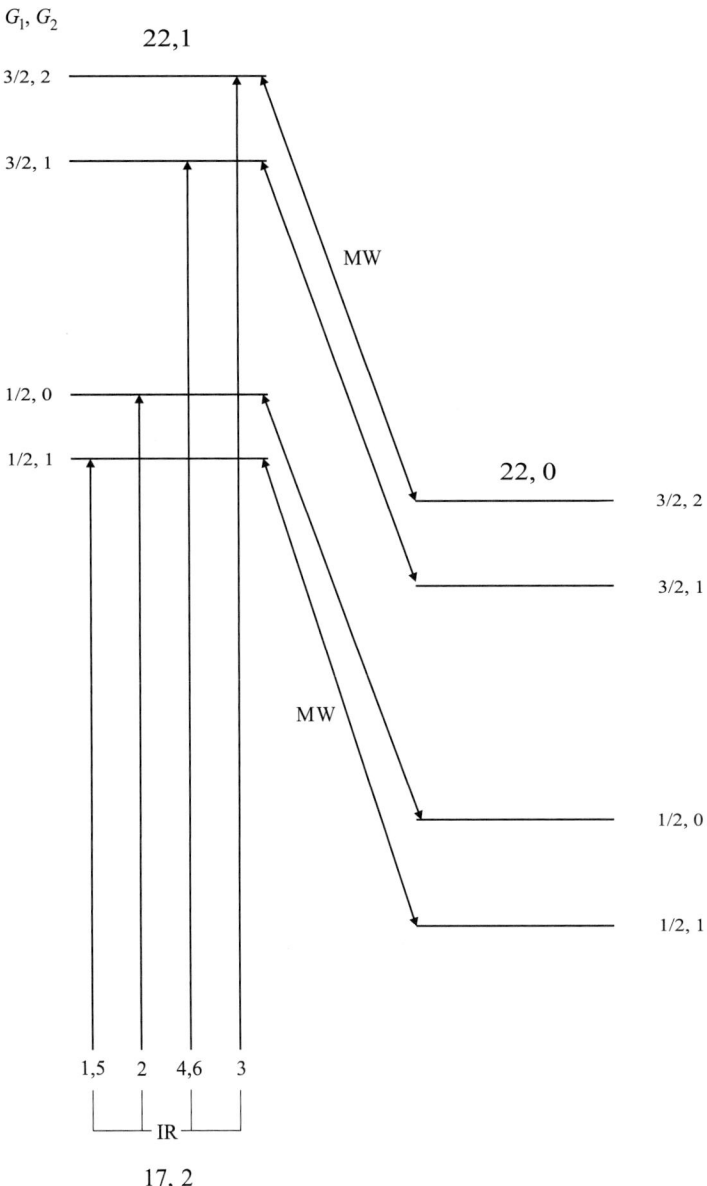

Figure 11.50. Energy level diagram (not to scale) showing the nuclear hyperfine structure of the HD$^+$ 22,1 and 22,0 vibration–rotation levels (labelled with the G_1 and G_2 quantum numbers described in the text). The infrared transitions which give rise to the six lines shown in figure 11.49 (a) are shown on the left-hand side of the figure, and the four observed microwave transitions are shown on the right-hand side.

Table 11.5. *Microwave transition frequencies (in MHz) for the hyperfine components of the 22,1 ← 22,0 rotational transition in HD$^+$, measured through double resonance studies*

22,1		22,0			
G_1	G_2	G_1	G_2	IR pump	frequency
1/2	0	1/2	0	2	9440.3
3/2	2	3/2	2	3	9442.2
1/2	1	1/2	1	1, 5	9442.8
3/2	1	3/2	1	4	9447.0

is the reason for the change of phase in the double resonance signals; note also that the signal-to-noise ratio is much larger when protons from 22,0 are detected, simply because the off-resonance proton fragment current is much smaller.

The four microwave transitions detected in this double resonance study are listed in table 11.5; the infrared pump transitions are also listed. The rotational transitions are all diagonal in G_1 and G_2, which meant that the experiment yielded only the *differences* in the proton and deuteron Fermi contact constants, and not their absolute values. However, a combination of the microwave and infrared studies [92] has given the absolute values, which are as follows:

$$22,0: \quad b_F(H) = 19.0 \, \text{MHz}, \quad b_F(D) = 217.7 \, \text{MHz}$$
$$22,1: \quad b_F(H) = 8.4 \, \text{MHz}, \quad b_F(D) = 218.2 \, \text{MHz}.$$

The dissociation energies of these levels have been calculated [93] to be 0.323 cm^{-1} for 22,0 and 0.057 cm^{-1} for 22,1.

The hyperfine constants listed above are remarkable. The free atom values are 1420.41 and 218.05 MHz for the proton and deuteron respectively, and in the simplest possible interpretation one might expect the values in HD$^+$ to be very close to one half of these values, namely, 710.2 and 109.0 MHz. In a more strongly bound vibration–rotation level, for example 17,1 with a dissociation energy of 994.551 cm^{-1}, the measured Fermi contact constants [94] are 711.9 and 111.1 MHz, very close to the simple expectations. As the lowest dissociation limit is approached, the proton constant decreases to almost zero, and the deuterium constant approaches its free atom value. This behaviour is, of course, completely at variance with the Born–Oppenheimer approximation, and in its last bound vibration–rotation level the HD$^+$ molecule is best regarded as a long-range complex, H$^+$...D. A quantitative treatment of the problem requires a non-adiabatic calculation [95, 93]; when this is performed the agreement between experiment and theory is excellent.

Microwave-microwave double resonance experiments similar to those described in this section have been carried out on H$_2^+$ [96], D$_2^+$ [97] and He$_2^+$ [98]. In these cases, however, the transitions studied are actually electronic transitions, despite being observed at microwave and millimetre-wave frequencies. We conclude, with regret, that they are beyond the scope of this book.

11.8.2. Radiofrequency/optical double resonance of N_2^+ in the $X\,^2\Sigma_g^+$ ground state

The N_2^+ ion is an important molecule, in the laboratory and in the atmosphere, and its electronic spectrum has been studied extensively. The ground state electron configuration may be written as

$$(1s\sigma_g)^2(1s\sigma_u)^2(2s\sigma_g)^2(2s\sigma_u)^2(2p\pi_u)^4(2p\sigma_g)^1:\ X\,^2\Sigma_g^+,$$

and the two lowest excited electronic states are

$$(1s\sigma_g)^2(1s\sigma_u)^2(2s\sigma_g)^2(2s\sigma_u)^2(2p\pi_u)^3(2p\sigma_g)^2:\ A\,^2\Pi_u(T_e = 9166.9\text{ cm}^{-1})$$
$$(1s\sigma_g)^2(1s\sigma_u)^2(2s\sigma_g)^2(2s\sigma_u)^1(2p\pi_u)^4(2p\sigma_g)^2:\ B\,^2\Sigma_u^+(T_e = 25461.4\text{ cm}^{-1}).$$

The electronic band systems arising from the $A \to X$ and $B \to X$ transitions have been extensively studied, and the second of these, occuring in the near-ultraviolet, was used by Berrah Mansour, Kurtz, Steimle, Goodman, Young, Scholl, Rosner and Holt [99] in an elegant radiofrequency double resonance investigation. They used a cw ring dye laser to pump transitions in the $v' = 0 \leftarrow v'' = 1$ vibrational band at 428 nm, and a small section of their electronic spectrum is shown in figure 11.51. The origin of the observed structure is explained in figure 11.52; since each ^{14}N nucleus has spin $I = 1$, the total nuclear spin I_T can be 2, 1 or 0. $I = 1$ (*para*) is associated with odd N rotational levels, whilst $I = 0$ and 2 (*ortho*) are associated with even N. Each P-branch rotational component ($\Delta J = -1$) is split into a doublet by the spin-rotation interaction, and there

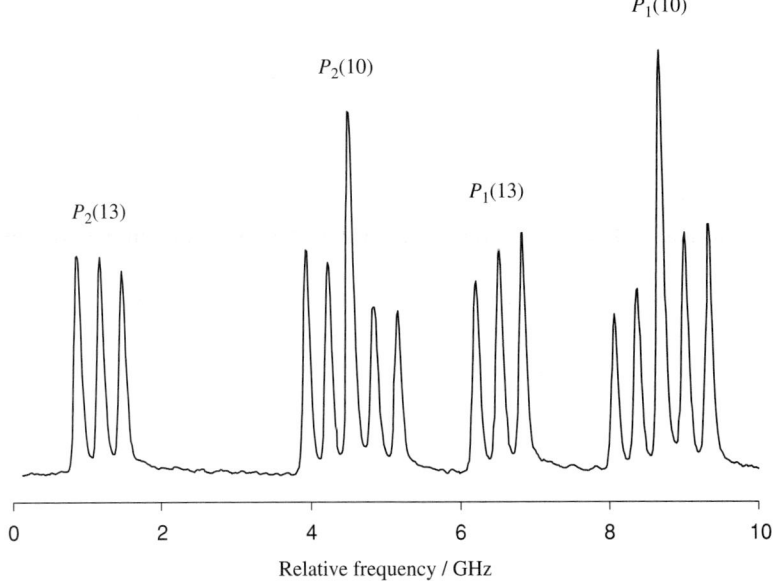

Figure 11.51. N_2^+ ion beam laser fluorescence spectrum near the bandhead of the $B\,^2\Sigma_u^+ \leftarrow X\,^2\Sigma_g^+$ (0,1) band, obtained by scanning the frequency of a cw dye laser [99].

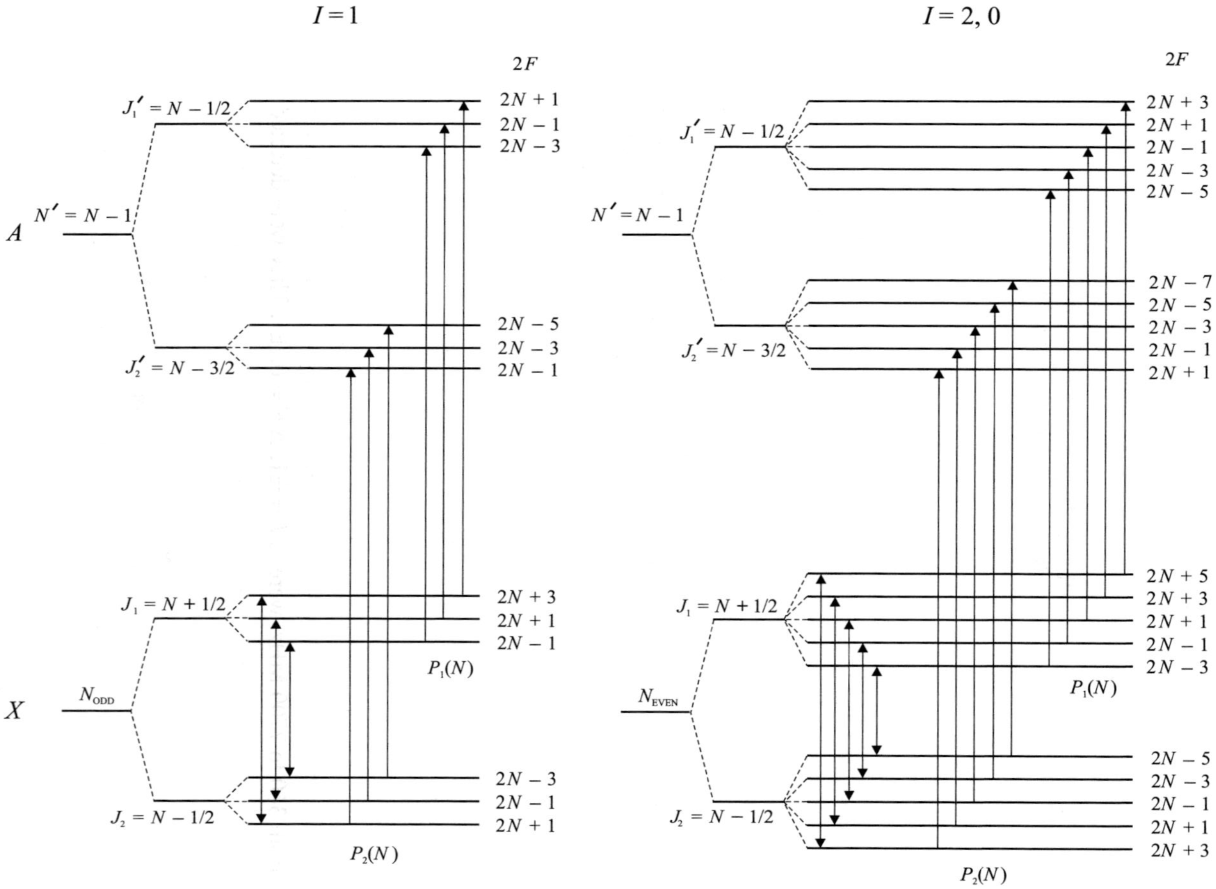

Figure 11.52. Energy level diagram for typical even-N and odd-N rotational levels of the $X\,^2\Sigma_g^+$ state of $^{14}N_2^+$. The principal P-branch optical pumping transitions to the $B\,^2\Sigma_u^+$ excited state, $P_1(N)$ and $P_2(N)$, are shown, as well as the principal radiofrequency double resonance transitions in the ground state. The $I = 0, F = J$ levels are not shown, but lie close to the $I = 2, F = J$ levels [99].

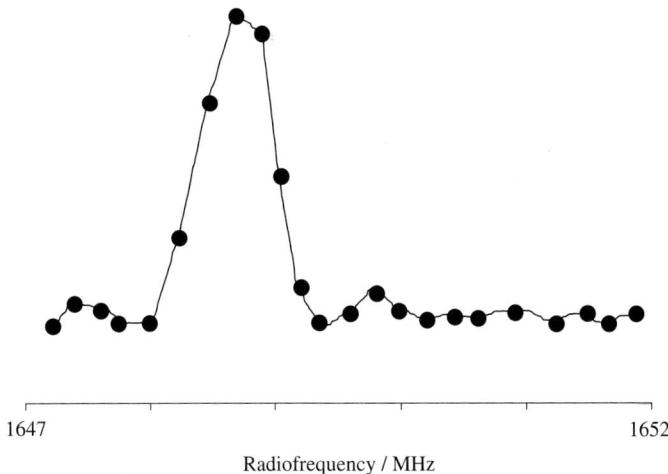

Figure 11.53. Radiofrequency resonance arising from the $J = 13/2, F = 9/2 \leftarrow J = 11/2, F = 7/2$ transition in the ground state of N_2^+ [99].

is then a further nuclear hyperfine splitting, giving a triplet structure for $I = 1$, or a quintet structure for $I = 2$, as illustrated in figure 11.51.

The apparatus used for this investigation was similar to that described in chapter 8 for studies of Na_2 [100]. N_2^+ ions were produced by a radiofrequency discharge in N_2, accelerated to form a beam, and mass analysed with a 90° magnetic sector. The ion beam was then exposed to a collinear laser beam; molecules excited at the 'pump' region A, were monitored through their 391 nm fluorescence at the 'probe' region B by means of a photon counting photomultiplier. In fact the experiments were performed by two different groups working at two different locations, using machines which differed somewhat in their details.

Between the A and B regions, a radiofrequency field was applied to induce fine-structure transitions within the $v'' = 1$ level of the *ground* electronic state, split by the nuclear hyperfine interaction. The selection rules for these transitions, which ranged in frequency from 360 to 7700 MHz, were $\Delta J = \pm 1, \Delta F = 0, \pm 1$. They were detected through resonant changes in the fluorescence intensity; an example of a radiofrequency double resonance line is shown in figure 11.53. The observed spectrum involved N values from 1 to 27.

The magnetic and electric hyperfine interactions for a molecule containing two equivalent ^{14}N nuclei were discussed in detail in chapter 8 for the $A\,^3\Sigma_u^+$ state of N_2. We summarise the results here, referring the reader to chapter 8 for a more thorough description.

For the Fermi contact interaction:

$$\langle \eta, \Lambda; N, S, J, I_T, F | b_F \mathbf{T}^1(\mathbf{S}) \cdot \mathbf{T}^1(\mathbf{I}_T) | \eta, \Lambda'; N', S, J', I_T, F \rangle$$
$$= b_F (-1)^{J'+F+I_T} \begin{Bmatrix} I_T & J' & F \\ J & I_T & 1 \end{Bmatrix} \langle N, S, J \| \mathbf{T}^1(\mathbf{S}) \| N', S, J' \rangle [I_T(I_T+1)(2I_T+1)]^{1/2}$$

$$= b_\text{F}(-1)^{J'+F+I_T} \begin{Bmatrix} I_T & J' & F \\ J & I_T & 1 \end{Bmatrix} (-1)^{J+N+1+S}[(2J+1)(2J'+1)]^{1/2} \begin{Bmatrix} J & S & N \\ S & J' & 1 \end{Bmatrix}$$
$$\times [S(S+1)(2S+1)I_T(I_T+1)(2I_T+1)]^{1/2}. \tag{11.69}$$

The magnetic dipolar hyperfine coupling also has familiar matrix elements:

$$\langle \eta, \Lambda; N, S, J, I_T, F | -\sqrt{10} g_S \mu_B g_N \mu_N (\mu_0/4\pi) \mathrm{T}^1(\boldsymbol{S},\boldsymbol{C}^2)\cdot\mathrm{T}^1(\boldsymbol{I}_T) | \eta, \Lambda'; N', S, J', I_T, F \rangle$$
$$= \sqrt{30} t (-1)^{J'+F+N+I_T+1} \begin{Bmatrix} I_T & J' & F \\ J & I_T & 1 \end{Bmatrix}$$
$$\times \begin{Bmatrix} J & J' & 1 \\ N & N' & 2 \\ S & S & 1 \end{Bmatrix} \begin{pmatrix} N & 2 & N' \\ 0 & 0 & 0 \end{pmatrix} [(2N+1)(2N'+1)I_T(I_T+1)$$
$$\times (2I_T+1)(2J+1)(2J'+1)S(S+1)(2S+1)]^{1/2}. \tag{11.70}$$

For the electric quadrupole interaction, generalising our earlier results to include matrix elements off-diagonal in the total nuclear spin, we have:

$$\langle \eta, \Lambda; N, S, J, I_1, I_2, I_T, F | -e \sum_{k=1,2} \mathrm{T}^2(\nabla\boldsymbol{E}_k)\cdot\mathrm{T}^2(\boldsymbol{Q}_k) | \eta, \Lambda'; N', S, J', I_1, I_2, I'_T, F \rangle$$
$$= \sum_k (-1)^{J'+I_T+F} \begin{Bmatrix} J & I_T & F \\ I'_T & J' & 2 \end{Bmatrix} (-1)^{J'+N+S}[(2J'+1)(2J+1)]^{1/2}$$
$$\times \begin{Bmatrix} J & N & S \\ N' & J' & 2 \end{Bmatrix} \langle \eta, \Lambda; N, \Lambda \| \mathrm{T}^2(\nabla\boldsymbol{E}_k) \| \eta', \Lambda'; N', \Lambda' \rangle$$
$$\times \langle I_1, I_2, I_T \| -e\mathrm{T}^2(\boldsymbol{Q}_k) \| I_1, I_2, I'_T \rangle. \tag{11.71}$$

The nuclear spin reduced matrix element is expanded as follows:

$$\langle I_1, I_2, I_T \| \sum_{k=1,2} \mathrm{T}^2(\boldsymbol{Q}_k) \| I_1, I_2, I'_T \rangle$$
$$= (-1)^{I_1+I_2}[(2I_T+1)(2I'_T+1)]^{1/2} \left[(-1)^{I'_T} \begin{Bmatrix} I_T & I_1 & I_2 \\ I_1 & I'_T & 2 \end{Bmatrix} \langle I_1 \| \mathrm{T}^2(\boldsymbol{Q}_1) \| I_1 \rangle \right.$$
$$\left. + (-1)^{I_T} \begin{Bmatrix} I_T & I_2 & I_1 \\ I_2 & I'_T & 2 \end{Bmatrix} \langle I_2 \| \mathrm{T}^2(\boldsymbol{Q}_2) \| I_2 \rangle \right]$$
$$= (-1)^{2I_N}[(2I_T+1)(2I'_T+1)]^{1/2} \begin{Bmatrix} I_T & I_N & I_N \\ I_N & I'_T & 2 \end{Bmatrix}$$
$$\times \langle I_N \| \mathrm{T}^2(\boldsymbol{Q}_N) \| I_N \rangle [(-1)^{I'_T} + (-1)^{I_T}]$$
$$= (-1)^{2I_N}[(2I_T+1)(2I'_T+1)]^{1/2} \begin{Bmatrix} I_T & I_N & I_N \\ I_N & I'_T & 2 \end{Bmatrix} \left(\frac{Q_N}{2}\right)$$
$$\times \begin{pmatrix} I_N & 2 & I_N \\ -I_N & 0 & I_N \end{pmatrix}^{-1} [(-1)^{I'_T} + (-1)^{I_T}]. \tag{11.72}$$

In the second line of (11.72) we have put $I_1 = I_2 = I_N$, and in the third line substituted the definition of the nuclear quadrupole moment.

The remaining matrix element in equation (11.71) is evaluated by putting $\Lambda = 0$ for the Σ state, and neglecting the mixing of excited electronic states by putting $q = 0$:

$$\langle \eta, \Lambda; N, \Lambda \| -T^2(\nabla E) \| \eta, \Lambda; N', \Lambda \rangle$$
$$= (-1)^N [(2N+1)(2N'+1)]^{1/2} \begin{pmatrix} N & 2 & N' \\ 0 & 0 & 0 \end{pmatrix} (q_0/2). \quad (11.73)$$

Consequently we combine equations (11.71), (11.72) and (11.73) to obtain the full result for the quadrupole matrix elements, which is

$$\langle \eta, \Lambda; N, S, J, I_1, I_2, I_T, F | -e \sum_{k=1,2} T^2(\nabla E_k) \cdot T^2(Q_k) | \eta, \Lambda; N', S, J', I_1, I_2, I'_T, F \rangle$$

$$= (eq_0 Q_N/4)(-1)^{2J'+2N+S+I_T+F} \begin{Bmatrix} J & I_T & F \\ I'_T & J' & 2 \end{Bmatrix} [(2J'+1)(2J+1)]^{1/2}$$

$$\times \begin{Bmatrix} J & N & S \\ N' & J' & 2 \end{Bmatrix} [(2N+1)(2N'+1)]^{1/2} \begin{pmatrix} N & 2 & N' \\ 0 & 0 & 0 \end{pmatrix} (-1)^{2I_N} [(2I_T+1)]$$

$$\times (2I'_T+1)]^{1/2} \begin{Bmatrix} I_T & I_N & I_N \\ I_N & I'_T & 2 \end{Bmatrix} \begin{pmatrix} I_N & 2 & I_N \\ -I_N & 0 & I_N \end{pmatrix}^{-1} [(-1)^{I'_T} + (-1)^{I_T}]. \quad (11.74)$$

Finally the electron spin–rotation interaction has matrix elements given by

$$\langle \eta, \Lambda; N, S, J, I_T, F | \gamma T^1(N) \cdot T^1(S) | \eta, \Lambda; N, S, J, I_T, F \rangle$$
$$= \gamma(-1)^{N+J+S} \begin{Bmatrix} S & N & J \\ N & S & 1 \end{Bmatrix} [N(N+1)(2N+1)S(S+1)(2S+1)]^{1/2}, \quad (11.75)$$

and the nuclear spin–rotation interaction is given by

$$\langle \eta, \Lambda; N, S, J, I_T, F | c_I T^1(N) \cdot T^1(I_T) | \eta, \Lambda; N, S, J', I_T, F \rangle$$
$$= c_I(-1)^{2J'+F+I_T+N+S+1} [(2J'+1)(2J+1)N(N+1)$$
$$\times (2N+1)I_T(I_T+1)(2I_T+1)]^{1/2} \begin{Bmatrix} I_T & J' & F \\ J & I_T & 1 \end{Bmatrix} \begin{Bmatrix} J & N & S \\ N & J' & 1 \end{Bmatrix}. \quad (11.76)$$

In the analysis of the spectrum, which involved high values of N, it was found necessary to include centrifugal distortion corrections to both the electron spin–rotation and nuclear dipolar constants, i.e.

$$t \to t + t_D N(N+1), \quad \gamma \to \gamma + \gamma_D N(N+1). \quad (11.77)$$

A matrix for each F value was set up and diagonalised with the aid of a least squares minimisation programme, and the spectra were fitted with seven constants, whose values (in MHz) were found to be:

$$\gamma = 276.922\,53, \quad \gamma_D = -3.9790, \quad b_F = 100.6040, \quad t = 28.1946,$$
$$t_D = -7.35, \quad eq_0 Q = 0.7079, \quad c_I = 0.011\,32.$$

It should be remembered that these constants are for the $v'' = 1$ level. A Hartree-Fock wave function for N_2^+ has been calculated by Cade, Sales and Wahl [101], from which the spin–rotation and magnetic hyperfine constants were calculated by Rosner, Gaily and Holt [102]; they were in excellent agreement with experiment. The dominant molecular orbital configuration, given at the beginning of this section, is

$$(1s\sigma_g)^2(1s\sigma_u)^2(2s\sigma_g)^2(2s\sigma_u)^2(2p\pi_u)^4(2p\sigma_g)^1: \quad X\,^2\Sigma_g^+.$$

Comparison can be made with the predictions of this simple molecular orbital model. The Fermi contact constant, for example, is expected to be small because the molecular orbital containing the unpaired electron has predominantly $2p$ atomic character. The observed contact interaction corresponds to only 6% s character. The dipolar interaction is also in good agreement with this simple description. The electric quadrupole coupling is much more difficult to describe in simple terms because it involves all electrons.

One might suppose that this pioneering application of radiofrequency/optical double resonance techniques to the study of an ion beam would soon be followed by others but, so far, this has not happened. This study involves a combination of different sophisticated techniques, not commonly found in the same laboratory.

11.8.3. Microwave/optical double resonance of CO$^+$ in the $X\,^2\Sigma^+$ ground state

Two different groups have described experiments aimed at obtaining microwave spectra of molecular ions in beams through double resonance experiments. Both groups have investigated the CO$^+$ ion, the first being Brown, Godfrey, McGilvery and Crofts [103] who used a rovibronic component of the $A\,^2\Pi_{3/2} \leftarrow X\,^2\Sigma^+$ band system to detect the two spin components of the lowest rotational transition, $N = 1 \leftarrow 0$, in the $v = 0$ level of the ground electronic state. Subsequently Johnson, Alexander, Hertel and Lineberger [104] described a somewhat different experiment applied to the same ion, which seemed likely to have more general applicability even though the subsequent sixteen years have been disappointingly quiet. We now describe these experiments [104] in more detail.

The first stage was the production of a pulsed free-jet molecular beam of helium containing 20% CO, which was then crossed with an electron beam to produce ionisation. The ions were produced close enough to the beam nozzle for cooling to occur downstream. Some 4 cm from the nozzle the beam entered a confocal Fabry–Perot cavity where it was exposed to millimetre wave radiation close to 120 GHz in frequency. Following microwave excitation, when on resonance, the beam was probed with a Nd:YAG pumped dye laser beam with the frequency chosen to drive rovibronic components of the $A\,^2\Pi$–$X\,^2\Sigma^+$ band system. Figure 11.54 shows two recordings of a spin component of the lowest rotational transition; the line shown in (a) is

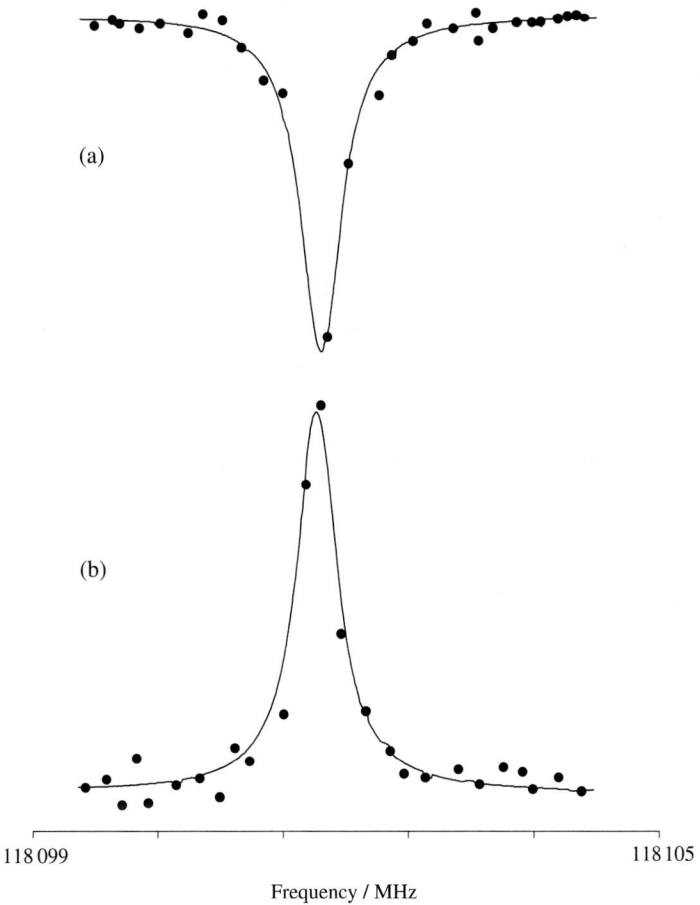

Figure 11.54. Microwave/optical double resonance line for CO^+, arising from one spin component of the lowest rotational transition in the $v = 0$ level of the $X\,^2\Sigma^+$ state [104]. Spectra (a) and (b) correspond to different laser pump transitions (see text).

obtained when the *lowest* level of the microwave transition is driven with the laser, so that resonance results in a decrease in the fluorescence intensity. Conversely, the line shown in (b) corresponds to an increase in fluorescence intensity observed when the *upper* level involved in the microwave transition is monitored by the laser probe. The sensitivity in the experiment is high.

Both spin components of the lowest rotational transition were observed, and an improved value of the rotational constant obtained. The technique was also applied to study the rotational spectrum of the CN radical, produced when methyl cyanide was added to the helium beam. In this case the rotational transition exhibited splitting due to the spin–rotation and nuclear hyperfine interactions. The results were essentially the same as those described in chapter 10, obtained from a conventional microwave absorption experiment.

11.9. Quadrupole trap radiofrequency spectroscopy of the H_2^+ ion

11.9.1. Introduction

It might be anticipated that since H_2^+ is the simplest of all molecules, and the subject of a huge number of theoretical investigations, its spectra would have been studied in great detail. This is, however, not the case because of certain particular characteristics of the molecule, which make it relatively inaccessible. Whilst H_2^+ is a respectably energetically stable molecule, it is highly reactive, particularly towards the parent molecule H_2. Consequently although electrical discharges or electron impact in H_2 readily produce H_2^+ ions, rapid reaction with H_2 leads to the production of H_3^+. H_2^+ ions are only preserved if they are either accelerated rapidly into a high-vacuum environment to form a beam, or produced in a very low-pressure environment such as that used in an ion trap. Even if a high concentration of H_2^+ can be produced, spectroscopic problems remain. The obvious problem of its homonuclear structure precludes vibrational or rotational spectroscopy, a problem that can only be removed by working with the heteronuclear species HD^+. We have described studies of the spectra of HD^+ in the previous section. Electronic spectroscopy also has its problems, which have been overcome in the rather special case involving transitions from the highest vibrational levels of the ground state. All of these spectroscopic investigations, electronic and vibrational, employed ion beam techniques. They were, however, preceded by the beautiful radiofrequency investigations, now to be described, which used an ingenious quadrupole trap method. This method has not been applied to any other molecular ion, although it would seem, at least in principle, to be more generally applicable, particularly with the advances in the underlying technology which have occurred during the past thirty years. It is not really a double resonance experiment, even though it does involve the simultaneous use of two different radiation sources. However it fits into this chapter perhaps better than any other.

11.9.2. Principles of photo-alignment

We first remind ourselves of some of the important features regarding the coupling of the angular momenta in the H_2^+ ion. In the electronic ground state ($^2\Sigma_g^+$ or $1s\sigma_g$ in the united atom nomenclature) the electronic spin angular momentum S may be coupled to the total nuclear spin angular momentum I to form a resultant G. For para-H_2^+ the total nuclear spin I is zero, so that the value of G can only be $1/2$. For ortho-H_2^+, however, the total nuclear spin has the value $I = 1$, so that G may take the values $3/2$ or $1/2$. The total spin vector G is now coupled to the rotational angular momentum vector N, to form the total angular momentum F; this coupling scheme is summarised in the vector model diagram shown in figure 11.55. So far as the rotational levels are concerned, para-H_2^+ has even N values only, whilst ortho-H_2^+ has odd values of N only.

Next we recall that the strongly allowed electronic transition between the $^2\Sigma_g^+$ ground state and the essentially repulsive $^2\Sigma_u^+$ first excited state, induced by the

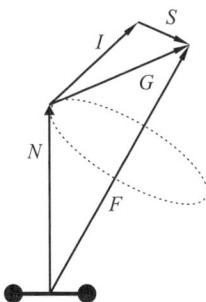

Figure 11.55. Vector model for the coupling of angular momentum vectors in the *ortho*-H_2^+ ion.

oscillating electric vector $E(t)$ of appropriate electromagnetic radiation, leads to photodissociation (except for special transitions involving the few bound long-range excited states). These electronic transitions are polarised along the direction of the internuclear axis; if an applied magnetic field B defines the direction of a space-fixed axis of quantisation, the relative orientation of B and $E(t)$ may be chosen. Dehmelt and Jefferts [105] showed that the individual photodissociation rates (R_{FM}) of different $|N, G, F, M_F\rangle$ levels of the ground state can be readily calculated. These rates depend only upon the ground state quantum numbers, and are listed for the first three rotational levels ($N = 0, 1, 2$) of H_2^+ in table 11.6. The population of each level decays exponentially in time as the photodissociation proceeds, according to the simple equation $N_{FM}(t) = N_{FM}(0) \exp(-R_{FM} t/\tau)$, where τ is the dissociation time, depending on the light intensity. The relative residual population factors after four dissociation time constants are also listed in table 11.6.

Table 11.6 reveals the interesting fact that as the photodissociation proceeds, a highly non-thermal population distribution of the levels is expected to develop; alternatively one may say that progressive preferential photo-alignment of the H_2^+ ions should occur. Before examining the consequences of this, and its experimental verification, we must give further thought to the wavelength requirements of the electromagnetic radiation producing photodissociation. It is known from studies of the electronic and vibrational spectra of the hydrogen molecular ion that electron impact ionisation of H_2 leads to the production of H_2^+ ions in which all of the vibrational levels of the electronic ground state ($v = 0$ to 19) are populated. The vibrational population factors are determined essentially by Franck–Condon overlap factors between the vibrational levels of the ground states of H_2 and H_2^+; these have been calculated by a number of authors. The maximum population occurs for $v = 2$, and slowly decreases for higher vibrational levels. Now H_2^+ ions in any vibrational level of the ground electronic state *can* be photodissociated, but the relative cross-sections for different vibrational levels depend upon the wavelength of the photodissociating radiation. The threshold wavelength for photodissociation out of $v = 3$ is 247 nm, for $v = 4$ it is 286 nm, for $v = 5$ it is 326 nm, and so on. Experimental verification of the predicted photo-alignment requires that the H_2^+ ions be contained in as nearly a collision-free environment as possible. Richardson, Jefferts and Dehmelt [106] employed a radiofrequency quadrupole

Table 11.6. *Calculated photodissociation rates R of the hyperfine sublevels of the three lowest rotational levels of the H_2^+ molecular ion, for the electric light vector aligned parallel to the static magnetic field. Relative residual population factors are listed after irradiation for four dissociation time constants τ*

N	G	F	$|M|$	R	Population
2	1/2	3/2	3/2	0.65	7.43
			1/2	1.35	0.45
		5/2	5/2	0.50	13.50
			3/2	1.10	1.23
			1/2	1.40	0.37
1	3/2	3/2	3/2	1.20	0.82
			1/2	0.80	4.08
		1/2	1/2	1.00	1.83
		5/2	5/2	0.75	4.98
			3/2	1.05	1.50
			1/2	1.20	0.82
	1/2	3/2	3/2	0.75	4.98
			1/2	1.25	0.67
		1/2	1/2	1.00	1.83
0	1/2	1/2	1/2	1.00	1.83

trap (see [107]), operated at an ambient hydrogen pressure of 3×10^{-10} Torr; effective trapping times of several seconds for 10^6 ions could be obtained with this system. For photodissociation they used a 500W high-pressure mercury arc lamp, coupled with Brewster-angle reflection from an assembly of quartz plates to produce linearly polarised light.

An important theoretical result derived by Dehmelt and Jefferts [105] was that the photodissociation rate is strongly dependent on the relative orientations of $E(t)$ and B. In particular, τ is four times larger for parallel alignment than for perpendicular alignment. Consequently all that is necessary to prove the presence of photo-alignment is to demonstrate the presence of this predicted anisotropy. Dehmelt and Jefferts achieved this by using a periodic sequence; the H_2^+ ions were irradiated, first in parallel alignment, and the numbers of H_2^+ and H^+ ions then counted. The sequence was repeated in perpendicular alignment, the results for parallel and perpendicular alignment being compared after a sufficient number of sequences had been recorded. The predicted difference was indeed recorded.

11.9.3. Experimental methods and results

The significance of the photo-alignment observations was that they pointed the way towards a possible scheme for detecting spectroscopic transitions between the F, M

levels, just as the prior preparation of aligned molecules was the key to detecting the molecular beam magnetic resonance spectrum of H$_2$. Transitions of the type $|G, F, M\rangle \rightarrow |G', F', M'\rangle$ (with $\Delta N = 0$) will occur in the radiofrequency region and will lead to equalisation of the populations of the two levels involved. An equivalent way of describing this situation is to say that saturation of the transition leads to effective photodissociation rate factors $R = 1.00$ for both levels. Consequently the occurrence of a spectroscopic transition can be detected by measuring a resonant change in the photodissociation rate.

The apparatus used by Jefferts [108] is shown in block diagrammatic form in figure 11.56. The ion trap was designed to allow simultaneous trapping of both H$_2^+$ and photoproduct H$^+$ ions, as well as individual extraction of both types of ion at the end of a measurement cycle. The latter consisted of a 50 ms electron impact burst to form the H$_2^+$ ions, a 120 ms irradiation period to produce photodissociation, and a

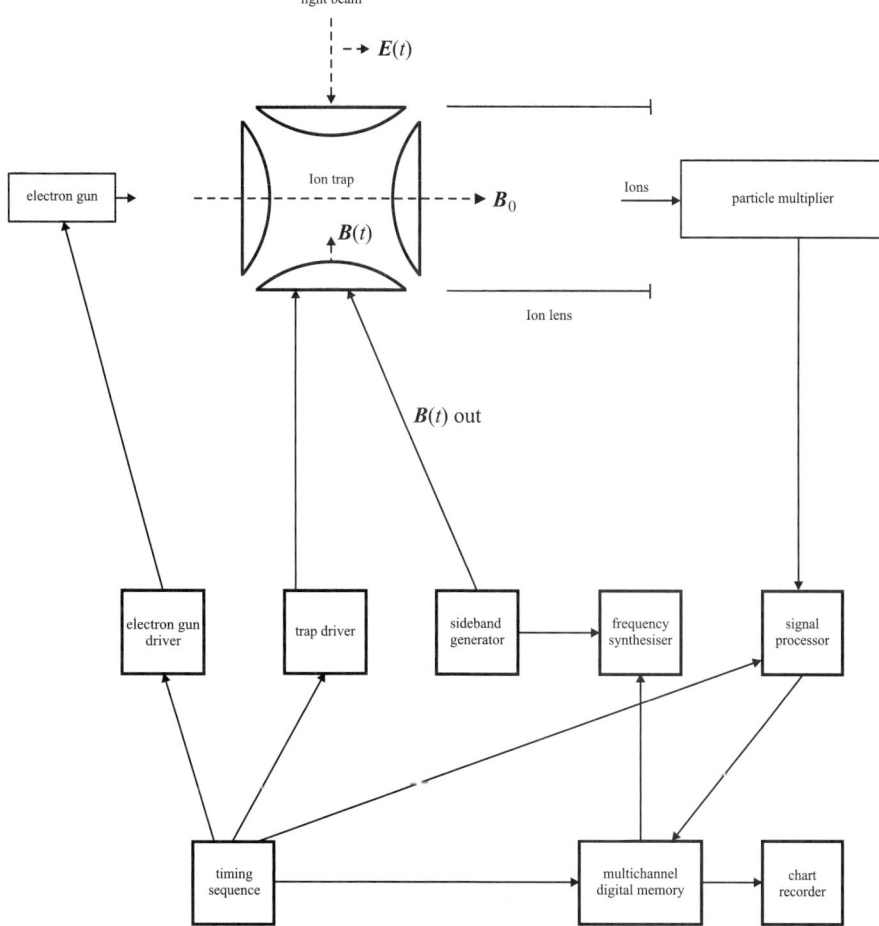

Figure 11.56. Block diagram of the ion quadrupole trap apparatus used to detect magnetic resonance and hyperfine spectra of the H$_2^+$ ion [108].

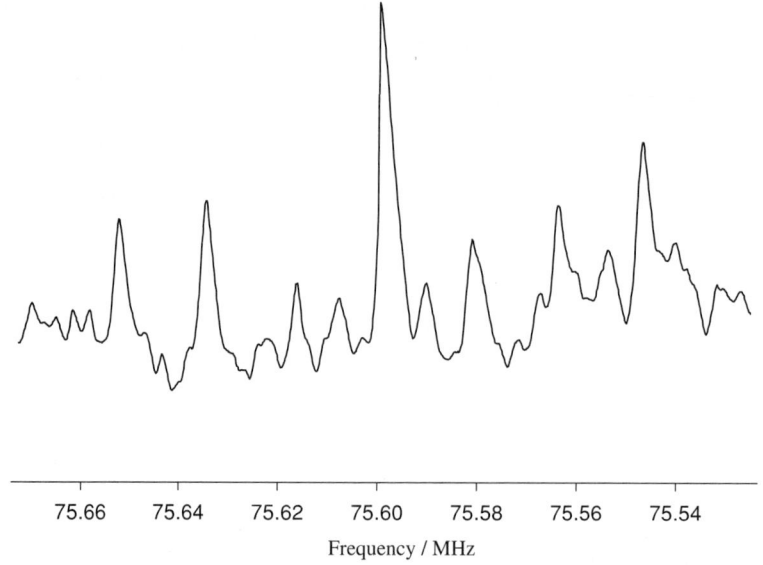

Figure 11.57. Radiofrequency hyperfine spectrum of the H_2^+ ion. The resonances arise from $\Delta F = \pm 1$ transitions within the $v = 5$ level [108].

30 ms period during which the ions were extracted and the H^+/H_2^+ ratio measured and stored in a multichannel memory. At the same time the frequency of a radiofrequency magnetic field, $B(t)$, was slowly swept. Different types of spectra were detected and recorded. In the earlier work [106] magnetic resonance spectra were recorded, but in later work [108, 109] transitions between hyperfine levels were observed in very small static magnetic fields; an example is shown in figure 11.57. The long residence time of the ions in the radiofrequency field enabled very high spectroscopic resolution to be achieved; line widths of 0.2 to 0.5 kHz were determined by inhomogeneities in the static magnetic field. Jefferts actually used a frequency sideband generator to ensure that all the Zeeman components of a hyperfine transition contributed simultaneously to its intensity. We now proceed to discuss the theory and analysis of these spectra in more detail.

11.9.4. Analysis of the spectra

The effective Hamiltonian used by Jefferts [108, 109] to analyse his spectra in zero magnetic field was expressed in the Frosch and Foley form,

$$\mathcal{H}_{\text{eff}} = b\boldsymbol{I} \cdot \boldsymbol{S} + cI_zS_z + \gamma \boldsymbol{S} \cdot \boldsymbol{N} + c_I \boldsymbol{I} \cdot \boldsymbol{N}. \tag{11.78}$$

Note that Jefferts used d for γ and f for c_I. The first two terms contain contributions from both the Fermi contact interaction and the axial component of the electron spin–nuclear spin dipolar interaction, z being along the direction of the internuclear axis.

The third and fourth terms describe the spin–rotation interactions for the electron and nuclear spins respectively. Equation (11.78) represents the Hamiltonian in zero magnetic field; additional terms would be needed to describe the Zeeman interactions in an applied static magnetic field.

We reformulate the effective Hamiltonian using irreducible tensors and obtain

$$\mathcal{H}_{\text{eff}} = b_\text{F} \mathbf{T}^1(\mathbf{I}) \cdot \mathbf{T}^1(\mathbf{S}) + \sqrt{6} g_S \mu_B g_N \mu_N (\mu_0/4\pi) \sum_p (-1)^p \mathbf{T}^2_{-p}(\mathbf{C}) \mathbf{T}^2_p(\mathbf{I},\mathbf{S})$$

$$+ \gamma \mathbf{T}^1(\mathbf{S}) \cdot \mathbf{T}^1(\mathbf{N}) + c_I \mathbf{T}^1(\mathbf{I}) \cdot \mathbf{T}^1(\mathbf{N}). \quad (11.79)$$

The four terms now represent uniquely the Fermi contact, dipolar, electron spin–rotation and nuclear spin–rotation interactions. We will examine the relationships between the parameters in (11.78) and (11.79) in due course.

We use a basis set $|\eta, \Lambda; I, S, G, N, F, M\rangle$ where η is taken to represent different vibrational levels of the ground electronic state; Jefferts' measurements involved the $v = 4$ to 8 levels, these being the ones with the optimum populations and photodissociation cross-sections. The matrix elements of each term in (11.79) are now readily calculated. The Fermi contact interaction is found to be diagonal in the chosen basis:

$$\langle \eta, \Lambda; I, S, G, N, F, M | b_\text{F} \mathbf{T}^1(\mathbf{I}) \cdot \mathbf{T}^1(\mathbf{S}) | \eta, \Lambda; I, S, G, N, F, M \rangle$$

$$= b_\text{F} (-1)^{S+I+G} \begin{Bmatrix} S & I & G \\ I & S & 1 \end{Bmatrix} \{ I(I+1)(2I+1)S(S+1)(2S+1) \}^{1/2}$$

$$= (b_\text{F}/2) \{ G(G+1) - S(S+1) - I(I+1) \}. \quad (11.80)$$

The matrix elements of the dipolar interaction are more complicated:

$$\langle \eta, \Lambda; I, S, G, N, F, M | \sqrt{6} g_S \mu_B g_N \mu_N (\mu_0/4\pi) \mathbf{T}^2(\mathbf{C}) \cdot \mathbf{T}^2(\mathbf{I},\mathbf{S}) | \eta, \Lambda'; I, S, G', N', F, M \rangle$$

$$= \sqrt{6} g_S \mu_B g_N \mu_N (\mu_0/4\pi) (-1)^{G'+F+N} \begin{Bmatrix} N' & G' & F \\ G & N & 2 \end{Bmatrix}$$

$$\times \langle \eta, \Lambda, N \| \mathbf{T}^2(\mathbf{C}) \| \eta, \Lambda', N' \rangle \langle I, S, G \| \mathbf{T}^2(\mathbf{I},\mathbf{S}) \| I, S, G' \rangle. \quad (11.81)$$

The spin part of this expression is readily evaluated:

$$\langle I, S, G \| \mathbf{T}^2(\mathbf{I},\mathbf{S}) \| I, S, G' \rangle = \sqrt{5} \{ (2G'+1)(2G+1)I(I+1)$$

$$\times (2I+1)S(S+1)(2S+1) \}^{1/2} \begin{Bmatrix} G & G' & 2 \\ I & I & 1 \\ S & S & 1 \end{Bmatrix}. \quad (11.82)$$

For the remaining part we restrict attention to matrix elements diagonal in η and Λ, noting that, since $\Lambda = 0$ for the $^2\Sigma_g^+$ ground state of H_2^+,

$$\langle \eta, N, \Lambda \| \mathbf{T}^2(\mathbf{C}) \| \eta, N', \Lambda \rangle$$

$$= (-1)^N \{ (2N+1)(2N'+1) \}^{1/2} \begin{pmatrix} N & 2 & N' \\ 0 & 0 & 0 \end{pmatrix} \langle \eta | C_0^2(\theta, \phi) r^{-3} | \eta \rangle. \quad (11.83)$$

Replacing the matrix element and its associated factors in (11.83) by the parameter t

and putting $S = 1/2$ and $I = 1$, we obtain the required general result,

$$\langle \eta, \Lambda, I, S, G, N, F, M | \mathcal{H}_{\text{dip}} | \eta, \Lambda, I, S, G', N', F, M \rangle$$
$$= 3\sqrt{30} t (-1)^{G'+F} \{(2N+1)(2N'+1)(2G+1)(2G'+1)\}^{1/2}$$
$$\times \begin{Bmatrix} N' & G' & F \\ G & N & 2 \end{Bmatrix} \begin{pmatrix} N & 2 & N' \\ 0 & 0 & 0 \end{pmatrix} \begin{Bmatrix} G & G' & 2 \\ 1 & 1 & 1 \\ 1/2 & 1/2 & 1 \end{Bmatrix}. \quad (11.84)$$

The electron spin–rotation interaction is straightforward,

$$\langle \eta, \Lambda, I, S, G, N, F, M | \gamma T^1(S) \cdot T^1(N) | \eta, \Lambda, I, S, G', N, F, M \rangle$$
$$= \gamma(-1)^{G'+F+N+G+I+S+1} \{(2G'+1)(2G+1)S(S+1)(2S+1)N(N+1)$$
$$\times (2N+1)\}^{1/2} \begin{Bmatrix} N & G' & F \\ G & N & 1 \end{Bmatrix} \begin{Bmatrix} G & S & I \\ S & G' & 1 \end{Bmatrix}, \quad (11.85)$$

and the nuclear spin–rotation interaction is given by a similar expression,

$$\langle \eta, \Lambda, I, S, G, N, F, M | c_I T^1(I) \cdot T^1(N) | \eta, \Lambda, I, S, G', N, F, M \rangle$$
$$= c_I (-1)^{2G'+F+N+I+S+1} \{(2G'+1)(2G+1)I(I+1)(2I+1)N(N+1)$$
$$\times (2N+1)\}^{1/2} \begin{Bmatrix} N & G' & F \\ G & N & 1 \end{Bmatrix} \begin{Bmatrix} G & I & S \\ I & G' & 1 \end{Bmatrix}. \quad (11.86)$$

The molecular parameters given in equation (11.78) can be readily related to those given in equations (11.80), (11.84), (11.85) and (11.86). The relationships are:

$$b_F = b + (1/3)c, \quad t = (1/3)c. \quad (11.87)$$

The hyperfine constants b and c used by Jefferts were those originally introduced by Frosch and Foley [9]; as we have discussed elsewhere, they have, to our minds, the disadvantage that they represent mixtures of the Fermi contact and dipolar interactions, whereas our b_F and t describe the magnitudes of these interactions separately.

We are now in a position to apply these results to the experimental studies of Jefferts [108, 109]. In the case of para-H_2^+ in the $N = 2$ level the situation is particularly simple because, since $I = 0$, only the electron spin-rotation interaction is involved. Consequently we have $G = 1/2$, so that $F = 5/2$ or $3/2$. Substitution in (11.85) gives the following energies:

$$\begin{aligned} N = 2, \quad G = 1/2, \quad F = 3/2: \quad & Energy = -3\gamma/2, \\ N = 2, \quad G = 1/2, \quad F = 5/2: \quad & Energy = \gamma. \end{aligned} \quad (11.88)$$

The $F = 3/2 \to 5/2$ transition frequency is therefore given by $5\gamma/2$, enabling Jefferts to determine the value of γ for five vibrational levels from $v = 4$ to 8. In table 11.8 the experimental and calculated transition frequencies are compared, and for these transitions the agreement between the two is good to 1 kHz. Note that equation (11.84) suggests that there are non-zero off-diagonal matrix elements of the dipolar interaction, but these have to involve the mixing of rotational levels differing in N by 2, which are widely separated in energy. The first-order analysis is obviously very accurate.

Table 11.7. *Matrix representation of the fine and hyperfine effective Hamiltonian (11.79) operating within the $|S, I, G, N, F, M\rangle$ basis set*

	$G = 3/2,$ $F = 5/2$	$G = 3/2,$ $F = 3/2$	$G = 3/2,$ $F = 1/2$	$G = 1/2,$ $F = 3/2$	$G = 1/2,$ $F = 1/2$
$G = 3/2, F = 5/2$	m_{11}	0	0	0	0
$G = 3/2, F = 3/2$	0	m_{22}	0	m_{24}	0
$G = 3/2, F = 1/2$	0	0	m_{33}	0	m_{35}
$G = 1/2, F = 3/2$	0	m_{42}	0	m_{44}	0
$G = 1/2, F = 1/2$	0	0	m_{53}	0	m_{55}

Turning now to *ortho*-H_2^+ in $N = 1$, the situation is considerably more complicated because all four interaction terms contribute, and more states are involved; specifically the appropriate five allowed hyperfine states are:

$$N = 1, \quad G = 3/2, \quad F = 5/2, 3/2, 1/2$$
$$N = 1, \quad G = 1/2, \quad F = 3/2, 1/2. \tag{11.89}$$

The matrix representation of the operator in equation (11.79) in this basis set is presented in table 11.7; the explicit forms of the matrix elements are as follows.

$$m_{11} = (1/2)b_F - (1/5)t + (1/2)\gamma + c_I$$
$$m_{22} = (1/2)b_F + (4/5)t - (1/3)\gamma - (2/3)c_I$$
$$m_{33} = (1/2)b_F - t - (5/6)\gamma - (5/3)c_I$$
$$m_{44} = -b_F - (1/6)\gamma + (2/3)c_I$$
$$m_{55} = -b_F + (1/3)\gamma - (4/3)c_I$$
$$m_{24} = m_{42} = -(1/2\sqrt{5})t + (\sqrt{5}/3)\gamma - (\sqrt{5}/3)c_I$$
$$m_{35} = m_{53} = (1/\sqrt{2})t + (\sqrt{2/3})\gamma - (\sqrt{2/3})c_I$$

Given values of the four constants for any particular vibrational level, it is a simple matter to calculate the energies of the five hyperfine levels, and hence the predicted transition frequencies. In practice, of course, the spectral analysis involves the inverse of this procedure.

Figure 11.58 shows an energy level diagram and the observed transitions. The transitions are, of course, magnetic dipole arising principally from coupling of the oscillating magnetic field with the electron spin magnetic moment,

$$\mathcal{H}'(t) = g_S \mu_B T^1(\boldsymbol{B}(t)) \cdot T^1(\boldsymbol{S}). \tag{11.90}$$

We now calculate the relative transition dipole moments for the transitions. The matrix

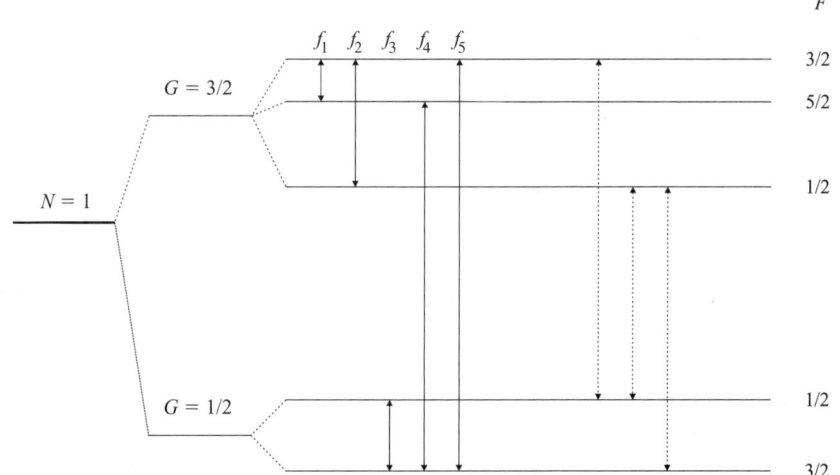

Figure 11.58. Hyperfine energy levels and magnetic dipole transitions for H_2^+ in the $N=1$ rotational level (not to scale). The dashed lines indicate magnetic-dipole allowed transitions which were not observed experimentally (see text).

elements of (11.90) within our chosen basis set are given by

$$\langle N, G, F, M | \mathcal{H}'(t) | N, G', F', M' \rangle$$
$$= g_S \mu_B \sum_p (-1)^p T_p^1(\boldsymbol{B}(t)) \langle N, G, F, M | T_{-p}^1(\boldsymbol{S}) | N, G', F', M' \rangle$$
$$= g_S \mu_B \sum_p (-1)^p T_p^1(\boldsymbol{B}(t))(-1)^{F-M} \begin{pmatrix} F & 1 & F' \\ -M & -p & M' \end{pmatrix}$$
$$\times \langle N, G, F \| T^1(\boldsymbol{S}) \| N, G', F' \rangle. \tag{11.91}$$

The reduced matrix element in (11.91) is given by

$$\langle N, G, F \| T^1(\boldsymbol{S}) \| N, G', F' \rangle$$
$$= (-1)^{F+N+1+G'} \{(2F+1)(2F'+1)\}^{1/2} \begin{Bmatrix} F & G & N \\ G' & F' & 1 \end{Bmatrix} \langle I, S, G \| T^1(\boldsymbol{S}) \| I, S, G' \rangle$$
$$= (-1)^{F+N+G'+G+I+S} \begin{Bmatrix} F & G & N \\ G' & F' & 1 \end{Bmatrix} \{(2F+1)(2F'+1)(2G+1)$$
$$\times (2G'+1)S(S+1)(2S+1)\}^{1/2} \begin{Bmatrix} G & S & I \\ S & G' & 1 \end{Bmatrix}. \tag{11.92}$$

Now the transition probabilities (TP) are given by the squares of the matrix elements (11.91). Noting that

$$\sum_{p,M'} \begin{pmatrix} F & 1 & F' \\ -M & p & M' \end{pmatrix}^2 = 1/3, \tag{11.93}$$

we obtain the required result:

$$TP = g_S^2 \mu_B^2 B^2 (2F+1)(2F'+1)(2G+1)(2G'+1)S(S+1)(2S+1)$$
$$\times \begin{Bmatrix} F & G & N \\ G' & F' & 1 \end{Bmatrix}^2 \begin{Bmatrix} G & S & I \\ S & G' & 1 \end{Bmatrix}^2. \quad (11.94)$$

Jefferts [109] observed five transitions for each vibrational level from $v = 4$ to 8, and gave values of the four constants for each vibrational level. One finds that the calculated transition frequencies, which are listed together with the experimental values in table 11.8 agree with experiment to an average deviation of about 80 kHz. Since the experimental values were accurate to ± 1.5 kHz, the analysis was not entirely satisfactory. This fact has been appreciated by Varshalovich and Sannikov [110] who have re-analysed the experimental data, derived a new set of constants, and recalculated the transition frequencies. We find that the average discrepancy in the calculated transition frequencies is now reduced to about 2.5 kHz (see table 11.8), which is a great improvement over the original analysis.

The transitions listed in the table are, in the notation $|G, F\rangle \to |G', F'\rangle$, as follows.

$f_1: 3/2, 5/2 \to 3/2, 3/2 \quad f_2: 3/2, 1/2 \to 3/2, 3/2 \quad f_3: 1/2, 3/2 \to 1/2, 1/2$
$f_4: 1/2, 3/2 \to 3/2, 5/2 \quad f_5: 1/2, 3/2 \to 3/2, 3/2 \quad f_6: 1/2, 5/2 \to 1/2, 3/2$

Their relative intensities, calculated from (11.94) are as follows.

$f_1: 0.4000, \quad f_2: 0.3704, \quad f_3: 0.1481,$
$f_4: 2.0000, \quad f_5: 0.5926, \quad f_6: 2.4000$

There are three other possible transitions, indicated by dashed lines in figure 11.58, which have comparable magnetic dipole intensities but were not observed experimentally. Table 11.6, however, shows that for the levels involved in these transitions, little or no photo-alignment is to be expected. The experimental and calculated transition frequencies are given in table 11.8, together with the determined values of the molecular constants.

This might appear to be a satisfactory conclusion, so far as the analysis of the observed spectrum is concerned. However, Carrington and Gammie [111] have re-examined the analysis and concluded that there does not appear to be any obvious reason why the nuclear spin–nuclear spin dipolar interaction should be neglected, since it is likely to be similar in magnitude to the nuclear spin–rotation interaction. This interaction was discussed for H_2 in chapter 8, where it was represented, in spherical tensor form, by the term

$$\mathcal{H}_{dip} = -g_H^2 \mu_N^2 (\mu_0/4\pi) \sqrt{6} T^2(C) \cdot T^2(I_1, I_2). \quad (11.95)$$

The second-rank tensors in this expression are defined, as shown elsewhere, by

$$T_p^2(I_1, I_2) = (-1)^p \sqrt{5} \sum_{p_1, p_2} \begin{pmatrix} 1 & 1 & 2 \\ p_1 & p_2 & -p \end{pmatrix} T_{p_1}^1(I_1) T_{p_2}^1(I_2), \quad (11.96)$$

$$T_q^2(C) = \langle \eta | C_q^2(\theta, \phi) r^{-3} | \eta \rangle. \quad (11.97)$$

Table 11.8. *Experimental and calculated transition frequencies, in MHz, for v = 4 to 8, and rotational levels with N = 1 and N = 2*

				$N=1$				$N=2$		$N=1$			$N=2$
v	Freq.	f_1	f_2	f_3	f_4	f_5	f_6	b_F	t	γ	c_I	γ
4	exp.	5.721	74.027	15.371	1270.550	1276.271	81.121	na	na	na	na	na
	calc.[a]	5.644	74.154	15.220	1270.575	1276.219	81.120	836.743	32.678	32.636	0.038	32.448
	calc.[b]	5.720	74.027	15.371	1270.549	1276.269	na	836.730	32.643	32.649	−0.034	na
5	exp.	5.258	68.933	14.381	1243.251	1248.509	75.601	na	na	na	na	na
	calc.[a]	5.183	69.053	14.238	1243.274	1248.457	75.600	819.239	30.393	30.421	0.036	30.240
	calc.[b]	5.259	68.933	14.382	1243.251	1248.510	na	819.229	30.361	30.432	−0.033	na
6	exp.	4.817	63.989	13.413	1218.154	1222.971	70.231	na	na	na	na	na
	calc.[a]	4.746	64.104	13.278	1218.176	1222.922	70.230	803.186	28.180	28.266	0.034	28.092
	calc.[b]	4.817	63.989	13.413	1218.154	1222.970	na	803.176	28.149	28.276	−0.031	na
7	exp.	4.395	59.164	12.461	1195.156	1199.551	64.977	na	na	na	na	na
	calc.[a]	4.330	59.273	12.332	1195.177	1199.507	64.978	788.519	26.025	26.156	0.032	25.991
	calc.[b]	4.394	59.164	12.460	1195.156	1199.551	na	788.509	25.995	26.167	−0.029	na
8	exp.	3.989	54.425	11.517	1174.169	1178.159	59.804	na	na	na	na	na
	calc.[a]	3.929	54.526	11.397	1174.188	1178.117	59.805	775.182	23.911	24.080	0.030	23.922
	calc.[b]	3.991	54.425	11.518	1174.169	1178.160	na	775.173	23.884	24.090	−0.028	na

Calc[a] uses the molecular constants derived by Jefferts [109].
Calc[b] uses the molecular constants derived by Varshalovich and Sannikov [110].
na means not applicable.

The matrix elements of (11.95) in our basis set may be developed as follows:

$$\langle \eta, \Lambda; I, S, G, N, F| - g_H^2 \mu_N^2 (\mu_0/4\pi)\sqrt{6}\, T^2(\boldsymbol{I}_1, \boldsymbol{I}_2) \cdot T^2(\boldsymbol{C})|\eta, \Lambda; I', S, G', N', F\rangle$$

$$= -g_H^2 \mu_N^2 (\mu_0/4\pi)\sqrt{6}(-1)^{G'+F+N} \begin{Bmatrix} N' & G' & F \\ G & N & 2 \end{Bmatrix} \langle I, S, G\|T^2(\boldsymbol{I}_1, \boldsymbol{I}_2)\|I', S, G'\rangle$$

$$\times \langle \eta, \Lambda, N\|T^2(\boldsymbol{C})\|\eta, \Lambda, N'\rangle. \tag{11.98}$$

The reduced matrix element involving the nuclear spins is evaluated as follows:

$$\langle I, S, G\|T^2(\boldsymbol{I}_1, \boldsymbol{I}_2)\|I', S, G'\rangle$$

$$= (-1)^{G'+I+1+S}\{(2G+1)(2G'+1)\}^{1/2} \begin{Bmatrix} G & I & S \\ I' & G' & 1 \end{Bmatrix} \langle I\|T^2(\boldsymbol{I}_1, \boldsymbol{I}_2)\|I'\rangle$$

$$= (-1)^{G'+I+1+S}\{(2G+1)(2G'+1)\}^{1/2} \begin{Bmatrix} G & I & S \\ I & G' & 1 \end{Bmatrix} \sqrt{5}\{(2I+1)(2I'+1)\}^{1/2}$$

$$\times \{I_1(I_1+1)(2I_1+1)I_2(I_2+1)(2I_2+1)\}^{1/2} \begin{Bmatrix} I_1 & I_1 & 1 \\ I_2 & I_2 & 1 \\ I & I' & 2 \end{Bmatrix}. \tag{11.99}$$

The second reduced matrix element in (11.98) is evaluated by noting that $\Lambda = 0$ for a Σ state, so that

$$\langle \eta, \Lambda, N|T^2(\boldsymbol{C})|\eta, \Lambda, N'\rangle = (-1)^N\{(2N+1)(2N'+1)\}^{1/2}$$

$$\times \begin{pmatrix} N & 2 & N' \\ 0 & 0 & 0 \end{pmatrix} \langle \eta|C_0^2(\theta, \phi)r^{-3}|\eta\rangle. \tag{11.100}$$

For the case of two point nuclear magnetic dipole moments, the remaining matrix element in (11.100) is calculated for $\theta = 0$, enabling us to define the dipolar parameter d_N as

$$d_N = g_H^2 \mu_N^2 (\mu_0/4\pi)\langle \eta|R^{-3}|\eta\rangle. \tag{11.101}$$

Here η contains the vibrational quantum number, so that d_N will have a different value for each vibrational level as the internuclear distance R changes.

We now have a problem in that we have five constants and five sets of data for each vibrational level. Perhaps the best approach to this problem is to determine the values of d_N from *ab initio* calculations, and use the experimental data to determine the values of the remaining four constants. Adiabatic calculations give the following values (in kHz) of d_N for the vibrational levels involved:

$$v = 4: d_N = 82.1, \quad v = 5: d_N = 77.7, \quad v = 6: d_N = 73.2,$$
$$v = 7: d_N = 68.6, \quad v = 8: d_N = 63.9.$$

Incorporation of these values in the analysis of the experimental data enables us to obtain the final sets of constants given in table 11.9.

Table 11.9. *Refitted values of the H_2^+ molecular constants (in MHz) using calculated values of the nuclear spin dipolar parameter d_N (in kHz)*

v		b_F	t	γ	c_I
4	$d_N = 82.1$	836.729	32.687	32.655	−0.036
	$d_N = 0$	836.731	32.643	32.649	−0.034
5	$d_N = 77.7$	819.227	30.402	30.437	−0.034
	$d_N = 0$	819.229	30.361	30.432	−0.032
6	$d_N = 63.2$	803.175	28.188	28.281	−0.032
	$d_N = 0$	803.176	28.149	28.276	−0.031
7	$d_N = 68.6$	788.508	26.032	26.171	−0.031
	$d_N = 0$	788.509	25.995	26.167	−0.029
8	$d_N = 63.9$	775.172	23.917	24.094	−0.028
	$d_N = 0$	775.173	23.883	24.090	−0.027

11.9.5. Quantitative interpretation of the molecular parameters

It will come as no surprise to the reader to learn that whilst the H_2^+ molecular ion is elusive to the experimental spectroscopist, it is a favourite molecule for the theorist, with hundreds of papers published. As the simplest one-electron molecule it does not have the complications arising from electron correlation. The difficulties that remain arise primarily from departures from the Born–Oppenheimer approximation which, although small for the homonuclear species, are nevertheless important in a fundamental sense. Within the Born–Oppenheimer approximation the Schrödinger equation for H_2^+ can be solved to very high accuracy by using series expansion methods; the very large literature existing prior to 1970 has been expertly summarised by Teller and Sahlin [112]. In general the theoretical treatments can be put into one of three main categories. In the *Born–Oppenheimer approximation* all terms coupling electronic and nuclear motions are neglected; in the *adiabatic approximation* terms coupling electronic and nuclear motions which are diagonal in the electronic state are included. In *non-adiabatic* calculations, coupling terms which are off-diagonal in the electronic state are treated by approximate methods.

For the reader who is interested in the quantitative aspects of these distinctions, the following analysis, due to Leach and Moss [113], may be helpful. The Hamiltonian for a one-electron molecule with nuclear masses m_1 and m_2 may be written

$$\mathcal{H} = \mathcal{H}_{BO} + (1/\mu)\mathcal{H}_{ad} + (1/\mu_a)\mathcal{H}_{gu}, \qquad (11.102)$$

where the reduced mass constants are

$$(1/\mu) = (1/m_1) + (1/m_2), \quad (1/\mu_a) = (1/m_1) - (1/m_2). \qquad (11.103)$$

The term involving μ_a is relevant only for the heteronuclear molecule, HD$^+$. If R is the internuclear distance and r_g is the position of the electron relative to the geometric centre of the nuclei, the different contributions to the Hamiltonian (11.102) may be written, in atomic units, as follows:

Born–Oppenheimer: $\mathcal{H}_{BO} = -(1/2)\nabla_g^2 - (1/r_1) - (1/r_2) + (1/R)$ (11.104)

adiabatic: $\mathcal{H}_{ad} = -(1/2)\nabla_R^2 - (1/8)\nabla_g^2$ (11.105)

non-adiabatic: $\mathcal{H}_{gu} = -(1/2)\nabla_g \cdot \nabla_R$. (11.106)

The Born–Oppenheimer potentials and wave functions, depending on R, are obtained by solving the problem

$$\mathcal{H}_{BO}\phi_t(r_g; R) = E_t(R)\phi_t(r_g; R). \quad (11.107)$$

If the eigenfunctions of the full Hamiltonian (11.102) are expressed as an expansion of the Born–Oppenheimer solutions,

$$\Psi(r_g; R) = \sum_t \phi_t(r_g; R) F_t(R), \quad (11.108)$$

the following set of coupled equations is obtained:

$$\left\{ E_s(R) - (1/2\mu)\nabla_R^2 - \int \phi_s^*(r_g; R)[(1/8\mu)\nabla_g^2 + (1/2\mu)\nabla_R^2]\phi_s(r_g; R)\,dr_g - E \right\} F_s(R)$$

$$= \sum_{t \neq s} \left\{ \int \phi_s^*(r_g; R)[-(1/8\mu)\nabla_g^2 - (1/2\mu)\nabla_R^2 - (1/2\mu_a)\nabla_g \cdot \nabla_R]\phi_t(r_g; R)\,dr_g \right.$$

$$\left. + \int \phi_s^*(r_g; R)[-(1/\mu)\nabla_R - (1/2\mu_a)\nabla_g]\phi_t(r_g; R)\,dr_g \cdot \nabla_R \right\} F_t(R). \quad (11.109)$$

The right-hand side of (11.109) contains terms off-diagonal in the electronic state (i.e. the non-adiabatic terms). If the right-hand side is set equal to zero, we obtain the adiabatic eigenvalue problem,

$$[-(1/2\mu)\nabla_R^2 + U_s(R)]F_s(R) = EF_s(R), \quad (11.110)$$

with the effective adiabatic potential

$$U_s(R) = E_s(R) + \int \phi_s^*(r_g; R)(1/\mu)\mathcal{H}_{ad}\phi_s(r_g; R)\,dr_g. \quad (11.111)$$

Exact solution of the non-adiabatic problem is not possible, and valuable concepts like the potential energy curve are lost. Nevertheless remarkable accuracy has been achieved in recent years, and some of the results obtained where experimental measurements are available for comparison are listed in table 11.10. The first set of calculated constants (1) are due to Ray and Certain [114] using an adiabatic potential of Bishop and Wetmore [115]. The second set (2) are given by McEachran, Veenstra and Cohen [116], also using an adiabatic potential. The final set are from calculations which include non-adiabatic terms; the Fermi contact parameter b_F is from Babb and Dalgarno [117], whilst the remaining constants are given by Babb and Dalgarno [118] and Babb [119]. Note that Babb [119] successfully calculates a negative value for c_I.

Table 11.10. *Comparison of experimental and calculated fine and hyperfine constants for the $v = 4$ vibrational level of H_2^+ (in MHz)*

	b_F	t	$\gamma\ (N=1)$	c_I	$\gamma\ (N=2)$
Experiment	836.730	32.643	32.649	-0.034	32.448
Calc. (adiab)(1)	837.113	32.653	32.65	0	32.44
Calc. (adiab)(2)	837.179	32.669	32.639	0.039	32.433
Calc. (non-ad)	836.747	32.643	32.658	-0.036	32.451

The experimental values are those derived by Varshalovich and Sannikov [110] from their re-analysis of Jefferts' original data.

In this section we have concentrated on calculations for H_2^+ only, which have particular relevance to the fine and hyperfine constants determined from Jefferts' experiments. Many other papers deal with calculations of the vibration–rotation level energies, for which there is much less experimental data. There are also many papers dealing with the heteronuclear molecule, HD$^+$, which is really a special case because the Born–Oppenheimer approximation collapses, particularly for the highest vibrational levels of the ground electronic state. Even the homonuclear species H_2^+ and D_2^+ exhibit some fascinating and unusual effects in their near-dissociation vibration–rotation levels. Finally we note that in order to match the accuracy of the experimental measurements for all the hydrogen molecular ion isotopomers, it is necessary to include radiative and relativistic effects.

One of the reasons why so much effort has been expended on highly accurate calculations of the constants given in table 11.10 is the desire to extrapolate the theoretical results to the ground vibrational level, $v = 0$, for which direct experimental data are not available, although studies of the Rydberg spectrum of H_2 [120] provide valuable but less accurate data. The efforts to improve the theory are motivated, at least in part, by radioastronomical searches for interstellar H_2^+, through observation of the magnetic dipole transitions discussed in this section. Several unsuccessful searches have been reported, the first being in 1968 [121], and since H_2^+ plays a fundamental role in extra-terrestrial chemistry, efforts to detect it directly are certain to continue.

References

[1] J. Brossel and F. Bitter, *Phys. Rev.*, **86**, 308 (1952).
[2] G.J. Ritter and G.W. Series, *Proc. R. Soc. Lond.*, **A238**, 473 (1957).
[3] H.E. Radford and H.P. Broida, *Phys. Rev.*, **128**, 231 (1962).
[4] K.M. Evenson, J.L. Dunn and H.P. Broida, *Phys. Rev. A*, **136**, 1566 (1964).
[5] K.M. Evenson, *Phys. Rev.*, **178**, 1 (1969).
[6] K.M. Evenson, *App. Phys. Lett.*, **12**, 253 (1968).
[7] D.W. Pratt and H.P. Broida, *J. Chem. Phys.*, **50**, 2181 (1969).
[8] H.E. Radford, *Phys. Rev. A*, **136**, 1571 (1964).

[9] R.A. Frosch and H.M. Foley, *Phys. Rev.*, **88**, 1337 (1952).
[10] K.R. German and R.N. Zare, *Phys. Rev. Lett.*, **23**, 1207 (1969).
[11] S.J. Silvers, T.H. Bergeman and W. Klemperer, *J. Chem. Phys.*, **52**, 4385 (1970).
[12] R.W. Field and T.H. Bergeman, *J. Chem. Phys.*, **54**, 2936 (1971).
[13] K.R. German, T.H. Bergeman, E.M. Weinstock and R.N. Zare, *J. Chem. Phys.*, **58**, 4304 (1973).
[14] R.W. Field, R.S. Bradford, H.P. Broida and D.O. Harris, *J. Chem. Phys.*, **57**, 2209 (1972).
[15] R.W. Field, R.S. Bradford, D.O. Harris and H.P. Broida, *J. Chem. Phys.*, **56**, 4712 (1972).
[16] R.W. Field, A.D. English, T. Tanaka, D.O. Harris and D.A. Jennings, *J. Chem. Phys.*, **59**, 2191 (1973).
[17] M. Takami, *Japan J. Appl. Phys.*, **15**, 1063 (1976); **15**, 1889 (1976).
[18] W. Lichten, *Phys. Rev.*, **120**, 848 (1960); **126**, 1020 (1962).
[19] O.W. Richardson, *Molecular Hydrogen and its Spectrum*, Yale University Press, 1934.
[20] G. Herzberg, *Spectra of Diatomic Molecules*, D.Van Nostrand Company, Inc., 1950.
[21] R.S. Freund and T.A. Miller, *J. Chem. Phys.*, **56**, 2211 (1972).
[22] M.L. Ginter, *J. Chem. Phys.*, **46**, 3687 (1967).
[23] T.A. Miller and R.S. Freund, *J. Chem. Phys.*, **56**, 3165 (1972).
[24] R.S. Freund and T.A. Miller, *J. Chem. Phys.*, **58**, 3565 (1973).
[25] T.A. Miller and R.S. Freund, *J. Chem. Phys.*, **58**, 2345 (1973).
[26] R. Jost, M.A. Marechal and M. Lombardi, *Phys. Rev.*, **A5**, 732, 740 (1972);
[27] P.R. Fontana, *Phys. Rev.*, **125**, 220 (1962).
[28] L.-Y. Chow Chiu, *J. Chem. Phys.*, **40**, 2276 (1964); *Phys. Rev.*, **137**, A384 (1965).
[29] T.A. Miller, *J. Chem. Phys.*, **59**, 4078 (1973).
[30] A.N. Jette and P. Cahill, *Phys. Rev.*, **160**, 35 (1967).
[31] R.S. Freund and T.A. Miller, *J. Chem. Phys.*, **59**, 5770 (1973).
[32] P.R. Brooks, W. Lichten and R. Reno, *Phys. Rev.*, **A4**, 2217 (1971).
[33] K.B. Jefferts, *Phys. Rev. Lett.*, **23**, 1476 (1969).
[34] M. Lombardi, *J. Chem. Phys.*, **58**, 797 (1973).
[35] T.A. Miller and R.S. Freund, *J. Chem. Phys.*, **59**, 4093 (1973).
[36] R.S. Freund, T.A. Miller and B.R. Zegarski, *Chem. Phys. Lett.*, **23**, 120 (1973).
[37] T.A. Miller, R.S. Freund and B.R. Zegarski, *J. Chem. Phys.*, **60**, 3195 (1974).
[38] P.J. Domaille, T.C. Steimle and D.O. Harris, *J. Mol. Spectrosc.*, **68**, 146 (1977).
[39] P.J. Domaille, T.C. Steimle and D.O. Harris, *J. Mol. Spectrosc.*, **66**, 503 (1977).
[40] J. Nakagawa, P.J. Domaille, T.C. Steimle and D.O. Harris, *J. Mol. Spectrosc.*, **70**, 374 (1978).
[41] T.C. Steimle, P.J. Domaille and D.O. Harris, *J. Mol. Spectrosc.*, **68**, 134 (1977).
[42] W.E. Ernst and T.Törring, *Phys. Rev.*, **A25**, 1236 (1982).
[43] W.E. Ernst, *Appl. Phys.*, **B30**, 105 (1983).
[44] W.E. Ernst and S. Kindt, *Appl. Phys.*, **B31**, 79 (1983).
[45] S.D. Rosner, R.A. Holt and T.D. Gaily, *Phys. Rev. Lett.*, **35**, 785 (1975).
[46] Y. Endo, S. Saito and E. Hirota, *Astrophys. J.*, **278**, L131 (1984).
[47] M.D. Allen, L.M. Ziurys and J.M. Brown, *Chem. Phys. Lett.*, **257**, 130 (1996).
[48] T. Kröckertskothen, H. Knöckel and E. Tiemann, *Mol. Phys.*, **62**, 1031 (1987).
[49] T. Kröckertskothen, H. Knöckel and E. Tiemann, *Chem. Phys.*, **103**, 335 (1986).
[50] A.W. Taylor, A.S.C. Cheung and A.J. Merer, *J. Mol. Spectrosc.*, **113**, 487 (1985).

[51] F. Ahmed, R.F. Barrow, A.H. Chojnicki, C. Dufour and J. Schamps, *J. Phys. B.*, **15**, 3801 (1982).
[52] C. Dufour, J. Schamps and R.F. Barrow, *J. Phys. B.*, **15**, 3819 (1982).
[53] T.C. Steimle, C.R. Brazier and J.M. Brown, *J. Mol. Spectrosc.*, **91**, 137 (1982).
[54] C.R. Brazier, J.M. Brown and T.C. Steimle, *J. Mol. Spectrosc.*, **97**, 449 (1983).
[55] T.C. Steimle, C.R. Brazier and J.M. Brown, *J. Mol. Spectrosc.*, **110**, 39 (1985).
[56] R.J. Saykally, K.M. Evenson, E.R. Comben and J.M. Brown, *Mol. Phys.*, **58**, 735 (1986).
[57] J.M. Brown, I. Kopp, C. Malmberg and B. Rydh, *Physica Scripta*, **17**, 55 (1978).
[58] J.M. Brown and A.J. Merer, *J. Mol. Spectrosc.*, **74**, 488 (1979).
[59] J. Hoeft, F.J. Lovas, E. Tiemann and T. Törring, *Z. Naturforsch.*, **25a**, 25 (1970).
[60] R. Ritschel, *Z. Phys.*, **79**, 1 (1932).
[61] M.C.L. Gerry, A.J. Merer, U. Sassenberg and T.C. Steimle, *J. Chem. Phys.*, **86**, 4754 (1987).
[62] T.C. Steimle, W.-L. Chang and D.F. Nachman, *Chem. Phys. Lett.*, **153**, 534 (1988).
[63] S.R. Langhoff and C.W. Bauschlicher, *Chem. Phys. Lett.*, **124**, 241 (1986).
[64] P.V. Madhavan and M.D. Newton, *J. Chem. Phys.*, **83**, 2337 (1985).
[65] R. Stringat, C. Athenour and J.L. Féménias, *Can. J. Phys.*, **50**, 395 (1972).
[66] K. Liu and J.M. Parson, *J. Chem. Phys.*, **67**, 1814 (1977).
[67] W.J. Childs and T.C. Steimle, *J. Chem. Phys.*, **88**, 6168 (1988).
[68] C.W. Bauschlicher and S.R. Langhoff, *J. Chem. Phys.*, **85**, 5936 (1986).
[69] K. Namiki, S. Saito, J. Scott Robinson and T.C. Steimle, *J. Mol. Spectrosc.*, **191**, 176 (1998).
[70] T.C. Steimle, J.E. Shirley, K.Y. Jung, L.R. Russon and C.T. Scurlock, *J. Mol. Spectrosc.*, **144**, 27 (1990).
[71] K.C. Namiki and T.C. Steimle, *J. Chem. Phys.*, **111**, 6385 (1999).
[72] D.A. Fletcher, K.Y. Jung and T.C. Steimle, *J. Chem. Phys.*, **99**, 901 (1993); **108**, 10 327 (1998).
[73] J.F. Harrison, *J. Phys. Chem.*, **100**, 3513 (1996).
[74] T.C. Steimle, J.S. Robinson and D. Goodridge, *J. Chem. Phys.*, **110**, 881 (1999).
[75] T. Nelis, S.P. Beaton, K.M. Evenson and J.M. Brown, *J. Mol. Spectrosc.*, **148**, 462 (1991).
[76] T.C. Steimle, D.F. Nachman, J.E. Shirley, D.A. Fletcher and J.M. Brown, *Mol. Phys.*, **69**, 923 (1990).
[77] J.M. Brown, A.S.C. Cheung and A.J. Merer, *J. Mol. Spectrosc.*, **124**, 464 (1987).
[78] D.A. Fletcher, K.Y. Jung, C.T. Scurlock and T.C. Steimle, *J. Chem. Phys.*, **98**, 1837 (1993).
[79] S.D. Rosner, R.A. Holt and T.D. Gaily, *Phys. Rev. Lett.*, **35**, 785 (1975).
[80] W.J. Childs and L.S. Goodman, *Phys. Rev.*, **A21**, 1216 (1980).
[81] W.J. Childs, O. Poulsen and T.C. Steimle, *J. Chem. Phys.*, **88**, 598 (1988).
[82] Y. Azuma and W.J. Childs, *J. Chem. Phys.*, **93**, 8415 (1990).
[83] B.E. Sauer, J. Wang and E.A. Hinds, *J. Chem. Phys.*, **105**, 7412 (1996).
[84] B.E. Sauer, J. Wang and E.A. Hinds, *Phys. Rev. Lett.*, **74**, 1554 (1995).
[85] R. Bacis, R. Collomb and N. Bessis, *Phys. Rev.*, **A8**, 2255 (1973).
[86] A. Bernard and A.M. Sibai, *Z. Naturforsch.*, **35a**, 1313 (1980).
[87] W.J. Childs, G.L. Goodman, L.S. Goodman and L. Young, *J. Mol. Spectrosc.*, **119**, 166 (1986).
[88] A. Carrington, I.R. McNab and C.A. Montgomerie, *Mol. Phys.*, **66**, 519 (1989).

[89] S.D. Rosner, T.D. Gaily and R.A. Holt, *Phys. Rev. Lett.*, **40**, 851 (1978).
[90] A. Carrington and R.A. Kennedy, *Mol. Phys.*, **56**, 935 (1984).
[91] A. Carrington, I.R. McNab, C.A. Montgomerie and J.M. Brown, *Mol. Phys.*, **66**, 1279 (1989).
[92] A. Carrington, I.R. McNab and C.A. Montgomerie, *Mol. Phys.*, **65**, 751 (1988).
[93] R.A. Kennedy, R.E. Moss and I.A. Sadler, *Mol. Phys.*, **64**, 165 (1988).
[94] A. Carrington, I.R. McNab and C.A. Montgomerie, *J. Phys. B:At. Mol. Opt. Phys.*, **22**, 3551 (1989).
[95] A. Carrington and R.A. Kennedy, *Mol. Phys.*, **56**, 935 (1985).
[96] A. Carrington, C.A. Leach, R.E. Moss, T.C. Steimle, M.R. Viant and Y.D. West, *J. Chem. Soc. Faraday Trans.*, **89**, 603 (1993).
[97] A. Carrington, C.A. Leach, A.J. Marr, R.E. Moss, C.H. Pyne and T.C. Steimle, *J. Chem. Phys.*, **98**, 5290 (1993).
[98] A. Carrington, P.J. Knowles and C.H. Pyne, *J. Chem. Phys.*, **102**, 5979 (1995).
[99] N.Berrah Mansour, C. Kurtz, T.C. Steimle, G.L. Goodman, L. Young, T.J. Scholl, S.D. Rosner and R.A. Holt, *Phys. Rev.*, **A44**, 4418 (1991).
[100] S.D. Rosner, R.A. Holt and T.D. Gaily, *Phys. Rev. Lett.*, **35**, 785 (1975).
[101] P.E. Cade, K.D. Sales and A.C. Wahl, *J. Chem. Phys.*, **44**, 1973 (1966).
[102] S.D. Rosner, T.D. Gaily and R.A. Holt, *J. Mol. Spectrosc.*, **109**, 73 (1985).
[103] R.D. Brown, P.D. Godfrey, D.C. McGilvery and J.G. Crofts, *Chem. Phys. Lett.*, **84**, 437 (1981).
[104] M.A. Johnson, M.L. Alexander, I. Hertel and W.C. Lineberger, *Chem. Phys. Lett.*, **105**, 374 (1984).
[105] H.G. Dehmelt and K.B. Jefferts, *Phys. Rev.*, **125**, 1318 (1962).
[106] C.B. Richardson, K.B. Jefferts and H.G. Dehmelt, *Phys. Rev.*, **165**, 80 (1968).
[107] W. Paul, O. Osberghaus and E. Fischer, *Forschungsber. Wirtsch. Verkehrsministeriums Nordrhein-Westfalen* **415** (1958); E. Fischer, *Z. Physik.*, **156**, 1 (1959); R.F. Wuerker, H. Shelton and R.V. Langmuir, *J. Appl. Phys.*, **30**, 342 (1958).
[108] K.B. Jefferts, *Phys. Rev. Lett.*, **20**, 39 (1967).
[109] K.B. Jefferts, *Phys. Rev. Lett.*, **23**, 1476 (1969).
[110] D.A. Varshalovich and A.V. Sannikov, *Astron. Lett.*, **19**, 290 (1993).
[111] A. Carrington and D.I. Gammie, unpublished work (1999).
[112] E. Teller and H.L. Sahlin, *Physical Chemistry: An Advanced Treatise*, Vol. 5, pp. 35–124, Academic Press, New York, 1970.
[113] C.A. Leach and R.E. Moss, *Ann. Rev. Phys. Chem.*, **46**, 55 (1995).
[114] R.D. Ray and P.R. Certain, *Phys. Rev. Lett.*, **38**, 824 (1977).
[115] D.M. Bishop and R.W. Wetmore, *Mol. Phys.*, **26**, 145 (1973).
[116] R.P. McEachran, C.J. Veenstra and M. Cohen, *Chem. Phys. Lett.*, **59**, 275 (1978).
[117] J.F. Babb and A. Dalgarno, *Phys. Rev. Lett.*, **66**, 880 (1991).
[118] J.F. Babb and A. Dalgarno, *Phys. Rev.*, **A46**, R5317 (1992).
[119] J.F. Babb, *Phys. Rev. Lett.*, **75**, 4377 (1995).
[120] Z.W. Fu, E.A. Hessels and S.R. Lundeen, *Phys. Rev.*, **A46**, R5313 (1992).
[121] A.A. Penzias, K.B. Jefferts, D.F. Dickinson, A.E. Lilley and H. Penfield, *Astrophys. J.*, **154**, 389 (1968).

General appendices

Appendix A
Values of the fundamental constants

A new best set of the fundamental constants has been compiled by NIST at Gaithersburg in 1999. A selected set is given in the following table.

quantity	symbol	value	unit
Planck constant	h	6.626 068 76(52)	10^{-34} Js
elementary charge	e	1.602 176 462(63)	10^{-19} C
electron rest mass	m_e	9.109 381 88(72)	10^{-31} kg
proton rest mass	m_p	1.672 621 58(13)	10^{-27} kg
atomic mass constant	$u = m_u$	1.660 538 73(13)	10^{-27} kg
Avogadro constant	L, N_A	6.022 141 99(47)	10^{23} mol^{-1}
Boltzmann constant	k	1.380 650 3(24)	10^{-23} J K^{-1}
Bohr radius	a_0	0.529 177 208 3(19)	10^{-10} m
Hartree energy	E_h	4.359 743 81(34)	10^{-18} J
Rydberg constant	R_∞	1.097 373 156 854 9(83)	10^7 m^{-1}
Bohr magneton	μ_B	9.274 008 99(37)	10^{-24} J T^{-1}
nuclear magneton	μ_N	5.050 783 17(20)	10^{-27} J T^{-1}
free electron g factor	g_S	2.002 319 304 373 7(82)	
speed of light (vacuum)	c	2.997 924 58	10^8 m s^{-1}
magnetic constant	μ_0	$4\pi \times 10^{-7}$ = 12.566 370 61...	10^{-7} N A^{-2}
vacuum permittivity	ε_0	8.854 187 817...	10^{-12} J^{-1} C^2 m^{-1}
	$4\pi \varepsilon_0$	1.112 650 056...	10^{-10} J^{-1} C^2 m^{-1}
proton-electron mass ratio	m_p/m_e	1836.152 667 5(39)	
electron volt	eV	1.602 176 462(63)	10^{-19} J

See http://physics.nist.gov/constants

Appendix B
Selected set of nuclear properties for naturally occurring isotopes

nucleus	isotope	atomic mass	natural abundance (%)	spin ($h/2\pi$)	magnetic moment (μ_N)	electric quadrupole moment (10^{-24} cm^2)
H	1	1.007 825	99.985	1/2	+2.792 85	—
	2	2.014 10	0.015	1	+0.857 44	+0.0028
	3	3.016 05	0	1/2	+2.978 96	—
He	3	3.016 03	0.000137	1/2	−2.127 62	—
	4	4.002 60	≈100	0	—	—
Li	6	6.015 121	7.5	1	+0.822 05	−0.0008
	7	7.016 003	92.5	3/2	+3.256 44	−0.041
Be	9	9.012 182	100.0	3/2	−1.1776	+0.05
B	10	10.012 937	19.9	3	+1.8006	+0.085
	11	11.009 305	80.1	3/2	+2.6886	+0.041
C	12	12	98.90	0	—	—
	13	13.003 355	1.10	1/2	+0.702 41	—
N	14	14.003 074	99.63	1	+0.403 76	+0.020
	15	15.000 108	0.37	1/2	−0.283 19	—
O	16	15.994 915	99.76	0	—	—
	17	16.999 131	0.04	5/2	−1.8938	−0.026
	18	17.999 160	0.20	0	—	—
F	19	18.998 403	100.0	1/2	+2.628 87	—
Ne	20	19.992 435	90.48	0	—	—
	21	20.993 843	0.27	3/2	−0.661 80	+0.103
	22	21.991 383	9.25	0	—	—
Na	23	22.989 767	100.0	3/2	+2.217 52	+0.101
Mg	24	23.985 042	78.99	0	—	—
	25	24.985 837	10.00	5/2	−0.855 45	+0.20
	26	25.982 593	11.01	0	—	—
Al	27	26.981 539	100.0	5/2	+3.641 51	+0.15
Si	28	27.976 927	92.23	0	—	—
	29	28.976 495	4.67	1/2	−0.5553	—
	30	29.973 770	3.10	0	—	—
P	31	30.973 762	100.0	1/2	+1.131 60	—

Appendix B (*cont.*)

nucleus	isotope	atomic mass	natural abundance (%)	spin ($h/2\pi$)	magnetic moment (μ_N)	electric quadrupole moment (10^{-24} cm^2)
S	32	31.972 070	95.02	0	—	—
	33	32.971 456	0.75	3/2	+0.643 82	−0.07
	34	33.967 866	4.21	0	—	—
Cl	35	34.968 852	75.77	3/2	+0.821 87	−0.08
	37	36.965 903	24.23	3/2	+0.684 12	−0.065
Ar	36	35.967 545	0.337	0	—	—
	38	37.962 732	0.063	0	—	—
	40	39.962 384	99.600	0	—	—
K	39	38.963 707	93.2581	3/2	+0.391 46	+0.049
	40	39.963 999	0.0117	4	−1.298	−0.061
	41	40.961 825	6.7302	3/2	+0.214 87	+0.060
Ca	40	39.962 591	96.941	0	—	—
	42	41.958 618	0.647	0	—	—
	43	42.958 766	0.135	7/2	−1.3173	−0.05
	44	43.955 480	2.086	0	—	—
	46	45.953 689	0.004	0	—	—
	48	47.952 533	0.187	0	—	—
Sc	45	44.955 910	100.0	7/2	4.756 49	−0.22
Ti	46	45.952 629	8.0	0	—	—
	47	46.951 764	7.3	5/2	−0.788 48	+0.29
	48	47.947 947	73.8	0	—	—
	49	48.947 871	5.5	7/2	−1.104 17	+0.24
	50	49.944 792	5.4	0	—	—
V	50	49.947 161	0.250	6	+3.345 69	+0.21
	51	50.943 962	99.750	7/2	+5.148 706	−0.04
Cr	50	49.946 046	4.345	0	—	—
	52	51.940 509	83.79	0	—	—
	53	52.940 651	9.50	3/2	−0.474 54	−0.15
	54	53.938 882	2.365	0	—	—
Mn	55	54.938 047	100.0	5/2	+3.4687	+0.32
Fe	54	53.939 612	5.9	0	—	—
	56	55.934 939	91.72	0	—	—
	57	56.935 396	2.1	1/2	+0.0906	—
	58	57.933 277	0.28	0	—	—
Co	59	58.933 198	100.0	7/2	+4.63	+0.41

Appendix B (*cont.*)

nucleus	isotope	atomic mass	natural abundance (%)	spin ($h/2\pi$)	magnetic moment (μ_N)	electric quadrupole moment (10^{-24} cm^2)
Ni	58	57.935 346	68.077	0	—	—
	60	59.930 788	26.223	0	—	—
	61	60.931 058	1.140	3/2	$-0.750\,02$	$+0.16$
	62	61.928 346	3.634	0	—	—
	64	63.927 968	0.926	0	—	—
Cu	63	62.929 598	69.17	3/2	$+2.2273$	-0.211
	65	64.927 793	30.83	3/2	$+2.3817$	-0.195
Zn	64	63.929 145	48.6	0	—	—
	66	65.926 034	27.9	0	—	—
	67	66.927 129	4.1	5/2	$+0.8755$	$+0.15$
	68	67.924 846	18.8	0	—	—
	70	69.925 325	0.6	0	—	—
Ga	69	68.925 580	60.108	3/2	$+2.016\,59$	$+0.17$
	71	70.924 700	39.892	3/2	$+2.562\,27$	$+0.11$
Ge	70	69.924 250	21.24	0	—	—
	72	71.922 079	27.66	0	—	—
	73	72.923 463	7.72	9/2	$-0.879\,467$	-0.17
	74	73.921 177	35.94	0	—	—
	76	75.921 401	7.44	0	—	—
As	75	74.921 594	100.0	3/2	$+1.439\,47$	$+0.31$
Se	74	73.922 475	0.89	0	—	—
	76	75.919 212	9.36	0	—	—
	77	76.919 912	7.63	1/2	$+0.535\,06$	—
	78	77.917 308	23.77	0	—	—
	80	79.916 520	49.61	0	—	—
	82	81.916 698	8.74	0	—	—
Br	79	78.918 336	50.69	3/2	$+2.106\,400$	$+0.331$
	81	80.916 289	49.31	3/2	$+2.270\,562$	$+0.276$
Kr	78	77.920 496	0.35	0	—	—
	80	79.916 380	2.25	0	—	—
	82	81.913 482	11.6	0	—	—
	83	82.914 135	11.5	9/2	$-0.970\,669$	$+0.253$
	84	83.911 507	57.0	0	—	—
	86	85.910 616	17.3	0	—	—
Rb	85	84.911 794	72.17	5/2	$+1.353$	$+0.23$
	87	86.909 187	27.83	3/2	$+2.7512$	$+0.13$

Appendix B (*cont.*)

nucleus	isotope	atomic mass	natural abundance (%)	spin ($h/2\pi$)	magnetic moment (μ_N)	electric quadrupole moment (10^{-24} cm^2)
Sr	84	83.913 430	0.56	0	—	—
	86	85.909 267	9.86	0	—	—
	87	86.908 884	7.00	9/2	−1.093 60	+0.34
	88	87.905 619	82.58	0	—	—
Y	89	88.905 849	100.0	1/2	−0.137 42	—
Zr	90	89.904 703	51.45	0	—	—
	91	90.905 644	11.22	5/2	−1.303 62	−0.21
	92	91.905 039	17.15	0	—	—
	94	93.906 314	17.38	0	—	—
	96	95.908 275	2.80	0	—	—
Nb	93	92.906 377	100.0	9/2	+6.1705	−0.32
Mo	92	91.906 808	14.84	0	—	—
	94	93.905 085	9.25	0	—	—
	95	94.905 840	15.92	5/2	−0.9142	−0.02
	96	95.904 678	16.68	0	—	—
	97	96.906 020	9.55	5/2	−0.9335	+0.26
	98	97.905 406	24.13	0	—	—
	100	99.907 477	9.63	0	—	—
Ru	96	95.907 599	5.54	0	—	—
	98	97.905 287	1.86	0	—	—
	99	98.905 939	12.7	5/2	−0.6413	+0.079
	100	99.904 219	12.6	0	—	—
	101	100.905 582	17.1	5/2	−0.7188	+0.46
	102	101.904 348	31.6	0	—	—
	104	103.905 424	18.6	0	—	—
Rh	103	102.905 500	100.0	1/2	−0.0884	—
Pd	102	101.905 634	1.02	0	—	—
	104	103.904 029	11.14	0	—	—
	105	104.905 079	22.33	5/2	−0.642	+0.66
	106	105.903 478	27.33	0	—	—
	108	107.903 895	26.46	0	—	—
	110	109.905 167	11.72	0	—	—
Ag	107	106.905 092	51.839	1/2	−0.113 57	—
	109	108.904 757	48.161	1/2	−0.130 56	—
Cd	106	105.904 176	1.25	0	—	—
	108	107.904 18	0.89	0	—	—
	110	109.903 005	12.49	0	—	—

Appendix B (*cont.*)

nucleus	isotope	atomic mass	natural abundance (%)	spin ($h/2\pi$)	magnetic moment (μ_N)	electric quadrupole moment (10^{-24} cm^2)
Cd	111	110.904 182	12.80	1/2	−0.594 886	—
	112	111.902 758	24.13	0	—	—
	113	112.904 400	12.22	1/2	−0.622 301	—
	114	113.903 357	28.73	0	—	—
	116	115.904 754	7.49	0	—	—
In	113	112.904 061	4.3	9/2	+5.529	+0.80
	115	114.903 880	95.7	9/2	+5.541	+0.81
Sn	112	111.904 826	0.97	0	—	—
	114	113.902 784	0.65	0	—	—
	115	114.903 348	0.36	1/2	−0.9188	—
	116	115.901 747	14.53	0	—	—
	117	116.902 956	7.68	1/2	−1.0010	—
	118	117.901 609	24.22	0	—	—
	119	118.903 310	8.58	1/2	−1.0473	—
	120	119.902 200	32.59	0	—	—
	122	121.903 440	4.63	0	—	—
	124	123.905 274	5.79	0	—	—
Sb	121	120.903 821	57.36	5/2	+3.363	−0.4
	123	122.904 216	42.64	7/2	+2.550	−0.5
Te	120	119.904 048	0.095	0	—	—
	122	121.903 054	2.59	0	—	—
	123	122.904 271	0.905	1/2	−0.736 95	—
	124	123.902 823	4.79	0	—	—
	125	124.904 433	7.12	1/2	−0.8885	—
	126	125.903 314	18.93	0	—	—
	128	127.904 463	31.70	0	—	—
	130	129.906 229	33.87	0	—	—
I	127	126.904 473	100.0	5/2	+2.8133	−0.79
Xe	124	123.905 894	0.10	0	—	—
	126	125.904 281	0.09	0	—	—
	128	127.903 531	1.91	0	—	—
	129	128.904 780	26.4	1/2	−0.7780	—
	130	129.903 509	4.1	0	—	—
	131	130.905 072	21.2	3/2	+0.691 86	−0.12
	132	131.904 144	26.9	0	—	—
	134	133.905 395	10.4	0	—	—
	136	135.907 214	8.9	0	—	—

Appendix B (*cont.*)

nucleus	isotope	atomic mass	natural abundance (%)	spin ($h/2\pi$)	magnetic moment (μ_N)	electric quadrupole moment (10^{-24} cm^2)
Cs	133	132.905 429	100.0	7/2	+2.582	−0.0037
Ba	130	129.906 282	0.106	0	—	—
	132	131.905 042	0.101	0	—	—
	134	133.904 486	2.42	0	—	—
	135	134.905 665	6.593	3/2	+0.838	+0.16
	136	135.904 553	7.85	0	—	—
	137	136.905 812	11.23	3/2	+0.9374	+0.245
	138	137.905 232	71.70	0	—	—
La	138	137.907 11	0.0902	5	+3.7136	+0.45
	139	138.906 347	99.9098	7/2	+2.7830	+0.20
Ce	136	135.907 140	0.19	0	—	—
	138	137.905 985	0.25	0	—	—
	140	139.905 433	88.43	0	—	—
	142	141.909 241	11.13	0	—	—
Pr	141	140.907 647	100.0	5/2	+4.275	−0.059
Nd	142	141.907 719	27.13	0	—	—
	143	142.909 810	12.18	7/2	−1.07	−0.6
	144	143.910 083	23.80	0	—	—
	145	144.912 570	8.30	7/2	−0.66	−0.33
	146	145.913 113	17.19	0	—	—
	148	147.916 889	5.76	0	—	—
	150	149.920 887	5.64	0	—	—
Sm	144	143.911 998	3.1	0	—	—
	147	146.914 895	15.0	7/2	−0.815	−0.26
	148	147.914 820	11.3	0	—	—
	149	148.917 181	13.8	7/2	−0.672	+0.075
	150	149.917 273	7.4	0	—	—
	152	151.919 729	26.6	0	—	—
	154	153.922 206	22.7	0	—	—
Eu	151	150.919 847	47.8	5/2	+3.472	+0.90
	153	152.921 225	52.2	5/2	+1.533	+2.41
Gd	152	151.919 786	0.20	0	—	—
	154	153.920 861	2.18	0	—	—
	155	154.922 618	14.80	3/2	−0.257	+1.30
	156	155.922 118	20.47	0	—	—
	157	156.923 956	15.65	3/2	−0.337	+1.36

Appendix B (*cont.*)

nucleus	isotope	atomic mass	natural abundance (%)	spin ($h/2\pi$)	magnetic moment (μ_N)	electric quadrupole moment (10^{-24} cm^2)
Gd	158	157.924 019	24.84	0	—	—
	160	159.927 049	21.86	0	—	—
Tb	159	158.925 342	100.0	3/2	+2.014	+1.43
Dy	156	155.925 277	0.06	0	—	—
	158	157.924 403	0.10	0	—	—
	160	159.925 193	2.34	0	—	—
	161	160.926 930	18.9	5/2	−0.480	+2.51
	162	161.926 795	25.5	0	—	—
	163	162.928 728	24.9	5/2	+0.673	+2.65
	164	163.929 171	28.2	0	—	—
Ho	165	164.930 319	100.0	7/2	+4.17	+3.49
Er	162	161.928 775	0.14	0	—	—
	164	163.929 198	1.61	0	—	—
	166	165.930 290	33.6	0	—	—
	167	166.932 046	22.95	7/2	−0.5639	+3.57
	168	167.932 368	26.8	0	—	—
	170	169.935 461	14.9	0	—	—
Tm	169	168.934 212	100.0	1/2	−0.2316	—
Yb	168	167.933 894	0.13	0	—	—
	170	169.934 759	3.05	0	—	—
	171	170.936 323	14.3	1/2	+0.493 67	—
	172	171.936 378	21.9	0	—	—
	173	172.938 208	16.12	5/2	−0.679 89	+2.80
	174	173.938 859	31.8	0	—	—
	176	175.942 564	12.7	0	—	—
Lu	175	174.940 770	97.41	7/2	+2.2327	+3.49
	176	175.942 679	2.59	7	+3.169	+4.92
Hf	174	173.940 044	0.162	0	—	—
	176	175.941 406	5.206	0	—	—
	177	176.943 217	18.606	7/2	+0.7935	+3.37
	178	177.943 696	27.297	0	—	—
	179	178.945 812	13.629	9/2	−0.6409	+3.79
	180	179.946 545	35.100	0	—	—
Ta	180	179.947 462	0.012	8	+4.77	—
	181	180.947 992	99.998	7/2	+2.370	+3.3
W	180	179.946 701	0.12	0	—	—
	182	181.948 202	26.3	0	—	—

Appendix B (*cont.*)

nucleus	isotope	atomic mass	natural abundance (%)	spin ($h/2\pi$)	magnetic moment (μ_N)	electric quadrupole moment (10^{-24} cm^2)
W	183	182.950 220	14.28	1/2	+0.117 784 8	—
	184	183.950 928	30.7	0	—	—
	186	185.954 357	28.6	0	—	—
Re	185	184.952 951	37.40	5/2	+3.1871	+2.18
	187	186.955 744	62.60	5/2	+3.2197	+2.07
Os	184	183.952 488	0.02	0	—	—
	186	185.953 830	1.58	0	—	—
	187	186.955 741	1.6	1/2	+0.064 651 9	—
	188	187.955 860	13.3	0	—	—
	189	188.958 137	16.1	3/2	+0.659 93	+0.86
	190	189.958 436	26.4	0	—	—
	192	191.961 467	41.0	0	—	—
Ir	191	190.960 584	37.3	3/2	+0.151	+0.82
	193	192.962 917	62.7	3/2	+0.164	+0.75
Pt	190	189.959 917	0.01	0	—	—
	192	191.961 019	0.79	0	—	—
	194	193.962 655	32.9	0	—	—
	195	194.964 766	33.8	1/2	+0.6095	—
	196	195.964 926	25.3	0	—	—
	198	197.967 869	7.2	0	—	—
Au	197	196.966 543	100.0	3/2	+0.145 75	+0.55
Hg	196	195.965 807	0.15	0	—	—
	198	197.966 743	9.97	0	—	—
	199	198.968 254	16.87	1/2	+0.505 885	—
	200	199.968 300	23.10	0	—	—
	201	200.970 277	13.18	3/2	−0.560 226	+0.39
	202	201.970 617	29.86	0	—	—
	204	203.973 467	6.87	0	—	—
Tl	203	202.972 320	29.524	1/2	+1.622 258	—
	205	204.974 401	70.476	1/2	+1.638 215	—
Pb	204	203.973 020	1.4	0	—	—
	206	205.974 440	24.1	0	—	—
	207	206.975 872	22.1	1/2	+0.582 58	—
	208	207.976 627	52.4	0	—	—
Bi	209	208.980 374	100.0	9/2	+4.111	−0.37

Appendix C
Compilation of Wigner 3-j symbols

k	Symbol	Analytical expansion
0	$\begin{pmatrix} J & 0 & J \\ -M & 0 & M \end{pmatrix}$	$(-1)^{J-M}(2J+1)^{-1/2}$
1	$\begin{pmatrix} J & 1 & J \\ -M & 0 & M \end{pmatrix}$	$(-1)^{J-M}\dfrac{M}{[J(J+1)(2J+1)]^{1/2}}$
	$\begin{pmatrix} J & 1 & J \\ -M\mp 1 & \pm 1 & M \end{pmatrix}$	$\pm(-1)^{J-M}\left[\dfrac{(J\mp M)(J\pm M+1)}{2J(J+1)(2J+1)}\right]^{1/2}$
	$\begin{pmatrix} J+1 & 1 & J \\ -M & 0 & M \end{pmatrix}$	$(-1)^{J-M+1}\left[\dfrac{2(J+M+1)(J-M+1)}{(2J+1)(2J+2)(2J+3)}\right]^{1/2}$
	$\begin{pmatrix} J+1 & 1 & J \\ -M\mp 1 & \pm 1 & M \end{pmatrix}$	$(-1)^{J-M}\left[\dfrac{(J\pm M+1)(J\pm M+2)}{(2J+1)(2J+2)(2J+3)}\right]^{1/2}$
2	$\begin{pmatrix} J & 2 & J \\ -M & 0 & M \end{pmatrix}$	$(-1)^{J-M}\dfrac{2[3M^2-J(J+1)]}{[(2J-1)(2J)(2J+1)(2J+2)(2J+3)]^{1/2}}$
	$\begin{pmatrix} J & 2 & J \\ -M\mp 1 & \pm 1 & M \end{pmatrix}$	$(-1)^{J-M}(1\pm 2M)\left[\dfrac{6(J\pm M+1)(J\mp M)}{(2J-1)(2J)(2J+1)(2J+2)(2J+3)}\right]^{1/2}$
	$\begin{pmatrix} J & 2 & J \\ -M\mp 2 & \pm 2 & M \end{pmatrix}$	$(-1)^{J-M}\left[\dfrac{6(J\mp M-1)(J\mp M)(J\pm M+1)(J\pm M+2)}{(2J-1)(2J)(2J+1)(2J+2)(2J+3)}\right]^{1/2}$

Appendix C (cont.)

k	Symbol	Analytical expansion
2	$\begin{pmatrix} J+1 & 2 & J \\ -M & 0 & M \end{pmatrix}$	$(-1)^{J-M+1} 2M \left[\dfrac{6(J+M+1)(J-M+1)}{2J(2J+1)(2J+2)(2J+3)(2J+4)} \right]^{1/2}$
	$\begin{pmatrix} J+1 & 2 & J \\ -M\mp 1 & \pm 1 & M \end{pmatrix}$	$\mp(-1)^{J-M} 2(J\mp 2M) \left[\dfrac{(J\pm M+1)(J\pm M+2)}{2J(2J+1)(2J+2)(2J+3)(2J+4)} \right]^{1/2}$
	$\begin{pmatrix} J+1 & 2 & J \\ -M\mp 2 & \pm 2 & M \end{pmatrix}$	$\pm(-1)^{J-M} 2 \left[\dfrac{(J\pm M+1)(J\pm M+2)(J\pm M+3)(J\mp M)}{2J(2J+1)(2J+2)(2J+3)(2J+4)} \right]^{1/2}$
	$\begin{pmatrix} J+2 & 2 & J \\ -M & 0 & M \end{pmatrix}$	$(-1)^{J-M} \left[\dfrac{6(J+M+2)(J+M+1)(J-M+2)(J-M+1)}{(2J+1)(2J+2)(2J+3)(2J+4)(2J+5)} \right]^{1/2}$
	$\begin{pmatrix} J+2 & 2 & J \\ -M\mp 1 & \pm 1 & M \end{pmatrix}$	$(-1)^{J-M+1} 2 \left[\dfrac{(J\mp M+1)(J\pm M+1)(J\pm M+2)(J\pm M+3)}{(2J+1)(2J+2)(2J+3)(2J+4)(2J+5)} \right]^{1/2}$
	$\begin{pmatrix} J+2 & 2 & J \\ -M\mp 2 & \pm 2 & M \end{pmatrix}$	$(-1)^{J-M} \left[\dfrac{(J\pm M+1)(J\pm M+2)(J\pm M+3)(J\pm M+4)}{(2J+1)(2J+2)(2J+3)(2J+4)(2J+5)} \right]^{1/2}$
3	$\begin{pmatrix} J & 3 & J \\ -M & 0 & M \end{pmatrix}$	$(-1)^{J-M+1} \dfrac{4M[3J(J+1)-5M^2-1]}{[(2J-2)(2J-1)(2J)(2J+1)(2J+2)(2J+3)(2J+4)]^{1/2}}$
	$\begin{pmatrix} J & 3 & J \\ -M\mp 1 & \pm 1 & M \end{pmatrix}$	$\mp(-1)^{J-M} \dfrac{[J(J+1)\mp 5M(\pm M+1)-2][12(J\pm M+1)(J\mp M)]^{1/2}}{[(2J-2)(2J-1)(2J)(2J+1)(2J+2)(2J+3)(2J+4)]^{1/2}}$
	$\begin{pmatrix} J & 3 & J \\ -M\mp 2 & \pm 2 & M \end{pmatrix}$	$\pm(-1)^{J-M} \dfrac{2(\pm M+1)[30(J\pm M+2)(J\pm M+1)(J\mp M)(J\mp M-1)]^{1/2}}{[(2J-2)(2J-1)(2J)(2J+1)(2J+2)(2J+3)(2J+4)]^{1/2}}$

Appendix C (cont.)

k	Symbol	Analytical expansion
3	$\begin{pmatrix} J & 3 & J \\ -M\mp 3 & \pm 3 & M \end{pmatrix}$	$\pm(-1)^{J-M}\left[\dfrac{20(J\pm M+3)(J\pm M+2)(J\pm M+1)(J\mp M)(J\mp M-1)(J\mp M-2)}{(2J-2)(2J-1)(2J)(2J+1)(2J+2)(2J+3)(2J+4)}\right]^{1/2}$
	$\begin{pmatrix} J+1 & 3 & J \\ -M & 0 & M \end{pmatrix}$	$(-1)^{J-M}\dfrac{[J(J+2)-5M^2][12(J-M+1)(J+M+1)]^{1/2}}{[(2J-1)(2J)(2J+1)(2J+2)(2J+3)(2J+4)(2J+5)]^{1/2}}$
	$\begin{pmatrix} J+1 & 3 & J \\ -M\mp 1 & \pm 1 & M \end{pmatrix}$	$-(-1)^{J-M}\dfrac{[(J+2)(J-5)-5(\mp M-1)(2J-1)](2J\pm 1)(J\pm M+1)(J\pm M+2)][(J\pm M+1)(J\pm M+2)]^{1/2}}{[(2J-1)(2J)(2J+1)(2J+2)(2J+3)(2J+4)(2J+5)]^{1/2}}$
	$\begin{pmatrix} J+1 & 3 & J \\ -M\mp 2 & \pm 2 & M \end{pmatrix}$	$-(-1)^{J-M}\dfrac{(J\mp 3M-2)[10(J\mp M)(J\pm M+3)(J\pm M+2)(J\pm M+1)(J\mp M-1)]^{1/2}}{[(2J-1)(2J)(2J+1)(2J+2)(2J+3)(2J+4)(2J+5)]^{1/2}}$
	$\begin{pmatrix} J+1 & 3 & J \\ -M\mp 3 & \pm 3 & M \end{pmatrix}$	$(-1)^{J-M}\left[\dfrac{15(J\pm M+4)(J\pm M+3)(J\pm M+2)(J\pm M+1)(J\mp M)(J\mp M-1)}{(2J-1)(2J)(2J+1)(2J+2)(2J+3)(2J+4)(2J+5)}\right]^{1/2}$
	$\begin{pmatrix} J+2 & 3 & J \\ -M & 0 & M \end{pmatrix}$	$(-1)^{J-M}2M\left[\dfrac{30(J+M+2)(J+M+1)(J-M+2)(J-M+1)}{(2J)(2J+1)(2J+2)(2J+3)(2J+4)(2J+5)(2J+6)}\right]^{1/2}$
	$\begin{pmatrix} J+2 & 3 & J \\ -M\mp 1 & \pm 1 & M \end{pmatrix}$	$\pm(-1)^{J-M}(J\mp 3M)\left[\dfrac{10(J\mp M+1)(J\pm M+3)(J\pm M+2)(J\pm M+1)}{(2J)(2J+1)(2J+2)(2J+3)(2J+4)(2J+5)(2J+6)}\right]^{1/2}$
	$\begin{pmatrix} J+2 & 3 & J \\ -M\mp 2 & \pm 2 & M \end{pmatrix}$	$\mp(-1)^{J-M}(J\mp 3M/2)\left[\dfrac{16(J\pm M+4)(J\pm M+3)(J\pm M+2)(J\pm M+1)}{(2J)(2J+1)(2J+2)(2J+3)(2J+4)(2J+5)(2J+6)}\right]^{1/2}$
	$\begin{pmatrix} J+2 & 3 & J \\ -M\mp 3 & \pm 3 & M \end{pmatrix}$	$\pm(-1)^{J-M}\left[\dfrac{6(J\pm M+5)(J\pm M+4)(J\pm M+3)(J\pm M+2)(J\pm M+1)(J\mp M)}{(2J)(2J+1)(2J+2)(2J+3)(2J+4)(2J+5)(2J+6)}\right]^{1/2}$

Appendix C (*cont.*)

k	Symbol	Analytical expansion
3	$\begin{pmatrix} J+3 & 3 & J \\ -M & 0 & M \end{pmatrix}$	$-(-1)^{J-M} \left[\dfrac{20(J+M+3)(J+M+2)(J+M+1)(J-M+3)(J-M+2)(J-M+1)}{(2J+1)(2J+2)(2J+3)(2J+4)(2J+5)(2J+6)(2J+7)} \right]^{1/2}$
	$\begin{pmatrix} J+3 & 3 & J \\ -M\mp1 & \pm1 & M \end{pmatrix}$	$(-1)^{J-M} \left[\dfrac{15(J\mp M+2)(J\mp M+1)(J\pm M+4)(J\pm M+3)(J\pm M+2)(J\pm M+1)}{(2J+1)(2J+2)(2J+3)(2J+4)(2J+5)(2J+6)(2J+7)} \right]^{1/2}$
	$\begin{pmatrix} J+3 & 3 & J \\ -M\mp2 & \pm2 & M \end{pmatrix}$	$-(-1)^{J-M} \left[\dfrac{6(J\mp M+1)(J\pm M+5)(J\pm M+4)(J\pm M+3)(J\pm M+2)(J\pm M+1)}{(2J+1)(2J+2)(2J+3)(2J+4)(2J+5)(2J+6)(2J+7)} \right]^{1/2}$
	$\begin{pmatrix} J+3 & 3 & J \\ -M\mp3 & \pm3 & M \end{pmatrix}$	$(-1)^{J-M} \left[\dfrac{(J\pm M+6)(J\pm M+5)(J\pm M+4)(J\pm M+3)(J\pm M+2)(J\pm M+1)}{(2J+1)(2J+2)(2J+3)(2J+4)(2J+5)(2J+6)(2J+7)} \right]^{1/2}$

Appendix D
Compilation of Wigner 6-j symbols

Note: in the following expressions, $s = a + b + c$.

k	Symbol	Analytical expansion
0	$\begin{Bmatrix} b & c & a \\ c & b & 0 \end{Bmatrix}$	$(-1)^s [(2b+1)(2c+1)]^{-1/2}$
1	$\begin{Bmatrix} b & c & a \\ c-1 & b-1 & 1 \end{Bmatrix}$	$(-1)^s \left[\dfrac{s(s+1)(s-2a-1)(s-2a)}{(2b-1)(2b)(2b+1)(2c-1)(2c)(2c+1)} \right]^{1/2}$
	$\begin{Bmatrix} b & c & a \\ c-1 & b & 1 \end{Bmatrix}$	$(-1)^s \left[\dfrac{2(s+1)(s-2a)(s-2b)(s-2c+1)}{2b(2b+1)(2b+2)(2c-1)(2c)(2c+1)} \right]^{1/2}$
	$\begin{Bmatrix} b & c & a \\ c-1 & b+1 & 1 \end{Bmatrix}$	$(-1)^s \left[\dfrac{(s-2b-1)(s-2b)(s-2c+1)(s-2c+2)}{(2b+1)(2b+2)(2b+3)(2c-1)(2c)(2c+1)} \right]^{1/2}$
	$\begin{Bmatrix} b & c & a \\ c & b & 1 \end{Bmatrix}$	$(-1)^{s+1} \dfrac{2[b(b+1)+c(c+1)-a(a+1)]}{[2b(2b+1)(2b+2)(2c)(2c+1)(2c+2)]^{1/2}}$
3/2	$\begin{Bmatrix} b & c & a \\ c-3/2 & b-3/2 & 3/2 \end{Bmatrix}$	$(-1)^s \left[\dfrac{(s-1)s(s+1)(s-2a-2)(s-2a-1)(s-2a)}{(2b-2)(2b-1)2b(2b+1)(2c-2)(2c-1)2c(2c+1)} \right]^{1/2}$
	$\begin{Bmatrix} b & c & a \\ c-3/2 & b-1/2 & 3/2 \end{Bmatrix}$	$(-1)^s \left[\dfrac{3s(s+1)(s-2a-1)(s-2a)(s-2b)(s-2c+1)}{(2b-1)2b(2b+1)(2b+2)(2c-2)(2c-1)2c(2c+1)} \right]^{1/2}$
	$\begin{Bmatrix} b & c & a \\ c-3/2 & b+1/2 & 3/2 \end{Bmatrix}$	$(-1)^s \left[\dfrac{3(s+1)(s-2a)(s-2b-1)(s-2b)(s-2c+1)(s-2c+2)}{2b(2b+1)(2b+2)(2b+3)(2c-2)(2c-1)2c(2c+1)} \right]^{1/2}$
	$\begin{Bmatrix} b & c & a \\ c-3/2 & b+3/2 & 3/2 \end{Bmatrix}$	$(-1)^s \left[\dfrac{(s-2b-2)(s-2b-1)(s-2b)(s-2c+1)(s-2c+2)(s-2c+3)}{(2b+1)(2b+2)(2b+3)(2b+4)(2c-2)(2c-1)2c(2c+1)} \right]^{1/2}$

Appendix D (cont.)

k	Symbol	Analytical expansion
3/2	$\left\{\begin{array}{ccc} b & c & a \\ c-1/2 & b-1/2 & 3/2 \end{array}\right\}$	$(-1)^s \dfrac{[2(s-2b)(s-2c)-(s+2)(s-2a-1)][(s+1)(s-2a)]^{1/2}}{[(2b-1)2b(2b+1)(2b+2)(2c-1)2c(2c+1)(2c+2)]^{1/2}}$
	$\left\{\begin{array}{ccc} b & c & a \\ c-1/2 & b+1/2 & 3/2 \end{array}\right\}$	$(-1)^s \dfrac{[(s-2b-1)(s-2c)-2(s+2)(s-2a)][(s-2b)(s-2c-2)+1]^{1/2}}{[2b(2b+1)(2b+2)(2b+3)(2c-1)2c(2c+1)(2c+2)]^{1/2}}$
2	$\left\{\begin{array}{ccc} b & c & a \\ c-2 & b-2 & 2 \end{array}\right\}$	$(-1)^s \left[\dfrac{(s-2)(s-1)s(s+1)(s-2a-3)(s-2a-2)(s-2a-1)(s-2a)}{(2b-3)(2b-2)(2b-1)2b(2b+1)(2c-3)(2c-2)(2c-1)2c(2c+1)}\right]^{1/2}$
	$\left\{\begin{array}{ccc} b & c & a \\ c-2 & b-1 & 2 \end{array}\right\}$	$(-1)^s \cdot 2 \dfrac{(s-1)s(s+1)(s-2a-2)(s-2a-1)(s-2a)(s-2b)(s-2c+1)}{(2b-2)(2b-1)2b(2b+1)(2b+2)(2c-3)(2c-2)(2c-1)2c(2c+1)}\right]^{1/2}$
	$\left\{\begin{array}{ccc} b & c & a \\ c-2 & b & 2 \end{array}\right\}$	$(-1)^s \left[\dfrac{6s(s+1)(s-2a-1)(s-2a)(s-2b-1)(s-2b)(s-2c+1)(s-2c+2)}{(2b-1)2b(2b+1)(2b+2)(2b+3)(2c-3)(2c-2)(2c-1)2c(2c+1)}\right]^{1/2}$
	$\left\{\begin{array}{ccc} b & c & a \\ c-2 & b+1 & 2 \end{array}\right\}$	$(-1)^s \cdot 2 \left[\dfrac{(s+1)(s-2a)(s-2b-2)(s-2b-1)(s-2b)(s-2c+1)(s-2c+2)(s-2c+3)}{2b(2b+1)(2b+2)(2b+3)(2b+4)(2c-3)(2c-2)(2c-1)2c(2c+1)}\right]^{1/2}$
	$\left\{\begin{array}{ccc} b & c & a \\ c-2 & b+2 & 2 \end{array}\right\}$	$(-1)^s \left[\dfrac{(s-2b-3)(s-2b-2)(s-2b-1)(s-2b)(s-2c+1)(s-2c+2)(s-2c+3)(s-2c+4)}{(2b+1)(2b+2)(2b+3)(2b+4)(2b+5)(2c-3)(2c-2)(2c-1)2c(2c+1)}\right]^{1/2}$
	$\left\{\begin{array}{ccc} b & c & a \\ c-1 & b-1 & 2 \end{array}\right\}$	$(-1)^s \dfrac{4[(a+b)(a-b+1)-(c-1)(c-b+1)][s(s+1)(s-2a-1)(s-2a)]^{1/2}}{[(2b-2)(2b-1)2b(2b+1)(2b+2)(2c-2)(2c-1)2c(2c+1)(2c+2)]^{1/2}}$
	$\left\{\begin{array}{ccc} b & c & a \\ c-1 & b & 2 \end{array}\right\}$	$(-1)^s \cdot 2 \dfrac{[(a+b+1)(a-b)-c^2+1][6(s+1)(s-2a)(s-2b)(s-2c+1)]^{1/2}}{[(2b-1)2b(2b+1)(2b+2)(2b+3)(2c-2)(2c-1)2c(2c+1)(2c+2)]^{1/2}}$
	$\left\{\begin{array}{ccc} b & c & a \\ c-1 & b+1 & 2 \end{array}\right\}$	$(-1)^s \dfrac{4[(a+b+2)(a-b-1)-(c-1)(b+c+2)][(s-2b-1)(s-2b)(s-2c+1)(s-2c+2)]^{1/2}}{[2b(2b+1)(2b+2)(2b+3)(2b+4)(2c-2)(2c-1)2c(2c+1)(2c+2)]^{1/2}}$
	$\left\{\begin{array}{ccc} b & c & a \\ c & b & 2 \end{array}\right\}$	$(-1)^s \dfrac{2[3X(X-1)-4b(b+1)c(c+1)]}{[(2b-1)2b(2b+1)(2b+2)(2b+3)(2c-1)2c(2c+1)(2c+2)(2c+3)]^{1/2}}$

In the final symbol listed, $X = b(b+1) + c(c+1) - a(a+1)$.

Appendix E
Relationships between cgs and SI units

The vacuum permeability μ_0 and vacuum permittivity ε_0 are related through the equation

$$\mu_0 = 1/\varepsilon_0 c^2.$$

We then have the results listed below.

quantity	cgs	SI
elementary charge	q	$q/(4\pi\varepsilon_0)^{1/2}$
electric field strength	E	$E(4\pi\varepsilon_0)^{1/2}$
electric potential	ϕ	$\phi(4\pi\varepsilon_0)^{1/2}$
magnetic flux density	$\boldsymbol{B}$	$\boldsymbol{B}/(\mu_0/4\pi)^{1/2}$
magnetic vector potential	$\boldsymbol{A}$	$\boldsymbol{A}/(\mu_0/4\pi)^{1/2}$
Bohr magneton	μ_B	$\mu_B(\mu_0/4\pi)^{1/2}$

Author index

Abragam, A., 29, 36
Abramowitz, M., 279, 285, 300, 301
Ahmed, F., 913, 976
Albritton, D.L., 282, 301, 343, 369
Alexander, M.L., 958, 959, 977
Aliev, M.R., 313, 369
Allen, M.D., 841–843, 845, 846, 848–850, 854, 856, 868, 868, 909, 912, 975
Alpert, B.D., 733, 864
Altman, R.S., 736, 864
Amano, T., 691, 726–728, 736, 756, 757, 759, 799–801, 862–868, 812, 814, 854, 900
Amiot, C., 787, 866
Anderson, C.D., 77, 122, 798, 866
Anderson, P.W., 275, 300
Anderson, T.G., 552, 578, 746–748, 835, 836, 865, 867
Appelbad, O., 364, 370
Arrington, C.A., 271, 300, 590, 680, 778, 779, 866
Asgharian, A., 744, 864
Athenour, C., 919, 976
Auty, A.R., 738, 740, 741, 864
Azuma, Y., 934, 935, 976

Babb, J.F., 973, 977
Bacis, R., 787, 859, 866, 869, 938, 976
Baird, J.C., 360, 370
Baker, M.R., 422, 489, 496, 575, 577
Balfour, W.J., 841, 868
Ball, J.A., 542, 578, 722, 863
Balle, T.J., 704, 708–710, 862
Bally, J., 722, 863
Bally, T., 218, 299
Barber, W.G., 291, 301
Bardeen, J., 123, 138
Barnes, R.G., 388, 574
Barrett, A.H., 713, 722, 790, 862, 863
Barrow, R.F., 913, 976
Bates, D.R., 224, 291, 300, 301
Bauschlicher, C.W., 918, 921, 976
Beaton, S.P., 666, 668–670, 672–677, 682, 927, 976
Beaudet, R.A., 542, 578
Becke, A.D., 219, 299

Beer, B.S., 552, 578
Begemann, M.H., 699–701, 736, 862
Belov, S.P., 733, 734, 742, 743, 759, 761, 779, 783, 784, 787, 864–866
Bendazzoli, G.L., 549, 578
Bender, C.F., 501, 577
Benedict, W.S., 506, 577
Benesch, W., 447, 461, 575
Bergeman, T.H., 876–878, 880, 883, 975
Beringer, R., 13, 36, 612, 681, 754, 759, 865
Bernadi, F., 549, 578
Bernard, A., 938, 976
Bernath, P.F., 32, 33, 37, 736, 836, 852, 854, 864, 868
Bernstein, H.J., 29, 37, 392, 575
Bernstein, R.B., 284, 301
Berrah Mansour, N., 953–955, 977
Bessis, N., 938, 976
Bethe, H.A., 109, 122, 123, 130, 138
Bird, R.B., 282, 301
Birge, R.T., 284, 301, 552, 578
Bishop, D.M., 219, 223, 224, 299, 300, 973, 977
Bitter, F., 870, 974
Blake, G.A., 722, 723, 790, 863, 866
Bleaney, B., 29, 36
Blinder, S.M., 507, 577
Bloch, C., 305, 369
Bogey, M., 742, 798, 837, 864, 866, 868
Bolef, D.I., 123, 138, 422, 575
Bonczyk, P.A., 487, 576
Borden, W.T., 218, 299
Born, M., 38, 59, 72, 213, 221, 299, 302, 316, 369
Boyd, T.L., 761, 865
Boys, S.F., 216, 299
Bradford, R.S., 883, 975
Bray, P.J., 388, 574
Brazier, C.R., 794, 796, 798, 805, 807, 810, 836, 866–868, 913–916, 976
Breit, G., 104, 122, 123, 125, 138
Breivogel, F.W. Jr., 487, 576
Brillouin, L., 277, 300
Brink, D.M., 32, 37, 139, 163, 175
Broida, H.P., 586, 680, 871–875, 883, 974, 975

Brom, J.M., 844, 846, 868
Brooks, P.R., 440, 444–446, 575, 900, 975
Brooks, R.A., 422, 575
Brossel, J., 870, 974
Brown, J.M., 169, 176, 251, 300, 303, 327, 331, 336, 338, 341, 343, 351, 352, 354, 356, 359, 360, 364, 368–370, 512, 526, 530, 537, 542, 545, 548, 552, 577, 578, 587, 589, 590, 608, 616, 618–624, 630, 631, 633, 634, 655–664, 666–670, 672–677, 680, 682, 728, 738, 740, 741, 790, 791, 794, 796–799, 804, 805, 807, 810, 820, 839, 841, 849, 854, 856, 864, 866–868, 909, 912–916, 927–930, 948–950, 975–977
Brown, R.L., 537, 578, 612, 613, 681
Brown, R.D., 958, 977
Budó, A., 552, 578
Bukowski, R., 224, 300
Buenker, R.J., 352, 368–370
Bunker, P.R., 32, 37, 40, 64, 72, 256, 300, 345, 369, 737, 823, 859, 864, 867, 869
Burke, B.F., 713, 722, 863
Burrus, C.A., 526, 537, 577
Busarow, K.L., 723, 863
Buxton, L.W., 710, 862
Byfleet, C.R., 359, 369, 596, 608, 680, 681

Cade, P.E., 501, 508, 577, 958, 977
Cahill, P., 428, 440, 444, 575, 799, 804, 867, 895, 898, 975
Campbell, E.J., 710, 862
Cao, J., 841, 868
Carballo, N., 835, 836, 868
Carrington, A., 29, 30, 32, 36, 37, 219, 220, 223, 228, 230, 288, 299–301, 352, 366, 368–370, 582–584, 587, 589, 596, 597, 599, 607, 608, 617, 641, 646, 648, 649, 652, 680, 681, 730, 736, 749, 777, 779, 811–817, 821, 824, 829, 832, 833, 858, 859, 864–868, 942–944, 946, 948–950, 952, 969, 976, 977
Carroll, P.K., 447, 575
Carson, T.R., 224, 300
Cashion, J.K., 280, 301
Casoli, F., 787, 866
Castle, J.G., 13, 36, 754, 759, 865
Cazzoli, G., 778, 866
Cecchi, J.L., 483, 485, 576
Cederberg, J.W., 487, 576
Cernicharo, J., 722, 836, 837, 839, 840, 863, 868
Certain, P.R., 973, 977
Chan, S.I., 422, 575
Chance, K.V., 728, 864
Chang, W.-L., 917, 976
Chen, D.-W., 733, 864
Cheung, A. S.-C., 676, 682, 841, 849, 868, 909, 928, 930, 975, 976
Cheung, L.M., 219, 223, 224, 299, 300
Cheval, G., 844, 868
Child, M.S., 66, 72, 276, 279, 300
Childs, W.J., 919–921, 931, 933–935, 938–941, 976

Chiu, C.C.W., 778, 866
Chojnicki, A.H., 913, 976
Chow Chiu, L.-Y., 432–436, 575, 894, 975
Chraplyvy, Z.V., 105, 122
Christy, A., 248, 300, 331, 369, 525, 529, 577, 618, 681, 807, 808, 867
Churassy, S., 859, 869
Civis, S., 742, 864
Clark, W.W., 759, 865
Clementi, E., 195, 298
Code, R.F., 388, 489, 503, 574, 577
Cohen, E.A., 367, 370, 778, 866
Cohen, M., 973, 977
Cohen, R.C., 723, 863
Colbourn, E.A., 303, 331, 354, 360, 369, 370, 657, 682
Cole, A.R.H., 526, 537, 577
Collomb, R., 938, 976
Comben, E.R., 552, 578, 624, 655–658, 681, 913, 976
Condon, E.U., 144, 175, 186, 298
Connelly, J.P., 738, 740–741, 864
Cook, H.M., 360, 370
Cook, R.L., 366, 370, 759, 865
Cooley, J.P., 13, 36
Cooley, J.W., 280, 301
Coolidge, A.S., 209, 299
Cooper, D.L., 525, 577
Corney, A., 258, 300
Cornwell, C.D., 496, 504, 577
Corkery, S.M., 666, 668, 682
Cote, R.E., 422, 575
Coulson, C.A., 209, 286, 299, 301, 858, 868
Cowan, M., 733, 864
Cox, S.G., 730, 864
Coxon, J.A., 607, 680
Critchley, A.D.J., 730, 856, 861, 864, 868
Crofton, M.W., 736, 864
Crofts, J.G., 958, 977
Crutcher, R.M., 551, 578
Curl, R.F., 668, 682
Curran, A.H., 591, 593, 596, 680
Currie, G.N., 584, 649, 652, 680, 681
Curtiss, C.F., 282, 301

Dabbousi, O.B., 489, 577
Dabrowski, I., 816, 867
Dailey, B.P., 366, 370
Dale, R.M., 537, 578
Dalgarno, A., 973, 977
Darwin, C.G., 85, 122
Davidson, E.R., 209, 299
Davidson, S.A., 799, 805, 867
Davies, P.B., 652, 681
Davis, R.E., 489, 576
Degli Esposti, C., 778, 866
Dehmelt, H.G., 961, 962, 964, 977
Delaval, J.M., 845, 868
Delcroix, B., 742, 864

de Leeuw, F.H., 487, 489, 496–500, 502, 508, 527, 576, 577
De Lucia, F.C., 743, 744, 759, 864, 865
Demtröder, W., 32, 37, 275, 300
Demuynck, C., 742, 798, 837, 864, 866, 868
Di Lauro, C., 352, 368–370
De Santis, D., 25, 36, 303, 369, 446–454, 459, 461, 462, 575
Destombes, J.L., 798, 837, 866, 868
Dickinson, D.F., 542, 578, 974, 977
Dickinson, J.T., 481, 576
Dilonardo, G., 728, 864
Dirac, P.A.M., 73, 74, 122
Dixon, R.N., 352, 358, 368–370
Dixon, T.A., 552, 578, 638, 681, 695, 697, 745–748, 750, 798, 835, 836, 862, 865–867
Doermann, F.W., 123, 125, 138
Domaille, P.J., 902–904, 975
Doose, J., 713, 862
Douglas, A.E., 713, 722, 862
Dousmanis, G.C., 29, 37, 525, 537, 538, 552, 577–579, 680, 692, 694, 788, 789, 862, 866
Dowling, J.M., 526, 537, 577
Drabbels, M., 723, 863
Dufour, C., 845, 868, 913, 976
Dulick, M., 364, 370
Dunham, J.L., 63, 65, 72, 243, 244, 282, 300, 301, 339, 345, 369, 501, 577
Dunn, J.L., 871–874, 974
Dunn, T.M., 232, 300
Durham, C.H., 844, 846, 868
Durie, R.A., 607, 680
Dyer, P.N., 597, 607, 608, 680, 812, 814, 867
Dymanus, A., 481, 487, 489, 496–500, 502, 508, 526–529, 533, 535, 537, 539, 542, 548, 549, 576–578, 723, 782, 787, 789, 811, 863, 866, 867

Eastman, D.P., 502, 577
Edmonds, A.R., 32, 37, 139, 157, 163, 175, 678, 680, 682
Ellder, J., 624, 681, 794, 866
Ellinger, Y., 787, 866
Endo, Y., 752, 763, 764, 767–770, 773, 775, 811, 814, 853, 865, 867, 868, 909, 912, 975
English, A.D., 883, 884, 975
English, T.C., 468, 469, 471, 473, 475, 481, 575, 576
Erickson, N.R., 746, 798, 865, 866
Ernst, W.E., 904–908, 975
Evenson, K.M., 338, 356, 364, 369, 370, 533, 536, 537, 544, 552, 577, 578, 584–586, 591, 609, 610, 622–624, 628–631, 633, 634, 655–658, 661–664, 666, 668–670, 672–677, 680–682, 726, 728, 729, 757, 778, 790, 791, 794, 799, 805, 810, 811, 863–867, 871–875, 913, 927, 974, 976
Ewen, H.I., 713, 862
Eyring, H., 207, 208, 299

Falick, A.M., 271, 300, 590, 680, 778, 779, 866
Farhoomand, J., 790, 866

Favero, P.G., 526, 537, 577, 778, 866
Féménias, J.-L., 844, 868, 919, 976
Fermi, E., 123, 138
Ferrari, C.A., 609, 611, 681
Field, R.W., 32, 37, 230, 276, 300, 313, 352, 368–370, 836, 868, 876–878, 880, 883, 884, 975
Fink, U., 722, 863
Fischer, C.F., 283, 301
Fischer, E., 962, 977
Fisk, G.A., 481, 482, 576
Fletcher, D.A., 677, 682, 927–933, 976
Flygare, W.H., 4, 32, 36, 37, 190, 298, 393, 504, 577, 704, 708–710, 862
Fock, V., 74, 122
Foldy, L.L., 77, 122
Foley, H.M., 123, 129, 138, 524, 577, 607, 672, 677, 680, 682, 810, 813, 840, 853, 867, 868, 875, 898, 919, 966, 975–977
Fontana, P.R., 432, 434, 435, 437, 575, 894, 975
Fournier, P.G., 736, 864
Fraser, G.T., 549–551, 578, 710, 738, 739, 842, 844, 845, 862, 864, 868
Freed, K.F., 553, 557, 578, 836, 868
Freeman, R.R., 745, 864
Frerking, M.A., 790, 866
Freund, R.S., 25, 36, 303, 369, 446–454, 459, 461, 462, 553, 575, 578, 885, 887–893, 895, 896, 898, 899–901, 975
Freund, S.M., 481, 482, 509, 511, 517, 523, 524, 526, 576, 577
Frisch, M.J., 218, 299
Freed, K., 553, 578
Frosch, R.A., 123, 129, 138, 524, 577, 607, 672, 677, 680, 682, 810, 813, 840, 853, 867, 868, 875, 898, 919, 966, 975–977
Fu, Z.W., 974, 977
Fujitake, M., 525, 577, 697–699, 814, 862, 867

Gaily, T.D., 416–421, 575, 906, 931, 942, 975–977, 955, 958
Gallagher, T.F., 483, 484, 486, 576
Gallagher, J.J., 521, 526, 577
Gammie, D.I., 969, 977
Gammon, R.H., 553, 555, 557, 558, 578, 749, 865
Gautier, T.N., 722, 863
Geiger, J.S., 624, 681
Geller, M., 367, 370
Gendriesch, R., 723, 863
Genzel, R., 734, 791, 792, 799, 866, 867
Gerig, J.T., 414, 575
Gerin, M., 787, 866
German, K.R., 876, 880, 882, 883, 975
Gero, L., 552, 578
Gerry, M.C.L., 741, 742, 864, 917–919, 976
Gillam, C.M., 286, 301
Gilliam, O.R., 733, 864
Gillies, C.W., 710, 842, 844, 845, 862, 868
Gillies, J.Z., 710, 862
Gilmore, W., 722, 863
Ginter, M.L., 892, 975

Godfrey, P.D., 958, 977
Goodman, G.L., 938–941, 953–955, 976, 977
Goodman, L.S., 931, 938–941, 976
Goodridge, D., 927, 976
Gold, L.P., 481, 487, 576
Gordon, R.J., 139, 175
Gordon, W., 74, 122
Gordy, W., 13, 36, 366, 370, 526, 537, 577, 695, 733, 743, 757, 759, 862, 864, 865
Goto, M., 745, 779, 781, 798, 865, 866
Gottlieb, C.A., 542, 578, 722, 836, 838–840, 863, 868,
Gottlieb, E.W., 722, 863
Grabner, L., 487, 576
Grabow, J.-U., 713, 862
Graff, G., 479, 481, 487, 576
Graham-Smith, F., 713, 863
Gray, B.F., 291, 301
Griffith, J.S., 303, 369
Gruebele, M.H.W., 655, 681, 701, 862
Guarnieri, A., 713, 862
Gudeman, C.S., 699–701, 736, 746–748, 798, 835, 836, 862, 865, 866, 868
Guelachvili, G., 787, 866
Guélin, M., 722, 836, 837, 839, 840, 863, 868

Haese, N.N., 736, 864
Hagen, J.P., 713, 862
Hahn, J., 713, 862
Hall, G.G., 214, 299
Hall, J.A., 361, 370
Hall, R.T., 526, 537, 577
Hannabuss, K., 32, 37, 73, 122
Hargreaves, W.A., 123, 138
Harrick, N.J., 385, 388, 574
Harris, D.O., 883, 884, 902–904, 975
Harris, F.E., 209, 299
Harrison, J.F., 927, 976
Hartmann, F., 859, 869
Hartree, D.R., 192, 298
Harvey, J.S.M., 363, 370
Hasse, H.R., 291, 301
Havenith, M., 723, 863
Hayashi, M., 759, 761, 865
Hehre, W.J., 216, 299
Heitler, W., 210, 299
Hebert, A.J., 487, 576
Henry, A.F., 612, 681, 759, 865
Henry, J.C., 713, 722, 790, 862
Hepp, M., 723, 863
Herbert, A.J., 481, 576
Herbst, E., 509, 511, 517, 523, 524, 526, 577, 743, 744, 810, 864, 867
Herman, R.M., 506, 577, 744, 864
Herschbach, D.R., 481, 482, 576
Hertel, I., 958, 959, 977
Herzberg, G., 12, 13, 32, 36, 201, 203, 213, 298, 299, 423, 552, 575, 578, 626, 681, 713, 722, 816, 859, 862, 867, 869, 885, 975

Hessels, E.A., 974, 977
Hilborn, R.C., 483, 484, 486, 576
Hill, E.L., 794, 866
Hillig, K.W., 778, 866
Hindermann, P.K., 496, 504, 577
Hinds, E.A., 936–938, 976
Hinkley, R.K., 361, 370
Hinz, A., 728, 790, 791, 864
Hinze, J., 805–808, 810, 867
Hirahara, Y., 759, 761, 865
Hirao, T., 854, 868
Hirota, E., 32, 33, 37, 525, 577, 652, 681, 697–699, 726, 752, 756, 757, 759, 763, 764, 767–770, 773, 775, 811, 812, 814, 853, 862, 863, 865, 867, 868, 909, 912, 975
Hirschfelder, J.O., 40, 72, 241, 282, 283, 300, 301
Hjalmarson, A., 794, 866
Hoeft, J., 739, 864
Hohenberg, P., 218, 299
Hollas, J.M., 32, 37, 706, 711, 862
Hollowell, C.D., 481, 487, 576
Holt, R.A., 416–421, 575, 906, 931, 942, 953–955, 958, 975–977
Honey, F.R., 526, 537, 577
Horani, M., 343, 369
Hougen, J.T., 251, 300, 320, 322, 336, 352, 368–370, 630, 659, 681, 682, 820, 867
Hovde, D.C., 609, 611, 681
Howard, B.J., 169, 176, 359, 369, 608, 680, 738, 740, 741, 811, 864, 867
Howard, R.A., 836, 868
Howells, M.H., 224, 300
Huang, K., 59, 72, 221, 299
Huber, K.P., 13, 32, 36, 251, 300, 423, 575, 659, 682, 820, 867
Hughes, A.N., 856, 861, 868
Hughes, H.K., 13, 15, 36, 463, 465, 467, 468, 481, 575, 576
Hughes, R.H., 690, 862
Hughes, V.W., 487, 576
Hulburt, H.M., 241, 300
Hund, F., 21, 36, 224, 300
Hunter, G., 291, 295, 301
Huo, W.M., 496, 501, 503, 508, 577
Hutson, J.M., 228, 230, 300, 814–817, 821, 824, 827, 829, 832, 833, 867
Hüttner, W., 352, 368–370
Hylleraas, E.A., 190, 243, 297, 298, 300, 301

Irvine, W.M., 624, 681, 787, 794, 866
Itoh, T., 89, 122

Jacobson, A.R., 487, 576, 745, 864
Jaffe, G., 292, 295, 301
James, H.M., 209, 299
Jarman, C.N., 852, 868
Jefferts, K.B., 713, 722, 749, 862, 863, 865, 900, 961–966, 970, 974, 975, 977

Jeffreys, H., 277, 300
Jennings, D.A., 630, 681, 726, 728, 729, 790, 863, 864, 884, 975
Jensen, Per 32, 37, 64, 72, 256, 300, 859, 869
Jette, A.N., 428, 440, 444, 575, 799, 804, 867, 895, 898, 975
Jeziorski, B., 224, 230
Johns, J.W.C., 251, 300, 537, 578, 659, 682, 820, 867
Johnson, B.R., 827, 867
Johnson, C.M., 521, 526, 577, 733, 864
Johnson, D.R., 691, 693, 812, 814, 862, 867
Johnson, D.W., 745, 864
Johnson, M.A., 958, 959, 977
Jost, R., 892, 975
Judd, B.R., 32, 37, 187, 298
Jung, K.Y., 922, 924, 927, 931–933, 976

Kagi, E., 759, 761, 865
Kaise, M., 351, 369, 512, 542, 545, 548, 577, 578, 616, 619–622, 656, 682
Kaiser, E.W., 489, 500–503, 507, 551, 576–578
Kanamori, H., 835, 868
Kasdan, A., 810, 867
Kaufman, M., 489, 576
Kauzmann, W., 189, 298
Kawaguchi, K., 652, 681, 759, 761, 768, 865
Kayama, K., 360, 370, 549, 578
Keenan, M.R., 710, 862
Keene, J., 722, 863
Kellogg, J.M.B., 13, 36, 123, 138, 375, 388, 391, 574
Kemble, E.C., 313, 369
Kennedy, R.A., 219, 223, 224, 288, 299–301, 736, 858, 864, 868, 946, 948, 952, 977
Kerr, C.L.M., 351, 369, 512, 542, 545, 548, 577, 578, 616, 619–624, 656, 682, 791, 866
Kewley, R., 695, 862
Khersonskii, V.K., 139, 175
Khoshla, A., 422, 489, 503, 575, 577
Kimball, G.F., 207, 208, 299
Kindt, S., 906, 908, 975
King, G.W., 283, 301
Kirchhoff, W.H., 691, 693, 862
Klaus, Th., 733, 734, 742, 743, 759, 761, 779, 864, 865
Klein, O., 74, 122, 280, 301
Kleinman, V.D., 139, 175
Klemperer, W., 481, 482, 487, 489, 490, 493, 494, 496, 509, 511, 517, 523, 524, 526, 549–551, 553, 555, 557, 576–578
Klisch, E., 783, 784, 787, 866
Knipp, J.K., 283, 301
Knöckel, H., 853, 868, 909–912, 931, 975, 976
Knowles, P.J., 952, 977
Kobayashi, K., 743, 779, 781, 864, 866
Koenig, S., 118, 122
Kohn, W., 218, 299

Kolos, W., 40, 72, 209, 210, 224, 299, 300, 507, 577
Kolsky, H.G., 376, 385, 574
Kopp, I., 251, 300, 618, 656, 659, 681, 682, 820, 867, 914, 976
Kotake, Y., 549, 578
Kovács, I., 32, 37
Kramers, H.A., 23, 36, 277, 300
Krauss, J.D., 713, 717, 719, 862
Krishnan, R., 218, 299
Kristiansen, P., 526, 533, 537, 577
Kröckertskothen, T., 853, 868, 909–912, 931, 975, 976
Krogdahl, M.K., 286, 301
Kronig, R. de L., 40, 72, 466, 575
Kroto, H.W., 32, 37, 358, 369
Krupnov, A.F., 723, 742, 863, 864
Kuchitsu, K., 352, 368–370
Kurtz, C., 953–955, 977
Kusch, P., 118, 122, 374, 422, 574, 575
Kutner, M., 722, 863
Kuwata, K., 549, 578
Kyrala, G.A., 736, 864

Lada, C.J., 722, 863
Lamb, W.E. Jr., 392, 575
Landau, L.D., 68, 72
Langhoff, S.R., 918, 921, 976
Langmuir, R.V., 962, 977
Larson, H.P., 722, 863
Latter, W.B., 722, 863
Laughlin, K.B., 723, 863
Law, M.M., 228, 230, 300, 813, 815–817, 821, 824, 829, 827, 832, 833, 867
Leach, C.A., 220, 228, 288, 299–301, 813, 815–817, 821, 824, 829, 858, 859, 867, 868, 952, 977, 972
Leavitt, J.A., 489, 496, 577
Ledsham, K., 291, 301
Lee, Y.T., 723, 863
Lefebvre-Brion, H., 32, 37, 230, 251, 276, 300, 313, 369, 659, 682, 820, 867
Legon, A.C., 704, 705, 710, 862
Leopold, K.R., 624, 681, 728, 864
Le Roy, R.J., 280, 284, 285, 301
Lesk, M.E., 553, 558, 578
Levy, D.H., 29, 37, 582–584, 587, 589, 596, 599, 607, 608, 641, 646, 648, 649, 652, 680, 681, 777, 779, 810, 812, 814, 866, 867
Lewen, F., 723, 725, 733, 863, 864
Lichten, W., 424, 428, 429, 436, 438–440, 444–446, 575, 885, 900, 975
Lide, D.R., 15, 36, 759, 865
Lie, G.C., 805–808, 810, 845, 846, 848, 849, 867, 868
Liedtke, M., 723, 725, 863
Lifshitz, E.M., 68, 72
Lilley, A.E., 974, 977
Lin, C.C., 612, 681

Lineberger, W.C., 810, 867, 958, 959, 977
Lipscomb, W., 501, 577
Liszt, H.S., 722, 863
Liu, B., 805–808, 810, 867
Liu, K., 919, 976
Loewenstein, E.V., 733, 734, 864
Logan, R.A., 422, 975
Lombardi, M., 892, 900, 975
London, F., 210, 299
Loren, R.B., 746, 798, 865, 866
Lovas, F.J., 487, 576, 710, 738, 739, 842, 844, 845, 862, 864, 868
Low, R.J., 738, 740, 741, 864
Lowe, R.S., 537, 578
Lubic, K.G., 609, 611, 681, 811, 867
Lucas, N.J.D., 366, 370, 617, 681
Luce, R.G., 487, 576
Lugten, J.B., 799, 867
Lundeen, S.R., 974, 977
Lurio, A., 25, 36, 303, 369, 446–454, 459, 461, 462, 575
Lutz, B.L., 722, 863

MacDonald, R.G., 591, 593, 596, 680
Madhavan, P.V., 917, 976
Maeda, A., 854, 868
Mahler, R.J., 586, 680
Malmberg, C., 618, 656, 681, 914, 976
Maloney, P.R., 722, 863
Mandl, F., 143, 175
Mantz, A.W., 282, 301, 733, 864
Marechal, M.A., 892, 975
Mariella, R.P., 509, 511, 517, 523, 524, 526, 577
Marple, D.T.F., 487, 576
Marr, A.J., 228, 288, 300, 301, 813, 815, 816, 817, 821, 824, 829, 867, 952, 977
Matcha, R.L., 487, 576
Matsumura, K., 738, 739, 864
Matsushima, F., 699, 701, 736, 862
McClain, E.F., 713, 862
McDowell, R.S., 733, 864
McEachran, R.P., 973, 977
McGilvery, D.C., 958, 977
McGonagle, D., 787, 866
McKellar, A.R.W., 537, 578
McKnight, J.S., 757, 865
McLachlan, A.D., 29, 36
McLay, D.B., 723, 863
McLean, A.D., 210, 299, 501, 508, 577, 836, 868
McLean, R.A., 417, 419–421, 575
McLeod, S., 33, 37
McNab, I.R., 220, 288, 299, 301, 730, 856, 858, 861, 864, 868, 942–944, 948–950, 952, 976, 977
Meath, W.J., 283, 301
Meeks, M.L., 713, 722, 790, 862, 863
Meerts, W.L., 489, 526–529, 533, 535, 537, 539, 542, 544, 548, 549, 577, 578, 616, 621, 681, 723, 782, 787, 789, 811, 863, 866, 867
Mehran, F., 422, 575

Mehring, M., 352, 368–370
Melendres, C.A., 487, 576
Merer, A.J., 251, 300, 352, 368–370, 530, 577, 656, 659, 676, 681, 682, 820, 839, 841, 844, 849, 867, 868, 909, 914, 917–919, 928, 930, 975, 976
Merritt, F.R., 13, 36
Meyer, D.M., 722, 863
Michael, E., 723, 863
Miller, C.E., 487, 576
Miller, R.C., 487, 576
Miller, S.L., 757, 758, 865
Miller, T.A., 25, 29, 36, 37, 303, 348, 351, 352, 368–370, 446–454, 459, 461, 462, 575, 582–584, 587, 589, 590, 596, 608, 641, 646, 648, 649, 652, 680, 681, 777, 779, 866, 885, 887–893, 895, 896, 898–901, 975
Miller, W.H., 281, 301
Millman, S., 374, 574
Milton, D.J., 336, 338, 341, 343, 351, 369, 512, 542, 545, 548, 577, 578, 616, 619–622, 656, 667, 682
Milverton, D.R.J., 749, 865
Minh, Y.C., 787, 866
Mirri, A.M., 526, 537, 577
Mizushima, M., 32, 37, 586, 591, 612, 680, 681, 757, 778, 865, 866
Moeller, C., 218, 299
Mollaaghababa, R., 836, 838, 840, 868
Montgomerie, C.A., 220, 299, 858, 868, 942–944, 948–950, 952, 976, 977
Moore, G.E., 506, 577
Moran, J.M., 722, 863
Moran, M.M., 624, 628, 629, 681, 794, 866
Morino, Y., 759, 812, 814, 865, 867
Morris, M., 722, 863
Morse, P.M., 66, 72, 238, 300
Morton, J.R., 775, 865
Moskalev, A.N., 139, 175
Moss, R.E., 32, 37, 73, 122, 220, 224, 288, 299, 301, 859, 869, 952, 972, 977
Moszynski, R., 224, 230
Mucha, J.A., 630, 681
Muenter, J.S., 489, 490, 493, 494, 496, 576, 577
Müller, R.P., 655, 681, 701, 862
Mulliken, R.S., 201, 228, 248, 299, 300, 331, 369, 525, 529, 577, 618, 681, 807, 808, 832, 867
Mundy, L., 746, 798, 865, 866
Myers, R.J., 271, 300, 590, 680, 778, 779, 866

Nachman, D.F., 677, 682, 917, 928–930, 976
Nair, K.P.R., 739, 864
Nakagawa, J., 903, 975
Namiki, K., 745, 761–763, 814, 850, 851, 865, 867, 868, 922–926, 976
Naude, S.M., 447, 575
Nelis, T., 338, 369, 661–664, 669, 674–677, 682, 810, 867, 927, 976
Nelson, H.M., 489, 496, 577
Nesbet, R.K., 215, 299
Nethercot, A.H. Jr., 733, 864

Neufeld, D.A., 722, 863
Neumann, R.M., 526, 577
Newton, M.D., 918, 976
Nichols, E.R., 733, 864
Nolt, I.G., 728, 790, 791, 864

O'Brien, L.C., 836, 868
O'Hare, P.A.G., 596, 680
Ohashi, N., 652, 681, 768, 865
Oike, T., 850, 868
Oka, T., 699, 701, 736, 862, 864
Okabayashi, T., 745, 776, 777, 845, 846, 850, 853, 865, 866, 868
Onda, M., 710, 862
Ono, M., 549, 578
Oppenheimer, R., 38, 72, 213, 299, 302, 369
O'Reilly, D.E., 496, 577
Osberghaus, O., 962, 977
Ozier, I., 422, 489, 503, 575, 577

Pack, R.J., 40, 72
Pack, R.T., 286, 301
Pak, I., 723, 863
Palke, W.E., 414, 575
Palmer, P., 722, 863
Palmieri, P., 549, 578
Park, H., 139, 175
Parr, R.G., 192, 210, 298
Parson, J.M., 919, 976
Paubert, G., 722, 863
Paul, W., 962, 977
Pauli, W., 55, 72, 123, 138
Pauling, L., 64, 72, 187, 237, 258, 298, 300
Pauzat, F., 787, 866
Paveliev, D.G., 723, 863
Pearson, E.F., 743, 864
Peek, J.M., 858, 868
Pekeris, C.L., 244, 300
Penfield, H., 722, 863, 974, 977
Penzias, A.A., 713, 722, 749, 862, 863, 865, 974, 977
Pesch, T.C., 841–843, 868
Petersen, F.R., 726, 736, 863
Peterson, K.I., 549–551, 578
Pfaff, J., 699–701, 736, 862
Phillips, T.G., 722, 863
Phipps, T.E., 376, 385, 574
Pickett, H.M., 367, 370, 761, 790, 865, 866
Piltch, N.D., 746–748, 798, 865, 866
Pinchemel, B., 853, 868
Pique, J.P., 859, 869
Plambeck, R.L., 746, 798, 865, 866
Plesset, M.S., 218, 299
Plummer, G.M., 713, 743, 744, 862, 864
Pople, J.A., 29, 37, 214–216, 218, 299, 392, 575
Porter, G., 607, 680
Poulsen, O., 933, 976
Powell, F.X., 691, 693, 759, 812, 814, 862, 865, 867
Poynter, R.L., 542, 578

Pradell, A.G., 118, 122
Prasad, C.V.V., 841, 868
Pratt, D.W., 875, 974
Preston, K.F., 775, 865
Pritchard, H.O., 291, 295, 301
Pryce, M.H.L., 303, 369
Purcell, E.M., 3, 36, 713, 862
Purnell, M.R., 364, 370
Pyne, C.H., 230, 288, 300, 301, 832, 833, 867, 952, 977

Qian, C.X.W., 841, 868
Quack, M., 352, 368–370
Quiguer, J., 742, 864

Rabi, I.I., 13, 36, 123, 138, 374, 375, 388, 391, 574
Racah, G., 165, 176
Radford, H.E., 29, 37, 537, 538, 542, 544, 578, 587, 591, 612–615, 617, 620–624, 628, 629, 652–655, 680, 681, 702, 703, 778, 791, 794, 862, 866, 871, 875, 974
Radom, L., 216, 299
Radostitz, J.V., 728, 864
Raftery, J., 364, 370
Ram, R.S., 852, 868
Ramsay, D.A., 251, 300, 352, 368–370, 607, 659, 680, 682, 820, 867
Ramsey, N.F., 10, 13, 32, 36, 37, 123, 138, 270, 300, 375, 376, 380–382, 384, 385, 388, 389, 391–393, 406–409, 415, 422, 471, 483–487, 489, 496, 503, 574–577, 745, 864
Rank, D.H., 502, 577
Rank, D.M., 551, 578
Rao, B.S., 502, 577
Rao, K.N., 282, 301, 697, 733, 862, 864
Rawson, E.B., 612, 681
Ray, D., 609, 611, 681, 973, 977
Raymonda, J.W., 489, 576
Read, W.G., 710, 778, 862, 866
Reinartz, K., 713, 862
Reinsch, E.-A., 551, 578
Renhorn, I., 364, 370
Reno, R., 440, 444–446, 575, 900, 975
Rees, A.L.G., 280, 301
Richard, E.G., 736, 864
Richards, W.G., 187, 298, 361, 364, 370, 525, 577
Richardson, C.B., 961, 964, 977
Richardson, O.W., 423, 575, 885, 975
Riggin, M., 537, 578
Ritschel, R., 917, 976
Ritter, G.J., 870, 974
Robinson, J.S., 922–924, 927, 976
Roetti, C., 195, 298
Rogers, A.E., 722, 863
Rogers, S.A., 836, 868
Roh, W.B., 733, 864
Rohlfs, K., 32, 37, 713, 716, 863
Rohrbaugh, J.H., 13, 36
Roothaan, C.C.J., 192, 209, 214, 298, 299

Rose, M.E., 32, 37, 73, 122, 139, 175
Rosenblum, B., 733, 864
Rosenfeld, L., 713, 722, 862
Rosmus, P., 551, 578
Rosner, S.D., 416–421, 575, 906, 931, 942, 975–977, 953–955, 958
Rostas, J., 251, 300, 343, 369, 659, 682, 820, 867
Roth, K.C., 722, 863
Runolfsson, O., 479, 481, 576
Russell, D.K., 359, 369, 596, 608, 652, 680, 681
Russon, L.R., 922, 924, 976
Rydbeck, O.E.H., 624, 681, 794, 866
Rydberg, R., 280, 301
Rydh, B., 618, 656, 681, 914, 976

Sadler, I.A., 224, 299, 952, 977
Sahlin, H.L., 207, 291, 299, 301, 972, 977
Saito, S., 590, 680, 691, 743, 745, 752, 761–764, 767, 768, 779, 781, 798, 811, 862, 864–868, 812, 814, 835, 845, 846, 850, 852, 853, 909, 912, 922–924, 975, 976
Sakamaki, T., 776, 777, 853, 865, 866, 868
Salek, A.H., 537, 577, 578, 723, 725, 759, 761, 782, 787, 863, 865, 866
Sales, K.D., 958, 977
Salpeter, E.E., 123, 130, 138
Salwen, H., 503, 505, 577
Sanders, T.M., 29, 37, 525, 538, 552, 577–579, 680, 692, 694, 788, 789, 862, 866
Sannikov, A.V., 969, 970, 977
Sarre, P.J., 749, 865
Sassenberg, U., 844, 868, 917–919, 976
Sastry, K.V.L.N., 695, 759, 862, 865
Satchler, G.R., 32, 37, 139, 163, 175
Sauer, B.E., 936–938, 976
Saykally, R.J., 364, 370, 552, 578, 591, 609–611, 655–658, 680, 681, 699–701, 723, 736, 778, 811, 835, 836, 862, 863, 866, 867, 913, 976
Scalabrin, A., 591, 680, 778, 811, 866, 867
Schafer, E., 609, 611, 681
Schamps, J., 845, 853, 868, 913, 976
Schawlow, A.L., 32, 37
Schieder, R., 723, 725, 733, 734, 863, 864
Schiff, L.I., 73, 122
Schilke, P., 722, 863
Schlapp, R., 23, 36, 759, 865
Schleyer, P.v.R., 216, 299
Schlier, C., 507, 577
Schmeltekopf, A.L., 282, 301
Schmid, R., 552, 578
Schneider, W.G., 29, 37, 392, 575
Scholl, T.J., 953–955, 977
Schrödinger, E., 74, 88, 122
Schubert, J.E., 364, 370, 624, 681
Schwarz, R., 713, 862
Scoles, G., 32, 37, 372, 574
Scott, P.R., 187, 298
Scurlock, C.T., 922, 924, 931–933, 976
Seeger, J.A., 581, 680

Segre, E., 123, 138
Segun, I.A., 279, 285, 300, 301
Series, G., 870, 974
Sham, L.J., 218, 299
Shaw, A.M., 228, 230, 300, 730, 813, 815–817, 821, 824, 829, 832, 833, 864, 867
Shea, J.A., 710, 862
Shelton, H., 962, 977
Shih, S., 224, 300
Shirley, J.E., 677, 682, 922, 924, 928–930, 976
Shkaev, A.P., 723, 725, 863
Shortley, G.H., 144, 175, 186, 298
Sibai, A.M., 938, 976
Silsbee, H.B., 376, 385, 574
Silver, B.L., 139, 175
Silvers, S.J., 876–878, 975
Silverman, S., 506, 577
Simmons, J.D., 836, 868
Siska, P.E., 816, 827, 867
Slater, J.C., 183, 298
Smith, D., 164, 176
Smith, D.R., 624, 652, 681
Smith, S.A., 414, 575
Smith, S.E., 730, 864
Snell, R.L., 746, 798, 865, 866
Softley, T.P., 187, 298, 736, 864
Soliverez, C.E., 312, 369
Solomon, P.M., 722, 863
Solunac, S.A., 223, 299
Souza, S.P., 722, 863
Sponer, H., 284, 301
Stacey, G.J., 799, 867
Steimle, T.C., 624, 656, 660, 677, 681, 682, 794, 796, 797, 804, 814, 866, 867, 902–904, 913–933, 952–955, 975–977
Steinbach, W., 757, 865
Stephenson, D.A., 481, 576
Stern, R.C., 553, 555, 557, 558, 578
Stevens, R.M., 501, 577
Stewart, A.L., 291, 301
Stone, A.J., 591, 593, 596, 680
Storey, J.W.V., 718, 720, 734, 791, 792, 863, 866
Story, T.L., 487, 576
Strahan, S.F., 609, 611, 681
Street, K. Jr., 481, 487, 576
Stringat, R., 919, 976
Stroh, F., 533, 536, 537, 577, 578
Strong, J., 712, 862
Stutzki, J., 723, 863
Stwalley, W.C., 285, 301
Suenram, R.D., 710, 738, 739, 842, 844, 845, 862, 864, 868
Sume, A., 794, 866
Swings, P., 713, 722, 862
Szanto, P.G., 552, 578, 746–748, 798, 835, 836, 865–867

Takagi, K., 699, 701, 736, 862
Takami, M., 768, 811, 865, 867, 884, 975

Tamassia, F., 356, 369
Tanaka, T., 883, 884, 975
Tang, K.T., 827, 867
Tanimoto, M., 745, 768, 776, 777, 811, 814, 845, 846, 850, 853, 865–868
Taylor, A.W., 909, 975
Taylor, S.M., 230, 300, 730, 832, 833, 864, 867
Teachout, R.R., 286, 301
Teller, E., 207, 291, 299, 301, 972, 977
Ter Meulen, J.J., 539, 542, 578, 723, 789, 863, 866
Thaddeus, P., 722, 836–840, 863, 868
Thornley, J.H.M., 164, 176
Thrush, B.A., 551, 578, 591, 593, 596, 652, 680, 681
Tiemann, E., 739, 743, 853, 864, 868, 909–912, 931, 975, 976
Tilford, S.G., 447, 461, 575, 836, 868
Toennies, J.P., 827, 867
Tolliver, D.E., 736, 864
Törring, T., 904, 905, 975
Townes, C.H., 13, 29, 32, 36, 37, 123, 138, 366, 370, 525, 538, 551, 552, 577–579, 680, 692, 694, 718, 720, 733, 734, 758, 788, 789, 791, 792, 862–866
Treffers, R.R., 722, 863
Tretyakov, M.Yu., 723, 725, 742, 863, 864
Trischka, J.W., 468–470, 476, 487, 503, 505, 575–577
Tromp, J.W., 285, 301
Turner, B.E., 624, 681, 722, 749, 752, 794, 798, 863, 865, 866

Uehara, H., 351, 369, 587, 589, 590, 620, 680

van den Heuvel, F.C., 537, 578, 811, 867
Vanderslice, J.T., 447, 461, 575
Vanek, M.D., 726, 728, 729, 863, 864
Van Esbroeck, P.E., 417, 419–421, 575
Vanesse, G.A., 712, 862
Van Vleck, J.H., 21, 30, 36, 37, 40, 72, 148, 169, 176, 248, 258, 275, 283, 300, 301, 303, 313, 346, 359, 369, 525, 530, 577, 622, 681, 794, 866
Van Wachem, R., 481, 487, 576
Varberg, T.D., 533, 536, 537, 577, 578, 624, 681, 733, 738, 740, 741, 791, 864, 866
Varshalovich, D.A., 139, 175, 969, 970, 977
Veenstra, C.J., 973, 977
Veillette, P., 537, 578
Verhoevre, P., 723, 863
Versluis, M., 723, 863
Veseth, L., 324, 352, 368–370, 611, 681, 820, 867
Viala, Y., 787, 866
Viant, M.R., 228, 288, 300, 301, 814–817, 821, 824, 829, 858, 859, 867, 868, 952, 977
Volokhov, S.V., 723, 725, 863
Vrtilek, J.M., 722, 836, 837, 839, 840, 863, 868

Wada, A., 835, 868
Wahl, A.C., 509, 577, 596, 680, 958, 977

Walker, C.K., 722, 863
Walker, K.A., 741, 742, 864
Walker, T.E.H., 361, 370
Walter, J., 207, 208, 299
Walters, A., 742, 864
Wang, J., 936–938, 976
Warner, H.E., 835, 836, 868
Watson, D.M., 718, 720, 734, 791, 792, 863, 866
Watson, J.K.G., 282, 301, 303, 313, 327, 331, 339, 341, 343, 345, 346, 354, 356, 357, 369, 657, 677, 682, 737, 743–745, 864
Watson, W.D., 551, 578
Wayne, F.D., 303, 331, 354, 360, 369, 370, 542, 578, 622–624, 652–655, 657, 681, 682, 791, 866
Weidner, R.T., 13, 36
Weinbaum, S., 210, 299
Weinreb, S.A., 713, 722, 790, 862
Weinstock, E.M., 880, 883, 975
Weiss, A., 210, 299
Weiss, R., 489, 493, 576
Weisskopf, V.F., 275, 300
Welch, W.J., 551, 578
Welch, W.M., 757, 865
Wells, J.S., 586, 622, 680, 681
Weltner, W. Jr., 674, 682, 844, 846, 868
Wentzel, G., 277, 300
Werner, H.-J., 551, 578
Werth, G., 481, 487, 576
West, B.G., 757, 865
West, Y.D., 288, 301, 952, 977
Wetmore, R.W., 223, 299, 973, 977
Wharton, L., 481, 487, 489, 576
Wick, G.C., 123, 138, 407, 575
Wicke, B.G., 558, 578
Wiggins, T.A., 502, 577
Wigner, E.P., 154, 176, 203, 299, 317, 369
Wilkinson, P.G., 447, 461, 575
Wilson, C., 360, 370
Wilson, E.B., 64, 72, 187, 237, 258, 298, 300, 690, 862
Wilson, R.W., 713, 722, 749, 862, 863, 865
Wilson, T.L., 32, 37, 713, 716, 863
Wind, H., 297, 301
Wing, W.H., 736, 864
Witmer, E.E., 203, 299
Winnewisser, G., 537, 577, 578, 713, 723, 725, 733, 734, 742, 743, 759, 761, 779, 782, 787, 862–866
Winnewisser, M., 695, 713, 759, 862, 865
Wolniewicz, L., 40, 72, 210, 224, 299, 507, 577
Woods, R.C., 552, 578, 638, 681, 695–697, 722, 745–748, 750, 798, 835, 836, 862, 863, 865–868
Woodward, D.R., 794, 796, 797, 804, 866, 867
Wouthuysen, S.A., 77, 122
Wright, B.D., 13, 36
Wuerker, R.F., 962, 977

Yamada, C., 525, 577, 697–699, 769, 770, 773, 775, 814, 862, 865, 867
Yamada, K.M.T., 537, 577, 578, 782, 787, 866

Yamamoto, S., 742, 779, 781, 835, 864, 866, 868
Yamazaki, E., 853, 868
Yardley, J.T., 733, 864
Yi, P.N., 422, 489, 503, 575, 577
Yoshimine, M., 210, 299, 501, 508, 509, 577
Yoshino, K., 816, 867
Young, L., 938–941, 953–955, 976, 977

Zacharias, J.R., 13, 36, 123, 138, 374, 375, 388, 391, 574
Zare, R.N., 32, 37, 45, 72, 139, 163, 175, 251, 281, 282, 300, 301, 343, 352, 368–370, 659, 682, 820, 867, 876, 880, 882, 883, 975
Zeiger, H.J., 123, 138
Zegarski, B.R., 900, 901, 975
Zieger, H., 422, 575
Zimmerer, R.W., 757, 865
Zink, L.R., 728, 790, 791, 864
Ziurys, L.M., 787, 798, 841–843, 845, 846, 848–850, 854, 856, 866, 868, 909, 912, 975
Zmuidzinas, J., 722, 863
Zorn, J.C., 468, 469, 471, 473, 475, 481, 487, 575, 576
Zuckerman, B., 624, 681, 722, 794, 863, 866
Zwart, E., 723, 863

Subject index

a, orbital hyperfine constant, 127, 573
a, A, symbols for electronic states, 200, 440
a_0, see Bohr radius
A, spin–orbit coupling constant, 357
α, Dirac matrix, 75
α, spin function, 181
α_e, vibration–rotation constant, 244
A coefficient, see Einstein coefficients of absorption and emission
ab initio methods, 213
absorption, 260, 683
abundance ratio of isotopes, 979
adiabatic approximation, 60, 223
AgH, see LiH $X^1\Sigma^+$
AgO $X^2\Pi$
   molecular parameters, 814
AlF $X^1\Sigma^+$
   rotational spectrum, 16
AlH, see LiH $X^1\Sigma^+$
alkali halides
   molecular beam electric resonance spectra, 487
   molecular beam magnetic resonance spectra, 421
alkaline earth oxides, 487
alkaline earth halides, 902
altitude–azimuth drive, see radio telescope
angular momentum
   addition of, 152
   electron orbital, L, 26, 144
   electron spin, S, 26, 77
   nuclear rotational, R, 26
   nuclear spin, I, 123
   total, J, 8, 397
anharmonic oscillator, 238, 283
anomalous commutation relations, 58, 148, 168
antibonding orbitals, see Bonding and antibonding orbitals
anticommutation, 106, 342, 546
ArHe$^+$, see HeAr$^+$
associated Laguerre functions, 179, 194
associated Legendre functions, 145, 179
asymmetric top, 150
atomic mass constant, 978
atomic masses, 979

atomic orbital, 180, 194
atomic states, 180, 184
atomic structure, 178
atomic units, 214
Auger detector, 373
Avogadro constant, 978

b, Frosch and Foley hyperfine constant, 573
b, B, electronic state symbols, 200
b_F, 24, 127
B, rotational constant, 235
B_e, rotational constant at equilibrium, 244, 356
B_v, rotational constant in vibrational level v, 244
B_{mn}, Einstein transition probability, 258
B coefficient, see Einstein coefficients of absorption and emission
β, constant in the Morse function, 66, 238, 267
β, Dirac matrix, 75
β, spin function, 181
β_e, vibration–rotation constant, 244
backward wave oscillator, 687, 723
BaO $X^1\Sigma^+$
   microwave/optical double resonance rotational spectrum, 883
BaO $A^1\Sigma^+$
   microwave/optical double resonance rotational spectrum, 883
Bethe–Salpeter equation, 107
Birge–Sponer extrapolation, 284
body-fixed axis system, 146
Bohr
   electron magneton, 348, 978
   frequency condition, 2
   nuclear magneton, 978
   orbit, 978
   radius (a_0), 180, 978
Bohr–Sommerfeld quantisation condition, 279
Boltzmann constant, 978
Boltzmann distribution law, 260
bonding and antibonding orbitals, 197
Born approximation, 8, 60
Born–Oppenheimer approximation, 9, 38, 60, 832
Born–Oppenheimer breakdown, 346, 357

bosons, 126, 255
Breit equation, 104, 130
BrO $X\,^2\Pi$
    microwave magnetic resonance spectrum, 597
    microwave rotational spectrum, 811

c, speed of light, 978
c, Frosch and Foley hyperfine constant, 573
c_I, nuclear spin–rotation constant, 378
C_{lm}, modified spherical harmonics, 145
CaCl $X\,^2\Sigma^+$
    microwave/optical double resonance rotational spectrum, 902
CaF $X\,^2\Sigma^+$, see SrF $X\,^2\Sigma^+$
carbon dioxide laser, 584
cartesian
    tensor, 14, 398, 561
    vector, 14
case (a), case (b), etc., see Hund's case (a), (b), etc.
CCl $X\,^2\Pi$
    molecular parameters, 814
CD $X\,^2\Pi$, see CH $X\,^2\Pi$
centre of mass
    molecular, 40, 41, 234, 396
    nuclear, 40, 43
centrifugal distortion, 242, 338, 546
CF $X\,^2\Pi$
    microwave magnetic resonance spectrum, 608
    molecular parameters, 768, 811
    rotational spectrum, 811
CFe, see FeC
cgs units, relationships to SI units, 4, 33, 993
CH $X\,^2\Pi$
    CD, far-infrared laser magnetic resonance spectrum, 633
    electronic structure, *ab initio* calculations, 805
    energy levels, 630
    far-infrared laser magnetic resonance spectrum, 628, 802
    Λ-doubling spectrum, 794, 802
    microwave/optical double resonance spectrum, 794
    millimetre wave rotational spectrum, 794
    molecular parameters, 805
    theory of Hund's case (a) to case (b) correlation, 627
CH $a\,^4\Sigma^-$
    far-infrared laser magnetic resonance spectrum, 661
    molecular parameters, 664
    rotational levels, 805
chemical shift, 378
Chraplyvy transformation, 105
circulator, 581
classification of electronic states, 26, 200
Clebsch–Gordan coefficients, 153, 157
Clebsch–Gordan series, 157
ClO $X\,^2\Pi$
    ^{35}Cl quadrupole interaction, 604
    microwave magnetic resonance spectrum, 597
    molecular parameters, 607, 813
    rotational spectrum, 811
    theory of the hyperfine interaction, 602
CN $X\,^2\Sigma^+$
    energy levels, 639
    interstellar spectrum, 749
    molecular parameters, 751
    rotational spectrum, 749
    Zeeman effect, 637, 639
CN $A\,^2\Pi_{3/2}$
    radiofrequency Λ-doubling transitions, 871
CN $B\,^2\Sigma^+$
    electronic transitions, 871
    ^{14}N hyperfine parameters, 875
CO $X\,^1\Sigma^+$
    far-infrared rotational spectrum, 732
    millimetre wave rotational spectrum, 732
    rotational constants of isotopomers, 733
CO $a\,^3\Pi$
    energy levels, 554, 656, 835
    far-infrared laser magnetic resonance spectrum, 655
    Λ-doubling theory, 556, 659
    microwave and millimetre wave rotational spectrum, 834
    molecular beam electric resonance spectrum, 555
    perturbation of $a\,^3\Pi$ state by $a'\,^3\Sigma^+$ state, 557
    Stark effect, 553
    Zeeman effect, 660
CO$^+$ $X\,^2\Sigma^+$
    interstellar spectrum, 722, 746
    microwave/optical double resonance spectrum, 958
    molecular parameters, 749
    rotational spectrum, 697, 745
CoH $X\,^3\Phi$
    far-infrared laser magnetic resonance spectrum, 669
    molecular parameters, 673
    rotational levels, 671
commutation relations
    anomalous, 148, 168
    normal, 142, 160, 322
composite systems, 165
Condon and Shortley convention, 144
configuration interaction, 196, 216
conservation of angular momentum, 79
contact transformation, 312, 352
core integrals, 191
Coriolis coupling, 329
correlation energy, 189, 196
correlation diagrams, 202, 231, 527, 598, 650, 755, 785
correlation rules, 203
Coulomb
    gauge, 102, 114
    integrals, 191
    potential, 178
coupled-cluster method, 218
coupled representation, 152, 173

coupling cases, *see* Hund's coupling cases, nuclear spin coupling cases
coupling
  of I_i to form I_T, 253, 417, 447
  of J and I to form F, 233, 743
  of J_a and R to form J, 229
  of L to the molecular axis, 225
  of L and R, 228, 425
  of Λ and R to form N, 226
  of L and S to form J_a, 228
  of λ_i to form L, 197
  of N and S to form J, 226, 425
  of s_i to form S, 199
  of S and I to form G, 233
CrCl, *see* CrF
CrF $X\,^6\Sigma^+$
  millimetre wave rotational spectrum, 851
  molecular parameters, 852
CrH $X\,^6\Sigma^+$
  far-infrared laser magnetic resonance, 666
  molecular parameters, 668
CrN $X\,^4\Sigma^-$
  microwave/optical double resonance rotational spectrum, 924
  molecular parameters, 927
CS $A\,^1\Pi$
  molecular parameters, 879
  radiofrequency/optical double resonance spectrum, 876
  Stark effect, 878
CsF $X\,^1\Sigma^+$
  molecular parameters, 481
  Stark effect, 465, 476
  strong field molecular beam electric resonance spectrum, 476
  weak field molecular beam electric resonance spectrum, 470
CSi, *see* SiC
CuBr $X\,^1\Sigma^+$, *see* CuCl $X\,^1\Sigma^+$
CuCl, CuBr $X\,^1\Sigma^+$
  rotational spectrum, 738
CuF $b\,^3\Pi$
  microwave/optical double resonance rotational spectrum, 913
  molecular parameters, 916
CuH $X\,^1\Sigma^+$, *see* LiH $X\,^1\Sigma^+$
CuO $X\,^2\Pi_{1/2}$
  microwave/optical double resonance rotational spectrum, 917
  molecular orbital theory, 918
  molecular parameters, 918
Curl's relationship, 668

d, Frosch and Foley hyperfine constant, 573
d orbitals, 182
D_e, rotational constant at equilibrium, 244
D_e, dissociation energy at equilibrium, 239
D_v, rotational constant in vibrational level v, 244
$^1\Delta$ states, 26, 587, 588, 591, 594, 778
$^2\Delta$ states, 665, 674, 927
$^3\Delta$ states, 841, 922
$^4\Delta$ states, 669
$^5\Delta$ states, 856, 909
$^6\Delta$ states, 846
D$_2$, *see* H$_2$
D$_2^+$, *see* H$_2^+$
Darwin term, 85, 94
DCl $X\,^1\Sigma^+$ electric resonance spectrum, *see* HCl $X\,^1\Sigma^+$
decoupled representation, 152, 226
density functional theory, 218
density matrix, 215
determinantal wave function, 183, 357
diamagnetism
  Hamiltonian, 116, 380, 403, 408, 500
  semiclassical theory for an atom, 392
  semiclassical theory for a diatomic molecule, 393
dipolar interaction
  electron–electron, 25, 430, 452, 563, 643, 661, 753
  electron–nuclear, 332, 441, 452, 561, 748, 765, 803
  nuclear–nuclear, 378, 492, 558
dipole electric field, 464
dipole–quadrupole interaction, 283
Dirac delta function, 24, 91
Dirac equation, 73
Dirac representation, 85
Dirac spin matrices, 78
dish, *see* radio telescope
dissociation energy, D_e, 66, 239
Doppler modulation, 699
Doppler shift, interstellar, 721
Doppler splitting, 721
Doppler width of spectral lines, 275
double zeta basis set, 195
dumbbell model, 233
Dunham variable, 63
Dunham expansion, 244, 282, 345, 501, 742

e, elementary electric charge, 978
e/f convention, 251
effective Hamiltonian, 29, 129, 302, 316
eigenfunctions
  of the anharmonic oscillator, 238
  of the harmonic oscillator, 64, 235
  of the rigid rotator, 233
  of the vibrating rotator, 243
Einstein coefficients of absorption and emission, 258, 718
electric dipole moment, 20, 116
electric dipole transitions, 261
electric field dissociation, 731
electric field gradient, 365
electric potential, 89, 99
electric resonance, *see* Molecular beam electric resonance
electromagnetic radiation, 1, 3, 35
electromagnetic spectrum, 2, 684

Subject index 1007

electronic states classification, 197, 200
electronic structure, 200, 213, 665
electron rest mass, 978
electron spin
  g-factor, 21, 80, 110
  magnetic moment, 21, 77, 80
electronvolt, eV, 978
electrostatic interaction potential, 365
elementary charge, 978
elliptical coordinates, 209, 289
emission spectra, 260
Euler angles, 46, 147, 245
exchange integrals, 191
exclusion principle, *see* Pauli exclusion principle

f orbitals, 182
$^3\Phi$ states, 669
$^4\Phi$ states, 854
$F(J)$, rotational term, 244
Fabry–Perot cavity, *see* Fourier transform
far-infrared
  laser frequencies, 586
  laser magnetic resonance, 584
  spectrometers, 585, 723
  telescope, 720
FeC $X\,^3\Delta$
  millimetre wave rotational spectrum, 841
FeCl $X\,^6\Delta$
  molecular parameters, 849
  rotational levels, 846, 848
FeF $X\,^6\Delta$
  electronic structure, 850
  millimetre wave rotational spectrum, 845, 847
  molecular parameters, 849
FeH $X\,^4\Delta$
  far-infrared laser magnetic resonance, 669
FeH $a\,^6\Delta$, 359
Felgett advantage, 712
FeO $X\,^5\Delta$
  microwave/optical double resonance spectrum, 909
  millimetre wave rotational spectrum, 856
  molecular parameters, 913
Fermi contact interaction, 24, 127, 332, 364, 440, 452, 748, 751, 763, 803
fermions, 126, 255
fine structure: higher order contributions, 327, 335, 661, 667, 852, 909
flop-out, flop in detection of molecular beams, 375, 463, 482, 555
Fock matrix, 215
Foldy–Wouthysen transformation, 80, 85
force constant k, 236
Fourier transform
  Fabry–Perot cavity, 708
  microwave spectrometer, 703, 708, 739
four-vector, 99
Franck–Condon integral, 269
free space cell, 698

fundamental constants, 978
fundamental transition, 238

g-factor
  electron orbital, g_L, 28, 351
  electron spin, g_S, 21, 351, 978
  nuclear spin, g_N, 20, 270, 351
  rotational, g_r or g_J, 20, 350, 406
g, gerade, 200, 245
γ, electron spin–rotation parameter, 21, 360
γ_e, vibration–rotation parameter, 244
gauge
  invariance, 101
  transformation, 102
Gaussian lineshape, 276
Gaussian orbitals, 195, 216
Gaussian quadrature, 279
GeF $X\,^2\Pi$
  microwave rotational spectrum, 810
  molecular parameters, 811
Grotrian diagram, 185

h, the Planck constant, 978
H, H_e, H_v, rotational constant, 243
H atom
  atomic orbitals, 178
  Schrödinger equation, 178
H$_2$ molecule, electronic states, 212, 224, 423, 424, 886
H$_2$ $X\,^1\Sigma_g^+$
  molecular beam magnetic resonance spectrum, 272, 372
  molecular orbital theory, 208
  molecular parameters, 416
  molecular quadrupole moment, 409, 416
  ortho and *para* forms, 254, 385, 425
  Zeeman effect, 384, 390, 391
H$_2$ $G\,^1\Sigma_g^+$
  microwave/optical double resonance (MOMRIE) spectrum, 885
  rotational levels and Zeeman effect, 889
H$_2$ $c\,^3\Pi_u$
  molecular beam magnetic resonance spectrum, 422, 439
  rotational levels, 437
H$_2$ $d\,^3\Pi_u$
  hyperfine interactions and parameters, 900
  microwave/optical double resonance (MOMRIE) spectrum, 892
  rotational levels and Zeeman effect, 894
H$_2$ $k\,^3\Pi_u$
  hyperfine interactions and parameters, 900
  microwave/optical double resonance (MOMRIE) spectrum, 900
  rotational levels and Zeeman effect, 901
H$_2^+$ $X\,^2\Sigma^+$
  adiabatic and non-adiabatic calculations, 207, 223, 972
  Born–Oppenheimer potential, 207, 221, 289, 858
  electronic wave function, 207
  hyperfine symmetry-breaking, 859, 860

H_2^+ $X\,^2\Sigma^+$ (Cont.)
  microwave electronic spectrum, 859
  microwave rotational spectrum, 856
  molecular parameters, 970
  quadrupole trap radiofrequency spectrum, 960
HD $X\,^1\Sigma^+$
  molecular beam magnetic resonance spectrum, 388
  Zeeman effect, 391
HD^+ $X\,^2\Sigma^+$
  hyperfine interaction parameters, 948
  microwave/infrared double resonance rotational spectrum, 948
  radiofrequency/infrared double resonance spectrum, 943
Hamiltonian
  electron kinetic energy, 7, 84
  nuclear kinetic energy, 7
harmonic oscillator
  potential curve, 241
  vibrational eigenfunctions, 237
  vibrational energy levels, 240
Hartree–Fock equation, 190, 192
Hartree unit of energy, 189, 978
HBr^+ $X\,^2\Pi_{3/2}$
  far-infrared laser magnetic resonance spectrum, 609
HCl $X\,^1\Sigma^+$
  molecular beam electric resonance spectrum, 500
  molecular parameters, 502, 508
  vibrational dependence of electric dipole moment, 506
HCl^+ $X\,^2\Pi_{3/2}$
  far-infrared laser magnetic resonance spectrum, 609
helium atom, 187
$HeAr^+$
  coupled-channel theory, 824
  electric field dissociation, 815
  Hund's case (c) rotational levels, 819
  long-range interaction potential, 827
  microwave spectrum in an ion beam, 816
  multimode Doppler pattern, 816
  near-dissociation energy levels, 818, 829
  potential energy curves, 815
  Zeeman effect (Hund's case (c) and case (e)), 817, 821, 828, 831
HeH^+ $X\,^1\Sigma^+$
  molecular parameters, 737
  rotational spectrum, 736
Heisenberg uncertainty principle, 273
$HeKr^+$
  electric field dissociation, 832
  Hund's case (e) rotational levels, 832
  microwave spectrum in an ion beam, 832
  near-dissociation energy levels, 833
Hermite polynomials, 65, 238
heteronuclear diatomic molecules, molecular orbitals, 204

HF $X\,^1\Sigma^+$
  magnetic shielding tensor, 500
  molecular beam electric resonance spectrum, 489
  Zeeman effect, 496
HF^+ $X\,^2\Pi_{3/2}$
  far-infrared laser magnetic resonance spectrum, 609
high frequency paramagnetism, 405
Hohenberg–Kohn theorem, 218
homonuclear diatomic molecules, molecular orbitals, 202
Hougen's isomorphic Hamiltonian, 320
Hund's case (a), 225, 649
Hund's case (b), 226, 627, 652
Hund's case (c), 228, 819
Hund's case (d), 228
Hund's case (e), 229, 819

i, molecule-fixed inversion operator, 245
I, nuclear spin angular momentum, 123
I, nuclear spin quantum number, 124
I, moment of inertia, 234, 399
indeterminacies, 352
interferometer
  far-infrared spectrometer, 713
  Michelson experiment, 711
intermediate coupling, 230
interstellar spectra, 722
inversion operator, 244, 250, 328
inversion symmetry of rotational levels, 244
ion beam spectrometer, 730, 942
ion trap, 963
irreducible spherical tensors, see Spherical tensor operators
isotopes, table of, 979
isotopic dependence of molecular parameters, 327, 344, 677, 737, 761, 788

J, total angular momentum, 152, 234
jj coupling in atoms, 186
J_a, resultant of coupling of L and S in Hund's case (c), 228, 819
Jacqinot advantage, 712
Jaffé expansion, 295
JWKB method, 277

κ, Born–Oppenheimer parameter, 38, 356
k, Boltzmann constant, 978
KH $X\,^1\Sigma^+$, see LiH $X\,^1\Sigma^+$
kinetic energy
  electrons in a molecule, 6, 40
  nuclei in a molecule, 6, 40
Klein–Gordon equation, 74
klystron, 686

l, orbital angular momentum of an electron, 183
l, azimuthal quantum number, 180, 201
L, orbital angular momentum, 144
λ, electron spin–spin coupling constant, 24, 360
Λ, resultant orbital angular momentum of the electrons about the internuclear axis, 26

Λ, quantum number of the resultant orbital angular momentum of the electrons about the internuclear axis, 26
Λ-doubling
  in Δ states, 659, 674
  in Π states, 26, 328, 362, 617, 807, 840
  parameters, 362
ladder operators, 143
Lagrangian, 68, 74, 103
Laguerre functions, 180
Lamb dips, 623
lamellar grating, *see* far-infrared spectrometers
LaO $X\,^2\Sigma^+$
  molecular parameters, 941
  radiofrequency/optical double resonance spectrum, 938
LaO $B\,^2\Sigma^+$
  molecular parameters, 941
  radiofrequency/optical double resonance spectrum, 938
Laplace operators, 6
Laplacian, 6, 178, 290
laser, 261
Legendre function, *see* associated Legendre functions
Legendre polynomials, 131, 264
Levi–Civita symbol, 143
LiBr $X\,^1\Sigma^+$
  molecular beam electric resonance spectrum, 483
  molecular parameters, 486
  Zeeman effect, 485
LiCl $X\,^1\Sigma^+$
  molecular beam electric resonance spectrum, 487
LiF $X\,^1\Sigma^+$
  molecular beam electric resonance spectrum, 487
LiH $X\,^1\Sigma^+$ state
  millimetre wave rotational spectrum, 743
linear combination of atomic orbitals (l.c.a.o.), 197, 206
line width
  collision, 275
  Doppler, 12, 274
  natural, 273
  predissociation, 286
  transit time, 273
LiO $X\,^2\Pi$
  construction of parity-conserved wave functions, 512
  molecular beam electric resonance spectrum, 509
  molecular parameters, 524
  nuclear hyperfine interactions, 517, 523
  theory of the Λ-doubling, 512
local oscillator, 701
long-range interactions, 282
Lorentz
  gauge, 102, 394
  invariance, 89, 99
  transformation, 98
Lorentzian line shape, 275

m, electron mass, 978
m_l, space-fixed component of l, 183
m_s, space-fixed component of s, 181
M_J, quantum number of the space-fixed component of J, 226
M_I, quantum number of the space-fixed component of I, 124
μ_e, electric dipole moment of the molecule, 116
μ, reduced mass, 178, 234
μ_B, electron Bohr magneton, 978
μ_I, magnetic dipole moment from nuclear spin angular momentum, 18, 270, 411
μ_L, magnetic dipole moment from electron orbital angular momentum, 269
μ_R, magnetic dipole moment from rotational angular momentum, 18, 270
μ_S, magnetic dipole moment from electron spin angular momentum, 21, 269
magnetic dipole transitions, 269, 754, 875
μ_N, nuclear magneton, 978
magnetic hyperfine Hamiltonian, 573
magnetic susceptibility
  tensor form, 405, 408
magnetic vector potential, 89, 94, 99, 125
maser, 261, 790
mass polarisation term, 6, 43, 317
Maxwell's equations, 33, 394
mercury atoms, 870
microwave
  absorption cells, 692
  magnetic resonance, 579
  radiation sources, 685
  resonant cavity, 582
  resonant cavity modes, 582
  resonant cavity Q factor, 582
  spectroscopy, 683
microwave/optical double resonance principles, 871, 884, 910, 931
microwave/optical double resonance polarisation spectroscopy, 870, 905
minimal basis set, 195
MnO $X\,^6\Sigma^+$
  millimetre wave rotational spectrum, 850
  molecular parameters, 852
modulation
  electric field (Stark), 583, 688
  magnetic field (Zeeman), 581, 692, 694
  source amplitude, 692
  source frequency, 692
  velocity (Doppler), 699
Moeller–Plesset perturbation theory, 218
molecular beams
  alignment, 373
  detectors, 373
  effusive, 372
  electric resonance, 463, 508
  magnetic resonance, 372
  supersonic (nozzle), 372
molecular orbital theory, 197
molecular parameters, definitions, 368
molecule-fixed coordinates, 51, 167

moment of inertia, 234
momenta
　electrons, 40
　nuclei, 40
momentum operator, 71
MoN $X\,^4\Sigma^-$
　microwave/optical double resonance rotational spectrum, 924
　molecular parameters, 927
Morse potential, 66, 238, 287
motional electric field, 487
motional Stark effect, 118

n, principal quantum number, 180, 201
N^2 form of the Hamiltonian, 343
Na atom, 185
$N_2\,A\,^3\Sigma_u^+$
　magnetic resonance spectrum, 446
　molecular parameters, 462
　nuclear statistical weights, 255
　rotational levels, 448
$N_2^+\,X\,^2\Sigma_g^+$
　nuclear hyperfine parameters, 957
　radiofrequency/optical double resonance, 953
$Na_2\,^1\Sigma_g^+$
　molecular beam magnetic resonance spectrum, 416
　molecular parameters, 420
NaH, $X^1\Sigma^+$, see LiH $X\,^1\Sigma^+$
NaO $X\,^2\Pi_{3/2}$
　microwave rotational spectrum, 811, 699
NbO X $^4\Sigma^-$
　electronic structure, 842, 846
　microwave rotational spectrum, 841
NBr $X\,^3\Sigma^-$
　microwave rotational spectrum, 763
NCl $X\,^3\Sigma^-$
　microwave rotational spectrum, 763, 769
　molecular parameters, 782
NCl $a\,^1\Delta$
　molecular parameters, 782
　rotational spectrum, 779
NCl $b\,^1\Sigma^+$
　microwave rotational spectrum, 741
negative energy solutions, 76
NF $X\,^3\Sigma^-$
　rotational spectrum, 763
NF $b\,^1\Sigma^+$
　rotational spectrum, 741
NF $a\,^1\Delta$
　electric quadrupole interaction for ^{14}N, 592
　magnetic hyperfine interaction for ^{14}N and ^{19}F, 593
　microwave magnetic resonance spectrum, 591
　molecular parameters, 594
　Stark effect, 594
　theory of the Zeeman effect, 594
NH $X\,^3\Sigma^-$
　far-infrared laser magnetic resonance spectrum, 652

nuclear hyperfine interactions, 654
NI $X\,^3\Sigma^-$
　microwave rotational spectrum, 763
NiH $X\,^2\Delta$
　far-infrared laser magnetic resonance, 674
　microwave/optical double resonance rotational spectrum, 927
　molecular parameters, 677, 930
NiO $X\,^3\Sigma^-$
　microwave rotational spectrum, 759, 762
NO $X\,^2\Pi$
　microwave magnetic resonance spectrum, 611
　molecular beam electric resonance spectrum, 526
　molecular parameters, 537, 788
　nuclear hyperfine interactions, 532, 537, 784, 946
　rotational spectrum, 782
　theory of the Λ-doubling, 527
node, 238
non-adiabatic effects, 67, 223
non-rigid rotor, 242
NS $X\,^2\Pi$
　molecular parameters, 609
nuclear
　dipole magnetic moment, 124, 979
　electric quadrupole interactions, 131, 604
　electric quadrupole moment, 124, 131, 162, 979
　gyromagnetic ratio, 124
　hyperfine parameters, 768, 844
　kinetic energy operator, 45, 70, 109
　magneton, 124, 978
　permutation operator, see permutation operator
　screening factor, 413
　shell theory, 124
　spin, 18, 123, 979
　spin coupling cases, 232
　spin–orbit coupling, 780, 803
　spin–rotation interaction, 129, 378, 413, 458, 491, 504, 535, 781
　spin–vibration interaction, 129
　Zeeman interaction, 128, 378

Ω, total angular momentum of the electrons about the internuclear axis, 26
Ω-doubling, 27
ω, vibrational frequency, 237
$\omega_e x_e,\ \omega_e y_e,\ \omega_e z_e$, vibrational constants, 241
$o,\ \Lambda$-doubling constant, 330, 618
O_2 molecule
　electron configurations, 200, 777
$O_2\,X\,^3\Sigma_g^-$
　microwave rotational spectrum, 272, 756
　molecular parameters, 758
$O_2\,a\,^1\Delta_g$
　far-infrared rotational spectrum, 778
　microwave magnetic resonance spectrum, 271, 587
　molecular parameters, 778
oblate symmetric top, 151
OD, see OH
OD$^+$, see OH$^+$

Subject index 1011

OH $X\,^2\Pi$
  electronic structure, 539, 622
  energy levels, 540, 614, 789, 791
  far-infrared laser magnetic resonance spectrum, 622
  far-infrared rotational spectrum, 790
  interstellar spectrum, 713, 792
  Λ-doubling and hyperfine parameters, 27, 542, 617
  Λ-doubling and centrifugal distortion, 27, 546
  microwave magnetic resonance spectrum, 613
  molecular beam electric resonance spectrum, 539
  molecular beam maser spectrum, 539
  molecular parameters, 544, 548, 622, 793
  Stark effect and electric dipole moment, 549
  theory of the Zeeman effect, 620
OH $A^2\Sigma^+$
  radiofrequency/optical double resonance spectrum, 880
  Zeeman effect, 881
OH$^+$ X $^3\Sigma^-$
  far-infrared laser magnetic resonance spectrum, 655
optical alignment, 416
optical detection, 416
optical state selection, 416
orbital hyperfine constant, 440, 28
orbit–orbit interaction, 91
origin transformations
  arbitrary to centre of mass of molecule, 41
  molecular to geometrical centre of mass, 44
  molecular to nuclear centre of mass, 43
ortho nuclear spin state, 253
overlap integral, 215
overtone transitions, 238

p orbitals, 182
p, Λ-doubling constant, 330
π-type molecular orbital, 197
P, total electronic angular momentum, 57
$^1\Pi$ states, 876
$^2\Pi$ states, 26, 328, 362, 509, 512, 527, 539, 597, 608, 609, 620, 627, 699, 782, 794, 810, 814, 871, 917
$^3\Pi$ states, 422, 437, 553, 555, 655, 659, 834, 836, 900, 913
para nuclear spin state, 253
parallel plate cell, 693
parity
  combinations of basis functions, 251, 512, 553, 651, 787, 820, 840, 887
  definition, 244
  selection rule, 266, 270
Pauli exclusion principle, 182, 317
Pauli spin matrices, 56, 75, 150
permittivity, 35
permutation operator
  case (a) and case (b) functions, 252
  definition, 200, 251
  ortho and *para* nuclear spin states, 253, 754

statistical weights, 254
perturbation theory, degenerate, 303
PF $X\,^3\Sigma^-$
  microwave rotational spectrum, 763
  molecular parameters, 768
phase convention, 144, 619
phase sensitive detection, 691
photo-alignment of H_2^+, 960
Planck constant, 978
Planck's radiation law, 259
PO $X\,^2\Pi$
  molecular parameters, 768
  rotational spectrum, 811
polarisation, 4, 5, 870
population inversion, 260
positron states, 77
potential curve, 276, 297, 807, 808
potential energy
  electron–electron, 44, 90
  electron–nuclear, 44, 90
  nuclear–nuclear, 44
  of the anharmonic oscillator, 65
  of the harmonic oscillator, 65
precession
  of L about the internuclear axis, 225
  of N and S about J, 227
  of S about the internuclear axis, 225
predissociation
  electronic, 288
  rotational, 286
projection operators, 305
prolate symmetric top, 151
proton rest mass, 978
pseudo contact interaction, 31
pulsed nozzle, 372, 704, 739, 742
pure precession hypothesis, 359, 622

q, electric field gradient, 464
q, Λ-doubling constant, 331
Q, molecular quadrupole moment, 409, 501
Q, nuclear quadrupole moment, 135, 568, 592
quadrupole
  asymmetry parameter, 137
  coupling constant, 18
  electric field, 464
  interaction, 17, 131, 138
quadrupole ion trap, 963
quadrupole–quadrupole interaction, 283
quantum electrodynamics correction to electron spin g-factor, 110
quantum mechanical tunnelling, 238
quasibound level, 287

Racah coefficients, 155
radial wave function, 180
radiation
  circularly polarised, 5
  density, 258, 261
  plane-polarised, 3
radiative transition probability, 258

radio astronomy, 713
radio telescope, 713, 714, 715
reduced mass, 61, 178, 234
reduced matrix element, 163, 174, 531, 663, 678
relativistic wave equation, 73
relativistic correction, 94
replacement theorem, 164
resonant cavity, 579
rest energy, 74
retarded interaction, 97
reversed angular momentum, 169
rigid rotor, 233
RKR inversion, 280
Roothaan–Hall equations, 215
rotational wave equation, 61, 146, 233
Rotation–vibration wave equation, 61, 243
rotation group
    irreducible representations, 143
rotational matrix, 55, 148, 174
rotation operators, 140
rotational
    angular momentum of the nuclei, 234
    constant, 243, 356
    Hamiltonian, 67, 319
    inversion symmetry, 250
    transition probabilities, 263
rotational Zeeman interaction, 20
Russell–Saunders coupling, 184
Rydberg constant, 214, 978
Rydberg states, 229

s orbitals, 182
s, spin vector of the electron, 80, 150
S, resultant electron spin angular momentum, 80
Σ, vector component of S along the internuclear axis, 225
Σ^+, Σ^- states, definition of, 200
$^1\Sigma^+$ states
    magnetic susceptibility, 408
    magnetic shielding tensor, 410, 414
    molecular quadrupole moments, 409
    nuclear spin–rotation tensor, 410, 415
    rotational magnetic moment tensor, 406
    other $^1\Sigma$ states, 16, 372, 388, 416, 465, 470, 483, 487, 489, 500, 732, 736, 738, 741, 743, 883, 885, 930
$^2\Sigma$ states, 697, 749, 859, 871, 880, 902, 919, 922, 930, 936, 938, 943, 953
$^3\Sigma$ states, 641, 649, 652, 655, 756, 759, 763
$^4\Sigma$ states, 661, 841, 924
$^6\Sigma$ states, 665, 851
S$_2$ $X\,^3\Sigma_g^-$
    microwave rotational spectrum, 759
saturation, 260
scalar coupling
    electron spin–nuclear spin, 24, 127
    nuclear spin–nuclear spin, 19
scalar product, 161, 172
Schrödinger equation
    diatomic molecule, 8
    H atom, 178
    harmonic oscillator, 237
    time-dependent, 256
    total wave function, 59
ScO $X\,^2\Sigma^+$
    microwave/optical double resonance rotational spectrum, 919
    molecular parameters, 921
    spin–rotation constant, 922
screening factor, 20, 378, 410
self-consistent field method, 192, 215
SeO, $X\,^3\Sigma^-$
    microwave magnetic resonance, 649
SeO $a\,^1\Delta$
    microwave magnetic resonance, 587
SF $X\,^2\Pi$, see CF $X\,^2\Pi$
SH, see OH
shift operators, 67, 143, 360
SI units, 33, 993
SiC $X\,^3\Pi$
    hyperfine interaction, 840
    interstellar spectrum, 839
    microwave and millimetre wave rotational spectrum, 836
    molecular parameters, 840
SiCl $X\,^2\Pi$
    molecular parameters, 814
SiF $X\,^2\Pi$
    microwave rotational spectrum, 810
    molecular parameters, 811
SiN $X\,^2\Sigma^+$, 752
Slater determinant, 183, 190, 214
Slater orbitals, 194
SO $X\,^3\Sigma^-$
    microwave magnetic resonance spectrum, 641
    microwave rotational spectrum, 760
    molecular parameters, 761
SO $a\,^1\Delta$
    microwave magnetic resonance spectrum, 587
    rotational spectrum, 779
SO $b\,^1\Sigma^+$
    millimetre wave rotational spectrum, 741
source modulation, 692
space-fixed axes, 7, 40
space-fixed inversion, 185, 244
spatial transformations, 53
special relativity, 73, 98
spherical harmonics, 23, 144, 179, 234
spherical polar coordinates, 7, 179
spherical tensor operators, 14, 159
spin Hamiltonian, 29, 44, 303
Spin-orbital, 183, 806
spin–orbit interaction, 30, 94, 186, 324, 357, 434, 667, 801
spin–other-orbit interaction, 92, 324
spinor, 52, 55, 76
spin–rotation interaction, 21, 323, 360, 747, 753, 803, 809

spin–spin interaction
  electron–electron, 23, 92, 360, 563
  electron–nuclear, 561
  nuclear–nuclear, 18, 378, 558
spin transformations, 52, 55
spin vector, 55
spontaneous emission, 259, 718
SrF $X\,^2\Sigma^+$
  microwave/optical double resonance rotational spectrum, 902
  molecular parameters, 904
Stark interaction, 20, 92
Stark Hamiltonian, 97, 114, 415
Stark modulation, 688
state selection, 374
statistical weights, 254
stimulated emission and absorption, 259, 720
superheterodyne detection, 701, 719
susceptibility tensor, 348, 351, 405
symmetric top wave functions, 150

t_0, axial dipolar hyperfine constant, 574
t_2, dipolar hyperfine constant, 574
tensor product, 171, 172
terahertz spectrometer, *see* far-infrared spectrometers
term symbols, 184
term values, 244
third-rank spin tensor, 336, 661, 678
TiCl $X\,^4\Phi$
  millimetre wave rotational spectrum, 854
TiN $X\,^2\Sigma^+$
  microwave/optical double resonance rotational spectrum, 922
  molecular parameters, 924
TiO $X\,^3\Delta$
  microwave/optical double resonance rotational spectrum, 922
  molecular parameters, 923
time-dependent perturbation theory, 256
transformations
  origin of coordinates, 40
  space-fixed to molecule-fixed, 7, 51, 52, 167
  cartesian to elliptical coordinates, 290
transition metal molecules, 906
transition moment, 256
transition between case (a) and case (b), 231
transition probabilities
  electronic, 268
  magnetic dipole, 269
  rotational, 263, 823
  vibrational, 267
translational
  motion, 40, 220
  Stark effect, 487
triangle rule, 154
tunable cavity microwave spectrometer, 702
tunable far-infrared spectrometer, 723
tunnelling, 238

u, ungerade, 200, 245
U_{kl}, vibration–rotation parameters, 347
united atom model, 201
unrestricted Hartree–Fock, 215

v, vibrational quantum number, 237
vacuum permeability, 35, 978, 993
vacuum permittivity, 35, 978, 993
valence bond function, 210
van Vleck pure precession hypothesis, 359
van Vleck transformation, 312
van Vleck–Weisskopf line shape, 275
variation method, 188, 207
vector diagrams, *see* Hund's case (a)–(e)
vector potential, 69, 75
velocity modulation, 699
vibrational
  averaging, 338
  constants, 66, 237, 241
  transition probabilities, 266
  wave equation, 63
vibrating-rotator
  eigenfunctions, 243
  energy levels, 66, 243
VO, *see* NbO
Voigt profile, 276

wave number unit, 1
Wigner–Eckart theorem, 163, 173, 335
Wigner symbols
  3-j symbols, 154, 987
  6-j symbols, 155, 991
  9-j symbols, 155
Wigner rotation matrix, 55, 148
Wigner–Witmer correlation rules, 203

X, symbol for the ground electronic state, 200

YbF $X\,^2\Sigma^+$
  radiofrequency/optical double resonance rotational spectrum, 936
YF $X\,^1\Sigma^+$
  microwave/optical double resonance spectrum, 930
YO $X\,^2\Sigma^+$
  microwave/optical double resonance rotational spectrum, 930
  molecular parameters, 933
YS $X\,^2\Sigma^+$
  microwave/optical double resonance rotational spectrum, 930
  molecular parameters, 936

Zeeman Hamiltonian, 19, 25, 96, 114, 347, 589, 605, 676
Zeeman modulation, 692
Zero point energy, 237
Zitterbewegung, 88, 92